TURBULENCE AND COHERENT STRUCTURES

# FLUID MECHANICS AND ITS APPLICATIONS
Volume 2

*Series Editor:* R. MOREAU
*MADYLAM*
*Ecole Nationale Supérieure d'Hydraulique de Grenoble*
*Boîte Postale 95*
*38402 Saint Martin d'Hères Cedex, France*

*Aims and Scope of the Series*

The purpose of this series is to focus on subjects in which fluid mechanics plays a fundamental role.

As well as the more traditional applications of aeronautics, hydraulics, heat and mass transfer etc., books will be published dealing with topics which are currently in a state of rapid development, such as turbulence, suspensions and multiphase fluids, super and hypersonic flows and numerical modelling techniques.

It is a widely held view that it is the interdisciplinary subjects that will receive intense scientific attention, bringing them to the forefront of technological advancement. Fluids have the ability to transport matter and its properties as well as transmit force, therefore fluid mechanics is a subject that is particulary open to cross fertilisation with other sciences and disciplines of engineering. The subject of fluid mechanics will be highly relevant in domains such as chemical, metallurgical, biological and ecological engineering. This series is particularly open to such new multidisciplinary domains.

The median level of presentation is the first year graduate student. Some texts are monographs defining the current state of a field; others are accessible to final year undergraduates; but essentially the emphasis is on readability and clarity.

***For a list of related mechanics titles, see final pages.***

# Turbulence and Coherent Structures

Selected Papers from "Turbulence 89: Organized Structures and Turbulence in Fluid Mechanics", Grenoble, 18–21 September 1989

edited by

O. METAIS

and

M. LESIEUR

*Institut de Mecanique de Grenoble,*
*Grenoble Cedex, France*

KLUWER ACADEMIC PUBLISHERS

DORDRECHT / BOSTON / LONDON

Library of Congress Cataloging-in-Publication Data

Turbulence and coherent structures / [edited] by O. Métais and M. Lesieur.
p. cm. -- (Fluid mechanics and its applications ; 2)
Includes index.

1. Fluid mechanics. 2. Turbulence. 3. Fluid dynamics.
I. Métais, O. (Olivier) II. Lesieur, Marcel. III. Series: Fluid mechanics and its applications ; v. 2.
QC145.2.T87 1990
620.1'064--dc20 90-5337

ISBN 978-90-481-4063-3

Published by Kluwer Academic Publishers,
P.O. Box 17, 3300 AA Dordrecht, The Netherlands.

Kluwer Academic Publishers incorporates
the publishing programmes of
D. Reidel, Martinus Nijhoff, Dr W. Junk and MTP Press.

Sold and distributed in the U.S.A. and Canada
by Kluwer Academic Publishers,
101 Philip Drive, Norwell, MA 02061, U.S.A.

In all other countries, sold and distributed
by Kluwer Academic Publishers Group,
P.O. Box 322, 3300 AH Dordrecht, The Netherlands.

*Printed on acid-free paper*

## Chairmen:

M. Lesieur (IMG) Y. Morchoisne (ONERA)

## Scientific Committee:

F.K. Browand (USA)
G. Duvaut (France)
J.C.R. Hunt (UK)
E. Krause (FRG)
J.C. Mc Williams (USA)
C.W. Van Atta (USA)
G.Comte-Bellot (France)
U. Frisch (France)
T. Kambe (Japan)
A. Linan (Spain)
Y. Pomeau (France)
A.M. Yaglom (USSR)

## Organizing Committee:

Chairman:
O. Leuchter (ONERA)

Scientific Secretary:
O. Métais (IMG)

J.P. Chollet
M. Lesieur
(IMG)

T.K. Dang
Y. Morchoisne
(ONERA)

M. Legrand
B. Masure
(DRET)

## Sponsoring Organizations and Companies:

- Aérospatiale
- Alliant Computer Systems France
- Avions Marcel Dassault
- Centre National d'Etudes Spatiales
- Centre National de la Recherche Scientifique
- Commissariat à l'Energie Atomique
- Crédit Lyonnais
- Institut National Polytechnique de Grenoble
- Ministère de la Défense (D.R.E.T.)
- Office National d'Etudes et de Recherches Aérospatiales
- Université Joseph Fourier
- U.S. Air Force (E.O.A.R.D.)
- Ville de Grenoble

# Contents

# Preface

In the last 25 years, one of the most striking advances in Fluid Mechanics was certainly the discovery of coherent structures in turbulence: laboratory experiments and numerical simulations have shown that most turbulent flows exhibit both spatially-organized large-scale structures and disorganized motions, generally at smaller scales. The development of new measurement and visualization techniques have allowed a more precise characterization and investigation of these structures in the laboratory. Thanks to the unprecedented increase of computer power and to the development of efficient interactive three-dimensional colour graphics, computational fluid dynamicists can explore the still mysterious world of turbulence. However, many problems remain unsolved concerning the origin of these structures, their dynamics, and their interaction with the disorganized motions. In this book will be found the latest results of experimentalists, theoreticians and numerical modellers interested in these topics.

These coherent structures may appear on airplane wings or slender bodies, mixing layers, jets, wakes or boundary-layers. In free-shear flows and in boundary layers, the results presented here highlight the intense three-dimensional character of the vortices. The two-dimensional large-scale eddies are very sensitive to three-dimensional perturbations, whose amplification leads to the formation of three-dimensional coherent vortical structures, such as streamwise, hairpin or horseshoe vortex filaments.

This book focuses on modern aspects of turbulence study. Relations between turbulence theory and optimal control theory in mathematics are discussed. This may have important applications with regard to, e.g., numerical weather forecasting. Attempts to utilize the dynamical systems theory to further the understanding of open turbulent flows such as boundary layers and jets are presented. It is also demonstrated that purely kinematic properties may be of importance for the flow dynamics and topology.

The large-scale vortices may break down into small-scale three-dimensional turbulence. The vorticity structures in the dissipative and inertial ranges of three-dimensional turbulence are also visualized in high-resolution computations. Their link with the intermittent nature of the turbulence is discussed. The multifractal nature of turbulence is investigated numerically in the dissipative range, and experimentally (through wavelet analysis) in the inertial range. A better understanding of the intermittent nature of inertial and dissipative scale is of great importance in the improvement of subgrid-scale models in the numerical simulations, where only the largest scales of motion are explicitly calculated (Large-Eddy Simulations or LES). The LES have fundamental applications to understanding the coherent-structure dynamics and constitute a first step towards the use of numerical simulations to study industrial flows. In this context, the spectral eddy viscosity concept applied to LES is a promising aspect.

The phenomenon of vortex breakdown is reviewed. The available experimental evidence suggests that vortex breakdown is a system instability or imperfect bifurcation of the vortex flow. Progress in numerical simulation of this phenomenon is presented. Other types of instabilities are studied. A nonlinear instability study is performed to explain the generation of secondary flow in rectangular ducts. Numerical simulations of internal gravity waves and vortical mode interactions in stably-stratified fluids are compared to a nonlinear perturbation analysis. Experiments on jets with large density differences show striking evidence of absolute instability.

Under the influence of strong thermal stratification, rapid rotation or strong magnetic field, the coherent structures tend to be more two-dimensional. The pioneering studies of two-dimensional turbulence were motivated by large-scale atmospheric and oceanic motions, which have a quasi-two-dimensional dynamics. In the past few years, the study of two-dimensional turbulence has undergone a renaissance, thanks to the possibility of high resolution two-dimensional computations: they have shown the emergence of isolated, coherent vortices and have allowed for the study of their dynamics. These vortices are of same nature as those encountered in wakes and mixing layers. This simple remark corresponds in fact to a major progress in meteorology or oceanography. It has allowed also the broadening of the concept of two-dimensional turbulence, from the earlier isotropic model to free-shear flows. A first step towards a *structured* two-dimensional turbulence theory is presented. Analogies and differences in the behaviour of two-dimensional turbulence and quasi-two-dimensional turbulence (like geostrophic turbulence) are discussed. Laboratory experiments can presently be designed in order to check numerical-simulation observations and statistical-theory results. The noise generation by turbulence and by coherent structures in two-dimensional turbulence is investigated.

In a geophysical environment, turbulent flows are often submitted to the stabilizing or destabilizing effects of a mean temperature (or density) gradient. New developments of classical statistical theories are compared to observations in the atmospheric convective boundary layer. In the stable case, the studies are focused on the influence of stratification on the flow structures, and on the generation of internal gravity waves by the coherent structures. In numerical simulations, a mathematical decomposition of the velocity field into wave and turbulent components is introduced.

Effects of compressibility, chemical reactions and magnetic fields on the coherent vortices are also examined. Statistical techniques of modelling are applied to compressible flows. A precise investigation of large scale coherent structures in MHD turbulence is performed through high resolution numerical simulations.

By focussing on these topics, this book aims at four goals: first, to achieve a better understanding of the organized structure dynamics. This constitutes a step towards the construction of improved statistical or deterministic theories of two-dimensional and three-dimensional *structured* turbulence. Second, to define laboratory experiments which can be studied in conjunction with numerical experiments. Third, the development of a unified point of view on the generation of these structures (instability or other mechanisms of spontaneous organisation).

An ultimate goal is to reach (for drag reduction purposes for instance) a higher *controllability* of turbulence.

*The Editors:*

*Olivier Métais*
*Marcel Lesieur*

# Remarks on Turbulence Theory

Jacques-Louis LIONS

*Collège de France and C.N.E.S.*

*Turbulence and Coherent Structures*
*O. Métais and M. Lesieur, eds.*
*Selected papers from "Turbulence 89: organized structures and turbulence in fluid mechanics"*
*Grenoble, 18-21 September 1989*

## 1. Introduction

The goal of this paper is to present some simple and somewhat preliminary remarks on Turbulence Theory *in Connection with Control Theory of Distributed Systems.*

The first remarks, presented in Section 2 below, address the question of using Control Theory to *derive* possibly "optimal", or at least "good", *numerical algorithms.* One has of course to make precise what is meant by "optimal" (or by "good"), i.e. to define a "cost function" in the terminology of Control Theory. In general terms, one wants to find algorithms which give the information we are looking for and which "suppress" the difficulties introduced by Turbulence or Chaos.

The other family of remarks, presented in Section 3 and 4, address the question of *controlling* turbulence. This situation arises if we can act on the system through *control functions.* Several approaches are, at least theoretically, concevable. They also depend on the "cost function". We consider two families of "cost functions". The first one to be considered (Section 3) expresses (or at least *should* express !) the "mixing" of scales which is *one* of the properties of Turbulence (cf. Lesieur, 1987). The second family of "cost functions" is connected with *Exact Controllability.* This will be presented in Section 4 below.

There is of course a large variety of models where the Remarks to follow would apply.

What we have particularly in mind is the classical set of Navier-Stokes equations.

We shall denote here by $y$ the velocity field and by $\Pi$ the pressure, a quite unusual notation in the "Turbulence" circle.

The reason is simply that in all what follows we think of $y$ as the state of our system, this state depending on control functions, these control functions being either "artificial" as in Section 2 or "natural" as in Sections 3 and 4.

We shall write therefore for the state equation :

$$\begin{aligned} &\frac{\partial y}{\partial t} + (y\nabla)y - \mu\Delta y = -\nabla\Pi \quad \text{in} \quad \Omega \times (0,T), \\ &div y = 0 \quad \text{in} \quad \Omega \times (0,T) \end{aligned} \qquad (1-1)$$

subject to *initial conditions* and to *boundary conditions.*

In (1.1) $\Omega$ is a bounded open set of $\mathbb{R}^n, n = 2$ or 3.

The control functions, when there are of the "natural" type, will be either in the "source terms", in the right hand side of (1-1), or on the boundary ; more precisely

$$\begin{aligned} &\frac{\partial y}{\partial t} + (y\nabla)y - \mu\Delta y = -\nabla\Pi + v\chi_\sigma \quad \text{in} \quad \Omega \times (0,T), \\ &divy = 0 \quad \text{in} \quad \Omega \times (0,T) \end{aligned} \qquad (1-2)$$

where

$$\begin{aligned} &\sigma = \text{open set contained in} \quad \Omega, \\ &\chi_\sigma = \text{characteristic function of} \quad \mathcal{O}, \\ &v = v(x,t) = \text{control function}(= \textit{distributed control}) \end{aligned}$$

Equations (1-2) express the fact that we can *act* on the system in the region $\mathcal{O}$ (not necessarily connected) and which can be "very small".

Another situation will be when the state equation is given by (1.1) with the control functions appearing on the boundary *(boundary control)*

$$y = v \quad \text{on} \quad \Gamma \times (0,T) \qquad (1-3)$$

where $\Gamma$ denotes the boundary of $\Omega$ and where the control function $v = \{v_1, ..., v_n\}$, $n = 2$ or 3, is subject to the compatibility condition

$$\int_\Gamma v\nu \quad d\Gamma = 0 \quad \text{for a.e.t} \qquad (1-4)$$

In (1-4), $\nu$ denotes the normal to $\Gamma$, chosen unitary and directed towards the exterior of $\Omega$ to fix ideas.

Another, more realistic case, is the situation where the boundary control $v$ is subject to (1-4) *and* to

$$v = 0 \quad \text{outside} \quad \Gamma_o \times (O,T), \quad \Gamma_o \subset \Gamma \qquad (1-5)$$

(in other words one can *act* on the part $\Gamma_o$ of the boundary).

As we said above, what we are going to say is "model independant", provided we assume it enjoys (so to speak) the "properties of turbulence".

Let us make now this point more precise. We consider a state equation given by

$$\frac{\partial y}{\partial t} + Ay + f(y) = 0 \quad \text{in} \quad \Omega \times (O,T) \qquad (1-6)$$

where $y = \{y_1, ..., y_N\}$ denotes the *state*.

In case we have a *distributed control*, (1-6) is replaced by

$$\frac{\partial y}{\partial t} + Ay + f(y) = v\chi_\sigma \qquad (1-7)$$

Both in (1-6) and (1-7) $A$ is a second order self adjoint elliptic system -or, more generally, an elliptic system in the general sense of Agmon-Douglis-Nirenberg and of Lopatinski, so as to cover the Navier-Stokes equations. The control function $v = \{v_1, ..., v_N\}$ may have some of its components equal to 0, so as again to cover the Navier-Stokes equations where no control appears in the last equation (1-2) $divy = 0$.

In order to fix ideas, we shall assume that

$$y = 0 \quad \text{on} \quad \Gamma \times (O,T) \qquad (1-8)$$

We *exclude* in this way "boundary control", but the necessary changes needed to cover also the situations (1-3) (and (1-5)) are simple at least theoretically. They are made explicit in the references given in the Bibliography.

We want now to express that (1.6) is some kind of "turbulent" system. According to Lesieur (1987) it means that the system has two properties:

"the" state y shoud de "unpredictable" (1 – 9)

the system satisfies the "increased mixing property" (the flow is able to mix transported quantities much more rapidly than if only molecular diffusion processes were involved). (1 – 10)

In order to make (1.9) more precise, we introduce *initial conditions* in a particular form. We shall assume that

$$y(x,0) = y^o(x) + \tau\hat{y}^o(x) \quad \text{in} \quad \Omega, \qquad (1-11)$$

where

$$y^o \quad \text{is } \textit{given} \text{ in an appropriate space, say} H, \qquad (1-12)$$

and where $\hat{y}^o$ is *arbitrary* in $H$, with norm 1, and where $\tau \in \mathbb{R}$ is "small".

We shall assume that "the" state $y$ is *"very sensitive"* to $\tau$. □
A few remarks are here in order.

*Remark 1.1*

Let us assume for a moment that $\tau$ and $\hat{y}^o$ are given. Then, given $v(v = 0$ in case of (1-6)), all data are known in (1-7), (1-9), (1-11). In the case of Navier-Stokes equations, it is known after Leray (1933, 1934) that in case *the space dimension equals 2* the state $y$ *is uniquely defined* in a suitable functional class (cf., also, Ladyzenskaya, 1958, for the case when $y(x, 0)$ is given in a smooth class and Lions and Prodi, 1959, for the case where $y(x, 0)$ is given in $(L^2(\Omega))^2$).

The problem for existence *and* uniqueness (in the same functional class) is *open* since Leray (1933, 1934).

But global (in time) existence of at least one solution is known, cf. Lions (1969) for a presentation of these results.

The view point which is adopted is then the following (cf. Lions, 1983, and the Bibliography therein) : one thinks of (1-7), (1-8) and (1-11) *as a set of constraints* which *do not* necessarily uniquely define $y$ as a function of $x, t$ and of $\tau\hat{y}^o$ and of $v$. □

*Remark 1.2*

In apparent contradiction with what is said in Remark 1.1 above, we shall *assume* that

equations (1-6) (or (1-7)), (1-8) and (1-11) uniquely define
(in a suitable functional class) the solution (1 − 13)
$y = y(x, t; \tau\hat{y}^o, v)$.

The "contradiction" is apparent since the strategy that we shall (try to) follow is to *derive* equations based on (1-13) and on control theory. But these equations are meaningful *independently* of (1-13) and we will (try to) study directly these equations. □

*Remark 1.3*

We have taken condition (1-8) to simplify the exposition. But "very sensitive" dependence applies as well to boundary conditions.

If we have the boundary conditions

$$y = g + \tau_1 \hat{g} \quad \text{on} \quad \Gamma \times (0, T) \tag{1-14}$$

where

$g$ is given $\hat{g}$ is arbitrary (with norm 1
in a suitable Banach space) and $\tau \in \mathbb{R}$ is small

then "the" solution is "very sensitive" to $\tau_1 \hat{g}$. □

*Remark 1.4*

It remains to make precise hypothesis (1-9), (1-10) using "the" state as in (1-13).

## 2. Approximation Methods

We consider here equations (1-6),(1-8) and the case where the initial condition is given :

$$y(x,0) = y^o(x) \quad \text{in} \quad \Omega \tag{2-1}$$

In all numerical algorithms, one *changes* equation (1-6) in order, among other things, to reduce the problem to a finite dimensional situation.

*One* way (of course not the only one !) to look at this is to rewrite (1-6) under the form

$$\frac{\partial y}{\partial t} + Ay + w = 0 \tag{2-2}$$

with the constraint

$$w = f(y). \tag{2-3}$$

The next step is to *relax* (2-3). Let us consider a functional $\phi(y)$ where we think of $y$, for the time being, as "the" solution of (1-6), (1-8), (2-1) and that we would like to keep "small".

**Exemple 2.1**

If we do not want to consider turbulence on $\omega \subset \Omega$ and during the time interval $(t_o, T)$, we could take

$$\phi(y) = \int\int_{\omega\times(t_o,T)} (curl\ y)^2 dxdt. \tag{2-4}$$

□

**Exemple 2.2**

We could also have "non oscillatory" functions in time on $\omega \times (t_o, T)$. Then

$$\phi(y) = \int\int_{\omega\times(t_o,T)} (\frac{\partial y}{\partial t})^2 dxdt. \tag{2-5}$$

□

**Exemple 2.3**

In a situation of the type "climatology" (cf. Lorenz, 1986), we do not want to take into account the "fast" variations related to gravity waves and we want to restrict the state to the "slow" variations related to Rossby waves. If $\{\varphi_j\}$ denote the set of (normalized) eigenfunctions of the (self-adjoint) operator $A$ * one divides the spectrum of $A$ into two parts $\{\lambda_\alpha\}$, $\{\lambda_\beta\}$, and we introduce

$$\phi(y) = \int_{t_o}^{T} \sum_{\beta} \left( \int_{\Omega} y \varphi_\beta dx \right)^2 dt. \tag{2-6}$$

Another approach is through the theory of *inertial varieties*. Cf. Foias *et al.* (1988). □

The *problem of optimal control* we introduce is now the following.

Let $y(w)$ be the solution of (2-2), (2-1), (1-8). We introduce

$$J_k(w) = \phi(y(w)) + k \int_{\Omega \times (0,T)} (w - f(y))^2 dx dt \tag{2-7}$$

where $k > 0$ plays the role of a *penalty term*.

We consider then the problem :

$$inf \cdot J_k(w). \tag{2-8}$$

*Remark 2.1*

One can write the *Optimality System* (cf. Lions,1968) corresponding to problem (2-8). One can add *constraints* of other types on $w$. □

*Remark 2.2*

Very many variants are of course possible. One can in particular *decompose* the solution $y$ in two parts, say $y = z + \zeta$ and apply the above ideas to *one* part, say $\zeta$. □.

*Remark 2.3*

Numerical computations based on ideas similar to the above are in progress. □.

*Remark 2.4*

In the above approach we have escaped the difficulty connected with the possibility of non uniqueness of the solution of Navier-Stokes equations in 3 dimensions. □

---

* Similar methods are used, at least formally, in cases when $A$ is *not* self-adjoint.

## 3. Control of Turbulence

Let us consider now (1-7), (1-8), (1-11) and let us assume *for the time being* that (1-13) holds true. We assume that we are in the situation (1-9), (1-10).

Let us consider a functional $\phi_1(y)$ which is, in principle, "inpredictable" if there are *small* variations in the initial condition (1-11).

We shall say (cf. Lions, 1989a, 1989b) that the control $v$ is *insensitive with respect to the functional* $\phi_1$ if

$$\frac{d}{d\tau}\phi_1(y(\tau,v)|_{\tau=0}=0, \quad \forall\hat{y}^o. \tag{3-1}$$

In (3-1), we have written $y(\tau,v)$ for $y(x,t;\tau\hat{y}^o,v)$ and we assume that the $\tau$ derivative does exist near the origin.

If $f$ and $\phi_1$ are smooth, then (3-1) is equivalent to

$$(\phi_1'(y(v)),\ y_\tau)=0, \quad \forall\hat{y}^o \tag{3-2}$$

where :

$$\phi_1' = \quad \text{Fréchet's derivative of} \quad \phi_1,$$

$$g(v) = \quad \text{solution of (1-7), (1-8) and (1-11) where} \quad \tau=0,$$

$$y_\tau = \frac{\partial}{\partial\tau}y(x,t;\tau\hat{y}^o,v)|_{\tau=0}.$$

If $f$ is smooth, $y_\tau$ is given by

$$\begin{aligned}&\frac{\partial}{\partial t}y_\tau + Ay_\tau + f'(y)y_\tau = 0 \quad (\text{where} \quad y=y(x)),\\ &y_\tau(0)=\hat{y}^o,\\ &y_\tau = 0 \quad \text{on} \quad \Gamma\times(0,T).\end{aligned} \tag{3-3}$$

We restrict the choice of $v's$ to those controls - *provided they exist* - which satisfy (3-2), the insensitive controls.

We consider next a second functional $\phi_2(y)$ which, in principle, is related to the "increased mixing property" (1-10).

*For instance*, in the case of Navier-Stokes equations, we could consider

$$\phi_2(y)=\phi(y) \quad \text{as given by (2-4).}$$

Then the problem of controlling turbulence *can* be expressed as :

$$inf \cdot \phi_2(y(v)) \tag{3-4}$$

where

$$\begin{aligned} &\frac{\partial}{\partial t}y(v) + Ay(v) + f(y(v)) = \chi_\sigma, \\ &y(v)|_{t=0} = y^o, \quad y(v) = 0 \quad \text{on} \quad \Gamma \times (O,T) \end{aligned} \tag{3-5}$$

*and where v is subject to* (3-2) (insensitivity). □

*Remark 3.1*

It has been implicitely assumed that there is existence and uniqueness of a solution. If this property is not proven (or not true) the necessary modifications are indicated below in Remark 3.4. □

We proceed by transforming (3-2). To this effect, we introduce as it is classical in sensitivity studies, the adjoint state $q$ defined by

$$\begin{aligned} &-\frac{\partial q}{\partial t} + Aq + f'(y)^* q = \phi_1'(y) \quad \text{where } y = y(v) \\ &q(x,T) = 0, \quad q = 0 \quad \text{on} \quad \Gamma \times (O,T). \end{aligned} \tag{3-6}$$

Then (3-2) becomes, using (3-3) and integration by parts

$$\int_\Omega q(x,0) = \hat{y}^o(x)dx = 0, \quad \forall \hat{y}^o \tag{3-7}$$

i.e.

$$q(x,0) = 0 \quad \text{in}\Omega. \tag{3-8}$$

*The insensitive controls with respect to* $\phi_1$ *are those which are such that* (3-5), (3-6), (3-8) *hold true.*

*Remark 3.2*

The *existence* of such insensitive controls seems to be in general a non trivial problem. Cf. Lions (1989a, 1989b, 1989c) for more details, variants and partial results. □

The problem can now be formulated : find

$$\inf_v \phi_2(y(v)), \quad v \quad \text{subject to (3-5),(3-6),(3-8).} \tag{3-9}$$

*Remark 3.3*

A quite realistic way to avoid the difficulty indicated in Remark 3.2 (but of course the question raised in this Remark remains of interest, at least we think !) is to *relax* (3-8) and to consider

$$\inf_v \cdot [\phi_2(y(v)) + k \int_\Omega q(x,0)^2 dx], \quad k > 0 \tag{3-10}$$

In this formulation one can add other constraints on $v$. □.

*Remark 3.4*

We can proceed one step further, in case existence and uniqueness is not proven (or not true). We consider

$$\begin{aligned} &\frac{\partial y}{\partial t} + Ay + f(y) = v\chi_\sigma, \\ &y(x,0) = y^o(x), \quad y = 0 \quad \text{on} \quad \Gamma \times (O,T) \end{aligned} \tag{3-11}$$

as a set of *constraints* and we consider

$$\inf[\phi_2(y) + k \int_\Omega q(x,0)^2 dx + \alpha \int\int_{\sigma\times(O,T)} v^2 dx dt] \tag{3-12}$$

where $y$ and $v$ are subject to (3-11), where $q$ is a function of $y$ defined by the solution of (3-6) and where $\alpha$ is a (small) parameter, $\alpha \geq 0$. The last term $\alpha \int\int_{\sigma\times(O,T)} v^2 dx dt$ is needed in situations where we could have many solutions corresponding to $v$. This term can be avoided for Navier-Stokes. □

*Remark 3.5*

If we do *not* take into account the insensitivity, which is motivated by (1-9), the problem reduces to a "classical" form :

$$\inf \cdot [\phi_2(y) + \alpha \int\int_{\sigma\times(O,T)} v^2 dx dt] \tag{3-13}$$

where $y$ and $v$ are subject to constraints (3-11). We refer to Fursikov (1981, 1982) and Lions (1983) for this point of view. □

*Remark 3.6*

Finding $v$ insensitive (with respect to $\phi_1$) amounts to find $v$ such that starting at time $t = 0$, with $y^o$, integrating (3-5) (if $n = 2$ in the case of Navier-Stokes equations) and then integrating (3-6) backward one is led to (3-8) at time $t = 0$. This is a question of the *exact controllability type*. Cf. Lions (1989a, 1989b, 1989c). □

## 4. Exact controllability

Let us now consider the equations

$$\frac{\partial y}{\partial t} + Ay + f(y) = v\chi_\sigma \tag{4-1}$$

subject to

$$y = 0 \quad \text{on} \quad \Gamma \times (0,T) \tag{4-2}$$

and where

$$y(0) = y^o \quad \text{is given.} \tag{4-3}$$

We formulate the problem of *exact controllability* in a way which makes sense even if existence *and uniqueness* of a solution of (4-1), (4-2), (4-3) *is not known.*

*The first formulation* is as follows : let $T$ be given and let $z^o$ be given in a suitable function space. We say that the system is *exactly controllable at time* $T$ if for every couple $y^o, z^o$, there exists a couple $\{v, y\}$ satisfying (4-1), (4-2), (4-3) such that

$$y(T) = z^o \quad , \quad (y(T) = y(x,T)). \tag{4-4}$$

Moreover, if there is exact controllability, one wants to find $v$ such that (4-4) holds true and *which minimizes* (for instance)

$$\int\int_{\sigma \times (O,T)} v^2 dxdt. \tag{4-5}$$

As we indicate in Lions (1989c), this formulation may be too restrictive and it can be replaced by the following :

Minimize

$$\int\int_{\sigma \times (O,T)} v^2 dxdt + k \int_\Omega |y(T) - z^o|^2 dx, \quad k > 0 \tag{4-6}$$

among all couples $\{v, y\}$ subject to (4-1), (4-2), (4-3).

Still another possibility is to consider

$$\inf \cdot \int_\Omega |y(T) - z^o|^2 dx \tag{4-7}$$

amont all couples $\{v, y\}$ subject to (4-1), (4-2), (4-3). □

*We conjecture* that for Navier-Stokes equations the inf · in (4-7) equals 0, i.e. that the *attaignable set* (the set spanned in $H = \{\varphi | \varphi \in$

$(L^2(\Omega))^n, div\varphi = 0, \varphi\nu = 0$ on $\Gamma\}$ by all possible states $(y(T))$ is *dense* in $H$. □

*Remark 4.1*

It seems possible that the inf $\cdot$ in (4-6) *decreases* as the viscosity $\nu$ decreases in Navier-Stokes equations.

A vague and fuzzy conjecture is that the more "chaotic" is a system, the closest it is from exact controllability. □

*Remark 4.2*

Some technical indications are given in Lions (1989c). For the exact controllability of *linear* systems, a general method has been introduced by the Author in 1988 (the so-called HUM method, Hilbert Uniqueness Method). For *nonlinear* systems, the situation is still largely open. Examples for nonlinear wave equations have been solved by Zuazua (1988, 1990) and by application of Zuazua, loc. cit. and further arguments, by Jinhai (1990).

## References

Foias, C., Sell, G. and R. Temam, 1988: Inertial manifolds for nonlinear evolutionnary equations. *J. Diff. Equations*, 73 , p. 309-353.

Fursikov, A.V., 1981: Problèmes de contrôle et résultats sur la résolution unique de problèmes mixtes pour les équations de Navier-Stokes et Euler tridimensionnelles. *Mat. Sbornik*, 115 (157), p. 281-306.

Fursikov, A.V., 1982: Propriétés des solutions de certains problèmes d'extremum liés au système de Navier-Stokes. *Mat. Sbornik*, 118 (160), (1982), p. 323-349.

Jinhai, Y., 1990, *Contrôlabilité exacte de l'équation des ondes non linéaires.* To appear.

Ladyzenskaya, O.A.,1958: Global solution of the boundary value problem for Navier-Stokes equations in 2 space dimensions. *Doklady Akad. Nauk. SSSR*, 123, p. 427-429.

Leray, J., 1933: Etude de diverses équations intégrales non linéaires et de quelques problèmes que pose l'hydrodynamique. *J.M.P.A. XII*, p. 1-82.

Leray, J., 1934: Essai sur le mouvement plan d'un liquide visqueux que limitent des parois. *J.M.P.A. XIII*, p. 331-418.

Lesieur, M., 1987: *Turbulence in fluids.* Martinius Nijhoff.

Lions, J.L., 1968: *Contrôle Optimal des Systèmes gouvernés par deséquations aux dérivées partielles.* Dunod. Gauthier-Villars, Paris.

Lions, J.L., 1969: *Quelques méthodes de résolution des problèmes aux limites non linéaires.* Dunod. Gauthier-Villars.

Lions, J.L., 1983: *Contrôle des systèmes distribués singuliers.* Gauthier-Villars.

Lions, J.L., 1988a: Exact Controllability, Stabilization and Perturbation for Distributed Systems. *J. Von Neumann Lecture. SIAM Rev.*, 30 , p. 1-68.

Lions, J.L., 1988b: *Contrôlabilité Exacte, Perturbation et Stabilisation de Systèmes Distribués.* Tome 1. Contrôlabilité exacte. Tome 2. Perturbations. Masson, Collection RMA.

Lions, J.L., 1989a: Insensibilization of functionals connected with distributed systems. *Conference Rambouillet*, June 1989.

Lions, J.L., 1989b: Insensitive controls. *Lecture at Franco-Italian-Soviet Meeting in Pavie*, October 1989. To appear.

Lions, J.L., 1989c: *San Servolo, Venise*, October 1989, To appear.

Lions, J.L. et Prodi, G., 1959: Un théorème d'existence et unicité dans les équations de Navier-Stokes en dimension 2. *C.R. Acad. Sci. Paris*, 248, p. 3519-3521.

Lorenz, E.N., 1986 On the existence of a slow manifold. *J. Atmos. Sci.*, 43 (17), p. 1547-1557.

Zuazua, E., 1988: Contrôlabilité exacte de systèmes d'évolution non linéaires. *C.R. Acad. Sci. Paris*, 306, p. 129-132.

Zuazua, E., 1990: Exact controllability for the semi-linear wave equation. *J.M.P.A.*.

# Hydrodynamic visualization of organized structures and turbulences in boundary layers, wakes, jets or propeller flows

Henri WERLÉ

*Office National d'Etudes et de Recherches Arospatiales*
*BP 72, 92322 Châtillon Cedex, France*

*Turbulence and Coherent Structures*
*O. Métais and M. Lesieur, eds.*
*Selected papers from "Turbulence 89: organized structures and turbulence in fluid mechanics"*
*Grenoble, 18-21 September 1989*

**Abstract**
The present paper reviews some typical examples of water tunnel visualizations, made at ONERA, revealing small instabilities and the organized structures that characterize turbulent flows. These visualisations demonstrate chiefly the time variation, in films of the tests (e.g. ONERA Film, no 1213, *Grosses structures et turbulence*; 1989), as well as the contribution of certain parameters such as the Reynolds number.

## 1. Introduction

Water tunnel visualizations are often an efficient, convenient and easy way to get a picture of the basic phenomena at play in a real flow, and to analyze and understand those phenomena better.

ONERA has acquired a broad experience in flow visualization techniques using liquid (Figs. 2b, 7, 9, 10) and gaseous (Figs. 2a, 4, 5, 6) tracers injected directly into the flows. These tracers preserve all of their efficiency through the laminar-turbulent transition. These methods have been used in the TH1 water tunnel first put into service in 1952 (Werlé (1974)), and then with the new TH2 tunnel installed in 1982 (Werlé et Gallon, 1982). With its extensive Reynolds number domain ($10^4 \leq R_{eL} \leq 10^6$), the latter tunnel has shown itself to be quite useful in studying turbulence and the flow instabilities related to the appearance of turbulence (Werlé, 1987).

## 2. Experimental technique

Let us recall that these are vertical, open circuit, gravity-driven water tunnels (Fig. 1).

An instanteous visualization of the flow is obtained by injecting filaments or sheets of dyes from the model wall that have the same density and viscosity as the water and that start out continuous and sticking more or less to the surface, depending on the rate of flow. These dye injections reveal the structure and character of the boundary layers (Fig. 2b), separations (Fig. 7cd), vortices (Figs. 9, 10), wakes (Figs. 4a, 5a) and jets (Figs 6a, 7b) or propeller flows (Fig. 8ac).

Fine air bubbles in suspension are also used to extend this visualization into the fluid : the paths followed by the air bubbles reveal the general appearance of the velocity field or of the mean flow, depending on the exposure time (Figs. 2a, 5b, 6b, 8b).

---

H. Werlé is a retired Senior Scientist from ONERA

Some improvements have recently been made in the test lighting (flood lighting with chronophotographic effect) and the way the pictures are taken (motorized camera and a stand consisting of a head with controlled-friction mount). The chief advantage of these techniques is that they generate views in absolute motion for the study of wakes and jets (Figs. 4cd, 5cd, 6cd), and that the visualizations can be analysed quantitatively (Werlé, 1988).

## 3. Results

**Boundary layer** along a flat plate without longitudinal pressure gradient in transitional regime. The parietal sheet of dye emitted upstream (Fig. 2bdf) reveals stages of transition in succession, i.e. laminar flow on the upstream side, then the appearance of Tollmien Schlichting waves followed by the formation of turbulence spots, then the development and generalization of the turbulence. The same phenomenon can be seen in a longitudinal section visualized by air bubbles (Fig. 2ace) (Werlé, 1988).

**Mixing layer after a flat plate**

In this test, the flow under a plate lying along the centerline of the test section is slowed down by a grid of low permeability (44 %). This results in the two parallel flows at different velocities, coming together downstream of the straight base of the model, with the formation of a mixing layer characterized by a vortex street, with all of the vortices rotating in the same direction and carried off by the current. In turbulent flows ($R_{eL} \approx 2.10^5$, in Fig. 3), the three-dimensional turbulence does not affect the organization of these large two-dimensional structures.

**Wake of a thick plate**

Testing the same plate, but without the grid, we get two parallel flows of equal velocity coming together downstream of the plate. After a small, fixed recirculation area at the base, we then observe the formation of a Karman alternating vortex street. In turbulent flow ($R_{eL} \approx 2.10^5$, Fig. 4), the turbulence of the plate boundary layers does not disturb the emission of the vortex rolls, so the cores -and even the whole field- can be visualized by air bubbles, photographied by a camera that follows the movement of the vortices (Figs. 4c and d).

**Wake of a thin plate**

The confluence of two laminar boundary layers at the trailing edge of a thin plate at zero incidence produces a particularly well-structured alternate vortex street (Fig. 5), but one which disorganizes periodically (Werlé, 1987). The air bubble visualizations, and particularly those taken along moving axes (Fig. 5c and d), confirm this wake structure and accurately position the vortices and measure their cores.

### Jet from a thin plate

Emitted from the trailing edge in the direction of the laminar flow, a turbulent two-dimensional jet gives to a mixing area. The dye identifying the jet visualizes the large structures - sometimes symmetrical and sometimes alternate - forming at the fluctuating boundaries of the jet (Fig. 6a). The air-bubbles show the outer streamlines curving back toward the jet (Fig.6, showing the mean flow observed along fixed axes) and the position of the vortex cores in the opposite direction of rotation from the one recorded in the case of a wake (Fig. 6c and d; observed along mobile axes). Finally, the visualizations obtained with a two-dimensional jet of small elongation confirms that the structures seen do become three-dimensional very fast (Fig. 7h) and also that their direction of rotation reverses when the velocity of the jet falls below $V_0$ (Fig.7i).

### Wake of axisymmetrical models

The separation of the turbulent boundary layer at the flat base of a cylinder set at zero incidence produces an emission of fine toric vortex structures that disorganize rapidly (Fig. 7a) around a recirculation area, the mean extent of which is revealed by the air bubble visualizations in figure 7b and which is the extension of an irregularly fluctuating three-dimensional wake.

On a cylinder with a curved afterbody of moderate slope, the separation is free. Located at the beginning of the retraction in laminar regime (Fig. 7c), it is retarded under the effect of the transition (Fig. 7d) and is absorbed in turbulent regime (Fig. 7e) (Werlé, 1987). In all three cases, the wake is three-dimensional fluctuating.

Finally, on a smooth sphere, the free separation occurs around a parallel located in the vicinity of the maximum cross section in subcritical regime ($R_{eD} = 10^5$ : Fig. 7f) or farther downstream in supercritical ($R_{eD} \geq 3.10^5$ : Fig. 7g). As $R_{eD}$ increases, the diameter of the toric vortices on the separation boundary decreases while their frequency increases; but these vortices rapidly break up downstream into small, isolated elements feeding a fluctuating three-dimensional wake.

### Flow around a high-speed aircraft propeller

One example of organized structures observed in unsteady regime is that of the tip vortices emitted by the blades of a tractor propeller (Werlé, 1989). Around a single propeller (Fig. 8a), these vortices remain organized in their helical path and move along the edge of the contracted outflow from the propeller. In a diameter section visualized by air bubbles (Fig.2b), we can see the trace of their cores, regularly spaced and all rotating in the same direction as they do in an axial jet.

Around a counterrotating propeller (Fig. 8c), the vortices from the first propeller are periodically cut by the blades of the second, which generate their own tip vortices farther out, and with their helical trajectories in the opposite direction. The core are found in another diameter cross section (Fig. 8d).

### Vortex breakdown in the flow and over a delta wing

One example of a three-dimensional vortex structure of unsteady type is that of the breakdown of an organized steady vortex under the effect of an adverse pressure gradient (Werlé, 1971).

In the case of an isolated vortex (Fig.9), this disorganization may be caused by an obstacle and the breakdown observed is of the spiral type, both in laminar and in turbulent flow.

In the case of the vortices from a delta wing, this phenomenon occurs above the upper surface at high angles of attack (Fig. 10) and climbs up to apex as the angle of attack increases. At low Reynolds, we then see a spiral breakdown preceded by a laminar bubble (Fig. 10a, c). The bubble disappears in transitional regime (Fig. 10b and d) or turbulent flow (Fig. 10e) ; but the spiral structure of the phenomenon appears only in the flash photographs (Fig. 10b) because of the longer mean exposure time of normal photographs and because of the diffusion of the dye by the turbulence.

## 4. Conclusion

Despite certain limitation due to the Reynolds number and the resulting qualitative character of the results, the water tunnel visualizations presented here reveal certain turbulence and organized structures mechanisms that occur in boundary layers, separations, wakes and jets.

It is hoped that they will be of some help in understanding these complex phenomena, being studied by so many researchers (see references), and guide, complement or confirm their research, whether they be calculations, tests or measurements.

### References

Werlé, H., 1971, Sur l'éclatement des tourbillons, *ONERA Tech. Rep.*, 175.

Werlé, H., 1974, Le tunnel hydrodynamique au service de la recherche aérospatiale, *ONERA Public. n156.*

Werlé, H., 1987, Flow instability and turbulence (visualizations in the ONERA water tunnels), *ONERA T.P.*, 94.

Werlé, H., 1987, Transition et turbulence, *ONERA Tech. Rep.*, 7.

Werlé, H., 1988, Visualisation des écoulements instationnaires dans les tunnels hydrodynamiques de l'ONERA á l'aide de procédés facilitant leur exploitation quantitative, *ONERA T.P.*, 46.

Werlé, H., 1989, Hydrodynamic visualization of the flow around a high-speed aircraft propeller, *ONERA T.P.*, 108.

Werlé, H. et M. Gallon, 1982, Le nouveau laboratoire de visualisation hydrodynamique de la Direction de l'aérodynamyque, *La Rech. Aérosp.*, 5,p.1-23.

A full bibliography would not be called in the context of this article, but the interested reader may fruitfully consult the references found in the following works, as well as those given by the participants to the International Conference on Turbulence, 1989, Grenoble, France.

W. Debler, 1989, *Experiences in Fluid Mechanics*, U. of Michigan, Ann Arbor (U.S.A.).

R. Granger, ed., 1988, *Experiments in Fluid Mechanics*, Vol. 2, Holt Rinehard and Winston Book Co.

M. Lesieur, 1987, *Turbulence in Fluids*, Martinus Nijhoff Publ., Dordrecht, Netherlands.

T.R. Moulden, W. Frost & A. Garner, 1977, *Handbook of Turbulence*, Vol. 1, Plenum Publ. Corp.

T. Okiishi, D. Young & B. Munson, 1990, *Fundamentals of Fluid Mechanics*, Publ. Jhon Wiley & sons.

L. Thomas, 1989, *Heat transfer*, Publ. Prentice-Hall.

M. Van Dyke, 1982, *An Album of Fluid Motion*, The Parabolic Press-Standford.

*Fig. 1 - General view of the water tunnel TH2 at the ONERA Châtillon center.*

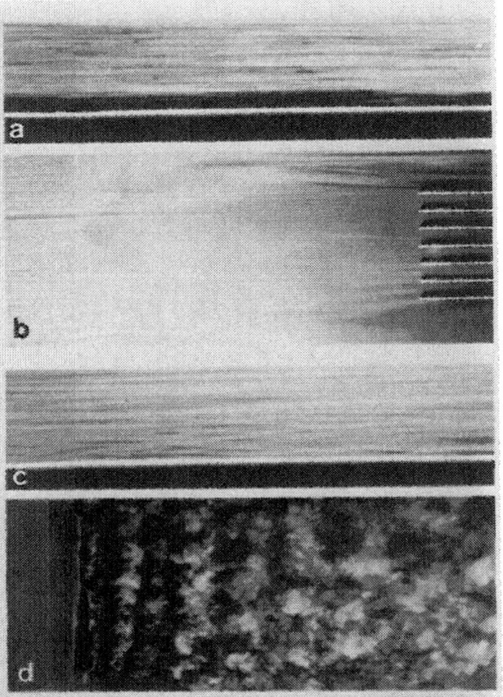


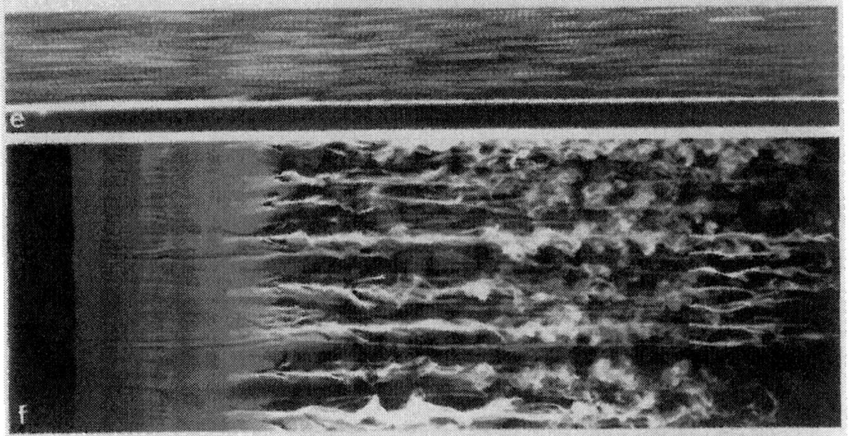


*Fig. 2 - Boundary layer along a flat plate without longitudinal pressure gradient. ab : laminar flow ; cd : turbulent flow. ef : transitional flow ($Re_{0.5L} = 10^5$). ace : visualization by air bubbles. bdf : visualization by dye emissions.*

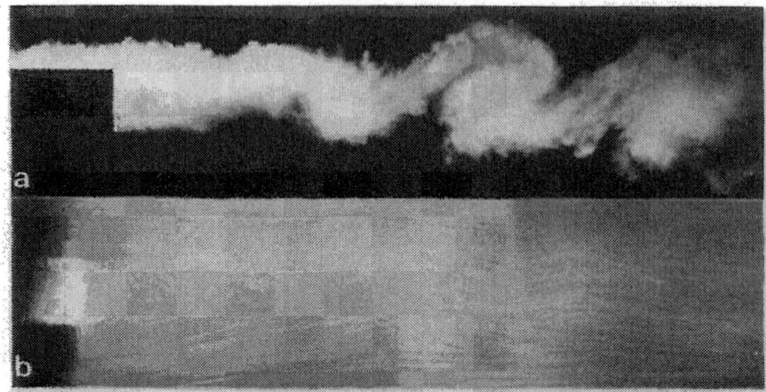


$\overline{Re}_L = 2 \cdot 10^5$

Upper flow : $V = 1.6\ V_0$

Lower flow : $V = 0.6\ V_0$

*Fig. 3 - Turbulent mixing layer issued from a flat plate separating two flows with different velocities.*

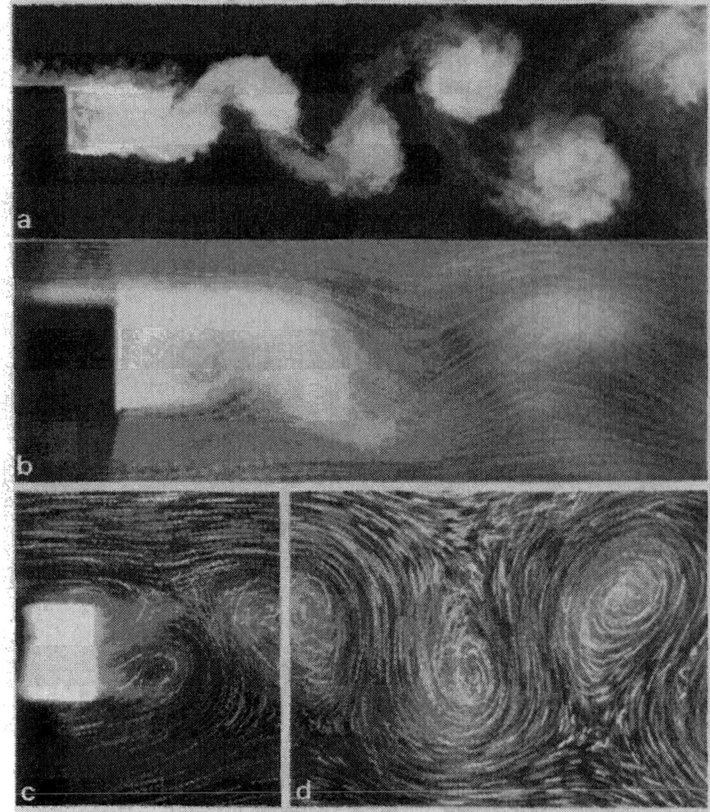


Views a visualization by dye emissions (taken with flash lighting)

Views bcd visualization by air bubbles (flood with chronophotographic effect).

Views cd taken with vortices following

*Fig. 4 - Karman vortices forming in the wake behind a thick plate without incidence.*

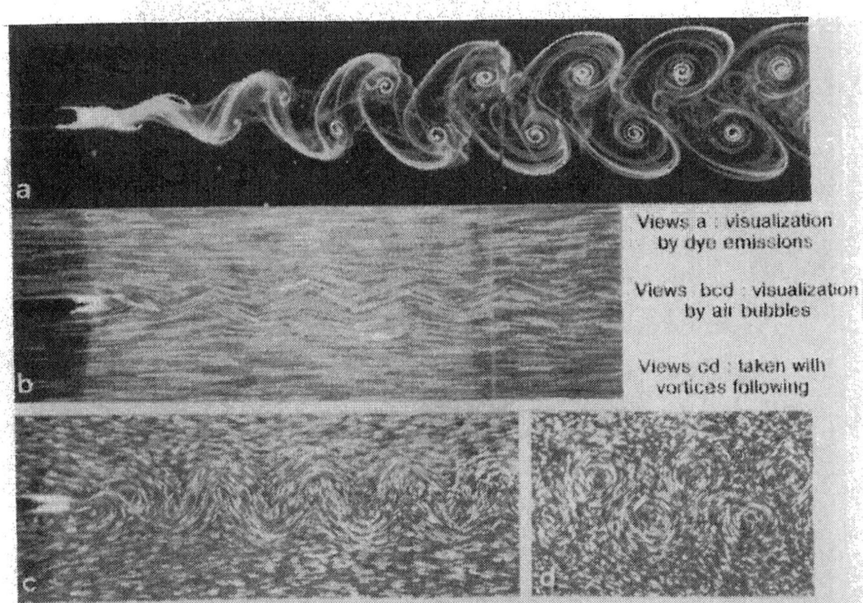


*Fig. 5 - Karman vortices forming in the wake behind a thin plate without incidence. ($Re_L \approx 10^4$).*

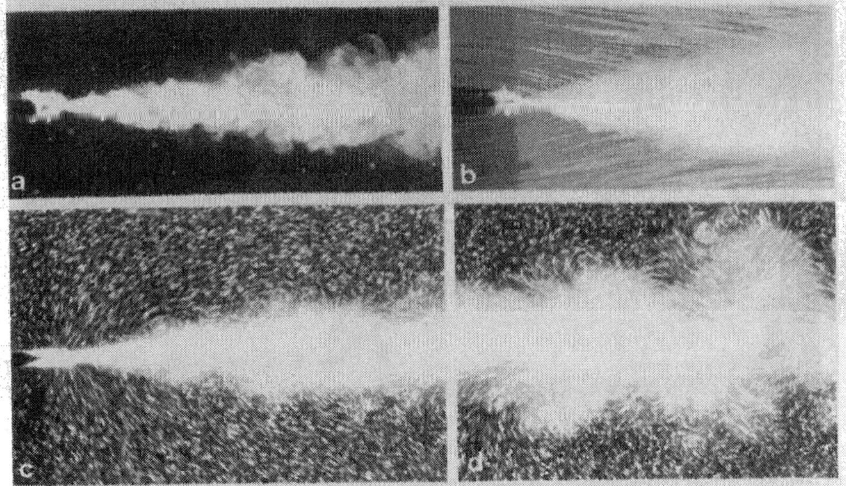


*Fig. 6 - Turbulent jet issued from the trailing edge of a thin plate without incidence. ($Re_L \approx 10^4$ - $\bar{V}_{jet} \approx 10\ V_0$) jet couloured by dye - short time exposure (a), long time exposure (b).*

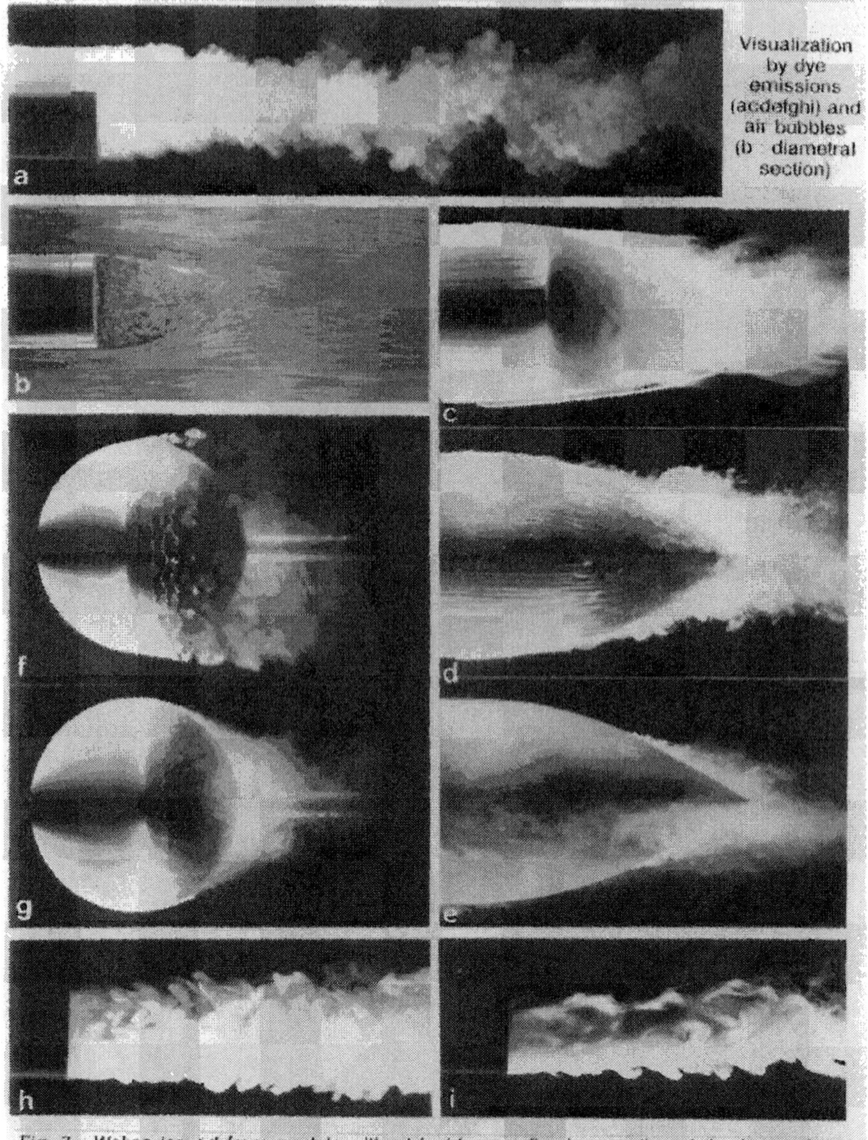


*Fig. 7 - Wakes issued from models without incidence : fixed separation at the base of a cylinder (ab), free separation at the curved afterbody (c : laminar, d : transitional, e : turbulent), free separation around a smooth sphere (f : subcritical, g : critical), jet issued from a flattened cylinder (h $\overline{V}_{jet} = 2 V_0$, i $\overline{V}_{jet} = 0.5 V_0$).*

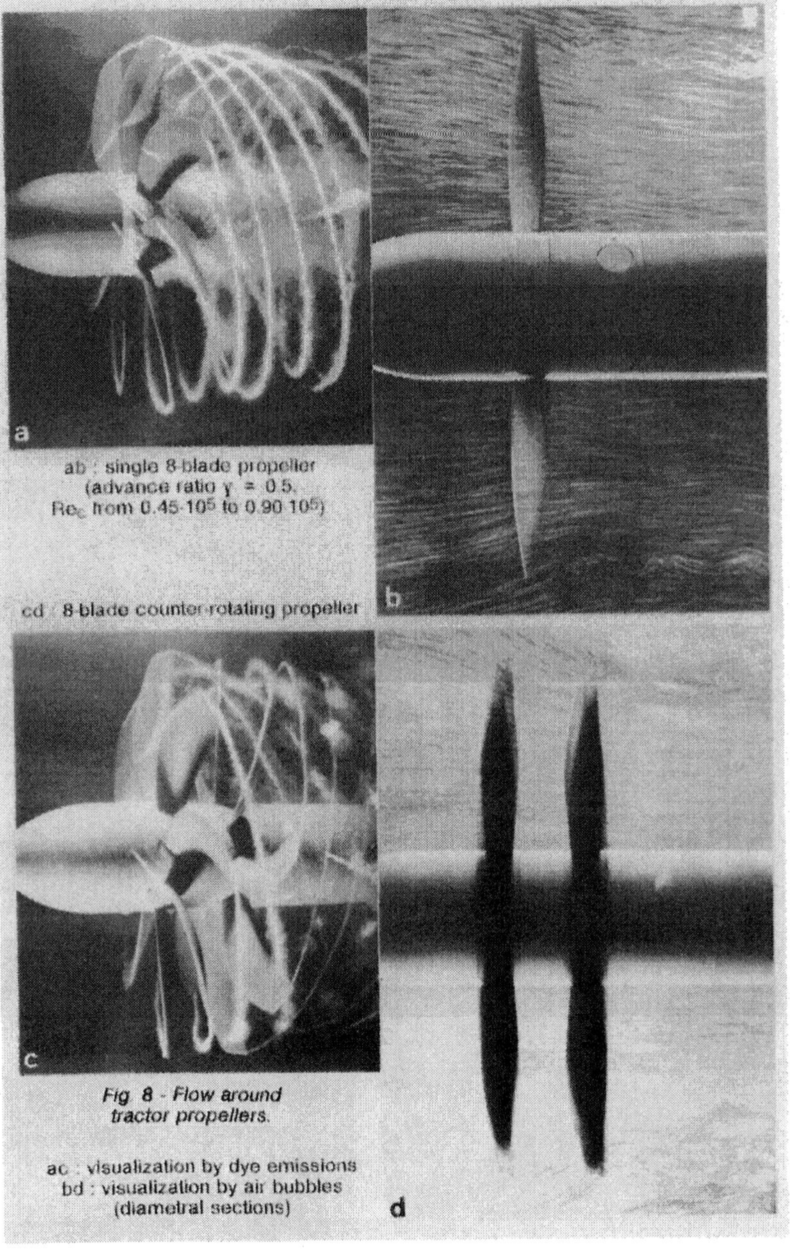


*Fig. 8 - Flow around tractor propellers.*

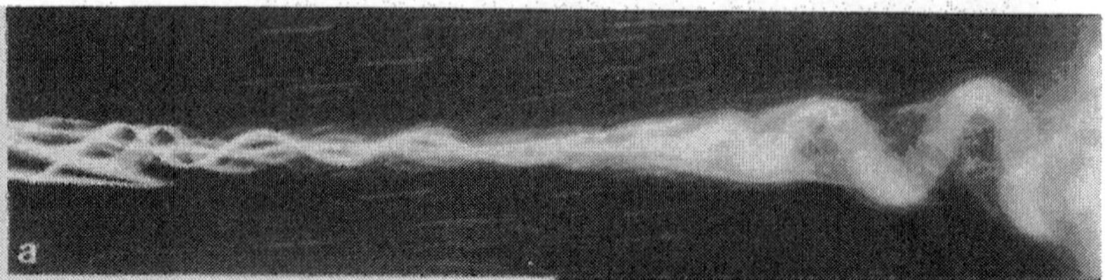


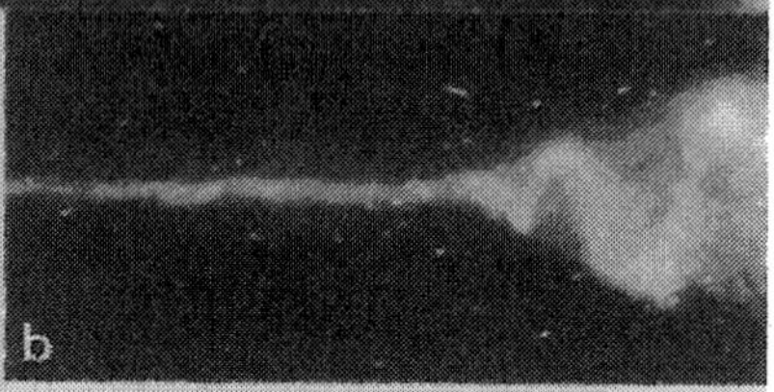


Fig. 9 - Spiral breakdown of an isolated vortex. a : laminar flow ($Re_c = 0.25 \cdot 10^5$). b : turbulent flow ($Re_c = 10^5$). (Artificial breakdown in front of an obstacle).

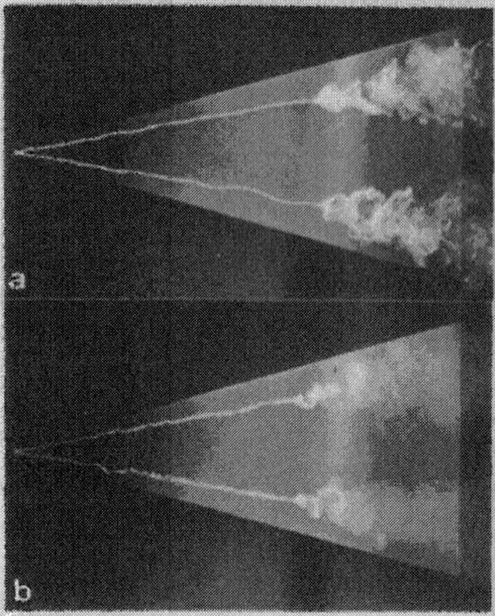


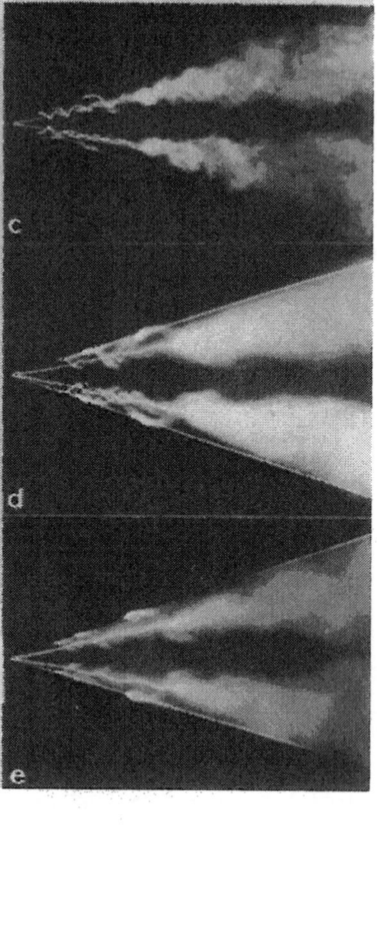


Fig. 10 - Breakdown of the delta wing vortices at high incidence ($\alpha$).
a $\alpha = 30°$ laminar flow ($Re_c = 0.2 \cdot 10^5$) $\alpha = 40°$ c
b $\alpha = 30°$ transitional flow ($Re_c = 10^5$) $\alpha = 40°$ d
turbulent flow ($Re_c = 4 \cdot 10^5$) $\alpha = 40°$ e

ab views taken by flash lighting

Sweep angle $\phi_{LE} = 75°$

# I. Free Shear Flows

# The emergence of characteristic (coherent?) motion in homogeneous turbulent shear flows

D.J. CARRUTHERS, J.C.H. FUNG, J.C.R. HUNT and R.J. PERKINS

*Department of Applied Mathematics and Theoretical Physics*
*University of Cambridge*
*Silver Street, Cambridge CB3 9EW, UK*

*Turbulence and Coherent Structures*
*O. Métais and M. Lesieur, eds.*
*Selected papers from "Turbulence 89: organized structures and turbulence in fluid mechanics"*
*Grenoble, 18-21 September 1989*

## 1. Introduction

In most turbulent flows there exist 'coherent structures' (CS) where the velocity and vorticity have a characteristic structures and where the large scale distributions of velocity and vorticity remain coherent – even as the structures move through and interact with the surrounding fluid. There appear to be two distinct approaches to describing coherent structures which reflect the different flows fields in which they occur, viz:

(a) Dynamical coherent structures are isolated regions where the vorticity is large and has a characteristic form and which have their own dynamical life and could exist in isolation (Hussain 1986). One approach to analyzing the dynamics of CS is to consider how typical forms of these vortex structures interact with each other and with the ambient shear flow. In the absence of a general theory for covering most kinds of vortex interaction it will be necessary to conduct many such computations to obtain general concepts from them (some aspect of these dynamical interaction were reviewed by Hunt 1987a,b).

(b) 'Kinematic' coherent structures are regions of characteristic turbulent motion with similar features and similar spatial distributions in all realisations of a particular turbulent flow. Different definitions have been given for these structures based on different properties such as orthogonality, peak velocity, Reynolds stress, strain rate versus vorticity, pressure, etc. (Lumley 1967, Adrian and Moin 1988, Blackwelder and Kaplan 1976, Wray and Hunt 1990). Such a classification has practical aspects, such as the identification of straining and recirculating regions which are significant for mixing, surface deformation, reaction, etc., as well as dynamical aspects, such as the characterisation of production and dissipation regions for energy and enstrophy.

In this paper Rapid Distortion Theory is used to give some new insights into the spectra of turbulent shear flows and the corresponding physical structure of the 'structures' generated by shear flow which are of type (b). Following Lee *et al.* (1987) random realisations of the distorted velocity field are also used to show some of the characteristic eddy structures including the simulation of the formation of vortex sheets or 'streaks'. It might be thought that in general the use of a linear theory which depends on the initial and boundary conditions would not allow detailed analysis of the structure of turbulence, since turbulent flows become decorrelated over space and time, and are intrinsically nonlinear. However there are certain features of turbulent flow structure which reach a steady state and change only very slowly even when the turbulence is being rapidly distorted so that some linear solutions may have certain properties that are independent of initial and boundary conditions ('eigensolutions') or are weakly dependent after statistical averaging ('statistical eigensolutions'). This provides the motivation for using our present approach in which we will obtain 'statistical eigensolutions'; it is discussed in greater detail by Hunt and Carruthers (1990) (subsequently referred to as HC).

Another theoretical approach to studying the structure of the turbulence is to use linear stability theory (Liu 1989) to calculate the eigenmodes of the mean velocity field (Lessen 1979, Gaster *et al.* 1985). The calculated structures are of type (a) above. In the early stages of shear flow distortions (Ho & Huerre 1984), the phase and amplitude distribution of large scale structures correspond approximately to the large-scale features of the nonlinear 'coherent structure', but the smaller-scale structures are not well modelled by these eigenmodes of the whole flow (Hussain 1986).

Finally, we objectively characterise these and other structures in an ensemble of realisations of the flow field. The flow is divided into eddy, streaming and convergence regions. We have compared direct numerical simulation with a 'kinematic' simulation of random Fourier modes, to show how the shape and distribution of these regions vary with the form of the spectrum and how they change when shear is applied.

## 2. Rapid distortion uniform shear

### 2.1. Analysis

Rapid Distortion Theory (RDT hereafter) is useful for understanding the structure of turbulence subjected to a suddenly imposed mean deformation and gives a good approximation to the most amplified components of turbulence over an arbitrary straining time (i.e. $T_{\mathcal{D}} > T_L$) if the linear strain rate is large compared with the nonlinear strain rate (i.e. $\mathcal{L}^* = (dU_1/dx_2)L/u_0 \gg 1$, where $L$ is the integral length scale and $u_0$ is the r.m.s. velocity), or if other disturbances are much larger than the distortion by turbulence acting on itself. Then the governing equation can be linearized by neglecting the nonlinear (turbulent inertial) terms. In homogeneous turbulence, analytic solutions for the Fourier components of the turbulent velocity can be obtained as functions of total strain (Moffatt 1965, Townsend 1970), and hence statistical quantities may be obtained directly.

Consider a uniform flow $\mathbf{U} = (\alpha x_2, 0, 0)$ rapidly distorting, over a period $t_0 < t < t_1$, a homogeneous turbulent velocity field $\mathbf{u}$ whose initial energy spectrum at $t = t_0$ is $E_0(k)$. Following Townsend (1970), and assuming local homogeneity (over the scale of the large eddies) the Fourier series for the velocity can be defined, i.e.

$$u_i(\chi, t) = \sum a_i(\chi, t) e^{i\chi \cdot \mathbf{x}}, \qquad (2-1)$$

and substituted into the general transport equation for velocity fluctuations to derive the rapid strain equations for the Fourier component amplitude $a_i(\chi, t)$:

$$\frac{\partial a_i(\chi, t)}{\partial t} = -a_j(\chi, t)\frac{\partial U_i}{\partial x_j} + \frac{2\kappa_i \kappa_j}{k^2}\frac{\partial U_j}{\partial x_l} a_l(\chi, t), \qquad (2-2)$$

and the local wavenumber $\chi$, defined by the constraint that wave fronts are conserved, i.e.

$$\frac{d\chi_i}{dt} + \chi_j \frac{\partial U_j}{\partial x_i} = 0, \qquad (2-3)$$

$d/dt$ being evaluated in a frame moving with the local mean flow, and coincident with a fixed frame initially. $\chi$ is related to $\boldsymbol{\kappa}$, the initial wavenumber by

$$\chi = (\kappa_1, \kappa_2 - \beta\kappa_1, \kappa_3) \qquad (2-4)$$

and $\beta = \alpha t$.

The solution (Townsend 1970; HC) to the rapid strain equations may be expressed in terms of initial conditions as

$$a_i(\chi, t) = A_{ij}(\chi, t) a_j(\boldsymbol{\kappa}, 0), \qquad (2-5)$$

i.e. the change in the amplitudes is linear and

$$A_{11}(\chi(\kappa, t), t) = \Bigg\{ \frac{\beta\kappa_1^2}{k_{13}^2}\left(\frac{k^2 - 2\kappa_2^2 + \beta\kappa_1\kappa_2}{\chi^2}\right) - \frac{k^2}{k_{13}^3}\frac{\kappa_3^2}{\kappa_1}\left[\tan^{-1}\left(\frac{\kappa_2}{k_{13}}\right) - \tan^{-1}\left(\frac{\kappa_2 - \beta\kappa_1}{k_{13}}\right)\right]\Bigg\} A_{22}(k, t_0), \qquad (2-6a)$$

$$A_{22}(\chi(\kappa, t), t) = \frac{k^2}{\chi^2} A_{22}(k, t_0), \qquad (2-6b)$$

$$A_{33}(\chi(\kappa, t), t) = \Bigg\{ \frac{\beta\kappa_1\kappa_3}{k_{13}^2}\left(\frac{k^2 - 2\kappa_2^2 + \beta\kappa_1\kappa_2}{\chi^2}\right) + \frac{k^2\kappa_3}{k_{13}^3}\left[\tan^{-1}\left(\frac{\kappa_2}{k_{13}}\right) - \tan^{-1}\left(\frac{\kappa_2 - \beta\kappa_1}{k_{13}}\right)\right]\Bigg\} A_{22}(k, t_0), \qquad (2-6c)$$

where $k$ is the wavenumber at $t = 0$ and $k_{13}^2 = \kappa_1^2 + \kappa_3^2$.

As in most computations of Rapid Distortion theory, it has been assumed that the initial turbulence was isotropic, so that $\Phi_{ij}(\boldsymbol{\kappa}, 0)$ could be expressed (Batchelor 1953) simply and uniquely as

$$\Phi_{ij}(\boldsymbol{\kappa}, 0) = \frac{E(k)}{4\pi k^4}\left(k^2\delta_{ij} - \kappa_i\kappa_j\right), \qquad (2-7)$$

where $E(k)$ is the energy spectrum.

The solutions for the Fourier components of the velocity may be used to obtain spectra. The solution for the form of velocity spectra at times $t > 0$ may be expressed in terms of the initial conditions as

$$\begin{aligned}\Phi_{ij}(\chi, t) &= A_{ip}A_{jq}\Phi_{pq}(\boldsymbol{\kappa}, 0) \\ &= \frac{E(k)}{4\pi k^4}\left[A_{ip}A_{jq}(k^2\delta_{pq} - \kappa_p\kappa_q)\right], \end{aligned}\qquad (2-8)$$

from which relevant statistics may be obtained by integration over all wave-numbers. For example

$$\langle u_i(\mathbf{x})u_j(\mathbf{x}+\mathbf{r})\rangle = \int_{\boldsymbol{\chi}} \Phi_{ij}(\chi, t)e^{i\boldsymbol{\chi}\cdot\mathbf{r}}\, d^3\chi. \qquad (2-9)$$

The distribution of turbulent energy over different scales can be obtained by integrating over the surfaces of spheres of which $dA$ is an element, i.e.

$$E_0(k) = \frac{1}{2}\int_{|\boldsymbol{\kappa}|=k} \Phi(\boldsymbol{\kappa}, 0)\, dA(\boldsymbol{\kappa}), \qquad (2-10)$$

Batchelor (1953).

The behaviour of the turbulence and its development are determined by the strain parameter, $\alpha$, which is the time integral of the mean shear. Notice that the quantity inside the square brackets of $(2-8)$ depends only on the direction of the wavenumber vector and on $\alpha$ and not its magnitude, so that the shear distortion does not discriminate between *high* and *low* wavenumbers.

An important controlling parameter in the study of turbulence structure is the shear-rate parameter

$$\alpha^* = \alpha L/u', \qquad (2-11)$$

which represents the ratio of the turbulence and mean field time-scales.

If $\alpha^* \gg 1$, this means that the mean deformation is very rapid compared to the turbulence time scale. If $\alpha^* \sim 1$, this relaxation time $t_r$ is of order $L/u'$, so the criterion for the importance of nonlinear terms is that $t > t_r \sim L/u'$ ('turnover time' or the Lagrangian time scale). But if $\alpha^*$ is large, the rate at which the anisotropy is growing by linear processes is so much greater than by nonlinear processes that, as Lee *et al.* (1987) have shown, the nonlinear redistributive processes are negligible even when $t > t_r$ (for spectra with a rapid cut-off at large wavenumbers $k$, e.g. the exponential decay).

### 2.2. Spectra

To investigate how the spatial structure of the turbulence changes with shear, 'two-point' spectra can be calculated using (2–6) and (2–10). We are particularly interested in the form of the spectra at high wavenumber to see whether there are any universal features of the small-scale turbulence in shear flow.

It is instructive to express the spectra in terms of the local wavenumber $\chi$. In a shear flow after a finite distortion the spherical surface of $\chi$ equals constant in wavenumber space corresponds to a surface $\chi^2 = \kappa_1^2 + (\kappa_2 - \kappa_1\beta)^2 + \kappa_3^2$, which in $\boldsymbol{\kappa}$-space is a spheroid flattened in the $\kappa_1$ direction and rotated and elongated in the direction $\kappa_2 = \kappa_1\beta$. The form of the components $E_{ii}(\chi)$ of the energy spectrum $E(\chi)$ are calculated (eq. 4.8) by HC.

HC (1990) derived the form of $E(\chi)$ when $\beta \gg 1$ by using asymptotic analysis of the integral (eq. 4.8 - HC) with two forms of the initial spectra

$$E_0(k) = u_0^2 L(kL)^N \exp(-k^2L^2) \qquad (2-12a)$$

and the von Kármán energy spectrum

$$E_0(k) = u_0^2 (kL)^4 / \left[1 + (kL)^2\right]^p, \qquad (2-12b)$$

where $N$ is an integer ($N \geq 0$) and $p > 1$. They found that whatever the initial spectrum, provided it decreases faster than $k^{-2}$, (which is characteristic of a spectrum of vortex sheets,) for large enough strain, over an increasing range of wavenumber, the energy spectrum tends to the limiting form of

$$E(\chi) \propto \widehat{\beta}\chi^{-2}, \qquad (2-13)$$

where $\widehat{\beta} = \beta \int_0^\infty kE_0(k)\, dk$ and $\beta$ is a constant. This results holds for broader classes of spectra than those specified in (2 – 12). However if the initial spectrum decreases slower than $k^{-2}$ as $k$ increases, in high Reynolds number turbulence ($Re_\lambda > 10^4$) the effect of shear is to maintain a $-5/3$ spectrum, which has been well established in many field experiments (e.g. see Monin & Yaglom 1971; Wyngaard & Cote 1972).

Computations of $E_{ii}(\chi)$ for the streamwise and vertical velocity components are displayed in figure 1 (for the simple spectrum $E(k) \propto$

$\exp\left(-k^2L^2\right)$. These show a tendency to the $\chi^{-2}$ spectrum, for the streamwise component and $\chi^{-4}$ for the vertical component when the initial spectrum decays faster than $k^{-2}$. (Non-linear processes tend to make components have some form as the most amplified – as shown in the DNS by Rogallo (1981))

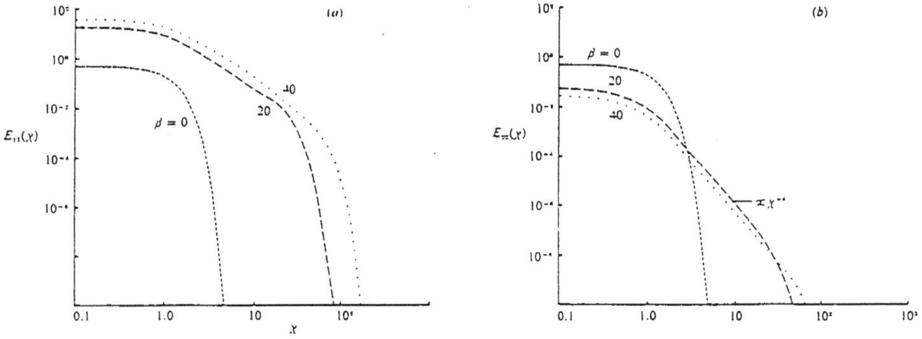


Fig.1. RDT calculation of $E_{ii}(\chi)$ for increasing shear rate $\beta$. The initial spectrum is $E_0(k) = u_0^2 L e^{-k^2L^2}$. (a) $E_{11}(\chi)$; (b) $E_{11}(\chi)$.

The similarity of $k^{-2}$ spectra is associated with the existence of vortex sheets (on a scale much less than $\chi^{-1}$), rather than being characteristic of any cascade process, which is found at much higher Reynolds numbers. In the former case, the 'vortex sheets' are highly localized gradients $\partial u_1/\partial x_2$, $\partial u_1/\partial x_3$, which surround long, narrow regions or 'streaks', where the streamwise fluctuations are positive or negative. These phenomena have been seen in DNS of individual flow realisations (Lee *et al.* 1987).

The results above imply that the essential mechanism responsible for the formation of the streaks in most turbulent shear flows is contained in the ***linear*** theory rather than being the effect of nonlinear interactions. The algebraic form of the spectra is consistent with the existence of discontinuities in velocity of velocity gradients on the scale of $k^{-1}$.

The spectra for the small scales of turbulent flows appear to be highly dependent on the Reynolds number (e.g. $Re_\lambda$) of the turbulence and the on mean velocity distribution. For fully developed turbulence generated by grids in wind tunnels, or obtained by DNS, without mean shear (where $Re_\lambda \leq 300$) the spectra decrease rapidly with $k$, typically $E(k) \propto e^{-k^2L^2}$ at the lowest values of $Re_\lambda$ (Champagne *et al.* 1970; Rogallo 1981), and $E(k) \propto k^{-n_H}$, where $n_H \geq 2$, at the higher values of $Re_\lambda$. However, in shear flows, even at these ranges of $Re_\lambda$, it is quite usual to find that $E_0(k)$ decays algebraically, i.e. $E_0(k) \propto k^{-n}$ when $-5/3 \leq n \leq 2$ – over a significant range. Rogallo (1981) found $n \to 2$ in his computations of turbulence in a homogeneous shear flow which extended to $\beta = 18$ and Rogers & Moin (1987) found that $n \approx 3$ for $\beta = 8$. In Champagne *et al.*'s (1970) wind-tunnel measurements they found that $n = 2$, even for quite small strain.

## 3. Kinematic simulation

In the following section, we show that in purely kinematic simulations which use random Fourier modes constrained to have a given energy spectrum (see Fung *et al.* 1990 and Fung 1990), similar 'streaks' were observed. By following the work of Hunt *et al.* (1988), we objectively characterise these and other structure in an ensemble of realisations of the flow field, and show how the shape and distribution of different regions change when shear is applied.

### 3.1. Methods of simulation

The numerical method of the Kinematic simulation of turbulence (KS) with mean shear distortion is essentially the same as in the case of homogeneous, isotropic turbulence simulation i.e. the technique of KS is to superpose linearly space-time Fourier modes with random coefficients and to allow the large scales to advect the small scales. The simulations are started from an initially-random velocity field with a prescribed Eulerian energy spectrum. The distortion of the velocity field by the mean shear is included by varying the wave-vectors $\boldsymbol{\kappa}$ and the Fourier coefficients $\mathbf{a}_n(\boldsymbol{\kappa})$ and $\mathbf{b}_n(\boldsymbol{\kappa})$ with time according to $(2-5$ and $2-6)$.

### 3.2. Detailed flow structures

In this section we shall investigate the detailed structure of the computed flow field in uniform shear. This will be done by examining contour plots of instantaneous velocity and vorticity fields. Also, to see how the structures changes with different initial spectra.

In order to examine the effect of low shear rate on the turbulence structure, contours of streamwise turbulent velocity $u$ from a homogeneous shear flow simulation with small $\alpha = 20.0$ and at $\beta = 9.4$ are plotted on a horizontal $(x, y)$-plane in figure 2 which indicates that at low shear rate, structures are not highly elongated in the streamwise direction.

In figures 3a and b, similar $u$-contour plots in a homogeneous turbulent shear flow field at high shear rate ($\alpha = 43.2$) are shown at different times $\beta = 4.7$ and 9.4 respectively. At the earlier time ($\beta = 4.7$), figure 3a shows structures somewhat elongated in the streamwise direction. The contour plot at a later time ($\beta = 9.4$ in figure 3b) clearly shows the existence of highly elongated high- and low-speed streak structures alternating in the spanwise direction similar to the streaks observed in the near-wall region of turbulent boundary layers (Kline *et al.* 1967). In contrast, the normal and spanwise velocity components do not exhibit

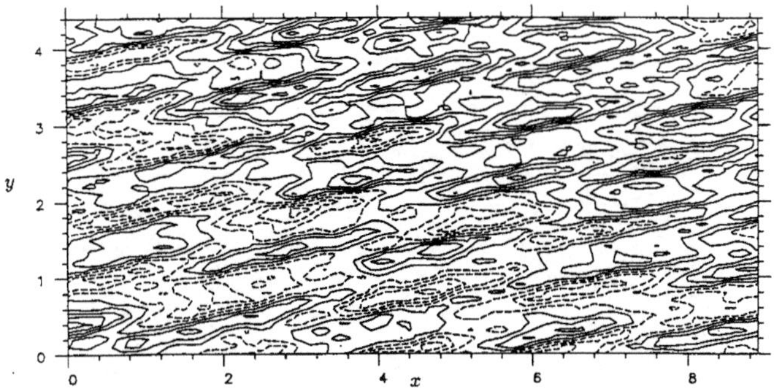


Fig.2. Contours of the streamwise turbulent velocity $u$ for a homogeneous turbulent field with von Kármán energy spectrum at low shear rate: $\alpha = 20.0$ and $\beta = 9.4$. Positive values are contoured by solid lines and negative values are contoured by dashed lines.

streaky structures. This means that, as shear is prolonged, the lateral extent of the streaks is mainly controlled by the effect of shear, and the presence of a solid boundary is not necessary, also that the essential mechanism responsible for the formation of the streaks is contained in the linear theory (Lee *et al.* 1987).

Figure 4a shows the contour plot of spanwise vorticity fluctuation, $\omega_z$ in the same $(x, z)$-plane as in figure 3b.

Streamwise streak pattern of $y$-vorticity can also be seen in Fig. 4b. Conditions are identical to those of Fig. 4a, and careful comparison with that figure shows that the vorticity spots reside in the relatively quiescent corridors between the high velocity streaks.

In order to focus on how the flow changes with different initial spectra, two other forms of initial spectra

$$E_1(k) = A_1 k^2 e^{-A_2 k^2} \quad \text{and} \quad E_2(k) = A_3 k^4 e^{-A_4 k^2} \quad (3-1)$$

have been used for our KS simulations together with RDT, where $A_1$, $A_2$, $A_3$ and $A_4$ are constants. Figures 5 and 6 show the contours of streamwise velocity components (with $\alpha = 40.0$ and $\beta = 23.7$) from the same homogeneous shear flow simulation with the above spectra. One can distinguish several localized regions of very high-speed fluid (large concentration of solid lines) that are located on the high-speed streak. It is apparent that the gross feature of the degree of streaky structures in figures 5 and 6 are higher than that of the von Kármán spectrum simulation. This is consistent with the results of RDT discussed in §2 when $E(k) \propto k^{-2}$ for initial spectrum with decay faster then $k^{-2}$.

Linear rapid distortion theory has shown how a velocity field of random Fourier components undergoing mean shear produces many of the

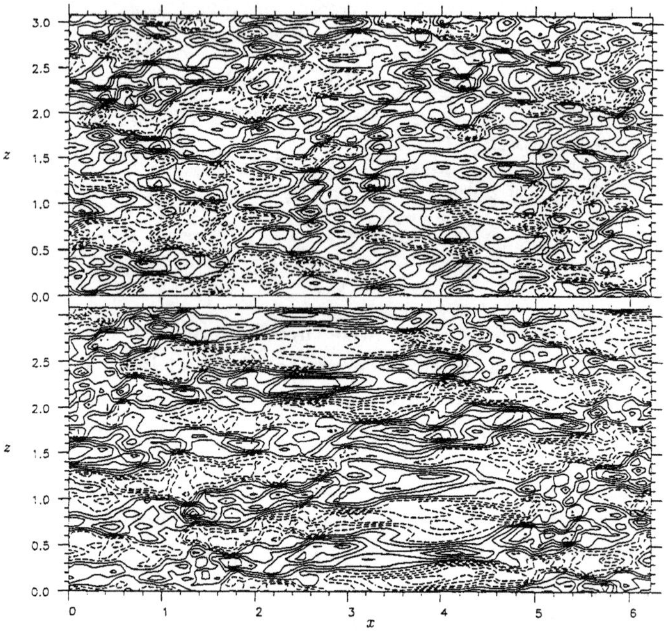


Fig.3. Contours of the streamwise turbulent velocity $u$ for a homogeneous turbulent field with von Kárman energy spectrum at high shear rate: $\alpha = 43.2$ and (a) $\beta = 4.7$; (b) $\beta = 9.4$.

characteristic structures in the flow that are observed in experiments and DNS at moderate Reynolds number. In the remaining part of this section, following the work by Hunt *et al.* (1988) and Wray and Hunt (1990), we objectively characterise the flow structure in terms of the local values of the deformation tensor $u_{i,j}$, specifically the second invariant II$= \dfrac{\partial u_i}{\partial x_j}\dfrac{\partial u_j}{\partial x_i}$ of the velocity field and the pressure $p$. The flow is divided into eddy regions ($E$), corresponding to regions of high swirl where $\mathrm{II} < -\mathrm{II}_0$, $p < -p_0$, convergence regions ($C$), where $\mathrm{II} > \mathrm{II}_0$, $p > p_0$, and streaming regions ($S$) where $-\mathrm{II}_0 < \mathrm{II} < \mathrm{II}_0$ and $u^2 > u_0^2$. Here $\mathrm{II}_0$, $p_0$ and $u_0$ are the r.m.s. values of II, $p$ and $u$ values. Plate 1 shows the velocity vectors superimposed on the pressure field, and there is good agreement between the pressure and the eddy and convergence regions. Also, there is good agreement (see Plate 2) between the observed structure and the regions determined on the zone algorithms, especially remembering that the criteria refer to 3-D space whilst the vectors are plotted in a 2-D plane. By using these criteria, we show how

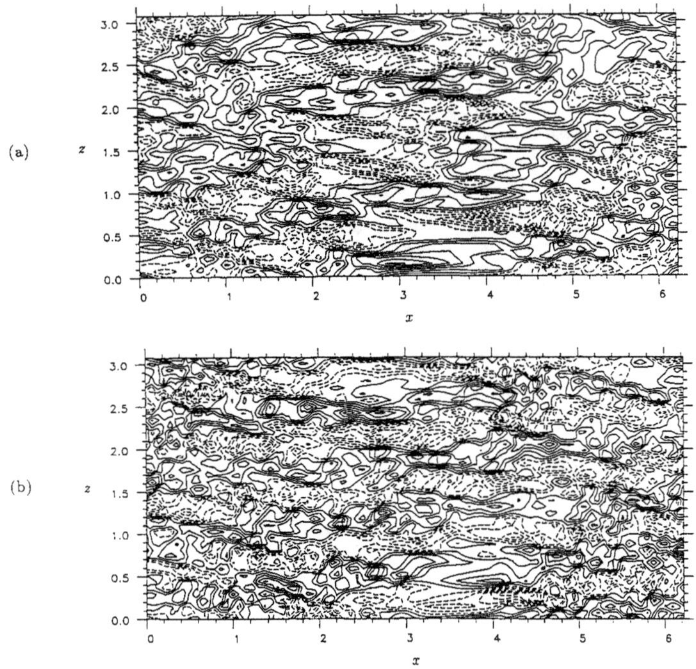


Fig.4. Contours of (a) spanwise vorticity fluctuation $\omega_z$; (b) vertical vorticity fluctuation $\omega_y$ in the $(x, z)$-plane.

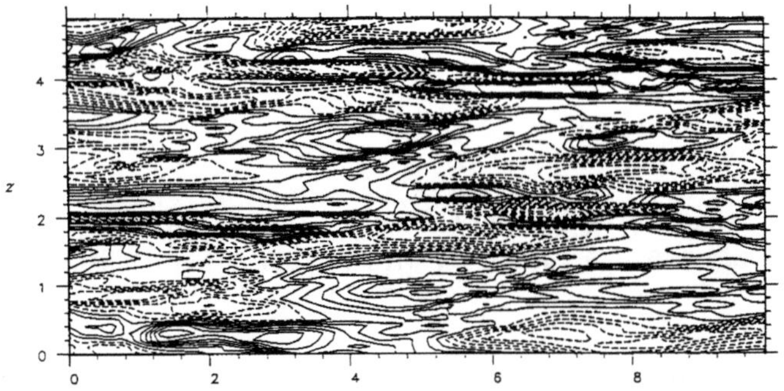


Fig.5. Contours of the streamwise turbulent velocity $u$ for a homogeneous turbulent field with energy spectrum $E_1$: $\alpha = 43.2$ and $\beta = 25.6$.

the shape and distribution of the $(E)$, $(S)$ and $(C)$ regions change when shear is applied.

Figure 7a shows the streaming regions of an initially isotropic velocity field distributed on the surfaces of cuboid. They seem to be connected,

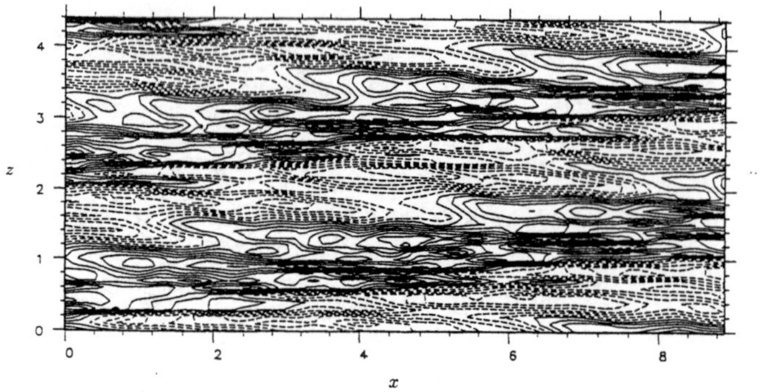


Fig.6. Contours of the streamwise turbulent velocity $u$ for a homogeneous turbulent field with energy spectrum $E_2$: $\alpha = 43.2$ and $\beta = 25.6$.

their shapes somewhat elongated with no preferred directions and they are uniformly distributed. However, after shear is applied at $\beta = 5.0$ (Fig. 7b), the distributions of streaming regions seem to be rather discrete and much more elongated in the direction of shear.

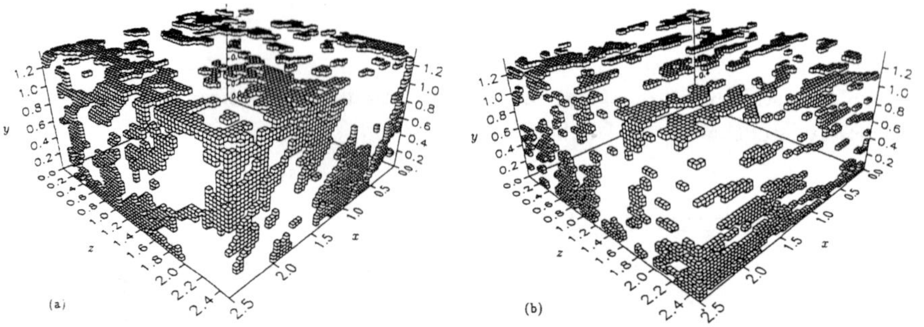


Fig.7. Streaming regions plotted only on the surface of a cuboid (a) initially homogeneous, isotropic turbulence with $\beta = 0.0$; (b) after shear is applied with $\beta = 5.0$.

Figures 8a and 9a show the eddy and convergence regions of an initially isotropic velocity field. It can be seen that both regions are distributed uniformly in space with some of the eddy regions connected and elongated, whilst the convergence regions are of round shapes. After shear is applied at $\beta = 5.0$, figure 8b and 9b show that both regions are elongated in the direction of shear. In order to find out the angle of inclination after distortion, we assume that individual eddy or convergence regions have elliptical shapes, and ellipses have been fitted to these regions, according to the following criteria: (i) equal position of centroid of area; (ii) equal

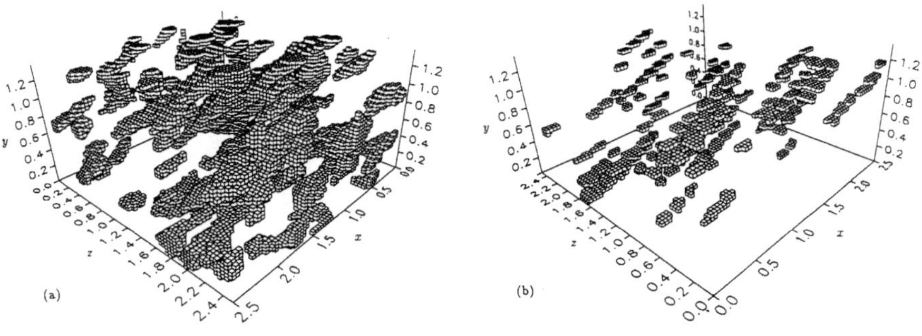


Fig.8. Eddy regions plotted only in the volume of a cuboid.

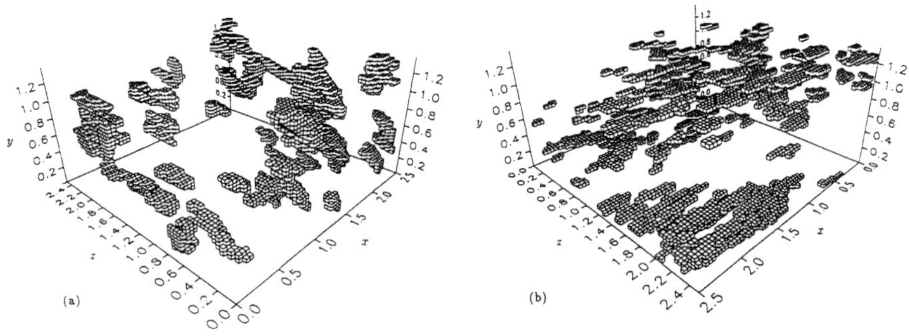


Fig.9. Convergence regions plotted in the volume of a cuboid.

area; (iii) equal 2nd moments of area. The results are shown in figure 10 and the average angle of inclination to the direction of the flow after an average of 10 realisations (of volume $(2L)^3$) is 9° for streaming regions, 12° for eddy regions and 13° for convergence regions. This is surprising since vorticity is at 45° to the flow direction (Rogers and Moin 1987). Obviously, these angles of inclination will be different for different values of $t(DU/dy)$.

## 4. Conclusions

(a) In strong shear flow, the turbulence tends to have a universal spectral form consistent with characteristic local kinematic structure, despite the Reynolds number not being large enough for the existence of a nonlinear cascade. (In fact, since shear changes the spectrum, it enhances the energy cascade processes.)

(b) Topological ('E', 'C' and 'S') zonal analysis is a good indicator of how turbulence changes – quantitative definition of coherent struc-

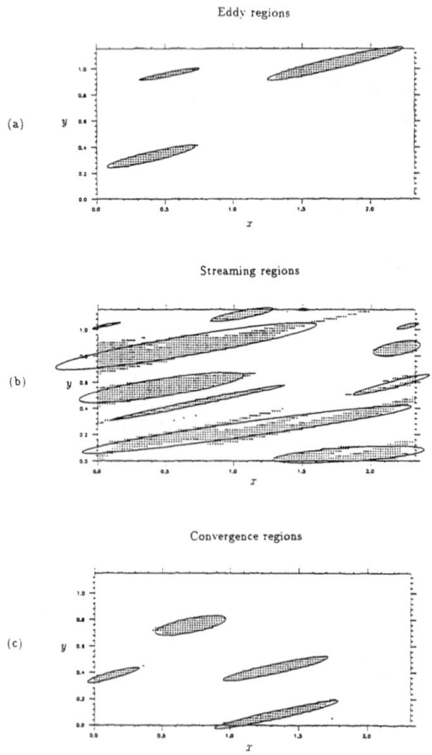


Fig.10. Elliptics have been fitted to (a) eddy regions; (b) streaming regions and (c) convergence regions.

tures.

(c) Linear calculations using Rapid Distortion Theory predict certain features of the eddy structures that are similar to Direct Numerical simulations. The form of these structures is not sensitive to initial anisotropy or spectra. This indicates that there are *common* features in the structure of large eddies in shear flows.

(d) Experiments in shear flows have also demonstrated that there are common features – $k^{-2}$ spectra, 'roller-eddies' (Townsend 1970) and streaks when $t(DU/Dy) \gg 1$.

**Acknowledgements**

We acknowledges helpful insight from many colleagues, especially M. Lee, P. Moin and A. Wray (C.T.R. NASA Ames), F. Hussain (Houston), C. Cambon (Lyon), M. Lesieur (Grenoble). D.J.C. acknowledges financial support from NERC and SERC. J.C.H.F. acknowledges financial support from Peterhouse and the use of computing resources at UKAEA Harwell and R.J.P. was supported by D.T.I. Warren Spring Laboratory.

## References

Adrian, R.J. & Moin, P. 1988 Stochastic estimation of organized turbulent structure: homogeneous shear flow. *J. Fluid Mech.* **190**, 531-559.

Batchelor, G.K. 1953 *The theory of homogeneous turbulence.* Cambridge University Press.

Blackwelder R.F. & Kaplan R.E. 1976 On the wall of the turbulent boundary layer. *J. Fluid Mech.* **76**, 89-112.

Champagne, F.H., Harris, V.G. & Corrsin, S. 1970 Experiment on nearly homogeneous turbulent shear flow. *J. Fluid Mech.* **41**, 81-139.

Fung, J.C.H. 1990 Ph.D dissertation University of Cambridge.

Fung, J.C.H., Hunt, J.C.R., Malik,N.A. and Perkins,R.J. 1990 Kinematic simulation of homogeneous turbulent flows generated by unsteady random Fourier modes. Submitted to *J. Fluid Mech.*

Hunt, J.C.R. 1987a Vorticity and vortex dynamics in complex turbulent flow. Proc. CANCAM. *Trans. Can. Soc. Mech. Engng.* **11**, 21-35.

Hunt, J.C.R. 1987b Coherent structures – comments on mechanisms. NASA Report CTR-S87.

Hunt, J.C.R., Moin, P. & Wray, A.A. 1988 Eddies, stream and convergence zones in turbulent flows. *NASA Report* CTR-S88.

Hunt, J.C.R. & Carruthers, D.J. 1990 Rapid distortion theory and the 'problem' of turbulence. To appear in *J. Fluid Mech.*

Gaster, M, Kit, E. & Wygnanski, I. 1985 Large-scale structures in a forced turbulent mixing layer. *J. Fluid Mech.* **150**, 23-39.

Ho, C.M. & Huerre, P. 1984 Perturbed free shear layers. *Ann. Rev. Fluid Mech.* **16**, 365-424.

Hussain, A.K.M.F. 1986 Coherent structures and turbulence. *J. Fluid Mech.* **173**, 303-356.

Lee, M.J., Kim, J. & Moin, P. 1987 Turbulent structure at high shear rate. In *Sixth Symposium on Turbulence Shear Flows, Toulouse, France* (ed. F. Durst *et al.*), pp. 22.6.1-22.6.6.

Lessen, M. 1979 On the power laws for turbulent jets, wakes and shearing layers and their relationship to the principle of marginal instability. *J. Fluid Mech.* **88**, 535-540.

Liu, J.T.C. 1989 Contributions to the understanding of large-scale coherent structures in developing free turbulent shear flows. *Adv. Appl. Mech.* **26**, 535-540.

Kline, S.J., Reynolds, W.C., Schraub, F.A. & Runstadler, P.W. 1967 The structure of turbulent boundary layers. *J. Fluid Mech.* **30** 741-773.

Lumley, J. 1965 The structure of inhomogeneous turbulent jets. In Atmospheric Turbulence and Radio Waves Propagation: Proc. of intern. Collq., Moscow, (eds. A.N. Yaglom & V.I. Tatarsky), pp. 166-178. Publishing House ⟨⟨NAUKA⟩⟩: Moscow.

Moffatt, H.K. 1965 The interaction of turbulence with strong wind shear. In Atmospheric Turbulence and Radio Waves Propagation: Proc. of intern. Collq., Moscow, (eds. A.N. Yaglom & V.I. Tatarsky), pp. 139-156. Publishing House ⟨⟨NAUKA⟩⟩: Moscow.

Monin, A.S.& Yaglom, A.M. 1971 *Statistical Fluid Mechanics.* Vol. I, MIT Press.

Rogallo, R.S. 1981 Numerical experiments in homogeneous turbulence. *NASA Tech. Memo.* 81315.

Rogers, N.M. & Moin, P. 1987 The structure of the vorticity field in homogeneous turbulent flows. *J. Fluid Mech.* **176**, 33-66.

Townsend, A.A. 1970 Entrainment and the structure of turbulent flow. *J. Fluid Mech.* **41**, 13-46.

Wray, A.A. & Hunt, J.C.R. 1990 Algorithms for classification of turbulent structures. In *Proc. IUTAM Symp. on Topological Fluid Mechanics.* Cambridge University Press.

Wyngaard, J.C. & Cote, O.R. 1972 Modelling buoyancy driven mixed layer. *J. Atmos. Sci.* **33** 1974-1988.

# Generation of coherent structures in free shear layers

Pierre COMTE, Yves FOUILLET, Marc-André GONZE, Marcel LESIEUR, Olivier METAIS and Xavier NORMAND

*Institut de Mécanique de Grenoble*
*BP 53X, 38041 Grenoble-Cedex, France*

*Turbulence and Coherent Structures*
*O. Métais and M. Lesieur, eds.*
*Selected papers from "Turbulence 89: organized structures and turbulence in fluid mechanics"*
*Grenoble, 18-21 September 1989*

## 1. Abstract

Direct numerical simulations are performed for the turbulent mixing layer and the turbulent plane wake. Coherent structures are qualitatively analyzed by means of graphic visualizations. In all cases, transition to turbulence is triggered by random noise superimposed onto a unidirectional basic flow. In the case of three-dimensional incompressible mixing layers, two regimes are found according to whether this noise is more two-dimensional or more three-dimensional. In the case of three-dimensional incompressible plane wakes, we first observe the formation of two staggered mixing-layers of opposite signs. Each of them features spanwise primary vortices connected together by streamwise secondary vortices. The interaction between the two layers then intensifies, until a Kármán street forms. At this moment, closed vortex loops bridging primary vortices of unlike signs are observed. Two-dimensional compressible mixing-layer simulations verify that compressibility effects inhibit the spreading of mixing layers. Eddy-shocklets are found when the convective Mach number reaches 0.8. A two-dimensional simulation of a spatially-growing plane wake is also presented.

## 2. Introduction

Free shear layers, such as mixing layers, jets and wakes, have been extensively investigated over the last two decades. In incompressible situations, the basically two-dimensional feature of the coherent structures involved by these free-shear flows has been established in most cases. In the mixing layer's case, Oster *et al.* (1977) performed correlation measurements in a situation close to that of Brown and Roshko (1974), *viz.*, at low ambient turbulence level. Wygnanski *et al.* (1979) came to similar conclusions in the presence of strong external disturbances. On the other hand, Chandrsuda *et al.* (1978) made plan-view smoke observations of mixing-layer coherent structures and found somewhat three-dimensional features such as 'helical vortex pairings', which limit their spanwise correlation. As reported by Breidenthal (1981), three-dimensionality also appears at smaller scale in the form of a 'wiggle disturbance' of the coherent structures, accompanied by streamwise 'lines' in between (see his Figure 8). These lines have been identified by Bernal and Roshko (1986) with the streamwise vortices previously found by Bradshaw (1966) and Miksad (1972). Bernal and Roshko (1986) also proposed a model for these secondary vortices, made of 'a single vortex line winding back and forth between the high-speed side of a primary vortex and the low-speed side of the following one'. Lasheras and Choi (1988) suggested that the latter was initially located at the stagnation line between the primary vortices, distorted by spanwise instability and stretched along the principal axis of positive strain rate. Metcalfe *et al.* (1987) shown numerically that the development of subharmonics of the most amplified streamwise mode, which induces succesive pairings of the primary vortices, was necessary to maintain the quasi-two-dimensional feature of the flow observed in the experiments. They thus suggested that 'persistent pairings could inhibit three-dimensional instability and, hence, keep the flow predominantly two-dimensional'. Starting with random perturbations, they obtained coherent structures which strongly resemble that observed by Bernal and Roshko (1986), *viz.*, quasi-two-dimensional Kelvin-Helmholtz rollers and streamwise counter-rotating secondary vortices (see their Figures 15 and 21). A similar flow pattern was obtained by Comte *et al.* (1989) by means of large-eddy simulation, using pseudo-spectral numerical methods similar to those employed by Metcalfe *et al.* (1987). We are now performing fully-turbulent simulations at higher resolution in a larger cubic domain, in such a way that four Kelvin-Helmholtz vortices can form and pair. The same numerical methods are then applied to study transition to turbulence of an incompressible plane wake. Preliminary results are presented in section 3.2.

We are also studying the effects of compressibility on the development of coherent structures in free shear flows. Two specific numerical codes have been developed recently by the team *Turbulence Modelling* in Grenoble. Our long-term motivation is to simulate three-dimensional spatially-growing mixing layers jets and wakes in sufficiently long a domain to reach a state of fully-developed turbulence. We are here presenting, in chapter 4, two-dimensional results of temporally-growing mixing-layer and spatially-growing plane wake simulations.

## 3. Three-dimensional incompressible simulations

Using a pseudo-spectral numerical method, the continuity, Navier-Stokes and passive-scalar transport equations are solved in the Fourier space in the form

$$\mathbf{k}.\hat{\mathbf{u}} = 0 \quad , \tag{3-1}$$

$$\frac{\partial \hat{\mathbf{u}}}{\partial t} = \Pi \left\{ F\ [F^{-1}(\hat{\mathbf{u}}) \times F^{-1}(\hat{\omega})] \right\} - \nu\ \mathbf{k}^2\ \hat{\mathbf{u}} \tag{3-2}$$

$$\frac{\partial \hat{\theta}}{\partial t} = -i\mathbf{k}.F\ [F^{-1}(\hat{\theta}).F^{-1}(\hat{\mathbf{u}})] - \kappa\ \mathbf{k}^2\ \hat{\theta} \quad , \tag{3-3}$$

where $\hat{\omega} = i\,\mathbf{k} \times \hat{\mathbf{u}}$ is the vorticity in the Fourier space ($i^2 = -1$), $F$ is the discrete Fourier transform operator, $\Pi$ is the projection on the plane normal to the wave vector $\mathbf{k}$, and $k = \sqrt{\mathbf{k}^2}$.

In all cases, periodic boundary conditions are applied in the streamwise $x$ and spanwise $z$ directions. The initial conditions result from the superposition of a small-amplitude random perturbation onto a one-directional basic profile, both for the velocity and the passive scalar fields. In the mixing layer's case, this basic velocity profile is defined by

$$u_0(y) = U \tanh\left(2y/\delta_i\right) \quad , \qquad (3-4)$$

$\delta_i$ being the initial vorticity thickness. The passive-scalar velocity profile is also proportional to $\tanh\,(2y/\delta_i)$. We then use free-slip boundary conditions in the transverse $y$ direction, by means of sine and cosine expansions.

In the plane wake's case, the initial basic profile is

$$u_0(y) = U \exp\left[-\ln 2\,(y/\delta_i)^2\right] \quad . \qquad (3-5)$$

In this case, $\delta_i$ corresponds to the half-velocity thickness ($\bar{u}(\delta_i) = U/2$), and the boundary conditions are periodic in all directions.

In all cases, the computational domain is cubic, with a side $L$ chosen equal to a multiple $m_i$ of the most amplified streamwise wavelength $\lambda_a$ predicted by the inviscid linear-stability theory, *viz.*, 7.07 $\delta_i$ and 7.96 $\delta_i$ for the mixing layer (Michalke 1964) and the plane wake (Sato and Kuriki 1961) respectively.

### 3.1. Mixing-layer simulations

We are studying the mixing layer's sensitivity to initial conditions by repeating the same simulation, but with initial perturbations of different spectral composition: in practice, the initial velocity and passive scalar fluctuations result from the superposition of two random perturbations whose spectra are broad-banded. The first perturbation is three-dimensional, of kinetic energy $\varepsilon_{3D}\, U^2$. The other one is two-dimensional, of energy $\varepsilon_{2D}\, U^2$.

All the mixing-layer simulations presented hereafter are performed with an initial Reynolds number $R = U\,\delta_i/\nu = 100$, with $m_i = 4$ and at a resolution of $64^3$ Fourier modes, except for the last run (cf. Plate 3), for which the resolution is of $128^3$. We present some spanwise $\omega_z$ and streamwise $\omega_x$ vorticity component iso-surfaces, the passive-scalar interface and some vortex lines at the same instant $t = 30\,\delta_i/U$.

The first simulation, whose pictures will not be presented here, is performed with $\varepsilon_{2D} = 10^{-4}$ and $\varepsilon_{3D} = 10^{-6}$. It evolves in an almost two-dimensional manner until the end of the calculation, at $t = 50\ \delta_i/U$, with formation of $m_i = 4$ Kelvin-Helmholtz vortices at about 15 $\delta_i/U$, which undergo a first pairing completed at 30 $\delta_i/U$. These values are in agreement with earlier two-dimensional simulations (Lesieur *et al.* 1988), where the eddies formed, on average, at 17 $\delta_i/U$ and terminated their first pairing at 35 $\delta_i/U$.

The second simulation, with $\varepsilon_{2D} = 10^{-4}$ and $\varepsilon_{3D} = 10^{-5}$, still involves four Kelvin-Helmholtz vortices which form at about 15 $\delta_i/U$ and merge at about 30 $\delta_i/U$. Plate 1 shows the two rollers resulting from this pairing; they are aligned along $z$ and slightly oscillate in phase, with a period equal to the domain's size $L = 4\ \lambda_a$. The vortex lines in between also oscillate with the same period, but 180° out of phase. Stretched along the braids by the strain field induced by the two rollers, these lines concentrate into vortex tubes which resemble the hairpin-like vortices found by Bernal and Roshko (1986) in a fully-developed turbulence situation. As proposed by Lasheras and Choi (1988), they might result from the growth of a spanwise oscillation about the stagnation line between the rollers. The undulation of the latter 180° out of phase might thus be due to the hairpin-like vortices which wrap around them, in contrast with the scenario suggested in Comte *et al.* (1989) who interpret the result of a simulation performed in a narrower domain, with $m_i = 2$ and $\varepsilon_{2D} = 0$. On the other hand, the in-phase snaking of the rollers reminds one of the translative instability investigated by Pierrehumbert and Widnall (1982). However, the spanwise wavelength 4 $\lambda_a$ found in the present calculation is too large to correspond to a strongly amplified mode of translative instability: indeed, the growth rate of the translative instability is maximal for spanwise wavenumbers larger than $2\pi/4\ \lambda_a$ and rapidly goes to zero when the wavenumber decreases [see Figure 7 of Pierrehumbert and Widnall (1982)].

The next simulation, performed with $\varepsilon_{3D} = 10^{-4}$ and $\varepsilon_{2D} = 0$, features at $t = 30\ \delta_i/U$ a vortex lattice of streamwise period 2 $\lambda_a$ and of spanwise period $L = 4\ \lambda_a$, as above (see Plate 2). This flow pattern gives the passive-scalar interface a streamwise-staggered $\Lambda$-shaped structure. By repeating the simulation at a resolution of $128^3$ Fourier modes with the same key-parameters, a similar vortex lattice is obtained, with the same periodicity. It can be conjectured from Plate 3 that this lattice results from helical pairing of four fundamental Kelvin-Helmholtz rollers. Indeed, the flow pattern somewhat resembles those observed experimentally in Chandrsuda *et al.* (1978) (see in particular their Figure 3). Moreover, cuts at constant $z$ of the passive-scalar interface where these helical pairings occur [see Plate 3 (c)] remind of pairings observed in two-dimensional simulations [see for example Lesieur *et al.* (1988)].

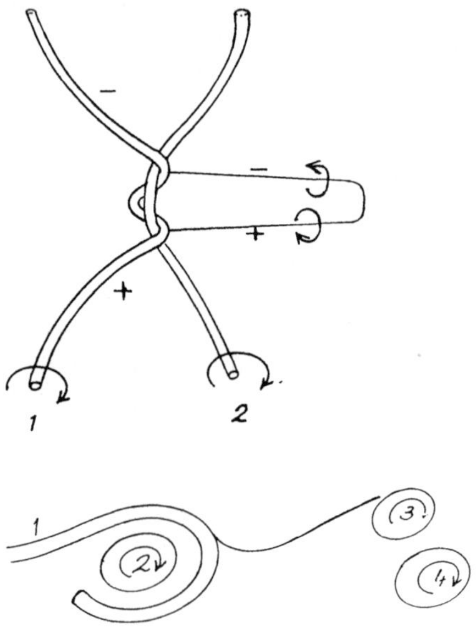


Fig.1. Sketch of the vortex pattern found in the $128^3$ simulation (see Plates 3 and 7 in colour): plan view and side view. The signs + and − near the vortex (1) and the hairpin vortex correspond to the sign of the streamwise vorticity component they carry locally.

In addition to the Kelvin-Helmholtz rollers which form the vortex lattice, Plate 3 (d) and (e) also show a single streamwise-stretched hairpin vortex of small section. It has to be noticed that one of its 'legs' carries more vorticity than the other one. If one numbers, as in Figure 1, the rollers from the left, it seems that this hairpin vortex results from one helical pairing of (1) and (2). Moreover, Plate 7 (b) [the colour counterpart of Plate 3 (d)] shows the sign of the streamwise vorticity component locally carried by each vortex[1]. One can thus check that the hairpin vortex rotates in the same direction as the primary vortices. Consequently, it is tempting to say that it is formed of vortex lines contained initially in (1) and streched towards (3) and (4) along the principal direction of positive strain rate, instead of being trapped by (2); the rollers' section being rather large, it seems reasonable to assume that, helped by self-induction effects, some vortex lines among the most external of (1) manage to escape from the attraction of (2).

An explanation for 'helical pairings' in terms of instability is proposed by Pierrehumbert and Widnall (1982), who noticed that the simulta-

[1] these signs are plotted in Figure 1

neous growth of a spanwise instability together with the subharmonic $k_x = 2\ \pi/2\ \lambda_a$ tends to make the rollers, which are $\lambda_a$ apart, undulate 180° out of phase one with respect to another. This tends to bring locally adjacent rollers closer together, which favours their pairing. In Pierrehumbert and Widnall (1982), the growth-rate corresponding to such pairings is maximal for $\beta = 0$, *viz.* for the purely two-dimensional case, and decreases when the spanwise wavenumber $\beta$ (or $k_z$) increases (see their Figure 5). This could be the reason why a spanwise period of 4 $\lambda_a$ is found in the present calculations, since it corresponds to the smallest non-zero $\beta$ which matches a Fourier mode involved in the simulation.

### 3.2. Plane-wake simulations

A plane wake of Gaussian initial velocity profile is simulated at a resolution of $48^3$ Fourier modes, with $m_i = 2$ and at an initial Reynolds number $R = U\ \delta_i/\nu = 160$.

The first simulation, whose figures will not be presented here, starts from purely two-dimensional initial conditions ($\varepsilon_{2D} = 2.\ 10^{-3}$ ; $\varepsilon_{3D} = 0$). A Kármán street forms at about 46 $\delta_i/U$ and remains two-dimensional until the end of the calculation.

The simulation is then resumed with $\varepsilon_{2D} = 0$ and $\varepsilon_{3D} = 2.\ 10^{-3}$. Plate 4 (a) shows, at $t = 46\ \delta_i/U$, two staggered rows of opposite-sign vortices. Each row closely resembles a three-dimensional mixing layer and features spanwise in-phase bulging rollers and also streamwise vortices connecting these rollers. At $t = 66\ \delta_i/U$, vortex-line plots [Plate 4 (e) and (f)], to be correlated with iso-vorticity surfaces [Plate 4 (b), (c) and (d)] show hairpin-like vortices together with clusters of interconnected vortex loops, as observed in many numerical simulations and experiments [see for example Chen *et al.* (1989), Lasheras and Meiburg (1990), Breidenthal (1980), Perry *et al.* (1980), Antonia *et al.* (1987)].

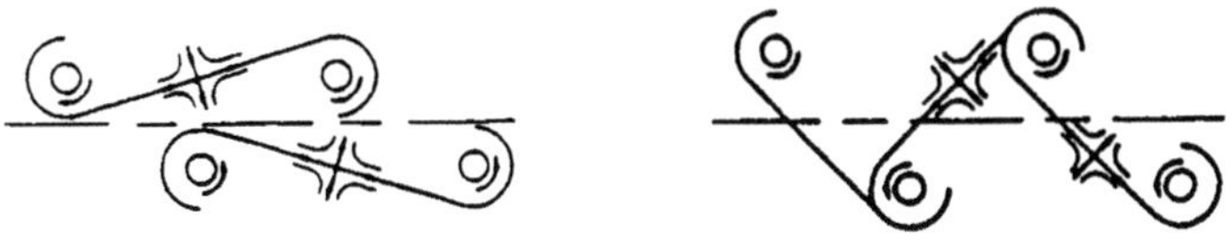

Fig.2. Sketch of the vortex patterns observed at $t = 46\ \delta_i/U$ (on the left) and $t = 66\ \delta_i/U$ (on the right) respectively.

The topological change which occured between $t = 46$ and $t = 66\ \delta_i/U$ have also been observed by Chen *et al.* (1989) in a compressible case (see their Figure 5-14).

## 4. Two-dimensional compressible simulations

Two finite-difference codes solving the complete compressible Navier-Stokes and energy equations for an idéal gas have been developped succesively in the team *Turbulence Modelling*, by Xavier Normand. We present first a temporally-growing mixing-layer simulation involving eddy shocklets, obtained with the most recent code. A spatially-growing wake simulation obtained with the other code is then presented.

### 4.1. Temporally-growing mixing layer

Let us now describe the most recent code. Although only two-dimensional results are presented here, it is a three-dimensional code. The variables are $\rho$, $p$, $\underline{u}=(u,v,w)$, $\mu$, $\varepsilon$, $e$, the density, static pressure, velocity vector, dynamic viscosity, internal energy per unit volume and total energy per unit mass respectively. The equations to be solved are in the following conservative form

$$\frac{\partial U}{\partial t} = \frac{\partial F}{\partial x} + \frac{\partial G}{\partial y} + \frac{\partial H}{\partial z} \qquad (4-1)$$

$$\text{with : } U = \begin{vmatrix} \rho \\ \rho u \\ \rho v \\ \rho w \\ e \end{vmatrix}$$

$$F = \begin{vmatrix} -\rho u \\ -\rho u^2 - \dfrac{1}{\gamma M^2} p + \dfrac{1}{Re}\tau_{xx} \\ -\rho u v + \dfrac{1}{Re}\tau_{xy} \\ -\rho u w + \dfrac{1}{Re}\tau_{xz} \\ -u(e+p) + \dfrac{\gamma M^2}{Re}(u\tau_{xx} + v\tau_{xy} + w\tau_{xz}) + \dfrac{\gamma\mu}{PrRe}\dfrac{\partial \varepsilon}{\partial x} \end{vmatrix},$$

$$G = \begin{vmatrix} -\rho v \\ -\rho u v + \dfrac{1}{Re}\tau_{xy} \\ -\rho v^2 - \dfrac{1}{\gamma M^2} p + \dfrac{1}{Re}\tau_{yy} \\ -\rho v w + \dfrac{1}{Re}\tau_{yz} \\ -v(e+p) + \dfrac{\gamma M^2}{Re}(u\tau_{xy} + v\tau_{yy} + w\tau_{yz}) + \dfrac{\gamma\mu}{PrRe}\dfrac{\partial \varepsilon}{\partial y} \end{vmatrix},$$

$$H = \begin{vmatrix} -\rho u \\ -\rho u w + \frac{1}{Re}\tau_{xz} \\ -\rho v w + \frac{1}{Re}\tau_{yz} \\ -\rho w^2 - \frac{1}{\gamma M^2}p + \frac{1}{Re}\tau_{zz} \\ -w(e+p) + \frac{\gamma M^2}{Re}(u\tau_{xz} + v\tau_{yz} + w\tau_{zz}) + \frac{\gamma\mu}{PrRe}\frac{\partial \varepsilon}{\partial z} \end{vmatrix},$$

where $\underline{\underline{\tau}}$ is the stress tensor with components:

$$\tau_{xx} = \frac{2}{3}\mu\left(2\frac{\partial u}{\partial x} - \frac{\partial v}{\partial y} - \frac{\partial w}{\partial z}\right)$$

$$\tau_{yy} = \frac{2}{3}\mu\left(-\frac{\partial u}{\partial x} + 2\frac{\partial v}{\partial y} - \frac{\partial w}{\partial z}\right)$$

$$\tau_{zz} = \frac{2}{3}\mu\left(-\frac{\partial u}{\partial x} - \frac{\partial v}{\partial y} + 2\frac{\partial w}{\partial z}\right)$$

$$\tau_{xy} = \mu\left(\frac{\partial u}{\partial y} + \frac{\partial v}{\partial x}\right)$$

$$\tau_{xz} = \mu\left(\frac{\partial u}{\partial z} + \frac{\partial w}{\partial x}\right)$$

$$\tau_{yz} = \mu\left(\frac{\partial v}{\partial z} + \frac{\partial w}{\partial y}\right)$$

The viscosity $\mu$ satisfies Sutherland's law

$$\mu = T^{3/2}\,\frac{1 + \frac{S_1}{T_0}}{T + \frac{S_1}{T_0}}$$

where $T$ is the temperature non-dimensioned by a reference temperature $T_0$ ($T = p/\rho$). In the case of air, $S_1 = 110\ K$.

In two dimensions, $\partial/\partial z = 0$ , which cancels $H$, the fourth component of $U$, $F$ and $G$, and also simplifies the expression for $\underline{\underline{\tau}}$.

The method of solution is derived from Bayliss *et al.* (1985). It involves the splitting of (4-1) into one-directional operators, each of them consisting of a predictor-corrector Mc-Cormack scheme which is second-order accurate in time and fourth-order accurate in space. The time-marching proceeds the following way

$$U^{n+2} = L_x\ L_y\ L_z\ L_z\ L_y\ L_x\ \ U^n,$$

the streamwise operator $L_x$ solving the equation

$$\frac{\partial V}{\partial t} = \frac{\partial F}{\partial x}$$

in the form $V^{n+1} = L_x V^n$ with

$$\bar{V}_i = V_i^n + \frac{\Delta t}{6\Delta x}\left[7(F_{i+1}^n - F_i^n) - (F_{i+2}^n - F_{i+1}^n)\right] \qquad (4-2)$$

$$V_i^{n+1} = \frac{1}{2}\left[V_i^n + \bar{V}_i + \frac{\Delta t}{6\Delta x}\left[7(F_i^n - F_{i-1}^n) - (F_{i-1}^n - F_{i-2}^n)\right]\right] \quad . \qquad (4-3)$$

The meshing is rectangular, constant in $x$ and stretched in the $y$-direction with the aid of the co-ordinate transformation

$$y = \frac{L_y}{2}\,\frac{(C_2-1)^{C_1}\,\eta}{(C_2-\eta^2)^{C_1}} \ , \quad \eta \in [-1,1] \quad ,$$

where $C_2$ controls the amount and $C_1$ the rate of stretching. The smaller the quantity $(C_2 - 1)$, the stronger the stretching.

The boundary conditions are periodic in the streamwise $x$ direction. For the other boundaries, we apply non-reflective conditions which consists, after neglecting viscous effects at the boundary, in solving a characteristic propagation equation $\partial k/\partial t + \lambda\ \partial k/\partial x = 0$. It is found that $\lambda = v - c$, $c$, or $v + c$. Positive values of $\lambda$ correspond to outgoing characteristics which are extrapolated from the values known within the computational domain. The incoming characteristics are calculated from the values at the boundary taken from the previous timestep.

A mixing layer between two streams of same temperature $T$, one of them being of velocity $U_1 = 2U$ and the other one being at rest ($U_2 = 0$) is simulated starting from a basic velocity profile

$$u_0(y) = U\ [1 + \tanh\,(2y/\delta_i)] \quad ,$$

onto which a small-amplitude random perturbation of wide spectrum is superimposed. The basic temperature profile

$$T_0(y) = T\left[1 + \frac{\gamma-1}{2}\ M_1^2\ \frac{1}{4}\left[1 - \tanh^2\left(\frac{2y}{\delta_i}\right)\right]\right]$$

is obtained from $u_0(y)$ by means of Crocco-Busemann's law.

The convective Mach number $M_c$ is here equal to half the Mach number $M_1$ of the stream at velocity $U_1$.

The effects of compressibility upon the development of the mixing layer have been studied in detail for several Mach numbers and temperature ratios. The results will not be presented here but it can be said that they are globally in good agreement with the results of the spatially-growing simulations performed by Lele (1989). Our aim is here to present something which somewhat expands upon the latter and brings an immediate validation for the code we are using: the simulation presented in Plate 5 corresponds to $M_c = 0.8$, as in Lele (1989), in a case eddy-shocklets were found. The resolution here is of $400 \times 100$ grid-points with a domain of size $L_x = 45\ \delta_i$ and $L_y = 15\ \delta_i$. Thus we have $L_x = 4\ \lambda_a(0.8)$, $\lambda_a(M_c)$ being the most amplified wavelength predicted by Blumen's (1970) inviscid linear stability analysis.

In order to obtain eddy-shocklets, Lele had to force the most amplified mode and its subharmonics at a high amplitude. Here, at a higher Reynolds number ($R = \rho_1 U_1 \delta_i / \mu = 5000$), there is no need for any deterministic forcing to obtain such shocklets, which appear on the Kelvin-Helmholtz vortices while they merge. The numerical code seems to behave in a satisfactory way since the steep pressure jumps due to the shocklets are solved without any noticeable numerical oscillation. However, it can be seen in Plate 5 that these shocklets induce small-amplitude vorticity oscillations.

### 4.2. Spatially-growing plane wake

The simulation presented hereafter corresponds to the final stage of development of the first compressible code written in the team *Turbulence Modelling* in Grenoble. This code, now made obsolete by the code described before (which would be able to perform the same simulation at a lower cost), solves the complete Navier-Stokes and energy equations with a constant viscosity in a non-conservative form, by means of fully-explicit finite-difference centered scheme. The reader is referred to Normand (1989) for description in details of the numerical method.

The basic velocity profile

$$u_0(y) = U_\infty - u_d \exp\left[-\ln 2\ (y/\delta_i)^2\right]$$

and the associated Crocco-Busemann basic temperature profile

$$T_0(y) = T_\infty \left[1 + \frac{\gamma - 1}{2} M_\infty^2 \left(1 - \frac{u_0^2(y)}{U_\infty^2}\right)\right]$$

are imposed every timestep at the upstream boundary $x = 0$ of the computational domain, together with small-amplitude time-dependent random noise. An open-boundary condition based upon the same characteristic expansion as in the code described above is applied at the downstream boundary. Free-slip boundary conditions are applied in the

$y$-direction. Plate 6 shows the vorticity and the temperature fields at the same moment, after a steady regime is obtained. The resolution is of $400 \times 100$, with an external Mach number $M_\infty = U_\infty/c_\infty = 2$ and a centerline Mach number of 0.47. As in the simulations performed by Chen *et al.* (1989), the velocity ratio $u_d/U_\infty$ is equal to 0.692 to meet Sato and Kuriki's (1961) experiments and stability calculations. The upstream Reynolds number $R = u_d\delta_i/\nu$, based upon the velocity deficit, is of 1200. It is tempting to interpret the basic flow as a double mixing layer between the external flow $(U_\infty, T\infty)$ and a *centerline flow* $[u_o(y=0) = U_\infty - u_d\ ,\ T_0(y=0)]$ and define a convective Mach number

$$M_c = \frac{U_\infty - u_o(y=0)}{c_\infty + c\ (y=0)} \qquad \text{with } c\ (y) = \sqrt{\gamma\ \Re\ T_0(y)}$$

as in the mixing layer's case. A value of 0.6 is found in the present case, but it is not clear that the arguments developed in the mixing layer case by Bogdanoff (1983) to introduce the convective Mach number as the only parameter accounting for compressibility effects (at least for $M_c \leq 0.6$) also apply in the wake case. Anyway, since the velocity deficit $u_d$ goes to 0 when $x$ goes to infinity, the value $M_c = 0.6$ suggests that the present compressible wake is not strongly affected by compressibility. This is partly confirmed by the linear-stability analysis performed by Chen *et al.* (1989): in two dimensions this analysis shows that, in the present case, the growth rate of the most amplified mode, which is antisymetric (sinuous), is lower than in the incompressible case by 30%, while its wavenumber is increased by 13%. The growth rate of the symmetric (varicose) mode is more affected by compressibility: its maximal growth rate is 10 times smaller than the maximal growth rate of the sinuous mode, whereas it is only 3 times smaller than in the incompressible case.

## 5. Summary and discussion

Transition to turbulence of different free-shear flows is obtained by means of small-amplitude broad-banded random perturbations.

In the case of the incompressible mixing layer, two regimes are found: when the initial disturbances are purely three-dimensional, a vortex lattice is obtained. Considering Plate 2 only, it seems that this lattice is made of oblique vortices arising from the growth of the oblique modes $(k_x = 2\pi/2\lambda_a\ ,\ k_z = \pm 2\pi/4\lambda_a)$. However, a similar vortex lattice is obtained at a higher resolution, and Plate 3 brings evidence that this lattice is made of initially-straight Kelvin-Helmholtz vortices which locally wind around each other. This was observed experimentally by Chandrsuda *et al.* (1978) who proposed the term 'helical pairing' to characterize this process. As proposed by Pierrehumbert and Widnall (1982), the latter is certainly initiated by the simultaneous growth of the sub-harmonic of the most amplified streamwise mode and a spanwise unstable mode of arbitrary wavelength, the longer this wavelength, the larger the growth rate. Plate 3 also shows a hairpin-like vortex which might have been ejected from a region where a helical pairing occurs.

When the initial disturbances are rather two-dimensional (but not completely two-dimensional, otherwise the flows remains two-dimensional), the same periods as before are found, but the development of the streamwise subharmonic here overwhelms that of the spanwise mode: the pairing is not helical any longer and gives rise to almost two-dimensional rollers of spacing 2 $\lambda_a$. Three-dimensionality appears as a spanwise oscillation of the vortex lines in between, in agreement with the scenario proposed by Lasheras and Choi (1988).

All this suggests that several mechanisms lead to three-dimensionalization of mixing layers and that they may coexist and interact in experiments or practical situations where the level of residual turbulence or perturbations is not low. Furthermore, some of these mechanisms, such as the ejection of haipin-like vortices where helical pairings occur, do not recur systematically.

Both in the case $\varepsilon_{2D} = 0$ and in the case $\varepsilon_{2D} \neq 0$, no visible spanwise periodicity is found, but the one imposed by the boundary conditions. Simulations in a domain of larger spanwise dimension will soon be done in both cases, for different Reynolds numbers. Indeed, the stability calculations performed by Metcalfe *et al.* (1987) pointed out the influence of the Reynolds number on the development of a small-amplitude spanwise mode superimposed onto a two-dimensional quasi-equilibrium flow involving a single Kelvin-Helmholtz vortex: for small Reynolds numbers ($\simeq$ 200), the curve showing the growth rate of the spanwise mode as a function of its wavenumber presents a sharp peak at low wavenumbers (see their Figure 8). When the Reynolds number increases, the peak becomes gradually less and less sharp and is shifted towards larger wavenumbers.

The wake simulations have shown two stages in the formation of a Kármán street, as found by Chen *et al.* (1989), with a vortex configuration involving closed loops, as observed in many experimental or numerical experiments.

The two-dimensional compressible mixing layer simulations have shown the decay of the Kelvin-Helmholtz instability when the convective Mach number $M_c$ increases, and the apparition of eddy-shocklets for $M_c \leq 0.8$, as found by Lele (1989). The existence of such shocklets in three-dimensional situations is still an open question, which could question the concept of convective Mach number beyond about 0.6.

As a general conclusion, it can be said that the use of random perturbations seems to be a good method to explore the routes leading to three-dimensionalization of free shear flows. This could be due to the fact that inflexional-shear instabilities are sufficiently amplified to emerge from noise rapidly enough not to require a deterministic forcing of the most amplified modes. Three-dimensional visualizations proved to be very useful to understand the vortex-structure dynamics, provided careful cross-checking of the different views is done.

## 6. Acknowledgements

This work was supported by DRET under contracts 88/150 and 87/808/11. Part of the computational time was allocated by the CCVR.

We are grateful to Pr. J. J. Riley for helpful comments on a draft of this manuscript.

## 7. References

ANTONIA, R., BROWNE, L. & BISSET, D. 1987 A description of the organized motion in the turbulent wake of a cylinder at low Reynolds number. *J. Fluid Mech.* **184**, 423-444.

BAYLISS, A., PARIKH, P., MAESTRELLO, L. & TURKEL, E. 1985 A fourth-order scheme for the unsteady compressible Navier-Stokes equations. AIAA Paper 85-1694, July 1985.

BERNAL, L. P. & ROSHKO, A. 1986 Streamwise vortex structure in plane mixing layers. *J. Fluid Mech.* **170**, 499-525.

BLUMEN, W. 1970 Shear layer instability of an inviscid compressible fluid. *J. Fluid Mech* **40**, 769-781.

BOGDANOFF, D. W. 1983 Compressibility effects in turbulent shear layers. *AIAA Journal* **21**, 926-927.

BRADSHAW, P. 1966 The effect of initial conditions on the development of a free shear layer. *J. Fluid Mech* **26**, 225-236.

BREIDENTHAL, R. 1980 Response of plane shear layers and wakes to strong three-dimensional disturbances. *Phys. Fluids* **23**, 1929-1934.

BREIDENTHAL, R. 1981 Structure in turbulent mixing layers and wakes using a chemical reaction. *J. Fluid Mech.* **109**, 1-24.

BROWN, G. L. & ROSHKO, A. 1974 On density effects and large structure in turbulent mixing layers. *J. Fluid Mech.* **64**, 775-816.

CHANDRSUDA, C., MEHTA, R. D., WEIR, A. D. & BRADSHAW, P. 1978 Effect of free-stream turbulence on large structure in turbulent mixing layers. *J. Fluid Mech.* **85**, 693-704.

CHEN, J. H., CANTWELL, B. J. & MANSOUR, N. N. 1989 Direct numerical simulations of a plane compressible wake: stability, vorticity dynamics and topology, Dept. Mech. Eng. Rep. No. TF-46, Stanford University , California.

COMTE, P., LESIEUR, M. & FOUILLET, Y. 1989 Coherent structures of mixing layers in large-eddy simulation. Proceedings of the IUTAM Symposium on *Topological Fluid Mechanics*, Cambridge, August 13-18, 1989, Cambrigde University Press, 648-658.

LASHERAS, J. C. & CHOI, H. 1988 Three-dimensional instability of a plane shear layer: an experimental study of the formation and evolution of streamwise vortices. *J. Fluid Mech.* **189**, 53-86.

LASHERAS, J. C. & MEIBURG, E. 1990 On the three-dimensional dynamics of coherent vortical structures forming in free shear flows. Present proceedings.

LELE, S. K. 1989 Direct numerical simulation of compressible free shear flows. AIAA-889-0374.

LESIEUR, M., STAQUET, C., LE ROY, P. & COMTE, P. 1988 The mixing layer and its point of view from the point of view of two-dimensional turbulence. *J. Fluid Mech.* **192**, 511-534.

METCALFE, R. W., ORSZAG, S. A., BRACHET, M. E., MENON, S. & RILEY, J. J. 1987 Secondary instability of a temporally growing mixing layer. *J. Fluid Mech.* **184**, 207-243.

MICHALKE, A. 1964 On the inviscid instability of the hyperbolic tangent velocity profile. *J. Fluid Mech.* **19**, 543-556.

MISKAD, R. W. 1972 Experiments on the nonlinear stages of free-shear layer transition. *J. Fluid Mech.* **56**, 645-719

NORMAND, X. 1989 Numerical simulations of compressible mixing layers. Proceedings of the *Instability and Transition Workshop* ICASE/LaRC, June 1989.

OSTER, D., WYGNANSKI, I. & FIEDLER, H. 1977 In *Turbulence in Internal Flows* (ed. S. N. B. Murthy). Washington: Hemisphere.

PERRY, A. E., LIM, T. T. & CHONG, M. S. 1980 The instantaneous velocity fields of coherent structures in coflowing jets and wakes. *J. Fluid Mech.* **101**, 243-256.

PIERREHUMBERT, R. T. & WIDNALL, S. E. 1982 The two- and three-dimensional insatbilities of a spatially periodic shear layer. *J. Fluid Mech.* **114**, 59-82.

SATO, H. & KURIKI, K. 1961 The mechanism of transition in the wake of a thin plate placed parallel to a uniform flow. *J. Fluid Mech* **11**, 321-351.

WYGNANSKI, I., OSTER, D., FIEDLER, H. & DZIOMBA, B. 1979 On the perseverance of a quasi-two-dimensional eddy-structure in a turbulent mixing layer. *J. Fluid Mech.* **93**, 325-335.

## 8. Plate captions

PLATE 1: Three-dimensional temporally-growing incompressible mixing layer at $t = 30\ \delta_i/U$. Resolution: $64^3$ Fourier modes in a cubic domain of side $L = 4\lambda_a \approx 28\ \delta_i$. Initial Reynolds number: $R = U\delta_i/\nu = 100$. Initial perturbations characterized by: $\varepsilon_{2D} = 10^{-4}$, $\varepsilon_{3D} = 10^{-5}$.

- passive-scalar interface: plan view (a), perspective view (b).
- vorticity iso-surfaces $\omega_x = \pm 0.15\ \omega_i$ and $\omega_z = 0.15\ \omega_i$, with $\omega_i = -2U/\delta_i$: plan view (c), perspective view (d).
- vortex lines: plan view (e), side view (f).

PLATE 2: Three-dimensional temporally-growing incompressible mixing layer at $t = 30\ \delta_i/U$. Resolution: $64^3$ Fourier modes in a cubic domain of side $L = 4\lambda_a \approx 28\ \delta_i$. Initial Reynolds number: $R = U\delta_i/\nu = 100$. Initial perturbations characterized by: $\varepsilon_{2D} = 0$, $\varepsilon_{3D} = 10^{-4}$.

- passive-scalar interface: plan view (a), perspective view (b).
- vorticity iso-surfaces $\omega_x = \pm 0.15\ \omega_i$ and $\omega_z = 0.15\ \omega_i$, with $\omega_i = -2U/\delta_i$: plan view (c), perspective view (d).

PLATE 3: Three-dimensional temporally-growing incompressible mixing layer at $t = 30\ \delta_i/U$. Resolution: $128^3$ Fourier modes in a cubic domain of side $L = 4\lambda_a \approx 28\ \delta_i$. Initial Reynolds number: $R = U\delta_i/\nu = 100$. Initial perturbations characterized by: $\varepsilon_{2D} = 10^{-4}$, $\varepsilon_{3D} = 10^{-5}$.

- passive-scalar interface: plan view (a), perspective view (b), cut at constant $z$ (c).
- vorticity iso-surfaces $\omega_x = \pm 0.15\ \omega_i$ and $\omega_z = 0.15\ \omega_i$, with $\omega_i = -2U/\delta_i$: plan view (d).
- vortex lines: plan view (e).

PLATE 4: Three-dimensional temporally-growing incompressible plane wake. Resolution: $48^3$ Fourier modes in a cubic domain of side $L = 2\lambda_a \approx 16\ \delta_i$. Initial Reynolds number: $R = U\delta_i/\nu = 160$.

- perspective view of the vorticity iso-surfaces $\omega_z = \pm 0.4\ |\omega_z\ |_{max}$ and $\sqrt{\omega_x^2 + \omega_y^2} = 0.5\ \left|\sqrt{\omega_x^2 + \omega_y^2}\ \right|_{max}$ at $t = 46\ \delta_i/U$ (a)
- perspective (b) and plan view (c) of the vorticity iso-surfaces $\omega_z = \pm 0.3\ |\omega_z|_{max}$ and $\sqrt{\omega_x^2 + \omega_y^2} = 0.25\ \left|\sqrt{\omega_x^2 + \omega_y^2}\ \right|_{max}$ at $t = 66\ \delta_i/U$.
- perspective view (d) of the vorticity iso-surfaces $|\omega| = 0.15\ |\omega|_{max}$ and $\sqrt{\omega_x^2 + \omega_y^2} = 0.35\ \left|\sqrt{\omega_x^2 + \omega_y^2}\ \right|_{max}$ at $t = 66\ \delta_i/U$.
- vortex lines at $t = 66\ \delta_i/U$. Side view (e) and left view (f).

PLATE 5: Two-dimensional temporally-growing compressible mixing layer at $M_c = 0.8$ and $R = 5000$. Resolution: $400 \times 100$.

- vorticity [(a) and (c)] and pressure [(b) and (d)] fields before and after the appartion of eddy-shocklets, *viz* at $t = 75$ and $80\ \delta_i/U$ respectively.

PLATE 6: Two-dimensional spatially-growing compressible plane wake at $M_\infty = 2$, $M(y = 0) = 0.47$ and $R = 1200$. Resolution: $400 \times 100$.

- vorticity (a) and temperature (b) fields.

PLATE 7: The two regimes found in the case of the three-dimensional temporally-growing incompressible mixing layer at a Reynolds number $R = U\delta_i/\nu = 100$. Perspective views of vorticity iso-surfaces at $t = 30\ \delta_i/U$. Initial perturbations characterized by:

- $\varepsilon_{2D} = 10^{-4}$, $\varepsilon_{3D} = 10^{-5}$ (a)
- $\varepsilon_{2D} = 0$, $\varepsilon_{3D} = 10^{-4}$ (b).

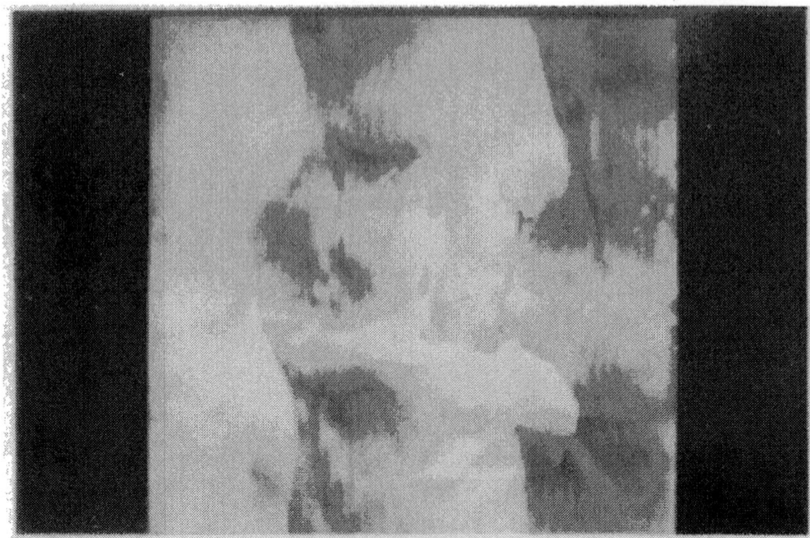

(a)

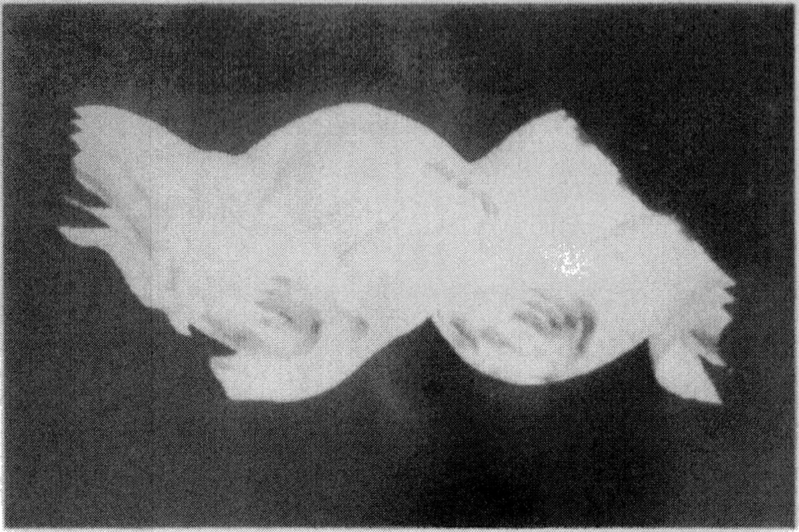

(b)

Plate 1

(c)

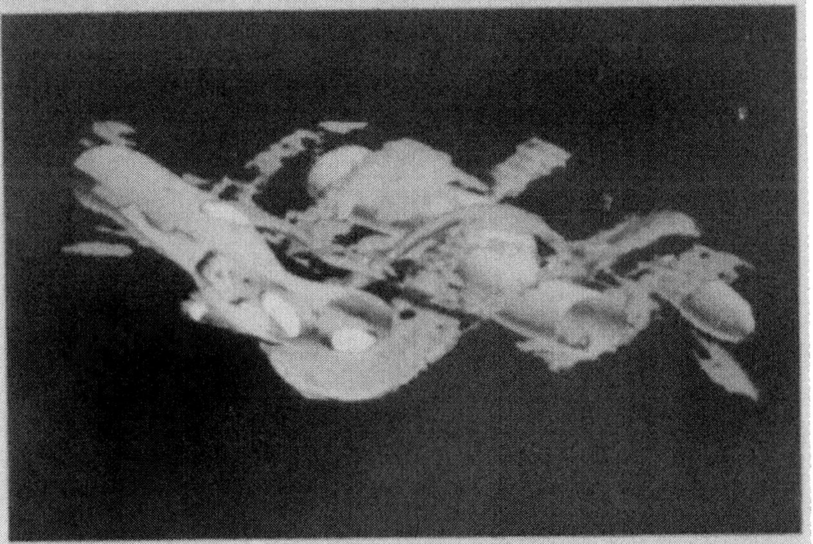

(d)

Plate 1 (cont'd)

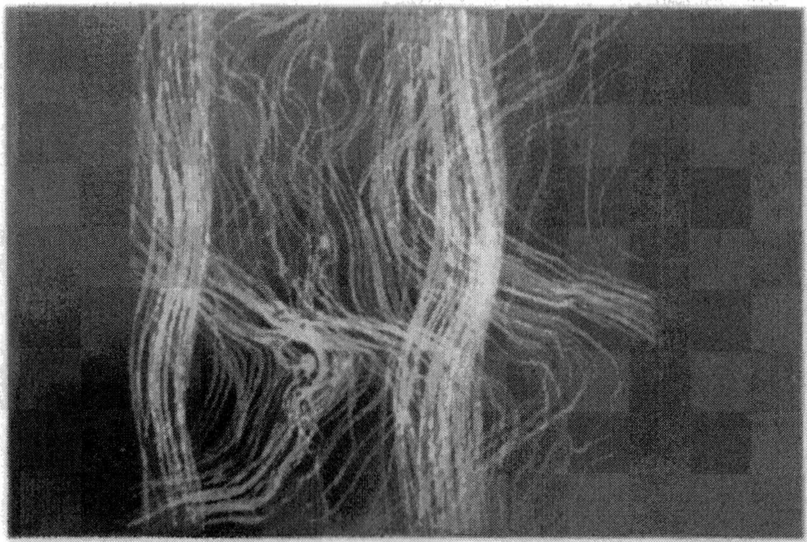

(e)

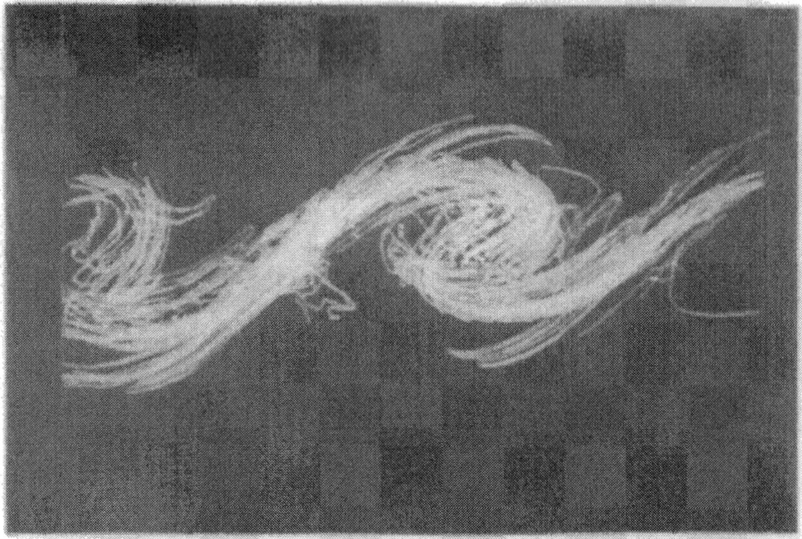

(f)

Plate 1 (cont'd)

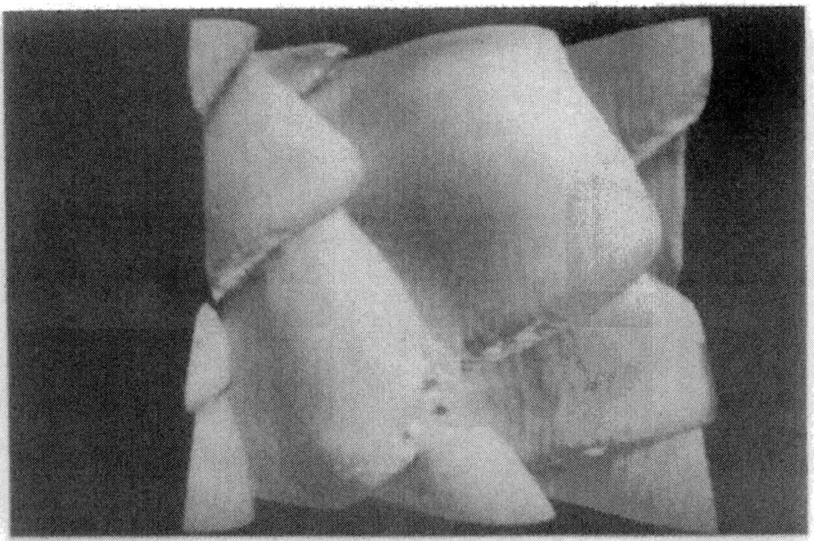

(a)

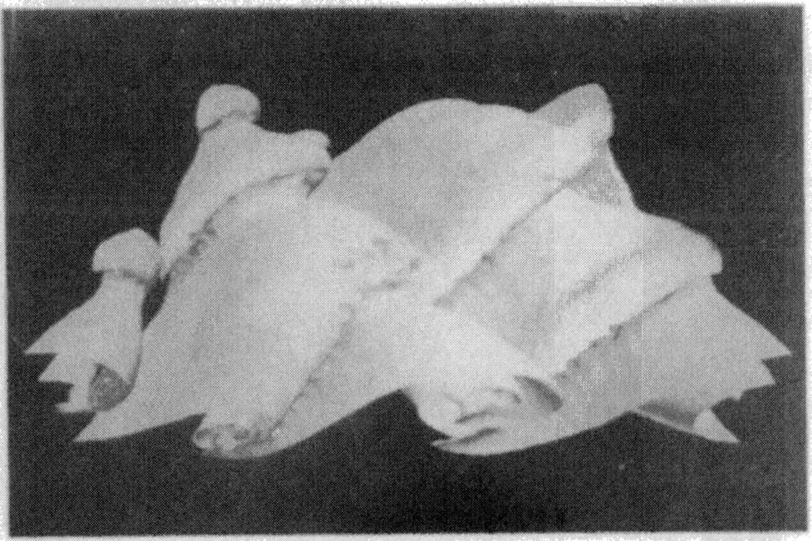

(b)

Plate 2

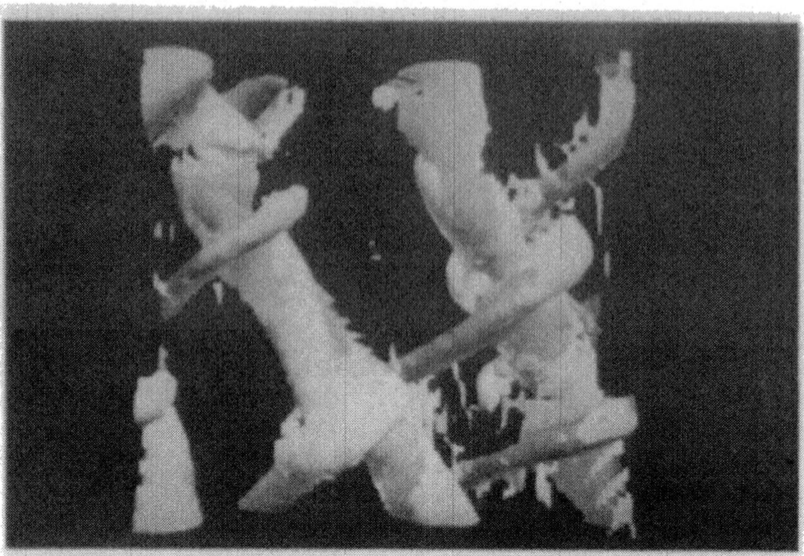

(c)

(d)

Plate 2 (cont'd)

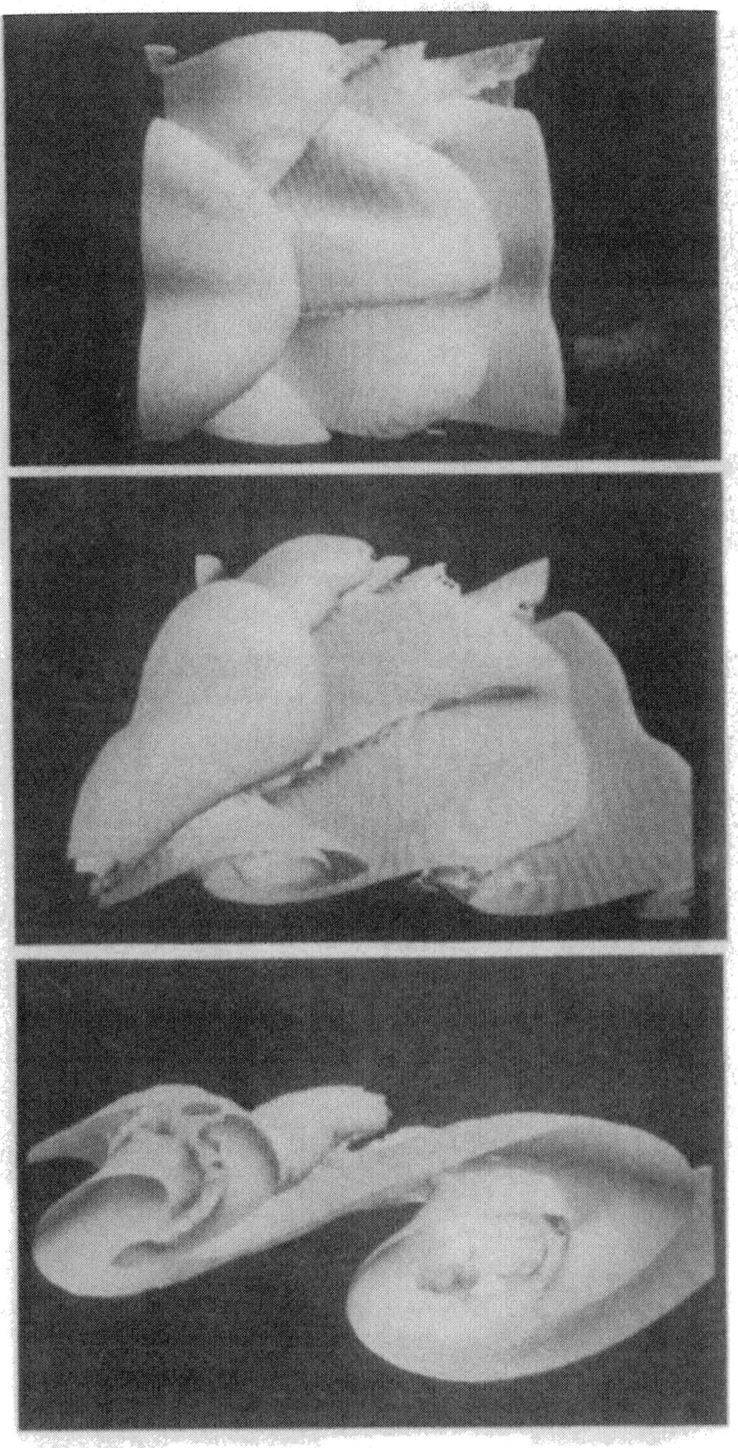

(a)

(b)

(c)

Plate 3

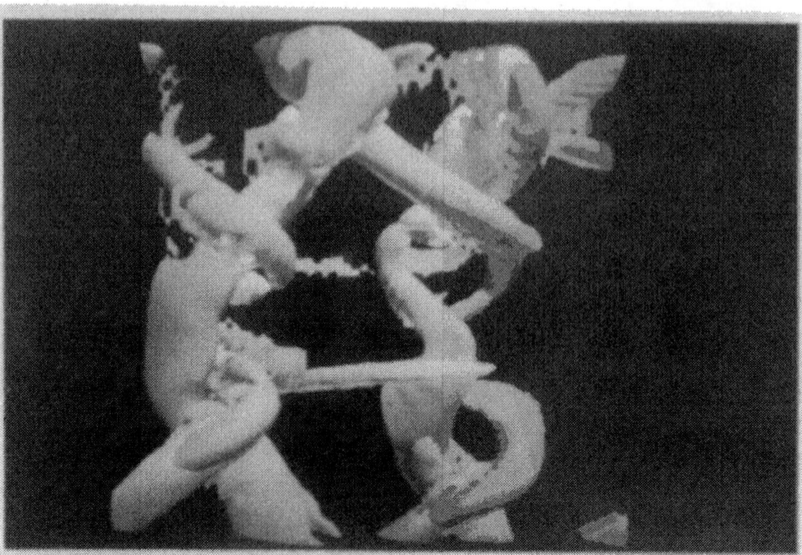

(d)

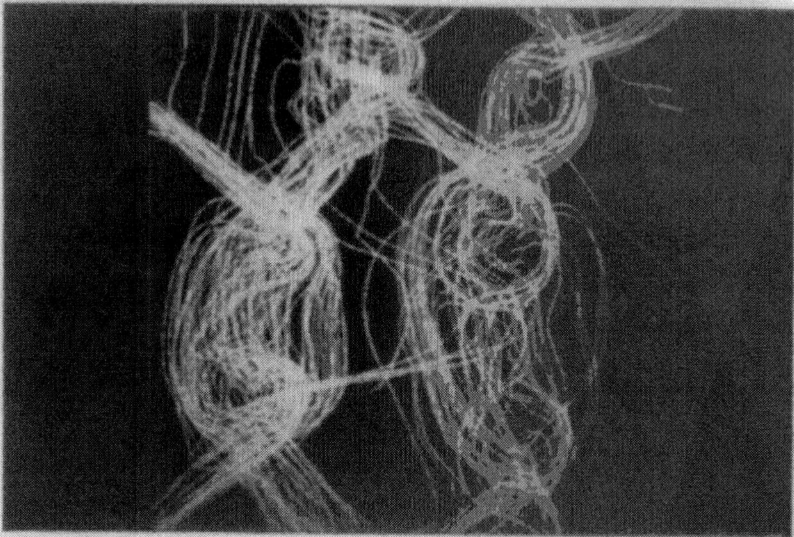

(e)

Plate 3 (cont'd)

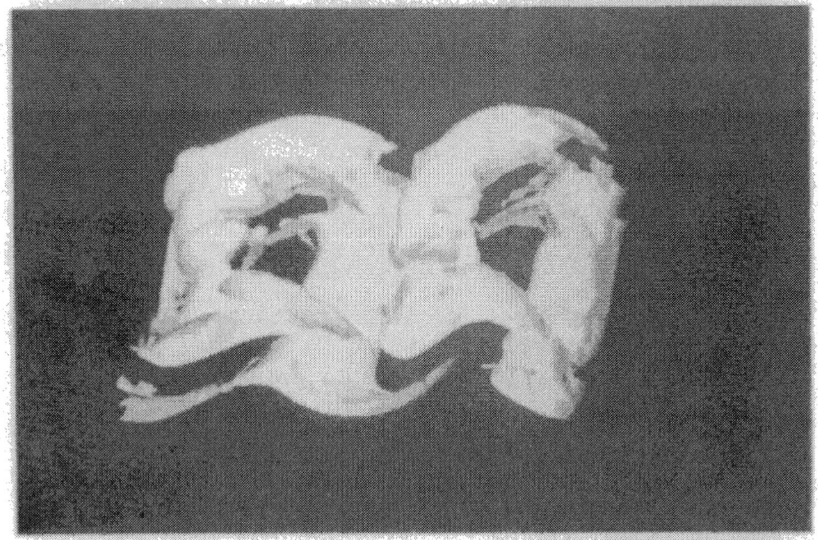

(a)

(b)

Plate 4

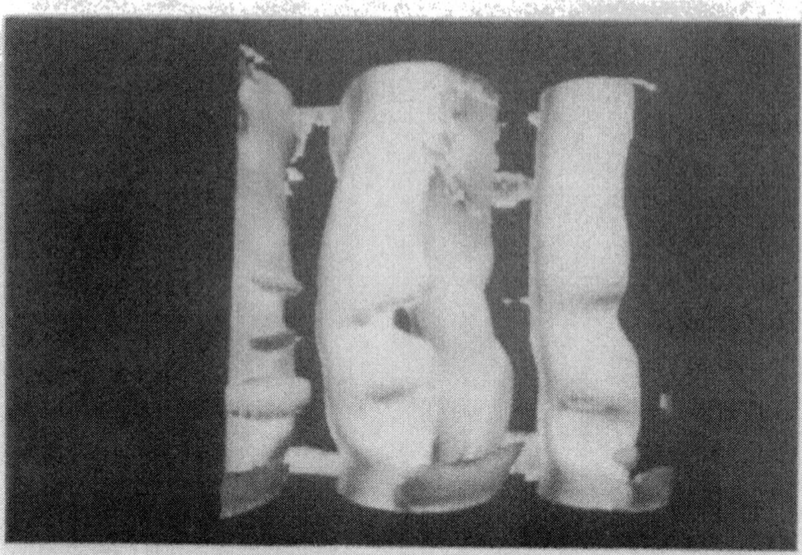

(c)

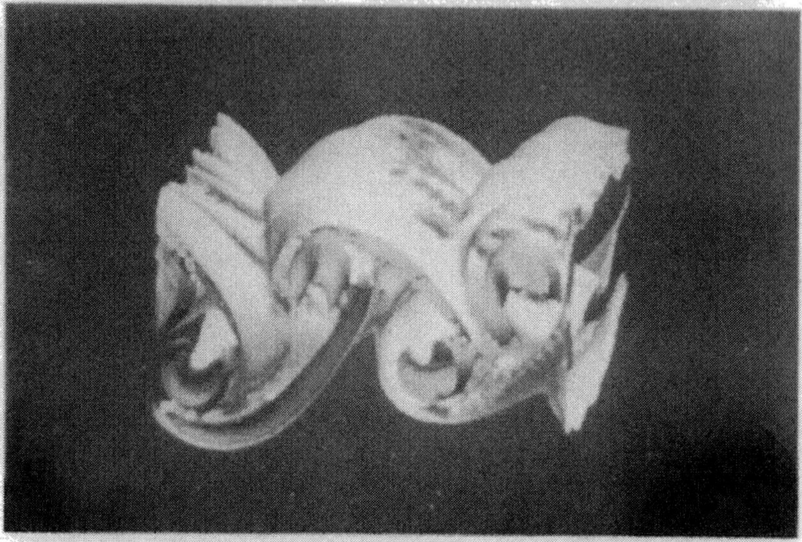

(d)

Plate 4 (cont'd)

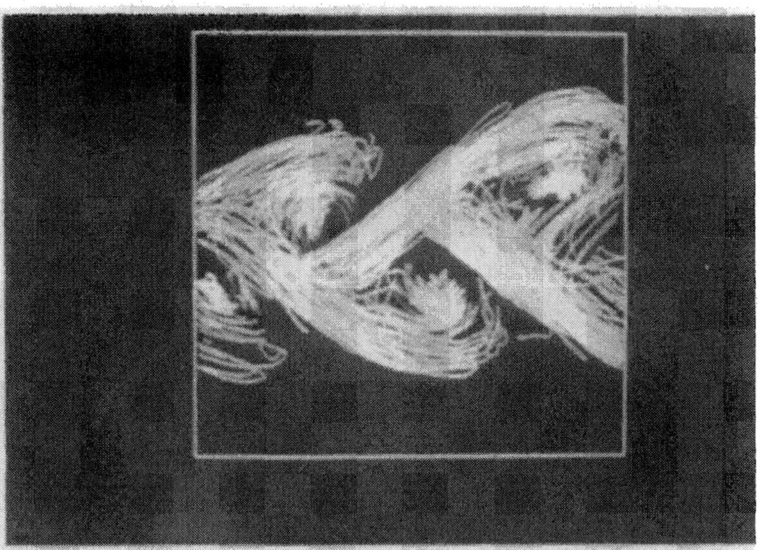

(e)

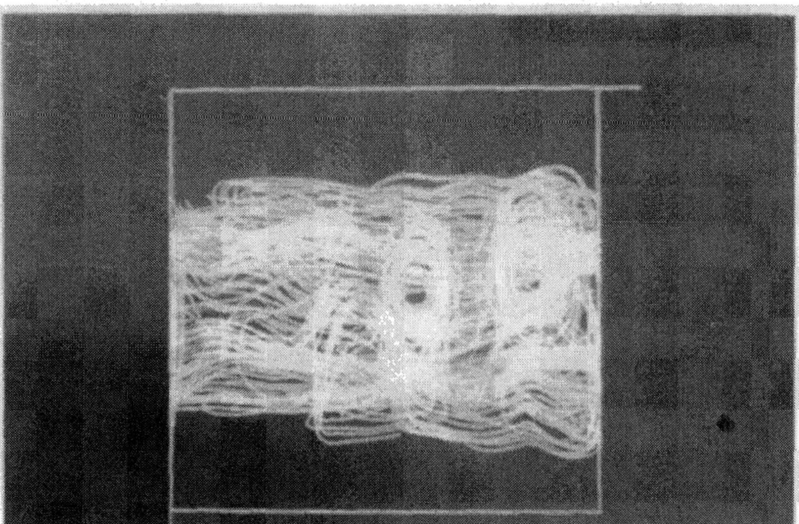

(f)

Plate 4 (cont'd)

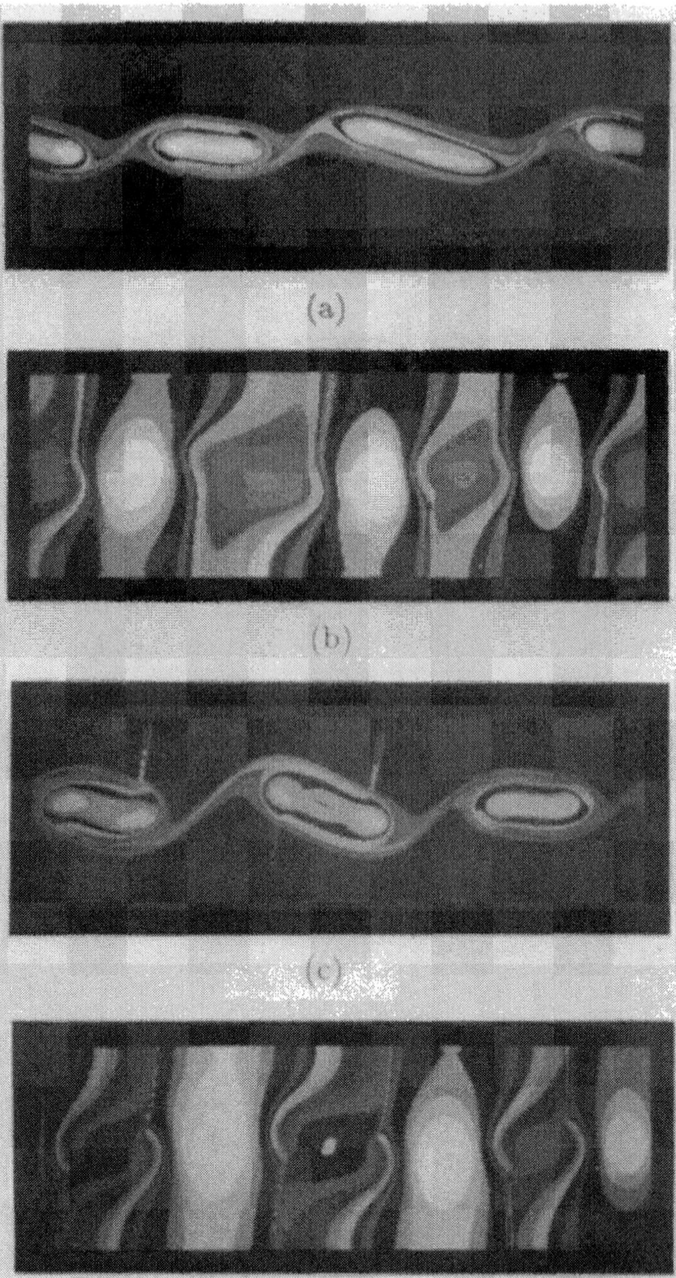

(a) (b) (c) (d)

Plate 5

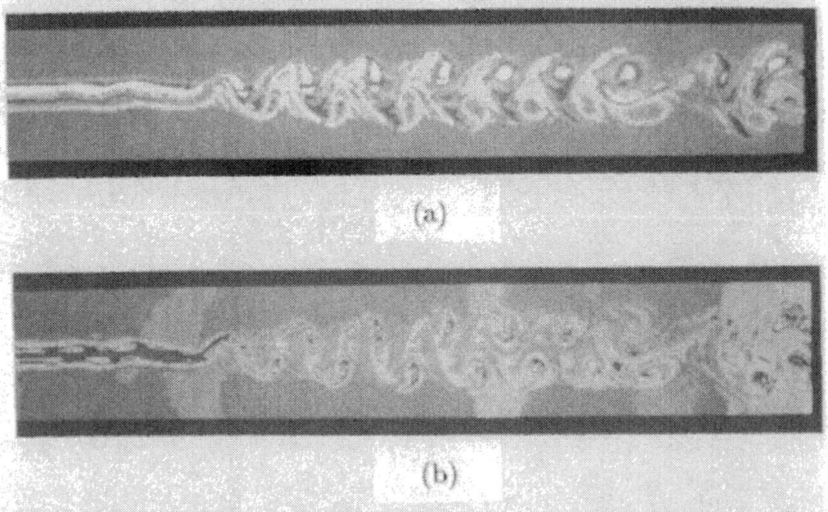

Plate 6

# Use of the proper orthogonal decomposition in a plane turbulent mixing layer.

Joël DELVILLE, Stéphane BELLIN and Jean Paul BONNET

*Centre d'Etudes Aérodynamiques et Thermiques L.E.A. U.A. CNRS 191*
*43 route de l'aérodrome, 86036 Poitiers-Cedex, France*

*Turbulence and Coherent Structures*
*O. Métais and M. Lesieur, eds.*
*Selected papers from "Turbulence 89: organized structures and turbulence in fluid mechanics"*
*Grenoble, 18-21 September 1989*

## 1. Introduction.

The existence of organized motions in turbulent shear flows has now been largely evidenced, since the early work of Townsend in 1956. During the last two decades, a large amount of studies has allowed to precise the characteristics of organized structures and to emphasis their significance in the turbulent mechanism. The techniques involved in most of these studies are essentialy based on flow-visualization and conditional sampling (see Cantwell 1981 or Hussain 1986). However it remains some limitations in the application of the actual knowledge of the coherent structures for the modelization of turbulence. First, visualizations are generally carried out at relatively low Reynolds numbers and then structures may generally only be "defined" as long as they remain not too chaotic. Second, the use of conditional sampling is limited by the need of the definition of an objective criterium. Lumley in 1967 proposed that the dominant coherent structure in a turbulent flow could be thought as the one that has the largest mean square projection on the velocity field. This approach, the Proper Orthogonal Decomposition (POD), is based on the extraction of mutually uncorrelated eigen-functions from the two-point correlation tensor. The dominant structure is then related to the higher eigen-value. Unfortunately this technique implies the experimental knowledge of the correlation tensor over a large number of points and therefore the use of large experimental data sets. Then only few early applications of the POD are found: Payne & Lumley in 1967 for the wake of a cylinder or Bakewell & Lumley 1967 for a wall flow. With the recent development of powerful experimental and computational means, the POD is more intensively used. Glauser et al. 1985 applied the POD to measurements in an axisymmetric jet mixing layer and recently Moin & Moser 1989 on the numerical simulation of a channel flow. The property of the POD to extract most of the energy of the turbulent field has allowed Aubry et al. 1988 to use the eigen-functions measured by Herzog 1988 to construct a low order system of ordinary differentiel equations that can efficiently describe the dynamics of the coherent structures in a wall flow. This has also been applied in the same way by Glauser et al. 1989 for the eigen-fuction determined for the jet case. Recently the POD has been extended to a forced plane mixing layer by Glezer et al. 1989.

## 2. Proper Orthogonal Decomposition.

### 2.1. General Equations.

The definition of Lumley for the dominant coherent structure leads to an integral eigen-value problem,

$$\int R_{ij}(\underline{X}, \underline{X}')\, \Phi_j^*(\underline{X}')\, d\underline{X}' = \lambda\, \Phi_i(\underline{X}),$$

where $R_{ij}$ is the two points space time correlation tensor:

$$R_{ij}(\underline{X}, \underline{X}') = \overline{u_i(\underline{X}) u_j(\underline{X}')},$$

and where $()^*$ denotes the complex conjugate. $\Phi_i$ is the deterministic candidate structure and $\underline{X} = (x_1, x_2, x_3; t)$, is a space time variable. Because of the symmetry of $R_{ij}$, it can be shown from the Hilbert-Schmidt theory that a discrete set of solutions may be found for this Fredholm equation which can be written as follows:

$$\int R_{ij}(\underline{X}, \underline{X}')\, \Phi_j^{*(n)}(\underline{X}')\, d\underline{X}' = \lambda^{(n)}\, \Phi_i^{(n)}(\underline{X}) \qquad n = 1, 2, 3, \cdots.$$

This discrete set can be chosen so that the eigen-functions $\Phi_i^{(n)}$ are orthonormal:

$$\int \Phi_i^{(p)}(\underline{X})\, \Phi_i^{*(q)}(\underline{X})\, d\underline{X} = \delta_{pq}. \tag{1}$$

The fluctuating random field $u_i(\underline{X})$ can then be rebuilt from the deterministic eigen-functions:

$$u_i(\underline{X}) = \sum_{n=1}^{\infty} a^{(n)} \Phi_i^{(n)}(\underline{X}), \tag{2}$$

the stochastic coefficients $a^{(n)}$ being calculated using :

$$a^{(n)} = \int u_i(\underline{X})\, \Phi_i^{*(n)}(\underline{X})\, d\underline{X}.$$

The $\lambda^{(n)}$ are real, positive, their sum is finite and their serie converges :

$$\lambda^{(1)} > \lambda^{(2)} > \lambda^{(3)}, \cdots. \tag{3}$$

### 2.2. Simplified Approach.

In the case of stationary, homogeneous or periodic directions, the POD reduces to harmonic decomposition for the corresponding variables. Then a Fourier analysis is preferred for these directions whilst the POD is applied in the inhomogeneous direction. In the present study a limited and simplified approach of the POD is used: only the streamwise component of the velocity, $u$, is taken into account, the homogeneous direction $Z$ is disregarded and invoking time stationarity, a Time Fourier Transform is used. Then the problem to solve may be written as follows:

$$\int \Psi(Y,Y';f)\,\widetilde{\Phi}^{*(n)}(Y;f)\,dY' = \lambda^{(n)}(f)\,\widetilde{\Phi}^{(n)}(Y;f) \qquad (4).$$

In this equation, $\Psi$ is the cospectrum of $u$ ( Time Fourier Transform of the conventional Space Time Correlation: $\Psi = \widetilde{R_{uu}}(Y,Y';\tau)$). The notation $\widetilde{(\;)}$ corresponds to the Time Fourier Transform and $Y$ is the inhomogeneous direction. Since $u$ is the only velocity component used, in the following, "energy" will be referred to the $\overline{u^2}$ participation.

### 2.3. Some Implications of this Decomposition.

The POD allows to extract structural information about the turbulent field in an objective way, in that it corresponds to an energy (or stastical occurence) based extraction. Therefore, the POD is optimal in the sense that for a given limited number of modes, no other expansion can be as representative of the turbulent energy. The proper values $\lambda^{(n)}$ are then related to the energy or probability of the mode of the decomposition, integrated over the whole inhomogeneous direction. By this technique, the more energetical mode is first extracted, then a less energetical one, etc.... As it will be seen in section 4, the modes of the decomposition can be related to a physical size or spatial partition (ie the higher the mode of POD is, the smaller the "corresponding size" is). Then the different modes of the decomposition may be thought as the statistical contribution of eddies of different sizes. On the other hand, as shown by Lumley (1970) for most of shear flows, a limited number of significant modes can be sufficient to be representative of the turbulent field. Endly, because of the linearity of the decomposition (equation 2) and of the stastical independency of the modes (equation 1), most of the statistics of the turbulent field (ie correlation, spectra,...) may be reconstructed from the participation of each mode of the POD. For example the cospectrum of $u$ measured for the two points $Y$ and $Y'$ is given by:

$$\Psi(Y,Y';f) = \sum_{n=1}^{\infty} \Psi^{(n)},$$

$$\text{where } \Psi^{(n)} = \lambda^{(n)}\,\widetilde{\Phi}^{*(n)}(Y;f)\,\widetilde{\Phi}^{(n)}(Y';f). \qquad (5)$$

## 3. Experimental Apparatus and Data Processing.

### 3.1. Flow Tested.

Experiments are performed in an open wind tunnel of square section 300x300 $mm^2$, with a 1.2m long test section. The plane mixing layer is created by a flat plate, splitting two air streams $U_a$=42.8m.$s^{-1}$ and $U_b$=25.2m.$s^{-1}$; the corresponding velocity ratio is r=0.6. The plate is bevelled with a 3% slope from 50mm upstream of the trailing edge. At this location, the boundary layers are fully turbulent with a thickness of about 7mm. The coordinates are $X$, $Y$ and $Z$ respectively in the streamwise direction, normal to the plane of the plate (inhomogeneous direction; $Y = 0$ at the center line of the mixing layer) and parallel to the trailing edge of the plate (homogeneous direction). The external level of turbulence is 0.3%. The measurements are performed in the asymptotic part of the flow at a distance $X$=600mm from the trailing edge of the plate. At this location, the local vorticity thickness $\delta_\omega$ is about 28 mm and the Reynolds number based on $\delta_\omega$ and on the velocity difference $\Delta U = U_a - U_b$ is $R_{\delta_\omega}$=33,000. The Strouhal number ($S_\omega = f_p.\delta_\omega/U_c$), based on the peak frequency $f_p$ of the u-spectrum measured in the vicinity of the mixing layer edges and on the mean velocity $U_c = (U_a + U_b)/2$, is found to be $S_\omega = 0.25$ (Delville et al. 1987).

### 3.2. Experimental Procedure.

A special home made rake of Hot Wires is used for this study. The rake is built of a set of up to 24 individual subminiature probes ingrooved in the thickness of a printed circuit board. The probes have a 1$mm^2$ squared section and are made of conical ended, golded Tungsten prongs of 0.2 mm diameter. The wires (W-Pt) are 0.5mm long with a 2.5 $\mu$m diameter. The probe separation is 3mm (of the same order as the Taylor microscale), then the rake extent is about twice the local vorticity thickness. The hot wires are driven by T.S.I 1750 Constant Temperature Anemometers. Velocity signals can be simultaneous sampled at a maximum rate of 100,000 samples/second per channel (12 bit analog digital converters). For each hot wire output, samples of 10,240 conversions are collected and then recorded on magnetic tape for further processing. Since POD is supposed to deal with "relatively large" scales, only a low band of frequency is retained for the present study. A maximum frequency of 5 kHz has been considered in the present situation to be sufficient, in the sense that the maximum relevant frequency is then 15 times the Strouhal frequency. For the computation of the spectra involved in this study, the sampling frequency is 10 kHz and 80 records are used. The total corresponding sampling time is then 82 seconds (ie about 25,000 time the Strouhal period). Spectra are evaluated by Fast

Fourier Transforms of size 512. All the present results are obtained in a configuration of 21 probes. Previous tests have been processed in order to validate the rake measurements: as well mean or RMS velocity profiles as space time correlation or spectra are found to be accurately measured by this way, when compared with conventional measurements.

## 4. Results and Discussion.

### 4.1. Energy distribution.

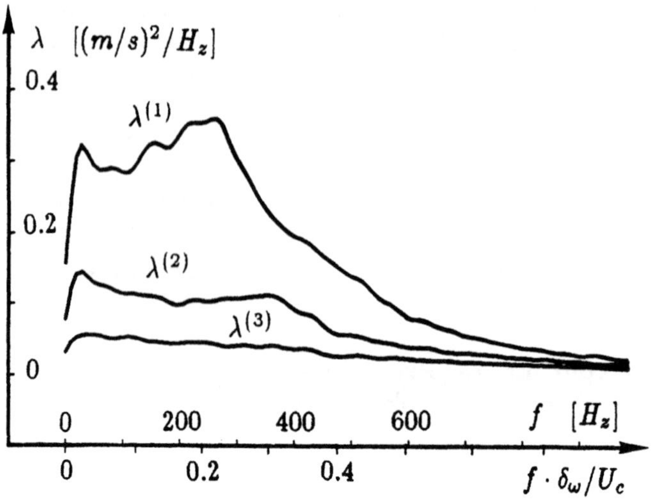


Fig.1. Eigen-values Distribution vs frequency.

In the POD, the eigen-functions $\widetilde{\Phi}^{(n)}(Y;f)$ are only representative of the preferred directions of the flow realizations. The level of energy corresponding to these directions is then found in the eigen-values. The figure 1 shows the frequency evolution of the eigen-values corresponding to the 3 first modes of the POD. For the first mode, a maximum of energy is located at a frequency close to the Strouhal frequency. The quick convergence of the eigen-values $\lambda^{(n)}(f)$ can be noticed. Globally, when integrated over all the frequencies, these eigen-values contain 43%, 19% and 9.5% for the mode 1, 2 and 3 respectively. The same behaviour of the eigen-values has been previously denoted by Glauser et al. 1985 for an axisymmetric jet mixing layer or by Herzog 1988 in the case of a wall flow, as well for the aspect as for the order of magnitude of the energy captured by the different modes.

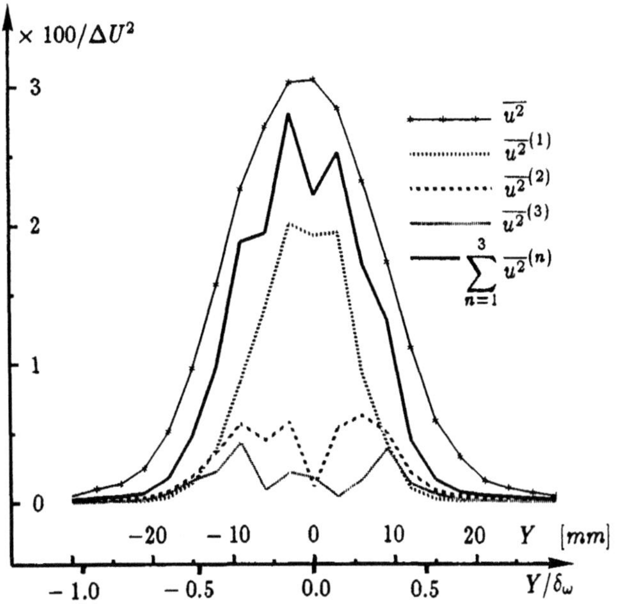


Fig.2. Contribution of modes to energy distribution over the $Y$ direction.

For each $Y$ location, an idea on the local contribution of each mode to energy may be given by considering equation (5). The total energy is then $\overline{u^2}(Y) = \int \Psi(Y,Y;f)\,df$ and the contribution of each mode $(n)$ is $\overline{u^2}^{(n)}(Y) = \int \Psi^{(n)}\,df$. The evolution of these contributions reflects the same hierarchy with the order of the mode as the proper values $\lambda$. However here, the energy is weighted by the "direction" of the modes via the eigen-functions $\tilde{\Phi}$ and it may exist some locations where the decrease with the order of the modes observed by the proper values (equation 3) is not verified by the energy. For example at $Y = 0$, the third mode is slightly more energetic than the second one; in the outer part of the mixing layer, the second and third modes become more representative of the energy distribution. An other interesting feature may be found in figure 2. The energy distribution of each mode exhibits a symmetrical behaviour about the $Y = 0$ position. This symmetry will be discussed in section 4.3.

### 4.2. Spectra reconstruction.

One of the consequences of the POD is that for each $Y$ location, spectra can be rebuilt from the eigen-functions $\tilde{\Phi}$ (equation 5). The figure 3 shows in a log-log representation, the spectrum $\Psi$ of $u$ measured on the mixing layer axis and the individual contribution $\Psi^{(n)}$ of the first

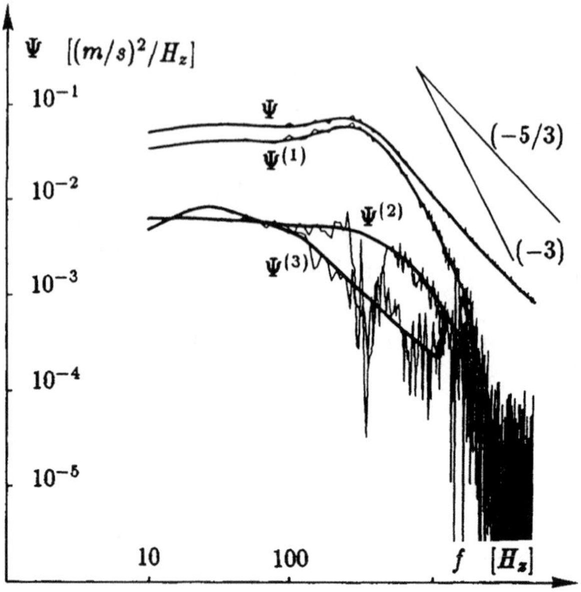


Fig.3. Contribution of modes to $u$-spectrum measured on mixing layer axis.

three modes. As for the eigen-values, the first mode captures for this position most of the energy on the entire frequency domain. Due to an effect equivalent to a "signal to noise ratio", the precision that can be obtained for each mode of the POD is clearly related to the effective energy of this mode in the frequency range considered. Then the less significative a given mode is, the less the precision of its estimation is. This feature can be noticed on figure 3 for frequencies in the range 100-1000 Hz, where the amplitude of the oscillations of $\Psi^{(n)}$ increases with the order $n$. The signal to noise ratio may be a consequence of the uncertainty of the cross-spectrum estimates. However, it may exist a relationship between the precision obtained for a given frequency range and the spacing of nodes used in the $Y$ direction.

The spectra of the figure 3 have been calculated with a spatial resolution of 21 nodes corresponding to the maximum spatial precision obtainable from this experiment, $\Delta_Y = 3mm$. The POD has also been performed on the same set of data but with larger spatial resolutions: 11 nodes and 5 nodes corresponding respectively to $\Delta_Y = 6mm$ and $12mm$. From these experiments, for the first mode of the POD, a cut off frequency $f_C$ has been defined corresponding to the frequency from where the spectrum of this mode begins to oscillate. $f_C$ may be considered as the maximum significative frequency for each decomposition. The evolution of $f_C$ with the spatial size $\Delta_Y$ (figure 4) shows that this

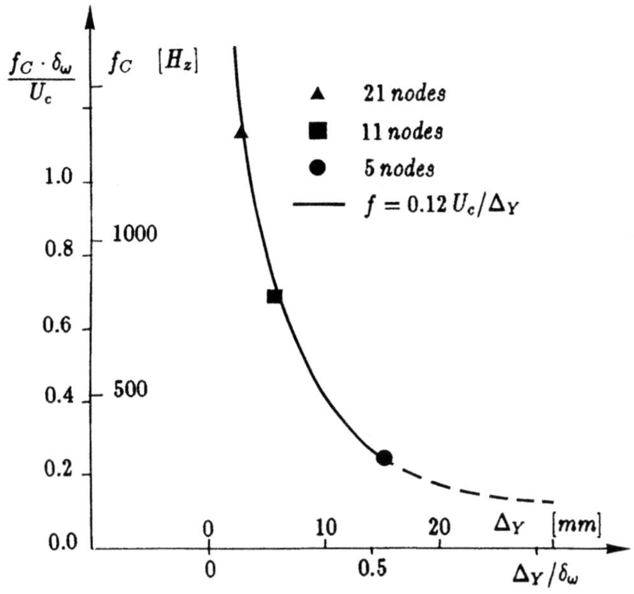


Fig.4. Evolution of the POD cut-off frequency for the first mode

frequency can be related to the probe separation. This can be precised by considering that the cutoff frequency $f_C$ can be related to the smallest discernable scale $d_X$ in the streamwise direction using a Taylor hypothesis: $f_C = 0.5\, U_c/d_X$. A spatial scale $d_Y$ in the $Y$ direction may be associated to $d_X$, with the same ratio as the large scales dimensions (ie $\delta_\omega$ for the $Y$ direction and $f_p.U_c$ for the $X$ direction). Then the corresponding smallest discernable scale in the $Y$ direction is $d_Y = S_\omega.d_X$. This scale has to be related to the mesh size $\Delta_Y$ and finally the cutoff frequency can be written as $f_C = S_\omega.U_c/(2.\Delta_Y)$. This law drawn on figure 4, matches with the experimental determination of $f_C$. The domain where the first eigen-spectra are significative is then located under the curve plotted on this figure.

For the most accurate case (21 nodes), when restricting to the domain of relevant frequencies, the spectrum of the first mode $\Psi^{(1)}$, measured on the mixing layer axis exhibits a special evolution (figure 3). In the usual inertial domain (corresponding to a -5/3 power law for the spectrum $\Psi$), the power law of the first mode is found to be close to -3. This particular, quite well defined behaviour of the first mode can be associated to the footprint of 2D turbulence. The same slope has been obtained for example by Jirka & Dracos 1989 from spectra measured conditionaly in large scales of a jet.

### 4.3. Correlations reconstruction.

As spectra, space time correlation $R_{uu}(Y, Y'; \tau)$ may be rebuilt from the eigen-functions by simply considering the inverse Time Fourier Transform of the cross-spectrum in equation (5). $R_{uu}$ may then be written as $R_{uu} = \Sigma R_{uu}^{(n)}$. Examples of the contribution of modes 1 to 3 to the space time correlation are shown on figure 5 and 6. For each sketch, the horizontal line indicates the position of the reference probe: on the mixing layer axis (figure 5) and in the vicinity of the mixing layer edge corresponding to the low velocity side (figure 6).

When the reference probe is located on the mixing layer axis, the first mode (5-b) is quite well representative of the space time correlation directly measured (5-a), as well for the shape of the contour plots as for the absolute levels of correlation. The addition of the other modes to the first one does not change the shape of the correlation but only allows a better restitution of the level of correlation (5-e and 5-f). All the contour plots of figure 5 exhibit the same mean typical inclination associated to the mean velocity difference of the mixing layer. As the first one, the third mode (5-d) is nearly symmetrical respective to the $Y = 0$ position, whilst the second mode is (5-c) antisymmetrical. The same topological behaviour has been noticed by Glezer [1] et al. 1989 where the even modes are found to be symmetrical and the odd ones antisymmetrical, when looking at the instantaneous decomposition. This characteristic indicates that the POD performs, in average, a geometrical partition of the field in the $Y$ direction. This partition is also observed when the reference probe is located near the edge of the mixing layer (figure 6): the odd modes are symmetrical whilst the even one is antisymmetrical. It should be emphazised that the original partition in the $Y$ direction is preserved, the maximum of the first mode being then not at the probe location but still on the mixing layer axis. The first mode (6-b) is then very different of the direct measurement (6-a) and is not able to reconstruct the maximum and minimum of energy corresponding to correlations measured for points located on both parts of the mixing layer. However when the higher modes (6-c) and (6-d) are added, the space time correlation is well rebuilt (6-e) and (6-f). These results show that the POD is suitable for capturing with very few modes, not only most of the statistics concerning the energy in the mixing layer but also the space time correlations which, in a certain way, are indicative of the large scale structuration of the turbulent field.

[1] In this case an other approach of the POD is used, where the velocity field is represented as a serie $U(X,t) = \sum_{n=1}^{\infty} a_n(t)\, U^{(n)}(X)$ ($X$ is the position within the field). $U^{(n)}(X)$ is a deterministic basis of the velocity field and $a_n$ are random coefficients.

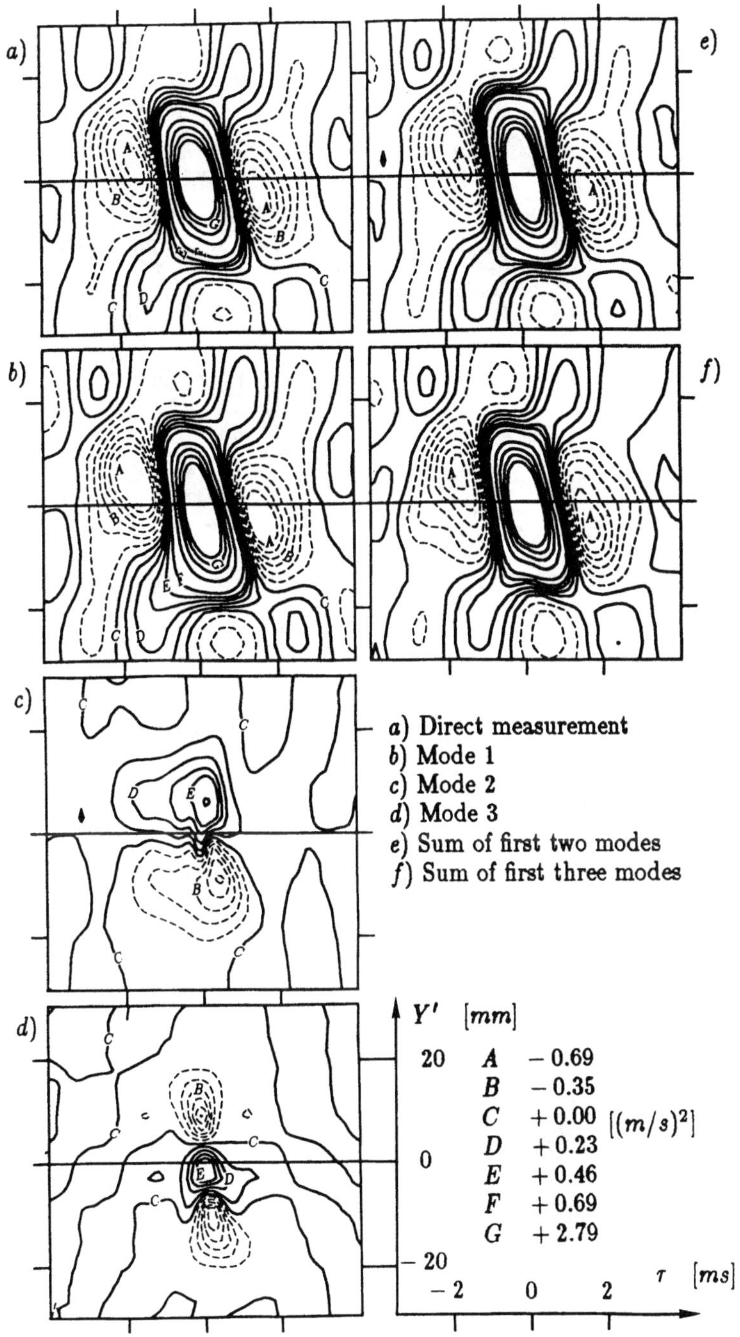


Fig.5. Contribution of modes to Space Time Correlation $R_{uu}(Y, Y'; \tau)$. Fixed probe on mixing layer axis: $Y = 0$

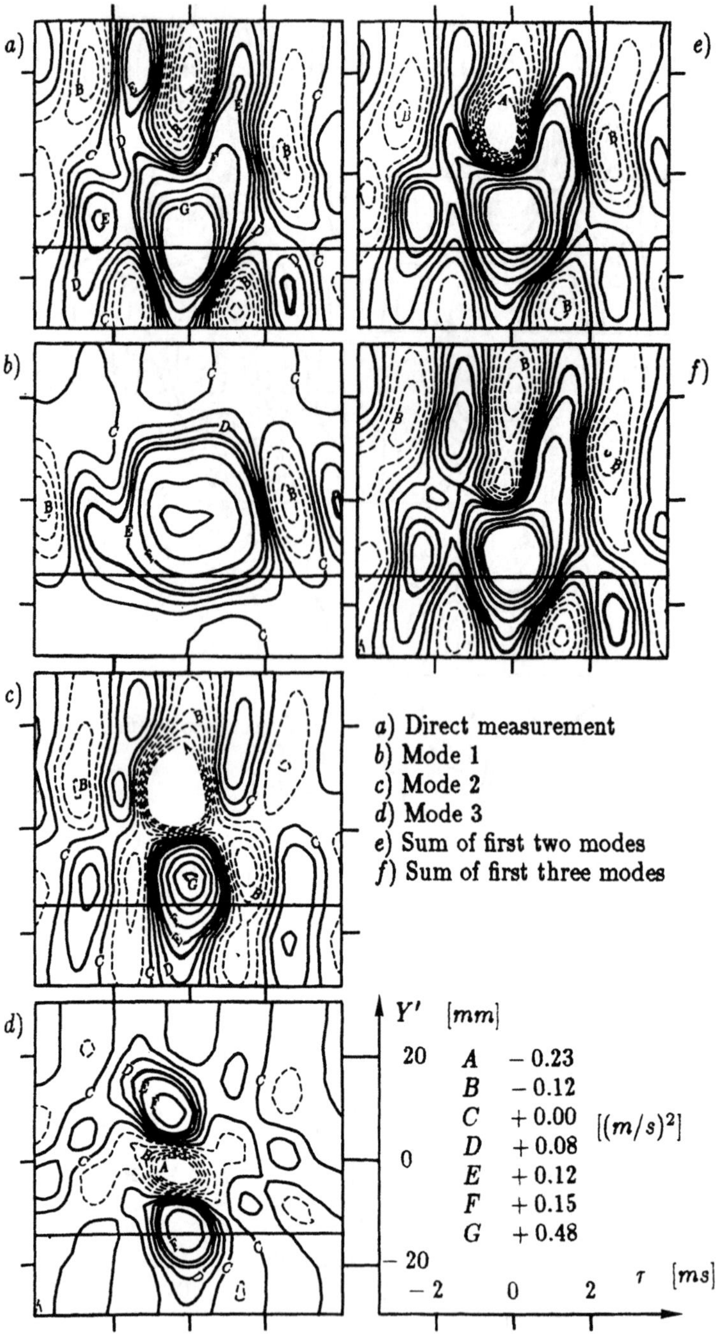


Fig.6. Contribution of modes to Space Time Correlation $R_{uu}(Y, Y'; \tau)$. Fixed probe on mixing layer edge: $Y = -\delta_\omega/2$

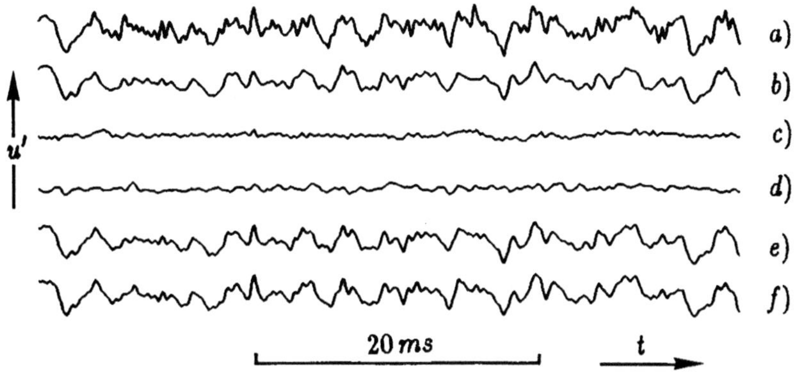


Fig.7. Contribution of modes to the instantaneous velocity measured on mixing layer axis. a) original velocity, b) 1st. mode, c) 2nd. mode, d) 3rd. mode, e) sum of first 2 modes, f) sum of first 3 modes).

**4.4. Reconstruction of the instantaneous velocity distribution.** The use of the POD alone is not sufficient to sort out the characteristic eddy in the physical domain ( i.e. in the present case: $\Phi(Y,t)$ ). A complementary theory must then be used in order to define the dominant eddy. Lumley (1967) suggests to make use of the shot noise expansion. Only second order correlations are used, then the time phase relationship is lost. To retrieve phase information, Lumley (1981) introduces third order statistics, but different techniques may be used (Moin 1989). However, even if the characteristic eddy cannot be easily reconstructed, the contribution of each mode to the instantaneous velocity may be obtained.

For a continuous time sample of velocities simultaneously measured, the instantaneous contribution of each mode to the velocity can be obtained from the deterministic eigen-functions. This contribution may be given by:

$$u^{(n)}(Y;t) = \int \Phi^{(n)}(Y;t-\tau)\, a^{(n)}(\tau)\, d\tau,$$

or in the frequency domain:

$$\widetilde{u}^{(n)}(Y;f) = \widetilde{a}^{(n)}(f)\, \widetilde{\Phi}^{(n)}(Y;f),$$

where

$$\widetilde{a}^{(n)}(f) = \int \widetilde{u}(Y;f)\, \widetilde{\Phi}^{*(n)}(Y;f)\, dY.$$

Then, if the velocity is known at the same time on the whole $Y$ extent (that is what is done with the rake of hot-wires), the contribution of each mode can be given. For this contribution, the phase information is then contained in the velocity signals themselves. This instantaneous reconstruction can be compared with the one that could be obtained by using the method of Sirovich (1987). The knowledge of these contributions may be of interest since, even if POD involves averaged contributions implying uncorrelated statistical eigen-functions, some interaction or triggering between modes may appear for short time durations.

An example of the reconstruction of the instantaneous velocity $u$ on the mixing layer axis is illustrated on figure 7. The first mode (7-b) is well representative of the large scale variations of the original signal (7-a). In order to get a better global estimation of how each mode participates to the velocity over the whole $Y$ extent, some Pseudo-Flow-Visualizations (PFV) have been performed. The technique of PFV used here has been described in Delville et al. (1989). In these PFVs, the vertical direction corresponds to a $Y$ extent of about $2\delta_\omega$, centered on the mixing layer axis, and the horizontal direction to a time period of about $2/f_p$. The instantaneous velocity distribution is related to colors through a Look Up Table (LUT). The colored plate 1 shows a typical sample of PFV extracted from a video movie realized from a long record of velocities. Of course the image obtained depends on the coloration rule used. Two LUTs have been used for this PFV. The LUT (I) is associated to the absolute velocity $u(Y,t)$. Then the instantaneous $Y$ extent of the mixing layer and the quasi-periodicity of the main pattern is exhibited (plate 1-a). An other LUT (II) is used for the fluctuating velocity: $u'(Y,t) = u(Y,t) - \overline{u}(Y)$. This LUT emphasizes both the turbulent intensity (cold-to-hot coloration rule) and the sign of the velocity fluctuation (sharp variation around $u' = 0$). Then, the phase shift between the upper and lower parts of the mixing layer is clearly exhibited (plate 1-b).

This LUT allows to get a visual estimation of the participation of the modes to signal reconstruction (plate 1-c to 1-e ). As for the correlations, the first mode tends to capture the energy where this energy is located: on the mixing layer axis (plate 1-c). For the outer parts of the mixing layer, the contribution of the first mode is very weak. The instantaneous contribution of the modes follows the same behaviour as the correlations: symmetry for the odd modes and antisymmetry for the second mode (plate 1-d, 1-e). On the other hand, the second and third modes are responsible of the phase opposition for the external parts of the mixing layer. The sum of the first four modes drawn on plate 1-f is then very well representative of the original velocity PFV for the entire $Y$ domain.

## 5. Conclusion.

The present study shows that, in the case of the turbulent plane mixing layer, this simplified version of POD is efficient as well for reconstructing the turbulent energy as for indicating the participation of the individual modes to the instantaneous realization of the flow. It has been proposed a direct relationship between the relevant frequency domain and the spatial mesh size through a simple law which ought to be confirmed for other flows. For the first mode, a special spectral behaviour can be found, corresponding to a bidimensional turbulence. On the other hand, the use of PFV allows to gain a global estimate of the participation of each mode and will be useful in the analysis of the interaction between modes.

### References

Aubry, N., Holmes, P., Lumley, J.L. & Stone, E. 1988 The dynamics of coherent structures in the wall region of a turbulent boundary layer. *J. Fluid Mech.* **192**, 115-173.

Bakewell, H.P. & Lumley, J.L. 1967 Viscous sublayer and ajacent wall region in turbulent pipe flow. Phys. Fluids **10**, 1880-1889.

Cantwell, B.J. 1981 Organized motion in turbulent flow. *Ann. Rev. Fluid. Mech.* **13**, 557-515.

Delville, J., Chahine, Z. & Bonnet, J.-P. 1987 Experimental study of an incompressible, plane mixing layer by temporal and spectral analysis. In *Advances in Turbulence* (eds. G. Comte-Bellot & J. Mathieu), pp. 435-444. Springer.

Delville, J., Bellin, S., Garem, J.H. & Bonnet, J.-P. 1989 Analysis of Structures in a turbulent, plane mixing layer by use of a pseudo flow visualization method based on hot-wire anemometry. In *Advances in Turbulence 2* (eds. H.-H. Fernholtz & H.E. Fiedler), pp. 251-256. Springer.

Glauser, M.N., Leib, S.J. & George, W.K. 1985 Coherent structures in the axisymmetric turbulent jet mixing layer. *Fifth Symp. on Turbulent Shear Flows 5, August 7-9, Ithaca, NY.*

Glauser, M.N., Zheng, X. & Doering, C. 1989 The dynamics of organized structures in the axisymmetric jet mixing layer. *Turbulence 89, September 18-21, Grenoble F.*

Glezer, A., Kadioglu, Z. & Pearlstein, A.J. 1989 Development of an extended proper orthogonal decomposition and its application to a time periodically forced plane mixing layer. *Phys. Fluids A* **1** (8), 1363-1373.

Herzog, S. 1986 The large scale structure in the near-wall region of turbulent pipe flow. Ph.D. Thesis, Cornell University.

Hussain, A.K.M.F. 1986 Coherent structures and turbulence. *J. Fluid Mech.* **173**, 303-356.

Jirka, G.H. & Dracos, T. 1989 Mixing in turbulent shear flows in a shallow fluid layer. *Turbulence 89, September 18-21, Grenoble F.*

Lumley, J.L. 1967 The structure of inhomogeneous turbulent flows. In *Atmospheric Turbulence and Radio Wave Propagation* (eds. A.M. Yaglom & V.I. Tatarsky), p. 166. Moscow: Nauka.

Lumley, J. 1970 *Stochastic Tools in Turbulence.* Academic.

Lumley, J. 1981 Coherent structures in turbulence. In *Transition and turbulence* (ed. R.E. Mayer) pp. 215-241. Academic.

Payne, F.R & Lumley, J.L. 1967 Large eddy structures of the turbulent wake behind a circular cylinder. *Phys. Fluids* S194.

Sirovich, L. 1987 Turbulence and the dynamics of coherent structures *Q. of Appl. Math.* **45** ,561-590.

Townsend, A.A. 1956 *The Structure of Turbulent Shear Flows.* Cambridge University Press.

# On the Three-Dimensional Dynamics of Coherent Vortical Structures Forming in Free, Shear Flows

J. C. LASHERAS[α] and E. MEIBURG[β]

[α] *University of Southern California*
*Los Angeles, California 90089-1453, USA*

[β] *Brown University*
*Providence, Rhode Island 02912, USA*

*Turbulence and Coherent Structures*
*O. Métais and M. Lesieur, eds.*
*Selected papers from "Turbulence 89: organized structures and turbulence in fluid mechanics"*
*Grenoble, 18-21 September 1989*

## 1. Abstract

The dynamics of the three-dimensional evolution of coherent vortical structures forming in moderate Reynolds number free, shear flows is discussed. Laboratory experiments of the evolution of transitional plane mixing layers, wakes and axisymmetric jets, developing from initially laminar conditions, subjected to temporal and spatially varying periodic perturbations are analyzed by means of an inviscid vortex dynamics model of the evolution of one or several layers of vortex filaments in which the initial vorticity is discretized. For planar and axisymmetric flow configurations, the formation and evolution of streamwise vortex tubes and their interaction with the primary, two-dimensional, coherent vortical structure is analyzed. The symmetry features and the topological transitions of the three-dimensional vorticity field resulting from the growth of the initial perturbations is studied for both the planar and axisymmetric cases.

Comparison between the experimental results and the numerical simulations indicates that important features of the three-dimensional evolution of the large scale vortical structures can be reproduced by inviscid vortex dynamics. Vortex stretching in the braid regions appears to be the primary mechanism for the formation of concentrated streamwise vorticity in free shear flows such as mixing layers, wakes and jets.

Several topological modes of the evolving three-dimensional vorticity field having different symmetry properties are shown to exist in both mixing layers and wake flows.

## 2. Introduction

Investigations performed over the past twenty years have greatly enhanced our understanding of the characteristic features of the development of turbulent, free shear flows (for reviews see Roshko 1976, Townsend 1979, Cantwell 1981, Ho and Huerre 1984, Liu 1989). Numerous flow visualization experiments, quantitative measurements, and more recently, numerical simulations have focused on the two-dimensional aspects of the large-scale structures which are known to contain a large portion of the turbulent kinetic energy and, that mantain certain degree of coherency over a substantial portion of the flow. They have shown that turbulent plane mixing layers and wakes, as well as axisymmetric jets, exhibit features which bear similarities to those of the corresponding laminar flow, namely the formation of coherent spanwise rollers, Karman vortices, and vortex rings, respectively. Both the experiments and the numerical simulations have also shown that under certain conditions these nominally two-dimensional flows are sensitive to three-dimensional perturbations. The amplification of these perturbation has been found to lead to the formation of three-dimensional coherent vortical structure. Furthermore the apearence of this streamwise vortical structure is believe to play an important role in the transitional process from the laminar stage to the fully turbulent regime.

The principal instability mode in mixing layers, wakes, as well as in jets, leads to the concentration of vorticity with spanwise coherence which induce strong strain fields between the developing vortical features. Under certain conditions, it was shown that the extensional component of the strain field can effectively amplify three dimensional perturbations leading to the stretching and reorientation of the predominantly spanwise or axysimmetric vortex structures into the streamwise direction (Corcos and Lin 1984, Lasheras and Choi 1988, Ashurst and Meiburg 1988, Meiburg, Lasheras and Martin 1989). Considering these common characteristics of the planar and axisymmetric free shear flows, it seems natural to look for generic features in their transitional and turbulent stages. Searching for potential ways to actively or passively control the flow, it is then important to elucidate how strong a link exists between the coherent two- and three-dimensional structures observed in the transitional stages and those that dominate the fully turbulent state. For plane boundary layers, Gilbert and Kleiser (1988) have recently demonstrated by means of numerical simulation that structures typical of wall turbulence, such as hairpin vortices and low- and high-speed streaks, exist already during the late stages of transition. If one assumes that free shear flows behave similarly, i.e. that the flow structures formed during the transitional process will affect the fully turbulent flow, it is tempting to analyze and classify the emergence of such structures as a function of specific initial or boundary conditions.

In this paper, we comparatively study the three-dimensional transition of plane shear layers and wakes, as well as axisymmetric jets, developing from initially laminar conditions. We will review the flow visualization experiments and three-dimensional vortex dynamics simulations of plane shear layers (Lasheras, Cho and Maxworthy 1986, Ashurst and Meiburg 1988, Lasheras and Choi 1988), plane wakes (Meiburg and Lasheras 1988, Lasheras and Meiburg 1990) and jets (Meiburg, Lasheras and Martin 1989) subjected to three-dimensional perturbations in their initial conditions. Because the question of possible generic aspects common to this class of free shear flows appears not only to be of fundamental interest for the understanding of the evolution of transitional and turbulent flows, but also to develop active or passive control mechanisms for mixing enhancement, we will investigate the common patterns in which the vorticity field undergoes topological changes.

A brief description of the numerical and experimental techniques is given in sections 2 and 3. We discuss plane shear layer in section 4. Section 5 will demonstrate the effect of introducing a second layer of vorticity of the opposite sign, i.e. the plane wake, and in section 6 we will discuss the case of the axisymmetric jet.

## 3. Experimental Techniques

The flow visualization experiments of plane shear layers and wakes were performed in water and, the axisymmetric co-flowing jet experiments were conducted in a wind tunnel. Detailed descriptions of the experimental facilities can be found in Lasheras, Cho and Maxworthy (1986), Lasheras and Choi (1988), Meiburg and Lasheras (1988), as well as in Meiburg, Lasheras and Martin (1989), here we will limit the discussion to a brief summary of the facility and the experimental techniques used.

### 3.1. Visualization Experiments of Shear Layers and Wakes

The experimental facility consists of a blow-down water channel where two laminar, horizontal streams are formed, initially separated by a flat plate. At the trailing edge of the splitter plate, the two streams meet and mix in a 2 m long test-section. The two laminar flows are created in a 80 cm long settling chamber followed by a three-dimensional, 40 cm long, contraction nozzle, with a contraction ratio of 9:1. The facility operates continuously in a closed loop by recirculating the fluid collected in the dump section back to the supply tanks through pumps.

We analyzed the response of the plane shear layers and wakes to several types of perturbations periodically distributed along the span. The first type consisted of a single-wave horizontal perturbation of the vorticity sheet emanating from the plate. This was achieved by modifying the trailing edge of the splitter plate to give it a sinusoidal "indentation" of wavelength $\lambda$ and amplitude $a$, (Figure 1a). The effect of the indentation in the plate is to cause a horizontal sinusoidal displacement of the geometrical origin of the vorticity layers, thus producing a periodic axial disturbance in the vortex sheets. The second type consisted of a single-wave sinusoidal vertical (cross-stream) perturbation in the vorticity sheets emanating from the plate. This is achieved by using a modified splitter plate containing a sinusoidal "corrugation" of wavelength $l$ and amplitude a (Figure 1b). The corrugation in the splitter plate causes a sinusoidal cross-stream displacement of the vorticity sheets emanating from the plate. The third type of perturbation studied consisted of a combined horizontal and vertical wave. By corrugating the indented splitter plate, both the axial (horizontal) and cross-stream (vertical) perturbation components can be created simultaneously. For this third case we limited our study to only the cases where the two perturbation components have the same wavelength. Furthermore, we studied only the cases where the indentation and the corrugation are in-phase or 180 degrees phase-shifted. Hereinafter these cases will be refered as "in-phase" and "out-of-phase" respectively.

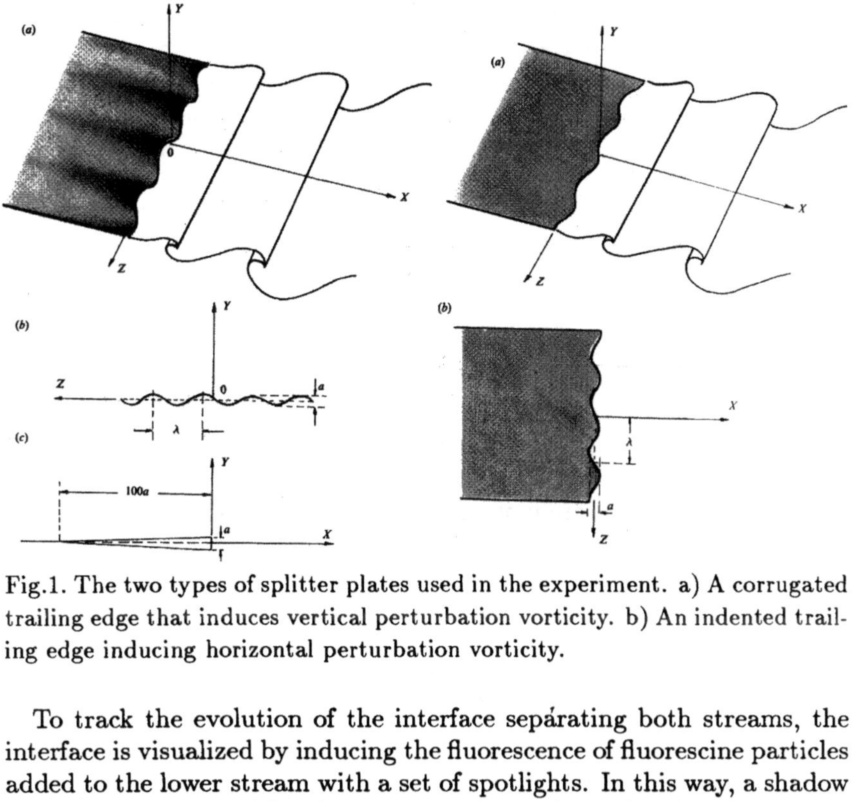


Fig.1. The two types of splitter plates used in the experiment. a) A corrugated trailing edge that induces vertical perturbation vorticity. b) An indented trailing edge inducing horizontal perturbation vorticity.

To track the evolution of the interface separating both streams, the interface is visualized by inducing the fluorescence of fluorescine particles added to the lower stream with a set of spotlights. In this way, a shadow effect can be created by the corrugated interface, thereby showing its three-dimensionality. A large concentration of fluorescine particles results in a solid, opaque effect at the interface. Additional information about the temporal and spatial evolution of the interface is obtained by using laser induced fluorescence (LIF) cross-cuts of the wake.

### 3.2. Visualization of Co-flowing Jets

Our base flow consists of two laminar coflowing jets of air. The primary jet instability is triggered through the addition of periodic pressure pulses in the axial direction along the central jet. Acoustic forcing of the central jet leads to a sinusoidal variation in the strength of its boundary layer, which in turn causes a strongly periodic axisymmetric roll-up when the two streams merge.

In addition, following a similar technique to the one used in plane shear flows described above, the three-dimensional instability is forced by introducing a sinusoidal corrugation at the nozzle lip. The number of wavelengths in the circumferential direction is varied from 4 to 9 in this study. The nozzle has an average diameter of 5 cm while the corrugation amplitude is 2 mm (an order of magnitude smaller than the thickness of the boundary layers). The central jet has velocities in the range of 40-50 cm/s while the outer jet, which merely serves to stabilize the flow pattern, typically has a velocity one order of magnitude smaller.

In order to visualize the three-dimensional evolution of the flow, we have modified a technique used to visualize flames which was developed by Chen and Roquemore (1986). The dry air stream that flows through the central jet is bubbled through a $TiCl_4$ solution until it becomes saturated with $TiCl_4$ vapors. Subsequently, this saturated dry air passes through a settling chamber equipped with the standard array of honeycomb and screens which is followed by an axisymmetric nozzle. The outer co-flowing air flow is intentionally kept saturated with the $H_2O$ vapor typically present in the laboratory environment. As the two co-flowing jets meet at the nozzle lip, the $TiCl_4$ vapors of the central jet react with the $H_2O$ vapors of the outer stream. Thus, only at the interface separating the two streams, the reaction

$$TiCl_4 + 2H_2O \longrightarrow TiO_2 + 4HCl$$

results in the generation of micron-size $TiO_2$ particles. The diffusivity of these particle is small enough to ensure that throughout the characteristic residence time in the test section they ocupy only the position of the interface. Therefore, the interface can subsequently be visualized by light scattering of these particles induced by planes of laser light that can be placed either at a given distance from the nozzle lip or traversing with the convective velocity of the vortex rings.

## 4. Numerical Technique

The non-divergent nature of the velocity field in incompressible flows, along with the definition of vorticity, allow for a complete description of the kinematics of the flow in the form of the Biot-Savart law. Employing the theorems of Kelvin and Helmholtz and following the general concepts reviewed by Leonard (1985), vortex filaments are used for the representation of the vorticity field. A detailed description can be found in Ashurst and Meiburg (1988), hence we will only review the main points of the technique here. Each filament is represented by a number of node points along its centerline, through which a cubic spline is fitted to give it a smooth shape. The Biot-Savart law is evaluated assuming a smooth invariant vorticity distribution around the filament centerline. Incorporation of this vorticity distribution into the Biot-Savart law then allows us to obtain the velocity at any position by integrating over the arclength of all filaments in the flowfield. For the numerical simulation, we limit ourselves to the temporally growing problem, i.e., our flow is periodic in the streamwise direction. The Biot-Savart integration is carried out with second order accuracy both in space and in time by employing the predictor-corrector time-stepping scheme and the trapezoidal rule. As the flow develops a three-dimensional structure, the vortex filaments undergo considerable stretching. To maintain an adequate resolution, additional nodes are introduced, based on a criterion involving distance and curvature. Furthermore, the time-step is repeatedly reduced as local acceleration effects increase. The filament core radius decreases as its arclength increases to conserve its total volume.

## 5. The Plane, Free Shear Layer

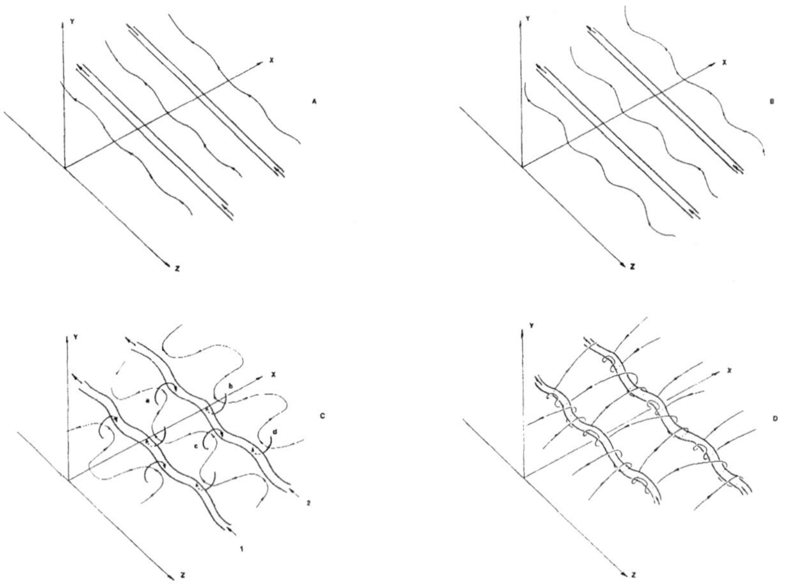


Fig.2. Schematic representation of the vorticity field evolving in a three-dimensional plane shear layer.

When the plane mixing layer is subjected to a combined spanwise and streamwise sinusoidal perturbation, its evolution is observed to qualitativelly undergo the following steps (Figure 2). Under the effect of the Kelvin-Helmholtz instability, the vorticity sheet emanating from the plate is redistributed into regions of strong vorticity (the cores of the spanwise vortices) and weak vorticity (the braids connecting consecutive spanwise vortices). As the instability amplifies, the evolving spanwise vortices begin to develop strain fields in the braids between them. In the topological structure of the vector field a saddle point (free stagnation point) appears in between two consecutive rolers. The perturbation component of the voticity present near the stagnation region is thereby subjected to to stretching along the principal direction of positive strain. The three-dimensional vorticity present in the braids then experience a progresive stretching evolving into counter rotating pairs of vortex tubes. The evolving axial loops finally envelop the large spanwise vortex tubes as shown in Figure 2d.

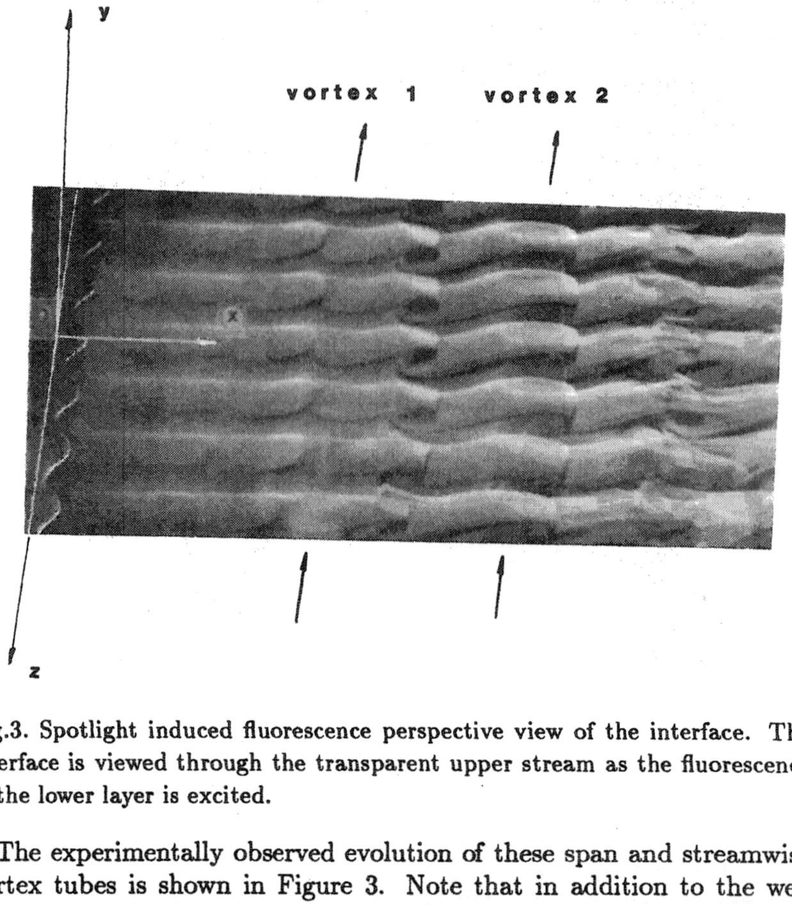


Fig.3. Spotlight induced fluorescence perspective view of the interface. The interface is viewed through the transparent upper stream as the fluorescence of the lower layer is excited.

The experimentally observed evolution of these span and streamwise vortex tubes is shown in Figure 3. Note that in addition to the well known concentration of vorticity in the spanwise rolers, a well-organized array of streamwise vortices appears superimposed onto the spanwise structure. A LIF end-view cross-cut of the convective evolution of a braid between to consecutive vortices is given in Figure 4. Note that a pair of counter-rotating streamwise vortices forms per each wavelength of the spanwise perturbation introduced at the trailing edge of the splitter

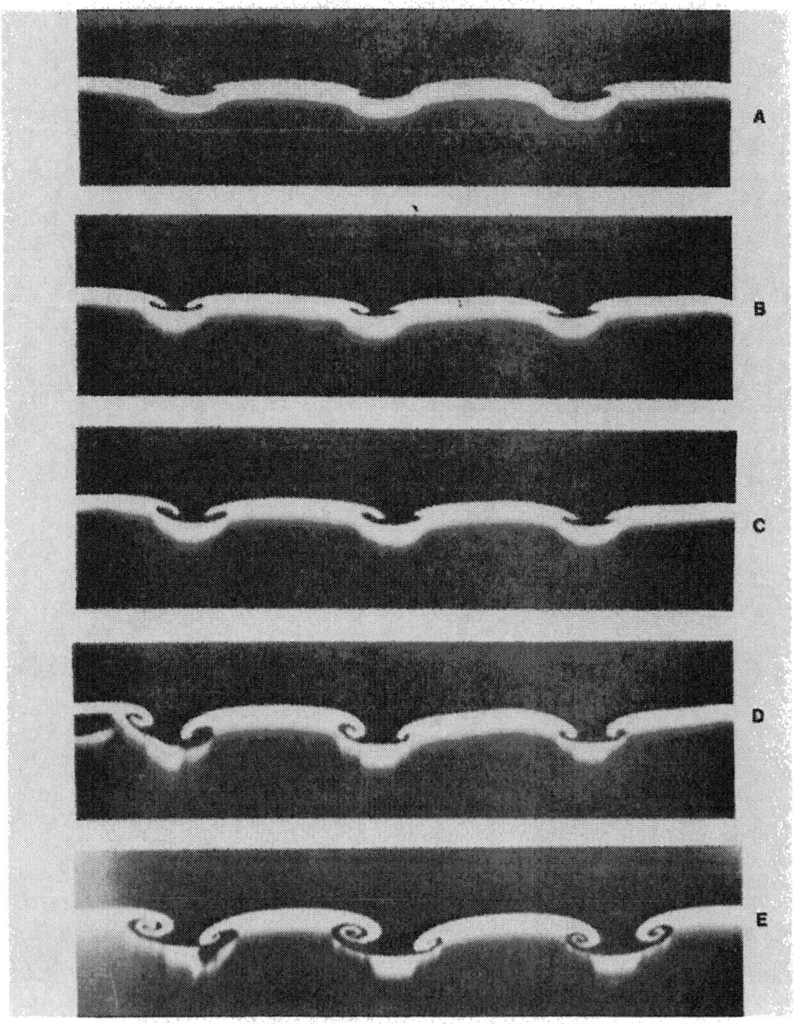


Fig.4. LIF end-view cross-cuts of the braids. The Downstream distance increasing from top to bottom of the figure.

plate. Note also the unequal spacing between the the evolving counter-rotating streamwise vortex pairs. The structure of the topology of the vorticity field after the axial vortex loops wrap around the spanwise vortices is shown in the plane view of the direct interface visualization of figure 5. Also shown in the figure is the topology of the vortex tubes between two consecutive spanwise vortices.

The same qualitative development observed in the experiments is precisely reproduced in the vortex dynamic simulations of Ashurst and Meiburg (1989) shown in Figure 6. The temporally developing shear

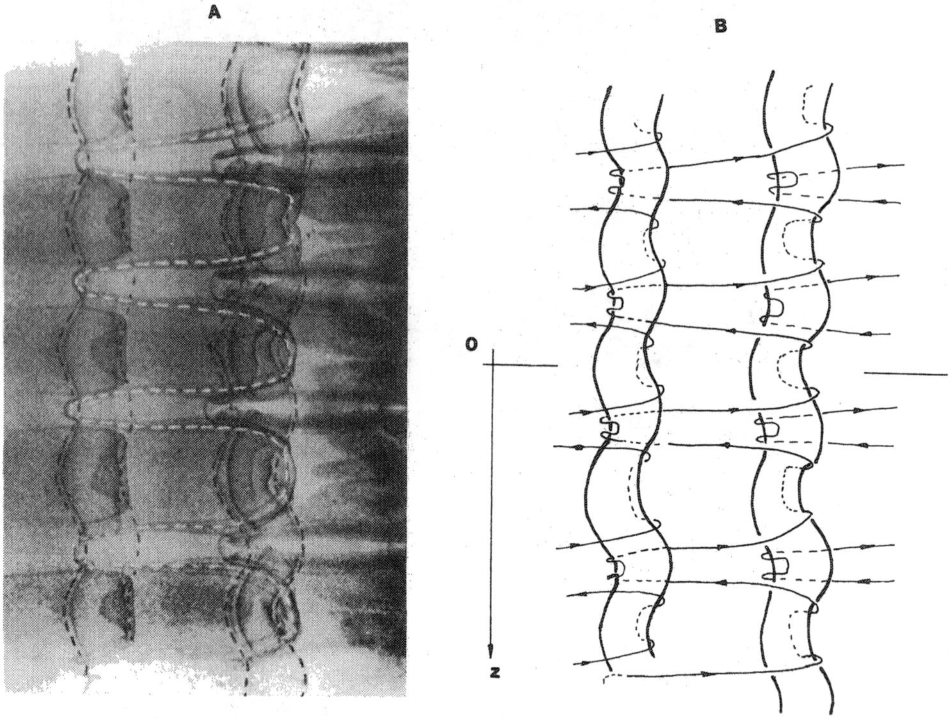


Fig.5. Structure of the vorticity field after the streamwise vortex loops have wraped around the spanwise vortices.

layer shows qualitativelly the sequence of events observed experimentally. As reinforced by the experiments, the vortex filament simulations show that the stretching in the braids is the driving mechanism for the concentration of streamwise vorticity in counter-rotating vortex tubes. Both the experiments and simulations supports the analysis of the three-dimensional motion performed earlier by Corcos (1979, 84). The simulation showed that the inclusion of two signs of vorticity (to model the wake component) leads to a more non-linear development of the three-dimensional structure. The above numerical results suggest that the non-uniform spacing between streamwise vortex tubes observed in the experiments (Figures 4 and 5) is the result of the interaction between the stronger (upper) and weaker (lower) vorticity emanating from the plate. Note in the cross-stream LIF cuts of figure 4 that near the origin of the mixing layer, wher the wake effect is still quite strong, the spacing between streamwise vortices is quite unequal, while as one moves downstream (and the wake component disappears) the distance among them becomes almost equal.

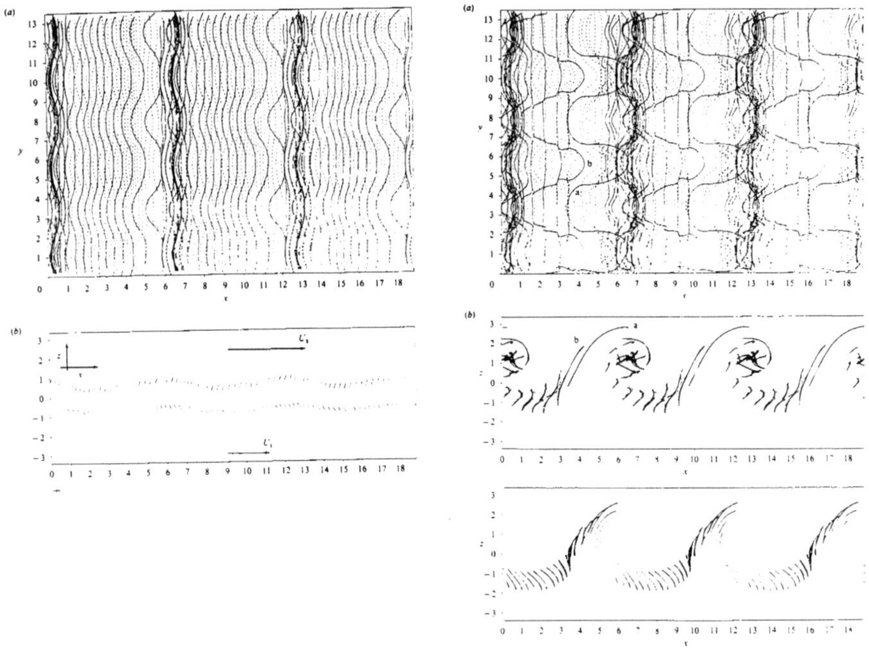

Fig.6. Inclusion of a second weaker layer of opposite vorticity in the shear layer simulation changes the symmetry properties and leads to better agreement with the experiment.

## 6. The Plane Wake

Under the combined effect of both a streamwise and a spanwise perturbation, the overall induction of the vorticity (concentration, reorientation, and stretching) results in the formation of counter-rotating pairs of streamwise vortices which are located in the braids connecting consecutive Karman vortices of opposite sign (planes inclined to the plane of the wake). Depending on the orientation of the initial spanwise perturbation, the wake was observed to develop two distinct, three-dimensional vorticity modes of different symmetry properties. Under the global induction of the vorticity, the perturbation vorticity in both layers is reoriented along the directions of positive strain in the evolving strain fields created by the two-dimensional spanwise vortices. When subjected to a vertical (cross-stream) initial spanwise perturbation this induction results in a $180^o$ phase shift between the perturbation-vorticity of both layers, leading to a loss of the vertical ($x$-$y$ plane) symmetry during the initial development of the wake. On the other hand, under the effect of a horizontal (plane of the wake) initial spanwise perturbation, the stretched perturbation vorticity reoriented along the directions of positive strain in both layers is found to remain in phase, preserving the vertical ($x$-$y$ plane) symmetry of the wake.

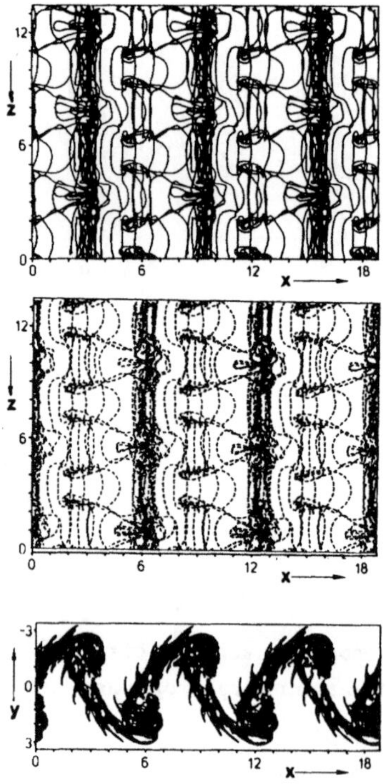


Fig.7. Vertical perturbation vorticity (mode 1). Numerical simulation of a three-dimensional plane wake. Shown are top, side, and front views of the vortex filament configurations at time 169 as well as the corresponding interface shapes.

The results of the numerical simulation at an inermediate time, for both the case of the vertical (cross-stream) and horizontal (plane of the wake) sinusoidal perturbations are given in Figures 7 and 8 respectively. The case selected for the study corresponds to a wavelength of the spanwise perturbation equal to 2/3 of the two-dimensional streamwise perturbation. At this time, in both cases, the formation of the Karman vortex street has been completed and the appearance in the braids of streamwise vorticity concentration is clearly seen. For the purpose of clarity, the vortex filaments evolving from the upper and lower layer are given in two separate plan views. Notice that while the side view of the

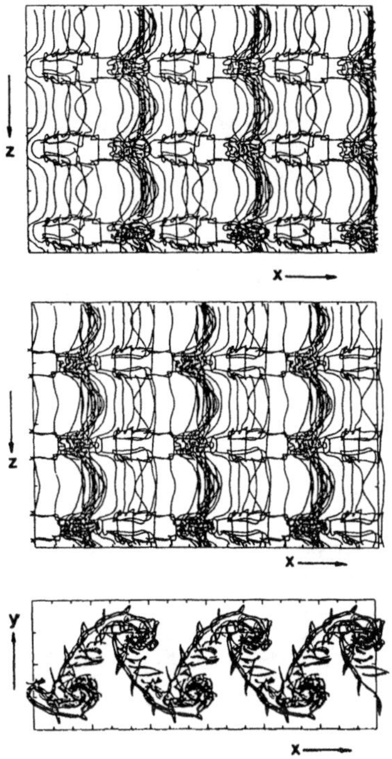


Fig.8. Horizontal perturbation vorticity (Mode 2). Numerical simulation of a three-dimensional plane wake. Shown are top, side, and front views of the vortex filament configurations at time 169 as well as the corresponding interface shapes.

vortex filaments in the two cases reveals a practically identical shape, the plan view of the three-dimensional vortex filaments already shows some important differences. In Mode 1 (resulting from the cross-stream perturbation, Figure 7), the counter-rotating pair of streamwise vortices formed in both layers acquires a lambda-shaped form, while in Mode 2 (resulting from the perturbation oriented in the plane of the wake, Figure 8) the counter-rotating pairs of streamwise vortices of both layers remain in the plan view always parallel to the $x$ direction, while their relative distance is not constant across the span.

In addition, the interaction of the evolving streamwise vortices with the spanwise ones leads to a vertical ($x$-$y$ plane) undulation in the axis of the spanwise vortices. Figure 9 shows the end-view of all the filaments composing the upper and lower layers. Observe that the axes of the Karman vortices are undulated in the vertical plane. In Mode 1, this undulation is in-phase between the Karman vortices of opposite sign,

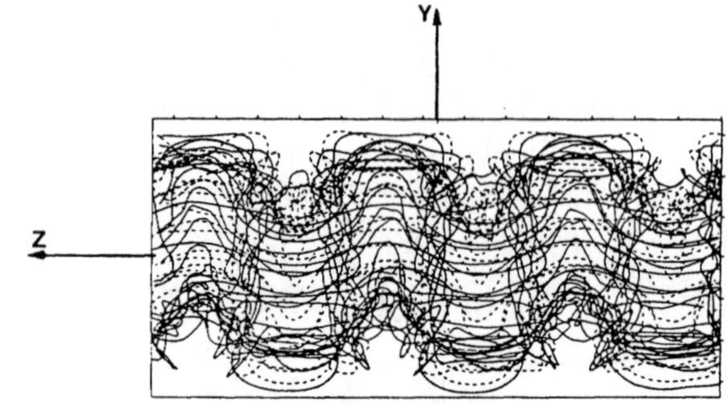


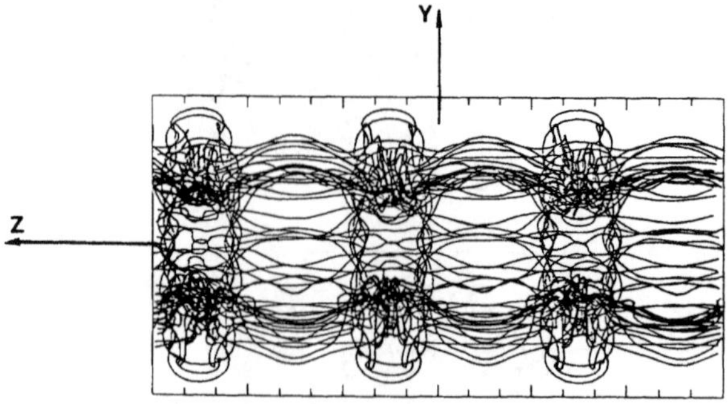


Fig.9. End-view of the filaments of the upper and lower layers. Modes 1 and 2.

while in Mode 2 it is 180° phase shifted, resulting in a varicose shape.

These undulations of the axes of the Karman-like vortices described above are clearly seen in both Modes 1 and 2 in the lateral perspective views of the visualized interface given in Figure 10. It should be pointed out that as was shown in our previous work (Meiburg and Lasheras, 1988), and observed in these interface visualizations, Mode 1 evolves from a vertically ($x$-$y$ plane) anti-symmetric configuration whereby the interaction of the streamwise and spanwise vortices leads to a modulation along the span in the strength of the Karman vortices. This modulation in the strength produced during the initial evolution of the three-dimensional instability is 180° phase shifted between the Karman vortices of opposite sign. However, the interaction of the streamwise vortices with the spanwise vortices of opposite sign leads to a loss of strength in the modulation, and eventually to a recovery in the vertical ($x$-$y$ plane) symmetry of the wake. On the other hand, in Mode 2, the modulation in the strength of the Karman vortices in each layer is in-phase, and thus, the initial evolution and the final stage always exhibits

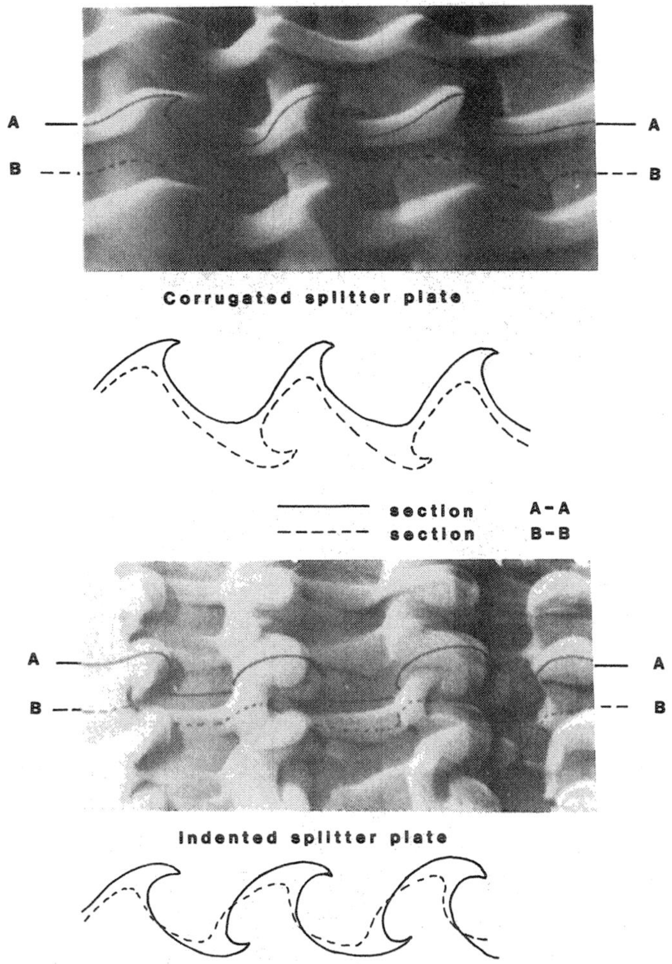


Fig.10. Flow visualization experiment of the plane wake. a) Corrugated trailing edge. b) Indented trailing edge.

a vertical symmetry at each $x$-$y$ plane.

Figures 11 and 12 show a three-dimensional perspective view of the two modes of vorticity resulting from the vertical and horizontal perturbation respectively. These schematic representations correspond to a late time step comparable to the one of Figures 7 and 8. Observe in Mode 1 that when the streamwise vortices generated through stretching in each layer wrap around two consecutive Karman vortices, the cancellation of vorticity of opposite sign leads to a recovery in the vertical symmetry of the wake. Notice also that in both Modes 1 and 2, at this late stage, the

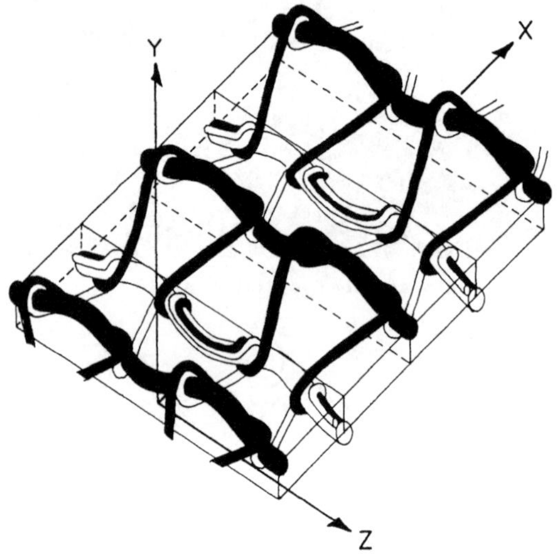


Fig.11. Vertical perturbation vorticity (Mode 1). Schematic representation of the vorticity field. Observe the formation of closed vortex loops.

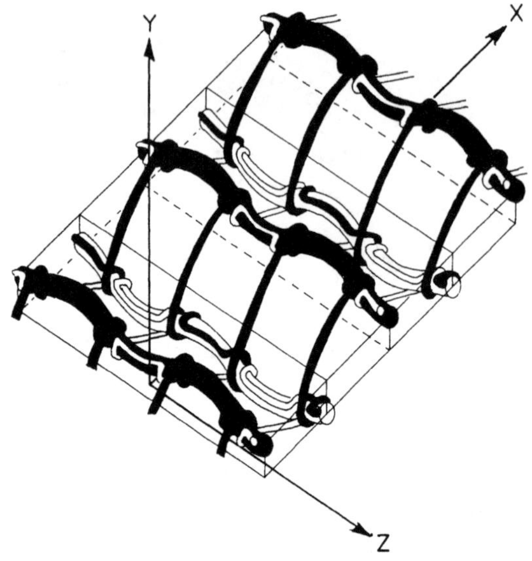


Fig.12. Horizontal perturbation vorticity (Mode 2). Schematic representation of the vorticity field. Observe the formation of closed vortex loops aligned in the direction of the flow.

strength of the Karman vortices becomes uniform along the span.

## 7. The Axisymmetric Jet

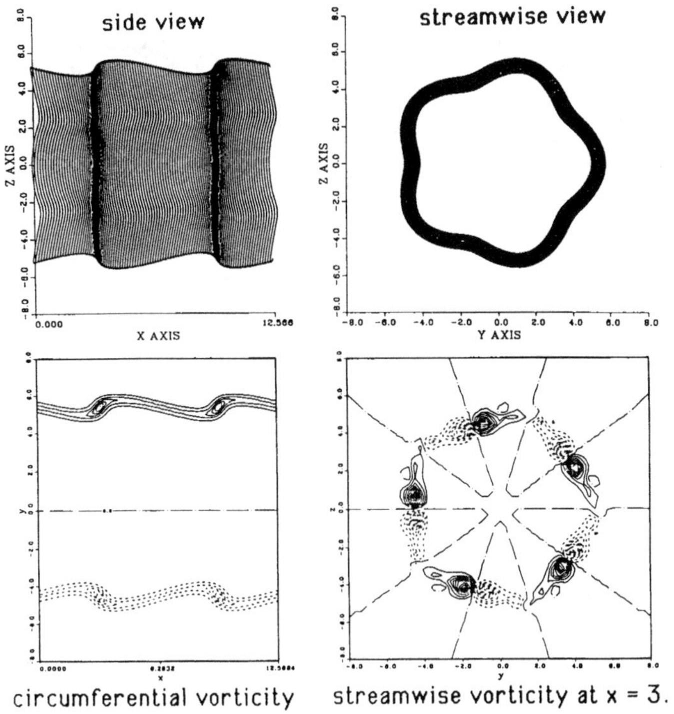


Figure 13: Numerical simulation of a three-dimensional axisymmetric jet. (t= 7.8)

Fig.13. Numerical simulation of a three-dimensional axisymmetric jet. Lateral view.

Figures 13, 14 and 15 show the results of a calculation in which we introduce both a perturbation in the circumferential direction as well as a single sinusoidal perturbation in the streamwise direction. This disturbance periodically displaces the filament centerline in the radial direction, thus duplicating the effect of the corrugated nozzle. Both the perspective views and the side views clearly show the roll-up of the axisymmetric shear layer into vortex rings. These vortex rings evolve developing a modulation in the circumferential direction (shown in the front views). Both from the side views and the perspective views, one can recognize that the portions of the vortex rings farthest away from the jet axis are retarded downstream, an effect caused by the strong self-induction of the vortex rings (as inferred from the curvature of the vortex rings). This self-induction effect tends to accelerate the outer sections of the vortex rings while slowing down the ones closer to the jet axis. The streamwise vorticity component that the vortex rings generate

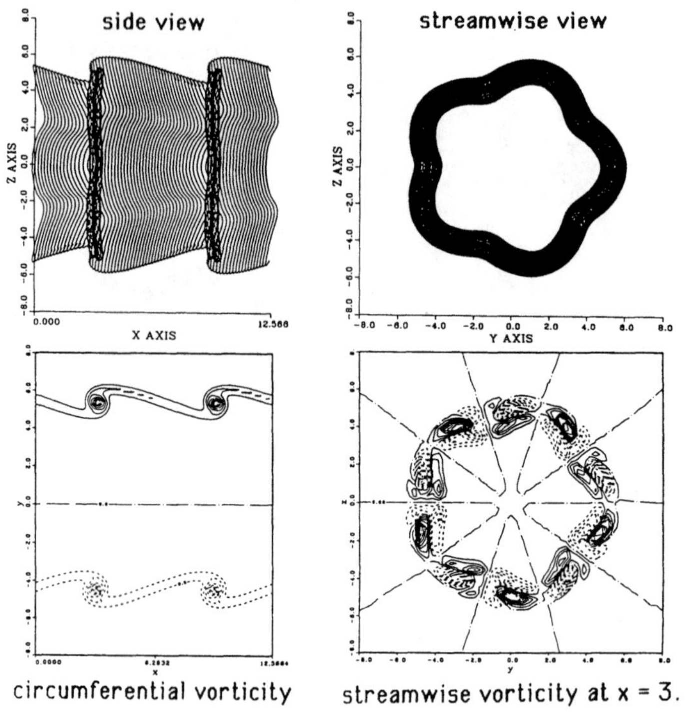


Fig.14. Numerical simulation of a three-dimensional axisymmetric jet. Perspective view.

in this way, is of opposite sign to that developing close to the free stagnation point in the braid regions. We furthermore recognize that the streamwise braid vorticity is oriented such that it will cause the wavy radial displacement of the vortex rings to be amplified. Hence it appear that there is a strong coupling between the vortex rings and the emerging streamwise braid vorticity: While the vortex rings set up the global strain field that amplifies small streamwise perturbations in the braid region, the streamwise vorticity created in this way amplifies wavy displacements of the vortex rings.

The above mentioned flow features become also manifested in the experimental flow visualization (Figure 16). A horizontal (cross-stream) laser sheet that cuts the jet clearly shows the periodic displacement of the jet interface (white in the figure) away from the axis, resulting in a lobe structure of the white targeted fluid ($TiO_2$ particles), while in between these lobes ambient fluid (dark) is entrained into the jet. The presence of the counterrotating pairs of streamwise vortex tubes, shown in Figures 14 and 15, manifested itself in the alternate-sign periodic fold-

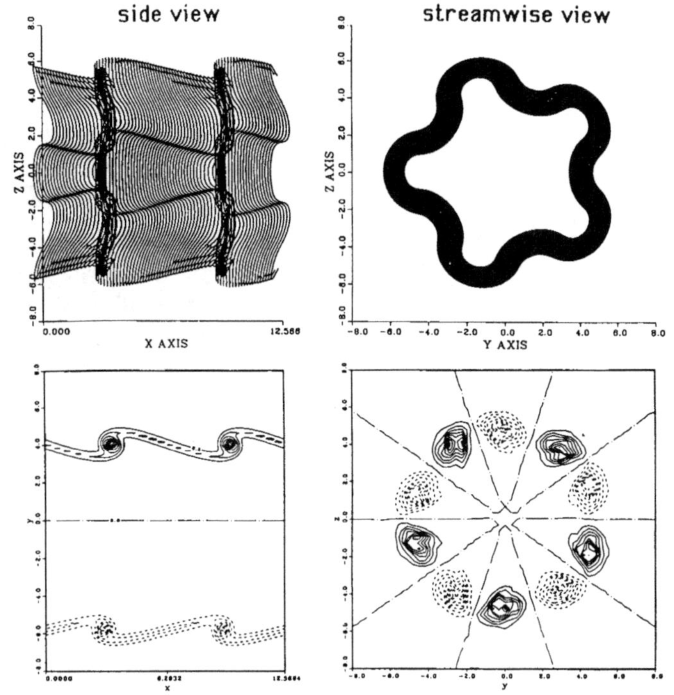


Fig.15. Numerical simulation of a three-dimensional axisymmetric jet. End view.

ing of the interface as one moves along the azimuthal coordinate along the cross-cut. This flow visualization corresponds to six wavelengths in the circumferencial direction.

## 8. Conclusions

On the basis of flow visualization experiments and three-dimensional vortex dynamics simulations we have studied various free shear flows with respect to their three-dimensional transition. By comparing plane shear layers and wakes, we observe that in addition to a strain field, which reorients the perturbed two-dimensional vorticity into streamwise direction, the presence of vorticity of the opposite sign may lead to the formation of closed vortex loops. The plane wake furthermore represents an example of a nominally two-dimensional flow that depending on its initial perturbations, can acquire three-dimensional states of different symmetry properties. The single axisymmetric shear layer behaves similarly to the plane shear layer in some respects. However, the axisymmetric nature of the extensional strain field leads to a vorticity configuration that through a cut-and-connect process may also result in the formation

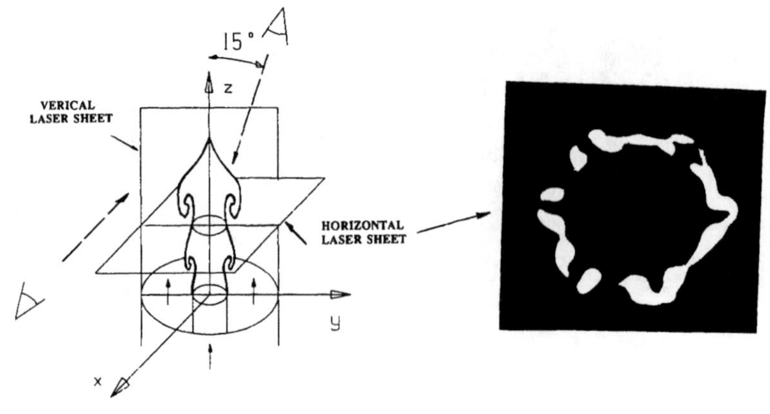


Fig.16. Flow visualization experiment for the three-dimensional axisymmetric jet.

of closed vortex loops with a self-induced velocity almost perpendicular to the axial direction.

This work is supported by the National Science Foundation under grant #MSM-8809438 (to EM), by DARPA under URI contract #N00014-86-K0754 to Brown University and by United Technologies Corporation (to JCL). The San Diego Supercomputer Center is providing computing time on its CRAY-X/MP.

## 9. References

ASHURST, W.T. and MEIBURG, E. Three-dimensional shear layers via vortex dynamics. J. Fluid Mech. 189, p. 87. 1988.

BROWAND, F.K. and TROUTT, T.R. The turbulent mixing layer: geometry of large vortices. J. Fluid Mech. 158, p. 489. 1985.

CANTWELL, B.J. Organized motion in turbulent flow. Ann. Rev. Fluid Mech. 13, p. 457. 1981.

CHEN, L.D. and ROQUEMORE, W.M. Visualization of jet flames. Combustion and Flame. 66, p. 81, 1986.

GILBERT, N. and KLEISER, L. Near-wall phenomena in transition to turbulence. To appear in Proceedings of the International Seminar on Near-Wall Turbulence, Dubrovnik, Yugoslavia, Hemisphere Press. 1988.

HO, C.M. and HUERRE, P. Perturbed free shear layers. Ann. Rev. Fluid Mech. 16, p. 365. 1984.

LASHERAS, J.C., CHO, J.S. and MAXWORTHY, T. On the origin and evolution of streamwise vortical structures in a plane free shear-layer. J. Fluid Mech. 172, p. 231. 1986

LASHERAS, J.C. and CHOI, H. Three-dimensional instability of a plane, free shear layer: an experimental study of the formation and evolution of streamwise vortices. J. Fluid Mech. 189, p. 53. 1988

LASHERAS, J.C. and MEIBURG, E. Three-dimensinal vorticity modes in the wake of a flat plate. (in press), Physics of Fluids A. 1990.

LEONARD, A. Computing three-dimensional incompressible flows with vortex elements. Ann. Rev. Fluid Mech. 17, p. 523. 1985.

LIN, S.J. and CORCOS, G.M. The mixing layer: deterministic models of a turbulent flow. Part 3: The effect of plane strain on the dynamics of streamwise vortices. J. Fluid Mech. 141, p. 139. 1984.

LIU, J.T.C. Coherent structures in transitional and turbulent free shear flows. Ann. Rev. Fluid Mech. 21, p.285. 1989.

MEIBURG, E., LASHERAS, J.C. and ASHURST, W.T. Topology of the vorticity field in three-dimensional shear layers and wakes. Fluid Dynamics Research 3, p. 140. 1987.

MEIBURG, E. and LASHERAS, J.C. Experimental and numerical investigation of the three-dimensional transition in plane wakes. J. Fluid Mech.190, p.1. 1988.

MEIBURG, E., LASHERAS, J.C. and MARTIN, J.E. Experimental and numerical analysis of the three-dimensional evolution of an axisymmetric jet. (in press) Turbulent Shear Flows 7. Springer-Verlag. F. Durst Ed. 1990.

PAREKH, D.E. and REYNOLDS, W.C. Bifurcating air jets at high subsonic speeds. Presentation at the Sixth Symposium on Turbulent Shear Flows, Toulouse, France. 1987.

ROSHKO, A. Structure of turbulent shear flows: a new look. AIAA J. 14, 1349. 1976

TOWNSEND, A.A. Flow patterns of large eddies in a wake and in a boundary layer. J. Fluid Mech. 95, 515. 1979.

# Visualizations of helium jets in air

R. RIVA, G. BINDER, S.TARDU and M. FAVRE-MARINET

*Institut de Mécanique de Grenoble*
*BP 53X, 38041 Grenoble Cedex, France*

*Université Joseph Fourier (Grenoble)*
*Institut National Polytechnique*
*Centre National de la Recherche Scientifique*

*Turbulence and Coherent Structures*
*O. Métais and M. Lesieur, eds.*
*Selected papers from "Turbulence 89: organized structures and turbulence in fluid mechanics"*
*Grenoble, 18-21 September 1989*

**Abstract**

Axisymmetric helium jets issuing into still air at low to medium Reynolds numbers are visualized by schlieren and differential interferometric techniques. It is shown that strong self sustaining sinusoidal oscillations set in as soon as $Re > 100$, a clear manifestation of absolute instability. This is confirmed by hot wire measurements. There is no evidence of pairing. The existence of previously observed side jets is confirmed. It is shown that these develop when the Reynolds number exceeds a critical value and become stronger with increasing $Re$. Large variations of the spread in the fully turbulent region are also observed. Finally it is shown that none of these phenomena occur in the jet generated from laminar tube flow.

## 1. Introduction

Flows with large density diffences have received increased attention in the past years because of their importance in combustion, exhaust jets of hot gases and heat transfer problems. A striking example of the latter is the surface heating of the space shuttle during reentry into the atmosphere. Even far from the nose, on the rear end of the fuselage the density difference due to the heat transfer to the surface may be several hundred percent.

In order to investigate the effects of such density differences on wall flows, an experiment is being undertaken in our laboratory where helium (or air) will be injected into a boundary layer of air (or helium). While the main facility -a small pressurized wind tunnel with a test section 300 × 300 mm - is being built we have started to study the response of hot wires and surface hot film gages in air-helium mixtures and to try out various visualization techniques based on density differences on a round jet.

These visualizations of helium jets revealed features which were so strikingly different from those observed on homogeneous jets of air into air that they were further investigated. Perusal of the litterature showed that only two other experiments with jets ligther than the ambiant fluid, carried out very recently (Monkewitz *et al.*, 1989a, 1989b), have disclosed similar observations . The aim of these investigations was to seek experimental support for the theoretical work on absolute instability developed since a few years by Huerre and Monkewitz (1985) and Monkewitz and Sohn (1988).

The existence of such instabilities was, in particular, predicted by these authors in the case of jets less dense than the ambiant medium. It is recalled that a flow is termed absolutely unstable if there exists a region in the flow field where some frequencies of a transient perturbation can grow with zero group velocity , in other words, where such a perturbation does not entirely move away from the source, in contrast to a convectively unstable flow where the perturbation will move away from the source everywhere. As a consequence, in an absolutely unstable flow a transient perturbation will not die out and will progressively contaminate the whole flow field. Moreover, the most amplified frequency will dominate all the others so that the flow behaves as an oscillator. The cited experiments have also shown the existence of side jets which break away from the main jet column and have a dramatic effect on the jet spread.

The present visualizations are evidence that absolute instability and some of its effects may occur in jets lighter than the ambiant fluid at quite low Reynolds numbers. The influence of the initial shear layer thickness is also illustrated.

## 2. Apparatus and experimental conditions

The jet facility consists of the usual settling chamber with screens and a honey-comb, a bell shaped 25:1 contraction ending in a 15 mm nozzle. Smaller diameter jets (dia. 6 and 3mm) where generated by placing contractions on the nozzle outlet. Observations were also made on a tube jet (dia. 8 mm).

Air-helium mixtures with velocities in the range 0 to 20 m/s and with molar fractions covering the full range 0 to 1 are continuously generated with an apparatus consisting of two sets of calibrated sonic throats mounted in parallel on the supply tubes (one set for each constituant, diameters 0.1 to 1.2 mm) and which may be selected in any combination by valves. Further changes of velocity and concentration may be obtained by changing the supply pressures of air and helium. This system has the advantages of producing flows with rigorously constant and reproducible velocities and concentrations and to block perturbations either from the regulating valve of the cylinder of compressed helium or from the air compressor.

Schlieren and differential interferometry photographs were taken with a continuous ligth source and a camera with a shutter speed of 1/1000s. Because of this limited shutter speed there is an increasing time integration effect with increasing jet velocity in addition to the unavoidable spatial integration on the photographs. Some hot wire measurements were also made in the potential core of the jet.

The Reynolds number $V_e D_e / \nu_e$ (index $e$ for exit conditions) of the 15mm jet with pure helium was varied between 100 and 2000 and could reach proportionaly higher values in the smaller jets. Monkewitz *et al.*

(1989a) use the viscosity evaluated at the avarage temperature between the hot jet and the ambiant fluid to compute $Re$. The present Reynolds numbers should, therefore, be multiplied by a factor 4 to 5 in order to compare similar flows in the two experiments. This factor is only approximate because values of the kinematic viscosity at the average temperature, at the average density or at the average mass fraction are all different, and these differences are larger at smaller values of the density ratio $S(=\rho_e/\rho_\infty)$. Since it is not clear at the present time which viscosity is physically the most significant it may be best to choose the exit value.

The densimetric Froude number $Fr = V_e(g'D_e)^{-1/2}$ ($g' = g(\rho_\infty - \rho_e)/\rho_e$) of the jet is an indication of the downstream distance $x/D_e$ where the flow changes from jet like to plume like, i.e. from momentum to buoyancy controlled. The Froude number varied here from 1 to 20 and reached much higher values in the smaller jets. The level of fluctuation in the exit section was approximatly 0.2% in air and in helium. The momentum thickness was $\theta/D_e = 0.01$ at $Re = 7000$ in air and $\theta/D_e = 0.036$ at $Re = 500$ in helium.

## 3. Results

### 3.1. Self-sustaining oscillations

The schlieren photographs Fig.2a, b, c and 3a, b, c respectively taken with the knife edge transverse to and aligned with the main flow direction and the interferograms Fig.4a, b, c (see colour plates) illustrate the existence of strong instability waves on the helium jet column. On these photographs attention is drawn onto the regular structures in initial jet region. These structures appear clearly as vortex rings on the schlieren pictures with the knife edge parallel to the axis especially on figure 3a. Although more faint, they are still visible on figures 3b and c. Caution must be exercised in the interpretation of the interferograms. Indeed, the interference fringes of a given colour are lines of constant $\partial E/\partial x$ ( or $\partial E/\partial y$), where $E$ is the optical thickness traversed by a light ray :

$$E(x,y) = (n_0 - 1)\int_L \rho dz$$

($x$: direction of flow, $y$: direction normal to jet axis and to direction of light, $z$: direction of light normal to the jet axis; miror parallel to $xy$ plane ; $n_0$ : index of refraction of ambiant medium). Thus, $E$ includes variations of both density and thickness in the $z$-direction across the jet. The vortex-like fringe pattern seen in the initial region of the jet on Fig.4a, b, c (see colour plates) reveal the existence of regular structures in the flow but they are not a direct representation of the ring vortices.

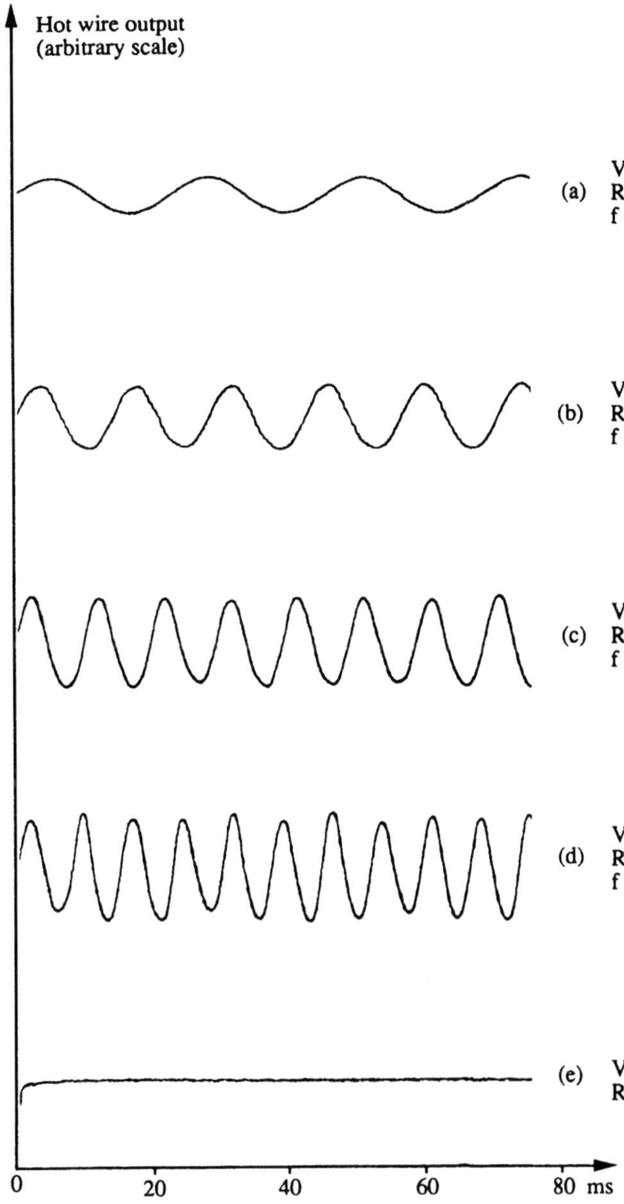


Figure 1 - Hot wire oscillograms in a helium jet (a,b,c,d) and in an air jet (e).Probe at 1D from the nozzle exit.

Hot wire oscillograms (Fig.1) in the same conditions show very stable, nearly purely sinusoidal signals repeatable in frequency as well as in amplitude up to Reynolds numbers of 600. At higher values of $Re$ the signals contain higher harmonics and become more noisy and it is then necessary to determine spectra as Monkewitz *et al.* (1989a, 1989b) in order to distinguish between the regimes of absolute and convective instability . Since the hot wire traces in the air jet at the same Re are perfectly flat (see the last trace on Fig.1) it is clear that the oscillations in the helium jet are due to the density difference. In order to insure that the hot-wire probes do not create appreciable interference with the flow, visualizations were performed with and without probes. There were no observable differences.

With the apparatus described earlier it is quite easy to repeat identical flow conditions since these are entirely determined by the selected combination of sonic throats and by the generating pressure and since the flow can quickly be turned on and off by opening and closing of a valve ( this also has the practical advantage to help save helium). For a given velocity , the jet immediatly enters into oscillations with a fixed frequency and a fixed saturation amplitude although no special precautions were taken to minimize perturbations from the environment. Clearly, the helium jet does indeed behave as a self exciting oscillator.

In the present experiments this regime of self-excited oscillations sets in at very low Reynolds numbers, approximatly when $Re > 100$, as soon as the jet becomes unstable. This is easily detected on the visualizations for when the flow is stable the jet column (rather a plume in this case) is perfectly smooth on schlieren pictures with the knife edge perpendicular to the jet axis and interferograms of $\partial E/\partial y$ have fringes aligned with the axis. This critical Reynolds number is considerably lower than the value of about 900 indicated by Sreenivasan *et al.* (1989) as the lower limit of absolute instability for $S = 0.14$. Because of limitations in flow rate, it was not possible to reach the upper R e limit for this density ratio where the flow becomes only convectively unstable.

The frequency which was simply measured on recorded oscillographs varies nearly as: $f = 0.35(V_e - 0.56)/D_e$ where $V_e$ is expressed in m/s and for $V_e < 10$m/s. Equivalently the Strouhal number is : $St = fD_e/V_e = 0.35(1 - 0.56/V_e)$. Since the frequency is nearly proportional to the velocity, the observed oscillations can not be caused by the cavity resonance of the settling chamber.

It is observed on figures 2 and 3 that the wave length of the instability structures and the streamwise extention of the transition region with regular structures decrease with $Re$. Random motions appear on these structures after about one or two wave lengths (also visible on the interferograms of Figure 4, see colour plates) and transition to fully turbulent flow follows rapidly. At $Re = 1610$ (Fig.3d), the transition region

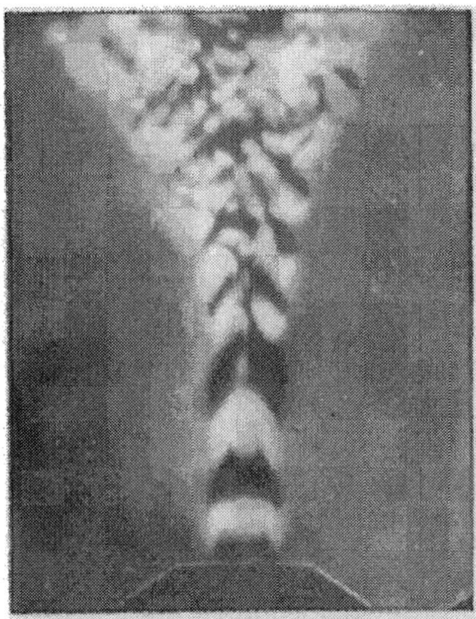

(a) Ve = 1 m/s Re = 130 Fr = 1

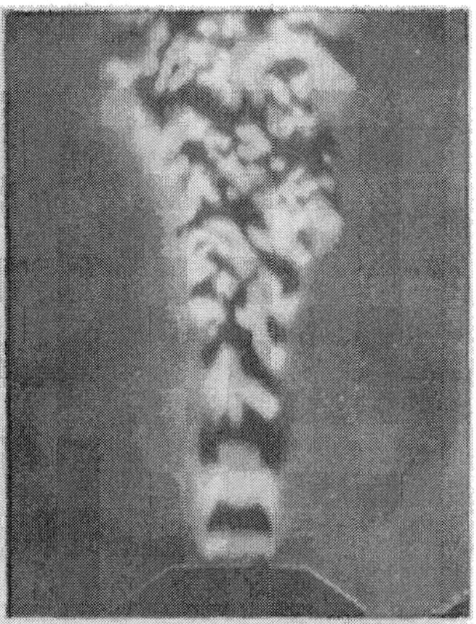

(b) Ve = 2 m/s Re = 260 Fr = 2.1

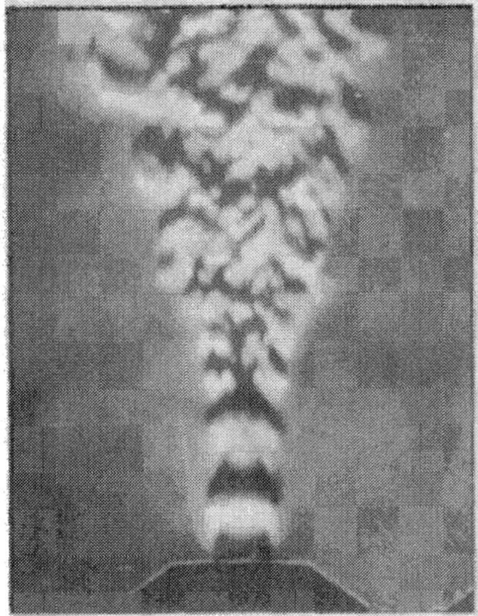

(c) Ve = 3.2 m/s Re = 400 Fr = 3.3

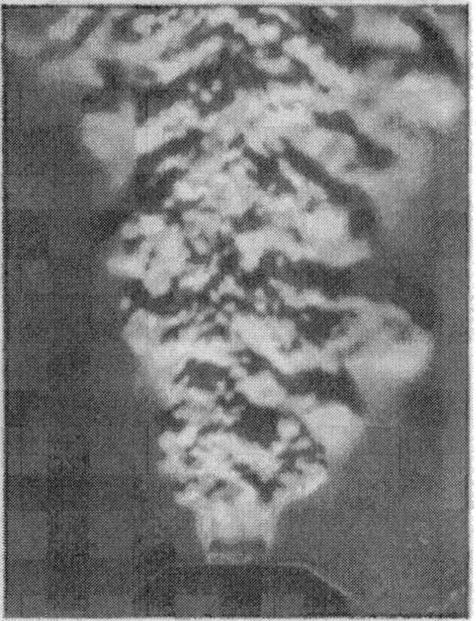

(d) Ve = 8.4 m/s Re = 1060 Fr = 8.8

Figure 2 - Schlieren visualizations of helium jets into air (horizontal knife edge) with a bell shaped nozzle.

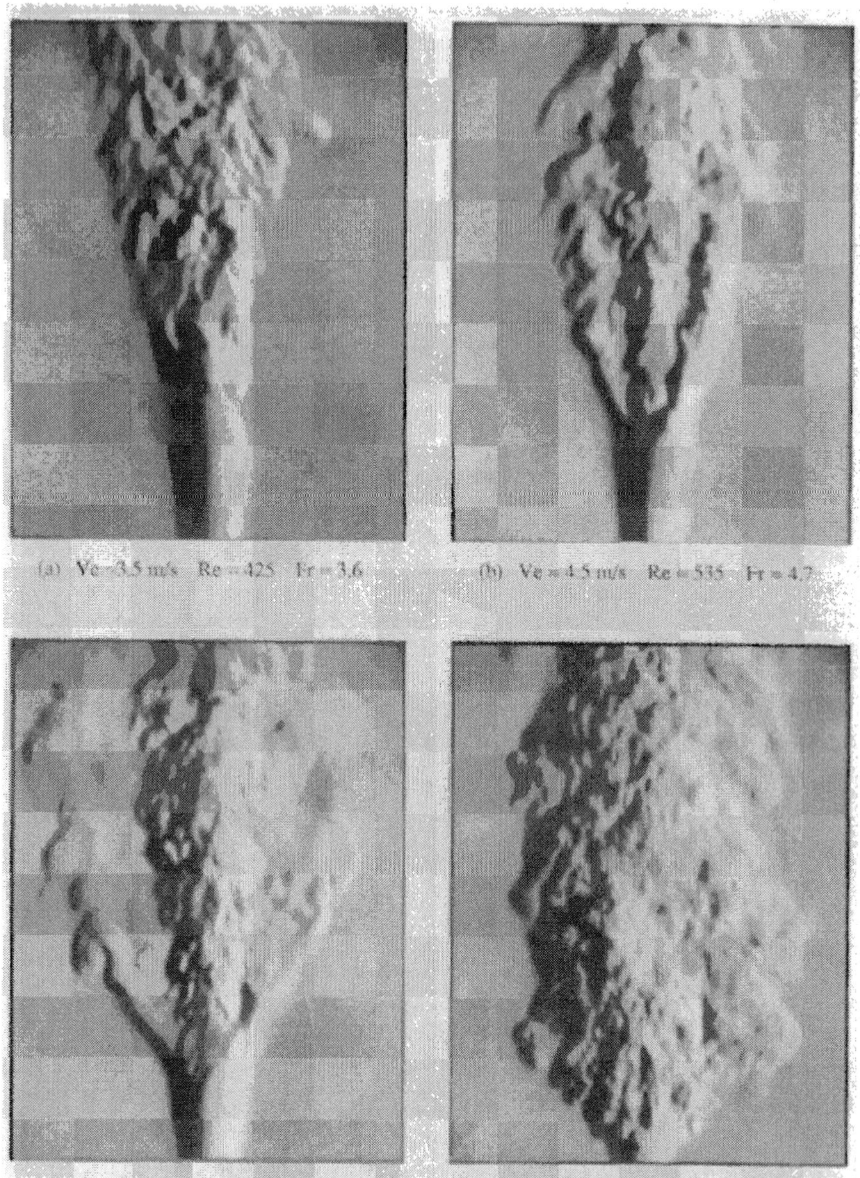

(a) Ve = 3.5 m/s Re = 425 Fr = 3.6

(b) Ve = 4.5 m/s Re = 535 Fr = 4.7

(c) Ve = 5.2 m/s Re = 625 Fr = 5.4

(d) Ve = 13.4 m/s Re = 1610 Fr = 14

Figure 3 - Schlieren visualizations of helium jets into air (vertical knife edge) with an end plate on the nozzle is quite short. Since in the two previous experiments (1,2) the smallest Reynolds numbers were larger than this value the instabilty structures could, therefore, not be be directly visualized. The decrease in wave length is in qualitative agreement with the trend of the theoretical results of Monkewitz and Sohn (1988). Quantitative comparisons are not possible so far because these computations were performed for $S > 0.5$ and our experiments were mainly done with pure helium ($S = 0.14$).

On all the visualizations performed so far -and not only on the representative sample presented here- no evidence of pairing was found. This observation is similar to that made in homogeneous jets forced at the preferred frequency (Hussain and Zaman, 1981; Favre-Marinet, 1986) where the dominant mode is so strong that transition to turbulence occurs without prior pairing.

The slant of the interference fringes on figure 4a (see colour plates) where the optical setup is oriented so as to visualize streamwise density gradients indicates that non-axisymmetric modes may be present. In order to see how the initial shear layer thickness influence the stability charateristics, experiments were performed with a jet issuing from a tube in which the flow was laminar and nearly fully developped. No instability waves could be observed in this jet as may be seen on Fig.5 and no oscillations could be detected by the hot-wire. The length of the laminar jet column decreases continuously with Re and at the downstream end of it there appears to be an abrupt transition to turbulence. These observations on the tube jet are in agreement too with the computations of Monkewitz and Sohn (1988) which show that the absolute growth rate of the pertubations for $S = 0.5$ decreases continuously with downstream distance i.e. with increasing shear layer thickness $\theta/D_e$, and eventually becomes negative.

### 3.2. Jet spread

The visualizations of the two previous experiments Monkewitz *et al.* (1989a, 1989b) and Sreenivasan *et al.* (1989) showed both the existence of side jets which break away nearly perpendicularly from the main jet column and which dramaticaly enhance the visual spread of the jet. Visualizations of cross-sections of the flow by Monkewitz *et al.* (1989a) show in addition that there may be several side jets on the periphery of - and that they rotate randomly about -the main jet column. These authors also suggest that these lateral ejections may be the result of the instability of the strong vortex rings.

The present observations confirm this very peculiar feature of jets less dense than the ambiant fluid. Figures 4d (see colour plates) and 6 are particularly striking examples.

Such a side jet is also visible on the left side of the jet on Fig.4c (see colour plates). What is further shown here is that these side jets only

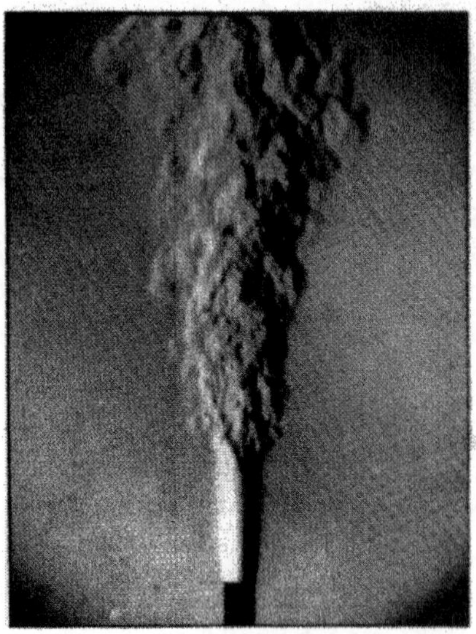

Figure 5 -Schlieren visualization of tube jet
Ve = 14.1 m/s
Re = 930
Fr = 20.1

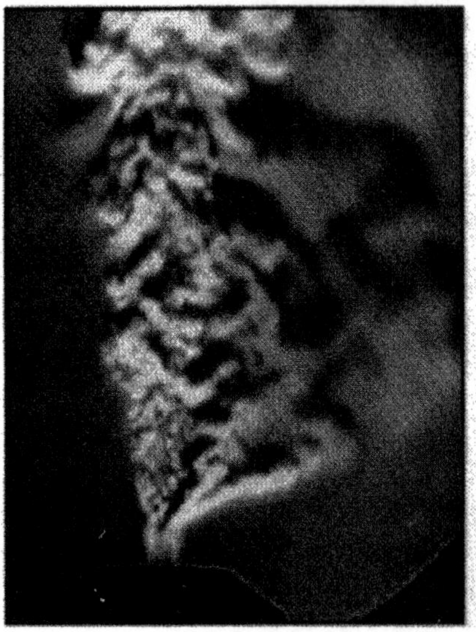

Figure 6 - Schlieren visualization of 6 mm jet
Ve = 12.5 m/s
Re = 620
Fr = 20.6

appear beyond a critical Reynolds number as illustrated by the sequences of figures 2 and 3. Absolute instability is, therefore, not sufficient to produce this phenomenon. In the ligth of the above observation on the decrease of the wave length of the instabilities with Reynolds number it seems that the side jets only occur when the spacing of the vortex rings is sufficiently small to allow strong interactions between them. The photographs 3b and c show in addition that the angle of these side jets with the main flow direction increase with $Re$ and that they originate at the end of the transition region. It is not clear whether some of the large bulges visible on the jets at higher $Re$ (Fig.2d and 3d) are the result of the convection of side jets from the initial region or whether they originated further donstream.

The critical Reynolds number for the occurrence of side jets and their strength depends no doubt on the nozzle geometry or equivalently on the ratio of shear layer thickness to diameter. Thus the side jets appear generally to be more violent on the 6 mm than on the 15 mm jet. In agreement with the absence of absolute instability in the tube jet, such side jets were never observed in this flow. An end plate as was the case in the experiments of figure 3, seems to facilitate the apparition of these jets.

The jet spread beyond the turbulent transition deduced from the visual observations depends upon the Reynolds number. At $Re \approx 300 - 400$, the expansion is notably slower than at lower or higher velocities (Fig.2) and even slower than for the homogeneous jet. This is particularly clear on the sequence a, b, c of Fig.4 (see colour plates). At the larger $Re$ (Fig.2d and 3d), the spread is first very rapid because of the side jets but then it is considerably slower, the width actually appears to be nearly constant. Since the plume has an expansion which is close to that of the homogeneous jet, this result cannot be explained by buoyancy effects.

More definite conclusions on this aspect must avait detailed velocity and concentration measurements.

## 4. Conclusion

This investigation shows that absolute instabilities exist in helium jets at Reynolds numbers as low as 100. They are revealed by strong self sustaining sinusoidal reproducible oscillations. Their Strouhal number is approximatly constant ($\sim 0.35$) but their wave lenght decreases with velocity. These instabilities are not observed on a tube jet. None of the visualizations show the occurence of pairing. It is shown that side jets develop beyong a critical Reynolds number and that their strenght increases with increasing velocity (or $Re$). The angle of spread in the fully turbulent region varies considerably with $Re$.

## Acknowledgement

This work is supported by the **Avions Marcel Dassault** and the **Centre National dEtudes Spatiales.**


## References


Favre-Marinet, M., Structures Cohérentes dans un Jet Rond Excité, *Doctoral Dissertation*, Univers. Scient. et Medicale & Inst. Nat. Polytechn. Grenoble, France (1986).

Huerre, P. and Monkewitz, P.A., 1985: *J. Fluid Mech.*, 159, pp. 151-168

Hussain, A.K.M.F. and Zaman, K.B.M.Q., 1981: *J. Fluid Mech* , 110, pp. 39-71

Monkewitz, P.A., Bechert, D.W., Barsikow, B. and Lehmann,B., 1989a: *Advances in Turbulence 2*, pp.455-460, Spinger.

Monkewitz, P.A., Lehmann, B., Barsikow, B. and Bechert, D.W., 1989b: *Phys. Fluids A 1*, 3, pp. 446-448

Monkewitz P.A. and Sohn K.D., 1988: *A.I.A.A. J.*, 26, pp. 911-916

Sreenivasan, K.R., Raghu, S. and Kyle,D., 1989: *Experiments in Fluids*, 7, pp. 309-317

# Thermal striping : structures in interacting jets

H.M. TSAI, D.C. LESLIE and P.R. VOKE

*The Turbulence Unit*
*Queen Mary and Westfield College*
*Mile End Road, London E1 4NS, GB*

*Turbulence and Coherent Structures*
*O. Métais and M. Lesieur, eds.*
*Selected papers from "Turbulence 89: organized structures and turbulence in fluid mechanics"*
*Grenoble, 18-21 September 1989*

**Abstract**

Turbulent jets at markedly different temperatures impinge on a steel structure above the cores of sodium-cooled fast nuclear reactors. Temperature fluctuations in the sodium induce fluctuations in the steel, and the thermal fatigue so produced can lead to premature failure.

A detailed description of this ***thermal striping*** phenomenon is beyond the reach of methods such as $k - \epsilon$. This paper reports a first attempt to predict it by Large Eddy Simulation, using a simplified geometry. It is found that LES can predict the probability distributions of the temperature field, and that it could therefore supplement the scarce and difficult experiments on this effect.

## 1. Introduction

The (stochastic) Numerical Simulation (NS) of turbulence is increasingly recognised as a valuable adjunct, and even an alternative, to experiment. So far attention has been concentrated on scientific results obtained from the Full version of NS, but this technique has another application which will also (in the opinion of the authors) be found to be of great value. This is its ability to predict engineering phenomena which are fundamentally beyond the reach of Reynolds- averaged closures such as $k - \epsilon$. The investigation reported in this paper is believed to be the first application of the Large Eddy Simulation version of NS to such a problem.

## 2. The problem

Figure 1 shows the primary circuit of a typical sodium-cooled fast nuclear reactor. A hot fast jet of sodium emerges from

the top of the active core. A slower colder jet comes out of the top of the breeding blanket adjoining the core, in which fresh plutonium is formed. A region of fluctuating temperatures is formed by the turbulent interaction of the two jets. These impinge normally onto the Above-Core Structure (ACS), a stainless steel component above the core, where they induce temperature fluctuations which may limit life by thermal fatigue: this is known as "thermal striping".

Figure 2 shows an idealisation of this thermal striping region.

On present understanding, this lifetime is determined by the magnitude of the largest temperature swings seen by critical regions of the structure. Quantities such as this are fundamentally beyond the reach of any Reynolds closure, but they should be predictable by Numerical Simulation. The Turbulence Unit (TU) of Queen Mary and Westfield College has therefore undertaken a feasibility study on behalf of the UK

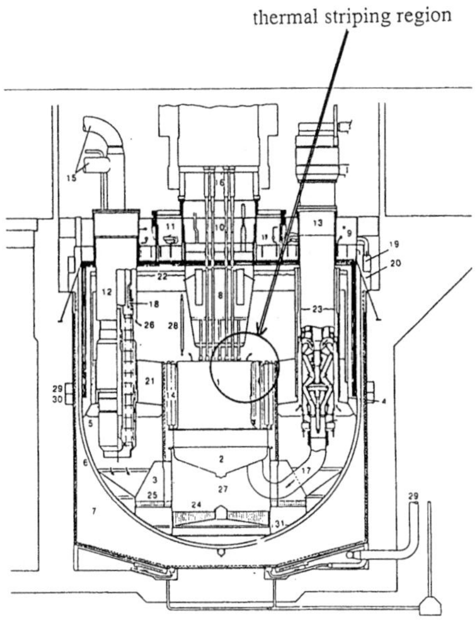


Fig.1. Primary circuit of commercial fast reactor from Mitchell, *J. Brit. Nuclear Energy Soc.*, **25**, 111, 1986

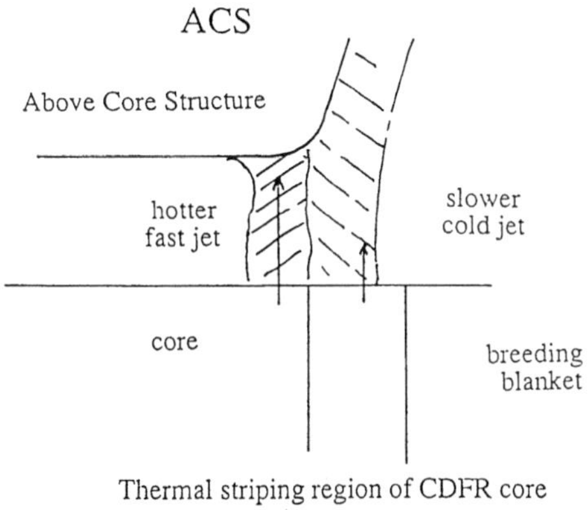


Fig.2. Idealisation of thermal striping region

Central Electricity Generating Board (CEGB), which is engaged in fast reactor research, to test the feasibility of predicting thermal striping by Large Eddy Simulation (LES): the Reynolds number *Re* is so high that it is not at present feasible to study this problem by Full Simulation.

The primary aim of this investigation has been to produce information on the swings in temperature in the region where the two jets interact.

The outputs of the simulations inevitably contain much additional data on the turbulent structures produced by this interaction, and it is this aspect which is of particular interest to Turbulence 89.

## 3. The style of simulation

At the start of this project the only TU code with sufficient speed was ECCLES (Gavrilakis *et al*, 1985). This code is periodic in the $x$ and $y$ directions; this implies that the mesh must be uniform and the flow periodic in these directions. The simulation has perforce been tailored to these limitations and in particular we have not attempted to represent the impingement of the jets on the steel of the ACS: figure 3, which may be

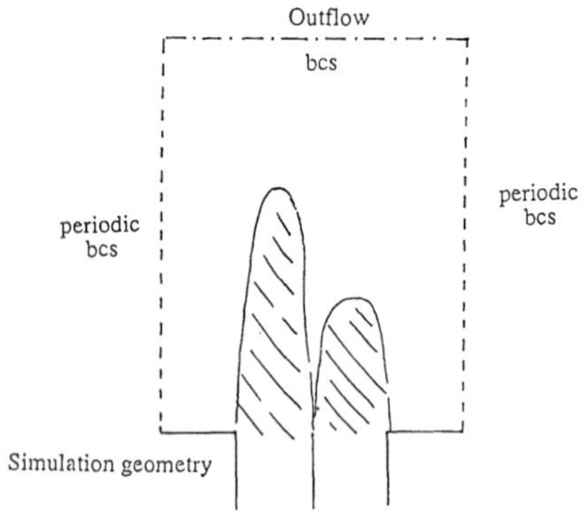


Fig.3. Comparison of reactor and simulation geometries

compared with figure 2 above, shows how the geometry of the thermal striping region has been simplified to accomodate the limitations of ECCLES. The work reported here is concerned only with the generation of temperature fluctuations by interaction between the jets.

Figure 4 illustrates the style of simulation, which is broken down into two parts. The ***precursor*** part is a normal simulation

of channel flow; synthetic boundary conditions (Schumann, 1975) are used because of the relatively high $Re$, which is not less than 20,000. We have used the SubGrid Model (SGM) of Moin and Kim (1982) with the inhomogeneous term set to zero.

The timescale of the large eddies is of order $h/u$, and this unit of time has been christened the ***letot*** (Large Eddy TurnOver Time). The precursor simulation is first ***conditioned*** (brought to a stationary state) by running for about 10 letot. It is then run for a further 6 to 12 letot, and the fields on the two outflow planes of the staggered mesh are now written to disc at each timestep: this does not slow the computation

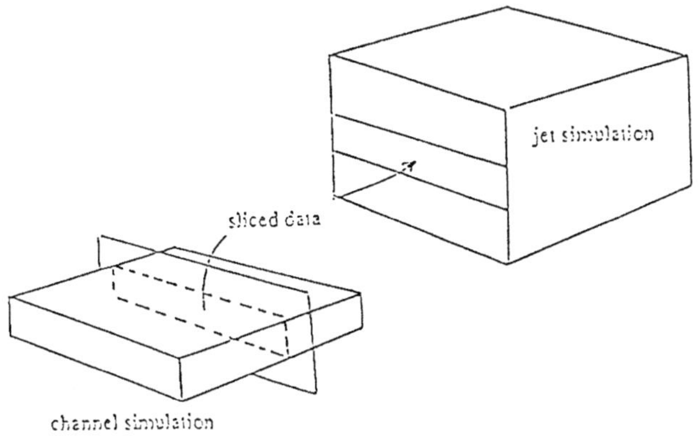


Fig.4. Division into precursor and successor simulations

appreciably. This process has been christened *slicing*: an account is given by Tsai *et al.* (1987).

Two different samples of these slices are used as inflow conditions for the two jets in the *successor* simulation, and zero gradient conditions are used at the outflow. Figure 5

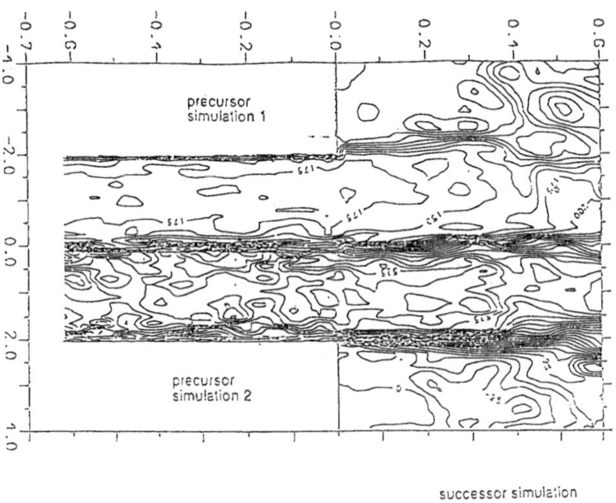


Fig.5. Computer composite of the two simulations

shows a computer composite of the precursor simulation together with the successor simulation which has been fed from it.

The only obvious objection to the slicing procedure is that the formation of the jet can have no reaction back into the downstream part of the channel flow: it is believed that this simplification does not cause any substantial error. The uniform mesh and periodic bcs in the cross-stream ($y$) direction is a more likely reason for the simulation to depart from the engineering situation. The uniform mesh coupled with the additional damping due to the subgrid model means that the simulation does not represent the inner wake. There is substantial evidence (see, eg, Andreopoulos and Bradshaw, 1980) that this will have little effect on the large structures well downstream of the jet efflux. The periodicity in the spanwise ($x$) direction agrees with the assumed nature of the flow.

There are no temperature fluctuations on the inflow plane of the successor simulation, and these fluctuations build up as the flow moves downstream: the CEGB has advised us that this is the best simple representation of conditions in the reactor. In consequence, the temperatures in the flow can only fluctuate between those of the hot and cold jets;

## 4. Simulation performed : analysis of results

4 simulations have been performed, as follows:

- jet speed ratio 1:1, $Pr = 1$ only
- jet speed ratio 2:1, $Pr = 1,\ 0.0656,\ 0.0043$

Comparing the two runs with $Pr = 1$ shows the effect of altering the speed ratio on the flow and temperature fields. Comparison of the runs with differing $Pr$ is important to the design engineer. The lowest $Pr$ value simulated is equal to that of the sodium in the core region of the fast reactor primary circuit, while experiments are often done in water ($Pr \simeq 1$) to save expense: the simulations should reveal whether conclusions drawn from the water experiments are valid for sodium. The intermediate $Pr$ value is the geometric mean of the other two; runs with this $Pr$ help to confirm trends.

These simulations have produced a total of about $5 \times 10^{10}$ items of data. We have processed this mass of information in three different ways: these are snapshots at a moment in time, averages over time and the spanwise dimension $x$, and samples taken as the simulation evolves in time. After processing, the data are presented either as a set of 2D contours or as 1D profiles of the dependence of one variable on one coordinate. A 2D dependence can be shown graphically as an isometric combination of 1D profiles: our plotting routine removes hidden lines.

### 4.1. Accumulation of probabilities

The temperature range $+1$, $-1$ together with a guard interval at either end is divided into 32 equal width bins. At each timestep the temperature in each mesh will lie in a particular bin, and this fact is logged. Counts are accumulated in the bins as the simulation proceeds, and are also averaged in the spanwise direction to improve statistical accuracy.

The probability that the temperature in a given mesh will lie in a particular bin is simply the ratio of the count in the bin for that mesh to the product of the number of timesteps and the number of spanwise meshes.

### 4.2. Logging of extremal events

The highest and lowest temperatures at a point are identified by the highest and lowest bin numbers with non-zero counts.

## 5. Comparison of averaged and instantaneous plots

Figure 6 shows profiles of the mean streamwise velocity for equal speed jets. The profiles on the inlet plane reflect the

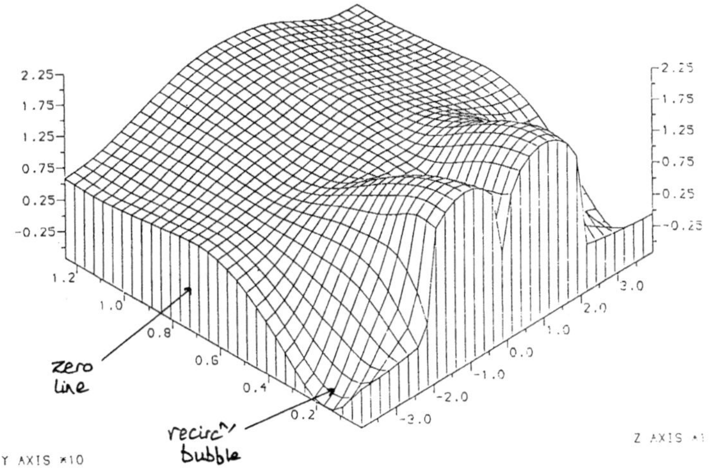


Fig.6. Mean streamwise velocity profiles for equal speed jets

nature of the data fed in from the precursor simulation. The profiles tend rather rapidly towards uniformity. Recirculation bubbles are to be expected on the side walls: it is surprising that they are so strong when the side walls are periodic rather than hard. Generally the flow seems very smooth, and there is no suggestion that it is turning through any very large angle.

The jets have been labelled with passive temperatures $\pm 1$, the left-hand jet being hot. The mean isotherms (not shown) confirm the impression given by figure 6, but the instantaneous isotherms shown in figure 7 give a very different picture.

The solid lines show hot fluid which has come from the left-hand jet: the irruption of hot fluid at the bottom of the picture is the periodic extension of the fluid leaving at the top of the diagram. It is evident that the fluid is turning through large angles. Figure 6 gives the opposite

Fig.7. Instantaneous isotherms for equal speed jets

impression because the jets are equally likely to turn either way: the averaging over time (and over the spanwise direction) also averages the turning angle to zero.

Figure 8 shows the effect of raising the jet speed ratio to 2:1 on the streamwise velocity profiles. The flow still seems smooth

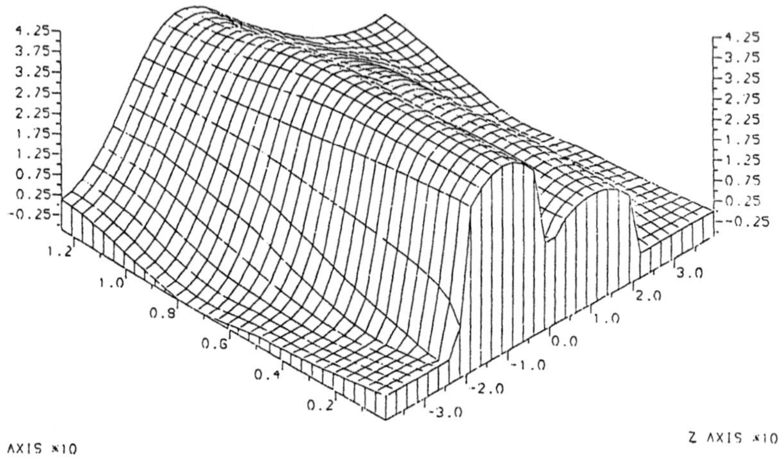


Fig.8. Mean streamwise velocity profiles for jet speed ratio of 2:1

but there is some suggestion that the fast left-hand jet is obstructing the streamwise movement of the slow right-hand jet. This is strongly confirmed by the instantaneous isotherms. The

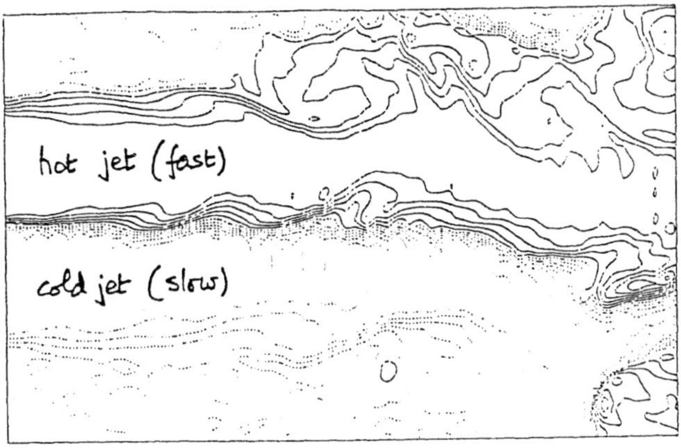


Fig.9. Instantaneous isotherms for jet speed ratio of 2:1

hot fast jet is free to move downstream, and is even expanding at the right-hand (downstream) edge of the picture; the slow cold jet (labelled by the dotted lines) is being pinched off.

Figures 10 and 11 show contours of constant temperature variance $< \theta'^2 >$. This quantity is zero on the inlet plane, since no

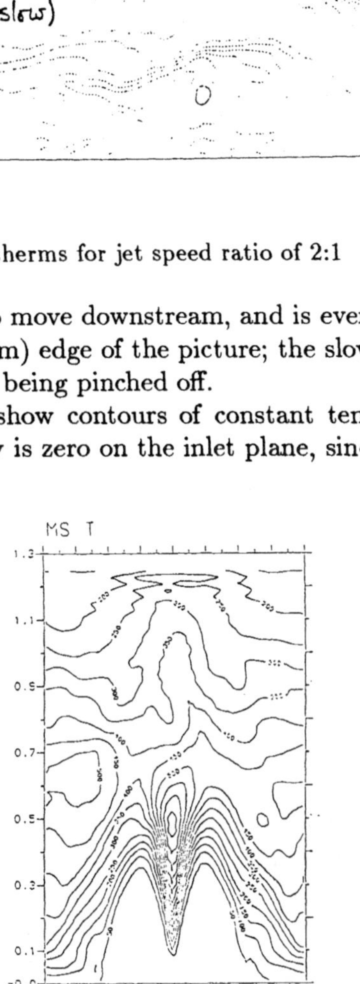


**Fig.10. Contours of constant $< \theta'^2 >$: equal speed jets, $Pr = 1$**

temperature fluctuations are passed from the precursor to the successor simulation. However there is a step change in $T$ from - 1 to +1 on crossing the median plane $y = 0$. This step change

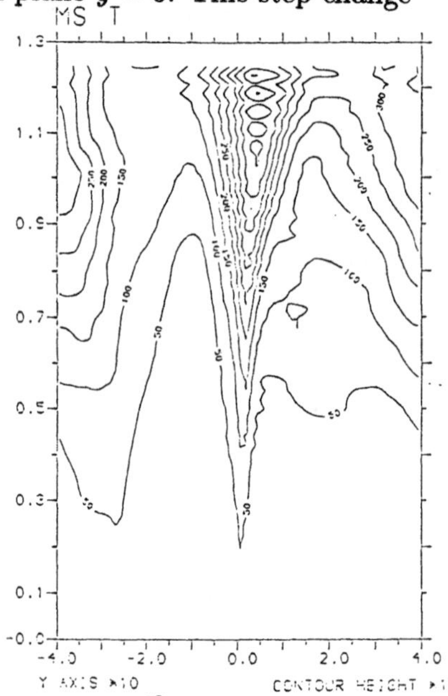


Fig.11. Contours of constant $< \theta'^2 >$: jet speed ratio 2:1, $Pr = 1$

is moved sideways in a random way by turbulent motions, and this is the main mechanism by which temperature fluctuations are built up. These are gradually destroyed by molecular and subgrid diffusion: the processes of build-up and destruction are clearly visible in the figure. The structures extend much further in the streamwise direction when the jet speed ratio is increased to 2:1, in agreement with conclusions drawn from the velocity plots.

## 6. Isopiths

An *isopith* is a contour of constant probability, from the Greek $\pi\iota\theta\alpha\nu o\tau\iota\zeta$ = probability. The accumulation of probability information is described in section 4 above: the information in this section is derivable only from numerical simulation or from

experiment. Figures 12 and 13 show profiles of the probability that the temperature will lie in the topmost active bin $1 > T > 0.963$, for both jet speed ratios: $Pr$ is again set equal to 1, so that $T$ can be used to mark the origin of the fluid. Consider for example the contour (not shown explicitly, but

inferrable from the figure) on which the probability defined above is 5%. This contour is the surface on which the probability that fluid from the hot jet has not been diluted by fluid from the cold jet is 5%. Because

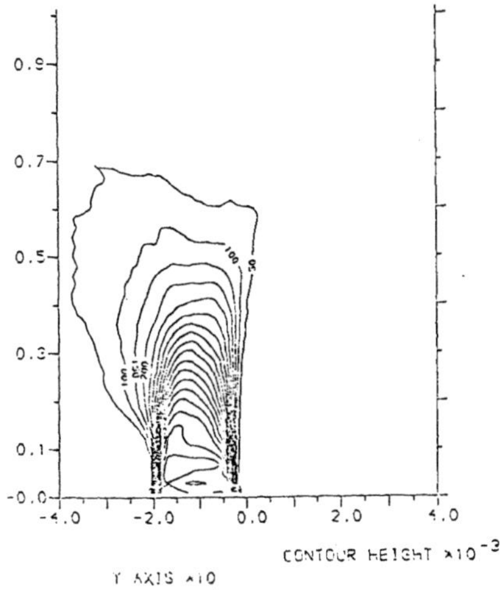


Fig.12. Isopiths for $1 > T > 0.963$: equal speed jets, $Pr = 1$

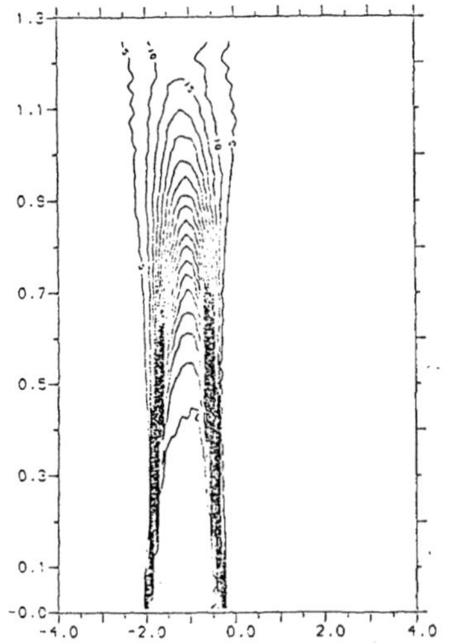


Fig.13. Isopiths for $1 > T > 0.963$; jet speed ratio 2:1, $Pr = 1$

the bins are of finite size, this contour marks the effective outer boundary of the hot jet. Comparison of these figures shows very clearly how much the hot jet is extended when the the jet speed ratio is increased from 1:1 to 2:1.

## 7. Maximum temperature swings

Figure 14 shows the maximum temperature swing experienced at each point in the $y - z$ plane. This is related to the quantities

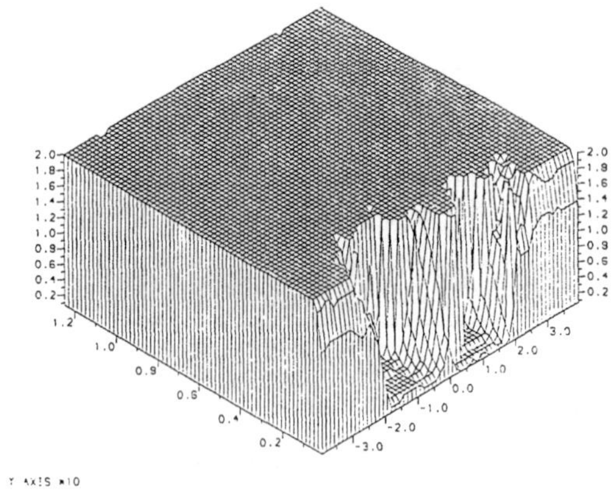

Fig.14. Maximum temperature swing: equal speed jets, $Pr = 1$

which the reactor designers need to assess the lifetime of the ACS (see section 2). [The designers actually wish to know the temperature swings, not in the fluid, but in the steel. This latter quantity cannot be calculated in the simulation

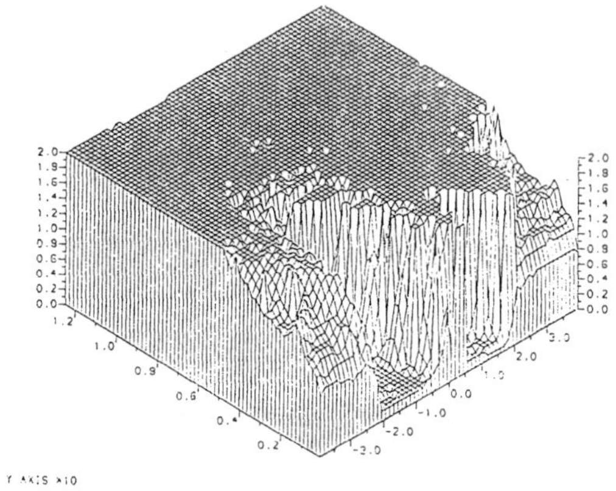

Fig.15. Maximum temperature swing: jet speed ratio 2:1, $Pr = 1$

reported here, but it is now being studied in a continuation project]. The maximum possible swing is 2, and figure 14 shows that when $Pr = 1$ and the jets have the same speed, this maximum swing is found over almost the whole of the plane. The region of less-than-maximum swing extends somewhat further downstream when the speed ratio is increased to 2:1. This reflects the streamwise extension of the structures produced by speeding up the hot jet, an effect which has been observed in every comparison of the two speed ratios.

The last figure shows the effect of reducing the Prandtl number from 1 to 0.0043, a value typical of the sodium coolant in a

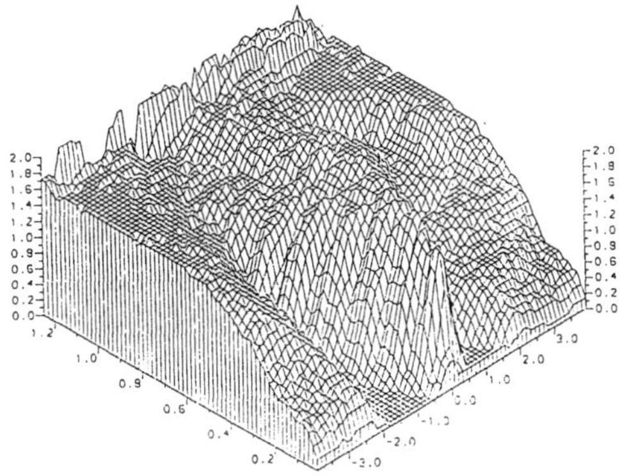

Fig.16. Maximum temperature swing: equal speed jets, $Pr = 0.0043$

(liquid metal cooled) fast reactor. The maximum possible swing is now found on the outflow plane only, and this is believed to be due to defects in the outflow boundary condition. Clearly the temperature swings are being damped very heavily by the high molecular conduction.

If the simulation is correct, then at these relatively low Reynolds numbers it will not be possible to infer temperature swings in liquid sodium from experiments in water. Such an infererence might still be justified at higher *Re*.

## 8. Conclusions and further work

The work shows that LES can be used to obtain extremal and other statistics which cannot be derived from any lesser method of prediction, and which are extremely hard to measure. The next step is to bring the simulation closer to the engineering situation, and in particular to study normal impingement onto a solid surface, and the consequent diffusion of heat into the surface. The QMC Turbulence Unit is now able to do this, thanks to the development of the fast singly periodic code DUCT, which is able to handle two pairs of arbitrary boundary conditions.

**Ackowledgements**

This work is funded jointly by the CEGB and the UK Science and Engineering Research Council (SERC). We are grateful for the interest and advice of Dr P Jones, the CEGB's Technical Correspondent for this project, and for the award of substantial time on the SERC XMP.

## References

Andreopoulos, J. & Bradshw, P. 1980. Measurements of interacting turbulent shear layers in the near wake of a flat plate. *J. Fluid Mech.* 100, 639-688.

Gavrilakis, S., Tsai, H.M., Voke, P.R. & Leslie, D.C. 1986 Large eddy simulation of low Reynolds number channel flow by spectral and finite difference methods in *Direct and Large Eddy simulation of turbulence*, (ed. U.Schumann & R.Friedrichs). Vieweg, Braunschweig.

Moin, P. & Kim, J. 1982 Numerical investigation of turbulent channel flow. *J. Fluid Mech.* 118, 341-377.

Schumann, U. 1975 Subgrid scale model for finite difference simulations of turbulent flows in plane channels and annuli. *J. Comp. Phys* 18, 376-404.

Tsai, H.M., Voke, P.R. & Leslie, D.C. 1986. Large edy simulation of turbulent free shear flows in *Proc 5th Tntl Conf on Numerical Methods in Laminar and Turbulent Flow, Montreal, Canada* (ed C.Taylor, W.G.Habashi & M.M. Hafez) Pineridge, Swansea, U.K.

# II. Boundary Layers

# Three-dimensional aspects of the organized motion in a turbulent boundary layer

R.A. ANTONIA and D.K. BISSET

*Department of Mechanical Engineering*
*University of Newcastle, N.S.W. 2308, Australia*

*Turbulence and Coherent Structures*
*O. Métais and M. Lesieur, eds.*
*Selected papers from "Turbulence 89: organized structures and turbulence in fluid mechanics"*
*Grenoble, 18-21 September 1989*

Velocity data have been obtained in a turbulent flat plate boundary layer with a rake of eight **X**-probes, first normal to the wall and then parallel at several heights. Identical detection procedures are applied to the lines of intersection of the planes of data, and sectional streamlines and vorticity contours are calculated throughout the planes from the conditioned velocities, giving different cuts through the same structures. Near the wall the revised, modified $u$-level approach is used to detect bursts. Further from the wall the WAG method is used to detect highly three-dimensional large structures, which appear as a series of spanwise vortices in the plane normal to the wall and as smaller, counter-rotating regions in planes parallel to the wall. Additional criteria are used to show that perfect symmetry is artificial in the latter planes, but other conclusions drawn from symmetrical results are not invalidated. A brief comparison is made with results in the far-wake of a circular cylinder.

## 1. Introduction

Certain features of organized motion in boundary layers about which there is general agreement have been summarised by Kline (1989), but Kline pointed out that there is much less agreement about how the features are interrelated, especially dynamically. One of the main problems is to obtain simultaneous, detailed quantitative measurements throughout a three-dimensional region of the flow for a sufficient period of time. No experimental method is available to solve this problem directly, and therefore the present work takes an indirect approach. It attempts to provide three-dimensional information about organized motion, at moderate Reynolds number, through conditional averaging procedures that are applied separately (but identically) to data taken in various planes either normal or parallel to the wall.

Full three-dimensional information is available (at fairly low Reynolds numbers) from databases generated from numerical solutions of the Navier-Stokes equations (e.g. Kim and Moin, 1986 for a turbulent duct; Robinson et al., 1989 for a boundary layer), but this highlights another difficulty. In one sense there is always too much data, and the problem is to extract basic information about fundamental features and dynamic processes from the vast amount of detail. Detection and conditional averaging are means to this end.

A working definition of organized motion, from the viewpoint of an experimentalist using conditional averaging procedures, is that detections of a certain feature in certain signal(s) should "organize" the conditional averages of other signals (different spatial location and/or physical property) recorded simultaneously. But there may be more than one kind of structure organising the flow, and each individual structure of a given kind may be of a different size and/or have a different orientation. Different detection criteria may sometimes focus on different features of the same underlying structure, but in other cases they may select an entirely different structure. These problems are serious when a three-dimensional description is desired, especially for the boundary layer where at least two distinct length scales exist. Also, conditional results obtained in planes parallel to the wall often show symmetry that is rarely present instantaneously.

The results reported here are based on velocity fluctuations recorded by a rake of eight **X**-probes in a turbulent, zero-pressure-gradient flat plate boundary layer at a momentum thickness Reynolds number $R_\theta \approx 2200$. This Reynolds number is below 5000, the value at which Reynolds number effects largely cease (Antonia et al., 1989 and references therein), but it is above the range for which large structural differences occur (Head and Bandyopadhyay, 1981; Antonia et al., 1989). The rake was initially in the $y$ direction (normal to the wall) with the **X**-probes recording velocity fluctuations $u$ and $v$. Further data were taken with the rake in the $z$ (spanwise) direction at several $y$ values, recording fluctuations $u$ and $w$. Taylor's hypothesis was invoked to convert the data time-series into data distributed throughout the $(x, y)$ and several $(x, z)$ planes.

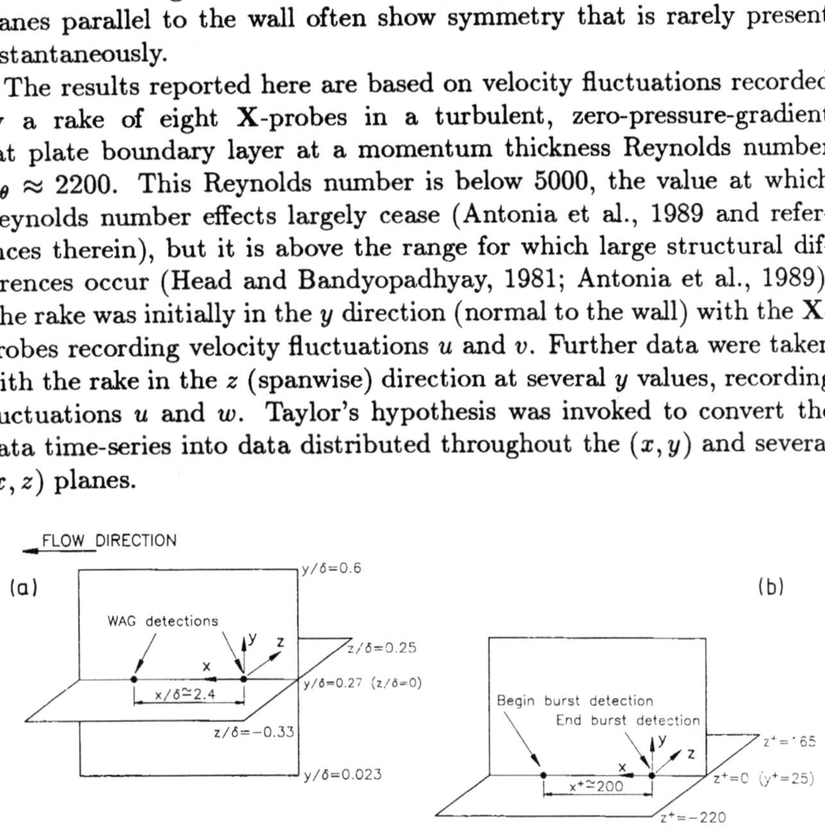


Fig.1. Results in $(x, y)$ and $(x, z)$ planes linked by identical detection procedures at the intersections. (a) intersection at $y = 0.27\delta$, WAG detections. (b) intersection at $y^+ \approx 25$, $u$-level detections.

Near the wall, the revised modified $u$-level detection method (Tiederman, 1989) is used to detect ejections and bursts. Further away from the wall, the window average gradient (WAG) detection method is used to focus on $\delta$-scale $u$-discontinuities (Kline, 1989). The sets of detections are used to generate conditional averages of velocities throughout $(x, y)$ and $(x, z)$ planes, which are then related to each other through their common detection procedure along the lines of intersection of the planes (see Figure 1). As expected, conditional velocities from probes other than the detection probe are "organized", but the results in $(x, z)$ planes are symmetrical about the $x$-axis. Non-symmetrical conditionally averaged burst detections are obtained by selecting bursts which have specified signs of $u$ near the end of the burst at $z^+ = \pm 80$ (a procedure used by Antonia and Bisset, 1990). Non-symmetrical $\delta$-scale $u$-discontinuities are obtained by splitting a set of WAG detections into two subsets according to the sign of $w$ at $x \approx -0.2\delta$ (relative to each detection). These additional detection conditions reduce the risk of oversimplification associated with single point detection methods.

Results from conditioning velocity components are presented primarily as sectional streamlines, i.e. streamlines calculated from two component velocity vectors in a given plane. Contour plots of components of the vorticity vector, both conditionally averaged and instantaneous, are also presented. The results are briefly compared with similar work in the far-wake of a circular cylinder.

## 2. Experimental Details

An open return blower type wind tunnel was used. The boundary layer was tripped by a 1.5 mm diameter rod and 150 mm of number 40 sandpaper mounted on one wall at the entrance to the working section (size 150 mm × 890 mm). The opposite wall was at a slight angle so that the pressure gradient was zero. Measurements were made 4.3 m downstream, where $\delta$, the boundary layer thickness, was about 68 mm, with a free stream velocity $U_1 = 4.2$ m/s. $R_\theta$, the momentum thickness Reynolds number, was 2180, and $U_\tau$, the friction velocity, was 0.173 m/s. The eight $90^o$ **X**-probes of the array were approximately equally spaced at 5 mm centres. The hot wires were 5 $\mu$m diameter Pt-10% Rh, with a length to diameter ratio of about 250. They were operated by constant temperature circuits with an overheat of 0.6, and the output signals were digitized for approximately 40 seconds at 2976 Hz per channel.

## 3. Detection and Conditional Averaging

The detection methods are both described in detail in Antonia et al. (1989), and will be outlined only briefly here. The revised, modified $u$-level method of Tiederman (1989), a revised version of the method of Luchik and Tiederman (1987), defines the beginning of an ejection as the point where the value of $u$ falls below $-L\overline{u^2}^{\frac{1}{2}}$, and the end of an ejection as the point where the value of $u$ rises above $-0.25L\overline{u^2}^{\frac{1}{2}}$, where $L$ is a positive threshold parameter. It is generally found that graphs of the logarithm of the cumulative probability of $\tau$, the interval between the end of an ejection and the beginning of the next, show two straight

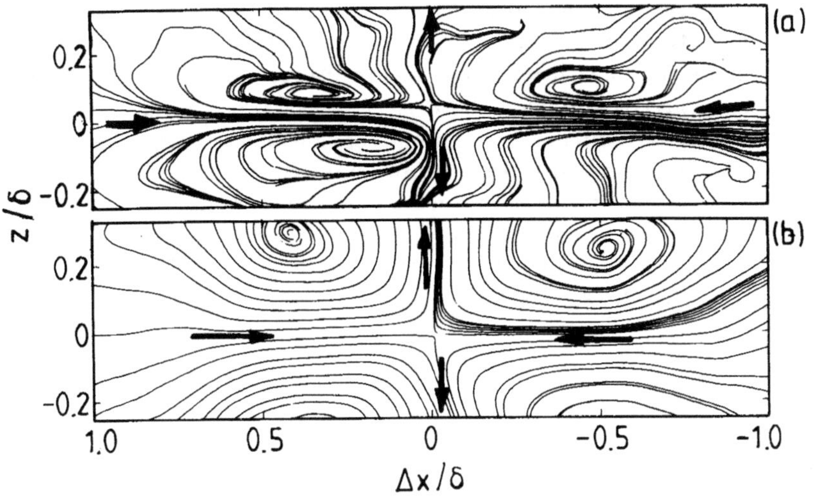


Fig.2. Sectional streamlines in $(x, z)$ planes with $U_c = \overline{U}(y)$, based on WAG detections in $u$ signals at (a) $y = 0.052\delta$, and (b) $y = 0.52\delta$.

line segments. The grouping time $\tau_B$ is defined as the value of $\tau$ at the intersection of the two lines. Ejections separated by less than $\tau_B$ are grouped into the same burst. Tiederman and co-workers have successfully calibrated the method against simultaneous flow visualization in a water channel. The mean period of bursts (after grouping) is constant at about $3\delta/U_1$ for a range of $L$ (where $L \sim 0.6$ for the present data).

The window average gradient (WAG) detection method, in its single point form, detects maxima in changes in the short time mean value of a signal. The direction of change depends on the value of *sign*, i.e. +1 or −1. The method moves a computation window of length $2\tau + 1$ samples through the digital data, and computes

$$WAG_j = \frac{sign}{2\tau}\left(\sum_{i=j+1}^{j+\tau} u_i - \sum_{i=j-\tau}^{j-1} u_i\right)$$

for every sample $j$. A detection region begins when $WAG_j$ first exceeds $k\overline{u^2}^{\frac{1}{2}}$, $k$ being a threshold parameter, and ends when $WAG_j$ becomes negative. The detection point $d$ within each region is the value of $j$ for which $WAG_j$ is largest. In the present work, $k$ is 0.45 and $2\tau + 1$ is equivalent to $1.6\delta/U_1$, and only $u$ signals, being common to both $(x, y)$ and $(x, z)$ planes, are used for detection.

The conditional average of a quantity f, based on a set of detections $d_i$, $i = 1, \ldots, n$, is $< f >_\kappa = \frac{1}{n}\sum_{i=1}^{n} f_{d_i+\kappa}$, where $\kappa$ is time (in samples; positive or negative) relative to the detection point. The subscript $\kappa$ is usually omitted.

In general, conditional averages are degraded (or "smeared") even at quite short times or distances away from detection points because of variation in size and orientation of structures, and, as noted before, symmetry may be implied where it is rarely present instantaneously. One method of reducing the amount of smearing present is to select subsets of detections based on additional criteria located at different points in space or time. Criteria used here are : (i) the length of a burst should fall within a certain range; (ii) the period between successive WAG detections should be close to the most commonly found value; and (iii) criteria to reduce artificial symmetry mentioned in the Introduction.

## 4. Results

### 4.1. WAG Detection Results

Sectional streamlines in $(x, z)$ planes at $y/\delta = 0.052$ and 0.52 are shown in Figure 2. The streamlines are calculated using the method of Bisset et al. (1989) from $<u>$ and $<w>$ conditioned on WAG detections at $\Delta x = z = 0$. Time has been converted to distance using $\Delta x = -0.8U_1\Delta t$ for this figure and all other figures based on WAG detections, because this conversion was found to be suitable for results in the $(x, y)$ plane for the present range of $y$ values (Antonia et al., 1989). The effective frame-of-reference convection velocity $U_c$ in Figure 2 is the local value of the mean velocity $\overline{U}$, however ($0.55U_1$ and $0.93U_1$ respectively). The streamlines form a pattern of four generally rotating regions (with adjacent regions in opposite directions) separated by a saddle point which marks the WAG detection point.

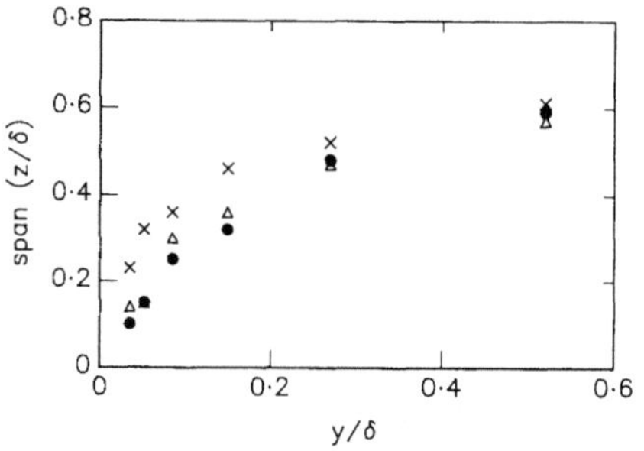


Fig.3. Spanwise scale of WAG detected regions, based on (•) the $z$ values at which detections cease to occur preferentially at the same time as at $z = 0$, and (Δ) the spans of low speed regions. Also shown (×) are estimates based on conventional correlation coefficients of $u$.

The rotating regions are larger at larger $y$, especially in the $z$ direction. The mean $z$ dimension of the region associated with a WAG detection (defined as the span of the low speed region downstream of the detection, approximately equal to the span of the upstream high speed region) is shown in Figure 3 as a function of $y/\delta$. Another estimate of the mean $z$ span of WAG detected regions was obtained as follows. It was found that, except for the lowest $y/\delta$, WAG detections from adjacent probes (separated in $z$) tend to occur at approximately the same time. For other probes separated more widely in $z$, detections tend not to occur at the same time. The distance between the two locations (at $z > 0$ and $z < 0$) where WAG detections ceased to occur preferentially at the same time as detections at $z = 0$, is shown in Figure 3. There is good agreement between the two estimates of $z$ span. On a semi-log plot, the results fall quite close to a straight line, the equation of which is $z/\delta = 0.43 \log_{10}(y/\delta) + 0.71$. The mean $z$ span of regions of low speed or high speed fluid obtained from conventional correlation coefficients [i.e. twice the distance to the zero crossing point in $\rho_{uu}(z)$] is also shown in Figure 3. The values are only a little larger than the spans of WAG detected regions and the trend is the same, suggesting that the organized motion giving rise to WAG detections is a major contributor to correlation coefficients. Contours of correlations between fluctuations in transverse vorticity $\omega_y$ throughout two $(x, z)$ planes are shown in Figure 4, and it is interesting that there is essentially no change at different $y$. Negative correlations are very small.

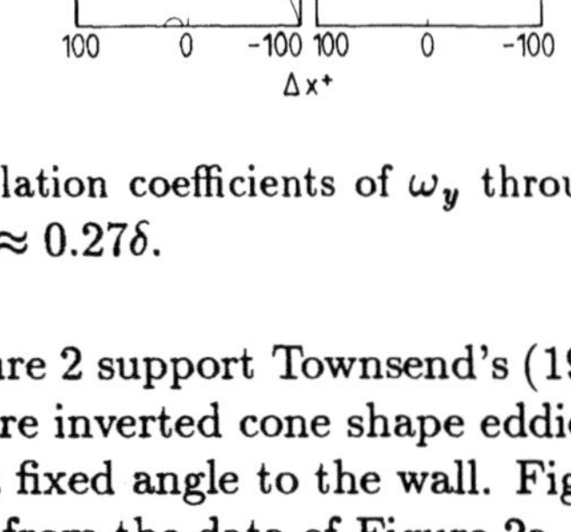


Fig.4. Contours of correlation coefficients of $\omega_y$ throughout $(x, z)$ planes at (a) $y^+ \approx 25$, and (b) $y \approx 0.27\delta$.

The patterns of Figure 2 support Townsend's (1976, p.157) suggestion that large structures are inverted cone shape eddies. If they exist, however, they are not at a fixed angle to the wall. Figure 5 shows sectional streamlines calculated from the data of Figure 2a, but this time with $U_c$ increased by $0.05U_1$. Streamlines at other $y$ values are affected similarly by changes in $U_c$. Clearly there is no single value of $U_c$ for which the rotating streamline patterns are apparent over a substantial range of $y$, and therefore, the axis of any eddy must itself be rotating quite rapidly about a spanwise axis. The predominant structure may actually have a more complex three-dimensional shape, however.

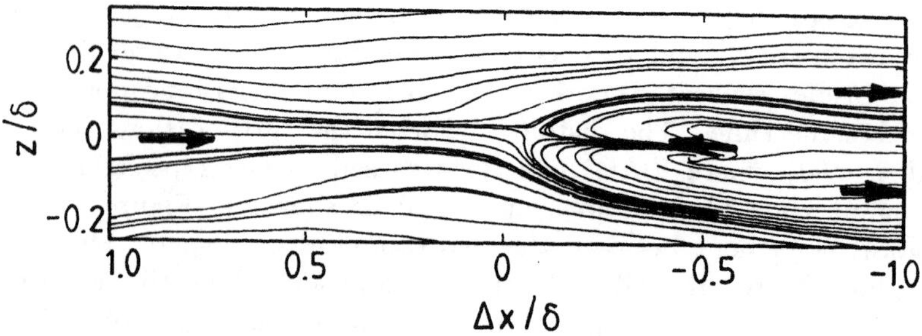


Fig.5. Sectional streamlines as for Fig. 2a, except $U_c = \overline{U} + 0.05U_1$.

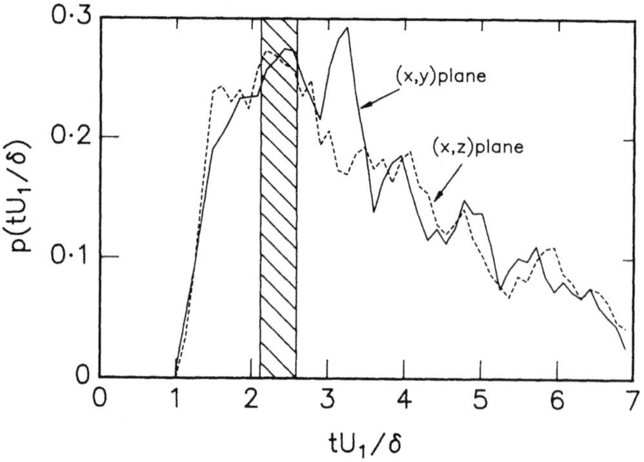


Fig.6. Probability density functions for periods between successive WAG detections in the $u$ signals from an **X**-probe at $y \approx 0.27\delta$ when the rake of probes was in either the $(x, y)$ or the $(x, z)$ plane.

Probability density functions of periods between successive WAG detections are shown in Figure 6 for the **X**-probe at $y = 0.27\delta$ in the $(x, y)$ plane and for one of the **X**-probes in the $(x, z)$ plane at the same $y$ value. The functions are quite similar to each other, as are the equivalent functions (not shown) from $u$-level burst detection (nearer the wall) in two planes. Subsets of detections separated by $\Delta t = 2.3\delta/U_1 \pm 10\%$ (the band indicated by crosshatching in Figure 6) were selected, and conditional averages $< u >$ and $< v >$ or $< u >$ and $< w >$ were computed throughout the respective planes. After adding $\overline{U}(y) - U_c$ to the $<u>$ components for the $(x, y)$ plane, the sectional streamlines shown in Figure 7 were computed. Corresponding contours of vorticity components $<\omega_z>$ and $<\omega_y>$ are shown in Figure 8.

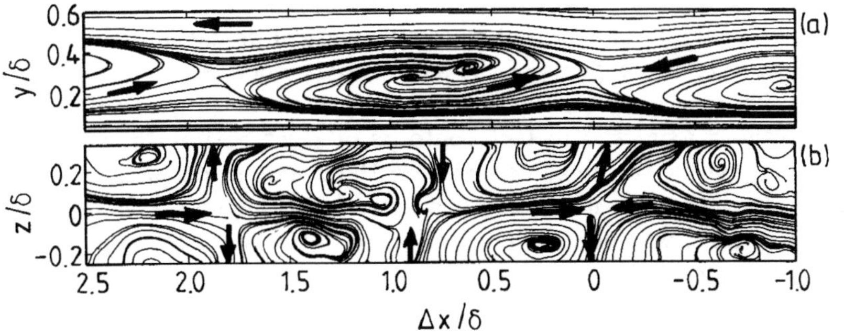


Fig.7. Sectional streamlines based on WAG detections separated by $\Delta t \approx 2.3\delta/U_1$, with $U_c = 0.8U_1$, in (a) the $(x, y)$ plane, and (b) the $(x, z)$ plane at $y \approx 0.27\delta$.

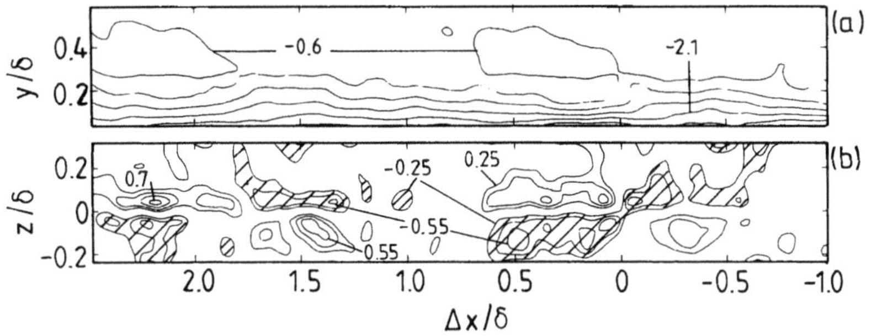


Fig.8. Contours of normalized $<\omega_y>$ and $<\omega_z>$ corresponding to Fig. 7. (a) $<\omega_z>$, contour increment $0.6U_1/\delta$; (b) $<\omega_y>$, contour increment $0.15U_1/\delta$, negative regions crosshatched.

In the $(x, y)$ plane (Figure 7a) the main structure appears to be a flattened rotating region of fluid, bounded upstream and downstream by saddle points located on the $\delta$-scale $u$-discontinuities. Spanwise vorticity $<\omega_z>$ is concentrated along the wall, Figure 8a, but there is also a region of significant $<\omega_z>$ well above the wall on the downstream edge of each $u$-discontinuity. In the $(x, z)$ plane (Figure 7b), there appear to be four smaller rotating regions between each pair of $u$-discontinuities (saddle points), with adjacent regions rotating in opposite directions. Contours of transverse vorticity $<\omega_y>$ (Figure 8b) follow the same pattern, and magnitudes of $<\omega_y>$ and $<\omega_z>$ associated with the WAG detected $u$-discontinuities are similar. The pattern in the $(x, z)$ plane is roughly antisymmetrical about the $x$ axis, with $<w>\approx 0$ along the axis. Although the streamline patterns in the two planes seem rather different, they are generated from an identical detection procedure and are therefore two aspects of the same large structures.

Unlike $<w>$, the instantaneous fluctuation $w$ is generally not zero along the $x$ axis, and in fact contours of $<w^2>$ (not shown) indicate that $|w|$ can be of the same order on the $x$ axis as elsewhere. The sign of $w$ along the $x$ axis, averaged over a short range of $x$ values relative to detections, was used as a criterion to divide sets of WAG detections into asymmetrical subsets. Figure 9 shows sectional streamlines resulting from such a division based on the sign of $w$ for $-0.3\delta < \Delta x < -0.1\delta$. Various other ranges of $x$ values were also tried, but there was no fundamental difference in results. The conditional averages are clearly asymmetrical, and certain rotating regions are now larger than neighbouring ones, but the general features of the pattern are unaltered. Asymmetry will be discussed further in the next section.

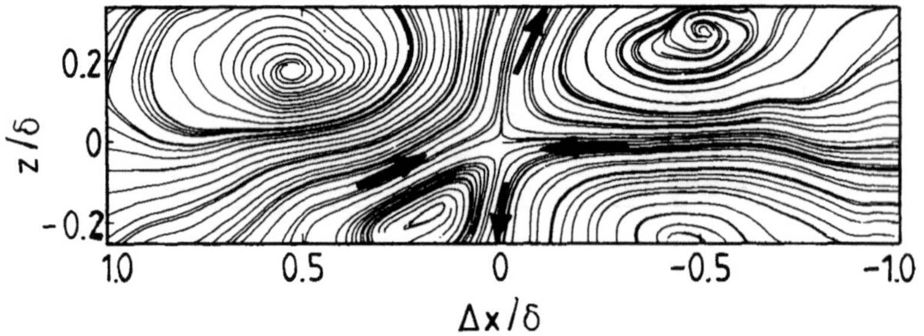


Fig.9. Sectional streamlines in the $(x, z)$ plane at $y = 0.27\delta$ based on WAG detections (at $\Delta x = z = 0$) for which $w > 0$ at $\Delta x \approx -0.2\delta$, with $U_c = \overline{U}$; compare with Fig. 2.

## 4.2. $U$-Level Burst Detection Results

Contours of $<u>$ from ten single hot wires conditioned on the ends of bursts (Antonia and Bisset, 1990) indicate that the length of the main low speed region of a burst is of order $\Delta x^+ = 200$. The "+" indicates normalization by $U_\tau$ and $\nu$, the kinematic viscosity. Bursts of approximately that length were selected from the set of all bursts from one **X**-probe in the $(x, z)$ plane at $y^+ = 25$ (for which $y = 0.035\delta$) and similarly from the **X**-probe at $y^+ = 17$ in the $(x, y)$ plane. Time is converted to distance using $\Delta x = -0.52 U_1 \Delta t$, where $0.52 U_1 \approx \overline{U}$ at $y^+ = 25$. Sectional streamlines calculated from the conditioned velocities (not using the full array of X-probes) are presented in Figure 10. The frame-of-reference velocity $U_c$ is $12U_\tau$ in both planes, slightly less than the $(x, z)$ plane mean velocity $12.7U_\tau$, in order to focus more specifically on the slow speed burst region. The equivalent contours of vorticity components $<\omega_y>$ and $<\omega_z>$ are shown in Figure 11. Contours of $<\omega_z>$ display the expected (anti)symmetry, but $(x, z)$ plane sectional streamlines (Figure 10b) are not quite symmetrical, probably because the spacing between probes and the spacing between the wires of each probe are rather large compared to the distance from the wall. It will be assumed that the lower (imperfect) rotating region of Figure 10b is the mirror image of the upper part, in which case there is good correspondence with contours of $<\omega_y>$ (Figure 11b).

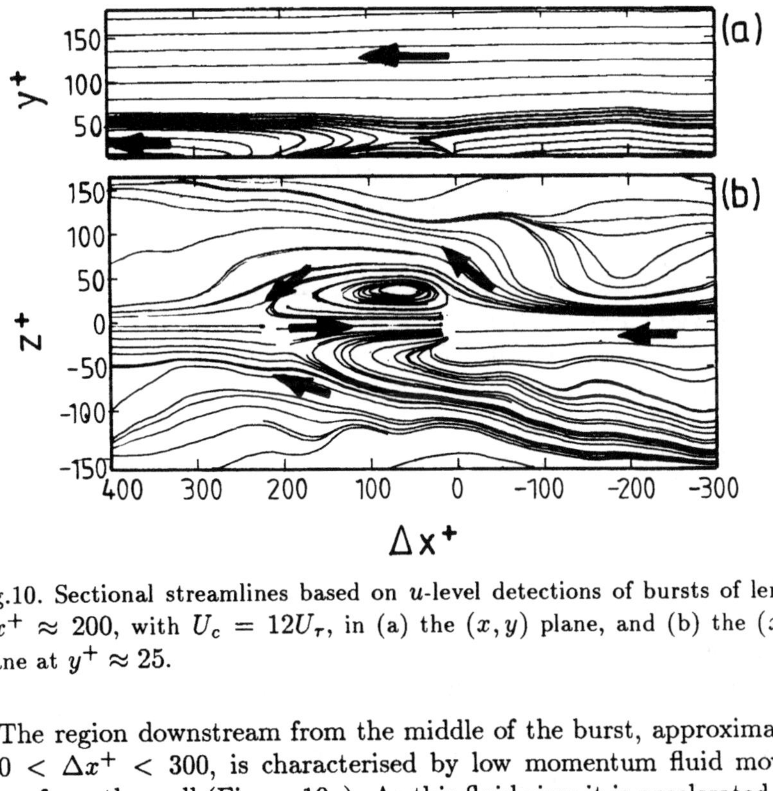


Fig.10. Sectional streamlines based on $u$-level detections of bursts of length $\Delta x^+ \approx 200$, with $U_c = 12U_\tau$, in (a) the $(x, y)$ plane, and (b) the $(x, z)$ plane at $y^+ \approx 25$.

The region downstream from the middle of the burst, approximately $100 < \Delta x^+ < 300$, is characterised by low momentum fluid moving away from the wall (Figure 10a). As this fluid rises it is accelerated and eventually carried away further downstream, as indicated by the sharp change of direction in the sectional streamlines. In the $(x, z)$ plane, however, fluid is moving *towards* this region of the burst from both sides (Figure 10b). Further upstream, say $25 > \Delta x^+ > -200$, fluid (mainly high momentum) is moving down towards the wall (Figure 10a). As the high momentum fluid reaches the low speed fluid within the burst, it

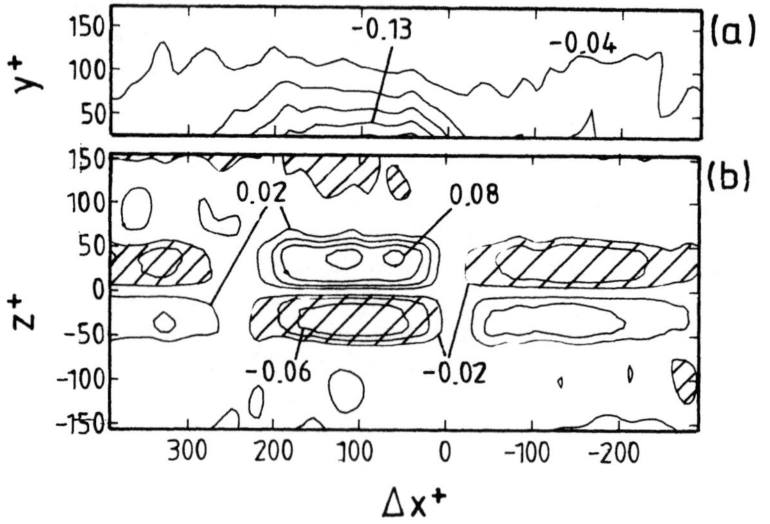


Fig.11. Contours of normalized $<\omega_y>$ and $<\omega_z>$ corresponding to Fig. 10. (a) $<\omega_z>$, contour increment $0.03U_\tau^2/\nu$; (b) $<\omega_y>$, contour increment $0.02U_\tau^2/\nu$, negative regions crosshatched.

forms a stagnation point (Figure 10b), and fluid at $z \neq 0$ is generally moving *away* from the burst.

Contours of $\omega_y$, generated from the data corresponding to one burst randomly chosen from the set used for Figures 10 and 11, are presented in Figure 12. This figure is not (anti)symmetrical about $z^+ = 0$, and peak values are considerably higher than in Figure 11, but there is no contradiction of the general impression given by $<\omega_y>$ contours in Figure 11.

Antonia and Bisset (1990) investigated the symmetry of $<u>$ distributions by splitting a set of end-of-burst detections into four subsets based on the signs of $\overline{\overline{u}}$ at $z^+ = 80$ and $z^+ = -80$, where $\overline{\overline{u}}$ is the mean value of $u$ for $-100 < \Delta x^+ < 100$ relative to each detection. In the present data, $|z^+| = 80$ falls half way between two **X**-probes, but $|z^+| = 80$ had been chosen as suitable after inspection of a considerable quantity of instantaneous $u$ contours, and also gave four detection subsets of roughly equal size. Therefore, a separation of $|z^+| = 80$, roughly equal to $1\frac{1}{2}$ **X**-probe spacings in the present data, was simulated by creating an artificial burst detection signal between two probes, i.e. the average of two $u$ signals from adjacent probes. After verifying that $u$-level detections in this signal behaved normally and produced approximately the usual sectional streamlines and contours of $<u>$, etc., the criteria on the signs of $\overline{\overline{u}}$ at $z^+ \approx \pm 80$ were applied to the set of $u$-level end-of-

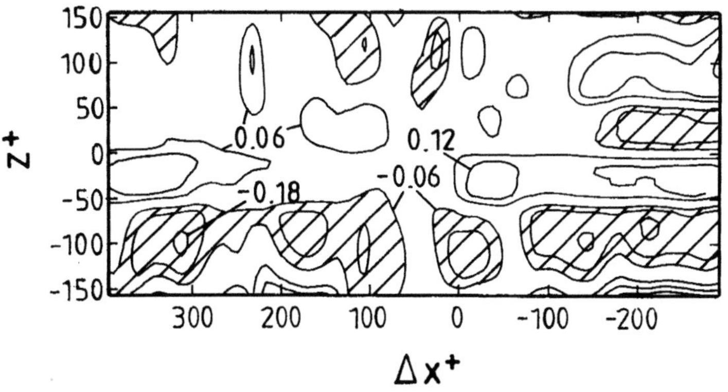


Fig.12. Contours of normalized $\omega_y$ from one burst randomly selected from the set of bursts used for Figs. 10b & 11b; contour increment $0.06U_\tau^2/\nu$, negative regions crosshatched.

burst detections. The detections of interest in the context of asymmetry are those for which the two $\overline{\overline{u}}$ values are different in sign, about 50% of all bursts. Sectional streamlines and contours of $<\omega_y>$ for the subset where $\overline{\overline{u}} < 0$ at $z^+ \approx 80$ and $\overline{\overline{u}} > 0$ at $z^+ \approx -80$ are presented in Figure 13 (the other subset produces the mirror image). In Figure 13a there are two rotating regions, but they rotate in the same direction and one is displaced considerably upstream (compare Figure 10b). The $<\omega_y>$ contours (Figure 13b) correspond well to the streamlines but are rather different from the almost perfectly (anti)symmetrical contours of Figure 11b. The general pattern of the movement of fluid within and around a burst is not greatly different from the symmetrical case, however, in the sense that fluid in the $(x, z)$ plane still moves towards the region around the beginning of the burst, and moves away from the region around the upstream end of the burst.

Johansson et al. (1987) and Guezennec et al. (1989) investigated artificial symmetry in conditional averages derived from numerical simulations of turbulent wall-bounded flows. Johansson et al. (1987) switched the $z$ axis according to the sign of $\partial u/\partial z$ at the upstream ends of VISA detected ejections, while Guezennec et al. (1989) used the sign of the net circulation in the $(y, z)$ plane to classify quadrant detections at $y^+ = 12$. The results from examination of asymmetry in both numerical simulations and the present experimental work, both near-wall and in the middle of the boundary layer, support these general conclusions: (a) that perfectly symmetrical (or antisymmetrical) structures are unlikely to occur, but (b) other general characteristics of structures are not misrepresented by the artificially symmetrical results of detection and conditional averaging.

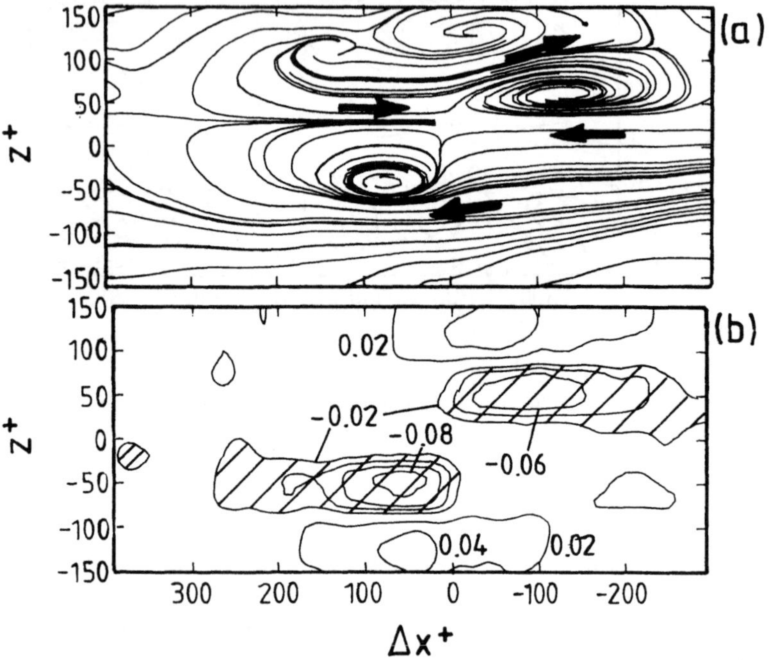


Fig.13. (a) Sectional streamlines at $y^+ \approx 25$ based on $u$-level bursts of length $\Delta^+ \approx 200$ for which $\overline{\overline{u}} < 0$ at $z^+ = 80$ and $\overline{\overline{u}} > 0$ at $z^+ = -80$ (compare Fig. 10b), with $U_c = 12U_\tau$. (b) Contours of the corresponding $<\omega_y>$ (compare Fig 11b), contour increment $0.02U_\tau^2/\nu$, negative regions crosshatched.

### 4.3. Comparison With a Cylinder Wake

The far-wake of a circular cylinder has been examined in a similar fashion with the same array of eight **X**-probes, and some results are briefly reported here for comparison with the boundary layer. The Reynolds number $R_d$ was about 1200, and the measurement station was $x = 420d$. WAG detections were obtained in $u$ signals from probes at $y \approx L$, where $L$ is the mean velocity halfwidth, in both $(x, y)$ and $(x, z)$ planes. Detections separated by the most probable spacing ($\Delta x \approx 3L$) were selected and used to condition the velocities from all probes. Sectional streamlines calculated from the conditioned velocities are presented in Figure 14, in a frame of reference moving at approximately the mean velocity of the $(x, z)$ plane. Bearing in mind that the wake is two-sided, these streamlines are reasonably similar to the boundary layer streamlines of Figure 7, although the rotational regions are more stretched out in the $x$ direction in the boundary layer. Contours of $<\omega_y>$ (not shown) follow the $(x, z)$ streamlines very closely.

Contours of correlations between $\omega_y$ values throughout the $(x, z)$ plane

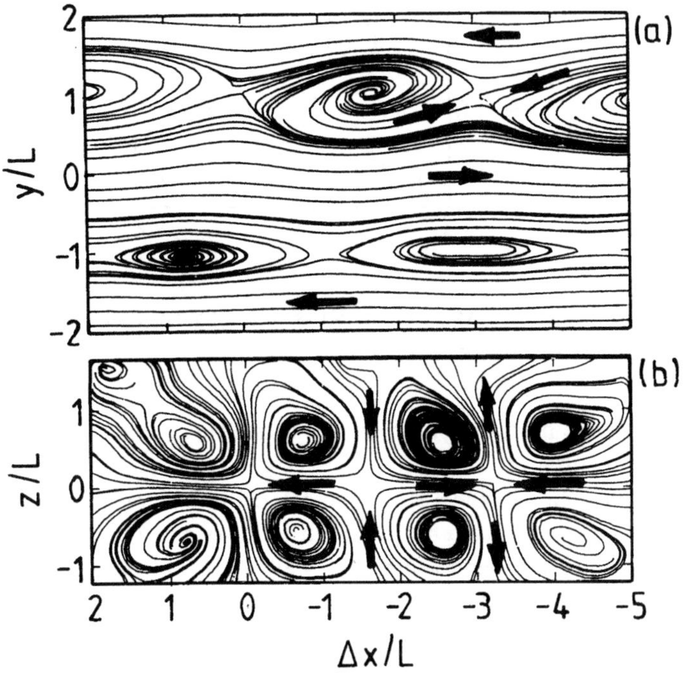


Fig.14. Sectional streamlines from WAG detections separated by $\Delta t \approx L/U_1$ in the far-wake of a circular cylinder, with $U_c \approx \overline{U}(y = L)$, (a) in the $(x, y)$ plane, and (b) in the $(x, z)$ plane at $y \approx L$.

are shown in Figure 15. Unlike the equivalent correlations for the boundary layer (Figure 4), the negative correlations are quite significant, and the scale of the contours is very close to that of the streamline patterns. Even though there are specific differences between the two flows, it is likely that there is the same general type of complex three-dimensional structure present in both, with the same relationship between its manifestations in $(x, y)$ and $(x, z)$ planes.

## 5. Conclusions

Detection procedures have been applied, separately but identically, to the $u$ signals at the lines of intersection of orthogonal planes of velocity vectors recorded by a rake of **X**-probes. Sectional streamlines and vorticity contours for different views of the same conditionally averaged turbulence structures have been presented. The results highlight the intense three-dimensionality and complex nature of the structures. No attempt has been made at a definitive description of the structures, however, particularly because of the lack of information about their development and decay as they move downstream. Such information is available

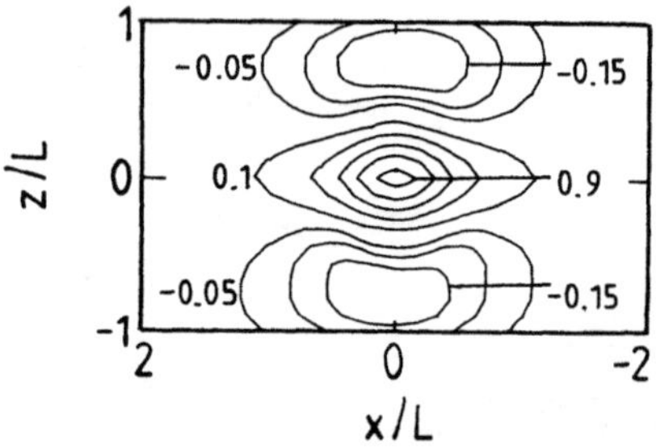


Fig.15. Contours of correlation coefficients of $\omega_y$ throughout the $(x, z)$ plane of the cylinder wake at $y \approx L$.

from numerical simulations, which could be studied in conjunction with experiment results.

In the buffer layer, the $(x, z)$ plane motion associated with a burst appears to correspond to a cut through the legs of a hairpin vortex. The $(x, y)$ plane motion does not agree so well with that concept, and in the $(x, z)$ plane it was shown that at least half of the bursts are not at all symmetrical. True hairpin vortices are probably very rare.

Beyond the buffer layer there is a complex large scale conditionally averaged motion that manifests as an elongated (in $x$) spanwise vortex in the $(x, y)$ plane, bounded by $\delta$-scale $u$-discontinuities. In the $(x, z)$ plane it manifests as a series of adjacent smaller counter-rotating regions. The $z$-span of the detected regions increases approximately logarithmically with distance from the wall, although the $x$ dimension is little changed. The rotating regions in the $(x, z)$ planes look like cuts through inverted cones, but in fact the cones (if they exist at all) have only a very small $y$ dimension, or else their axes are themselves rotating about the $z$ axis — they cannot be seen from a single frame-of-reference velocity over a significant range of $y$ values. Symmetry in $(x, z)$ planes was shown to be artificial, but other general features of the motion are not misrepresented in the symmetrical results. The probable connection between this motion and bursting nearer the wall was considered in Antonia et al. (1989) for the $(x, y)$ plane.

A broadly similar organized motion exists in the far-wake of a circular cylinder, but in this case there is a closer relationship between the detected motion and transverse vorticity $\omega_y$.

**Acknowledgements** The support of the Australian Research Council is gratefully acknowledged. We wish to thank Drs. L.W.B. Browne and D.A. Shah for their contributions to the experimental work.

**References**


Antonia, R. A. & Bisset, D. K. 1990 Spanwise Structure in the Near-Wall Region of a Turbulent Boundary Layer. *J. Fluid Mech.* **210**, 437-458.

Antonia, R. A., Bisset, D. K. & Browne, L. W. B. 1989 Effect of Reynolds Number on the Organized Motion in a Turbulent Boundary Layer, in S. J. Kline and N. H. Afgan (eds.) *Near Wall Turbulence.* New York, Hemisphere.

Bisset, D. K., Antonia, R. A. & Browne, L. W. B. 1989 Spatial Organisation of Large Structures in the Turbulent Far-Wake of a Cylinder. *J. Fluid Mech.* [submitted].

Guezennec, Y. G., Piomelli, U. & Kim, J. 1989 On the Shape and Dynamics of Wall Structures in Turbulent Channel Flow. *Phys. Fluids A* **1**, 764-766.

Head, M. R. & Bandyopadhyay, P. 1981 New Aspects of Turbulent Boundary- Layer Structure. *J. Fluid Mech.* **107**, 297-2338.

Johansson, A. V., Alfredsson, P. H. & Kim, J. 1987 Shear-Layer Structures in Near-Wall Turbulence. *Center for Turbulence Research Proceedings of the Summer Program 1987*, 237-251.

Kim, J. & Moin, P. 1986 The Structure of the Vorticity Field in Turbulent Channel Flow. Part 2. Study of Ensemble-Averaged Fields. *J. Fluid Mech.* **162**, 339-363.

Kline, S. J. 1989 Quasi-Coherent Structures in the Turbulent Boundary Layer. Part I : Status Report on a Community-Wide Summary of the Data, in S. Kline and N. H. Afgan (eds.) *Near Wall Turbulence*, New York, Hemisphere.

Luchik, T. S. & Tiederman, W. G. 1987 Time Scale and Structure of Ejections and Bursts in Turbulent Channel Flow. *J. Fluid Mech.* **174**, 529-552.

Robinson, S. K., Kline, S. J. & Spalart, P. R. 1989 Quasi-Coherent Structures in the Turbulent Boundary Layer. Part II : Verification and New Information from a Numerically Simulated Flat-Plate Layer, in S. Kline and N. H. Afgan (eds.) *Near Wall Turbulence*, New York, Hemisphere.

Tiederman, W. G. 1989 Eulerian Detection of Turbulence Bursts, in S. Kline and N. H. Afgan (eds.) *Near Wall Turbulence*, New York, Hemisphere.

Townsend, A. A. 1976 *The Structure of Turbulent Shear Flow*, 2nd edition, Cambridge, Cambridge University Press.

# Large eddy simulation of boundary layers with a step change in pressure gradient

Rainer FRIEDRICH and Friedemann UNGER

*Lehrstuhl für Strömungsmechanik der Technischen Universität München*
*Arcisstr.21, 8000 München 2, Federal Republic of Germany*

*Turbulence and Coherent Structures*
*O. Métais and M. Lesieur, eds.*
*Selected papers from "Turbulence 89: organized structures and turbulence in fluid mechanics"*
*Grenoble, 18-21 September 1989*

## 1. Summary

Statistically two-dimensional and steady high Reynolds number boundary layers over a flat plate with a sudden change in the freestream longitudinal pressure gradient are simulated based on the Navier-Stokes equations for an incompressible fluid. Low-pass filtering of these equations via integration over the mesh volume of a cartesian equidistant staggered grid leads to a second order accurate finite difference scheme in space which conserves mass exactly. The explicit time integration scheme applied is of second order accuracy with respect to convective terms. The use of the projection method results in a Poisson equation for the pressure which is evaluated with the help of a vectorized direct Poisson solver. Three different flow situations are simulated corresponding to zero-, negative- and positive pressure gradient boundary layers. In all three cases the same inflow boundary conditions are used which are Dirichlet conditions for the velocity vector at each time step, taken from a separate large eddy simulation of a turbulent boundary layer. Thus, a direct comparison of the effects of acceleration and deceleration on the turbulence structure is possible. It is found that the statistically observed inhibition of turbulence fluctuations in an accelerated boundary layer is due to a decrease in the intensity of coherent (shear stress producing) structures whereas the characteristic enhancement of fluctuating activity results from an increase in peak fluctuations of these structures.

## 2. Introduction

The last twenty years of turbulence research have led to a 'new mental picture of turbulence structure' (Bradshaw 1987). According to this picture turbulence is highly organized, producing coherent structures which are large contributors to shear stress and therefore to mixing. Hussain (1986) has given a useful definition of these structures as open volumes with distinctive large-scale vorticity distributions.

Fiedler (1987) defines coherent structures as recurrent, appearing at best quasi-periodically and having a characteristic life-span. They are comparable in size with the global dimensions of the flow field and therefore amenable to prediction methods like the large eddy simulation (LES) technique.

In a recent paper (Friedrich & Richter 1989) we have discussed different ways of deriving transport equations for LES via spatial low-pass filtering of the basic continuum equations. Among these, Schumann's

(1975) approach which consists in a filtering over the mesh volume itself and thus leads to a semi-discrete system of transport equations in a natural way, is the one we adopt here. Other techniques generally use twice the mesh spacing in the three coordinate directions as proper filter widths and have the possibility to control the smoothing process independently of the current discretization.

We start from the integral form of the continuity and momentum equations for an incompressible fluid:

$$\int_S \vec{u} \cdot \vec{n} dS = 0, \qquad (2-1)$$

$$\partial_t \int_V \vec{u} dV + \int_S [\vec{u}(\vec{u} \cdot \vec{n}) - \overline{\overline{T}} \cdot \vec{n}] dS = 0. \qquad (2-2)$$

V is a closed and time-independent region in space surrounded by a surface $S$ . $\vec{n}$ denotes the outwardly directed unit normal vector to $dS$. $\overline{\overline{T}}$ is the stress tensor and $\vec{t}=\overline{\overline{T}} \cdot \vec{n}$ the stress vector which not only depends on space and time, but also on the orientation of the surface element $dS$, i.e. on $\vec{n}$ itself. Applying (2-1), (2-2) to the finite volume $\Delta V$ of a cartesian grid system with surface elements $\Delta S_1 = \Delta x_2 \Delta x_3, \Delta S_2 = \Delta x_1 \Delta x_3$, $\Delta S_3 = \Delta x_1 \Delta x_2$, we get the integral conservation equations for a grid volume:

$$\sum_{j=1}^{3} \int_{\Delta S_j^*} \vec{u} \cdot \vec{n}_j dS = 0, \qquad (2-3)$$

$$\partial_t \int_{\Delta V} \vec{u} dV + \sum_{j=1}^{3} \int_{\Delta S_j^*} (\vec{u}(\vec{u} \cdot \vec{n}_j) - \overline{\overline{T}} \cdot \vec{n}_j) dS = 0. \qquad (2-4)$$

Note that $\Delta S_j^*$ stands for $\Delta S_j^+ \cup \Delta S_j^-$, namely the sum of surface elements lying on opposite sides of $\Delta V$ and having unit normal vectors $\vec{n}_j$ which point in positive and negative coordinate directions. The stress tensor in (2-4), when expressed in cartesian coordinates for a Newtonian fluid, has the form:

$$T_{ij} = -p/\rho\, \delta_{ij} + \nu D_{ij}, \qquad D_{ij} = \partial_j u_i + \partial_i u_j. \qquad (2-5)$$

Equations (2-3),(2-4) suggest the definition of discrete variables as grid surface and grid volume averages,viz:

$${}^{i}\overline{u_i} = \int_{\Delta S_i} u_i dS/\Delta S_i, \qquad {}^{v}\overline{u_i} = \int_{\Delta V} u_i dV/\Delta V, \qquad (2-6a,b)$$

$$^{j}\overline{u_i u_j} = \int\limits_{\Delta S_j} u_i u_j dS/\Delta S_j, \qquad (2-7)$$

$$^{i}\overline{p} = \int\limits_{\Delta S_i} p\, dS/\Delta S_i, \qquad (2-8)$$

$$^{j}\overline{D_{ij}} = \int\limits_{\Delta S_j} D_{ij} dS/\Delta S_j. \qquad (2-9)$$

This time, integration in (2-6a,-7,-8,-9) is carried out over $\Delta S_j^+$ alone. Inserting (2-6) to (2-9) into (2-3) and (2-4) leads to an unclosed set of equations. The most serious closure problem results from the surface averaged convection terms. If we split the instantaneous velocity vector into a surface average and a deviation from it, according to

$$u_i =^j \overline{u_i} + u_i', \qquad ^{j}\overline{u_i} = \int\limits_{\Delta S_j} u_i\, dS/\Delta S_j \qquad (2-10)$$

and substitute this into (2-7), we end up with an unknown subgrid scale (SGS) stress $^{j}\overline{u_i' u_j'}$ :

$$^{j}\overline{u_i u_j} =^j \overline{u_i}\,^j\overline{u_j} + {}^{j}\overline{u_i' u_j'}. \qquad (2-11)$$

Solving (2-4) on a staggered grid allows for the following assumptions:

$$^{v}\overline{u_i} \approx^i \overline{u_i} \qquad (2-12)$$

$$^{i}\overline{p} \approx^v \overline{p}. \qquad (2-13)$$

The volume averaged pressure is defined according to (2-6b). Components of $^{j}\overline{u_i}$ for which $j \neq i$ are linearly related to $^{i}\overline{u_i}$. Conformity conditions are used for discrete variables and gradients in grid planes where cell surfaces of adjacent grid volumes coincide.

## 3. Filtered N-S equations and time-integration

The finite volume equations (2-3),(2-4) for the cell $\Delta V$ together with definitions and assumptions (2-5) to (2-13) can be put in the more compact form of the *filtered Navier-Stokes equations*:

$$\delta_j\, {}^j\overline{u_j} = 0, \qquad (3-1)$$

$$\partial_t\, {}^i\overline{u_i} = -\delta_j({}^j\overline{u_i}\,{}^j\overline{u_j} + {}^j\overline{u_i'u_j'}) - \delta_i\, {}^v\overline{p}/\rho + \nu\delta_j\, {}^j\overline{D_{ij}}, \qquad (3-2)$$

$\delta_j$ stands for the central finite difference opperator

$$\delta_j\phi = (\phi(x_j + \Delta x_j/2) - \phi(x_j - \Delta x_j/2))/\Delta x_j. \qquad (3-3)$$

The SGS stress is related to the filtered deformation tensor following Schumann's (1975) suggestion:

$${}^j\overline{u_i'u_j'} = -\mu_{iso}({}^j\overline{D_{ij}} - < {}^j\overline{D_{ij}} >) - \mu_{inh} < {}^j\overline{D_{ij}} > + \frac{1}{3}\delta_{ij}\, {}^k\overline{u_k'u_k'}. \qquad (3-4)$$

The right-hand side contains a fluctuating part (in Reynolds' sense) and a statistical part which accounts for inhomogeneities in the SGS turbulence. These are certainly important close to walls and in the case of coarse grids. The last term ensures equality on both sides for $i=j$. The triangular brackets in (3-4) indicate statistical averages. The eddy viscosities $\mu_{iso}$, $\mu_{inh}$ contain length scales which are related to the grid spacing (in the case of $\mu_{iso}$) or are empirical. The velocity scales are computed either from the fluctuating or statistically averaged deformation tensor ${}^j\overline{D_{ij}}$. For more details see Schmitt (1988).

We use an explicit scheme for *integration* of (3-2) *in time*. It is a variant of Chorin's (1968) fractional-step (or time-splitting) method. The pressure plays the role of a projection operator which maps an arbitrary velocity field into one which is divergence free. Writing (3-2) in a symbolic form, neglecting overbars and using $C_i$ for convection terms and $D_i$ for diffusion terms, the difference scheme reads:

$$u_i^* = f_1 u_i^{n-1} + f_2 u_i^n + f_3(C_i^n + D_i^{n-1}), \qquad (3-5)$$

$$u_i^{n+1} = u_i^* - f_3\delta_i p^{n+1}, \qquad (3-6)$$

$$\delta_i u_i^{n+1} = 0. \qquad (3-7)$$

The integration proceeds in cycles of 49 leapfrog-steps ($f_1=1, f_2=0, f_3=2\Delta t$) and one averaging step ($f_1=0.5, f_2=0.5, f_3=1.5\Delta t$) in order to avoid $2\Delta t$-oscillations which might result from the leapfrog scheme. The initial integration step is an Euler step ($f_1=0, f_2=1, f_3=\Delta t$). During each time step an approximative velocity field $u_i^*$ is first computed neglecting the pressure gradient (3-5). It is then corrected, using (3-6). The pressure field needed in (3-6) results from solving the discrete Poisson equation

$$\delta_i u_i^* = f_3 \delta_i \delta_i p^{n+1}, \qquad (3-8)$$

which follows from taking the divergence of (3-6) and considering (3-7). The algebraic system (3-8) is evaluated with a fast elliptic solver. It uses a fast fourier transformation in the homogeneous direction in which there is no mean flow (periodic boundary conditions) and the cyclic reduction technique (Schumann & Sweet 1976) in a vectorized version. Time steps $\Delta t$ are chosen such that the Courant numbers are well below one. The integration scheme is mass-, momentum- and energy-conserving and of second order accuracy with respect to convection terms. Boundary conditions for $p$ can either be of Dirichlet or von Neumann type. Because of (3-6) they may be expressed in terms of velocity components and it is important to note that the numerical solution is independant of the boundary value of $u_i^*$ (Peyret & Taylor 1983).

The question of *initial values* for the 3D velocity vector in the whole computational domain may be addressed quickly. Since $u_i$ is composed of a statistical and a fluctuating part, we use experimental or empirical data for the first part and add Gaussian random numbers which are weighted with $(2k/3)^{1/2}$ ($k$= mean turbulence energy) and made divergence -free with a single use of the Poisson solver. This procedure is relevant for the very first simulation only. Succeeding simulations are started from end data of previous runs by simply adjusting mean velocities to the new flow conditions, if necessary.

## 4. Specification of flows and boundary conditions

It is useful to describe the geometry of the flow field and its global characteristics before discussing numerical boundary conditions.

In units of the boundary layer thickness $\delta_0$ (measured at the inflow plane) the computational domain has the size $24 \times 4 \times 2(4)$ in x,y,z i.e. in main flow, spanwise and wall normal directions. The value 4 in brackets refers to a retarded boundary layer which grows much faster than that in the other cases considered. From $x$=8 onwards the flow is subjected to a constant pressure gradient by specifying the freestream velocity $u_\infty(x, y, z{=}2(4))$ accordingly. Three different values of $dp_\infty/dx$ are investigated in three separate large eddy simulations, namely $-10, 0, 10$ (nondimensionalized with $\delta_0, u_\tau^2$ ). We refer to these simulations as Cases 1,2,3. Upstream of $x$=8 $u_\infty$ is constant in all the three cases according to a zero pressure gradient. The Reynolds number at the entrance plane ($x$=0) of the computational domain is $Re_\tau = u_\tau \delta_0/\nu$ =3240 or 90000 when based on $u_\infty$ and $\delta_0$.

In order to have proper and equal ***inflow boundary conditions*** for Cases 1,2,3, we have performed a fourth LES in a $8 \times 4 \times 2$ domain for a zero-pressure gradient boundary layer with the same Reynolds number as above (Case 4). The question of inflow boundary conditions arises here too, of course. The 3D instantaneous velocity vector has to be specified in the entrance plane in this case also during the whole simulation process. Again we split $u_i$ into its statistical, $< u_i >$, and fluctuating part,$u_i''$, and specify $< u_i >$ according to Klebanoff's (1955) data. For this case, the three components of $u_i''$ are taken from plane $x$=6.25 and transferred to $x$=0 after weighting them with the local ratios of the rms-values at $x$=0 (Klebanoff data) and $x$=6.25 (simulation). This is also referred to as ***structure periodic boundary conditions***. A more detailed description is given in Richter et al. (1987) and Friedrich & Richter (1989). All the remaining boundary conditions are analogous to those for Cases 1,2,3 and will be discussed below. As mentioned, Case 4 provides inflow data for the other 3 simulations; i.e. the velocity vectors from plane $x$=4 are stored on magnetic tape over 12000 time steps $\Delta t(= 10^{-3}\delta_0/u_{\tau_0})$ and used as Dirichlet inflow conditions for Cases 1,2,3.

Since the flow field is 2D in the mean and has a spanwise extent of 4 boundary layer thicknesses, the assumption of periodicity of all computational variables in ***y-direction*** makes sense.

The ***upper plane*** of the computational domain ($z$=2(4)) allows the turbulence-free flow to pass through. The longitudinal velocity component is constant in the range $0 \leq x < 8$ and decreases or is constant or increases according to $\delta_x p_\infty = -10, 0, 10$ for $8 \leq x \leq 24$. The spanwise velocity component is set equal to zero whereas the vertical one is extrapolated, fulfilling continuity.

Boundary conditions for the ***outflow plane*** ($x$=24) may have a remarkable effect on the upstream flow field if not designed carefully. We found extrapolating statistically averaged velocity components and solving convection equations for the velocity fluctuations, viz.

$$\delta_x(\delta_x < \bar{u} >) = 0, \quad \delta_x < \bar{v} >= 0, \quad \delta_x < \bar{w} >= 0, \qquad (4-1)$$

$$\partial_t u_i'' + < \bar{u} > \delta_x u_i'' = 0, \qquad (4-2)$$

to provide satisfactory results.

A comparison of the performance of these conditions with that of simpler conditions is made in Richter et al. (1987).

It follows from the filtered equations (3-1,-2) that ***wall boundary conditions*** can be specified for the normal velocity and the shear stress components. According to Schumann (1975) and Grötzbach (1986) the wall shear stress is in phase with the tangential velocity component, thus:

$$\overline{w}|_w = 0, \qquad (4-3)$$

$$\bar{\tau}_{zx}|_w = <\bar{\tau}_{zx}>_w / <\bar{u}>_{k=1} \cdot \bar{u}_{k=1}, \qquad (4-4)$$

$$\tau_{zy}|_w = \nu \bar{v}_{k=1}/(\Delta z/2). \qquad (4-5)$$

$k$=1 refers to the first grid volume, close to the wall. The statistical values of the shear stress and longitudinal velocity incorporate empiricism insofar as the logarithmic law is used to relate them:

$$<\bar{u}>_{k=1} = u_\tau/\kappa \cdot (\ln(Eu_\tau \Delta z/\nu) - 1), \qquad (4-6)$$

$$\kappa = 0.4, \quad E = 9.025.$$

(4-6) is obtained from averaging

$$<u> /u_\tau = 1/\kappa \ln(Ezu_\tau/\nu) \qquad (4-7)$$

between 0 and $\Delta z$. We are aware of the fact that assumptions (4-3,-4) might be improved by correlating the wall shear stress with the velocity along a typical flow structure, i.e. to take $\bar{u}_{k=1}$ from a position suitably displaced in downstream direction (See, Piomelli 1989). We intend to test such conditions next.

## 5. Results

In a former study (Friedrich & Richter 1989) we have presented LES results for the same flow with a very coarse grid of 96 × 16 × 16 cells and were able to predict mean and some of the instantaneous flow phenomena remarkably well. The present simulations provide improved data on a 192 × 32 × 32(64) grid of a $(x, y, z)$-coordinate system.

We start by discussing mean flow quantities which are obtained from instantaneous variables by averaging in $y$-direction and over time (12000 respectively 18000 time samples have been used).

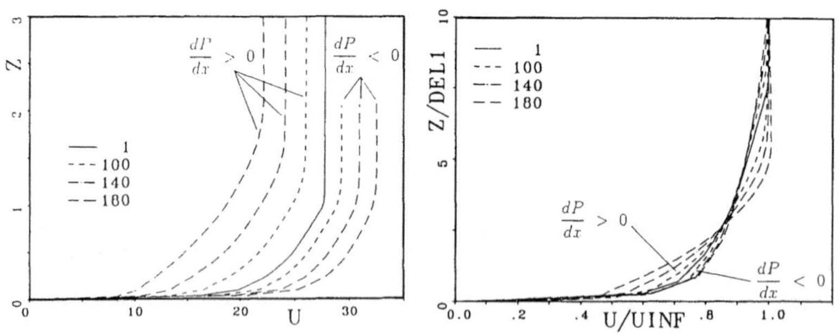


Fig.1. a,b Mean $u$-velocity profiles, Cases 1,3

Figure 1a shows the mean longitudinal velocity (nondimensionalized with $u_\tau$) for accelerated and retarded flow at three different positions $x$=12.4, 17.4, 22.4 (the corresponding mesh index is $I$=100, 140, 180). The solid line represents the inflow profile. It is that of a zero-pressure gradient boundary layer with $Re_\tau$=3240. In the fully turbulent region where viscous terms are unimportant the mean pressure gradient primarily affects the $< u >$-field and depends only weakly on $z$ except for a short region around $x$=8 where the flow looses its boundary layer character. Especially in the accelerated case the velocity profiles are approximately shifted without alteration of $\partial_z < u >$ (Townsend 1962). This is also obvious from figure 1b where $< u > /u_\infty$ is plotted against $z/\delta_1$, $\delta_1$ being the displacement thickness. A fairly big change in the shape of profiles occurs in the case of retarded flow.

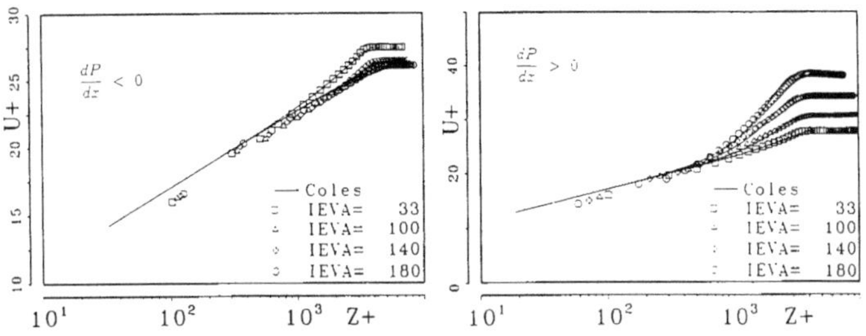


Fig.2. Mean $u$-velocity profiles, comp. with Coles' law of the wake, Cases 1,3

A semi-logarithmic plot of the velocity profiles is given in figure 2 for favourable and adverse pressure gradients. The symbols from squares to circles correspond to the positions $x$=4, 12.4, 17.4, 22.4. The profiles are compared with Coles' law of the wake:

$$< u > /u_\tau = 1/\kappa\left[\ln(Ez^+) + \Pi(1 - \cos(\pi z/\delta_c))\right], \qquad (5-1)$$

where $E$=9.025, $\kappa$=0.4 and $\Pi(x)$, $\delta_c(x)$ were determined such that (5-1) leads to the simulated displacement thickness. The effects of inhibition respectively enhancement of turbulent fluctuations due to acceleration/deceleration of the flow are apparent from the total shear stress (sum of Reynolds and viscous stress) profiles in figure 3.

When the local wall shear stress is used for scaling, the typical maxima appear away from the wall for retarded flow. The solid line represents the inflow profile which is in good agreement with Klebanoff's experiment. Since the mean pressure gradient does not appear in the Reynolds stress transport equation, it affects the turbulence fluctuations only through the mean velocity field, i.e. via the production terms. The friction coefficients for the three cases are plotted in figure 4. Triangular symbols represent the acceleration case. $c_f$ seems to approach its maximum

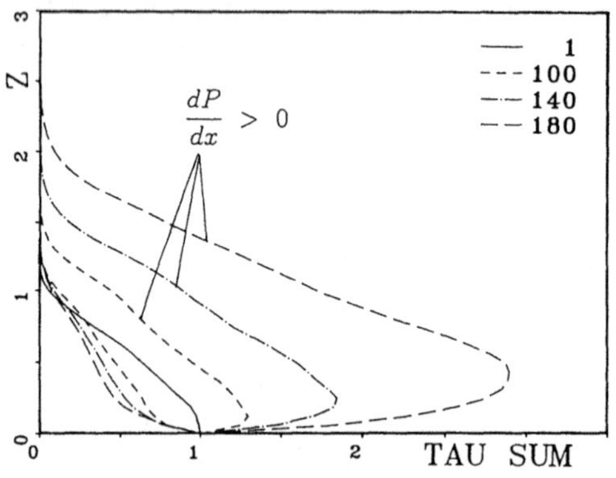


Fig.3. Total shear stress profiles, Cases 1,3

value at the end of the computational domain. A longer domain would have been necessary to predict the decrease in $c_f$ demonstrated e.g. by Deissler (1974). The retarted flow (circles) is still far from separation.

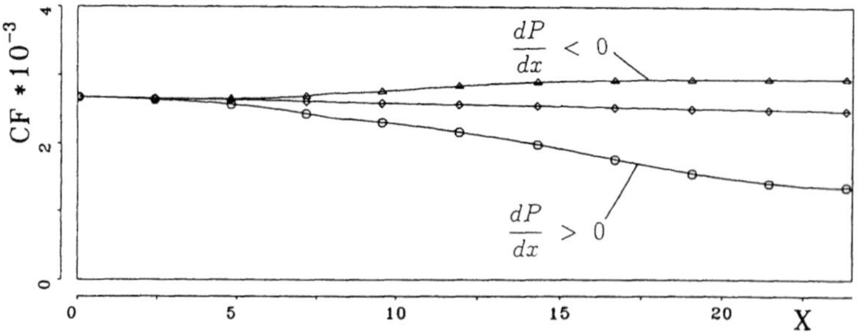


Fig.4. Friction coefficient, Cases 1,2,3

The aim of the following figures is to demonstrate the pressure gradient effect on the instantaneous turbulence structure. As a first example we show instantaneous u- and w-profiles in figures 5a,b, 6a,b, 7a,b for the cases 1,2,3 in a vertical plane aligned with the mean flow ($y$=1.31). Only a portion of the computational domain is presented. The position where the pressure gradient suddenly changes is further upstream ($x$=8). The differences in the first profiles at $x$=10.5 are minor due to the fact that the inflow conditions at $x$=0 are the same in all the three flows. Further downstream, however, the flow fields develop in different ways. The dashed-dotted lines represent mean profiles.

The information contained in the $u$- and $w$-profiles together with their statistical means suffices to detect local regions from which large contributions to the Reynolds stress can be expected. These regions are either characterized by negative $u$- and positive $w$-fluctuations (Q2-event)

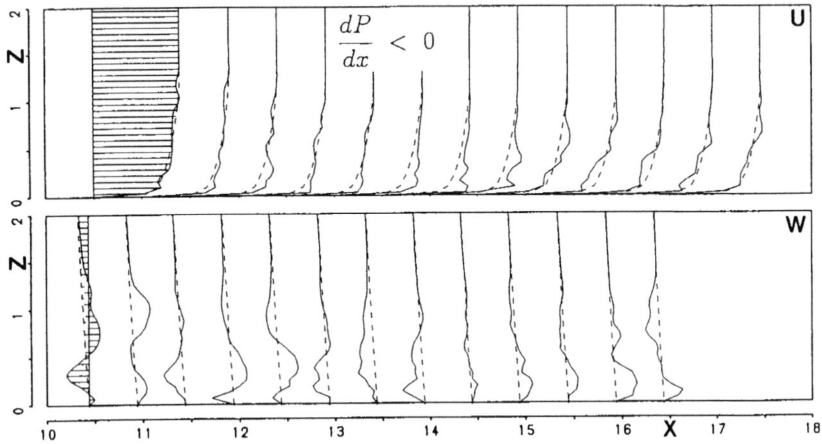


Fig.5. a,b Instantaneous longitudinal and vertical velocity profiles, Case 1

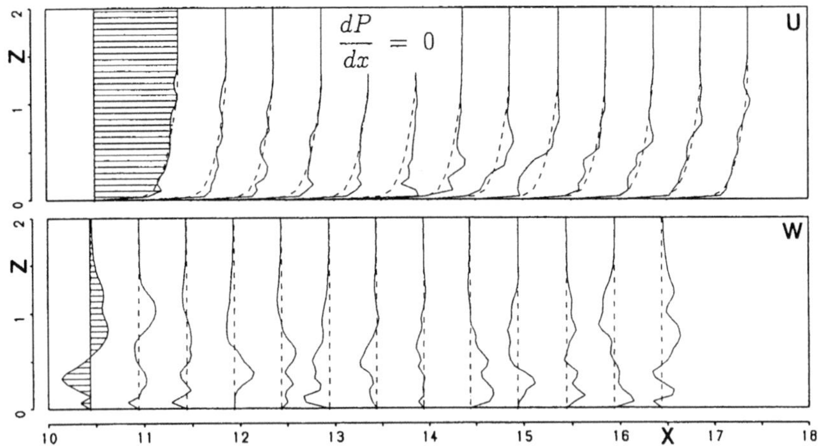


Fig.6. a,b Instantaneous longitudinal and vertical velocity profiles, Case 2

or by positive $u''$- and negative $w''$-values (Q4-event). In the figures below we have plotted contour lines of $-u''w''$-products, taking solid lines representative of positive values (Q2 $\cup$ Q4) and dashed lines representative of negative values ($Q1 := \{-u''w''|u'' > 0, w'' > 0\}$ and $Q3 := \{-u''w''|u'' < 0, w'' < 0\}$). Simultaneous plots of $(u'', w'')$-vectors in the same planes ($y$=1.31) allow to distinguish Q2- from Q4-events which we may also call ejection- and sweep-type events, respectively. Figures 8a,b, 9a,b and 10a,b demonstrate what happens to large scale Reynolds stress producing (or coherent) structures in the case of negative, zero and positive pressure gradients.

Acceleration of the flow not only increases the spacing between these

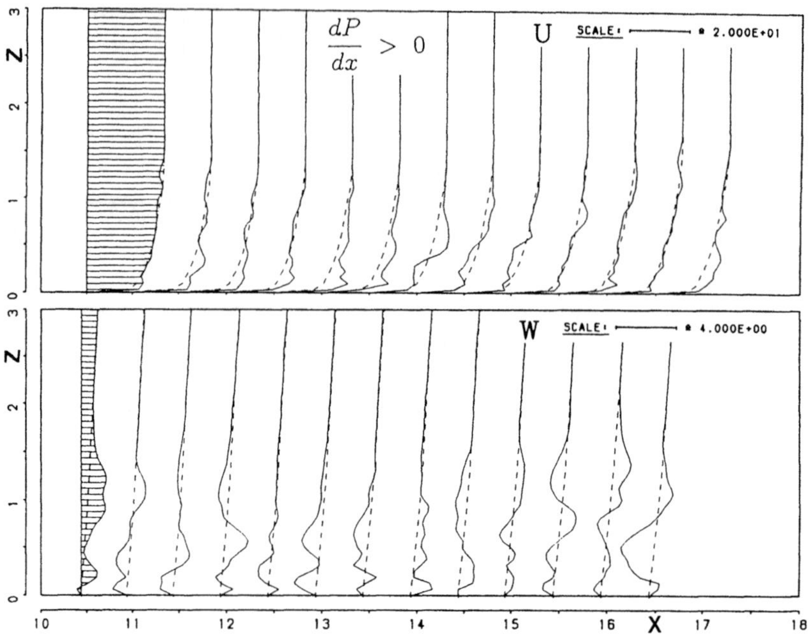


Fig.7. a,b Instantaneous longitudinal and vertical velocity profiles, Case 3

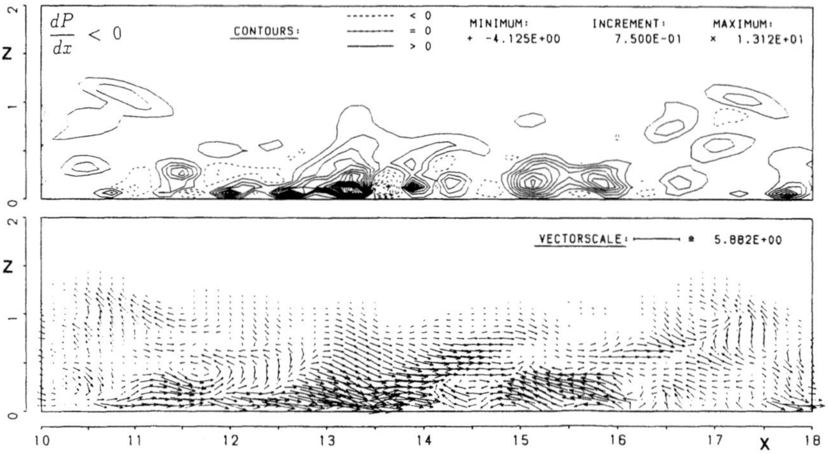


Fig.8. a,b Contour lines of $-u''w''$ and $(u'', w'')$-vectors, Case 1

structures (compare the positions of the Q2/Q4-structures in figure 9 and 10) it also decreases the peak values of $u''w''$. This is perhaps more apparent from figure 11a,b which shows contour lines for Cases 1,2 ($dp/dx \leq 0$) with $-u''w'' > 2$ ($u_\tau^2$ being the reference value). De-

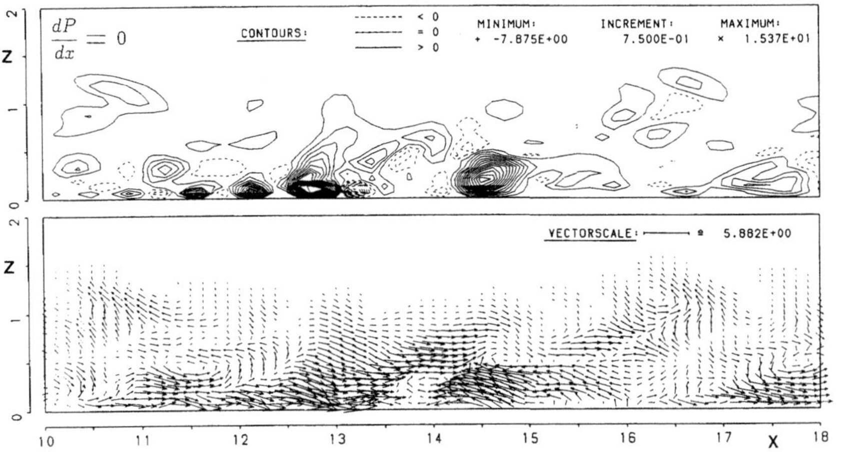


Fig.9. a,b Contour lines of $-u''w''$ and $(u'', w'')$-vectors, Case 2

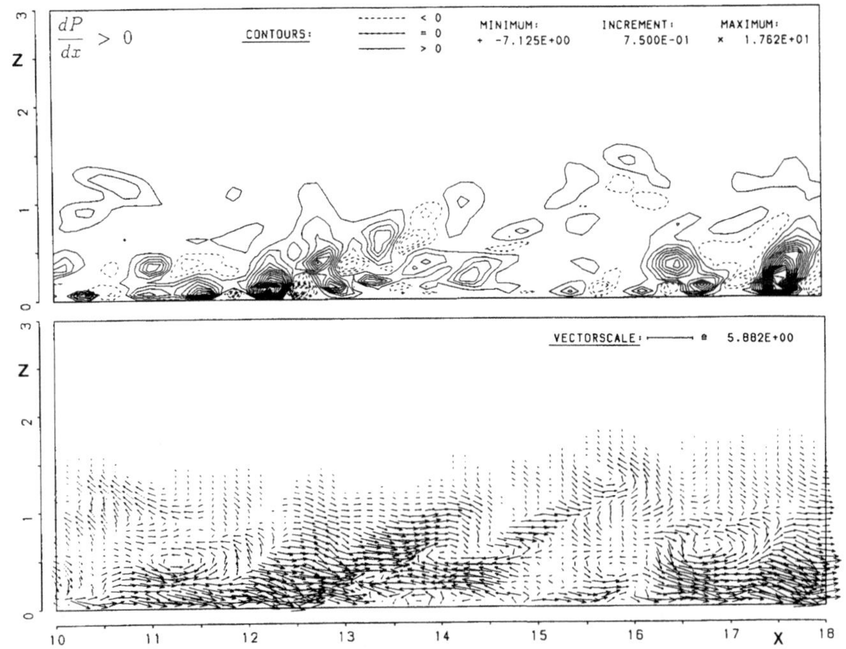


Fig.10. a,b Contour lines of $-u''w''$ and $(u'', w'')$-vectors, Case 3

celeration of the flow on the other hand decreases the spacing between large scale structures and increases their intensity. It is difficult to say whether the different spacing between the structures in the two cases explains the inhibition or enhancement of turbulence activity in the field because the accelerated flow convects large scale structures faster than the decelerated flow, so that the combination of different spacing and convection speed might perhaps compensate. This question can only be clarified with the help of the time series of $\{u''w''| - u''w'' > 2\}$.

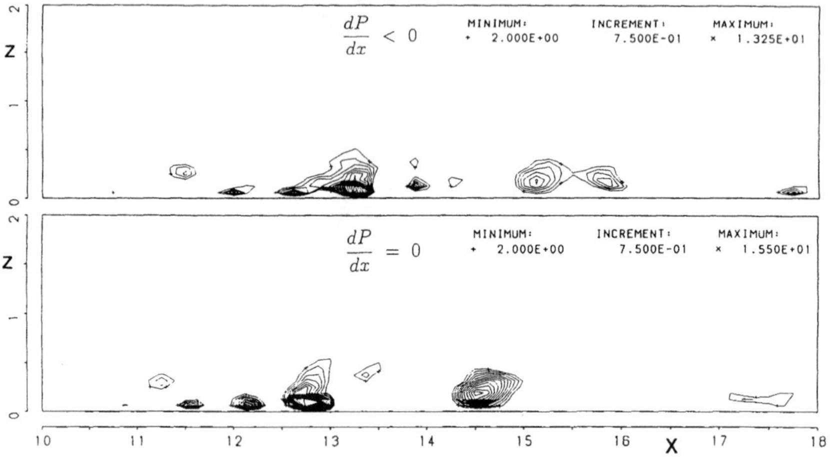


Fig.11. a,b Contour lines of $-u''w''$, minimum value is 2, Cases 1,2

One pressure gradient effect, however, remains for sure. It is the change in intensity of 'coherent' structures and can be illustrated on the basis of contour surfaces of $-u''w''$ taking an upper threshold value of 3.5. Figures 12a,b and 13a,b present perspective as well as top views of these structures, for the acceleration and deceleration cases.

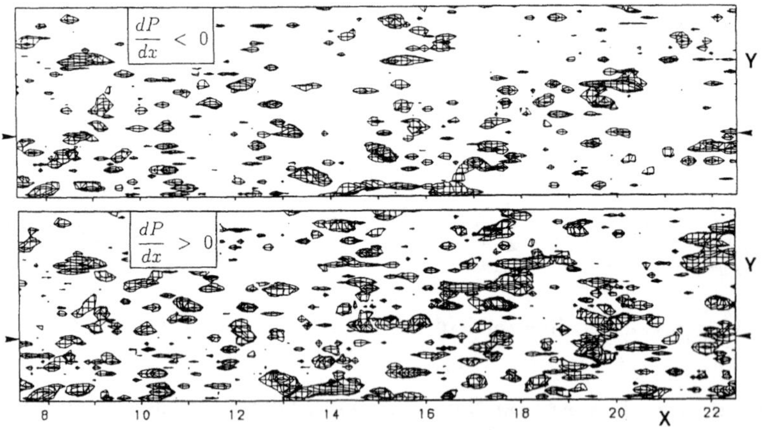


Fig.12. a,b Top view of the stress producing structures, Cases 1,3

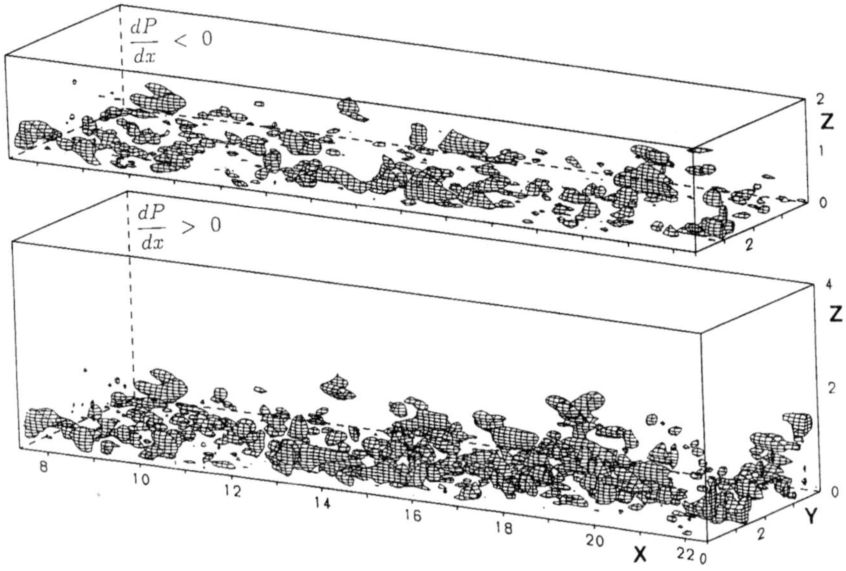


Fig.13. a,b Perspective view of the stress producing structures, Cases 1,3

Cartographic representations of velocity fluctuations and of the squared fluctuating deformation tensor are given in plates 1 and 2. Each figure contains three panels corresponding to favourable, zero and positive pressure gradient boundary layers (from top to bottom). Negative values are displayed by the blue end of the spectrum and positive values by the red end. The flow is from left to right and meets the step change in pressure gradient at $x$=8. At this position the flow structures have still the same shape and intensity. Further downstream they are deformed and their downstream spacing is changed compared to the zero-pressure gradient case. Especially from the spanwise velocity fluctuations we draw the conclusion that acceleration of the flow stretches the large scale structures and makes them more parallel to the wall whereas deceleration increases their inclination. Both effects are certainly a consequence of the changed vertical velocity field. We refer to photographs by R.E.Falco in Van Dyke (1982) concerning a visualization of boundary layers with strong pressure gradients. The squared fluctuating deformation tensor in figure 18 which partly contributes to the dissipation rate, gives a good impression of the instantaneous edge of the boundary layers.

Acknowledgement. The authors wish to thank the German Research Association (DFG) for supporting these studies in one of its Priority Research Programmes.

## References

Bradshaw, P. 1987 *Coherent Structure.* Proc. CTR Summer Progam. Rept CTR-S87. Stanford University.

Chorin, A.J. 1968 *Numerical solution of the Navier -Stokes equations.* Mathematics of Computations **22**, p. 745–762.

Deissler, R.G. 1974 Int. J. Heat Mass Transfer **17**, p. 1079.

Fiedler, H.E. 1987 *Coherent Structures.* In Advances in Turbulence (ed. G. Comte-Bellot & J.Mathieu), Proc. 1st Europ. Turbulence Conf., July 1986, p. 320–336.

Friedrich, R. & Richter, K. 1989 *Effect of rapid change in pressure gradient on turbulent momentum and heat transport in flat plate boundary layers.* In Notes on Numerical Fluid Mechanics, Vol.25, p.108–124, Vieweg Verlag, Braunschweig.

Grötzbach, G. 1986 In Encyclopedia of Fluid Mechanics (ed. N.P. Cheremisinoff), Gulf Publishing, West Orange, New Jersey.

Hussain, A.K.M.F. 1986 *Coherent Structures and Turbulence.* J. Fluid. Mech. **173** , 303–356.

Klebanoff,P.S. 1955 *Characteristics of turbulence in a boundary layer with zero pressure gradient.* NACA–TR–1247.

Peyret, R. & Taylor, Th.D. 1983 *Computational methods for fluid flow.* Springer Series in Computational Physics. Springer, New York, Heidelberg.

Piomelli,U., Ferziger,J. & Moin, P. 1989 *New Approximate Boundary Conditions For Large Eddy Simulations of Wall-Bounded Flows.*,to appear in Phys. Fluids.

Richter, K., Friedrich, R. & Schmitt,L. 1987 In Proc. 6th Turbulent Shear Flow Conference, Toulouse, France, Sept. 7–9.

Schmitt, L. 1988 *Grobstruktursimulation turbulenter Grenzschicht-, Kanal- und Stufenströmungen.* Dissertation TU München.

Schumann, U. 1975 *Subgrid scale model for finite difference simulations of turbulent flows in plane channels and annuli.* J. Comp. Physics **18**, p. 376–404.

Schumann, U. & Sweet, R.A. 1976 *A direct method for the solution of Poisson's equation with Neumann boundary conditions on a staggered grid of arbitrary size.* J. Comp. Physics **20**, p. 171–182.

Townsend, A.A. 1962 J. Fluid Mech. **12**, p.536.

Van Dyke, M. 1982 *An Album of Fluid Motion. The Parabolic Press*, p. 96.

# III. Fundamentals

# Intermittent structure of dissipation in isotropic turbulence viewed from a direct simulation

Iwao HOSOKAWA[α] and Kiyoshi YAMAMOTO[β]

[α] *University of Electro-Communications,*
*Chofu, Tokyo 182, Japan*

[β] *National Aerospace Laboratory*
*Chofu, Tokyo 182, Japan*

*Turbulence and Coherent Structures*
*O. Métais and M. Lesieur, eds.*
*Selected papers from "Turbulence 89: organized structures and turbulence in fluid mechanics"*
*Grenoble, 18-21 September 1989*

**Abstract**

The multifractal nature of isotropic turbulence is numerically investigated. A decaying isotropic turbulence, realized on a $128^3$ computergrid by the spectral method, fully develops after a proper time, revealing the Kolmogorov energy spectrum in a certain inertial range of wavenumber. At this state of turbulence, the generalized dimensions $D_q$ with respect to dissipation measure in space and the corresponding intermittency exponents $\mu_q$ are calculated for $-30 \leq q \leq 30$, based on the 3D dissipation field. Through a contact transformation of the continuous $D_q$, such an $f - \alpha$ spectrum is obtained that is well known in the theory of chaos. Our result shows a multifractal distribution in space of the scaling index $\alpha$ , comparable with, but somewhat distinct from, the recent experiment and model of Meneveau and Sreenivasan. A similar investigation of a passive scalar field is added.

## 1. Introduction

With the recognition of intermittency in isotropic turbulence, the concept of scale similarity of the velocity structure functions or moments of dissipation in scale $r$, which belongs to the inertial range, was introduced. With this, the lognormal (See Monin & Yaglom 1975), $\beta$ (Frisch, Sulem & Nelkin 1978), and other models came into play to characterize the scale similarity in each way. These models have been studied in comparison with experiments (Antonia, Satyaprakash & Hussain 1982; Anselmet, Gagne, Hopfinger & Antonia 1984) but none of them are completely satisfactory. Recently, Frisch and Parisi (1985) suggested the multifractal distribution in space of a scaling index for velocity difference in $r$. On the other hand, Meneveau and Sreenivasan (1987a) treated another scaling index for dissipation averaged over a domain of scale $r$, verified its multifractal distribution by experiment and proposed the so-called $p$ model (Meneveau & Sreenivasan 1987b) to explain the result. The multifractal theories of these two groups of authors are theoretically equivalent, as is described later. It is to be noted here that the latter authors utilized the concept of generalized dimensions $D_q$, which is now essential for the theory of chaotic systems, with respect to a measure of dissipation. Their method, i.e. first deciding $D_q$ from data according to the definition of Hentschel and Procaccia (1983), is effective in insuring the necessary monotonic behavior of $D_q$. Here we calculate $D_q$ for $-30 \leq q \leq 30$ from our data of a directly simulated isotropic turbulence at the fully-developed state (Yamamoto & Hosokawa 1988; Hosokawa & Yamamoto 1989), and hence give the intermittency exponents $\mu_q$ and the corresponding $f - \alpha$ spectrum. (We note that this process is somewhat different from what we announced in Abstracts of Turbulence 89. This is because we changed the process to the simpler and more effective one in the sense described above after the abstract was written.) As a result, it is no doubt that the scaling index $\alpha$ , and then the velocity and dissipation in scale $r$, disrtibute in space in a multifractal way. Finally, some problematic points left on the multifractal structure of isotropic turbulence are discussed.

## 2. Direct numerical simulation

We numerically solved the 3D Navier-Stokes equation under a periodic boundary condition on a $128^3$ computergrid, starting from an isotropic random velocity field with an energy spectrum like $k^4 exp(-2k^2)$($k$: normalized wave number). The scheme of our direct simulation was an alias-free spectral Galerkin method [basically the same as Orszag & Patterson's (1972) method]. But we used only the solenoidal components of the velocity field in the basic equation to guarantee the imcompressibility condition, and for time integration, the (4-th order) Runge-Kutta-Gill scheme was taken. The Taylor-scale Reynolds number $R_\lambda$ reached at the fully-developed state of the decaying turbulence was almost 100. This value is not high enough to produce a very wide inertial range in the energy spectrum where we must see the Kolmogorov $-5/3$ law appear; in fact, the width of the inertial range of wavenumber achieved is only about a half decade. Nevertheless, what we can deduce from this result seems to be considerably non-trivial. For details, the reader is referred to our publications cited in Introduction. Only a typical energy spectrum is shown here in Figure 1, in which $R = 500$ mean $\nu$ (kinematic viscosity) $= 0.002$. It is seen that the fully-developed state is reached at time $t = 10$. (And then enstrophy arrives at its maximum.) All the data discussed here is taken at this state of the velocity field.

As an interesting example which shows a feature of our turbulence, we present Figure 2 (see colour plates). High-strain regions in space taking charge of about 14% of the total dissipation are indicated in red in Figure 2(a), while vorticity-concentrated regions taking charge of about the same dissipation are in black. The two kinds of regions seem to be strongly interrelated with each other, but the former is more diffusive than the latter, which has clearly sharp forms like worms and spots. Iso-norm-of-strain surfaces, which contain about 75% of the total dissipation and then along which there exists relatively low strain, are seen in Figure 2(b) as much as can be viewed by the reader. They are just like a sponge. Do these two pictures not indicate that there are various dissipative regions in space with various fractals? Perhaps so, but not exactly. As will be described in Chap. 4, real fractals are to reveal themselves when dissipation is treated in scales of the inertial

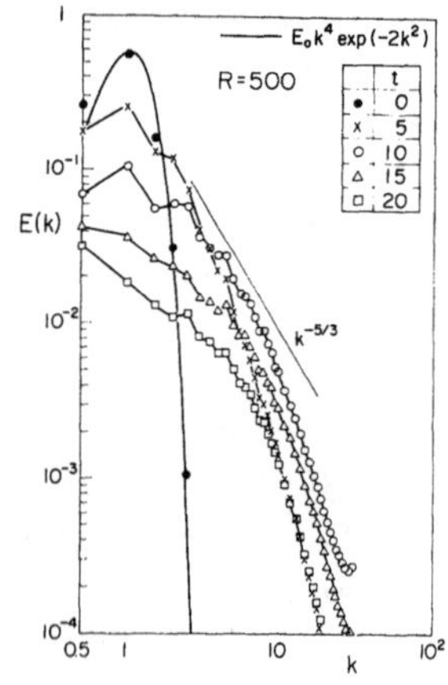


Fig.1. Energy spectrum of our turbulence for $R = 500$ at various times. (Transcription from Yamamoto & Hosokawa 1988.)

range.

## 3. Generalized dimensions and intermittency exponents

If the probability measure of events in the $i$-th disjoint subbox of scale $r$ in space is assigned as $p_i = (E_r)_i/E_L$, where $E_r$ is the total dissipation (per unit mass) of the velocity field within the subbox and $L$ $(= 4\pi)$ is the macroscale (the largest scale in our turbulence), then the $q$-th order generalized dimension $D_q$ is defined as

$$\sum_i^N p_i^q = (r/L)^{(q-1)D_q} \tag{3-1}$$

according to Hentschel & Procaccia (1983); $N$ is the total number of subboxes of scale $r$. It is important to note that the subbox with $p_i = 0$, if any, should be taken out of account in the sum in (3-1) in order for the formula to be still meaningful for $q < 0$.

On the other hand, the $q-$th order intermittency exponent $\mu_q$ is defined as

$$\langle \varepsilon_r^q \rangle / \varepsilon_L^q \sim (r/L)^{-\mu_q} \tag{3-2}$$

where $\varepsilon_r$ is the dissipation averaged over a domain of scale $r$, and the angular bracket means the ensemble average. [For details, see Monin & Yaglom (1975); the concept of $\varepsilon_r$ was introduced by Obukhov and Kolmogorov in 1962.] In (3-2), the case of $q < 0$ has not generally been considered; since (3-2) totally loses its meaning if $\varepsilon_r$ vanishes anywhere in the flow field. This point makes a big contrast between (3-1) and (3-2).

Obviously, both (3-1) and (3-2) represent the scale similarity of dissipation in the inertial range of $r$. Since $E_r \sim \varepsilon_r r^d, E_L \sim \varepsilon_L L^d$ and $N = L^d/r^d$ ( $d$: the spatial dimension), $D_q$ and $\mu_q$ can be related with each other as

$$-\mu_q = (q-1)(D_q - d). \qquad (3-3)$$

For $q < 0$, this equation is meaningful only when (3-2) has its meaning, and when not, it would give rise a strange behavior of $\mu_q$. We note that the lognormal and $p$ models have a meaningful $\mu_q$ for $q < 0$, while the $\beta$ and other models fail to, because $\varepsilon_r$ can vanish there. Our numerical simulation has a symptom that there is no probability of $\varepsilon_r = 0$, at least, within the resolution of our calculation, as is understood in Hosokawa & Yamamoto(1989). It is well known that the exponent of the $p$−th order velocity structure function $\zeta_p$ is related with $\mu_q$ as $\zeta_p = p/3 - \mu_{p/3}$ as a result of scaling analysis. But the calculation of $\zeta_p$ is omitted here.

In Figure 3 is seen the log-log plot of $(\sum_i^N p_i^q)^{1/(q-1)}$ vs. $r/L$ for various values of $q$. The dashed lines bound the inertial range of $r$, which corresponds to the inertial range of wavenumber.

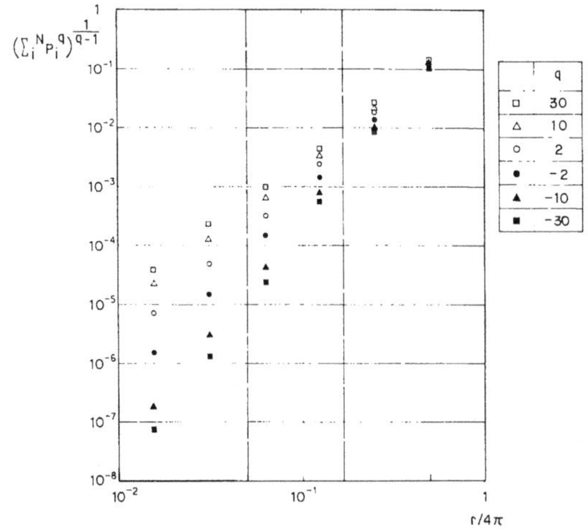


Fig.3. $(\sum_i^N p_i^q)^{1/(q-1)}$ vs. $r/4\pi$. The dashed lines bound the inertial scale range. The main box containing flow with side $L = 4\pi$ is split into $8^n$ subboxes with side $r = L/2^n$; $n$ runs from 1 to 6. (Transcription from Hosokawa & Yamamoto 1990.)

It is surprising that the data points are almost on straight lines much beyond the inertial range. (The seventh points to be on the left-hand side were omitted, because they deviate appreciably from the straight lines.) Therefore, the $D_q$ values were calculated as the gradients decided by the least-square fit to the six points. This is just the first approach and never intended to be finally fixed. It may be comparable on exactness with the approach in which the $D_q$ values are to be the gradients decided by only the two points inside the inertial range. We are planning to take much more points inside the inertial range by changing the method of subbox production; the result will be published elsewhere. The circles in Figure 4 indicate the $D_q$ curve obtained. The solid line is the $p$ model of Meneveau & Sreenivasan(1987b). Thus, our first approach to $D_q$ is close to the $p$ model for $|q| \leq 4$, but gradually deviates from it as $|q|$ increases. However, it may be noted that both curves are rather comparable on the qualitative trend; indeed, they both seem to behave far more reasonably than the lognormal and $\beta$ models (with $\mu = \mu_2 = 0.2$ ) drawn in the Figure.

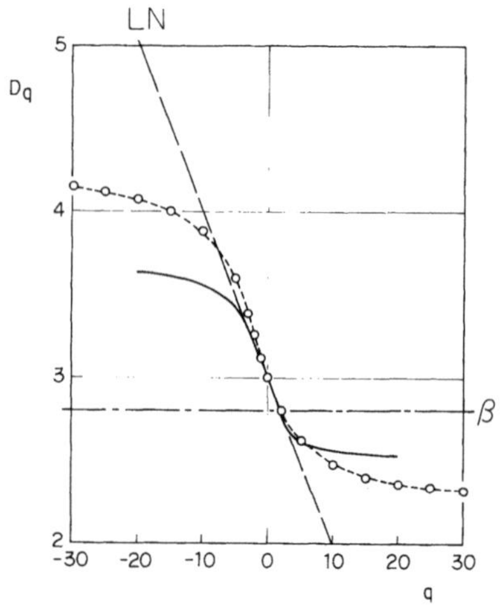


Fig.4. Generalized Dimensions $D_q$ vs. $q$. The circles denote the calculated values. The dotted line is an analytical fitting $[D_q = D_{-1.8} - 0.647\mathrm{Tan}^{-1}0.191(q+1.8)]$. The solid line is the curve of the 3D version of the $p$ model. The dashed line denotes the lognormal model (with $\mu = 0.2$), and the dot-and-dash line the $\beta$ model (with $\mu = 0.2$).

Next, the $\mu_q$ curve obtained through (3-3) is plotted by circles in Figure 5. Figure 5(a) is particularly interesting in showing that $\mu_q$ is meaningful for $q < 0$ in our result as well as in the lognormal and $p$ models, as has already been described. The $\mu_q$ there may be interpreted to characterize intermittency of very weak dissipation. In Figure 5(b) are seen the behaviors of $\mu_q$ for $q \geq 0$ gathered from many sources.

Our result is situated there in a very reasonable manner in comparison with the others, even though it is the first approach supposed to contain some error.[1] Also, we point out that our result satisfies Novikov's (1971) inequality: $\mu_q \leq \mu + q - 2$ *for* $q \leq 2$ as well as the $p$ and $\beta$ models.

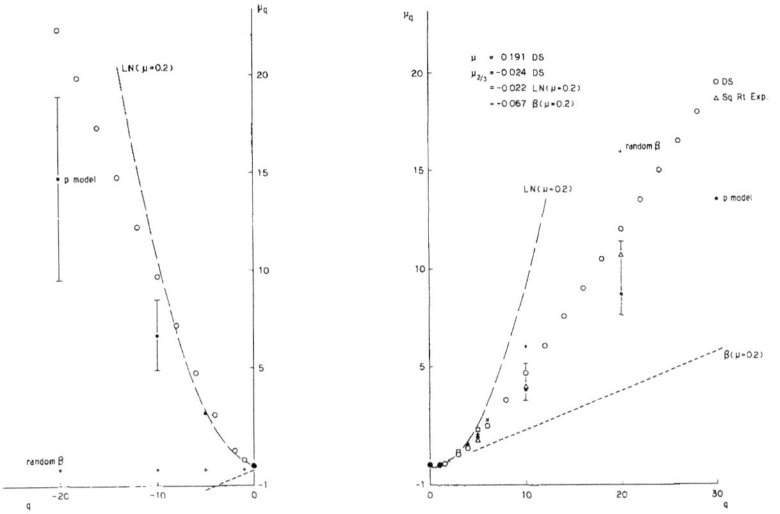


Fig.5. (a) Intermittency exponents $\mu_q$ vs. $q \leq 0$. All symbols are the same as in (b). (b) Intermittency exponents $\mu_q$ vs. $q \geq 0$. The dashed and dotted lines, respectively, denote the lognormal and $\beta$ models with $\mu = 0.2$. Our direct simulation is plotted by circles, the random $\beta$ model (Benzi, Paradin, Parisi & Vulpiani 1984) by +, the $p$ model by small solid circles, the square-root exponential model (Hosokawa 1989) by triangles, Anselmet et al's (1984) experiment by ×, Antonia et al's (1982) experiment by squares, and the error bars of Meneveau et al's (1987a) by ⊢⊣ . (Transcription from Hosokawa & Yamamoto 1990.)

It may be interesting that our result is close to the square-root exponential model (Hosokawa 1989), at least, for $0 < q < 30$, although this

[1] Later, we have investigated the error of our result under the assumption that the data points in Figure 3 distribute Gaussianly around the least-square straight line with the average standard deviation given by the least square error. If three times the square root of the local variance of $D_q$ is considered as the local error, we can obtain the local error bars on Figires 4, 5 and 6 which are slightly less in magnitude than (and have the same trend as) those in Meneveau et al's (1987a) experimental result. And then the negative parts of f in Figure 6 are within the error thus obtained.

model fails to be meaningful for $q < 0$ as badly as some other models (Novikov 1971; Frisch et al 1978; Benzi et al 1984; Nakano et al 1985; Andrews et al 1989; the reason why these models do so has already been explained.) The values of $\mu(=\mu_2)$ and $\mu_{2/3}$ in our result are given on Figure 5(b), the latter of which is compared with those of the lognormal and $\beta$ models with $\mu = 0.2$.

## 4. Scaling law and the $f - \alpha$ spectrum

It is well known (for example, Frisch 1985; Meneveau & Sreenivasan 1987a) that the Euler equation:

$$\frac{\partial \mathbf{v}}{\partial t} = -\mathbf{v} \cdot \nabla \mathbf{v} - \nabla p, \qquad (4-1)$$

where $\mathbf{v}$ is the velocity field, $\nabla = \partial/\partial \mathbf{r}$, $p$ the pressure field, $t$ the time variable and $\mathbf{r}$ the distance vector, is invariant to the scale change:

$$\begin{aligned} \mathbf{r}' &= \lambda \mathbf{r} \\ \mathbf{v}' &= \lambda^{\alpha/3} \mathbf{v} \\ t' &= \lambda^{1-\alpha/3} t \end{aligned} \qquad (4-2)$$

This allows the existence of self-similar solutions in the inertial range of scale $|\mathbf{r}| = r$ in which viscosity is supposed to be ineffective and the effect of a macroscale boundary condition to be negligibly small. As a result of the scaling law in (4-2), we have the following dependence of energy and dissipation on scale $r$:

$$v_r^2/2 \sim r^{2\alpha/3}, \qquad (4-3)$$

$$\varepsilon_r \sim v_r^3/r \sim r^{\alpha-1}, \qquad (4-4)$$

where $v_r$ is considered to be the velocity difference over scale $r$ and $\varepsilon_r$ has appeared already in (3-2). It should be pointed out here that $\varepsilon_r = 0$ can never happen in the form of (4-4) (i.e. intrinsically within the scaling law), unless $\alpha \to \infty$. Obviously, $\alpha = 1$ gives the Kolmogorov law, and $\alpha = D - 2$ the $\beta$ model ($D$: fractal dimension). Generally, $\alpha$ may range over some (probably) finite domain, and an iso-$\alpha$ set in space may have a fractal dimension $f(\alpha)$ (Mandelbrot 1977).

We know from (3-1)

$$N\langle E_r^q \rangle \sim r^{(q-1)D_q} \qquad (4-5)$$

if $r$ is sufficiently small that $N$ is large enough. The left-hand side may be calculated in an alternative way, if the probability density of distribution of $\alpha$ is assumed as $\rho(\alpha)$. That is

$$(L^d/r^d)\int \rho(\alpha)(\varepsilon_r r^d)^q r^{d-f(\alpha)}\mathrm{d}\alpha \sim r^{(q-1)D_q}, \qquad (4-6)$$

where $r^{d-f(\alpha)}$ is the space fraction of an iso-$\alpha$ set. With the substitution of (4-4) and the assumption of $f"(\alpha) < 0$, the steepest descent method (for $r \to 0$) directly leads to a contact transformation:

$$\begin{aligned} f(\alpha) &= \alpha q - (q-1)(D_q - d + 1) + d - 1 \\ \alpha &= \frac{\mathrm{d}}{\mathrm{d}q}[(q-1)(D_q - d + 1)], \end{aligned} \qquad (4-7)$$

which gives the $f - \alpha$ spectrum suitable for turbulence in some sense. [Note that this $\alpha$ is different from $\alpha$ in the $f - \alpha$ spectrum often used in the theory of strange attractors (Jensen, Kadanoff & Libchaber 1985; Halsey, Jensen, Kadanoff, Procaccia & Shraiman 1986; Jensen 1987); if $\alpha$ in the latter case is denoted as $\alpha_s$, we should have $\alpha_s = \alpha - 1 + d$.] Frisch & Parisi (1985) derived a spectrum euivalent to (4-7) (in terms of $h = \alpha/3$) in a similar way but by directly relating the spectrum with $\zeta_p$. But they appear to have skipped a consideration of the case of $p < 0$, which corresponds to $q < 0$. (We note that there can be a branch with $f'(\alpha) < 0$ in the spectrum corresponding to $q < 0$, since $D_q$ is meaningful also then.)

Our result is indicated by circles in Figure 6. The left branch corresponds to $D_q$ for $q > 0$, which represents dimensions of strongly dissipative sets in space and, mostly, on which $\varepsilon_r$ is singular, as is known from (4-4). It is remarkable that our result, the lognormal and the $\beta$ model almost coincide in the range of $0.6 < \alpha < 1.2$ which includes the location ($\alpha \approx 1.13$) of the summits of the curves. It may be interesting to see that the set in space with $\alpha = 1$ (to give the exact Kolmogorov law) has a fractal dimension, a little bit deviated from 3. Since a negative dimension cannot be considered, the small negative parts of $f$ observed in the Figure should be neglected as errors. It is natural to consider that the branch points accumulate, at most, on the $\alpha$ axis. Since all $\alpha$ stay positive, energy in scale $r$ in (4-3) cannot be singular, at least, so far as we judge from our data presented. The lognormal model breaks down for $\alpha < 0$ and $\alpha > 2.2$ because $f < 0$, while the $p$ model accumulates at $f = 2$ in both branches; this means that the strongest and the weakest dissipative region are surfacelike in this model.

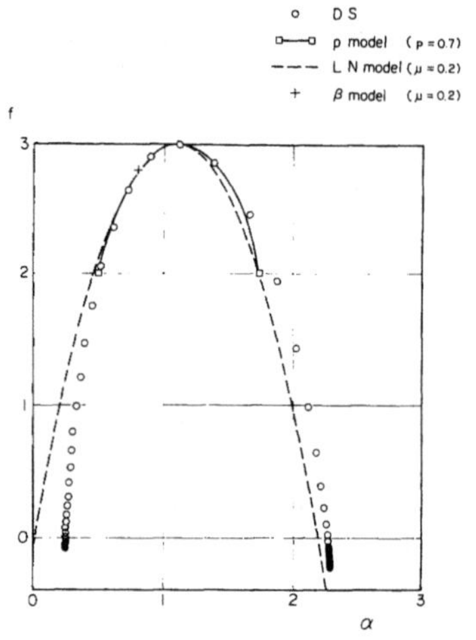


Fig.6. The $f - \alpha$ spectrum. Our direct simulation is plotted by circles, the $p$ model by a solid line terminated by squares, the $\beta$ model (with $\mu = 0.2$) by +, and the lognormal model (with $\mu = 0.2$) by a dashed line.

## 5. The $f - \alpha$ spectrum of a passive scalar field

A passive scalar field, say temperature, with Prandtl number 1 was investigated, which develops together with the above-described turbulence from the same type of initial state. In Figure 7 we can see the temperature spectrum. The temperature field reveals much more rapidly the Kolmogorov spectrum than the velocity field. In fact, it fully develops at $t = 5$. At this time, it gives the Batchelor constant as 1.5.

Therefore, the temperature field at $t = 5$ is used for calculating its own $D_q$ and $f - \alpha$ spectrum based on moments of the temperature dissipation in scale $r$ in the same way as (3-1). The $f - \alpha$ spectrum thus obtained is shown in Figure 8, in comparison with the recent experimental result of Prasad, Meneveau & Sreenivasan (1988). Although our result is considered to have an appreciable error, a main feature is common in both results. $\alpha$ in this spectrum occupies a wider range than that of the velocity field.

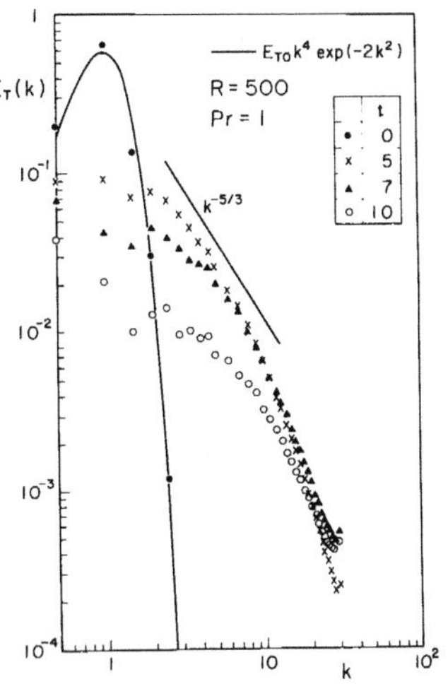


Fig.7. Temperature spectrum for $R = 500$ and Prandtl number 1 at various times.

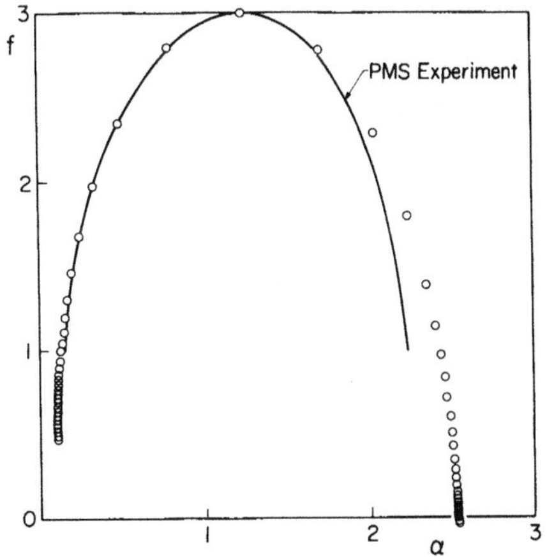


Fig.8. The $f - \alpha$ spectrum of the temperature field. Our result is indicated by circles, and Prasad et al's (1988) experiment by a solid line.

## 6. Problematic points

6.1. Our data is based on only one realization of the flow field because of limited computer time. We expect that the universal feature of isotropic turbulence is invariant to a particular change of initial condition, but some statistical variation will be unavoidable. Also, the resolution of the $128^3$ grid is limited. In these views, our result remains plausible rather than precise. It is hoped that a more refined calculation will produce a more precise result; in connection, the work of Vincent & Menegucci (1989) is to be noticed.

6.2. The explicit form of $\rho(\alpha)$ is still undiscovered. It would be found by calculating $\alpha$ at all points in space [through (4-4)] to make up the histogram of it. This is a next work to be done, although it might be difficult within the present resolution. Interestingly, Bacry, Frisch, Gagne & Hopfinger (1989) reported that they did probably the same thing (in terms of $h = \alpha/3$) by wavelet analysis of a real high-Reynolds number turbulence($R_\lambda = 2720$).

6.3. There remains a question of where the end points of the two branches in the $f - \alpha$ spectrum should be. They are sensitive to the statistical variation of data. However, it is natural to think that the coordinates of the end points are such that $0 < \alpha < 3$ and $0 \leq f \leq 2$. In fact, only the value of $\alpha$ in this range guarantees that both of the proper time and velocity of the eddy of scale $r$, which are $t_r \sim r^{1-\alpha/3}$ and $v_r \sim r^{\alpha/3}$, respectively, can be non-singular as $r$ tends to 0; this tendency of $v_r$ was expected by Frisch et al (1978). The same question exists for the case of a passive scalar field, too.

However, Bacry et al (1989) mentioned a new fact that the minimum of $\alpha$ can be negative. If it were true, $v_r$ could not stay non-singular any more and our result, as badly as some models, should have a big error in the left branch in the spectrum. This is a very subtle point, at the present time, in the problem of isotropic turbulence with a multifractal structure.

## References


Andrews, L. C., Phillips, R. L., Shivamoggi, B. K., Beck, J. K. & Joshi, M. L. 1989 *Phys. Fluids* **A1**, 999.

Anselmet, F., Gagne, Y., Hopfinger, E. J. & Antonia, R.A. 1984 *J. Fluid Mech.* **140**, 63.

Antonia, R. A., Satyaprakash, B. R. & Hussain, A. K. M. 1982 *J. Fluid Mech.* **119**, 55.

Bacry, E.,Frisch, U., Gagne, Y. & Hopfinger, E. 1989 Spoken at Turbulence 89; *Abstracts*, p. 145.

Benzi, R., Paradin, G., Parisi, G & Vulpiani, A. 1984 *J. Phys.* **A17**, 3521.

Frisch, U. 1985 In *Turbulence and Predictability in Geophysical Fluid Dynamics and Climate Dynamics* (ed. M. Ghil), p. 71. North-Holland.

Frisch, U & Parisi, G. 1985 *ibid*, p. 84.

Frisch, U., Sulem, P. L. & Nelkin, M. 1978 *J. Fluid Mech.* **87**, 719.

Halsey, T. C., Jensen, M. H., Kadanoff, L. P., Procaccia, I. & Shraiman, B. I. 1986 *Phys. Rev.* **A33**, 1141.

Hentschel, H. G. E. & Procaccia, I. 1983 *Physica* **8D**, 435.

Hosokawa, I. 1989 *Phys. Fluids* **A1**, 186.

Hosokawa, I. & Yamamoto, K. 1989 *J. Phys. Soc. Japan* **58**, 20.

Hosokawa, I. & Yamamoto, K. 1990 *J. Phys. Soc. Japan* **59**, 401.

Jensen, M. H. 1987 In *International Conference on the Physics of Chaos and System far from Equilibrium* (ed. M. Duang-Van), p. 487. North-Holland.

Jensen, M. H., Kadanoff, L. P. &Libchaber, A. 1985 *Phys. Rev. Lett.* **55**, 2798.

Mandelbrot, B.1977 *Fractals: Form, Chance and Dimension.* W. H. Freeman.

Meneveau, C. & Sreenivasan, K. R. 1987a *Nucl. Phys. B(Proc. Suppl.)* **2**, 49.

Meneveau, C. & Sreenivasan, K. R. 1987b *Phys. Rev. Lett.* **59**, 1424.

Monin, A. S. & Yaglom, A. M. 1975 *Statistical Fluid Mechanics*, Vol. 2., Chap. 8. MIT Press.

Nakano, T. & Nelkin, M. 1985 *Phys. Rev.* **A31**, 1980

Novikov, E. A. 1971 *Appl. Math. Mech.* **35**, 231.

Orszag, S. A. & Patterson, Jr., G. S. 1972 *Phys. Rev. Lett.* **28**, 76.

Prasad, R. P., Meneveau, C. & Sreenivasan, K. R. 1988 *Phys. Rev. Lett.* **61**, 74.

Vincent, A. & Menegucci, M. 1989 Spoken at Turbulence 89; *Abstracts*, p. 231.

Yamamoto, K. & Hosokawa, I. 1988 *J. Phys. Soc. Japan* **57**, 1532.

# The spatial structure of homogeneous turbulence at Reynolds Numbers around 1000

A. VINCENT and M. MENEGUZZI

*CERFACS, 42 av. G. Coriolis*
*31000 Toulouse, France*

*Turbulence and Coherent Structures*
*O. Métais and M. Lesieur, eds.*
*Selected papers from "Turbulence 89: organized structures and turbulence in fluid mechanics"*
*Grenoble, 18-21 September 1989*

**Abstract**

A direct numerical simulation is used to obtain a fully developed statistically stationary three-dimensional homogeneous and isotropic turbulent field at a Reynods number around 1000 ($R_\lambda \approx 150$). The energy spectrum displays an inertial subrange. The velocity derivative distribution, known to be strongly non gaussian, is found to be close to but not exponential. Visualization of the flow confirms the previous finding that the vorticity is organized in very elongated thin tubes. The width of these tubes is of the order of the Kolmogorv scale, while their length can reach the integral scale of the flow.

## 1. Introduction

We report here some results of a simulation of homogeneous turbulence obtained by direct solution of the incompressible Navier-Stokes equation with periodic boundary conditions on a $240^3$ grid. A forcing term at low wavenumbers was added to the equation in order to produce a statistically steady flow. Early attempts at direct simulation of homogeneous three-dimensional turbulence were restricted to modest Reynolds numbers (Orszag and Patterson (1972), Siggia and Patterson (1978)). They have allowed a study of the dissipation scale properties of a turbulent flow. It becomes now feasible, on computers of the last generation, to reach Reynolds numbers at which there exists an inertial subrange in the energy spectrum (Kerr (1985), Hosokawa and Yamamoto (1988), She et al. (1990)). One of the main objectives of our simulation was to obtain a more extended inertial subrange.

## 2. The calculation

We solve the Navier-Stokes equation for incompressible fluids in rotational form

$$\frac{\partial \mathbf{v}}{\partial t} = \mathbf{v} \times \omega - \nabla(p + v^2/2) + \nu \nabla^2 \mathbf{v} + \mathbf{f} \tag{1}$$

with the continuity equation:

$$\nabla \cdot \mathbf{v} = 0$$

$\mathbf{v}$ is the velocity field, $\omega = \nabla \times \mathbf{v}$ the vorticity, $p$ the pressure, $\nu$ the kinematic viscosity and $\mathbf{f}$ a force field. Since we are interested in (statistically) homogeneous turbulent flows, we take periodic boundary conditions in all directions. In Fourier space the two equations can be combined to give

$$\frac{\partial \mathbf{v_k}}{\partial t} = \mathbf{P(k)} \cdot (\mathbf{v} \times \omega)_\mathbf{k} - \nu k^2 \mathbf{v_k} + \mathbf{f_k} \tag{2}$$

where the tensor **P** is the projector on the space of solenoidal fields, defined as

$$P_{ij}(\mathbf{k}) = \delta_{ij} - \frac{k_i k_j}{k^2}$$

A pseudospectral method is used to compute the right-hand side of this equation (see Gottlieb and Orszag 1977). The time marching is done using a second order finite-difference scheme. An Adams-Bashfort scheme is used for the non linear term while the dissipative term is integrated exactly. The resulting numerical scheme is

$$\frac{\mathbf{v_k}^{n+1} - \mathbf{v_k}^{n} e^{-\nu k^2 \delta t}}{\delta t} = \mathbf{P}(\mathbf{k}) \cdot \left[ \frac{3}{2} (\mathbf{v} \times \omega)_{\mathbf{k}}{}^{n} e^{-\nu k^2 \delta t} - \frac{1}{2} (\mathbf{v} \times \omega)_{\mathbf{k}}{}^{n-1} e^{-2\nu k^2 \delta t} + \mathbf{f_k}{}^{n} e^{-\nu k^2 \delta t} \right]$$

To start (or restart) the calculation, we use a second order Runge-Kutta scheme.

The simplest way to force the flow in this geometry would be to take a force field which is a white noise in time and space. This would allow us to control the energy injection rate, but would prevent us from asking questions about the intrinsic chaotic behaviour of the turbulent flow, for instance by studying time series of the velocity or its derivatives at a given point. We therefore chose to force the field at low wavenumbers in a deterministic way. All Fourier modes with $k < 2$ are forced with a constant amplitude **f** independent of **k**. This results in large fluctuations of the energy injection rate $\epsilon_i = < \mathbf{f} \cdot \mathbf{v} >$. With a deterministic forcing, one can only hope to reach a steady-state regime in the sense that $\epsilon_i$ fluctuates in time around a constant value.

The calculation presented here was done with $240^3$ Fourier components, and a viscosity of $10^{-3}$. One time-step takes 12 seconds on a Cray-2, using the 4 processors. Integration over one turnover time (defined below) takes of the order of 4 hours of Cray-2.

Let us recall the definition of some characteristic quantities used in the following. Three characteritic lengths are used, the integral scale

$$l_0 = \frac{\int_0^\infty k^{-1} E(k) dk}{\int_0^\infty E(k) dk},$$

the Taylor microscale

$$\lambda = \left[ \frac{\int_0^\infty E(k) dk}{\int_0^\infty k^2 E(k) dk} \right]^{\frac{1}{2}},$$

and the Kolmogorov dissipation scale

$$l_D = \left(\frac{\nu^3}{\epsilon}\right)^{\frac{1}{4}},$$

where $\epsilon$ is the mean energy dissipation rate per unit mass. The two characteristic time-scales of homogeneous turbulence are the eddy turnover time

$$\tau_0 = l_0/v_0,$$

where $v_o$ is the root mean square velocity, and the dissipation time

$$\tau_\nu = l_0^2/\nu.$$

With these quantities one can define two Reynolds numbers, the integral scale Reynolds number $R = v_0 l_0/\nu$ and the Taylor microscale Reynolds number $R_\lambda = \lambda v_0/\nu$.

## 3. The spectra of the flow

After integration over 10 turnover times we obtain a statistically stationary regime in the above sense. The energy spectrum, shown in figure 1, displays a power law range for $k < 20$, with an exponent a little larger than -5/3.

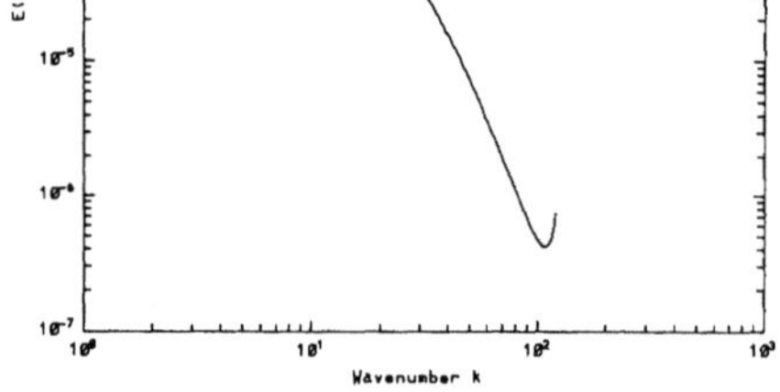


Fig.1. Energy spectrum averaged over 10 different times

The fact that this is an inertial range is confirmed by inspection of the energy flux spectrum, shown in figure 2. The energy flux $\Phi(k)$ is defined as

$$\Phi(k) = \int_k^\infty T(k)dk \tag{3}$$

where $T(k)$ is the energy transfer, which appears in the energy equation

$$\frac{dE}{dt}(k) = F(k) + T(k) - \nu k^2 E(k) \tag{4}$$

($F(k)$ is the energy injection spectrum). In other words, $\Phi(k)$ is the rate of energy transfered per unit mass by all modes with $k' < k$ to modes with $k' > k$. From equation (2), the energy transfer is

$$T(k) = \int \mathbf{v_k} \cdot \mathbf{P(k)} \cdot (\mathbf{v} \times \omega)_{\mathbf{k}} d\Omega. \tag{5}$$

where the integration is over the angles in Fourier space. By definition, the energy flux is constant in the inertial range, which is seen in figure 2 to occupy a little more than one decade in wavenumber space.

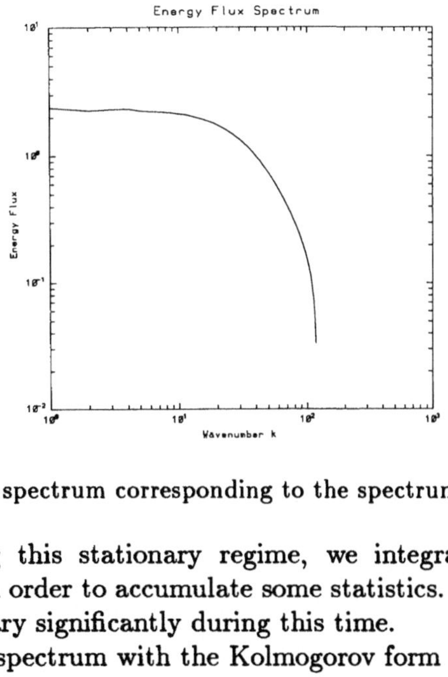


Fig.2. Energy flux spectrum corresponding to the spectrum of Fig. 1

After reaching this stationary regime, we integrate for ten more turnover times in order to accumulate some statistics. The energy spectrum does not vary significantly during this time.

By fitting the spectrum with the Kolmogorov form

$$E(k) = C_K \epsilon^{2/3} k^{-5/3}, \tag{6}$$

we obtain a value of 2 for the Kolmogorov constant $C_K$, which does not vary by more than 3% in the statistically steady-state period. This value of $C_K$ is a little larger than the experimental value 1.7 (see Monin and Yaglom 1975 for references). Our Reynolds numbers are $R \approx 1000$ and $R_\lambda \approx 150$.

## 4. The statistical distribution of v and its derivatives

It is generally assumed that turbulent flows are ergodic, and experimentalists use time distributions of a velocity component $v_i$ or its derivatives $\partial_i v_j$ at a given space location to obtain probability distributions. Our time-series are too short to apply the same method to our results. It seems more efficient to use space distributions, for which we have $\approx 1.4 \times 10^7$ grid point values at each time. Using approximately 70 different times, we reach a total of $\approx 10^9$ values. The resulting statistical distribution of $v_x$ is shown in figure 3, after being normalized so that $\sigma^2 =< v^2 >= 1$. It is compared to a gaussian distribution, shown with a dashed line. The distribution we obtain is close to gaussian, in agreement with what is known from experimental data (see Monin and Yaglom 1975 for a review).

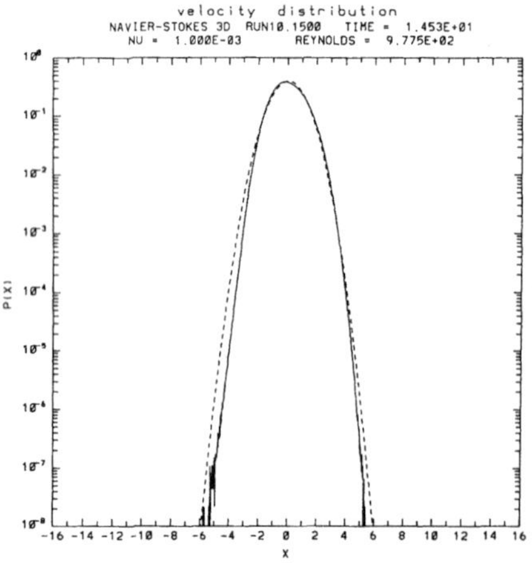


Fig.3. Probability distribution of a velocity component, normalized so that $\sigma = 1$, shown together with a gaussian distribution (dashed line)

Figure 4 shows the distribution of $\partial_x v_x$, known from experiments to be strongly non gaussian. Also shown is a gaussian, again normalized so that $\sigma = 1$. It is seen that the distribution we obtain is closer to an exponential than to a gaussian, but our statistical sample allows one to see a departure from an exponential law. The wings are indeed not straight lines in this linear-logarithmic plot. The same is true for the distribution of a vorticity component $\omega_x$, shown in figure 5. A detailed analysis of these statistical distributions will be published elsewere (Vincent and Meneguzzi 1990).

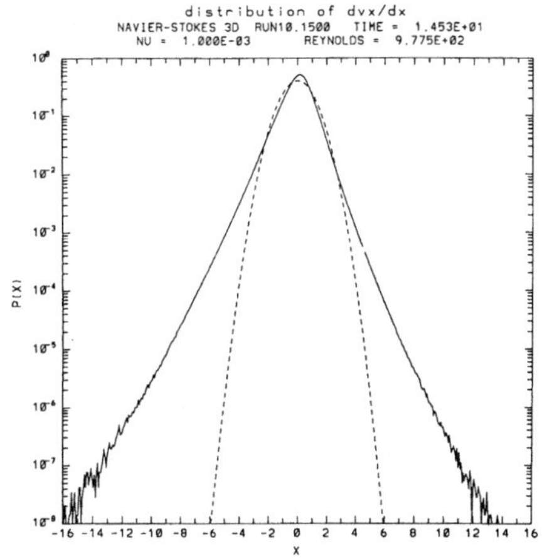


Fig.4. Probability distribution of $\frac{\partial v_x}{\partial x}$ normalized so that $\sigma = 1$, shown together with a gaussian distribution (dashed line)

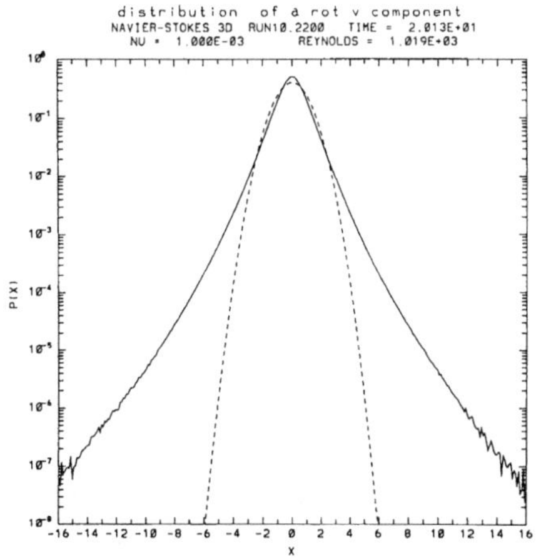


Fig.5. Probability distribution of a vorticity component, normalized so that $\sigma = 1$, shown together with a gaussian distribution (dashed line)

## 5. The structure of the flow

Figure 6 shows a 3D picture of the vorticity field. The vorticity vector at each grid point is represented by a vector, here so small that individual vectors can hardly be seen. Vectors are only plotted if their modulus is larger than a given threshold. By varying this threshold and rotating the figure on a graphics screen one can explore the structure of the field in detail. One can see that the vorticity is organized in thin elongated tubes, as previously reported by Siggia and Patterson (1978), Kerr (1985), She et al. (1990).

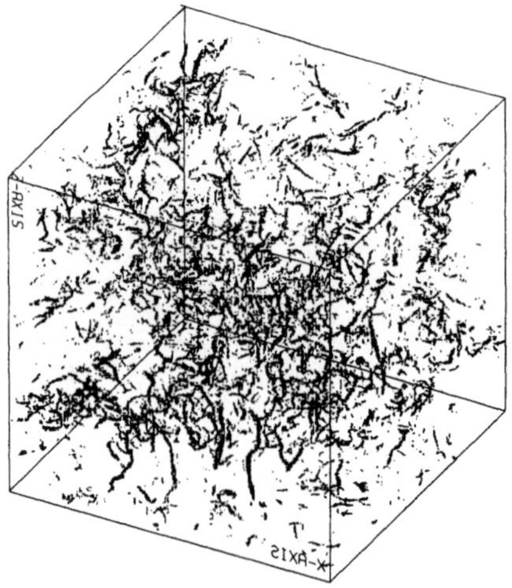

Fig.6. View of the vorticity field, represented by a vector of length proportional to the vorticity amplitude at each grid point. Only vectors larger than a given threshold value are shown

Figures 7 shows the effect of lowering the threshold, and therefore letting smaller amplitude vorticity vectors appear. The thickness of these tubes is of the order of a few dissipation scales, here a few grid points. Their length seems to be of order 1 , i.e. of the same order as the integral scale (the cube size is $2\pi$). Figure 8 (see colour plates) shows a detailed view of one of these vorticity tubes. It displays a sub-cube one sixth the size of the complete one, with 40 grid points on each side. Similar plots for the velocity field show mainly the forcing field if one uses a high threshold. But when one eliminates the highest velocities, the tubes begin to appear.

We have examined the shape of many of these tubes in order to see wether they are in fact rolled-up vorticity sheets, but this does not seem to be the case. Figure 9 gives an example of the projection of the velocity field on a plane perpendicular to a vorticity tube, while figure 10 shows the curves of constant vorticity on such a plane. As one can see on these figures, there is little evidence of spiral structure, or any other two dimensional structure.

Space correlations between vorticity and stress tensor were reported by Kerr (1985). At each grid point, we compute the eigenvalues $\lambda_1, \lambda_2, \lambda_3$ of the stress tensor $S_{ij} = \frac{1}{2}(\partial_i v_j + \partial_j v_i)$, with $\lambda_1$ always negative and

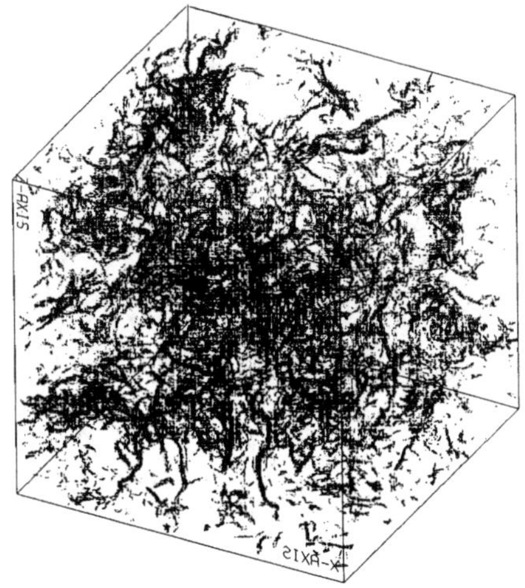

Fig.7. Same as figure 6, but with a lower threshold value, and therefore more vectors represented

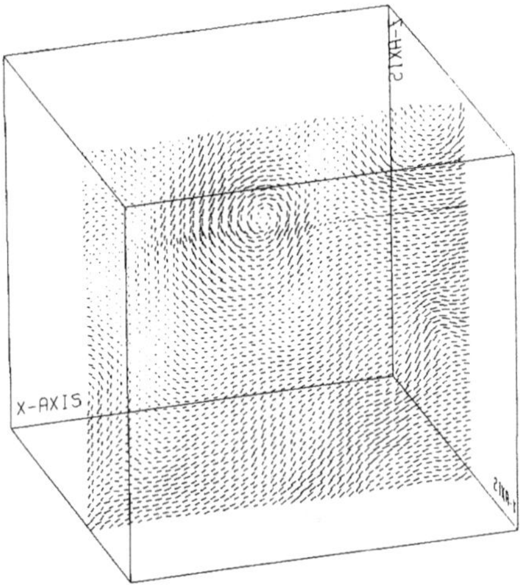


Fig.9. Projection of the velocity field on a plane perpendicular to a particular vorticity tube

$\lambda_3$ always positive. Indeed, $\nabla \cdot \mathbf{v} = 0$ implies that $\lambda_1 + \lambda_2 + \lambda_3 = 0$. The intermediate eigenvalue $\lambda_2$ is found to be positive in approximately two thirds of the cases, in agreement with the results of the analysis of the Kerr (1985) and Ashurst et al. (1987). Three-dimensional plots of the eigenvector fields, are used to study the correlations with vorticity. Let us call $\mathbf{e_1}, \mathbf{e_2}, \mathbf{e_3}$ eigenvectors associated with $\lambda_1, \lambda_2, \lambda_3$, each eigenvector being given the amplitude and sign of its associated eigenvalue. Figure

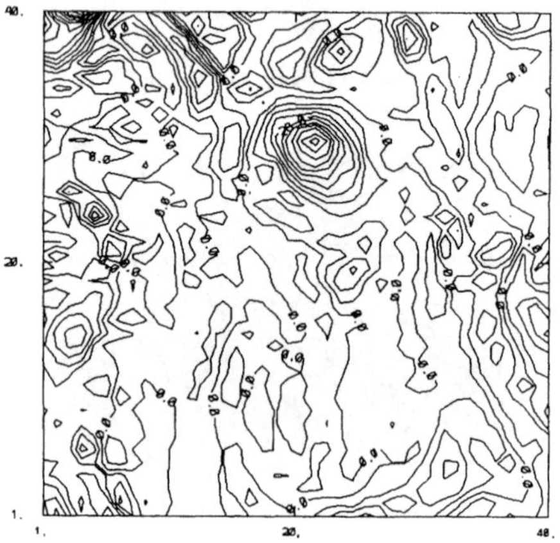

Fig.10. Curves of constant vorticity in a plane perpendicular to a selected vorticity tube

11 (see colour plates) shows the field $\mathbf{e_3}$ (in green) together with the vorticity field (in red). Again, only vectors above a certain amplitude are displayed, and therefore the two fields are not generally shown at the same grid points. The highest rate of strain is seen to be generally in the vicinity of vorticity tubes. The eigenvector $\mathbf{e_3}$, associated with the positive eigenvalue, i.e. the stretching direction, is generally perpendicular to the vorticity. Note in the upper part of the picture, a vorticity tube seen along its axis, and the perpendicular field $\mathbf{e_3}$ in its vicinity. Fig. 12 (see colour plates) is analogous to Fig. 11 (see colour plates) but for $\mathbf{e_1}$, the eigenvector associated with the negative eigenvalue of the strain, which is also seen to be perpendicular to the vorticity. The same tube as before is seen along its axis, with the field $\mathbf{e_1}$ perpendicular to it. Figure 13 (see colour plates) shows again the same view of the vorticity, with this time the intermediate eigenvector $\mathbf{e_2}$. This eigenvector is the one which is often aligned with vorticity. It is the case, for instance, for the previously mentioned vorticity tube seen along its axis. This is not easy to see on figure 13, but becomes clear after rotating the fields (Figure 14, upper right corner, see colour plates). A discussion of this correlation of vorticity with the intermediate rate of strain direction is presented in Ashurst et al. (1987), where an explanation is proposed using a model due to Vieillefosse (1982, 1984). The results presented here are preliminary. A more thorough examination of the velocity and vorticity structure in space, as well as their detailed statistical properties, is under way.

## References


Ashurst W.T., Kerstein A.R., Kerr R.M., Gibson C.H. (1987), *Phys. Fluids* **30**, 2343.

Anselmet F., Gagne Y., Hopfinger E.J., Antonia R.A.(1984), *J. Fluid Mech.* **140**, 63.

Basdevant C., Legras B., Sadourny R., Beland M. (1981), *J. Atmos. Sci.* **38**, 2305.

Benzi R., Patarnello S., Santangelo (1987) ICE-0015 to appear in *J.Phys. A*.

Brachet M.E., Meiron D.I., Orszag S.A., Nickel B.G., Morf R.H., Frisch U (1983), *J. Fluid Mech.* **130**, 411.

Brachet M.E., Meneguzzi M., Politano H., Sulem P.L. (1988) *J. Fluid Mech.* **94**, 333.

Frisch U., Sulem P.L., Nelkin M (1978), *J. Fluid Mech* **87**, 719.

Herring J.R., McWilliams J.C. (1985), *J. Fluid Mech.* **153**, 229.

Kerr R. (1985), *J. Fluid Mech.* **153**, 31.

Kolmogorov A.N. (1962) *J. Fluid Mech.* **13**, 82.

Kraichnan R.H. (1974), *J. Fluid Mech.* **62**, 305.

Legras B., Santangelo P, Benzi R. (1988), *Europhys. Lett.* **5**, 37.

McWilliams J.C.(1984), *J. Fluid Mech.* **146**,21

Monin A.S., Yaglom A.M. (1975), *Statistical Fluid Mechanics*, Vol 2, MIT Press.

Novikov E.A., Stewart R.W. (1964), *Isv. Geophys.* **3**, 408

Oboukhov A.M. (1962), *J. Fluid Mech.* **13**, 77

She Z.S., Jackson E., Orszag S.A. (1990) to appear in *J. Sci. Comp.*

Siggia E.D., Patterson G.S. (1978) *J. Fluid Mech.* **86**, 567.

Van Atta C.W., Chen W.Y. (1970), *J. Fluid Mech.* **44**, 145.

Vieillefosse P. (1982), *J. Phys. Paris* **43**, 837.

Vieillefosse P. (1984), *Physica* **125 A**, 150.

Vincent A., Meneguzzi M. (1990), submitted to *J. Fluid Mech.*

# Wavelet analysis of fully developed turbulence data and measurement of scaling exponents.

Emmanuel BACRY[α,β], Alain ARNEODO[α], Uriel FRISCH[β], Yves GAGNE[γ] and Emil HOPFINGER[γ]

[α] *CNRS, Centre de Recherches Paul Pascal*
*Domaine Universitaire, 33405 Talence-Cedex, France*

[β] *CNRS, Observatoire de Nice*
*BP 139, 06003 Nice-Cedex, France*

[γ] *CNRS, Institut de Mécanique de Grenoble*
*BP 53X, 38041 Grenoble-Cedex, France*

*Turbulence and Coherent Structures*
*O. Métais and M. Lesieur, eds.*
*Selected papers from "Turbulence 89: organized structures and turbulence in fluid mechanics"*
*Grenoble, 18-21 September 1989*

**Abstract**

Wavelet analysis can be used to measure directly the scaling exponents characterizing the local multifractal behaviour of the velocity field at inertial-range scales, without recourse to dissipation-type quantities. Preliminary results using data from the Modane S1 wind tunnel indicate that the most frequent exponents are close to the Kolmogorov 1941 value of 1/3. Violent rare events are however found with negative exponents of $-0.1$ or less.

## 1. Multifractals and fully developed turbulence.

In 1983, Parisi and Frisch [1984] introduced the *multifractal* model of fully developed turbulence in order to account for previously available experimental data of Anselmet *et al.* [1984]. These results concerned the structure functions of order $p$, defined as $S_p(\ell) = \langle (\delta v_\ell)^p \rangle$, where $\delta v_\ell$ is a (longitudinal) velocity increment over a distance $\ell$. It was reported in Anselmet *et al.* [1984] and later confirmed in Gagne [1987] and Gagne *et al.* [1989] that (i), $S_p$ scales in the inertial-range like a power law $\ell^{\zeta_p}$ (except for oscillations interpreted by Smith *et al.* [1986]) and (ii), the exponent $\zeta_p$ has a nonlinear dependence on $p$. The interpretation of the nonlinearity proposed in Parisi and Frisch [1984] was to assume that a very high Reynolds number turbulent flow has singularities (actually, almost singularities given the smoothing action of viscosity). These singularities have a range of possible scaling exponents $h$. For each $h$ there is a set in $\mathbf{R}^3$ of Hausdorff dimension $D(h)$ near which velocity increments behave as $\ell^h$. The contribution of singularities of exponent $h$ to the structure function of order $p$ will be proportional to $\ell^{ph}$ times the probability of intersecting a fractal set of dimension $D(h)$ with a ball of diameter $\ell$. The latter is proportional to $\ell^{3-D(h)}$. ( If not familiar with fractal dimensions, check it for integer-dimension objects such as a point, a curve and a surface.) Hence, the singularities of exponent $h$ contribute a term proportional to $\ell^{ph+3-D(h)}$. For $\ell \to 0$ the contribution with the smallest exponent dominates. It follows that $\zeta_p = \min_h (ph + 3 - D(h))$, i.e. is given by a Legendre transform. Nonlinear dependence of $\zeta_p$ on $p$ is equivalent to the assumption that there is more than a single value for the scaling exponent.

Earlier models of Mandelbrot [1974], who pioneered the use of fractals in turbulence, did also include nonlinear dependence of $\zeta_p$ and are now described by him as "multifractal". His approach did not involve Legendre transforms. From the view-point of fully developed turbulence it presents an important difference with the Parisi-Frisch model (and also with the $\beta$-model of Frisch *et al.* [1978]). Following a tradition going back to Kolmogorov [1962], Mandelbrot was working not with velocity increments but with the "local dissipation" $\epsilon_\ell$, defined as local spatial average over a ball of size $\ell$ of the dissipation function

$$\epsilon(\mathbf{x}) = \nu \sum_{ij} \left(\partial_i u_j(\mathbf{x}) + \partial_j u_i(\mathbf{x})\right)^2 ,$$

where $\nu$ is the kinematic viscosity and $\mathbf{u}(\mathbf{x})$ is the velocity field. In such work it is assumed that a relation of the following form holds (which is suggested by dimensional analysis)

$$\epsilon_\ell \sim \frac{(\delta v_\ell)^3}{\ell}, \qquad (1-1)$$

at least in a scaling sense (i.e. for evaluation of moments). If we accept eq.( 1-1), it follows that wherever the velocity has a scaling exponent $h$, the "local dissipation" will have an exponent (traditionally denoted by $\alpha$)

$$\alpha = 3h - 1 .$$

Eq. (1-1) is however questionable because it has a consequence contradicting experimental data. It was shown in Gagne [1987], Gagne *et al.* [1989] and Castaing *et al.* [1990] that the probability distribution function of $\delta v_\ell$, for $\ell$'s in the inertial range, has a finite nonvanishing value for $\delta v_\ell = 0$ (nothing special is happening for vanishing velocity increments). This implies that structure functions exist only for $p > -1$ (an absolute value is needed on velocity increments for noninteger $p$). In contrast, the dissipation moments $< (\epsilon_\ell)^q >$ exist for negative orders $q$ far less than the lower limit $-1/3$ implied by (1-1) if it were true. Indeed, the local dissipation is the spatial integral of a positive quantity; hence its probability distribution function vanishes (probably very fast) for the argument zero. Thus, there is no obvious relation between the dimensions $D(h)$ and the distribution of scaling exponent of the dissipation, traditionally denoted by $f(\alpha)$ in the context of dynamical systems (Halsey *et al.* [1986]). The former is a pure inertial-range quantity while the latter is a mixed, inertial-range and dissipation-range quantity.

Measurements of $f(\alpha)$ based on a simplified version of the local dissipation have been reported by Meneveau and Sreenivasan [1987]. Such measurements involve difficulties in high Reynolds number flow because squared time-derivatives of the velocity may not be fully representative of the local dissipation; furthermore, rare large excursions of the velocity derivative signal are not well captured by existing probes.

We shall show that the new technique of ***wavelet analysis*** can be used for direct measurements of the scaling exponents $h$. The main ideas of wavelet analysis of turbulence, already presented in Argoul *et al.* [1989], are summarized in Section 2. The measurement of scaling exponents is presented in Section 3. Concluding remarks are made in Section 4.

## 2. The wavelet transform

The wavelet transform of a signal (Grossman and Morlet [1987]) permits an analysis both in physical space and in scale space (scales are, roughly, inverse wave numbers). The wavelet transform consists in expanding functions in terms of "wavelets" which are constructed from one single function $g$ by means of dilatations and translations.

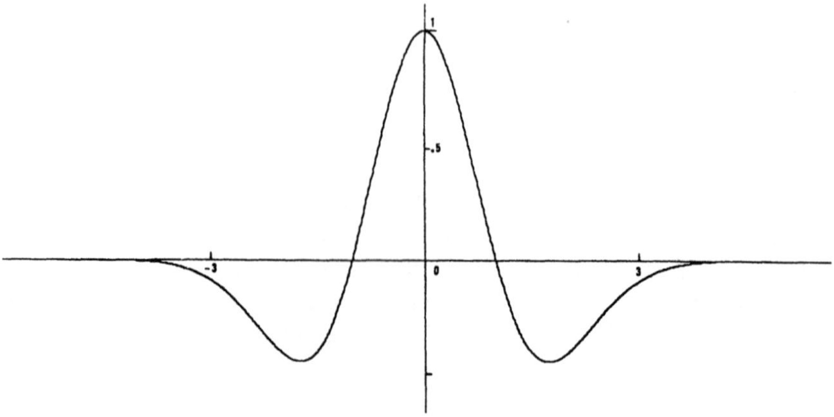


Fig.1. Analysing wavelet: the Mexican hat.

Figure 1 gives an exemple of such a function $g$, which is called analysing wavelet. It is necessary, in order to perform wavelet transforms, that the analysing wavelet should be localized around $x = 0$ and have a vanishing integral. Further conditions may be needed. From the single analysing function $g$, we construct a set of wavelet functions

$$g_{a,x_0}(x) = \frac{1}{a} g(\frac{x - x_0}{a})$$

by translations and dilatations. The wavelets $g_{a,x_0}(x)$ are then used as "mathematical microscopes", in which $g$ characterizes the optics, while $x_0$ is the position analysed and $1/a$ the magnification. The parameter $a$ which has dimension of a length, will henceforth be called the (analysing) scale. The wavelet transform of a signal $s(x)$ is defined as

$$T_g(a, x_0) = \frac{1}{a} \int_{-\infty}^{+\infty} s(x) g(\frac{x_0 - x}{a}) dx \qquad a > 0.$$

In others words, $T_g(a, x_0)$ is simply the convolution product of the signal and the scaled analysing function $g_a(x) = \frac{1}{a} g(\frac{x}{a})$. The wavelet transform may also be regarded as a local Fourier transform: for a given position $x_0$, the amplitude of $T_g(a, x_0)$ will be largest when the scale $a$ is roughly equal to the inverse local wavenumber of the signal.

For most of our study of developed turbulence data, we choose as analysing wavelet the second derivative of a Gaussian function; this is called the "Mexican hat". Occasionally, we also used a stepwise constant approximation of the Mexican hat, called the "French top hat". The corresponding function is
$g(x) = 1(\mid x \mid < 1), -1/2(1 \leq \mid x \mid \leq 3), 0(\mid x \mid > 3)$. This speeds up the numerical evaluation of wavelet transforms and is useful for revealing certain large-scale structures (Grasseau [1989]). We have also used as analysing function $g$, the first,rather than the second, derivative of a Gaussian in order to provide further check that our results are not sensitive to the particular choice of analysing function.

The turbulent velocity signal analysed comes from the wind tunnel S1 of ONERA at Modane (For details on the experiment, see Gagne [1987] and Castaing *et al.* [1990]). Figure 2 shows the energy spectrum of the streamwise component of the velocity. The Taylor scale based Reynolds number is $R_\lambda = 2720$ and the extent of the inertial range following (approximately) the Kolmogorov $-5/3$ law is almost three decades. The Kolmogorov dissipation scale $\eta = 0.3mm$ is too small to permit accurate analysis in the dissipation range. All the results reported here concern inertial-range scales.

The wavelet transform permits a, geometrically very revealing, display of a one-dimensional signal as a two-dimensional unfolding of space and scale. Figure 3 (see colour plates) shows the wavelet transform obtained with the "French top hat" wavelet applied to the velocity signal. The coordinates are linear (abcissa: position $x$, ordinate: scale $a$, rainbow colours for the amplitude $T(a, x)$ ). It is important to note that only positive values of $T(a, x)$ are coded; thus black regions correspond to both zero and negative values of $T(a, x)$. The colour is defined for each value of the scale $a$ by using 256 levels; so, for a given scale $a_0$ the colours changes from purple to red when the amplitude $T(a_0, x)$ increases from zero to its maximum value.

It is seen that the patterns revealed by figure 3a are drastically different from those displayed on the others figures 3b and 3c. Figure 3a displays only "large scales", ranging from $\ell_0/10$ to $28\ell_0$, where $\ell_0$ is the integral scale of the turbulent field. It reveals structures of size of or-

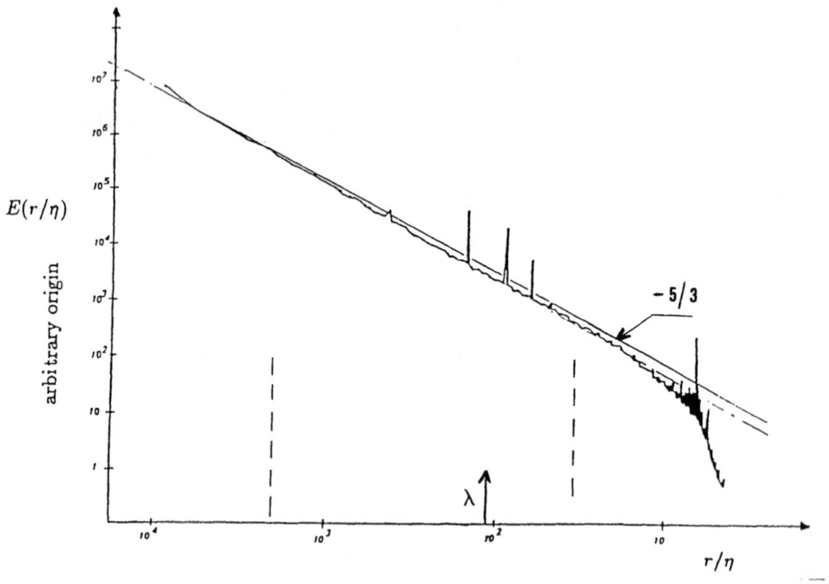


Fig.2. Energy spectrum of Modane turbulence data (log-log plot)

der $\ell_0$, randomly distributed in space. The cone-like patterns seen point towards the strongest structures of scale $\simeq \ell_0$ and are generated by the edges (discontinuites) of the "French top hat". Figures 3b and 3c are obtained by analysing much smaller scales $a$, reduced by a factor 20 (b) and $20^2$ (c). The position variable is scaled by the same factor. The analysis shown on figure (b) is made near the arrow on figure (a); similarly for (c) with respect to (b). The full range of scales displayed in figures 3b and 3c covers the inertial range from the integral scale down to about one quarter of the Taylor scale. These figures show a hierarchical branching process. In Argoul *et al.* [1989] the reader will find versions of figures 3b and 3c using the same signal and processing, but with a Mexican hat analysing wavelet: the features are then identical but a bit smoother.

In order to test the sensitivity of this branching process to the particular representation chosen, another representation is displayed in figures 4a and 4b (see colour plates) for the same signal.( Figure 4b is an enlargement of figure 4a near the upper right branch.) We still use linear scales for position $x$ and analysed scales $a$. However, the rainbow colours are now applied to the whole picture and not independently for each analysing scale $a$, as before; furthermore, the scale of the amplitude of $T(a, x)$ is logarithmic and the analysing wavelet is the first derivative of a Gaussian. Note that the self-similarity of the branching process is now particularly conspicuous.

The successive branchings observed in figures 3b and 3c are suggestive of a hierarchical cascade process with fractal (or multifractal) branching. When the wavelet analysis is applied to chaotic processes presenting multifractal attractors (in their phase-space), the latter structure is clearly revealed by wavelet analysis (Arnéodo *et al.* [1988], [1989]). For high Reynolds number turbulence data the situation is not so clear. Following a suggestion of Kraichnan [1988] we have performed a wavelet analysis on artificial Gaussian signals having the same spectrum as the Modane signal. We also observed some form of fractal branching in the Gaussian case. Similar observations were made by Everson and Sirovich [1989]. A possible explanation is the following. A Gaussian process with a power-law spectrum at high wavenumbers has a small-scale behaviour locally isomorphic to that of "fractional Brownian motion" of exponent $H$ close to 1/3 (Mandelbrot [1965]). The times when such a process $B_H(t)$ crosses a given value form a fractal set of dimension $1 - H$. This is seen for example in the work of Everson and Sirovich, where fractional Brownian motion produces hierarchical branching in displays of its wavelet transform. Indeed, we observe that the wavelet transform puts the emphasis on small scales and the colour coding emphasizes level-crossing. Of course, the small-scale statistics of the true turbulent signal is far from Gaussian, so that the fractal branching observed in its wavelet transform is quite different and may to some extent reflects *dynamically significant* structures. With the representation used for figure 3, dynamically significant events having strong localized gradients appear as red tongues (and are not seen in Gaussian signals). With the representation of figure 4, dynamically significant events are harder to detect. We also note that a recent study of two-dimensional wavelet transforms for concentration data in a jet reveals considerable structure not present in Gaussian processes (Everson *et al.* [1989]).

Clearly, the problem of interpreting geometrical structures observed in display of wavelet transforms requires further study. We shall now concentrate on more quantitatives studies of the wavelet transforms which are independent of such considerations.

## 3. Determination of scaling exponents

An immediate consequence of the definition of the wavelet transform is the following: if the signal $s(x)$ has the scaling exponent $h(x_0)$ at the point $x_0$, that is $s(x_0+\ell)-s(x_0)\sim \ell^{h(x_0)}$ for $\ell\to 0$, then $T_g(a,x_0)\sim a^{h(x_0)}$ for $a\to 0$ (Holschneider [1988]).

Thus, we can extract the local scaling exponent by analysing the behaviour of the wavelet transform as a function of scale $a$, for fixed position $x_0$. A log-log plot of the wavelet amplitude versus the scale $a$, for fixed position $x_0$, will reveal the exponent. An example of such a plot is given in figure 5.

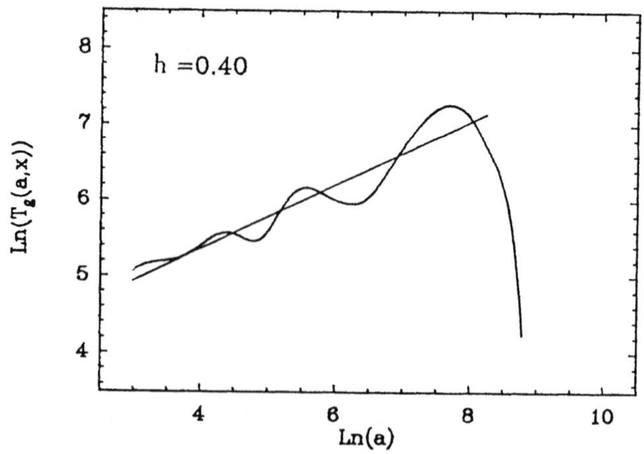


Fig.5. Local scaling exponent revealed in log-log plot of amplitude $T(a,x)$ vs. scale $a$.

We see that at scales smaller than the integral scale, the amplitude of the wavelet coefficients $T(x_0,a)$ has a power law behaviour with superimposed oscillations. Such oscillations are unavoidable in any situation where fractal branching occurs. For example, if the analysis were done for a measure concentrated on the standard Cantor set, the oscillations would be perfectly periodic in $\log a$ and correspond to the exact replication of the set after rescaling by a factor $1/3$ ( Smith *et al.* [1986]). Using a least square fit to a power law over inertial-range scales, we can extract the local scaling exponent. For the case shown in figure 5, the local scaling exponent is $h(x_0)\simeq 0.4$ not very far from the Kolmogorov 1941 value of $1/3$.

There are instances where oscillations are so large as to make the determination of exponents very uncertain. An example is shown in figure 6a: the oscillation is so strong that the wavelet amplitude reverses its sign.

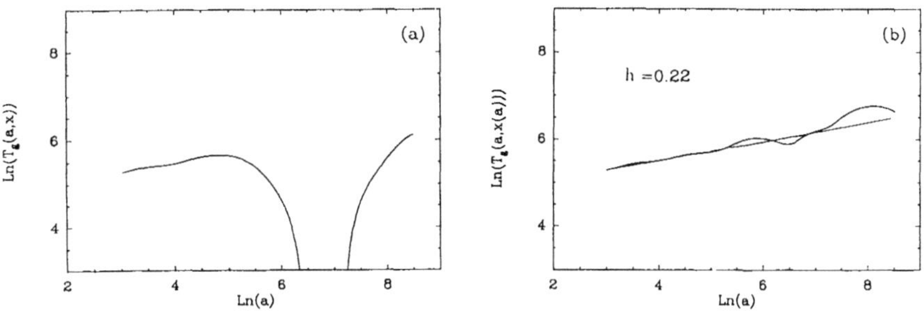


Fig.6. (a) Same as 5 near a point presenting a strong oscillation; (b) Same as 6a using the maximum-tracking technique.

We have thus begun to explore an alternative technique suggested by Mallat [1989]. This technique consists in tracking local maxima and minima of the wavelet transform in the neighborhood of the point $x_0$ under investigation. These obey the same scaling properties as the wavelet transform. Figure 7 shows a "backbone" of the wavelet transform where only relative maxima and minima are shown, respectively as large dots and small dots. Here, maxima and minima are determined by varying the position variable $x$ at fixed analysing scale $a$. For the same case as shown in figure 6a, figure 6b shows a log-log plot of the local maximum value of the wavelet amplitude versus the scale $a$, following a local maximum curve emanating from $x_0$. The determination of the exponent is now much cleaner. Of course, the method of tracking maxima/minima requires more computer resources.

In figure 8a, we show a log-log plot of amplitude versus scale for a point $x_0$ having a very strong singularity with a negative exponent $h(x_0) \simeq -0.13$. The signal itself (figure 8b) shows a violent event with very sharp velocity gradients and very large signal excursion.

The above types of events were obtained from a systematic investigation involving 2000 points $x_0$ selected at random among the $4.10^6$ points in our experimental sample. The histogram of the measured scaling exponents is shown in figure 9. We can observe that the most probable value of the local scaling exponents is $h = 1/3$ corresponding to the Kolmogorov 1941 value. The values of the measured exponents $h(x_0)$ range beetwen $-0.3$ and $0.8$. We believe that the data are statistically significant for negative exponents down to $-0.1$ or beyond.

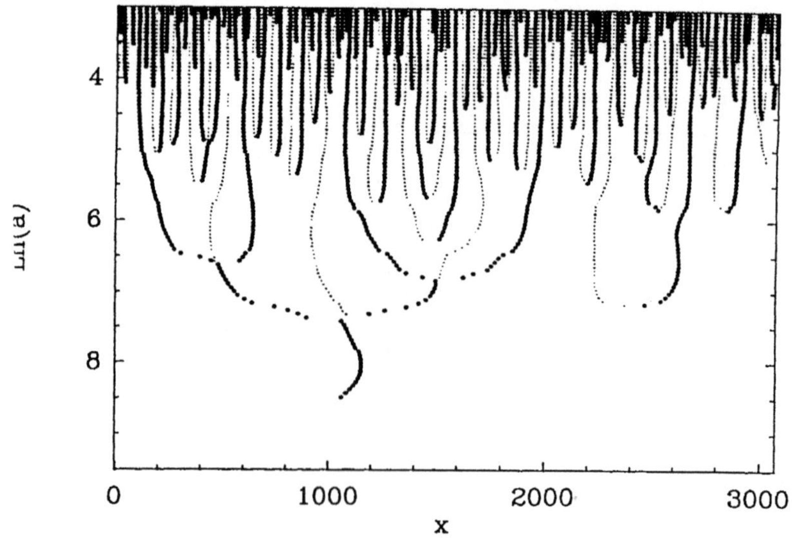


Fig.7. Backbone of wavelet transform showing local maxima and minima.

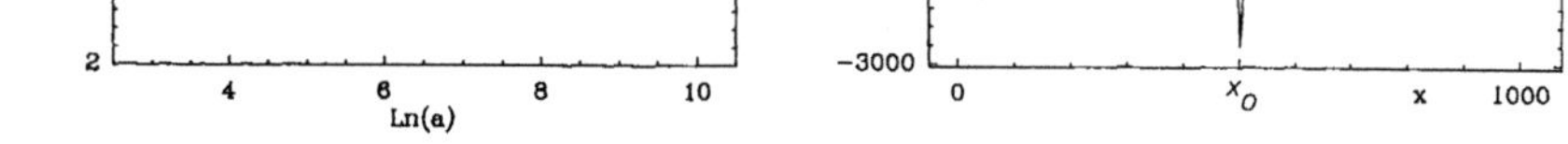


Fig.8. (a) Same as 5 near a point $x_0$ having exponent $h = -0.13$; (b) Velocity signal near $x_0$ corresponding to 8a.

## 4. Concluding remarks

We have found that the wavelet transform gives us direct access to local scaling exponents at inertial-range scales. A wide range of scaling exponent is obtained, thereby confirming that the multifractal picture

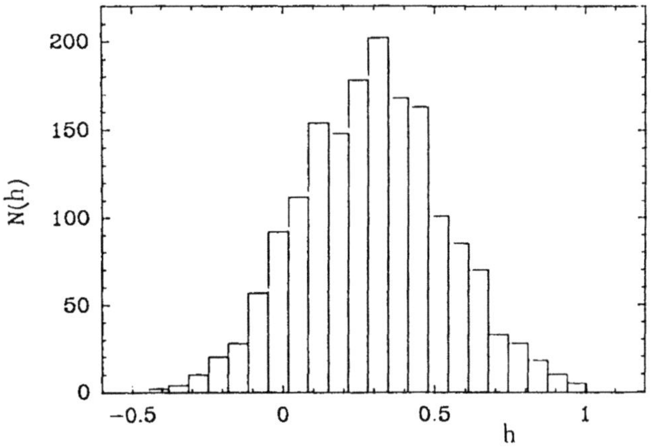


Fig.9. Histogram of scaling exponents (from 2000 data points out of 4 million).

of Parisi-Frisch [1984] is appropriate. Actually, the range of measured exponents is surprisingly large. Negative exponents do not seem to have been previously reported. One tentative interpretation could be the occasional passage near the probe of slender vortex filaments of the sort observed in numerical simulations by Siggia [1981], She *et al.* [1989] and Meneguzzi and Vincent [1989]. We also note that Kuo and Corrsin [1972] reported evidence for rod-like structures in fully developed turbulence. At first sight it seems that this interpretation should be rejected on the ground that the probe is measuring along a line (in the reference frame of the mean flow) and such a line has an almost surely empty intersection with vortex filaments which, on inertial- range scales, appear as one dimensional objects. Actually, the vortex filaments can approach the probe arbitrarily close and there is a finite probability for a sphere of radius $\ell$ (an inertial-range scale) to have non-empty intersections with both. This probability is approximately proportional to $\ell^4 = \ell^{3-(-1)}$; this led Mandelbrot [1989a] to attribute a *negative* dimension (here approximately $-1$) to such a situation.

As for the negative scaling exponents, they do not imply that the velocity becomes infinite in the limit of zero viscosity $\nu$. Indeed, the scaling law holds only over inertial-range separations $\ell$ and, as $\nu \to 0$, the constant in front of the power laws with negative exponents may well go to zero . We note also that a negative exponent in the sense we are using (for the velocity) does not pose any particular difficulty as is the case for negative $\alpha$'s in $f(\alpha)$ functions, which require a special interpretation (Mandelbrot [1989b]).

It is clear that considerable further work is needed for reliable determination of the dimensions $D(h)$ associated to the distribution of singularities. Here we have only obtained the distribution of exponents corresponding to a rather small subset of the experimental sample. It is too small to permit verification that the probability of encountering a singularity of given exponent in a set of size $\ell$ scales like $\ell^{3-D(h)}$, as predicted by the multifractal model.

**Acknowledgements** This work was supported by DRET (contracts 86/1336 and 89/201) and by EEC SCIENCE program (SC1-0212-C).

## References

Anselmet, F., Gagne, Y., Hopfinger, E.J. & Antonia, R.A. 1984 *J. Fluid. Mech.*, **140**, 63-89.

Argoul, F., Arnéodo, A., Grasseau, G., Gagne, Y., Hopfinger, E.J. & Frisch, U. 1989 *Nature*, **338**,no 6210, 51-53.

Arnéodo, A;, Argoul, F., Elezgaray, J. & Grasseau, G. 1989 *in Nonlinear Dynamics* (ed. G. Turchetti), p.130. World Scientific, Singapore.

Arnéodo, A., Grasseau, G. & Holschneider, M. 1988 *Phys. Rev. Lett.*, **61**, 2287-2289.

Castaing, B., Gagne, Y. & Hopfinger, E.J. 1990, preprint.

Everson, R. & Sirovich, L. 1989, Report 89-182, Center for Fluid Mechanics, Brown University.

Everson, R., Sirovich, L. & Sreenivasan, K.R. 1989, to appear in *Phys.Lett. A*.

Frisch, U., Sulem, P.L. & Nelkin, M. 1978 *J. Fluid. Mech.*, **87**, 719-736.

Gagne, Y. 1987, Thèse. Université de Grenoble.

Gagne, Y., Hopfinger, E.J. & Frisch, U. 1988 *in New Trends in Non Linear Dynamics and Pattern Forming Phenomena: The Geometry of Non Equilibrium* (ed. P. Huerre and P. Coullet). Plenum Press.

Grasseau, G. 1989, Thèse. Université de Bordeaux.

Grossman, A. & Morlet, J. 1987 *Mathematics and Physics, Lecture on Recent Results* (ed. L. Streit) World Scientific, Singapore.

Halsey, T.C., Jensen, M.H., Kadanoff, L.P., Procaccia, I. & Shraiman, B.I. 1986 *Phys. Rev.*, **A33**, 1141-1151.

Holschneider, M., 1988 *J. Stat. Phys.*, **50**, 963.

Kolmogorov, A.N. 1962 *J. Fluid. Mech.*, **13**, 82-85.

Kraichnan, R.H. 1988 Private Communication.

Kuo, A.Y-S., & Corrsin, S. 1972 *J. Fluid. Mech.*, **56**, 447- 479.

Mallat, S. 1989 presented at the 3rd Conference on *Wavelets* Marseille, May-June 1989.

Mandelbrot, B.B. 1965 *C. R. Acad. Sci. Paris*, **260**, 3274

Mandelbrot, B.B. 1974 *J. Fluid. Mech.*, **62**, 331-358.

Mandelbrot, B.B. 1989a *in STATPHYS***17** (Proceedings of the Rio de Janeiro meeting, 1989), ed. C. Tsallis, North-Holland (Special volume of Physica A and Book).

Mandelbrot, B.B. 1989b *in Frontiers of Physics: Landau memorial conference* (ed. E. Gotsman) Pergamon Press, New-York.

Meneguzzi, M. & Vincent, A. 1989 submitted to *J. Fluid. Mech.*.

Meneveau, C. & Sreenivasan, K.R. 1987 *Nuclear physics B*.**2** (Suppl.), 49-67. North-Holland.

Parisi, G. & Frisch, U. 1984 *Turbulence and Predictability in Geophysical Fluid Dynamics and Climate Dynamics* (ed. M. Ghil, R. Benzi and G. Parisi), p.84. North-Holland.

She, Z.S., Jackson, E. & Orszag, S.A. 1989, submitted to *Nature*.

Siggia, E. 1981 *J. Fluid. Mech.*.**107**, 375-406.

Smith, L.A., Fournier, J.D. & Spiegel, E.A. 1986 *Phys. Lett.*, **A114**,465-473.

# How Solenoidal can be the Lamb Vector in Turbulent Flow? Consequences and Implications

A. TSINOBER[α]

[α]*Department of Applied Mathematics and Theoretical Physics*
*Silver Street, Cambridge CB3 9EW, U.K.*

*On summer leave from the Faculty of Engineering*
*Tel-Aviv University, Tel-Aviv 69978, Israel*

*Turbulence and Coherent Structures*
*O. Métais and M. Lesieur, eds.*
*Selected papers from "Turbulence 89: organized structures and turbulence in fluid mechanics"*
*Grenoble, 18-21 September 1989*

**Abstract**
It is shown that the potential part of Lamb vector $\lambda = \omega \times \mathbf{u}, \omega = \nabla \times \mathbf{u}$, where $u$ is an isotropic Gaussian solenoidal random field, is much larger than its solenoidal part. This result shows that purely kinematic properties of turbulent flow can lead to essential reduction of its nonlinearity and thereby to impose considerable constraints on flow dynamics and topology.

## 1. Introduction

The Lamb vector $\lambda = \omega \times \mathbf{u}$ is known to be of profound importance for fluid dynamics generally and for the dynamics of turbulence in particular, since the nonlinear term in the Navier-Stokes equations $(\mathbf{u} \cdot \nabla)\mathbf{u} = \omega \times \mathbf{u} + \nabla(\frac{1}{2}u^2)$ so that the last term can be incorporated into the pressure gradient term, while the vorticity velocity cross-product is intimately related to gradients of Reynolds stresses in turbulent shear flows.

Recent studies have demonstrated the importance of studying the properties of the Lamb vector in the context of topological flow properties, fixed points of the Euler equations and reduction of nonlinearity in turbulent flows (Arnold (1974), Herring and Kerr (1989), Kraichnan and Panda(1988), Moffat(1985, 1986a,b), Shtilman and Polifke(1989)). Moffatt has suggested that turbulent flow may spend a large proportion of its time in a neighbourhood of fixed points of Euler equations, i.e. solutions of $\omega \times \mathbf{u} = \nabla h$ , $h = -p - \frac{1}{2}u^2$. This implies that in turbulent flow the potential part of the Lamb vector can be expected to be larger than its solenoidal part. Indeed, direct numerical computations performed by Kraichnan and Panda(1988) showed that the quantity $< (\omega \times \mathbf{u} + \nabla h)^2 >$ is only 0.57 of the corresponding value for a Gaussian field with the same energy spectrum. Shtilman and Polifke(1989) corroborated their results and showed that the Lamb vector contains a substantial potential part by calculating the probability density function of the angle between the Fourier image of the Lamb vector $\lambda(\mathbf{k})$ and the wave vector $\mathbf{k}$. They found that there was a high probability that $\lambda(\mathbf{k})$ and $\mathbf{k}$ were aligned, which implies that in physical space the Lamb vector should have a large potential part. Moreover, they found also, that this property is characteristic for an initial Gaussian field and that the flow dynamics is such that it enhances this tendency. Finally, some clear experimental indications were found, that the Lamb vector has a considerable potential part Dracos et al. (1989).

In view of the above research it is of interest to look into the *kinematic* properties of the Lamb vector *analytically*.

## 2. Kinematic properties of a cross product of two vectors

Our purpose is to find the relation between the potential and the solenoidal parts of a vector of the type $\lambda = \mathbf{a} \times \mathbf{b}$, i.e. between $\nabla\alpha$ and $\nabla \times \beta$ such that $\lambda = \nabla\alpha + \nabla \times \beta$ (Helmholz decomposition) and

$$\nabla^2\alpha = \nabla \cdot \lambda \qquad \nabla^2\alpha = \frac{\partial\lambda_i}{\partial x_i}, \qquad (2-1)$$

$$\nabla^2\beta = -\nabla \times \lambda \qquad or \qquad \nabla^2\beta_j = -\epsilon_{jem}\frac{\partial\lambda_m}{\partial x_e} \qquad (2-2)$$

We will use a technique used by Batchelor(1951) and Tsinober et al. (1987) in different contexts and calculate correlations of the type $< \frac{\partial\alpha(\mathbf{x})}{\partial x_k}\frac{\partial\alpha(\mathbf{x+r})}{\partial x_k} >$ etc. in terms of statistical properties of solenoidal vector fields **a, b** which are assumed to be random and isotropic. Using the property of homogeneity it is straightforward to obtain from (2-1)

$$\nabla^2 < \frac{\partial\alpha(\mathbf{x})}{\partial x_k}\frac{\partial\alpha(\mathbf{x}+\mathbf{r})}{\partial x_k} >= \frac{\partial^2}{\partial r_i \partial r_j} < \lambda_i(\mathbf{x})\lambda_j(\mathbf{x}+\mathbf{r}) > \qquad (2-3)$$

In the following we consider three cases:

**Case A.** $\lambda = \omega \times \mathbf{u}$, where $\omega$ is an isotropic solenoidal vector field and $\mathbf{u}$ is a constant vector.

**Case B.** $\lambda = \omega \times \mathbf{u}$ where $\omega$ and $\mathbf{u}$ are isotropic random and statistically independent solenoidal vector fields, $\omega \neq \nabla \times \mathbf{u}$. $F(r)$ and $f(r)$ are the longitudinal correlation functions of the vector fields $\omega$ and $\mathbf{u}$ correspondingly, i.e.

$$< u_i(\mathbf{x})u_j(\mathbf{x}+\mathbf{r}) >= -\frac{1}{2r}f'(r)r_ir_j + \left\{ f(r) + \frac{1}{2}rf'(r) \right\}\delta_{ij},$$

**Case C.** $\lambda = \omega \times \mathbf{u}$ where $\omega = \nabla \times \mathbf{u}$, $\mathbf{u}$ is an isotropic Gaussian solenoidal random vector field in the sense that its fourth moments are expressed through its second moments as for a Gaussian field and $f(r)$ is the longitudinal correlation function of the vector field $\mathbf{u}$ *Case A*. This case is primarily of interest as an illustration. Here the results from Tsinober et al (1987) can be used directly to obtain that $< (\nabla\alpha)^2 >= < (\nabla \times \beta)^2 >= \frac{1}{2} < \lambda^2 >$, i.e. the potential and the solenoidal parts of the cross product $\omega \times \mathbf{u}$ contribute equally to $< \lambda^2 >$. This case can be considered also as a particular of the *Case B*. The calculation here involves the representation of the right hand side of (2-3) in terms of second order derivatives of

$$< \omega_e(\mathbf{x})u_m(x)\omega_p(\mathbf{x}+\mathbf{r})u_q(\mathbf{x}+\mathbf{r}) >=$$

$$< \omega_e(\mathbf{x})\omega_p(\mathbf{x}+\mathbf{r}) >< u_m(\mathbf{x})u_q(\mathbf{x}+\mathbf{r}) >,$$

After some tedious algebra one obtains the following results

$$< \frac{\partial\alpha(\mathbf{x})}{\partial x_k}\frac{\partial\alpha(\mathbf{x}+\mathbf{r})}{\partial x_k} >= \frac{1}{2} < \lambda_k(\mathbf{x})\lambda_k(\mathbf{x}+\mathbf{r}) >$$

$$+\frac{1}{4}r^2 f'(r)F'(r) - \int_r^\infty pf'(p)F'(p)dp$$

$$< \lambda_k(\mathbf{x})\lambda_k(\mathbf{x}+\mathbf{r}) >= 6f(r)F(r) + 2r\{f(r)F(r)\}' + \frac{1}{2}r^2 f'(r)F'(r)$$

In particular

$$< (\nabla\alpha)^2 >= \frac{1}{2} < \lambda^2 > - \int_o^\infty pf'(p)F'(p)dp, \qquad < \lambda^2 >= 6f(0)F(0).$$

For example if $f(r) = f(0)e^{-\alpha r^2}, F(r) = F(0)e^{-\beta r^2}$,

$$< (\nabla\alpha)^2 >= (\frac{1}{2} - \frac{1}{6}\frac{\alpha\beta}{(\alpha+\beta)^2}) < \lambda^2 >,$$

$$i.e. \frac{11}{24} < \lambda^2 > \leq < (\nabla\alpha)^2 > \leq \frac{1}{2} < \lambda^2 > .$$

One sees that in this case again the contribution of the potential and solenoidal parts to $< \lambda^2 >$ is almost equal. It is therefore of particular interest to consider the *Case C*. Here the right hand side of (2-3) is represented in terms of fourth order derivatives of expressions of the type

$$< u_i(\mathbf{x})u_j(\mathbf{x})u_k(\mathbf{x}+\mathbf{r})u_\ell(\mathbf{x}+\mathbf{r}) >=< u_i(\mathbf{x})u_j(\mathbf{x}) >< u_k(\mathbf{x})u_\ell(\mathbf{x}) >$$

$$+ < u_i(\mathbf{x})u_k(\mathbf{x}+\mathbf{r}) >< u_j(\mathbf{x})u_\ell(\mathbf{x}+\mathbf{r}) >$$

$$+ < u_i(\mathbf{x})u_\ell(\mathbf{x}+\mathbf{r}) >< u_j(\mathbf{x})u_k(\mathbf{x}+\mathbf{r}) >,$$

i.e. the velocity field is supposed to be Gaussian (for the discussion of departures from Gaussianity see Herring and Kerr(1989)). The algebra in this case is much more lengthy than in the previous one with the following result

$$< \frac{\partial\alpha(\mathbf{x})}{\partial x_k}\frac{\partial\alpha(\mathbf{x}+\mathbf{r})}{\partial x_k} >= \frac{1}{2} < \lambda_k(\mathbf{x})\lambda_k(\mathbf{x}+\mathbf{r}) > +12\int_r^\infty \frac{1}{p}f'^2 dp$$

$$-11f'^2 + 4rf'f'' - \frac{r^2}{4}f''^2 - \frac{r^2}{4}f'f'', \qquad (2-4)$$

$$< \lambda_k(\mathbf{x})\lambda_k(\mathbf{x}+\mathbf{r}) >= -\frac{16}{r}ff' - 14ff'' - 9f'^2$$

$$-8rf'f'' - \frac{r^2}{2}f''^2 - 2rff''' - \frac{r^2}{2}f'f''' \qquad (2-5)$$

In particular (note that $f''(0) < 0$)

$$< (\nabla\alpha)^2 >= \frac{1}{2} < \lambda^2 > +12\int_0^\infty \frac{1}{p}f'^2 dp,$$

$$< \lambda^2 >= -30f(0)f''(0).$$

For example if $f = f(0)e^{-\alpha r^2}$ then

$$< (\nabla\alpha)^2 >\approx 0.7 < \lambda^2 >$$

$$< (\nabla \times \beta)^2 >\approx 0.3 < \lambda^2 >$$

It is remarkable that in the case when $\omega = \nabla \times \mathbf{u}$ the potential part of the Lamb vector $\lambda = \omega \times \mathbf{u}$ is considerably larger than its solenoidal part. Relations between the power spectra of $\nabla\alpha, \nabla\times\beta$ and $\lambda$ and the power spectrum of $\mathbf{u}$ can be obtained from (2-4) and (2-5). It is also of interest to find the correlation coefficient between $\lambda$ and $\nabla\alpha$. Since $< \nabla\alpha \cdot \nabla \times \beta >= 0$ it follows immediately that

$$< \lambda \cdot \nabla\alpha > / < \lambda^2 >^{\frac{1}{2}}< (\nabla\alpha)^2 >^{\frac{1}{2}} = (< (\nabla\alpha)^2 > / < \lambda^2 >)^{\frac{1}{2}} \approx 0.84$$

$$< \lambda \cdot \nabla \times \beta) > / < \lambda^2 >^{\frac{1}{2}}< (\nabla \times \beta)^2 >^{\frac{1}{2}} \simeq 0.55$$

In the dynamical context, since $\nabla \cdot \mathbf{u} = 0$ and $\nabla^2(p + \frac{1}{2}u^2) = -\frac{\partial\lambda_i}{\partial x_i}$ the potential part of $\lambda$ is identified with $(p + \frac{1}{2}u^2)$, i.e. $\alpha \equiv -(p + \frac{1}{2}u^2)$. Therefore, the potential part of the total nonlinear term $\lambda + \nabla p$ in the Navier-Stokes equations vanishes as can be checked also by direct calculations as above. In other words, the mean square of the total nonlinear term $< (\lambda + \nabla p)^2 >=< \lambda^2 > - < (\nabla p)^2 >$ (c.f. eq.(2) in Shtilman & Polifke (1989)).

## 3. Concluding Remarks and Implications

The results for the case C show that the Lamb vector $\lambda = \omega \times \mathbf{u}, \omega = \nabla \times \mathbf{u}$ for a random Gaussian solenoidal vector field $\mathbf{u}$ is 70% potential in the sense that 70% of the mean square $< \lambda^2 >$ is contributed by its potential part. The first implication of this result is that there is a substantial reduction of nonlinearity in turbulent flows due to purely kinematic constraints in agreement with computations by Shtilman and Polifke (1989) in the sense that the potential part of the Lamb vector can be included into the pressure gradient term. We see this reduction can be as large as 70%. As we already mentioned in the introduction, the dynamics of the flow is such that this tendency is amplified further (Kraichnan & Panda (1988), Shtilman and Polifke (1989)). It is remarkable that the "remaining" 30% (or even less in real turbulent flows) are able to produce such strong effects as they do (vortex stretching, high level of turbulent dissipation, etc.). Another aspect is that, again for purely kinematic reasons, the turbulent flow is quite close (in the above sense) to the set of fixed points of the Euler equations, which imposes certain constraints on the topology of the flow field, i.e. vectors $\omega$ and $\mathbf{u}$ are not "far" from being tangent to $\alpha$-surfaces, where $\alpha$ is the potential of the potential part of the Lamb vector $\lambda$. This property, enchanced by the flow dynamics, strengthen the belief that studying of steady-state Euler flows (i.e. such that $\omega \times \mathbf{u} = \nabla\alpha$ and in particular Beltrami flows) are of special importance for the analysis of the development of spatial turbulence(Zaslavsky et al. (1988)). The results reported here show that purely kinematic properties (e.g. Bio-Savart law (Pumir and Siggia (1989)), can be important in understanding of turbulent flows. It is also noteworthy that, as follows from results for the case C, flows with the property $\omega \times \mathbf{u} = \nabla\alpha$ (for discussion of such and similar flows see Tsinober (1989)) should be anisotropic and/or nongaussian, while it is easy to show that Beltrami flows can't be isotropic. Finally, a quite speculative implication is that the above result provides a feeling as to how nonlinear turbulent flow can be. It is likely that further progress in understanding of turbulence can be achieved via studying the *structure* (temporal, spatial, spectral) of the fields $\alpha, \beta$ and their derivatives in actual turbulent flows, e.g. the relation between $< (\nabla\alpha)^2 >$ and $< (\nabla \times \beta)^2 >$ in the so called coherent structures, etc.


### Acknowledgements

The author is grateful to E. Kit and L. Shtilman for stimulating discussions and to G.K. Batchelor and H.K. Moffatt for their interest and comments. The final stage of this work was supported by SERC during the summer stay of the author in DAMTP, Cambridge, UK.

## References

Arnold, V.I. 1986 The asymptotic Hopf invariant and its applications, *Sel. Math. Sov.* **5**, 327-345; translation from a Russian paper originally published in 1974.

Batchelor, G.K. 1951 Pressure fluctuations in isotropic turbulence. *Proc. Cambridge Philos. Soc.*, **47**, 359-374.

Dracos, T., Kholmyansky, M., Kit, E. and Tsinober, A. 1989 Some experimental results on velocity-velocity gradients measurements in turbulent grid flows. To be published in the *Proceedings of the IUTAM Symposium on Topological Fluid Mechanics*, held in Cambridge, August 13-18, 1989, eds. H.K. Moffatt and A. Tsinober, Camb. Univ. Press.

Herring, J.R. and Kerr, R.M. 1989 Numerical simulation of turbulence. *Theoretical and Applied Mechanics*, eds. Germain, P., Piau, M. and Caillerie, D., Elsevier (North-Holland), Amsterdam - New York - Oxford - Tokyo, pp. 101-116.

Kraichnan, R.H. and Panda, R. 1988 Depression of nonlinearity in decaying isotropic turbulence. *Phys. Fluids* **31**, 2395-2397.

Moffatt, H.K. 1985, 1986a Magnetostatic equilibria and analogous Euler flows or arbitrary complex topology. Part I. Fundamentals. *J. Fluid Mech.* **159**, 359-378. Part 2. Stability considerations. *J. Fluid Mech.* **166**, 359-378.

Moffatt, H.K. 1986b On the existence of localized rotational disturbances which propagate without change of structure in an inviscid fluid. *J. Fluid Mech.* **173**, 289-292.

Moffatt, H.K. 1989 Fixed points of turbulent dynamical systems and suppression of nonlinearity. *Whither turbulence workshop*, Cornell University, March 22-24 1989.

Pumir A. and Siggia, E. 1989 Collapsing solutions to the 3-D Euler equations. To be published in the *Proceedings of the IUTAM Symposium on Topological Fluid Mechanics*, held in Cambridge, August 13-18, 1989, eds. H.K. Moffatt and A. Tsinober.

Shtilman, L. and Polifke, W. 1989 On the mechanism of the reduction of nonlinearity in incompressible turbulent flows. *Phys. Fluids* **A1**, 778-780.

Tsinober, A., Kit, E. and Teitel, M. 1987 On the relevance of potential difference methods for turbulence measurements. *J. Fluid Mech.*, **175**, 447-461.

Tsinober, A. 1989 Turbulent drag reduction versus structure of turbulence. To be published in the *Proceedings of the 2nd Symposium on*

*Structure of Turbulence and Drag Reduction*, held in Zurich, July 25-28, 1989, ed. A. Gyr, Springer.

Zaslavsky, G.M., Sagdeev, R.Z. and Chernikov, A.A. 1988 Stochastic nature of streamlines in steady-state flows. *Soviet Phys. JETP*, **67**(2), 270-274; Chernikov, A.A., Sagdeev, R.Z. and Zaslavsky, G.M. 1988 Chaos: How regular can it be? *Physics Today*, Nov. 27-30; Beloshapkin, V.V. et al 1989 Chaotic streamlines in pre-turbulent states, *Nature*, **337**, 133-137.

# IV. Chaos and Instability

# Bifurcations and bursting of streaks in the turbulent wall layer

N. AUBRY and S. SANGHI

*Benjamin Levich Institute and*
*Department of Mechanical Engineering*
*City College of the City University of New York, New York, U.S.A.*

*Turbulence and Coherent Structures*
*O. Métais and M. Lesieur, eds.*
*Selected papers from "Turbulence 89: organized structures and turbulence in fluid mechanics"*
*Grenoble, 18-21 September 1989*

**Abstract**

The dynamics of infinitely long steamwise rolls in the wall region of a turbulent boundary layer have been described by a ten dimensional representation of Aubry *et al.* (1988). In this model, the rolls undergo intermittent cross stream dynamics which qualitatively resemble the cyclic bursting event experimentally observed. The explosion phenomenon in this model is due to the presence of a heteroclinic cycle in the phase space, a connection between saddle points, where each saddle represents streamwise rolls. We demonstrate here that the dynamics of streamwise rolls of finite length is also strongly intermittent. This intermittency is still of the heteroclinic type. Streamwise modes, absent in Aubry *et al.*'s model and included in the present study, participate to the explosion events, corresponding to a streamwise burst in the physical space. A stability analysis of the systems linearized at the fixed points which represent steady rolls is presented in dynamical systems of dimensions as high as 32 and 54 so that bifurcations to intermittency are clearly identified. We show that all non zero streamwise Fourier modes eventually burst, resulting in a streamwise energy cascade during the bursts. Although the explosion of the zero and first non zero streamwise Fourier modes occur through bifurcations of the system linearized about the fixed points (i.e. the rolls), the activation of all higher modes is a result of non linear phenomena.

## 1. Introduction

Despite the contribution of many researchers, the mechanism of turbulence production in near wall turbulence is poorly understood. It seems, however, well established that a large fraction of the production occurs during violent explosions (or "bursts") of otherwise quiescent streamwise streaks (see for example Cantwell 1981). These streaks have been first visualized by Kline *et al.* (1967) using hydrogen bubbles and they have been associated with streamwise vortices (Bakewell and Lumley 1967, Blackwelder and Eckelman 1979, Hogenes and Hanratty 1982, Kim 1985, Adrian *et al.* 1987, Moin and Moser 1989, Guezennec 1989). The structure of near wall turbulence has recently received considerable interest and significant progress has been made with the generation of highly detailed numerical simulation data (Moin and Kim 1982, Spalart 1988). The complexity of the data, however, has made their interpretation extremely difficult in terms of coherent structures and their dynamics so that the formation of the streaks together with their dynamics are not well described. There is a need for simplified numerical simulations and models. Jimenez and Moin (1989) carried out an extremely reduced simulation in which the spanwise periodic length is about 100 wall units so that one streak can be isolated and studied in detail. The striking result is that this coarse simulation generates approximately right first and second order statistics (see also Moin at this conference).

One of the most detailed studies of the structure of the wall region is that of Robinson *et al.* (1988) who analyzed Spalart's numerical data. The different instabilities, however, are rather absent there because they are not well described by direct numerical simulations. The generally accepted idea is "a violent small-scale secondary instability on a large-scale primary disturbance" (Landahl 1975). Jang *et al.* (1986) proposed that streak formation or "primary" instability is due to a resonnance between the vertical vorticity (Squire) mode and the Orr-Sommerfeld mode occurring at a spanwise wavenumber of 90 wall units, which is in agreement with experimental observations. We will not address this question in the present paper. The streak breaking or "secondary instability" is still an open question. One postulates that it is generated by an inflectional instability of the streamwise velocity profile observed in visualization experiments (Kline *et al.*, 1967) and numerical data (Moin and Kim, 1982). Jimenez (1987), while carrying out a two dimensional (streamwise/vertical) numerical simulation of a channel flow, found that ejection of vorticity (necessarily $\omega_z$) away from the wall was driven by the updraft induced by the Tollmien-Sclichting wave. Jimenez *et al.* (1988) examined data generated from three dimensional direct numerical simulations and found similar inclined shear layers of $\omega_z$. They argue that "the mechanism for the generation of ejections in the sublayer may be essentially two dimensional although there are undoubtedly three dimensional effects present".

Another model is that of Aubry *et al.* (1988) who derived a set of ten ordinary differential equations for the infinitely long streamwise rolls close to a wall. The rolls, extracted from experimental two point correlations and identified with the first proper orthogonal (or Kharhunen-Loève) eigenmode, exhibit intermittency whose nature is clearly identified: it is due to the presence of a heteroclinic cycle connecting hyperbolic fixed points. The solution spends a relatively long time near a fixed point where it stays quasi-steady, oscillates in a growing oscillatory motion and suddenly bursts through a brief jump along a heteroclinic orbit to another fixed point (only translated in the cross-stream direction). In the physical space, streamwise rolls, quiescent for a long time, starts to oscillate, suddenly burst into other structures in the cross-stream direction and get reformed. This strongly intermittent dynamics repeats itself an infinite number of times. The influence of noise of any kind was clearly identified: it triggers the eruption event earlier and the time between two events, which otherwise keeps increasing, equilibrates. In the theoretical model as well as in the boundary layer, a large fraction of the noise comes from the outer layer whose role then becomes clear.

Until now, there was no distinction in the literature between the large scale breakup and the small scale generation, all was called "secondary instability". In this paper, we will make a clear distinction since the instability mechanisms in both phenomena are found to be quite different.

## 2. The Model

Our model is an extension of that of Aubry *et al.* (1988) in which the streamwise series is no longer truncated to the zero Fourier mode alone. While it is now clear that streamwise variations are not necessary for the appearance of the first destabilization of the rolls, they undoubtedly participate in the overall dynamics and statistics. For the latter, we can refer to the streamwise broad band spectrum of Moin and Kim (1982). For the former, visualization experiments have shown that fluid particles are carried downstream along the streaks over a distance of about 1000 wall units until they suddenly get ejected and "spread out in all directions" (Corino and Brodkey, 1969).

Another issue of Aubry *et al.* (1988) is the robustness of the proposed intermittency mechanism to the order of truncation. One knows that in many cases the behavior of a finite or low dimensional dynamical system is a consequence of the low order of truncation. Curry *et al.* (1984) who extended the Lorenz system to a full simulation of the Boussinesq equations did not observe any strong chaos at high resolution. This is a typical example in which qualitative dynamics is a product of inadequate spatial resolution. Although Aubry *et al.*'s low dimensional model was able to reproduce qualitatively the basic bursting event of the near wall region, it showed a certain number of deficiencies among which we can mention the improper distribution of energy in the three velocity components, the weak turbulence production during the bursts and the quasi-steadiness of the motion between the bursts (see Aubry 1987).

The derivation of the model is similar to that of Aubry *et al.* (1988). The set of ordinary differential equations derived is considerably more complex and has been presented in Aubry (1987) and briefly in Aubry and Sanghi (1989). The flow studied is the wall region: $0 \leq y^+ \leq 40$ ($^+$ denotes wall variable normalization with kinematic viscosity $\nu$ and friction velocity $u_\tau$) in a pipe flow of Reynolds number 8750 based on the centerline velocity and the diameter of the pipe. The spatial two point correlation tensor has been measured by Herzog (1986). As Aubry *et al.* (1988), we are using proper orthogonal (or Karhunen-Loève) eigenfunctions $\phi$ extracted from these measurements in the wall normal direction and Fourier modes in the other directions to derive sets of dynamical equations for time dependent coefficients. More specifically, the fluctuating velocity is expanded as follows:

$$u_i(x_1, x_2, x_3, t) = \frac{1}{\sqrt{L_1 L_3}} \sum_{lkn} a_{lk}^{(n)}(t)\, e^{2\pi i(lx_1 + kx_3)} \phi_{i_{\mathbf{k}}}^{(n)}(x_2), \quad (1)$$

where $l, k$ are the streamwise and spanwise wavenumbers and $\mathbf{k}$ is the wavenumber vector. The index $n$ denotes the order of the eigenmode. Only six (positive) spanwise Fourier modes ($L_3^+ = 333$) and one proper orthogonal $n = 1$ are considered here as in Aubry *et al.*(1988). The eigenfunctions are extracted from the equation:

$$\int \Phi_{ij_{\mathbf{k}}}(x_2, x_2')\phi_{j_{\mathbf{k}}}^{(n)}(x_2')dx_2' = \lambda_{\mathbf{k}}^{(n)} \phi_{i_{\mathbf{k}}}^{(n)}(x_2), \quad (2)$$

where $\Phi_{ij}$ the Fourier transform of the autocorrelation tensor in the $x_1, x_3$ directions at zero time lag: $R_{ij}(x_1 - x_1', x_2 - x_2', x_3 - x_3')_{t-t'=0}$. The energy transfer to the unresolved scales is represented by a simple generalization of Heisenberg spectral model. Such a representation is fairly crude but it has been used in numerical simulations with considerable success. The loss of energy is adjusted upward and downward by introducing a coefficient $\alpha$ ($\alpha$ is referred as the "Heisenberg" parameter) which is varied. This model is similar to that of the ten dimensional system of Aubry *et al.* (1988) but the numerical values of the computed terms are different due to the addition of the extra modes (see Aubry and Sanghi 1989).

As recalled above, Aubry *et al.* (1988) identified the role of the fluctuating pressure at the top of the layer. Because this role, although not negligible, does not affect the qualitative nature of the intermittent solutions, the fluctuating pressure term will be omitted in this study.

## 3. Properties of the ODEs

The properties of the new set of ODEs can be found in Aubry (1987) and Aubry and Sanghi (1989). Let us mention two fundamental properties which will be used in the next section for the description of the solutions: 1. the existence of $S^1$ families of solutions due to the physical statistical reflexion symmetry about a $x_3 = 0$ plane (contained in the experimentally measured two point correlation tensor) 2. the existence of invariant subspaces, in particular the $l = 0$ subspace and the even-$k$ subspace.

The invariance of the former subspace ($l = 0$) is particularly important. It guarantees that the intermittent behavior exhibited by Aubry *et al.* (1988).'s ten dimensional system is still present together with its intermittent dynamics. However, this subspace could be globally unstable so that these solutions would no longer be observed in the long time dynamics. It seems necessary to emphasize that by no mean this subsystem represents a 2 1/2 dimensional real fluid flow with no streamwise dependency. Our real fluid flow, the one measured by Herzog, is of course fully three dimensional and thus the extracted proper orthogonal eigenfunctions $\phi_{lk}^{n}$ are characteristic of a three dimensional velocity field. In particular, the coefficients $a^{(n)}$ do not depend on the components $u_i$ $(i = 1, 2, 3)$. The restriction of Aubry *et al.* (1988) to the first streamwise Fourier modes in the velocity expansion was only a strategy to retain the least number of modes possible. This was justified physically by the large streamwise length scale of the streaks and mathematically by the invariance of the $l = 0$ subspace so that solutions of larger systems with initial conditions in this subspace would remain in this subspace. Analyses restricted to invariant subspaces is a usual procedure in dynamical systems theory. Neither the dynamical equations of Aubry *et al.* (1988) nor the $l = 0$ subsystem of our present models is representative of a flow without streamwise variations. If this was the case, as pointed out by Moffatt (1990), the first component of the Navier Stokes equations should decouple from the second and third components and turbulence would simply decay. This would be possible only if components 1 and 3 of the eigenfunctions decoupled (in equation (2)) at zero streamwise wavenumber as they do at zero spanwise wavenumber for symmetry reasons (see Aubry 1987). Because the experimentally measured correlation tensor is that of a fully three dimensional turbulent field in which there is no streamwise reflection symmetry, this does not happen and the velocity field is expanded in a "three dimensional way" as shown in equation (1) where all components are coupled, even in the $l = 0$ subspace (except at the wavenumbers $(l = 0, k = 0)$). In a decaying (thus non stationary) 2 1/2 dimensional flow, both the extraction of the eigenfunctions from experimental data and the derivation of the dynamical system would not be applicable without major modifications of these techniques. We are not considering this case here. The physical interpretation of this is that infinitely long streamwise rolls, whose plane of circulation is inclined upstream such that the Reynolds shear stress is negative, are able to maintain themselves. Turbulent energy production $-< u_1 u_2 > U_{,2}$ is injected into the $u_1$ equations and redistributed into the $u_2$ and $u_3$ components by coupling imposed by the experimental eigenfunctions.

## 4. Specific sets of equations

Addition of a single non zero streamwise Fourier mode (and its oposite) in the 10 dimensional system of Aubry et al. leads to 32 dimensions. Addition of two, three, four, five and six non zero Fourier modes gives 54, 76, 98, 120 and 142 dimensions respectively. Aubry (1987) showed that addition of two streamwise scales or more gave velocity variances and Reynolds shear stress approximating fairly well the original experimental data (Herzog, 1986). Improvement can be then achieved by addition of more proper orthogonal modes (which is not considered here). Although more systems have been derived and numerically integrated, we are reporting here in detail results obtained with 32 and 54 dimensional dynamical systems in a periodic cell of streamwise length $L_1^+ = 667$.

If the series in (1) was infinite or if the effect of smaller scales was ignored, the invariant zero streamwise mode subsystem would be identical to the Aubry et al.'s ten dimensional system. However, the energy transfer model which accounts for the loss of energy towards the unresolved modes makes the systems slightly different. Numerical simulations, however, show that the qualitative nature of the solutions is identical in both systems but the bifurcations occur at different values of the Heisenberg parameter $\alpha$. This is a trivial result. Recall that this representation changes, on the one hand, the linear terms by a modified viscosity, and, on the other hand, the quadratic terms, by the addition of a pseudo-pressure term. The non-dimensional transport coefficient $1/Re_T$ is computed from the first neglected modes in the three spatial directions. When the first neglected mode is higher than the last mode of the experimental series, the loss of energy in this particular direction is neglected since it cannot be computed (the last modes of the experimental series are sufficiently high enough to make it effectively negligible). Because the transport coefficient $1/Re_T$ represents the global action of unresolved modes, it is an increasing function of the neglected energy and a decreasing function of the cut-off wavenumber/eigenmode. In the zero streamwise mode model, the energy contained in all the experimental streamwise Fourier modes was injected into the zero streamwise wavenumber by simple adjustment of the periodic streamwise length. In our model, the value of the streamwise length $L_1$ is imposed by the streamwise cut-off which is smaller than the experimental one. The energy included in the explicit modes is thus lower and the transport coefficient larger than in the zero streamwise mode model (in which it was 6.28). The latter decreases as more modes get included in the same periodic box and takes the values 9.92 and 7.00 in the 32 and 54 dimensional systems respectively. The values of the coefficients of the additional quadratic terms introduced by the pseudo-pressure term in the Heisenberg representation are also affected. We report in table 1 the critical values of $\alpha$ at which bifurcation branches emerge from the trivial branch. These values correspond to the critical values $\alpha^c$ of each linear term taken independently. They increase as more modes get included (since $1/Re_T$ decreases).

| l | k | $\alpha^c_{l,k}$ (32 dim.) | $\alpha^c_{l,k}$ (54 dim.) |
|---|---|---|---|
| 0 | 1 | 1.48 | 2.08 |
| 0 | 2 | 1.53 | 2.16 |
| 0 | 3 | 1.39 | 1.97 |
| 0 | 4 | 1.23 | 1.74 |
| 0 | 5 | 1.03 | 1.45 |
| 1 | 0 | 0.87 | 1.28 |
| 1 | 1 | 1.31 | 1.91 |
| 1 | 2 | 0.85 | 1.19 |
| 1 | 3 | 0.54 | 0.74 |
| 1 | 4 | 0.55 | 0.76 |
| 1 | 5 | 0.56 | 0.78 |

Table 1. Bifurcation values of $\alpha$ from the trivial solution in the 32 and 54 dimensional systems.

The critical values of $\alpha^c$ for the zero streamwise modes are in the same order as in the ten dimensional model, that is the (0,2), (0,1), (0,3), (0,4), (0,5) modes bifurcate successively; this is due to the fact that the $l = 0$ subsystem is very similar to the ten dimensional system.

## 5. 32 dimensional dynamical system

### 5.1. Results of Numerical Integrations

We now outline the behavior of typical solutions as the Heisenberg parameter $\alpha$ is reduced.

When $\alpha$ is larger than 1.53, the trivial solution is the global attractor. For $1.52 \leq \alpha \leq 1.51$ we observe a non zero fixed point in the ($l = 0$, even $k$) subspace. As $\alpha$ decreases from 1.50 to 1.22, the solution becomes intermittent in the zero streamwise mode subspace. This intermittency is similar to window I of intermittency present in the ten dimensional model of Aubry *et al.* (1988) and rather simple. For $1.22 \leq \alpha \leq 1.06$, the solution is steady again and is a combination of the modes (0,2) and (0,4) only. Another window (II) of intermittency appears for $1.05 \leq \alpha \leq 0.88$. This intermittency is of the same type as that of window II of intermittency in the ten dimensional model. It occurs in the zero streamwise mode subspace, the non zero streamwise modes decaying very fast. As in the window II of intermittency of the ten dimensional model, the (0,2)/(0,4) mixed modes are destabilized in the (0,1)/(0,3)/(0,5) directions. This solution encounters bursts involving growing oscillatory motions of the spanwise even and odd modes (at constant modal phases), corresponding to cross-stream oscillations of the streamwise rolls followed by a sudden breakup. Until now, the sequence and nature of the solutions observed have been very similar to those observed in the ten dimensional model (see Aubry *et al.* (1988) for time series), indicating that the zero streamwise mode subspace is attracting for the whole system. In these solutions, the non zero streamwise modes are inactive after a very short transient during which they decay to zero. As a decreases more, the invariant zero streamwise subspace becomes unstable. For $0.80 \leq \alpha \leq 0.89$, we observe a third window of intermittency. Between the bursts, the solution is the same as in the first and second windows, almost steady since it stays in a neighborhood of one equilibrium in the zero streamwise/even spanwise subspace. However, it changes significantly during the bursts since *all* the modes then participate to the dynamics. Figure 1 reproduces some time series.

Different projections of the phase space clearly show presence of a heteroclinic cycle connecting fixed points in the (0,2)/(0,4) subspace (see Aubry and Sanghi, 1989). For $0.80 \leq \alpha \leq 0.50$, the solution is attracted to the zero streamwise subspace where it becomes much more complex, apparently chaotic. The heteroclinic cycle starts travelling and then becomes unstable. At this time, the non zero streamwise modes collapse. For lower values of $\alpha$, the $(l = 0)$-subspace is unstable again and a fourth window of intermittency arises. In the intermittent behavior previously described, the solution between the bursts was always quasi steady due to the proximity of a fixed point. The new solution is different in nature: it is very disorganized, apparently chaotic between the bursts in the zero streamwise subspace and this degree of complexity increases during each burst due to the sudden excitement of the non zero streamwise modes of high frequency motions (see figure 2).

We speculate that the cause of the intermittency is the presence of an attracting heteroclinic cycle (in a generalized sense) which connects hyperbolic, apparently chaotic, objects. The saddle is no longer a fixed point but the complicated region of the zero streamwise subspace, the attractor (perhaps strange) of the solution when the initial condition lies in the zero streamwise subspace. For an initial condition outside of any invariant subspace, the solution leaves the ten dimensional zero streamwise subspace, starts a brief excursion in the full 32 dimensional phase space and comes back to the zero streamwise subspace.

We summarize in table 2 the different solutions observed. **0** denotes the zero complex number.

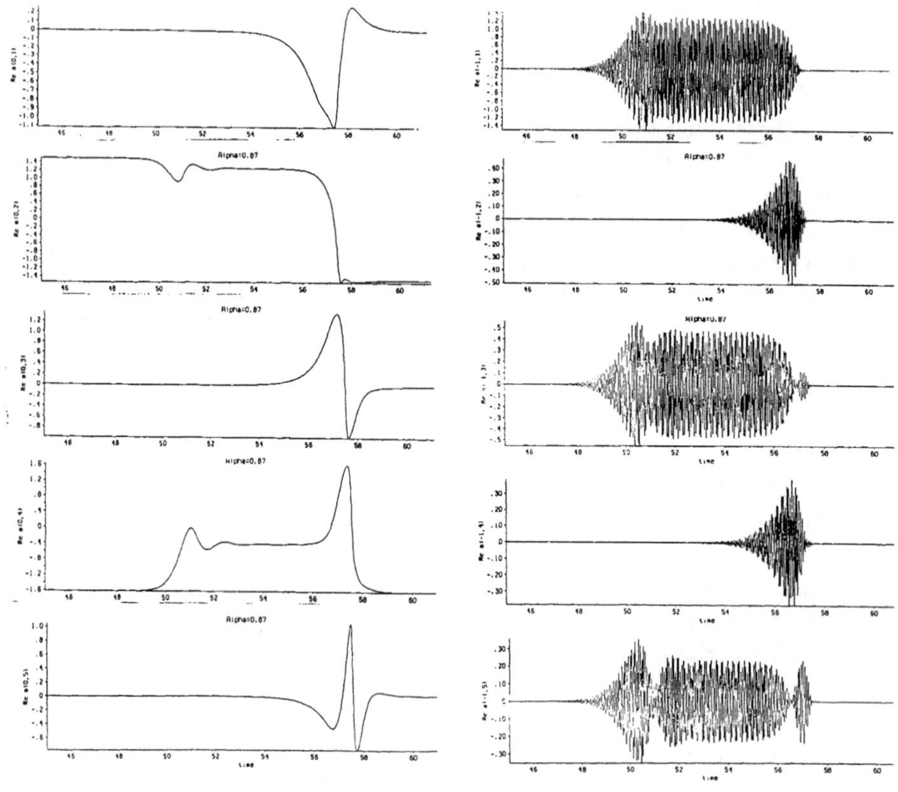

Fig.1. Typical solution in window III of intermittency of the 32 dimensional system. a. Time series of the real parts of $a_{01}, a_{02}, a_{03}, a_{04}, a-11, a-12, a-13, a-14, a-15$ for $\alpha = 0.87$. b. Projection of the phase plane on the (Re($a_{02}$), Re($a04$)) plane showing heteroclinic cycle involving bursting of non zero streamwise modes (window III of intermittency).

Vectorial representations of the velocity field in cross-sections of the flow at zero streamwise location are displayed in figures 3 and 4 for typical solutions in windows III and IV of intermittency respectively.

In window III, before and after the burst, the field is steady and represents two pairs of streamwise rolls in the periodic box of cross-stream length $L_3^+ = 333$. The rolls are symmetric with respect to a $(x_1, x_2)$ plane. The first and last pictures are shifted in the spanwise direction by $L_3^+/4$. During the burst, we can observe variations of positions and

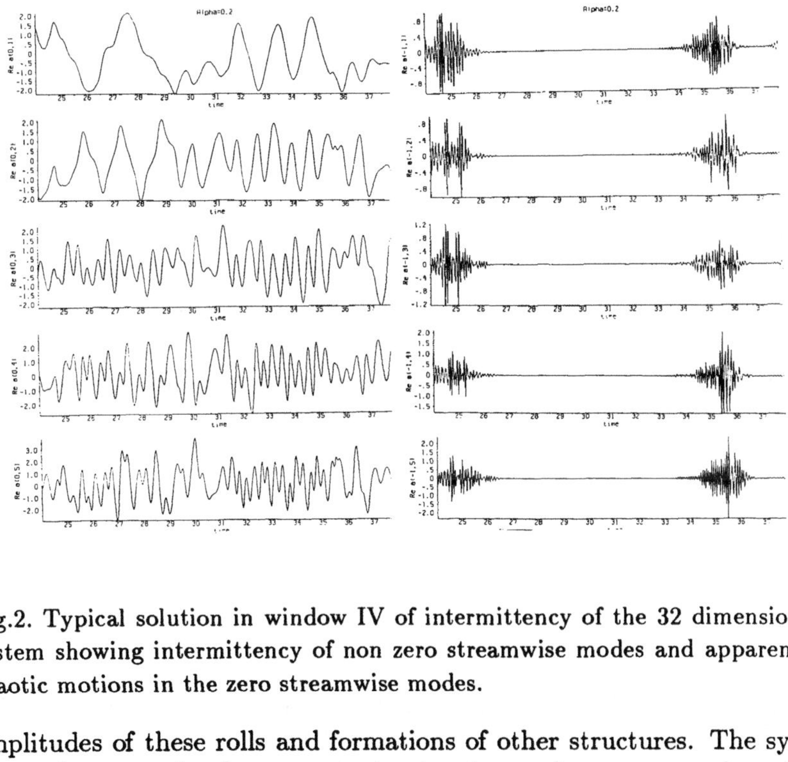


Fig.2. Typical solution in window IV of intermittency of the 32 dimensional system showing intermittency of non zero streamwise modes and apparently chaotic motions in the zero streamwise modes.

amplitudes of these rolls and formations of other structures. The symmetry, however, is always present: structures always appear in pairs. The evolution of the flow cross-section in the fourth window of intermittency shows that the solution is very disorganized and asymmetric even between the bursts (see figure 4).

## 5.2. Linearized stability of the fixed points

Since the first three types of intermittency appear through bifurcations from fixed points, we are investigating the nature of these bifurcations. This was done in Aubry *et al.* (1988) for the ten dimensional dynamical system in which the second window of intermittency appeared through a

| Heisenberg parameter $\alpha$ | behaviour |
|---|---|
| $\alpha \geq 1.53$ | trivial solution $a_{l,k} = \mathbf{0}, \forall(k,l)$ is the global attractor |
| $1.52 \geq \alpha \geq 1.51$ | stable fixed point $a_{02}, a_{04} \neq 0$,other modes are zero |
| $1.50 \geq \alpha \geq 1.22$ | intermittency in the $a_{0,k}$ subspace: $a_{1,k} = \mathbf{0}, \forall k$; $a_{02}, a_{04} \neq \mathbf{0}$; $a_{0,k} = 0, \forall k \neq 2,4$ between bursts; $a_{0,k} \neq \mathbf{0}, \forall k$ during bursts (no oscillations) |
| $1.22 \geq \alpha \geq 1.06$ | stable fixed point $a_{02}, a_{04} \neq 0$,other modes are zero |
| $1.05 \geq \alpha \geq 0.88$ | intermittency in the $a_{0,k}$ subspace: $a_{1,k} = \mathbf{0}, \forall k$; $a_{02}, a_{04} \neq \mathbf{0}$; $a_{0,k} = 0, \forall k \neq 2,4$ between bursts; $a_{0,k} \neq \mathbf{0}, \forall k$ during bursts ($a_{01}, a_{03}, a_{05}$ oscillate) |
| $0.87 \geq \alpha \geq 0.80$ | intermittency in the full space $a_{02}, a_{04} \neq \mathbf{0}$; $a_{0,k} = 0, \forall k \neq 2,4$ between bursts, $a_{l,k} \neq \mathbf{0}, \forall l,k$ during bursts |
| $0.80 \geq \alpha \geq 0.50$ | complex behavior in the $a_{0,k}$ subspace: $a_{1,k} = \mathbf{0}, \forall k$; $a_{0,k} \neq \mathbf{0}, \forall k$ |
| $0.50 \geq \alpha$ | intermittency in the full space; complex behavior in the $a_{0,k}$ subspace between bursts; $a_{l,k} \neq \mathbf{0}$, $\forall(l,k)$ during bursts |

Table 2. Qualitative dynamical behavior in the 32 dimensional system.

subcritical Hopf bifurcation. As in their study, we can limit our stability calculation to one of the two real fixed points: $a_{02} = -r_2, a_{04} = -r_4$ where $r_2$ and $r_4$ are strictly positive values and all other components are zero without loss of generality. The result will be the same for any element of the symmetric $S^1$ circle of fixed points. Linearization of the full system at this fixed point leads to diagonalization of the $32 \times 32$ block matrix as follows : ($Re(a_{02})$,$Re(a_{04})$), a $2\times2$ matrix, ($Im(a_{02})$,$Im(a_{04})$), a $2 \times 2$ matrix, ($Re(a_{01})$, $Re(a_{03})$, $Re(a_{05})$), a $3 \times 3$ matrix, ($Im(a_{01})$, $Im(a_{03})$, $Im(a_{05})$), a$3 \times 3$ matrix, ($a_{-11}$, $a_{-13}$, $a_{-15}$, $a_{11}$, $a_{13}$, $a_{15}$), a $12 \times 12$ matrix and ($a_{-12}$,$a_{-14}$, $a_{10}$, $a_{12}$, $a_{14}$), a $10 \times 10$ matrix. The eigenvalues of each of these matrices are given in Appendix A. One eigenvalue of the imaginary part of $a_{02}$ and $a_{04}$ has always zero real part (as

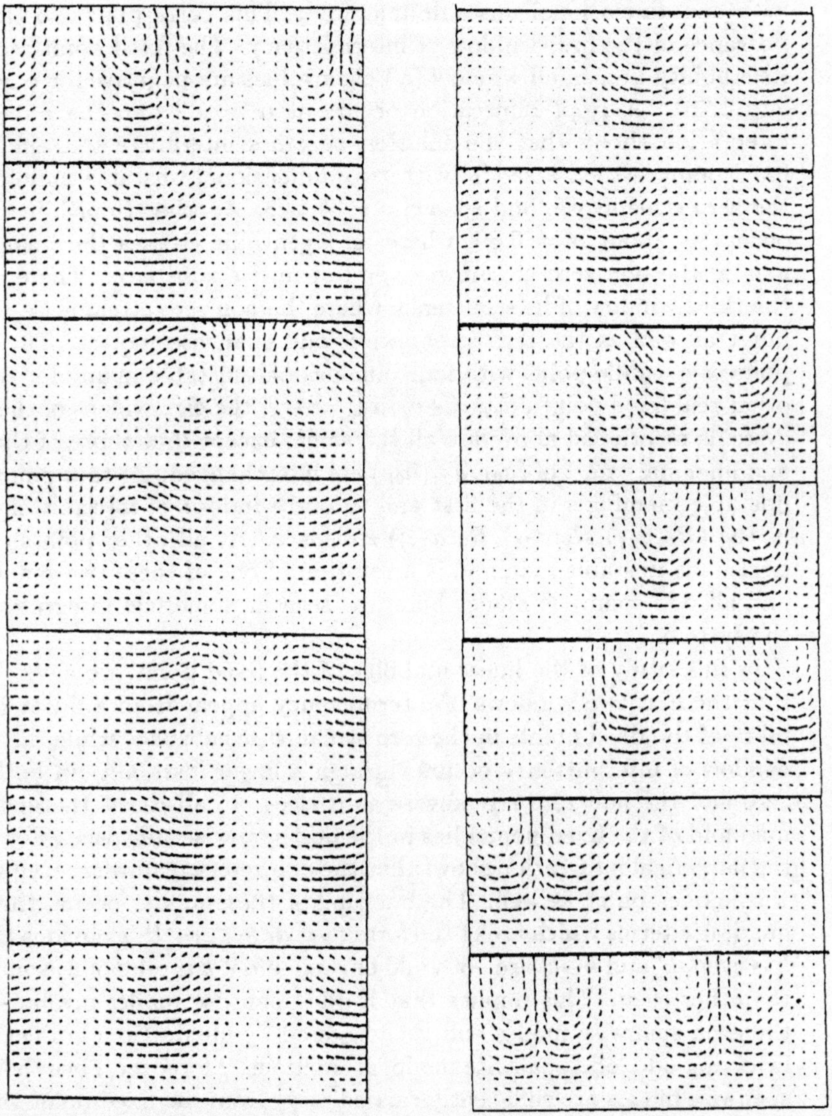

Fig.3. Vector representation of $u_2, u_3$ in a $(x_2, x_3)$-plane during a burst at $\alpha = 0.87$ as time increases.

in the ten dimensional system) which corresponds to variations along the $S^1$ circle of fixed points. Its eigenvector is tangent to this circle. The eigenvalues of the matrices involving the non zero streamwise wave number are always complex conjugates. For parameter values smaller than 1.49, all eigenvalues have negative real parts, showing that the fixed point is stable. The first bifurcation occurs at $\alpha = 1.50$ at which an eigenvalue becomes positive in the imaginary part of zero streamwise, odd spanwise mode subspace. The fixed point becomes a saddle with one dimensional unstable manifold. This corresponds to the appearance of the first window of intermittency. The fixed point recovers its stability in a small window (all eigenvalues are negative for $\alpha = 1.10$ and 1.20). A Hopf bifurcation occurs at $\alpha = 1.7$ and the numerical integration shows that the solution becomes intermittent again. The fixed point is then a saddle with two dimensional unstable manifold in the zero streamwise, odd spanwise subspace. Another Hopf bifurcation occurs at about $\alpha = 0.89$, where an eigenvalue crosses the imaginary axis in the non zero streamwise, odd spanwise subspace. There starts the third window of intermittency where the non zero streamwise modes burst as well as the zero streamwise, odd spanwise modes. The fixed point is a saddle point with four dimensional unstable manifold. At the other real fixed point , $a_{02} = +r_2, a_{04} = -r_4$, the eigenvalues of the subsystems mentioned above are all the same, except that those of the real and imaginary part of $(a_{01}, a_{03}, a_{05})$ are interchanged. At that point, the unstable manifolds in the first and second windows of intermittency lie in the $(Re(a_{01}), Re(a_{03}), Re(a_{05}))$ subspace. At all other points of the circle, the unstable manifold is a rotation of one of these two manifolds and all odd spanwise modes burst, as seen in numerical integrations of the system.

In this study of the linear stability of the fixed point, we have shown that the first two windows of intermittency appear from a linear instability of the fixed points in the zero streamwise subspace while the third window of intermittency occurs through a linear instability in both the zero and the non zero streamwise subspaces. In all cases, the unstable manifold of the fixed points lies in the odd spanwise subspace. However, the numerical integration shows that the non zero streamwise, even spanwise modes burst as well. The bursting of these modes occurs through quadratic terms which are of the form $a_{l'k'}\, a_{l-l',k-k'}$. It is obvious that if $k$ is even, $k'$ and $k-k'$ are both odd or even and if $k$ is odd, one is odd and the other even. This implies that if odd spanwise modes are bursting, the even spanwise modes will get triggered by quadratic non linearities involving odd/odd spanwise mode interactions. Moreover, once the even spanwise modes are sufficiently carried away from the fixed point values, the non linear even/even spanwise modes interactions also contribute to the burst. This is a non linear instability which is delayed in time (see

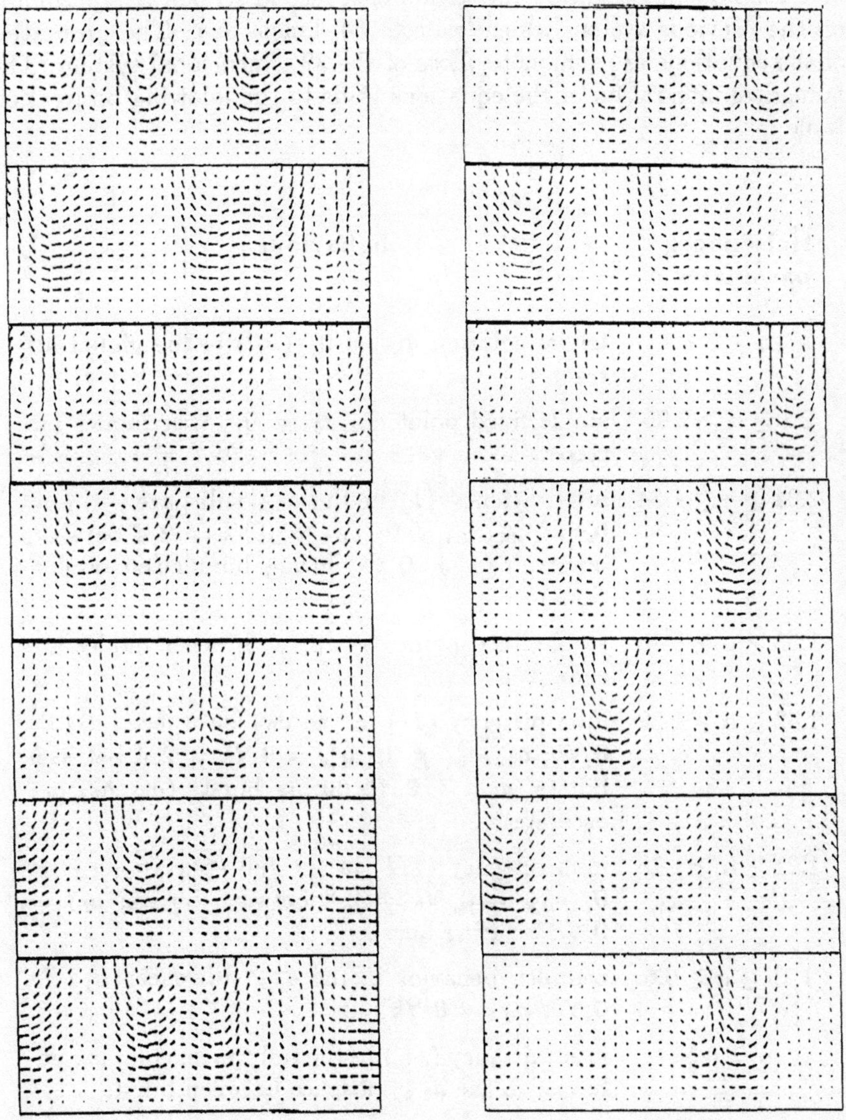

Fig.4. Vector representation of $u_2, u_3 ina(x_2, x_3)$-plane as time increases. Typical solution in window IV of intermittency of the 32 dimensional.

time series in figures 1 and 2).

## 6. 54 Dimensional Dynamical System

### 6.1. Numerical Integrations

We are now investigating the action of a second streamwise harmonic on the previous system which leads to 54 dimensions. The properties of the equations are similar to those of the 32 dimensional system. The numerical integration of the equations leads to the folowing bifurcation table :

| **Heisenberg parameter $\alpha$** | **behaviour** |
|---|---|
| $\alpha \geq 2.1$ | trivial solution $a_{l,k} = \mathbf{0}, \forall(k,l)$ is the global attractor |
| $2.1 \geq \alpha \geq 2.05$ | stable fixed point $a_{02}, a_{04} \neq 0$, other modes are zero |
| $2.04 \geq \alpha \geq 1.84$ | intermittency $(I)$ in the $a_{0,k}$ subspace: $a_{1,k} = \mathbf{0}, \forall k$; $a_{02}, a_{04} \neq \mathbf{0}$; $a_{0,k} = 0, \forall k \neq 2,4$ between bursts; $a_{0,k} \neq \mathbf{0}, \forall k$ during bursts (no oscillations) |
| $1.83 \geq \alpha \geq 1.43$ | stable fixed point $a_{02}, a_{04} \neq 0$, other modes are zero |
| $1.42 \geq \alpha \geq 1.22$ | intermittency $(II)$ in the $a_{0,k}$ subspace: $a_{l,k} = \mathbf{0}, \forall k$; $a_{02}, a_{04} \neq \mathbf{0}$; $a_{0,k} = 0, \forall k \neq 2,4$ between bursts; $a_{0,k} \neq \mathbf{0}, \forall k$ during bursts ($a_{01}, a_{03}, a_{05}$ oscillate) |
| $1.21 \geq \alpha \geq 1.18$ | intermittency $(III)$ in the full space $a_{02}, a_{04} \neq \mathbf{0}$; $a_{0,k} = 0$, $\forall k \neq 2,4$ between bursts; $a_{l,k} \neq \mathbf{0}, \forall l, k$ during bursts |
| $1.15 \geq \alpha \geq 0.6$ | complex behavior in the $a_{0,k}$ subspace: $a_{1,k} = \mathbf{0}, \forall k$; $a_{0,k} \neq \mathbf{0}, \forall k$ |
| $\alpha \cong 0.5$ | intermittency $(IV)$ in the full space; complex behavior in the $a_{0,k}$ subspace between bursts |

Table 2. Qualitative dynamical behavior in the 54 dimensional system.

Time series of typical solutions in window III are displayed in figure 5. Reconstructions of the velocity field are similar to those obtained with the 32 dimensional system for the same qualitative solutions.

### 6.2. Stability analysis of the fixed points

Similar to our stability analysis in the 32 dimensional dynamical system, we compute the eigenvalues of the Jacobian matrix at the fixed point $a_{02} = -r_2, a_{04} = -r_4$. Linearization of the full system leads to diagonalization of the $54 \times 54$ block matrix as follows: $(Re(a_{02}), Re(a_{04}))$, a $2 \times 2$ matrix, $(Im(a_{02}), Im(a_{04}))$, a $2{\times}2$ matrix, $(Re(a_{01}), Re(a_{03}), Re(a_{05}))$, a $3{\times}3$ matrix, $(Im(a_{01}), Im(a_{03}), Im(a_{05}))$, a $3{\times}3$ matrix, $(a_{-12}, a_{-14}, a_{10}, a_{12}, a_{14})$, a $10 \times 10$ matrix, $(a_{-11}, a_{-13}, a_{-15}, a_{11}, a_{13}, a_{15})$, a $12 \times 12$ matrix, $(a_{-21}, a_{-23}, a_{-25}, a_{21}, a_{23}, a_{25})$, a $12 \times 12$ matrix and $(a_{-22}, a_{-24}, a_{20}, a_{22}, a_{24})$, a $10 \times 10$ matrix. The eigenvalues of each of these matrices are given in Appendix B. As in the 32 dimensional dynamical system, one can see the bifurcations in which the first and second windows of intermittency appear in the $(Im(a_{01}), Im(a_{03}), Im(a_{05}))$ subspace. The third window emerges through a (subcritical) Hopf bifurcation occurring in the $(a_{-11}, a_{-13}, a_{-15}, a_{11}, a_{13}, a_{15})$ subspace. The eigenvalues of the second non zero streamwise mode subspace stay negative for all these parameter values. However, numerical integrations show that the second non zero streamwise mode bursts as well as the first one. It is easy to see that the second mode will get activated by non linear, quadratic interactions (this argument is similar to that used earlier for the excitation of even spanwise modes). This non linear instability is delayed. As the second non zero streamwise mode starts oscillating, the odd spanwise modes get destabilized first followed by the even spanwise modes (for the reason mentioned above).

### 6.3. Bursting of higher order streamwise modes

In view of the hierarchy of instabilities presented above, one can ask whether higher order streamwise harmonics will burst as well. Experimental visualizations have shown that sublayer bursting events are characterized by generation of fine scale turbulence. It is obvious that this cascade of energy towards finer scales is not limited to two harmonics. The study of the 54 dimensional dynamical system showed that once the first non zero streamwise mode bursts through a subcritical Hopf bifurcation, the second non zero streamwise mode bursts also by non linear interactions. Would higher harmonic be triggered in the same manner? Examination of the quadratic terms shows that all modes such that $l = 2$ is destabilized by the $l = 1/l = 1$ interactions for example. In the same way, the modes such that $l = 3$ are destabilized by $l = 1/l = 2$ interactions, the modes such that $l = 4$ are destabilized by $l = 1/l = 3$ and $l = 2/l = 2$ interactions, etc. This process, however, has its limit. Such

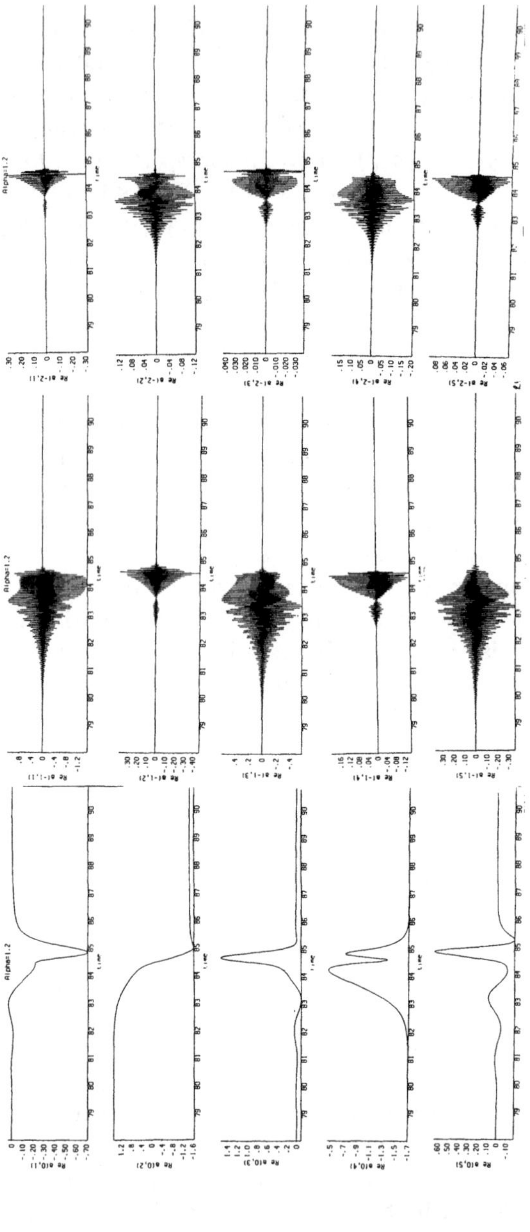

**Fig.5. Typical solution in window III of intermittency of the 54 dimensional system. Time series of the real parts of $a_{01}, a_{02}, a_{03}, a_{04}, a_{-11}, a-12, a-13, a-14, a-15, a-21, a-22, a-23, a-24, a-25$ for $\alpha = 0.87$.**

behavior occurs for ranges of the parameter $\alpha$ at which the larger scales destabilize. These values of $\alpha$ make the negative part of the linear terms in the small scale equations very large. When the non linear quadratic terms are strong enough in the bursting phenomenon to counteract these stabilizing terms, small scales start to burst as well. This destabilization is more and more difficult as scales get smaller. At smaller values of $\alpha$, higher modes bifurcate from the trivial solution. For these (possibly negative) values of $\alpha$ given by linear terms (see above), intersections of a multitude of bifurcation branches make the problem extremely difficult. We are not addressing this problem here.

The non linear triggering of higher streamwise Fourier modes has been demonstrated above in a 54 dimensional system, including 3 (positive) streamwise modes. Integrations of systems including higher Fourier and proper orthogonal modes have been performed, showing that intermittency persists for some range of the bifurcation parameter. Details on these higher dimensional dynamical systems will be reported elsewhere.

## 7. Concluding Remarks

It is now clear that the intermittent behavior displayed by the ten dimensional dynamical system of Aubry *et al.* (1988) in which streamwise variations were negected persists as non zero streamwise Fourier modes are included. As the bifurcation parameter (which adjusts the loss of energy towards the unresolved modes) decreases, four types of intermittency are observed. For large parameter values, the system displays intermittent behavior in the zero streamwise mode subspace only. As in the ten dimensional system of Aubry *et al.* (1988), infinitely long streamwise rolls burst in the cross-stream direction. For lower values of the parameter, intermittency is observed in the full space so that streamwise modes, mostly inactive, get excited during the explosion events, resulting into a streamwise burst in the physical space. The fourth window of intermittency occurs as the parameter decreases more, showing that the motions in the zero streamwise subspace is not necessarily close to a fixed point but can already be of a more complex, i.e.turbulent nature. There is thus no doubt that in this dynamical system intermittency is generic. It is identified with the presence of a heteroclinic cycle connecting saddles. In the first three windows, these saddles are fixed points; in the fourth window, they are more complex (apparently chaotic) manifolds. A linear stability study of the fixed points in the first three windows proved that intermittency appeared through clearly identified bifurcations. In the first window, the stable fixed point becomes unstable with one dimensional unstable manifold. In the second window, the stable fixed point becomes unstable through a subcritical Hopf bifurcation in the $(a_{01}, a_{03}, a_{05})$ subspace with two dimensional unstable manifold. In the third window, the already unstable fixed point gains unstable directions through a (subcritical) Hopf bifurcation in the $(a_{-11}, a_{-13}, a_{-15}, a_{11}, a_{13}, a_{15})$ subspace. The unstable manifold of the fixed point becomes four dimensional. The excitation of higher Fourier modes occurs through (quadratic) non linearities for this range of the parameter $\alpha$.

The quasi-steadiness of the rolls motions between the bursts is of course the result of the extreme simplicity of the model. Undoubtedly, a fully developed turbulent flow cannot be steady, even for small periods of time, but keeps fluctuating randomly. The fourth type of intermittency shows that this more realistic situation appears in our model: The velocity field is apparently chaotic, involving only a certain number of scales and suddenly bursts into even more complex behavior involving the totality of the modes included in the model. Here, the orbit connects apparently chaotic, globally unstable regimes. This chaotic regime had been induced from perturbation of the heteroclinic cycle present for lower parameter values (Silnikov, 1970). In this type of intermittency, the rolls do not appear necessarily as a pair. In other words, they are not symmetric, i.e. invariant under reflexion a $x_1/x_2$ plane and single rolls are frequent, as observed experimentally.

**Aknowledgements**

This work was supported by NSF under grant number MSS-8957462 and by the City University of New York under a PSC-CUNY grant. Computer resources were provided by the University Computer Center of the City University of New York. These sources of support are gratefully aknowledged.

# Appendix A

**Eigenvalues of the 32 dimensional system linearized about a fixed point for the different subsystems (see text)**

The first column contains the bifurcation parameter values. (a,b) denotes the complex conjugate pair a+ib, a-ib.

| | | |
|---|---|---|
| 1.50 | [illegible] | [illegible] |
| 1.40 | -13.[illegible] | -6.[illegible] |
| 1.30 | (-11.5[illegible],1.[illegible]) | |
| 1.20 | -14.772 | -11.442 |
| 1.10 | -20.566 | -8.926 |
| 1.00 | -26.555 | -7.024 |
| 0.95 | -30.236 | -6.172 |
| 0.90 | -33.927 | -5.365 |
| 0.89 | -34.617 | -5.336 |
| 0.88 | -35.413 | -5.[illegible] |

Eigenvalues of the subsystem: Re $a_{02}$, $a_{04}$

| | | |
|---|---|---|
| 1.50 | -15.824 | 0.000 |
| 1.40 | -13.793 | 0.000 |
| 1.30 | -11.657 | 0.000 |
| 1.20 | -9.392 | 0.000 |
| 1.10 | -6.956 | 0.000 |
| 1.00 | -4.358 | 0.000 |
| 0.95 | -2.968 | 0.000 |
| 0.90 | -1.565 | 0.000 |
| 0.89 | -1.302 | 0.000 |
| 0.88 | -1.019 | 0.000 |

Eigenvalues of the subsystem: Im($a_{02}$, $a_{04}$

| | | | |
|---|---|---|---|
| 1.50 | -37.335 | -4.634 | -1.573 |
| 1.40 | -33.639 | (-3.337,1.391) | |
| 1.30 | -29.970 | (-3.365,2.222) | |
| 1.20 | -26.119 | (-3.314,2.604) | |
| 1.10 | -22.010 | (-3.195,2.739) | |
| 1.00 | -18.031 | (-3.018,2.493) | |
| 0.95 | -15.993 | (-2.920,2.078) | |
| 0.90 | -13.875 | (-2.839,1.050) | |
| 0.89 | -13.437 | (-2.829,0.495) | |
| 0.88 | -12.991 | -3.662 | -1.954 |

Eigenvalues of the subsystem: Re($a_{01}$, $a_{03}$, $a_{05}$)

| | | | |
|---|---|---|---|
| 1.50 | -27.336 | -4.694 | 0.004 |
| 1.40 | -33.640 | -3.731 | 0.453 |
| 1.30 | -29.576 | -2.654 | 0.420 |
| 1.20 | -26.042 | -0.987 | -0.321 |
| 1.10 | -22.155 | (-0.216,1.475) | |
| 1.00 | -18.267 | ( 0.221,2.089) | |
| 0.95 | -16.415 | ( 0.450,2.316) | |
| 0.90 | -14.612 | ( 0.693,2.502) | |
| 0.89 | -14.266 | ( 0.744,2.534) | |
| 0.88 | -13.925 | ( 0.795,2.565) | |

Eigenvalues of the subsystem: Im($a_{01}$, $a_{03}$, $a_{05}$

| | | | | | |
|---|---|---|---|---|---|
| 1.50 | (-67.,17.) | (-56.,135.) | (-56.,135.) | (-31.,100.) | (-31.,100.) |
| 1.40 | (-63.,19.) | (-52.,131.) | (-52.,131.) | (-29.,96.) | (-29.,98.) |
| 1.30 | (-59.,21.) | (-49.,127.) | (-48.,127.) | (-27.,92.) | (-26.,96.) |
| 1.20 | (-54.,22.) | (-45.,122.) | (-45.,123.) | (-25.,88.) | (-24.,93.) |
| 1.10 | (-50.,24.) | (-41.,118.) | (-41.,118.) | (-23.,84.) | (-21.,91.) |
| 1.00 | (-45.,25.) | (-37.,113.) | (-37.,114.) | (-21.,79.) | (-19.,89.) |
| 0.95 | (-42.,26.) | (-35.,111.) | (-35.,112.) | (-20.,76.) | (-17.,88.) |
| 0.90 | [illegible] | (-33.,[illegible]) | (-32.,[illegible]) | (-19.,73.) | (-16.,87.) |
| 0.89 | (-39.,[illegible]) | (-33.,[illegible]) | (-32.,[illegible]) | (-18.,72.) | (-15.,87.) |
| 0.88 | (-39.,[illegible]) | (-32.,107.) | (-32.,108.) | (-18.,72.) | (-15.,87.) |

Eigenvalues of the subsystem: ($a_{-12}$, $a_{-14}$, $a_{10}$, $a_{12}$, $a_{14}$

| | | | | | | |
|---|---|---|---|---|---|---|
| 1.50 | (-69.,149.) | (-69.,102.) | (-49.,123.) | (-49.,123.) | (-16.,94.) | (-12.95,100.) |
| 1.40 | (-65.,145.) | (-64.,96.) | (-47.,[illegible]) | (-46.,119.) | (-16.,88.) | (-10.02,[illegible]) |
| 1.30 | (-60.,140.) | (-58.,90.) | (-44.,[illegible]) | (-43.,[illegible]) | (-16.,83.) | (-7.73,99.) |
| 1.20 | (-55.,136.) | (-53.,85.) | (-41.,114.) | [illegible] | (-16.,78.) | (-5.61,97.) |
| 1.10 | (-50.,131.) | (-48.,79.) | (-38.,[illegible]) | (-37.,106.) | (-15.,73.) | (-3.55,95.) |
| 1.00 | (-45.,126.) | (-43.,73.) | (-34.,[illegible]) | (-33.,102.) | (-15.,68.) | (-1.52,92.) |
| 0.95 | (-43.,123.) | (-41.,71.) | (-32.,106.) | (-32.,100.) | (-14.,65.) | (-0.53,90.) |
| 0.90 | (-40.,120.) | (-38.,68.) | (-30.,95.) | (-30.,104.) | (-14.,62.) | ( 0.41,88.) |
| 0.89 | (-40.,119.) | (-38.,68.) | (-30.,97.) | (-29.,103.) | (-13.,61.) | ( 0.58,85.) |
| 0.88 | (-39.,118.) | (-38.,67.) | (-30.,97.) | (-28.,103.) | (-13.,61.) | ( 0.76,87.) |

Eigenvalues of the subsystem: ($a_{-11}$, $a_{-13}$, $a_{-15}$, $a_{11}$, $a_{13}$, $a_{15}$)

# Appendix B

**Eigenvalues of the 54 dimensional system linearized about a fixed point for the different subsystems (see text)**

The first column contains the bifurcation parameter values. (a,b) denotes the complex conjugate pair a+ib, a-ib.

| | | |
|---|---|---|
| 2.10 | -15.617 | -2.101 |
| 2.00 | -14.109 | -5.701 |
| 1.90 | -11.838 | -10.031 |
| 1.85 | (-11.503,1.797) | |
| 1.75 | (-12.608,1.591) | |
| 1.65 | -17.030 | -10.517 |
| 1.55 | -21.143 | -8.936 |
| 1.45 | -25.372 | -7.567 |
| 1.40 | -27.616 | -6.931 |
| 1.35 | -29.952 | -6.327 |
| 1.30 | -32.501 | -5.757 |
| 1.25 | -35.204 | -5.219 |
| 1.22 | -36.926 | -4.912 |
| 1.21 | -37.518 | -4.812 |
| 1.20 | -38.110 | -4.714 |
| 1.19 | -38.731 | -4.616 |

Eigenvalues of the subsystem: Re($a_{02}$, $a_{04}$)

| | | |
|---|---|---|
| 2.10 | -15.622 | 0.000 |
| 2.00 | -14.184 | 0.000 |
| 1.90 | -12.696 | 0.000 |
| 1.85 | -11.931 | 0.000 |
| 1.75 | -10.352 | 0.000 |
| 1.65 | -8.699 | 0.000 |
| 1.55 | -6.962 | 0.000 |
| 1.45 | -5.133 | 0.000 |
| 1.40 | -4.184 | 0.000 |
| 1.35 | -3.215 | 0.000 |
| 1.30 | -2.228 | 0.000 |
| 1.25 | -1.229 | 0.000 |
| 1.22 | -0.626 | 0.000 |
| 1.21 | -0.425 | 0.000 |
| 1.20 | -0.224 | 0.000 |
| 1.19 | -0.026 | 0.000 |

Eigenvalues of the subsystem: Im($a_{02}$, $a_{04}$)

| | | | |
|---|---|---|---|
| 2.10 | -36.671 | -4.480 | -2.357 |
| 2.00 | -34.062 | (-3.553,1.309) | |
| 1.90 | -31.417 | (-3.568,1.975) | |
| 1.85 | -30.080 | (-3.555,2.195) | |
| 1.75 | -27.374 | (-3.550,2.513) | |
| 1.65 | -24.624 | (-3.452,2.701) | |
| 1.55 | -21.828 | (-3.393,2.757) | |
| 1.45 | -18.982 | (-3.258,2.630) | |
| 1.40 | -17.538 | (-3.189,2.457) | |
| 1.35 | -16.074 | (-3.121,2.160) | |
| 1.30 | -14.577 | (-3.062,1.627) | |
| 1.25 | -13.018 | -3.519 | -2.531 |
| 1.22 | -12.020 | -4.651 | -1.432 |
| 1.21 | -11.659 | -4.929 | -1.181 |
| 1.20 | -11.304 | -5.203 | -0.948 |
| 1.19 | -10.920 | -5.483 | -0.729 |

Eigenvalues of the subsystem: Re($a_{01}$, $a_{03}$, $a_{05}$)

| | | | |
|---|---|---|---|
| 2.10 | -36.671 | -4.672 | -0.164 |
| 2.00 | -34.062 | -3.973 | 0.110 |
| 1.90 | -31.420 | -3.221 | 0.143 |
| 1.85 | -30.035 | -2.804 | 0.087 |
| 1.75 | -27.389 | -1.749 | -0.238 |
| 1.65 | -24.659 | (-0.595,1.032) | |
| 1.55 | -21.909 | (-0.382,1.630) | |
| 1.45 | -19.164 | (-0.069,2.047) | |
| 1.40 | -17.810 | (0.091,2.219) | |
| 1.35 | -16.480 | (0.255,2.371) | |
| 1.30 | -15.189 | (0.426,2.504) | |
| 1.25 | -13.950 | (0.607,2.619) | |
| 1.22 | -13.239 | (0.720,2.678) | |
| 1.21 | -13.008 | (0.759,2.696) | |
| 1.20 | -12.781 | (0.799,2.713) | |
| 1.19 | -12.556 | (0.839,2.726) | |

Eigenvalues of the subsystem: Im($a_{01}$, $a_{03}$, $a_{05}$)

# Appendix B (cont.)

| | | | | | |
|---|---|---|---|---|---|
| 2.10 | (-67., 24.) | (-55.,141.) | (-55.,141.) | (-31.,107.) | (-31.,107.) |
| 2.00 | (-64., 25.) | (-53.,138.) | (-53.,138.) | (-29.,104.) | (-29.,105.) |
| 1.90 | (-61., 26.) | (-50.,135.) | (-50.,135.) | (-28.,101.) | (-27.,104.) |
| 1.85 | (-60., 27.) | (-49.,133.) | (-49.,134.) | (-27.,100.) | (-27.,103.) |
| 1.75 | (-57., 28.) | (-46.,131.) | (-46.,131.) | (-26., 97.) | (-25.,101.) |
| 1.65 | (-53., 29.) | (-44.,128.) | (-44.,128.) | (-24., 94.) | (-23., 99.) |
| 1.55 | (-50., 30.) | (-41.,124.) | (-41.,125.) | (-23., 90.) | (-21., 98.) |
| 1.45 | (-47., 31.) | (-38.,121.) | (-38.,122.) | (-21., 87.) | (-19., 96.) |
| 1.40 | (-45., 32.) | (-37.,119.) | (-36.,120.) | (-21., 85.) | (-18., 96.) |
| 1.35 | (-43., 32.) | (-35.,117.) | (-35.,119.) | (-20., 83.) | (-17., 95.) |
| 1.30 | (-42., 32.) | (-34.,115.) | (-33.,117.) | (-19., 81.) | (-16., 94.) |
| 1.25 | (-40., 33.) | (-33.,113.) | (-32.,115.) | (-18., 79.) | (-15., 94.) |
| 1.22 | (-39., 33.) | (-32.,112.) | (-31.,114.) | (-16., 77.) | (-15., 93.) |
| 1.21 | (-38., 33.) | (-31.,112.) | (-31.,113.) | (-18., 77.) | (-14., 93.) |
| 1.20 | (-38., 33.) | (-31.,111.) | (-30.,113.) | (-17., 77.) | (-14., 93.) |
| 1.19 | (-37., 33.) | (-31.,111.) | (-30.,112.) | (-17., 76.) | (-14., 93.) |

Eigenvalues of the subsystem: $(a_{-12}, a_{-14}, a_{10}, a_{12}, a_{14})$

| | | | | | | |
|---|---|---|---|---|---|---|
| 2.10 | (-69.,155.) | (-68.,103.) | (-48.,130.) | (-48.,129.) | (-17.,100.) | (-13.70,107.) |
| 2.00 | (-65.,152.) | (-65., 99.) | (-47.,128.) | (-46.,126.) | (-18., 96.) | (-11.64,107.) |
| 1.90 | (-62.,149.) | (-61., 95.) | (-45.,125.) | (-44.,123.) | (-18., 92.) | (-9.90,107.) |
| 1.85 | (-60.,147.) | (-59., 93.) | (-44.,124.) | (-43.,122.) | (-18., 90.) | (-9.08,106.) |
| 1.75 | (-57.,144.) | (-55., 89.) | (-42.,122.) | (-41.,119.) | (-17., 87.) | (-7.51,105.) |
| 1.65 | (-53.,141.) | (-52., 85.) | (-40.,120.) | (-39.,116.) | (-17., 83.) | (-5.97,104.) |
| 1.55 | (-50.,137.) | (-48., 81.) | (-38.,118.) | (-36.,113.) | (-16., 80.) | (-4.46,102.) |
| 1.45 | (-47.,134.) | (-45., 77.) | (-35.,115.) | (-34.,110.) | (-16., 76.) | (-2.97,100.) |
| 1.40 | (-45.,132.) | (-43., 75.) | (-34.,114.) | (-33.,109.) | (-16., 74.) | (-2.24, 99.) |
| 1.35 | (-43.,130.) | (-41., 73.) | (-32.,113.) | (-32.,107.) | (-15., 72.) | (-1.52, 98.) |
| 1.30 | (-41.,128.) | (-39., 71.) | (-31.,106.) | (-30.,111.) | (-15., 70.) | (-0.83, 96.) |
| 1.25 | (-39.,125.) | (-38., 69.) | (-30.,104.) | (-29.,110.) | (-14., 68.) | (-0.17, 95.) |
| 1.22 | (-38.,124.) | (-37., 65.) | (-29.,103.) | (-27.,109.) | (-14., 67.) | ( 0.20, 94.) |
| 1.21 | (-38.,124.) | (-36., 68.) | (-29.,103.) | (-27.,109.) | (-14., 66.) | ( 0.32, 94.) |
| 1.20 | (-38.,123.) | (-36., 68.) | (-29.,102.) | (-27.,109.) | (-14., 66.) | ( 0.44, 93.) |
| 1.19 | (-37.,123.) | (-36., 67.) | (-28.,102.) | (-26.,108.) | (-14., 65.) | ( 0.55, 93.) |

Eigenvalues of the subsystem: $(a_{-11}, a_{-13}, a_{-15}, a_{11}, a_{13}, a_{15})$

| | | | | | |
|---|---|---|---|---|---|
| 2.10 | (-102.,234.) | (-102.,234.) | (-71.,203.) | (-71.,204.) | (-45.,304.) |
| 2.00 | (-97.,231.) | (-97.,231.) | (-68.,199.) | (-68.,201.) | (-43.,296.) |
| 1.90 | (-92.,227.) | (-91.,227.) | (-66.,194.) | (-66.,198.) | (-41.,287.) |
| 1.85 | (-89.,226.) | (-89.,226.) | (-64.,192.) | (-64.,196.) | (-40.,282.) |
| 1.75 | (-84.,222.) | (-84.,223.) | (-62.,193.) | (-62.,187.) | (!38.,273.) |
| 1.65 | (-79.,219.) | (-78.,219.) | (-59.,190.) | (-59.,182.) | (-36.,265.) |
| 1.55 | (-75.,215.) | (-73.,215.) | (-56.,187.) | (-56.,177.) | (-34.,256.) |
| 1.45 | (-70.,211.) | (-68.,211.) | (-54.,184.) | (-53.,171.) | (-31.,247.) |
| 1.40 | (-68.,208.) | (-65.,209.) | (-52.,183.) | (-51.,168.) | (-30.,242.) |
| 1.35 | (-66.,205.) | (-63.,207.) | (-51.,181.) | (-50.,164.) | (-29.,238.) |
| 1.30 | (-64.,202.) | (-60.,204.) | (-50.,180.) | (-48.,161.) | (-28.,233.) |
| 1.25 | (-62.,199.) | (-57.,201.) | (-49.,178.) | (-46.,157.) | (-27.,228.) |
| 1.22 | (-60.,197.) | (-56.,199.) | (-48.,177.) | (-45.,155.) | (-26.,225.) |
| 1.21 | (-60.,196.) | (-55.,198.) | (-48.,177.) | (-45.,154.) | (-26.,224.) |
| 1.20 | (-59.,196.) | (-55.,197.) | (-48.,177.) | (-45.,154.) | (-26.,223.) |
| 1.19 | (-59.,195.) | (-54.,197.) | (-48.,176.) | (-45.,153.) | (-25.,222.) |

Eigenvalues of the subsystem: $(a_{-22}, a_{-24}, a_{20}, a_{22}, a_{24})$

| | | | | | | |
|---|---|---|---|---|---|---|
| 2.10 | (-118.,171.) | (-118.,248.) | (-94.,226.) | (-94.,227.) | (-36.,232.) | (-32.51,243.) |
| 2.00 | (-112.,165.) | (-111.,245.) | (-90.,221.) | (-89.,224.) | (-36.,224.) | (-30.14,242.) |
| 1.90 | (-107.,158.) | (-105.,242.) | (-87.,216.) | (-85.,222.) | (-36.,217.) | (-28.06,240.) |
| 1.85 | (-104.,155.) | (-102.,240.) | (-85.,214.) | (-82.,220.) | (-36.,214.) | (-27.06,239.) |
| 1.75 | (-98.,149.) | (-96.,237.) | (-80.,209.) | (-78.,217.) | (-36.,207.) | (-25.08,236.) |
| 1.65 | (-93.,143.) | (-90.,233.) | (-76.,204.) | (-73.,214.) | (-37.,200.) | (-23.09,232.) |
| 1.55 | (-87.,136.) | (-84.,230.) | (-72.,200.) | (-68.,211.) | (-37.,193.) | (-21.04,228.) |
| 1.45 | (-82.,130.) | (-78.,225.) | (-67.,195.) | (-62.,208.) | (-37.,186.) | (-18.92,223.) |
| 1.40 | (-79.,127.) | (-75.,223.) | (-65.,193.) | (-60.,206.) | (-37.,181.) | (-17.82,220.) |
| 1.35 | (-76.,124.) | (-73.,220.) | (-63.,191.) | (-57.,205.) | (-37.,177.) | (-16.70,217.) |
| 1.30 | (-74.,121.) | (-70.,217.) | (-61.,188.) | (-54.,203.) | (-37.,172.) | (-15.57,214.) |
| 1.25 | (-71.,118.) | (-68.,213.) | (-59.,186.) | (-51.,201.) | (-37.,168.) | (-14.43,211.) |
| 1.22 | (-70.,116.) | (-66.,211.) | (-58.,184.) | (-50.,200.) | (-37.,165.) | (-13.75,208.) |
| 1.21 | (-69.,115.) | (-66.,210.) | (-58.,184.) | (-49.,199.) | (-37.,164.) | (-13.52,207.) |
| 1.20 | (-69.,115.) | (-65.,209.) | (-58.,183.) | (-48.,199.) | (-36.,163.) | (-13.30,207.) |
| 1.19 | (-68.,114.) | (-65.,208.) | (-58.,183.) | (-48.,199.) | (-36.,162.) | (-13.05,206.) |

Eigenvalues of the subsystem: $(a_{-21}, a_{-23}, a_{-25}, a_{21}, a_{23}, a_{25})$

**References**

Adrian, R. J., Moin, P. and Moser R. D. 1987 Stochastic estimation of conditional eddies in turbulent channel flow. In Proceedings of the 1987 Summer Program of Center for Turbulence Research (NASA Ames/Stanford Univ., Stanford, CA 1987)

Armbruster, D., Guckenheimer, J. and Holmes, P. 1988 Heteroclinic cycles and modulated travelling waves in systems with 0(2) symmetry. Physica 29D, 257-282.

Aubry. N. 1987 A dynamical system / coherent structure approach to the fully developed turbulent wall layer. Ph. D. thesis, Cornell University.

Aubry, N., Holmes, P., Lumley, J.L., & Stone, E. 1988 The dynamics of coherent structures in the wall Region of a turbulent boundary layer. *J. Fluid Mech.* 192, 115-173.

Aubry, N. & Sanghi, S. 1989 Streamwise and spanwise dynamics in the turbulent wall layer. In *Forum on chaotic Flow*, ed. K.N. Ghia. New York: ASME.

Aubry, N., Lumley, J.L., & Holmes, P. 1990 The effect of drag reduction on the wall region. *Theoret. comput. Fluid Dynamics.* In press.

Bakewell, P & Lumley, J.L. 12967 Viscous sublayer and adjacent wall region in turbulent pipe flow. *Phys. Fluids* 10, 1880-1889.

Blackwelder, R. F. & Eckelman, H. 1979 Streamwise vortices associated with the bursting phenomenon. *J. Fluid Mech.* 94 (3), 577-594.

Cantwell, B.J. 1981 Organized motion in turbulent flow. *Ann. Rev. Fluid Mech.* 13, 457-515.

Corino, E.R. & Brodkey, R.S. 1969 A visual investigation of the wall region in turbulent flow. *J. Fluid Mech.* 37, 1-30.

Curry, J. H., Herring, J. R., Loncaric, J. & Orszag, S.A. 1984. Order and disorder in two- and three-dimensional Benard c onvection. *J. Fluid Mech.* 147, 1-38.

Doedel, E.J. & Kernevez, J.P. 1985 Software for continuation problems in ordinary differential equations with applications. *Applied Mathematics Rep.* California Institute of Technology.

Foias, C and Temam, R. 1977 Structure of the set of stationary solutions of the Navier Stokes equations. *Comm. Pure Appl.*, 30, 149-164.

Guezennec, Y. G. 1989 Stochastic estimation of coherent structures in turbulent boundary layers. *Phys. Fluids A* 1 (6), 1054-1060.

Herzog, S. 1986 The large scale structure in the near-wall region of turbulent pipe flow. Ph.D. thesis, Cornell University.

Hogenes, J.H.A. & Hanratty, T. J. 1982 The use of multiple wall probes to identify coherent flow patterns in the viscous wall region. *J. Fluid Mech.* 95, 655-679.

Jang, P. S., Benney, D. J. & Gran, R. L. 1986 On the origin of streamwise vortices in a turbulent boundary layer. *J. Fluid Mech.* 169, 109-123.

Jimenez, J. 1987 Bifurcations and bursting in two-dimensional Poiseuille flow. *Phys. Fluids* 30 (12), 3644-6.

Jimenez, J., Moin, P. , Moser, R.& Keefe, L. 1988 Ejection mechanisms in the sublayer of a turbulent channel. *Phys. Fluids* 31 (6), 1311-3.

Jimenez, J.& Moin, P. 1989 A minimal turbulent channel. American Physical Society/Division of Fluid Mechanics, 42nd annual meeting, DA3. Palo Alto, CA.

Kline, S.J., Reynolds, W.C., Schraub, F.A. & Runstadler, P.W. 1967 The structure of turbulent boundary layers. *J. Fluid Mech.* 30, 741-73.

Kim, J. 1985 Turbulence structures associated with the bursting event. *Phys. Fluids* 28, 52-58.

Landahl, M. 1975 Wave breakdown and turbulence. *SIAM J. Appl. Math.* 28 (4), 735-56.

Loève, M. 1955 *Probability Theory.* Van Nostrand.

Lumley, J.L. 1967 The structure of inhomogeneous turbulent flows. In *Atmospheric Turbulence and Radio Wave Propagation* (ed. A.M. Yaglom & V.I. Tatarski), pp. 166-178. Moscow: Nauka.

Moffatt, H.K. 1989 Fixed points of turbulent dynamical systems and suppression of nonlinearity. In *Whither Turbulence* (ed. J.L. Lumley), 22-24 March, Cornell University, NY. Springer-Verlag.

Moin, P. & Kim, J. 1982 Numerical investigation of turbulent channel flow. *J. Fluid Mech.* 118, 341-377.

Moin, P. & Moser, R.D. 1989 Characteristic-eddy decomposition of turbulence in a channel, *J. Fluid Mech.* 200, 471-509.

Robinson, S.K., Kline, S.J. and Spalart, P.R. 1988 Quasi-coherent structures in the turbulent boundary layer: Part II. Verification and new information from a numerically simulated flat-plate layer. In *Near Wall Turbulence*, eds. S. J. Kline *et al.*, Washington DC: Hemisphere.

Silnikov, L. P. 1970 A contribution to the problem of the structure of an extended neighbourhood of a rough equilibrium state of saddle-focus type. *Math. USSR Sbornik* 10, 91-102.

Spalart, P. R. 1988 Direct simulation of a turbulent boundary layer up to Rt = 1410. *J. Fluid Mech.* 187, 61-98.

Temam, R. 1988 *Infinite-Dimensional Dynamical Systems in Mechanics and Physics.* Applied Mathematical Sciences 68. Springer-Verlag.

Willmarth, W. W. & Lu, S. S. 1972 Structure of the Reynolds stress near the wall. J. Fluid Mech. 55, 65-92.

# The Dynamics of Organized Structures in the Axisymmetric Jet Mixing Layer


Mark GLAUSER[α], Xiaowei ZHENG[β] and Charles R. DOERING[γ]

[α] *Department of Mechanical Engineering and Institute for Nonlinear Studies Clarkson University, Potsdam, New York, USA 13676*

[β] *Department of Mechanical Engineering, Clarkson University Potsdam, New York, USA 13676*

[γ] *Department of Physics and Institute for Nonlinear Studies Clarkson University, Potsdam, New York, USA 13676*



*Turbulence and Coherent Structures*
*O. Métais and M. Lesieur, eds.*
*Selected papers from "Turbulence 89: organized structures and turbulence in fluid mechanics"*
*Grenoble, 18-21 September 1989*

## ABSTRACT

The near field jet mixing layer is modeled in a manner similar to that applied by Aubry et al. (1988) to the near wall region of the turbulent boundary layer. In this work the instantaneous velocity field is expanded in terms of the empirical eigenfunctions obtained by Glauser and George (1987). These eigenfunctions were extracted from the measured cross-spectral tensor by application of the proper orthogonal decomposition theorem (POD) suggested by Lumley (1967). Galerkin projection is then applied to the Navier Stokes equations in conjunction with this representation, resulting in low-dimensional sets of ordinary differential equations. The methods of dynamical systems theory are then used to analyze these equations. This work consists of an attempt to utilize the dynamical systems model to further our understanding of the transfer of turbulent energy between various azimuthal modes and streamwise wavenumbers and relate this to the turbulence production phenomena in the jet mixing layer. With this model the sequence by which the various modes contribute in time can be examined.


## 1. INTRODUCTION

In recent years, two separate developments have altered the basic statistical framework of turbulence. There is an abundance of experimental evidence implying the existence of coherent structures and, on the theoretical side, recent applications of dynamical systems theory to turbulence suggest that such flows reside on relatively low dimensional manifolds or attractors(v. Aubry et al. 1988, Bartuccelli et al. 1989 and references therein).

Aubry et al. (1988) appear to be among the first to apply low dimensional dynamical systems theory to a turbulent open flow system. They used the POD to provide basis functions to obtain low dimensional sets of ordinary differential equations from the Navier–Stokes equations. Used in conjunction with Galerkin projection, the POD yields an optimal set of basis functions in the sense that the resulting truncated system of ordinary differential equations captures the maximum amount of kinetic energy among all possible truncations in the same order over a specific domain. Sirovich and Rodriguez (1987) have shown, for the Ginzburg–Landau system, that these basis functions are fairly robust and can be used over a wide range of the bifurcation parameter. The basis functions that Aubry et al. (1988) used were those obtained experimentally by Herzog (1986). The equations that they derived, which exhibit intermittent behavior, where then analyzed using dynamical systems theory. Their results are not inconsistent with events seen in experimental work.

The axisymmetric jet is a good candidate for a similar approach. This was demonstrated by Glauser et al. (1987), where the instantaneous signals were almost completely reconstructed with only 3 proper orthogonal modes. In this work the instantaneous velocity field of the axisymmetric jet mixing layer is expanded in terms of the empirical eigenfunctions obtained by Glauser and George (1987). These eigenfunctions were extracted from the measured cross spectral tensor by application of the POD. In the streamwise directions the flow is approximately homogeneous; so, as a first step, Fourier modes were used. In the azimuthal direction the flow is periodic, hence the use of the Fourier modes in this direction as well. In the radial direction the flow is strongly inhomogeneous; so eigenfunctions obtained from the POD were utilized. Galerkin projection is then used in conjunction with the POD to obtain a truncated system of ordinary differential equations. The modes neglected in the truncation are accounted for by a Heisenberg model. This work appears to be the first attempt at utilizing low dimensional dynamics in conjunction with Galerkin projection and the POD to examine the temporal dynamics of coherent structures in a high Reynolds number axisymmetric jet free shear layer.

## 2. PROPER ORTHOGONAL DECOMPOSITION

In 1967 Lumley suggested that a coherent structure should be that structure which has the largest mean square projection on the velocity field. This process involves maximizing the mean square energy via the calculus of variations and leads to the integral eigenvalue problem

$$\lambda \phi_i(\tilde{x}) = \int R_{ij}(\tilde{x}, \tilde{x}')\phi_j(\tilde{x}')d\tilde{x}'. \qquad (2-1)$$

The symmetric kernel of this Fredholm integral equation is the two-point correlation tenser $R_{ij}$ defined by

$$R_{ij}(\tilde{x}, \tilde{x}') = \overline{u_i(\tilde{x})u_j(\tilde{x}')}, \qquad (2-2)$$

where $\tilde{\phi}$ is the candidate structure; $\tilde{x}$ and $\tilde{x}'$ represent different spatial points in the inhomogeneous directions and different times if the flow is non-stationary.

From the Hilbert-Schmidt theory it can be shown that the solution of a Fredholm integral equation of the first kind for a symmetric kernel is a discrete set, hence equation (2-1) can be written as

$$\lambda^n \phi_i^n(\tilde{x}) = \int R_{ij}(\tilde{x}, \tilde{x}')\phi_j^n(\tilde{x}')d\tilde{x}'. \qquad (2-3)$$

The eigenfunctions of the Fredholm equation are orthogonal over the interval and

$$\int \phi_i^n(\tilde{x})\phi_i^m(\tilde{x})d\tilde{x} = \delta_{nm} \qquad (2-4)$$

for normalized eigenfunctions. The random coefficients are calculated from

$$a^n = \int \tilde{u}_i(\tilde{x})\phi_i^{n*}(\tilde{x})d(\tilde{x}) \qquad (2-5)$$

where the $\phi_i^n$ are the eigenfunctions obtained from equation (2-3). The eigenvalues of the Fredholm equation with a real symmetric kernel are all real and uncorrelated,

$$\overline{a^n a^m} = \lambda^n \delta_{nm}, \qquad (2-6)$$

and the fluctuating random field $\tilde{u}_i$ can be reconstructed from the eigenfunctions and the random coefficients via

$$\tilde{u}_i(\tilde{x}) = \sum_{n=0}^{\infty} a^n \phi_i^n(\tilde{x}). \qquad (2-7)$$

The turbulent kinetic energy is the sum over n of the eigenvalues $\lambda^n$, and each structure makes an independent contribution to the kinetic energy and Reynolds stress.

If the random field is homogeneous or periodic in one or more directions or stationary in time, the eigenfunctions become Fourier modes, so that the POD reduces to the harmonic orthogonal decomposition in these directions. In the jet studied the flow is periodic in the azimuthal direction and almost homogeneous in the streamwise direction. We will not transform over time in this case because we are keeping time in the random coefficients. Under these conditions the spectral tensor may be defined by

$$S_{ij}(r,r',m,k_1) = \int R_{ij}(r,r',\theta,x)e^{-2\pi i(k_1 x+m\theta)}dxd\theta, \qquad (2-8)$$

so that equation (2-3) becomes

$$\lambda^n(m,k_1)\phi_i^n(r,m,k_1)$$
$$= \int_\Omega S_{ij}(r,r',m,k_1)\phi_j^n(r',m,k_1)dr'. \qquad (2-9)$$

In the above expressions $r$ and $r'$ represent different spatial locations in the radial direction ( the inhomogeneous direction), $x$ and $\theta$ are the separations in the streamwise and azimuthal directions respectively, $k_1$

is the streamwise wavenumber and $m$ is the azimuthal mode number. It should be noted that in this case the eigenvalues and eigenfunctions are now a function of $m$ and $k_1$. Equation (2-9) is then solved numerically using the values of $S_{ij}(r, r', m, k_1)$ obtained by Glauser and George (1987). The continuity equation must be utilized with this data base to obtain the full cross-spectral tensor because only the radial and streamwise velocity components were measured. Taylor's hypothesis is used to convert the temporal frequency to streamwise spatial wavenumber because the cross-spectral tensor data of Glauser and George was obtained only at one streamwise location. The continuity equation can be written in terms of $R_{ij}$ as

$$\frac{\partial R_{ij}}{\partial x_i} = 0. \qquad (2-10)$$

The correlation tensor can be represented in the term of Fourier inverse transform of the spectral tensor as

$$R_{ij}(r, r', \theta, x) = \sum_{m=0}^{\infty} \int_{-\infty}^{\infty} e^{2\pi i(k_1 x + m\theta)} S_{ij}(r, r', m, k_1) dk_1, \qquad (2-11)$$

so that the continuity equation becomes, in terms of the spectral tensor,

$$-k_1 i S_{1j} + \frac{\partial S_{2j}}{\partial r} + \frac{S_{2j}}{r} - \frac{2\pi m}{ri} S_{3j} = 0. \qquad (2-12)$$

We now solve for $S_{3j}$ and obtain

$$S_{3j} = \frac{ri}{2\pi m}[-k_1 i S_{1j} + \frac{\partial S_{2j}}{\partial r} + \frac{S_{2j}}{r}] \qquad (2-13)$$

where $j = 1, 2, 3$ and $S_{1j}$, $S_{2j}$ for $j = 1, 2$ are given from the experimental measurements of Glauser and George (1987).

## 3. THE MOMENTUM EQUATION

The Navier–Stokes equations for an incompressible flow in cylindrical coordinates, after application of Reynold's decomposition and substitution of a relationship between the divergence of Reynold's stress and the mean pressure and velocity can be written as (v. Zheng 1990)

$$\frac{\partial}{\partial t}\tilde{u}_i + M\tilde{u}_i + (-\frac{\tilde{u}_\theta^2}{r} + \overline{U}_z \frac{\partial \tilde{u}_r}{\partial z})\delta_{ir} + (\frac{\tilde{u}_r \tilde{u}_\theta}{r} + \overline{U}_z \frac{\partial \tilde{u}_\theta}{\partial z})\delta_{i\theta}$$

$$+(\overline{U}_z \frac{\partial \tilde{u}_z}{\partial z} + \tilde{u}_r \frac{\partial \overline{U}_z}{\partial r})\delta_{iz} + \overline{M\tilde{u}_i} - \frac{\overline{\tilde{u}_\theta^2}}{r}\delta_{ir} + \frac{\overline{\tilde{u}_r \tilde{u}_\theta}}{r}\delta_{i\theta}$$

$$= -\frac{r(\delta_{ir} + \delta_{iz}) + \delta_{i\theta}}{r\rho}\frac{\partial \tilde{p}}{\partial i} + \nu[\nabla^2 \tilde{u}_i - (\frac{\tilde{u}_r}{r^2} + \frac{2}{r^2}\frac{\partial \tilde{u}_\theta}{\partial \theta})\delta_{ir}$$

$$+(-\frac{\tilde{u}_\theta}{r^2} + \frac{2}{r^2}\frac{\partial \tilde{u}_r}{\partial \theta})\delta_{i\theta}], \qquad (3-1)$$

where

$$\nabla^2 = \frac{\partial^2}{\partial r^2} + \frac{1}{r}\frac{\partial}{\partial r} + \frac{1}{r^2}\frac{\partial^2}{\partial \theta^2} + \frac{\partial^2}{\partial z^2},$$

$M = \tilde{u}_r\frac{\partial}{\partial r} + \frac{\tilde{u}_\theta}{r}\frac{\partial}{\partial \theta} + \tilde{u}_z\frac{\partial}{\partial z}$, $i = r, \theta, z$, $\delta$ is the Kronecker delta, $\tilde{p}$ is the fluctuating pressure , $\tilde{u}_r$, $\tilde{u}_\theta$ and $\tilde{u}_z$ are the fluctuating velocities in the radial, azimuthal and streamwise direction respectively, and $\overline{U}_z$ is the mean streamwise velocity. The overbar denotes an average, which in this case is defined as a spatial average over the homogeneous and periodic directions (v. Aubry et. al. 1988). In the above equation it is assumed that $\rho$ and $\mu$ are constant and that the body forces are zero.

One change is made to equation (3-1) before we perform a Fourier Transform and the Galerkin projection on it. This involves deriving a relationship between the mean streamwise velocity and the fluctuating velocity and pressure from the Navier–Stokes equations. This change is needed because the actual measured mean velocity profile will be incorrect for the severely truncated system of equations to be studied. This relationship, which expresses the mean velocity $\overline{U}_z$ in terms of the Reynolds stress $\overline{\tilde{u}_r\tilde{u}_z}$ can be shown to be (v. Zheng 1990)

$$\overline{U}_z = \frac{1}{\nu}\int \overline{\tilde{u}_r\tilde{u}_z} dr \qquad (3-2)$$

where, as a first step, we have neglected the terms which come from the pressure (i.e., an isobaric jet). Equation (3-2) is then used to eliminate $\overline{U}_z$ from equation (3-1). In the axisymmetric jet this gives some feedback to the system of equations as the fluctuation changes. It should be noted that this feedback increases as the Reynolds stress gets stronger. This is similar to the relationship derived by Aubry et al. (1988) for the turbulent boundary layer.

## 4. GALERKIN PROJECTION

The Galerkin method is well known and has been used extensively to study turbulence and the instability of various fluid flows (v. Lin et al 1987 and references therein). The essential idea of the method is to expand the dependent variables in terms of a finite series of independent basis functions. The basis functions form a complete basis for the relevant class of functions and they satisfy the relevant boundary conditions. In this work we use a Galerkin projection in conjunction with the POD (to supply the basis functions) which minimizes the error (measured in terms of mean square energy) due to the truncation and yields a set of ordinary differential equations for the coefficients (v. equation 2-5).

The Galerkin projection is performed on the Fourier Transform (in the azimuthal and streamwise directions) of the Navier–Stokes equations so that is is useful to define the following equations:

$$\tilde{u}_i(z,\theta,r,t) = \sum_m \int_{-\infty}^{\infty} e^{-2\pi j(k_1 z + m\theta)} \hat{u}_i(k_1,m,r,t)dk_1, \qquad (4-1)$$

and

$$\hat{u}_i(k_1,m,r,t) = \int \int_{-\infty}^{\infty} e^{2\pi j(k_1 z + m\theta)} \tilde{u}_i(z,\theta,r,t)dzd\theta. \qquad (4-2)$$

We expand $\hat{u}_i(k_1,m,r,t)$ in terms of the coefficients and eigenfunctions in the following manner (v. equation (2-7)),

$$\hat{u}_i(k_1,m,r,t) = \sum_n a^n_{k_1,m}(t)\phi^n_{i,k_1,m}(r). \qquad (4-3)$$

We then substitute these equations into the Fourier Transform of the Navier-Stokes equations and perform the Galerkin projection. This is written as

$$(\phi^l, N) = \int_0^R N_i(t,r)\phi_i^{l*}(r)dr = 0 \qquad (4-4)$$

where $N_i(t,r)$ represents the Fourier Transform of the Navier–Stokes equations.

Finally, after performing the above and utilizing the orthogonality condition (v. equation (2-4)) we obtain a set of ordinary differential equations for the coefficients:

$$\frac{da^l_{k_1,m}}{dt} = \nu \sum_n a^n_{k_1,m}$$

$$\{-4\pi^2 k_1^2\delta_{nl} + \int_0^R [(\frac{d^2}{dr^2} + \frac{1}{r}\frac{d}{dr} - \frac{4\pi^2 m^2}{r^2})\phi^n_{i,k_1,m}$$

$$-(\frac{1}{r^2}\phi^n_{r,k_1,m} + \frac{4\pi jm}{r^2}\phi^n_{\theta,k_1,m})\delta_{ir}$$

$$+(-\frac{1}{r^2}\phi^n_{\theta,k_1,m} + \frac{4\pi jm}{r^2}\phi^n_{r,k_1,m})\delta_{i\theta}] \cdot \phi^{l*}_{i,k_1,m} dr\} \qquad (4-5a)$$

$$-\sum_{k_1' m'}\sum_{pq} a^p_{k_1',m'} a^q_{k_1-k_1',m-m'}$$

$$\{\int_0^R [\phi^p_{r,k_1',m'} \frac{d}{dr}\phi^q_{i,k_1-k_1',m-m'} + \frac{2\pi j(m-m')}{r}$$

$$\phi^p_{\theta,k_1',m'}\phi^q_{i,k_1-k_1',m-m'} + 2\pi j(k_1-k_1')\phi^p_{z,k_1',m'}\phi^q_{i,k_1-k_1',m-m'}$$

$$-(\frac{1}{r}\phi^p_{\theta,k_1',m'}\phi^q_{\theta,k_1-k_1',m-m'})\delta_{ir}$$

$$+(\frac{1}{r}\phi^p_{r,k_1',m'}\phi^q_{\theta,k_1-k_1',m-m'})\delta_{i\theta}] \cdot \phi^{l*}_{i,k_1,m} dr\} \qquad (4-5b)$$

$$-\frac{1}{\nu}\sum_{k_1' m'}\sum_{pqr} a^p_{k_1',m'} a^{q*}_{k_1',m'} a^r_{k_1,m}$$

$$\{\int_0^R [(\int_0^{R'} \phi^p_{r,k_1',m'}\phi^{q*}_{z,k_1',m'} dr')\phi^r_{i,k_1,m} 2\pi j(k_1)$$

$$+\phi^p_{r,k_1',m'}\phi^{q*}_{z,k_1',m'}\phi^r_{r,k_1,m}\delta_{iz}] \cdot \phi^{l*}_{i,k_1,m} dr\}$$

$$-\int_0^R \frac{r(\delta_{ir}+\delta_{iz})+\delta_{i\theta}}{\rho r}\frac{\partial \hat{p}_{k_1,m}}{\partial i} \cdot \phi^{l*}_{i,k_1,m} dr\}. \qquad (4-5c)$$

In our ordinary differential equation there are four parts. The term on the left hand side is the time derivative of the coefficient $a$. The first term on the right hand side (RHS) (v. equation 4-5a) is a linear term that comes from diffusion term of the Navier–Stokes equations. The second term on the RHS (v. equation 4-5b) is a quadratic term that is a consequence of the fluctuation–fluctuation interactions and exhibits the energy transfer between the different eigenfunctions(from the POD) and the Fourier modes. The last term on the RHS (v. equation 4-5c) is a cubic term that is a direct result of the mean velocity–fluctuating velocity interaction. Equation (4-5) can be written in a more compact form as:

$$A\frac{da^l}{dt} = Bl + Cq + Dc \qquad (4-6)$$

where A, B, C and D are all matrices and $l$, $q$ and $c$ denote the linear, quadratic and cubic terms respectively. A in this case is an an identity matrix because of the orthogonality of the eigenfunctions. It should be noted that the pressure term in equation (4-5) will vanish if the integration covers the whole domain. In the jet shear layer we cover enough of the domain to neglect the pressure term (unlike Aubry et al. 1988 who had to account for the pressure).

## 5. ENERGY TRANSFER MODEL

When we truncate at some cutoff point we need to account for the energy transfer between the included and neglected modes. The effect of the neglected modes will be accounted for by utilizing a Heisenberg model (v. Aubry et al. 1988). The assumption is that the neglected modes withdraw energy from the modes that are kept, as if a certain turbulence viscosity were present.

The equation for $\nu_T$, our turbulence viscosity, can be shown to be (v. Zheng (1990))

$$\nu_T = \frac{\alpha \sum_{k_1,m,n} \lambda^n_{k_1,m}}{R \sum_{k_1,m,n} \lambda^n_{k_1,m} \overline{G}} \qquad (5-1)$$

where

$$\overline{G} = \overline{\int_0^R \left[\frac{d\phi^n_{i,k_1,m}}{dr}\frac{d\phi^{n*}_{i,k_1,m}}{dr} - \frac{1}{r^2} 4\pi^2 m^2 \phi^n_{i,k_1,m}\phi^{n*}_{i,k_1,m}\right] dr - 4\pi^2 k_1^2}$$

and $\alpha$ is a dimensionless parameter. In equation (5-1), the sum denotes a sum over the first neglected modes. We now substitute $\nu_T$ into our ordinary differential equations. The equations then have the form

$$A\frac{da^l}{dt} = B(1 + \nu_T/\nu)l + Cq + Dc. \qquad (5-2)$$

The effect of the energy transfer model is to introduce the parameter $\alpha$ which becomes the bifurcation parameter in our system of ODE's. The larger $\alpha$ the more energy the neglected modes take from our system so that the flow system should be laminar. As $\alpha$ decreases less energy is extracted so that we expect the system to become chaotic.

## 6. TRUNCATIONS AND DISCUSSION

At the present time various truncations are being explored. A basic guide in the selection of which terms to keep is to retain a minimum number of terms but, to keep as much energy in the system as is necessary to retain the essential dynamics of the flow phenomena (v. Aubry et al. 1988). Figure 1 is a plot of the dominant eigenvalue obtained from the application of the POD, plotted versus streamwise wavenumber and azimuthal mode number.

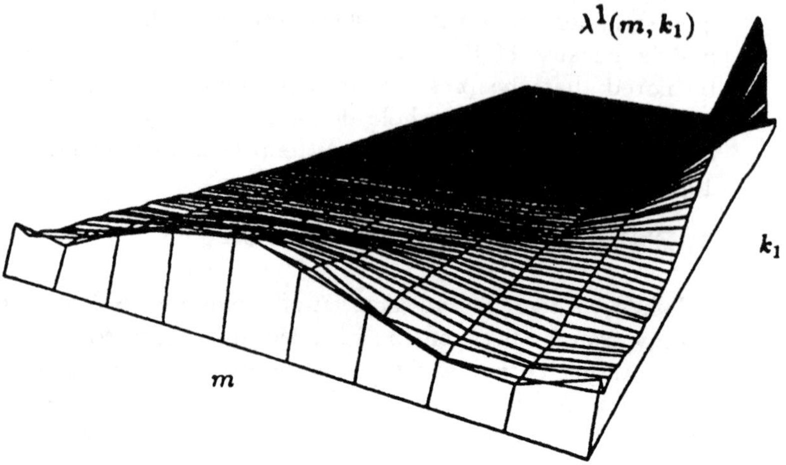


Fig.1. Dominant eigenvalue plotted as a function $k_1$ and $m$.

This plot indicates that there is an exchange of energy between streamwise wavenumbers and azimuthal mode numbers, and in particular, that there is a maximum energy path (remember that the eigenvalues are energy integrated across the jet shear layer) between the two peaks. The peak in the wavenumber direction corresponds to the Strouhal frequency of the jet. The peak in the azimuthal mode number direction is approximately at mode 6. Several azimuthal modes around the peak are kept as well as several around the peak in the streamwise wavenumber direction in order to retain the essential dynamics. For the initial model presently being studied, only the dominant eigenfunction will be used in the inhomogeneous direction. This corresponds to setting $l = n = p = q = r = 1$ in equation (4-5). This is justified because the dominant eigenvalue is significantly larger than the next smallest eigenvalue and in fact, typically contains 50 percent or more of the energy. All of the higher modes exhibit this same dominance of the first eigenvalue (v. Glauser and George 1987). The specific streamwise wavenumbers and azimuthal modes used in the initial study are shown in figure 2. For $k_1 = 0$, six azimuthal mode numbers are used and for $m = 0$ six values of $k_1 = 0$ are utilized. In addition to these, 6 other combinations are also used as can be seen from figure 2.

These were selected to try and capture the maximum energy path shown in figure 1. This particular truncation results in a set of 18 complex or 36 real, ordinary differential equations. The actual equations can be found in Zheng (1990) and are not included here.

As was discussed earlier, figure 1 indicates that there is an exchange of energy between streamwise wavenumbers and azimuthal mode num-

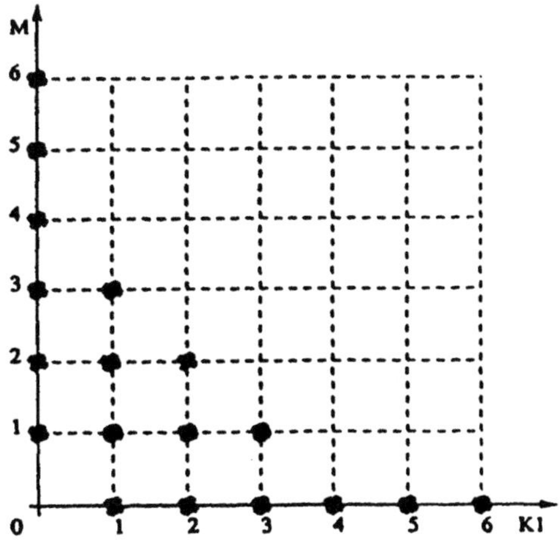


Fig.2. Modes retained in the dynamical systems model.

bers, and in particular, there is a maximum energy path between the two peaks. This is consistent with the mechanism for turbulence production suggested by Glauser and George (1987) and it is anticipated that the dynamical systems model will show this as well. This proposed mechanism is based on extensive azimuthal correlations taken in the near field jet shear layer at $x/D = 3$ and their subsequent breakdown into azimuthal modes. The proposed mechanism consists of four stages which are shown in figure 3.

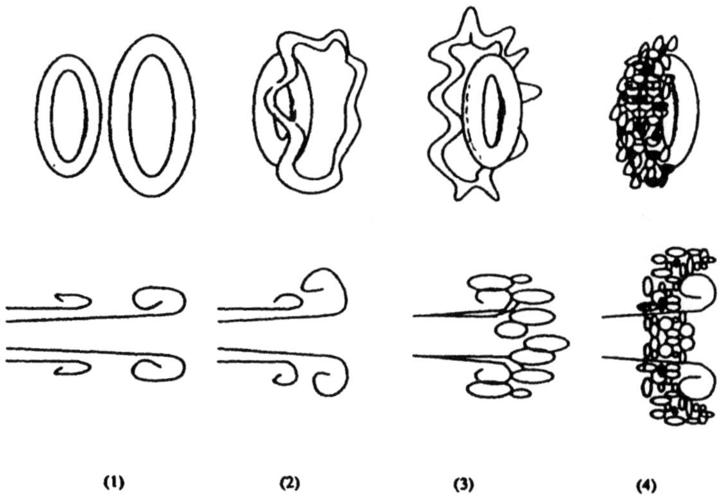


Fig.3. The *proposed* four stages of turbulence production.

Briefly, vortex ring-like concentrations arise from an instability of the base flow, the induced velocities from vortices which have already formed providing the perturbation for those which follow. This feedback mechanism has been suggested by others (v. Hussein 1986). These pairs of rings then behave like the textbook examples or inviscid rings, the rear vortex ring overtaking the vortex ring ahead of it, the rearward vortex being reduced in radius and the forward being expanded by their mutual interaction. The rearward ring is stabilized by the reduction in its vorticity (by compression) thus the predominance of the 0th mode (seen from the correlations, v. Glauser and George 1987) on the high speed side (core region). The forward ring has its vorticity increased by stretching as it expands in radius. This narrowing of its core while the radius is expanding causes the leading vortex to become unstable (as in the Widnall-Sullivan mechanism (v. Widnall and Sullivan 1973 and Yamada and Matsui 1978), thus the predominance of the 4-6 modes (seen from the azimuthal correlations of Glauser and George 1987) from the center of the shear layer outwards. This growing wavy deformation of the leading vortex ring causes it to acquire a streamwise component of vorticity consistent with the numerical results of Meiburg et al. (1989) and the experimental results of Hussain (1986). The continued effect of the rearward vortex on the forward one, and the now highly distorted rings interaction with itself, accelerates the instability until its vorticity is now entirely in small scale motions, in effect an energy cascade from modes 4-6 all the way to dissipative scales. This incoherent turbulence is swept from the outside where it has been carried, back to the center of the mixing layer as the still-intact rearward vortex passes. It is this collection of the debris, both small-scale vorticity and fluid material, which has been recognized as "pairing." The entire process is repeated as a new rearward vortex overtakes and destabilizes the one ahead of it.

## CONCLUSIONS

Dynamical equations have been derived for the random coefficients in the axisymmetric jet shear layer. These equations exhibit, as expected the same type of nonlinearities (quadratic and cubic) as those of Aubry et al. (1988). The equations are currently being solved numerically.

## ACKNOWLEDGEMENTS

This work was supported by NSF under grant number MSM–8808872 and Pratt and Whitney/UTRC. These sources of support are gratefully acknowledged.

## REFERENCES


Aubry, N., Holmes, P., Lumley, J.L. and Stone, E. (1988) The Dynamics of Coherent Structures in the Wall Region of a Turbulent Boundary Layer, *J. Fluid Mech.* **192**, *115-173.*

Bartuccelli, M., Constantin, P., Doering, C.R., Gibbon, J.D., and Gisselfalt, M. (1989) Hard Turbulence in a Finite Dimensional Dynamical System?, in press *Physics Letters A - Nonlinear Science.*

Glauser, M.N. and George, W.K. (1987) An Orthogonal Decomposition of the Axisymmetric Jet Mixing Layer Utilizing Cross-Wire Velocity Measurements. *Proc. 6th Symp. Turbulent Shear Flows, Toulouse, France, Sept. 7-9.*

Glauser, M.N., Leib, S.J. and George, W.K. (1987) Coherent Structures in the Axisymmetric Jet Mixing Layer. *Turbulent Shear Flows 5, Springer-Verlag.*

Herzog, S. (1986) The Large Scale Structure in the Near-Wall Region of Turbulent Pipe Flow. PH.D. Dissertation, Cornell University.

Hussain, A.K.M.F., (1986) Coherent Structures and Turbulence. *J. Fluid Mech.* **173**, *303-356*

Lin, S.P., Chen, H.B., and Tobak, M.(1987) Instability of A Nonparallel Flow, *Proceedings of 1st International Conference of Fluid Mechanics, Beijing.*

Lumley, J.L. (1967) The Structure of Inhomogeneous Turbulent Flows, In *Atm. Turb. and Radio Wave. Prop. (Yaglom and Tatarsky, ed.) Nauka, Moscow, 166-1,8.*

Meiburg, E., Lasheras, J.C., and Martin, J.E. (1989) Experimental and Numerical Analysis of the Three-Dimensional Evolution of an Axisymmetric Jet, *Seventh Symposium on Turbulent Shear Flows, Stanford University, August 21-23, 1989*

Sirovich, L. and Rodriguez, J.D. (1987) Coherent Structures and Chaos: A Model Problem, *Physics Letters A,* **120**, *Number 5.*

Widnall, S.E. and Sullivan, J.P. (1973) *Proc. R. Soc. London Ser. A332, 335.*

Yamada, H. and Matsui, T. (1978), Preliminary study of mutual slip–through of a pair of vortices. *Physics of Fluids,* **21**, *Number 2.*

Zheng,X. (1990), A Low Dimensional Description of the Axisymmetric Jet Mixing Layer. PH.D. Thesis , Dept Mechanical Engineering, Clarkson University.

# Instability of the rectangular duct flow and generation of the secondary flow

Tomomasa TATSUMI[α] and Takahiro YOSHIMURA[β]

[α] *Kyoto Institute of Technology*
*Kyoto 606, Japan*

[β] *Department of Physics, Faculty of Science, Kyoto University*
*Kyoto 606, Japan*

*Turbulence and Coherent Structures*
*O. Métais and M. Lesieur, eds.*
*Selected papers from "Turbulence 89: organized structures and turbulence in fluid mechanics"*
*Grenoble, 18-21 September 1989*

A prominient feature of the turbulent flows in ducts with non-circular cross-sections is the "secondary flow", which is the earliest known coherent structure of turbulent flows (Prandtl, 1926). It is our aim to derive this coherent vortex structure in the rectangular duct flow as the equilibrium state of the finite-amplitude disturbances in the flow.

First, the stability of the laminar flow in a rectangular duct with an arbitrary aspect ratio $A$ is investigated. The velocity components of both the basic laminar flow and the disturbance are expanded into the Legendre polynomials and a matrix equation for the expansion coefficients is derived. By solving the eigenvalue problem of the matrix equation, the stability characteristics of the basic flow at a Reynolds number $R$ to the disturbance of the wavenumber $\alpha$ are determined.

The flow is found to be unstable for $A \geq A_c = 3.2$ and stable otherwise, with the critical Reynolds number $R_c$ decreasing with increasing $A$: $R_c =$ 36600, 18400, 10400, 8200 and 6800 for $A = 3.5, 4, 5, 6$ and 8 respectively. This trend is in accordance with the known result that $R_c = 5772$ for the limit of $A \to \infty$ or the plane Poiseuille flow (Orszag, 1971). The flow field of the disturbance clearly shows the existence of thin high shear layers along the critical layer at which the velocity of the basic flow equals the phase-velocity of the disturbance.

Lastly, the generation of the secondary flow is briefly discussed. The nonlinear growth of the disturbance of the fundamental wavenumber $\alpha$ is considered and the evolution of the zero-wavenumber harmonic which is responsible to the secondary flow is examined. The contribution to this harmonic from the nonlinear inertia term is worked out numerically, and the flow field of this contribution is found to be qualitatively similar to Prandtl's picture of the secondary flow but more concentrated to the region around the critical layer.

## 1. Introduction

It is well known that turbulent fluid motions have two-fold aspects: the apparent randomness and the intrinsic determinism. In fact, fluid motions are genuinly deterministic in the sense that they are governed by certain autonomous equations of motion, but on the other hand, they show random behaviour when they become unstable against external disturbances at supercritical state.

In a fully developed turbulent state, which is far from the critical state, the flow field is composed of fully randomized small-scale eddy motions and relatively coherent large-scale motions. It is indeed an unexpected finding of organized vortex-row structures in a turbulent mixing layer by Brown & Roshko (1974) that has attracted interests of researchers to the coherent large-scale structures in turbulent flows.

Among various coherent structures, including the above-mentioned vortex-rows and the horse-shoe vortices in turbulent boundary layers, those having the most definite identity are the stationary vortex structures such as the Taylor-Görtler vortices which are formed along a streamwise curved solid wall. These vortices can be derived theoretically as the solution of the linearlized disturbance equation to the basic laminar flow and their evolution in time can also be predicted theoretically. In fact, this type of flows constitute the turbulent flows whose dynamical properties are best clarified.

The equally familiar but less understood coherent structures in turbulent flows are the so-called "secondary flow" of the turbulent flow in a straight duct of non-circular cross-section. This is the subject of our study beginning with the present work.

## 2. Secondary flow

The existence of the secondary flow was first suggested by Prandtl (1926) in order to explain the experimental contours of the mean axial velocity of turbulent flows through a straight duct of non-circular cross-section. The contours for the flow in a rectangular duct which were measured by Nikuradse (1926) are shown in Fig.1($a$). It may be noticed that, instead of rapidly assuming an oval shape as shown in Fig.2 for laminar flows, the contours near the walls show marked indentations near the corners. These indentations may be explained if the secondary flow of the type shown in Fig.1($b$) exists. The marked curvature of the velocity contours implies a mean velocity component normal to the velocity contours from the concave to the convex side. Experimental verification of the existence of the secondary flows was made by Nikuradse (1930) using a photographic method for triangular cross-sections, but no such a verification has been given for rectangular cross-sections. The secondary flow is nothing but a cross-section of longitudinal vortex structures in turbulent duct flow, which are the earliest known coherent motions in turbulent shear flows.

The secondary flow in turbulent duct flows has been dealt with theoretically by many authors mostly using statistical models for the Reynolds stress, but the results obtained so far are more or less model dependent and may not be taken as rigorous theoretical prediction. Another way of approach to this flow is to investigate the instability of the basic laminar flow and derive the secondary flow as the result of unstable disturbances. Unlike the Taylor-Görtler vortices, the secondary flow is not obtainable as the solution of the linearized disturbance equation but can be derived as the second-order solution of the nonlinear disturbance equation.

The theory of hydrodynamical stability has so far been applied to relatively simple flows whose velocity distributions depend only upon a single space variable, at least in approximate sense. Two-demensional and axi-symmetric flows including the plane Poiseuille flow, the Hagen-Poiseuille flow, the Blasius boundary layer, free mixing layers, jets and wakes belong to this category of flows. On the other hand, those flows which have more complex velocity distributions were left untouched since the semi-analytical methods which were useful for the simple flows become cumbersome for the complex flows. These flows may be treated more conveniently by numerical method combined with the orthogonal function expansion of the velocity field. In what follows the Legendre polynomials are used as the orthogonal functions in view of their rapid convergence and feasibilty of evaluating integrals.

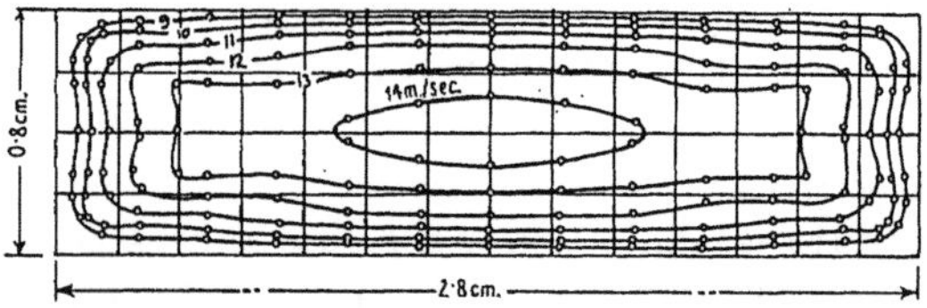


(a)

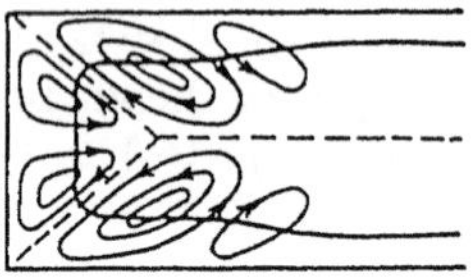

(b)

Fig.1. Secondary flow in a rectangular duct. (*a*) Contours of constant axial velocity (after Nikuradse, 1926). (*b*) Streamlines of the secondary flow (after Plantdl, 1926).

## 3. Rectangular duct flow

The velocity field of the laminar flow through a rectangular duct of an arbitrary aspect ratio $A(\geq 1)$ can be obtained as the solution of two-dimensional potential problem in the rectangular domain, and the problem was already solved by Saint-Venant (1855) as a mathematically equivalent torsion problem of an elastic column of rectangular cross-section. Although his solution was expressed in terms of a series of trigonometric and hyperbolic functions, here the solution is expanded in the Legendre polynomials for convenience of dealing with the stability problem as well.

The velocity $\mathbf{u}$ and the pressure $p$ of the flow are governed by the following equations of motion and continuity:

$$\frac{\partial \mathbf{u}}{\partial t} + (\mathbf{u} \cdot \mathrm{grad})\mathbf{u} = -\mathrm{grad}\, p + \frac{1}{R}\Delta \mathbf{u}, \qquad (3-1)$$

$$\mathrm{div}\, \mathbf{u} = 0, \qquad (3-2)$$

where all variables have been non-dimensionalized using a characteristic length $L_0$ and velocity $U_0$ of the flow, and

$$R = U_0\, L_0/\nu, \qquad (3-3)$$

$\nu$ being the kinetic viscosity of the fluid, represents the Reynolds number characterizing the flow.

Take the rectangular coodinate $\mathbf{x} = (x, y, z)$ with the $x$-axis along the central axis of the duct and the $y$- and $z$-axes each parallel to the narrower and wider sidewalls respectively. If we express the width of the narrower and wider walls as $2L_0$ and $2AL_0$ respectively, the walls are expressed in non-dimensional form as $y = \pm 1$, $z = \pm A$, where $A \geq 1$ represents the aspect ratio. Then, the boundary conditions for (3-1) and (3-2) are given by

$$\mathbf{u} = 0 \quad \text{at} \quad y = \pm 1 \quad \text{and} \quad z = \pm A. \qquad (3-4)$$

In order to solve the boundary-value problem related to (3-1),(3-2) and (3-4), we expand the velocity $\mathbf{u}$ and the pressure $p$ into series of the Legendre polynomials with respect to $y$ and $z$. Concerning the detailed expressions for the cases of the stationary laminar flow and its stability, reference may be made to Tatsumi & Yoshimura (1990).

If we asume the stationary laminar flow in a recangular duct to be pararell to the axis of the duct and unchanged in the axial direction, the velocity $\mathbf{u}$ and the pressure $p$ are expressed in the following form:

$$\left.\begin{aligned} \mathbf{u}(\mathbf{x}, t) &= \mathbf{U}(\mathbf{x}) = (U(y, z), 0, 0), \\ p(\mathbf{x}, t) &= P(\mathbf{x}). \end{aligned}\right\} \qquad (3-5)$$

Substituting (3-5) into (3-1) and (3-2), we find that (3-2) is automatically satisfied and (3-1) is reduced to the following equations:

$$\frac{\partial P}{\partial x} = -\kappa = \text{const.}, \qquad \frac{\partial P}{\partial y} = \frac{\partial P}{\partial z} = 0, \tag{3-6}$$

$$\left(\frac{\partial^2}{\partial y^2} + \frac{\partial^2}{\partial z^2}\right) U(y,z) = -R\kappa, \tag{3-7}$$

where $\kappa$ is a positive constant representing the pressure drop downstream. It also follows from (3-4) that

$$U(y,z) = 0 \quad \text{at} \quad y = \pm 1 \quad \text{and} \quad z = \pm A, \tag{3-8}$$

which gives the boundary condition for (3-7).

The velocity distribution $U(y,z)$ is calculated numerically by solving the equation (3-7) under the boundary condition (3-8) for $A = 1, 2, \ldots,$ 10, 20 and 30, and the results for $A = 3$ and 4 are shown graphically in Fig.2 as examples. It may be seen that the contour of constant velocity takes an oval shape for large constants unlike the indented shape for the turbulent flow. For large aspect ratios, $A \geq 5$ say, the velocity distribution becomes to be composed of three regions: a central region which is equivalent to the plane Poiseuille flow with parabolic velocity profile and a pair of terminal regions which have a universal velocity distribution irrespective to the aspect ratio.

The flux of the flow and its asymptotic formula for large aspect ratio are immediately derived from the above result and confirmed to be in agreement with classical results up to the digits calculated.

## 4. Stability of rectangular duct flow

The velocity $\hat{\mathbf{u}}$ and the pressure $\hat{p}$ of the external disturbance are decomposed into the Fourier series in the $x$ coordinate and a component with the wavenumber $\alpha$ may be expressed in the form,

$$\left.\begin{aligned} \begin{pmatrix} \hat{\mathbf{u}} \\ \hat{p} \end{pmatrix}(\mathbf{x},t) &= \begin{pmatrix} \tilde{\mathbf{u}} \\ \tilde{p} \end{pmatrix}(y,z) exp\left[i\alpha(x-ct)\right], \\ \tilde{\mathbf{u}} &= (\tilde{u}, \tilde{v}, \tilde{w}). \end{aligned}\right\} \tag{4-1}$$

where $c$ is a complex constant $c = c_r + ic_i$, $c_r$ representing the phase velocity and $\alpha c_i$ the logarithmic amplification rate of the disturbance. Substituting the expressions (4-1) into the equations of motion and continuity, and neglecting nonlinear terms with respect to $\tilde{\mathbf{u}}$ and $\tilde{p}$, we obtain a set of linear equations for the disturbance. These equations are further simplified by eliminating $\tilde{u}$ and $\tilde{p}$ and written in the form of a pair of linear equations for $\tilde{v}$ and $\tilde{w}$ including $U$ as the coefficient function.

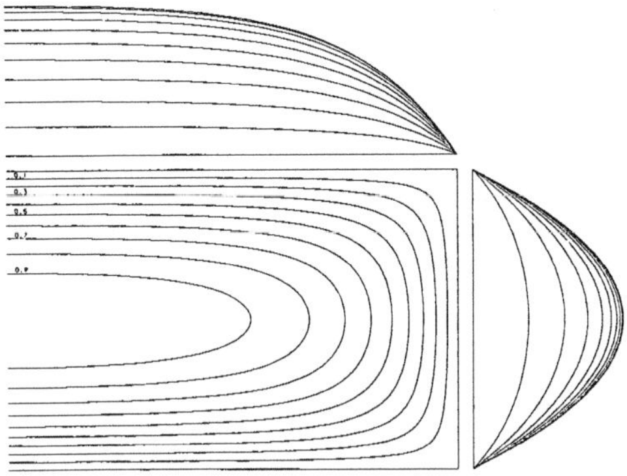

(a)

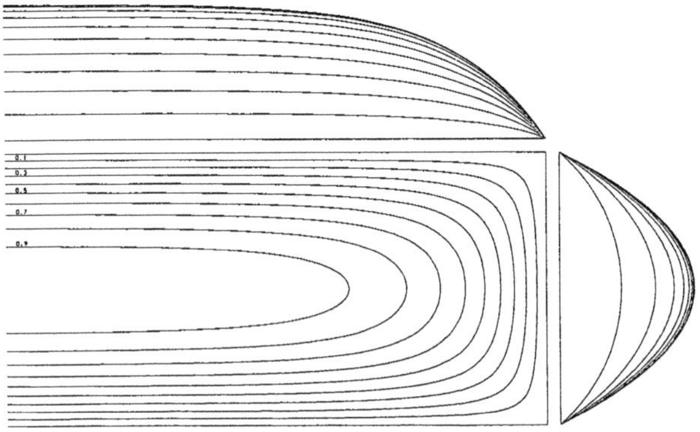

(b)

Fig.2. Contours of constant axial velocity for $U = 0, 0.1, \ldots, 1.0$ and velocity profiles at sections $y$ or $z = 0, 0.1, \ldots, 1.0$. (*a*) $A = 3$. (*b*) $A = 4$.

The stability of the basic flow (3-5) is examined by substituting the Legendre polynomial expansion of the velocities of both the basic flow (3-5) and the disturbance (4-1) into the disturbance equations and the boundary conditions and then solving an eigenvalue problem of the resulting matrix equation. The positive, negative and zero eigenvalues of $c_i$ represent the unstable, stable and critical states of the basic flow respectively.

For the basic flow whose velocity $U(y, z)$ is an even function of $y$ and $z$, as is the case for (3-5), the disturbance equations allow the following four modes of solution with different parity:

$$\text{I:}\quad (\tilde{v}(\mathrm{e},\mathrm{e}), \tilde{w}(\mathrm{o},\mathrm{o})), \qquad \text{II:}\quad (\tilde{v}(\mathrm{e},\mathrm{o}), \tilde{w}(\mathrm{o},\mathrm{e})),$$
$$\text{III:}\quad (\tilde{v}(\mathrm{o},\mathrm{e}), \tilde{w}(\mathrm{e},\mathrm{o})), \qquad \text{IV:}\quad (\tilde{v}(\mathrm{o},\mathrm{o}), \tilde{w}(\mathrm{e},\mathrm{e})),$$

where, for instance, $\tilde{v}(\mathrm{e},\mathrm{o})$ means that $\tilde{v}$ is an even and odd function of $y$ and $z$ respectively. The general features of the velocity field of each mode may be recognized from Fig.3 which shows the instantaneous velocity fields of the four modes for $A = 1$, $R = 1000$ and $\alpha = 1$ with unspecified initial phase.

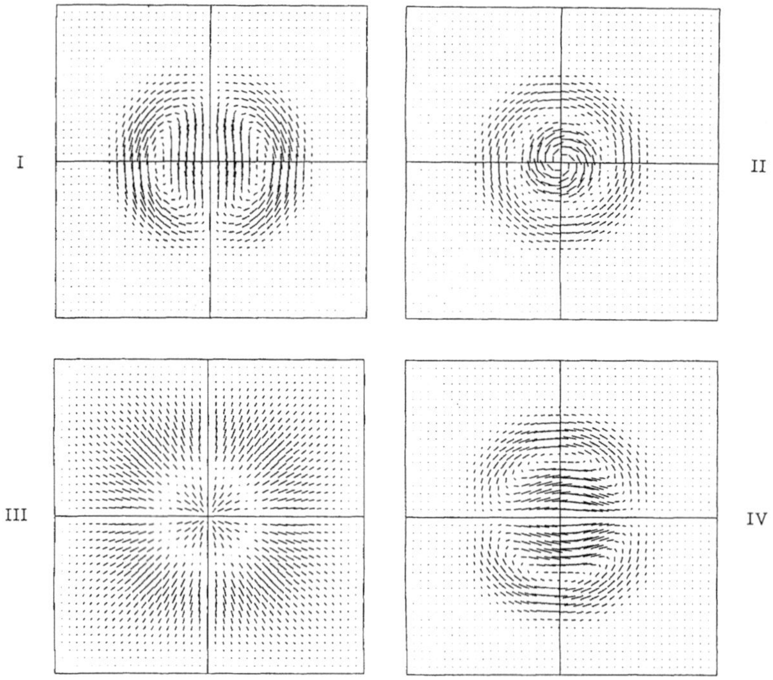


Fig.3. Velocity fields of the modes I, II, III and IV of the disturbance in the $(y, z)$-plane. $A = 1$, $R = 1000$, $\alpha = 1.0$, the phase arbitrary.

For the square cross-section $A = 1$, the modes I and IV become identical and there remain only three different modes. In the limit of $A \to \infty$, on the other hand, the modes I and II tend to the symmetric $\tilde{v}$ disturbance and the mode III and IV tend to the antisymmetric disturbance, the $\tilde{w}$ component disappearing in either case. The plane Poiseuille flow corresponding to the limit $A \to \infty$ is known to be unstable with the critical Reynolds number $R_c = 5772$ against the symmetric $\tilde{v}$ disturbance and stable against the antisymmetric disturbance (Orszag, 1971). Thus, the modes I and II are expected to have finite Reynolds numbers for certain ranges of finite aspect ratio $A$, whereas there is no such a guarantee for the modes III and IV.

## 5. Numerical results

The stabilty of the basic flow has been examined against the four modes for the Reynolds numbers up to 50000 and the wavenumbers below 1.1. It is found that there is no particular mode which is most unstable or least stable throughout all ranges of the parameters, but the mode I is the most unstable in the sense that it becomes unstable first with increasing values of the aspect ratio and the Reynolds number. Thus, according to the eigenvalues of this mode, the flow is stable for small aspect ratios $A \leq 3$ and unstable for large ones $A \geq 4$ with the following values for the critical Reynolds number $R_c$ and the wavenumber $\alpha_c$ :

| $A$ | 3.5 | 4 | 5 | 6 | 8 | $\infty$ |
|---|---|---|---|---|---|---|
| $R_c$ | 36600 | 18400 | 10400 | 8200 | 6800 | 5772 |
| $\alpha_c$ | 0.71 | 0.80 | 0.91 | 0.94 | 0.98 | 1.02 |

The value of $R_c$ for each aspect ratio $A$ has been determined as the minimum Reynolds number on the neutral stability curve in the $(R, \alpha)$ plane as shown in Fig.4, and the critical wavenumber $\alpha_c$ is defined as the wavenumber corresponding to $R = R_c$. The values of $R_c$ and $\alpha_c$ for $A = \infty$ have been calculated separately by solving the $\tilde{v}$ disturbance equation in this limit, and they are in perfect agreement with the values given by Orszag (1971) for the plane Poiseuille flow, up to the indicated digits.

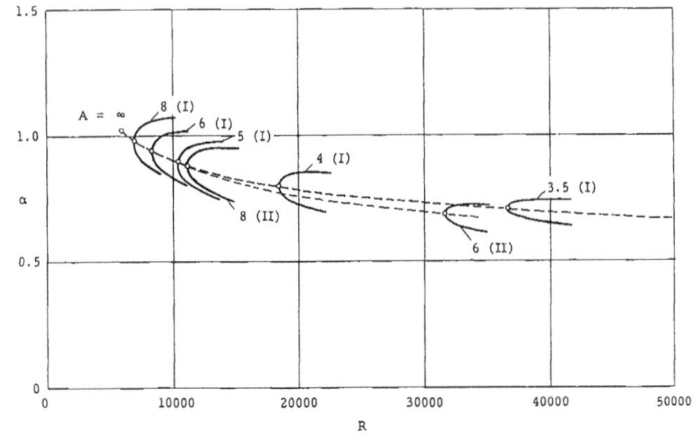


Fig.4. Neutral curves near the critical points for various aspect ratios: $A = 8, 6, 5, 4$ and 3.5 for the mode I; and $A = 8$ and 6 for the mode II.

Fig.4 shows that the critical point $(R_c, \alpha_c)$ for the mode I moves along a smooth line with increasing $A$ and seems to approach the point for the plane Poiseuille flow in the limit of $A \to \infty$. With decreasing $A$, the critical point moves along the line toward larger values of $R_c$ and seems to tend to infinity for a certain critical value of $A$. It is difficult to determine accurately the critical aspect ratio $A_c$ which separates the stability and instability of the rectangular duct flow, but extraporation from the above table gives $A_c = 3.2$ as a rough estimate. Contours of constant $c_i$ for the stable and unstable disturbances can be obtained in similar manner as the neutral curve and the result for the mode I disturbance at $A = 5$ is shown in Fig.5.

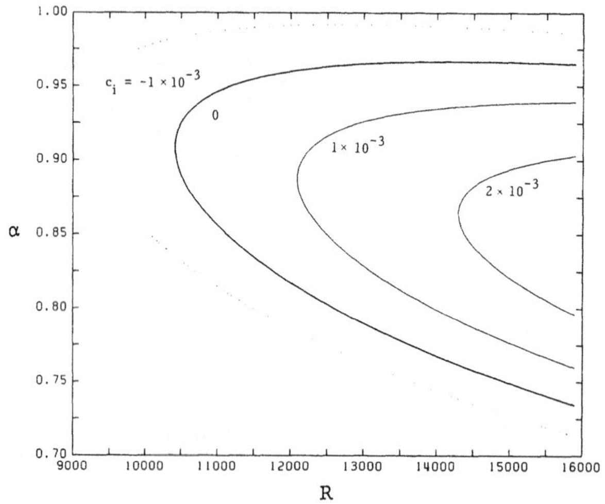


Fig.5. Contours of constant $c_i$ for the mode I disturbance. $A = 5$.

The mode II of the disturbance, which also reduces to the symmetric $\tilde{v}$ disturbance in the limit of $A \to \infty$, is found to be unstable, but with higher value of $R_c$ for each aspect ratio $A$ than the mode I as shown in the following table and in Fig.4:

| $A$ | 6 | 8 |
|---|---|---|
| $R_c$ | 31500 | 11000 |
| $\alpha_c$ | 0.69 | 0.89 |

The modes III and IV, both of which reduce to the stable antisymmetric $\tilde{v}$ disturbance in the limit of $A \to \infty$, are found to be stable for the range of the parameters examined.

## 6. Comparison with experiments

A number of experimental works have so far been done on the instability and transition of the rectangular duct flows of various aspect ratios. The earlier works by Schiller (1923, $A = 1$ and 3), Davies & White (1928, $A = 38$ to 710), Cornish (1928, $A = 2.92$) and others were, however, all hydraulic experiments which detect the transition by measuring the pressure drop along the flow direction. The critical Reynolds numbers obtained by these authors are considerably lower than theoretical values, but in view of the nature of their measurements it is likely that these experimental values are related with some sort of nonlinear instability rather than the linear instability of present concern.

Direct measurement of the instability of the flow was first made by Kao & Park (1970, $A = 8$) using modern technique of examining growth or decay of the controlled artificial disturbances. Their results expressed as the neutral stability curve in the $(R, \alpha)$ plane, gives the critical Reynolds number $R_c = 2600$ and $\alpha_c = 1.5$. Since their definition of the Reynolds number is based on the hydraulic diameter $d$ $(= 4AL_0/(1+A))$ and the mean velocity $U_m$ $(= Q/4AL_0^2$, $Q$ being the volume flux), their value of $R_c$ for $A = 8$ corresponds to $R_c = 1191$ in the present definition. These experimental values of $R_c$ and $\alpha_c$ do not agree with the corresponding theoretical values for either modes I or II.

At present we have no clear explanation for this discrepancy, but there are some reasons to believe that these experimental values are not related with the linear stability of the flow. First, the flow with a finite aspect ratio $A$ is considered to be stabilized by the presence of the end walls compared with that of infinite $A$ or the plane Poiseuille flow, so that the critical Reynolds number for $A = 8$ should be higher than $R_c = 5772$ for the latter. Theoretical result that gives higher critical Reynolds number with decreasing aspect ratio $A$ for both the modes I and II actually supports this conjecture. Secondly, the experimental neutral curve covers much a wider wavenumber range than the corresponding theoretical curves in Fig.4 and the unstable region includes the zero wavenumber range which is known to be linearly stable. According to these reasons the experimental instability obtatined by Kao & Park (1970) using controlled excitation could be attributed to a sort of nolinear subcritical instability. More detailed and precise measurements are needed in order to establish experimentally the linear stability characteristics of the flow.

The velocity field and the isovorticity contours of the disturbance of the mode I in a cross-sectional plane are displayed in Fig.6($a$) and ($b$) respectively for $A = 5$, $R = 12000$ and $\alpha = 0.9$. As easily seen from the neutral curve in Fig.4, this disturbance is slightly unstable at $R = 12000$ and stable at $R = 10000$, but the velocity and vorticity fields for these Reynolds numbers are found to be almost identical to each other. It may be observed in Fig.6($a$) and ($b$) that the disturbance fields contain thin regions of very high shear along the critical layer $U = c_r = 0.24$ which is indicated by a broken-line contour in Fig.6($c$).

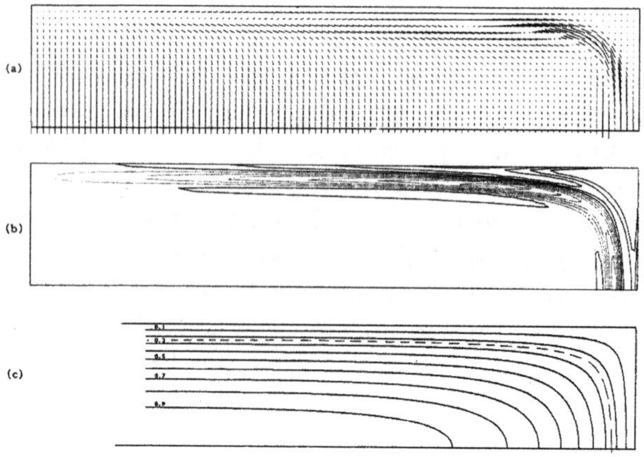


Fig.6. Flow patterns of the disturbance in the $(y, z)$-plane. $A = 5$, $R = 12000$, $\alpha = 0.9$, the phase arbitrary. ($a$) Velocity vector field. ($b$) Isovorticity contours. ($c$) Isovelocity contours of the basic flow. Broken line indicates the critical layer, $U = c_r = 0.24$.

## 7. Generation of secondary flow

Lastly, a preliminary discussion will be made on the generation of the secondary flow in a rectangular duct flow. Let us consider the temporal evolution of an initially small unstable disturbance of a certain wavenumber $\alpha$. As the disturbance grows up to a finite magnitude, a number of higher harmonics are generated due to the nonlinearity of the equation of motion, and the velocity and the pressure of the disturbance are now expressed in the form,

$$\begin{pmatrix} \hat{\mathbf{u}} \\ \hat{p} \end{pmatrix} (\mathbf{x}, t) = \sum_{n=-\infty}^{+\infty} \begin{pmatrix} \tilde{\mathbf{u}} \\ \tilde{p} \end{pmatrix} (y, z; t) exp\,[in\alpha x], \qquad (7-1)$$

where the $\tilde{\mathbf{u}}_n$ and $\tilde{p}_n$ are complex in general and satisfy the conditions, $\tilde{\mathbf{u}}_{-n} = \tilde{\mathbf{u}}_n^*$ and $\tilde{p}_{-n} = \tilde{p}_n^*$, $*$ denoting the complex conjugate, which follow from the reality of the left-hand side of (7-1).

The disturbance of finite magnitude does no longer grow up exponentially in time like a small disturbance as expressed by (4-1), but its growth is slowed down since the energy supplied to the fundamental of the disturbance is consumed more effectively by the nonlinear transfer of energy to its higher harmonics. Eventually, the disturbance attains an equilibrium magnitude under the balance between the energy supply from the basic laminar flow and the viscous dissipation of energy of the whole disturbance (7-1). Turbulent flow is thus realized as such an equilibrium state of the basic flow and the disturbance, and the secondary flow is observed as the transverse cross-section of the stationary part of such a disturbance.

In order to work out the equilibrium state of the disturbance we have to deal with the nonlinear problem concerning the full system of harmonics of the disturbance (7-1). Such a problem will be dealt with in a separate paper and here we shall only discuss the first nonlinear effect to the growth of the disturbance in its earlier stage.

In a turbulent flow which is composed of the basic flow (3-5) and the disturbance of the form of (7-1), the mean velocity field is given by $(U + \tilde{u}_0, \tilde{v}_0, \tilde{w}_0)$, where the component $\tilde{u}_0$ represents the reaction of the disturbance to the basic flow and the component field $(\tilde{v}_0, \tilde{w}_0)$ is nothing but the secondary flow in a transverse cross-section. The equations governing the amplitude functions $\tilde{\mathbf{u}}_n$ and $\tilde{p}_n$ of (7-1) are derived from substituting (7-1) into the equations (3-1) and (3-2) and making use of the orthogonality relations of the Fourier series. The linear parts of the equations for the fundamental $\tilde{\mathbf{u}}_1$ and $\tilde{p}_1$ are identical to the equations for the small disturbance (4-1), and the equations for the $n$-th harmonic $\tilde{\mathbf{u}}_n$ and $\tilde{p}_n$ include the corresponding linear terms and also the nonlinear terms which arise from the inertia term and the pressure term of the equation of motion (3-1).

At earlier stage of growth of the disturbance, there exist only the fundamentals $\tilde{\mathbf{u}}_{\pm 1}$ and $\tilde{p}_{\pm 1}$ of small magnitude, $O(\varepsilon)$ say, and then the zero-wavenumber harmonic $\tilde{\mathbf{u}}_0$ and $\tilde{p}_0$ and the $2\alpha$ wavenumber harmonic $\tilde{\mathbf{u}}_{\pm 2}$ and $\tilde{p}_{\pm 2}$, arise with the magnitudes of $O(\varepsilon^2)$. The contribution to the zero-wavenumber harmonic $\tilde{\mathbf{u}}_0$ from the nonlinear terms may be evaluated approximately by replacing $\tilde{\mathbf{u}}_1$ and $\tilde{p}_1$ by the linear solution $\tilde{\mathbf{u}} exp\,[i\alpha ct]$ and $\tilde{p} exp\,[i\alpha ct]$ respectively. Among such contributions to $\tilde{\mathbf{u}}_0$ from the nonlinear terms, the one from the inertia term has been worked out numerically for the initial disturbance of the mode I, $A = 5$, $R = 12000$ and $\alpha = 0.9$. Fig.7($a$) and ($b$) show the contributions to the velocity field and the isovorticity contours of the $(\tilde{v}_0, \tilde{w}_0)$ components respectively and Fig.7($c$) the contribution to the axial component $\tilde{u}_0$. It may be seen from Fig.7($a$) that the velocity field is qualitatively similar to Prandtl's picture of the secondary flow as shown in Fig.1($b$) but more concentrated to the region around the critical layer. Fig.7($c$) also shows that the component $\tilde{u}_0$ gives rise to the indentation of the isovelocity contours in the same sense as the experimental contours by Nikuradse as shown in Fig.1($a$). The above comparisons suggest that the nonlinear effects considered here produce qualitatively favourable results for explaining the secondary flow and the indented contours of the axial velocity in a rectangular duct flow, but to work out the equilibrium state of the turbulent duct flow remains as the subject of future works.

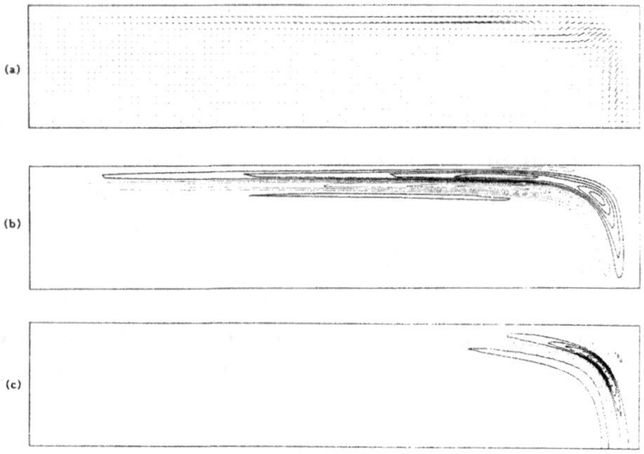


Fig.7. Flow patterns of the contributions to the secondary flow from the inertia term in the $(y, z)$-plane. Initial disturbance of the mode I, $A = 5$, $R = 12000$ and $\alpha = 0.9$. $(a)$ Velocity vector field. $(b)$ Isovorticity contours. $(c)$ Isovelocity contours of the axial flow.

**Acknowledgements**
The authors wish to express their hearty thanks to Prof. Takuji Kawahara, Prof. Michio Yamada and Dr. Masako Seki for their helpful advice. This work has been partially supported by the Grant in Aid for Scientific Research from the Ministry of Education, Science and Culture.

## References

Brown, G.L. & Roshko, A. 1974 On density effects and large structures in turbulent mixing layers. *J. Fluid Mech.* **64**. 775–814.

Cornish, R.J. 1928 Flow in a pipe of rectangular cross-section. *Proc. R. Soc. Lond.* A **120**. 691–700.

Davies, S.J. & White, C.M. 1928 An experimental study of the flow of water in pipes of rectangular section. *Proc. R. Soc. Lond.* A **119**. 92–107.

Kao, T.W. & Park, C. 1970 Experimental investigations of the stability of channel flows. Part 1. Flow of a single liquid in a rectangular channel. *J. Fluid Mech.* **43**. 145–164.

Nikuradse, J. 1926 Geschwindigkeitsverteilung in Strömungen. *Forschungsarbeiten des Ver. deuitsch. Ing.* No.281, pp.13–14.

Nikuradse, J. 1930 Untersuchungen über turbulente Strömungen in nicht kreisförmigen Rohren. *Ingenieur-Archiv.* **1**, 306–332.

Orszag, S. 1971 Accurate solution of the Orr-Sommerfeld stability equation. *J. Fluid Mech.* **50**. 689–703.

Prandl, L. 1926 Uber die ausgebildete Turbulenz. *Verhandl. 2 intern. Kongr. techn. Mech. Zürich*, pp.70–74.

Saint-Venant, B. de 1855 Mémoire sur la torsion de prismes. *Mémoire de l'Academie des Sciences des Savants Etrangers.* **14**. 233–560.

Schiller, L. 1923 Uber den Strömungswiderstand von Rohren vershiedenen Querschnits und Rauhigkeitsgrades. *Z. Angew. Math. Mech.* **3**, 2–13.

Tatsumi, T. & Yoshimura, T. 1990 Stability of the laminar flow in a rectangular duct. To be published in *J. Fluid Mech.* **212**.

# V. Vortex Breakdown

# Vortex breakdown: a coherent transition trigger in concentrated vortices

S. LEIBOVICH

*Cornell University*
*Ithaca, N.Y., U.S.A.*

*Turbulence and Coherent Structures*
*O. Métais and M. Lesieur, eds.*
*Selected papers from "Turbulence 89: organized structures and turbulence in fluid mechanics"*
*Grenoble, 18-21 September 1989*

## 1. Introduction

Vortex breakdowns are organized structures that usually lead to turbulent flow in vortices. They are among the many kinds of large amplitude disturbances that are possible in concentrated vortices having a streaming component of velocity along the vortex axis. While there is no universal agreement on a definition of "vortex breakdown", most observers associate the term with a condition in which there is a rapid axial deceleration of the fluid leading to a stagnation point and region of reversed axial flow. By rapid, I mean that the fluid on the vortex axis is brought to rest in distances comparable to the vortex core diameter. Since this distance is small compared to the length scales of interactions generally taking place in concentrated vortices, the occurrence of vortex breakdown appears as a dramatic event.

Vortex breakdown events in flows in closed geometries can remain completely laminar. The only example of this of which I am aware is the Vogel-Ronnenberg or VR flow (Vogel, 1968; Ronnenberg, 1977; Escudier, 1984; Lugt and Abboud, 1987) which takes place in a stationary cylinder with one rotating lid. Such flows are exceptional and have not yet proved of technological interest. While the the occurrence of breakdown in closed flows may prove useful in the future in establishing underlying physical mechanisms, the vortex flows that have been of technological interest to date are open flows.

When vortex breakdown occurs in open flows the process always seems to be nonaxisymmetric and unsteady, and to trigger a transition to turbulence leaving turbulent flow in its wake, whether the breakdown structure is itself laminar or turbulent. Vortex breakdown therefore falls in the category of a transitional flow.

The complexities of this phenomenon are such that no generally accepted view of the underlying mechanisms responsible for it has emerged, despite the importance of the problem and the efforts expended on it over a period of more than thirty years. Several reviews (Hall, 1966, 1972; Leibovich, 1978, 1984; Escudier, 1988) outline the few theoretical developments (all of which, as as might be expected, concentrate on highly idealized situations) which have been advanced. In this paper, I will not survey the various competing theoretical notions; instead, the reader is referred to the existing reviews. I will, however, argue that the experimental observations imply that vortex breakdown arises as system instability, and is jointly controlled by local effects, due to the detailed structure of the vortex core, and by global effects, due to the conditions imposed on the vortex on length scales much larger than the vortex core size by the flow in which the vortex resides.

Most attention has been devoted to the phenomenon as it arises in aerodynamics, because of its dramatic effects on the forces and moments on therefore on the controllability of aircraft (see, for example, Lambourne and Bryer, 1961, for early consideration of this issue, and Peake and Tobak, 1983, or Polhamus, 1986, for more recent discussions). The paper by Peake and Tobak shows a number of photographs of vortex breakdowns on aircraft, as does that by Erickson and Gilbert (1983).

Vortex breakdown arises in other contexts: is believed to occur in nature, in tornados and dust devils (see, for example, Maxworthy, 1973, or Fiedler and Rotunno, 1986), in flows over automobiles, and in internal flow devices, like gas turbine combustors (Gouldin et al. 1985). While breakdown generally leads to undesirable effects in external aerodynamics, its promotion would be desirable in trailing vortices at landing and takeoff to reduce the dangers trailing vortices pose to aircraft in their wake; and breakdown is promoted in swirling combustors because the low speed recirculating region functions there as a fluid mechanical flameholder.

Another intriguing occurrence of vortex breakdown has been noted by Hopfinger, Browand, and Gagne (1984), in an experiment on the effect of rotation on turbulence. In the experiment, an oscillating grid placed at the bottom of a rotating tank locally generates turbulence like that in a nonrotating system. Away from the immediate neighborhood of the grid, the turbulence is organized by the rotation, the most remarkable aspect of the organization being a series of vortices with core vorticity as much as 50 times the tank background. Their axes are parallel to the rotation axis with their "feet" rooted in the turbulent layer near the grid. Axial motion occurs in these vortices, and vortex breakdown. A subsequent study by Maxworthy, Hopfinger, and Redekopp (1986) provides added experimental detail. These experiments suggest other possibilities for the appearance of vortex breakdown in geophysical contexts, and this question deserves further attention.

Some of the important consequences of vortex breakdown arise from the turbulence it generates downstream. In the context of aerodynamic flows, for example, the mean structure of the vortex core in the wake of a breakdown event is substantially altered by turbulent mixing, and these structural changes, which include enlargement of the vortex core, affect the downstream development of the flow and its interaction with solid structures.

To study the fundamental aspects of vortex breakdown, it is preferable to isolate a single vortex, rather than deal with the complications of flows such as those shown in practical aerodynamic configurations. This has been done by a number of experimentalists (see the review articles cited for references), usually by creating a vortex in one of several ways, and then confining it to a tube. The data to be reviewed in the next section was obtained by Faler and Leibovich (1977, 1978), and by Garg and Leibovich (1979) in an apparatus modeled after Sarpkaya's (1972). In this apparatus, swirl is imparted to water drawn radially inward past a set vanes, and the flow is then turned so that it passes into a tube. The tube is in three sections, the inner walls of the first and third are cylindrical, and the central section has conical walls that diverge slightly. It is in the latter section where the vortex breakdown events to be discussed are stabilized in position.

## 2. Mean Flow Fields

Figure 1 is a photograph of vortex breakdown: the fluid enters at the left and exits to the right, and a thin filament of dye is introduced on the upstream axis. Two morphologically

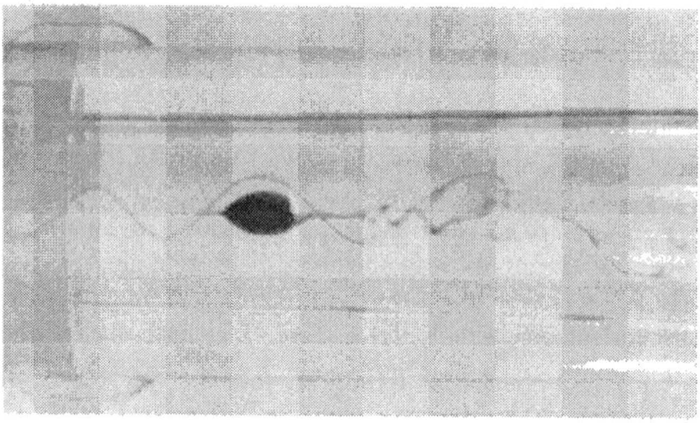

Fig.1. Photograph showing the simultaneous occurrence of "bubble" and "spiral" forms of vortex breakdown.

distinct forms of vortex breakdown typical of flows of aerodynamic type both appear in the picture. The first to occur as one proceeds downstream has been called the "bubble", or B- form of breakdown. It is made visible by dye which fills a recirculation region resembling a solid body of revolution with an aft portion that is not quite axisymmetric. A stagnation point marks the upstream end of the bubble. When the flow is averaged over azimuth (or over time, which has the same effect in these flows), then the structure is of course accurately axially-symmetric, and terminates at a downstream (mean) stagnation point. The dye enters the recirculation region across the downstream section of the structure, where the symmetry of the instantaneous flow is broken. Speeds in the bubble are much smaller than those typifying the flow upstream, or anywhere well outside the bubble, so the residence time of the dyed fluid is much greater in the bubble than it is at other fixed locations. Eventually, the dyed fluid elements leave the bubble, again through the same downstream section. In the flow in Figure 1, the ejected dye regroups near the axis, and then suffers another breakdown, this time of "spiral", or S- form. Here the instantaneous flow again is brought to rest at the upstream end of the breakdown. Reverse axial flow follows the stagnation point, but the dye does not collect in a recirculation zone, but rather continues along in a corkscrew fashion until instabilities occuring in the wake cause transition to turbulent flow.

The interior of the recirculation region of a B-type of breakdown is complicated. The mean streamlines in the interior, as reconstructed by Faler and Leibovich (1978) from measured velocity components, is shown in figure 2, and illustrates the complexity. In addition to the front and rear stagnation points of the main (mean) recirculation zone, there are two other interior stagnation points on the axis, marking a second (mean) recirculation zone nested within the main bubble. In the instantaneous flow, the downstream terminus of the interior bubble is the site of what appears to be an embryonic S-type of breakdown, which may prove to be the genesis of nonaxisymmetric motion within the bubble. It is as if a spiral heart beats within the breast of a B-breakdown.

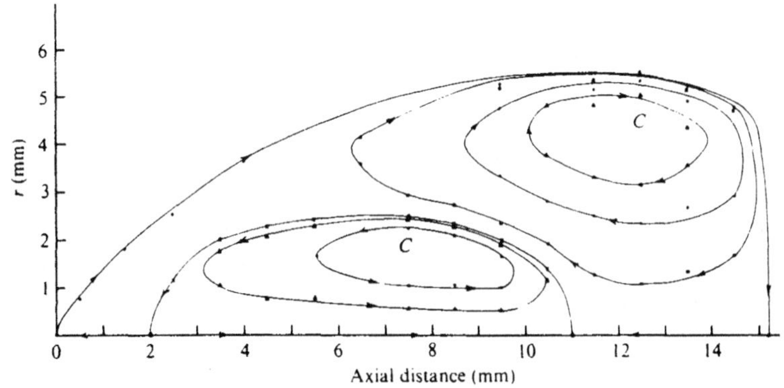


Fig.2. Mean streamlines in the recirculation zone of a B-breakdown.

In both B and S forms of breakdown, the radial extents of the coherent structures - which we associate with the recirculation zone, if there is an identifiable one, or the axial interval containing detectable reversed axial flow otherwise - are close to that of the local upstream vortex core, and the axial extents are one to perhaps three vortex core diameters. Both structures are azimuthally asymmetric and unsteady. The main form of unsteadiness is in azimuthal structure, but there is also a positional unsteadiness in which the entire structure wanders in the axial direction. For fixed experimental conditions, the wandering for a given type of breakdown is within a fixed axial range. The wandering, which appears to be chaotic, is to some extent controllable, and in any case generally takes place on a time scale large compared to fluctuations associated with azimuthal variations.

In flows in this apparatus, and typically in flows arising in aerodynamic contexts, the axial flow upstream of breakdown is jet-like. The maximum axial speeds occur on the vortex axis and are several (up to 3 or more) times the axial speeds outside the vortex core. Fluid at the jet axis is decelerated to rest at the front stagnation point of a vortex breakdown within an axial distance of one or two vortex core diameters. The flow downstream from the breakdown structure is turbulent and wake-like, with a mean velocity deficit resembling that found in the wake of a solid bluff body. Furthermore, the vortex core in the wake is decidedly larger than that upstream.

Figure 3 shows laser Doppler traverses of the axial and azimuthal velocity components both upstream of breakdown (the "approach flow") and at several locations in the wake of a B- type breakdown which occurs at the 2.3 cm station. Data for a range of experiments in this facility is summarized in Leibovich (1984).

There is no qualitative difference between the shapes of the velocity profiles in the approach flows, nor in the wake flows, of B and S-forms of breakdown, although the vortex core expansion in the wake is systematically greater for B-forms than for S-forms.

## 3. Wave Propagation Characteristics

Concentrated vortices admit dispersive waves which propagate along the vortex axis: the vortex core acts like a waveguide. These waves are similar in some waves to surface waves in water, or to internal waves. They are not self-excited, even though co-existing self-excited waves arise in vortices which are unstable. Qualitative explanations and predictions of the motions of vortex breakdown structures following transients caused by imposed alterations of flow conditions sometimes can be given using simple wave properties, as has been shown by Faler and Leibovich (1977).

As in the case of water waves on a moving stream, or gasdynamics, flows may be categorized as supercritical or subcritical (in gasdynamics, supersonic or subsonic) according to whether infinitesimal waves are capable of propagating upstream or not. The point of division is the critical flow. There are many families of waves in vortices, in contrast to waves on a moving stream of water of finite depth. The analog of water of (undisturbed) constant depth is the columnar vortex, which in

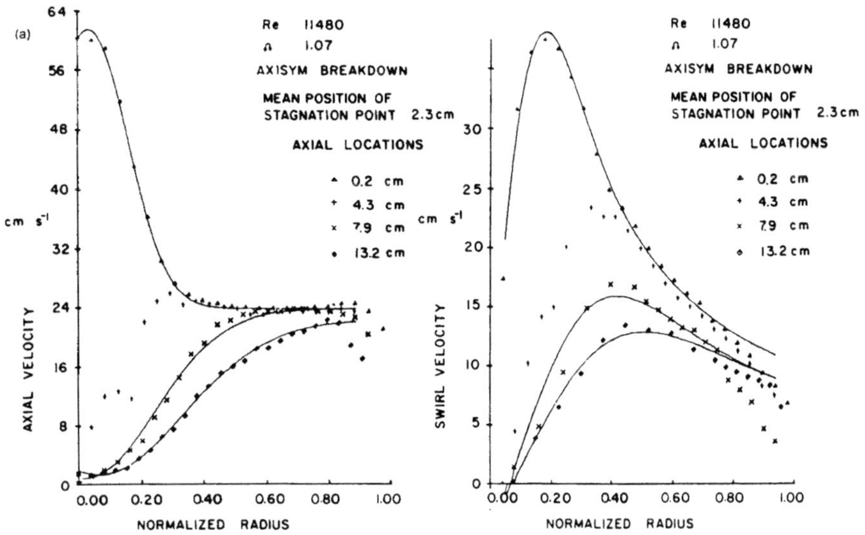


Fig.3. Typical time-averaged profiles of axial and swirl velocity components upstream and in the wake of a vortex breakdown. A B-type breakdown existed in this flow at a mean axial position of 2.3 cm.

its undisturbed state is axially symmetric, with a structure independent of distance along the vortex axis. For example, waves may be either axially symmetric or asymmetric. One really robust experimental feature of flows of vortices of aerodynamic type which subsequently suffer breakdown has been identified: approach flows are supercritical and (mean) wake flows are subcritical when the waves considered are axisymmetric. As a practical matter, to test this assertion given experimentally determined (mean) velocity profiles, one must first assume that the flows in question are columnar. A columnar vortex has an axially symmetric undisturbed state with structure independent of distance along the vortex axis, and is the analog of water of (undisturbed) constant depth.

### 3.1. Axisymmetric waves

Figure 4 summarizes the results of my analysis (Leibovich, 1984) of the data of Garg and Leibovich (1979), showing the transition "jumps" for both B and S-forms of breakdown. The "Froude number" $N$ has been calculated for upstream and downstream mean data assuming axial symmetry: $N > 1$ and $N < 1$ are supercritical and subcritical, respectively. Each flow is characterized by a dimensionless parameter $q$ that describes its relative level of swirl; alternatively, $q$ measures the ratio of the axial to the azimuthal vorticity. Each upstream flow is shown by a triangle, each downstream flow by a circle, and data for a given experiment is connected by a line with an arrow showing the direction of the transition. Transitions associated with B-type breakdowns cluster together, and those with S-types cluster in a different region. The figure also summarizes the stability characteristics of these flows. Here "strong" instability means the instability criterion derived by Keith Stewartson and me (Leibovich and Stewartson, 1983) is satisfied, leading to broadband three-dimensional inviscid instability with sizeable growth rates. The region simply marked "unstable" has been found numerically to be unstable for these experimental flows, and the growth rates are smaller. Notice that, at least according to the stability considerations accounted for in the figure, all approach flows are stable to three- dimensional (and axisymmetric) perturbations. All wake flows are unstable, and those following B-breakdowns are more violently unstable, according to the diagram. This suggests that the larger core expansions involved in B-breakdowns (in the data sets shown, core expansion ratios range from 2 to nearly 3 for B-types, and from 1.6 to slightly over 2 for S-types) is associated with their enhanced instability.

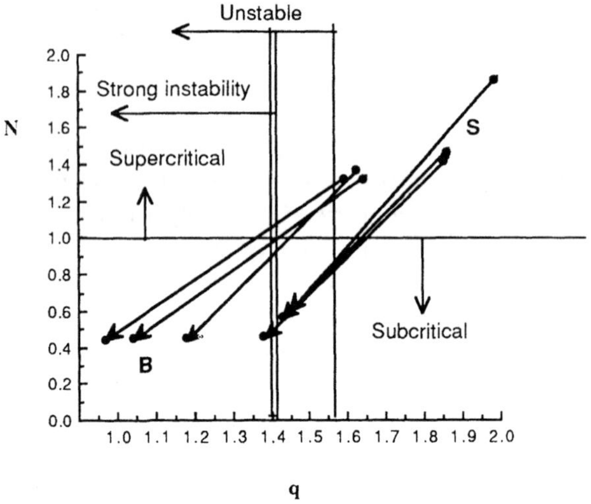


Fig.4. Summary of transitions for flows with bubble (B) types and spiral (S) types of vortex breakdowns. $N$ is the Froude number, and $q$ is a measure of the ratio of axial to azimuthal vorticity in the flow states analyzed.

Benjamin (1962) proposed a well-known theory of vortex breakdown in which approach flows are supercritical. In this theory, the approach flow undergoes a transition leading to a downstream axisymmetric subcritical flow "conjugate" to the upstream approach flow. Here conjugate refers to columnar flows sharing the same radial boundaries, mass flow, angular momentum distribution, and stagnation pressure distribution.

The transition connecting the two flow states is itself outside the scope of the theory. This clearly has parallels with the theory of shock waves in gases. In the latter case, however, the transition is fully determined by the Rankine-Hugoniot conditions independent of the shock wave structure. This is not the case in vortex breakdown. Here the analog of the shock structure is the recirculation region and contiguous zones with substantial axial accelerations, and the transition between approach and wake flows (the analog of the shock jump condition) cannot be determined without knowledge of the complete flow. Furthermore, although the transition region is of negligible extent compared to the variation of external conditions imposed on the vortex, it is not small compared to the vortex core, which is the only length scale available to the presumably columnar vortices joined by the transition.

### 3.2. Nonaxisymmetric waves

It may seem arbitrary to describe a flow as supercritical or subcritical based only on the way axially-symmetric perturbations behave. The theory of nonaxisymmetric waves is much richer than that for axisymmetric waves - there are many more possibilities and consequently fewer general statements can be made. It is not clear when, or indeed whether, nonaxisymmetric perturbations can propagate upstream to disturb a flow that is supercritical on axisymmetric grounds. Some information is known, however, about the data sets of Garg and Leibovich (1979) described above. Tsai and Widnall (1980) have numerically computed the dispersion relations and group velocities for "bending" modes, that is, those with azimuthal wavenumber $m = \pm 1$, for these data sets. For these examples, they find that when the flow is supercritical to axisymmetric waves, it is also supercritical to bending waves. A flow subcritical to axisymmetric waves, however, may be supercritical to bending waves.

The computational results are of course necessarily limited to the special velocity profiles assumed in the numerical work. There is information available concerning the long-wave dispersion relations for columnar vortices of arbitrary core structure in concentrated vortices in an unbounded fluid. Thus, Leibovich, Brown, and Patel (1986) (or LBP) show that an infinite number of branches of the dispersion relation exist. Those found fall into two categories: "slow" bending waves, and fast center modes of arbitrary azimuthal wavenumber $m$. In the limit of indefinitely long waves, the group and phase velocities on the slow dispersion branch tend to zero in a reference frame in which the axial velocity vanishes at large distances from the vortex centerline. Thus, any flow having a nonzero axial velocity away from the centerline is supercritical at least with respect to these long bending waves. Furthermore, LBP find that long center modes have a group velocity exceeding the axial speed on the axis by an amount proportional to $W''(0)\Omega(0)/\Omega''(0)$, where the proportionality factor is a positive function of the azimuthal wavenumber and mode number. Here $W(r)$ is the axial velocity and $\Omega(r)$ the angular velocity of the undisturbed columnar vortex, and the prime represents differentiation. Typical vortices have $\Omega(0)/\Omega''(0) < 0$. Therefore, vortices with jet-like axial velocities (where typically $W''(0) < 0$) are supercritical with respect to long center modes of all azimuthal wavenumber $m$. This additional information suggests that it may be generally true that flows supercritical to axisymmetric waves are also supercritical to nonaxisymmetric waves. The question, however, remains an open one.

### 3.3. Relation between criticality and stagnation

If a vortex changes its structure very slowly, then its propagation characteristics may be reasonably educed on a station- by- station basis, where at each axial location the vortex is treated locally as columnar. This is analogous to the common parallel-flow assumption for two-dimensional shear flows. Suppose a vortex is supercritical at some station, with a positive velocity on the axis. If this flow leads to vortex breakdown, the flow on the axis will decelerate as one proceeds downstream, eventually forming a stagnation point. The axial decelerations are forced by the variation of the waveguide. Imagining that this is so gradual that the locally columnar assumption can be made, we can ask whether the flow at the station where stagnation occurs is still supercritical, or whether it is subcritical to axisymmetric waves.

This question can be given a fairly general answer. The flow must have first become subcritical to axisymmetric disturbances before the station at which the stagnation point is located. For flows confined in radial extent, this may be seen from the variational arguments in Leibovich (1979), pp 406-409. It is proved there that if the minimum of the axial velocity $W(r)$ in the vortex occurs in the interior of the flow, or on a boundary where $W' = 0$, then there is a branch of axisymmetric waves with phase speeds less than the minimum of $W$ over $r$. It is a kinematic requirement that $W'(0) = 0$ for all flows. Therefore, if there is an incipient stagnation point on the axis, this will correspond to a global minimum speed, and the result cited means that there are waves with phase velocity directed upstream. Further, from section 4 of the same paper, this implies that the group speed of long waves is directed upstream, and so the flow is subcritical. The same conclusion concerning the upstream directed phase speed was arrived at by Cowley (private communication) for radially unbounded flow by a completely different argument. It is not immediately evident how this conclusion that flows with an incipient stagnation point are necessarily subcritical may be reconciled with the work of Keller et al. (1985, 1988), which suggests that such flows need not be subcritical to axisymmetric disturbances.

We close this section with the sketch in figure 5, which summarizes the characteristics of the mean flow field of a vortex breakdown.

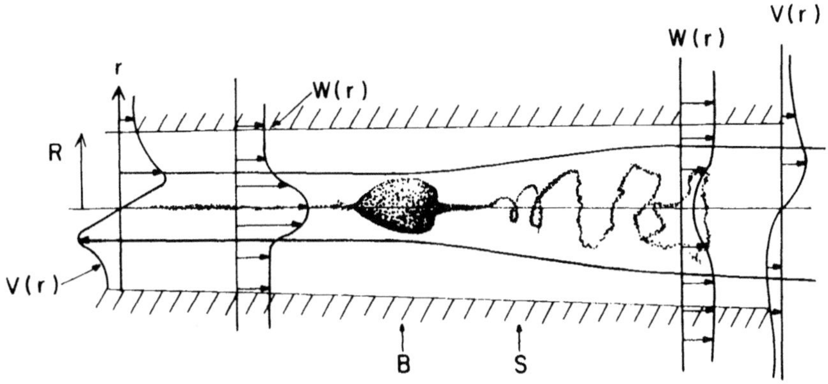


Fig.5. Sketch summarizing the features of a vortex breakdown. "B" marks the location of a bubble type of breakdown, and "S" that of a spiral type: these need not both occur in the same flow. The flow upstream is supercritical and stable, the flow downstream is subcritical and unstable.

## 4. Fluctuating Velocity Field

The velocity fields in the recirculation regions of vortex breakdowns have a strong time period component. This periodic flow persists in the near wake, weakening and being augmented by broadband fluctuations further downstream. This behavior is demonstrated by the data collected by Garg and Leibovich (1979) and shown in figure 6. Each part shows the power spectra of the axial velocity component for selected radial locations at a particular axial station in the wake of the breakdown. These show the temporal coherence in the near wake, and turbulence in the wake further downstream.

Garg and Leibovich (1979) show that the strong spectral peaks in the near wake can be quantitatively related to instability of the local mean velocity profiles to bending waves. The further downstream in the wake, the larger the growth rates, and when strong instability arises, that is, when the instability criterion of Leibovich and Stewartson (1973) is satisfied, not only are the growth rates sizeable, but an infinite number of azimuthal modes are simultaneously excited, which by itself indicates that strong instability in this sense is likely to be associated with transition to turbulence.

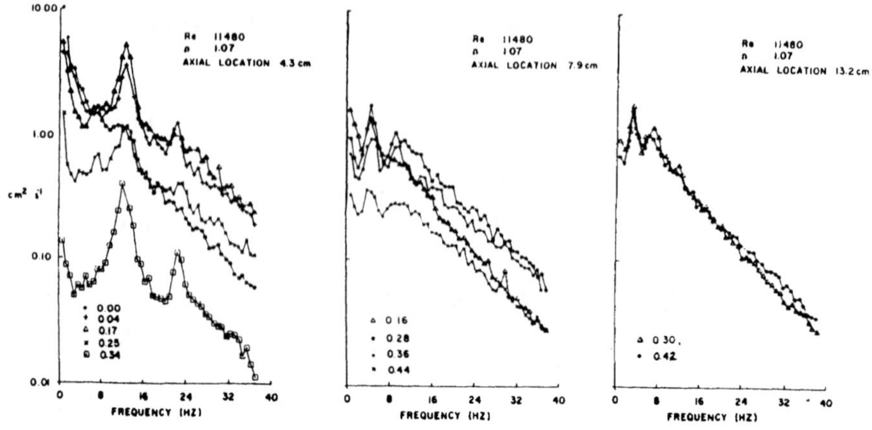


Fig.6. Power spectra of axial velocity fluctuations at several wake positions following the B-breakdown found in the flow of fig. 3. Each panel shows spectra at several radial locations.

## 5. System instability, or imperfect bifurcation

The linear axisymmetric waves discussed earlier have finite amplitude counterparts in columnar vortices. The weakly nonlinear problem is described by amplitude equations, and when the fluid is bounded radially, as in a tube, the long wave version of the amplitude equation is the Korteweg-deVries equation as discussed in Benjamin (1967), Pritchard (1970), and Leibovich (1970). The weakly nonlinear waves permitted are cnoidal and solitary waves. Leibovich and Kribus (1990) show that these weakly nonlinear waves have strongly nonlinear counterparts, and how the various families of waves are related to each other. As in the weakly nonlinear case, strongly nonlinear solitary waves necessarily have supercritical approach flows. Thus the supercritical barrier provided to upstream propagation of infinitesimal waves is not present for a nonlinear wave of sufficient amplitude, a situation again reminiscent of gasdynamics.

If the underlying waveguide is slowly varying, the bifurcation is imperfect, and a weakly nonlinear theory of wave-like perturbations to the unperturbed waveguide flow has been given by Leibovich and Randall (1973) and Randall and Leibovich (1973). In this case, the waves are forced by the variation of the waveguide, which we may think of as the axial pressure gradient associated, for example, with the axial variation of waveguide geometry. As shown in these papers, a long, small amplitude axisymmetric perturbation is amplified by the forcing imposed by the waveguide if this forcing is associated with an increase in the unperturbed pressure in the downstream direction. Growth of the wave is associated with an increase in the speed of the wave in the upstream direction, and a shortening of its axial length scale. This process continues unless and until some wave energy loss mechanism comes into play to absorb the work done by the pressure gradient on the wave, at which time the finite amplitude wave can equilibrate in amplitude and position. Leibovich (1983, and its more accessible version published in 1984) suggests that instability of the axisymmetric wave to nonaxisymmetric perturbations is a likely mechanism to cause a transfer of axisymmetric wave energy out of the system. This possibility is capable of explaining both the dominate axisymmetric portion seen experimentally in all vortex breakdowns, as well as the symbiotic and much smaller nonaxisymmetric part that is observed.

If this theory can be extended to strongly nonlinear motions, in the way that Leibovich and Kribus show that the columnar waveguide disturbances can be extended, then a physical mechanism for breakdown is suggested (Leibovich, 1983). Downstream sources initiate weak disturbances, the longest axisymmetric components of which have the greatest phase and group speeds and therefore are to be found furthest upstream. These are amplified by their interaction with the forcing imposed by the waveguide, causing accelerated upstream motion, wave growth and shortening, and ultimate loss of stability of the waves to nonaxisymmetric disturbances when the wave amplitude exceeds some threshold, and subsequent equilibration of the combined structure in both size and position. On the basis of their experimental observations in flows very different from those already cited, Maxworthy, Mory, and Hopfinger (1983) also suggest, in a paper published in the same 1983 volume as my cited 1983 paper, that breakdown is associated with loss of stability of large axially symmetric waves to nonaxisymmetric perturbations.

A movie shown at the Turbulence 89 conference showed this process in the Cornell vortex tunnel, a very similar frame sequence is presented by Sarpkaya (1972). A sketch in figure 7 shows the sequence of events seen in the film to lead to the birth and stabilization of a vortex breakdown, and the caption describes the process according the to mechanistic scenario suggested above.

## 6. Numerical simulation

In this paper, I have given only slight attention to theory, despite the fact that a great part of the efforts which have been expended on this problem to date have been of a theoretical nature. I have so far given no attention at all to numerical simulation of vortex breakdown, although there also has been activity in this arena, and the level of activity is growing as computational fluid mechanics generally grows in capability and confidence. To numerically simulate flows containing vortex breakdown is a very difficult task, however, and most of the computations that have been carried out have been restricted to be axially symmetric.

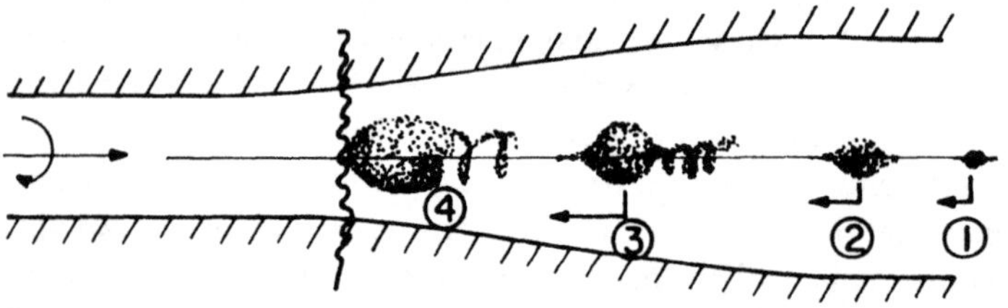


Fig.7. Birth of vortex breakdown. A slight swelling of streamsurfaces near the axis begins the process (1), and propagates upstream. It grows, while continuing to be axisymmetric (2), as it continues to propagate upstream. This upstream propagation can penetrate a supercritical region, if the wave amplitude is sufficiently large. Further wave growth leads to a loss of stability to asymmetric perturbations (3), and the transfer of axisymmetric energy permits stabilization of the structure (4). The flow upstream of the wavy line is supercritical.

Because of a propensity for axisymmetric swirling flows to lose stability to nonaxisymmetric perturbations, the axisymmetric flows which have been computed are not realized in experiment. In point of fact, in my view all attempts at numerical simulation, at least until now, have been unsatisfactory for one reason or another.

None of the review articles cited gives more than cursory attention to numerical issues as they relate to vortex breakdown, yet there is a clear need for careful and critical assessment of the current situation and future prospects of numerical simulation of vortex breakdown. This cannot be done here, but one or two general remarks may be appropriate.

In transition problems in shear flows, there are "standard" models to study. For example, in boundary layers, the flat plate with zero pressure gradient and associated Blasius flow serves as a standard starting point for study, and for the development of numerical treatments of transition. The vortex breakdown problem is intrinsically more complicated: at a minimum, it involves two velocity profiles rather than one to describe the undisturbed state, or entrance conditions, and furthermore, the effects of axial pressure gradient are central rather than a complication. One of the genuine inconveniences of the problem is the lack of a standard model.

Perhaps the remarks of the previous paragraph are better rephrased. To prepare a problem for computation, one must isolate a real-world phenomenon from its surroundings, which is done mathematically by specifying boundary conditions. There is no agreed-upon standard laboratory experiment to define a vortex breakdown problem to model by numerical experiment.

The problem of specifying boundary conditions is much more troublesome, however, than just settling on a given physical experiment to model. To capture a vortex breakdown event, a computational domain must include a region that is supercritical, as well as a region that is subcritical. More precisely, the inflow boundary must be supercritical, and a subcritical region must exist downstream, most likely terminating at the outflow boundary. In early computations, the inflow boundary was not restricted to be supercritical, and this led to large amplitude structures which propagated upstream as far as possible, that is to say, to the inflow boundary, corrupting the results.

The outflow boundary in the computations of which I am aware have been the terminus of a subcritical region, but apparently no attention has been given to the significance of this fact. This general situation is reminiscent of transonic flow, and there great efforts are made to try to ensure that subsonic boundaries do not cause reflections of disturbances. Apparently, even in that problem, which is easier, there is no clear way to impose appropriate nonreflecting boundary conditions except in one space dimension, where consideration of characteristic directions helps to settle the matter. In the case of vortex breakdown, the subcritical wake region not only permits upstream influence the outflow boundary, it is also a region of instability and therefore any waves reflected from the outflow boundary will be amplified, leading to a substantial artificial boundary effect with the potential of serious contamination of the results. Since three-dimensional perturbations in vortex flows tend to be more highly unstable than axisymmetric ones, this problem is likely to be more pressing in fully three dimensional simulations, but the dangers of contamination even from reflections of neutrally propagating waves would appear to be significant.

## 7. Summary

Vortex breakdown induces a transition to turbulence. The vortex core undergoes a structural change in the laminar-turbulent transition process, leading to core enlargement, and reduction of peak vorticity and swirl in the wake relative to the approach flow. Consequently, the location of breakdown separates a vortex flow into two distinct regions with different core structures.

Only a few general conclusions seem to be supported by experimental data. Experiments indicate that approach flows are supercritical and stable to infinitesimal disturbances and (mean) wake flows are subcritical.and unstable to three-dimensional disturbances. Key ingredients for vortex breakdown are levels of swirl somewhat above those leading to critical conditions, and adverse axial pressure gradients imposed on the core.

I have argued here that the available experimental evidence suggests that vortex breakdown is a system instability or imperfect bifurcation of the vortex flow. This process depends on more than one distinctly identifiable physical processes: neutral axisymmetric wave propagation, wave growth caused by boundary forcing, and energy transfer from large axisymmetric waves to three-dimensional perturbations by flow instability. These processes, when taken together, describe the observed qualitative aspects of the vortex breakdown. This picture is the basis of an on-going research program at Cornell, in which we are attempting to determine if it can provide a consistent theoretical framework for the phenomenon.

The subject has proceeded, no doubt for too long, without the aid of reliable numerical experiments capable of capturing the three-dimensional and time-dependent behavior exhibited in the laboratory. We may hope that numerical experiment will, in the near future, assume a more prominent role. To this end, it is to be hoped that serious attention be given to boundary conditions, and particularly to outflow conditions.

**Aknowledgements**

This work was supported by the U.S. Air Force Office of Scientific Research under grants AFOSR-89-0346, monitored by Dr. L. Sakell.

## References

Benjamin, T. B. 1962 Theory of the vortex breakdown phenomenon, *J. Fluid Mech.*, **14**, 593-629.

Benjamin, T. B. 1967 Some developments in the theory of vortex breakdown, *J. Fluid Mech.*, **28**, 65-84.

Erickson, G.E. and Gilbert, W.P. 1983 Experimental investigation of forebody and wing leading-edge vortex interactions, In: *Aerodynamics of Vortical Type Flows in Three Dimensions*, A. D. Young, ed., AGARD Conference Proceedings No. **342** (NATO), Paper 11.

Escudier, M.P. 1984 Observations of the flow produced in a cylindrical container by a rotating endwall, *Exp. Fluids*, **2**, 189-196.

Escudier, M.P. 1988 Vortex breakdown: Observations and explanations, in *Prog. Aero. Sciences*, **25**, 189-229.

Faler, J.H. and Leibovich, S. 1977 Disrupted states of vortex flow and vortex breakdown, *Phys. Fluids*, **20**, 1385-1400.

Faler, J.H. and Leibovich, S. 1978 An experimental map of the internal structure of a vortex breakdown, *J. Fluid Mech.*, **86**, 313-335.

Fiedler, B.H.and Rotunno, R. 1986 A theory for the maximum windspeeds in tornado-like vortices, *J. Atmos. Sci.*, **43**, 2328-2340.

Garg, A.K.& Leibovich, S. 1979 Spectral characteristics of vortex breakdown flowfields, *Phys. Fluids*, **22**, 2053-2064.

Gouldin, F.C., Depsky, J.S. and Lee, S.L. 1985 Velocity field characteristics of a swirling flow combustor, *AIAA J.*, **23**, 95-102.

Hall, M. G. 1966 The structure of concentrated vortex cores, in *Prog in Aero Sci.*, **7**, ed.by D. Küchemann, 53-110.

Hall, M. G. 1972 Vortex breakdown, *Ann. Rev. Fluid Mech.*, **4**, 195-218.

Hopfinger, E.J., Browand, F.K. and Gagne, Y. 1982 Turbulence and waves in a rotating tank, *J. Fluid Mech.*, **125**, 505-534.

Keller, J.J., Egli, W. and Exley, J. 1985 Force- and loss-free transitions between flow states, *ZAMP*, **36**, 854-889.

Keller, J.J., Egli, W. and Althaus, R.. 1988 Vortex breakdown as a fundamental element of vortex dynamics, *ZAMP*, **39**, 404-439.

Lambourne, N.C. and Bryer, D.W. 1961 The bursting of leading edge vortices; Some observations and discussion of the phenomenon, *Aero. Res. Coun.*, **R&M** 3282.

Leibovich, S. 1970 Weakly nonlinear waves in rotating fluids, *J. Fluid Mech.*, **42**, 803-822.

Leibovich, S. 1978 The structure of vortex breakdown, *Ann. Rev. Fluid Mech.*, **10**, 221-246.

Leibovich, S. 1979 Waves in parallel or swirling stratified shear flows, *J. Fluid Mech.*, **93**, 401-412.

Leibovich, S. 1983 Vortex stability and breakdown. In: *Aerodynamics of Vortical Type Flows in Three Dimensions*, A. D. Young, ed., AGARD Conference Proceedings No.**342** (NATO), Paper 23.

Leibovich, S. 1984 Vortex stability and breakdown: Survey and extension, *AIAA J.*, **22**, 1192-1206.

Leibovich, S., Brown, S.N., & Patel, Y. 1986 Bending waves on inviscid columnar vortices, *J. Fluid Mech.*, **173**, 595-624.

Leibovich, S. and Kribus, A. 1990 Large amplitude wavetrains and solitary waves in vortices, *J. Fluid Mech.* (In press)

Leibovich, S. and Randall, J.D. 1973 Amplification and decay of long nonlinear waves, *J. Fluid Mech.*, **53**, 481-493.

Leibovich, S. and Stewartson, K. 1983 A sufficient condition for the instability of columnar vortices, *J. Fluid Mech.*, **126**, 335-356.

Lugt, H.J. & Abboud, M. 1987 Axisymmetric vortex breakdown in rotating fluid within a container, *J. Fluid Mech.*, **179**, 179-200.

Maxworthy, T., Mory, & Hopfinger, E. 1983 Waves on vortex cores and their relation to vortex breakdown. In: *Aerodynamics of Vortical Type Flows in Three Dimensions*, A. D. Young, ed., AGARD Conference Proceedings No. **342** (NATO), Paper 29.

Maxworthy, T. 1973 A vorticity source for large-scale dust devils and other comments on naturally occurring columnar vortices, *J. Atmos. Sci.*, **30**, 1717-1722.

Maxworthy, T., Hopfinger, E.J., and Redekopp, L.G. 1985 Wave motions on vortex cores, *J. Fluid Mech.*, **151**, 141-165.

Polhamus, E.C. 1986 Vortex lift research: Early contributions and some current challenges, in *Vortex Flow Aerodynamics I*, ed. by J. F. Campbell, R. F. Osborn, and J.T. Foughner, Jr., **NASA CP-2416**, pp. 1-30.

Pritchard, W.G. 1970 Solitary waves in rotating fluids, *J. Fluid Mech.*, **42**, 61-83.

Randall, J. D. & Leibovich, S. 1973 The critical state: A trapped wave model of vortex breakdown, *J.Fluid Mech.*, **53**, 481-493.

Ronnenberg, B. 1977 Ein selbstjustierendes 3-Komponenten-Laser-doppleranemometer nach dem Vergleichsstrahlverfahren, angewandt für Untersuchungen in einer stationären zylinder-symmetrischen Drehströmung mit einem Rüchstromgebiet. *Max- Planck Inst. für Strömungsforschung*, Göttingen, Bericht **20**.

Sarpkaya, T. 1971 On stationary and travelling vortex breakdowns, *J. Fluid Mech.*, **45**, 545-559.

Peake, D.J. and Tobak, M. 1983 On issues concerning flow separation and vortical flows in three dimensions. In: *Aerodynamics of Vortical Type Flows in Three Dimensions*, A. D. Young, ed., AGARD Conference Proceedings No. 342 (NATO), Paper 1.

Tsai, C-Y., and Widnall, S.E.1980 Examination of group-velocity criterion for breakdown of vortex flow in a divergent duct, *Phys. Fluids*, **23**, 864-870.

Vogel, H.U. 1968 Experimentelle Ergebnisse über die laminare Strömung in einem zylindrischen Geh use mit darin rotierender Scheibe, *Max-Planck Inst. für Strömungsforschung*, Göttingen, Bericht **6**.

# Numerical simulation of vortex breakdown via 3-D Euler equations

Philippe MEGE[α], Thien Hiep LE[α], Yves MORCHOISNE[β,α]

[α] *Office National d'Etudes et de Recherches Aérospatiales*
*BP 72, 92322 Châtillon Cedex, France*

[β] *Laboratoire d'Analyse Numérique*
*Université Paris VI, 75230 Paris Cedex 05, France*

*Turbulence and Coherent Structures*
*O. Métais and M. Lesieur, eds.*
*Selected papers from "Turbulence 89: organized structures and turbulence in fluid mechanics"*
*Grenoble, 18-21 September 1989*

*Contents*

## 1. Introduction

The long term goal of this study is the modelling of vortex breakdown that occurs in some aerodynamic configuration at high angles of attack, (fighters highly swept delta wings or missiles for example). Experimental and theoretical work done at ONERA [1,2] and elsewhere [3] has not yet provided a full understanding of the phenomenon.

Our contribution to this study consists in a numerical simulation based on solving the 3-D Euler equations for an unsteady incompressible flow. Preliminary results were obtained using a pressure - velocity formulation with periodic boundary conditions, the Euler equations being discretised by 2nd order finite difference schemes [4]. In this paper, we will present the continuation to this work by implementing more realistic boundary conditions and 4th order finite difference discretisation schemes.

## 2. Continuous formulation

The problem consists in solving the unsteady incompressible flow Euler equations in the domain $:\Omega = [-a,a]X[-b,b]X[c,d]$. With variables made dimensionless (using a reference length L and velocity $U_r$, the equations are :

$$\forall x = (x_1, x_2, x_3) \epsilon \Omega, t \epsilon [O, T]$$

$$\frac{\partial U}{\partial t} + C + \nabla(P + \frac{1}{2}U^2) = 0 \qquad (1)$$

$$\nabla . U = 0 \qquad (2)$$

where $U$ is the velocity, $P$ the pressure, $C = (\nabla X U) X U$ represents the remainder to $\nabla(\frac{U^2}{2})$ of the convective term.

To these equations, we must add:

- initial conditions: $U(x, t = 0) = U_0(x)$
- periodic boundary conditions for velocity and pressure in the transverse directions $x_1$ and $x_2$:

$$U(x_1 + 2a, x_2, x_3) = U(x_1, x_2, x_3)$$

$$U(x_1, x_2 + 2b, x_3) = U(x_1, x_2, x_3)$$

$$P(x_1 + 2a, x_2, x_3) = P(x_1, x_2, x_3)$$

$$P(x_1, x_2 + 2b, x_3) = P(x_1, x_2, x_3)$$

- an upstream Dirichlet boundary condition for $x_3 = c$ :

$$U(x_1, x_2, c, t) = f(x_1, x_2)$$

- a downstream boundary condition for $x_3 = d$, which may be :
  - either a Neumann condition on total pressure:

$$\frac{\partial Q}{\partial n}(x_1, x_2, d) = 0.$$

where $Q = P + \frac{1}{2}U^2$ is the total pressure and $\frac{\partial Q}{\partial n}$ its normal derivative on the surface $x_3 = d$.
  - or Neumann conditions on the three components of the velocity:

$$\frac{\partial U_1}{\partial n}(x_1, x_2, d) = 0.$$

$$\frac{\partial U_2}{\partial n}(x_1, x_2, d) = 0.$$

$$\frac{\partial U_3}{\partial n}(x_1, x_2, d) = 0.$$

For sufficiently large computational domain , these boundary conditions will allow the study of an isolated vortex, without disturbing too much its evolution to breakdown.

The continuity equation is taken into account by minimising a norm of the velocity divergence as function of the pressure (seen as a Lagrange multiplier) :

$$\frac{\partial}{\partial Q}\left|\frac{\partial U_i}{\partial x_i}\right| = 0$$

## 3. Discretisation

Discretisation is done by space and time finite difference schemes, similar to those developed for direct simulation of turbulence [5]. Let $\delta x_l$ and $\delta\xi_l$ be the grid steps respectively in physical and reference space, and $x_l$ and $\xi_l$ the corresponding coordinates.

### 3.1. Space discretisation

The computational domain is discretised by a stretched cartesian grid. Space derivatives of the variable f are obtained by the following formulae:

$$\frac{\partial f}{\partial x_l} = \frac{\partial f}{\partial \xi_l}\frac{\partial \xi_l}{\partial x_l} \qquad l = 1, 2, 3$$

$$\text{with} \qquad \frac{\partial \xi_l}{\partial x_l} = 1/\frac{\partial x_l}{\partial \xi_l}$$

4th order finite difference schemes (or 2nd order schemes, not reported here) are used to approximate derivatives.

- at a current point:

$$\left(\frac{\partial f}{\partial \xi_l}\right)_m = -\frac{f_{m-2} - 8f_{m-1} + 8f_{m+1} - f_{m+2}}{12\delta\xi_l} \qquad m = i, j, k$$

- for points near boundaries :

In a non periodical direction , the derivative is computed near the boundaries in the following manner:

- for a point on the upstream boundary :

$$\left(\frac{\partial f}{\partial \xi_l}\right)_1 = \frac{1}{\delta\xi}\left(-\frac{25}{12}f_1 + 4f_2 - 3f_3 + \frac{4}{3}f_4 - \frac{1}{4}f_5\right)$$

- for a point next the upstream boundary :

$$\left(\frac{\partial f}{\partial \xi_l}\right)_2 = \frac{1}{\delta\xi}\left(-\frac{1}{4}f_1 - \frac{5}{6}f_2 + \frac{3}{2}f_3 - \frac{1}{2}f_4 + \frac{1}{12}f_5\right)$$

- for a point on the downstream boundary :

$$\left(\frac{\partial f}{\partial \xi_l}\right)_M = \frac{1}{\delta\xi}\left(\frac{25}{12}f_M - 4f_{M-1} + 3f_{M-2} - \frac{4}{3}f_{M-3} + \frac{1}{4}f_{M-4}\right)$$

- for a point next the downstream boundary :

$$\left(\frac{\partial f}{\partial \xi_l}\right)_{M-1} = \frac{1}{\delta\xi}\left(\frac{1}{4}f_M + \frac{5}{6}f_{M-1} - \frac{3}{2}f_{M-2} + \frac{1}{2}f_{M-3} - \frac{1}{12}f_{M-4}\right)$$

Terms such as $\frac{\partial x_l}{\partial \xi_l}$ are approximated by finite difference schemes similar to those used for the variable f derivatives.

### 3.2. Time discretisation

The discrete equations are computed at time $(n+\frac{1}{2})\Delta t$, e.g.:

$$A(U^{n+1}-U^n)+\Delta t A_1 A_2 D Q^{n+\frac{1}{2}}+\Delta t C^* = 0 \qquad (3)$$

with :
- $U^n = U(n\Delta t)$, the velocity evaluated at time $t^n = n\Delta t$,
- $Q^{n+\frac{1}{2}} = Q((n+\frac{1}{2})\Delta t)$, the total pressure evaluated at time $t^{n+\frac{1}{2}} = (n+\frac{1}{2})\Delta t$,
- $C^* = C(U^*)$ with $C(U) = (DXU)XU$, function of the velocity $U^*$
- $U^* = 1.5U^n - 0.5U^{n-1}$

where $D$ approximates the $\nabla$ operator by centered differences.
$A$ is the factorised operator $A_1A_2A_3$, with:

$$A_i = 1 - \eta\frac{\partial^2}{\partial {x_i}^2} \qquad i = 1,2,3$$

$$\text{and} \qquad \eta = 0(\Delta t^2)$$

The boundary conditions (periodic, Neumann, or Dirichlet) are taken into account in these operators:$A_1$ and $A_2$ insure of the periodicity condition while $A_3$ deals with upstream and downstream conditions.

In operator $A$, the second order derivative is discretised by a 2nd order finite difference scheme. An unconditionnally stable scheme is thus obtained. The $\eta$ parameter chosen as $O(\Delta t^2)$, the scheme is 2nd order in time [6].

Applying the discretised divergence operator $D.$ to equation (3), the total pressure $Q$ satisfies the following Poisson equation:

$$-D.{A_3}^{-1}DQ^{n+\frac{1}{2}} = \frac{D.U^{n+1}}{\delta t} - \frac{D.U^n}{\delta t} + D.A^{-1}C^*$$

Pressure is computed by minimising the discrete norm of $D.U^{n+1}$, the resulting linear system is solved by the iterative Multigradient method [7].

Matrices $A_1, A_2$ and $A_3$ are $L.U.$ factorised, reducing the solution of the associated linear systems to a double sweep at each time step.

Once total pressure is computed, $U^{n+1}$ is given by the following equation :

$$U^{n+1} = U^n - \delta t({A_3}^{-1}DQ^{n+\frac{1}{2}} + A^{-1}C^*)$$

To evaluate the influence of the downstream boundary condition, a Neumann condition was imposed on the velocity then on the pressure.

Both conditions gave highly similar results away from the immediate vicinity of the downstream boundary.

## 4. Numerical results

### 4.1. Accuracy of the numerical scheme

*4.1.1. Test problem for accuracy study for the 2nd order scheme*
Accuracy studies were done on a 3-D convection equation with a prescribed solution giving rise to a right hand term :

$$\frac{\partial U}{\partial t} + (U.\nabla)U = R.H.T.$$

The velocity field was:
$U = (u, v, w)$ where

$$u = a_x sin\left(\frac{2\pi x}{a_x}\right) cos\left(\frac{2\pi y}{a_y}\right) f(t)$$

$$v = a_y sin\left(\frac{2\pi y}{a_y}\right) cos\left(\frac{2\pi x}{a_x}\right) f(t)$$

$$w = 1.55 f(t)$$

where $f(t) = (t + 100)/100$, $a_x = 10.$, $a_y = 12$. The computed numerical solution with a space step equal to 0.2 and a time step of 0.05 showed that the error on the solution is around $10^{-3}$ for a 87x63x9 space discretisation.

*4.1.2. Comparison between 2nd order and 4th order schemes*
Space accuracy was estimated for the same velocity field and equation. Accuracy versus discretisation and scheme order was measured by the $l_2$ norm of solution error to be seen in table below.

| GRID | DF2 | DF4 |
|---|---|---|
| 21x15x9 | $2.3\ 10^{-2}$ | $1.2\ 10^{-3}$ |
| 45x33x9 | $4.5\ 10^{-3}$ | $5\ 10^{-5}$ |
| 87x63x9 | $1.26\ 10^{-3}$ | $3.8\ 10^{-6}$ |

Although total computation time being about 30% greater for 4th order than for 2nd order scheme for any given grid, it appears for a required accuracy, that 4th order calculation is 15 times cheaper than a 2nd order calculation. Following results will be given for a 4th order discretisation.

## 4.2. Vortex breakdown

### *4.2.1. Test case and initialisation*

The study concerns the evolution of an isolated vortex the characteristics of which, in the initial undisturbed state, are given by the axisymmetrical velocity field written below:

$$V_r = 0$$

$$V_\theta = 0.825\frac{(1 - e^{-10.9r^2})}{r}$$

$$V_z = 0.95 + 1.55e^{-10r^2}$$

where $V_r, V_\theta, V_z$ are the radial, tangential, and axial component of the velocity and r the radial distance. Velocity is made dimensionless with a reference value $U_r$, defined as 1.73 $\overline{r_c}$, where $\overline{r_c}$ is the radial distance corresponding to a maximum of $V_\theta(\overline{r_c} = 0.34)$.

The given velocity profiles are those described by Batchelor [8]. The chosen parameters have the values used by Nakamura [9]. These conditions were obtained by experimental measurements [10], the value of $V_z$ on the axis exceeding 1. corresponds to an overspeed typical of delta wing vortices.

In cartesian coordinates ($x = x_1, y = x_2, z = x_3$), the azimuthal velocity splits into:

$$u = \frac{-0.825y\left(1 - e^{-10.9(x^2+y^2)}\right)}{x^2 + y^2}$$

$$v = \frac{-0.825x\left(1 - e^{-10.9(x^2+y^2)}\right)}{x^2 + y^2}$$

$$\text{with} \qquad \sqrt{x^2 + y^2} = r$$

Figure 1 shows axial and azimuthal velocity profiles versus radius and their respective analitycal representations.

In order to obtain a three-dimensional breakdown, an initial pertubation $u_p$ of intensity $\epsilon$ was added to the $U$ component:

$$u_p = \epsilon e^{-\beta z^2}$$

$$\text{with} \qquad \epsilon = 0.015 \qquad \text{and} \qquad \beta = 2$$

*4.2.2. Grid generation*

Due to the initial velocity field, the vortex is concentrated around the z axis. The stretched cartesian grid is designed so as to concentrate points in the breakdown area. Grid point coordinates are consequently defined as functions of the coordinates $\xi_1$ and $\xi_2$ in reference space by :

$$x = f(\xi_1)$$

$$y = f(\xi_2)$$

with $$f(\xi_l) = \frac{\alpha_1(\xi_l - \beta_1)}{1 - \gamma_1(\xi_l - \beta_1)^2} \qquad l = 1, 2$$

The $z$ coordinate is given by the following geometrical progression:

$$z^k = z^{k-1} + \Delta z^0 \alpha$$

with $\Delta z^0 = 0.05$ and $\alpha = 1.11$

Figure 2 and 3 show the computational grid in a plane $Z = Cst$, normal to the vortex axis and in a plane $\theta = Cst$ containing the vortex axis, where there are respectively 21x21 and 21x49 discretisation points. Computational domain dimensions are: $[-8, 8]$ x $[-8, 8]$ x $[-2, 14]$. The constant values of $\alpha_1, \beta_1, \gamma_1$ are : $\alpha_1 = 0.23$ $\beta_1 = 11.$ $\gamma_1 = 6.9.10^{-3}$ and give minimal and maximal mesh size :

$$\Delta x_{min} = \Delta y_{min} = 0.2; \Delta z_{min} = 0.05$$

$$\Delta x_{Max} = \Delta y_{Max} = 2.7; \Delta z_{Max} = 1.2$$

*4.2.3. Numerical results*

Computations were done on a CRAY 2 with a 21x21x49 space discretisation. The computation cost is $9.10^{-6}$ second per point and per time step for a 4th order scheme, with an average of 6 iterations per time step for the Multigradient method to obtain a divergence norm less than $10^{-3}$ . When an downstream boundary condition is imposed on the pressure in the implicit phase , cost rises slightly due to the fact that velocity divergence is not minimised on the downstream boundary.

Results presented concern axial velocity $V_z$, total pressure $Q$ and vorticity $\omega$.

Figure 4 shows the axial velocity evolution versus the longitudinal distance Z for three different times. Constant at initial time, the velocity suddenly drops at the breakdown. Then, this velocity increases downstream and causes a second instability. These result and those that follow have been obtained with the Neumann outflow boundary condition on the total pressure ($\frac{\partial Q}{\partial n} = 0$).

On figure 5 are displayed axial velocity and total pressure distributions along the Z-axis at the same time $t = 300\Delta t$. As can be seen, total pressure and axial velocity are strongly correlated. We can observe after a very little drop, an important raise of the total pressure. This abrupt variation is probably the sign of the desorganisation beginning.

Figure 6 compares results obtained for the axial velocity distribution along the Z-axis (with 21x21x403 points and a constant meshsize in axial direction) to experiments of Uchida et al. and to the simulation made by Nakamura et al.. We observe the same behaviour of the axial velocity and similar types of structures appear in the three cases. Without further information on Nakamura's initial disturbance shape, it is difficult to achieve more quantitative comparisons, because the initial disturbance shape may influence the vortex breakdown process.

Figures 7 and 8 (see colour plates) give iso-vorticity evolutions in a yz plane for 0 and 300 time steps. Initially, vorticity distribution is constant along the flow axis. At time $t = 15$ (corresponding to 300 time steps) vorticity contours have to a slight extent changed. Whereas at initial time, the vorticity is highly confined and distribution is axisymmetric, the vortex expands with a non axisymmetric behaviour.

Total pressure contours evolution is also studied at same times as above. Figures 9 and 10 (see colour plates) indicate clearly a sudden drop in total pressure followed by a high positive gradient immediatly downstream. The pressure field is totaly desorganised corresponding to the vortex breakdown.

Figure 11 (see colour plates) represents at time $t = 15$ (300 time steps) the trajectories of particles which can be compared with experimental streaklines. We can see the presence of a well-formed helix structure. On figure 12 (see colour plates), corresponding to 500 time steps ($t = 25$), two areas are to be noticed : upstream, total breakdown of the structure and further downstream the formation of a bubble.

A qualitative comparison can be made especially concerning the helicoidal structure of the breakdown between figure 11 and figure 13 (see coulour plates) which is an experimental visualisation obtained by M. Werlé in the water tunnel of the ONERA.

## 5. Conclusion and perspectives

A 3-D unsteady inviscid incompressible flow method has been developed and applied to vortex breakdown. Results so far concerning simulation of a breakdown are very encouraging:

- the main known characteristics of the phenomenon (3-D structures, unsteadiness) showed up satisfactorily,
- the choosen method, the most robust among several studied, allows the analysis of the breakdown phenomenon,
- the scheme is accurate and thus allows limited discretisation which minimises the computing cost.

The next step will be developing Large Eddy Simulation (L.E.S.) taking into account small structures by implementing a subgrid model into Navier-Stokes equations [5]. A detailed parametric study on core size, rotation rate and pertubation will give better understanding of the physical phenomenon and we suggest to find breakdown criteria, based on pressure correlation in time, less empirical than those actually used.

**Acknowledgments**

This research was sponsored by the Direction des Recherches , Etudes et Techniques (DRET) of the Délégation générale de l'Armement (DGA).

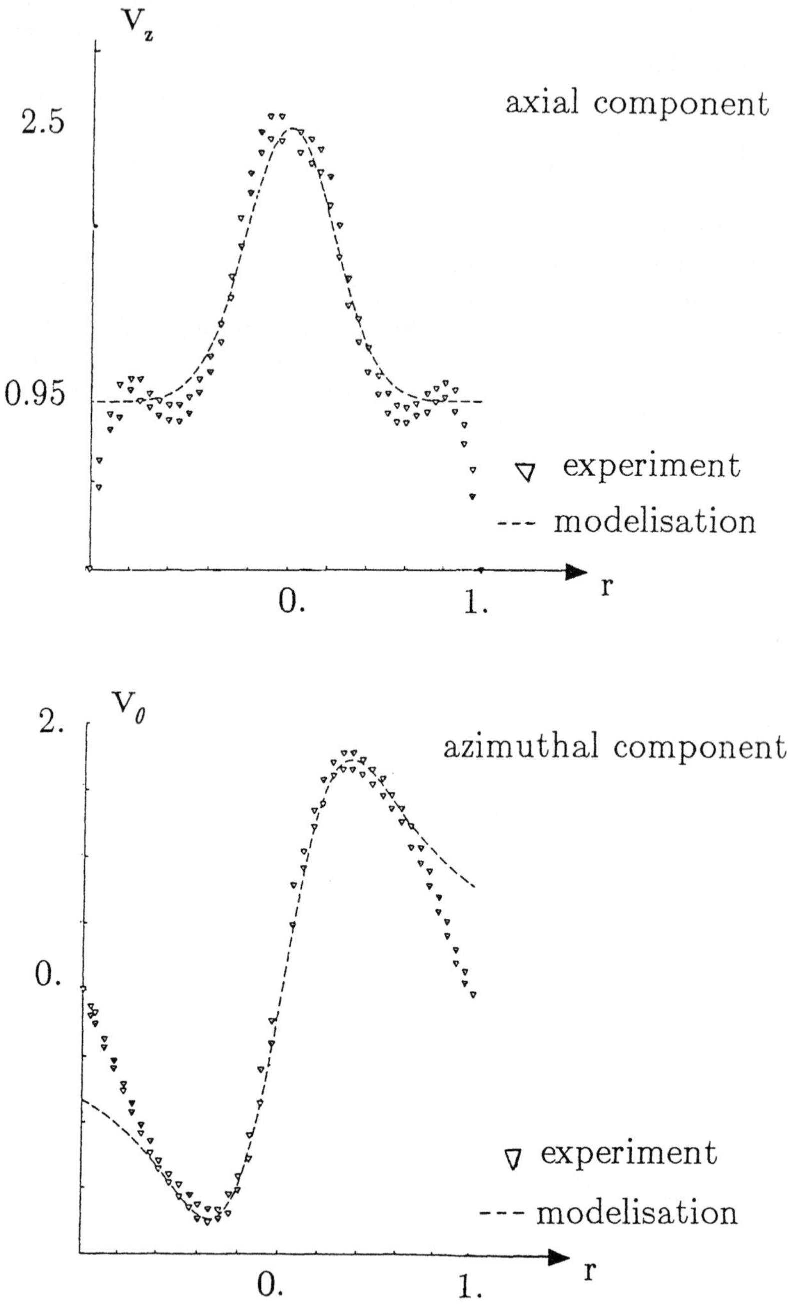


**Figure 1 : Initial velocity profiles**

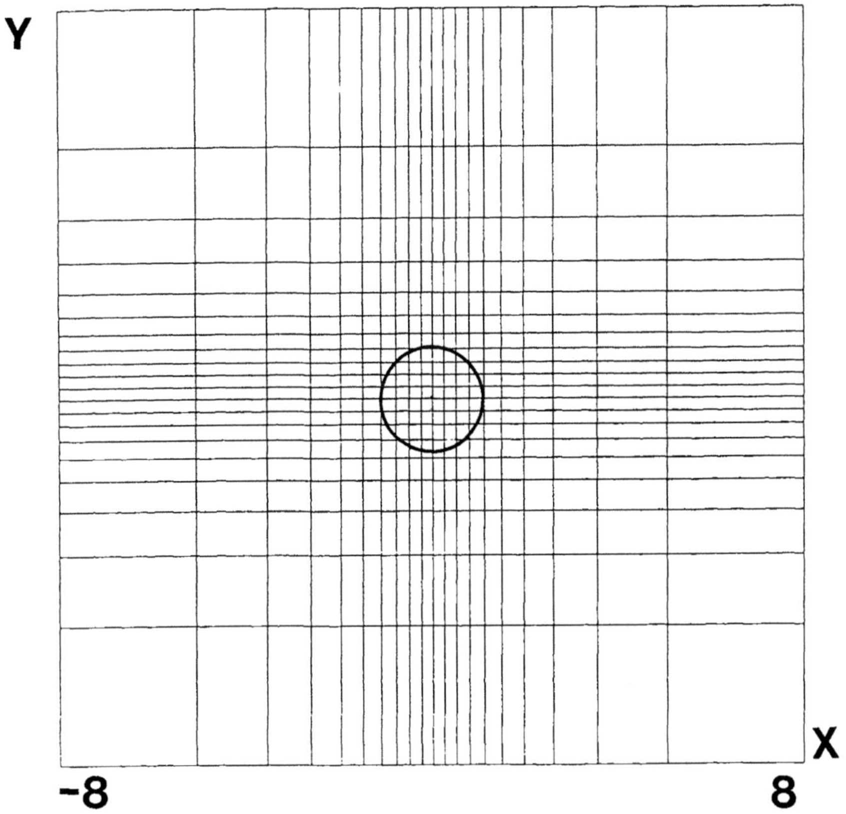


**Figure 2 : Grid in the XY plane (21×21)**

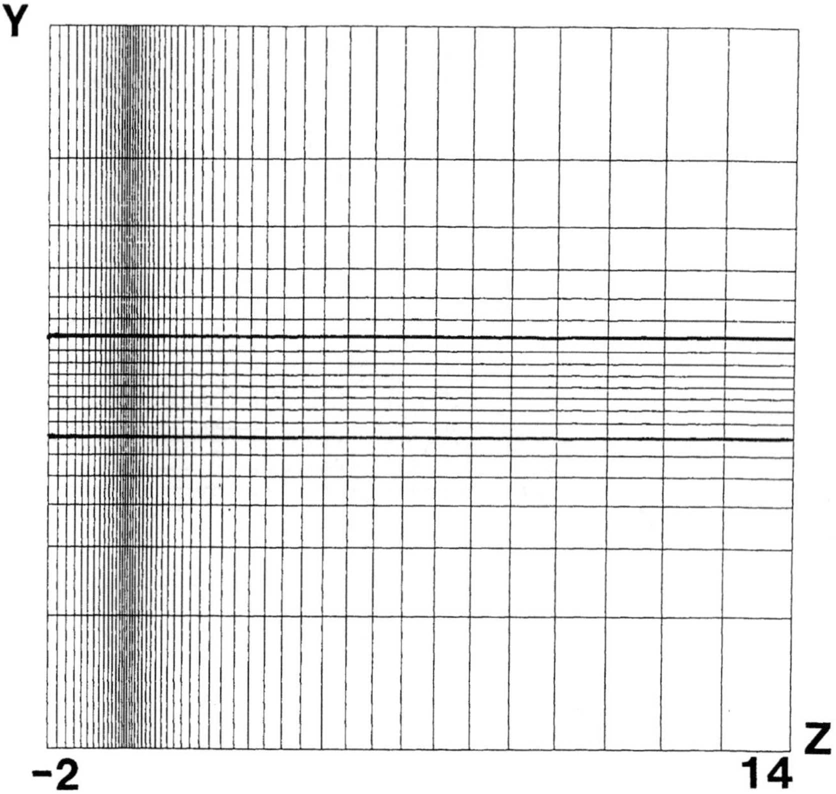


Figure 3 : Grid in the YZ plane (21×49)

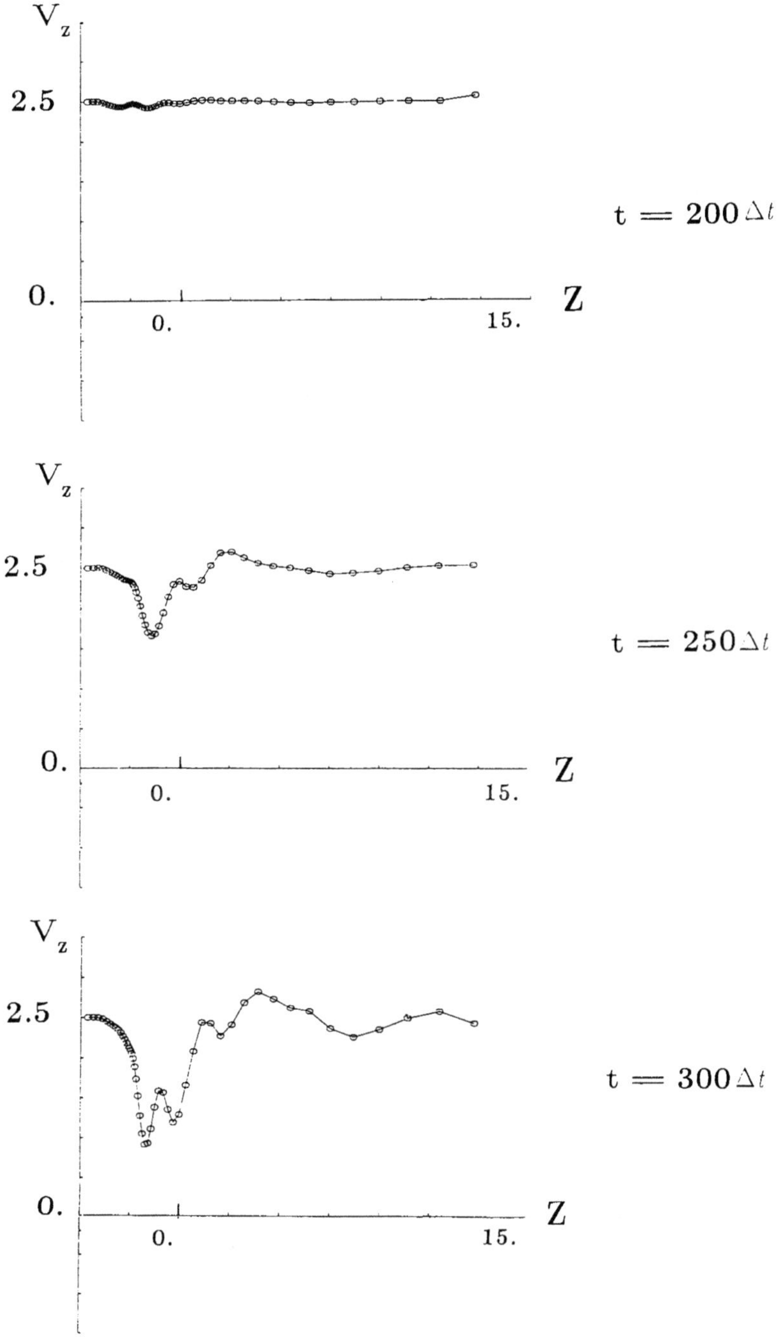


**Figure 4 : Axial velocity evolution**

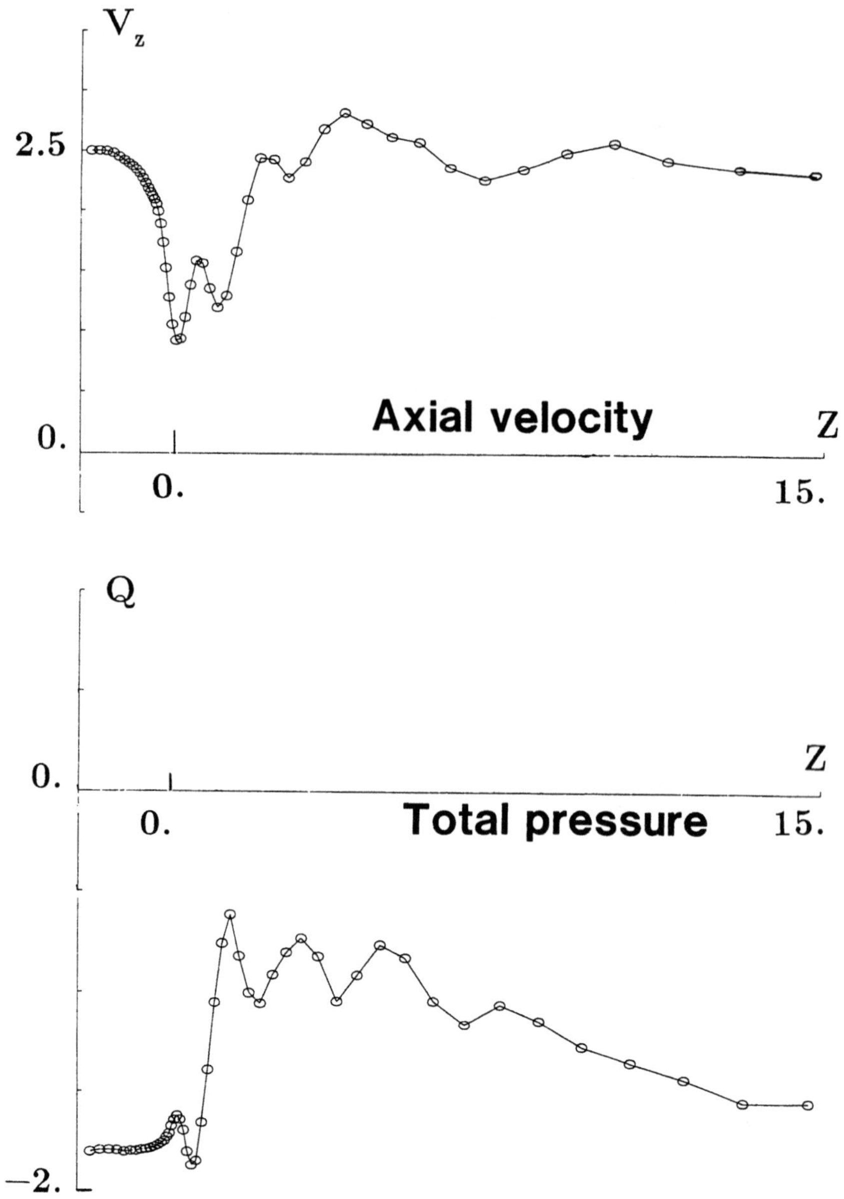


**Figure 5 : Longitudinal distribution**

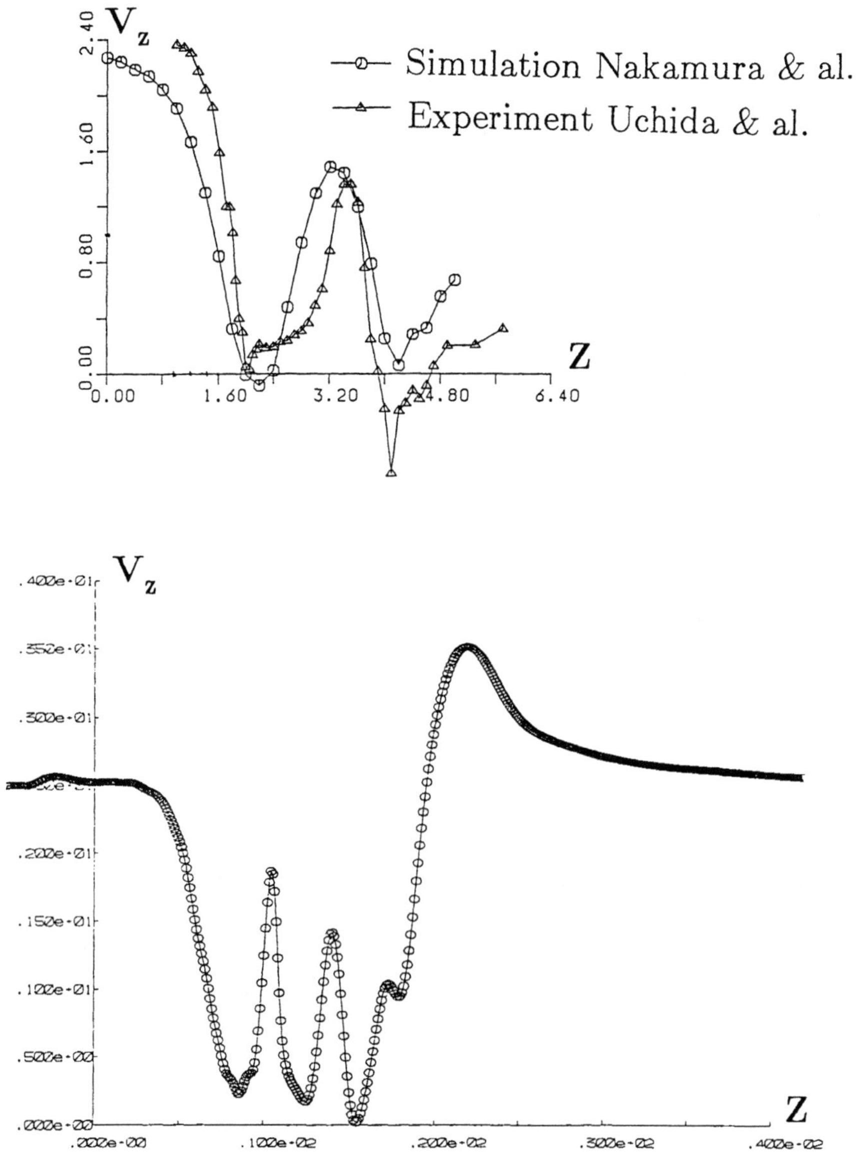


**Figure 6 : Axial velocity distribution along the Z-axis**

## References

[1] J. Délery : "Etat des études sur les écoulements tourbillonnaires effectués par la division d'Aérodynamique Fondamentale. ONERA-R.T. n° 21/7252 AN - Mars 1988.

[2] Benay R., Pagan D. : "Etude numérique de l'éclatement de tourbillons soumis à des gradients de pression". La Recherche Aérospatiale n° 1988-1, p.15-27 (English and French edition).

[3] Newsome R.W., Kandil O.A. : "Vortical flow aerodynamics. Physical aspects and numerical simulation". AIAA Paper 87-0205, AIAA 25th Aerospace Science Metting. Reno, Nevada. January 12-15, 1987.

[4] Lê T.H., Mège Ph., Morchoisne Y. : "Numerical simulation of vortex breakdown by solving the Euler equations for an incompressible fluid". La Recherche Aérospatiale n° 1989-5 (English and French edition).

[5] Dang K., Deschamps V. : "Numerical simulations of transitional channel flow". 5th International conference on numerical methods in laminar and turbulent flow. Montréal 6-10 Juillet 1987. TP ONERA n° 1987-98.

[6] Mège Ph. : "Simulation numérique de l'éclatement tourbillonnaire". Rapport de stage de DEA de Mécanique (Paris VI) Juin 1988.

[7] Ryan J., Lê T.H., Morchoisne Y. : "Panels code solvers". 7th GAMM, Louvain (Belgique), 9-11 Septembre 1987.

[8] Batchelor G.K. : "Axial flow in trailing line vortices". J.Fluid Mech., Vol 20 p.645, 1964.

[9] Nakamura Y., Leonard A., Spalart P.R. : "Vortex breakdown simulation". AIAA Paper 85-1581, Cincinnati, Ohio. Juillet 1985.

[10] Uchida S., Nakamura Y., Ohsawa M. : "Experiments on the axisymetric vortex breakdown in a swirling air flow". Transaction of the Japan society for aeronautical and space sciences.Vol 27 1985.

[11] Pagan D., Solignac J.L. :"Etude expérimentale de l'éclatement d'un tourbillon engendré par une aile delta". La Recherche Aérospatiale n°1986-3 (English and French edition).

# VI. Two-dimensional Turbulence

# The Coherent Vortices of Two-dimensional and Geostrophic Turbulence

James C. McWILLIAMS[α]

[α]*National Center for Atmospheric Research*
*P.O. Box 3000, Boulder, CO 80307, USA*

*Turbulence and Coherent Structures*
*O. Métais and M. Lesieur, eds.*
*Selected papers from "Turbulence 89: organized structures and turbulence in fluid mechanics"*
*Grenoble, 18-21 September 1989*

# 1. Introduction

## 1.1. Two-dimensional turbulence and vortices

The study of two-dimensional turbulence has undergone a renaissance in recent years, largely as a result of the discovery that isolated, coherent (i.e., long-lived and having a characteristic circular flow pattern) vortices spontaneously emerge from random initial conditions or forcing at large Reynolds number. Furthermore, once they emerge, the vortices dominate the subsequent flow evolution. The primary focus in these studies—as it will be for me today—has been on spatially homogeneous, temporally decaying solutions (e.g., Babiano et al., 1987; Basdevant et al., 1981; Benzi et al., 1986, 1988; Brachet et al., 1986, 1988; Herring and McWilliams, 1985; McWilliams, 1984, 1989c, 1990a; Santangelo et al., 1990).

Vortex dynamics in two-dimensional flow—their characteristic shapes, stabilities, and interactions—is one of the oldest subjects in fluid dynamics, in the work of Kirchhoff, Rayleigh, and others. This subject also has been rejuvenated by the realization of its relevance to turbulent flows. (A recent survey of vortex dynamics is presented in McWilliams, 1989b.)

## 1.2. Three-dimensional turbulence and vortices?

Two-dimensional turbulence has long been viewed as a paradigm for large-scale, geophysical motions (McWilliams, 1983). It seems to me increasingly plausible that it may quite generally be a useful paradigm for three-dimensional turbulence in flow environments which induce a significant degree of anisotropy in the sense that one velocity component is substantially weaker than the other two. There are even some hints from recent numerical simulations of isotropic, three-dimensional turbulence that this anisotropic view may be a locally useful one in a reference frame oriented with the vorticity vector near extrema in its magnitude (e.g., Meneguzzi, personal communication), although much more work is needed to investigate this possibility.

## 1.3. Survey of topics

The topics I wish to discuss today are the following:

a) an overview of the behavior of two-dimensional turbulence, as I presently understand it from the studies cited above;

b) some modest steps towards a theory of structured turbulence (i.e., turbulent flows with coherent vortices or coherent structures of perhaps some other configuration); and
c) the extension of our understanding of two-dimensional turbulence to at least some types of anisotropic, three-dimensional turbulence, in particular geostrophic turbulence (i.e., in a rotating, stably stratified fluid).

## 2. An overview of the behavior of two-dimensional turbulence

### 2.1. Evolution from smooth initial conditions

We begin with an illustration of the evolution of two-dimensional flow from smooth initial conditions, defined by a random-phase realization of an isotropic wavenumber spectrum of moderate bandwidth. (The solution is described more fully in McWilliams, 1990a.)

A sequence of images of the vorticity field $\zeta(x,y)$ is shown in the accompanying plates with a gray-scale shading that makes a sharp transition from black for quite small amplitudes (less than 10% of the peak extremum even at late times) to white for larger amplitudes. This shading exhibits quite small black zones for random fields of the type specified as initial conditions at $t = 0$, which has the pattern of grout between the tiles in a mosaic. In these images, one can see the evolution away from a random field to a highly structured one (i.e., with coherent vortices).

The times here can be interpreted by comparison with an initial circulation time, $t_c = \frac{1}{\text{rms } \zeta} \approx 0.02$, and a time of maximum enstrophy (i.e., vorticity variance) dissipation, $t_z \approx 0.25$. The latter can be interpreted as $t_z = t_c \cdot \ln(\text{Re})$, where Re is a Reynolds number (Lesieur, 1987). As the flow evolves, both $t_c$ and $t_z$ increase as enstrophy is dissipated and $Re$ increases. Thus, on a time characterized by many $t_z$ in these images, we see an expansion of the black area and an organization of the white areas into nearly circular structures (the coherent vortices). This evolution occurs through intermediate stages of extrusion of $\zeta$ into filaments by the local strain field, which are then sites of large dissipation.

### 2.2. The competition between vorticity and strain

The dynamics of two-dimensional turbulence is in essence a competition between vorticity and strain. For a two-dimensional fluid with velocity components $(u, v)$, we define

$$\zeta = v_x - u_y, \ \sigma_1 = v_x + u_y, \ \sigma_2 = u_x - v_y, \ u_x + v_y = 0. \qquad (2-1)$$

Quadratic measures of vorticity and strain are

$$Z = \zeta^2, \ S = \sigma_1^2 + \sigma_2^2. \qquad (2-2)$$

For spatially homogeneous conditions, one can show that the competitors are well matched; that is,

$$\langle Z \rangle = \langle S \rangle, \quad \langle \zeta \sigma_i \rangle = 0 \ (i = 1, 2), \tag{2-3}$$

where the angle brackets denote area averages over the domain.

For smooth initial conditions, one can partition the domain into a curvilinear checkerboard pattern in which the distorted squares surround vorticity centers or extrema where $Z > S$ (these regions we denote by $\mathcal{Z}$) and the interstices ($\mathcal{S}$) are zones of high strain with $S > Z$. The areas of $\mathcal{Z}$ and $\mathcal{S}$ are initially equal.

The flow evolution is quite different in $\mathcal{Z}$ and $\mathcal{S}$ (Weiss, 1981). In $\mathcal{S}$, vorticity gradients grow with time. From smooth initial conditions, vorticity frontogenesis by straining advection occurs on a time $t_c$ within $\mathcal{S}$, which results in a kinetic energy spectrum $\propto k^{-4}$ for large wavenumbers $k$. On a time of many $t_c$, chaotic advection (i.e., the chaos of point vortices) repeatedly folds the fronts and shapes them into closely packed, elongating filaments; this folding results in a spectrum which becomes shallower until it is $\propto k^{-3}$, as in a classical enstrophy inertial range. On a time of $t_z$, the initial set of filaments become as thin as the dissipation scale, and the enstrophy they contain is dissipated.

In $\mathcal{Z}$, the amplitude of vorticity gradients does not have a systematic trend. On a time $t_c$, the process of axisymmetrization is initiated about the vorticity centers: vorticity gradients locally increase on the edges of $\mathcal{Z}$, and filaments are discarded into $\mathcal{S}$, and these events may occur through several cycles. On a time $t_z$, the local vorticity distribution remaining within $\mathcal{Z}$ approaches a smooth, axisymmetric, monopole (i.e., with $\zeta$ monotonically decreasing from the extremum). A circular flow pattern is a steady solution of the two-dimensional Euler equations, and, by the Rayleigh criterion, a circular monopole is stable.

Thereafter, for $t > t_z$, coherent vortices exist within $\mathcal{Z}$, and elongating, weakening (dissipating) vorticity filaments exist within $\mathcal{S}$. The spectrum progressively steepens from $\propto k^{-3}$ even for wavenumbers well below the dissipation range. The vortices move chaotically by mutual advection, and they usually stay within $\mathcal{Z}$ since vorticity extrema are preserved following fluid parcels except for dissipation. The vortices occasionally form transient (reversible) aggregates of opposite sign (dipoles or tripoles). Like-sign vortices irreversibly merge when they come close enough to each other, while casting off some of their vorticity into $\mathcal{S}$ as new filaments. Finally, vortices are deformed under the influence of the strain field from other vortices, either reversibly (if they remain sufficiently within $\mathcal{Z}$) or irreversibly (if they move into $\mathcal{S}$ for a sufficient period); for example, the latter can occur if a weak vortex comes close to strong one and is deformed by the surrounding annulus of strong strain.

Meanwhile, the filaments within $\mathcal{S}$ continue to elongate and weaken, and, on the rare occasions when they become sufficiently distant from other vortices so that they move into $\mathcal{Z}$ due to their own vorticity, they become unstable, roll up, and cause a secondary generation of vortices on a smaller scale.

With time these processes lead to fewer, sparser, larger vortices; an increase in the area of $\mathcal{S}$ and a decrease in the area of $\mathcal{Z}$; and a decrease of the enstrophy within $\mathcal{S}$. Together these trends effect an inverse cascade of energy and a forward cascade and dissipation of enstrophy. They also create an increasingly intermittent character to the flow evolution as the vortex density decreases and their close encounters (during which non-conservative behavior is initiated through mergers and straining deformations) become increasingly rare. That is, the turbulent cascade is significantly slowed down by the presence of the coherent vortices; in particular, the enstrophy dissipation rate steadily declines for $t > t_z$, and it does so to a far greater degree than a flow with an equivalent spectrum that lacks vortex structures (McWilliams, 1989c). For $t \gg t_z$, the time required for appreciable spectrum evolution increases as $\ln t$: for $2^N$ vortices to combine through successive mergers to become 2 vortices of opposite sign—the usual end state of two-dimensional turbulence, after which the only significant evolution is by vorticity diffusion—requires a time which is exponentially dependent on $N$ since the rate of vortex encounter, hence merger, is a positive function of vortex density.

## 3. An analysis of the coherent vortices of two-dimensional turbulence

### 3.1. Towards a theory of structured turbulence

It is, of course, natural to seek a theory for the evolution of structured turbulence. It is now quite clear that no theory will be adequate, even for such gross properties as the evolution of the energy spectrum, that does not include the effects of vortices. It may be true that vortex dynamics are so strongly implicit in the Navier Stokes equations that many different theoretical approaches could capture their effects, including, for example, the traditional statistical moment closures (but present evidence is negative: Herring and McWilliams, 1985). It is more direct, and perhaps more attractive, to base a theory on vortices as elemental entities. However, in contrast to assuming a population of point or line vortices within a framework of Hamiltonian dynamics, the theory must be an essentially non-conservative one in order to represent changes in such quantities as vortex number, enstrophy, and the circulation trapped within the vortices—all of which happen in ways that involve strong shape deformations for the vortices.

I am in no position yet to present such a theory, nor is anyone else as far as I know, although I believe several people are interested in the possibility. Instead, I will present the results of an analysis of the evolution of the vortex population in a numerical solution for two-dimensional turbulence (the same one as shown above). This can be viewed as part of the answer which the as yet unfound theory must provide.

### 3.2. An automated vortex census

We perform a vortex census that identifies the coherent vortices in a solution and measures their properties of interest. Specifically, for each time, we consider all $N_e$ extrema of $\zeta(x, y)$ that exceed a specified (small) threshold amplitude and test them for their similarity to an archetypal vortex structure. About each extremum, $\zeta = \zeta_v$, a curve $\mathcal{C}$ is defined such that $\zeta = 0.2\zeta_v$. For acceptance as a coherent vortex we require that $\mathcal{C}$ be simply connected and that $\zeta$ inside $\mathcal{C}$ have a value in the range $[0.2, 1.0] \cdot \zeta_v$ and be close to axisymmetric about the extremum (as determined by several shape measures: the first and second spatial moments and the arc length of $\mathcal{C}$). (See McWilliams (1990a) for more details.) Thus, $N_v$ vortices are selected.

For each of the selected vortices, we calculate an elliptical fit to $\mathcal{C}$, with major and minor semi-diameters $a$ and $b$. A mean radius for the vortex is defined by $r = \sqrt{ab}$. The most common shape deformation from axisymmetry is elliptical—the response to a spatially uniform strain (Moore and Saffman, 1971)—and a mean ellipticity is defined by $\epsilon = \sqrt{(\frac{a}{b})^2 - 1}$. Finally we define a vortex circulation $\Gamma$ as the integral of $|\zeta|$ within $\mathcal{C}$.

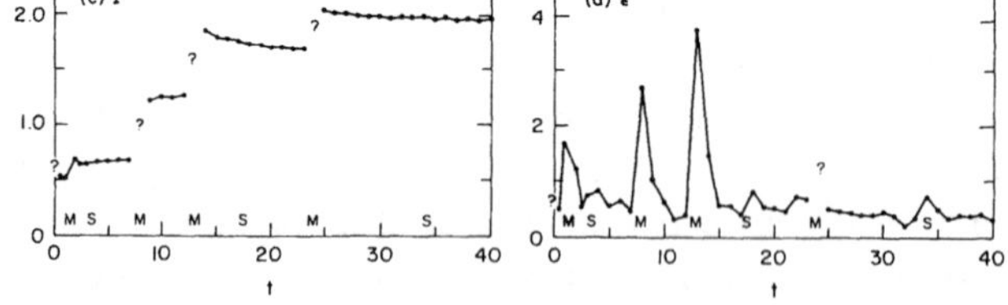


Fig.2. Vortex properties for the vortex whose trajectory is in Fig. 1. "?" indicates the times when the vortex not selected due to strong shape deformations.

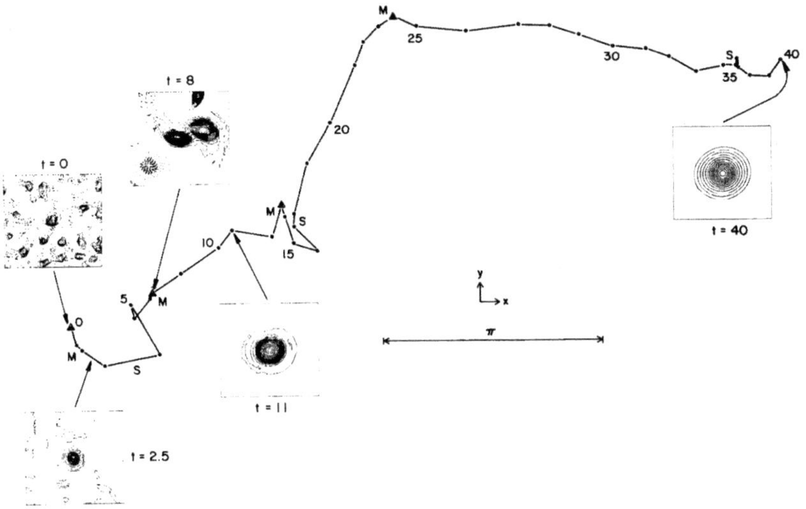


Fig.1. The trajectory of a single vortex. Positions of $\zeta_v$ are plotted at unit times. Events of merger (M) and strong straining deformation (S) are marked beside the trajectory; "$\Delta$" indicates the times when the vortex not selected due to strong shape deformations. In addition, there are selected local plots of $\zeta(x,y)$ in subdomains magnified by a factor of 5 relative to the trajectory scale.

Before examining the vortex population as a whole, consider the life-history of a single vortex over a time interval of $\Delta t \approx 10^3 t_c$. Its trajectory (Fig. 1) is irregular, but its Lagrangian autocorrelation time increases as the vortex density decreases. This particular vortex experiences four mergers and three additional episodes of strong, but reversible, straining deformation during this interval; the mergers lead to abrupt increases in vortex size and circulation (Figs. 1 and 2). The vortex amplitude slowly declines due to diffusion, and its rate is somewhat enhanced during merger or strong deformation. Of course, not all vortices survive for such a long interval; this particular vortex is among the strongest ones at later times, and on average stronger vortices have longer lifetimes than weaker ones (see below).

Initially only a small fraction of the vorticity extrema are selected as vortices, and this fraction further declines during the interval $0 < t < t_z$ as the closely packed vorticity centers experience strong mutual straining and shape deformation (Fig. 3a). Thereafter, the selection ratio increases until, at late time, almost all extrema are accepted as vortices. Strong extrema are more likely to be accepted than weak ones (Fig. 3b) since the effectiveness of axisymmetrization increases with

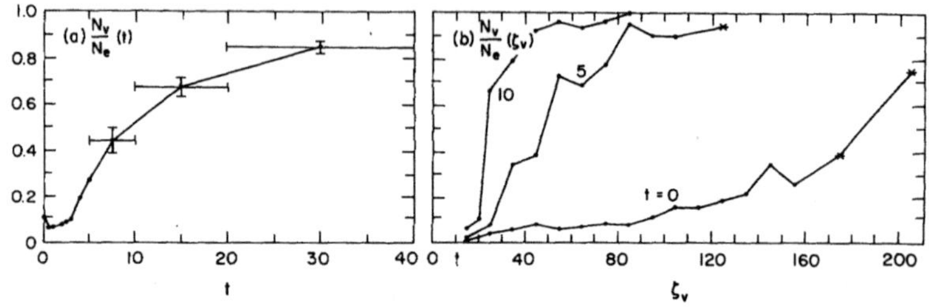


Fig.3. Selection ratio, $\frac{N_v}{N_e}$, as a function of (a) time and (b) vortex amplitude, $\zeta_v$.

vorticity amplitude for a given strain environment.

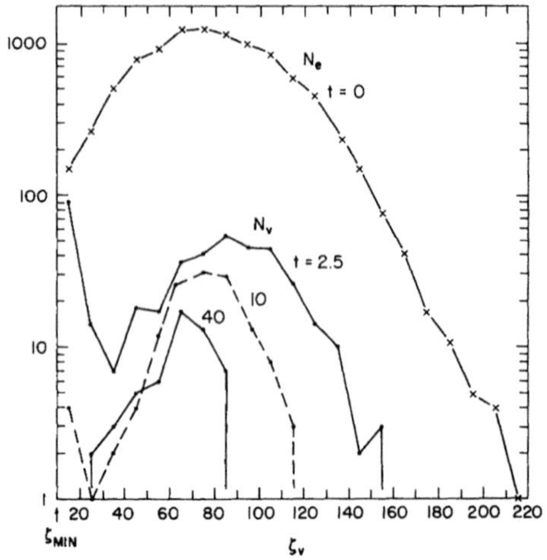


Fig.4. Amplitude distributions of initial extrema ($\times$) and selected vortices at subsequent times ($\cdot$), with amplitude bins $\Delta\zeta_v = 10$.

The vortices have a broad distribution in amplitude, which reflects the broad distribution of initial extrema (Fig. 4). With time the population declines (Fig. 5) due to mergers and irreversible straining episodes. In addition the center of the distribution shifts to lower amplitude due to diffusive amplitude decay. The population also decreases more rapidly at small amplitude since, in a pair interaction between vortices of unequal strength, the weaker one is more likely to be destroyed by straining or to have its extremum be absorbed into the flanks of the stronger one through merger. An exception can be seen in Fig. 4 at early times, where there is secondary generation of vortices at small amplitude from filaments (see Chapter 2). The population decrease has a power-law form, $N_v \propto t^{-\alpha_N}$, where the best-fit exponent is $\alpha_N = 0.71$ for $5 \leq t \leq 40$.

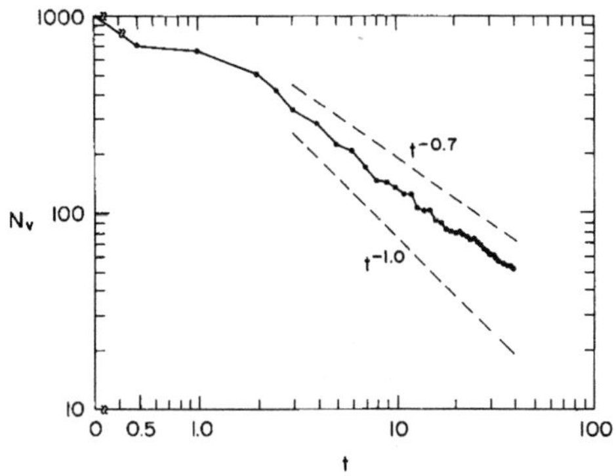


Fig.5. Evolution of the vortex population.

For comparison, a model of vortex motion as independent random walks plus destruction through close approaches has the following population equation:

$$\frac{dN_v}{dt} \propto -N_v^2, \qquad (3-1)$$

whose power-law solution is $N_v \propto t^{-1}$. Thus, the actual population decrease is slower than for random encounters, probably because the mutual advection of (point) vortices yields a less frequent rate of encounter for a given population density.

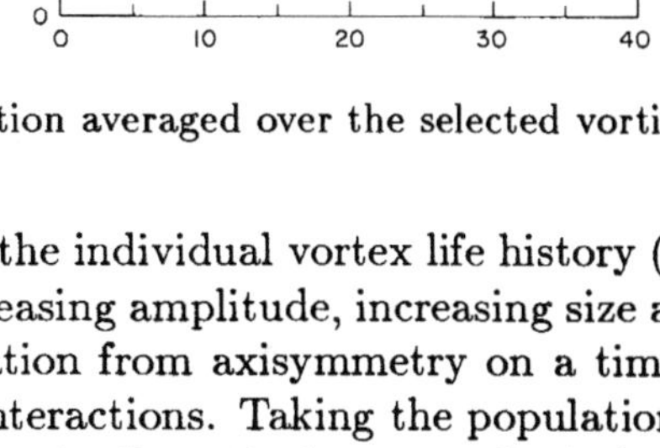


Fig.6. Vortex circulation averaged over the selected vortices.

As illustrated in the individual vortex life history (Figs. 1-2), the vortices also have decreasing amplitude, increasing size and circulation, and decreasing deformation from axisymmetry on a time scale longer than individual vortex interactions. Taking the population average, however, acts to smooth the episodic variations seen for individual vortices (e.g., $\langle \Gamma \rangle$ in Fig. 6, where the angle brackets now denote a population average). For most properties, the tendencies have approximately a power-law form, as does $N_v(t)$ in Fig. 5. Over the same time interval as before, we obtain the fits

$$\begin{aligned}\langle r\rangle &\propto t^{+.20} \equiv t^{\alpha_r}\\ \langle\Gamma\rangle &\propto t^{+.35} \equiv t^{\alpha_\Gamma}\\ \langle\zeta_v\rangle &\propto t^{-.09} \equiv t^{\alpha_\zeta}.\end{aligned} \qquad (3-2)$$

For a simple, inviscid model of merger of equal vortices with constant vorticity profiles, these exponents satisfy

$$2\alpha_r = \alpha_\Gamma, \ \alpha_\zeta = 0, \qquad (3-3)$$

which roughly correspond to the best-fit exponents in (3-2). If we also include the effects of weak (but strain-enhanced) diffusion, then the expected relations are

$$2\alpha_r > \alpha_\Gamma \gg -\alpha_\zeta, \qquad (3-4)$$

which have an even better correspondence with (3-2). If the population evolved only through mergers which conserve all the area and vorticity inside the participating vortices, then total circulation within the vortices, $\langle\Gamma\rangle N_v$, would be conserved. From (3-2), however, we see that

$$\begin{aligned}\langle\Gamma\rangle N_v &\propto t^{\alpha_\Gamma+\alpha_N}\\ &= t^{-.36} = e^{-.32\tau},\end{aligned} \qquad (3-5)$$

where $\tau$ is a transformed time that increases by a unit amount for every halving of the vortex population; in the present solution, this occurs three times between $t \approx 2.5$, 5.7, 13.2, and 38.5, and the best-fit logarithmic relation is

$$\tau(t) = 1.10 \ln t - 0.94. \qquad (3-6)$$

Thus, as the population evolves, there is a fractional loss of circulation in the vortices, which is $1 - e^{-.32} = .27$ on a population halving time. This loss results from straining destruction events and leaky mergers.

Finally, we consider the behavior of the average vortex deformation. A fit of the form

$$\langle\epsilon\rangle \propto t^{\alpha_\epsilon} \qquad (3-7)$$

is not particularly apt over the whole time interval of vortex dominance. For example, we obtain the following partial-interval fits:

$$\begin{aligned}\alpha_\epsilon &= -.34 \text{ for } 2.5 < t < 15,\\ \alpha_\epsilon &= -.15 \text{ for } 25 < t < 40.\end{aligned} \qquad (3-8)$$

Moore and Saffman (1971) found steady solutions for elliptical vortices with uniform vorticity profiles in a uniform strain field. In these solutions, the departure of the the vortex aspect ratio from unity is approximately proportional to the strain amplitude and inversely proportional to the vorticity of the vortex. For two-dimensional turbulence, the strain field is due to the distant vortices. The far-field strain from a vortex is equal to the ratio of its circulation and the square of the distance from it. The average distance between vortices is inversely proportional to the square root of the vortex density. The density is proportional to the population. Together these statements imply

$$\frac{a}{b} - 1 \propto \frac{\langle \Gamma \rangle N_v}{\langle \zeta_v \rangle}. \qquad (3-9)$$

$\epsilon$ has a nonlinear relation to $\frac{a}{b}$, and its asymptotic forms are

$$\begin{aligned} \epsilon &\sim \frac{a}{b} \quad \text{for } \frac{a}{b} \gg 1, \\ &\sim \sqrt{\frac{a}{b}} \quad \text{for } 0 < (\frac{a}{b} -)1 \ll 1. \end{aligned} \qquad (3-10)$$

Thus, from this crude model of vortex deformation, we expect

$$\alpha_\epsilon = c(\alpha_\Gamma + \alpha_N - \alpha_\zeta) = .28c, \qquad (3-11)$$

where $c$ decreases from 1 at early time (large deformations) to 0.5 at late time (small deformations). The correspondence with the measured exponents in (3-8) is reasonably good, and it is even slightly better if partial-range fits are used for the right-hand-side exponents in (3-11).

In summary, through the recent efforts of many scientists, I believe that much progress has been made on achieving at least a qualitative understanding of the importance and behavior of coherent vortices in two-dimensional turbulence. I am therefore hopeful that progress can soon be made on developing a quantitative theory.

## 4. Coherent vortices in three-dimensional turbulence

### 4.1. Anisotropic flow environments

Two-dimensional turbulence is a popular subject of investigation, in part, because it is relatively accessible, both computationally and theoretically (e.g., vortex dynamics). However, two-dimensional flow is also physically unrealizable in a three-dimensional universe, even though some special situations come close. Therefore, two-dimensional turbulence is most interesting, in relation to natural fluid phenomena in either laboratory or geophysical or astrophysical environments, insofar as it serves as a paradigm for anisotropic turbulence rather generally. (Anisotropy here is defined as $w \ll (u, v)$.)

The sources of anisotropy, even at small scales, can be found in the large-scale environment of the turbulence. Examples of conducive environments are stable stratification (with buoyancy frequency $N(z) > 0$ and Richardson number $Ri \gg 1$), unstable stratification ($Ri < 0$), rotation (with Coriolis frequency $f$ and Rossby number $Ro \ll 1$), mean shear ($\overline{U}(y)$), and certain configurations of electromagnetic fields.

There can be no doubt that each of these regimes will exhibit many departures from two-dimensional flow, but I believe it is likely to be a fruitful hypothesis that they may also exhibit essential similarities. The exploration of each of these regimes at large $Re$ is still in preliminary stages. Furthermore, their investigation with respect to coherent vortices can be quite difficult insofar as it involves $\zeta$, intermittency, Lagrangian-frame coherence, large $Re$, and relatively small spatial scales.

In this conference we have heard a great deal about coherent structures in shear flows. As my final topic, I will report on another regime where coherent vortices can be important. It is that of rotating, stably stratified flow ($Ri \gg 1$; $Ro \ll 1$), also called geostrophic turbulence since the flow is primarily along isobars at a speed proportional to the horizontal pressure gradient. This regime is characteristic of much of the large-scale flow on the earth and other planets. Here the association with two-dimensional turbulence is found to be quite strong indeed.

### 4.2. Geostrophic turbulence

We consider an analogous problem to the two-dimensional one discussed above: the decay at large $Re$ from random, broad-band initial conditions. We assume, for simplicity, that $f$ and $N$ are spatially uniform. The domain is horizontally homogeneous (i.e., periodic) and vertically bounded (with boundary conditions of zero normal flow and tangential stress).

I do not wish to focus here on the statistical behavior of the solution (for that, see McWilliams, 1989a). As in two-dimensional turbulence, isolated, coherent vortices do spontaneously emerge and dominate the subsequent evolution. Their characteristic structure is a one-signed distribution in $\zeta(x, y, z)$, which has the shape of an axisymmetric monopole in horizontal planes and has its horizontal extrema along an axis that is aligned with the vertical and extends a distance typically much less than the domain height. There may be several three-dimensional $\zeta$ extrema along the axis. These features are illustrated in Figs. 9-10 of McWilliams (1989a).

The important processes of two-dimensional vortex dynamics—axisymmetrization, merger, straining deformation—are also important for geostrophic vortices. In addition, there are analogous processes that are essentially related to the vertical structure of the vortices: alignment (the development of phase correlations in the vertical within a volume with $\zeta$ of one sign), attachment (the joining of two vortices whose central extrema are at different levels, leading to multiple three-dimensional extrema on the axis), and vertical straining deformation (tilting of phase lines from vertical alignment in response to the field of $S_V = u_z^2 + v_z^2$ from other vortices).

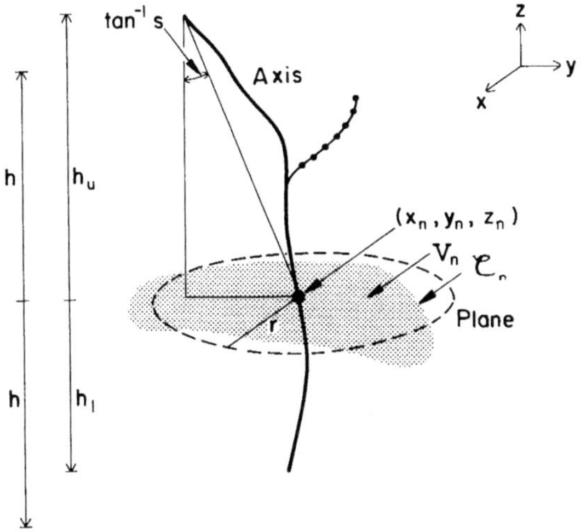


Fig.7. Axis and plane census scheme. The large black dot denotes the strongest three-dimensional vorticity extremum, the stippled region denotes the vortex interior in the plane of this extremum, the solid line denotes the axis, and the dotted line denotes a secondary axis branch discarded during the elimination of redundancy.

We perform an automated vortex census for this geostrophic turbulence solution as well. The procedure is an extension of the two-dimensional one described above (see McWilliams, 1990b, for more information). It is schematically shown in Fig. 7. The vortex properties associated with the horizontal plane of the central extremum are defined as before. The vortex height $h$ is the average of the upper and lower half-heights (excluding half-heights that intersect the boundary); vertical extent is determined by the same criterion as for $\mathcal{C}$ in the horizontal plane, $\zeta = 0.2\zeta_v$. The vortex slope $s$ measures the tilt of the axis from vertical alignment.

We now present some of the results of the census. This will be done with only brief comments, both to match the schedule of this conference and to reflect our as yet much less complete understanding of geostrophic, as compared to two-dimensional, vortex dynamics (see also McWilliams, 1989b).

Fig. 8 shows the decline in vortex population. It is significantly steeper than in two-dimensional turbulence (Fig. 5), suggesting that the attachment interaction increases the rate of vortex amalgamation. (The rates

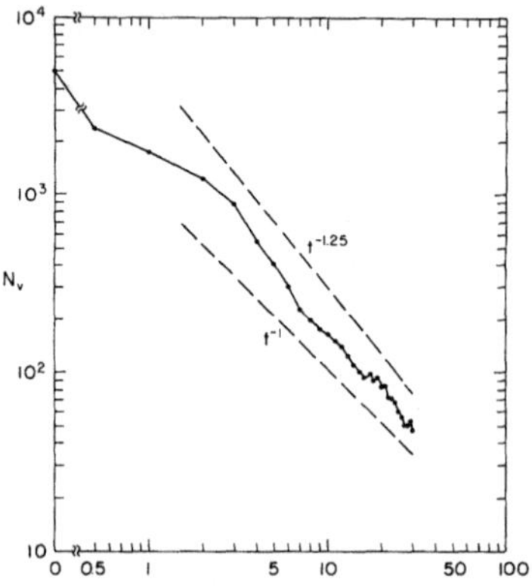


Fig.8. Evolution of the vortex population.

of close encounter between vortices may also be different in the two regimes, of course.)

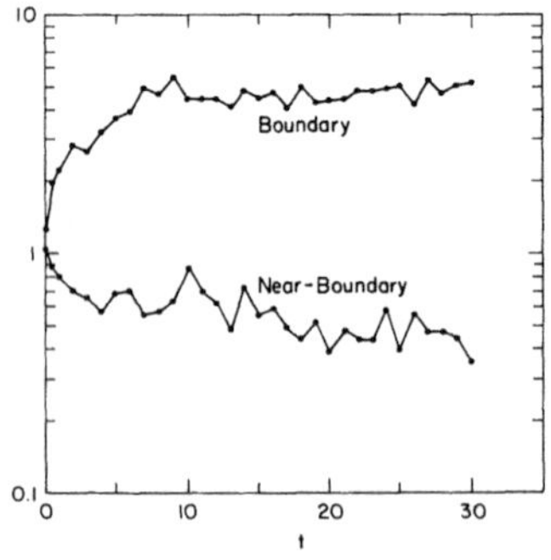


Fig.9. Relative frequency of occurrence of vortex central extrema in the levels either closest to the vertical boundaries or in the adjacent interior. The normalization is such that the average over all levels is unity.

The presence of vertical boundaries leads to the development of a vertical inhomogeneity in the turbulence. In particular, there is a tendency for vortex extrema near the boundary to migrate towards it during straining interactions with other vortices (Fig. 9).

Vortices in geostrophic turbulence tend to grow in size, both vertically and horizontally, through merger and attachment and the selective destruction of weaker vortices through irreversible straining interactions (Fig. 10-11). In spite of the fact that individual merger or attachment events are more likely to increase either $r$ or $h$ alone, rather than both together, there is little change in the population-averaged aspect ratio after an early time interval of $t < t_z$ (Fig. 12; $t_z \approx 0.25$ in this

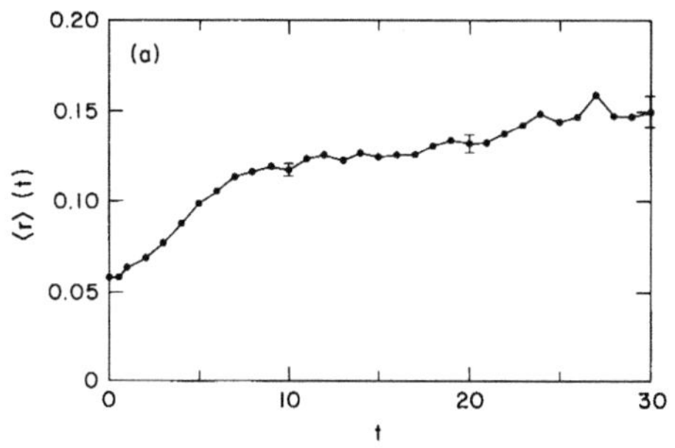


Fig.10. Average vortex radius.

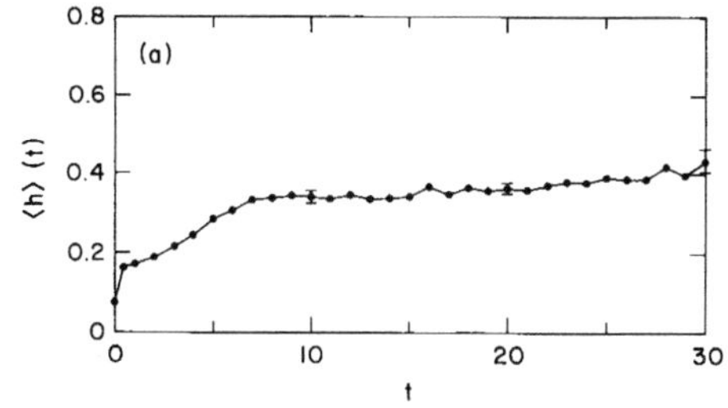


Fig.11. Average vortex height.

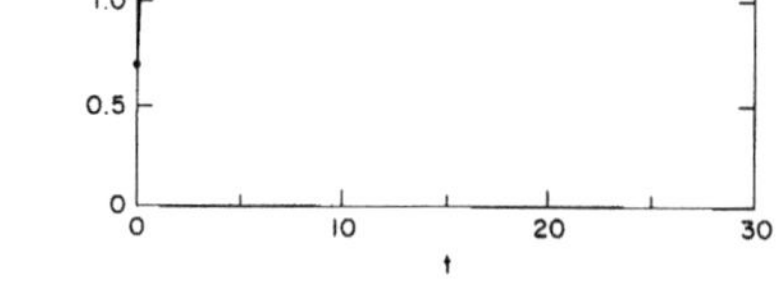


Fig.12. Average, normalized vortex aspect ratio.

solution as well). This aspect ratio is normalized by $\frac{N}{f}$, since the total energy and potential enstrophy spectra are expected to be approximately three-dimensionally isotropic in the stretched coordinate frame $(x, y, \frac{N}{f}z)$ (Charney, 1971).

Another consequence of attachment interactions is that the average number of distinct, three-dimensional extrema on a vortex axis tends to increase with time (Fig. 13).

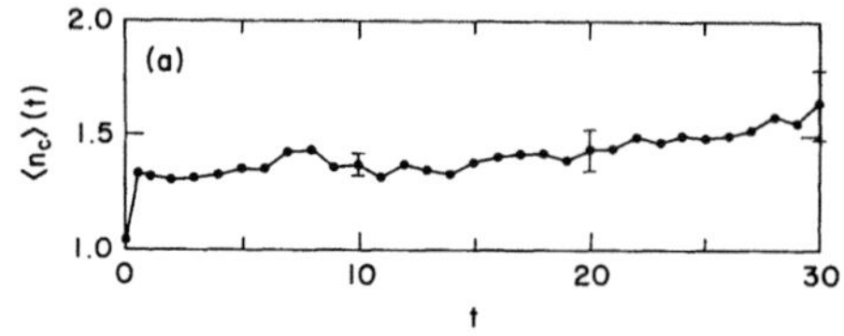


Fig.13. Average number of three-dimensional vorticity extrema on a vortex axis, $n_c$.

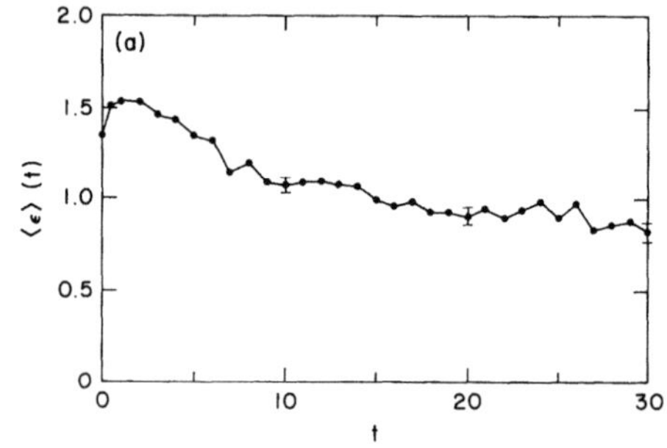


Fig.14. Average elliptical deformation in the central plane.

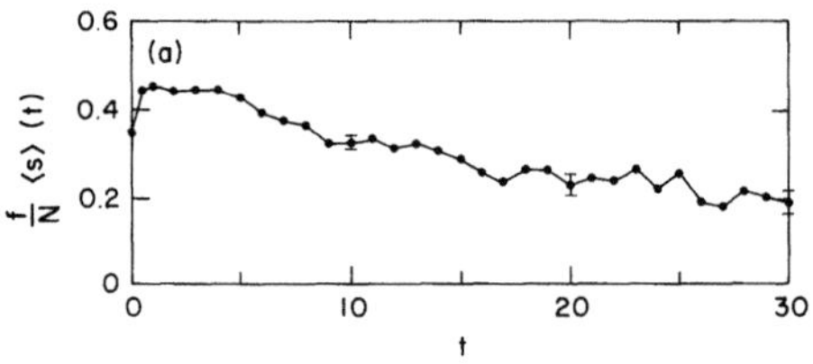


Fig.15. Average, normalized axis slope.

Finally, both horizontal and vertical shape deformations tend to decrease with time in geostrophic turbulence as the vortices become sparser and the mutually straining becomes weaker (Figs. 14-15). The average elliptical deformation is larger here than in two-dimensional turbulence, by about 50% at late times. This is most likely due to the presence of an additional source of strain, $S_V$, in three dimensional flow that induces horizontal as well as vertical deformations in the vortices.

I hope these results convey some sense of the properties of the coherent vortices of geostrophic turbulence. Of course, there is still much that remains to be understood about their dynamics, as well as their frequency of occurrence in the great variety of geostrophic regimes that occur in nature.

## References


Babiano, A., C. Basdevant, B. Legras & R. Sadourny 1987 Vorticity and passive-scalar dynamics in two-dimensional turbulence. *J. Fluid. Mech.* **183**, 379-397.

Basdevant, C., B. Legras, R. Sadourny & M. Beland 1981 A study of barotropic model flows: intermittency, waves, and predictability. *J. Atmos. Sci.* **38**, 2305-2326.

Benzi, R., G. Paladin, S. Patarnello, P. Santangelo & A. Vulpiani 1986 Intermittency and coherent structures in two-dimensional turbulence. *J. Phys. A: Math. Gen.* **19**, 3771-3784.

Benzi, R., S. Patarnello & P. Santangelo 1988 Self-similar coherent structures in two-dimensional decaying turbulence. *J. Phys. A: Math. Gen.* **21**, 1221-1237.

Brachet, M., M. Meneguzzi & P. Sulem 1986 Small-scale dynamics of the high Reynolds number two-dimensional turbulence. *Phys. Rev. Lett.* **57**, 683-686.

Brachet, M., M. Meneguzzi, H. Politano & P. Sulem 1988 The dynamics of freely decaying two-dimensional turbulence. *J. Fluid Mech.* **194**, 333-349.

Charney, J. 1971 Geostrophic turbulence. *J. Atmos. Sci.* **28** 1087-1095.

Herring, J. & J. McWilliams 1985 Comparison of direct numerical simulation of two-dimensional turbulence with two-point closure: the effects of intermittency. *J. Fluid Mech.* **153**, 229-242.

Lesieur, M. 1987 *Turbulence in Fluids.* Kluwer.

McWilliams, J. 1983 On the relevance of two-dimensional turbulence to geophysical fluid motions. *J. de Méchanique, Num. spéc.*, 83-97.

McWilliams, J. 1984 The emergence of isolated coherent vortices in turbulent flow. *J. Fluid Mech.* **146**, 21-43.

McWilliams, J. 1989a Statistical properties of decaying geostrophic turbulence. *J. Fluid Mech.* **198**, 199-230.

McWilliams, J. 1989b Geostrophic vortices. In *Proc. International School of Physics Enrico Fermi*, Italian Physical Society, in press.

McWilliams, J. 1989c A demonstration of the suppression of turbulent cascades by coherent vortices in two-dimensional turbulence. *Phys. Fluids*, in press.

McWilliams, J. 1990a The vortices of two-dimensional turbulence. *J. Fluid Mech.*, submitted.

McWilliams, J. 1990b The vortices of geostrophic turbulence. *J. Fluid Mech.*, submitted.

Moore, D., & P. Saffman 1971 Structure of a line vortex in an imposed strain. *Aircraft Turbulence and its Detection.* Plenum Press.

Santangelo, P., R. Benzi & B. Legras 1990 The generation of vortices in high-resolution, two-dimensional, decaying turbulence and the influence of initial conditions on the breaking of self-similarity. Submitted.

Weiss, J. 1981 The dynamics of enstrophy transfer in two-dimensional hydrodynamics. *La Jolla Inst. Tech. Rpt.*, LJI-TN-81-121.

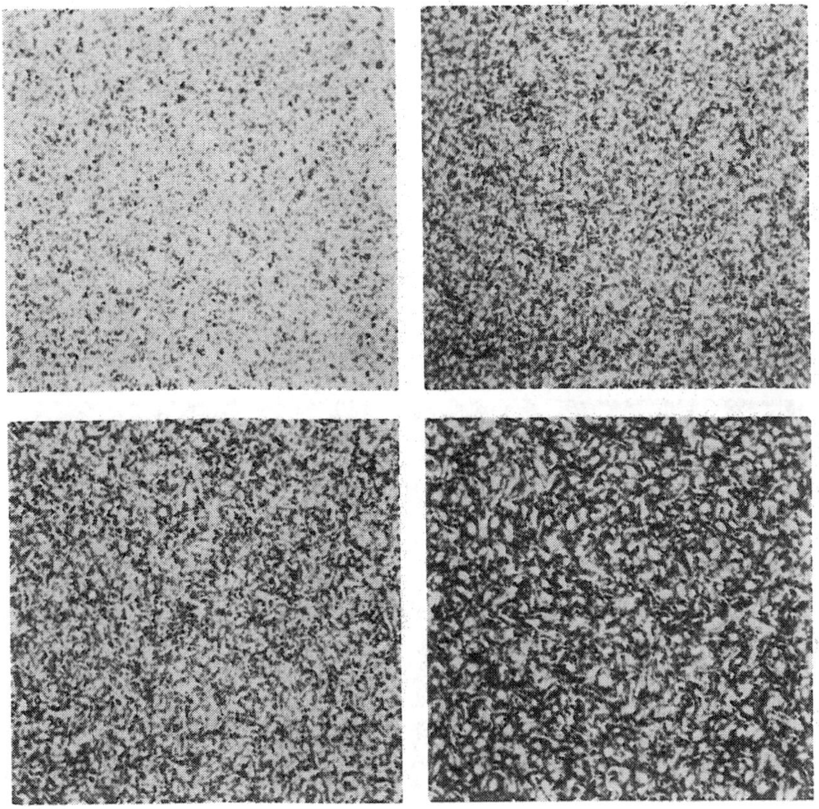

Platel. Vorticity at $t = 0, 0.25, 0.5$, and 1.2.

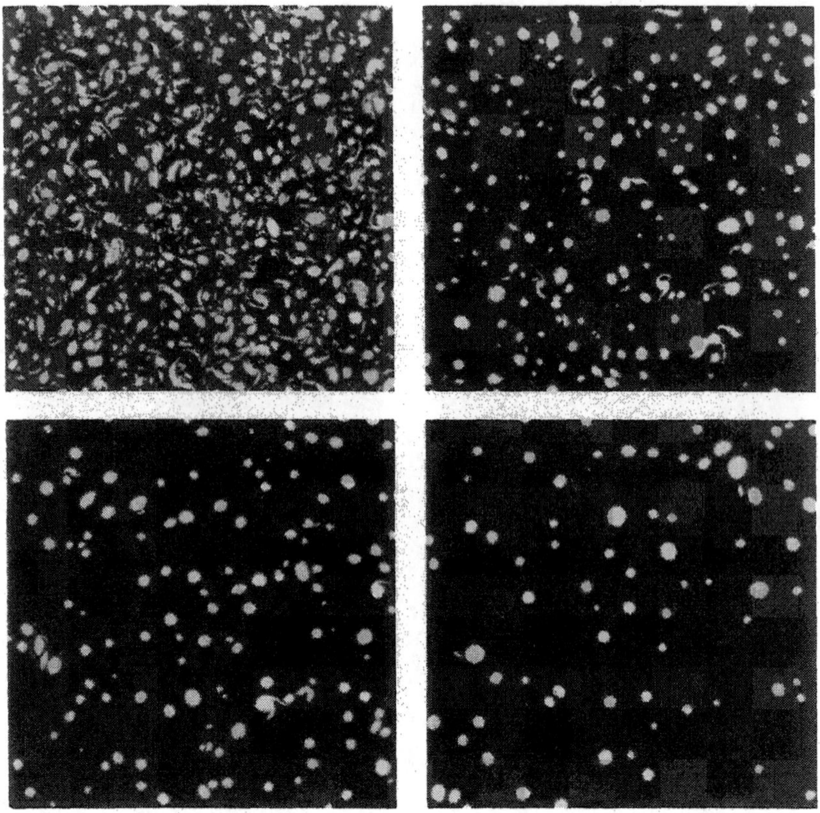

Plate2. Vorticity at $t = 2.4$, 5.6, 10.0, and 20.0.

# Experiments On Forced, Quasi-Two-Dimensional Turbulence at Upwelling Fronts

S. NARIMOUSA[α], T. MAXWORTHY[β], and G. R. SPEDDING[γ]

[α] *University of Southern California*
*Los Angeles, California 90089-1453, USA*

[β] *University of Southern California*
*Los Angeles, California 90089-1453, USA*

[γ] *University of Southern California*
*Los Angeles, California 90089-1453, USA*

*Turbulence and Coherent Structures*
*O. Métais and M. Lesieur, eds.*
*Selected papers from "Turbulence 89: organized structures and turbulence in fluid mechanics"*
*Grenoble, 18-21 September 1989*

## 1. Abstract

Coastal upwelling fronts have been generated around the outer edge of a cylindrical tank, filled with a two-layer fluid system and driven by a surface stress. Initially, an axisymmetric front was observed which subsequently became unstable to small baroclinic eddies. These eddies continued to grow until, eventually, they reached an equilibrium size. Under some circumstances, cyclonic eddies pinched-off from the fully developed front and moved away from the mean position of the front into the fluid interior. Streak photographs of the fully-developed, flow field were digitized to generate a velocity field interpolated onto a regular grid. A direct two-dimensional FFT was performed on the turbulent kinetic energy deduced from such images and one-dimensional energy ($E$) wavenumber ($K$) spectra were extracted. Consistently, $K^{-5/3}$ energy spectra were found at lower $K$ and approximately $K^{-5}$ spectra at higher $K$. In any given experiment, the two spectral slopes meet close to a wavenumber $K_w = 2\pi/\lambda_w$ (where $\lambda_w$ is the mean diameter of a frontal eddy). According to classical theories, $K_w$ would be the input wavenumber and the range of $K$ with a $K^{-5/3}$ spectrum the inverse energy cascade which yielded a Kolmogorov constant ($C$) that varied within the limits $3.6 \leq C \leq 6.3$. The approximately $K^{-5}$ range, which is much steeper than that predicted by the theories, is consistent with those found frequently in numerical experiments.

## 2. Introduction

The pioneering studies of 2D turbulence were concerned with the structure of atmospheric meso and large scale eddies, which were thought to have 2D structure. In the case of three-dimensional turbulence , Kolmogorov's (1941) cascade hypothesis indicated an equilibrium inertial subrange in which kinetic energy cascaded from low to higher wavenumber ($K$), and the energy spectra took the form $E \sim K^{-5/3}$. In case of 2D turbulence there is a second invariant of motion, the enstrophy, which is defined as half of the squared vorticity $\omega^2/2$. This led Kraichnan (1967) and Leith (1968) to propose, both a direct enstrophy and an inverse energy cascade inertial range for the case of forced 2D turbulence. They found, that $E \sim K^{-5/3}$ in the inertia range in which energy cascaded to lower $K$ (an inverse cascade) and an $E \sim K^{-3}$ inertial range in which enstrophy cascaded to higher $K$. Batchelor (1969) found the $K^{-3}$ range for the case of a freely-decaying 2D turbulence, which was consistent with Kraichnan's and Leith's energy spectrum for the enstrophy cascade range. Studies of three-dimensional, quasi-geostrophic turbulence (Charney 1971, Salmon 1978, Herring 1980 and Heyer and Sadourny 1982) indicated the existence of two inertial ranges $K^{-3}$ and $K^{-5/3}$, which was similar to spectral characteristics of 2D turbulence (also see Lesieur (1987) section 5.5).

One important motivation that triggered this study were observations of thin, long off-shore extending, cold filaments in satellite (IR) images taken of the west coast of the USA during coastal upwelling episodes (Narimousa and Maxworthy (1989)). Intense turbulent jets or rivers which are believed to transport the cold upwelled water far off-shore, have been recognized as a possible source for generating these filaments. Narimousa and Maxworthy (1985, 1986 and 1987a,b) observed, in the laboratory, that the turbulent jets were often associated with off-shore eddies which could be generated in a variety of ways: (a) large amplitude instability of standing waves generated by coastal perturbations; (b) large amplitude baroclinic instability of the front itself; and (c) production of eddies due to the presence of an off-shore curl in the surface stress. In all cases, interaction between the off-shore eddies produced the off-shore jets, which then transported the upwelled water off-shore. If the structure of such eddy fields obey the laws of 2D or geostrophic turbulence, then the intense turbulent rivers which seem to thread through such eddy fields, and are responsible for transport of material across the eddy field, would seem to be an important characteristic of such flows.

In the present study we employed the experiments of Narimousa and Maxworthy (1987a) to investigate the structure of mesoscale turbulence generated via baroclinic instabilities at upwelling fronts. Since the flow field was maintained by a constant supply of energy (see section 3), the turbulent flow produced was forced and statistically steady. The case of freely-decaying mesoscale turbulence generated at a sharp density front in the laboratory has already been reported by Griffiths and Hopfinger (1984). From measurements of the dispersion of neutrally buoyant particles they obtained an energy spectrum of the form $\sim K^{-2.5}$. Their measured spectral slope was smaller than the classical $K^{-3}$ law, and much less than those found in numerical simulations of 2D turbulence. Later Mory and Hopfinger (1986), Maxworthy et al. (1987) and Narimousa et al. (1987) employed the same particle dispersion methods and again they found energy slopes close to those reported by Griffiths and Hopfinger (1984) for the enstrophy cascade range. It appeared that independently of how one generated 2D turbulence in the laboratory, the measured spectral slope was always close to $\sim K^{-2.5}$, if particle dispersion methods were used. The value of the slope did not change as a result of dramatic increase in the number of particles, considerable changes in values of turbulent Reynolds number or rotation rate (Maxworthy et al. (1987), Narimousa et al. (1987c)). However, recent studies (summarized in Maxworthy 1989) have shown that such methods are in error when the actual spectral slope is larger than -3. In this case Babiano et al. (1985) and Bennet (1984) argue that the lower spectral slopes found by particle dispersion methods were due to nonlocality of the particle dynamics while the arguments which lead to the classical relationship between particle diffusion and spectral slope are based on local dynamics. As in Maxworthy (1989) we decided to employ other means to calculate the energy spectra. In (section 4) we show graphs of $E$-$K$ inferred from taking a direct 2D, FFT of mean square turbulent velocities.

## 3. The Experiment

Since the present study employs the experiments of Narimousa and Maxworthy (1987a), the reader is referred to to that paper for a detail discussion of the apparatus and the range of parameters used, only a brief description will be given here.

Two layers of salt water of slightly different densities were brought to solid body rotation inside a cylindrical tank (90 cm in diameter and 20 cm in depth) having a conical bottom with a slope of about 0.27. The surface of the upper fluid was in contact with a circular, smooth disc, which, as it rotated, applied a stress to this surface producing an alongshore shear flow and an offshore Ekman flux in the top layer. As a result, an upwelling front was formed around the outer edge of the tank. The velocity vectors of the current and the upwelling frontal movement were observed by placing small neutrally buoyant particles in the front and recording their motion by streak photography. The direction of motion of these particles was found by firing an electronic flash at the beginning of the time exposure. To reveal the structure of the eddies, the camera was set to rotate at a rate close to the drift velocity of the eddies. In this way a frozen flow field was recorded (figure 1) and the resulting streaks represented the turbulent velocities.

## 4. Results

The digitized turbulent velocity data of a given flow field (figure 1) was interpolated into a regular $32^2$ grid (see figure 2). Then a direct 2D, FFT was performed on the mean squared turbulent velocities deduced from the interpolated data. This resulted in a 2D energy spectra, which consisted of surfaces of constant $K$ that are centered at the origin of a rectangular ($k_x$, $k_y$) coordinate system ($k_x$ is the wavenumber in the $x$-direction, $k_y$ is the wavenumber in the $y$-direction and $K = \left(k_x^2 + k_y^2\right)^{1/2}$). In order to extract a 1D energy spectra, the energy was averaged around each surface of the associated wavenumber.

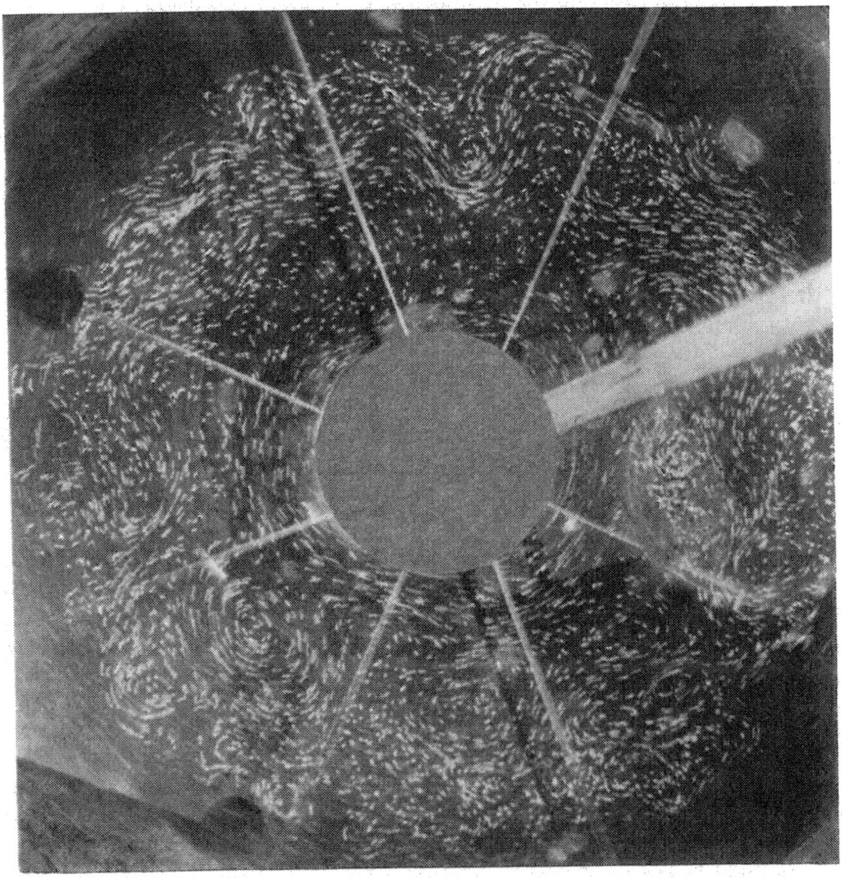

Fig.1. A streak photograph of frozen flow field showing an example of long-time behavior of meso-scale eddy structures at an upwelling front and their immediate vicinity.

In this way an average energy level was calculated for each $K$ and a 1D wavenumber-energy spectra was obtained (see figure 3).

At lower values of $K$, figure 3 shows a consistent $K^{-5/3}$ spectra slope for a rather wide range of values of $\theta_*$, where, $\theta_* = (g'h_0)/(u_* f \lambda_s)$ (here $g' = g\delta\rho$, $\delta\rho$ is the density difference between the two layers, $f$ is the

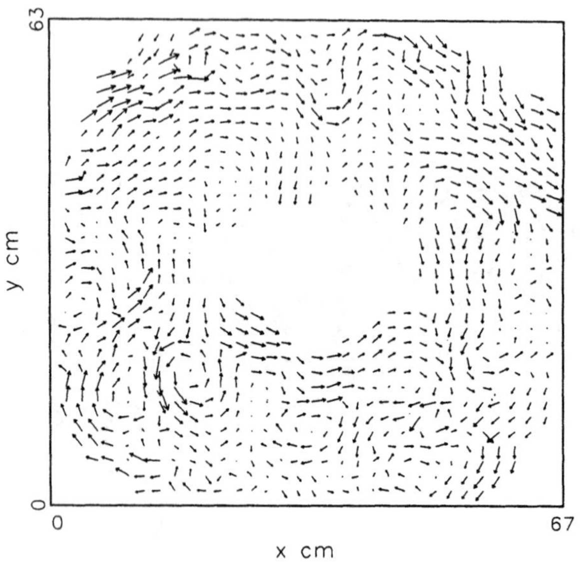


Fig.2. The original particle streak data (figure 1) as interpolated on to a regular $32^2$ grid by convolution with an adaptive Gaussian window.

Coriolis parameter, $h_0$ depth of the top layer, $u_*$ is the friction velocity and $\lambda_s$ is the width of the upwelled water at the surface). At higher values of $K$, the direct method, however, produces a second spectral range with a slope much steeper than that obtained at lower $K$. This spectral form changed from $K^{-5.4}$ at values of $\theta_* \approx 1.7$, to $K^{-4.32}$ at values of $\theta_* \approx 20$, indicating a noticeable decrease in value of the slope $\alpha$ as the value of $\theta_*$ increased.

The most important result is that the energy spectra consists of two distinct ranges. The $K^{-5/3}$ range at lower values of $K$, and the range with much steeper slope ($\sim K^{-5}$) at higher values of $K$. According to the classical theories of forced 2D turbulence (see section 1), the $K^{-5/3}$ is the inverse energy cascade inertial range. This result is consistent also with numerical simulations of Herring and McWilliams (1985) and Frisch and Sulem (1984) and experimental results of Sommeria (1986). Classical theories predict the $K^{-3}$ for the enstrophy cascade inertial range at higher $K$, also. At such wavenumbers the slope of kinetic energy ($\sim K^{-5}$) is considerably steeper than the ($K^{-3}$) predicted by the theories. The much steeper slope found in the present study is consistent with those predicted frequently by numerical experiments. High resolution numerical experiments of Brachet et al. (1988) showed that energy spectra is not unique at $K^{-3}$ in the so called enstrophy cascade range. They observed a $K^{-4}$ range for a short time which changed sharply to the

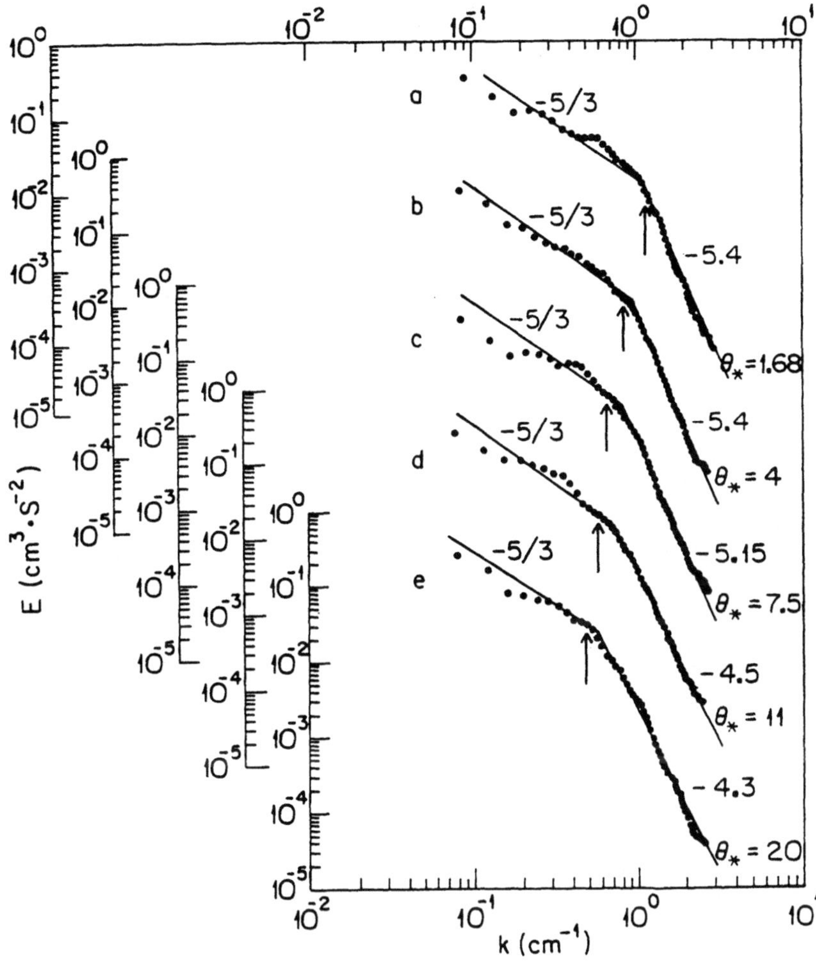


Fig.3. One-dimensional energy spectra $(E)$ as a function of wavenumber $(k)$, inferred from a direct 2D, FFT of the mean square turbulent velocities deduced from the interpolated data such as that shown in figure 2.

$K^{-3}$ when the enstrophy dissipation was maximum and at a later time, due to isolated concentration of vorticity, coherent vortices dominated the flow field. McWilliams (1984) showed that such coherent structures last for a long time and he identified them with an approximate $K^{-5}$ energy spectrum. His isovorticity contours clearly revealed the structure of these isolated, coherent vortices.

In the present study, the recorded turbulent velocity field represents the fully developed, long time behavior of the flow.

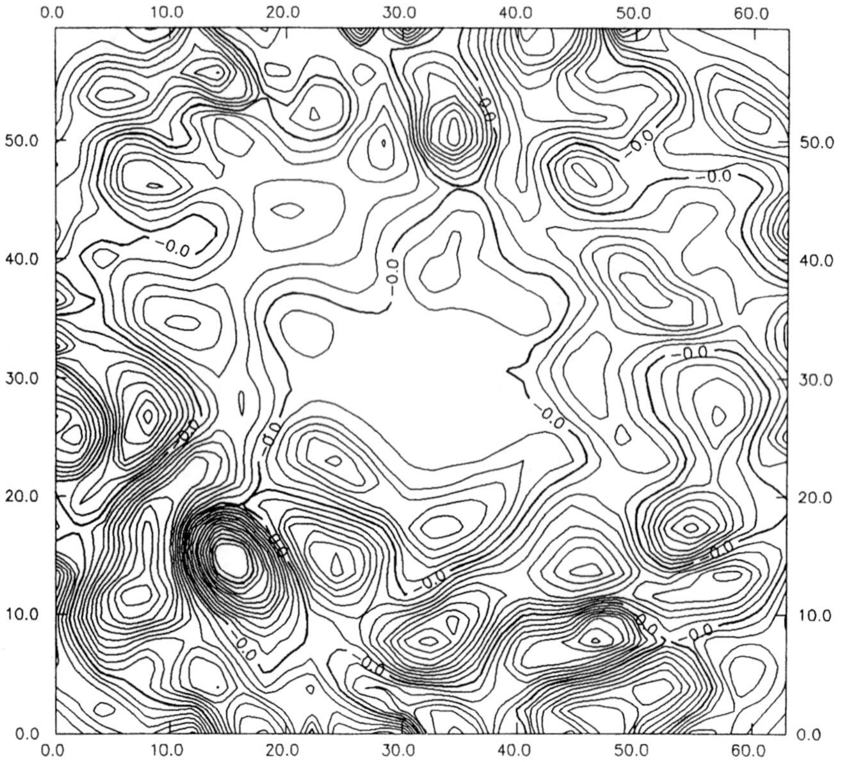


Fig.4. Iso-vorticity contours for figure 1.

Figure (4) shows an example of contour map of the vorticity field in which many coherent vortex structure can be seen. These structures are closely packed as in Brachet et al. (1988) and Santangelo et al. (1989) for example, and not isolated as found by McWilliams (1984). However, we have found that such coherent vortex structures produce an approximate

$K^{-5}$ energy spectra also.

In addition, figure (2) shows that in any given experiment, the two spectral slopes meet close to the wavenumber $K_w = 2\pi\lambda_w$ where, according to the classical theories, $K_w$ (the wavenumber of the frontal eddies), is the input wavenumber. Since $\lambda_w \sim (g'h_0)/(u_*f)$ is a function of both the Rossby deformation radius $R_0 = (g'h_0)^{.5}/f$ and the Richardson number $R_0 = (g'h_0)/u_*^2$, we can write

$$K_w = \frac{2\pi}{\lambda_w} \sim 2\pi\left(\frac{u_*f}{g'h_0}\right) \sim 2\pi R_i^{-.5}R_0^{-1}.$$

Furthermore it is possible to estimate the two-dimensional Kolmogorov constant ($C$) according to

$$E(k) = C\epsilon^{2/3}(k)^{-5/3},$$

The calculated values of $3.6 < C < 6.3$ are within the range of those estimated by Lilly (1969), 4.3-6.2 Herring and Kraichnan (1975), 6.5, Herring and McWilliams (1985), 2.9-4.8 and Sommeria (1986), 3-7, but smaller than those of Frisch and Salem (1984), 9, and much smaller than those of Siggia & Aref (1981), 14.

## 5. Summary and Conclusions

When we performed a direct 2D, FFT on the mean square turbulent velocities, the induced 1D energy spectra produced two distinct inertia ranges also; a $K^{-5/3}$ spectrum at lower $K$ and one approximately varying as $K^{-5}$ at higher $K$. The $K^{-5/3}$ result is consistent with that predicted for the inverse energy cascade inertia range and the $K^{-5}$ is in agreement with those obtained in numerical simulations for the enstrophy cascade in stationary flow. The two inertia ranges meet close to a wavenumber that is associated with the wavenumber of the frontal eddies ($K_w \sim 2\pi(u_*f)/(g'h_0)$) and therefore according to classical theories $K_w$ is the input wavenumber. We have estimated the 2D Kolmogorov constant ($C$), within the $K^{-5/3}$ inertia range. The values of $C$ varied within the range $3.6 < C < 6.3$.

According to classical theories and numerical simulations of 2D or geostrophic turbulence, the spectral slopes ($K^{-5/3}$ at lower $K < K_w$ and $K^{-5}$ at higher $K > K_w$) found in the present study suggest that turbulence in our model is two dimensional. This in turn suggests that the prototype system we are attempting to model can also be considered to be a 2D turbulent field. Since jet-like structures or rivers evolve in such a field in the laboratory then presumably they do also in the natural flow, this helps to explain the filamentary structures which transport passive markers, e.g. heat, as observed in satellite images.

## 6. Acknowledgment

The authors wish to thank the Office of Naval Research for supporting this research under Grant No. N00014-87-K-0809.

## 7. References

BABIANO, A., BASDEVANT, C. & SADOURNY, R. 1985 Structure functions and dispersion laws in two-dimensional turbulence. *J. Atmos. Sci.* **42**, 941-949.

BATCHELOR, G. K. 1969 Computation of energy spectrum in homogeneous two-dimensional turbulence. *Phys. Fluid Suppl.* **12**, II 233.

BRACHET, M. E., MENEGUZZI, M., POLITANO, H. & SULEM, P. L. 1988 The dynamics of freely decaying two-dimensional turbulence. *J. Fluid Mech.* **194**, 333-349.

CHARNEY, J. G. 1971 Geostrophic turbulence. *J. Atmos. Sci.* **28**, 1087-1095.

FRISCH, U. & SULEM, P. L. 1984 Numerical simualtion of inverse cascade in two-dimensional turbulence. *Phys. Fluids* **27**, 1921-1923.

GRIFFITHS, R. W. & HOPFINGER, E. J. 1986 The structure of mesoscale turbulence and horizontal spreading at ocean fronts. *Deep-Sea Research* **31**, 245-269.

HERRING, J. R. 1980 Statistical theory of quasi-geostrophic turbulence. *J. Atmos. Sci.* **37**, 969-977.

HERRING, J. R. & McWILLIAMS, J. C. 1985 Comparison of direct numerical simulation of two-dimensional turbulence with two-joint closure: the effects of intermittency. *J. Fluid Mech.* **153**, 229-242.

HEYER, J. M. & SADOURNY, R. 1982 Closure modeling of fully developed baroclinic instabilities. *J. Atmos. Sci.* **39**, 707-721.

KOLMOGOROV, A. N. 1941 Dispersion of energy in locally isotropic turbulence. *C. R. Acad. Sci. URSS*, **32**, 16.

KRAICHNAN, R. H. 1967 Inertial ranges in to-dimensional turbulence. *Phys. Fluids* **10**, 1417-1423.

KRAICHNAN, R. H. 1975 Statistical dynamics of two-dimensional flow. *J. Fluid Mech.* **67**, 155.

LEITH, C. E. 1968 Diffusion approximation for two-dimensional turbulence. *Phys. Fluids* **11**, 671-673.

LESIEUR, M. 1987 *Turbulence in fluids.* Martinus Nijhoff Publishers.

LILLY, D. K. 1969 Numerical simulation of two-dimensional turbulence. High-speed computing in fluid dynamics. *Phys. Fluid Suppl.* **12**, II 240.

MAXWORTHY, T. 1989 The dynamics of two-dimensional turbulence. *Proc. of Conf. on the "Oceanography of Sea Straits"*, les Arcs, France.

MAXWORTHY, T., CAPERAN, P. & SPEDDING, G. R. 1987 Two-dimensional turbulence and vortex dynamics in a stratified fluid. *Proc. Third Inter. Conf. on Stratified Flows*. Pasadena, California.

McWILLIAMS, J. C. 1984 The emergence of isolated coherent vorticies in turbulent flow. *J. Fluid Mech.* **146**, 21-43.

MORY, M. & HOPFINGER, E. J. 1986 Structure functions in a rotationally dominated turbulent flow. *Phys. Fluids* **29**, 2140-2146.

NARIMOUSA, S. & MAXWORTHY, T. 1985 Two-layer model of shear driven coastal upwelling in the pressence of bottom topography. *J. Fluid Mech.* **159**, 503-531.

NARIMOUSA, S. & MAXWORTHY, T. 1986 Effects of a discontinuous surface stress on a model of coastal upwelling. *J. Phys. Oceanogr.* **16**, 2071-2083.

NARIMOUSA, S. & MAXWORTHY, T. 1987a Coastal upwelling on a sloping bottom: The formation of plumes, jets and pinched-off cyclones. *J. Fluid Mech.* **176**, 169-190.

NARIMOUSA, S. & MAXWORTHY, T. 1987b Effects of coastline perturbations on coastal currents and fronts. *J. Phys. Oceanogr.* **17**, 1296-1303.

NARIMOUSA, S., MAXWORTHY, T. & SPEDDING, G. R. 1987c Structure of meso-scale baroclinic turbulence. *Sixth Symposium on Turbulent Shear Flows*. Toulouse, France, September 7-9.

NARIMOUSA, S. & MAXWORTHY, T 1989 Application of a laboratory model to the interpretation of satellite and field observations of coastal upwelling. *Dyn. Atmos. Oceans* **13**, 1-46.

SALMON, R. 1978 Two-layer quasi-geostrophic turbulence in a simple case. *Geophys. Astrophys. Fluid Dyn.* **10**, 25-51

SANTANGELO, R., BENZI, R. & LEGRAS, B. 1989 The generation of vortices in high-resolution, two-dimensional decaying turbulence and the influence of initial conditions on the breaking of self-similarity. *Phys. Fluids* **1**, 1027-1034.

SIGGIA, E. D. & AREF, H. 1981 Point vortex simulation of the inverse cascade in two-dimensional turbulence. *Phys. Fluids* **24**, 171-173.

SOMMERIA, J. 1986 Experimental study of the two-dimensional inverse energy cascade in a square box. *J. Fluid Mech.* **170**, 139-168.

# Merging of two-layer baroclinic vortices

Jacques VERRON[α], Emil HOPFINGER[α] and James C. McWILLIAMS[β]

[α]*Institut de Mécanique de Grenoble*
*BP 53X, 38041 Grenoble-Cédex, France*

[β]*National Center for Atmospheric Research*
*PO BOX 3000, Boulder 80307 CO, USA*

*Turbulence and Coherent Structures*
*O. Métais and M. Lesieur, eds.*
*Selected papers from "Turbulence 89: organized structures and turbulence in fluid mechanics"*
*Grenoble, 18-21 September 1989*

## 1. Introduction

Studying the interaction of isolated vorticity regions is of primary importance for the understanding of turbulence in general and the behaviors of many geophysical situations in particular. In the context of shear flows it has received much attention from experimentalists (Winant and Browand, 1974, Brown and Roshko, 1974) as well as from theoretical and computational points of view (Zabusky and Deem, 1971, Christiansen, 1973, Christiansen and Zabusky, 1973). Isolated vortex structures have also been studied in detail theoretically and numerically (Kida, 1981, Overman and Zabusky, 1982). However it is only recently that progress has been made in the understanding as to why and how merging occurs. Axisymmetrization due to filament formation appears to be the chief merger mechanism (Melander, Zabusky and McWilliams, 1989). The problem of the interaction of baroclinic vortex structures is even more difficult. Point vortices in the simplest configuration, namely a two-layer stratified fluid, assuming geostrophy, have been recently investigated by Hogg and Stommel (1985a). In this case there is no merging but there is cluster formation formation (Hogg and Stommel, 1985b). Griffiths and Hopfinger (1987, hereafter GH) studied experimentally the interaction and merging conditions of real, finite core vortices in a two layer stratified rotating fluid. A strong dependance of merging conditions on stratification was observed. Contour dynamics calculations of the same problem by Polvani, Zabusky and Flierl (1989, hereafter PZF) indicated however that merging conditions are not altered by stratification except when the two layers were not of equal depth.

The physical experiments by GH were conducted in a rotating circular tank where a two-layer stratification is created with the two layers of equal depth. Two-like sign vortices were initiated by injection (resulting in cyclones) or suction (anticyclones) of fluid in the upper layer of the flow. Vortex interaction was then studied for different configurations. In particular merging conditions were studied as a fonction of stratification. Figure 1 shows the curve they obtained for the critical distance $d_c$ as a function of the internal Rossby radius of deformation $\lambda$, below which merging occurs and above which stable conditions were encountered for times long compared with the orbital time . Both quantites, $d$, the initial distance between the vortex centers and $\lambda = \sqrt{g'H}/f_0$, the Rossby radius are adimensionalized by the inital vortex radius $R$. $g'$ measures the reduced gravity $g\Delta\rho/\rho$ where $\Delta\rho$ is the stratification jump between the two layers and $\rho$ the mean stratification. $H$ is the layer depth and $f_0$ the basic rotation. In the limiting barotropic situation it is known from theoretical developments (i.g. Zabusky et al., 1979) and computations that the critical distance for merging should be around $d_c/R \approx 3.3$. GH recover this results in a barotropic situation. By varying the stratification merging was found to be strongly dependent on $\lambda/R$ with a decrease of $d_c/R$ for smaller $\lambda/R$ ($\lambda/R < 2$) and an important increase of $d_c/R$ for larger $\lambda/R$ ($2 < \lambda/R < 4$). The range of $\lambda/R > 4$ was not investigated in the experiments. However as $\lambda/R$ gets larger the interface will become more and more rigid and in the limit of $\lambda/R$ going to infinity the surface layer must behave as in a barotropic situation as the layers become completely uncoupled. The critical distance $d_c/R$ is then expected to decrease back to 3.3.

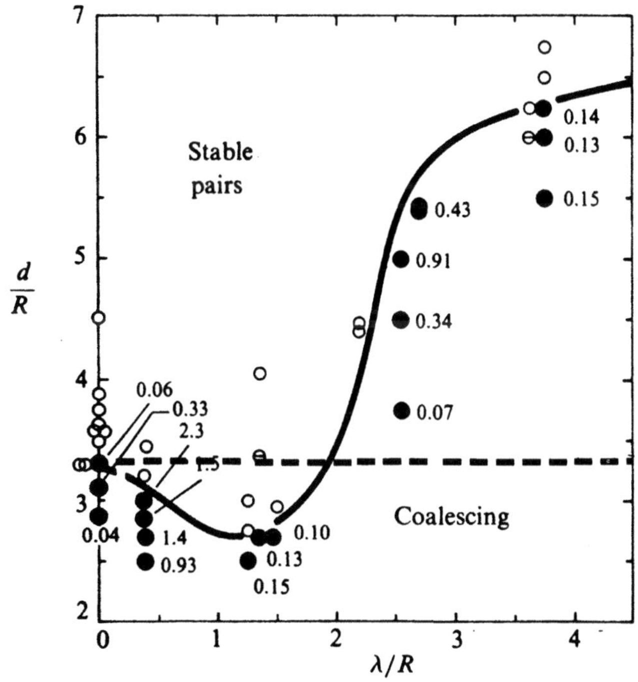


Fig.1. Classification of stable and unstable vortex pairs expressed by the relative initial distance $d/R$ as a function of the relative Rossby radius of deformation $\lambda/R$ as obtained by Griffiths and Hopfinger (1987). The dotted line corresponds to the results by Polvani, Zabusky and Flierl (1989) and also the present PVI results.

In this paper our general motivation is to explore how stratification may affect vortex merging but a more immediate motivation was to try to clarify the dramatic discrepancy between the results by GH and PZF as seen form Figure 1.

## 2. Relative vortex initialization

The problem is reconsidered by using direct numerical simulations of the complete set of potential vorticity equations including dissipation (Pedlosky, 1979). Defining the potential vorticity $Q_i$ as:

$$Q_1 = \Delta\psi_1 + f_0\frac{\eta}{H_1} = \omega_1 + f_0\frac{\eta}{H_1}$$

$$Q_2 = \Delta\psi_2 - f_0\frac{\eta}{H_2} = \omega_2 - f_0\frac{\eta}{H_1}$$

these equations write:

$$\frac{DQ_1}{Dt} = -V_1$$
$$\frac{DQ_2}{Dt} = -V_2$$

where :

$$\frac{DQ_i}{Dt} = \frac{\partial Q_i}{\partial t} + J(\psi_i, Q_i)$$

The indices 1 and 2 designates respectively the upper and lower layers. The interface deviation $\eta$ is written as :

$$\eta = \frac{f_0}{g'}(\psi_2 - \psi_1)$$

where $\psi_i$ are the streamfunctions related with the velocity by $u = -\frac{\partial\psi}{\partial y}$, $v = \frac{\partial\psi}{\partial x}$ and the relative vorticites $\omega_i$ by $\omega_i = \Delta\psi_i$. $H_i$ is the layer depth. Rotation is assumed to be constant at a rate $f_0$ (f-plane approximation).

The coupling coefficient is $\lambda^{-2} = f_0^2/g'H$ when $H_1 = H_2 = H$ . $\lambda$ is the internal Rossby radius of deformation. The lateral friction is parameterized with a high order viscosity of the biharmonic type and the bottom friction effect with a simple linear drag :

$$V_i = -A_4\nabla^6\psi_i - r\Delta\psi_i$$

The calculation domain is a square fluid tank with horizontal dimension $L$ and vertical thickness $H_1 + H_2$. Discarding dissipation leads to the model used by PZF. Assuming stationarity and dropping dissipation terms one can recover the simplified model GH derived analytically in their paper. Boundary conditions are free slip along the lateral walls.

The rigid lid approximation is also included for the purpose of filtering external gravity waves and therefore gaining on the computional time step. The model uses finite difference method with a pseudo-spectral solver for the Helmholtz equation. The parameters for the simulations are the ones of the GH experiment.

The first calculations of merging conditions were performed with the following initial conditions: two vortices are created in the upper layer at time $t = 0$. Their shape is defined by:

$$\omega = \omega_0[1 - 0.5(tanhX + 1)]$$

with $X = \alpha(\frac{r}{R} - 1)$. If not otherwise indicated $\alpha = 4$. The deep layer flow is at rest.

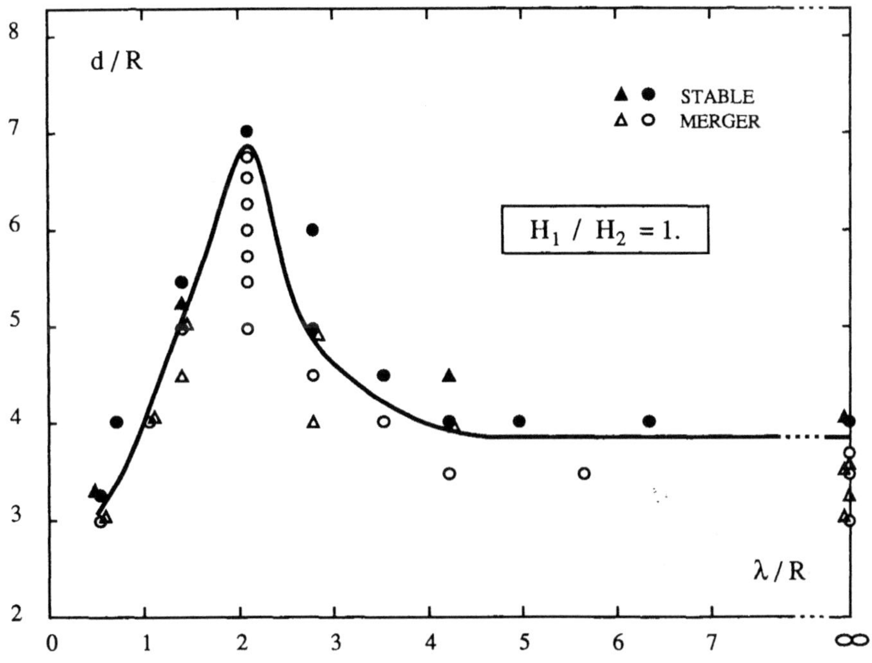


Fig.2. Merger stability curve obtained with RVI calculations when $H_1 = H_2$ (same notation as Figure 1).

The results for the stability curve are shown in Figure 2. Three main regimes can be identified. When $\lambda/R < 1$ the baroclinic effect stabilizes the vortex pair and the critical distance $d$ for merging is less than the corresponding barotropic critical distance $d_c/R \approx 3.3$. On the contrary, in the range $1 < \lambda/R < 3$, merging is very much favored and the critical distance increases to twice the barotropic value. Beyond $\lambda/R = 3$ there is a rapid decrease in critical distance and the barotropic conditions are recovered when $\lambda/R > 4$ .

The most striking result is the sharp peak at $\lambda/R \approx 2$ in the merging boundary where the critical distance is twice the barotropic value. When compared with the experimental merging boundary it is seen that the increase is similar in both cases with nearly identical maximum values but the experimental curve reaches a maximum at $\lambda/R \approx 3$ instead of 2 in the numerical simulations. In part this shift may be due to differences in the definition of vortex radius but this does not account for everything. When $\lambda/R$ is less than one, the critical merging distance is 3 or less which means that vortices nearly touch each other before merging. Actually the regime of smallest $\lambda/R$ is difficult to predict because baroclinic instability will start to compete with merging stability. Complex situations are encountered there for which stability may no longer be described in the simple $(d/R, \lambda/R)$ parameter space.

It should also be noted that on Figure 2, the barotropic limit ($\lambda \longrightarrow \infty$) is not exactly 3.3 but $d_c/R \approx 3.7$, which is due to the vortex initial shape which is a smoothed representation of a Rankine vortex.

Varying the layer depth ratio may lead to some change in merging stability. Figure 3 relates to a case representative of mid-latitude oceanic conditions. The ocean thermocline is placed at 1000 m in a 5000 m "deep ocean". Stratification and rotation are chosen to be typical of mid-latitudes : $g' = 0.02 m/s^2$ and $f_0 = 10^{-5} s^{-1}$. This corresponds to a Rossby radius of deformation $\lambda$ of about 43 km. The stability curve is similar to the previous one, except that, in the range of parameters being investigated, less favorable situations than barotropic seem not be exist. On the other hand "Peak" conditions for merging reach somewhat larger values ($d_c/R \approx 7.5$). This implies a strong scale selectivity for vortices having a size of about 30 km.

By looking at the flow patterns and their evolution in time, some features are worthwhile to be mentioned. When $\lambda/R$ is more than about 5 the flow structure is not different from the barotropic situation. However when $\lambda/R$ decreases the filamentation process seems to be reduced compared with the barotropic situation. Vortex straining results in more compact vorticity patches than elongated filaments. The axisymmetrization process by filamentation as being understood from purely 2D situations, appears to be less active. This may be related to the creation of vorticity of opposite sign which arises more and more strongly in the vortex interaction process (Figure 4). For low $\lambda/R$, final merger (assuming the word "merger" may be still relevant here) is quite far from axisymmetry and, for example, tripoles are being found(Figure 5).

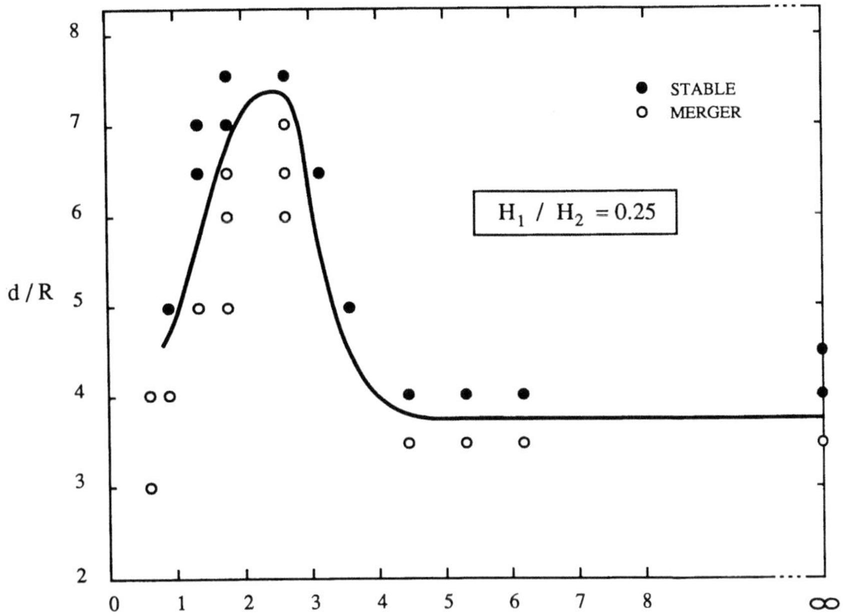


Fig.3. Merger stability curve for the present calculations with RVI initialization when $H_1 = 0.25H_2$ (same notation as Figure 1).

## 3. Potential vortex initialization

In the second numerical experiments of merging initial conditions were chosen similar to PZF. Two upper layer potential vortices were created at t=0. Their shape is defined as previously except $Q$ is now replacing $\omega$ in the form:

$$Q = Q_0[1 - 0.5(tanhX + 1)]$$

The lower layer flow is no longer at rest but of constant potential vorticity.

The results obtained were identical to PZF, that is to say stratification does not have any influence on merging as long as layers have equal depths.

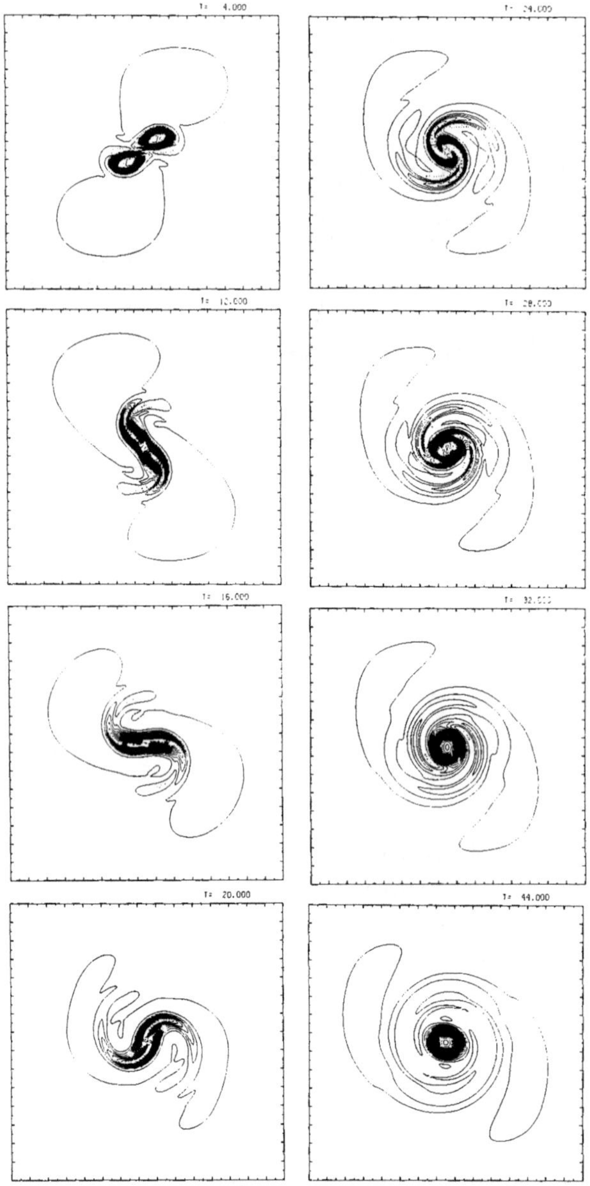

Fig.4. Vorticity field for a time sequence of merging with $H_1/H_2 = 1$, $\lambda/R = 5.66$, $d/R = 3.5$. Time are given in seconds.

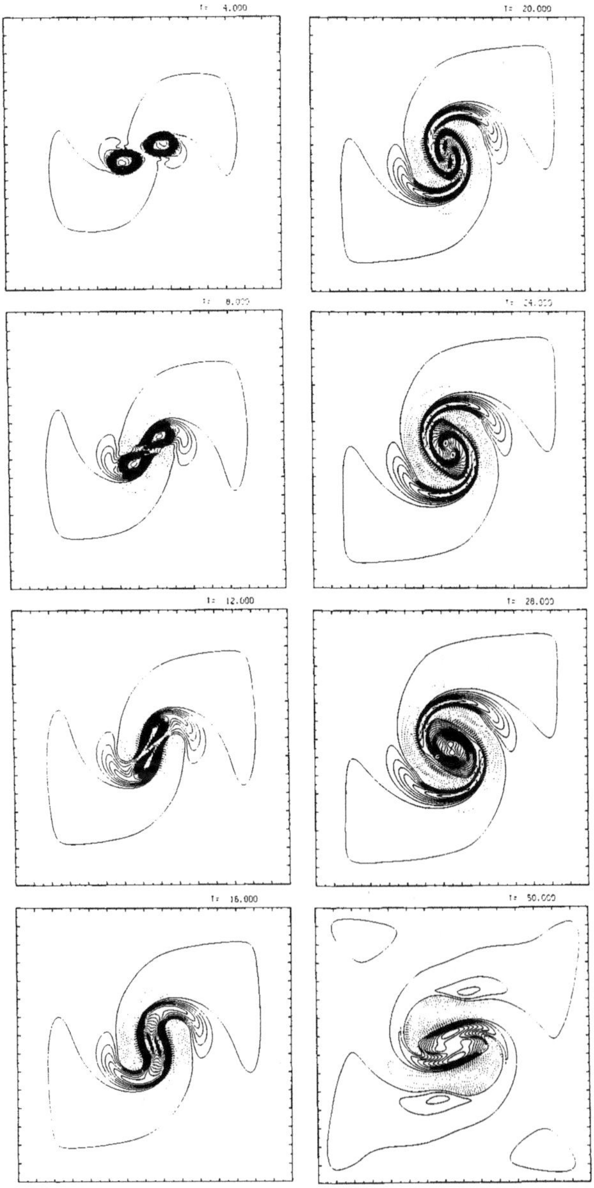

Fig.5. Vorticity field for a time sequence of merging with $H_1/H_2 = 0.25$, $\lambda/R = 0.89$, $d/R = 4$. Time are given in seconds.

## 4. Discussion

The numerical simulations presented reproduce the results of PZF when potential vorticity initialization (PVI) is used. Most remarkable is however that results similar to GH experiments are obtained when the initialization is of relative vorticity (RVI) type. This emphasizes for the first time the apparent importance of initial conditions on the flow behavior, at least as far as merging is concerned. The contradiction between GH and PZF results could therefore be attributed to a difference in initial conditions.

The problem that initialization as performed in GH experiments is more of the PVI type rather than RVI as the final merging results would suggest. By creating vortices by injection or suction of water in the upper layer of a rotating two-layer fluid tank, the interface is deviated and one would expect to have some relative vorticity in the deep layer according to the potential vorticity conservation. To some extent this is confirmed by the velocity profiles measured by GH (Figure 5 of GH paper). Azimuthal motion is clearly present in the bottom layer and velocity profiles look like what would be expected from PVI calculations. In the RVI case, the deep flow is at rest and some vertical shear is present even at large distances which does not exist in the PVI case.

In the search for other explanations, further simulations have been done to assess the role of non-conservative effects (horizontal viscous effects, Ekman bottom dissipation) or the approximation of a Rankine vortex profile. Tests of the effects of these three particular aspects revealed that these do, as expected, play a role in the merging process. But, these effects are not altered in an appreciable way by stratification. At this time these seem therefore, not to be good candidates for bringing clarification to the problem. The role of interfacial friction has not been explored so far.

The observed dependency of merging on initial conditions is fascinating in itself in a fundamental and practical context. In particular the localisation of a sharp peak of the stability boundary at $\lambda/R \approx 2$ requires considerable effort to explain physically why this occurs. All models with various initial conditions are schematization of realistic flows. The fundamental question is then to know which model or combination of models is the most relevant one for of the "real world". If the RVI case is part of the solution the strong scale selective effect of stratification raises questions concerning the inverse cascade in geostrophic turbulence.

The calculations have been made with the numerical facilities of the Centre de Calcul Vectoriel pour la Recherche in Palaiseau.

## References

Brown G.L. & A. Roshko, 1974: On density effects and large struture in turbulent mixing layers. *J. of Fluid Mechanics*, **64**, 775-816.

Christiansen J.P., 1973: Numerical simulation of hydrodynamics by the method of point vortices. *J. of Comp. Physics*, **13**, 363-379.

Christiansen J.P. & N.J. Zabusky, 1973: Instability coalescence and fission of finite area vortex structures. *J. of Fluid Mechanics*, **61**, 219-243.

Griffiths R.W. & E.J. Hopfinger, 1986: Experiments with baroclinic vortex pairs in a rotating fluid. *J. of Fluid Mechanics*, **173**, 501-518.

Griffiths R.W. & E.J. Hopfinger, 1987: Coalescing of geostrophic vortices. *J. of Fluid Mechanics*, **178**, 73-97.

Hogg N. & H. Stommel, 1985: The heton, an elementary interaction between discrete baroclinic geostrophic vortices and its implications concerning eddy heat-flow. *Proc. R. Soc. Lond.*, **A 397**, 1-20.

Kida S., 1981: Motion of elliptic vortex in a uniform shear flow. *J. Phys. Soc. Japan*, **50**, 3517-3520.

Melander M.V., Zabusky N.J. & J.C.McWilliams, 1988: Symmetric vortex merger in two dimensions : causes and conditions. *J. of Fluid Mechanics*, **195**, 303-340.

Overman E.A. & N.J. Zabusky, 1982: Evolution and merger of isolated vortex structures. *Phys. Fluids*, **25**, 1297-1305.

Pedlosky J., 1979: Geophysical Fluid Dynamics, Springer-Verlag, Berlin.

Polvani L.M., Zabusky N.J. & G.R. Flierl, 1989: Two-layer geostrophic vortex dynamics. Part 1. Upper layer V-states and merger. *J. of Fluid Mechanics*, **205**, 215-242.

Winant C.D. & F.K. Browand, 1974: Vortex pairing: the mechanism of turbulent mixing layer growth at moderate Reynolds number. *J. of Fluid Mechanics*, **63**, 237-255.

Zabusky N.J., Hughes M.H. & K.V. Roberts, 1979: Contour dynamics for the Euler equations in two dimensions. *J. of Comp. Phys.*, **30**, 96-106.

Zabusky N.J. & G.S. Deem, 1971: Dynamical evolution of two-dynamical unstable shear flows. *J. of Fluid Mechanics*, **47**, 353-379.

# Transmission of acoustic waves through mixing layers and 2D isotropic turbulence.

D. JUVÉ, P. BLANC-BENON and G. COMTE-BELLOT

*Laboratoire de Mécanique des Fluides et d'Acoustique, URA CNRS 263*
*Ecole Centrale de Lyon, BP 163, 69131 ECULLY Cedex, France*

*Turbulence and Coherent Structures*
*O. Métais and M. Lesieur, eds.*
*Selected papers from "Turbulence 89: organized structures and turbulence in fluid mechanics"*
*Grenoble, 18-21 September 1989*

**Abstract**

Ray tracing and parabolic equation methods have been used to study the properties of acoustic waves transmitted through turbulent velocity fields. A numerical simulation permits individual realizations of the turbulent field, which then allow, if desired, an ensemble averaging of the fields. Two flows have been considered, 2D isotropic turbulence, and a 2D mixing layer .The following complementary aspects are developed: the occurrence of caustics, the reinforced or weakened zones of the acoustic field, the eigenrays between a source and a receiver, and the associated travel times, variances, and scintillation index.

## 1. Introduction

The propagation of acoustic waves through inhomogeneous and/or random media has attracted a great deal of interest and its applications are numerous. Noise annoyance contours in the vicinity of industrial plants or airports have to include the pressure level changes due to the turbulence (wind or temperature) existing in the real atmospheric surroundings. In underwater acoustics, wave front distortions or phase changes have a direct effect on the detection ability of fixed or towed arrays used to locate noise sources. Remote sensing devices, such as sodars, used to detect high altitude turbulence as a complement to optical techniques, are based on the efficiency of the interaction between the acoustic waves and the flow. In all these cases there is a need for numerical models in order to estimate all the details of the interaction.

The classical method of handling the propagation of waves in random media relies on a statistical approach. As a first step, making use of several assumptions, one obtains an equation with random coefficients (Helmholtz or parabolic equation). Then other equations are deduced for second and higher moments (intensity, correlations...). The latter can only be closed using a hypothesis which implies, in essence, a very short correlation along the mean direction of the wave propagation (delta-correlated hypothesis). Initially this theory was developed for optical applications and the turbulent medium was characterized by a random refraction index. However, in acoustics, the presence of a flow introduces an additional effect of wave convection (ovalization of wave fronts) which is neglected in the classical approach.

We have developed a different approach in order to avoid purely statistical theory, not well-founded hypothesis, and inadaptation to fluctuating velocity fields. We use a deterministic propagation of an acoustic wave through numerically simulated turbulent fields. In this way not only one realization of the turbulent field can be created, but many and this allows for the obtaining of ensemble averages. Similar approaches were introduced very recently in optics (Martin and Flatte, 1988; Hesselink and Sturtevant, 1988). Our work presents several noteworthy differences : i) a consideration of velocity fluctuations instead of a refraction index such as in temperature changes (at least for part of the study) ; ii) a specifically tailored random field ; iii) simultaneous use of two complementary methods for describing acoustic waves (geometrical theory and the parabolic equation method).

In this article we will explain the non-averaged approach that we have developed and indicate its potentialities. We emphasize, in particular, the visualization of the acoustic field in order to illustrate the considerable deformation that it undergoes due to the cumulative effect of very weak turbulent fluctuations along the propagation path. These visualizations are presented, on the one hand for isotropic homogeneous turbulence and, on the other hand, for two steps of the downstream development of a 2 D mixing layer (Comte, Lesieur, Laroche and Normand, 1989). We also provide some quantitative statistical data in the case of isotropic homogeneous turbulence, not only in order to offer a comparison with existing theories, but also to demonstrate that new results can be generated.

Although the present work deals with two dimensional fields, its extension to three dimensional fields presents no essential difficulty if not a marked increase in calculating time. In fact 2 D and 3 D behaviors do not seem to be very different qualitatively (Blanc-Benon, Juvé, Karweit and Comte-Bellot, 1990) and this justifies our approach.

## 2. Modelling of acoustic wave propagation

The following two complementary descriptions are used : i) the geometrical acoustics approach which gives simple and clear visualizations of the focusing properties of a medium in random motion ; it is well suited to the computation of quantities linked to the transit time along the rays (wavefront distortion, phase fluctuations...) ; ii) The parabolic equation method which is well adapted to the determination of the energetics of the acoustic field (mean intensity, fluctuations, correlations...)

### 2.1. Geometrical acoustics

In this high frequency approximation, the acoustic variables, such as pressure $p(\vec{x},t)$, are written in the form

$$p(\vec{x},t) = A(\vec{x}).e^{iS(\vec{x})}.e^{-i\omega t}$$

where $\omega$ is the angular frequency. The amplitude A and the local wave vector $\vec{K}(\vec{x}) = \vec{\nabla}(S)$ are assumed to vary slowly on the scale of a wavelength $\lambda = 2\pi c_0/\omega$ ($c_0$ being the speed of sound ). An asymptotic expansion for $\omega \to \infty$ of the exact linearized fluid mechanics equations gives the dispersion relation of acoustic waves propagating in an arbitrary velocity field $\vec{V}(\vec{x})$(Candel, 1977) :

$$\omega = Kc_0 + \vec{K}.\vec{V} \quad K = \| \vec{K} \|$$

This non linear first order equation (for S ) can be solved by the method of characteristics. The rays which are the lines tangent to the group velocity can be determined through the following Hamiltonian system :

$$\frac{dx_i}{dt} = \frac{c_0 p_i}{p} + v_i$$
$$\frac{dp_i}{dt} = -p_j \frac{\partial v_j}{\partial x_i}$$
$$p^2 = (1 - \frac{\vec{v}.\vec{p}}{c_0})^2$$

where $\vec{p} = \vec{K}/k_0 \quad (k_0 = \omega/c_0)$ is a non dimensional wave vector. The rays have been parametrized by the transit time from the source to a given point.

Plotting the rays permits a clear visualization of the trajectories followed by the acoustic energy radiating from an initially isotropic source. The spatial distribution of rays is a qualitative indicator of the local intensity of the field, since the square of the amplitude is inversely proportional to the cross-section of a ray tube. Quantitative information can be obtained by solving two additional differential equations (in 2 D) which govern the evolution along each ray of infinitesimal cross-section elements (Candel, 1977). It should be emphasized that the computation of acoustic intensity is, however, a difficult task at the crossing points of several rays. The amplitudes and travel times along each of these eigenrays have to be carefully determined in order to consider the multiple interferences existing at a given frequency. It is in fact more convenient to use a wave formulation such as the parabolic equation method which is described below. An important piece of information can be obtained from the evolution in space of an infinitesimal ray tube and that is the location of caustics (envelope of a family of rays) where the ray tube section vanishes. The geometrical approximation predicts infinite levels at the caustics (since diffraction is neglected) and, in practice, for high frequency waves this corresponds to a considerable reinforcement of the field. Prediction of the occurrence of caustics in random media is therefore an important research topic which has attracted a lot of theoretical work in recent years (White, 1984; Kulkarny and White, 1982).

### 2.2. The parabolic equation method

Unlike the geometrical approach, the parabolic equation method cannot be directly derived from the fluid mechanics equations in a linearized form. An "equivalent" inhomogeneous quiescent medium has to be defined and characterized by an index of refraction $n(\vec{x}) \simeq 1 - v_1(\vec{x})/c_0$ where $v_1$ is the component of the velocity field along the initial (unperturbed) propagation direction of the acoustic wave (Tatarski, 1971 ; Candel, 1979). This implicitly takes for granted that a "preferred " direction exists and that energy is scattered for the most part in a narrow conical region around this direction. Indeed this is the case when the ratio of the acoustic wavelength to the characteristic scale of the velocity field L is much smaller than one ($\lambda/L \ll 1$). When this condition is met, it is also possible to reduce the Helmholtz equation (which describes the propagation of waves in an inhomogeneous medium) to a parabolic form suitable for numerical treatment as well as for theoretical analysis ( Tatarski,1971). Use of the envelope transformation $p(\vec{x}) = \Phi(\vec{x})e^{ik_0 x}$, and suppression of the small term $\partial^2\Phi/\partial x^2$ gives :

$$2ik_0\frac{\partial\Phi}{\partial x} + \frac{\partial^2\Phi}{\partial y^2} + (n^2 - 1){k_0}^2\Phi = 0$$

This equation can be solved in an efficient way by using a spatial Fourier transform in the transverse direction. The field at abscissa $x+h$ is related to its value at $x$ by the following algorithm (introduced in underwater acoustics by Tappert, 1977) :

$$\Phi(x+h,y) = e^{ik_0 h(n^2-1)/2}\ \mathcal{F}^{-1}\left\{e^{is^2h/2k_0}\ \mathcal{F}\left\{\Phi(x,y)\right\}\right\}$$

Here, $\mathcal{F}$ denotes the direct Fourier Transform and s is the conjugate variable of y. The choice of the step $h$ depends both on the wavelength and on the inhomogeneities of the medium (scales, fluctuation levels). In practice $h$ can often be made to be much greater than $\lambda$ rendering the algorithm very efficient (a detailed analysis of the errors inherent in this method can be found in DiNapoli and Deavenport,1979).

Two more questions remain to be solved : how do we choose the initial distribution of the pressure in the plane $x = 0$ ? and how do we suppress the artificial reflexions at the limits of the computation domain in the y direction ?

Our goal in the initialization of the field is to model a point source emitting in a limited angular sector ,typically $\pm 20°$ around the x-axis. This choice corresponds to our intention to compare the parabolic equation results with those of ray tracing. This spherical type of source is also closer to the ones used in experimental work than the plane wave frequently encountered in other numerical simulations. As shown by

Tappert (1977) a good initial field consists of a Gaussian distribution centered on the position of the source, its width being determined by the acoustic wavelength and the desired aperture angle.

The second question is solved by using a complex refraction index ; the small imaginary part induces a progressive artificial attenuation of the wave in the last 10% of the sampling points in the positive and negative y directions, so that the field is virtually reduced to zero at the two boundaries.

## 3. Propagation through 2D homogeneous isotropic turbulence

Most of the theoretical studies on wave propagation through random media have been developed for statistically isotropic and homogeneous fields. For these fields many results are available using various theoretical approaches such as geometrical acoustics, method of smooth perturbations, strong fluctuations theory, .... This offers opportunities to test our numerical scheme as well as some of the hypotheses used in theoretical studies.

### 3.1. Modelling of the random field

During the transit time of the acoustic wave the velocity field is, as usual, considered to be frozen. The medium can then be modelled by a sequence of independent realizations of a random field. Following Kraichnan (1970), the velocity at a given point $\vec{x}$ is simulated by the sum of a limited number N of random incompressible Fourier modes :

$$\vec{V}(\vec{x}) = \sum_{l=1}^{N} \vec{U}(\vec{K}^l) \cos(\vec{K}^l.\vec{x} + \Psi^l)$$

$$\vec{U}(\vec{K}^l).\vec{K}^l = 0$$

The direction of the wave vector $\vec{K}^l$ and the phase $\Psi^l$ are independant random variables with uniform distributions. In our simulations, we have considered the amplitude $\| \vec{U}(\vec{K}^l) \|$ to be a deterministic variable whose value is set according to the energy spectrum $E(K)$ under consideration. In this paper we have used a field with a Gaussian correlation function $f(r) = e^{-r^2/L^2}$, where L is related to the integral scale $L_f$ by $L_f = L\sqrt{\pi}/2$ . This form is often encountered in theoretical investigations due to its analytical simplicity. It is also convenient for numerical simulations because it covers a limited range of energetic wave numbers. In 2D the energy spectrum is related to f(r) by the general formula :

$$E(K) = \frac{v'^2}{2} K \int_0^{\infty} \frac{\partial}{\partial r}(r^2 f(r)) \, J_0(Kr) dr$$

$J_0$ is the Bessel function of the first kind of zero order and $v'^2$ is the mean square value of the velocity fluctuations. With the Gaussian correlation function $f(r)$ indicated above one obtains :

$$E(K) = v'^2 \frac{K^3 L^4}{8} \; e^{-K^2 L^2/4}$$

In our simulations this spectrum has been sampled with $N = 50$ modes linearly distributed between $K_{min} = 0,1/L$ and $K_{max} = 10/L$ . In order that the mean properties of the simulated field be close to the desired ones (homogeneity, isotropy, correlation length), the minimum number of realizations of the field has to be around 500 (Blanc-Benon *et al.*, 1990). For acoustic quantities, however, the global trends are obtained with only 100 realizations, because of a further mean along the propagation path.

It is important to note that the description in terms of Fourier modes presents a definite advantage for ray tracing : the differential equations to be solved contain first order (geometry of rays) and second order (ray tube cross-sections) derivatives of the velocity field. In our simulation these derivatives are obtained analytically, avoiding the usual finite-difference approximations. Numerical errors are then reduced and computation time is saved.

### 3.2. Acoustic field visualizations

The first figures provided illustrate the large distortions that acoustic waves undergo when travelling through one realization of a turbulent velocity field even though the fluctuation rate is weak (turbulent Mach number$M_t = v'/c_0$ less than 0.01). These visualizations also present an additional interest, which is to allow at least a qualitative confrontation between the two approaches used. The obtaining of figures with similar tendancies is a proof of the validity of our results, since each of the approaches is based on different approximations : high frequencies for geometrical acoustics ; definition of an equivalent refraction index for the parabolic method.

In geometric acoustics the only characteristic length scale is the parameter L introduced in the correlation function $f(r)$ . The various visualizations are therefore presented in non dimensional variables $x/L$ and $y/L$ . The turbulent Mach number is equal to $6.10^{-3}$. The maximum propagation distances are $x/L = 30$ and $x/L = 60$. For the parabolic approximation, three values of $\lambda/L$ were retained : $\lambda/L = 0.05; 0.01; 0.2$. The source emission is within an angle of $\pm 20°$ in respect to the x-axis. These values have been selected according to two criteria : on the one hand, it is necessary to respect the application conditions of the theories ($\lambda/L \ll 1$, small angles, weak fluctuations) ; on the other hand,

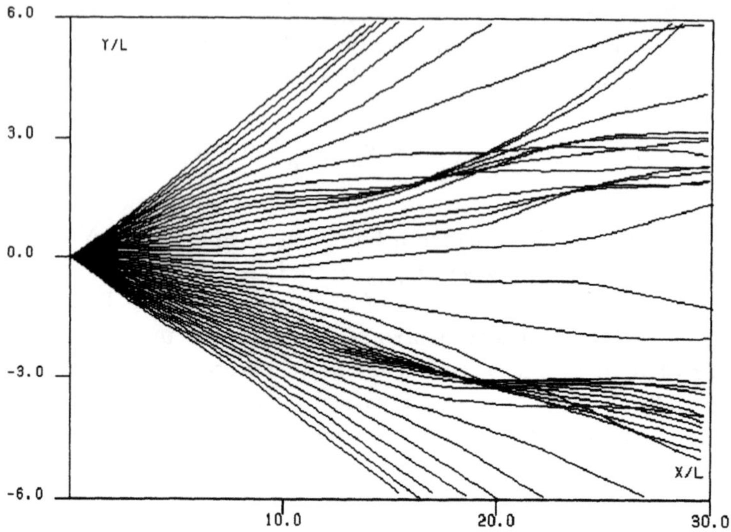


Fig.1. Propagation of acoustic rays in a single realization of a 2D isotropic velocity field ($v_1'/c_0 = 0.003$, $L = 0.1m$). ote the distortion of rays and the occurrence of caustics

the observed effects should correspond to realistic situations (in spite of the 2D character of the modelisation) such as those found during propagation in the atmosphere or in laboratory experiments simulating it ( Blanc-Benon and Juvé 1987).

Figure 1 shows a typical example of ray tracing through a turbulent field realization. It can be noted that the initially linear trajectories launched at regularly spaced angular intervals ($\Delta\theta = 0.5°$) are highly deformed. When two neighboring rays are observed, large increases or decreases of the cross-section of elementary tubes can be seen. Accordingly, large local variations of the acoustic intensity are obtained and have to be connected to the occurrence of caustics as soon as a sufficient distance from the source is reached. In Figure 1, two strong concentrations of rays occurred around $\theta = -9°$ and $x/L = 20$ and around $\theta = 5°$ and $x/L = 15$ ; a third weaker caustic occurs around $\theta = 3°$ and $x/L = 25$.

Figure 2 (see colour plates) corresponds to the same realization but depicts a distribution of acoustic levels using the parabolic equation method with $\lambda/L = 0.1$. In order to eliminate the geometrical spreading of the wave, we expressed the results in terms of a ratio of the acoustic intensity obtained with turbulence to the acoustic intensity in a homogeneous medium. This ratio was then expressed in color levels according to a logarithmic scale . The dark red, for example, corresponds to a reinforcement of the initial level equal to or above 10 dB, which is highly significant. The similar behavior of this figure and Figure 1 is striking. In particular, the high increase zones of the level directly correspond to the caustics predicted by the geometrical theory. The decrease zones of the level are between caustics and associated with regions where few rays pass. It is also interesting to observe the elongated filament structure in the direction of local propagation. This structure corresponds to an aforementioned suggestion that, for $\lambda/L \ll 1$ , scattering mostly occurs within small angles. As a result, the field evolution is much more gradual in the axial direction than in the transverse one, thus permitting a rather large step. For Figure 2 we chose $\Delta x/L = 0.1$ ( i.e. $\Delta x = \lambda$ and $\Delta y/L = 0.025$) which are rather conservative values.

Figure 3 (see colour plates) depicts the effect of a wave frequency multiplied by 2 (that is $\lambda/L = 0.05$) for the same turbulent field realization. All the characteristic traits of Figure 2 are found, but this time with a greater sharpness of details especially perceptible in the transverse direction. This result corresponds to the decrease of the low frequency filtering effect : in the acoustic field (not too close to the source) no detail less than the wavelength can be detected. The wave frequency effect is therefore relatively modest. The effect of the characteristic length scale L of the turbulent field is much larger as seen in Figure 4 (see colour plates) where L is half the size of that in Figure 2 (with $\lambda/L = 0.2$ ). The spatial evolution of the intensity is clearly more rapid than previously. This can bee seen for example in the region of weak intensity around the $x$-axis. These evolutions correspond rather well to the idea according to which the characteristic length scale of the axial variation of intensity fluctuation would be $k_0L^2$ or $L^2/\lambda$, showing a greater sensitivity to a modification of $L$ rather than to a modification of $\lambda$ (Spivack and Uscinski, 1988).

### 3.3. Some additional results

The following results deal with average acoustic field properties obtained using both the ray and the parabolic equation methods. As a whole, these results are in agreement with theoretical prevision ; this validates our simulation technique. However, in certain cases, the theory and the simulation separate and it would seem that the differences noted are due to a deficiency of certain assumptions of theoretical developments. The latter constitutes a justification for using the deterministic approach.

i) Travel time along the rays

One of the most interesting parameters that the geometric theory can provide is the travel time between a source and a receiver in an effort to reconstruct wave fronts perturbed by turbulence. Fluctuations in this travel time were studied theoretically as early as the 1960's by Chernov

(1960) and Keller (1962). The essential result deals with the variance of the travel time fluctuation which is expressed by the following formula:

$$\overline{\Delta t^2} = \frac{2\mu^2}{{c_0}^2}\, l_c\, x$$

where $\mu = v_1^{'}/c_0$ , $l_c$ is the correlation length scale of the random field, $x$ the distance between the source and the receiver.

In Figure 5 we show the variances obtained for two values of the rms velocity fluctuation ($v_1^{'}/c_0 = 0.003; 0.006$). For the weaker value, the agreement between theory and simulation is very good: the evolution of $\overline{\Delta t^2}$ with $x$ is in particular very linear up to around $x/L = 30$ ; the simulated variance increases afterwards slightly more quickly than predicted. For the second rms value, the difference between theory and simulation rapidly increases. The linear increase of $\overline{\Delta t^2}$ with $x$ occurs only up to $x/L = 15$ ; afterwards $\overline{\Delta t^2}$ exhibits an almost quadratic growth, with a simulated value at $x/L = 40$ which is twice as big as that predicted by the Chernov theory. The reason for this behavior is not yet entirely clear, but it would seem to be directly linked to the existence of caustics. The higher the turbulent intensity, the shorter the distances at which the first custic occurs. The focalisation effect therefore becomes essential. As long as no caustic occurs, only one eigenray links the source to the receiver. If one caustic occurs, three eigen-rays coexist with different travel times and, qualitatively, this can lead to an increase in the variance of real travel times in relation to the theoretical predictions which implicitly assume the existence of a unique eigenray. To consolidate this point of view, Codona and al (1985) have recently demonstrated the importance of the occurrence of caustics when evaluating the average travel time in a random scalar field.

The probability of the appearance of the caustics in a random scalar field has recently been theoretically explored by Kulkarny and White (1982). In our numerical simulation applied to the evolution of the cross section of an elementary ray tube, we determined the point at which a given ray touches a caustic and then we estimated the probability density of occurrence of the first caustic (Blanc-Benon and Juvé, 1990). In Figure 6 we report the results obtained for the two values of the r.m.s. velocity fluctuation. The curves present the shapes predicted by Kulkarny and White (1982). The probability of meeting a caustic is very small up to a certain distance. Afterwards the probability increases rapidly, reaching a maximum before slowly decreasing in a quasi exponential way. However, the probability density maximum which should theoretically occur at $s = 3.5$ according to Kulkarny and White (1982) occurs instead at 2 ($s$ is defined as $s = 12^{1/3}\pi^{1/6}\mu^{2/3}\ x/L$ ). This difference can no doubt be explained by the vectorial character of our random velocity

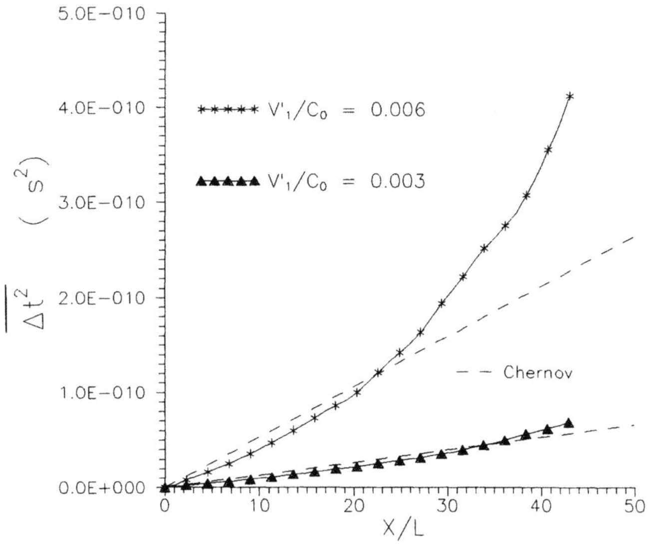


**Fig.5.** Variances of arrival time on a circle of radius $x$ from an acoustic point source (2D isotropic field, $v_1'/c_0 = 0.003$ and 0.006, L = 0.1 m ; 250 realizations). Continuous curves are Chernov predictions

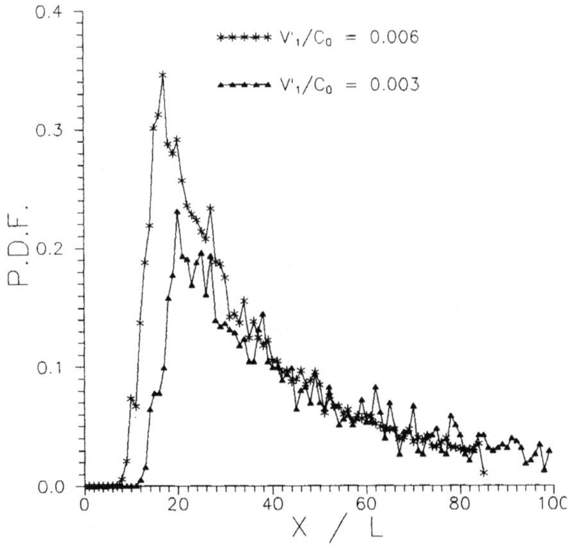


**Fig.6.** Probability density function for the occurrence of the first caustic (2D isotropic field, $v_1'/c_0 = 0.003$ and0.006, L = 0.1 m ; 250 realizations)

field. For the same characteristic length scale L introduced in the correlation function f(r), the kinetic energy spectrum is richer in spatial high frequencies than in the scalar case studied by Kulkarny and White (1982). The existence of these more energetic small structures leads to a wave focalisation at a shorter distance, which is precisely what is seen in the simulations. In addition, we note that the maximum of the probability distribution is obtained at $x/L = 32$ when $v_1'/c_0 = 0.003$ and earlier at $x/L = 20$ when $v_1'/c_0 = 0.006$ and that, at the same respective distances, the variances obtained by simulation begin to slightly differ from the theoretical predictions.

ii) Intensity fluctuations

To obtain quantitive information on the acoustic intensity at a given point is difficult in geometrical acoustics. On the contrary, the parabolic equation method is ideally suited for this. The most significant parameter of intensity fluctuation is the standard deviation $\sigma_I = \sqrt{\left\langle (I - <I>)^2 \right\rangle / <I>^2}$ (in optics $\sigma_I^2$ is known as the scintillation index). In the domain of small fluctuations the Rytov approximation gives results for the log amplitude fluctuations $\chi = \ln(A/A_0)$; $A_0$ is the field amplitude in the absence of perturbation, and the standard deviation $\sigma_\chi$ is approximately linked to $(\sigma_I)_{Rytov}$ by $(\sigma_I)_{Rytov} = 2\sigma_\chi$. In the case of a spherical wave in a Gaussian random medium, $(\sigma_I)_{Rytov}$ is then found by Ishimaru (1978) to be :

$$(\sigma_I)^2_{Rytov} = 2\sqrt{\pi} \left( \frac{v_1'}{c_0} \right)^2 L{k_0}^2 \int_0^x 1 \Big/ \left( 1 + \frac{k_0^2 L^4 x^2}{16\eta^2 (x-\eta)^2} \right) d\eta$$

A saturation phenomenon is also known to occur in the scintillation index for values of $\sigma_I$ close to 1 . This behavior has been experimentally observed and predicted by ad hoc asymptotic theories ( Tatarski, 1971 ; Tatarski and Zavorotnyi ,1980) which even indicate a slight decrease of the standard deviation for $(\sigma_I)_{Rytov} \gg 1$. No theory currently exists to predict behavior in the transitional region between the Rytov calculation and the asymptotic expression mentionned above. It is in this region, which is perhaps the most interesting, at least in acoustics, that our numerical simulation furnishes the most new elements.

Our results on that topic are presented in Figure 7.

Two frequency and two scale $L$ values are used (corresponding to $\lambda/L = 0.05; 0.1; 0.2$ when $v_1'/c_0 = 0.006$). In these situations the strength parameter $\Gamma = k_0^3 \mu^2 x^2 l_c$ defined by Spivack and Uscinski( 1988) ranges from around 2 to 120. In the weak fluctuation region ($\sigma_I$ less than around 0.3 to 0.5) the agreement with the Rytov formula is excellent. The computations then show the transition towards a $\sigma_I$ plateau culminating slightly above 1 and depending on $\Gamma$ . However, it is not possible

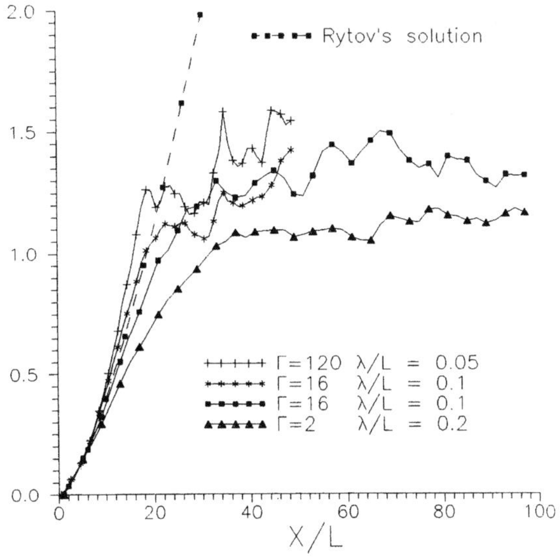


Fig.7. Standard deviation of acoustic intensity fluctuation (2D isotropic field, $v_1'/c_0 = 0.003$ and0.006, L =0.1 m; $\lambda/L = 0.05, 0.1, 0.2$ ; 250 realizations)

to detect any decrease tendancy whatever for $\sigma_I$. It must nevertheless be noted that, at a great distance, the acoustic level is strongly reduced due to the beam divergence, and moreover, a greater number of realizations would be required to reduce the residual curve oscillations. Finally, let us also add that Spivack and Uscinski(1988), using a different approach and an equation governing the 4th order moment, also obtained a plateau for $\sigma_I$, after a maximum whose position and height depend on the strength parameter.

## 4. Propagation through a 2D mixing layer

The mixing layer which has been the subject of numerous investigations, both experimental and numerical, is also of great interest to our study. Its principal characteristic is the formation of vortices which can coalesce (Corcos and Sherman, 1984; Lesieur *et al.*, 1988). We have chosen to show the influence of such a flow on the propagation of acoustic waves emitted perpendicular to the mean flow direction. A possible application could be the use of the modulation of the transmitted acoustic beam as a means to detect the existence of well-defined turbulence structures. The most interesting stage of the merging of two structures has been retained. Two computations were carried out, immediately before and immediately after the coalescence. The velocity fields corresponding to these two time steps were graciously furnished by P. Comte from the Fluid Mechanics Laboratory in Grenoble (Comte *et al.*, 1989). His computations were made for a 2D mixing layer increasing in time. The initial field has a stream function $\Psi = U\delta_0 \log(cosh(y/\delta_0))$, the vorticity thickness is given as $\delta_i = 2\delta_0$. The computation domain is square, with a side equal to four times the most amplified wave length predicted by the inviscid linear stability theory ($\lambda_{st} = 7\delta_i$) . The initial Reynolds number based on $\delta_i$ and U, the half velocity difference, is 1000. Accordingly, our acoustic wavelength $\lambda$ was chosen in such a way that $\lambda \sim .2\delta_i$ , the Mach number $U/c_0$ being 0.03.

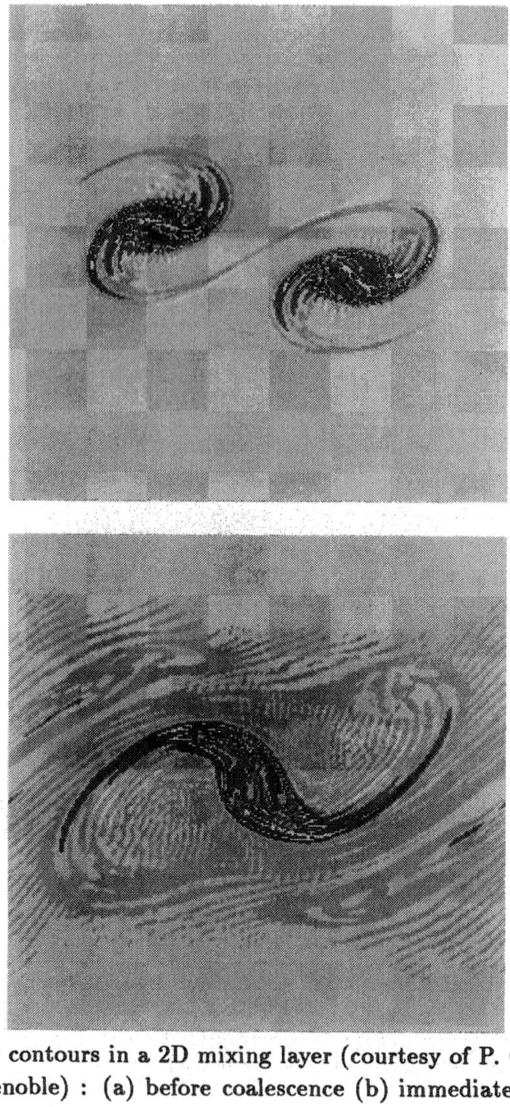

**Fig.8. Vorticity contours in a 2D mixing layer (courtesy of P. Comte, Institut Mécanique Grenoble) : (a) before coalescence (b) immediately after coalescence**

In Figure 8 we provide the vorticity distribution given by P. Comte which allows a clear visualization of the position of the vortices. From an acoustic point of view the most important characteristic is the velocity component in the direction of propagation, that is to say, component v. Therefore in Figure 9 we give the v distribution in gray levels. The dark gray regions correspond to positive components and the light gray to negative ones.

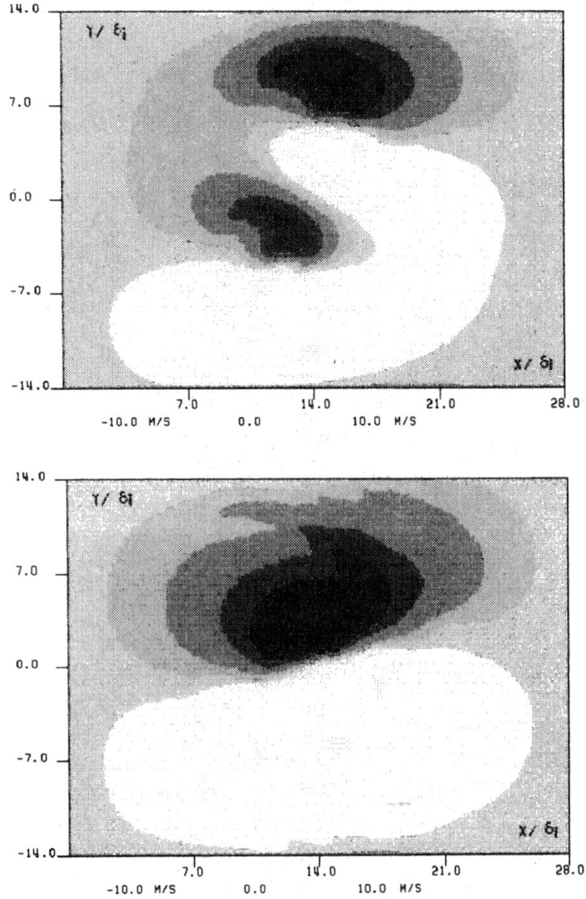


Fig.9. Iso-contours of the lateral velocity v for the flow pictured in Figure 8 : (a) before coalescence (b) immediately after coalescence. Dark gray regions are positive v and light gray regions are negative $v$

The acoustic fields we obtained using the parabolic equation method are given in Figure 10. The similarity with the maps of Figure 9 is striking. Before coalescence takes place, we see four regions where weakening and strengthening of the acoustic field alternate. These regions correspond to the four high velocity regions of Figure 9. After coalescence, we note a region of strong strengthening and a region of strong weakening associated with regions where the absolute magnitude of v is very large.

Despite the absence of ray tracing, which is difficult here because of the interpolation needed to express the derivatives, it is possible to obtain a qualitative idea of the behavior of an acoustic wave. Indeed, focusing and divergence of rays are essentially linked to the presence of intense velocity gradients between vortices as well as inside the vortices. Illustrations are provided in Figure 11.

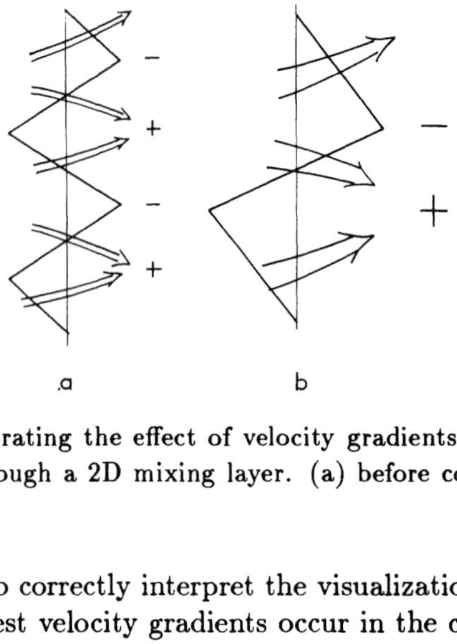


Fig.11. Sketch illustrating the effect of velocity gradients on th propagation of acoustic rays through a 2D mixing layer. (a) before coalescence (b) after coalescence.

This allows us to correctly interpret the visualizations of the acoustic fields. The strongest velocity gradients occur in the case (b) producing focusing and diverging effects which are stronger than in case (a).

## 5. Conclusion

When applied to individual realizations, the ray theory and the parabolic equation methods can provide complementary and original results. For example, ray paths, caustics, reinforced or weakened zones of the transmitted acoustic field can be clearly visualized.

Averaging over an ensemble of such realizations permit us to obtain statistical results which have been compared with theoretical predictions in the case of 2D isotropic turbulence. For the fluctuations of acoustic intensity, similar results are obtained using both these approaches in the range of weak fluctuations described by the Rytov theory, and in the range of large fluctuations where the saturation phenomenon takes place. In addition, our simulations permits us to cover the intermediate region where no theory is currently available. For the fluctuations of propagation time, which can be interpreted as phase changes or wave front distortions, our numerical results show the limits of the classical approaches (e.g. Chernov) as soon as caustics occur.

The representation of a (vectorial) velocity field by a (scalar) refraction index ($-u^{'}/c_0$ versus $-t^{'}/2.T_0$) also requires special scrutiny. The distribution of kinetic energy among the different scales of the inhomogeneous field has to be taken into account in the r.m.s. level $u^{'}$ as well as in the length scale L used to make a non dimensional representation of the results.

Using the same technique it could be possible to create inhomogeneous and/or random media involving non negligible backscattering due to strong inhomogeneities and to compute the resulting acoustic fields by solving forward and backward parabolic equations.

**Aknowledgements**
The authors would like to thank R. Lambrecht for her very kind editorial assistance.

## References

Blanc-Benon, Ph. and Juvé, D., 1987: Statistical properties of acoustic waves propagating through turbulent thermal fields. *AIAA Paper 87-2727*, AIAA 11th Aeroacoustics Conf., Palo-Alto.

Blanc-Benon, Ph., Juvé, D., Karweit, M. and Comte-Bellot, G., 1990: Simulation numérique de la propagation des ondes acoustiques à travers une turbulence cinématique, *J. d'Acoustique 3*, 1, 1-8.

Blanc-Benon, Ph. and Juvé, D., 1990: Simulation numérique de la propagation d'une onde acoustique dans une turbulence bidimensionnelle par une méthode de rayons. 1er Congrès Français d'Acoustique, *Supplément au Journal de Physique*, Editions de Physique, C2, 1149-1152.

Candel, S.M., 1977: Numerical solution of conservation equations arising in linear wave theory : application to aeroacoustics, *J. Fluid Mech.*, **83**, 465-493.

Candel, S.M., 1979: Numerical solution of wave scattering problems in the parabolic approximation, *J. Fluid Mech.*, **90**, 465-507.

Chernov, L.A., 1960: *Wave propagation in a random medium.* Mc Graw-Hill, New York.

Codona, J.L., Creamer, D.B., Flatte, S.M., Frelich, R.G. and Henyey, F.S., 1985: Average arrival time of wave pulses through continuous random media, *Phys. Rev. Letters 55*, (1), 9-12.

Comte, P., Lesieur, M., Laroche, H. and Normand, X., 1989: Numerical simulations of turbulent plane shear layers - in *Turbulent Shear Flows 6*, Springer-Verlag, 360-380.

Corcos, G.M. and Sherman, F.S. 1984 The mixing layer : deterministic models of a turbulent flow. Part 1. Introduction and the two-dimensional flow, *J. Fluid Mech.*, **139**, 29-65.

DiNapoli, F.R. and Deavenport, R.L., 1979: Numerical models of underwater acoustic propagation. In *Ocean Acoustics*, ed. by J.A. De Santo, Topics in Current Physics, Springer Verlag. 79, 157.

Hesselink, L. and Sturtevant, B., 1988: Propagation of weak shocks through a random medium, *J. Fluid Mech.*, **196**, 513-553.

Ishimaru A., 1978: *Wave propagation and scattering in random media*, Vol. 2. Academic Press.

Keller, J.B., 1962: Wave propagation in random media. Proc. Symp. *Appl. Math. 13*, Amer. Math. Soc. Providence R.I.

Kraichnan, R.H., 1970: Diffusion by a random velocity field, *Phys. Fluids*, 13, 22-31.

Kulkarny, V.A. and White, B.S., 1982: Focusing of waves in turbulent inhomogeneous media, *Phys. Fluids 25*, 10, 1770-1784.

Lesieur, M., Staquet, C., Le Roy, P. and Comte, P., 1988: The mixing layer and its coherence examined from the point of view of two-dimensional turbulence, *J. Fluid Mech.*, **192**, 511-534.

Martin, J.M. and Flatte, S.M., 1988: Intensity images and statistics from numerical simulation of wave propagation in 3-D random media. *Applied Optics 27*, (11), 2111-2126.

Spivack, M. and Uscinski, B.J., 1988: Accurate numerical solution of the fourth-moment equation for very large values of $\Gamma$ . *J. Modern Optics*, 35, 11, 1741-1755.

Tappert, F.D. 1977: The parabolic approximation method - in *Wave propagation and underwater acoustics*, ed. by J.B. Keller and J.S. Papadakis. Lecture Notes in Physics, Springer-Verlag. 70, 224-287.

Tatarski, V.I. 1971: *The effects of the turbulent atmosphere on wave propagation*. IPST Keter Press, Jerusalem.

Tatarski, V.I. and Zavorotnyi V.U., 1980: Strong fluctuations in light propagation in a randomly inhomogeneous medium. *Progress in Optics*, edited by E. Wolf, Vol. 18, North Holland, New York .

White, B.S., 1984: The stochastic caustic. *SIAM J. Appl. Math. 44*, 1, 127-149.

# VII. Geophysical Fluid Dynamics

# Spectra and correlation functions of surface layer atmospheric turbulence in unstable thermal stratification

B.A. KADER, A.M. YAGLOM

*Institute of Atmospheric Pysics*
*USSR Academy of Sciences, Moscow, USSR*

*Turbulence and Coherent Structures*
*O. Métais and M. Lesieur, eds.*
*Selected papers from "Turbulence 89: organized structures and turbulence in fluid mechanics"*
*Grenoble, 18-21 September 1989*

## 1. Introduction

A surface layer of the Earth's atmosphere above flat homogeneous ground on hot summer days is the most easily accesible example of an unstably stratified boundary layer. The mean air temperature $T = T(z)$ is here decreasing with height z and the mean density $\rho = \rho(z)$ is increasing with z. As it is customary in the micrometeorology the atmospheric surface layer (ASL) will be defined as the lower air layer of the thickness $z_s \leq 0.1z_i$ (where $z_i$ is the height of the lower edge of the inversion layer where $T(z)$ is increasing with z) satisfying the condition that the turbulence fluxes of momentum and temperature $\tau = \rho\langle -uw\rangle$ and $Q = \langle wt\rangle$ can be considered as independent of z. (In this paper the turbulent fluctuations of longitudinal, i.e. directed along the mean wind, lateral, and vertical velocity components, and temperature are denoted by $u = u_1$, $v = u_2$, $w = u_3$, and $t = u_4$, and the angular brackets symbolize averaging). It is known that in the middle latitudes on a hot summer day $z_s \approx 50m$ and $z_i \approx 1km$. The thin lowest sublayer, where the molecular transfers and ground roughness affect directly the turbulence structure, will be always excluded from the consideration. Then the statistical characteristics of the surface-layer turbulence can be assumed to depend only on the following three dimensional parameters: $\tau/\rho$ (or equivalently, $u_* = (\tau/\rho)^{1/2}$), $Q$, and the buoyancy parameter $\beta = g\gamma$, where g is the acceleration due to gravity and $\gamma$ is the cœfficient of thermal expansion (wich is equal to $1/T_0$ for an ideal gas, $T_0$ being the mean absolute temperature of the surface layer).

The theoretical analysis of the ASL begins with the classical papers (Prandtl, 1932; Obukhov 1946; Monin and Obukhov, 1954) ; its history is sketched by Monin and Obukhov (1954). According to the Monin-Obukhov similarity theory of the surface-layer turbulence (see Monin and Oboukhov, 1954; Monin and Yaglom,1971, chap.4) the vertical profile of any one-point dimensionless moment of turbulent fluctuations is given within the ASL by a universal function of dimensionless height $\zeta = z/L$, where $L = u_*^3/Q\beta$ is the so-called Obukhov length. (The moments are made dimensionless by division by the appropriate product of the powers of velocity and temperature scale $u_*$ and $T_* = Q/u_*$). Note also that the length L is usually replaced in the meteorological literature by the length $L_0 = -L/\kappa \approx -2.5L$, where $\kappa \approx 0.4$ is the von Karman constant. $L_0$ is just the length introduced by Obukhov (1946) ; however, the length L, which is positive in unstable stratification and independent of the empirical value of $\kappa$, is more convenient for our aims than $L_0$. Similar results are also valid for two-point statistical characteristics of turbulence. For example, the longitudinal (in the mean wind direction) autospectra (for $i = j$) and cospectra (for $i \neq j$) $E_{ij}(k)$ of fluctuations $u_i$, $u_j$ can be represented within the ASL as:

$$E_{ij}(k) = u_*^{(i)} u_*^{(j)} z F_{ij}(kz, \zeta), \quad i,j = 1,2,3,4, \tag{1}$$

where $u_*^{(1)} = u_*^{(2)} = u_*^{(3)} = u_*$, $u_*^{(4)} = T_*$, and $F_{ij}(q,\zeta)$ are universal functions of two variables. However, equations (1) are inconvenient for practical applications since the functions of two variables cannot be reliably determined from scattered micrometeorological data. Therefore much consideration is given in the micrometeorological literature (starting from Prandtl, 1932; Obukhov, 1946; Monin and Obukhov, 1954) to the search of the ASL sublayers where the general similarity equations can be simplified. In this paper the suggestion by Betchov and Yaglom (1971) based on the reconsideration of Zilitinkevich's (1971) recommendation will be widely used. According to this suggestion within the ASL there are three special sublayers that have a self-preserving turbulence structure. They are i) **dynamic sublayer** at $z \ll L$, where the thermal stratification does affect the air flow and the buoyancy parameter $\beta$ can be excluded from the list of relevant dimensional parameters, ii) **dynamic-convective sublayer** at moderate values of $\zeta = z/L$, where the three parameters Q, $u_*$ and $\beta$ are relevant but the vertical and horizontal motions are energetically uncoupled (i.e. there is no energy exchange between them) and therefore, by applying the dimensional analysis, the vertical and horizontal lengths can be assumed to have different (and independent) dimensions $L_x$ and $L_z$, and iii) **free convection sublayer** at $z \gg L$, where the friction velocity $u_*$ can be excluded from the list of relevant physical parameters. (These three sublayers can also be called the **forced convection sublayer, mixed convection sublayer** and **free convection sublayer**, but then it should be taken into account that the term 'mixed convection' has here some special meaning). In these three sublayers there is no combination of the relevant physical parameters having the same dimension as the distance from the ground $z$ ; this is just the reason why the sublayer turbulence is self-preserving when $z$ is varying.

In 1971 there were no data suitable for the verification of the suggestion formulated by Betchov and Yaglom (1971). Therefore it was indicated by Betchov and Yaglom (1971) that possibly the free convection sublayer does not exists within the ASL since the sublayer lower edge would be located above the upper edge of the ASL. (The same assumption was also stated by Wyngaard *et al.* (1971), Yaglom (1971) and Monin and Yaglom (1975, p. 853-854). However, the new data collected from the recent micrometeorological observations (in particular, from those at the Tsimlyansk Field Station of the Moscow Institute of Atmospheric Physics during the summers of 1981-1987) have changed the situation. According to these data all the three aforementioned sublayers exist quite often within the ASL on hot summer days, the upper edge of the dynamic sublayer being located near the level $z = 0.1L \approx 0.04 \mid L_0 \mid$, while the lower and upper edges of the dynamic-convective sublayer are in the vicinity of the levels $z = 0.3L \approx 0.12 \mid L_0 \mid$ and $z = 3L \approx 1.2 \mid L_0 \mid$ and the level $z = 5L \approx 2 \mid L_0 \mid$ can be taken as the lower edge of the free convection sublayer (see Kader and Yaglom, 1989, where the measurements of mean wind velocity and temperature profiles and profiles of one-point turbulent moments are analyzed). In this paper the assumption of the existence of the three indicated sublayers within the ASL is applied to the analysis of the measurements of longitudinal autospectra, cospectra, and autocorrelation and cross-correlation functions. Some of the results given below are also shortly considered by Kader and Yaglom (1984), Kader (1987), Kader (1988), Kader and Yaglom (1987), Kader *et al.* (1989).

## 2. Spectra of Turbulent Fluctuations in an Unstable Surface Layer

The longitudinal autospectra and cospectra can easily be determined from the observations at a fixed point with the aid of Taylor's frozen turbulence hypothesis. Therefore there is much experimental data concerning such spectra. We shall exclude from our consideration both the range of very high wave numbers $k \geq k_0$ (where $k_0 = (\epsilon/\nu^3)^{1/4}$ is the Kolmogorov wave number, $\epsilon$ is the mean rate of energy dissipation, and $\nu$ - the kinematic viscosity of air) and the range of very low wave numbers $k < k_1$ corresponding to the disturbances affected by the mesoscale ground inhomogeneities and/or the heights $z_i$ and $z_c = u_*/2\omega_z$ (where $\omega_z$ is the vertical component of the angular velocity of the Earth's rotation). The spectra in the wave number range $k_1 < k \ll k_0$ satisfy the Monin-Obukhov similarity theory which implies equations (1).

### 2.1. Spectra within the dynamic sublayer

The thermal stratification does not affect the air flow within the dynamic sublayer and therefore here the stratification can be assumed to be neutral. The neutral stratification quite often occurs in the atmosphere and there are many observations made in neutral conditions. Moreover, many laboratory measurments are related to the logarithmic sublayer of a non-stratified turbulent boundary layer and this sublayer can be assumed to be identical to the ASL in neutral stratification. Note, however, that according to all the data the intensity of temperature fluctuations within the dynamic sublayer of the ASL is much higher than the same intensity in the laboratory boundary layers. (This difference is apparently due to the ever present thermal inhomogeneity of the ground ; Kader and Yaglom, 1989).

Since the buoyancy forces play no role within the dynamic sublayer, we can assume that $\beta = 0$, i.e. $L = \infty$. Hence $z/L = 0$, i.e. the last argument of the functions $F_{ij}(q, \zeta)$ is constant here and can be omitted. Thus

$$E_{ij}(k) = u_*^{(i)} u_*^{(j)} z F_{ij}(kz) \quad (2)$$

within the dynamic sublayer, where $F_{ij}(q)$ are universal functions of one variable.

Let k belong to the inertial subrange, i.e. $z^{-1} < k < k_0$. Then, according to the theory of locally isotropic turbulence, the spectra $E_{ii}(k)$, $i = 1, 2, 3, 4$, must be proportional to $k^{-5/3}$ (Monin and Yaglom, chap.8). Hence

$$F_{ii}(q) = C_{ii} q^{-\frac{5}{3}}, \ i = 1, 2, 3, 4, \quad (3)$$

for $b_{ii} < q < a_{ii}$ where $C_{ii}$ and $b_{ii}$ are universal constants and $a_{ii}$ is proportional to $k_0 z$. The theory of locally isotropic turbulence also implies that $C_{22} = C_{33} = 4C_{11}/3$ and $C_{11} u_*^2 z^{-2/3} = C_K \epsilon^{2/3}$, $C_{44} T_*^2 z^{-2/3} = C_{OC} N \epsilon^{-1/3}$, where N is the rate of dissipation for $\langle t^2 \rangle /2$ and $C_K$ and $C_{OC}$ are the Kolmogorov and Obukhov-Corrsin constants (universal cœfficients of the longitudinal velocity and temperature spectra in the inertial subrange). It is known that within the dynamic sublayer $\epsilon = A_U u_*^3/z$, $N = A_T T_*^2 u_*/z$ where $A_U \approx 2.5, A_T \approx 2.12$ (see, e.g., Kader and Yaglom, 1989). Therefore,

$$C_{11} \approx 1.84 C_K, C_{22} = C_{33} \approx 2.45 C_K, C_{44} \approx 1.56 C_{OC} \quad (4)$$

As to the cospectra $E_{ij}(k)$, where $(i, j) = (1, 3), (1, 4)$ or $(3, 4)$, they fall off in the inertial subrange more rapidly than $k^{-5/3}$ - approximately as $k^{-\alpha_{ij}}$, where in the first approximation, we can take that $\alpha_{13} = \alpha_{14} = \alpha_{34} = 7/3$. (According to Kaimal, Wyngaard, Izumi and Coté (1972) the estimates $\alpha_{13} = \alpha_{34} = 7/3$, $\alpha_{14} = 5/2$ are slightly more acccurate. However, the difference between 7/3 and 5/2 is slight and it is possible to assume that $\alpha_{ij} = 7/3$ also for $(i, j) = (1, 4)$.)

Large-scale turbulent disturbances that correspond to wave numbers $k < 1/z$ are always anisotropic and the spectral shapes at such values of k differ from those in the inertial subrange. It seems reasonable to assume that the statistical regime of very large disturbances, whose length scales are considerably greater than the distance from the wall $z$, does not change at small and moderate variations in the value of $z$. This implies that the spectra and cospectra $E_{ij}(k)$ do not depend on $z$, if $kz$ is small enough. According to (2) it follows from this conclusion that

$F_{ij}(kz) = G_{ij}(kz)^{-1}$ and $E_{ij}(k)$ is proportional to $k^{-1}$ for any $i$ and $j$, if $k$ is small enough. (Without a reference to the data it is, of course, impossible to say what means here "small enough", but it is known that asymptotic relations have often a broad range of applicability). The validity of the $-1$ power law for spectra of the shear-flow turbulence was predicted by Tchen (1953). Later several experimenters found that the data on the velocity and temperature spectra of some laboratory wall flows showed the existence of a wave-number range where the spectral curves had (in the logarithmic representation) the slope $-1$. Perry and Abell (1975) and Korotkov (1976) derived the $-1$ power law for the spectrum of the longitudinal velocity component in tube and channel flows from the assumption of the independence of the spectrum on $z$ (without indicating the range of validity of the stated assumption). Later Kader (1987) and (1984) showed that this assumption was valid with good accuracy for all the longitudinal spectra and cospectra of velocity components and temperature fluctuations in both the laboratory logarithmic sublayers of turbulent boundary layers and the dynamic sublayers of the ASL. It was also found by Kader that the transition zone between the subrange, where a spectrum satisfies the $-5/3$ power law and $-1$ power law, is usually very narrow so that it is possible to assume that these two subranges adjoin each other. The same is also true for the longitudinal cospectra of turbulent fluctuations but with the replacement of the theoretical $-5/3$ power law by purely empirical $-7/3$ power law (see Figs.1-4).

According to figs. 1 and 2 the spectra and cospectra of the surface-layer turbulence within the dynamic sublayer can be described accurately enough in a wide wave-number range by the equations :

$$F_{ii}(q) = \begin{cases} C_{ii}q^{-\frac{5}{3}} & \text{for } b_{ii} < q < a_{ii}, \\ G_{ii}q^{-1} & \text{for } d_{ii} < q < b_{ii}, \end{cases} \tag{5}$$

$$F_{ij}(q) = \begin{cases} C_{ij}q^{-\frac{5}{3}} & \text{for } b_{ij} < q < a_{ij}, \\ G_{ij}q^{-1} & \text{for } d_{ij} < q < b_{ij}, \end{cases} \tag{6}$$

where $C_{ii}, C_{ij}, G_{ii}, G_{ij}, b_{ii}$ and $b_{ij}$ are universal constants, $a_{ii}$ and $a_{ij}$ are proportional to $k_0 z$ and $d_{ii}$ and $d_{ij}$ can depend on the observation place and weather conditions. The Tsimlyansk data imply that

$$\begin{gathered} C_{11} \approx 0.95, C_{33} \approx 1.2, C_{44} \approx 1.1; C_{13} \approx -0.5, C_{14} \approx -0.6, C_{34} \approx 0.5; \\ G_{11} \approx 0.95, G_{33} \approx 0.35, G_{44} \approx 0.9; G_{13} \approx -0.2, G_{14} \approx -0.6, G_{34} \approx 0.2; \\ b_{11} \approx 1, b_{33} \approx 6.3, b_{44} \approx 1.4; b_{13} \approx 2, b_{14} \approx 1, b_{34} \approx 2 \end{gathered} \tag{7}$$

(atmospheric measurements of $E_{22}(k)$ are strongly affected by wind direction variations and are not reliable enough ; therefore the constants

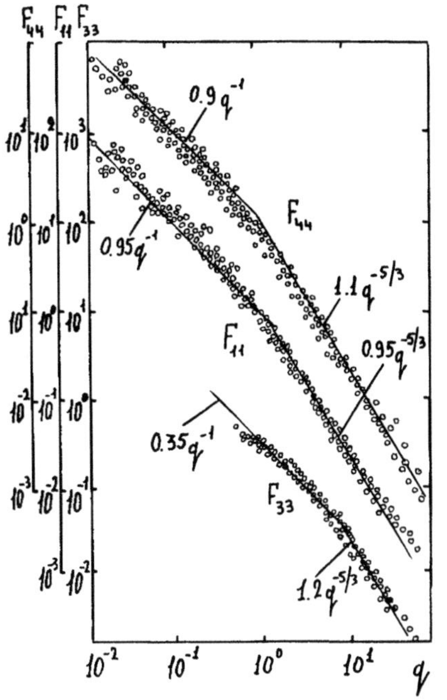


Fig.1. Normalized longitudinal autospectra of surface-layer turbulence in neutral stratification

related to these spectra are not included in (7)). The laboratory data on the velocity spectra and cospectra are quite close to those obtained in the atmosphere (see, e.g., fig.3) but the comparison of figs. 1 and 4 shows that the value of $G_{44}$ in the laboratory flows is three times smaller than the atmosphere. (Note that till now no reliable data on the laboratory values of $G_{14}$ and $G_{34}$ have been obtained). Strong difference between the temperature spectra in the laboratory flows and ASL is apparently due to the influence of the thermal inhomogeneities of the ground which intensifies considerably the temperature fluctuations. According to (4) the values (7) of $C_{11}, C_{33}$ and $C_{44}$ lead to the estimates $C_K \approx 0.52, C_K \approx 0.49$ and $C_{OC} \approx 0.7$ which agree well with the other available estimates of the Kolmogorov and Obukhov-Corrsin constants (see Monin and Yaglom, 1975; Yaglom, 1981).

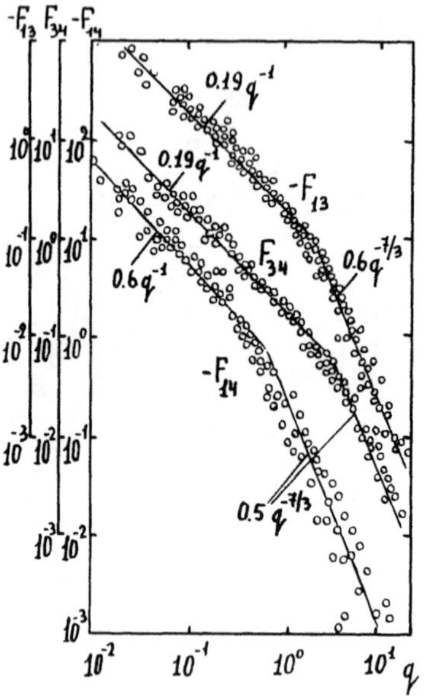


Fig.2. Normalized longitudinal cospectra of surface-layer turbulence in neutral stratification

## 2.2. Spectra within the free convection sublayer

Within the free convection sublayer (at $z \geq 5L$) the friction velocity $u_*$ must be excluded from the list of relevant physical parameters. According to (1) this implies that $F_{ij}(q,\zeta) = \zeta^{2/3} F_{ij}^{(1)}(q)$ for $i,j = 1,2,3$, $F_{i4}(q,\zeta) = F_{i4}^{(1)}(q)$ for $i = 1,3$, and $F_{44}(q,\zeta) = \zeta^{-2/3} F_{44}^{(1)}(q)$. Hence

$$E_{ij}(k) = (Q\beta)^{\frac{2}{3}} z^{\frac{5}{3}} F_{ij}^{(1)}(kz), \quad i,j = 1,2,3,$$

$$E_{i4}(k) = QzF_{i4}^{(1)}(kz), \quad i,j = 1,3, \tag{8}$$

$$E_{44}(k) = Q^{\frac{4}{3}} \beta^{-\frac{2}{3}} z^{\frac{1}{3}} F_{44}^{(1)}(kz)$$

within the free convection sublayer where $F_{ij}^{(1)}(q)$, $i,j = 1,2,3,4$, are universal functions of one variable.

In the inertial wave-number subrange the spectra $E_{ii}(k)$, $i = 1,2,3,4$, must satisfy the $-5/3$ power law. Therefore $F_{ii}^{(1)}(q) = C_{ii}^{(1)} q^{-5/3}$ in the inertial subrange, i.e.

$$E_{ii}(k) = C_{ii}^{(1)} (Q\beta)^{\frac{2}{3}} k^{-\frac{5}{3}}, \; i = 1,2,3; \; E_{44}(k) = C_{44}^{(1)} Q^{\frac{4}{3}} \beta^{-\frac{2}{3}} z^{-\frac{4}{3}} k^{-\frac{5}{3}} \tag{9}$$

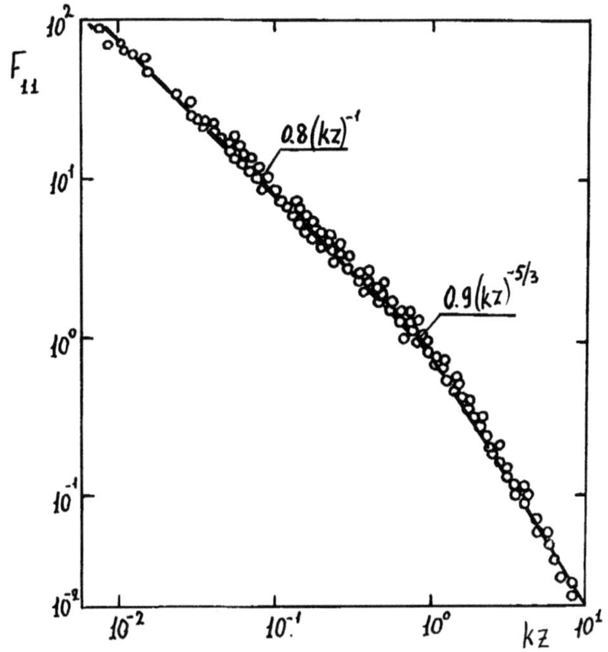


Fig.3. Normalized spectrum $F_{11} = E_{11}/u_*^2 z$ within the logarithmic sublayers of laboratory boundary layers.

As for the cospectra $E_{13}(k)$, $E_{14}(k)$ and $E_{34}(k)$, according to the experimental data they fall off in the inertial subrange approximately as $k^{-7/3}$. Therefore it is possible to assume in the first approximation that $F_{ij}^{(1)}(q) = C_{ij}^{(1)} q^{-7/3}$ for $i \neq j$, i.e.

$$E_{13}(k) = C_{13}^{(1)}(Q\beta)^{\frac{2}{3}} z^{-\frac{2}{3}} k^{-\frac{7}{3}}; \; E_{i4}^{(1)}(k) = C_{i4}^{(1)} Q z^{-\frac{4}{3}} k^{-\frac{7}{3}}, \; i = 1, 3, \quad (10)$$

in the inertial subrange of the free convection turbulence.

If $k$ is smaller than $1/z$, then, as in the case of the dynamic sublayer, it is reasonable to assume that the spectral shapes for such large-scale disturbances do not depend on $z$. Then, according to (7), $F_{ij}^{(1)}(q) = G_{ij}^{(1)} q^{-5/3}$, $i, j = 1, 2, 3$; $F_{i4}^{(1)}(q) = G_{i4}^{(1)} q^{-1}$, $i = 1, 3$; $F_{44}^{(1)}(q) = G_{44}^{(1)} q^{-1/3}$, where $G_{ij}^{(1)}$ are constant cœfficients, i.e.

$$E_{ij}(k) = G_{ij}^{(1)}(Q\beta)^{\frac{2}{3}} k^{-\frac{5}{3}}, \; i, j = 1, 2, 3;$$

$$E_{i4}(k) = G_{i4}^{(1)} Q k^{-1}, \; i = 1, 3; E_{44}(k) = G_{44}^{(1)} Q^{\frac{4}{3}} \beta^{-\frac{2}{3}} k^{-\frac{1}{3}} \quad (11)$$

for small $k$.

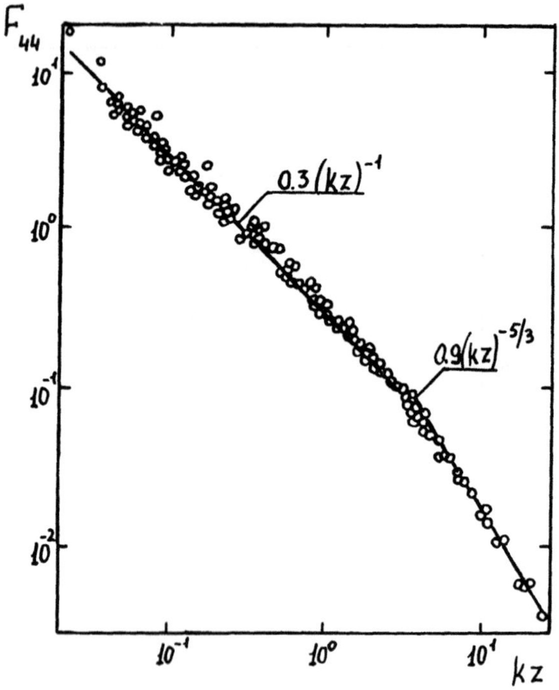


Fig.4. Normalized temperature spectrum $E_{44}/T_*^2 z$ within the logarithmic sublayers.

According to (9) and (11) the spectra $E_{ii}(k)$, $i = 1, 2, 3$, must include two $-5/3$ power subranges where the same $-5/3$ power law is valid with the possibly different coefficients $C_{ii}^{(1)}$ and $G_{ii}^{(1)}$. Of course, only the 'right' $-5/3$ power subrange that includes large wave numbers k is really the inertial subrange corresponding to small-scale isotropic turbulent disturbances. Therefore $C_{11}^{(1)} = 3C_{22}^{(1)}/4 = 3C_{33}^{(1)}/4 = C_K\epsilon^{2/3}(Q\beta)^{-2/3}$, $C_{44}^{(1)} = C_{OC}N\epsilon^{-1/3}Q^{-4/3}\beta^{2/3}z^{4/3}$. Note that within the free-convection sublayer $\epsilon = C_\epsilon Q\beta$, $N = C_N Q^{5/3}\beta^{-1/3}z^{-4/3}$ where $C_\epsilon$ and $C_N$ are constants (see Monin and Yaglom, 1971,sec.7.5). The experimental data collected at the Tsimlyansk Field Station and analysed by Kader and Yaglom (1989) show that $C_\epsilon \approx 1.2, C_N \approx 0.65$; hence

$$C_{11} \approx 1.1C_K, C_{22} = C_{33} \approx 1.45C_K, C_{44} \approx 0.61C_{OC} \qquad (12)$$

Note also that the air flow within the free convection sublayer of the ASL is not identical to the idealized shear-free convection where the mean velocity vanishes and the turbulence is axisymmetric. It is shown by Kader and Yaglom (1989) that in the atmospheric free-convection sublayer the wind shear $dU/dz$ is small but different from zero ; therefore $\langle -uw \rangle$ and $E_{13}(k)$ do not vanish here.

Let us now consider the results of spectral measurements within the free-convection sublayer (i.e at $z > 5L$). The condition $z > 5L$ is fulfilled not very often within the ASL but, if the measurements are made at the heights of the order of several tens of meters on hot summer days with weak but stable wind, the validity of this condition is also not a very rare exception.

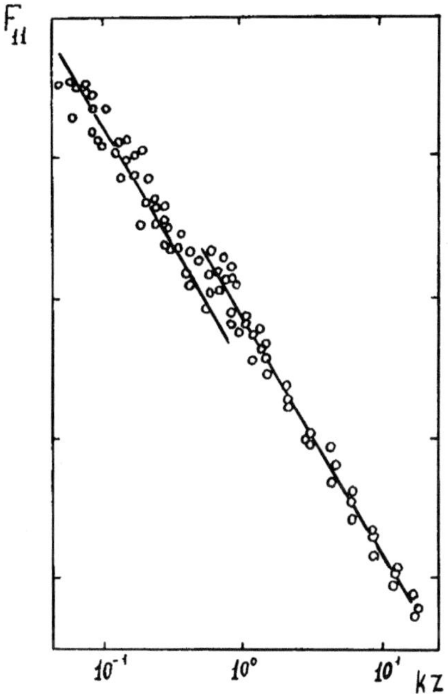


Fig.5. Normalized spectrum $F_{11} = E_{11}/w_*^2 z$ at the height $z = 50m$ above the ground on hot days derived from aircraft measurements.

In fig.5 the measured values of the spectrum $E_{11}(k)$ of the longitudinal wind velocity component are shown ; these values were derived from the aircraft measurements when the plane was flying along the mean wind direction at the height $z = 50$ m over the steppe on hot summer days. No scale is given on the ordinate axis since the results obtained during four flights are collected in fig.5 and all the data are normalized to make them fit each other in the high-wave-number range. We see that the data for low wave numbers also agree well with each other in fig.5 and all the data are clustered near two lines with the same slope corresponding to the $-5/3$ power law. More detailed results were obtained during 1981-1987 at the Tsimlyansk Field Station where all the spectra and cospectra of the fluctuations $u_i$, $i = 1, 2, 3, 4$, where measured from a 40 m mast simultaneously with the values of $Q = \langle wt \rangle$ and $u_* = \langle -uw \rangle^{1/2}$ (cf. Kader and Yaglom, 1989). Much attention was given to the conditions of hot sunny weather and weak but stable wind during these measurements ; these conditions correspond to small values of $L$, i.e. to large values of $z/L$. The spectral data for $z > 5L$ are shown in figs. 6 and 7.

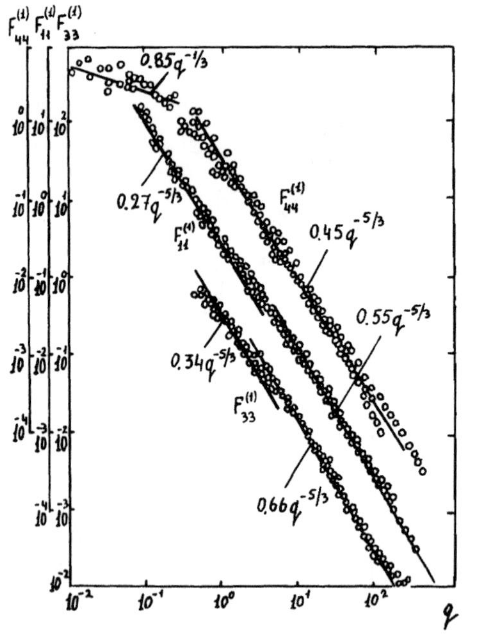


Fig.6. Normalized longitudinal autospectra of surface-layer turbulence for $z > 5L$.

These data prove to be quite scattered (especially for small values of $k$) but they do not contradict the theoretical equations (9)-(11). The transition zones between the power subranges at high and low values of $k$ are here wider than those for the dynamic sublayer but, nevertheless, if we exclude quite inaccurate measurements of $E_{22}(k)$, then all the other spectra and cospectra can be again described in the first approximation by simple two-line equations :

$$F_{ii}^{(1)}(q) = \begin{cases} C_{ii}^{(1)} q^{-\frac{5}{3}} & \text{for } b_{ii}^{(1)} < q < a_{ii}^{(1)}, \\ G_{ii}^{(1)} q^{-\frac{5}{3}} & \text{for } d_{ii}^{(1)} < q < b_{ii}^{(1)}, \end{cases} \tag{13}$$

for $i = 1, 3$,

$$F_{44}^{(1)}(q) = \begin{cases} C_{44}^{(1)} q^{-\frac{5}{3}} & \text{for } b_{44}^{(1)} < q < a_{44}^{(1)}, \\ G_{44}^{(1)} q^{-\frac{1}{3}} & \text{for } d_{44}^{(1)} < q < b_{44}^{(1)}, \end{cases} \tag{14}$$

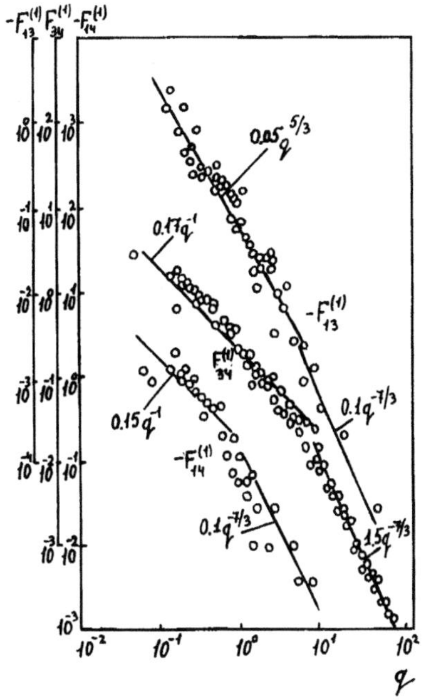


Fig.7. Normalized longitudinal cospectra for $z > 5L$.

$$F_{13}^{(1)}(q) = \begin{cases} C_{13}^{(1)} q^{-\frac{7}{3}} & \text{for } b_{13}^{(1)} < q < a_{13}^{(1)}, \\ G_{13}^{(1)} q^{-\frac{5}{3}} & \text{for } d_{13}^{(1)} < q < b_{13}^{(1)}, \end{cases} \tag{15}$$

$$F_{i4}^{(1)}(q) = \begin{cases} C_{i4}^{(1)} q^{-\frac{7}{3}} & \text{for } b_{i4}^{(1)} < q < a_{i4}^{(1)}, \\ G_{i4}^{(1)} q^{-1} & \text{for } d_{i4}^{(1)} < q < b_{i4}^{(1)}, \end{cases} \tag{16}$$

for $i = 1, 3$, where

$$\begin{gathered} C_{11}^{(1)} \approx 0.55, C_{33}^{(1)} \approx 0.66, C_{44}^{(1)} \approx 0.45; \\ C_{13}^{(1)} \approx -0.1, C_{14}^{(1)} \approx -0.1, C_{34}^{(1)} \approx 1.5; \\ G_{11}^{(1)} \approx 0.27, G_{33}^{(1)} \approx 0.34, G_{44}^{(1)} \approx 0.45; \\ G_{13}^{(1)} \approx -0.05, G_{14}^{(1)} \approx -0.15, G_{34}^{(1)} \approx 0.17; \\ b_{11}^{(1)} \approx 2, b_{33}^{(1)} \approx 3.5, b_{44}^{(1)} \approx 0.4, b_{13}^{(1)} \approx 3, b_{14}^{(1)} \approx 0.4; b_{34}^{(1)} \approx 5. \end{gathered} \tag{17}$$

According to (13) functions $F_{ii}^{(1)}(q)$, $i = 1, 3$, are discontinuous at the points $q = b_{ii}^{(1)}$. It is clear, however, that these discontinuities are not real but are due only to the simplified model description of functions

$F_{ii}^{(1)}(q)$. Note also that according to (12) the values $C_K \approx 0.5$, $C_K \approx 0.46$ and $C_{OC} \approx 0.74$ of the kolmogorov and Obukhov-Corrsin constants correspond to the values (17) of $C_{11}^{(1)}, C_{33}^{(1)}$ and $C_{44}^{(1)}$. These estimates of $C_K$ and $C_{OC}$ agree well with the other available experimental values of these constants.

### 2.3. Spectra within the dynamic-convective sublayer

In the flow region above the dynamic sublayer but below the free-convection sublayer both parameters $u_*$ and $\beta$ affect the turbulence. Therefore the general equations (1) must be used here for spectra and cospectra. However in the middle part of this region (apparently at $0.3L \leq z \leq 3L$ ; see Kader and Yaglom, 1989) the so-called dynamic-convective sublayer is located where equation (1) can be simplified substantially. Within this sublayer the energy exchange between horizontal and vertical motions is very small and can be neglected in the first approximation so that these two types of motion can be considered to be energetically uncoupled. It is explained by Betchov and Yaglom (1971) and Kader and Yaglom (1989) that this implies the possibility to apply the directional dimensional analysis to the study of turbulence within the dynamic-convective sublayer, i.e. to assume that the lengths of horizontal and vertical segments have different dimensions $L_x$ and $L_z$. Therefore $u_* = \langle -uw \rangle^{1/2}$ has dimension $L_x^{1/2} L_z^{1/2} T^{-1}$ (where T is the time dimension) in the framework of the directional dimensional analysis and the combinations $w_* = (Q\beta z)^{1/3}$ and $u_{**} = u_*^2/w_* = u_*^2(Q\beta z)^{-1/3}$ (of dimensions $L_z T^{-1}$ and $L_x T^{-1}$) must be used as vertical and horizontal velocity scales. Similarity $\Theta_* = Q/w_* = Q^{2/3}(\beta z)^{-1/3}$ and $l_* = (u_*^2/w_*^2)z = u_*^2(Q\beta)^{-2/3} z^{1/3}$ must be used as scales for the temperature and horizontal length.

Directional dimensional analysis implies the following equations for the spectra and cospectra $E_{ij}(k)$, $i, j = 1, 2, 3, 4$,

$$E_{ij}(k) = w_*^{(i)} w_*^{(j)} l_* F_{ij}^{(2)}(kl_*) \tag{18}$$

where $w_*^{(1)} = w_*^{(2)} = u_{**}$, $w_*^{(3)} = w_*$, $w_*^{(4)} = \Theta_*$. These equations are special cases of (1) which include only the functions of one variable $q_1 = kl_* = (u_*^2/w_*^2)kz$. We should take into account, however, that equations (18) are valid only for some subrange of the wide range of wave numbers where equations (1) are valid. In fact, equations (18) are valid only if the sinusoidal disturbances of the horizontal and vertical velocities with the wave length $2\pi/k$ are energetically uncoupled. The last condition is clearly untrue within the inertial wave-number subrange since the disturbances corresponding to this subrange are isotropic and isotropy of the velocity field cannot be achieved without the energy exchange between horizontal and vertical motions. Therefore general equations (1) must be used for the wave numbers in the inertial subrange and only beyond the large-scale limit of this subrange (i.e. whithin the energy subrange) equations (18) have chances to be true.

Equations (1) for autospectra $E_{ii}(k)$ can be simplified significantly within the inertial subrange. The $-5/3$ power law must be valid in this subrange and therefore

$$F_{ii}(kz,\zeta) = C_{ii}^{(2)}(\zeta)(kz)^{-\frac{5}{3}}, \ \zeta = \frac{z}{L}, \ i = 1,2,3,4, \tag{19}$$

for $1 < kz \ll k_0 z$ where $C_{11}^{(2)}(\zeta) = C_K(\epsilon z/u_*^3)^{2/3}$, $C_{22}^{(2)}(\zeta) = C_{33}^{(2)}(\zeta) = 4C_K(\epsilon z/u_*^3)^{2/3}/3$, $C_{44}^{(2)}(\zeta) = C_{OC}(Nz/u_*T_*^2)(\epsilon z/u_*^3)^{-1/3}$. The analysis of the results of Tsimlyansk measurements of the autospectra (based on the assumption that $C_K = 0.5$, $C_{OC} = 0.7$) shows that $\epsilon z/u_*^3$ is pratically independent of $\zeta = z/L$ and is close to 2.5 within the dynamic-convective sublayer while $Nz/u_*T_*^2 = Nzu_*/Q^2$ falls off with $\zeta$ and can be described in the first approximation by equation $Nz/u_*T_*^2 \approx 0.85\zeta^{-1/3}$ for $0.3 < \zeta < 3$ (see Kader and Yaglom, 1989). This implies that

$$C_{11}^{(2)}(\zeta) \approx 0.92, C_{22}^{(2)}(\zeta) = C_{33}^{(2)}(\zeta) \approx 1.2, C_{44}^{(2)}(\zeta) \approx 0.44\zeta^{-\frac{1}{3}}$$

$$\text{for } 0.3 < \zeta < 3, \tag{20}$$

The cospectra $E_{ij}(k)$, $i \neq j$, fall off approximately as $k^{-7/3}$ within the inertial subrange ; hence in the first approximation

$$F_{ij}(kz,\zeta) = C_{ij}^{(2)}(\zeta)(kz)^{-\frac{7}{3}}, (i,j) = (1,3), (1,4), \text{ or } (3,4), \tag{21}$$

in this subrange.

In the range of small wave numbers $k < 1/z$ corresponding to relatively large-scale turbulent disturbances the equation (18) is apparently valid. However, we must take into account that (18) is justified by the directional dimensional analysis and in the framework of this analysis the horizontal wave length $2\pi/k$ and distance from the ground $z$ have different dimensions and cannot be compared with each other. Therefore we cannot conclude that $E_{ij}(k)$ must be independent of $z$ for $kz < 1$. Instead of this we can conclude that $E_{ij}(k)$ must be independent of $l_*$ for $kl_* < 1$ but this conclusion cannot be used so easily, as that about the independence on $z$, to derive the form of the functions $F_{ij}^{(2)}(q_1)$. Therefore we start from quite another reasoning. Let us remind that equation (18) is based on the assumption that the vertical and horizontal motions are energetically uncoupled. This implies that statistical characteristics of $u_1$ and $u_2$ fluctuations are independent of the parameter $\beta$ which enters only the dynamic equation for the vertical velocity component $u_3$. Since

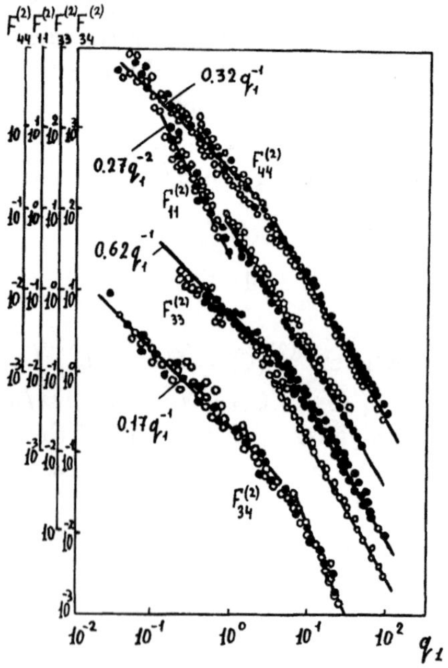

Fig.8. Normalized spectra of surface-layer turbulence for $z = 0.35L$ (•) and $z = 1.4L$ (∘).

$w_*^{(1)} = w_*^{(2)}$ is proportional to $\beta^{-1/3}$ and $l_*$ is proportional to $\beta^{-2/3}$, the independence of $E_{11}(k)$ and $E_{22}(k)$ on $\beta$ implies that

$$F_{11}^{(2)}(q_1) = G_{11}^{(2)} q_1^{-2}, F_{22}(q_1) = G_{22}^{(2)} q^{-1}. \tag{22}$$

The forms of the functions $F_{33}^{(2)}(q_1)$, $F_{44}^{(2)}(q_1)$ and $F_{34}^{(2)}(q_1)$ can be derived from the requirements of the independence of $E_{ij}(k)$ on $l_*$ for $kl_* < 1$. In fact, $u_*$ enters equations (18) for $E_{33}(k)$, $E_{44}(k)$, and $E_{34}(k)$ only through $l_*$ and therefore the independence on $l_*$ is equivalent here to the independence on $u_*$. It follows from this that

$$F_{33}^{(2)}(q_1) = G_{33}^{(2)} q_1^{-1}, F_{44}(q_1) = G_{44}^{(2)} q^{-1}, F_{34}^{(2)}(q_1) = G_{34}^{(2)} q^{-1} \tag{23}$$

for $kl_* < 1$. The situation with the cospectra $E_{13}(k)$ and $E_{14}(k)$ is more complicated and we shall not consider these cospectra within the dynamic-convective sublayer in the present paper. Note, however, that the available experimental data make the impression that the functions $F_{13}^{(2)}(q_1)$ and $F_{14}^{(2)}(q_1)$ are also proportional to $q_1^{-1}$.

Experimental values of the functions $F_{ii}^{(2)}(q_1)$, $i = 1, 3, 4$, and $F_{34}^{(2)}(q_1)$ derived from the measurements at the Tsimlyansk Field Station at $0.3L < z < 3L$ are shown in Fig.8.

We see that the experimental data for small values of $q_1$ cluster along the lines described by equations (22) and (23) with the cœfficients

$$G_{11}^{(2)} \approx 0.27, G_{33}^{(2)} \approx 0.62, G_{44}^{(2)} \approx 0.32, G_{34}^{(2)} \approx 0.17. \qquad (24)$$

The small-wave-length limits of the subranges where the equations (22) and (23) are valid can be crudely estimated as follows :

$$b_{11}^{(2)} \approx 0.6, b_{33}^{(2)} \approx 2.7\zeta^{-5/3}, b_{44}^{(2)} \approx 2, b_{34}^{(2)} \approx 5 \qquad (25)$$

according to the experimental data. Beyong these limits the functions $F_{ij}^{(2)}(q_1)$ begin to fall off approximately as $q_1^{-5/3}$ and $F_{34}^{(2)}(q_1)$ as $q_1^{-7/3}$ but the experimental points become differentiated according to the value of $\zeta = z/L$ in the subrange of large $q_1$-values and this differentiation is best seen in the data on $F_{33}^{(2)}(q_1)$. This can be easily explained by substituting (19) and (21) into (1) and comparing the results with (18), since this comparison shows that within the inertial subrange $F_{11}^{(2)}(q_1) = C_{11}^{(2)}(\zeta)\zeta^{2/9}q_1^{-5/3}, F_{33}^{(2)}(q_1) = C_{33}^{(2)}(\zeta)\zeta^{-10/9}q_1^{-5/3}, F_{44}^{(2)}(q_1) = C_{44}^{(2)}(\zeta)\zeta^{2/9}q_1^{-5/3}, F_{34}^{(2)}(q_1) = C_{34}^{(2)}(\zeta)\zeta^{-8/9}q_1^{-7/3}$.

According to (20) this implies that

$$F_{11}^{(2)}(q_1) \approx 0.92\zeta^{2/9}q_1^{-5/3}, F_{33}^{(2)}(q_1) \approx 1.2\zeta^{-10/9}q_1^{-5/3},$$

$$F_{44}^{(2)}(q_1) \approx 0.44\zeta^{-1/9}q_1^{-5/3} \qquad (26)$$

and this shows that $F_{33}^{(2)}(q_1)$ varies much faster with $\zeta$ than $F_{11}^{(2)}(q_1)$ and $F_{44}^{(2)}(q_1)$.

## 3. Correlation Functions of Turbulent Fluctuations

The longitudinal (in the direction of mean wind) correlation functions of turbulent fluctuations are the Fourier transforms of the corresponding longitudinal spectra, i.e.

$$\langle u_i(x+r,y,z)u_i(x,y,z)\rangle = B_{ii}(r) = \int_0^\infty \cos kr E_{ii}(k)dk \qquad (27)$$

Hence, if the values of $E_{ii}(k)$ are known in the broad wave-number range, then the values of $B_{ii}(r)$ can be evaluated from (27) if the known values of $E_{ii}(k)$ make the main contribution to the right-hand part of (27) for considered values of $r$. Similarly the measured values of the cospectrum $E_{ij}(k), i \neq j$, can be used to evaluate the symmetrized cross-correlation function

$$B_{ij}(r) = \frac{1}{2}\{\langle u_i(x+r,y,z)u_j(x,y,z)\rangle + \langle u_i(x,y,z)u_j(x+r,y,z)\rangle\} \quad (28)$$

since the function (28) is just the Fourier transform of $E_{ij}(k)$ :

$$B_{ij}(r) = \int_0^\infty \cos kr E_{ij}(k) dk \tag{29}$$

### 3.1. Correlation functions within the dynamic sublayer

We already know that within the dynamic sublayer of the unstable ASL (and hence within the whole ASL in neutral stratification) the spectra $E_{ii}(k), i = 1,2,3,4$, are given within the broad wave-number range by equations (2) with $i = j$ and (5). In the dissipation subrange beyond the small-scale limit of the inertial subrange, the spectra and cospectra fall off with $k$ very rapidly (apparently at the exponential rate). Moreover, the behaviour of spectra and cospectra beyond the large-scale limit of the subrange where the $-1$ power law is poorly known since the spectral measurements become unreliable when the wave numbers are very small. The main contribution to the variances $\sigma_i^2 = \langle u_i^2 \rangle = \int_0^\infty E_{ii}(k)dk$ and covariances $\langle u_i u_j \rangle = \int_0^\infty E_{ij}(k)dk$ is apparently due to the spectral subrange where the $-1$ power law is valid for $E_{ii}(k)$ and, respectively, $E_{ij}(k)$ (Kader, 1987; Kader, Yaglom and Zubkovskii, 1989; Kader, 1984). The main contribution to the values of the correlation functions (27) and symmetrized cross-correlation functions (28) for the values of $r$ that will be of main interest below is due to the inertial subrange and the subrange where the $-1$ power law is valid. Therefore it is possible to neglect often the contribution of the dissipation subrange and large-scale subrange where $kz < d_{ii}$ or $kz < d_{ij}$ to the integral on the right-hand sides of (27) and (29). (The same results will be, in fact, obtained, if we take into account all the values of $k$ and assume that $E_{ii}(k)$ and $E_{ij}(k)$ decrease exponentially for $kz > a_{ii}$ or $kz > a_{ij}$ and are constant or proportional to $k^{-\alpha}$, where $\alpha < 1$, for $kz < d_{ii}$ or $kz < d_{ij}$).

If $r \leq k_0^{-1}$, then the dissipation subrange cannot be neglected but here the main contribution to the integral on the right-hand side of (27) is made by wave numbers $k$ such that $kr < 1$. Therefore $\cos kr$ can be replaced here by the first two terms of its Taylor's expansion. Then we obtain easily that

$$B_{ii}(r) = \sigma_i^2 (1 - A_i r^2) \text{ for } r \leq k_0^{-1} \tag{30}$$

where $A_1 = \epsilon/30\nu\sigma_1^2$, $A_i = \epsilon/15\nu\sigma_i^2$ for $i = 2,3$, $A_4 = N/6\chi\sigma_4^2$, $\chi$ being the cœfficient of thermal diffusivity (see, e.g., Monin and Yaglom, 1975, eqs. (21.16) and (21.86), or Kader, Yaglom and Zubkovskii, 1989).

For larger values of r we can neglect the contributions of the small-scale and large-scale subranges where $E_{ii}(k)$ cannot be represented in the form (2). Then equation (27) can be rewritten as

$$B_{ii}(r) = \sigma_i^2 - \int_0^\infty (1 - \cos kr) E_{ii}(k) dk$$
$$= \sigma_i^2 - \{1 - a_i^{-2} \int_0^\infty (1 - \cos qr/z) F_{ii}(q) dq\} \qquad (31)$$

where $a_i = \sigma_i / u_*^{(i)}$, $q = kz$, and this implies, in particular, that $R_{ii}(r) = B_{ii}(r)/\sigma_i^2$ is the universal function of $r/z = \xi$. It is easy to see that if $k_0^{-1} \ll r < z$ (i.e. if $r$ belongs to the inertial subrange of lengths), then $a_{ii}\xi \gg 1$, $b_{ii}\xi \ll 1$ and the inertial subrange $b_{ii} < q < a_{ii}$ of $q$-values makes the main contribution to the integral on the right-hand side of (31). Therefore the function $F_{ii}(q)$ can be replaced by $C_{ii}q^{-5/3}$ on the right-hand side of (31) and this replacement leads to the well-known 2/3 power law of Kolmogorov and Obukhov

$$B_{ii}(r) = \sigma_i^2 \{1 - B_i (\frac{r}{z})^{\frac{2}{3}}\}, B_i = 3C_{ii} \frac{\Gamma(\frac{1}{3})}{4a_i^2}, \qquad (32)$$

where $\Gamma$ is the $\Gamma$ -function (see, e.g., Monin and Yaglom, 1975, Secs. 21.4 and 21.6, or Kader, Yaglom and Zubkovskii, 1989).

Let now $r > z$ (i.e. $\xi = r/z > 1$) and $b_{ii}^{-1} < \xi < d_{ii}^{-1}$. In this case (considered in detail by Kader, Yaglom and Zubkovskii, 1989) both the intervals $d_{ii} < q < b_{ii}$ and $b_{ii} < q < a_{ii}$ contribute significantly to the integral on the right-hand side of (31). Note that the inequalities $d_{ii} \ll b_{ii} \ll a_{ii}$ are always true in the ASL (for laboratory turbulent flows at moderate Reynolds numbers the ratio $b_{ii}/a_{ii}$ is usually not too small and depends on Re). Let us assume for simplicity that $b_{ii}\xi \gg 1, d_{ii}\xi \ll 1$ (this is not strong restriction since the asymptotic relations usually have a rather broad range of applicability). If we substitute (5) into (31) and use equations 9.10(8), 9.10(10), 9.8(9) and 9.8(11) from Erdélyi, Magnus, Oberhettinger and Tricomi (1953) together with the conditions $C_{ii} = G_{ii}(b_{ii})^{2/3}$ (this follows from te continuity of $F_{ii}(q)$ at $q = b_{ii}$) and $b_{ii}/a_{ii} \ll 1$, we find that

$$B_{ii}(r) \approx \sigma_i^2 \{1 - D_i - d_i ln \frac{r}{z}\}, D_i = \frac{G_{ii}(1.5 + \mathbf{C} + ln(b_{ii}))}{a_i^2}, d_i = \frac{G_{ii}}{a_i^2} \qquad (33)$$

for $b_{ii}^{-1} < r/z < d_{ii}^{-1}$ where $\mathbf{C} \approx 0.58$ is Euler's constant.

Similar results can be obtained for the symmetrized longitudinal cross-correlation functions $B_{ij}(r)$. It is easy to see that

$$B_{ij}(r) = B_{ij}(0) - A_i r^2 \quad \text{for } r \leq k_0^{-1} \qquad 34$$

where $A_{ij} = \langle(\partial u_i/\partial x)(\partial u_j/\partial x)\rangle/2$. Moreover, equation (6) implies that

$$B_{ij}(r) = B_{ij}(0) - H_{ij}(\frac{r}{z})^{\frac{4}{3}}, \qquad (35)$$

where $H_{ij} = u_*^{(i)}u_*^{(j)}\{\Gamma(-4/3)/2\}C_{ij}$. Finally, for $b_{ij}^{-1} < r/z < d_{ij}^{-1}$ the following equation

$$B_{ij}(r) = B_{ij}(0) - u_*^{(i)}u_*^{(j)}\{D_{ij} - G_{ij}ln\frac{r}{z}\}, \qquad (36)$$

where $D_{ij} = G_{ij}(0.75 + \mathbf{C} + ln(b_{ij}))$, can be derived by the arguments similar to those used in the derivation of (33). (The replacement of the exponent -7/3 by -5/2 on the right-hand side of (6) implies only the replacement of the constant 0.75=3/4 in the expression for $D_{ij}$ by $2/3 \approx 0.67$).

The above equations for the functions $B_{ii}(r)$ and $B_{ij}(r)$ are compared with the results of direct measurements (headed by S.L. Zubkovskii) of these functions at the Tsimlyansk Field Station. Five sets of instruments were used, each of them containing a three-component sonic anemometer and a low-inertia resistance thermometer. The sets were placed in a line along the mean wind direction at the same height of 3.2 m and the values of the wind components and temperature fluctuations were recorded on magnetic tape during 150-minute runs. The distance between the first set and the others were selected to be 3, 10, 30, and 100 m. The measured values of $B_{11}(r), B_{33}(r), B_{44}(r)$ and $B_{13}(r), B_{14}(r)$, obtained when the stratification was pratically neutral and the mean wind speed, wind direction and temperature did not change appreciably during the observation run, are compared in figs. 9 and 10 with the equations (33) and (36) with the cœfficients computed from the results of spectral measurements at the same place.

We see that the empirical data agree quite satisfactorily with the theoretical equations for all values of r in the interval from 3 to 100 m. This agreement can be also considered as a confirmation of Taylor's frozen turbulence hypothesis since the spectra $E_{ii}(k)$ and $E_{ij}(k)$ are found here from the measurements at a fixed point with the aid of this hypothesis while the spatial correlation functions $B_{ii}(r)$ and $B_{ij}(r)$ were measured directly by the instruments located at a number of points.

### 3.2. Correlation functions within the free convection sublayer

Equations (8), (13)-(16), (27) and (29) allow one to determine the shapes of correlation functions $B_{ii}(r)$ and $B_{ij}(r)$ of turbulence within the free convection sublayer. We shall consider only the autocorrelation functions $B_{ii}(r)$ since the derivation of the results for the cross-correlation functions $B_{ij}(r), i \neq j$, differs only slightly from that for autocorrelation functions.

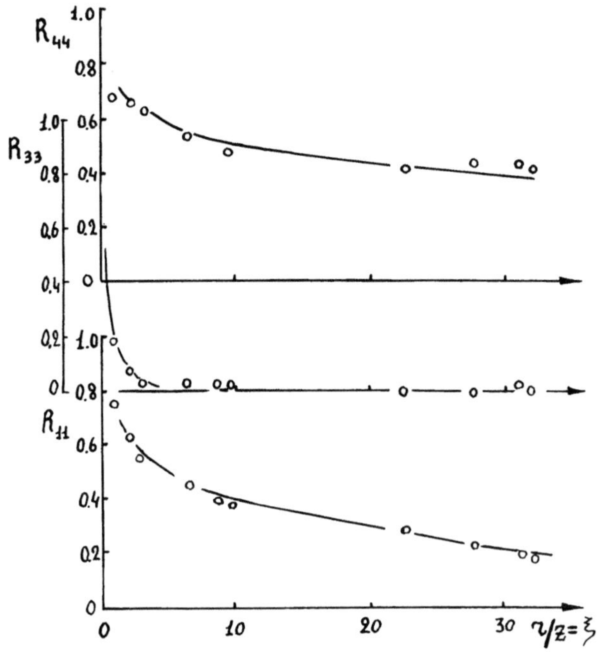


Fig.9. Normalized longitudinal autocorrelation functions $B_{ii}(r)/B_{ii}(0) = R_{ii}(r)$ of surface-layer turbulence in neutral stratification.

It is clear that all the arguments used above to derive (30) are also valid for the free-convection turbulence. Equation (32) is also valid for the free-convection correlation functions in the inertial subrange of lengths $r$ (but, of course, $C_{ii}$ and $a_i = \sigma_i/u_*^{(i)}$ must be replaced here by $C_{ii}^{(1)}$ and $a_i^{(1)} = \sigma_i/w_*^{(i)}$). Let us now consider the case of greater values of $r$ satisfying the inequalities $r > z$, $(b_{ii}^{(1)})^{-1} < r/z < (d_{ii}^{(1)})^{-1}$. Equation (31) is clearly valid also in the free convection sublayer, if $a_i$ is replaced by $a_i^{(1)}$ and $F_{ii}(q)$- by the function $F_{ii}^{(1)}(q)$ given by (13) or (14). It is easy to show that the significant contribution to the integral on the right-hand side of (31) is made only by the intervals $d_{ii}^{(1)} < q < b_{ii}^{(1)}$ and $b_{ii}^{(1)} < q < a_{ii}^{(1)}$ of q-values. Let us assume for simplicity that $b_{ii}^{(1)}\xi \gg 1$, $d_{ii}^{(1)}\xi \ll 1$, where $\xi = r/z$. Then, as in the derivation of (33), it is easy to show that the contribution of the latter of these two intervals is close to $1.5C_{ii}^{(1)}(b_{ii}^{(1)})^{-2/3}$ (cf. Kader, Yaglom and Zubkovskii, 1989). Let now $i = 1, 2$, or $3$ ; then for the estimation of the contribution of the former interval we can use the same arguments which have led us to equation (32) and supplement these arguments by applications of the appropriate equations from Sec. 9.10 of Erdélyi, Magnus, Oberhettinger

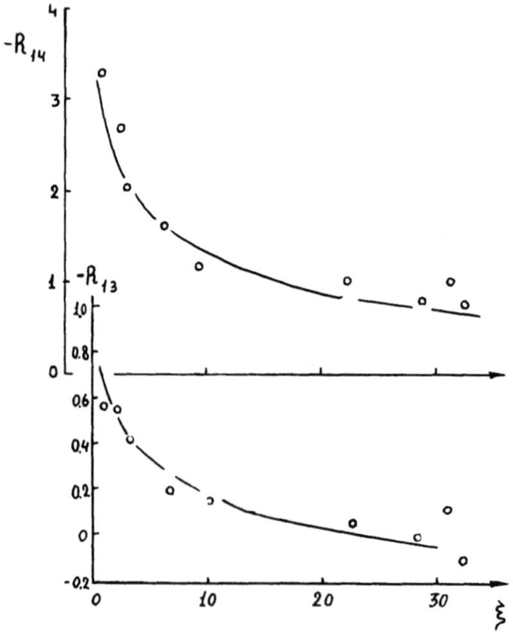


Fig.10. Normalized longitudinal cross-correlation functions $B_{13}(r)/u_*^2 = R_{13}(r)$ and $B_{14}(r)/Q = R_{14}(r)$ in neutral stratification.

and Tricomi (1953). Then we easily obtain that this contribution is close to $0.75\Gamma(1/3)G_{ii}^{(1)}\xi^{2/3}$. Hence

$$B_{ii}(r) \approx \sigma_i^2\{1 - \frac{w_*^2}{\sigma_i^2}[1.5C_{ii}^{(1)}(b_{ii}^{(1)})^{-\frac{2}{3}} + 0.75\Gamma(1/3)G_{ii}^{(1)}(\frac{z}{r})^{\frac{2}{3}}]\} \quad (37)$$

for $i = 1, 2$, or 3 and $(b_{ii}^{(1)})^{-1}$. If $i = 4$, then one must use equation (14) for $F_{44}^{(1)}(q)$ ; in this case it is easy to show that the contribution of the interval $d_{44}^{(1)} < q < b_{44}^{(1)}$ to the integral on the right-hand side of (31) is close to $G_{44}^{(1)}[1.5(b_{44}^{(1)})^{2/3} - 0.5\Gamma(2/3)\xi^{-2/3}]$. Thus

$$B_{44}(r) \approx \sigma_4^2\{1 - \frac{\Theta_*^2}{\sigma_4^2}[1.5C_{44}^{(1)}(b_{44}^{(1)})^{-\frac{2}{3}} + 1.5G^{(1)}(b_{44}^{(1)})^{\frac{2}{3}} - 0.5\Gamma(\frac{2}{3})G_{44}^{(1)}(\frac{r}{z})^{-\frac{2}{3}}]\} \quad (38)$$

for $(b_{44}^{(1)})^{-1} < r/z < (d_{44}^{(1)})^{-1}$.

Unfortunately, till now no experimental data have been obtained on the values of correlation functions of the turbulent fluctuations within the free convection sublayer. Therefore there is no possibility to compare equations (37) and (38) with the experimental results.

### 3.3. Correlation functions within the dynamic-convective sublayer

The above arguments used to derive equations for the correlation functions within the dynamic and free convection sublayers of an unstable ASL can be applied, with slight modifications, to the evaluation of the same functions within the dynamic-convective sublayer. It is clear that equations (30) and (32) are also valid in this sublayer, if the constant $C_{ii}^{(2)}(\zeta)$. Hence only the behaviour of the correlation functions for $(b_{ii}^{(2)})^{-1} < r/l_* < (d_{ii}^{(2)})^{-1}$ needs investigation.

Let us again limit ourselves to the study of autocorrelation functions $B_{ii}(r)$. The similarity of equations (5) for $F_{ii}(q)$ to equations (23) for $F_{33}^{(2)}(q_1)$ and $F_{44}^{(2)}(q_1)$ implies the complete similarity of the evaluations of functions $B_{33}(r)$ and $B_{44}(r)$ within the dynamic-convective sublayer to the evaluation of autocorrelation functions within the dynamic sublayer. In particular, the forms of $B_{33}(r)$ and $B_{44}(r)$ for $(b_{ii}^{(2)})^{-1} < r/l_* < (d_{ii}^{(2)})^{-1}$, where $i = 3$ or 4, are given by the following equations quite similar to (33) :

$$B_{33}(r) \approx \sigma_3^2\{1-\frac{w_*^2}{\sigma_3^2}[1.5C_{33}^{(2)}(b_{33}^{(2)})^{-2/3}+G_{33}^{(2)}(\mathrm{C}+ln\frac{b_{33}^{(2)}w_*^2}{u_*^2}+ln\frac{z}{r})]\}, \quad (39)$$

$$B_{44}(r) \approx \sigma_4^2\{1-\frac{\Theta_*^2}{\sigma_4^2}[1.5C_{44}^{(2)}(b_{44}^{(2)})^{-2/3}+G_{44}^{(2)}(\mathrm{C}+ln\frac{b_{44}^{(2)}w_*^2}{u_*^2}+ln\frac{z}{r})]\}, \quad (40)$$

Since, however, the shapes of $E_{11}(k)$ and $E_{22}(k)$ beyond the large scale limit of the inertial subrange in the case of the dynamic-convective sublayer are not the same as in the case of dynamic sublayers, the functions $B_{11}(r)$ and $B_{22}(r)$ for the dynamic-convective sublayer must be determined independently. It is easy to see that the replacement of the spectral -1 power law by the -2 power law simplifies the evaluation of functions $B_{ii}(r)$ corresponding to spectra $E_{ii}(k)$. In particular, for $(b_{11}^{(2)})^{-1} < r/l_* < (d_{11}^{(1)})^{-1}$ we now obtain the result

$$B_{11}(r) \approx \sigma_1^2\{1 - \frac{3u_*^4}{2w_*^2\sigma_1^2}C_{11}^{(2)}(b_{11}^{(2)})^{-2/3} - \frac{\pi}{2}G_{11}^{(2)}\frac{u_*^2}{\sigma_1^2}\frac{r}{z}\}, \quad (41)$$

and similar equation is valid also for $B_{22}(r)$.

In fig. 11 the functions $B_{ii}(r)/B_{ii}(0) = R_{ii}(r)$, $i = 1, 3, 4$, given by equations (39), (40) and (41) are compared with the results of the direct measuremens of the longitudinal correlation functions $R_{ii}(r)$ of the surface layer turbulence in the moderately unstable stratification. These measurments were performed at the Tsimlyansk Field Station in the summer of 1985. The functions $B_{ii}(r)$ were measured for $3m \leq r \leq 100m$

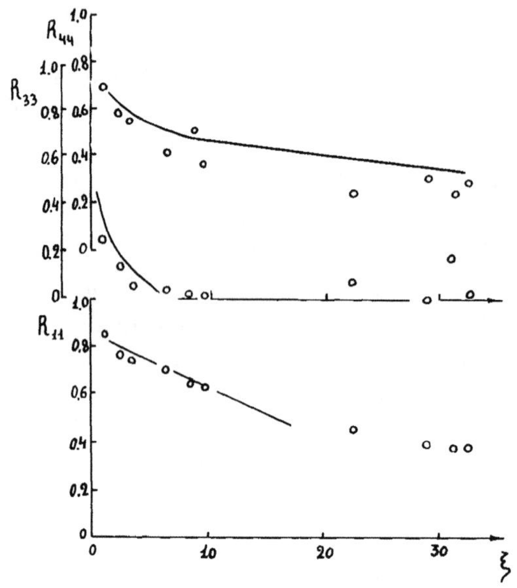


Fig.11. Normalized longitudinal correlation functions $R_{ii}(r)$ of surface-layer turbulence for $z \simeq 0.5L$

at the height $3.1m$ in the conditions corresponding to $L \approx 6.5m$ (i.e. $\zeta = z/L \approx 0.5$). We see that empirical data agree quite satisfactorily with the theoretical equations.

The authors are grateful to Dr. L.R. Tsvang, Dr. S.L. Zubkovskii and M.M. Fedorov for the permission to use the experimental data obtained by them and for numerous discussions of the results of this work.

## References

Betchov, R. & Yaglom, A.M., 1971: Comments on the theory of similarity as applied to turbulence in an unstably stratified fluid, *Izv. Akad. Nauk SSSR, Ser. Fiz. Atmosf. i Okeana*, **7**, 1270-1279 (829-832 in Engl. transl. of the journal).

Erdélyi, A., Magnus, W., Oberhettinger, F. & Tricomi, F.G., 1953: *Higher Transcendental Functions*, Vol. 2, McGraw-Hill, New York, 396pp.

Kader, B.A., 1984: Structure of anisotropic velocity and temperature fluctuations in a developed turbulent boundary layer, *Izv. Akad. Nauk*

*SSSR, Ser. Mekh. Zhidk. i Gaza*, N° 4, 47-56.

Kader, B.A., 1987: Anisotropic fluctuations of wind velocity and temperature in neutrally stratified atmospheric surface layer. In *Meteorological Researches*, N° 28 (eds. L.R. Tsvang & B.A. Kader), pp. 26-35, Acad. Sci. USSR, Soviet Geophysical Committee, Moscow (in Russian).

Kader, B.A., 1988: Three-layer structure of unstably stratified atmospheric surface layer, *Izv. Akad. Nauk SSSR, Ser. Fiz. Atmosf. i Okeana*, **24**, 1235-1250.

Kader, B.A. & Yaglom, A.M., 1984: Turbulent structure of an unstable atmospheric surface layer. In *Nonlinear and Turbulent Processes in Physics* (ed. R.Z. Sagdeyev), vol.2, pp. 829-845, Harwood Acad. Publ., New York.

Kader, B.A. & Yaglom, A.M., 1987: Spectra of anisotropic velocity and temperature fluctuations in turbulent wall flows. In *Problems of Turbulent Flows* (ed. V.V. Struminsky), pp. 65-74, Nauka, Moscow (Transl. in *Fluid Mech. - Soviet Research*, **16**, N°6, 89-102, 1987).

Kader, B.A. & Yaglom, A.M., 1989: Mean fields and fluctuation moments in unstably stratified turbulent boundary layers. *J. Fluid. Mech.*, **212** (in press).

Kader, B.A., Yaglom, A.M. & Zubkovskii, S.L., 1989: Spatial correlation functions of surface layer atmospheric turbulence at neutral stratification, *Boundary-Layer Meteor.* , **47**, 233-249.

Kaimal, J.C., Wyngaard, J.C., Izumi, Y. & Coté, O.R., 1972: Spectral characteristics of surface layer turbulence, *Quart. J. Roy. Meteor. Soc.*, **98**, 563-589.

Korotkov, B.N., 1976: Some types of local self-similarity of the velocity field of wall turbulent flows, *Izv. Akad. Nauk SSSR, Ser. Mekh. Zhidk. i Gaza*, N° 6, 35-42.

Monin, A.S. & Obukhov, A.M., 1954: Basic laws of turbulent mixing in the atmospheric surface layer, *Trudy Geofiz. Inst. Akad. Nauk SSSR*, N°24 (151), 163-187.

Monin, A.S. & Yaglom, A.M., 1971: *Statistical Fluid Mechanichs*, Vol.1, MIT-Press, Cambridge, Mass., 769pp.

Monin, A.S. & Yaglom, A.M., 1975: *Statistical Fluid Mechanics*, Vol. 2, MIT-Press, Cambridge, Mass. 874pp.

Obukhov, A.M., 1946: Turbulence in thermally inhomogeneous atmosphere, *Trudy Inst. Teor. Geofiz. Akad. Nauk SSSR*, N° 1, 95-115 (Transl. in *Boundary-Layer Meteor.*, **3**, 7-29, 1971).

Perry, A.E. and Abell, S.J., 1975: Scaling laws for pipe-flow turbulence, *J. Fluid Mech.*, **67**, 257-271.

Prandtl, L., 1932: Meteorogische Anwendung der Strömungslehre, *Beitr. Phys. fr. Atmosph.*, **19**, 188-202.

Tchen, C.M., 1953: On the spectrum of energy in turbulent shear flow, *J. Res. Nat. Bur. Standards*, **50**, N°1, 51-62.

Wyngaard, J.C., Coté, O.K. & Izumi, Y., 1971: Local free convection, similarity, and the budgets of shear stress and heat flux, *J. Atmos. Sci.*, **28**, 1171-1182.

Yaglom, A.M., 1974: Data on turbulence characteristics in the atmospheric surface layer, *Izv. Akad. Nauk SSSR, Ser. Fiz. Atmosf. i Okeana*, **10**, 566-586 (341-352 in Engl. transl. of the Journal).

Yaglom, A.M., 1981: Laws of small-scale turbulence in atmosphere and ocean, *Izv. Akad. Nauk. SSSR, Ser. Fiz. Atmosf. i Okeana*, **17**, 1235-1257 (919-935 in Engl. transl. of the journal).

Zilitinkevich, S.S., 1971: On the turbulence and diffusion under free convection conditions,*Izv. Akad. Nauk SSSR, Ser. Fiz. Atmosf. i Okeana*, **7**, 1263-1269.

# Organized Structures in Strongly Stratified Flows

James J. RILEY[α], Marie-Pascale G. LELONG[β], and Donald N. SLINN[α]

[α] *University of Washington*
*Seattle, Washington, USA*

[β] *National Center for Atmospheric Research*
*Boulder, Colorado, USA*

*Turbulence and Coherent Structures*
*O. Métais and M. Lesieur, eds.*
*Selected papers from "Turbulence 89: organized structures and turbulence in fluid mechanics"*
*Grenoble, 18-21 September 1989*

We report on numerical simulations of internal wave and vortical mode interactions in stably-stratified fluids. Two problems were considered: a wave/vortical mode resonance, and the Taylor-Green problem. In both cases the results of the simulations supported the perturbation theory of Lelong and Riley (1989) for Froude numbers of order one of less. Furthermore, in each case computed in this range of Froude number, the vortical mode exhibited strong instabilities, transferring energy to larger horizontal scales.

## 1. Introduction

Visual observations from laboratory experiments indicate that stable density stratification can totally alter the dynamics of a turbulent flow. For example, Lin and Pao (1979) observed from visualization studies that, in the later stages of decay of a turbulent wake in a stably-stratified fluid, the wake exhibited large-scale, quasi-horizontal meanders superimposed on the undulations of an internal wave field. (See Figure 1, taken from Hopfinger, 1987). Liu (1980) has observed similar behavior in the later stages of decay of grid turbulence in a stably stratified fluid. More recently, van Heijst and Flór (1989) observed the long-time behavior of a short duration jet in a strongly-stratified fluid and again noted the quasi-horizontal character of the motion.

Riley et al. (1981) have suggested that the presence of quasi-horizontal structures, for which we will use the term vortical modes, is due to the stratification domination of the flow. Introducing a Froude number defined by $F = u'/NL$, where $u'$ is an rms velocity, $N$ is the buoyancy frequency, and $L$ is an integral scale (characterizing the larger-scale structure), they argued that the vortical modes appear when $F$ is small. Furthermore, they offered scaling arguments to explain why the flow should consist of both vortical modes and internal waves when $F$ becomes small, and gave a mathematical decomposition of the velocity field into wave and vortical mode components. Subsequently Lilly (1983) extended these ideas to large-scale geophysical flows (with rotation), and suggested that vortical modes might explain recent atmospheric spectral data (Gage, 1979) at intermediate scales. Müller (1988; see also Müller et al., 1988) has suggested that vortical modes explain the vertical fine-structure observed in the ocean internal wave field. Recently Métais and Herring (1989; Herring and Métais, 1989) performed direct numerical

Fig.1. Wake collapse of a towed sphere in a continuously stratified fluid (from Hopfinger, 1987).

simulations of both decaying homogeneous turbulence and also of forced homogeneous turbulence in strongly-stratified fluids and found the decomposition suggested by Riley et al. to be useful in interpreting the results of the simulations.

In an effort to understand the vortical mode dynamics and their interaction with the internal wave field, Lelong and Riley (1989; see also Lelong, 1989) have recently examined the nonlinear interaction of simple monochromatic internal waves and vortical modes. Taking advantage of the small Froude number, they have formulated the problem using perturbation methods to examine these interactions. In particular, they have concluded the following:

i wave-wave resonant interaction theory extends intact when applied in the presence of weak vortical modes. It is also unaffected by the generalizing assumption that the wave vectors constituting the resonant triad not lie in a vertical plane;

ii a potentially important wave-vortical mode resonant triad exists, in which the vortical mode plays a catalytic role as energy is exchanged between two internal waves of equal frequency;

iii Any two vortical modes can excite a third vortical mode to form a resonant vortical triad. To lowest order, vortical modes satisfy the two-dimensional flow equations in each horizontal plane, while varying in the vertical direction. This verifies rigorously the heuristic suggestions of Riley et al.

The purposes of the work presented in this paper are to (i) examine the validity of the weak interaction theory of Lelong and Riley, and to (ii) investigate further the interactions identified by them. Direct numerical simulations are used to achieve these purposes, with computations being performed for cases predicted by the theory, giving 'exact' solutions to the fully-nonlinear problem. In the next section, we briefly review the theory of Lelong and Riley and discuss their principal results. In the third section, calculations for wave-vortical mode interactions are presented and compared with theory. In addition, simulations are presented for initial conditions consisting only of vortical modes. The behavior of the resulting flows is examined, and the results compared to the scaling analysis suggested by the theory. In the final section, the results are summarized and discussed.

## 2. Results from Perturbation Analysis

As mentioned in the previous section, vortical modes are visually observed in laboratory experiments when the Froude number based upon the energy containing range, i. e., $F = u'/NL$, is small. This Froude number can be interpreted as the ratio of two time scales: $N^{-1}$, the buoyancy period, and $L/u'$, an advective time scale. In the experiments, the buoyancy period has thus become small compared to the advection time. $F$ being small is also the usual requirement for linear or weakly nonlinear internal wave theory to be valid.

This suggests the use of multiple time scale perturbation analysis to analyze the dynamics of the internal waves and vortical modes. Lelong and Riley have carried out such an analysis, and begin by writing the velocity field $\vec{u}(\vec{x}, t)$ in the form, as suggested by Lilly (1983):

$$\vec{u} = \nabla \wedge \psi \vec{i}_z + \nabla_H \phi + w \vec{i}_z \, . \qquad (2-1)$$

Here the subscript $H$ denotes the horizontal component. To lowest order the stream function $\psi$ completely determines the vortical mode, while the vertical velocity $w$ (or $\phi$, where $w = -\int \nabla^2 \phi \, dz$) specifies the internal wave field. When these expressions are substituted into the Navier-Stokes equations subject to the Boussinesq approximation, an equation for $\psi$, namely the vertical vorticity equation is obtained while an equation for $w$ results from manipulation of the vertical and horizontal momentum equations.

Velocities are nondimensionalized by $u'$, lengths by $L$, and time by $N^{-1}$. (It is assumed that $N$ is uniform in space.) All nondimensionalized dependent variables are expanded in power series of $F$, and terms of like powers of $F$ are equated. Furthermore, the buoyancy and advective times are now both treated as independent variables. To lowest order (on the short time scale $N^{-1}$), linear theory is recovered. On this time scale, the vortical modes are steady. The equations at the next order give an $O(F^2)$ correction to these solutions unless a resonance exists, in which case the form of the lowest order solutions has to be reassessed in order for the series expansion to remain valid. The elimination of secular terms from the second order equations yields the evolution of the solutions on the slow, advective time scale. The results depend on the specific problem considered, i. e., on the initial conditions.

### 2.1. Wave-Wave Interactions

For this case, the initial conditions are taken to be two linear internal waves and a linear vortical mode. The internal waves are arbitrarily oriented, while the vortical mode is assumed to be harmonic in space and is given in terms of a stream function as in Equation (2-1) above. The results of the analysis are an extension of resonant wave interaction theory in the presence of a vortical mode, and the application of resonance theory to triads which are out of the vertical plane. Wave resonant triads are unaffected by the presence of the vortical mode, unless there exists a wave-vortical mode resonance (see the next subsection below). Furthermore, resonant triad analysis is readily extended to wave triads not lying in a vertical plane. The only deviation from the vertical plane case is that the gravity is effectively reduced by the cosine of the angle of the triad-containing plane with respect to the horizontal.

## 2.2. Wave-Vortical Mode Interactions

In this case, the initial conditions consist of one internal wave (with wave number vector $\vec{\kappa}_1$ and frequency $\omega$) and one vortical mode (with wave number vector $\vec{\kappa}_2$). The analysis predicts that a resonance can occur with a second internal wave (with wave number vector $\vec{\kappa}_3$) of the same frequency if the wave numbers satisfy

$$\vec{\kappa}_1 \pm \vec{\kappa}_2 \pm \vec{\kappa}_3 = 0\,. \qquad (2-2)$$

The wave number vectors of the internal waves lie on a vertical cone whose surface is at angle $\theta = \cos^{-1}\omega$. (See Figure 2.) Note that this interaction could act to redistribute energy broadly in wave number space.

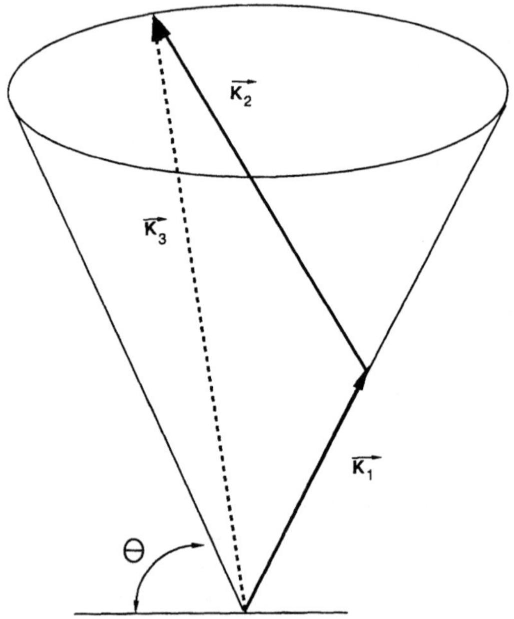


Fig.2. Wave/vertical mode resonant triad.

The nonlinear amplitude equations for this case can be solved analytically, and it is found that the role of the vortical mode is catalytic, i. e., it is needed for the interaction to occur, but it does actively participate in the energy exchange. (This mechanism is reminiscent of the elastic scattering interaction discovered by Phillips, 1968, although the dynamics for the present case are significantly different.) The two wave modes exchange energy harmonically on the slow time scale, the vertical velocity being given by

$$w = \cos \Gamma F t \sin(\vec{\kappa}_1 \cdot \vec{x} - \omega t) - \sin \Gamma F t \cos(\vec{\kappa}_3 \cdot \vec{x} - \omega t) . \qquad (2-3)$$

$2\pi/\Gamma F$ is the interaction period, where $\Gamma$ is given by

$$\Gamma = \frac{B \kappa_1 \kappa_3 \sin^2 \theta \sin \Delta\gamma}{2} \left\{ \cos^2 \theta \cos \Delta\gamma + \sin^2 \theta \right\} . \qquad (2-4)$$

Here $B$ is the amplitude of the vortical mode and $\Delta\gamma$ is the angle between the horizontal components of the wave number vectors of the two waves.

### 2.3. Vortical Mode Interactions

In this final case, Lelong and Riley have considered initial conditions consisting of two vortical modes. It is found that all interactions are resonant because the resonance condition imposed on the frequencies is identically satisfied, the frequency of any vortical mode being zero regardless of its wavenumber. As a consequence, the vortical mode equations are fully-nonlinear on the vortical mode time scale ($L/u'$). It is found that, to lowest order, the vortical mode velocity and pressure satisfy:

$$\frac{\partial}{\partial t} \vec{u}_H + \vec{u}_H \cdot \nabla \vec{u}_H = -\nabla p + R^{-1} \nabla^2 \vec{u}_H \qquad (2-5a)$$

$$\nabla \cdot \vec{u}_H = 0 . \qquad (2-5b)$$

Here $R = u'L/\nu$ is the Reynolds number. Note that these equations describe horizontal motion in each horizontal plane, but with vertical variation retained. This result was suggested by Riley et al. (1981) using heuristic arguments. In addition to the conclusion that the vortical modes satisfy Equation (2-5), it was found that the vortical modes excite internal waves at higher order, with the amplitude of the internal waves scaling as $F$ (or the energy as $F^2$).

## 3. Simulations Results

In this section we present results of direct numerical simulations and comparisons with theory for both wave-vortical mode interactions and vortical mode self-interactions. The simulations employ pseudo-spectral numerical methods with leap-frog time-stepping (smoothed every 25 time steps). The wave-vortical mode simulations were performed on $32 \times 32 \times 32$ point computational grids, while the vortical mode interactions used $64 \times 64 \times 64$ point grids. All calculations were performed on an Ardent Titan Mini-Supercomputer Server.

### 3.1. Wave-Vortical Mode Interactions

As discussed in the previous section, a resonance was found to exist between a vortical mode and two internal waves if the two waves are of the same frequency and the three wave number vectors form a triangle on the surface of a vertically-oriented cone. In order to examine these interactions we have performed direct numerical simulations for a number of different resonance conditions, and for a variety of different Froude numbers. The following case is typical of the results from these simulations, and represents the interaction of an internal wave with a horizontally-shearing current. The latter is a special case of a vortical mode. The initial vertical velocity of the internal wave satisfies

$$w(\vec{x}, 0) = \cos(\vec{\kappa}_1 \cdot \vec{x}),$$

while the vortical mode is initially given by

$$\vec{u}(\vec{x}, 0) = \left(\kappa_{2_2} B \sin(\vec{\kappa}_2 \cdot x), -\kappa_{2_1} B \sin(\vec{\kappa}_2 \cdot x), 0\right).$$

We choose

$$\vec{\kappa}_1 = (2, 0, 2),$$

giving a wave propagating at 45° to the horizontal, and

$$\vec{\kappa}_2 = (-2, 2, 0).$$

Therefore the resonant wave should be found at

$$\vec{\kappa}_3 = (0, 2, 2).$$

$B$ is taken to be 2.0, and the Froude number $F$ to be 0.02. From the scaling analysis, this implies that the wave frequency $\omega$ in the simulations is $2\pi/F$, while the interaction frequency $\Gamma$ is 2.0. The perturbation theory indicates that both the kinetic and the potential energy in the initial internal wave ($\vec{\kappa}_1$) should vary with time as $0.5 \cos^2(2t)$, and that in the forced wave ($\vec{\kappa}_3$) as $0.5 \sin^2(2t)$.

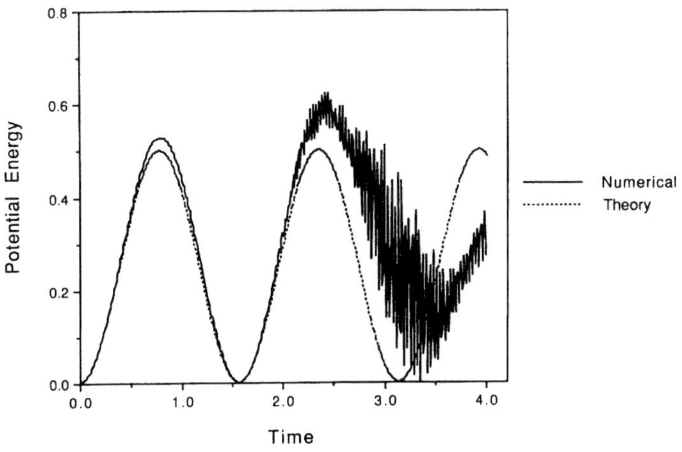


Fig.3. Potential energy in the forced wave $\vec{\kappa}_3$; $F_t = 0.02$.

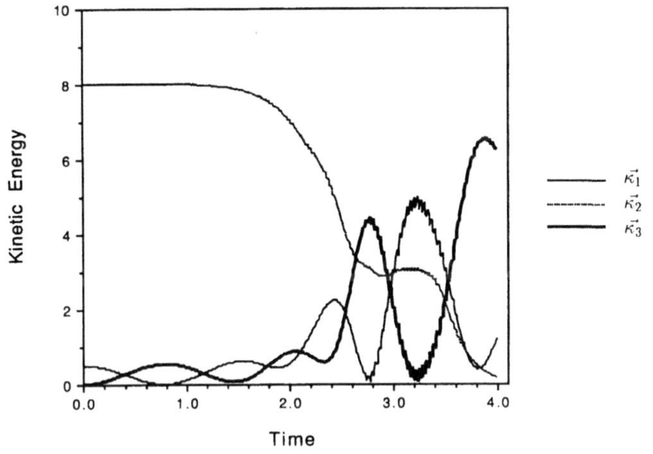


Fig.4. Kinetic energy in all three modes; $F_t = 0.02$.

Figure 3 contains a plot of the potential energy at the forced wave number $\vec{\kappa}_3$ versus time taken from the simulations. Also plotted is the theoretical prediction. We see that the simulations follow the theory fairly closely for a time of about 2 to 2.5, at which point the potential energy in this wave mode begins to oscillate (near the buoyancy frequency), and the results begin to deviate significantly from the theory. More insight into this problem is given by examining the kinetic energy of each mode for this case, as shown in Figure 4. According to the theory, the kinetic energy in the vortical mode should remain uniform in time. Again we see that the simulation results follow the predicted oscillation in the wave kinetic energy up to a time of about 2 to 2.5, at which point the computed solutions rapidly diverge from the theoretical predictions. Furthermore, the kinetic energy in the vortical mode remains approximately constant up to this time, and it then begins to exhibit a large deviation from its initial value.

Visual analysis of the results of this case determined that the horizontally shearing motion was subject to a shear instability which arose dramatically at about a time of 2. Furthermore this instability transferred energy mainly into the vortical mode components at wave numbers $\vec{\kappa}_1$ and $\vec{\kappa}_3$. Prior to this instability, however, the perturbation theory gave accurate predictions of the interactions.

### 3.2. Vortical Mode Interactions

In order to examine nonlinear vortical mode interactions we have considered initial conditions consisting of a sum of spatially-harmonic vortical modes. The initial velocity field is given by

$$\vec{u}(\vec{x},0) = \cos x_3(\cos x_1 \sin x_2, -\sin x_1 \cos x_2, 0), \qquad (3-1)$$

while the initial perturbation density field is identically zero. For the nonstratified case this initial condition defines the Taylor-Green problem (Taylor and Green, 1937), with the velocity field oriented such that it is initially horizontal. This problem has received much attention in the literature (e. g., Orszag, 1971). The perturbation analysis predicts that as $F \to 0$, the flow field should satisfy Equation (2-5) to lowest order. The exact solution to these equations, satisfying the above initial conditions, is

$$\vec{u}(\vec{x},0) = e^{(-3R^{-1}t)} \cos x_3(\cos x_1 \sin x_2, -\sin x_1 \cos x_2, 0). \qquad (3-2)$$

The nonlinear terms in the momentum equation [Equation (2-5a)] are exactly balanced by the pressure gradient, so that the equations become linear. The streamline pattern remains independent of time as the velocity field decays due to viscosity. This solution is an extension of a well-known solution for the two-dimensional Taylor-Green problem (see, e. g., Staquet, 1985).

We performed a series of simulations for a number of different Froude numbers, using the above initial conditions for each case. For all the cases presented the initial Reynolds number $R$ was fixed at 200. Figure 5 gives a plot of the volume integrated kinetic energy of the horizontal velocity (normalized by its initial value) for these different cases. The case with $F = \infty$ corresponds to the Taylor-Green problem, and agrees with previously published simulation results. We see that increasing the stratification decreases the decay rate, as might be expected. Furthermore as $F$ becomes small, roughly $F \leq 1.0$, the computed solutions

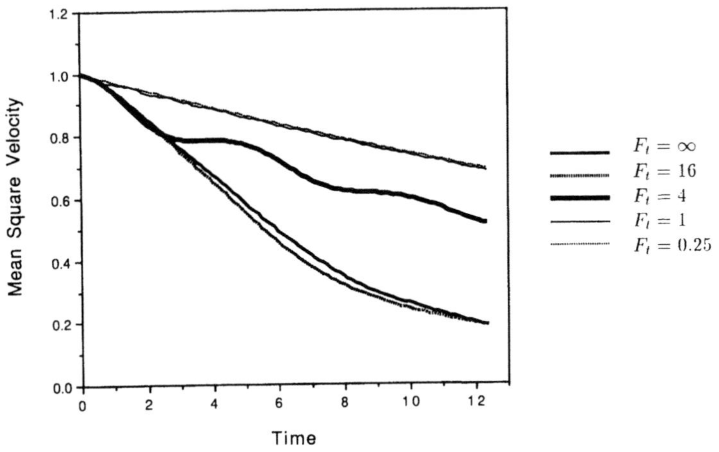


Fig.5. Taylor-Green problem: normalized kinetic energy of horizontal velocity for various Froude numbers.

coincide, as expected from the perturbation theory. Figure 6 gives a plot of the same data on an expanded scale, for the cases with stronger stratification. Also plotted is the result from the theoretical solution, Equation (3-2). We see that the solution to the predicted asymptotic equations agrees very well with the simulation results for $F \leq 1.0$.

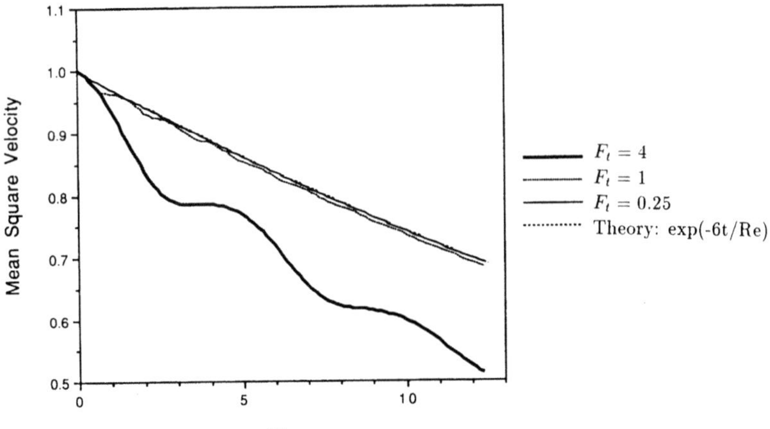


Fig.6. Taylor-Green problem: normalized kinetic energy of horizontal velocity for several Froude numbers – expanded scale.

Figure 7 contains plots of the wave energy for these same cases. The wave energy contains the potential energy plus the wave part of the kinetic energy, as defined by Equation (2-1). Note that the wave energy

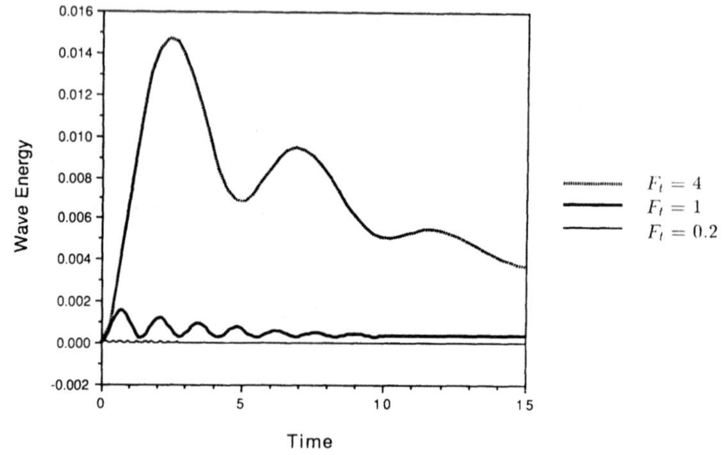


Fig.7. Taylor-Green problem with white noise: wave energy for three different Froude numbers.

decreases significantly as $F$ is decreased. When these results are replotted, scaled with $F^{-2}$, as suggested by the theory, then the wave energy collapses well for $F \leq 4.0$ (Figure 8). Therefore, for $F \leq 1.0$, the perturbation analysis is consistent with the results for both the horizontal kinetic energy, which consists mainly of the vortical mode, and the wave energy.

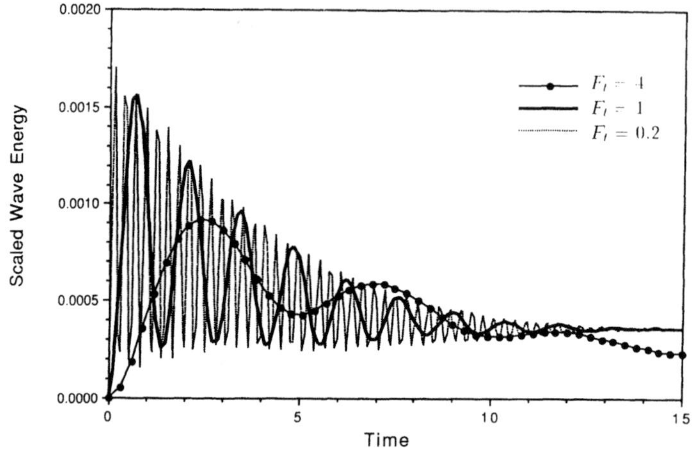


Fig.8. Wave energy scaled by $1/F_t^2$ for three different Froude numbers.

This present case is somewhat degenerate because, as $F$ becomes small, the nonlinear and horizontal pressure gradient terms come into balance, leading to the simple viscous decay given by Equation (3-2). As

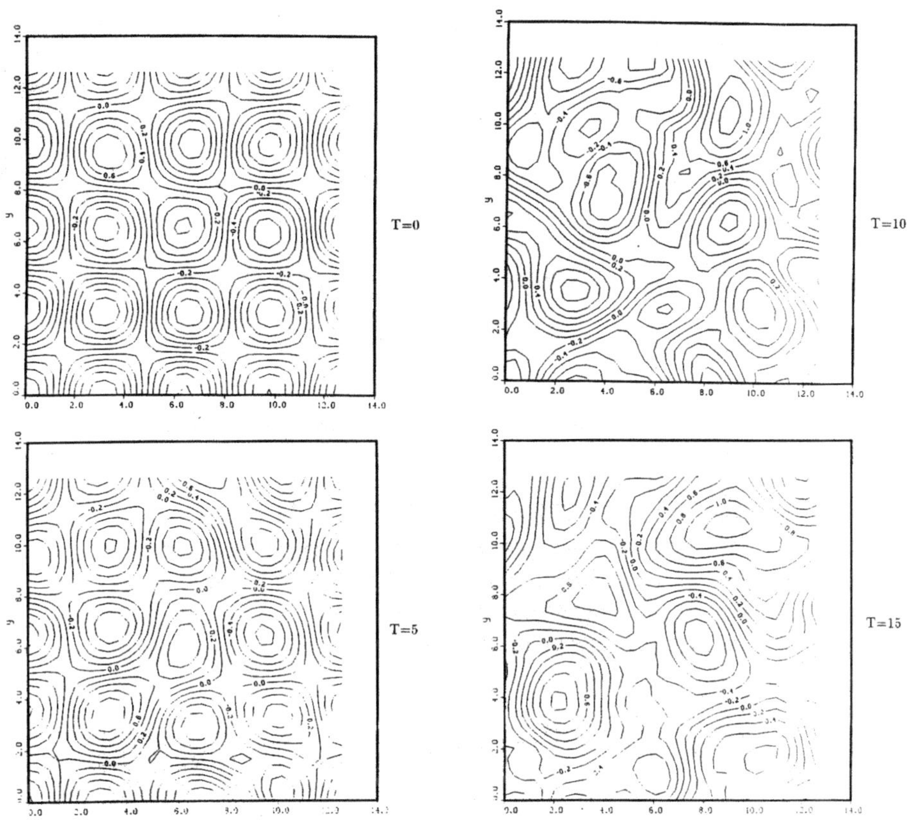


Fig.9. Sequence of constant contours of the stream function in a horizontal plane for $F_t = 0.2$ at $T = 0$, $T = 5$, $T = 10$, and $T = 15$.

mentioned, this is similar to the result for the two-dimensional Taylor-Green problem. It is well-known for the two-dimensional problem, however, that the solution is unstable to subharmonic perturbations. The length scales of the flow continually grow larger as energy is nonlinearly transferred to lower wave numbers. Therefore we also performed a series of simulations for different $F$ using the above initial conditions, Equation (3-1), but with white noise added. Figure 9 gives a sequence of constant contours of $\psi$ in a horizontal plane for the case $F = 1.0$. At this value of stratification these contours approximate streamlines in a horizontal plane. We see that the flow develops nonlinearly in time, as energy is continually transferred to larger scales, reminiscent of the two-dimensional problem. Plate 1 contains plots of $\psi$ in two different horizontal planes at a later time. We see that the two layers have become decoupled, the flow in the two planes being very different. Finally, Figure 10 has plots of the total energy versus time for the different cases computed. Again the results converge for $F \leq 1.0$, consistent with the predictions of the perturbation theory.

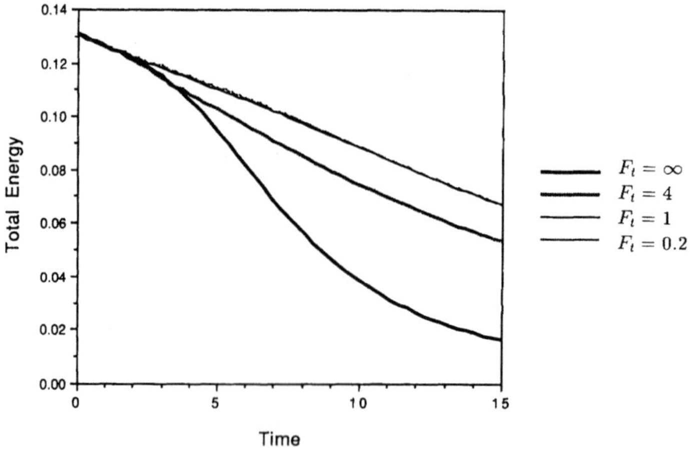


Fig.10. Taylor-Green problem: total energy for several Froude numbers.

## 4. Conclusions and Discussion

We have reported on direct numerical simulations of internal wave and vortical mode interactions in strongly stratified flows. The objectives are to examine the validity of the weak interaction theory of Lelong and Riley (1989), and to further investigate the interactions identified by them. One of the interactions examined is the resonance of two internal waves and a vortical mode. We have found that the perturbation analysis predicts the interaction very well if the Froude number is small enough, approximately $F \leq 1.0$. From the numerical simulations we have also found, however, that the vortical modes considered were highly unstable, and ultimately experienced breakdown as the fluctuations in the flow increased. In all cases considered for this interaction the vortical mode consisted at least partially of horizontally varying currents with multiple inflection points. These flows satisfy both Rayleigh's and Fjortoft's necessary conditions for instability (Drazin and Reid, 1981). Thus, it is not surprising that they were unstable. The instabilities appear to feed energy into the vortical modes at the same wave number as the internal waves. Once the unstable fluctuations grew to an appreciable amplitude, then the results of the simulations deviated strongly from the perturbation theory predictions.

For the case of vortical mode interactions we considered the Taylor-Green problem. An exact solution to the perturbation equations was found for this case, an extension of a well-known two-dimensional result. The results of the numerical simulations agreed well with this solution for approximately $F \leq 1.0$. Furthermore, the scaling of the potential energy predicted by the theory was also consistent with the simulation results when this condition was satisfied. When white noise was added to the initial Taylor-Green field, for small $F$ the flow exhibited subharmonic instabilities similar to those observed for the two-dimensional Taylor-Green case. Again the perturbation equations as well as the predicted scaling of the potential energy were found to hold for approximately $F \leq 1.0$.

In both cases the simulations emphasize the importance of nonlinearity in the dynamics of the vortical modes. Furthermore, especially in the Taylor-Green case, it is clear that upscale transfer of energy to larger horizontal scales occurs, a phenomena suggested by Lilly (1983), and observed in the laboratory (Lin and Pao, 1979) and in other numerical simulations (Riley et al., 1981; Herring and Métais, 1989). It was also clear that the different horizontal layers tend to become uncorrelated, since the horizontal flow dynamics differ in each layer. How the layers remain weakly coupled, whether through viscous effects or shear instabilities, must surely depend strongly on the Reynolds number.

If vortical modes exist on geophysical scales then their spatial characteristics and dynamical properties become of interest. Clearly, if vortical modes somewhat similar to those considered here were prevalent, then the vortical mode field would be rapidly evolving in time and probably continually be subject to shear instabilities. However if vortical modes existed as two-dimensionally (in the horizontal) stable flows, e. g., in stable rotation satisfying Rayleigh's circulation criterion (Lelong, 1989; McWilliams, 1985), then perhaps they would be more persistent dynamically.

## References


Drazin, P. G., and Reid, W. H. . 1981. *Hydrodynamic Stability*. Cambridge University Press.

Gage, K. S. 1979. "Evidence for a $k^{-5/3}$ Law Inertial Range in Mesoscale Two-Dimensional Turbulence", *J. Atmos. Sci.*, **36**, 1950-1954.

Herring, J. R., and Métais, O. 1989. "Numerical Experiments in Forced Stably-Stratified Turbulence", *J. Fluid Mech.*, **202**, 97-115.

Hopfinger, E. J. 1987. "Turbulence in Stratified Fluids: A Review", *J. Geophys. Res.*, **92**(C5), 5287-5304.

Lelong, M.-P. G. 1989. *Weakly-Nonlinear Internal Wave/Vortical Mode Interactions in Stably-Stratified Flows*, Ph. D. Thesis, University of Washington.

Lelong, M.-P. G., and Riley, J. J. 1989. "Internal Wave/Vortical Mode Interactions in a Stably-Stratified Flow", in preparation.

Lilly, D. K. 1983. "Stratified Turbulence and the Mesoscale Variability of the Atmosphere", *J. Atmos. Sci.*, **40**, 749-761.

Lin, J.-T., and Pao, Y.-H. 1979. "Wakes in Stratified Fluids", *Ann. Rev. Fluid Mech.*, **11**, 317-338.

Liu, H.-T. 1980. Private communication.

McWilliams, J. C. 1985. "Submesoscale Coherent Vortices in the Ocean", *Rev. Geophys.*, **23**, 165-182.

Métais, O., and Herring, J. R. 1989. "Numerical Simulations of Freely Evolving Turbulence in Stably-Stratified Fluids", *J. Fluid Mech.*, **202**, 117-148.

Müller, P. 1988. "Vortical Motions", in *Small-Scale Turbulence and Mixing in the Ocean*, ed. by J. C. J. Nihoul and B. M. Jamart, Elsevier, 285-302.

Müller, P., Lien, R.-C., and Williams, R. 1988. "Estimates of Potential Vorticity at Small Scales in the Ocean", *J. Phys. Ocean.*, **18**(3), 401-416.

Orszag, S. A. 1971. "Numerical Simulation of Incompressible Flows Within Simple Boundaries: Accuracy", *J. Fluid Mech.*, **49**(1), 75-112.

Phillips, O. M. 1968. "The Interaction Trapping of Internal Gravity Waves", *J. Fluid Mech.*, **34**, 407-416.

Riley, J. J., Metcalfe, R. W., and Weissman, M. A. 1981. "Direct Numerical Simulations of Homogeneous Turbulence in Stably-Stratified Fluids", in *Proc. AIP Conf. on Nonlinear Properties of Internal Waves*, ed. by B. J. West, 79-112.

Staquet, C. 1985. *Étude Numérique de la Turbulence Bidimensionnelle Homogène et Cisaillée.* Thèse Docteur de l'Institut National Polytechnique de Grenoble.

Taylor, G. I., and Green, A. E. 1937. "Mechanisms of the Production of Small Eddies From Large Ones", *Proc. Roy. Soc.*, **A 158**, 499-521.

van Heijst, G. J. F., and Flór, J. B. 1989. "Dipole Formation and Collisions in a Stratified Fluid", *Nature*, **340**(6230), 212-215.

# Fossil and Active Turbulence

Robert C. SANDERSON[α], Andy D. LEONARD[β], Jackson R. HERRING[γ], and James C. HILL[δ]

[α] *APTEK, Inc.*
*Colorado Springs, Colo., U.S.A. 80906-3578*

[β] *CFD Research Corporation*
*Huntsville, Ala., U.S.A. 35805*

[γ] *National Center for Atmospheric Research*
*Boulder, Colo., U.S.A. 80307-3000*

[δ] *Iowa State University*
*Ames, Iowa, U.S.A. 50011-2230*

*Turbulence and Coherent Structures*
*O. Métais and M. Lesieur, eds.*
*Selected papers from "Turbulence 89: organized structures and turbulence in fluid mechanics"*
*Grenoble, 18-21 September 1989*

## 1. Introduction

Stably stratified turbulence is commonly encountered in geophysical flows and can dramatically alter the evolutionary characteristics of the turbulence by introducing oscillatory behavior in both the scalar and velocity transport fields. Although only velocities in the vertical direction are subject to buoyancy forces, the horizontal components are also affected through nonlinear interaction with the vertical velocity components. If the length scales of the velocity field are small relative to those at which the buoyancy forces become strong, the turbulence is classified as "active" by Gibson [1986]. In buoyancy-dominated turbulence, some or all motions have inertial forces less than or equal to buoyant forces. There is also reduced turbulent scalar transport and dissipation compared with active turbulence of similar total energy. These ideas have been used by Gibson [1986] in his physical modeling of this problem to derive criteria for discriminating stably stratified turbulent flow which is buoyancy-dominated from inertially-dominated (active) turbulence. His results have received support from the experimental studies of Stillinger et al. [1983] and Itsweire et al. [1986].

Fossil turbulence has been defined in several ways, as discussed by Gibson [1986]. One definition is that it is a passive scalar field containing spatial deviations from the mean which remain after the turbulent motions of the fluid have disappeared due to viscous dissipation. The scalar field fluctuations are thus the remains or fossilized remnants created by the defunct turbulent velocity field. This situation is most easily envisioned in high Prandtl or Schmidt number fluids (e.g. sea water) in which scalar fluctuations diffuse very slowly. This definition is subject to some objections, as turbulence is generally considered to be a dynamic process, not a static scalar field. Secondly, it is somewhat restrictive in that velocity fields which occur in stably stratified fluids may become dominated by buoyancy forces to the point that they no longer possess the mixing and eddy-dissipative properties normally associated with nonstratified flows having the same energy content. In this sense they may be considered to be the dynamic remnants or fossils of the original turbulence (Gibson [1986]). This paper will use this latter definition in a theoretical and numerical investigation of the characteristics of dynamic turbulence which is strongly buoyancy-dominated.

## 2. Problem Statement and Linear Theory

The problem to be addressed is incompressible homogeneous axisymmetric turbulence with a uniform mean scalar gradient in the vertical direction, leading to stable stratification. The buoyancy forces are assumed to obey the linear Boussinesq approximation. The resulting equations governing the fluctuating velocity and scalar fields are

$$\left(\frac{d}{dt}+\nu k^2\right)u_i(\mathbf{k},t)-\beta g n_i\vartheta(\mathbf{k},t)=-\frac{\imath}{2}P_{imn}(\mathbf{k})\sum_{\mathbf{p}+\mathbf{q}=\mathbf{k}}u_m(\mathbf{p},t)u_n(\mathbf{q},t) \tag{2-1}$$

$$\left(\frac{d}{dt}+\alpha k^2\right)\vartheta(\mathbf{k},t)+an_iu_i(\mathbf{k},t)=-\imath\mathbf{k}_m\sum_{\mathbf{p}+\mathbf{q}=\mathbf{k}}u_m(\mathbf{p},t)\vartheta(\mathbf{q},t) \tag{2-2}$$

with

$$P_{imn}(\mathbf{k})=k_mP_{in}(\mathbf{k})+k_nP_{im}(\mathbf{k})\ ;\ P_{im}(\mathbf{k})=\delta_{im}-\frac{k_ik_m}{k^2}. \tag{2-3}$$

In this paper we define

$$N\equiv\sqrt{a\beta g} \tag{2-4}$$

as the Brunt-Väisälä frequency, with $a$ the magnitude of the scalar gradient, $\beta$ the volumetric expansion coefficient, and $g$ the gravitational constant. The unit vector $n_i$ is in the direction of the mean scalar gradient $\mathbf{a}$. The mean scalar gradient, represented by the $an_iu_i$ term in equation (2-2), is assumed to be unaffected by the turbulence, thus leading to a constant Brunt-Väisälä frequency. It should be noted here that the Boussinesq buoyancy term in equation (2-1) is an additional (linear) force which is added to the standard Navier-Stokes equations to describe the buoyancy forces, and the $an_iu_i$ term in equation (2-2) comes from the nonlinear scalar convection term but is linear because of the uniformity of the mean scalar gradient. The energy interchange between the kinetic and potential energy fields in this problem is primarily a linear process with the nonlinearities confined to the inertial convection of energy from one wavenumber to another.

Due to the linearity of the forcing terms in equations (2-1) and (2-2), insight into the behavior of the system may be obtained using (linear) rapid distortion theory (RDT). The linearized nondimensional two-point single-time correlation equations for nonzero initial total kinetic energy are

$$\left(\frac{d}{dt}+\frac{2k^2}{R_\lambda}\right)U_{ij}(\mathbf{k},t)-N^{*2}n_i\left[U_{\theta j}(\mathbf{k},t)+U_{j\theta}(-\mathbf{k},t)\right]=0 \tag{2-5}$$

$$\left(\frac{d}{dt}+\frac{(1+Pr)k^2}{PrR_\lambda}\right)U_{i\theta}(\mathbf{k},t)+n_jU_{ij}(\mathbf{k},t)-N^{*2}n_iU_{\theta\theta}(\mathbf{k},t)=0 \tag{2-6}$$

$$\left(\frac{d}{dt} + \frac{2k^2}{PrR_\lambda}\right) U_{\theta\theta}(\mathbf{k},t) + n_i \left[U_{i\theta}(\mathbf{k},t) + U_{\theta i}(-\mathbf{k},t)\right] = 0 \qquad (2\text{–}7)$$

with the equation for $U_{\theta i}(\mathbf{k},t)$ identical to equation (2-6) and where the following definitions are used:

$$U_{\iota\kappa}(\mathbf{k},t) \equiv \langle \hat{u}_\iota(\mathbf{k},t)\hat{u}_\kappa(-\mathbf{k},t)\rangle; \; \iota,\kappa \;=\; 1,2,3,\theta \qquad (2\text{–}8)$$

with $\hat{u}_\kappa(\mathbf{k},t)$ or $\hat{u}_\iota(\mathbf{k},t)$ given by

$$\hat{u}_\kappa(\mathbf{k},t) = \begin{cases} u_\kappa(\mathbf{k},t)/u_\text{o}, & \text{for } \kappa = 1,2, \text{ or } 3 \\ \vartheta(\mathbf{k},t)/a\lambda_\text{o}, & \text{for } \kappa = \theta. \end{cases} \qquad (2\text{–}9)$$

Equivalently, the potential energy spectrum $P(\mathbf{k},t) \equiv U_{\theta\theta}(\mathbf{k},t)N^{*2}/2$ may be used in place of $U_{\theta\theta}$. In equations (2-5) through (2-7), $R_\lambda \equiv u_\text{o}\lambda_\text{o}/\nu$, $Pr \equiv \nu/\alpha$, and $N^* \equiv N\lambda_\text{o}/u_\text{o}$, with $u_\text{o}$ the initial turbulence intensity and $\lambda_\text{o}$ the initial Taylor velocity microscale. The eigenvalues $\xi_i$ corresponding to equations (2-5) through (2-7) and the equation for $U_{\theta i}(\mathbf{k},t)$ are

$$\xi_1 = \xi_2 = -\frac{(1+Pr)k^2}{PrR_\lambda} \qquad (2\text{–}10)$$

$$\xi_3, \xi_4 = -\left(\frac{1+Pr}{PrR_\lambda}\right)k^2 \mp \sqrt{\left(\frac{Pr-1}{PrR_\lambda}\right)^2 k^4 - 4\left(\frac{N\lambda_\text{o}}{u_\text{o}}\right)^2}. \qquad (2\text{–}11)$$

If $N=0$ the last two eigenvalues reduce to their isotropic form:

$$\xi_3 = -\frac{2}{R_\lambda}k^2 \text{ and } \xi_4 = -\frac{2}{PrR_\lambda}k^2. \qquad (2\text{–}12)$$

The eigenvalues $\xi_1$ and $\xi_2$ are associated with the scalar transport field $\langle \mathbf{u}\theta\rangle$, $\xi_3$ corresponds with the velocity field and $\xi_4$ with the scalar or potential energy field. The eigenvalues $\xi_1$ and $\xi_2$ are always real and thus indicate purely dissipative behavior of the $\langle \mathbf{u}\theta\rangle$ field, while the nature of $\xi_3$ and $\xi_4$ depends upon the sign of the discriminant in equation (2-11). A non-negative discriminant indicates behavior which is dissipative but modified from the isotropic form. Negativity of the discriminant introduces oscillatory behavior into the problem.

The negativity of the discriminant is wavenumber dependent. As the wavenumber $k$ increases from zero, the discriminant approaches zero, which occurs at a wavenumber value $k_\text{o}$ of

$$k_\text{o} = \sqrt{\frac{2N^* PrR_\lambda}{|1-Pr|}}. \qquad (2\text{–}13)$$

For wavenumbers greater than $k_o$ the discriminant will be positive and the integrating factors are strictly dissipative (although changed from their passive scalar values), while for wavenumbers smaller than $k_o$ the integrating factors will consist of a dissipative real part and an oscillatory imaginary part. The dissipative part (the first term in equation (2-11)) is the arithmetic mean of the eigenvalues for the passive scalar case. The imaginary part results in oscillatory terms with frequency dependent upon wavenumber.

Figure 1 compares the wavenumber dependence of the magnitude of the imaginary part of the eigenvalues (equation (2-11)), or the radial frequency of the resulting sinusoidal (and cosinusoidal) terms, with a typical initial kinetic energy spectrum. The hydrodynamic conditions are similar to the initial conditions of Métais and Herring [1989] but with $Pr = 0.7$ and 200. Because the oscillatory frequency is close to the $k=0$ value over the energy-containing range of the spectrum, it appears that any activeness of the turbulence in the sense of Gibson [1986] would be due to the initial conditions and tend to disappear in time. Itsweire et al. [1986] reported this to be the case for their experiments.

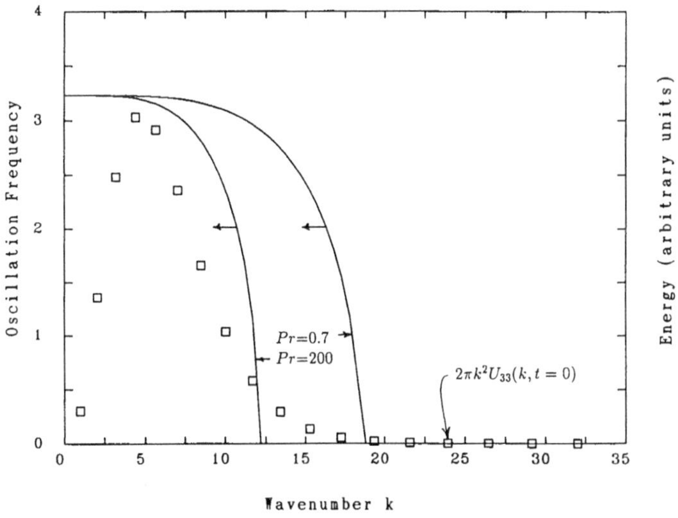


Fig.1. Wavenumber dependence of the oscillatory part of the linear eigenvalues.

The linear integrating factors are formed by taking the exponential of the eigenvalues multiplied by the time at which the solution is desired. The part of the integrating factors corresponding to the real parts of the eigenvalues we designate the dissipative part, while that due to the imaginary portions we call the oscillatory part. Figure 2 shows the equality of the scalar and velocity dissipation integrating factors for wavenumbers up to $k_o$, after which they diverge and approach their passive scalar values with increasing $k$. The consequences of this behavior (for $Pr < 1$) are that energy dissipation in the velocity field is, for given spectral levels, increased from passive scalar levels in the lower wavenumber range, while that for the scalar or potential energy field is decreased. For Prandtl numbers greater than unity this behavior is reversed, as is evident in the figure.

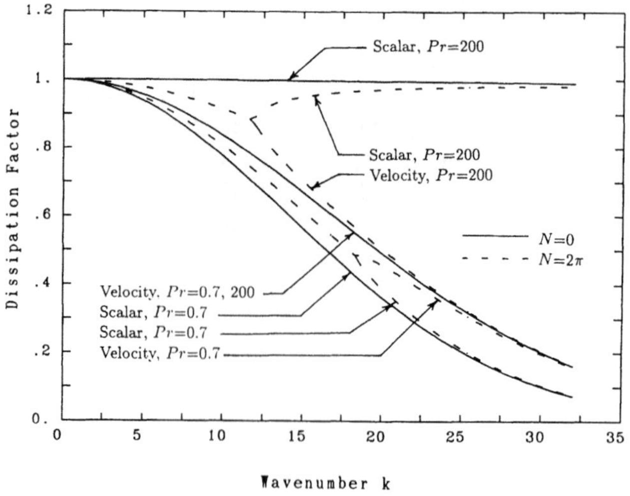


Fig.2. Dependence of dissipative part of linear integrating factors upon wavenumber for $Pr=0.7$, $N=2\pi$.

Understanding the interaction of these linear effects with nonlinear advection requires the solution of the full nonlinear equations. Sanderson et al. [1986] used the direct interaction approximation (DIA) of Kraichnan [1959, 1964b] to study stably stratified turbulence and make qualitative comparisons with the work of Stillinger et al. [1983]. The DIA was also compared by Sanderson and Hill [1988] with the direct numerical simulations (DNS) of Métais and Herring [1989] and the experiments of Itsweire et al. [1986]. Based upon the success of this previous work the present paper compares the DIA predictions with those of DNS for the cases of active and strongly buoyancy-dominated ("fossil") turbulence.

## 3. Numerical Methodology

The full nonlinear governing equations were solved numerically using direct numerical simulation and the direct interaction approximation. The code used for the DNS is a modified version of the pseudo-spectral code developed by R.M. Kerr which employs a third order Runge-Kutta time integration scheme. DNS was used only for computations at Prandtl numbers near unity, while the DIA was used for these conditions as well as at high $Pr$.

The DIA was derived following Kraichnan [1964b] for thermally driven turbulence. The tensor correlations were reexpressed in terms of their eigenfunctions. Variation with angle $\theta$ between the wavenumber vector $\mathbf{k}$ and the direction $\mathbf{n}$ of the mean scalar gradient was represented by a Legendre series expansion, following Craya [1958] and Herring [1974]. For the velocity-velocity correlation, e.g., we have

$$U_{ij}(\mathbf{k},t,t') = \sum_{\lambda=1}^{2}\sum_{l=0}^{\infty} \varphi_l^{\lambda}(k,t,t')P_l(\mu_k)e_i^{\lambda}(\mathbf{k})e_j^{\lambda}(\mathbf{k}), \tag{3–1}$$

where $i,\ j\ =\ 1,2,$ or 3 and

$$\mathbf{e}^1(\mathbf{k}) \equiv \frac{\mathbf{k}\times\mathbf{n}}{|\mathbf{k}\times\mathbf{n}|}\ ;\ \mathbf{e}^2(\mathbf{k}) \equiv \frac{\mathbf{k}\times\mathbf{e}^1(\mathbf{k})}{|\mathbf{k}\times\mathbf{e}^1(\mathbf{k})|}\ ;\ \mu_k \equiv \frac{\mathbf{n}\cdot\mathbf{k}}{|\,\mathbf{k}\,|} \tag{3–2}$$

and $P_l(\mu_k)$ is a Legendre polynomial. The first eigenfunction $\varphi^1(k,t,t')$ contains only horizontal velocity components and is thus denoted the vortical component, while the second eigenfunction $\varphi^2(k,t,t')$ contains vertical and horizontal velocity components and is denoted the wave component of the velocity field (see Métais and Herring [1989]). The velocity-velocity and scalar-scalar correlations are tensors of even order and so were represented by Legendre polynomials of even order (i.e. $P_l(\mu_k)$ with $l=2m,\ m=0,1,2,...$), and the velocity-scalar and scalar-velocity correlations (vectors) by odd order Legendre functions of the first kind ($P_l^1(\mu_k)$, with $l=2m+1,\ m=0,1,2,...$). The complete listing of the DIA equations will appear in a future paper.

The geometric coefficients were evaluated using the method of residues (Schumann and Herring [1976]) while the wavenumber convolution integrals were evaluated using two dimensional tensor B-splines. The time integration of the differential equations was effected by the use of the exact integrating factors (determined by use of the eigenvalues from equations (2-10) and (2-11)) for the linear portion of the equations, analogous to the method outlined in Kraichnan [1964a].

Both numerical methods employed wavenumbers between 1 and 32. The simulations were performed with a box size of $64^3$ while the DIA computations used 18 wavenumbers spaced logarithmically. In keeping with the exploratory nature of this study—as well as to remain within the memory limits of the available computational equipment—the Herring-Craya representations for the correlation and response functions were limited to one Legendre coefficient each.

| Name | $E_{\text{horiz.}}$ or $\varphi_{HH}$ | $E_{\text{vert.}}$ or $\varphi_{VV}$ | $E_{\text{potential}}$ |
|---|---|---|---|
| 3-D | Nonzero | Nonzero | Zero |
| 2-D | Nonzero | Zero | Nonzero |
| 0-D | Zero | Zero | Nonzero |

Table 1: Initial Energy Distributions.

The initial energy distributions used for the DIA/DNS comparisons are summarized in Table 1 with $\varphi_{HH} \equiv \int_0^\infty 2\pi k^2 \varphi_0^{\lambda=1}(k,t{=}0)dk$, $\varphi_{VV} \equiv \int_0^\infty 2\pi k^2 \varphi_0^{\lambda=2}(k,t{=}0)dk$, $E_{\text{horiz.}} \equiv \int_0^\infty \int_{-1}^1 \pi k^2 [U_{11}(\mathbf{k},t{=}0) + U_{22}(\mathbf{k},t{=}0)]d\mu_k dk$, $E_{\text{vert.}} \equiv \int_0^\infty \int_{-1}^1 \pi k^2 U_{33}(\mathbf{k},t{=}0)d\mu_k dk$, and with $E_{\text{potential}} \equiv \int_0^\infty \int_{-1}^1 2\pi k^2 P(\mathbf{k},t{=}0)d\mu_k dk$. The 3-D case may be active or fossil (buoyancy-dominated) depending upon the hydrodynamic conditions, while the 2-D and 0-D cases are used as idealizations of possible fossil turbulence situations. Although the value of $\varphi_{VV}$ depends, in general, upon both $E_{\text{horiz.}}$ and $E_{\text{vert.}}$, the lowest order truncation used in this study requires $\varphi_{VV}$ to be identically zero when $E_{\text{vert.}}$ is zero, i.e. for the 2-D and 0-D initial conditions, thus enabling a direct comparison between the DIA and DNS results for these cases. All nonzero initial energy spectral shapes were of the form $ak^3e^{-bk}$ with $a$ and $b$ constant and peak wavenumber $k = 4.76$, except for the value 3.60 used with the scalar field in the 2-D case. A viscosity value of $\nu = 0.007$ was employed. All results are for Brunt-Väisälä frequencies $N$ of $2\pi$ and a Prandtl number of 0.7 unless otherwise noted.

## 4. Results and Discussion

The results of the (truncated) DIA theory were assessed by comparing its performance in the three cases of Table 1 with those of the DNS. The evolution of total energy (kinetic plus potential) for each of the cases outlined in Table 1 is plotted in figure 3 as calculated by the DNS and the DIA computations.

Both techniques yield similar results, although the DIA results are not quite as sensitive to initial conditions as are the DNS results. Both predict decreased rates of total energy dissipation as the turbulence field is reduced in initial importance, i.e. as first the vertical component and then all components are initialized with zero kinetic energy. This occurs despite the fact that, with $Pr = 0.7$, the scalar diffusivity is higher than the kinematic viscosity. This attests to the importance, even under strongly buoyancy-dominated conditions, of the nonlinear energy cascade in determining total energy dissipation.

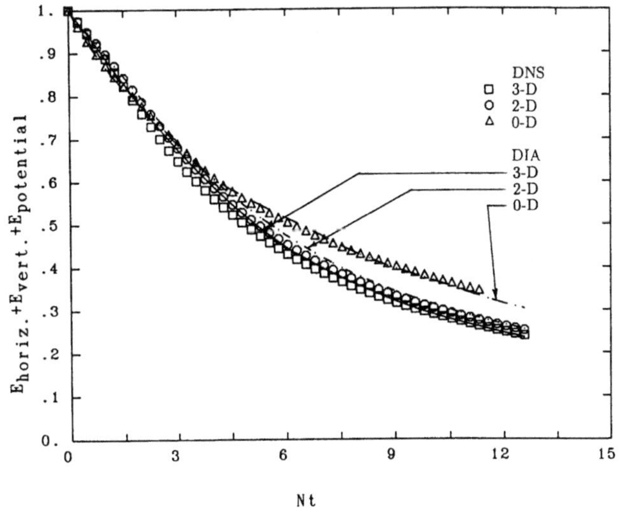


Fig.3. Total energy histories for DIA and DNS computations.

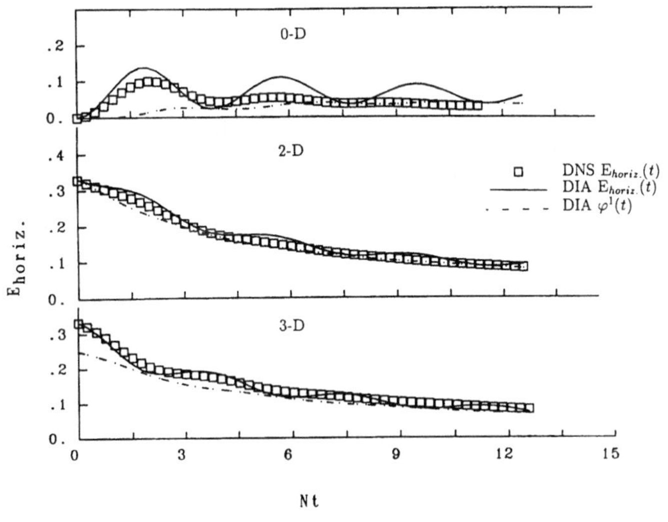


Fig.4. Half-horizontal DNS and DIA energy histories for the cases of Table 1. For comparison the vortical DIA energy $\varphi_{\mathbf{HH}}$ is indicated by the dashed lines.

Figure 4 compares the predicted evolutions of the half-horizontal energy fields for the DNS and DIA. The 3-D and 2-D cases display reasonably good agreement between the two methods, but significant disagreement occurs in the 0-D case. Although the horizontal velocity field is isotropic in all cases, the 0-D represents an especially severe test of the DIA treatment of the nonlinear terms, since the growth of the horizontal kinetic energy is due solely to nonlinearities. Thus, while some errors in the DIA may not be significant in the 3-D and 2-D cases, they will be quite obvious in the 0-D situation. It is apparent that the truncated Legendre series expansion performs most poorly in the 0-D case. The question of to what degree the inclusion of higher order Legendre functions can improve this performance cannot be answered at this time.

Figures 5 and 6 compare the results of the two techniques for the potential and vertical kinetic energy histories of the 0-D problem. They are representative of all the cases tested. Again the DIA displays shortcomings in its treatment of nonlinearities, as it tends to oscillate more strongly than the DNS, behaving like the linear problem in that the energies become very nearly zero at their minima. The difference in phase is unexplained at this time but may also be due to the nonlinearities.

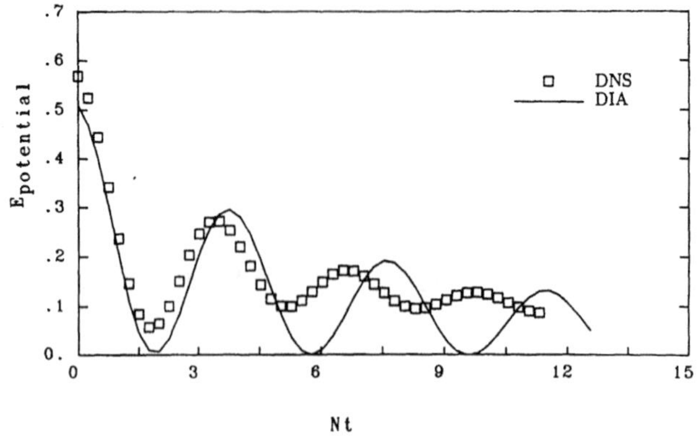


Fig.5. Potential energy histories for the 0-D case.

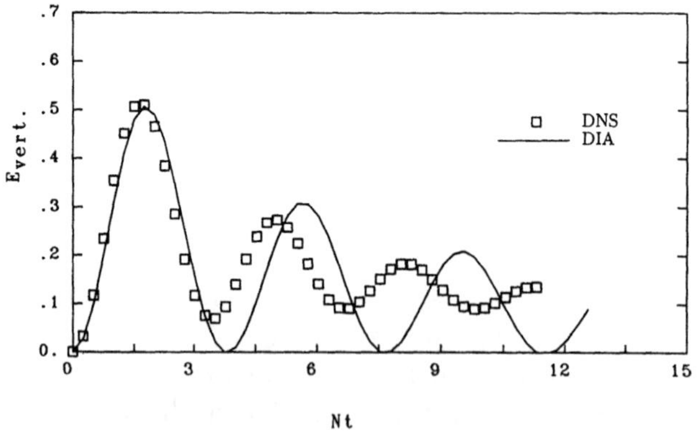


Fig.6. Vertical kinetic energy histories for the 0-D case.

Figures 7 and 8 compare the evolution of the scalar transport eddy diffusivity ratio ($\varepsilon_h/\alpha \equiv -\langle u_3\vartheta\rangle/\alpha a$) and the scalar transport correlation coefficient ($\rho_{u\theta} \equiv \langle u_3\vartheta\rangle/u_3'\vartheta'$), respectively, as computed by the DNS for the three cases of Table 1. The large $\rho_{u\theta}$ magnitudes observed in air flow experiments by Lienhard and Van Atta [1990] and numerically by Métais and Herring [1989] and by Sanderson and Hill [1988] are evident in figure 8. Although oscillations of $\rho_{u\theta}$ for the 3-D and 2-D cases are opposite in phase, the differences in amplitude between the two cases are relatively small, apparently due to the presence in both cases of significant amounts of horizontal kinetic energy contributing to the nonlinearities of the problem. The 0-D case displays somewhat stronger oscillations, perhaps due to reduced nonlinearities caused by the low levels of the horizontal velocity field compared with the other cases. The DIA results for these quantities behave in a similar manner but display the same shortcomings as discussed earlier with respect to the half-horizontal kinetic energy field in the 0-D problem.

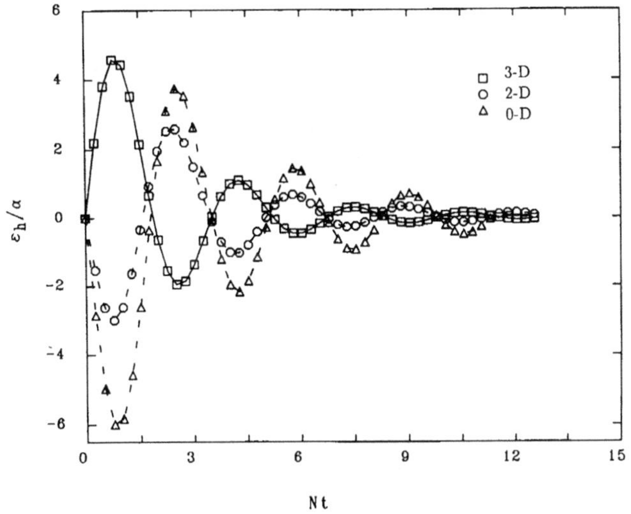


Fig.7. DNS eddy diffusivity ratio histories for the cases of Table 1.

The DIA also shows some interesting characteristics of the anisotropy of the turbulence. Figure 9 shows the evolution of the ratio of the vertical and horizontal DIA eigenfunctions at two wavenumbers in the 3-D case for conditions similar to those used by Métais and Herring [1989] ($Pr$= 1, $N = 0.98$), in which the initial potential energy field was small but nonzero. The highest wavenumber ($k = 29.145$ or $k^* = k\lambda_o = 12.15$) is well into the dissipation range, while the lower one ($k = 10.029$ or $k^* = 4.18$) is in the upper end of the energy containing range. Although

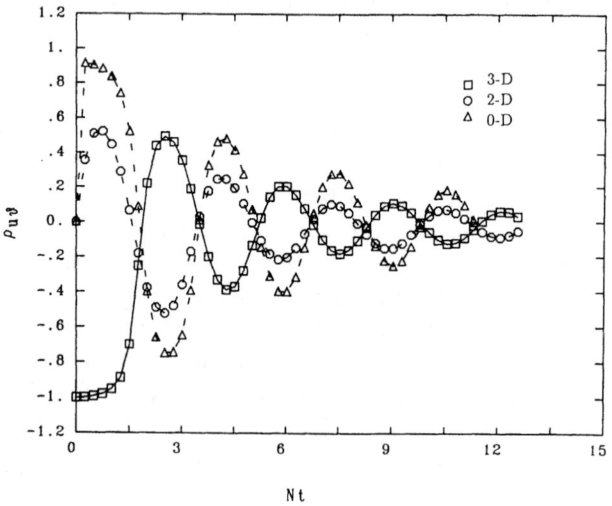


Fig.8. DNS histories of scalar transport correlation coefficients for the cases of Table 1.

these experiments showed a transition to buoyancy domination, neither wavenumber exhibits significant anisotropy. To obtain more history with the DIA (which requires all historical values of the variables due to its two time structure), subsequent studies used the higher value of $N = 2\pi$.

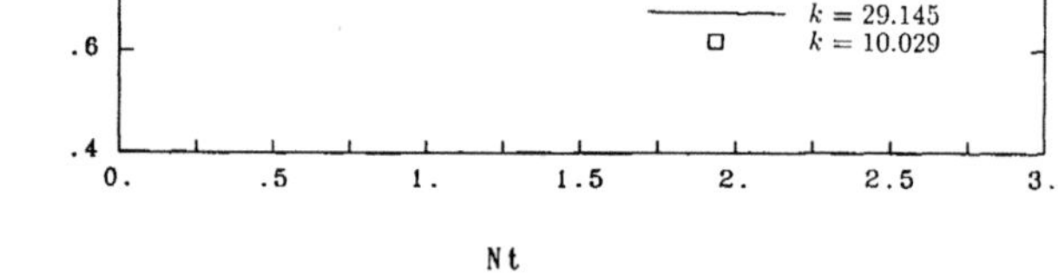


Fig.9. Evolution of the DIA eigenfunction ratio $\varphi_0^{\lambda=2}(k,t,t)/\varphi_0^{\lambda=1}(k,t,t)$, indicated as $\varphi_{VV}/\varphi_{HH}$ on the figure, for the 3-D case ($N = 0.98$).

Figure 10 illustrates the DIA eigenfunction ratio for the same two wavenumbers at the higher $N$ for a 3-D problem. Significantly more anisotropy, indicated by the deviation of the eigenfunction ratio from unity, is evident at both wavenumbers, especially for $k$ in the energy-containing range, due to the stronger stratification. The corresponding behavior for the high wavenumber under 0-D conditions (all initial energy in the potential field) for Prandtl numbers of 0.7 and 200 is shown in figure 11 and the same for the lower wavenumber in figure 12. Both wavenumbers exhibit significantly stronger oscillations than for the 2-D problem. Although the low Prandtl number case ($Pr=0.7$) has oscillatory integrating factors over a wider wavenumber range than does the high Prandtl number case ($Pr=200$), the latter case displays stronger high wavenumber oscillations even though oscillations are absent from the linear problem at this $Pr$.

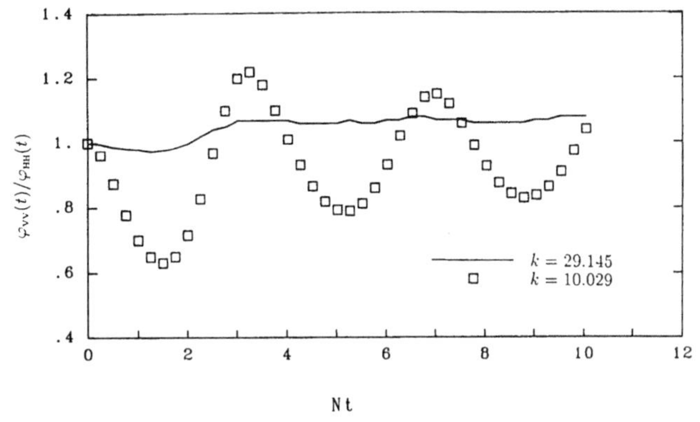


Fig.10. Evolution of the DIA eigenfunction ratio for the 3-D case ($N=2\pi$).

The enhanced anisotropy of the kinetic energy field in the case of 0-D initial conditions is due to the efficient (linear) transfer of energy from the potential field to the vertical velocity field and the relatively slower nonlinear transfer from the vertical to horizontal velocity components. This is illustrated in figure 13 where the vertical energy, predicted by DNS and normalized with respect to total energy, is plotted as a function of time for the three cases of Table 1. The 0-D case experiences oscillations of the largest magnitude. The effectiveness of the linear energy transfer is also reflected in the comparison of the scalar transport correlation coefficients in figure 8. Although the differences are not so pronounced here, the 0-D case exhibits the highest oscillation peaks. With reference to figure 13, it should be pointed out that differences in initial potential energies could account for these variations. However, the differences in initial potential energies are less by a factor of two, while the two vertical energy fields differ in peaks by a factor of approximately three. It appears that the presence of significant horizontal kinetic energy may act to damp the oscillations in the vertical field by nonlinear energy transfer to and from the vertical modes.

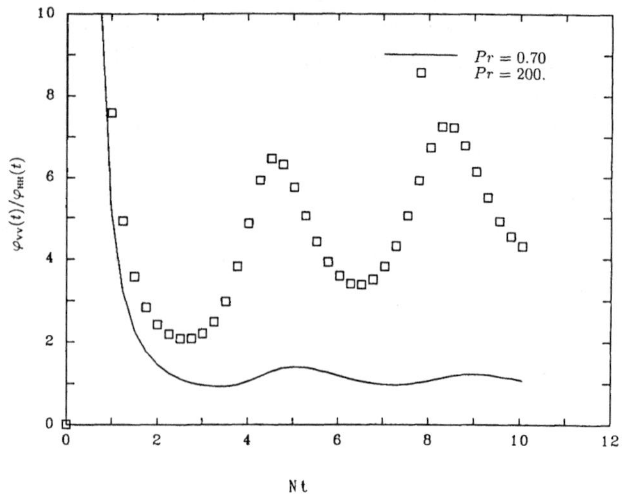


Fig.11. Evolution of the DIA eigenfunction ratio for the 0-D case ($N = 2\pi$, $k = 29$).

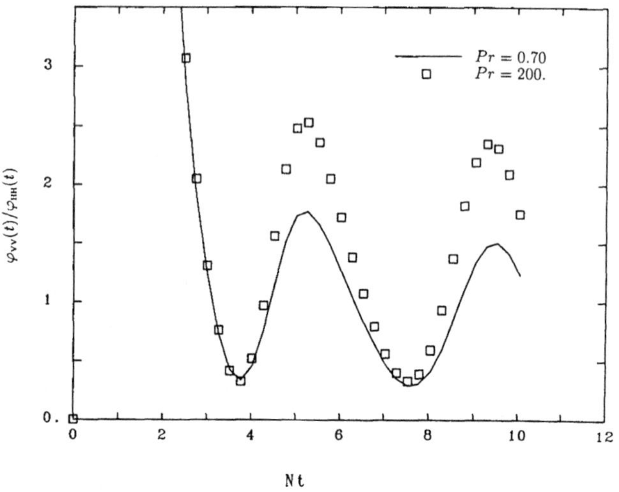


Fig.12. Evolution of the DIA eigenfunction ratio for the 0-D case ($N = 2\pi$, $k = 10$).

Figures 14–16 compare typical DNS three-dimensional spectra of the vertical kinetic, half-horizontal, and potential energy fields for the 0-D, 2-D, and 3-D cases at a Prandtl number of 0.7. The figures show that there is more scalar energy at low $k$ in the 0-D and 2-D cases than in the 3-D case. Also, the energy spectra for the 2-D case resemble those

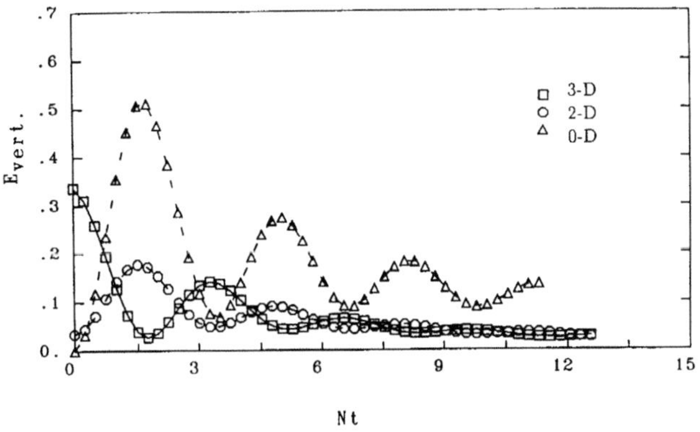


Fig.13. Evolution of the vertical kinetic energy (predicted by DNS) for stably stratified turbulence ($N = 2\pi$).

of the 0-D case at low $k$ and the 3-D case at high $k$, probably because of the smaller nonlinear transfer rate in the 0-D case.

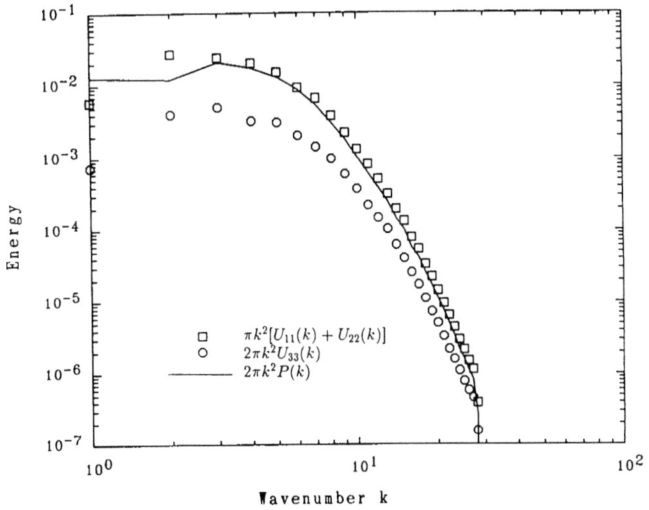


Fig.14. DNS energy spectra for the 0-D case ($N = 2\pi, Pr = 0.7, Nt = 8.55$).

Figures 17 and 18 compare typical spectra from DIA computations for the 0-D case and Prandtl numbers of 0.7 and 200. The vortical eigenfunction ($\varphi_{\mathrm{HH}}$) spectrum displays a pronounced upward shift in peak wavenumber compared to the wave eigenfunction ($\varphi_{\mathrm{VV}}$) and potential energy spectra. The shift is not nearly as pronounced with the half-horizontal and vertical kinetic energy spectra, probably due to the clearer distinction between vortical and wave modes in the eigenfunction

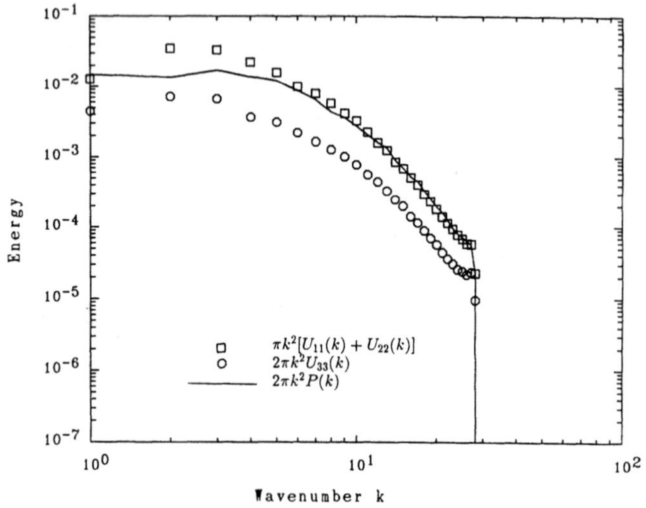


Fig.15. DNS energy spectra for the 2-D case ($N=2\pi$, $Pr=0.7$, $Nt=8.55$).

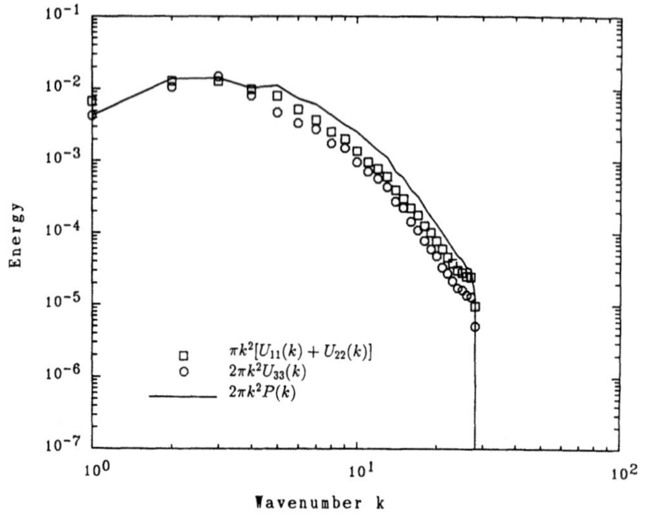


Fig.16. DNS energy spectra for the 3-D case ($N=2\pi$, $Pr=0.7$, $Nt=8.55$).

representation. There is a reversal of relative magnitudes of vertical and potential energies between the DNS and DIA results, possibly because of a phase shift in the energy exchange between these fields, as suggested by figures 5 and 6. Increasing the Prandtl number in figure 18 does not significantly affect the low $k$ spectral shapes or relative peaks. This suggests that inferences from the studies of Stillinger, et al. [1983] and

Itsweire, et al. [1986], concerning the large scales of the velocity and scalar fields, may be applied to Prandtl numbers near unity.

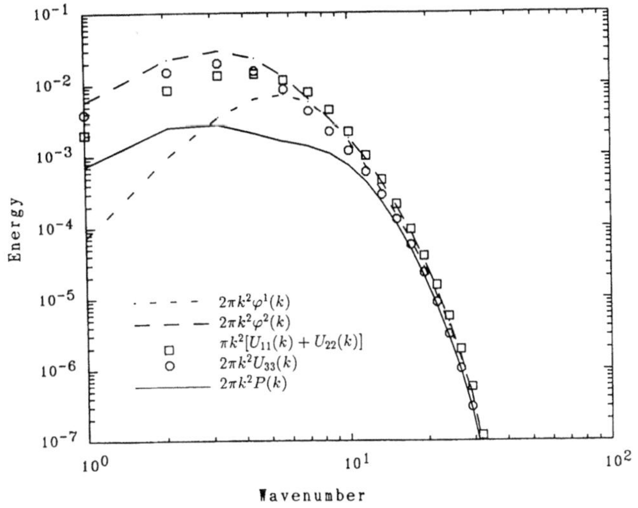


Fig.17. DIA energy spectra for the 0-D case ($N = 2\pi$, $Pr = 0.7$, $Nt = 8.55$).

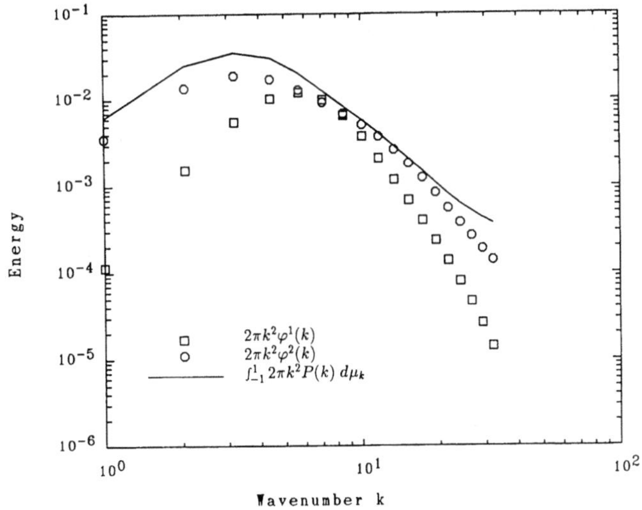


Fig.18. DIA energy spectra for the 0-D case ($N = 2\pi$, $Pr = 200$, $Nt = 8.55$).

Figure 19 compares the spectra for the horizontal, vertical and potential energies of the nonlinear DIA and the linear (RDT) problems for $N = 2\pi$. The linear RDT analysis displays no spectral shift; rather, all

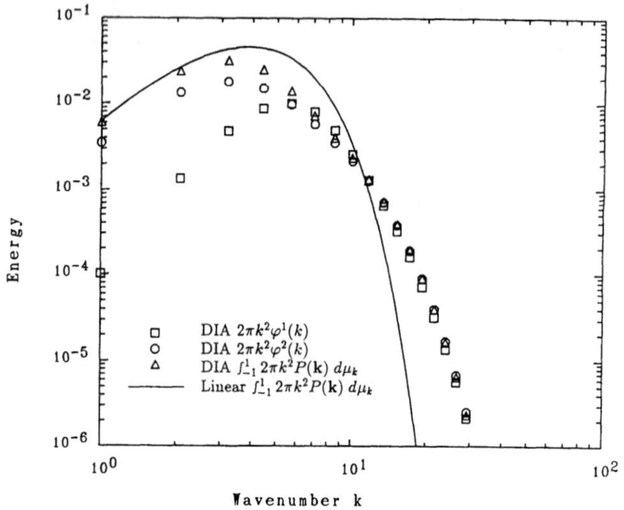


Fig.19. Comparison of nonlinear and linear energy spectra for the 0-D case ($N=2\pi, Pr=0.7, Nt=6.79$).

of the spectra evolve to the same shape. The shift in the DIA spectrum appears to be a result of stronger nonlinear energy transport up the wavenumber spectrum for the horizontal kinetic energy than for the vertical kinetic and potential energies. Examination of the exact kinetic and potential energy equations suggests a reason for this difference. The equations for a vertically oriented gradient ($n_i=\delta_{i3}$) are

$$\left(\frac{d}{dt}+\frac{2k^2}{R_\lambda}\right)U_{11}(\mathbf{k},t) = -\imath P_{1mn}(\mathbf{k}) \sum_{\mathbf{p}+\mathbf{q}=\mathbf{k}} \langle u_m(\mathbf{p},t)u_n(\mathbf{q},t)u_1(-\mathbf{k},t)\rangle \tag{4-1}$$

$$\left(\frac{d}{dt}+\frac{2k^2}{R_\lambda}\right)U_{33}(\mathbf{k},t) - N^{*2}\left[U_{3\theta}(\mathbf{k},t)+U_{\theta 3}(\mathbf{k},t)\right] = -\imath P_{3mn}(\mathbf{k}) \sum_{\mathbf{p}+\mathbf{q}=\mathbf{k}} \langle u_m(\mathbf{p},t)u_n(\mathbf{q},t)u_3(-\mathbf{k},t)\rangle \tag{4-2}$$

$$\left(\frac{d}{dt}+\frac{2k^2}{PrR_\lambda}\right)U_{\theta\theta}(\mathbf{k},t) + \left[U_{\theta 3}(\mathbf{k},t)+U_{3\theta}(\mathbf{k},t)\right] = -\imath k_m \sum_{\mathbf{p}+\mathbf{q}=\mathbf{k}} \langle u_m(\mathbf{p},t)\vartheta(\mathbf{q},t)\vartheta(-\mathbf{k},t)\rangle, \tag{4-3}$$

with the equation for $U_{22}(\mathbf{k},t)$ similar to that for $U_{11}(\mathbf{k},t)$. The nonlinear terms in equations (4-1) through (4-3) are formed from the product of three dynamical quantities. The third variable is associated with the energy variable for which the equation is derived: $u_1(-\mathbf{k},t)$ for the equation governing $U_{11}(\mathbf{k},t)$, $u_3(-\mathbf{k},t)$ for the equation governing $U_{33}(\mathbf{k},t)$, and $\vartheta(-\mathbf{k},t)$ for the equation governing $U_{\theta\theta}(\mathbf{k},t)$. The stratification reduces the magnitude of the vertical velocity components and scalar variables to values which are generally less than those of the horizontal velocity components. Thus the vertical kinetic and potential energy equations have nonlinear terms whose third factors are reduced in magnitude by the stratification, while the third factors of the horizontal kinetic energy nonlinear terms are the relatively undamped horizontal velocity components. The nonlinear terms in the potential energy equation contain one velocity factor and two scalar factors and thus are doubly reduced by the stratification. Consequently, the potential and vertical kinetic energies can be expected to behave in a somewhat more linear fashion than does the horizontal kinetic energy.

## 5. Conclusions

The truncated DIA tested here appears to predict qualitatively correct behavior of buoyancy-dominated or fossil turbulence but is inadequate for producing reliable quantitative results. It is possible that the angular dependence of the eigenfunctions requires more accurate representation by retaining higher order Legendre functions than used here. The numerical studies of fossil turbulence have shown some interesting dependencies upon the relative energy distributions. These include (1) enhanced oscillations of the scalar transport correlation coefficient and the vertical kinetic energy field, (2) a higher degree of anisotropy (especially at high Prandtl numbers) for the cases with zero initial vertical kinetic energy compared to the case of an isotropic initial velocity field, and (3) a stronger spectral shift of energy to low wavenumbers for the scalar and vertical velocity fields than for the horizontal velocity field. This last effect appears to be due to the repression of the nonlinearities of the scalar and vertical velocity fields by the mean density gradient. The high $Pr$ results indicate little effect of $Pr$ on this peak wavenumber shift, implying a small $Pr$ effect on the large scale motions.

## 6. Acknowledgements

The authors are grateful for the support of the Iowa State University Computation Center, the National Center for Supercomputing Applications, the IBM Palo Alto Scientific Center (through the IBM Research Assistance Program), and the National Center for Atmospheric Research which is funded by the National Science Foundation. The authors also appreciate the use of the turbulence simulation code provided by Dr. R.M. Kerr and the assistance of Ms. M. Chakrabarti in carrying out the simulations.

## References


Craya, A. 1958 *Contribution a l'Analyse de la Turbulence a des Vitesse Moyennes*, P.S.T. Ministere de l'Air (Fr.).

Gibson, C.H. 1986 Internal waves, fossil turbulence, and composite ocean microstructure spectra. *J. Fluid Mech.* **168**, 89–117.

Herring, J.R. 1974 Approach of axisymmetric turbulence to isotropy. *Phys. Fluids* **17**, 859–872.

Itsweire, E.C., Helland, K.N., & Van Atta, C.W. 1986 Evolution of grid-generated turbulence in a stably stratified fluid. *J. Fluid Mech.* **162**, 299–338.

Kraichnan, R.H. 1959 The structure of isotropic turbulence at very high Reynolds numbers. *J. Fluid Mech.* **5**, 497–543.

Kraichnan, R.H. 1964a Decay of isotropic turbulence in the direct-interaction approximation. *Phys. Fluids* **7**, 1030–1048.

Kraichnan, R.H. 1964b Direct-interaction approximation for shear and thermally driven turbulence. *Phys. Fluids* **7**, 1048–1062.

Lienhard V, J.H. & Van Atta, C.W. 1990 The decay of turbulence in thermally stratified flow. *J. Fluid Mech.* **210**, 57–112.

Métais, O. & Herring, J.R. 1989 Numerical simulations of freely evolving turbulence in stably stratified fluids. *J. Fluid Mech.* **202**, 117–148.

Sanderson, R.C., Hill, J.C. and Herring, J.R. 1986 Transient behavior of a stably stratified homogeneous turbulent flow. *Advances in Turbulence*, G. Comte-Bellot & J. Mathieu, eds. Springer-Verlag, Berlin, 184–190.

Sanderson, R.C. and Hill, J.C. 1988 Dynamics of decaying stably stratified turbulent flow. *Eighth Symposium on Turbulence and Diffusion* (Preprint Volume), American Meteorological Society, Boston, Mass., 53–56.

Schumann, U. & Herring, J.R. 1976 Axisymmetric homogeneous turbulence: a comparison of direct simulations with the direct-interaction approximation. *J. Fluid Mech.* **76**, 755–782.

Stillinger, D.C., Helland, K.N., & Van Atta, C.W. 1983 Experiments on the transition of homogeneous turbulence to internal waves in a stratified fluid. *J. Fluid Mech.* **131**, 91–122.

# Coherent Structures in Stratified Turbulent Shear Flows Deduced from Direct Simulations

Thomas GERZ

*DLR, Institut für Physik der Atmosphäre*
*D-8031 Oberpfaffenhofen, FRG*

*Turbulence and Coherent Structures*
*O. Métais and M. Lesieur, eds.*
*Selected papers from "Turbulence 89: organized structures and turbulence in fluid mechanics"*
*Grenoble, 18-21 September 1989*

**Abstract**

The character of coherent turbulence structures at the presence of mean shear and stable stratification is investigated using databases of three-dimensional velocity and temperature perturbation fields. The databases have been obtained by two direct numerical simulations of flows with gradient-Richardson numbers $Ri = 0.13$ and 1.32 in a discretized domain with 160 gridpoints in each direction. The horseshoe like vortex structures known as a feature in unstratified shear flows also show up in stratified flows with low (subcritical) values of the Richardson number $Ri = 0.13$. They control the dynamics of the flow like heat and momentum transport. In supercritical flows with $Ri = 1.32$, no three-dimensional vortex structures are found but the vorticity is organized in either medium-sized sheets or narrow but long streaks which both are almost horizontally oriented. The former represent vertically decoupled shear layers; the latter can be associated with internal gravity waves which propagate in the spanwise direction.

## 1. Introduction

Horseshoe and hairpin like structures of the fluctuating vorticity field are known to exist in neutrally stratified heterogeneous as well as in homogeneous shear flows. Head and Bandyopadhyay (1981) showed pictures of inclined vortex structures in a turbulent boundary layer by flow visualization techniques. They found largely elongated ***hairpin*** vortices or vortex pairs at large Reynolds numbers and much less elongated ***horseshoe*** or $\Omega$-shaped vortex loops at low Reynolds numbers. Rogers, Moin and Reynolds (1986) and Rogers and Moin (1987) studied the structure of the vorticity field in homogeneous unstratified turbulent shear flows by direct numerical simulation with a resolution of $128^3$ gridpoints. They observed similar structures as in the heterogeneous boundary layer flow. The inclination angle against the mean flow direction $\theta$ is 45° initially. When the flow is fully developed $\theta$ drops to final values of 35° – 40°. The average hairpin leg spacing is found to decrease with increasing Reynolds number, which follows the picture that the structure changes form a horseshoe shape to a hairpin shape with increasing turbulence intensity. The inclination results from the rate of deformation of the mean flow $\partial U_i/\partial x_j$, which can be decomposed into a symmetric (strain-rate) part $(\partial U_i/\partial x_j + \partial U_j/\partial x_i)/2$ and an anti-symmetric (rotation-rate) part $(\partial U_i/\partial x_j - \partial U_j/\partial x_i)/2$. The angle of the principal axes of the main strain rate is 45°. The vortices are formed by roll-up or deformation of sheets of the mean vorticity $\mathbf{\Omega} = \nabla \times \mathbf{U}$. They are maximum stretched by the mean strain at an angle of 45°. When they become more elongated due to stretching the vortices finally are rotated by the rotational part of the mean flow deformation leading to the reduced inclinations which are observed. At these values of $\theta$ a balance has been reached between mean rotation and production of newly stretched vortex filaments by the mean strain.

Here, the question will be addressed if and which kind of vortex structures exist when the flow is sheared and thermally stably stratified. Gerz *et al.* (1989) performed direct numerical simulations of stably stratified homogeneous turbulent shear flows for a large variety of gradient-Richardson numbers $Ri$. They studied the temporal evolution and the statistical behaviour of the turbulence quantities as a function of $Ri$. Gerz and Schumann (1990) executed similar calculations of stably stratified sheared and unsheared turbulent flows to investigate differences in the flow behaviour at absence and presence of a mean shear. In flows with shear, $Ri$ has been varied in a range from zero to 1.32. Statistical analysis of the data shows a separation into subcritical and supercritical flow regimes depending on the value of $Ri$. *Subcritical* regimes are recognized by increasing turbulent kinetic and available potential energy in time whereas all perturbations are damped out if the flow is *supercritical.* The critical value of the Richardson number was found to be 0.13 approximately in those flows with increasing tendency for higher Reynolds-number flows (Holt *et al.*, 1989). A temporally persistent counter-gradient heat-flux has been found at small scales in the flow due to an imbalance of dynamic and thermal dissipation rates. All these simulations used $64^3$ gridpoints in the disretized domain. In order to resolve all turbulent scales properly they were restricted to moderatly turbulent flows with a Reynolds number based on root-mean-square (rms) velocity $v$ and Taylor-micro length $\lambda$ of $Re_\lambda = 25$. In subcritical flows shear dominates stratification and hence vortex structures are expected to develop. Histograms of the inclination angle of the vortices indeed showed maxima between 35° and 40° for subcritical values of $Ri$. It was further found that these angles became smaller (25° − 35°) when $Ri$ was supercritical. However, the resolution was too low to reveal coherent vortex structures. Only elongated bands of equally oriented spanwise velocity could be distinguished (Gerz, 1988). Recently, Holt *et al.* (1989) performed simulations of stratified and sheared homogeneous turbulent flows with $128^3$ gridpoints. First evaluations showed the existence of horseshoe vortex structures and ring-like vortices in a subcritical ($Ri = 0.08$) and in a supercritical flow ($Ri = 0.5$), respectively.

| | | |
|---|---|---|
| Rms velocity | $\upsilon = (\overline{u_i u_i}/3)^{1/2}$ | 0.02086 |
| Rms temperature | $T' = (\overline{TT})^{1/2}$ | 0. |
| Initial energy-density spectrum | $\hat{E}(k) = 16(2/\pi)^{1/2}(2\pi)^{-1}\upsilon^2 k^4 k_p^{-5}\, e^{-2(k/k_p)^2}$ | |
| Peak wave number | $k_p$ | 4 |
| Shear number | $Sh = (\ell/\upsilon)\,(\Delta U/L)$ | 3.56 |
| Reynolds number (Taylor-micro) | $Re_\lambda = \upsilon\,\lambda/\nu$ | 26.8, 80.4 |
| Reynolds number (box) | $Re = \Delta U\, L\, /\, \nu$ | 28220 |
| Prandtl number | $Pr = \nu\, /\, \gamma$ | 1 |
| Richardson number | $Ri = (\alpha g\, dT_R/dz)\, /\, (dU/dz)^2$ | 0.13, 1.32 |
| Grid points | $M^3$ | $160^3$ |
| Integration time | $St$ | 12 |
| Memory space (in $10^6$ words) | | 55.3 |
| CPU time (in hours) | | 22.75 |

Table1. Initialization parameters, characteristic numbers and used computer resources for direct simulation of stratified and sheared homogeneous turbulence. The angular brackets <> denote the ensemble average over the entire domain.

## 2. Flow parameters

In the present study databases of instantaneous three-dimensional velocity and temperature fields of a subcritical ($Ri = 0.13$) and a strongly supercritical ($Ri = 1.32$) flow will be examined at two time levels. The databases have been obtained by integrating the incompressible Navier-Stokes and temperature equations numerically in a cubic domain and time with a very high resolution of $160^3$ gridpoints and up to time $St = 12$, a time scale normalized by the mean shear rate $S$. The physical properties are basically identical with the respective low-resolution simulations performed by Gerz and Schumann (1990). We consider non-dimensional equations and variables which have been normalized by reference quantities like the differences of velocity $\Delta U$ and temperature $\Delta T$ between bottom and top of the domain, by the box border length $L$ and by the constant density. The resulting characteristic numbers are the non-dimensional shear $S = (L/\Delta U)(dU/dz) = 1$ and stratification $s = (L/\Delta T_R)(dT_R/dz) = 1$, the Reynolds number $Re = \Delta U L/\nu$, the Richardson number $Ri = (\alpha g\; dT_R/dz)/(dU/dz)^2$ and the Prandtl number $Pr = \nu/\gamma$ where $T_R$, $\alpha$, $g$, $\nu$, and $\gamma$ are the reference temperature, the constant isobaric volumetric expansion coeffcient, the gravitational acceleration, the kinematic viscosity, and the thermal conductivity, respectively. A detailed description of the mathematical and numerical approach which is used for the presented studies has been given in Gerz *et al.* (1989). The initial values and characteristic numbers as well as the memory space and the CPU time needed for one run on a CRAY-XMP 2/16 computer are summarized in table 1.

## 3. Energies and fluxes

Figure 1 illustrates the temporal evolution of both flows with $Ri = 0.13$ and 1.32 by means of kinetic energy $E_{kin} =< u_i u_i > /2$, available potential energy $E_{pot} = Ri < TT > /2$, total energy $E_{tot} = E_{kin} + E_{pot}$, and correlation coefficients of the negative vertical momentum flux $- < uw > /(u'w')$, the horizontal heat flux $< uT > /(u'T')$ and the negative vertical heat flux $- < wT > /(w'T')$. The primes denote the rms values, $\phi' = (< \phi\phi >)^{1/2}$. Whereas the kinetic energy grows slightly and the potential energy remains constant in time for $Ri = 0.13$, all fluctuation intensities decay strongly if the flow is supercritical with $Ri = 1.32$. Both flows, however, have reached quasistationarity at $St = 4$ when the correlation coefficients are fairly constant in time (Gerz *et al.*, 1989). These statistics are very close to the corresponding lower resolution statistics of Gerz and Schumann (1990), which guarantees the independence of numerical results from the number of grid points as long as the resolution constraints for a direct simulation are satisfied as pointed out by Reynolds (1989).

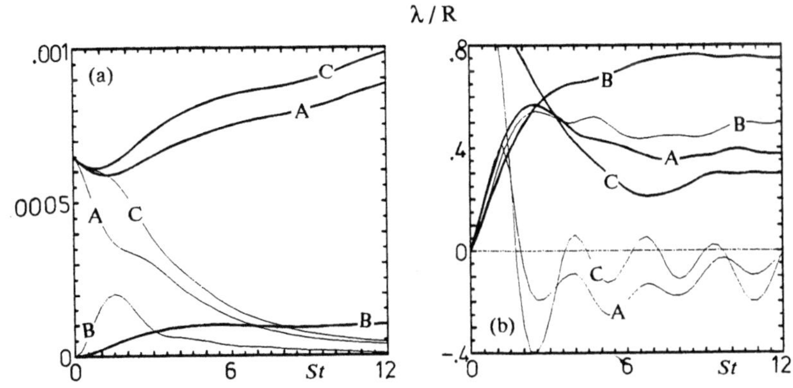


Fig.1. Turbulence mean values versus time for the subcritical ($Ri = 0.13$, thick lines) and the supercritical flow ($Ri = 1.32$, thin lines). (a) Kinetic energy $E_{kin}$ (A), available potential energy $E_{pot}$ (B), total energy $E_{tot}$ (C); (b) the correlation coefficients of the negative vertical momentum flux (A), the downstream heat flux (B) and the negative vertical heat-flux (C).

In the subcritical case strong vertical mixing occurs at all times as indicated by the negative fluxes of momentum and heat (thick lines in Fig.

1b): The flow is fully turbulent. Analysing typical length scales which describe the vertical extension of eddies Gerz and Schumann (1990) found that buoyancy only affects the largest scales for $Ri = 0.13$. For $Ri = 1.32$, vertical mixing is entirely suppressed and energy is transferred between kinetic and potential reservoirs at large scales by the vertical heat flux with a oscillation period of $\pi/N$ where $N = S(Ri)^{1/2}$ is the Brunt-Visll frequency. These oscillations are predicted by rapid-distortion theory (Hunt *et al.*, 1988). The fact, however, that the temporal mean value of both momentum and heat flux are not zero but persistently positive lies in the imbalance of dissipation between velocity and temperature fluctuations at small scales and is a molecular Prandtl-number effect (Gerz *et al.*, 1989; Gerz and Schumann 1990). Buoyancy influences all scales now and, hence, also gravity waves are expected to exist at a broad range of wave numbers.

## 4. Vorticity and enstrophy

| | | $Ri = 0.13$ | | $Ri = 1.32$ | |
|---|---|---|---|---|---|
| | $St = 0$ | $St = 6$ | 12 | $St = 6$ | 12 |
| $\omega'_x$ | 1.00 | 1.11 | 1.16 | 0.60 | 0.24 |
| $\omega'_y$ | 1.01 | 1.07 | 1.20 | 0.57 | 0.25 |
| $\omega'_z$ | 1.01 | 0.93 | 1.00 | 0.28 | 0.11 |
| $e$ | 1.52 | 1.62 | 1.89 | 0.38 | 0.07 |
| $I_1$ | 0.004 | 0.123 | 0.132 | 0.066 | 0.008 |
| $I_2$ | 0.811 | 0.728 | 0.989 | 0.038 | 0.0007 |
| $I_3$ | 0. | -0.051 | -0.058 | -0.021 | -0.003 |
| $I_4$ | -0.875 | -1.062 | -1.260 | -0.235 | -0.027 |

Table2. Vorticity and enstrophy statistics.

In this section we evaluate the vorticity fields at $St = 6$ and 12 statistically. The vorticity has been calculated from the instantaneous velocity fields at these times. In a vertically sheared mean flow $U(z)$ the dimensionless vorticity reduces to a lateral mean component $\Omega_y = S$ and to fluctuations $\omega_i = \epsilon_{ijk}\partial u_k/\partial x_j$ ($i = 1, 2, 3$ or $x, y, z$) where $\epsilon_{ijk}$ represents the unit alternating tensor which is 1 if $i, j, k$ are in cyclic order, -1 if $i, k, j$ are in cyclic order or zero if any two of the indices are equal. In a sheared and stratified flow the equation for $\omega_i$ reads in the Boussinesq system

$$\frac{\partial\omega_i}{\partial t}+\frac{\partial}{\partial x_j}(u_j\omega_i)+Sx_3\frac{\partial\omega_i}{\partial x_1}=S(\omega_3\delta_{i1}+\frac{\partial u_i}{\partial x_2})+\frac{\partial}{\partial x_j}(\omega_j u_i)$$
$$+|Ri|\;\epsilon_{ij3}\frac{\partial T}{\partial x_j}+\frac{1}{Re}\frac{\partial^2\omega_i}{\partial x_j^2},\quad i=1,2,3 \qquad (4-1)$$

where $\delta_{ij}$ is the Kronecker$-\delta$. The first term on the right-hand side of the equation is the production due to mean rotation or mean shear followed by a nonlinear interaction between vorticity and velocity gradients, the so-called twisting term. The third term introduces baroclinicity due to horizontal temperature fluctuations and is non-zero only for the horizontal vorticity components. The last term decribes the molecular diffusion of $\omega_i$. A measure for the intensity of vorticity fluctuations is the turbulent enstrophy of the flow. Its statistical mean (volume averaged) is defined in analogy to the kinetic energy as $e=<\omega_i\omega_i>/2$. Its balance equation stems from multiplying eq. (4-1) with $\omega_i$ and ensemble averaging. It reads

$$\frac{\partial e}{\partial t}=S<\epsilon_{1jk}u_j\frac{\partial\omega_z}{\partial x_k}>+<\omega_i\frac{\partial}{\partial x_j}(\omega_j u_i)>$$
$$+|Ri|\;<\epsilon_{ij3}\omega_i\frac{\partial T}{\partial x_j}>-\frac{1}{Re}<\frac{\partial\omega_i}{\partial x_j}\frac{\partial\omega_i}{\partial x_j}> \qquad (4-2)$$
$$=I_1+I_2+I_3+I_4$$

where again all four types of interaction (here labeled as $I_1$ to $I_4$) exist. In table 2 the values of rms vorticities $\omega'_x$, $\omega'_y$, $\omega'_z$, of enstrophy $e$, and of the interaction terms are listed for both runs with $Ri=0.13$ and 1.32 and times $St=0$, 6 and 12.

Starting from isotropically distributed vorticity at $St=0$ the enstrophy and all its components increase (decrease) in time for $Ri=0.13$ (1.32) according to the subcritical or supercritical character of the flow. Anisotropy shows up above all in the vertical rms vorticity component $\omega'_z$. While it is only little smaller than $\omega'_x$ or $\omega'_y$ for $Ri=0.13$, it has less than half of the magnitude of the horizontal vorticities for $Ri=1.32$. But when the flow is quasistationary it decays at approximately the same rate between $St=6$ and 12. Looking at the tabulated values of the terms in the enstrophy equation (4-2) it appears that diffusion $I_4$ is the most dominant contribution for all Richardson numbers and times reported. Next comes the nonlinear twisting term $I_2$. It is dominant for $Ri=0.13$ whereas it loses influence for $Ri=1.32$. But still it plays a considerable role also in strongly supercritical flows ($St=6$) and is negligible only late at $St=12$. It is remarkable that the baroclinity $I_3$ which acts as a sink for $e$ is the smallest contribution in the enstrophy balance and comes into play only for $Ri = 1.32$ where the other terms (except $I_4$) undergo a strong decay. In stratified mixing shear layers Staquet and Riley (1989) observed a strong decay of $\omega'_x$ and $\omega'_y$ compared to $\omega'_z$ which they interpreted as due to the baroclinicity effect. However, baroclinity does not play this role in the homogeneous stratified flow which we consider here although the Richardson number is large. A more detailed study is necessary to evolve the character of the baroclinic term in the enstrophy balance.

## 5. Inclination angle and two-point correlations of vorticity

Hairpin or horseshoe vortex structures which build up in shear flows are inclined against the mean shear stream by angle $\theta$. Hence, the legs of those vortices consist of downstream and vertical vorticity $\omega_x$ and $\omega_z$ such that $\theta = \arctan(\omega_z/\omega_x)$. Figure 2 depicts the vorticity-weighted distribution of $\theta$ for both stratifications at $St = 6$ and 12. For $Ri = 0.13$, the maxima lie at 25° for $St = 6$ and 23° for $St = 12$ with a maximum-to-minimum ratio of 7.1:1 ($St = 6$) and 5.7:1 ($St = 12$). The magnitude of $\omega' = (<\omega_i\omega_i>)^{1/2}$ is $1.76S$ initially and increases to $1.82S$ ($1.97S$) at $St = 6$ (12). Rogers at al. (1986) found that hairpin vortices can form in a shear flow only when $\omega' = 1.5S$ or larger. Hence, although the flow with $Ri = 0.13$ is stably stratified we expect coherent horseshoe or hairpin like vortex structures since $\omega' > 1.5S$ at all times. In contrast to Rogers and Moin (1987)'s data we observe that the peak in the $\theta$-distribution is about 10° lower and that it is strongly skewed. The former difference can be explained by stratification effects which limit the vertical extension of the largest eddies and also the roll-up of sheets of mean vorticity. The skewed distribution of $\theta$ which forms to a secondary maximum between 80° and 90° at time $St = 12$ indicates that at some positions of the flow the downstream vorticity is much weaker compared to the vertical vorticity. This implies that in a few regions of the flow the motion circulates horizontally. In order to decide whether this is a buoyancy effect and how it may be related to gravity waves which can exist at the largest scales, more informations on the changing of structures with varying Richardson numbers are needed.

For $Ri = 1.32$, the maxima of $\theta$ range from 13° at $St = 6$ to 8° at $St = 12$. The distribution is not skewed and with an almost constant maximum-to-minimum ratio of 14:1 it is much more pronounced than in the former case. Most of the vorticity is organized in almost horizontal planes, $\omega_x >> \omega_z$, i.e. horizontal gradients of vertical velocity and vertical gradients of horizontal velocity are larger than horizontal gradients of horizontal velocity components. This coincides with observations of increasing horizontal Taylor-micro length scales and decreasing vertical

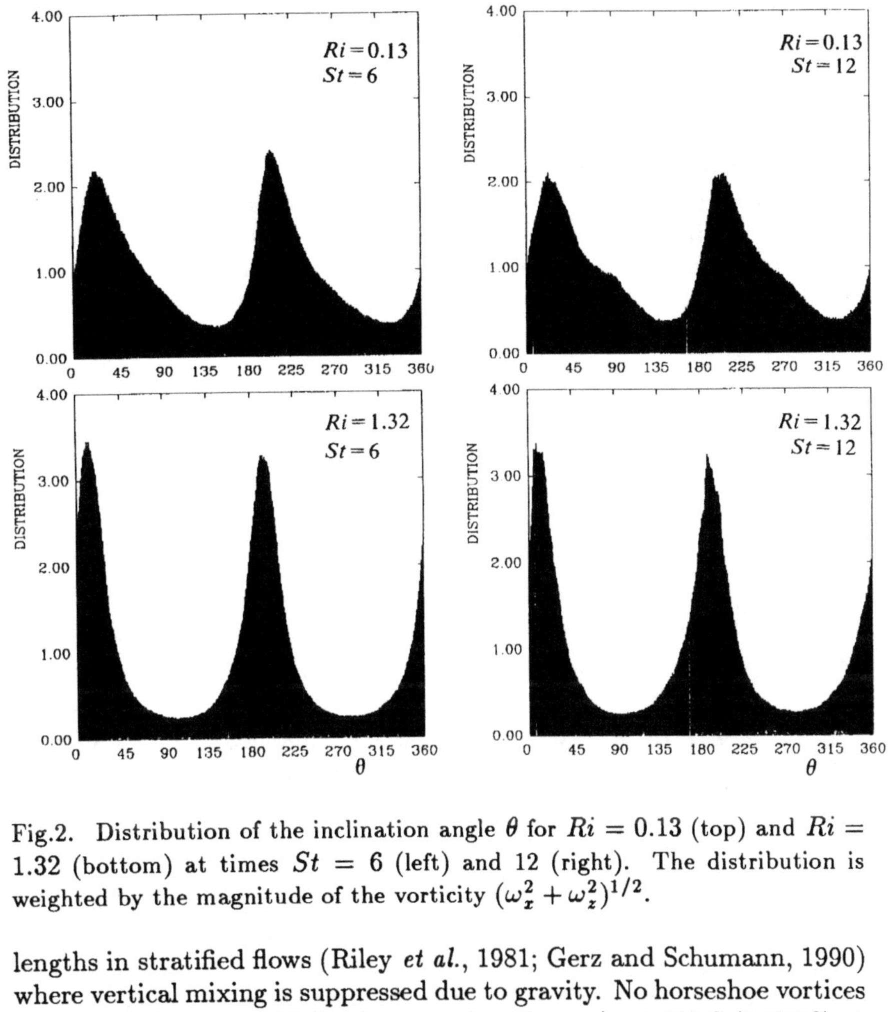


Fig.2. Distribution of the inclination angle $\theta$ for $Ri = 0.13$ (top) and $Ri = 1.32$ (bottom) at times $St = 6$ (left) and 12 (right). The distribution is weighted by the magnitude of the vorticity $(\omega_x^2 + \omega_z^2)^{1/2}$.

lengths in stratified flows (Riley *et al.*, 1981; Gerz and Schumann, 1990) where vertical mixing is suppressed due to gravity. No horseshoe vortices are expected for that Richardson number since $\omega' = 0.872S$ ($0.365S$) at $St = 6$ (12) and hence considerably smaller than $1.5S$.

We now focus on projected vorticity in a plane which is inclined against the $x$-direction of the flow by angle $\theta$. We define a new coordinate system $(\xi, \eta, \zeta)$ which results from the original system $(x, y, z)$ by rotation around the $y = \eta$-axis. The vector $(\omega_x, \omega_y, \omega_z)(x, y, z)$ then transforms into the rotated sytem as

$$(\omega_\xi, \omega_\eta, \omega_\zeta)(\xi, \eta) = (\omega_x cos\theta + \omega_z sin\theta, \omega_y, -\omega_x sin\theta + \omega_z cos\theta) \quad (5-1)$$

where $\omega_\xi$ and $\omega_\eta$ describe the projected portion of $\omega$ in the plane and $\omega_\zeta$ its component perpendicular to the plane. We consider two-point correlation coefficients of the components $\omega_\xi$ and $\omega_\eta$ which lie in the inclined plane. The two-point correlation coefficient for two quantities $a(x_i)$ and $b(x_i)$ at a given time is defined as

$$R_{ab}(r_i) = \frac{< a(x_i)\, b(x_i + r_i) >}{a'b'} \qquad (5-2)$$

where $x_i$ defines the position of each point and $r_i$ is the separation vector which is normalized by the length scale $(\nu/(dU/dz))^{1/2}$ or $(SRe)^{-1/2}$ in the non-dimensionalized system. This allows comparison with data from flows with different viscosity or Reynolds number. The separation varies between zero and half the box size (restricted by periodicity). The brackets <> again average over all three space directions.

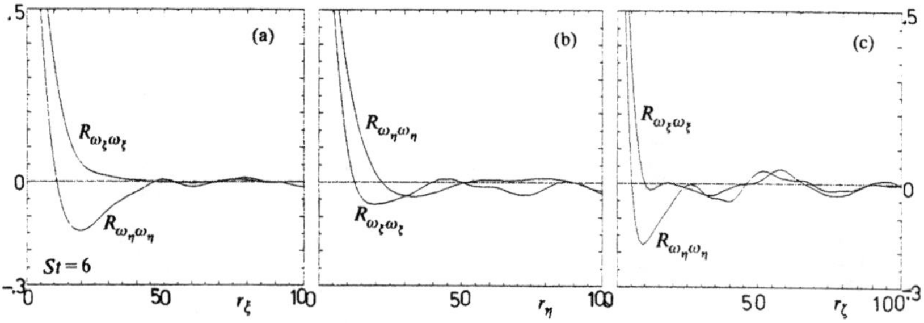


Fig.3. $Ri = 0.13$, $\theta = 36°$ : Two-point correlation coefficients of vorticity $R_{\omega_\xi\omega_\xi}$, $R_{\omega_\eta\omega_\eta}$ as functions of $r_\xi$ (a) $r_\eta$ (b) and $r_\zeta$ (c) at $St = 6$.

Figures 3 and 4 depict plots of $R_{\omega_\xi\omega_\xi}$ and $R_{\omega_\eta\omega_\eta}$ in the directions $r_\xi$, $r_\eta$ and $r_\zeta$ at $St = 6$ for $Ri = 0.13$ with an inclination of $\theta = 36°$ and at $St = 6$ and 12 for $Ri = 1.32$ with $\theta = 10°$. The selected values of $\theta$ are representative for angles with large vorticity-weighted frequency (fig. 2) where therefore we expect most of the structures. Provided that vortices which are inclined at 36° belong to hairpin structures we expect that the corresponding two-point correlation coefficients at distances $r_\xi$, $r_\eta$ and $r_\zeta$ reveal informations on the typical length, width and thickness of the vortex structures, respectively, lying in that plane.

For the subcritical flow with $Ri = 0.13$, we observe that in the $\xi$-direction, i.e. along the assumed structures, the longitudinal correlation $R_{\omega_\xi\omega_\xi}(r_\xi)$ decays slowly and remains positive throughout the plane (fig. 3a) showing that the inclination of the legs of the vortices has about the value of 36°. The typical distance between the centres of both legs of a hairpin vortex can be read from the distance between the maximum and minimum value of lateral correlation $R_{\omega_\xi\omega_\xi}(r_\eta)$ (fig. 3b) which is $r_\eta = 20$ or $r_\eta(SRe)^{-1/2} = 0.12\%L$. $R_{\omega_\xi\omega_\xi}(r_\zeta)$ (fig. 3c) decays very fast to zero indicating that vortex legs are rather thin in the direction perpendicluar

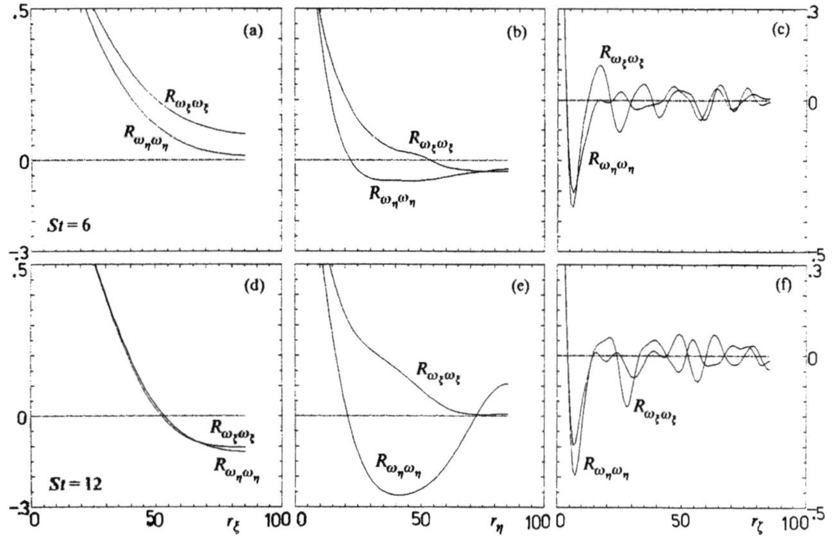


Fig.4. $Ri = 1.32$, $\theta = 10°$ : Two-point correlation coefficients of vorticity $R_{\omega_\xi \omega_\xi}$, $R_{\omega_\eta \omega_\eta}$ as functions of $r_\xi$, $r_\eta$ and $r_\zeta$ at $St = 6$ (a-c) and at $St = 12$ (d-f).

to the inclined plane. Due to the fact that hairpin vortices develop by deformation of the spanwise mean vorticity $\Omega_y$, the tips of those structures must be oriented in the positive spanwise direction $\eta$. From a hairpin or horseshoe like shape of a structure one can conclude that the typical length of a tip (in $\eta$-direction) can not exceed the distance between the legs. Hence, we expect that the longitudinal correlation of the spanwise vorticity $R_{\omega_\eta \omega_\eta}(r_\eta)$ drops quicker than $R_{\omega_\xi \omega_\xi}(r_\xi)$ and approaches zero when $R_{\omega_\xi \omega_\xi}(r_\eta)$ is minimum which is indeed verified in fig. 3b. The distance between the maximum and vanishing values of $R_{\omega_\eta \omega_\eta}(r_\xi)$ (fig. 3a) reveals the typical width of a tip with $r_\xi = 10$. The rather strong negative value of $R_{\omega_\eta \omega_\eta}$ at $r_\xi = 20$ indicates some coherent vorticity in the $-\eta$-direction which, however, cannot belong to horseshoe vortices if our picture of positively oriented vortex tips is right. Also $R_{\omega_\eta \omega_\eta}(r_\zeta)$ (fig. 3c) drops quickly and reaches rather strong negative values which seems to be appropriate in stratified flows. We further observed that all correlations have changed only insignificantly at $St = 12$ (not shown here) which means that also the vortex structures have almost equal size at both times $St = 6$ and 12.

Supercritically stratified shear flows behave similar as unsheared stratified flows (Britter, 1988) where vertical lengths decrease and horizontal length scales increase in time indicating a decoupling of horizontal layers in the vertical direction (Gerz and Schumann, 1990). Plots of the two-point correlation coefficients in fig. 4 corroborate this result: In the almost horizontal plane ($\theta = 10°$) the values of both longitudinal and lateral correlations are much larger than in the previous case and they still increase in time. On the other hand the correlations in the vertical $\zeta$-direction are small without considerable change in time. They drop quickly and change sign alternatingly. Hence we can assume that stable stratification causes horizontally oriented thin sheets of spanwise and downstream vorticity. Their horizontal structure, however, is not uniform in the downstream and spanwise direction. Vorticities $\omega_\xi$ and $\omega_\eta$ have similar correlations in directions $\xi$ and $\zeta$ but they differ largely in the spanwise coordinate. In the last section we will see that this asymmetry is due to spanwise propagating internal gravity waves.

## 6. Coherent structures

Instantaneous fields of various perturbation quantities are depicted for both simulations with $Ri = 0.13$ and 1.32 to illustrate typical flow and structural properties and to emphasize typical structural differences between a subcritically and a supercritically stratified flow. All subsequent figures depict cross-sections through field quantities which have been derived from the instantaneous fields of velocity and temperature perturbations at times $St = 6$ and 12. The cross-sections lie in the $(\xi, \eta)$-plane at the position $(x, z) = (0, 0.49)$.

The velocity field is isotropically initialized by random numbers and, hence, also the vorticity at $St = 0$ is isotropically distributed in space and cannot reveal structural features. This situation drastically changes for $Ri = 0.13$ at times $St = 6$ and 12 (fig. 5). Both, hairpin and horseshoe vortices can be clearly identified. The maximum magnitude of the vorticity in the plane is 7.07%$S$ (9.14%$S$) and the average vorticity of the structures has a magnitude of approximately 2.83%$S$ (3.66%$S$) at $St = 6$ (12). Hence, the vorticity of the structures is about 1.5 (1.9) times larger than the respective values of the rms vorticity $\omega'$. The spacing between the legs of the discernible vortices turns out to be about 10 % to 15 % of the box length $L$. This confirmes very well the statistical mean distance deduced from the lateral two-point correlation $R_{\omega_\xi \omega_\xi}(r_\eta)$ in fig. 3b. Also in subcritically stratified shear flows hairpin vortices are important for the efficient transport of momentum and heat. Fig. 6 shows contour plots of the fluxes of momentum $m_\zeta$ and heat $h_\zeta$ perpendicular through the plane at $St = 12$. It is obvious from th plots that the fluxes are strongest between the legs of the hairpin vortices which can be identified in fig. 5. In a homogeneous incompressible

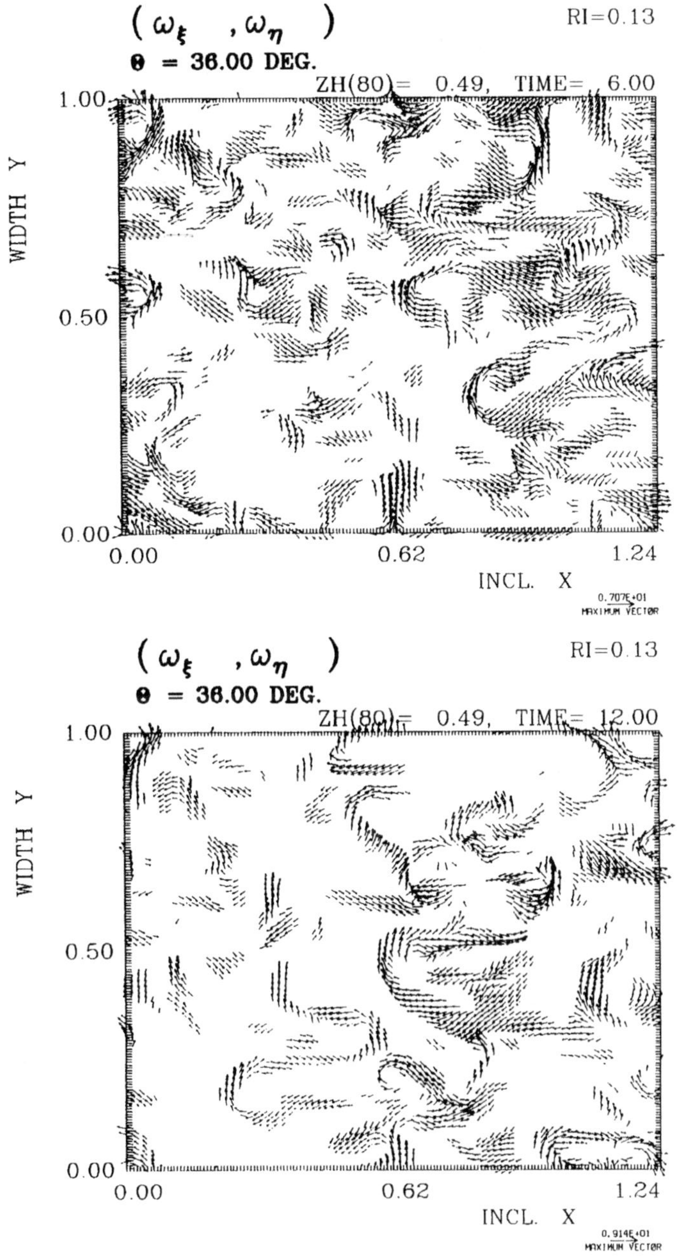


Fig.5. $Ri = 0.13$, $\theta = 36°$. Projection of vorticity vectors at $St = 6$ (top) and 12 (bottom) on the inclined plane. Vectors are shown if their magnitude exceeds a fifth of the magnitude of the maximum vector in the plane.

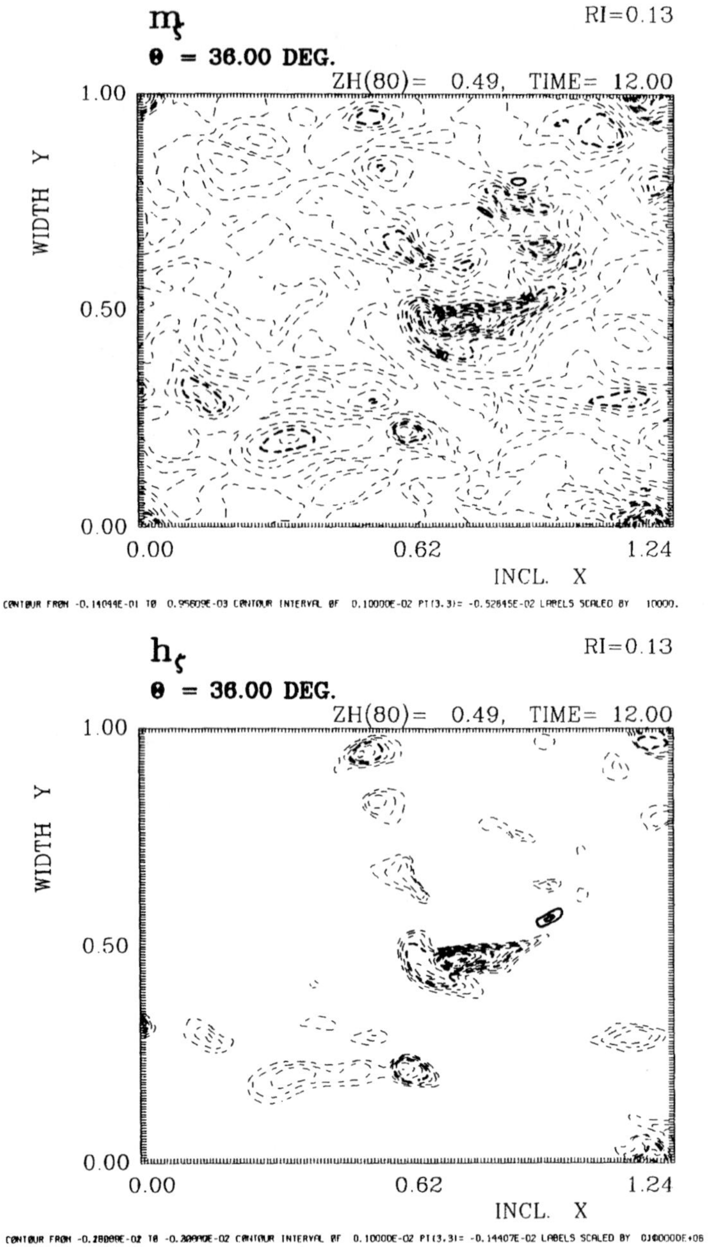


Fig.6. $Ri = 0.13$, $\theta = 36°$. Contour plots of the instantaneous fluxes perpendicular through the inclined plane of momentum $m_\zeta$ (top) and heat $h_\zeta$ (bottom) for $St = 12$. The contour interval is 0.001. Solid/dashed lines indicate positive/negative fluctuation values.

flow the square of rms vorticity is proportional to the dissipation rate of kinetic energy, $\omega'^2 = < \omega_i \omega_i > = \epsilon/\nu$. Fig. 7 depicts a contour plot of $\epsilon$ in the flow with $Ri = 0.13$ at $St = 12$. It appears that $\epsilon$ is strong where the vorticity is strong and often also the vortex structures can be re-detected by structures in the dissipation contours.

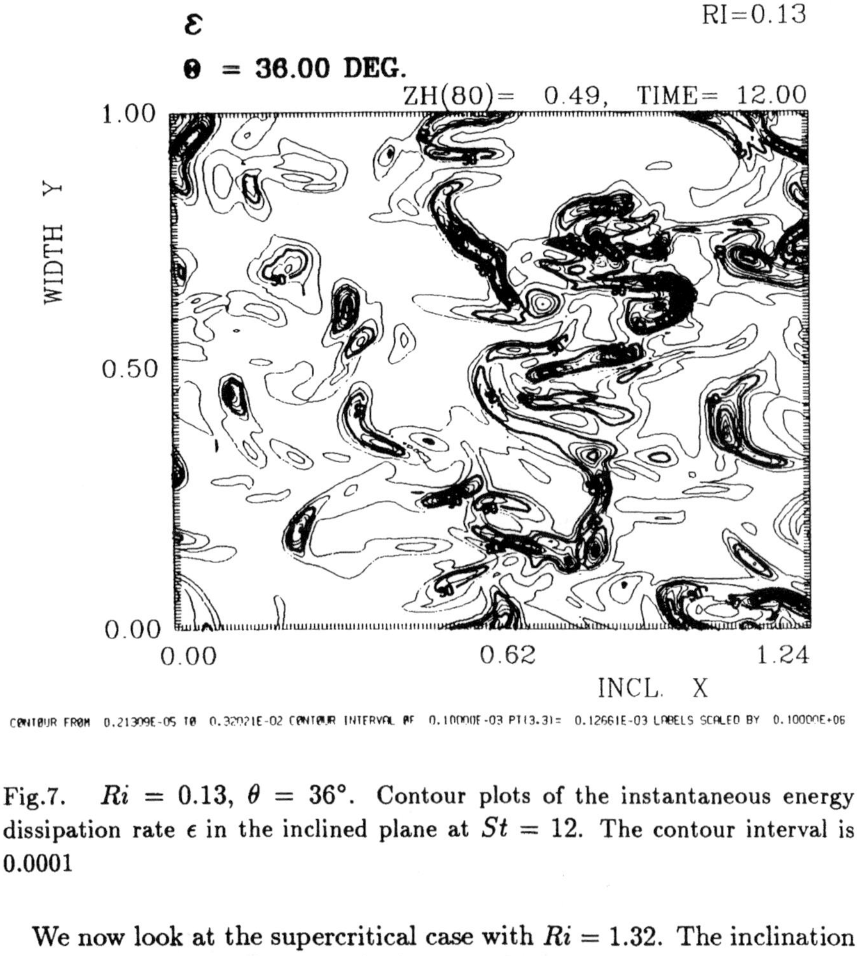


Fig.7. $Ri = 0.13$, $\theta = 36°$. Contour plots of the instantaneous energy dissipation rate $\epsilon$ in the inclined plane at $St = 12$. The contour interval is 0.0001

We now look at the supercritical case with $Ri = 1.32$. The inclination angle of the plane is $\theta = 12°$ which is a computationally convenient value and close to the peak angle in the histogram (fig. 2). Figure 8 depicts projected vorticity vectors $(\omega_\xi, \omega_\eta)$ at $St = 6$ and 12 if their magnitude exceeds one third of the maximum magnitude in that plane. In contrast to fig. 5 no horseshoe or hairpin like structures can be recognized. Instead we observe areas of medium size of equally oriented vorticity which often form spots of divergence and convergence. This observation corroborates the interpretations made above by analysing the two-point correlations in fig. 4: Vorticity is organized in almost horizontal planes.

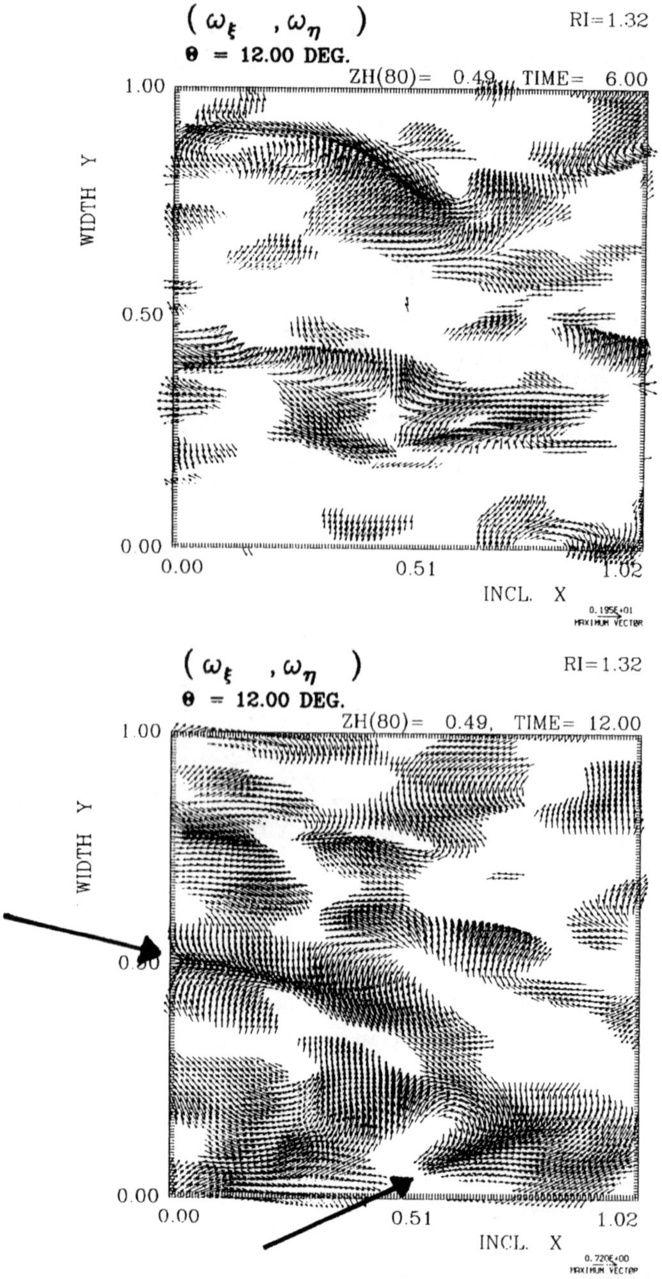


**Fig.8.** $Ri = 1.32$, $\theta = 12°$. Projection of vorticity vectors at $St = 6$ (top) and 12 (bottom) on the inclined plane. Vectors are shown if their magnitude exceeds a fifth of the magnitude of the maximum vector in the plane.

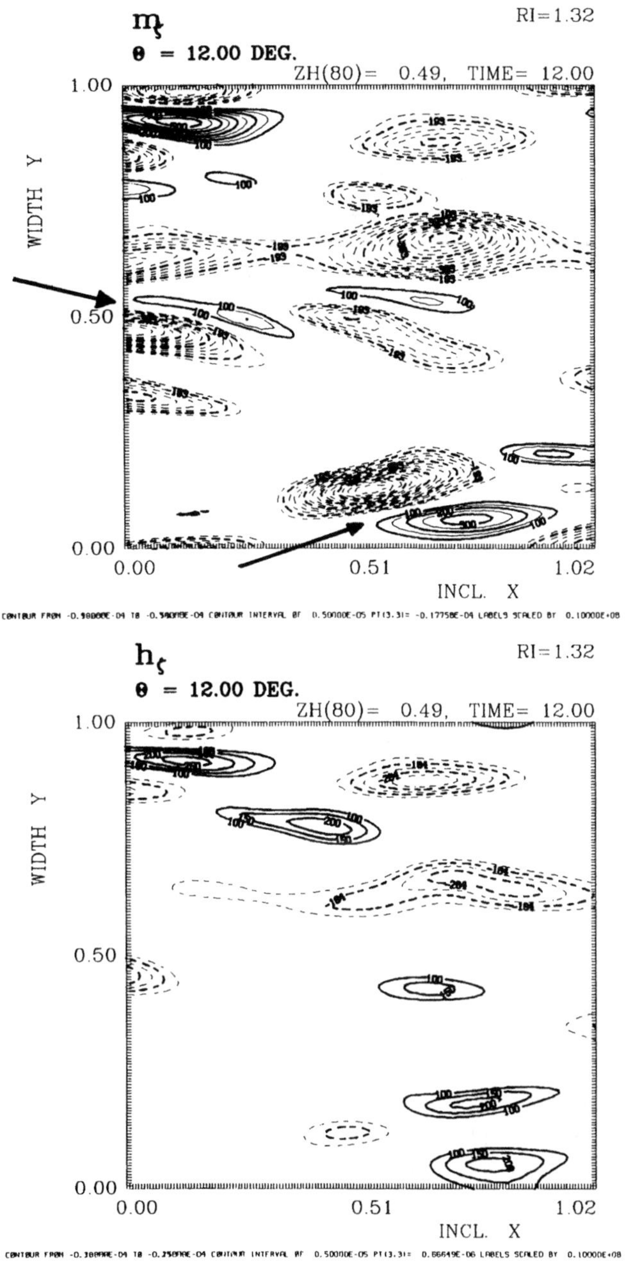


**Fig.9.** $Ri = 1.32$, $\theta = 12°$. **Contour plots of the instantaneous fluxes perpendicular through the inclined plane of momentum $m_\zeta$ (top) and heat $h_\zeta$ (bottom) for $St = 12$. The contour interval is 5. $10^{-4}$. Solid/dashed lines indicate positive/negative fluctuation values.**

We analysed the structures at $St = 12$ in a second plane which lies about $\Delta z = 0.045L$ above the first one. This is the distance where $R_{\omega_\xi \omega_\xi}(r_\zeta)$ and $R_{\omega_\eta \omega_\eta}(r_\zeta)$ have strongest negative values, see fig. 4f. We found indeed similar sheets of vorticity (not depicted here) but the vectors were oriented antiparallelly to the vectors in the lower plane.

Different from sheets are features with the largest vorticity and the largest length scales in the plane. These are rather narrow but long streaks of vorticity combined with broad zones of convergence of vorticity perpendicular to the streaks. The maximum magnitude of the vorticity in the plane is $1.95\%S$ ($0.72\%S$) and the average vorticity of the streaks has been estimated to $1.5\%S$ ($0.6\%S$) at $St = 6$ (12) which is still about 1.8 (1.6) times larger than the respective values of the rms vorticity $\omega'$ in that flow. Two of those streaks at time $St = 12$ are identified by thick arrows pointing at them. These streaks indicate spanwise propagating gravity waves as can be seen from fig. 9 by means of the fluxes of momentum and heat at $St = 12$ : In contrast to the flow with $Ri = 0.13$ where $m_\zeta$ and $h_\zeta$ are negative, i.e. down gradient, we find for $Ri = 1.32$ both negative and positive, i.e. counter-gradient contributions at large scales in alternating sequence due to the exchange of energy between the kinetic and potential reservoirs, see section 3. As the arrows in fig. 9 indicate, $m_\zeta$ is negative on the right side of a streak, positive on its left side and zero exactly on the ridge of the streak where the perpendicular component vanishes. Although the plots depict only one arbitrarily chosen cross-section through the fields these streaks make the difference plausible which occurs between the downstream and spanwise two-point correlations in fig. 4e. It is mostly the component $\omega_\eta$ which converges laterally towards the streaks leading to strongly negative correlations $R_{\omega_\eta \omega_\eta}(r_\eta)$. On the other hand, $\omega_\xi$ is always equally oriented in the spanwise direction (at a given $\xi$-position) which may explain the positive values of $R_{\omega_\xi \omega_\xi}(r_\eta)$. Hence we can associate those streaks with a circular movement of flow particles through the plane similar to movements of particles in water waves.

## 7. Conclusion

Coherent vorticity structures in sheared homogeneous turbulence under subcritical and supercritical stratification have been analysed at two times during flow evolution. The database of instantaneous velocity and temperature fields has been obtained by integrating the full Navier-Stokes and temperature equations numerically in a cubic domain with $160^3$ gridpoints.

It has been found that under subcritical conditions with $Ri = 0.13$ horseshoe vortices develop similar as in neutrally stratified flows studied by Rogers and Moin (1987). They control the dynamics of the flow like heat and momentum transport. Gravity waves are likely to exist at the largest scales but they do not change the turbulent character, i.e. strong mixing, of the flow. In supercritical flows with $Ri = 1.32$, no three-dimensional vortex structures are found but the vorticity is oriented almost horizontally and is organized in medium-sized sheets or in narrow but long streaks. The former represent vertically decoupled shear layers and the latter can be associated with spanwise propagating gravity waves.

It is an intention for future work to analyse the history of vortex structures in weakly as well as in strongly stably stratified shear flows in order to understand the physics of gravity-wave creation out of isotropic turbulence movements. We believe that the different mechanisms which balance the enstrophy (as well as the potential enstrophy) equation should be studied carefully by means of both statistical and structural informations in order to understand turbulence-wave interaction better.

*Acknowledgement.* I wish to thank Kilian Krettenauer and Dr. Thomas Hauf for many fruitful and illuminating discussions during the creation of this work.

## References

Britter, R.E. 1988 Laboratory experiments on turbulence in density-stratified fluids. Proc. 8th Conf. on *Turbulence and Diffusion,* April 26-29, 1988 at San Diego.

Gerz, T. 1988 Direct simulation of stably stratified homogeneous turbulent shear flows; Proc. 8th Conf. on *Turbulence and Diffusion,* April 26-29, 1988 at San Diego, 66-69.

Gerz, T., Schumann, U. & Elghobashi, S. 1989 Direct numerical simulation of stratified homogeneous turbulent shear flows. *J. Fluid Mech.* **200**, 563-594.

Gerz, T. & Schumann, U. 1990 Length scales of sheared and unsheared stratified homogeneous turbulence deduced from direct simulations. *Turbulent Shear Flow 7,* Springer-Verlag, in press.

Head, M.R. & Bandyopadhyay,P. 1981 New aspects of turbulent boundary-layer structure. *J.Fluid Mech.* **107**, 297-338.

Holt, S.E., Koseff, J.R. & Ferziger, J.H. 1989 The evolution of turbulence in the presence of mean shear and stable stratification. Proc. *7th Symp. on Turbulent Shear Flows,* 21.-23. Aug. 1989, Stanford Univ., 12-2.

Reynolds, W.C. 1989 The potential and limitations of direct and large eddy simulations. Position Paper. *Whither Turbulence Conf.,* Cornell Univ., 22.-24. 3. 1989.

Riley, J.J., Metcalfe, R. & Weissman, M. 1981 Direct numerical simulations of homogeneous turbulence in density-stratified fluids. In *Nonlinear Properties of Internal Waves,* West,B.J.(ed.), **76**, 79-112.

Rogers, M.M., Moin, P. & Reynolds, W.C. 1986 *The structure and modeling of the hydrodynamic and passive scalar fields in homogeneous turbulent shear flow.* NASA, Rep.TF-25, Dep. Mech. Engin., Stanford University.

Rogers, M.M. & Moin, P. 1987 The structure of the vorticity field in homogeneous turbulent flows. *J. Fluid Mech.* **176**, 33-66.

Staquet, C. & Riley, J.J. 1989 A numerical study of a stably-stratified mixing layer. *Turbulent Shear Flow 6,* Springer-Verlag, 381-397.

# Influence of a shear on a stably-stratified flow

Chantal STAQUET

*Laboratoire de physique, Ecole Normale Supérieure de Lyon*
*46 allée d'Italie, 69364 Lyon Cédex 07, France*

*Turbulence and Coherent Structures*
*O. Métais and M. Lesieur, eds.*
*Selected papers from "Turbulence 89: organized structures and turbulence in fluid mechanics"*
*Grenoble, 18-21 September 1989*

**Abstract**

The dynamics of a stably-stratified mixing layer exhibit the major features of small scale dynamics in stratified turbulence. In order to get some insight into those dynamics, a stably-stratified mixing layer is studied by + means of direct numerical simulations of resolution up to $128^3$. The three-dimensional Boussinesq equations are solved using a pseudo-spectral method.

The fundamental feature exhibited by our calculations is the organization of the flow into thin and quasi-horizontal layers of spanwise vorticity (vorticity of the mean flow and associated fluctuations). Depending upon the initial value of the Reynolds and Richardson numbers, two regimes have to be distinguished: a regime of "diffusive layers" (for low values of these parameters), and a regime of "baroclinic layers" (for higher values). In the former regime, the layer dynamics are driven by the shear and diffusive effects, which eventually balance. In the regime of baroclinic layers (upon which we shall focus in this paper), horizontal density variations and the mean shear lead to the development of a secondary instability of the Kelvin-Helmholtz + type in the layers. We show that this instability does not develop in a two-dimensional layer and propose an argument to explain this behavior.

## 1. Introduction

A number of experiments of homogeneous (unsheared) stably-stratified flows have been performed (see Hopfinger(1987), for a review), either by numerical or laboratory experiments. These studies all show the same trend, namely an organization of the flow into quasi-horizontal layers (of velocity, vorticity or density) after the vertical turbulent motions of the flow have collapsed. The quasi-horizontal layers are decorrelated along the vertical and their relative motions generate velocity shears. The presence of a velocity shear and the existence of a stable density gradient may therefore be responsible for the + major features of small scale dynamics in stratified turbulence. A study of the stably-stratified mixing layer could then help to understand those dynamics.

A stably-stratified mixing layer develops at the interface of two parallel streams of fluid moving horizontally at different velocities and having different densities, the upper stream being lighter than the lower one. A few laboratory experiments of stably-stratified mixing layers have already been conducted. In Thorpe's studies, a temporal mixing layer is generated by tilting up a closed tube containing a stratified liquid fluid away from the horizontal. Thorpe studied the turbulent aspect of the flow, focusing upon its two-dimensional character (e.g. Thorpe, 1973) as well as the dynamics of internal and interfacial waves (e.g., Thorpe, 1981). Scotti and Corcos (1973) created a stably-stratified mixing layer in a wind tunnel in order to test the linear stability theory of such a flow, and compared their results with the numerical calculations of Hazel (1972). The most extensive experimental study of a stably-stratified mixing layer has been performed by Koop and Browand (1973). They studied all stages of the flow, from the early development of instability waves to the regime where the mixing layer reaches a maximum amplitude due to stratification effects and finally collapses. Experiments were carried out for different Richardson numbers (a ratio of buoyancy to shear forces), covering the range of values for which the layer is (linearly) unstable.

Measurements of fine scale vorticity structures are very difficult in laboratory experiments, while these structures can be localized and studied using numerical experiments. Hence, numerical experiments may be a fruitful tool to explore such flows. This tool has however some limitations: because the resolution cannot be very high, the Prandt number has to be of order 1 (in order to properly resolve all density scales) while the Reynolds number cannot exceed a few hundred.

Two-dimensional numerical experiments have been performed, in which a weakly stably-stratified mixing layer consisting of one Kelvin-Helmholtz billow is studied. Patnaik et al. (1976), using $64^2$ simulations, focused in detail on + the stage of the flow for which the layer reaches a maximum amplitude ("climax state"); Corcos and Sherman (1976) complemented the numerical study by proposing theoretical models for the braid dynamics and the establishment of the climax state. Klaassen and Peltier (1985), using $128^2$ simulations, examined the influence of the Reynolds number upon the energetics of the flow.

A numerical study of a three-dimensional stably-stratified mixing layer is presented here. The striking feature exhibited by our calculations, once two-dimensional and three-dimensional instabilities have developed, consists of the formation of quasi-horizontal layers of spanwise vorticity stretched by the mean shear. (The spanwise vorticity is the vorticity of the mean flow and associated fluctuations). Depending upon the initial values of the Reynolds and Richardson numbers, two different regimes have to be distinguished: for low values of these parameters, we shall show that the layer dynamics are dominated + by the shear and diffusive effects, while for higher values of these two + parameters, the dynamics are driven by the shear and baroclinic effects (due to horizontal density variations). The former regime will be denoted the "regime of diffusive layers" while the latter regime will be referred to as the "regime of baroclinic layers". In this paper, we shall focus upon the latter regime.

## 2. Equations of motion and numerical model

We assume that the fluid motion is described by the Navier-Stokes equations in the Boussinesq approximation

$$\frac{\partial \mathbf{u}}{\partial t} + (\mathbf{u} \cdot \nabla)\mathbf{u} = -\nabla p - Ri\ \rho\ \mathbf{i}_z + \frac{1}{Re}\ \Delta \mathbf{u} \qquad (2-1a)$$

$$\frac{\partial \rho}{\partial t} + (\mathbf{u} \cdot \nabla)\rho = \frac{1}{Pr Re}\ \Delta \rho \qquad (2-1b)$$

$$\nabla \cdot \mathbf{u} = 0. \qquad (2-1c)$$

Equations (2-1) are written in dimensionless form: the length and velocity + scales have been chosen to be, respectively, $\delta_i$, the initial vorticity + thickness of the flow, and $\Delta U$, the initial mean velocity difference across the mixing layer. The parameter $\delta_i$ is a measure of the initial mixing layer thickness and is defined by $\Delta U / (d\overline{U_i}/dz)_{\max}$, $\overline{U}_i(z)$ being the initial mean velocity profile + and $z$ the vertical coordinate. The density has been scaled by $\sqrt{Pr}\Delta\rho$, where $Pr = \nu/\kappa$ is the Prandtl number and $\Delta\rho$ is the density difference across the layer. We decompose the density field as $\rho = \rho_0 + \overline{\rho}(z) + \rho'$, where $\rho_0$ is + a constant density reference, $\overline{\rho}(z)$ is the ambient (or mean) + density and $\rho'$ is the deviation from the ambient. The initial mean density + profile will be denoted $\overline{\rho_i}(z)$. The results will be expressed in + terms of a stratification time scale, $T_{BV} = 2\pi/N_i$, where $N_i{}^2 = (-g/\rho_0)(d\overline{\rho_i}/dz)_{z=0} = (g/\rho_0)\Delta\rho\sqrt{Pr}/\delta_i$ is the maximum of the Brunt-Väisälä frequency at $t = 0$. We shall also use in the last section an advective time scale $\delta_i/\Delta U$. With this scaling, the nondimensionalized parameters which come into play in the equations are: $Pr$, the Prandtl number; $Re$, the Reynolds number, initially equal to $\delta_i \Delta U/\nu$; and the minimum of the gradient Richardson number, $Ri_{min} = (-g/\rho_0)\min\left[(d\overline{\rho_i}/dz)/(d\overline{U_i}/dz)^2\right] = (N_i\delta_i/\Delta U)^2$ initially, which is inversely proportional to the square of a Froude number.

Equations (2-1) are solved numerically using a pseudo-spectral method to compute the spatial derivatives and nonlinear terms, and a third order Adams-Bashforth scheme for time stepping. We study a temporal mixing layer (i.e. the flow develops in time, as opposed to a spatial evolution) and this allows the use of periodic boundary conditions along the streamwise ($x$) direction. The boundary conditions are also periodicity along the spanwise ($y$) direction and of free- + slip type along the vertical.

The initial mean velocity and density profiles are those that would be brought about by molecular diffusion from initial coplanar discontinuities of density + and horizontal velocity at $z = 0$. These profiles, written in nondimensionalized + form, are expressed by:

$$\overline{\mathbf{U}}(z, t = 0) = \overline{U}_i(z)\ \mathbf{i}_x \quad \text{with} \quad \overline{U}_i(z) = \operatorname{erf}(\sqrt{\pi}\ z)$$

$$\overline{\rho}_i(z) = -\frac{1}{\sqrt{Pr}} \operatorname{erf}(\sqrt{\pi Pr}\ z)$$

and are represented in Figure 1.

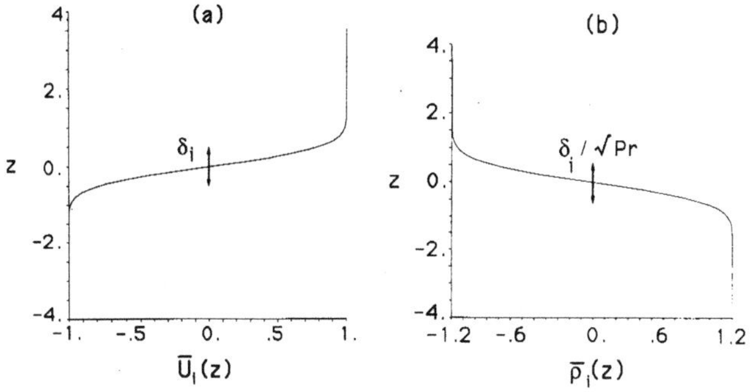


Fig.1. Initial mean velocity (a) and mean density (b) profiles. The thickness of each profile is also plotted.

N density perturbation is superimposed initially upon $\overline{\rho}_i(z)$, + while both two-dimensional and three-dimensional perturbations are added to $\overline{\mathbf{U}}$. The two-dimensional perturbations are solutions of the linear stability equation (Taylor-Goldstein) for the most amplified wavenumber $k_f$ + and its first subharmonic $k_f/2$, and have an amplitude equal to 2.5% of $+triangleU$. These two perturbations promote respectively the development of the Kelvin-Helmholtz instability and vortex pairing when the size of the + computational domain in the streamwise direction is taken large enough so as to allow the formation of two Kelvin-Helmholtz vortices. The initial three-dimensional perturbation is narrow-banded, with wavelength in the spanwise direction equal + to the size of the computational domain in that direction, and with amplitude equal to 1% of $\Delta U$. In addition, this perturbation has no initial vertical vorticity.

Two calculations are presented here. In the first one, where the regime of diffusive layers is found, the initial Richardson number is taken small enough so that vortex pairing, which mainly involves vertical motions, can occur: $Ri = 0.1\sqrt{Pr}$, that is, here $Ri = 0.084$. (This value stems from Koop and Browand's (1979) experimental study). To study the merging process, we choose the size of the numerical domain so that two Kelvin-Helmholtz billows can develop. The resolution is 128x64x129 (in the $x$, $y$, and $z$ directions respectively) and the initial Reynolds number is equal to 200. In the second calculation, where the regime of baroclinic layers is found, the initial + Richardson number is taken high enough so that pairing does not occur: $Ri = 0.2\sqrt{Pr} = 0.167$. It therefore suffices to examine the development of one Kelvin-Helmholtz billow only.

Because the flow is strongly-stratified (but still unstable) and a fine resolution is used (128x128x129), a higher value for the initial Reynolds number than in the previous calculation can be used: $Re = +560$. The Prandtl number is equal to 0.7 for both calculations.

To determine the maximum Reynolds number for which a reliable computation can be done, we have used the equation for the potential vorticity evolution. In the Boussinesq approximation, the potential vorticity reduces to: $\pi = \omega \cdot \nabla\rho/\rho_0$ and satisfies the following equation for a viscous fluid:

$$\frac{D\pi}{Dt} = \frac{1}{Re}\,\Delta\omega\cdot\nabla\rho + \frac{1}{Pr\,Re}\,\omega\cdot\Delta(\nabla\rho) \qquad (2-2)$$

where $D/Dt$ denotes the material derivative. The right-hand side of equation (2-2), involving third-order spatial derivatives of velocity + or density, is accurately computed if the smallest scales of the flow are properly resolved. The criterion we have chosen so as to have this condition satisfied consists of computing independently the two sides of equation (2-2), written for the volume-averaged potential enstrophy $(1/V)\int\int\int \pi^2\, dV$ (in order to remove space dependency), and adjusting the Reynolds number till + the two sides are equal (or the two corresponding curves coincide). Though this criterion involves mean values, it is still rather accurate because the sources and sinks of potential vorticity are very localized in space. The sensitivity of this criterion to the initial Reynolds number is demonstrated in fig. 2. The two sides of equation (2-2) written for the volume-averaged potential enstrophy are compared for two different values of the initial Reynolds number: Re=640 (fig. 2a) and Re=560 (fig. 2b), in the case of the strongly-stratified mixing layer calculation ($Ri = 0.167$). Fig. 2a shows that the two curves strongly depart, + while they coincide (fig. 2b) when a lower Reynolds number is chosen. A comparison of these two curves for the calculation of the weakly-stratified mixing layer ($Ri = 0.084$ and $Re = 200$) is displayed in fig. 2c; the two curves coincide, + hence attesting that the Reynolds number has been properly chosen.

As we shall deal throughout the paper with the spanwise component of vorticity $\omega_y$, let us write the Boussinesq equation for this component:

$$\frac{D\omega_y}{Dt} = \omega\cdot\nabla\mathbf{u} - Ri\,\frac{\partial\rho}{\partial x} + \frac{1}{Re}\,\Delta\omega_y \qquad (2-3)$$

Equation (2-3) shows that streamwise variation of the density field modifies (produces or destroys) spanwise vorticity; this important effect will be hereafter referred to as baroclinic effect, because $-Ri\ \partial\rho/\ +partialx$ is the spanwise component of the baroclinic torque, whose expression reduces to $Ri\ (\nabla\rho\ \mathbf{x}\ \mathbf{i_s})$ in the Boussinesq approximation.

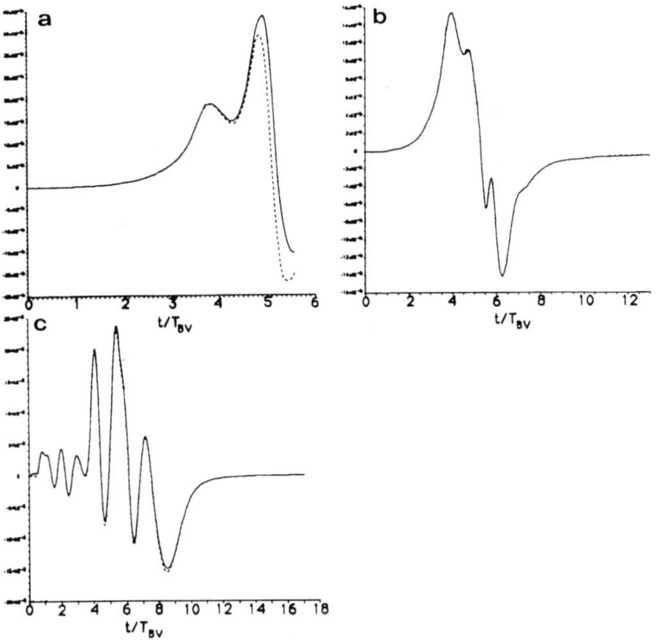


Fig.2. Criterion to determine the right value for the + Reynolds number. The l.h.s of equation (2-2) written for the volume-averaged potential enstrophy (straight line) is compared to the r.h.s (dashed line) for the following initial parameters: (a) Ri=0.167 and Re=640; (b) Ri=0.167 and Re=560; (c) Ri=0.083 and Re=200. Pr=0.7 in all cases.

## 3. The regime of diffusive layers

The mixing layer evolution is presented in fig. 3, via surfaces of constant spanwise vorticity.

The initial spanwise vorticity is confined into a thin two-dimensional layer. At $t = 0.8\ T_{BV}$, two coherent structures develop (fig. 3a), which would merge if the mixing layer were not stratified. Fig. 3b actually shows that vortex pairing reduces here to the sliding of two layers of vorticity the one on top of the + other. Two mechanisms are responsible for this behavior: (i) vertical motions are weakened by the stable stratification and (ii) baroclinic effects, through streamwise variations of the density field, tend to destroy the vorticity of the vortex cores while they reinforce the vorticity in the region between the cores (also called "braid"). These two effects eventually lead to the formation of + three layers of vorticity, the top and bottom layers being quasi-two-dimensional (fig. 3c). Notice that the intermediate layer, which results

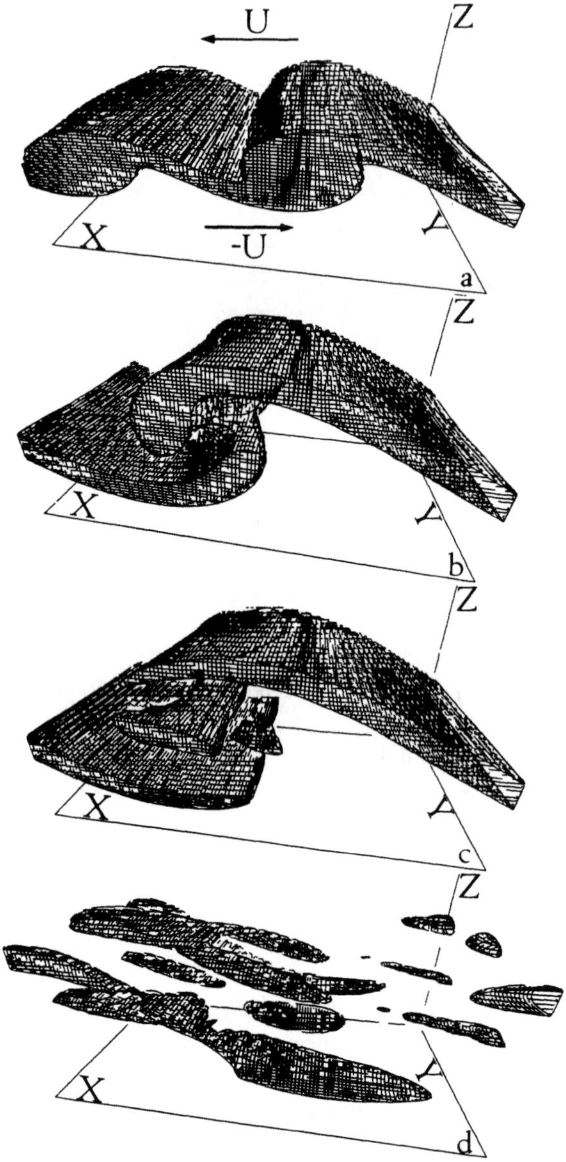


Fig.3. Evolution of the weakly-stratified + mixing layer ($Ri$ = 0.084). Surfaces of constant spanwise vorticity $\omega_y$, w absolute value equal to 50% of the maximum of $|\omega_y|$, are + plotted at successive times: (a) $t = 0.8\ T_{BV}$; (b) $t = 2.9\ T_{BV}$; (c) $t = 3.4\ T_{BV}$; (d) $t = 7.3\ T_{BV}$.

from the merging of the two vortex core residuals, is already marked by the presence of the three-dimensional perturbation. From this time on, the growth of the three-dimensional perturbation highly modifies the flow dynamics, in strongly enhancing the spanwise variation of the flow (fig. 3d). To sum up, the combined effects of the pairing process, stratification effects and three-dimensional perturbation growth redistribute the initial vorticity, which was confined into + a thin two-dimensional layer, into a number of thinner mixing layers of finite spanwise extent.

We shall now focus upon those layer dynamics. Though, as already said, stratification effects have led to the formation of these thin layers, they do not play any significant role in their dynamics, once these layers have formed. This can be seen in fig. 4, where vertical cross-sections of spanwise vorticity and corresponding density are plotted at $t = 7.3\ T_{BV}$. In the region where + the vorticity layer exists in fig. 4a, we see in fig. 4b that the density field displays very small streamwise variations. Hence, the layer vorticity should be affected very weakly only by baroclinic effects. It follows that the layer + dynamics should be dominated by the shear and diffusive effects. This regime will be denoted "regime of diffusive layers".

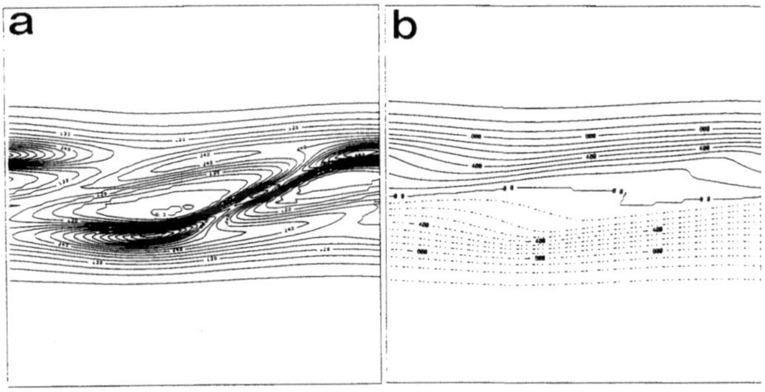


Fig.4. Vertical cross-section + of the flow. Contours of constant spanwise vorticity (a) and corresponding density (b) are plotted at $t = 7.3\ T_{BV}$.

Diffusive effects and shear have opposite actions upon the layers and should eventually balance. As an example, the shear stretches the layers and makes + their thickness decrease while diffusive effects enhance layer thickness. The layer thickness should therefore reach a constant value. As a consequence, stretching and rotation of the layers by the shear should lead to a transfer of energy towards large scales (fig. 5), which is indeed observed (fig. 6). Also, the peak value of the layer vorticity decreases exponentially with time during this regime of diffusive layers. We have developed a simple theoretical model which predicts quantitatively this exponential decreaese of the vorticity together + with the constant layer thickness. Further detail will be given in a subsequent paper.

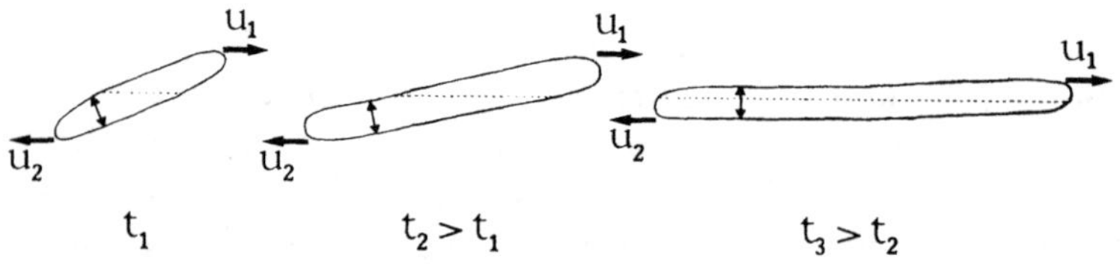


Fig.5. Sketch of the action of the shear upon the vorticity layer. Supposing its thickness (⟷) to be constant, the layer stretches and rotates; hence, its streamwise extent (— — —) increases.

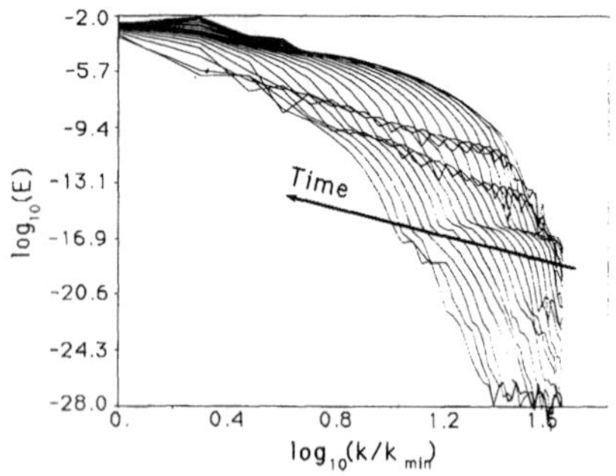


Fig.6. One-dimensional longitudinal energy spectrum of the + streamwise velocity component. The spectrum is computed across a layer and plotted at successive times during the regime of diffusive layers. The transfer of ener towards larger scales is interpreted as being due to the action of the shear + upon the layer (the wavenumber where a peak of energy is observed at a given ti would correspond to the streamwise extent of the layer -see fig. 5).

## 4. Regime of baroclinic layers

### 4.1. The flow dynamics are driven by the shear and baroclinic effects

The mixing layer evolution is visualized at successive times in fig. 7 through vertical cross-sections of the spanwise vorticity.

The initial Richardson number is now twice as high as in the previous calculation. The effect of such a strong stratification upon the instability development clearly appears in fig. 7b: the vortex core cannot form anymore because of baroclinic effects, while these effects continuously

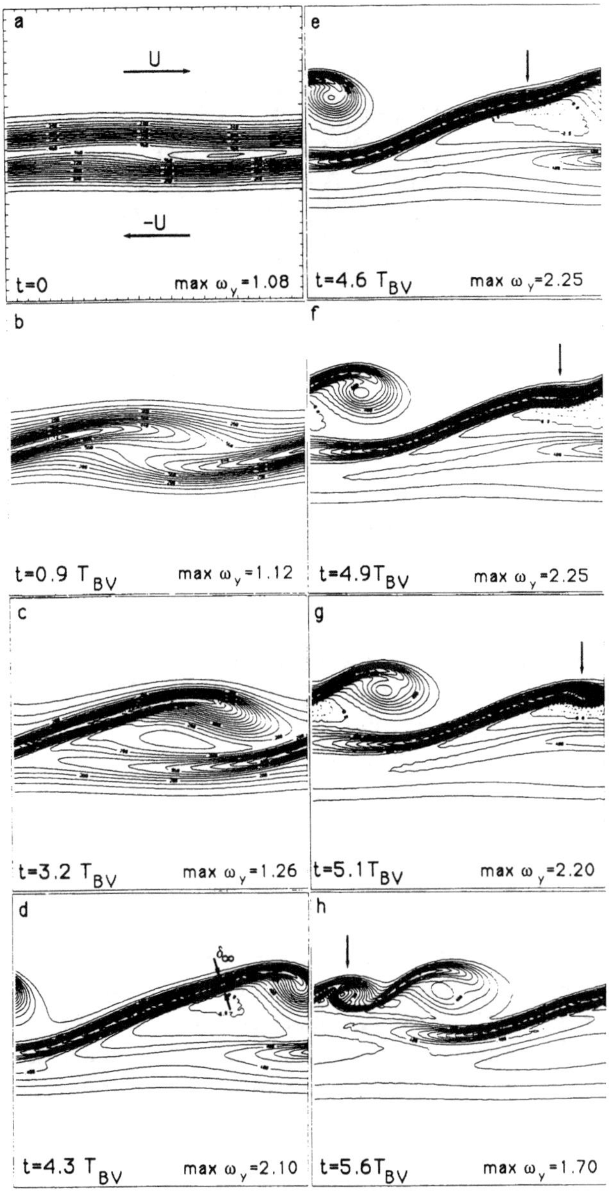


Fig.7. Evolution of the strongly-stratified mixing layer ($Ri = 0.167$). A vertical cross-section of the spanwise vorticity is plo successive times. The arrows in fig. 7e to 7h indicate the location where the secondary instability of Kelvin-Helmholtz type develops.

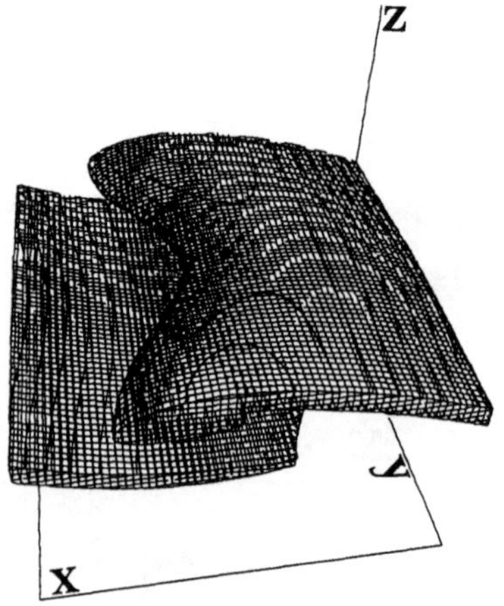


Fig.8. Surfaces of + constant spanwise vorticity $\omega_y$, with absolute value equal to 50% of the maximum of $|\omega_y|$, are plotted a $t = 3.2\ T_{BV}$. Notice the strong distortion of the vorticity field along the spanwise direction, as a result of the three-dimensional perturbation growth; this should produce spanwise vorticity.

reinforce the braids, which move in opposite directions. The following scenario can be proposed to interpret fig. 7c and 7d: the braid vorticity is also enhanced (by stretching) by the growth of the three-dimensional perturbation, as shown by figure 8 where surfaces of constant spanwise vorticity are represented at $t = 3.2\ T_{BV}$. The spanwise vorticity therefore increases. As a result, a vortex forms at the end of the + braid and induces a velocity field of pure strain type along the braid. Consequently, the braid aligns with the direction of maximum strain and its angle with the horizontal increases; its length also increases while its thickness decreases. Hence, in this region of the flow, the vorticity eventually is confined in a + very thin and intense tilted shear layer, as a result of the three-dimensional perturbation growth and baroclinic effects. Notice that the thickness of this layer is much thinner than the initial layer thickness and that the vorticity maximum is at the center of the layer.

The crucial role of baroclinic effects is demonstrated in fig. 9, where contours of constant density (still in the vertical cross-section we are considering) are plotted at the same time as in fig. 7d. We see that, in the upper part of the flow where the spanwise vorticity layer exists, density contours are aligned with the tilted layer, therefore leading to a nonzero streamwise gradient.

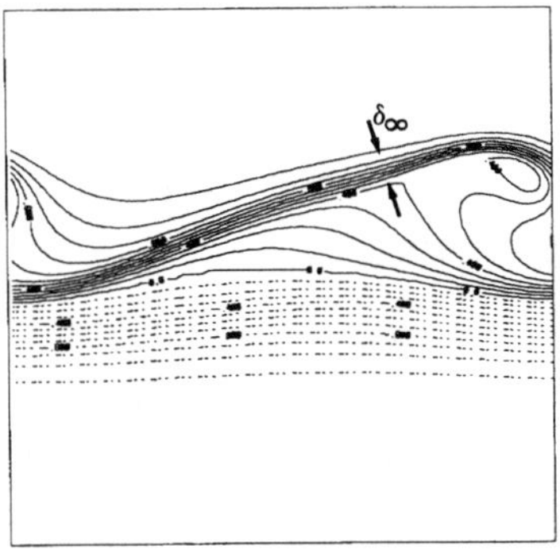


Fig.9. Contours of constant density corresponding to fig. 7d.

## 4.2. The baroclinic layer becomes unstable

The spanwise vorticity layer displayed in fig. 7d is intense and very thin, and subjected to a moderate normal density gradient. (Fig. 9 actually shows that the density gradient normal to the layer is about four times weaker that the one existing at $t = 0$.) The layer can therefore become unstable. Fig. 7e to 7h show indeed that a new instability of the Kelvin-Helmholtz type develops in the + layer. Notice that this instability develops as if the flow were not, or only weakly-stratified, as a strong vortex forms (while this vortex could not form in the earlier stage of the flow development, because of baroclinic effects). This observation is consistent with the Richardson number's being small (equal to 0.03) at the location where the vortex develops in fig. 7e. The same vertical cross-section at later times shows that the vortex core progressively disappears because of baroclinic effects.

A three-dimensional view of the flow is presented in fig. 10, where surfaces of constant spanwise vorticity are plotted at $t = 5.6\ T_{BV}$. Fig. 10 shows that the secondary instability of Kelvin-Helmholtz type has a finite + extent along the spanwise direction. This spanwise extent is imposed by the spanwise wavelength of the initial three-dimensional perturbation.

The calculation presented in this section had first been run with a slightly higher Reynolds number ($Re = 640$). That run was stopped after a few Brunt-Väisälä periods because the criterion we had chosen to determine + whether the smallest scales of the flow were properly resolved was not satisfied (see + fig. 2a). However, because of its higher Reynolds number, this calculation still is of interest as shown by fig. 11. A vertical cross-section of the spanwise vorticity (in the same vertical plane as in fig. 7) is presented at $t = 5.6\ T_{BV}$ in this figure. We see that the growth of the secondary Kelvin-Helmholtz instability + now leads to the formation of two vortices. This result suggests the possible

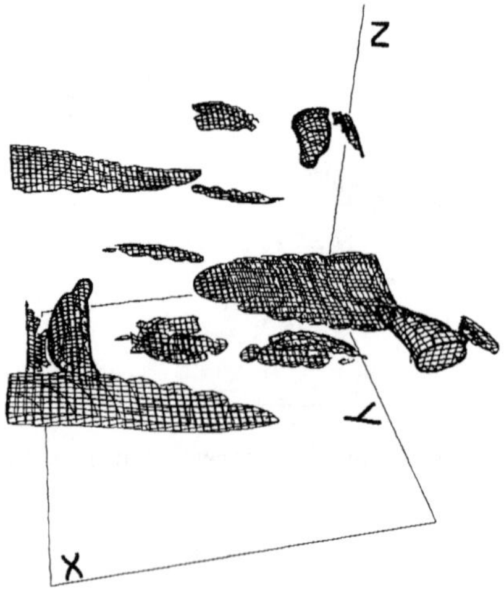


Fig.10. Surfaces of constant + spanwise vorticity $\omega_y$, with absolute value equal to 50% of the maximum + of $|\omega_y|$, are plotted at $t = 5.6\ T_{BV}$. The outline of the Kelvin- + Helmholtz billows resulting from the secondary instability growth has been shown in + boldface for clarity.

existence, in geophysical flows for example where Reynolds numbers are very + high, of a sort of cascade of Kelvin-Helmholtz instabilities towards smaller scales: + as the braids are reinforced by baroclinic effects (while these effects + progressively destroy vortex cores), the braids could in turn become unstable, and this instability would occur at a smaller scale than the first secondary instability. The braids resulting from this (second) secondary instability could in turn + become unstable and so on. This mechanism could be responsible for the properties of mixing at small scales in geophysical flows.

Such a secondary instability has been predicted theoretically by Corcos and Sherman (1976). These authors proposed a two-dimensional dynamical model for the density and vorticity fields which describes the evolution of these two fields + in a frame of reference attached to the layer and in the neighboorhood of the stagnation point. The velocity field acting upon the layer in this region is of pure strain type. As shown by numerical simulations (Patnaik et al., 1976), the rate of strain of this outer velocity field is approximately constant once the Kelvin-Helmholtz instability has developed. Assuming a self-similar behavior + for the two fields, an asymptotic form (as time goes to infinity) for the vorticity and density fields was obtained, from which an expression for the Richardson number at the center of the layer was derived. For sufficiently high values of the Reynolds number, Corcos and Sherman thus showed that

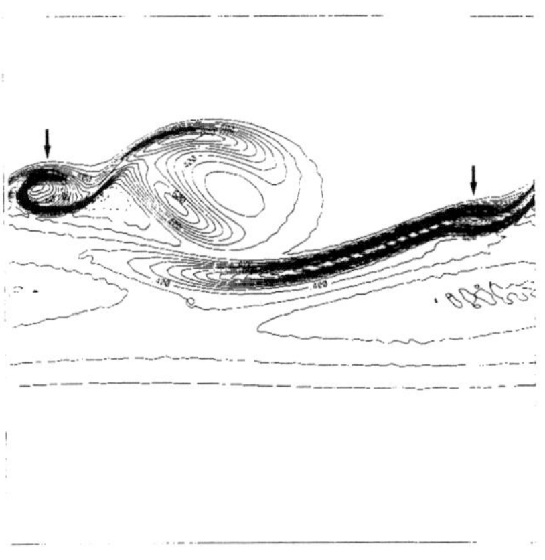

Fig.11. Same as fig. 7h but for Re=640.

this parameter could be much smaller than 0.25, that is, according to linear stability theory, the + layer could become unstable.

The model also predicts that the thickness of the density and vorticity layers should reach an asymptotic expression, which we write here in nondimensional form:

$$\delta\ (t=\infty) = (\pi/2)^{1/2} (\gamma\ Re)^{-1/2}, \qquad (4-1)$$

$\gamma$ being the rate of strain of the outer velocity field. $\delta\ (t = +\infty)$, denoted $\delta_\infty$ hereafter, is the same for the two layers if the Prandtl number is equal to 1. Let us compute $\delta_\infty$ for our three- + dimensional calculation (for which the Prandtl number is close to 1). This will allow us + make comparison with the thicknesses of the vorticity and density layers represented + in fig. 7d and 9 respectively. We first have to compute the rate of strain $\gamma$ and use for that the asymptotic expression for the vorticity maximum predicted + by the model

$$\max \omega_y\ (t=\infty) \ = \ (2/\pi)^{1/2}\ Ri\ (Re/Pr)^{1/2} \sin\ + \theta\ \gamma^{-1/2}, \quad (4-2)$$

written again in nondimensional form. $\theta$ is the angle that the braid makes with the horizontal at the stagnation point. Assuming that infinite time is the time just preceeding the secondary + instability development, we use fig. 7e to get max $\omega_y = 2.25$ and $\theta = +21$ degrees. This yields $\gamma = 0.09$, the time unit being the advective time scale $\delta_i/\Delta U$. Using (4-1), we find $\delta_\infty = 0.36$. This + value is represented in figures 7d and 9 respectively. An excellent agreement is + found. Though this value has been obtained by a two-dimensional model, such a good agreement is not so surprising if one considers that the braid dynamics + primarily depend upon two-dimensional effects, namely shear stretching and streamwise variation of the density. The influence of three-dimensional effects would then come into play through the value of the rate of strain $\gamma$.

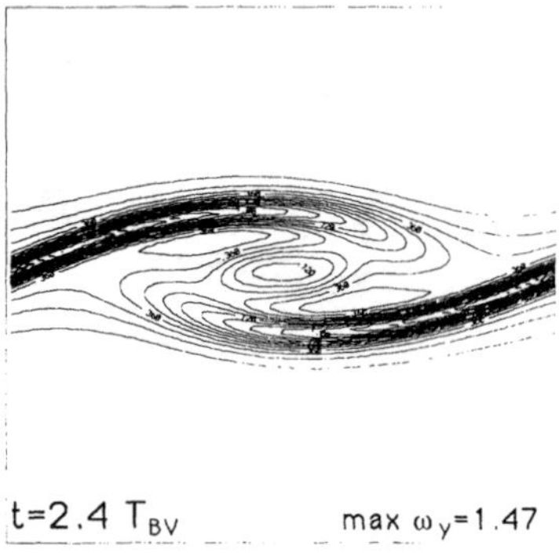


Fig.12. Two-dimensional numerical simulation of the strongly- + stratified mixing layer (resolution: 128x129, Ri=0.167, Re=800). The brai stable.

## 4.3. The two-dimensional flow is stable

Corcos and Sherman did not observe the development of the secondary instability in their two-dimensional numerical calculations (Patnaik et al., 1976); neither did we when running the two-dimensional version of our three-dimensional calculation (fig. 12). The value of the Richardson number at the stagnation point, denoted $Ri_B$, is nevertheless equal to 0.18 in our two-dimensional calculation. This means that the layer could be unstable, but on a very long time scale. One reason for such a large value of + the local Richardson number is that the braid vorticity is only reinforced by baroclinic effects in two dimensions, while nonlinear effects (through vorticity stretching) may also enhance this vorticity in a three-dimensional flow. The + maximum of the braid vorticity is indeed equal to 2.25 in the three-dimensional + calculation (fig. 7e) while max $\omega_y = 1.47$ in the two-dimensional version ( + fig. 12). In order to get a braid (at least) as strong as in the three-dimensional calculation, we increased the resolution of the two-dimensional calculation as well as the Reynolds number. A 256x257 numerical simulation was thus performed with initial Reynolds number equal to 4000. But here again the braid remains stable (fig. 13), though max $\omega_y = 3.93$ and $Ri_B = 0.07$. Because this two-dimensional flow is highly symmetric (while this symmetry is broken by the growth of the three-dimensional perturbation, see fig. 7c), we tried to disrupt that symmetry using various + ways. The braid still remained stable.

We then propose the following argument to explain the stability of the two-dimensional braid. As already said, the braid + is subjected to an outer velocity field of pure strain type, whose component normal to the

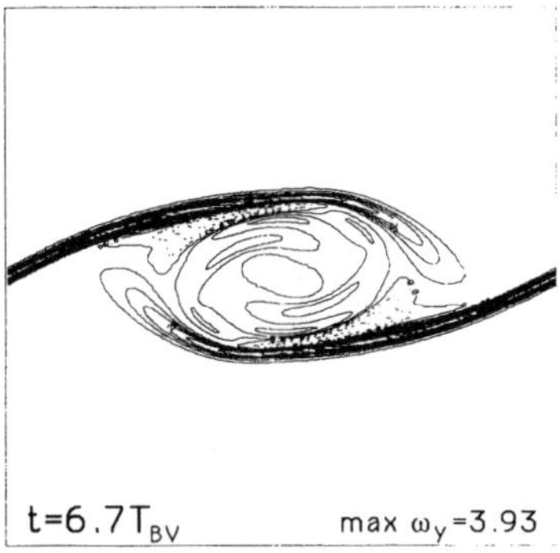


Fig.13. Same as fig. 12 but for + higher resolution (256x257) and higher Reynolds number (Re=4000). + The initial Richardson number is unchanged. The braid remains stable.

layer, of the form

$$u_\eta = -\gamma\eta \qquad (4-3)$$

converges towards the layer; $\eta$ is the coordinate across the layer. This velocity component therefore opposes the instability development. Indeed, (4-3) yields $d\eta/dt = -\gamma\eta$ ; that is, $\eta \sim e^{-\gamma t}$ . Hence, any quantity perpendicular to the layer will decrease exponentially at a rate equal + to $\gamma$. It follows that the growth rate of the instability, say $\alpha$ when the outer velocity field is uniform and along the streamwise direction (standard mixing layer), is now equal to $\alpha - \gamma$. If $\gamma$ is greater than $\alpha$, the instability cannot develop.

In the three-dimensional calculation, we have noticed that the secondary instability develops as if no stratification effect were present; $\alpha$ is therefore close to the growth rate of a neutral mixing layer (0.19), which is greater than the value of $\gamma$ we found ($\gamma$=0.09). Hence, $\alpha - \gamma > 0$.

Let us now compute $\alpha$ and $\gamma$ for the two-dimensional calculations mentionned above. We use again equation (4-2) to compute $\gamma$. Taking max $\omega_y$ = 1.47 and $\theta$ = 23 degrees from fig. 12, we find $\gamma$ = 0.36: this value is larger than the largest growth rate of the Kelvin-Helmholtz instability (equal to 0.19 for a neutral mixing layer); + hence, $\alpha - \gamma < 0$. The same conclusion can be drawn for the calculation with + higher resolution and Reynolds number, where a value of 0.32 for $\gamma$ is computed (with max $\omega_y$ = 3.93 and $\theta$ = 26 degrees from fig. 13). + Finally, in the calculation performed by Patnaik et al., $\gamma$ is equal to 0.24 (according to their fig. 5b), which leads again to a negative value for $\alpha - \gamma$.

The fact that $\gamma$ is equal to 0.09 in our three-dimensional calculation + and 0.36 in the two-dimensional version suggests that three-dimensional effects weaken the rate of strain of the outer velocity field acting upon the vorticity layer. Hence the secondary instability, of Kelvin-Helmholtz type and therefore two-dimensional, could only develop in a three-dimensional flow.

## 5. Conclusion

In the description of the flow we have proposed, we have focused upon its "turbulent" part only (we should better say: "nonlinear, nonpropagating" part, + as the values of the Reynolds numbers we have used are quite moderate), hence neglecting its wave component. Though a complete description of the flow would require the study of this part of the flow, it can be argued that the internal wave component has at least been taken into account implicitly: indeed, in the atmosphere for example, internal wave motion is mainly responsible for the vertical variability of the velocity profile and is therefore at the origin of velocity shears.

According to our calculations, a behavior for the stably-stratified mixing layer can be proposed, which consists of the organization of the flow into quasi-horizontal layers of spanwise vorticity. When the initial Reynolds and Richardson numbers are high enough, we have shown that a secondary instability of Kelvin-Helmholtz type develops in the layers, provided the flow + is three-dimensional. A two-dimensional layer (that is, developing in a two-dimensional flow) is indeed stable; the argument we have proposed as an attempt to elucidate this latter behavior would need to be refined, for example, by doing a linear stability analysis in the neighbourhood of the stagnation + point of a two-dimensional vorticity layer subjected to a pure strain velocity field.

Our results display two phenomena associated with energy transfers: in the regime of diffusive layers, the stretching and rotation of the layers by the local shear result in an energy transfer towards larger scales (supposing the layer thickness is constant), while in the regime of baroclinic layers, a + cascade of Kelvin-Helmholtz instabilities towards smaller scales has been proposed. + These two mechanisms, which require shear to be present, could provide alternate explanations for the origin of the spectra measured in geophysical flows.

This regime of *spanwise* vorticity layers, which exists in the earlier stage of the flow when nonlinear and stratification effects still compete, is followed by a quite different regime. Staquet and Riley (1989a) have indeed shown that quasi-horizontal patches of *vertical* vorticity (imbedded in a field of internal waves) remain in the later stage of the flow, when its dynamics are dominated by stratification effects. Staquet and Riley (1989b) have suggested that these patches could be characterized by their containing all of the potential vorticity of the flow; the waves, by contrast, would contain no potential vorticity. The present study, where one focuses upon the stage of the flow existing before the patches form, might provide some + insight into the mechanism of the patch generation.

**Acknowledments**

We wish to thank Dr J. Sommeria for his comments and suggestions regarding this work, as well as Dr J.J. Riley, who read the manuscript and made fruitful comments. Computations have been performed on the Cray 2 of CCVR (Centre de + Calcul Vectoriel pour la Recherche). Computation time has been allocated by the + Scientific Committee of CCVR.

**References**

Corcos, G.M. & Sherman, F.S. 1976 Vorticity concentration and the dynamics of unstable free shear layers. *J. Fluid Mech.* **73**, 241.

Hazel, P. 1972 Numerical studies of the stability of inviscid stratified shear flows. *J. Fluid Mech.* **51**, 39.

Hopfinger, E.J. 1987 Turbulence in stratified fluids: a review *J. Geophys. Research* **92-C5**, 5287.

Klaassen, G. P. & Peltier, W. R. 1985 Evolution of finite amplitude Kelvin-Helmholtz billows in two spatial dimensions *J. Atm. Sci.* 1321.

Koop, C. G. & Browand, F. K. 1979 Instability and turbulence in a stratified fluid with shear. *J. Fluid Mech.* **93**, 135.

Patnaik, P. C., Sherman, F.S. & Corcos, G.M. 1976 A numerical simulation of Kelvin-Helmholtz waves of finite amplitude. *J. Fluid Mech.* **73**, 215.

Scotti, R. S. & Corcos, G. M. 1972 An experiment on the stability of small disturbances in a stratified shear layer. *J. Fluid Mech.* **52**,

Staquet, C. & Riley, J. J. 1989a A numerical study of a stably-stratified mixing layer. In *Selected Papers from the Sixth Symposium on Turbulent + Shear Flows*, Springer, Berlin, 381.

Staquet, C. & Riley, J. J. 1989b On the velocity field associated with potential vorticity. *Dyn. Atmos. and Oceans* **14**, 93.

Thorpe, S. A. 1973 Experiments on instability and turbulence in a stratified shear flow. *J. Fluid Mech.* **61**, 731.

Thorpe, S. A., 1981 An experimental study of critical layers. *J. Fluid Mech.* **103**, 321.

# Gravity wave patterns in the wake of a sphere in a stratified fluid

J.M. CHOMAZ[α], P. BONNETON[α], A. BUTET[α], E.J. HOPFINGER[β], and M. PERRIER[α]

[α]*Météorologie Nationale CNRM Toulouse*
*42 avenue Coriolis ,31057 Toulouse ,France*

[β]*Institut de Mécanique de Grenoble*
*BP 53X, 38041 Grenoble-Cedex, France*

*Turbulence and Coherent Structures*
*O. Métais and M. Lesieur, eds.*
*Selected papers from "Turbulence 89: organized structures and turbulence in fluid mechanics"*
*Grenoble, 18-21 September 1989*

**Abstract**
We investigate the entire internal gravity wave field generated by towing a sphere in a linearly stratified fluid. We are able to discriminate the lee waves, generated by the virtual object composed of the sphere and the turbulent wake, from the gravity waves emited by the turbulent wake. A sharp transition is observed between the field dominated by the first type of waves and the field dominated by the second process . This transition occurs at a well defined Froude number $F_c = 4$ and is associated with the appearance of coherent structures in the wake of the sphere. These structures, when they collapse, generate waves. To avoid boundary interactions some experiments were done in the large stratified towing tank facility of the french meteorology.

## 1. Introduction

It is essentiel, in the context of oceanographical and meteorological problems, to understand stratified turbulence and in particular its interactions with internal gravity waves. Such waves propagate energy vertically and horizontally, and invalidate local parameterization. Furthermore, stratification affects vertical motion, hence introducing anisotropy. Recently, a number of experimental and numerical results concerning stratified turbulence have been published (see Hopfinger (1987) for an exhaustive review, Browand and al (1987), Lin and Pao (1979) for experiments and Riley, Metcalfe and Weissman (1981), Riley (1985) and Métais, Herring (1989) for numerical simulations). The majority of experiments deals with decaying grid turbulence in the presence of stable stratification , but wakes and jets have also been investigated. In grid turbulence, when the turbulence is strong near the grid , it evolves as if there were no stratification. Further downstream the turbulent integral lenght scale , $l$, increases and the turbulent velocity decreases so that at one stage the turbulence will feel the stratification effect which will tend to limit the vertical growth. This limitation is estimated by the Ozmidov (1965) scale $L_R$ defined by equating inertial time $({L_R}^2/\epsilon)^{1/3}$ and buoyant time $N^{-1}$

$$L_R = \left(\frac{\epsilon}{N^3}\right)^{1/2}$$

where $N$ is the Brunt-Visl frequency $N = \sqrt{-g/\rho_0 \, d\rho/dz}$, and $\epsilon$ the kinetic energy dissipation rate. When the integral scale $l$ reaches the Ozmidov scale $L_R$, internal waves are generated which interact with the turbulence .

The physical interpretation of this result is straightforward in terms of excitation frequency. For gravity waves, $N$ is a cut-off frequency. As long as the turbulence time scales have a smaller value than $N^{-1}$, there is no coupling, and the turbulence is not affected by stratification. On the other hand , when the time scale $(l^2/\epsilon)^{1/3}$ exceeds $N^{-1}$ a strong interaction occurs, converting energy from turbulence into waves. It is to be expected that the wave field will have a wave length near the maximum vertical scale, and an excitation frequency close to $N$.

In the present experiment, we test these ideas by observing the behavior of the wake of a sphere towed in a stratified tank. The visualization technique used for waves can resolve a wave field, crossing a horizontal plane, of less than 1 $mm$ amplitude. At the same time, the turbulent flow and also the wave field in a vertical plane were observed by a shadowgraph technique. The behavior of the flow is quite complex because the stratification affects all stages of the wake development. As shown by the numerical simulations of Hanazaki (1988), the geometry of the flow around the sphere and flow separation are strongly dependent on the Froude number $F$,which is the ratio of the advection frequency $U/R$ to the Brunt-Visl frequency :

$$F = \frac{U}{RN},$$

where $R$ is the radius of the sphere, U its velocity , and $N$ is in $rad\ s^{-1}$.

In this article, we will not treat in detail the full range of the Froude number from 0.3 to 12.2 but focus on the transition we have discovered between flows dominated by lee waves and flows governed by coherent structures. As we will see, this transition occurs around $F_c = 4$ and the difference of behavior below and above $F_c$ will be illustrated by three typical runs : two below $F_c$, at $F = 1.6$ and 3.2, the other at $F = 12.2$. For the other experiments, between $F = 0.3$ and $F = 12.2$, only wavelenghts as a function of $F$ are given .

## 2. Experimental Technique

### 2.1. Apparatus

Two different tanks, equiped the same way, were used. A "small" one, made entirely of glass, of dimensions 50 cm deep by 50 cm wide by 400 cm long. It was used to observe the near wave field and the turbulence. A "big" one whose dimensions are 100 cm by 300 cm by 2200 cm. This size was required for the study of the far wake and wave field which are very sensitive to wall reflections and interactions.

The sphere was 5 cm in diameter. It was towed respectively between 1.0 cm s$^{-1}$ and 65 cm s$^{-1}$ in the small tank and between 9 cm s$^{-1}$ and 40 cm s$^{-1}$ in the big one, giving Reynolds numbers of $500 < Re < 32000$.

It was very difficult to find a way to support the sphere because two opposite conditions had to be satisfied :

- absence of sphere vibration ;
- absence of disturbances in the flow.

After many trials we ended up with a very efficient set-up presented in Fig. 1. The sphere was filled with lead sand and suspended by three stainless steel wires (0.1 mm) going through three fixed pulleys and attached together . No vibrations or disturbances were visible or mesurable.

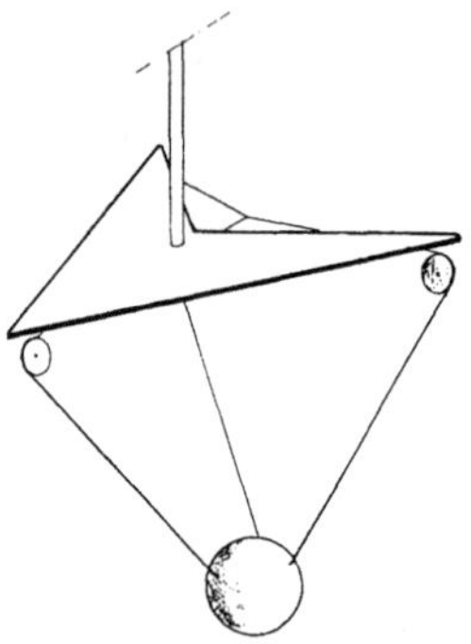

Fig.1. Schematic of the support of the sphere.

Both tanks were linearly stratified with a salt solution. The small tank was filled by gravity flow from two tanks, one full of salt solution and the other one contained initialy pure water in which a mixer was placed. The big tank used computer controled flow rate pumps, mixing directly saturated salt water with pure water. Any value of $N$ can be established in this way. In the present experiments $N = 1.25$ rad $s^{-1}$ was used in the big tank.

Several techniques were used to visualize the turbulent and wave motions. In the case of turbulence, only shadowgraph pictures were used. The wave fields was visualized by a technique discussed in the following chapter. All these images were recorded by a video camera and processed in real time or defered, by an image processing bord connected with a P.C.

## 2.2. Visualization of gravity waves

We have adopted the classical technique of fluorescence induced by laser to visualize an internal wave field in a non turbulent region. This method, associated with an image processing device, allows to monitor the vertical displacements in a whole horizontal plane. These motions of small amplitude (a few mms) and of small velocity (0.1 mm $s^{-1}$) are difficult to capture by a point probe. The energy contained in waves is nevertheless important because of their spacial extend.

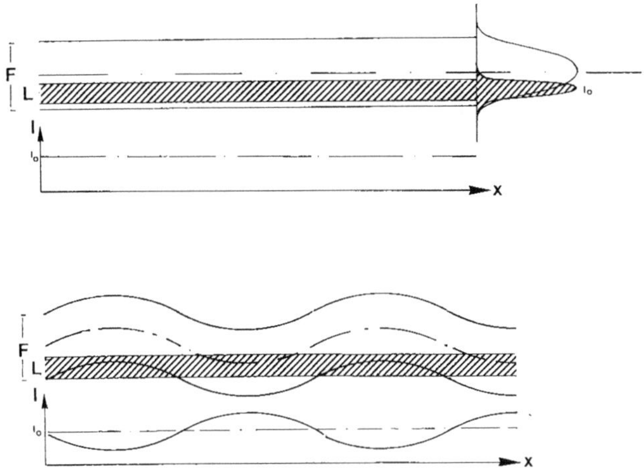


Fig.2. Gravity wave visualization : Side view of the fluoresceine plane F and laser sheet L, and the corresponding light intensity I .(a)respective position at rest; (b) respective position when a wave deforms the fluoresceine plane

The experimental trick to achieve such a precision consists in dying with fluoresceine a single thin sheet of fluid. This sheet is horizontal and lit with an even thinner laser plane (Fig. 2a). At rest, the whole plane is uniformly bright. When an internal wave passes by, it displaces particules of fluid up and down, making brightness to vary. Iso-intensity of the perturbation characterise the isophases of the internal gravity wave field (Fig. 2b) .

The experimental set up is sketched in Figure 3. The sequence of the manipulations is as follow : A coton thread (5), soaked in fluoresceine dye and stretched on a frame (4) is towed horizontally through the fluid. The motion is sufficiently slow and regular, to avoid vortices behind the thread ($Re < 1$).

We obtain a homogeneous horizontal fluoresceine plane, the thickness of which can be controlled by the towing speed and the characteristics of the thread. The thickness will slowly vary and increase by molecular diffusion. To get quantitative measurements, we had to take into account this thickness variation or run the experiments under the same conditions.

The second step consists in creating a laser sheet using, in our case, a vibrating mirror and to make it coincide with the horizontal fluoresceine plane. This adjustment is very precise and needs a certain know-how

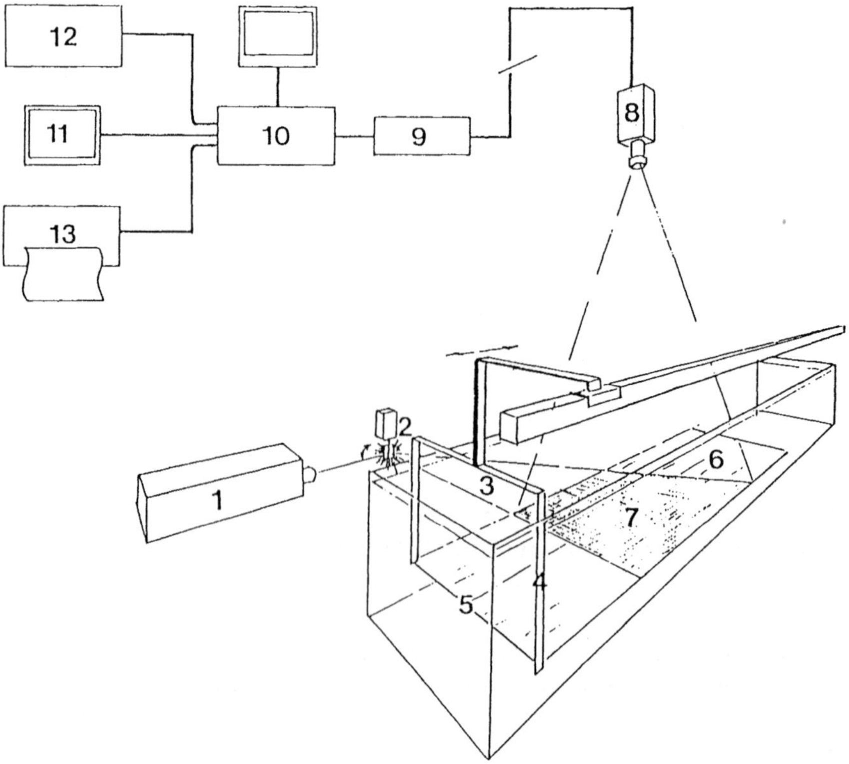


Fig.3. Schematic diagram of the wave visualization technique : 1. power laser , 2. oscillating mirror , 3. horizontal laser sheet , 4. slowly moving frame , 5. coton yarn soaked in fluoresceine dye , 6. horizontal fluoresceine plane , 7. observation plane ( coplanar intersection of 3. and 6.) , 8. TV camera , 9. to .13. image processing system , 10. PC with an image processing board , 11. image display , 12. laser printer output , 13. color prints.

when used in the 22x3x1 $m^3$ facility. With a Brunt-Visl frequency of 1.25 rad $s^{-1}$, the radius of curvature of the laser ray is 20 m, and for a 3 m long trajectory, it becomes really tricky. In order to discriminate upward and downward motions, the laser sheet should be off center and thinner than the fluoresceine one (figure 2). The visualization will be efficient only if the experiment preserves a non-turbulent zone where it will be possible to observe the gravity wave field. Had the flow been turbulent at the location of the plane, the fluoresceine would disperse.

## 3. Gravity waves

### 3.1. Linear Theory

Internal gravity waves are a specific feature of stratified flows. Basic understanding involves a good knowledge of linear wave theory (see for instance Lighthill, 1978). The description may be limited to the dispersion relation (relation between $\mathbf{k}$ and $\omega$) for a planar wave in $\exp(i(\mathbf{k}.\mathbf{x}-\omega t))$

$$\omega^2 = N^2 \left(\frac{|\mathbf{k_H}|}{|\mathbf{k}|}\right)^2, \tag{1}$$

where $\mathbf{k_H}$ is the horizontal projection of the wave vector $\mathbf{k}$. From (1) we can see that the value of $\omega$ fixes the angle of $\mathbf{k}$ with the horizontal, but not the modulus. This degeneracy explains why in each experiment, a length scale selection mechanism had to be specified.

From (1) we deduce also that $N$ is a cut-off frequency for gravity waves. For these two reasons the interpretation of the wave-turbulence coupling in terms of time scales seems more physical than an interpretation in terms of length scales.

Figure 4 presents the main characteristics of a linear wave. Equation (1) gives the way the energy is propagated through the stratified medium: the group velocity, $\mathbf{v_G} = d\omega/d\mathbf{k}$ is orthogonal to the phase velocity $\mathbf{v}_\varphi = (\omega/|\mathbf{k}|^2)\mathbf{k}$ . Their vertical projections are in opposite directions and their horizontal projections are colinear and in the same direction.

From a side view, the propagation of energy follows the isophase. On the contrary, in a horizontal plane, the energy propagates itself in a direction normal to the isophases.

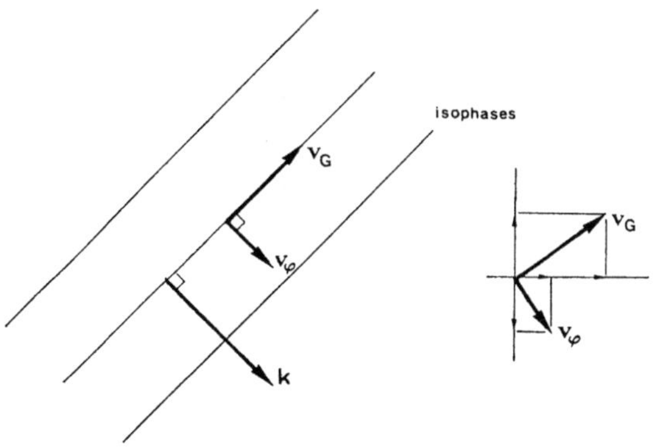


Fig.4. Representation of $\mathbf{v_G}$ and $\mathbf{v}_\varphi$ in a vertical plane

As we observe the waves crossing a horizontal plane away from the excitation zone, we need to discuss the dependence of the information propagation speed on **k**. In a set of waves associated with the same length scale, $1/k$, the fastest upward moving wave, corresponds to a propagation at 45° because :

$$v_{G_z} = -N\frac{k_z|\mathbf{k_H}|}{|\mathbf{k}|^3} \tag{2}$$

We also observe from (2) that waves with small values of $k$ (large scale) propagate faster than those of short wave length .

For an excitation frequency, $\omega$, close to the Brunt-Visl frequency $N$, the wave vector tends to be horizontal (vertical isophases). The corresponding group velocity points vertically and vanish when $\omega$ equals $N$. Thus, excitations close to the Brunt-Visl frequency propagate nearly vertically and very slowly. They characterize the behavior of the wave field a long time after it has been emitted.

From these considerations, we conclude that :

- oscillating a material point at a certain frequency will select the angle of wave propagation but not the length scale, thus a local excitation must be characterized by a frequency and a lenght scale made to correspond to a well set physical problem.

- moving a material point at a velocity **U**, gives a length scale. The two quantities **U** and $N$ are sufficient to characterize the whole wave field, except the amplitude. The amplitude is specified when the material point is replaced by an object of finite dimensions (a sphere in our case).

## 3.2. Lee waves generated by a moving material point P

We looked for the wave field moving with the object. The shape of the isophases is fixed in the moving frame. At each location M in the moving frame, the asymptotically selected **k** vector is such that

$$\mathbf{PM}//\mathbf{v_G} - \mathbf{U}, \text{ and of same direction.}$$
$$\mathbf{v_G} \text{ computed for } : \omega = \mathbf{k.U}$$

Taking the geometric optic approximation for computation of isophase surface $\varphi = \mathbf{k.PM}$ , the equation scaled by $U/N$ becomes:

$$\begin{cases} x = \varphi\left(1-(z/\varphi)^2\right)\dfrac{1}{\sin\theta} \\ y = \pm\, z\sqrt{\cos^2\theta-(z/\varphi)^2}\,\dfrac{1}{\sin\theta} \end{cases} \tag{3}$$

where $\theta$ is the angle of **k** with the vertical.

## 4. Results

### 4.1. Lee waves regime

The experimental technique described in section 2 allowed us to obtain an image of the wave field in a horizontal plane ($z = z_o$). For small Froude number, we observe a strong lee wave which we may compare with linear theory developped previously. Figure 5a shows the image for $F = 1.6$ and $z_0 = 4.5$ cm measured from the center of the sphere of radius 2.5 cm ; the Brunt-Visl frequency is here fixed at $N = 1.25$ $rad\ s^{-1}$.

The large amplitude of the waves makes the visualization strongly non linear. The dye zone passes above and below the lighted zone, making the isophases $\varphi = n\pi$ visible as dark fringes. Figure 5b compares the observed isophases with linear theory. The agreement is very good especialy if one keeps in mind that the theory is derived for a material point and the experiments were done with a 2.5 cm radius sphere.

The sphere creates a mean flow which affects the lee wave evolution (Figure 6). At a late stage, the lee wave corresponds to a perturbation at the Brunt-Visl frequency with a phase velocity equal to U and a vertical, nearly zero group velocity. Therefore, a particle of fluid keeps its oscillating behavior but moves with the mean flow. This remark explains the deformed shape of the lee wave observed at long time. This deformation is slow, and we may still define the lee wave at any time and any location as the gravity wave field moving with the sphere.

For $F = 3.2$ we display in Plate 1 the full sequence of pictures obtained at different times . The first picture of the lee wave on top of the sphere ($tN = 0$) is not processed. The others are obtained by subtracting two images delayed by $\Delta t.N = \pi$ in order to remove fixed echos and to increase the signal to noise ratio. Figure 7 shows a side view shadowgraph picture of the same flow. The turbulent wake of the sphere is greatly affected by the stratification. Its vertical excursion is limited to the size of the sphere. The spiral vortex, present in homogeneous fluids, is inhibited by stratification at this Froude number. From shadowgraph pictures as well as turbulence measurements made by O. Thual et al. for the same conditions, we observe that the turbulent wake collapses between $Nt = 2$ and $Nt = 10$. In Plate 1, we note the existence of small scale waves superimposed on the lee wave. Their typical size correspond to $\lambda/D \sim 1$. These waves are generated by the largest and most energetic structures present in the turbulent zone. The size of these structures is, as observations show, limited to D. The same is true for $F = 1.6$ where the wake is limited to a fraction of the diameter ($\sim 0.7D$) and the longest wave emited by the turbulent zone is of the same size ($\lambda/D \sim 0.7$).

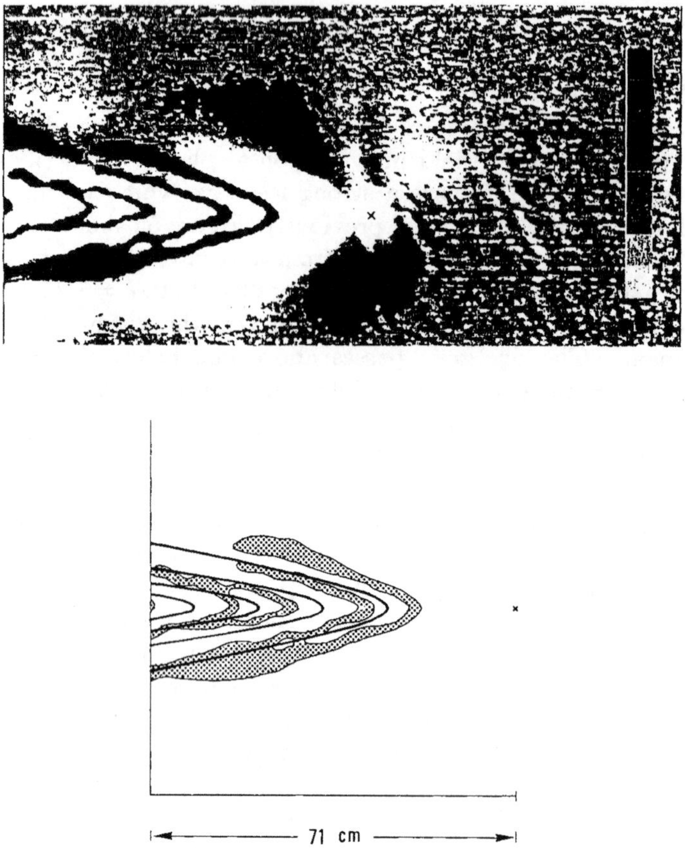


Fig.5. Phase lines of lee waves generated by a sphere of radius 2.5 cm for $F = 1.6$. (a) Visualization in a horizontal plane, 4.5 cm above the center of the sphere. (b) Comparison with linear theory.

### 4.2. Coherent structures regime

When $F > 4$ the lee waves disappear (within the precision of the visualization) and the turbulent wake of the sphere starts spiraling and generates, fairly coherent, upward and downward turbulent motions. These coherent structures have a vertical development limited by buoyancy effects. Their collapse generates gravity waves and also reactivates the small scales of the turbulence. This sequence is illustrated in Plate 2 for $F = 12.7$. The set of pictures are shadowgraph side views, where the non uniformity of the lighting is corrected and a histogram equalization algorithm is applied in order to make the gravity waves visible. The gravity wave field looks like the one generated by an impulsively moved

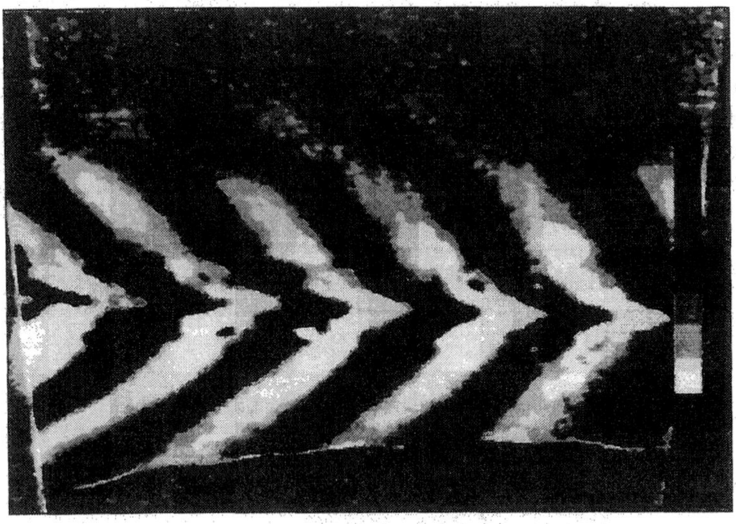

Fig.6. Horizontal visualization of lee waves, for $F = 1.6$, at 30 seconds ($Nt = 37.5$) after passage of the sphere.

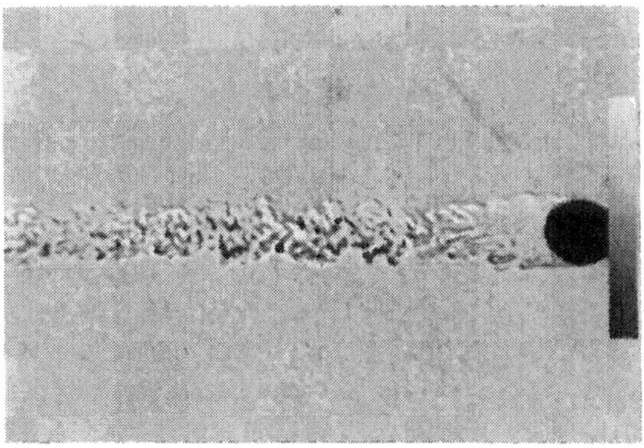

Fig.7. Shadowgraph side view of the wake of the sphere, for $F \simeq 3.2$.

object which has the size of the coherent structures (impulse response to a perturbation of typical size $3D$).

Plate 3 presents the horizontal wave field 12.5 cm above the center of the sphere, obtained in a similar experiment in the big tank. Because of light absorption problems, only one half of the horizontal field is completely visible. The full sequence is recorded by a video camera ; no signal moving at the speed of the sphere is measurable. Thus, at all time, the wave field is free from lee waves. Waves are a result of the collapse of the large scale turbulent eddies. After $Nt = 36$, a wave field with a nearly coherent pattern and of frequency $N$ emerges. Its wave length decreases very slowly and is about $4.2D$. The associated horizontal phase speed is $0.09U$.

In Figure 8 are ploted the measured mean values of $\lambda/D$ versus time. The first part of the curve corresponds to the propagation of waves, emited by each single collapse, with a frequency different from $N$. The waves of smaller wave length cross the plane later. The maximum wave length is equal to the size of the helicoidal vortex structures. The second part is the asymptotic wave field near to the Brunt-Visl frequency, generated by the coherent effect of all the collapses happening with a phase defined by the shedding frequency of the sphere .

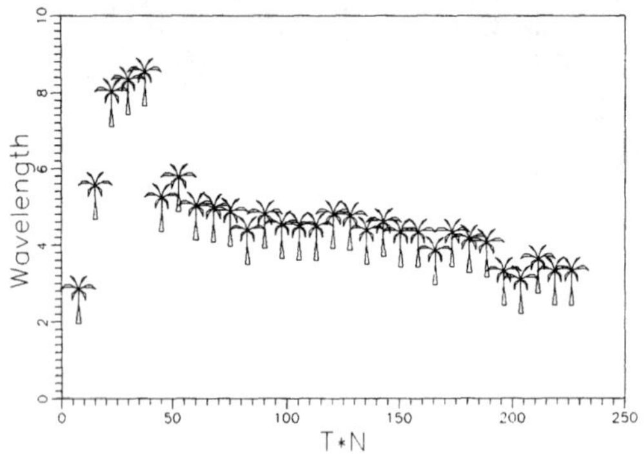


Fig.8. Internal wave length, normalized by $D$ , as a function of nondimensional time $Nt$, for $F = 12.7$.

### 4.3. Global View

Figure 9 summarizes the characteristics of the observed horizontal wave fields. It presents the measured asymptotic wave length $\lambda/D$ as a function of Froude number. The Reynolds number is proportional to the Froude number in the present set of experiment ; for $F = 3.2$ we have $Re = UD/\nu = 5000$. Figure 9 is composed of several branches. The lee wave branch corresponds to waves moving at the horizontal wave speed equal to U. It exists up to $F \sim 4$ and is well described by linear theory. When $F > 4$, another branch takes over and is representative of the waves generated by the collapse of the spiral vortex structures generated behind the sphere. In these conditions the associated phase speed is small compared to U. The change from lee waves to a wave field generated by the coherent structures of the turbulent wake seems well defined and characterized by a Froude number of $F_c = 4$ which we refer to as critical Froude number for the wave field. Other critical Froude numbers may be associated with changes in flow separation (Hunt and Snyder 1980).

In figure 9 is also plotted the smallest mean value of wave length which appears during the evolution. Before the transition it seems linked to the largest energetic structures of the turbulent wake (of about the vertical extent of the wake).

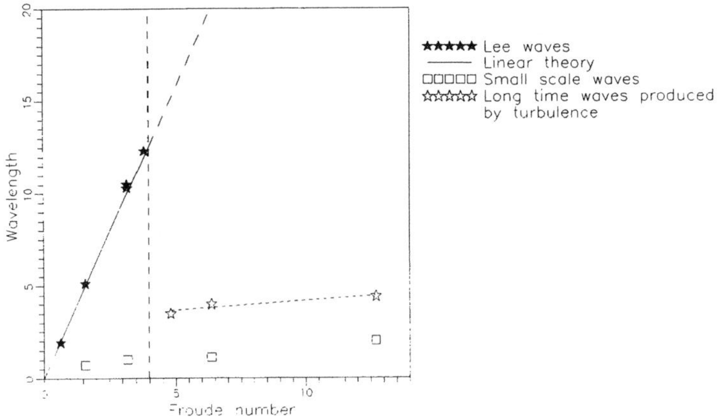


Fig.9. Internal wave length, normalized by $D$, as a function of $F$.

## 5. Conclusion

We have presented results concerning the internal wave field generated by a sphere moving through a stratified fluid. Reynolds numbers are in the range $500 < Re < 32000$ and Froude numbers based on the radius range from 0.5 to 13. The results show a clear transition at $F_c = 4$ from the lee wave dominated wave field to a field governed by the collapse of the turbulent coherent structures shed behind the sphere. In spite of the relatively large amplitude of the lee waves, comparison of the lee wave pattern with linear theory is found to be remarkably good. The turbulent wake wave field which dominates when $F > F_c$ is qualitatively similar to that generated by an impulsively started obstacle at size $3 - 4.5D$.

First observations of the far wake dominated by quasi two dimensional motion indicate that this transition alters radically its evolution. In the coherent structure regime, the flow stays uncorrelated on the vertical during a very long time ($Nt > 1500$).

This work was supported by the Météorologie Nationale Française and contract DRET number 88-126. It was made possible by the team SPEA and we wish to thank for their kind help, enthousiasm and efficiency : J.C BOULAY, C. NICLOT, M. NICLOT, S. LASSUS-PIGAT, B. BEAUDOIN and H. SCHAFFNER. We have benefitted from discussions with J.B. FLOR.

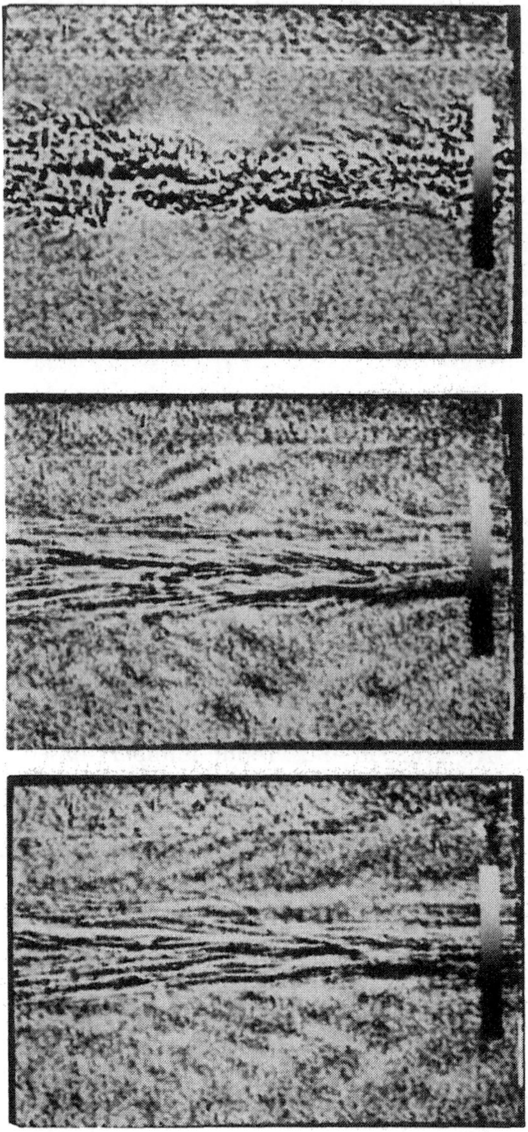

Plate2. Shadowgraph side view of the wake of the sphere , for $F = 12.7$ (a) $Nt = 7.5$ ; (b) $Nt = 18.8$ ; (c) $Nt = 50.3$. The sphere is moving from left to right.

**References**

Browand, F.K, D. Guyomar, & S.C. Yoon 1987 . The behavior of a turbulent front in a stratified fluid : Experiments with an oscillating grid. *J. Geophys. Research*,**92**, pp 5329-5341 .

Dickey, T.D., & G. L. Mellor 1980 . Decaying turbulence in neutral and stratified fluids. *J. Fluid Mech.*, **99**, pp 13-31.

Hanazaki, H. 1988 . A numerical study of three-dimensional stratified flow past a sphere. *J. Fluid Mech*, **192**, pp 393-419.

Hopfinger, E.J. 1987 . Turbulence in a stratified fluids : A review, *J. Geophys. Research*, **92**, pp 5287-5303 .

Hunt,J.C.R & Snyder,W.H 1980 . Experiments on stably and neutrally stratified flow over a model three dimensional hill. *J. Fluid Mech*, **96**, pp 671-704 .

Itsweire, E.C., K.N. Helland, & C. W. Van Atta 1986. The evolution of a grid-generated turbulence in a stably stratified fluid. *J. Fluid Mech.*, **162**, 299-338 .

Lightill J. 1978 *Waves in fluids.* Cambridge University Press.

Lin, J.T., & Y.H. Pao 1979 : Wakes in stratified fluids. *Ann. Rev. Fluid Mech.*, **11**, 317-338 .

Métais, O.J., & J.R. Herring 1989 : Numerical simulations of freely evolving turbulence in stably stratified fluids. *J.Fluid Mech.*, **202**, 117-148.

Riley, J.J., 1985 : A review of turbulence in stably stratified fluids, Seventh Symposium on Turbulence and Diffusion, Boulder, Colorado, USA, Amer. Meteorol. Soc., 377 pp.

Riley, J.J., R.W. Metcalfe, & M.A. Weissman, 1981 : Direct numerical simulations of homogeneous turbulence in density stratified fluids.*Proc. AIP Conf. Nonlinear properties of internal waves.* Bruce J. West, Ed., 79- 112.

Thual.O. 1987 : Questions about an experiment on the wake of a sphere in a stratified fluid. Proc. of the workshop, les Houches, FRANCE, March 10-18 .

# VIII. Compressible and Reacting Flows

# The compressible mixing layer

Dany VANDROMME[α,γ], Hieu HAMINH[β,γ]

[α] *URA CNRS 230, INSA de Rouen*
*BP 8, 76131 Mont Saint Aignan-Cedex, France*

[β] *URA CNRS 005, IMF Toulouse*
*Av du Prof. C. Soula, 31400 Toulouse, France*

[γ] *C.E.R.F.A.C.S*
*42, Av. Gustave Coriolis, 31057 Toulouse Cedex, France*

*Turbulence and Coherent Structures*
*O. Métais and M. Lesieur, eds.*
*Selected papers from "Turbulence 89: organized structures and turbulence in fluid mechanics"*
*Grenoble, 18-21 September 1989*

## 1. Motivation of the study.

Despite important efforts of many scientists over the decades, turbulence continues to persist as one of the least understood field of fluid mechanics, and therefore much improvements in fluid flow systems are still expected from a better understanding of this physics.

Turbulence research has experienced two deep transformations in the past two decades: *i*) the first is the discovery of organized structures in turbulent flows (BROWN and ROSHKO, 1974), *ii*) the second is the integration of fluid motion partial differential equations, using powerful supercomputers.

Indeed, because of new visualization techniques and efficient spectrum analysis methods, it has been found that the unsteady features are quite often present in complex turbulent flows, even in compressible flows, and more precisely in complex situations of industrial interest.

These natural unsteady features, called also *organized structures* have two important consequences: *i*) significant modifications of *mixing properties* and *ii*) reduced confidence into classical statistic turbulence modelling because of *unsteadiness.*

Typical example of such natural unsteadiness exists in the mixing layers. Since the famous experiment of BROWN and ROSHKO (1974), various studies have been devoted to the free shear layer, experimentally and also by direct numerical simulations. But most of them have been devoted to incompressible flows.

Interest is growing again for the compressible turbulent mixing layer, because of today's increasing interest for supersonic combustion systems. Although experiments and direct simulations are still needed to improve our knowledge of the physics and also our prediction capabilities, it seems that in the near future, the use of the turbulence modelling will be the only way to predict mean and turbulent quantities in complex flows.

However, a raw extension of classical statistical turbulence models to unsteady flows is not straightforward: many physical hypothesis on which the classical models rely, are only valid in very restrictive cases. They must be reconsidered again in perspective of different identified characteristic time scales present in the flows.

The compressible mixing layer study presented here has two objectives:

*i*).- Testing the capability of turbulence modelling to handle correctly natural unsteadiness appearing in compressible turbulent flows.

*ii*).- Analyzing physical aspects linked to the unsteadiness problem and the role of various flow parameters.

In a next step, the objective of the study will be to improve eventually the classical turbulence models for numerical predictions of compressible unsteady turbulent flows, by comparison with: *i*) direct numerical simulation data, and *ii*) experimental results.

## 2. General context of turbulence in fluids.

Nowadays, it is well accepted that all turbulent flows contain *both* **ordered** and **disordered** features. Their respective proportions depend on the Reynolds number, on the flow geometry, on the boundary and initial conditions ... The **organized structures** and the **random structures** are more or less perceivable according to the analysis facilities. The basic question for engineers is: does the numerical prediction of real flows need to distinguish between these different kinds of structures ? The answer depends on the current status of the classical turbulence model capability in the numerical predictions of the complex turbulent flows.

Although we believe that there exist no "**universal turbulence models**" to describe *all* industrial **complex** flows, we must also recognize that current models can already produce satisfactory results, even in complex three dimensional flows. Up to what limit are the classical statistic turbulence models valid?

## 3. Back to the statistical modelling validity

Statistical modelling lead to relatively satisfactory results when spectral equilibrium is obtained for the turbulence energy, according to the Kolmogorov theory. Therefore, it is only true for the fully-developped turbulence or "**old turbulence**" !

Unfortunately, in finite size industrial facilities the flows are rather concerned by ... the "**young turbulence**" ! This "just born" turbulence, following the *transition*, is characterized by: *i*).- The simultaneous presence of both regimes: laminar and turbulence; *ii*).- A non-equilibrium energy spectrum.

Consequently, it is more judicious to redefine the physical variable decomposition for modelling.

**4. Organized structures and decomposition problem.**
According to HUSSAIN (1983), a coherent structure is a *connected, large-scale quantity of turbulent fluid with a phase-correlated vorticity over its spatial extent.* Thus, the coherent (organized) structures would occur at *random phases* and their description must involve *phase-dependent* information. Let us recall the definition of *phase averaging*: it is an *ensemble average* of any property of fluid at a *particular phase* and a *particular location* with respect to the boundary of the structure:

$$< f(x_k, \phi) > \;=\; \lim_{N\to\infty} \frac{1}{N} \sum_{i=1}^{N} f_i \left[x_k, \phi + \phi_i(t)\right] \tag{1}$$

where $\phi_i(t)$ denotes the random phases of occurence of the successive structures at the location $(x_k)$ toto When these structures occur at regular intervals (a situation that essentially occurs in Karman vortex shedding for instance or when structures are induced by a controlled excitation), this definition reduces to the *periodic phase average*:

$$< f(x_k, \phi) > \;=\; \lim_{N\to\infty} \frac{1}{N} \sum_{i=1}^{N} f_i \left[x_k,\; t + iT\right] \tag{2}$$

where $T$ is the period of occurence of the structure and $t$ is the instant corresponding to the *reference phase.*[1]

Because the phase-average provides time-dependent description, it gives more information than what is looked for in most engineering situations, where one is interested in *time-average*:

$$\bar{f}(x_k) \;=\; \lim_{T\to\infty} \frac{1}{T} \int_{-T/2}^{+T/2} f(x_k, \tau) d\tau \tag{3}$$

If $\bar{f}$ is considered as a time-averaged quantity, the starting time is irrelevant.

Then, let us use the triple decomposition, suggested by REYNOLDS and HUSSAIN (1971):

$$\underbrace{f(x_k, t)}_{Instantaneous\ quantity} \;=\; \underbrace{\bar{f}(x_k)}_{a} + \underbrace{f_c(x_k, t)}_{b} + \underbrace{f_r(x_k, t)}_{c} \tag{4}$$

[1] In a context of separated flows (the flow around a circular cylinder for example), this reference phase must be determined experimentally, without special difficulty. Otherwise, on a theoretical point of view, a more mathematical thorough analysis is desirable.

Part $(a)$ representing the time-averaged value, which does not depend on time, is a quantity easily accessible by experiments, provided that the measurements are made during a time long enough, compared to characteristic turbulence time scales.

Part $(b)$ specifies the *coherent* or *organized unsteadiness.* We believe that this part $(b)$ have a *deterministic character*: meanning that it could be *numerically predicted* under some conditions. Coherent unsteadiness should come from geometrical properties of the flow, of the initial and boundary conditions. The signatures of these organized structures are more or less sharp according to the physics of the problem: vortex shedding behind a bluff body should be more organized than a simple free shear layer.

Finally, the last part $(c)$ indicates all *random fluctuations.* In fact, this "*randomness*" denomination is used as an "apparently chaotic" sense and not with a "by chance" meaning. Let us recall that turbulent flows are characterized by the simultaneous presence of eddies of different sizes and frequencies, and their interactions appear as "disordered". It is generally admitted that if computers exist with enough power in speed and in memory, it should be possible, in principle, to simulate numerically these eddies, even those which are "apparently" random.

If we regroup now the two components parts which can "reasonably" be numerically predicted in a deterministic sense, such as the time-averaged part $(a)$ and the coherent unsteady features $(b)$, then we obtain a new quantity called **phase averaging** (which is also the classical **ensemble average**), which supports all statistical processes in turbulence modelling. Therefore, *phase average* are the *ensemble average* of all (successive) structures at the "same age" or "phase" in their evolution. Then:

$$< f(x_k,t) > \;=\; \bar{f}(x_k) \;+\; f_c(x_k,t) \tag{5}$$

Consequently:

$$f(x_k,t) \;=\; < f(x_k,t) > \;+\; f_r(x_k,t) \tag{6}$$

This relationship (6) leads to some remarks:

- If the part $f_r(x_k,t)$ is "minimized", we obtain:

$$f(x_k,t) \;=\; < f(x_k,t) > \;+\; \underbrace{f_r(x_k,t)}_{\longrightarrow 0}$$

At the extreme limit, when $f_r(x_k,t) = 0$, the decomposition leads to a *direct simulation.* If this random part $f_r(x_k,t)$ is not equal to zero but is still smalll, then we have a Large Eddy Simulation (the classical L.E.S.), which must be complemented by some *sub-grid scale modelling* for the random part. But in that case, the cut-off frequency of the filter (splitting between random and deterministic motion) is imposed by the mesh size.

- Otherwise, if the unsteady coherent part $f_c(x_k, t)$ is "minimized" (no phase information in the averaging process), the deterministic part reduces to the time-mean quantity $\bar{f}(x_k)$ (which is only possible for stationnary processes), then:

$$f'(x_k, t) \;=\; f_c(x_k, t) \;+\; f_r(x_k, t) \;=\; f(x_k, t) \;-\; \bar{f}(x_k) \quad (7)$$

This quantity $f'(x_k, t)$ is nothing else than the classical fluctuations (abusively called the *random fluctuations*), and we are back to the classical statistical modelling for turbulence, with all weaknesses whixh have been identified previously.

Consequently, the decomposition (6) appears as an intermediate approach, between the classical turbulence modelling, rather simple to handle, but insufficient, and the direct simulation, a more "formal" approach, but unrealistic or even impossible for complex industrial flows. Although this approach is similar in spirit to the Large Eddy Simulation method, it is different because of the cut-off filter which does not depend uniquely on the mesh cell size. For this reason, this intermediate approach is called the Semi-Deterministic Modelling (S.D.M.).

In the S.D.M. approach, all quantities $< f(x_k, t) >$ will be determined from the numerical solution of a partial differential equation set, while the quantity $f_r(x_k, t)$ effects will be included in *turbulence models* appearing in partial differential equations belonging to the previous set.

It is worth to point out that in this decomposition, we have also the same property:

$$< f_r(x_k, t) > \;=\; 0 \quad (8)$$

which is equivalent to the ensemble average of the fluctuating part in the classical Reynolds decomposition.

By using an **ensemble averaging**, the Navier-Stokes equations lead to the *same* formal Reynolds equation set containing the *same unknown correlations* to be modeled.

The *transport equations* for unknown correlations can be obtained by the *same* way than for the classical time-mean correlations. But, classical closures need to be remodelled, including unsteady features.

## 5. Turbulence modelling for compressible flows

Although that mass-weighted average (FAVRE decomposition) is not able to describe all effects of variable density, for instance the turbulent mass fluxes, i.e. the density-velocity correlations, unsteady features due to density variations, mainly by the mean density, are correctly included in this decomposition.

Mass weighted average leads *formally* to the same equations to be solved, including turbulence quantity transport equations. Since the pressure is still splitted according to Reynolds decomposition, extra terms representing compressibility effects may be present in the equation set, since $\overline{u''} \neq 0$ and $\overline{p' \frac{\partial u_i''}{\partial x_i}} \neq 0$.

In the present work, a transport equation turbulence model, including compressibility effects (terms CT), is used: equation is written now as:

$$\frac{\partial}{\partial t}(\bar{\rho} k) + \frac{\partial}{\partial x_k}(\bar{\rho}\, \widetilde{U_k}\; k) = -\bar{\rho}\, \widetilde{u_i u_j} \frac{\partial \widetilde{U_i}}{\partial x_j} + \frac{\partial}{\partial x_k}\left(\frac{\mu_t}{\sigma_k}\frac{\partial k}{\partial x_k}\right) - \bar{\rho}\epsilon + \mathrm{CT} \quad (9)$$

$$\frac{\partial}{\partial t}(\bar{\rho}\epsilon) + \frac{\partial}{\partial x_k}(\bar{\rho}\, \widetilde{U_k}\; \epsilon) = C_1 P_k \frac{\epsilon}{k} - C_2\, \bar{\rho}\, \frac{\epsilon^2}{k} + \frac{\partial}{\partial x_k}\left(\frac{\mu_t}{\sigma_\epsilon}\frac{\partial \epsilon}{\partial x_k}\right) + \mathrm{CT} \quad (10)$$

For compressible turbulent flows, a specific energy equation is needed. Finally, neglecting the fluctuations of the laminar viscosity, this energy equation can be written as:

$$\begin{aligned}\frac{\partial}{\partial t}(\bar{\rho}\widetilde{E}) + &\frac{\partial}{\partial x_k}((\bar{\rho}\widetilde{E} + \bar{p} + \frac{2}{3}\bar{\rho}k)\widetilde{U_k} - \widetilde{U_j}(\mu + \mu_t)\widetilde{S_{kj}}) \\ &+ \frac{\partial}{\partial x_k}(-\mu_k \frac{\partial k}{\partial x_k} - (\gamma\, C_v\, \delta_{kj}\, \frac{\mu_t}{Pr_t} + \delta_{kj}\, K)\frac{\partial \widetilde{T}}{\partial x_j}) = 0\end{aligned} \quad (11)$$

More details about turbulence modelling in compressible flows are given by HAMINH and VANDROMME (1989) and VANDROMME (1989).

## 6. Numerical technique.

For these numerical simulations, the explicit-implicit method of MACCORMACK (1981) (2nd order accuracy in space and in time) is used. It is a general Navier-Stokes code for unsteady compressible turbulent flows previously applied to several complex flow cases (supersonic compression corner, supersonic expansion, transonic shock interaction...), see for instance HAMINH *et al.* (1985) and VANDROMME-HAMINH (1985, 1986). The mixing layer is treated as a two-dimensional problem on which is living a three-dimensional turbulence (a three-dimensional version of the code is currently developped for the same flow geometry).

## 7. Application to a compressible turbulent free shear-layer

Since it is possible to simulate unsteady turbulent flows by dividing these motions into an "**organized**" part for which equations of motion are solved and a remaining "**incoherent**" part represented by a turbulence model, this method is applied to calculate the mixing layer originated by a backward-facing step (without the bottom wall). Although both two-equation and second-order turbulence models are able to yield reasonnable results, we present here only preliminary results obtanined by a two-equation model $\bar{k}, \bar{\epsilon}$.

External Mach number scale from 0.8 to 3, internal Mach number being zero (convective Mach number is ranged from 0.4 to 1.5). Reynolds number ranged from $7.42 \times 10^6$ to $1.41 \times 10^7$. $(190 \times 80)$ grid is used for all calculations presented.

## 8. Preliminary results.

Preliminary results show that global characteristics of the simulated flows, for instance the spreading rate of the mixing layer, depend on characteristic Mach number.

Figure 1 presents density and pressure contours for the external Mach number 1.4 case. The results shows the existence of low pressure zone in the core of the structures where the vorticity is constant, and high pressure zone in the streching parts between vortices. It is also interesting to point out that graphic presentations of different quantities may reveal complementary physical features.

Figures 2, 3 and 4 show respectively the time evolution of large structures in Mach=1.4 case, by their density, pressure and iso-Mach contours. It is interesting to point out the formation of a strong compression zone on the back of the large structure.

Figures 5, 6 and 7 present the influence of the Mach number. They show that unsteady features appear naturally without any artificial perturbation or excitation, for external Mach number lower than 1.6, corresponding to a convective Mach number equal to 0.8. For larger Mach numbers, unsteadiness seems to be progressively but strongly damped. This fact agrees with the experimental work of PAPAMOSCHOU and ROSHKO (1988). Previous results presented concern mainly the earliest structures. Longer (in time) calculations are needed to obtained more complete description and understanding of the flow.

## 9. Concluding remarks.

- For General Unsteady Turbulent Flows:
  - → It is possible to simulate unsteady turbulent flows by dividing these motions into an "**organized**" part for which equations of motion are solved and a remaining "**incoherent**" part represented by a turbulence model.
  - → Two-equation turbulence models and second-order turbulence models are able to yield reasonnable results.
  - → Preliminary results show that global characteristics of the simulated flows, for instance the spreading rate in mixing layers, depend on the turbulence model used.
  - → Improving of the turbulence modelling needs more experimental data (conditional sampling) and direct simulation data (higher Reynolds numbers).
- For Specific compressible unsteady turbulent flows:
  - → Graphic presentations of different quantities may reveal complementary physical features.
  - → Strong compression zones are observed in rapid flow part, but shocklets don't yet occur.
  - → Preliminary results show that: Unsteadiness is observed only for Mach numbers (high speed flow) under 1.8 (corresponding to convective Mach number $\simeq 0.9$. For larger Mach numbers, unsteadiness seems to be pregressively but strongly damped.
  - → Previous results presented concern mainly the earliest structures. Longer (in time) calculations are needed to obtained more complete description and understanding of the flow.

**Acknowledgments.**

This work has been possible with a grant of NASA-Ames Research Center to the University of California Davis (NCC2-186) monitored by Dr J.R. Viegas, Dr J.G. Marvin and Dr R.W. Rubesin. The authors are indebted for their continous collaboration in this work. They express also their many thanks to Pr. R.W. MacCormack and to Pr W. Kollmann for their continual advises.

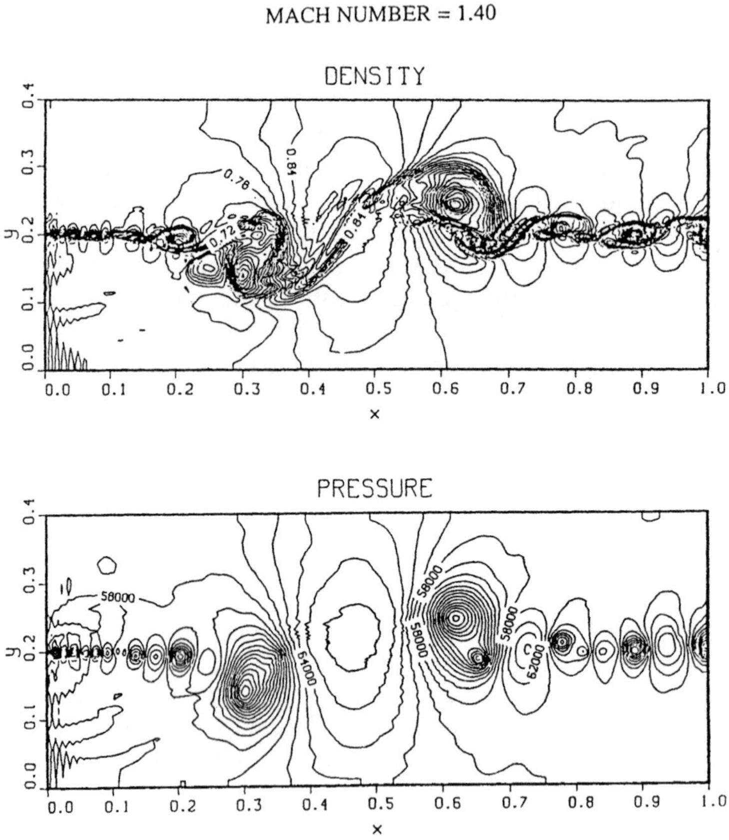


Fig.1. Density and Pressure Contours for Mach = 1.4 case.

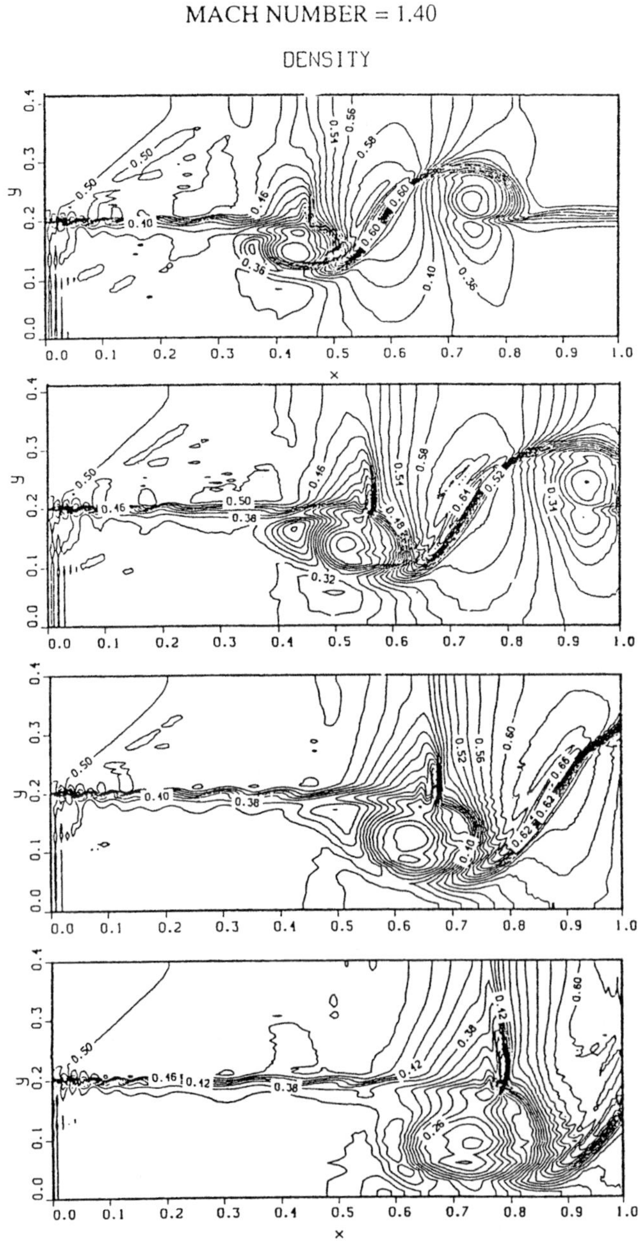


Fig.2. Time-evolution of density contour for Mach = 1.4 case.

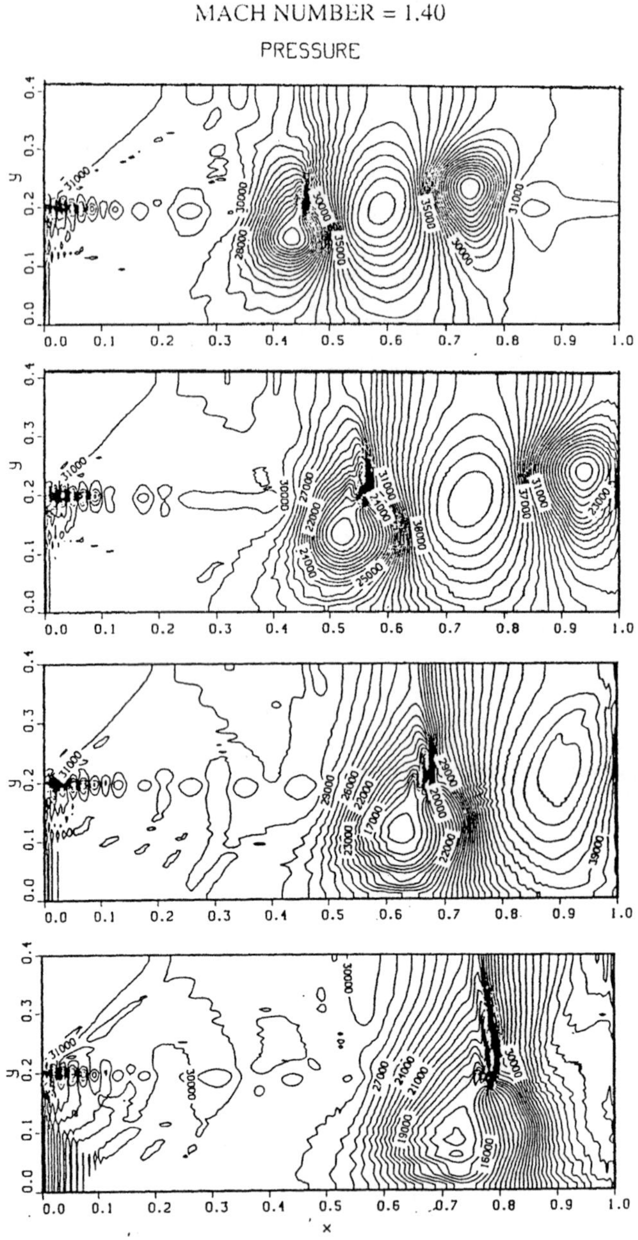


Fig.3. Time-evolution of pressure contour for Mach = 1.4 case.

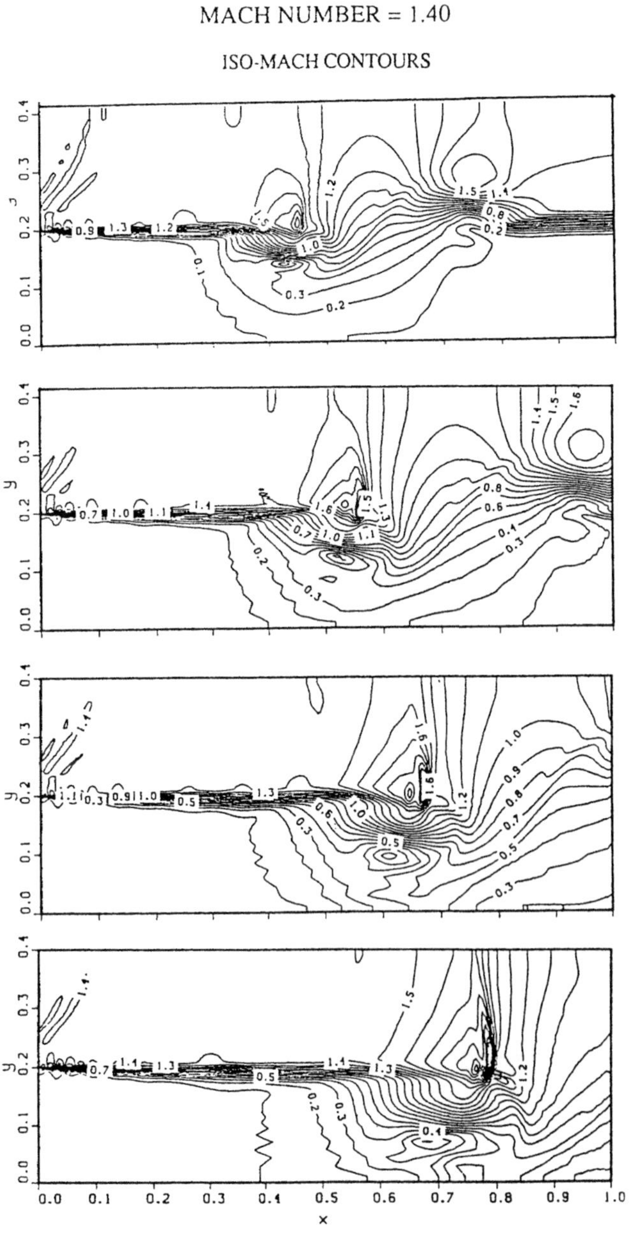


Fig.4. Time-evolution of iso-Mach contour for Mach = 1.4 case.

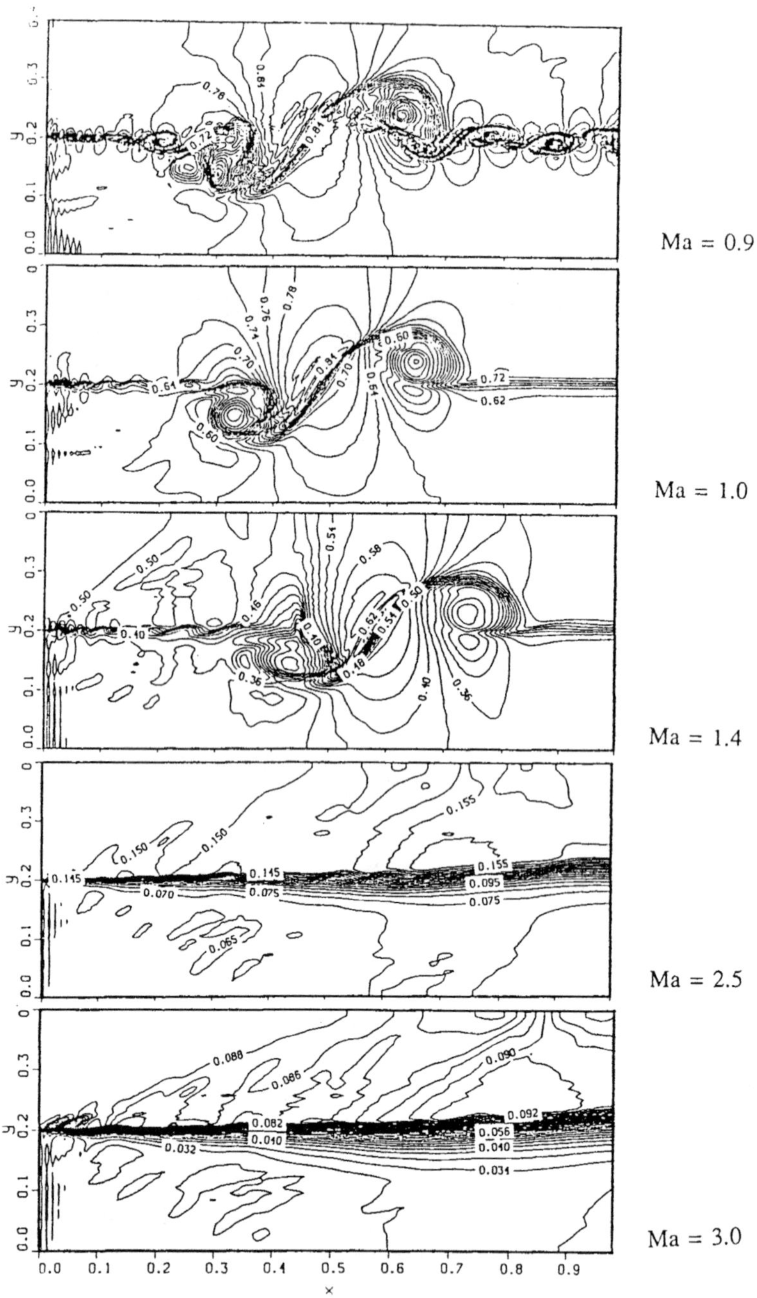


Fig.5. Influence of Mach number: Density contours.

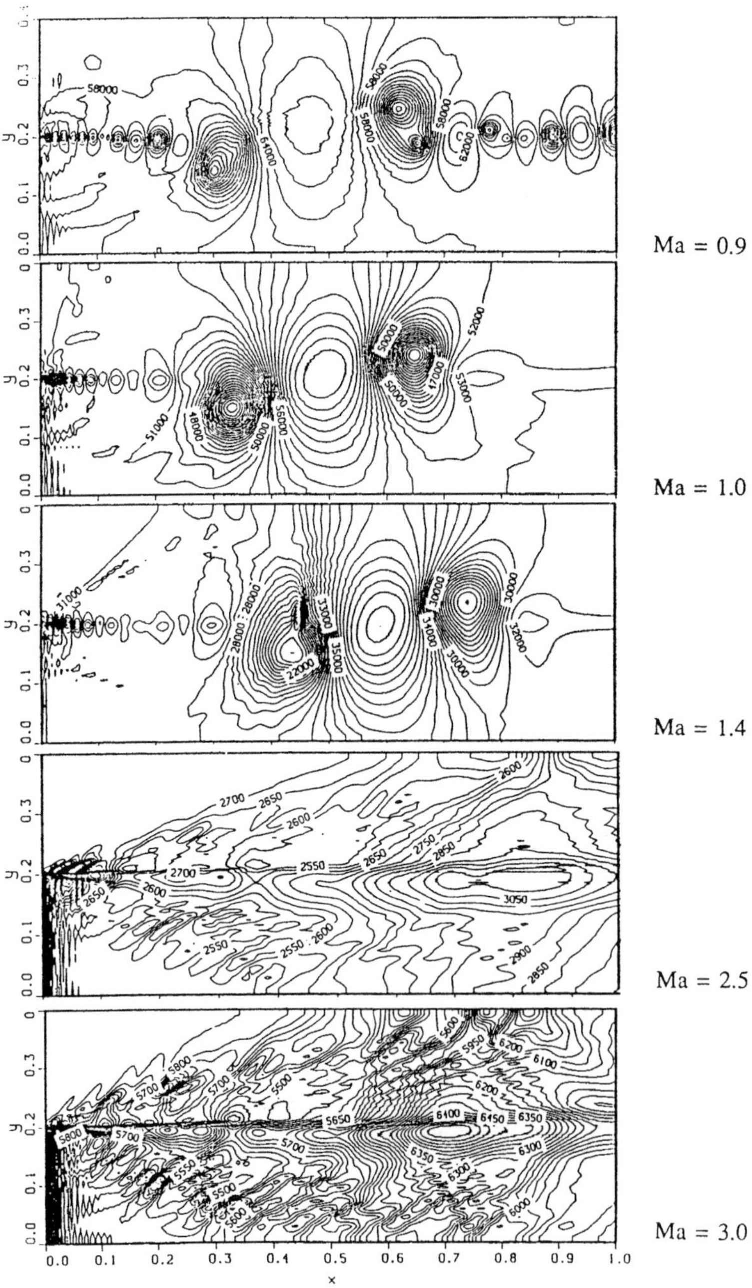


Fig.6. Influence of Mach number: Pressure contours.

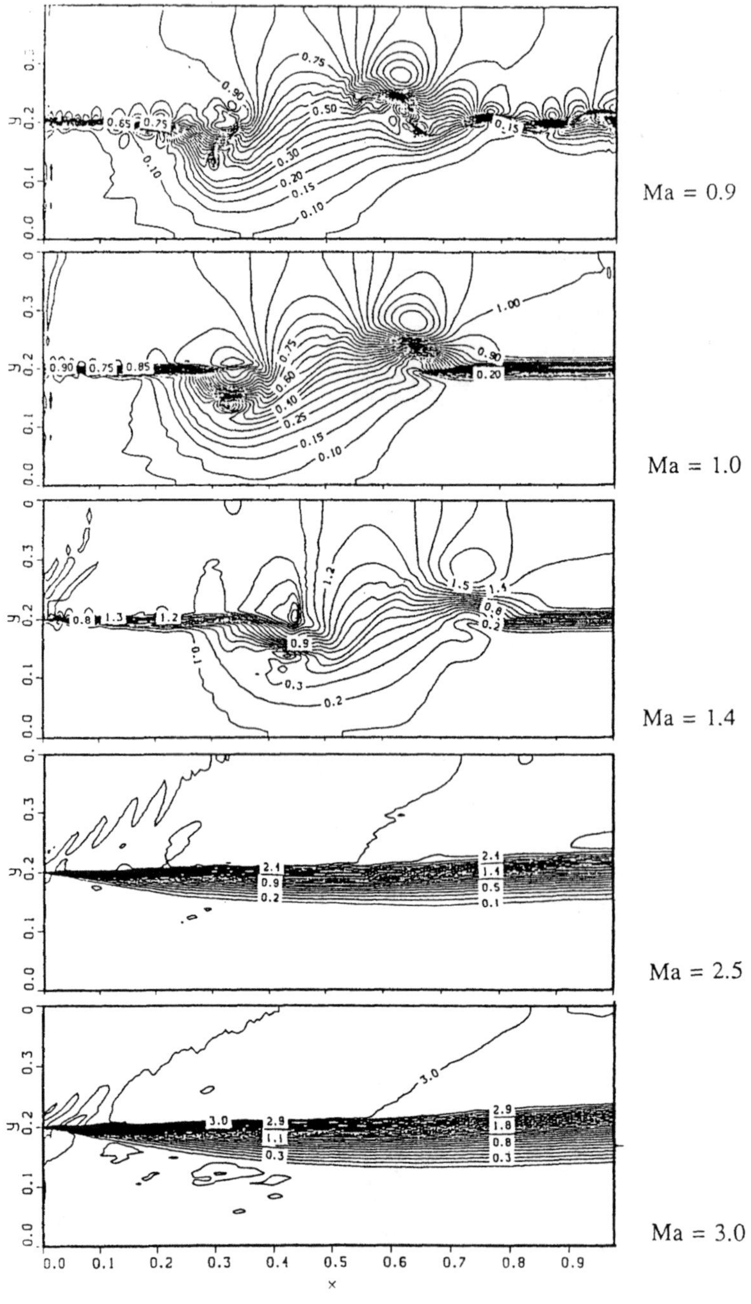


Fig.7. Influence of Mach number: Iso-Mach contours.

**References.**

Brown, G.L. & Roshko, A. 1974 On density effects and large structure in turbulent mixing layers. *J. Fluid Mech.* **64**, 775-816.

Ha-Minh, H., Rubesin, M.W., Vandromme, D.D. & Viegas, J.R., 1985 On the Use of Second Order Closure Modelling for the Prediction of the Turbulence Boundary-Layer/Shock Wave Interaction: Physical and Numerical Aspects.*Proceeding of the International Computational Fluid Dynamic Symposium*, Tokyo.

Ha-Minh, H. & Vandromme, D.D., 1986, Modelling of Compressible Turbulent Flows: Present Possibilities and Perspectives. *IUTAM Symposium, Ed: J. Délery*,13-16, Sringer Verlag.

Ha-Minh, H. & Vandromme, D.D. 1989, Compressibility effects on turbulence in high speed flows, *Summer workshop on Theory and Modelling of Turbulent Flows, Ecole Centrale de Lyon*, July 3-13.

Hussain, A.K.M.F. 1983 Coherent Structures-Reality and Myth. *Phys. Fluids*, **26** (10), 2816-2850.

Mac Cormack, R.W. 1981 A Numerical Method for Solving the Equations of Compressible Viscous Flow. *A.I.A.A. Paper* No 81-0110.

Papamoschou, D & Roshko, A 1988 The Compressible Turbulent Shear Layer: An Experimental Study. *J. Fluid Mech.* **197**, 453-477.

Reynolds, W.C. & Hussain, A.K.M.F. 1971 The Mechanics of an Organized Wave in Turbulent Shear Flow. Part 3: Theoretical models and Comparaison with Experiments, *J. Fluid Mech.* **54**, 263-288.

Vandromme, D.D. & Ha-Minh, H. 1985 Introduction to the Modelling of Turbulence: Turbulence Modelling for the Compressible Flows. *S.C.1985-1, Von Karman Institute for Fluid Dynamics*, June 1985.

Vandromme, D.D. & Ha-Minh, H. 1986 About the Coupling of Turbulence Closure Models with Averaged Navier-Stokes Equations.*Journal of Computational Physics*, **65**, No 2, 386-409, August.

Vandromme D., 1989, Turbulence modelling and implementation in Navier-Stokes solvers, *Von Karman Institute Lecture Serie 1989-8, "Numerical methods for flows in turbomachines", Rhode Saint Genese, Belgium.*

# Numerical simulations of supersonic plane Couette flow instability

N. DOLEZ[α] and J. LÉORAT [β]

[α] *Observatoire Midi–Pyrénées and CERFACS*
*42, Avenue G. Coriolis, 31057–Toulouse, France*

[β] *Observatoire de Meudon*
*92195-Meudon, France.*

*Turbulence and Coherent Structures*
*O. Métais and M. Lesieur, eds.*
*Selected papers from "Turbulence 89: organized structures and turbulence in fluid mechanics"*
*Grenoble, 18-21 September 1989*

## 1. Introduction

Due to the importance of gravitational forces, flows of astrophysical interest are often observed in a supersonic regime, which gives rise for their description to various analytical or numerical difficulties. Let us briefly mention as examples the supersonic turbulence observed in interstellar clouds, stellar or galactic jets and accretion disks where gravitational energy is thought to be converted into radiation. The interested reader may consult for instance an account of recent works concerning disk stability in Narayan & Goodman (1989) and the references therein. However, despite the central role played by such dynamical phenomena in the description of cosmic objects, the basic problems associated with matter inflow and angular momentum outflow remain still unsolved. The complexity of the problem urges on studying first the stability properties of supersonic flows in the simpler case of plane shearing and examining in a second step questions related to rotation and gravitation.

Study of the compressibility effects for the plane mixing layer of an inviscid unbounded fluid indeed shows that the the linear stability problem exhibits peculiar features in the supersonic regime(Miles 1958, Gerwin 1968, Ray 1982, Choudhury & Lovelace 1984). We do not attempt here to review analytical or numerical works related to this question, which deals with linear, sinusoidal or hyperbolic tangent profiles, but we note that, even in this simple geometry, the identification of growing modes is not a straightforward task (see for example, Blumen 1970, Blumen & al. 1975, Artola & Majda 1987).

To study the properties of real supersonic flows,in either an astrophysical or an aeronautical context, numerical simulations are needed to describe the non-linear regime. For dissipative fluids, an external forcing is necessary to avoid viscous decay and keep, say, Reynolds and Mach numbers at fixed values. Couette flow, where energy injection is performed not by ad-hoc volume forces but through boundary conditions, appears in fact as the simplest case where non-linear saturation could be observed in numerical simulations of supersonic flows. Moreover, in a cylindrical geometry, it represents a convenient model as well for astrophysical flows as for experimental configurations.

The linear stability properties of supersonic plane Couette flow have been recently studied in some detail by Glatzel, in the inviscid (1988) and in the viscous case (1989, henceafter called G89). Instability bands related to resonant interactions between modes are found. For sonic modes, the width of the bands is close to unity and the first two bands are centered near the Mach numbers 5 and 7.5. At a Reynolds number of 1000, the first viscous-sonic resonance occurs for Mach numbers between about 1.5 and 2. At smaller speeds the flow is linearly stable as for incompressible fluids (the issue for finite amplitude instability seems to be still not known). The lowest critical Reynolds number is below one hundred, so that the first source of numerical difficulties is not the resolution, but the rather high value of the first instability band of critical Mach numbers.

We present in this paper results obtained at a Reynolds number of 160, which is sufficient to show the fundamental instability and saturation of supersonic Couette flow. After introducing the equations, their parameters and the numerical scheme in the next Section, instability in the linear regime is examined in Section 3 and preliminary results concerning the nonlinear saturation are given in Section 4. The last concluding Section includes a short discussion of work in progress and planned on the subject.

## 2. Physical assumptions and numerical method

For a preliminary investigation, a polytropic equation of state allows to treat easily the two limit cases of isothermal (infinite thermal conductivity) and isentropic flows as well as the case of infinitesimal fluctuations corresponding to the linear analysis.

Let $\rho_0$ be the mean density, $P$ the pressure and $a_0$ the sound velocity corresponding to mean density, then

$$P = (\rho_0 {a_o}^2/\gamma)(\rho/\rho_0)^\gamma, \qquad (2-1)$$

where the polytropic index $\gamma$ is equal to 1 for isothermal flows, 5/3 for adiabatic flows of monoatomic gases, 2 for shallow water simulations.

We assume that the bulk viscosity is zero and that the kinematical shear viscosity is constant, which is convenient for the implicit step of the integration in time, so that the viscous stress tensor $\mathbf{S}$ is:

$$S_{ij} = \nu\rho(\partial_i u_j + \partial_j u_i - (2/3)\delta_{ij}\nabla.\mathbf{u}) \qquad (2-2)$$

To adimensionalize the dynamical equations, the density unit is $\rho_0$, the half separation $D$ of the two horizontal boundaries is taken as length unit, while the speed $U$ at the boundaries gives the velocity unit and thus the time unit, $D/U$ is based on large scale characteristics of the flow. The numerical procedure integrates the following set of adimensionalized equations:

$$\begin{aligned} \partial_t\rho + \nabla\rho\mathbf{u} &= 0, \\ \partial_t\mathbf{u} + \mathbf{u}.\nabla\mathbf{u} &= -Ma^{-2}(\gamma\rho)^{-1}\nabla\rho^\gamma + Re^{-1}\rho^{-1}\nabla\mathbf{S}, \end{aligned} \qquad (2-3)$$

where the parameters $Ma$ and $Re$ are respectively the convective Mach and Reynolds numbers,

$$\begin{aligned} Ma &= U/a_0, \\ Re &= UD/\nu. \end{aligned} \tag{2-4}$$

The flow is assumed to be periodic with respect to $x$ coordinate. The aspect ratio is chosen to be equal to $\pi$ (the period is $L = 2\pi D$ as in G89) and the boundary conditions are

$$\begin{aligned} u(x+2\pi, y) &= u(x,y), \\ u(x,-1) = -u(x,+1) &= 1. \end{aligned} \tag{2-5}$$

Spectral methods (Gottlieb & Orszag 1977) have proven their ability to treat supersonic flows with simple boundary conditions (see for example, Léorat & al.1984, Passot & Pouquet 1987, Canuto & al. 1988 and references therein). Since the velocity is fixed on the plane $y = 0$ and $y = 2$, it is expanded on the basis of Tchebychev polynomials $T_n$, while periodicity in the $x$ direction allows to use the Fourier basis, so that if the field $e(x, y, t)$ stands for the density or a velocity component,

$$e(x,y,t) = \sum_{m=-M/2}^{M/2-1} \sum_{n=0}^{N} \hat{e}_{mn} exp(imx) T_n(y), \tag{2-6}$$

where M and N are the (even) dimensions of the Fourier and Tchebychev basis. A Fourier-Tchebychev code, which was in fact first developed to study the compressible cylindrical Couette flow, has been used with minor modifications for the plane case presented in this paper. The boundary conditions (2.5) for the velocity are fullfilled by using the standard tau-collocation method. The density is obtained from the continuity equation (2.3), where the last two Tchebycheff coefficients are kept equal to zero. If this filtering is not done, a numerical instability is encountered quite suddenly after about 30000 time steps. Advance in time is implicit (Crank-Nicholson scheme) only for second order derivatives in the y direction. A second order Adams-Bashforth scheme is used for the other terms. Note that the second order derivatives in the x direction (involving Fourier space) do not introduce more stability constraints than the first order derivatives in the y direction, where Tchebychev space is used.

Stability of the explicit part of the time scheme asks for a time step dt beeing smaller than some local advection time (at speed close to $u + a_0$) over the smallest grid scale, $l_{min}$, which is given by the highest rank $N$ of Tchebychev polynomials ($l_{min} = \pi^2/(2N^2)$). One can thus write

$$dt < A\pi^2/(2N^2(1 + Ma^{-1})), \tag{2-7}$$

where the coefficient A is of order one. According to this rough estimate, highl y supersonic flows would be easier to simulate than subsonic ones, where acoustic waves introduce a shorter time scale. For Mach numbers greater than one, shock waves however let small scales in, which are controlled by the Mach and Reynolds number based on molecular viscosity (numerical viscosity in negligible). It is found by numerical experience that the thickness of shock waves has to be covered by at least three grid points. For Reynolds numbers below 200 and Mach numbers below 10, a resolution of 32*32 grid points may be used, with a time step dt=.004 (dt=0.002 for 48*48 and dt=.001 for 64*64 grid points), compared to a large scale turn-over time of order unity. Note that low resolutions are useful since non linear saturation is reached at Re=160 only after t=200(see Section 4 below). With an Alliant FX80 computer, integration of a time step needs 0.08s of CPU time in the 32*32 case, while on a Cray II, it is 0.039 s for 48*48 and 0.062 s for 64*64 points.

## 3. Simulations of the linear growth

We study numerically in this Section the instability of the stationary solution of (2.3)

$$\begin{aligned} \rho &= 1, \\ u_x &= y - 1, \\ u_y &= 0, \end{aligned} \tag{3-1}$$

which is linearly stable for incompressible fluids. Recall that direct numerical simulations of two-dimensional flows without symmetries are constrained to Reynolds numbers below a few hundreds. The Glatzel's linear analysis (G89) gives the growth rate of the unstable modes at a Reynolds number of 1000 and shows also that the lowest critical Reynolds number, which is reached at a critical Mach number of 4.86 for the mode k= 0.971, is 83.54 (using the same adimensionalized units as in Section 2).

We choose as initial pertubations to (3-1), a divergenceless velocity pertubation restricted to the strip $0.8 \leq y \leq 1.2$, without initial density fluctuations. To measure the growth of the instability at different spatial scales, we introduce the Fourier energy spectrum $E^y_m$ of the $y$ velocity-component, integrated between the moving walls $y = 0$ and $y = 2$:

$$E^y_m = \sum_{n=0}^{N} \hat{u}^y_{mn} \hat{u}^{y*}_{mn} \tag{3-2}$$

We have assumed here that the Tchebychev polynomials are normalized. The pertubation has initially a small constant amplitude ($10^{-8}$) and random phases in the 8 lowest Fourier modes . We have first coarsely explored the parameter space by varying the Mach number at the fixed

Reynolds number of 160, using the results o f G89 as a guide. No instability is observed at Mach number 4 and below, whereas at Ma= 4.9 , growth of $E^y_m$ occurs on wave number $m = 1$. At greater Mach numbers, the flow (3-1) recovers stability. The second instability band close to $Ma = 7.5$ predicted for $Re = 1000$ is not observed numerically at$Re = 160$ or 320. This is plausibly a consequence of too low a Reynolds number ((see for example figure 12 of G89) and will be taken again using improved numerical resources.

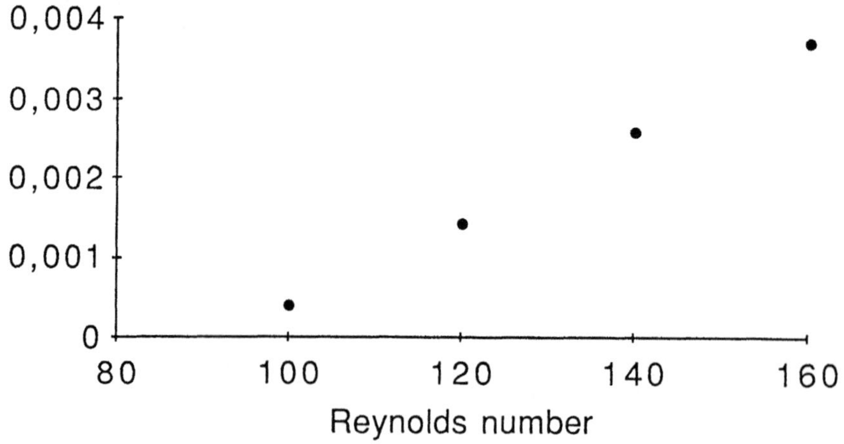


Fig.1. Growth rates and critical Reynolds number at Ma=4.9

Measuring the variation of $E^y_m$ growth rate $\sigma$ with Reynolds number, one obtains the critical Reynolds number Re* at a given Mach number, assuming a power law dependence, $\sigma \sim (Re - Re*)^\alpha$. The results for Reynolds numbers 160, 140, 120 and 100, concerning the growth rates of $E^y_1$ are summarized on fig. 1, which shows that the index $\alpha$ is close to 1/2, and the critical Mach number is about 93 at Mach number 4.9. There is a small difference with the value given by G89, which is 83.54, for slightly different parameters (Ma=4.86 and m=0.971).

The growth rate at greater Reynolds numbers may be obtained keeping the same low resolution of 32*32 grid points, since during the linear regime, the truncat ion at small scales ($m \leq N/2$) has no influence on the growth around the modes m close to 1. We have thus found that the growth rate of the *amplitude* of mode m=1 at $Re = 1000$ is 0.062,

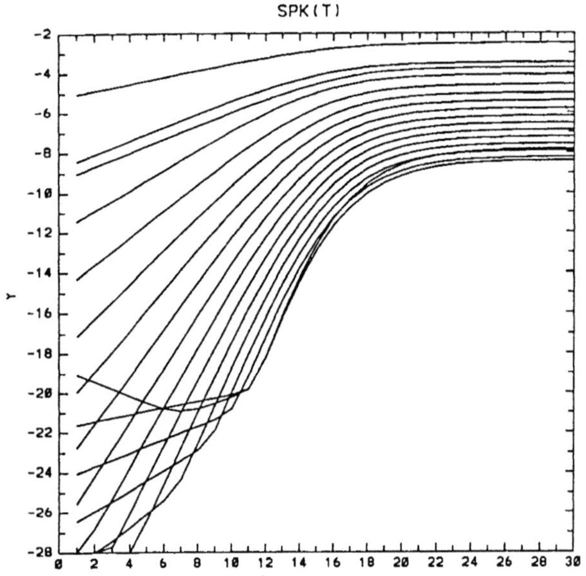


Fig.3. Evolution of $E_m^y$ (t=64 to 256)

which is very close to the value which is read on figure 4 of G89. The main objective of the numerical study is the description of the non-linear regime, so that we do not give here more details on the eigenmodes of the linearized dynamical equations. Notice however that the stream lines of the pertubed velocity in the linear phase are similar to the ones obtained at saturation (see figure 4 below).

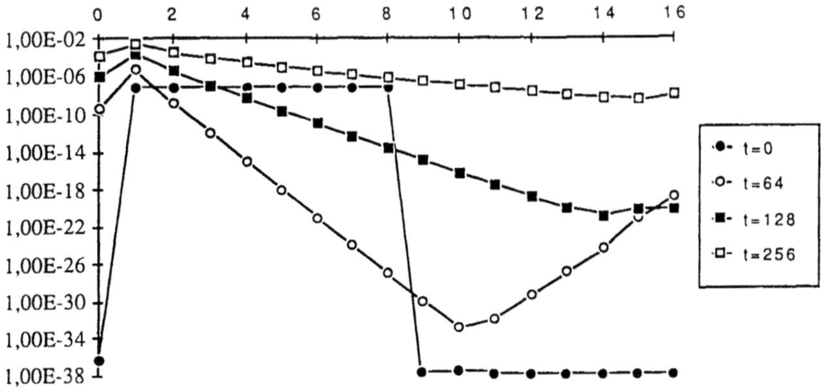


Fig.2. Evolution of Fourier energy spectra

In the early stage, the energy spectrum $E_m^y$ has a positive slope at large wave-numbers, as is seen on figure 2, where the ordinate axis has been extended on purpose down to $10^{-38}$: this is likely due mainly to the numerical errors introduced by the second order time integration scheme. As energy rises in this domain of wave-numbers, the relative importance of this error vanishes and figure 2 shows that most of the energy spectrum is close to an exponential.

The evolution curves of the 16 Fourier modes of $E_m^y$ from time 64 to 256 are shown on figure 3: the smallness of the growth rate delays the saturation to times greater than 200.

## 4. The non-linear stable state at $Re = 160$

We give now a short account of the steady state reached at t=256. The stream lines of the perturbed velocity (difference between the actual velocity and the unstable solution (3-1) are shown on figure 4. There are four rolls, such that the y component of velocity dominates only in two thin jet-like domains forming an angle of about 15 degrees with the moving walls. If this were a Mach angle, it would correspond to a Mach number of 3.9. The maximum speed of the pertubed velocity is O.25.

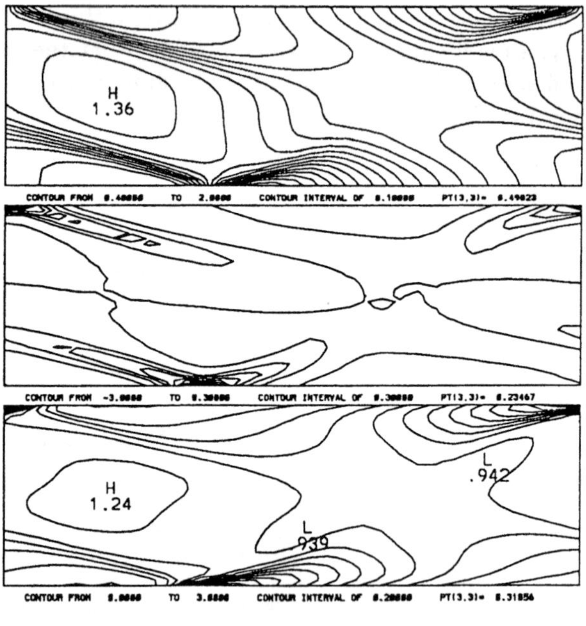


Fig.5. Density, divergence and curl contours

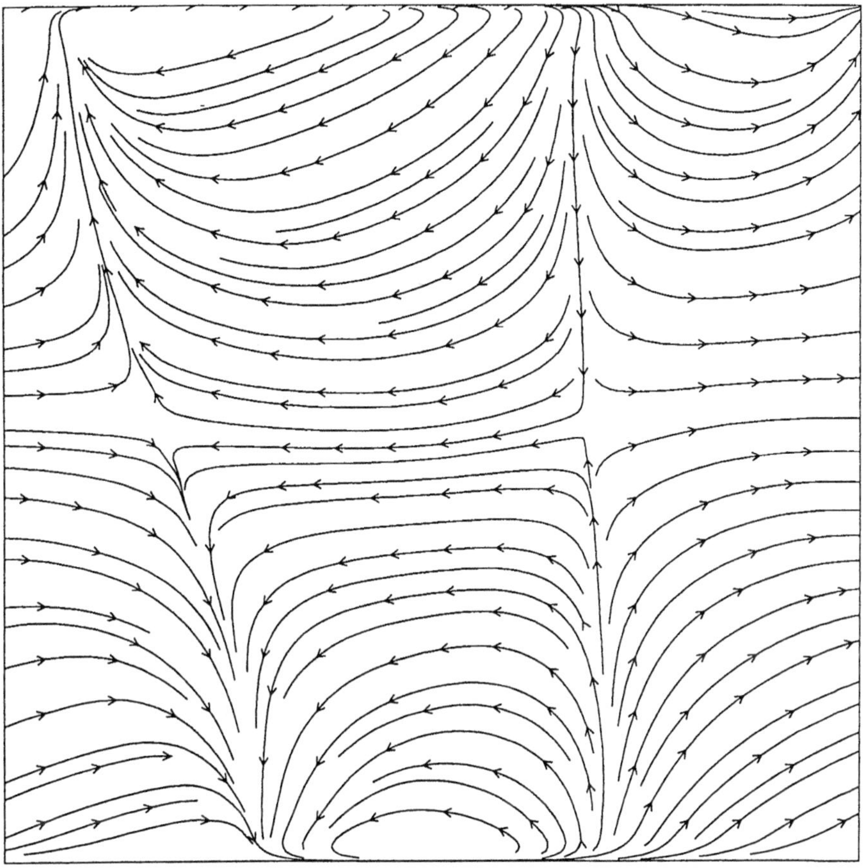

Fig.4. Stream lines of the velocity pertubations at saturation

There are two centers of symetry (grid coordinates:M=6,N=16 and M=22, N=16) separated by half the wave length: the periodicity prescribes that the centers of symetry be even in number.

The density contours at the same time (t=256) are given on figure 5, with the divergence and curl contours. As could be inferred form the stream lines, there are two regions of negative divergence in each half of the flow (y¡1 and y¡1), corresponding also to strong density gradients. The first is a linear strip at the angle given above, and may turn into a shock at greater Reynolds numbers while the second is concentrated on a point close to the wall.

The density field is shown with greater detail in the last figure, which shows successively two streamwise profiles (on the wall y=0 and in the

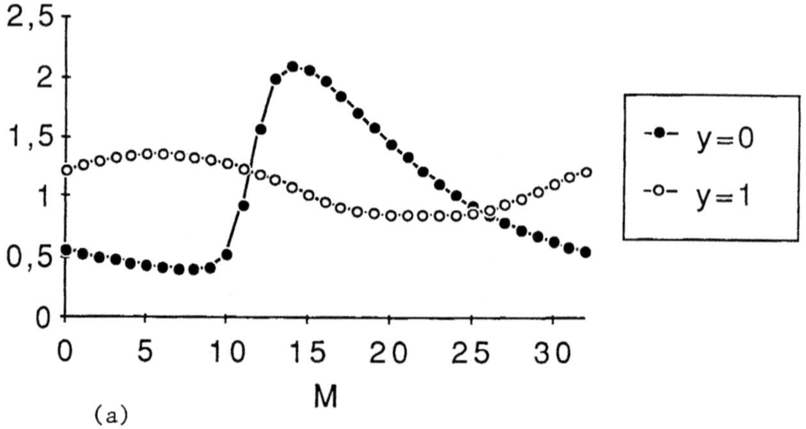


(a)

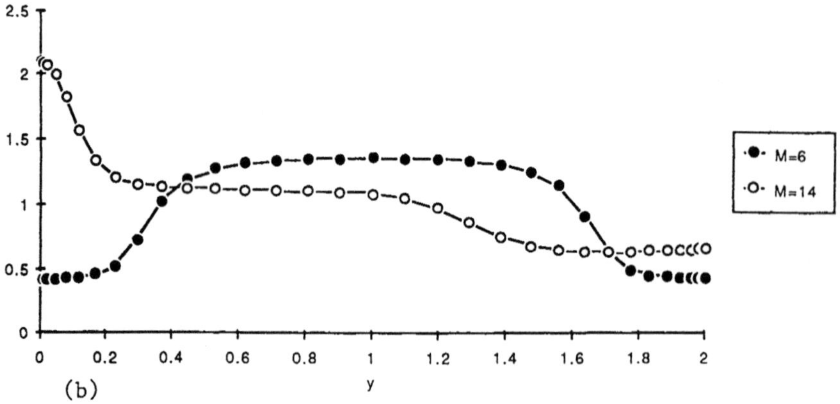


(b)

Fig.6. Density profiles

center y=1), two spanwise profiles (the x grid point M=6 corresponds to the center of symetry and M=14 passes trough the density maximum. The density jump is of the order of 5 at wall, and extends over three grid meshes. These gradients are responsible of the numerical difficulties associated with supersonic flows: the maximal Reynolds number which could be reached with 32*32 grid points was 180 and an increase of the resolution is needed to ge t higher Reynolds numbers.

The last figure shows the density surface viewed from above (with x=0 and y=0 as nearest point from observer).

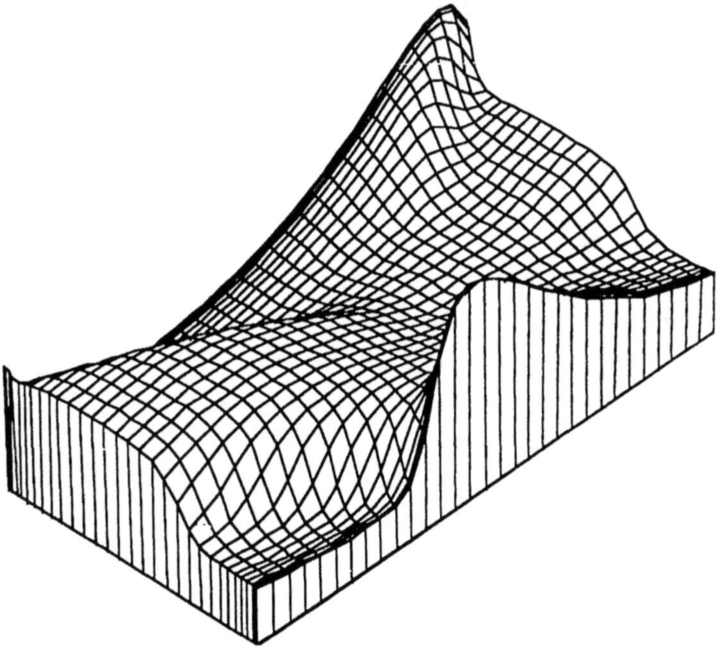

Fig.7. Density surface

## 5. Conclusion

At a Reynolds number of 160 and Mach number of 4.9, the supersonic plane Couette flow reaches a steady state with strong density fluctuations. The linear analysis of Glatzel(1989) has been a particularly valuable guide for the selection of parameters for the direct numerical simulations of such a flow and helped also to validate the numerical method.

At greater Reynolds numbers, one expects an enhancement of the density gradients and formation of shocks. The main question to be answered numerically is now the following: will the flow always be in a steady state or will non-linear interactions give rise to time dependent flows and supersonic turbulence? this last case would obviously lead to interesting developments for astrophysical or experimental situations. Another very important situation which is studied with the same numerical code is the influence of rotation on shear instabilities of compressible flows.

The numerical approach will thus contribute to a better understanding of the various instabilities occuring in compressible flows, for which experimental difficulties have prevented to get an amount of data comparable to the one existing for incompressible flows.

*Acknowledgments* We want to thank particularly L. Valdettaro for a useful discussion on the numerical treatment of boundary conditions. We have benefited from the computing facilities of the *CERFACS*(Toulouse, France) and of the *Centre Calcul Vectoriel pour la Recherche* (Palaiseau, France), and also from CNRS grants (RCP *fluides astrophysiques en régime supersonique*).

## References

Artola M. and Majda A.J., 1987, *Physica D*, **28**, 253.

Blumen W., 1970, *J. Fluid Mech.*, **40**, 769.

Blumen W., Drazin P.G. & Billings D.F., 1975, *J. Fluid Mech.*, **40**, 769.

Canuto C., Hussaini M.Y., Quarteroni A., Zang T.A., 1988, *Spectral methods in fluid dynamics*, Springer Verlag, New York.

Choudhury S.R. & Lovelace R.V.E., 1984, *Ap.J.*, **283**, 331.

Gerwin R.A., 1968, *Rev. Mod. Phys.*, **40**, 652.

Glatzel W., 1989, *J. Fluid Mech.*, **202**, 515.

Glatzel W., 1988, *Mon. Not. R. astr. Soc.*, **231**, 795.

Gottlieb D.,& Orszag S.A., 1977, *Numerical analysis of spectral methods*, SIAM, Philadelphia.

Léorat J., Poyet J.P. and Pouquet, A., 1984, in *Proceedings of Colloquium on problems of Collapse and numerical Relativity*, p.287, eds Bancel M. & Signore M., Reidel, Dordrecht.

Miles J.W., 1958, *J. Fluid Mech.*, **4**, 538.

Narayan R. & Goodman J., *Non-axisymmetric shear instabilities en thick accretion disks*, to appear in the Proceedings of the NATO Workshop on *Theory of accretion disks*, March 1989, Garching, West Germany.

Passot T.& Pouquet A., 1987, *J. Fluid Mech.*, **181**, 441.

Ray T.P., 1982, *Mon. Not. R. astr. Soc.*, **198**, 617.

# Reactive Scalar as a Tracer of Organized Structures in Turbulent Mixing Layers

Jean-Pierre CHOLLET[α], Maria-Roser VALLCORBA[α] and Ralf GATHMANN[α]

[α]*Institut de Mécanique de Grenoble*
*BP 53X, 38041 Grenoble-Cedex, France*

*Turbulence and Coherent Structures*
*O. Métais and M. Lesieur, eds.*
*Selected papers from "Turbulence 89: organized structures and turbulence in fluid mechanics"*
*Grenoble, 18-21 September 1989*

# 1. Introduction

## 1.1. Transport, diffusion and eddy structures

The mixing layer can be viewed as a good prototype of real flows where instability mechanisms, associated with mean velocity shears, produce large coherent structures which strongly influence transport, mixing and diffusion of scalar quantities such as temperature or any tracer concentration (Lesieur 1987). The behaviour of a passive scalar in mixing layers which has been widely documented (e.g. Comte *et al.* 1987, Comte *et al.* 1989, Normand *et al.* 1988) is significantly altered when the scalar is the concentration of species which undergo chemical reactions. The strongly coherent structures of the mixing layer influence dramatically mixing properties of the flow field, and then the efficiency of chemical reactions which can be boosted by streams which help to put reacting species closer ones to the others and drain products away. Turbulent mixing acts through a continuous distribution of length and time scales ranging from small scales where diffusive processes dominate to large scales which are ruled by the most prominent coherent structures.

Interactions between turbulence and chemical reactions can be considered in order to: *(i)* take advantage of species properties to trace coherent structures, *(ii)* study the influence of structures upon the chemical reaction efficiency. A scalar field exhibits some typical signature due to the coherent flow structures when taking advantage of its dying or reactive properties. Such visualisation techniques have been used by experimentalists, (e.g. Broadwell & Breidenthal 1982, Koochesfahani & Dimotakis 1986, Mungal & Dimotakis 1984). In other respects, numerical simulations of chemical mixing layers are attractive because the global efficiency of a reaction is strongly determined by species mixing through fine scale turbulence and large scale flow structures; these features have a drastic importance in industrial combustion problems.

Institut de Mécanique de Grenoble: Unité Mixte de Recherche 101: C.N.R.S., Université J.Fourier-Grenoble 1, Institut National Polytechnique de Grenoble

### 1.2. Turbulent reactive flows

Real chemically reacting turbulent flows depend on many phenomena related one to the other in an intricate way, and associating chemical reactions, thermodynamics and nonlinear evolutions of turbulent fields. These phenomena involve a wide range of scales, from very small ones where chemical interactions take place between molecules to large scales corresponding to the coherent structures of the turbulent flow field.

In order to focus on a reliable description of turbulence as it interacts with chemical reactions, no "turbulence model" is used here, velocities and scalar quantities are computed at each time and space step. In order to make this direct numerical simulation tractable, hypotheses, primarily non compressibility and two dimensionality, are used to provide a well suited environment in order to develop detailed analysis of basic trends of reaction-turbulence interactions, especially as far as largest scales are concerned.

### 1.3. Chemical reactions

Chemical phenomena cannot be taken into detailed account since they depend on scales much smaller than the Kolmogorov dissipative scales of the turbulent field; also, they can be viewed as very "local" in space, contrary to turbulent interactions whose largest ones, featuring the eddy structures, have the size of the computation domain. Chemistry formulation can be considered as a parameterization of very small scales, the simulation being direct only as long as turbulent fields are concerned.

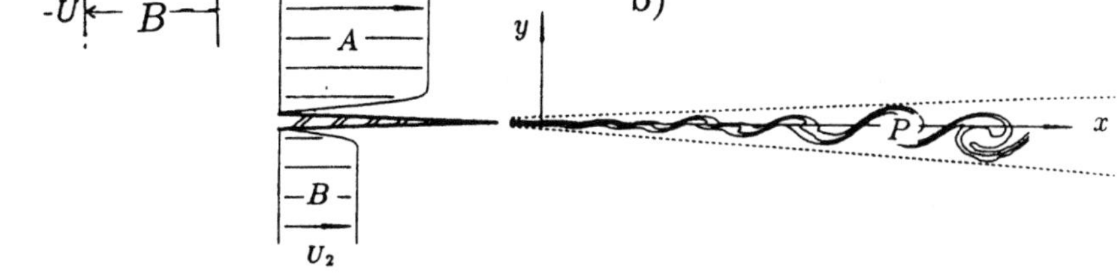


Fig.1. a) Temporal mixing layer (with $U = \frac{U_1 - U_2}{2}$ and an initial layer thickness $\delta_i$, b) Spatial mixing layer

The chemical reaction under consideration is a binary, one step, non-reversible reaction $\nu_A A + \nu_B B \rightarrow \nu_P P + Q$. Chemical species $A$, $B$ and $P$ are handled through their concentrations (or molar fractions) $C_A$, $C_B$ and $C_P$ and are consumed (or produced) with the reaction rate $\dot{w}$. The only reactions considered here keep the number of moles unchanged ($\nu_P = \nu_A + \nu_B$), this restriction is in good agreement with flow incompressibility. $Q = c_p T_q$ is heat released by the reaction. Two different formulations of the reaction rate $\dot{w}$ are used:
(i) either an isothermal reaction :

$$\dot{w} = -\frac{C_A{}^{\nu_A} C_B{}^{\nu_B}}{\tau_c} \qquad (1-1)$$

where $\tau_c$ is a characteristic time of the chemical process whose value is constant, which is helpful when analysing phenomena with various values of the Damkhöler number $Da = \tau_t/\tau_c$. The time scale $\tau_t = \delta_i/U$ is associated to the large scales of the turbulent field since it is based on the initial thickness $\delta_i$ and the velocity lag $U$ ( figure 1).
(ii) or an Arrhenius law to take account of the dependence of the reaction upon temperature $T$:

$$\dot{w} = -BC_A{}^{\nu_A} C_B{}^{\nu_B} e^{-\frac{T_A}{T}} \qquad (1-2)$$

The activation temperature $T_A$ and the constant $B$ depend on the chemical reaction. Formulation (1-1) is a limit of (1-2) for $\frac{T_A}{T} \rightarrow 0$

## 2. Time-evolving turbulent fields

### 2.1. Velocity and passive scalar fields

Assuming flow incompressibility and newtonian fluid behaviour, the velocity field $\mathbf{u}(\mathbf{x},t)$is non divergent ($\nabla.\mathbf{u} = 0$) and evolves according to the Navier Stokes equation :

$$\frac{\partial \mathbf{u}}{\partial t} + \mathbf{u}.\nabla\mathbf{u} = -\frac{\nabla p}{\rho} + \nu\nabla^2\mathbf{u} \qquad (2-1)$$

where $\nu$ is the viscosity, $\rho$ the constant density of the fluid, and $p$ the pressure field.

The various scalar quantities $\theta(\mathbf{x},t)$ which have to be considered in reacting flow numerical simulations are treated as passive, with velocity $\mathbf{u}$ evolving according to (2-1) independently of $\theta(\mathbf{x},t)$:

$$\frac{\partial \theta}{\partial t} + \mathbf{u}.\nabla\theta = D_\theta \nabla^2\theta + S_\theta \qquad (2-2)$$

where $D_\theta$ is the molecular diffusivity and $S_\theta$ a "source"; $S_\theta$ is negative when $\theta$ is the concentration of species consumed by a chemical reaction; $S_\theta = 0$ in the case of the inert (non reactive) scalar.

Restricting our interest to two dimensional flows, the vorticity $\omega = \nabla \wedge \mathbf{u}$ is a scalar quantity which evolves according to equation (2-2) (with $D_\theta = \nu$). The equations which control the evolution of the various scalar quantities differ only through the diffusivity coefficients $D_\theta$ and the source terms $S_\theta$. Of particular interest is the case where Schmidt, Prandtl and Lewis numbers are unity ($D = \nu = \kappa = D_A = D_B = D_P$); moreover stochiometric coefficients are chosen to be ($\nu_A = \nu_B = 1$, $\nu_P = 2$). With these conditions, scalar fields evolve according to :

$$(\frac{\partial}{\partial t} + \mathbf{u}.\nabla) \begin{pmatrix} \omega \\ C_A \\ C_B \\ C_P \\ T \\ \Phi_i \end{pmatrix} = D\nabla^2 \begin{pmatrix} \omega \\ C_A \\ C_B \\ C_P \\ T \\ \Phi_i \end{pmatrix} + \begin{pmatrix} 0 \\ \dot{w} \\ \dot{w} \\ -2\dot{w} \\ -T_q\dot{w} \\ 0 \end{pmatrix} \qquad (2-3)$$

where $\Phi_i$ is an inert (non reacting) scalar; except for $\omega$, the formulation can be used for both two and three dimensional problems. Scalars $C_i$ are constrained to range from 0 to 1.

## 2.2. Time evolutions of scalar fields

The passive inert scalar, such as $\Phi_i$ which was used in previous studies (Normand *et al.*, 1988, Comte *et al.* 1989) is the assymptotic behaviour of a very slowly reacting scalar ($\tau_c \to \infty$, $Da \to 0$). From (2-3), some fields, for instance $\Phi_i$ and $\omega$, are observed to evolve according to the same equation; nevertheless, as generally being submitted to different initial and boundary conditions they do not collapse to the same solution. Inert (non reactive) scalars can be built from concentrations, like $\Phi_1 = C_A - C_B$, and used in numerical codes in order to improve their efficiency (as in section 4), taking advantage of the fact that species concentrations do not evolve independantly when they are not diluted in a neutral fluid, since continuity prescribes $C_A + C_B + C_P = 1$.

When attempting to model real flows with a two dimensional approximation, the viscosity $\nu$ should not be considered as the molecular viscosity but viewed as a model of three dimensional phenomena of energy transfer towards the small scales where the dissipation takes place through the genuine molecular viscosity $\nu^{(m)}$. Two dimensional numerical simulations, according to (2- 1), are then used to describe large "quasi" two dimensional scales which are damped by $\nu$ which acts as an eddy viscosity. Therefore, Schmidt numbers, $Sc_i = \nu/D_i$, of the numerical simulation are not necessarily equal to Schmidt numbers

$Sc_i^{(m)} = \nu^{(m)}/D_i^{(m)}$ associated to molecular viscosity and diffusivities. As a result, twodimensional simulations with $Sc_i \approx 1$ can be considered when computing large scales of liquid reacting flows although $Sc_i^{(m)}$ are usually high in such fluids. $Sc_i^{(m)}$ must be used in direct three dimensional simulations in order to take a complete account of all turbulent scales.

### 2.3. Numerical scheme, initial and boundary conditions

In order to improve the computation efficiency, a stream function ($\psi$) is defined and the calculation of $(\mathbf{u},\omega)$ is changed into $(\psi,\omega)$; the general features of the numerical code are described by Lesieur *et al.* 1988, Comte *et al.* 1989. The computation domain is rectangular with a regular mesh grid : $x \in [0, L_x]$, $y \in [-L_y/2, +L_y/2]$. In addition to source terms $S_\theta$, boundary and initial conditions make the difference between the evolutions of the scalar fields.

$$
\begin{aligned}
C_A(x,\frac{L_y}{2},t) = 1,\ & C_B(x,\frac{L_y}{2},t) = 0,\\
C_A(x,-\frac{L_y}{2},t) = 0,\ & C_B(x,-\frac{L_y}{2},t) = 1,\\
T(x,\pm\frac{L_y}{2},t) = T_0,\ & C_P(x,\pm\frac{L_y}{2},t) = 0
\end{aligned}
\qquad (2-4)
$$

$(\psi,\omega)$ flow field is calculated with: *(i)* a free slip condition for $y = \pm L_y/2$, *(ii)* an initial condition which is based on a basic mean tanh velocity profile with noise added and random triggering of the most instable mode.

Two kinds of calculations which complement each other are presented, these are the **temporal** and the **spatial** approximations of the mixing layer. With periodical streamwise conditions (at $x = 0, L_x$), the temporal mixing layer can give a detailed description of the interactions between eddies when following the mean flow. [1] The spatial mixing layer is computed with inflow (at $x = 0$) and outflow (at $x = L_x$) conditions and, as being closer to real situations, is better able to provide reliable averaged quantities such as the layer thickness $\delta(\mathbf{x},t)$. Moreover, spatial calculations provide mean stationary solutions whose statistics do not depend on initial conditions. The outflow ($x = L_x$) boundary conditions is designed to let momentum and various scalars escape freely from the computation domain (Normand *et al.* 1988).

---

[1] the mean advecting velocity corresponds to $\frac{U_1+U_2}{2}$ of figure 1.b)

## 3. Chemical reaction and coherent flow structures

### 3.1. Temporal simulations

The temporal appoximation is well suited to detailed studies of the influence of eddy structures on the reaction, since spatial resolution, here up to $256^2$, can be devoted to the description of few interacting structures with initial conditions giving rise, in the present case, to 2 or 4 eddies. When focusing on the dependance of the species fields $C_i(\mathbf{x}, t)$ upon the Damkhöler number $Da = \tau_t/\tau_c$, the isothermal reaction (1-1) provides a convenient frame of approximation since the value of $Da$ is well defined for each test and remains constant.

In order to display the evolutions of velocity and chemical species, in space and time, a video film was made; it shows how large scale structures interact one with the other, inducing intrusions of large quantities of $A$ (or $B$) species into the domain intially filled with $B$ (or $A$) species and how deep these intrusions are, depending on the Damkhöler number.

Two fields, at time $t_1 \approx 35\delta_i/U$, selected from the results are shown on plates 1 and 2. With $Da \approx 1$, the concentration of one of the reactive species $C_A$, on plate 1, [1] develops a "thin tongues" pattern which is less pronounced than in the inert scalar case corresponding to $Da \to 0$ (as in Comte *et al.* 1987). Still for $Da \approx 1$, similar pictures of vorticity $\omega$ and product concentration $C_P$ would have shown that such a $C_P$ is a very good tracer of the vorticity. Species spatial distribution is quite different for very fast reactions ($Da \approx 10^4$), [2] the eddy parttern of the $C_P$ field, on plate 2, is scrambled yielding a wide strip of high $C_P$ values, which suggests to use such $C_P$ as a mean of visualising the layer thickness. In this same fast reaction case, the reactive $C_A$ tongues, are completely eroded, when compared to plate 1, with a drastic smoothing of the inerface between $A$ and $B$ regions.

### 3.2. Reactions rates, iso value contour lengths

Reaction rates $\dot{w}(\mathbf{x}, t)$ are calculated at each time step since they need to be known in order to compute the evolution of $C_i$ fields. Studying the regions of high $\dot{w}$ is the most direct method to get the features which characterize the interactions between turbulence and reaction. The evolutions of $C_P$ and $\dot{w}$, as they are observed on the video film, exhibit significant differences, with rapidly varying values of $\dot{w}$ along the braids, especially for fast reactions.

---

[1] concentration varies from 1 to 0 as colour evolves from red to blue

[2] in this latter case, the interactions between turbulence and chemistry are little ruled by the large scale structures, which makes the two dimensional approximation probably less appropriate when dealing with real problems

Various mean quantities have been calculated at each time step in order to characterize the evolution of the reaction: p.d.f. of $C_i$, mean reaction rate $< \dot{w}(t) >$, $< C_i(t) >$ and $< \dot{w}(y,t) >$ profiles averaged along the streamwise $x$ direction. Beside these usual ways of analysing results, another field characterisation is the measure of iso-value contour lengths [3] The length depends of course on the threshold value $\theta_0$ chosen to draw the iso-line; it increases as $\theta_0$ decreases . Nevertheless, the evolutions of this length with time appear to be quite similar and almost proportional one to the other, for a whole range of threshold values. Measuring the contour length of an iso- $C_P$ line corresponding to the threshold $C_0$ yields the length $L_{P,C_0}(t)$ which is presented in figure 2 for various values of the $Da$ number. A quantity similar to $L_P(t)$ can be estimated by experimentalists by taking pictures of flows with a dye triggered by the product $P$.

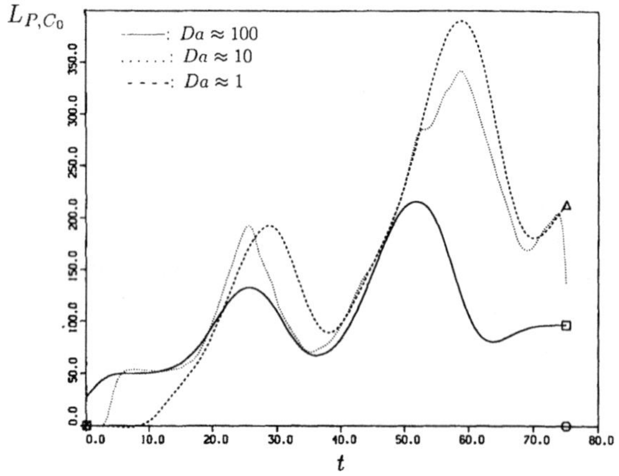


Fig.2. Evolution with time of the length of an iso-$C_P$ contour, $C_P$ being the concentration of species produced by the reaction

$L_P(t)$ has been drawn for $C_0 = 0.6$; the evolution strongly depends on the eddy pairing which occur at $t_{p1} =\approx 35\delta_i/U$ and $t_{p2} =\approx 75\delta_i/U$. The contour is observed to lengthen, up to the first pairing (at $t_{p1}$), this first stage is driven by the velocity field since it appears not to depend on the $Da$ value; after reaching its maximum value, $L_P(t)$ decreases because of the eddy merging; the peaks occur before eddy pairing, when structures are close enough one to the other with a lot of interpenetration of their peripheral regions. With a faster reaction ($Da$ $\approx$, 100), the contour is shorter since $P$ is rapidly produced before any significant turbulent

[3] this length measured from the results of a two dimensional simulation corresponds to a specific surface area in three dimensional real flows

mixing takes place, which creates relatively small high $C_P$ regions. The higher values of $L_P$ are obtained for $Da \approx 1$.

$L_P(t)$ helps to understand better ways of tracing flow structures with chemical species, but a measure of an isovalue of the reaction rate $\dot{w}$ would be more meaningful when turning to combustion problems since this iso-$\dot{w}$ contour can be interpreted as a reaction front or a flame length. The threshold $\dot{w}_0$ is selected in order to get results whose general behaviour is not too sensitive to its value. The length $L_{\dot{w},\dot{w}_0}(t)$ calculated for $\dot{w}_0 = 0.06$, is plotted on fig.3.

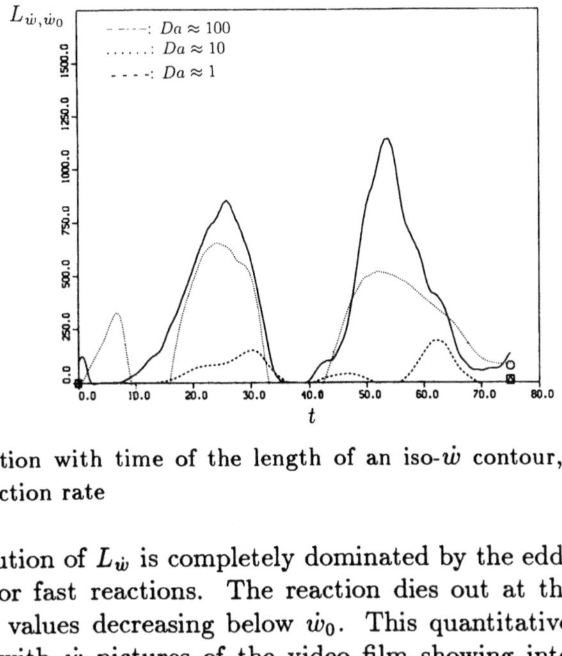


Fig.3. Evolution with time of the length of an iso-$\dot{w}$ contour, $\dot{w}$ being the chemical reaction rate

The evolution of $L_{\dot{w}}$ is completely dominated by the eddy structures, especially for fast reactions. The reaction dies out at the eddy pairing with $\dot{w}$ values decreasing below $\dot{w}_0$. This quantitative result is in agreement with $\dot{w}$ pictures of the video film showing intermittent intense activity in the braid regions. Fast reactions ($Da \approx 100$) enable $L_{\dot{w}}$ to decrease very rapidly from high values, before the eddy pairing.

## 4. Spatial mixing layer to model a diffusion flame

### 4.1. Spatial mixing layer with temperature dependent reaction rate

Spatial approximation seems to be closer to real cases since it can provide more realistic averaging procedures, for instance to compute the thickness $\delta(x,t)$ growth rate, and deal with many practical cases where the mean flow is stationary. The requested boundary conditions are *(i)* an inflow condition: the velocity shear between the two streams carrying

$A$ and $B$ respectively *(ii)* a radiative outflow condition (at $x = L_x$) which let eddies and scalars escape freely from the computation domain. This configuration can be viewed also as corresponding to a crude model of basic mechanisms of diffusion flames. The only reaction rate formulation considered in this section is the temperature dependent reaction rate according to Arrhenius law (1-2).

Taking advantage of the fact that the fluid is made only of species $A$, $B$ and $P$, which yields $C_A(\mathbf{x},t)+C_B(\mathbf{x},t)+C_P(\mathbf{x},t) = 1$ and restricting to the the case where the quantity of moles remains unchanged through the reaction ($\nu_A + \nu_B = \nu_P$), an inert scalar $\mathbf{\Phi}_s(\mathbf{x},t)$ is considered in such a way that the species concentrations are formulated as :

$$C_i(\mathbf{x},t) = \epsilon_i^{(1)} \frac{1+\epsilon_i^{(2)}\mathbf{\Phi}_s(\mathbf{x},t)}{2} - \epsilon_i^{(3)}\nu_i \frac{T(\mathbf{x},t)-T_0}{T_q} \qquad (4-1)$$

with $\epsilon_A^{(1)} = \epsilon_B^{(1)} = 1$, $\epsilon_P^{(1)} = 0$, $\epsilon_A^{(2)} = -\epsilon_B^{(2)} = 1$, $\epsilon_A^{(3)} = \epsilon_B^{(3)} = -\epsilon_P^{(3)} = 1$

Evolutions of ($\mathbf{u}$, $T$, $C_A$, $C_B$, $C_P$, $T$) can be calculated from the evolutions of ($\omega$, $\psi$, $T$, $\mathbf{\Phi}_s$). The product concentration $C_P$ and the temperature variation ($T(\mathbf{x},t)-T_0$) remain proportional because they are ruled by the same equation (in (2-3)) and the same boundary and initial conditions (in (2-4)): 0 at $y = \pm L/2$ and $x = 0$. $\omega$ and $\mathbf{\Phi}_s$ are governed by the same equation but their boundary conditions differ. This formulation helps to reduce the computation effort which is always heavy in spatial calculations which require numerous grid points along the streamwise direction. The evolution of the four components of ($\psi, \omega, \mathbf{\Phi}_s, T$) vector are calculated in space and time by numerical methods which are similar to those used in section 3, with some improvements: non uniform mesh grid in the $y$ direction, step size refinement to handle fast reactions. Assuming constant density in reactive flows allows to treat temperature and all species concentrations as passive scalars which do not act on the velocity field. Consequently, the value of the time step $\Delta_\omega t$ under use to integrate the ($\psi,\omega$) flow field can be chosen to be significantly different from the value of the time step $\Delta_\theta t$ used to integrate the scalar fields ($\mathbf{\Phi}_s, T$); faster reactions can be simulated by using smaller values of $\Delta_\theta t/\Delta_\omega t$.

### 4.2. Vorticity, temperature and product concentrations

Various realisations $\Omega_i$ of the flow are obtained through changing the white noise which is superimposed on the velocity profile at the inflow condition. Results presented on figure 4 and 5, are obtained at time $t_2 \approx 2.5t_s$ where $t_s$ is the time $L_x/U$, with a spatial resolution of $64 \times 256$. The upstream part of the domain where the instability expands is not represented and values are plotted only for $128\Delta x < x < 256\Delta x$.

These figures must be viewed as snaphots at time $t_2$ since picturing the non stationary behaviour around mean stationay values would require to show the film.

Flow field is well described by vorticity $\omega$ which reachs its peak value in the eddy cores; exact amplitude and locations would, of course, change depending on the realisation $\Omega_i$ and on time $t_2$. The temperature (or $C_P$) structures develop in the same locations as the vorticity structures; slight shifts which possibly appear are resorbed through terms which depend on $\nu$, $D_i$, $\nabla C_i$ and $\nabla\omega$.

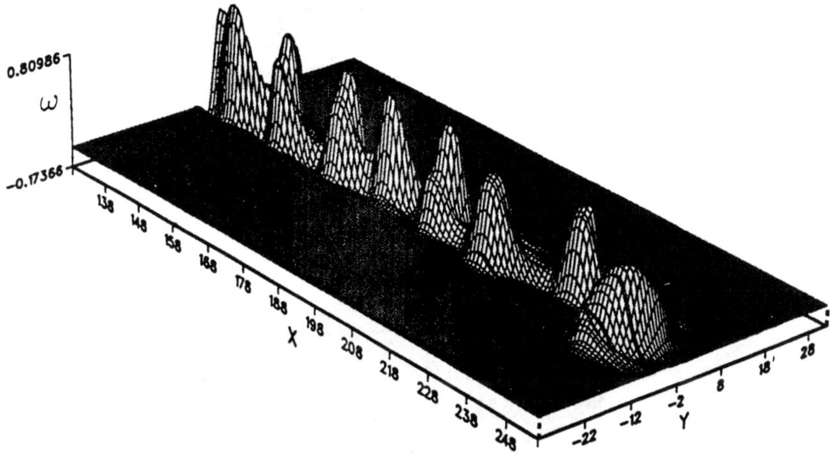


Fig.4. Vorticity field $\omega(x, y, t_2)$

Various cases of chemical reaction according to Arrhenius law (1-2) were tested; a fast reaction with $B = 5000, T_q = 5, T_A = 14$ produces the temperature field of figure 5. Flame gets ignited from the inlet and the fluid is hot everywhere along the interface between the $A$ and $B$ domains.

For a slower reaction with $B = 300, T_q = 2, T_A = 10$, the temperature is represented, at time $t_3 = 1.5t_s$, on figure 6; fluid is hot inside the eddies an cold in the braid regions. Variations of temperature along the streamwise direction is rather strong with minimal (cold) values between eddies especially when the eddies are far enough one from the other. This simulation and its results can be considered as a first step towards a direct simulation of a diffusion flame, featuring flame ignition and extinction, and any other transient regime.

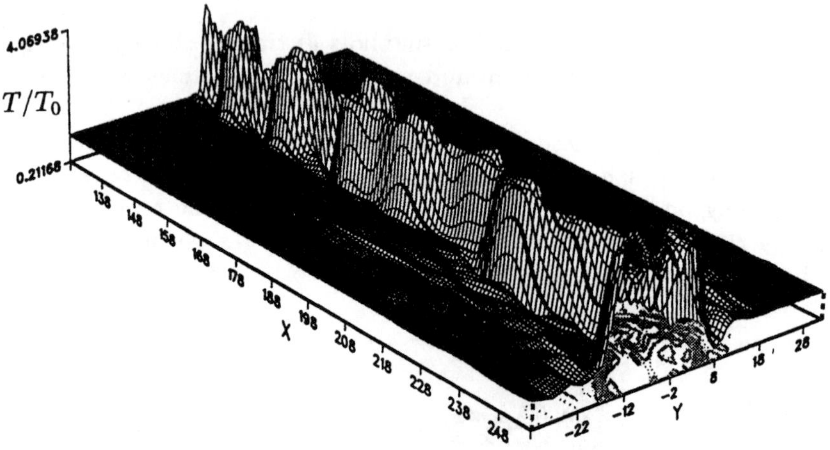


Fig.5. Temperature field $T(x, y, t_2)$

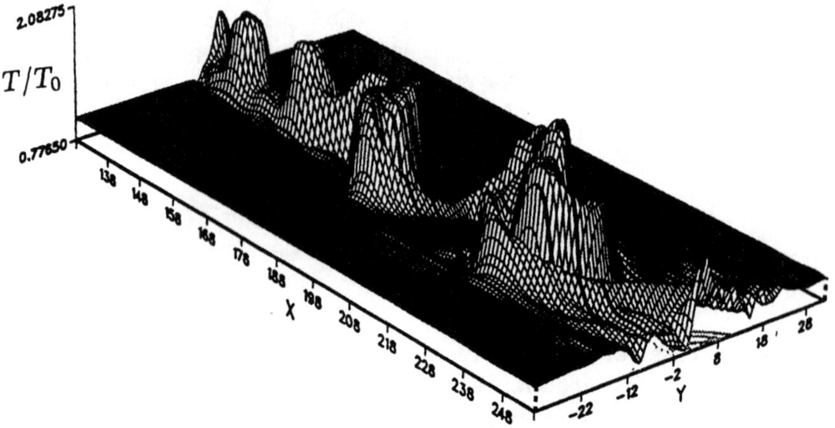


Fig.6. Temperature field $T(x, y, t_3)$

## 5. CONCLUSION

Although vorticity appears to be the relevant quantity to study turbulent velocity fields and flow structures, no method of direct visualisation of vorticity distribution is avalaible. Tracing at least one of the flow with a scalar does not always give a clear picture of the eddy structures. The proper way to get scalar tracing of the flow structure is then to observe the concentration of species $P$ produced by the reaction of species $A$ and $B$ carried by each of the two streams of the layer; the reaction must be slow enough ($Da$ of the order of 1). Faster reactions have some averaging properties which are helpful to get statistical quantities, for instance when defining a "visual" thickness of the mixing layer. Tracing the vorticity field with the $C_P$ product could be improved by designing purely artificial chemical species $C_t$ for the purpose of analysing results from numerical simulations. The reaction rate $\dot{w}$ would then be tailored to: *(i)* maintain high $C_t$ values with positive values of $\dot{w}$ in order to balance the diffusion $D_t \nabla C_t$ *(ii)* strongly damp low $C_t$ by consuming the scalar with a negative $\dot{w}$. The initial conditions for the self-sustained $C_t$ would be a peak at $y = 0$ in order to trace the evolution of the interface between the two streams of the mixing layer.

Mixing layers with temperature dependent chemistry, especially with spatial calculations, are an appropriate frame to study the basic mechanisms of a diffusion flame which can ignite by itself. Temperature, though strongly correlated with flow structures, depends on the characteristics of the chemical reaction, with various regimes possibly resulting in flame extinction.

As for homogeneous flows Picart *et al.* 1987, real mixing layers need three dimensional simulations (as in Riley et al. 1986) to reliably describe phenomena from dissipative conductive scales to large eddy structures. Nevertheless, due to the limitations in computer memory size and processing time, two dimensional calculations have still to be used to: *(i)* simulate spatial streamwise development of the flow in a domain whose size is long enough, in the $x$ direction, to let chemistry and dynamics evolve and interact, *(ii)* derive and tune relevant methods to analyse the computed flow and scalar fields, for instance the contour calculation which is presently extended to surface aera analysis of three dimensional simulation results.

ACKNOWLEDGEMENTS

This work has been partly supported by D.R.E.T. under contract 86/199. Contour calculation of $C_P$ fields was initiated jointly with A.Picart and R.Borghi, when working on homogeneous reactive flows.

## References

Broadwell, J.E. & Breidenthal, R.E., 1982, A simple model of mixing and chemical reaction in a turbulent shear layer, *J. Fluid Mech.*, **125**, 397-410 .

Comte, P., Lesieur, M. & Chollet, J.P., 1987, Simulation numérique d'un jet plan turbulent, *C.R. Acad. Sci., Paris*, **205**, série II, 1037-1044.

Comte, P., Lesieur, M., Laroche, H. & Normand, X. , 1989, Numerical simulations of turbulent plane shear layers, *Turbulence Shear Flows 6*, Springer-Verlag, 360-380.

Koochesfahani, M.M. & Dimotakis, P.E. , 1986, Mixing and chemical reactions in a turbulent liquid mixing layer, *J. Fluid Mech.*, **170**, 83-112.

Lesieur, M., 1987, *Turbulence in Fluids*, Kluwer.

Lesieur, M., Staquet, C., Le Roy, P. & Comte, P. , 1988, The mixing layer and its coherence examined from the point of view of two- dimensional turbulence, *J. Fluid Mech.*, **192**, 511-534.

Mungal, M.G. & Dimotakis, P.E. , 1984, Mixing and combustion with low heat release in a turbulent shear layer, *J. Fluid Mech.*, **148**, 349-382.

Normand, X., Comte, P. & Lesieur, M., 1988, Siumlation numérique d'une couche de mélange se développant spatialement, *La Recherche Aérospatiale*, **6**, 45-52.

Picart, A., Borghi, R. & Chollet, J.P. , 1987, Numerical simulation of turbulent reactive flows : new results and discussion Proc. 6th TSF, Toulouse.

Riley, J., Metcalfe, R.W. & Orszag, S.A. , 1986, Direct Numerical Simulations of Chemically Reacting Turbulent Mixing Layers, *Phys. Fluids*, **29**(2), 406-422.

# IX. MHD Turbulence

# Large-eddy simulation of MHD turbulent channel flow under a uniform magnetic field

Yutaka SHIMOMURA

*Department of Physics, Faculty of Science, University of Tokyo*
*Bunkyo-ku, Tokyo, 113, Japan*

*Turbulence and Coherent Structures*
*O. Métais and M. Lesieur, eds.*
*Selected papers from "Turbulence 89: organized structures and turbulence in fluid mechanics"*
*Grenoble, 18-21 September 1989*

## 1. Introduction

Liquid metals such as mercury are weakly electrically-conducting fluids, and their motions are described by magnetohydrodynamics. These magnetohydrodynamic (MHD) fluids have many applications to engineering devices: their dynamics are very important in designing liquid metal pumps, electromagnetic flow meters, MHD power generators (Yakhot & Branover 1982), liquid metal heat exchangers in nuclear fusion reactors (Shercliff 1979), and some kinds of crystal pullers (Hjellming & Walker 1987). For pure scientific research, these MHD fluids are known to be good candidates to make two-dimensional turbulence in laboratory experiments (Schumann 1976; Sommeria & Moreau 1982).

In most of realistic cases, the flow (the velocity field) is turbulent with large Reynolds number. On the other hand, the magnetic Reynolds number, which is the dimensionless parameter in the induction equation for magnetic fields, is very small. For, the magnetic Reynolds number $R_m$ is defined by using the electrical conductivity $\sigma$, the magnetic permeability $\mu$, the characteristic velocity $U_0$, and the characteristic length scale $L$ as

$$R_m = \sigma \mu U_0 L. \qquad (1-1)$$

So, weakly conducting fluid generally has small magnetic Reynolds number. For $R_m \ll 1$, the diffusion term is dominant in the induction equation for the magnetic field, so that the induced magnetic field is very small $(O(R_m))$ compared with the imposed field $\mathbf{B_0}$ $(O(R_m^0))$ , and negligible in a first approximation of $O(R_m^0)$.

For non-conducting turbulent shear flows, the streaky structures (Kline et al. 1967) and the hairpin vortices (Willmarth & Tu 1967) are important organized structures associated with the turbulent energy production and transfer in the wall region of turbulent boundary layers. Are these organized structures present under a magnetic field? How do they, if present, change from those obeserved in the usual non-conducting turbulent shear flows?

In order to answer these questions, it could be the easiest way to do the "direct"numerical simulation of realistic high-Reynolds number turbulence. Actually, however, this approach is unpractical now: in spite of the great advance of the speed and memory size of the computing hardware, high-Reynolds number turbulence remains unsolved by the "direct"numerical simulation because it involves untractablly huge number of degrees of freedom. Of course, the global turbulence modelings such as the two-equation model (Bradshaw, Cebeci & Whitelaw 1981; see for MHD version Shimomura 1988) are helpless in explaining the detais of the flow structures.

The appoach lying between the "direct"numerical simulation and the global turbulence modeling is large-eddy simulation (LES) (see for example Rogallo & Poin 1984) where the large-scale motions are explicitly computed and only the subgrid-scale ones are modeled. The advantage of LES is that it can compute the comparatively detaild structure of turbulence by introducing a small number of the modeling parameters. In fact, Moin & Kim (1982) studied non-conducting turbulent channel flow by using LES, and succeeded in capturing the streaky structure in the vicinity of the wall. Later, the existence of hairpin vortices was also confirmed (Moin & Kim 1985) using their LES database.

The objective of the present paper is to analyse how magnetic fields change these organized structures in MHD turbulent channel flows at low $R_m$ under a uniform magnetic field $\mathbf{B_0}$ which is perpendicular to the insulated walls. We study it numerically by using LES technique.

In §2, we formulate LES in MHD at low $R_m$ version and describe the numerical method for channel flow configuration. In §3, results on the effects of magnetic field on the streaky and vorticity structures are presented. Finally, our conclusions are summarized in §4.

## 2. Formulation

### 2.1. Governing equations for the large-scale fields

The dynamics of an incompressible MHD flow at low $R_m$ under a uniform magnetic field is described in the first approximation of $R_m$ expansion by a system of the following equations (see for example Schumann 1976):

$$\frac{\partial}{\partial t}\mathbf{u} + \mathbf{u}\cdot\nabla\mathbf{u} = -\nabla p + \nu\nabla^2\mathbf{u} + \frac{1}{\rho}\mathbf{j}\times\mathbf{B_0}, \qquad (2-1)$$

$$\nabla\cdot\mathbf{u} = 0, \qquad (2-2)$$

$$\mathbf{j} = \sigma(-\nabla\varphi + \mathbf{u}\times\mathbf{B_0}), \qquad (2-3)$$

$$\nabla^2\varphi = \nabla\cdot(\mathbf{u}\times\mathbf{B_0}) = \mathbf{B_0}\cdot\omega, \qquad \omega = \nabla\times\mathbf{u}. \qquad (2-4)$$

Here, $\mathbf{u}, \mathbf{B_0}, \mathbf{j}, p, \varphi$ denote the velocity, the applied uniform magnetic field, the electric current density, and the kinematic pressure, the electrostatic potential, respectively. The material constants $\rho, \nu, \sigma$ are called the fluid density, the kinematic viscosity, and the electrical conductivity, respectively. The equation (2-1) is the Navier-Stokes equations including Lorentz force $\mathbf{j}\times\mathbf{B_0}$, (2-2) is the incompressibility condition, (2-3) is

Ohm's law, and (2-4) is the Poisson equation satisfied by the electrostatic potential.

In LES, we introduce a spatial filtering operation to decompose a flow field $f$ into a large-scale or grid-scale (GS) component $\bar{f}$ and a small-scale or subgrid-scale (SGS) component $f''$ as follows:

$$f = \bar{f} + f'', \tag{2-5}$$

$$\bar{f}(\mathbf{x}) = \int\int\int G(\mathbf{x},\mathbf{y})f(\mathbf{y})d\mathbf{y}, \tag{2-6}$$

where $G(\mathbf{x},\mathbf{y})$ is a filter function which averages small-scale variation. For a plane channel geometry, Mion and Kim (1982) used the Gaussian filter function in planes parallel to the walls, and the top-hat filter function in the direction normal to the walls. Now, applynig the filtering process to our system of governing equations (2-1)-(2-4), we get the filtered equations for GS flow fields:

$$\frac{\partial}{\partial t}\bar{u}_i + \frac{\partial}{\partial x_j}(\bar{u}_j\bar{u}_j) = -\nabla\bar{p} + \nu\nabla^2\bar{u}_i + \frac{\sigma}{\rho}\left((-\nabla\bar{\varphi} + \bar{\mathbf{u}} \times \mathbf{B_0}) \times \mathbf{B_0}\right)_i +$$

$$\frac{\partial}{\partial x_j}(R_{ij} + L_{ij} + C_{ij}), \tag{2-7}$$

$$\nabla \cdot \bar{\mathbf{u}} = 0, \tag{2-8}$$

$$\nabla^2\bar{\varphi} = \nabla \cdot (\bar{\mathbf{u}} \times \mathbf{B_0}) = \mathbf{B_0} \cdot \bar{\omega}, \tag{2-9}$$

where the SGS Reynolds stress $R_{ij}$, the Leonard stress $L_{ij}$, and the cross correlation $C_{ij}$ are given by the same definitions as in the usual non-conducting case since Lorentz force is linear with regard to the velocity $\mathbf{u}$ at low $R_m$; namely,

$$R_{ij} = -\overline{u_i''u_j''}, \tag{2-10}$$

$$L_{ij} = \bar{u}_i\bar{u}_j - \overline{\bar{u}_i\bar{u}_j}, \tag{2-11}$$

$$C_{ij} = -\left(\overline{\bar{u}_i u_j''} + \overline{u_i''\bar{u}_j}\right). \tag{2-12}$$

In order to close the filtered equations (2-7)-(2-9), we need SGS modeling for $R_{ij}$, $L_{ij}$, $C_{ij}$. For non-conducting flows, the Smagorinsly model is usually adopted for $R_{ij}$, while both $L_{ij}$ and $C_{ij}$ are often neglected. This modeling is shown to work well, and conserves Galilean invariance as pointed out by Speziale (1985). In MHD case studied here, the validity of this modeling is not obvious. In fact, Shimomura (1988) showed theoretically by using the two-scale direct interaction approximation (Yoshizawa 1984) combined with the double expansion with regard to $R_m$ and the interaction parameter that at the level of ensemble average, the usual eddy-viscosity representation of the Reynolds stress should be modified by the effect of magnetic field. It was also found by comparing the numerical and experimental data that a modified two-equation model incorporating this effect is better than the existing one for non-conducting flow in explaining the laminarization by the strong magnetic field. In LES, however, it is not straightforward to construct a SGS model including the magnetic effects since the length scale is given *a priori* if we are based on the Smagorinsky model. In the present study, expecting that it still works when the applied magnetic field $\mathbf{B_0}$ is weak, we adopt the SGS modeling mentioned above; namely ,

$$R_{ij} = -\frac{1}{3}\overline{u''_a u''_a}\delta_{ij} + \nu_e \bar{e}_{ij}, \qquad \bar{e}_{ij} = \left(\frac{\partial \bar{u}_i}{\partial x_j} + \frac{\partial \bar{u}_j}{\partial x_i}\right), \tag{2-13}$$

$$L_{ij} = 0, \qquad C_{ij} = 0, \tag{2-14}$$

where the SGS eddy viscosity $\nu_e$ is given in the Smagorinsky model by

$$\nu_e = (C_s\Delta)^2\sqrt{\frac{1}{2}\bar{e}_{ab}\bar{e}_{ab}}. \tag{2-15}$$

In the expression (2-15), $C_s$ is the model constant chosen as 0.1, and $\Delta$ is the representative length scale of the form

$$\Delta(x, y, z) = (\Delta x \Delta y \Delta z)^{1/3}, \tag{2-16}$$

where $\Delta x$, $\Delta y$, and $\Delta z$ represent the computational mesh sizes in the $x$, $y$, and $z$ directions, respectively. Here, we should note that the SGS model (2-13)-(2-16) can not explain low-Reynolds-number effect on the SGS turbulence near the wall. So, as is usually done for the LES of non-conducting turbulence, the same modification is introduced in the present study; $\Delta$ is multiplied by the wall damping function $f$ of Van Driest type, which is given in case that the wall is located on the $z = 0$ plane by

$$f\left(z^+\right) = 1 - \exp\left(-\frac{z^+}{A^+}\right), \qquad A^+ = 25, \tag{2-17}$$

In (2-17), $z^+$ is the distance to the nearest wall in the wall units and it is defined with the friction velocity $u_\tau$ as

$$z^+ = \frac{z u_\tau}{\nu}, \qquad u_\tau = \sqrt{\nu \frac{\partial \langle \bar{u}_1 \rangle}{\partial z}\bigg|_{z=0}}, \tag{2-18}$$

where $\langle \bar{u}_1 \rangle$ denotes an ensemble mean of the streamwise GS velocity.

The equations (2-7)-(2-9) for the GS fields with the modeling (2-13)-(2-15), and (2-16) multiplied by the wall damping function (2-17) constitute the basic system of equations to be solved in the present numerical simulation of MHD turbulent channel flow at low $R_m$ under a uniform and not so strong applied magnetic field.

### 2.2. MHD channel flow under a uniform magnetic field

We study a MHD turbulent plane channel flow at low $R_m$ under a uniform magnetic field which is normal to the walls.

The flow configuration and the coordinate system used in this paper are shown in figure 1.

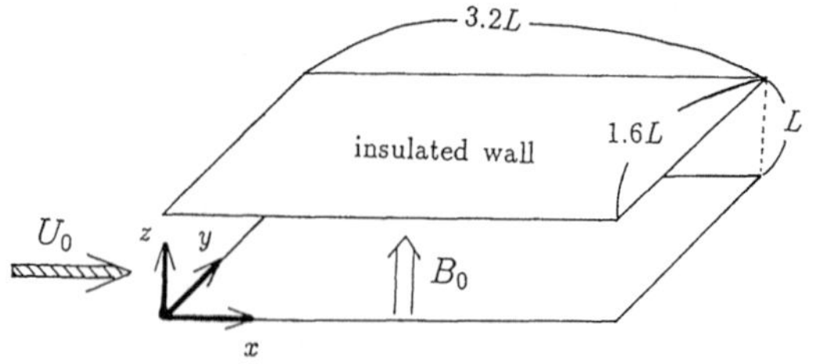


Fig.1. Flow configuration

Here, $x$, $y$, and $z$ denote the streamwise, spanwise, and normal ( to the wall ) directions respectively; $u_1$ , $u_2$, and $u_3$ are the velocity components along $x$, $y$, and $z$ respectively. The origin $O$ of the normal coordinate, $z = 0$, is located on the bottom wall. A magnetic field of magnitude $\mathbf{B_0}$ is uniformly applied in the direction of positive $z$ to the conducting fluid flow between two horizontal walls, which are separated each other at a distance $L$. We suppose that the walls are insulated and its dimensions $L_x$ and $L_y$ are much larger than the channel whole width $L$ ( $L_x, L_y \gg L$ ).

Next, let us state the boundary conditions. On the insulated walls, all the velocity components and the electric current density normal to the wall vanish:

$$\mathbf{u} = 0, \qquad j_3 = 0 \qquad (z = 0, L). \qquad (2-19)$$

Combining (2-3), and (2-19), we get the non-slip condition on the GS components of velocity and the Neumann condition on the GS component of electrostatic potential on the walls:

$$\bar{\mathbf{u}} = 0, \qquad \frac{\partial \bar{\varphi}}{\partial z} = 0 \qquad (z = 0, L). \qquad (2-20)$$

In the $x$, $y$ directions, we introduce the finite computational domain in real calculation to impose the periodic boundary conditions on $\mathbf{u}$, $\nabla p$,

$\nabla\varphi$, or, $\bar{\mathbf{u}}$, $\nabla\bar{p}$, $\nabla\bar{\varphi}$. This artificial condition can mimic the assumed flow between the infinitely extended horizontal walls only if the computational domain is large enough: the lengths, $L_x$ and $L_y$, of the computational domain should be larger than the two-point correlation lengths between flow variable fluctuations in $x$ and $y$ directions, respectively. In the present calculation, we employ

$$L_x = 3.2L, \qquad L_y = 1.6L, \tag{2-21}$$

which gives enough sizes for non-conducting turbulent channel flow (Horiuti 1987).

In order to specify the flow, we need one more condition, which is also required to obtain the laminar solution or Hartmann flow. This condition is determined by the state of conducting channel sidewalls in a real experiment. Here, having correspondence with the experiment by Brouillette & Lykoudis (1967), we suppose that the sidewalls constitute an open circuit. Then, the net spanwise current in the flow domain should disappear:

$$\int_0^{L_x} dx \int_0^{L_y} dy \int_0^{L} dz\, j_2 = 0. \tag{2-22}$$

If we apply filtering to (2-22) and denote the average over a $x-y$ plane by $\langle \quad \rangle$, we get

$$\int_0^{L} \langle \bar{j}_2 \rangle\, dz = 0, \tag{2-23}$$

which establishes by virtue of (2-3) the average of spanwise derivative of the GS electrostatic potential as

$$\left\langle \frac{\partial}{\partial y}\bar{\varphi} \right\rangle = -B_0 U_0, \tag{2-24}$$

where $U_0$ is the average velocity in the channel defined as

$$U_0 = \frac{1}{L}\int_0^{L} \langle \bar{u}_1 \rangle\, dz. \tag{2-25}$$

The condition (2-24) is utilized in the method described in the next subsection **2.3.**

### 2.3. Numerical method

For a channel flow explained in the foregoing subsection **2.2**, the equation (2-7) and (2-9) are recast with the help of (2-13), (2-14), and (2-24) in the dimensionless form

$$\frac{\partial}{\partial t}\bar{u}_i + \frac{\partial}{\partial x_j}(\bar{u}_j\bar{u}_j) = -\frac{\partial}{\partial x_i}\bar{P}' + \frac{\partial}{\partial x_j}\left(\left(\frac{1}{R_e} + \nu_e\right)\bar{e}_{ij}\right)$$

$$+N\left(-\epsilon_{ij3}\frac{\partial}{\partial x_j}\bar{\varphi}' + \bar{u}_3\delta_{i3} - \bar{u}_i\right) + \left(-\left\langle\frac{\partial \bar{P}}{\partial x}\right\rangle + N\right)\delta_{i1}, \qquad (2-26)$$

$$\frac{\partial^2}{\partial x_j \partial x_j}\bar{\varphi}' = \frac{\partial \bar{u}_2}{\partial x} - \frac{\partial \bar{u}_1}{\partial y} = \bar{\omega}_3, \qquad (2-27)$$

where

$$\bar{P} = \bar{p} + \frac{1}{3}\overline{u''_a u''_a}, \qquad (2-28)$$

and the quantity with prime means the deviation from the average over a $x-y$ plane:

$$\bar{P}' = \bar{P} - \langle\bar{P}\rangle, \qquad \bar{\varphi}' = \bar{\varphi} - \langle\bar{\varphi}\rangle. \qquad (2-29)$$

The channel whole width $L$, the average velocity $U_0$, and the applied magnetic field $B_0$ are used to normalize the equations (2-26) and (2-27) where we have two dimensionless parameters; the Reynolds number $R_e$ and the interaction parameter $N$, which are defined as

$$R_e = \frac{U_0 L}{\nu}, \qquad N = \frac{\sigma}{\rho}\frac{B_0^2 L}{U_0}. \qquad (2-30)$$

For later convenience, let us introduce one more dimensionless parameter called the Hartmann number $H_a$ given by

$$H_a = \sqrt{R_e \cdot N} = \sqrt{\frac{\sigma}{\rho\nu}}B_0 L. \qquad (2-31)$$

In the present calculation, the equation (2-26) and (2-27), coupled with (2-8), (2-15)-(2-18), and (2-20), are numerically solved by a finite difference method at three Hartmann numbers $H_a = 0, 30, 40$ with the Reynolds number $R_e$ fixed at 29000.

As a computational grid, we adopt a full staggered system where each flow variable is defined at different node. The equation of continuity (2-8) is enforced at node points for $\bar{P}'$ through the Poisson equation for $\bar{P}'$ while the momentum equations (2-26) and the Poisson equation (2-27) for $\bar{\varphi}'$ are evaluated at node points for $\bar{u}_i$ and $\bar{\varphi}'$, respectively. Note, especially, that the location of node point for $\bar{\varphi}'$ allows for a compact evaluation of the quantities related to $\bar{\varphi}'$. In $x$ and $y$ directions, $N_x = N_y = 64$ grid points are distributed uniformly, whereas $N_z = 65$ grid points with non-uniform spacings are placed in $z$ direction. The location $z_i$ $(0 \le z_i \le 1)$ of the mesh is given by

$$z_i = \frac{1}{2}\left(1 + \frac{1}{a}\tanh\left(\frac{1}{2}\xi_i \ln\frac{1+a}{1-a}\right)\right), \qquad (2-32)$$

where

$$\xi_i = -1 + \frac{2i}{N_z} \qquad (i = 0, 1, \cdots, N_z), \qquad a = 0.98346. \qquad (2-33)$$

Partial derivatives in space are approximated by the second-order central difference formulae. The Poisson equations for $\bar{P}'$ and $\bar{\varphi}'$ are first discretized in the physical space, second Fourier-transformed in $x$ and $y$ directions, and finally solved by the inversions of tri-diagonal matrices. The time is advanced by a second-order Adams-Bashforth scheme with the total flow quantity conserved, which specify the value of $-\langle \partial \bar{P} / \partial x \rangle$ in (2-26). The time step is set at 0.01 in the dimensionless unit. Computations were done with HITAC S-810 Model 20 system of the Computer Center of the University of Tokyo and with the FACOM VP-100 system of the Institute of Industrial Science.

## 3. Results

In this section, we show the results of three cases at $H_a = 0$, 30, and 40 with the fixed $R_e = 29000$, and discuss the effect of magnetic field on the turbulence by comparing them. For each case, the governing equations are integrated forward in time until the solutions reach statistically steady states. This stationary states are identified by approximate periodicity of horizontally averaged GS component of streamwise velocity. After the statistically steady states, horizontal averages of various quantities are averaged in time period $t = 3.2$ for which the equations are further integrated. In the following, $\langle \cdot \rangle$ means time and horizontal average.

### 3.1. Streaky structures

Figures 2 and 3 show the streamwise and spanwise two-point correlation functions of $\bar{u}_1'$ at $z = 0.0078$, respectively. They are defined as

$$R_{11}(r_1; z) = \frac{\langle \bar{u}_1'(x, y, z)\bar{u}_1'(x + r_1, y, z)\rangle}{\langle \bar{u}_1'^2(x, y, z)\rangle},$$

$$R_{11}(r_2; z) = \frac{\langle \bar{u}_1'(x, y, z)\bar{u}_1'(x, y + r_2, z)\rangle}{\langle \bar{u}_1'^2(x, y, z)\rangle}.$$

In figures 2 and 3, the solid, the dashed, and the broken lines represent the results of $H_a = 0, 30$, and 40, respectively.

At $H_a = 0$, the slow decay of the streamwise correlation function $R_{11}(r_1; z)$ in figure 2 indicates the high elongation of the $\bar{u}_1'$ contour in the streamwise direction, and the negative value of the spanwise correlation function $R_{11}(r_2; z)$ in figure 3 at $r_2 \sim 0.05$ implies the alternating structures of well-known high- and low-speed streaks. The negative peak

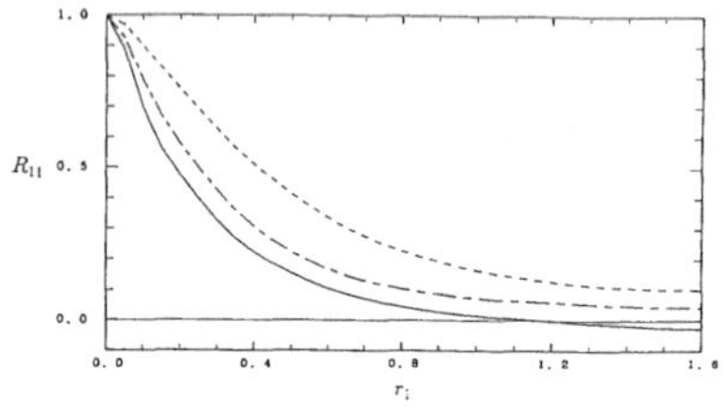

Fig.2. Streamwise two-pioint correlation functions $R_{11}(r_1; z)$ at $z = 0.0078$: $H_a = 0$ ; $H_a = 30$ ; $H_a = 40$

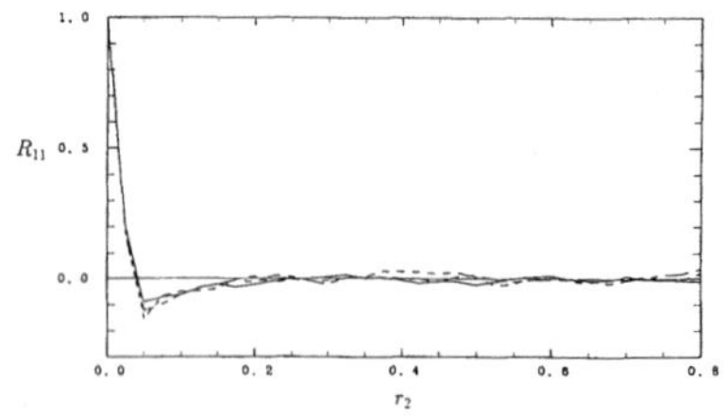

Fig.3. Spanwise two-point correlation functions $R_{11}(r_2; z)$ at $z = 0.0078$: $H_a = 0$ ; $H_a = 30$ ; $H_a = 40$

of $R_{11}(r_2; z)$ occurs at $r_2 \sim 0.05$ ($r_2^+ \sim 75$). This estimates the mean streak spacing to be 150 in the wall unit, which is close to the generally accepted value 100.

We notice that as the Hartmann number is raised, all the streamwise two-point correlations extend over much longer distances. On the other hand, there is no major difference between the spanwise two-point correlations at $H_a = 0, 30$, and 40.

These results on $R_{11}(r_1; z)$ and $R_{11}(r_2; z)$ are consistent with the pictures in figure 4 which shows the contour plots of $\bar{u}'_1$ in the x-y plane at $z = 0.0078$ ($z^+ \sim 10$). In three cases (a), (b), and (c) in figure 4, both the levels and the number of contour lines are same. Positive values are contoured by solid lines and negative values are contoured by dashed lines. In addition, the regions of positive values are shaded by dots.

At $H_a = 0$, the streaky structures are observed in (a): the highly elongated regions of high-speed fluid ($\bar{u}'_1 > 0$) and low-speed fluid ($\bar{u}'_1 < 0$) are alternating in the spanwise direction. When a magnetic field is applied, the streaky structures are more elongated in the streamwise direction while the mean spacing in the spanwise direction between two adjacent high-speed streaks (or low-speed ones) are not so changed. In other words, the magnetic field change coherent structures anisotropically.

If we carefully compare three cases of different Hartmann numbers in

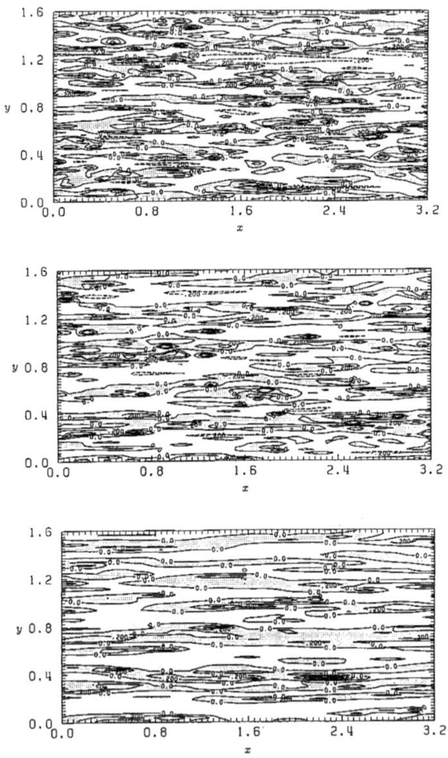


Fig.4. Contour lines of $\bar{u}_1'$ in the $x - y$ plane at $z = 0.0078$: (a) $H_a = 0$; (b) $H_a = 30$; (c) $H_a = 40$

figure 3, it is found that the amplitude of the negative peak of $R_{11}(r_2; z)$ is slightly larger at large Hartmann number than at small one. This means that the streaky structures are more striking under a magnetic field. However, this is a result of the longer streamwise correlation length of $R_{11}(r_1; z)$ under a magnetic field, as shown in figure 2 and 4: the area of negative spanwise correlation are automatically increased by the streamwise streching of streaks. Here, it should be noted that the velocity gap between high- and low-speed streaks is decreased by a magnetic field as seen in figure 4.

### 3.2. Vorticity structures

In this subsection, we present the statistical distributions of the inclination angle of the vorticity vector at $H_a = 0$, 30, and 40. For non-conducting flow or at $H_a = 0$, Moin & Kim (1985) showed by using a LES database (Moin & Kim 1982) that away from the wall, the distribution of the inclination angle of vorticity vector projected on to the $x - z$ plane gains its maximum at about 45° (and the complement angle, −135°) to the wall. We follow their technique to study how the distribution is changed by a magnetic field.

The inclination angle of the projection of the instant vorticity vector on $x - z$ planes is defined as

$$\theta = \tan^{-1}\left(\bar{\omega}_3/\bar{\omega}_1\right).$$

We calculate $\theta$ at each grid point and gather samples in the time period $t = 3.2$ at fixed $z$ positions to generate histograms describing the distributin.

To obtain the better statistical sample, the contribution of each grid point is weighted with the (normalized) magnitude of the projected vorticity vector

$$\bar{\omega}_{13} = \frac{\bar{\omega}_1^2 + \bar{\omega}_3^2}{\langle \bar{\omega}_1^2 + \bar{\omega}_3^2 \rangle}.$$

This procedure is used to detect the distributions of the velocity of high strength. We have checked that the result of the calculation without weighting is not qualitatively changed from the present one. In the following, we show the result on the lower half of the channel, but the histograms are appropriately averaged from both sides of the channel centerline for further improvement of the staistical sample. We have also checked that the data from only one side of the channel centerline is qualitatively same as the present result.

Figure 5 shows the histograms describing the distribution of $\theta$ for $H_a = 0$, 30, 40 at several distances $z$ from the wall. For $H_a = 0$, the overall distributions are same as that of Moin & Kim (1985); in the vicinity of the wall or (a), the distribution attains its maximum at slightly negative angle $\theta_{max}$ around 0°, while away from the wall (from $z = 0.1$ to 0.3), they peak at about $\theta_{max} = 45°$. The 180° period of the histograms in figure 5 indicates the adequacy of the present statistical sample.

Let us focus on the change of the distribution caused by magnetic fields. At $z = 0.0021$, the magnitude of the negative angle $\theta_{max}$ becomes gradually larger and the histogram shows a broader distribution as the Hartmann number is increased. At $z = 0.0166$, three distributions are quite different from one another; the distributions at $H_a = 30$ and 40 peak at about 90° and the maximum-to-minimum ratios are higher than that at $H_a = 0$. This difference is gradually lessened as the distance $z$ from the wall is increased. At $z = 0.1224$ where most of the vorticity vectors are inclined at $\theta_{max} = 45°$ for $H_a = 0$, $\theta_{max}$ for $H_a = 30$, and 40 are attracted to the value 45°. The angles $\theta_{max}$ at $z = 0.2892$ are almost same between three cases and slightly larger than 45°.

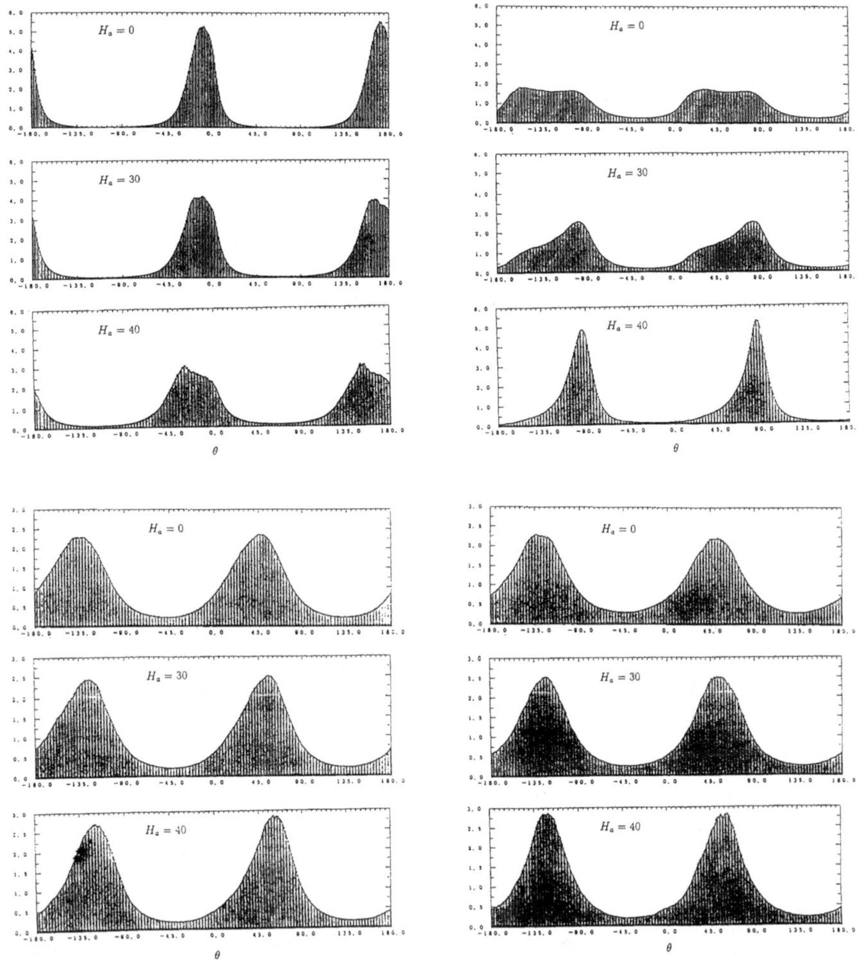


Fig.5. Distributions of the inclination angle $\theta$ of the projection of vorticity vectors in $x-y$ planes ($\theta = \tan^{-1}(\bar{\omega}_z/\bar{\omega}_x)$): (a) $z = 0.0021$; (b) $z = 0.0166$; (c) $z = 0.1224$; (d) $z = 0.2892$

## 4. Summary and conclusions

It is found that the magnetic field makes MHD turbulent channel flow quiet and organized, and its process is very anisotropic. The conclusions on the streaky and vorticity structures are summarized as follows.

The streaky structures are more elongated in the streamwise direction by the magnetic field while the mean spacing in the spanwise direction between two adjacent high-speed streaks (or low-speed ones) is not so changed. However, the velocity gap between high- and low-speed streaks is decreased by the magnetic field.

Away from the wall, the angle, at which the distribution of the inclination angle of vorticity vector projected on to a $x - z$ plane gains its maximum, gets larger (than 45°) to the wall under the magnetic field.

## References

Bradshaw, P., Cebeci, T. & Whitelaw, J.H. 1981 *Engineering Calculation Methods for Turbulent Flow.* Academic.

Brouillette, E.C. & Lykoudis, P.S. 1967 Magneto-fluid-mechanic channel flow. I. Experiment. *Phys. Fluids* **10**, 995-1001.

Hjellming, L.N. & Walker, J.S. 1987 Melt motion in a Czochralski crystal puller with an axial magnetic field: motion due to buoyancy and thermocapillarity. *J. Fluid Mech.* **182**, 335-368.

Horiuti, K. 1987 Comparison of conservative and rotational forms in large eddy simulation of turbulent channel flow. *J. Comput. Phys* **71**, 343-370.

Kline, S.J., Reynolds, W.C., Schraub, F.A. & Runstadler, P.W. 1967 The structure of turbulent boundary layers. *J. Fluid Mech.* **30**, 741-773.

Moin, P. & Kim, J. 1982 Numerical investigation of turbulent channel flow. *J. Fluid Mech.* **118**, 341-377.

Moin, P. & Kim, J. 1985 The structure of the vorticity field in turbulent channel flow. Part 1. Analysis of instantaneous fields and statistical correlations. *J. Fluid Mech.* **155**, 441-464.

Rogallo, R.S. & Moin, P. 1984 Numerical simulation of turbulent flows. *Ann. Rev. Fluid Mech.* **16**, 99-137.

Schumann, U. 1976 Numerical simulation of the transition from three- to two-dimensional turbulence under a uniform magnetic field. *J. Fluid Mech.* **74**, 31-58.

Shercliff, J.A. 1979 Thermoelectric magnetohydrodynamics. *J. Fluid Mech.* **91**, 231-251.

Shimomura, Y. 1988 Statistical analysis of magnetohydrodynamic turbulent shear flows at low magnetic Reynolds number. *J. Phys. Soc. Jpn.* **57**, 2365-2385.

Sommeria, J. & Moreau, R. 1982 Why, how, and when, MHD turbulence becomes two-dimensional. *J. Fluid Mech.* **118**, 507-518.

Speziale, C.G. 1985 Galilean invariance of subgrid-scale stress models in the large-eddy simulation of turbulence. *J. Fluid Mech.* **156**, 55-62.

Willmarth, W.W. & Tu, B.J. 1967 Structure of turbulence in the boundary layer near the wall. *Phys. Fluids* **10**, S134-S137.

Yakhot, A. & Branover, H. 1982 An analytical model of a two-phase liquid metal magnetohydrodynamic generator. *Phys. Fluids* **25**, 446-451.

Yoshizawa, A. 1984 Statistical analysis of the deviation of the Reynolds stress from its eddy viscosity representation. *Phys. Fluids* **27**, 1377-1387.

# Coherent Structures in MHD Turbulence and Turbulent Dynamo

Shinichiro YANASE[α], Jiro MIZUSHIMA[β] and Shigeo KIDA[γ]

[α]*Engineering Mathematics, Faculty of Engineering*
*Okayama University, Okayama 700, Japan*

[β]*Department of Information Science*
*Sagami Institute of Technology, Fujisawa 251, Japan*

[γ]*Applied and Computational Mathematics*
*Princeton University, Princeton, NJ 08544, U.S.A.*
*On leave from Research Institute for Mathmatical Science*
*Kyoto University, Kyoto 606, Japan*

*Turbulence and Coherent Structures*
*O. Métais and M. Lesieur, eds.*
*Selected papers from "Turbulence 89: organized structures and turbulence in fluid mechanics"*
*Grenoble, 18-21 September 1989*

## 1. Introduction

The mechanism of magnetic field generation by self inductive action in MHD fluid is called 'dynamo effects' and has been studied extensively by many researchers (see Moffatt, 1978). Successfull dynamo theories have a common feature of the crucial importance of a lack of reflectional symmetry in fluid motions.

An kinematic theory of the turbulent dynamo was proposed by Steenbeck, Krause and Rädler (1966) and it was developed further by Moffatt (1970) and others. The basic method they used for the study of the turbulent dynamo is a two-scale analysis, where the turbulent field is assumed to consist of a magnetic field of large scales and velocity and magnetic fields of much smaller scales. A turbulent magnetic field of small scales is excited by a turbulent velocity field as a form of Alfvén waves in the presence of an weak initial magnetic field of large scales. Then, turbulent velocity and magnetic fields of small scales strengthen and maintain the magnitude of the large-scale magnetic field. This process is called as the '$\alpha$ effect'. In order that the effect may work, it is crucial that the helicity of a turbulent velocity field should have a finite value, which is equivalent of a lack of reflectional symmetry in the velocity field.

Kraichnan (1965) proposed a phenomenological theory by extending Kolmogorov's theory and concluded that $E_K(k) = E_M(k) \propto k^{-3/2}$ in non-helical case, where $E_K(k)$ and $E_M(k)$ are the kinetic and the magnetic energy spectrum, respectively. Pouquet, Frisch and Léorat (1976) confirmed Kraichnan's result for the non-helical case and argued the mechanism of the turbulent dynamo in a helical MHD turbulence by EDQNM approximation. Saito, Mizushima and Futami (1985) derived another conclusion that $E_K(k) = E_M(k) \propto k^{-5/3}$ for a non-helical MHD turbulence by using the modified 0-4 th order cumulant approximation. The critical condition for the turbulent dynamo was obtained using EDQNM theory by Léorat, Pouquet and Frisch (1981). They obtained a critical magnetic Reynolds number above which the magnetic energy is generated and showed that it is independent of the kinetic Reynolds number.

Meneguzzi, Frisch and Pouquet (1981) made a numerical simulation of MHD turbulence. In their simulation, the kinetic energy and the maximal kinetic helicity are injected into the turbulence. Their results show that the magnetic energy is generated near the injection wavenumber at first, and then the peak of the magnetic energy spectrum drifts to smaller wavenumbers until it reaches to the minimum wavenumber. It then keeps growing linearly without saturation. The linear growth was shown to be consistent with a two-scale model by Kraichnan (1979). They also found that even if the total helicity is zero, the turbulent dynamo is still effective. Since no large-scale magnetic field was assumed as the initial condition, the large-scale magnetic field does not seem to be essential for the turbulent dynamo. It can be said that their work is the first to have demonstrated the turbulent dynamo effect by numerical simulation.

Recently Kida, Yanase and Mizushima (1989) made a numerical simulation of a three-dimensional MHD turbulence by using the spectral method. The MHD turbulence is assumed to be maintained by an external force and have the high-symmetry (Kida, 1985). The critical condition of the turbulent dynamo was investigated and the statistical properties of the MHD turbulence was studied in detail. They showed the way how the large-scale magnetic field develops, the process of which has been explained by closure theories as the inverse-cascade of the magnetic energy or the magnetic helicity. A statistically equilibrium state of the MHD turbulence sustained by an external force was traced for a long period.

We use the same numerical data obtained by Kida, Yanase and Mizushima (1989) and investigate the large-scale coherent structures in MHD turbulence. Especially, the relation between the vorticity concentration in the velocity field and the generation of the large-scale magnetic field is discussed in this paper.

## 2. Basic equations

Suppose that an MHD fluid is filled in a cubic box with the length of side $2\pi$ and is exerted by an external force. Periodic boundary conditions are taken in three orthogonal directions. We assume that both the velocity field and the vector potential of the magnetic field have the high symmetry and that the fluid has a hyper viscosity and a hyper magnetic diffusivity instead of the usual viscosity and the magnetic diffusivity in order to increase Reynolds numbers which can be simulated with a given resolution. It may be found that the large-scale dynamics remain essentially unaffected by the choice of the modified viscosity and diffusivity (Meneguzzi et al., 1981; Passot and Pouquet, 1988; Farge, 1988).

If the magnetic field is rescaled to give it the dimension of velocity, we have the following basic equations.

$$\frac{\partial \mathbf{u}}{\partial t} + \mathbf{u}.\nabla\mathbf{u} = \mathbf{b}.\nabla\mathbf{b} - \nabla p - \nu\Delta^2\mathbf{u} + \mathbf{f}, \qquad (2-1)$$

$$\frac{\partial \mathbf{b}}{\partial t} + \mathbf{u}.\nabla\mathbf{b} = \mathbf{b}.\nabla\mathbf{u} - \lambda\Delta^2\mathbf{b}, \qquad (2-2)$$

$$\nabla.\mathbf{u} = 0, \qquad (2-3)$$

$$\nabla.\mathbf{b} = 0, \qquad (2-4)$$

where $\mathbf{u}$ is the velocity, $\mathbf{b}$ the magnetic field, $p$ the total pressure, $\nu$ the hyper viscosity, $\lambda$ the hyper magnetic diffusivity, and $\mathbf{f}$ a force which drives the flow.

As for the external force we assume that it is applied in such a way that several Fourier components always have prefixed amplitudes as follows:

$$u_x = \frac{1}{2}\sin x(\cos 3y\cos z - \cos y\cos 3z)$$

$$+\frac{\sqrt{0.12}}{2}[\sin 4x(\cos 2y + \cos 2z) - 2\sin 2x(\cos 4y + \cos 4z)], \qquad (2-5)$$

where $u_x$ is the $x$ -component of the velocity. The other two-components, $u_y$ and $u_z$, are given by replacing $x$, $y$ and $z$ in (2-5) cyclically. Velocity field (2-5), which satisfies the high symmetry condition, is determined as to maximize the local helicity. It is noted that information for a single component of the velocity in the fundamental box, $(0 \le x, y, z \le \pi/2)$, is sufficient to describe the whole velocity field.

The kinetic energy $E_K(t)$ and the magnetic energy $E_M(t)$ per unit mass are defined by

$$E_K(t) = \frac{1}{2}\sum_{k_x}\sum_{k_y}\sum_{k_z}|\hat{\mathbf{u}}(\mathbf{k})|^2, \qquad (2-6)$$

$$E_M(t) = \frac{1}{2}\sum_{k_x}\sum_{k_y}\sum_{k_z}|\hat{\mathbf{b}}(\mathbf{k})|^2, \qquad (2-7)$$

where $\hat{\mathbf{u}}(\mathbf{k})$ and $\hat{\mathbf{b}}(\mathbf{k})$ are the Fourier coefficients of $\mathbf{u}(\mathbf{x})$ and $\mathbf{b}(\mathbf{x})$, respectively, $\mathbf{k}$ is the wavenumber vector. The kinetic energy spectrum $E_K(k)$ and the magnetic energy spectrum $E_M(k)$ are obtained respectively by summing up the Fourier coefficients $\hat{\mathbf{u}}(\mathbf{k})$ and $\hat{\mathbf{b}}(\mathbf{k})$ within a band of a shell containing $k = |\mathbf{k}|$. We also define the total energy spectrum as $E_T(k) \equiv E_K(k) + E_M(k)$.

The kinetic and the magnetic energy dissipations, $D_K(t)$ and $D_M(t)$, are defined by

$$D_K(t) = 2\nu\sum_k k^4 E_K(k), \qquad (2-8)$$

$$D_M(t) = 2\lambda\sum_k k^4 E_M(k). \qquad (2-9)$$

The kinitic and magnetic helicities are zero due to the high symmetry condition if the average is taken over the whole domain. However, if

| Case | $\nu$ | $\lambda$ | $t_{max}$ | $t_{eq}$ | $\Delta t$ | $\beta$ |
|---|---|---|---|---|---|---|
| I | $1 \times 10^{-5}$ | $1 \times 10^{-5}$ | 170 | 100 | 0.01 | 0.142 |
| II | $5 \times 10^{-6}$ | $5 \times 10^{-6}$ | 100 | 50 | 0.02 | 0.256 |
| III | $2 \times 10^{-6}$ | $2 \times 10^{-6}$ | 200 | 30 | 0.02 | 0.381 |
| IV | $2 \times 10^{-6}$ | $1 \times 10^{-5}$ | 310 | 210 | 0.01 | 0.158 |
| V | $1 \times 10^{-5}$ | $2 \times 10^{-6}$ | 70 | 30 | 0.01 | 0.397 |

Table1. Parameters adopted for five cases.

the average is taken in the fundamental box, both helicities can have finite values. We shall call them the local kinetic helicity and the local magnetic helicity, and denote them by $H_K(t)$ and $H_M(t)$, respectively:

$$H_K(t) = \frac{1}{2} < \mathbf{u}.\omega >, \qquad (2-10)$$

$$H_M(t) = \frac{1}{2} < \mathbf{a}.\mathbf{b} >, \qquad (2-11)$$

where $<>$ means the spatial average over the fundamental box. We expect these local helicities to play important roles as well as the usual helicities.

## 3. Numerical simulation

The numerical simulation is made by using the spectral method with $128 \times 128 \times 128$ collocation points. The aliasing interactions in the calculation of the nonlinear terms are suppressed by the 2/3 rule. The time marching is made by the Runge-Kutta-Gill scheme.

The initial condition for the velocity field is taken as a fully developed turbulence, which is created by solving (2-1) and (2-3) without the magnetic field for sufficiently large period in which a statistically equilibrium state is attained. The initial condition for the magnetic field is given as a small random noise which has the following energy spectrum:

$$E_M(k) = Ak^{10}\exp(-Bk^2), A = 9.00 \times 10^{-39}, B = 5.50 \times 10^{-6}. \quad (3-1)$$

The ratio of $E_M(t)$ to $E_K(t)$ at $t = 0$ is $1.50 \times 10^{-4}$.

we adopt five pairs of the values of the hyper viscosity and the hyper magnetic diffusivity as listed in Table 1. We also list $\Delta t$, the time increment used for the time advancement, $t_{max}$, the final time of the numerical simulation, and $t_{eq}$, the time when the MHD turbulence attains a statistically equilibrium state for each case in the same table. It is noted that it takes very long time ($t_{eq} > 200$) for the MHD turbulence to attain its stationary state for case IV. It seems that the values of $t_{eq}$ becomes very large if $\lambda/\nu \gg 1$.

## 4. Statistical properties of MHD turbulence

In this section, we investigate the statistical properties of the MHD turbulence mainly for case III, where $\nu$ and $\lambda$ take the smallest values in our five cases.

### 4.1. Developments of energies and energy dissipations

The time developments of the kinetic energy $E_K(t)$ and the magnetic energy $E_M(t)$ for $0 \leq t \leq 200$ are shown in Fig.1. The kinetic energy $E_K(t)$ does not change significantly throughout the whole time, but only fluctuates around the mean value. On the other hand, the magnetic energy $E_M(t)$ increases in the initial period $0 \leq t \leq 30$, and then it also fluctuates around its own mean value. We should make a notice that the mean value of $E_M(t)$ for $t > 150$ becomes smaller than that for $30 < t < 150$ (about 2/3) and that this change of the mean value may be related to the difference in the structure of the turbulent magnetic field before and after $t = 150$, which will be shown later. The ratio of the averaged values of $E_M(t)$ to $E_K(t)$ during $t_{eq} \leq t \leq t_{max}$ is about 0.09. Since $E_K(t)$ and $E_M(t)$ depend upon large scale motions, they may be affected by the periodic boundary or the high symmetry condition.

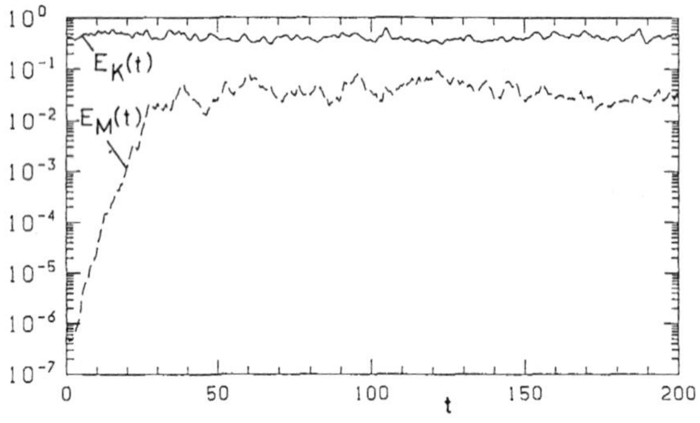


Fig.1. Time developments of the kinetic and magnetic energies.

The time developments of the kinetic and the magnetic energy dissipations, $D_K(t)$ and $D_M(t)$, are plotted in Fig.2. It is interesting to note that the magnitudes of $D_K(t)$ and $D_M(t)$ are nearly equal after $t = 30$. If we denote the time averages of $D_K(t)$ and $D_M(t)$ by $\bar{D}_K$ and $\bar{D}_M$, respectively, then $\bar{D}_M/\bar{D}_K = 0.8$, where the time average is taken for $t_{eq} \leq t \leq t_{max}$. The values of $\bar{D}_M/\bar{D}_K$ for the other four cases are shown in Table 2.

Only the kinetic energy is supplied to the MHD turbulence in the present case. The present result shows that 55% of the kinetic energy is transferred to higher wavenumber region of the kinetic energy spectrum,

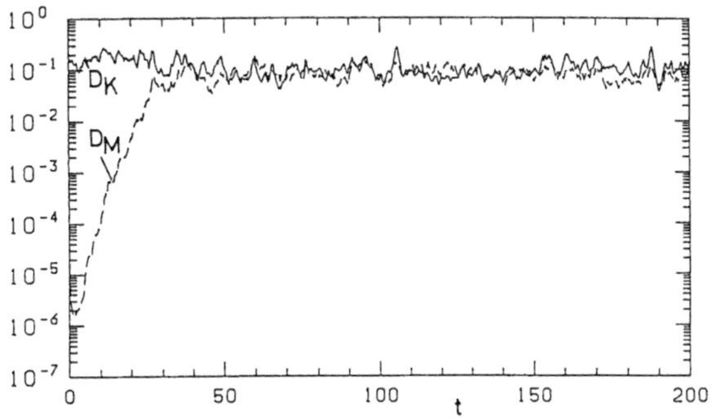


Fig.2. Time developments of the kinetic and magnetic energy dissipations.

where it is dissipated by the kinematic viscosity and 45% is transformed into the magnetic energy by the nonlinear interaction between velocity and magnetic fields and dissipated by the magnetic diffusivity in the higher wavenumber region.

| Case | $\bar{E}_K$ | $\bar{E}_M$ | $\bar{D}_K$ | $\bar{D}_M$ | $\bar{E}_M/\bar{E}_K$ | $\bar{D}_M/\bar{D}_K$ |
|---|---|---|---|---|---|---|
| I | $4.57 \times 10^{-1}$ | $1.81 \times 10^{-2}$ | $1.52 \times 10^{-1}$ | $4.01 \times 10^{-2}$ | $3.96 \times 10^{-2}$ | $2.64 \times 10^{-1}$ |
| II | $4.98 \times 10^{-1}$ | $1.66 \times 10^{-2}$ | $1.69 \times 10^{-1}$ | $4.76 \times 10^{-2}$ | $3.33 \times 10^{-2}$ | $2.82 \times 10^{-1}$ |
| III | $4.14 \times 10^{-1}$ | $3.80 \times 10^{-2}$ | $9.95 \times 10^{-2}$ | $8.22 \times 10^{-2}$ | $9.18 \times 10^{-2}$ | $8.26 \times 10^{-1}$ |
| IV | $4.84 \times 10^{-1}$ | $7.00 \times 10^{-3}$ | $1.65 \times 10^{-1}$ | $2.38 \times 10^{-2}$ | $1.45 \times 10^{-2}$ | $1.44 \times 10^{-1}$ |
| V | $4.15 \times 10^{-1}$ | $3.73 \times 10^{-2}$ | $1.00 \times 10^{-1}$ | $8.31 \times 10^{-2}$ | $8.99 \times 10^{-2}$ | $8.31 \times 10^{-1}$ |

Table2. Time averaged values of energies and energy dissipations.

### 4.2. Critical condition of turbulent dynamo

It is shown that the magnetic energy $E_M(t)$ decreases abruptly at first due to the strong dissipative effect at high wavenumber region. Then, it increases almost exponentially as $E_M(t) \propto \exp(\beta t)$, for cases I, II, III and V. For case IV, the $E_M(t)$ increases and decreases repeatedly before $t \cong 120$ and then begins to grow exponentially. The growth rate $\beta$ is listed in Table 1, where they are evaluated from the data in the period of $0 \leq t \leq t_{eq}$ for cases I, II, III and V, and from those in the period of $120 \leq t \leq t_{eq}$ for case IV. As seen in Table 1, the growth rate $\beta$ hardly depends on $\nu$ within the present parameter range, but it depends on $\lambda$. We obtain an approximate linear relation between $\beta$ and $\log\lambda$ by using the least squares method:

$$\beta = A \log \lambda + B, A = -0.342, B = -1.56. \qquad (4-1)$$

If we put $\beta = 0$ in eqs.(4-1), we obtain the critical value of the hyper magnetic diffusivity as $\lambda_c = 2.660 \times 10^{-5}$. However, it is expected that the value of $\beta$ deviates from relation (4-1) toward smaller values around the critical state as indicated by Léorat, Pouquet and Frisch (1981), so $\lambda_c$ should be less than $2.660 \times 10^{-5}$, that is, $1 \times 10^{-5} < \lambda_c < 2.660 \times 10^{-5}$. It is also confirmed from our additional numerical simulation that $\lambda_c \simeq 2 \times 10^{-5}$.

### 4.3. Kinetic and magnetic helicities

Helicities are regarded as ones of the most important quantities in the dynamo theory. We show the time development of $H_K(t)$ in Fig.3. The time averaged value of $H_K(t)$ is about 2.3. It is interesting to note that it is always positive though its magnitude changes very much.

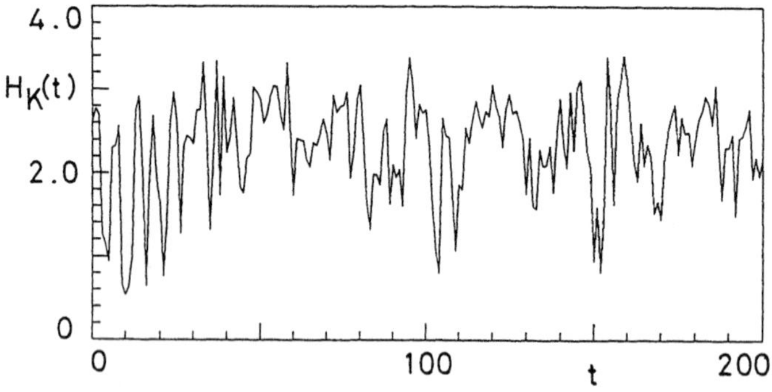


Fig.3. Time developments of the local kinetic helicity.

This is because the Fourier components of the velocity field (2-5) are held constant in amplitudes which gives a positive helicity in $0 \leq x, y, z \leq \pi/2$. The continuous injection of the positive kinetic helicity to the turbulence may maintain some coherent structures in the vorticity fields.

Next, we show the time development of $H_M(t)$ in Fig.4, which gives us the most important information. It is seen that for $t \leq 22$, $H_M(t) \approx 0$, but for $t \geq 30$ it starts to oscillate and takes negative values almost all time later then. The averaged value is $-4.30 \times 10^{-4}$. Unlike $H_K(t)$, $H_M(t)$ sometimes becomes positive. It should be remarked that after $H_M(t)$ takes large negative values around $t = 120$, the amplitude of the oscillation decreases to nearly 30% of that before $t = 120$. The striking variation of $H_M(t)$ around $t = 120$ suggests a qualitative change of the structure of the MHD turbulence after $t = 150$. After $t = 150$, less energy is transferred from the kinetic component to the magnetic component and the intermittent variations of the kinetic and the magnetic energies become milder. It suggests a formation of some coherent structures in the magnetic fields that $H_M(t)$ usually takes negative values and rarely becomes positive.

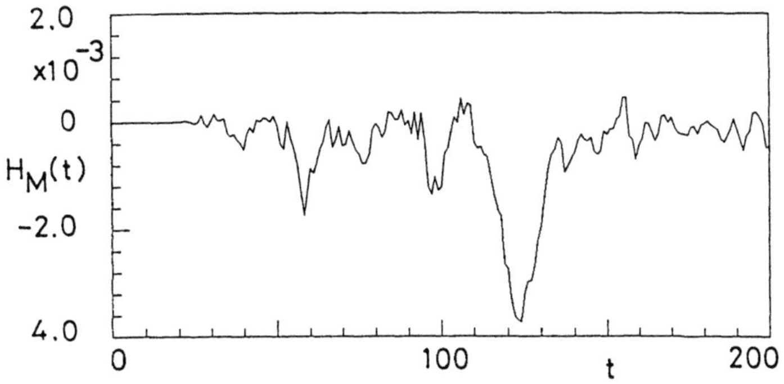


Fig.4. Time developments of the local magnetic helicity.

From the data obtained for energies and helicities, we can discriminate three characteristic periods of the evolution of the MHD turbulence. The first one is the initial period where the magnetic energy $E_M(t)$ increases exponentially during $0 \leq t \leq 30$. After a transitional time, the MHD turbulence arrives at an equilibrium state and the magnetic energy oscillates around the mean value. A careful observation of the data makes it clear that there are two distinctive periods in the equilibrium state. The earlier one is the dynamic equilibrium period for $50 \leq t \leq 130$. In this period, $E_M(t)$ oscillates with a large amplitude and the magnetic helicity varies greatly, taking negative values. Then the turbulence enters into the second transitional time. After $t = 150$, the quiescent equilibrium period, we may call, comes out where $E_M(t)$ becomes a little smaller and $H_M(t)$ oscillates with a small amplitude.

### 4.4. Energy spectra

Delicate properties of MHD turbulence is expressed better by the energy spectra of velocity and magnetic fields than integrated quantities discussed in 4-1 and 4-3. In this section, only the principal results of Kida, Yanase and Mizushima (1989) are summarized. In Fig.5, the kinetic energy spectrum $E_K(k)$, the magnetic energy spectrum $E_M(k)$, and the total energy spectrum $E_T(k)$ averaged over $50 \leq t \leq 70$ are plotted. Three straight lines represent the $k^{-5/3}$, $k^{-2}$ and $k^0$ power spectra, respectively.

It is seen that there exists a wavenumber region where $E_T(k) \propto k^{-5/3}$. By use of the least squares method, we calculate the Kolmogorov constant $Ko$ of the total energy spectrum averaged over $50 \leq t \leq 70$:

$$E_T(k) \cong Ko(D_K(t) + D_M(t))^{2/3} k^{-5/3}, Ko = 2.08, \qquad (4-4)$$

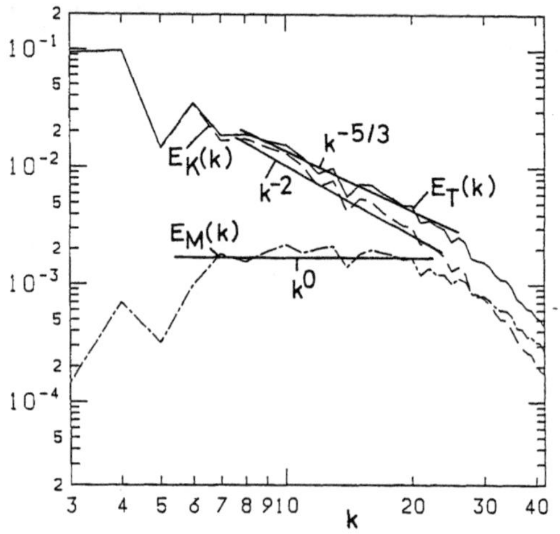


Fig.5. Averaged energy spectra for $50 \leq t \leq 70$.

where the total dissipation $D_K(t) + D_M(t)$ is also evaluated by averaging it over $50 \leq t \leq 70$. It is very interesting to note that the magnitude of the Kolmogorov constant $Ko$ for the MHD turbulence is close to that obtained experimentally and numerically in fluid turbulence (Kida and Murakami, 1987). It is also found that $E_K(k) \propto k^{-2}$ and $E_M(k) \propto k^0$ in the middle wavenumber region. Therefore, the magnetic energy spectrum $E_M(k)$ has the similar form as the enstrophy spectrum $k^2 E_K(k)$. These results agree with Batchelor's conjecture that the vorticity field behaves similarly as the magnetic field if the magnetic energy is small enough (Batchelor, 1950).

In the high wavenumber region, however, Batchelor's prediction does not hold, because a vital energy transfer from the velocity field to the magnetic field occurs and the velocity field should be strongly affected by the presence of the magnetic field. It should be noted that $E_M(k) > E_K(k)$ at the highest wavenumber region, which agrees with the results obtained by Meneguzzi et at. (1981). This shows that there is no equipartition between the kinetic and the magnetic components.

**4.5. Energy transfer between kinetic and magnetic fields**

It is very interesting to know how the kinetic energy, which is injected to the velocity field in the low wavenumber region, is transferred to the magnetic field. The total energy transfer, $E_{tr}$, from the velocity field to the magnetic field is defined as

$$E_{tr} \equiv \sum_{i,j=1}^{3} < b_i b_j \partial u_i / \partial x_j >, \qquad (4-5)$$

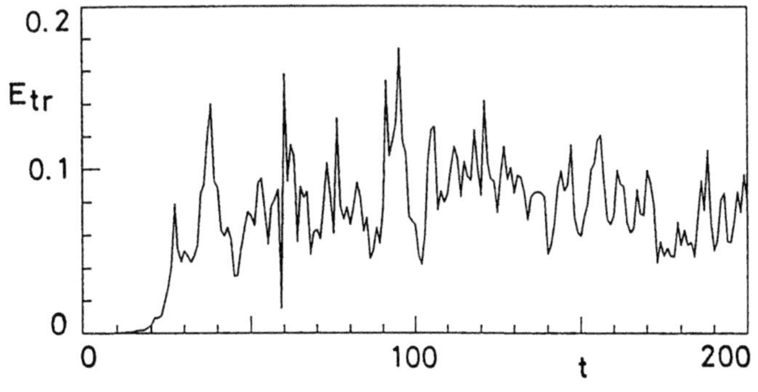


Fig.6. Time developments of the total energy transfer.

where the bracket means the space average per unit mass. As shown in Fig.6, $E_{tr}$ is always positive. Thus, the energy is continuously transferred from the velocity field to the magnetic field. The detailed discussion of the energy transfer between the velocity and the magnetic fields is made better by investigating the energy transfer spectra. However, the discussion on the energy transfer spectra is omitted due to the restricted space available here (see Kida, Yanase and Mizushima, 1989).

## 5. Coherent structures in MHD turbulence

In Fig.7, we plot an iso-surface of $|\omega|^2$ in the fundamental box, where $\omega$ is the vorticity calculated from (2-5). The surface level is 62.5% of the maximum. The point denoted by $O$ is the origin $(0,0,0)$ and the one denoted by $\Delta$ is $(\frac{\pi}{2}, \frac{\pi}{2}, \frac{\pi}{2})$. The axis of the torus of the iso-surface of $|\omega|^2$ is directed from the origin to $(\frac{\pi}{2}, \frac{\pi}{2}, \frac{\pi}{2})$. We call this direction as (1,1,1) direction hereafter. We can imagine the structure of the external force from this figure.

In the initial period, $0 \leq t < 30$, the magnetic field repeats gathering and scattering, making rather small structures, while its averaged magnitude grows exponentially. The structure of the magnetic field is affected by its initial condition very much in this period. The peak value of $|\omega|^2$ oscillates with a large amplitude and the structure of the velocity field is almost determined by the external force. Sometimes there appears a weak large-scale structure in (1,1,1) direction, and then it dissapears.

At $t \simeq 30$, there appears a magnetic ring whose axis is in (1,1,1) direction (see Fig.8(b)). At the same time, the magnitude of the magnetic field saturates and attains its almost equilibrium value. So, it is said that the formation of the magnetic ring is an indication of attaining the equilibrium state of magnetic field. On the other hand, the peak value of $|\omega|^2$ become very small (about a half). Fig.8(a) shows the iso-surface of $|\omega|^2$. We can see a twisted long and slender structure in (1,1,1) direction.

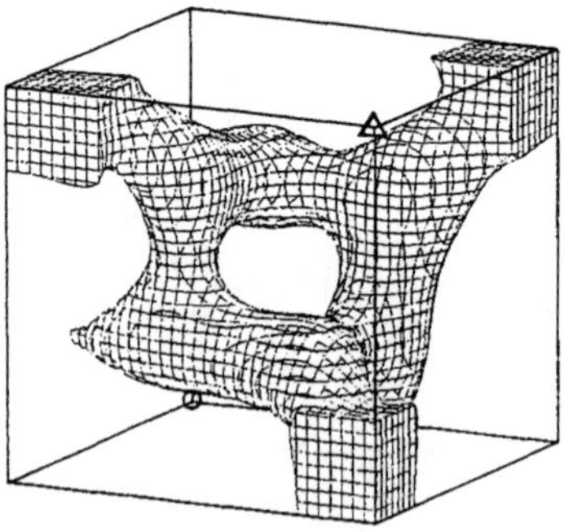

Fig.7. Iso-surface of $|\omega|^2$ of the maintained vorticity field. Surface level is 62.5% of the maximum. The point denoted by $O$ is the origin $(0,0,0)$ and the one denoted by $\triangle$ is $(\frac{\pi}{2},\frac{\pi}{2},\frac{\pi}{2})$.

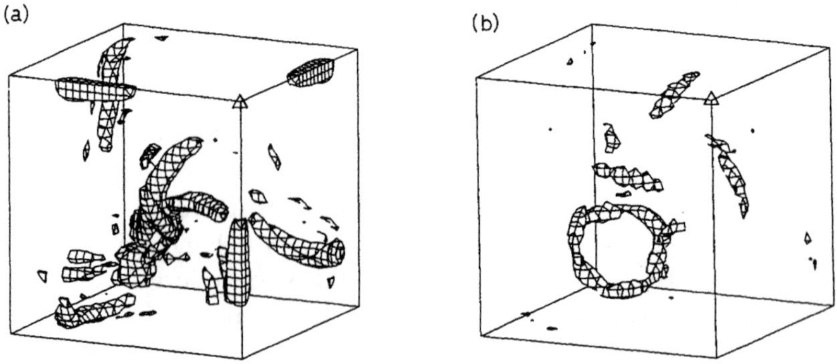


Fig.8. (a).Iso-surface of $|\omega|^2$. (b). Iso-surface of $|\mathbf{b}|^2$. $t=30$.

In the transion period, $30 < t < 50$, we can not find any definite large-scale structures neither in the magnetic nor the velocity fields.

At $t \simeq 50$, large-scale structures are created both in the magnetic and velocity fields. Fig.9(a) shows the iso-surface of $|\omega|^2$ at $t=52$. It is seen that there is a long structure like a cigar in (1,1,1) direction. Fig.9(b) shows the structure of the magnetic field at the same instant. It must be noted that both the fields have the large-scale structures which are oriented in the same direction with each other. The iso-surface of $|\omega|^2$ is surrounded by that of the magnetic field. We have confirmed this disposition of the structures of the magnetic and the velocity fields by ploting contours of iso-surface of $|\omega|^2$ and $|\mathbf{b}|^2$ on several cross-sections parallel to $x$-$y$ plane. It is interesting to note that there is a cylindrical hole in the center of the magnetic field, that is, the magnetic field has a tube-like structure. The vorticity, on the other hand, has no hole and is concentrated in a cigar-like region. Plate 1 is a graphical output of the magnetic field at $t = 52$.

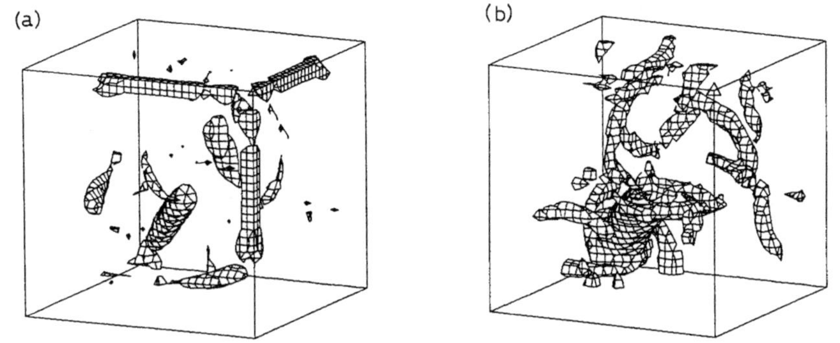


Fig.9. (a).Iso-surface of $|\omega|^2$. (b). Iso-surface of $|\mathbf{b}|^2$. $t = 52$.

After $t \simeq 50$, the velocity and the magnetic fields have a strong correlation with each other. Occasionally, the large-scale structures in both the fields resemble with each other. A typical example is seen at $t = 70$. Fig.10(a) shows the iso-surface of $|\omega|^2$ at $t = 70$, and Fig.10(b) the structure of the magnetic field at the same instant. The iso-surface of $|\mathbf{b}|^2$ is surrounded by that of $|\omega|^2$, which is in contrast with the structure at $t = 52$.

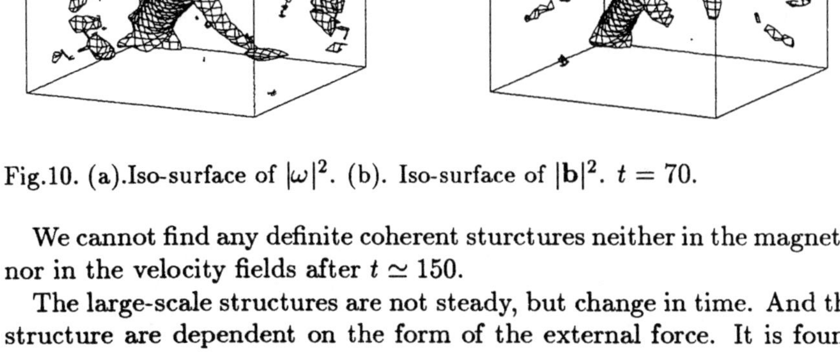


Fig.10. (a).Iso-surface of $|\omega|^2$. (b). Iso-surface of $|\mathbf{b}|^2$. $t = 70$.

We cannot find any definite coherent sturctures neither in the magnetic nor in the velocity fields after $t \simeq 150$.

The large-scale structures are not steady, but change in time. And the structure are dependent on the form of the external force. It is found that there is a difference in the mechanism of the formations of the large-scale coherent structures between the magnetic field and the velocity field. The coherent structure in the magnetic field appears as a form of ring. Then the radius of the ring shrinks so that a local concentration of the magnitude of the magnetic field occurs. On the other hand, the large-scale structure of the velocity field is formed by twists of vortex filaments.

## References

Batchelor, G. K. 1950 On the spontaneous magnetic field in a conducting liquid in turbulent motion. *Proc. Roy. Soc.* **A201**, 405-416.

Farge, M. 1988 Vortex motion in a rotating barotropic fluid layer. *Fluid Dyn. Res.* **3**, 282-288.

Kida, S. 1985 Three-dimensional periodic flows with high-symmetry. *J. Phys. Soc. Jpn* **54**, 2132-2136.

Kida, S. & Murakami, Y. 1987 Kolmogorov similarity in freely decaying turbulence. *Phys. Fluids* **30**, 2030-2039.

Kida, S., Yanase, S. & Mizushima, J. 1989 Statistical properties of MHD turbulence and turbulent dynamo. submitted to *Phys. Fluids.*

Kraichnan, R. H. 1965 Inertial-Range spectrum of hydromagnetic turbulence. *Phys. Fluids* **8**, 1385-1387.

Kraichnan, R. H. 1979 Consistency of the $\alpha$ -Effect turbulent dynamo. *Phys. Rev. Lett.* **42**, 1677-1680.

Léorat, J., Pouquet, A. & Frisch, U. 1981 Fully developed MHD turbulence near critical magnetic Reynolds number. *J. Fluid Mech.* **104**, 419-443.

Meneguzzi, M., Frisch, U. & Pouquet, A. 1981 Helical and nonhelical turbulent dynamos. *Phys. Rev. Lett.* **47**, 1060-1064.

Moffatt, H. K. 1970 Turbulent dynamo action at low magnetic Reynolds number. *J. Fluid Mech.* **41**, 435-452.

Moffatt, H. K. 1978 *Magnetic Field Generation in Electrically Conducting Fluids.* Cambridge Univ. Press.

Monin, A. S. & Yaglom, A. M. 1975 *Statistical Fluid Mechanics of Turbulence.* MIT Press.

Pouquet, A., Frisch, U. & Léorat, J. 1976 Strong MHD helical turbulence and the nonlinear dynamo effect. *J. Fluid Mech.* **77**, 321-354.

Saito, Y., Mizushima, J. & Futami, N. 1985 Statistical properties of strong MHD turbulence without helicity. *J. Phys. Soc. Jpn.* **54**, 134-145.

Steenbeck, M., Krause, F. & Rädler, K. -H. 1966 A calculation of the mean electromotive force in an electrically conducting fluid in turbulent motion, under the influence of Coriolis force. *Z. Naturforsch.* **21**a, 369-376. [English translation: Roberts & Stix 1971, 29-47].

# List of Participants

P. AGOSTINI
STCPMVN
2 Bd Gambetta
B. P. 53
60321 COMPIEGNE Cedex
tel: 44 20 37 23

P. ALBAREDE
Laboratoire de recherche en combustion
Université de Provence
252 Faculté Saint-Jérome
13397 MARSEILLE Cedex 13
tel: 91 28 83 66
fax: 91 02 05 50

T. ALZIARY DE ROQUEFORT
Laboratoire d'Etudes Aérodynamiques
CEAT
43 Route de l'Aérodrome
86000 POITIERS Cedex
tel: 49 53 70 40
fax: 49 53 70 01
E-mail: UMDF007 @FRORS31

F. ANSELMET
IMST
12 av. du Géneral Leclerc
13003 MARSEILLE
tel: 91 50 54 39
E-mail: FRAUNIE@FRMOP11

R. A. ANTONIA
Dept of Mechanical Engineering
University of Newcastle
Rankin Drive
NEWCASTLE, NSW 2308
AUSTRALIE
tel: 049 68 54 08
fax: 049 60 22 28
E-mail: munnari!cc.nu.oz.au!meraa@uunet.uu.ne

R. ARINA
Centro Ricerche FIAT
Politecnico di Torino
C.so Duce degli Abruzzi,24
TORINO, 10129
ITALIE
tel: 011 5566812
fax: 011 5566899
E-mail: arina@itopoli . bitnet

P. ATTEN
Directeur de Recherches CNRS
CNRS
Laboratoire d'Electrostatique
Av. des Martyrs
GRENOBLE

G. AUBERT
Directeur
Ecole Normale Supérieure de Lyon
48 Allée d'Italie
69364 LYON Cedex 07
tel: 72 72 80 00
fax: 72 72 80 80

N. AUBRY
Benjamin Levich Institute
City College of New-York
NEW-YORK, N.Y. 10031
U.S.A
tel: (212) 690 81 57
fax: (212) 690 83 74
E-mail: MENNA@CCNYVME

B. AUPOIX
ONERA/CERT
2, avenue Edouard Belin
B. P. 4025
31055 TOULOUSE Cedex
tel: 61 55 70 37
fax: 61 55 71 72

E. BACRY
Ecole Normale Supérieure
45 Rue d'Ulm
75005 PARIS
E-mail: bacry@frulm63

C. BASDEVANT
Laboratoire de Météorologie Dynamique
Ecole Normale Supérieure
24, rue Lhomond
75231 PARIS Cedex 05
tel: (1) 43 31 72 73
fax: (1) 45 87 34 89
E-mail: Claude @FRULM11

S. BELLIN
Laboratoire d'études Aérothermiques
CEAT
43 route de l'aérodrome
86036 POITIERS Cedex
tel: 49 53 70 00
fax: 49 53 70 01

M. BELORGEY
Laboratoire de Mecanique des Fluides
Université du Havre
25 Rue Philippe Le Bon
B. P. 540
76058 LE HAVRE Cedex
tel: 35 22 49 29
fax: 35 21 49 59

M. BERTHIER
SETH
CENG
17 rue des Martyrs
B. P. 85 X
38041 GRENOBLE Cedex
tel: 76 88 35 93

J.- P. BERTOGLIO
Laboratoire de Mécanique des Fluides et d'Acoustique
Ecole Centrale de Lyon
36, avenue Guy de Collongues
B. P. 163
69131 ECULLY Cedex
tel: 78 33 81 27 poste 47 20
fax: 78 33 34 75

G. BINDER
Institut de Mécanique de Grenoble
Domaine Universitaire
B. P. 53 X
38041 GRENOBLE Cedex
tel: 76 82 50 45
fax: 76 82 50 01

J .- P. BONNET
Laboratoire d'Etudes Aérodynamiques
C.E.A.T
43 Route de l'Aérodrome
86036 POITIERS Cedex
tel: 49 53 70 31
fax: 49 53 70 01
E-mail: umdf008@FRORS31

M. BOUTHIER
Laboratoire de Modélisation en Mécanique
Université P. et M.Curie
T66 Place Jussieu
75005 PARIS

J. BROSSARD
Laboratoire de Mécanique des Fluides
Université du Havre
25 rue Philippe Le Bon
B. P. 540
76058 LE HAVRE Cedex
tel: 35 22 49 29
fax: 35 21 49 59

F. K. BROWAND
Department of Aerospace Engineering
University of Southern California
University Park
LOS ANGELES, CA 90089-1191
USA
tel: (213) 743 2035
fax: (213) 748 8728
E-mail: BROWAND@GROGHE

C. CAMBON
Laboratoire de Mécanique des Fluides et d'Acoustique
Ecole Centrale de Lyon
36, avenue Guy de Collongues
69131 ECULLY Cedex
tel: 78 33 81 27

D.J. CARRUTHERS
University of Cambridge
D.A.M.T.P.
Silver Street
CAMBRIDGE, CB3 9EW
GRANDE-BRETAGNE
tel: (44) 223 337 858
fax: (44) 223 337918
E-mail: dc31@phoenix.cambridge.ac.uk

C. CAULFIELD
DAMTP
University of Cambridge
Silver Street
CAMBRIDGE, CB39EW
GRANDE-BRETAGNE
tel: (44) 223 33 79 00
fax: (44) 223 33 79 18
E-mail: CPC12.PHX.CAM@EARN.U.K.

P. CHASSAING
IMFT
Av. du Pr. Camille Soula
31400 TOULOUSE
tel: 61 28 58 32
fax: 61 62 09 76

M.- P. CHAUVE
IMST
12 Avenue du Gal Leclerc
13003 MARSEILLE

J.- P. CHOLLET
Institut de Mécanique de Grenoble
Domaine Universitaire
B. P. 53 X
38041 GRENOBLE Cedex
tel: 76 82 51 19
fax: 76 82 50 01
E-mail: Chollet@FRGREN81

J.- M. CHOMAZ
Météo Nationale
CNRM
42 Av G. de Coriolis
31057 TOULOUSE Cedex
tel: 61 07 93 87
fax: 61 07 96 00

C. M. COATS
University of Leeds
Dept. of Fuel and Energy
LEEDS, LS29JT
GRANDE-BRETAGNE
tel: (44) 532 332505
fax: (44) 532 336017

S. COLLINEAU
Centre de Recherches de Bordeaux
INRA
33140 PONT-DE-LA-MAGE
tel: 56 84 32 77

G. COMTE-BELLOT
Lab. de Méc. des Fluides et d'Acoustique
Ecole Centrale de Lyon
36, avenue Guy de Collongues
B. P. 163
69131 ECULLY Cedex
tel: 78 33 81 27
fax: 78 43 39 62
E-mail: GCB@FRFLU51

P. COMTE
Institut de Mécanique de Grenoble
Domaine Universitaire
B. P. 53 X
38041 GRENOBLE Cedex
tel: 76 82 51 21
fax: 76 82 50 01
E-mail: Comte@frgren81

P. CORBET
Conseil Générale de l'Isère
TETRAPOLE
Hôtel du Département , rue Fantin-Latour
B. P. 1096
38022 GRENOBLE Cedex
tel: 76 60 38 38

O. R. COTE
Chief, Geophysics and Space
E.O.A.R.D.
223/231 Old Marylebone Road
LONDON, NW1 5th
GRANDE-BRETAGNE
tel: (01) 409-4437

J.- C. COURTY
Aérodynamique Théorique
Avions Marcel Dassault
78 Quai Marcel Dassault
92214 SAINT CLOUD
tel: (1) 47 11 47 34

D. D'HUMIERES
Laboratoire de Physique Statistique
Ecole Normale Supérieure
24, rue Lhomond
75005 PARIS
tel: (1) 45 87 02 84 poste 3354 (secr.)
fax: (1) 45 87 34 89
E-mail: DOMINIQ@FRULM62

D. DANG TRAN
Direction de l'Aérodynamique
ONERA
29 Av. de la Division Leclerc
B. P. 72
92322 CHATILLON Cedex
tel: (1) 46 57 11 60
fax: (1) 46 56 25 23

F. DARVE
Directeur de l'ENSHMG
INPG
ENSHMG
Domaine Universitaire
SAINT-MARTIN D'HERES

E. DAVID
Institut de Mécanique de Grenoble
Domaine Universitaire
B. P. 53X
38041 GRENOBLE Cedex

J. DE LEIRIS
Maire-Adjoint
COVEG
30 Bd Gambetta
38 000 GRENOBLE

B. DELHAYE
Laboratoire EM2C
Ecole Centrale de Paris
Grande voie des vignes
92290 CHATENAY MALABRY Cedex
tel: (1) 47027056

J. DELLA DORA
INPG
Laboratoire TIM3
46 Av. Félix Viallet
38000 GRENOBLE

J. DELVILLE
Laboratoire d'Etudes Aérodynamiques
C.E.A.T
43 Route de l'Aérodrome
86036 POITIERS Cedex
tel: 49 53 70 57
fax: 49 53 70 01
E-mail: umdf008@FRORS31

E. DERIAT
ONERA
29 Av. de la Division Leclerc
B. P. 72
92322 CHATILLON Cedex
tel: (1) 46 57 11 60
fax: (1) 46 56 25 23

T. DRACOS
Institut für Hydromechanik und Wasserwirtshaft
Federal Institute of Technology Zurich
ETH-Höngerberg
HONGERBERG, 8093
SUISSE
tel: 01 377 30 66 poste 30 65
fax: 01 371 55 48

B. DUBRULLE
CERFACS
42 Av. Coriolis
31000 TOULOUSE
tel: 61 07 96 96
E-mail: dubrulle@fromp51

D. DUTOYA
O.N.E.R.A.-Energétique
Avenue de la Division Leclerc
B. P. 72
92322 CHATILLON Cedex
tel: (1) 46 57 11 60
fax: (1) 46 56 25 23

G. DUVAUT
Directeur Scientifique Général
ONERA
29 av. de la Division Leclerc
B. P. 72
92322 CHATILLON Cedex
tel: (1) 46 57 11 60 poste 2448
fax: (1) 46 56 25 23

M. FANDRE
Administrateur Délégué
AD11
CNRS
25, Av. des Martyrs
B. P. 166X
38042 GRENOBLE Cedex
tel: 76 88 10 60
fax: 76 88 11 61

M. FARGE
Laboratoire de Météorologie Dynamique
Ecole Normale Supérieure
24, rue Lhomond
75231 PARIS Cedex 05
tel: (1) 43 31 72 73
fax: (1) 45 87 34 89
E-mail: MFARGE @FRULM11

Y. FAUTRELLE
ENSHMG
Laboratoire MADYLAM
Domaine Universitaire
B. P. 95
38 402 GRENOBLE Cedex
tel: 76 72 52 05

S. FAUVE
Laboratoire de Physique
E.N.S. Lyon
48, allée d'Italie
69364 LYON Cedex 07
tel: 72 72 81 48
fax: 72 72 80 80
E-mail: fauve@frensl61.bitnet

M. FAYOL
STPA
4 avenue de la Porte d'Issy
75996 PARIS-ARMEES
tel: (1) 45524361
fax: (1) 45524435

H. H. FERNHOLZ
Herman Föttinger Institut
Technische Universität Berlin
Strasse des 17 Juni 135
D-1000 BERLIN 12
RFA
tel: (030) 314 23359

Y. FOUILLET
Institut de Mecanique de Grenoble
Domaine Universitaire
B. P. 53X
38041 GRENOBLE Cedex
tel: 76 82 51 21
fax: 76 82 50 01

W. FRANK
Institut für Strömungslehre
Universität Karlsruhe
Kaiserstrasse 12
KARLSRUHE, D-7500
R.F.A.

R. FRIEDRICH
Lehrstuhl für Strömungsmechanik
Techniche Universität München
Arcistraße 21
8000 MUNCHEN 2
RFA
tel: (089) 21052508
fax: (089) 21052000
E-mail: T784103@DM0LRZ01

U. FRISCH
Observatoire de Nice
Le Mont Gros
B. P. 139
06003 NICE Cedex
tel: 92 00 30 35
fax: 92 00 30 58
E-mail: URIEL @FRONI51

Y. GAGNE
Institut de Mécanique de Grenoble
Domaine Universitaire
B. P. 53 X
38041 GRENOBLE Cedex
tel: 76 82 50 35
fax: 76 82 50 01

A. GARINO
IMFM
1, rue Honorat
MARSEILLE, 13003
tel: 96 08 16 90
fax: 91 08 58 91

P. GASCON
Premier Adjoint au Maire
Hôtel de Ville
11 Bd Jean-Pain
38021 GRENOBLE Cedex

R. GATHMANN
Institut de Mécanique de Grenoble
Domaine Universitaire
B. P. 53 X
38041 GRENOBLE Cedex
tel: 76 71 50 00
fax: 76 82 50 01

M. GERMANO
Politecnico Torino
Corso Duca Abruzzi 24
TORINO,
ITALIE
tel: 11-5566814

T. GERZ
Institut für Physik der Atmosphäre
DLR
D-8031 OBERPFAFFENHOFEN
RFA
tel: 08153 / 28586
E-mail: PA67@DLRVM

M. GLAUSER
Dept of Mechanical and Industrial Eng.
Clarkson University
211W Old Main
POTSDAM, NY 13676
USA
tel: (315) 268-6683
fax: (315) 268-4475
E-mail: glauser@sun.soe.clarkson.edu

A. GLEZER
Dept of Aerospace and Mechanical Eng.
University of Arizona
Aero Buildind 16
TUCSON, Arizona 85721
USA
tel: (602) 621-6135
E-mail: glezer@arizjvax

M.- A. GONZE
Institut de Mécanique de Grenoble
Domaine Universitaire
B. P. 53X
38041 GRENOBLE Cedex
tel: 76 82 51 22
fax: 76 82 50 01

D. GRAND
Service des transferts thermiques
Centre d'Etudes Nucléaires de Grenoble
Avenue des Martyrs
B. P. 85X
38 041 GRENOBLE Cedex
tel: 76 88 46 28
fax: 76 88 51 61

R. GUILLEMIER
Colonel
DRET/SDR G7
26,Bv.Victor
75 996 PARIS
tel: (16-1) 45 52 46 74

J.- P. GUIRAUD
Université Pierre et Marie Curie
27 Av du 11 Novembre
92190 MEUDON

H. HA MINH
IMFT
Avenue du Professeur Camille Soula
31400 TOULOUSE
tel: 61 28 58 34
fax: 61 07 96 13
E-mail: Haminh@FRIMFT11

A. HAMDOUNI
C.E.A.T.
43 route de l'Aérodrome
86036 POITIERS Cedex
tel: 49 53 70 00

E. HERVIEU
SETH/LEF
Centre d'Etudes Nucléaires de Grenoble
B. P. 85 X
38041 GRENOBLE Cedex
tel: 76 88 45 33
fax: 76 88 51 77

J. C. HILL
Chemical Engineering Department
Iowa State University
231 Sweeney Hall
AMES, Iowa 50011-2230
USA
tel: (515) 294-4959
fax: (515)294-9273
E-mail: n4.jch@isumvs

N. HOLLINGWORTH
Science&Technology Div.
Kluwer Academic Publishers
Spuiboulevard 50
DORDRECHT, 3300 AA
PAYS-BAS
tel: (central)(0)78 334911
fax: (0)78 334254

E. HOPFINGER
Institut de Mécanique de Grenoble
Domaine Universitaire
B. P. 53 X
38041 GRENOBLE Cedex
tel: 76 82 50 43

I. HOSOKAWA
University of Electro-communications
Chofu
TOKYO, 182
JAPON
tel: 0424 83 2161
fax: 0424 84 3327
E-mail:
c31973%tansei.cc.u-tokyo.junet@japan.csnet

B.L HUA
Centre Océanologique de Bretagne
IFREMER
B. P. 70
29 263 PLOUZANE Cedex
tel: 98 224287
E-mail: Lien@heol2.ifremer.fr

J. C. R. HUNT
D.A.M.T.P.
University of Cambridge
Silver Street
CAMBRIDGE, CB3 9EW
GRANDE-BRETAGNE
tel: (44) 223 337870
fax: (44) 223 337918

F. HUSSAIN
Department of Mechanical Engineering
University of Houston
University Park, 4800 Calhoun rd
HOUSTON, Texas 77004
U.S.A
tel: (713) 749-4444
fax: (713) 741 2614
E-mail: MECE1W@UHVAX1

L. JACQUIN
ONERA
8 RUE DES VERTUGADINS
92190 MEUDON
tel: (1) 45 34 75 01
fax: (1) 45 34 82 16

J. JIMENEZ
School of Aeronautics
Pl. Cardenal Cisneros 3
MADRID, E 28040
ESPAGNE
tel: (34) 1 734 7512
E-mail: c0130001@emdupm11

F. JOHNSON
Boeing Company
P. O. Box 3707
SEATTLE, WASHINGTON,
U.S.A.
tel: 206 865 6275
fax: (206) 865 2134

T. KAMBE
Department of Physics
University of Tokyo, Faculty of Sciences
7-3-1 Hongo; Bunkyo-Ku
TOKYO, 113
JAPON
tel: (81) 03 812 2111 poste 41 88
fax: (81) 03 814 9717

L. KLEISER
Institut für Theoretische Strömungsmechanik
D.L.R.
Burtsgenstraße 10
D-3400 GOTTINGEN
RFA
tel: (05 51) 709-2295
fax: (05 51) 709-2101
E-mail: TS22@DLRVMGO

P. LALLEMAND
Laboratoire de Physique Statistique
Ecole Normale Supérieure
24 rue Lhomond
75231 PARIS Cedex 05
tel: (1) 45 52 46 05 (DRET)
E-mail: Lalleman@FRULM62

M. LANCE
Laboratoire de Mécanique des Fluides
Ecole Centrale de Lyon
36 avenue Guy de Collongues
B. P. 163
69131 ECULLY Cedex

M. LARCHEVEQUE
Laboratoire de Météorologie Dynamique
Ecole Normale Supérieure
24, rue Lhomond
75231 PARIS Cedex 15
tel: (1) 43 29 12 25 poste 3469

H. LAROCHE
Institut de Mécanique de Grenoble
B. P. 53X
38041 GRENOBLE CEDEX

J. C. LASHERAS
Department of Mechanical Engineering
University of Southern California
University Park
LOS ANGELES, CA 90089-1453
USA
tel: (213) 743-0156
fax: (213) 749 1271
E-mail: lasheras@groghe ou jlasher@castor

A. LATROBE
Directeur régional
CISI Ingénierie
57 rue Pierre Sémard
38 000 GRENOBLE
tel: 76 88 50 50
fax: 76 88 50 75

S. LE GARREC
DEMT/SMTS/TTMF
Commissariat à l'Energie Atomique
Bâtiment 458
91191 GIF SUR YVETTE
tel: (1) 69 08 20 46

L. LE PENVEN
Laboratoire de Mécanique des Fluides
Ecole Centrale de Lyon
36 avenue Guy de Collongues
B. P. 163
69131 ECULLY Cedex

P. LE QUERE
LIMSI - CNRS
B. P. 133
91403 ORSAY Cedex
tel: 16 (1) 69 85 80 76
E-mail: ULUU008 @ FRORS31

M. LE RAY
Lab. d'Hydrodynamique
Université de Valenciennes
d'Aérodynamique et d'Energeetique
59326 VALENCIENNES
tel: 27 42 41 00

M. LEGRAND
Centre d'études de Vaujours
Commissariat à l'Energie Atomique
B. P. 7
77 COURTRY
tel: 49 36 86 19

M. LEGRAND
Groupe 6
DRET
26 Bd. Victor
75996 PARIS-ARMEES
tel: (1) 45 52 49 26

B. LEGRAS
Laboratoire de Météorologie Dynamique
Ecole Normale Supérieure
24, rue Lhomond
75231 PARIS Cedex 05
tel: (1) 45 35 99 57
fax: (1) 43 36 83 92
E-mail: LEGRAS@FRULM11

S. LEIBOVICH
Sibley School of Mechanical and Aerospace Engineering
Cornell University
ITHACA, NY 14 853-7501
USA
tel: (607) 255-3477
fax: (607) 254-4588
E-mail: leibov@flow.tn.cornell.edu

C. E. LEITH
Lawrence Livermore National Laboratories
L-16
P. O. Box 808
LIVERMORE, CA 94550
USA
tel: (415) 423-1612
fax: (415) 422-1370

H. LEMONNIER
SETh/LEF
Centre d'Etudes Nucléaires de Grenoble
B. P. 85 X
38041 GRENOBLE Cedex
tel: 76 88 45 40
fax: 76 88 51 77

A. D. LEONARD
Chemical Engineering Department
Iowa State University
231 Sweeney Hall
AMES, IA 50021
USA
tel: (515)-294-6766
E-mail: N2.ADL@ISUMVS

A. LEONARD
California Institute of Technology
Graduate Aeronautical Laboratories
M/C: 301-46 E. California
PASADENA, CA 91 125
USA
tel: (818) 356-4465
fax: (818) 449-2677
E-mail: Tony@wdproc.caltech.edu

J. LEORAT
Directeur de Recherches CNRS
DAEC
Observatoire de Meudon
Place Janssen
92195 MEUDON Cedex
tel: (1) 45 07 74 21
fax: (1) 45 07 74 69
E-mail: LEORAT@FRMEU51

L. LESIEUR
Mathématiques
Université Paris-Sud
5 Allée des Sophoras
92330 SCEAUX

M. LESIEUR
Directeur du GDR MFN
Institut de Mécanique de Grenoble
Domaine Universitaire
B. P. 53 X
38041 GRENOBLE Cedex
tel: 76 82 50 64
fax: 75 82 50 01
E-mail: Lesieur@FRGREN81

D. C. LESLIE
Department of Nuclear Engineering
Queen Mary College
Mile End Road
LONDON, E14NS
GRANDE-BRETAGNE
tel: (44) 01-975-5195
fax: (44) 01-975-5500
E-mail: leslie%turbulence.unit@qmc.ac.uk

G. LESPINARD
Président de l'INPG
Institut National Polytechnique de Grenoble
46, avenue Félix Viallet
38 031 GRENOBLE Cedex
tel: 76 47 98 55
fax: 76 46 88 24

O. LEUCHTER
Adjoint Scientifique
ONERA
29 Av. de la Division Leclerc
B. P. 72
92322 CHATILLON Cedex
tel: (1) 46 57 11 60

A. J. LIBCHABER
James Franck Institute
University of Chicago
5640 S. Ellis Avenue
CHICAGO, Illinois 60637
U.S.A.
tel: (312) 702-1919
fax: (312) 702-5863
E-mail: albert@control.uchicago.edu

J.- L. LIONS
Président du C.N.E.S.
C.N.E.S.
2, place Maurice Quentin
75001 PARIS
tel: (1)45 08 75 00
fax: (1) 45 08 76 76

L. LLIBOUTRY
Laboratoire de Glaciologie
CNRS
B. P. 53X
38041 GRENOBLE Cedex
tel: 76 42 58 72

J. LYS
Direction de la Recherche
Régie Nationale des Usines Renault
9-11 Av. du 18 Juin 1940
92 500 RUEIL-MALMAISON
tel: (1) 47 32 02 53

H. MAEKAWA
Department of Mechanical Engineering
Kagoshima University
1-21-40, Koorimoto
KAGOSHIMA, 890
JAPON
tel: 0992-54-7141
fax: (81) 992-50-3181

J. MAESTRATI
Avions Marcel Dassault
Aéroport de Mérignac
B. P. 24
33701 MERIGNAC
tel: 56 55 23 19

P. MAHOUDEAU
O.N.E.R.A.
29 Avenue de la division Leclerc
B. P. 72
92322 CHATILLON Cedex
tel: (1) 46 57 11 60
fax: (1) 46 56 25 23

B. MALRAISON
LEMD/CNRS
Av.Martyrs
38042 GRENOBLE Cedex
tel: 76 88 11 71

J.- D. MARION
Aerodynamique Theorique
Avions Marcel Dassault
78 Quai Marcel Dassault
92214 SAINT-CLOUD
tel: (1) 47 11 47 34

T. MARTIN
Agence de Saclay
CISI - INGENIERIE - DCI
B. P. 28
91192 GIF SUR YVETTE
tel: (1)69 08 59 36

B. MASURE
ICA
Direction des Constructions Navales
D.G.A., S.T.
8, boulevard Victor
75732 PARIS Cedex 15
tel: (1) 40 59 20 00

J. MATHIEU
Laboratoire de Mécanique des Fluides
Ecole Centrale de Lyon
36, avenue Guy de Collongues
B. P. 163
69131 ECULLY Cedex
tel: 78 33 81 27
fax: 78 33 34 75
E-mail: bataille@frflu51

T. MAXWORTHY
Dept of Mechanical Engineering
University of Southern California
University Park
LOS ANGELES, CA 90089-1453
USA
tel: (213) 743-6240
fax: (213) 749-1271

J. C. MCWILLIAMS
N.C.A.R.
P. O. Box 3000
BOULDER, Colorado 80307
USA
tel: (303) 497-1352
fax: (303) 497-1137
E-mail: jcm@ncar.ucar.edu

P. MEGE
ONERA
29 Av. de la Division Leclerc
B. P. 72
92320 CHATILLON
tel: (1) 46 57 11 60 poste 3135
fax: (1) 46 56 25 23

E. MEIBURG
Div. of Applied Mathematics
Brown University
37 Manning Street
PROVIDENCE, RI 02912
U.S.A.
tel: (401) 863-2507

M. MENEGUZZI
CERFACS
42 Avenue G. Coriolis
31057 TOULOUSE Cedex
tel: 61 07 96 90
fax: 61 07 96 13
E-mail: mngz@frtls12

O. METAIS
Institut de Mécanique de Grenoble
Domaine Universitaire
B. P. 53 X
38041 GRENOBLE Cedex
tel: 76 82 51 22
fax: 76 82 50 01
E-mail: METAIS@FRGREN81

M. MICHARD
Laboratoire de Mécanique des Fluides
Ecole Centrale de Lyon
36 avenue Guy de Collongues
B. P. 163
69131 ECULLY Cedex

J. MIZUSHIMA
Sagami Institute of Technology
1-1-25 Tsujido Nishikaigan
FUJISAWA,
JAPON
tel: 0466-34-4111 (296)
fax: 0466-33-7009
E-mail: A33040@jpnkudpc

P. MOIN
Director, Center for Turbulence Research
Dept of Mechanical Eng.
Stanford University
STANFORD, CA 94305-3030
USA
tel: (415) 725-2081
fax: (415) 725-4862
E-mail: moin@eul.span.nasa.gov

Y. MORCHOISNE
Division d'Aérodynamique
ONERA
29 Av. de la Division Leclerc
B. P. 72
92322 CHATILLON Cedex
tel: (1) 46 57 11 60
fax: (1) 46 56 25 23

R. MOREAU
ENSHMG
Laboratoire MADYLAM
Domaine universitaire
B. P. 95
38402 SAINT MARTIN D'HERES Cedex
tel: 76 82 52 06

N. NEMOZ
Président de l'UJF
Université Joseph Fourier
Domaine Universitaire
B. P. 53X
38041 GRENOBLE Cedex

J.- L. NICOLAS
Dept de Mathématiques
Université Claude Bernard
1 Place Louis Chazette
69001 LYON

A.- M. NICOLAS
Dept de Mathématiques
Université Claude Bernard
1 Place Louis Chazette
69001 LYON

E. NIENALTOWSKA
Direction de la Recherche
Régie Nationale des Usines Renault
9-11 Av du 18 Juin 1940
92500 RUEIL-MALMAISON
tel: (1) 47 32 02 53
fax: (1) 47 08 32 59

X. NORMAND
Institut de Mécanique de Grenoble
Domaine Universitaire
B. P. 53 X
38041 GRENOBLE Cedex
tel: 76 82 51 21

M. S. OGGIANO
Dip. Ing. Aeronautica
POLITECNICO di Torino
Centro Studi Dinamica Fluidi - C.N.R.
TORINO, 10129
ITALIE
tel: (11) 5566803
fax: (11) 5566899
E-mail: Tordella@ITOPOLI

A. ORMANCEY
ISC
DRET/SDR G7
26, Bd.Victor
75996 PARIS
tel: (16-1) 45 52 49 30

S. A. ORSZAG
Program in Applied and Comp. Mathematics
Princeton University
218 Fine Hall, Washington Road
PRINCETON, New Jersey 08544
USA
tel: (609) 452-6206
E-mail: sao@acm.princeton.edu

T. OTAGURO
Institute of Computational Fluid Dynamics
2-1-4 Haramachi, Meguro-Ku
TOKYO,
JAPAN
tel: 81-3-711-0411
fax: 81-3-714-6343

D. OUESLATI
L.M.D.
Ecole Normale Supérieure
24, rue Lhomond
75231 PARIS Cedex 05
tel: (1) 43 31 72 73
fax: (1)45 87 34 89

J. PAUCHET
Centre d'Etudes et de Recherches de Grenoble
ACB - Groupe ALSTHOM
Voie Privée Rhône-Poulenc
38800 LE PONT DE CLAIX
tel: 76 40 90 40
fax: 76 98 08 81

P. PERRIER
Avions Marcel Dassault
78 Quai Marcel Dassault
B. P. 300
92914 SAINT-CLOUD Cedex
tel: (16 1)47 11 32 58

L. PHAM NGUYEN
ALLIANT
2 place de l'Equerre - Silic 243
94568 RUNGIS Cedex
tel: 33(1)46 75 03 54
fax: 33(1)49 78 07 01

M. PIAU
Institut de Mécanique de Grenoble
Domaine Universitaire
B. P. 53X
38041 GRENOBLE Cedex

A. PICART
CNRS 230
Faculté des Sciences de Rouen - C.O.R.I.A.
B. P. 118
76134 MONT-SAINT-AIGNAN Cedex
tel: 35 14 65 73
fax: 35 70 83 84
E-mail: UTHD008 @FRORS31

S. PIGNY
ENSHMG
Laboratoire MADYLAM
Domaine Universitaire
B. P. 95
38402 SAINT-MARTIN D'HERES Cedex

E. PRIEUR
CRAY Research France
7 rue de Tilsitt
PARIS, 75017
tel: (1)47 66 01 55
fax: (1)43 80 03 65

M. PROVANSAL
Laboratoire de Recherche en Combustion
Université de Provence
252 Faculté Saint Jérome
13397 MARSEILLE Cedex13
tel: 91 28 81 27

A. PUMIR
Laboratoire de Physique Statistique
Ecole Normale Supérieure
24, rue Lhomond
75231 PARIS Cedex 05
tel: (1) 43 29 12 25
fax: (1) 45 87 34 89
E-mail: PUMIR @Frulm62

L. QUARTAPELLE
Istituto di Fisica
Istituto Politecnico di Milano
Piazza Leonardo da Vinci, 32
MILANO, 20139
ITALIA
tel: 02 23996139
E-mail: RFISIC01@IMIPOLI

J. M. REDONDO
D.A.M.T.P.
University of Cambridge
Silver Street
CAMBRIDGE, CB3 9EW
GRANDE-BRETAGNE
tel: (44) 223 33 79 11
fax: (44) 223 33 79 18
E-mail: jmr10@phoenix.cambridge.ac.uk

J. J. RILEY
Dept of Mechanical Engineering, FU10
University of Washington
SEATTLE, WA 98195
USA
tel: (206) 543-5347
fax: (206) 545-8047
E-mail: rileyj@max.u.washington.edu

A. ROSHKO
Graduate Aeronautical Laboratories
California Institute of Technology
Mail Code 105-50
PASADENA, CA 91 125
USA
tel: (818) 356-4484
fax: (818) 449-2677
E-mail: cf. LEONARD Tony

G. ROUAS
STCPMVN
2 Bd. Gambetta
B. P. 53
60321 COMPIEGNE
tel: 44 20 37 23

E. RUIZ
O.N.E.R.A. - Energétique
29 Avenue de la division Leclerc
B. P. 72
92322 CHATILLON Cedex
tel: (1) 46 57 11 60
fax: (1) 46 56 25 23

M. SAGNER
Bertin et Cie
B. P. 3
78 373 PLAISIR Cedex
tel: 34 81 87 48

J. P SALVODELLI
Directeur Commercial
ALLIANT
2 place de l'Equerre-Silic 243
94568 RUNGIS Cedex
tel: 33(1)46 75 03 54
fax: 33(1)49 78 07 01

R. SCHIESTEL
IMST
12, Avenue du Général Leclerc
MARSEILLE, 13003
tel: 91 50 54 39

R. SEMERIA
Adjoint au Chef du Dép. Transf. d'Energie du CEA
CENG
CEA
B. P. 85X
38 041 GRENOBLE
tel: 76 88 39 11
fax: 76 88 51 61

Y. SHIMOMURA
Department of Physics
University of Tokyo, Faculty of Science
Bunkyo-Ku
TOKYO, 113
JAPON
tel: 03-812-2111 poste 4216
fax: 81-3-814-9717
E-mail: d32759@tansei.cc.u-tokyo.junet

A. SILVEIRA
Service des Tranferts Thermiques
Centre d'Etudes Nucléaires de Grenoble
Avenue des Martyrs
B. P. 85X
38041 GRENOBLE

J. SOMMERIA
ENSHMG
Laboratoire MADYLAM
B. P. 95
38402 SAINT-MARTIN D'HERES
tel: 76 82 52 04
E-mail: DNGUYEN at FRGREN81

C. STAQUET
Laboratoire de Physique
Ecole Normale Supérieure de Lyon
48, allée d'Italie
69364 LYON Cedex 07
tel: 72 72 8193
fax: 72 72 80 80
E-mail: STAQUET@FRENSL61. bitnet

P. SULLIVAN
Department of Mechanical Engineering
Clarkson University
POTSDAM, NEW YORK,
U.S.A.
tel: 315 264 2203
E-mail: pierre@sun.soe.clarkson.edu

B. M. SUMER
Institute of Hydrodynamics and Hydraulic Eng.
Techn. University of Denmark
b. 115
LYNGBY, 2800
DANEMARK
tel: (45) (42) 88 22 22
fax: 45 42 88 22 39

L. TA PHUOC
Directeur de Recherches CNRS
LIMSI
B. P. 133
91403 ORSAY Cedex
tel: (1) 69 85 80 78

J.- M. TANGUY
STCPMDN
2 Bd. Gambetta
B. P. 53
60321 COMPIEGNE
tel: 44 20 37 23

S. TARDU
Institut de Mécanique de Grenoble
Domaine Universitaire
B. P. 53X
38041 GRENOBLE Cedex
tel: 76 82 51 31

T. TATSUMI
President, Kyoto Institute of Technology
Kyoto Institute of Technology
Matsugasaki Sakyoku
KYOTO, 606
JAPON
tel: 075-791-3211
fax: 075 781 0190
E-mail: D53640 @ JPNKUDPC

V. TAVERNE
CERFACS
42 Avenue G. Coriolis
31057 TOULOUSE Cedex
tel: 61 07 96 96
fax: 61 07 96 13

C. TENAUD
Laboratoire D'étude Aérothermique
CEAT
43 Rue de l'Aérodrome
86036 POITIERS Cedex
tel: 49 53 70 40
fax: 49 53 70 01

D. TORDELLA
Dip. Ing. Aeronautica
POLITECNICO di TORINO
Corso Duca degli Abruzzi 24
TORINO, 10129
ITALIE
tel: (11) 5566812
fax: (11)5566899
E-mail: Tordella@itopoli

D. N. TRAN
Aérodynamique Théorique
Avions Marcel Dassault
78 Quai Marcel Dassault
92214 SAINT-CLOUD
tel: (1) 47 11 47 34

A. TROUVE
Laboratoire EM2C
Ecole Centrale de Paris
92295 CHATENAY MALABRY Cedex
tel: (1) 47 02 70 56

A. TSINOBER
University of Tel-Aviv
Faculty of Engineering
TEL-AVIV, 69978
ISRAEL
E-mail: D78@TAUNOS (or tsinober@taunos)

J. M. USSEGLIO
Directeur de la Division Numérique
Laboratoire d'Hydraulique de France
6 rue de Lorraine
38 130 ECHIROLLES
tel: 76 33 42 08
fax: 76 33 42 39

M. R. VALLCORBA
Institut de Mécanique de Grenoble
Domaine Universitaire
B. P. 53 X
38041 GRENOBLE Cedex
tel: 76 82 50 00
fax: 76 82 50 01

J. VERRON
Institut de Mécanique de Grenoble
Domaine Universitaire
B. P. 53 X
38041 GRENOBLE Cedex
tel: 76 82 50 66
fax: 76 82 50 01
E-mail: VERRON@FRGREN81

R. VERSTAPPEN
University of Twente
P. O. Box 217
ENSCHEDE, 7500 AE
PAYS-BAS
tel: 053-893416
fax: 053-356695

C. VIGNERON
Département de Recherche Spatiale
Observatoire de Paris
92195 MEUDON-PRINCIPAL Cedex
tel: (1) 45 07 76 75
fax: (1) 45 07 28 06
E-mail: catala@frmeu51

A.- P. VINCENT
CERFACS
42 Avenue de Coriolis
31057 TOULOUSE Cedex
tel: 61 07 96 27
fax: 61 07 96 13
E-mail: VINCENT@FRMOP11

H. WERLE
ONERA
B. P. 72
92322 CHATILLON Cedex
tel: (1) 46 57 11 60

D. WILDERMUTH
Institute of Hydromechanics and Water Res.
ETH Hönggerberg
ZURICH, 8049
SUISSE
tel: 0041 (1) 377 30 65

A. M. YAGLOM
Institute of Atmospheric Science
Academy of Sciences of USSR
Pyzhevsky 3
MOSCOW, 109017
USSR

N. YURCHENKO
Ukrainian Academy of Sci.
Institute of Hydromechanics
ulitsa Zheliabova, 8/4
KIEV-57, 252057
U.R.S.S.
tel: (7) 044 446 4313

N. J. ZABUSKY
Dept of Mechanical Engineering
Rutgers University
Busch Campus
PISCATAWAY, NJ 08855
U.S.A
tel: (201) 932-5869
fax: (201) 932-5313
E-mail: nzabusky@jove.rutgers.edu

# Author Index

# COLOUR PLATES

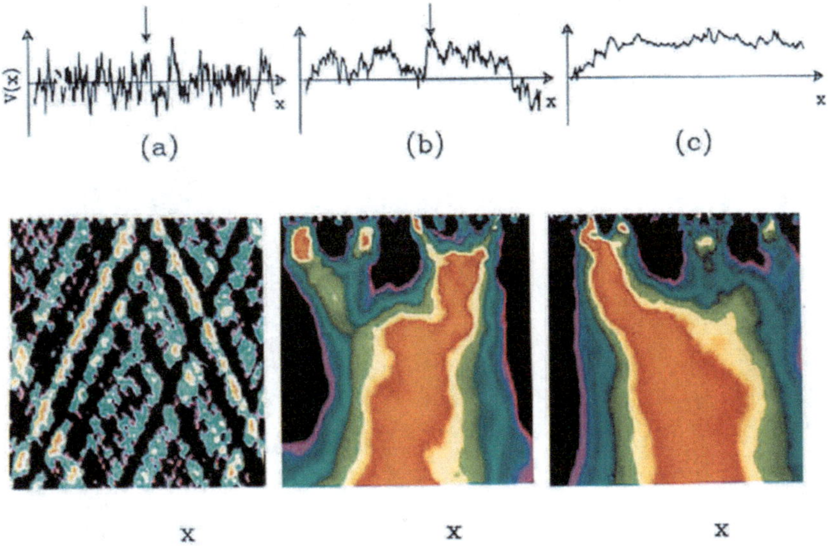


**Fig.3.** (a) Wavelet transform of turbulent signal schown in upper part (French top hat): large scales; (b) Same as 3a at intermediate scales (near arrow on 3a); (c) Same as 3a at small scales (near arrow on 3b).

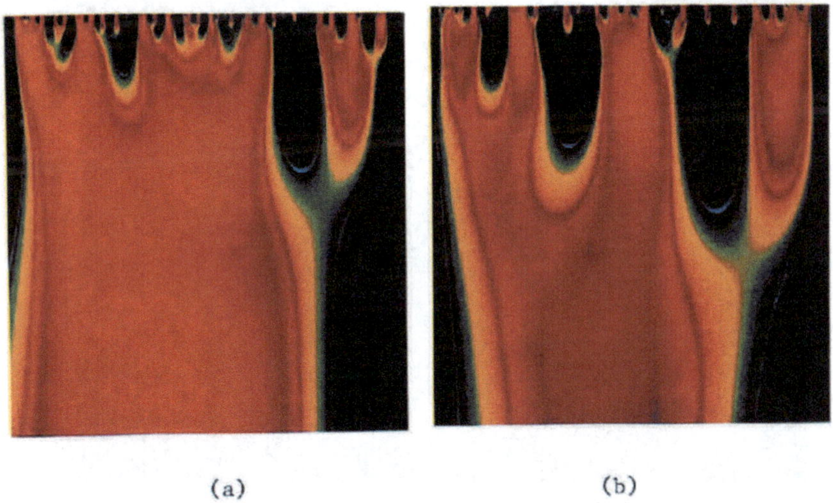


**Fig.4.** (a) Wavelet transform of turbulent signal (analysing wavelet and colour coding different from case of Fig.3., see text); (b) Enlargement of part of Fig.4a. near upper right branch.

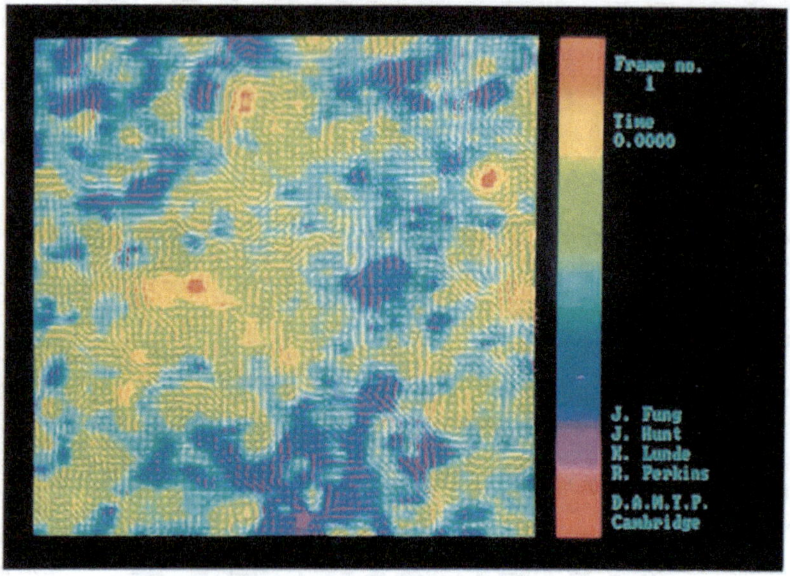


**Plate 1.** Pressure field on top of the velocity field.

Colour key: red – yellow – green – light blue– dark blue – pink – purple
for pressure: $p < 0$ $p = 0$ $p > 0$

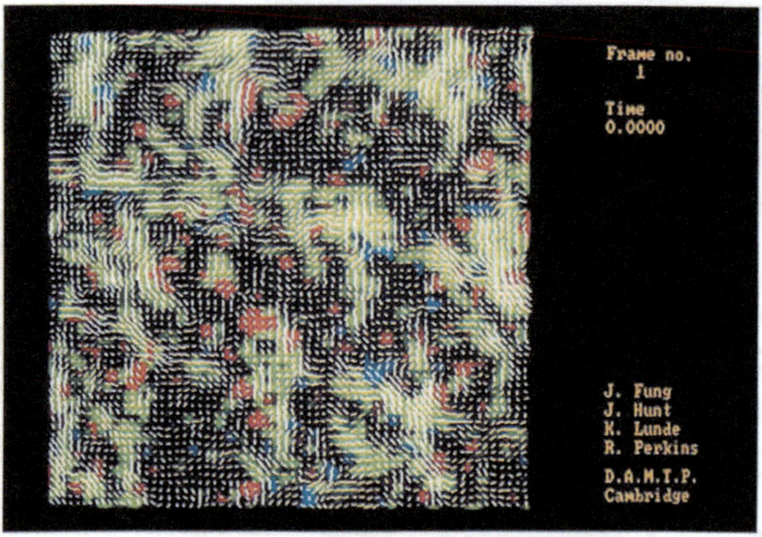


**Plate 2.** Eddies (red), streaming (green) and convergence (blue) regions on top of the velocity field.

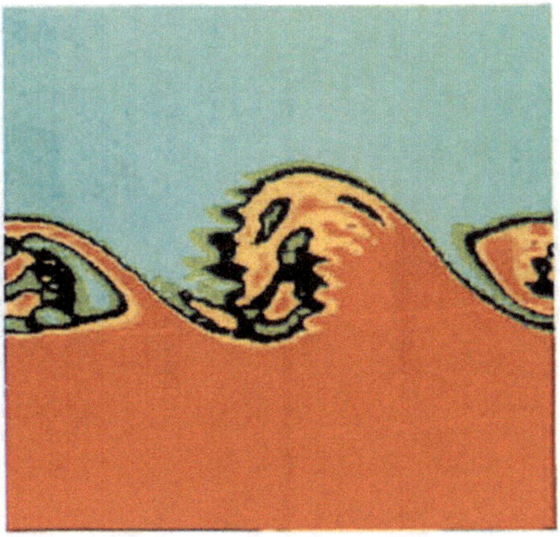

**Plate 1.** Reactive species concentration $C_A(x,y)$ field at time $t_1$ for $Da \approx 1$

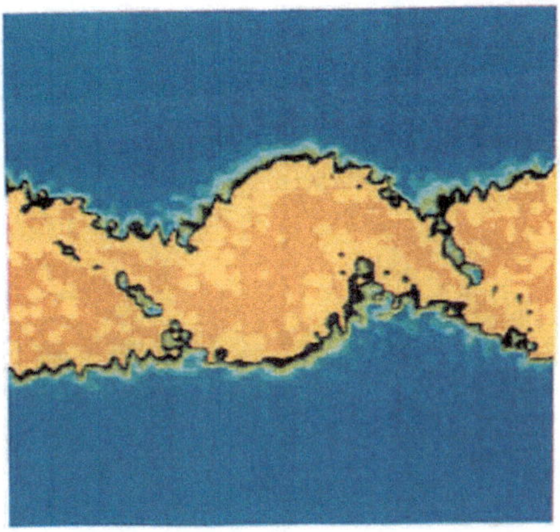

**Plate 2.** Product concentration $C_P(x,y)$ field at time $t_1$ for $Da \approx 10^4$ (fast chemical reaction)

**Plate 1**

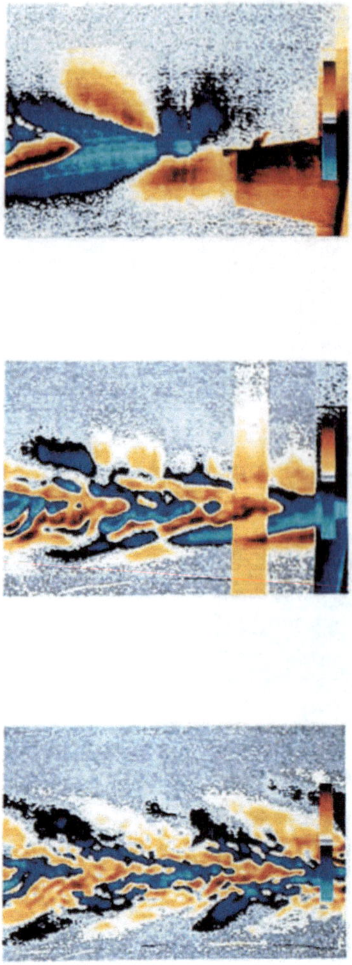

**Plate 3**

**Plate 1.** Visualization of lee waves in an horizontal plane, 7.5 cm above the center of the sphere, for $F = 3.2$ (a) $Nt = 0$ ; (b) $Nt = 12.5$ ; (c) $Nt = 37.7$. The sphere is moving from left to right.

**Plate 3.** Visualization of internal waves in an horizontal plane, 12.5 cm above the center of the sphere, for $F = 12.7$. (a) $Nt = 6.3$ ; (b) $Nt = 12.5$ ; (c) $Nt = 37.7$. The sphere is moving from right to left.

(a)

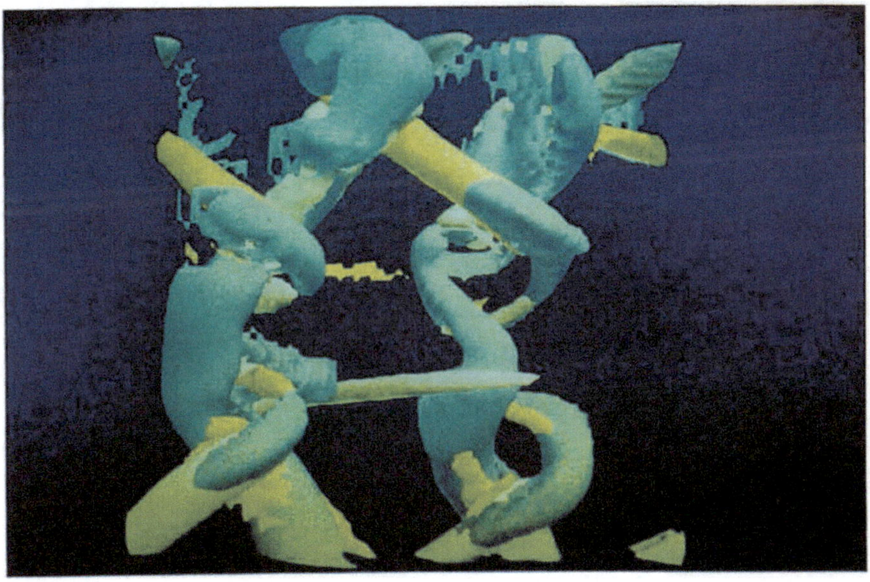

(b)

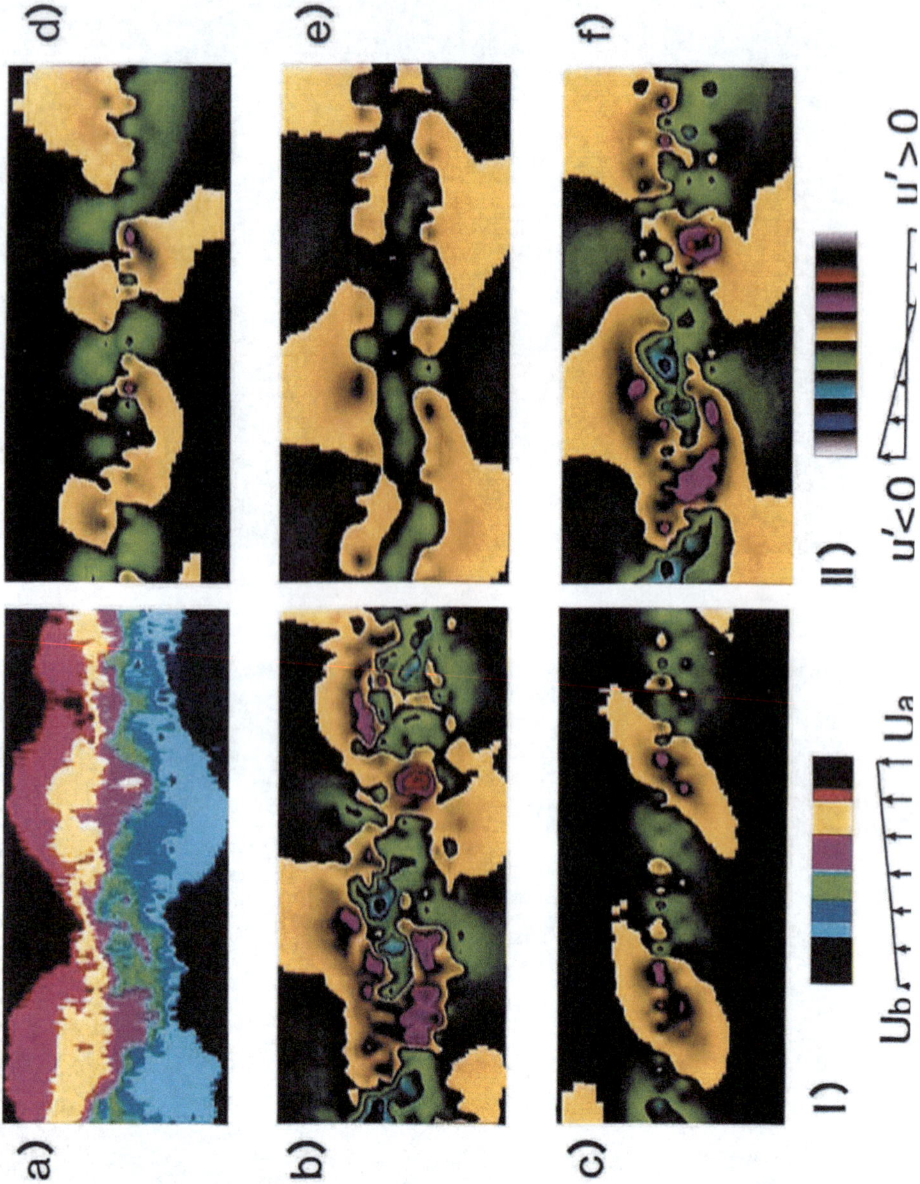


**Plate 1.** Instantaneous Contribution of individual modes : *a), b) original velocity, c) first mode, d) second mode e) third mode f) sum of first four modes.*

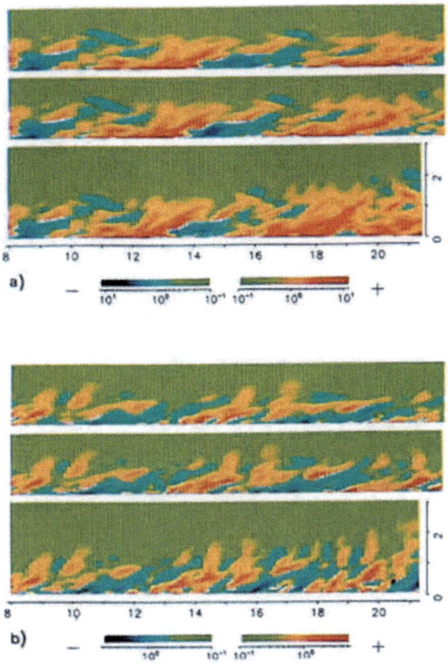

**Plate 1.** a,b Cartographic representations of the $u$-velocity fluctuations (a) and the $v$-velocity fluctuations (b).

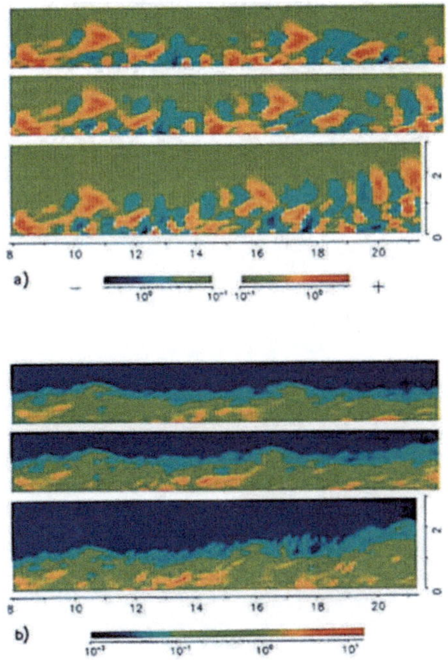

**Plate 2.** a,b Cartographic representations of the $w$-velocity fluctuations (a) and the squared fluctuating deformation tensor (b).

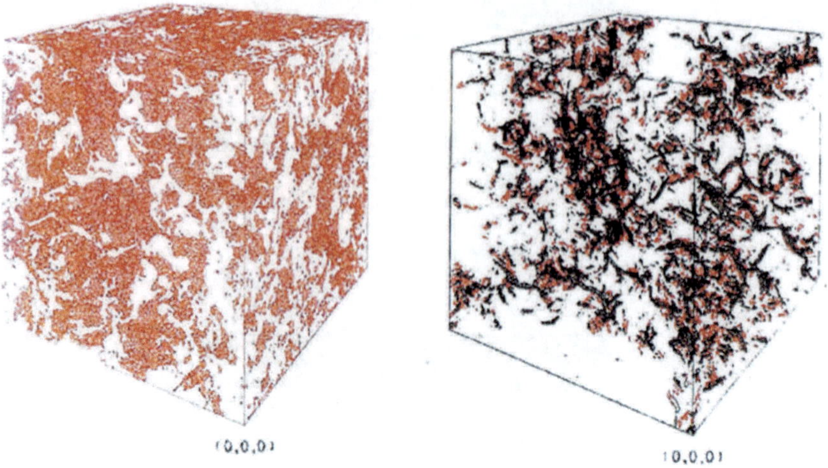


**Fig.2.** (a) Vorticity-concentrated regions (black) with the normalized vorticity magnitude greater than 0.3 and high-strain regions (red) with the normalized strain norm greater then 0.44 in the fully-developed turbulence. (b) Iso-norm-of strain surfaces with the normalized norm of strain equal to 0.15 in the fully-developed turbulence, which can be viewed from the reader. (Transcription from Hosokawa & Yamamoto 1989.)

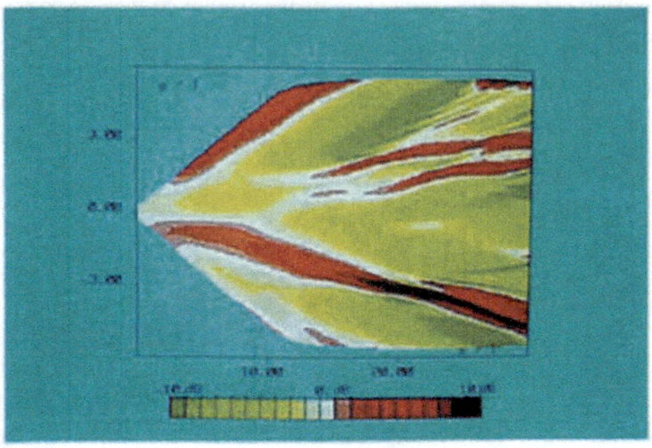

**Fig.2.** Acoustic pressure levels for the same turbulent realization used in Figure 1 (parabolic equation and $\lambda/L = 0.1$). Reinforced and weakened zones are clearly visible and can be associated with the caustics depicted in Figure 1.

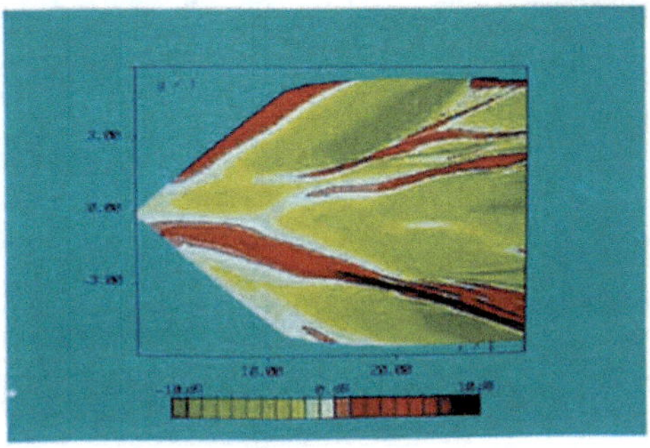

**Fig.3.** Acoustic pressure levels for the same turbulent realization as above, except that the wave frequency is now multiplied by 2 ($\lambda/L = 0.05$). Note the similarity with Figure 2.

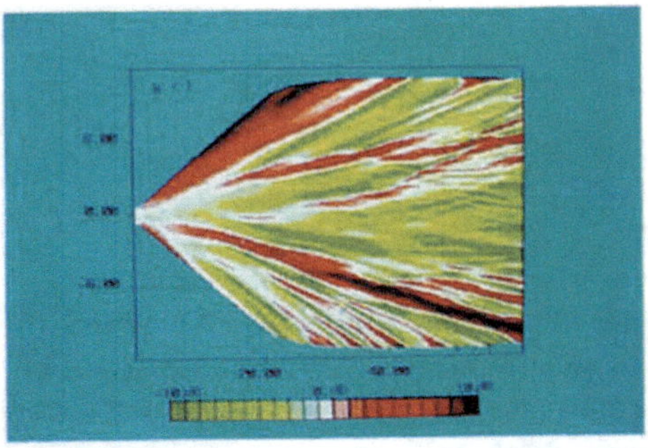

**Fig.4.** Acoustic pressure levels for a turbulent field whose length scale is half the size of that in Figure 2, with $\lambda/L = 0.2$). Observe the much finer spatial structure of the field in comparison with Figures 2 and 3.

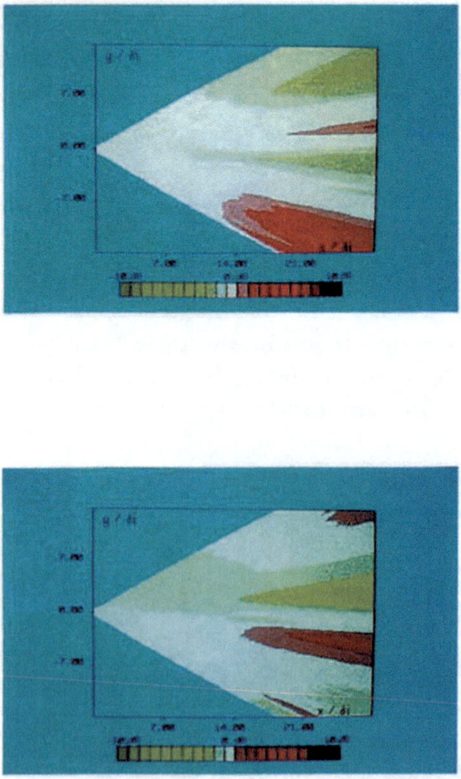

**Fig.10.** Acoustic pressure levels for waves transmitted through the mixing layer perpendicularly to the mean flow (parabolic equation and $\lambda/\delta_i = 0.2$). Note the four alternating regions of weakening and strengthening for case (a),before coalescence ; with only two for case (b), immediately after coalescence.

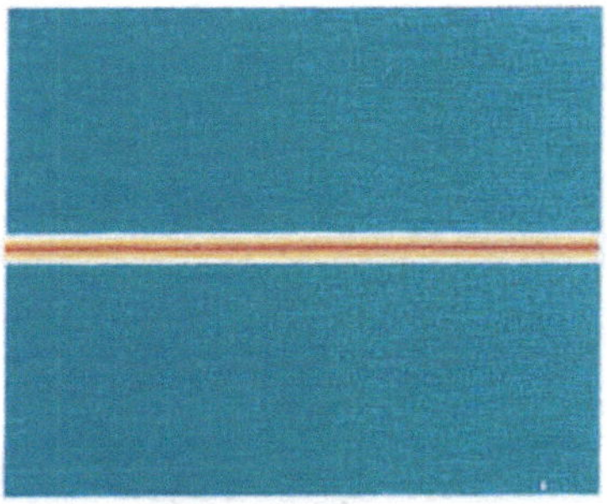

Figure 7: Initial vorticity contours

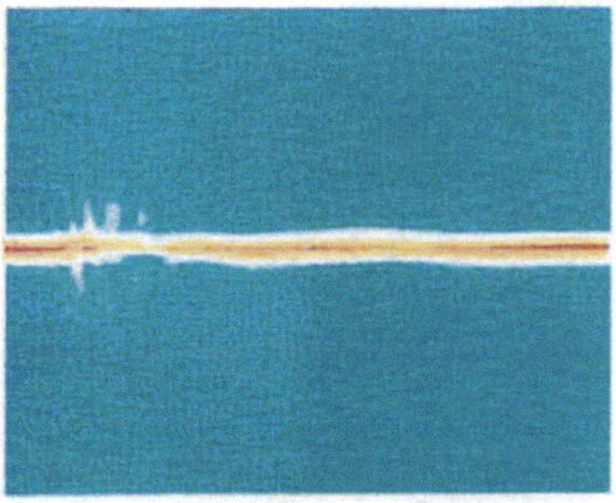

Figure 8: Vorticity contours at t = 15

Figure 9: Initial total pressure contours

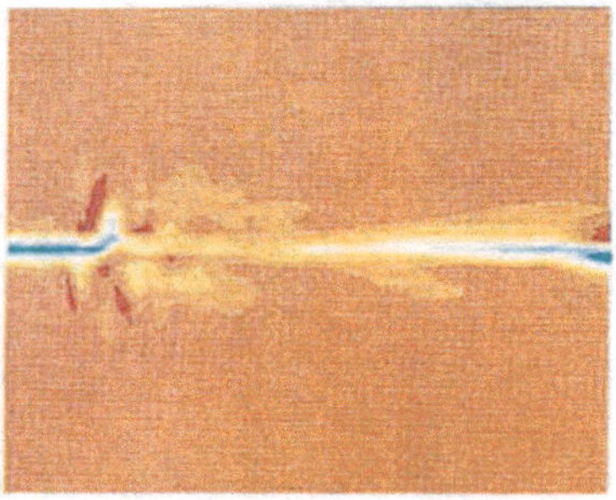

Figure 10: Total pressure contours at t = 15

Figure 11: Trajectories of particles at t = 15

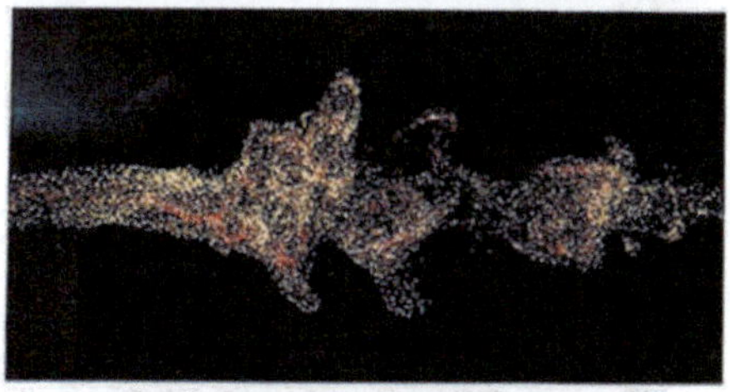

Figure 12: Trajectories of particles at t = 25

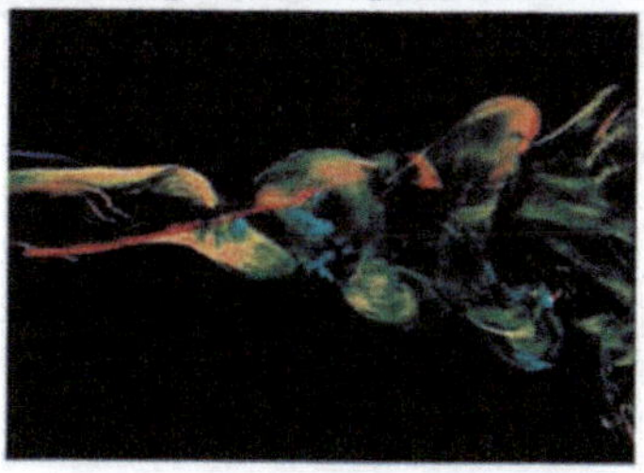

Figure 13: Visualisation of experimental streaklines obtained by M. Werle (ONERA)

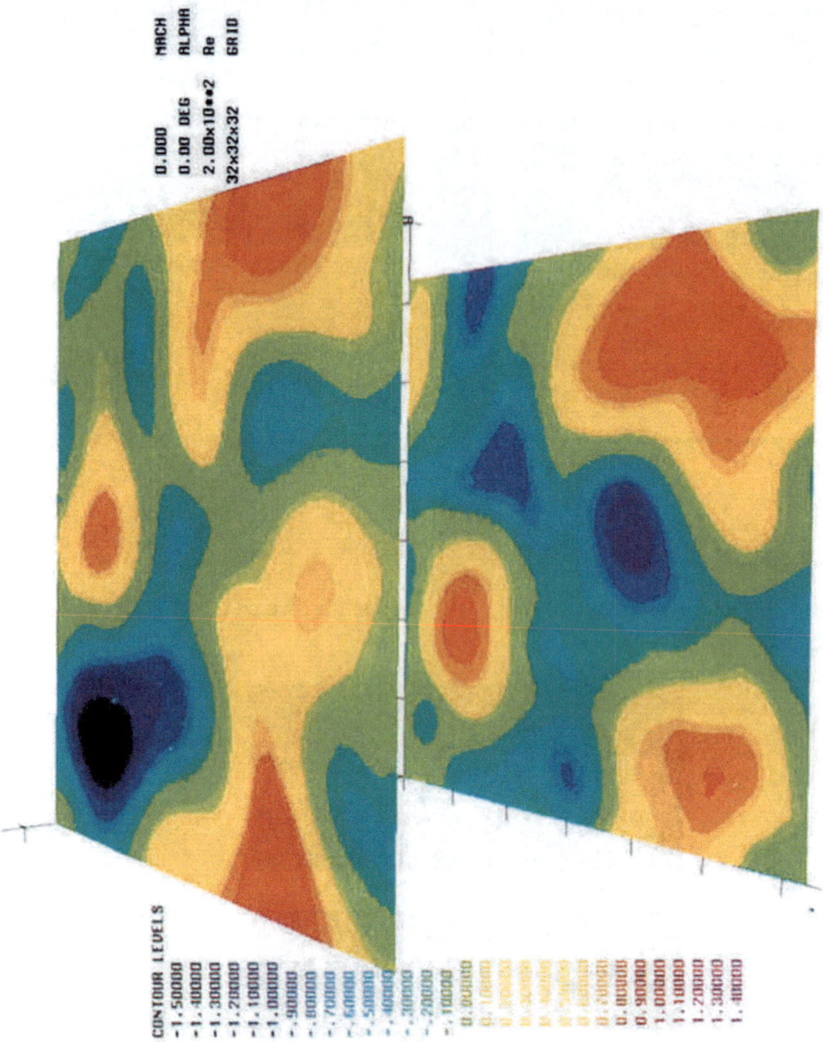


**Plate 1.** Constant contours of the stream function in two different horizontal planes for $F_t = 0.2$ at $T = 15$.

(a) Ve = 1.6 m/s Re = 200 Fr = 1.7

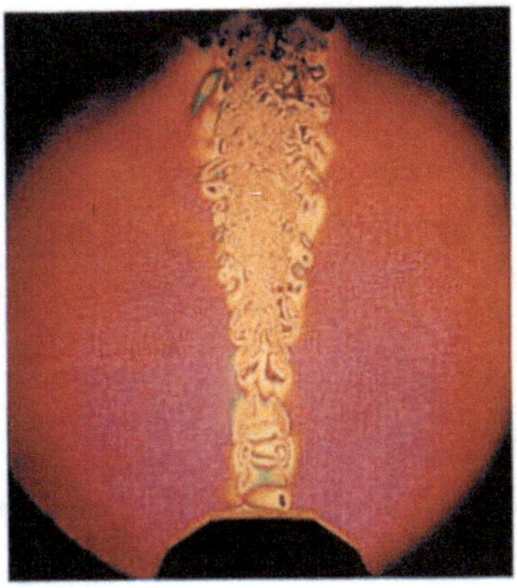

(b) Ve = 2.5 m/s Re = 315 Fr = 2.6

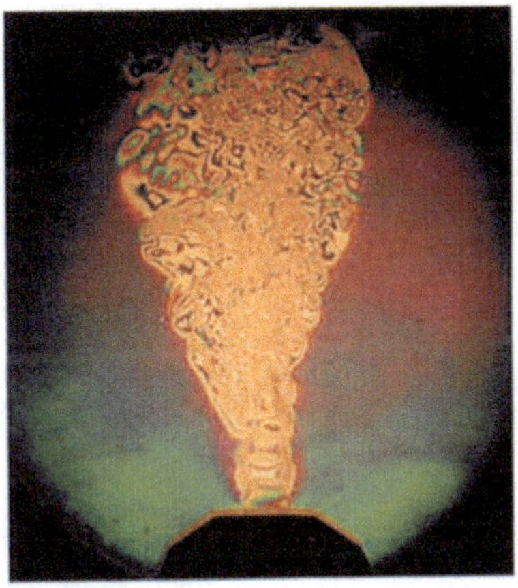

(c) Ve = 4.3 m/s Re = 532 Fr = 4.5

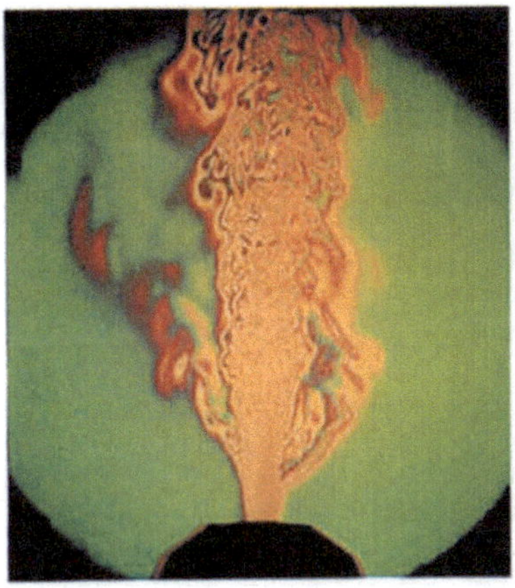

(d) Ve = 13.1 m/s Re = 630 Fr = 21.6

Figure 4 – Interferometric visualizations of helium jets into air: a,b,c 15 mm nozzle
d 6 mm nozzle

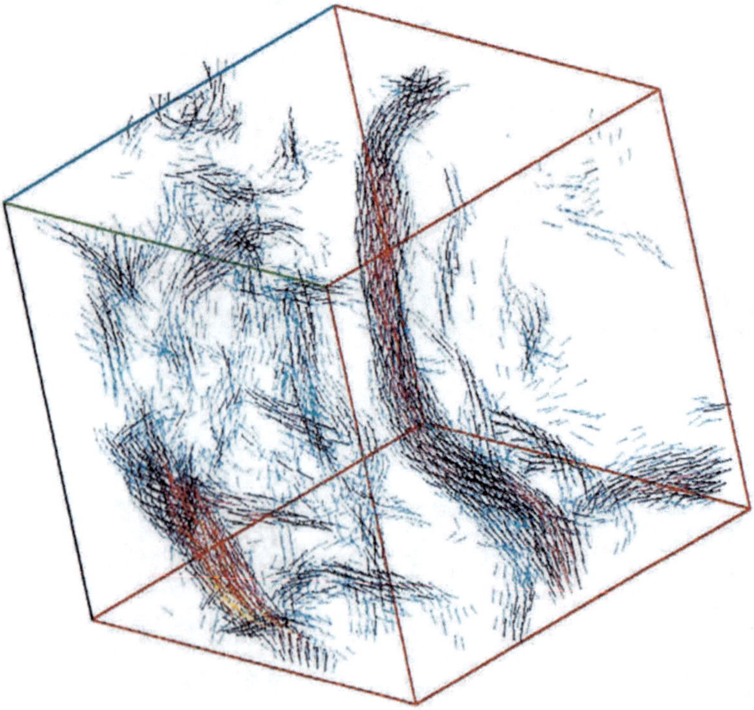

**Fig.8.** Detail of the vorticity field, showing a sub-cube of size one sixth of the complete cube, i.e. 40 grid points in each direction.

**Figure 11**

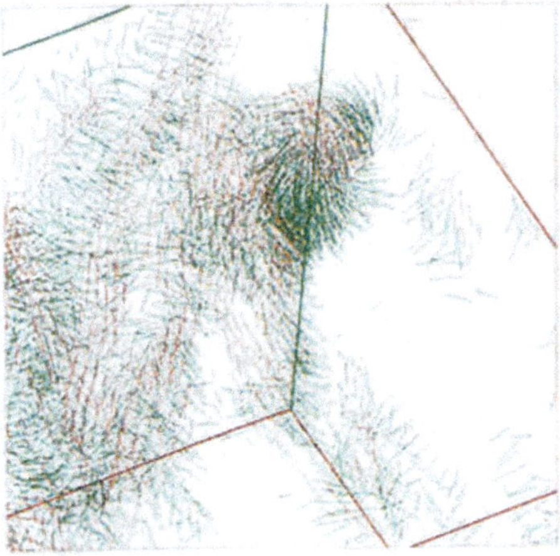

**Figure 12**

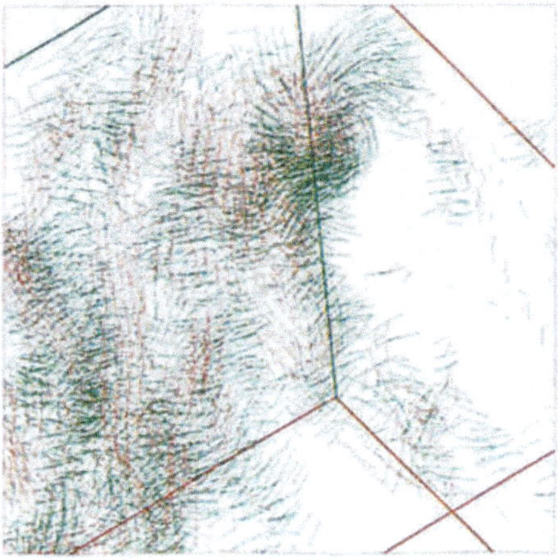

**Figure 13**

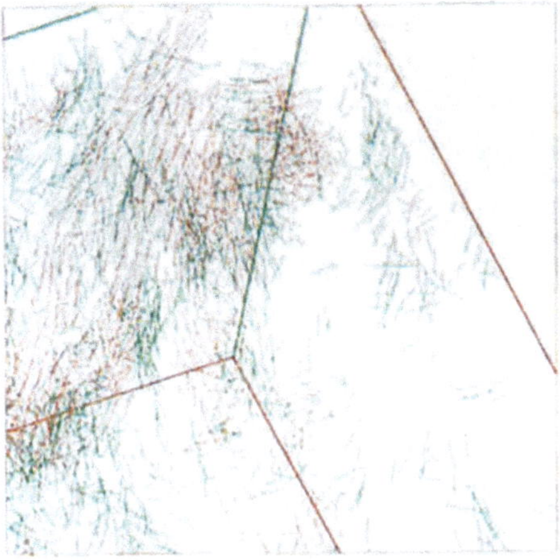

**Figure 14**

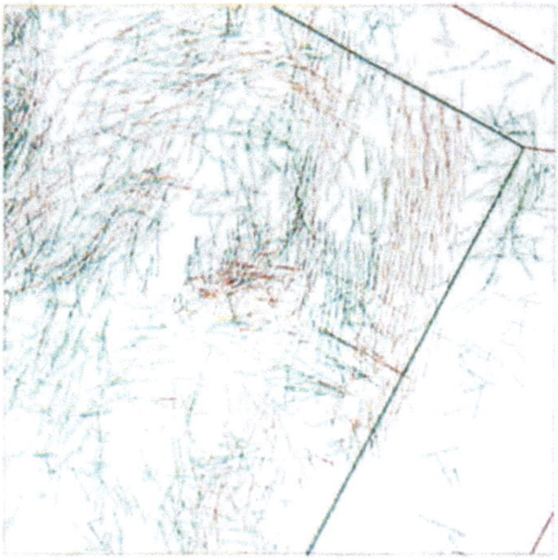

**Fig.11.** Vorticity field, in red, and eigenvector $\mathbf{e_3}$ associated with the largest (positive) eigeinvalue, in green. Note in the upper part of the figure the vorticity tube seen along its axis.

**Fig.12.** Vorticity field, in red, and eigenvector $\mathbf{e_1}$ associated with the smallest (negative) eigeinvalue, in green. Note in the upper part of the figure the same vorticity tube as in Fig 11.

**Fig.13.** Vorticity field, in red, and eigenvector $\mathbf{e_2}$ associated with the intermediate eigeinvalue, in green. Note in the upper part of the figure the same vorticity tube as in Figs 11 and 12.

**Fig.14.** Vorticity field, in red, and eigenvector $\mathbf{e_2}$ associated with the intermediate eigeinvalue, in green. The figure has been rotated respect to figures 11-13. The vorticity tube seen along its axis in those figures is now seen from the side in the upper right area of the picture.

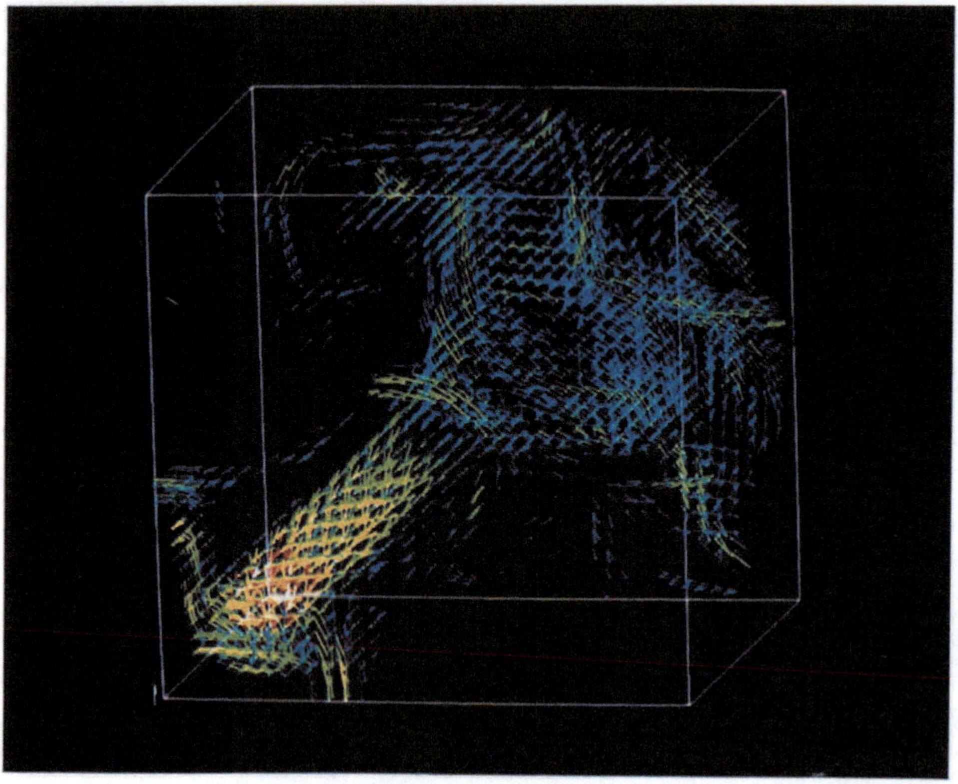

**Plate 1.** Graphical output of the magnetic field at $t = 52$.

# Mechanics

From 1990, books on the subject of *mechanics* will be published under two series.

## *FLUID* MECHANICS AND ITS APPLICATIONS

*Series Editor:* R. Moreau

*Aims and Scope of the Series*

The purpose of this series is to focus on subjects in which fluid mechanics plays a fundamental role. As well as the more traditional applications of aeronautics, hydraulics, heat and mass transfer etc., books will be published dealing with topics which are currently in a state of rapid development, such as turbulence, suspensions and multiphase fluids, super and hypersonic flows and numerical modelling techniques. It is a widely held view that it is the interdisciplinary subjects that will receive intense scientific attention, bringing them to the forefront of technological advancement. Fluids have the ability to transport matter and its properties as well as transmit force, therefore fluid mechanics is a subject that is particularly open to cross fertilisation with other sciences and disciplines of engineering. The subject of fluid mechanics will be highly relevant in domains such as chemical, metallurgical, biological and ecological engineering. This series is particularly open to such new multidisciplinary domains.

1. M. Lesieur: *Turbulence in Fluids*. 2nd rev. ed., 1990 ISBN 0-7923-0645-7
2. O. Métais and M. Lesieur (eds.): *Turbulence and Coherent Structures*. 1991 ISBN 0-7923-0646-5
3. R. Moreau: *Magnetohydrodynamics*. 1990 ISBN 0-7923-0937-5
4. E. Coustols (ed.): *Turbulence Control by Passive Means*. 1990 ISBN 0-7923-1020-9

## *SOLID* MECHANICS AND ITS APPLICATIONS

*Series Editor:* G.M.L. Gladwell

*Aims and Scope of the Series*

The fundamental questions arising in mechanics are: *Why?*, *How?*, and *How much?* The aim of this series is to provide lucid accounts written by authoritative researchers giving vision and insight in answering these questions on the subject of mechanics as it relates to solids. The scope of the series covers the entire spectrum of solid mechanics. Thus it includes the foundation of mechanics; variational formulations; computational mechanics; statics, kinematics and dynamics of rigid and elastic bodies; vibrations of solids and structures; dynamical systems and chaos; the theories of elasticity, plasticity and viscoelasticity; composite materials; rods, beams, shells and membranes; structural control and stability; soils, rocks and geomechanics; fracture; tribology; experimental mechanics; biomechanics and machine design.

1. R.T. Haftka, Z. Gürdal and M.P. Kamat: *Elements of Structural Optimization*. 2nd rev.ed., 1990 ISBN 0-7923-0608-2
2. J.J. Kalker: *Threedimensional Elastic Bodies*. 1990 ISBN 0-7923-0712-7

Kluwer Academic Publishers – Dordrecht / Boston / London

# Mechanics

From 1990, books on the subject of *mechanics* will be published under two series:
***FLUID* MECHANICS AND ITS APPLICATIONS**
*Series Editor:* R.J. Moreau
***SOLID* MECHANICS AND ITS APPLICATIONS**
*Series Editor:* G.M.L. Gladwell

Prior to 1990, the books listed below were published in the respective series indicated below.

MECHANICS: DYNAMICAL SYSTEMS
Editors: L. Meirovitch and G.Æ. Oravas

1. E.H. Dowell: *Aeroelasticity of Plates and Shells.* 1975 ISBN 90-286-0404-9
2. D.G.B. Edelen: *Lagrangian Mechanics of Nonconservative Nonholonomic Systems.* 1977 ISBN 90-286-0077-9
3. J.L. Junkins: *An Introduction to Optimal Estimation of Dynamical Systems.* 1978 ISBN 90-286-0067-1
4. E.H. Dowell (ed.), H.C. Curtiss Jr., R.H. Scanlan and F. Sisto: *A Modern Course in Aeroelasticity.* *Revised and enlarged edition see under Volume 11*
5. L. Meirovitch: *Computational Methods in Structural Dynamics.* 1980 ISBN 90-286-0580-0
6. B. Skalmierski and A. Tylikowski: *Stochastic Processes in Dynamics.* Revised and enlarged translation. 1982 ISBN 90-247-2686-7
7. P.C. Müller and W.O. Schiehlen: *Linear Vibrations.* A Theoretical Treatment of Multi-degree-of-freedom Vibrating Systems. 1985 ISBN 90-247-2983-1
8. Gh. Buzdugan, E. Mihăilescu and M. Radeş: *Vibration Measurement.* 1986 ISBN 90-247-3111-9
9. G.M.L. Gladwell: *Inverse Problems in Vibration.* 1987 ISBN 90-247-3408-8
10. G.I. Schuëller and M. Shinozuka: *Stochastic Methods in Structural Dynamics.* 1987 ISBN 90-247-3611-0
11. E.H. Dowell (ed.), H.C. Curtiss Jr., R.H. Scanlan and F. Sisto: *A Modern Course in Aeroelasticity.* Second revised and enlarged edition (of Volume 4). 1989 ISBN Hb 0-7923-0062-9; Pb 0-7923-0185-4
12. W. Szemplińska-Stupnicka: *The Behavior of Nonlinear Vibrating Systems.* Volume I: Fundamental Concepts and Methods: Applications to Single-Degree-of-Freedom Systems. 1990 ISBN 0-7923-0368-7
13. W. Szemplińska-Stupnicka: *The Behavior of Nonlinear Vibrating Systems.* Volume II: Advanced Concepts and Applications to Multi-Degree-of-Freedom Systems. 1990 ISBN 0-7923-0369-5
    Set ISBN (Vols. 12–13) 0-7923-0370-9

MECHANICS OF STRUCTURAL SYSTEMS
Editors: J.S. Przemieniecki and G.Æ. Oravas

1. L. Frýba: *Vibration of Solids and Structures under Moving Loads.* 1970 ISBN 90-01-32420-2
2. K. Marguerre and K. Wölfel: *Mechanics of Vibration.* 1979 ISBN 90-286-0086-8

# Mechanics

3. E.B. Magrab: *Vibrations of Elastic Structural Members.* 1979 ISBN 90-286-0207-0
4. R.T. Haftka and M.P. Kamat: *Elements of Structural Optimization.* 1985
   *Revised and enlarged edition see under* Solid Mechanics and Its Applications, Volume 1
5. J.R. Vinson and R.L. Sierakowski: *The Behavior of Structures Composed of Composite Materials.* 1986 ISBN Hb 90-247-3125-9; Pb 90-247-3578-5
6. B.E. Gatewood: *Virtual Principles in Aircraft Structures.* Volume 1: Analysis. 1989
   ISBN 90-247-3754-0
7. B.E. Gatewood: *Virtual Principles in Aircraft Structures.* Volume 2: Design, Plates, Finite Elements. 1989 ISBN 90-247-3755-9
   Set (Gatewood 1 + 2) ISBN 90-247-3753-2

---

MECHANICS OF ELASTIC AND INELASTIC SOLIDS

Editors: S. Nemat-Nasser and G.Æ. Oravas

---

1. G.M.L. Gladwell: *Contact Problems in the Classical Theory of Elasticity.* 1980
   ISBN Hb 90-286-0440-5; Pb 90-286-0760-9
2. G. Wempner: *Mechanics of Solids with Applications to Thin Bodies.* 1981
   ISBN 90-286-0880-X
3. T. Mura: *Micromechanics of Defects in Solids.* 2nd revised edition, 1987
   ISBN 90-247-3343-X
4. R.G. Payton: *Elastic Wave Propagation in Transversely Isotropic Media.* 1983
   ISBN 90-247-2843-6
5. S. Nemat-Nasser, H. Abé and S. Hirakawa (eds.): *Hydraulic Fracturing and Geothermal Energy.* 1983 ISBN 90-247-2855-X
6. S. Nemat-Nasser, R.J. Asaro and G.A. Hegemier (eds.): *Theoretical Foundation for Large-scale Computations of Nonlinear Material Behavior.* 1984 ISBN 90-247-3092-9
7. N. Cristescu: *Rock Rheology.* 1988 ISBN 90-247-3660-9
8. G.I.N. Rozvany: *Structural Design via Optimality Criteria.* The Prager Approach to Structural Optimization. 1989 ISBN 90-247-3613-7

---

MECHANICS OF SURFACE STRUCTURES

Editors: W.A. Nash and G.Æ. Oravas

---

1. P. Seide: *Small Elastic Deformations of Thin Shells.* 1975 ISBN 90-286-0064-7
2. V. Panc: *Theories of Elastic Plates.* 1975 ISBN 90-286-0104-X
3. J.L. Nowinski: *Theory of Thermoelasticity with Applications.* 1978
   ISBN 90-286-0457-X
4. S. Łukasiewicz: *Local Loads in Plates and Shells.* 1979 ISBN 90-286-0047-7
5. C. Fiřt: *Statics, Formfinding and Dynamics of Air-supported Membrane Structures.* 1983 ISBN 90-247-2672-7
6. Y. Kai-yuan (ed.): *Progress in Applied Mechanics.* The Chien Wei-zang Anniversary Volume. 1987 ISBN 90-247-3249-2
7. R. Negruţiu: *Elastic Analysis of Slab Structures.* 1987 ISBN 90-247-3367-7
8. J.R. Vinson: *The Behavior of Thin Walled Structures.* Beams, Plates, and Shells. 1988
   ISBN Hb 90-247-3663-3; Pb 90-247-3664-1

---

# Mechanics

MECHANICS OF FLUIDS AND TRANSPORT PROCESSES

Editors: R.J. Moreau and G.Æ. Oravas

1. J. Happel and H. Brenner: *Low Reynolds Number Hydrodynamics*. With Special Applications to Particular Media. 1983 ISBN Hb 90-01-37115-9; Pb 90-247-2877-0
2. S. Zahorski: *Mechanics of Viscoelastic Fluids*. 1982 ISBN 90-247-2687-5
3. J.A. Sparenberg: *Elements of Hydrodynamics Propulsion*. 1984 ISBN 90-247-2871-1
4. B.K. Shivamoggi: *Theoretical Fluid Dynamics*. 1984 ISBN 90-247-2999-8
5. R. Timman, A.J. Hermans and G.C. Hsiao: *Water Waves and Ship Hydrodynamics*. An Introduction. 1985 ISBN 90-247-3218-2
6. M. Lesieur: *Turbulence in Fluids*. Stochastic and Numerical Modelling. 1987 ISBN 90-247-3470-3
7. L.A. Lliboutry: *Very Slow Flows of Solids*. Basics of Modeling in Geodynamics and Glaciology. 1987 ISBN 90-247-3482-7
8. B.K. Shivamoggi: *Introduction to Nonlinear Fluid-Plasma Waves*. 1988 ISBN 90-247-3662-5
9. V. Bojarevičs, Ya. Freibergs, E.I. Shilova and E.V. Shcherbinin: *Electrically Induced Vortical Flows*. 1989 ISBN 90-247-3712-5
10. J. Lielpeteris and R. Moreau (eds.): *Liquid Metal Magnetohydrodynamics*. 1989 ISBN 0-7923-0344-X

MECHANICS OF ELASTIC STABILITY

Editors: H. Leipholz and G.Æ. Oravas

1. H. Leipholz: *Theory of Elasticity*. 1974 ISBN 90-286-0193-7
2. L. Librescu: *Elastostatics and Kinetics of Aniosotropic and Heterogeneous Shell-type Structures*. 1975 ISBN 90-286-0035-3
3. C.L. Dym: *Stability Theory and Its Applications to Structural Mechanics*. 1974 ISBN 90-286-0094-9
4. K. Huseyin: *Nonlinear Theory of Elastic Stability*. 1975 ISBN 90-286-0344-1
5. H. Leipholz: *Direct Variational Methods and Eigenvalue Problems in Engineering*. 1977 ISBN 90-286-0106-6
6. K. Huseyin: *Vibrations and Stability of Multiple Parameter Systems*. 1978 ISBN 90-286-0136-8
7. H. Leipholz: *Stability of Elastic Systems*. 1980 ISBN 90-286-0050-7
8. V.V. Bolotin: *Random Vibrations of Elastic Systems*. 1984 ISBN 90-247-2981-5
9. D. Bushnell: *Computerized Buckling Analysis of Shells*. 1985 ISBN 90-247-3099-6
10. L.M. Kachanov: *Introduction to Continuum Damage Mechanics*. 1986 ISBN 90-247-3319-7
11. H.H.E. Leipholz and M. Abdel-Rohman: *Control of Structures*. 1986 ISBN 90-247-3321-9
12. H.E. Lindberg and A.L. Florence: *Dynamic Pulse Buckling*. Theory and Experiment. 1987 ISBN 90-247-3566-1
13. A. Gajewski and M. Zyczkowski: *Optimal Structural Design under Stability Constraints*. 1988 ISBN 90-247-3612-9

# Mechanics

MECHANICS: ANALYSIS

Editors: V.J. Mizel and G.Æ. Oravas

1. M.A. Krasnoselskii, P.P. Zabreiko, E.I. Pustylnik and P.E. Sbolevskii: *Integral Operators in Spaces of Summable Functions.* 1976 ISBN 90-286-0294-1
2. V.V. Ivanov: *The Theory of Approximate Methods and Their Application to the Numerical Solution of Singular Integral Equations.* 1976 ISBN 90-286-0036-1
3. A. Kufner, O. John and S. Pučík: *Function Spaces.* 1977 ISBN 90-286-0015-9
4. S.G. Mikhlin: *Approximation on a Rectangular Grid.* With Application to Finite Element Methods and Other Problems. 1979 ISBN 90-286-0008-6
5. D.G.B. Edelen: *Isovector Methods for Equations of Balance.* With Programs for Computer Assistance in Operator Calculations and an Exposition of Practical Topics of the Exterior Calculus. 1980 ISBN 90-286-0420-0
6. R.S. Anderssen, F.R. de Hoog and M.A. Lukas (eds.): *The Application and Numerical Solution of Integral Equations.* 1980 ISBN 90-286-0450-2
7. R.Z. Has'minskiĭ: *Stochastic Stability of Differential Equations.* 1980 ISBN 90-286-0100-7
8. A.I. Vol'pert and S.I. Hudjaev: *Analysis in Classes of Discontinuous Functions and Equations of Mathematical Physics.* 1985 ISBN 90-247-3109-7
9. A. Georgescu: *Hydrodynamic Stability Theory.* 1985 ISBN 90-247-3120-8
10. W. Noll: *Finite-dimensional Spaces.* Algebra, Geometry and Analysis. Volume I. 1987 ISBN Hb 90-247-3581-5; Pb 90-247-3582-3

MECHANICS: COMPUTATIONAL MECHANICS

Editors: M. Stern and G.Æ. Oravas

1. T.A. Cruse: *Boundary Element Analysis in Computational Fracture Mechanics.* 1988 ISBN 90-247-3614-5

MECHANICS: GENESIS AND METHOD

Editor: G.Æ. Oravas

1. P.-M.-M. Duhem: *The Evolution of Mechanics.* 1980 ISBN 90-286-0688-2

MECHANICS OF CONTINUA

Editors: W.O. Williams and G.Æ. Oravas

1. C.-C. Wang and C. Truesdell: *Introduction to Rational Elasticity.* 1973 ISBN 90-01-93710-1
2. P.J. Chen: *Selected Topics in Wave Propagation.* 1976 ISBN 90-286-0515-0
3. P. Villaggio: *Qualitative Methods in Elasticity.* 1977 ISBN 90-286-0007-8

# Mechanics

## MECHANICS OF FRACTURE

Editors: G.C. Sih

1. G.C. Sih (ed.): *Methods of Analysis and Solutions of Crack Problems*. 1973 ISBN 90-01-79860-8
2. M.K. Kassir and G.C. Sih (eds.): *Three-dimensional Crack Problems*. A New Solution of Crack Solutions in Three-dimensional Elasticity. 1975 ISBN 90-286-0414-6
3. G.C. Sih (ed.): *Plates and Shells with Cracks*. 1977 ISBN 90-286-0146-5
4. G.C. Sih (ed.): *Elastodynamic Crack Problems*. 1977 ISBN 90-286-0156-2
5. G.C. Sih (ed.): *Stress Analysis of Notch Problems*. Stress Solutions to a Variety of Notch Geometries used in Engineering Design. 1978 ISBN 90-286-0166-X
6. G.C. Sih and E.P. Chen (eds.): *Cracks in Composite Materials*. A Compilation of Stress Solutions for Composite System with Cracks. 1981 ISBN 90-247-2559-3
7. G.C. Sih (ed.): *Experimental Evaluation of Stress Concentration and Intensity Factors*. Useful Methods and Solutions to Experimentalists in Fracture Mechanics. 1981 ISBN 90-247-2558-5

## MECHANICS OF PLASTIC SOLIDS

Editors: J. Schroeder and G.Æ. Oravas

1. A. Sawczuk (ed.): *Foundations of Plasticity*. 1973 ISBN 90-01-77570-5
2. A. Sawczuk (ed.): *Problems of Plasticity*. 1974 ISBN 90-286-0233-X
3. W. Szczepiński: *Introduction to the Mechanics of Plastic Forming of Metals*. 1979 ISBN 90-286-0126-0
4. D.A. Gokhfeld and O.F. Cherniavsky: *Limit Analysis of Structures at Thermal Cycling*. 1980 ISBN 90-286-0455-3
5. N. Cristescu and I. Suliciu: *Viscoplasticity*. 1982 ISBN 90-247-2777-4

Kluwer Academic Publishers - Dordrecht / Boston / London

GPSR Compliance

*The European Union's (EU) General Product Safety Regulation (GPSR) is a set of rules that requires consumer products to be safe and our obligations to ensure this.*

*If you have any concerns about our products, you can contact us on ProductSafety@springernature.com*

In case Publisher is established outside the EU, the EU authorized representative is:

Springer Nature Customer Service Center GmbH
Europaplatz 3
69115 Heidelberg, Germany

**Batch number: 10815968**

Printed by Printforce, the Netherlands

Wilhelm Schäberle

# Ultrasonography in Vascular Diagnosis

## A Therapy-Oriented Textbook and Atlas

Second Edition

Springer

*Author*
Dr. Wilhelm Schäberle
Dept. of Visceral, Vascular, Thoracic, and Pediatric Surgery
Klinik am Eichert
Eichertstr. 3
73035 Göppingen
Germany
ach@kae.de

*Translator*
Bettina Herwig
Hauptstr. 4H
10317 Berlin
Germany
herwigb@zedat.fu-berlin.de

ISBN 978-3-642-02508-2 e-ISBN 978-3-642-02509-9
DOI 10.1007/978-3-642-02509-9
Springer Heidelberg Dordrecht London New York

Library of Congress Control Number: 2010937909



*Cover design*: eStudioCalamar, Fiqueres/Berlin

Printed on acid-free paper

Springer is part of Springer Science+Business Media (www.springer.com)

*To my wife Solange*
*and my children Jan and Philip*

# Preface to the Second English and Third German Edition

The longer and the more intensively one has been working with medical imaging, the more questions of a broader, more general kind one is confronted with: How well does the image represent the truth? Can our interpretation of the imaging findings explain the patient's disease? Which imaging appearances mean that the patient requires treatment, and if so, which treatment? When does imaging (including incidental findings) lead to unnecessary interventions – due to users not being aware of the intrinsic problems of a diagnostic method or failing to take its inherent limitations into account? These issues are relevant for all diagnostic modalities, including the traditional gold standard of angiography and more recent developments such as magnetic resonance and computed tomography angiography. Applied to diagnostic ultrasonography, the more specific question that arises is how we misinterpret echo patterns or ultrasound features and consequently make erroneous treatment decisions. These problems become particularly manifest when dealing with the morphology of internal carotid artery plaque, where the sonographic appearance of the plaque may be used as a criterion for making treatment recommendations. In the name of scientific rigor investigators sometimes end up focusing too heavily on a single aspect of a complex problem, in turn giving rise to specific assumptions and hypotheses that affect the study design and ultimately lead to wrong, contradictory, and biased results, as well as to the wrong therapeutic conclusions. Despite these cautionary remarks, however, there is good scientific evidence that vascular duplex ultrasonography – as long as both the morphologic appearance and hemodynamic findings are taken into account and as long as the examiner remains critically aware of the methodological basis – comes very close to depicting the true clinical situation in patients with vascular disease. Although somewhat neglected by some "schools of ultrasound," where color flow images (which are more angiography-like) are preferred, spectral Doppler analysis can provide some very valuable information. In particular it can depict the hemodynamic situation (in both normal and diseased vessels) with excellent sensitivity, making it highly useful in the diagnostic assessment of vascular disease and in solving problems of differential diagnosis.

In terms of method and didactic approach this second English edition continues in the tradition of the earlier German editions and of the first English edition and emphasizes the therapeutic relevance of the diagnostic measures being taken. For details on this approach please refer to the earlier prefaces. Staying in the same pedagogical vein but seeking to advance this method further, this extensively revised edition incorporates even more diagrams and tables. The hope is that this will help make examination protocols and complex diagnostic procedures even easier to visualize and understand. This new edition presents the most recent scientific insights as well as new developments in ultrasound technology, which are discussed with regard to their role in providing therapeutically relevant diagnostic information for treating patients with vascular disease. On points where there is no clear consensus regarding the diagnostic status of certain ultrasound features and findings, these controversies are discussed. The atlas sections of the individual chapters have also been expanded to include even more examples of ultrasound findings obtained in the routine clinical setting, along with examples of less common vascular diseases. A focus here is on showing the reader how to interpret Doppler waveforms and how to use the hemodynamic information to help make a diagnosis. As a little cultural aside, color flow ultrasound can also be counted on to produce images with highly artistic color compositions. My 4-year-old son's comment, upon seeing the proofs of the book, was, "Your new art book is really beautiful."

The author would like to thank Ms Zorn, Ms Rieker, Ms Mehlbeer, and Ms Lietz for secretarial assistance. My thanks are also due to Ms Mütschele for her support in preparing the diagrams and figures. Thank you also to Ms Herwig for the translation and excellent support through all stages of preparing this English edition. Further, I would like to thank the staff of Springer-Verlag for their excellent support in preparing this new edition, particularly Ms Heilmann and Mr Bachem. Most of all, however, I would like to thank my family for their patience and understanding and for the humor that is necessary and makes it easier to pull off a project like this.

Göppingen
November 2010

Wilhelm Schäberle

# Preface to the First English Edition

Vascular ultrasonography becomes increasingly valuable the more the diagnostic query to be answered is based on the clinical findings and the more the examination is performed with regard to its therapeutic consequences. As with other specialties that make use of ultrasound findings, the diagnostic yield of vascular ultrasound relies crucially on the close integration of the examination into the routine of the clinician or physician treating the patient. That is why in the German-speaking countries, vascular ultrasound is chiefly performed by angiologists and vascular surgeons. Duplex ultrasound can indeed be regarded as an integral component of the angiologic examination or an extension of the clinical examination by fairly simple technical means. Thus the sonographic findings do not simply supplement other imaging modalities but, together with the clinical findings, provide the basis for deciding whether medical therapy, a radiologic intervention,or surgical reconstruction is the most suitable therapy for an individual patient. This means that in a patient with atherosclerotic occlusive disease of the leg arteries, the patient's clinical presentation determines whether or not surgical repair is necessary, while the duplex sonographic findings serve to plan the kind of repair required and to confirm the localization and extent of the vascular pathology suspected on clinical grounds. Up to this point, no invasive diagnostic tests are needed. Angiography continues to have a role in planning the details of the surgical procedure, i.e., identification of a suitable recipient vessel for a bypass graft. Some surgical procedures such as thromboendarterectomy of the carotid arteries or femoral bifurcation can be performed without prior angiography, which does not provide any additional information that would affect the surgical strategy. Duplex sonography has evolved into the gold standard for answering most queries pertaining to venous conditions (therapeutic decision-making in thrombosis, planning of the surgical intervention for varicosis, chronic venous insufficiency).

Special emphasis is placed on the therapy-oriented presentation of indications for vascular ultrasonography, including the sonographic differentiation of rare vascular pathology and the role of the ultrasound examination in conjunction with the patient's clinical findings. The abundant images provided are intended to facilitate morphologic and hemodynamic vascular evaluation and put the reader in a position to become more confident in identifying rare conditions as well, which often have a characteristic appearance and are thus recognizable at a glance. The high acceptance of the diagnostic concept advocated here as reflected in the success of the first two editions of the book in the German-speaking countries led to the decision to have an English edition. I would like to thank Springer-Verlag, in particular Dr. Heilmann, for making this English edition possible.

Göppingen
August 2005

Wilhelm Schäberle

# Preface to the Second German Edition

Vascular duplex sonography is the continuation of the clinical examination of vascular disease by fairly simple technical means. A sonographic examination relies on interaction with the patient and is guided by the clinical findings, therapeutic relevance, and treatment options available. It is highly examiner-dependent and does not easily lend itself to full documentation of the results, which are thus difficult to communicate and verify. For these reasons, sonographers require thorough training, both to avoid inaccurate findings with disastrous consequences for patients and in order not to discredit the method.

The format of the first edition with a text section and an atlas for each vascular territory has been retained as has the subdivision of the individual chapters into sections on sonoanatomy, examination technique, normal findings, abnormal ultrasound findings, and diagnostic role of the sonographic findings. Given the special focus of this textbook on clinically and therapeutically relevant aspects of vascular ultrasonography, each of the main chapters (peripheral arteries and veins, extracranial arteries supplying the brain, hemodialysis shunts, and abdominal and retroperitoneal vessels) has been supplemented with a section on the clinical significance of ultrasound examinations in the respective vascular territory. This addition was considered necessary in order to do justice to the expanding and changing role of diagnostic ultrasound since the first German edition six years ago. While until only a few years ago vascular ultrasound was used for orientation or served as a supplementary diagnostic test only, it has since evolved into a key modality in this field. It has since even become a kind of gold standard in the diagnostic evaluation of veins, in particular in patients with thrombosis and varicosis. In this setting, venography has lost its significance and its use is now restricted to exceptional cases where it serves to obtain supplementary information to answer specific questions.

In patients with arterial disease, duplex ultrasound is an integral part of the step-by-step diagnostic workup. The sonographic findings provide the key to adequate therapeutic management (medical therapy, radiologic intervention, or vascular surgical repair). Together with the patient's clinical status, duplex sonography is thus decisive for establishing the indication for medical therapy or invasive vascular reconstruction. Duplex sonography has replaced angiography in the localization of a vascular obstruction and the evaluation of its significance. The invasive radiologic modality is used only to identify a suitable recipient segment in patients scheduled for a bypass procedure or in combination with a catheter-based intervention (PTA and stenting). The morphologic information provided on the vessel lumen and wall as well as on perivascular structures makes nonatherosclerotic vascular disorders a domain of ultrasound.

Ultrasonography is the method of first choice in evaluating carotid artery stenoses for stroke prevention by identifying those patients who would benefit from surgical repair on the basis of hemodynamic parameters but also taking into account morphologic information. Ultrasound can retain its central role in therapeutic decision-making only if its advantages are fully exploited, which means that the examination should be performed by the angiologist or vascular surgeon who is also treating the patient. This is why this second edition is again intended mainly for angiologists and vascular surgeons.

The revised edition also describes recent developments such as the use of ultrasound contrast media, or echo enhancers, in angiology and the B-flow mode although their role in the routine clinical setting is small from the angiologist's and vascular surgeon's perspective. The use of ultrasound contrast media in differentiating liver tumors is not dealt with in detail since it is mainly of interest to gastroenterologists and visceral surgeons and would therefore go beyond the scope of this textbook.

As in the first edition, great care was taken in selecting illustrative ultrasound scans of high quality for the atlas, following the motto "an ultrasound image must speak for itself ". The sonomorphologic context is important for didactic purposes; that is why the pathology of interest is not shown in a magnified view (zoom) but presented in the constellation in which it appears in the course of a routine examination. In those settings where the sonication conditions are poor but an ultrasound examination nevertheless appears to be indicated from a clinical perspective as in postoperative patients, the examples shown were not selected specifically but are such as illustrate this fact. Angiograms, and in individual cases graphic representations, are intended to clarify the situation.

The abundant images contained in the atlas sections reflect the intention not only to present abnormal finding as such but to illustrate more clearly situations that are relevant from a therapeutic perspective and to also show the development of vascular pathology. Adhering to the ultrasound convention of depicting cranial on the left side of the image and caudal on the right, the blood flow direction is color coded in accordance with the defaults settings of the ultrasound equipment. This means, for instance, that the internal carotid artery is coded in blue, indicating arterial blood flow away from the transducer. Following this convention, it is thus not necessary to first have to look for the color key, and orientation is facilitated when complex vascular territories such as the abdominal and retroperitoneal vessels are examined.

The detailed introduction to the fundamental physical principles of diagnostic ultrasound and basic hemodynamics under normal and abnormal conditions as well as the detailed description of vascular anatomy, examination protocols, and of the interpretation of the findings aim at providing the beginner with an introduction to vascular ultrasound. It is hoped that the richly illustrated atlas sections will facilitate the first steps for the beginner. For experienced sonographers, the detailed illustrations also of rare vascular pathology are expected to broaden their knowledge and help them diagnose rare disorders with greater confidence. To this end the role of ultrasound examinations is compared with that of other diagnostic modalities and tips and tricks are described that facilitate the examination and provide a basis for tackling more difficult diagnostic tasks. That is why all diseases in which ultrasonography is indicated and that are of relevance for angiologists and vascular surgeons are represented by images in the atlas. Rare vascular conditions can often be identified sonographically at a glance. Where appropriate, additional angiograms illustrate the role of the respective modality in comparison, and occasionally the situation is further clarified by an intraoperative photograph.

The constant support I received from Professor R. Eisele is gratefully acknowledged. My special thanks are due to the co-workers of Springer-Verlag for their excellent cooperation in preparing the second edition and to Ms. R. Mütschele for her assistance in preparing the graphics.

Göppingen
February 2004

Wilhelm Schäberle

# Preface to the First German Edition

Conventional and color duplex ultrasonography has evolved into an indispensable tool for the diagnostic evaluation of vascular pathology. As a noninvasive test that can be repeated any time, sonography is increasingly replacing conventional diagnostic modalities that cause more discomfort to the patient. The combination of gray-scale sonographic information for evaluating topographic relationships and morphologic features of vessels with the qualitative and quantitative data obtained with the Doppler technique enables fine diagnostic differentiation of vascular disorders. In particular, the hemodynamic Doppler information is a useful supplement to the findings obtained with radiologic modalities. Being noninvasive and easy to perform any time, duplex sonography precedes more invasive, stressful, and expensive diagnostic tests in the step-by-step diagnostic workup of patients with vascular disease. It provides crucial information for optimal therapy and will replace invasive modalities such as angiography and venography as examiners gain skills and experience and ultrasound equipment becomes more sophisticated.

The significance duplex ultrasonography has gained in the hands of angiologists and vascular surgeons is also reflected in the further education programs for these specialties. This book therefore aims at providing a detailed description of the diagnostic information that can be obtained by (color) duplex sonography in those vascular territories that are relevant to angiologists and vascular surgeons. Each of the main chapters introduces beginners to the relevant vascular anatomy and scanning technique while at the same time offering detailed discussions of the parameters involved and a thorough review of the pertinent scientific literature to help experienced sonographers become more confident in establishing their diagnoses.

The first chapter presents the basic hemodynamic concepts that are relevant to vascular sonography and the fundamental physical and technical principles of vascular ultrasonography. This introductory chapter is intended to help readers grasp the potential and limitations of the method.

The situation in Germany is different from that in many other countries in that duplex sonography is performed primarily by angiologists, internists, and increasingly by vascular surgeons rather than by radiologists. On the basis of a patient's clinical findings, it is thus possible to specifically address therapeutically relevant questions in performing the sonographic examination. Besides general assessment of the vascular status, ultrasound can thus serve to acquire additional diagnostic information important for therapeutic decision-making in general and for planning the surgical procedure in particular. Duplex sonography in the hands of the clinician who is also treating the patient is seen as the continuation of the clinical examination by technical means. That is why the emphasis in this book is on the clinical and therapeutic role of ultrasound findings, and the individual chapters are organized according to such pragmatic aspects.

Each of the six main chapters deals with a specific vascular territory and consists of a text section as well as an atlas section with ample illustrations and detailed descriptions of normal findings, variants, and abnormal findings.Whenever considered appropriate for better illustration of complex pathology, the sonographic images have been supplemented with angiograms or CT scans. The comparison also illustrates the advantages and disadvantages of the respective radiologic modalities. As many rare vascular disorders are diagnosed at a glance by an experienced sonographer, their appearance is shown in numerous figures. Series of ultrasound scans document the course of the examination and complex hemodynamic changes in vascular disorders as well as their clinical significance and changes under therapy. The legends provide detailed descriptions allowing the reader to use the atlas sections independently for reference when looking for information on specific vascular conditions.

Different ultrasound modes are described in detail and their respective merits and shortcomings are discussed for the benefit of readers using different equipment. Gray-scale sonography alone (compression ultrasound) is quite sufficient for the diagnostic assessment of thrombosis while conventional duplex ultrasonography is a valid modality for diagnosing therapeutically relevant abnormal changes of the femoropopliteal territory. In most instances, color flow images are shown together with the Doppler waveform but occasionally "only" the conventional duplex scan is presented to illustrate the fact that many abnormalities can be identified by conventional duplex ultrasound alone. Despite the additional diagnostic information obtained with the color-coded technique, quantitative evaluation relies on the Doppler frequency spectrum. The color duplex mode can facilitate the examination procedure (identification of small vessels, recanalization, differential diagnosis) but the sonographer needs some basic knowledge of the conventional Doppler technique for the proper interpretation of color flow images.

My special thanks are due to Professor R. Eisele for promoting the use of diagnostic ultrasound in the department of vascular surgery at our hospital and for his valuable advice. I thank Ms. G. Rieker and Ms. E. Stieger and Mrs. B. Sihler for typing the manuscript and Ms. R. Uhlig for the photographic work in preparing the figures.

Finally I would like to express my thanks to the publishers, Springer-Verlag, and in particular to Ms. Zeck and Dr. Heilmann, for their excellent cooperation and constructive support.

Göppingen
December 1997

Wilhelm Schäberle

# Contents

# Fundamental Principles 1

## 1.1 Technical Principles of Diagnostic Ultrasound

### 1.1.1 Gray-Scale Ultrasonography (B-Mode)

#### 1.1.1.1 Historical Milestones

The potential for using the reflection of ultrasound in the visualization of the internal organs of the human body was recognized about 70 years ago. The first attempts at using ultrasound in medical diagnosis were made in the late 1930s by the Austrian neurologist K.T. Dussik. He developed what he referred to as hyperphonography, a sonographic transmission technique for the visualization of the cerebral ventricles. In the 1940s as well, American scientists began experimenting with ultrasound reflection to examine biological objects. Among the early pioneers were Ludwig and Struthers, who used this new technique for detecting gallstones. Other important milestones in the history of diagnostic ultrasound were the development of B-mode imaging by Howry and Bliss and the introduction of the echo pulse method by Leksell in Sweden, which he used to determine the position of midline brain structures in the intact skull, thus marking the start of echoencephalography. There was also the original description of M-mode echocardiography by Edler and Herz in the early 1950s.

The Japanese physicist Satomura is credited with implementing the first medical applications of the Doppler principle. He and his colleagues investigated the use of Doppler frequency shifts to evaluate moving cardiac structures and to measure the velocity of red blood cells. The advent of the first real-time scanner, developed by Krause and Soldner, completely changed the practice of medical ultrasound scanning and marks yet another important step in the success story of diagnostic ultrasound.

Modern ultrasound offers excellent image quality and diagnostic capabilities, with its outstanding position among radiologic imaging techniques being due to its versatility, low cost, flexibility, and safety.

This chapter introduces the physical and technical fundamentals of medical ultrasound, which will help readers to make optimal use of the diagnostic capabilities of ultrasound; it also outlines the range of techniques for different applications that are available today.

#### 1.1.1.2 Sound Waves

When a particle is activated to vibrate around its equilibrium position, the vibration is transmitted to a neighboring molecule in the medium and from there to the next molecule and so on. In this way, kinetic energy is propagated from one molecule to the next, spreading through the medium in a sine wave pattern. This pattern of the spreading of kinetic energy is known as a continuous wave or an acoustic wave (sound wave). A sound wave alternately compresses (positive pressure) and expands (negative pressure) the medium it travels through (Fig. 1.1). Particles can vibrate parallel or perpendicular to the direction of energy propagation, giving rise to longitudinal waves (along the direction of travel) and transverse waves (perpendicular to the direction of travel). Particles

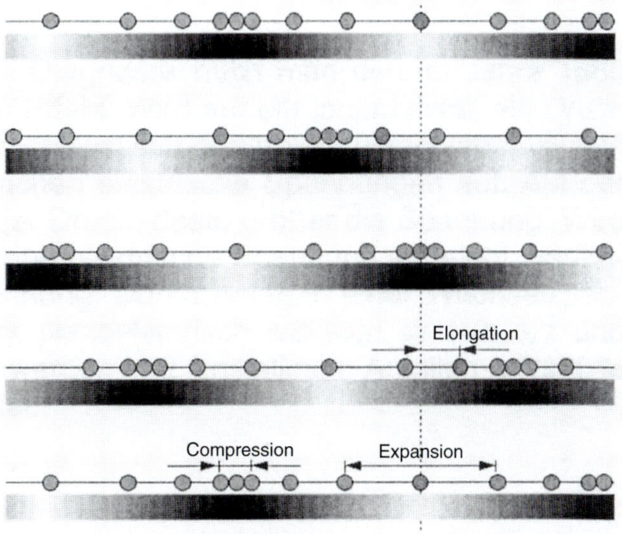


**Fig. 1.1** Diagram of the propagation of a longitudinal wave illustrating cyclic compression and expansion (courtesy of Hitachi Ltd., who also provided the historical material presented in Sect. 1.1.1.1)


W. Schäberle, *Ultrasonography in Vascular Diagnosis*,
DOI: 10.1007/978-3-642-02509-9_1, © Springer-Verlag Berlin Heidelberg 2011

excited in the ultrasound range vibrate around their resting positions at a rate of 20,000 to 1 billion times per second.

In gases and liquids, only longitudinal wave propagation is possible, as the shear forces necessary for the spread of transverse vibration are absent. In physical terms, biological tissues can be viewed as viscous fluids, which is why the effect of transverse waves is negligible. In such a medium the speed of sound increases with density, which in turn is defined by the force of molecular cohesion (Table 1.1). The average speed of sound in biological tissue is about 1,540 m/s.

Waves can be described with reference to several properties. Wavelength $\lambda$ is the distance between two consecutive points of maximum compression, and frequency $f$ is the number of vibrations of a molecule per unit time, given in hertz (Hz). One hertz corresponds to one cycle per second, or 1 Hz = 1/s. The frequency range of diagnostic ultrasound is 2–30 MHz. The speed of a sound wave, $C$, is the product of wavelength and frequency:

$$C = \lambda \cdot f$$

The wavelengths occurring in diagnostic ultrasound are determined by the frequency emitted by the transducer (carrier frequency) and range from 0.8 to 0.1 mm over the frequency range of 2–15 MHz that is typically used in medical imaging (Table 1.2). The properties defining a sound wave are summarized in Table 1.3.

**Table 1.1** Typical sound velocities, densities, and attenuation values in some important biological tissues and other media in the body

| Medium | Sound velocity (m/s) | Density (g/cm$^3$) | Attenuation (dB/MHz/cm) |
|---|---|---|---|
| Fat | 1,470 | 0.97 | 0.5 |
| Bone marrow | 1,700 | 0.97 | |
| Muscle | 1,568 | 1.04 | 2 |
| Liver | 1,540 | 1.055 | 0.7 |
| Brain | 1,530 | 1.02 | 1 |
| Bone (compact) | 3,600 | 1.7 | 4–10 |
| Water (20°C) | 1,492 | 0.9982 | 0.002 |
| Air | 331 | 0.0013 | |

**Table 1.2** Properties of diagnostic ultrasound over a typical frequency range

| Transmit frequency (MHz) | Wavelength (mm) | Penetration depth (cm) | Lateral resolution (mm) | Axial resolution (mm) |
|---|---|---|---|---|
| 2 | 0.78 | 25 | 3 | 0.8 |
| 3.5 | 0.44 | 14 | 1.7 | 0.5 |
| 5 | 0.31 | 10 | 1.2 | 0.35 |
| 7.5 | 0.21 | 6.7 | 0.8 | 0.25 |
| 10 | 0.16 | 5 | 0.6 | 0.2 |
| 15 | 0.1 | 3.3 | 0.4 | 0.15 |

**Table 1.3** Parameters defining a sound wave

| | |
|---|---|
| Period | Duration of a complete vibration |
| Wavelength | Spatial extension of a period |
| Frequency | Number of periods per second |
| Amplitude | Measure of sound energy |

#### 1.1.1.3 Generating Ultrasound Waves

In most ultrasonic transducers for medical imaging, the piezoelectric effect discovered by Pierre and Jacques Curie in 1880 is used to generate ultrasound waves. When mechanical stress is applied to piezoelectric materials such as ionic crystals, they experience an elastic deformation which results in a shift in internal charge distribution. In this way, electric voltages are generated at the surfaces – which are negative on one side and positive on the other. As the degree of stress increases so does the voltage. Conversely, when a positive or negative voltage is applied to the surface of a piezoelectric crystal, the material expands or contracts, depending on the direction of the current. When an alternating current is applied, the piezoelectric crystal is activated and begins to vibrate. Materials possessing strong piezoelectric properties are quartz and tourmaline. State-of-the-art transducers use semicrystalline polymers such as polyvinylidene fluoride (PVDF).

#### 1.1.1.4 Physical Factors Affecting the Ultrasound Scan

An ultrasound image is created by processing the echoes returning to the transducer from various depths of the body upon emission of an ultrasound pulse of a specific frequency. A two-dimensional (2D) image is generated from adjacent ultrasound lines. Two-dimensional morphologic images are acquired by applying short pulses of energy using only a small number of wavelengths to optimize spatial resolution. The *round trip time* is *the time delay between the emission of an ultrasound pulse and the return of the reflected echo* and is a function of the *distance* between the transducer and reflector. Reflection occurs at the boundaries between media that differ in their sound propagation properties. Each tissue structure in the human body has a specific sound resistance, or *acoustic impedance*. The impedance is equal to the *speed of sound propagation* multiplied by the *density* of the medium. Hence, an *ultrasound scan does not visualize tissue structures directly but rather interfaces between tissues of different acoustic impedance. The greater the difference in impedance, the greater the reflection of the ultrasound wave* and *the smaller its transmission into deeper tissue*. Other physical processes besides reflection and scattering that affect the ultrasound scan are refraction, interference, diffraction, attenuation, and absorption.

#### 1.1.1.4.1 Reflection and Refraction

The propagation of sound waves in biological tissues is governed by the laws of wave optics. Tissues vary in density and hence differ in acoustic impedance. Impedance $Z$ is the product of the density of a medium and the speed of sound in it. At an acoustic interface in the body, an incident ultrasound beam is partially reflected and partially refracted. Refraction means that the wave passes through the interface, changing its direction of travel (Fig. 1.2).

The reflection gradient, $R$, is given by the following equation for incident angles perpendicular to an interface:

$$R = \left(\frac{Z1 - Z2}{Z1 + Z2}\right)^2$$

For an ultrasound beam striking the interface between liver tissue ($Z1 = 1.66 \times 10^5$) and renal tissue ($Z2 = 1.63 \times 10^5$), the equation yields a reflection gradient of 0.000008, meaning that this boundary reflects less than one hundred thousandth of the incident energy. In contrast, nearly all of the incident energy (over 99%) is reflected from the interface between fatty tissue and air ($Z1 = 1.42 \times 10^5$, $Z2 = 43$, $R = 0.9987$), leaving virtually no ultrasound energy to travel deeper into the tissue. This is why the lungs or bowel loops containing air cannot be examined by ultrasonography, and is also the reason that contact gel is used to eliminate air intervening between the ultrasound probe and the skin surface.

The echo pulses reflected back from an interface between media of different acoustic impedance can be processed for image generation only if the interface is relatively perpendicular to the ultrasound beam (angles of incident and reflected beam). For this reason, structures such as vessel walls perpendicular to the beam appear fairly bright compared to vessel walls tangential to the beam since most echo pulses are reflected back to the transducer by the former. Reflection occurs at the surfaces of particles that are larger than the wavelength, while scattering predominates when they are smaller.

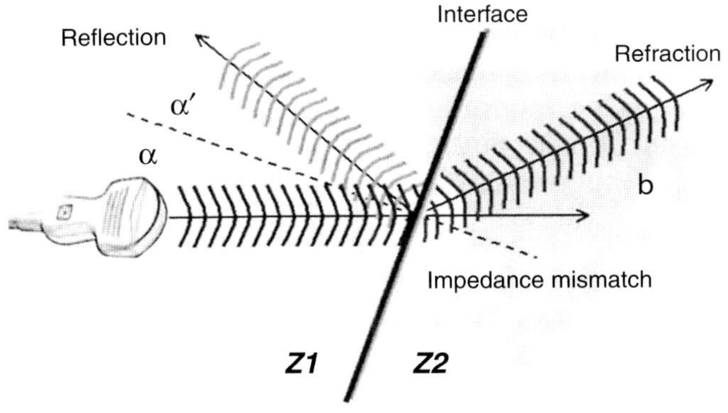


**Fig. 1.2** Interaction of ultrasound with interfaces in the body according to the laws of wave optics (for details see text) (courtesy of Hitachi Ltd.)

#### 1.1.1.4.2 Scattering

The interface between tissues of different acoustic impedance is typically not smooth but rough. A sound wave interacting with a rough surface will be scattered in all directions in the form of a spherical wave rather than along one path (Fig. 1.3). An incident ultrasound wave is also mostly scattered when it strikes an object that is much smaller than the wavelength, and it is reflected when it strikes an object much larger than its wavelength. Scattering gives rise to the characteristic echotexture of parenchymal organs in ultrasound images.

Since structures perpendicular to the beam are rare in clinical ultrasound examinations, an ultrasound image is chiefly generated from a mixture of reflected and scattered echoes. Aggregations of tissue cells scatter the beam diffusely in all directions but with a low energy only. Therefore, a structure appears bright and is clearly defined when it is perpendicular to the ultrasound beam because the image information is mainly derived from reflected echoes; its visualization is weaker and less bright when the ultrasound beam strikes tangentially and only diffusely reflected echoes are available to generate the image, although the impedance is identical in both cases.

The intensity of the returning echoes is dependent on the emitted frequency. Attenuation increases with frequency and limits the penetration depth of the ultrasound pulse. The emitted intensity decreases exponentially with distance and is influenced by an *attenuation coefficient* that varies with

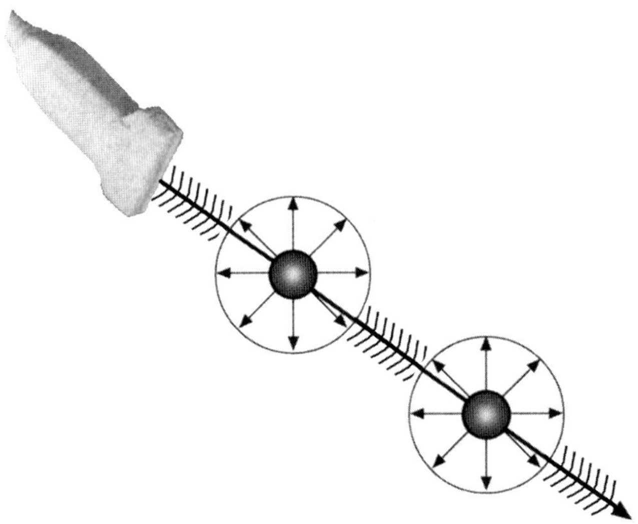

**Fig. 1.3** Scattering of ultrasound: most ultrasound beams do not strike reflecting structures in the body at a right angle, which is why the incident beam is scattered in all directions. As a result, only a small proportion of the emitted energy is backscattered to the transducer and available for generating the ultrasound image. An ultrasound beam reflected from an interface between tissues of different acoustic impedance will yield much stronger echoes than a beam scattered at the same boundary (resulting in poorer visualization) (modified from Widder 2004)

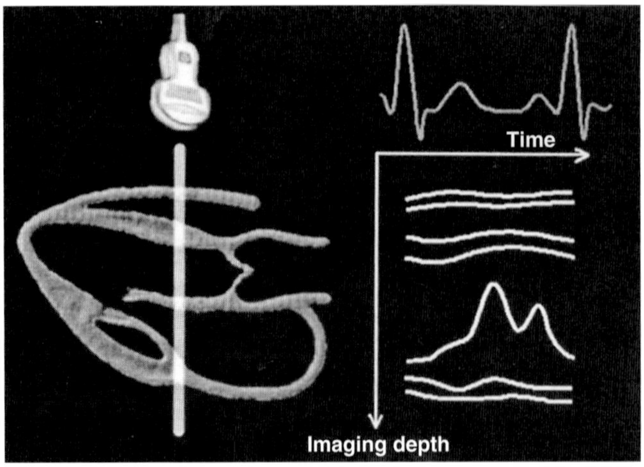


**Fig. 1.6** In M-mode scanning, the temporal changes in returning echoes are displayed, thereby recording the motion of reflecting interfaces toward and away from the transducer over time (courtesy of Hitachi Ltd.)

precise measurement of the cardiac chambers and walls and quantitative evaluation of cardiac motion.

#### 1.1.1.6 Resolution

*Image resolution*, which is given in millimeters, is defined as the smallest distance between two structures that is necessary to represent them as separate entities on a monitor. When applied to ultrasound scans, *resolution* describes the spatial discrimination between two structures differing in acoustic impedance. A distinction is made between *axial resolution* (resolution in the direction of sound propagation) and *lateral resolution*.

*Axial resolution* depends on the length of the excitation pulse and is typically one or a few wavelengths. It increases with the frequency of the emitted pulse due to the consequent shortening of wavelength. Attenuation, however, also increases with frequency, limiting the depth from which echoes can be reflected. The user must therefore strike a balance between spatial resolution and imaging depth. Axial resolution depends on wavelength alone and improves as the wavelength decreases (or the frequency increases). It ranges from 0.2 to 1 mm, depending on the transducer frequency.

*Lateral resolution* is the ability to separate two closely spaced echoes that lie in a plane perpendicular to the direction of the sound wave. It is also influenced by the transmit frequency but is mainly determined by the focusing properties of the ultrasound system and the resulting beam properties.

Lateral resolution is proportional to the beam width (Fig. 1.7). The beam profile changes along the beam path, consisting of a well-focused, narrow near field and a divergent far field. As lateral resolution is proportional to the beam diameter, it is best in the near-field area. *The ultrasound beam can be focused to improve image quality*. The slow speed of sound in human tissue (1,540 m/s) and the aim of achieving a high frame rate (real-time imaging) limit the number of scan lines per image. In order to relate the echoes to a specific depth, it is necessary to wait for the arrival of the returning echo from the respective depth of the preceding pulse before emission of the next ultrasound pulse. The transmitted or received pulse is focused in a longitudinal direction relative to the transducer, and focusing of the returning pulse in the scan plane is optimized in smaller steps (dynamically, almost continuously with the arrival time of the pulse).

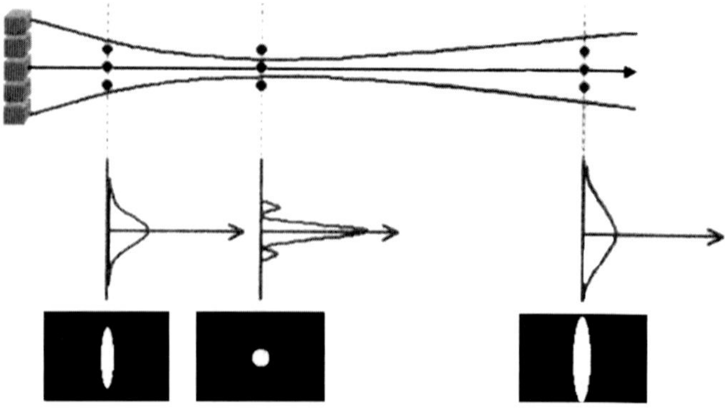

**Fig. 1.7** Effect of beam focusing on lateral resolution (courtesy of Hitachi Ltd.)

The possible resolution is proportional to the wavelength. It is ½ $\lambda$ (wavelength) for axial resolution and much poorer for lateral resolution with a value of 4 $\lambda$. Depending on the wavelength, axial and lateral resolution increase with the ultrasound frequency. However, greater penetration depths are achieved only by using transducers with lower transmit frequencies (attenuation). When deeper vessels are scanned, a compromise must be found at the expense of an adequate discrimination of the vessel structures of interest (poorer spatial resolution resulting from a lower transmit frequency).

The *scanning depth* encoded in the B-mode image is calculated from the round trip time. As attenuation increases with the penetration depth (round trip time), echo signals arriving from deeper within the body are progressively more strongly amplified in order to visualize them with the same intensity in the resulting image (see "Time-Gain Compensation" in the preceding section). Overall gain and depth gain are adjusted according to the distance of the vessel of interest from the body surface. The gain is crucial for the amplitude or intensity of the signal, and along with output energy and signal-to-noise limit, it must be set properly when assessing vascular structures.

#### 1.1.1.7 Beam Focusing

There are several techniques for focusing an ultrasound beam. The simplest option is to use an acoustic lens, which has the same effect as a glass lens for visible light. A concave acoustic lens placed in front of the transducer provides weak focusing at a fixed depth. The site of maximum focusing is referred to as the focal point or focal zone. Alternatively, the crystal in the transducer can be made concave, providing internal focusing. This technique is used in single-element mechanical sector scanners.

More flexible beam forming, with a variable depth of the focal point, is accomplished with electronic beam focusing. Modern ultrasound machines use array transducers, which consist of multiple crystal elements placed side by side. Depending on the scanner type, the number of individual elements ranges from 60 to 256. Variable numbers of elements can be activated simultaneously to form an ultrasound beam. If the elements forming the beam are excited at slightly different times, a concave wavefront is generated, causing the beam to converge at the focal point. The site of the focal point can be manipulated by varying the number of active elements and the pattern of excitation of individual elements. The user can thus adjust the beam to achieve maximum resolution at the anatomic site of interest. Modern scanners use multiple zone focusing, which reduces the frame rate, as several consecutive beams with different focal points are transmitted to generate a scan line. A technique known as dynamic focusing allows the focus of the beam to be altered during reception by imposing variable delays on signals from different depths. With this technique, the reception focus can be optimized without compromising the frame rate. Groups of 8–128 elements are used for focusing the beam (Fig. 1.8).

Lateral resolution is limited by the proximity of the transducer elements activated to emit an ultrasound pulse. Resolution along the longitudinal axis can be improved by exciting only a limited number of elements at a time and not the whole array. A more focused beam is achieved by later excitation of the transducer elements in the center. Dynamic focusing is accomplished by applying small time delays to the excitation pulses driving the individual transducer elements. Resolution in the third direction, or slice thickness, depends on the position in the image.

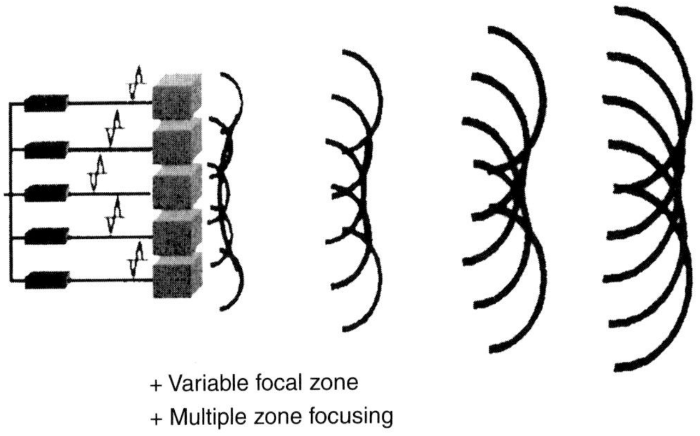


**Fig. 1.8** Beam focusing in modern array probes. By delaying the firing of the central element after the firing of the outer elements a curved wavefront is produced, resulting in a focused beam (courtesy of Hitachi Ltd.)

### 1.1.1.8 Types of Transducers

#### 1.1.1.8.1 Principle of Operation

Most electronic ultrasound transducers used today contain a number of individual piezoelectric elements for transmitting and receiving ultrasonic waves. To create a complete image, the ultrasound beam has to pass through adjacent areas of tissue. Parallel ultrasound beams are generated by varying the groups of elements within the array that are simultaneously active. A group of elements is excited to generate the first scan line. The next adjacent scan line is formed by shifting the group of active elements along the transducer array, one element position from the first group – for example, elements 1–5 produce the first beam, 2–6 the second, 3–7 the third, and so on (Fig. 1.9). The second ultrasound beam generated in this way is said to be shifted by the width of one element. The number of scan lines used to generate an image can be increased by varying the number of elements activated simultaneously to generate each beam. For instance, if the second beam is generated using the same group of elements as for the first beam plus one additional element on the left side (and no element on the right side is switched off), then the axis of the second beam is shifted by half an element width relative to the first beam. The third beam is generated by removing one element on the right side without adding an element to the left side. In this way, the number of lines scanned to produce an image is doubled. A higher line density is desirable for improving image quality; however, it also reduces frame rate.

#### 1.1.1.8.2 Linear Arrays

In a linear array transducer, the individual crystal elements are arranged in a straight row (Fig. 1.10) and can be pulsed to generate adjacent parallel ultrasonic beams, producing a rectangular image with nearly constant resolution over the entire scan depth. A linear array is made up of 60–196 elements, with an element width of 1–4 $\lambda$, and operates at a frequency of 5–13 MHz. An acoustic lens can be used for focusing perpendicular to the direction of beam propagation.

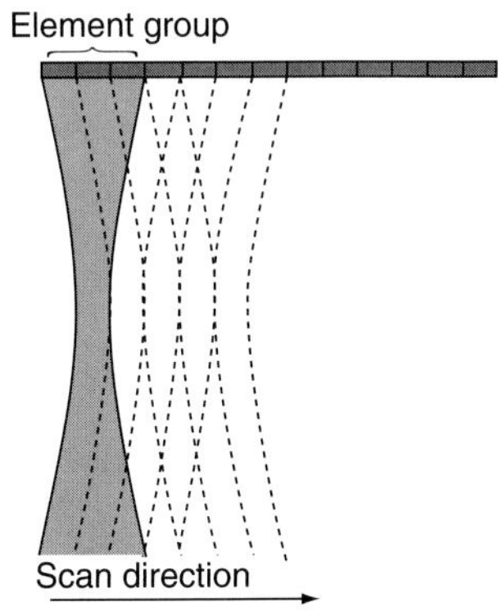


**Fig. 1.9** Emission and reception of a series of parallel ultrasound beams by successive excitation of groups of transducer elements for generation of an ultrasound image (courtesy of Hitachi Ltd.)

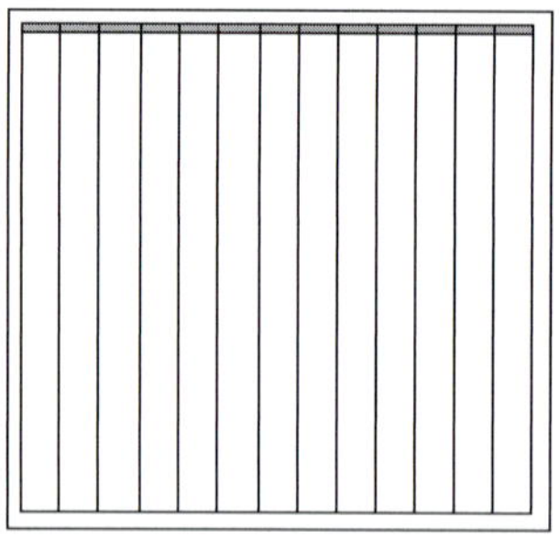

**Fig. 1.10** Diagram of a linear array with the crystal elements arranged in a straight row (courtesy of Hitachi Ltd.)

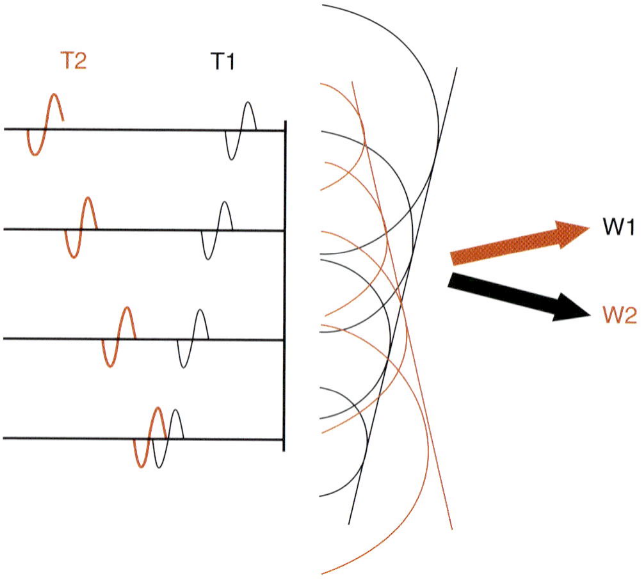


**Fig. 1.12** Generation of a pie-shaped image by the successive excitation of groups of elements in a phased-array probe (courtesy of Hitachi Ltd.)

#### 1.1.1.8.3 Curved or Convex Arrays

A curved or convex array transducer is a linear array, with the individual elements arranged along a curved line, to produce a sector image (Fig. 1.11). As the lines fan out with increasing distance from the transducer, lateral resolution decreases with depth. A typical curvilinear array consists of at least 96 elements and has a radius of 25–80 mm and a frequency range of 3–7 MHz. Most curvilinear scanners produce sector images ranging in size from 60° to 90°.

#### 1.1.1.8.4 Sector Scanners

Sector scanners have a smaller radius (<25 mm) than curved arrays and also have a small footprint, resulting in a narrow near field. With a beam-steering angle >90°, these probes are especially useful where access is difficult, such as in the imaging of the heart through the intercostal spaces (echocardiography), or for endoluminal applications such as transvaginal ultrasound.

#### 1.1.1.8.5 Phased Arrays

In a phased-array transducer, the elements are also arranged in a linear array. The difference is that all elements are excited to generate a scan line. However, time delays are introduced between pulsing consecutive elements to produce a wavefront that is no longer perpendicular to the transducer face (Fig. 1.12). By choosing appropriate delays between the excitation of individual elements, it is possible to direct the beam at a desired angle. Using this method, the beam can be steered through a range of angles to produce a sector image. Phased-array transducers use a smaller array of elements (64–128), resulting in a small footprint of 12–20 mm. The beam covers a sector of 80–90° with frequency ranging from 2 to 7 MHz. Since they require complex electronic circuitry, phased-array devices are expensive and are used mainly for cardiac and transcranial imaging.

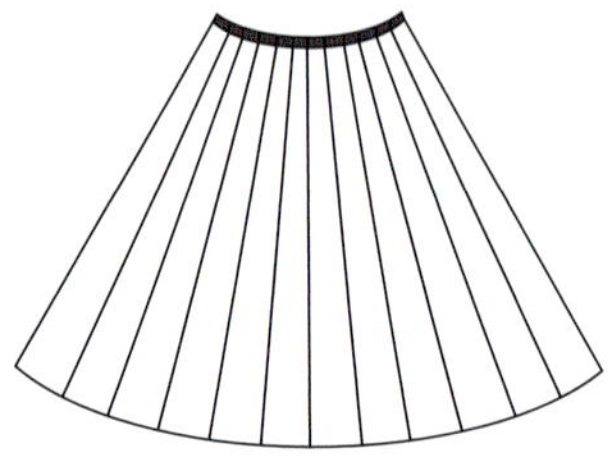

**Fig. 1.11** Diagram of a curved array with the crystal elements arranged along a curved line (courtesy of Hitachi Ltd.)

#### 1.1.1.8.6 Mechanical Sector Scanners

Compared with phased arrays, mechanical systems are fairly simple regarding the control of transducer elements and signal processing. There are basically two designs of mechanical devices: the rotating wheel transducer and the wobbler transducer.

- *Rotating wheel transducer*. This type usually comprises three to five transducer elements mounted 120–72° apart on a wheel. A motor housed in the handle turns the wheel at a constant rate in one direction. One of the crystal elements at a time is activated as it rotates past an acoustically transparent window. The active element scans a sector-shaped region. Then the next crystal rotates past the window, generating a second image.
- *Wobbler transducer*. In this type of mechanical sector scanner, a single crystal oscillates about a pivotal point, producing a beam that covers a sector of 60–100°. Since the wobbler transducer consists of a single crystal element, no complex adjustment is required. Another advantage it has over the rotating wheel transducer is that the sector angle is variable. Both mechanical devices are limited, however, by the fact that only a single element is used to produce the ultrasound beam and thus only fixed beam focusing is possible.

#### 1.1.1.8.7 Annular Phased Arrays

An annular phased array is an oscillating transducer combining features of mechanical and electronic devices. Instead of a single element, the transducer consists of several concentric rings (annuli). Each ring can be excited separately, allowing variable focusing in two dimensions.

#### 1.1.1.8.8 Disadvantages of Mechanical Transducers

Regardless of their design, mechanical probes are subject to wear and require maintenance. Moreover, they are relatively slow, not allowing rapid switching between different scan modes (B-mode, M-mode, Doppler). Real-time display of B-mode/M-mode or B-mode/Doppler information is generally not possible.

### 1.1.1.9 Ultrasound Artifacts

Artifacts play a much greater role in diagnostic ultrasound compared with other imaging modalities such as computed tomography (CT) or magnetic resonance imaging (MRI). One fundamental issue is that several simplifying assumptions are made, namely that parameters such as the speed of sound in tissues, the propagation of ultrasound, and the attenuation are constant. Another important source of artifacts in the ultrasound image is the use of inadequate instrument settings. At the same time, however, some of the major artifacts described below can be exploited to advantage because they may provide additional diagnostic information on the target tissue.

#### 1.1.1.9.1 Posterior Shadowing

Acoustic shadowing is the occurrence of hypoechoic areas behind certain objects due to loss of energy; it is one of the most commonly encountered ultrasound artifacts. These artifacts can occur deep to a strong reflector such as air, which is difficult to penetrate by the ultrasound beam because of a strong acoustic mismatch, or behind highly attenuating structures such as bone or calculi, which absorb much of the ultrasound energy (Fig. 1.13).

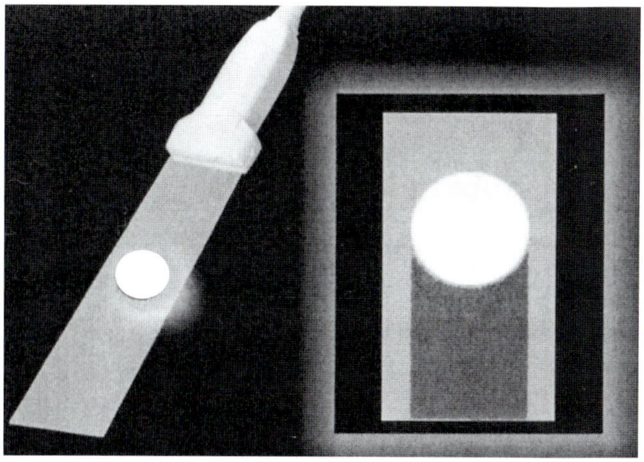

**Fig. 1.13** Posterior shadowing occurs behind interfaces of high acoustic impedance or highly attenuating objects (courtesy of Hitachi Ltd.)

#### 1.1.1.9.2 Acoustic Enhancement

Acoustic enhancement is an increase in brightness behind a low-attenuating area, in particular fluid-filled spaces such as cysts. An ultrasound beam passing through fluid is nearly unchanged because fluid reflects and attenuates only little of the ultrasound energy. With optimal time-gain compensation setting, echoes returning from behind a low-attenuation region are amplified more than necessary. Acoustic enhancement can be useful in distinguishing a fluid-filled lesion such as a cyst from a solid mass (Fig. 1.14).

#### 1.1.1.9.3 Edge Effect

The edge effect is a form of acoustic shadowing that is observed at the margins of curved, fluid-filled spaces such as cysts and is assumed to be caused by a combination of refraction and reflection. When a parallel ultrasound beam passes through the lateral border of such a space, sound is diverted into the surrounding tissue. This phenomenon also

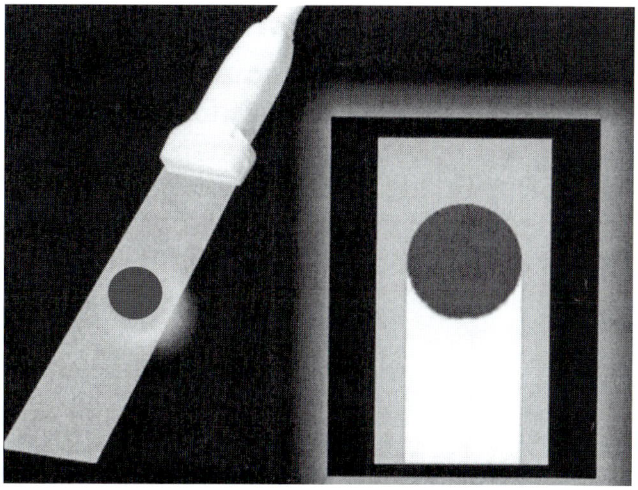

**Fig. 1.14** Acoustic enhancement occurs behind low-attenuating areas (courtesy of Hitachi Ltd.)

**Fig. 1.15** Edge effects are caused by a combination of refraction and reflection when a parallel ultrasound beam passes through the lateral border of a curved, fluid-filled space. *Right section:* Transverse image of an artery showing the effect of an ultrasound beam tangentially hitting the arterial wall. The beam is refracted, giving rise to an acoustic shadow posteriorly, where no ultrasound energy is present that can be reflected (courtesy of Hitachi Ltd.)

explains the incomplete display of the margins of certain structures such as the fetal head or a blood vessel depicted in cross-section (Fig. 1.15).

#### 1.1.1.9.4 Side Lobes

A transducer transmits not only the main beam (also called the main lobe) but also some weaker beams, or side lobes, on either side of the primary beam in the near field. When a side lobe strikes a strong reflector, the obliquely deflected echoes are misrepresented in the resulting image because they are processed as if they had originated from the main beam (Fig. 1.16). Modern ultrasound systems use various techniques, such as delay time calculation or suppression of echoes not returning along a path perpendicular to the transducer face, to minimize side lobe effects.

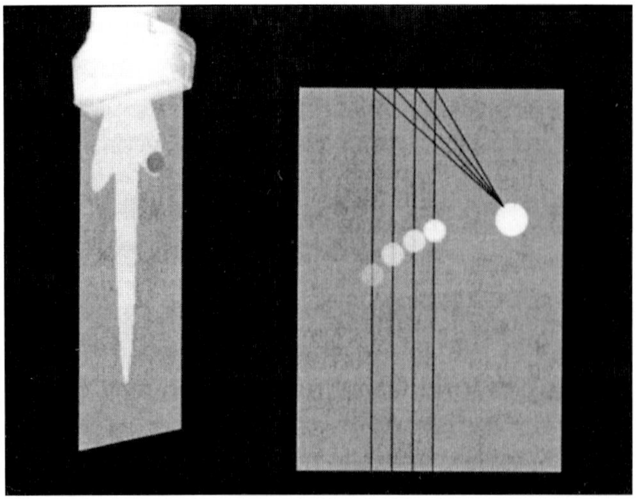

**Fig. 1.16** Diagram of side lobe artifacts (courtesy of Hitachi Ltd.)

#### 1.1.1.9.5 Reverberation Artifact

This type of artifact is also known as multiple reflection artifact and occurs when ultrasound is reflected back to the transducer from a strongly reflective surface in the near field. Part of the returning echo is properly processed by the transducer, while another part is reflected back into the body. Sound can thus bounce back and forth between the reflector and the transducer face. The resulting reverberation artifact is seen in the display as several equidistant echoes decreasing in brightness with depth. This artifact typically arises when there is a large acoustic impedance mismatch near the transducer (soft tissue/air interface) (Fig. 1.17).

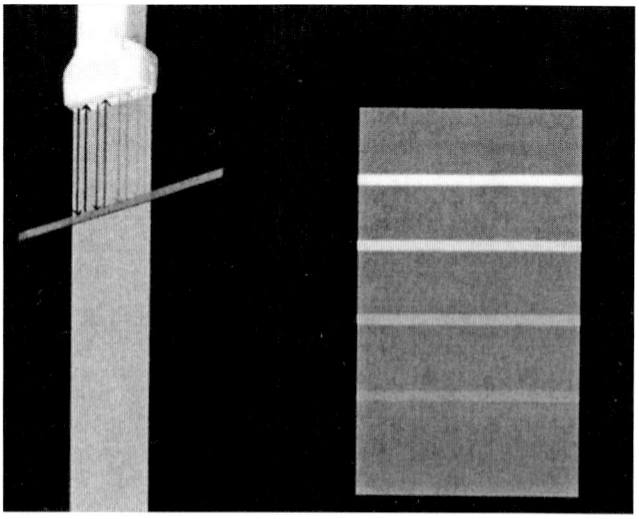

**Fig. 1.17** Diagram of reverberation artifact. This is the repeat reflection of an ultrasound beam hitting a strong reflector near the transducer. In this situation, sound will bounce back and forth between the reflector and the transducer. Echoes from multiple reflections return to the transducer later than the direct echoes and appear in the image as interfaces deeper in the body (courtesy of Hitachi Ltd.)

1.1.1.9.6 Geometric Distortion

In processing returning echoes and creating an image, the ultrasound system relies on certain assumptions, for example, that ultrasound travels in a straight line or at a constant speed in the body. In fact, however, an ultrasound beam can be deflected from its straight path, and the speed of sound varies slightly with the tissue. As a result, the ultrasound image may not reflect the exact anatomic location of a feature.

## 1.1.2 Basic Physics of Doppler Ultrasound

In 1842, the Austrian physicist and mathematician Christian Johann Doppler described what is now called the Doppler effect or Doppler shift. This phenomenon refers to the change in frequency of a wave resulting from relative movement between the source of the wave and an observer. A familiar example is an ambulance siren: although the emitted frequency remains the same, the siren has a higher pitch when the ambulance is approaching and a lower pitch when the vehicle is receding. The pitch changes abruptly at the moment the ambulance passes the observer. Thus, the pitch of the siren perceived by the human ear depends on the direction of motion relative to the observer and remains consistently high while the vehicle is approaching and consistently low while it is receding. This is different from the intensity of the sound, or the loudness of the siren, which increases gradually as the vehicle approaches and then decreases gradually after the vehicle has passed the observer. The Doppler effect occurs when the source or the observer is moving toward or away from the other or when both are moving relative to each other. For a vehicle traveling at a speed of 100 km/h, the difference in pitch due to the Doppler effect is almost two whole tones.

In diagnostic ultrasound, the Doppler effect is used to calculate blood flow velocity from the differences in frequency between the emitted and reflected waves; this was first reported by Satomura in 1959. The signals reflected by moving red blood cells have a different frequency than the emitted beam. In this case, the transducer transmitting and receiving the signals is stationary and the frequency shift is caused by the motion of the reflector (red blood cells). Also in this case, the Doppler shift occurs twice – when the ultrasound beam emitting from the stationary transducer strikes the red blood cells and when the blood cells backscatter the signal, now acting as a moving source with the transducer becoming a stationary receiver. The Doppler shift frequency depends on the frequency of the transmitted ultrasound waves, the velocity of the moving red blood cells, and the angle at which the Doppler beam intersects the vessel. This angle is known as the Doppler angle.

The Doppler effect can be used to determine blood flow velocity because the Doppler shift frequency depends on the

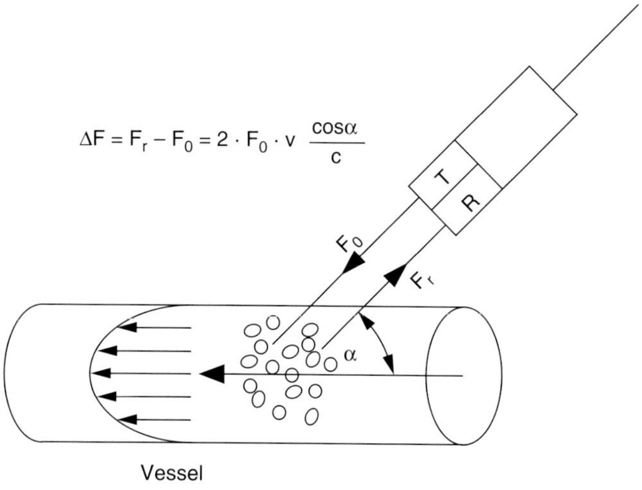


**Fig. 1.18** Diagram of Doppler interrogation of a vessel with laminar blood flow. The *arrows* in the vessel are vectors representing different flow velocities. Blood flow is fastest in the center and decreases toward the wall. The drawing illustrates the effect of the angle of incidence on the Doppler measurement. In the equation for calculating the Doppler shift, this angle is represented by the cosine function. The Doppler shift increases with the acuity of the angle (cosine of 90° = 0) (*T* transmitter, *R* receiver, $F_0$ emitted frequency, $F_r$ reflected frequency)

direction of blood flow and is proportional to the speed of the moving red blood cells. The shift is detected by the Doppler probe. The direction of blood flow relative to the transducer determines whether the returning echoes have a higher or lower frequency, and the flow velocity determines the magnitude of the frequency shift (Fig. 1.18). This relationship is expressed in the Doppler equation:

$$F_d = F_r - F_0 = \frac{2F_0 \cdot v \cdot \cos\alpha}{c}$$

$F_d$ Doppler frequency shift
$F_0$ emitted frequency
$F_r$ reflected frequency
$v$ mean flow velocity of the reflecting red blood cells
$c$ speed of sound in soft tissue (about 1,540 m/s)
$\alpha$ angle between ultrasound beam and direction of blood flow

In the transcutaneous measurement of blood flow by Doppler ultrasound, *angle correction* is necessary to calculate the flow velocity since the axis of the ultrasound beam is not in line with the longitudinal axis of the vessel or the direction of flow. The transformation with representation of the different velocity vectors is expressed mathematically as a cosine function of the angle between the sound beam and the blood vessel (cosα).

$F_d$ is proportional to the velocity of blood flow, cosα, and the carrier frequency of the ultrasound beam.

For angles of about 90°, the cosine function yields values around 0, at which there is no Doppler frequency shift, and the Doppler shift increases as the angle decreases (with a maximum cosα of 1 at $\alpha = 0°$).

The blood flow velocity is calculated by solving the Doppler shift equation for V:

$$V = (F_r - F_0) \cdot \frac{c}{\cos\alpha \cdot 2F_0}$$

This formula allows one to calculate the *blood flow velocity* from the measured *Doppler frequency shift* at a given *transmit frequency* and *angle of incidence*. The accuracy of the velocity estimation increases with the acuity of the angle. Ideally, a Doppler angle of less than 60° should be used to avoid unacceptably large errors in the calculation of flow velocity. At angles above 60°, even minor errors in determining the Doppler angle (which are unavoidable in the clinical setting, especially when curved vessels are interrogated) unduly distort the velocity calculation. At angles around 90°, a Doppler shift is no longer detectable and the flow direction cannot be determined. This is reflected in the color duplex scan by the absence of color-coded flow signals although flow is present.

Table 1.4 lists the Doppler shift frequencies for different angles of incidence, illustrating how the percentage error in estimating blood flow velocity increases with the Doppler angle. The values were calculated for a transmitted frequency of 6 MHz and a blood flow velocity of 1 ms/1.

It is apparent from the examples listed in Table 1.4 that no Doppler shift occurs at a 90° angle of incidence. The reason is that when the ultrasound beam is perpendicular to the direction of blood flow, there is no relative movement between the Doppler probe and the red blood cell. Optimal results can be obtained when the Doppler beam is aligned parallel to the blood flow. If this is not possible, accurate velocity estimates can only be made if the Doppler angle is measured using angle correction. The Doppler angle is measured by placing the angle correction cursor parallel to the direction of flow in the B-mode image. For precise calculation, a correction factor of 1/cosα is used. Table 1.5 lists the correction factors for different Doppler angles and the overestimation or underestimation of blood flow velocities resulting from errors in the alignment of the angle correction cursor.

The data in Table 1.5 illustrate how the error in calculating blood flow velocities increases with the Doppler angle. The examiner must therefore try to minimize the insonation angle for Doppler interrogation.

**Table 1.4** Dependence of the Doppler shift frequency on the Doppler angle

| Angle $\alpha$ | 0° | 30° | 45° | 60° | 90° |
|---|---|---|---|---|---|
| Cosα | 1 | 0.866 | 0.707 | 0.5 | 0 |
| $\Delta f_{(MHz)}$ | 7.79 | 6.75 | 5.51 | 3.90 | 0 |
| Percentage error | 0 | 13 | 29 | 50 | 100 |

**Table 1.5** Errors in blood flow velocity estimates in relation to the Doppler angle

| Angle α (deg) | Correction factor 1/cosα | Error in calculated blood flow velocity (%) |
|---|---|---|
| 30 | 1.15 | ±3 |
| 45 | 1.41 | ±6 |
| 60 | 2.00 | ±9 |
| 70 | 2.92 | ±14 |
| 75 | 3.86 | ±21 |
| 80 | 5.76 | ±30 |

Doppler shift frequencies are extracted by the demodulator of the ultrasound system based on a comparison between the backscattered Doppler-shifted signal and the transmitted frequency. The Doppler shift frequencies occurring in medical imaging are in the audible range and can be output to a loudspeaker. Information about the direction of flow relative to the transducer can also be extracted from the Doppler signal; this, however, requires more sophisticated demodulation techniques. Blood flow toward the transducer produces a positive frequency shift, and blood flow away from the transducer a negative shift.

Blood flow velocity varies across a vessel lumen. Blood cells move faster in the center and slower near the wall due to friction, giving rise to a *laminar flow profile*. Other factors affecting the flow profile include the pulsatility of blood flow and the elasticity of the vessel wall or changes in flow resulting from bends in the vessel, branching, and narrowing. The Doppler signal derived from flowing blood thus contains a range of frequencies, which can be extracted using a mathematical algorithm called the *fast Fourier transform (FFT). This spectral analysis enables changes in blood flow velocity to be displayed over time.* In the resulting Doppler spectrum or waveform, the magnitudes of positive and negative shifts are displayed above and below the baseline, respectively. The image intensity or brightness indicates the amplitude, which corresponds to the number of red blood cells moving at a particular velocity, thus reflecting the volume of blood flow.

#### 1.1.2.1 Continuous Wave Doppler Ultrasound

Continuous wave (CW) Doppler (Fig. 1.19) uses two transducer elements – one continuously transmitting and the other continuously receiving ultrasound. Blood flow velocity is calculated from the frequency shift of the signal reflected by the moving red blood cells.

CW Doppler systems may be directional or nondirectional. Nondirectional systems cannot discriminate between positive and negative flow directions. In a directional system, information on the flow direction is extracted from the phase shift. As ultrasound is continuously transmitted and

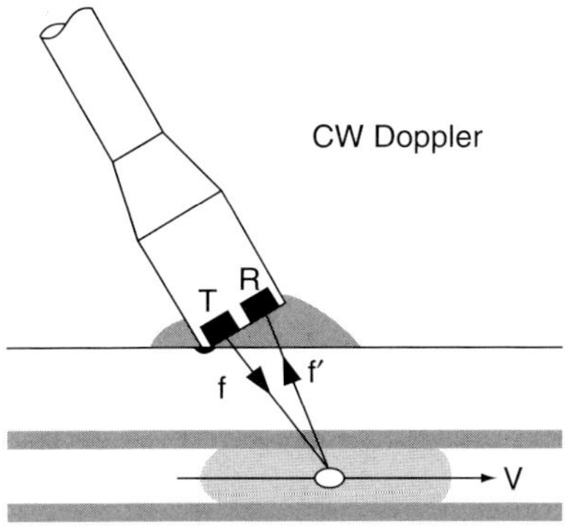


**Fig. 1.19** Diagram of continuous wave (CW) Doppler ultrasound. Ultrasound pulses are continuously emitted by the transmitter (*T*) and the frequency-shifted signals reflected by red blood cells moving at different velocities return to the receiver (*R*)

received, CW Doppler cannot assign the returning Doppler signal to a specific depth. Hence, the returning signal contains flow information from all vessels along the beam path. With arteries and veins often lying close together, the CW Doppler signal simultaneously represents arterial and venous flow. When performed with a high transmit frequency, CW Doppler allows sensitive examination of superficial vessels.

The advantage of CW Doppler lies in the detection of high flow velocities without aliasing, which is accomplished by the use of separate transmit and receive crystals for the simultaneous emission and reception of ultrasound signals.

### 1.1.2.2 Pulsed Wave Doppler Ultrasound/Duplex Ultrasound

Pulsed wave (PW) Doppler (Fig. 1.20) is similar to conventional B-mode scanning in that the same piezoelectric elements alternately emit ultrasound pulses and receive the incoming echoes.

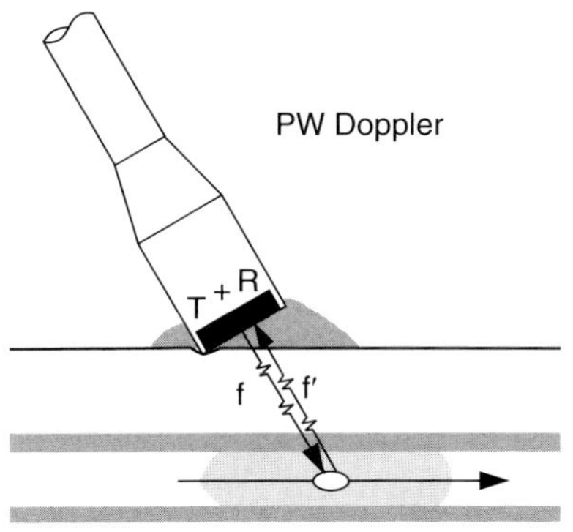


**Fig. 1.20** Diagram of pulsed wave (PW) Doppler ultrasound. The transducer alternately emits short ultrasound pulses (*T* transmitter) and records the reflected echoes at defined intervals (*R* receiver)

The depth at which a returning signal originated can be determined by calculating the round trip time (based on knowledge of the speed of sound in tissue) as follows: a short pulse is emitted, and the system is switched off for some time before the receive mode is switched on. In this way, only echoes arriving at the transducer face with the system in the receive mode are processed, ignoring echoes arriving during the off-mode. The time during which the transducer is in the receive mode is the *range gate*. By changing the range gate, the operator can define the sample volume or Doppler window. A typical sample volume encompasses the entire diameter of the target vessel. The number of pulses emitted per second is the *pulse repetition frequency* (*PRF*). The maximum PRF that can be used decreases with the depth of the vessel interrogated, as it then takes longer for the echoes to return to the transducer.

Sound waves travel through the human body at a fairly constant speed of about 1,540 m/s. Hence, the *round trip time* varies with the distance between the reflector and the transmitter, and the operator can define a scan depth by means of a time filter. An electronic gate then opens briefly, allowing only signals from this site to pass, while discarding all echoes coming in earlier or later. It is thus possible to selectively record Doppler signals from the specified depth. The combination of PW Doppler with real-time gray-scale imaging is the basis for *duplex ultrasonography*. PW Doppler has the advantage of providing axial resolution (discrimination of vessels along the ultrasound beam), but is limited by the fact that it fails to adequately record high-velocity signals (depending on the transmit frequency and penetration depth). Using a single crystal for transmitting and receiving signals requires a delay between pulses for the processing of returning echoes. The longer the pulse delay, the lower the peak flow velocity that can be detected.

*Duplex ultrasound* combines 2D real-time imaging with pulsed Doppler and thus provides flow information from a sample volume at a defined depth. Duplex scanning enables calculation of blood flow velocity from the Doppler frequency shift as the angle of incidence between the ultrasound beam and the vessel axis can be measured in the B-mode image.

### 1.1.2.3 Frequency Processing

In a blood vessel, blood components move with different velocities, which are represented in the Doppler spectrum by a range of frequencies with different amplitudes reflecting the distribution of flow velocities in the vessel. The spectrum is analyzed using FFT, which breaks down the waveform into a series of sinusoidal waveforms. For the individual frequency values, the corresponding amplitudes

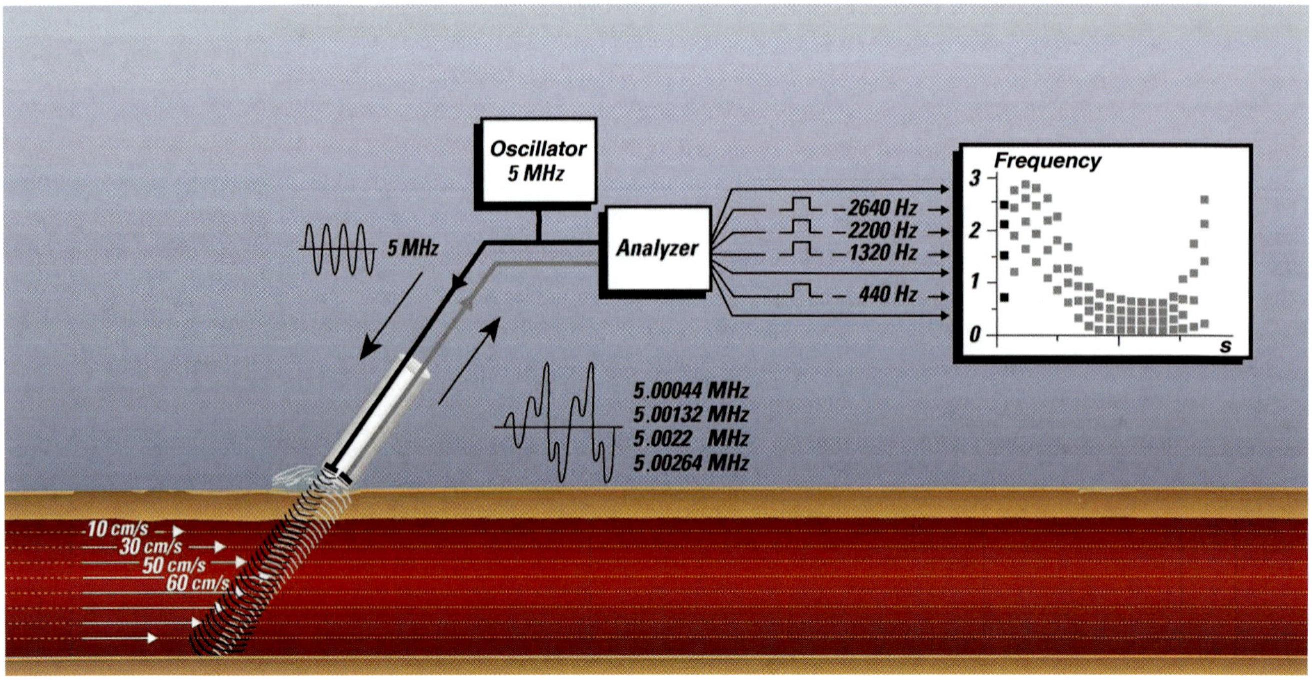


**Fig. 1.21** Function of a Doppler transducer. Ultrasound waves are emitted by an oscillator and reflected by red blood cells moving through the vessel at different velocities. The signal is reflected with a shifted frequency, or Doppler shift, which is determined by the speed and relative direction of the moving reflectors. The received Doppler signal is composed of a range of frequencies, which have to be sorted by fast Fourier transform (FFT) before they can be displayed over time in the form of a Doppler frequency spectrum or waveform (diagram courtesy of GE Healthcare)

are calculated and displayed in different shades of gray (Fig. 1.21).

According to Fourier's theorem, any periodic waveform can be reconstructed from its component waveforms. Conversely, in spectral analysis, a complex waveform of a given frequency (Doppler shift frequency) is decomposed into its frequency components. In this case, the FFT yields the amplitudes of the individual frequencies of the respective sine and cosine functions, which together make up the waveform. The individual frequencies thus separated are continuously displayed over time in the *Doppler frequency spectrum* (*spectral waveform*). The Doppler spectrum contains the following information on blood flow (Fig. 1.22a):

- The vertical axis representing different flow velocities as Doppler frequency shifts
- The horizontal axis representing the time course of the frequency shifts
- Density of points, or color intensity, on the vertical axis representing the number of red blood cells moving at a specific velocity (may also be plotted in the form of a histogram)

Flow toward and away from the transducer is processed simultaneously and respectively represented above and below the baseline (zero flow velocity line).

**Table 1.6** Spectral displays

| | |
|---|---|
| *Power spectrum* | Quantitative frequency distribution in terms of amplitude |
| *Frequency spectrum* | Temporal distribution of shifted frequencies or blood flow velocities |
| *Usual mode of display* | Frequency spectrum |
| Levels of brightness or color represent the density of a given frequency in the frequency band | |

Alternatively, some ultrasound devices display the magnitudes of the different velocity components in a separate *power spectrum*. This is done by measuring the signal intensities of the individual Doppler frequencies at a specific time in the cardiac cycle and displaying the spectral distribution in a histogram (Fig. 1.22b and Table 1.6).

#### 1.1.2.4 Blood Flow Measurement

The most important parameters for evaluating blood flow that can be derived from the Doppler frequency spectrum are:

- The peak systolic frequency (mainly relevant for quantifying stenosis)
- The peak end-diastolic frequency (stenosis, flow character)

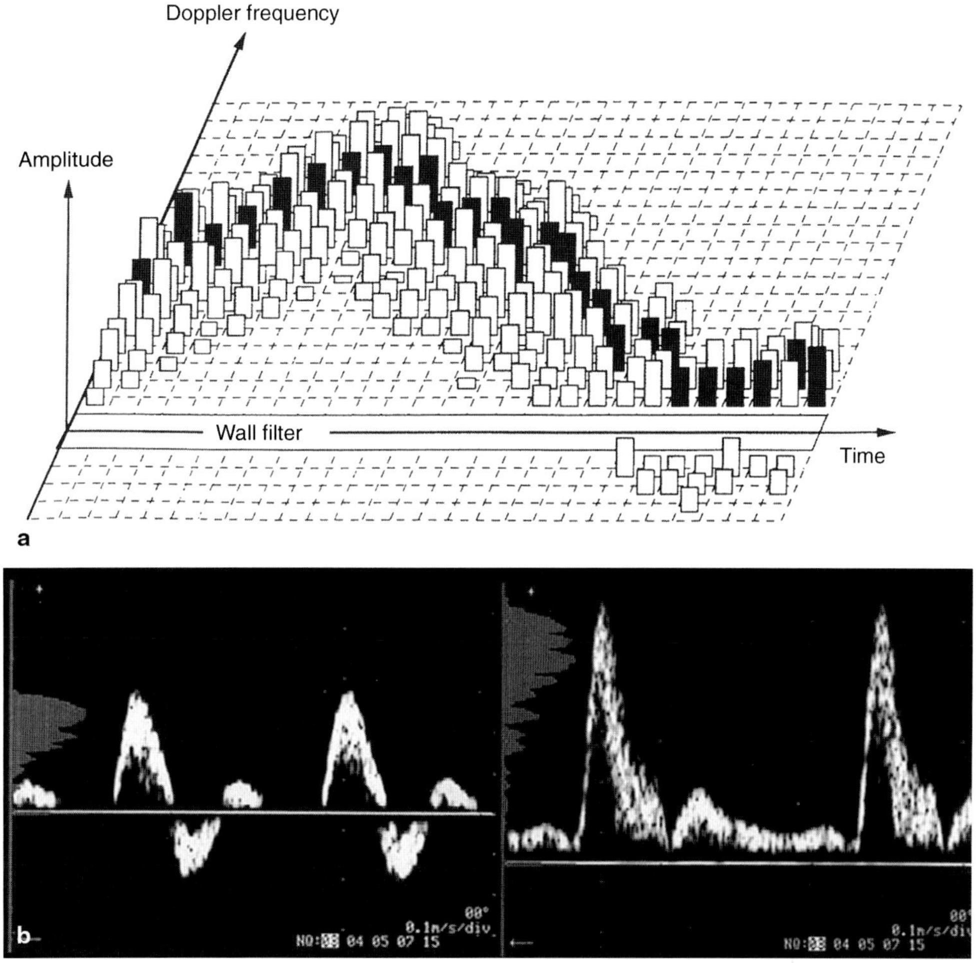


**Fig. 1.22** **a** Three-dimensional Doppler frequency spectrum showing the distribution of individual Doppler shifts (amplitudes), flow directions (above and below the time axis), and flow velocities (computed from Doppler frequency shifts). The heights of the boxes correspond to the amplitudes of the respective Doppler frequencies. A Doppler frequency spectrum represents amplitudes by different levels of brightness. In color-coded duplex ultrasound (black boxes), the averaged flow velocity at a specific point in time is displayed in color according to the flow direction and superimposed on the 2D gray-scale image in real time (according to P.M. Klews, in Wolf and Fobbe 1993). **b** Doppler frequency spectrum of the superficial femoral artery (*left section*). The histogram plotted on the vertical axis on the left represents the distribution of the different Doppler frequency shifts during systole. In the Doppler waveform, this distribution is represented by different levels of brightness (laminar flow). The *right section* shows the corresponding distribution during systole in the common carotid artery, which has less pulsatile flow

- The average flow velocity
- The mean or intensity-weighted flow velocity (which is the basis for calculation of the volume flow rate)
- The variance (spectral broadening due to disturbed flow)

Based on these parameters, the following quantitative determinations can be performed:

- Angle-corrected peak systolic and end-diastolic velocities can be calculated from the Doppler waveform. Mean flow velocity is calculated on the basis of the signal intensities.
- The volume flow rate is calculated from the *mean flow velocity* and the *cross-sectional area* according to the following equation:

$$Q(mL/min) = 60 \cdot mean\ flow\ velocity\ (cm/s) \cdot cross\text{-}sectional\ area\ (cm^2)$$

Quantitative evaluation of blood flow requires estimation of the Doppler angle. The Doppler shift alone does not provide this information. Knowledge of the Doppler angle is necessary to calculate the angle-corrected blood flow velocity. To minimize errors in the calculation of blood flow velocity and other quantitative determinations, the angle must be as small as possible and should not exceed 60°.

An *angle setting error* of ±5° causes a 20% error in the velocity measurement at a Doppler angle of 60°. Higher angles of incidence lead to an overproportional increase in the magnitude of the error (Fig. 1.23a–d).

Various measures are available to optimize the angle of incidence for Doppler measurement:

- Use of a unilateral waterpath (linear-array transducer)
- Activation of beam steering: successive firing of the elements in the linear-array transducer produces an ultrasound wave that is emitted from the transducer at a specific angle, which makes the color box appear tilted
- Shifting and angling of the transducer (sector and curved-array transducers): a curved-array transducer with a small radius enables a wide range of motion including angulation for optimization of the Doppler angle. However, the examiner must be aware that the color coding may change as a result of a change in the flow direction relative to the sector-shaped ultrasound beam. In this case, the area of transition between red and blue is black (while it is yellow in aliasing). No Doppler frequency shift information is obtained from this black zone because the ultrasound beam is at a 90° angle to the vessel axis

In vitro waterbath experiments in which two precision pumps generated different flow profiles demonstrated good

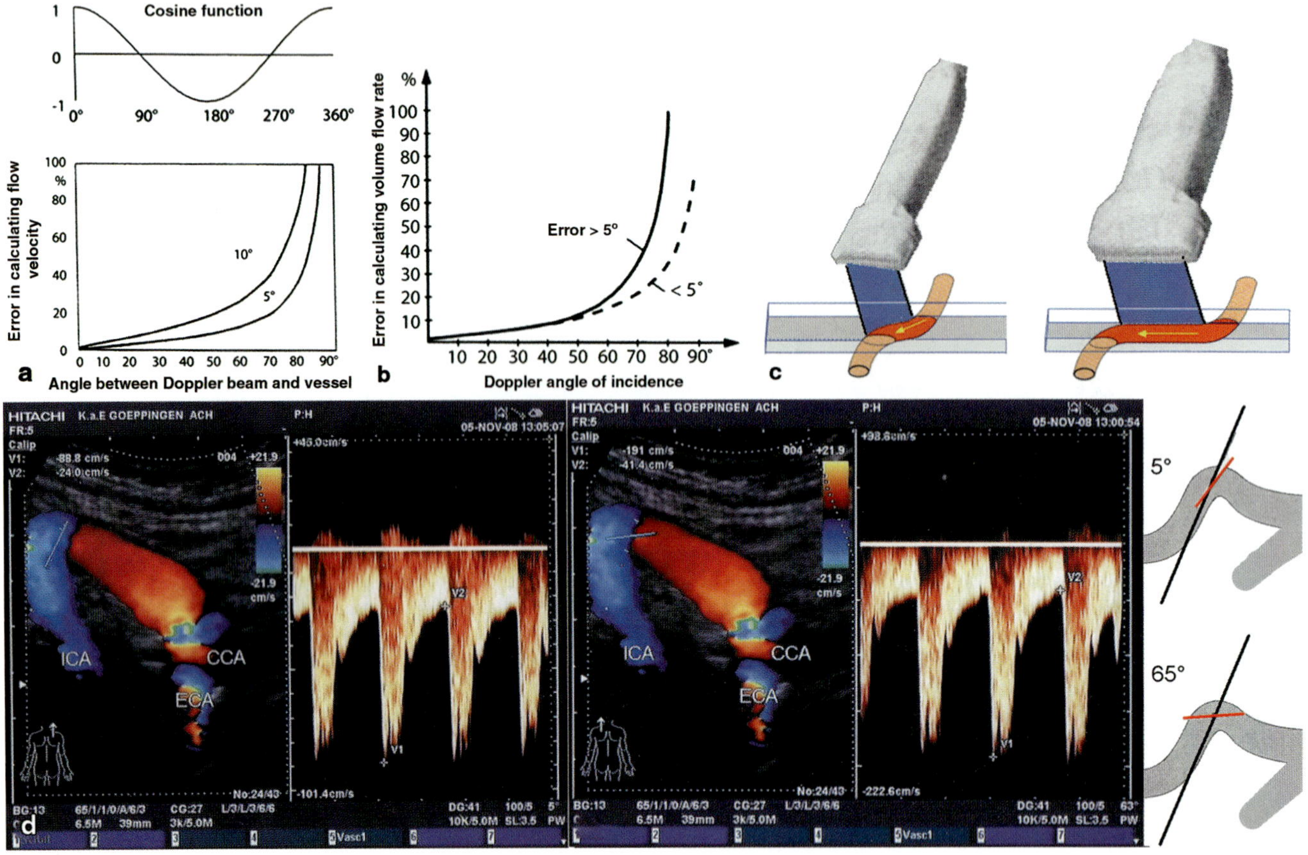


**Fig. 1.23** **a** The Doppler equation incorporates the angle between the ultrasound beam and the flowing blood in the form of the cosine function (cosα), with the shift being highest when the beam strikes the vessel tangentially (cosine of 0° = 1) and lowest when the beam is perpendicular to the vessel (cosine of 90° = 0). The larger the Doppler angle, the greater the resulting error in the velocity estimate in case of inaccurate placement of the angle correction cursor (graphically shown for errors of 5° and 10°). Such errors are unavoidable, particularly when aligning the angle cursor with the vessel wall in curved vessel segments. **b** The graph illustrates the angle-dependent error in flow measurement for a misalignment of ±5°. Overestimation of the Doppler angle results in greater error in the velocity estimate than underestimation. **c** Inaccurate velocity estimation resulting from misalignment of the angle correction cursor in vessels running obliquely through the scan plane. Alignment of the angle correction cursor is more difficult if a blood vessel passes obliquely through the scan plane in the B-mode image (*left drawing*). An oblique course is suggested if only a short segment of a long straight vessel is depicted. In such a case, the transducer should be turned to obtain a B-mode scan visualizing a long straight vessel segment (*right drawing*) for optimal positioning of the angle correction cursor. **d** Uncertainty concerning the Doppler angle of incidence in a tortuous vessel. In a curved vessel segment, the angle of insonation varies through a range of 5–65° over a short stretch, making it difficult to accurately determine the Doppler angle for calculating flow velocity. *Left color flow image and corresponding Doppler waveform:* Velocity measurement in a very tortuous ICA. With the sample volume positioned in the curved segment (to confirm or rule out clinically suspected kinking stenosis), a maximum peak systolic velocity (PSV) of 88 cm/s and a peak end-diastolic velocity of 24 cm/s were calculated with a Doppler angle of 5° (*top drawing*). *Right color flow image and waveform:* With an assumed Doppler angle of 65°, a PSV of 191 cm/s and an end-diastolic velocity of 41 cm/s were calculated in the curvature of the vessel (*bottom drawing*)

correlation ($r = 0.98$) between the volume flow rates measured by duplex ultrasound and volumetry (Schäberle and Seitz 1991; Fig. 1.24).

Even in deeper vessels, highly reproducible measurements can be obtained by performing Doppler interrogations at angles as close to 0° as possible to minimize the effects of errors in angle setting. Repeated ultrasound measurement of flow in the superior mesenteric artery performed in 28 fasting subjects in the morning revealed a day-to-day variation of 11% in peak systolic velocity (PSV) and of 9.7% in peak end-diastolic velocity (Fig. 1.25). Repeated diameter measurement using the leading-edge method showed a day-to-day variation of 2.2% (Schäberle and Seitz 1991).

Another source of error is the inadequate setting of transmit and receive gain, resulting in over- or underestimation of mean flow velocity (Fig. 1.26).

The *main source of errors in flow measurement*, however, is the inaccurate determination of the vascular diameter and the resulting inaccuracy in calculating the cross-sectional area (Fig. 1.27 and Table 1.7). In B-mode images, the vessel walls appear larger than their true anatomic size. This is due

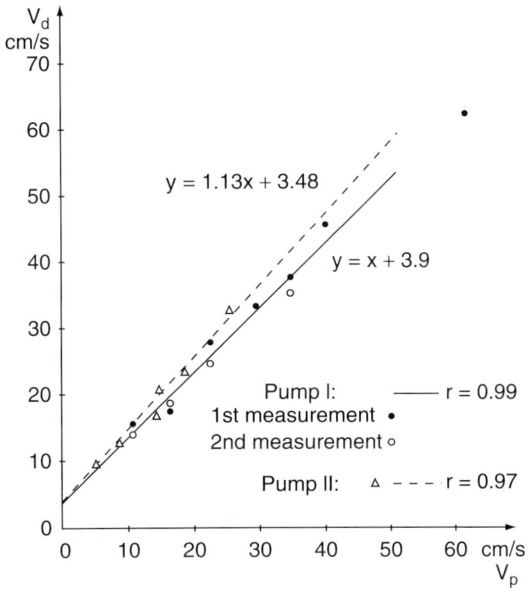


**Fig. 1.24** In vitro flow measurement by duplex ultrasound. Comparison of mean flow velocity determined by duplex ultrasound ($V_d$) and volumetry ($V_p$). Different flow profiles were generated by two precision pumps (I and II). The mean axis shift of 3.75 cm/s with shift of the zero line was due to a software error and was corrected by the manufacturer following these experiments. $V_p$ = mean actual flow velocity calculated from volumetrically determined flow rate/cross-sectional area of the tube; $V_d$ = mean flow velocity determined by duplex ultrasound (mean of five individual measurements) (Schäberle and Seitz 1991)

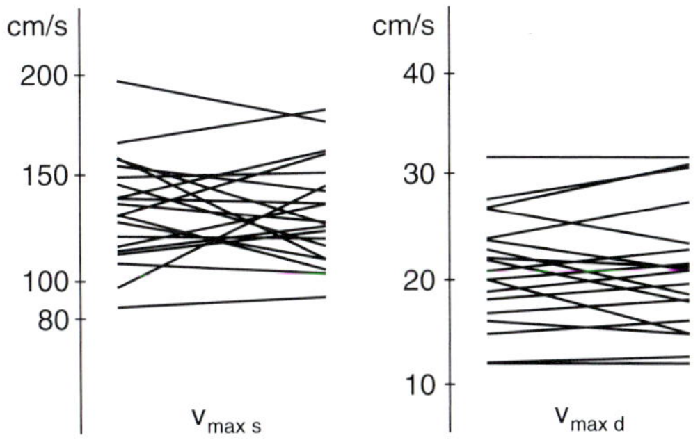


**Fig. 1.25** Peak systolic ($V_{max\ s}$) and late diastolic ($V_{max\ d}$) velocities measured in the superior mesenteric artery of fasting subjects on two successive days ($n = 28$)

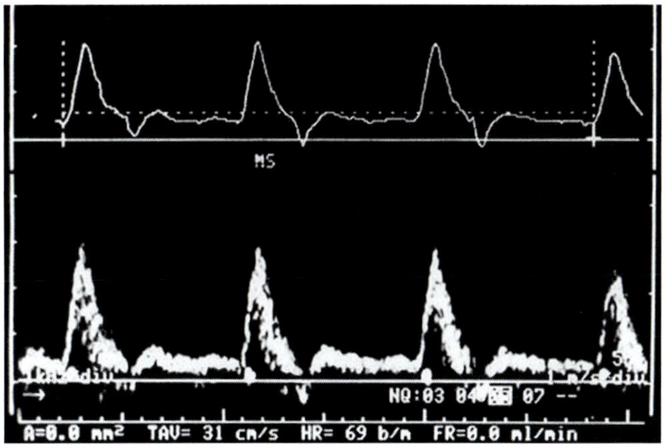


**Fig. 1.26** Doppler waveform sampled in the superior mesenteric artery with adequate instrument settings (*bottom*) with the corresponding mean flow velocities calculated by the device (*top*). The mean velocity averaged over three cardiac cycles is 31 cm/s

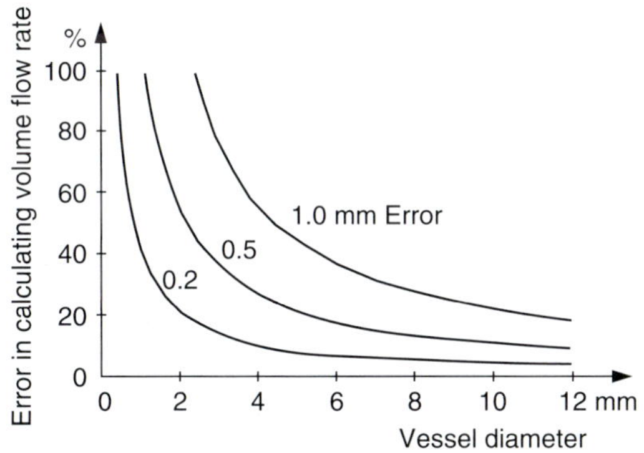


**Fig. 1.27** Errors in flow rate estimation resulting from different measurement accuracies in determining vessel diameter (for errors ranging from 0.2 to 1.0 mm)

**Table 1.7** Sources of error in blood flow rate determination with duplex ultrasound

- Determination of mean flow velocity
- Doppler angle error
- Measurement of vessel cross-section
  – Measuring accuracy (blooming effect)
  – Assumption of a circular cross-section
  – Variation in cross-section during cardiac cycle
  – Respiratory variation in vessel cross-section (veins)

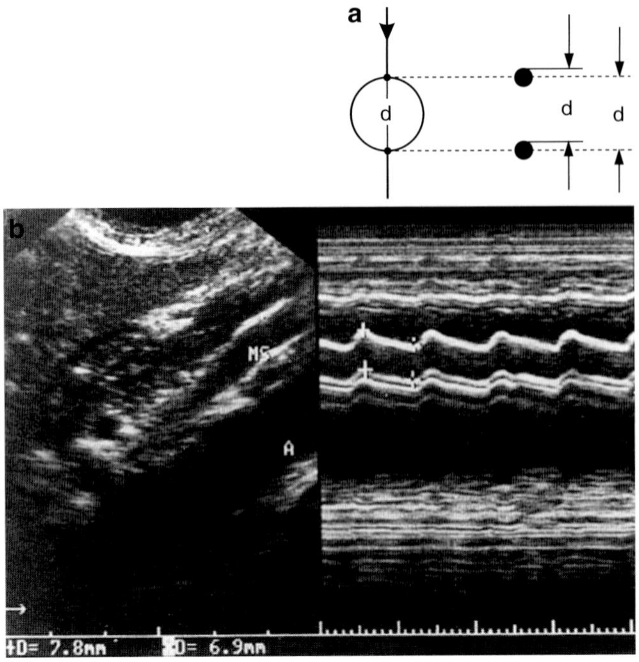


**Fig. 1.28** **a**, **b** Determination of vessel diameter. **a** Diagram of the leading-edge method. The vessel diameter is measured from the beginning of the reflection from the outer vessel wall to the beginning of the echo reflected from the opposite inner wall. **b** Measurement of the diameter of the superior mesenteric artery (*MS*) using the leading-edge method: The figure illustrates how the vessel wall is overemphasized as a result of the blooming effect. Systolic–diastolic variation in diameter: the gray-scale scan on the *left* coincidentally depicts the maximum systolic extension of 7.8 mm, while the time–motion display shows the variation in diameter from 7.8 mm in systole to 6.9 mm in diastole

to the so-called *blooming effect* resulting from the strong reflection of the ultrasound beam at the boundary between tissues of different acoustic impedance (Fig. 1.28b).

The measurement errors described above can be minimized and systematized by using the leading-edge method and a low gain.

Using the *leading-edge method*, the diameter is measured from the reflection of the outer wall to that of the opposite inner wall (Fig. 1.28a). In vitro experiments demonstrated a greater accuracy for diameters below 13 mm and showed the overestimation of diameters to be less severe than the underestimation associated with diameter measurement from inner wall to inner wall (Smith 1984). Moreover, use of the leading-edge method systematizes the unavoidable measurement error, thereby improving the reproducibility of measurements.

Fluctuations in diameter during the cardiac cycle are taken into account by measurement of both systolic and diastolic diameters, which enter the equation for calculating flow volume with different weightings.

The resistance indices – the *pulsatility index* (*PI*) and the *resistance index* (*RI*) according to Pourcelot – can be calculated independently of the Doppler angle. The resistance indices, in particular the Pourcelot index, also reflect *wall elasticity* as well as the *peripheral resistance* of the organ supplied (Fig. 1.29).

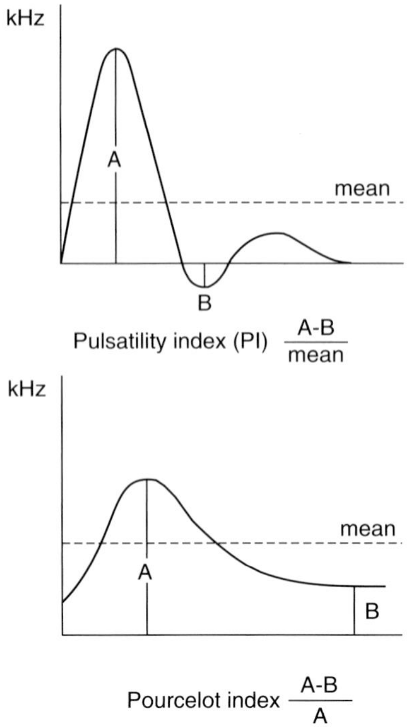


**Fig. 1.29** Diagrams of the resistance indices. The Pourcelot index is calculated from peak systolic and end-diastolic velocities, while the pulsatility index (PI) can only be calculated when the ultrasound equipment allows determination of mean flow velocity

In vessels with greater peripheral resistance, the Pourcelot index is higher and end-diastolic velocity decreases. Stenosis or occlusion in peripheral arteries with triphasic flow alters the Doppler waveform and the PI. The index can thus serve as a semiquantitative parameter for estimating the degree of stenosis. In an artery supplying a parenchymal organ, a relevant decrease in the Pourcelot index between the prestenotic and the poststenotic segment can be interpreted as indicating hemodynamically significant stenosis, for instance, when examining a patient with suspected renal artery stenosis.

### 1.1.3 Physical Principles of Color-Coded Duplex Ultrasound

#### 1.1.3.1 Velocity Mode

Color duplex ultrasound combines the presentation of 2D morphologic information with superimposed flow data of a defined area displayed in color. The frame rate is much lower for the color-coded 2D display of flow information than for the conventional (black-and-white) display because it takes much longer to compute the 2D distribution of flow.

In conventional duplex ultrasound, a small gate (sample volume) is defined in the real-time gray-scale image for which Doppler frequency shift data are obtained by analyzing separate scan lines. This information is displayed in the form of a Doppler waveform. To simultaneously measure flow velocities at different sites, several sample volumes are placed along adjacent beam paths. The distribution of sites defines a region of interest (ROI) for which flow velocity data are sampled. The Doppler signals from multiple sites cannot be analyzed using FFT because it would take too much time and also because it is not possible to simultaneously display all Doppler spectra generated using this approach. Consider that if we have 20 active scan lines, each with 50 sampling sites, this would mean data from 1,000 sample volumes! To cope with this amount of data, most systems use a technique known as autocorrelation, which compares two consecutive pulses returning from the sample sites of a given color scan line for phase shifts in order to estimate the mean Doppler shift frequency. Four sampling sites are usually sufficient for autocorrelation, compared with 128 sites for Fourier analysis. The phase shift information extracted by *autocorrelation* is a direct measure of the mean velocity distribution in a sample volume. The flow information is superimposed on a B-mode scan using red and blue to encode blood flow toward and away from the transducer, respectively. Different *degrees of brightness encode blood flow velocity*, with brighter colors indicating faster flow. As with PW Doppler, color Doppler ultrasound is also limited by angle dependence and aliasing. Aliasing is indicated by a color reversal in the color display.

Hence *multigate pulsed Doppler* uses several sample volumes, from which the respective Doppler frequency shift information is gathered and analyzed simultaneously using several independent channels. The flow information can thus be analyzed and displayed without delay compared with the conventional duplex scan. This technique is used in color-coded M-mode echocardiography and is the basis of color-coded duplex sonography. An ultrasound line with several measuring gates is swept over the B-mode scan or a defined part thereof in 50–150 ms. The number of scan lines is limited by the geometric arrangement of the crystals in the transducer, and the available lines must be divided into those for generation of the gray-scale image (B-mode image lines) and those for blood flow velocity measurement (Doppler lines). Most scan lines are reserved for generation of the B-mode image with only every third or fourth line being available for acquiring Doppler information. Due to the lower number of Doppler lines, the missing information between two Doppler lines must be interpolated. Using the multigate technique with placement of several sample volumes along the Doppler beam path, a 2D display of blood flow distribution is generated. About ten pulse packets are necessary for each color Doppler line to obtain precise information (as opposed to only one pulse packet for each B-mode scan line). Before emission of each new pulse packet, all echoes from the preceding pulse must have returned from the maximum scan depth in order to be able to assign them correctly. Hence, the time required for generating a color Doppler scan line is ten times that needed for a B-mode scan line. The information from the 50–250 measuring gates along each scan line is processed and analyzed simultaneously by separate channels.

Based on a mean ultrasound propagation velocity in the human body of 1,540 m/s, an echo signal reflected from a depth of 10 cm has a round trip time of 130 μs, which is the time required for the generation of a B-mode scan line. It takes ten times longer, that is, 1.3 ms, to generate a color Doppler scan line. If 50 color Doppler scan lines are used to generate a color Doppler image, the overall time required is about 65 ms, resulting in a frame rate of 15 images per second. *The frame rate decreases when more color Doppler scan lines are required*, which in turn depends on the width of the color box selected. A frame rate of at least 20/s is necessary for a smooth display with good temporal resolution.

When deeper structures in the body are imaged, as in abdominal ultrasound, the longer delay of the echo pulses makes it necessary to use lower pulse repetition frequencies (PRFs) for precise spatial resolution of the echo pulses, which likewise decreases the frame rate.

In *triplex scanning*, a Doppler waveform from a defined site in the B-mode scan that provides detailed information on flow velocity is obtained in addition to the gray-scale morphologic data and 2D flow velocity information of the duplex technique. The PRF for sampling of the color flow information must be kept constant, which severely restricts the arrangement of the pulse sequences. The frame rate is reduced further in the triplex mode because of the scan lines required for generating the Doppler spectrum. The maximum PRF that can be used in triplex scanning is markedly lower than that for the generation of a Doppler frequency spectrum, while the B-mode scan or color flow display is frozen. Therefore, at higher flow velocities, the Doppler frequency spectrum should be sampled in the mono mode following color duplex scanning for orientation and placement of the sample volume in the B-mode scan.

Some improvement is possible with the use of the *interleave technique*, in which a Doppler pulse echo cycle is processed after each second to fourth B-mode scan line. This technique thus enables a rather fast frame rate at a low PRF, for example, when scanning deep body structures. A much higher frame rate is possible with this technique compared with the analysis of one Doppler scan line after each B-mode line.

With the use of a conventional processing technique, the color-coded flow image is generated from the Doppler frequency shift of the ultrasound echoes reflected by flowing red blood cells. The frequency shift information can be analyzed and displayed in relation to velocity (velocity mode) or in relation to power (power mode).

In contrast, the *color-velocity imaging technique*, which is a so-called time-domain technique, determines flow velocity from the round trip times. This is done by comparing changes in the patterns of the reflected echoes between two successive B-mode scan line pulses and deriving blood flow information from changes in the echo pattern over time.

Quantitatively, the direction and velocity of flow can be determined using a *cross-correlation technique*. However, this procedure makes high demands on computing capacity and has not established itself although it is superior to shift-based techniques because there is no aliasing and angle dependence and it enables higher frame rates. The cross-correlation technique does not determine the Doppler frequency or phase shift of an echo compared to the transmitted pulse but compares two successive pulse echo cycles in a defined space and at a predetermined time delay. In other words, the positional change of two pulse echo cycles is determined in relation to time. The velocity of the moving medium with its characteristic echo pattern is then calculated from the temporal shift. This technique relies on the recognition of the echo pattern, which requires an excellent signal-to-noise ratio. The only factor that limits the PRF is the scanning depth.

The *autocorrelation technique* compares demodulated Doppler signals. The accuracy of determining the mean Doppler shift is higher when blood flow is constant and decreases when flow becomes more turbulent. The bandwidth of the different flow components is given by the variance, which can be displayed in color by adding a green shade. Analysis of several pulse cycles will yield a more accurate value of mean flow velocity.

The amount of Doppler information collected from the large number of sample volumes along the scan line (in contrast to a single, circumscribed gate in conventional duplex scanning) is too large to be processed by spectral analysis using the FFT with decomposition of the spectrum into its component parts and display of the proportions of different flow velocities at a given location. Instead, the faster autocorrelation technique is used to *calculate* the mean frequency shift and the corresponding *mean velocity*. Averaging of the frequencies *reduces the spectral information to a color pixel* that represents an intensity-weighted mean Doppler frequency shift *in a direction-dependent manner*. Blood flow toward the transducer is displayed in red, and flow away from the transducer in blue. Less saturated hues indicate higher flow velocities. The instrument alternately displays the gray-scale scan and the color flow information whenever a Doppler frequency shift occurs.

In the inverse color mode, veins can be displayed in blue and arteries in red regardless of the true flow directions relative to the transducer. The use of this mode will cause problems when there is abnormal reversal of flow or when scanning abdominal regions with a complex vascular anatomy. The ultrasound convention of displaying cranial portions of the vascular anatomy on the left side of the monitor and peripheral segments on the right allows direct identification of the flow direction (toward the heart or toward the periphery) from the color coding when the transducer position is known (without first having to look for the information that indicates the inverse color display and varies from one manufacturer to the next).

#### 1.1.3.2 Power (Angio) Mode

*Amplitude-based flow processing* has recently gained attention as an additional color flow imaging technique. The principle was first reported in the literature in 1994, and this technique is now available for state-of-the-art ultrasound systems. Various names have been used by manufacturers including power Doppler (imaging), color amplitude imaging, and color power or color energy Doppler.

While Doppler techniques processing frequency or phase information use a high-pass filter (HPF) to extract the blood signal and to distinguish it from tissue echoes, the amplitude-based flow techniques additionally process the intensity or amplitude of the received echoes to differentiate signals from stationary and moving reflectors. Tissue echoes typically have 1,000 times higher intensity than echoes reflected by blood cells, allowing a much better separation of the two signal types. While conventional color Doppler techniques require a frequency analysis to process blood flow information, amplitude-based techniques assign colors directly from the echo intensities, similar to the assignment of gray-scale values in B-mode ultrasound.

In a power Doppler image, the *sum of the Doppler signal intensities* reflected by *moving particles* is represented by *levels of brightness*, while the magnitude of the flow velocity and different velocities as well as flow direction (in older devices) are ignored (Fig. 1.30a, b). The color intensity in the power Doppler mode is determined by the density of the moving reflectors. The power mode is more sensitive to slow flow and flow with only a few reflectors as compared with the velocity mode. The signal-to-noise ratio improves with the number of gates placed along each color Doppler line. While the power mode is similar to other techniques in that it only registers reflected echo signals within a certain range of Doppler frequency shifts, it is largely independent of the angle between the ultrasound beam and the blood vessel. Since blood does not flow strictly in one direction, there will always be some reflection of echo pulses even at an unfavorable insonation angle, but color intensity is reduced at an angle around 90°. The power Doppler mode is particularly suitable for evaluating slow flow in small

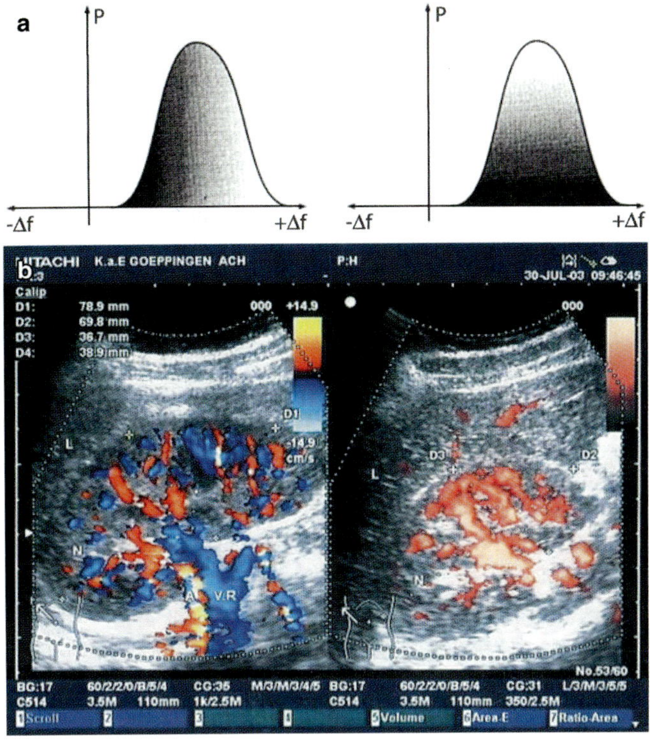


**Fig. 1.30** **a** Types of color coding used to display flowing blood: In the velocity mode (*left*), *red* and *blue* represent the flow direction and lighter shades represent a higher Doppler shift. In the power/angio mode (*right*), higher amplitudes of the reflected ultrasound echoes are displayed in lighter shades irrespective of the frequency and flow direction. **b** Depiction of the right kidney by color duplex ultrasound in the velocity mode on the *left* and in the power Doppler mode on the *right*. In the velocity mode, the color coding of the vessels indicates the blood flow direction, with veins being shown in *blue* (flow away from the transducer) and arteries in *red* (toward the transducer). Vessels to the level of the interlobar vessels are depicted. The power Doppler mode provides no information on the flow direction of blood but enables evaluation of slow flow and is less angle-dependent, which is why this mode is superior in depicting parenchymal blood flow in small vessels. The kidney (*N*) is marked with *plus signs*. Part of the liver (*L*) is depicted near the abdominal wall with visualization of peripheral vessels in the power Doppler mode

vessels and can thus be used to examine peripheral perfusion as well as perfusion in small tumor vessels or in parenchymal organs (Table 1.8).

The power mode is limited by the fact that it does not provide qualitative or semiquantitative information on flow velocity. Moreover, it is more susceptible to artifacts induced by organ movement and has a poorer temporal resolution. There is no aliasing because the power mode is independent of the magnitude of the Doppler frequency shift. The main advantage of the power mode lies in the fact that it uses very low PRFs (in the range of some 100 Hz), which in turn enables the resolution of very small Doppler frequency shifts (slow flow). Such low PRFs would be highly susceptible to aliasing in velocity-dependent color coding (Table 1.8).

Since power Doppler techniques process intensities in addition to frequency shifts induced by moving red blood cells, there is no color reversal due to changes in flow direction or aliasing. Power Doppler imaging provides no information on relative flow velocities or flow direction. An important *advantage* is the *higher sensitivity of power Doppler to much slower blood flow* compared with conventional color Doppler techniques. Moreover, the signal-to-noise ratio is much better, and the technique is virtually independent of the angle of insonation.

The power mode represents the intensities of the signals reflected by moving blood as different levels of brightness based on their amplitudes. The so-called bidirectional power mode offered by the manufacturers of state-of-the-art scanners enables simultaneous color coding of flow direction (blue and red). This is achieved by the activation of some additional ultrasound beams which solely serve the purpose of sampling and processing flow direction information using the autocorrelation technique.

#### 1.1.3.3 B-Flow Mode (Brightness Flow)

The B-flow mode is not a Doppler technique based on the processing of the Doppler shift but a *B-mode scanning technique* that *compares gray-scale scans over time* to identify changes in the spatial positions of reflectors (blood cells) by the successive emission of coded pulse packets. If the echoes returning along the same scan line upon transmission of two successive ultrasound beams give an identical gray-scale

**Table 1.8** Comparison of the two methods of flow encoding: velocity mode and power Doppler (angio) mode

| Mode | Advantages | Disadvantages |
|---|---|---|
| Velocity mode | Detection of flow velocity and direction, high temporal resolution | Angle-dependent display of flow (which in turn affects color filling of lumen), aliasing |
| Power Doppler mode (angio mode) | Evaluation of flow largely angle-independent (resulting in good color filling), depiction of slow flow, sensitivity to small flow volumes, few artifacts, good color filling improves discrimination of patent lumen from mural structures | No information on flow velocity and direction, more difficult differentiation of arteries and veins, no hemodynamic information |

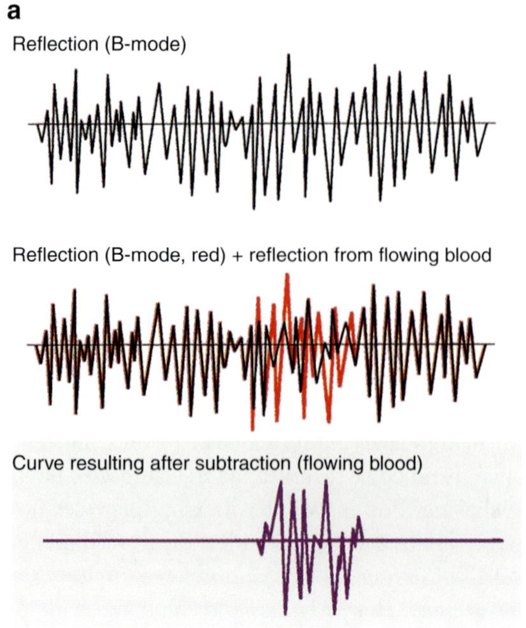


**Fig. 1.31** a Signal generation in B-flow ultrasound. Two reflections, I and II, produced by two successive ultrasound beams emitted along the same scan line give identical signals if the echoes are returning from stationary objects. Conversely, if signals are reflected from moving targets (red blood cells), there will be a circumscribed change in the signal pattern (superimposition). A subtraction image representing flowing blood can be generated by subtracting the echoes from stationary tissue, improving the differentiation between blood and surrounding stationary tissue (see Fig. A5.36)

pattern, the echoes have been reflected by stationary tissue. Conversely, if the echoes are reflected by stationary tissue and moving red blood cells, a slight difference in the pattern can be seen (Fig. 1.31a). The echo signals are subtracted from one another, and the brightness is determined by the number of reflectors and partly also by their velocity. The conventional B-mode image generated from the stationary echoes is displayed around the flow information. Since the successive pulses are emitted at defined intervals and in digitally encoded form, it is possible to eliminate interfering echoes and use only the encoded echoes in the subtraction procedure. Hence, only the amplitude signal reflected by moving particles is processed in the interval between two pulses. As the signal strength increases not only with the number of reflecting particles (volume flow) but also with flow velocity, a jet within a stenosis is depicted with a higher signal intensity.

B-flow images depict blood flow with *high spatial resolution* and allow *good delineation of flowing blood from the vessel wall*. Echoes from stationary tissue are either suppressed or displayed with reduced gain to provide anatomic orientation. The advantage of this technique lies in the simultaneous display in a single image of blood flow information (comparable to angiography) and morphologic details of the vessel wall with high resolution while showing no or only little angle dependence. Disadvantages of B-flow imaging are the occurrence of artifacts in highly pulsatile, atherosclerotic vessels, the susceptibility to artifacts caused by wall motion, and the still limited scanning depth.

Therefore, B-flow ultrasound can only be used to image the carotid arteries and other superficial vascular beds. Another limitation is that it does not allow reliable differentiation of blood flow direction or quantitative estimation of flow velocity. A good signal-to-noise ratio is important, rendering the method unreliable in patients with a somewhat deeper vertebral artery or for scanning the carotid circulation in patients with large necks.

With further technical advancement, B-flow imaging will, in principle, enable morphologic quantification of stenosis and differentiation of the vessel wall from the patent lumen even if only slow flow is present (e.g., in ulceration). The B-mode provides a high-resolution display of the vessel wall contour with separate representation of blood flow in the B-flow mode. There is no superimposition of blood flow information as in color duplex scanning (see Fig. A5.23).

### 1.1.3.4 Intravascular Ultrasound

With technical advances and the development of miniaturized ultrasound probes, it is now possible to examine a vascular ROI from within the vessel. Intravascular ultrasound (IVUS) is performed using percutaneously inserted intravascular probes, which are advanced to the target site by means of very small catheter systems and thin guidewires under radiologic guidance. IVUS is used to assess atherosclerotic lesions, other vessel wall conditions such as dissection, and perivascular structures close to the vessel wall including tumor infiltration. The technique is most suitable for evaluating catheter-based interventions of both coronary and peripheral arteries. In this setting, the ultrasound probe can be inserted through the introducer sheath, already placed for the interventional procedure, allowing evaluation for wall changes and postinterventional complications with high spatial resolution.

Various mechanical and electronic high-frequency phased-array probes are available for single use. A higher transmit frequency improves axial resolution but limits penetration depth. On a 360° IVUS image, the *normal arterial wall* has a *three-layer appearance* resulting from the reflection of the ultrasound beam at the boundaries between layers of differing acoustic impedance. The sonographic layers do not correspond to the histologic layers of the intestinal wall. The histologic thickness of the normal intima is below the axial resolution of IVUS probes. The

b

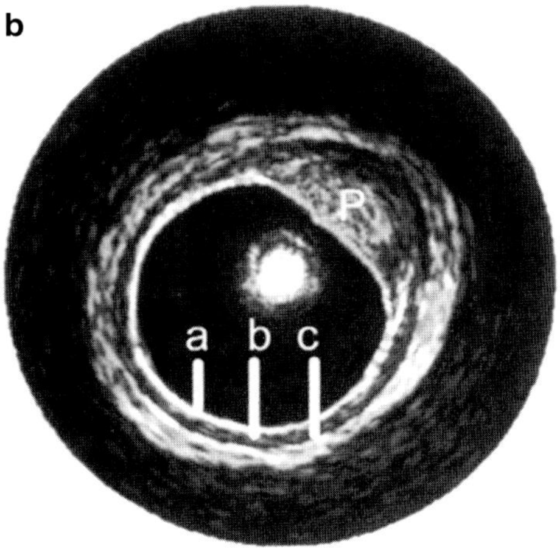


**Fig. 1.31** **b** Three-layer appearance of the arterial wall in intervascular ultrasound (IVUS). The layers distinguished by ultrasound do not represent the tissue layers that make up the arterial wall. The bright inner ring (*a*) is the interface between blood and the intima (which gives a high signal because of the large acoustic mismatch). The second, darker ring (*b*) roughly corresponds to the media, and the third ring, which is also bright (*c*), is the transition from the adventitia to the perivascular connective tissue. The normal intima is not visualized by IVUS; only a thickened intima or plaque will be seen (*P*)

*bright inner ring* thus represents the *interface between blood and intima*. The *intima* is seen only when it is *thickened by atherosclerosis*. The outer hyperechoic ring is the reflection of the interface between the adventitia and periadventitial tissue and is clearly distinct from a markedly more hypoechoic middle layer. This layer corresponds to the muscularis and results in the characteristic three-layer appearance of arteries of the muscular type. With its high resolution, IVUS is superior to percutaneous ultrasound in terms of demonstrating intimal thickening and plaques; it also improves plaque characterization as it allows better differentiation of plaque types such as fibrotic or necrotic plaques (Fig. 1.31b).

*Examination procedure.* With the patient positioned supine, the femoral artery is punctured and a 6F to 8F introducer sheath is placed using a Seldinger technique. A guidewire serves to advance the ultrasound probe to the target region under fluoroscopic guidance.

IVUS is used to identify and characterize atherosclerotic and other wall lesions and to evaluate perivascular changes, for instance, in patients with suspected tumor infiltration. Following an interventional treatment, IVUS can be performed to identify residual stenosis or complications such as an intimal flap or dissection. IVUS is an invasive and technically demanding procedure, and the single-use probes are expensive, which is why IVUS is confined to specialized centers and is typically performed in conjunction with a vascular intervention.

#### 1.1.3.5 Three-Dimensional/Four-Dimensional Ultrasound

A three-dimensional (3D) ultrasound display is reconstructed from a series of 2D images acquired by manually or automatically moving the ultrasound probe across the body surface perpendicular to the transducer plane. There are several options for displaying the 3D information: either as a composite, as a transparent volume block, or as a display in which the observer can select a point, for which the information will then be presented in all three dimensions. These techniques enable a 3D display of wall lesions or plaques as well as their relationship to perivascular tissue. These rendering techniques, however, are most useful in the documentation or demonstration of findings. During the examination, the sonographer is able to grasp spatial relationships more rapidly, making use of the flexibility of the ultrasound probe and switching to different scan planes. Compared with the fixed planes available in a 3D rendering, this flexibility is superior when it comes to evaluating challenging anatomy. It also provides more freedom in circumventing artifacts. Three-dimensional vascular ultrasound is degraded by vascular pulsation. With the computational capacity afforded by state-of-the-art ultrasound systems, complete volumes can be displayed in fractions of seconds. If the frame rate for volume generation is high enough to track motion, this is known as four-dimensional ultrasound (with the time course of the changes observed representing the fourth dimension).

### *1.1.4 Factors Affecting (Color) Duplex Imaging – Pitfalls*

The pitfalls of duplex ultrasonography are summarized in Table 1.9.

#### 1.1.4.1 Scattering, Acoustic Shadowing

Air (in the intestine and lungs) and calcified structures (bones, calcified plaque) produce scattering and acoustic shadowing. These structures are not penetrated by the ultrasound beam and thus prevent collection of morphologic and Doppler flow information from body regions behind them. Bowel gas can be pushed out of the way by firmly pressing the transducer against the bowel, thereby enabling visualization of retroperitoneal structures and flow. In all other cases, the examiner must try and circumvent such structures by changing the transducer position.

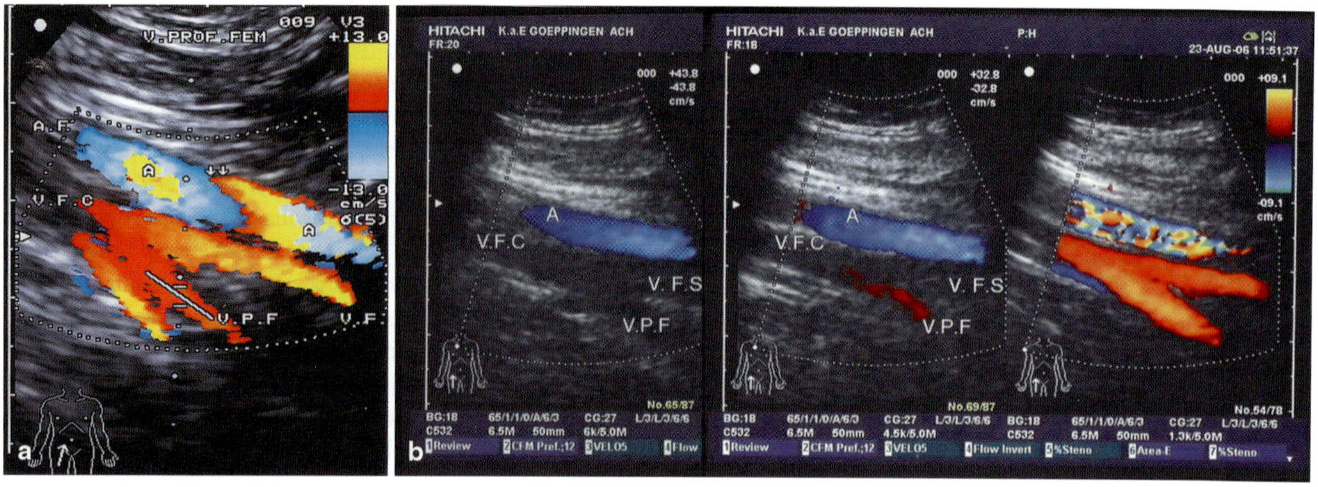


**Fig. 1.36** **a** The pulse repetition frequency (PRF) is adequately set for evaluating venous flow (confirmed here in the deep femoral vein). The figure illustrates the pitfalls of color duplex scanning in the superficial femoral artery. Color-coded flow signals are depicted throughout the lumen of the deep femoral vein (duplicated in the example shown, *V.P.F.*) and in the superficial femoral vein. Flow is toward the transducer (*red*). Flow in the superficial femoral artery (*A.F*) is in the opposite direction (*blue*). The PRF is too low for arterial blood flow, resulting in aliasing indicated by the color change through *yellow* to *red*. As a result of the low frame rate (proportional to the PRF), the change in flow direction from systole to early diastole is seen in the *center* of the image (indicated by *arrows*). The changed flow direction is reflected by the color change from *dark blue* to *dark red*. In the further course of the vessel, there is a second area of aliasing with color changing from *red* to *yellow* to *blue*. The increasingly smaller Doppler angle also contributes to these phenomena and is also responsible for the lighter color in the superficial femoral vein running posteriorly. **b** Effects of different PRFs on the depiction of arterial and venous flow. A lower PRF is required to detect the slower venous flow. The *leftmost image* depicts the artery with good color filling (*blue*), while no flow is detected in the vein, which may be misinterpreted as thrombosis. With increasingly lower PRFs (from *left* to *right*), there is increasingly severe aliasing in the artery despite normal blood flow, which must not be misinterpreted to indicate stenosis. In the vein, more flow signals are depicted (from *left* to *right*) as the PRF is increased. In the *middle section*, flow is only detected in the profunda vein (which is insonated at a smaller angle). The *rightmost* image shows good color filling (*red*, flow toward transducer) of the entire venous lumen, ruling out thrombosis

The options available to overcome these limitations can be summarized as follows:

- Use of a higher PRF to register higher Doppler shift frequencies, but this will reduce the scan depth (longer pulse delays with greater depth; Fig. 1.37).
- Use of a lower transmit frequency, but this will reduce the spatial resolution of the gray-scale scan.
- Shifting of the baseline, both in the spectral display and on the color scale. The PRF cannot be increased indefinitely as this would introduce range ambiguity. Alternatively, aliasing can be overcome by manipulating the measurement range as defined by the positive and negative Nyquist limits. The baseline is normally midway between these two limits, allowing equal forward and backward flow velocities to be detected. By shifting the baseline in either direction, the measurement range for one direction is expanded at the expense of the other. The total measurement range for nonaliased flow detection remains constant.
- Some ultrasound machines have a high PRF option for the detection of faster flow. The higher frequency is achieved by emitting ultrasound pulses before the preceding signals have been processed. Processing is done using additional Doppler lines. However, this approach introduces

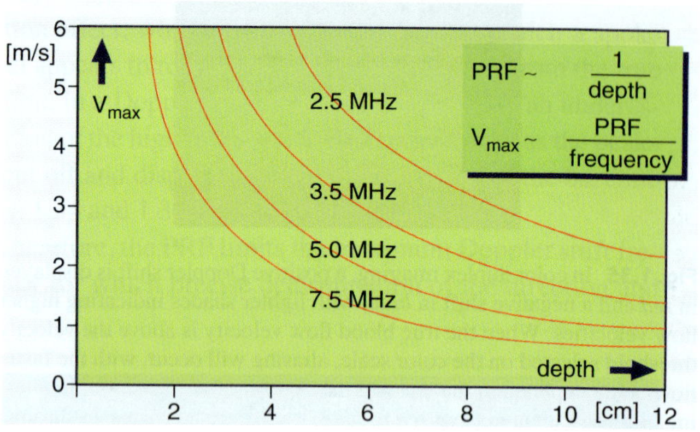


**Fig. 1.37** The extent to which the pulse repetition frequency (PRF) can be increased to solve the aliasing problem is limited by the distance of the target vasculature from the transducer. The round trip time of a pulse increases with the depth of the target vessel, which means that the PRF has to be lowered to ensure that all echoes from the previous pulse have been received before the next pulse is emitted – otherwise the range ambiguity problem may occur. Therefore, the maximum flow velocity ($V_{max}$) that can be recorded without aliasing is lower when a deeper vessel is examined. The relationships are illustrated in the graph. When a 6-cm-deep artery is interrogated with a 5-MHz transducer, aliasing will occur at a peak systolic velocity (PSV) of 2 m/s. With a lower transmit frequency of 3.5 MHz, aliasing does not occur unless flow velocity in the artery increases to 3 m/s. This is because the Doppler shift frequency is lower when a smaller transmit frequency is used, enabling higher flow velocities to be measured without aliasing (drawing courtesy of GE Healthcare)

some unreliability in terms of spatial resolution as Doppler information from a larger number of gates must be processed to generate the spectral display.

- CW Doppler using separate crystals for continuously transmitting and receiving signals is not limited by an upper Doppler shift threshold. This technique does not allow local differentiation of the reflected signals since all frequencies reflected back from along the beam path are processed to generate the Doppler waveform.

#### 1.1.4.4 Minimum Flow Velocity Detectable – Wall Filter, Frame Rate

Doppler-shifted echoes return from flowing blood but also from the moving vessel wall. Shifts due to wall motion typically have a high amplitude but a low frequency and affect the velocity spectrum. In contrast, the Doppler shifts caused by the flowing blood are of higher frequency and lower amplitude. To eliminate the interfering signals from the Doppler spectrum, a wall filter is used. The *wall filter* serves to eliminate low-frequency noise artifacts and vascular wall motion from the frequency spectrum. It is a high pass filter, meaning that high frequencies pass the filter, while low-frequency noise artifacts and vascular wall motion are filtered out and discarded. The user can select a filter cutoff to define the minimum frequency range to be displayed. Most filters have a cutoff range of 100–400 Hz, while a few ultrasound systems use cutoffs of up to 1,600 Hz. The filter eliminates all frequencies below the selected cutoff, including Doppler shift frequencies caused by slow flow (Fig. 1.38a, b). When the filter choice is inappropriate, the elimination of signals from slow-flowing venous blood may suggest absence of flow. Also, the mean flow velocity may be overestimated if slow flow components with low Doppler shifts are filtered out. To overcome these problems, devices have been introduced that eliminate only low frequencies with high amplitudes (typical of wall motion) or that identify and eliminate typical patterns of tissue motion.

The *frame rate* is the number of color-coded ultrasound scans that can be generated per second and determines the minimum flow velocity that is detectable. It is mainly dependent on the width of the color box and scan depth selected by the examiner. A larger color box requires more color Doppler lines to be processed for image generation. Moreover, image generation is slower when deeper structures are scanned due to the longer round trip time (Table 1.10). Lower frame rates result in a poorer temporal resolution with failure to detect rapid changes or short flow phenomena. Cardiac perfusion and flow in peripheral arteries can only be examined with high frame rates (at least 20 images per second). Lower frame rates, on the other hand, improve the spectral Doppler display and the sensitivity for low Doppler shift frequencies because more Doppler information can be sampled. Despite the poorer temporal resolution, low frame rates yield better results in scanning veins and small arteries (Table 1.11).

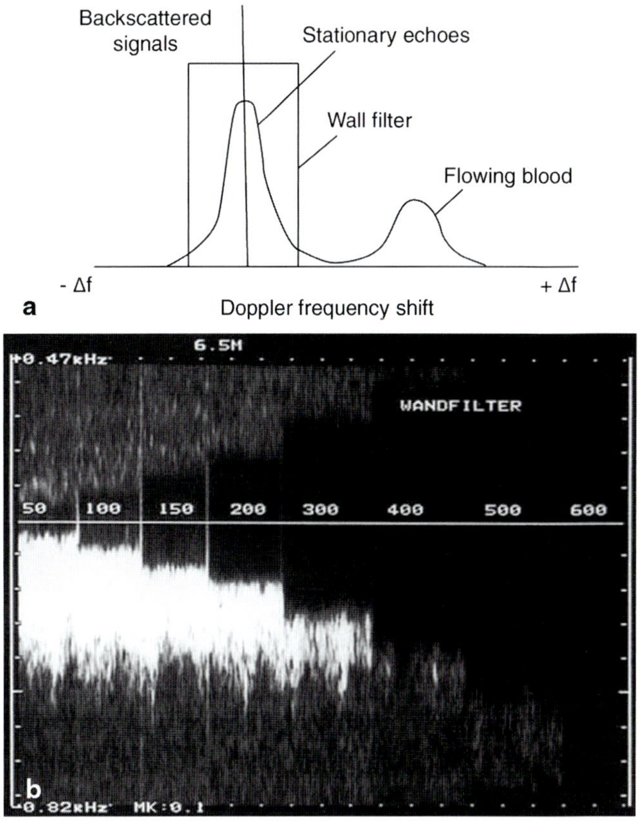


**Fig. 1.38** **a** A wall filter is used to eliminate the low-frequency signals caused by motion of the vessel wall. The cutoff should be selected so as not to inadvertently remove flow-related low-frequency signals, for instance, when evaluating slow venous flow. **b** When the cutoff is gradually increased, at first only the low Doppler shift frequencies are eliminated, while further increments will eliminate venous flow signals as well. A wall filter of 500 Hz eliminates venous flow signals up to 450 Hz, while slower flow is already eliminated at lower values

**Table 1.10** Factors affecting the frame rate in color Doppler imaging

| | Frame rate | |
|---|---|---|
| | Higher | Lower |
| Size of color box | Smaller | Larger |
| Number of scan lines | Fewer | More |
| Scan depth | Smaller | Greater |
| Lower frequency limit | Higher | Lower |

**Table 1.11** Options for improving color filling of the vessel lumen

- Lower Doppler transmit frequency
- Lower PRF
- Lower wall filter
- Optimization of insonation angle
- Higher color gain

#### 1.1.4.5 Transmit and Receive Gain

The receive gain must be set so as to ensure that all informwation from the reflected echo pulses is displayed, while at the same time preventing artifacts due to overmodulation. The gain is adjusted separately for B-mode, Doppler, and color duplex ultrasound because different pulses are used for each of these modes. Readjustment during the examination may be required, for instance, when measuring intima-media thickness. Correct measurement requires careful tuning of the receive gain so that boundary lines are visible and overestimation resulting from the blooming effect is avoided.

Receive gain setting for the acquisition of Doppler information must ensure that, in the presence of laminar flow, the clear systolic window in the display of the Doppler spectrum contains no artifacts while not eliminating signals from slow flow. Conversely, some overmodulation of the receive gain may be necessary in the presence of subtotal occlusion to determine the correct PSV (high-frequency Doppler shifts with low amplitudes due to the small number of reflecting red blood cells in the stenosis jet), even if this results in more artifacts. In color duplex imaging, overmodulation of the gain will lead to color overflow and may obscure atherosclerotic lesions in arteries or thrombus in a partially occluded vein (Fig. 1.39a). This phenomenon contributes to the unreliability of planimetric stenosis grading, which is based on the cross-sectional vessel area depicted in color duplex scans.

#### 1.1.4.6 Doppler Angle

The evaluation of an artery or vein in the B-mode image combined with the spectral Doppler information obtained with a defined ultrasound beam and known angle of insonation enables calculation of the true flow velocity from the Doppler shift. To select a proper Doppler angle relative to the flow direction, *a long stretch of the vessel of interest must be depicted in* the longitudinal plane (parallel vessel walls along the width of the monitor). *The smaller the angle* between the Doppler beam and the direction of the flowing blood, the higher the Doppler frequency shift of the reflected echo and the higher the sensitivity for flow detection. At angles of 60–90°, the Doppler frequency shift decreases and flow velocity measurement becomes progressively more unreliable, and no flow signals are depicted when the Doppler beam is perpendicular to the vessel wall (Fig. 1.39b, c). Conversely, *perpendicular insonation of a reflecting surface provides optimal B-mode information.* In the clinical setting, the examiner must find a compromise between optimal B-mode imaging and optimal flow evaluation. The usual procedure is to first try and achieve a perpendicular beam angle for morphologic evaluation of the vessel wall and to then sample the Doppler information either by activation of *beam steering* (lateral beam deflection) or with a smaller angle through *shifting and angling of the curved-array or sector probe and placement of the Doppler gate at the edge of the imaging field.*

In color duplex imaging with a curved-array transducer, different shades of color that do not reflect different flow velocities are displayed when interrogating a vessel coursing parallel to the body surface (e.g., carotid or femoral artery) with the transducer held perpendicular to the body surface. Under these conditions, flow near the wall is displayed in lighter shades due to higher Doppler shifts and in increasingly darker colors toward the center, where the frequency shift decreases as the insonation angle increases. This phenomenon is relevant only in the velocity mode and not in the power mode, as the latter is virtually independent of the Doppler angle.

#### 1.1.4.7 Physical Limitations of Color Duplex Ultrasound

As a result of the vast amount of information to be processed, color duplex scanning has a much poorer spatial and temporal resolution than pure B-mode imaging. Axial resolution is proportional to the wavelength in B-mode imaging, while it is dependent on the number of sample volumes placed along the color Doppler scan line in the color duplex mode. The *use of smaller sample volumes improves axial resolution but at the expense of sensitivity and accuracy in Doppler shift evaluation* as the signal-to-noise ratio deteriorates. Lateral resolution in color-coded duplex scanning is determined by the number of color Doppler lines processed per centimeter. The frame rate decreases as the number of Doppler lines increases, resulting in a lower temporal resolution, in particular at greater scan depths.

As a result of these limitations, the axial resolution of color duplex ultrasound is on the order of 0.4–1.0 mm with a lateral resolution of only 1.0–2.0 mm, which is four to ten times lower than the B-mode scan resolution.

The *frame rate* in the color duplex mode ranges from 50 to 200 ms (corresponding to an image repetition rate of 5 Hz), depending on the scanning depth and size of the color box. When a low PRF is selected, the speed at which the color Doppler scan lines sweep the sector is similar to or slightly below the mean flow velocity in arteries. Therefore, a single ultrasound scan may simultaneously depict systolic flow and early diastolic flow (e.g., displayed in red and blue, respectively, see Fig. 1.36). Due to the low temporal resolution, however, the color coding does not fully reflect the pulsatile character of flow.

Slow flow produces smaller Doppler frequency shifts, which have to be extracted from short echo pulse packets for each scan line consisting of a number of individual pulses. The scan lines must be processed successively.

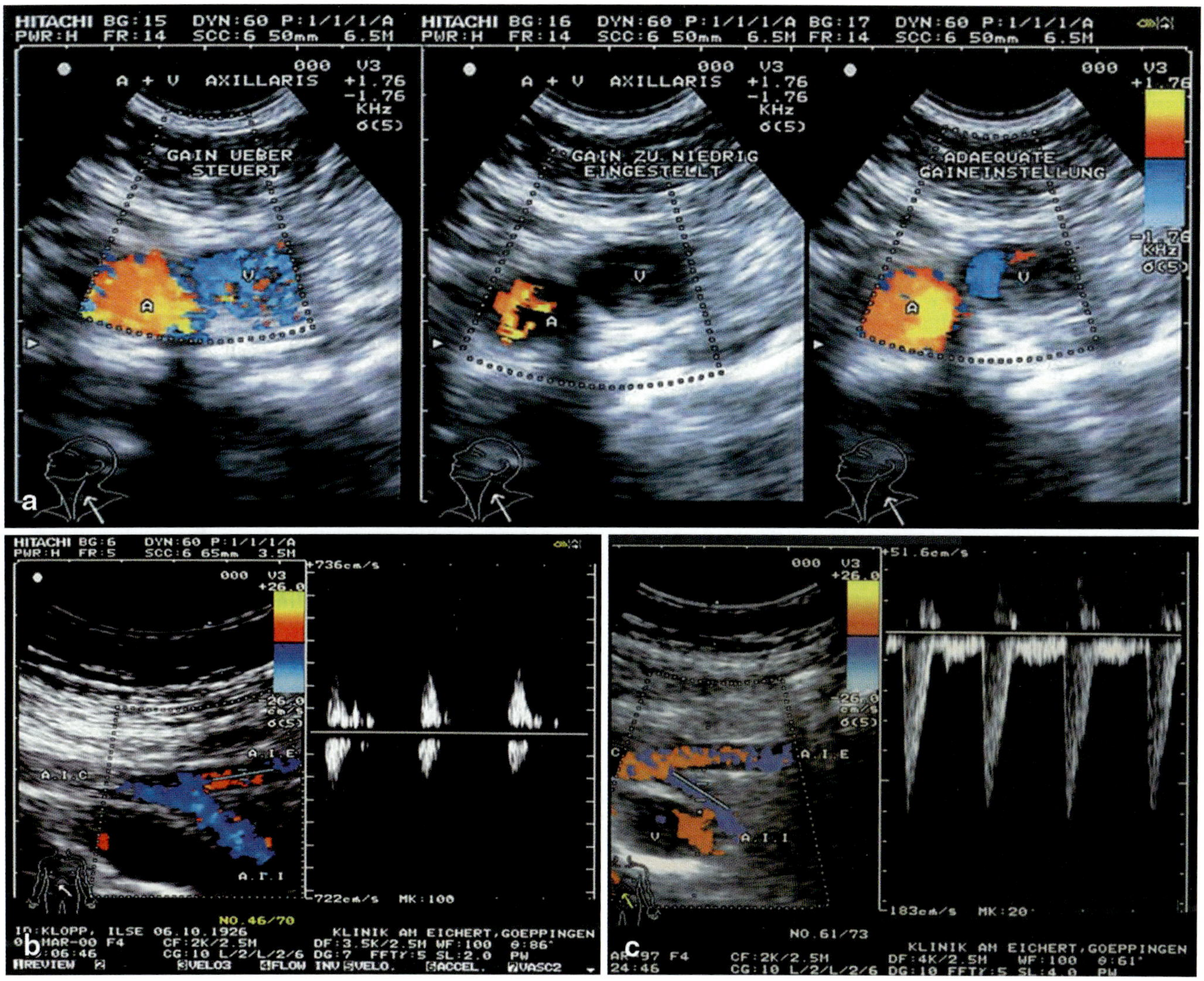


**Fig. 1.39** **a** Effects of different gain settings on the visualization of mural thrombosis in the axillary vein, as illustrated by a series of images obtained with the same pulse repetition frequency (PRF). The scan on the *left* obtained with gain overmodulation shows the mural thrombus to be largely obscured by artifacts extending beyond the patent lumen (*blue*). In the next scan (*middle section*), no flow signals are depicted along the thrombus in the axillary vein because the gain is too low. In addition, there is incomplete color filling of the axillary artery (*red*) cranial to the vein. The scan on the *right* illustrates adequate gain setting. There is proper color coding of the artery (*red*) and the thrombus in the axillary vein is depicted with adjacent *blue* venous flow signals. **b** An optimal gray-scale scan is obtained with a beam perpendicular to an interface between tissues of different acoustic impedance. In contrast, angles of 80–90° yield the lowest frequency shifts according to the Doppler equation and thus provide poor (color) duplex information. Therefore, interrogation of the iliac bifurcation at the preferred site of stenosis of the external iliac artery (*A.I.E*) poses a problem as adequate Doppler angles can be achieved only by tilting the transducer and moving it up or down (see Fig. 2.2b). In the color duplex scan, the change between *red* and *blue* at the origin of the external iliac artery illustrates the problem of determining the flow direction. The Doppler waveform is not helpful either, showing flow in both directions. As a result of the poor Doppler angle, an iliac artery stenosis with a peak systolic velocity (PSV) of 250 cm/s is suggested. **c** The posterior origin of the internal iliac artery (*A.I.I*), on the other hand, can be interrogated with a Doppler angle of less than 60° from the same transducer position, yielding a diagnostic Doppler waveform, from which a correct PSV of 120 cm/s is calculated

Though the insonation angle should ideally be as small as possible for optimal velocity measurement, this is not always practical because there will be a *longer delay when the color box is tilted* (beam steering). This is why one must find a compromise, in particular when examining deeper vessels. Tilting the color box by 20° and 30° prolongs the round trip time by 13% and 31%, respectively.

Color duplex imaging, like all diagnostic ultrasound techniques, is impaired by *scattering* and *acoustic shadowing* caused by bowel gas or calcified structures (bone or calcified plaques in the lumen). The examiner can circumvent such interfering structures by moving the transducer, but this often increases the distance between the transducer and the target anatomy, and hence the round trip time.

A strong reflector in the beam path can act like a mirror and generate a phantom image in another area of the scan. Such mirror images can be identified by angling the transducer, which will make the mirror artifacts disappear or appear in a different location. When the ultrasound beam strikes interfaces of high acoustic impedance at a right angle, reverberations (repeat echoes) may occur with the ultrasound pulses being reflected to and fro, resulting in a kind of ping-pong effect. Slight angulation of the transducer prevents reverberations but will also reduce reflection from the interface and thus degrade image quality.

### 1.1.5 Ultrasound Contrast Agents

Color duplex scanning with a high-resolution transducer usually allows adequate visualization of the peripheral arteries in the gray-scale mode; the evaluation of blood flow, however, may be impaired, either by the presence of sclerotic vascular lesions or by scattering due to edema or other localized soft tissue changes. In these situations, a microbubble contrast agent, or echo enhancer, can be used to improve blood flow imaging. For vascular ultrasound examinations in the clinical setting, however, a contrast agent is rarely needed as there are only a few specific diagnostic tasks (identification of a suitable target vessel for crural bypass grafts for example) where a poor acoustic window prevents color duplex and spectral Doppler from providing therapeutically relevant information.

*Ultrasound contrast agents typically consist of encapsulated microbubbles and enhance the contrast between blood and surrounding tissue by producing strong reflections in the vascular compartment during their lifetime of 3–5 min after intravenous administration.* Microbubbles thus have the potential to improve the localization of vessels with slow flow that are notoriously difficult to identify with conventional ultrasound techniques. Nevertheless, ultrasound contrast agents are mainly used to evaluate organ and lesion perfusion, to characterize focal liver lesions for example. They play no role in the examination of patients with angiologic or vascular surgical conditions in the routine clinical setting. Notable exceptions are small peripheral vessels and lower leg arteries with slow, postocclusive flow, transcranial duplex scanning, and the search for endoleaks after stenting of the aorta. With the use of contrast media, however, ultrasound to some extent gives up the crucial advantages it normally has over other imaging modalities – low cost, short examination time, and noninvasiveness. Moreover, in those body regions where diagnostic improvement is achieved through the administration of an echo enhancer, competing imaging modalities are used because they provide high accuracy and often enable better documentation of the findings. Since the focus of this book is on the role of ultrasonography in vascular surgery and angiology, the basic principles, advantages, and limitations of ultrasound contrast agents will only be discussed briefly below.

Ultrasound contrast agents with different compositions and properties are commercially available from various manufacturers. Basically, the contrast agents consist of microbubbles composed of a gas core encapsulated by a thin shell or stabilized by a carrier medium.

The stability of the microbubbles is determined by the composition of the gas core and the shell. Microbubble shells are made of surface stabilizers (palmitic acid, phospholipids) or substances that form a capsule on the molecular level (albumins, polymers). The stability of the core can be improved by using a heavier gas instead of air (sulfur hexafluoride, perfluoropropane). Such gas-filled bubbles persist in the bloodstream longer as they have a lower solubility and diffusivity. Bubbles containing a high-molecular-weight gas are used in newer contrast agents. These bubbles can pass the pulmonary circulation because they are more resistant to destruction by exposure to changing pressures.

**Approved ultrasound contrast agents** A variety of microbubble contrast agents differing in core and shell composition are undergoing preclinical and clinical testing. While a number of agents have been approved for echocardiography, there are basically only two agents that are currently on the market and widely used for vascular ultrasound in the clinical setting.

Levovist, the first ultrasound contrast agent approved in Europe, consists of a suspension of galactose-based air bubbles coated with a stabilizing palmitic acid layer. The suspension is prepared immediately before administration and can be given as a bolus or short infusion. Following intravenous bolus injection, it takes the microbubbles approximately 10–20 s to arrive in the arteries. After about 1 min and an initial phase of blooming (artifactual flow outside the vessel), a period of uniform enhancement occurs which lasts for about 2 min and then recedes within 3–4 min. The blooming effect can be avoided by administering the suspension as a short infusion with an injection rate that is adjusted according to diagnostic need (typically between 1 and 4 mL/min). Depending on the target anatomy, bolus injection of Levovist will increase the Doppler signal intensity by 10–15 dB.

SonoVue is an echo enhancer approved in Europe for cardiac and vascular applications. The microbubbles contain sulfur hexafluoride and are predominantly stabilized by a phospholipid coating (mean bubble size of 2–3 μm). Bolus injection (at a rate of 1 mL/s) produces maximum enhancement after 10–20 s, followed by a gradual decrease in intensity over the next 10–12 min (imaging window). When given as a short infusion, a 15–20 dB signal intensity increase can be observed after 1–2 min, depending on the injection rate. As with Levovist, the problem of initial blooming can be corrected by lowering the color gain. The slow linear decrease in enhancement following the initial peak results in adequate enhancement persisting long enough for most vascular applications (using bolus injection at a rate of 1 mL/s).

The behavior of contrast microbubbles in an ultrasound field depends on their properties and stability as well as the acoustic energy applied. Microbubbles are strong reflectors as a result of the acoustic mismatch that exists between the gas and the surrounding blood or tissue. The amount of backscatter is determined by the bubble diameter and compressibility as well as the transmitted ultrasound frequency.

Ultrasound contrast media thus act by increasing the proportion of scattered and reflected ultrasound pulses from the blood, thereby improving the Doppler signal and the signal-to-noise ratio. The *gas-filled microbubbles* used in most ultrasound contrast agents enhance backscatter as a result of the large difference in acoustic impedance that exists between the gas and the liquid blood. The extent of backscatter depends on the microbubble concentration and the reflection capacity of the individual bubbles, which is a function of their scatter cross-section. The maximum bubble size is limited by the fact that they must pass the lungs (bubble size <8 μm). Studies have shown that backscatter enhances the echo signal intensity of blood by 15–25 dB (Kaps and Seidel 1999). Low ultrasound beam power induces linear oscillation of the microbubbles, transforming the bubbles into small "ultrasound transmitters." This is another mechanism by which microbubbles enhance the ultrasound signal. The resonance frequency is inversely proportional to the bubble diameter.

At a higher power, the microbubbles oscillate in a nonlinear fashion, producing strong signals at fundamental and nonfundamental frequencies. The nonfundamental or *harmonic frequencies* are multiples of the transmitted frequency. The second harmonic has the highest energy and is therefore most relevant for diagnostic ultrasound (Fig. 1.40). Because the resonance frequencies of the 2–7-μm gas bubbles are within the range of the transmitted frequencies of 2.5–10 MHz typically used in diagnostic ultrasound, the bubble vibrations produce an additional signal amplification. The corresponding frequencies are received and processed in addition to the Doppler-shifted frequencies. The resonance effect of the microbubbles improves the signal-to-noise ratio by a further 30–35 dB, provided that the bandwidth extends over a sufficient range of frequencies to enable the generated harmonics to be detected (Correas et al. 1997).

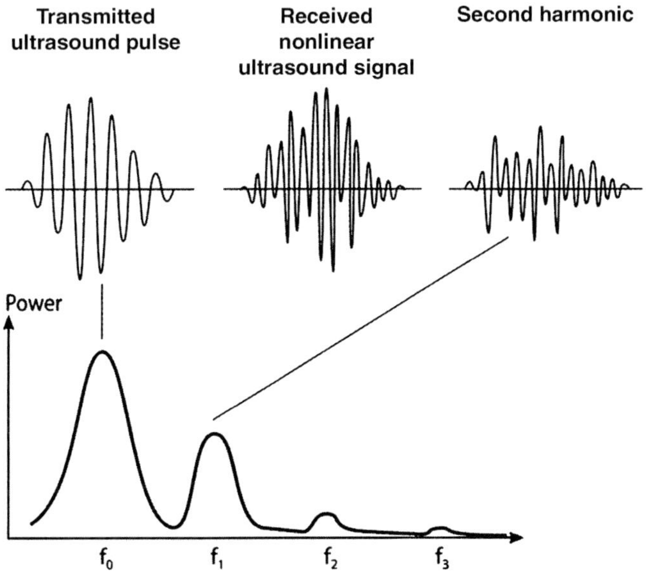


**Fig. 1.40** The Doppler-shifted signal returning to the transducer contains both the fundamental and harmonics, which can be extracted by comparison with the emitted pulse

Short pulses of high energy can be applied to make the microbubbles burst, producing ultrasound signals that are detected with high sensitivity. In contrast to the enhancing mechanisms outlined in the preceding sections, bursting is independent of flow and the resulting signals merely show the distribution of the collapsed microbubbles at a certain time point. This technique was introduced as *sonoscintigraphy* (Cosgrove et al. 1998) and is increasingly being used to differentiate and characterize focal liver lesions (now known as stimulated acoustic emission, see below).

High microbubble concentrations must be avoided because they may obscure deeper structures by backscattering most of the incident ultrasound energy (contrast shadowing).

Both Sonovist and Levovist are taken up and eliminated by the reticuloendothelial system of the liver. In the liver, the Sonovist bubbles burst when exposed to a high-energy beam and thus contribute to the differentiation of liver metastases, which do not have a reticuloendothelial system. In addition, focal liver lesions may be differentiated on the basis of their blood supply (predominantly portal venous versus arterial) and differences in contrast medium arrival times after bolus injection. Other echo enhancers are available as suspensions (some of which contain human albumin) of particles stabilized with other materials for specific applications.

The different effects and properties of microbubble contrast agents outlined above result in complex interactions with the ultrasound field, which can be exploited by a variety of contrast-specific ultrasound techniques. There are basically three strategies for visualizing the effects of contrast agents (Fig. 1.41).

**Contrast-enhanced duplex sonography** The reflection of ultrasound by microbubbles present in the blood selectively enhances the vascular system (in color duplex and power Doppler), thus improving the delineation of arteries and veins from surrounding tissue. The velocity of flowing blood is not affected by the microbubbles, and both quantitative and semiquantitative spectral Doppler measurements can be performed (CW/PW Doppler); in the spectral Doppler mode, however, the gain must be lowered. Using a frequency-based technique such as Doppler imaging, the effect of contrast enhancement can only be exploited in large and medium-sized vessels as the signals returning from slow microcirculatory flow in small vessels cannot be adequately separated from the echoes produced by moving tissue.

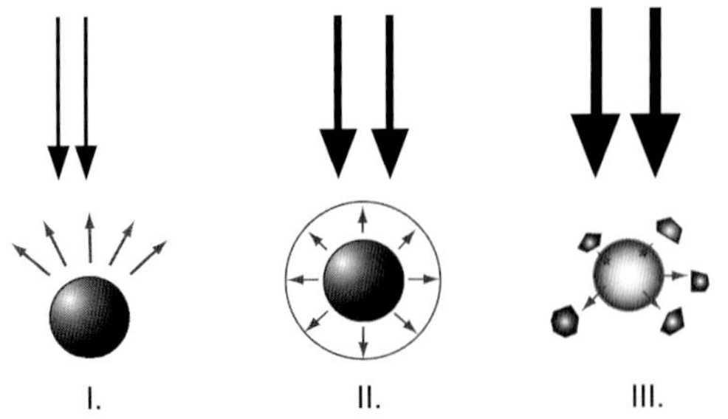


**Fig. 1.41** Ultrasound techniques using contrast agents. The amount of ultrasound energy applied determines how the microbubbles interact with the ultrasound beam, giving rise to different techniques of image generation. Exposure of microbubbles to low energy (I) causes an increase in reflection and backscatter. Intermediate ultrasound energy levels (II) induce nonlinear oscillation of the microbubbles, resulting in the emission of second harmonics, which selectively enhance the blood signal (contrast harmonic imaging). High ultrasound energy (III) destroys the microbubbles, releasing frequencies that provide information on the distribution of the contrast bubbles in the circulation (stimulated acoustic emission)

Microbubble contrast agents have been used to evaluate slow flow in peripheral vessels, including postocclusive flow, and carotid artery stenosis, and studies have shown that echo enhancers improve the accuracy of color duplex ultrasound in identifying pseudo-occlusion (Fürst et al. 1999; Ferrer et al. 2000). The sonographic evaluation of the vertebral territory can be improved in particular in patients with poor scanning conditions or with a hypoplastic vertebral artery.

Echo enhancers have also been advocated to improve the signal-to-noise ratio in the examination of deep vessels such as the pelvic or renal arteries. Using state-of-the-art ultrasound equipment, however, these vascular areas rarely pose diagnostic problems. On the other hand, a study of renal artery stenosis investigating duplex scanning with echo enhancer administration demonstrated the diagnostic yield to increase from 64 to 84%, while the improvement in the sensitivity and specificity for identifying high-grade stenosis was negligible (Claudon et al. 2000).

**Contrast harmonic imaging** This ultrasound technique can improve vascular imaging by selectively displaying the harmonics specific to microbubbles. A broadband transducer is used to detect the harmonics generated by the contrast microbubbles, especially the second harmonic. The backscatter from microbubbles allows better separation of the echoes from stationary tissues, thus offering advantages in the sonographic evaluation of slow flow compared with the fundamental techniques.

**Stimulated acoustic emission imaging** Contrast agent microbubbles are destroyed when exposed to high sound energy (high mechanical index). The bursting bubbles emit transient ultrasound signals, which can be detected with a broadband transducer. The system registers the variation in signals from pulse to pulse. This information is displayed in conjunction with the spatial information of the B-mode image to show the contrast agent distribution in the macro- and microcirculation at a certain point in time.

**Specific instrument settings using echo enhancers** The administration of echo enhancers in color duplex ultrasound leads to overmodulation of the color Doppler signal with color overflow obscuring perivascular structures and parts of the vessel wall, in particular when larger vessels are examined. This effect can be eliminated by downregulating the receive gain. On the other hand, the strong reflection produced by the contrast bubbles leads to sound attenuation in structures farther away from the transducer than the enhanced vessel. In Doppler scanning after echo enhancer administration, there will be spectral broadening with almost complete filling-in of the systolic window. This effect can likewise be counteracted by downregulating the receive gain.

In summary, all sonographic techniques exploiting the different effects of ultrasound contrast agents in the vascular compartment aim at improving the sensitivity to blood flow phenomena. With *the recent advances in ultrasound technology*, however, for most applications this is not necessary. Moreover, in those rare cases where an adequate diagnostic evaluation with color duplex imaging is not possible, the use of a microbubble contrast agent is often limited as well and offers no benefits. Despite these cautionary remarks, *a few specialized uses have emerged* including transcranial duplex imaging, evaluation for renovascular disease, vein mapping prior to crural bypass surgery, and the search for endoleaks following aortic stenting. It is expected that ultrasound contrast agents will be used increasingly in the evaluation of microcirculation, e.g., for perfusion studies (Seidel et al. 2006; Claudon et al. 2008).

## 1.1.6 Safety of Diagnostic Ultrasound

Ever since the early 1960s when this technique was first used for diagnostic imaging, the potential biological hazards of medical ultrasound have been discussed. Ultrasound traveling through the human body can have two effects known to cause changes in biological systems. Firstly there are the thermal effects resulting from the conversion of ultrasound energy into heat and secondly there are the mechanical effects arising from pressure changes associated with the propagation of sound waves in a medium.

### 1.1.6.1 Thermal Effects

The use of diagnostic ultrasound can give rise to tissue heating because sound energy is absorbed and converted to heat. The ability to absorb energy varies with the tissue; it is low in body fluids (amniotic fluid, blood, urine) and high in bones. Adult bones absorb 60–80% of the incident ultrasound energy. In addition, absorption in the body is also affected by technical parameters, most notably the frequency transmitted by the transducer. Higher frequencies are absorbed more rapidly. A temperature increase of 2.5°C can severely damage biological tissues, while an increase of 1°C is generally considered harmless. Experimental evidence suggests that the thermal effects of diagnostic ultrasound procedures pose no health hazard.

#### 1.1.6.2 Mechanical Effects

Most of the mechanical effects of ultrasound that are potentially harmful to living tissues are related to the formation, growth, and possible collapse of tiny gas bubbles in the ultrasound field, a process known as cavitation. Recall that ultrasound propagates through tissue in waves of alternating high and low pressure. Bubble formation or the expansion of existing bubbles occurs when the negative pressure is large enough. The development of cavitation and its effects depend on the frequency and intensity of the transmitted ultrasound waves as well as on the focus of the acoustic field. Two types of cavitation are commonly described: stable and inertial (or transient).

Stable cavitation refers to the continuous oscillation of gas-filled bodies in response to the alternating positive and negative pressures to which they are exposed in an ultrasound field. These cyclic expansions and contractions result in an increased flow in the fluid-like medium surrounding the vibrating bubbles. This is known as microstreaming and can generate very high pressures with disruption of cell membranes. In the other form of cavitation, inertial cavitation, existing bubbles or cavitation nuclei expand during the low-pressure phase and then collapse violently. Microbubble collapse is a highly localized process occurring on the order of microseconds. Collapsing bubbles can produce extremely high temperatures and pressures but these dissipate rapidly. Bursting bubbles therefore have the potential to destroy cells and tissues. There is scientific evidence to suggest that inertial cavitation is a threshold phenomenon and will only occur if microbubbles already present in the acoustic field are exposed to excessive sound pressures and frequencies. Pressure below the cavitation threshold will never by itself lead to cavitation, not even during extremely long exposure to ultrasound. Inertial cavitation induced by diagnostic ultrasound procedures therefore remains a mere theoretical possibility and has never been reported in vivo. However, one must also be aware that in vivo evidence of potential bioeffects of ultrasound is very difficult, if not impossible, to obtain: cavitation can occur anywhere in the body, and the damage it produces may be very local, involving only a few cells. Modern ultrasound equipment incorporates safety mechanisms allowing the user to limit the average acoustic output, thus avoiding peak pressures that could theoretically lead to cavitation or other mechanical bioeffects.

#### 1.1.6.3 Specific Risks of Individual Ultrasound Techniques

**B-Mode** B-mode imaging is generally performed at very low acoustic output, resulting in intensities below 10 mW/$cm^2$. The individual pulses are very short (<1 ms) and are emitted at a PRF of less than 5 kHz to achieve high spatial resolution. As the energy transmitted into the body is dissipated over a large volume, the resulting rise in tissue temperature is so small that it remains below the limit of detection. Diagnostic B-mode ultrasound is considered absolutely safe in terms of potential hazards to patients.

**M-Mode** This technique uses higher energies and may theoretically cause tissue heating. In this mode, a stationary beam is emitted repeatedly to evaluate moving structures. The scan volume is smaller than in B-mode imaging, but the PRF is much lower (only about 1 kHz). M-mode sonography is also considered safe.

**CW Doppler** As with M-mode techniques, the scan volume is small and there is continuous exposure. The power output can reach up to 100 mW, and some procedures have the potential to produce biologically significant temperature rises. Possible mechanical effects are much less of a concern than in B-mode or M-mode scanning, despite the higher acoustic output. The transmit power depends on the depth of the target anatomy. It is the operator's responsibility to keep the overall examination time as short as is consistent with achieving diagnostically useful results.

**PW Doppler** Again, the exposed volume is relatively small and the PRF is high. The individual pulses are often twice as long as with B-mode or M-mode techniques. Taken together, the machine settings used in PW Doppler applications can result in considerable exposure, and the risk of tissue heating is far greater. Conversely, mechanical effects are negligible because the intensity of the emitted pulses is the same as in B-mode and M-mode scanning.

**Color Doppler** The acoustic output of color Doppler equipment is intermediate between that of B-mode and PW Doppler. Mechanical effects are negligible. The emitted ultrasound pulses are distributed over a relatively large tissue volume. Temperature rises are higher than with B-mode imaging but lower than with PW Doppler techniques.

Current uses of diagnostic ultrasound expose the body to intensities that do not exceed 100 mW/$cm^2$. There is no evidence that these intensities damage living tissues.

Nevertheless, the examiner should always seek to minimize exposure by limiting both the power output and the duration of scanning to what is absolutely necessary to obtain the desired diagnostic information. This approach is known as the ALARA principle (as low as reasonably achievable) and applies to all diagnostic imaging modalities.

## 1.2 Hemodynamic Principles

### 1.2.1 Laminar Flow

Although blood flow is subject to specific conditions due to the solid components in plasma and the elasticity of the vessel wall, it basically follows the laws of hemodynamics. These laws govern the flow of a fluid in tubes and apply to watery or oily solutions of a constant viscosity (Newtonian fluid) and assume that flow velocity under these conditions is primarily a function of the pressure difference that exists between the two ends of the tube. These ideal conditions for continuous laminar flow are typically not met in a living organism because various factors like elasticity of the vessel wall, pulsatility resulting from cardiac activity, curving of vessels, and branching affect blood flow, resulting in changing velocity distributions in the different layers. Moreover, blood is not a watery or oily solution of constant viscosity but a suspension of solid blood cells in plasma. The viscosity of blood is primarily dependent on its hematocrit and is only constant at levels below 10, increasing exponentially at higher values. Other factors affecting blood viscosity are plasma viscosity and vessel diameter. In the terminal capillary bed, viscosity is additionally influenced by the deformation of red blood cells. Despite these specific features of blood flow, some basic hemodynamic terms and laws are useful and will make it easier to understand normal and abnormal flow in arteries and veins. In addition, in vitro experiments and in vivo blood flow measurements using duplex scanning have provided new insights into the flow behavior in specific vessels under normal and abnormal conditions as well as under the influence of pharmacologic agents.

*Continuous* or *laminar flow* is characterized by a constant velocity over time. Flow in a tube is brought about by a pressure difference between the two ends of the tube. The pressure difference ($P_1$–$P_2$) is proportional to the *volume flow rate*. The volume flow rate ($I$) is proportional to the tube diameter ($r$) and inversely proportional to its length ($l$) and the viscosity of the fluid ($\eta$). Mathematically, this relationship is expressed in the Hagen–Poiseuille law:

$$I = \frac{(P_1 - P_2) \cdot \theta \cdot r^4}{8 \cdot l \cdot \eta} = \frac{(P_1 - P_2)}{R}$$

By analogy with Ohm's law, *flow resistance* can be calculated from the Hagen–Poiseuille equation:

$$R = \frac{8 \cdot l \cdot \eta}{\theta \cdot r^4}$$

It follows that resistance is proportional to the length of the tube ($l$) and the viscosity of the liquid ($\eta$). Overall resistance is most strongly affected by the radius ($r$), which enters the equation in the fourth power. This means that decreasing the vessel radius by one-half, for example, increases the flow resistance by a factor of 16. Peripheral resistance in the vascular system is regulated according to demand, primarily by the tone of the arterioles, and affects the pulsatility depicted in the Doppler waveform of large vessels.

The *flow profile* of laminar flow is determined by inertial and frictional forces. Friction produces a laminar or, in the 3D model, parabolic flow profile. Flow is fastest in the center of a vessel and decreases toward the wall, where it approximates zero.

Color duplex ultrasound reflects this flow profile by lighter color shades in the center (fast flow) and darker shades near the wall (slow flow) (Fig. 1.42a). The following factors determine the shape of the flow profile of blood:

- Velocity
- Viscosity (internal friction)
- Adhesion of the blood to the vessel wall (external friction)
- Cohesion (forces that occur between adjacent molecules of like composition)

Blood differs from Newtonian fluid in that its viscosity is not an inherent property that only varies with temperature but is mainly determined by the hematocrit level and other factors such as plasma viscosity (which in turn is predominantly dependent on the fibrinogen concentration), red blood cell deformability, and the degree of shearing. In an artery or vein with laminar flow, shear forces (e.g., thrust) are weakest in the center and strongest near the wall (Fig. 1.42b).

According to the continuity law, a decrease in the cross-sectional area in the course of a vessel segment leads to an increase in mean flow velocity. Blood flow through a vessel segment with an abrupt change in caliber becomes flattened (plug flow) upon entering the narrower vessel segment. In plug flow inertial forces are stronger than frictional forces, resulting in the same flow velocity of all fluid layers in the vessel except for a thin layer near the wall. So-called turbulent flow results when the inertial forces become even stronger than the frictional forces, which bring order to the course of flow. Turbulent flow is characterized by an irregular flow pattern with flow in different directions. The typical parabolic flow pattern develops after a certain stretch along which frictional forces predominate. Physiologically, this occurs when blood leaves the left ventricle and enters the ascending aorta. The sharp velocity gradient between central portions and the thin boundary layer near the wall in plug flow is associated with strong shear forces.

The abrupt decrease in the cross-sectional area in a stenotic vessel segment and the resulting increase in flow velocity are associated with progressive disturbance of laminar flow, which will finally become turbulent. Turbulent flow above a critical velocity is characterized on color duplex ultrasound by a mosaic of colors reflecting

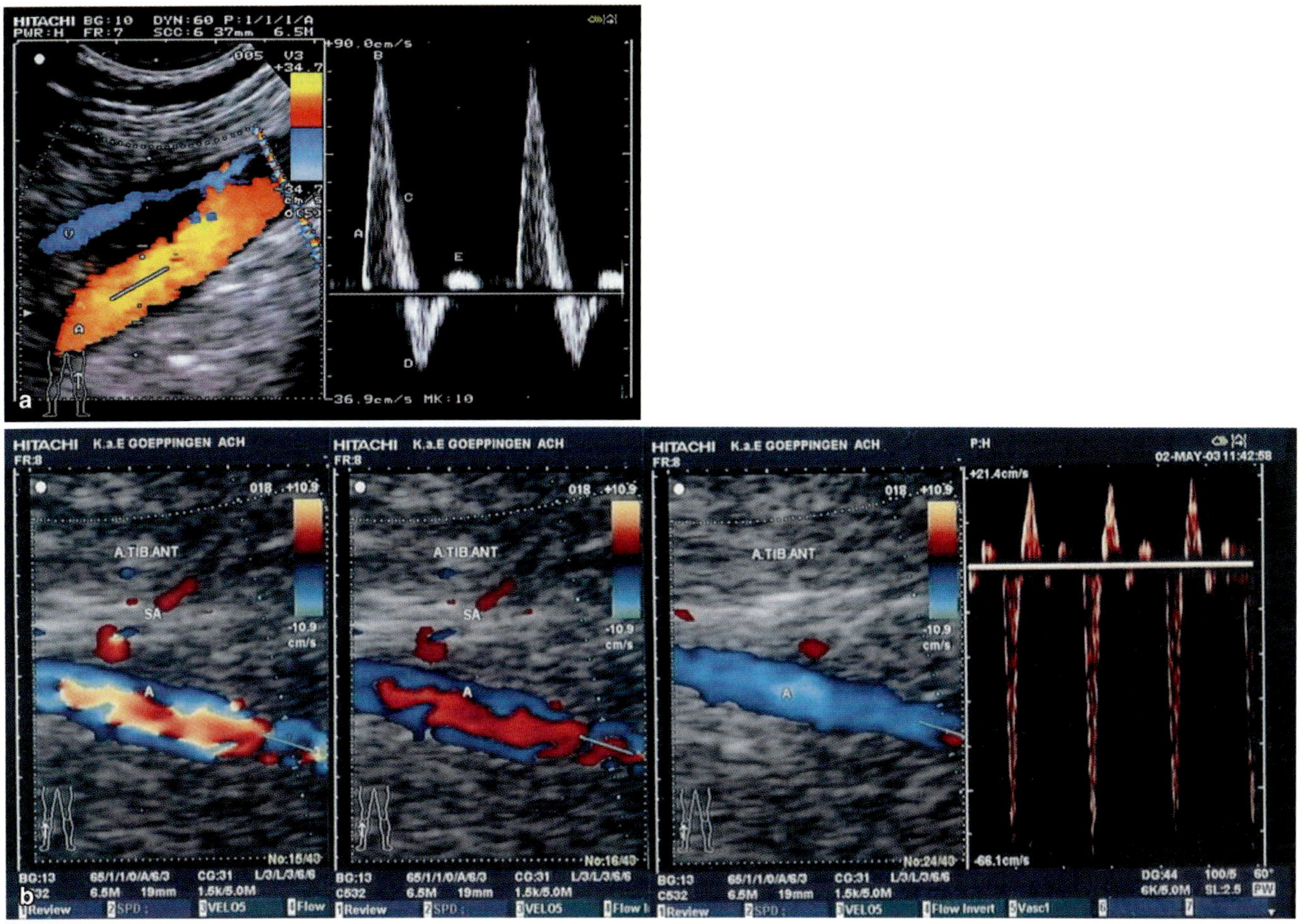


**Fig. 1.42** **a** Typical triphasic Doppler waveform of the popliteal artery. In the color duplex image, laminar flow is characterized by lighter coloring in the center with darker colors representing slow flow near the wall. *Red* indicates blood flow toward the transducer. A vein closer to the transducer is displayed in *blue*, indicating blood flow away from the transducer. The triphasic waveform consists of a steep upslope (*A*), a deceleration phase (*C*) following peak systolic flow (*B*), a short phase of early diastolic backward flow (*D*), and forward flow from mid-diastole to end diastole (*E*). The magnitude and duration of diastolic forward flow (*E*) depend on peripheral resistance (sympathetic tone) and the thrust generated by the compliant aorta (windkessel effect). Blood flow toward the transducer is displayed above the baseline, and flow away from it below the line. The Doppler angle of insonation is 59°, and peak systolic velocity (PSV) is 85 cm/s. The different intensities of the individual pixels in the Doppler waveform reflect the number of red blood cells at a given point in time. The amplitude can also be represented in the form of a histogram. **b** Use of a low pulse repetition frequency (PRF) to ensure good color filling of arteries with slow flow in the lower leg. In the first color image (*leftmost section*), the anterior tibial artery (*blue*, flow toward the periphery, away from the transducer) is depicted with central aliasing (color change from *blue* to *yellow* to *red*). This example illustrates a laminar flow profile with fast flow in the center and lower flow velocities near the arterial wall due to friction. The second color scan simultaneously depicts early diastolic blood flow at the same site, which is due to the low frame rate. In the center, arterial reflux due to high peripheral resistance is seen as a superimposed wave (*red*, toward transducer) while flow toward the periphery predominates nearer the walls (*blue*, away from transducer). The view illustrates true flow reversal relative to the ultrasound beam rather than aliasing. True flow reversal is characterized by a color change from *blue* to *black* to *red*. The third color scan depicts blood flow (*blue*, away from transducer) toward the periphery and without aliasing in mid-diastole. The blood flow in a peripheral artery during one cardiac cycle as depicted by these three color duplex scans is additionally represented in the Doppler frequency spectrum (*rightmost section*), which presents the flow changes over time

the different flow directions. The transition from laminar to turbulent flow can be calculated by means of the dimensionless *Reynolds number*, which depends on mean flow velocity ($v$), vessel diameter ($d$), density of the fluid ($p$), and viscosity ($\eta$):

$$Re = \frac{v \cdot d \cdot p}{\eta}$$

Turbulent flow occurs at Reynolds numbers above 2,000. Blood flow becomes turbulent when the vessel lumen is narrowed by a stenosis. The details are highly complex but need not be taken into account by the sonographer in determining the degree of stenosis.

In turbulent flow, part of the kinetic energy is converted into sound energy, which can be auscultated as a characteristic bruit.

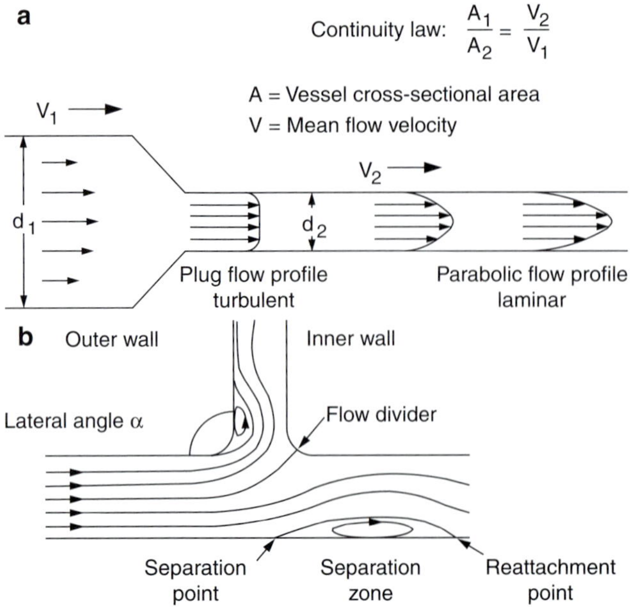


**Fig. 1.43** **a** When a fluid such as blood enters a narrower lumen, parabolic flow changes into plug flow and returns to its original profile only after having traveled some distance under the influence of shear forces. According to the continuity law, flow velocity increases in proportion to the decrease in diameter. **b** Flow pattern at a vessel branching. Thrust and shearing are highest at the inner wall of the branching. Separation occurs at the outer wall, where thrust is rather low. Physiologic flow separation is depicted sonographically in the carotid bulb (see Fig. A5.1) (according to Lanzer and Yoganathan 1991)

In vessels with pulsatile flow, which normally is laminar, turbulent flow may occur at specific phases of the cardiac cycle under physiologic conditions. This phenomenon depends on the flow profile (high pulsatility) and pulse rate.

A sudden increase in the vessel diameter results in a longer flow profile and greater velocity gradient across the vessel lumen. If the difference between a narrow and wide (poststenotic) segment exceeds a certain value, *flow separation* and eddy currents will occur near the wall. Flow separation near the wall is also observed in branching blood vessels (Fig. 1.43).

Flow separation is associated with recirculation zones in which relative stasis of flow in conjunction with shear forces induces platelet aggregation with release and adhesion of procoagulative agents, which in turn can trigger local atherogenic processes. This is a possible mechanism contributing to the preferred occurrence of atherosclerotic lesions in dividing and branching vessel segments.

The phenomenon of flow separation is depicted most impressively by color duplex scanning at the origin of the internal carotid artery (ICA), where it occurs as a result of both widening in the bulb area and branching (see Fig. A5.1). Flow separation may also be observed in poststenotic vessel segments as a result of the abrupt increase in the cross-sectional area (see Fig. 1.46).

## *1.2.2 Flow Profiles and Perfusion Regulation*

Unlike laminar flow, *pulsatile flow* changes periodically over time. Phases of acceleration and deceleration vary in relation to changes in pressure. The *pressure amplitude* generated by the left ventricle is reduced by the *compliance of the aorta and other large elastic arteries* (windkessel effect), resulting in a more steady flow. Another factor affecting the flow profile is the *peripheral resistance.*

Flow is highly pulsatile in the extremity arteries because peripheral resistance is high at rest, giving rise to the characteristic triphasic Doppler waveform. An increase in the peripheral blood demand leads to dilatation of the arterioles, and the resulting decrease in peripheral resistance changes the Doppler waveform. *Peripheral resistance may decrease under normal* (muscle activity) *or abnormal conditions* (local inflammation, postocclusive ischemia, tumor perfusion) and is associated with an *increase in the diastolic flow component* in the spectrum. Moreover, the character of the waveform is affected by central regulatory processes (increase in heart rate, blood pressure) and vessel wall elasticity (diabetes mellitus).

As the peripheral resistance is a crucial factor affecting the waveform, a distinction is made between low-resistance flow and high-resistance flow (Fig. 1.44).

### 1.2.2.1 Low-Resistance Flow

Arteries supplying parenchymal organs and the brain are characterized by a fairly steady blood flow as a result of *low peripheral resistance*. In these arteries, a moderate systolic rise is followed by a steady flow that persists throughout diastole. This flow profile is typical of the renal, hepatic, splenic, internal carotid, and vertebral arteries.

The windkessel function of large vessels including the aorta leads to a *continuous diastolic flow* in arteries supplying parenchymal organs due to a lower peripheral resistance of the terminal vascular territories supplied by these arteries (Fig. 1.44).

The windkessel effect thus ensures a more continuous flow than would result from the action of the left ventricle and aortic valve alone. As a result, flow will become more pulsatile when this function and the normal elasticity of the vessels are lost.

### 1.2.2.2 High-Resistance Flow

*High peripheral resistance* results in a more pulsatile flow with a steep systolic upslope during the acceleration phase, followed by deceleration and a significant reflux in early

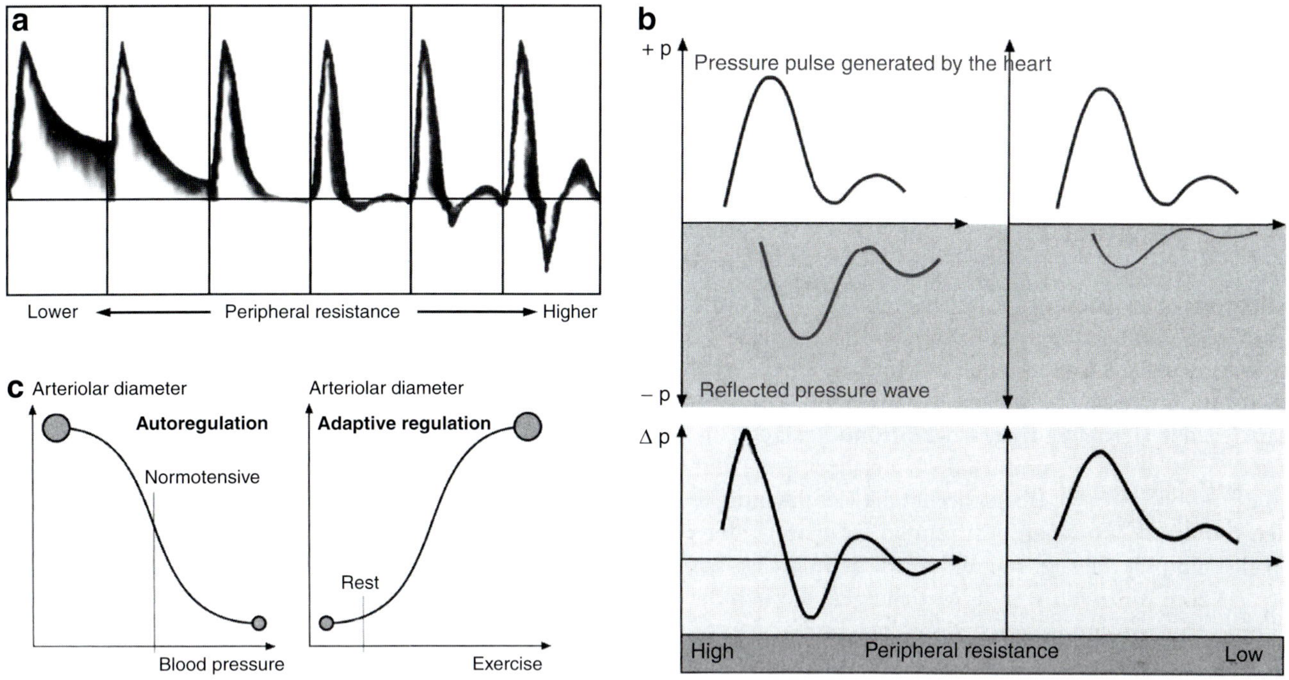


**Fig. 1.44** Effect of peripheral resistance on the Doppler waveform. **a** Pulsatility increases with peripheral resistance. **b** Flow pulse curves resulting from superimposition of the pressure pulse generated by the heart and the pressure waves reflected by the distal vascular bed in arteries with high peripheral resistance (*left*) and low peripheral resistance (*right*). For an identical pressure pulse generated by the heart (above the baseline), the pulse curve resulting from interaction with the reflected pulse (below the baseline) is highly pulsatile if peripheral resistance is high and less pulsatile if peripheral resistance is low. The pressure gradient (ΔP) calculated by subtracting the reflected pulse wave from the pulse wave emitted by the heart is directly proportional to blood flow velocity according to the Hagen–Poiseuille law. **c** Changes in arteriolar diameter in response to blood pressure changes (autoregulation, *left*) and in response to exercise (adaptive regulation, *right*)

diastole and short backward flow in mid-diastole. Zero flow is typically seen in end diastole. This pattern is referred to as *triphasic flow*.

The systolic pulse wave is in part reflected by the high peripheral resistance and thus moves backward through the arterial system until the flow is again redirected toward the periphery by the influx of blood during the next cardiac cycle. This flow component is small due to the high peripheral resistance.

As a result of the high pressure in the arterioles supplied by the limb arteries, significant blood flow in these vessels occurs only during systole when systemic pressure is higher than peripheral pressure. The pressure during diastole is too low to produce blood flow toward the periphery.

High-resistance flow is typical of the arteries supplying the muscles and the skin. These are the leg and arm arteries and the external carotid artery. The ratio of skin to muscle supply determines the extent of diastolic forward flow. When the peripheral demand increases (muscle activity, inflammation), the arterioles reduce the peripheral resistance, and forward flow, primarily during diastole, increases.

*Transitions between these two flow patterns* may occur *under normal and abnormal conditions*. Besides, there are vessels with mixed flow. A typical example is the superior mesenteric artery, which has pulsatile flow like a limb artery but also has a significant end-diastolic flow component. A surgically created arteriovenous (AV) fistula will turn high-resistance flow into low-resistance flow. A change from low-resistance to high-resistance flow in a transplant renal artery indicates graft rejection.

#### 1.2.2.3 Blood Flow Regulation

Blood flow throughout the body is regulated to ensure adequate perfusion of organs and tissues and to adjust the blood supply to a body region in response to varying demand (which is activity-dependent in the extremity arteries or increases after a meal in the mesenteric arteries). Local blood flow is regulated by arterioles and capillaries, which can dilate or constrict selectively and are therefore referred to as resistance vessels. Two types of regulation are distinguished: *autoregulation* maintains a constant blood flow (cerebral perfusion, renal vessels), while *adaptive regulation* varies blood supply according to demand (Fig. 1.44c).

Normal resting perfusion is maintained by constricted arterioles, resulting in high peripheral resistance and pulsatile flow. Widening of the arterioles in response to an increase in demand during exercise leads to augmented blood flow. The decreased peripheral resistance results in less pulsatile flow and a larger diastolic component.

In autoregulation, the ability of the arterioles to adjust their diameter serves to maintain constant blood flow and compensates for changes in arterial blood pressure or other systemic factors. *A drop in perfusion pressure leads to a compensatory widening of the resistance vessels*, and an increase in perfusion pressure to a compensatory narrowing.

### 1.2.3 Determining the Degree of Stenosis

According to the *continuity law*, the volume flow rate (cross-sectional area multiplied by mean flow velocity) is identical in all vessel segments. Therefore, an *abrupt decrease in arterial diameter* is associated with an *increase in blood flow velocity* (a 50% decrease in diameter, corresponding to a 75% decrease in cross-sectional area, will result in a four times higher flow velocity). The flow profile flattens out (plug flow) when the blood enters a narrower segment. If the increase in flow velocity is known, it is possible, in principle, to calculate the degree of stenosis according to the continuity equation:

$$X = 100 \cdot \left(1 - \frac{V_1}{V_2}\right)$$

$X$ percentage stenosis grade (cross-sectional area reduction)
$V_1$ prestenotic velocity
$V_2$ intrastenotic velocity

This equation for estimating the degree of stenosis does not take into account other systemic factors (blood pressure, wall elasticity, peripheral resistance) that may affect the ultrasound measurement. A sudden decrease in the cross-sectional area is associated with strong acceleration forces, which turn laminar flow into *plug flow* along the constricted segment. The increase in kinetic energy resulting from the increase in mean flow velocity in the stenotic segment is associated with a decrease in static (lateral) pressure. This conservation of energy is expressed in the *Bernoulli equation* as the sum of lateral pressure energy and kinetic energy:

$$P_1 + \frac{1}{2} \cdot \rho \cdot V_1^2 = P_2 + \frac{1}{2} \cdot \rho \cdot V_2^2$$

$P_1$ prestenotic lateral pressure
$P_2$ intrastenotic lateral pressure
$V_1$ prestenotic flow velocity
$V_2$ intrastenotic flow velocity
$\rho$ density of blood

Using the Bernoulli equation, the conversion of static pressure energy into kinetic energy resulting from a reduction in the vessel cross-sectional area is expressed as follows:

$$P_1 - P_2 = \frac{1}{2} \cdot \rho \cdot \left(V_2^2 - V_1^2\right)$$

The poststenotic increase in the cross-sectional flow area leads to turbulent flow, flow separation, and vortexing. The pronounced turbulence occurring in the presence of high-grade stenosis is associated with eddy currents and backward flow, resulting in the irreversible loss of most of the kinetic energy, while the loss due to inertial and frictional forces is negligible (Weber et al. 1992). The *pressure drop across a stenosis* (substantial portion of the measured Doppler blood pressure) thus predominantly *reflects the lost kinetic energy* in the stenosis. Therefore, the kinetic energy expressed as PSV in the stenosis can serve as a measure of the pressure drop over the stenosis (according to the simplified Bernoulli equation: $P_1 - P_2 = 4 \times V_2^2$; neglecting prestenotic velocity $P_1$ in favor of intrastenotic velocity $V_2$) and the degreee of stenosis (see Sect. 2.1.6.1.1).

Blood flow resumes its laminar profile only farther downstream of the stenosis as a result of decreasing turbulence and the influence of mural friction. The conversion of much of the remaining kinetic energy into static pressure energy promotes the dilatation of an atherosclerotically damaged vessel wall behind the stenosis.

In determining the degree of stenosis, the examiner must be aware that, due to friction losses inside the stenosis, the measured grade may be lower than expected. In high-grade stenosis with a reduced volume flow rate, flow velocity is already decreased in the prestenotic vessel segment, resulting in an intrastenotic increase in velocity that is lower than expected from the diameter reduction. The peak intrastenotic velocity may thus be lower in high-grade stenosis with a reduced volume flow rate than in moderate stenosis causing less pronounced flow reduction.

In the presence of two or more sequential stenoses, the respective pressure drops reduce PSV upstream of the next (second or third) stenosis, so that no absolute flow velocity levels can be given as thresholds for stenosis grading. In these cases, one should use quotients of flow velocities measured inside a stenosis and in the prestenotic segment. Velocity ratios greater than 2, corresponding to an increase in PSV inside the stenosis of over 100%, suggest a hemodynamically significant stenosis (>50%). This method cannot be used for grading stenosis occurring immediately downstream of a vessel division, which is a common site of stenosis (origins of profunda femoris or renal arteries, carotid bifurcation). For these vessel segments, empirical threshold velocities (reference value, angiography) must be determined to identify hemodynamically significant stenosis. Such thresholds are affected by systemic factors (hypercirculation, hypertension).

Doppler angle correction and cursor alignment are crucial in measuring blood flow velocity for stenosis grading. Ideally, the Doppler angle should be 60° or less; angles above 70° may result in unreliable flow velocity measurement. Alignment of the angle correction cursor may be technically challenging in curved or branching vessels.

PSV is measured in the stenosis or the jet seen in the color duplex image. The intrastenotic increase in flow velocity in high-grade stenosis results in a higher Reynolds number and is associated with turbulent flow. Turbulence is characterized on color duplex images by a typical mosaic pattern around the central stenosis jet with backward flow near the wall. Depending on the increase in flow velocity, eddy currents may be seen over a length of several centimeters downstream of the stenosis (Fig. 1.45a–f).

The jet tapers off distal to the stenosis, while the zone of turbulent flow widens until it occupies the entire lumen. Further downstream, flow becomes laminar again. *The pressure drop across a stenosis is determined by its length and degree*. These two parameters, together with poststenotic turbulence, govern the loss of kinetic energy. The magnitude of the intrastenotic pressure drop correlates with the kinetic pressure energy present in the stenosis jet. Based on measurement of the stenosis jet, the pressure drop across the

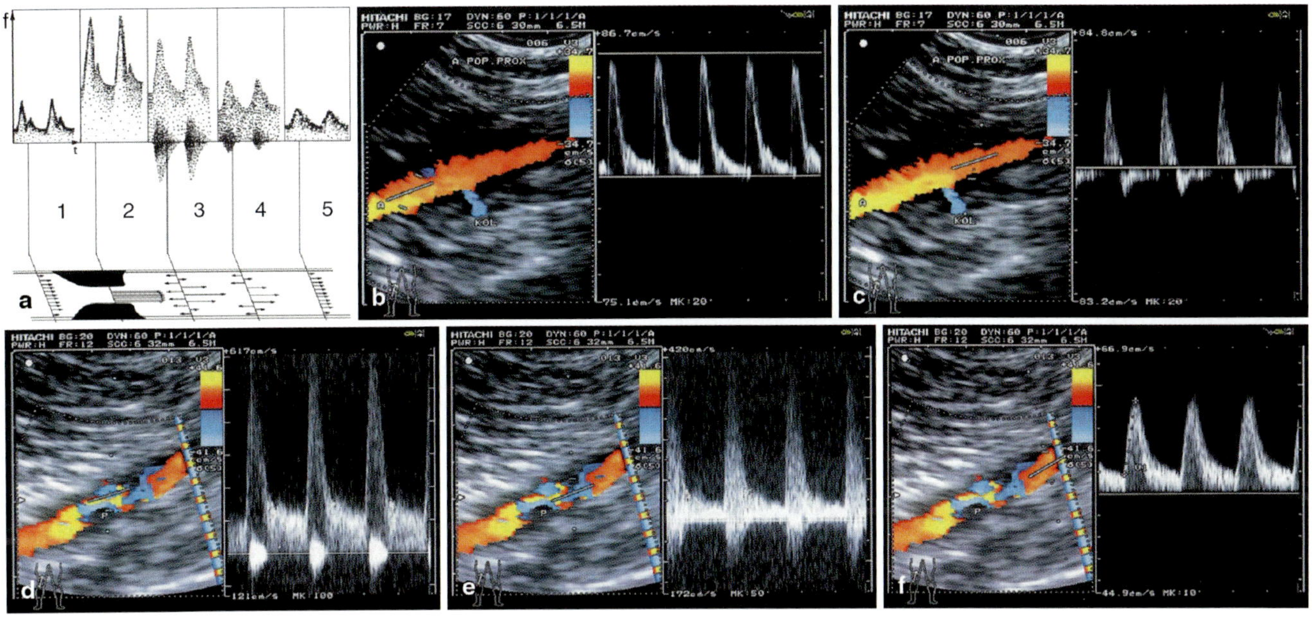


**Fig. 1.45** **a** Diagrams of Doppler waveforms obtained at different sites in and around a stenosis (illustrated here for internal carotid artery (ICA) stenosis):

1. prestenotic (laminar, pulsatile);
2. intrastenotic (plug profile, maximum increase in peak systolic velocity (PSV), increase varies with diameter reduction);
3. immediately poststenotic (marked turbulence, increased PSV);
4. poststenotic (return to lower flow velocity, residual turbulence);
5. further downstream from stenosis (relamination of flow but decreased pulsatility, flattened waveform with larger diastolic component).

**b-f** The diagrams in **a** show stenosis-related flow changes in the carotid artery, but basically the same patterns will be seen in a stenotic peripheral artery (see Table 2.9), except that flow is more pulsatile. **b** Popliteal artery stenosis: Proximal to the origin of hemodynamically significant collaterals, the normal triphasic prestenotic waveform may show reduced pulsatility. The extent to which pulsatility is decreased depends on the amount of collateral flow and compensatory ischemic widening of arteries and arterioles in the periphery. When the sample volume is placed upstream of the origin of a collateral (*KOL*), there is only a subtle incisure in the Doppler waveform in early diastole and continuous blood flow throughout diastole. Farther away from the transducer, a collateral arising from the popliteal artery is depicted, which provides significant compensatory circulation. The hemodynamic effects of collateralization are rarely depicted in such an impressive manner. **c** Downstream of the collateral, the increased pulsatility reflects the flow resistance in the popliteal artery immediately before the stenosis. The abnormal diastolic flow is not only a function of the stenosis-related peripheral resistance but also of vessel elasticity. Proximal to the sample volume, the hemodynamically significant collateral (*KOL*) arising from the popliteal artery is displayed in *blue*. The stenotic popliteal artery segment is not depicted but is likely to be located to the right of the view presented. **d** The high-grade stenosis caused by a hypoechoic plaque produces aliasing in the color duplex scan. The Doppler spectrum shows the flow velocity to be increased to over 600 cm/s (waveform 2 in **a**). In the segment depicted in the B-mode scan, the popliteal artery stenosis appears to be less severe because the plaque predominantly involves the lateral vessel wall as compared with the spectral display, which reflects the hemodynamic significance of the stenosis (see Fig. 5.7a, b). Like angiography, the B-mode scan reduces the three-dimensional lumen to a two-dimensional, longitudinal plane. Depending on the plane of the gray-scale image, the severity of the stenosis caused by plaque may be underestimated and may differ from the hemodynamic degree of stenosis. **e** In the immediate poststenotic segment, turbulent flow is predominant both in the color duplex image and in the Doppler waveform. PSV is still increased at 300 cm/s (waveform 3 in **a**). **f** Three centimeters downstream of the stenosis, flow is monophasic with a delayed systolic rise and reduced PSV (waveform 5 in **a**) stenosis can be estimated using the simplified Bernoulli equation (neglecting prestenotic flow velocity) as

$$P_1 - P_2 = 4 \cdot V_2^{\,2}\,(jet\ velocity)$$

The jet axis is generally not parallel to the vessel wall, especially when the stenosis is eccentric. *Visualization of the stenosis jet in the B-mode image can help to align the angle correction cursor to the direction of flow* for reliable measurement of flow velocity in the jet (Fig. 1.46).

A study correlating the invasively measured mean catheter pressure gradient across stenotic vessel segments with the pressure gradient determined from the stenosis jet by means of duplex ultrasound (Strauss et al. 1993) found a correlation for iliac artery stenoses of $R = 0.77$. This investigation was based on preceding model calculations and neglected viscous friction losses and energy losses due to poststenotic turbulence.

In vitro measurements in a model of pulsatile peripheral flow demonstrated good agreement for the pressure decrease across a stenosis measured invasively and that determined by Doppler ultrasound. The correlation found for different grades of stenosis was $R = 0.98$ (Strauss et al. 1990; Weber et al. 1992). Still, quantitative evaluation of stenosis using only the absolute values of peak intrastenotic frequency shifts or peak velocities is discouraged, as the magnitude of intrastenotic flow velocity is also affected by various other factors such as central regulatory mechanisms (blood pressure), collateral pathways, and peripheral resistance.

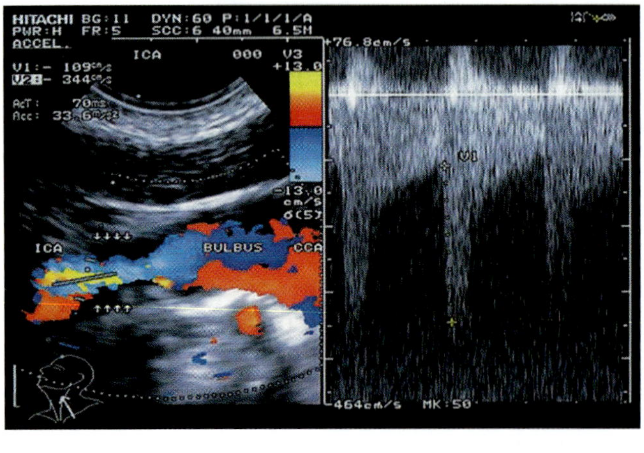


**Fig. 1.46** In the eccentric high-grade stenosis of the internal carotid artery (ICA) shown, the Doppler angle adjusted relative to the vessel wall differs by about 10° from the angle relative to the stenosis jet. Flow in the vessel is displayed in *blue* (away from transducer) with aliasing due to the stenosis jet. Distal to the hypoechoic plaque, blood flow is depicted in *red*, indicating flow reversal resulting from flow separations and eddy currents. The color change in the common carotid artery and the bulb (*red* to *blue*) is due to the change in flow direction relative to the ultrasound beam (flow toward and away from transducer) and flow separation in the bulb (*blue*). Proper alignment of the Doppler angle with the stenosis jet would have yielded a higher flow acceleration. The turbulent flow components are reflected more adequately in the Doppler waveform (see Fig. 5.16)

Empirical data suggest that a stenosis becomes hemodynamically significant and causes clinical symptoms when the vessel diameter is reduced by at least 50% (corresponding to a cross-sectional area reduction of more than 75%). The pressure drop across a stenosis increases with its degree and length and is reflected by a decrease in the peripheral blood pressure determined by Doppler sonography.

Because absolute values of PSV using thresholds for different degrees of stenosis, as defined by ROC analysis, are influenced by systemic factors (blood pressure during the examination, vessel wall elasticity), many investigators have explored stenosis grading using the so-called *systolic velocity ratio* (SVR; calculated as intrastenotic PSV divided by prestenotic PSV). It has been shown that a ratio >2 corresponds to a hemodynamically significant stenosis (>50%) and a ratio >4 to a severe stenosis (>75%).

These stenosis grades are calculated according to the following formula:

$$\left(1 - \frac{1}{PSV\ ratio}\right) \cdot 100 = stenosis\ grade$$

The PSV ratio is only reliable if prestenotic and intrastenotic velocities are measured in the same artery. Problems may arise, for example, when grading a stenosis in a bifurcation if the prestenotic PSV is measured in the segment proximal to the bifurcation, where the blood flow velocity may be influenced by other factors (e.g., difference in arterial diameter or flow profiles). Therefore, the PSV ratio cannot be used to grade stenosis of the renal artery origins or bifurcations (carotid, femoral, and iliac arteries), the most common sites of stenosis. To overcome this limitation, empirical data and graphic interpolations were used to establish nomograms, which enable stenosis grading on the basis of intrastenotic PSV in relation to PSV proximal or distal to the stenotic segment (Ranke et al. 1995). However, measuring the reference velocity distal to the stenosis may introduce new sources of error (hemodynamic effects of collaterals). Nomograms were originally developed for the peripheral arteries but have since been used to grade stenoses of the cerebral arteries as well. The use of nomograms for stenosis grading is based on the assumption that the nonstenotic vascular diameter in the stenotic segment is identical to the diameter of the prestenotic or poststenotic segment in which the reference velocity is measured (Fig. 1.47). It follows that the intrastenotic to poststenotic PSV ratio is used to grade a bifurcation stenosis, at the origin of the ICA, for instance (see Fig. 5.6b). Here, no collaterals enter the poststenotic segment, and in accordance with the continuity principle, identical flow velocities can be assumed proximal and distal to the stenosis. As described above, the differing hemodynamic situation in the common carotid artery precludes the use of the prestenotic velocity. This is why stenosis at the ICA origin is graded according to

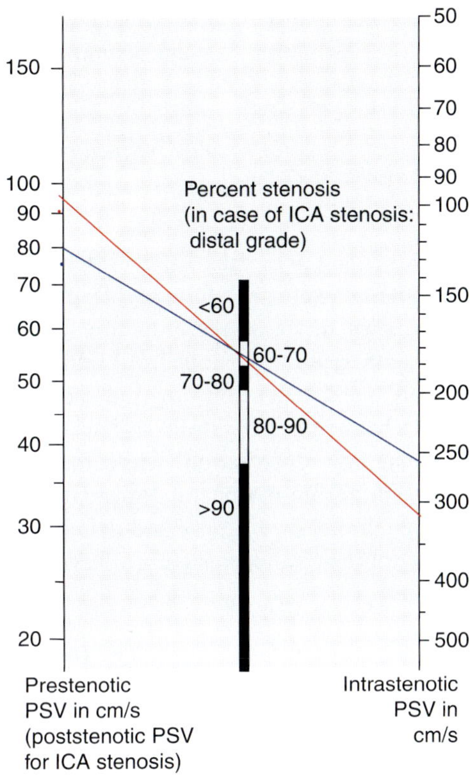


**Fig. 1.47** Nomogram for grading stenosis (Modified from Ranke et al. 1995). Using graphic interpolation, the degree of stenosis is determined from the ratio of intrastenotic peak systolic velocity (PSV) to the proximal or distal PSV. Example 1: 60–70% stenosis of the superficial femoral artery (*red line*) with a PSV of 320 cm/s in the stenosis and a prestenotic PSV of 95 cm/s. Example 2: internal carotid artery (ICA) stenosis (*blue line*) with a PSV of 260 cm/s in the stenosis and a poststenotic PSV of 80 cm/s, yielding a distal stenosis grade of 60–70%. This can be converted to a local stenosis grade of 70–80% (see Sect. 5.4)

the NASCET criteria (based on the appearance of the distal ICA), which are based on the measurement of intra- and poststenotic PSV and graphic interpolation but which do not take plaque thickness in the carotid bulb into account. Conversion tables (see Sect. 5.1.1) are used to derive the local stenosis grade.

Section 5.6.1.2 discusses the factors that may influence the PSV; these factors must be considered in order not to over- or underestimate stenosis severity.

One important factor is that a decrease in peripheral resistance, for example, during muscle activity, is associated with a relative increase in the degree of stenosis. The increased blood volume required in the periphery per unit time leads to a relatively greater reduction of the blood flow through the stenosis above a certain degree of stenosis, resulting in a greater discrepancy between the flow volume required in the periphery and the volume that can pass the stenotic vessel segment. As a result, the peripheral dilatation associated with muscle activity can reduce the stenosis-related perfusion pressure to such an extent that relative or absolute ischemia may occur. The hemodynamic effects of an exercise-induced, hemodynamically significant perfusion reduction in the presence of a stenosis that is not hemodynamically significant at rest are also reflected in the Doppler waveform: there is a more pronounced increase in the diastolic component during exercise but, above all, a longer rest after exercise is required before the postocclusive Doppler waveform returns to its normal triphasic pattern (as compared with the contralateral side). In addition to the local degree of stenosis, the reduction in peripheral perfusion is also affected by other occlusive processes and above all by cardiac function (in particular systolic pressure) and the extent of collateralization.

The decrease in pulsatility is primarily due to the high pressure gradient associated with luminal narrowing. The changes in the spectral waveform proximal to a vessel obstruction vary with collateral perfusion and the distance between the site of sampling and the vessel lesion. Close to the lesion, pulsatility increases as a result of the high resistance. When the Doppler information is sampled proximal to the origin of relevant collateral vessels, peripheral resistance causes a less pulsatile flow profile (see Fig. 1.45b–f). The hemodynamic changes resulting from widening of the arterioles, which decrease their tone as the blood supply drops, affect the flow pattern in the prestenotic vessel segment through the collateral pathways.

Grading of stenosis at arterial origins (ICA, profunda femoris, and renal arteries) relies on empirical data as the continuity equation does not apply to vessel divisions. In the clinical setting, it is not generally necessary to determine an exact percentage as the therapeutic management of a hemodynamically significant stenosis is guided by the patient's clinical symptoms and the vessel segment affected.

On color duplex images acquired with adequate settings, aliasing will already suggest a stenosis. Nevertheless, quantitative evaluation must be performed by analysis of the Doppler waveform with angle-corrected velocity measurement using the criteria outlined above.

Again and again it has been proposed to measure the degree of a stenosis planimetrically by determining the residual patent lumen, visualized in the color flow mode, in relation to the vessel lumen (wall). However, this approach is often impaired or yields unsatisfactory results due to inaccuracies resulting from color overflow (few color scan lines with interpolation) and scattering or acoustic shadowing due to intrastenotic structures such as calcified plaques. Under ideal conditions with complete direct visualization of the stenotic segment, absence of aliasing, and localization of the stenosis outside a bifurcation, determination of the residual lumen by color duplex ultrasound was found to have a satisfactory diagnostic accuracy of 85% compared with angiography (Steinke et al. 1990). Planimetric measurement appears to be most suitable for estimating the degree of mild to moderate stenosis (see Figs. A5.5, A5.19, and 5.12) but should not be used unless plaque echogenicity enables reliable definition of the patent lumen on B-mode images. Planimetric stenosis grading on the basis of the cross-sectional area reduction is justified only because these stenoses have no hemodynamically relevant effect and therefore will not be detected by spectral Doppler. The use of color duplex images for defining the patent lumen for stenosis grading has inherent methodological limitations and is discouraged.

With the angle of incidence perpendicular to the vessel (i.e., $\alpha = 90°$) in the transverse plane, the Doppler equation predicts Doppler-shifted frequencies approximating zero, resulting in poor or very inadequate visualization of blood flow. These limitations can be overcome to some extent, but the remedies are likewise subject to error.

Slightly tilting the transducer to obtain a Doppler angle <90° will improve the Doppler signal, but this comes at the expense of introducing errors into the measurement of the vascular diameter and cross-sectional area, which becomes an oblique plane not showing the exact cross-section but rather an ellipsoid vessel section. The second remedy is to increase the color gain to make up for the inadequate color display of the patent lumen, but, as described above, this may result in color spillover obscuring the vessel wall and plaque edge.

**Acceleration time** Another important Doppler parameter is the systolic acceleration time (also known as the systolic rise time) or acceleration index. Normal arterial blood flow is characterized by a rapid systolic upstroke with a short rise time of less than one hundredth of a second before PSV is reached. Distal to a high-grade stenosis or occlusion, the upstroke is delayed, resulting in a longer acceleration time and a diminished acceleration index ($\delta V/\delta T$) – mainly because flow through collaterals is slower.

Acceleration time measurements are mainly used in the sonographic evaluation of the renal arteries, where a prolonged acceleration time measured at the hilum is considered a sign of a severe proximal stenosis.

The *delayed systolic upstroke* in the *poststenotic Doppler* waveform is due to the slower systolic pressure build-up in the poststenotic segment. A longer acceleration time in the poststenotic waveform is an indirect sign of a hemodynamically significant stenosis. Behind high-grade stenosis, the delayed equalization of central and peripheral (prestenotic and poststenotic) pressure during the cardiac cycle also contributes to the persistent diastolic flow. *This postocclusive increase in diastolic pressure is due to the decreased peripheral resistance resulting from arteriolar widening secondary to reduced blood flow.* A stenosis may be identified faster by obtaining Doppler spectra at selected sites and evaluating them for prestenotic and poststenotic criteria. If a stenosis is suspected in a vessel segment that is difficult to interrogate directly, the so-called damping factor can be calculated from the resistance indices (pulsatility indices) proximal and distal to the suspected stenosis in order to estimate the significance of the flow obstruction.

$$Damping\ factor = \frac{Proximal\ pulsatility\ index}{Distal\ pulsatility\ index}$$

Stenosis of less than 60% has little effect on the poststenotic Doppler waveform. Only higher-grade stenoses are associated with an increasing reduction in poststenotic PSV, a less steep systolic upslope, and delayed diastolic decrease with persistent flow to the periphery. The decreased PSV and delayed systolic rise are primarily due to the proximal flow obstruction, while the monophasic flow profile results from peripheral vasodilatation secondary to a mismatch of blood supply and demand. The latter can thus also influence the prestenotic waveform via the collateral vessels.

## 1.3 Machine Settings

Proper selection of scanning parameters is of utmost importance for vascular examinations using color-coded duplex ultrasound (Table 1.12). Adjustment is done individually for the respective vessel to be interrogated, in particular when spectral Doppler information is collected. The most important aspects are summarized below.

**Transducer**

- Selection of a suitable frequency
- Adequate presetting (magnification, gain, focusing, PRF, wall filter, etc.)

**Table 1.12** Optimal machine settings for specific diagnostic tasks

| Parameter | Evaluation of flow pattern in stenosis | Evaluation of small vessels with slow flow |
|---|---|---|
| | Depiction of fast flow | Depiction of slow flow |
| *Color duplex ultrasound* | | |
| PRF | As high as possible | Low |
| Color box | Small | Fairly large |
| Doppler angle | Intermediate (30–60°) | As small as possible |
| Wall filter | Intermediate | Low |
| Color gain amplification | Intermediate | High |
| *Pulsed Doppler ultrasound* | | |
| PRF | As high as possible | As low as possible |
| Wall filter | Intermediate | Low |
| Doppler angle | <70° | As small as possible |
| Transducer | Low frequency | Higher frequency |

**Adjustment of B-mode scanning parameters**

- Identification of the target vessel in the transverse plane
- 90° clockwise rotation of the transducer to depict the vessel in the longitudinal plane
- Optimization of the image size
- Optimization of gain and time-gain compensation: black vessel lumen, walls sharply delineated

**Color Doppler**

- Angling of the transducer to interrogate the vessel in the longitudinal plane (to improve angle of incidence)
- Positioning of the color box, possibly tilting it (parallel scanner), to ensure the smallest angle possible in relation to the vessel axis
- Selecting the size of the color box in such a way as to ensure a high enough frame rate
- Optimizing gain
- Adjusting PRF (Fig. 1.48a, b):
  - Higher PRF when aliasing occurs
  - Lower PRF when no flow is detected but slow flow is expected
- Adjusting high-pass filter (HPF) (rarely necessary as mostly coupled to PRF):
  - Lower cutoff to detect very slow flow
  - Higher cutoff when motion artifacts occur
- If color filling of the vessel is insufficient, a *lower-frequency transducer* should be selected to improve penetration depth. The poorer spatial resolution is negligible in the color mode as it is compensated for by the superior color signal.

**Pulsed Doppler**

- Placing the Doppler gate in the center of the lumen in the color-mode viewfinder at an angle <60° relative to the longitudinal vessel axis
- Angle correction
- Optimizing the Doppler gate width to cover the entire lumen
- If the Doppler signal is poor, the color mode should be frozen in triplex scanning to improve Doppler resolution
- Optimizing the spectral waveform:
  - Gain:
    - Absent or sketchy curve: *gain* ↑
    - Overmodulation (complete filling-in of systolic window, mirror image): *gain* ↓
  - PRF:
    - Peak frequency/velocity cutoff (aliasing): *PRF* ↑
    - Spectrum too "small": *PRF* ↓
  - Filter:
    - Depiction of slow flow: *HPF* ↓
    - Elimination of interfering low frequencies: *HPF* ↑

An optimal waveform can be sampled from the site interrogated by minimally changing the transducer position using for orientation the Doppler tracing but above all the *acoustic Doppler signal*. An abnormal signal is often heard before the transducer provides the corresponding graphic information (in particular when there is scattering due to plaque).

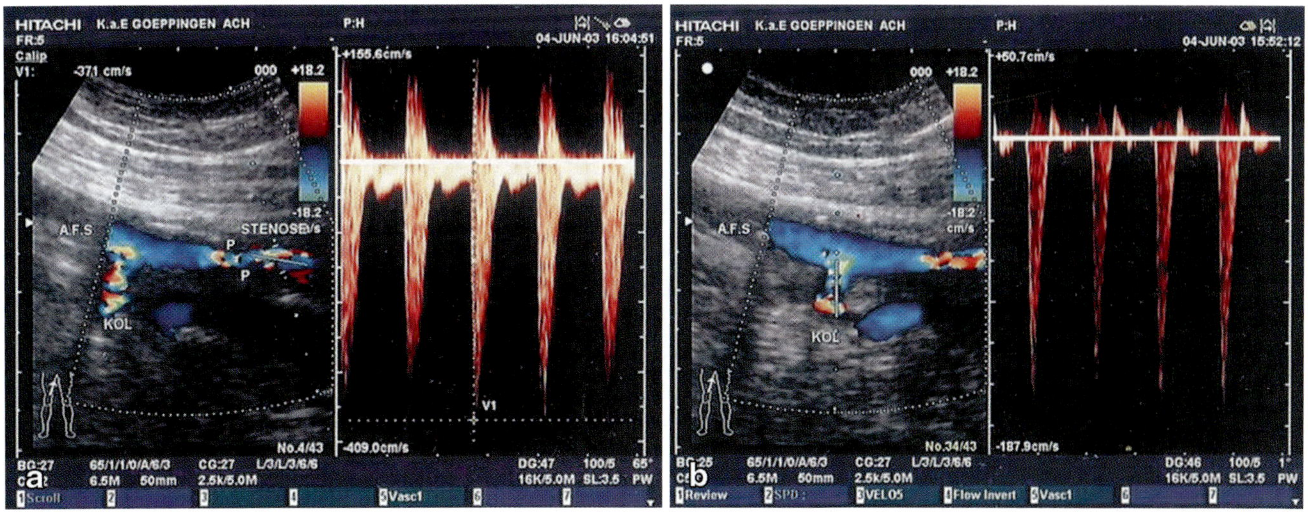


**Fig. 1.48** **a** Semiquantitative measurement of flow velocity by color duplex ultrasound requires proper adjustment of the pulse repetition frequency (PRF) and other scan parameters. Aliasing should not occur in vessels with normal flow. A compromise between different requirements may occasionally be necessary. In color duplex scanning, aliasing occurs if the PRF is too low to register high Doppler shifts. The Doppler frequency is affected by the flow velocity of the reflecting particles and the size of the Doppler angle. The scan shows the superficial femoral artery in *blue* (flow away from transducer toward periphery) with aliasing (on the right side of the scan), which suggests a stenosis unless instrument settings are inadequate. The Doppler waveform obtained with an angle of 65° confirms the stenosis with a peak systolic velocity (PSV) of 371 cm/s. **b** Aliasing due to a small angle of insonation. The collateral (*KOL*) arising from the posterior aspect of the superficial femoral artery (*A.F.S*) has a normal Doppler waveform with an angle-corrected PSV of 170 cm/s (very small Doppler angle of 1°)

Note: In vessel segments with high-grade stenosis and very turbulent flow, high-energy frequencies from slow flow are predominant. For stenosis grading, however, it is important to register the fast flow that often has a low energy and is concentrated in a thin jet. The following measures are recommended *to determine peak flow velocity in high-grade stenosis*:

1. Selection of a high PRF
2. Overmodulation of the spectrum by a higher gain to also depict the low-amplitude, high Doppler frequencies
3. Localization of the jet by minimally and carefully changing the transducer position under acoustic and visual guidance

# Peripheral Arteries 2

As the population ages, the number of people who develop atherosclerotic occlusive disease will increase as well. Atherosclerosis affects not only the coronary arteries and extracranial cerebral vessels but the arteries of the extremities as well. The prevalence of symptomatic arterial occlusive disease (AOD) in men and women aged 55–75 is 4.5% and, when asymptomatic disease is included, over 20%. Atherosclerosis impairs an individual's quality of life, causing immobility and disability. Life expectancy is reduced by about 10 years in male AOD patients. The main causes of death are coronary heart disease (55% of patients with AOD vs. 36% without AOD) and cerebral conditions (11% with AOD and 4% without). While AOD is the manifestation of generalized atherosclerosis affecting all vascular territories, there is a high rate of concomitant involvement of the coronary and extracranial cerebral vessels, in particular in AOD of the pelvic arteries. Patients in whom peripheral AOD (PAOD) has been diagnosed require individual therapeutic management based on the stage of the disease and the suitability of their vascular system for surgical repair as well as further prophylactic diagnostic and therapeutic measures (coronary heart disease, carotid stenosis). The wider range of therapeutic options available today, percutaneous interventions in particular, can prevent the loss of a limb by means of reconstruction or recanalization of the occluded vessel segments. Timely interventions restore perfusion and improve quality of life in patients with steno-occlusive disease. Early diagnosis is thus essential for the initiation of proper treatment, and duplex ultrasonography is a well-suited noninvasive modality for an efficient and low-risk primary diagnostic workup.

## 2.1 Pelvic and Leg Arteries

### *2.1.1 Vascular Anatomy*

#### 2.1.1.1 Pelvic Arteries

At the level of L4–L5, or the umbilical level when scanning from the anterior approach, the abdominal aorta divides into the two *common iliac arteries*. They descend into the true pelvis taking an arched course along which they bifurcate at about the most posterior point. The common and external iliac veins course behind the respective arteries. The *internal iliac artery* arises at the level of the sacroiliac joint, coursing in a posterior direction to supply the pelvic organs, pelvic wall, and buttocks. The *external iliac artery* is the continuation of the common iliac artery and arches into the lacuna vasorum under the inguinal ligament. It runs medial to the iliopsoas muscle and gives off the inferior epigastric and deep circumflex iliac arteries shortly before it reaches the inguinal ligament. These two arteries can serve as collaterals in pelvic artery occlusion.

Diameters range from 0.6 to 1.4 cm in the common iliac artery, 0.5–1.0 cm in the external iliac artery, and 0.4–0.8 cm in the internal iliac artery.

#### 2.1.1.2 Leg Arteries

The *common femoral artery* is about 2–4 cm long and, below the inguinal ligament, divides into the *profunda femoris artery*, which typically arises from the posterolateral aspect, and the *superficial femoral artery* (Fig. 2.1a).

The origin of the *profunda femoris artery* is variable, and several branches may arise directly from the common femoral artery. Typically, one branch arises posteriorly, while the main branch arises posterolaterally or, in rare cases, posteromedially. Branches of the medial and lateral circumflex arteries open into the femoral bifurcation and form the main collateral pathways in pelvic artery stenosis or occlusion. The profunda femoris artery is the most important collateral route in occlusion of the femoropopliteal segment. Somewhat distal to the femoral bifurcation, the *deep femoral veins* converge and unite with the *superficial femoral vein*. The deep veins cross the bifurcation. The superficial femoral vein runs behind the artery on its course to the distal thigh. The plexus passes posteriorly through the vasoadductor membrane at the level of the adductor canal (Hunter's canal), and the superficial femoral artery continues as the *popliteal artery*.

Interventional radiologists and vascular surgeons subdivide the popliteal artery into three segments with the P1

W. Schäberle, *Ultrasonography in Vascular Diagnosis*,
DOI: 10.1007/978-3-642-02509-9_2, 

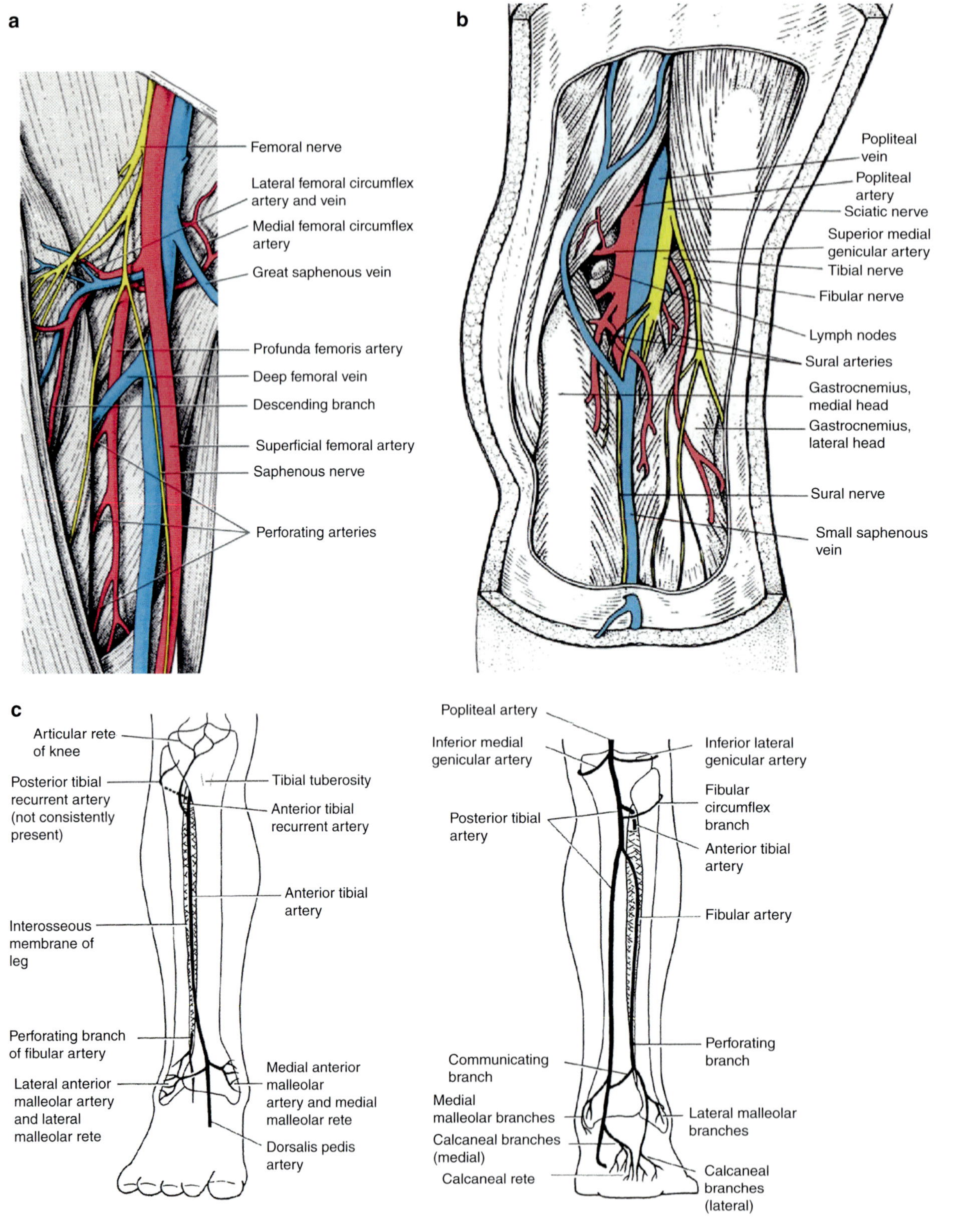


**Fig. 2.1** **a** Arteries and veins in the groin (from Heberer and van Dongen 1993). **b** Arteries and veins in the popliteal fossa (from Heberer and van Dongen 1993). **c** Anterior and posterior views of the lower leg arteries

segment in angiography extending from the origin to the upper edge of the patella and the P2 segment to the knee joint cleft. The P3 segment extends to the origin of the *anterior tibial artery*, which passes anteriorly through the interosseous membrane at the lower edge of the popliteal muscle. It continues through the extensor compartment anterior to the interosseous membrane to the upper edge of the ankle joint with its proximal segment coursing relatively close to the fibula. The popliteal artery continues as the highly variable *tibiofibular trunk* with a length of about 1–5 cm (Fig. 2.1b), branching into the posterior tibial and fibular arteries. The *posterior tibial artery* is the main artery of the lower leg and passes through the deep crural fascia between the superficial and deep flexors. It continues posterior to the medial ankle to the sole of the foot, where it divides into the larger lateral plantar artery and the smaller medial plantar artery. The lateral plantar artery extends to the deep plantar arch, which completes the arch of foot via its collaterals to the dorsalis pedis artery, establishing a connection to anterior tibial branches. The *fibular artery* courses posteromedial to the fibula, also at the level of the deep crural fascia. It ends at the distal lower leg and gives off several arteries supplwwying the muscles and serving as collaterals when other calf arteries become occluded.

The arteries below the knee vary widely in caliber. In cases of hypoplasia or, very rarely, aplasia, the other arteries act as collaterals. The anterior tibial artery is absent or shorter in 2% of the population, the posterior tibial artery in about 5%. In about 7% of individuals, the fibular artery is the main calf artery, communicating with the distal posterior tibial artery or dorsalis pedis artery via large collaterals arising at the level of the ankle joint. Collateralization through the rete malleolare at the ankle joint plays an important role in atherosclerotic occlusion or hypoplasia.

## 2.1.2 Examination Protocol and Technique

### 2.1.2.1 Pelvic Arteries

The scanning depth required for an examination of the pelvic arteries makes it necessary to use a *convex transducer* with a frequency of 3.5–5 MHz. The proper pulse repetition frequency (PRF) (no aliasing) and gain are selected in a nondiseased arterial segment. The examination is performed with the patient in the supine position and after an adequate period of rest to prevent false-positive results due to reactive hyperemia. Exercise-induced hyperemia takes longer to return to normal in patients with atherosclerotic occlusive lesions.

Hemodynamically significant stenosis above the inguinal ligament can be identified quickly by obtaining a spectral Doppler waveform from the common femoral artery. A triphasic waveform with a peak systolic velocity (PSV) of at least 70 cm/s (comparison with contralateral side) rules out higher-grade stenosis at the pelvic level with a high degree of confidence.

For a closer evaluation, the aortic bifurcation is identified in the transverse plane at the umbilicus, and the common and external iliac arteries are scanned for stenotic lesions in longitudinal orientation (Fig. 2.2a). Bowel loops, in particular when filled with air, produce marked scattering and *attenuation*, impairing continuous scanning of the pelvic vessels. They can be pushed out of the way by exerting pressure on them with the transducer or can be avoided by moving the transducer around them.

The arched course of the iliac arteries through the true pelvis makes it more difficult to achieve a small angle of incidence, especially at the deepest point, the *iliac bifurcation*

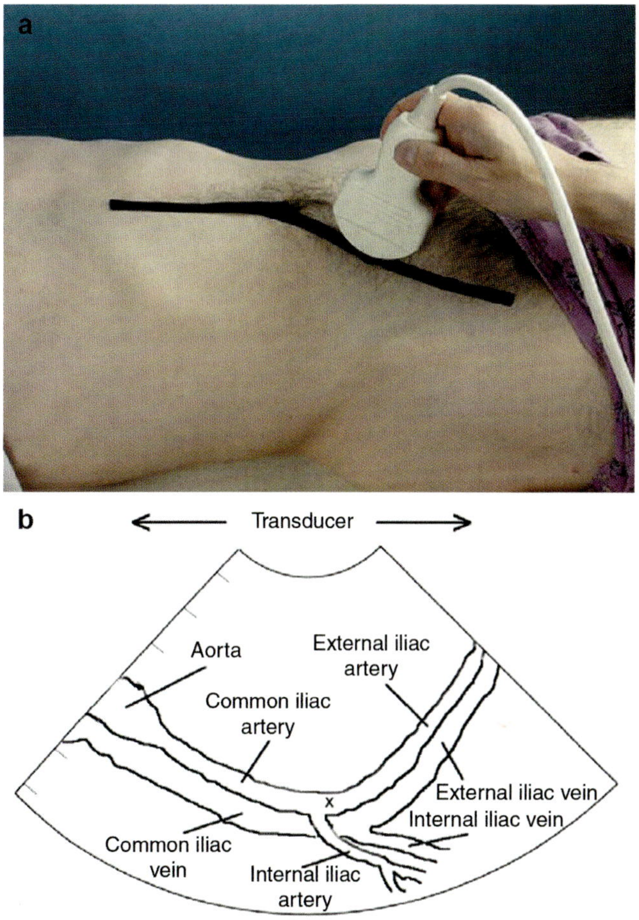


**Fig. 2.2** **a** Transducer positioning for pelvic artery examination (aortic bifurcation at the umbilical level). **b** Diagram of the arched course of the iliac vessels through the true pelvis. The iliac bifurcation is situated at the most posterior point, where the common iliac artery gives off the internal iliac artery. The common and external iliac veins run posterior to the arteries of the same name. The origin of the external iliac artery is a common site of atherosclerotic stenosis (*x*). From the normal anterior transducer position, however, stenosis cannot be graded due to an extremely poor Doppler angle of about 90°. To obtain a better Doppler angle, the transducer must be moved upward or downward (*arrows*) and angled (see Fig. 1.39b, c)

(Fig. 2.2b), where the origin of the external iliac artery is particularly prone to the development of stenosis. The Doppler angle is optimized by moving the probe along the course of the artery and angling it (see Figs. 1.39b, c and 2.2b). If the examination is technically challenging, an attempt should be made to at least evaluate the proximal segments of the common and external iliac arteries (including Doppler measurement), as they are *predilection sites for pelvic artery stenosis.* The Doppler spectra are analyzed, with an eye to the established criteria for stenosis, and then compared, as differences between spectra from two different sites may provide indirect clues about stenotic lesions between the two measurement sites.

#### 2.1.2.2 Leg Arteries

The superficial course of the leg arteries makes it possible to use a higher-frequency transducer of 5–7.5 MHz, depending on the thickness of the intervening soft tissue layer. Linear transducers require the activation of beam steering and careful alignment with the course of the vessel to achieve an *adequate Doppler angle. Curved-array transducers* with a small radius yield B-mode scans with somewhat poorer detail resolution; they have, however, the advantage of enabling fast Doppler angle correction (<60°) by angling the transducer and placing the sample volume at the lateral edge of the imaging field. In contrast, beam steering in most ultrasound devices enables a maximum deflection of the emitted beam of only 20°, resulting in a Doppler angle of 70° when interrogating vessels running parallel to the skin surface. The evaluation of spectral tracings for indirect signs of occlusive lesions in the vicinity of the sample volume requires an angle of less than 60°.

The femoral (Fig. 2.3a) and anterior tibial arteries are examined in the supine position, all other lower leg arteries and the popliteal artery in the prone position with slight elevation of the distal calf by a support placed under the ankles.

In general, vessels are visualized in two planes. First, the artery is identified in the *transverse plane*. The PRF and gain are set so as to ensure good color filling of the arterial lumen while avoiding aliasing. For a preliminary overview in transverse orientation, the transducer must be angled distally or cranially to achieve a small angle of insonation relative to the vessel. A survey in the transverse plane has the advantage of enabling rapid identification of dilated vessel segments, high-grade stenosis (aliasing), and occlusions including those at the origins of collaterals once settings are optimized (Fig. 2.4, Tables 2.1 and 2.2). Abnormal findings need to be confirmed and quantified in the *longitudinal plane*. An adequate angle of insonation and Doppler angle correction in the B-mode scan are prerequisites for accurate stenosis grading. As with grayscale sonography, the monitor display of vascular images in

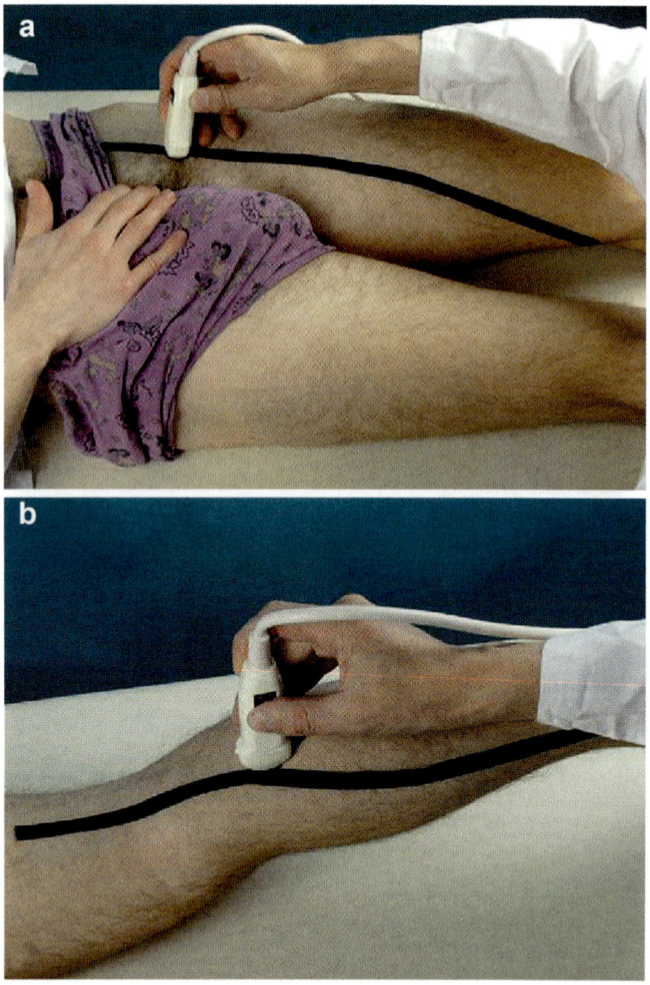


**Fig. 2.3 a** Transducer positioning for examination of the femoral arteries (transverse plane for identification of the vessel, longitudinal plane for measuring blood flow velocity). **b** Transducer positioning for examination of the distal popliteal artery (at the junction of the P3 segment and tibiofibular trunk)

the longitudinal plane depicts the cranial vessel segment on the left and the distal segment on the right.

With the patient in the supine position and after an adequate period of rest (>5 min), the common femoral artery is identified in the transverse plane and followed along its length to the bifurcation. The transducer is positioned (on the inner thigh in most patients) to view the bifurcation in such a way that the profunda femoris artery, which typically arises from the posterolateral aspect, comes to lie exactly behind the superficial femoral artery. In this way, in longitudinal orientation, the bifurcation is depicted as a fork, which facilitates vessel identification and proper angle correction for evaluation of a stenosis at the origin of the profunda femoris. Especially when occlusions have been identified in the femoropopliteal segment, the profunda femoris artery should be followed to the level of second-order branches to check for the presence of more distal stenosis. The superficial femoral

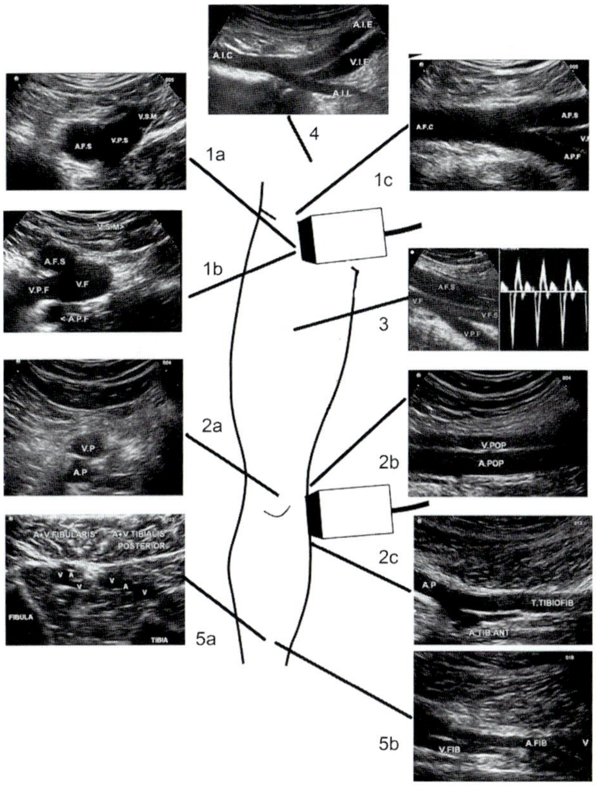


**Fig. 2.4** Sonoanatomy of the leg arteries at representative sites (transverse views on the *left* for identification of the target arteries; longitudinal views on the *right* for evaluation and detection/characterization of stenosis with Doppler measurement)
1a and 1b – transverse views of femoral bifurcation in the groin
1c – longitudinal view of femoral bifurcation
2a – transverse view of popliteal fossa
2b – longitudinal view of popliteal artery
2c – longitudinal view of tibiofibular trunk
3 – longitudinal view of superficial femoral artery
4 – longitudinal view of iliac bifurcation
5a – transverse view of fibular and posterior tibial arteries from posterior approach
5b – longitudinal view of fibular artery and vein

artery is scanned in longitudinal orientation down the inner thigh. Scattering and attenuation due to connective tissue structures within the adductor canal may require readjustment of the gain in the B-scan and Doppler modes.

Within the adductor canal, a vessel is easier to follow with the leg turned outward and the knee slightly bent. The popliteal artery and vein are best examined with the patient lying prone. The vein runs posterior to the artery. Alternatively, the popliteal artery can be scanned with the patient supine, the knee at a 30°–60°angle of flexion, and the transducer placed in the popliteal fossa. As the anterior tibial artery arises from the anterolateral aspect of the popliteal artery, its origin appears farther away from the transducer (Figs. 2.3b and 2.2e). The origin may be relatively high in individuals with a short P3 segment or very low when P3 is long. In about 4% of the population, all three lower leg arteries jointly arise in a trifurcation (Lippert and Pabst 1985). After having pierced the interosseous membrane, the anterior tibial artery is traced distally along its anterolateral course in longitudinal orientation.

The tibiofibular trunk varies in length from 1 to 5 cm, depending on the level of origin of the anterior tibial artery. Its division into the posterior tibial and fibular arteries is identified in the transverse plane. In the longitudinal view, the vessels are continuously scanned for stenosis or occlusion by following their courses distally (Fig. 2.5). If a vessel disappears from the scanning plane, it can easily be identified again by changing to a transverse view. The tibia and fibia cast acoustic shadows and can serve as landmarks. Further orientation is provided by the hyperechoic band of the deep crural fascia. Under good insonation conditions, the lower leg arteries can be visualized down to the ankle region. Time can be saved by obtaining two spectral tracings, one in the proximal and the other in the distal segment of the respective lower leg artery, which in general excludes a hemodynamically significant obstruction between the two sampling sites when both show the same flow profile and normal PSV.

The dorsalis pedis artery and the posterior tibial artery behind the medial ankle (Fig. 2.6) are examined in the supine position using a high-frequency transducer (7.5–10 MHz). These arteries are identified in the transverse plane to then perform Doppler measurements in the longitudinal plane. From there, the plantar artery can be scanned in the transverse plane to the level of the interdigital arteries.

### 2.1.3 Specific Aspects of the Examination from the Perspective of the Angiographer and Vascular Surgeon

Vascular duplex ultrasound of the lower extremities is done for three purposes – *screening, evaluation for treatment planning, and follow-up after surgical and interventional procedures* or medical treatment. Screening ultrasound is performed to identify individuals in need of treatment either in a group of the population or among patients presenting with vascular disease. Such a test must have a high sensitivity and a low rate of false negatives. A false-positive diagnosis can be corrected by the subsequent (invasive) vascular examination. Screening tests should be inexpensive and not put patients at risk.

The German Epidemiological Trial on Ankle-Brachial Index (GetABI; Lange et al 2007) confirms earlier investigations (Neuerburg-Heusler 1984), showing that Doppler measurement combined with determination of the ankle-brachial index (ABI) is an ideal screening test for PAOD. This test has an accuracy of over 90% in differentiating between patients with pain caused by PAOD and patients with pain of other etiologies when clinical examination and pulse measurement are inconclusive (except for pelvic artery occlusion with good collateral circulation).

**Table 2.1** Duplex examination of the pelvic and leg arteries (sequence of steps; see Figs. 2.4, 2.9, 2.10)

| Site (level) | Ultrasound technique | Purpose, diagnostic tasks |
|---|---|---|
| I Groin | *B-mode*<br>Transverse<br>Longitudinal (rotate transducer over the common femoral artery from transverse to longitudinal plane) | Anatomic orientation (major branches and junctions), origin of profunda femoris, vessel walls (plaque?), femoral bifurcation |
| | *(Optional: color duplex*, longitudinal*)* | Stenosis? (suggested by aliasing) |
| | *PW Doppler: waveform*<br>(a) Longitudinal: curved array transducer tilted cranially or linear transducer with beam steering cranially; sample volume in distal external iliac artery/common femoral artery junction<br>(b) Longitudinal: tilt transducer or steer beam peripherally; Doppler waveform in profunda femoris and superficial femoral artery/origin<br>(c) Longitudinal (suspected common femoral artery stenosis): optional continuous examination of the common femoral artery with continuous spectral Doppler recording (moving the tilted transducer toward the inguinal ligament) | Exclusion/suggestion of pelvic artery stenosis (triphasic/monophasic)<br>Interpretation of Doppler waveform (compare with contralateral side)<br>Stenosis of profunda femoris origin: PSV >180 cm/s; stenosis of femoral bifurcation |
| *Optional evaluation of pelvic level (depending on Doppler waveform at groin)* | In case of abnormal common femoral Doppler waveform (monophasic, reduced PSV compared with contralateral side): continuous mapping of iliac artery in longitudinal plane (3.5–5 MHz transducer) | Detection and grading of stenosis<br>Color duplex and spectral Doppler: PSV ratio >2/>4 or PSV >180 cm/s in bifurcation |
| II Popliteal fossa (Popliteal artery) | *B-mode*<br>Transverse | Identification of popliteal artery to evaluate wall and perivascular structures: nonatherosclerotic vascular disease – aneurysm? |
| | Longitudinal | Course of the artery, perivascular structures, evaluation of wall (nonatherosclerotic disease – aneurysm?) |
| | (*Optional longitudinal color duplex*) | Stenosis? (suggested by aliasing) |
| | *PW Doppler: waveform*<br>(a) Longitudinal: curved array transducer tilted cranially or linear transducer with beam steered cranially<br>(b) Longitudinal: tilt transducer or steer beam peripherally<br>(c) Mapping of popliteal artery if waveform from distal segment is abnormal | Interpretation of Doppler waveform; indirect criteria for stenosis/occlusion of superficial femoral artery (triphasic/monophasic)<br>Comparison of Doppler waveforms from proximal superficial femoral artery and proximal popliteal artery<br>Popliteal artery stenosis? Mapping of popliteal artery may be indicated if abnormal Doppler waveform is obtained from distal popliteal artery (monophasic, reduced PSV compared with contralateral side)<br>Popliteal artery stenosis: abrupt doubling of PSV |
| *Optional evaluation of thigh level (depending on abnormalities of Doppler waveform in popliteal fossa)* | If popliteal artery exhibits monophasic flow or unilaterally reduced PSV: *continuous longitudinal examination of superficial femoral artery* with the tilted transducer (or with beam steering in the case of a linear array being used; color duplex as needed) and continuous spectral Doppler recording | Stenosis criteria: abrupt doubling of PSV for >50% stenosis (PSV ratio >2/>4); possibly measurement of occlusion length (color duplex) |
| III Anterior and posterior tibial arteries at the ankle in stage III and IV AOD | *B-mode*, transverse<br>*Duplex ultrasound*, longitudinal | Identification of arteries<br>Spectral Doppler analysis: is the waveform abnormal? |
| If clinically relevant | In case of abnormal tibial Doppler waveforms:<br>Continuous examination of calf arteries<br>Transverse: localization of vessels<br>Longitudinal: color duplex and Doppler spectrum to detect stenosis | Stenosis criteria: PSV ratio >2/>4<br>Occlusion: no flow (using low PRF)<br>Search for target vessel for crural bypass graft |

**Table 2.2** Diagnostic algorithm in PAOD (key points)

| Question | | Criteria |
|---|---|---|
| I | Is something to be done? | • Clinical presentation!<br>• Duplex ultrasonography?<br>• Angiogram unnecessary (obsolete) |
| II | What is to be done? | • (Color) duplex ultrasonography<br>• Angiogram unnecessary (obsolete) |
| III | How is it to be done? | • Angiography with PTA<br>• Angiography, MRA, or duplex ultrasound to select a target segment for crural bypass procedure<br>• Supplementary invasive diagnostic tests |

Ad I: Is the patient's pain even caused by PAOD? Consider clinical presentation, pulses, ABI (highly valid screening test for PAOD). The therapeutic strategy in PAOD is solely guided by clinical necessity (i.e., the patient's symptoms)

Ad II: Color duplex to identify the level of obstruction (pelvis, thigh, calf). Individual treatment approach based on clinical necessity as well as on therapeutic measures possible in a patient and their prognosis. Length of occlusion determines whether percutaneous transluminal angioplasty (PTA) can be attempted or bypass graft surgery is necessary. In occlusion of a pelvic or thigh artery, the decision for bypass graft surgery and the selection of a target segment can be made on the basis of duplex findings

Ad III: Selective angiography with PTA without prior diagnostic angiogram. Selection of the crural target segment in multilevel obstruction or in combined popliteal/crural obstruction using angiography or dedicated magnetic resonance angiography (MRA). A pedal target artery is identified using color duplex ultrasound combined with angiography or magnetic resonance imaging (MRI). The step-by-step diagnostic approach demands that no additional diagnostic tests (e.g., more invasive and more expensive) be performed unless the results may affect the therapeutic decision

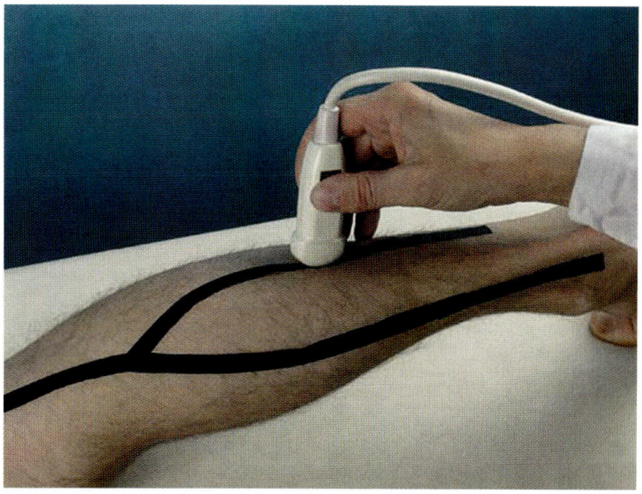

**Fig. 2.5** Transducer position for examination of fibular and posterior tibial arteries (course indicated by thick black line)

As the *therapeutic approach* to PAOD is *symptom-oriented* (vs. prognosis-oriented as is the case for the carotid territory), the extent of the ultrasound examination depends on the clinical question to be answered; no comprehensive vascular mapping is done (Table 2.3). A step-by-step algorithm is recommended to ensure an efficient ultrasound examination of the lower extremity (Fig. 2.7). Noninvasive duplex ultrasound is performed if AOD is suggested by the patient's history, clinical examination with full evaluation of peripheral pulses, and ABI measurement. The ultrasound findings serve to help decide

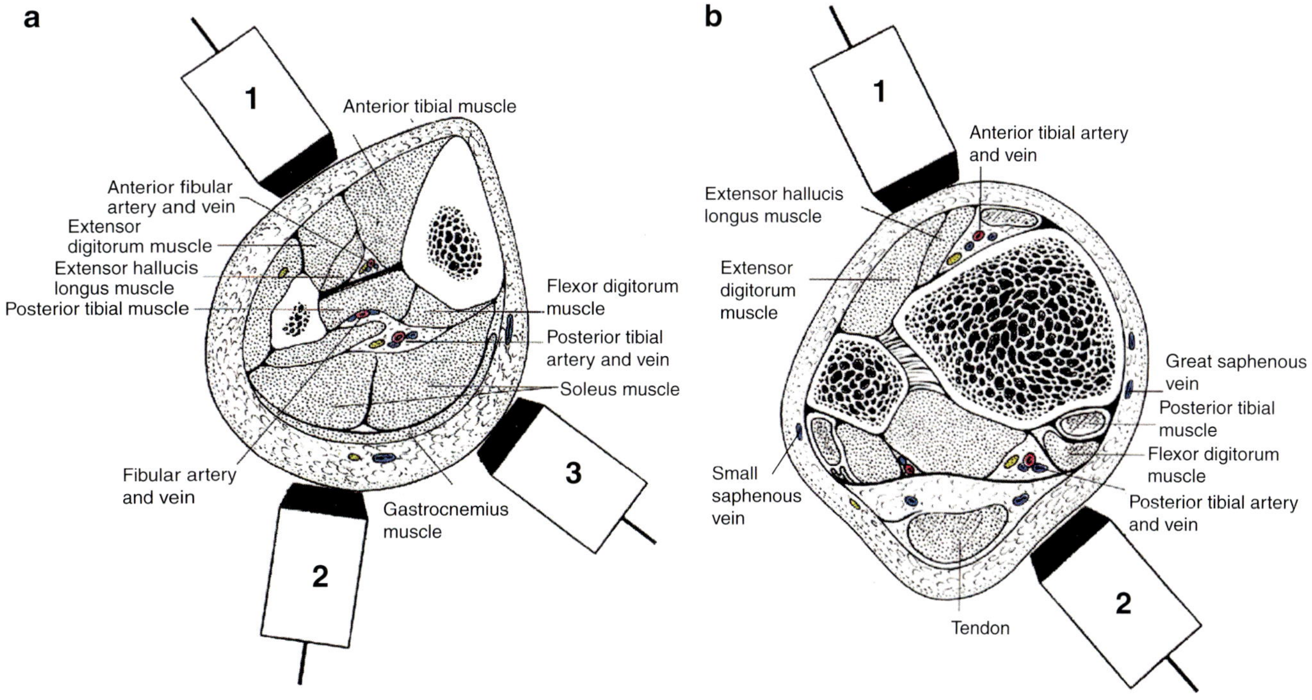


**Fig. 2.6 a, b** Anatomy of the anterior and posterior tibial arteries at the mid-calf level (**a**) and at the ankle joint (**b**). At the mid-calf level, the arteries can be scanned from an anterior (*1*), posterior (*2*), or posteromedial transducer position (*3*). At the ankle, the transducer positions for scanning the anterior tibial artery (*1*) and posterior tibial artery (*2*) are shown

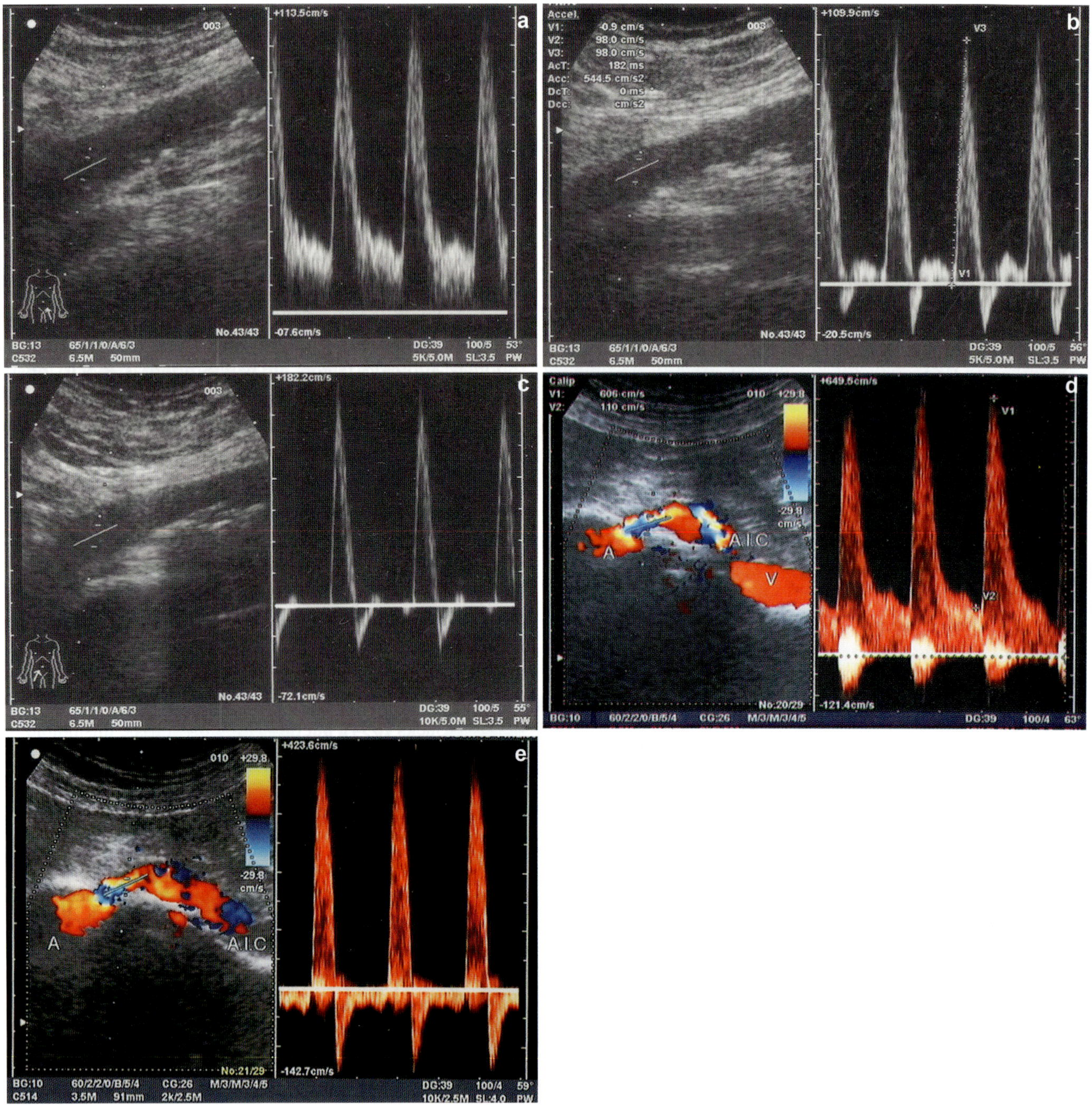

**Fig. 2.10** Common iliac artery stenosis, grading at rest. **a** The Doppler waveform recorded in the *left* groin (common femoral artery) shows monophasic flow. Doppler measurement was performed immediately after positioning of the patient, who had walked from the waiting room to the examination room. **b** After 5 min of rest, a normal triphasic waveform with a slightly delayed systolic upstroke (acceleration time of 182 ms, PSV of 96 cm/s) was obtained. However, the PSV here is markedly different from the PSV measured on the contralateral side (PSV of 170 cm/s), which should prompt continuous duplex imaging of the pelvic segment despite the triphasic waveform. **c** The Doppler spectrum from the contralateral common femoral artery is triphasic with a PSV of 170 cm/s. **d** In this patient, the monophasic waveform with delayed return to normal after 5 min of rest and the reduced PSV were found to be caused by a stenosis of the common iliac artery at its origin from the aorta. The waveform recorded immediately after the positioning of the patient for the examination shows criteria of high-grade stenosis (>90%; peak velocities > 6 m/s during systole and > 1 m/s during diastole, monophasic flow). **e** The correct stenosis grade can be estimated from the Doppler waveform obtained in the stenotic segment after 5 min of rest and is about 70% (PSV of 380 cm/s, triphasic flow). This example illustrates the importance of performing Doppler measurements at rest to ensure accurate stenosis grading by spectral analysis (PSV, indirect criteria)

vasodilatation in response to a mismatch of blood supply and demand. This peripheral situation in turn also affects the prestenotic waveform via the collaterals.

The ABI decreases after exercise, and flow becomes less pulsatile, which may result in a monophasic waveform. In the absence of vascular disease, the ABI and Doppler waveform will return to normal after a short rest. This is why a short waiting period following positioning of the patient on the couch (>3 min) is necessary to obtain accurate quantitative measurements and Doppler information. Alternatively, the recovery phase can be used for spectral Doppler measurement, which may help in differentiating between absence of stenosis (normal waveform at rest) and high-grade proximal stenosis with good collateralization (delayed return to normal after exercise).

Studies using conventional angiography as the gold standard reported a surprisingly high accuracy for ultrasound in diagnosing hemodynamically relevant pelvic artery stenosis, when pulsatility as reflected in the Doppler waveform from the groin was used as the only criterion. One group found 98% sensitivity and 81% specificity (kappa, 0.81) with a positive predictive value (PPV) of 86% and negative predictive value (NPV) of 97% (Eiberg et al. 2001); other groups reported 87% sensitivity and 88% specificity (Walter et al. 1984), 81% sensitivity and 98% specificity with a kappa of 0.82 (Cossman et al. 1989), 95% sensitivity and 80% specificity with a kappa of 0.74 (Sensier et al. 1998), and 95% sensitivity with 89% specificity (Skaalan et al. 2003). However, other investigators reported disappointing results for the method, with 82% sensitivity and 39% specificity and a kappa of 0.22 (Rosfors 1993). The discrepancies are attributable to how different investigators define and interpret triphasic flow in the Doppler waveform. Moreover, the authors provide no details on how they took into account possible effects of proximal stenosis (e.g., of the femoral bifurcation) on pulsatility. The most reliable method is to compare pulsatility (triphasic/monophasic waveform), PSV, and acceleration time on the affected side with the contralateral side to make allowance for the fact that good collateralization is possible in the pelvis. This is how the author's group achieved 95% sensitivity and 98% specificity in the detection of >60% stenoses in 85 patients with suspected pelvic artery stenosis (intermittent claudication, pulses, ABI). Stenosis was confirmed by angiography in 32 of the patients.

A hemodynamically significant, higher-grade stenosis can be ruled out indirectly by calculating the pulsatility index (see Fig. 1.29) for the common femoral artery. A significant stenosis of the aortoiliac segment can be excluded if the pulsatility index is smaller than 5.5 (Johnson et al. 1983; Neuerburg et al. 1991). The following pulsatility indices have been determined: 8.5 ± 3.5 in a normal population, 2.8 ± 1.6 in isolated stenosis at the pelvic level, 2.3 ± 1.0 in concomitant pelvic and upper leg occlusion, and 6.3 ± 2.6 in isolated femoral artery occlusion. The pulsatility index increases with the extent of collateralization (see Fig. 2.18), giving rise to false-negative results. A pulsatility index with a cutoff of 4 was found to have 94% sensitivity and 82% specificity for identifying isolated aortoiliac obstruction (Thiele et al. 1983).

Indirect stenosis criteria can be used when the insonation conditions in the true pelvis are poor. Whenever abnormal findings are encountered, however, an attempt should also be made to identify the stenosis directly. Under normal scanning conditions, state-of-the-art (color) duplex ultrasound equipment often allows faster direct identification of a stenosis or occlusion than is possible with indirect criteria.

When the external iliac artery is occluded and the lateral circumflex artery provides collateral flow, backward flow occurs in the proximal profunda femoris and common femoral arteries. This is seen in the Doppler examination as reversed flow with a monophasic character.

Strauss et al. (1995), investigating stenoses of the external iliac artery, found the following correlations between the flow velocity measured by duplex ultrasound and various other parameters:

- Cross-sectional area reduction determined densitometrically and the hemodynamic stenosis grade calculated as the PSV ratio (calculated from intrastenotic and prestenotic PSV): $R = 0.64$
- PSV and densitometrically determined cross-sectional area reduction: $R = 0.56$
- Pressure drop calculated from the flow velocity determined by color duplex ultrasound using the Bernoulli equation (Fig. 2.11) and the pressure drop measured by intra-arterial catheter: $R = 0.86$

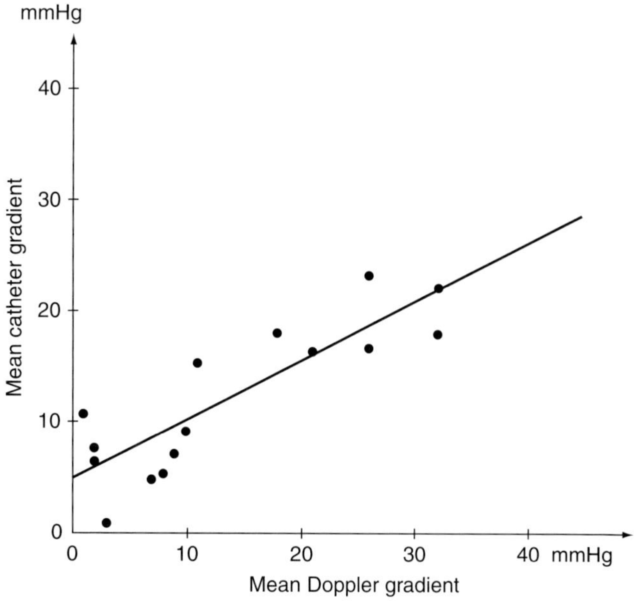


**Fig. 2.11** Correlation ($R = 0.86$) of the mean pressure drop over a pelvic artery stenosis calculated from color duplex ultrasound using the simplified Bernoulli equation ($p = 4 \times PSV^2$) and the drop measured by intra-arterial catheter (Strauss et al. 1995)

The high correlation ($R = 0.86$) between the invasively and noninvasively measured pressure drop (see Fig. 2.11) is attributable to the fact that the change in pressure across a stenosis also incorporates the effects of collateralization. The drop in pressure across a stenosis is smaller when it is bridged by collateral pathways than when there is no collateralization.

#### 2.1.6.1.2 Leg Arteries

Preferred sites of femoral artery stenosis due to atherosclerosis are the origins of the superficial and profunda femoris arteries at the bifurcation and the adductor canal.

Isolated stenosis or occlusion of the common femoral artery is uncommon (about 4%); most patients with common femoral artery stenosis have concomitant obstructions of the superficial femoral and lower leg arteries. Occlusion of the common femoral artery or femoral bifurcation is of considerable clinical significance and should be treated by surgical repair (TEA) whenever possible; collateralization here is poor, as all collateral pathways (via the iliac and profunda femoris arteries) comprise the femoral bifurcation, and auxiliary collaterals have a low capacity. The profunda femoris artery supplies the thigh muscles and is the most important collateral in all arterial obstructions distal to the femoral bifurcation. As a phylogenetically old vessel, the profunda femoris artery is rarely affected by sclerotic changes distal to its origin. All isolated obstructions of the profunda femoris are due to embolism or occur in patients with diabetes mellitus. Stenosis at the origin of the profunda femoris is more common in patients with atherosclerosis of the femoral bifurcation and is clinically relevant due to the key role of the profunda femoris as a collateral in obstruction of the femoropopliteal circulation. Surgical repair of the profunda femoris artery is the treatment of choice. The superficial femoral artery is the preferred site of atherosclerotic lesions and is the most common site of isolated occlusions, which have an incidence of 27%. Occlusion of both the femoral and popliteal arteries occurs in 40–45% of cases. In all cases of isolated popliteal artery occlusion, thrombosed popliteal aneurysm and nonatherosclerotic vascular disorders (which preferably affect the popliteal artery) must be ruled out in the differential diagnosis.

The treatment of femoropopliteal artery occlusion depends on the clinical presentation, cause, site, and length of the occluded segments. These frequently affected and hence clinically significant vessels are easily accessible to duplex scanning as they lie close to the surface and there are no intervening scatterers. Many studies have confirmed the diagnostic accuracy of duplex ultrasound in evaluating femoropopliteal occlusive disease (Table 2.8). The precise information on the site and length of an occlusion provided by duplex ultrasound is necessary for therapeutic decision making, which is ultimately dictated by what is required clinically (Fig. 2.11).

##### *Gray-Scale Imaging*

Atherosclerotic wall lesions are already apparent on the gray-scale scan as irregularities of the wall contour, intimal thickening, or plaques. In case of echogenic, noncalcified plaques in larger vessels, the B-mode image also allows a rough estimate of the extent of luminal narrowing; the quantitative assessment of the hemodynamic significance, however, is always done by Doppler assessment.

An atherosclerotic occlusion is suggested if extensive intraluminal plaques are depicted and the arterial wall can no longer be seen. Lesions caused by atherosclerosis can be differentiated from luminal narrowing caused by external structures.

*Medial sclerosis in diabetics* is characterized by diffuse calcification of the middle layer of the arterial wall. Sonographically, the calcifications produce irregular and inhomogeneous wall thickening with scattering and acoustic shadowing and impair flow imaging in the (color) duplex mode.

##### *(Color) Duplex Imaging*

In the limb arteries, the high peripheral resistance produces pulsatile, nearly laminar flow; this is reflected in the Doppler spectrum by a narrow bandwidth with a clear systolic window. The typical triphasic waveform is characterized by a steep systolic rise and subsequent decrease, followed by a short early diastolic reflux and forward flow, with the magnitude and duration depending on the vascular territory supplied. Physiologic changes in the laminar flow profile can occur at vessel origins and in curved segments.

An obstruction caused by stenosis or external compression leads to flow acceleration in proportion to the cross-sectional area reduction (see Fig. 1.43, continuity law), and the flow becomes turbulent (see Fig. 1.45). A slight increase in flow velocity can be observed with 40–50% luminal narrowing, but a relevant drop in peripheral arterial blood pressure (Doppler pressure) is unlikely or will only occur during exercise. Mild stenosis (<50%) has only little effect on the flow profile. With increasing narrowing, however, flow becomes less pulsatile with turbulence and eddy currents downstream of the stenosis. A reduction of the cross-sectional area exceeding 75% (>50% diameter reduction) is associated with an intrastenotic increase in PSV of more than 100% compared with the prestenotic arterial segment (Jäger et al. 1985a, b; Moneta et al. 1992). Flow becomes less and less pulsatile, ultimately resulting in a monophasic waveform typical of flow within and behind a high-grade stenosis (Cossman et al. 1989; Polak et al. 1991). The changes depicted by (color) duplex ultrasound within a stenosis are referred to as direct stenosis criteria and the poststenotic changes in the flow profile as indirect stenosis criteria (Table 2.9). The examiner must be aware that the indirect criterion of monophasic flow merely indicates peripheral

**Table 2.8** Sensitivity, specificity, and diagnostic accuracy of duplex ultrasonography compared with angiography in the diagnosis of hemodynamically significant stenosis (>50%), occlusion, and aneurysm of the pelvic and leg arteries

| Author | Vascular territory | Duplex technique | Reference method | Sensitivity (%) | Specificity (%) | Accuracy (%) |
|---|---|---|---|---|---|---|
| Kohler et al. (1987) | Femoropopliteal | Conventional | Conventional angio | 82 | 92 | – |
| Legemate et al. (1991) | Aortoiliac | Conventional | IA DSA | 89 | 92 | 91 |
| Allard et al. (1994) | Aortoiliac<br>Femoropopliteal | Conventional | Conventional angio | 83<br>87 | 96<br>93 | 92<br>90 |
| Cossman et al. (1989) | Iliac<br>Common femoral<br>Superficial femoral<br>Profunda femoris<br>Popliteal | Color | Conventional angio | 81<br>70<br>87<br>71<br>85 | 98<br>97<br>85<br>95<br>97 | 92<br>93<br>87<br>93<br>93 |
| Mulligan et al. (1991) | Femoropopliteal | Color | Conventional angio | 89 | 91 | |
| Moneta et al. (1992) | Iliac<br>Common femoral<br>Superficial femoral<br>Profunda femoris<br>Popliteal | Color | Conventional angio or IA DSA | 89<br>76<br>87<br>83<br>67 | 99<br>99<br>98<br>97<br>99 | |
| Strauss et al. (2001) | Iliac<br>Common femoral<br>Superficial femoral<br>Profunda femoris<br>Popliteal | Color | Conventional angio or IA DSA | 87<br>75<br>94<br>79<br>94 | 73<br>91<br>72<br>96<br>92 | 83<br>86<br>88<br>86<br>93 |
| Schäberle (1998) | Femoropopliteal, iliac, proximal segments of crural arteries | Color | Conventional angio or IA DSA; intraoperative | 97 | 98 | 97 |
| Polak et al. (1990) | Femoropopliteal | Color | Angiography or IA DSA | 88 | 95 | 93 |
| Landwehr et al. (1990) | Femoropopliteal | Color | Angiography or IA DSA | 92 | 99 | 96 |
| Koennecke et al. 1989 | Femoropopliteal | Color | Angiography or IA DSA | 97 | 97 | 97 |
| Legemate et al. (1991) | | Color | Angiography | 84 | 96 | |
| Ranke et al. (1992) | | Color | Angiography | 87 | 94 | |
| Katsamouris et al. (2001) | Aortoiliac<br>Femoropopliteal<br>Tibial | Color | Angiography | 86<br>99<br>80 | 90<br>94<br>91 | 88<br>96<br>83 |
| Aly et al. (1998) | Aortoiliac<br>Femoropopliteal<br>Crural | Color | Angiography | 89<br>100<br>82 | 99<br>99<br>99 | |

*IA DSA* intra-arterial digital subtraction angiography

dilatation, which may also be caused by other factors such as peripheral infection (Table 2.10).

Duplex ultrasound using the direct stenosis criteria has an *accuracy ranging from 83% to 97%* in identifying hemodynamically significant stenosis and occlusion of the aortoiliac and femoropopliteal arteries compared with angiography (see Table 2.8).

On color duplex images, the subtle flow acceleration associated with mild to moderate stenosis is identified by lighter color shades (primarily within the stenosis jet). With increasing stenosis severity, retrograde flow components associated with eddy currents and flow separations are depicted as color changes. High-grade stenosis with turbulent flow is characterized by a mosaic of colors and aliasing. Color duplex

**Table 2.9** Duplex ultrasound of peripheral arteries – stenosis criteria and factors affecting PSV

| | |
|---|---|
| **Direct stenosis criteria** | Turbulent flow<br>Flow acceleration PSV > 180 cm/s<br>Abrupt doubling of PSV (PSV ratio)<br>Perivascular tissue vibration |
| **Indirect stenosis criteria** | Flow profile:<br>• Damping (triphasic/monophasic)<br>• Delayed systolic rise |
| **Factors affecting PSV** | Systolic blood pressure<br>Vessel wall elasticity (increased pulsatility, medial sclerosis)<br>Sympathetic tone, outflow resistance, peripheral widening<br>PSV measurement in poststenotic segment (grade of stenosis, collaterals)<br>Collateral function<br>• Artery in which PSV is measured acts as a collateral (increase in PSV)<br>• Artery in which PSV is measured is bridged by a collateral (decrease in PSV) |

**Table 2.10** Factors damping the triphasic waveform (change from high-resistance to low-resistance flow)

| | |
|---|---|
| **Physiologic:** | Muscle activity |
| **Abnormal:** | Fever<br>Hypercirculation<br>Downstream infection<br>Postocclusive dilatation |

scanning performed with adequate settings thus enables rapid localization of a stenosis and semiquantitative estimation of its grade.

For precise stenosis quantification, however, a *Doppler frequency spectrum* must be recorded, which is highly sensitive in depicting the hemodynamic changes occurring in the prestenotic, intrastenotic, and poststenotic arterial segments (Table 2.11).

*Proximal to a high-grade stenosis*, flow may become less pulsatile due to changes in peripheral resistance. In the spectral display, however, the steep systolic rise remains unchanged (in contrast to the postocclusive waveform). The closer the sample volume is placed to a high-grade stenosis or occlusion, the less it is affected by collateral flow. When no hemodynamically significant collaterals arise between the sample volume and the flow obstruction, there may be very pronounced pulsatility or even *to-and-fro flow* (thump pattern; see Fig. 1.45).

*Distal to a stenosis* with flow acceleration (according to the continuity law) and higher-grade stenosis with loss of the triphasic flow profile, the monophasic pattern persists, and flow additionally becomes turbulent. Depending on the degree of stenosis, the poststenotic Doppler spectrum will show a decreased PSV, a delayed upstroke, and reduced pulsatility or even monophasic flow (Table 2.11). In larger arteries such as the iliac and femoral arteries, a high-grade stenosis may also be indicated by the so-called confetti phenomenon outside the blood vessel (due to tissue vibration) or by high-frequency signals in the Doppler waveform, the so-called seagull's cry.

The loss of pulsatility distal to a high-grade stenosis (Fig. 2.12) or occlusion is due to a decrease in peripheral resistance (widening of collateral vessels, reduced arteriolar tone) and a pressure drop across the stenosis. The pressure difference between the heart and the periphery is no longer equalized during a single cardiac cycle and there may be flow throughout diastole.

Most study groups comparing duplex or color duplex ultrasound and angiography in patients with PAOD found good agreement, with sensitivities and specificities ranging from 85% to 99% (see Table 2.8). More recent studies report values of over 90%, but earlier studies using only conventional duplex ultrasound show surprisingly good results as well: as early as 1986 Jäger et al. found a sensitivity of 96% and a specificity of 81% for the demonstration of abnormal changes of the pelvic and leg arteries by duplex ultrasound compared with angiography. It is noteworthy that the sensitivity was the same and the specificity higher compared with the agreement between two radiologists interpreting the same angiograms (97% sensitivity, 68% specificity).

The comparison of two examination modalities based on different criteria will necessarily yield slightly discrepant results. Angiography (but also DSA and X-ray densitometry) relies primarily on morphologic features, while duplex ultrasound uses functional parameters to assess the hemodynamic significance of a stenosis. Selective angiography continues to be the gold standard against which a newly introduced method is measured. Note, however, that angiographic methods also have their limitations in stenosis grading, as they depict only the perfused lumen and not the vessel wall. Moreover, they reduce the three-dimensional lumen to the two dimensions of the film. In particular, the posterior wall plaques frequently occurring in the pelvic arteries may be underestimated on anteroposterior views (see Fig. 5.21).

To diagnose hemodynamically significant stenosis (>50%), some study groups use mainly qualitative, indirect criteria, but the majority of investigators use either a PSV of 1.8 or 2 m/s as a cutoff or a doubling of PSV relative to the prestenotic segment (Fig. 2.13).

In summary, published data indicate that duplex ultrasonography has a sensitivity and specificity of over 90% in detecting hemodynamically relevant stenosis (>50%) compared with angiography. When state-of-the-art ultrasound equipment is used, sensitivity and specificity increase to over 95%. These results mainly apply to the aortoiliac and femoropopliteal segments. For duplex ultrasound of the arteries below the knee, sensitivities and specificities of 82/74%, 90/93%, and 90/92% have been reported (compared with

**Table 2.11** Grading of peripheral artery stenosis (see Fig. 1.45). The grade is defined as the percentage reduction in cross-sectional area. The criteria are not fully applicable in branching vessels. There are no strict boundaries between the different grades as the hemodynamic effects of a stenosis depend on a complex interaction of different factors (modified according to Wolf and Fobbe 1993; Cossman et al. 1989; Polak et al. 1991)

| Stenosis grade | (Color) duplex (intrastenotic) | (Color) duplex (just distal to stenosis) | Waveform far distal to stenosis | Waveform proximal to stenosis | PSV ratio[a] |
|---|---|---|---|---|---|
| No stenosis | Triphasic waveform (PSV < 150 cm/s) | Clear spectral window<br>Markedly pulsatile flow<br>Steep systolic upslope | Unchanged | Unchanged | <1.5 |
| 20–50%<br>Low-grade stenosis | Increase in PSV (150–200 cm/s) | Only mild turbulence<br>Moderate spectral broadening may occur | Same as prestenotic | Normal | 1.5–2 |
| 51–75%<br>Intermediate-grade stenosis | Further increase in PSV (200–350 cm/s)<br>Slight reduction in pulsatility | Eddy currents<br>Possibly slight turbulence<br>Partial filling-in of systolic window | Slightly reduced pulsatility | Normal | 2–4 |
| 76–95%<br>High-grade stenosis | Very pronounced increase in PSV (>350 cm/s)<br>Reduction in pulsatility<br>Monophasic | Considerable turbulence<br>Complete filling-in of systolic window<br>Monophasic flow | Lengthened systolic acceleration time<br>Reduced pulsatility | Amplitude normal or slightly reduced (compared to other side)<br>Pulsatility may be reduced before sites of origins of collaterals | >4 |
| >95%<br>Subtotal occlusion | More marked reduction in pulsatility Marked increase in PSV and end-diastolic velocity<br>Monophasic | Pronounced turbulence<br>Completely filled-in systolic window<br>Monophasic | Flattened systolic peak<br>Considerably reduced pulsatility<br>Monophasic | Reduced amplitude<br>Increased pulsatility immediately before stenosis<br>Prestenotic pulsatility reduced before sites of origins of collaterals | >4 |
| Occlusion | No flow signal detectable | Very reduced flow in distal segment<br>Marked damping of waveform<br>Monophasic | Very flat systolic peak<br>Monophasic | Low amplitude<br>Thump pattern immediately before occlusion: increased pulsatility, small complex with large negative component<br>Decreased pulsatility before sites of origins of collaterals | |

[a]PSV ratio: intrastenotic PSV divided by proximal PSV

angiography and using a ratio of intrastenotic to prestenotic PSV of >2 as a criterion for >50% stenosis) (Moneta et al. 1991/1992; Aly et al. 1998; Collins et al. 2007).

In a study by our group (Schäberle 1998) comparing (color) duplex imaging with angiography in 125 patients with typical symptoms of AOD (grades II–IV), the ultrasound examination was found to have a sensitivity of 96%, a specificity of 98%, and a diagnostic accuracy of 97%. The sites of steno-occlusive disease were the femoropopliteal arteries in 31%, the pelvic level in 12%, the lower leg arteries in 18%, and multiple levels in 39% of the cases. The criteria for hemodynamically significant stenosis were PSV of over 180 cm/s at sites of branching (e.g., origin of profunda femoris; Fig. 2.13a) and a focal 100% increase in PSV in straight arterial segments. The indirect criteria listed in Table 2.9 (prestenotic and poststenotic waveform changes) served as supplementary criteria.

The *main trunk of the profunda femoris artery* is of particular significance in the diagnostic evaluation as it is the most important collateral pathway in occlusion or stenosis of the superficial femoral artery. It is often involved in obstruction of the superficial femoral artery; however, blood supply to the lower leg can be improved by a minor surgical intervention (profunda femoris repair, TEA). Lesions at the origin of the profunda femoris artery are important to identify, but on angiograms they may be obscured by superimposed vessels. Reliable assessment is thus only possible if an *additional oblique projection* is obtained.

In another study by our group (Strauss et al. 1989), investigating the hemodynamics at the origin of the profunda femoris with determination of the degree of stenosis from PSV, the PPV and NPV were 86% and 91%, respectively, compared with angiography as the reference method. ROC analysis identified a PSV of 180 cm/s as the optimal cutoff

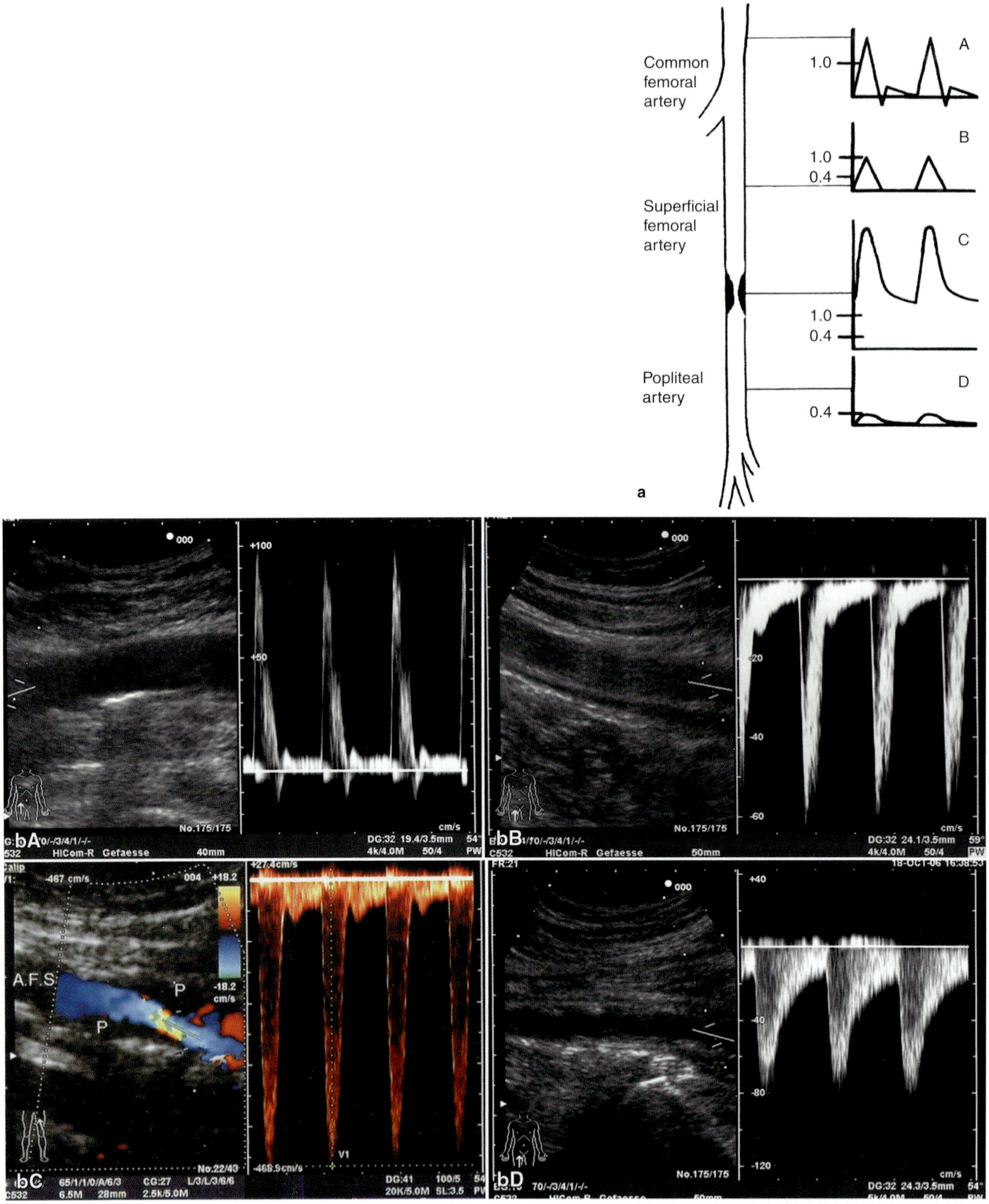


**Fig. 2.12** **a** Doppler waveform changes induced by a superficial femoral artery stenosis in different segments of the peripheral circulation. **bA** Nearly normal triphasic waveform from common femoral artery (far away from stenosis). **bB** Prestenotic waveform from proximal superficial femoral artery with slightly reduced PSV and reduced or absent early diastolic dip but steep systolic upstroke. **bC** Monophasic flow profile with intrastenotic PSV increased to more than twice the prestenotic velocity. **bD** Monophasic poststenotic waveform from popliteal artery with delayed systolic upstroke and low PSV (damped waveform)

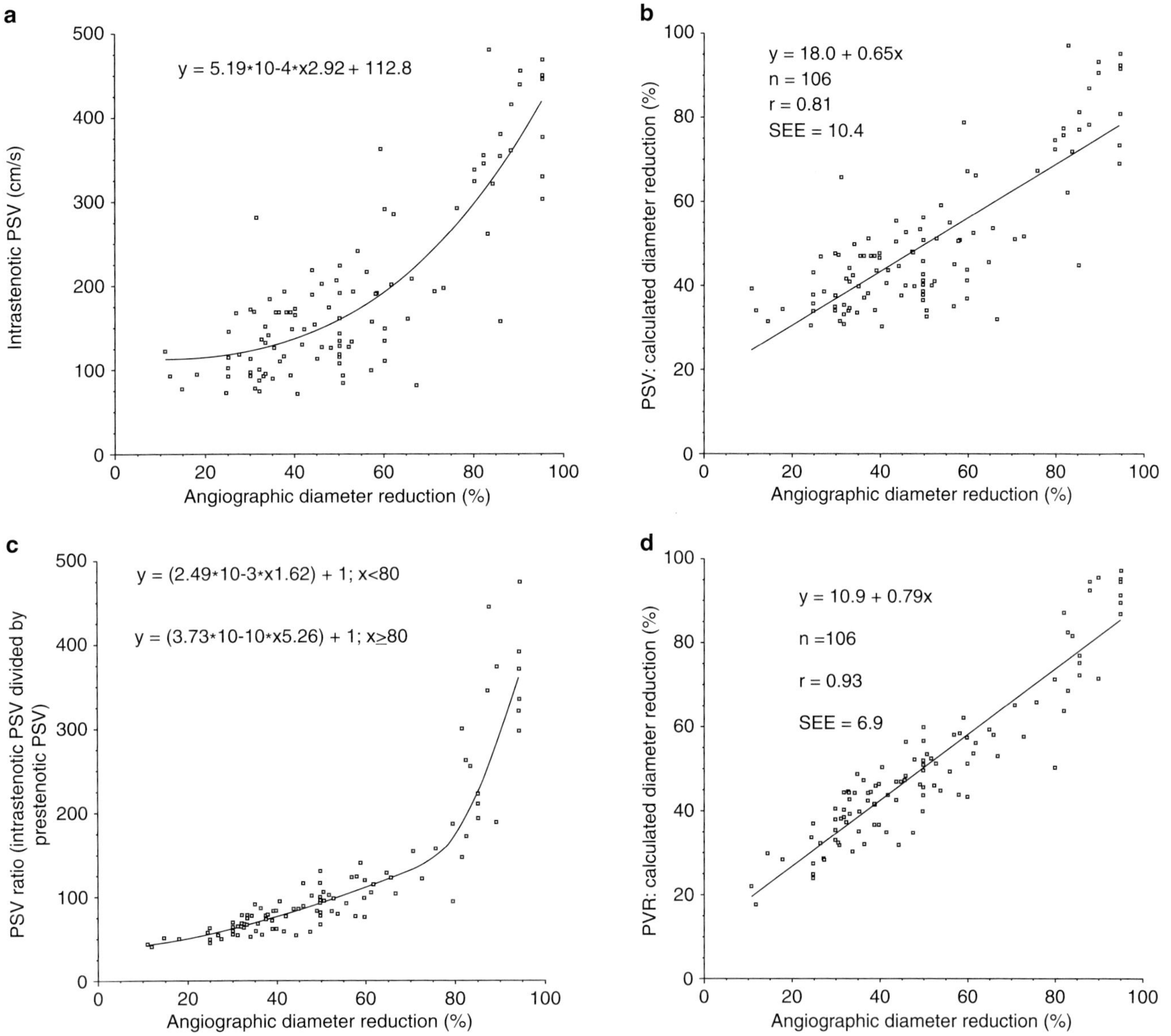


**Fig. 2.13 a–d** Correlation of intrastenotic peak systolic velocity (PSV) (**a, b**) and of PSV ratio (intrastenotic PSV divided by prestenotic PSV) (**c, d**) with percent diameter reduction at angiography (graphs from Ranke et al. 1992). **a** Correlation of intrastenotic PSV and angiographic diameter reduction in 106 femoral artery stenoses using a PSV cutoff of 180 cm/s for >50% stenosis (correlation $r = 0.81$). **b** Linear regression analysis of percent diameter stenosis calculated from PSV versus angiographic diameter stenosis. **c** The PSV ratio correlates better with angiographic diameter reduction ($r = 0.93$) because it is less susceptible to variations in systemic factors (systolic blood pressure, vessel wall elasticity). According to this analysis, the best results were achieved using a PSV ratio of 2.4 as a cutoff, which identified >50% stenosis with 87% sensitivity and 94% specificity. **d** Linear regression analysis of percent diameter reduction calculated from PSV ratio versus angiographic diameter reduction

for differentiating normal arteries and low-grade stenosis from higher-grade stenoses (>50%) (see Fig. 2.14).

The more distal an occlusion, the more the examiner has to take into account the collateral function of the vessel examined. For instance, the profunda femoris artery as the main collateral in occlusion of the superficial femoral artery may show an increase in mean flow velocity of over 100% at its origin without itself being stenosed. Moreover, flow in a collateral artery bridging an occluded segment is less pulsatile because peripheral resistance is decreased. In this situation, only an increase in PSV above a defined threshold (see Table 2.9) and a monophasic flow profile are valid criteria for diagnosing a stenosis (Fig. 2.15).

Plaque configuration (concentric versus eccentric) is another factor influencing the degree of increase in PSV within a stenosis. A concentric plaque causing the same degree of stenosis in terms of diameter reduction as an eccentric plaque leads to a greater increase in flow velocity across

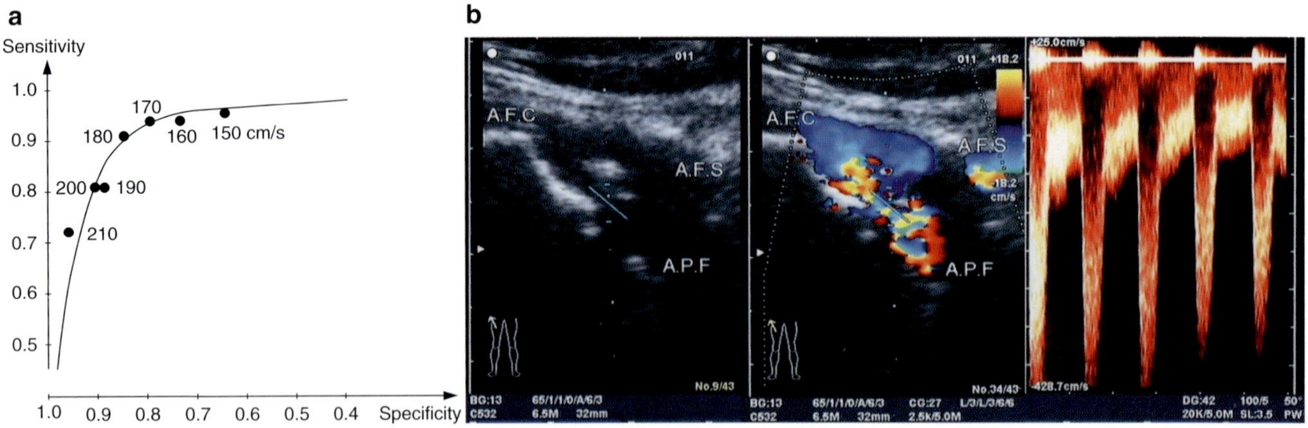


**Fig. 2.14** **a** Profunda femoris artery stenosis: ROC curve for determining the sensitivity and specificity at different velocity cutoffs in the profunda femoris artery measured by duplex ultrasound in comparison with angiography. **b** Stenosis at profunda femoris origin with a PSV of 403 cm/s, monophasic flow, and aliasing. The images show the femoral bifurcation with the common femoral artery (*A.F.C.*), superficial femoral artery (*A.F.S.*), and profunda femoris artery (*A.P.F.*) in one scan plane. Acoustic shadowing due to plaque in the superficial femoral artery; stenosis in the obscured segment is suggested by aliasing (blue indicates flow away from the transducer, toward the periphery)

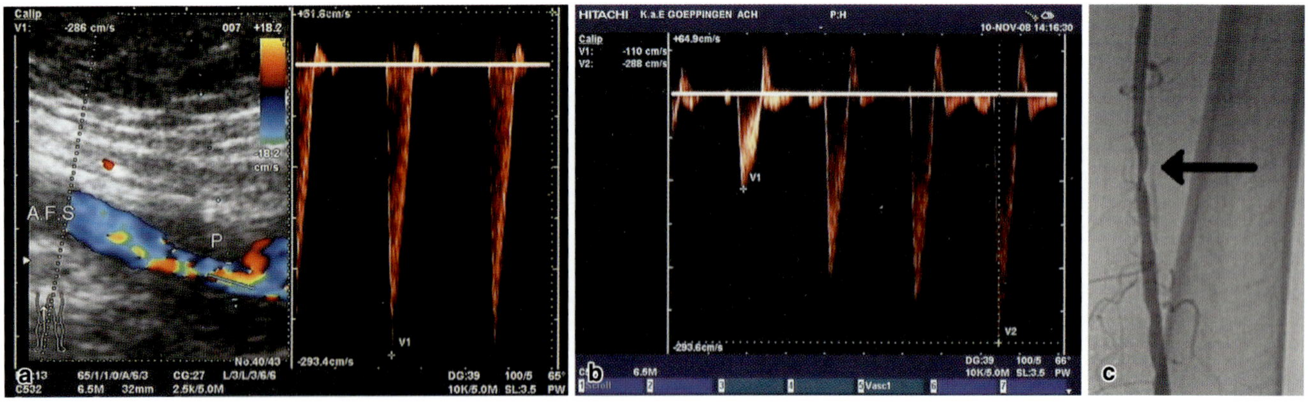


**Fig. 2.15 a–c** Grading of superficial femoral artery stenosis. **a** Hypoechoic plaque (*P*) causes 50–70% stenosis of the superficial femoral artery with a spectral Doppler-derived PSV of 290 cm/s. Aliasing in the stenotic segment; sample volume placed in the stenotic jet, antegrade flow away from the transducer (coded blue). **b** A continuous Doppler trace is obtained from the superficial femoral artery (*A.F.S.*) by moving the angled transducer (acute Doppler angle) across the skin starting 2 cm proximal to the stenosis. This Doppler measurement reveals a PSV of 290 cm/s in the stenotic jet versus 110 cm/s in the prestenotic segment. The PSV ratio calculated from these values (intrastenotic to prestenotic PSV) is more than 2 but less than 4, corresponding to a stenosis grade of more than 50% but less than 70%. The stenosis is not a high-grade stenosis, which is why the Doppler waveform shows normal triphasic flow (no arteriolar dilatation). **c** Angiogram confirms 50–70% stenosis

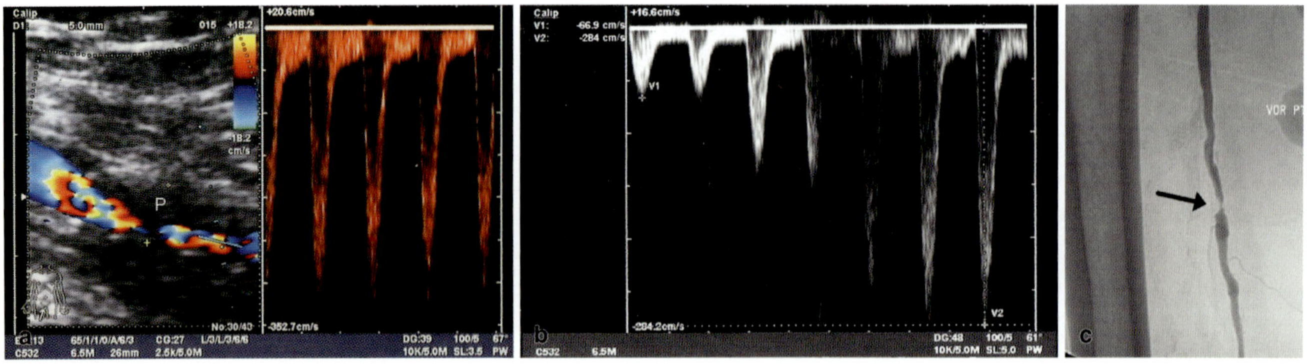


**Fig. 2.16 a–c** Grading of high-grade stenosis. **a** Hypoechoic plaque (*P*) causes high-grade stenosis with a PSV of almost 4 m/s and monophasic flow. Similar constellation as in Fig.2.15, except that the stenosis is high-grade. **b** Continuous measurement as described in Fig. 2.15 reveals an increase in PSV from 60 cm/s to over 3 m/s in the stenosis, corresponding to a PSV ratio > 4, which indicates high-grade stenosis. **c** Angiogram confirms high-grade stenosis of the superficial femoral artery

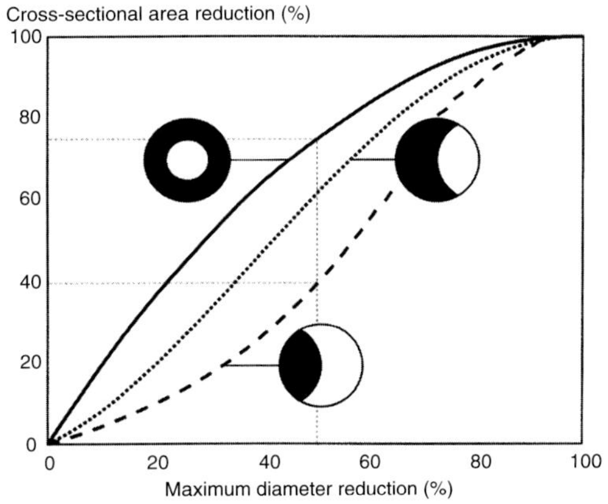


**Fig. 2.17** Effects of different plaque configurations (concentric – eccentric) on cross-sectional area reduction for different angiographic stenosis grades. Because of these differences, a stenosis caused by an eccentric plaque will lead to a lower increase in PSV (as measured by duplex ultrasound) compared with a stenosis of the same grade caused by a concentric plaque. The hemodynamic effect of a stenosis is determined by the cross-sectional area reduction (see Figs. A2.9 and 5.12)

the stenosis (angiographic stenosis grading; Fig. 2.16). The reason is that a 50% diameter reduction of the arterial lumen will produce a cross-sectional area reduction of 75% when a concentric plaque is present as opposed to only 50% when caused by an eccentric plaque (Fig. 2.17). The hemodynamic stenosis severity determined by duplex ultrasound is thus a more adequate measure of plaque-related blood flow obstruction than is morphologic appearance (angiography). The increase in PSV within the stenosis reflects the cross-sectional area reduction.

### *Spectral Doppler Assessment of Calf Arteries*

The indirect stenosis criteria can also facilitate and shorten the sonographic examination of the lower leg arteries, which are less amenable to ultrasound evaluation. The search for steno-occlusive lesions or the evaluation of potential bypass targets below the knee begins with a Doppler interrogation of the dorsalis pedis and posterior tibial arteries. The Doppler spectra from these sites are compared with a waveform from the popliteal artery. The examiner then proceeds to measure Doppler spectra from the proximal calf arteries for comparison with the waveforms from the ankle area in order to narrow down sites of obstruction. Finally, if relevant for treatment planning, the examiner can try and localize individual lesions (stenosis or occlusion). If the calf arteries are being examined to identify the site of distal anastomosis for a crural bypass graft once occlusive disease of the popliteal artery and trifurcation has been confirmed, the examiner first identifies the artery with the highest blood flow in the ankle area. This artery is then continuously scanned from the ankle upward using low-flow settings to detect the slow flow in the calf arteries (similar to venous flow), searching for lesions that might preclude its use as a bypass target and identifying the most suitable site for the distal anastomosis. At the same time, the candidate artery is screened for a greater than 100% increase in PSV, which indicates a hemodynamically relevant stenosis, even in vessel segments distal to an occlusion, possibly rendering it unsuitable for use as a bypass target.

In patients with extensive atherosclerotic disease or longstanding diabetes with severe medial sclerosis, sonographic evaluation of the arteries below the knee is impaired. In these patients, calcified lesions may produce acoustic shadowing, precluding long segments of the arteries from being evaluated for the presence of stenosis or occlusion. When acoustic shadowing occurs, stenosis grading becomes inaccurate and the length of an occluded segment can be misinterpreted. Acoustic shadowing is a problem that cannot be overcome by the use of ultrasound contrast agents. Good knowledge of the sonoanatomy of the calf vessels is important to accurately assess vascular disease in this territory and to minimize the risk of misinterpretation that may result from mistaking a collateral for a main calf artery.

Duplex scanning can supplement angiography in identifying a target site for the distal anastomosis of a pedal bypass graft. The excellent spatial resolution afforded by high-frequency probes (10 MHz) is superior to angiography, which, in patients with proximal occlusion, is often limited by poor opacification of calf or pedal arteries. A potential bypass target artery is searched for plaques using the B-mode, and a Doppler spectrum is obtained to establish patency of the pedal arch (Hofmann et al. 2004). Only a few studies have investigated the role of duplex imaging in diabetics with primary peripheral occlusion or diabetic foot syndrome (Boström et al. 2002; Dyet et al. 2000; Schneider and Ogawa 1998). The results are difficult to compare because the patient populations investigated are very heterogeneous in terms of clinical stage and severity of macroangiopathy. What is noteworthy about these studies is that in a large proportion of patients, results for the calf arteries were inconclusive (29%), the calf arteries were not examined systematically (22%), or ultrasound failed to visualize these arteries (13%). The fibular artery, often the only patent calf artery in diabetics, was found to be least accessible to sonographic evaluation.

Detection of an isolated stenosis in the plantar arch remains a problem because continuous evaluation of these arteries is not always possible. This is the reason that duplex imaging alone cannot be used to decide whether the posterior tibial artery or the dorsalis pedis artery is more suitable to receive the bypass graft. The hope of overcoming these limitations by using ultrasound contrast agents has not been fulfilled. Contrast agents produce overenhancement of collateral vessels (blooming effect), impairing the identification of the main calf arteries (Dyet et al. 2000; Ubbink et al. 2002).

*Role of Ultrasound Contrast Agents*

Few scientific data are available on how contrast agents can improve the sonographic diagnosis. In a recent multicenter study (Sidhu et al. 2006), the percentage of poorly visualized vascular segments decreased from 40.7% to 7.4% when SonoVue was given at a dose of 2.4 mL. A subgroup analysis of agreement with different reference methods (angiography, CT angiography, magnetic resonance imaging) showed that the diagnostic accuracy of color duplex imaging increased from 30.7% to 68.9% after administration of the contrast agent. A limitation of this study is the use of different ultrasound equipment and the diversity of vascular territories investigated (ranging from carotid to peripheral arteries) in a total of 82 patients. The subgroups are not well defined with regard to results and ultrasound accuracy in different body regions. Most notably, it would be desirable to have separate results for the calf arteries, as this is the only vascular territory for which the supplementary use of a contrast agent appears to have some justification. Using the scanning strategy presented above (Table 2.1), an ultrasound contrast agent is only necessary in those cases where the standard technique fails to unequivocally identify a suitable target vessel for a planned bypass onto a calf artery. This is typically the case in long-standing diabetes with medial sclerosis, where acoustic shadowing obscures long vessel segments. However, medial sclerosis can be expected to impair contrast-based ultrasound evaluation as well.

*Effects of Collaterals on the Doppler Waveform*

*Postocclusive perfusion pressure in the periphery is affected by preocclusive systemic pressure and, above all, by flow resistance in the collateral circulation* (Fig. 2.18). *Collateral resistance*, in turn, *depends on the number and size of collateral vessels*, the length of the occluded segment to be bridged, and blood viscosity. When collateral resistance is low, the situation in the periphery is less dramatic, and there is only moderate postocclusive vasodilatation; the Doppler waveform has a fairly normal appearance and pulsatility. Hence, *postocclusive pulsatility* is determined not only by the grade of stenosis or length of obstruction but also by the *extent of collateralization*. When there is good collateralization (e.g., in pelvic artery stenosis), the Doppler waveform is less abnormal, with only little damping of the triphasic flow profile. The magnitude of the postocclusive Doppler pressure is likewise a function of both the occlusive process and collateral resistance. Moreover, good collateralization with division of blood flow can lead to a lower PSV in the stenotic segment than would be expected based on the degree of stenosis.

This complex hemodynamic situation is reflected in the waveform and must be taken into account in interpreting spectral Doppler traces. With the development of collateral pathways in patients with chronic occlusive disease, postocclusive

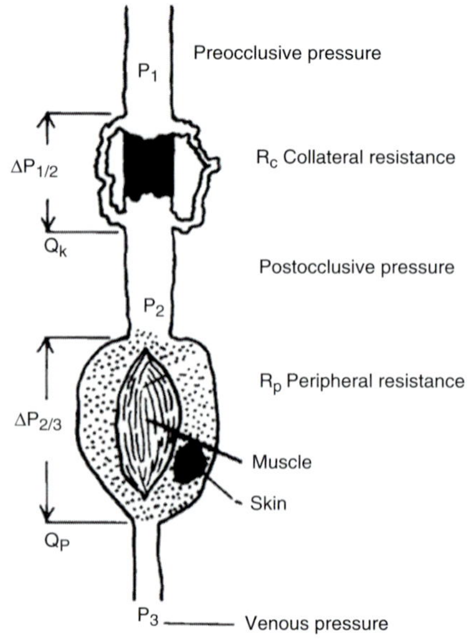


**Fig. 2.18** Postocclusive arterial pressure depends on the degree of stenosis or length of occlusion as well as on collateral resistance. Pressure increases when there is low collateral resistance and extensive collateral flow. In the presence of good collateralization, the spectrum is less abnormal with a higher PSV and low end-diastolic velocity. A nearly normal triphasic waveform may be seen in patients with pelvic artery occlusion and good collateralization (from Rieger and Schoop 1998)

flow becomes more pulsatile (Fig. 2.19). Collaterals also affect flow velocity and the flow profile in the immediate vicinity of a stenosis (see Table 2.9 and Fig. 2.20). *How collaterals alter the waveform configuration depends on where they arise and enter relative to the site of Doppler measurement.* In a prestenotic segment proximal to the origin of collaterals, loss of peripheral resistance results in a monophasic waveform with persistent diastolic flow through the collaterals. Systolic flow velocity is high with a rapid upsweep to peak velocity. A prestenotic Doppler spectrum recorded between the site of origin of collaterals and a high-grade stenosis will show more pulsatile flow because of the high resistance resulting from the stenosis. The flow volume in the immediate prestenotic segment decreases with increasing collateral flow. As a consequence, PSV within the stenosis is lower than in a stenosis of similar degree with little or no collateralization (See Table 2.9 and Fig. 2.20), and its severity may be underestimated.

*Multilevel Obstruction*

In patients with multilevel occlusive disease, the *poststenotic Doppler spectrum* recorded behind the most proximal obstruction is also influenced by *flow alterations caused by lesions distal to the sampling site*. For instance, the waveform will be more pulsatile if there is strong resistance due to subtotal occlusion of the artery more distally (Fig. 2.21).

In patients with *sequential stenoses or occlusions*, the usual stenosis criteria may thus lead to misinterpretation. In grading

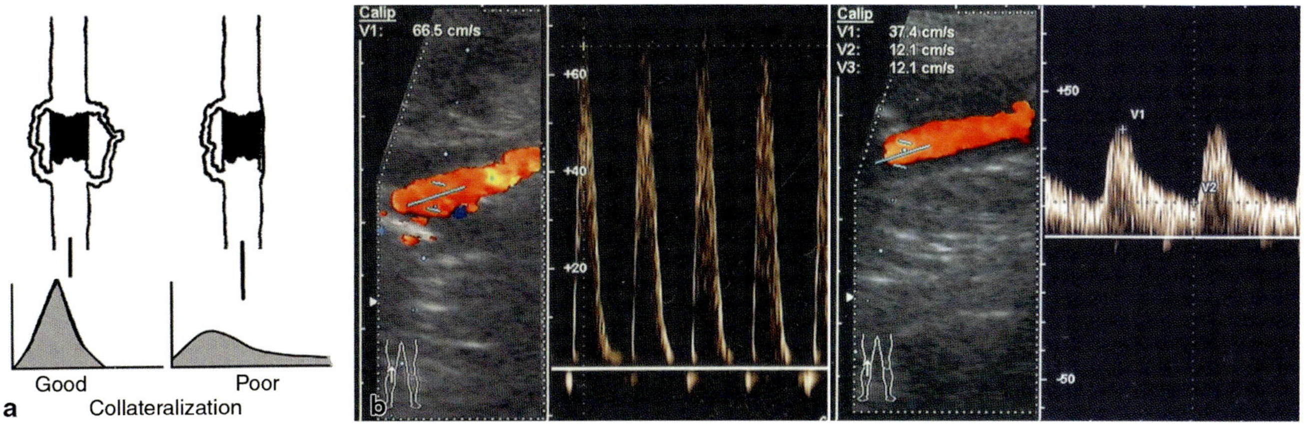


**Fig. 2.19** **a** With poor collateralization, the Doppler waveform obtained in the postocclusive segment (*right*) is damped with a large diastolic component, resulting from distal dilatation in response to chronic peripheral ischemia. With increasing collateralization, blood flow distal to an occluded segment becomes more pulsatile (*left*), approaching normal flow when there is optimal collateralization and postocclusive pressure approximates preocclusive pressure. **b** The pulsatility of the Doppler spectrum recorded distal to an occlusion is determined by the magnitude of collateral flow. Good collateral pathways can compensate for an occluded main artery and ensure adequate perfusion, at least at rest. For instance, if an isolated occlusion of a pelvic artery or the superficial femoral artery develops slowly over years, one may occasionally record a triphasic Doppler waveform over the popliteal artery, but with reduced PSV and delayed acceleration. Conversely, the poorer the collateral situation, the more monophasic the waveform and the lower the PSV (relative to end-diastolic flow velocity) become. The first example illustrates the findings in superficial femoral artery occlusion with good collateralization (*left*): pulsatile Doppler waveform from the popliteal artery with short retrograde flow in early diastole and zero diastolic flow (ABI of 0.8). The second example (*right*) shows a monophasic waveform from a patient with poor collateralization of superficial femoral artery occlusion: persistent diastolic flow and a low PSV (ABI of 0.5).

a second stenosis, the examiner has to take into account the *hemodynamic changes* (change in pulsatility and pressure drop) *produced by the preceding stenosis*: the postocclusive decrease in velocity after the first stenosis will result in a lower absolute PSV in the second stenosis (and the PSV of 180 cm/s proposed as a criterion for isolated stenosis does not apply in sequential stenoses; Table 2.9). Therefore, only the criterion of a 100% increase in PSV can be used to classify a sequential stenosis as hemodynamically significant.

Stenosis grading by comparison of intrastenotic and prestenotic flow velocities is rather unreliable for an origin stenosis because the prestenotic segment has a different diameter and different hemodynamics. In cases where direct color duplex scanning of a stenotic segment is not possible due to acoustic shadowing by calcified plaques, the proximal and distal Doppler measurements have to be compared (Table 2.12). If the PSV and waveform are similar, the plaque does not produce hemodynamically significant narrowing.

### Vascular Occlusion

Color duplex scanning and Doppler ultrasound demonstrate vascular occlusion as an absence of flow signals. Note, however, that the failure to demonstrate flow may also be due to inadequate instrument settings (gain, PRF) or technical limitations of (color) duplex ultrasound in the area of calcified plaques (see Table 2.12). The problem of posterior acoustic shadowing due to calcifications in the vessel wall is mainly encountered in diabetic patients with medial sclerosis and can be overcome by comparing the Doppler waveforms before and after the calcified segment (monophasic profile downstream of occlusion) and searching for collaterals arising upstream of the obstruction and emptying downstream (see Fig. 2.22a–d).

Duplex scanning is highly accurate in determining the *length of occlusion.* A study of our group including 40 legs with femoropopliteal occlusion demonstrated a correlation of 0.96 between angiography and duplex ultrasound. The length of the occluded segment was less than 5 cm in 21%, 5–10 cm in 54%, and over 10 cm in 25% cases. Pelvic artery occlusion ($n = 30$) was correctly identified by duplex ultrasound in all patients; however, due to the poorer scanning window, the peripheral extent was sometimes overestimated by several centimeters (“dead water zone”). A similar correlation ($R = 0.95$ in 98 extremities) between the sonographically and angiographically measured length of occluded segments was reported by other authors (Karasch et al. 1993).

Slow postocclusive flow may lead to overestimation of the occluded segment, in particular when collateralization is poor. Further downstream, evaluation by duplex scanning is better due to the increased flow resulting from emptying collaterals. If visualization is poor, blood flow detection can be improved by intravenous administration of ultrasound contrast medium (Langholz et al. 1992). In the routine clinical setting, however, ultrasound contrast media are used in individual cases only.

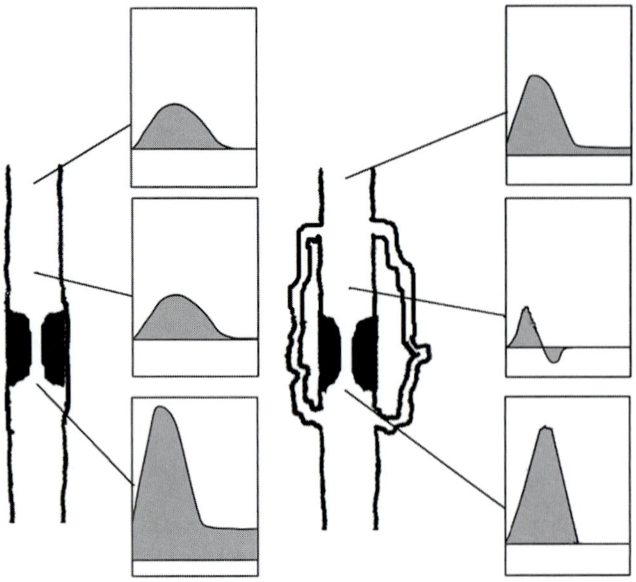

**Fig. 2.20** Diagrams illustrating the effects of collateralization on pre- and intrastenotic waveforms (flow character and velocity). With little or no collateralization (or in bypass stenosis), the prestenotic waveform will show a reduced PSV. In this situation, flow in the prestenotic segment is less affected by compensatory peripheral dilatation, and a knocking waveform (thump pattern) may be obtained in very high-grade stenosis (*left* example). In case of good collateralization, PSV distal to the origin of a collateral pathway is relatively normal; however, there will be some diastolic flow transmitted through the collateral (as a result of reduced perfusion, leading to peripheral widening). Flow is more pulsatile between the origin of a collateral and the stenosis; some to-and-fro flow may be seen immediately proximal to a high-grade stenosis (*right* example). If there are no collaterals, intravascular pressure leads to a higher intrastenotic flow velocity (*left*). The lower pressure in a collateral vessel diverts flow from the stenotic artery. The resultant decrease in pressure in the main artery reduces intrastenotic flow velocity compared with the velocity expected in a stenosis of the same degree without collateral circulation (*right*). This is why the severity of a stenosis may be underestimated if the effects of collateralization are ignored. The ratio of intrastenotic to prestenotic PSV is independent of the extent of collateralization

**Table 2.12** Duplex ultrasound of the peripheral arteries – intrinsic limitations of the method

| | |
|---|---|
| **B-scan:** | Calcified plaque: posterior acoustic shadowing<br>Edema: scattering |
| **Doppler:** | Calcified plaque: posterior acoustic shadowing<br>Maximum flow velocity detectable: limited by PRF |

A low PRF and a high gain are needed to detect the slow flow distal to an occlusion and to correctly identify the distal end of the occluded segment.

An *occlusion*, like a high-grade stenosis, *affects the preocclusive and postocclusive Doppler waveforms*. When using a duplex ultrasound system without a color option, the examiner can approach the occluded zone by sampling Doppler information at both ends. Signals from collaterals coursing parallel to the occluded artery may be misinterpreted as patency shortly before the refilled segment of the artery is actually reached, giving rise to underestimation of the length of the occlusion. Collaterals entering the main artery can be identified by an apparent sudden flow acceleration resulting from the different insonation angle and above all by the change in flow direction indicated by the Doppler signal. Once a site of origin or re-entry of a collateral has been identified, a Doppler tracing obtained with angle correction will enable diagnosis of a stenosis. Evaluation of the postocclusive flow also contributes to therapeutic decision making (medical therapy or repair).

The flow rate downstream of multilevel occlusions with poor collateralization may occasionally drop below the limit of detection of (color) duplex scanning – even when a high-resolution transducer with proper settings is used. In these cases, spectral Doppler measurement with a high gain and a low PRF can often detect any residual flow that may still be present.

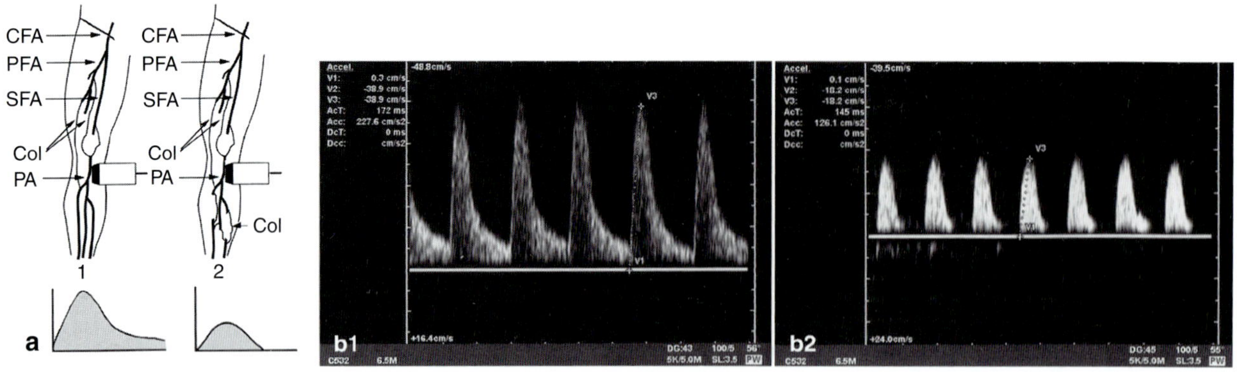


**Fig. 2.21** **a** Diagrams illustrating the effects of peripheral outflow on popliteal artery spectral Doppler measurements in superficial femoral artery occlusion. When outflow is poor due to occlusion of a calf artery (*right*), higher outflow resistance leads to a more pulsatile postocclusive waveform; the waveform is monophasic when the calf arteries are patent (*left*) (*CFA*, common femoral artery; *PFA*, profunda femoris artery; *SFA*, superficial femoral artery; *PA*, popliteal artery; *Col*, collaterals). **b** Illustration of the effects of differences in peripheral outflow on popliteal artery Doppler recordings in two patients with superficial femoral artery occlusion and comparable collateralization. **b1** In the first case, the calf arteries are patent and there is good peripheral outflow. There is an acceleration time of 172 ms, a PSV of 39 cm/s, and an end-diastolic velocity of 8 cm/s (corresponding to *left* drawing in **a**. **b2** In the second case, all three calf arteries are occluded, and the foot is supplied through collaterals. Here, a knocking waveform (thump pattern) is obtained from the popliteal artery. In this case, acceleration time is 145 ms with a PSV of 18 cm/s. Flow is more pulsatile due to higher outflow resistance (corresponding to *right* drawing in **a**

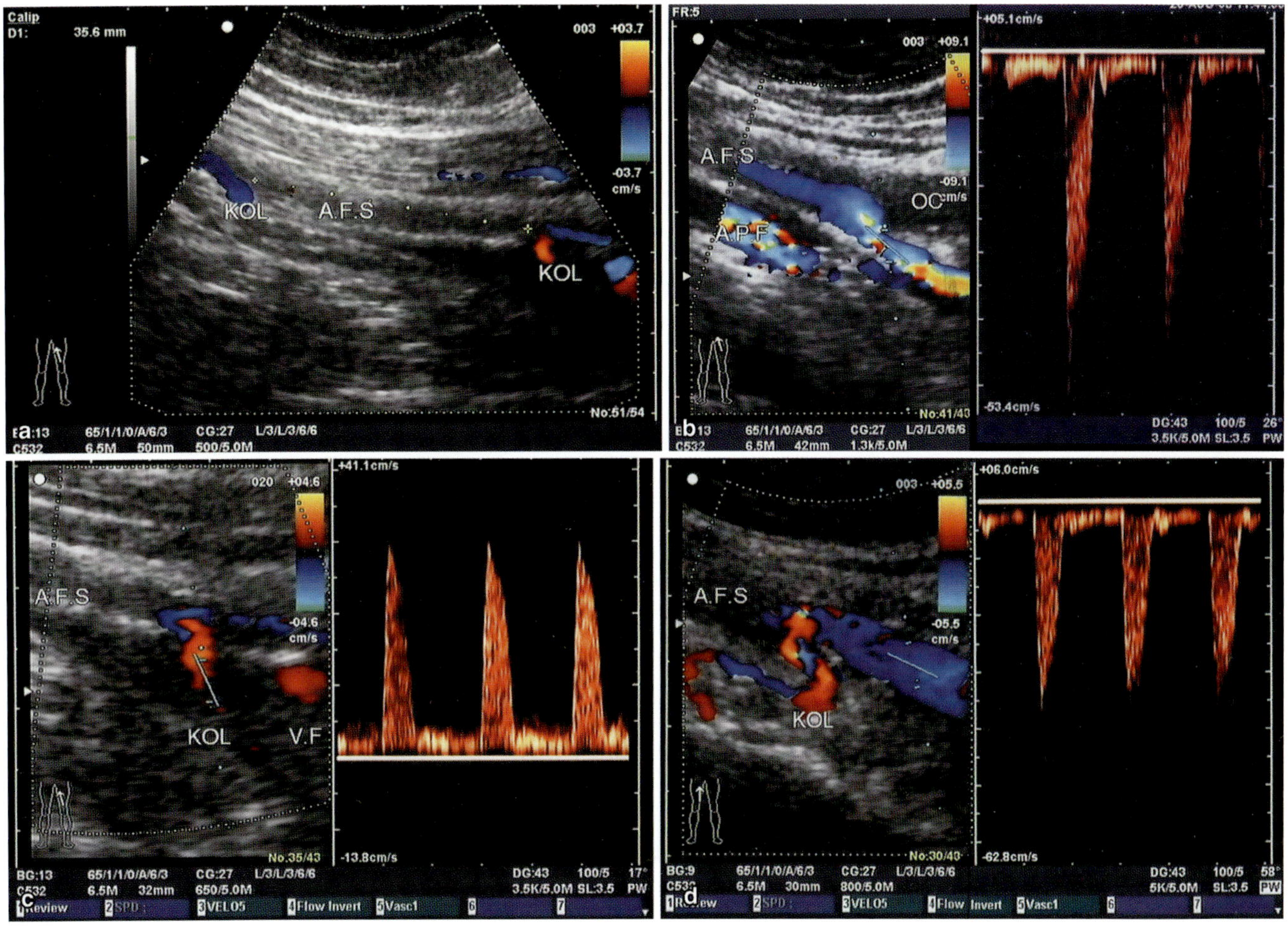


**Fig. 2.22 a–d** Superficial femoral artery occlusion. **a** Exact determination of the length of an occluded segment is important for therapeutic decision making (PTA vs. bypass grafting). First, the length is determined by duplex scanning using a low PRF to also detect slow flow (3.5 cm in the example shown). Additional evaluation of collaterals is recommended to confirm the measured length, especially when calcified plaques cause acoustic shadowing and impair assessment of the main artery. The image shows a dilated collateral segment (*KOL*) proximal to the occlusion (flow away from transducer, coded blue, *left*) and another collateral segment refilling the superficial femoral artery (flow toward transducer, coded red, *right*). **b** Detailed evaluation of collaterals: the dilated collaterals indicate the beginning and end of the occluded segment (transducer moved to focus on the sites of origins of collaterals). The Doppler waveform from the origin of the collateral shows pulsatile flow with a velocity of 50 cm/s, indicating good inflow into the collateral system (aliasing in the color flow image is due to small Doppler angle and does not indicate stenosis in this case). **c** Detail showing the collateral resupplying the superficial femoral artery distal to the occluded segment. Flow is toward the transducer (red) with a PSV of 30 cm/s. **d** Doppler spectrum from the superficial femoral artery segment (*A.F.S.*) resupplied by the collateral (*KOL*) distal to the occluded segment. The postocclusive pulsatility suggested by the waveform with a small diastolic flow component and early diastolic decrease in flow velocity (resulting from the reflected pressure wave), a PSV of almost 40 cm/s, and a rather steep systolic upstroke indicate adequate compensatory collateral circulation. The good collateral flow in this case maintains nearly normal pressure in the postocclusive segment, which ensures adequate peripheral blood supply at rest without a need for arteriolar dilatation. The findings (sonographic length of occlusion) would theoretically justify an attempt at PTA (if clinically indicated), but in this case favor a conservative strategy: the Doppler measurements indicate good collateral circulation, while the relationship between the occluded segment and the collateral resupplying the main artery distal to the occlusion suggests that there is a risk that the collateral may become occluded during PTA

### 2.1.6.2 Arterial Embolism

Arterial embolism with ischemia is typically of cardiac origin (80–90%) with the remaining cases being accounted for by arterioarterial emboli, chiefly arising from a partially thrombosed aneurysm and rarely from an atherosclerotic lesion.

The site and length of an embolic occlusion are identified by Doppler measurement or by the absence of flow signals in the color duplex mode. In the less common cases of subtotal occlusion, some residual flow will be detected along the hypoechoic thromboembolus near the wall (see Fig. A2.39). An embolic occlusion is suggested by the depiction of a hypoechoic and homogeneous mass in the vessel lumen, good delineation of the wall with preservation of its smooth contour, and the absence of plaques.

Embolic occlusions typically occur at bifurcations, where the embolus creates a nidus for the formation of appositional thrombi that may extend proximally to the site of the nearest hemodynamically significant branching. The distal end of the occlusion is identified with a low PRF and a high gain in order to depict the slow flow resulting from poor collateralization. In addition to identifying and characterizing the embolic occlusion, searching for the source of the embolus is an integral part of the examination (echocardiography, duplex ultrasound of the aorta and peripheral arteries focusing on the identification of a possible popliteal artery aneurysm).

#### 2.1.6.3 Aneurysm

Aneurysms most commonly affect the aorta and the popliteal artery but may also occur in the femoral and iliac vessels, in particular in patients with dilated angiopathy (Schuler et al. 1993). An aneurysm is seen on gray-scale images as a *saccular or spindle-shaped dilatation* of the vessel lumen. *Mural thrombi* in the aneurysm are often identified by their slightly higher echogenicity relative to flowing blood and are confirmed by the absence of color in the color duplex mode. Thrombotic deposits can cause stenosis, in particular when they occur at the distal end of an aneurysm. No flow signals will be depicted in a completely thrombosed aneurysm. Angiography is not the method of reference for assessing a partially thrombosed aneurysm while *computed tomography (CT)* depicts the morphology and extent of an aneurysm but provides no hemodynamic information. Patients with an isolated occlusion in the popliteal territory should undergo an ultrasound examination to rule out a thrombosed aneurysm or compression syndrome prior to a radiologic intervention.

Aneurysms of the popliteal artery are often bilateral, and 25–30% of patients have a concomitant aneurysm of the abdominal aorta. Aneurysms can occlude or rupture or cause embolic occlusion of peripheral vessels – making amputation necessary.

Surgery is indicated for all aneurysms with a diameter greater than 2 cm and also for smaller ones when they are saccular or contain thrombotic deposits (Fig. 2.23a, b). Thrombotic aneurysms in the knee area carry a higher risk of

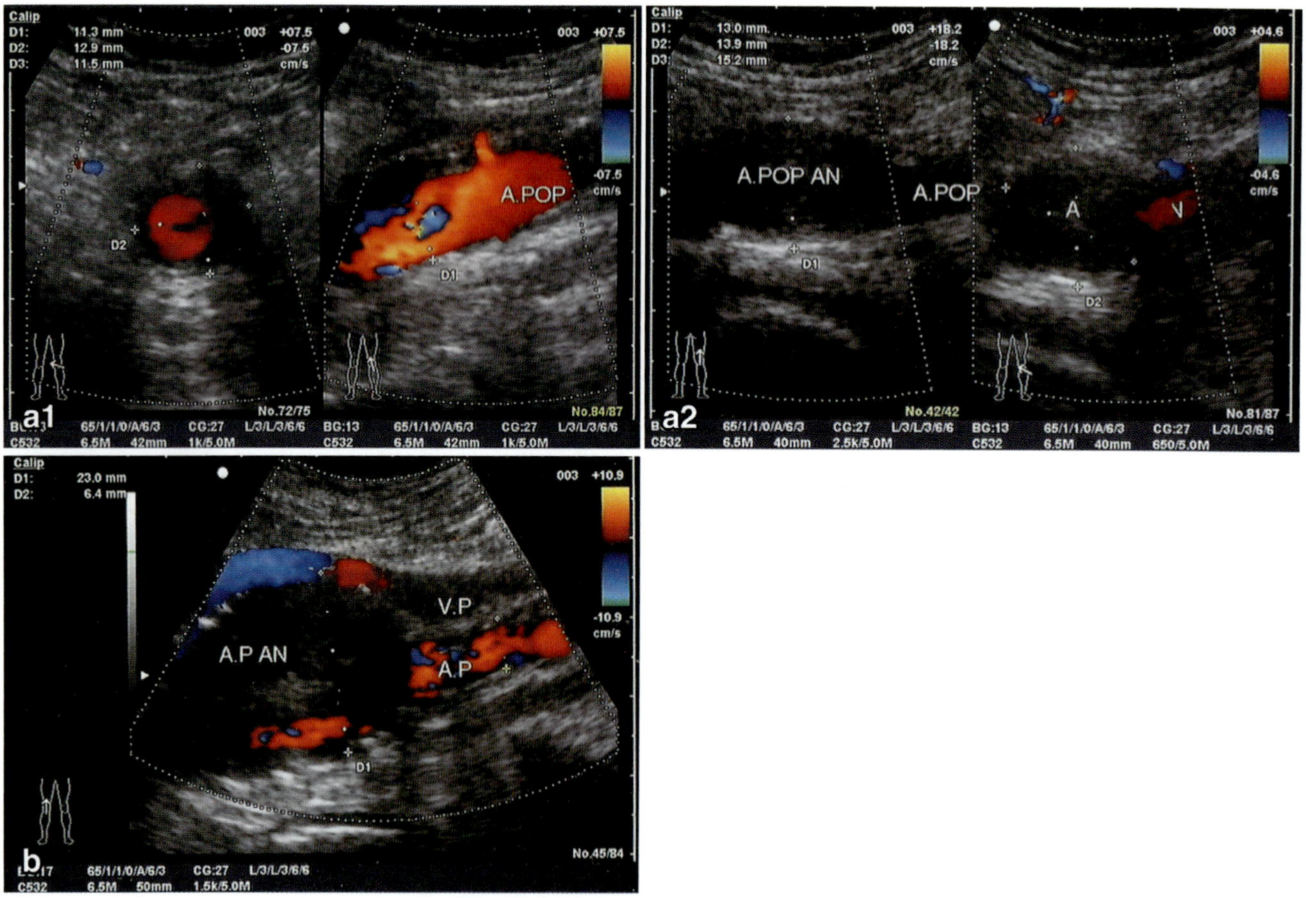


**Fig. 2.23 a** Serial ultrasound of small popliteal artery aneurysm. **a1** Small, partially thrombosed aneurysm of the popliteal artery with a maximum diameter of 13 mm. Such an aneurysm can cause embolism with occlusion of a lower leg artery (due to the shear forces and kinking of the artery when the knee is bent). For this reason, surgical repair is indicated. Color duplex imaging with a low PRF allows good delineation of the patent lumen in transverse and longitudinal planes. Plus signs indicate the total luminal diameter. **a2** Same popliteal artery aneurysm half a year later (patient refused surgery). There is a slight increase in aneurysmal diameter to 15 mm, and complete thrombosis of the aneurysm (*A.POP.AN*) and trifurcation has occurred. The longitudinal image (*left*) shows the transition from the normal arterial lumen to the aneurysm. **b** Large, partially thrombosed popliteal artery aneurysm (*A.P.AN*) (2.5 cm in diameter); the normal artery diameter distal to the aneurysm is 6.4 mm. The accompanying popliteal vein (*V.P*) is displaced and compressed (longitudinal view)

arterial embolism even when they are small because they are exposed to the shear forces of the bending knee.

This is why, in patients with popliteal artery aneurysm, the risk of rupture is less relevant than the risk of peripheral embolism arising from thrombus, and *diameter is not the main criterion in identifying candidates for surgical repair. Detection of mural thrombus is thus the primary diagnostic task* in these patients. Even a small popliteal artery aneurysm is operated on if partial mural thrombosis is demonstrated.

*Duplex ultrasound is the method of first choice, yielding reliable information on the diameter of the aneurysm, its shape, and the presence of thrombosis* (see Fig. A2.42) – thus providing the foundation for planning the surgical procedure. An aneurysm is superficial and can be examined with a high-resolution transducer. The diameter is measured, and thrombotic material in the lumen is identified (absence of color coding) in transverse orientation, while the shape is assessed in the longitudinal plane.

*Pseudoaneurysm (*also known *as false aneurysm)* is a typical complication of arterial puncture performed for diagnostic angiography or interventional procedures and is observed in up to 4% of persons after percutaneous transluminal angioplasty (PTA) and cardiac catheterization (Hust and Schuler 1992; Moll et al. 1991). A suture aneurysm is a pseudoaneurysm developing after vascular surgery, in particular after bypass operations.

A false aneurysm must be differentiated from a perivascular hematoma with transmitted pulsation, but this is difficult on clinical grounds (Thomas et al. 1989). Using duplex ultrasound, an aneurysm can be differentiated from hypoechoic, perivascular structures such as hematoma, seroma, or lymphocele by the depiction of *to-and-fro flow*, which is diagnostic of a pseudoaneurysm and requires no angiographic confirmation. To-and-fro flow occurs in the neck of a pseudoaneurysm due to changing pressures: at the high intraluminal pressure during systole, blood flows through the narrow neck into the aneurysm at a rather high velocity. Under the reversed pressure conditions during diastole, the blood flows back into the artery at a slightly lower flow rate. Reflux is typically turbulent.

Before ultrasound enabled the precise localization of the aneurysmal neck relative to the skin surface, the treatment of choice was surgical closure. With advances in ultrasound equipment, it is now possible to occlude over 90% aneurysms by *compression of the neck using color duplex imaging for guidance* (Fellmeth et al. 1991; Hust et al. 1993). Thrombosis occurs after 10–30 min of compression (see Fig. A2.49).

Success rates of 70–90% have been reported for ultrasound-guided compression of pseudoaneurysm (Krumme et al. 1995; Lange et al. 2001) compared with 95–100% for induction of thrombosis by sonographically guided thrombin injection into the aneurysmal sac (Table 2.13; Vicente and Kazmers 1999; Wixon et al. 2000). Thrombin injection must be performed with great caution to prevent the highly concentrated thrombin solution (e.g., 5,000 IU in 5 mL) from entering the circulation during diastole. Inadvert outflow can be prevented by slow instillation starting in the periphery of the aneurysmal sac. The neck is identified by demonstration of forward and reverse flow components as blood enters the aneurysm during systole and exits during diastole. This flow pattern causes a characteristic audible Doppler signal (steam engine sound). Although visualization is impaired by hematoma and edema, the needle can be reliably placed because it is easily recognized by its high echogenicity within the hypoechoic or anechoic pseudoaneurysm. Rapidly moving the needle tip back and forth will help in locating the needle and check for correct positioning. Injection results in *instantaneous thrombosis around the needle tip and, if the solution is injected slowly, will prevent peripheral thrombin outflow*. Spilling into distal arteries can occur only if the needle is mistakenly placed in the neck or if thrombin is injected as a rapid bolus. Thromboembolic complications leading to amputation have been reported. Some authors therefore recommend starting injection near the wall; however, areas near the neck will thrombose spontaneously once the aneurysmal sac has been obliterated.

**Table 2.13** Ultrasound-guided vascular interventions

| | |
|---|---|
| False aneurysm (typically iatrogenic): | Compression of neck, thrombin injection |
| Postoperative fluid collection around implants: | Drainage (infection, abscess?) |

Thrombin injection has the advantage of rapidly inducing thrombosis, while compression therapy is less expensive and has the added advantage of reducing the aneurysm volume, leaving a smaller hematoma that produces less swelling and pressure (see Fig. 2.24c).

### 2.1.6.4 Rare Stenosing Arterial Diseases of Nonatherosclerotic Origin

The popliteal artery is a common site not only of atherosclerotic stenosis and occlusion or embolism but also of rare vascular disorders, in particular *compression syndromes*. Though venography or angiography remains the gold standard, these modalities may be limited in identifying the cause of a compression syndrome, especially when occlusion has already occurred (Table 2.14).

Duplex imaging provides information on the degree and hemodynamic relevance of luminal narrowing and enables evaluation of the vessel wall and perivascular structures, thus allowing identification of the cause of nonatherosclerotic vascular disease (Table 2.15).

Suspected compression of an artery by muscular structures (entrapment syndrome) can be confirmed by functional tests, and its hemodynamic significance can be determined by spectral Doppler (Figs. A2.27, A2.49, A2.50). Another domain of duplex imaging is the evaluation of vascular complications in compression syndromes

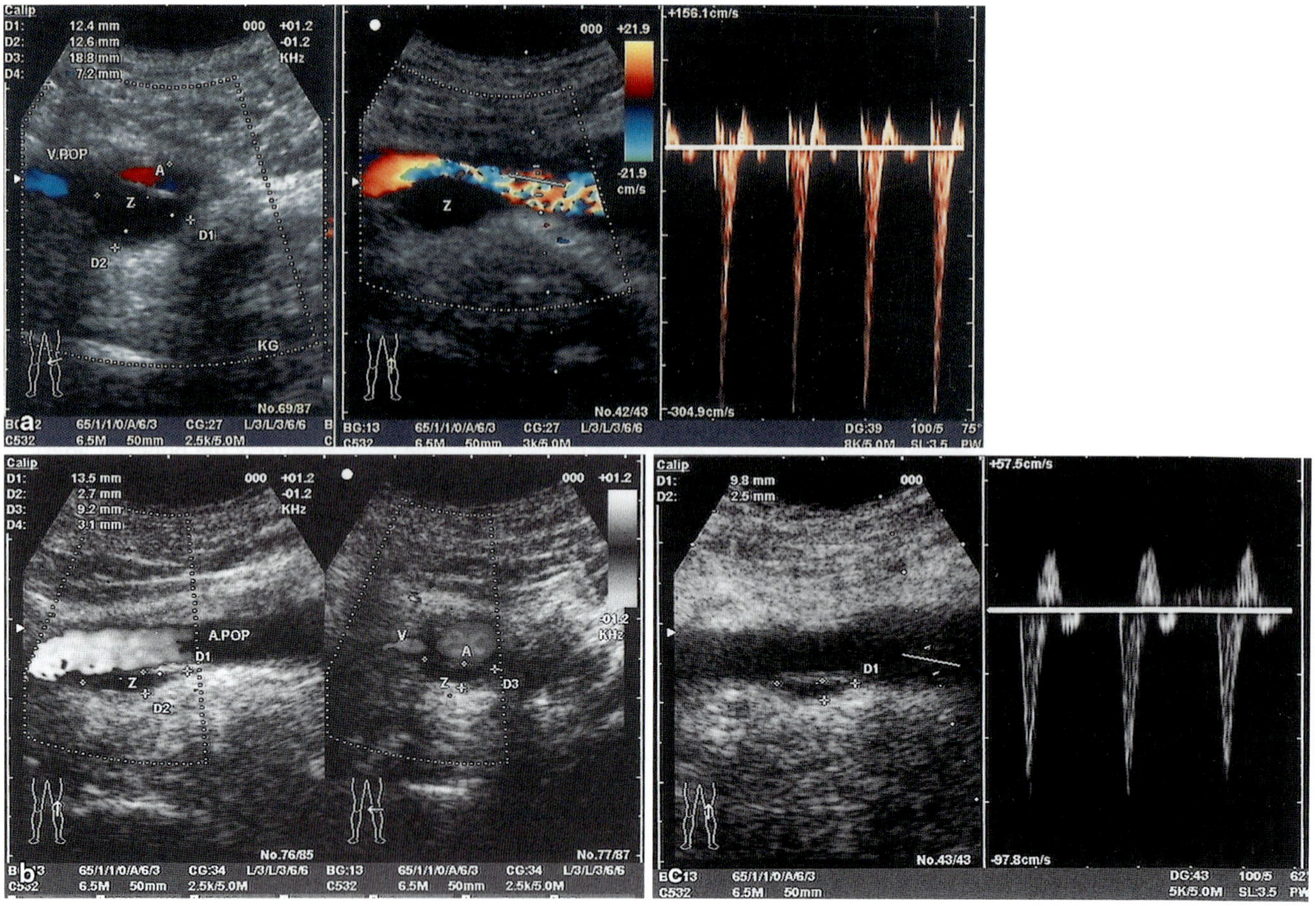


**Fig. 2.25 a–c** Adventitial cystic disease – variable cyst size. **a** This patient with intermittent claudication of variable severity due to adventitial cystic disease (*Z*) shows a highly variable cyst size over a period of 2 weeks. The transverse view (*left*) demonstrates compression of the popliteal artery with a residual lumen of 20–30%. Aliasing (*middle section*) and a Doppler-derived flow velocity of 3 m/s confirm stenosis due to cystic compression. **b, c** The cyst has become so small that it is easily overlooked on routine ultrasound if one is unaware of the earlier findings (**b**, longitudinal view on the *left*, transverse view on the *right*). The diameter of the cyst has decreased from 1 cm to 2.7 mm, and the arterial lumen is no longer compromised (not seen angiographically). The Doppler waveform is normal

**Table 2.16** Differential diagnosis of sonographically hypoechoic vascular and perivascular structures in the popliteal fossa

| |
|---|
| • Aneurysm of the popliteal artery (true/false) |
| • Hematoma, seroma, abscess |
| • Hemangioma |
| • Baker's cyst |
| • Adventitial cystic disease |
| • Dissection with thrombosis of false lumen |
| • Tumor |
| • Venous aneurysm |

Surgical enucleation of the cyst or resection of the involved vessel segment with replacement by a venous bypass graft is the treatment of choice. If an operation is not possible or refused by the patient, the cystic fluid can be aspirated under ultrasound guidance (Schäberle and Eisele 1996).

### 2.1.6.4.2 Popliteal Entrapment Syndrome

Entrapment of the popliteal artery was first described in 1879 by a medical student in Edinburgh. Few data are available on the incidence of this syndrome, but it seems to be more common than assumed in the past. A study performed in members of the Greek army reported an incidence of 0.17% (Bouhoutsos and Daskalakis 1981), while an autopsy study found an incidence of 3.5% (Gibson 1977). In the above-quoted analysis by our group (Sect. 2.1.6.4), the incidence was 0.1% in symptomatic patients with stage II or III disease. The higher incidence of popliteal entrapment in asymptomatic patients appears to be attributable to the fact that malformation of the medial head of gastrocnemius, which causes the entrapment constellation, may occur without causing symptoms. An entrapment constellation is occasionally seen in patients examined for other reasons (e.g., suspected thrombosis, preoperative vein mapping prior to varicosis surgery) (Fig. A2.51).

Intermittent claudication, chiefly associated with walking uphill, is the cardinal clinical symptom. Paresthesia and rest pain or trophic disorders have been observed but are uncommon. Both our results and published data indicate that thrombosis or segmental arterial occlusion is already present at the time of diagnosis in 50–70% of patients. Bilateral involvement was reported to occur in 30–50% of cases but was seen in only one patient (9%) of our series (of symptomatic patients).

The popliteal artery courses through the center of the intercondylar fossa together with the popliteal vein and the tibial nerve and gives off a variable number of branches along this course (sural arteries). An atypical course of the popliteal artery, and possibly of the popliteal vein as well, or abnormal attachment of the medial head of the gastrocnemius muscle can lead to compression of the vessels during muscle contraction.

Plantar flexion compresses the popliteal artery, leading to intermittent reduction of lower leg perfusion and claudication with secondary intimal proliferation of the vessel wall. Intimal damage may give rise to the formation of mural thrombi with subsequent complete occlusion, while compression of the artery may be associated with poststenotic dilatation. The mural thrombi developing in the aneurysm may cause arterial embolism with occlusion of peripheral vessels. These complications can be prevented by early surgical division of the compressing structure.

Insua et al. (1970) distinguish four types of popliteal entrapment syndrome according to the relationship between artery and muscle (Fig. 2.10a–f):

- In *types I and Ia*, the popliteal artery courses on the medial side of the medial head of the gastrocnemius muscle. Type I refers to a malformation of the artery, type Ia to the malformation of the medial head of the gastrocnemius, which attaches to the femur more laterally and cranially than under normal conditions, thereby displacing the artery from its normal path. Sonographically, these two types are suggested by the depiction of muscle tissue between the artery and vein, which usually course through the popliteal fossa together.
- In *types II und IIa*, the artery and vein have a normal course but are compressed by structures crossing the popliteal fossa (abnormal attachment of a lateral extension of the medial gastrocnemius head, abnormal course of plantar muscle).

Rarely, a well-developed gastrocnemius can cause intermittent claudication. During contraction, the hypertrophied heads compress the popliteal artery and occasionally the vein as well.

Younger persons presenting with typical claudication should be examined for the presence of popliteal entrapment. This is done sonographically by carefully following the course of the popliteal artery, evaluating its relationship to muscular structures (Fig. 2.26) and performing the plantar flexion test (Table 2.17; Fig. 2.27).

The appearance of stenosis signals in the popliteal artery during plantar flexion confirms muscle-induced arterial compression. Angiography does not yield any relevant additional information that may affect therapeutic management (Table 2.18), especially since ultrasound also depicts the vascular complications such as occlusion and poststenotic dilatation (Table 2.19). The examiner should bear in mind that isolated popliteal occlusion may be due to popliteal entrapment, particularly in younger persons, and that aneurysmal changes of the popliteal artery may be due to poststenotic dilatation (Fig. A3.67a).

Therapeutic management consists in division of the structure compressing the popliteal artery and repairing the artery in case of secondary occlusion (Steckmeier et al. 1989). In Insua type I popliteal entrapment syndrome, the atypically attaching medial gastrocnemius head is divided.

#### 2.1.6.4.3 Raynaud's Disease

Raynaud's disease is characterized by *intermittent attacks of ischemia of the fingers and toes*, typically brought on by cold and enhanced by emotional stress, local compression, and conditions that are associated with an increased sympathetic tone. The vasospasm is relieved by heat or drug treatment. Primary or idiopathic Raynaud's disease (no underlying disease; no occlusion of finger arteries) is distinguished from a secondary form (e.g., in patients with scleroderma or simultaneous finger artery occlusion). Ischemic attacks often occur bilaterally, affecting the second to fifth fingers while in most cases sparing the thumb. The toes are involved in only about 2% of cases. Women are affected two to five times more commonly than men, primarily between the ages of 20 and 50.

The diagnosis of Raynaud's disease chiefly relies on the typical clinical presentation, while further diagnostic procedures are only required to identify vasospasm as the underlying cause of the clinical symptoms. Again, Doppler and duplex ultrasound with spectral analysis provide useful information. The examination is performed with exposure to cold to provoke the vasospasm and exposure to heat to relieve the spasm. Under exposure to cold, the systolic finger artery pressure in Raynaud's disease drops markedly by 20–50% compared with only up to 10% in healthy persons. Duplex scanning typically demonstrates residual perfusion in the common digital arteries, while there is no or only partial flow in the distal finger arteries during spasm. The *vasospasm* produces *a markedly pulsatile flow pattern* with a short systolic peak and absence of diastolic flow in the hand arteries and the common digital arteries. In contrast to the flow signal

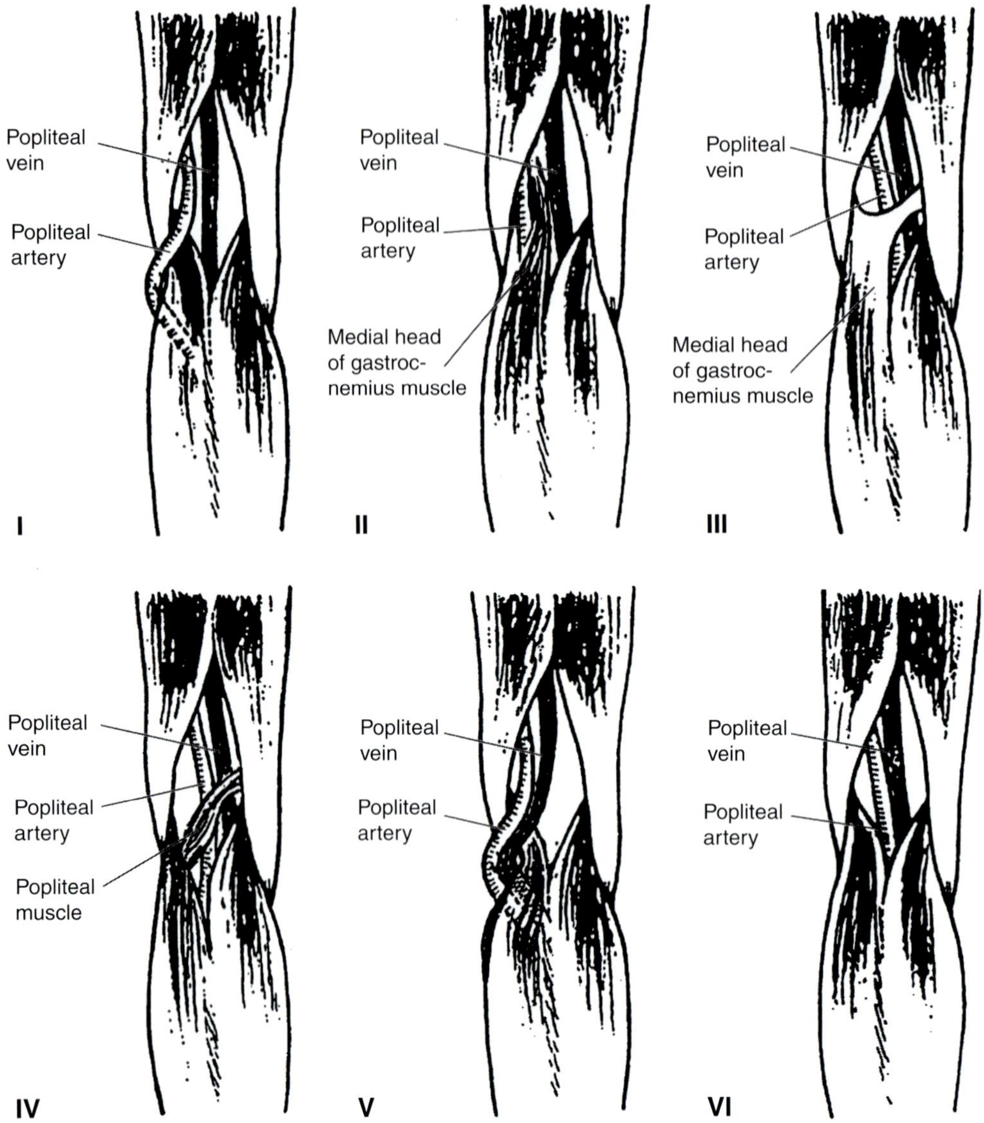


**Fig. 2.26** Classification of popliteal entrapment syndrome (modified from Insua). *I* The popliteal artery courses medially over the posterior aspect of the normal attachment of the medial head of the gastrocnemius to return to its normal course in front of the muscle (corresponding to Insua type I). *II* The medial head of the gastrocnemius attaches more cranially and laterally, thus forcing the popliteal artery to take an abnormal course around the head (see Figs. 2.17 and A2.49; corresponding to Insua type Ia); the popliteal vein may be compressed as well. *III* The attachment of the medial head of the gastrocnemius has an accessory lateral extension, or the plantar muscle takes an abnormal course. The path of the popliteal artery is normal, but the artery and vein may be compressed to variable degrees, depending on the strength of the muscle fibers (Fig. A3.67a) coursing to the lateral femoral condyle (corresponding to Insua types II and IIa). *IV* The popliteal artery and vein are compressed by the popliteal muscle, an abnormal branch of the tibial nerve, or a fibrous ligament (according to Rich). *V* In rare cases, the popliteal vein follows the artery along its abnormal path and is compressed as well. Only one case of an isolated abnormal course of the popliteal vein has been reported so far. *VI* Normal course of the popliteal artery and vein through the popliteal fossa with compression of both vessels by a well-developed gastrocnemius muscle during muscle contraction, giving rise to intermittent claudication or venous congestion (see Figs. A2.50 and A3.67b, c)

**Table 2.17** Differential diagnosis – nonatherosclerotic vessel diseases

| | |
|---|---|
| **Vessel wall:** | • Adventitial cystic disease: cysts in adventitia<br>• Arteritis: concentric wall thickening |
| Morphology – (hemodynamics) | |
| **Vessel compression:** | • Popliteal entrapment syndrome<br>• Compression syndrome of upper thoracic aperture<br>(Compression by perivascular structures) |
| Hemodynamics in functional tests – (morphology) | |

seen in the presence of distal finger artery occlusion, *heat exposure induces vasodilatation* with hyperemia, resulting in pronounced diastolic flow in the proximal finger arteries. Other helpful diagnostic procedures are oscillography and pressure measurement in the finger arteries. Raynaud's disease must be differentiated from vascular disorders of the larger proximal vessels as well as from arterial embolism (popliteal artery aneurysm, aortic aneurysm, thoracic outlet syndrome with poststenotic subclavian aneurysm). This

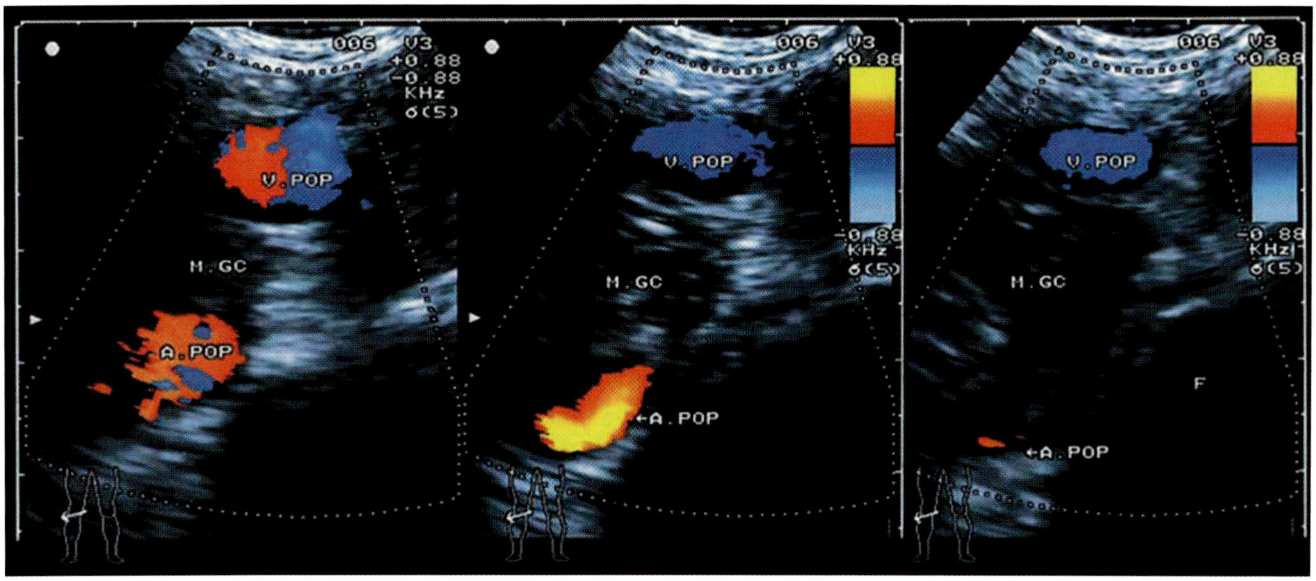


**Fig. 2.27** Popliteal entrapment syndrome – functional test. 38-year-old athletic man with entrapment syndrome caused by a gastrocnemius muscle malformation (Insua type I). The transverse views from *left* to *right* show progressive compression of the popliteal artery by the gastrocnemius muscle (*M.GC*) with increasing plantar flexion. The muscle attachment between the popliteal artery and vein (*V.POP*) forces them apart, thereby compressing the artery (subtotal occlusion on *rightmost scan*)

**Table 2.18** Diagnostic workup of nonatherosclerotic vessel disease prior to surgical repair

| | |
|---|---|
| Compression syndrome, adventitial cystic disease: | Mandatory: duplex ultrasound (morphology and hemodynamics during provocative maneuver) |
| | Optional: angiography (not necessary in: adventitial cystic disease and popliteal entrapment syndrome without occlusion) |
| | May be supplemented by MRI, CT |
| Vascular inflammation: | Duplex ultrasound to confirm the diagnosis and prevent unnecessary vessel repair |

**Table 2.19** Vascular complications of compression syndrome

| |
|---|
| • Mural thrombi secondary to local vessel wall lesions |
| • Poststenotic aneurysm |
| • Thrombotic occlusion secondary to vascular damage or aneurysmal dilatation |
| Method of choice for diagnosing and evaluating complications: duplex ultrasound (morphology and hemodynamics) |

is especially the case if color duplex imaging demonstrates occlusion of interdigital arteries and if the clinical picture of trash foot is present.

### 2.1.6.4.4 Paraneoplastic Disturbance of Acral Perfusion

A tumor can compromise blood flow by the following pathomorphologic and pathophysiologic mechanisms:

1. Local displacement and compression (tumors of soft tissue, nerves, vessels, and bones as well as metastases) or tumor infiltration of the vessel wall, which may be associated with arterial tumor embolism
2. Paraneoplastic vasculitis
3. Paraneoplastic hyperviscosity of the blood and hypercoagulable state

Duplex ultrasound can identify the site of *external compression or infiltration* of the arterial wall by the *tumor* and provides information on the hemodynamic significance of the narrowing that is useful for therapeutic decision making. Arteries, with their strong muscle coat and intramural pressure, are much less susceptible to local tumor compression than veins.

### 2.1.6.4.5 Buerger's Disease

Buerger's disease, or thromboangiitis obliterans, is a vascular condition with an intermittent course and is characterized by segmental and multiple occlusions of small and medium-sized arteries of the extremities. As a panangiitis, it can be differentiated from atherosclerosis and other inflammatory vascular diseases. Though the etiology of Buerger's disease is unknown, there is an association between the onset and progression of the disease and cigarette smoking – 93–99% of the patients smoke. Remission is seen in most patients who quit for good. The severity and frequency of disease episodes correlate with the patient's smoking habits. The clinical symptoms depend on the extent and site of occlusions. Pain at rest and acral necrosis occur at an early stage, while typical intermittent claudication of the calf muscles is rare. The duplex findings can exclude other disorders such as arterial embolism, aneurysms of the aorta and popliteal artery, popliteal entrapment syndrome, atherosclerosis, and macroangiopathy. In Buerger's disease, the walls of the large arteries appear normal without hyperechoic atherosclerotic plaques or thickening on B-scan ultrasound performed with a high-resolution transducer. Occlusions primarily involve the arteries below the knee including the pedal arteries, and disseminated occlusion of the interdigital arteries may also be seen. The occluded lumen has low echogenicity. Vein involvement is seen as segmental phlebitis. *Duplex ultrasound often depicts corkscrew-like collaterals or revascularization characterized by short segments in which repeated flow reversal can be seen* (Fig. 2.28a). This flow pattern is also reflected in the Doppler waveform.

#### 2.1.6.4.6 Inflammatory Conditions

Inflammatory vascular disease may be localized or generalized. Primary vasculitis arises in the vessel wall, while secondary forms refer to inflammatory vascular involvement in systemic diseases (rheumatoid arthritis, collagen disease). The following primary inflammatory vascular diseases are distinguished according to the vessels affected:

- Inflammation of large vessels (giant cell arteritis, Takayasu's arteritis)
- Inflammation of medium-sized vessels (polyarteritis nodosa, Kawasaki's disease)
- Inflammation of small vessels (Wegener's granulomatosis, microscopic polyangiitis, Schoenlein–Henoch purpura, Churg–Strauss syndrome)

The clinical diagnosis is suggested by the symptoms and changes in the organ or body region supplied by the inflamed arteries. These can range from pathognomonic local skin lesions with palpable purpura to organ loss (kidney) or limb ischemia when peripheral arteries are involved. During the active stage of vascular inflammation, most patients have a markedly elevated erythrocyte sedimentation rate (typically above 100 during the first hour) with only a slight increase in C-reactive protein. Additional findings are anemia, mild to moderate leukocytosis, and marked thrombocytosis. Supplementary tests include protein electrophoresis, complement determination, and antibody serology. Duplex ultrasound allows localization and quantification of vascular constriction, but its foremost role is to noninvasively demonstrate the typical inflammatory thickening of the vessel wall. This is done using a high-resolution transducer (7–10 MHz), which will depict the *characteristic "macaroni sign" consisting of a higher-level echo from the lumen/intima interface surrounded by a concentric, homogeneous tube-like structure* (Fig. 2.28b) of lower echogenicity (intima-media complex) (Maeda et al. 1991). If the scanner resolution is not high enough, the macaroni sign may be visualized only in large and medium-sized vessels. With disease progression, thickening will lead to the development of concentric and rather long stenoses with subsequent obliteration.

Polyarteritis nodosa is segmental inflammation of medium-sized arteries and is characterized by circumferential wall thickening with luminal narrowing or constricted arterial segments alternating with dilated segments (Fig. 2.28b).

The sonomorphologic differentiation of intimal thickening and sclerotic changes (calcification) from circumferential thickening of long vessel segments (macaroni sign) in

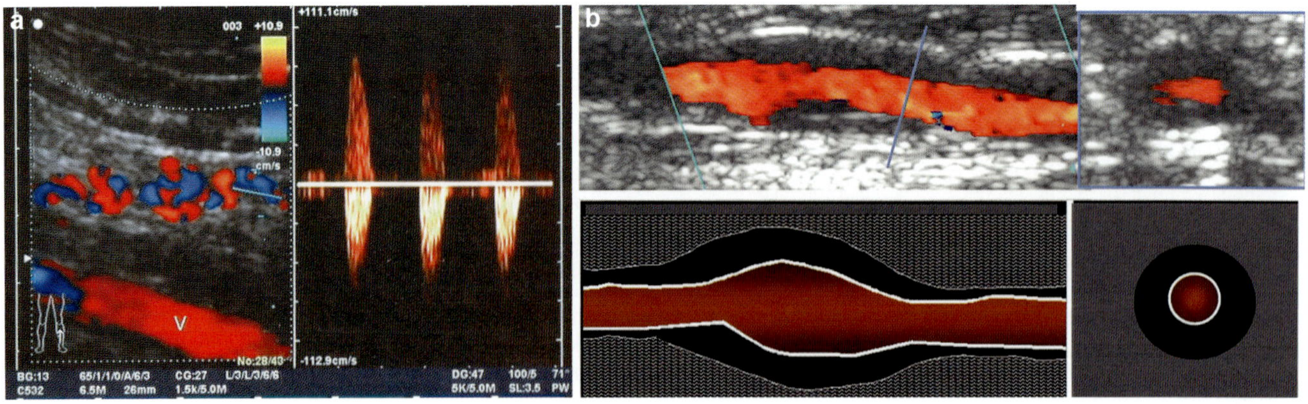


**Fig. 2.28** Thromboangiitis obliterans and Polyarteritis nodosa. **a** Occlusion of the posterior tibial artery in thromboangiitis obliterans (hypoechoic artery without plaque). The patent vein (*V*) with flow coded red is seen deep to the occluded artery. A tortuous, recanalized collateral (with sample volume) is depicted close to the transducer with red and blue indicating flow in both directions; this is also reflected in the spectral Doppler trace obtained from a short segment of the artery. These findings are characteristic of a corkscrew collateral and indicate recanalization in thromboangiitis obliterans. **b** In polyarteritis nodosa, inflammatory thickening of the wall causes circumferential luminal narrowing, which may at times alternate with dilated segments. The diagrams illustrate circumferential arterial wall thickening in longitudinal and transverse orientation (Fig. 2.28b courtesy of K. Amendt)

inflammatory disease has important clinical implications. It is highly accurate in superficial vessels (carotid, subclavian, axillary, and femoral arteries), which can be examined with high-frequency transducers.

Computed tomography (CT) and magnetic resonance imaging (MRI) are the reference methods for the diagnostic evaluation of the vessel wall. Angiography can identify the segments affected by inflammatory constriction but does not demonstrate wall thickening (see Table 2.18). Duplex ultrasound thus makes an important contribution to the correct diagnosis and treatment of steno-occlusive disease caused by vasculitis, thereby providing a sound basis for initiating proper treatment and sparing patients unnecessary interventions or surgery.

#### 2.1.6.4.7 Chronic Recurrent Compartment Syndrome of the Calf

Chronic recurrent compartment syndrome of the lower leg most commonly involves the anterior compartment. It is an overuse condition causing a rise in intracompartmental pressure with subsequent microvascular compromise. Patients present with swelling, tension, and severe pain. Foot pulses are palpable. Muscle damage can lead to an increase in creatine kinase; intracompartmental pressure, which even during activity is normally below 20 mmHg, may increase two- to fourfold. When duplex imaging is performed after activity to reproduce increased compartmental pressure, the calf veins of patients with chronic compartment syndrome will appear compressed or may be collapsed (at pressures >30 mmHg). At pressures >50 mmHg, arterial flow becomes more pulsatile. With a further increase in pressure, the orthograde diastolic flow component disappears, and finally a line-like artifact will appear in the waveform at late diastole. In patients with compartment syndrome of the anterior lower leg, the B-mode scan may show intrafascial edema.

## *2.1.7 Follow-up After Surgical and Interventional Treatment*

With its proven validity compared with the gold standard and intraoperative findings, duplex ultrasound is an excellent imaging modality for treatment planning in patients with steno-occlusive disease of the pelvic, femoral, and popliteal arteries including the trifurcation and also for surveillance after surgery or endovascular interventions. The duplex ultrasound information on the sites and severity of stenosis or occlusion, in conjunction with the clinical manifestation, helps the physician to decide whether conservative management with walking exercises, an interventional radiologic procedure, or surgical repair (TEA or bypass grafting) is the most suitable treatment for the patient. The surgical procedure and revisions in case of complications can be planned in the infrainguinal arteries with the same accuracy as with angiography (Wain et al. 1999; Ligush et al. 1998). In patients with adequate scanning conditions, no pretherapeutic angiography is necessary for planning reconstructive procedures in this territory including the P1 segment. Evaluation of the recipient artery is important both in the preoperative workup and in the examination of patients with restenosis or reocclusion after surgery. As this can be done very accurately by duplex ultrasound including evaluation of the trifurcation and segmental evaluation of lower leg arteries, the indication for a P1 femoral artery bypass can be established without preoperative angiography on condition that an inflow obstruction (pelvic artery) has been excluded.

### 2.1.7.1 Thromboendarterectomy

An important indication for TEA is stenosis of the femoral artery bifurcation, which is easily accessible to ultrasound examination. The duplex findings alone can serve to identify candidates for TEA and plan the procedure.

Duplex imaging enables very detailed planning of TEA or profundaplasty for obstructive lesions at the origin of the profunda femoris as well as of repeat interventions in patients with complications or recurrent stenosis. Moreover, hemodynamic evaluation by duplex ultrasound overcomes the limitations of angiography in this region (superimposition and posterior wall plaque).

Ultrasound can also be used to assess the outcome of treatment, for example, after surgical elimination of stenosis in the profunda femoris artery performed to improve its collateral function in superficial femoral artery occlusion. In this setting, the success of surgery can be confirmed by the sonographic identification of improved flow with higher peak systolic and diastolic velocities in the refilled popliteal artery (Fig. A2.11).

### 2.1.7.2 Percutaneous Transluminal Angioplasty and Stenting

*Angioplasty* aims at improving peripheral perfusion by dilatation of a narrowed vessel segment. The therapeutic procedure in patients with occlusion (PTA or bypass surgery) can be planned beforehand once the length of the occluded segment has been determined. In angioplasty, atherosclerotic plaques are fragmented and pressed into the arterial wall, often resulting in intimal or medial tears. The irregular surface is susceptible to the deposition of thrombotic material. A complication of angioplasty is recurrent stenosis due to fragmented plaques extending into the lumen, dissection, elastic rebound forces, intimal hyperplasia, or progression of atherosclerosis. Residual stenosis persisting after intra-arterial thrombolytic therapy with dissolution of thrombotic material by administration of plasmin activators is also treated by dilatation. Other interventional procedures, apart from standard PTA, include arterectomy, rotational angioplasty, and laser angioplasty.

*Duplex scanning is the first-line diagnostic modality to follow up patients after PTA* (with or without stent implantation) or bypass surgery for the early identification of those who require reintervention. A study showed that 85% of patients with a postinterventional PSV ratio above 2 had restenosis (Mewissen et al. 1992).

The role of *duplex ultrasound* in patient who underwent interventional procedures is to identify complications (dissection, aneurysm, perforation) and to diagnose residual or recurrent stenosis caused by thrombotic deposits or fragmented plaques protruding into the lumen. The morphologic appearance of the wall in the dilated segment and hemodynamic information are important criteria for detecting restenosis. Subintimal hemorrhage due to intimal or medial tears may be identified as hypoechoic wall thickening in the dilated segment while thrombotic deposits appear as hypoechoic intraluminal areas without color coding.

A stent is identified by its serrated or mesh-like appearance. A focal increase in flow velocity is the most important criterion for residual or recurrent stenosis after PTA, stenting (where special attention must be paid to the stent ends), and bypass grafting (primarily at the anastomotic sites) (see Fig. A2.4e).

In most cases, flow evaluation within a stent requires a higher color gain. Eddy currents and turbulent flow at the proximal and distal ends suggest that the stent does not fit snugly to the wall, which can promote restenosis.

The accurate diagnosis of complications such as arteriovenous (AV) fistula, pseudoaneurysm, and hematoma and the timely identification of residual or recurrent stenosis are crucial for post-PTA patency. *A hemodynamically significant residual or recurrent stenosis is suggested by a doubling of the flow velocity within the dilated segment.* The detection of a hemodynamically significant stenosis by duplex sonography is a predictor of 1-year patency. One study reported a 1-year patency rate of 83% in the absence of stenosis as opposed to only 15% when a functional stenosis was diagnosed (Mewissen et al. 1992).

Several studies have shown *duplex imaging to be more sensitive than angiography in detecting residual stenosis or residual flow disturbance* following PTA. In one study, 20% of residual stenoses > 50% based on duplex ultrasound were classified as causing <30% diameter reduction at angiography. The sonographic stenosis criteria were PSV > 180 cm/s (Fig. 2.29) and an intrastenotic to prestenotic PSV ratio > 2.5 (Kinney et al. 1991; Mewissen et al. 1992). The presence of residual stenosis classified as causing >50% diameter reduction by duplex scanning was found to predict late failure (15% success rate) while late patency was observed for <50% diameter reduction (84% success rate). Based on these results, it is recommended to perform a follow-up duplex scan within 1 month of PTA to identify candidates for reintervention (>180 cm/s PSV or >2.5 PSV

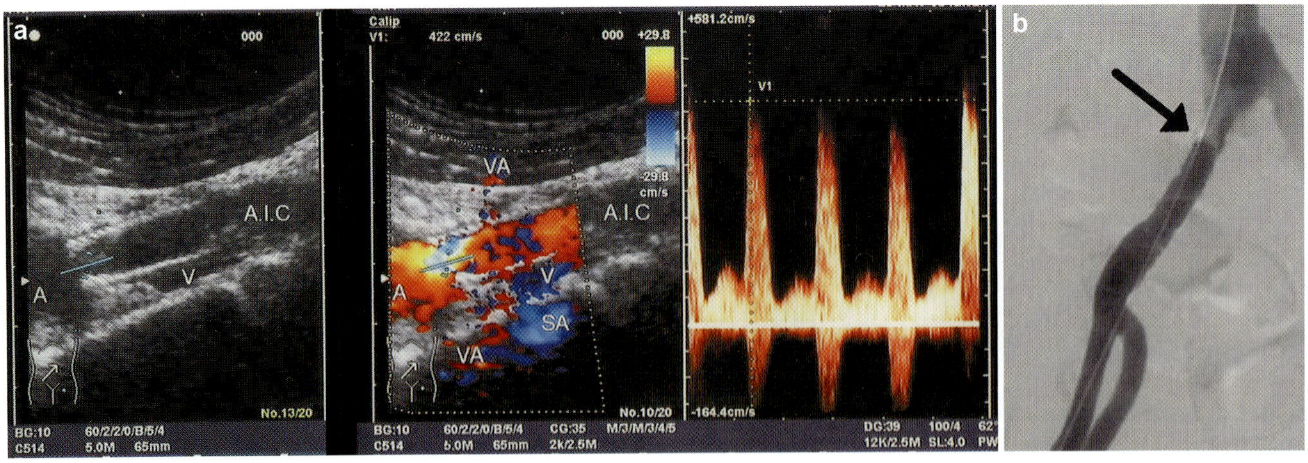


**Fig. 2.29** **a** Patient after PTA and stenting of high-grade common iliac artery stenosis. In the gray-scale image (*left*), the stent is identified by its serrated appearance. The color duplex scan shows aliasing at the proximal stent end and disturbed flow within the stent (red and blue flow signals). *A* aorta; *A.I.C* A. common iliac artery; *V* common iliac vein; *VA* vibration artifact; *SA* mirror artifact. Spectral Doppler interrogation of the segment with aliasing demonstrates a PSV of 422 cm/s and very turbulent flow, consistent with high-grade stenosis. **b** Angiogram fails to adequately show the stenosis or its cause. Dilatation was performed on the basis of the duplex ultrasound findings and led to resolution of clinical symptoms (patient's walking distance before reintervention was limited to 180 m); color duplex confirmed the elimination of the stenosis, and the ABI normalized from 0.8 to 1.1

ratio). Note that, because the arterial wall is rigid in a stented segment, a 20–30% higher PSV than in native arteries should be considered normal (based on studies of stented carotid arteries; study data for peripheral arteries are not available).

### 2.1.7.3 Bypass Graft Surveillance

The appearance of a bypass in gray-scale imaging depends on the material used. The thin wall of an *autologous* venous bypass graft is very difficult to delineate in case of occlusion. Such a bypass is easier to identify, in particular in older occlusion, if the sonographer has information on its course (anatomic, extra-anatomic). In patients with a venous bypass graft, the entire length must be scanned because the former valves are common sites of stenosis, especially in an in situ bypass with residual valve cusps. An AV fistula developing from a perforating vein that has not been ligated can be identified by the presence of perivascular tissue vibration artifacts in the color duplex mode.

In contrast, the wall of a *synthetic bypass graft* is always clearly seen. A PTFE (polytetrafluoroethylene) prosthesis has a typical echogenic double contour and a Dacron bypass a sawtooth-like appearance.

In the postoperative evaluation and surveillance of a synthetic graft, special attention must be given to possible anastomotic stenoses. Narrowing within the bypass is due to neointimal hyperplasia and occurs later. About 20–30% of venous bypass grafts develop strictures on the basis of neointimal hyperplasia within the first year of surgery.

Different factors can cause occlusion of a bypass at different times after surgery:

- *Immediate postoperative occlusion* within the first days after surgery may be due to an inadequate surgical technique resulting in anastomotic stenosis or a poor outflow tract. The latter can be identified by assessing flow in the recipient artery.
- *Early occlusion*, within the first year, chiefly results from neointimal hyperplasia, predominantly causing stenosis at the anastomotic sites, or from flow disturbance due to progression of atherosclerosis distal to the bypass. If the occlusion is due to an impaired inflow secondary to atherosclerotic lesions of the proximal artery with loss of the triphasic waveform, the examiner must carefully evaluate the native artery to identify the site of the lesion.
- *Late occlusion* is predominantly caused by progression of atherosclerosis.

Abnormal fluid around a bypass graft should be punctured under ultrasound guidance for microbiologic testing, in particular in patients with clinical signs of infection. Before puncture, a suture aneurysm should be excluded by color duplex imaging (see Fig. A2.25).

Hematoma, seroma, and suture aneurysm appear as pulsatile masses at the site of anastomosis, each having a characteristic color duplex appearance, which allows it to be differentiated at a glance.

#### 2.1.7.3.1 Methodological Considerations and Stenosis Criteria

*The criteria for stenosis grading* are based on those for the native peripheral arteries. It must be noted, however, that the hemodynamic changes in a bypass graft may occasionally lead to a monophasic waveform that does not suggest abnormal flow. Eddy currents at the sites of anastomoses cause spectral broadening, which is likewise normal (see Figs. A2.33 and A2.28–A2.31).

The normal peak velocity is a function of the relative cross sections of the bypass and the proximal and distal native arteries. The complex relationships make it difficult to give a reliable general threshold velocity. Still, one can rule out a hemodynamically significant stenosis with some confidence if PSV at the site of anastomosis is below 2 m/s and if there is no evidence of a size mismatch between the graft and the native vessels (Table 2.20).

The *Doppler frequency spectrum* recorded in a bypass is affected by *several factors* that have to be taken into account, especially in patients with severe atherosclerosis and in assessing bypass grafts onto a calf artery. Pulsatility is physiologically dependent on the demand-oriented widening of the arterioles (monophasic flow). In a bypass, pulsatility is further affected by *differences in elasticity* (depending on the *material* used for grafting) and an *increase in outflow resistance if there*

**Table 2.20** Duplex ultrasound surveillance for diagnosis of complications after bypass grafting: suture aneurysm, abscess, imminent occlusion (failing bypass), stenosis

| Ultrasound technique | Stenosis criterion/failing bypass |
|---|---|
| Single PSV measurement in bypass graft (indirect criterion of flow obstruction/stenosis in the graft) | Proximal to stenosis: reduced PSV<br>Distal to stenosis: dampened flow, reduced PSV, delayed systolic rise<br>PSV < 45 cm/s: failing bypass |
| Analysis of representative spectral waveform | Triphasic: good graft function<br>Monophasic: impaired drainage, reduced peripheral resistance |
| Mapping of bypass and anastomoses<br>Increased PSV: direct stenosis criterion | PSV ratio > 2: moderate stenosis<br>PSV ratio > 4: high-grade stenosis<br>PSV > 2–2.5 m/s//>3–3.5 m/s: stenosis |

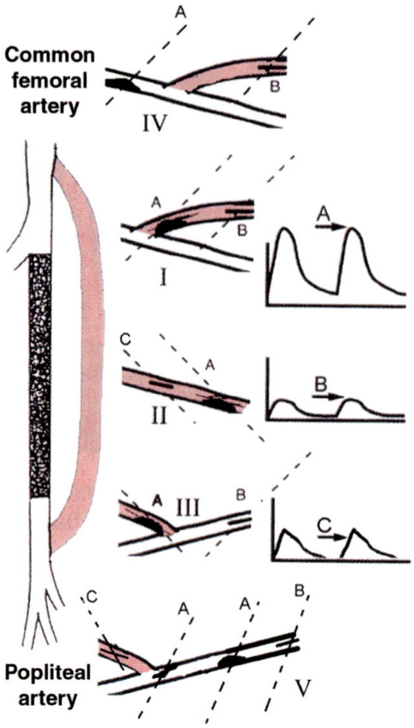


**Fig. 2.31** Common sites of bypass graft stenosis and spectral Doppler changes (*A*, *B*). Femoropopliteal bypass graft bridging occluded superficial femoral artery. *I* Stenosis of proximal bypass anastomosis (*A*) with poststenotic waveform (*B*). *II* Stenosis in the main body of the graft (at valve site) with focal doubling of flow velocity (*A*) compared with prestenotic waveform (*C*) and postocclusive flow profile distal to the stenosis (*B*). *III* Stenosis of distal bypass anastomosis (*A*) with poststenotic flow profile in the popliteal outflow tract (*B*). *IV* Stenosis of the native artery proximal to the bypass anastomosis, which is due to progressive atherosclerosis. Stenotic waveform (*A*) in the arterial segment proximal to the stenosis and poststenotic flow profile in the artery distal to the stenosis and in the bypass graft (*B*). *V* Stenosis of the artery distal to the lower bypass anastomosis. Stenotic waveform (*A*) at the site of stenosis with poststenotic flow profile distally (*B*) and prestenotic waveform (*C*) in the proximal arterial segment and in the bypass graft

27 patients with bypass occlusion showed no abnormalities in the preceding duplex examination (low flow manifested by PSV < 45 cm/s or threefold focal increase in PSV compared with adjacent segment).

#### 2.1.7.3.2 Controversy About the Benefit of Duplex Bypass Graft Surveillance Programs

Despite a lack of scientific consensus, many authors advocate duplex ultrasound surveillance after bypass grafting, at least in patients with venous grafts (Tables 2.20 and 2.21) and in the first postoperative year when the risk of occlusion is highest and the prognosis of bypass revision is good (Passman et al. 1995; Taylor et al. 1990).

In serial examinations, the sonographer should compare the pulsatility and flow velocity in the bypass graft with the

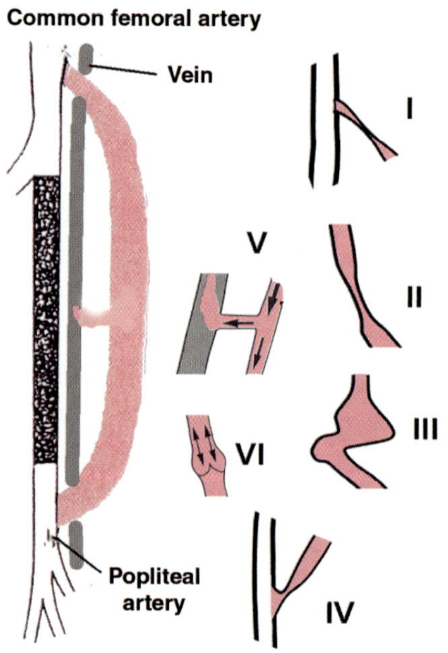


**Fig. 2.32** A venous bypass graft is susceptible to a number of specific complications (*I–VI*), which must be taken into account in bypass surveillance in addition to the common sites of stenosis described in Fig. 2.31. *I* A reversed vein graft is prone to stenosis just distal to the proximal stenosis, the narrowest part of the graft. *II* Scar formation with narrowing at valve sites. *III* Dilatation with elongation and kinking of the graft. *IV* In situ vein graft with a narrow distal end can result in luminal narrowing just proximal to the distal anastomosis. *V* Failure to ligate all side branches of an in situ vein can give rise to an AV fistula (between the bypass graft and the venous system). *VI* Residual valve in an in situ vein graft giving rise to graft stenosis or occlusion

early postoperative results. If there is a decrease in PSV in the bypass or triphasic flow becomes monophasic, the examiner must look for proximal or distal stenoses, applying the criteria described above. Even if a stenosis can be identified and localized with duplex ultrasound, in certain situations (e.g., older bypass grafts) preoperative DSA is recommended to plan the extent of the revision and to detect additional stenosis proximal or distal to the graft, unless this can be done by duplex imaging.

In a cost-effectiveness analysis, infrainguinal venous bypass graft surveillance with revision for duplex-detected stenoses resulted in a 1-year patency rate of 93% versus 57% for grafts revised after thrombosis (Wixon et al. 2000). The amputation rate was also lower (2% vs. 33%). Especially patients with critical leg ischemia at the time of bypass grafting were shown to benefit from sonographic surveillance and revision for stenosis (Visser 2001). In this study, patients in the duplex surveillance group had a major amputation rate of 1.7% compared with 7.7% in patients undergoing clinical monitoring with ABI. The cost of diagnosis and treatment was only half as high in the duplex group. The subgroup of patients with intermittent claudication at the time of bypass surgery benefited less from sonographic surveillance.

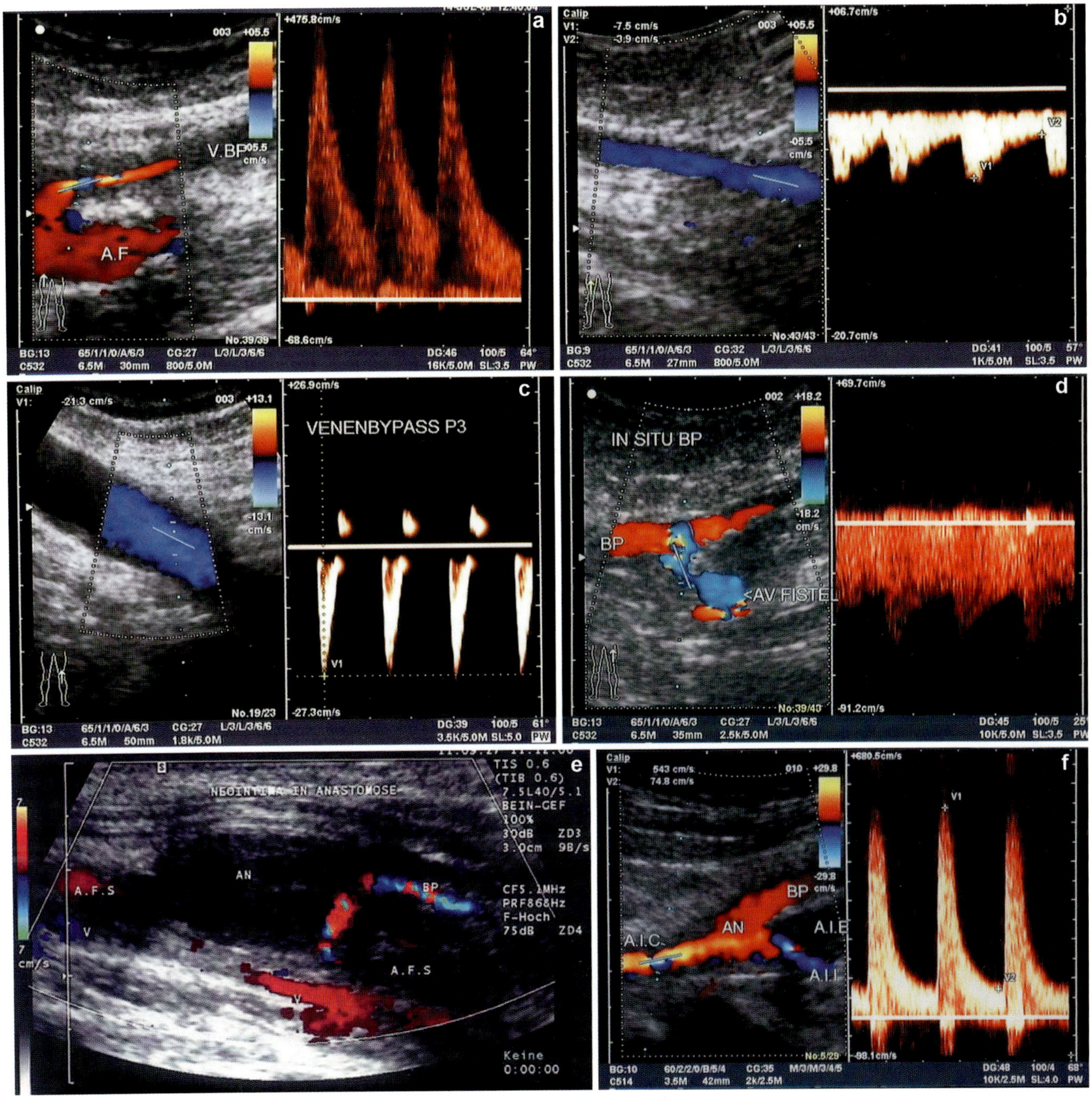


**Fig. 2.33** **a** Stenosis in a femorocrural vein bypass graft just distal to the proximal anastomosis (long stenosis due to use of a small-caliber vein segment for grafting): the Doppler waveform obtained with the sample volume placed in the stenotic area (indicated by aliasing) demonstrates monophasic flow with a PSV of 4.5 m/s and an end-diastolic velocity of 1.25 m/s. **b** The Doppler measurement in the graft distal to the stenosis shows very slow, monophasic flow (7.5 cm/s during systole and 3.9 cm/s during diastole), which is typical of a poststenotic waveform. **c** If no flow obstruction is present above, within, or below a bypass graft, an altered flow velocity within the graft may be due to a size mismatch between graft (synthetic or venous) and recipient artery (e.g., calf arteries). Very slow flow may occur in a synthetic graft with too large of a diameter or in a dilated vein graft (as in the example, where PSV is 21 cm/s). A triphasic waveform obtained in a bypass graft indicates good graft function with adequate peripheral perfusion. In the example, the vein graft anastomosed onto the tibiofibular trunk is dilated to 1.3 cm, and the fibular artery is the only patent calf artery. **d** An in situ vein graft must be scrutinized for the presence of an AV fistula (arising from a nonligated branch or perforating vein), especially if peripheral pulses are poorer than would be expected after a bypass procedure. There is monophasic flow with a large diastolic component in the graft segment proximal to and within the AV fistula due to direct outflow into the venous system (*BP* bypass). The site of the AV fistula is marked for ligation. **e** High-grade stenosis at the proximal bypass anastomosis (*AN*) due to neointimal hyperplasia (hypoechoic). **f** Iliacofemoral bypass graft (*BP*) with high-grade proximal stenosis in the common iliac artery (*A.I.C.*) giving rise to a monophasic Doppler waveform with a PSV of 550 m/s. A poststenotic waveform is obtained in the bypass graft (as in **b**). *A.I.I.* internal iliac artery, *A.*I.E. occluded external iliac artery

**Table 2.21** Indications for duplex ultrasound surveillance of bypass grafts. Recommendations based on controversial results of published studies

| | |
|---|---|
| I | Routine duplex ultrasound surveillance appears to have no benefit in:<br>• Infrainguinal synthetic bypass graft<br>A failing bypass (clinical symptoms, decrease in ABI) should prompt a duplex ultrasound scan to search for possible causes |
| II | Routine duplex ultrasound surveillance (except for one examination 3–6 months after implantation) unlikely to be beneficial in patients with venous bypass graft (in situ or reversed graft) who meet the following conditions:<br>• Large-caliber bypass graft vein (>5 mm), normal graft vein<br>• Normal intra-/postoperative completion study<br>• Bypass graft procedure performed in patients with stage II disease (intermittent claudication)<br>• Good patient compliance |
| III | Routine duplex ultrasound surveillance of venous bypass grafts is likely to be beneficial in the following situations:<br>• Thin bypass vein<br>• Nonsaphenous vein bypass grafts and composite grafts<br>• Patients with poor intra-/postoperative results (increased outflow resistance, low flow, poor venous drainage)<br>• Bypass recipient with chronic critical ischemia (stage III and IV AOD)<br>• Distal-origin bypass grafts in the presence of proximal atherosclerotic lesions<br>(Study by Davies et al. 2005 does not present a subgroup analysis for a differentiated approach) |

Although some discrepancy exists among published results, patients with autologous vein grafts appear to benefit from routine duplex surveillance, while for those with synthetic grafts the results are much poorer. Many studies are limited by the fact that they investigated mixed populations of patients with venous and synthetic bypass grafts. The poorer predictive value of routine surveillance in patients with synthetic grafts seems to be attributable to the complexity of factors that can cause occlusion in these cases.

In patients with failing grafts, bypass patency and limb salvage rates can be improved if graft-threatening lesions are recognized in time and surgical revision is done before occlusion occurs. For this reason, investigators are interested in defining *reliable duplex criteria to predict graft failure*. The potential of ultrasound to identify imminent graft failure, using either PSV measurement within the graft (PSV < 45 cm/s) or graft mapping, was investigated in numerous studies, but *no consistent picture emerges*. A large meta-analysis (Golledge et al. 2009) comparing 2,680 duplex surveillance and 3,969 nonsurveillance vein grafts showed that routine duplex surveillance improved bypass patency rates but not the (long-term) limb salvage rate. Still, most scientific evidence seemed to suggest that patients with venous bypass grafts benefit from routine sonographic surveillance. The results of the Vein Graft Surveillance Randomised Trial (VGST), however, have led to a reappraisal of the role of routine duplex surveillance in patients with venous grafts as well. In this trial, no difference was found between clinical and duplex surveillance in terms of primary patency, primary assisted patency, secondary patency, and amputation rates.

Thus, while there is *no added benefit of routine duplex monitoring of leg bypass grafts* (as shown by Kaplan–Meier analysis), *a duplex scan is necessary whenever serial clinical examination, pulse status, or measurement of ABI show deterioration*.

No subgroup analysis was done, which is a limitation of the VGST. Current experience and the evidence from other studies seem to suggest that patients who have a poorer prognosis from the start will benefit from being enrolled in a duplex surveillance program. These include patients in whom the bypass is created using a thin-caliber vein, a nonsaphenous vein, or a composite graft; patients with poor intra-/postoperative results; patients with chronic critical ischemia and no alternative options for restoring blood flow; and all patients with distal-origin bypass grafts in the presence of severe atherosclerotic lesions of the inflow tract (Table 2.20).

If, for cost-efficiency reasons, not all patients are monitored, at least grafts with poor intraoperative flow and increased outflow resistance (bypasses onto calf arteries) should be subjected to duplex surveillance (Mercer et al. 1999). A critical venous bypass with a poorer prognosis should be examined 3 and 6 months after surgery and every 6 months thereafter. This recommendation is based on the observation that despite a normal postoperative angiogram, a considerable proportion of venous bypass grafts (25–37%) develop stenosis within the first 6 postoperative weeks, resulting in a significantly increased 3-month occlusion rate (Ihnat et al. 1999). The decision for bypass revision is based on the severity of stenosis and poststenotic flow rates; clinical symptoms alone may be misleading as they vary with the patient's disease stage at the time of the bypass procedure. Finally, the proposed threshold velocities must be applied flexibly, taking into account the diameters of the graft and recipient artery (diameter mismatch). Future studies should try and establish different threshold values for different subgroups of grafts (autologous vein versus synthetic material, recipient vessel above or below the knee, status of outflow tract, bypass diameter).

#### 2.1.7.4 Ultrasound Vein Mapping Prior to Peripheral Bypass Surgery

Autologous venous bypass grafts having higher short-term and long-term patency rates are superior to other procedures in peripheral bypass surgery. However, vein preparation may be time-consuming in patients with variants such as an aberrant path or duplication and in obese patients. When the great or small saphenous vein is considered, the superficial course can be identified with a high-resolution transducer (6.5–10 MHz) and marked on the skin before surgery. Moreover, duplicated veins can be localized and the most suitable branch selected for grafting. When an in situ bypass is planned, perforating veins can also be marked for intraoperative ligation to prevent development of an AV fistula. The vein diameter is measured in transverse orientation with the great saphenous vein normally having a diameter of 3–4 mm in the lower leg; very small veins (<2 mm) are unsuitable for grafting. If two branches are present, the larger one is selected. Moreover, unnecessary dissection can be avoided by discarding varicose or postthrombophlebitic veins with thickened walls and sclerosis. Altogether, *preoperative sonographic vein mapping for selection of a suitable graft shortens the length of surgery and helps prevent unnecessary incisions and extensive exposure*.

### 2.1.8 Role of (Color) Duplex Ultrasound Compared with Other Modalities: Problems and Pitfalls

In the step-by-step diagnostic workup of PAOD, the patient's history, clinical examination with evaluation of pulses, and Doppler pressure measurement come before noninvasive duplex scanning, which may then be followed by invasive angiography (Fig. 2.7). The clinical stage of AOD and the sonographic findings are the basis for further patient management, either initiation of treatment or ordering of specific supplementary diagnostic tests (Tables 2.22 and 2.23).

For example, patients with iliac or femoropopliteal stenosis can undergo diagnostic angiography with PTA standby. In contrast, patients with longer occlusions of the pelvic or thigh arteries can be scheduled for bypass surgery without angiography if ultrasound has demonstrated adequate peripheral outflow as well as patency of the popliteal and calf arteries. Angiography is likewise not necessary in patients in whom an aneurysm has been confirmed sonographically.

**Table 2.22** Advantages and disadvantages of duplex imaging

| Advantages | Disadvantages |
|---|---|
| Noninvasiveness<br>Multilevel evaluation<br>Evaluation of<br>• wall morphology<br>• surrounding structures<br>• intraluminal structures<br>• plaque<br>Stenosis grading based on<br>• morphology<br>• hemodynamics<br>Low cost | Documentation of findings<br>Evaluation of collateral pathways<br>Evaluation of terminal vascular bed<br>Specific methodological limitations (calcification, air, obesity, edema) |

**Table 2.23** Advantages and disadvantages of angiography

| Advantages | Disadvantages |
|---|---|
| Documentation of findings<br>Visualization and evaluation of collateral pathways<br>Evaluation of terminal vascular bed<br>Fairly short training period | Invasiveness and complications (false aneurysm, embolism, bleeding, local thrombosis, AV fistula)<br>Visualization of patent lumen only<br>Projection-related problems:<br>• Stenosis grading<br>• Evaluation of bifurcation<br>Nonvisualization:<br>• Vessel wall<br>• Surrounding structures<br>High cost<br>Radiation exposure and contrast medium administration |

Patients in whom ultrasound reveals external compression (popliteal entrapment syndrome, cystic adventitial disease) can also be operated on without prior angiography, which provides no additional information and merely serves to document the vascular status. Depicting only the vessel lumen, angiography is inferior to ultrasound in evaluating perivascular structures. A further drawback of angiography is the reduction of the three-dimensional vessel lumen to the two-dimensional plane of the film (see Fig. 5.21).

With this limitation, the diameter reduction randomly depicted in the imaging plane does not necessarily represent the true cross-sectional area reduction, as the wall changes may vary along the length of the stenosis (concentric – eccentric; regular – irregular). The angiographic stenosis grade may thus differ from the grade determined by Doppler, which reflects the hemodynamic effects of the stenosis. Duplex and Doppler ultrasound may also yield discrepant information about plaque because the former shows the extent of a plaque in the form of color gaps within the lumen while the latter assesses the hemodynamic effects in terms of flow acceleration. For accurate and reproducible quantification, it is thus necessary to consistently depict a stenosis in different planes both sonographically and angiographically (Table 2.23).

This is especially important when assessing the pelvic arteries and femoral trifurcation, where eccentric plaques of the posterior wall are common. The mere morphologic assessment of an eccentric plaque afforded by angiography may overestimate the resulting stenosis compared with its hemodynamic effects, even when evaluated in different planes. If only an anteroposterior angiogram is available, however, an eccentric stenosis may be overlooked or underestimated.

On the other hand, plaque configuration may not receive adequate attention in assessing the hemodynamic relevance of a stenosis, and an examiner must bear in mind that a concentric plaque produces more marked flow effects than an eccentric plaque; for instance, a concentric stenosis with a

50% diameter reduction causes a 75% cross-sectional area reduction as opposed to 50% or less when an eccentric stenosis with the same diameter reduction is present.

Evaluation of stenoses at the origin of the profunda femoris artery by angiography may be impaired by superimposed vessels. Ultrasound is superior when performed with an adequate angle of insonation. In a study of 30 patients scheduled for TEA of the profunda femoris to improve collateralization of superficial femoral occlusion, the high-grade stenosis demonstrated by ultrasound and confirmed intraoperatively was identified definitely by angiography in only 70% of the cases, and there was considerable interobserver variability in stenosis grading.

Catheter-based digital subtraction angiography (DSA) via a transfemoral or transbrachial approach is the gold standard if views in two or three planes are obtained and if adequate opacification of the distal arteries is ensured. Proper timing taking into account the longer contrast agent transit time to the thigh and foot is important to avoid misinterpreting angiographically occult distal runoff vessels. Magnetic resonance angiography (MRA) using dedicated coils is superior to DSA in evaluating the thigh and foot arteries in patients with reduced contrast agent inflow due to occlusive disease of more proximal arteries (Fellner et al. 1999; Owen et al. 1992; Kreitner et al. 2000). Several investigators have shown that MRA allows good evaluation of distal arteries, including the pedal arch in patients with foot ischemia, and reliable identification of patent runoff vessels in cases where DSA fails to reveal a target vessel for pedal bypass grafting (Dorweiler et al. 2002; Kreitner et al. 2000).

For a complete lower extremity arterial evaluation, an imaging technique optimized for visualization of the calf and foot arteries needs to be supplemented by a modality best suited to imaging the iliofemoral arteries. Noninvasive duplex ultrasound is ideal for the proximal leg arteries and has the added advantage of providing information on the flow effects of steno-occlusive lesions of the iliac and femoropopliteal arteries. It is therefore conceivable that a diagnostic strategy combining (color) duplex imaging of the proximal leg arteries with MRA of the thigh and pedal arch may in the future replace invasive DSA, which carries a number of risks (use of contrast medium, radiation exposure, vascular puncture, and catheterization).

The clinical usefulness of a diagnostic depends not only on its diagnostic accuracy and the relevance of the results for therapeutic decision making but also on how *efficiently the examination can be performed*. It is theoretically possible to perform a complete duplex scan from the pelvic level down to the pedal arteries including identification of a *target artery for a bypass procedure* (which may require a contrast-enhanced examination if a bypass onto a thigh artery is contemplated). Few scientific data are available on the time requirement for sonographic evaluation of the lower limb arteries. In a study performed by the author (as yet unpublished), 220 patients with claudication, forefoot lesions, or rest pain were selected for a duplex examination of the leg arteries on the basis of prior Doppler pressure measurement. In this preselected population, duplex scanning identified hemodynamically relevant stenosis or occlusion in 93% of cases. Following the above-described protocol (Table 2.1), the examination started with the recording and interpretation of Doppler waveforms from the groin (both external iliac/common femoral artery junctions for comparison), followed by the profunda femoris origin and superficial femoral artery. Next, spectral Doppler measurement of the proximal and distal popliteal segments (P1 and P3) of the symptomatic leg was performed. Only if spectral analysis or comparison of the waveforms from the proximal and distal popliteal segments revealed any abnormalities or relevant changes was the proximal arterial segment continuously mapped in the B-mode with continuous spectral Doppler recording (longitudinal plane, sample volume slightly larger than the arterial lumen). Occlusion length was determined taking into account the sites of origin and entry of collaterals. Using this algorithm in the study population of 220 patients, it took 5.2 min to establish a diagnosis and to decide on the therapeutic approach (TEA or bypass graft including identification of the segment for the distal anastomosis). Patients with a long history of diabetes were not included in this study because severe medial sclerosis with acoustic shadowing makes the sonographic examination more time-consuming. This sonographic method allows reliable treatment decisions to be made, including selection of a bypass target artery for surgical management of relevant stenosis or occlusion, in patients with stage II disease and a patent popliteal artery. In patients with stage III or IV AOD and occlusion at the pelvic or thigh level but a patent popliteal segment (from P1 to the tibiofibular trunk), the inflow obstruction above the popliteal artery is eliminated (PTA or bypass), which can be done without examination of the calf arteries. Imaging of the calf arteries is necessary only if the popliteal artery is stenosed or partially occluded and an appropriate artery has to be identified for the distal anastomosis of a crural bypass. This is possible but time-consuming, as several authors have demonstrated (Boström et al. 2002; Hofmann et al. 2004). Alternatively, angiography or magnetic resonance angiography with dedicated coils can be used. Angiography has several disadvantages including invasiveness, contrast medium administration, and poor visualization of calf arteries in patients with long proximal occlusions. When duplex ultrasound is performed to identify a potential crural bypass target vessel, Doppler recordings are obtained from the dorsalis pedis and posterior tibial arteries at the ankle level (and possibly also from the distal fibular artery) and are then compared with the waveforms from the P3 popliteal segment or proximal segments of the calf arteries. The artery in which the Doppler spectra from the proximal and distal sites are least different is the most suitable runoff vessel for the planned bypass and should subsequently be mapped to confirm its suitability.

Sonographic follow-up examinations can be performed to monitor the outcome of surgical repair and can help initiate timely revision in case of residual or anastomotic stenosis (see Table 2.20).

Calcified plaque impairs sonographic assessment as *acoustic shadowing* obscures affected vessel segments and other structures lying behind them. Long calcified lesions may prevent the acquisition of an adequate Doppler signal even at a higher gain. In such cases, the degree of a stenosis is determined by comparing waveforms obtained proximal and distal to the lesion. An unchanged PSV and a triphasic flow profile suggest a plaque without hemodynamic significance.

Proper angulation of the ultrasound beam for quantifying a stenosis may be difficult to achieve despite changing the transducer position. This is especially the case with the arched segment of the common iliac artery at the junction with the external iliac artery. Aliasing may occur in arteries interrogated with a very small Doppler angle. As a consequence, high velocity peaks can be cut off. Deep arteries such as the internal iliac artery are most susceptible to aliasing.

In patients with long-standing stenosis, extensive collateral pathways may have developed. The reduced blood flow and altered flow in the main branch resulting from a division of the blood volume between the collaterals and the artery can lead to *underestimation* of the stenosis because duplex ultrasound solely relies on hemodynamic parameters (see Fig. 2.18).

*Color duplex scanning* with a high-resolution transducer enables assessment of collaterals, in particular when the occluded segment is short, and will detect stenosis at the reentry sites of collateral pathways (see Fig. A2.14). Nevertheless, ultrasound is inferior to angiography in providing a full overview of the collateral circulation of an occlusion. The hemodynamic situation distal to an occluded segment can merely give hints. A relatively high PSV in the postocclusive segment distal to the emptying of the collaterals suggests good collateralization. However, PSV must be interpreted with caution and is also increased in diabetics with rigid arterial walls due to medial sclerosis.

If two or more sequential stenoses are present, the more distal ones may be underestimated or difficult to assess, especially in *diabetic macroangiopathy*. Moreover, the *hemodynamic changes* associated with diabetic medial sclerosis and peripheral occlusions further impair stenosis grading by duplex ultrasound. Angiographically, such sequential stenoses are identified by their typical goose throat appearance.

When diagnostic ultrasonography and vascular repair are performed by different persons or even by different departments, the lack of *continuous documentation* of the findings – in particular in patients with complex patterns of vascular lesions – limits the communication of the findings to the surgeon. Ultrasonography is an invaluable tool in the hands of the radiologist or vascular surgeon who also treats the patient and benefits from the additional hemodynamic information. The diagnostic gain also depends on the skills of the examiner, as ultrasound is highly examiner-dependent.

A systematic review of 48 studies investigating the accuracy of different imaging modalities for detecting significant arterial stenosis (>50%) in patients with symptomatic peripheral arterial occlusive disease was presented by Collins et al. (2007). The studies included in the review used DSA of the whole leg as the gold standard. The synthesized data showed CT angiography (five studies) to have a median sensitivity of 97% (range, 89–100%) and a median specificity of 99.6% (range, 99–100%); time-of-flight (TOF) magnetic resonance imaging (four studies) had a median sensitivity of 86% (range, 77–100%) and a median specificity of 93.8% (range, 85–98%). MRA using a gadolinium-based contrast agent was found to have a median sensitivity of 94% (range, 85–100%) and a median specificity of 99.2% (range, 97–99.8%). Color duplex ultrasound (7 studies) had a similar diagnostic performance with a median sensitivity of 90% (range, 74–94%) and a median specificity of 99% (range, 89–100%). Duplex imaging and contrast-enhanced MRA had similar sensitivities of about 90% in detecting hemodynamically significant stenosis below the knee. Compared with the other noninvasive imaging modalities, duplex scanning causes no patient discomfort and has a low complication rate. CT is limited by a high radiation exposure and the use of iodine-based contrast media; stenosis grading is limited by the presence of calcified plaques or stent-related artifacts. The quality of contrast-enhanced MRA depends on the use of state-of-the-art equipment; general limitations include poor visualization of below-knee arteries unless dedicated coils are used, the susceptibility to artifacts, overestimation of stenosis severity, and venous superimposition, which is most relevant in the periphery. MRI is contraindicated in patients with pacemakers and defibrillators. Gadolinium-based contrast agents should not be used in patients with impaired renal function because they may induce life-threatening nephrogenic systemic fibrosis (acute renal failure rate of up to 2%) (Collins et al. 2007).

An experienced examiner is able to assess the lower leg arteries throughout their course using a high-resolution transducer. Extra time is required for the separate evaluation of an occlusion or stenosis. What ultrasound fails to provide is an overview of complex patterns of disturbed perfusion with multiple occlusions of the lower leg arteries that collateralize each other, especially with regard to evaluation of peripheral outflow (Table 2.24).

Acoustic shadowing from calcified plaques may impair the detection and grading of stenosis in small arteries. In such cases, the sonographic search for a patent target artery for a distal bypass is very time-consuming if the presence of

**Table 2.24** Different indications for angiography and color duplex ultrasonography. More plus signs indicate a more favorable rating

| Diagnostic task | Angiography | Color duplex |
|---|---|---|
| Atherosclerosis screening | Not possible | +++ |
| Therapeutic decision making (medical, PTA, surgery) | ++ | +++ |
| Planning of surgical procedure | +++ | + |
| Follow-up: PTA, surgery | + | +++ |
| Critical ischemia | Mandatory | +++ |
| Vessel wall evaluation (aneurysm/inflammatory conditions) | (+) | +++ |
| Evaluation of surrounding structures | Not possible | +++ |
| Analysis of occlusion (cause) | (+) | +++ |
| Evaluation of collaterals | +++ | + |
| Analysis of plaque morphology, progression, regression | (+) | +++ |
| Atypical arteries | +++ | + |
| Diabetic foot | +++ | + |
| Documentation of findings | +++ | – |

proximal occlusion or multilevel stenosis precludes the use of indirect stenosis criteria. Despite these limitations, some investigators reported accuracy rates for the detection of steno-occlusive lesions below the knee that would allow surgical planning without prior angiography. According to these studies, preoperative duplex ultrasound alone allows reliable identification of a patent runoff vessel for a planned bypass graft below the femoropopliteal arteries. These patients were found to have the same intraoperative results as well as primary and secondary bypass patency and limb salvage rates as patients undergoing preoperative angiography.

The studies comparing duplex imaging with DSA in the assessment of the cruropedal arteries report widely divergent results (Karacagil et al. 1996; Koelemay et al. 1997) with sensitivities and specificities for the detection of occlusion ranging from 50% to 90%. However, the investigators used 5 MHz transducers, which are inadequate for the pedal vessels, and the low resolution may explain the poor results. Other study groups (Boström et al. 2002; Hofmann et al. 2001) describe limitations of nonselective IA DSA in the visualization of the outflow vessels of the foot (see Fig. A2.18). In a consecutive series of 49 patients reported by Hofmann et al. (2001), the pedal arteries in 32 cases (65.3%) were hardly or not at all visualized by DSA. Based on the sonographic evaluation using a high-resolution transducer (13 MHz), all 32 patients underwent pedal bypass grafting with a 2-year patency rate of 69.5%. Boström et al. (2002) compared 157 vascular surgical procedures (32 inguinal TEA, 91 femoropopliteal bypass, 34 femorocrural bypass) performed solely on the basis of the ultrasound findings and 172 procedures (28 inguinal TEA, 144 femoropopliteal and femorocrural bypass) planned on the basis of angiography. The second group included patients with inadequate ultrasound results (in particular also patients with femorofibular bypass procedures). Cruropedal bypass grafting (ratio of 1:2 [ultrasound versus angiography group]) and femorocrural bypass grafting (ratio of 1:1) were also performed using only the ultrasound findings for planning. The primary patency rate of the bypass grafts was 59% in the ultrasound group versus 64% in the angiography group. In 98% of the patients examined by preoperative ultrasound, no intraoperative revision of the planned procedure was necessary (not even on the basis of intraoperative DSA).

Selective DSA with positioning of the catheter in the superficial femoral artery or more distally continues to be the gold standard. When technically adequate DSA is not possible or visualization of the cruropedal arteries is poor, a supplementary high-resolution duplex ultrasound examination is required. Provided that the necessary infrastructure and ultrasound equipment are available and that there is adequate communication between the sonographer and the operator (ideally the same person), patients with good insonation conditions can undergo TEA or femoropopliteal and femorocrural bypass procedures on the basis of the duplex ultrasound findings alone.

## 2.2 Arm Arteries

### 2.2.1 Vascular Anatomy

The innominate artery (brachiocephalic trunk) arises from the aortic arch on the right and behind the sternoclavicular joint divides into the subclavian and common carotid arteries. On the left, the *subclavian artery* arises directly from the aortic arch as does the common carotid artery at a more proximal site. Along its course, the subclavian artery first gives off the *vertebral artery* cranially. Further along the course, the thyrocervical trunk arises, likewise from the posterior aspect, and at once divides into a branch supplying the thyroid and other branches supplying the skin and soft tissue. Together with the brachial plexus, the subclavian artery (Fig. 2.34) passes through the scalene triangle (between the anterior and medial scalene muscles and cranial to the first rib) and arches over the pleural dome, crossing under the clavicle, to continue as the *axillary artery*. In individuals with a cervical rib, the subclavian artery is displaced cranially and anteriorly. Distal to the thyrocervical trunk, the internal thoracic (mammary) artery arises and descends behind the anterior chest wall and about one fingerbreadth lateral to the sternum.

The branches of the axillary artery have extensive collateral connections to the branches of the subclavian artery and supply the region of the shoulder girdle. The axillary artery

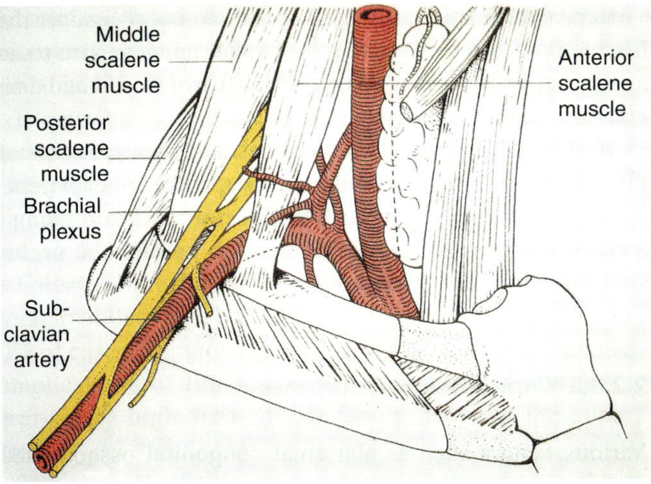


**Fig. 2.34** Course of the subclavian artery through the scalene triangle. The subclavian artery runs between the first rib, medial scalene muscle, and anterior scalene muscle (from Heberer and van Dongen 1993)

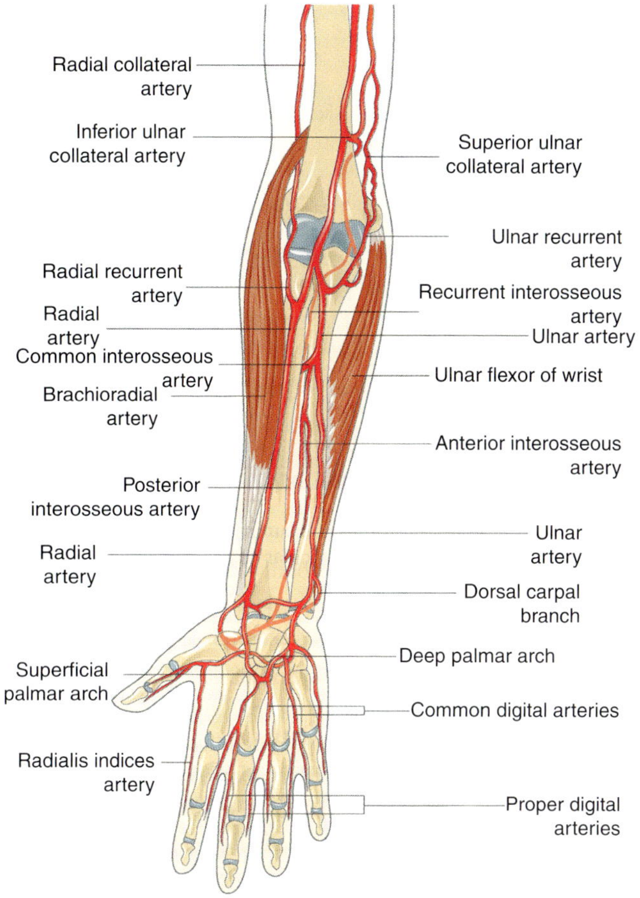


**Fig. 2.35** Anatomy of the arm vessels

courses along the lower border of the pectoralis muscle through the axilla and continues as the *brachial artery*. The latter runs through the medial bicipital groove near the humerus to the elbow and divides into the radial and ulnar arteries at the level of the joint space. There are anatomic variants in which the brachial artery already gives off the radial artery in the upper arm (in about 15%) or arises directly from the distal axillary artery (1–3%). The ulnar artery may likewise originate from the axillary artery (in about 1%).

The radial artery continues through the forearm on the ulnar side of the radius to the wrist where it unites with the deep branch of the ulnar artery to form the deep palmar arch. The radial artery primarily feeds the deep arch and the ulnar artery the superficial arch. The main branches of the superficial arch give off the common palmar digital arteries, which in turn give rise to the proper palmar digital arteries, the main vessels supplying the fingers (Fig. 2.35). A complete connection between the superficial and deep palmar arches is present in only about 80–90% of individuals.

## 2.2.2 Examination Protocol and Technique

The subclavian and axillary arteries are scanned at *5–7.5 MHz* while a higher-frequency transducer is appropriate more distally where the arteries lie closer to the surface. The finger arteries are examined with a *7.5–10 MHz transducer*. Especially in the supraclavicular fossa, curved array or sector transducers are superior to linear transducers. The subclavian and axillary arteries are best imaged in the supine position with the examiner behind the patient's head, as in the examination of the carotid arteries. The forearm and finger arteries are examined in the sitting patient with the hand supinated.

The arm arteries are traced along their course from the supraclavicular area to the palmar arch. In the upper arm, the arteries are easily identified by B-mode scanning based on sonoanatomic knowledge. The vessels of the palmar arch and fingers are localized with the aid of color duplex imaging. Spectral Doppler information is sampled in longitudinal orientation with a smaller angle of insonation.

The proximal portion of the subclavian artery is interrogated with the transducer in the supraclavicular position. In evaluating the supra-aortic branches, the examiner should pay special attention to the origin of the vertebral artery, which must be differentiated from the thyrocervical trunk. Rhythmical tapping of the vertebral artery suboccipitally will be transmitted and appear in the Doppler waveform from the proximal segment of the artery.

The axillary artery is identified cranial to the axillary vein with the transducer placed in the infraclavicular fossa and followed along its path to the axilla (Fig. 2.36). The brachial artery is examined in the upper arm from a medial position.

diagnose this condition. A study in a German population shows that patients consulted an average of 6.5 specialists before the syndrome was finally diagnosed – after a mean of 4.3 years (Gruss et al. 1989; Gruss and Geissler 1997).

Prior to apparative tests, a vascular examination is performed with determination of pulses, auscultation, and bilateral Doppler pressure measurement. Unilateral pulse reduction or obliteration with elevation or abduction of the arm is not a very specific symptom and is seen in 30–60% of asymptomatic young adults. The *AER* (abduction external rotation) *test* continues to have a significant clinical role: arms in 90 degree AER position with fist clenching every 2 s for 3 min. Most subjects will experience fatigue, pain, and heaviness before the end of the 3-min test period. Formication suggests compression of the upper plexus. Additional pain and pallor of the fingers indicate arterial compression. The diagnosis of a thoracic outlet syndrome further requires measurement of the *nerve conduction velocity* of the ulnar and median nerves. A definite delay in proximal ulnar and median nerve conduction in the plexus/axilla segment is diagnostic of compression, but normal conduction does not exclude the thoracic outlet syndrome. Ulnar motor velocities above 65 m/s are normal and velocities below 45 m/s indicate damage to the plexus (Urschel 1976).

In cases where the (color) duplex findings confirm interdigital artery occlusion, the examiner must look for the source of embolism, primarily a partially thrombosed *aneurysm*.

Such aneurysms primarily occur *secondary to compression-induced wall damage* in thoracic outlet syndrome or as a result of traumatic damage to the distal ulnar artery in hypothenar syndrome. The intermittent compression in thoracic outlet syndrome may also give rise to intraluminal thrombus formation. The specific type of thoracic outlet syndrome (Table 2.25) is diagnosed by ultrasonography of the respective sites of compression as suggested by the patient's history and clinical symptoms using the following provocative tests:

- *Scalenus muscle test* (*Adson test*) in combination with the hyperabduction test to identify arterial compression in the scalene triangle: the hyperextended head is turned toward the affected side (Schoop 1988) with the neck muscles tensed and possibly with additional hyperabduction and rotation of the arm.
- *Costoclavicular* or *hyperabduction test* to identify venous compression in the costoclavicular space: gliding of the clavicle over the first rib with the arm hyperabducted narrows the passage, thereby inducing venous compression. However, venous compression in this area is more commonly caused by weak shoulder muscles, which are better identified by a downward pull on the posteriorly turned arm (with the shoulder drawn back, inspiration).

During the *provocative maneuvers*, the *Doppler spectrum is recorded from the target site* or distal to it if the vessel is not accessible to scanning. The test is positive if there is flow acceleration in the compressed artery or if an altered flow profile is obtained distal to the compressed segment. The examiner can also move the scanner toward the compressed segment from the periphery, intermittently recording the Doppler information. As clinical symptoms are often unspecific, provocative tests are necessary to diagnose the specific type of compression syndrome (Table 2.26). If there is occlusion of the forearm or finger arteries, it is crucial to identify the partially thrombosed aneurysm of the subclavian artery, which typically develops on the basis of a scalene muscle or cervical rib syndrome. In rare cases, emboli may arise from thrombotic deposits of the damaged wall of the axillary artery in hyperabduction syndrome. These changes are caused by intermittent compression of the axillary artery and will be identified with the transducer placed in the axilla (see Fig. A2.63).

Patients with venous compression and clinical symptoms should initially be treated by physical therapy to strengthen the shoulder muscles or by resection if bony abnormalities such as a cervical rib or exostosis are present.

In the thoracic outlet syndrome with compression or mechanical irritation of the arteries or nervous plexus, the first rib should be resected before secondary damage to the vessel wall with development of aneurysm occurs. If secondary damage has already occurred, the affected arterial segments must be resected as well.

*Inflammatory conditions* primarily involve the subclavian artery (Takayasu's arteritis) and are characterized by *hypoechoic circumferential wall thickening* of long segments. If the clinical presentation and patient's history suggest *Raynaud's syndrome*, this must be confirmed by *provocative tests* with heat and cold exposure after interdigital artery occlusion has been ruled out (see above, Sect. 2.1.6.4.3).

**Table 2.25** Compression syndromes of the upper thoracic aperture and arm vessels

| |
|---|
| • Cervical rib syndrome |
| • Scalenus anterior syndrome (arterial: Adson test) |
| • Scalenus minimus syndrome |
| • Costoclavicular compression syndrome (venous: hyperabduction test) |
| • Pectoralis minor syndrome |
| • Compression syndrome of the brachial artery |

**Table 2.26** Compression syndromes of the upper thoracic aperture. Duplex scanning with confirmation by angiography, venography, intraoperative findings. Tentative clinical diagnosis: arterial and/or venous compression ($n = 680$)

| Compression syndrome | Number |
|---|---|
| Cervical rib syndrome (arterial) | 3 |
| Scalenus anterior syndrome (arterial)<br>– with poststenotic aneurysm | 6<br>2 |
| Costoclavicular compression syndrome (venous)<br>– with venous thrombosis | 8<br>5 |
| Pectoralis minor syndrome (arterial) | 2 |

## 2.3 Atlas: Peripheral Arteries

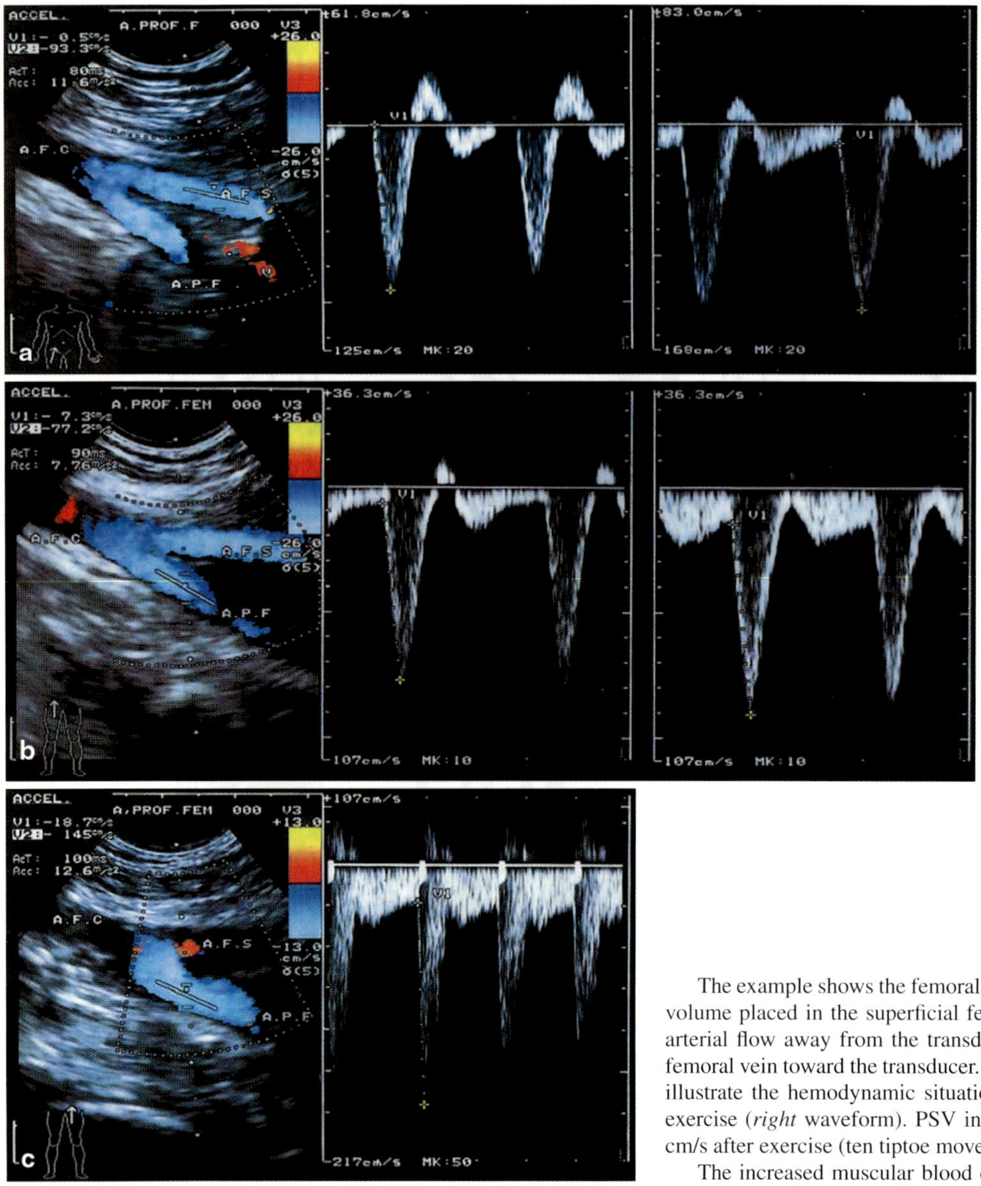


**Fig. A2.2 a–f**
**Femoral bifurcation – normal perfusion**

**a** Gray-scale scan and color flow information supplement each other: along the course of an artery, some segments may be better appreciated in the B-mode scan, others in the color Doppler mode. In the example, the superficial femoral artery (*A.F.S*) and profunda femoris artery (*A.P.F*) are insonated with a smaller angle, improving their visualization in the color mode whereas wall structures perpendicular to the ultrasound beam (here: common femoral artery, *A.F.C*, *left* side of scan) are seen more clearly in the B-scan. Ultrasound pulses striking the vessel wall at a perpendicular angle produce a detailed image of the wall, which is a strong reflector. In contrast, a smaller angle between the direction of flowing blood and the beam is necessary to ensure accurate spectral Doppler measurement and reliable evaluation of flow.

Although all extremity arteries have a triphasic pulsatile flow profile under normal conditions resulting from the high peripheral resistance at rest, different spectral waveforms may be obtained, depending on the territory supplied by the vessel interrogated. High-resistance flow as in the superficial femoral artery, which mostly supplies skin and subcutaneous tissue and only some muscle tissue, gives rise to a pulsatile, triphasic waveform with zero flow in end diastole.

The example shows the femoral bifurcation with the Doppler sample volume placed in the superficial femoral artery (*A.F.S*). Blue indicates arterial flow away from the transducer, red the flow in the superficial femoral vein toward the transducer. The corresponding Doppler tracings illustrate the hemodynamic situation at rest (*left* waveform) and after exercise (*right* waveform). PSV increases from 90 cm/s at rest to 141 cm/s after exercise (ten tiptoe movements).

The increased muscular blood demand during exercise is met by a decrease in peripheral resistance and is reflected in the Doppler wave form by an increase in peak end-diastolic velocity from 0 (*left* waveform) to 16 cm/s (*right* waveform).

**b** Femoral bifurcation (longitudinal scan): The profunda femoris (*A.P.F*) supplying more muscle tissue has a slightly less pulsatile flow but the profile is still triphasic. Peak velocity at rest (*left* waveform) is 77 cm/s in systole and 7 cm/s at end diastole. After exercise (*right* waveform), PSV increases to 90 cm/s with end-diastolic velocity doubling to 15 cm/s. The color change in the color duplex image from red, to black, to blue reflects the change in flow direction relative to the ultrasound beam (toward transducer: red; away from transducer: blue). (*A.F.S* superficial femoral artery, *A.F.C* common femoral artery).

**c** When there is occlusion of the superficial femoral artery (*A.F.S*), the profunda femoris is the main collateral to bridge the occluded segment. The increased blood flow in the profunda femoris circulation is reflected by a higher flow velocity, which may increase by 50–80% even if there is no stenosis at the origin of the profunda femoris.

The Doppler waveform from the profunda femoris artery (*A.P.F*) illustrates peak velocity in case of occlusion of the superficial femoral, which is 145 cm/s in systole and 18 cm/s in diastole. Reversed flow due to eddy currents at the origin of the occluded superficial femoral is displayed in red.

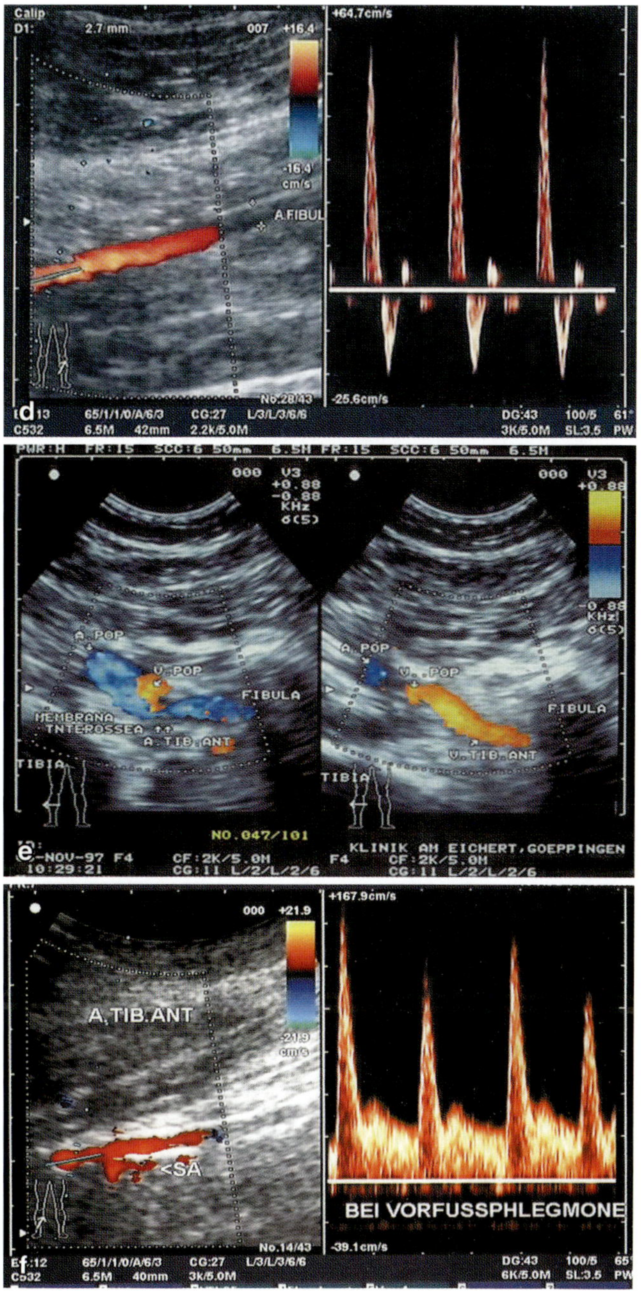


**Fig. A2.2 (continued)**
**d** The leg arteries have a triphasic waveform down into the periphery, as illustrated here by the normal fibular artery. The artery has a diameter of 2.7 mm.

**Sonoanatomy of the anterior tibial artery origin**
**e** The popliteal artery gives off the anterior tibial artery, which courses anteriorly to pierce the interosseous membrane, in front of which it descends, initially taking a course close to the fibula. The scan depicts the anterior tibial artery insonated from a posterior transducer position (popliteal fossa) in blue (flow away from transducer) downstream of its origin from the popliteal artery (*A.POP*) as it pierces the interosseous membrane (hyperechoic structure between tibia and fibula). With the transducer slightly angled, the anterior tibial vein is depicted in blue (flow toward transducer) along its course parallel to the artery and as it enters the popliteal vein.

**Hyperemia**
**f** Peripheral inflammation is another factor affecting the character of the Doppler waveform besides an increased flow resulting from exercise-induced hyperemia or collateral flow through an artery. In the example, a phlegmon of the foot results in a monophasic waveform with reduced pulsatility and a rather high end-diastolic flow velocity of 22 cm/s. An upstream stenosis can be ruled out because the steep systolic upslope is preserved. The normal anterior tibial artery has a triphasic flow profile with short diastolic forward flow after the notch and zero end-diastolic flow (see **d**). The variation in PSV seen here is attributable to absolute arrhythmia with a variable ventricular filling phase (*SA* mirror artifact).

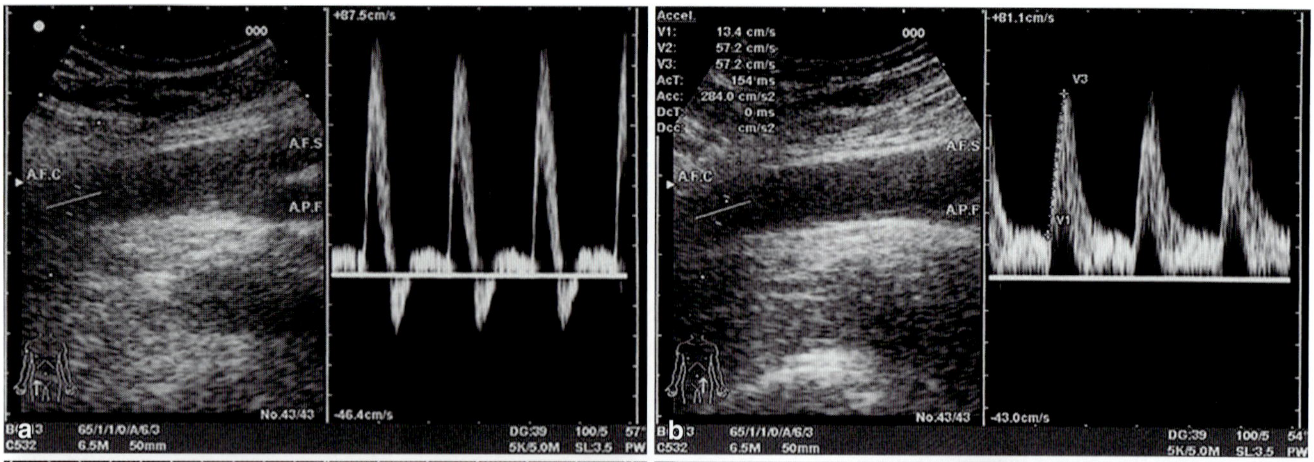


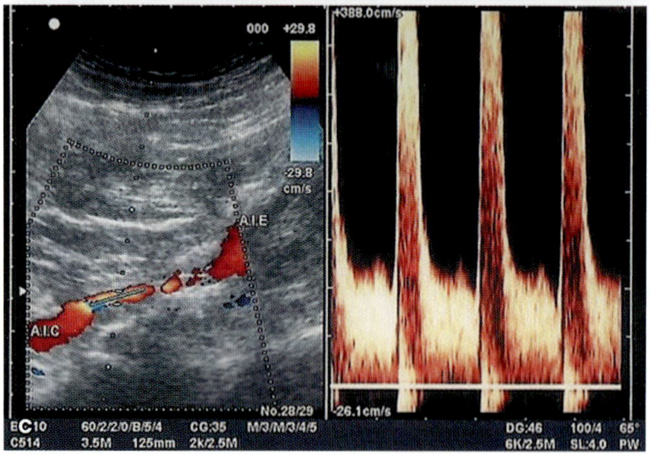


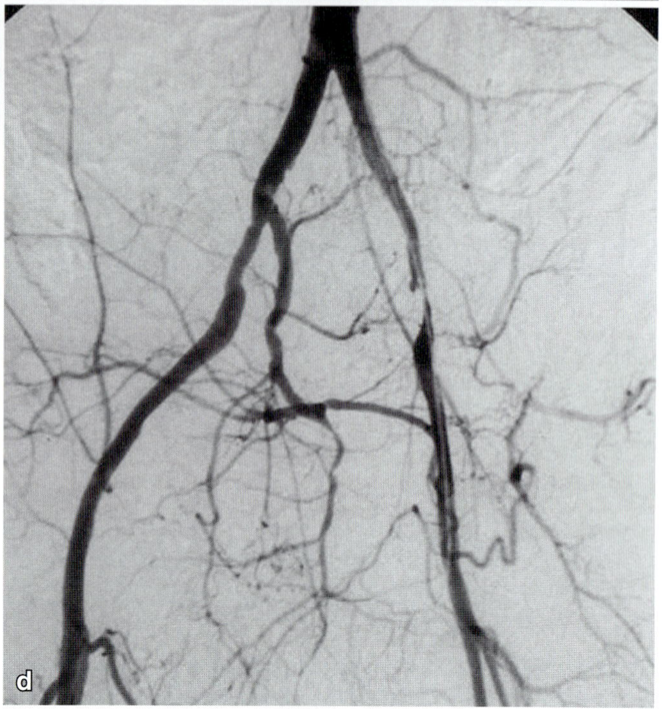


**Fig. A2.3 a–d**
**Pelvic artery stenosis**

**a** In evaluating a patient with suspected pelvic artery obstruction, the examiner first obtains Doppler tracings from both common femoral arteries to compare these with regard to triphasic flow, steep systolic upslope, and magnitude of PSV. Reliable Doppler shift analysis requires an insonation angle below 60°. In this example, the angle is 50° on the *right* and 54° on the *left*. The Doppler waveform from the *right* groin shows triphasic flow with a systolic upslope and a PSV > 80 cm/s.

**b** The waveform from the *left* common femoral artery illustrates postocclusive flow with a monophasic profile, reduced PSV (57 cm/s), and delayed systolic rise.

**c** Monophasic flow profile caused by a high-grade stenosis of the common iliac artery (*A.I.C*) due to plaque, mainly of the posterior wall. Sonographic criteria for stenosis are aliasing in the color duplex image and a Doppler-derived PSV of over 4 m/s. Due to aliasing, the PSV is cut off and must be interpolated (about 4.5 m/s). The simplified Bernoulli equation, $P = 4 \cdot PSV^2$, yields a maximum pressure drop of 81 mmHg over the stenosis, resulting in a poststenotic decrease in systolic velocity.

**d** Angiogram demonstrates the high-grade iliac stenosis as a filling defect in the lumen.

**Fig. A2.4 a–f**

**External iliac artery stenosis – diagnosis by distal Doppler measurement**

**a** Under good insonation conditions, the B-mode scan reveals stenosing plaques (*P*) in the pelvic artery territory on the basis of an irregular vessel contour, acoustic shadowing due to calcification, and protrusion of hyperechoic structures into the lumen.

**b** Color duplex imaging is necessary for qualitative estimation of the degree of stenosis on the basis of aliasing, and spectral Doppler measurement enables quantitative hemodynamic grading. The example shows a stenosis in the external iliac artery (*A.I.E*) immediately distal to the bifurcation. The internal iliac artery (*A.I.I*) is depicted farther away from the transducer. The high PSV suggests a higher-grade narrowing with the pulsatile flow profile (early diastolic dip) indicating only little impairment of peripheral hemodynamics at rest. Taken together, these findings are consistent with good collateralization and/or stenosis < 75%.

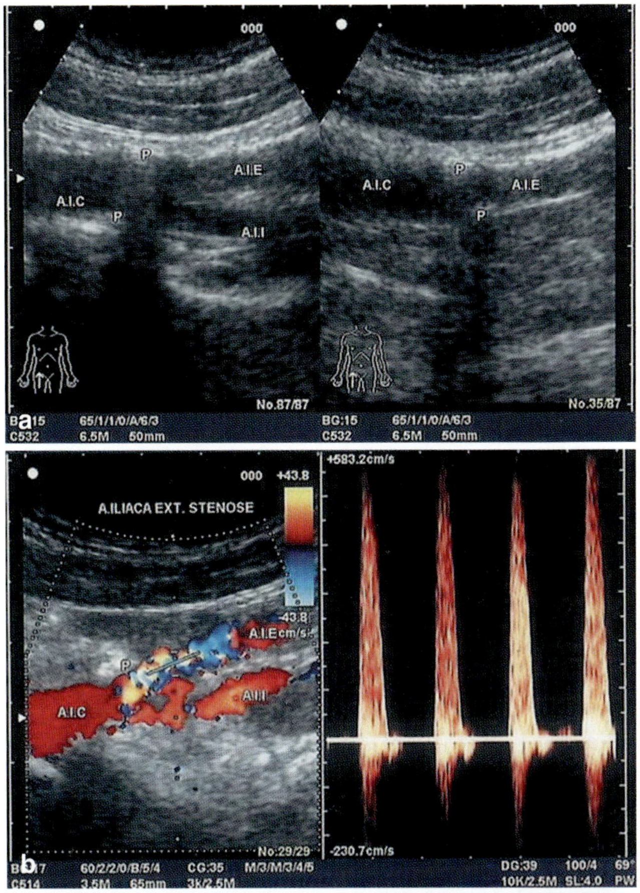


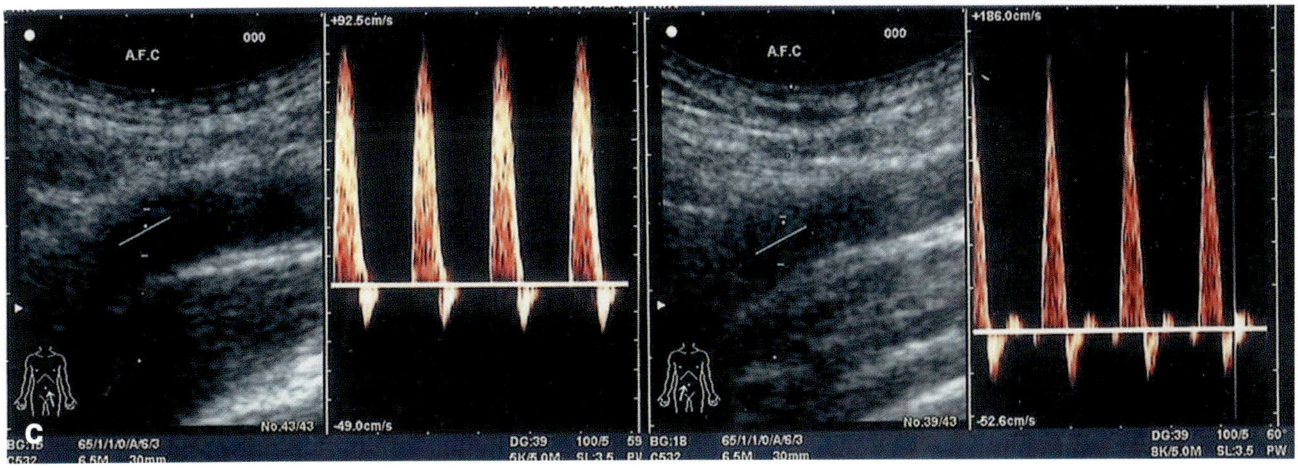


**c** Such a stenosis will be noted in the Doppler tracing from the groin and a more peripheral site only if the examination is performed carefully, the Doppler information is acquired with a small angle, and a meticulous spectral analysis is done with comparison of both sides. In this case, the waveform from the stenotic side shows an early diastolic retrograde flow component, indicating that high peripheral resistance is preserved, while PSV is markedly decreased compared with the unaffected side (80 vs. 150 cm/s). Moreover, the waveform from the unaffected side displays the preserved diastolic flow resulting from the windkessel effect of the aorta.

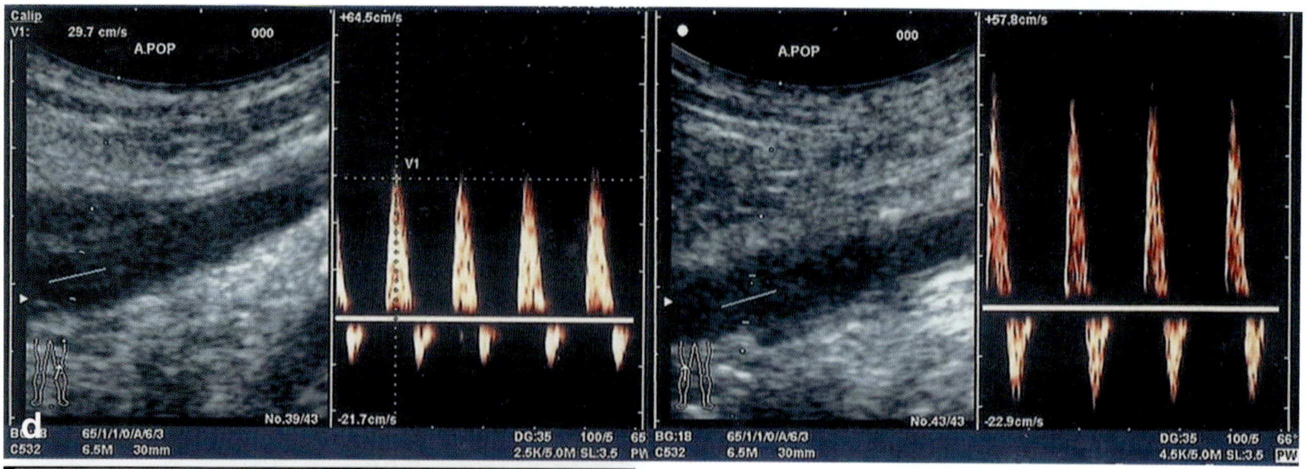


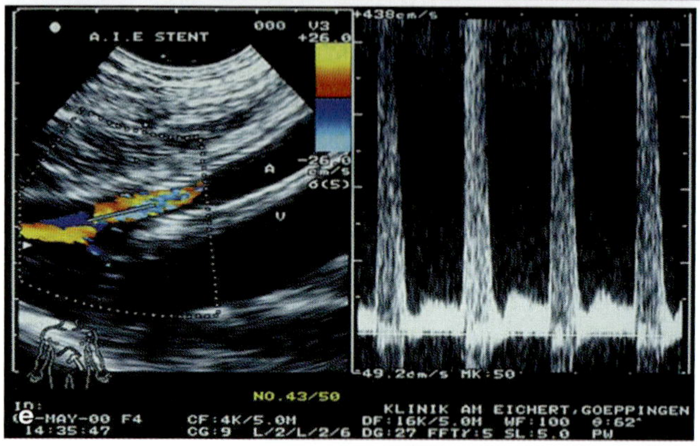


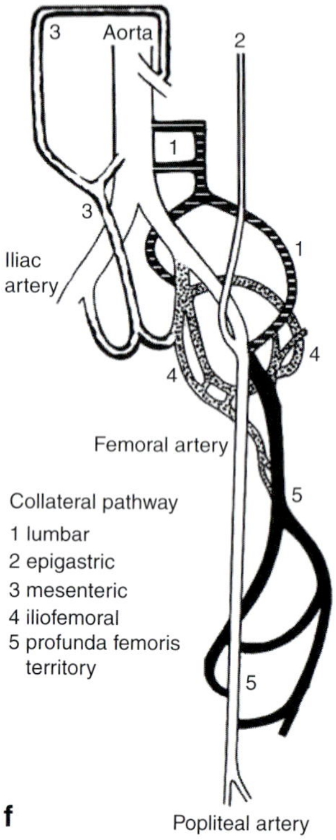


f

**Fig. A2.4 (continued)**

**d** The upstream stenosis on the *left* is apparent in the popliteal artery waveform as a dampened flow signal that is also audible and in a postocclusive PSV of 29 versus 45 cm/s on the normal side.

Another way of demonstrating a stenosis is to sample the Doppler information after exercise: after ten knee bends, the waveforms from the groin and popliteal artery are monophasic on both sides (due to exercise-induced reduction of peripheral resistance). On the unaffected side, the waveform returns to its normal triphasic shape within a minute. A delayed return to this normal profile indicates a stenosis, which prevents adequate peripheral perfusion during and after exercise. The length of the delay varies with the stenosis grade and collateralization.

**e** Status after dilatation and stenting of a proximal external iliac artery stenosis with high-grade stenosis at the proximal stent end (aliasing, PSV > 4 m/s). The mesh-like structure enables good estimation of stent length.

**Collateral pathways in stenosis and occlusion of the aortoiliac and iliofemoral territories**

**f** The better the collateralization, the less pronounced the postocclusive changes in the Doppler waveform.

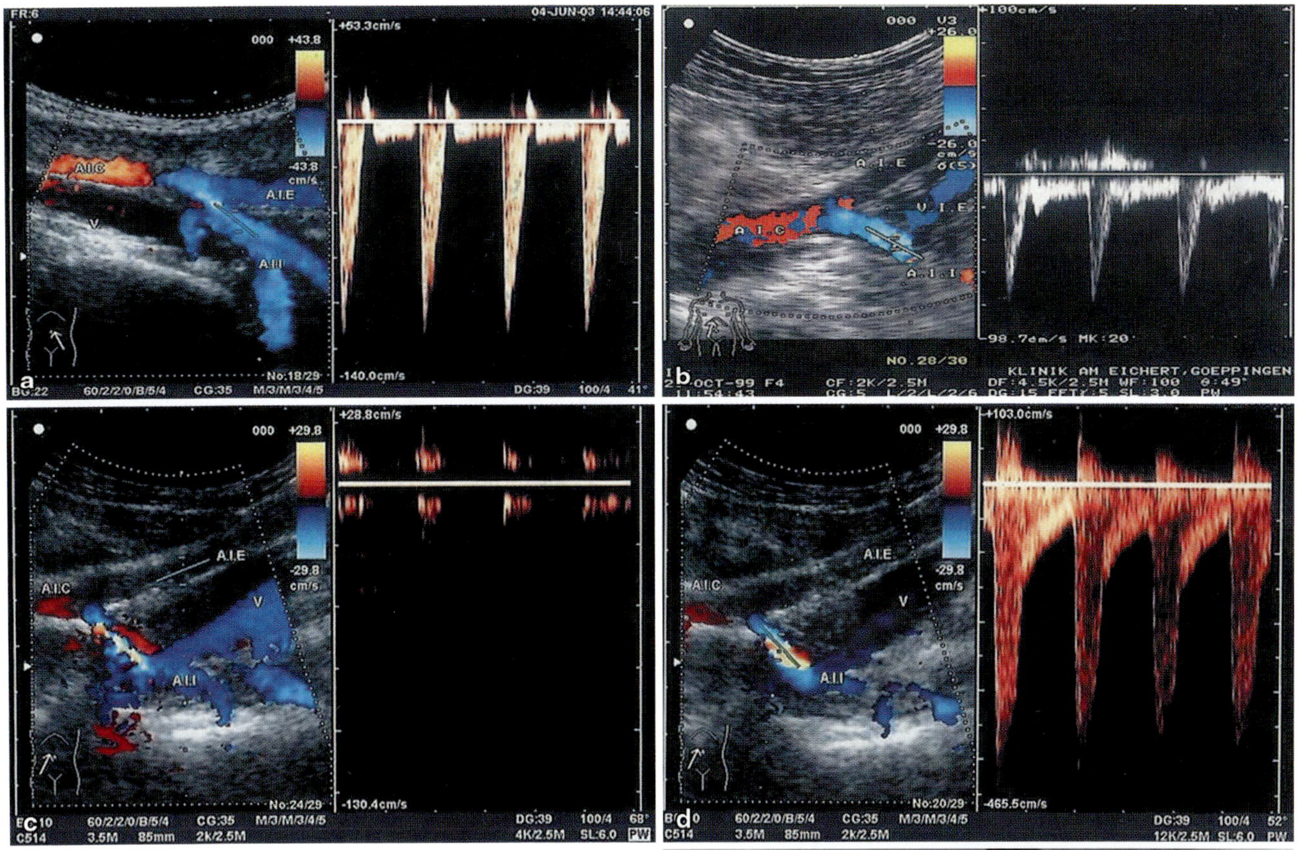


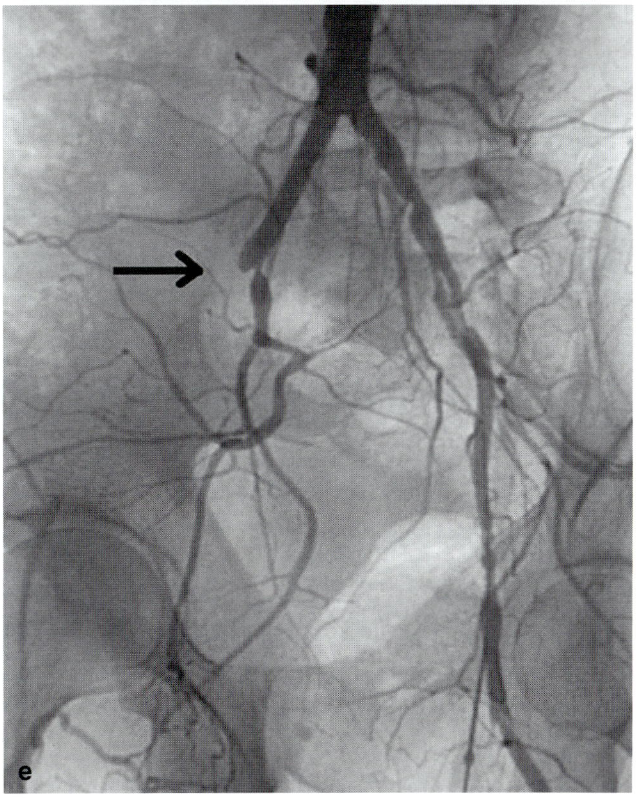


**Fig. A2.5 a–h**
**Iliac artery – stenosis – occlusion – collateral pathways**
**a** The iliac bifurcation with the origin of the internal iliac artery (*A.I.I*) is situated at the deepest point of the true pelvis. The internal iliac artery courses posteriorly (blue, away from transducer, toward periphery). The waveform shows a pulsatile profile but with diastolic flow because the internal iliac artery empties into the pelvic vessels. The color change from red to blue in the bifurcation is due to the changed flow direction relative to the ultrasound beam. With the high PRF selected to depict fast arterial flow, no flow signals are obtained from the iliac vein (*V*) posterior to the artery. (*A.I.E* external iliac artery, *A.I.C* common iliac artery).
**b** Exact localization of an occlusion is crucial for planning a bypass procedure. In case of external iliac artery occlusion (*A.I.E*), the internal iliac artery (*A.I.I*; blue, away from transducer) is an important collateral pathway.
**c**, **d** 54-year-old patient with intermittent claudication with a short walking distance and erectile dysfunction (see Chap. 7) due to external iliac artery occlusion (Doppler waveform with wall pulsation but no flow signals) and concomitant high-grade internal iliac stenosis (aliasing and PSV of 4 m/s).
**e** Oblique angiographic projection depicting right-sided occlusion of the external iliac artery and internal iliac stenosis. The internal iliac artery stenosis on the *left* is obscured by superimposed structures.

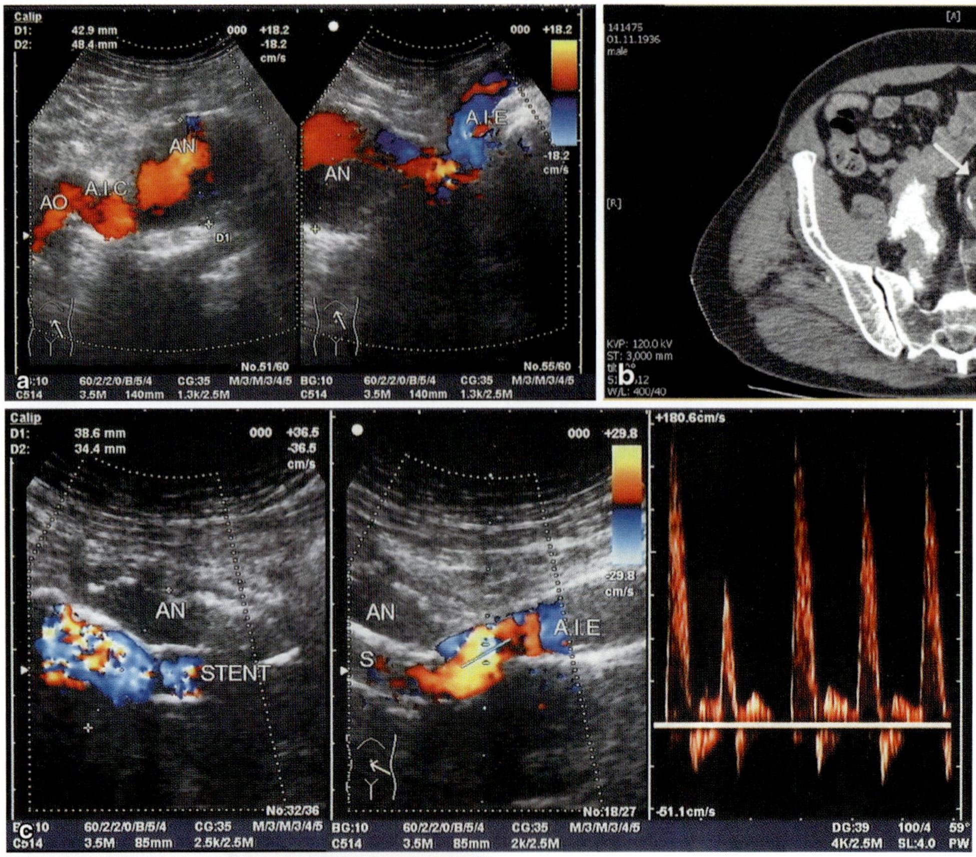


**Fig. A2.8 a–f**
**Iliac artery aneurysm – stent**
**a** Partially thrombosed common iliac artery aneurysm. The *left* section shows the origin of the common iliac artery from the aorta (*Ao*) and the aneurysm (*An*). The *right* section shows the partially thrombosed aneurysm (*An*) and the iliac bifurcation (*A.I.E.* external iliac artery).
**b** CT scan of the partially thrombosed iliac aneurysm.
**c** Color duplex imaging after endovascular repair with a covered stent demonstrates normal flow in the stent. No signs of endoleak or stenosis. The waveform shown is from the distal stent end.

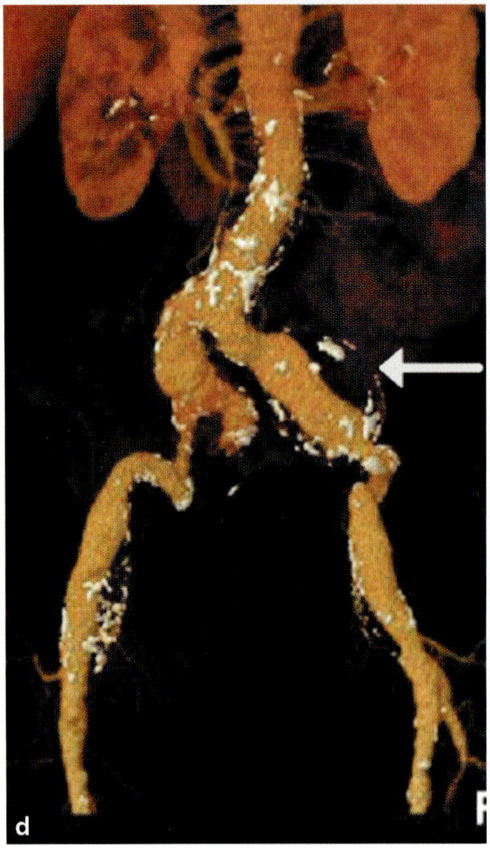

**d** CT angiography (3D reconstruction) confirms elimination of the iliac artery aneurysm (*arrow*) after stent placement.

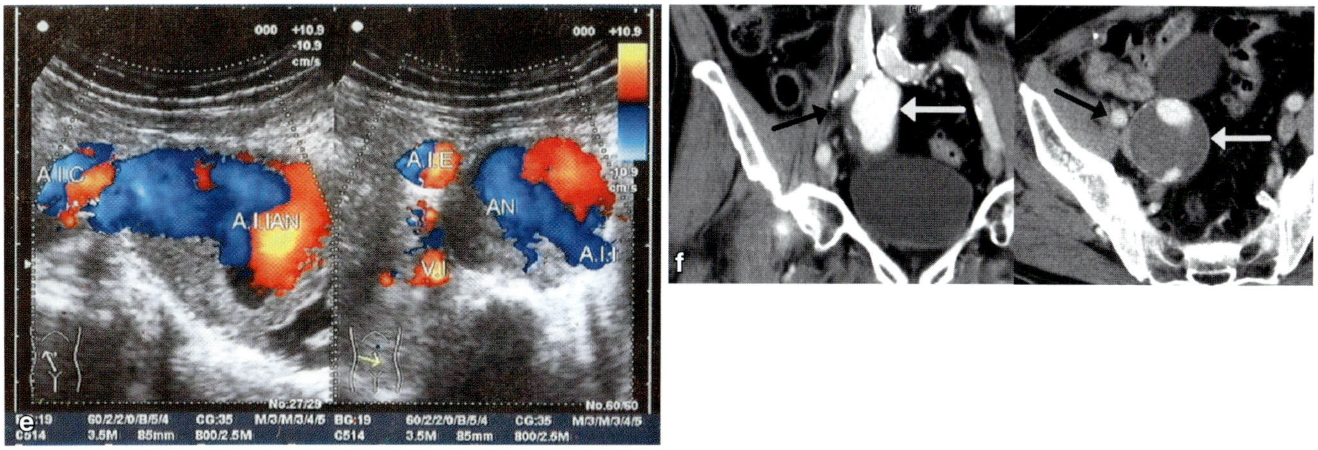


**Fig. A2.8 (continued)**
**Internal iliac artery aneurysm**
**e** Longitudinal (*left*) and transverse (*right*) images of a long, partially thrombosed aneurysm (*AN*) close to the origin of the internal iliac artery (*A.I.I*) with a diameter of 4 cm. The longitudinal image also depicts the origin of the internal iliac artery from the common iliac artery (*A.I.C*). The transverse image shows the aneurysm just distal to the bifurcation; it is adjacent to the external iliac artery (*A.I.E*) but does not communicate with it. The external and internal iliac veins (*V.I.E* and *V.I.I*) are seen deep to the external iliac artery.
**f** Axial and longitudinal CT scans showing the partially thrombosed internal iliac artery aneurysm (*white arrow*) adjacent to the external iliac artery (*black arrow*).

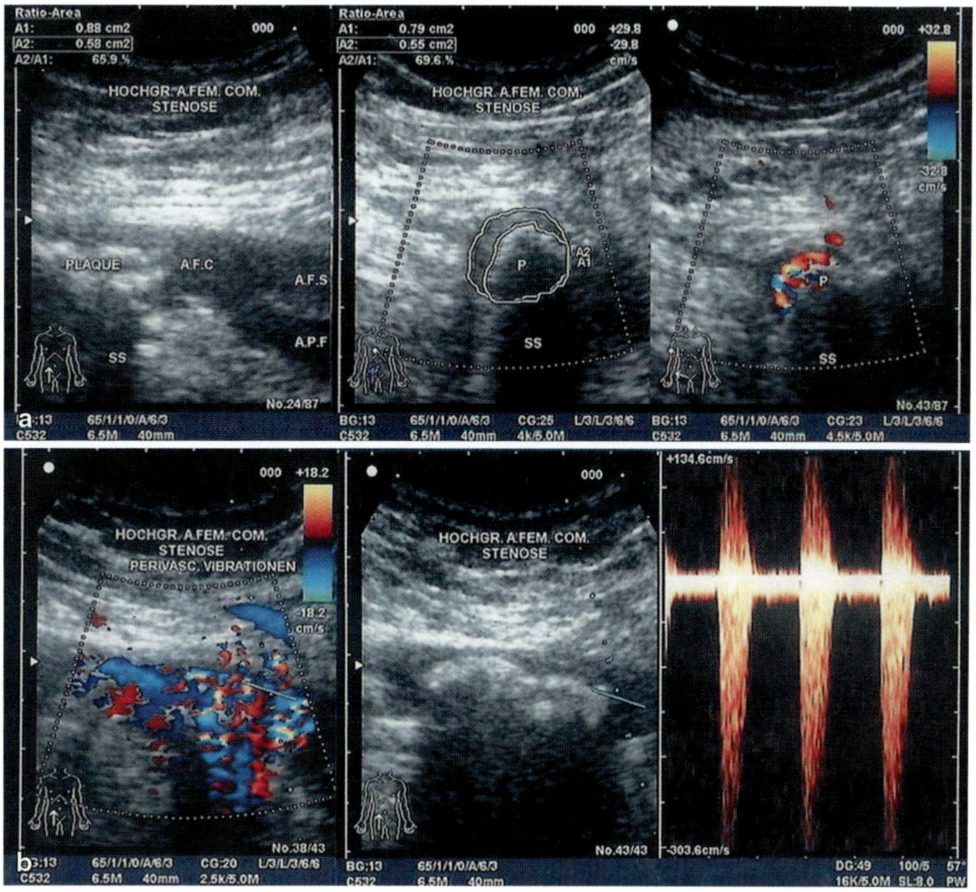


**Fig. A2.9 a–d**
**Common femoral artery stenosis**
**a** In the common femoral artery, stenosing plaque chiefly involves the posterior wall. The B-scan (right leg) depicts the extent and morphology of the plaque with acoustic shadowing (*SS*). The transverse scans (B-scan in the *middle*, color duplex scan on the *right*) illustrate the problem of estimating the degree of stenosis on the basis of cross-sectional area reduction (vessel cross section/plaque cross section; planimetric measurement: 69.9%). Acoustic shadowing limits evaluation of the artery and plaque contour in the B-mode, precluding accurate differentiation of the atherosclerotic wall lesion from the patent lumen. In the color duplex mode, plaque can be obscured by vibration artifacts and color overflow.
**b** Perivascular vibration artifacts may preclude the differentiation of the patent lumen from plaque by color duplex scanning in large arteries with high-grade stenosis (*left section*: longitudinal color scan; *middle section*: corresponding B-scan). Using the B-mode information for plaque localization, in conjunction with color duplex visualization of the stenosis jet, facilitates placement of the sample volume in the jet for recording a representative Doppler spectrum. Here, PSV > 3.5 m/s, pronounced turbulence, and monophasic flow suggest high-grade stenosis.

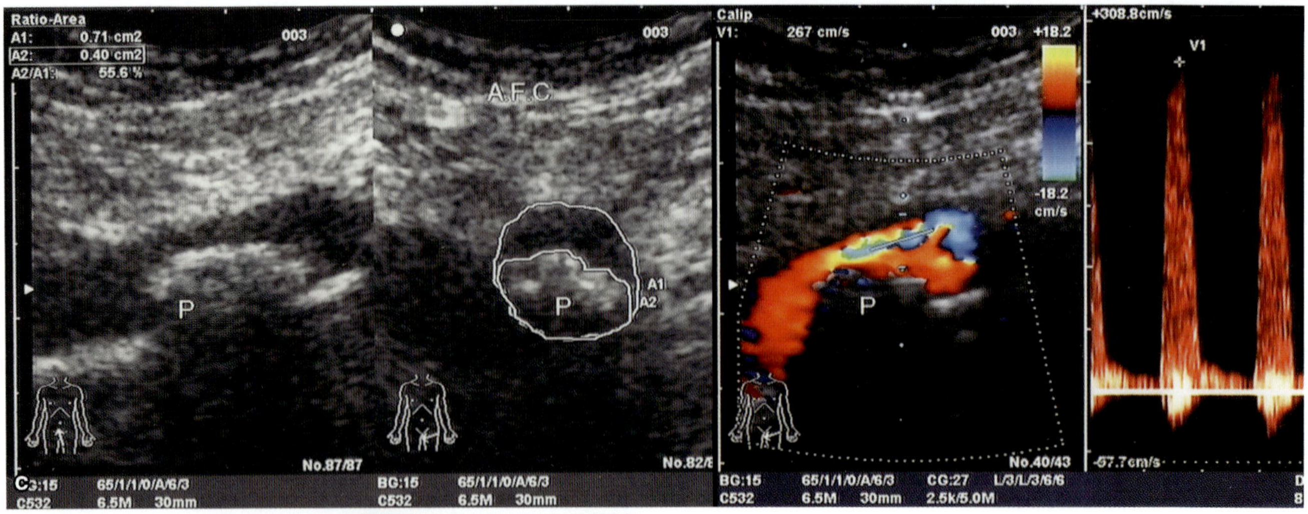


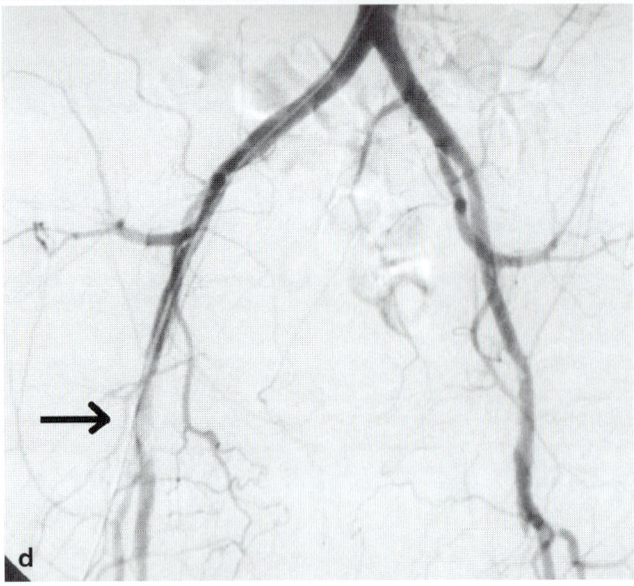


**Fig. A2.9 (continued)**

**c** On the left side, a posterior plaque (*P*) protrudes into the lumen of the distal external iliac artery at the junction with the common femoral artery (longitudinal and transverse views in the *left part* of the figure). The calcified plaque produces strong echoes and causes intermediate-grade stenosis (with roughly 56% luminal narrowing based on planimetric measurement). The color duplex image and the Doppler waveform (*right part*) confirm nearly 60% luminal narrowing with a PSV of 260 cm/s.

**d** Angiogram showing luminal narrowing of the common femoral arteries, which is more severe on the *right* side (*arrow*). Posterior plaque is common in this segment and, in the example, precludes angiographic quantification of the stenosis in the *left* common femoral artery (unless the anteroposterior view is supplemented by additional projections).

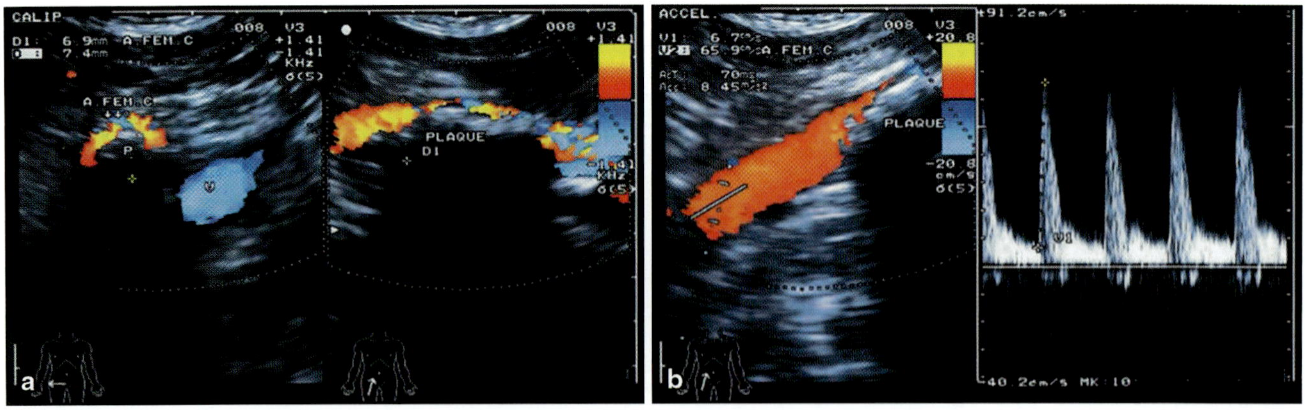


**Fig. A2.10 a–h**

**Common femoral artery stenosis/occlusion**

**a** High-grade stenosis of the common femoral artery involving the bifurcation. The stenosis is caused by a calcified plaque (*P*) with posterior shadowing and protrusion into the lumen. Transverse view of the artery on the *left* and longitudinal view on the *right*. The common femoral vein (*V*) is seen in blue on the medial side of the artery. The high grade of the stenosis is reflected by the color change from red to yellow to blue (aliasing).

**b** The effects of the high-grade stenosis on the prestenotic flow pattern (pulsatility) in the external iliac artery are chiefly caused by the collaterals arising upstream of the stenosis and by peripheral dilatation.

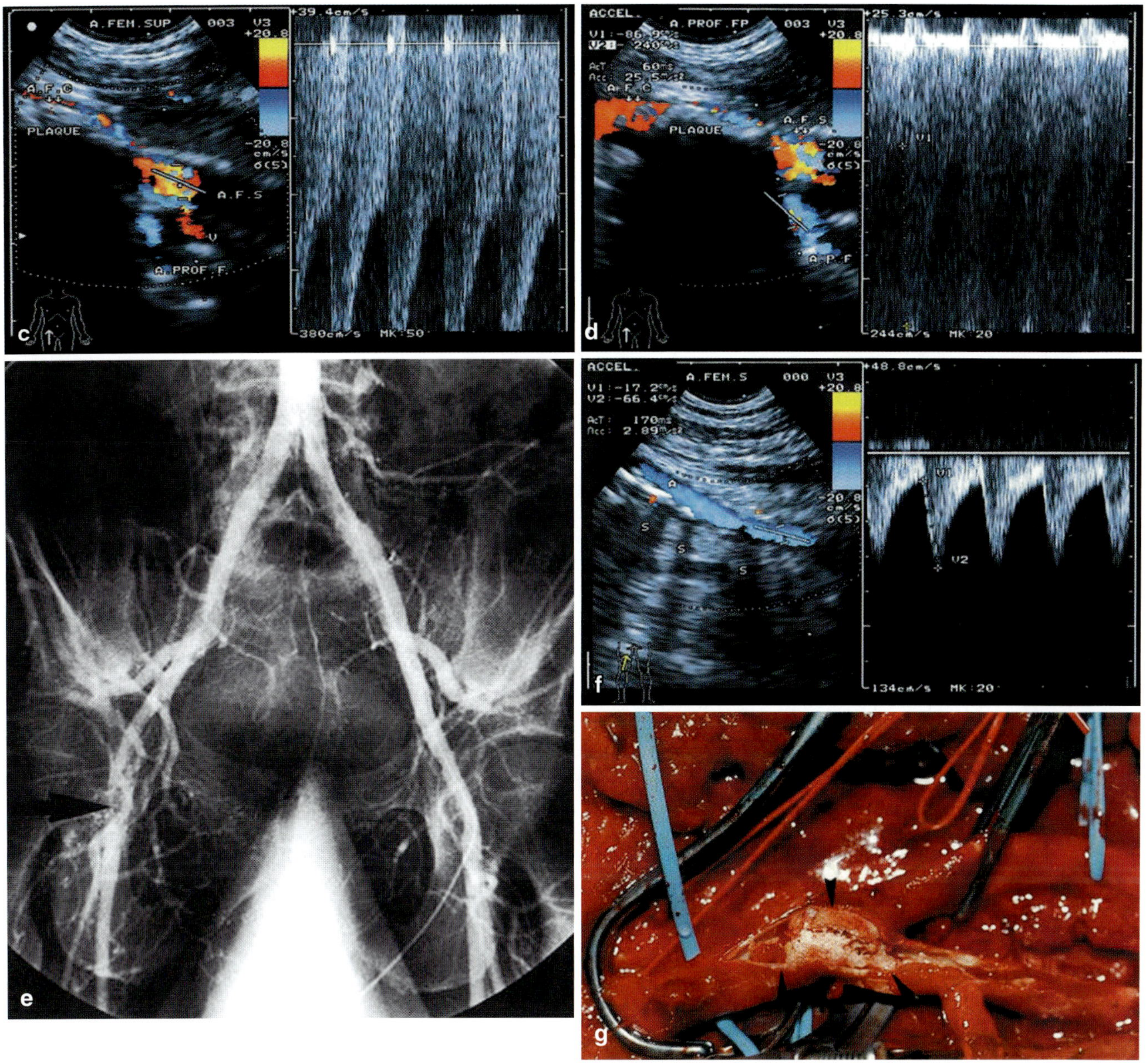


**Fig. A2.10 (continued)**

**c** The Doppler waveform from the proximal superficial femoral artery (*A.F.S*) immediately distal to the plaque reflects the high-grade stenosis of the bifurcation. The velocity peaks are cut off (aliasing). PSV is >380 cm/s with an end-diastolic velocity >200 cm/s. Posterior to the artery, the vein (*V*) is seen in red and the profunda femoris in blue (*A.PROF.F*). **d** Acoustic shadowing due to the calcified plaque makes it very difficult to assess the stenosis involving the origin of the profunda femoris in both the gray-scale scan and the color image. A very high gain is required to obtain a Doppler measurement that confirms the stenosis. There is monophasic flow with a PSV of 240 cm/s and an end-diastolic velocity of 86 cm/s. **e** Angiogram: The high-grade stenosis of the *right* common femoral artery has led to extensive collateralization compared with the unaffected side. Because it is caused by posterior plaque, the stenosis is difficult to grade in the anteroposterior projection alone and would require a second imaging plane.

**f** The superficial femoral artery shows the typical features of poststenotic flow with markedly reduced pulsatility, delayed and reduced systolic upslope, and monophasic flow. The gray-scale scan on the *left* depicts atherosclerotic changes in the form of hyperechoic plaques with acoustic shadowing (*S*), which do not cause stenosis. **g** With duplex scanning ruling out pelvic artery stenosis and demonstrating patency down to the popliteal artery, a planned TEA with patching of the femoral bifurcation could have been performed without preoperative angiography. As illustrated by this example, angiography does not add any relevant therapeutic information and merely serves to document the findings. The intraoperative site confirms the high-grade stenosis caused by calcified plaques (*arrowheads*). The artery is exposed from the common femoral to the profunda femoris artery. The origin of the superficial femoral artery with the lumen still constricted by the plaque is clamped off with straight forceps (*right side* of photograph). A round clamp is fixed around the proximal common femoral artery.

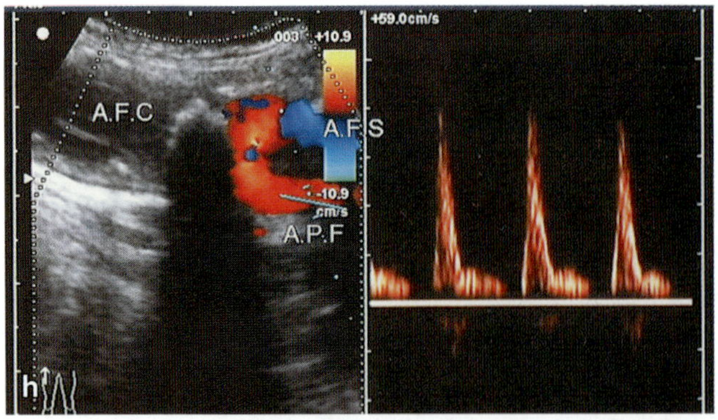


**Fig. A2.10 (continued)**

**Common femoral artery occlusion with collateralization**

**h** Occlusion (absence of flow signals) of the common femoral artery (*A.F.C*) with refilling of the superficial femoral artery (*A.F.S*; forward flow coded in blue, away from transducer) through the profunda femoris artery (*A.P.F*), which exhibits backward flow at its origin (red, toward transducer) and is supplied by the femoral circumflex artery. These findings indicate good collateralization. The flow profile is postocclusive (monophasic waveform with delayed systolic upstroke). The occlusion (with plaque and acoustic shadowing) is caused by a thrombosed aneurysm (2 cm) seen distal to the occluded common femoral artery segment.

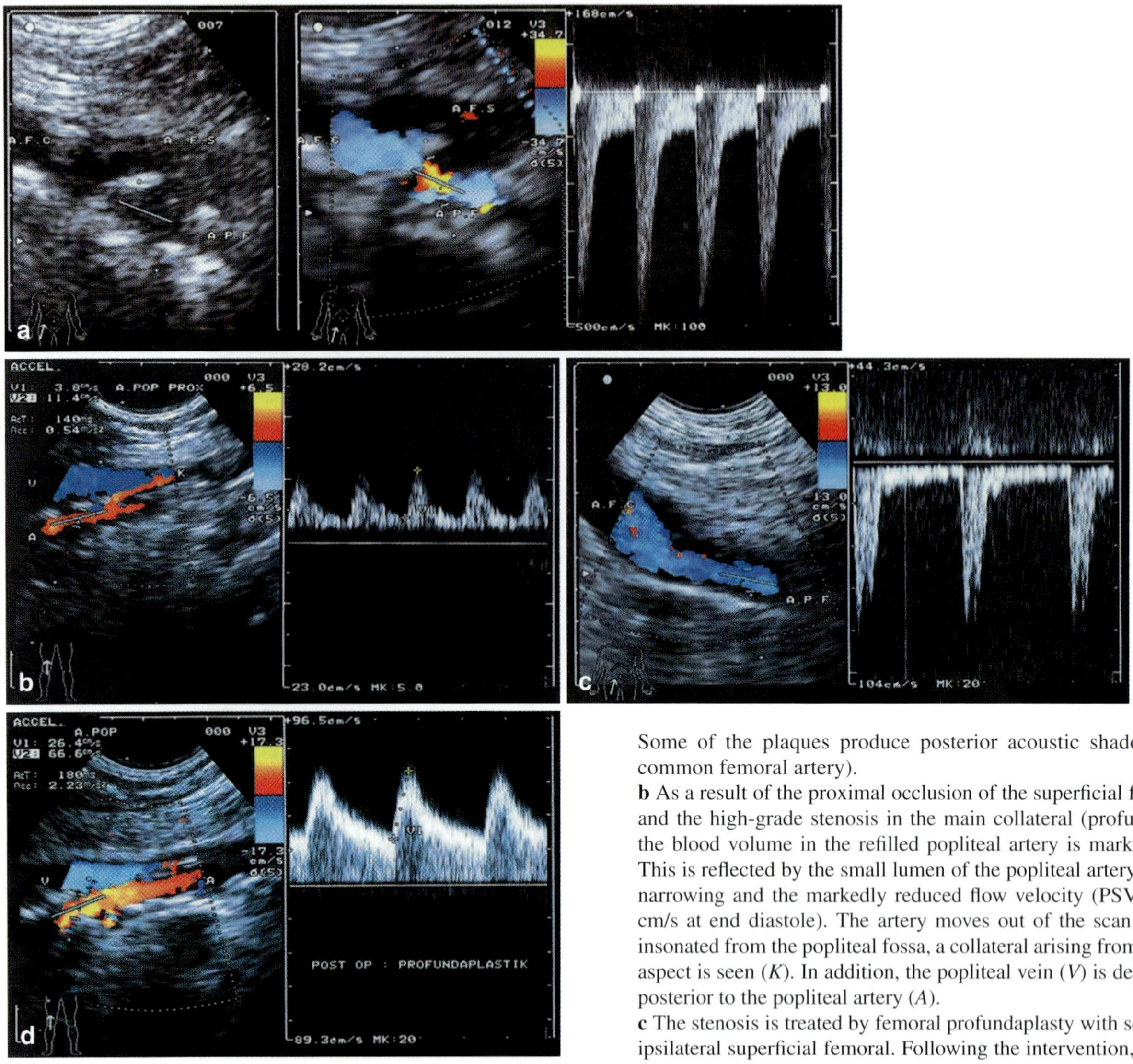


**Fig. A2.11 a–d**

**Stenosis at origin of profunda femoris artery – TEA**

**a** High-grade stenosis of the profunda femoris artery (*A.P.F*) with a monophasic flow profile and a peak velocity of 480 cm/s in systole and 90 cm/s in end diastole. In the color duplex image (*middle section*), the flow acceleration produces aliasing. The superficial femoral artery (*A.F.S*) is occluded; only the distal end (about 1 cm) is patent, but the slow flow is not detected with the high PRF used, and only some retrograde flow (red) is recorded. The gray-scale scan (*left section*) depicts plaques of different echogenicity with marked wall irregularities. Some of the plaques produce posterior acoustic shadowing (*A.F.C* common femoral artery).

**b** As a result of the proximal occlusion of the superficial femoral artery and the high-grade stenosis in the main collateral (profunda femoris), the blood volume in the refilled popliteal artery is markedly reduced. This is reflected by the small lumen of the popliteal artery with chronic narrowing and the markedly reduced flow velocity (PSV of 11 and 3 cm/s at end diastole). The artery moves out of the scan plane. When insonated from the popliteal fossa, a collateral arising from the posterior aspect is seen (*K*). In addition, the popliteal vein (*V*) is depicted in blue posterior to the popliteal artery (*A*).

**c** The stenosis is treated by femoral profundaplasty with severing of the ipsilateral superficial femoral. Following the intervention, the profunda femoris artery (*A.P.F*) shows a peak velocity in the area of the patch of 80 cm/s during systole and 10 cm/s at end diastole. The diastolic flow component and the reduced pulsatility are due to collateral flow in the profunda femoris and the difference in wall elasticity in the area of the patch. (*A.F.C* common femoral artery).

**d** Improved perfusion, following profunda femoris repair with persistent superficial femoral occlusion is reflected in the waveform obtained from the refilled femoral artery at about the same site as the preoperative waveform presented above, now showing a peak velocity of 66 cm/s in systole and 26 cm/s at end diastole. The postocclusive flow character is due to persistent superficial femoral occlusion.

**Fig. A2.12 a–e**

**Stenosis at origin of profunda femoris artery (recurrence)**

**a** Duplex ultrasound has become the method of choice for diagnosing and grading stenosis of the profunda femoris artery, as anteroposterior angiograms are limited for various reasons: the femoral bifurcation may be obscured by overlying vessels, variants in the course of the artery may not be assessable, and stenosis caused by posterior wall plaque is difficult to grade. In this patient with prior TEA of the common femoral artery, aliasing in the color duplex scan and a PSV > 3 m/s with a monophasic flow profile indicate recurrent high-grade stenosis.

**b** The corresponding angiogram depicts the stenosing plaque at the origin of the profunda femoris. Recurrent stenosis and wide lumen of the common femoral artery following TEA.

**Distal profunda femoris stenosis**

**c** Distal profunda femoris stenosis only requires treatment if it involves the main branch of the artery, which courses parallel to the superficial femoral artery and is thus relevant as a potential collateral in case of superficial femoral artery occlusion. The color duplex image (*left*) shows high-grade stenosis of the profunda femoris artery approximately 4 cm from its origin with a Doppler-derived PSV > 5 m/s. The proximal segment of the superficial femoral artery (*A.FEM.S.*) is patent. The angiogram confirms the stenosis in the distal profunda femoris artery and a patent proximal superficial femoral artery with an occlusion in the lower thigh. The angiogram also allows clear differentiation between the main trunk of the profunda femoris, which is relevant as a collateral in superficial femoral artery occlusion, and a second branch of the profunda femoris arising posteriorly. The latter plays no role as a collateral in superficial femoral artery occlusion; it supplies the upper thigh muscles and receives the circumflex artery (providing arterial flow in case of occlusion of the common femoral or external iliac artery). In a patient with superficial femoral artery occlusion, the sonographic examination cannot be confined to the origin of the profunda femoris but must include a length of 7–8 cm to identify any therapeutically relevant stenosis.

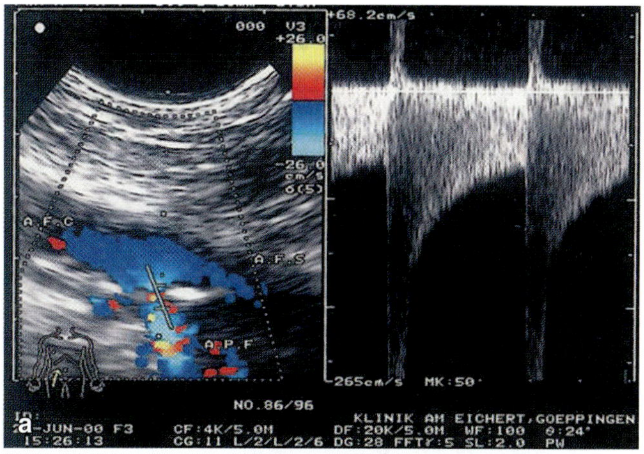


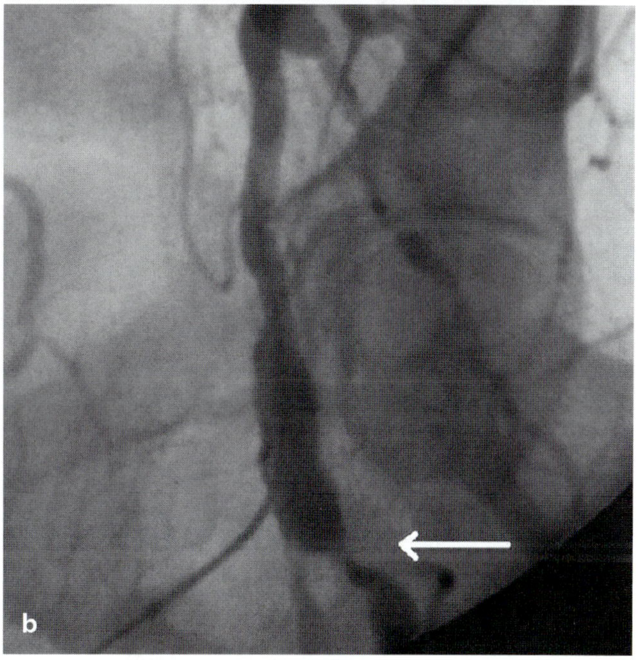


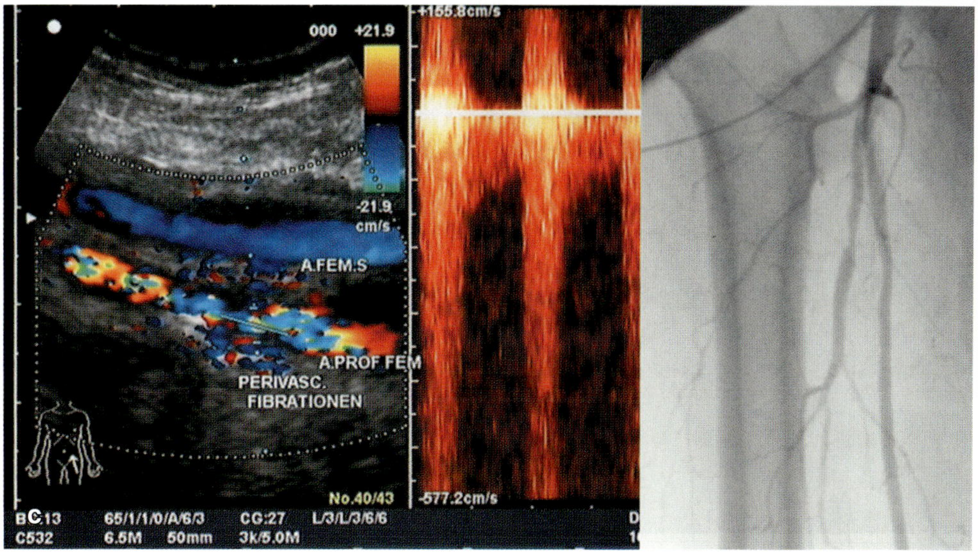

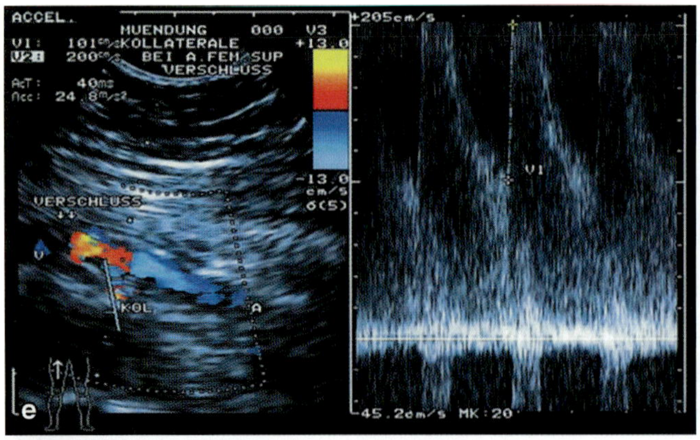


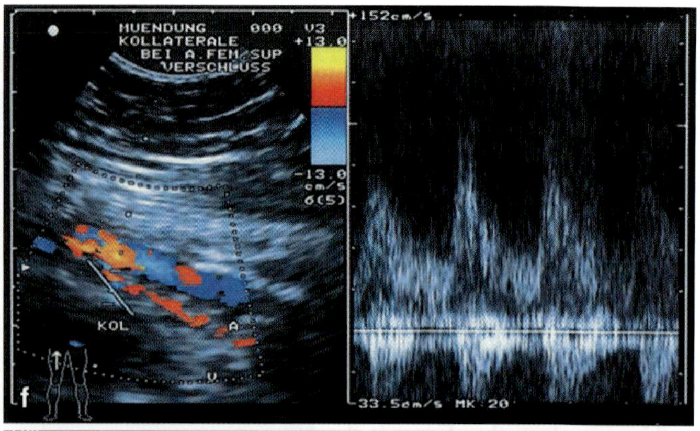


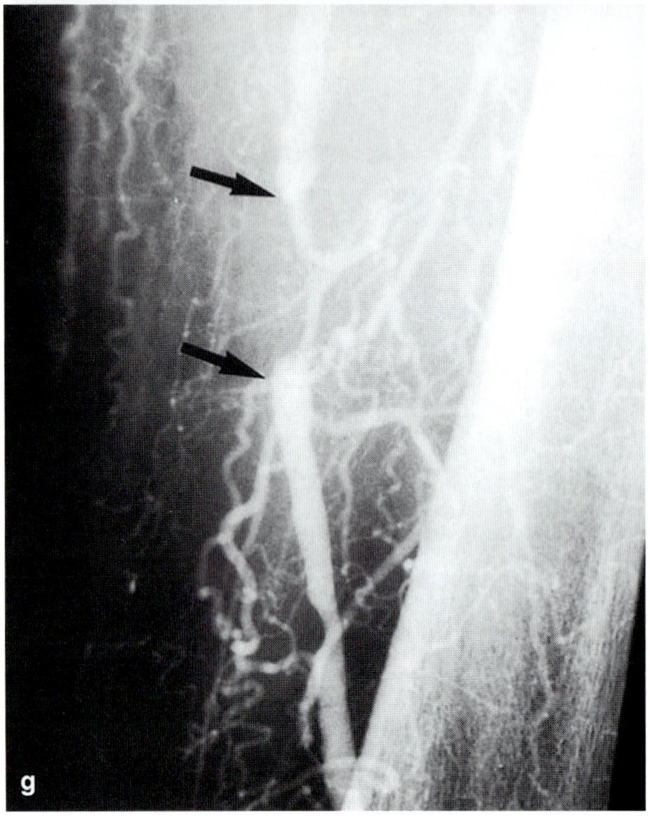


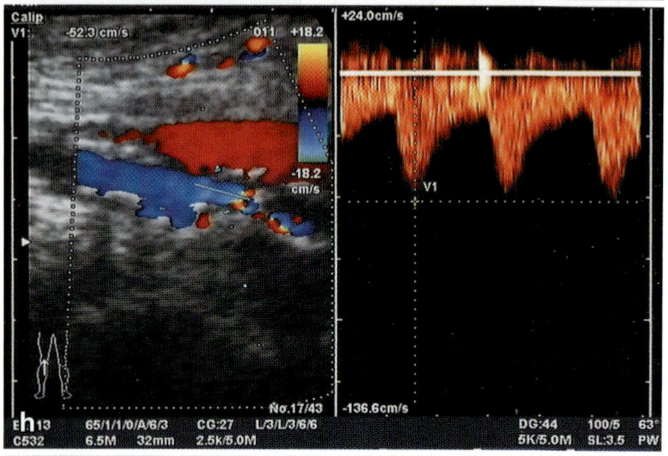


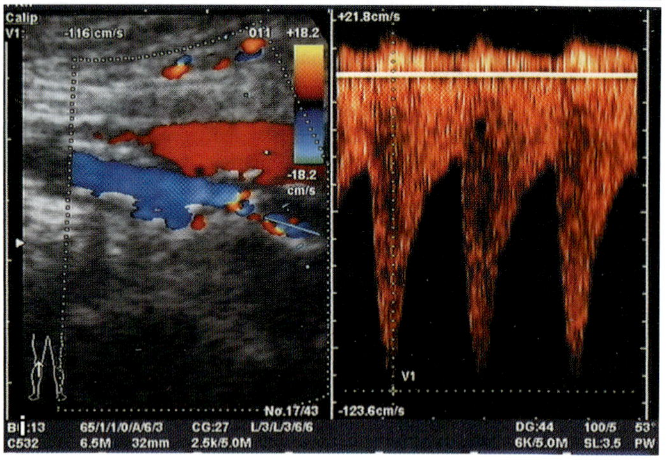


**Fig. A2.14 (continued)**

**e, f** Just proximal to the refilled segment, two further collaterals (*KOL*) empty into the superficial femoral on its posterior aspect with flow toward the transducer. In **f** a long segment of the collateral is depicted in red while the superficial femoral is shown in blue (flow away from transducer). With a PSV of 95 cm/s, this collateral is not stenosed whereas the second collateral **e** entering the artery more proximally shows criteria of stenosis at its site of entry on duplex ultrasound and in the Doppler waveform (aliasing, PSV of 100 cm/s at end diastole and over 250 cm/s in systole). The occlusion is indicated by *arrows* in **e**. Retrograde flow components in the superficial femoral artery are displayed in red.

**g** Angiogram: Confirmation of the 2-cm occlusion of the superficial femoral artery. Also seen are the anterior collateral pathway and the two collaterals emptying into the artery on its posterior aspect (*distal arrow*). The latter are supplied by profunda femoris collaterals.

**h** Downstream stenosis in the distal popliteal artery (P3 segment). The Doppler waveform obtained distal to the entry of the collaterals bridging the occlusion shows postocclusive flow (delayed systolic upstroke, monophasic flow) with a PSV of 52 cm/s. The downstream stenosis is indicated by aliasing.

**i** Doppler interrogation of the suspected stenosis (aliasing) reveals a focal increase in PSV to 116 cm/s. This increase alone does not indicate a relevant stenosis; however, in conjunction with the postocclusive decrease in flow velocity to 50 cm/s between the occluded femoral artery segment and the popliteal stenosis, a 50–60% stenosis is suggested. A PSV ratio (intrastenotic to prestenotic PSV) > 2 was calculated.

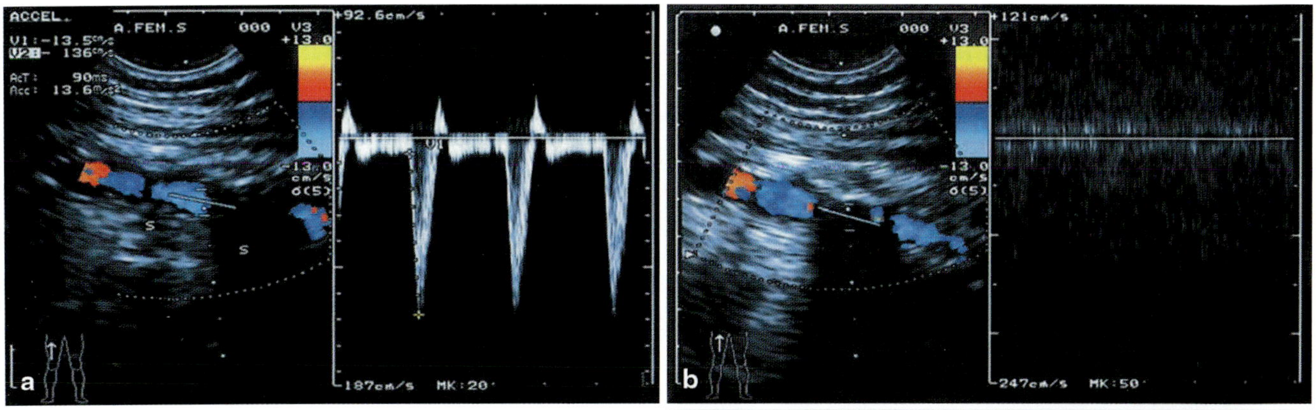


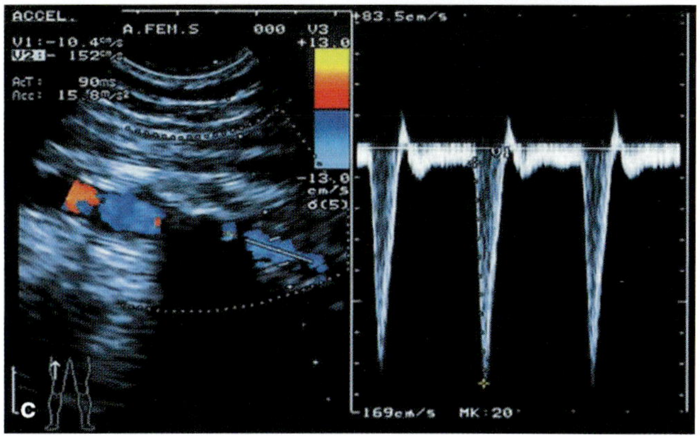


**Fig. A2.15 a–c**
**Artifact due to acoustic shadowing**

**a** In contrast to the example presented in Fig. A2.14a–g, the absence of flow signals along a 1-cm segment of the superficial femoral artery in this case is not due to occlusion but to acoustic shadowing produced by a calcified plaque. Just proximal to this segment, there is pulsatile, triphasic flow with a PSV of 136 cm/s.

**b** Neither color duplex scanning nor the Doppler waveform depicts flow in the segment obscured by acoustic shadowing.

**c** The Doppler waveforms distal and proximal to the obscured segment are identical, excluding a higher-grade stenosis or occlusion of the nonvisualized segment. The slightly higher flow velocity of 152 cm/s may be due to a moderate lumen reduction or an angle-related error.

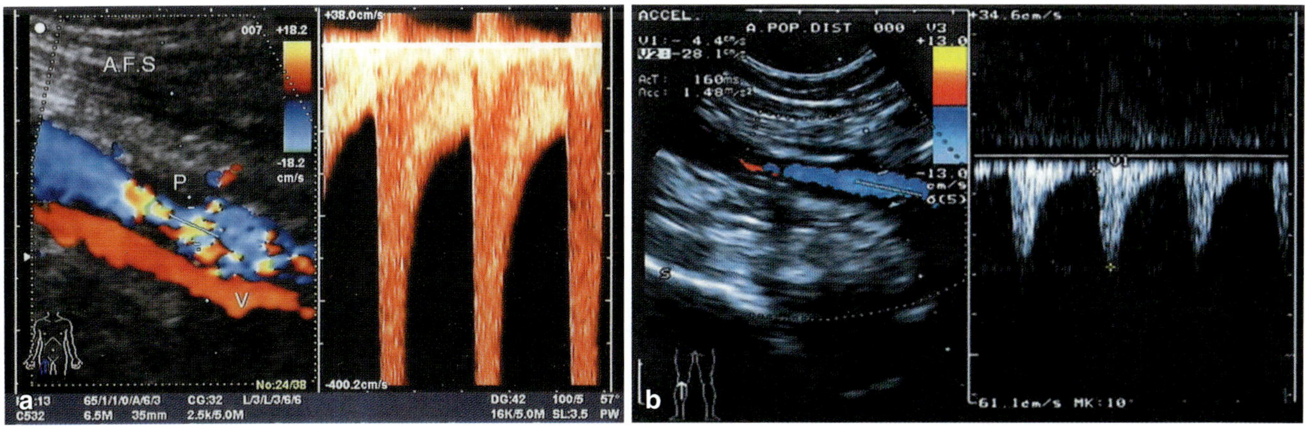


**Fig. A2.16 a–h**
**Follow-up after PTA**

**a** Patient with a long history of diabetes mellitus and stage III AOD. Besides medial sclerosis, there is a high-grade stenosis at the junction of the superficial femoral and popliteal arteries. The stenosis is associated with monophasic flow with a PSV of over 400 cm/s in the Doppler waveform and aliasing.

**b** In the distal popliteal artery downstream of the stenosis, the Doppler waveform shows the typical changes consisting of delayed systolic rise, reduced PSV (28 cm/s), and monophasic flow profile.

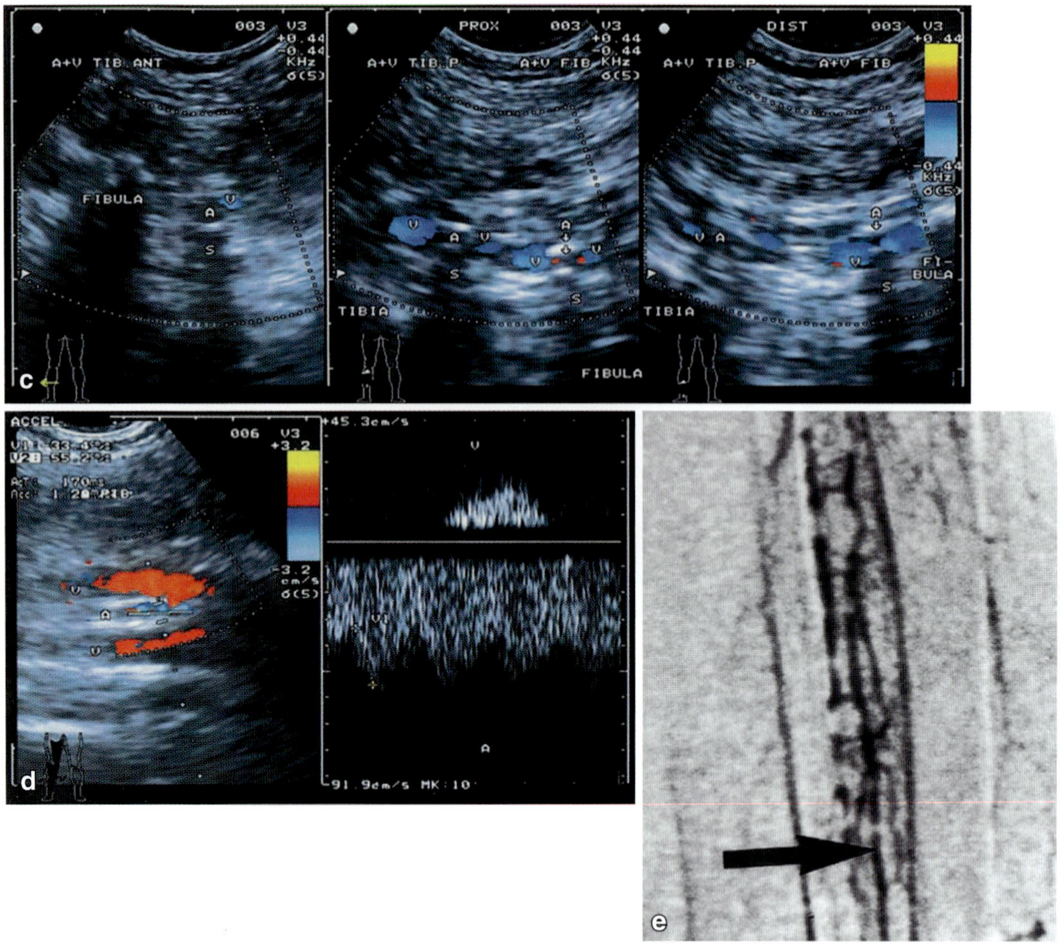


**Fig. A2.16 (continued)**

**c** The fibular artery is the only patent calf artery. The anterior view (*left section*) depicts the anterior tibial artery (*A*) on the medial side of the fibula; there is acoustic shadowing posterior to the artery due to medial sclerosis with calcifications. The good color filling of the vein (*V*, blue) confirms adequate instrument setting for the detection of the slower poststenotic flow. The posterior view (*middle section*) depicts the posterior tibial and fibular arteries in the distal lower leg in transverse orientation. With the transducer tilted, the veins are shown with flow away from the transducer (blue). Only residual flow near the wall, displayed in red, is detected in the fibular artery; this is due to plaque-induced acoustic shadowing of the middle segment. As a result of acoustic shadowing, the proximal fibular artery (*right section*) appears to be occluded in the color duplex mode.

**d** To confirm patency of the fibular artery, it must be scanned in longitudinal orientation to obtain a Doppler tracing with an adequate angle of incidence. Here, an attempt was made to avoid the acoustic shadows by moving and rotating the transducer. It was thus possible to also depict flow signals in the distal fibular artery. There is only sparse coloring of the arterial lumen despite a low PRF and high gain. In this view of the distal fibular artery, the accompanying veins anterior and inferior to the artery are displayed in red. The poststenotic Doppler waveform displays the arterial flow below the baseline (away from transducer) with an almost venous poststenotic character. The flow signal represented above the baseline is from an accompanying vein with flow toward the transducer.

**e** Angiogram: The patent distal segment of the fibular artery is indicated by an *arrow*. The posterior tibial artery is occluded, and the anterior tibial artery is filled with blood only along a short stretch of its middle segment, supplied by collaterals from the fibular artery.

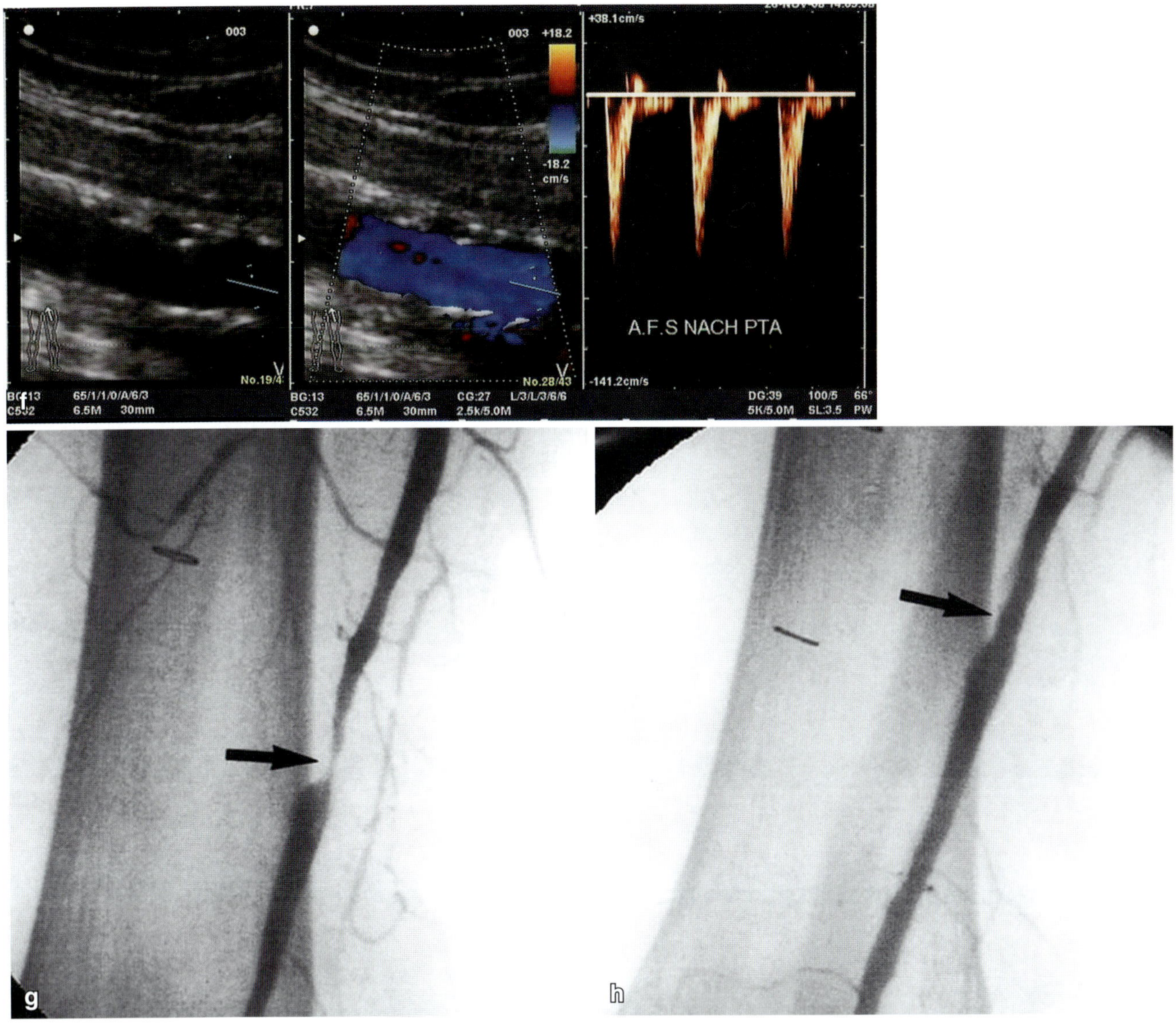


**Fig. A2.16 (continued)**
**f** Duplex scanning is an excellent follow-up modality after vascular interventions. After angioplasty, a PSV of 75 cm/s is measured at the site of the former stenosis (see **a**); the plaques are fragmented. The hypoechoic wall thickening may be due to subintimal hemorrhage

**g** Angiogram: High-grade stenosis at the femoropopliteal junction (see **a**). **h** Angiogram after angioplasty of the stenosis at the femoropopliteal junction (*arrow*) demonstrates successful dilatation. The angiogram corresponds to the duplex ultrasound findings in **f**.

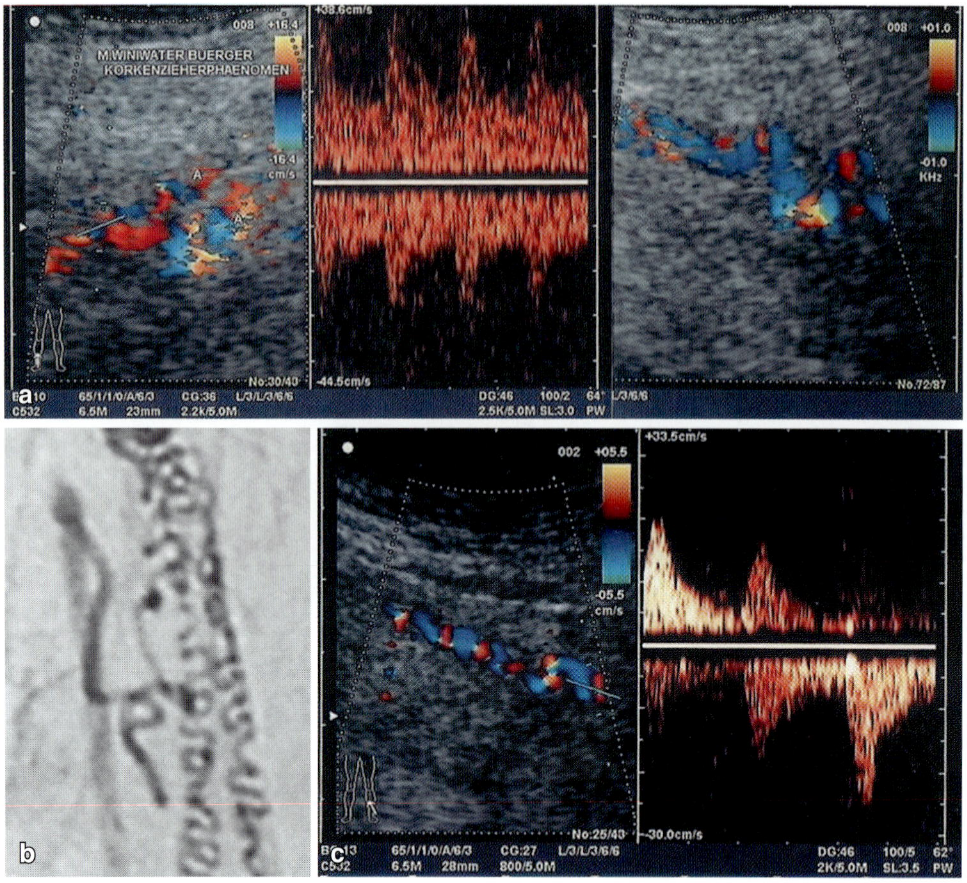


**Fig. A2.17 a–c**
**Buerger's disease (thromboangiitis obliterans)**
**a** Typical color duplex ultrasound appearance of corkscrew collaterals in Buerger's disease (*left* and *right* sections with vessel branching). The changes from red to blue reflect the changing flow directions of the blood relative to the ultrasound probe. The Doppler waveform obtained with placement of the sample volume in this segment depicts the typical pulsatile flow with flow components toward and away from the transducer (*middle section*).
**b** Angiographic appearance of corkscrew collaterals.
**c** Doppler waveform from a short stretch of a small collateral artery illustrates the characteristic changes in flow direction. This is pathognomonic and should not confuse the sonographer during the examination.

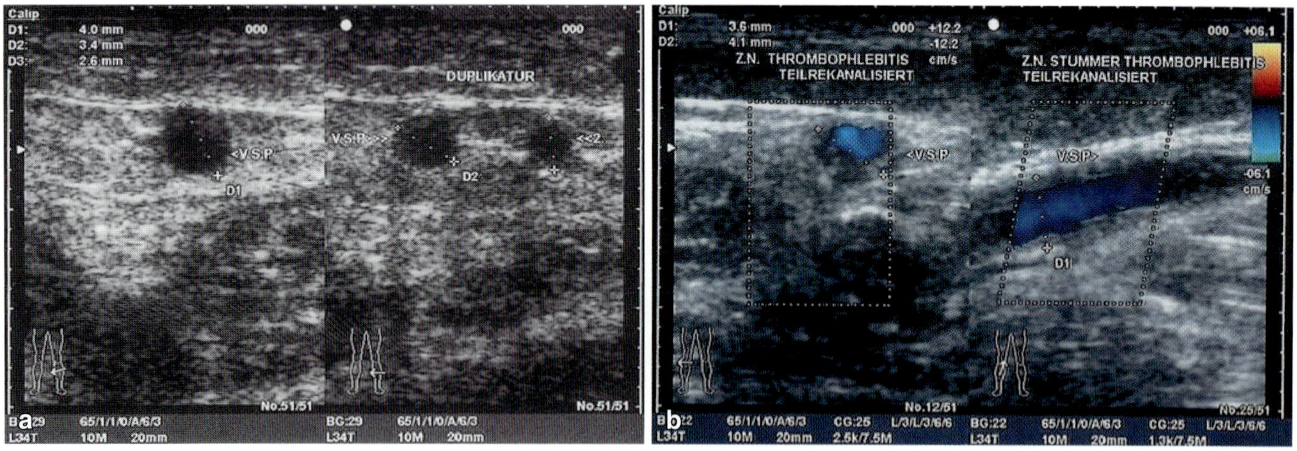


**Fig. A2.18 a, b**
**Planning of bypass procedure, graft selection (vein mapping), target vessel**
**a** Suitable veins for grafting can be selected preoperatively. This includes measurement of the diameter, which should be over 2 mm for a crural bypass. Preoperative marking of the course of the selected vein on the skin reduces the length of incision and shortens operation time. In case of duplication, the most suitable branch in terms of diameter and course is selected sonographically. The transverse view on the *left* shows a suitable small saphenous vein with a diameter of 4 mm and the scan on the *right* obtained more distally a duplicated vein with a thicker (3.4 mm) and a thinner branch (2.6 mm).
**b** Veins with postthrombophlebitic changes are unsuitable for grafting. These can be identified by sonography, which will demonstrate a patent lumen with sclerotic wall thickening, as illustrated here for the small saphenous vein. A recanalized thrombophlebitic vein shows the same features as a postthrombotic deep vein: wall sclerosis and thickening, residual thrombi, and valve incompetence. In the example, transverse and longitudinal views (*left* and *right*, respectively) depict flow in the patent lumen of the small saphenous vein (blue) and hypoechoic wall thickening.

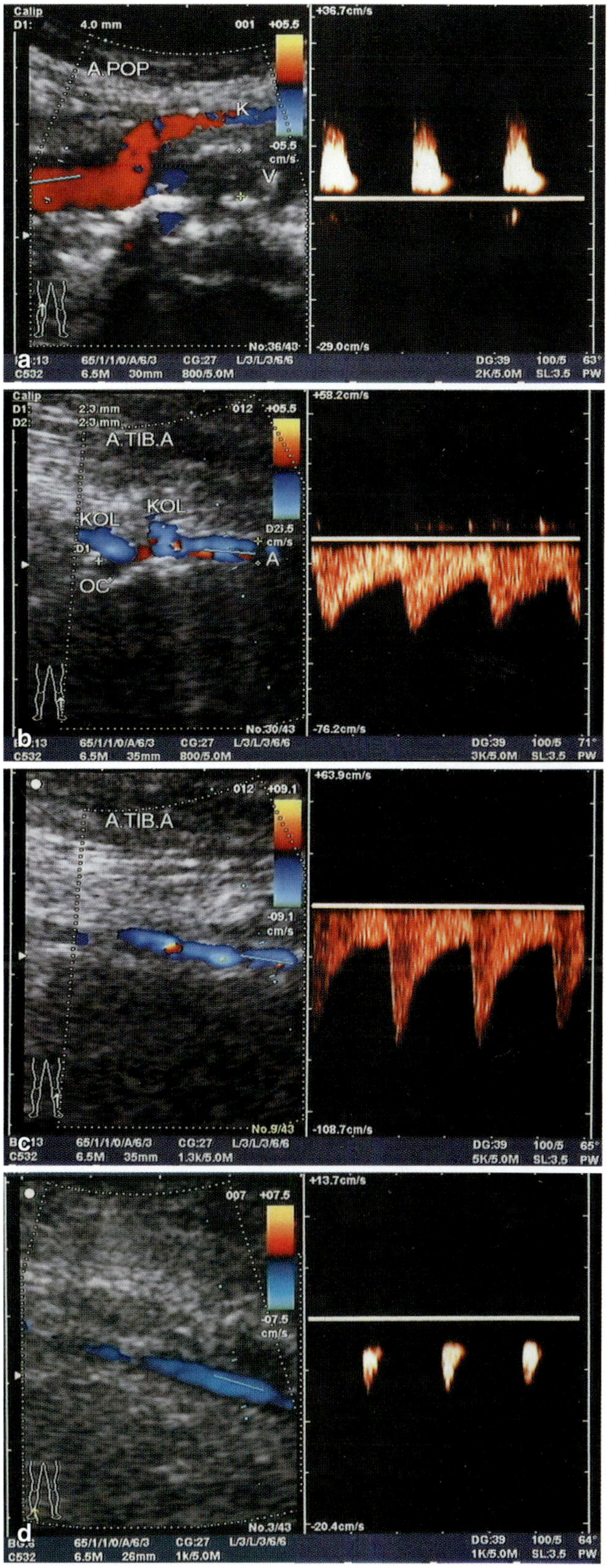


**Fig. A2.19 a–f**

**Selecting target vessel for distal bypass procedure**

**a** Color duplex imaging demonstrates occlusion of the P2 and P3 popliteal segments in a patient with grade IV AOD. A sural artery provides collateral flow (*K*). The Doppler waveform shows a preocclusive thump pattern.

**b** The search for a runoff vessel to position the distal anastomosis of a femorocrural bypass reveals collaterals (*KOL*) resupplying flow to the proximal tibial artery (*A*).

**c** The anterior tibial artery is then followed distally down to the ankle, where a Doppler spectrum is obtained. There is no stenosis, and the waveform pattern suggests good peripheral runoff, confirming the anterior tibial artery to be an ideal target vessel for a bypass procedure.

**d** Doppler measurements show the fibular artery and posterior tibial artery (with multiple stenoses and only a short patent segment) to be inadequate target vessels for the planned bypass: high pulsatility and low PSV indicate poor peripheral runoff. In this case, the Doppler findings already show these two arteries to be unsuitable to receive the bypass, and a complete examination is not necessary.

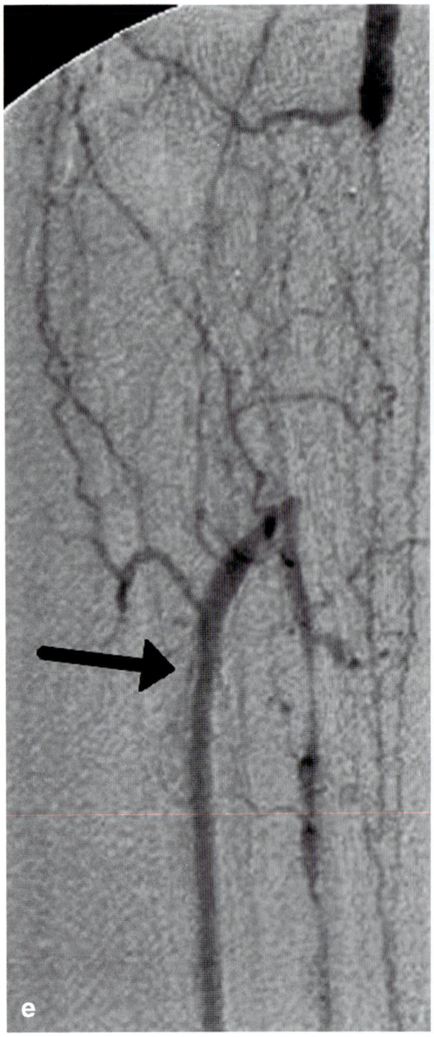


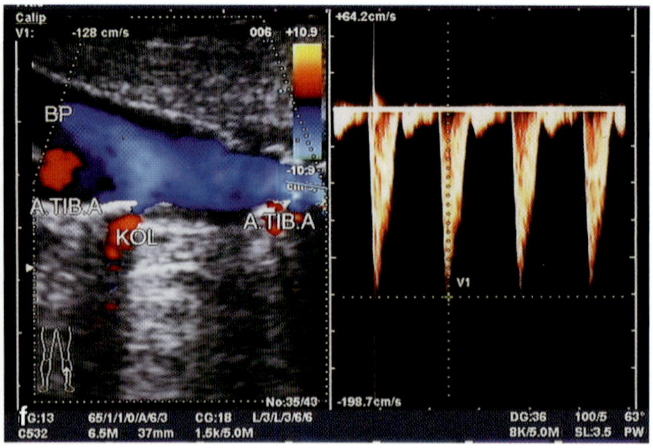


**Fig. A2.19 (continued)**

**e** Angiogram confirms popliteal artery occlusion and suitability of the anterior tibial artery for use in a distal bypass procedure.

**f** Doppler findings after bypass grafting onto the anterior tibial artery suggest restoration of peripheral perfusion: the anterior tibial artery waveform recorded just distal to the bypass anastomosis shows pulsatile flow and a PSV of 128 cm/s with a short systolic rise time.

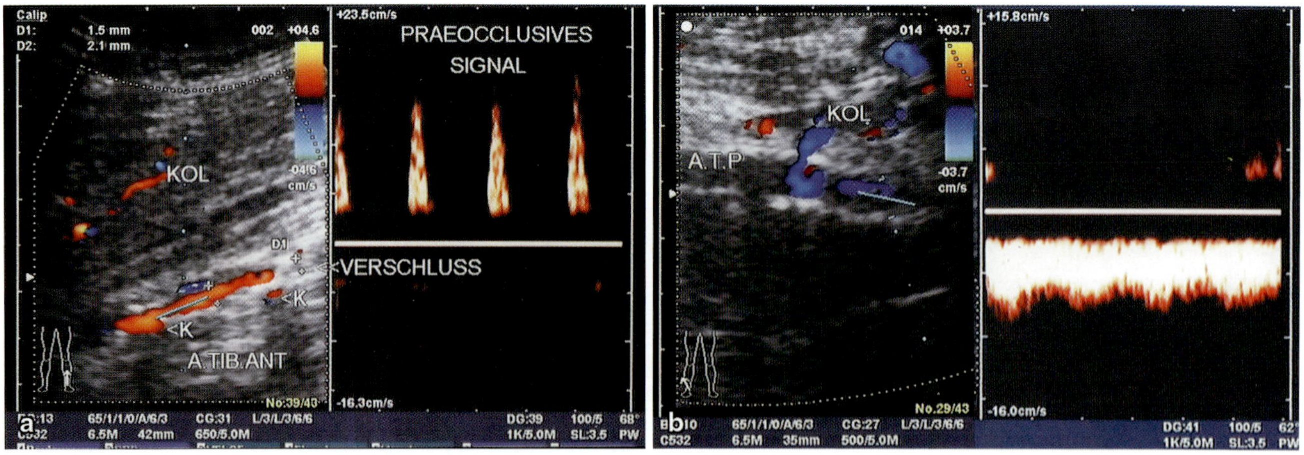


**Fig. A2.20 a–e**

**a** Duplex examination before a planned bypass procedure in stage IV AOD with occlusion of calf arteries. The peripheral arteries are patent down to and including the popliteal artery. Spectral Doppler analysis reveals absent diastolic flow (preocclusive thump) in the proximal anterior tibial artery between the origin of a collateral (*K*) and an occlusion demonstrated by color duplex. The diameter (2 mm) of the occluded artery is marked with plus signs. The occlusion extends down to the ankle joint.

**b** The posterior tibial artery is also occluded proximally (3.5 cm in length) with a collateral vessel (*KOL*) providing flow distal to the occlusion. Flow is diminished and very slow (PSV of 8 cm/s); the flow pattern resembles that in a vein, but the direction is away from the probe, indicating that an artery is being interrogated. In a situation like this, with occlusion and collateral refilling of the main artery, it is important to continuously scan the artery for another 4–5 cm to decide whether distal runoff can be evaluated by Doppler measurement, which will be the case if there is adequate resupply of blood through the collateral pathway.

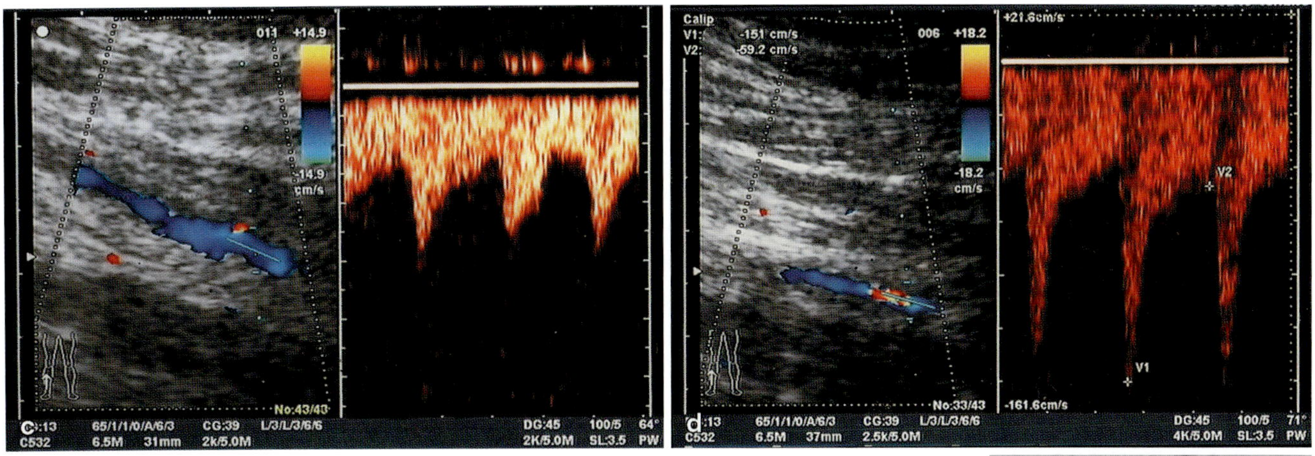


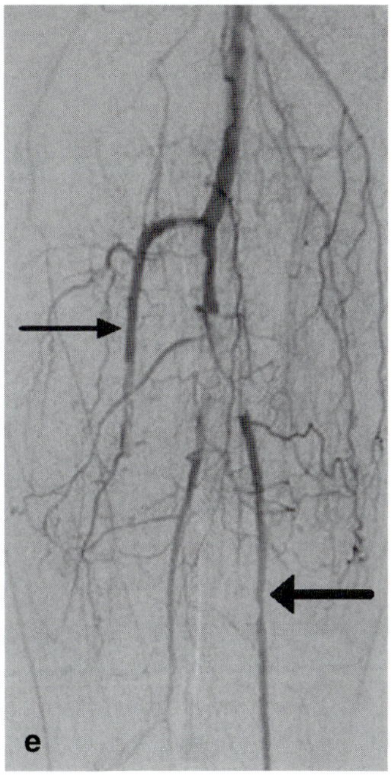

**Fig. A2.20 (continued)**

**c** More distally, flow is again increased due to inflow from additional collaterals (systolic velocity of 60 cm/s and diastolic velocity of 25 cm/s; delayed systolic upstroke).

**d** Two centimeters distal to the waveform presented in **c**, there is aliasing in the posterior tibial artery and Doppler interrogation reveals an increase in PSV of slightly more than 100% (150 cm/s), corresponding to 50–60% stenosis.

**e** Angiogram confirms occlusions of the anterior tibial artery and proximal posterior tibial artery. The latter is supplied via collaterals distal to the occluded segment. The *thick arrow* indicates a mildly narrowed segment of the posterior tibial artery (see **d**) and the *thin arrow* the anterior tibial artery

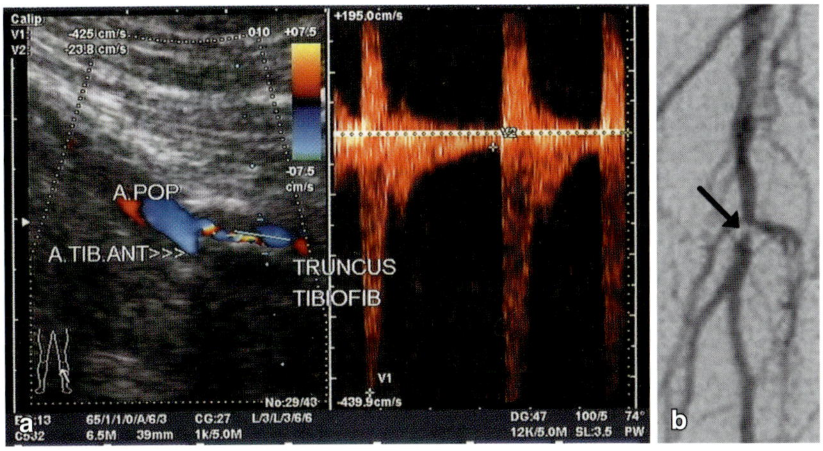


**Fig. A2.21 a–h**

**Stenosis of lower leg arteries**

**a** High-grade stenosis of the tibiofibular trunk just below the origin of the anterior tibial artery. The tibiofibular trunk can be visualized with the angled probe in the popliteal fossa. High-grade stenosis is suggested by aliasing in the color mode and confirmed by spectral Doppler analysis with a PSV > 4 m/s, monophasic flow pattern, and marked turbulence. In such cases, a clinically indicated PTA can be performed without prior diagnostic angiography.

**b** Angiogram confirms high-grade stenosis of the tibiofibular trunk.

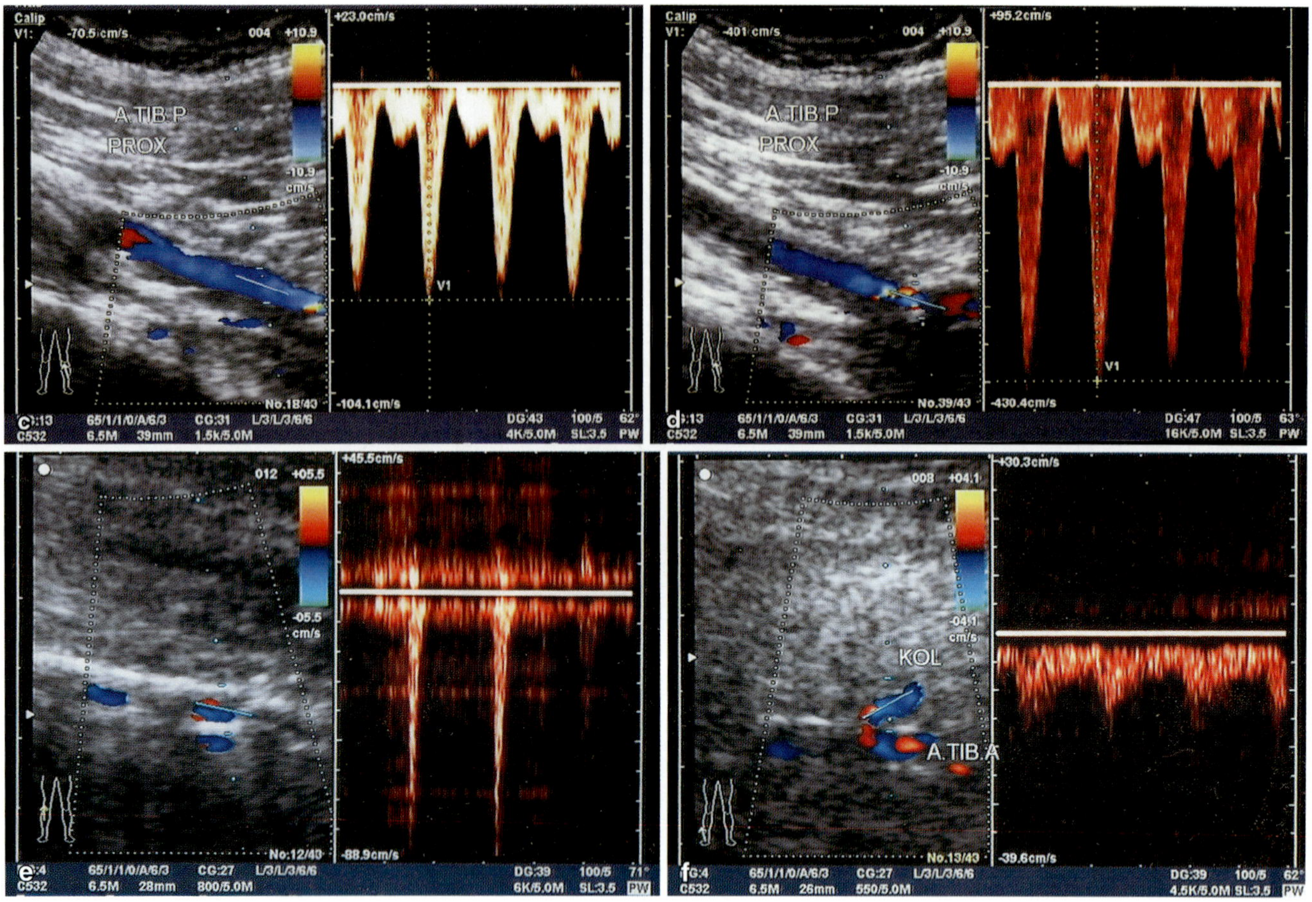


**Fig. A2.21 (continued)**

**Posterior tibial artery stenosis**

**c** High-grade stenosis of the posterior tibial artery. A stenosis in a calf artery cannot be localized on the basis of plaque demonstrated by B-mode ultrasound. Instead, the length of the artery must be evaluated by color duplex with a low PRF, focusing on the occurrence of aliasing (in the *right* margin of the image shown).

**d** The stenosis is graded by calculating the PSV ratio. The Doppler spectrum obtained with the sample volume placed in the posterior tibial artery shows a PSV of 400 cm/s, which, together with the PSV of 70 cm/s measured in **a**, yields a PSV ratio of >4, confirming the presence of a high-grade stenosis in the posterior tibial artery.

**Occlusion of lower leg arteries – Doppler waveforms**

**e** The anterior tibial artery is the only calf artery that is still patent proximally; the waveform obtained from the proximal anterior tibial artery does not show postocclusive flow (triphasic, not shown). Scanning downward, distal to the origins of several collaterals, the shape of the waveform more and more resembles a preocclusive thump pattern (shortened systolic flow and reduced PSV as signs of impaired outflow). The absence of color-coded flow in the patent tibial artery (diameter of 1.5 mm) in the color duplex mode is due to calcified plaques and acoustic shadowing.

**f** In contrast, the distal anterior tibial artery shows a postocclusive flow pattern (monophasic, delayed systolic upslope). The course of the anterior tibial artery is followed upward to identify the site of occlusion and determine the length of the occluded segment.

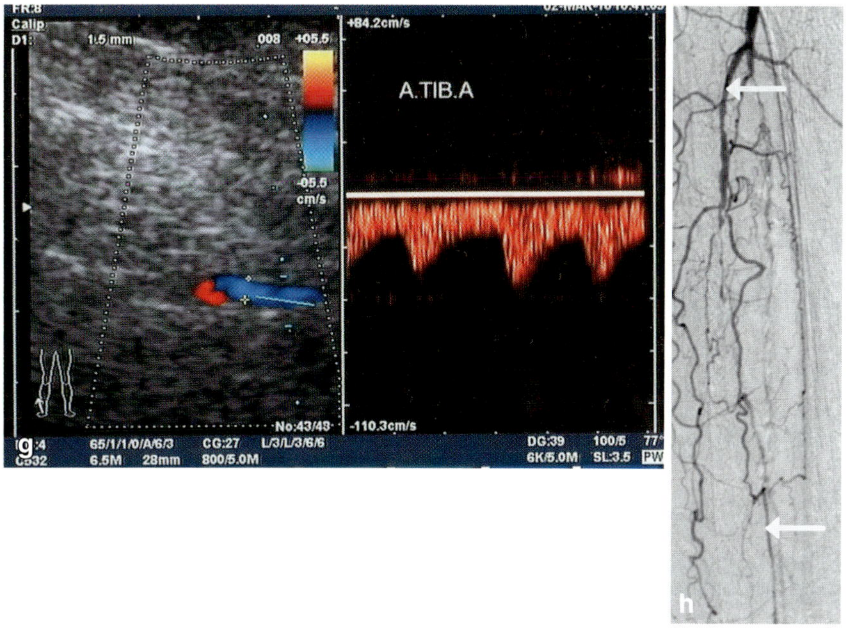


**Fig. A2.21 (continued)**
**g** The identification of collaterals and their sites of entry relative to the occluded segment and information on the flow direction are helpful in identifying an occlusion and its length. A vessel with retrograde flow entering the main artery indicates postocclusive refilling.
**h** Angiogram confirming occlusion of the anterior tibial artery, about 7 cm in length, and refilling through collaterals. The other calf arteries are occluded. The anterior tibial artery segments above and below the occlusion are indicated by *arrows*.

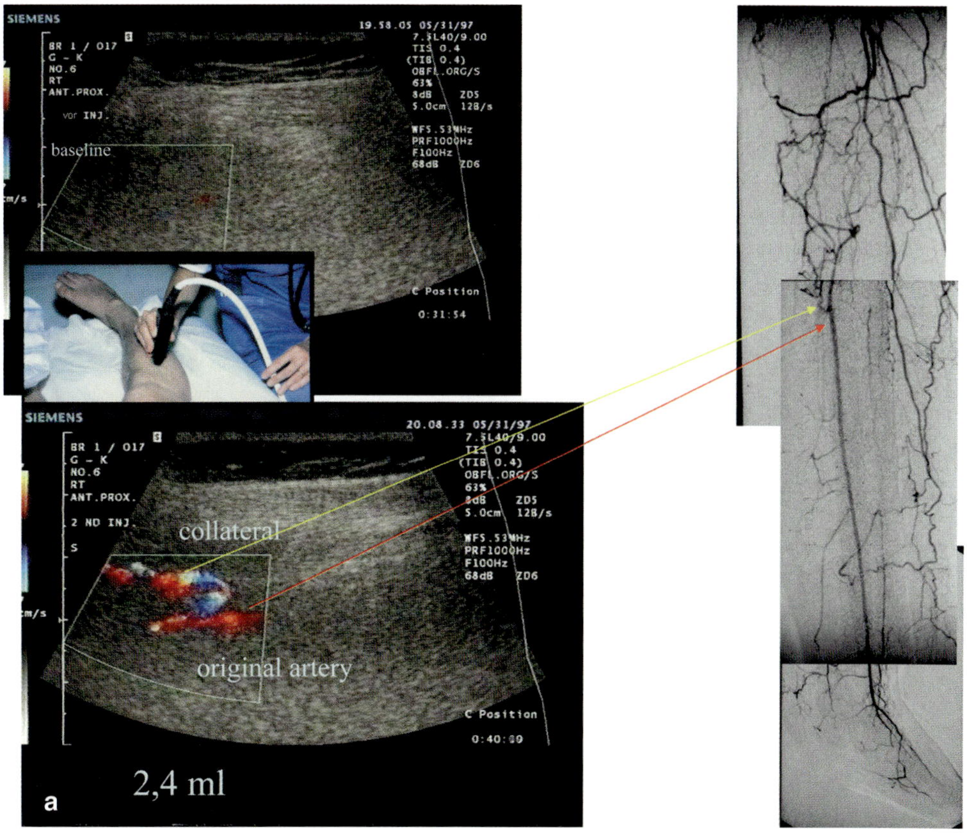


**Fig. A2.22 a–c**
**Contrast-enhanced ultrasonography of lower leg arteries** (courtesy of J. Langholz)
**a** Color duplex imaging may not enable complete evaluation of the lower leg arteries in search for a suitable target vessel for a crural bypass (as shown for the anterior tibial artery in the *top left* image). The scan obtained after administration of SonoVue (*bottom left*) reveals a large collateral resupplying the anterior tibial artery. The enhancement persisted for 9 min, which was long enough to evaluate all lower leg arterial segments. The angiogram confirms the findings of the contrast-enhanced examinations (the red and yellow *arrows* connect corresponding sites in the sonogram and angiogram).

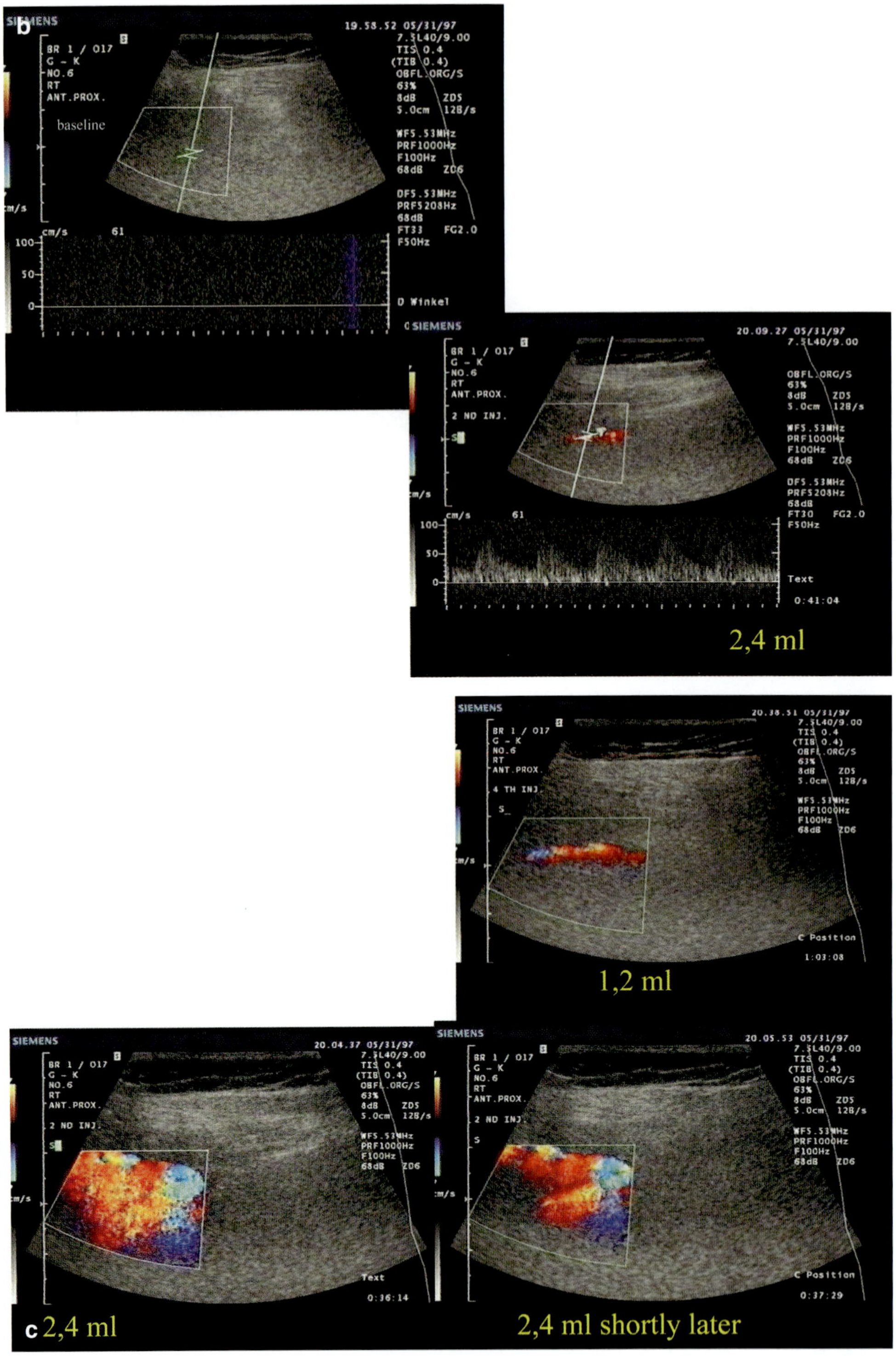


**Fig. A2.22 (continued)**

**b** Both the color duplex scan and the Doppler spectrum fail to show blood flow in the anterior tibial artery (*left*). The second scan shows the improved visualization of flow in the anterior tibial artery after administration of ultrasound contrast agent, which also improves spectral analysis by enhancing signal amplitudes. After administration of the contrast agent, the artery can be continuously followed distally to identify hemodynamically significant stenosis, which are then graded (B-scan image quality degraded by lower leg edema).

**c** Blooming effect. Overenhancement of the color display may be observed (*bottom left*) if the contrast agent dose is too high or immediately after inflow of the intravenous bolus into the target vessel (despite adequate dose). Dilution leads to good color visualization of blood flow shortly after injection of the agent (*bottom right*). Conversely, if the dose is too small, there will be incomplete enhancement of flowing blood in the patent lumen (*top right*) (see **a**).

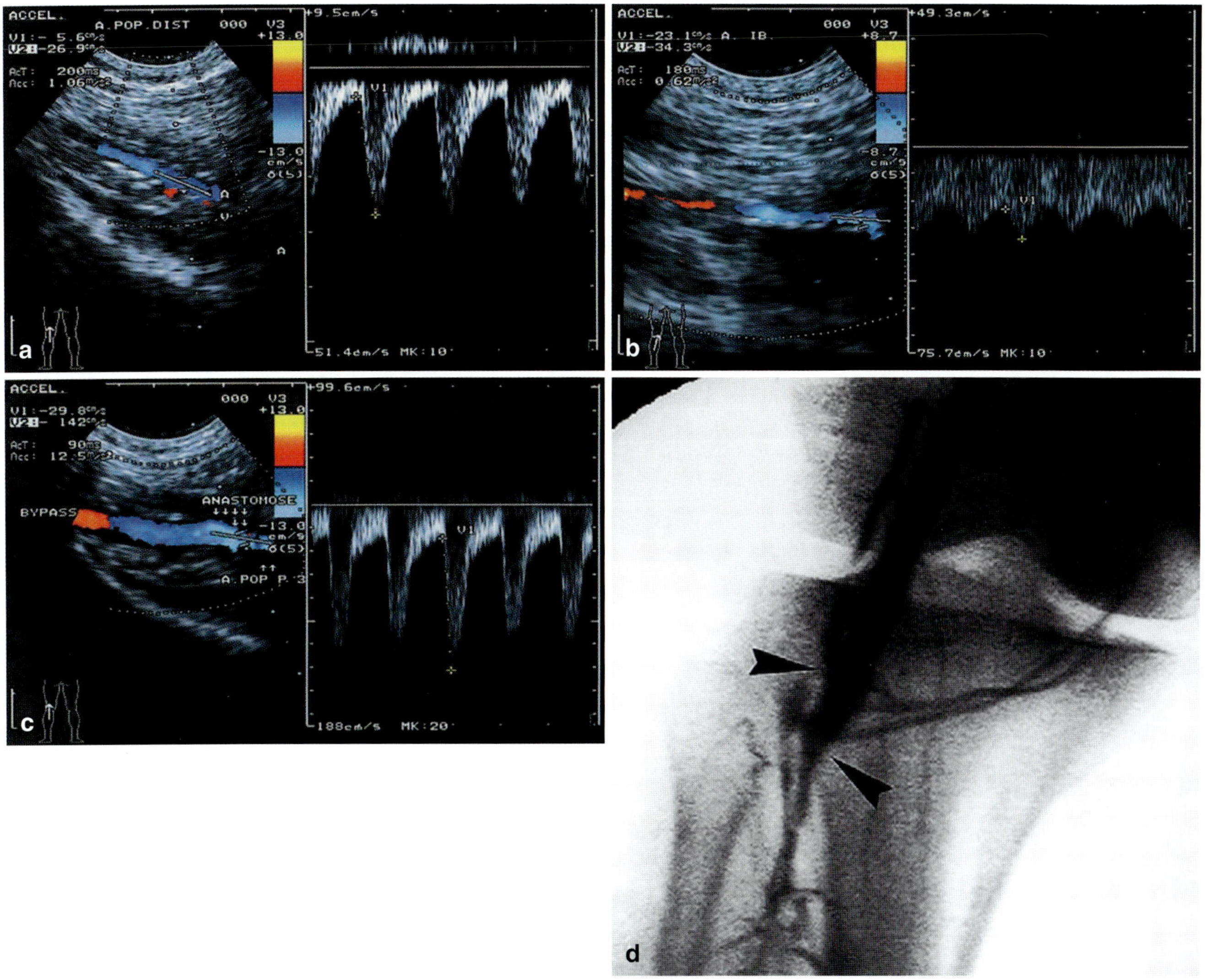


**Fig. A2.23 a–f**

**Bypass target vessel – follow-up after bypass procedure**

**a** Refilled distal popliteal artery with only a minor calcified plaque downstream of the occlusion. This gray-scale appearance suggests that segment 3 is suitable for the connection of the femoropopliteal bypass graft.

**b** Continuous scanning of the vessels down the leg reveals that only the fibular artery is patent. The color duplex scan shows color reversal from red, to black, to blue, which is due to the changing flow direction relative to the transducer. The PSV measured down to the distal lower leg is still fairly high (34 cm/s in a nonstenotic segment) in view of the multiple proximal occlusions. Together with a diastolic velocity of 23 cm/s, these findings suggest satisfactory blood supply to the foot. Evaluation of the hemodynamic situation in the lower leg arteries is crucial for estimating the prognosis if a bypass procedure is contemplated.

**c** Duplex ultrasound is the ideal modality for follow-up after bypass surgery. In the example, the femoropopliteal PTFE bypass connected onto the P3 segment is suggested by the just barely visible double contour (*left section*); the color change is again due to the change in flow direction relative to the ultrasound beam. There is no anastomotic stenosis (PSV of 142 cm/s). The steep systolic rise rules out an upstream stenosis. The monophasic flow profile is attributable to the rigidity of the graft wall and the changed peripheral resistance resulting from the presence of only a single patent lower leg artery providing collateral flow. The anastomosis is indicated by *arrows*. Intraoperative inspection of the arterial segment confirmed the presence of only little plaque and hence suitability as a recipient segment for the bypass.

**d** Intraoperative angiogram: The large-caliber bypass is depicted on the *right* and the thin native popliteal artery on the *left*. The artery is discontinuous with multiple occlusions. The occlusion at the level of the knee joint cleft is depicted over a length of 1 cm. No anastomic stenosis (length of anastomosis indicated by *arrowheads*). The band-like structure is a surgical sling and not an artifact.

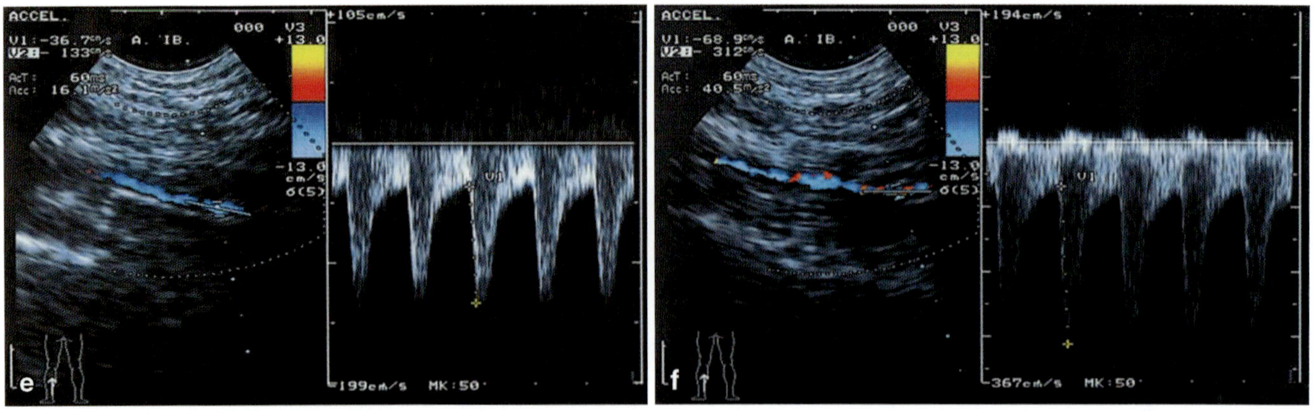


**Fig. A2.23 (continued)**

**e** The Doppler tracing obtained in the fibular artery after bypass surgery nicely depicts the improved perfusion (see **b**) with a peak velocity of 133 cm/s during systole and 36 cm/s at end diastole (waveform similar to that recorded at the anastomosis and depicted in **c**). The relatively fast flow is due to the diameter mismatch between the bypass graft and the fibular artery, the only patent vessel of the lower leg in this patient.

**f** There is a circumscribed flow acceleration to 312 cm/s during systole and 68 cm/s at end diastole 20 cm above the ankle joint. The intrastenotic to prestenotic PSV ratio is 2.3, corresponding to a hemodynamically significant stenosis (50–70%).

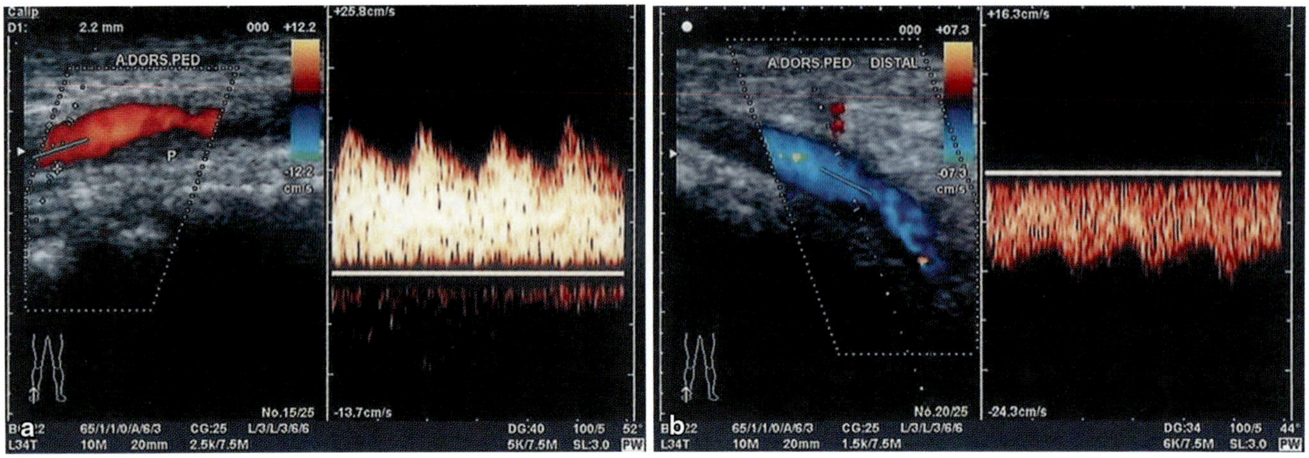


**Fig. A2.24 a–d**

**Target vessel for pedal bypass**

**a** Slow flow in the superficial pedal arteries is visualized by high-resolution duplex imaging using a high-frequency transducer and a low PRF. Plaques and stenoses are depicted, and a Doppler waveform showing the typical postocclusive monophasic flow, often with an almost venous profile, indicates upstream occlusion. Mean flow velocity or PSV and the diastolic flow component are other important parameters for determining whether an artery would provide adequate outflow when used as the recipient segment of a planned bypass. This information is important to estimate bypass patency prior to surgery. The patient presented has stage IV AOD with occlusion of all lower leg arteries. The dorsalis pedis artery shows monophasic, postocclusive flow just above the ankle joint. There is some luminal narrowing from a hypoechoic plaque (*P*).

**b** Further down, shortly before it enters the arch of foot, the dorsalis pedis artery has an unchanged monophasic flow profile with good perfusion, suggesting that the artery is a suitable candidate for connection of a pedal bypass graft. In this patient, with multiple upstream occlusions, a more pulsatile flow profile would suggest poor outflow.

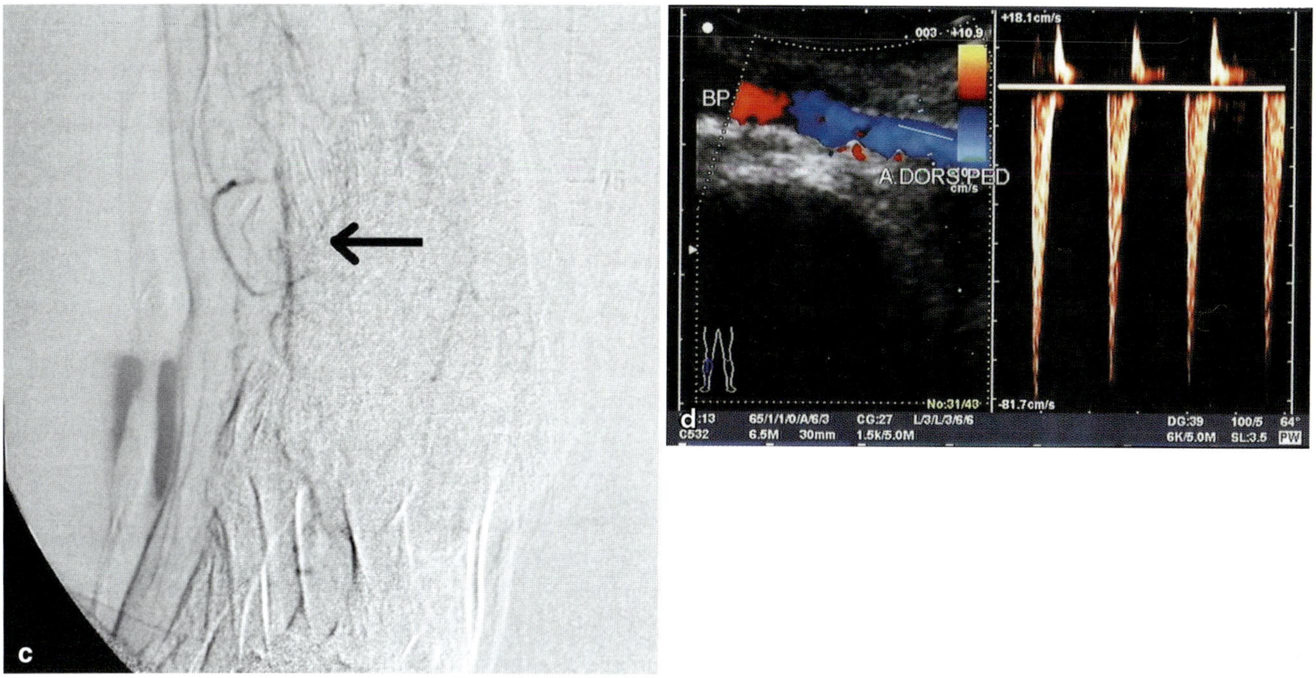


**Fig. A2.24 (continued)**

**c** Angiogram of the pedal vessels demonstrates patency of the artery though there is poor opacification due to the proximal occlusions. Angiography is inferior to color duplex in predicting whether this artery will ensure adequate runoff for a pedal bypass.

**d** The venous graft anastomosed onto the dorsalis pedis artery has triphasic flow with a PSV of nearly 80 cm/s, indicating adequate perfusion of the foot without ischemic vasodilatation in the periphery.

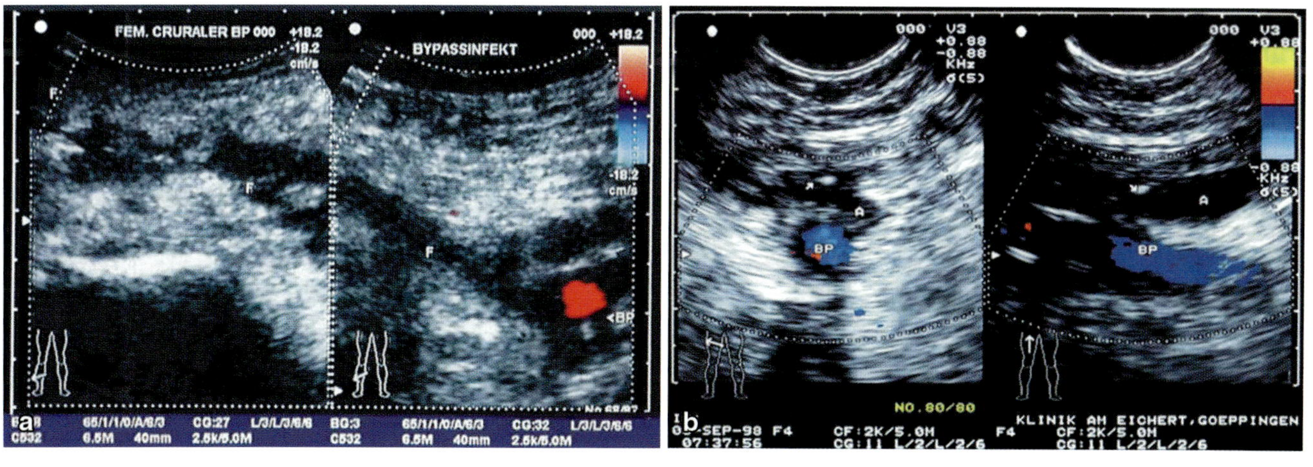


**Fig. A2.25 a–c**

**Follow-up after bypass procedure – graft infection**

**a** A hypoechoic fistula some centimeters in length extends from below the skin to the distal anastomosis of a P2 bypass, indicating graft infection, although the initial clinical appearance of the wound suggested only a superficial, subcutaneous infection.

**b** If gray-scale ultrasound depicts elongated hypoechoic to anechoic areas around a bypass graft, an infection of the bypass has to be excluded, in particular if the respective clinical signs are present. The simplest test is ultrasound-guided aspiration with placement of the hyperechoic needle tip (*N*) in the hypoechoic zone adjacent to the graft. The needle may have to be moved about a bit under suction to reach a fluid collection.

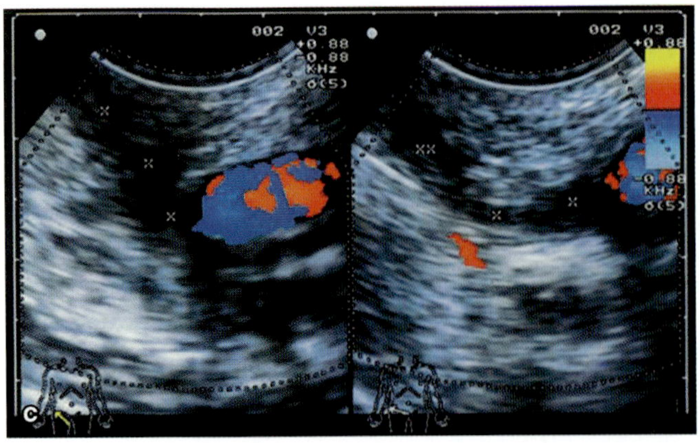

**Fig. A2.25 (continued)**
**c** Six days after crossover bypass grafting, a hypoechoic fistula (*x*) is seen extending from the site of the anastomosis to the subcutaneous tissue. There is only mild reddening of the skin at the scar site. In this patient, Staphylococcus aureus was isolated from the cloudy fluid aspirated under ultrasound guidance. The slightest clinical suspicion of bypass graft infection should prompt a sonographic examination to rule out complications and initiate timely graft revision.

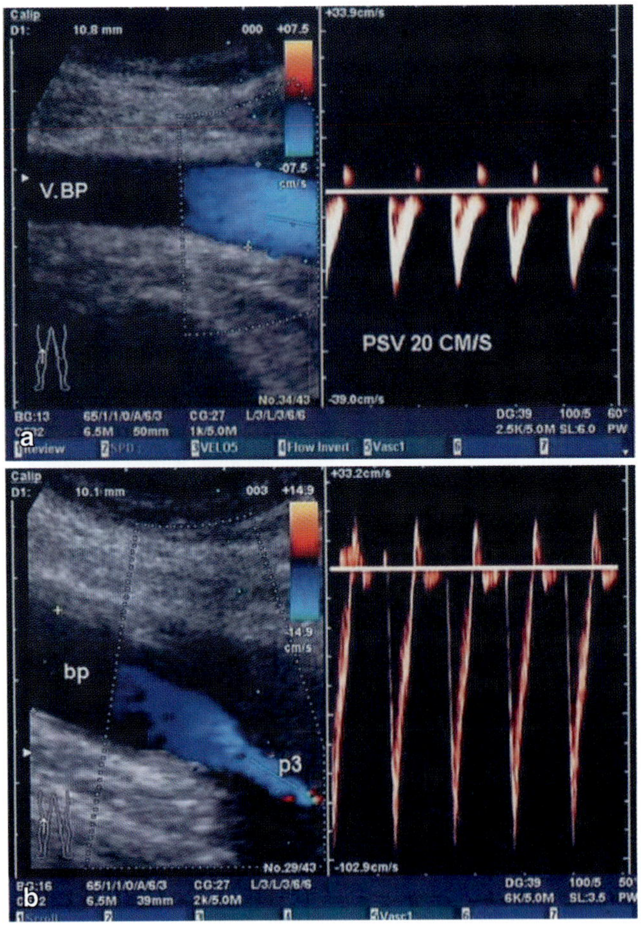


**Fig. A2.26 a, b**
**Interpretation of Doppler waveforms from within bypass grafts**
**a** The blood flow velocity within a bypass graft is largely determined by its diameter and that of the distal recipient artery. In the example shown, the PSV in the dilated venous bypass graft (*V.BP*; diameter of 11 mm) is only 20 cm/s although there is no stenosis distal to the measurement site. The waveform is pulsatile and exhibits a steep systolic upstroke.
**b** There is no stenosis at the distal anastomosis with the distal popliteal artery (P3). The increase in PSV to 102 cm/s is due to the size mismatch between the dilated graft (*bp*; see **a**) and the normal-caliber distal popliteal artery. The triphasic and pulsatile waveform recorded in the popliteal artery distal to the anastomosis is that of a normal peripheral artery. A sonographer examining a bypass graft should compare the pulsatility and flow velocity with the baseline values determined sonographically within the first 3 months of the bypass procedure.

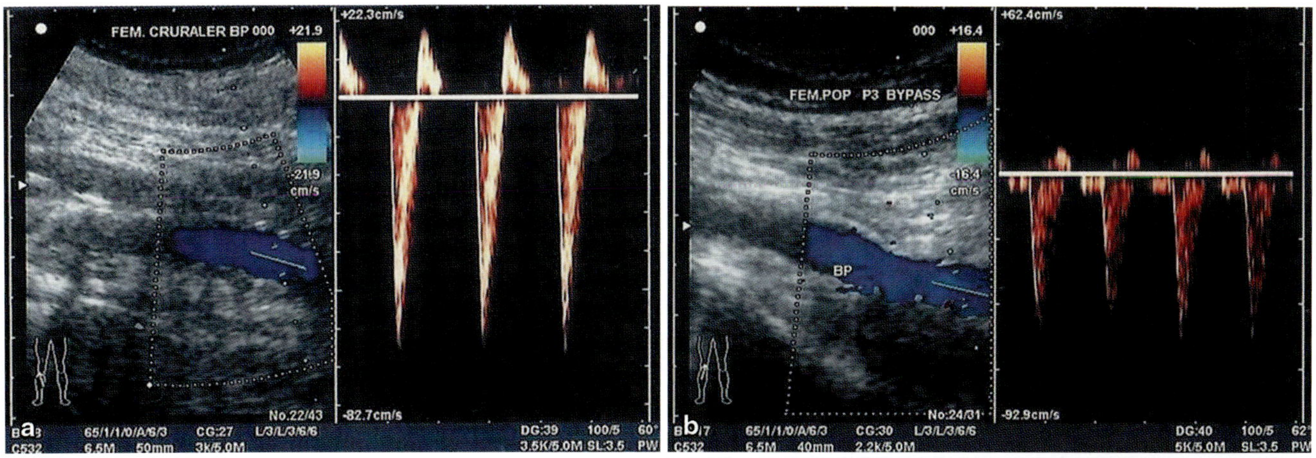


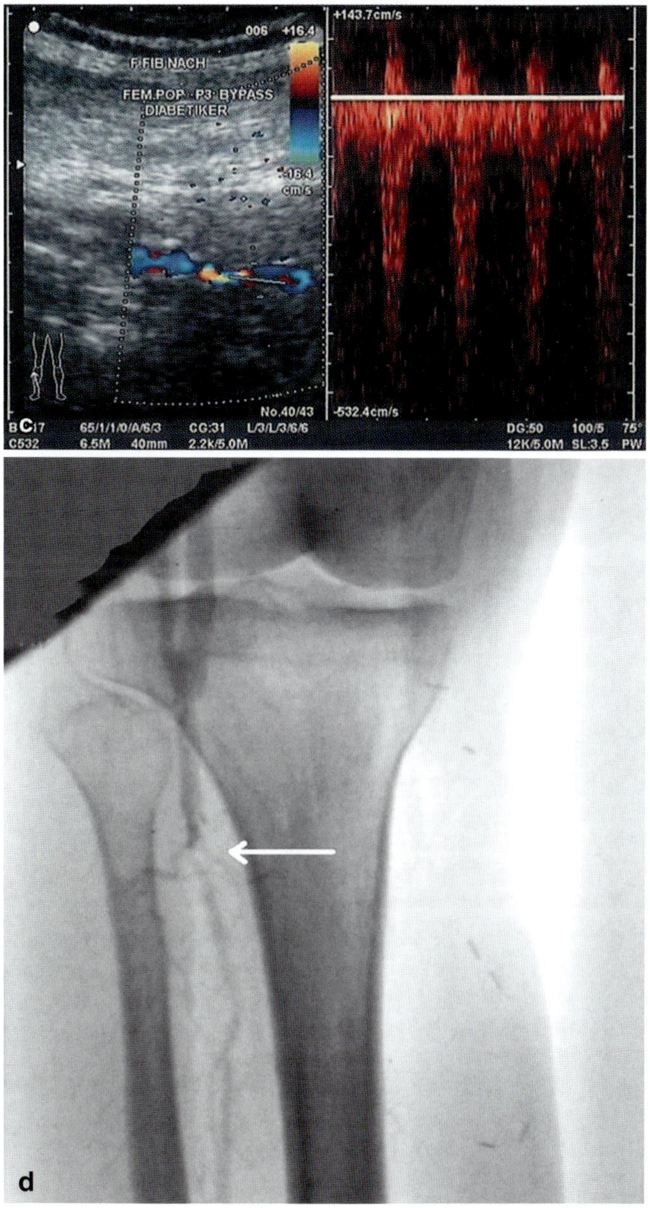


**Fig. A2.27 a–d**
**Flow through bypass graft – outflow obstruction**

**a** A triphasic flow profile with a steep systolic upslope and PSV of over 60 cm/s (in the present case 65 cm/s) suggest good bypass function. Nevertheless, the examiner should check the anastomoses, in particular the distal one, for the presence of stenosis and should also evaluate run-off in the recipient artery.

**b** In contrast, the 1-year ultrasound follow-up of this femoropopliteal bypass graft at the P2 level demonstrates a reduced PSV of only 45 cm/s while the flow profile is still triphasic. These findings may be indicative of an outflow obstruction. The patient is a diabetic with a necrotic great toe, which should produce a monophasic flow profile due to peripheral dilatation.

**c** In the lower leg, only the fibular artery is patent, and color duplex scanning demonstrates a high-grade stenosis at the junction between the tibiofibular trunk and the fibular artery with aliasing and a PSV of about 5 m/s. The stenosis obstructs bypass outflow.

**d** Control angiogram confirms high-grade stenosis (*arrow*) of the tibiofibular trunk at the junction with the fibular artery.

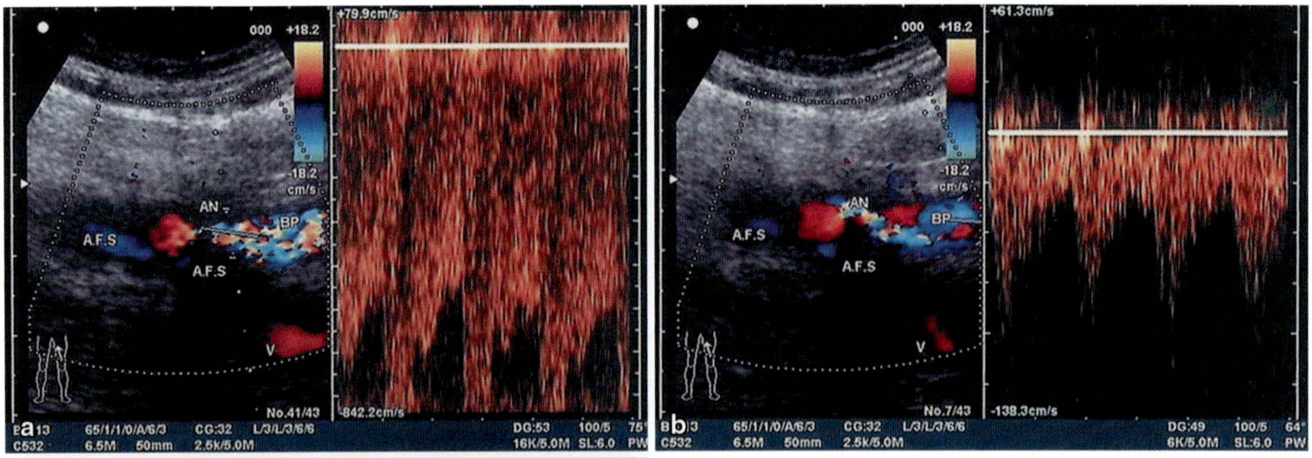


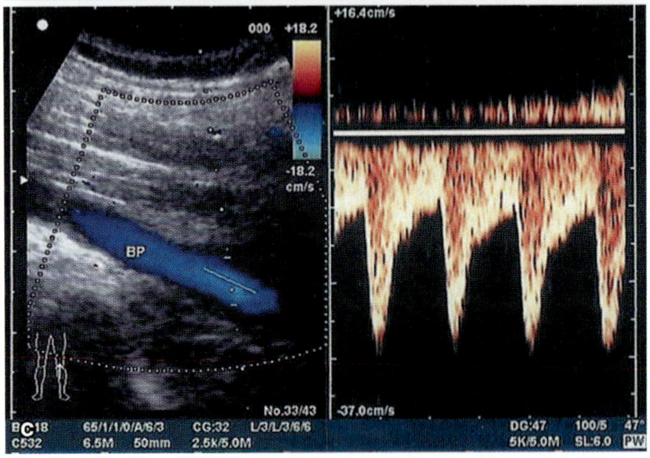


**Fig. A2.28 a–c**
**Stenosis of bypass anastomosis**

**a** Routine 6-month follow-up of a femorocrural bypass graft in a patient with stage IV AOD demonstrates high-grade stenosis at the proximal anastomosis (aliasing, PSV > 700 cm/s) due to neointimal proliferation (hypoechoic wall thickening at anastomotic site).
**b** Flow slightly distal to the anastomosis is highly turbulent, but flow velocity is normal again.
**c** The bypass is patent but exhibits a postocclusive flow profile (monophasic, delayed systolic rise, PSV < 30 cm/s). The slow flow in the bypass due to proximal anastomotic stenosis suggests that there is a risk of imminent bypass occlusion.

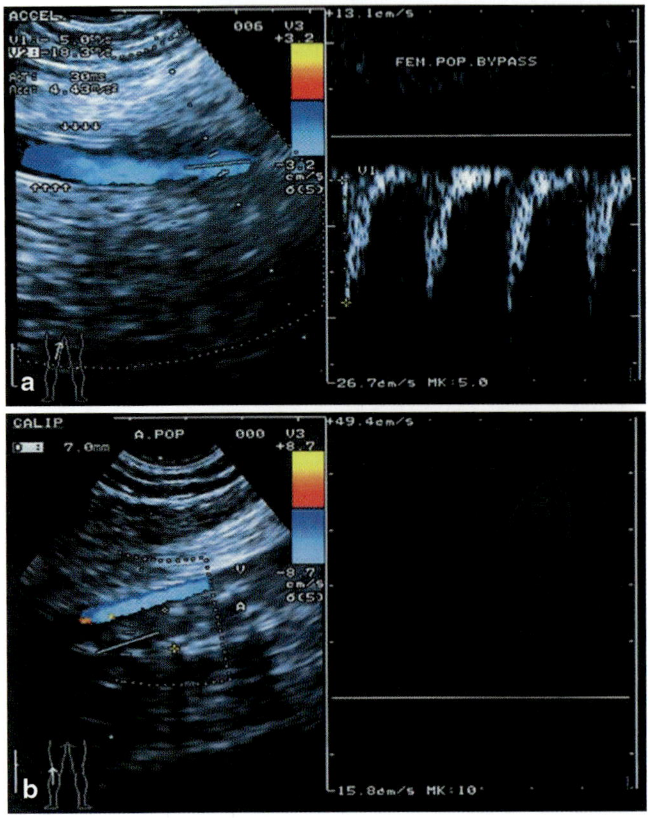


**Fig. A2.29 a–j**
**Bypass graft occlusion**

**a** The markedly reduced PSV of 18 cm/s in the PTFE bypass graft (double contour) is an indicator of imminent occlusion (threshold velocity: 35–40 cm/s). In interpreting flow velocities measured in a bypass, the examiner must take into account a possible size mismatch between graft and recipient artery.
**b** The slow flow in the femoropopliteal bypass anastomosed to the proximal popliteal artery is due to newly occurring occlusion of the popliteal artery (*A*) distal to the anastomosis 3 years after bypass grafting. The poor delineation of the artery from surrounding tissue due to plaques with acoustic shadowing suggests progression of atherosclerosis as the cause of the occlusion. Thrombectomy is not promising in this case; instead, a suitable recipient segment should be identified for an extension of the bypass.

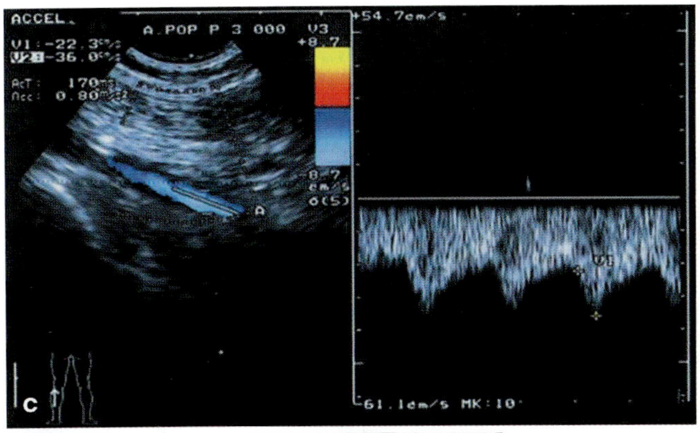


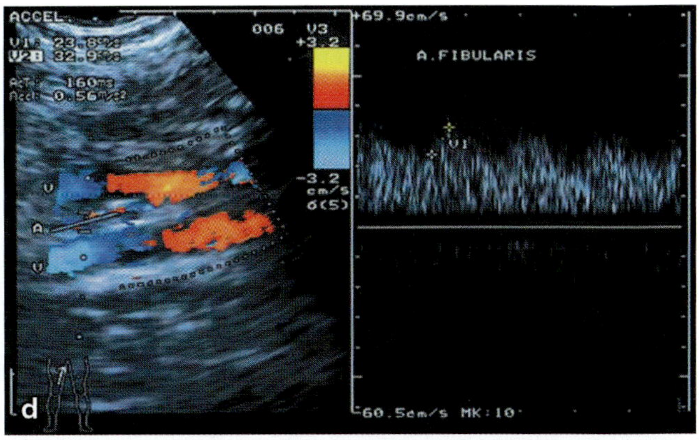


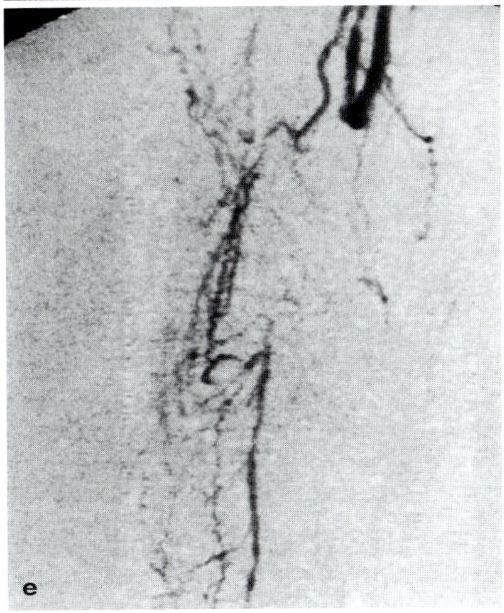


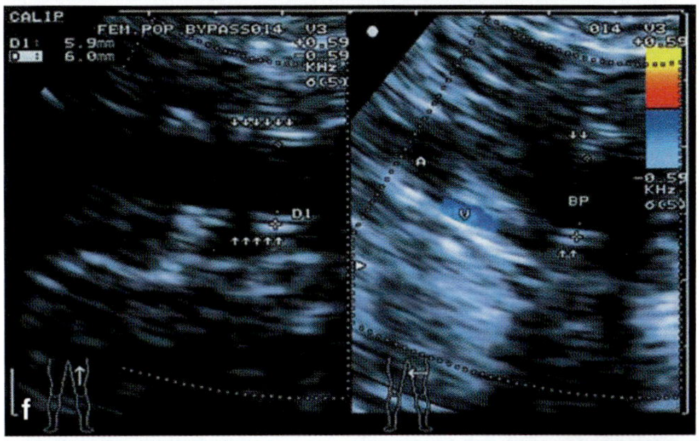


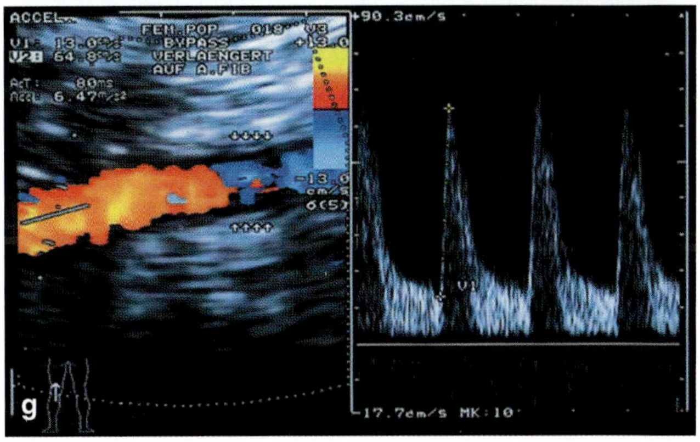


**Fig. A2.29 (continued)**

**c** The P3 segment of the popliteal artery is refilled and shows the typical postocclusive Doppler waveform (markedly delayed systolic rise and PSV decreased to 36 cm/s).

**d** Duplex evaluation of the outflow situation shows the fibular artery to be the only patent major artery. Only sparse flow signals are obtained from the artery despite adequate instrument setting, as shown by the good color filling of the accompanying veins. Besides the low PSV of 32 cm/s, the Doppler waveform shows a low amplitude despite a rather high gain. The waveform is much like a venous waveform but is typical of an artery with flow to the periphery toward the transducer (70° Doppler angle). Blue indicates that flow in the veins is away from the transducer at the level of the sample volume (color change resulting from flow reversal relative to the transducer).

**e** Control angiogram: With the popliteal artery occluded, the bypass (*top right*) supplies the superficial femoral artery (*to the left of it*) retrogradely. Collaterals arising from the superficial femoral artery bridge the occluded popliteal artery and refill the distal popliteal and the fibular arteries.

**f** A few days later, occlusion of the bypass occurs. The longitudinal view on the *left* shows no flow signals; the transverse view on the *right* depicts the vein in blue (*V*) to the *left* of the occluded bypass (*BP*) and, anterior to it, the proximal superficial femoral, which is also occluded (*A*).

**g** After thrombectomy of the graft and extension to the tibiofibular trunk, both color duplex and the Doppler waveform demonstrate good flow in the graft with a PSV of 64 cm/s. The monophasic flow profile is due to peripheral dilatation rather than a bypass stenosis as the steep systolic upslope is preserved.

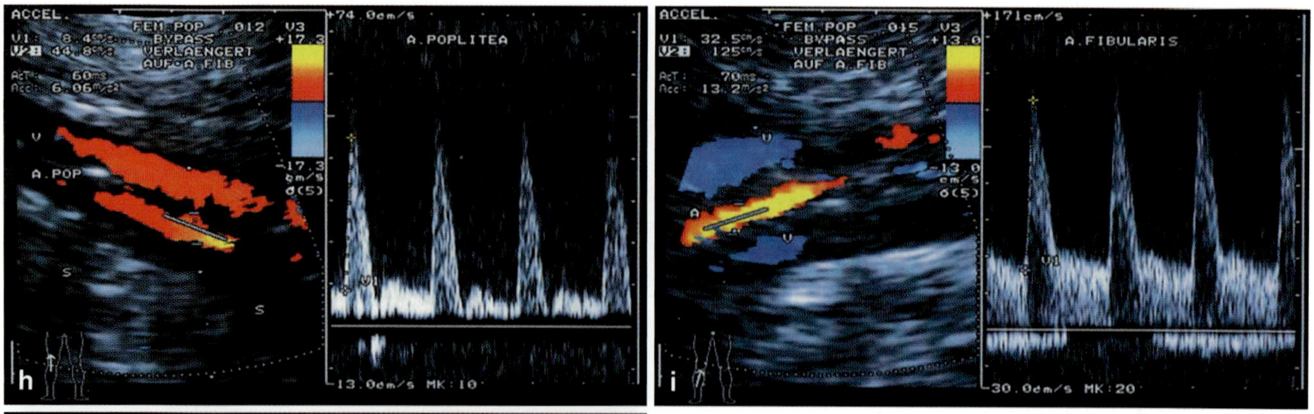


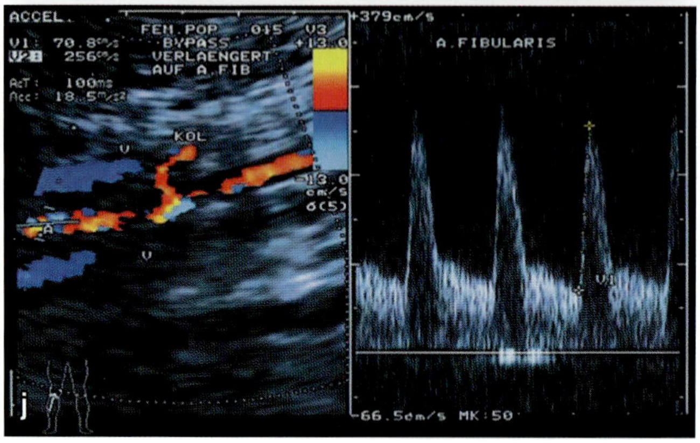


**Fig. A2.29 (continued)**

**h** Following bypass extension to the tibiofibular trunk, there is retrograde flow coded red in the distal popliteal artery (flow toward transducer as in the accompanying artery). Flow is triphasic and pulsatile.

**i** A rather high flow signal with a PSV of 125 cm/s and an end-diastolic flow of 30 cm/s is obtained from the proximal fibular artery after bypass extension. As this flow velocity is not circumscribed but detectable over a longer graft segment, it is not caused by stenosis but by a diameter reduction at the junction between the graft and the fibular artery, which is the only patent lower leg artery providing collateral flow in this case.

**j** However, there is aliasing in the distal fibular artery with a color change from red, to yellow, to blue in the duplex scan and flow acceleration with a PSV of 256 cm/s. The circumscribed doubling of the flow velocity compared with the prestenotic velocity suggests a hemodynamically significant peripheral stenosis. Next to the fibular vein displayed in blue (flow away from transducer), a collateral (*KOL*) with flow toward the transducer is depicted in red, supplying blood to the distal territory of the occluded posterior tibial artery.

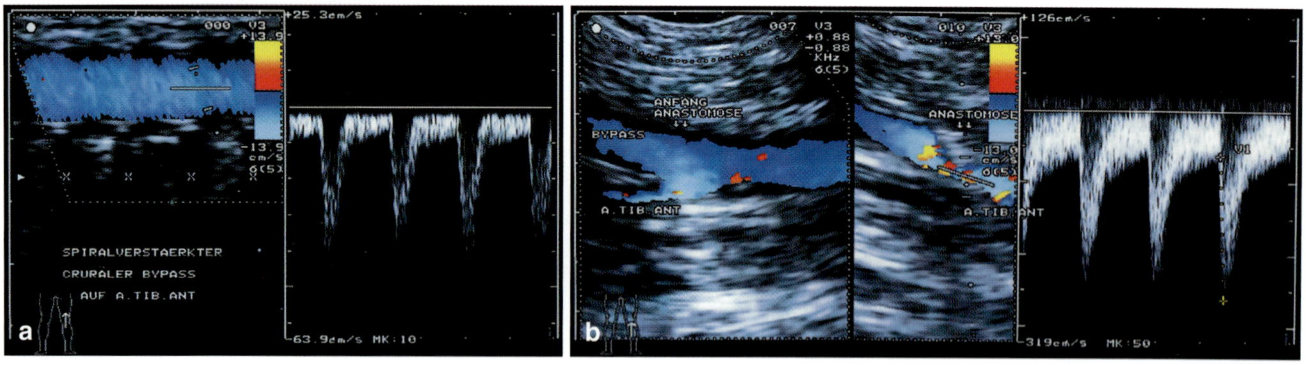


**Fig. A2.30 a, b**

**Stenosis of anastomosis**

**a** A coiled crural bypass graft anastomosed onto the anterior tibial artery shows rather slow flow of 45 cm/s during systole and 8 cm/s at end diastole. The coil is identified by the absence of signal (*X*). The slow flow may be due to the pronounced mismatch in diameter between the bypass (6 mm) and the rather thin anterior tibial artery. The sonographer must exclude a distal stenosis.

**b** The bypass graft was connected to the anterior tibial artery over a long segment. In the *left section*, the proximal end of the anastomosis to the anterior tibial artery is indicated by *arrows*. The anterior tibial artery arises posterior to the bypass graft and shows retrograde flow in its proximal portion. In a crural bypass graft with a long end-to-side anastomosis, the distal end of the anastomosis is the chief site of stenosis. The transducer is tilted to optimize the Doppler angle for interrogation of the junction of the bypass with the anterior tibial artery (*middle section*). There is aliasing at the distal end and flow acceleration with a peak velocity of 260 cm/s during systole and 63 cm/s at end diastole. The magnitude of the flow acceleration suggests a hemodynamically significant anastomotic stenosis

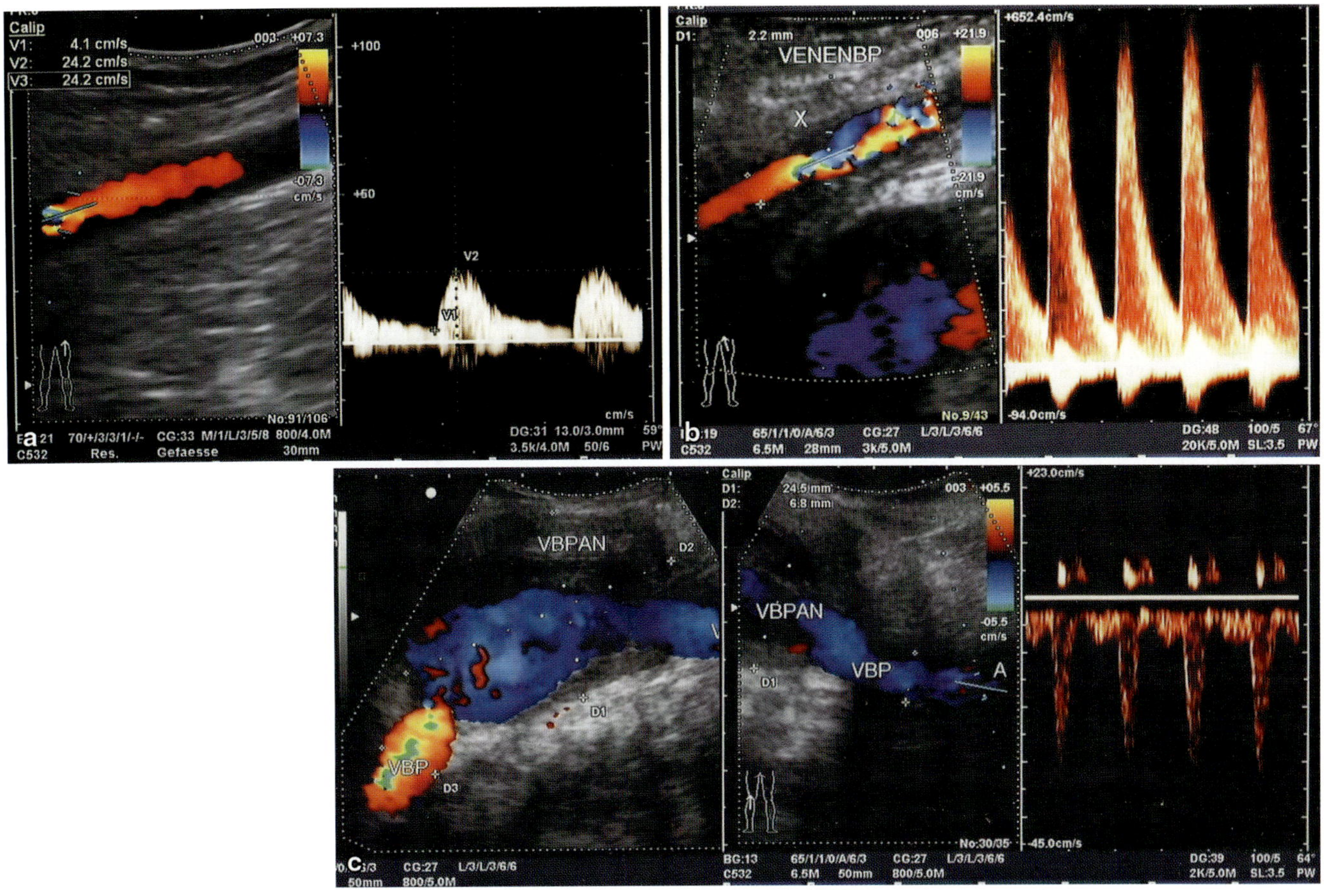


**Fig. A2.31 a–c**

**Saphenous vein bypass graft – stenosis at the site of venous valve**

An autologous bypass graft (great saphenous vein) is more difficult to identify, especially when it is occluded, due to the thin venous wall and the frequent extra-anatomic course. Color duplex helps identify the graft, but spectral Doppler measurement is necessary for quantitative evaluation.

**a** A postocclusive waveform with a PSV of 24 cm/s and an end-diastolic velocity of 4.1 cm/s obtained in the main body of the graft indicates proximal stenosis.

**b** While stenosis is rare within a synthetic bypass, the entire length of a venous graft must be carefully scrutinized for the presence of stenosis. In an in situ vein graft, stenosis tends to develop at sites of retained valves. In the example, the color flow image and Doppler measurement reveal a short, high-grade stenosis with a PSV of 6 m/s at the site of a valve leaflet, confirming the stenosis suggested by the postocclusive waveform presented in **a**.

**Venous bypass graft aneurysm**

**c** Aneurysmal dilatation is a late complication of bypass surgery and is often associated with elongation of the graft (*V.B.P*). The overview (*leftmost* image) shows a partially thrombosed aneurysm of a venous graft (measuring 2.5 × 3.8 cm) 2 cm above the distal anastomosis with the P3 segment of the popliteal artery (*V.B.PAN*). The second scan shows the site of anastomosis (*A*), from which the Doppler waveform was obtained.

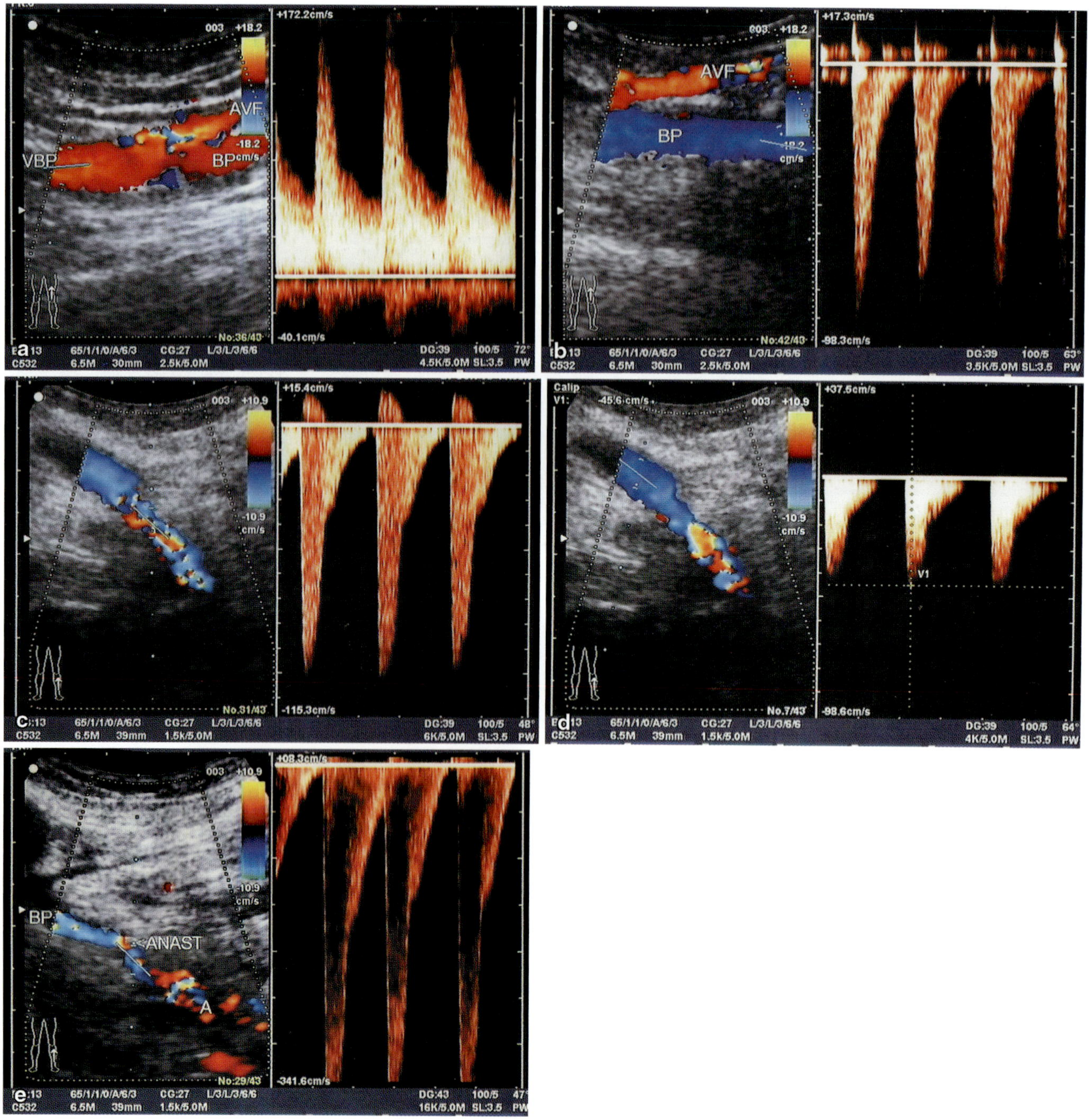


**Fig. A2.32 a–f**

**In situ vein graft – AV fistula and stenosis**

**a** Waveform from an in situ vein graft with a steep systolic upstroke but monophasic flow pattern and large diastolic component. The high flow volume in the graft with a PSV of 150 cm/s and an end-diastolic velocity of 50 cm/s is attributable to a larger arteriovenous fistula (*AVF*).

**b** Distal to the high-flow fistula (*AVF*), the flow velocity in the graft (*BP*) is much lower. Doppler interrogation shows a PSV of 70 cm/s and a monophasic pattern, but with some end-diastolic flow. The waveform is characteristic of low-resistance flow.

**c** Moreover, there is a stenosis 4 cm proximal to the distal anastomosis at the site of a retained valve leaflet. Stenosis is suggested by a focal increase in PSV to 1 m/s and by the monophasic flow.

**d** A PSV ratio >2 is calculated (prestenotic PSV of 45 cm/s), corresponding to approximately 50% stenosis. The color duplex image shows aliasing at the site of stenosis. The position of the AV fistula identified by ultrasound was marked on the skin for ligation, while the 50% stenosis was *left* untreated.

**e** Over the next 3 months, the patient developed a second, high-grade stenosis at the distal anastomosis (*ANAST*) with a PSV of >3.5 m/s.

**Fig. A2.32 (continued)**

**f** Angiogram showing the anastomosis stenosis and relative luminal narrowing about 3 cm proximal to the stenosis; the degree of stenosis is difficult to estimate.

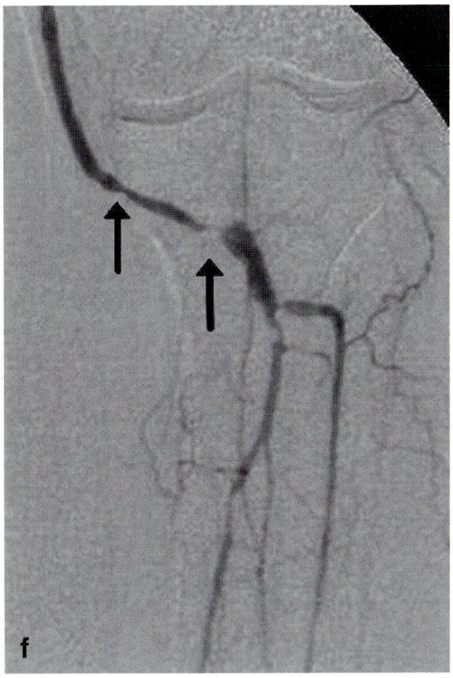


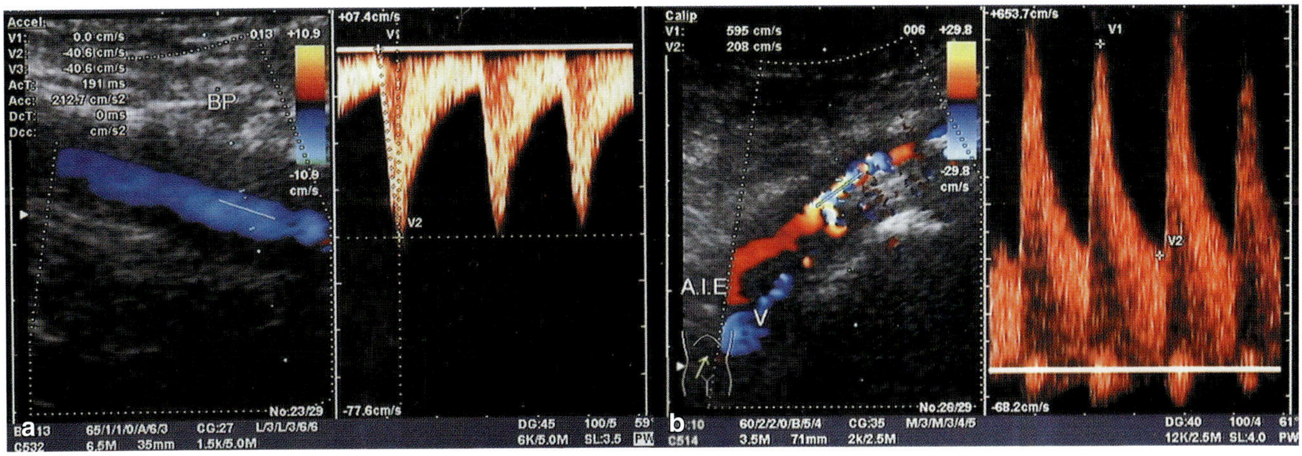


**Fig. A2.33 a, b**

**Bypass graft – inflow stenosis**

**a** There is controversy about the significance of stenosis of the inflow artery for bypass prognosis. Inflow stenosis is suggested if spectral Doppler examination of the bypass demonstrates the characteristic features of poststenotic flow including a monophasic waveform with a delayed systolic upstroke, reduced PSV, and persistent diastolic flow. The graft proximal to the measurement site and the inflow artery should then be followed cranially to identify the stenosis.

**b** High-grade external iliac artery stenosis caused by posterior plaque. Stenosis is suggested by aliasing and confirmed by Doppler measurement (monophasic flow, PSV of 550 cm/s, end-diastolic velocity of 220 cm/s).

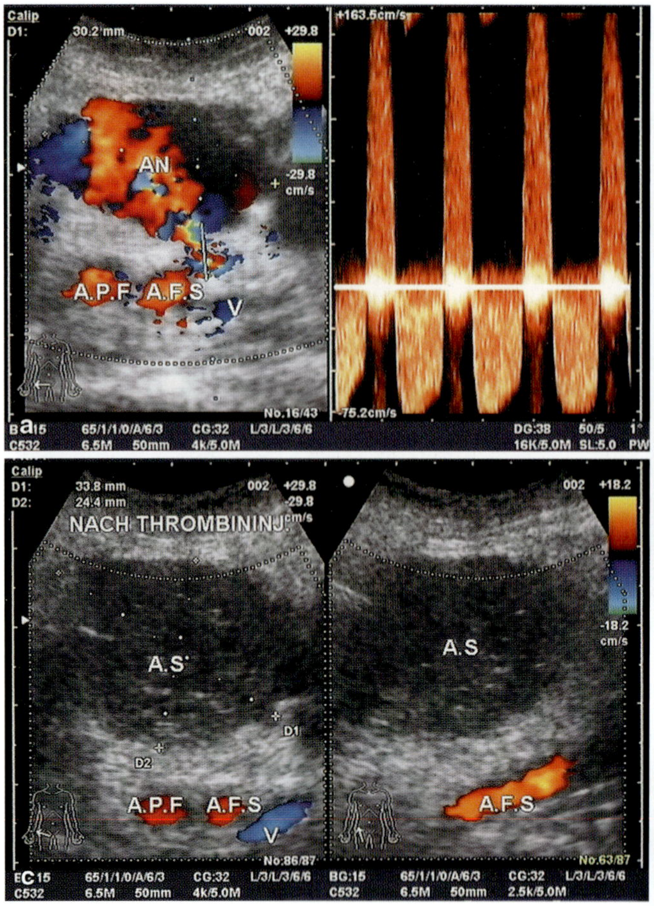


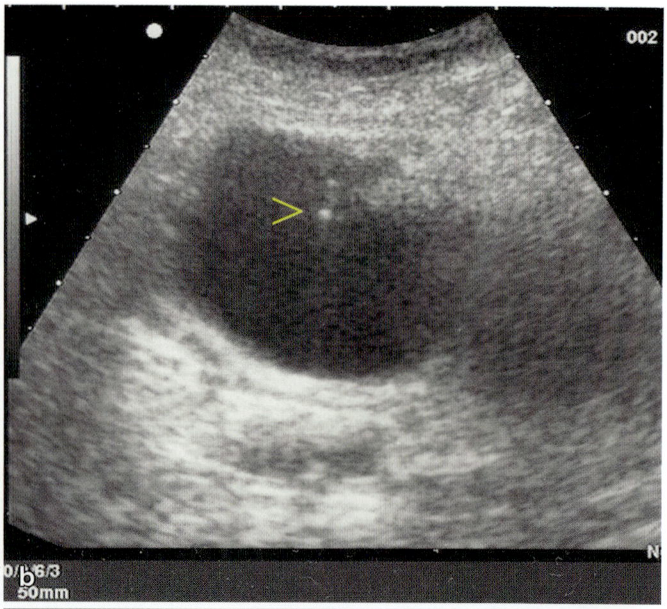

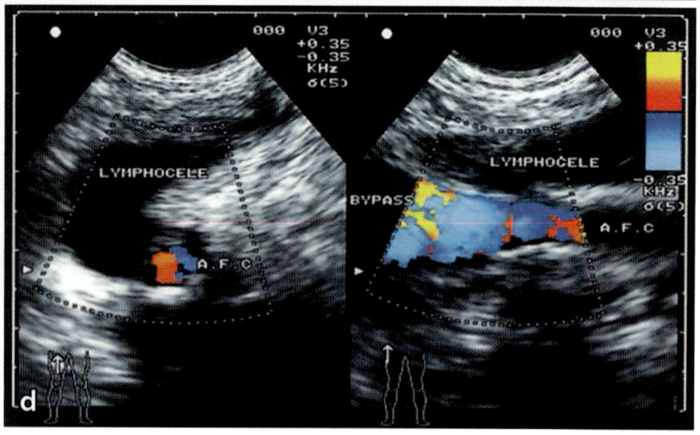


**Fig. A2.34 a–f**
**Pseudoaneurysm**

**a** Transverse view of the thigh reveals a pseudoaneurysm (*AN*) arising from the superficial femoral artery (*A.F.S*). With the sample volume placed in the neck, Doppler measurement depicts the characteristic to-and-fro flow with high-frequency flow into the aneurysm in systole and backward flow into the artery throughout diastole.

**b** For treatment of the aneurysm by thrombin instillation, a needle is advanced into the aneurysm and the tip positioned between the center of the cavity and the near wall under ultrasound guidance (needle tip identified by bright echo).

**c** Thrombin is instilled at a dose of 5,000 IU dissolved in 2 mL of saline solution. Complete thrombosis of the aneurysm (*AN*) has occurred after instillation of one to two drops as demonstrated by cessation of flow within the cavity in the color duplex mode; shown in transverse orientation on the *left* and in longitudinal orientation on the *right* (*A.F.S* superficial femoral artery, *A.P.F* profunda femoris artery, *V* femoral vein).

**d** Alternatively, a pulsatile mass developing in the groin after bypass grafting may be a seroma, hematoma, lymphocele, or abscess. All these entities show transmission of arterial pulsation to the skin surface and can be differentiated by ultrasound. The transverse view on the *left* depicts a lymphocele extending to the anastomosis between the common femoral artery and the iliofemoral bypass graft. A lymphocele differs from an aneurysm in that no flow signals can be obtained from the hypoechoic or anechoic lumen.

The longitudinal view (*right section*) depicts the bypass graft on the *left* side and the common femoral artery on the *right* with the anastomosis roughly in the middle. The lymphocele is located near the transducer. Since the differential diagnosis also includes infected fluid or an abscess, with extension to the bypass anastomosis in this case, ultrasound-guided puncture for bacteriological testing is indicated. At the same time, an attempt should be made to drain all the fluid.

**Fig. A2.34 (continued)**

**Suture aneurysm**

**e** In patients who have undergone an iliacofemoral bypass procedure, palpation of a mildly pulsatile mass at one of the anastomoses suggests a suture aneurysm. A transverse sonogram shows hypoechoic fluid extending laterally at the site of anastomosis. Color duplex imaging demonstrates flow in a portion of the lesion adjacent to the bypass graft. This appearance is also consistent with vibration artifacts. The suspected suture aneurysm is confirmed by Doppler measurement at the site of communication between the mass and the anastomosis, demonstrating the characteristic to-and-fro flow and steam engine sound. This sound is produced by high systolic inflow into an aneurysm and pandiastolic flow reversal.

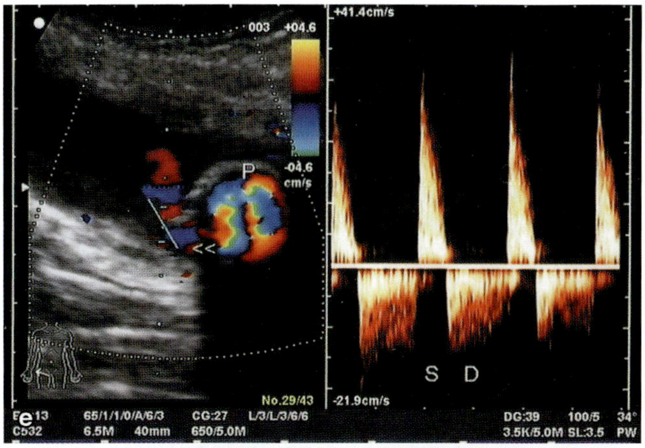


**f** Seroma at an aortofemoral bypass anastomosis with similar color duplex findings as described in **e**. However, the Doppler waveform recorded at the site of apparent flow (coded red) does not show to-and-fro flow (as in the suture aneurysm) but a signal generated in the seroma by wall motion of the vessel prosthesis. The example nicely illustrates that spectral Doppler measurement can differentiate true flow signals in a pseudoaneurysm from transmitted pulsation (which is also important when examining patients with suspected endoleaks after aortic stenting).

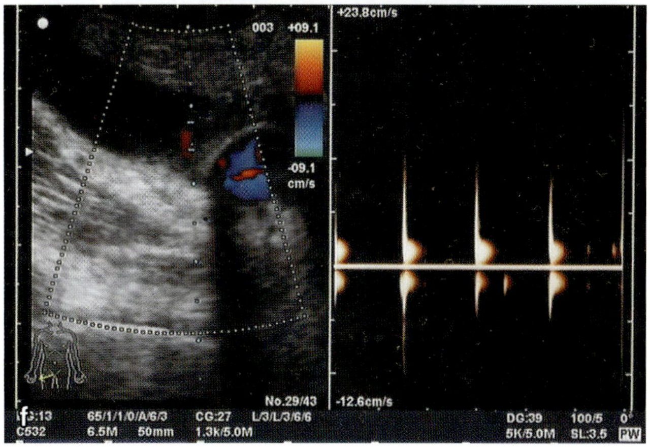


**Fig. A2.35 a–c**

**Pseudoaneurysm – compression therapy/thrombin injection**

**a** In the color duplex mode, the examiner identifies the neck connecting the pseudoaneurysm to the femoral artery and then occludes it by exerting pressure with the transducer. During the procedure, which may take up to half an hour, adequate compression is indicated by the absence of flow signals in the neck and cavity. Following the procedure, absence of flow in the cavity indicates that complete thrombosis has been accomplished. If only partial thrombosis is apparent after the procedure, it is often easier to induce complete thrombosis in a second session on the next day (compression bandage), or complete thrombosis may occur spontaneously. Alternatively, thrombosis of a pseudoaneurysm may be induced by thrombin injection. However, thrombin injection often leaves a larger residual hematoma, which may cause persistent symptoms. Thrombin injection is indicated if the site of the aneurysm precludes compression or in patients with perforated aneurysm or suture aneurysm (which may be infected).

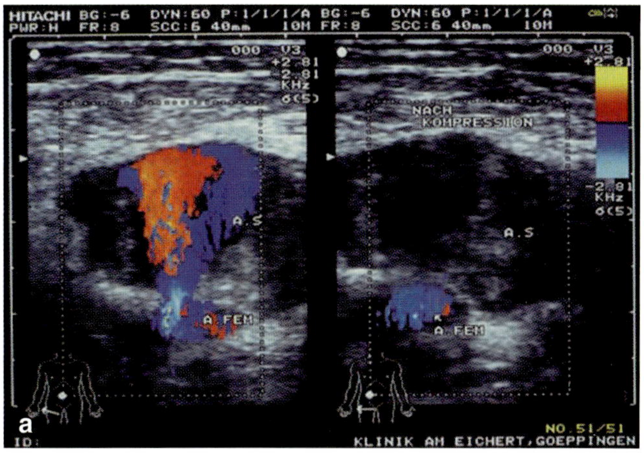

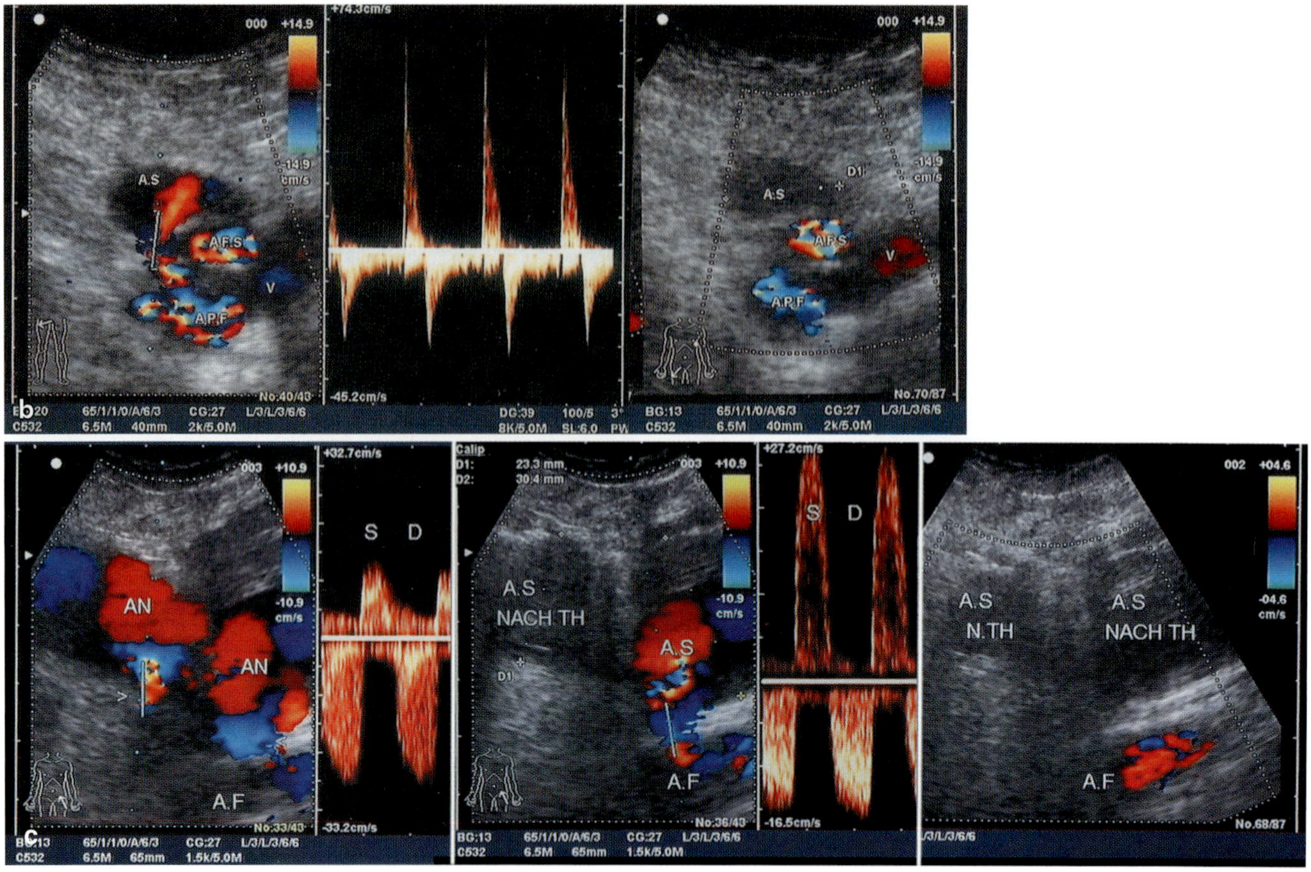


**Fig. A2.35 (continued)**

**Pseudoaneurysm – compression therapy**

**b** A small pseudoaneurysm (*A.S*) measuring only 2 cm but not occluding spontaneously arises somewhat atypically from the profunda femoris artery (*A.P.F*) about 2 cm distal to the femoral bifurcation (*left section*). With the sample volume placed in the neck, the typical systolic–diastolic to-and-fro flow is recorded. On the medial side of the neck, the superficial femoral artery (*A.F.S*) and vein (*V*) are depicted in the transverse plane. Compression of the neck with the transducer in a more lateral position brings about complete thrombosis of the aneurysm after 15 min (*right section*).

**Large pseudoaneurysm with multiple perforation – thrombin injection**

**c** A very obese patient developed a large hematoma extending from the *left* groin to the lower abdomen following angiography with cannulation of the femoral artery (*A.F.*). Pseudoaneurysm is suggested by the demonstration of flow (*AN*). The *leftmost scan* shows the sample volume placed in the neck (*arrow*) with the characteristic to-and-fro flow in the corresponding Doppler spectrum. There is a second aneurysm with a separate communication with the femoral artery (probably due to repeated puncture). The total length of both aneurysms is over 6 cm. The scan obtained after thrombin treatment of the upper aneurysm (*A.S. NACH TH – middle section*) shows the remaining second aneurysm (*A.S.*) arising from the femoral artery (*A.F.*). The Doppler waveform from the neck of the second aneurysm also shows the typical to-and-fro flow. Blood flow in the neck is very slow (30 cm/s during systole and 16 cm/s at end diastole), suggesting a large perforation defect. A total dose of 5,000 IU thrombin was required to induce closure of both aneurysms, which is very high. Very slow injection was started in the margin to minimize the risk of thrombin escape into the femoral artery. The *rightmost image* confirms complete thrombosis of both aneurysms and the patent femoral artery (*A.F.*) posteriorly. The poor color filling of the femoral artery despite a low PRF is due to scatter by the hematoma. Leg perfusion was normal, and foot pulses were palpable.

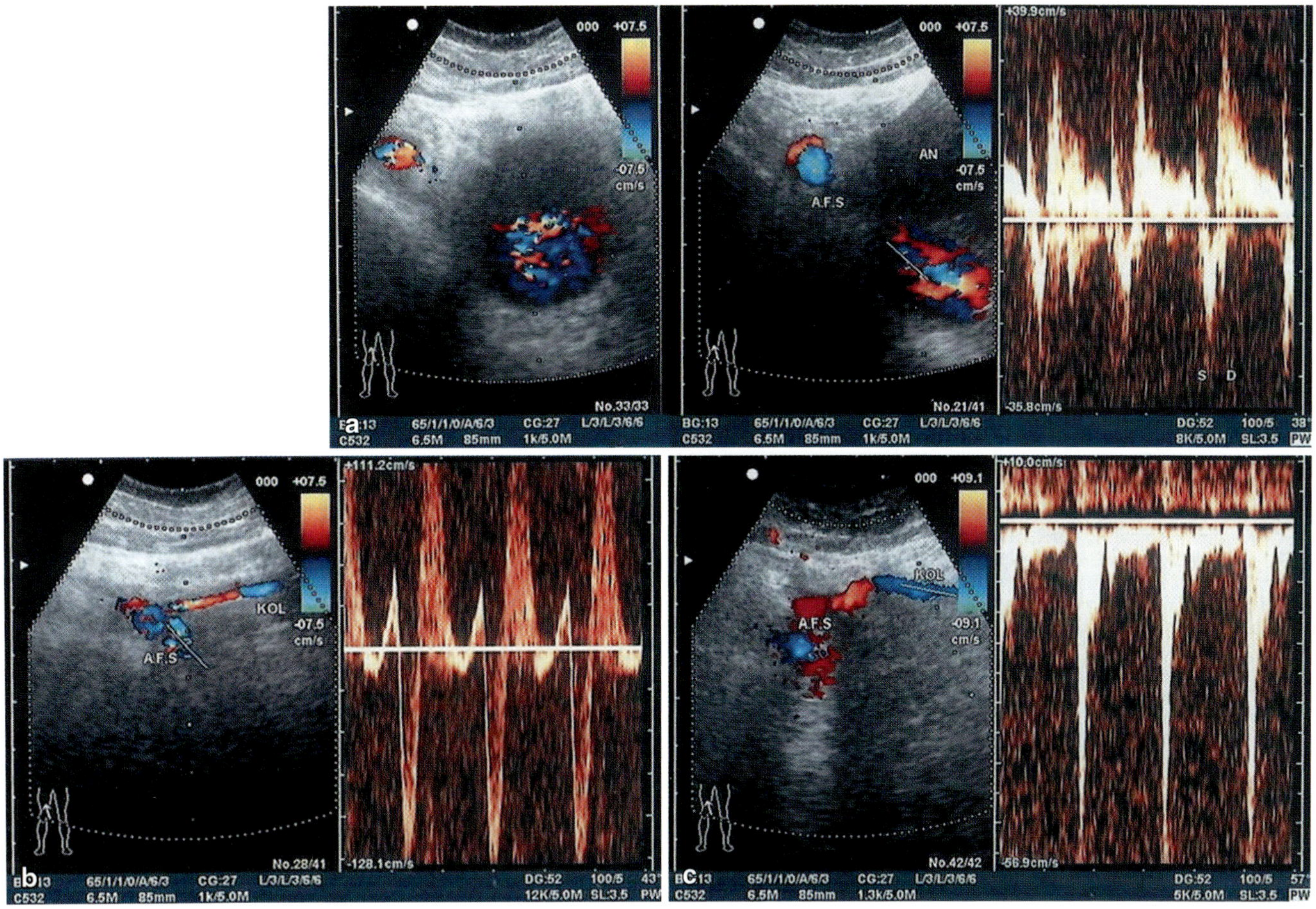


**Fig. A2.36 a–f**

**Contained perforation of superficial femoral artery aneurysm – thrombin injection**

**a** In this case of a pulsatile tumor at the level of the adductor canal, color duplex scanning demonstrates a contained perforated aneurysm at the femoropopliteal junction. The popliteal artery distal to the aneurysm is occluded and the femoral segment just above the aneurysm exhibits to-and-fro flow in the Doppler waveform.

**b** The to-and-fro flow can be followed to the site of origin of a large collateral (*KOL*) in the distal superficial femoral artery (*A.F.S*). Systolic inflow (toward the periphery, away from transducer) is followed by early diastolic outflow (toward transducer) and a further diastolic to-and-fro movement.

**c** The Doppler waveform from the main collateral (which presumably developed to bridge an older popliteal occlusion) has a just barely visible incisure but persistent diastolic flow (peripheral dilatation). The patient with considerable comorbidity and clinical stage II AOD did not require vascular repair; therefore, the perforated and only partially thrombosed aneurysm was treated with thrombin injection to induce complete thrombosis.

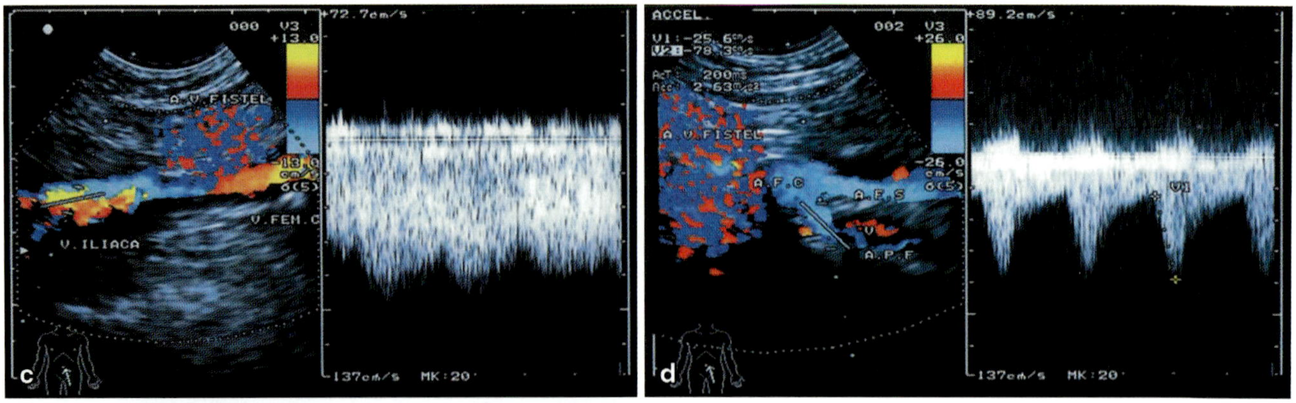


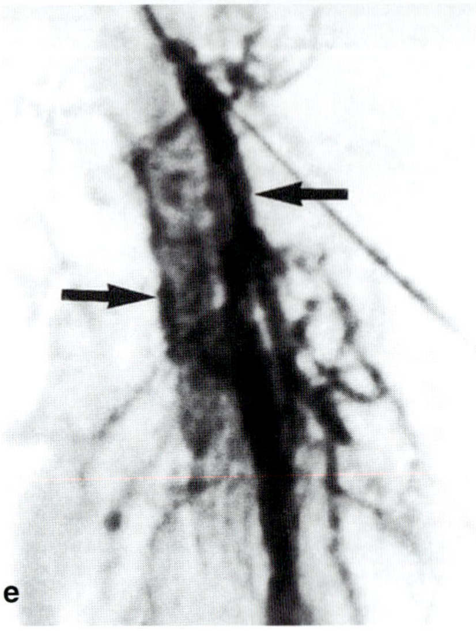

**Fig. A2.38 (continued)**

**c** The iliac vein exhibits the venous flow signal typical of an AV fistula: high-frequency flow (angle-corrected: 90 cm/s) with pulsatile variation. Adjustment of the PRF to venous flow leads to aliasing (*left* side of color scan).

**d** The Doppler waveform from the profunda femoris artery distal to the AV fistula has a delayed and flattened systolic upslope and a monophasic profile with a fairly large diastolic flow component. This is a typical poststenotic profile, caused by the puncture-induced AV fistula and the high-grade stenosis resulting from the plaques shown in **a**. For differentiation of the cause of the perivascular vibration artifacts, the downstream circulation must be evaluated (fistula: venous; stenosis: arterial). As illustrated here, vessel manipulation by puncture may not only induce fistula formation but may also cause stenosis through detachment of a plaque from the vessel wall.

**e** Angiogram: Contrast medium outflow in the iliac vein typical of a fistula. Angiography does not allow precise localization of the fistula, nor does it provide definitive evidence for the stenosis in this segment (superimposition). *Left arrow* indicates the femoral vein, *right arrow* indicates the femoral artery.

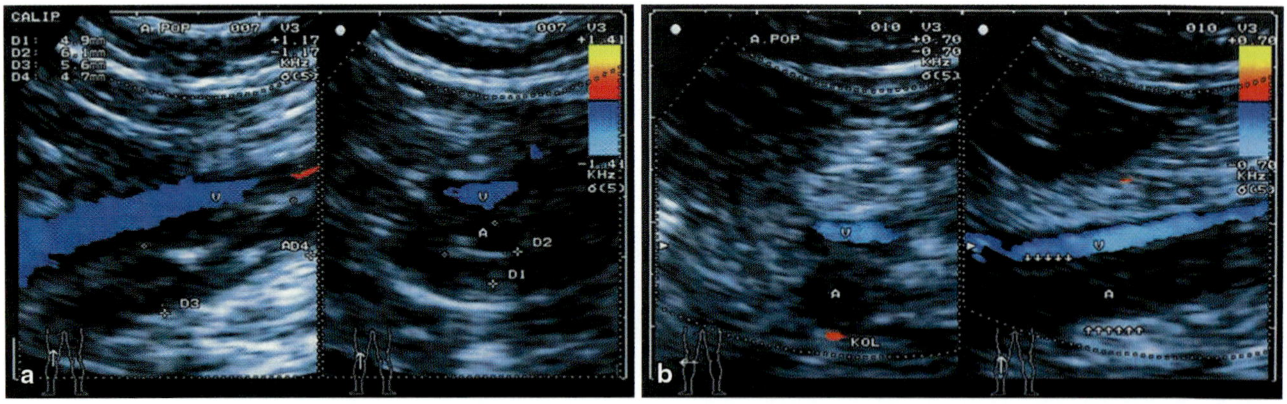


**Fig. A2.39 a–d**

**Popliteal artery occlusion – atherosclerosis versus embolism**

**a** Atherosclerotic occlusion of the popliteal artery, longitudinal view on the *left*, transverse view on the *right*: Popliteal vein displayed in blue close to transducer. Extensive plaque throughout the artery with poor demarcation of the wall contour, in conjunction with the inhomogeneous and partially very hyperechoic vessel lumen, suggests an atherosclerotic process. Based on these ultrasound findings, catheter thrombolysis, possibly with PTA, is not promising. Instead, bypass grafting is indicated, if clinically necessary.

**b** Embolic occlusion: The lumen of the popliteal artery is filled with a hypoechoic, homogeneous thrombus or embolus. There is good delineation of the vessel wall without signs of plaque. Anterior to the popliteal artery, the popliteal vein is depicted in blue; posterior to it, a red arterial collateral (*KOL*) is seen.

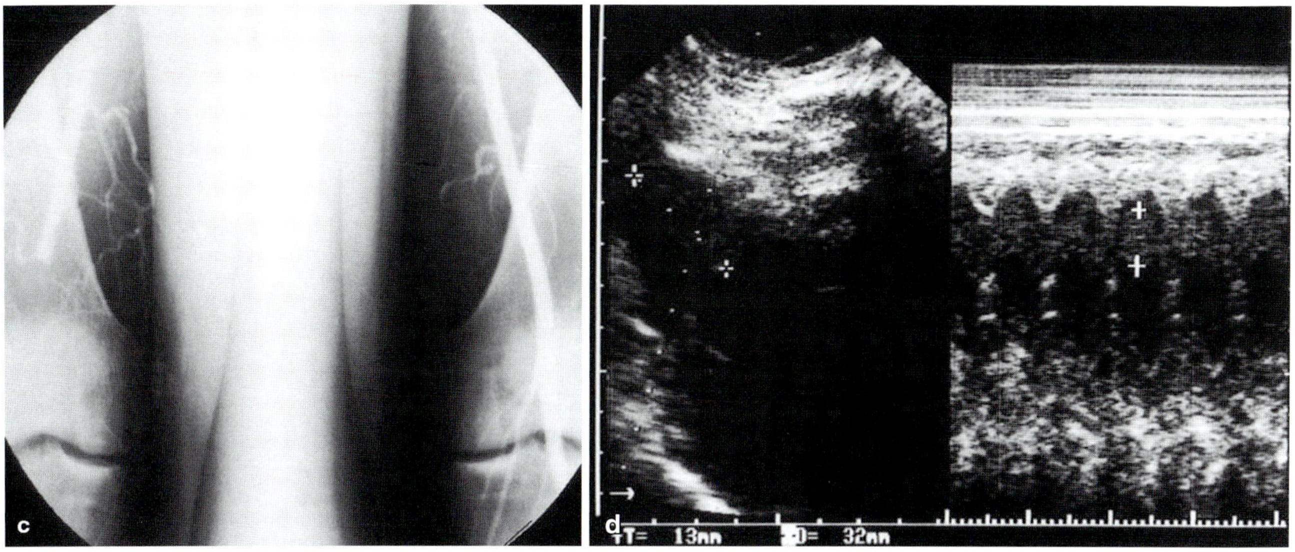

**Fig. A2.39 (continued)**

**c** Angiogram: Popliteal artery occlusion.

**d** Further diagnostic workup reveals a thrombus at the tip of the *left* ventricle as the source of embolic occlusion. The floating thrombus is indicated by *plus signs* in the two-dimensional view on the *left* and in the time-motion display on the *right*.

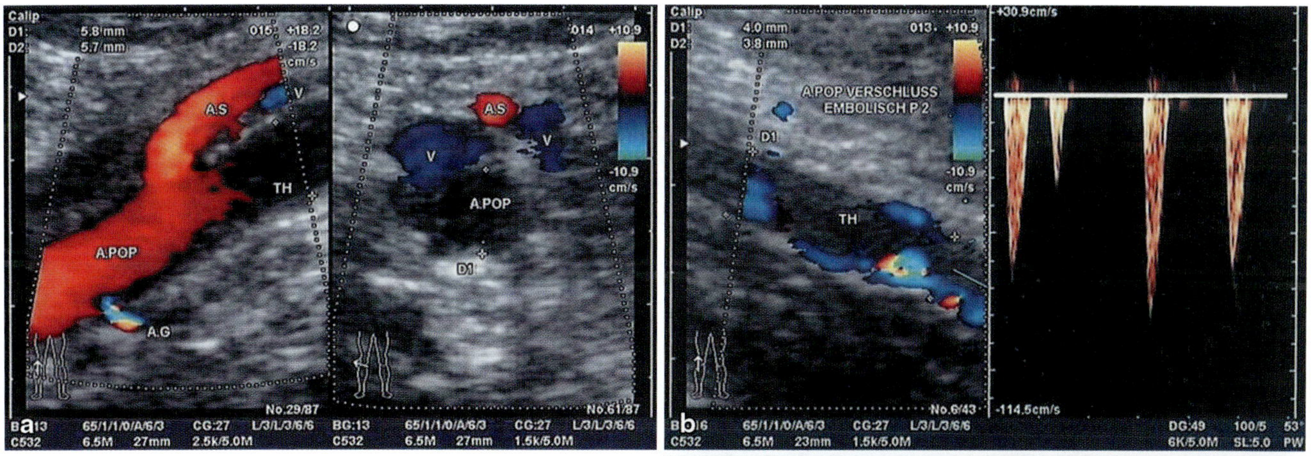


**Fig. A2.40 a–c**

**Embolic occlusion**

**a** Emboli grow in a cranial direction by thrombotic apposition up to the next branching of a hemodynamically significant collateral or become lodged in a bifurcation. In the case of embolic popliteal artery occlusion presented here (longitudinal view on the *left* and transverse view on the *right*), the artery is patent down to the origin of the sural artery while the distal portion is occluded (*TH*). The vessel wall is smoothly delineated and shows no atherosclerotic changes.

**b** When there is spontaneous partial or complete recanalization of thromboembolic occlusions, the Doppler waveform at follow-up will show flow signals near the arterial walls. In the example, flow (blue, away from transducer) along the intraluminal thromboembolic material is demonstrated in the distal popliteal artery. The thrombus (*TH*) is homogeneous and clearly delineated from the wall, which shows no atherosclerotic changes.

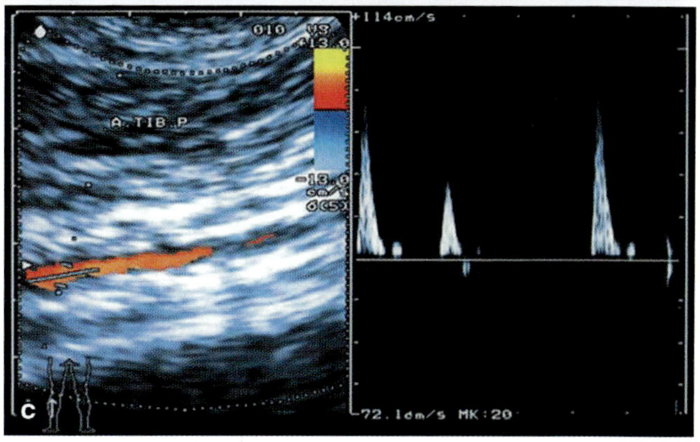


**c** Despite flow obstruction through the popliteal artery thrombus, the Doppler tracing (arrhythmia) from the patent lower leg arteries shows triphasic flow (as illustrated here in the distal posterior tibial artery). With compensation through collateral perfusion, the flow obstruction in the popliteal artery has only little effect on peripheral perfusion. Complete recanalization of the popliteal artery was observed after another 2 days of heparin therapy.

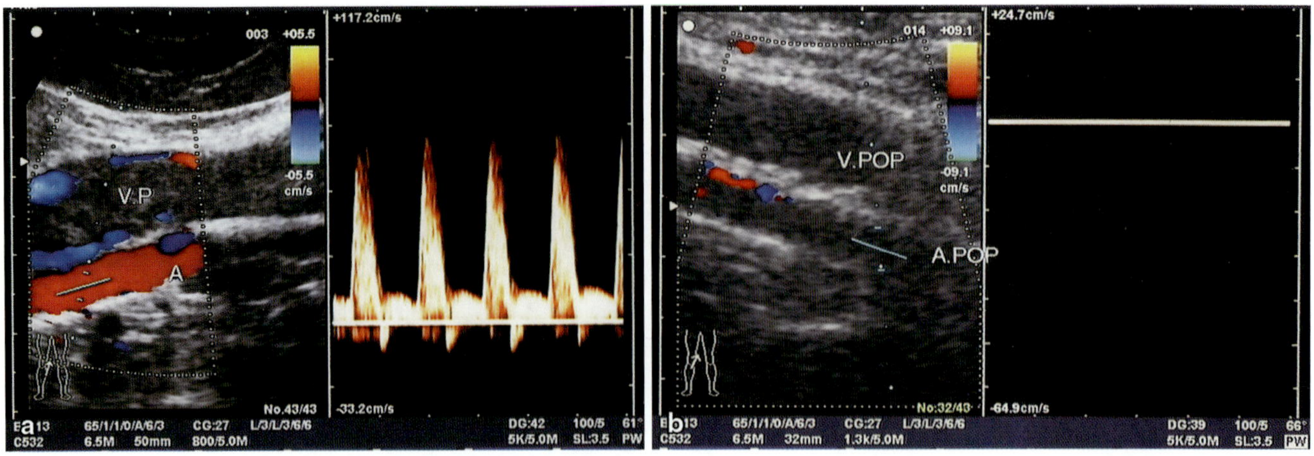


**Fig. A2.41 a, b**

**Arterial occlusion in deep leg vein thrombosis and patent foramen ovale**

**a** Deep vein thrombosis of the leg and ipsilateral arterial embolism in a patient with a patent foramen ovale presenting with a 1-week history of calf swelling and acute-onset forefoot ischemia. There is thrombosis of the calf veins and of the popliteal vein with a free-floating thrombus (*V.P*). The proximal popliteal artery (P1 segment) is patent with high diastolic flow due to low peripheral resistance; regular heartbeat.

**b** The popliteal artery is occluded distal to the origins of sural branches with residual flow around the thrombus; no plaque is demonstrated. Suspected patent foramen ovale was confirmed by echocardiography.

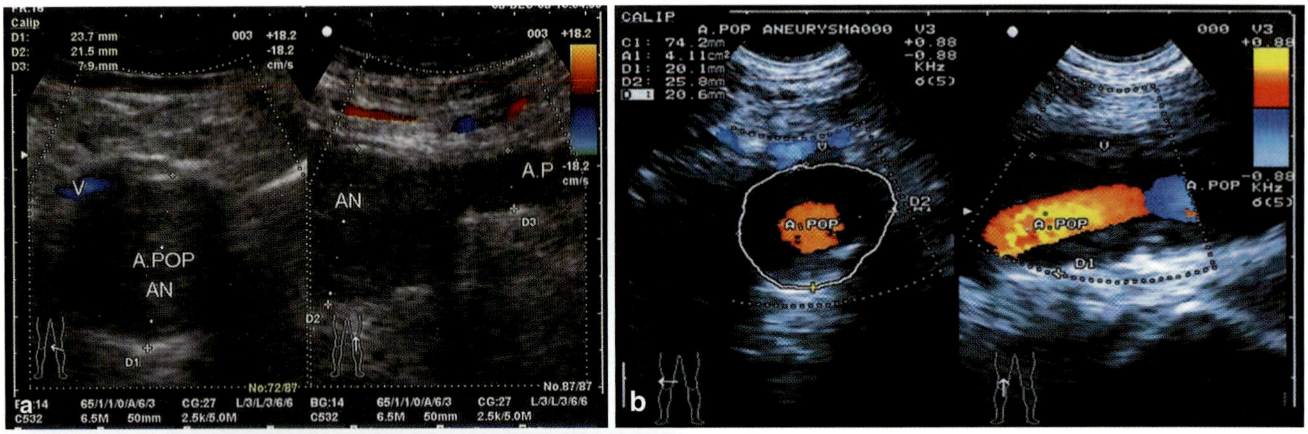


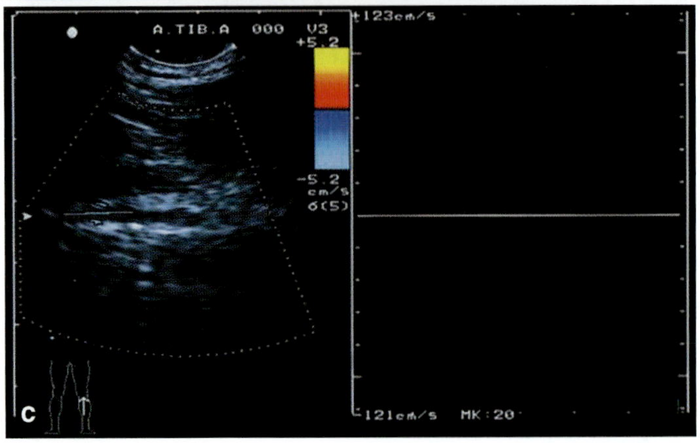


**Fig. A2.42 a–d**

**Aneurysm – popliteal artery**

**a** Patient with ischemic rest pain due to popliteal artery occlusion caused by a completely thrombosed aneurysm. Segments of the compressed vein displayed in blue are seen near the transducer. No flow signals are obtained from the lumen of the popliteal aneurysm. Transverse view of the aneurysm on the *left* (*A.POP*) and longitudinal view on the *right*.

**b** Apart from the risk of perforation, a partially thrombosed popliteal artery aneurysm tends to embolize into distal vessel segments. Chronic recurrent emboli can lead to occlusion of a main lower leg artery, as illustrated here for the mid-third of the anterior tibial artery. Neither color duplex nor the Doppler waveform shows flow. In this example, a femoropopliteal or crural bypass graft to bridge the aneurysm has an unfavorable prognosis due to embolic impairment of the outflow tract.

**c** On the contralateral side, the patent lumen of the popliteal artery (red) is surrounded by the hypoechoic thrombotic deposits of the partially thrombosed aneurysm. The patent lumen of the aneurysm roughly corresponds to the lumen of a normal popliteal artery, as shown on the transverse scan on the *left* and the longitudinal scan on the *right*. The white line indicates the extent of the aneurysm (*left section*).

**Fig. A2.42 (continued)**

**d** Angiogram: Popliteal artery with occlusion on the *left* and aneurysmal dilatation on the *right*. An estimate of the length and diameter of the aneurysm is not possible.

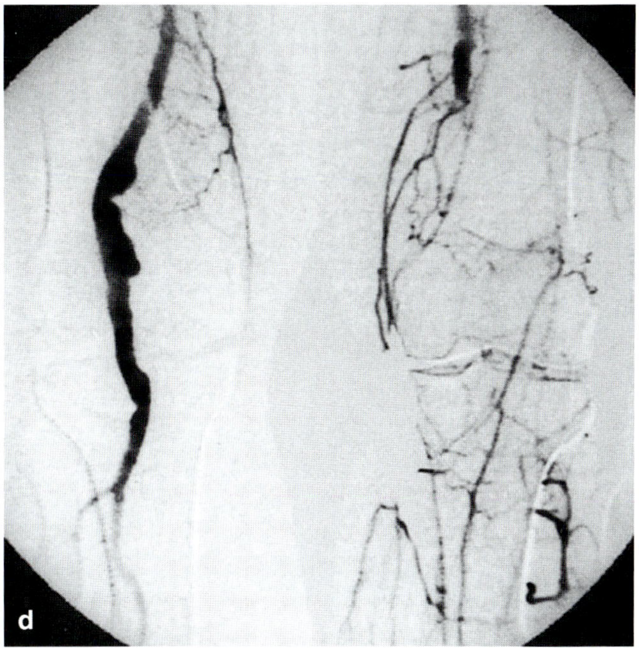


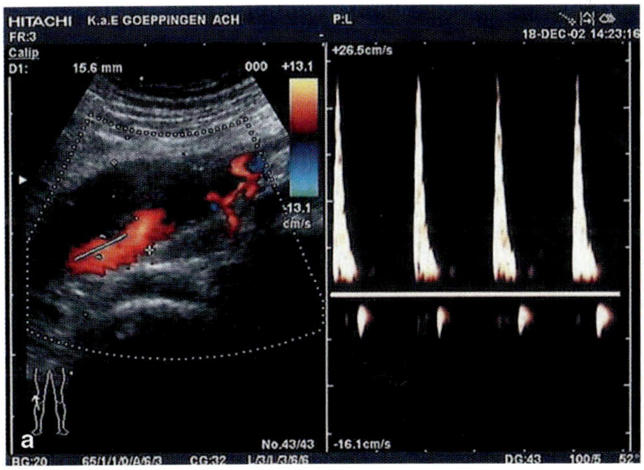


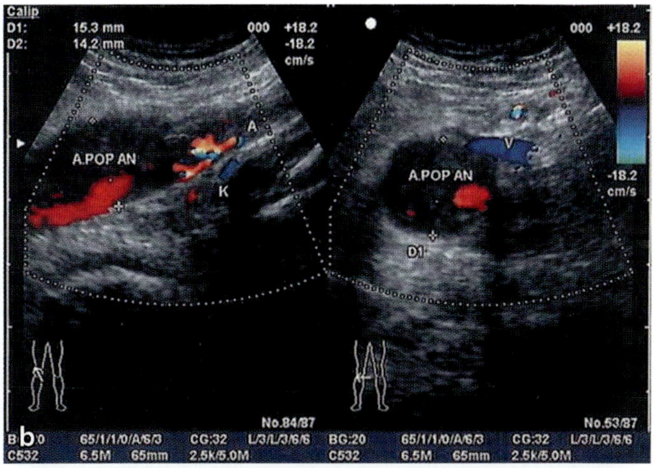


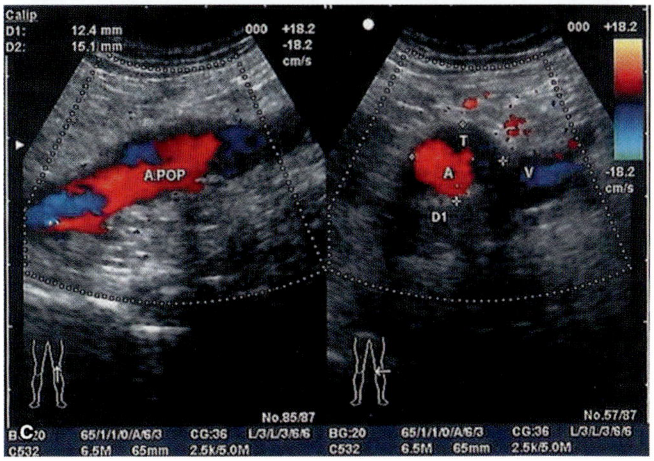


**Fig. A2.43 a–d**

**Small popliteal artery aneurysm with arterioarterial embolism**

**a, b** Patient with small popliteal aneurysms on both sides. Ultrasonography demonstrates occlusion of the popliteal artery distal to the aneurysm on the *right*. The aneurysm is partially thrombosed and has a diameter of 1.5 cm. There is only residual flow through the aneurysm via collaterals (arising from the popliteal artery in the distal aneurysm). These are patent but outflow is obstructed. This situation is reflected by a thump pattern and a PSV reduced to 22 cm/s.

**c** Scans of the left popliteal artery (longitudinal view on the *left*, transverse view on the *right*) depict the small aneurysm (diameter of 1.5 cm) with only little thrombosis (clearly seen on the transverse view only) and a patent residual lumen of normal width. The lower leg arteries are still patent. The control examination performed prior to elective aneurysm resection showed an unchanged configuration of the aneurysm, but lower leg artery occlusions due to arterial embolism.

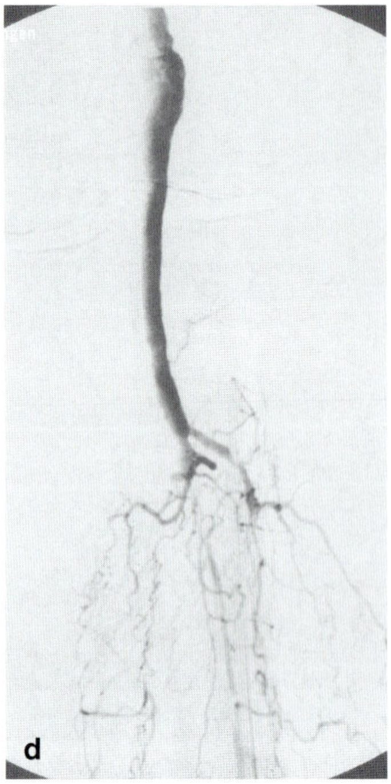


**Fig. A2.43 (continued)**

**d** Left-sided angiogram showing occlusions of lower leg arteries without significant dilatation of the popliteal artery. Only at the upper margin of the image does the popliteal artery appear somewhat ectatic (corresponding ultrasound scans in **c**).

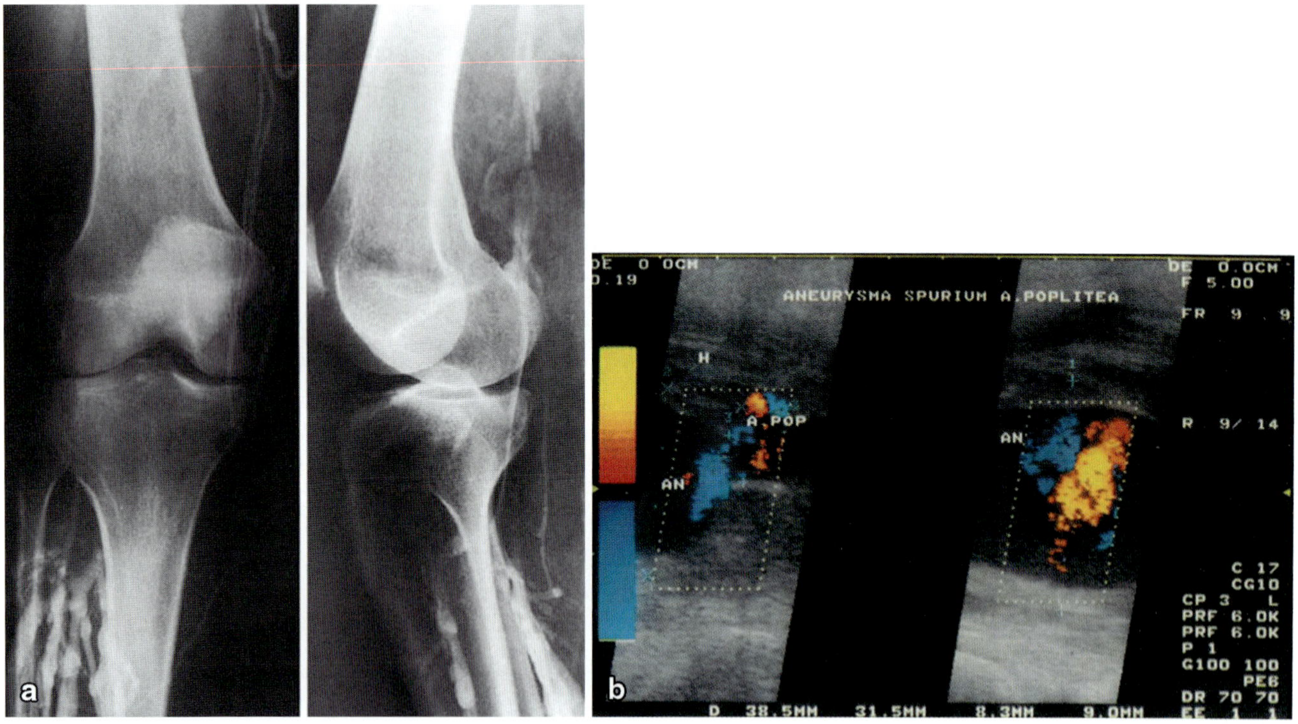


**Fig. A2.44 a–d**

**Pseudoaneurysm following arthroscopy**

**a** Iatrogenic damage to the vessels in the popliteal fossa is a rare but serious complication of knee arthroscopy. In the case presented, a large pseudoaneurysm developed after outpatient arthroscopy with partial resection of the medial meniscus. Venography performed for swelling of the calf showed contrast defects in the popliteal vein, which were misinterpreted as popliteal vein thrombosis.

**b** Duplex imaging after initiation of anticoagulation treatment demonstrates the pseudoaneurysm. In the aneurysm, there is flow toward and away from the transducer (*right section*). Black areas without flow signals either indicate stasis in the aneurysm or are due to the failure to obtain flow signals at an angle of 90° (cos 90° = 0).

The scan on the *left* depicts the canal connecting the popliteal artery (*A.POP*) and the aneurysm (*AN*) in blue, indicating flow from the artery into the aneurysm. The aneurysm is surrounded by a hematoma (*H*). Ultrasound shows the popliteal vein to be compressed by the aneurysm rather than thrombosed.

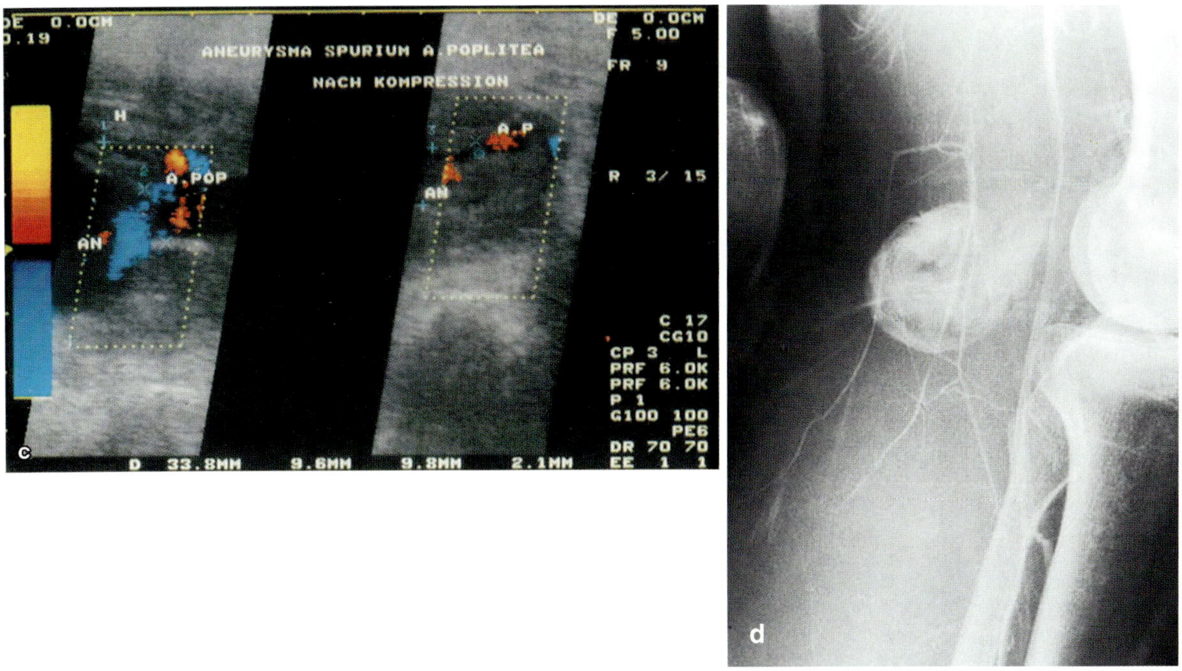


**Fig. A2.44 (continued)**

**c** The attempt to induce thrombosis of the aneurysm by compression failed because the neck is too wide and there is no adequate structure against which to compress it. The *right section* shows persistent flow after attempted compression. Thrombin injection would have been an alternative in this case but experience with this therapy was still limited when this patient was seen 15 years ago.
**d** Angiogram: Pseudoaneurysm of the popliteal artery.

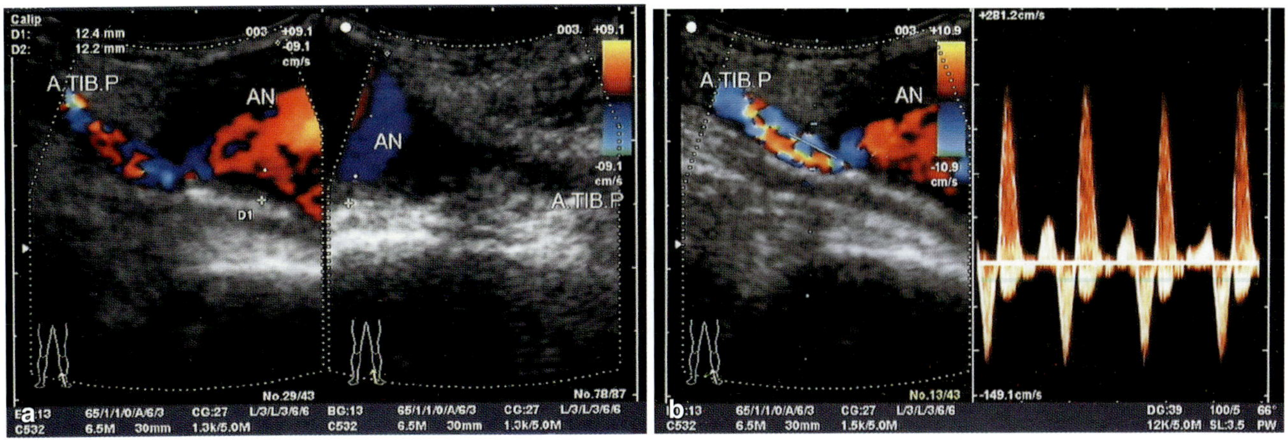


**Fig. A2.45 a, b**

**Aneurysm of posterior tibial artery**

**a** Traumatic aneurysm (13 mm in diameter) of the posterior tibial artery just above the ankle joint. There is an abrupt increase in diameter from 2.5 to 13 mm (montage of two adjacent scans showing the aneurysm in the center). The posterior tibial artery is patent proximal to the aneurysm and occluded distal to it (*A.TIB.P*). A collateral artery arises from the aneurysm.

**b** The posterior tibial artery has a triphasic flow pattern just proximal to the aneurysm (*AN*). The distal segment is occluded, and flow is maintained through a collateral arising from the aneurysm. The resulting higher outflow resistance leads to a diastolic to-and-fro flow pattern (normal mid-diastolic flow with reversed early and end-diastolic flow).

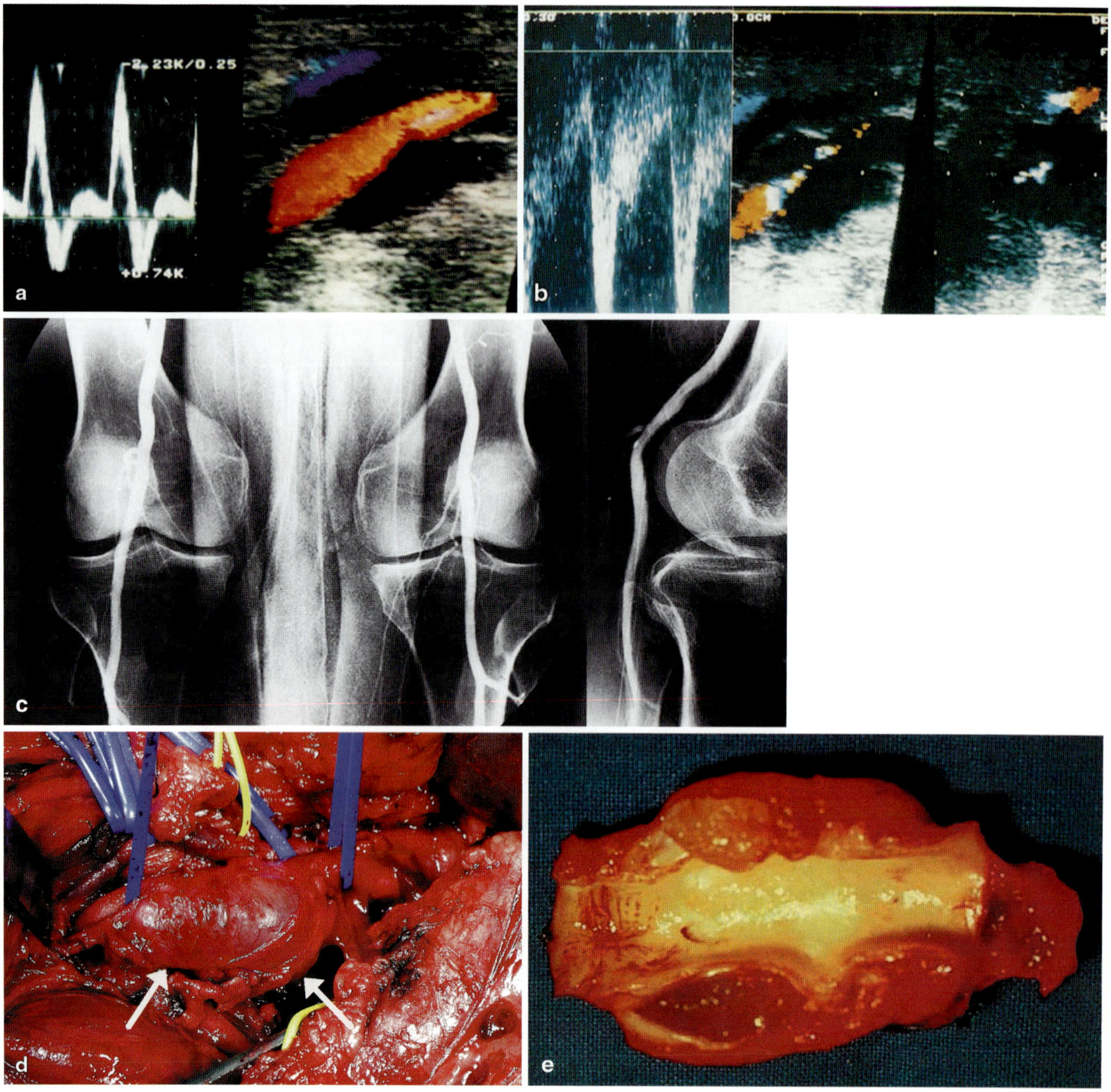

**Fig. A2.46 a–e**
**Adventitial cystic disease**

**a** The popliteal artery (red) is surrounded by hypoechoic cystic lesions, which produce slight indentation of the patent lumen but no hemodynamically significant narrowing. The Doppler waveform shows triphasic flow. The patient reports intermittent claudication with a highly variable walking distance.

**b** Seven days after the first examination, the patient presents with severe claudication and a maximum walking distance of 30 m. Ultrasound shows a markedly increased cyst volume with high-grade stenosis of the popliteal artery (*middle section*: longitudinal view; *right section*: transverse view). Color duplex ultrasound depicts a small residual lumen between the cysts with accelerated flow and aliasing. The Doppler tracing is presented in the inverted mode with arterial flow displayed below the baseline. The waveform indicates stenosis with monophasic flow and a flow velocity of over 3 m/s.

**c** Angiography performed 2 weeks later: Fairly inconspicuous popliteal artery with only slight anterior indentation, identified on a lateral view. Duplex scanning performed at this time shows a marked decrease in cyst size without hemodynamically significant stenosis, similar to the situation depicted in **a**.

**d** Intraoperative view of cystic adventitial degeneration (*arrow*). Blue slings are placed around the popliteal artery proximal and distal to the diseased arterial segment.

**e** The therapy of choice is surgical resection of the cyst-bearing arterial segment or enucleation of the cysts if the intima is still intact. In the example, gross inspection of the surgical specimen shows the adventitial cysts to be filled with gelatinous material.

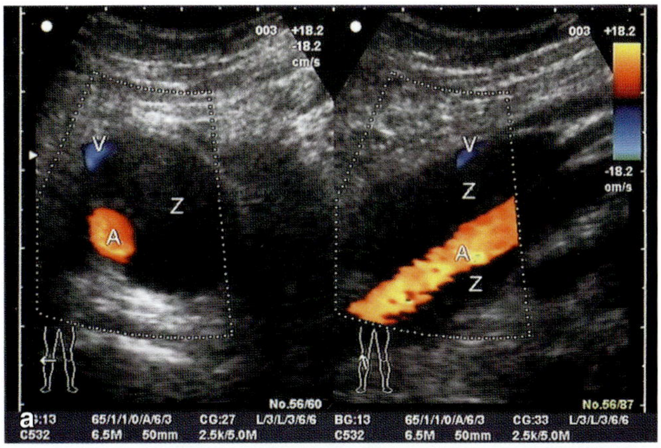


**Fig. A2.47 a, b**

**a** 40-year-old patient with intermittent foot pain resembling that of polyneuropathy. Arterial duplex imaging of the popliteal fossa reveals large cysts causing only mild luminal narrowing of the popliteal artery without significant hemodynamic effects. The patient reports no episodes of typical intermittent claudication but variable neurologic signs and symptoms. The neurologic examination revealed slightly reduced peripheral nerve conduction velocity.

In patients with adventitial cystic disease, the symptoms vary with the number, size, and location of cysts within the narrow confines of the popliteal fossa. An occasional patient may present with (intermittent) pain due to nerve compression by a large cyst, while the popliteal artery is not compromised. The patient shown has a large cyst (*Z*), but neither the color duplex images (transverse view on the *left*, longitudinal view on the *right*) nor the spectral Doppler measurement (not shown) suggest significant narrowing of the arterial lumen.

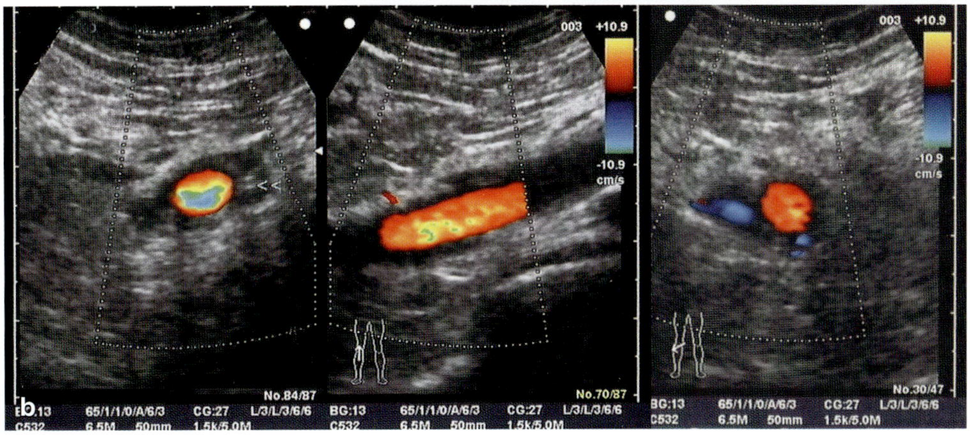


**b** Because the patient refused an operation, the cyst was drained and sclerosed under ultrasound guidance (transverse and longitudinal views on the *left*); histologic examination of the gelatinous cyst fluid confirmed adventitial cystic disease.

Following ultrasound-guided drainage using a 1.8-mm needle, the cyst was sclerosed with 1 mL of 95% ethyl alcohol to prevent recurrence (needle tip identified by bright echo). The patient's symptoms disappeared after treatment.

*Right* section: Follow-up after 1 month reveals no recurrent or residual cyst; popliteal vein with blue-coded flow lateral to the artery.

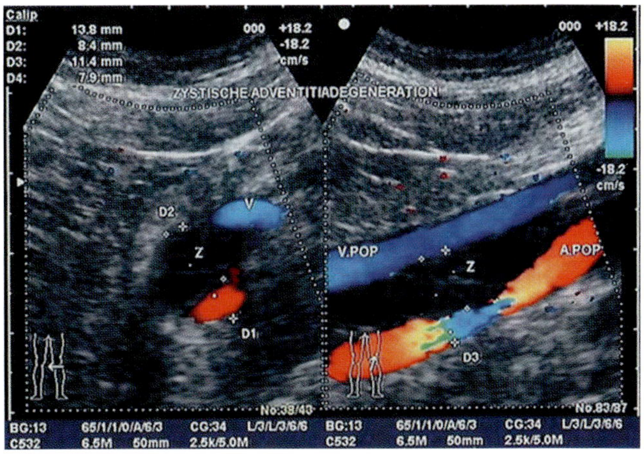


**Fig. A2.48**

**Adventitial cystic disease – differentiation from dissection**

Single or multiple cysts as well as cysts extending over a long vessel segment may occur in adventitial cystic disease. A cyst involving a long popliteal segment as in this example may be difficult to differentiate from dissection with complete thrombosis of the false lumen (see Figs. A2.52a and A5.24a, b). The popliteal artery (*A.POP*) is shown in transverse orientation on the *left* and in longitudinal orientation on the *right* with the cyst (*Z*) narrowing a long segment of the artery (confirmed intraoperatively). There is aliasing as a result of cystic luminal narrowing. The popliteal vein (*V.POP*) is depicted closer to the transducer with flow coded in blue.

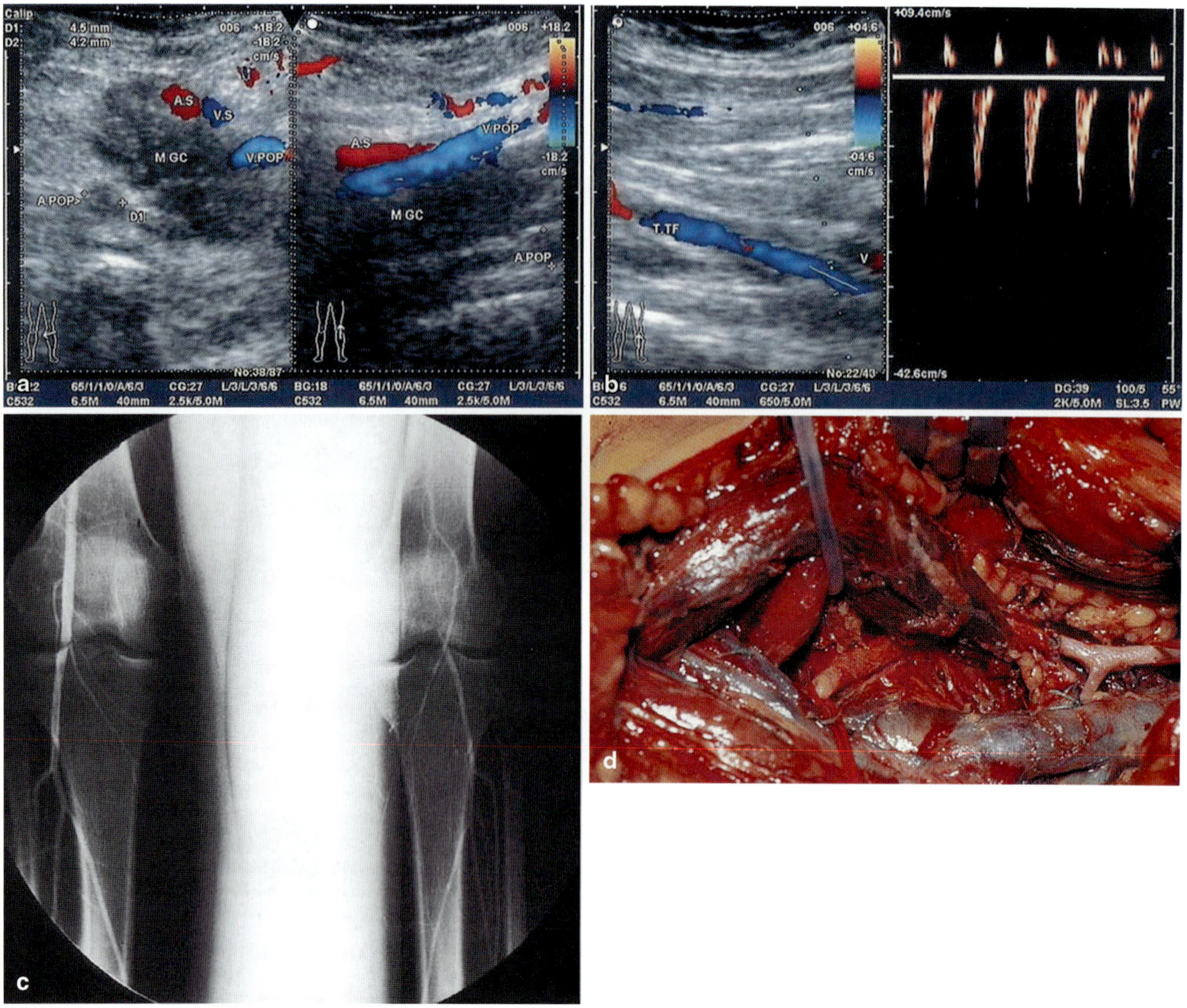


**Fig. A2.49 a–d**
**Entrapment syndrome**

**a** Isolated popliteal artery occlusion due to malformation of the medial head of the gastrocnemius muscle forcing the artery to course around the head on the medial side. In this type of malformation, the medial head of the muscle is located between the popliteal artery and vein – which thus do not pass through the popliteal fossa together – and compresses the artery against the femur with each plantar flexion. Intermittent compression damages the vessel wall with deposition of thrombotic material, which may ultimately progress to occlusion. No color duplex signal is obtained from the popliteal artery (*A.POP*). Posterolateral to the head of the gastrocnemius, the patent popliteal vein (*V.POP*) is depicted closer to the transducer with a blue flow signal. Posterior to it, the artery (red) supplying the soleus muscle and serving as a collateral and the vein are shown. The arteries acting as collaterals are markedly dilated due to the chronic occlusion process and may thus be confused with the popliteal artery. The sonoanatomic situation (transverse section on the *left* and longitudinal section on the *right*) is as follows: The popliteal artery courses anterior to the popliteal vein and is depicted farther away when scanning is performed from the posterior approach. The muscle-supplying arteries acting as collaterals arise from the posterior aspect of the popliteal artery and course posterior to the popliteal vein and are thus closer to the transducer than the vein.

**b** In this case with good collateralization of a chronic occlusive process, the flow profile in the refilled tibiofibular trunk does not show the typical postocclusive monophasic flow but is triphasic, though damped. Peak systolic velocity is just under 20 cm/s. Additional collaterals enter distally. There is no postocclusive peripheral dilatation at rest.

**c** Angiogram: Short occlusion of the *left* popliteal artery with refilling at the level of the knee joint cleft (lateral collateral).

**d** The intraoperative site confirms the ultrasound findings. The popliteal artery and vein do not pass through the popliteal fossa together because the medial gastrocnemius head (encircled with transparent vessel loop) attaches between the artery (distal segment encircled with red vessel loop) and the vein (at *lower margin*). The proximal popliteal artery (on the *right*) gives off the collateral already identified sonographically and accompanying the vein.

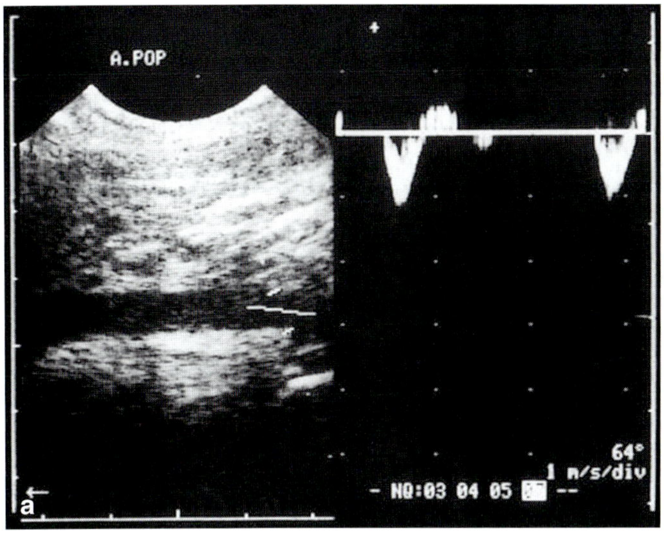


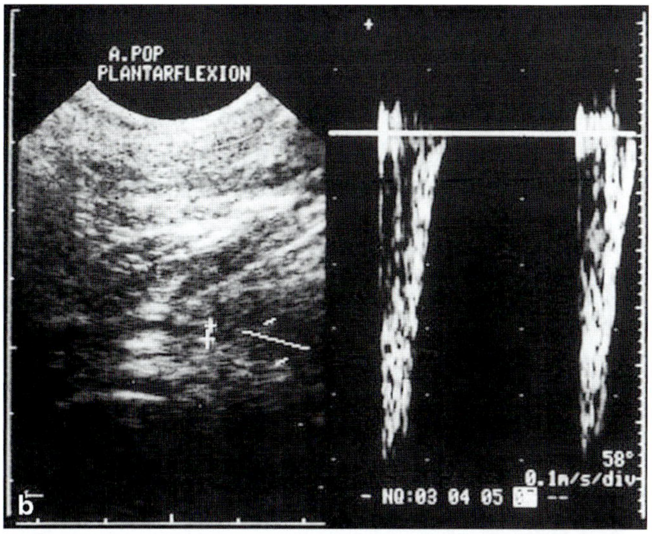


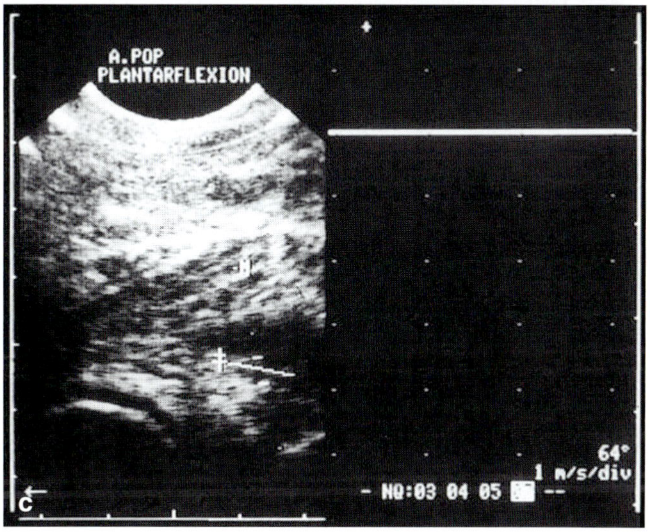


**Fig. A2.50 a–c**
**Entrapment syndrome**
**a** Calf swelling with occasional pain in a young patient caused by external compression of the vessels in the popliteal fossa due to a hypertrophied head of the gastrocnemius with normal attachment. The popliteal artery and vein pass through the popliteal fossa together and the vein is already compressed by the relaxed muscle (see Fig. A3.67b, c). The popliteal artery is not stenosed and shows triphasic flow in the Doppler measurement.
**b** Progressive compression of the popliteal artery occurs with increasing plantar flexion, producing a stenosis signal in the Doppler waveform with loss of triphasic flow and a PSV of 300 cm/s.
**c** Further plantar flexion leads to complete occlusion of the popliteal artery through muscular compression (see Fig. A3.67a–c: Popliteal entrapment syndrome with arterial and venous compression). This form of entrapment syndrome (type IV, see classification in Fig 2.26) occurs without malformation and is solely due to a well-developed gastrocnemius muscle (which may result from anabolic intake).

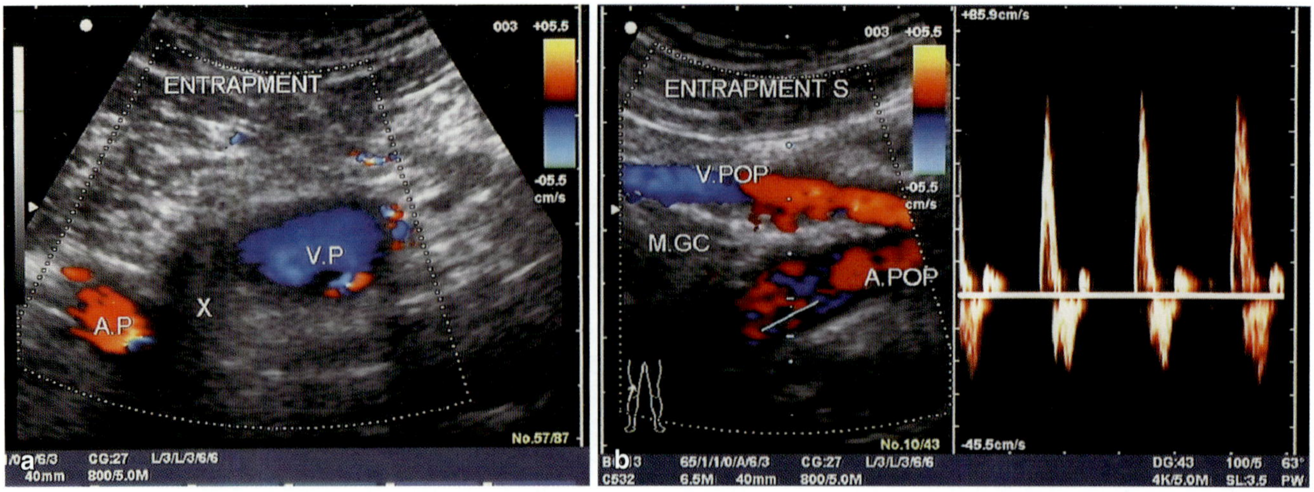


**Fig. A2.51 a, b**
**Entrapment constellation**
**a** The scan shows the characteristic abnormality of popliteal fossa anatomy predisposing an individual to popliteal entrapment: muscle structures (*X*) lying between the popliteal artery (*A.P*) and vein (*V.P*). This anatomic constellation may be present even if no compression of vascular structures can be elicited by plantar flexion of the ankle. In the literature, only little attention has been paid to this anatomic deviation in asymptomatic individuals, but it explains why popliteal entrapment is much more commonly encountered at autopsy than in the clinical setting. An examiner may see this anatomic constellation during a careful sonographic evaluation of the popliteal fossa in patients examined for other reasons (e.g., suspected venous thrombosis, chronic venous insufficiency). The identification of musculotendinous structures (attachment of medial head of gastrocnemius) between the artery and vein in the popliteal fossa is pathognomonic of this constellation.
**b** In this case, neither color duplex imaging nor Doppler interrogation shows popliteal artery (*A.POP*) narrowing during provocative maneuvers (plantar flexion of the ankle). There is normal triphasic flow and PSV is not increased. Longitudinal view showing the head of gastrocnemicus (*M.GC*) between the popliteal artery (*A.POP*) anteriorly (closer to transducer) and the popliteal vein (*V.POP*) posteriorly (scan obtained during maximum plantar flexion).

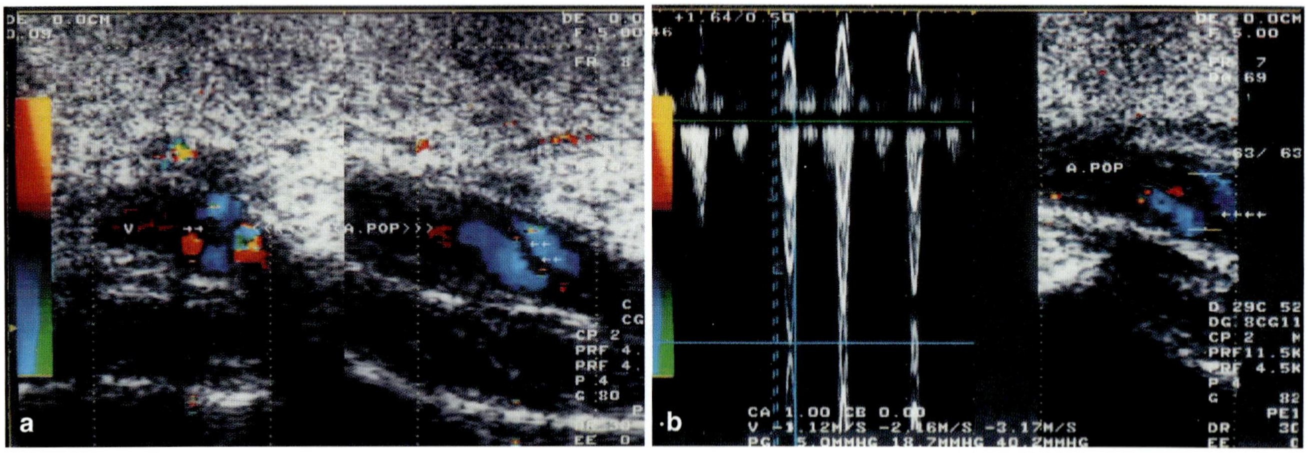


**Fig. A2.52 a, b**
**Dissection**
**a** Popliteal trauma can cause arterial wall dissection with ischemia-related symptoms of varying severity. A tear of the intima with detachment is depicted as a disruption of the color-coded arterial flow column in the color duplex mode. In the arteries below the inguinal ligament, dissection rarely leads to the development of two lumina over a long segment but gives rise to intimal flaps that can block the lumen. The transverse view of the artery is shown on the *left* and the longitudinal view on the *right*; *arrows* indicate the dissection. The transverse view of the popliteal artery depicts at least two lumina with different color codings.
**b** The Doppler waveform from the dissected popliteal artery simultaneously depicts flow signals of different velocities and directions. There is proper blood flow to the periphery, but a "knocking" waveform consistent with several blood columns with different flow velocities. Intraoperative inspection revealed a partial tear of the intima and two lumina with forward flow and flow-obstructing intimal flaps more distally. The resected arterial segment was replaced by a venous graft.

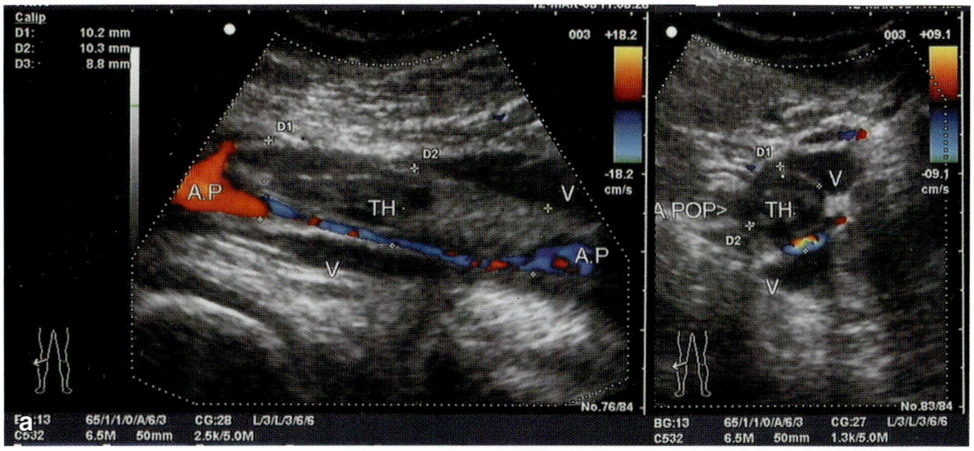


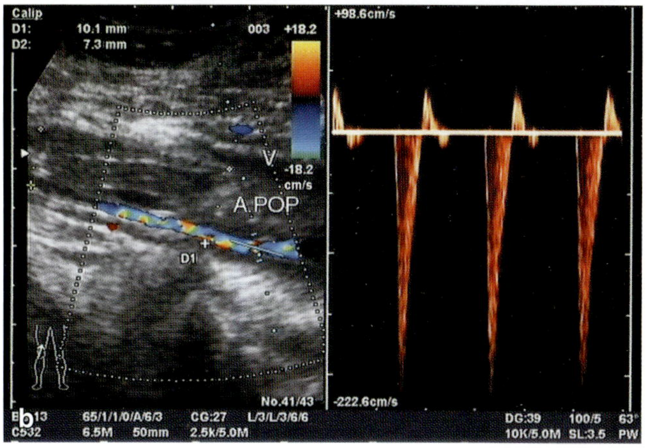


**Fig. A2.53 a, b**
**Dissection**

**a** Dissection of the popliteal artery with complete thrombosis of the false lumen (*TH*) following blunt trauma to the popliteal fossa. Longitudinal view (*left*) and transverse view (*right*) show the patent residual lumen of the artery (*A.P*), which is narrowed by the thrombosed false lumen.
**b** PSV in the compromised popliteal artery segment is increased to 220 cm/s. Plus signs indicate the popliteal artery lumen and the thrombosed false lumen (*left* section of the figure).

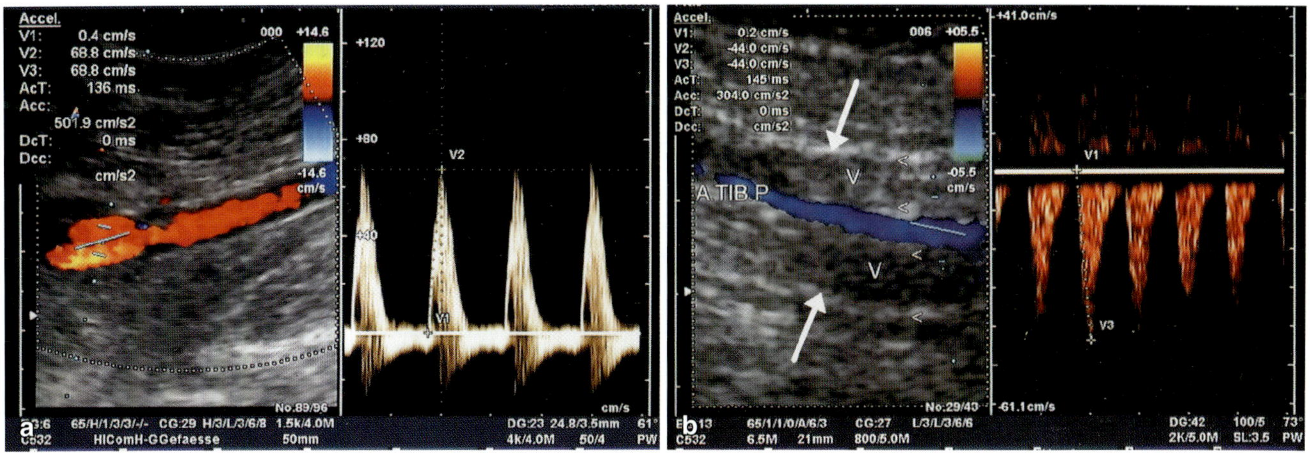


**Fig. A2.54 a, b**
**Progressive ischemia due to venous outflow obstruction (extensive venous thrombosis)**

**a** 75-year-old woman with a history of AOD and occlusion of the superficial femoral artery with very good collateral circulation. Doppler recording shows triphasic flow with a PSV of 68 cm/s in the popliteal artery. The most salient feature of postocclusive flow seen in this case is a delayed systolic upstroke with a rise time of 136 ms.
**b** The patient developed secondary peripheral thrombosis ascending to the common femoral vein (level of the inguinal ligament) and presenting with swelling and acute ischemic pain in the forefoot and classic signs of ischemia as well as early ischemic toe necrosis.

Duplex scanning reveals no macroangiopathic changes in perfusion compared with her status prior to the onset of thrombosis. In the lower leg, the posterior tibial and dorsalis pedis arteries are patent to just below the ankle joint. The Doppler waveform from the posterior tibial artery (same as in the dorsalis pedis artery) is presented, confirming a largely normal PSV (44 cm/s). The longer rise time of 145 ms is consistent with postocclusive flow. However, in a patient with foot ischemia, the Doppler recording should also reflect the flow effects of peripheral dilatation; the pulsatile flow profile seen in this case is due to venous outflow obstruction caused by extensive venous thrombosis. For illustration, the two thrombosed veins (*V*) are depicted above and below the arteries (venous wall indicated by *arrow*); also seen is the posterior tibial artery (*A.TIB.P*). The veins are dilated and no flow signals are obtained despite a low PRF. The dorsalis pedis vein was also thrombosed (not shown).

The ultrasound and Doppler findings show that disease progression with toe necrosis in this patient is attributable to venous obstruction caused by extensive thrombosis also involving the arterioles. This condition cannot be remedied by a femoropopliteal bypass graft. Nevertheless, a bypass procedure was performed in the acute situation, but no improvement ensued. In summary, in this case of stage IIa PAOD, extensive thrombosis led to the clinical and sonographic picture known as phlegmasia coerulea dolens.

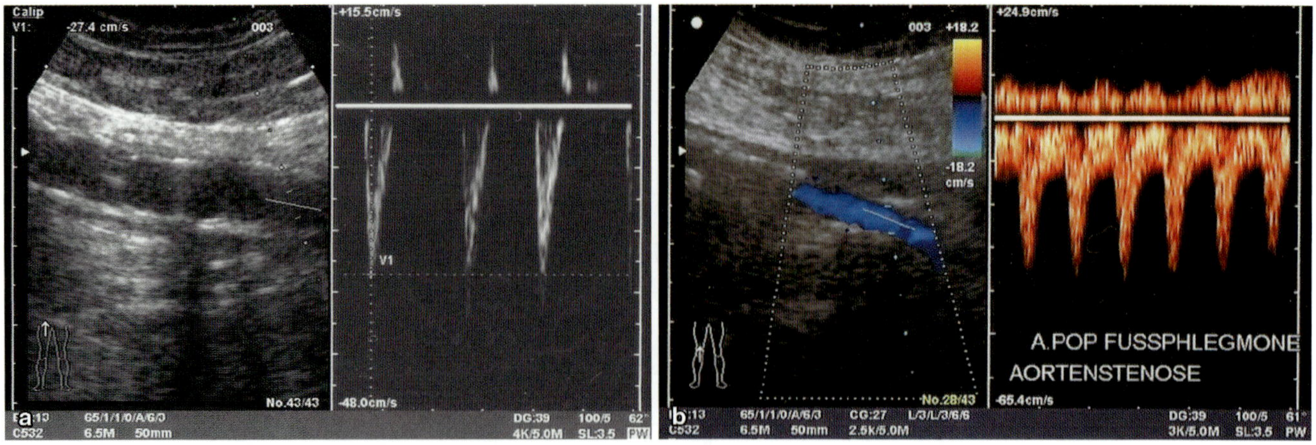


**Fig. A2.55 a, b**

**Cardiac causes of abnormal spectral Doppler findings**

**a** Patient with low peak systolic velocities at multiple Doppler sampling sites in the leg. As no stenosis is identified, one should consider cardiac insufficiency with reduced cardiac output as a possible underlying cause. If this is the case, PSV will be reduced in all arterial segments. In the example, a PSV of 25 cm/s is measured in the proximal superficial femoral artery; plaques are identified in one artery, while no stenosis or occlusion is detectable down to the ankle joint.

**b** In a patient with a higher-grade aortic stenosis, a Doppler tracing from a peripheral artery will show the same poststenotic pattern as distal to a stenosis of a peripheral artery: delayed systolic upstroke, reduced PSV (32 cm/s in the case shown), and monophasic flow. In this case, a foot phlegmon further contributes to the changes in the spectral waveform from the popliteal artery.

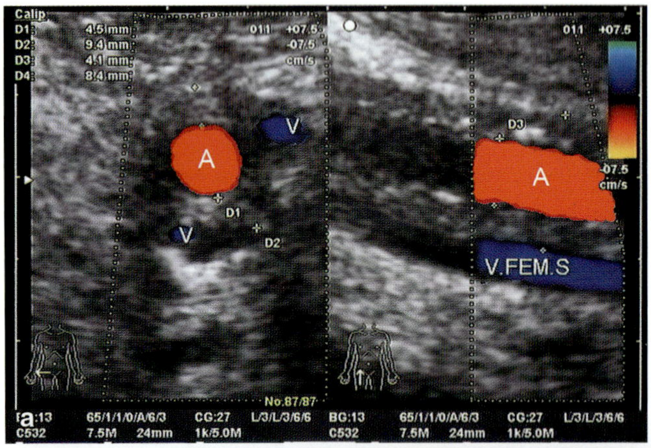


**Fig. A2.56 a–c**

**Vasculitis**

**a** Vasculitis of the femoral artery (longitudinal view on the *left*, transverse view on the *right*) with concentric hypoechoic wall thickening in a patient with concomitant atherosclerosis. The atherosclerotic plaques on the luminal side of the media are seen as hyperechoic deposits on the thickened wall.

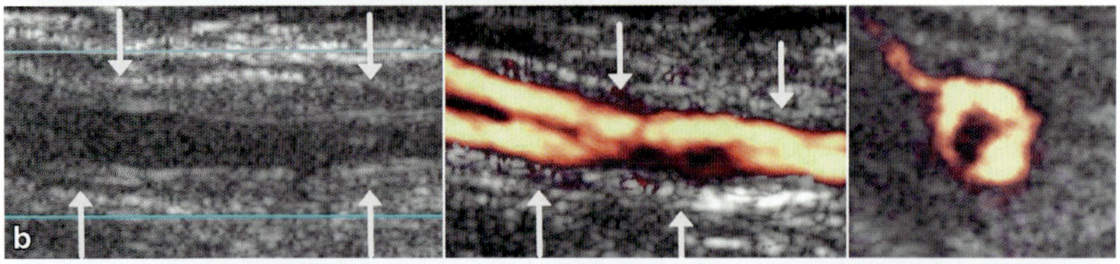

**b** Calf artery (posterior tibial artery) in polyarteritis nodosa with circumferential wall thickening. *Left* section shows longitudinal view, *middle* section a longitudinal power Doppler image, and *right* section a transverse power Doppler image.

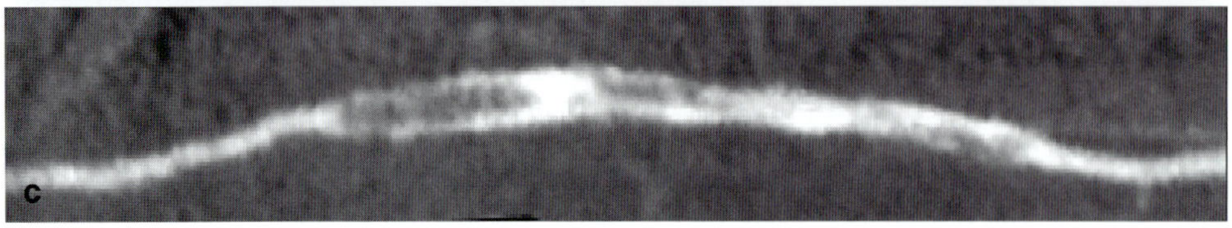

**c** Angiogram of the same artery as in **b** (Figs. b and c courtesy of K. Amendt).

**Fig. A2.57**

**Inflammatory vascular disease**

Vascular inflammation – Takayasu's arteritis of the subclavian and common carotid arteries or polyarteritis nodosa – leads to concentric wall thickening with a centrally perfused lumen in extremity vessels as in the case presented here. It is identified on ultrasound by the macaroni sign. There is a normal echo reflected from the wall interface while the remainder of the arterial wall is depicted as a concentric, hypoechoic structure (wall thickening) over a long segment without signs of atherosclerotic plaques. Progressive inflammatory wall thickening may ultimately lead to occlusion of the affected vessel. Aneurysmal changes may also occur. The longitudinal view on the *left* and the transverse view on the *right* show the concentric wall thickening of a lower leg artery in a patient with polyarteritis nodosa. (Due to reflux caused by postthrombotic venous changes, the veins depicted to the *left* and *right* of the artery are likewise displayed in blue).

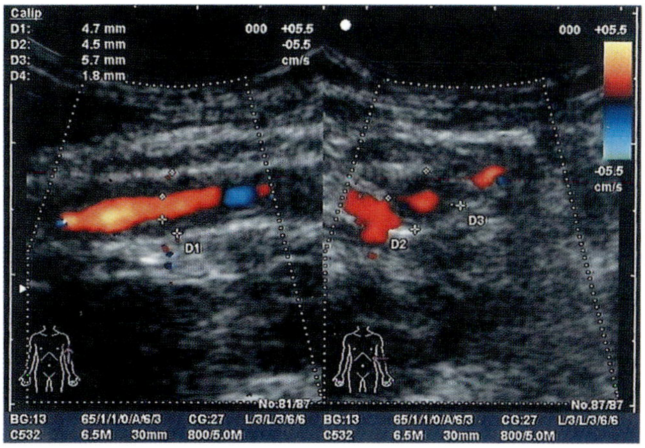


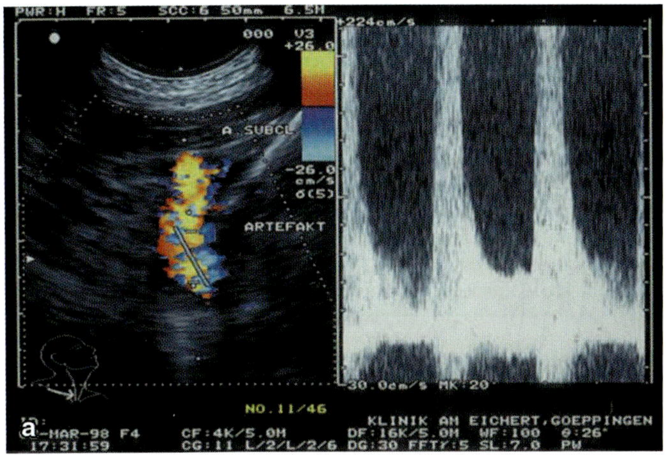


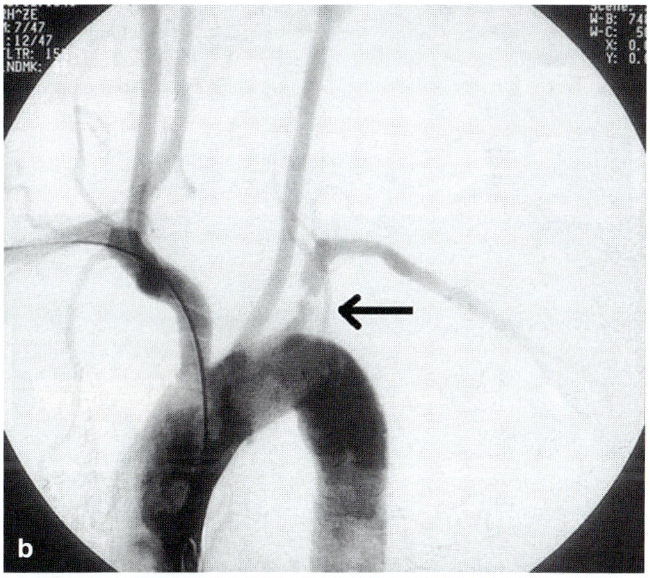


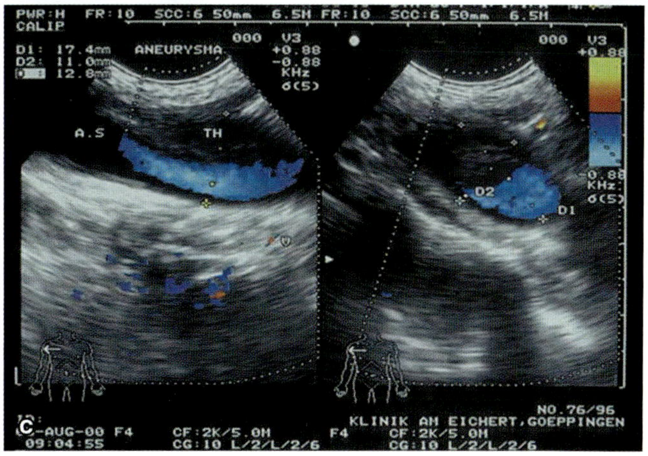


**Fig. A2.58 a–c**

**Subclavian artery stenosis due to atherosclerosis**

**a** Scanning of the subclavian artery from the supraclavicular position demonstrates direct signs of stenosis: increased peak systolic flow, aliasing, and perivascular vibration artifacts. Atherosclerotic stenoses of the arm arteries typically affect the origin of the subclavian artery and cannot be identified on the basis of direct criteria in all cases. Instead, the diagnosis has to rely on indirect criteria such as monophasic postocclusive flow.

**b** Angiogram showing stenosis of the subclavian artery on the *left*.

**c** The aneurysm (*AN*) of the *right* subclavian artery (longitudinal view on the *left*, transverse view on the *right*) is not depicted angiographically (see **b**) due to thrombosis.

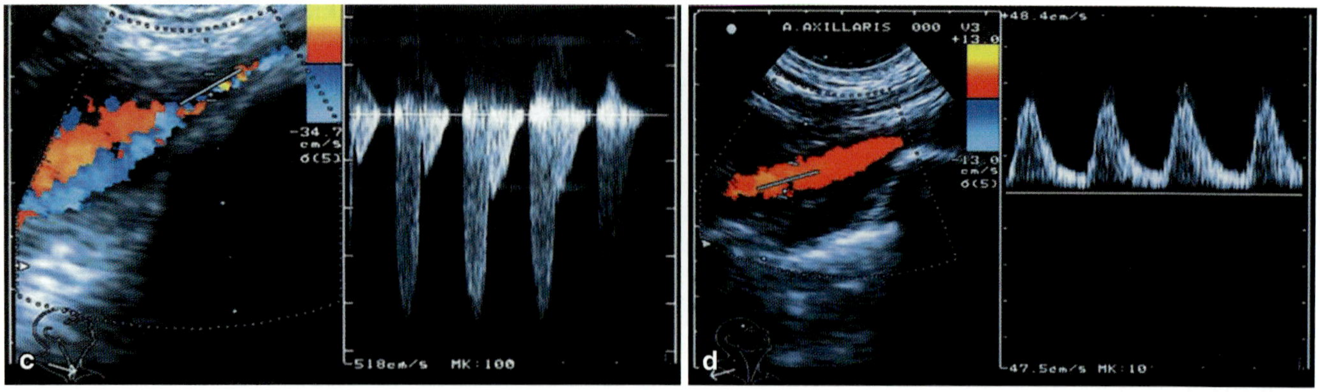


**Fig. A2.62 (continued)**

**c** Examination during Adson's test reveals compression of the subclavian artery with aliasing on color duplex scanning and a PSV > 400 cm/s in the Doppler waveform (stenosis signal). The test is positive for a compression syndrome with poststenotic dilatation.

**d** Specific anatomic conditions (obesity and short neck) may prohibit proper placement of the transducer during Adson's test. In such cases, compression during provocation can be demonstrated by the presence of the typical poststenotic changes in the Doppler waveform sampled in the axillary artery with the transducer placed in the infraclavicular fossa.

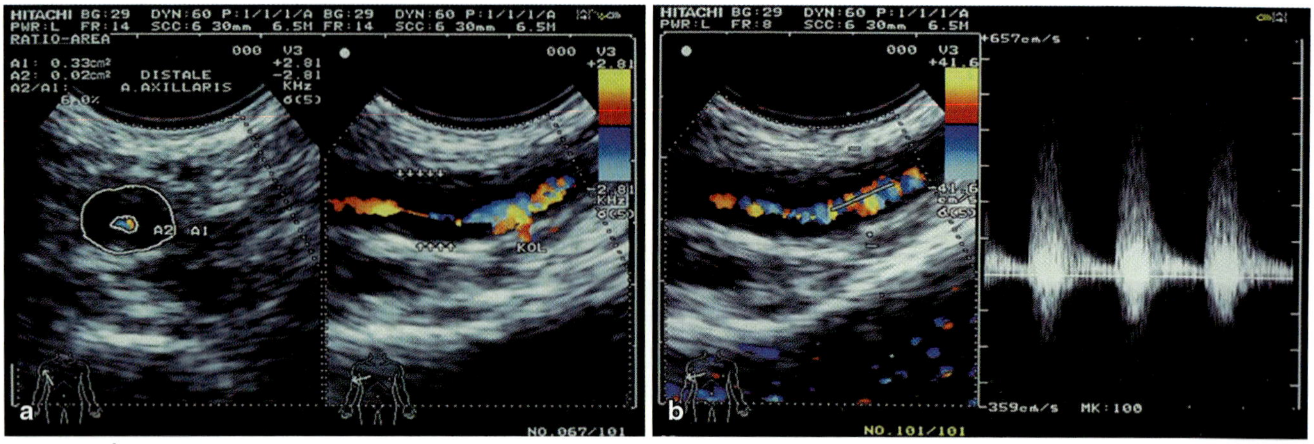


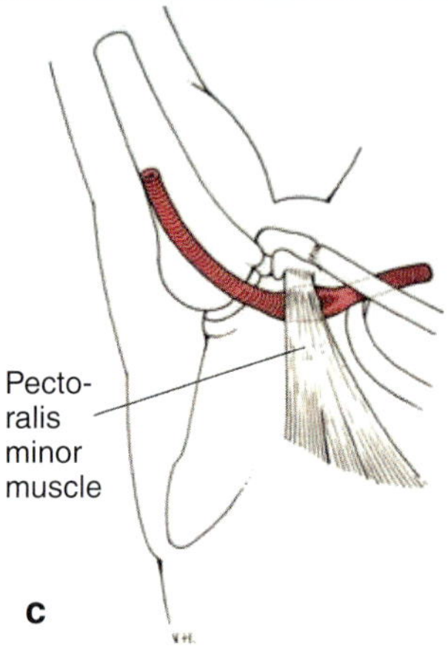


**Fig. A2.63 a–c**

**Pectoralis minor syndrome**

**a** The scan of the axillary artery obtained during hyperabduction with the transducer in the armpit shows the compression-induced stenosis as well as the complications of long-standing compression syndrome: extensive, though circumscribed, wall damage with thickening and local thrombus formation. In the transverse view on the *left*, the normal vessel diameter is marked; aliasing in the color duplex mode enables differentiation of the perfused lumen from mural thrombus.

A low echogenicity and concentric wall thickening as in this case may also occur in vasculitis (which must be considered in the differential diagnosis of patients presenting with elevated inflammatory parameters).

**b** Doppler waveform showing high-grade stenosis with a flow velocity of over 3 m/s, monophasic flow, and turbulence.

**c** Diagram of compression of the axillary artery between the pectoralis minor muscle and the coracoid process during hyperabduction (from Heberer and van Dongen 1993).

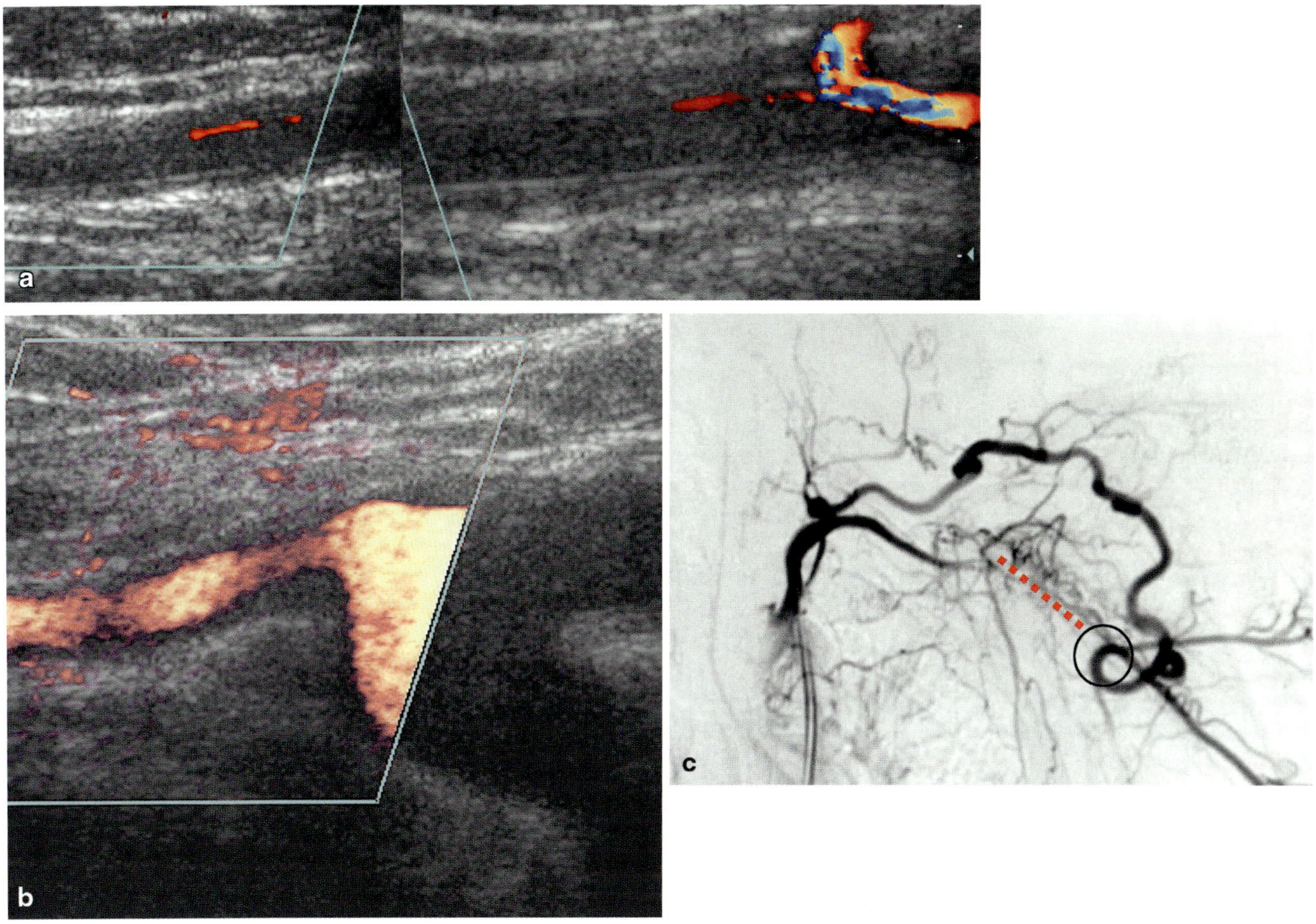


**Fig. A2.64 a–c**
**Takayasu's arteritis with subclavian artery occlusion**
**a** Long occlusion of the axillary artery and distal subclavian artery. There is conspicuous circumferential wall thickening of low echogenicity.
**b** Resupply of the axillary artery through dilated collaterals (*right*) and inflammatory wall thickening of the axillary artery (*left*).
**c** Angiogram showing occlusion of the subclavian and axillary arteries with good collateralization (indicating a chronic process). The *circle* indicates the site of entry of the collateral into the artery and corresponds to the detail shown in **b**. The *dotted red line* corresponds to the occluded arterial segment visualized in **a** (courtesy of K. Amendt).

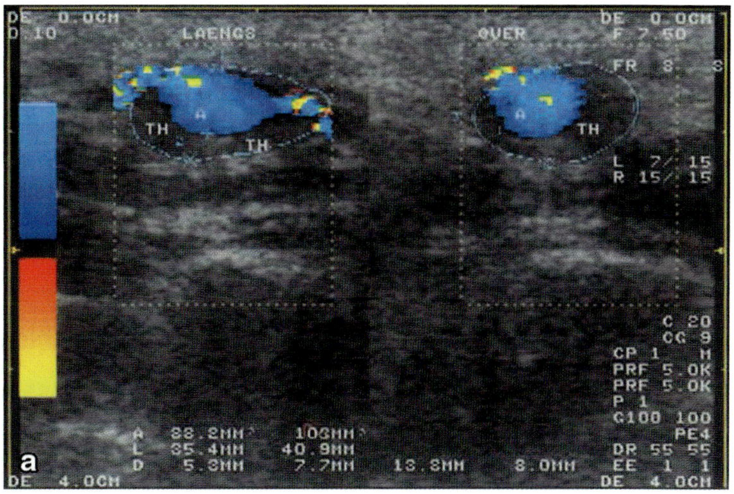


**Fig. A2.65 a, b**
**Aneurysm of the ulnar artery (hypothenar syndrome)**
**a** Patient with ischemia of the pads of fingers 4 and 5 due to arterial emboli from an aneurysm of the distal ulnar artery proximal to the palmar arch. The extent of the aneurysm is outlined in the color duplex scans (longitudinal on the *left*, transverse on the *right*) to illustrate the difference between the overall size of the partially thrombosed aneurysm (20 × 18 mm) and the perfused lumen.

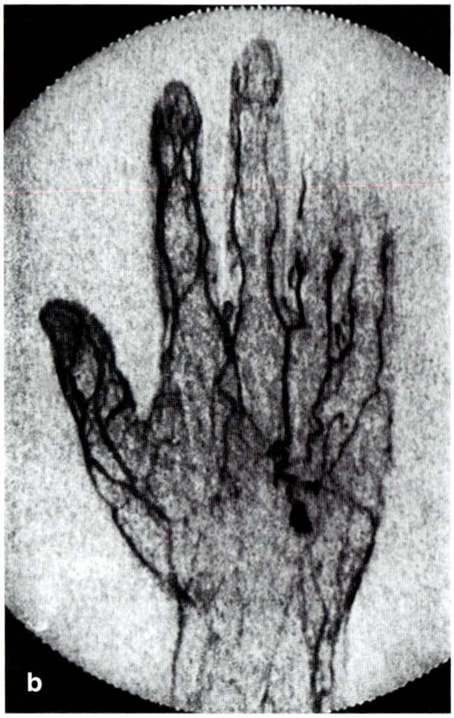


**Fig. A2.65 (continued)**

**b** Angiogram: Aneurysmal dilatation with a rather small caliber of the distal ulnar artery at the junction with the palmar arch and peripheral occlusions of the digital arteries of fingers 4 and 5. The largely thrombosed aneurysm of the ulnar artery was confirmed intraoperatively.

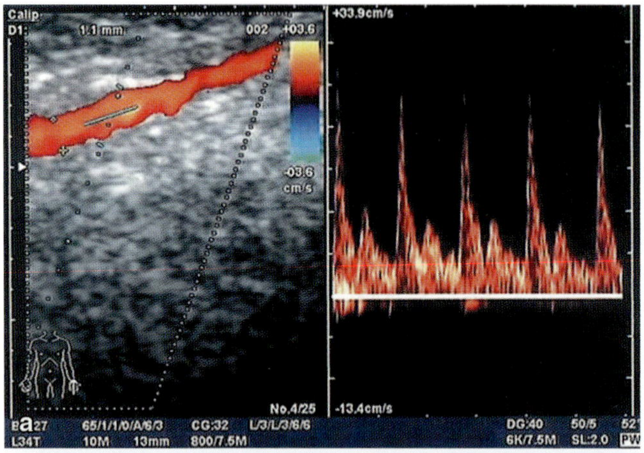


**Fig. A2.66 a–e**

**Interdigital artery occlusion – Raynaud's disease**

**a** Interdigital arteries to the *right* and *left* of the metacarpal bones scanned from the palm show pulsatile flow (pulsatility dependent on sympathetic tone).

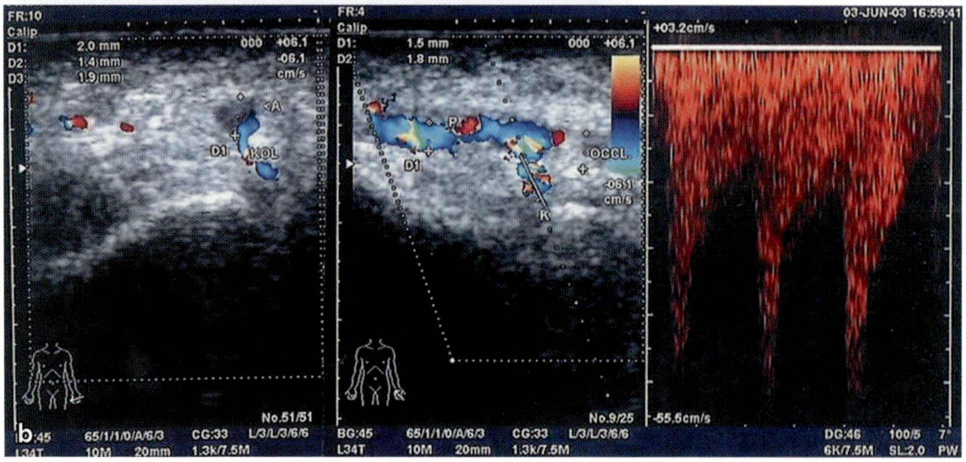


**b** In case of interdigital artery occlusion, the small collateral vessels show monophasic flow due to peripheral dilatation. The occluded interdigital artery with the origin of a collateral is depicted in transverse orientation in the *left section* and in a longitudinal plane in the *middle section*. It has a diameter of 2 mm and a plaque (*P*) is depicted.

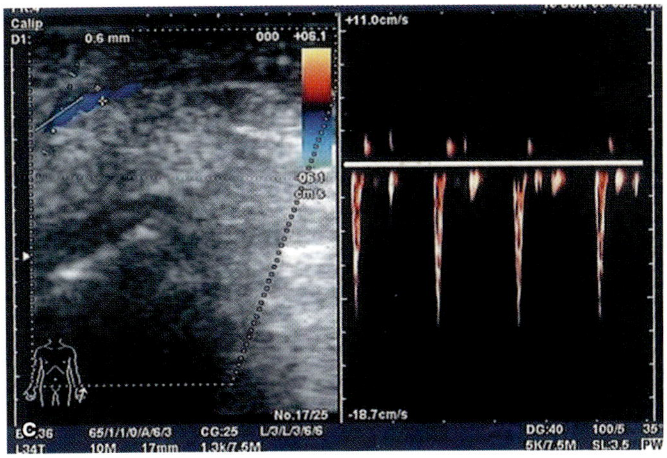

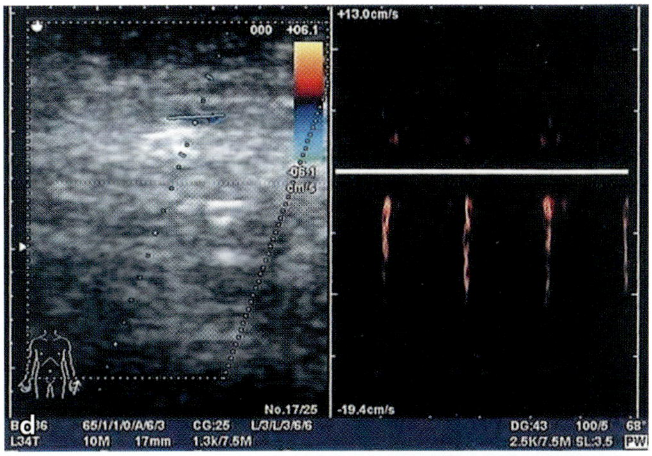

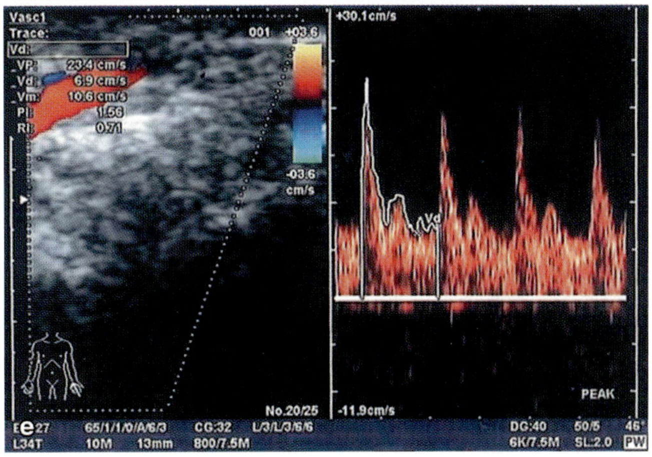

**Fig. A2.66 (continued)**
**c** A common digital artery in Raynaud's disease with a diameter of 0.6 mm and very pulsatile flow due to vasospasm (atypically displayed in blue because the transducer had to be rotated to visualize the artery).
**d** A "knocking" waveform is recorded from the affected distal interdigital artery due to peripheral spasms associated with Raynaud's disease.
**e** Arterial dilatation induced by bathing of the hand leads to less pulsatile flow with a large diastolic component.

# 3 Peripheral Veins

## 3.1 Pelvic and Leg Veins

### 3.1.1 Vascular Anatomy

Three groups of leg veins that are affected by different clinical conditions can be distinguished:

- Epifascial (superficial) veins
- Subfascial (deep) veins
- Transfascial (perforating) veins

The epifascial veins belong to the superficial venous system of the leg and the subfascial veins to the deep venous system with the transfascial or perforating veins establishing connections between these two venous systems. The deep veins accompany the arteries of the same name (Fig. 3.1).

The iliac vein runs through the true pelvis posterior to the iliac artery, pierces the inguinal ligament, and then immediately passes to the medial side of the artery, where it continues as the common femoral vein. Just below the inguinal ligament, the great saphenous vein enters the common femoral vein on its anteromedial aspect. The common femoral vein receives the deep femoral vein just after the division of the common femoral artery into the deep and superficial branches. The deep femoral vein runs between the arterial branches of the femoral bifurcation. Distally, the superficial femoral vein courses along the posterior aspect of the artery of the same name. In most individuals, a second, large branch of the deep femoral vein opens into the superficial femoral vein. Different variants exist as to where, how, and how many deep femoral vein branches enter the superficial femoral vein.

A single superficial femoral vein is present in 62% of individuals only, 21% have a duplicated vein, and in another 14%, even three or more branches are present. If there is more than one vein, these may vary in caliber and course lateral or anterior to the artery rather than posterior to it. While the iliac vein has no valves, the superficial femoral vein has about four to five (Weber and May 1990). After its passage through the adductor canal, the superficial femoral vein becomes the popliteal vein, which runs posteriorly along the artery of the same name (closer to the transducer when scanning from the popliteal fossa). The small saphenous vein enters the proximal popliteal vein (see Fig. 3.1) on its posterior aspect at a highly variable level. Just below the saphenopopliteal junction, the small saphenous vein perforates the deep fascia and descends along the back of the calf. The distal popliteal vein receives the calf muscle veins (soleus and gastrocnemius veins) at various levels around the cleft of the knee joint. Just before flowing into the popliteal vein, the proximal small saphenous vein gives off a connecting branch to the deep muscle veins of the thigh, the femoropopliteal vein (Fig. 3.2a).

The popliteal vein may be present as a single or paired vessel and arises from the union of the posterior tibial and the fibular veins. It receives the anterior tibial vein as the first lower leg vein at a variable level. The main lower leg veins typically follow the arteries of the same name. The anterior tibial veins penetrate the interosseous membrane and course along its anterior aspect. The fibular veins run close to the fibula in the deep crural fascia between the superficial and deep flexors, as do the tibial veins, but on the posteromedial aspect of the tibia.

The *superficial (epifascial) venous drainage system* consists of two subsystems, that of the great saphenous vein and that of the small saphenous vein, which receive the larger arch veins and side branches. The great saphenous vein extends from the back of the foot to the medial malleolus and takes a medial course through the lower and upper leg to about 2–3 cm below the inguinal ligament, where it opens into the popliteal vein. There is variation in the tributaries to the great saphenous vein in the lower leg, but these are mainly the following: the posterior arch vein, which is connected to the major deep veins, in particular the posterior tibial vein, through the perforating veins (Cockett I, II, and III); the great saphenous branch from the back of the foot; and the anterior tributary vein. In the thigh, connections to the deep venous system are established by Dodd's perforators. Just before its junction with the common femoral vein, the great saphenous vein receives tributary veins from the thigh and lateral branches (lateral and medial accessory great saphenous vein), which then establish connections to the abdominal (epigastric) veins and become important as collaterals in pelvic vein thrombosis (Fig. 3.2b).

W. Schäberle, *Ultrasonography in Vascular Diagnosis*,
DOI: 10.1007/978-3-642-02509-9_3, © Springer-Verlag Berlin Heidelberg 2011

**Patient Positioning** The inferior vena cava and iliac vein are examined with the patient in the supine position. If there is overlying air, improvement may be achieved by repositioning the patient on the right or left side; bowel gas can be pushed aside by applying pressure with the transducer. The femoral vein is scanned in the supine patient with the knee slightly bent and a slight outward rotation of the leg. An experienced examiner can scan the popliteal vein and lower leg veins with the patient in the supine or semilateral position and the knee slightly bent. Alternatively, the popliteal vein can be examined with the patient in the prone position. However, to avoid collapse of the veins due to hyperextension of the knee, the ankle should be slightly elevated by placing a cushion underneath. When the patient is sitting or standing, the filling of the veins in the lower leg is better, which makes it easier to localize them and evaluate their compressibility. Valve incompetence in the popliteal vein, the superficial lower leg veins (varicosis), and the perforating veins is best evaluated in the sitting patient. The proximal great saphenous vein and the femoral vein are examined with the patient supine (like the femoral artery, see Fig. 2.3a).

**Examination Technique** (Table 3.1) In the diagnostic evaluation of thrombosis, the deep veins are continuously scanned from the groin to the ankle and checked for the absence of intraluminal thrombi by intermittent compression

**Table 3.1** Ultrasound examination of the veins

| | | |
|---|---|---|
| **B-mode ultrasound** | Transducer position | Transverse (except for external iliac vein) |
| | Criteria: | • Compressibility<br>• Lumen width<br>• Wall morphology<br>• Internal structures |
| | Note: | Reversed compression maneuver in adductor canal |
| | Documentation: | • Split scan: without/with compression<br>• Normal findings as outlined in the text, abnormal findings according to the situation |
| **(Color) duplex ultrasound** | Transducer position | Longitudinal plane, overview in transverse plane |
| | Criteria: | • Spontaneous flow, augmented flow (Valsalva, valve function test)<br>• Luminal filling (defects?)<br>• Wall contour abnormalities, perivascular structures |
| | Note: | Doppler interrogation always in longitudinal orientation |
| | Documentation: | B-scan with corresponding waveform, color Doppler scan as needed |

(see Figs. A3.3a, b and A3.17). First, the common femoral vein is identified on the medial side of the common femoral artery below the inguinal ligament and followed in transverse orientation down to its junction with the superficial femoral vein. Along the course of the common femoral vein, the terminations of the great saphenous vein and of the deep femoral veins from the upper leg muscles are tested for compressibility as well.

In evaluation of the pelvic veins, compression ultrasound does not yield valid results because a continuous structure against which to compress the veins is not available, and the abdominal organs and fatty tissue preclude reliable compression, in particular in obese patients. Nevertheless, compression ultrasound can be performed, especially in slender patients. The arched iliac veins in the true pelvis are tested with the transducer in transverse orientation with additional longitudinal scanning as required. If adequate evaluation of compressibility is not possible in this way, patency must be evaluated by spectral Doppler.

If the scanning conditions are poor, occlusive thrombosis of the *pelvic veins* can be ruled out by obtaining a Doppler spectrum from the junction of the common femoral and external iliac veins, where the insonation window is good. In the presence of an obstruction, respiratory phasicity of flow is absent or reduced compared to the unaffected side. This is evaluated by analyzing the Doppler waveform recorded in the external iliac vein (posterior to the artery) somewhat above the inguinal ligament in the longitudinal plane and with a low PRF. With the patient stretched in the supine position, the common femoral vein segment passing under the inguinal ligament may be compressed, especially in slender patients. In such cases, visualization can be improved by slight outward rotation of the hip joint.

Next, with the patient in the supine position, the *superficial femoral vein* is followed down the leg in transverse orientation and is intermittently compressed (every 1–2 cm). The termination of the deep femoral vein is examined by color duplex ultrasound in the longitudinal plane (Fig. 3.3). In the distal segment of the superficial vein, at the level of the adductor canal, compression is difficult due to the absence of a bony structure and the interfering connective tissue. Instead, the examiner must press the muscle and vessels against the transducer from below with his or her other hand to achieve adequate compression (Fig. 3.4).

Below the adductor canal, the *popliteal vein* is scanned from posteriorly. This part of the examination is performed with the patient supine and the knee slightly bent or in the prone position with a support under the ankles. The bent knee ensures better filling and hence improved visualization. With the knee stretched or even overstretched in the flat position, the popliteal vein is often collapsed or compressed by the surrounding connective tissue structures, pushing the vein against the artery and bony structures.

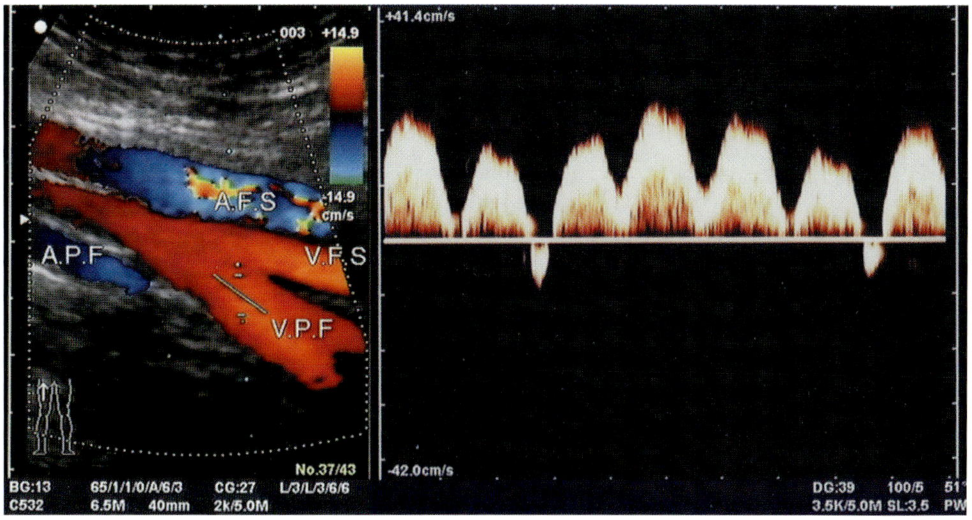


**Fig. 3.3** Sonographic anatomy of the junction of the superficial (*V.F.S.*) and deep (*V.P.F.*) femoral veins. There are usually two main branches of deep thigh veins that join with the superficial femoral vein to form the common femoral vein. One passes under the superficial femoral artery just below the femoral bifurcation, and the second (the one seen in the image) enters the femoral vein slightly more distally. Rare muscle vein thrombosis of the thigh nearly always involves this more distal branch. The Doppler waveform from the deep femoral vein shows respiratory phasicity and sometimes also cardiac pulsatility (as seen here)

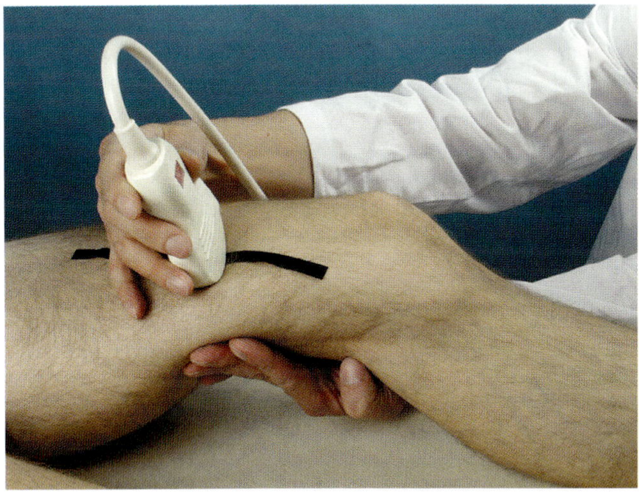

**Fig. 3.4** Compression ultrasound applying pressure with the transducer alone is inadequate at the level of the adductor canal. Instead, the examiner must additionally push the vein against the transducer from below with the flat hand

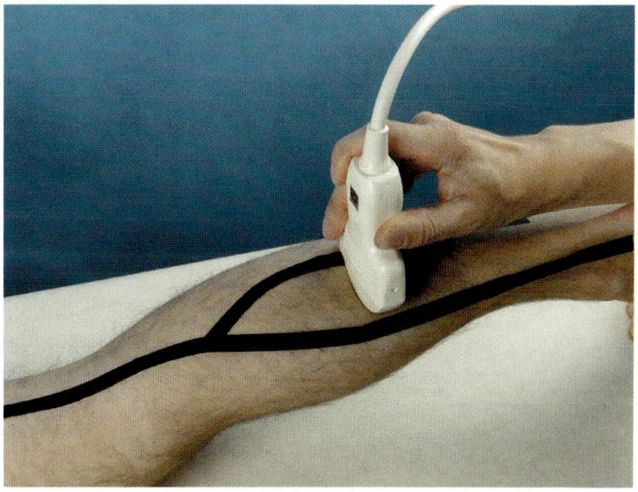

**Fig. 3.5** Compression ultrasound of the lower leg veins (course marked). The transducer is positioned on the calf such that the ultrasound beam is perpendicular to the interosseous membrane between the tibia and fibula

Following evaluation of the popliteal vein for compressibility in transverse orientation, it is traced downward to the confluence of the fibular and posterior tibial veins. The anterior tibial vein entering at a higher level is often identified at its point of entry by means of color duplex only. After having been identified, the vein is evaluated for compressibility from an anterior approach. The accompanying veins are then continuously scanned downward and intermittently compressed using the anterior tibial vein as a landmark.

For scanning of the anterior tibial vein, the transducer is placed on the extensors and then moved so as to achieve a beam direction roughly perpendicular to the interosseous membrane between the tibia and fibula. The procedure is the same for scanning of the fibular and posterior tibial veins, except that the transducer is in a posterior position on the gastrocnemius muscle. The two veins are traced distally with intermittent compression (Fig. 3.5).

While the popliteal and femoral veins are reliably identified by B-mode ultrasound, the lower leg veins may have to be localized using the arteries of the same name as landmarks, which are visualized by color duplex. The hyperechoic interosseous membrane is an anatomic landmark for identifying the anterior tibial artery and vein coursing in it, whereas the deep crural fascia between the deep flexors and the soleus and gastrocnemius muscles is not always depicted well enough to serve as a landmark for identifying the

a

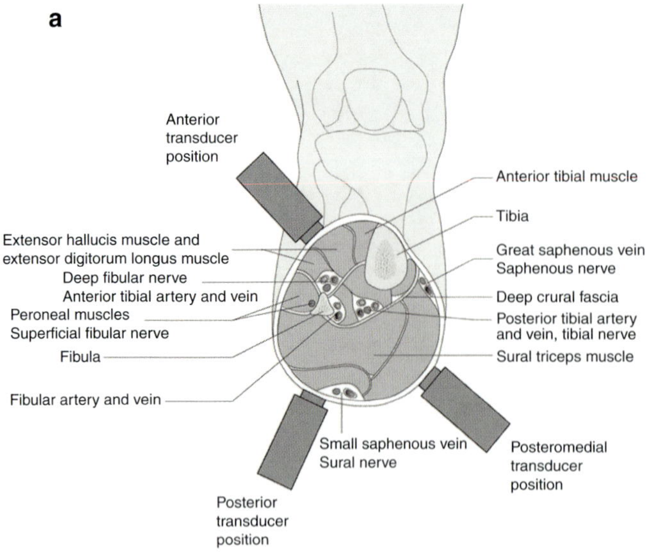


b

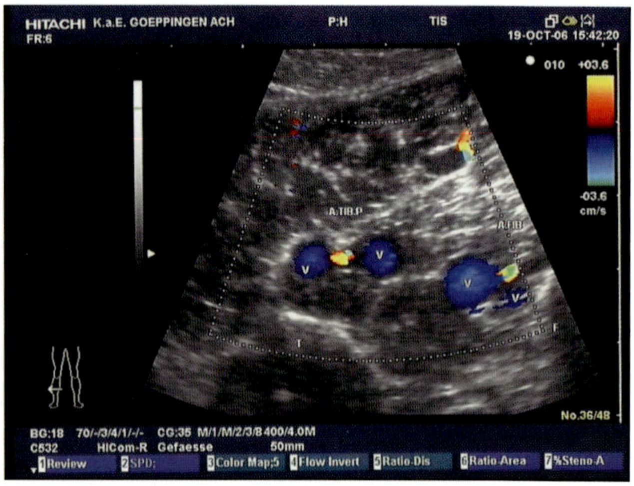

**Fig. 3.6** **a** Cross-sectional anatomy of the lower leg and transducer positions. **b** Sonoanatomy of the calf veins with the transducer in the posteromedial position. The transverse view shows the posterior tibial artery and vein (*left* part of image) somewhat posterior to the tibia (*T*) and the fibular artery and vein posteromedial to the fibula (*F*). Blue indicates flow in the veins, red flow in the artery (with some aliasing resulting from the low PRF chosen to depict slow venous flow). The artery lies within the slightly hypoechoic deep crural fascia and is accompanied by two veins on the left and right

posterior tibial and fibular veins coursing in it (Fig. 3.6). The fibular vein is easier to identify from the posterior approach (and the proximal anterior tibial vein from the anterior approach) as it courses close to the fibula. Diagnostic assessment of thrombosis can be performed with the patient in the supine or prone position, but better filling facilitates visualization of the veins in the sitting patient.

In addition to the *main veins of the lower leg*, evaluation of patients with suspected thrombosis also includes compression scanning of the *muscle veins* (the gastrocnemius veins opening into the popliteal vein) and the soleus veins entering the main veins of the lower leg to rule out muscle vein thrombosis (the same applies to the deep femoral vein in the thigh). The muscle veins can be traced in a distal direction from their sites of entry into the main vein, especially when there is adequate venous filling with the patient sitting.

The ultrasound examination in thrombophlebitis serves to determine the extent, in particular the cranial extent, and involvement of the deep venous system (inflow into deep venous system). This is done by performing compression ultrasound of the great or small saphenous vein in transverse orientation after identification of the clinically inflamed segment and using the same criteria as in the diagnostic assessment of thrombosis.

In thrombophlebitis, special attention must be paid to the sites of entry of the small and great saphenous veins into the popliteal and common femoral vein, respectively, which are checked for compressibility in transverse plane.

In patients examined for thrombosis, compression is always performed in transverse orientation: it is easier to identify a vein and follow its course down the leg, thus preventing false-negative results. Longitudinally, when pressure is applied, a noncompressible vein may be displaced and disappear from the scanning plane, thereby mimicking compressibility.

#### 3.1.2.2 Chronic Venous Insufficiency and Varicosis

In patients with chronic venous insufficiency of the deep veins or varicosis of the superficial veins, the affected venous segments are scanned longitudinally for the presence of venous reflux, which is induced by applying the respective provocative tests. For the demonstration of valve incompetence of the deep veins, the tests for reflux diagnosis are performed with Doppler recording in longitudinal orientation at selected sites (key sites) in the common femoral, superficial femoral, and popliteal veins.

Function of the proximal valves is evaluated during Valsalva's maneuver in the recumbent patient and Doppler sampling in the common and superficial femoral veins during increased abdominal pressure. Valve incompetence is demonstrated by persistent backward flow to the periphery, indicated by a corresponding color change in the color flow image. If this test is positive for proximal valve incompetence, the test is progressively extended to the popliteal vein and lower leg veins to determine the distal extent of the valve incompetence.

In patients with competent proximal valves (common femoral and proximal superficial femoral veins), distal insufficiency is diagnosed by the demonstration of persistent flow reversal (over 1 s) in the Doppler waveform (Fig. 3.7) or on color duplex scanning in the popliteal vein (longitudinal scan) during compression and release with the patient sitting or standing.

Incompetence of the terminal valve of the great saphenous vein is assessed longitudinally during Valsalva's maneuver (Fig. 3.8a–c). In case of truncal varicosis, the great saphenous vein is tracked distally to identify the *lowest point of incompetence* through intermittent Valsalva maneuvers (grading according to Hach). In case of sufficiency of the proximal segment, the extent of distal varicosis of the great saphenous vein is determined by *intermittent compression testing* along the vein in the cranial direction (Fig. 3.8b: compression with the thumb distal to the transducer) to identify the proximal and distal points of insufficiency (transition from persistent reflux to absent reflux upon compression with subsequent release) in the sitting or standing patient. This valve function test is also used to diagnose reflux in the small saphenous vein (Fig. 3.9a, b).

To assess valve competence of the perforating veins, these are first identified in their typical locations (e.g., Cockett's group in the distal medial lower leg, Boyd's group in the proximal lower leg, or Dodd's group in the upper leg) (Fig. 3.10).

On B-mode images, the perforating veins are identified as hypoechoic, tubular structures passing through the

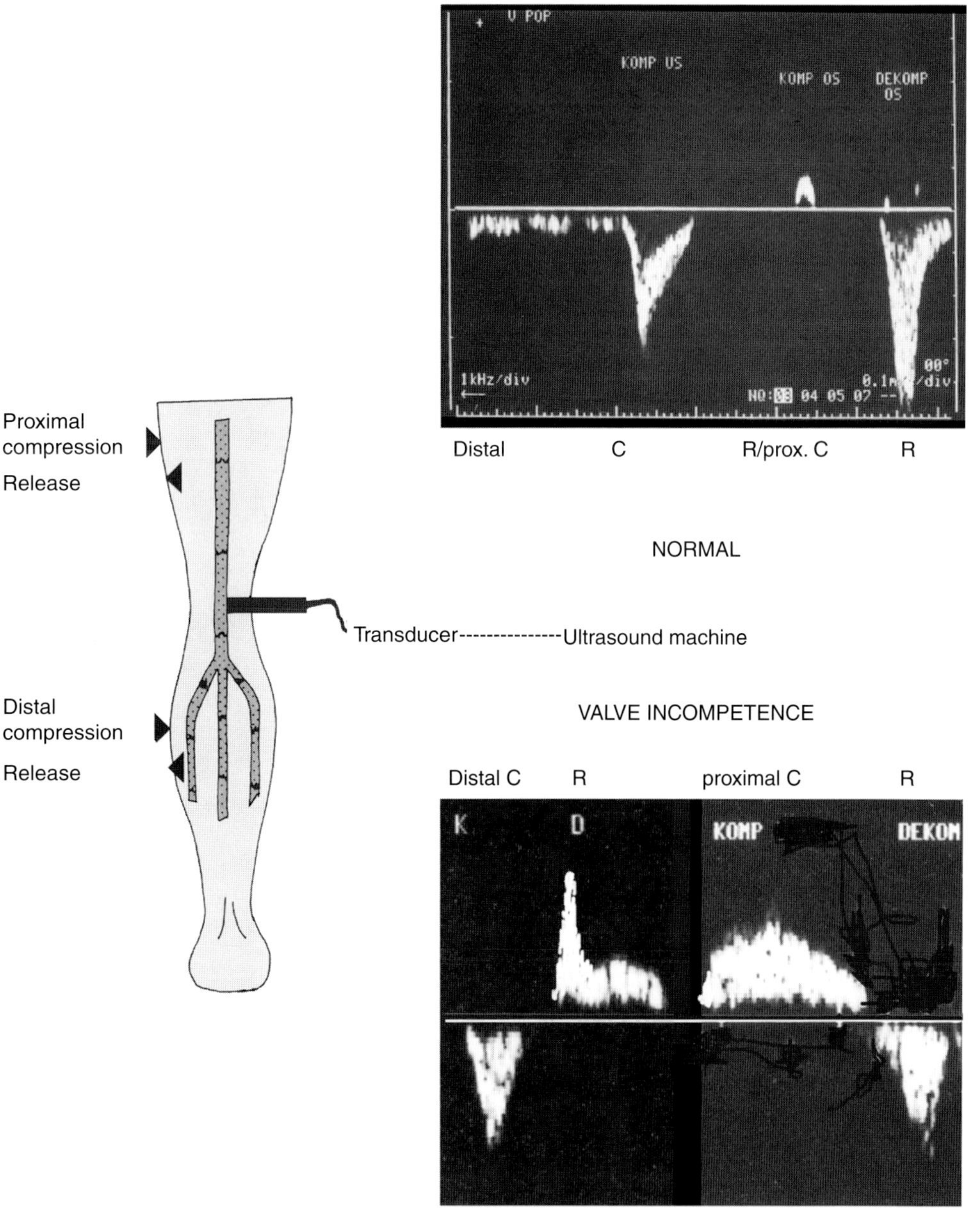


**Fig. 3.7** Illustration of the proximal and distal valve function test with Doppler measurement in the popliteal vein. The *top* waveform represents the normal findings obtained when there is proper valve closure (distal compression and release on the *left*; proximal compression and release on the *right*); the *bottom* waveform shows reflux due to valve incompetence (*C* compression, *R* release of compression)

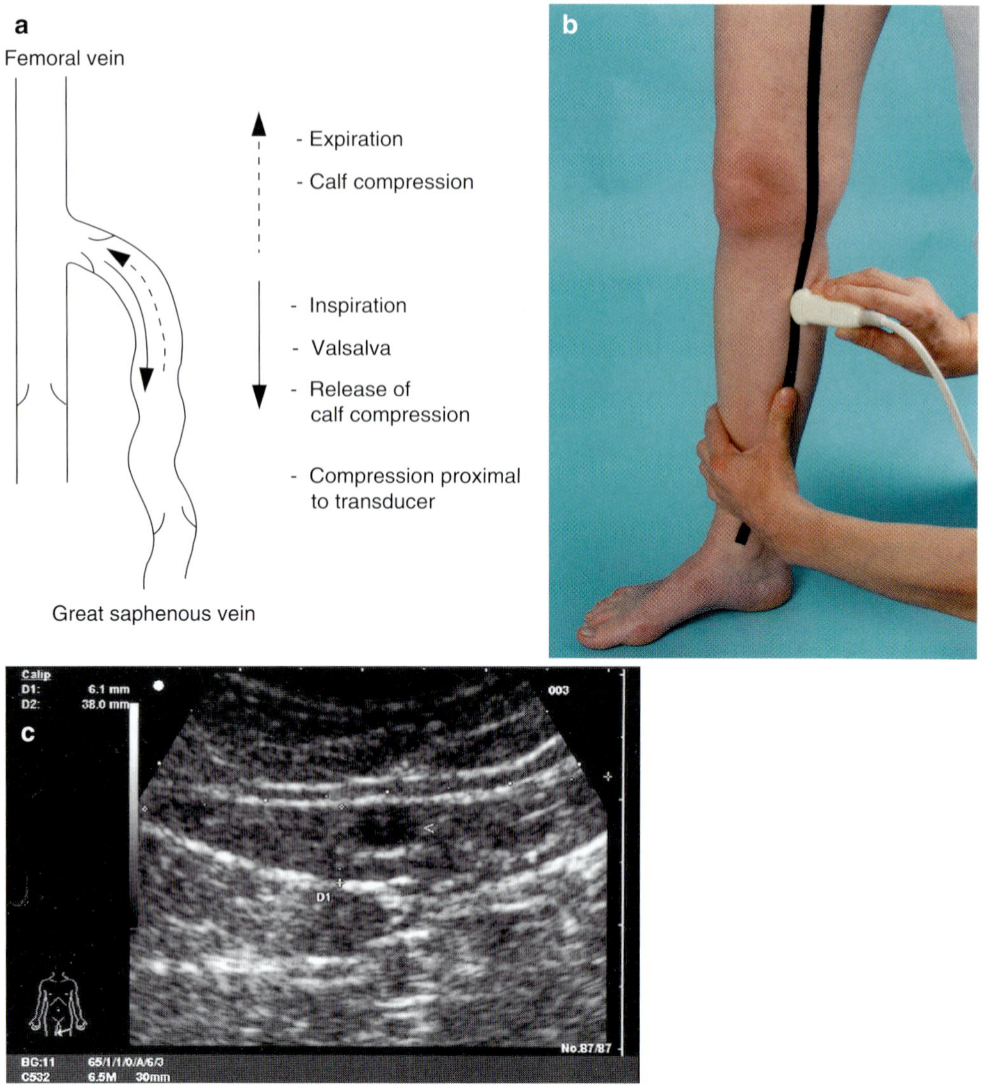


**Fig. 3.8** **a** Valve function test in the great saphenous vein. **b** Transducer position for evaluating valve competence of the distal great saphenous vein using compression and release. Alternating distal compression of the vein and release during recording of the Doppler spectrum is performed with the left thumb. **c** Transverse image of the great saphenous vein (*arrowhead*) in the saphenous compartment enclosed by the bright saphenous fascia anteriorly and the muscle fascia posteriorly (an appearance known as the "saphenous eye"). The great saphenous vein enters the common femoral vein from anteromedially. Within the saphenous compartment, it can be distinguished from branch varices coursing outside this compartment

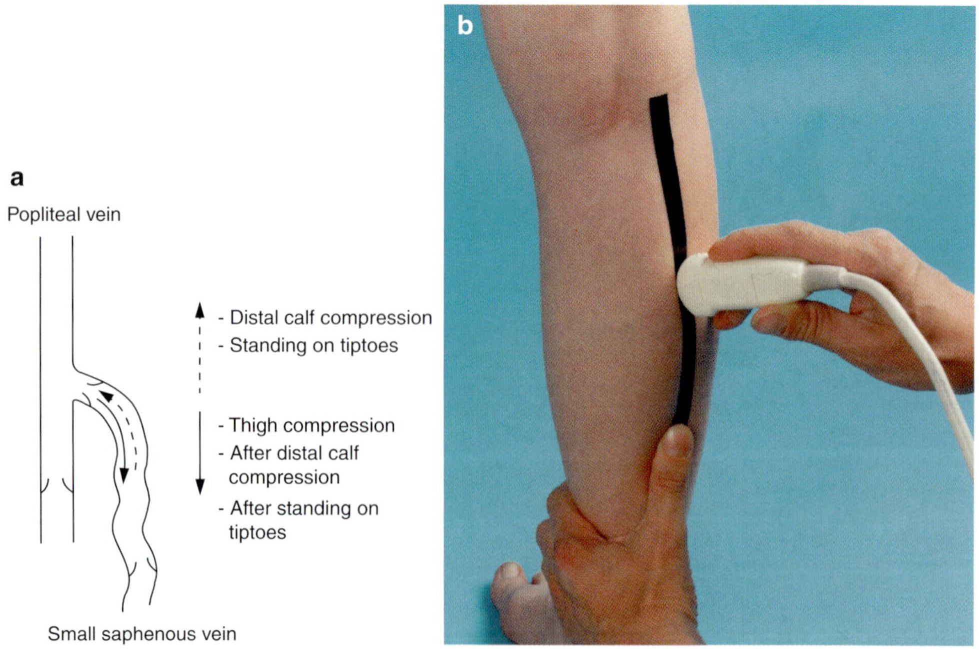


**Fig. 3.9** **a** Valve function test in the small saphenous vein. **b** Transducer position for evaluating valve competence of the small saphenous vein (see legend to Fig. 3.8b)

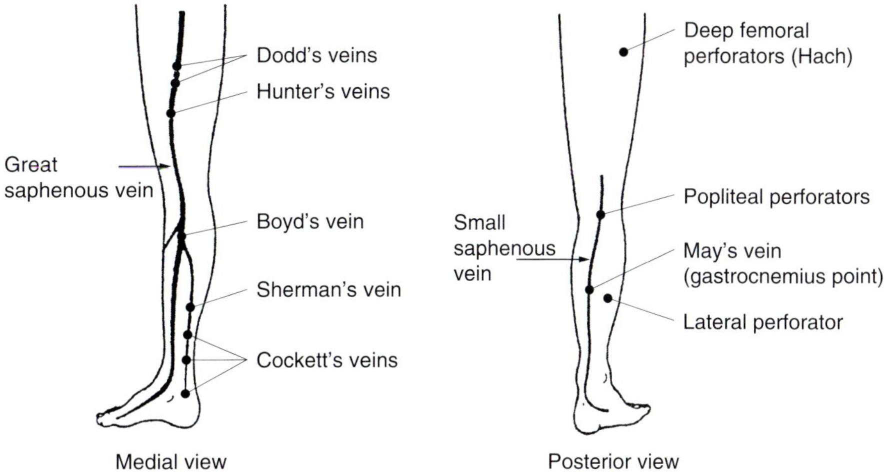


**Fig. 3.10** Typical locations of clinically relevant perforating veins

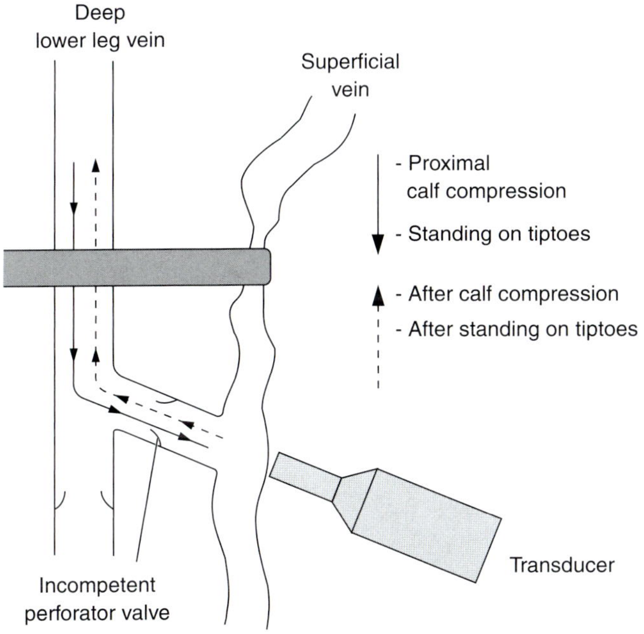


**Fig. 3.11** Valve function test in the perforating veins

deep fascia from the superficial to the deep veins. Once identified, the compression test is performed as described in Fig. 3.11.

If there is incompetence, color duplex or duplex scanning with the sample volume placed in the perforating vein depicted on the B-scan will demonstrate reflux (retrograde flow from the deep into the superficial system) during compression of the calf just proximal to the Doppler sample volume. When the valves function properly, there is no backward flow from the deep to the superficial veins. Compression of the calf will cause stoppage of flow but no reversal. Application of a tourniquet proximal to the site of scanning can prevent interference from flow in insufficient superficial veins.

### *3.1.3 Normal Findings*

The leg veins, with their *delicate walls* and *low intraluminal pressure*, can be *completely compressed with the transducer* so that normal veins become nearly invisible on ultrasound, or only a hyperechoic reflection indicating the wall but no lumen is seen. The breathing-related intra-abdominal pressure changes lead to *respiratory fluctuation* of venous flow with faster flow during expiration due to lower intra-abdominal pressure (upward movement of diaphragm) and slower flow during inspiration due to higher intra-abdominal pressure (downward movement of diaphragm). This pressure-dependent flow pattern is transmitted through the upper leg veins into the major deep veins in the distal lower leg and into the major superficial veins (great and small saphenous veins) in the recumbent patient. Respiratory fluctuation of venous flow may be overridden by *cardiac pulsatility* (changes in atrial pressure) in the iliac and proximal femoral veins, especially in young patients (Table 3.2).

The pocket-like valves ensure undisturbed flow from the periphery to the center. When there is pressure reversal with backward flow, flow ceases upon closure of the valves (after a short reflux of 0.3 s on average). Proper valve closure is demonstrated using Valsalva's maneuver or the valve function test (see Fig. 3.7).

**Table 3.2** Factors affecting venous reflux

| |
|---|
| • Vis-a-tergo |
| • Changes in intra-abdominal and intrathoracic pressure (suction pump) |
| • Cardiac suction pump (systole, early diastole) |
| • Musculovenous pump (requires competent valves): |
| – Competent perforating veins prevent blood flow into superficial veins |
| – Competent valves distal to the contracting muscle prevent backward flow |

### 3.1.4 Documentation

As with the examination protocol, the documentation of findings is dictated by the clinical question to be answered.

In Germany, the Deutsche Gesellschaft für Ultraschall in der Medizin (DEGUM; German Society of Ultrasound in Medicine) issued guidelines for documenting the findings of venous ultrasound of the legs based on clinical experience and requirements. The routine sites for which the duplex ultrasound findings (B-scan and Doppler waveform) should be documented in all cases are the superficial femoral, deep femoral, common femoral, and popliteal veins. Additional results are documented according to abnormal findings in the individual case.

#### 3.1.4.1 Deep Vein Thrombosis of the Leg

Transverse scans of the common femoral vein at about the level of the termination of the great saphenous vein, the superficial femoral vein somewhat distal to the site of entry of the deep femoral vein, and the popliteal vein are documented with and without compression.

Supplementary scans of the major lower leg veins (posterior tibial and fibular veins) obtained in the proximal lower leg should be documented, again with and without compression, together with a Doppler spectrum recorded at the junction of the external iliac vein with the common femoral vein to show unobstructed venous return of the pelvic veins. If the duplex ultrasound findings are documented, these should comprise longitudinal scans with the corresponding Doppler spectra confirming preserved respiratory phasicity of blood flow in the common femoral, superficial femoral, and deep femoral veins near their sites of entry, and in the popliteal vein.

In addition, if venous thrombosis is diagnosed, the abnormal findings must be documented (noncompressible venous segments) in transverse views with and without compression, or by means of the Doppler waveform obtained in longitudinal orientation, showing absence of flow or an abnormal flow profile. If the absence of flow is documented on a color duplex scan, the scan must contain information to the effect that proper instrument settings including a low PRF and adequate gain were used.

#### 3.1.4.2 Chronic Venous Insufficiency and Varicosis

Longitudinal scans of the terminal portions of the common femoral, superficial femoral, and great saphenous veins are documented with the corresponding Doppler spectra recorded during normal breathing and Valsalva's maneuver. Terminal valve function of the popliteal vein and the small saphenous vein is documented on longitudinal scans with the corresponding Doppler spectra obtained during compression and release. Color duplex scans alone are inadequate for documenting reflux because the duration must be quantified to differentiate abnormal reflux from the short backward flow that is normal before valve closure.

### 3.1.5 Clinical Role of Duplex Ultrasound

#### 3.1.5.1 Thrombosis and Postthrombotic Syndrome

The incidence of deep vein thrombosis of the leg is 1–2% per year and increases with age. Various noninvasive diagnostic tests have been proposed for the diagnosis of this common condition, which often takes an asymptomatic or unspecific clinical course but has serious early (pulmonary embolism) and late complications (chronic venous insufficiency in about 50% of cases). These include plethysmography, thermography, iodine fibrin test, and Doppler ultrasonography (Bollinger and Franzeck 1982; Hull et al. 1984; Kakkar 1972; Lepore et al. 1978; Neuerburg-Heusler and Hennerici 1995; Sandler et al. 1984; Strandness 1977).

The methods are either very time consuming or yield reliable results only in specific vessel segments. CW Doppler ultrasound used to be the noninvasive modality of first choice in the diagnostic assessment of valvular incompetence of the superficial and deep veins and, as a functional modality, showed good results in diagnosing pelvic and upper leg venous thrombosis including popliteal artery thrombosis with reported accuracies of up to 90%. However, isolated lower leg venous thrombosis and central thrombi surrounded by flowing blood are difficult to detect. A review of 2,060 patients who underwent additional venography yielded a sensitivity of 84% and a specificity of 88% for CW Doppler ultrasound in demonstrating venous thrombosis (Wheeler 1985).

Combining morphologic information (B-scan) and functional information (Doppler), duplex ultrasound has evolved into the central noninvasive modality for venous diagnosis.

Stasis is an important risk factor for the development of deep vein thrombosis of the leg in addition to a hypercoagulable state and a damaged vessel wall. Thus, immobilization plays a crucial role in the pathogenesis of thrombosis of the deep veins, which primarily arises in the muscle veins in bedridden patients or patients with cast immobilization of the leg. The risk of thrombosis without heparin prophylaxis is 10–30% in general surgery and up to 54% in hip surgery (Lippert and Pabst 1985). Venous thrombi are ascending in over 90% of cases and have an annual incidence of 160/100,000 inhabitants in Germany with pulmonary embolism occurring in 60/100,000 inhabitants per year. Thrombosis of the deep pelvic and leg veins is the source of pulmonary embolism in over 90% of cases. The importance of isolated lower leg venous thrombosis should not be underestimated as it may extend cranially and cause pulmonary embolism, though often asymptomatic, in 15–26% of cases (Kroegel

2003). In contrast, iliofemoral thrombosis has a 56–85% incidence of pulmonary embolism. The mortality of pulmonary embolism ranges from 0.1% to 5%, depending on the risk group (Polak 1992).

Data in the literature on the distribution of venous thrombosis of the legs are not very consistent. In a study of 1,084 lower extremities with acute venous thrombosis, the thrombosis was localized above the knee in 51%, below the knee in 32%, and in a superficial vein in 17% (Kerr et al. 1990). A venographic study (Schmitt et al. 1977) of deep venous thrombosis of the leg showed concomitant involvement of the common iliac vein in 16%, external iliac vein in 33%, common femoral vein in 46%, deep femoral vein in 45%, superficial femoral vein in 65%, popliteal vein in 66%, anterior tibial vein in 73%, posterior tibial vein in 82%, and fibular vein in 77%.

A study investigating 189 venograms in the early 1990s (Cogo et al. 1993) identified isolated calf vein thrombosis in 18% of cases. The vast majority of the 82% of patients with proximal vein thrombosis had popliteal vein involvement, while only 8% were found to have isolated pelvic vein thrombosis. No case of isolated thrombosis of the superficial femoral vein was reported in this study.

The generous use of diagnostic ultrasound in patients with clinically suspected deep vein thrombosis of the leg can help reduce the incidence of thrombosis of the pelvic and femoral veins.

A retrospective analysis of thrombosis distribution in patients who were consecutively examined in 2008 with a generous application of ultrasonography (18% thrombosis prevalence) showed that – with this approach – most patients can be identified when thrombosis is still confined to the lower leg arteries (indication for sonography: swelling of the leg or calf pain for which no other cause was apparent).

Isolated lower leg thrombosis was present in 63% of cases (including 8% isolated calf muscle vein thrombosis), 23% had extension to the popliteal vein and 11% involvement of the femoral and popliteal vein, while only 3% of patients had isolated or concomitant pelvic vein thrombosis. With one exception, isolated pelvic vein thrombosis extended down to the level of the saphenofemoral junction. There was one case of isolated superficial femoral vein thrombosis, which was seen in a patient with duplication of this vein. In 1.5% of patients with popliteal vein thrombosis, the site of origin was a muscle vein (soleus or gastrocnemius), as thrombosis was absent in the other major calf veins. Thrombosis of the deep femoral vein with thrombus extension into the common femoral vein accounted for 0.7% of cases. The analysis included a total of 280 cases of lower extremity deep vein thrombosis.

The high proportion of isolated calf vein thrombosis in this population may be attributable to the fact that symptoms indicative of thrombosis following trauma or surgery of the leg prompted a sonographic examination in all cases, frequently revealing calf vein thrombosis (in particular of the fibular vein or muscle veins).

Also contributing to this distribution is the policy of early diagnosis and treatment of deep vein thrombosis of the leg pursued by the ultrasound laboratory at the author's institution. This helps reduce the number of cases with thrombosis of the popliteal and distal superficial femoral veins, which always arise from ascending calf thrombosis. Our observations therefore underscore the need for always including the calf arteries when examining patients for vein thrombosis.

Most thrombi arise in the venous sinusoids of the lower leg muscles or in regions of relatively stagnant blood flow behind the pocket-like valves of the popliteal and femoral veins (Figs. 3.12a and A3.24). In the majority of patients, *deep vein*

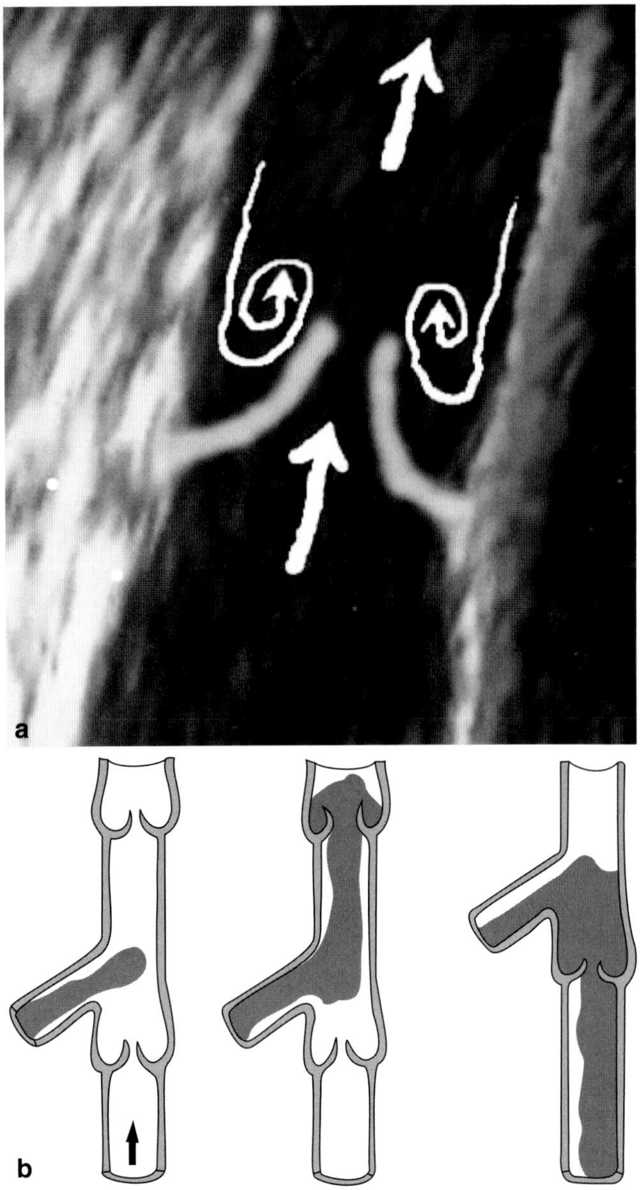

**Fig. 3.12** **a** Turbulent flow and eddy currents in the pocket-like valves can lead to local stasis with release of procoagulant substances. **b** Diagram of a thrombus (*left*) extending from a tributary (e.g., lower leg muscle vein) into the main vein, where it can ascend (*center*) or descend (*right*)

*thrombosis of the legs develops in the valves of the calf muscle veins* (soleus or gastrocnemius veins) (> 50%) or in the valves of the fibular vein. Recirculation in the cusps induces platelet activation and the release of procoagulant substances, which may lead to the formation of a red thrombus. It is estimated that 20–30% of such thrombi undergo spontaneous thrombolysis through the simultaneous activation of the fibrinolytic system and that about 50% of the thrombi become organized and thus remain clinically asymptomatic. About 20–30%, however, exhibit appositional growth with extension into the deep venous system, and from there may continue to grow cranially. With increasing growth into the major deep veins, a thrombus may become free-floating and lead to pulmonary embolism without causing any severe local clinical symptoms such as swelling or pain. Local clinical symptoms leading to the initiation of diagnostic measures may thus not occur – unless a thrombus occludes a main vein or interferes with blood flow by protruding from a muscle or superficial vein into a major deep vein (Figs. A3.26–A3.28).

Depending on flow in the partially thrombosed tributary vein and the flow obstruction caused by the growing thrombus in the main vein, *further growth is ascending or descending* (Fig. 3.12b). Descending venous thrombosis is less common, and in primary pelvic vein thrombosis, it is twice as common on the left side. This is attributed to a pelvic vein spur, a connective tissue structure producing chronic wall trauma with luminal narrowing, as the vein is compressed against the spur by the pulsation of the common iliac artery. The spur is not easily accessible to imaging procedures.

In the case of complete thrombotic occlusion of the deep leg veins, blood drains through superficial veins, chiefly the great saphenous vein. The increased blood flow from the great saphenous vein stops the growth of most ascending thrombi at the saphenofemoral junction. In the superficial femoral vein, blood from the deep femoral vein stops ascending thrombus growth or surrounds a thrombus extending more proximally (Fig. 3.13a). In isolated descending pelvic vein thrombosis, the blood is drained through epigastric collaterals or suprapubic pudendal veins (Fig. 3.13b). On duplex scanning, this collateralization is identified by retrograde flow at the saphenofemoral junction (see Fig. A3.8).

The incidence of early deep vein thrombosis of the legs is much higher than suspected on clinical grounds due to its fairly asymptomatic course. There is a risk of serious early (pulmonary embolism) and late complications (chronic venous insufficiency with crural ulceration). For these reasons, diagnostic measures to identify deep venous thrombosis of the legs should be applied liberally and should also be used in patients presenting with unspecific symptoms. This is underscored by the fact that compression duplex ultrasound offers an inexpensive, noninvasive, and accurate diagnostic modality for the evaluation of these patients and that anticoagulation treatment can effectively reduce the risk of pulmonary embolism and prevent further thrombus growth.

*With most venous thromboses arising from the lower leg veins, assessment of the muscular and main veins in this area is an integral part of diagnostic sonography in these patients.*

Ultrasound has the advantage of "*illuminating the blind spots*" *of venography*. In the elderly, stasis due to degenerative ectasia of gastrocnemius and soleus veins is a common source of ascending thrombosis. For technical reasons

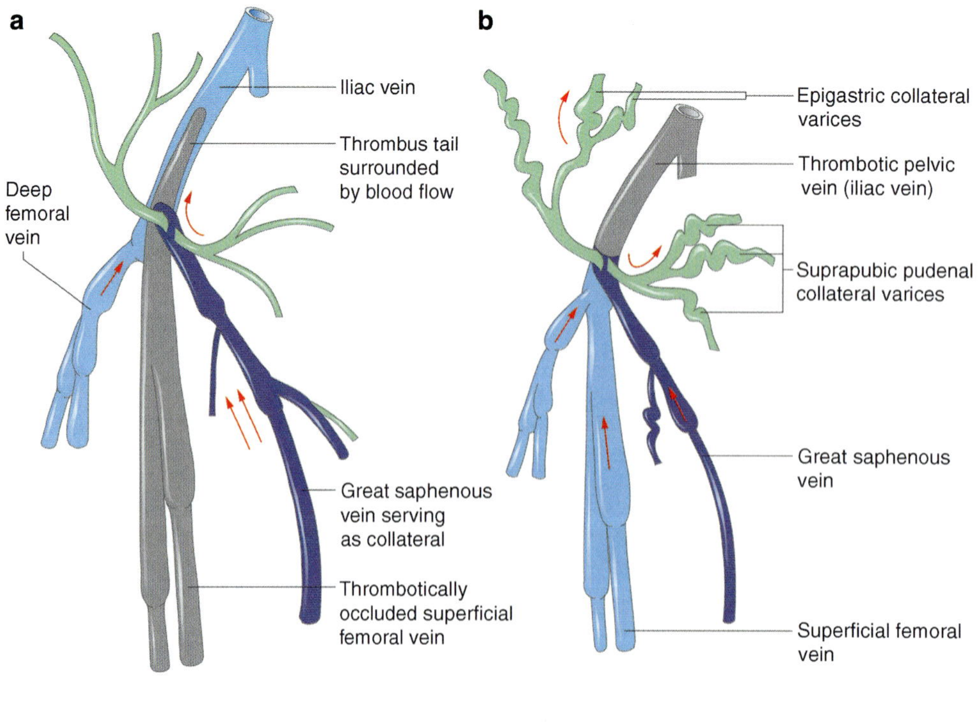


**Fig. 3.13** **a** Venous blood flow in thrombotic femoral vein occlusion with venous return from the periphery occurring primarily through the great saphenous vein (*indicated by arrows*). **b** Collateral pathways in descending (isolated) pelvic vein thrombosis: suprapubic pudendal and epigastric collaterals

(valve function), this form of calf muscle vein thrombosis and the less common deep femoral vein thrombosis cannot be identified by venography; these veins are not opacified or take up the contrast medium only through retrograde flow. Sonographic data suggest that ascending thrombophlebitis with thrombus extension from muscle or perforating veins into the deep venous system is a much more common cause of deep venous thrombosis of the leg than has been assumed thus far.

The fibular vein is a typical source of error in venography, as nonvisualization of this vein may indicate the presence of thrombus, or it may simply be due to limitations of the method. At the same time, the fibular vein is the most common site of isolated venous thrombosis of the calf with ascending thrombus growth. In an analysis of 105 cases of isolated venous thrombosis of the lower leg (without involvement of the popliteal vein) by our group, the fibular vein alone was affected in 48 instances, the posterior tibial vein in 36, and the posterior tibial and fibular veins in 21. Only one case of anterior tibial vein thrombosis was attributable to a large traumatic hematoma of the anterior compartment. Spontaneous thrombosis of the anterior tibial vein is always caused by descending thrombus growth from the popliteal vein.

*Soft tissue lesions* such as abscess, hematoma, or perforated Baker's cyst cause similar clinical symptoms but usually have distinct sonographic features allowing them to be differentiated from deep venous thrombosis or to be confirmed by ultrasound-guided biopsy.

Thrombus organization begins on day 3 or 4 with attachment to the venous wall, and ingrowth of capillaries occurs after 8–12 days (Leu 1973). At the end of the first week, lipoblasts and fibrocytes start to induce the formation of collagen fibrils that fill the hollow and intercapillary spaces left after liquefaction and absorption (Rotter 1981). As cellular infiltration is an ongoing process, a thrombus is composed of layers reflecting the different stages of development. Further *organization* is associated with *shrinkage of the vein*, which can be seen with ultrasound. Hemolysis with partial degradation of fibrin occurs after days to weeks.

Complete thrombus organization may transform superficial and small veins into strands of fibrous scar tissue. In most cases, however, there will be recanalization of the lumen through the ingrowth of capillaries. The latter dilate and become merged, thereby re-establishing patency over a course of several months. However, recanalization is associated with shrinkage and destruction of the valves as well as fibrosis and thickening of the wall. The main mechanism involved in the recanalization of a thrombosed vein is the high fibrinolytic potential of the venous wall.

The duration of thrombus organization depends on the vessel diameter and intraluminal pressure and may additionally be affected by external factors such as application of compression bandages. Attachment of the thrombus to the wall by collagen fibers will invariably have occurred by day 8–10. Thrombolytic therapy (e.g., streptokinase) performed at this time or later will recanalize the vessel but cannot prevent venous valve destruction in most cases. When surgical thrombectomy is performed at this stage, only central thrombus portions can be removed, while mural residues remain and may give rise to the postoperative development of recurrent thrombi through appositional growth. Residual thrombotic material near valves induces valve incompetence. Late sequelae are calcifications of the venous wall.

Collateralization and recanalization following acute deep venous thrombosis lead to the more or less complete reconstitution of venous drainage. Long segments of an occluded vein are recanalized in most cases (endogenous thrombolysis). Recanalization is a highly variable process: it may begin after 3–4 weeks in small vessels and may take 3–9 months in large veins such as the popliteal and femoral veins (Irninger 1963). The patency of a deep vein is re-established 3 months after the onset of thrombosis in about half of all cases (Killewich et al. 1989).

Venographic studies show that, within 1 year, complete recanalization occurs in up to 35% of all venous thromboses and partial recanalization in another 55%, with only 10% of patients showing persistent occlusion. The clinical severity of the *postthrombotic syndrome* mainly depends on the *degree of valve incompetence*, in particular of the *popliteal vein*, while a persisting lumen reduction after thrombosis has only a minor effect. The insufficient venous return is further compromised by secondary damage (widening with subsequent valve incompetence) to superficial and perforating veins resulting from the higher pressure and volume overload due to collateral flow (secondary varicosis).

Anticoagulation and compression therapy are major components in the management of thrombosis. The latter serves to limit the extent of progressive dilatation of the collateral veins induced by the increased outflow resistance, in particular during the first 3 months. In summary, chronic venous insufficiency (disturbed venous return from peripheral veins) can have the following causes:

- Obstruction of deep veins
- Valve incompetence of deep veins
- Valve incompetence of superficial veins
- Valve incompetence of perforating veins
- Calf muscle pump dysfunction (Fig. 3.14)

The superficial system (varicosis) and deep system (chronic venous incompetence) may develop secondary changes in response to disease of the respective other system. These secondary changes result from the compensatory increase in pressure and volume and may worsen the state of the already compromised venous return.

The morphologic features associated with the postthrombotic syndrome can be demonstrated in part by B-mode

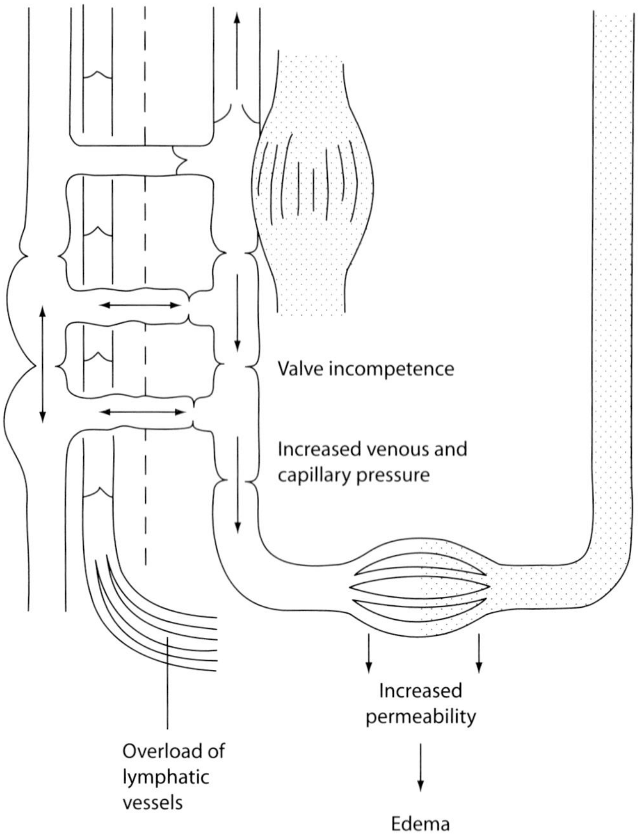


**Fig. 3.14** Diagram of the pathophysiological changes occurring in chronic venous incompetence. The drawing represents the superficial venous system on the *left* and the deep system on the *right*. When there is valve dysfunction in a major deep vein, the calf muscle pump (compression of the veins) fails to propel the blood toward the center (centripetal), resulting in at least partial reversal of flow toward the periphery (centrifugal). The ensuing recirculation via incompetent perforators and dilated varicose superficial veins further contributes to inefficient drainage. The increase in venous and capillary pressure results in a higher fluid infiltration and permeability of the damaged capillary wall. Interstitial edema in turn can lead to an overload of the lymphatic system with lymphatic microangiopathy in severe cases. Extensive and partially indurated edema in severe chronic venous insufficiency is not due to insufficient venous drainage alone but mainly to secondary lymphatic drainage insufficiency (according to Rieger and Schoop 1998)

ultrasound but above all by venography. The functional parameters reflecting the severity of reflux are reliably determined by duplex ultrasound and play a crucial role in planning treatment (type, extent, and duration of compression therapy).

Other causes of calf swelling besides acute thrombosis and chronic venous insufficiency include edema of different etiology (cardiac, lymphedema, lipedema). After exclusion of thrombosis and incompetent valves by duplex imaging, sonography can also provide important clues for differentiating lymphedema and lipedema. *Lymphedema* is characterized by the presence of primarily longitudinal, anechoic clefts (due to fluid collections) in the thickened subcutaneous tissue, while in lipedema such clefts are absent, and the subcutaneous layer appears rather uniform.

### 3.1.5.2 Varicosis

Varicosis of the great or small saphenous vein is caused by valve incompetence. In the primary form this is due to constitutional or external factors. In secondary varicosis, on the other hand, the valves of the superficial system fail due to pressure and volume overload resulting from disease (e.g., thrombosis) of the deep venous system (Table 3.3).

Four grades of varicosis of the great saphenous vein are distinguished according to Hach, depending on the length of involvement from its termination to its origin (Fig. 3.15). Grade I is varicosis of the terminal valve, grade II extension to the distal upper leg, grade III to the proximal lower leg, and grade IV complete incompetence of the vein down to the ankle.

*Primary superficial varicosis* can in turn lead to *pressure and volume overload of the deep venous system with secondary damage* resulting from the formation of *pathways of venous reflux*. Under these conditions, the blood draining through the deep veins reaches a proximal point of insufficiency in the superficial system (typically the terminal valve of the great saphenous vein) and then flows back down to the distal point of insufficiency (defined as the highest competent valve or the deepest incompetent valve). At this point, the blood flows back into the deep system and toward the heart (Fig. 3.16a). The resulting overload of the perforating and deep veins induces secondary valve incompetence in these segments. When the deep vein valves are competent, this condition is referred to as compensated recirculation and when they become incompetent as decompensated recirculation.

*Complete truncal varicosis* of the great saphenous vein is characterized by reflux in the saphenofemoral junction (i.e., the terminal valve of the great saphenous vein is incompetent). The ensuing pressure buildup causes secondary failure of distal valves (peripheral varicosis, grading according to Hach, see Fig. 3.15).

In *incomplete varicosis* of the great saphenous vein, the proximal valves initially tend to be competent, while the first dysfunctional valve (i.e., the proximal point of insufficiency) is more distal. Below this point, the valves of the great

**Table 3.3** Causes of valve incompetence

| |
|---|
| • Destruction (postthrombotic) |
| • Dilatation with incomplete coaptation |
| – Weakness of the venous wall (acquired, congenital) |
| – Pressure overload |
| – Volume overload (secondary, varicosis) |
| • Anomalies |

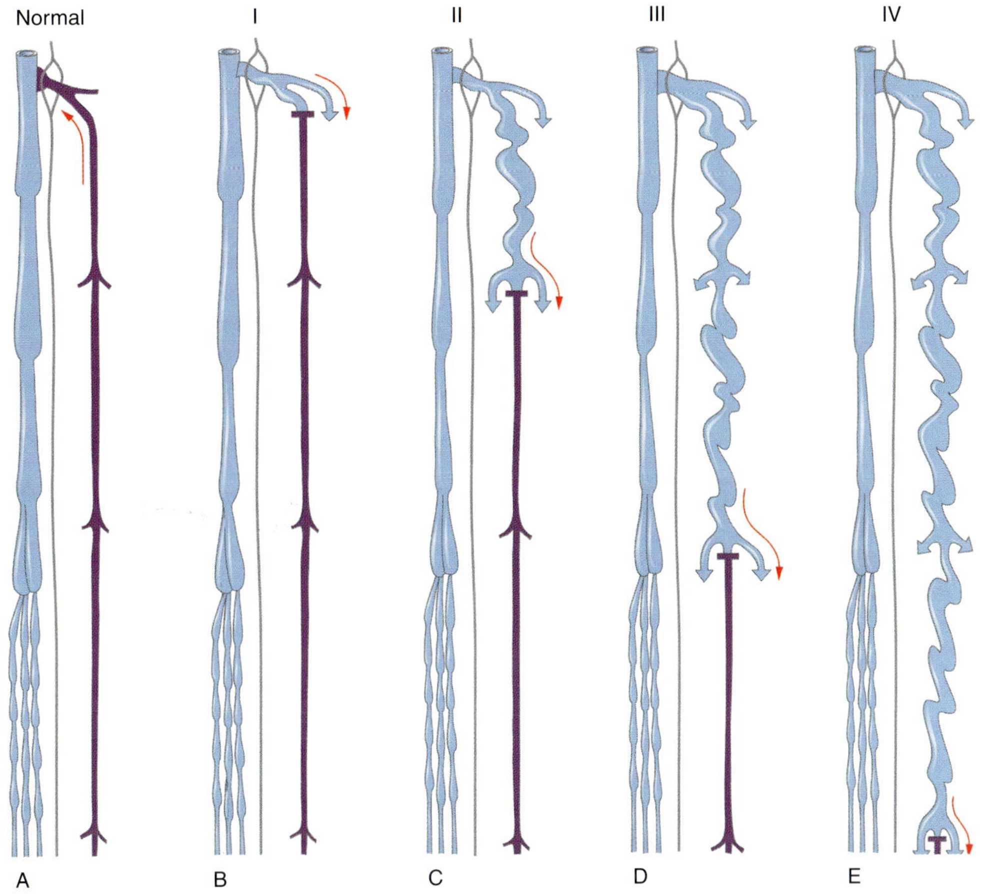


**Fig. 3.15** Grades of truncal varicosis of the great saphenous vein according to Hach:
*A* Normal blood flow toward the heart in the great saphenous vein;
*B* Grade I: incompetent terminal valve of great saphenous vein, possibly with concomitant lateral branch varicosis of accessory veins;
*C* Grade II: varicosis of great saphenous vein in upper leg, possibly with concomitant lateral branch varicosis;
*D* Grade III: varicosis of great saphenous vein extending to proximal lower leg, possibly with concomitant varicosis of anterior or posterior tributary vein of lower leg;
*E* Grade IV: varicosis of great saphenous vein extending down to ankle region with more or less severe lateral branch varicosis

saphenous vein are incompetent. In incomplete varicosis, the insufficient junction between the superficial and the deep venous system may involve the perforating veins, a branch of the great saphenous vein, or both. Incompetence of one or more perforating veins is the most common form. In this case, some of the blood entering the superficial system through the incompetent perforator flows into the distal portion of the great saphenous vein below this junction to then re-enter the deep system through a distal perforator.

In the second type, a varicose branch is responsible for reflux between the deep venous system and the proximal point of incompetence of the great saphenous vein (Fig. 3.16b). In most cases (55%), the incompetent branch is the lateral accessory saphenous vein (anterior variant); less commonly it is the medial accessory saphenous vein. In the posterior variant, the medial accessory saphenous vein establishes an incompetent venous communication with the proximal small saphenous vein via the femoropopliteal vein (Giacomini anastomosis). In this form of incomplete distal great saphenous vein varicosis, the incompetent Giacomini anastomosis connects the great and small saphenous veins.

Careful sonographic evaluation (see Tables 3.4 and 3.12) of the extent of varicosis with identification of the upper and lower points of insufficiency, secondary involvement of the deep venous system, and the presence of recirculation pathways is crucial for selecting the most suitable treatment (obliteration, surgery, compression). Surgery is performed to remove the insufficient portion of the affected deep vein between the upper and lower insufficiency points, sparing uninvolved venous segments for later arterial reconstruction. Incompetent superficial segments and perforating veins will invariably lead to *recurrent varicosis* if they are not removed. This is why precise determination of the distal point of insufficiency and the identification of insufficient perforating veins is crucial for the outcome of therapy. Duplex ultrasound is the method of choice and gold standard for this indication.

*Thrombophlebitis* is a typical complication of varicosis and is diagnosed by B-mode sonography using the same criteria as in the assessment of deep venous thrombosis. Since thrombophlebitis often extends beyond its clinically apparent boundaries, identification of the proximal thrombus end by imaging modalities is clinically relevant to exclude involvement of the deep system.

Moreover, further progression of thrombophlebitis into the deep system must be prevented by high ligation of the saphenofemoral junction in cases where the disease process already extends close to the deep veins. Alternatively, transient anticoagulation in combination with local symptomatic measures can be performed to prevent further progression in such cases.

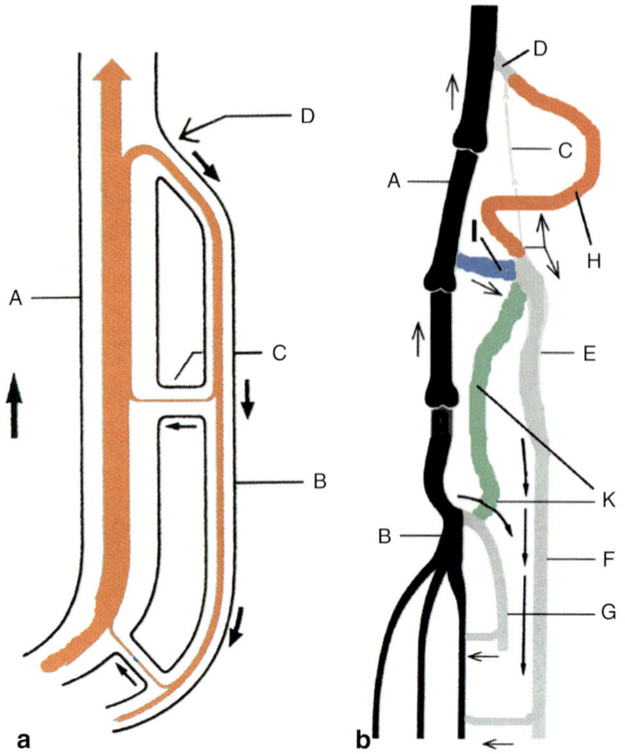


**Fig. 3.16** **a** Diagram of the recirculation pathway in superficial venous insufficiency. Part of the blood draining toward the heart in the deep venous system (*A*) flows back to the periphery through an incompetent terminal valve (*D*) or through incompetent superficial veins (great or small saphenous vein; *B*). The blood is then recirculated from the superficial to the deep venous system via perforating veins (*C*), resulting in volume overload of the deep (*A*) leg veins (according to Rieger and Schoop 1998). **b** Diagram of the different forms of incomplete varicosis of the great saphenous vein (*red*: lateral branch type, anterior variant; *blue*: perforator type; *green*: lateral branch type, posterior variant). *A* femoral vein; *B* popliteal vein; *C* competent (intact) proximal portion of the great saphenous vein; *D* terminal valve of great saphenous vein; *E, F* varicose segments of the great saphenous vein in the distal thigh and lower leg; *G* small saphenous vein; *H* lateral accessory saphenous vein; *I* Dodd perforator (incompetent); *K* medial accessory saphenous vein and Giacomini anastomosis (reflux through an incompetent connection between the small saphenous vein, femoropopliteal vein, medial accessory saphenous vein, and great saphenous vein). The proximal point of insufficiency is the transition from the competent to the incompetent great saphenous vein segment (*arrows*)

B-mode ultrasonography is the most suitable imaging modality both to identify the upper end of the thrombus for the initiation of adequate therapeutic management and to follow up therapy.

A retrospective analysis of the ultrasound findings in 363 patients with thrombophlebitis demonstrated growth of the thrombus into the deep venous system over an observation period of 10 days in 11% of the cases. Seventy percent of these cases were accounted for by great saphenous vein thrombophlebitis with thrombus growth into the common femoral vein (Foley et al. 1989).

**Table 3.4** Components of the sonographic evaluation of patients with varicosis

| |
|---|
| • Evaluation of major superficial veins (great and small saphenous veins), terminations, reflux pathways (truncal insufficiency) |
| • In the case of incomplete truncal varicosis: |
| – Determination of the upper point of insufficiency |
| – Determination of the lower point of insufficiency |
| • Identification of incompetent perforating veins |
| • Demonstration of secondary major vein insufficiency/valve incompetence of major deep veins |
| • Identification of variants in terminations of superficial veins. Morphologic variants |
| • Detection of thrombus or residual thrombus in the superficial and deep venous systems |
| • Quantification of impaired venous drainage |

More recent ultrasound studies of thrombophlebitis show thrombotic involvement of the deep venous system in 11–44% of the cases, which is a much higher rate than suspected on the basis of the clinical appearance (Blättler 1993; Blättler et al. 1996; Gaitini 1990; Gaitini et al. 1988; Lutter et al. 1991; Jorgensen et al. 1993; Ascer et al. 1995). Since therapeutic management must be extended to the deep veins in these patients, the indication for ultrasonography of the deep leg veins should be established generously.

In the routine clinical setting, the question to be answered is whether the chronic venous insufficiency the patient has developed is secondary to great or small saphenous vein varicosis or whether it exists in the context of the postthrombotic syndrome. As the therapeutic consequences are different, adequate diagnostic workup always includes evaluation of the morphologic and functional status of the major deep veins. *Primary valve incompetence of the superficial veins* (varicosis) without involvement of the deep veins is treated by surgical removal of the affected superficial vein segments to prevent dermatologic damage as well as secondary involvement of the deep leg veins due to pressure and volume overload (so-called Trendelenburg private circulation; Hach and Hach-Wunderle 1994). In *secondary valve incompetence of the superficial veins* with simultaneous deep vein involvement (postthrombotic), on the other hand, excision of the varices will not provide much improvement with regard to venous return. With few exceptions, surgery is not indicated in this situation. Instead, patients, including those operated on, are treated by a rigorous compression regimen (which must also be continued after surgery).

*Incomplete recanalization* or nearly complete postthrombotic occlusion of deep veins is a contraindication to the surgical removal of incompetent superficial vein segments. Tailoring therapeutic procedures to the individual patient relies on precise information regarding the localization and extent of morphologic and hemodynamic abnormalities.

Duplex scanning is superior to all other imaging modalities in providing this information (Table 3.4).

If sclerotherapy is planned for the treatment of varicosis of a side branch or mild truncal varicosis, the ultrasound examination can additionally serve to guide the puncture for injection of the sclerosing agent, particularly in obese patients, allowing immediate assessment of outcome.

## 3.1.6 Duplex Ultrasound – Diagnostic Criteria, Indications, and Role

### 3.1.6.1 Thrombosis

The chief sonographic criterion for the presence of venous thrombosis (of the deep and superficial systems) is the inability to compress the affected vein with the transducer in transverse orientation (Table 3.5; Figs. 3.17–3.19 and 3.21).

Full compressibility of a vein is indicated by complete obliteration of its lumen on ultrasound. Only a high-resolution transducer will depict the thin venous wall as an echogenic line within the muscle tissue. Incomplete compressibility indicates a thrombus surrounded by flowing blood (adherent to wall, floating) or partial recanalization after thrombosis with residual thrombus or severe wall sclerosis preventing full compression (Figs. 3.18 and 3.19).

**Table 3.5** Criteria for identifying fresh deep venous thrombosis of the leg by compression ultrasound in the B-mode

- Compressibility of the vein
- Dilatation of the vein (excluding breathing-related lumen variation)
- Hyperechoic intraluminal structure
- Perivascular structures (to identify extravascular causes of disturbed drainage)

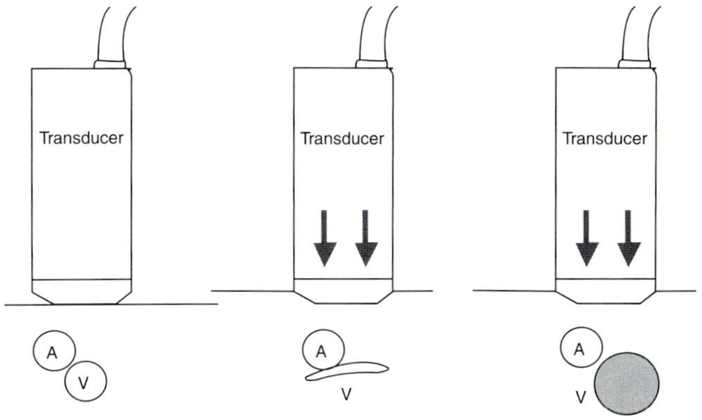


**Fig. 3.17** Diagram of compression ultrasound. A patent vein is completely compressible and virtually disappears when pressure is exerted with the transducer (*second drawing*). A thrombosed vein retains its shape upon compression (*right drawing*), while a partially thrombosed or partially recanalized vein can be compressed to some degree. A very fresh thrombus in a large vein may also be compressible to some extent, while the vein itself often has a wider lumen than the unaffected vein or the accompanying artery. A more or less hyperechoic intraluminal structure may be visualized

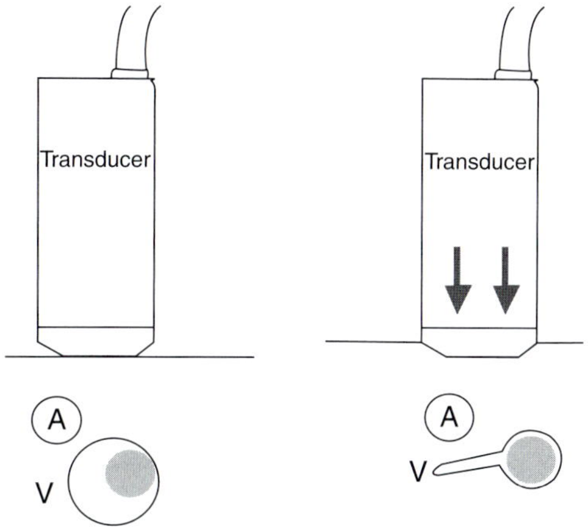


**Fig. 3.18** When pressure is applied to a vein containing a mural thrombus surrounded by flowing blood (*right drawing*), only the nonthrombosed vein segment is compressible. The delineation of the thrombus within the lumen depends on its composition, which determines its echogenicity

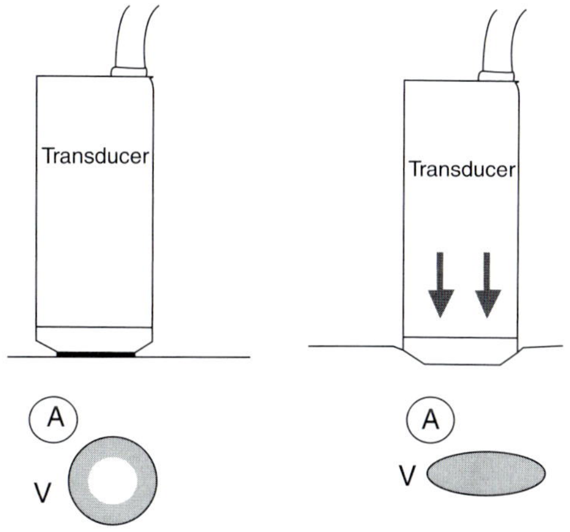


**Fig. 3.19** Partially recanalized vein. The extent to which the vein can be compressed depends on the degree of recanalization. Residual mural thrombi prevent full compression

The examination is usually performed with the patient lying on the examination table. Having the patient sit or stand may augment blood flow and improve evaluation of the calf veins. In the calf, the presence of a fresh thrombus improves visualization because the hypoechoic dilated vein is more conspicuous than a collapsed vein or small, thin-walled vein with normal blood flow. *Positive ultrasound findings are nearly 100% specific for deep vein thrombosis of the leg. Compression ultrasound can rule out thrombosis in the thigh and in the popliteal fossa with acceptable accuracy*, while some uncertainty remains in below-knee thrombosis, even with color

Doppler imaging. If the findings are equivocal and the clinical presentation is highly indicative of thrombosis (high pretest likelihood), additional phlebography or possibly a D-dimer test or repeat ultrasound after 5 days should be performed.

Study data indicate that color duplex imaging does not improve diagnostic accuracy in deep vein thrombosis, and color duplex is required only to diagnose isolated pelvic thrombosis. Color depiction of flow improves the diagnostic evaluation at the pelvic level, especially in obese patients in whom compression maneuvers are difficult to perform.

*Color duplex imaging* improves the visualization of *residual blood flow around a thrombus* or *floating thrombus* and also of *recanalized veins*, which often have a small lumen.

Many studies have demonstrated the high accuracy of B-scan ultrasound in the detection of thrombosis in comparison with venography (Table 3.6). In symptomatic patients, compression ultrasound has a sensitivity of 95% and a specificity of nearly 100% in demonstrating fresh thrombosis of the upper leg and popliteal vein. Incompressibility of a vein is both a necessary and sufficient condition for the diagnosis of deep venous thrombosis of the leg. Additional color duplex scanning does not improve the accuracy, and the use of color duplex alone tends to have a poorer specificity than compression ultrasound due to false-positive findings as a result of slow venous flow or a poor acoustic window in the lower leg.

In the rare cases of *isolated pelvic vein thrombosis*, compression ultrasound alone is often unreliable due to the lack of an adequate structure against which to compress the vein and is also impaired by interfering overlying structures. In these cases, a thrombosis is diagnosed on the basis of absent flow or an abnormal flow signal (compared to the unaffected side) in the Doppler waveform or a gap in color filling in color duplex scanning.

Careful scrutiny of the pelvic axis is indicated if an *abnormal Doppler waveform* with reduced respiratory phasicity and slower flow compared to the contralateral side is obtained *in the distal external iliac vein* in the longitudinal plane. The patient must lie supine with the thigh slightly abducted and externally rotated to ensure undisturbed venous outflow under the inguinal ligament. Flat positioning with the thigh stretched will compress the vein as it courses under the inguinal ligament, reducing or even eliminating respiratory phasicity in the Doppler waveform obtained from this site. However, since even isolated pelvic vein thrombosis typically involves the entire external iliac vein (including outflow through veins of the saphenofemoral junction and abdominal wall), the thrombosis can be demonstrated by B-mode and compression ultrasound above the inguinal ligament. The only pitfalls are thrombi protruding from the internal iliac vein into the common iliac vein, which escape detection by indirect hemodynamic flow analysis in the groin because they do not obstruct flow. This also applies to incomplete mural thrombi in a partially patent pelvic vein.

The small-caliber vessels below the knee are less well demarcated from the inhomogeneous echotexture of muscle tissue. Still, the criteria for *isolated lower leg venous thrombosis* are the same as in the upper leg: dilatation of the vessel lumen, noncompressibility, and possibly depiction of an inhomogeneous, hypoechoic intraluminal structure. Better filling of the veins is achieved if the examination is performed in the sitting or standing patient. *Since a tubular structure distended by fresh thrombosis can be identified more easily than a normal vein*, nonvisualization can be interpreted to indicate absence of fresh thrombosis. Note, however, that this only holds true for fresh venous thrombosis, whereas *older thrombi shrink* and often become more hyperechoic and inhomogeneous with the venous lumen returning to its normal diameter. Hence, the vein is again more difficult to differentiate from surrounding muscle tissue, rendering the method less accurate in identifying older thrombosis below the knee.

In the diagnosis of fresh (clinically relevant) calf thrombosis in patients with clinical symptoms, compression ultrasound has a sensitivity of 85–90% and a specificity of over 95% (see Table 3.6). Besides inspection of the main lower leg veins, which are identified on B-mode or color flow images using the arteries of the same name as landmarks, the examiner must pay special attention to the gastrocnemius and soleus veins, which are a common site of origin for deep vein thrombosis, especially in immobile patients. Stasis of flow may occur as the muscle veins become ectatic with age. The diagnostic criteria are the same as for the main veins (dilated, incompressible tubular structure within the muscle).

There is some remaining diagnostic uncertainty in the lower leg, and some patients may require venography, especially when the acoustic window is poor. Nevertheless, venography has its limitations as well and does not achieve 100% certainty either because some veins of the lower leg, especially the fibular vein group, are notoriously difficult to image.

*Different algorithms have been proposed* to compensate for the *diagnostic uncertainty of compression ultrasound in the calf* (85–90% sensitivity) and thus to optimize the diagnostic and therapeutic management of these patients. The options available for reducing diagnostic uncertainty are the use of the highly sensitive but rather unspecific D-dimer test, repeat ultrasound after 1 week, and venography in high-risk patients. Based on empirical and clinical experience, patients with suspected venous thrombosis can be assigned to a high-probability or a low-probability group on the basis of their risk factors, the severity of clinical signs, and the likelihood of alternative conditions that may explain the symptoms. These factors are assessed by means of scores that assign patients to different pretest likelihood groups, which determine further diagnostic measures if the ultrasound findings are inconclusive. For instance, high-risk patients will undergo supplementary venography or a D-dimer test, while no further diagnostic measures will be taken in patients with a low risk (Fig. 3.21).

**Table 3.6** Studies investigating the role of compression ultrasound, duplex ultrasound, and color duplex ultrasound in larger patient populations with deep leg vein thrombosis (using venography as the gold standard)

| Author/Year | Patients [n] | Thrombosis [n] | Sensitivity [%] | Specificity [%] |
|---|---|---|---|---|
| *Compression ultrasound* | | | | |
| Appelman 1987 | 112 | 52 | 96 | 97 |
| Dauzat 1986[b] | 145 | 100 | 94 | 100 |
| Elias 1987[b] | 430 | 303 | 98 | 95 |
| Habscheid 1990[a] | 238 | 153 | 96 | 99 |
| Hobson 1990 | 209 | – | 99 | 100 |
| Krings 1990 | 182 | – | 95 | 97 |
| Lensing 1989[a] | 220 | 66 | 99 | 100 |
| Pederson 1991 | 215 | 113 | 89 | 97 |
| Herzog 1991[a] | 113 | 57 | 88 | 98 |
| Langholz 1991 | 64 | 25 | 76 | 88 |
| *Compression ultrasound: analysis of lower leg veins only (thrombosis)*[a] | | | | |
| Habscheid 1990 | 37 | – | 89 | 99 |
| Elias 1987 | 92 | – | 91 | 96 |
| *Duplex ultrasound* | | | | |
| De Valois 1990 | 180 | 61 | 92 | 90 |
| Comerota 1990 | 103 | 44 | 96 | 93 |
| Killewich 1989[a] | 47 | 38 | 92 | 92 |
| Van Ramshorst 1991 | 117 | 64 | 91 | 95 |
| Schäberle 1991[a,c] | 125 | 56 | 97 | 98 |
| Betzl 1990 | 66 | – | 97 | 72 |
| *Color duplex ultrasound* | | | | |
| Schindler 1990 | 97 | 54 | 98 | 100 |
| Grosser 1990[a] | 180 | 154 | 94 | 99 |
| Van Ramshorst 1991 | 117 | 64 | 91 | 95 |
| Schönhofer 1992 | 100 | 63 | 97 | 98 |
| Miller 1996 | 216 | 98 | 99 | 100 |
| Fürst 1990 | 102 | 39 | 95 | 99 |
| Persson 1989[a] | 264 | 16 | 100 | 100 |
| Rose 1990[a] | 69 | 32 | 79 | 88 |
| Van Gemmeren 1991 | 114 | 74 | 96 | 97 |
| Langholz 1991 | 116 | 65 | 100 | 94 |
| Fobbe 1989 | 103 | 58 | 96 | 97 |
| Lensing 1989 | 220 | – | 91 | 99 |
| Krings 1990 | 235 | – | 93 | 96 |
| Schweizer 1993 (with ultrasound contrast medium) | 78 | 70 | 96 | 100 |

Note that the lower leg veins were not included in the examination in all cases
[a]Lower leg included in examination and analysis
[b]Compression ultrasound, in part, supplemented by CW Doppler
[c]Primary compression ultrasound with optional duplex scanning (primarily for pelvic vein assessment and inconclusive lower leg findings)

**Controversies About the Extent of the Examination** Some protocols do not include the calf veins in the evaluation by compression ultrasound but are instead restricted to the area from the distal external iliac artery, just above the inguinal ligament, to the distal popliteal vein or tibiofibular junction (possibly with repeat examination after 3–5 days). They are assumed to enable adequate diagnosis of thrombosis even when only two key sites – the femoral bifurcation and the patients with longer-standing thrombosis (dilated veins with spontaneous flow signals in the deep and superficial compartments). The presence of collateral pathways is an additional criterion for differentiating between an older and a more recent thrombosis and recurrence (Table 3.9; Figs. A3.17 and A3.18).

Documentation of the patency of all relevant veins by duplex ultrasound is time-consuming and would prohibit the

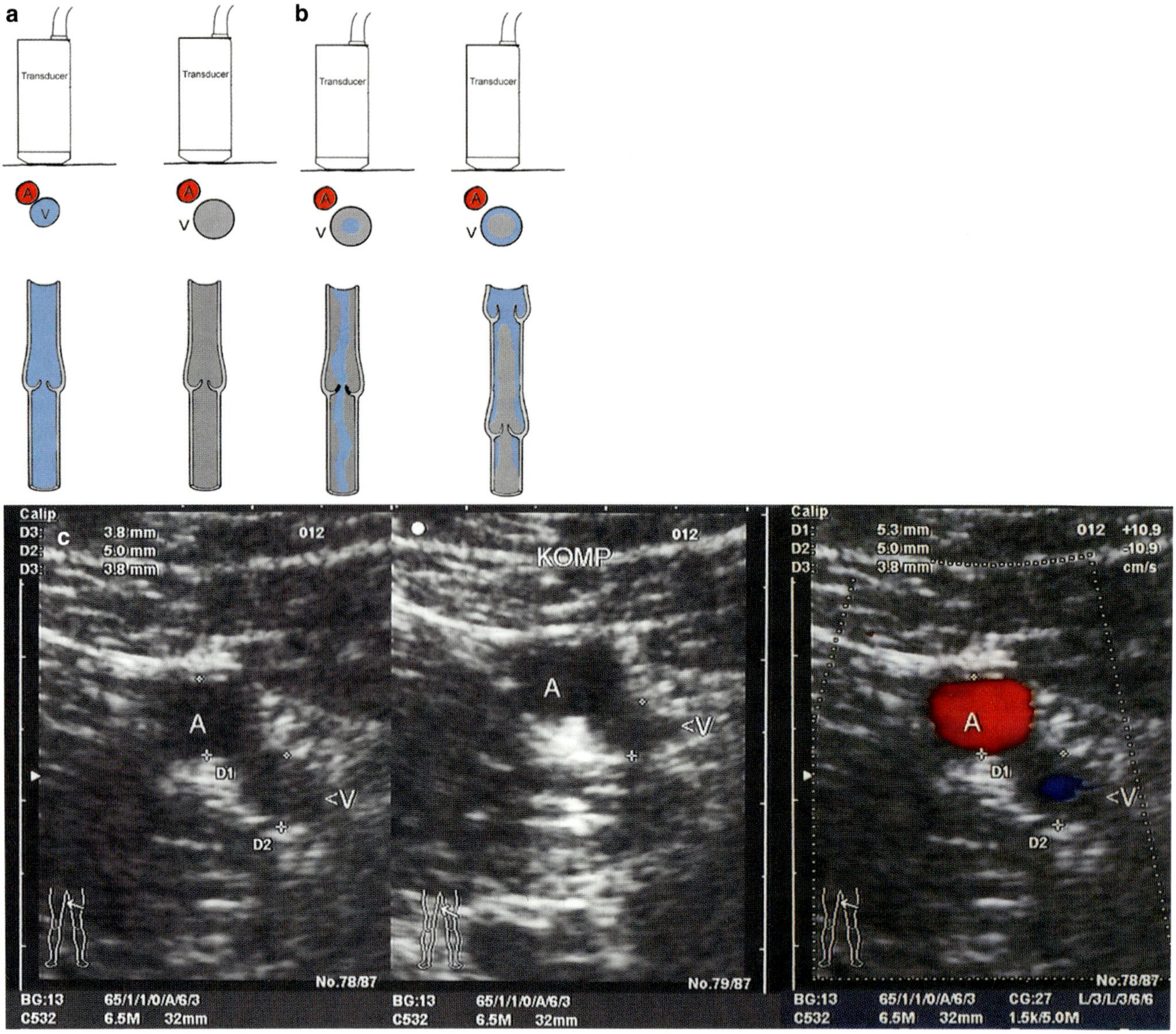


**Fig. 3.22 a–e** Diagnostic role of color duplex imaging in venous thrombosis. **a** Color duplex imaging does not significantly improve diagnostic accuracy compared with compression ultrasound, providing no additional information for ruling out thrombosis (*left*) or detecting occlusive thrombosis (*right*) (see Fig. 3.17). **b** Color duplex offers advantages in detecting recanalization and estimating its degree (*left*) because it depicts flow (which may involve imaging during Valsalva maneuver or with compression); it also offers advantages in identifying nonocclusive mural thrombus or free-floating thrombus (*right*) by detecting flow signals around the thrombus (which may require a provocative maneuver). Settings must be adjusted to depict slow flow (low PRF). **c** Half a year after acute deep vein thrombosis of the leg, the shrunken lumen of the femoral vein (*left section*; compare the lumen of the artery, indicated by *A*) is less clearly delineated from surrounding muscle and connective tissue. The second scan, obtained during compression with the transducer, shows incomplete compression of the vein (see Fig. 3.19) with a diameter reduction from 5 to 3.8 mm (*plus signs*). The color duplex scan (*right section*) reveals the cause of poor compressibility in **b**: there is a small recanalized lumen with flow coded in blue in the center of the superficial femoral vein (*V*). **d** Longitudinal view (*left*) and transverse view (*right*) of thrombus in the popliteal vein (*V.POP*) partially surrounded by flowing blood. The thrombus has all criteria of acute thrombosis: hypoechoic, markedly dilated vein, good demarcation from perivascular connective tissue, and marginal flow. These features differentiate this case from older thrombosis with partial recanalization. **e** Transverse view (*left*) and longitudinal view (*right*) of popliteal vein thrombosis (of 6 weeks duration) with beginning recanalization: there are flow signals in the center of the lumen (meandering flow and multiple recanalized lumina are also present). The Doppler waveform from this segment reflects the flow obstruction due to extensive residual thrombosis: flow is slow and respiratory phasicity is lost

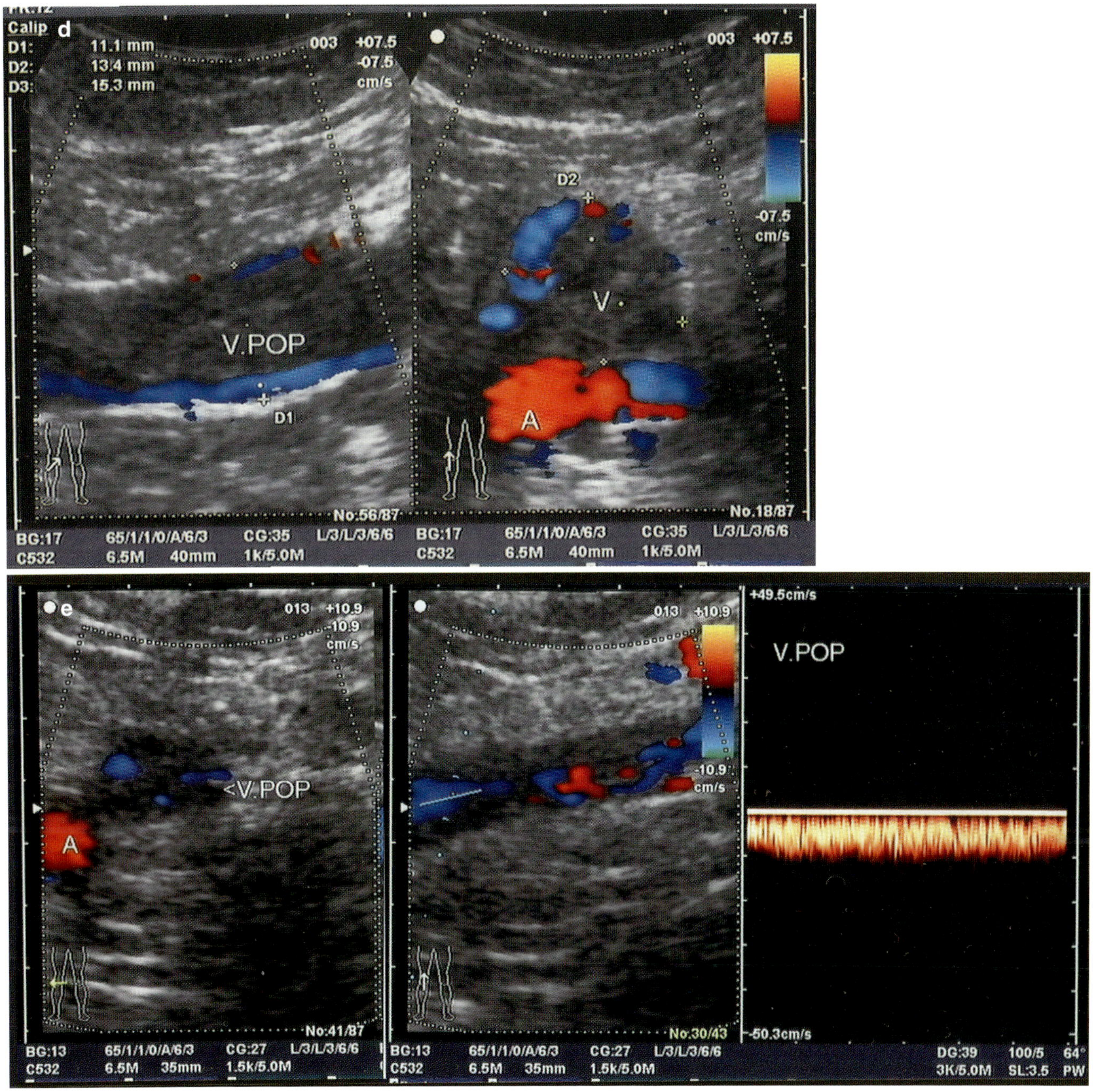


**Fig. 3.22** (continued)

**Table 3.8** Added diagnostic benefit of color duplex ultrasound in acute venous thrombosis of the leg

| |
|---|
| • Detection of residual flow near the wall |
| • Identification of collateral vessels |
| • Evaluation of the pelvic veins |
| • Demonstration of recanalization |
| • Patency of calf veins |

liberal use of ultrasound advocated by the author. Therefore, all ultrasound laboratories should implement a standardized and efficient algorithm for the diagnostic management of patients with suspected acute venous thrombosis of the legs. This can be done using compression ultrasound, which enables examination of both legs in about 10–15 min.

**Spectral Doppler Analysis** Obstruction of venous return (thrombosis, external compression) leads to increased intravascular pressure distal to the obstruction, demonstrated by duplex ultrasound as slower flow (Table 3.9).

The increased venous flow resistance associated with occlusion, compression, or persisting postthrombotic obstruction eliminates the respiratory phasicity of venous drainage (resulting from changes in intra-abdominal pressure).

**Table 3.9** Venous ultrasound

| |
|---|
| **Indications for gray-scale ultrasound/compression ultrasound** |
| • Evaluation of thrombosis (exclusion, confirmation, extent, age) with localization and differentiation (main veins, muscle veins) |
| • Thrombophlebitis (extent, thrombus protrusion into major deep vein) |
| • Follow-up (spontaneous resolution, thrombolysis, thrombectomy) |
| • Differential diagnosis of perivascular structures compressing the vein (Baker's cyst, soft tissue tumor, hematoma, abscess, wall tumor) |
| **Indications for color duplex ultrasound** |
| • Follow-up after thrombosis (spontaneous or thrombolysis-induced recanalization) |
| • Pelvic vein thrombosis |
| • Floating thrombi |
| • Chronic venous insufficiency/postthrombotic syndrome (valve incompetence of deep leg veins: severity of reflux, extent, degree of recanalization) |
| • Varicosis (extent, severity of reflux, secondary incompetence of main veins; preoperative evaluation: determination of upper and lower points of insufficiency, identification of incompetent perforating veins) |
| • Vein mapping prior to bypass surgery (suitability of saphenous vein for venous bypass grafting) |
| • Venous aneurysm (extent; characterization: spindle-shaped, saccular, intraluminal thrombosis) |

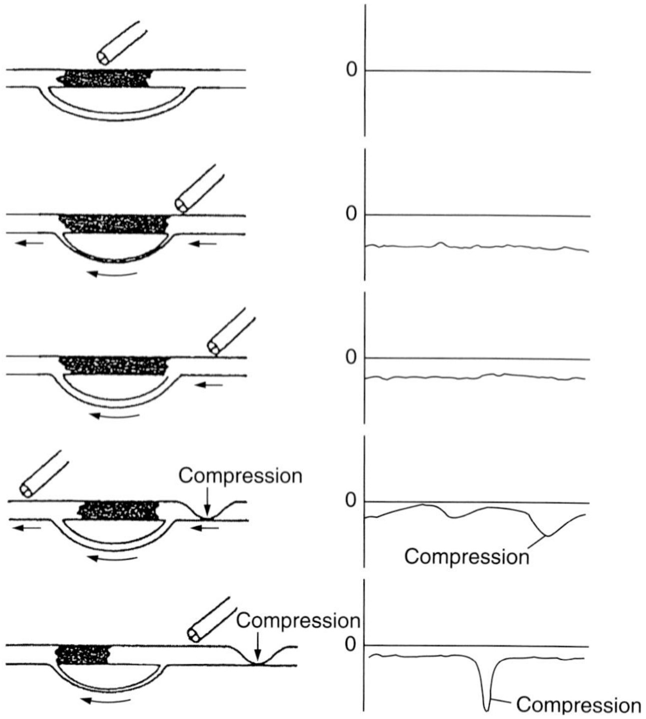


**Fig. 3.23** Diagram of Doppler waveforms obtained in flow-obstructing thrombosis – functional tests

This loss is reflected in the Doppler examination as a constant flow velocity in the vein distal to the obstructed segment and as a reduced flow velocity, which is best appreciated by comparison with the other leg (Fig. 3.23; Table 3.10).

**Thrombus Age** Initial hopes of determining thrombus age by means of sonomorphologic criteria and thus of selecting the most promising therapeutic option (surgery, thrombolysis, anticoagulation) have been disappointed. What is possible though is to differentiate very recent thrombi from much older ones (Fig. 3.24 and Table 3.16), which are characterized by increasing inhomogeneity and echogenicity with a decrease in luminal diameter (Fig. 3.25). In general, however, there is wide interindividual variation in thrombus development, and *the criteria are not reliable enough for therapeutic decision-making*. This holds true especially for the clinically relevant identification of thrombi that are still amenable to recanalization measures, i.e., thrombi not older than 1 week. Nevertheless, the sonomorphologic criteria can contribute to the therapeutic decision in individual cases. For instance, a homogeneous thrombus with lower echogenicity in a markedly dilated vein with good demarcation of the wall and thrombus portions surrounded by flowing blood is more likely to respond to thrombolytic therapy and will undergo rapid recanalization.

Older thrombosis is characterized by *progressive narrowing of the venous* lumen and a *loss of wall conspicuity* (Fig. 3.25a and Table 3.16). The poorer delineation from surrounding muscle tissue and increasing echogenicity of the thrombus contribute to a *lower accuracy* of ultrasound in diagnosing thrombosis, especially in the calf. Following the acute stage, early recanalization can also be demonstrated by color flow imaging. When the recanalized lumen is still small, sparse and slow venous flow may be depicted after augmentation (calf compression, Valsalva maneuver) even if spontaneous flow is not detectable (with the transducer set to detect slow flow). Both gray-scale ultrasound and color duplex imaging allow evaluation of venous drainage, which may be compromised by poor recanalization or *postthrombotic changes*. The findings may range from complete recanalization with a

**Table 3.10** Evaluation of flow obstruction by duplex scanning using criteria known from CW Doppler ultrasound

| |
|---|
| • Zero flow in the thrombotically occluded vein |
| • Decreased flow velocity with reduction or elimination of respiratory phasicity due to proximal thrombus or venous compression by surrounding structures |
| • Flow signal unaffected by respiration, possibly of high frequency, along partially thrombosed or compressed vessel segments or along thrombi surrounded by flowing blood (differential diagnosis: flow signal from collateral vein not subject to respiratory phasicity) |
| • Augmented flow abnormally reduced in the valve function test distal or proximal to thrombosis or flow obstruction |

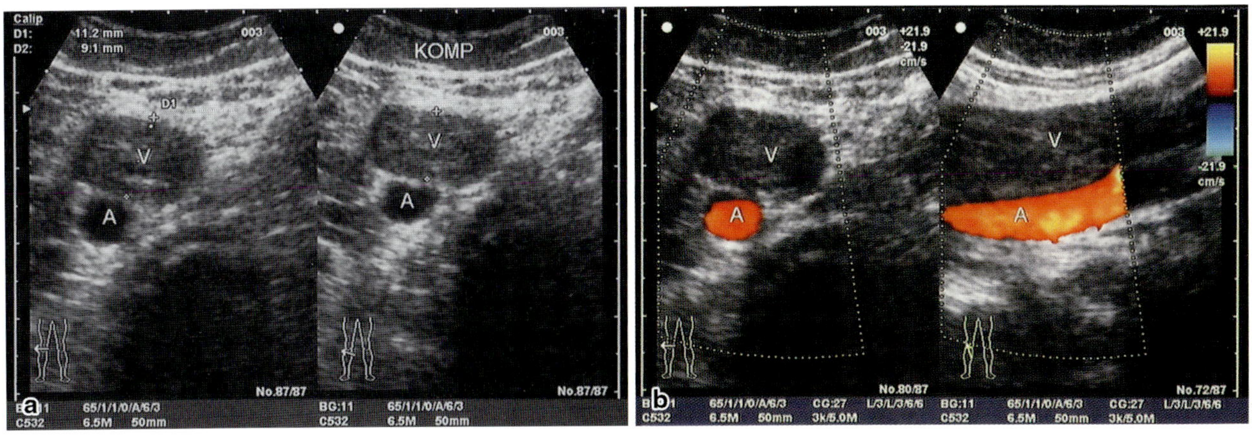


**Fig. 3.24** **a** Acute thrombosis of the popliteal vein (*V*), which is dilated by the thrombus (compare arterial diameter indicated by *A*). The wall of the thrombosed vein is sharply demarcated from surrounding tissue. The thrombus is identified on the gray-scale image as a hypoechoic, homogeneous lesion. The *right scan* obtained during compression with the transducer shows that the soft clot is still somewhat compressible (with a diameter reduction from 11 to 9 mm). **b** Color duplex imaging (transverse view on the *left*, longitudinal view on the *right*) fails to depict flow signals (even with calf compression), indicating occlusive thrombosis

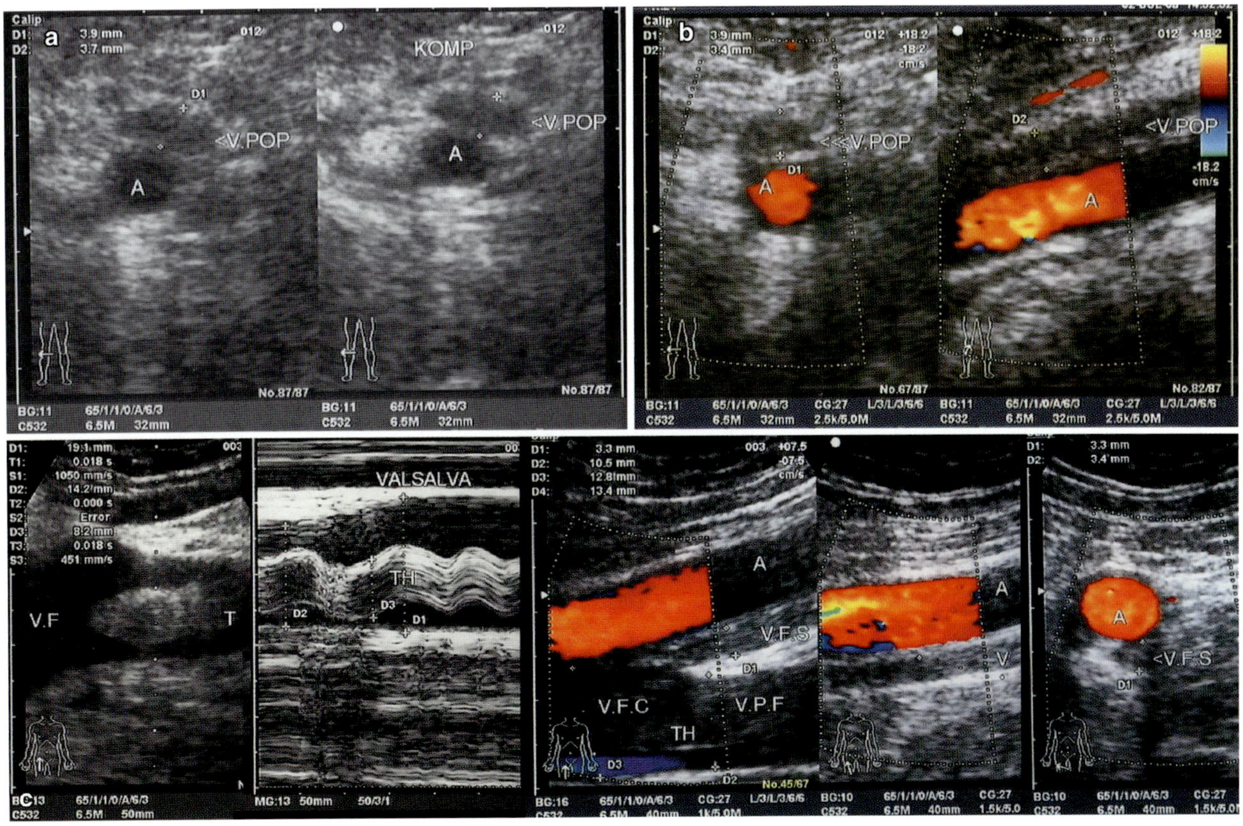


**Fig. 3.25** **a** Older thrombosis (>3 months) of the popliteal vein. Transverse views of the vein without compression (*left*) and with compression (*right*). The wall of the thrombosed popliteal vein (*V.POP*) is blurred and difficult to delineate from surrounding tissue. The lumen has shrunken (and is smaller than that of the accompanying artery). At this stage (after fibroblast invasion), the thrombus cannot be compressed with the transducer (diameter of 3.9 and 3.7 mm without and with compression, respectively). **b** Transverse (*left*) and longitudinal (*right*) color duplex images fail to depict flow in the popliteal vein (*V.POP*), confirming occlusive thrombosis. The blurred wall and poor delineation from surrounding tissue are most obvious on the longitudinal image. Red-coded flow indicates the accompanying popliteal artery (*A*). **c** Patient with mild swelling of the leg one year after deep vein thrombosis. The older thrombosis causes complete occlusion of the superficial femoral vein (thin vein with poorly differentiated wall and higher echogenicity on the color flow images, *V.F.S*, *V.*) The first color flow image additionally shows a fresh appositional thrombus (*TH*) extending from the old thrombosis into the common femoral vein (*V.F.C.*) and deep femoral vein (*V.P.F.*). The appositional portion has low echogenicity and is attached to the wall anteriorly with flow posteriorly (blue). It is well delineated from the wall and causes marked dilatation of the vein. The *leftmost scan* shows a free-floating component (3 cm in length), and the time-motion scan next to it shows the floating thrombus to be highly mobile within the dilated vein (during Valsalva maneuver). There is a risk of pulmonary embolism, and prompt anticoagulation treatment should be resumed

normal sonographic appearance of the venous wall to persistent thrombotic occlusion with luminal narrowing, residual thrombi of various size, wall sclerosis, and synechia. The severity of reflux due to incompetent valves is assessed by spectral Doppler sampling with Valsalva's maneuver or compression testing. If the sonographic examination shows not only postthrombotic changes but also newly obstructed venous segments (hypoechoic thrombus and dilatation of the vein compared with the accompanying artery), this is a sign of recurrent thrombosis (Fig. 3.25c) – especially if the proximal thrombus end is surrounded by flow. In contrast, flow signals in the center of a thrombosed vein depicted by color duplex indicate older thrombosis with beginning recanalization. Impaired venous drainage due to thrombus surrounded by flow or residual thrombus in a recanalized vein with a narrow lumen is identified by the absence of respiratory phasicity in the Doppler waveform (Fig. 3.19).

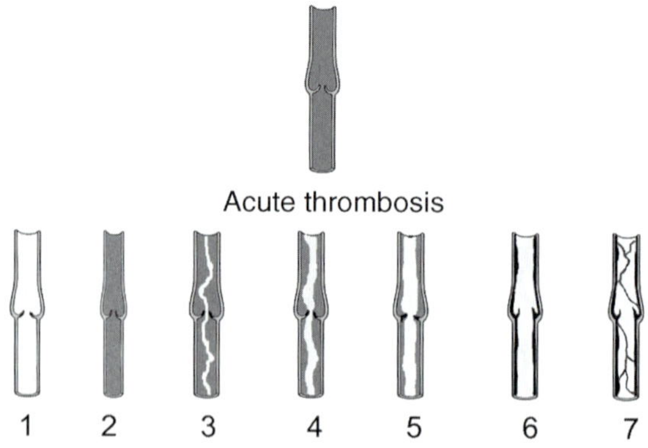


**Fig. 3.26** Persistent venous changes following deep vein thrombosis. *1* Complete recanalization, only valve damage. Normal gray-scale and compression ultrasound findings of the vein. *2* Persistent occlusive thrombosis. Sonographically, the vein typically has a decreased lumen (diameter is no larger than that of the accompanying artery) with higher echogenicity and cannot be compressed. *3* Narrow recanalized lumen. The recanalized vein is often missed by B-mode or compression ultrasound; color duplex imaging is usually required to demonstrate flow in the vein (and Valsalva maneuver or compression may be necessary to elicit flow). *4* Recanalization with residual mural thrombus. Visualization of the recanalized vein may be poor on gray-scale images, and the vein is not fully compressible. *5* Recanalization with marked wall thickening. In most cases, the patent lumen is sonographically delineated from the thickened wall, and the vein can be compressed, but full compression is prevented by the thickened wall. *6* Recanalization with wall sclerosis. Hyperechoic wall with focal posterior shadowing on B-mode imaging and incomplete compressibility due to thickened, sclerotic wall. *7* Intraluminal synechia and membranes. Hyperechoic and slightly mobile when the vein is compressed; membranes, typically with concomitant wall sclerosis, preclude complete compression of the vein

### 3.1.6.2 Chronic Venous Insufficiency

While thrombosis can be evaluated by B-mode imaging alone, Doppler ultrasound is necessary for the hemodynamic assessment of the severity of insufficient venous return in chronic venous incompetence. A high-resolution transducer nicely depicts venous wall sclerosis by an increase in echogenicity and thickening of the wall in the presence of a patent lumen; it also serves to identify valves damaged and immobilized by the sclerotic process. B-mode imaging alone is insufficient in evaluating the postthrombotic patient because a recanalized vein will have a normal sonomorphologic appearance in about 30% of cases. The other 70% show wall irregularities and thickening, a strand-like vein with a smaller lumen (Fig. 3.26), or a vein that has become dilated after recanalization due to the pressure and volume overload resulting from incompetent valves (see Figs. A3.34 and A3.39–A 3.41). If recanalization is delayed, serial ultrasound will show residual mural thrombi or a thickened wall, from which residual thrombus cannot be differentiated by ultrasound. The affected vein cannot be fully compressed, and color flow images will depict a narrow lumen with flow signals surrounded by an inhomogeneous area of mixed low and high echogenicity extending to the perivascular connective tissue (wall thickening and residual thrombi cannot be distinguished; Fig. 3.27).

*Incomplete compressibility* of the postthrombotic vein is due to *wall thickening*, and the degree to which a vein can be compressed depends on the extent of recanalization and the presence of intraluminal structures (Fig. 3.28a, b), which are inhomogeneous and mostly hypoechoic relative to the surrounding connective tissue. Especially the postthrombotic femoral vein may be difficult to identify since all that may remain is a cord-like structure visible on B-scan ultrasound. For this reason, it is helpful to use the accompanying artery as a landmark. A recanalized femoral vein with incompetent valves will dilate during Valsalva's maneuver (Tables 3.11 and 3.16).

Hence, B-mode ultrasound alone is insufficient to evaluate the recanalization process as it does not depict flow in sclerotic segments, and wall sclerosis may preclude compression of the vein (Table 3.11).

The clinical manifestation of the postthrombotic syndrome is chiefly influenced by the degree to which venous return is compromised, which in turn varies with the degree of thrombosis and *recanalization* as well as with the *presence of collateral pathways*. *Valve incompetence in the main veins* determines the extent of reflux, which correlates well with the extent of the initial thrombosis. Duplex ultrasound studies demonstrate abnormal reflux after recanalization of thrombotic deep vein segments in about 45–70% of cases after 1–3 years, while normal findings with preserved valve function are seen in 12–30% after recanalization (Johnson et al. 1995; Markel et al. 1992) and 10–20% of vessel segments remain completely obstructed (Johnson et al. 1995).

*Venous reflux* is measured with a Valsalva maneuver to evaluate proximal valve function and the compression test to evaluate distal insufficiency. The increase in intra-abdominal pressure induced by the Valsalva maneuver produces a short, physiologic backward flow with a mean duration of 0.3 s (Fig. 3.28c–f). Reflux persisting for over 1 s is abnormal.

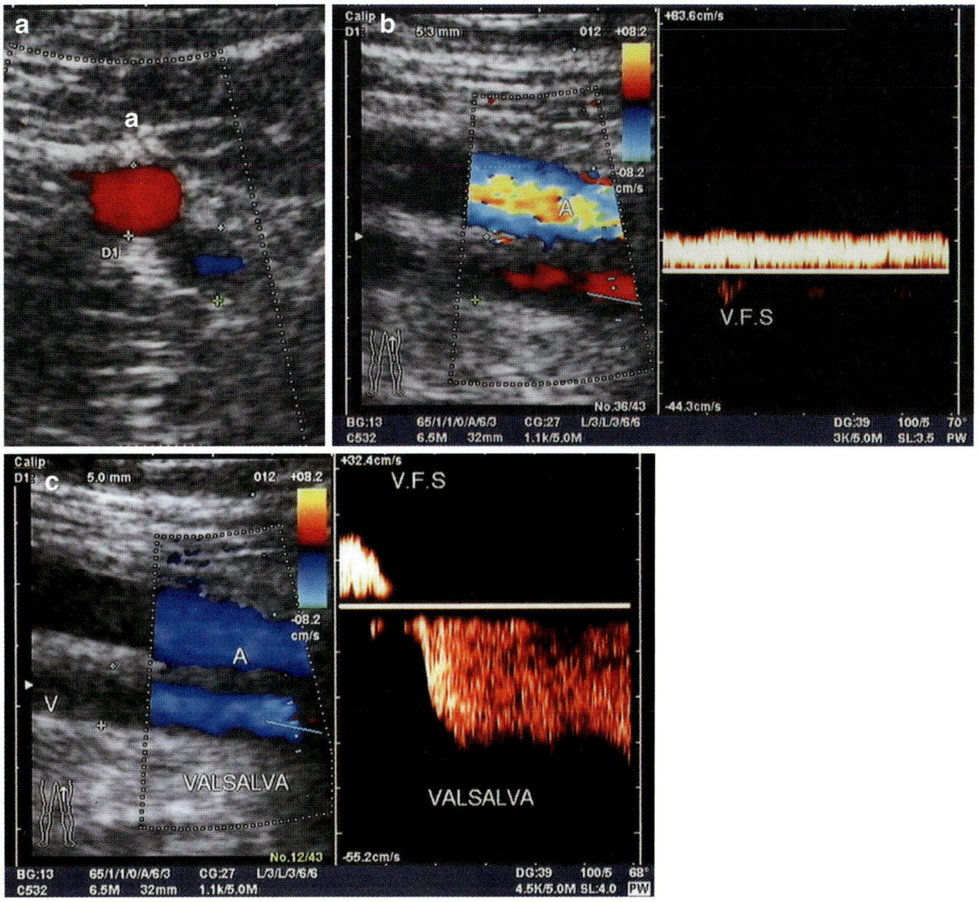


**Fig. 3.27** **a** Transverse image shows the artery, with flow coded red, and the femoral vein posterolaterally (indicated by *plus signs*). The center of the vein is recanalized (blue-coded flow signals) and surrounded by an inhomogeneous, thickened wall with areas of high and low echogenicity. Residual mural thrombus cannot be differentiated from the wall, which is also poorly delineated from surrounding connective tissue on the gray-scale image. **b** The longitudinal color flow image shows flow in the center of the recanalized venous lumen (red, toward transducer) and extensive residual mural thrombosis. The low PRF chosen to depict slow venous flow causes aliasing in the accompanying popliteal artery (blue, flow away from transducer). **c** Flow reversal in the vein elicited by Valsalva's maneuver (blue, away from transducer) indicates incompetent valves. In the Doppler waveform (*right*), the flow reversal is seen as persistent flow below the baseline

Studies in patients with stage II or III chronic venous insufficiency demonstrated sensitivities of 77–91% and specificities of 85–100% for reflux assessment by duplex ultrasound (Araki et al. 1993; Neglen and Raju 1992). Moreover, the duplex findings correlated better with the clinical stage than did ascending venography.

The *Valsalva test* can be performed in the recumbent patient, while the *compression test* should be performed with the patient sitting or standing, which increases the diagnostic accuracy by 10%. The compression test can be performed with manual compression distal to the ultrasound probe or in a standardized manner using a cuff for compression. Standardized spectral Doppler recordings of reflux in the popliteal vein are obtained upon sudden deflation of a blood pressure cuff placed around the calf and inflated to 100 mmHg. The use of such standardized provocative maneuvers is a prerequisite for the interindividual comparability of results and for the conducting of scientific studies. The compression test reproduces the flow variations resulting from the intermittent compression of the muscle veins by the calf muscle pump.

Because the main veins are also embedded in the muscle, contraction not only brings about drainage to the center but also induces flow toward the periphery, which is prevented by competent valves. In patients with incompetent perforators, the muscle pump is also responsible for abnormal reflux from the deep into the superficial veins.

Therefore, incompetent valves reduce the efficiency of the muscle pump because the venous pressure, normally decreasing upon muscular activity, remains unchanged or drops only a little. These complex interactions must be taken into consideration when selecting a site for placing the Doppler sample volume and also in interpreting the flow data obtained.

In severe valvular incompetence, as in the postthrombotic syndrome, reflux can be induced not only by Valsalva's maneuver but also by normal inspiration or deep inspiration in the horizontal position. Persisting reflux throughout inspiration indicates valve incompetence. Under normal conditions, the craniocaudal pressure gradient ensures rapid valve closure and thus prevents reflux.

Pressure and volume overload occurring distal to postthrombotically altered veins can lead to secondary damage through hyperextension of the valvular rings in the formerly unaffected vessel segments (Killewich et al. 1989). The same pathomechanism leads to secondary, nonpostthrombotic valvular incompetence of the deep veins in patients with a long history of truncal varicosity (Trendelenburg private circulation; Fig. 3.29a–c). In this secondary form, as in primary chronic venous insufficiency of the deep leg veins, the B-scan shows dilatation of the affected vein but no wall

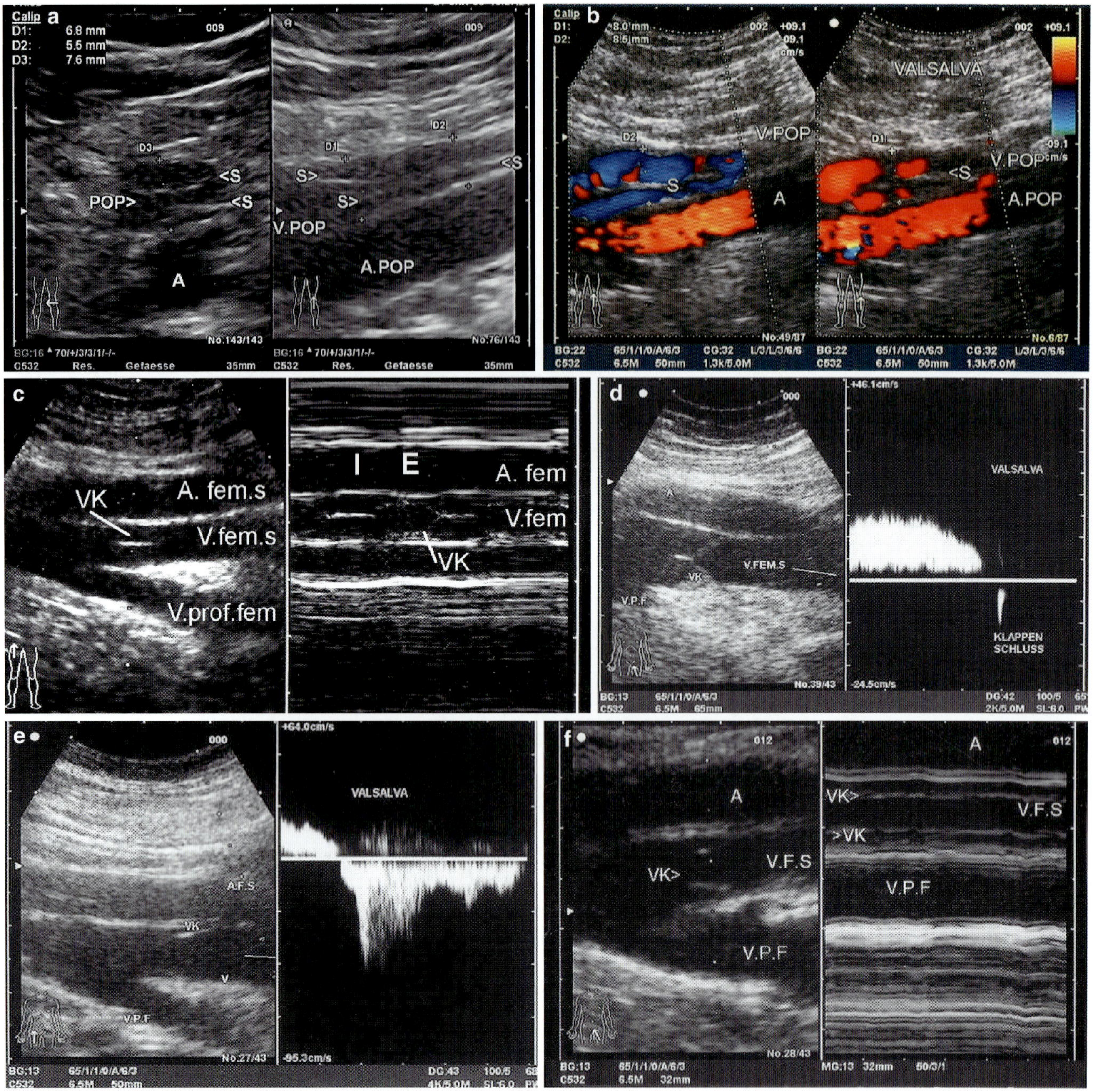


**Fig. 3.28** **a** Transverse (*left*) and longitudinal (*right*) views of the recanalized popliteal vein (*POP*) with postthrombotic wall sclerosis and strands (*S*) (see Fig. A3.39c). **b** Longitudinal color duplex scans during Valsalva's maneuver demonstrate postthrombotic flow reversal in the recanalized popliteal vein (*V.POP*): the *left* scan shows venous flow between intraluminal strands (*S*); the scan during Valsalva's maneuver (*right*) shows flow reversal in the vein, resulting in the same flow direction as in the accompanying popliteal artery (red-coded flow toward the transducer in both vessels). Gaps in the color coding indicate the strands, appearing as membrane-like structures within the lumen. **c** The time-motion scan obtained with the ultrasound beam through the valve leaflets allows good evaluation of valve motion: the leaflets (*VK*) are open during expiration (*E*) and close to the wall, and they are closed during inspiration (*I*) as a result of the increased intra-abdominal pressure, which prevents reflux of the venous blood (closed leaflets seen in the center of the lumen). **d** Valsalva's maneuver shows adequate valve closure (*VK*) in the superficial femoral vein (*V.FEM.S*), indicated by a short flow reversal (flow toward the periphery) just before valve closure and by cessation of reversed flow upon complete closure. **e** Postthrombotic damage, with sclerotic fixation of the valve leaflets to the wall, results in failure of the valve to close (*VK*) and is indicated by persistent reflux (flow away from transducer, toward the periphery) during Valsalva's maneuver. **f** In the corresponding time-motion scan, failure of the damaged valve to close (*VK*) is seen as adherence of the leaflets to the wall

**Table 3.11** Ultrasound criteria in evaluating postthrombotic syndrome

| | |
|---|---|
| **B-scan** (normal in 30–40%) | • Narrowing of vessel lumen<br>• Blurred wall structure<br>• Thickened wall<br>• Wall sclerosis, intramural calcifications<br>• Intraluminal connective-tissue strands, synechia<br>• Echogenic vessel lumen<br>• Not fully compressible |
| **Color duplex imaging** | • Degree of recanalization<br>• Incompetence of major veins<br>• Better visualization of wall sclerosis and of postthrombotic wall lesions<br>• Identification of collaterals<br>• Incompetence of superficial veins (secondary)<br>• Incompetence of perforating veins |

thickening or inhomogeneous structures within the lumen. Moreover, the vein can be completely compressed. Under good insonation conditions, mobility of the valve can be demonstrated, distinguishing this condition from the postthrombotic syndrome, where valve immobility is due to fibrotic thickening (see Figs. 3.28 and A3.5).

The *cause of valvular incompetence influences the character of reflux* (Evers and Wuppermann 1995, 1997). Reflux in postthrombotic valve incompetence sets in immediately with the provocative maneuver (without signs of valve movement), increases rapidly, peaks during the first seconds, and then decreases (type B). In contrast, in primary chronic venous insufficiency and primary varicosis, the abnormal reflux occurs with a short delay after physiologic reflux (which normally leads to valve closure) and is continuous (see Fig. A3.41) and slower (type A). In severe venous dilatation without any residual valve function, however, even primary chronic venous insufficiency of the superficial or deep venous system results in immediate high-frequency reflux.

*Reflux velocity can serve as a semiquantitative measure of postthrombotic valve damage*. The velocity increases during the first year and then reaches a plateau. Moreover, the reflux velocity and duration may be increased by secondary postthrombotic changes caused by pressure and volume overload.

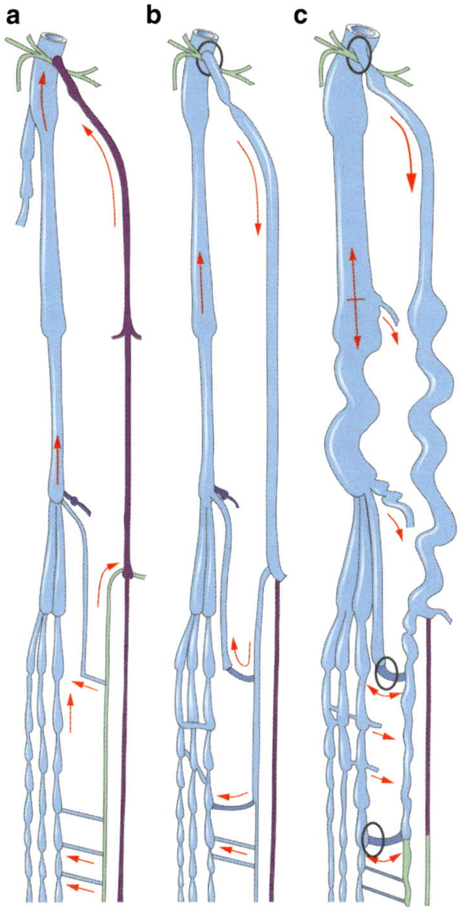


**Fig. 3.29 a–c** Development of chronic venous insufficiency with valvular incompetence of the deep veins secondary to truncal varicosity of the great saphenous vein. **a** Physiologic blood flow direction in deep and superficial veins (*red arrows*). Perforating veins drain the blood from the superficial to the deep venous system. **b** Truncal varicosis of the great saphenous vein is associated with retrograde flow through the perforators back into the superficial venous system. This hypercirculation (Trendelenburg private circulation) is compensated as long as the valves of the deep veins remain competent, but it leads to dilatation of superficial side branches and perforators as well as volume overload of the deep veins. **c** Hypercirculation eventually becomes decompensated as a result of volume overload, leading to dilatation and subsequent valve failure of the deep veins. This valve failure in turn leads to pressure-induced dilatation and secondary valve incompetence of further perforating veins (Cockett's, Boyd's, and Dodd's perforating veins). Finally, the muscle pump becomes insufficient, resulting in full-blown drainage insufficiency

### 3.1.6.3 Varicosis

The variability of venous reflux patterns makes it necessary to establish the following data to identify candidates for varicose surgery and to select the most suitable approach and extent of the intervention: competence/incompetence of the saphenous junctions, proximal and distal points of insufficiency in patients with superficial vein varicosis, identification of incompetent perforating veins, and state of the deep venous system (Wong et al. 2003). The type and extent of venous reflux including the vein segments involved are identified using Valsalva's maneuver and need to be known for surgical planning (Fig. 3.16a). A compression test with the patient standing is performed to determine the upper and lower points of insufficiency, which is especially important to spare competent vein segments for future bypass graft procedures in patients with incomplete superficial varicosis. Also with the patient standing, perforating veins are identified in their typical locations and evaluated using the compression test. Identifying incompetent perforators for elimination is important as they might otherwise give rise to recurrent varicosis.

A short reflux before valve closure is normal in the main veins; reflux with backward flow persisting for more than 0.5 s is abnormal according to the criteria of the Union Internationale de Phlébologie (UIP). Investigators meanwhile may advocate different cutoff values for different vein segments (Coleridge-Smith et al. 2006).

It is also very important to evaluate the deep venous system before surgery, which will help prevent secondary post-thrombotic incompetence of a superficial vein with renewed deterioration of venous drainage. Because the level of the saphenopopliteal junction is highly variable, it should be identified and marked in the preoperative examination. In patients who have undergone stripping of the great saphenous vein, a remaining stump can be identified with Valsalva's maneuver, but several years after surgery, it cannot always be reliably differentiated from neoreflux or neovascularization (Turton et al. 1999).

The procedure for evaluating valve incompetence of the superficial veins is the same as for the deep veins, namely Doppler measurement during provocative tests. Although flow reversal induced by provocative maneuvers is also documented in color flow images, only spectral Doppler enables accurate measurement of the duration of reflux. Clinically, it is important to identify the *proximal extent of venous insufficiency*. In complete varicosis of the great saphenous vein, which is the more common type, the proximal point is identified by repeated Doppler sampling beginning just below the saphenofemoral junction. Valve incompetence is indicated by reflux persisting for more than 1 s (several seconds in severe incompetence) upon increased intra-abdominal pressure during Valsalva's maneuver. Short reflux (<0.5 s) before valve closure is normal. This test is performed intermittently along the course of the great saphenous vein until the distal point of insufficiency is identified (transition from reflux to normal flow), which determines the grade of great saphenous vein varicosis according to Hach (Figs. 3.15 and 3.30). In Hach grades I to III, varicose side branches often enter the vein in the area of the distal point of insufficiency.

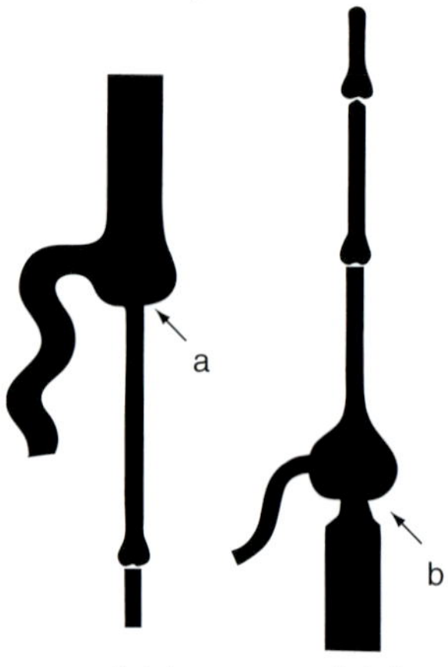


**Fig. 3.30** Valve incompetence of major superficial veins. *Left drawing*: distal point of insufficiency (*a*) with pressure-induced dilatation of a side branch or perforating vein (varicose degeneration) originating from the insufficiency point. Normal vein segment with functioning valves distal to this point. *Right drawing*: proximal point of insufficiency (*b*). Normal vein with competent valves proximal to the point of insufficiency and valve incompetence distal to it

Clinically apparent varicose changes of the great saphenous vein are often restricted to the lower leg. Nevertheless, duplex scanning frequently also reveals incompetence above the knee, where dilatation of the great saphenous vein is less pronounced due to the lower intravascular pressure. A study by our group including 103 patients with great saphenous vein varicosis demonstrated involvement of the saphenofemoral junction in 66% of the cases with clinically normal findings of the upper leg. Therefore, evaluation of the junction is mandatory in patients undergoing duplex scanning for lower leg varicosis.

In incomplete great saphenous vein varicosis, only the distal valves are incompetent while the terminal and preterminal valves are intact.

Four main types of *incomplete truncal varicosis* are distinguished:

- In incomplete varicosis of the perforator type with an intact saphenofemoral junction, distal incompetence of the great saphenous vein originates from an insufficient perforating vein, e.g., a Dodd vein in the upper leg (see Fig. A3.46).
- In incomplete varicosis of the lateral branch type with proximal competence, the distal great saphenous vein varicosis is maintained by varicosis in a side branch, e.g., an insufficient lateral accessory vein in the thigh (see Fig. A3.43d–f).
- In the posterior type, venous incompetence involves the small saphenous vein and femoropopliteal vein, from where it extends to the great saphenous vein via an accessory branch of the latter (Giacomini anastomosis) (see Fig. A3.30b).
- Distal varicose lateral branches such as veins communicating between the great and small saphenous veins can maintain secondary varicose changes of the distal saphenous vein.

Preoperative identification of the upper point of insufficiency in distal varicosis by duplex scanning is important to ensure complete surgical removal of the incompetent vein segments, thereby preventing recurrence and sparing competent venous segments for possible later bypass procedures. The dilated varicose segments are identified in the standing patient, followed to the point of insufficiency, and marked on the skin. Surgeons will benefit most from real-time imaging of venous morphology and function by duplex scanning if they perform the examination themselves.

In incomplete distal great saphenous insufficiency, the Valsalva test yields no valid results because the proximal valves still function properly. Instead, the compression test is performed with the patient standing. The incompetent great saphenous vein depicted by B-mode imaging is followed upward to identify the proximal point of insufficiency for tailoring the surgical procedure. Divisions into arch veins can thus be identified by duplex scanning as well and evaluated for incompetent valves. The ultrasound examination thus enables accurate identification of all incompetent venous segments prior to surgical removal.

The *perforating veins* can also be identified sonographically (Figs. 3.10 and 3.31) in their typical locations (Cockett's, Boyd's, and Dodd's perforators) and evaluated for competence. Demonstration of reflux from the deep into the superficial veins in the compression test, with elimination of flow in the superficial veins by means of a tourniquet, indicates perforator incompetence (Fig. 3.11). Incompetent perforating veins arising from the great saphenous vein below the knee are identified by B-mode imaging as slightly tortuous, tubular structures coursing to the posterior tibial territory in the transfascial compartment (Fig. A3.45). Competent perforating veins are usually so small that they will not be detected unless a thorough search is done with a high-resolution transducer and the patient in the standing position.

In a comparative study, 95.5% of a total of 252 insufficient perforating veins diagnosed by color duplex ultrasound were confirmed intraoperatively. In comparison, venography identified only 65% of these insufficient perforating veins (Stiegler et al. 1994). The accuracy of palpation alone is 49% and that of CW Doppler 75%.

The ultrasound examination for *incompetence of the small saphenous vein* is also performed with the patient standing. If the deep veins are still competent, however, insufficiency can only be evaluated by using the compression test and not with Valsalva's maneuver (Fig. 3.31). After identification of the highly variable saphenopopliteal junction, spectral Doppler is obtained while performing the compression test using the other hand or a cuff to compress the vein distal to the transducer. The course of the small saphenous vein is then traced downward with repeated compression to identify the distal point of insufficiency. Valve incompetence is suggested by dilatation – primarily in the popliteal fossa – and a tortuous course resulting from elongation of the vein.

During *endovenous interventions*, such as radiofrequency ablation (VNUS method) or intravascular laser therapy, ultrasound is used to monitor correct positioning of intraluminal probes. In foam sclerotherapy, ultrasound allows real-time monitoring of the spread of the sclerosant, and foam migration into the deep venous system can be prevented by compressing the terminal portion of the vein being treated.

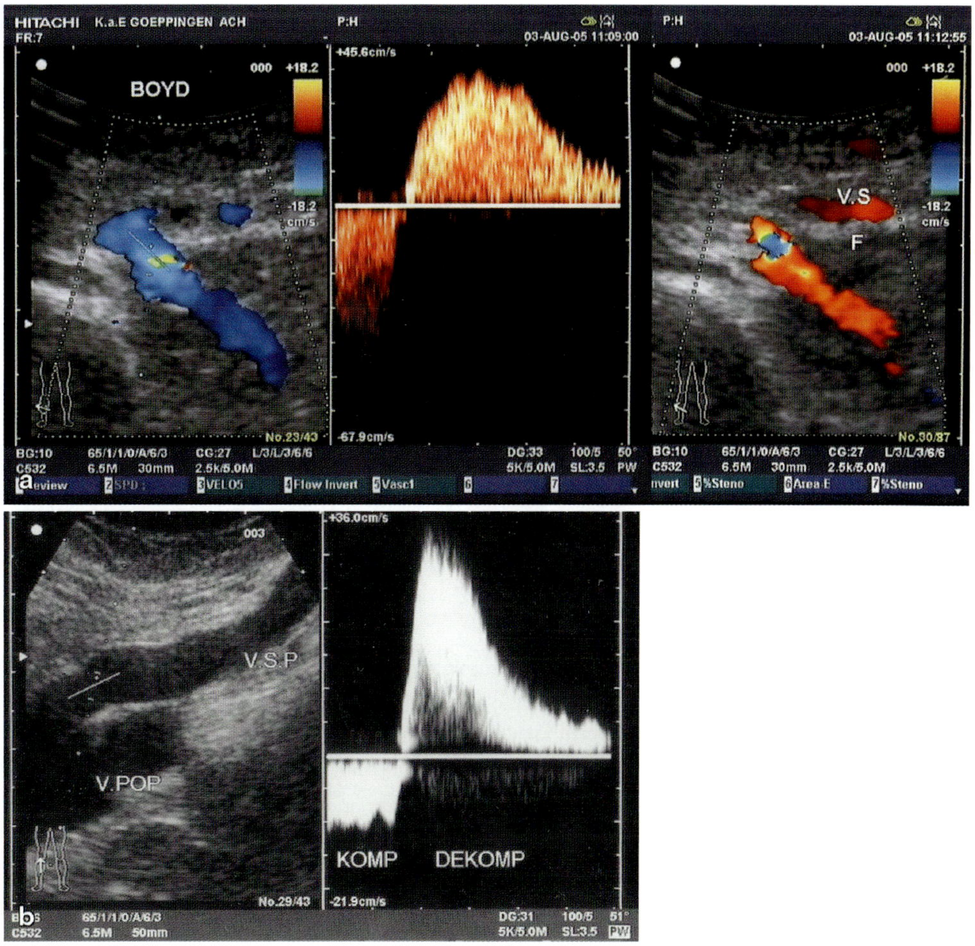


**Fig. 3.31** **a** Valve incompetence of a Boyd perforator in the calf. The valve function test reveals normal flow from the superficial to the deep venous system (*leftmost* scan, blue-coded flow) and reversal upon release of compression (*rightmost* scan, red-coded flow) (*F* fascia, *V.S* great saphenous vein). The Doppler waveform shows persistent reflux from the deep into the superficial system (flow above the baseline, toward the transducer). **b** The Doppler waveform from the saphenopopliteal junction (sample volume as indicated in the longitudinal view obtained with the transducer in the popliteal fossa). There is flow toward the heart during compression of the calf with the hand (away from transducer, below the baseline) and persistent retrograde flow upon release (above the baseline), indicating severe valve incompetence

In endovenous *radiofrequency obliteration*, the introducer sheath is advanced from a distal access *using ultrasound guidance* and placed in the great saphenous vein just below the site of entry of the superficial epigastric vein (Fig. A3.48) to prevent occlusion of this vein during treatment. Ultrasound enables good visualization of the superficial epigastric vein in this area as it extends cranially from the saphenofemoral junction. The saphenous side branches arising more distally, in particular the lateral and medial accessory veins, will be obliterated by the thermal energy applied during radiofrequency ablation. The intervention is performed under *tumescence anesthesia*, which is applied under sonographic guidance and serves to compress the target vein, thereby ensuring a good energy transfer to the vein wall, and also to protect the surrounding structures from heat.

Tumescence anesthesia is performed by injecting tumescence fluid (modified Klein's solution) around the vein once the catheter is in place. This is done under ultrasound guidance and serves to create a circumferential fluid layer of 4–5 mm around the vein and compress it (target diameter for the great saphenous vein: 4–6 mm). These conditions have been shown to ensure nearly pain-free endovascular laser treatment. Although the tumescence solution spreads within the saphenous compartment, it is necessary to inject the fluid along the course of the vein every few centimeters, starting proximally and using ultrasound for guidance.

The so-called CHIVA technique aims at hemodynamic correction by eliminating reflux pathways, while preserving the saphenous trunk as a route of venous drainage. Sites of reflux from the deep system into the superficial veins are identified by duplex imaging and marked for open surgical ligation. The result is evaluated by duplex ultrasound to ensure that all reflux pathways have been eliminated and that flow in all patent veins is from the superficial into the deep system.

*Recurrent varicosis* is the occurrence of varicosis at a new site or in a varicose vein segment previously treated with sclerotherapy or an endovascular intervention. Recurrent varicosis in the strict sense cannot occur after varicose vein resection or crossectomy with great saphenous vein stripping. In this situation, varicosis can recur if the residual saphenous stump is too long and veins terminating proximally connect to this stump or to a saphenous branch. Thus, when a patient presents with recurrent varicosis, the first diagnostic step is to evaluate the saphenofemoral junction with duplex imaging. Valsalva's maneuver is performed to identify any incompetent lateral branches arising from the residual great saphenous vein, and then connecting to a more distal superficial vein, or a persisting lateral or medial accessory branch that has undergone varicose degeneration (Table 3.12).

A limitation of duplex imaging is that it does not always allow reliable differentiation of a residual saphenous vein trunk from neoreflux or *neovascularization* (Fig. A3.50).

**Table 3.12** Varicosis: relevant findings and therapeutic consequences

| Duplex ultrasound findings | Therapeutic consequences |
|---|---|
| Incompetent great saphenous vein/saphenofemoral junction vs. peripheral/lateral-branch varicosis | Stripping/resection vs. local treatment, sclerotherapy |
| Secondary varicosis with (postthrombotic) incompetence of the deep leg veins | Compression treatment, surgery in exceptional cases only; ulcer: perforator dissection, division of fascia |
| Upper and lower points of incompetence in incomplete saphenous vein varicosis | Preservation of relevant competent vein segments for possible later bypass graft operations |
| Incompetent perforating veins | Ligation of incompetent perforators identified by ultrasound for prevention of recurrent varicosis |
| Postthrombotic syndrome with partial vs. complete occlusion by residual old thrombus | Contraindication to varicose surgery (impaired venous drainage would deteriorate further) |

### 3.1.6.4 Varicophlebitis

Varicophlebitis is defined as *thrombotic inflammation of a varicose superficial vein* and is a typical complication of varicosis. Thrombophlebitis is an inflammation of a previously healthy vein and typically occurs as a paraneoplastic complication. The clinical relevance of varicophlebitis has long been underestimated. The most serious complication is *thrombus extension into the deep venous system* (Fig. 3.32). The risk of progression of superficial thrombophlebitis into the deep venous system is 16% with 70% of cases accounted for by thrombus extension via the saphenofemoral junction and 20% by progression through perforators (Chengelis et al. 1996).

The complications of deep vein thrombosis with a relatively high risk of pulmonary embolism (caused by free-

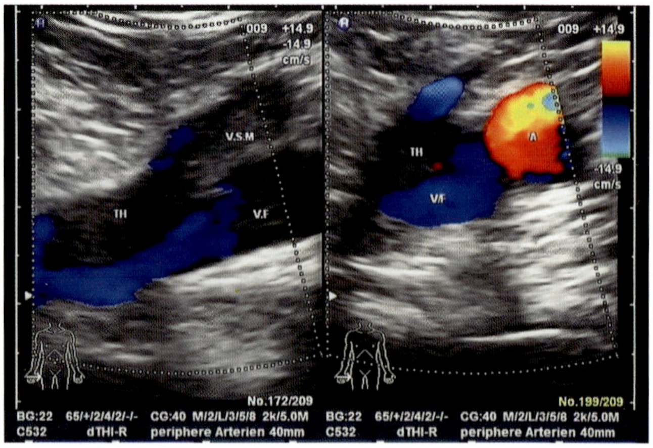


**Fig. 3.32** Longitudinal and transverse images of thrombophlebitis of the great saphenous vein (*V.S.M.*) with thrombus (*TH*) adhering to the wall and extending 1.5 cm into the common femoral vein (*V.F.*)

floating components; Bergquist 1986) have led to adoption of *a more aggressive therapeutic strategy* in patients with superficial varicophlebitis extending close to the saphenous junction. The foremost aim in the management of varicophlebitis is to prevent deep vein thrombosis and the attendant risk of embolic complications, which is why a therapy-oriented staging system is proposed (Table 3.13).

The *need to perform ultrasound* in all patients with clinically manifest disease arises from the fact that patients with clinical signs of calf varicophlebitis (classic signs of inflammation such as redness or pain, tumor) often have above-knee extension even when the thigh appears normal. Ultrasound is performed to *define the proximal extent of thrombophebitis*, and compression ultrasound of the deep venous system is done to identify possible extension via incompetent perforating veins. Surprisingly, up to 20% of patients with thrombophlebitis have been reported to have concomitant calf vein thrombosis; however, the degree of coincidence between the two depends on the patient population investigated; concomitant calf thrombosis is also more common in *paraneoplastic* thrombophlebitis. The exact extent of thrombophlebitis as defined by compression ultrasound dictates the therapeutic consequences (see Table 3.13).

The diagnostic criteria for thrombophlebitis (compressibility) are the same as for deep vein thrombosis of the leg (Blättler et al. 1996).

**Table 3.13** Classification of varicophlebitis (compression ultrasound) and treatment options

| Extent of thrombophlebitis | Treatment |
|---|---|
| Varicophlebitis of a lateral branch or of the great saphenous vein below the knee | Local anti-inflammatory therapy |
| Varicophlebitis of the great saphenous vein in the thigh but far below the saphenofemoral junction or ascending small saphenous vein thrombosis with extension above the level of the May perforator | 2 weeks of anticoagulation with low-dose heparin |
| Thrombophlebitis ascending to the junction (proximal 5 cm of great saphenous vein or proximal 3 cm of small saphenous vein) | Crossectomy, severing of the great saphenous vein or small saphenous vein; anticoagulation treatment alone in exceptional cases only |
| Thrombus extends beyond the saphenofemoral or saphenopopliteal junction with deep vein thrombosis of variable severity | Anticoagulation with low-molecular-weight heparin and overlapping change to phenprocoumon (6 months) |
| Varicophlebitis with secondary deep vein thrombosis resulting from thrombus extension through an incompetent perforator ("collar stud thrombosis") | Anticoagulation with low-molecular-weight heparin and overlapping change to phenprocoumon (6 months) |

## 3.1.7 Rare Venous Abnormalities

### 3.1.7.1 Venous Aneurysm

The term aneurysm has for a long time been used to refer to the local widening of arteries only, while the venous counterpart was described as ectasia or aneurysmal dilation. The term venous aneurysm is now commonly used to designate a marked, localized sacculated or spindle-shaped dilation of a venous segment (at least 2.5–3 times the normal luminal diameter).

Histologically, a *venous aneurysm is a true aneurysm* with a wall consisting of all venous layers but with thinning of the muscle coat. Medial sclerosis may occasionally be present. The etiology is unknown, but various mechanisms have been proposed including embryonal defects, excessive local pressure in narrow anatomic spaces, and trauma (Fischer et al. 1996; Smets et al. 1997; Aldridge and Comerota 1993). Venous aneurysms are very rare and mainly occur in the legs, in particular the *popliteal vein*. Aneurysms of leg veins account for about 65% of all venous aneurysms, 17% occur in neck veins, and 14% in the arms (overview in Ritter 1993). Leg aneurysms mainly affect the major deep veins but may also occur in the superficial compartment, where they have to be differentiated from varicosis and regional or diffuse phlebectasia, which, when occurring in the calf veins, may take the form of long tubular ectasias. Most venous aneurysms go undetected unless there is thrombus formation or *pulmonary embolism* secondary to *local thrombosis*. Other complications are due to compression of surrounding structures, which typically manifest as abnormal sensations, and very rarely rupture and hemorrhage. The popliteal vein is the most common site of venous aneurysm in the leg, and a popliteal aneurysm typically presents with pulmonary embolism caused by thrombus arising in the aneurysm cavity.

More asymptomatic aneurysms of the popliteal vein have been discovered since compression ultrasound and duplex imaging became widely accepted as the first-line modalities for the evaluation of deep venous thrombosis and valve incompetence of the leg.

As with arterial aneurysms, fusiform or spindle-shaped (Fig. 3.33a, b) and saccular aneurysms are distinguished. Experimental investigations in models have shown that quasilaminar flow is predominant in fusiform aneurysms while turbulent flow with flow separations occurs in saccular aneurysms (Brunner et al. 1997; Haaverstad et al. 1995). Stasis in dead water zones in an aneurysm can give rise to thrombus formation (see Fig. A3.51a). Only anecdotal cases of pulmonary embolism caused by venous aneurysms have been reported (Biesseaux et al. 1994; Seino et al. 1994).

One case of paradoxical embolism has been described (Manthey et al. 1994). Most venous aneurysms are incidentally detected in patients examined for exclusion of deep

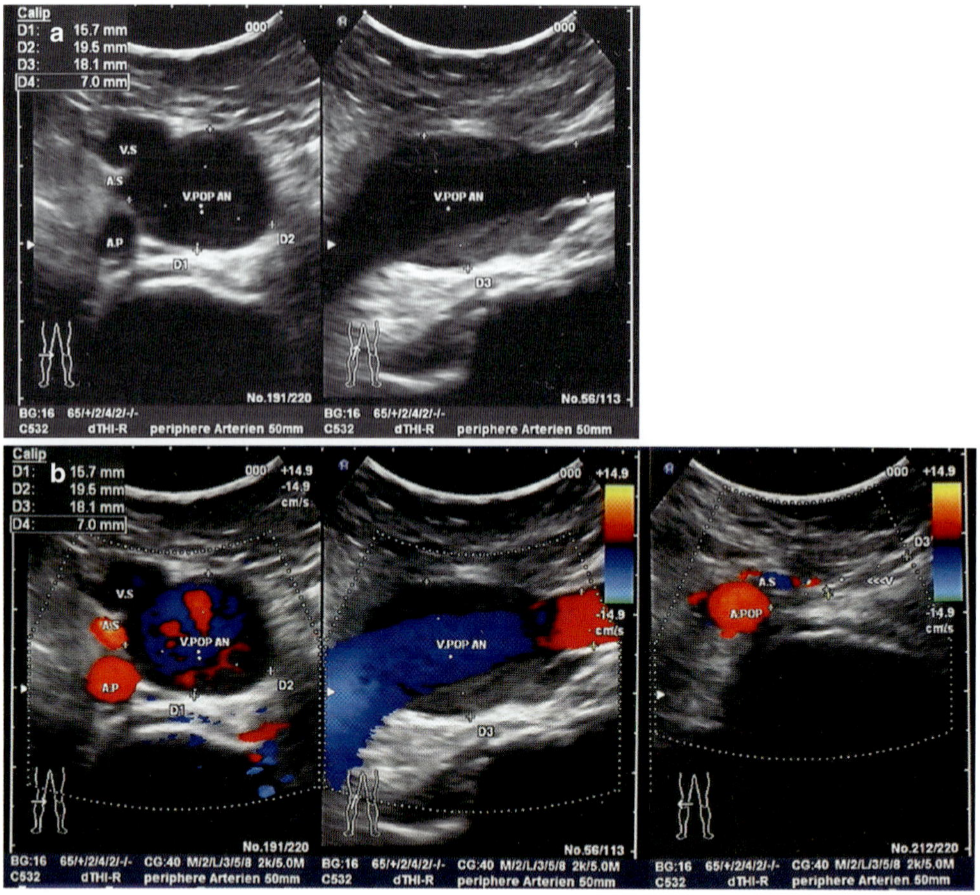


**Fig. 3.33** **a** Spindle-shaped aneurysm of the popliteal vein (transverse view on the *left* and longitudinal view on the *right*) with an increase in venous diameter from 7 to 18 mm just below the saphenopopliteal junction (*V.S*). **b** There is slow flow due to stasis in the color duplex mode, which may mimic mural thrombosis (despite a low PRF). Thrombosis of the aneurysm is ruled out by compression ultrasound, which reveals complete compressibility of the dilated popliteal segment (*right* section, transverse plane)

venous thrombosis of the lower extremity. The patients typically report pain and swelling. In a study by our group, four saccular aneurysms and four spindle-shaped aneurysms of the popliteal vein were detected in 11,500 ultrasound examinations of the deep leg veins performed in patients with suspected thrombosis (Schäberle and Eisele 2001). Two of the saccular aneurysms were partially thrombosed and the patients suffered pulmonary embolism. The other two saccular aneurysms were incidental findings without pulmonary embolism, and one of them was associated with deep venous lower leg thrombosis. The four spindle-shaped aneurysms were not thrombosed and there was no evidence of prior pulmonary embolism in these patients (see Figs. A3.51–A3.55).

Venography is limited in assessing the size and extent of mural thrombosis of an aneurysm in the popliteal fossa due to interference from overlying muscle veins and the small saphenous vein. Real-time ultrasound, on the other hand, provides definitive information on the relationship of an aneurysm to the veins terminating in it, the aneurysm diameter, its shape, and intra-aneurysmal thrombus formation.

A review of the literature suggests that the majority of investigators advocate surgical repair of popliteal vein aneurysm, the preferred technique being tangential resection with lateral plication of the venous wall (Bergqvist et al. 2006). This may be attributable to the fact that the majority of venous aneurysms included in this review were found in patients presenting with pulmonary embolism. An analysis conducted by the author and other sonographic studies, however, suggest that the wider use of ultrasound for the examination of the veins reveals popliteal vein aneurysms before complications such as pulmonary embolism or thrombosis make them apparent. In light of this new situation, other criteria besides the occurrence of pulmonary embolism or thrombosis are needed to decide when a popliteal vein aneurysm should be operated on.

The shape (saccular or spindle-shaped) and thrombus formation are crucial criteria for therapeutic management. Based on sonographic findings, the four saccular aneurysms in our study were resected and the ultrasound diagnosis confirmed. The spindle-shaped aneurysms were followed up and treated by combined anticoagulation and compression or compression alone according to their size.

The conclusion to be drawn from these observations is that venous aneurysm resection is indicated after episodes of thrombosis or pulmonary embolism and when the shape is markedly saccular. Color duplex imaging nicely depicts vortexing and relative stasis side by side in the aneurysm sac, which are the nidus for local thrombus formation (Figs. A3.51–A3.55). Conversely, incidentally detected spinde-shaped aneurysms require no surgical repair and, especially when they are small (<2.5 cm), can simply be monitored sonographically.

These cases underscore the ability of noninvasive duplex findings to provide the basis for a differentiated approach to the treatment of the rare popliteal vein aneurysms (incidence of 0.07% of all patients examined for suspected deep venous leg thrombosis in our study). The intraoperative findings confirm the validity of duplex scanning, which is the method of first choice in the diagnostic evaluation of venous aneurysm.

The incidences reported in the literature range from 0.2% in a total of 2,500 ultrasound examinations (Rubin et al. 1990) to 0.8% in 3,380 ultrasound examinations performed in patients with suspected deep vein thrombosis and venous insufficiency (Labropoulos et al. 1996).

#### 3.1.7.2 Tumors of the Venous Wall

Venous stasis or obstruction of inflow with unilateral leg edema of unclear origin may be caused by a benign or malignant tumor of a venous wall. Such tumors may give rise to the development of appositional thrombi as wall compression or infiltration progresses. With ultrasound, the tumor can be demonstrated directly as a circumscribed wall thickening and differentiated from venous thrombosis, thus serving as a basis for establishing the indication for surgical resection. Benign venous wall tumors include papillary endothelial hyperplasia, hemangioma, leiomyoma, and fibroma. Malignant tumors are angiosarcoma and leiomyosarcoma. Sonomorphologic criteria for differentiating benign and malignant tumors are invasive growth and tumor vascularization (see Fig. A3.66).

#### 3.1.7.3 Venous Compression

The venous wall has only a thin muscle layer and is therefore easily compressed by lymphoma (predominantly in the true pelvis and groin), *perivascular tumors*, hematoma, abscess, and arterial aneurysm (above all of the popliteal artery), causing flow obstruction with the clinical signs of thrombosis. With its ability to visualize both the vein itself and the surrounding structures, ultrasound will either demonstrate the cause of disturbed drainage directly or give indications for further, more specific diagnostic measures such as ultrasound-guided aspiration or biopsy.

In rare cases, *popliteal entrapment syndrome involves both the artery and the vein*, for instance in the presence of an ectopic popliteal muscle or pronounced hypertrophy of the heads of the gastrocnemius muscle. In such cases, outflow obstruction can be elicited by active plantar flexion (see Fig. A3.67).

A vein compressed by external structures will show only augmented flow (A-sound) upon distal compression while spontaneous flow is absent. Normal respiratory phasicity is lost distal to the flow obstruction. When a small residual lumen is still patent, spectral Doppler depicts a high-frequency flow signal resembling a stenosis signal (see Figs. A3.62 and A3.63).

#### 3.1.7.4 Adventitial Cystic Disease of Veins

Venous adventitial cystic disease is very rare, occurring 20–50 times less commonly than its arterial counterpart. As in the arteries, the lesions are histologically true ganglia (in terms of cyst contents and wall composition) in the adventitial layer of the diseased vein.

The synovial cysts in the adventitia compromise the venous lumen (Paty 1992; Schraverus 1997; Chakfe 1997; Hach-Wunderle 2003).

As with arterial adventitial degeneration, compression of the underlying vein and the associated clinical symptoms vary with the filling of the cysts, and the lesions always develop close to a joint. Occasionally, the surgeon will encounter a communication between an adventitial cyst and the joint capsule. Venous adventitial cystic degeneration most commonly affects the common femoral vein and popliteal vein. Depending on the degree of luminal compression, there may be swelling of the leg with a sensation of congestion distal to the lesion, which intensifies during physical activity. Edema usually recedes over night. As with the arterial counterpart, there will be *cystic lesions in the venous wall* (which may be multiple), with a *narrowing of the lumen* on B-mode ultrasound. The degree of luminal narrowing and the hemodynamic relevance of the outflow obstruction can be evaluated using color duplex imaging. Venous obstruction is indicated by loss of respiratory phasicity and reduced flow velocity in the Doppler waveform obtained distal to the degenerative lesion.

Luminal narrowing varies with the size of the cyst (Fig. A3.65) and is reflected by a variable increase in flow velocity (Doppler waveform) in the diseased vein segment as well as by a flow signal similar to that seen with a stenosis.

Inconclusive findings should be resolved by computed tomography or magnetic resonance imaging; ascending venography will only show external compression.

#### 3.1.7.5 Differential Diagnosis: Lymphedema, Lipedema

Once duplex imaging has ruled out chronic venous insufficiency as the underlying cause of edema, edema of cardiac origin must be differentiated from lymphedema and lipedema. Lymphedema has rather characteristic sonomorphologic features including increased echogenicity of the thickened subcutaneous layer and sound scattering with demonstration of anechoic clefts. These clefts tend to be longitudinal in orientation, distinguishing them from clefts in cardiac edema.

compression of a vein by a lymphoma can be identified, and the severity of the outflow obstruction estimated from spectral Doppler recording (spontaneous and augmented flow) distal to the compressed vein.

The accuracy of compression and duplex ultrasound must be measured against that of venography, the traditional gold standard. It is noteworthy that venography itself was never validated in comparison with another modality. Opacification of the veins by the contrast medium enables direct assessment of the lumen, but no extravascular information is provided. Moreover, filling defects may give rise to misinterpretation because they may be due to occlusion or methodological limitations, especially in the lower leg. The lower leg veins are depicted incompletely in 15% of cases and inadequately in 4% (Schmitt 1974, 1977). Complete visualization of the venous segments of the lower extremity varies from 61–96%. Repeat venography in patients with clinical signs of venous thrombosis of the leg performed due to progression of symptoms within 5 days of a negative initial venography enabled definitive diagnosis of thrombosis in 1.3% of cases (Hull et al. 1981).

These results call into question the suitability of venography as the gold standard in the diagnosis of thrombosis. In a study performed by our group in 159 patients with clinically suspected deep vein thrombosis who were independently examined by duplex scanning and venography (Schäberle and Eisele 1991), venography yielded inconclusive or false results or failed to identify the cause of leg swelling in 21 cases due to methodological limitations. In these cases, duplex scanning demonstrated an AV fistula, thrombophlebitis, muscle vein thrombosis, and venous compression by pelvic tumors, Baker's cysts, or arterial aneurysms. In five instances, venography yielded false-negative results compared to ultrasound (deep femoral vein thrombosis; complete thrombosis of one branch of a paired superficial femoral vein; lower leg venous thrombosis; mural thrombosis of venous aneurysm). False-positive venograms were seen in three cases (compression of the popliteal vein and proximal lower leg veins by a ruptured Baker's cyst and a large false aneurysm; misinterpretation of filling defect). The ultrasound findings were confirmed intraoperatively or through additional examinations. In this study, venography had an accuracy of 95% compared to duplex ultrasound.

In a large study of 430 consecutively examined patients, the 5% discrepancy between venography and ultrasound (including the lower legs) is attributed to false-negative venograms, which are represented in the study as false-positive ultrasound findings (Elias et al. 1987). Part of the false-negative venograms was retrospectively accounted for by sonographically identified venous thrombi in the lower leg and misinterpretation of filling defects.

Complete evaluation of all deep lower leg veins is often not possible by venography because thrombi producing complete occlusion of small veins are missed as these are not apparent as filling defects (Table 3.15).

**Table 3.15** Ultrasonography supplementing venography

| | |
|---|---|
| 1. | Evaluation of the deep femoral vein |
| 2. | Identification of muscle vein thrombosis |
| 3. | Thrombosis of veins entering the saphenofemoral junction |
| 4. | Demonstration of thrombus in an unopacified calf vein |
| 5. | Visualization of the thrombus end (poor contrast medium flow/ long thrombotic segment) |
| 6. | Concomitant evaluation of soft tissues |

Several studies conducted since the late 1980s found sensitivities and specificities of compression ultrasound ranging from 87% to 100% compared with venography as the gold standard (see Table 3.6). In one of these studies, performed by our group and including 131 legs with clinically suspected deep venous thrombosis (confirmed by venography in 73 legs) in 125 patients (72 women, 53 men; mean age 55 ± 18.5 years), ultrasound had a sensitivity of 97% and a specificity of 98% (Schäberele and Eisele 1991). Discrepancies mainly involved the lower leg veins.

Despite the controversy about the clinical importance of lower leg venous thrombosis, symptomatic patients should undergo early diagnostic assessment and treatment to prevent appositional thrombus growth. Compression ultrasound is also a valid diagnostic modality in isolated calf vein thrombosis, which is detected with sensitivities and specificities of 85–90% (see Table 3.6; Elias et al. 1987; Habscheid et al. 1990). Even isolated muscle vein thrombosis is highly conspicuous on compression ultrasound and can be detected as an accessory finding with a high degree of accuracy.

*Conventional duplex ultrasound* was not found to lead to a statistically significant improvement in the diagnosis of thrombosis, but flow assessment by spectral Doppler facilitates diagnostic evaluation in venous segments notoriously difficult to scan such as the pelvic veins. The demonstration of flow signals adjacent to central thrombi or in early recanalization can contribute to therapeutic decision making.

*Valve function* is assessed using gray-scale ultrasound supplemented by Doppler measurement (duplex or color duplex). The postthrombotic assessment of valve function in the superficial veins is the key to proper therapeutic management, apart from evaluation of spontaneous or thrombolysis-induced recanalization. Scanning begins at the saphenous junctions and proceeds down the leg to determine the distal extent of reflux.

As a supplementary tool in the hands of the surgeon, *color duplex imaging* was found to have an accuracy of 95.4% in identifying incompetent perforating veins as confirmed by the intraoperative findings (Stiegler et al. 1994). Moreover, duplex ultrasound can be used to assess the outcome of therapy, as in sclerotherapy of varices.

The risk of *appositional thrombus growth* with propagation into a major deep vein secondary to superficial thrombophlebitis was first noted and documented by ultrasonography. Patients with local thrombophlebitis treated conservatively should undergo sonographic follow-up to identify appositional growth, if indicated on the basis of their symptoms. High ligation of the saphenofemoral junction aimed at preventing further appositional growth and pulmonary embolism is indicated in thrombosis extending to the terminal portion of the saphenous vein or if a thrombus protrudes from a superficial vein into a major deep vein. Unlike venography, ultrasound enables a very reliable evaluation of the saphenofemoral and saphenopopliteal junctions (Barrelier 1993; Schuler et al. 1995; Schönhofer et al. 1992). Since superficial thrombophlebitis additionally correlates with occult deep vein thrombosis in the lower leg, this should be confirmed or ruled out by ultrasound (Jorgensen et al. 1993).

Precise determination of thrombus age is crucial for optimizing the therapeutic approach (thrombolysis, surgery, or conservative treatment). Thrombus age is often underestimated if based on the clinical findings and history alone. Venography chiefly relies on indirect criteria (collateralization) to estimate thrombus age, while the morphologic appearance is of limited value in venographic age determination. The ultrasonographic evaluation of thrombus morphology provides the most reliable data for estimating thrombus age despite the variations that exist in echotexture and extent

**Table 3.16** Sonographic findings in the diagnostic evaluation of venous thrombosis and estimation of thrombus age

| Thrombosis | Findings |
|---|---|
| Absent | • Complete compressibility of vein<br>• Thin wall<br>• Breathing, Valsalva's maneuver, and distal compression elicit identical changes in flow on both sides<br>• No elicitation of retrograde flow (indicating adequate valve closure) |
| Fresh thrombus (<8 days) | • Noncompressibility of vein<br>• Venous diameter at least twice that of accompanying artery<br>• Flow signals near the wall if thrombus is still surrounded by blood or in the presence of a floating thrombus<br>• Thrombus tends to be homogeneous and hypoechoic<br>• Good delineation of vessel wall, in part with hypoechoic halo<br>• Rare and small collateral veins are occasionally demonstrated by color duplex scanning |
| Older thrombus (>2–3 weeks) | *Total occlusion:*<br>• Noncompressibility of vein<br>• Diameter shrinks to less than twice that of accompanying artery<br>• No flow signals, thrombus tends to become more hyperechoic and inhomogeneous<br>• Poor delineation of vessel wall, hyperechoic halo may still be present<br>*Partial occlusion:*<br>• Partial compressibility of vein<br>• Vessel lumen comparable in diameter to that of accompanying artery<br>• Signs of marginal and central recanalization<br>• Collaterals begin to dilate |
| Postthrombotic lesions | *Persistent occlusion:*<br>• Reduced venous lumen (same as or smaller than the diameter of accompanying artery)<br>• Vessel wall poorly demarcated against surrounding soft tissue<br>• Fully developed collateral vessels<br>• *Partial recanalization:*<br>• Meandering flow signal in center of vein<br>• Little or no respiratory fluctuation of flow<br>• Short residual occlusions<br>• Sclerotic thickening and rigidity of vessel wall; incomplete compressibility<br>*Recanalization:*<br>• Flow signal throughout lumen<br>• Sclerotically altered segments and sonographically normal wall segments<br>• Incompetent valves demonstrated with Valsalva's maneuver or valve function test (compression and release)<br>• Variable lumen with widened and narrowed segments |

of venous dilatation in relation to thrombus genesis and localization (Table 3.16; Fobbe et al. 1991).

Early recanalization is suggested by the presence of spontaneous or augmented flow in the venous lumen. Incompetent valves in recanalized thrombotic veins are identified by functional tests (Valsalva's maneuver, compression test). Wall irregularities and hyperechoic deposits as well as a rigid vessel wall in the compression test already suggest the postthrombotic syndrome in the gray-scale scan. Noninvasive color duplex scanning can be repeated any time and is thus an excellent modality for monitoring the response to thrombolysis and deciding when treatment can be discontinued. Moreover, this modality can be used to evaluate the outcome of thrombectomy and to quantitatively estimate the flow volume after the creation of a bucket handle shunt, which is done by determining blood flow in the common femoral artery proximal to the shunt and comparing it with the contralateral side.

Close sonographic follow-up of the *natural history* (in patients undergoing heparinization with subsequent phenprocoumon administration) in comparison with thrombolysis or thrombectomy provides data for a critical appraisal of these therapies and may lead to changes in the currently established therapeutic approaches. The fact that patients undergoing thrombolytic therapy for more than 10 days develop valve incompetence casts doubt on this approach, although it leads to recanalization (but fairly late). Moreover, one must take into account the increasing bleeding risk associated with longer thrombolytic therapy. Valve damage occurs with increasing thrombus organization; the valves immobilized by residual thrombotic material can be depicted in the recanalized veins if the scanning conditions are good. Comparative studies of the late outcome of thrombolytic therapy by thrombus age, treatment duration, and thrombus localization in comparison with the natural history would be desirable but would require a complex study design.

Our preliminary experience suggests that *isolated pelvic vein thrombosis* has a high rate of spontaneous recanalization. Twelve of 15 patients with isolated pelvic vein thrombosis treated conservatively (full-dose heparin with subsequent phenprocoumon and compression therapy) showed complete recanalization after 6 months. Because the iliac veins have no valves, these patients are unlikely to develop a postthrombotic syndrome as long as the thrombosis does not descend into the femoral vein or such extension is prevented by heparin and compression therapy.

The sonographic examination yields very detailed information on the *extent and localization of thrombosis*. It provides direct evidence as to whether a thrombus is localized in a calf muscle vein or a main vein or if a thrombus protruding into the femoral vein originates from the great saphenous vein or from the deep femoral vein. Pulmonary emboli may arise from the proximal deep femoral vein, especially if an appositional thrombus protruding into the common femoral vein is dislodged. Venographically, the deep femoral vein is often not depicted or only after retrograde contrast filling, which does not occur in most patients with competent valves. Evaluation of the distal segment of the deep femoral vein (4–8 cm) is an integral part of the ultrasound examination, which enables good visualization of this segment. Another source of embolism is thrombosis at the termination of the internal iliac vein. This area should be assessed by duplex scanning for the presence of a floating thrombus, particularly in patients with pulmonary embolism. However, sonographic assessment of this area is limited in obese patients or when there is scattering and acoustic shadowing due to overlying bowel gas. If visualization of the pelvic veins is limited, occluding pelvic vein thrombosis can be excluded indirectly by comparing a Doppler waveform from the common femoral vein with a waveform from the contralateral vein.

Ultrasound, not involving radiation exposure and not requiring contrast medium administration, is the method of choice in pregnant women, children and adolescents, and in patients allergic to contrast medium. Another advantage of this noninvasive modality is the short examination time of 5–10 min per leg in the diagnostic assessment of thrombosis. Some extra time is required for the valve function tests in patients with the postthrombotic syndrome and varicosis.

While the detection and evaluation of thrombosis by ultrasound is highly examiner-dependent, the learning curve is steep, and performance improves rapidly. In a study of 99 patients using venography as reference standard, the initial sensitivity of duplex ultrasound was 67% in the thigh and 57% in the calf, increasing to 100% and 79%, respectively, after the first 50 examinations. Specificity was already over 95% when the study began, indicating that the pathognomonic criteria for thrombosis are highly accurate (Leutz et al. 1994).

As with arterial aneurysms, the extent, localization, and presence of mural thrombosis of *venous aneurysms* can be assessed more reliably using ultrasound compared with radiographic techniques that rely on the opacification of the patent lumen (angiography, venography). The decision for conservative management or surgical resection is mainly based on the presence of partial thrombosis, the shape of the aneurysm, and its extent.

Venous compression syndromes of the upper and lower extremities can be assessed sonographically by analyzing the flow patterns elicited by provocative tests.

#### 3.1.9.2 Chronic Venous Insufficiency

In patients with the postthrombotic syndrome, duplex scanning is a highly valid modality both for depicting morphologic changes of the venous wall (B-scan) and for identifying insufficiency of the main vein (Doppler). Recanalization is visualized with a high degree of accuracy and drainage insufficiency is evaluated semiquantitatively by measuring flow velocity and the duration and extent of reflux. It is important to differentiate reflux from persisting occlusion because the latter has important prognostic implications with regard to insufficient venous drainage. Careful documentation of the findings is crucial for comparison and diagnosis of recurrent thrombosis at follow-up. In the occasional patient, venography is required to confirm morphologic wall changes and recurrent thrombosis.

Noninvasive duplex imaging would also be a suitable modality for determining functional parameters in monitoring the effectiveness of *pharmacologic treatment*. However, for adequate evaluation and monitoring of chronic venous insufficiency, it would be desirable to have threshold values, e.g., of the vessel diameter above which abnormal venous dilatation should be assumed.

To overcome this lack, various parameters were determined by duplex scanning in 30 subjects with normal vessels (18 men, 12 women; mean age 34.7 ± 7.3 years). Mean values were calculated for each leg from five individual measurements. The measurements were performed with the subjects in the flat supine position and the feet lowered 10°. The following values were measured in the common femoral vein just above the saphenofemoral junction ($n = 60$ legs):

- Diameter: 11.7 ± 2.1 mm during expiration, 12.4 ± 2.2 mm during inspiration
- Planimetrically determined cross-sectional area: 1.07 ± 0.28 $cm^2$ during expiration, 1.16 ± 0.31 $cm^2$ during inspiration
- Peak flow velocity: 23.5 ± 8.3 cm/s during expiration; mean flow velocity averaged over 3 respiratory cycles with normal inspiration depth and abdominal breathing: 7.7 ± 1.9 cm/s.

During Valsalva's maneuver the femoral vein diameter increased to 1.82 ± 0.6 $cm^2$. The mean intraindividual diurnal variation in the cross-sectional area from morning to evening was 19.2% in expiration and 17.7% in inspiration. Peak expiratory flow velocity varied by 18.9% and mean flow velocity by 17.3%. Measurements performed on different days demonstrated a variation in the cross-sectional area of 24.6% during expiration and of 27.8% during inspiration while peak expiratory flow velocity varied by 24.7% and mean flow velocity by 21.4%. Other diameters determined in the 30 subjects were 8.9 ± 1.8 mm in the superficial femoral vein just after the origin of the deep femoral vein and 8.7 ± 1.6 mm in the popliteal vein at about the level of the knee joint cleft. The mean diameter of the great saphenous vein measured just below its termination in standing subjects was 5.6 ± 1.9 mm.

The parameters show wide intraindividual variation in repeat measurements performed on the same day or from day to day. Blood flow velocity as well as the diameter and cross-sectional area of large veins vary with breathing. Our results are confirmed by Marshall (1990), Hirschl and Bernt (1990), and Ludwig (1991), who report similar variations in diameter and cross-sectional area. Taking these factors into account, the mean flow rate determined for the common femoral vein just above the saphenofemoral junction is 503 ± 137 mL/min calculated as mean cross-sectional area multiplied by mean flow velocity ($V_{mean}$).

The respiration-induced variations in cross-sectional area are incorporated in this calculation as follows:

$$\textit{Mean cross-sectional area} = 1/3 \times (2 \times \textit{venous cross-section in expiration} + \textit{cross-section in inspiration})$$

Arterial blood flow shows less variation. In experimental measurement series performed in waterbaths by our group, the correlation between the average arterial flow velocity calculated from mean arterial flow velocities repeatedly measured by duplex scanning and the actual flow velocity was $R = 0.98$ (Schäberle and Seitz 1991). In the calculation, the pulsatile diameter variation is accounted for according to the formula:

$$\textit{Radius} = 1/3 \times \left(2\, R_{diastolic} + R_{systolic}\right)$$

Measurements in the superior mesenteric artery showed good reproducibility with a day-to-day variation of 10% for blood flow velocity and of 2.2% for vessel diameter. In contrast, quantitative determination of venous flow is much less precise because venous diameter measurement is subject to errors that are difficult to control. The diameters of large veins vary with breathing and the proximal iliac vein even shows pulsatile diameter variation. These variations are difficult to quantify and respiratory maneuvers do not easily lend themselves to standardization. The large veins are typically not circular but elliptical.

Another source of error is the use of high-pass filters that eliminate slow venous flow components from the Doppler spectrum, resulting in overestimation of mean blood flow velocity. In the above-described measurement series in 30 subjects with normal vessels, comparison of flow in the common femoral artery and in the common femoral vein just proximal to the saphenofemoral junction surprisingly showed mean venous flow to be 21% faster than arterial flow. In a series by Ludwig (1991) venous blood flow in the common femoral vein was about 32% higher than in the artery. These results are attributable to the above-described sources of error.

Cross-sectional areas or diameters markedly exceeding the above-described normal values suggest chronic venous incompetence, but the wide variation already seen under normal conditions does not allow reliable separation by corresponding angle-corrected time-velocity spectra. If venous thrombosis has been demonstrated, additional transverse views of the diseased vessel segment without and with compression are documented.

### 3.2.5 Clinical Role

Thrombosis of the arm veins is rare compared with venous thrombosis of the legs and is most commonly caused by

- Paraneoplasia
- Obstructed drainage in the narrow costoclavicular space (Fig. A3.73)
- Tumor compression in the thoracic outlet or
- Thrombogenic effects of portal catheters, central venous catheters, or pacemaker probes in the axillary and subclavian veins (Fig. 3.35)

The *narrow costoclavicular space* through which the vein must pass is formed by the first rib, the clavicle, and the subclavian muscle. Venous drainage may be obstructed when the already narrow passage is constricted further by weak shoulder muscles, rib callus, or exostosis. Obstruction can be reproduced during the examination by *hyperabduction* of the arm. Following recanalization after thrombolytic therapy, resection of the first rib is indicated if an outflow obstruction in the costoclavicular space is demonstrated in the hyperabduction test.

Ultrasound can be used to evaluate *patency* prior to insertion of a *pacemaker probe* or central venous catheter, especially in patients with clinical signs or a history of previous venous thrombosis of the upper extremity. Prior to placement of a central venous catheter in the jugular vein, ultrasound can additionally serve to identify the correct vessel in order to preclude misplacement or complications in patients with an abnormal course of the vein.

The risk of pulmonary embolism in venous thrombosis of the arm is very low. Clinically relevant postthrombotic damage is unlikely because there is good collateralization. Nevertheless, early examination by noninvasive duplex is indicated in patients with clinical signs of thrombosis in order to stop further progression by full-dose heparin therapy. Thrombophlebitis is typically caused iatrogenically and is rarely associated with thrombus extension into major deep veins.

### 3.2.6 Duplex Ultrasound Findings and Their Diagnostic Significance

The ultrasound findings in thrombosis are the same as in deep vein thrombosis of the legs: the vein appears markedly dilated and contains homogeneous or inhomogeneous structures, it cannot be compressed, and (color) duplex imaging depicts either no flow signals or, in the case of central thrombus, only flow near the wall. Any *internal or external obstruction of venous drainage eliminates cardiac pulsatility and eliminates or reduces respiratory phasicity of flow*. Therefore, in cases where methodological limitations prohibit adequate evaluation of the subclavian vein from the supraclavicular position, a Doppler waveform recorded in the axillary vein (from the infraclavicular fossa) can be used to exclude significant flow obstruction. Apart from the subclavian vein, the easily accessible axillary vein may be affected by thrombosis because blood from the arms is emptied into this vein through thoracic wall collaterals. Compression ultrasound is used to

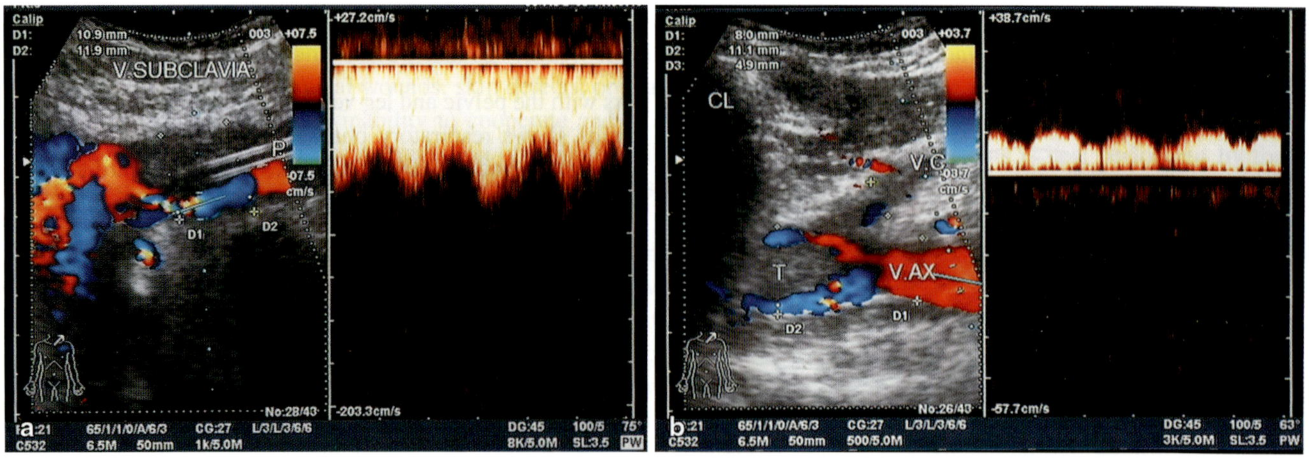


**Fig. 3.35** **a** Thrombus (*T*) surrounded by flowing blood in the subclavian vein; the thrombus in this patient is due to the thrombogenic effect of an implanted venous access port (*P*). **b** The port tube inserted through the cephalic vein (*V.C.*, occluded by suture) extends to the site of the thrombus, which is surrounded by flowing blood and extends into the axillary vein (*V.AX.*). The blood flow direction distal to the thrombus is normal (red, toward transducer), but flow is markedly disturbed by the thrombus: reduced flow velocity (10 cm/s) and absent cardiac pulsatility and respiratory phasicity. The cephalic and axillary veins are indicated by plus signs; *CL* clavicle

exclude thrombosis of the *axillary vein* (with the transducer *in the infraclavicular fossa*) and of the peripheral veins. As already mentioned, the compression test is unreliable in the subclavian vein, where flow has to be demonstrated directly by means of a Doppler spectrum. Sonographically, thrombophlebitis can be distinguished from deep thrombosis on the basis of the course of the vessel affected and its relationship to anatomic landmarks (accompanying artery of the same name).

We analyzed 610 patients with arm swelling examined by duplex ultrasound, which demonstrated arm vein thrombosis in 96 cases. In 61 of the patients, thrombosis was attributable to paraneoplasia, a central venous catheter, or a pacemaker probe. Nine of the remaining 35 patients, in particular younger ones, underwent thrombolytic therapy, which led to recanalization in seven cases. Subsequently, flow obstruction was demonstrated in the costoclavicular space by duplex scanning with provocative maneuvers in five of the patients.

In the *hyperabduction test*, the arm on the affected side is raised above the head while the Doppler spectrum is recorded in the axillary vein with the transducer in the infraclavicular fossa. Note, however, that there may be some physiologic flow obstruction during extreme hyperabduction, especially in slender patients. A more reliable way of demonstrating clinically significant narrowing of the passage between the clavicle and the first rib is to use a modified "apron grip" test, in which the Doppler spectrum in the axillary vein is recorded while pulling down the arm while it is turned backward and rotated to the side. Flow reduction with elimination of cardiac and respiratory fluctuation is abnormal (see Figs. A3.73 and A3.74).

The underlying mechanism is complex. Typically, however, the clinical symptoms are caused by weakness of the shoulder muscles. Normal muscle tone elevates the clavicle above the neurovascular bundle. If the muscles are too weak, however, the bundle is compressed by the movement of the clavicle over the first rib during abduction with rotation.

Obstruction in the costoclavicular space primarily affects the vein while the scalenus and cervical rib syndromes do not affect venous drainage as the subclavian vein passes in front of the scalenus anterior muscle. The costoclavicular compression syndrome is primarily treated by physical therapy aimed at strengthening the muscles of the shoulder girdle. The outcome can be followed up by duplex scanning. If the symptoms persist after physical therapy, or after recanalization following thrombolytic therapy of compression-induced thrombosis, resection of the first rib is indicated.

Color duplex scanning provides precise information on both spontaneous and thrombolytic *recanalization* after arm vein thrombosis as well as on lesions persisting after thrombosis such as luminal variations, residual mural thrombi, or wall sclerosis (see Fig. A3.75). Postthrombotic valvular incompetence is of little clinical relevance in the arm. Ultrasound is superior to all other imaging modalities in providing the information on morphologic wall lesions and their hemodynamic effects that is necessary to initiate the most effective treatment.

### 3.2.7 Diagnostic Role of Duplex Ultrasound Compared with Other Modalities

Duplex ultrasound is equal to the current gold standard, venography, in evaluating the arm veins for thrombosis, as their superficial course provides good insonation conditions. It is superior to venography in evaluating the wall of the axillary vein for postthrombotic lesions persisting after recanalized thrombosis. Moreover, ultrasound yields highly accurate information on the hemodynamic effects of vein compression by tumor or in the costoclavicular compression syndrome, especially by means of the Doppler frequency spectrum recorded distally during provocative maneuvers. Since imaging of the thoracic outlet is limited, CT is superior in assessing tumor-related inflow obstruction. As in the legs, venography is superior in evaluating the extent of collateralization. In controlled studies using venography as the reference modality, duplex ultrasound was found to have a sensitivity of 94% and a specificity of 96% in detecting arm vein thrombosis (Koksoy et al. 1995; Haire et al. 1991).

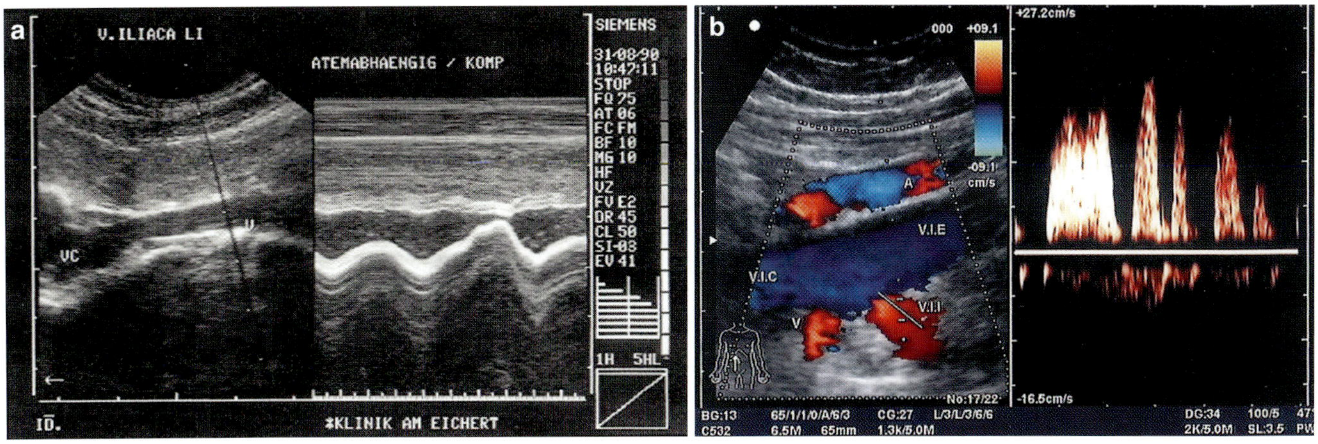


**Fig. A3.1 a, b**

**a** Respiratory variation in diameter, as seen in the vena cava (*VC, left*), also occurs in the iliac vein (illustrated on the *right* in the time-motion mode); in slender individuals, the iliac vein is also compressible (*Komp*).

**Course of the pelvic veins**

**b** The pelvic veins (common iliac vein/*V.I.C* and external iliac vein/*V.I.E*) take an arched course through the true pelvis posterior to the arteries (see Fig. 2.2b) of the same name (*A*). The internal iliac vein (*V.I.I*) enters the common iliac vein on its posterior aspect at its lowest point (flow toward transducer, coded in red). There is respiratory phasicity of the flow profile, and younger individuals may additionally show cardiac pulsatility. In the example, a second leg vein (red) enters the common iliac vein (blue) slightly above the internal iliac vein.

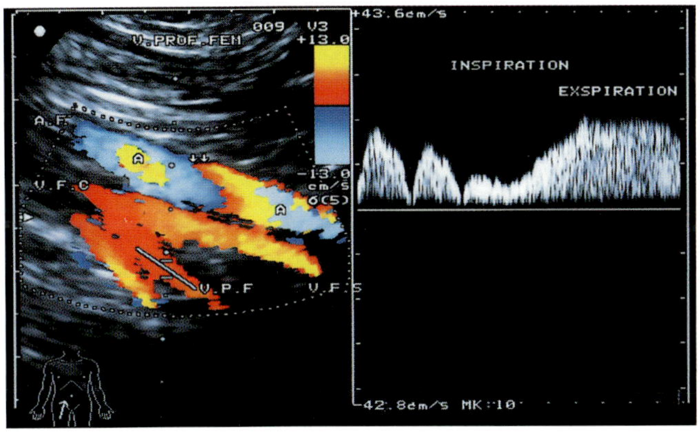


**Fig. A3.2**

**Venous Doppler waveform**

Just below the inguinal ligament, the common femoral vein (*V.F.C*) divides into the deep femoral vein (*V.P.F*) and the superficial femoral vein (*V.F.S*). Flow in the veins is characterized by respiratory fluctuation when no proximal obstruction (thrombus, compression) is present. The Doppler waveform on the *right* illustrates the respiratory variation in flow velocity in the deep femoral vein. Venous flow velocity is decreased due to an increase in intra-abdominal pressure during abdominal inspiration. There is cardiac modulation even during expiration (waveform obtained in a young woman). With the PRF adjusted to venous flow, the femoral artery close to the transducer shows aliasing (*A*) and flow reversal at diastole (*arrows*).

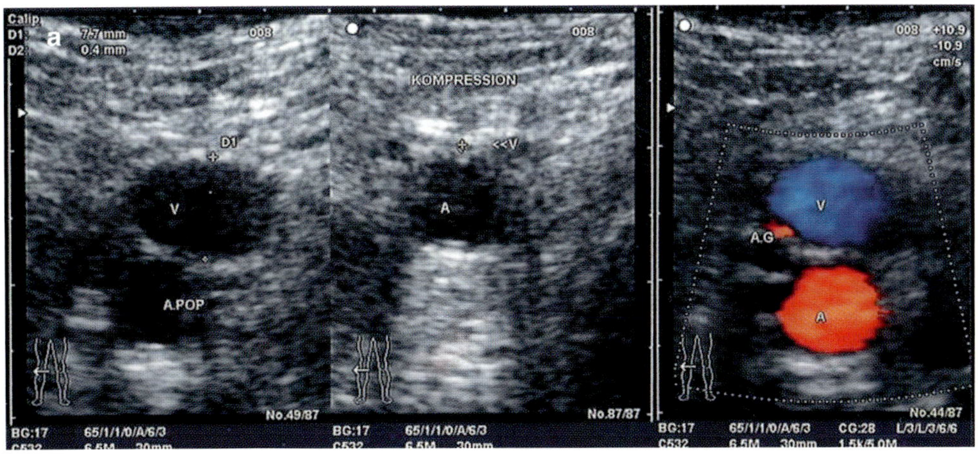


**Fig. A3.3 a, b**

**Popliteal vein thrombosis: evaluation by compression/color duplex ultrasound**

**a** The transverse gray-scale scan on the *left* depicts the popliteal artery (farther away from transducer, which is placed in the popliteal fossa) anterior to the popliteal vein. In the *middle section*, the vein becomes invisible as it is completely compressed by the transducer, indicating absence of an intraluminal thrombus. The transverse scan on the *right* obtained at the same level in the popliteal fossa confirms patency of the popliteal artery (red) and vein (blue).

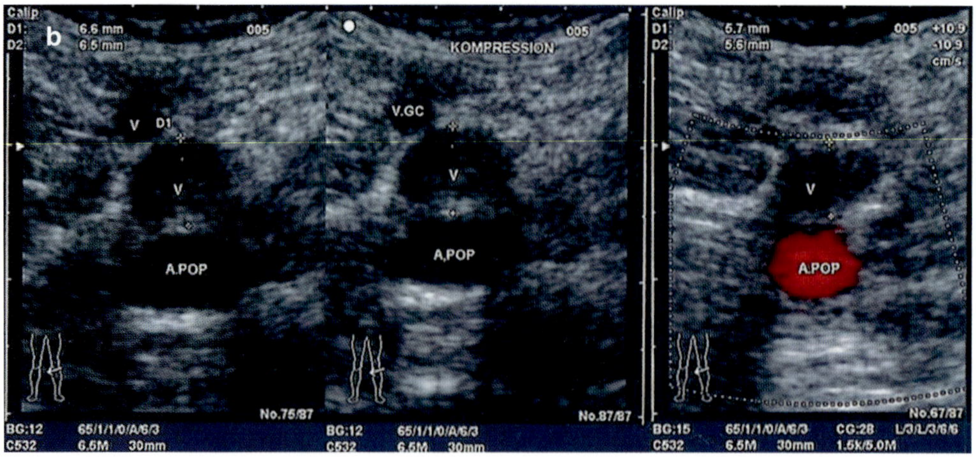


**Fig. A3.3 (continued)**

**b** In this example, a somewhat older thrombus is already depicted in the gray-scale scan (*left section*) as an intraluminal structure, more hyperechoic than the artery. Applying pressure with the transducer does not compress the vein (*middle section*). The color duplex scan (*right section*) depicts flow only in the popliteal artery (red) and no flow signals in the vein (despite a low PRF).

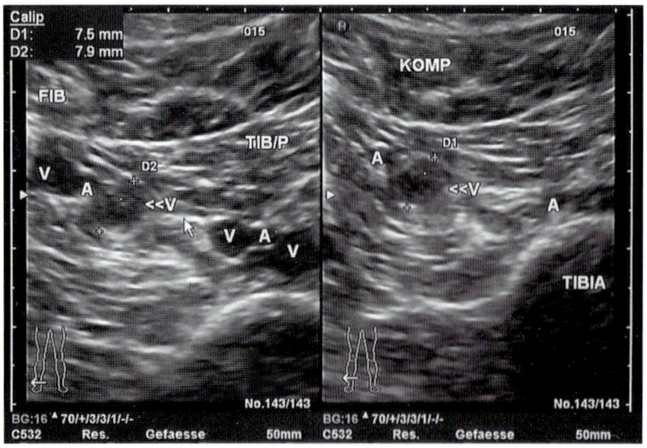


**Fig. A3.4**

**Lower leg veins**

The main veins in the lower leg are identified using the corresponding arteries as landmarks. The veins typically occur in pairs that run on both sides of the respective artery with color coding indicating flow in the opposite direction. Taking the popliteal vein as a point of departure, the examiner follows the veins toward the periphery in transverse orientation. Below the confluence, the courses can be followed continuously, that of the fibular vein close to the fibula and that of the posterior tibial vein along the more hyperechoic deep crural fascia (*arrow*).

In patients examined for suspected thrombosis, the examiner evaluates the tubular structure (*V*) adjacent to the artery (*A*, which serves as landmark) for complete compressibility (*right section*). In the example shown, fibular vein thrombosis is suggested by noncompressibility of one of the fibular veins (*V*) and the higher echogenicity of its lumen. The other tubular structures (*V*) seen adjacent to the corresponding arteries (*A*) in the *left image* can be compressed, becoming invisible while pressure is being applied with the transducer (*right image*).

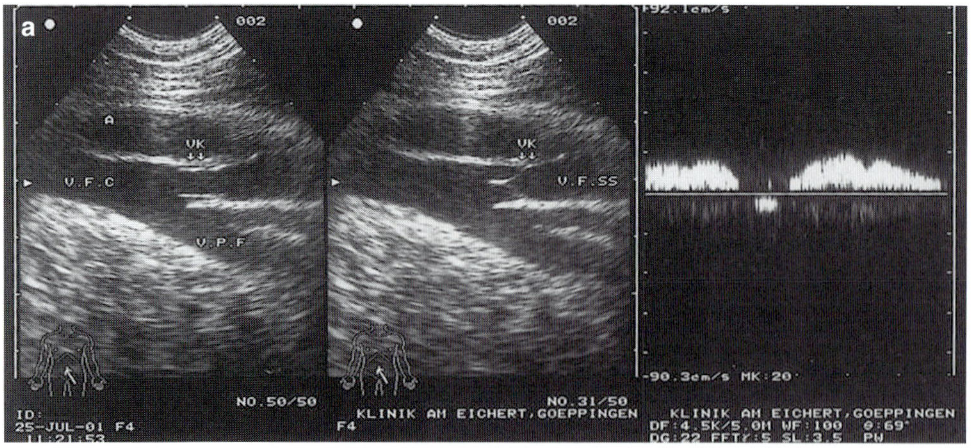


**Fig. A3.5 a, b**

**Normal venous valve function**

**a** Under good insonation conditions, normal valve function is apparent in the gray-scale scan. The valve (*VK*) depicted in the superficial femoral vein (*V.F.S*) just before it receives the deep femoral vein (*V.P.F*) is shown during expiration (valve open) on the *left* and during inspiration (valve closed) in the *middle section*. The cusps prevent flow reversal during the inspiration-induced increase in intra-abdominal pressure.

The respiratory variation in blood flow is documented in the Doppler waveform. Flow toward the transducer is increased in expiration and decreased in inspiration. Flow may even decrease to zero, and the short transient reflux until valve closure may be absent if the sample volume is placed near a valve.

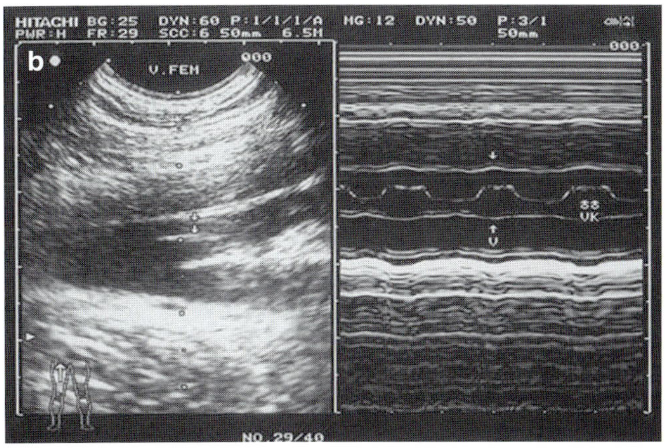


**Fig. A3.5 (continued)**

**b** The time-motion display documents valve motion from inspiration to expiration: the hyperechoic line in the lumen represents the closed valve during inspiration with subsequent expiratory opening being shown as movement of the valve leaflet toward the vessel wall.

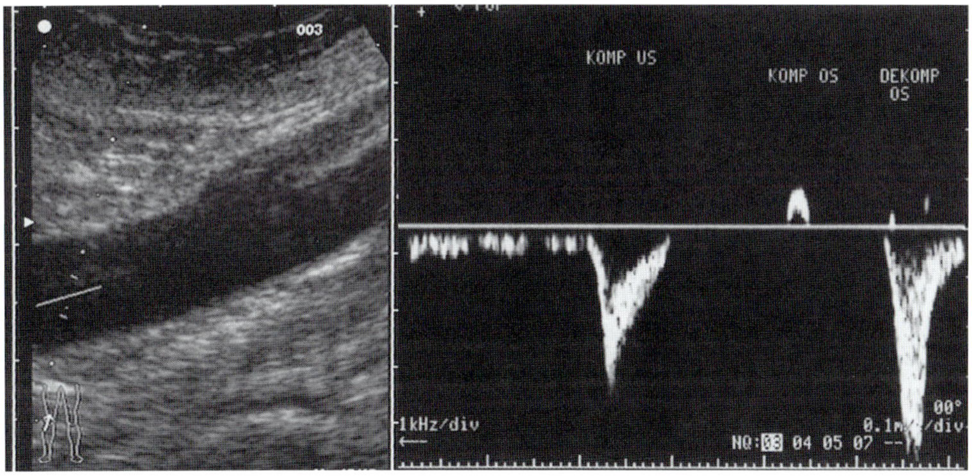


**Fig. A3.6**

**Valve function test**

Doppler frequency data sampled during functional tests provide information on proximal or distal flow obstruction and valve function. The femoral vein is commonly evaluated using Valsalva's test.

Illustration of the effect of alternating compression and release in the popliteal vein. Flow with normal respiratory phasicity is followed by augmented flow (upon compression of the lower leg soft tissue distal to the sample volume) toward the heart. No flow signal is recorded upon release of compression. Compression of the muscle and vein in the upper leg (*KOMP OS*) induces a short signal indicating distal flow that persists until valve closure.

Valve incompetence would be associated with persistent reflux. The example shows an induced signal indicating flow toward the heart (A-sound) after release of compression of the upper leg (*DEKOMP OS*), i.e., proximal to the sampling site. In the absence of flow obstruction, there is a steep upstroke in the Doppler waveform. The elicited flow increase is less pronounced when a flow obstruction is present between the site of compression and the site of Doppler measurement (see Figs. A3.62a, b and 3.23).

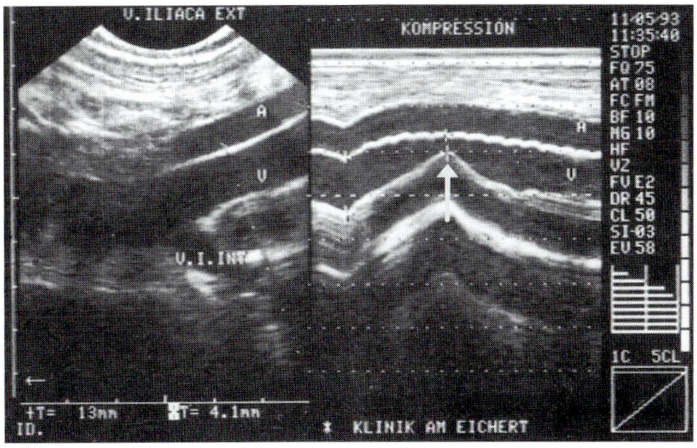


**Fig. A3.7**

**Compression ultrasound of pelvic veins**

The veins at the pelvic level can be compressed with the transducer only in slender patients. Therefore, the absence of compressibility is not a reliable indicator of thrombosis in this territory; the presence of compressibility however does rule out thrombosis. The example illustrates compressibility in the time-motion mode (*arrow*).

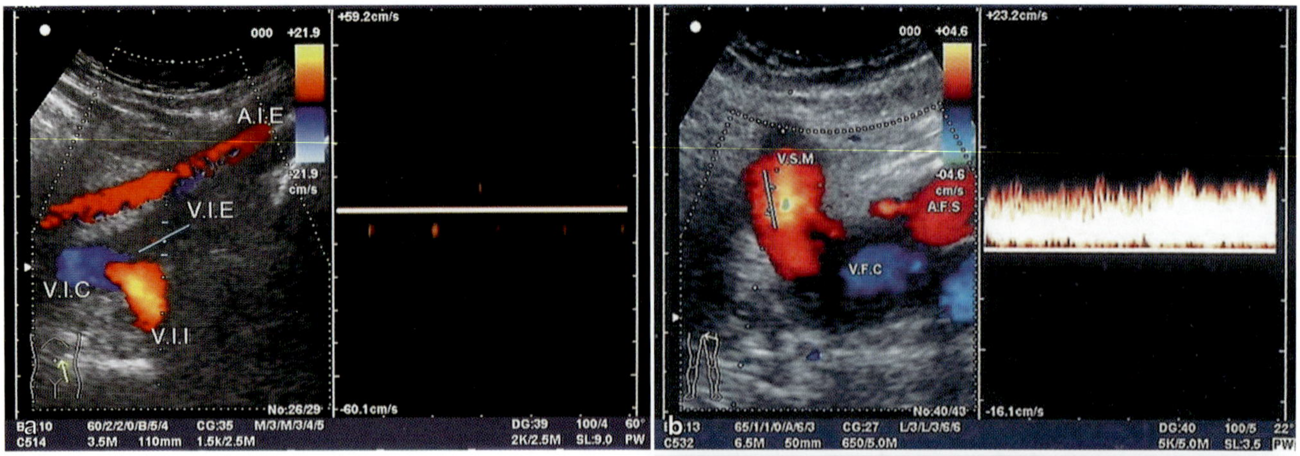


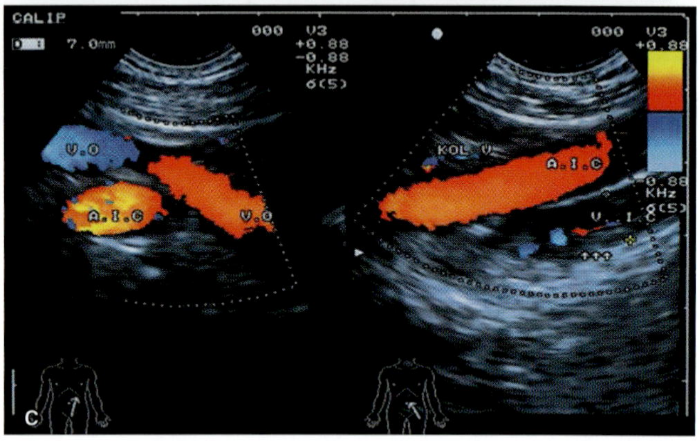


**Fig. A3.8 a–c**
**Pelvic vein thrombosis – collateral circulation**
**a** Compression ultrasound is unreliable, especially in obese patients, because an adequate structure against which to compress the vein is not available and a greater scanning depth is required. Sonographically, the vein is identified posterior to the iliac artery and thrombosis is demonstrated by failure to obtain a flow signal in the color flow mode. Occlusion is confirmed by the absence of flow signals in the Doppler frequency spectrum. In the example, neither the color flow image nor the Doppler waveform shows flow in the external iliac vein (*V.I.E*), confirming thrombosis of this vein (assuming that instrument settings are correct). There is flow in the internal iliac vein (*V.I.I*; red, toward transducer) and in the common iliac vein (*V.I.C*, blue, away from transducer, toward the heart), showing patency of the venous system at this level, which will ensure drainage toward the heart. Also seen is the external iliac artery (*A.I.E*) anteriorly.
**b** In pelvic vein occlusion, venous return must be ensured by collaterals. One important collateral pathway includes the saphenofemoral junction (termination of the great saphenous vein in the groin), which shows a retrograde flow signal (displayed in red, toward transducer, no respiratory phasicity) and drains into abdominal wall collaterals (epigastric vein). The latter are not subject to intra-abdominal pressure changes.
**c** Early recanalization 3 months later is difficult to identify by conventional duplex scanning in the pelvic territory. The scan on the *right* demonstrates central flow signals in the common iliac vein (*V.I.C*), which is still markedly thrombosed. Anteriorly, the common iliac artery is depicted with flow coded red (*A.I.C*). The scan on the *left* depicts the dilated ovarian vein (*V.O*) coursing over the common iliac artery (*A.I.C*). The color change from red to blue is due to the change in flow direction relative to the ultrasound beam. While thrombosis can be diagnosed by gray-scale or conventional duplex ultrasound, the detection of early signs of recanalization falls in the domain of color duplex.

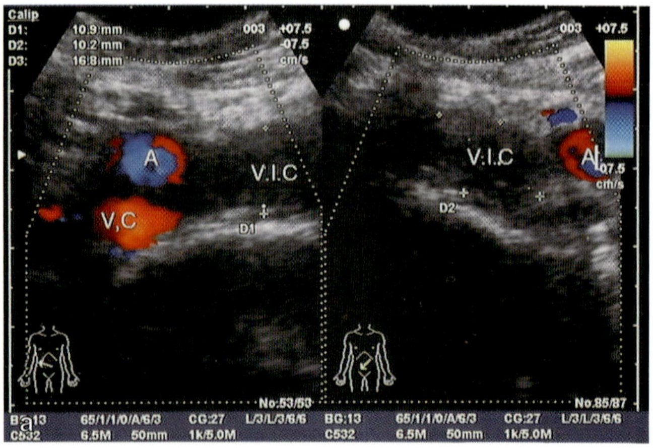


**Fig. A3.9 a–c**
**Pelvic vein thrombosis – early recanalization**
**a** Descending thrombosis of the left common iliac vein (*V.I.C*; longitudinal scan on the *left*, transverse scan on the *right*). When seeing a patient with this form of thrombosis, the examiner should always think of a venous spur as a possible cause. Color duplex imaging depicts no flow; vena cava (*V.C*) with red-coded flow.

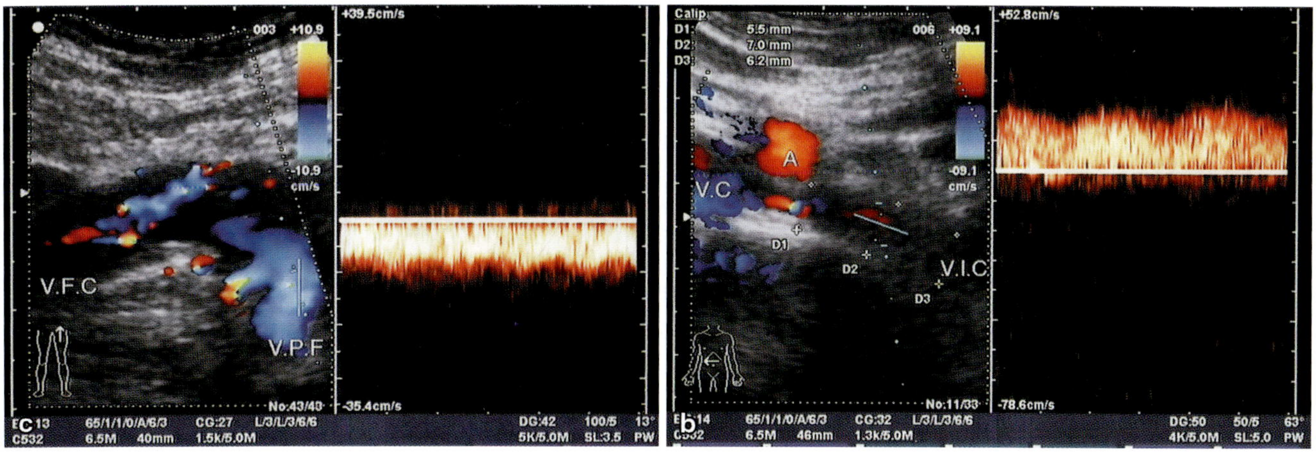


**Fig. A3.9 (continued)**

**b** Follow-up ultrasound examination shows a shrunken lumen of the common iliac vein as well as flow signals, indicating early recanalization (smaller diameter compared with recent thrombosis). Transducer tilted slightly differently, which is why the vena cava (*V.C*) is depicted with flow in the opposite direction relative to the transducer (blue; *A* aorta).

**c** Retrograde flow in the deep femoral vein (*V.P.F*; blue, away from transducer; below the baseline in the Doppler waveform) indicates inadequate drainage through the pelvic veins due to poor recanalization. Flow signals in the center of the partially recanalized common femoral vein (*V.F.C*).

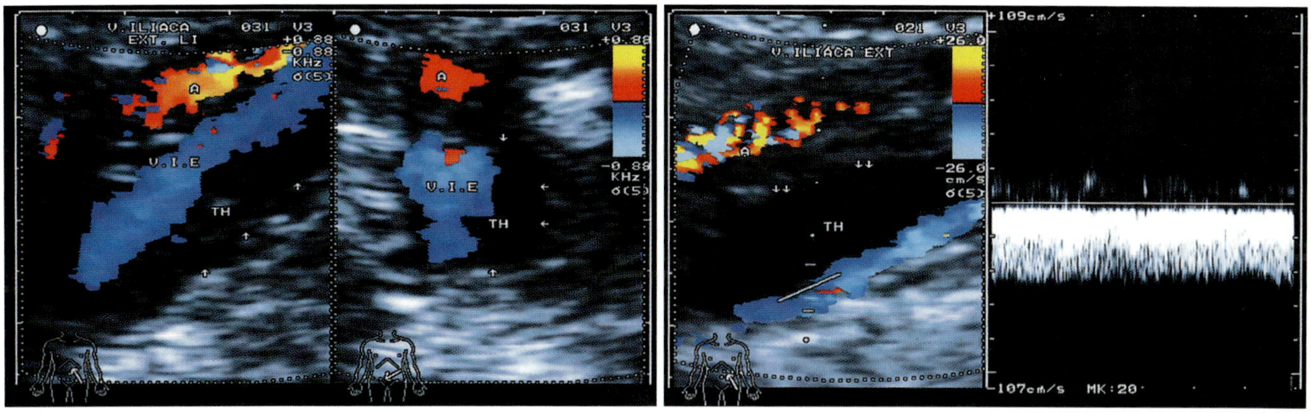


**Fig. A3.10 a, b**

**a** In this young woman with scintigraphically proven pulmonary embolism, ultrasound showed patent deep leg veins, while there were thrombi in the iliac vein. In contrast to recanalization with central flow signals as shown in **c**, the fresh thrombus (*TH*) in the external iliac vein (*V.I.E*) is attached to the wall and surrounded by flowing blood.

**b** More proximally, close to the junction with the internal iliac vein, the thrombus is surrounded by flowing blood posteriorly. The flow obstruction produces a high-frequency, continuous signal (with loss of respiratory phasicity) typical of venous stenosis. The peak flow velocity is 50 cm/s. With the PRF adjusted to venous flow, there is aliasing in the artery anterior to the vein. The thrombus is indicated by *arrows*. Not shown is the Doppler waveform obtained in the uninvolved common femoral vein distal to the thrombus. The patent lumen of the common femoral vein is relatively wide, and there is only little reduction of respiratory fluctuation. A thrombus in an otherwise patent lumen as in this case may be overlooked if only the Doppler frequency spectrum from the common femoral vein is analyzed, even if the spectrum is compared with that of the other side and functional tests are done. Routine venography may likewise fail to identify such mural thrombi in an otherwise patent iliac vein or to differentiate them from flow phenomena.

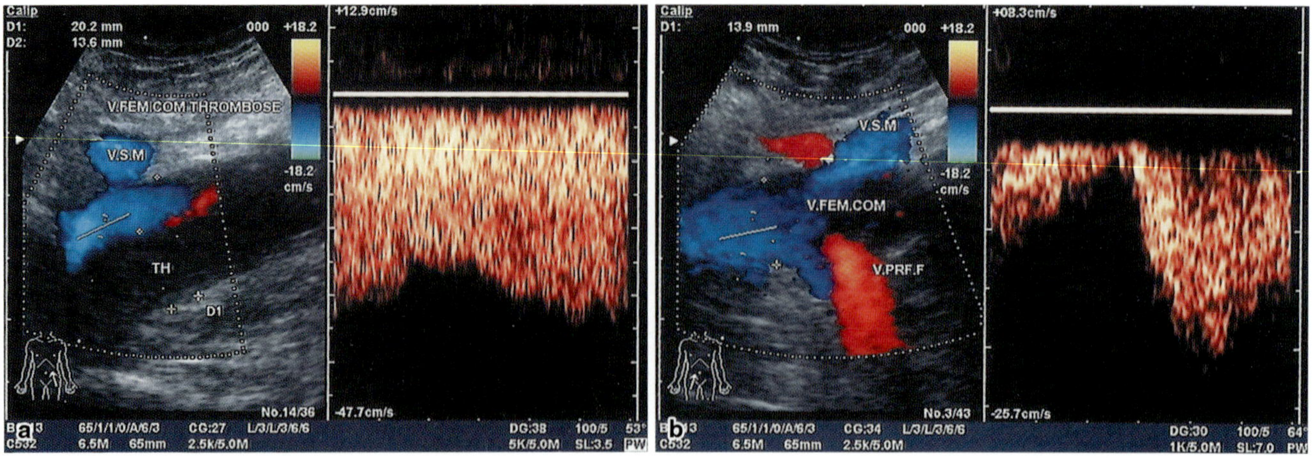


**Fig. A3.11 a, b**
**Elimination of respiratory phasicity by flow obstruction**
**a** Flow obstruction along a thrombus in a partially thrombosed vein eliminates or reduces respiratory phasicity and results in a higher-frequency flow signal in the Doppler waveform, unless the blood is drained through collaterals (according to the continuity law, as in arterial stenosis).
**b** Since some residual respiratory variation may still be depicted in the Doppler waveform, it must be compared with the other side. In the example, respiratory fluctuation is much less pronounced on the left than on the unaffected right side. Common femoral vein (*V.FEM.COM*) and saphenofemoral junction (*V.S.M*) coded in blue, and a deep femoral vein (*V.PRF.F*) coded in red (flow toward transducer).

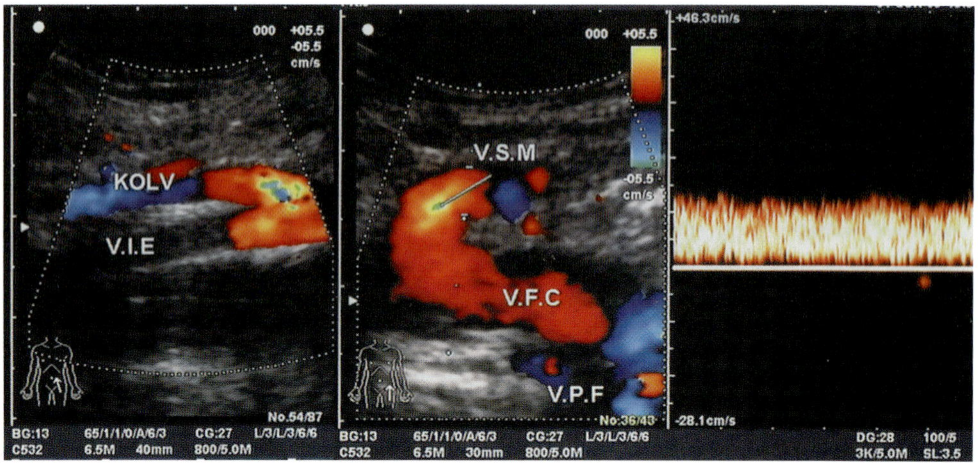


**Fig. A3.12**
**Collateral circulation in pelvic vein thrombosis**
The degree of recanalization is not always easy to estimate in older thrombosis of the external iliac vein. The shrunken vein is more difficult to evaluate on its course into the true pelvis and must be differentiated from collaterals such as the epigastric veins, which arise at the level of the inguinal ligament and, when dilated, may be of similar size as the shrunken external iliac vein (*leftmost* image).
Retrograde flow at the saphenofemoral junction (*middle section, V.S.M*) indicates outflow obstruction of the pelvic veins; the course of the abdominal wall collaterals can be followed using color flow imaging. The Doppler waveform (*right section*) also shows retrograde flow toward the transducer in the great saphenous vein.

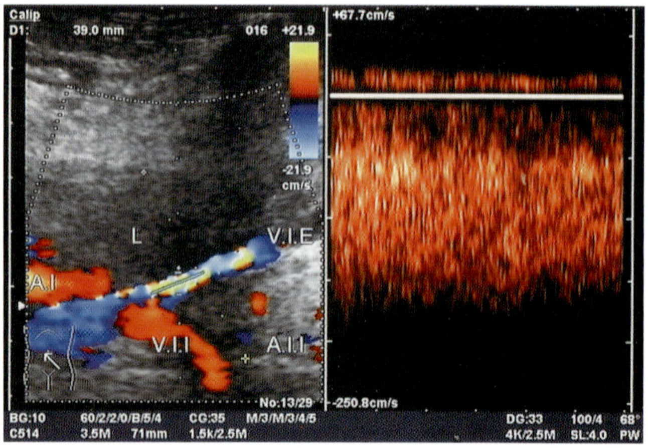


**Fig. A3.13**
**External pelvic vein compression by lymphoma**
Leg swelling and widening of major veins secondary to outflow obstruction at the pelvic level due to external compression of the external iliac vein (*V.I.E*) by lymphoma (*L*). (*V.I.I* = internal iliac vein). The Doppler spectrum shows high flow velocity of 170 cm/s in the external iliac vein and absence of respiratory phasicity.

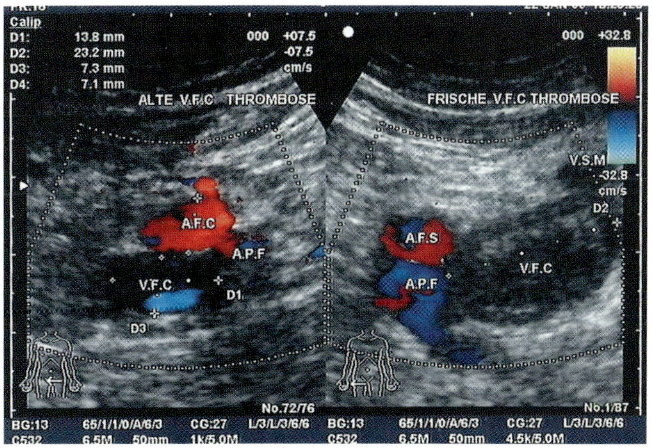


## Fig. A3.14

**Criteria for estimating thrombus age**

The major signs of fresh deep venous thrombosis of the leg are pronounced dilatation of the vein and good delineation of the homogeneous, often hypoechoic, thrombosed venous lumen from perivascular connective tissue. The two scans obtained in the same patient show fresh deep venous thrombosis with the venous lumen dilated to well over twice that of the accompanying artery in the right leg (*right section*) and an old thrombosis of the common femoral vein already partially recanalized in the area of the femoral bifurcation at the same level on the left (flow in the *V.F.C* displayed in blue). (*A.F.C* common femoral artery, *A.P.F* deep femoral artery, *A.F.S* superficial femoral artery).

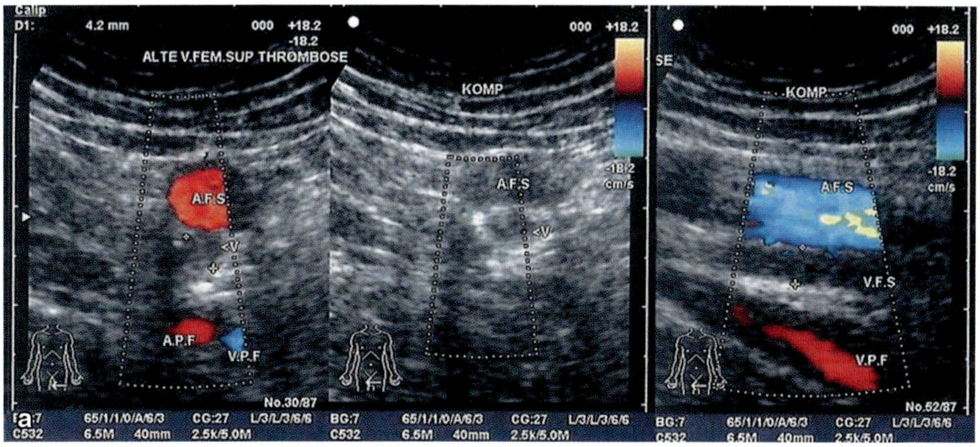


## Fig. A3.15 a, b

**Older femoral vein thrombosis**

**a** Older thrombosis is associated with shrinkage of the lumen (relative to the corresponding artery). The scan on the *left* shows the thrombotically occluded superficial femoral vein (*V*) posterior to the red superficial femoral artery (*A.F.S*). Demarcation is much poorer than in fresh thrombosis. The compression scan (*middle section*) shows noncompressibility of the vein depicted posterior to the artery. The longitudinal scan (*right section*) of the occluded superficial femoral vein (*V.F.S*) shows absence of flow in the vein posterior to the artery shortly before receiving the deep femoral vein (*V.P.F*). The deep femoral vein is recanalized with flow toward the transducer coded in red. However, marginal hypoechoic thrombosis still persists along the patent venous lumen.

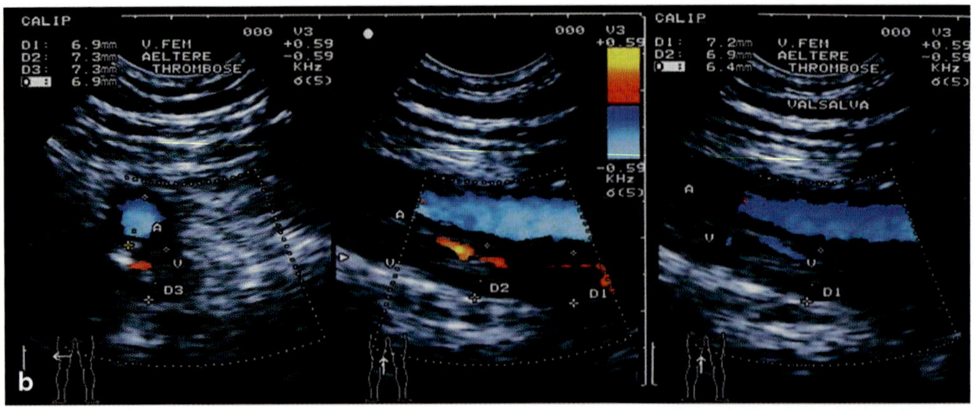


**Fig. A3.15 (continued)**

**Femoral vein thrombosis, early recanalization**

**b** While color duplex ultrasound contributes only little to the diagnostic evaluation of thrombosis and thrombus age, it has its role in demonstrating early recanalization. In the older femoral vein thrombosis presented here (poor demarcation of the wall, venous diameter identical to that of the artery), the color duplex image demonstrates meandering flow within the thrombosed lumen as a sign of early recanalization. Flow in the vein posterior to the artery is depicted in red (directed toward the heart) on the transverse scan (*left section*) and longitudinal scan (*middle section*). The scan on the *right* demonstrates flow reversal after Valsalva's maneuver, indicating impaired valvular function. Early recanalization in patients on thrombolytic therapy is identified in the same way. The depiction of flow in the center of the vessel differentiates recanalization in older thrombosis from fresh thrombus with residual flow near the wall.

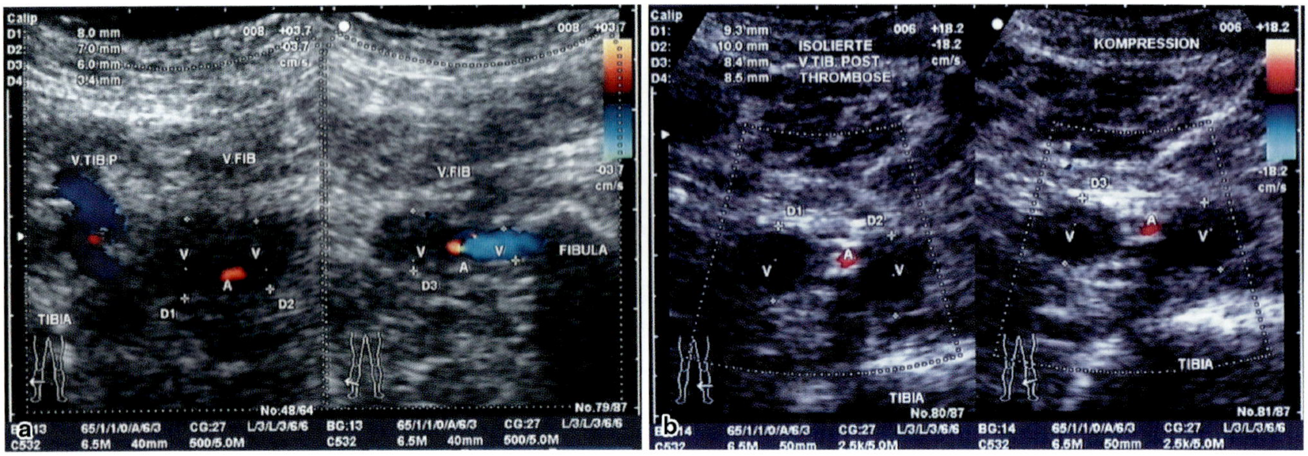


**Fig. A3.16 a, b**

**Lower leg venous thrombosis (fresh – older)**

**a** With a good acoustic window, fresh venous thrombosis of a lower leg vein is easily identified as a hypoechoic tubular structure within the soft tissue in the typical location of the vein. The lumen is wider than that of the corresponding artery. The fibular artery and vein course along the medial aspect of the fibula. Compression ultrasound and color duplex are highly accurate in differentiating between patent and thrombotic segments. The scan on the *right* shows one patent vein and one thrombosed vein, and the *left* scan shows thrombosis of both fibular veins, while both adjacent posterior tibial veins are patent with flow coded in blue.

**Fresh isolated thrombosis of the posterior tibial vein**

**b** Paired posterior tibial vein to the left and right of the artery. The vein is not compressible (right) and the lumen is increased to more than twice that of the artery, indicating fresh thrombosis. The thrombus is hypoechoic and has a homogeneous echotexture; color duplex depicts no flow signals in the vein.

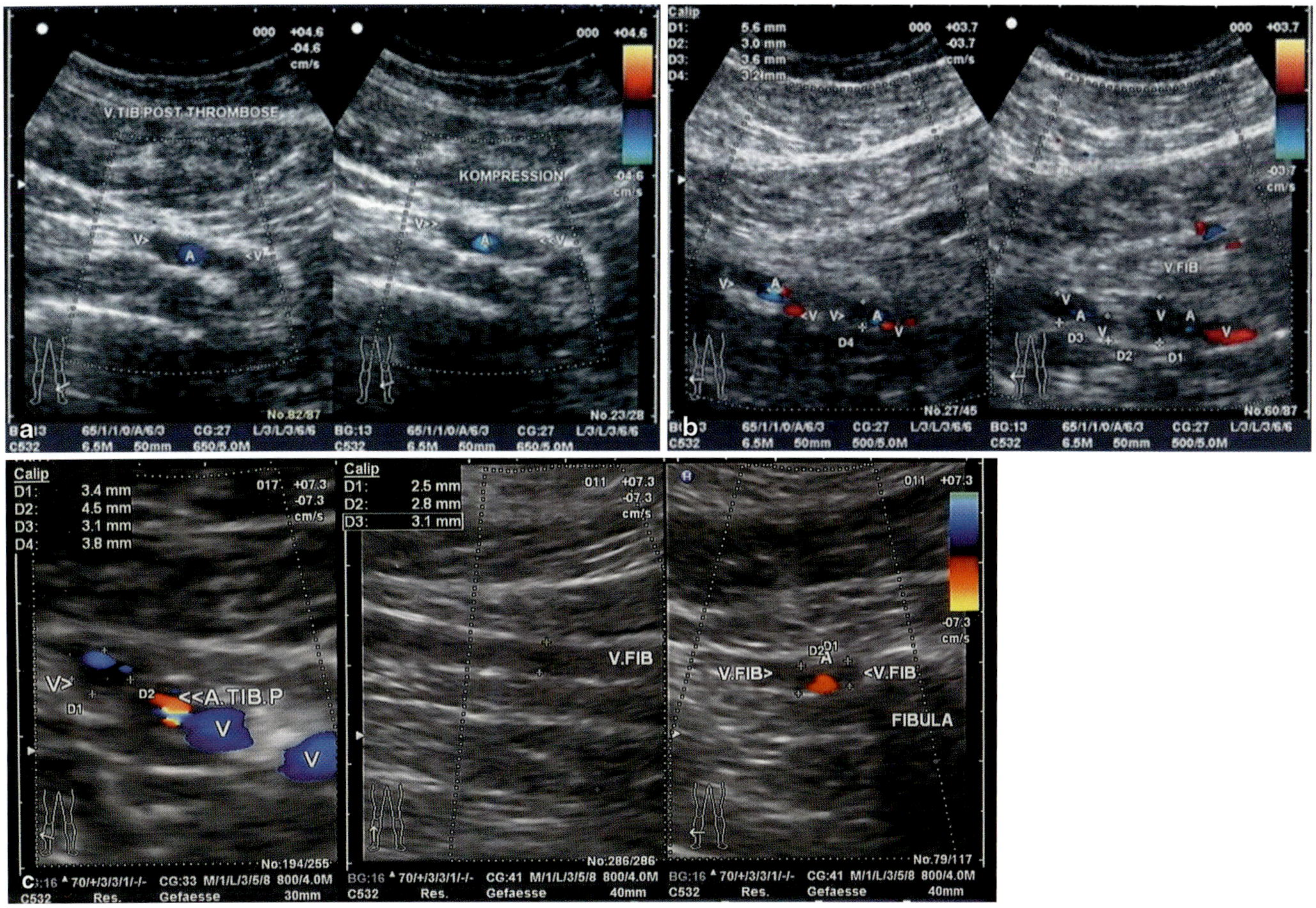


**Fig. A3.17 a–c**

**Older isolated thrombosis of the posterior tibial vein**

**a** Hypoechoic tubular structures are seen to the left and right of the posterior tibial artery. The structures cannot be compressed (compression scan on the *right*) and have about the same diameter as the artery, indicating older thrombosis with shrinkage of the lumen. Color duplex ultrasound depicts flow in the artery in blue, while there are no flow signals in the thrombosed vein, suggesting that recanalization has not yet occurred.

**b** Older thrombosis is more difficult to demarcate from surrounding soft tissue, as the diameter of the vein decreases with thrombus age and the thrombotic lumen may become more hyperechoic. Residual thrombosis is difficult to detect if parts of the vessel have already undergone recanalization and can be compressed or show flow signals in the color duplex mode. It is not clear, however, whether the identification of such residual thrombosis is clinically relevant. The scan on the *left* shows red venous flow to the right of the fibular and posterior tibial arteries (blue in this example). The hypoechoic tubular structures to the left reflect older venous thrombosis. Occluded lower leg veins that have undergone permanent fibrotic transformation are very difficult to identify (*right scan*). In such cases, the examiner must very thoroughly search the vicinity of the artery for tubular structures and try to elicit a response to flow augmentation by distal compression in the standing patient.

**Older calf thrombosis**

**c** Following acute thrombosis, recanalization may set in early as indicated in the example by partial recanalization (blue) of the left posterior tibial vein (*left section*; diameter of the vein indicated by *plus signs*). On the other hand, recanalization may occur late or not at all, which is indicated by a shrunken lumen without flow signals, shown here for the fibular vein (*V.FIB*) on longitudinal and transverse views (*middle* and *right sections*). Older thrombosis is characterized by more echogenic thrombus and shrinkage of the venous lumen, which is more difficult to distinguish from surrounding muscle and fatty connective tissue. The examples illustrate the difficulties encountered in diagnosing older thrombosis both in partial recanalization and in completely occluded postthrombotic veins with shrunken lumina.

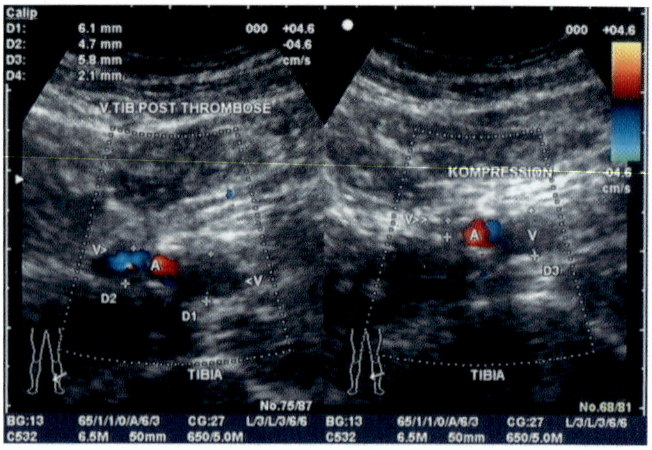


**Fig. A3.18**

**Recurrent thrombosis after recanalization**

The posterior tibial vein to the right of the artery (*A*, red) is markedly dilated (6.1 mm) and shows no flow, suggesting acute thrombosis. The vein to the left exhibits flow coded in blue with a surrounding hypoechoic margin corresponding to wall thickening as a sign of recanalized thrombosis. The compression scan (*right section*) shows only little compressibility of the thrombosed vein.

No venous flow signals are depicted to the left of the artery during compression. However, the vein is not fully compressed. There is a hypoechoic area in the connective tissue corresponding to the post-thrombotically thickened walls. The vein is surrounded by hyperechoic connective tissue of the deep crural fascia, and the tibia is depicted farther away from the transducer.

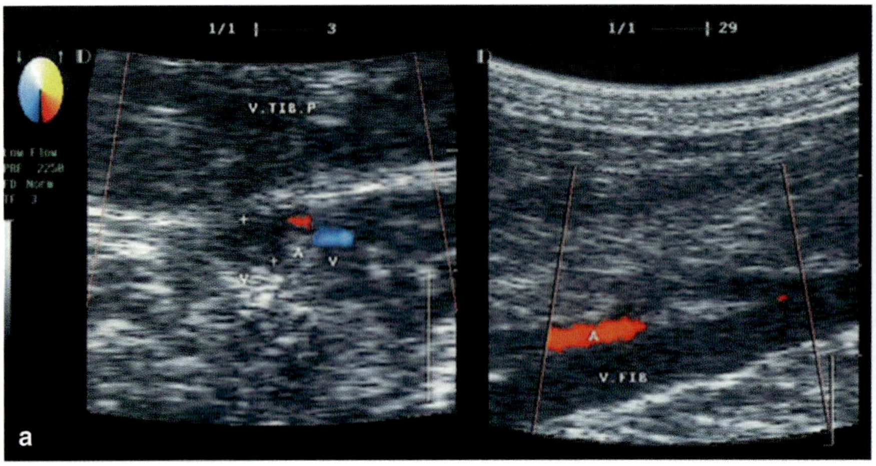


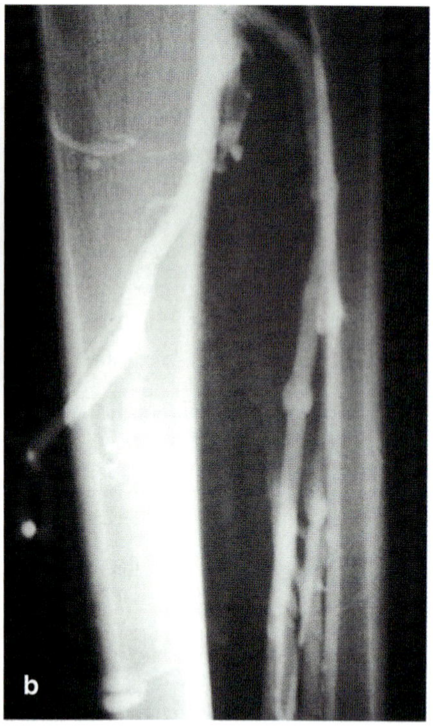


**Fig. A3.19 a, b**

**Lower leg venous thrombosis – ultrasound and venography**

**a** Isolated lower leg thrombosis of a single vein group may be overlooked or misinterpreted on venography. Moreover, small filling defects are difficult to locate to a muscle vein or a major vein. B-scan sonography identifies fresh thrombosis of a lower leg vein as a hypoechoic tubular structure along the respective artery.

Color duplex scanning corroborates the diagnosis by the failure to demonstrate flow when performed with a low PRF. Residual flow may be indicated by a signal elicited by muscle compression distal to the transducer (A-sound). The scan on the *left* shows a patent posterior tibial vein (*V*) with blue-coded flow to the right of the red artery (*A*), while the second vein (*V*) to the left of the artery is thrombosed. The marked dilatation of the vein (> twice arterial lumen) and the low-level echo of the thrombus suggest acute thrombosis. The longitudinal scan on the *right* shows the fibular vein to be thrombosed as well. The lumen is much wider than that of the corresponding artery (*A*) with flow depicted in red. The thrombotic vein is hypoechoic and homogeneous, clearly demarcating it from the surrounding soft tissue.

**b** Venogram: Filling defect in the posterior tibial vein. The fibular vein is not depicted.

**Fig. A3.20 a–c**

**Isolated fibular vein thrombosis – venography**

Venography is limited in the evaluation of the fibular vein. A filling defect in this vein may be due to a technical limitation or thrombosis.

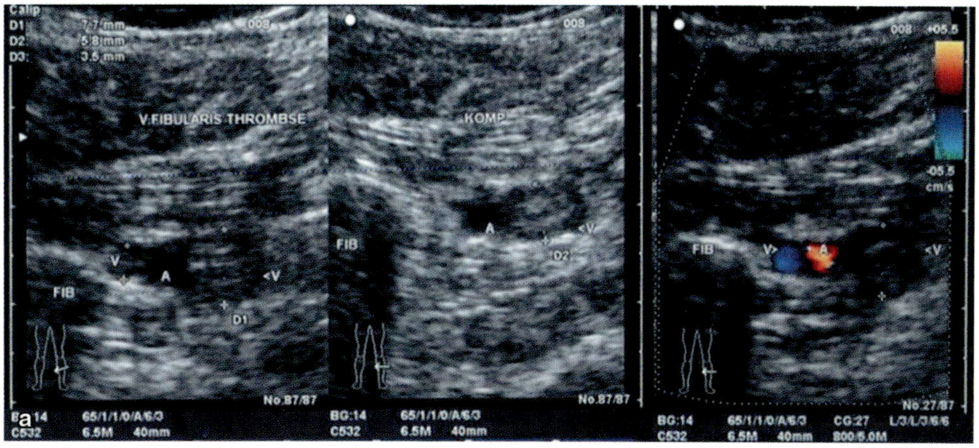


**a** Sonographic depiction of fresh thrombosis of a fibular vein with patency of the second fibular branch. The transverse view (*left section*) depicts round, tubular structures to the left and right of the artery and the fibula (*FIB*) to the left. The veins are indicated by *plus signs*; the thrombosed vein (right) is markedly dilated (7.7 mm), while the patent vein (left) has a diameter of 3.5 mm. The compression scan (*middle section*) fails to depict the vein to the left of the artery (*A*) because its lumen is completely compressed. The vein to the right shows only little compressibility (diameter reduced from 7.7 to 5.8 mm), indicating fresh thrombus. The color duplex scan (*right section*) depicts the patent vein in blue to the left of the artery (red), the thrombosed vein (*V*) to the right (marked with *plus signs*). The thrombosed vein is hypoechoic, markedly dilated, and shows no flow.

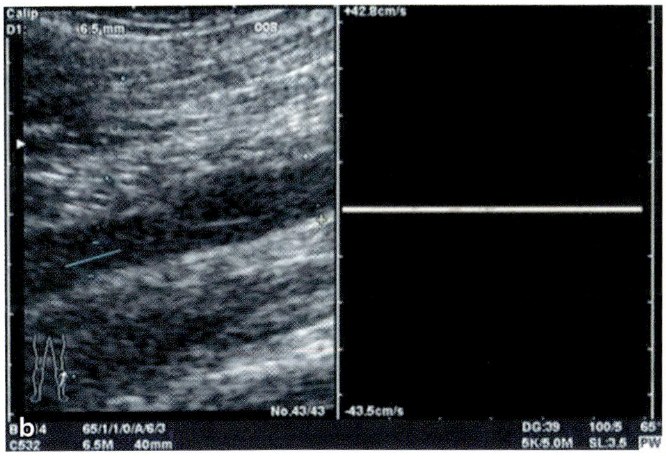


**b** Longitudinal duplex scan of the markedly dilated vein without flow signals in the Doppler waveform. The B-mode view shows a valve immobilized by clot.

**c** The venogram fails to depict the fibular veins. Based on the duplex sonographic demonstration of one patent and one thrombosed branch of the fibular vein, this example illustrates that absence of opacification may be due to thrombosis or technical limitations.

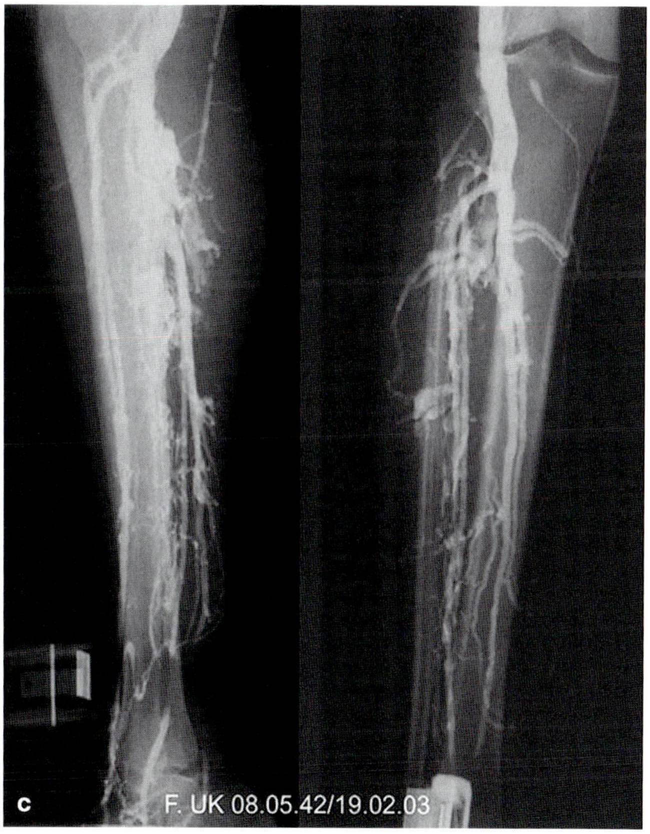

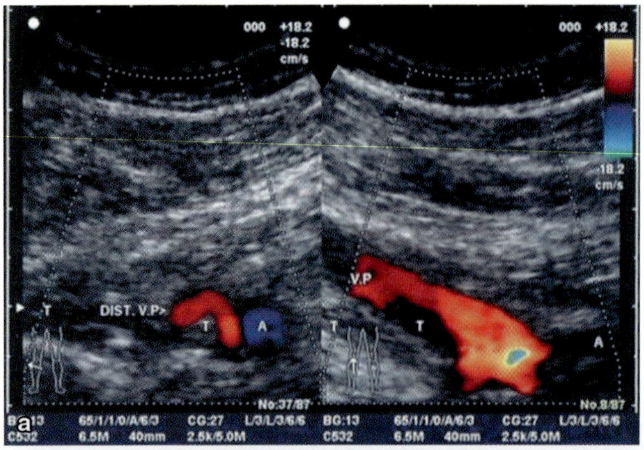


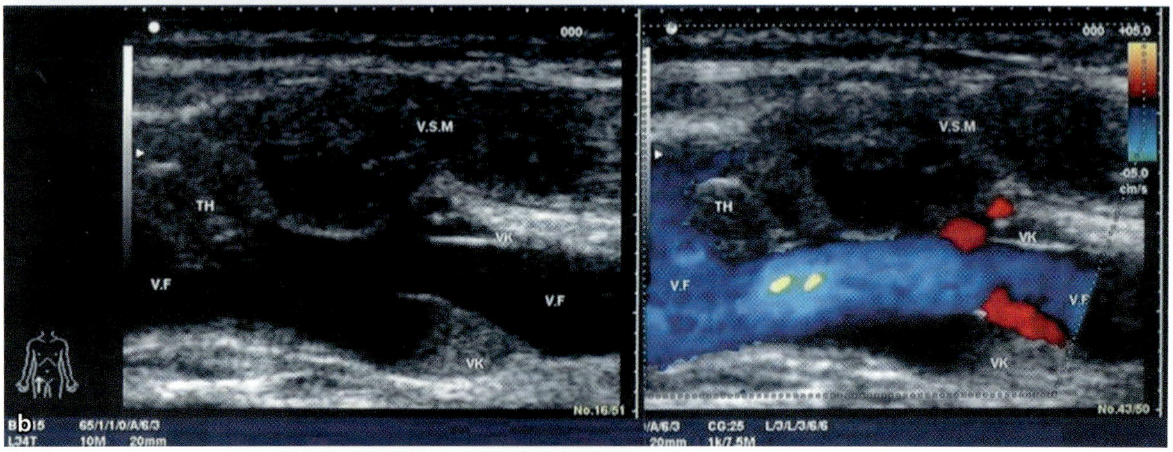


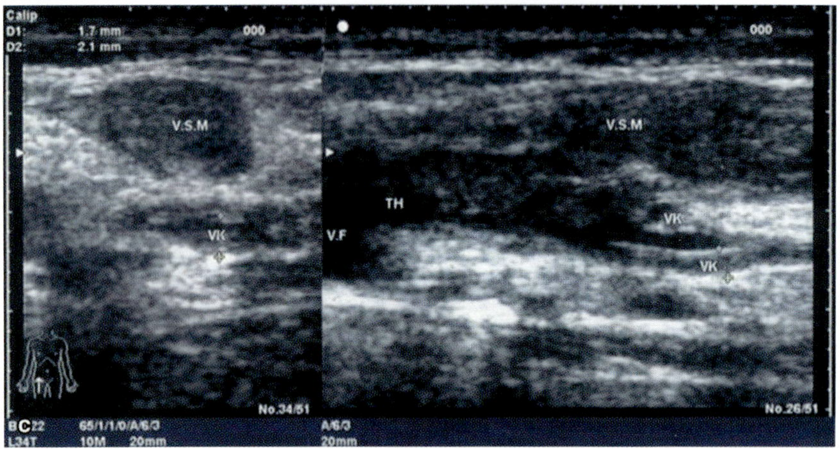


**Fig. A3.24 a–e**

**Asymptomatic venous thrombosis developing in valve cusps**

**a** Ultrasound has a markedly lower sensitivity in asymptomatic thrombosis than in symptomatic thrombosis. This is due to the fact that thrombus surrounded by flowing blood may be overlooked in calf vein segments notoriously difficult to scan, especially if there is only little dilatation and partial compressibility, or if clot is confined to valve pockets. The transverse (*left*) and longitudinal views (*right*) depict the distal popliteal vein with a patent lumen (red, flow toward transducer) but with absent color coding at the valve. Color duplex scanning facilitates the identification of such subtle abnormalities in problematic areas. However, to exclude flow phenomena as a possible cause of the filling defect, the thrombus must be confirmed by a compression scan of this vein segment.

**b** Duplex scanning performed in a clinically asymptomatic patient prior to stripping of varicose veins demonstrates thrombophlebitis of the great saphenous vein (*V.S.M*) with thrombus (*TH*) protruding into the common femoral vein (*V.F*). The gray-scale scan (*left*) shows a hyperechoic structure in a valve (*VK*) somewhat distal to the saphenofemoral junction. On the color duplex scan, absence of color coding indicates the thrombus (*TH*) including its valvular component, which prevents proper opening of the valve (despite augmented flow elicited by compression of the thigh/A-sound). Red color in the valve area indicates eddy flow (see Fig. 3.12a), particularly in the pocket of the valve (*VK*) depicted closer to the transducer. To rule out a flow-related cause of this subtle change in the color coding, the thrombus must be confirmed by a compression scan.

**c** The compression scans (transverse view on the *left* and longitudinal view on the *right*) show noncompressibility of the great saphenous vein (*V.S.M*) and incomplete compression of the femoral vein at the level of the thrombotic valve (residual diameter of 2 mm, see *markings*). The example illustrates two main sources of thrombosis of the major deep veins: thrombus development in a valve pocket (for its pathogenesis see Fig. 3.12a) and extension of thrombi from superficial or muscle veins.

**Fig. A3.24 (continued)**

**d** Thrombus in a venous valve pocket (illustrated for the great saphenous vein in the thigh) can lead to stasis of blood flow and thus become a nidus for venous thrombosis or thrombophlebitis. Absence of flow signals due to stasis can be differentiated from true thrombus using compression ultrasound or using color duplex imaging with a very low PRF (aliasing in the vein in *left section*) during Valsalva's maneuver or distal compression. In case of thrombosis, the valve leaflets (*VK*) will not move and Valsalva's maneuver will not elicit flow between the venous wall and the leaflet of the incompetent valve (*right section*).

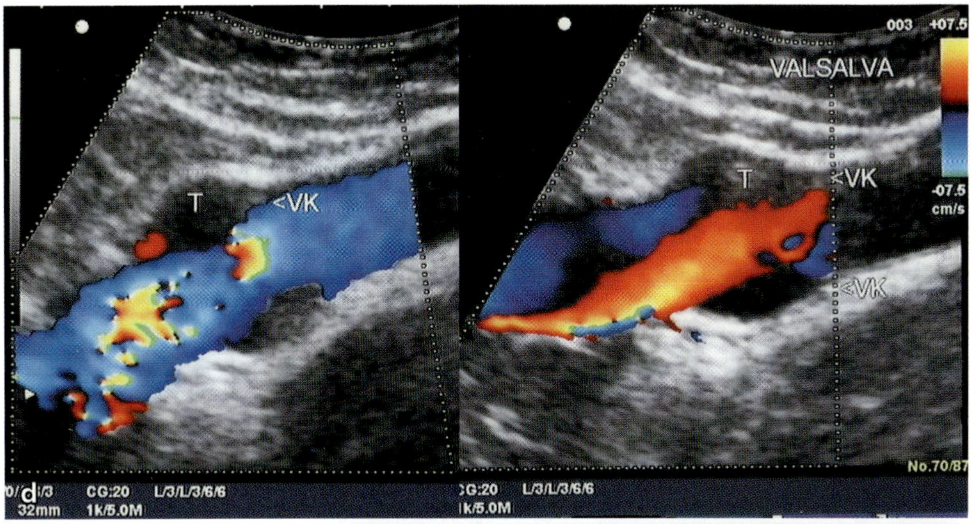


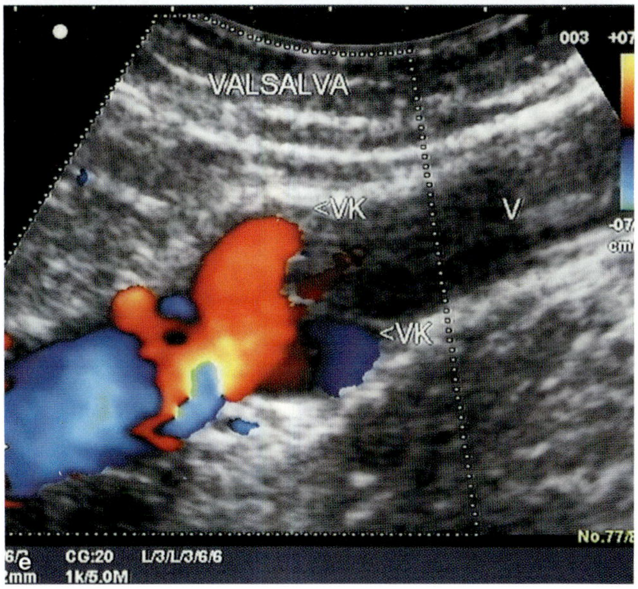


**e** Adequate valve closure. The example shows an incompetent saphenofemoral junction with an incompetent arch vein, while function of the great saphenous vein in the thigh is normal. The proximal valves are incompetent, and the scan depicts the first competent valve, indicated by adequate closure with Valsalva's maneuver. Retrograde flow causes flow signals extending into the valve pockets during closure (distal point of insufficiency). There is no flow distally, except for a minimal, thin stream coded in red and indicating minimal leakage of the valve; this is no evidence of relevant valve incompetence (*VK* = valve leaflet).

**Fig. A3.25 a–d**

**Deep femoral vein thrombosis**

**a** The deep femoral veins (*V.P.F*) join the superficial femoral vein on its posterolateral aspect. Deep femoral vein thrombosis is rare but can cause pulmonary embolism if there is ascending thrombus protrusion into the common femoral vein. The color duplex scan depicts the superficial femoral vein (*V.F.S*) with flow toward the transducer on the *right* (red) and with flow away from it on the *left* (blue). Several branches of the deep femoral vein (*V.P.F*) enter the superficial femoral vein posteriorly. Neither color duplex nor the Doppler spectrum depicts flow in the deep femoral vein joining the superficial femoral vein distally. The mural thrombus protrudes into the superficial femoral vein at the junction with the common femoral vein. The second deep femoral vein entering more proximally is not thrombosed (at *left* margin).

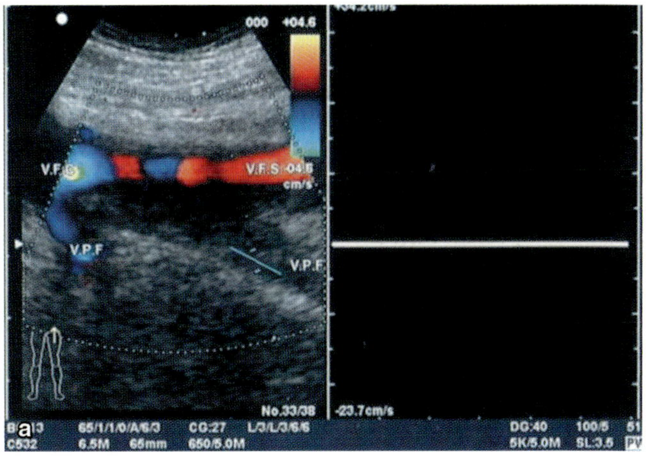

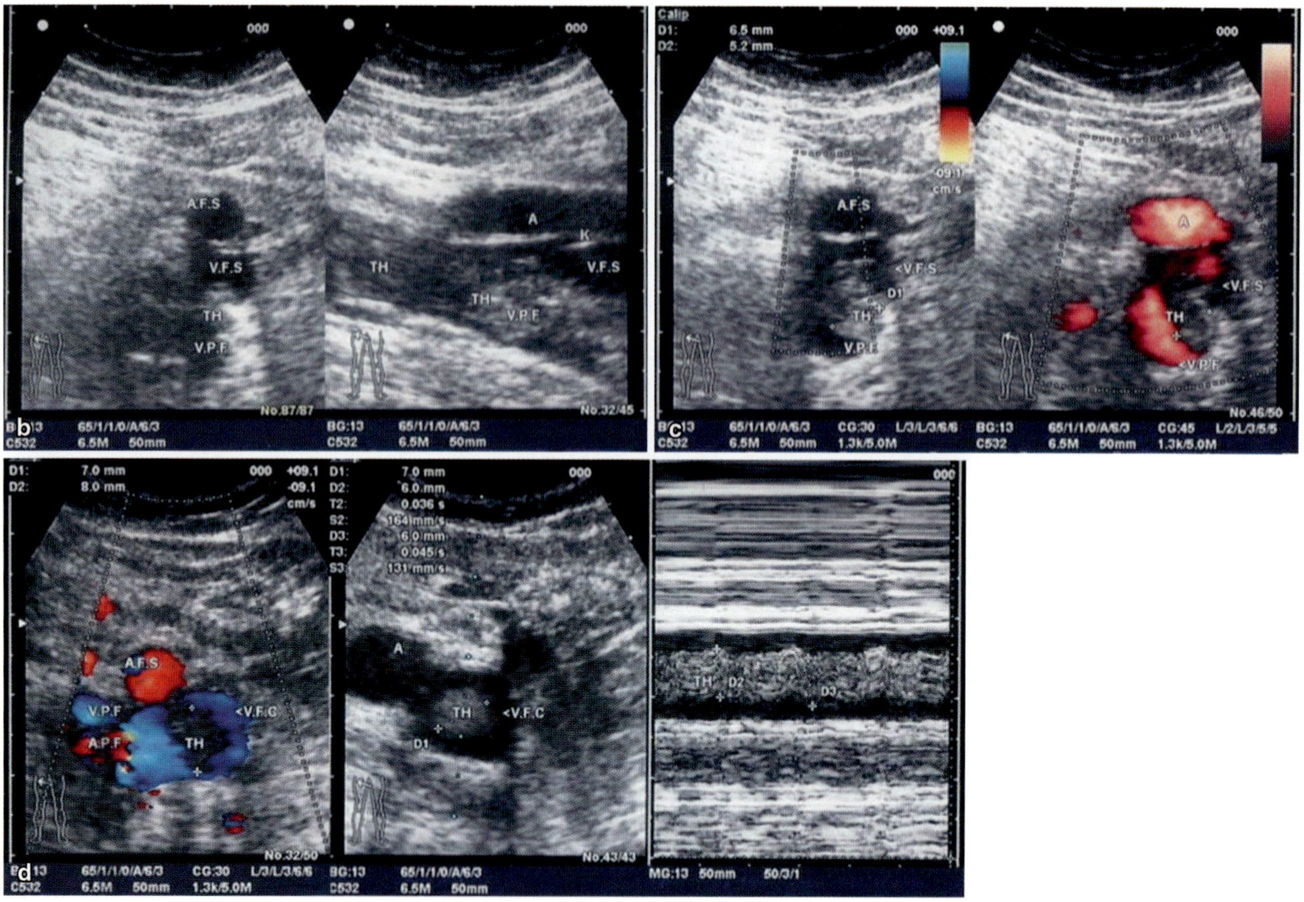


**Fig. A3.25 (continued)**

**Deep femoral vein thrombosis with thrombus growth into the common femoral vein**

**b** The transverse scan (*left*) and longitudinal scan (*right*) depict the thrombus (*TH*) growing appositionally from the deep femoral vein (*V.P.F*) into the common femoral vein. The superficial femoral vein (*V.F.S*) is not thrombosed. (*A.F.S* superficial femoral artery; *K* venous valve).

**c** Transverse view of the femoral venous confluence at a slightly higher level. The power mode (*right scan*) shows the thrombus extending from the deep femoral vein (*V.P.F*) to be partially surrounded by flowing blood. In the common femoral vein, the blood from the patent superficial femoral vein (*V.F.S)* flows along the free-floating thrombus tail anteriorly. The superficial femoral artery (*A*) is depicted above.

**d** Depiction of the free-floating thrombus in the common femoral vein at the level of the femoral artery bifurcation (*left section*) and 1 cm above it (*middle section*). The color duplex scan depicts a rim of blue-coded flow signals around the thrombus (*TH*).

The time-motion display (*right section*) nicely shows the floating movement of the thrombus in the bloodstream. (*A.F.S* = superficial femoral artery, *A.P.F* = deep femoral artery, *V.F.C* = common femoral vein, *V.P.F* = cranial branch of deep femoral vein joining the common femoral vein between the deep and superficial femoral arteries).

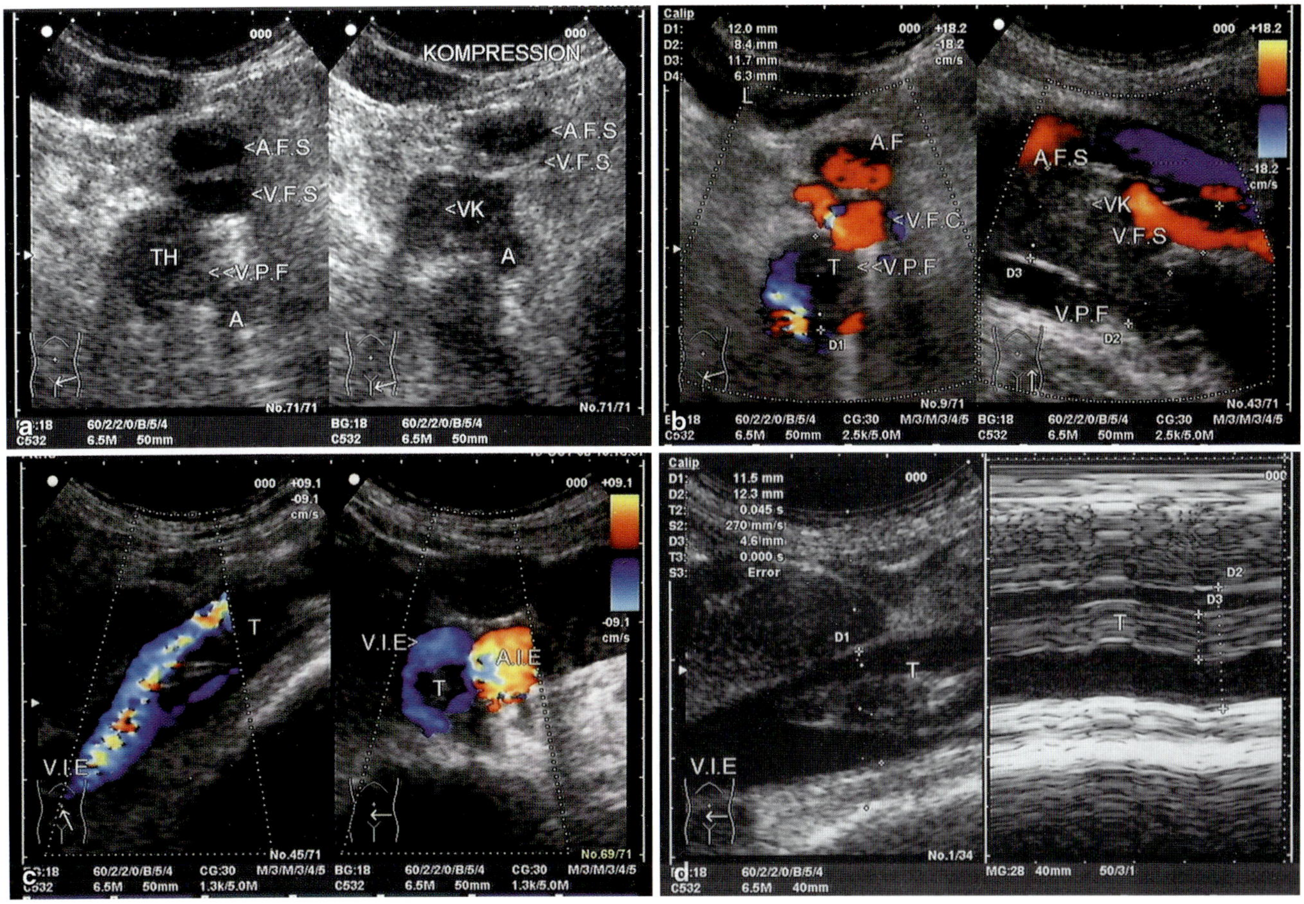


**Fig. A3.26 a–d**

**Pelvic vein thrombosis secondary to deep femoral thrombosis**

**a** Compression ultrasound (transverse plane) shows compressibility of the superficial femoral vein (*V.F.S*; *right section*), while the deep femoral vein is not compressible (*TH* in *V.P.F*).

**b** Color flow images (transverse section on the *left* and longitudinal section on the *right*) show flow in the superficial femoral vein (red) and no thrombosis; the deep femoral vein (*V.P.F*) joins the superficial vein posteriorly. Also depicted are the superficial femoral artery (*A.F.S*) anterior to the vein and the profunda femoris artery posteriorly.

**c** The thrombus (*T*) extends into the external iliac vein (*V.I.E*) and is shown to be surrounded by flowing blood on the color flow image.

**d** The time-motion display documents floating of the long thrombus (*T*) in the external iliac vein.

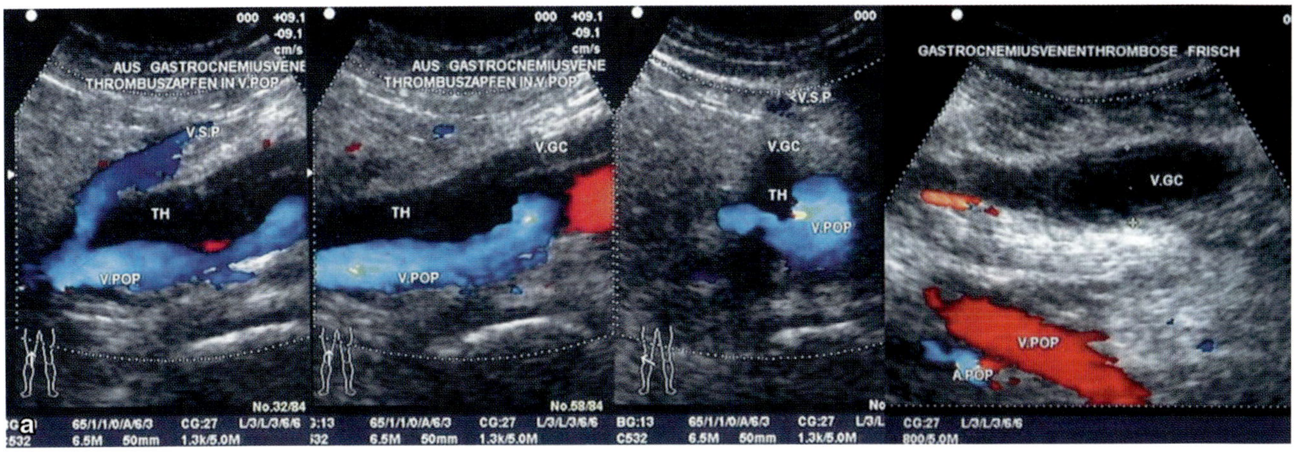


**Fig. A3.27 a, b**

**Calf muscle vein thrombosis**

**a** The two longitudinal views (*leftmost* and *left center*) and the transverse view (*right center*) show a gap (*TH*) in the color-coded flow in the popliteal vein (*V.POP*). An ascending thrombus (*TH*) protrudes into the popliteal vein from a thrombosed gastrocnemius vein (*V.GC*). More cranially, the small saphenous vein (*V.S.P*) is depicted with blood flow in blue. The mural thrombosis ascending from the gastrocnemius vein into the popliteal vein ends at the saphenofemoral junction (*leftmost* and *left center*). The gastrocnemius vein thrombosis cannot be traced further distally (*rightmost section*).

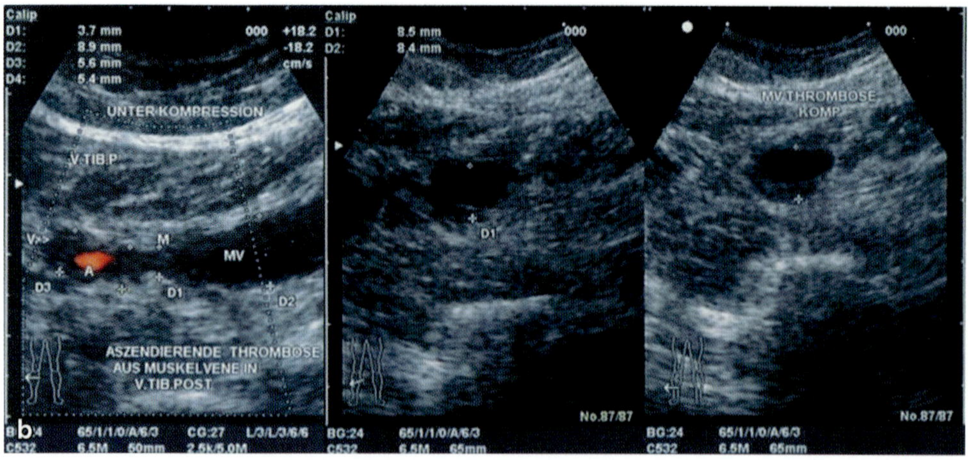


**Fig. A3.27 (continued)**

**b** Muscle vein thrombosis of the lower leg is suggested by the depiction in the soleus or gastrocnemius muscle of hypoechoic tubular structures that cannot be compressed. The veins are markedly dilated, which makes them much easier to identify by ultrasound than normal muscle veins.

The distinction between muscle vein thrombosis and thrombosis of a major calf vein is made sonoanatomically. The major veins run parallel to the lower leg arteries of the same name. The transverse scan (*middle section*) depicts a hypoechoic structure in the soleus muscle. Noncompressibility of the vein confirms muscle vein thrombosis (*right section*). The oblique color duplex scan on the left depicts the thrombosed soleus vein (*MV*, labeled as D2) on its course from the mid-calf to the knee, where it enters (labeled as D1) the posterior tibial vein. The latter is likewise thrombosed up to the tibiofibular junction with appositional thrombus growth, but somewhat distal to the entry site of the muscle vein it is compressible. The scan on the *left* was obtained during compression and depicts the hypoechoic, noncompressible posterior tibial veins (labeled as D3 and D4) to the left and right of the posterior tibial artery (red).

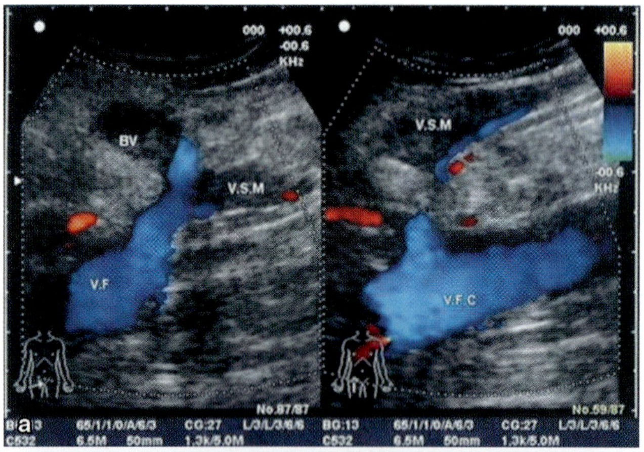


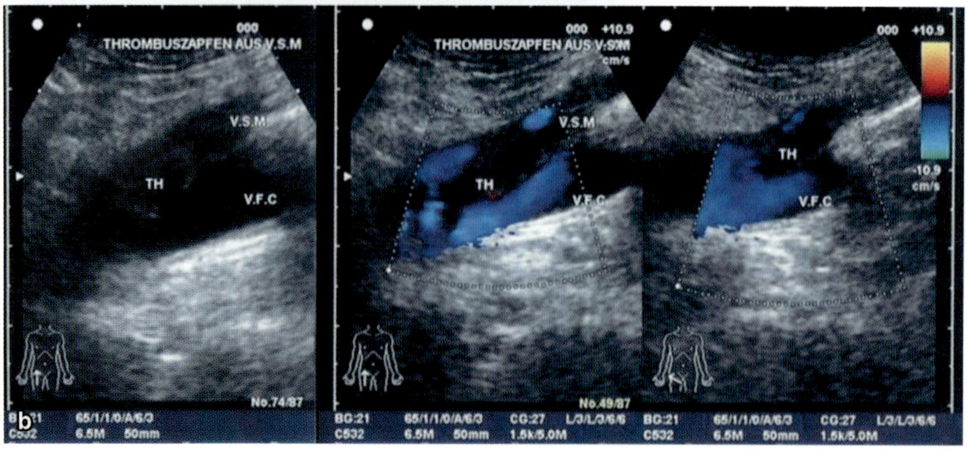


**Fig. A3.28 a–c**

**Thrombophlebitis**

**a** The proximal extent of thrombophlebitis may be greater than suggested by the clinical findings. The patient shown presented with reddening along the course of the great saphenous vein up to the mid-thigh, while color duplex scanning (transverse view on the *left* and longitudinal view on the *right*) demonstrates gaps in the color coding extending up to 1.5 cm below the saphenofemoral junction. The longitudinal view depicts flow in blue along the thrombus. Ultrasound also demonstrates thrombophlebitic involvement of the clinically normal anterior tributary vein (*BV*). Surgical ligation is indicated to prevent further thrombus growth into deep veins.

**b** Ascending thrombophlebitis can extend into a deep vein in the form of a cone-shaped thrombus. The B-scan (*left section*) already depicts a slightly more hyperechoic thrombus (*TH*) protruding into the anechoic lumen of the common femoral vein (*V.F.C*) from the great saphenous vein (*V.S.M*). On the color duplex scan (*right*), the thrombus (*TH*) protruding into the common femoral vein is identified by the absence of color in the blue-coded lumen.

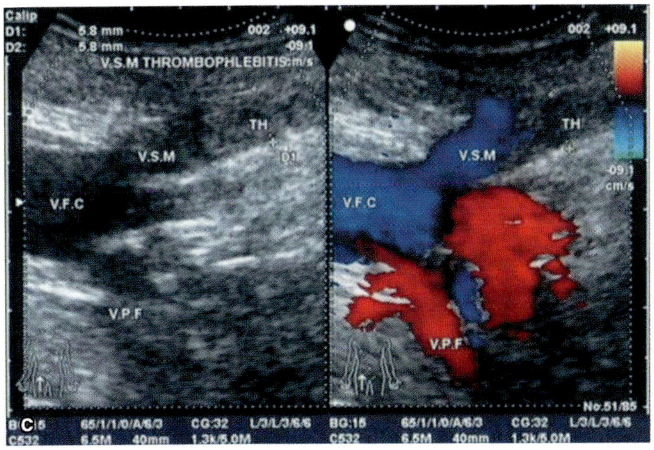


**Fig. A3.28 (continued)**

**c** Based on the duplex findings, high ligation of the great saphenous vein is indicated but was refused by the patient. In this case, the course of endogenous thrombolysis under heparin therapy can thus be followed. After three weeks, the thrombus in the great saphenous vein has receded to 1 cm below the junction. The scan on the *left* demonstrates the thrombus (*TH*) in the lumen of the great saphenous vein (*V.S.M*). The color scan on the *right* depicts flow in the great saphenous vein in blue (away from transducer, toward center) and a branch of the deep femoral vein (*V.P.F*) coming from posteriorly with flow toward the transducer coded in red.

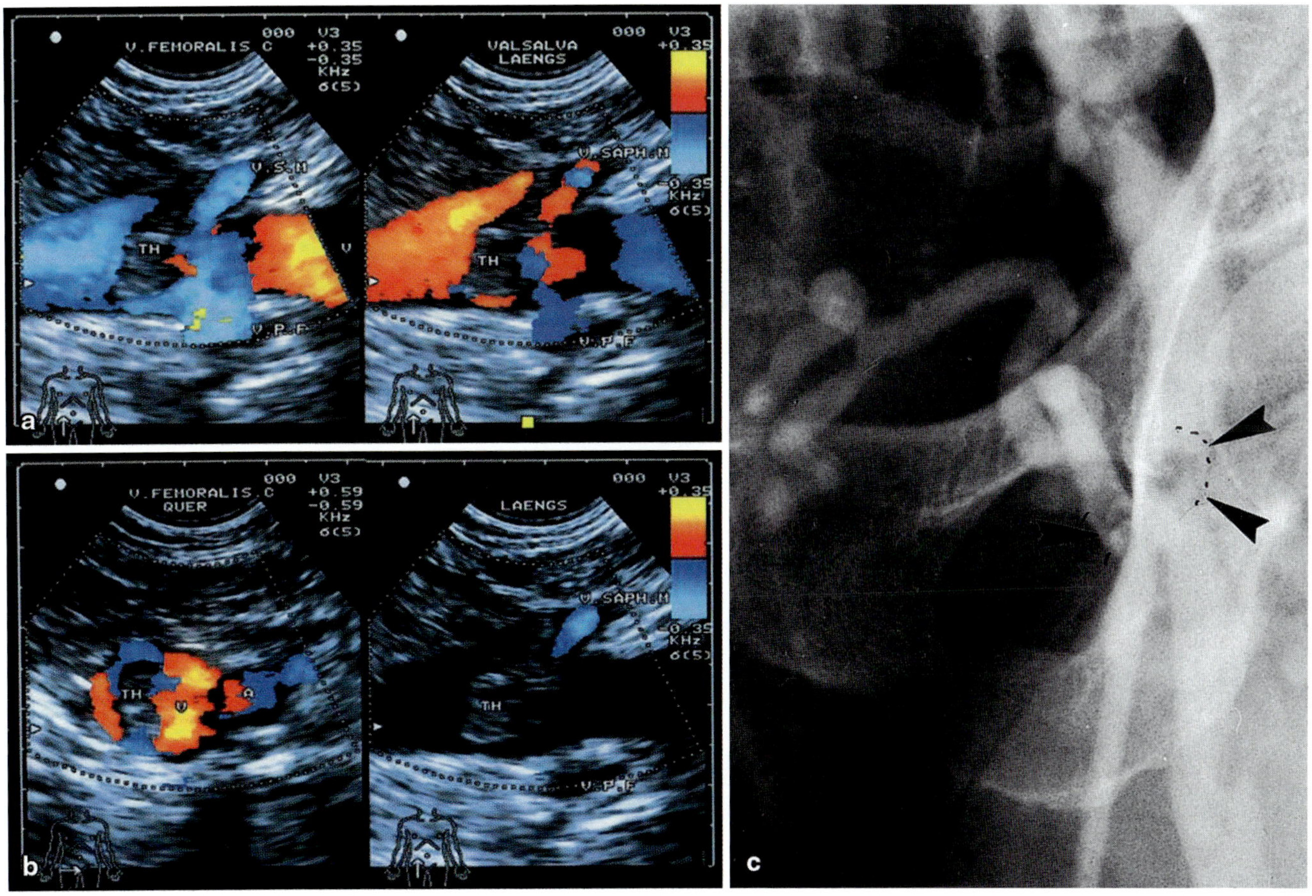


**Fig. A3.29 a–c**

**Thrombus protrusion from superficial vein**

**a** Free-floating thrombus (*TH*) protruding into the common femoral vein from a tributary vein entering the saphenofemoral junction. The scan on the *left* depicts the common femoral vein with flow toward the heart (blue). The scan on the *right* obtained during Valsalva's maneuver shows retrograde flow extending into the common femoral and great saphenous veins (color change from blue to red). The great saphenous vein is patent below the terminal valve (depicted with flow in blue close to the transducer at the right margin of both scans).

**b** The longitudinal (*right*) and transverse (*left*) views obtained during Valsalva's maneuver show the thrombus protruding into the common femoral vein to be surrounded by flowing blood on all sides. During inspiration, there is zero flow without flow signals in the common femoral vein and only some residual, obstructed flow in the patent great saphenous vein. The extent of the floating tail is clearly seen as there is no color spillover.

**c** Venogram: Confirmation of the protruding thrombus.

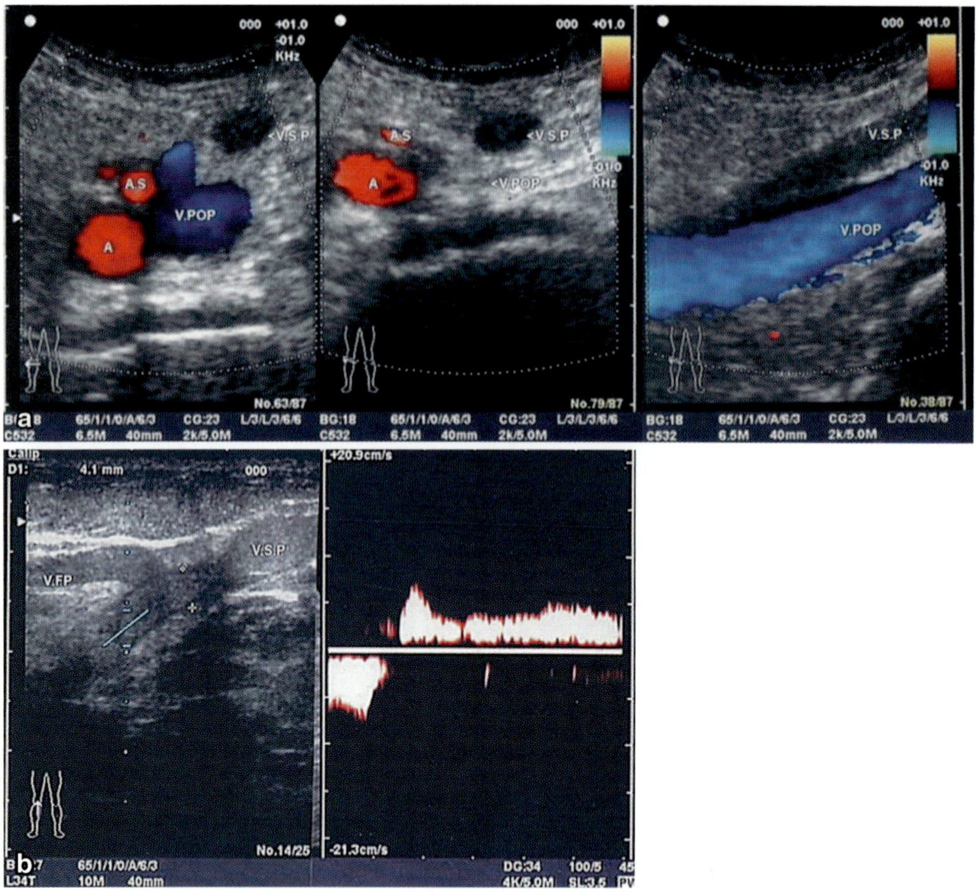


**Fig. A3.30 a, b**

**Thrombophlebitis of small saphenous vein**

**a** Patients with thrombophlebitis of the small saphenous vein often present with unspecific clinical symptoms that may mimic deep vein thrombosis. For this reason, diagnostic evaluation of patients for exclusion of deep vein thrombosis must also include the small saphenous vein. The transverse view on the left depicts the small saphenous vein (*V.S.P*) as a nonperfused hypoechoic tubular structure posterior to the popliteal vein (*V.POP*). The compression scan (*middle section*) shows noncompressibility of the vein. The longitudinal scan (*right section*) depicts the small saphenous vein (*V.S.P*) without flow to the level of the saphenopopliteal junction. There is no thrombus extension into the popliteal vein (*V.POP*).

**Femoropopliteal vein**

**b** The femoropopliteal vein (*V.FP*) passes posteriorly from the small saphenous vein (*V.S.P*, dilated by fresh thrombus) just below the saphenopopliteal junction. Despite thrombophlebitis of the small saphenous vein distal to the site of entry of the femoropopliteal vein, the proximal valve function test demonstrates reflux at the saphenopopliteal junction due to femoropopliteal valve incompetence.

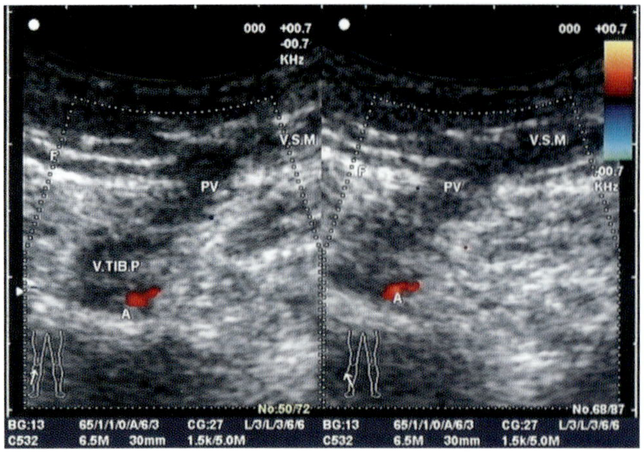


**Fig. A3.31**

**Thrombosis arising from thrombophlebitis extending through perforator**

Extension of thrombophlebitis into the deep venous system can also occur through a perforating vein. In the case presented, extensive thrombophlebitis of the great saphenous vein (*V.S.M*) gives rise to a thrombus extending through a perforating vein (*PV*) into the posterior tibial vein (*V.TIB.P*), where it causes a circumscribed thrombosis 3 cm in length. Next to the vein, the artery is depicted with flow in red. The great saphenous, perforating, and posterior tibial veins are markedly dilated by the thrombus and not compressible (*right section*). The hyperechoic reflection indicates the site at which the vein pierces the fascia (*F*).

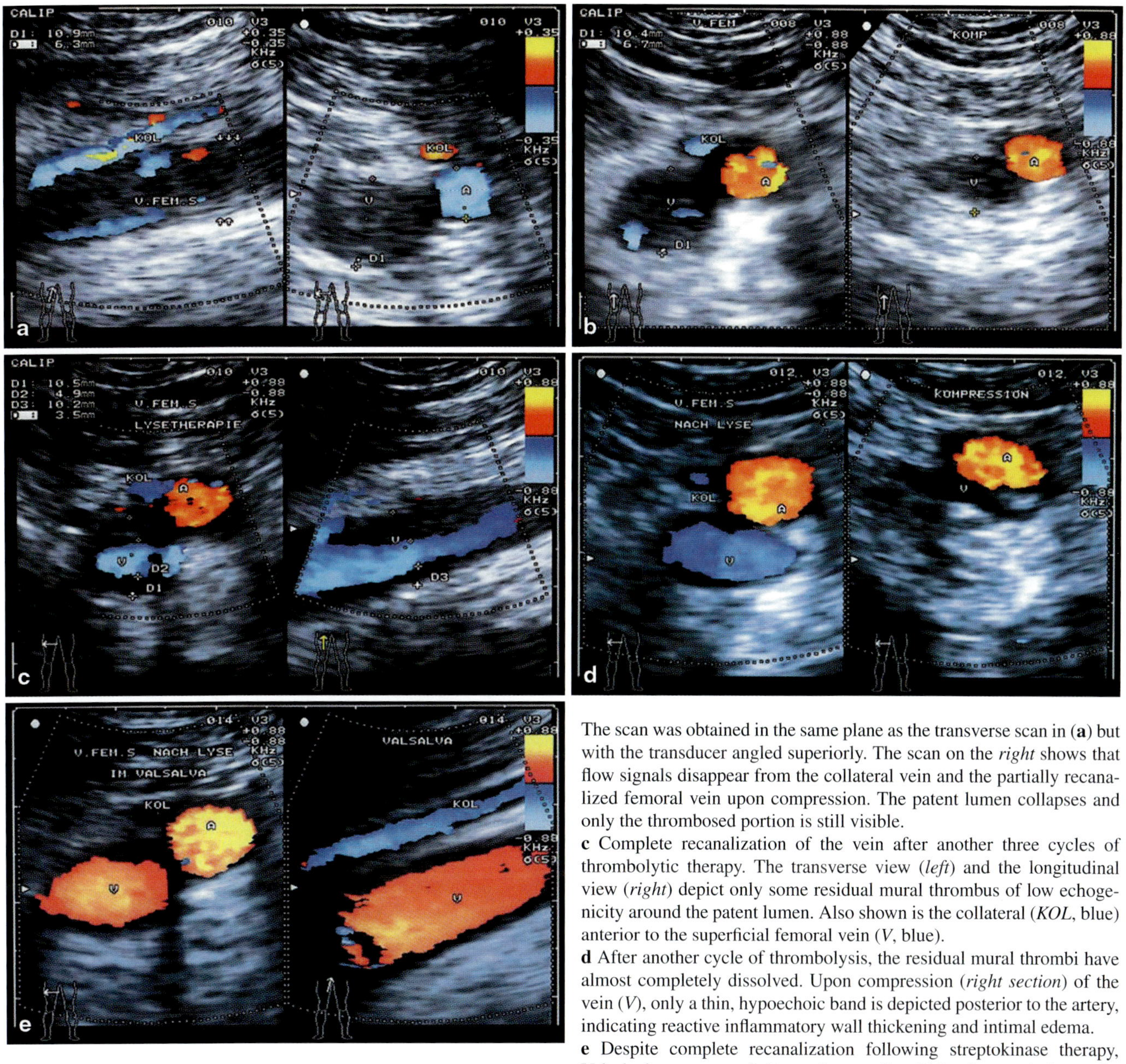


**Fig. A3.32 a–e**
**Monitoring of thrombolytic therapy**
**a** Marked dilatation of the superficial femoral vein (compared with the corresponding artery) and the hypoechoic, homogeneous thrombus with a just barely visible hypoechoic halo suggest acute thrombosis. The transverse view depicts a collateral (*KOL*) with flow in red anterior to the superficial femoral artery (*A*). The longitudinal view on the *left* shows a more proximal segment of the superficial femoral vein (*V*). The thrombus proximal to the site of entry of a collateral vein is surrounded by residual flow near the walls (blue).
**b** Both central and peripheral segments of the thrombosed superficial femoral vein (*V*) show flow coded in blue as a sign of beginning recanalization after three cycles of thrombolytic therapy with streptokinase. The scan was obtained in the same plane as the transverse scan in (**a**) but with the transducer angled superiorly. The scan on the *right* shows that flow signals disappear from the collateral vein and the partially recanalized femoral vein upon compression. The patent lumen collapses and only the thrombosed portion is still visible.
**c** Complete recanalization of the vein after another three cycles of thrombolytic therapy. The transverse view (*left*) and the longitudinal view (*right*) depict only some residual mural thrombus of low echogenicity around the patent lumen. Also shown is the collateral (*KOL*, blue) anterior to the superficial femoral vein (*V*, blue).
**d** After another cycle of thrombolysis, the residual mural thrombi have almost completely dissolved. Upon compression (*right section*) of the vein (*V*), only a thin, hypoechoic band is depicted posterior to the artery, indicating reactive inflammatory wall thickening and intimal edema.
**e** Despite complete recanalization following streptokinase therapy, Valsalva's maneuver elicits persistent reflux. Valve damage is due to the delay of more than 10 days between the onset of thrombosis and complete recanalization. The scan on the *left* demonstrates blood flow in the same direction (coded red) in the vein (*V*) and the corresponding artery (*A*). The scan on the *right* shows flow toward the heart (blue) in the competent collateral vein (*KOL*) during Valsalva's maneuver. This forward flow in the competent collateral is induced by the calf muscle pump because some patients inadvertently also contract their muscles during the Valsalva maneuver.

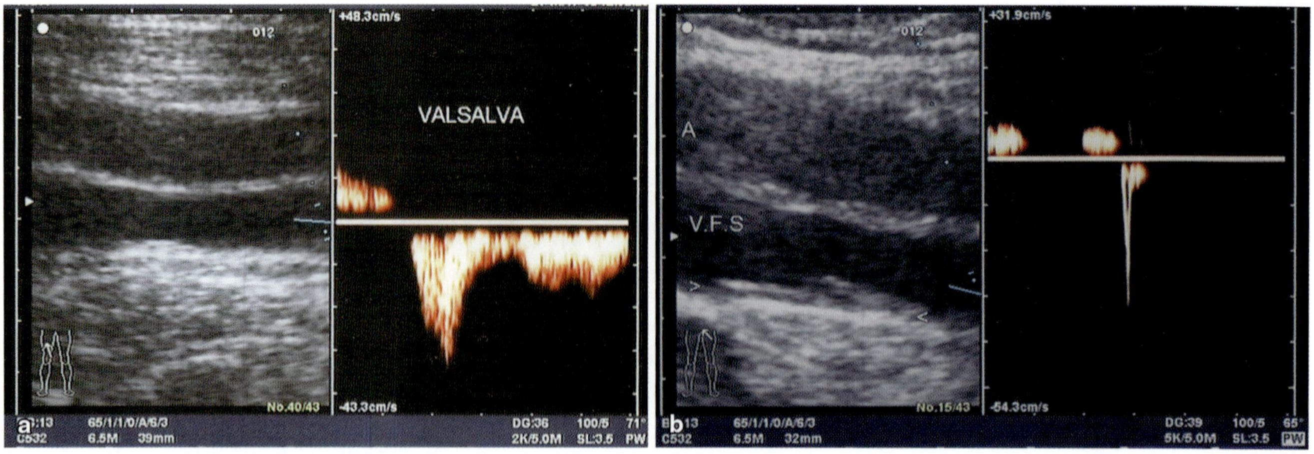


**Fig. A3.33 a, b**

**Postthrombotic syndrome – valve function**

**a** The severity of insufficient venous drainage depends on the degree of recanalization and the development of postthrombotic valve incompetence of major veins. If there is complete recanalization, the veins may appear perfectly normal on B-mode ultrasound with valvular dysfunction being the only sequela of postthrombotic valve damage.

**b** Conversely, there may be normal function of individual venous segments, which will prevent reflux, even if B-mode scans show vascular wall changes (sclerosis, thickening). In the example, Valsalva's maneuver elicits only a short reflux before valve closure (Doppler waveform) although the B-mode scan demonstrates postthrombotic wall thickening.

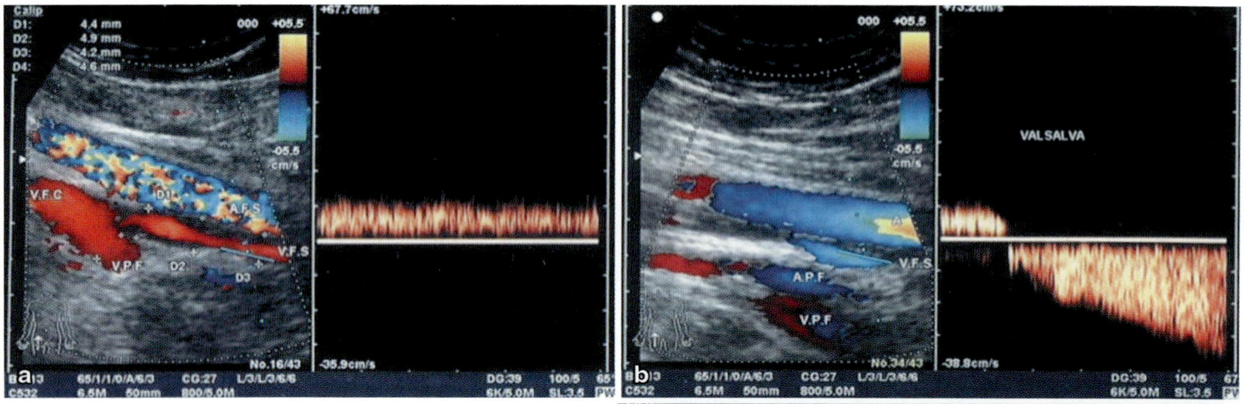


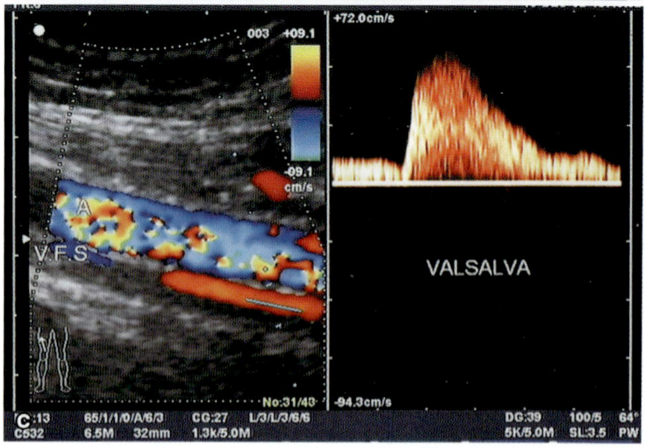


**Fig. A3.34 a–e**

**Postthrombotic syndrome – recanalized lumen**

**a** In about 10% of cases, thrombosis leads to permanent damage of the vein, depicted sonographically as a hypoechoic, tubular strand with a thin caliber adjacent to the artery (as in Fig. A3.15). In most cases, however, there is postthrombotic recanalization but often with a smaller lumen. In the example shown, the superficial femoral vein is patent 4 months after thrombosis, but only trickling flow is present. Hypoechoic thrombotic wall deposits and sclerotic wall lesions persist. Aliasing in the superficial femoral artery closer to the transducer confirms the PRF to be adequate for the detection of slow venous flow. The continuous flow signal resulting from the loss of respiratory phasicity indicates persistent flow obstruction in the recanalized vein.

**Postthrombotic syndrome – paradoxical flow during Valsalva's maneuver**

**b** Flow in the superficial femoral vein (*V.F.S*) during Valsalva's maneuver is coded in blue (away from transducer), and the Doppler waveform shows flow reversal.

**c** When Valsalva' maneuver produces increased flow rather than flow reversal in a recanalized vein (here the superficial femoral vein), this indicates flow through dilated collaterals. In the example, Valsalva's maneuver induces blood flow from the incompetent great saphenous vein into the femoral vein via incompetent perforating veins. The resulting flow increase in the superficial femoral vein shown in the Doppler waveform suggests that the vein is only partially recanalized and that there is persistent severe obstruction of venous drainage toward the heart.

**Fig. A3.34 (continued)**
**d** In more distal, partially recanalized vein segments such as the popliteal vein (distal to the Dodd perforators, through which the blood enters the deep system), Valsalva's maneuver induces typical to-and-fro flow with flow reversal.
**e** Valsalva's maneuver reveals severe terminal valve incompetence of the great saphenous vein (reflux in the Doppler waveform).

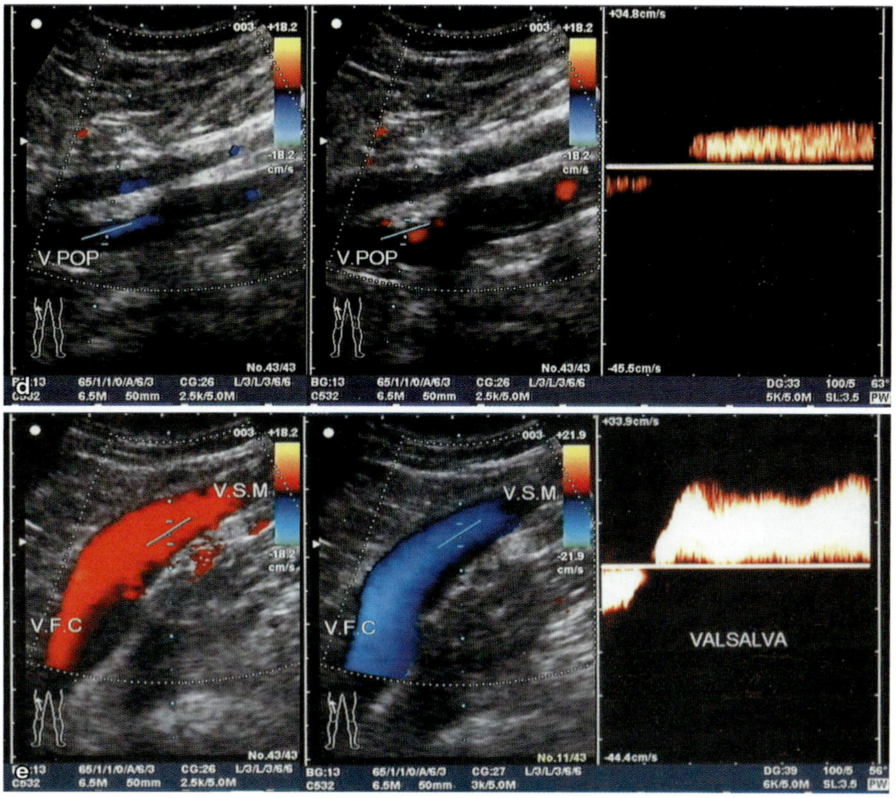


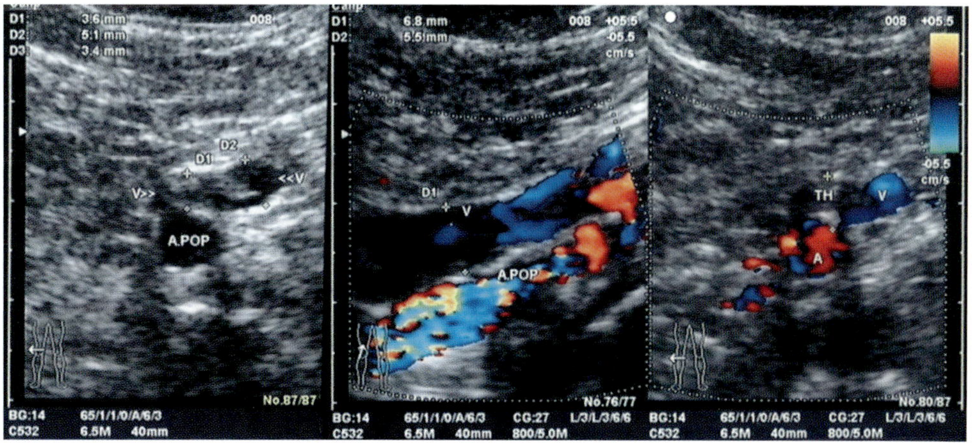


**Fig. A3.35**
**Partial recanalization**
Only meandering flow is detectable between residual thrombi (*TH*) in the popliteal vein (*V*) in the longitudinal color flow scan in the *middle* (with aliasing in the popliteal artery due to the low PRF). The transverse gray-scale scan (*left*) depicts hyperechoic residual thrombi adjacent to the anechoic recanalized lumen, which shows flow displayed in blue in the color duplex image (*right*).

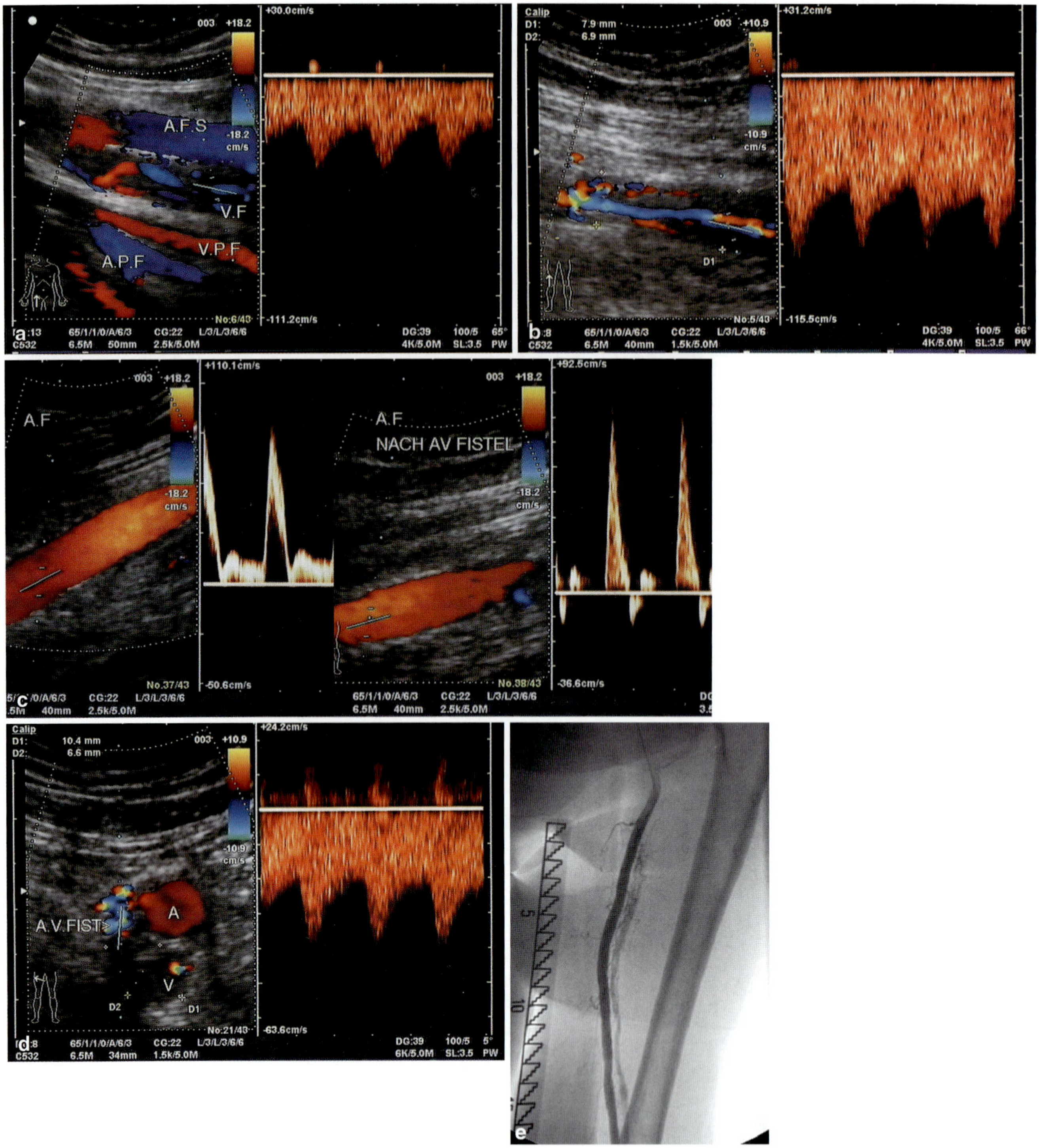


**Fig. A3.36 a–e**
**Postthrombotic recanalization of vein with arteriovenous fistula**
**a** Patient with venous thrombosis of the thigh showing the typical signs of early recanalization on color flow imaging after 4 months: meandering flow and flow signals mostly confined to the center of the vein. The Doppler waveform obtained from the partially recanalized vein shows retrograde pulsatile flow. A possible cause is an arteriovenous (AV) fistula; in this patient, the flow toward the periphery is due to occlusive thrombosis proximally.
**b** In the distal femoral vein, color duplex ultrasound also shows signs of recanalization with residual mural thrombus, with the Doppler waveform demonstrating high-frequency flow toward the periphery.
**c** To search for the AV fistula, the length of the femoral artery is scanned from proximal to distal with continuous Doppler recording. A sudden change to more pulsatile flow indicates the site where to look for the AV fistula. The Doppler waveform on the *left* was obtained in the femoral artery, upstream of the AV fistula and the one on the *right* downstream of the fistula.
**d** Transverse scan of the AV fistula between the superficial femoral artery (*A*) and the femoral vein (*V*). The sample volume is placed in the fistula, and the Doppler waveform shows the typical pulsatile flow pattern of a fistula; however, the frequency is lower than expected. The femoral vein is still largely thrombosed, but some flow is present, suggesting recanalization (next to the *V*). The communication between the AV fistula and the recanalized venous lumen is not visualized because it does not lie in the scan plane.
**e** Angiogram simultaneously depicts the artery and a thin stream in the vein with flow directed toward the periphery. The preceding color duplex examination provides the explanation for this phenomenon.

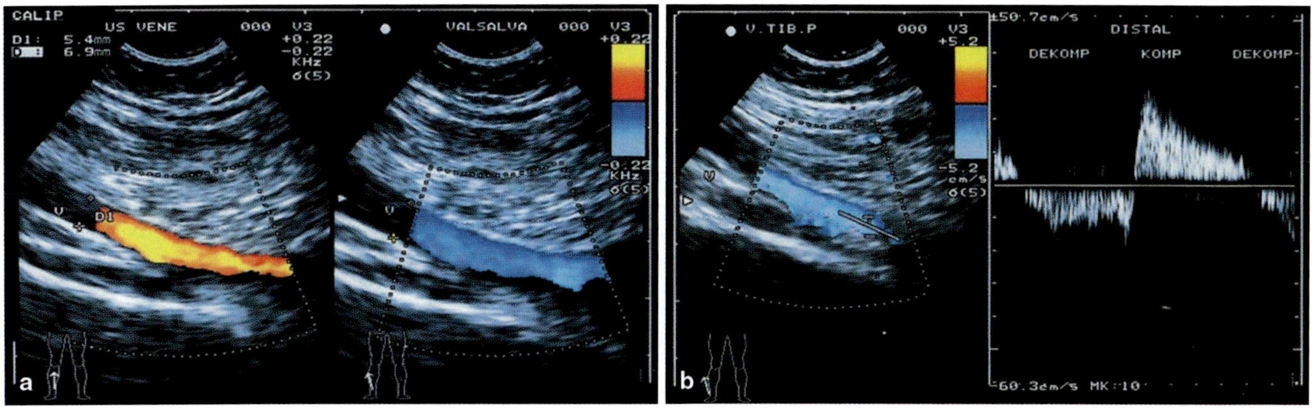


**Fig. A3.37 a, b**

**a** The postthrombotic syndrome is distinguished from primary chronic venous insufficiency of the deep leg veins, in which valve failure is due to venous dilatation. The walls are delicate and free of deposits and therefore easy to compress. In the example shown, valve incompetence in the proximal posterior tibial vein is associated with persistent reflux during Valsalva's maneuver, indicated by the color change from red to blue.

In case of severe dysfunction of all venous valves proximal to the transducer, even deep abdominal inspiration can induce a reflux, and normal rhythmical inspiration and expiration may induce to-and-fro flow.

**Valve incompetence of calf veins**

**b** The duration of reflux can be determined in the Doppler frequency spectrum, which thus enables differentiation of short physiologic reflux prior to valve closure from persistent reflux due to incompetent valves. Blue indicates reflux in the posterior tibial vein away from the transducer. Repeated and somewhat longer compression and release lead to alternating flow toward the transducer during compression (*KOMP*) and away during release (*DEKOMP*).

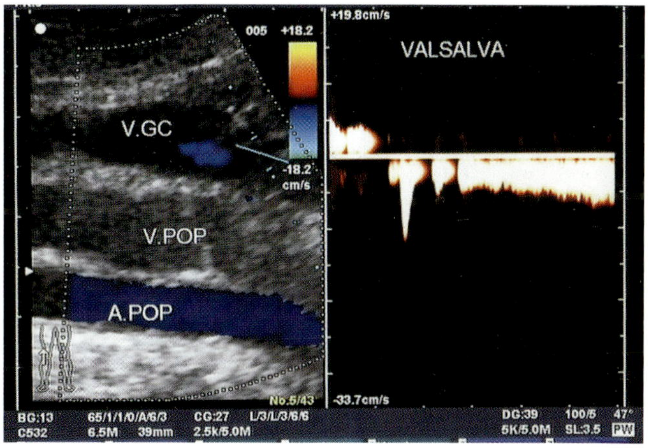


**Fig. A3.38**

**Dilated muscle veins**

Patient with crural ulcer but without signs of insufficiency of the great saphenous vein in the thigh. There is valve incompetence of the superficial femoral vein and the proximal popliteal vein with tight valve closure in the major veins distally. Valsalva's maneuver reveals valve incompetence in the form of persistent reflux (Doppler waveform) in a dilated gastrocnemius vein (*V.GC*). Color flow imaging (*left*) shows no flow in the popliteal vein (*V.POP*) with Valsalva's maneuver, indicating competent valves. The crural ulcer in this patient was caused by incompetent indirect perforating veins (not shown) and healed after elimination of the incompetent perforators identified by ultrasound (several weeks of prior compression therapy had no effect).

Such dilated gastrocnemius and soleus veins can cause stasis of blood flow, giving rise to calf thrombosis with extension into the popliteal vein.

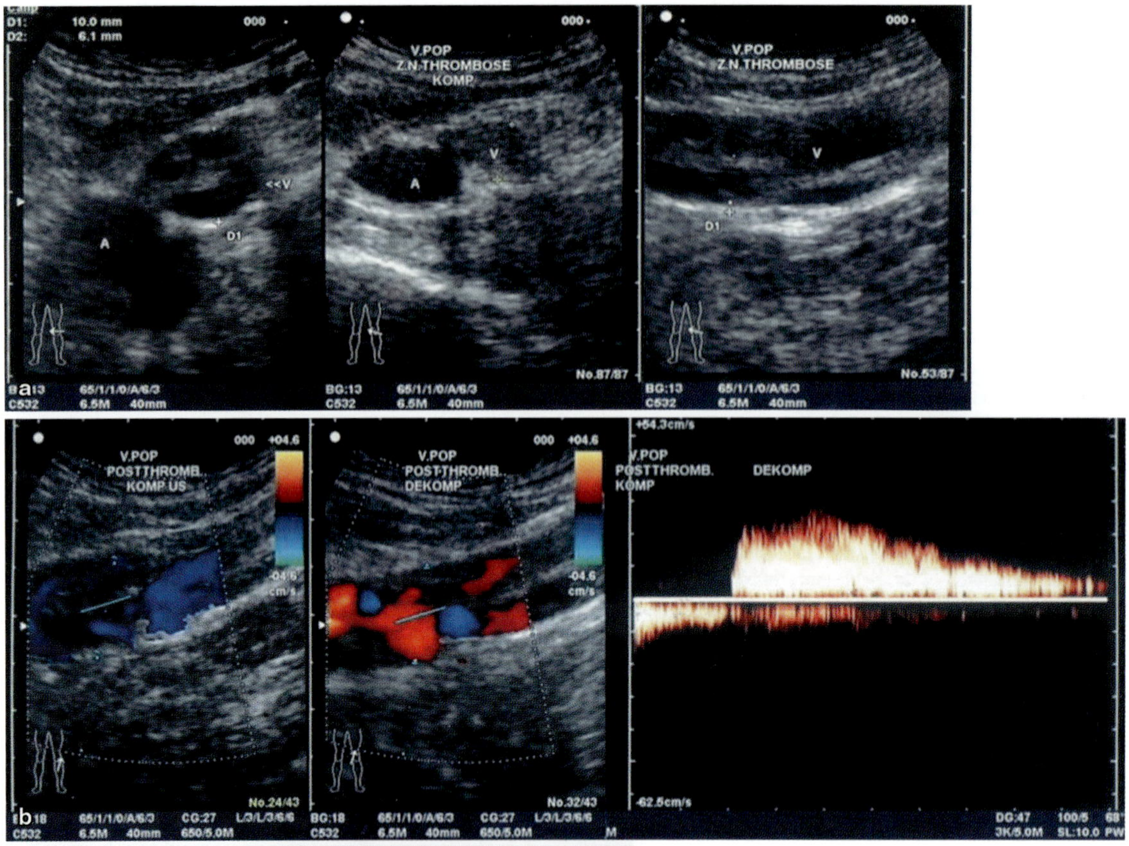


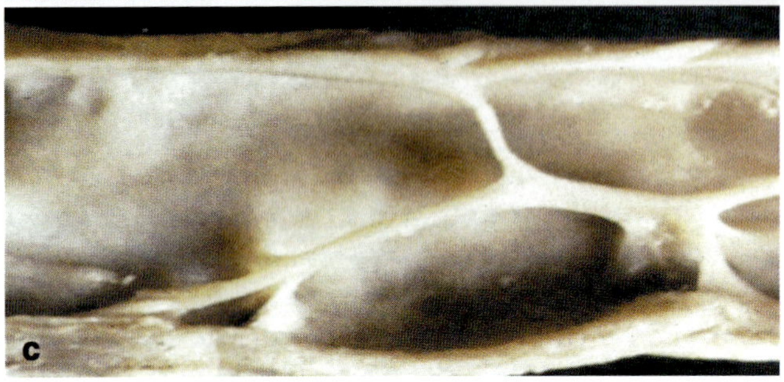


**c** Postthrombotic recanalization of the vein with extensive wall sclerosis and residual strands.

**Fig. A3.39 a–g**
**Postthrombotic syndrome – residual lesions/synechia**
**a** The compression scan (*middle section*) shows an inhomogeneous, mostly hypoechoic structure in the soft tissue at the site of the incompletely recanalized vein. The scan on the *left* depicts hyperechoic thread-like structures in the partially recanalized (more hypoechoic) lumen. These structures, which may occasionally have a honeycomb appearance, are sclerotic strands persisting after thrombosis. The scan on the *right* shows these structures in longitudinal orientation.
**b** Color duplex scanning depicts spontaneous or augmented postthrombotic flow among such fibrotic intraluminal structures. In this case, color-coded flow signals in the vein appear only upon peripheral compression (*left scan*). The second scan shows reflux after release of compression (red, toward transducer), indicating incompetent valves. The Doppler spectrum recorded during compression and release confirms persistent reflux.

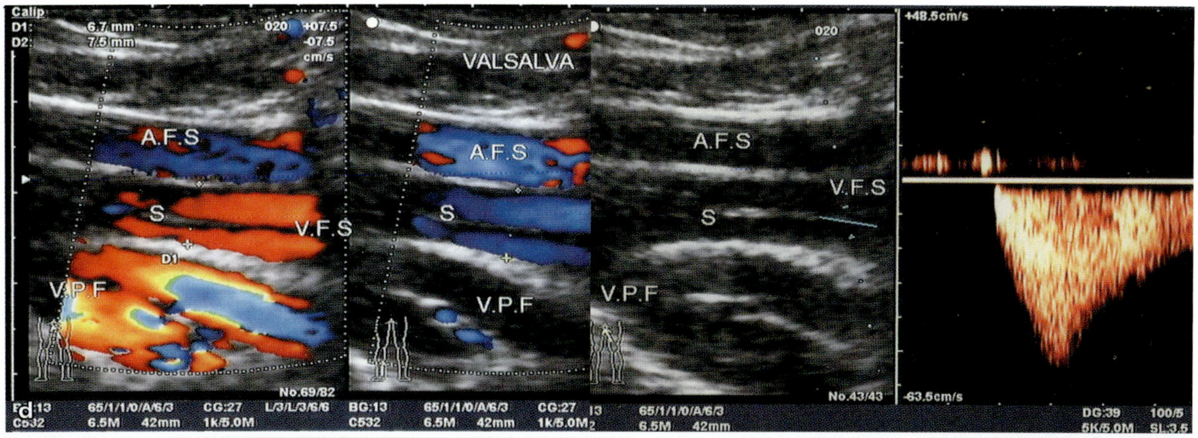


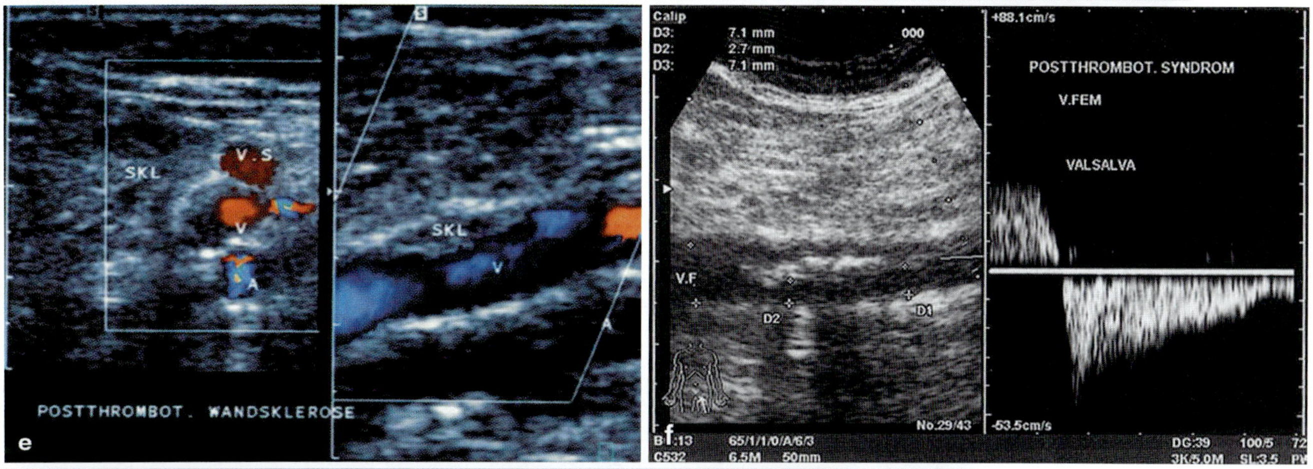


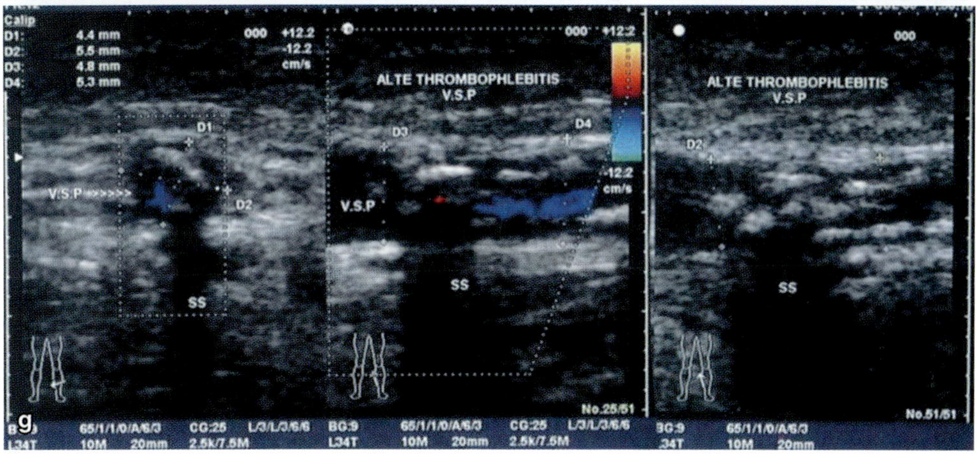


e The popliteal vein is completely patent, but there is postthrombotic wall sclerosis depicted as hyperechoic thickening of the wall (*SKL, longitudinal view on the right*). The transverse scan on the *left* also depicts more hypoechoic areas in the lumen, corresponding to residual thrombotic deposits on the wall or wall thickening. These abnormalities appear to the left of the recanalized patent lumen displayed in red and farther away from the transducer. The more superficial small saphenous vein appears normal shortly before it joins the popliteal vein.

**Fig. A3.39 (continued)**

**d** Intraluminal synechia (*S*) extend to the proximal superficial femoral vein (*V.FS*). *Leftmost section*: Recanalized superficial vein (red-coded flow, toward the heart). The *middle section* shows blood flow along the strands (*S*) toward the periphery with Valsalva's maneuver (blue flow away from transducer). Normal valve closure in the deep femoral veins (*V.PF*) without reflux (*A.FS* = superficial femoral artery). The gray-scale scan depicts synechia in the recanalized lumen, and the Doppler waveform shows slow flow due to obstruction by the strands and marked reflux elicited by Valsalva's maneuver (flow away from transducer, toward the periphery).

**f** Postthrombotic wall lesions can lead to wall sclerosis and calcifications with acoustic shadowing on ultrasound. The Doppler waveform shows reflux due to incompetent valves.

**g** Vasosclerotic changes with wall thickening and calcification may also occur after thrombophlebitis. In the example, the longitudinal view on the *right* shows the hyperechoic sclerotic wall lesions with intraluminal deposits that are partially calcified, as demonstrated by posterior acoustic shadowing (*SS*). The longitudinal scan in the *middle* and the transverse scan on the *left* also depict the thin recanalized lumen of the postthrombophlebitic small saphenous vein with flow in blue.

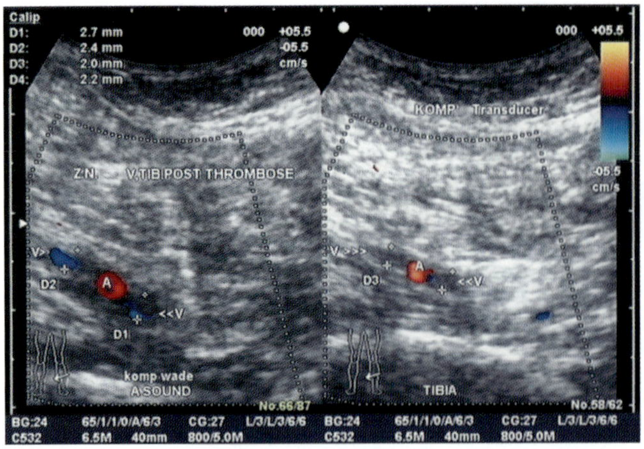


## Fig. A3.40

**Postthrombotic lower leg vein**

The diagnosis of older lower leg vein thrombosis or residual postthrombotic lesions relies on the examiner's sonoanatomic knowledge, as the affected veins will no longer be visualized as markedly dilated tubular structures. The corresponding arteries serve as landmarks. It is often difficult to detect spontaneous flow in (partially) recanalized lower leg veins. In the case presented, a flow signal (A-sound) can be recorded with adequate instrument settings in the duplicated posterior tibial vein (depicted to the left and right of the artery of the same name) following distal compression (lower calf). A vein with a normal lumen would have become invisible when compressed. The hypoechoic small-caliber structures persisting upon compression in the soft tissue along the course of the veins to the left and right of the artery indicate residual thrombotic lesions or fibrotic wall changes.

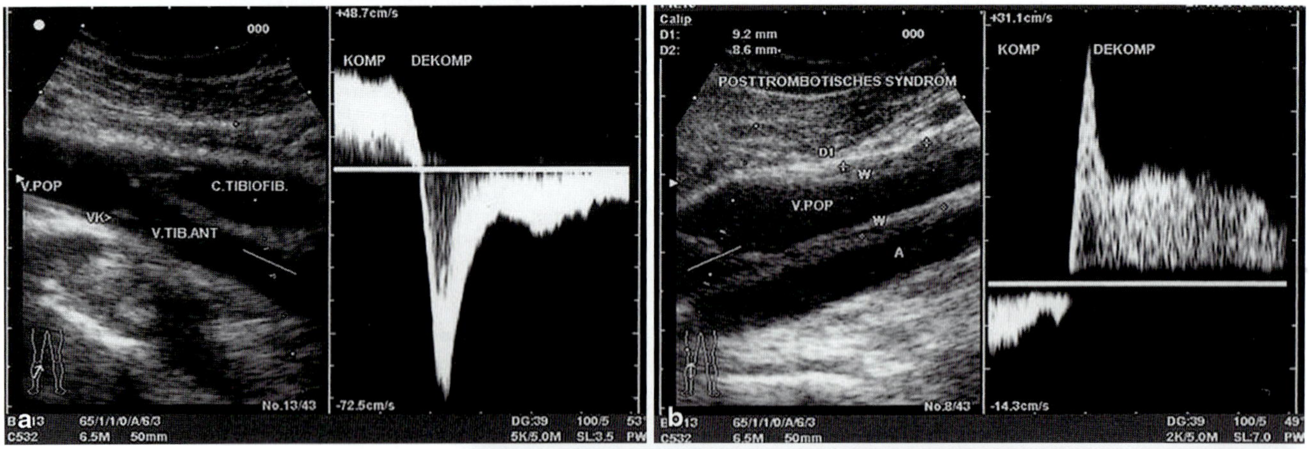


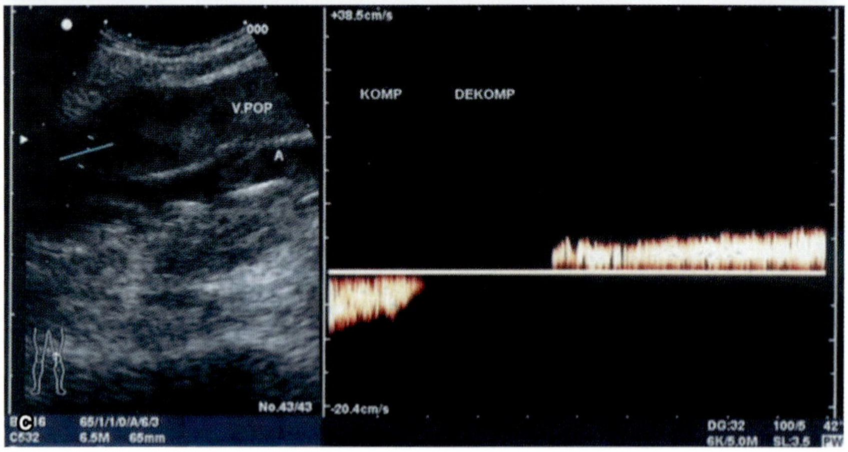


## Fig. A3.41 a–g

**Degrees of valve incompetence**

**a** Postthrombotic thickening of a venous valve (*VK*) with adhesion to the wall prevents closure, which is indicated by reflux during Valsalva's maneuver or the valve function test (compression and release). Here, postthrombotic sclerosis and valve incompetence of the popliteal and lower leg veins lead to immediate backward flow with a high signal upon compression of the calf (*KOMP*) with subsequent release (*DEKOMP*). This finding indicates complete valve failure. Once the blood column expulsed from the calf has flowed back, a decrease in the reflux signal induced by release of compression is noted. The example illustrates valve incompetence of the anterior tibial vein (*V.TIB.ANT*) before it enters the popliteal vein (*V.POP*).

**b** The popliteal vein also shows postthrombotic valvular adhesion and wall sclerosis (*W*), reflected sonographically as hyperechoic wall thickening (wall near transducer). The Doppler waveform from the popliteal vein (*V.POP*) depicts the prompt and pronounced reflux (toward transducer) upon release of compression (*DEKOMP*) as a sign of complete valve failure.

**c** Primary chronic venous insufficiency with dilatation but with some residual valve function is indicated by delayed reflux upon Valsalva's maneuver or release of compression. The example shows the delayed reflux (toward transducer) with a lower but constant flow in the popliteal vein.

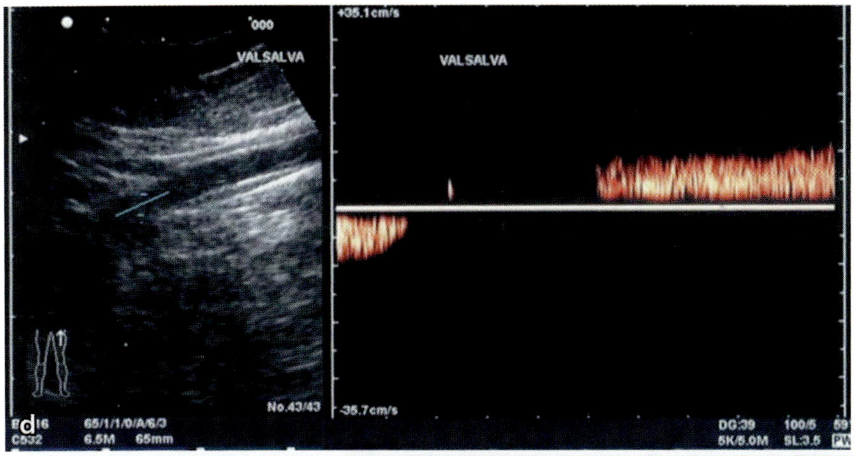


**Fig. A3.41 (continued)**

**d** A similar pattern of backward flow is seen in this case of varicosis of the great saphenous vein with early, mild valvular incompetence. There is delayed but constant reflux through the leaking valve after Valsalva's maneuver.

**e** Marked varicose dilatation produces severe valvular incompetence without residual function, as in the postthrombotic syndrome, resulting in immediate and pronounced reflux with high-velocity flow toward the periphery in Valsalva's test.

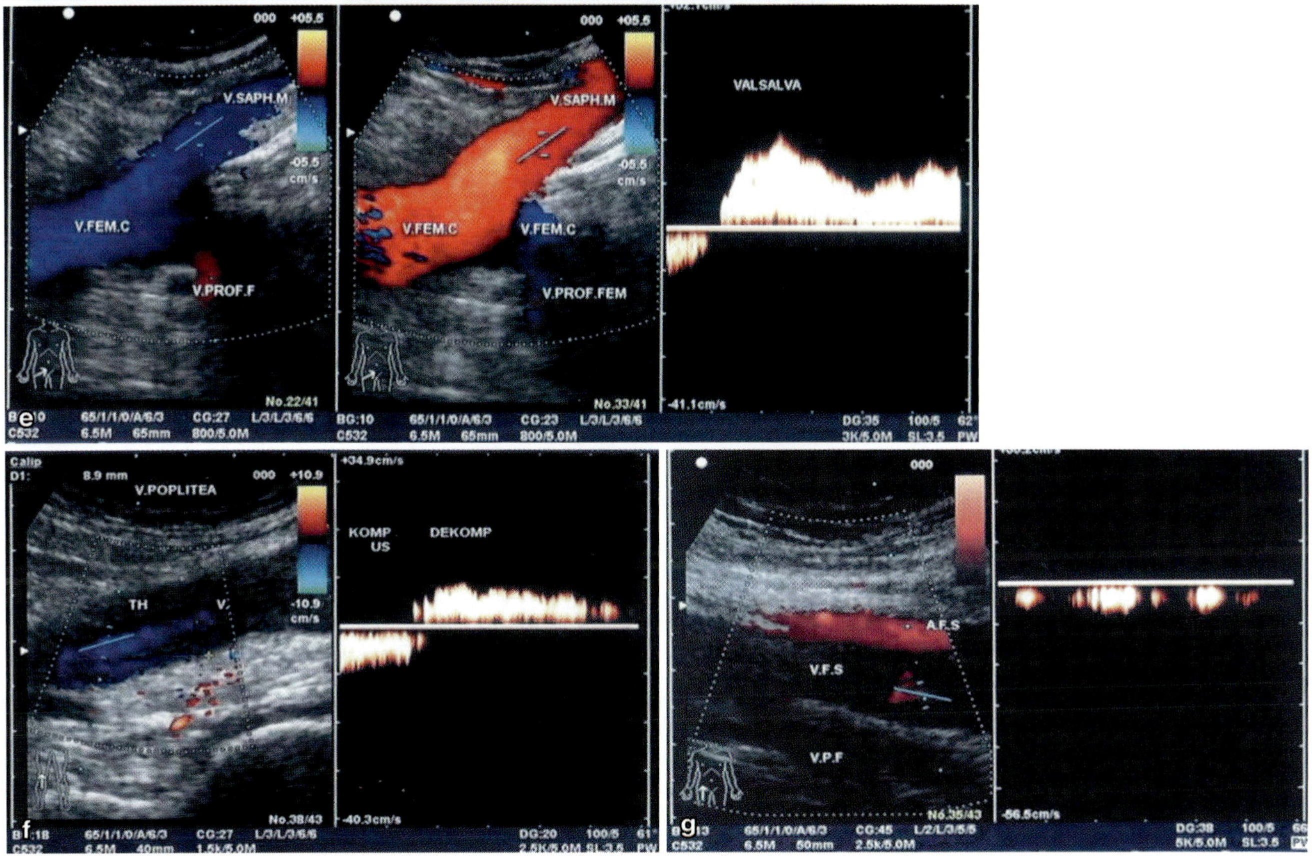


**f** The reflux resulting from postthrombotic valve incompetence is additionally influenced by flow obstruction due to residual thrombus. In the example, the popliteal vein is still partially thrombosed (*TH*) with only slow spontaneous flow. Compression (*KOMP*) of the calf induces constant flow from the periphery to the heart, while the backward flow occurring upon release of compression (*DEKOMP*) is less pronounced and less persistent than would be expected in extensive, recanalized thrombosis. The reduced backward flow is due to flow obstruction by residual thrombus.

**g** Slight reflux through a small leak in a valve leaflet during prolonged Valsalva's maneuver can be detected using a high-resolution transducer with a low PRF or in the power mode (for detection of slow flow). The power-mode scan on the *left* shows only little flow (red) directly behind the leaking valve leaflet in the proximal superficial femoral vein. If there is only little leakage, the sample volume must be placed close to the valve in order to depict the slight reflux during Valsalva's maneuver (flow toward periphery, away from transducer). As only little blood leaks back into the vein, no flow signals are detectable in the remaining venous segments. Such slight leakage as in this case should not be overinterpreted as valve incompetence but merely illustrates the high sensitivity of high-resolution ultrasound scanners for the detection of small flow volumes. However, repeat Doppler sampling along the course of the vein with provocative maneuvers is necessary to definitely exclude any clinically relevant reflux. Anterior to the vein, the superficial femoral artery (*A.F.S*, red) is depicted; and posterior to it, the deep femoral vein (*V.P.F*, without flow signals during Valsalva's maneuver).

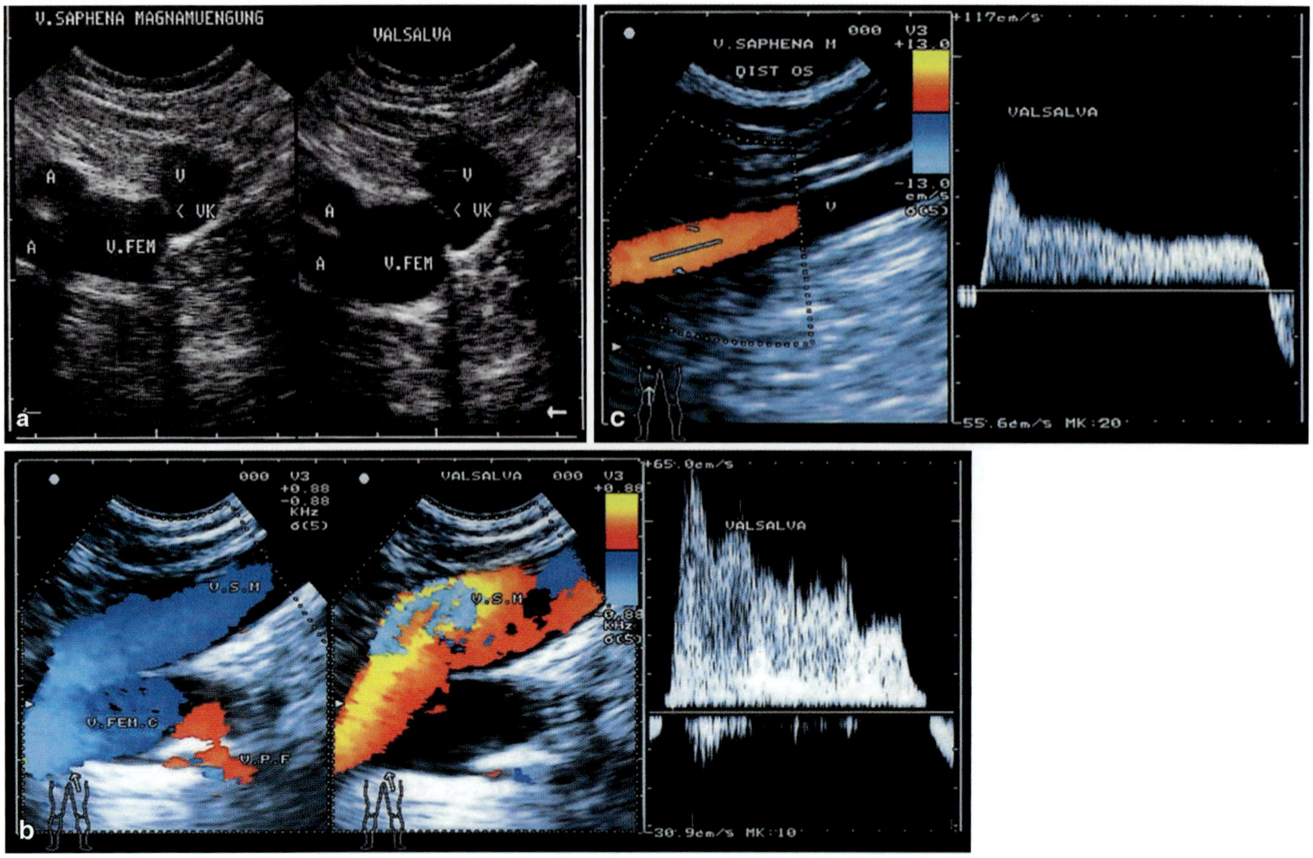


**Fig. A3.42 a–c**
**Truncal varicosis of great saphenous vein (distal extent)**
**a** The transverse B-mode scans show dilatation of the proximal great saphenous vein during Valsalva's maneuver with the incompetent valve leaflet turning distally and thus becoming visible (*VK*).
**b** Incompetent terminal valve of the great saphenous vein. The scan on the left shows blood flow toward the heart (blue). The great saphenous vein (*V.S.M*) courses close to the transducer, and a deep femoral vein (*V.P.F*) with flow displayed in red is seen entering the common femoral vein (*V.FEM.C*) posteriorly. Valsalva's maneuver (*middle section*) induces reflux (red) with aliasing due to the low PRF adjusted to slow venous flow. Proper valve closure in the common femoral vein prevents reflux into the deep venous system. The Doppler frequency spectrum (*right*) recorded at the saphenofemoral junction during Valsalva's maneuver shows flow to the periphery (toward transducer).
**c** The distal point of insufficiency of the great saphenous vein for grading according to Hach is identified by determining reflux during Valsalva's maneuver (toward transducer) in the color duplex mode or in the Doppler tracing obtained along the course of the vein from above the knee (*V*) to the lower leg.

**Fig. A3.43 a–f**
**Incomplete truncal varicosis of great saphenous vein**
**a** Transverse scan on the *left* and longitudinal scan on the *right* (or rather oblique scan) show reflux in the lateral accessory saphenous vein (*V.BV* = arch vein) in the right groin, induced by Valsalva's maneuver. Absence of flow in the proximal great saphenous vein (*V.S.M*, indicated by *plus signs*) during Valsalva's maneuver indicates competent valves in this segment. Directly below the saphenofemoral junction and the incompetent terminal valve, the vein divides into an incompetent accessory saphenous vein and the competent great saphenous vein (*V.f* = femoral vein).

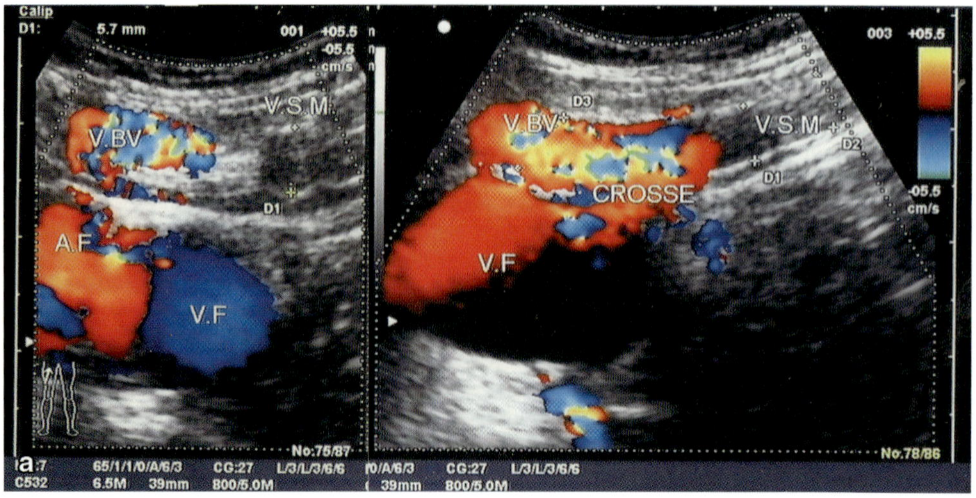

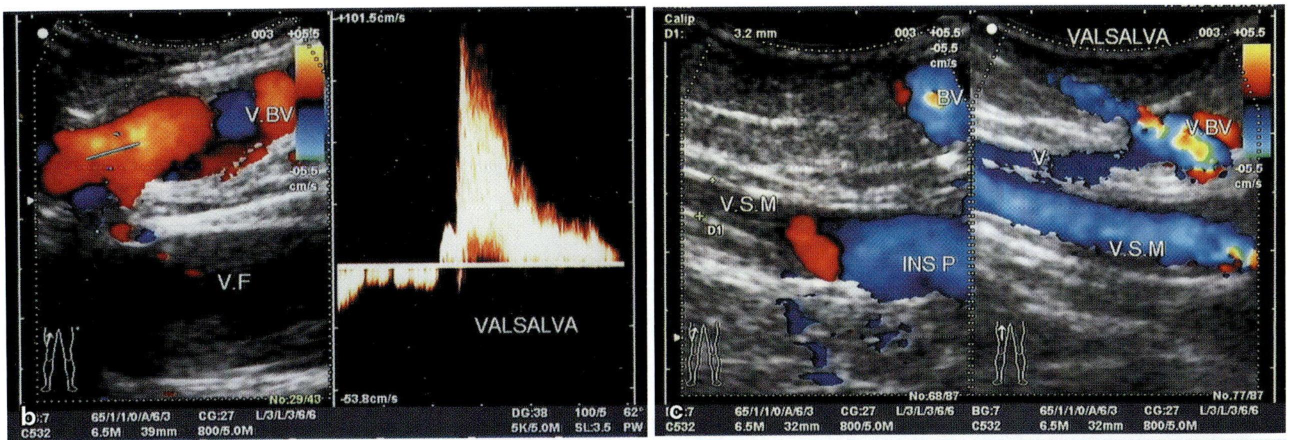


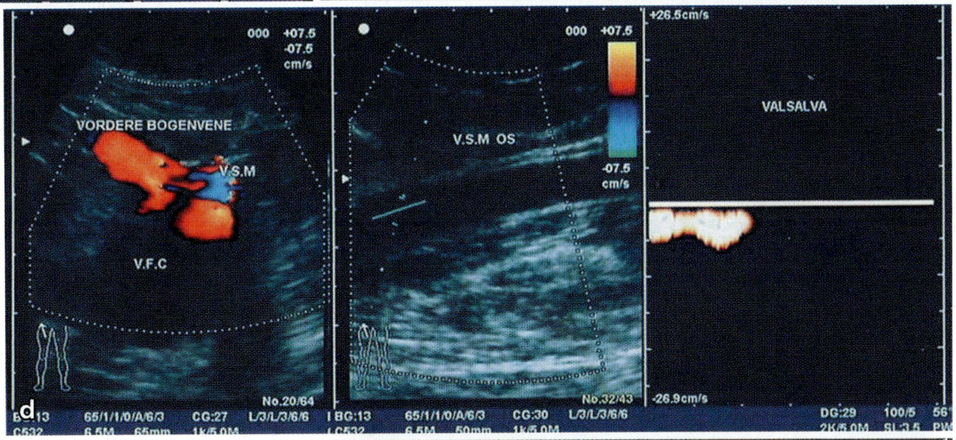


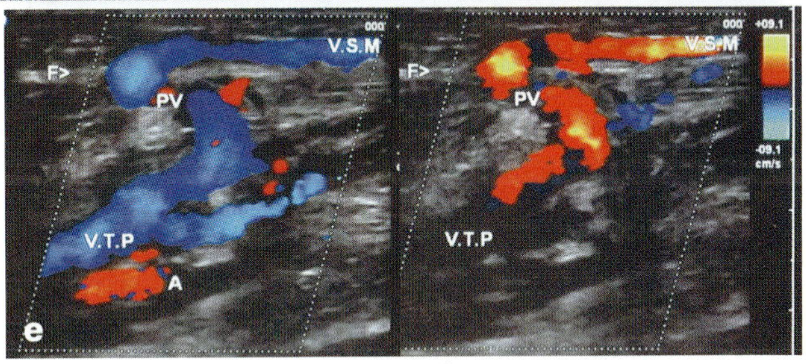


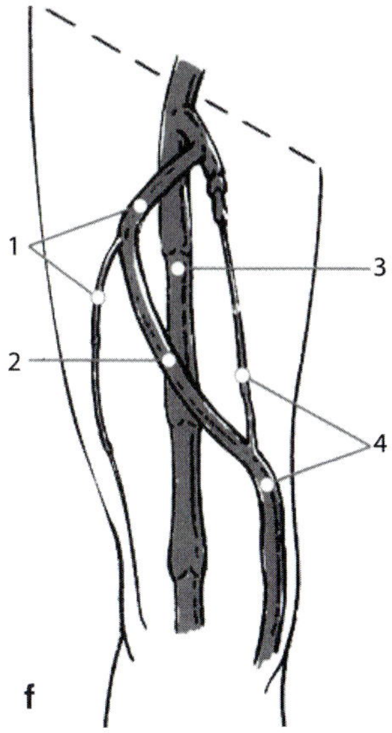


**Fig. A3.43 (continued)**

**b** Flow into the periphery (toward the transducer) induced by Valsalva's maneuver is seen in the lateral accessory saphenous vein (arch vein) on the color flow image (red) and in the Doppler waveform.

**c** Longitudinal scan depicting the accessory saphenous vein (*V.BV* = arch vein) and great saphenous vein (*V.S.M*) including the vein connecting the two (*V* = bucket handle anastomosis) in one plane. The *right* image shows reflux in this venous system upon Valsalva's maneuver: flow toward the periphery coded in blue (away from transducer). There is reflux from the arch vein into the great saphenous vein via the connecting vein. The valves of the great saphenous vein below this site (proximal point of insufficiency) are incompetent. The *left* image (composite image of proximal segment) also shows the upper point of insufficiency of the great saphenous vein (*V.S.M*, indicated by *plus signs*); there is no flow in the competent proximal segment of the great saphenous vein.

**d** Neither color duplex nor the Doppler waveform shows flow reversal just below the saphenofemoral junction with Valsalva's maneuver, confirming competence of the proximal great saphenous vein.

**e** Insufficient Cockett I perforating vein in the calf. The perforating vein establishes a transfascial (*F*) connection between the great saphenous vein (*V.S.M*) and the posterior tibial vein (*V.T.P*). When the calf is compressed (*left* section), there is flow toward the center (blue) in the great saphenous vein, posterior tibial vein, and perforating vein (from the superficial into the deep venous system). *Right* section shows reversed flow (from deep into superficial system, encoded in red) upon release of compression, indicating incompetence of the perforating vein. The great saphenous vein is also incompetent distal to the incompetent perforator (reflux, encoded in red), while the posterior tibial vein is competent, as indicated by the absence of flow reversal upon release of compression.

**f** Incomplete truncal varicosis of the great saphenous vein of the lateral branch type: *1* lateral accessory saphenous vein, *2* bucket handle anastomosis, *3* superficial femoral vein, *4* great saphenous vein.

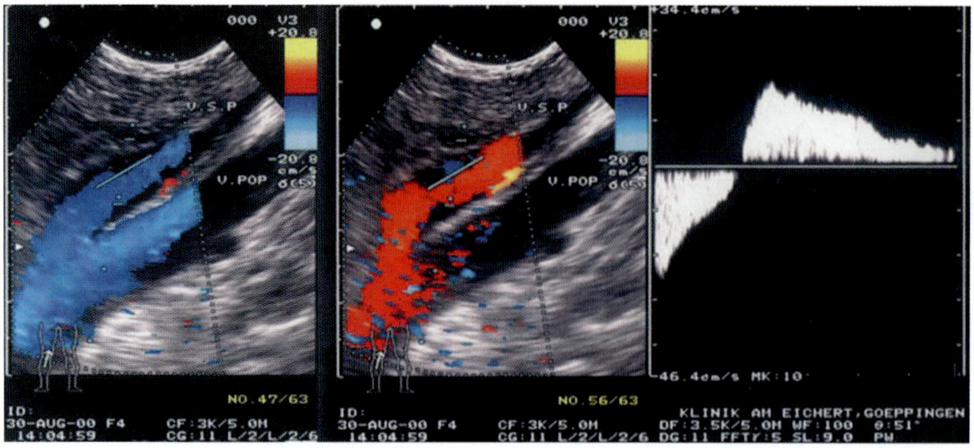


**Fig. A3.44**
**Truncal varicosis of small saphenous vein**
The Doppler spectrum recorded in the saphenopopliteal junction (*V.S.P* = small saphenous vein, *V.POP* = popliteal vein) shows normal flow toward the heart during calf compression and high-velocity reversed flow upon release of compression. In the color flow images, flow reversal in the small saphenous vein is indicated by red color coding (flow toward the periphery, *middle section*), while the absence of flow in the popliteal vein suggests competent valves here.

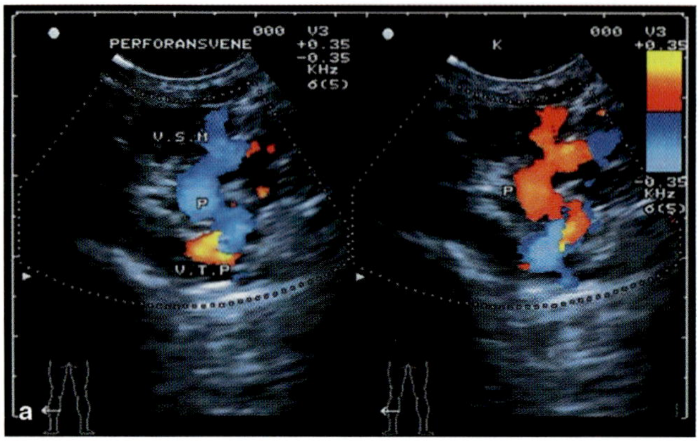


**Fig. A3.45 a–c**
**Valve incompetence of perforating vein**
**a** Incompetent perforating veins are identified by looking for transfascial tubular structures originating from branches of the great or small saphenous vein in transverse orientation using a high-frequency transducer. In the example, compression of the calf proximal to the probe with elimination of flow in the superficial veins by means of a tourniquet induces retrograde flow (displayed in red) from the posterior tibial vein (*V.T.P*) into the great saphenous vein (*V.S.M*) with a return to forward flow (blue, away from transducer) upon release of compression. Reflux from the deep venous system into the superficial system in this test confirms an incompetent perforating vein (see Fig. A3.43e).

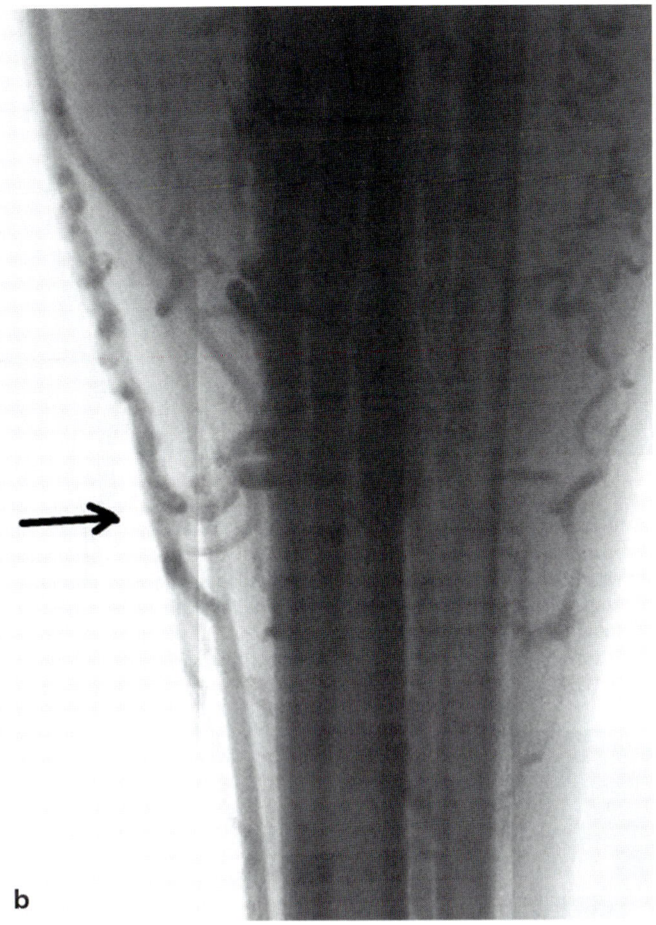


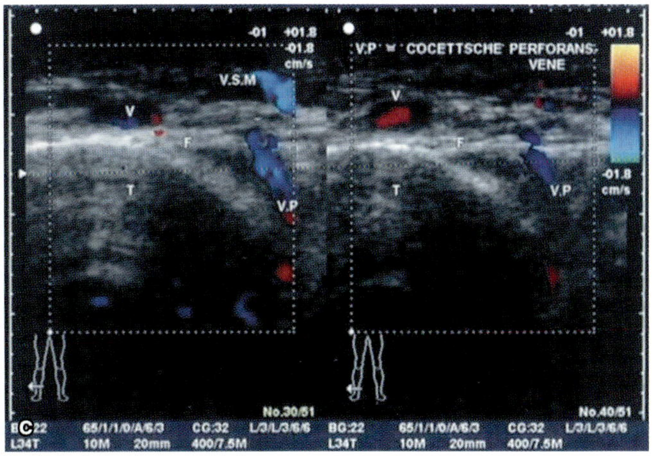


**Fig. A3.45 (continued)**

**b** Venogram depicting an incompetent perforating vein between the great saphenous vein and the posterior tibial vein.

**c** Widening associated with valve incompetence makes it much easier to identify an abnormal perforating vein than a normal one. A very small perforating vein (*V.P*) of the lower leg is depicted along its transfascial course (*F*) with flow.

During compression, there is flow in the perforating vein (*V.P*), coded in blue, from the superficial into the deep system and no reflux upon release of compression. The scan on the *left* depicts a perforator, the one on the *right* a Cockett perforator. Between the fascia (*F*) and the skin, the great saphenous vein and lateral branch veins (*V*) are depicted.

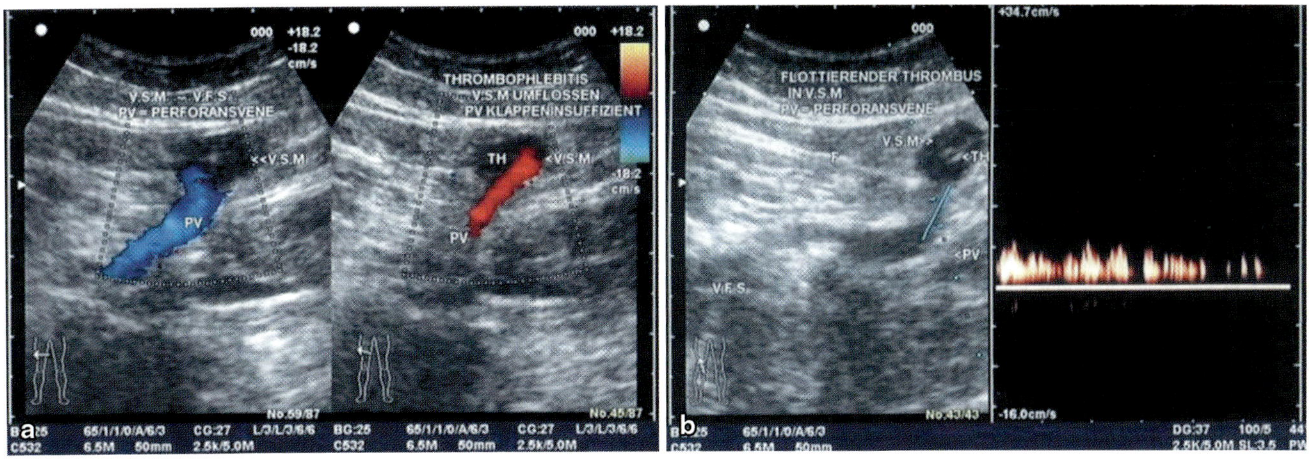


**Fig. A3.46 a–d**

**Thromboembolism from great saphenous vein and Dodd perforator incompetence**

**a** Patient with thrombophlebitis clinically extending to the knee and sonographic demonstration of a thrombus in the great saphenous vein with proximal extension to the level of the mid-thigh. The proximal end (3 cm) is surrounded by flowing blood. At this level, the transverse view depicts a Dodd perforator (*PV*) with normal flow into the deep venous system and an increase in flow velocity upon compression of the great saphenous vein just above the thrombophlebitic segment. Release induces reflux into the superficial system (displayed in red, scan on the *right*), indicating valve incompetence of the perforating vein. The thrombus in the great saphenous vein (*TH*) is identified by the absence of color coding.

**b** B-mode scan depicting the thrombus (*TH*) in the great saphenous vein (*V.S.M*). The Doppler spectrum recorded in the perforating vein demonstrates reflux from the superficial femoral vein (*V.F.S*) upon release of compression.

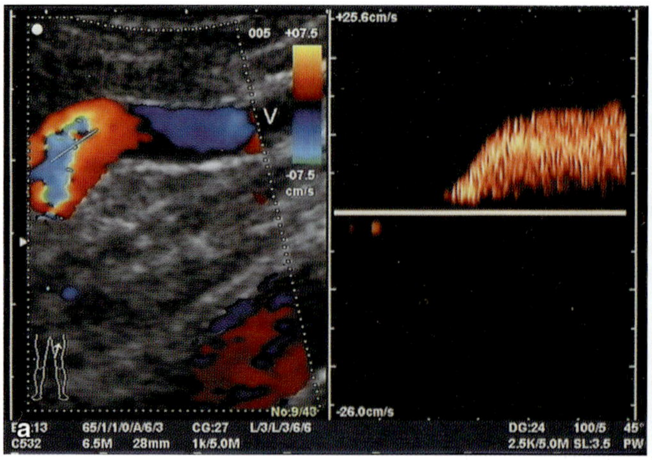


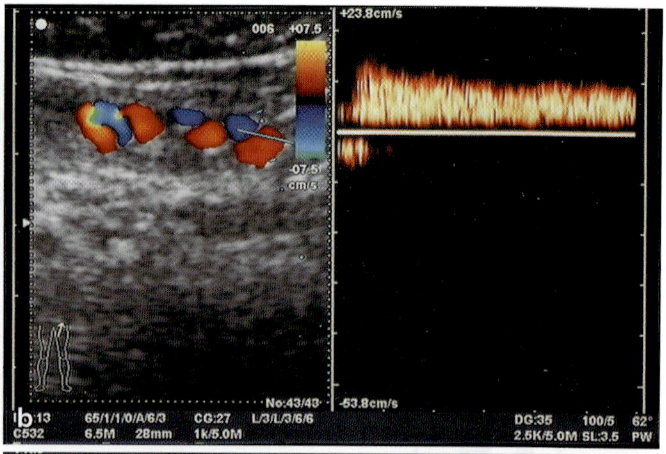

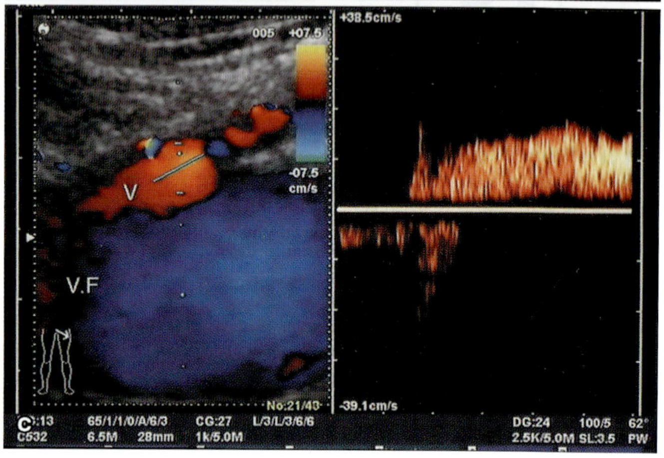


**Fig. A3.50 a–c**
**Recurrent varicosis**

**a** If there is visible recurrent varicosis, the course of the affected vein must be examined with color duplex sonography from distal to proximal to search for incompetent valves. The example shows a dilated and elongated varix in the medial thigh with peripheral flow (toward transducer) upon Valsalva's maneuver in a patient who underwent stripping of the great saphenous vein.

**b** A therapeutically relevant diagnostic task is to show whether a recurrent varix arises from a lateral branch or a perforating vein or whether it communicates with the saphenofemoral junction. A varix communicating with the former saphenofemoral junction (following crossectomy) may have a very thin lumen and show a very tortuous course over a short distance (similar in appearance on color flow image to arterial corkscrew collaterals in thromboangiitis obliterans). Nevertheless, such a varix is clinically relevant and will show reflux upon Valsalva's maneuver; its tortuous course appears on color flow imaging as repeated color reversal due to the changing flow direction relative to the ultrasound beam (and indicates neovascularization).

**c** A thin vein (*V*) is seen arising from the femoral vein (*V.F*) in the area of the former saphenofemoral junction. This vein shows retrograde flow upon Valsalva's maneuver (flow toward transducer indicated by red color; flow above the baseline in the Doppler waveform).

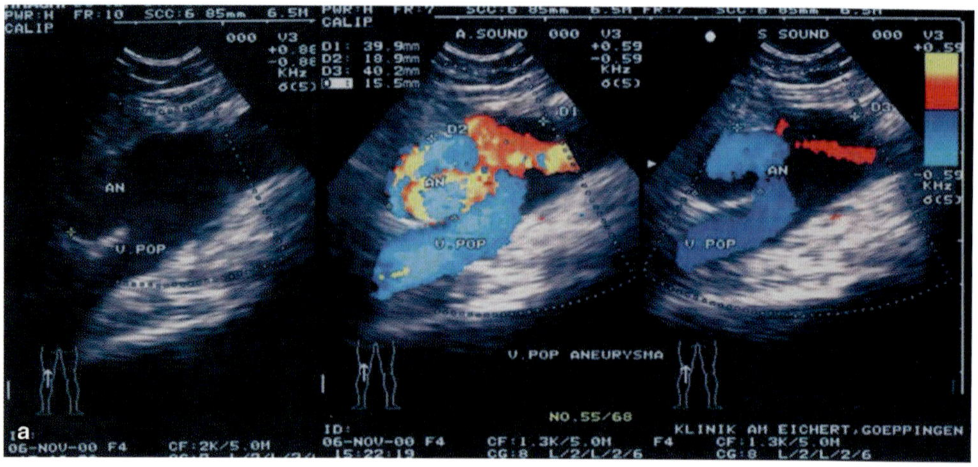


**Fig. A3.51 a–d**
**Venous aneurysm**

**a** Gray-scale scan depicting a saccular aneurysm (*AN*) as a distended sac at its preferred site, the popliteal vein (*V.POP*). Color duplex evaluation (*right section*) of spontaneous flow demonstrates zones of nearly complete stasis (S-sound). Compression of the calf (A-sound) induces pronounced eddy currents in the popliteal vein aneurysm (*middle section*).

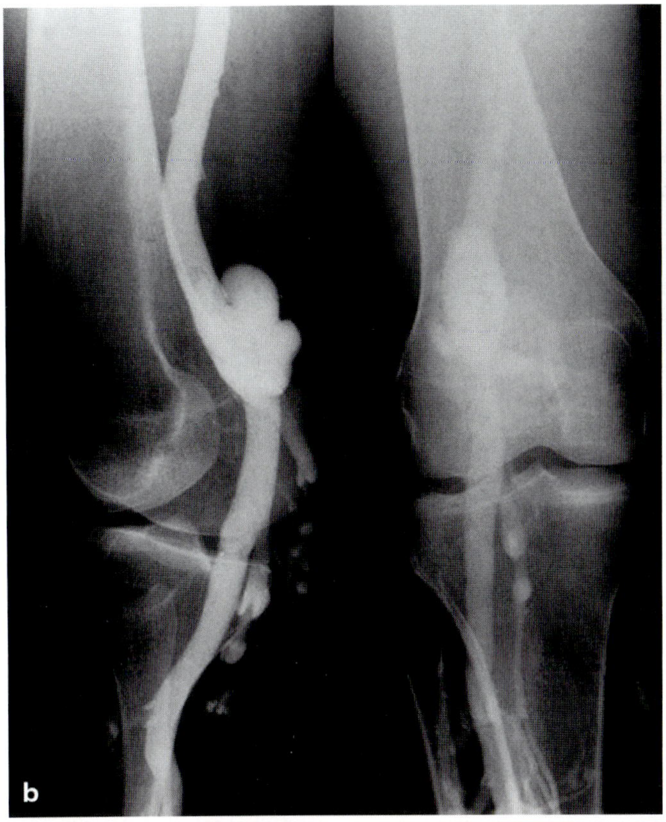


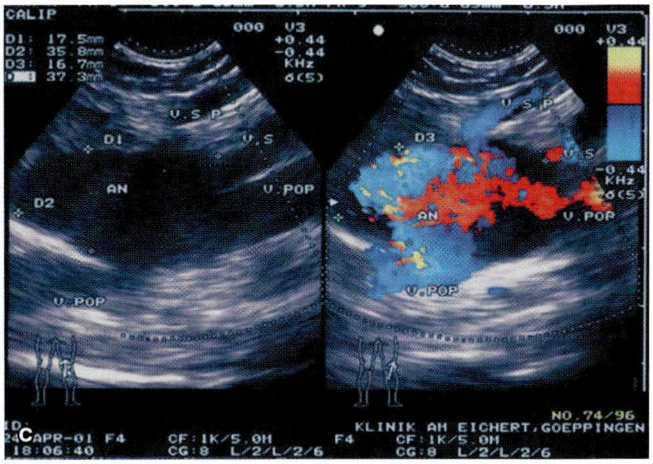


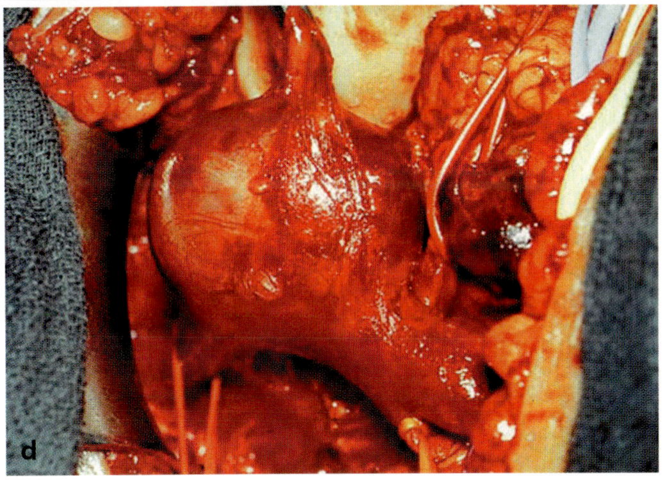


**Fig. A3.51 (continued)**

**b** Venogram demonstrating saccular aneurysm of the popliteal vein. In a nonthrombosed aneurysm, as in the case shown, opacification corresponds to the sonomorphologic shape (see *left section* in **a**).

**c** With a different transducer orientation, the small saphenous vein (*V.S.P*) and a gastrocnemius vein (*V.S*) entering the aneurysm are depicted. The maximum cross-sectional diameter of the aneurysm is 2.5 cm.

**d** The intraoperative site confirms the sonomorphologic appearance of the saccular aneurysm. The saccular cranial end is exposed on the left, and the two veins (gastrocnemius vein and small saphenous vein) entering the aneurysm are seen in the center. The popliteal vein (*left margin*) and a vessel entering the distal popliteal vein (*right*) are encircled with red vessel loops.

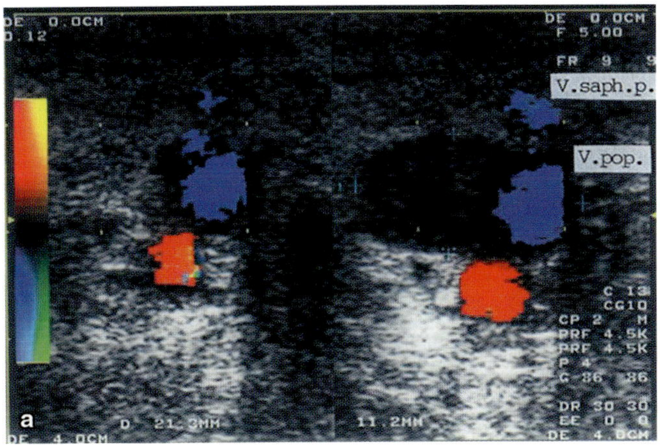


**Fig. A3.52 a–c**

**Venous aneurysm with thrombus**

58-year-old patient with scintigraphically proven pulmonary embolism. Saccular popliteal vein aneurysm extending to the terminal segment of the sural vein with complete thrombosis of the aneurysmal sac; only the normal popliteal lumen is patent.

**a** The scan on the *left* depicts the popliteal vein with flow in blue proximal to the aneurysm; the scan on the *right* shows the dilated segment of the popliteal vein (*V.POP*) with mural thrombosis.

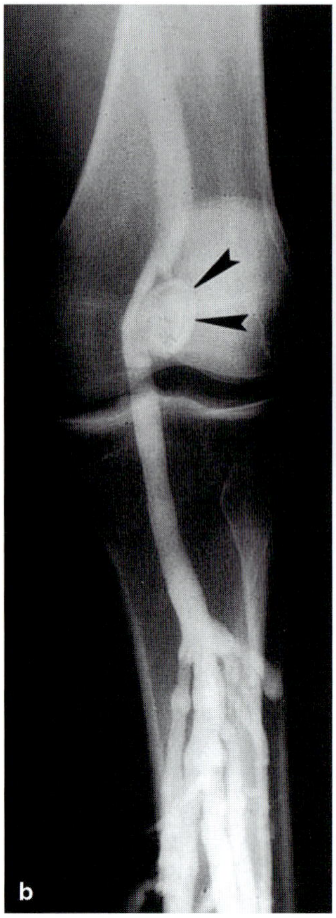


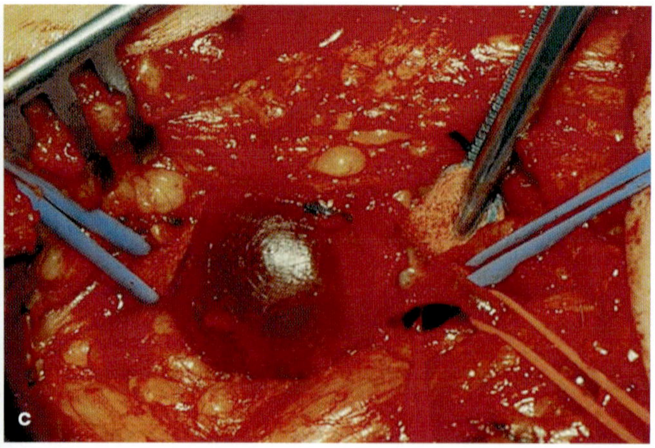


**Fig. A3.52 (continued)**

**b** Venogram: Mural thrombosis precludes visualization of the popliteal vein aneurysm and only aneurysmal dilatation at the entry site of a tributary vein is demonstrated (above knee joint cleft).

**c** The intraoperative site confirms the ultrasound findings of popliteal vein aneurysm (*center*) with mural thrombosis of the saccular portion and aneurysmal dilatation of the terminal sural vein. Popliteal vein encircled with blue vessel loops proximally and distally and sural vein with red vessel loop.

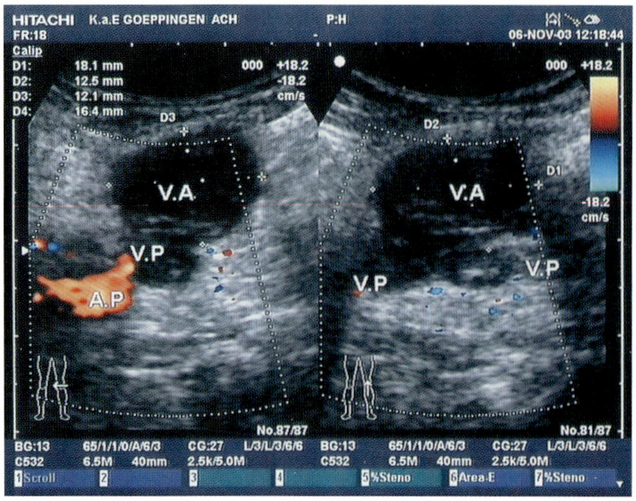


**Fig. A3.53**

**Venous aneurysm and deep venous thrombosis of leg**

There is complete thrombosis of the popliteal vein (*V.P*). Both the transverse scan (*left*) and the longitudinal scan (*right*) additionally demonstrate a saccular venous aneurysm (*VA*) with a diameter of nearly 2 cm. The aneurysm is thrombosed as well. This young patient had no other risk factors for venous thrombosis, suggesting that thrombosis from the venous aneurysm caused secondary popliteal vein thrombosis. Venous aneurysm must be differentiated from an ectatic terminal segment of a varicose small saphenous vein or an ectatic gastrocnemius vein.

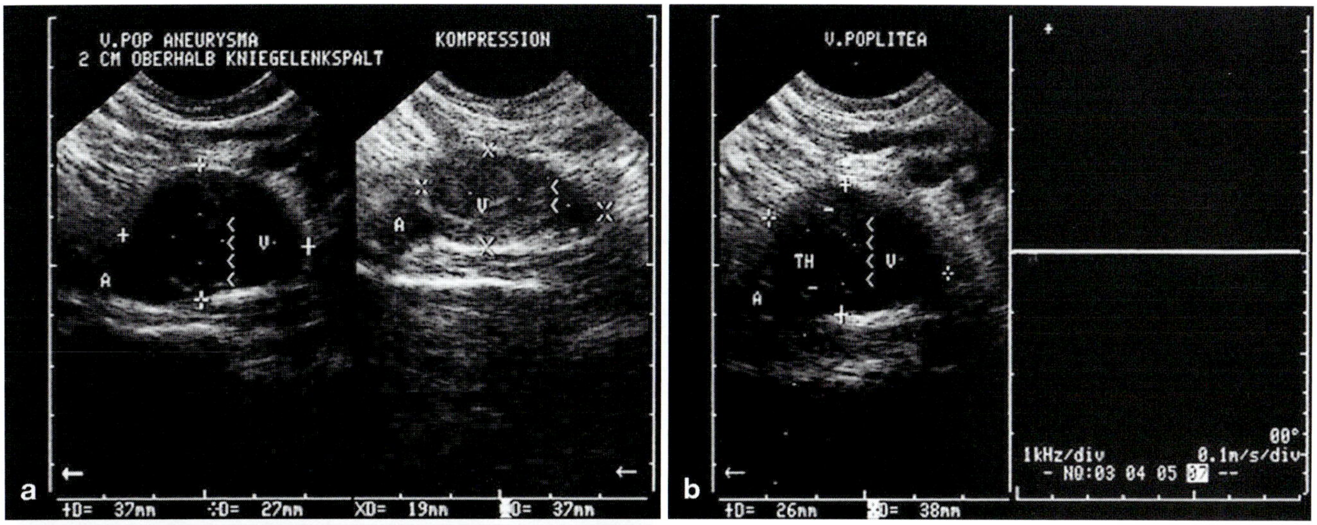


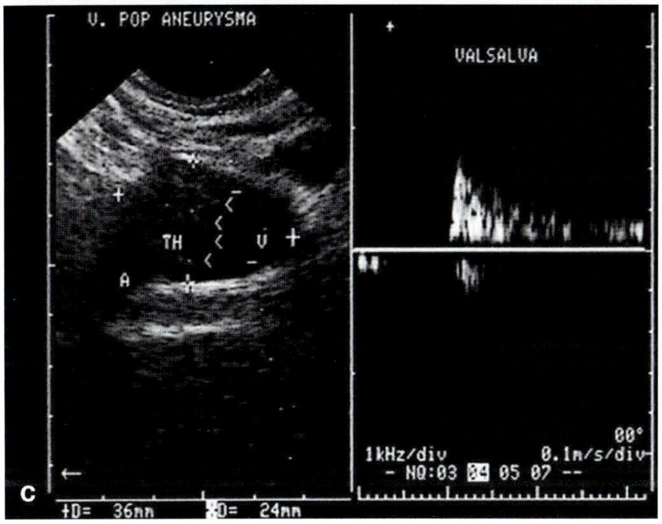


**Fig. A3.54 a–c**
**Saccular popliteal vein aneurysm**
**a** 45-year-old patient with recurrent pulmonary embolism; saccular popliteal vein aneurysm with nearly complete thrombosis, leaving only a small residual lumen, demonstrated by sonography and venography.
**b, c** The aneurysm has a maximum cross-sectional extent of 38 mm. Duplex ultrasound enables differentiation of the thrombotic portion (**b**) from the nonthrombotic residual lumen. Flow is depicted in the patent lumen, and there is reflux during Valsalva's maneuver, indicating valve incompetence (**c**). The patient had concomitant femoral vein incompetence and therefore underwent ligation of the superficial femoral vein to prevent further pulmonary embolism.

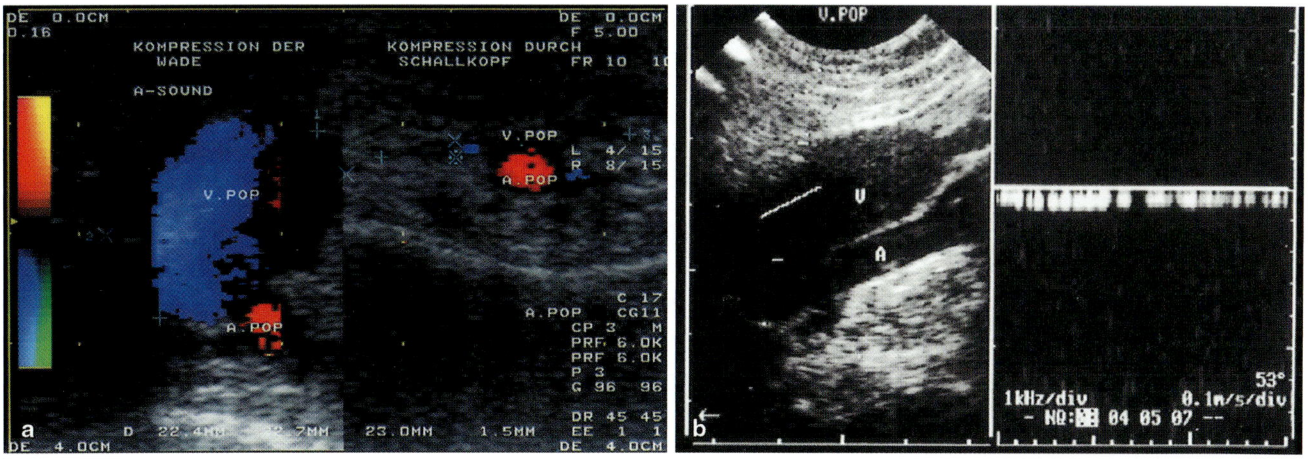


**Fig. A3.55 a, b**
**Spindle-shaped popliteal vein aneurysm**
**a** Spindle-shaped aneurysm in a patient presenting with a tendency for calf swelling. The aneurysm has a maximum width of 28 mm and can be completely compressed with the transducer. Due to the slow venous flow, there is no rich color filling throughout the patent lumen; however, the complete compressibility of the lumen rules out partial thrombosis.
**b** The spindle shape of the aneurysm is apparent from the longitudinal view; calf swelling is caused by valve incompetence in this area. Surgery was not indicated; the patient underwent compression therapy. Serial ultrasound examinations over 5 years showed an unchanged aneurysm without mural thrombosis.

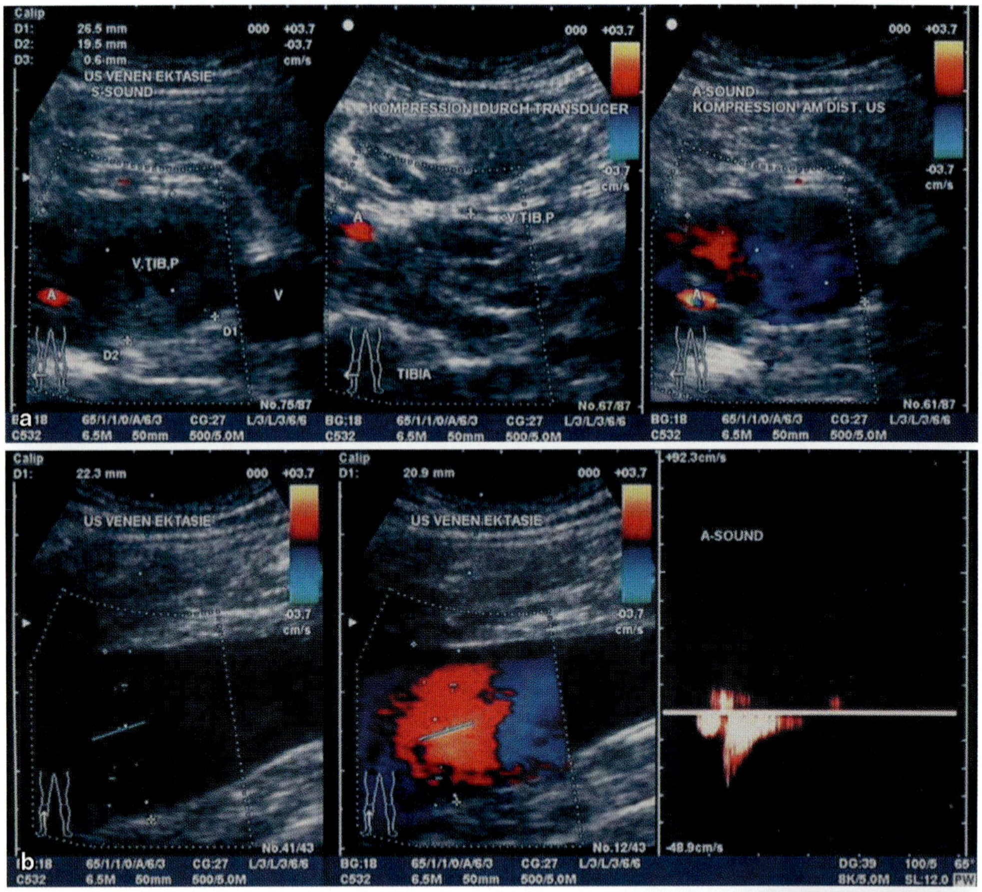


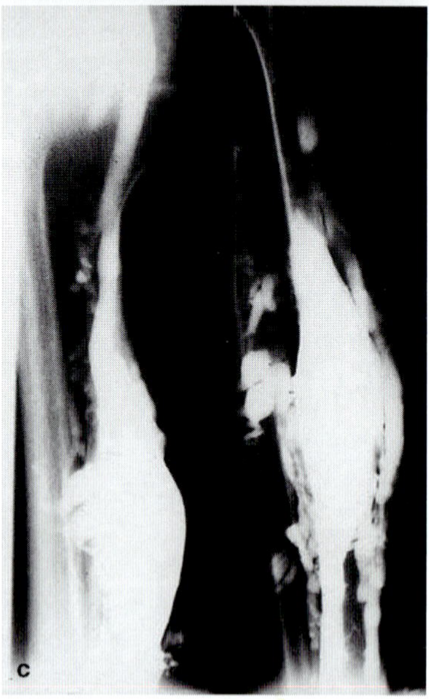

**Fig. A3.56 a–c**
**Venous ectasia of the lower leg**

**a** Ectatic degeneration chiefly involves the muscle veins of the gastrocnemius group, while severe ectasia of the major lower leg veins is rare. In the 50-year-old patient presented here, spindle-shaped ectatic changes of the posterior tibial vein (*V.TIB.P*) were the source of scintigraphically proven pulmonary embolism. The B-scan appearance suggests thrombosis. The ectatic veins have a diameter of up to 2.5 cm and can be completely compressed (*middle section*); the lumen of the posterior tibial vein is indistinguishable (marked). To the left of the vein, the posterior tibial artery is depicted with flow in red. There is no spontaneous flow in the vein (S-sound, *left section*), but augmented flow signals can be obtained upon distal compression of the calf (A-sound).

**b** The longitudinal scan likewise fails to depict flow in the spindle-shaped ectatic posterior tibial vein (*middle section*). Augmented flow (A-sound) is demonstrated by color duplex scanning and in the Doppler waveform following compression distal to the transducer.

**c** Venogram: Spindle-shaped ectatic dilatations of muscle veins and lower veins in the calf.

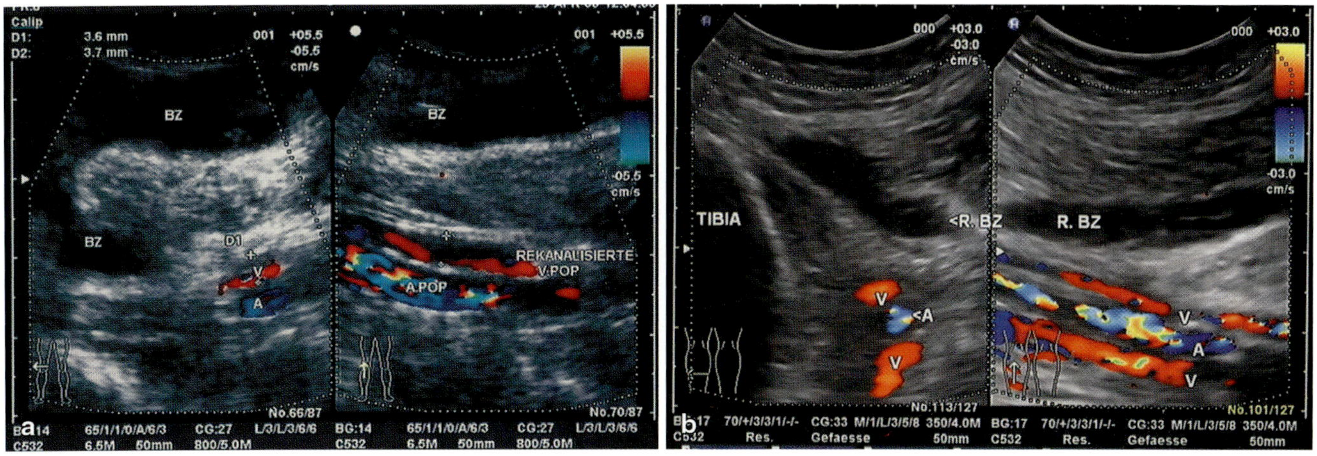


**Fig. A3.57 a, b**
**Differential diagnosis of venous thrombosis – Baker's cyst, hematoma, abscess**
**a** Leg pain and acute swelling in this patient is not caused by the postthrombotic changes in the popliteal vein (*V*) or by recurrent thrombosis, but by a large Baker cyst (*BZ*). The transverse view on the *left* and longitudinal view on the *right* depict the recanalized vein, but the walls are still markedly thickened. The low PRF adjusted to slow venous flow produces aliasing in the popliteal artery (*A.POP*).
**b** Ruptured Baker's cysts present the classic symptoms of lower leg venous thrombosis. They are typically seen as hypoechoic or anechoic, subfascial leaking structures with cystic residues in the popliteal fossa. Ultrasound-guided aspiration alleviates or completely eliminates the symptoms.

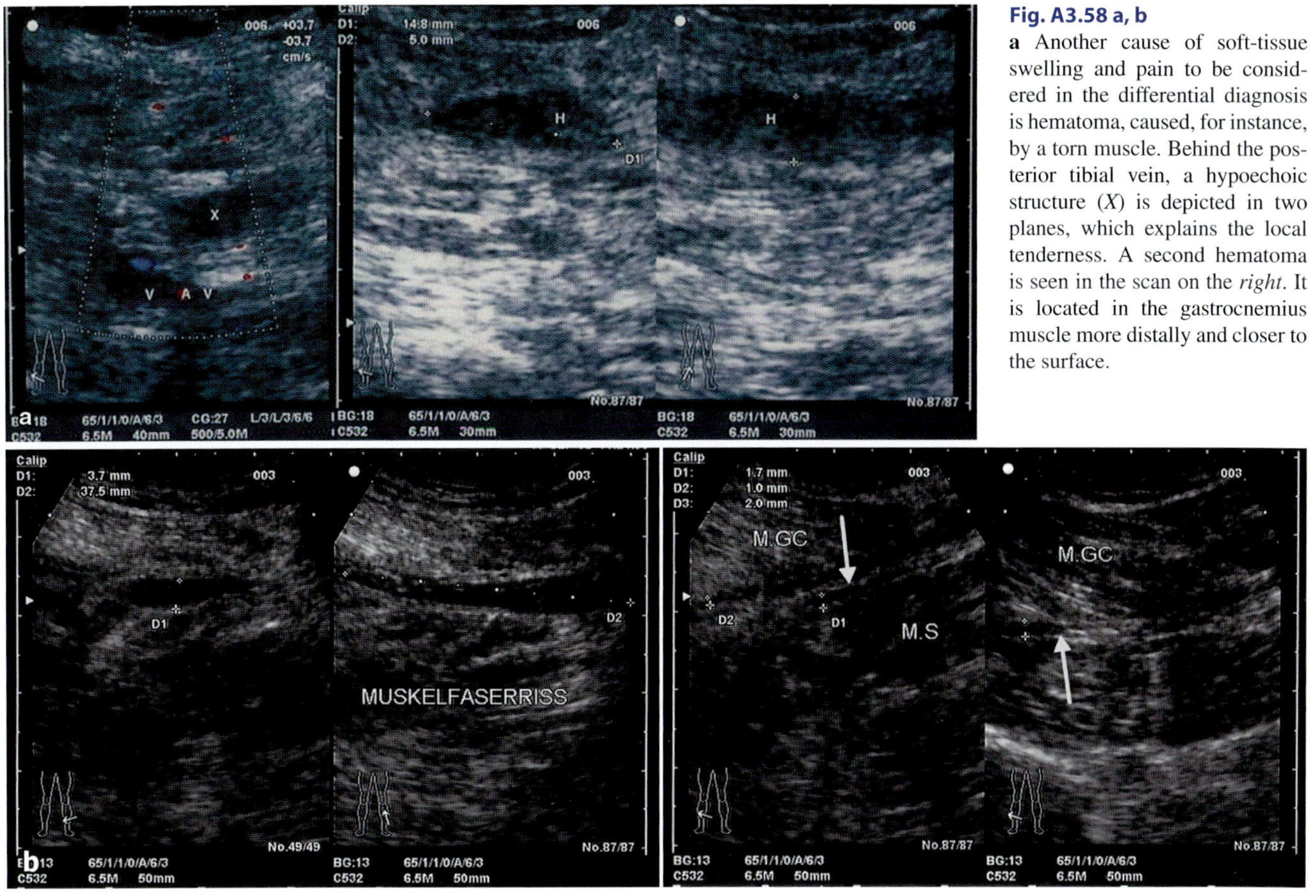


**Fig. A3.58 a, b**
**a** Another cause of soft-tissue swelling and pain to be considered in the differential diagnosis is hematoma, caused, for instance, by a torn muscle. Behind the posterior tibial vein, a hypoechoic structure (*X*) is depicted in two planes, which explains the local tenderness. A second hematoma is seen in the scan on the *right*. It is located in the gastrocnemius muscle more distally and closer to the surface.

**Calf swelling due to torn muscle**
**b** Free fluid (blood) secondary to a muscle strain may be very inconspicuous in patients presenting with symptoms of calf thrombosis. The examiner must look for bands of low echogenicity at the sites of muscle fasciae, in particular between the gastrocnemius and soleus muscle. The *left section* shows a hematoma secondary to a torn muscle. The hematoma is indicated by *plus signs*. The *right sections* show a second case with very little free fluid (*arrow*) between the gastrocnemius (M.GC) and soleus muscle (M.S).

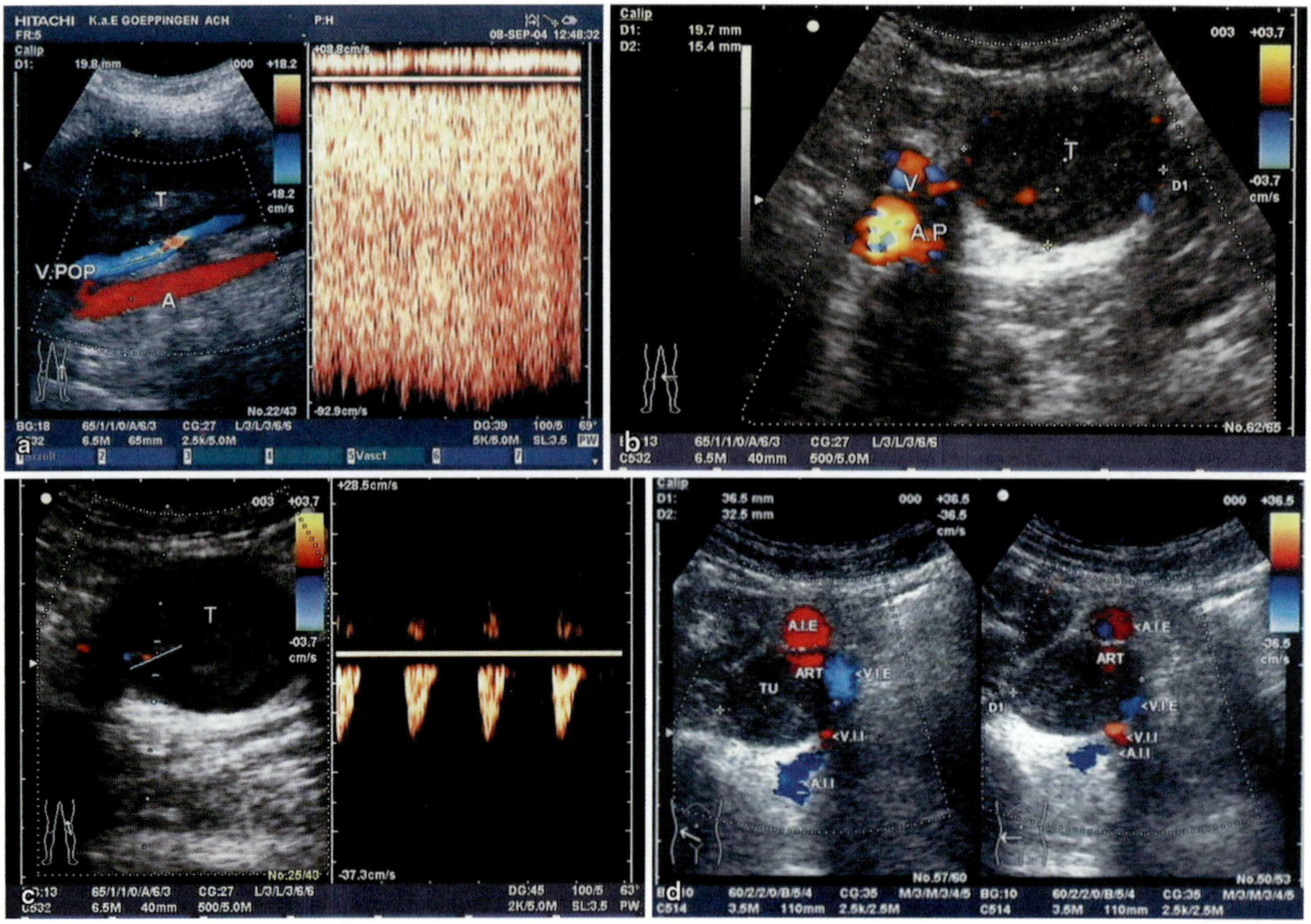


**Fig. A3.59 a–d**
**Calf swelling due to popliteal fossa tumor**

**a** External compression of the popliteal vein (*V.POP*) by a sarcoma in the popliteal fossa, reflected in the Doppler waveform as a high-frequency signal (flow velocity of 90 cm/s, absence of respiratory phasicity).
**b** 45-year-old woman with calf swelling; differential diagnosis: thrombosis. The detection of flow (low PRF) can help differentiate hypoechoic tumorous lesions from cysts with internal echoes due to intralesional hemorrhage.
**c** The Doppler waveform shows arterial flow as evidence of a solid tumor (sample volume placed in the area with flow signals in the color duplex scan). Schwannoma was diagnosed after removal of the tumor.
**d** Painful leg swelling caused by a tumor in the iliac bifurcation. Transverse views of the lower abdomen depict the external iliac vein (*V.I.E*, blue, flow away from transducer) and artery (*A.I.E*, red, toward transducer) anterior to the tumor and the internal iliac vein (*V.I.I*, red, toward transducer) and artery (*A.I.I*, blue, away from transducer) posterior to it. The hypoechoic tumor lies in the bifurcation and primarily compresses the external iliac vein (scan on the *right* obtained at a higher level). Posterior to the external iliac artery, a mirror artifact, caused by the interface of high acoustic impedance, is visible.

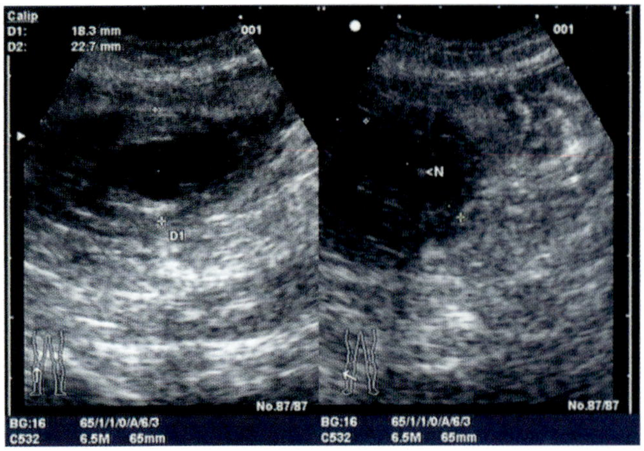


**Fig. A3.60**
**Calf swelling due to subfascial abscess**

An intramuscular abscess is not always associated with inflammation of the skin but may be diagnosed incidentally in patients undergoing ultrasonography for suspected venous thrombosis. It is seen on gray-scale scans as a hypoechoic, inhomogeneous structure and is confirmed by ultrasound-guided aspiration (*N* = needle tip).

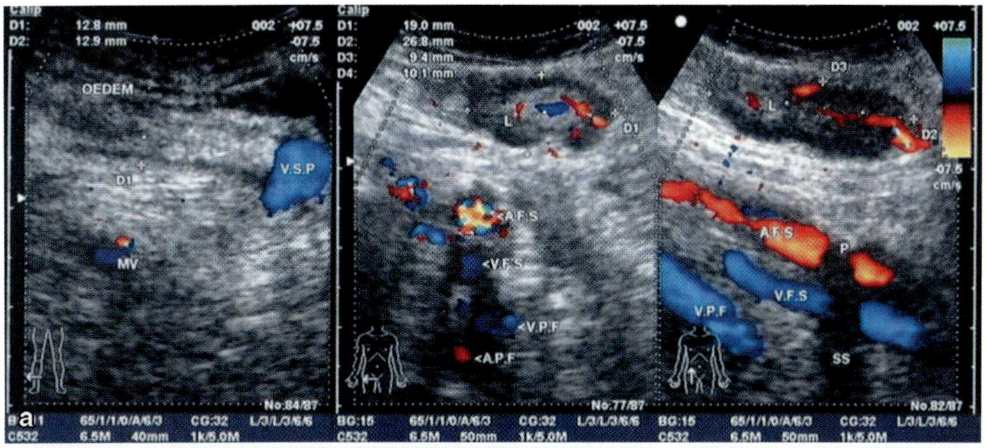


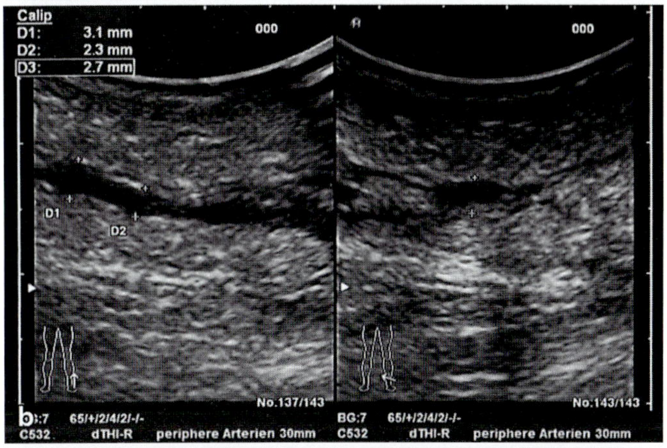


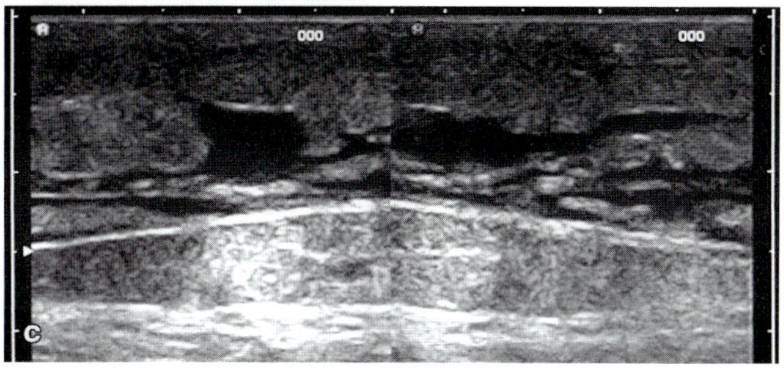


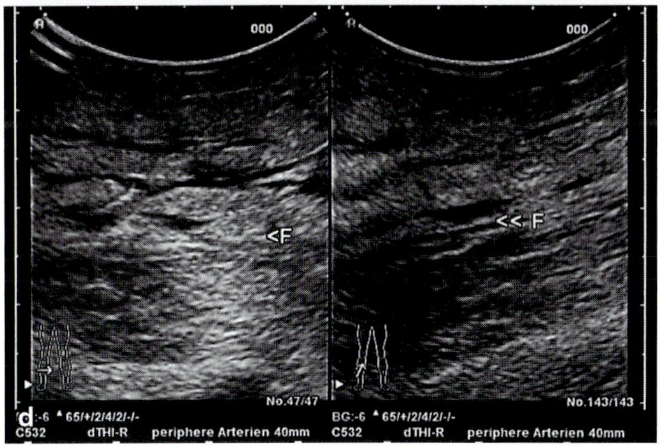


**Edema in the differential diagnosis of calf swelling**

**b** Channel-like structures in the subcutaneous fatty connective tissue (which tend to be near fasciae) on gray-scale scans are pathognomonic of lymphedema. For reliable differentiation from edema due to other causes, the dilated lymphatics must be visualized as tubular structures on longitudinal (*left section*) and transverse scans (*right section*). A thin, wall-like structure is occasionally identified by its higher echogenicity between the lumen and connective tissue (*left section*, adjacent to plus signs). Diameter 2–3 mm.

**c** Transverse and longitudinal images of lymphedema with markedly dilated lymphatic vessels.

**d** Edema due to other causes (cardiac, secondary to chronic venous incompetence) has a honeycomb-like appearance on both longitudinal and transverse scans, indicating fluid collections in connective tissue clefts of the subcutaneous fatty tissue (*F* = fascia with underlying muscle tissue).

**Fig. A3.61 a–d**

**Edema of different etiology, lymphoma, lymphedema**

**a** Another cause of leg swelling is cardiac, inflammatory, or lymphogenic edema with epifascial fluid collections in fatty or connective tissue clefts. Edema causes scattering and thus impairs evaluation of deeper subfascial areas and detection of thrombosis in the lower leg. In the case shown, there is edematous subcutaneous thickening (indicated by *plus signs*, 12 mm). The small saphenous vein (*V.S.P*) and a gastrocnemius vein (*MV*) are seen in transverse orientation.

Inflammatory edema is associated with reactively enlarged lymph nodes in the groin. They are depicted as hypoechoic, inhomogeneous structures that can be differentiated from thrombophlebitis by gray-scale ultrasound in two planes, which will demonstrate their round shape. Color duplex scanning with a low PRF depicts the supply and perfusion of the lymph node. Atherosclerosis with wall irregularities and calcified plaque (*P*) with acoustic shadowing (*SS*) is seen as an accessory finding.

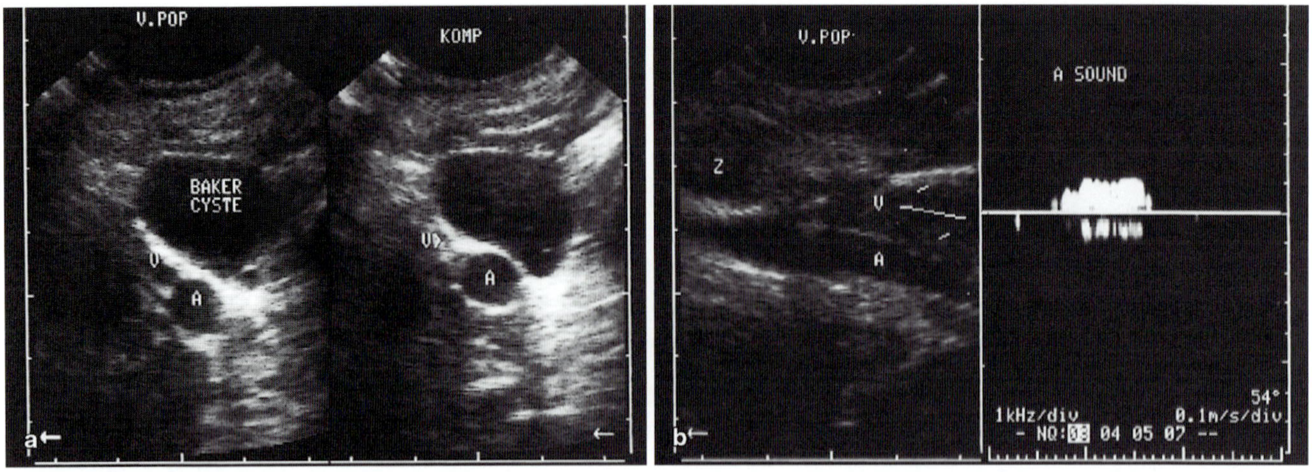


**Fig. A3.62 a, b**
**Vein compression by Baker's cyst**
Patient with calf swelling caused by a large Baker's cyst compressing the vein. The lumen of the vein is still patent but reduced. Pressure applied with the transducer causes complete collapse of the vein as seen on the transverse scan (*right*) in **a**. A month later, the cyst has increased in size, now not only compressing the vein but also displacing the artery. The Doppler waveform (**b**) shows no spontaneous flow signals but only a reduced A-sound upon strong compression of the calf muscles. The sample volume is placed in the vein (longitudinal scan); *Z* indicates the cyst.

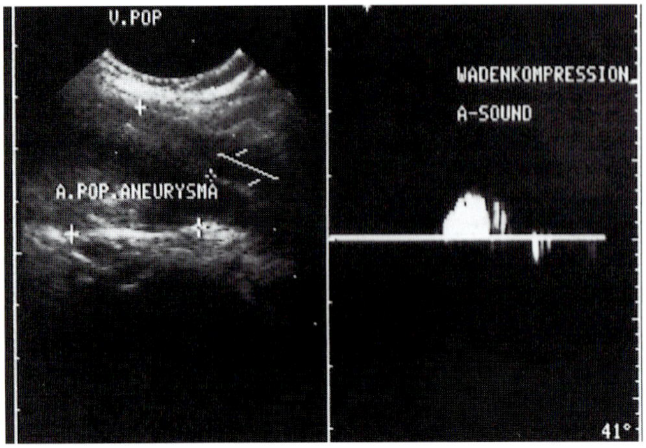


**Fig. A3.63**
**Vein compression by aneurysm**
Swelling of the lower leg caused by proximal vein compression as a differential diagnosis of vein thrombosis. In this case, the popliteal vein is compressed by an aneurysm of the popliteal artery. The Doppler waveform shows no spontaneous flow in the popliteal vein distal to the aneurysm. Calf compression elicits only a reduced signal (A-sound) due to the downstream flow obstruction. The sample volume placed in the venous lumen is shown above the popliteal artery with the *plus signs* indicating the extent of the aneurysm.

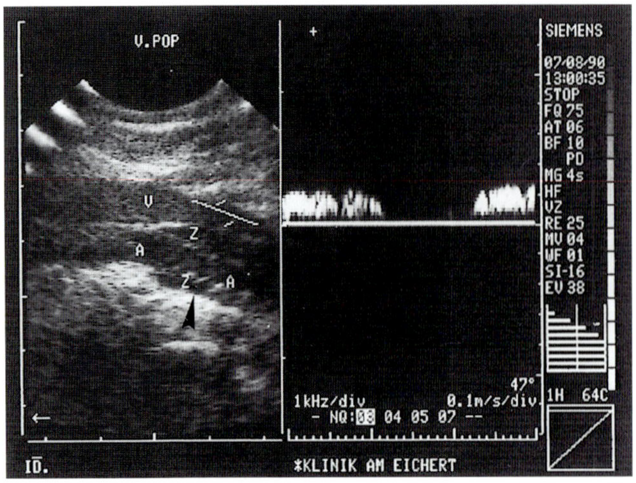


**Fig. A3.64**
**Venous wall indentation due to adventitial cysts**
Adventitial cystic disease of the popliteal artery causes indentation of the popliteal vein, evident on venography and ultrasonography. The Doppler waveform shows preserved respiratory phasicity and normal flow velocity (no stenosis signal).

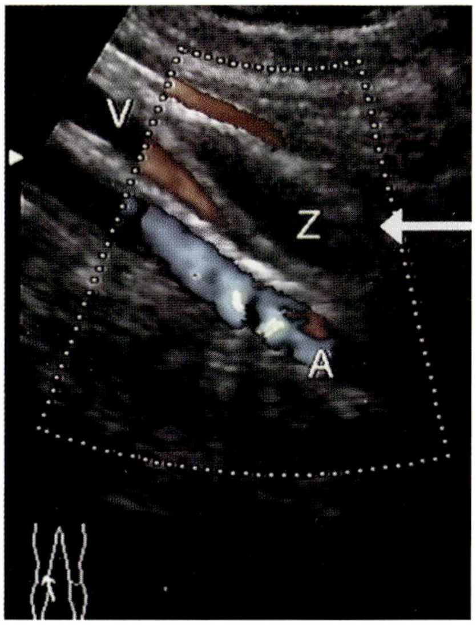


**Fig. A3.65**
**Adventitial cystic disease of the popliteal vein**
A cyst (*Z*) in the venous wall narrows the lumen of the distal popliteal vein (*V*). Variable filling of the cyst causes intermittent calf swelling with symptom-free intervals. Adventitial cystic disease of the popliteal vein was confirmed intraoperatively.

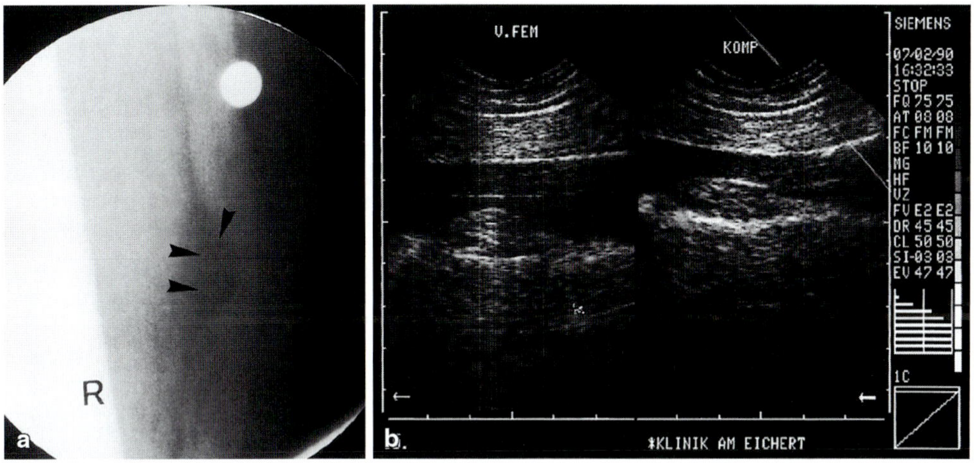


**Fig. A3.66 a, b**
**Venous wall tumor**
The contrast medium filling defect in the venogram (**a**) is caused by a tumorous lesion of the venous wall depicted by ultrasound (**b**). The wall is delineated and not disrupted (longitudinal scan). Scanning from an anteromedial position depicts the artery near the transducer and adjacent to the vein, which is compressible (*KOMP*, *right section* in **b**). Histologic workup of the surgical specimen yielded the diagnosis of a venous wall fibroma.

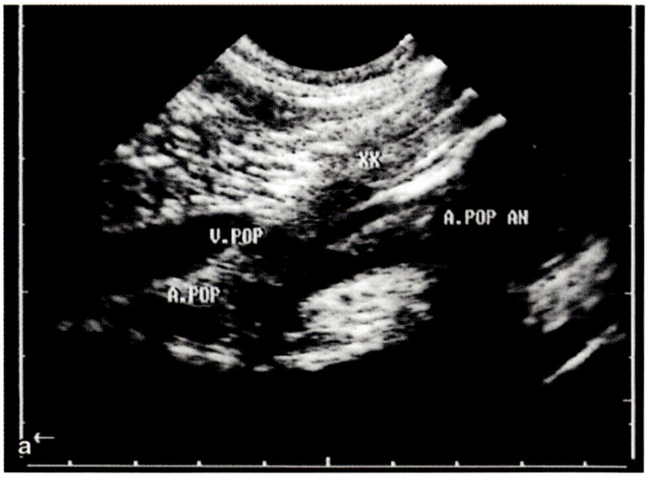


**Fig. A3.67 a–c**
**Entrapment syndrome**
**a** An entrapment syndrome of the popliteal artery very rarely involves the popliteal vein as well (see popliteal entrapment syndrome in Sect. 2.1.6.4.2). In a 45-year-old patient, malformation of the medial head of the gastrocnemius with a lateral extension (*XX*) to the lateral condyle of the femur (Insua type II) causes stenosis of the popliteal artery with poststenotic, thrombotic dilatation (*A.POP AN*). The atypical lateral gastrocnemius extension in this case also impairs venous return in the popliteal vein (*V.POP*), which is compressed between the dilated artery and the lateral muscle extension (*XX*).

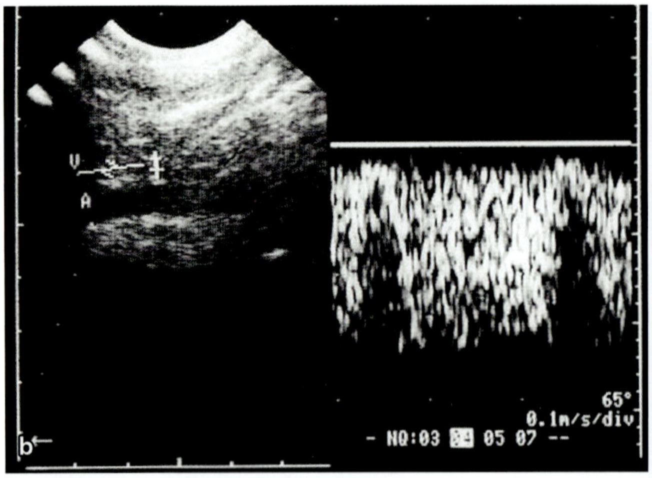


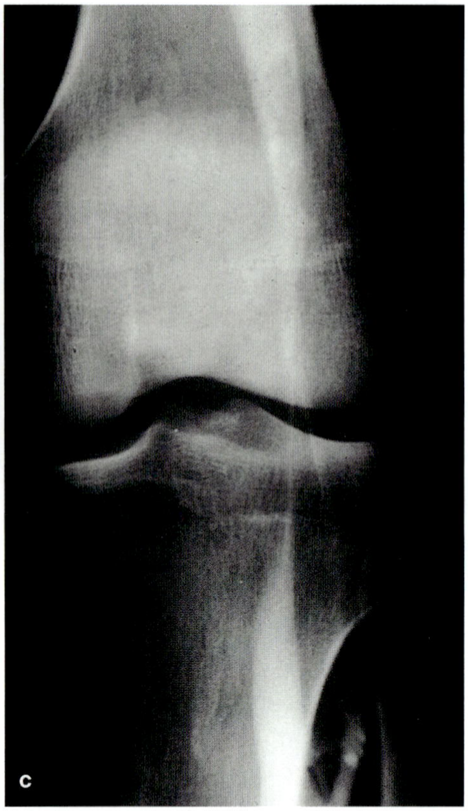


**Fig. A3.67 (continued)**
**Entrapment syndrome caused by hypertrophied gastrocnemius heads**
**b** 35-year-old athletic patient (well-developed calf muscles) presenting with calf swelling and exercise-induced pain. Ultrasound demonstrates compression of the popliteal vein by a hypertrophied gastrocnemius muscle with two strong heads but normal courses in the popliteal fossa. The Doppler spectrum sampled in the compressed popliteal vein with the patient lying in a relaxed position shows a stenosis signal interrupted by arterial pulsation. The vein has a lumen of 2 mm. The angle-corrected flow velocity is over 100 cm/s and respiratory variation is lost (same patient as in Fig. A2.50 a–c).
In this patient with the rare combination of arterial and venous compression, calf swelling was caused by compression of the popliteal vein at rest and exercise-induced pain by compression of the artery during plantar flexion.
**c** Venogram: The vein appears compressed. A large popliteal artery aneurysm or a large Baker cyst may have a similar venographic appearance.

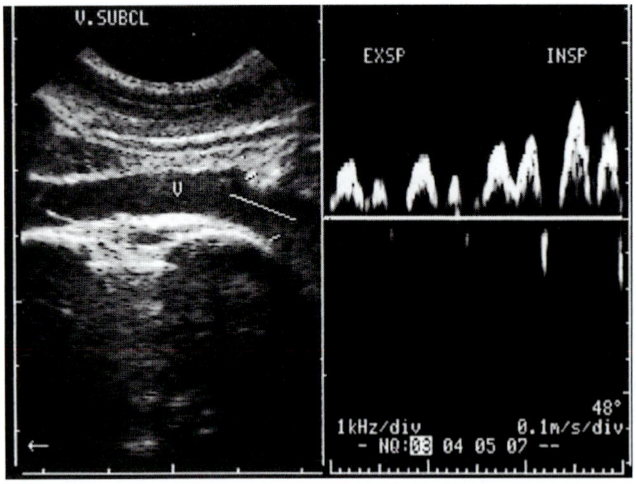


**Fig. A3.68**
**Axillary vein – normal findings**
Junction of axillary and subclavian veins with respiratory and typical cardiac fluctuation of blood flow velocity. The B-mode scan on the *left* depicts a venous valve.

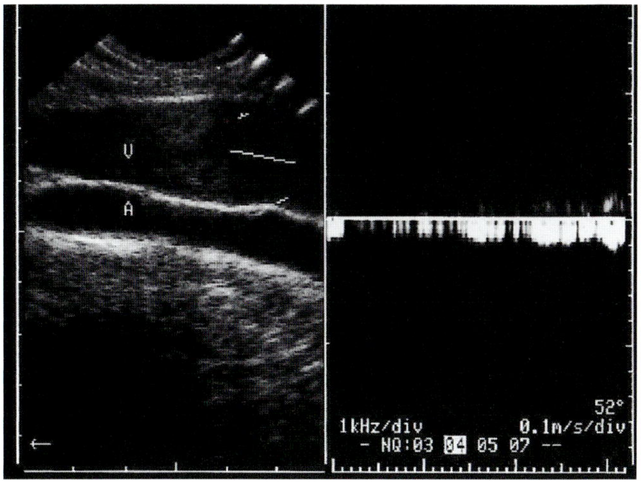


**Fig. A3.69**

**Obstruction of thoracic inflow**

Obstruction of venous inflow in the upper thoracic aperture by a mediastinal tumor is seen on the B-mode scan as a dilatation of the veins (shown here for the jugular vein). Blood flow is slower and cardiac pulsatility is eliminated.

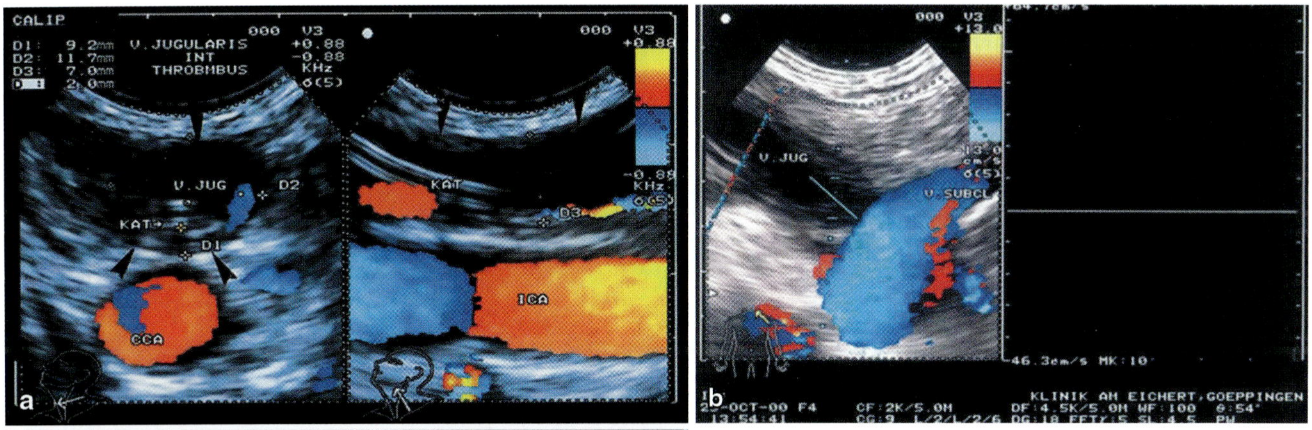


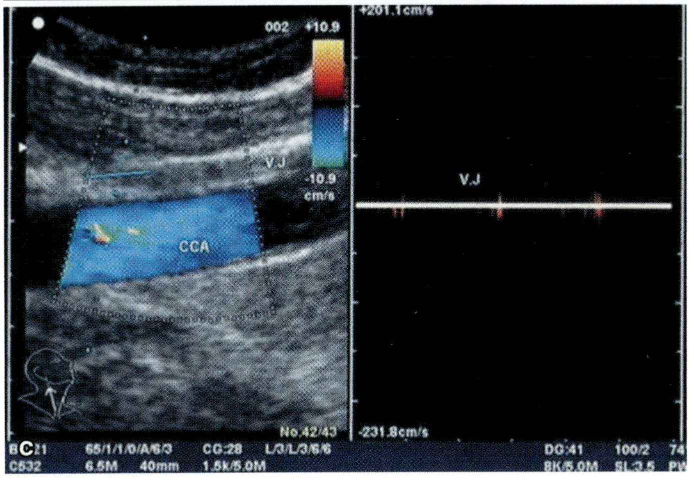


**Fig. A3.70 a–e**

**Jugular vein thrombosis (central venous catheter)**

**a** Foreign bodies in a vein (pacemaker, central venous catheter) have thrombogenic effects. In the example, the double contour indicates the central venous catheter (*KAT*) in the thrombosed jugular vein with residual flow near the wall depicted in blue on the transverse scan (*left*). The common carotid artery is seen medial to the thrombosed jugular vein (*arrowheads*; transverse scan on the *left*, longitudinal scan on the *right*). Flow in the artery is in the opposite direction. The color change from red, to black, to blue in the artery is due to a change in flow direction relative to the ultrasound beam.

**b** Due to the thrombosis, flow signals are absent from the jugular vein, which is scanned along its course toward the center to the level of its junction with the subclavian vein (*V.SUBCL*); the subclavian vein is patent.

**Older jugular vein thrombosis**

**c** In older jugular vein thrombosis, there may be partial recanalization or persistent obstruction with depiction of the vein as a connective tissue strand with a rather thin lumen adjacent to the carotid artery. An ultrasound examination of patients in whom implantation of a central venous catheter is planned can thus prevent unnecessary puncture.

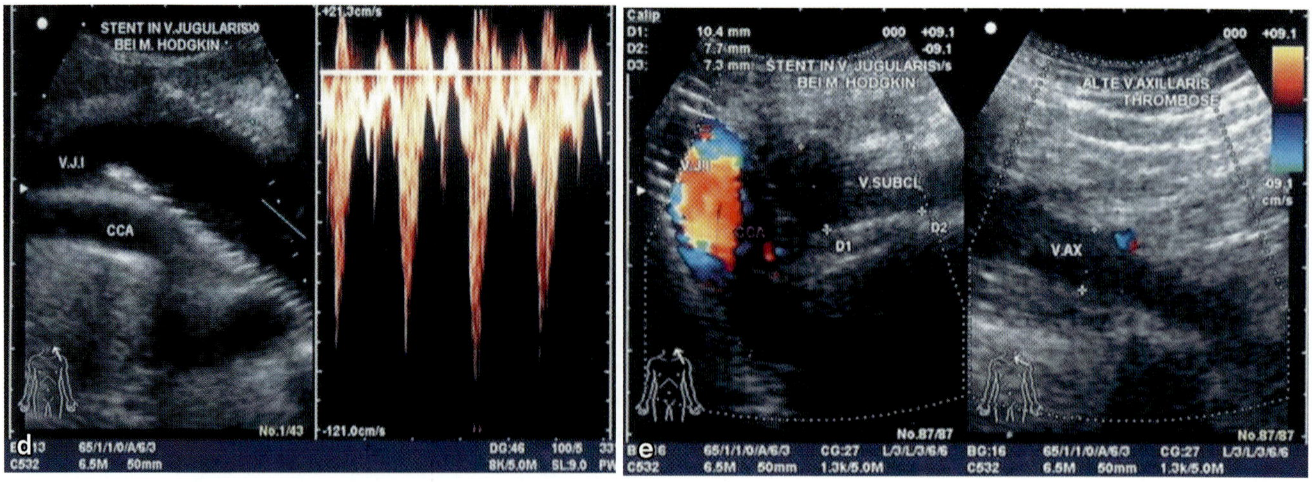


**Fig. A3.70 (continued)**

**Jugular vein compression treated by stenting**

**d** The mesh-like pattern indicates a stent placed to maintain patency of the jugular vein obstructed in the upper thoracic aperture by Hodgkin lymphoma.

**e** After stenting of the compressed jugular vein (*left section*), the patient developed thrombosis of the subclavian (*V.SUBCL*) and axillary veins (*V.AX*, *right section*).

**Fig. A3.71 a–c**

**Axillary vein thrombosis (thrombolytic therapy)**

**a** Hypoechoic and homogeneous thrombi in the axillary and subclavian veins with clear demarcation from the wall indicate acute thrombosis. The vein is markedly dilated compared to the artery posterior to it. Collateral veins are depicted anteriorly.

**b** Following two cycles of thrombolytic therapy with ultrahigh-dose streptokinase administration, color duplex imaging demonstrates beginning recanalization. There is complete recanalization of the distal axillary vein (flow depicted in red, toward transducer). In the proximal axillary vein (scan on the *left*), there is flow along one side of the thrombus (blue, due to change in flow direction relative to transducer). A chest wall collateral is seen anteriorly. The transverse view (*middle section*) depicts a second, larger hypoechoic mural thrombus in an otherwise patent axillary vein with flow in blue. The compression test confirms a thrombus and excludes a flow phenomenon due to inadequate instrument settings (scan on the *right*; see Fig. 1.39a). The patent lumen is collapsed and only the thrombosed, noncompressible portion is still identifiable as a hypoechoic structure. The transverse scans depict the axillary artery (*A*) posterocranially (*CL* = clavicle).

**c** There is full recanalization of the vein after another cycle of thrombolytic therapy. The Doppler waveform demonstrates respiratory as well as cardiac fluctuation of flow (M-shaped profile). The presence of cardiac pulsatility indicates that the thrombus was still very fresh when thrombolytic therapy was initiated; an older thrombus would have caused inflammatory changes and rigidity of the venous wall.

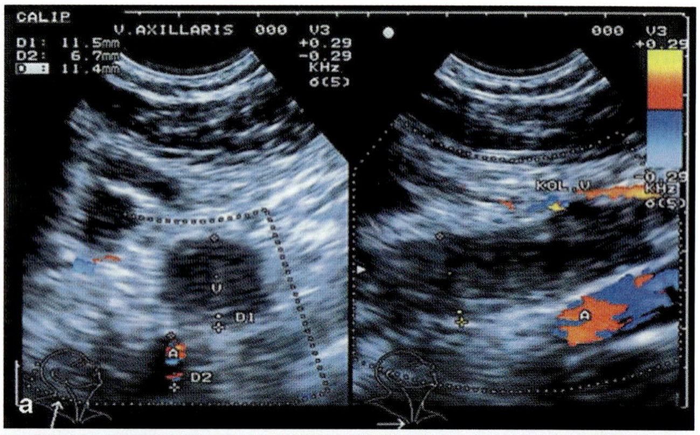


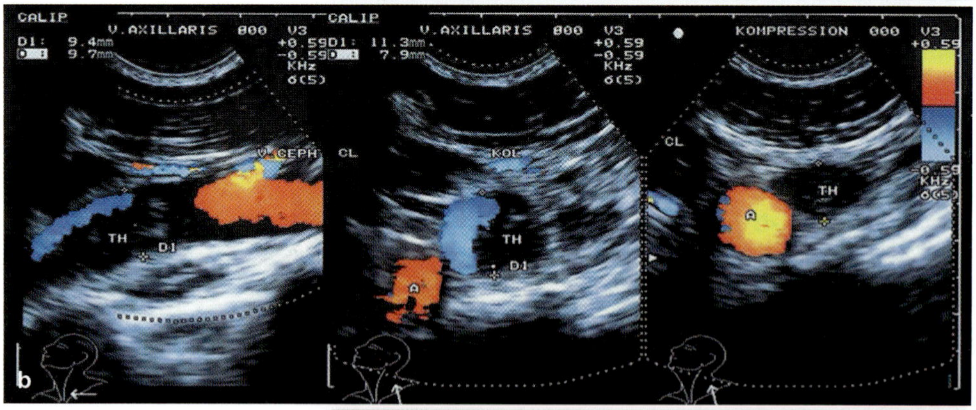


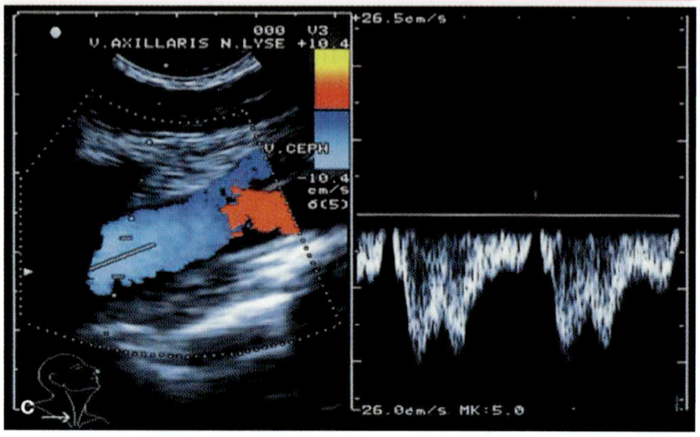

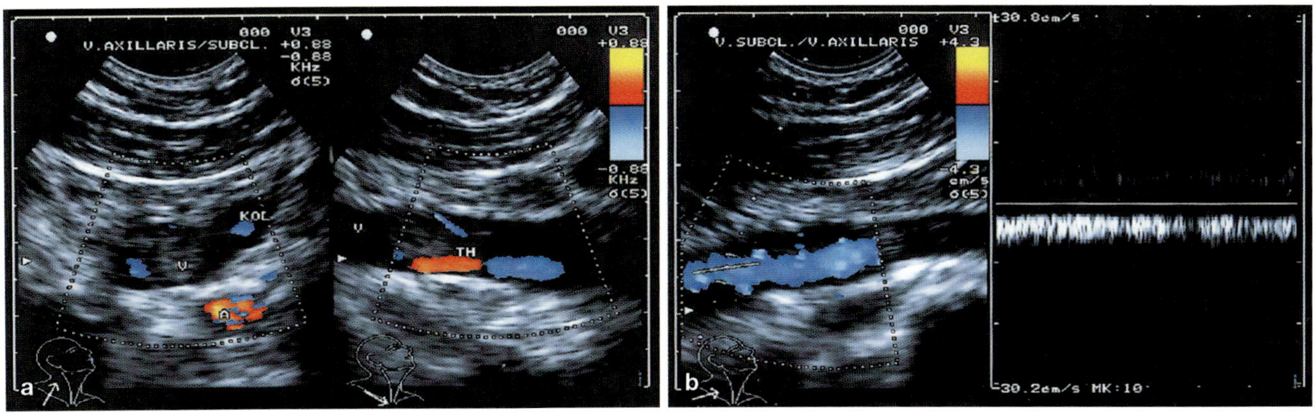


**Fig. A3.72 a, b**
**Recanalization**

**a** Thrombosis of the axillary vein as in the case presented in Fig. A3.71a–c; however, five cycles of thrombolytic therapy are necessary before signs of recanalization appear (transverse scan on the *left*, longitudinal scan on the *right*).

**b** Recanalization of the axillary vein is complete after another three cycles, but the wall is still markedly thickened as indicated by the hypoechoic structure surrounding the patent lumen (blue). The Doppler waveform shows no cardiac variation. Postthrombotic inflammatory changes, possibly associated with thrombotic deposits, lead to rigidity of the wall. In the case shown, the thrombogenic wall lesions carry a high risk of early recurrence of thrombosis. In this patient, recurrent thrombosis of the axillary vein with occlusion was seen 2 days later despite adequate heparinization.

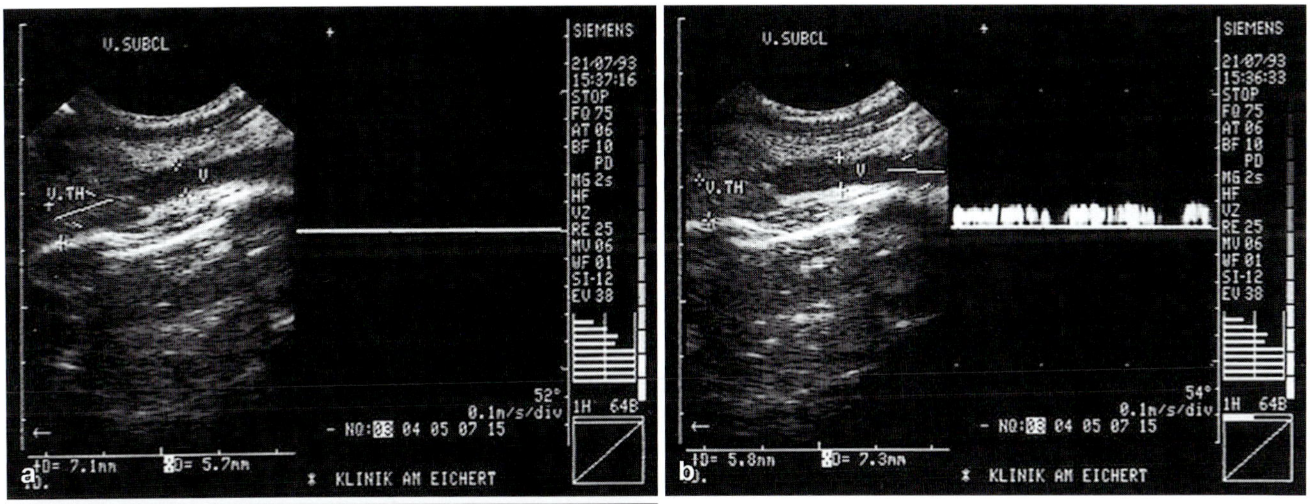


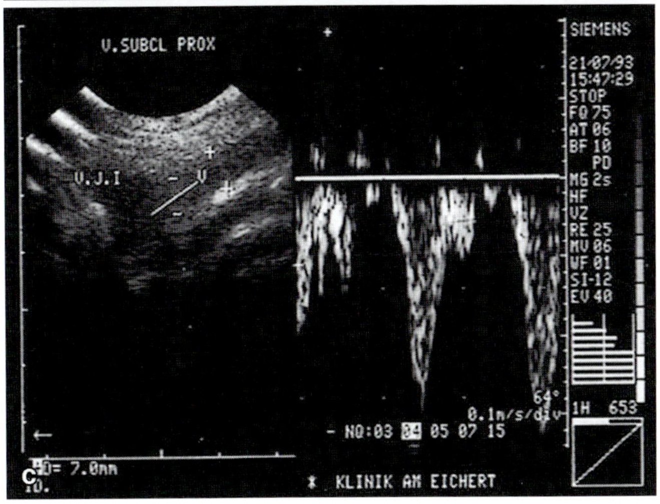


**Fig. A3.73 a–e**
**Costoclavicular compression syndrome – thrombosis, thrombolytic therapy**

**a** 17-year-old patient presenting with a 5-day history of swelling of the right arm and lividity of the hand and lower arm. She reported recurrent transient but very mild swelling of the arm. Ultrasound identified a short thrombus at the junction of the subclavian vein with the axillary vein immediately distal to the costoclavicular space.

**b** Upstream of the thrombus, the axillary vein is patent and the Doppler waveform indicates disturbed drainage with a continuous flow signal showing neither respiratory phasicity nor cardiac pulsatility.

**c** Proximal to the clavicle, the subclavian vein is patent and flow is normal flow with respiratory and cardiac variation.

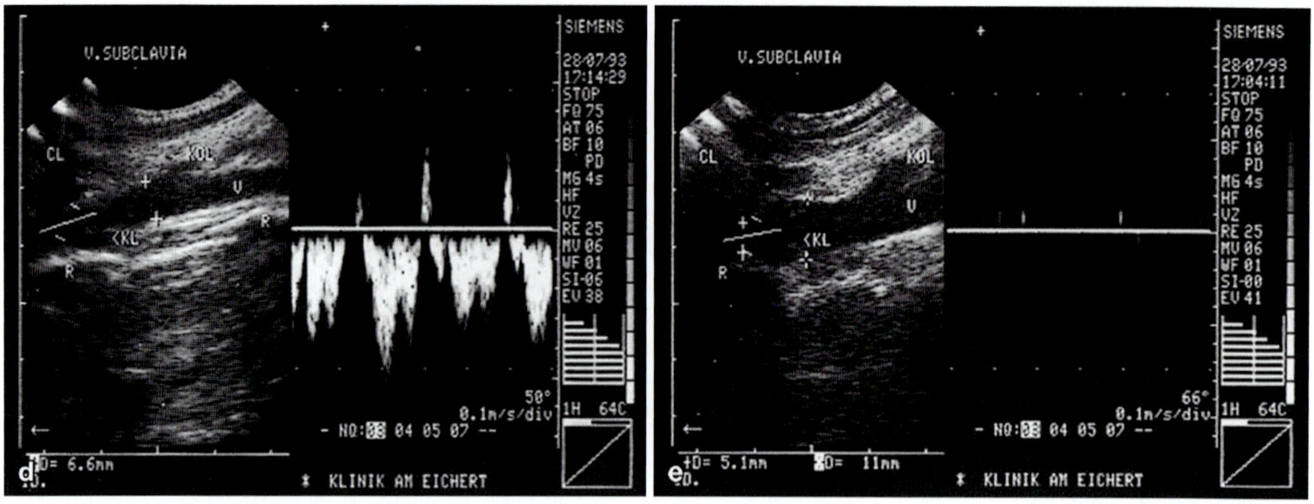


**Fig. A3.73 (continued)**

**d** After 3 cycles of ultrahigh-dose streptokinase, recanalization of the vein was observed and the suspected costoclavicular compression syndrome was confirmed as the cause of thrombosis by duplex scanning. The Doppler waveform sampled in the supine position with the arm relaxed shows a normal flow profile.

**e** Upon strong pulling of the arm in the posteroinferior direction, the vein becomes dilated distal to the costoclavicular space due to congestion. With the transducer in the infraclavicular fossa, the dilated subclavian and axillary veins as well as collateral veins (*KOL*) are seen. No flow signal is detected immediately distal to the costoclavicular space, indicating compression-induced occlusion of the subclavian vein. Valves (*KL*) are seen in the dilated lumen. (*CL* = clavicle).

A Doppler frequency spectrum should be obtained to document the costoclavicular compression syndrome because the color duplex findings are difficult to quantify and are more susceptible to artifacts as a result of the maneuvers performed to induce compression.

**Fig. A3.74**

**Costoclavicular compression syndrome**

The passage of the vein through the costoclavicular space between the clavicle and the first rib is difficult to depict due to bone-induced acoustic shadowing. Scanning of the vein in this area is possible only if tangential beam orientation can be achieved in slender patients. Under these conditions, a continuous high-frequency stenosis signal will be obtained from this vein segment with increasing abduction of the arm, possibly until complete occlusion of the subclavian vein has occurred. The findings presented were obtained in a 29-year-old patient with costoclavicular compression syndrome. As in most cases of this syndrome, the subclavian artery was not compressed and showed triphasic flow in the duplex examination.

In this patient, the same duplex findings could be elicited when the outwardly rotated arm was pulled in the posteroinferior direction. Extreme hyperabduction can induce compression of the subclavian vein in the costoclavicular space with demonstration of disturbed venous return in the Doppler waveform also in subjects without clinical symptoms of compression syndrome. For this reason, the hyperabduction test must be interpreted with caution. An abnormal Doppler waveform sampled while the outwardly rotated arm is being pulled posteroinferiorly is a more specific sign of the costoclavicular compression syndrome.

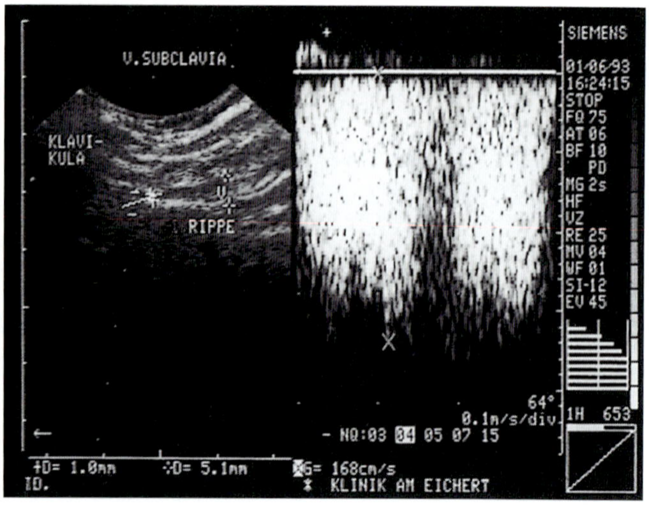

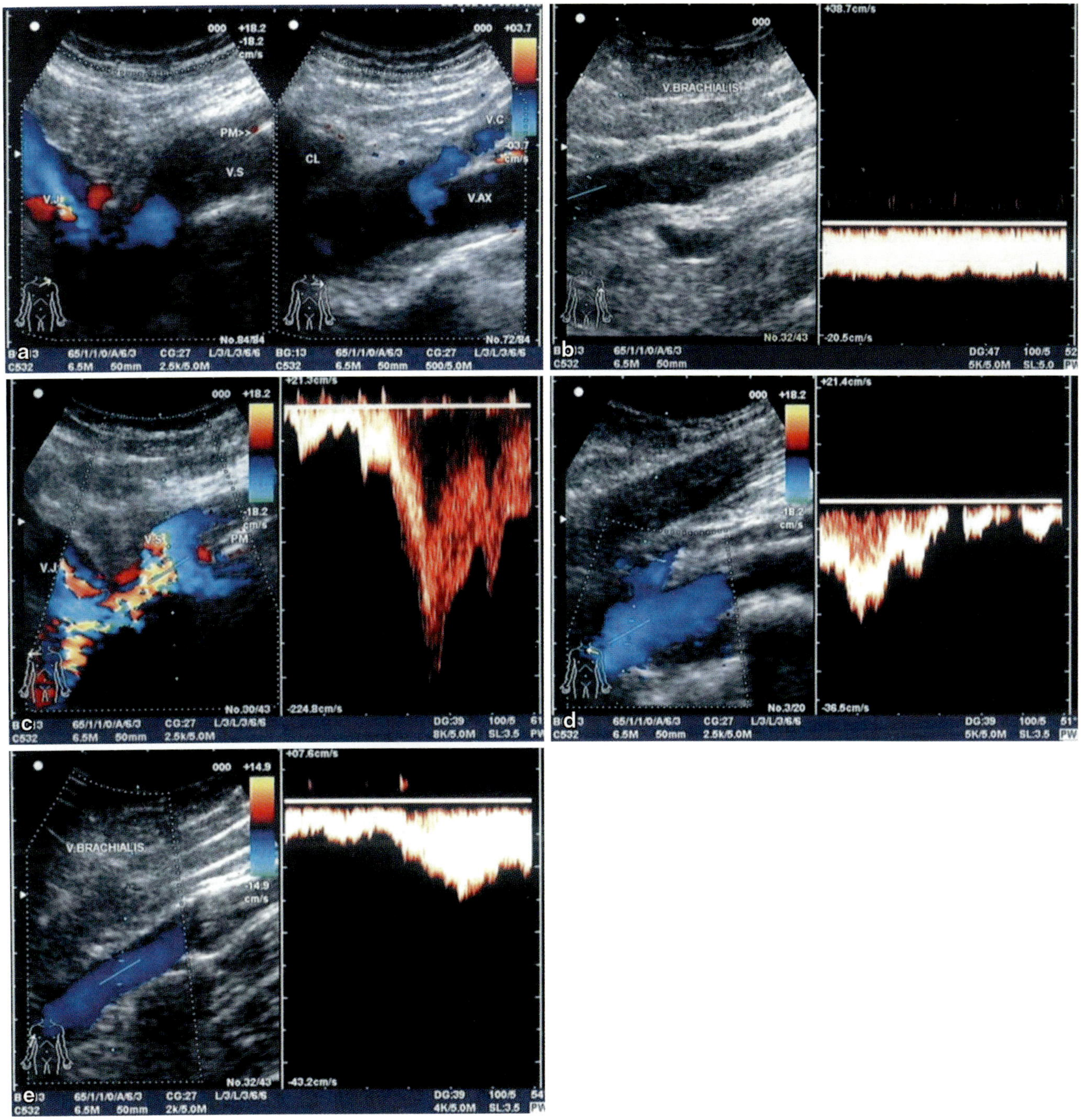


**Fig. A3.75 a–e**

**Follow-up of subclavian vein thrombosis after pacemaker implantation**

**a** One week after pacemaker implantation, ultrasound demonstrates thrombosis of the subclavian vein (*left section*) and of the axillary vein (*right section*). Only isolated segments of the partially thrombosed axillary vein show flow signals when scanned with a low PRF. The subclavian vein (*V.S*) is completely thrombosed to the level of entry of the jugular vein (*V.J*). The pacemaker probe (*PM*) is identified by the hyperechoic double reflection in the lumen.

**b** The Doppler waveform of the brachial vein shows the band-like flow profile with absence of respiratory phasicity typical of upstream flow obstruction (thrombosis).

**c** After only 2 days of low-molecular heparin (weight-adjusted therapeutic dose), the patient shows surprisingly early spontaneous recanalization. Residual thrombi are seen only around the pacemaker probe (*PM*). Moreover, there is narrowing of the subclavian vein as it enters the confluence (aliasing, but without demonstration of flow obstruction in the Doppler waveform).

**d** The axillary vein is completely recanalized with restoration of respiratory and cardiac flow variation (no central flow obstruction).

**e** The brachial vein now exhibits respiratory phasicity of flow with slight cardiac pulsatility, indicating absence of a flow obstruction (spectral Doppler recording at the same site as in **b**).

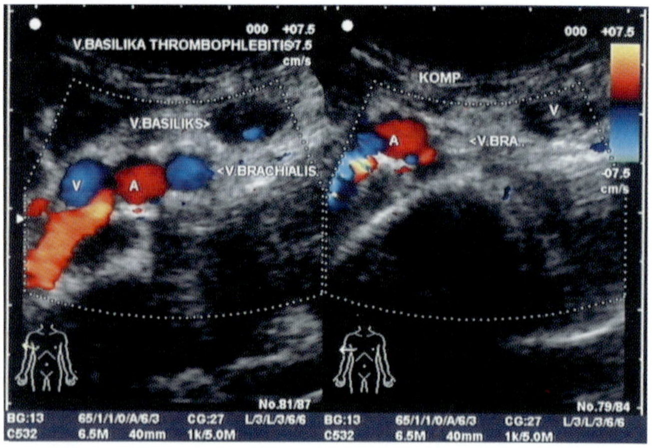


**Fig. A3.76**

**Thrombophlebitis of arm veins**

The arteries serve as landmarks to distinguish the deep veins from the superficial veins, thereby enabling differentiation of thrombosis and thrombophlebitis. In the example, the findings rule out venous thrombosis (brachial vein with flow displayed in blue, compressible as shown on the *right*) and confirm thrombophlebitis (basilic vein not compressible, superficial course, no accompanying artery).

# Shunts

# 4

## 4.1 Clinical Role of Shunt Diagnosis

### 4.1.1 Background

Of the 50,000 patients with end-stage renal failure in Germany, each year some 15,000 become candidates for creation of a hemodialysis access. A native arteriovenous (AV) fistula has a better prognosis with longer patency and fewer complications such as infections and is therefore preferred to the creation of a dialysis fistula with use of a synthetic implant (Tordoir et al. 2007). An advantage of a synthetic shunt is that it can be used earlier, while a native fistula needs time to mature before it can be used for hemodialysis access. A synthetic shunt is the second option in patients whose native vein (typically the cephalic vein) is deemed unsuitable because of a small lumen or because it has undergone thrombotic or fibrotic degeneration as a result of frequent puncture. A minimum flow volume is necessary to ensure adequate dialysis treatment. Protocols in the USA require a flow volume of at least 350 mL/min, while some European countries accept smaller volumes, for example, 200–300 mL/min in Germany. This requirement informs the preoperative search for a shunt vein or makes use of a synthetic loop necessary. Preoperative vascular mapping contributes important information for selecting the most suitable hemodialysis access for each patient.

### 4.1.2 Diagnostic Tasks in Patients with Spontaneous and Therapeutic Shunts

An AV fistula is a direct communication between an artery and a vein that bypasses the capillary bed. Such short circuits may be *congenital* or *acquired*. Acquired fistulas usually develop after trauma or as complications following puncture procedures or catheter examinations (Table 4.1).

In addition, an AV fistula or shunt may be created for *therapeutic* purposes, chiefly to provide *access for hemodialysis* tubing or to produce a transient connection in the groin after thrombectomy for pelvic vein thrombosis.

An AV fistula established for hemodialysis access must ensure an adequate flow volume without causing steal-induced ischemia or signs of cardiac insufficiency and provide a large enough segment for puncture. The site should be chosen with a view to minimizing patient discomfort.

The wrist or the bend of the elbow is the best site for a dialysis shunt, both in terms of surgical technique and ease of access. A minimum shunt volume of 300 mL/min is required for dialysis, while a volume exceeding 15–20% of the cardiac output may lead to cardiac insufficiency.

In cases where the vein is not adequate for creating a direct AV connection, a synthetic graft (polytetrafluoroethylene/PTFE or Gore-Tex) may be interposed.

The classic AV connection for hemodialysis is the Brescia-Cimino fistula, which is an end-to-side anastomosis between the cephalic vein and radial artery at the level of the wrist (Fig. 4.1a) or between the cephalic vein and brachial artery in the bend of the elbow. Synthetic shunts are established as loops from the brachial artery at the elbow to the basilic or brachial vein, most commonly as a U-shaped loop implanted subcutaneously in the lower arm. Alternatively, a shunt can be established with interposition of a straight graft between the brachial artery and the cephalic, axillary, or jugular vein (Fig. 4.1b).

The standard diameter of a PTFE prosthesis is 5 or 6 mm. It can be used for hemodialysis immediately after implantation. A direct AV fistula on the other hand requires 3–4 weeks of maturation, undergoing sufficient dilatation and wall thickening before it can be punctured.

Natural AV shunts have a better prognosis than synthetic implants, which may be affected by various functional problems requiring repeat revision (see Table 4.5 in Sect. 4.8).

Both the unphysiologically high flow rates and repeat puncture of the access vein induce intimal proliferation, frequently leading to stenosis and occlusion. Published shunt patency rates range widely, depending on the patient population, inclusion criteria, and type of shunt investigated. The patency rate reported for Brescia-Cimino shunts is 80–90%


DOI: 10.1007/978-3-642-02509-9_4, © Springer-Verlag Berlin Heidelberg 2011

**Table 4.1** Types of arteriovenous (AV) fistulas

| | |
|---|---|
| Congenital AV fistula: | Direct anatomic linkage between the arterial and venous system or indirect communication through short circuits in the soft tissue |
| Acquired AV fistula: | Iatrogenic: complication of arterial catheter examinations or renal transplant biopsy<br>Trauma<br>Spontaneous |
| Therapeutic AV fistula: | Temporary: after thrombectomy for pelvic vein thrombosis<br>For hemodialysis access in renal failure |

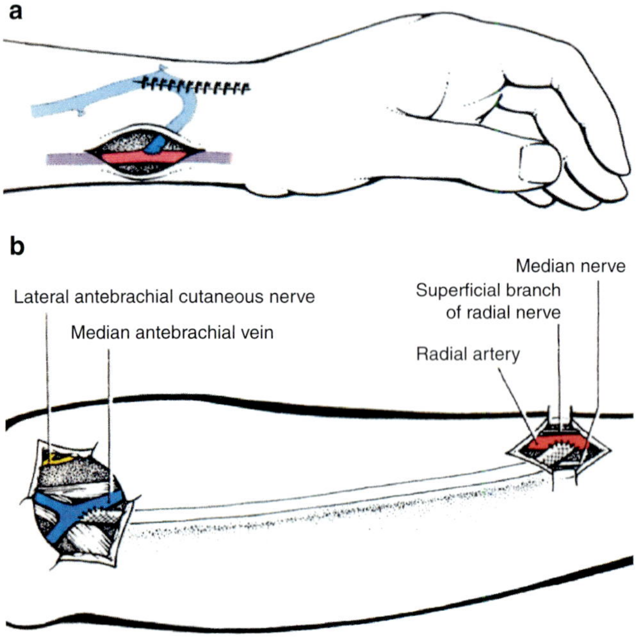


**Fig. 4.1** **a** Distal cephalic vein fistula of the lower arm: side-to-end anastomosis of the radial artery and distal cephalic vein. **b** Example of a synthetic shunt with an alloplastic graft (from Heberer and van Dongen 1993)

after 1 year, 63–87% after 2 years, and about 65% after 4 years (Ahmad et al. 1998; Brittinger et al. 1966; Harnoss et al. 1991; Keller et al. 1991, 1988). In contrast, synthetic AV shunts have patency rates of 62–90% after 1 year, 50–79% after 2 years, and about 40% after 4 years (Haimov 1979; Munda 1983; Tellis 1979). Since the morbidity and quality of life of patients on chronic hemodialysis depend on good shunt function, the timely recognition and proper interpretation of shunt problems is an important clinical task. Noninvasive modalities such as color duplex ultrasonography are the most suitable diagnostic tests, enabling early identification of the underlying cause of a reduced flow rate through the fistula or other shunt complications and prompt initiation of adequate therapeutic measures.

*Clinically*, a fistula is recognized by a palpable thrill and a more or less persistent high-frequency bruit that is present throughout the cardiac cycle and varies with shunt flow.

The following types of congenital, acquired, and therapeutic AV fistulae are distinguished:

- Congenital AV fistulae: Direct anatomic communications between the arterial and venous system (malformation) or blood-conducting structures between the two systems such as aneurysmal changes or multiple short circuits in the soft tissue or bone.
- Acquired AV fistulae:
  - Spontaneous fistulae without an apparent cause in the patient's history or clinical presentation
  - Traumatic
  - Iatrogenic as a complication of arterial puncture or biopsy of a parenchymal organ such as a kidney graft
- Therapeutic AV fistulae:
  - Temporary AV fistula after thrombectomy for pelvic vein thrombosis
  - For improved patency of femorocrural bypass grafts with a poor outflow tract (controversial)
  - AV fistula for hemodialysis access

Color duplex ultrasound has the following indications in assessing patients with a suspected AV fistula:

- Congenital or acquired AV fistula:
  - Demonstration of the fistula
  - Localization
  - Identification of the feeding artery and draining vein
  - Estimation of shunt flow
- Therapeutic AV fistula:
  - Estimation of shunt flow
  - Evaluation of shunt complications:
    - Shunt occlusion
    - Stenosis (at site of anastomosis or within the fistula)
    - Stenosis of proximal artery or of shunt vein
    - Peripheral ischemia (steal phenomena) in high-flow shunts and follow-up after shunt banding
    - Puncture aneurysm
    - Perivascular complications: abscess, hematoma

On color duplex ultrasound, a congenital or acquired (non-therapeutic) AV fistula is characterized by a color Doppler bruit (*mosaic of colors*) resulting from *highly turbulent flow* in the fistula and *perivascular vibration*. Moreover, the higher flow velocity will cause aliasing if the scan parameters are set for depicting normal venous flow.

The *Doppler waveform* of the feeding artery shows a *monophasic flow profile* with a *large diastolic component*, which is due to lower peripheral resistance. The *draining vein* has an *arterialized flow profile* with severe turbulence.

The (color) duplex examination allows identification and evaluation of the feeding artery and draining vein.

In therapeutically established shunts, ultrasonography enables noninvasive evaluation of shunt complications and is useful for estimating shunt flow. Assessment of blood flow in the feeding artery has been found to be the most reliable method of determining shunt flow rate, as measurement within the shunt or draining vein is degraded by turbulence and variability of the shunt diameter. The measurement in the inflow artery can be used to calculate the shunt volume by comparing it with the contralateral side or by subtracting the blood flow in the artery distal to the shunt. The flow volume is calculated from the mean velocity and the cross-sectional area of the vessel.

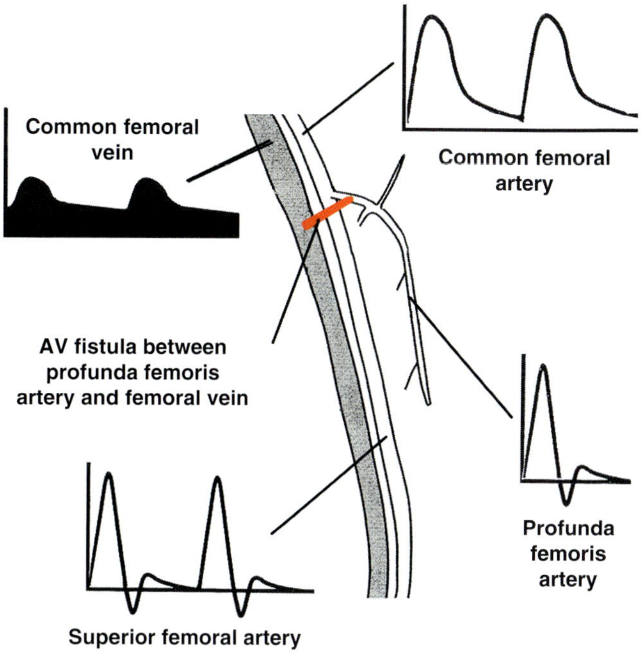


**Fig. 4.2** An iatrogenic arteriovenous (AV) fistula commonly develops between the proximal profunda femoris artery and the femoral vein, typically when the puncture is made too far peripherally. The diagrams show the Doppler waveform changes proximal to an AV fistula (common femoral artery, monophasic pattern) and distal to an AV fistula (superficial femoral and profunda femoris arteries, both triphasic). The Doppler waveform from the draining vein proximal to the AV fistula shows pulsatile flow (common femoral vein). The magnitude of the diastolic flow component proximal to the AV fistula reflects the flow volume within the fistula

## 4.2 Examination Protocol, Technique, and Diagnostic Role

### 4.2.1 Congenital and Acquired Fistulae

Accurate information on the site of a fistula with its feeding artery and draining vein is helpful for planning the surgical procedure.

When a fistula is suspected on clinical grounds, this information serves to guide the sonographic search (in the color duplex mode) for the arteriovenous communication, the inflow artery, and the draining vein. The transducer frequency must be adjusted to the required scanning depth. The high flow velocities in a fistula and the occurrence of perivascular vibration artifacts make it necessary to use a high pulse repetition frequency (PRF).

The choice of transducer, patient positioning, and the procedure depend on the body region in which the fistula is clinically suspected (e.g., palpable thrill or limb swelling due to disturbed venous drainage). Using the color mode, the examiner can look for the prominent color bruit caused by tissue vibration around the fistula.

In the periphery, where flow resistance is typically high, the *feeding artery* is identified by an *abnormal Doppler waveform* with low-resistance flow and *increased diastolic flow*. Distal to the fistula, the artery resumes its normal triphasic flow profile (Fig. 4.2).

An AV fistula with a hemodynamically significant shunt volume is identified by intermittent sampling of the arterial Doppler frequency spectrum in front of and behind the suspected site. This is possible because there is persistent diastolic flow in the inflow artery of an AV fistula, while the typical pulsatile and triphasic flow of extremity arteries predominates distal to the fistula. In this way, the site of the fistula can be identified as the point of transition between these two arterial flow patterns. At the same time, venous return proximal to this point will show pulsatile variation. In addition to precise localization, which is relevant for planning the surgical procedure, the shunt volume may have to be calculated as well. This is done on the basis of the diameter of the feeding artery, determined from the B-mode image, and the mean flow velocity (angle-corrected Doppler measurement), from which the normal blood flow volume of the artery is subtracted. The normal flow volume is determined in the artery of the same name on the contralateral side. Veins draining a fistula are characterized by an arterialized flow profile.

### 4.2.2 Dialysis Access

In the sonographic evaluation of therapeutic AV shunts and their clinical complications, it is not the morphologic or hemodynamic changes as such that are crucial for therapeutic decision making, but rather the clinical manifestations they produce. The indication for therapy is chiefly established on the basis of the clinical problems, while the choice of treatment is made on the basis of duplex ultrasound or the results of other imaging modalities (shunt

PTA, creation of a new shunt, shunt revision, ligation of collateral veins, resection of aneurysm, banding of fistula, shunt closure).

The clinical question to be answered determines the extent of the sonographic examination. In patients presenting with signs of peripheral ischemia or cardiac insufficiency, for instance, it is necessary to quantify the shunt volume. If dialysis flow has become insufficient, the examiner must look for stenosis of the shunt artery or vein.

A *shunt of the lower arm* is best examined in the sitting patient with the elbow slightly bent and the lower arm resting on a support. The superficial course of the arm vessels enables their examination with a high-frequency transducer (7.5–10 MHz). Use of a linear transducer has the advantage of providing better contact with the arm. An upper arm dialysis fistula is examined in the supine patient with the arm comfortably positioned on a support for optimal exposure of the fistula site for transducer maneuvers. A high PRF is necessary to depict the fast flow resulting from the high shunt volume, whereas the gain must be downregulated to eliminate vibration artifacts. A lower PRF is necessary to depict slow postocclusive flow. First, the examiner evaluates the length of the fistula in transverse orientation, followed by hemodynamic evaluation in the longitudinal plane with sampling of Doppler spectra in the feeding artery, the shunt vein (along the length of the access segment) or the synthetic loop graft, and in the draining vein. Additional Doppler measurements can be performed at the anastomotic sites. Based on the clinical query and the hemodynamic findings obtained in the shunt (flow character), the examiner then narrows down the site of the abnormality by continuously scanning the length of the fistula. Due to the high shunt-related diastolic flow, the indirect criteria of arterial obstruction (monophasic flow profile) are not reliable. Therefore, in inconclusive cases, the feeding limb must be scanned continuously from the subclavian artery to the brachial or radial artery in longitudinal orientation. A Doppler spectrum is obtained to detect flow acceleration as a sign of stenosis.

In patients with a Brescia-Cimino dialysis access, the shunt veins must be followed along their course toward the center to identify dilatation or narrowing (Fig. 4.3a–c).

The veins must be examined with very light pressure to avoid compression, which may be misdiagnosed as a stenosis. This is achieved by placing several fingers or the edge of the hand holding the transducer on the arm outside the course of the shunt. The transducer can thus be moved with very subtle pressure. Moreover, undue pressure is avoided by placing the transducer somewhat lateral to the apex of the vein wall and then angling it to interrogate the vein. Synthetic grafts are less susceptible to compression.

Apart from palpation, the course of a shunt is most easy to track sonographically in transverse orientation. Spectral

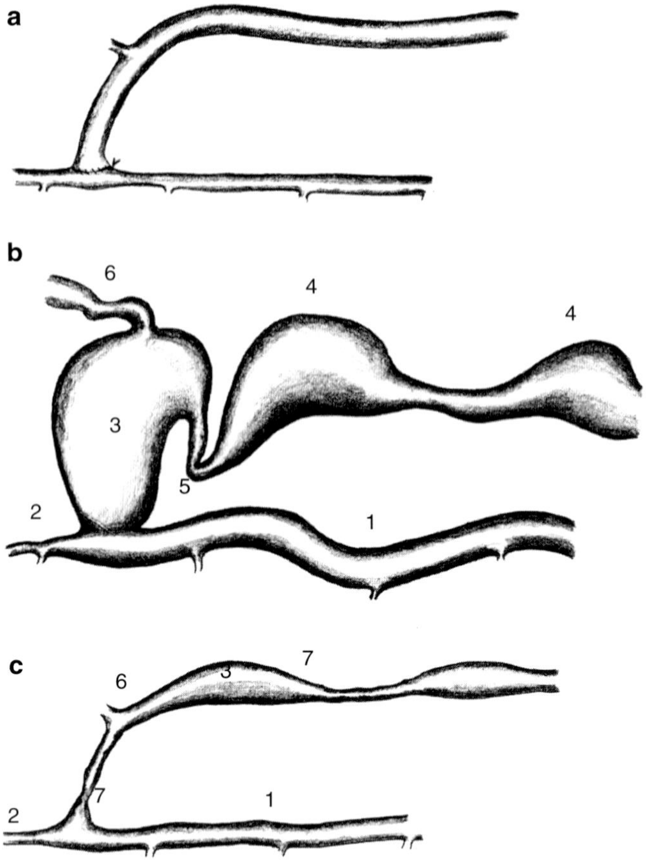


**Fig. 4.3** **a–c** Diagram of the morphologic changes that can occur in a hemodialysis access: **a** shortly after creation of the AV shunt; **b** dilatation; **c** stenosis (*1* feeding artery, *2* draining artery, *3* shunt vein, *4* dilatation or stricture at site of frequent venipuncture, *5* kinking-induced stenosis, *6* side branch of the shunt vein, *7* stenosis) (from Scholz 1995)

Doppler measurement is performed in the longitudinal plane at sites suspicious for stenosis. The known thrill in the shunt area may produce perivascular vibration and thus impair assessment. This problem can be overcome by carefully compressing the area next to the transducer with the flat hand, while at the same time avoiding excessive compression of the vein. When an intricate shunt is examined, an overview of flow directions in the different venous limbs can be obtained in the color duplex mode.

When a narrowing is encountered in the B-mode examination or when aliasing or perivascular tissue vibration (mosaic of colors) appears in the color duplex mode, a Doppler waveform is obtained from the respective site in the shunt in longitudinal orientation to confirm a stenosis. Artifacts caused by perivascular vibration can be avoided by slight soft-tissue compression. Proper positioning is verified in the B-mode by slightly changing the pressure exerted with the transducer. Tortuous veins are better appreciated transversely, but the Doppler waveform is nevertheless recorded in longitudinal orientation.

The sonographer also evaluates perivascular structures using longitudinal and transverse sections (and color duplex as needed) to differentiate hematoma, abscess, and shunt aneurysm.

## 4.3 Typical Shunt-Related Changes in the Doppler Waveform

The low peripheral resistance associated with an AV short circuit results in a monophasic waveform with continuous systolic and diastolic flow and a large diastolic flow component in the feeding artery (Table 4.2). A return to pulsatile flow in a therapeutic AV shunt indicates a low shunt volume due to disturbed venous drainage, shunt stenosis, or shunt occlusion.

Perivascular tissue vibration around the shunt, especially during systole, may be conspicuous as a mosaic of colors. The smaller the shunt caliber and the larger the jet, the more pronounced the perivascular vibration artifact. Turbulent flow in the shunt is identified by a mix of colors and by considerable spectral broadening in the Doppler waveform; there may even be retrograde flow components during systole.

The vein draining the shunt is dilated and, due to arterialization, shows pulsatile flow and spectral broadening (primarily close to the fistula) due to turbulence. Vibration of the vessel wall and adjacent soft tissue in the color duplex scan can be reduced by slight manual throttling of arterial inflow in order to minimize artifacts, which is especially important when recording Doppler spectra for comparison of flow velocities before and after a suspected stenosis.

All draining veins have arterialized flow. Lateral venous branches that divert blood away from the access vein, but are unsuitable for hemodialysis, can thus be identified and ligated.

The flow changes observed in a hemodialysis shunt that has been used for many years may lead to intricate flow patterns in arteries that are only indirectly, through collaterals, connected to the feeding arteries (e.g., steal phenomena, supply of a radial artery shunt by the palmar arch and ulnar artery). Evaluation of color-coded blood flow directions allows correct interpretation and identification of shunt problems (such as ischemia of the fingers and reduced shunt flow) under such complex flow conditions as well.

**Table 4.2** Color duplex findings in the presence of an AV fistula

| |
|---|
| Monophasic flow profile due to continuous systolic and diastolic flow with a large diastolic component in the feeding artery |
| Pulsatile flow in the arterialized draining vein |
| Very turbulent flow across the fistula (over a long stretch in a dialysis access) |
| Perivascular tissue vibration around the fistula |
| Dilatation of the inflow artery and draining vein after several years of use for hemodialysis access |

The flow velocities measured by spectral Doppler vary widely with fistula age and dilatation. Peak systolic velocity (PSV) in the inflow artery may be up to twice as high as in the contralateral counterpart with a large diastolic flow component, resulting in a Pourcelot resistance index of 0.7–0.4. Depending on the diameter, even greater variability in flow velocity may be seen in the draining vein, which is arterialized with flow velocities of 50–150 cm/s. Flow velocity in a synthetic AV access graft varies with arterial inflow and venous outflow resistance. Depending on the graft diameter, systolic velocities range from 100 to 400 cm/s with 60–200 cm/s at end diastole (Lockhart et al. 2001).

## 4.4 Shunt Maturation and Flow Volume Measurement

A decreased flow through the access fistula, due to complications such as stenosis, impairs hemodialysis function. Only a high-grade shunt stenosis becomes functionally relevant, which, according to Kathrein (1988, 1991), is defined as a decrease in the volume flow rate below 250 mL/min. Although this would seem to be the most obvious thing to do, blood flow is not measured directly in the affected access vein. This is because abrupt changes in diameter, especially in older shunts, and changes in the lumen shape (elliptical) give rise to errors. Determination of mean velocity within the fistula is also impaired by turbulent flow (spectral broadening). For these reasons, the *flow volume in an AV hemodialysis access can be determined most reliably* by measuring the mean flow velocity *in the main feeding artery* (typically the brachial artery). The flow volume is the mean velocity (calculated from the Doppler waveform of the feeding artery by means of the implemented software) multiplied by the cross-sectional area (see Sect. 1.1.2.4). Flow volume measurements performed on different ultrasound machines may vary by up to 30%. One reason is the use of different methods *for determining the cross-sectional area* (direct planimetric measurement or calculation from diameter, leading-edge method). Another is the way in which flow velocity is determined: it may be calculated as the mean velocity in the vessel cross section or as the median velocity. An inadequate receive gain can thus produce measurement errors. Calibration measurements are rarely done before flow volumes are measured. The discrepancies are less relevant as long as serial measurements are performed with the same equipment.

Grosser et al. (1991) investigated volume flow measurements performed in the brachial artery, radial artery, and shunt vein and found *the best reproducibility for measurements in the brachial artery*. Pitfalls are the calculation of the cross-sectional areas of small arteries (blooming effect) and the determination of mean flow velocity when flow is very turbulent.

The most accurate results are obtained by calculating the difference in arterial blood volumes before and after the shunt anastomosis. Alternatively, the volume in the inflow artery can be determined relative to that of the contralateral artery. It should be noted, however, that the blood flow to the arm is negligible in most cases compared with the high shunt flow.

Measurement of flow volumes in the draining vein is discouraged, especially in older shunts, as it is subject to the above-mentioned errors. However, when there is complex branching of veins, an attempt can be made to determine the functionally relevant shunt flow of the access vein. To do so, the Doppler sample volume must be placed in a straight segment of vein with little turbulence and without major caliber variation.

The flow volume can serve to evaluate the maturity of a newly created fistula. A wide range of flow rates from 500 to 1,200 mL/min make an AV fistula suitable for hemodialysis.

A retrospective study investigated the accuracy of ultrasound in the evaluation of AV fistula maturity in 69 patients using different criteria such as minimum venous diameter (Robbin et al. 2002). The fistula was adequate for dialysis when the diameter was at least 4 mm in 89% of cases versus 44% for diameters of less than 4 mm. A flow volume of 500 mL/min or greater enabled adequate dialysis in 84% of cases versus only 43% if flow volume was less than 500 mL/min. Inadequate shunt maturation should prompt a color duplex search for stenosis, focusing on the inflow artery and the anastomosis. Venous outflow obstruction is more likely to induce the formation of a collateral pathway circumventing the shunt vein. Such a collateral is inadequate for dialysis and can cause swelling of the arm. Arterial or venous stenosis can be treated by PTA and followed up by serial ultrasonography; however, long-term patency is poor (Clark et al. 2007). An anastomotic stenosis requires surgical revision. Poor shunt flow increases the risk of occlusion, which is over 50% when the flow volume drops below 300 mL/min (Lockhart et al. 2001; Bay et al. 1998).

## 4.5 Documentation

The documentation of findings depends on the diagnostic task and the type of shunt complication suspected. Individual images often give only a poor representation of the intricate vascular patterns that may be encountered, especially in hemodialysis patients who have undergone repeat shunt revision. To facilitate serial examinations, the following views should be documented in addition to the shunt complication (such as degree of stenosis):

- A longitudinal view and Doppler waveform of the feeding artery
- A longitudinal view and Doppler waveform of the anastomosis
- A longitudinal view with Doppler waveform of the shunt vein

Further documentation depends on the clinical question to be answered. Prior to shunt revision, all vessel segments involved must be assessed and the findings documented, for example, the axillary or jugular vein if placement of a synthetic graft is planned. A drawing of the vascular anatomy around the fistula site, where the parameters determined are also marked in, may be helpful in documenting complex vascular relationships for subsequent follow-up examinations and for reporting the findings to colleagues.

## 4.6 Vascular Mapping Prior to AV Fistula Creation

Creation of a direct, native AV fistula for dialysis access is always preferable to a synthetic graft. The site should be as far distal as possible so as to minimize ischemic complications and to avoid excessively high flow volumes. Note, though, that a Brescia-Cimino fistula connecting the radial artery and cephalic vein in the forearm has an early failure rate of 15.3% with a 1-year patency rate of only 62.5% (Rooijens et al. 2004). Failure or early occlusion can be minimized by performing preoperative vascular mapping, especially if the creation of a forearm fistula is planned. In most cases, a careful preoperative clinical evaluation will identify the most suitable shunt procedure for dialysis access (Table 4.3); however, preoperative color duplex imaging has been shown to facilitate the decision and improve the patency rate (Silva et al. 1998; Huber et al. 2002). The radial artery diameter should be at least 2–2.5 mm (Korten et al. 2007); a diameter of <1.5 mm is associated with low flow volumes and an early occlusion rate

**Table 4.3** Preoperative assessment of candidate vessels for creation of an AV hemodialysis access: prerequisites for adequate shunt flow and good shunt function

| Preoperative color duplex | Minimum | Waveform |
|---|---|---|
| Velocity in inflow artery | > 50 cm/s | Triphasic arterial waveform |
| Arterial diameter | > 2.0 mm | – |
| Venous drainage | – | Venous flow with respiratory phasicity and, toward the center, cardiac pulsatility |
| Venous diameter (possibly with placement of a tourniquet) | > 2.5 mm | – |

of up to 45% (Parmar et al. 2007). Atherosclerosis may also prevent adequate maturation of the draining vein. The cephalic vein should be >2.5 mm in diameter to ensure good venous return. A tourniquet can be placed to determine the diameter of the vein with maximum filling (Lockhart et al. 2006). A cephalic vein diameter <2 mm was reported to result in inadequate shunt maturation in 24% of cases (Mendes et al. 2002). Preoperative diameter measurement has high intra- and interobserver agreement, confirming the validity of the method (Planken et al. 2006). Preoperative color duplex mapping should rule out flow obstruction or sclerosis of the cephalic vein, which may have developed as a result of thrombosis from repeat prior cannulation. Doppler measurements should be obtained to confirm unimpaired central venous drainage through the axillary vein (respiratory phasicity and cardiac pulsatility of flow). For ease of cannulation, a superficial vein is preferred. Ideally, the candidate vein should not be more than 0.5 cm from the skin surface. In patients with obese arms though, an alternative vein may have to be used; the vein will be mobilized and tunneled superficially, or the placement of a synthetic loop will have to be considered.

## 4.7 Abnormal Findings (Inadequate Dialysis)

A color duplex scan is indicated whenever the clinical examination yields findings that warrant further workup or if shunt flow is inadequate for dialysis. Obstructed venous drainage may present with edema, swelling of the arm, livid discoloration, and secondary bleeding after hemodialysis. Obstruction may also cause an increased venous pressure at the dialyzer. Arterial problems may cause ischemia with pain and necrosis of the fistula hand (steal phenomenon) or may lead to inadequate shunt flow.

*Poor shunt flow and inadequate dialysis* have many causes with typical changes on duplex scanning. These are:

- Decreased inflow due to stenosis or occlusion of the feeding arteries
- Decreased drainage due to obstruction of runoff veins (thrombus, stenosis)
- Cardiac insufficiency
- Insufficient blood flow in the access segment due to diversion of blood into venous branches
- Partial shunt thrombosis with reduction of patent lumen
- Anastomotic stenosis

In addition, the function and prognosis of a dialysis fistula can be impaired by the following factors:

- Shunt aneurysm (true or false)
- Steal phenomena and peripheral ischemia
- Local infection, hematoma
- Difficult access due to deep or small vessels

### 4.7.1 Shunt Stenosis

The clinical presentation and palpation findings as well as problems encountered during hemodialysis (needle clinging – insufficient inflow; increased venous pressure – outflow obstruction) suggest the type of shunt complication and the site of obstruction that can be expected. Based on these clues, the suspected shunt segment is identified in the color duplex mode and evaluated for the presence of stenosis. This is done by means of the Doppler frequency spectrum sampled in longitudinal orientation, which is analyzed using basically the same criteria as in the peripheral arteries. Direct criteria include local flow acceleration, turbulent flow, and perivascular vibration artifacts. Changes in the flow profile are of limited value as monophasic flow predominates in a fistula due to the low resistance resulting from the venous short circuit. Still, a high-grade constriction will induce increased upstream pulsatility and decreased downstream pulsatility. Localized obstructions can be identified in the B-mode if a high-resolution transducer is used. Their hemodynamic significance is subsequently evaluated in the Doppler frequency spectrum. Moreover, the B-mode information enables differentiation of intramural and extramural causes of luminal narrowing. An example of an intramural process is intimal proliferation. Other obstructing lesions are local thrombotic deposits. These can be differentiated from extramural structures such as hematomas. Early intimal proliferation is seen as a hypoechoic wall deposit or a color filling defect. With further progression, the proliferating intima becomes inhomogeneous and may calcify.

Because normal flow velocity in a fistula is already high, a higher PSV of 2.5 m/s is used as a cutoff to identify hemodynamically significant stenosis of an AV fistula. Moreover, a doubling of the PSV compared with the upstream velocity can serve as a criterion for stenosis in a recently established fistula, but, due to caliber irregularities, is unreliable in older fistulas. Indirect criteria for shunt stenosis include a triphasic flow profile in the feeding artery and a drop in shunt flow below 250 mL/min (Table 4.4).

**Table 4.4** Criteria for identifying relevant hemodialysis access stenosis

| | |
|---|---|
| **Direct criteria** | |
| Luminal narrowing (B-mode) | |
| Systolic velocity ratio | Cutoff |
| • Arterial inflow | > 2.0 |
| • Venous outflow | > 3.0 |
| • Arteriovenous anastomosis | > 3.0 |
| **Indirect criteria** | |
| Reduced shunt volume (< 300 mL) | |
| Prestenotic waveform: high-resistance flow (triphasic) instead of low-resistance flow | |
| Delayed systolic upstroke distal to suspected stenosis | |

(Modified from Tordoir et al. 1989; Kathrein 1991; Grosser et al. 1991)

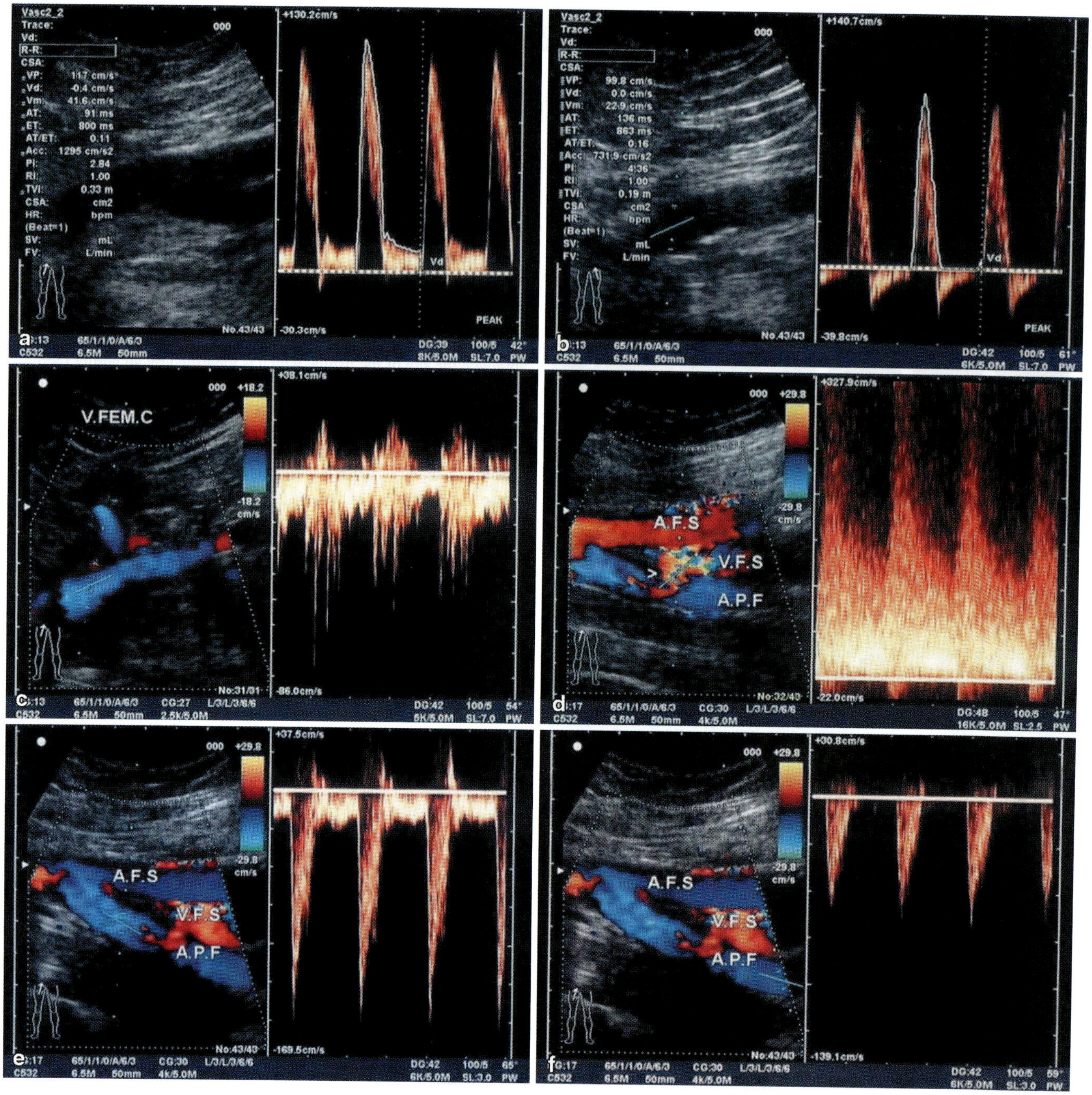


**Fig. A4.2 a–f**

**Iatrogenic arteriovenous fistula**

**a** There is continuous diastolic flow in the common femoral artery on the right compared to the contralateral side. The peak velocity is 117 cm/s during systole and 10 cm/s at end diastole with a mean flow velocity of 47.6 cm/s.

**b** Comparison with the unaffected side shows flow in the left common femoral artery to be triphasic with a PSV of 99.8 cm/s and a mean peak velocity of 22.9 cm/s. The common femoral artery diameter is the same on both sides.

**c** The common femoral vein on the right has a pulsatile flow profile (with flow toward the center displayed in blue) characteristic of an arterialized vein draining an AV fistula.

**d** As illustrated by the case presented, an iatrogenic AV fistula (as a complication of cardiac catheterization) nearly always develops between the superficial femoral vein and the profunda femoris artery. This complication typically occurs when the access site in the groin is chosen too low. The search for the fistula reveals a connection extending from the profunda femoris artery (*A.P.F*; blue flow away from transducer) to the superficial femoral vein (*V.F.S*) with a high-frequency flow signal (aliasing, red) and a flow velocity of over 3.5 m/s in the Doppler frequency spectrum. Anteriorly, the superficial femoral artery is depicted (*A.F.S*; red, toward transducer).

**e** The Doppler frequency spectrum sampled in the profunda femoris artery (*A.P.F*) proximal to the AV fistula shows a large diastolic flow component and the same flow profile as the common femoral artery.

**f** Distal to the AV fistula (see **d**), the profunda femoris artery (*A.P.F*; coded in blue) shows a triphasic profile without end-diastolic flow. This change in flow pattern proves that the AV fistula is located between the two sampling sites (in **e** and **f**).

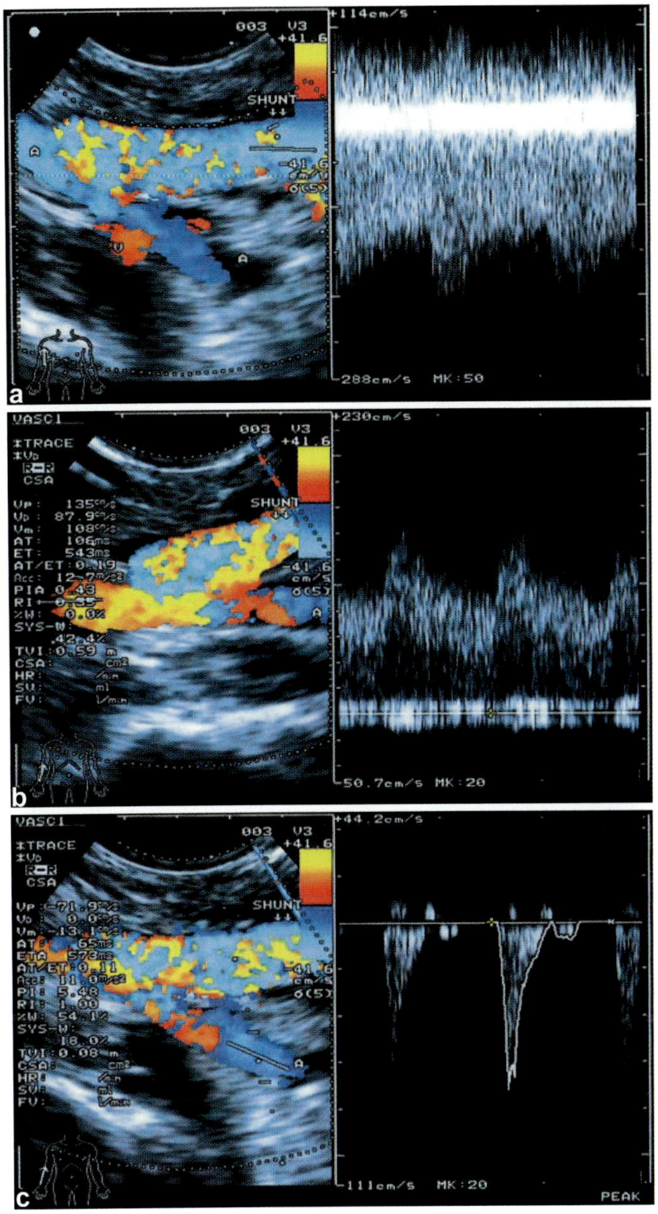


**Fig. A4.3 a–c**

**Hemodialysis access – flow measurement**

**a** Oblique image of the anastomosis of a Brescia-Cimino shunt (end-of-vein-to-side-of-artery anastomosis) in the bend of the elbow with marked turbulence at the anastomosis. Stretched brachial artery coursing posterior to the anastomosis.

**b** The color scan (*left section*) shows the brachial artery with flow in the cranial direction coded in red and mild aliasing on the *left* and the distal brachial artery on the *right* (coded blue). The sharp transition from red to blue appears to indicate flow reversal but is due to a change in flow direction relative to the transducer. The faster blood flow in the feeding artery is indicated by the lighter color.

The Doppler frequency spectrum sampled from the feeding artery (*right section*) shows shunt flow with a large diastolic flow component (peak end-diastolic flow velocity of 95 cm/s). With a calculated mean flow velocity of 108 cm/s and brachial artery diameter of 4.8 mm, the flow volume in the feeding artery is 1,170 mL/min.

**c** The brachial artery distal to the shunt has the typical flow profile of arm arteries: triphasic waveform without an end-diastolic component. The flow volume calculated for the brachial artery segment immediately after the shunt origin is 129 mL/min (0.16 $cm^2$ × 60 × 13 cm/s). The resulting shunt flow volume (calculated as the difference in flow volumes in the brachial artery before and after the shunt origin) is 1,040 mL/min.

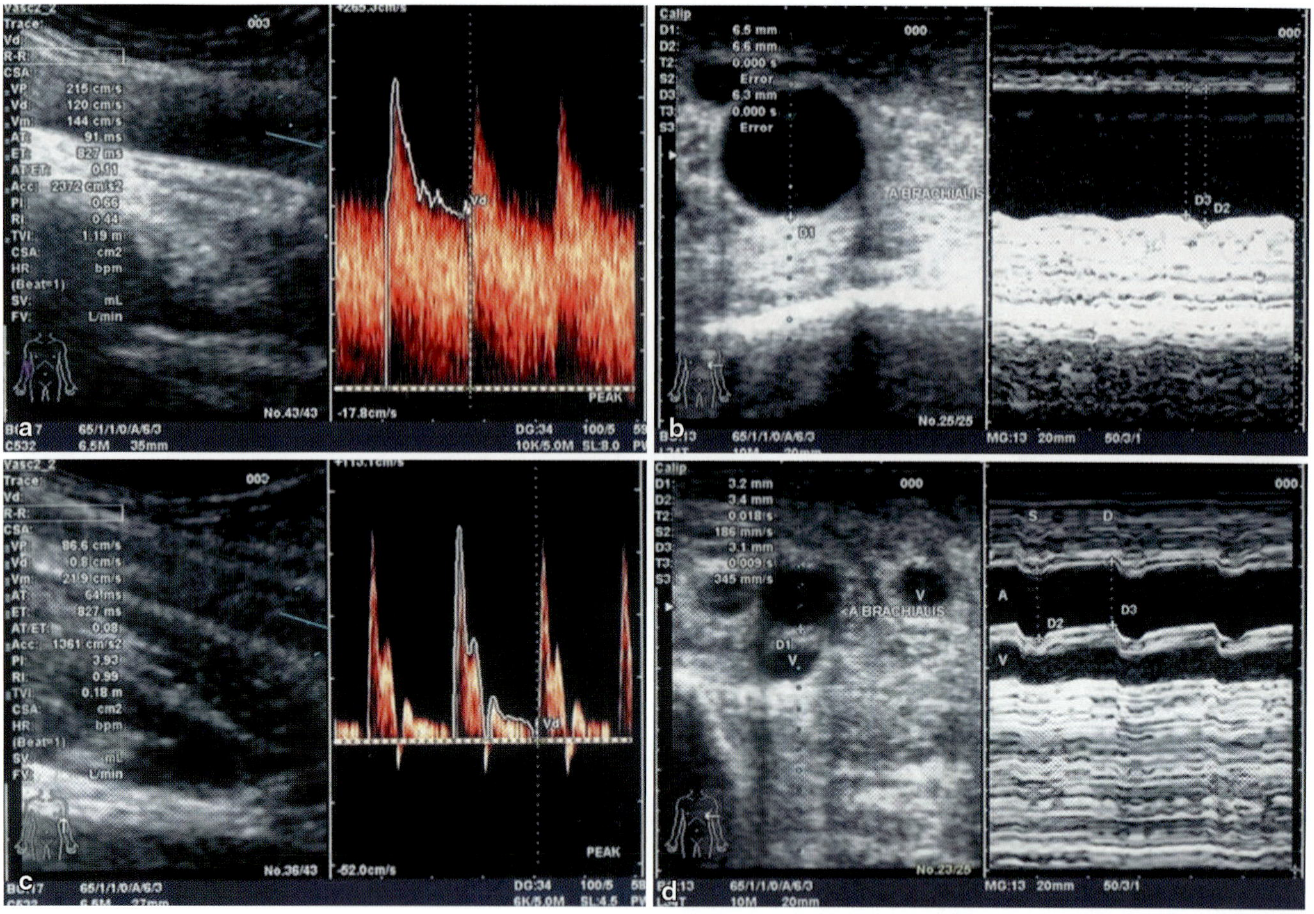


**Fig. A4.4 a–d**

**Complications of hemodialysis access – flow measurement**

Excessive shunt flow can lead to steal phenomena with subsequent ischemia of the hand or cardiac insufficiency. Since hemodialysis patients often have considerable comorbidity, the shunt must be examined as a possible cause of newly occurring signs of cardiac insufficiency. Duplex ultrasound is the simplest and most reliable method for measuring the flow volume in the AV fistula. As an alternative to the method presented in Fig. A4.3c, the flow volume in the brachial artery can be determined on both sides and the shunt volume calculated as the difference between the shunt-bearing arm and the non-shunt-bearing arm.

**a** Mean flow velocity is calculated with the integrated software using a Doppler frequency spectrum sampled with an angle of less than 60° (here: 144 cm/s).

**b** At the same site, the vessel diameter is measured in the B-mode scan (6.5 mm). For precise measurement, the systolic and diastolic diameters have to be determined and weighted at a ratio of 1:2. This is done in the time-motion mode with scanning of the vessel at an angle as close to 90° as possible. In the example, a flow volume of 2,778 mL/min is calculated from the mean flow velocity and cross-sectional area.

**c** The same measurements are performed in the brachial artery of the non-shunt-bearing arm, where a mean flow velocity of 21.9 cm/s is calculated. The flow profile is triphasic.

**d** After determination of the mean cross-sectional area from the systolic and diastolic diameters, a mean flow volume of 108 mL/min is calculated The example also illustrates the flow-induced dilatation of the arterial vessels as a cause of increased flow in long-standing shunts [the diameter differences between the views with spectral recordings (**a**, **c**) and those with time-motion displays (**b**, **d**) are due to the use of different scales].

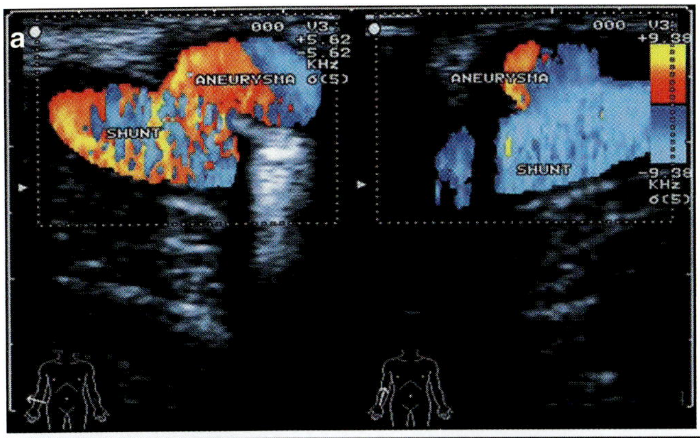


**Fig. A4.5 a, b**
**Puncture aneurysm**

**a** Puncture aneurysm of a shunt vein is seen on ultrasound as a circumscribed dilatation with flow signals. It is unlikely to produce spontaneous occlusion of a synthetic shunt. The scan on the *left* depicts the synthetic shunt with hyperechoic walls, from which a puncture aneurysm arises. The corresponding longitudinal scan on the *right* documents a fairly small puncture aneurysm.

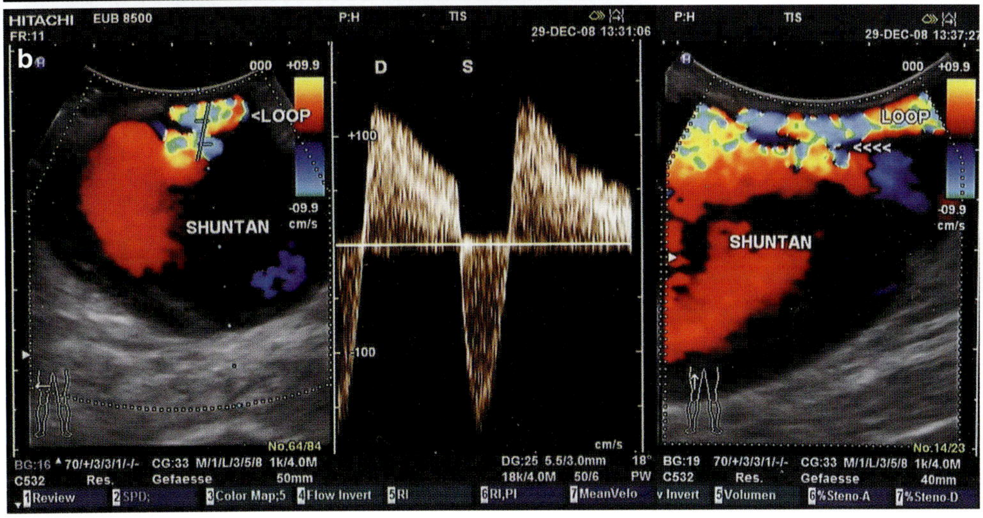


**b** Woman with a long history of hemodialysis and a loop in the thigh following loss of hemodialysis shunts in both arms due to multiple complications. A posterior puncture aneurysm was suspected, due to piercing of the far wall of the shunt (*SHUNTAN*). The Doppler waveform obtained with the sample volume placed in the leak between the loop and the shunt aneurysm reveals the changes characteristic of a false aneurysm: flow into the aneurysm (below the baseline, away from transducer) during systole (*S*) and back into the loop as a result of the changed pressure during diastole (above the baseline, toward transducer). In inconclusive cases, a spectral Doppler measurement can thus help to differentiate between severe ectasia of the fistula (only in a direct AV fistula without an interposed conduit) and puncture aneurysm (false aneurysm).

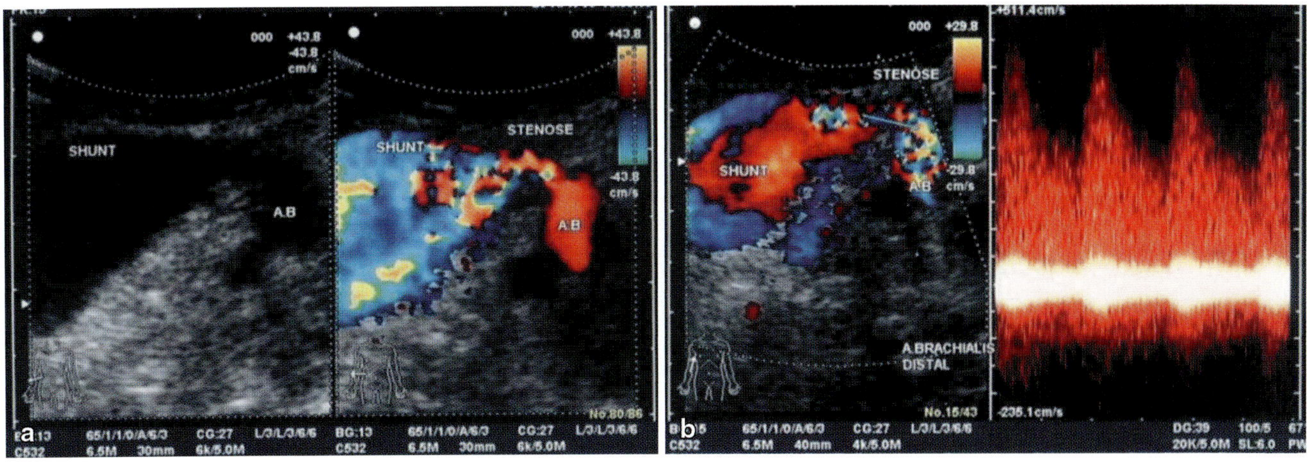


**Fig. A4.6 a, b**
**Stenosis of dialysis access with poststenotic ectasia**

**a**, **b** A Brescia-Cimino shunt established in the bend of the elbow many years before shows marked variations in caliber with high-grade stenosis upstream of an aneurysmal dilatation. Color duplex scanning demonstrates turbulence and aliasing; the PSV measured by Doppler is more than 500 cm/s. The red flow signals in the aneurysm are due to eddy currents.

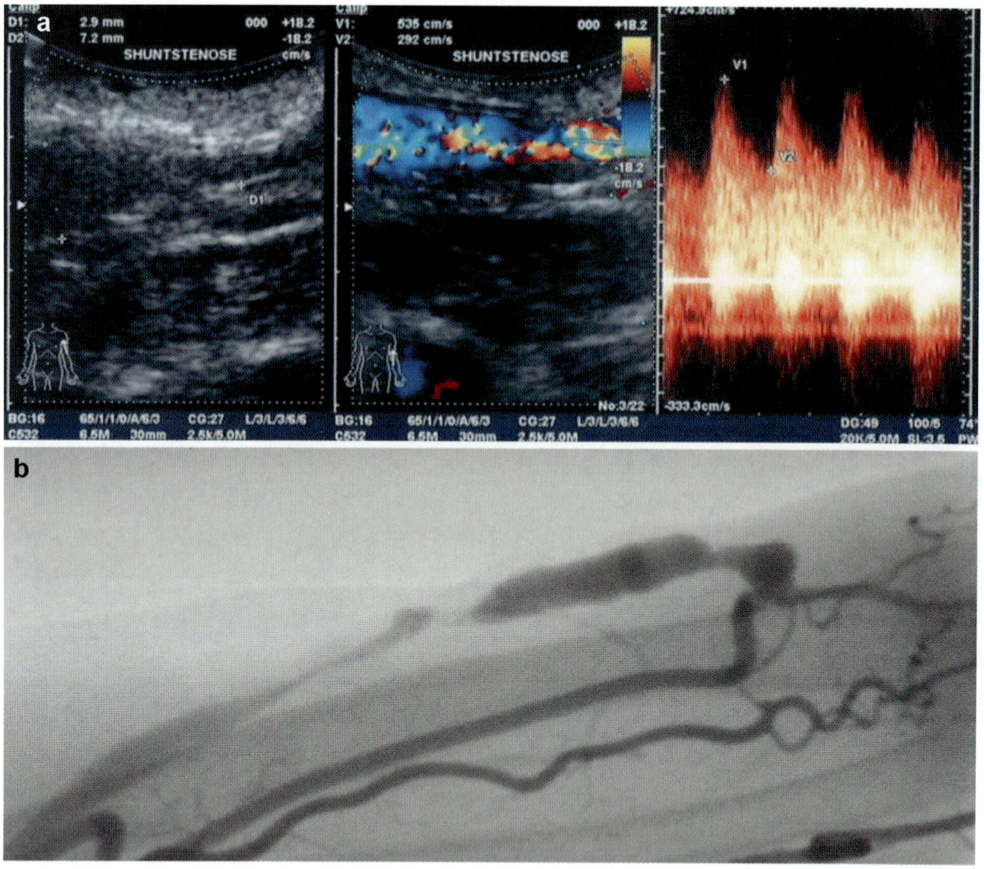


**Fig. A4.7 a, b**
**Stenosis of dialysis access**
**a** Perivascular tissue vibration impairs morphologic as well as hemodynamic assessment of a shunt stenosis by color duplex scanning. Therefore, morphologic assessment is done on the basis of the B-mode scan without activation of the color mode (*left section*). Assessment in the color duplex mode may be improved by controlled compression of the feeding artery to reduce shunt flow. A high-grade stenosis is indicated by the Doppler frequency spectrum with a peak flow velocity of 536 cm/s in systole and 292 cm/s at end diastole.
**b** Angiogram demonstrating shunt stenosis (after revision and synthetic graft interposition).

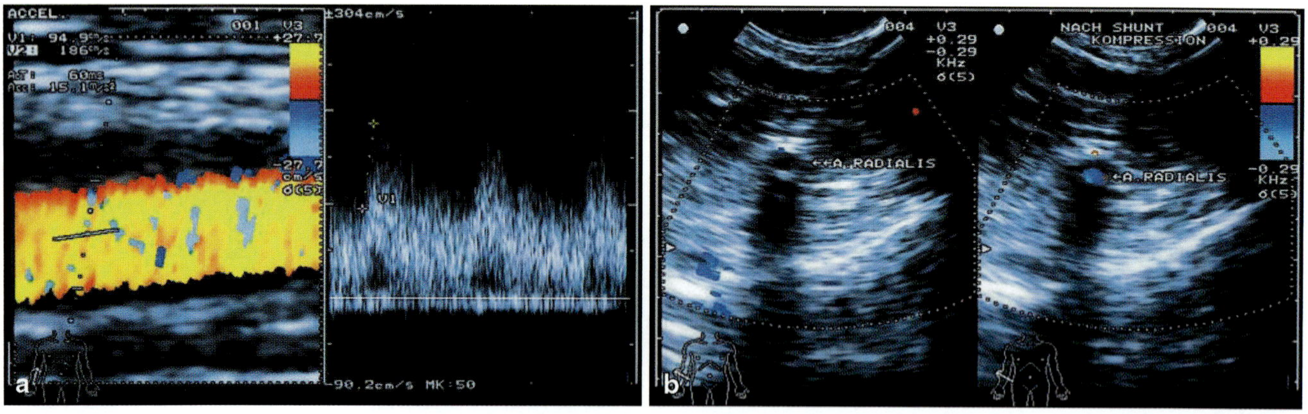


**Fig. A4.8 a–e**
**Steal phenomenon**
**a** A patient with a hemodialysis shunt in the bend of the elbow which functioned for many years developed ischemic necrosis of the finger pads. The shunt was found to be patent and showed a high flow rate with peak velocities of 186 cm/s in systole and 94 cm/s at end diastole over a long stretch.

**b** With the shunt patent, no flow is detected in the radial artery by color duplex or in the Doppler frequency spectrum. The transverse view of the radial artery on the *left* demonstrates marked medial sclerosis with posterior acoustic shadowing obscuring flow. The scan on the *right* shows blood flow coded in blue in the radial artery upon compression of the shunt. Angiogram likewise visualizes the distal radial artery only while the shunt is being compressed in the bend of the elbow.

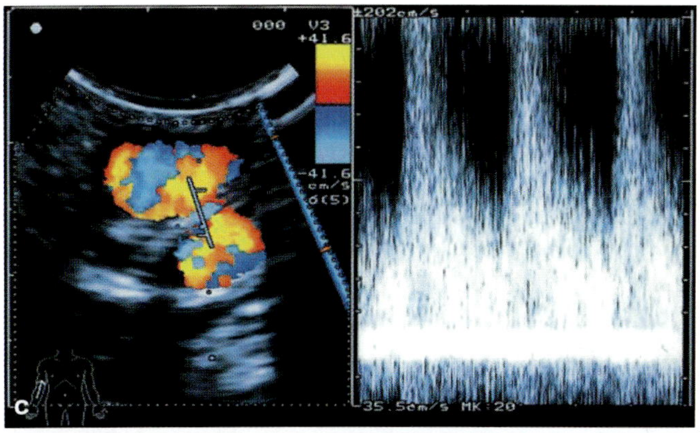


**Fig. A4.8 (continued)**

**c** After banding of the shunt, stenosis is demonstrated in this area with a peak flow velocity of 280 cm/s in systole and 100 cm/s at end diastole.

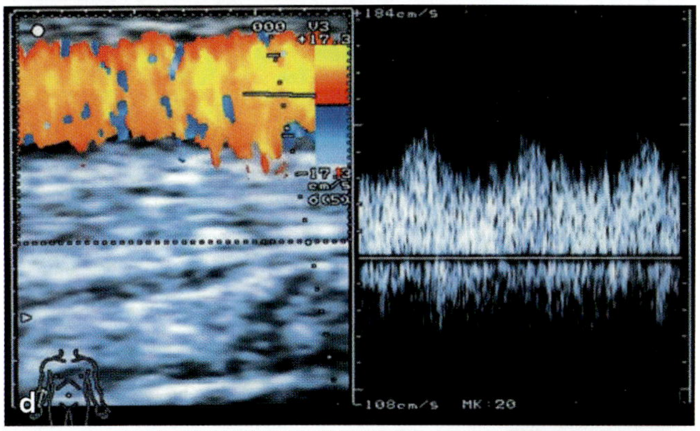


**d** The peak flow velocity in the shunt is reduced to 95 cm/s in systole and 60 cm/s at end diastole.

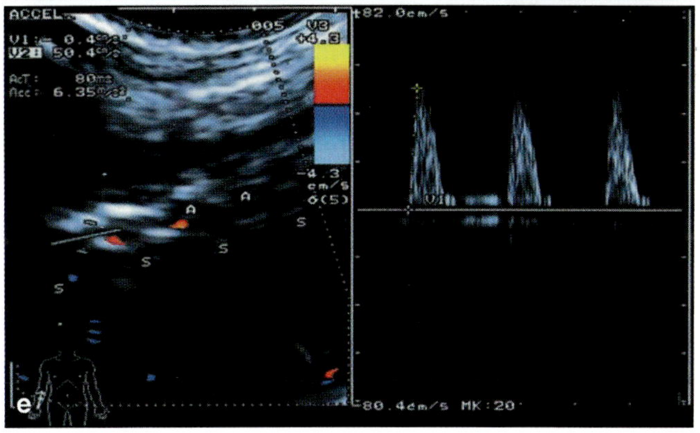


**e** Although visualization is impaired by medial sclerosis, flow with a systolic velocity of 50 cm/s is detectable in the radial artery (*A*) after banding. However, only isolated spot-like flow signals are depicted in the radial artery despite a high gain (indicated by posterior artifacts due to overmodulation) and a low PRF. The calcified plaques and medial sclerosis induce acoustic scattering and shadowing (*S*).

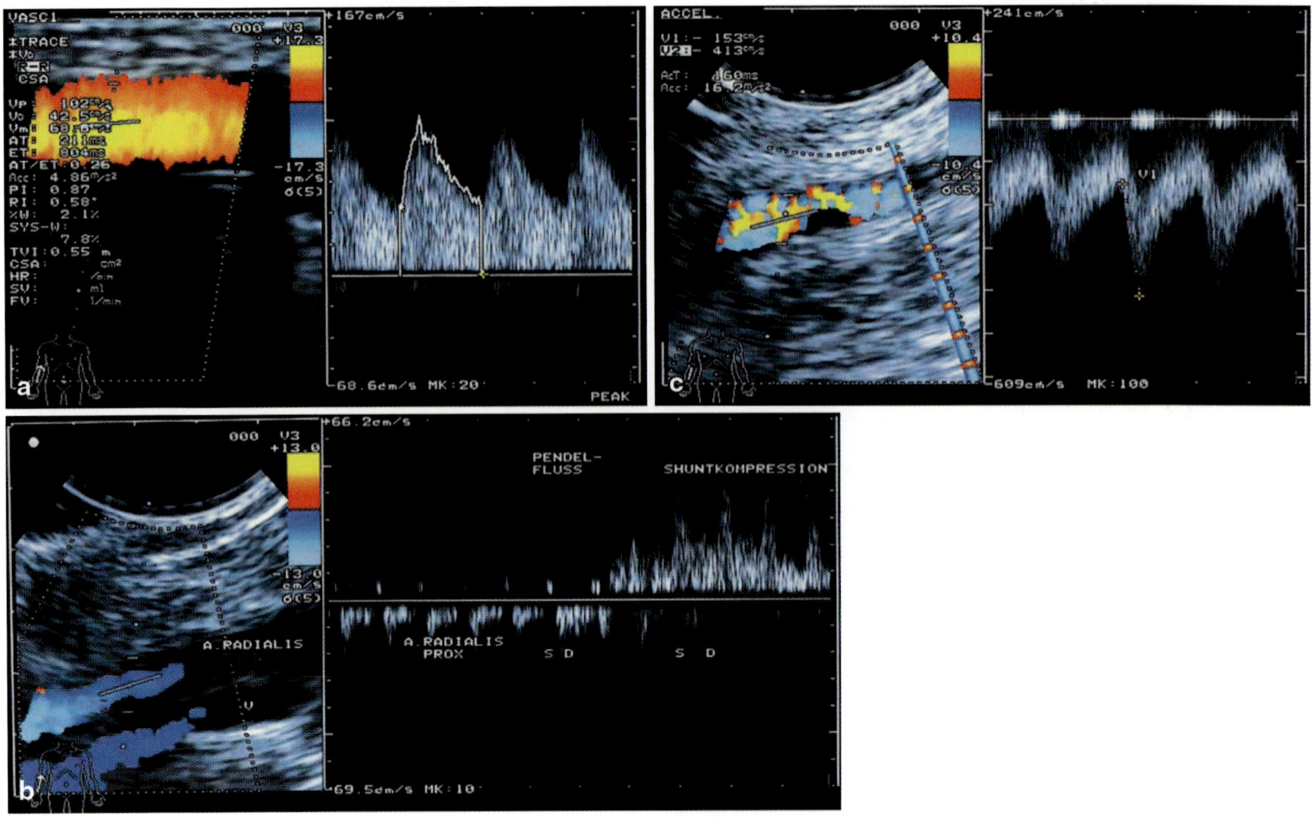


**Fig. A4.9 a–c**

**Stenosis of shunt artery**

**a** Synthetic loop with a diameter of 6 mm implanted for hemodialysis between the brachial artery and the basilic vein in the bend of the elbow. Shunt flow is in the upper normal range: flow volume of 1,153 mL/min calculated from a mean flow velocity of 68.8 cm/s derived from the Doppler frequency spectrum. The patient has ischemic pain of the fingers.

**b** Color duplex scanning demonstrates retrograde flow in the proximal radial artery (depicted in blue, away from transducer, same flow direction as in the parallel vein depicted posteriorly). The shunt diverts blood away from the periphery, which is why there is retrograde flow in the brachial artery distal to the shunt anastomosis and in the radial artery (proximal is on the *left*). These arteries are filled through cutaneous and muscle collaterals.

The Doppler waveform (*right section*) shows the flow pattern associated with the steal phenomenon: there is to-and-fro flow with a short forward flow during systole (*S*) and persistent retrograde flow during diastole (*D*) with flow reversal upon compression of the shunt. The flow reversal leads to forward flow in the distal brachial artery and in the radial artery. The abnormal diastolic flow is due to ischemia-induced dilatation of the arterioles.

**c** The retrograde flow in the proximal radial artery distal to the shunt is due to stenosis of the brachial artery just proximal to the shunt anastomosis (distal is on the *left*). In this case, there is a normal shunt flow volume and flow through collateral arteries from the proximal brachial artery to the lower arm arteries with shunt flow causing steal with retrograde flow in the proximal lower arm arteries.

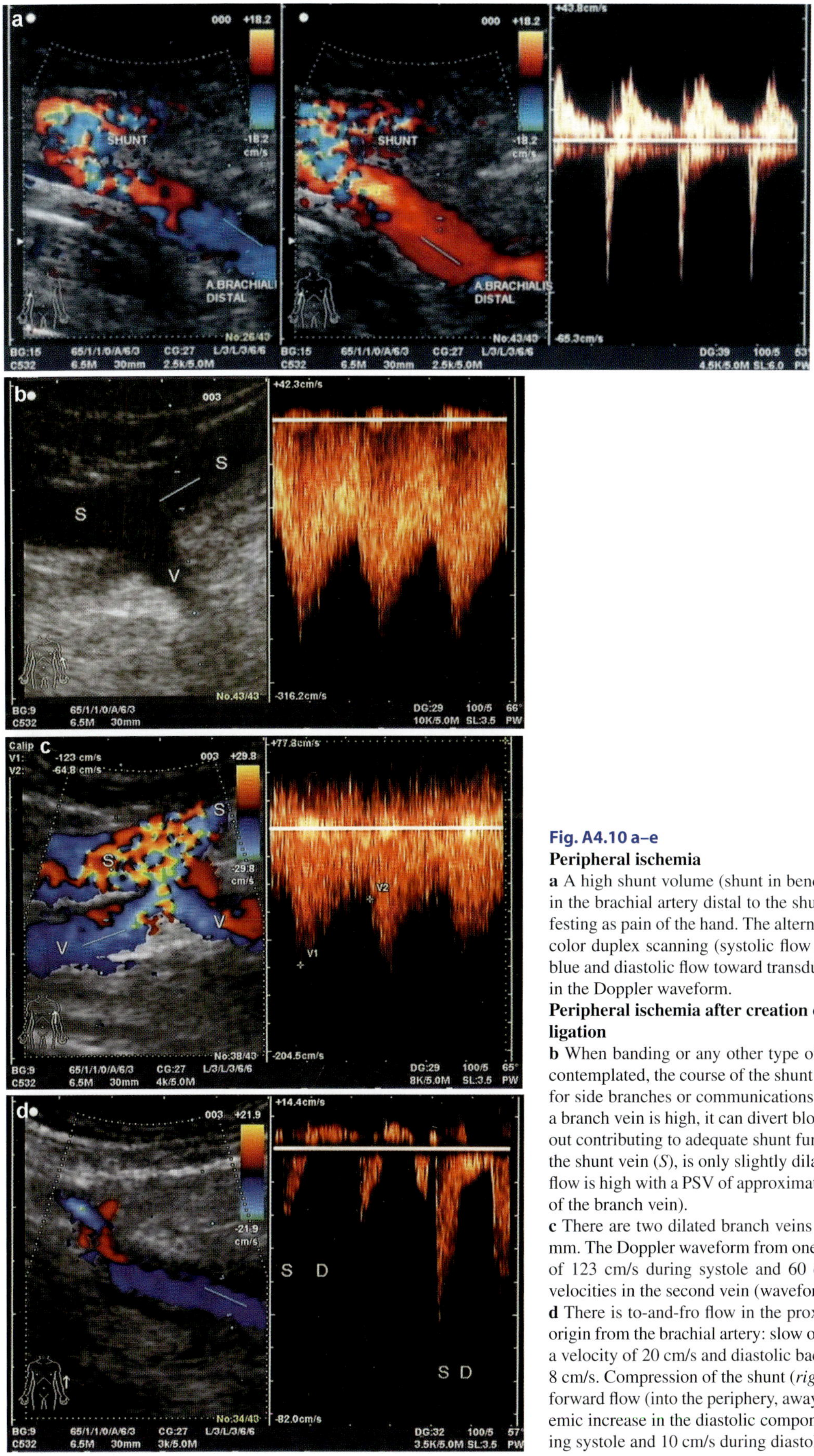


**Fig. A4.10 a–e**

**Peripheral ischemia**

**a** A high shunt volume (shunt in bend of elbow) causes to-and-fro flow in the brachial artery distal to the shunt with peripheral ischemia manifesting as pain of the hand. The alternating flow is demonstrated both on color duplex scanning (systolic flow away from transducer depicted in blue and diastolic flow toward transducer and shunt depicted in red) and in the Doppler waveform.

**Peripheral ischemia after creation of dialysis access – branch vein ligation**

**b** When banding or any other type of shunt revision or even closure is contemplated, the course of the shunt vein should be evaluated to search for side branches or communications with deeper veins. If flow in such a branch vein is high, it can divert blood away from arm perfusion without contributing to adequate shunt function. The cephalic vein, which is the shunt vein (*S*), is only slightly dilated with a diameter of 1.2 cm, but flow is high with a PSV of approximately 2.5 m/s (proximal to the origin of the branch vein).

**c** There are two dilated branch veins (*V*) with diameters of 0.8 and 0.7 mm. The Doppler waveform from one of the veins shows flow velocities of 123 cm/s during systole and 60 cm/s at end diastole with similar velocities in the second vein (waveform not shown).

**d** There is to-and-fro flow in the proximal radial artery shortly after its origin from the brachial artery: slow orthograde flow during systole with a velocity of 20 cm/s and diastolic backward flow (*D*) with a velocity of 8 cm/s. Compression of the shunt (*right*) results in systolic and diastolic forward flow (into the periphery, away from transducer) with a postischemic increase in the diastolic component. Flow velocity is 40 cm/s during systole and 10 cm/s during diastole.

**Fig. A4.10 (continued)**

**e** The lateral branches described in **c** were sonographically marked and exposed for ligation in order to improve hand perfusion and salvage the dialysis access. Following revision, the improved hemodynamic situation is demonstrated by repeat spectral Doppler measurement at the same site as in **d**: orthograde flow is restored (without backward flow) and a PSV of 35 cm/s is measured (compare the waveform in **d**). The patient's symptoms resolved after the intervention.

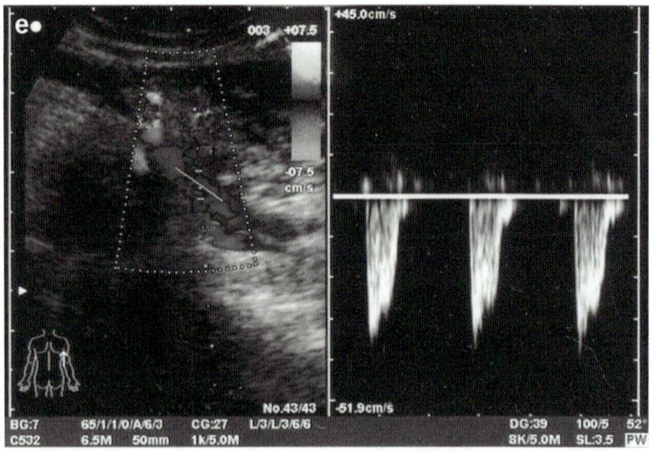


**Fig. A4.11 a, b**

**Peripheral ischemia, dilated fistula, steal effect**

**a** Patient with a dilated Brescia-Cimino fistula in the wrist (12 mm diameter) and ischemic pain in the finger pads but with otherwise good shunt function (shunt not shown). There is high flow in the proximal radial artery (not shown) with retrograde flow in the distal segment (coded in red, toward transducer). The Doppler waveform confirms retrograde flow with reduced systolic flow velocity (*S*) and a high diastolic flow velocity of 75 cm/s (*D*). Compression of the shunt elicits flow reversal (*KOMP SHUNT*) with an orthograde flow direction (away from transducer) and a large diastolic flow component (postischemic) in the radial artery.

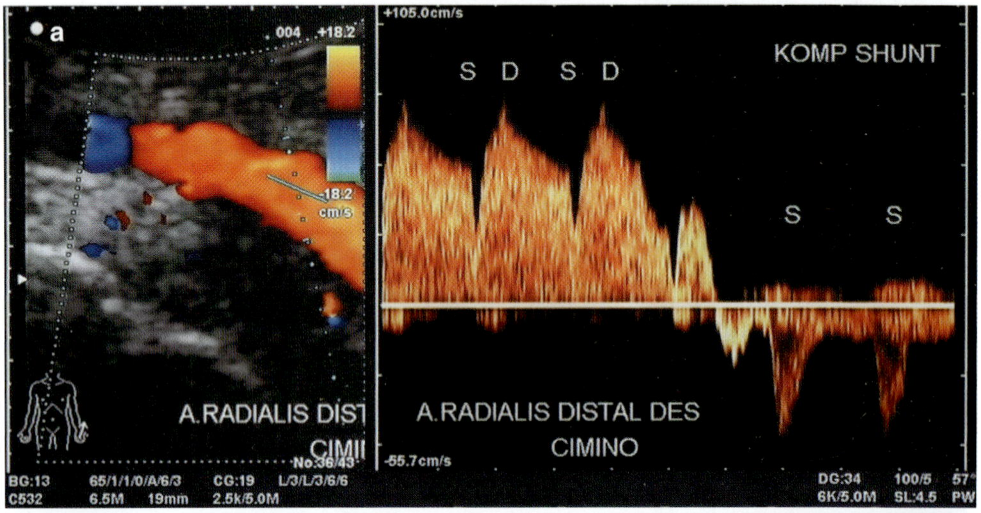


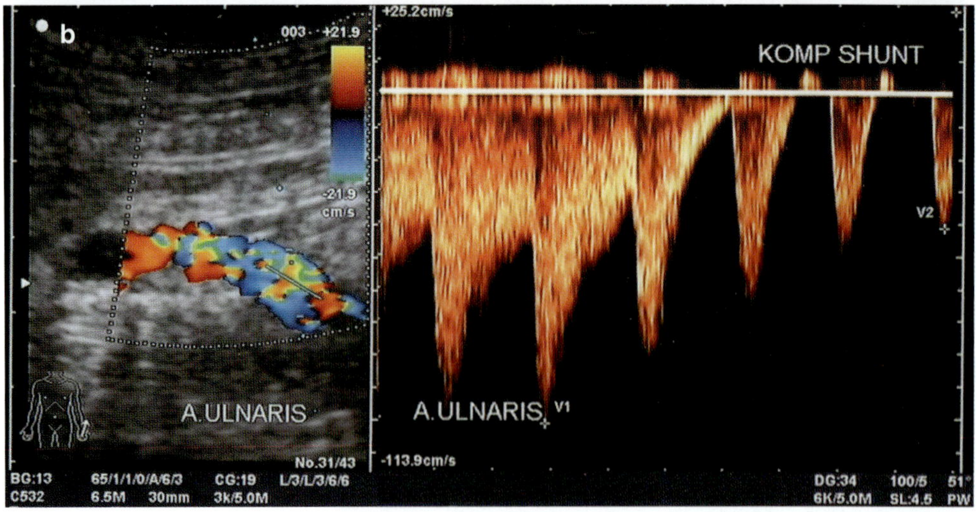


**b** The ulnar artery shows high orthograde flow (aliasing in the color flow image) toward the periphery (coded in blue, away from transducer; below the baseline in the waveform). The Doppler waveform is that of an artery supplying an AV fistula with a large diastolic component (44 cm/s at end diastole) and a high PSV of 100 cm/s. Upon compression of the fistula (*KOMP SHUNT*), the flow pattern normalizes (triphasic flow characteristic of peripheral arteries) with a PSV of 45 cm/s.

These findings are consistent with a steal effect in a high-flow shunt; arterial blood flow is insufficient, and the ulnar artery is recruited to also supply the shunt via the palmar arch, which explains the retrograde flow in the radial artery distal to the shunt. Based on these sonographic findings, the radial artery is ligated distal to the fistula. This measure eliminates the steal effect and restores adequate blood supply to the hand through the ulnar artery.

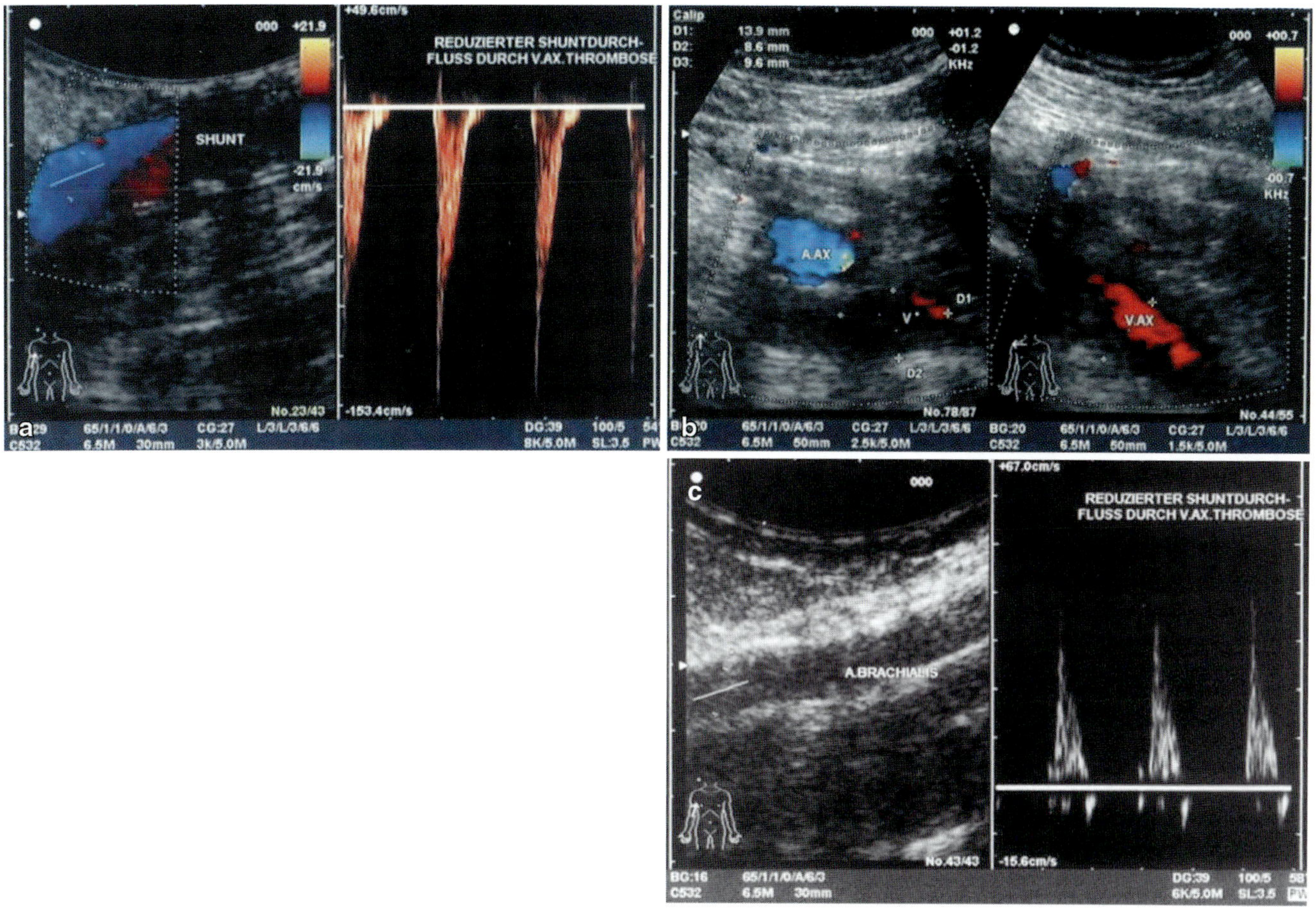


**Fig. A4.12 a–c**

**Outflow obstruction – central venous thrombosis proximal to dialysis access**

In stenotic disease proximial to a dialysis access, the Doppler waveform from the fistula is less pulsatile with a delayed systolic rise and an increased diastolic flow component (resembling venous flow). Conversely, impaired venous drainage (thrombosis, stenosis, compression) results in a more pulsatile flow profile.

**a** The abnormal Doppler waveform from the hemodialysis access lacks a diastolic component, suggesting an increased flow resistance (flow obstruction) distal to the sample volume.

**b** Color duplex imaging demonstrates thrombosis of the axillary vein with little residual flow near the walls coded in red. The lumen of the vein (*V*) is nearly completely filled by the thrombus.

**c** The outflow obstruction leads to high-resistance flow as indicated by the typical triphasic arterial flow profile in the brachial artery feeding the shunt. In cases of inadequate shunt flow during hemodialysis, impaired drainage can be diagnosed from the Doppler frequency spectrum sampled in the brachial artery or in the shunt. The cause of the obstruction can then also be identified sonographically. In the case presented, thrombosis of the axillary vein was revealed. In patients with a synthetic loop graft, the same flow profile may indicate stenosis of the venous anastomosis.

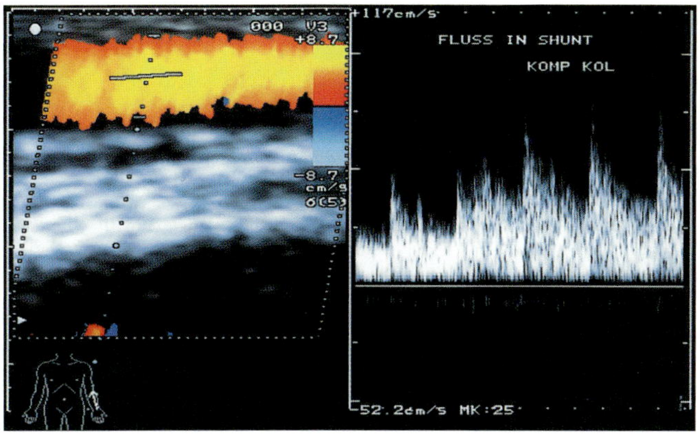


**Fig. A4.13**

**Inadequate shunt flow due to branch veins**

Brescia-Cimino fistula at the wrist with inadequate flow for hemodialysis. Once stenosis has been excluded, the examiner must search for venous collaterals that may reduce the fistula flow available for dialysis, thus requiring surgical ligation. In this case, compression of a sonographically identified collateral with relevant flow leads to a doubling of flow in the main shunt vein. The increase in flow velocity is documented in the Doppler frequency spectrum, which shows that this increase is chiefly due to the larger diastolic flow component.

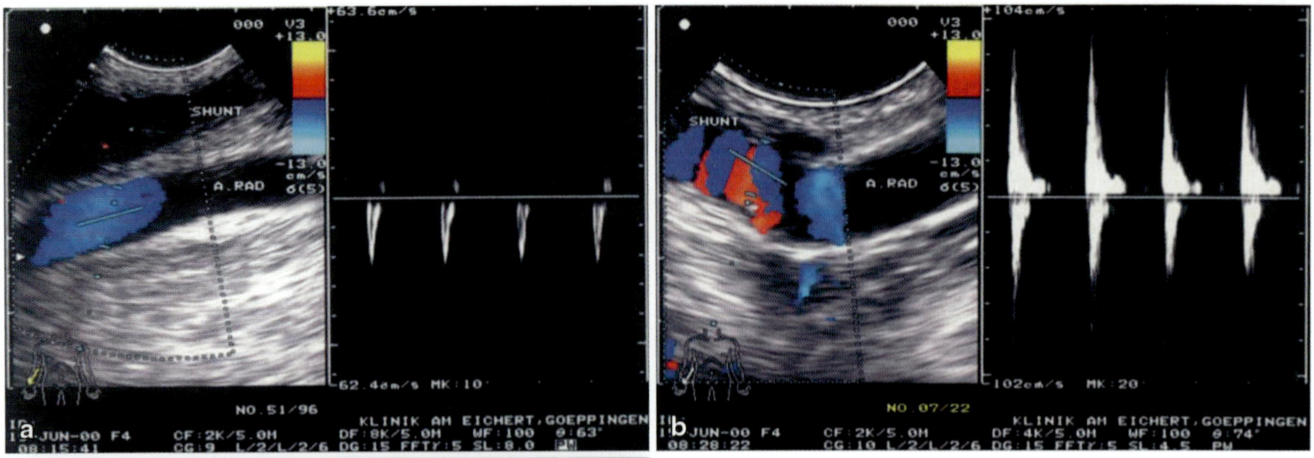


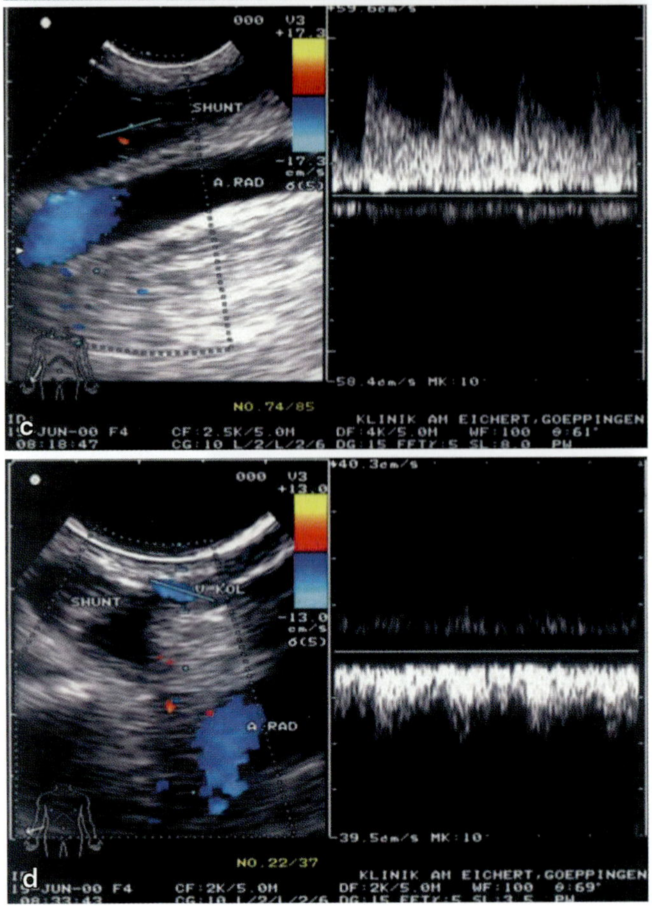


**Fig. A4.14 a–d**

**Partial occlusion of dialysis access**

Progressive intimal hyperplasia and thrombi in the shunt reduce flow, which is identified indirectly as high-resistance flow in the Doppler spectrum sampled in arterial segments upstream of the shunt.

**a** Image of the lower arm depicting the radial artery and the nearly completely thrombosed Brescia-Cimino fistula (closer to transducer) in one scanning plane. The artery has a triphasic flow profile with a low PSV.

**b** The shunt is patent at the anastomosis, but a thump pattern is seen (knocking waveform)

**c** Inside the shunt, color duplex demonstrates meandering flow with a typical shunt profile but markedly reduced velocity.

**d** Along the course of the shunt, there are thrombi surrounded by flow signals proximal to the origin of a small venous branch (*V.KOL*). Distal to it, the shunt is occluded by thrombosis.

# 5 Extracranial Cerebral Arteries

Cardiovascular disease constitutes the most common cause of death in Western industrialized countries. The most serious cerebrovascular manifestation is stroke with its complications, which are fatal in one third of cases. Patients who survive cerebral infarction often suffer from irreversible damage and paralysis and require permanent care. With atherosclerosis of the carotid system becoming more common in old age, cerebral infarction gains relevance as the population ages (Fabres et al. 1994; Mannami et al. 1997; Roederer et al. 1984). Over 60–70% of all ischemic cerebral infarctions are caused by arterial embolism, typically arising from the carotid artery (Bock et al. 1993; Evans 1999; Roederer et al. 1984).

Carotid thromboendarterectomy (TEA), first performed by De Bakey in 1953, is a highly effective surgical procedure for reducing the risk of stroke in patients with atherosclerosis of the carotid system. This has been confirmed in several large trials in individuals with symptomatic carotid stenosis performed in Europe (European Carotid Surgery Trial (ECST)) and the USA (North American Symptomatic Carotid Endarterectomy Trial (NASCET)) as well as in an asymptomatic population (Asymptomatic Carotid Atherosclerosis Study (ACAS)) (Table 5.1). These studies compared the natural history with the morbidity and mortality after carotid surgery stratified by clinical stage and grade of carotid stenosis. The results of all three studies suggest that carotid reconstruction is beneficial in individuals with symptomatic high-grade stenosis (>70%) and in selected cases of 60–70% symptomatic stenosis. In high-grade asymptomatic stenosis on the other hand, the only advantages surgical repair has over the natural history are in individuals with a low risk of perioperative morbidity and plaque morphology predictive of a high risk of embolism.

Suitable diagnostic tests are necessary for identifying those patients who will benefit from the therapeutic measures confirmed in these large trials to be advantageous. More specifically, this involves identifying subjects with carotid stenosis who also have a high risk of embolism and will benefit from carotid TEA. Color duplex ultrasound is a noninvasive method that can be repeated at any time and has evolved into a highly accurate method of grading carotid stenosis (the risk of embolism increases with the degree of stenosis). Moreover, sonography also provides information on plaque morphology, the second major factor affecting the risk of embolism.

The superficial course of the carotid arteries, without interfering structures, enables detailed sonographic evaluation of the arterial segment accounting for the majority of cerebral infarctions. Given these ideal scanning conditions and the fact that the vast majority of carotid stenoses occur at the origin of the internal carotid artery (ICA), continuous wave (CW) Doppler ultrasound was already highly accurate in detecting higher-grade carotid stenosis.

Duplex ultrasonography enables precise morphologic evaluation of arterial lesions and their locations, and angle-corrected spectral Doppler measurement yields information on the hemodynamic significance of stenosis. In addition, assessment of plaque morphology helps estimate the risk of embolism. Using these ultrasound techniques, it is possible to identify candidates for surgery or medical management of carotid artery stenosis without the need for additional invasive evaluation.

## 5.1 Normal Vascular Anatomy and Important Variants

The brain derives its blood supply from the two carotid arteries and the two vertebral arteries. The latter unite at the inferior border of the pons to form the basilar artery. In over 70% of individuals, the left *common carotid artery (CCA)* arises directly from the aortic arch before the origin of the subclavian artery (Fig. 5.1a). The right CCA originates from the *innominate artery* (*brachiocephalic trunk)*, which arises from the aortic arch and additionally gives off the

W. Schäberle, *Ultrasonography in Vascular Diagnosis*,
DOI: 10.1007/978-3-642-02509-9_5, © Springer-Verlag Berlin Heidelberg 2011

**Table 5.1** Results of randomized multicenter trials comparing surgical versus medical treatment of symptomatic (NASCET, ECST) and asymptomatic carotid artery stenosis (ACAS)

| | NASCET | ECST | ACAS |
|---|---|---|---|
| No. of patients | 659 | 778 | 1,659 |
| – Surgical | 328 | 455 | 825 |
| – Medical | 331 | 323 | 834 |
| Perioperative stroke rate | 2.1% | 6.6% | 1.4% |
| Morbidity/mortality rate (natural history) | 5.8% | 7.5% | 2.3% |
| Risk reduction (relative) | 65% | 43% | 53% |
| Men | | | 66% |
| Women | | | 17% |

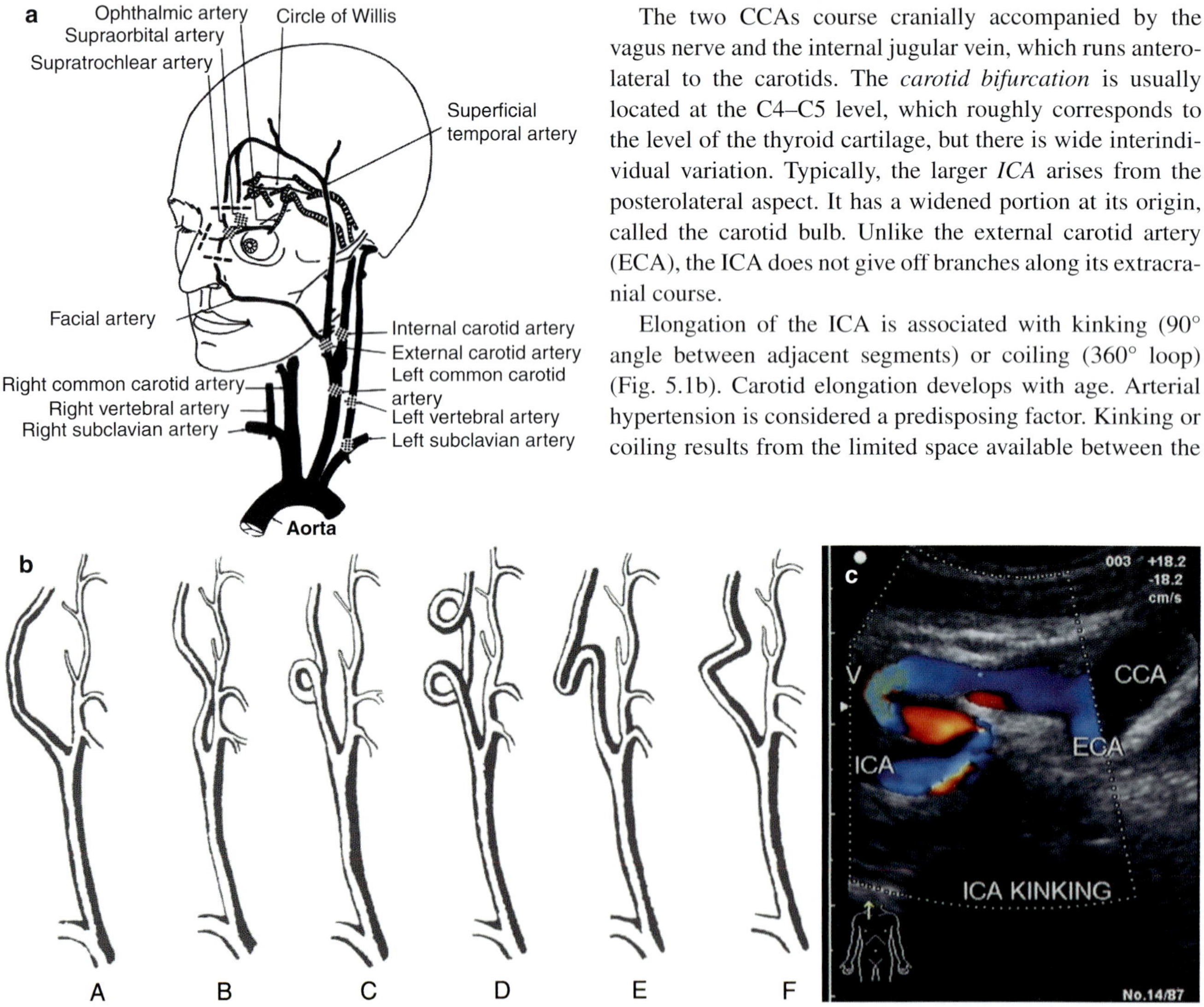


**Fig. 5.1** **a** Diagram of the arteries supplying the brain (marked are the sites for taking representative measurements and documenting results). **b** Variants resulting from elongation of the internal carotid artery (ICA) (shown for the left artery): *A* C-shaped course, *B* S-shaped course, *C* coiling, *D* double coiling, *E* kinking, *F* double kinking. **c** Color flow image showing severe kinking of the *ICA*, indicated by a change in flow direction relative to the transducer (change from *blue* to *red* color coding)

subclavian artery. The most important variants of the supra-aortic arteries, which originally developed from the branchial arches, are:

- Common origin of the innominate artery and left CCA from the aortic arch (13%)
- Persisting communicating trunk arising from the aortic arch and giving off first the left CCA and then the innominate artery (9%)
- Bilateral innominate artery dividing into the CCA and the subclavian artery (1%)
- Situs inversus (very rare)

The normal innominate artery on the right has a length of 4–5 cm. It crosses under the brachiocephalic vein and, behind the right sternoclavicular joint, divides into the right subclavian artery and the right CCA.

The two CCAs course cranially accompanied by the vagus nerve and the internal jugular vein, which runs anterolateral to the carotids. The *carotid bifurcation* is usually located at the C4–C5 level, which roughly corresponds to the level of the thyroid cartilage, but there is wide interindividual variation. Typically, the larger *ICA* arises from the posterolateral aspect. It has a widened portion at its origin, called the carotid bulb. Unlike the external carotid artery (ECA), the ICA does not give off branches along its extracranial course.

Elongation of the ICA is associated with kinking (90° angle between adjacent segments) or coiling (360° loop) (Fig. 5.1b). Carotid elongation develops with age. Arterial hypertension is considered a predisposing factor. Kinking or coiling results from the limited space available between the

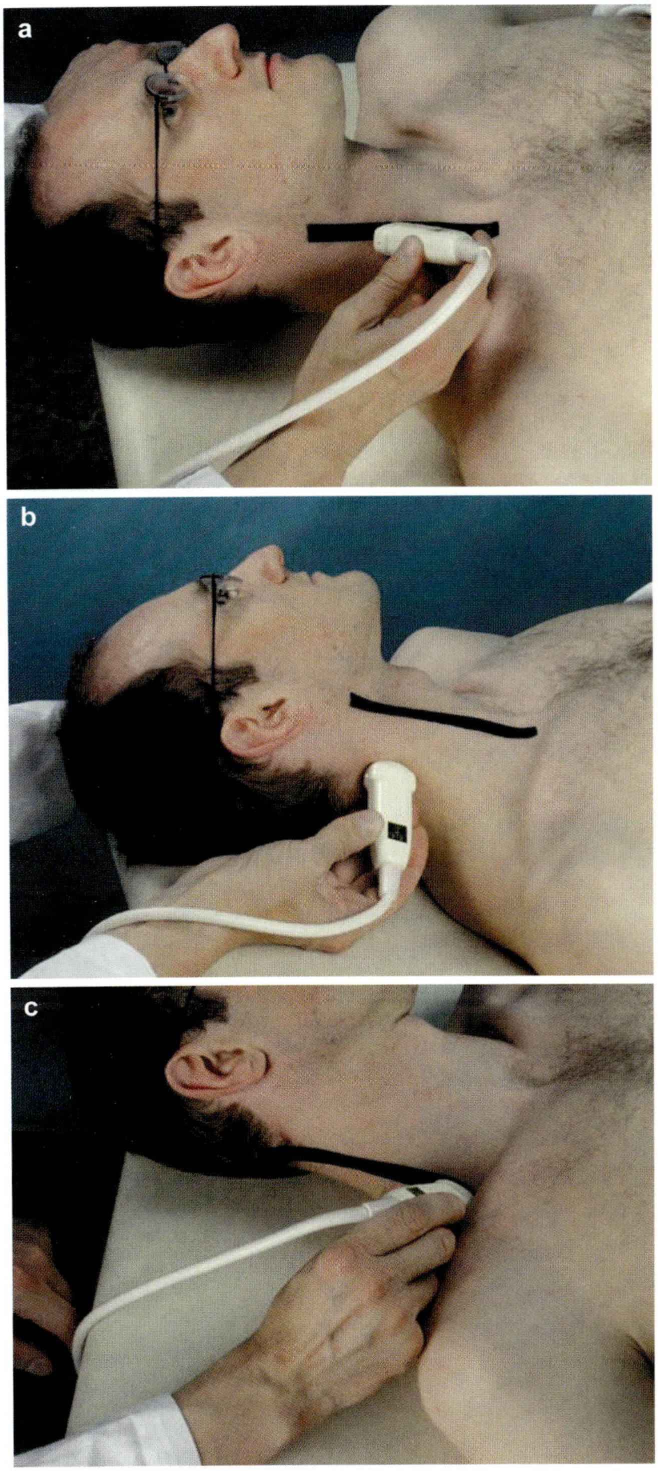


**Fig. 5.2** **a** Scanning of extracranial carotid arteries from anterolateral view (in front of sternocleidomastoid muscle). **b** Scanning of the carotid artery (course indicated by *black line*) from the posterolateral probe position (behind sternocleidomastoid muscle). **c** Transducer position for scanning of vertebral artery (at its origin; course marked)

two points of fixation, the bifurcation and the base of skull, but even severe kinking rarely causes hemodynamically significant stenosis (see Fig. A5.2).

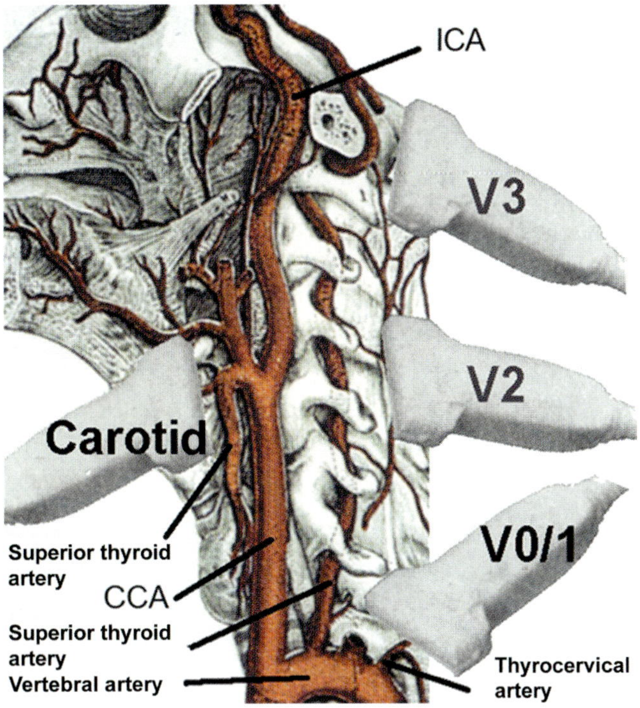


**Fig. 5.3** Vascular anatomy of the extracranial cerebral arteries. Transducer positions for imaging of vertebral segments V0/1, V2, and V3 and for imaging of the carotid bifurcation

The ECA arises from the anteromedial aspect of the ICA; in about 10% of cases its point of origin is lateral or posterolateral. On its course, it first gives off the superior thyroid artery (STA) and then branches to supply the skin and extracranial organs.

The two *vertebral arteries* originate from the ipsilateral subclavian arteries at the C6 level and then pass through the transverse foramina of the corresponding vertebrae, thus taking a partially intraosseous course on their way to the skull base. They often differ in caliber and may exhibit unilateral hypoplasia or aplasia, which is compensated for by contralateral hypertrophy. The left vertebral artery typically has a larger caliber and, in up to 4% of cases, arises directly from the aortic arch.

Somewhat distal to the vertebral artery, the thyrocervical trunk arises from the subclavian artery. The differentiation is significant in the duplex ultrasound examination. For a precise description of the site of lesions, the vertebral artery is divided into four segments (Fig. 5.3):

- The V0/V1 segment, which extends from the origin to the site of entry into the transverse foramen of C6
- The V2 segment, which is the part coursing through the cervical vertebral foramina
- The V3 segment, which takes an arched course around the atlas and is therefore also referred to as the atlas loop
- The V4 segment, which is the intracranial part of the vertebral artery

## 5.2 Examination Technique and Protocol

The superficial location of the cervical arteries enables their examination with a high-frequency transducer (5–7.5 MHz or even 10 MHz), yielding B-mode images with high spatial resolution. The patient is in the supine position with the head slightly hyperextended, and the examiner sits at the patient's head. The course of the arteries and the carotid bifurcation are identified in the transverse section, and the Doppler spectrum is sampled longitudinally. The spatial orientation is the same as in the examination of other body regions with display of cranial on the left side of the image and caudal on the right.

### 5.2.1 Carotid Arteries

The examination begins by obtaining a survey of the *carotid bifurcation in transverse orientation* to determine the location and course of the ICA and ECA in relation to each other. The following variants may be encountered:

- In about 90% of the population, the ICA courses posterolateral to the ECA.
- In about 10% of cases, the ICA is seen at the same level and medial to the ECA.
- In rare cases, the ICA lies in front of the ECA.

For adequate angulation of the beam and precise identification of stenosis or plaque, the examiner must move the transducer around to obtain a view depicting the carotid bifurcation as a tuning fork. There are three standardized approaches for longitudinal imaging:

- Positioning of the transducer between the larynx and sternocleidomastoid muscle for sagittal anteroposterior sections (Fig. 5.2a)
- Lateral approach through the sternocleidomastoid muscle
- Posterolateral approach with the transducer behind the sternocleidomastoid muscle (Fig. 5.2b)

The posterolateral transducer position will enable good visualization of the bifurcation in most patients whose ICA follows a normal course. In this position the ICA is depicted near the transducer.

*B-mode ultrasound* serves to identify the course of the carotid arteries and also provides *initial information* on the vessel wall *in transverse orientation* and in the above-described longitudinal views (Fig. 5.4).

Sonographically, three layers of the normal arterial wall can be distinguished: an inner layer depicted as a hyperechoic line next to the lumen; a middle zone seen as a somewhat broader, hypoechoic layer; and an outer layer of slightly higher echogenicity, which is poorly demarcated from the perivascular fatty tissue. Since ultrasound does not visualize tissues or tissue layers directly but rather the echoes reflected by interfaces between zones of different acoustic impedance, the three layers seen do not exactly match the three anatomic wall layers – the intima, media, and adventitia. The intima and media are sonographically indistinguishable, which is why it is not possible with ultrasound to evaluate the intima alone. It is therefore common practice to measure the thickness of the intima–media complex instead.

The sonographic thickness of the intima–media complex is increasingly being used as a surrogate marker for subclinical atherosclerosis and as a measure of therapeutic outcome in interventional studies (e.g., to monitor statin therapy). It is therefore desirable that a standardized method for measuring carotid *intima–media thickness* (IMT) be used to minimize interobserver variability. A perpendicular angle of incidence ensures optimal evaluation of the vessel wall, which is the case if the target vessel courses parallel to the skin surface. If the angle is smaller, the examiner should move the transducer back and forth or rotate it slightly. IMT is measured in the far wall of the CCA to exploit the blood-filled lumen as an acoustic window for optimal visualization of the two echogenic lines demarcating the intimal and medial layers. The leading-edge method (see Sect. 1.1.2.4 and Fig. 5.5) is recommended to minimize blooming artifacts (which appear at boundaries where there is a large mismatch in acoustic impedance). Serial IMT measurements should always be performed at the same site; most investigators prefer the far wall 2–3 cm proximal to the carotid bifurcation. Use of a high-frequency transducer (>10 MHz) with high axial resolution is recommended for evaluation of the wall as measurement accuracy increases with transducer frequency (see Table 1.2, Sect.1.1). Note, however, that although it is technically feasible, differentiation of structures smaller than 0.01 mm is beyond the resolution capacity of the human eye (and may introduce measurement errors, blooming, etc.). Finally, it is recommended that the measurement of IMT be performed at end diastole to minimize variations through the cardiac cycle (Meyer and Strobel 2008).

No agreement exists regarding the need for detailed sonomorphologic characterization of plaque in routine clinical examination. While most patients with over 70% stenosis are candidates for surgery based on this degree of stenosis alone, plaque morphology is relevant for therapeutic decisions regarding patients with 60–70% stenosis.

Following evaluation of the vessel wall and identification of plaque, *blood flow is assessed in the Doppler mode* in the longitudinal plane. In color duplex imaging with proper settings, stenosis is suggested by the occurrence of aliasing and an occlusion by the absence of color filling in the lumen. Moreover, the color duplex mode can help identify the course of a kinked or coiled ICA.

While color duplex imaging is optional for initial orientation, Doppler spectra with qualitative *determination of*

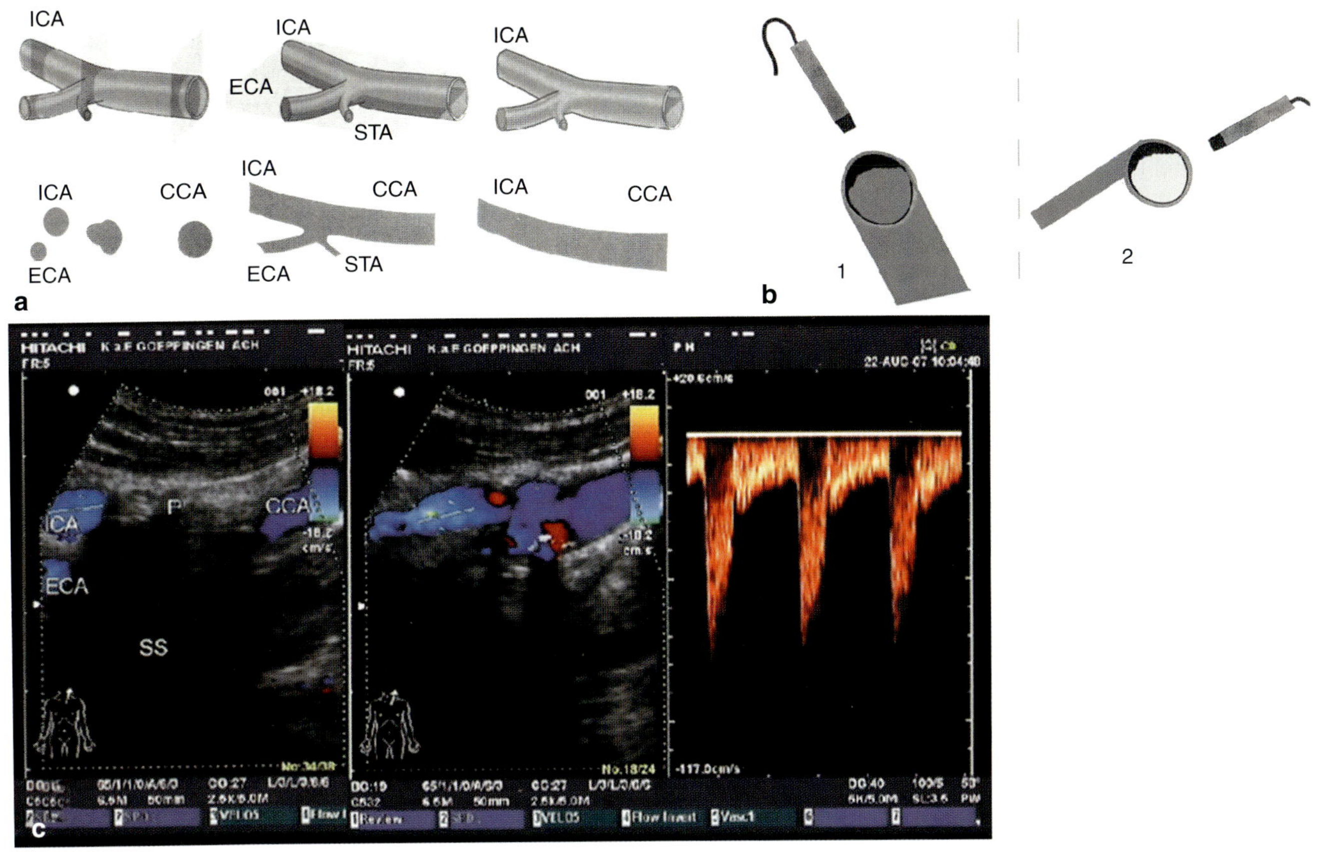


**Fig. 5.4** **a** Diagrams illustrating ultrasound examination of the carotid bifurcation. The *leftmost drawing* illustrates the sites of transverse examination for an overview and identification of the carotid arteries. The *second drawing* illustrates the posterolateral transducer position, which usually depicts the carotid bifurcation as a fork with the internal carotid artery (ICA), which runs posteriorly, appearing closer to the transducer and the external carotid artery (ECA) appearing farther from it. Often, this transducer position allows sonoanatomic identification of the ICA by demonstrating its wider bulb and also of the ECA by visualizing the superior thyroid artery (STA) arising from it; this position also enables evaluation of plaque morphology. The *third drawing* illustrates the anterior transducer position, which is used for plaque evaluation or Doppler recording in cases where acoustic shadowing impairs imaging in the posterolateral position. **b** Diagram illustrating how posterior acoustic shadowing obscuring the lumen can be circumvented by turning the transducer from position 1 (e.g., posterolateral position) to position 2 (e.g., anterior position) for evaluating plaque morphology/surface and for assessing the severity of stenosis. Position 2, unlike position 1, will also allow recording of a Doppler spectrum. The drawings illustrate how a small, calcified plaque can impair evaluation of the vascular lumen if the vessel is examined in only one plane. **c** *Left section* (obtained with the transducer in a posterolateral position) shows how acoustic shadowing from a calcified plaque completely eliminates flow signals from the ICA and ECA and obscures vascular structures in the B-mode as well. *Middle and right sections* illustrate how the sickle-shaped calcified plaque is circumvented and the bulb and ICA can be evaluated. There are no signs of hemodynamically relevant luminal narrowing. No flow acceleration is demonstrated by color duplex or spectral Doppler, precluding relevant stenosis caused by the plaque

*angle-corrected flow velocities in longitudinal orientation must be obtained* from the CCA, ICA, and ECA (Table 5.2). Spectral Doppler sampling should be performed in the ICA at short intervals. Use of a larger sample volume will often enable continuous examination of the CCA and ICA in the duplex mode, especially from the posterolateral approach. In this way, a continuous spectrum can be obtained and analyzed throughout the CCA and ICA, similar as with CW Doppler ultrasound. The ECA is scanned only at its origin for differentiation from the ICA and for the identification of possible stenosis.

The posterolateral transducer position is usually preferred for Doppler measurement as it visualizes the bifurcation as a fork, along with the CCA, the bulb, and long portions of the ICA and ECA in a single scan. Moreover, angle correction is easier, and the intervening soft tissue improves visualization in this transducer position. However, when kinking or coiling is present, different scanning planes are necessary to follow a maximum segment of the carotid.

*To minimize errors in flow velocity measurement* in the ICA, the examiner should try to achieve an *insonation angle of less than 60°*. This requires selection of an adequate transducer. Scanning with a linear transducer is performed using beam steering with maximum cranial deflection. Linear transducers provide the best resolution for morphologic assessment of the vessel wall in the B-mode, while a *curved-array transducer with a small radius is superior* for spectral Doppler recording with an adequate angle, especially when the transducer must be tilted along the arched course of the

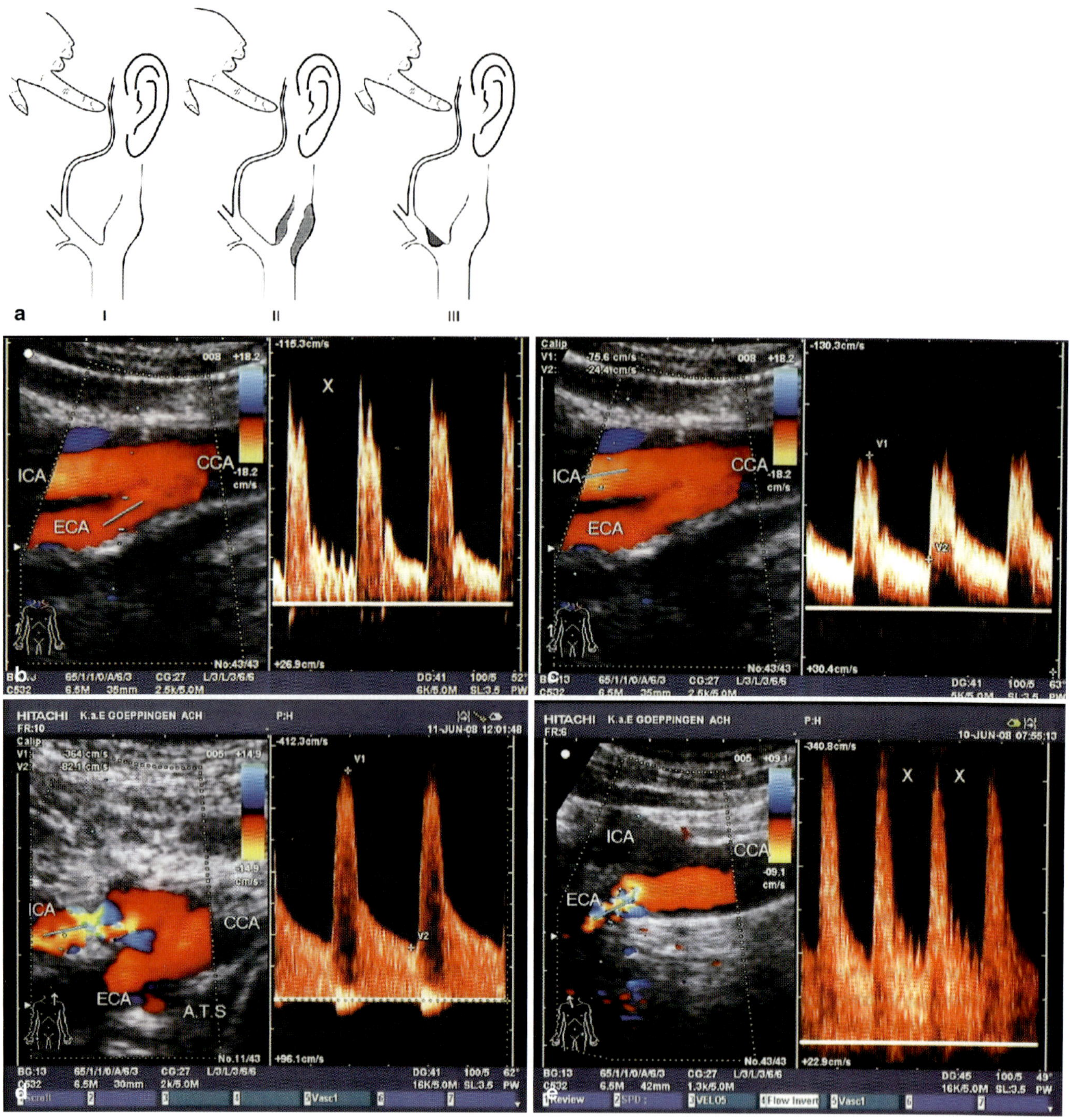


**Fig. 5.6** Duplex imaging of the carotid bifurcation and differentiation between the internal carotid artery (*ICA*) and external carotid artery (*ECA*). Flow in the ECA is more pulsatile and is modulated by the signal transmitted upon rhythmical tapping of the temporal artery (**a I** and **b**), especially during diastole (*X*); this modulation is absent in the ICA (**a I** and **c**). When stenosis is present, differentiation between the ICA and ECA on the basis of pulsatility is difficult, and plaque may impair identification of the superior thyroid artery (*A.T.S*) arising from the ECA or of the wider bulb of the ICA. The tapping test, however, still provides a clearcut differentiation, as the signal is not transmitted into the spectrum from a stenotic ICA (**a II** and **d**). The Doppler waveform from a stenotic ECA (**a III** and **e**) shows similar pulsatility of flow as the waveform from a stenotic ICA, and transmitted tapping (*X*) into the ECA allows differentiation of the two arteries in this situation

forward vertebral artery flow and suspected exercise-induced subclavian steal syndrome due to subclavian stenosis or occlusion, the clinical situation can be reproduced during the examination. With continuous spectral Doppler recording in the vertebral artery, a blood pressure cuff around the upper arm is inflated to over 250 mm Hg and then released after 3–5 min to induce reactive hyperemia in the arm. If high-grade stenosis or occlusion of the proximal subclavian artery with a subclavian steal phenomenon and vertebrovertebral crossover is present, this maneuver will induce flow reversal in the ipsilateral vertebral artery and an increase in flow velocity in the contralateral vertebral artery.

## 5.3 Documentation

Normal findings should be documented in longitudinal views (B-mode) of both CCAs, ICAs, and ECAs with the corresponding Doppler frequency spectra or angle-corrected velocity spectra. Abnormal findings are documented in additional B-mode images and the corresponding Doppler spectra recorded in longitudinal orientation from the affected sites. The angle-corrected velocity measurement must reflect the degree of stenosis. Moreover, the report should contain information on the localization of nonstenotic plaques as well as on the vascular segments (according to the documentation sites given in Fig. 5.1a) in which the systolic and end-diastolic Doppler shift frequencies or angle-corrected velocities were obtained.

## 5.4 Normal Findings

### 5.4.1 Carotid Arteries

The CCA has a constant luminal diameter of about 7 mm to the level of the bulb. Flow is pulsatile with a large diastolic component. Peak systolic velocity (PSV) ranges from 60 to 100 cm/s (Table 5.4). The wider lumen and vessel branching lead to eddy currents in the carotid bulb even under normal conditions, which is reflected in color duplex images and Doppler waveforms. PSV is decreased in the bulb, and flow separation may lead to retrograde flow on the side opposite the ECA, seen as changes in the color coding in color duplex images (see Figs. A5.1 and 1.43b).

The normal thickness of the *intima–media complex* measured in the B-mode (from the lumen–intima interface, the first bright line, to the media-adventitia interface, the second bright line) is *0.5–0.6 mm and increases somewhat with age.*

Supplying the brain, the *ICA* has low-resistance flow with a Doppler waveform that is characterized by a steep systolic upslope followed by *monophasic flow* with a fairly large diastolic component. In contrast, flow in the ECA is more pulsatile with a smaller diastolic component. The common carotid supplying both vessel territories has a mixed waveform. As in all other vascular territories, the pulsatility in the carotid system is determined by peripheral resistance and vessel elasticity and therefore increases with age (Table 5.5).

**Table 5.4** Average diameters (D) and flow velocities in the extracerebral arteries (meta-analysis)

| Artery | PSV (cm/s) | EDV (cm/s) | D (mm) |
|---|---|---|---|
| CCA | 50–80 | 15–30 | 6.0–7.5 |
| ICA | 60–90 | 20–40 | 4–6 |
| ECA | 60–100 | 10–20 | 3.5–4.5 |
| Vertebral artery | 20–70 | 5–35 | 3–5 |

*PSV* peak systolic velocity, *EDV* end-diastolic velocity, *CCA* common carotid artery, *ICA* internal carotid artery, *ECA* external carotid artery

**Table 5.5** Causes of reduced and increased pulsatility in the Doppler waveform from the extracranial cerebral arteries

| Change | Cause |
|---|---|
| Reduced pulsatility | • High-grade proximal flow obstruction<br>• AV fistula or angioma in distal segment<br>• Hyperperfusion (e.g., in hyperthyroidism)<br>• Aortic stenosis |
| Increased pulsatility | • High-grade distal flow obstruction<br>• Increased intracranial pressure<br>• Severe cerebral microangiopathy<br>• Aortic insufficiency<br>• Low heart rate |

*AV* arteriovenous

### 5.4.2 Vertebral Arteries

The flow velocity in the vertebral artery reported in different studies varies widely from 19 to 98 cm/s in systole and from 6 to 30 cm/s at end diastole. The resistance index (RI) ranges from 0.62 to 0.75 (Tratting et al. 1992). Assessment of the artery, especially at its origin, is impaired by caliber variation and the occurrence of congenital hypoplasia. At the same time, evaluation of the origin is important because it is the most common site of stenosis.

The normal diameter of the vertebral artery is 3–5 mm but it is common for one vertebral artery to be larger than the other, resulting in right–left diameter differences of over 2 mm.

In case of lateral differences in flow velocity and poor visualization of the origin, the examiner must differentiate a *proximal stenosis* from *hypoplasia.* Hypoplasia is suggested by an unchanged waveform, possibly with diminished diastolic flow, and pronounced widening of the contralateral artery (see Figs. 5.27 and A5.37).

In contrast, stenosis is indicated by less pulsatile flow, a smaller systolic component, and more pronounced diastolic flow.

## 5.5 Clinical Role of Duplex Ultrasound

### 5.5.1 Carotid Arteries

The aim of sonographic assessment of the extracranial cerebral arteries is to prevent cerebral infarction with its harmful sequelae and permanent deficits (Table 5.6).

While diagnostic ultrasound and treatment of peripheral vessels are symptom-oriented, the purpose of carotid artery evaluation is prognosis-oriented in that it aims at identifying patients at risk for stroke and initiating adequate preventive measures in these high-risk patients (Tables 5.7, 5.8, and 5.9).

**Table 5.6** Updated classification of extracranial carotid artery stenosis

| Stage | Clinical presentation/symptoms |
|---|---|
| **Stage I** | Asymptomatic carotid artery stenosis |
| IA | Asymptomatic stenosis without high-grade contralateral carotid stenosis or contralateral occlusion |
| IB | Asymptomatic stenosis with high-grade contralateral carotid stenosis or contralateral occlusion |
| **Stage II** | Symptomatic carotid stenosis: ipsilateral transient deficit within the preceding 6 months |
| IIA | Amaurosis fugax |
| IIB | Hemispheric symptoms reversible within 24 h (TIA) |
| **Stage III** | Indications for emergency carotid TEA |
| IIIA | Crescendo TIAs |
| IIIB | Progressive/Acute stroke |
| **Stage IV** | Symptomatic carotid stenosis: ipsilateral stroke within the preceding 6 months |
| Rankin 0 | Stroke with fully reversible neurologic deficits (duration > 24 h, PRIND) |
| Rankin 1 | Stroke without significant disability |
| Rankin 2 | Mild stroke with slight disability and/or slight aphasia |
| Rankin 3 | Moderate stroke with moderate disability with preserved ability to walk and/or moderately severe aphasia |
| Rankin 4 | Severe stroke, unable to walk without assistance, and/or complete aphasia |
| Rankin 5 | Stroke with severe disability: patient bedridden or requiring wheelchair (exceptional indication) |

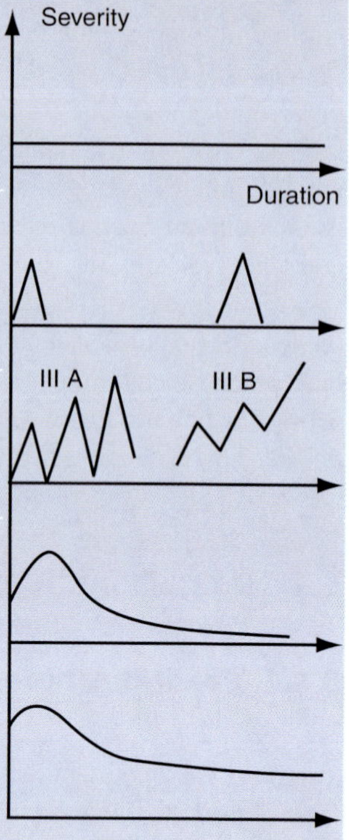


High-grade carotid stenosis ≥70% (NASCET criteria) or ≥80% (ECST criteria), determined angiographically or sonographically. Graphic representation of the duration (horizontal axis) and severity (vertical axis) of the respective neurologic deficits

*TIA* transient ischemic attack, *TEA* thromboendarterectomy, *PRIND* prolonged reversible ischemic neurologic deficit

**Table 5.7** Indications for duplex ultrasound of the extracranial cerebral arteries

- After TIA, PRIND, stroke: identification of underlying cause; additional CT and echocardiography
- Diagnostic evaluation for asymptomatic carotid stenosis suspected on clinical grounds (auscultation, risk factors, atherosclerosis with coronary heart disease or pelvic artery stenosis)
- Indication for operative treatment of carotid stenosis: plaque morphology, hemodynamic stenosis grading
- Diagnostic evaluation of pulsatile neck mass (aneurysm, transmission of pulsation through extravascular tumor)
- Diagnostic evaluation of traumatic intimal dissection
- Diagnostic evaluation of inflammatory disease (Takayasu's arteritis, temporal arteritis)
- Diagnostic evaluation of disturbed perfusion in the posterior circulation (vertebral artery stenosis, subclavian steal syndrome due to subclavian artery occlusion)
- (Diagnosis of brain death)
- Follow-up after surgical repair (TEA) or PTA with stenting (immediately after intervention, then at 6-month intervals)

*TIA* transient ischemic attack, *PRIND* prolonged reversible ischemic neurologic deficit, *CT* computed tomography, *TEA* thromboendarterectomy, *PTA* percutaneous transluminal angioplasty

Over the last decades, TEA has evolved into a suitable method for treating high-grade ICA stenosis, which accounts for the majority of cerebral infarctions. The main drawback of carotid TEA is that it may cause what it is supposed to prevent, namely stroke or death. This is why the surgical risk must be weighed against the risk of untreated stenosis. Numerous prospective randomized multicenter studies *compared the natural history and the surgical risk for symptomatic and asymptomatic carotid stenoses of different grades* (see Table 5.1). Endarterectomy in symptomatic carotid stenosis aims at eliminating the vascular source of emboli and/or the flow obstruction in individuals having suffered cerebral infarction with mild residual deficits.

The *ECST* and the *NASCET* compared antiplatelet therapy versus endarterectomy in patients with symptomatic carotid artery stenosis. A recent reanalysis of these data demonstrates that carotid TEA statistically highly significantly reduces the risk of ipsilateral stroke by 16% after 5 years in individuals with 70–99% stenoses. This means that six operations have to be performed to prevent one ipsilateral stroke over a 5-year period (number needed to treat (NNT)). In individuals with 50–69% stenosis, the absolute risk reduction (ARR) diminishes to 4.6%. TEA has no

**Table 5.8** Comparison of the risk of stroke for surgical versus medical management. Perioperative risk (stroke/death) and absolute risk reduction (ARR) of ipsilateral stroke over a 5-year period in patients with symptomatic carotid artery stenosis[a]

| Degree of stenosis (%) | Operative risk[b] (%) | Risk of stroke | | ARR[c] (%) | P | NNT |
|---|---|---|---|---|---|---|
| | | Surgical (%) | Medical (%) | | | |
| <30 | 6.7 | 12 | 10.0 | –2.2 | 0.05 | – |
| 30–49 | 8.4 | 15 | 18.2 | 3.2 | 0.6 | 31 |
| 50–69 | 8.4 | 14 | 18.6 | 4.6 | 0.04 | 22 |
| 70–99 | 6.2 | 10 | 26.0 | 15.9 | <0.001 | 6 |

[a]Summary of results obtained in 6,029 randomized patients from the ECST (*n* = 3,018), the VA (Veterans Affairs) trial 309 (*n* = 189), and the NASCET (*n* = 2,885); stenosis grade according to NASCET criteria
[b]All strokes/deaths occurring within 30 days; a total of 3,248 patients operated on
[c]Including perioperative stroke/death. *NNT* number needed to treat

advantage in individuals with stenoses <50% and is harmful in those with <30% stenosis compared to the natural history. The rate of severe perioperative complications (stroke, death) was 6.2% for stenoses greater than 70% and 8.4% for 50–69% stenoses (Table 5.8).

The wider use of validated noninvasive diagnostic modalities such as Doppler and duplex sonography as well as the known relationship between coronary heart disease and carotid stenosis and their associated risk of stroke make it more and more important to decide which patients with subclinical carotid artery stenosis would also benefit from surgery. However, in this population with a lower risk of spontaneous stroke (annual rate of less than 1% in stenosis <70% versus 2–5% in those with >70% stenoses, depending on comorbidity), it is more difficult to demonstrate a statistical benefit of therapeutic measures. While earlier studies revealed no benefit of operative treatment, the ACAS demonstrated an advantage for the patients operated on for carotid stenoses of 60–99% compared to patients undergoing medical treatment. The 5-year stroke risk was 5.1% in the surgical group versus 11% in the medical care group. The perioperative risk of stroke and death was 2.3% including the rate of 1.2% of preoperative angiography. Based on these data, the American Heart Association (AHA) recommends surgery for asymptomatic carotid artery stenosis >60% if the center performing the intervention has a perioperative risk of less than 3%. Patient management is dictated by the degree of stenosis determined by duplex imaging and the clinical stage (Table 5.9).

Apart from the degree of stenosis (Fig. 5.9), *plaque morphology* is the other major determinant of stroke risk. Ulcerated plaques with superimposed thrombi, intraplaque hemorrhage, and atheromatous plaques are associated with a higher risk of stroke as compared with smooth, fibrous plaques. However, no imaging modality exists that enables a reliable estimate of the risk of embolism on the basis of plaque morphology.

Yet, in individual cases, the sonomorphologic appearance may provide information as to the risk of embolism. Some studies suggest that hypoechoic plaques have a two to five times greater tendency to embolize than hyperechoic ones.

Arterial emboli from atherosclerotic plaques in carotid stenosis account for 55–60% of all strokes (territorial infarction; Fig. 5.7a). Another 30–35% are due to cardiogenic embolism, and less than 5% are due to hemodynamically reduced perfusion, especially in multiple-vessel disease (borderzone infarction). Other rare causes accounting for less than 5% of cases are inflammatory vessel disease, microangiopathy, and dissection.

#### 5.5.1.1 Defining the Degree of Stenosis

Despite the clinical relevance of ICA stenosis, there is no agreement about how it should be quantified, and various methods have been proposed. The difficulty in grading ICA stenosis is chiefly attributable to the greater width of the carotid bulb, the preferred site of ICA stenosis. Pronounced plaques in this slightly dilated portion of the carotid artery may cause considerable luminal narrowing without having any hemodynamic effects.

Basically, two methods exist for grading ICA stenosis. The *local degree of stenosis* is defined as the ratio of the patent residual lumen to the local vessel lumen without the plaque and gives the best estimate of plaque thickness (which is relevant for the ensuing risk of embolism) and the true extent of vascular obstruction. However, angiography enables only a rough and indirect estimate of the local degree of a stenosis because, unlike duplex ultrasound, it does not depict the original vessel diameter.

The *distal degree of stenosis* is calculated from the diameter of the residual lumen of the stenosed segment and that of the distal ICA (which is fairly constant up to the base of the skull). This method allows accurate assessment of the *reduction in blood supply to the brain* caused by a stenosis and classifies mild to moderate stenosis of the carotid bulb as hemodynamically nonsignificant (Figs. 5.7b and 5.8).

**Table 5.9** Extracranial cerebral arteries – duplex ultrasound findings and therapeutic consequences

| Diagnosis | Ultrasound findings and clinical presentation | Therapy |
|---|---|---|
| Plaques | No hemodynamic stenosis, asymptomatic or symptomatic | Medical management |
| ICA stenosis | Hemodynamically significant stenosis <70%, asymptomatic | Medical management |
| | Stenosis >70%, asymptomatic | Surgical reconstruction (TEA) acceptable but only proven if perioperative risk is low (according to ACAS study):<br>• Dependent on plaque morphology<br>• If perioperative morbidity/mortality rate <3%<br>• Annual stroke rate of 2% in medical care group versus 1% in surgical group<br>Surgery only if life expectancy >5 years |
| | 50–70% stenosis, symptomatic | Surgical reconstruction (TEA):<br>• Acceptable but not proven<br>• In TIA <6 months and plaque morphology suggesting a high risk of embolism (ulceration, hypoechogenicity, irregular surface) |
| | >70% stenosis, symptomatic | Indication for surgery (TEA):<br>• Proven<br>• Risk reduction relative to natural history increases as the perioperative morbidity and mortality rate decreases (target: <5%; Table 5.8) |
| | >70% stenosis, stage IV | Surgery only after nearly complete resolution of symptoms about 2–6 weeks after acute event. Prophylactic surgery of asymptomatic side may be indicated if there is stenosis on this side as well |
| ICA occlusion | Stages IV | Usually no operation, emergency operation may be contemplated only immediately after the event (mortality of up to 9%); otherwise medical management |
| Subclavian artery stenosis/occlusion | Steal syndrome, symptomatic | PTA, extrathoracic bypass procedure or transposition |
| ECA stenosis | High-grade | External carotid angioplasty indicated only in multiple-vessel disease (occlusion of ICA) with borderzone ischemias and proven extracranial and intracranial collateralization |
| Carotid artery dissection | Mostly due to trauma, asymptomatic, patent or thrombosed false lumen | Medical management, anticoagulation (intimal flap becomes attached or false lumen undergoes obliteration or thrombosis in most cases). Fixation or resection of intimal flaps only in exceptional cases with pronounced neurologic deficits and floating flaps |
| Kinking or coiling | Asymptomatic if nonstenotic<br>Symptomatic if associated with stenosis | Medical management<br>Resection |
| Inflammatory vessel disease (Takayasu's arteritis, temporal arteritis) | Wall thickening (macaroni sign) with or without hemodynamically significant stenosis | Cortisone theapy, no surgical reconstruction |
| Carotid body tumor | Well perfused tumor in the carotid bifurcation (color duplex) | Complete tumor resection, embolization only in patients with a high risk of morbidity |
| Vertebral artery stenosis | High-grade stenosis, asymptomatic<br>High-grade stenosis, symptomatic | Medical management<br>Chiefly located at origin, surgical reconstruction or PTA |

*ICA* internal carotid artery, *TEA* thromboendarterectomy, *ACAS* Asymptomatic Carotid Atherosclerosis Study, *TIA* transient ischemic attack, *PTA* percutaneous transluminal angioplasty, *ECA* external carotid artery, *CCA* common carotid artery

The distal grading method is primarily used in the USA, while determination of the local degree is more common in Europe. Hence the NASCET used the former and the ECST the latter. This must be borne in mind when the results of studies are compared. As the relation between the diameter of the carotid bulb and that of the distant internal carotid is fairly constant, the two degrees of stenosis can be easily converted into each other using the following formulae:

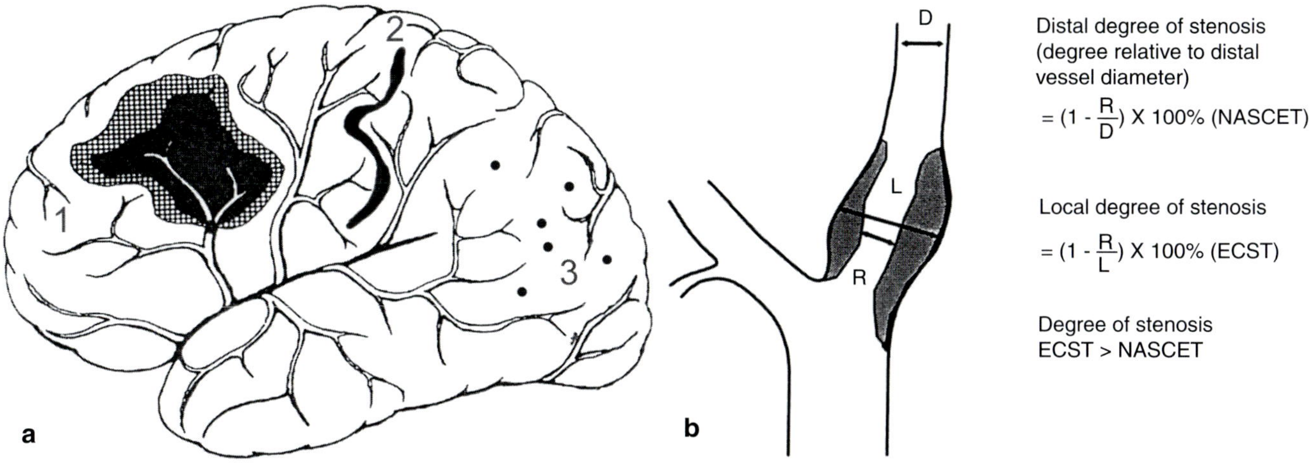


**Fig. 5.7** **a** Types of cerebral infarction. *1* Territorial infarction: due to arterioarterial embolism (carotid, cardiac). *2* Borderzone infarction: hemodynamic origin, reduced perfusion in terminal vascular bed, chiefly in patients with multiple-vessel disease. *3* Lacunar infarction: microangiopathy. **b** Methods of stenosis grading (local versus distal degree of stenosis). Due to the larger vessel diameter in the bulb, a stenosis classified as mild to moderate using the local grading method may not be graded as a stenosis by the distal grading method. Since the stenosis-induced decrease in perfusion only has a minor role in the development of cerebral ischemia, whereas plaque thickness is crucial for the associated risk of embolism, the local degree of stenosis is of greater clinical significance. For instance, eccentric plaques causing only moderate stenosis of the bulb may already carry a considerable risk of embolism based on their thickness

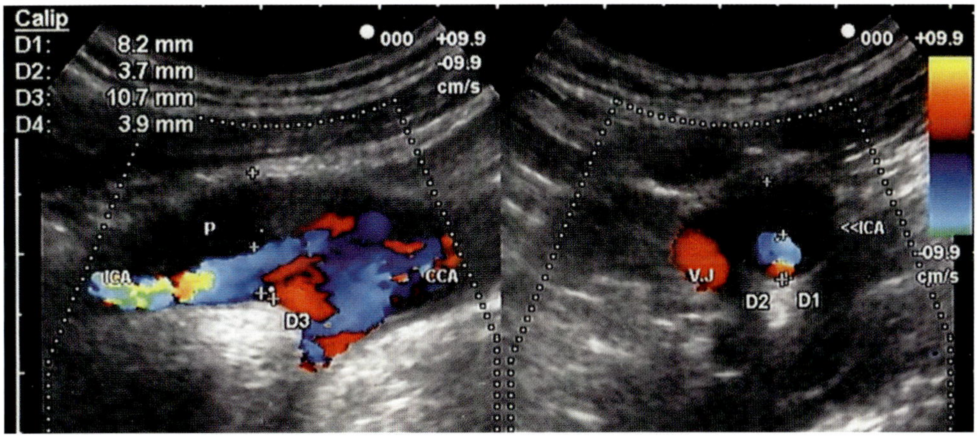


**Fig. 5.8** Color duplex images (longitudinal scan on the *left* and transverse scan on the *right*) demonstrating hypoechoic eccentric plaque of the carotid bulb. Calculation using the local grading method yields a 65% diameter reduction (60–70% stenosis). According to the distal stenosis grade (diameter of the distal ICA in the longitudinal scan: almost 5 mm), this is a 20–30% stenosis and surgery is not indicated. Conversely, the local stenosis grade establishes an indication for surgery, especially when additionally considering the plaque morphology (hypoechoic) and configuration (very eccentric and thickness >5 mm: high shearing forces). The final decision for surgery also depends on the patient's age and concomitant diseases. The example illustrates how the use of the two methods of stenosis grading (local versus distal) can lead to different therapeutic consequences

$$\text{Local (ECST) degree of stenosis (\%)} = 0.6 \times \text{distal (NASCET) degree (\%)} + 40\%$$

$$\text{Distal degree of stenosis (\%)} = \text{local degree of stenosis (\%)} - 40\%/0.6$$

These equations yield the following correspondences between the distal and local degrees of stenosis:

| | | | | | | | | | |
|---|---|---|---|---|---|---|---|---|---|
| Distal (%) | 0 | 50 | 60 | 67 | 70 | 75 | 85 | 90 | (NASCET criterion) |
| Local (%) | 40 | 70 | 75 | 80 | 82 | 85 | 90 | 95 | (ECST criterion) |

In Germany, the local grading method was recommended as the standard by the German Society of Ultrasound in Medicine (*Deutsche Gesellschaft für Ultraschall in der Medizin, DEGUM*) many years ago (Widder et al. 1986). Plaque causing luminal narrowing of up to 40% in the carotid bulb is classified as nonstenosing according to the distal quantification method because stenosis with a local grade of up to 30% reduces the bulbous lumen only to the diameter of the distal carotid artery. Note, however, that *hemodynamic alterations are less relevant for the risk of cerebral infarction*

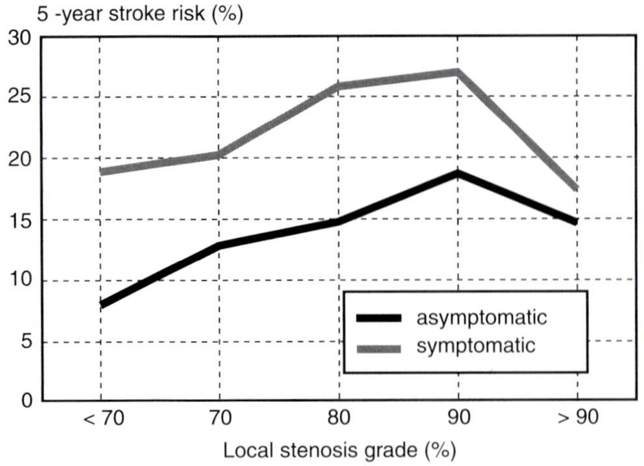


**Fig. 5.9** Risk of ipsilateral cerebral infarction by degree of ICA stenosis in symptomatic and asymptomatic stenosis (according to Widder 2004)

*than the risk of embolism, which increases with the plaque thickness*. Therefore, *eccentric plaque in the bulb may already pose a considerable risk of embolism before it causes hemodynamic effects*.

Angiography, the traditional gold standard for carotid assessment, has methodological limitations as it grades a stenosis on the basis of purely morphologic criteria. It is an invasive procedure that involves radiation exposure and contrast-medium-related side effects as well as the risk of minor stroke in 1.3–4.5% of cases and major stroke in 0.6–1.3% (Davies et al. 1993; Dion et al. 1987; Hankey et al. 1990; Moore 2003). The risk of angiography is higher in symptomatic stenoses than in asymptomatic ones, and the risk of inducing stroke may increase to stenosis up to 12.5% in patients with bilateral high-grade carotid stenosis (Theodotou et al. 1987). The ACAS provides the most detailed analysis. According to this study, *angiography* performed at radiologic centers is associated with a *combined neurologic morbidity and mortality of 1.2%* in asymptomatic patients, which is only slightly lower than the 1.52% risk associated with carotid TEA in the same patient population. Given the fact that diagnostic angiography has a stroke risk similar to the therapeutic intervention, *it is becoming more and more common to perform carotid TEA* (Fig. 5.10) *without prior diagnostic angiography* (Chervu et al. 1994). This is made possible in part by the use of high-resolution ultrasound, which has been shown in comparative studies with histologic workup to be superior to angiography in assessing plaque morphology and the risk of embolism (Fontenelle et al. 1994; O'Donnell et al. 1985).

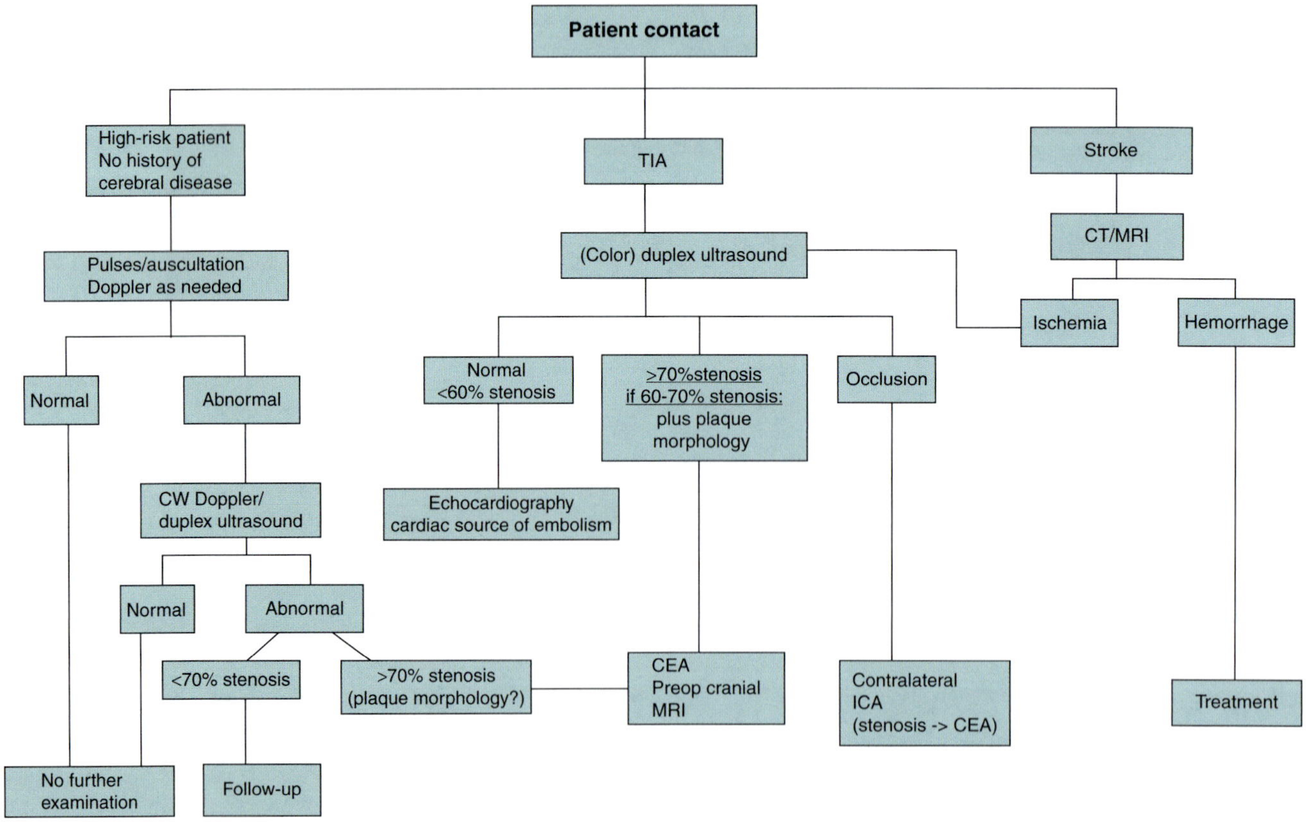


**Fig. 5.10** Diagnostic algorithm in patiens with suspected ICA stenosis. If >70% stenosis has been diagnosed by duplex ultrasound, the patient can proceed to surgery without further preoperative imaging of the carotid arteries. In patients with 60–70% stenosis, plaque morphology on B-mode imaging is considered as an additional criterion in identifying those for whom carotid endarterectomy (CEA) is recommended

#### 5.5.1.2 Plaque Morphology

In recent years, *ultrasound measurement of carotid intima-media thickness (IMT) has become an established technique for estimating the risk of cardiovascular morbidity and mortality* and has been used as a surrogate marker for pre- or subclinical atherosclerosis and for monitoring the outcome of treatment (e.g., statins) in interventional studies.

Risk factors such as long-standing hypertension or hyperlipoproteinemia damage the intima, first becoming manifest as thickening of the intima–media complex. Thickening above 1 mm is considered abnormal and a thickness of 2 mm or more is defined as plaque (Li et al. 1996).

However, thickening of the intima–media complex is an aging phenomenon. While IMT is below 0.6 mm in young healthy individuals, an average increase of 0.1 mm per decade of life is regarded as normal after the age of 40(Homma et al. 2000). Serious arterial wall changes should be expected when the increase in thickness exceeds 1.5 mm. Individuals with an IMT >1.5 mm or small focal plaques often have aortic plaques, which have recently been implicated as a possible cause of embolic cerebral infarction. Measurement of IMT therefore provides a general estimate of the total atherosclerotic burden, and patients with marked thickening of the intima–media complex have an increased embolic risk arising from atherosclerotic plaques in the aortic arch. Once a *plaque* has reached a certain thickness, it disturbs the nutrition of the intima, which is not supplied by vessels of its own but through diffusion from the vessel lumen. The initial plaque continues to grow through the accumulation of lipids, lipoproteins, and cholesterol. The interruption of the nutrient supply can lead to central necrosis (Fig. 5.11) with formation of an atheroma, which may become organized through fibroblast invasion and thus develop into a stable lesion. Alternatively, there may be rupture of the covering intimal layer with discharge of degenerative atheromatous debris into the bloodstream and embolization to the brain. As a result of *lipid inclusion* and central necrosis, a plaque can increase in size to such an extent that it represents a considerable *obstacle to pulsatile blood flow*. Sonographically, such a plaque is identified by pulsatile longitudinal movement with the blood flow. Fibroblast invasion leads to sclerosis, ultimately resulting in calcification of the plaque.

A rapid *increase in plaque size* may also be due to internal hemorrhage, which is attributed to very minute, vulnerable vessels growing in from the adventitia. Exposure to flowing blood can lead to rupture of the thin plaque cap (intima) with embolization to the brain of necrotic or thrombotic plaque components (Fig. 5.11a–f). Plaque rupture triggers repair processes with re-endothelization of the former plaque area resulting in a rather *smoothly covered niche* that poses no risk of embolization. Unfortunately, this fairly harmless state may be difficult to differentiate from ulceration by angiography and ultrasound alike.

Less harmless sequelae are *ulcerative defects* with incomplete re-endothelialization that may still release thrombotic material into the bloodstream.

The turbulent flow occurring in stenotic segments can induce the deposition of thrombotic material, especially at the distal end of a plaque, with ultimate progression to occlusion of the ICA.

The risk of embolism is determined not only by the degree of stenosis but also by plaque morphology as such. The following types of plaques can be distinguished in the carotid system on the basis of their macroscopic appearance:

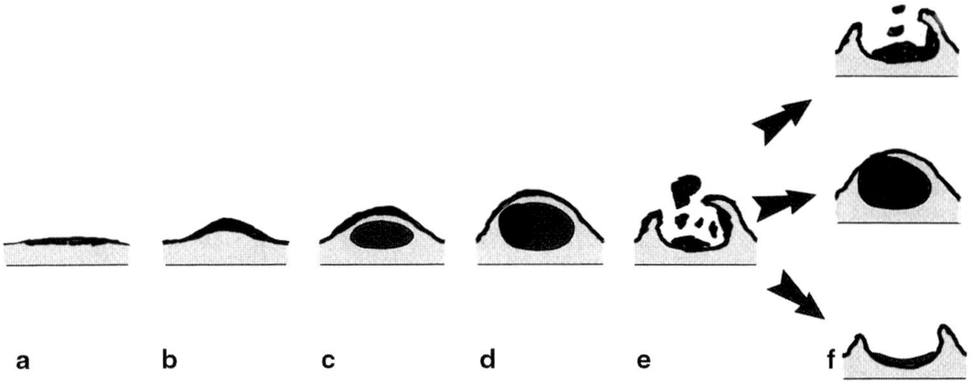


**Fig. 5.11 a–f** Stages of plaque development. **a** Initial atherosclerotic wall thickening (intima–media complex thickening). **b** Further increase in wall thickness with development of plaque. **c** Plaque increases in size through lipid inclusion and may undergo central necrosis (atheroma); disturbed nutrition of the plaque. **d** Intramural hemorrhage through rupture of ingrowing vessels. **e** Rupture of the plaque cap induced by pulsatile blood flow (longitudinal pulsation) with ulceration mainly of proximal portions. **f** Re-endothelialization of the ulcer with formation of a washed-out niche as a fairly stable residue (*bottom*), re-endothelialization of the ulcerative plaque (*middle*), or persisting ulcerative plaque with recurrent embolism and only partial repair of the vulnerable surface (*top*)

- Flat, fibrous plaque
- Atheromatous or soft plaque
- Calcified or hard plaque
- Ulcerative plaque
- Hemorrhagic plaque

A large study compared the carotid TEA specimens with the clinical symptoms in 1,252 consecutive patients. The incidence of plaque ulceration was 77% in patients with transient ischemic attacks (TIAs) and 79% in those with prior stroke, which was significantly higher than in asymptomatic patients with an ulceration rate of 60%. The incidence of intraplaque hemorrhage did not differ significantly between symptomatic and asymptomatic patients but was significantly higher in patients with greater than 90% carotid stenosis (Park et al. 1998).

For estimation of the risk of embolism, it would be desirable to have an imaging modality that provides reliable information on plaque morphology. This is difficult, however, since most atherosclerotic lesions are chiefly composed of variable amounts of atheromatous material with high lipid content and fibrous material rich in collagen. The inhomogeneous composition of plaques is reflected in their ultrasound appearance, but it is not possible to identify individual plaque components on the basis of their different echo levels and to thus make use of this information for predicting the risk of embolism. *Ulcerated plaques* are *difficult to differentiate from washed-out cavities that have become re-endothelialized.*

## 5.5.2 Vertebral Arteries

Transitory ischemic attacks or strokes involving the vertebrobasilar system are much less common than those arising from the carotid territory. Stenosis at the origin of the vertebral artery rarely requires surgical or interventional treatment, in particular because the risk of embolism is lower. In patients with multiple-vessel disease and a global reduction in cerebral perfusion, repair is mainly done in the carotid territory.

While lesions in the carotid system present with highly specific hemispheric symptoms, the clinical manifestation is much less specific when the vertebrobasilar system is affected. Dizziness is the chief symptom, but may also be caused by numerous nonvascular conditions. Apart from atherosclerotic lesions, acute symptoms of vertebrobasilar insufficiency may be due to dissection, typically occurring after trauma.

In patients with subclavian artery occlusion, the vertebral artery is scanned to evaluate its collateral function in subclavian steal syndrome (complete–incomplete).

Ultrasonography is the method of choice for morphologic assessment as well as demonstration of atherosclerotic lesions and dissection. Published data suggest that the vertebral artery is amenable to sonographic assessment in over 80–90% of cases, depending on the segments included in the analysis. Before treatment is initiated, however, the sonographic examination should be supplemented by other imaging modalities such as computed tomography (CT) angiography, magnetic resonance imaging (MRI), and possibly angiography.

# 5.6 Ultrasound Criteria, Measurement Parameters, and Diagnostic Role

## 5.6.1 Carotid Arteries

### 5.6.1.1 Plaque Evaluation and Morphology

**Intima–Media Thickness** There has been a long controversy regarding the role of B-mode plaque evaluation in estimating the risk of embolism, and recent studies have not clarified this issue. What is undisputed is that B-mode sonomorphologic criteria allow a detailed description and classification of plaques with good interobserver and intraobserver agreement. Technical developments and the use of high-resolution transducers (>10 MHz) have improved the detection and evaluation of small plaques as well as the measurement of IMT. The latter is therefore increasingly being used to identify individuals with an increased cardiovascular risk.

The thickness of the intimal and medial layers can be most reliably measured on a longitudinal scan, which shows two bright lines in the far wall. These two lines represent the lumen–intima interface and the interface between the adventitia and surrounding tissue. The combined thickness of the intimal and medial layers is measured because the two cannot be distinguished by ultrasound.

The intima–media complex can be measured with good interobserver agreement, and the age-related increase in thickness is well established. The normal IMT is <0.7 mm, with a thickness >1 mm being abnormal and >2 mm representing plaque. Homma et al. (1997, 1999, 2000) found a linear increase in IMT from a mean of 0.49 mm before age 40 to 1.02 mm in subjects older than 100 and proposed the following formula for calculating the age-related normal thickness: (0.009 × age) + 0.116.

Interventional studies (e.g., of statin treatment; Hedblad et al. 2001; Kang et al. 2004) used serial sonographic measurements of IMT to monitor treatment outcome. A measurement accuracy with an error of less than 0.1 mm has been reported (Meyer and Strobel 2008), but this requires an axial resolution afforded only by transducers with a very high frequency of >15 MHz, which may not provide the penetration necessary for imaging the CCA in all patients. Probes with a frequency of 10 MHz or less provide a maximum axial resolution of about 0.2 mm and are unlikely to detect changes of less than 0.1 mm in serial measurements. In addition, deviations of 0.1–0.2 mm result from interobserver variability and the use of different ultrasound systems (Baldassarre et al. 2000; Kanters et al. 1997). Despite these limitations, high-resolution transducers provide the best option available for monitoring IMT.

Various sites in the CCA and ICA have been explored to measure IMT, and the distal CCA 2–3 cm proximal to the bifurcation has been identified as the ideal site for IMT measurement. Areas of plaque should be excluded, but once plaque has been demonstrated, IMT measurement is no longer required to estimate the cardiovascular risk (Poli et al. 1988; Bond et al. 1989; Ebrahim et al. 1999; Sun et al. 2002; Homma et al. 2001; Sakaguchi et al. 2003; Sutton-Tyrrell et al. 1992; Meyer and Strobel 2008).

Long circumferential thickening of the arterial wall, especially when homogeneous and hypoechoic, could point to vasculitis. Suspected vasculitis should be ruled out or confirmed by additional clinical and laboratory examinations and sonographic evaluation of the vascular territories most susceptible to this condition (subclavian artery).

Small plaques in the ICA become more frequent in the normal population after age 50 with a prevalence of up to 80% in those over 80. Because they are so common and their natural history is unclear, the significance of small internal carotid plaques and their therapeutic relevance remain unclear.

**Plaque Features** Plaques and stenoses mainly involve the *bifurcation* and the first 2 cm of the ICA and ECA. The superficial location of these carotid segments enables scanning with high-resolution, *high-frequency transducers* that also allow evaluation of *plaque morphology*. The morphologic description of a plaque comprises the following features:

- Localization:
  - Anterior/posterior wall
  - Proximal/distal
- Extension:
  - Circular/semicircular
  - Plaque diameter
- Plaque configuration:
  - Concentric
  - Eccentric
- Plaque surface:
  - Clearly delineated/poorly delineated/not delineated
  - Smooth/irregular (0.4–2.0 mm deep); ulcer (>2.0 mm deep)
- Plaque composition:
  - Homogeneous/inhomogeneous
- Echogenicity:
  - Echogenic (with or without acoustic shadowing)/echolucent/cannot be visualized

The great flexibility of transducer positioning and the possibility of imaging in different planes afforded by duplex ultrasound facilitate the evaluation of plaque in a way that other cross-sectional imaging modalities do not. Nevertheless, the individual ultrasound scan reduces the three-dimensional (3D) plaque to a two-dimensional (2D) representation (Fig. 5.12), which limits the usefulness of B-mode imaging for evaluating plaque evolution over time. This is because serial measurements of plaque thickness, an important predictor of embolism risk, are unreliable. For these reasons, it is recommended that the plaque be insonated from different directions and that its greatest thickness be measured instead of using standardized planes for measurement. Plaque configuration also contributes to the risk of embolism and must not be neglected. *The risk is higher for an eccentric plaque because it is thicker on one side and the shear forces acting on this thicker plaque portion protruding into the lumen are greater than those acting on a concentric plaque* – even when the two are causing the same degree of stenosis (Fig. 5.13a).

Using a high-resolution transducer, the examiner should first obtain an unbiased impression of plaque morphology without any gross pathologic criteria or prognostic factors in mind. Plaque appearance on gray-scale images provides no direct information whatsoever about plaque composition – whether fibrous, atheromatous, stable, unstable, or ulcerated. Instead, the examiner must always bear in mind that the ultrasound image is a display of differences in acoustic impedance between tissues and does not reflect tissue properties directly.

The *plaque surface*, which is the boundary between flowing blood and the plaque components, is described in terms of visibility and irregularity or the presence of ulceration. Note, however, that the visibility of a reflecting structure such as the plaque boundary is primarily determined by the *angle of incidence of the ultrasound beam* (i.e., the intensity with which the boundary is depicted depends on whether the returning echoes have been reflected or scattered by the interface; see Figs. 1.2 and 1.3).

B-flow ultrasound is a new technique that allows highly accurate separation of vessel wall structures from the perfused lumen, providing high-resolution images of the extracranial carotid territory. This territory, being superficial, can be examined with a high-frequency transducer. The technique affords superior discrimination of plaque and flowing blood because – unlike Doppler imaging – it is independent of the angle of insonation (Fig. A5.36). For precise evaluation of the plaque surface and the degree of luminal narrowing for stenosis grading, the target vessel needs to be visualized in different planes. B-flow ultrasound can only be used to examine superficial vessels, and, like all ultrasound techniques, its diagnostic yield is degraded by scattering and acoustic shadowing from calcified plaque. The studies performed so far underscore the good differentiation between the patent lumen and wall structures, pointing out the potential for evaluating plaque contour and for grading stenosis on the basis of morphologic criteria (Jung et al. 2007). Comparative data on the sensitivity and specificity of B-flow ultrasound in the evaluation of plaque morphology and stenosis grading compared with color duplex imaging or angiography in a large study population are still lacking.

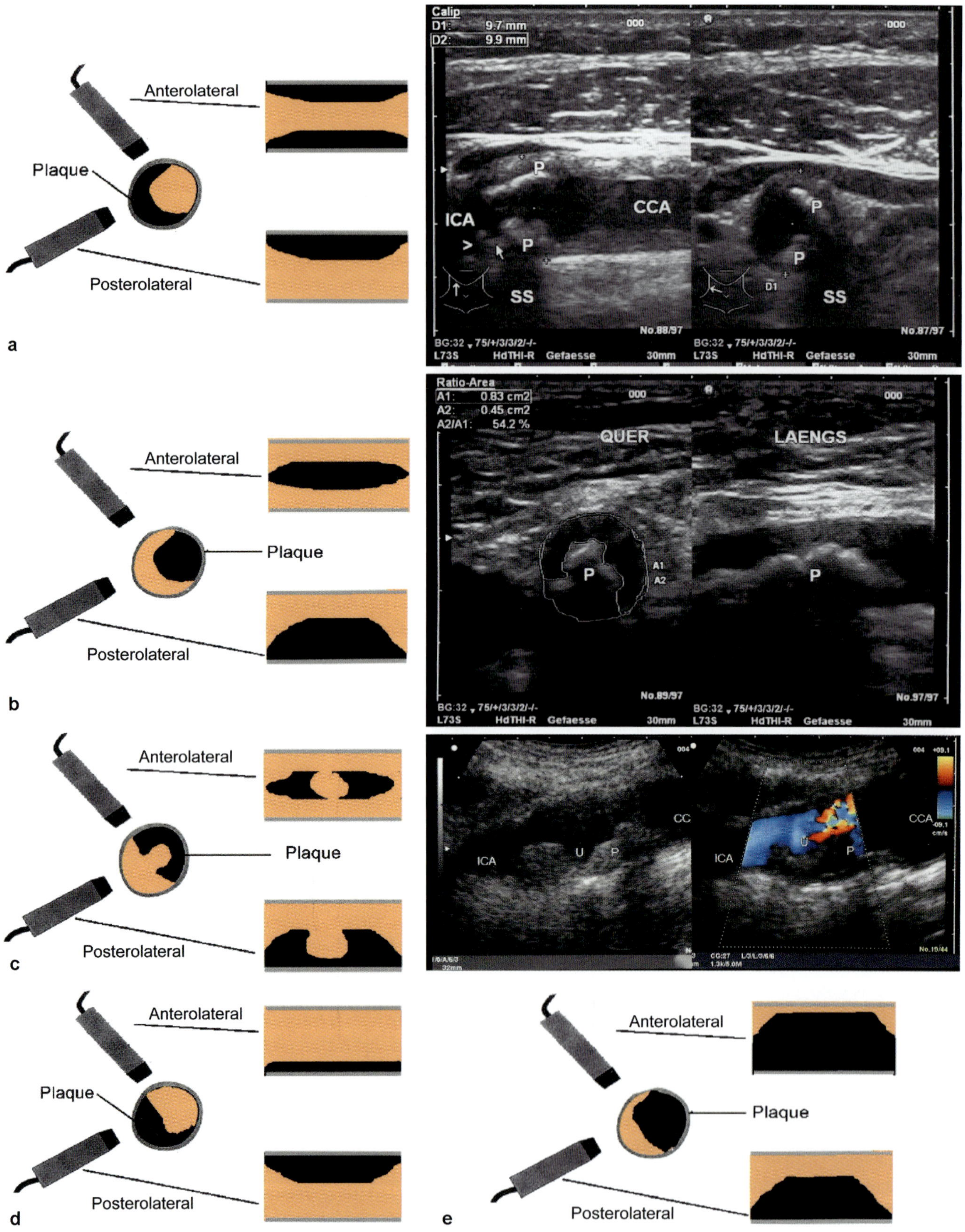
Anterolateral
Plaque
Posterolateral
a
Calip
D1: 9.7 mm
D2: 9.9 mm
P
CCA
ICA
P
SS
P
P
D1
SS
No.88/97
No.87/97
BG:32 75/+/3/3/2/-/-
L73S HdTHI-R Gefaesse 30mm
BG:32 75/+/3/3/2/-/-
L73S HdTHI-R Gefaesse 30mm
Anterolateral
Plaque
Posterolateral
b
Ratio-Area
A1: 0.83 cm2
A2: 0.45 cm2
A2/A1: 54.2 %
QUER
LAENGS
P
A1
A2
P
No.89/97
No.97/97
BG:32 75/+/3/3/2/-/-
L73S HdTHI-R Gefaesse 30mm
BG:32 75/+/3/3/2/-/-
L73S HdTHI-R Gefaesse 30mm
Anterolateral
Plaque
Posterolateral
c
CC
ICA
U
P
CCA
ICA
U
P
Anterolateral
Plaque
Posterolateral
d
Anterolateral
Plaque
Posterolateral
e

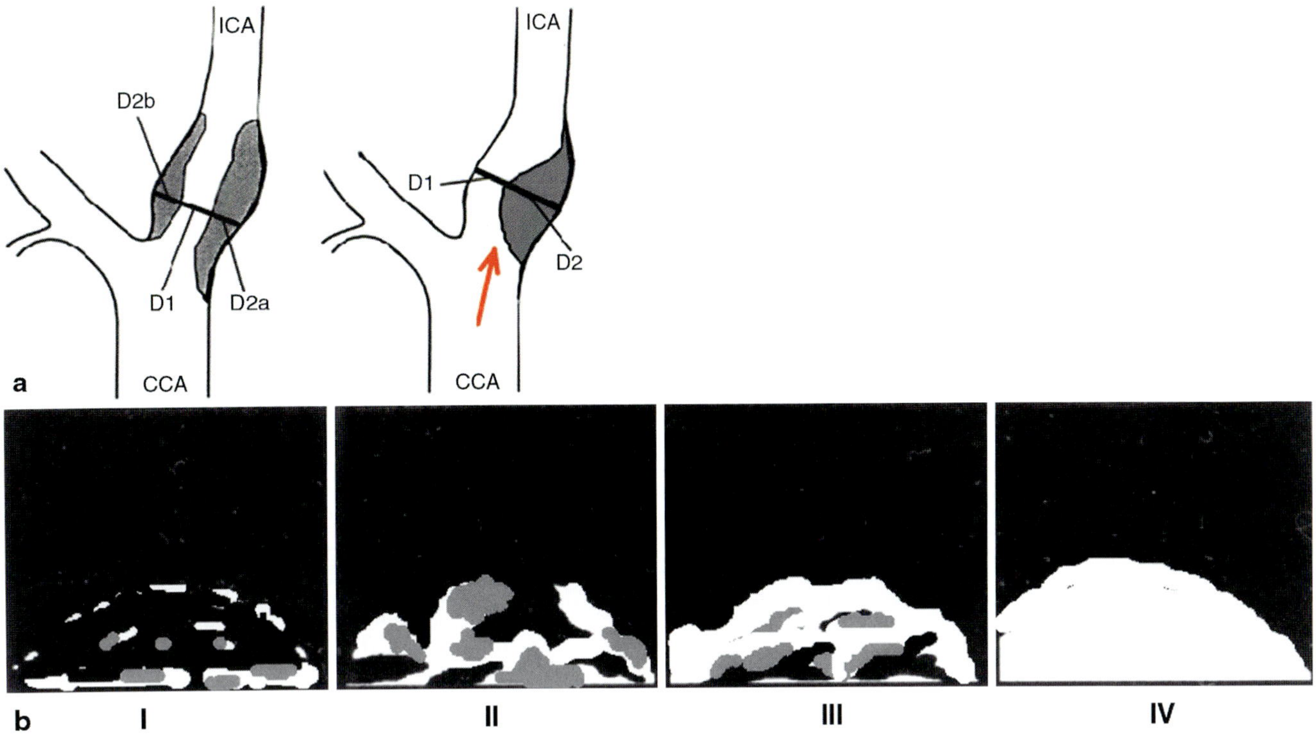


**Fig. 5.13** **a** Diagrams illustrating internal carotid artery (ICA) stenosis caused by concentric (*left* drawing) versus eccentric plaque (*right* drawing). Although the degree of stenosis is the same (approx. 65%), the eccentric plaque is thicker and therefore poses a higher risk of embolism: the shear forces acting on it (*red arrow*) are greater, and the plaque is therefore more likely to rupture.
**b** Sonomorphologic types of carotid artery plaque (based on the Gray-Weale classification; see Figs. A5.9 and A5.10):
Type *I* – predominantly echolucent lesions with a low gray-scale value, similar to that of the lumen; the surface is interrupted and not consistently visible.
Type *II* – mixed, substantially echolucent lesions with small areas of echogenicity and interrupted, irregular surfaces.
Type *III* – mixed, substantially echogenic lesions with mostly regular and clearly delineated surfaces.
Type *IV* – predominantly echogenic lesions of uniform density with mostly smooth and clearly delineated surfaces

**Plaque Differentiation** The way in which a gray-scale ultrasound image is formed also plays a role when evaluating plaque makeup and echotexture. Echodensity is described in shades of gray ranging from very dark to very bright (echolucent to echogenic). The reference values used are those of the hypoechoic flowing blood (lowest gray-scale value) and the hyperechoic boundary (high gray-scale value) between the adventitia and surrounding connective tissue in the far wall. The echotexture can be described as homogeneous (uniform appearance) or inhomogeneous (irregular distribution of bright pixels or absence of echoes). If a plaque is heterogeneous, areas close to the lumen, which are most relevant for the risk of embolism, are most important in terms of echogenicity and homogeneity. Acoustic shadowing is the only ultrasound phenomenon that provides direct information on a histopathologic tissue feature, as it indicates total reflection of the incident ultrasound beam by a calcified structure. It is a sign of a calcified plaque, which is more stable. The Gray-Weale classification was proposed to provide a unified description on the basis of the many criteria of plaque morphology used in the literature and distinguishes four types of plaques based on echogenicity (Gray-Weale 1988; Fig. 5.13b).

Supplementing this classification with other important criteria including plaque surface characteristics (Geroulakos et al.

**Fig. 5.12 a–e** Scanning in at least two planes (as in angiography) is necessary to characterize plaque (thickness, morphology) (anterior and posterolateral transducer positions). If, for instance, a small bowl-shaped plaque is imaged in only one plane, the stenosis grade can be overestimated or underestimated (see Fig. 5.21). Diagrams *a* – *e* illustrate different sonomorphologic plaque shapes (transverse plane on the *left* and anterolateral and posterolateral longitudinal sections on the *right*). The drawings depict eccentric concave (**a, d**) and convex (**b, e**) plaques and ulceration (**c**) and illustrate how eccentric plaques convexly protruding into the lumen can be overestimated in certain scan planes, while concave eccentric plaques may be underestimated. The plaques in **a** and **b** (diagrams and corresponding ultrasound images) cause roughly the same cross-sectional area reduction (approx. 50%) but differ in thickness and in the amount of diameter reduction they cause (how these parameters are evaluated depends on the scan plane). With the transducer in the posterolateral position, the degree of stenosis caused by a large eccentric plaque (examples **b** and **e**) may be overestimated; with an anterior approach, it may be slightly underestimated. **c** Ulcer in a large eccentric plaque. In this example, only the posterolateral transducer position allows adequate evaluation (as indicated by diagrams)

1994; Langsfeld et al. 1989; Lusby 1993; Widder 1995), one can distinguish the following types of plaques on ultrasound:

- *Type IV*: echogenic and homogeneous plaque with a clearly delineated, smooth surface
- *Type III*: plaque of mixed echogenicity with predominantly echogenic portions and an irregular surface
- *Type II*: predominantly echolucent or heterogeneous plaque with a poorly delineated surface
- *Type I*: plaque not visualized or suggested only by isolated echogenic spots in an otherwise echolucent lesion: plaque only indirectly identified as a filling defect on color flow images

Assigning a plaque to one of these four categories faces two fundamental problems. First, most plaques are very heterogeneous, with components belonging to different categories, while other portions are not visualized at all and simply cannot be categorized. Here, one also has to take into account that the echoes used to form the ultrasound image are affected by the interaction of ultrasound with structures in the body and are not a direct representation of the target tissue (see Sect. 1.1.1.4). Second, an intraoperative analysis has shown that echolucent plaques are fibrous or atheromatous with surprisingly similar frequency (Widder et al. 1990).

Despite these discouraging remarks, the following assumptions are valid regarding the modified Gray-Weale classification of plaques.

Some studies suggest that *predominantly echolucent* plaque (type I) corresponds to atheroma with lipid inclusions and intraplaque hemorrhage, which make the plaque *unstable* and have been shown to be associated with a *significantly increased risk of stroke. Echogenic and homogeneous plaques with a smooth surface* (type IV), on the other hand, are associated with a *low risk of embolization* (see Figs. A5.10–A5.12). Unfortunately, the most common plaques (types II and III) are difficult to assess in terms of prognosis, and sonomorphologic risk assessment has a disappointingly low accuracy of 50–70%, comparable to that of angiography (Estol et al. 1991; Friedrich et al. 1988; Streitzler et al. 1994). Eccentric plaques carry a higher risk of embolism than concentric plaques because they are thicker, yet cause the same degree of stenosis (diameter reduction).

Despite its rather disappointing overall accuracy, under certain circumstances, sonographic plaque classification can provide useful information for selecting patients for surgery. These include in particular patients with asymptomatic high-grade stenosis or symptomatic 50–70% stenosis and type I or IV plaque. Hopes were set on *ultrasound contrast media*, but they have not been able to overcome the difficulties in differentiating between low-risk cavities resulting from defect healing and recent ulceration.

**Plaque Echogenicity** The difficulties outlined here have not discouraged attempts to distinguish between "uncomplicated" and "complicated" plaques, the latter believed to pose a higher risk of embolism. Uncomplicated or stable plaques are assumed to be homogeneously fibrous or partially calcified and to have an intact fibrous cap. Conversely, unstable plaques predominantly consist of atheromatous material and may contain necrotic areas and blood: a plaque cap is either absent or appears thinned or visibly interrupted. Several authors assume that degenerative processes induced by inflammation or bacterial infection play an important role in the development of high-risk lesions (Libby 2002; O'Leary et al. 1991; O'Donnell 1985; Bassiouny et al. 1977). Several studies reported in the literature, including investigations using 3D ultrasound, suggest that larger plaque volumes are associated with complications such as ulceration (Schminke et al. 2000; AbuRahma et al. 2002; Pedro et al. 2002). Many authors focus on plaque echogenicity and the appearance of the surface, assuming these to be the most crucial features for predicting the risk of embolism.

Some investigators assume that echolucency indicates high lipid content, a known predictor of an increased embolism risk. However, evaluation of echogenicity is very subjective and also depends on the ultrasound scanner settings. To overcome these limitations, the *adjacent sternocleidomastoid muscle* can be *used as the standard of reference* with the plaque being classified as echolucent if it is *less echogenic than the muscle* (see Fig. A5.9). Very echolucent plaques are indistinguishable from flowing blood on B-mode ultrasound. Bright spots within the lesion may be the only sign that an atheromatous plaque is present at all. In these cases, color duplex imaging is required to indirectly reveal the plaque border as a filling defect. Increasing deposition of collagen and cellular matrix is believed to be associated with higher echogenicity. Very bright echoes are characteristic of dense fibrous or calcified areas and, in concjunction with acoustic shadowing, indicate calcification. Calcification in a plaque may be focal or diffuse. If acoustic shadowing obscures the vessel wall or precludes Doppler measurement, the transducer can be moved to avoid the inferring structure (Fig. 5.14).

A *heterogeneous ultrasound appearance* suggests a *mixed plaque*, and inhomogeneous components in a predominantly echolucent lesion appear to be associated with a higher risk of embolism.

One study showed that 80% of patients with neurologic deficits scheduled for internal carotid disobliteration had very heterogeneous plaques (Banafsche et al. 1995). In another study, more heterogeneous plaques were found to be associated with higher stenosis grades (Banafsche 1998).

Plaques with areas of lower echogenicity near the luminal surface appear to carry a higher risk of embolism. Published studies, however, provide no clear answer as to whether the risk is higher for homogeneous versus heterogeneous plaques (Szajzel 2006; El-Barghouty et al. 1994; Wijeyaratne et al. 2003).

For these reasons, caution is advised when applying histologic terms to the description of ultrasound appearances of carotid plaques, and it is emphasized again that a gray-scale image is a display not of tissues but of interfaces between areas of different acoustic impedance. To stay on the safe side, as

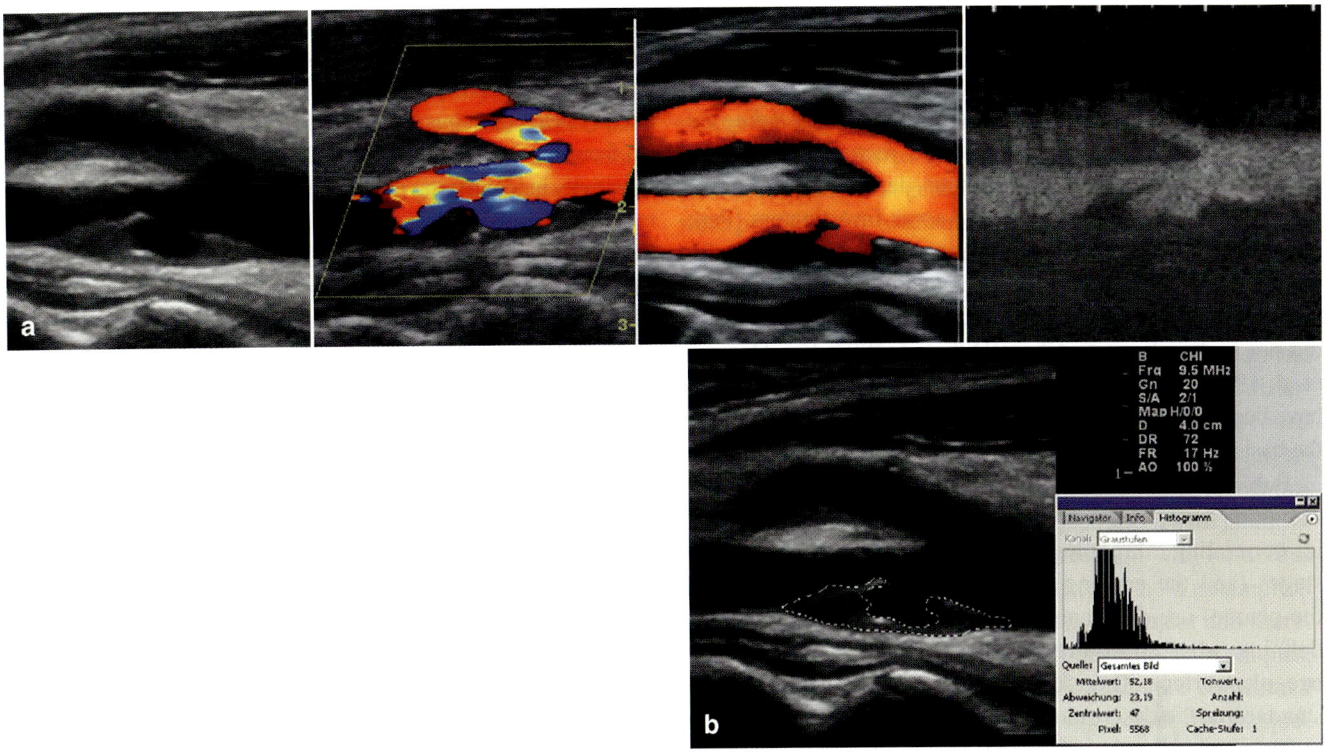


**Fig. 5.14** **a** Echolucent eccentric plaque at the origin of the internal carotid artery with deep ulcer (>2 mm) and no hemodynamic significance following plaque rupture. *Images from left to right*: B-mode, color duplex, power Doppler, and B-flow mode. The ulcerated portion is clearly differentiated from the patent lumen on the B-mode image, but the power Doppler and B-flow images allow the most accurate evaluation of the ulcer contour and delineation from flowing blood. Sonographic plaque morphology (echogenicity and plaque configuration) is increasingly being used as a criterion in identifying candidates for surgery or in deciding on the best reconstructive approach (carotid stenting or TEA). New ultrasound techniques are used to improve plaque imaging, and both B-flow and contrast-enhanced color duplex techniques allow more accurate evaluation of plaque configuration and better identification of bowl-shaped defects. However, even with these new techniques, it remains difficult to differentiate a plaque ulcer with a high risk of embolism from a washed-out niche, which, in terms of embolism risk, is considered rather harmless (see Fig. 5.11). **b** Computation of the gray-scale median (GSM) of the echolucent carotid plaque in **a**. The plaque is outlined in a normalized B-mode image (linear scaling using input and output values of two reference points: blood, 0–5; adventitia, 185–195), and the computer program (Adobe Photoshop CS) generates a histogram representing its composition and a median value (GSM of 47 in this case) (figure courtesy of Werner Lang)

long as the correlation between the sonographic appearance and histologic makeup of a plaque is still vague, it is recommended that the ultrasonographer simply describe the plaque (Woodcock et al. 1992).

**Plaque Thickness** *Longitudinal pulsatile movement* of a plaque also appears to *contribute to the risk of embolism. Shear forces* acting on the plaque *increase with plaque thickness* and may cause rupture of the vulnerable cap with discharge of embolic material. Progressive lumininal narrowing due to plaque thickening leads to higher flow velocities in the stenotic segment (>350 cm/s), which is associated with an increased risk of ulceration (Beach 1992). The risk of necrosis and ulceration with embolization also increases with the occurrence of necrosis, which in turn results from an insufficient nutrient supply as the plaque increases in length and thickness.

Rupture of the plaque cap can lead to ulceration, which is characterized on the B-mode scan by a heterogeneous echotexture with disruption of the surface or a crater-like defect.

Shear forces are higher when a plaque protrudes into the lumen, which is why an eccentric plaque poses a greater risk of embolism compared with a concentric plaque causing the same degree of stenosis. A concentric plaque occupies the entire luminal vessel circumference, while an eccentric plaque protrudes into the lumen on one side (Fig. 5.13a).

**Prognostic Relevance of Plaque Morphology** A number of prognostic criteria have been derived from the sonographic appearance of plaques to identify patients with a favorable or unfavorable prognosis:

| | |
|---|---|
| **Favorable prognosis:** | • Fairly echogenic and homogeneous plaque<br>• Plaque surface smooth and clearly delineated<br>• Calcification<br>• Short plaque (<1 cm)<br>• Thin plaque (<4 mm) |
| **Unfavorable prognosis:** | • Predominantly echolucent with plaque suggested only by echogenic spots; surface poorly delineated<br>• Long plaque (>1 cm)<br>• Plaque diameter >4 mm<br>• Longitudinal pulsatile movement of the plaque in a cranial direction |

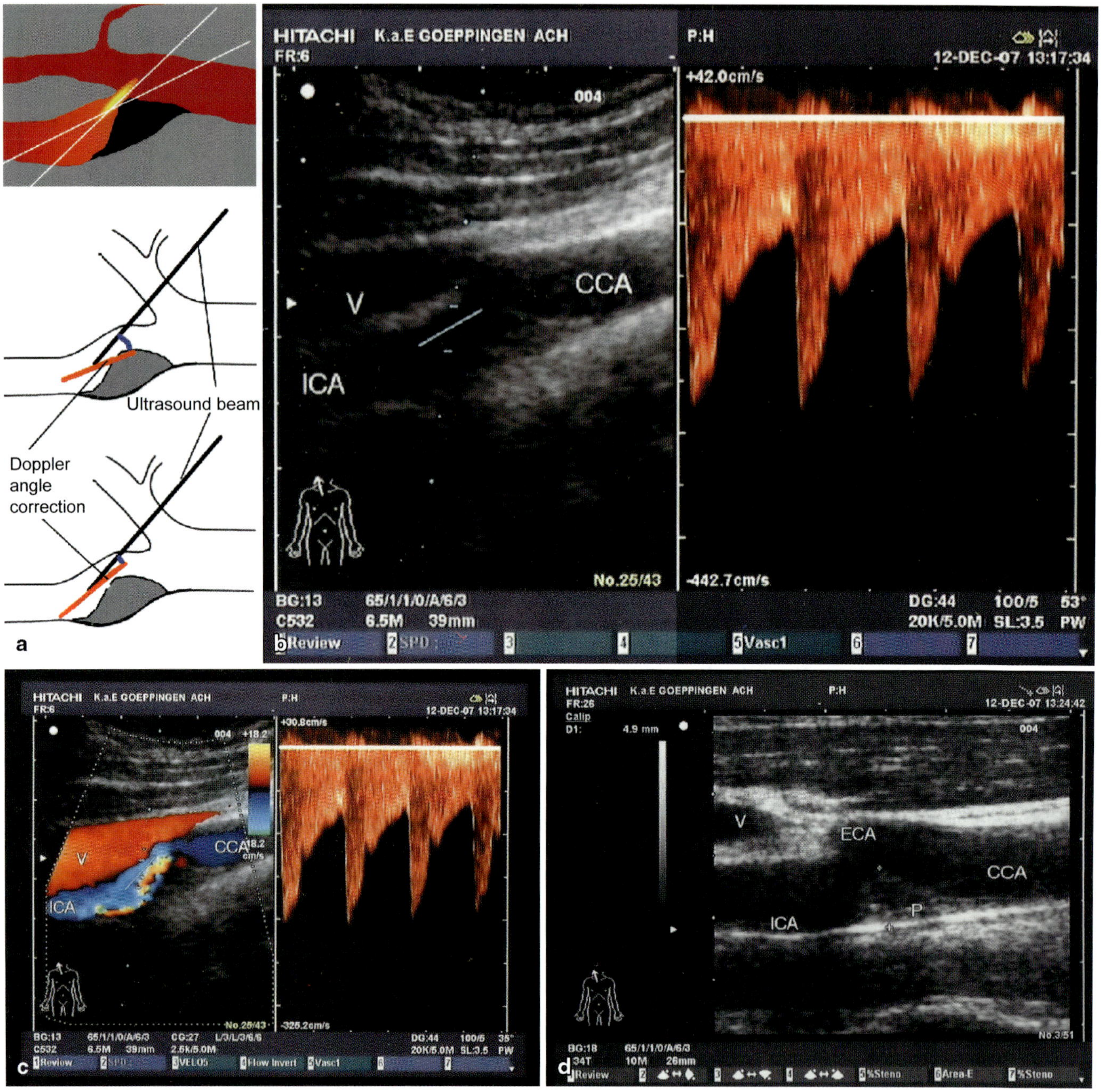


**Fig. 5.16** **a** The Doppler angle correction cursor (*red*) is usually aligned parallel to the vessel wall (*middle* diagram) because it is assumed that blood flow is parallel to the wall (*black line* represents the ultrasound beam and *blue* the Doppler angle). When an eccentric plaque is present, the flow jet visualized by color duplex imaging is frequently not parallel to the wall but is deviated toward the plaque (*yellow* in the *top* diagram). Alignment of the Doppler angle correction cursor parallel to the stenosis jet depicted in the duplex scan can result in a discrepancy of 5–10%, when compared with correction parallel to the vessel wall (*bottom* diagram). **b** Eccentric stenosis at the origin of the internal carotid artery (*ICA*): measured with Doppler angle correction parallel to the vessel wall, a peak systolic velocity (PSV) of 280 cm/s is calculated, corresponding to >70% stenosis (53° Doppler angle relative to arterial wall). **c** The flow jet depicted in the color duplex image (aliasing indicated by yellow color coding) is at a smaller angle to the ultrasound beam (see **a**). With the corrected Doppler angle of 35° (relative to stenosis jet), a PSV of 210 cm/s is calculated from the (same) Doppler shift frequency, corresponding to <70% stenosis (according to ECST criteria). The divergent results with regard to PSV and stenosis grade illustrate that the standard method of Doppler angle correction does not yield adequate results when the assumption on which it is based does not hold. **d** Sonomorphologically, the plaque has two features of vulnerability: low echogenicity and marked eccentricity with longitudinal pulsation in the real-time B-mode (high shear force acting on the plaque, see Fig. 5.13a; plaque thickness of 4.9 mm). These plaque features justify an indication for surgery, even in patients with stage I stenosis. Note the clearer definition of the plaque (*P*) in **d** compared with **b**. Plaque conspicuity decreases with the angle between the ultrasound beam and the wall (due to increased scatter)

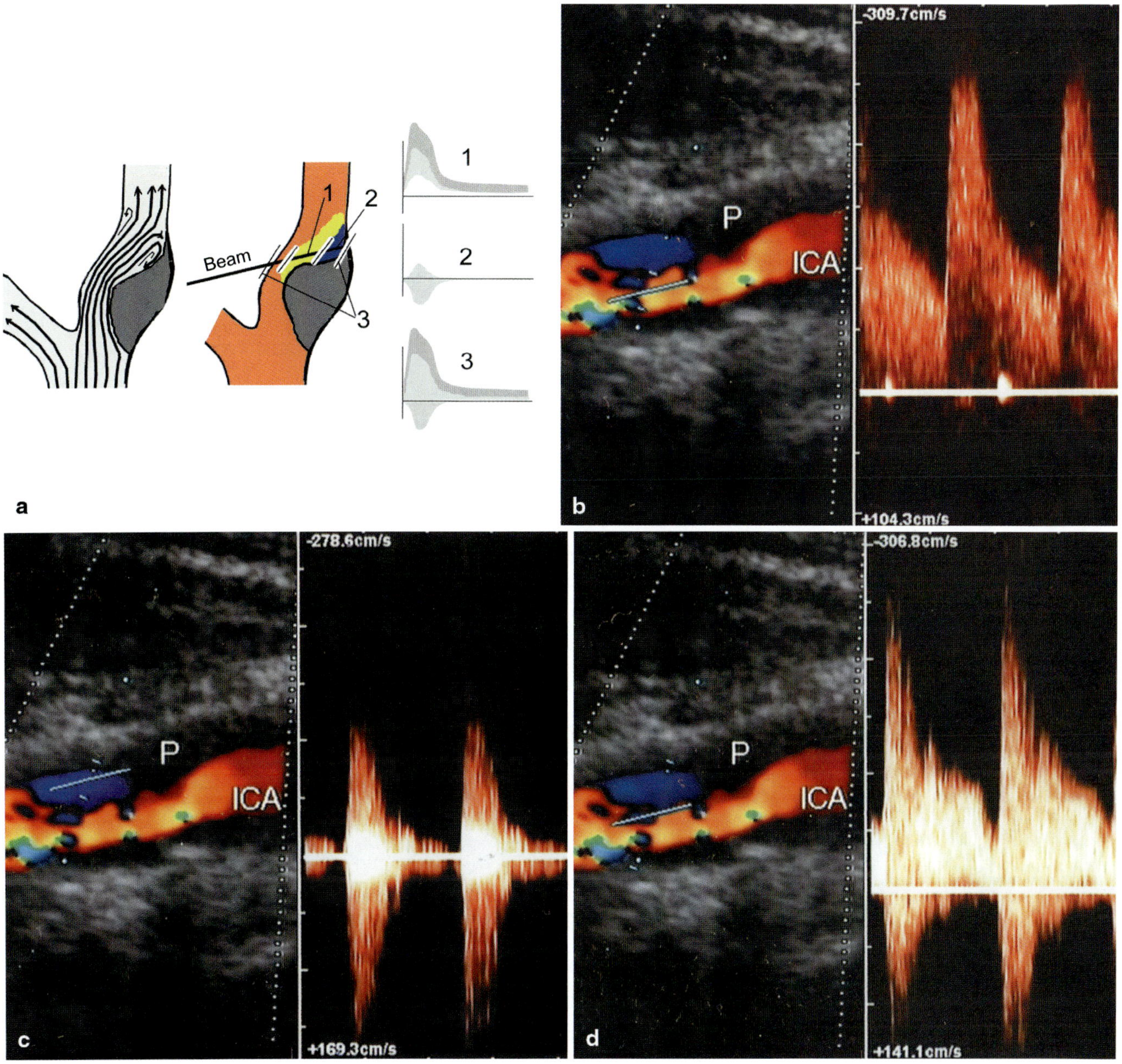


**Fig. 5.17** **a** The *left* drawing shows the course of flow in a high-grade stenosis caused by eccentric plaque in the internal carotid artery (ICA) with eddy currents and retrograde flow components directly behind the plaque. The *middle* drawing shows the duplex color coding (based on mean velocity) of blood flow velocity and direction in a stenosis. Flow velocity is highest in the flow jet, which is indicated by aliasing (color change from red to yellow); change in flow direction at sites of eddy currents is indicated by red and blue color. The Doppler waveforms (diagrams on the *right*) represent blood flow velocity and direction at different sampling sites in the area of stenosis (indicated by the numbers 1–3). The waveform from the center of the stenosis (*1*) reflects the intrastenotic flow velocity in the stenotic jet. The waveform measured in a small sample volume placed directly behind the plaque (*2*) represents the flow in the area of eddy currents, seen as predominantly retrograde flow (represented below the baseline in the waveform). Adequate selection of a sample volume covering the entire width of the arterial lumen (*3*) depicts both maximum peak systolic velocity (PSV) and the eddy currents distal to the plaque. **b** Color duplex image and Doppler waveform from a sample volume placed in the stenotic jet (aliasing with color change from *red* to *yellow* to *light blue*) in an eccentric ICA plaque. **c** The waveform obtained with the sample volume placed at the margin directly behind the plaque predominantly represents eddy currents (color change from *red* to *black* to *blue*), which appear predominantly as retrograde flow (below the baseline in the waveform). **d** With a sample volume covering the entire width of the lumen, the waveform reflects both the high PSV and the eddy currents with retrograde flow components

**Table 5.10** Duplex ultrasound criteria for the classification of internal carotid artery (ICA) stenosis. (Modified according to Arning et al. 2010; Jacobs et al. 1985; Bluth et al. 1988; Neale et al. 1994; Faught et al. 1994; Moneta et al. 1995; Abu Rahma et al. 1998; Grant et al. 2003)

| Stenosis grade (distal, NASCET criteria) (%) | | 10 | 20–40 | 50 | 60 | 70 | 80 | 90 | Occlusion |
|---|---|---|---|---|---|---|---|---|---|
| Stenosis grade (local, ECST criteria) (%) | | 45 | 50–60 | 70 | 75 | 80 | 90 | 95 | Occlusion |
| Main criteria | 1. B-mode image | +++ | + | | | | | | + |
| | 2. Color duplex image | + | +++ | + | + | + | + | + | +++ |
| | 3. PSV at the site of maximum stenosis (cm/s) | | 120–160 | 200 | 250 | 300 | 350–400 | 200–500 | No flow signal |
| | 4. PSV in the poststenotic segment (cm/s) | | | | >50 | >50 | <50 | <30 | No flow signal |
| | 5. Collaterals and developing collaterals (periorbital arteries/ACA) | | | | | (+) | ++ | +++ | +++ |
| Additional criteria | 6. Decrease in diastolic velocity in the prestenotic segment (CCA), increased pulsatility | | | | | (+) | ++ | +++ | +++ |
| | 7. Abnormal flow in the poststenotic segment, turbulence | | | + | + | ++ | +++ | (+) | |
| | 8. End-diastolic velocity at the site of maximum stenosis (cm/s) | <40 | <50 | <90 | <100 | >100 | >120 | | |
| | 9. Confetti sign | | | | (+) | ++ | ++ | | |
| | 10. ICA/CCA PSV ratio | | | >2 | >3 | >4 | >4.5 | | |

Notes on criteria 1–10: Stenosis grading according to NASCET: grades given in 10% intervals (±5%). Local stenosis grading (according to ECST) is preferred by the German Society of Ultrasound in Medicine (DEGUM) because it better takes into account plaque thickness (risk of embolism). Criterion 1: The lumen may appear hyperechoic in occlusion. Criterion 2: Differentiation of mild stenosis (local aliasing) from nonstenotic plaque; depiction of flow directions in moderate and high-grade stenosis; demonstration of occlusion. Criterion 3: Cutoff velocities for stenosis grades identified by ROC curve analysis. Criteria apply to stenosis 1-2 cm in length and are limited when multiple vessels are affected. Criterion 4: Measured far distal to the stenosis, i.e., outside the jet and the zone of disturbed flow. Criterion 5: May affect only one collateral pathway; diagnostic accuracy is lower when only extracranial arteries are examined. Criterion 9: Confetti sign only when a low PRF is used. Abbreviations: *PSV* peak systolic velocity, *ACA* anterior cerebral artery, *CCA* common carotid artery, *ICA* internal carotid artery; *+ signs* indicate diagnostic relevance (compared with other criteria)

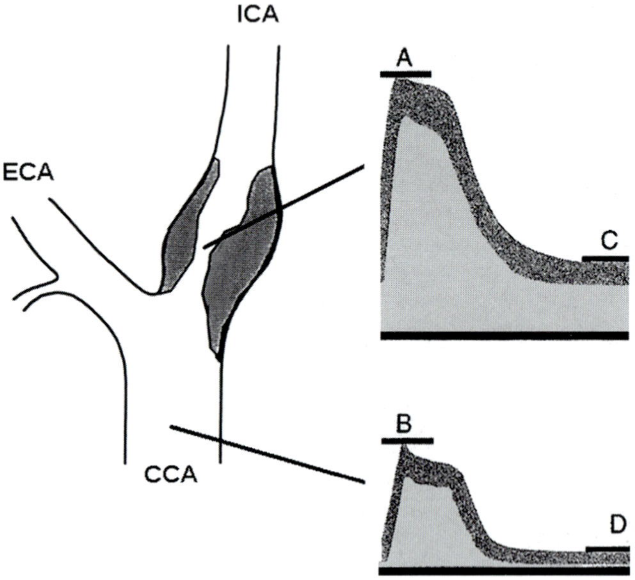


**Fig. 5.18** Stenosis criteria for grading internal carotid artery (ICA) stenosis. *A* = peak systolic velocity (PSV) within the stenosis; *C* = peak end-diastolic velocity (EDV) in the stenosis; *A/B* = ratio of intrastenotic PSV in the ICA and PSV in the common carotid artery (CCA); *C/D* = ratio of peak EDV in the stenosis and peak EDV in the CCA

**Table 5.11** Pitfalls in stenosis grading resulting from other factors that affect peaks systolic velocity (PSV)

| Stenosis grading | Factors affecting PSV/source of error |
|---|---|
| Underestimation of stenosis | Low-blood pressure phase<br>Proximal stenosis, e.g., aortic stenosis<br>Very-high-grade stenosis, subtotal occlusion<br>Long stenosis<br>Tandem stenosis (additional distal/intracranial stenosis) |
| Overestimation of stenosis | Current hypertension<br>Hyperperfusion (contralateral carotid occlusion or very-high-grade stenosis)<br>Very pulsatile flow (medial sclerosis in diabetes mellitus)<br>Small or contracted vessels<br>Very short stenosis<br>Very eccentric stenosis |

- *Minimum end-diastolic velocity (EDV)*: This parameter also increases with the stenosis grade for stenosis causing at least 50% diameter reduction. The EDV threshold for >50% stenosis is 40 cm/s with velocities >80–100 cm/s suggesting high-grade stenosis. This parameter is especially useful

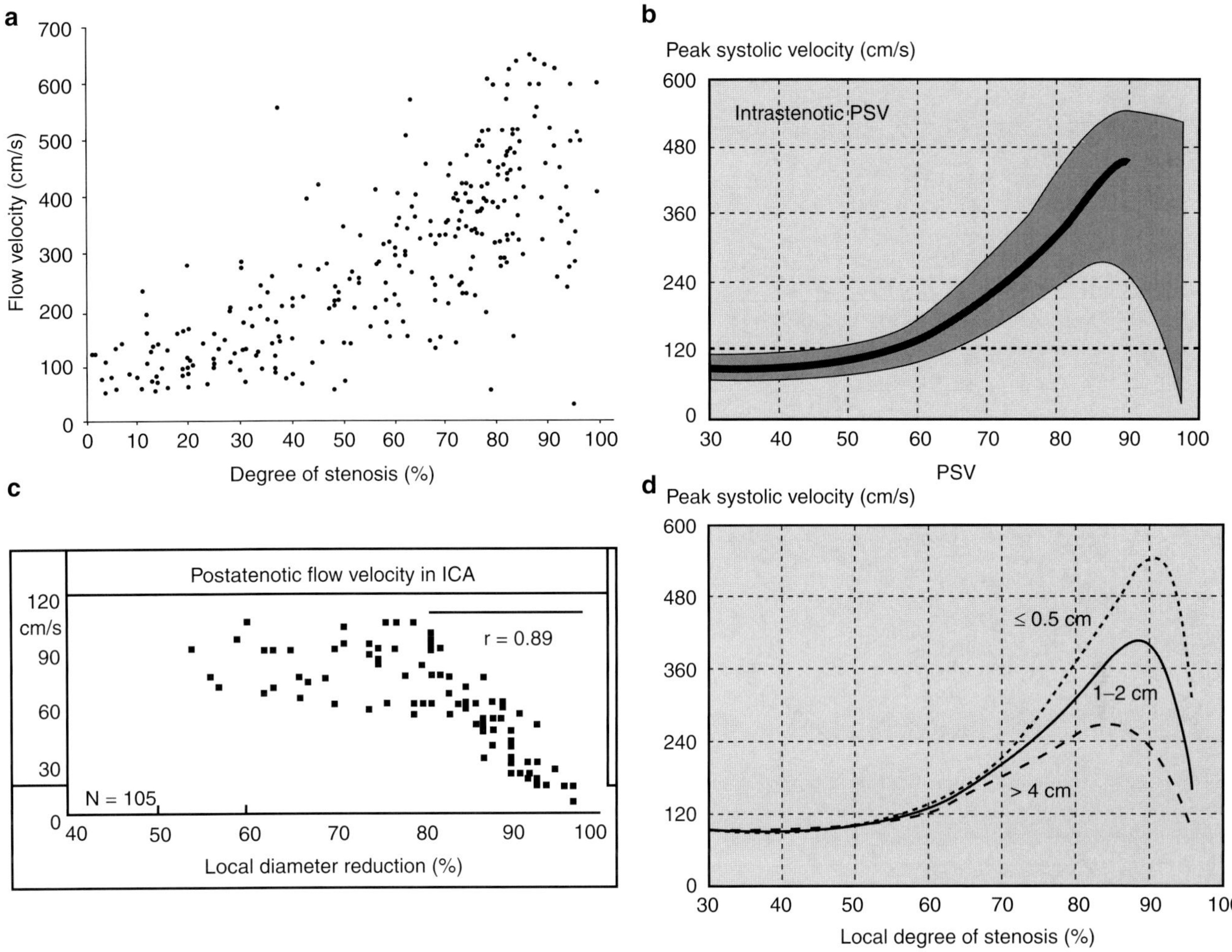


**Fig. 5.19** **a** Internal carotid artery (ICA) stenosis: relationship between calculated angiographic stenosis and peak systolic velocity (PSV) measured in the stenosis (from Moneta et al. 1995). **b** Flow velocity (PSV) as a function of stenosis grade. Intrastenotic PSV increases continuously with luminal diameter reduction (according to the continuity law) and degreases again in very-high-grade stenosis/subtotal occlusion with friction loss; this is especially the case when a long segment is stenosed. **c** Distribution of poststenotic PSV as a function of local diameter reduction. There is a decrease above a stenosis grade of about 85% (according to Görtler 1998). **d** Relationship between the length of stenotic plaque (length of stenosis) and expected maximum PSV as a function of local stenosis grade. When a long high-grade stenosis is present, friction loss may lead to a lower PSV in the stenotic jet than would be expected based on the degree of stenosis. This association is especially important when luminal narrowing is due to dissection. However, the actual decrease in PSV to be expected also depends on the plaque configuration and has not been investigated systematically (according to Widder 2004)

for quantifying subtotal occlusion, in which PSV is more difficult to determine (Carpenter et al. 1996).

- *Ratio of PSV in the ICA to that in the ipsilateral CCA (ICA/CCA PSV ratio)*: Absolute flow velocities are influenced by physiologic and abnormal systemic factors (hypertension, aortic valve stenosis, medial sclerosis, contralateral occlusion, extent of intracranial collateralization). The effect of these factors can be reduced by calculating the ratio of PSV in the stenosis to PSV in the prestenotic segment (Moneta et al. 1993; Howton et al. 2008; Carpenter et al. 1995). However, as the ECA also arises from the CCA, the ratio will be altered if the ECA is stenosed or acts as a collateral in the case of ICA stenosis.
- *Ratio of intrastenotic systolic velocity (PSV)* in the ICA and velocity measured distally (see Sect. 1.2.3 and Fig. 1.47). Identical stenosis grades (risk of embolism) can be associated with different intrastenotic and above all poststenotic velocities, depending on the length of the stenosis (Fig. 5.19d). The resulting error in stenosis grading can be minimized by calculating the ratio of intrastenotic to poststenotic velocity, preferably using intensity-weighted mean flow velocities rather than PSV. Using this method, a mean velocity ratio of >5 has been reported to be highly sensitive and specific for detecting higher-grade stenosis (Ranke et al. 1999; Table 5.12).

These parameters have accuracies of 83–97% using intra-arterial angiography as the gold standard. Several studies have shown good interobserver agreement both for grading ICA stenosis (kappa = 0.7) and for identifying candidates

**Table 5.12** Sensitivity, specificity, and accuracy of duplex ultrasound in internal carotid artery (ICA) stenosis grading using different sonographic criteria (angiography as reference standard)

| Author | Year | n | Method (stenosis criterion) | | | | Results | | | | |
|---|---|---|---|---|---|---|---|---|---|---|---|
| | | | Stenosis grade (%) | PSV (cm/s) | EDV (cm/s) | PSV ratio | Sensitivity (%) | Specificity (%) | PPV (%) | NPV (%) | Accuracy (%) |
| Faught et al. | 1994 | | 70 | | 130 | | | | | | 93 |
| Polak et al. | 1992 | | 50 | 125 | | | | | | | 83 |
| Huston et al. | 2000 | 915 | 50 | 130 | 70 | 1.6 | 92 | 90 | 90 | 91 | 91 |
| | | 915 | 70 | 230 | | 3.2 | 86 | 90 | 83 | 92 | 89 |
| Ranke et al. | 1999 | | 70 | $V_{meanICAst/ICA}$ $d > 5$ | | | 94 | 98 | | | 97 |
| Soulez et al. | 1999 | | 70 | | | | 94 | 81 | 62 | 98 | |
| | | | 60 | | | 2.9 | 94 | 80 | 72 | 96 | |
| Abu Rahma et al. | 1998 | | 50 | 140 | | | 92 | 95 | 97 | 89 | 93 |
| | | | 60 | 150 | 65 | | 82 | 97 | 96 | 86 | 90 |
| | | | 70 | 180 | 96 | | 85 | 95 | 91 | 92 | 92 |
| Grant et al. | 2000 | | 70 | 225 | | | | | | | 90 |
| Carpenter et al. | 1996 | 110 | 70 | 210 | | | 94 | 77 | 68 | 96 | 83 |
| | | 110 | 70 | | 70 | | 92 | 60 | 73 | 86 | 77 |
| | | 110 | 70 | | | 3.3 | 100 | 65 | 65 | 100 | 79 |
| Hood et al. | 1996 | 457 | 70 | 230 | 100 | | 78 | 97 | 88 | 94 | 93 |
| Carpenter et al. | 1995 | | 60 | 230 | | | 98 | 87 | 88 | 98 | 92 |
| | | | 60 | | 40 | | 97 | 52 | 86 | 86 | 86 |
| | | | 60 | | | 2.0 | 97 | 73 | 78 | 96 | 76 |
| | | | 60 | 230 | 40 | 2.0 | 100 | 100 | 100 | 100 | 100 |
| Browman et al. | 1995 | 75 | 70 | 175 | | | 91 | 60 | | | |
| Moneta et al. | 1995 | 176 | 60 | 260 | 70 | 3.2–3.5 | 84 | 94 | 92 | 88 | 90 |
| Neale et al. | 1994 | 60 | 70 | 270 | 110 | | 96 | 91 | | | 93 |
| Moneta et al. | 1993 | | 70 | 325 | 130 | 4 | 83 | 90 | 80 | 92 | 88 |
| Finkenzeller et al. | 2008 | 21 | 50<br>70 | 120<br>200 | 50<br>100 | 1.5<br>2.0 | Correlation with IADSA, $r = 0.852$ (Pearson) | | | | |

*PSV* peak systolic velocity, *EDV* end-diastolic velocity, *PPV* positive predictive value, *NPV* negative predictive value, *IADSA* intra-arterial digital substraction angiography

for surgery (kappa = 0.72; Griffiths et al. 2001). When the stenosis grade determined using absolute parameters provides no definitive basis for recommending surgery (threshold stenosis), effects of systemic conditions such as hypertension or hypercirculation (fever, hyperthyroidism) should be considered (Table 5.11). Medial sclerosis in long-standing diabetes leads to pulsatile flow with a larger systolic component and a smaller diastolic component. Contralateral carotid occlusion (Fig. 5.20), high-grade stenosis, or multi-vessel disease with vertebral artery involvement may also lead to artificially elevated flow velocities in the carotid system, depending on collateralization (see Figs. A5.18 and A5.19). To avoid overestimation in patients with significant contralateral stenosis, a higher PSV cutoff of 140–150 cm/s should be used to discriminate between low-grade and hemodynamically significant (>50%) carotid stenosis (modified according to AbuRahma et al. 1995). Not taking these factors into account will lead to false-positive results and overestimation of carotid stenosis (Horrow et al. 2000; Busuttil et al. 1996).

In the study by Busuttil et al. (1996), duplex ultrasound overestimated the degree of carotid stenosis in 27% of patients with severe contralateral disease (using angiography as the reference). Following unilateral carotid endarterectomy (CEA), PSV on the unoperated side decreased on average by 36 cm/s. Other investigators reported a 20–40% higher PSV due to compensatory flow in patients with over 90% contralateral stenosis (Henderson et al. 2000).

*The compensatory increase in PSV* is determined not only by the *severity of the contralateral stenosis* but also by the *amount of collateral flow through other pathways* (ipsilateral ECA and supratrochlear artery, posterior circulation).

In another study including 107 patients with asymptomatic 50–99% contralateral carotid stenosis by standard criteria (PSV >125 cm/s), an early postoperative duplex examination showed a mean decrease in PSV of 48 cm/s

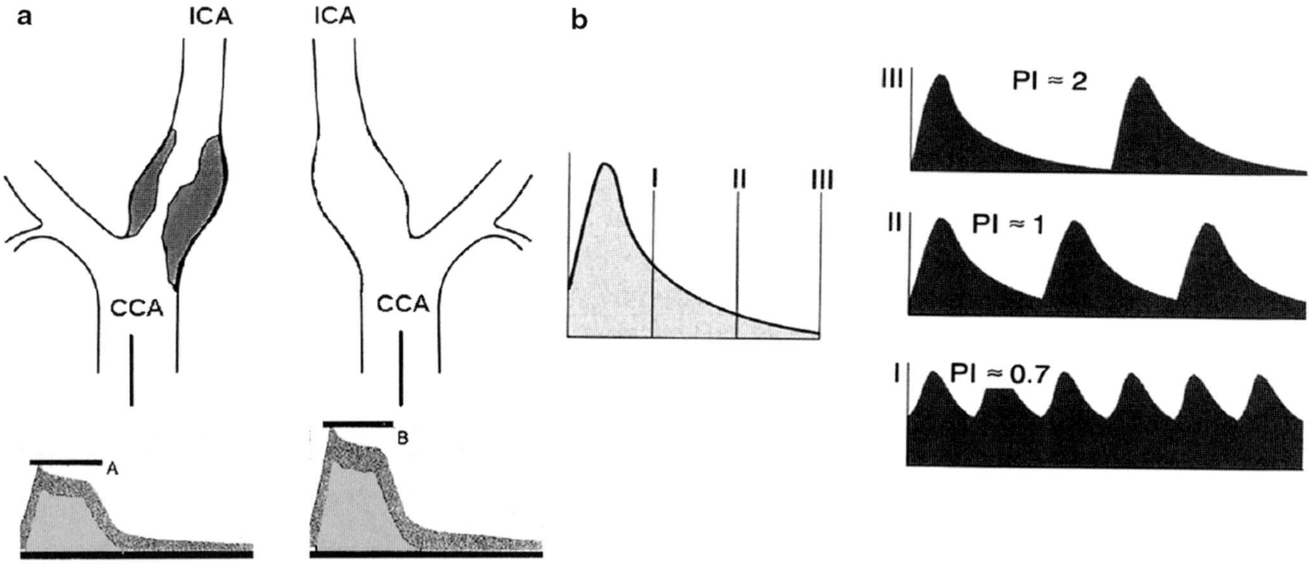


**Fig. 5.20** **a** Unilateral high-grade ICA stenosis or occlusion (*left* diagram) is associated with an increase in flow velocity in the contralateral common and internal carotid arteries if they function as the primary collateral pathway; this must be borne in mind in stenosis grading. **b** Because EDV decreases with the duration of diastole, stenosis grading on the basis of EDV is dependent on a patient's heart rate (as well as other systemic factors such as current blood pressure, contralateral occlusion, vessel elasticity). As a consequence, resistance indices incorporating EDV also vary with the heart rate

(10%) and a mean decrease in EDV of 36 cm/s (19%) in 38 patients (Abou-Zamzam et al. 2000). The authors conclude that patients with severe bilateral carotid stenosis should be restudied with duplex scanning after the first operation before undergoing CEA of the contralateral side.

Another point to be considered in *estimating the degree of carotid stenosis* is that a *long high-grade stenosis* (in particular a stenosis of >3 cm) will cause a less pronounced increase in PSV than a shorter stenosis of the same degree. The thresholds defined in investigations using angiography as the gold standard are usually based on the most common stenosis length of 1–2 cm. According to the Hagen-Poiseuille law, flow resistance also depends on the length of the narrowed segment. A very short stenosis will cause a more marked increase in PSV, and this is why the length of the stenosis has to be considered in stenosis grading as well (although no study-based data exist).

A *stenosis* of the intracranial internal carotid system occurring *in tandem* with an extracranial stenosis will reduce flow velocity in the extracranial carotid artery, resulting in a less marked increase in PSV across the extracranial stenosis (stenosis mismatch). In-tandem lesions with high-grade intracranial stenosis may be suspected if the flow velocity in the extracranial ICA distal to the extracranial stenosis is markedly lower and if flow is more pulsatile than would be expected on the basis of the degree of luminal narrowing (see Table 5.11).

Normal blood flow velocity is higher in the thinner vessels of slim patients or arteries narrowed by other factors such as temporary vascular contraction compared with the velocities determined in a general population. This will then translate into higher peak velocities in a stenotic segment.

In light of what has just been outlined, it is clear that a stenosis should never be graded simply on the basis of the measured PSV but that the above factors must always be considered in a critical appraisal to avoid over- or underestimation. Additional aspects that may contribute to a comprehensive diagnostic assessment include blood flow in the supratrochlear artery, flow character and velocity in the CCA, and parameters of the intracranial circulation derived from transcranial imaging.

Some authors prefer measurement of *end-diastolic velocity* for stenosis grading, arguing that this parameter is less affected by the patient's blood pressure at the time of the examination. Nevertheless, data from comparative studies still show PSV to be the most reliable velocity parameter for stenosis grading since *EDV varies with the heart rate* and other systemic factors (Fig. 5.20b). Repeated measurement of constant blood flow in the same vessel at increasing heart rates would yield increasingly higher EDV due to shortening of the cardiac cycle, resulting in an artificially lower RI.

The *effects of systemic factors* such as *hypertensive episodes* or greater pulsatility due to reduced wall elasticity can be minimized by using the ICA-to-CCA velocity ratio (ICA/CCA PSV ratio) for stenosis grading. This ratio can be calculated as a supplementary parameter whenever the absolute PSV yields a threshold stenosis and it is assumed that the measurement was influenced by systemic factors.

The compilation of studies in Table 5.12 shows that accuracy rates of over 90% can be achieved in stenosis grading on the basis of the hemodynamic parameters derived from duplex sonography (using angiography as the gold standard). Hence, the accuracy of duplex imaging is comparable to the

**Table 5.13** Agreement between two independent radiologists in identifying and classifying hemodynamically significant carotid stenosis on angiography

| Author/year | Agreement between two independent radiologists (%) |
|---|---|
| Croft et al. 1980 | 88 |
| Moneta et al. 1993 | 93 |

interobserver variability between two radiologists evaluating the same angiograms (Table 5.13). A closer look at the published data reveals that ultrasound achieves consistently good results with sensitivities and specificities of approximately 90% in identifying greater than 70% carotid artery stenosis (which is relevant for identifying surgical candidates). For stenosis grades of 50–70%, several studies and a meta-analysis of 41 studies investigating different imaging modalities in comparison with intra-arterial digital subtraction angiography (IADSA) revealed sensitivities for duplex sonography that were 5–30% lower but specificities of over 90% (Wardlaw et al. 2006). An explanation for these results is not apparent from the meta-analysis, but the use of different criteria (threshold velocities) for defining hemodynamically relevant stenosis (50% stenosis or greater) and systemic factors (blood pressure, wall elasticity) appear to be contributing factors. There is an ongoing debate about the most suitable velocity parameter for the grading of carotid artery stenosis – PSV, EDV, or the ratio of PSV in the stenotic segment of the ICA to the prestenotic PSV in the CCA (ICA/CCA PSV ratio). The lack of agreement is due to the fact that published results yield no uniform picture and that the heterogeneity of study designs makes the results difficult to compare.

A study investigating *ICA/CCA PSV ratios* determined by duplex scanning in over 300 carotid artery examinations with receiver operating characteristic (ROC) analysis revealed that a ratio of at least 4.0 provided the best accuracy for the detection of 70–99% stenosis according to the NASCET criteria (Moneta et al. 1993). Another study using the same criteria in 158 carotid examinations found 91% sensitivity, 90% specificity, and 90% accuracy in detecting 70–99% ICA stenosis compared with angiography (Table 5.12).

Nevertheless, PSV has turned out to be the most reliable velocity parameter in detecting and quantifying high-grade carotid artery stenosis, showing consistently high accuracy in many studies (Arning et al. 2003; Lal et al. 2004; Lewis et al. 2002) (Table 5.12). If the ultrasound examination is technically adequate, patients in whom higher-grade stenosis is diagnosed need no additional imaging study for stenosis grading before undergoing CEA (Grant et al. 2003; Lewis et al. 2002). The decision to recommend CEA always involves weighing the predicted risk of vascular events against the perioperative and postoperative morbidity/mortality rates. Various *attempts at defining an ideal PSV cutoff* for identifying hemodynamically relevant stenosis (>50%) by means of *ROC curve analysis* in studies using angiography as the reference standard show that a *higher PSV improves specificity, albeit at the cost of sensitivity*; conversely, a lower cutoff velocity improves sensitivity but lowers the specificity of the method (see Fig. 6.9: ROC analysis for determining the cutoff velocity for renal artery stenosis; Fig. 2.14a: for profunda femoris stenosis). This means that one has to compromise and define the PSV cutoff according to whether good sensitivity or good specificity is more important. Moneta et al. (1995) found a combination of PSV >260 cm/s and EDV >70 cm/s to have 84% sensitivity, 94% specificity, 92% positive predictive value, and 90% accuracy for discriminating 60–99% stenosis. Similar results were obtained with an ICA-to-CCA PSV ratio >3.2 (Table 5.12). In asymptomatic patients the statistical benefit of prophylactic CEA is smaller, and the number needed to treat to prevent one stroke is higher than is the case for patients with symptomatic carotid stenosis. For this reason it has been proposed that velocity thresholds with a higher positive predictive value be used in asymptomatic patients. In the same patient population, Moneta et al. achieved a 95% positive predictive value for 60–99% asymptomatic ICA stenosis when using an intrastenotic PSV of 290 cm/s combined with an EDV of 80 cm/s (NASCET grade).

Technical advances in recent years and the advent of high-resolution transducers have led to improved sensitivities and specificities of 90–95% in correctly identifying hemodynamically significant carotid artery stenoses. The correlation of intra-arterial angiography and color-coded duplex imaging is 0.8– 0.9 (Faught et al. 1994; Sitzer et al. 1993). Doppler waveforms sampled distal to high-grade carotid stenosis may show reduced PSV and filling of the spectral window (see Table 5.11 and Fig. 1.45a–f).

Atherosclerotic elongation of the ICA leads to tortuosity, kinking, and coiling due to the limited space available between the bulb and the base of the skull. Such changes typically do not require treatment and are often incidental findings that impair duplex evaluation. Only kinking stenosis, especially when symptomatic, should be operated on (see Fig. A5.2).

Even severe kinks or coils will produce a stenosis only if the artery takes a sharp turn; they may however impair flow velocity measurements due to the difficulty of achieving an adequate Doppler angle. Therefore, indirect criteria such as turbulent flow must be considered as well. If the kinking is pronounced, the constriction may vary with different functional positions of the cervical spine.

Most stenoses at the origin of the ICA (bulb), the most common site of stenosis of the carotid system, are due to atherosclerosis. Distal carotid stenosis is rare and typically has other underlying causes such as fibromuscular dysplasia, wall dissection (usually due to trauma), and kinking (as described above).

If the acoustic window is poor or the depth of penetration is inadequate for scanning the extracranial ICA segments

**Table 5.14** Accuracy of angiography in comparison with pathologic workup of surgical specimens

| Author/year | Accuracy of angiography relative to pathology (%) |
|---|---|
| Croft et al. 1980 | 79 |

near the base of the skull, the examiner should switch to a lower frequency (5 MHz) and use a curved-array transducer with a small radius.

Despite the range of flow velocities associated with a given stenosis grade, as determined by angiography, and despite the pitfalls in measuring PSV there is good agreement between duplex ultrasound grading on the basis of peak velocity and angiographic grading (Tables 5.11–5.14).

Another aspect to be mentioned here is that ultrasound equipment from different manufacturers may yield different flow velocity measurements – suggested by discrepancies of 5–10% reported in a phantom study (Fillinger et al. 1996). This must also be borne in mind when interpreting and comparing the different threshold velocities that have been proposed for diagnosing high-grade stenosis.

Finally, it must be mentioned that the extent of collateralization also contributes to the individual risk of embolism. The intrastenotic PSV in an 80–90% stenosis is lower when there is good collateralization as opposed to the same degree of stenosis in a patient with poor collateral pathways. Lower intrastenotic velocities reduce the risk of embolism because of the smaller shear forces acting on the plaque.

*Role of poststenotic diameter as a prognostic factor*: Slow development of high-grade ICA stenosis is accompanied by increasing collateral supply to the brain by way of the contralateral carotid system and the vertebral artery.

As a result, carotid blood flow on the side of stenosis is decreased. A subgroup analysis of the ECST shows that a reduced poststenotic ICA lumen has important prognostic implications for patients with high-grade carotid stenosis (Rothwell and Warlow 2000). In this analysis, patients with poststenotic narrowing of the ICA defined as an ICA-to-CCA ratio <0.42 (which usually means <3 mm) had a two thirds lower risk of stroke than patients without narrowing at 5-year follow-up. The authors assume that poststenotic narrowing may be protective as blood flow distal to the stenosis is insufficient for carrying emboli to the brain.

For the reasons outlined above, color flow imaging (color duplex or power mode) is inaccurate for stenosis quantification (diameter reduction on longitudinal view or cross-sectional area reduction on transverse view) on the basis of the width of color-coded flow relative to the lumen diameter at the site of maximum stenosis. The inherent methodological limitations (angle dependence of the flow signal) are eliminated using the B-flow technique, which allows better differentiation of flowing blood from the vessel wall or plaque because of its superior discrimination of echoes from mobile and stationary reflectors. A small study (21 patients) has recently evaluated the performance of different imaging techniques in discriminating 50–95% stenosis compared with intra-arterial DSA, which served as the gold standard. In this study, B-flow imaging showed the highest correlation (Pearson) with DSA ($R = 0.94$), followed by contrast-enhanced MRI (1.5 Tesla, standard coil; $R = 0.9117$), color duplex sonography, and reconstructed contrast-enhanced CT (both $R = 0.85$) (Finkenzeller et al. 2008). Note, however, that stenoses in this study were graded using the NASCET criteria (distal stenosis grade) and that plaque evaluation in both B-mode and B-flow imaging crucially depends on plaque morphology, in particular on the presence of calcification. The accuracy of stenosis grading with B-flow imaging decreases with the amount of calcification.

The gold standard, angiography, also has pitfalls – even when the stenosis is evaluated in different projections and quantification is done in the plane with the most severe luminal narrowing. Independent interpretations of the same angiograms by two radiologists show variation in accuracy ranging from 80% to 93%. Errors in angiographic stenosis grading arise from the fact that the 3D plaque is projected onto the 2D film, where the resulting degree of narrowing varies with the imaging plane (Fig. 5.21a, b). Moreover, plaques of different configuration which produce the same diameter reduction produce different reductions in the cross-sectional area.

When comparing duplex imaging and angiography, one must also consider whether the angiographic findings are based on *local or distal stenosis grading*. As mentioned above, the distal degree of stenosis was used in the NASCET and the local degree in the ECST. The distal degree of stenosis is easier to determine angiographically than the local degree, as the latter requires estimation of the bulb diameter through interpolation. Distal quantification would be more appropriate if the hemodynamic effect of a stenosis had a role in the development of stroke (the hemodynamic significance, or flow reduction, depends on the luminal narrowing relative to the (distal) normal ICA). Since this is not the case and the majority of strokes are caused by arterial emboli arising from plaques, *the German Society of Ultrasound in Medicine recommends use of the local quantification method (ECST).*

The risk of embolism associated with a plaque increases with its thickness, and in the bulb an eccentric plaque may reach a dangerous thickness, with a higher risk of rupture due to longitudinal pulsatile movement, even if no high-grade stenosis is present according to NASCET criteria. Due to the greater width of the bulb, circumscribed narrowing here of about 30% has no stenotic effect because it merely reduces the lumen diameter to that of the more distal portion of the artery (see Fig. 5.8). The local degree of stenosis can be calculated from the distal degree according to the formula: local degree (ECST criteria, in %) = 0.6 × distal degree (NASCET criteria, in %) + 40% (see Sect. 5.5.1.1).

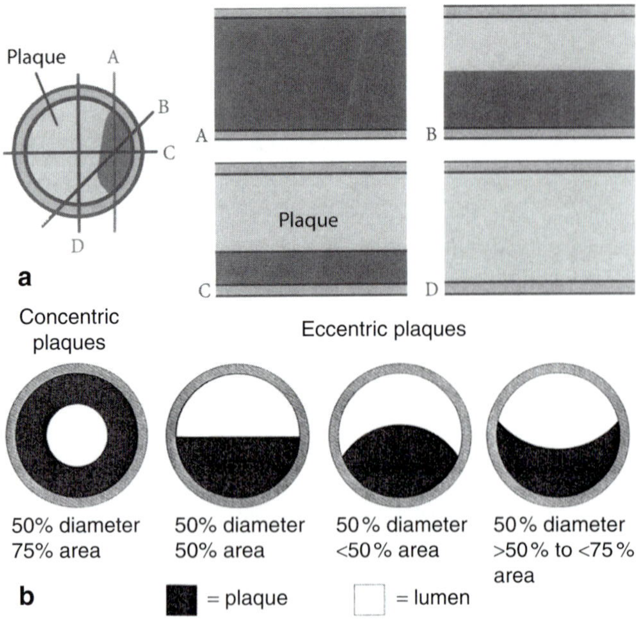


**Fig. 5.21** **a** Eccentric luminal narrowing can vary widely in appearance and suggest different degrees of stenosis, depending on the imaging plane (angiography and B-mode ultrasound). The diagrams illustrate how the same plaque can lead to different estimates of the degree of stenosis caused by it: A: 0%, B: about 50%, C: about 70%, D: 100%. This source of misinterpretation results from the reduction of the three-dimensional vascular lesion to the two-dimensional imaging plane. Color duplex facilitates evaluation of eccentric plaques, which, if possible, should be assessed in transverse and longitudinal orientation. **b** Plaque morphology (concentric–eccentric) can lead to variable degrees of cross-sectional area reduction despite identical diameter reduction. Eccentric plaques are thicker than concentric ones even when they produce identical stenosis grades. They are associated with a higher risk of stroke, as they are exposed to greater shear forces (longitudinal pulsation) and are thus more susceptible to rupture.These two factors (reduction to two planes and differences in plaque configuration) explain the discrepancies in stenosis quantification between duplex ultrasound and angiography in case of eccentric plaques. Differences in plaque configuration lead to problems when stenosis grading on the basis of hemodynamic criteria with duplex ultrasound is compared with angiography using diameter reduction as a morphologic criterion. For instance, a 50% diameter reduction measured angiographically and indicating a 50% stenosis corresponds to a 50% stenosis based on hemodynamic grading (expressed in terms of cross-sectional area reduction) when the plaque is eccentric, but to a 75% stenosis when the plaque is concentric. The stenosis grade determined hemodynamically by spectral Doppler and expressed as area reduction reflects the extent of blood flow reduction more adequately. This is also relevant in evaluating peripheral or renal artery stenosis

In an analysis of 1,001 angiograms of the ICA, 34% of stenoses were classified as 70–99% stenoses according to the local stenosis grade versus only 16% using the distal stenosis grade (Rothwell et al. 1994).

Compared with the ICA, the *ECA* has a more pulsatile flow profile with a smaller diastolic component. Stenosis typically occurs at its origin but becomes clinically relevant only in patients with concomitant ICA occlusion, where blood must be supplied to the brain by way of extracranial branches such as the supratrochlear artery.

Stenosis of the CCA is rare. The most common sites of CCA stenosis are the origin of the artery from the aortic arch or from the innominate artery and the distal CCA segment just below the bifurcation. In case of subtotal or total CCA occlusion, the ICA can be supplied through branches of the ECA (such as the superior thyroid artery). Depending on the course of the collateral pathways, the ECA will show variable retrograde filling. The hemodynamic stenosis criteria for the CCA are the same as for the ICA.

The pulsatility of blood flow varies with the peripheral resistance and wall elasticity. A stenosis causes a more pulsatile flow in the prestenotic segment and a less pulsatile flow in the poststenotic segment. Other causes of decreased pulsatility include arteriovenous (AV) fistula, hyperperfusion due to hyperthyroidism, and tachycardia. Higher pulsatility may be due to loss of wall elasticity as in medial sclerosis, increased intracranial pressure, aortic insufficiency, and bradycardia (see Table 5.5).

### 5.6.1.3 Occlusion

Occlusions of the carotid territory are chiefly due to local thrombus formation secondary to stenosing atherosclerosis at the origin of the ICA. As there are no arteries emptying into or arising from the extracranial portion of the ICA, an occlusion will extend to the level of the next branching in the petrous bone or intracranially into the ophthalmic artery. Embolic occlusions mainly affect the distal, intracranial segments of the internal carotid system. In the most severe form, there will be to-and-fro flow in the Doppler waveform (thump pattern) and the thrombus may extend cranially to the level of the bifurcation. Differentiation between subtotal and total ICA occlusion is important as surgery is usually only recommended for subtotal occlusion. Unfortunately, the ingrowth of vessels into the connective tissue of an organized occlusion makes it difficult to distinguish the two. Thus, the depiction of color-coded flow signals in the ICA segment near the base of the skull indicating a patent lumen is the decisive criterion for establishing the differential diagnosis.

*Adjusting the scanner settings* (low wall filter, *low pulse repetition frequency*, high gain) is important for detecting slow flow and small blood volumes in *pseudo-occlusion*. Assuming settings are adequate, the absence of flow distal to the atherosclerotic lesion is the most reliable evidence of an occlusion. Since no vessels arise from the ICA extracranially, the Doppler waveform sampled far distal to the bulb (near base of skull) provides the most reliable information for differentiation, especially since there are no structures interfering with the Doppler signal. Blood flow distal to a subtotal occlusion often assumes a venous character with a slow systolic velocity. Using these criteria, color duplex

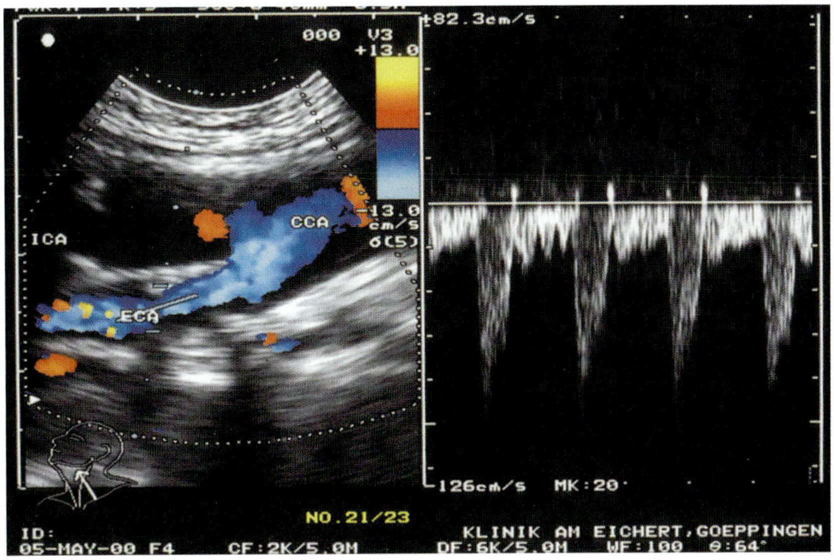


**Fig. 5.22** Image of the carotid bifurcation showing occlusion of the internal carotid artery (*ICA*) (no color-coded flow signals). The Doppler waveform from the external carotid artery (*ECA*) shows large diastolic flow and pulsatility as in the ICA, indicating that the ECA supplies the brain from externally to internally by way of the supratrochlear artery (see Fig. A5.13). Tapping of the temporal artery with transmission of this effect into the waveform proves that the waveform is actually from the ECA although the pulsatility appears similar to that of the ICA

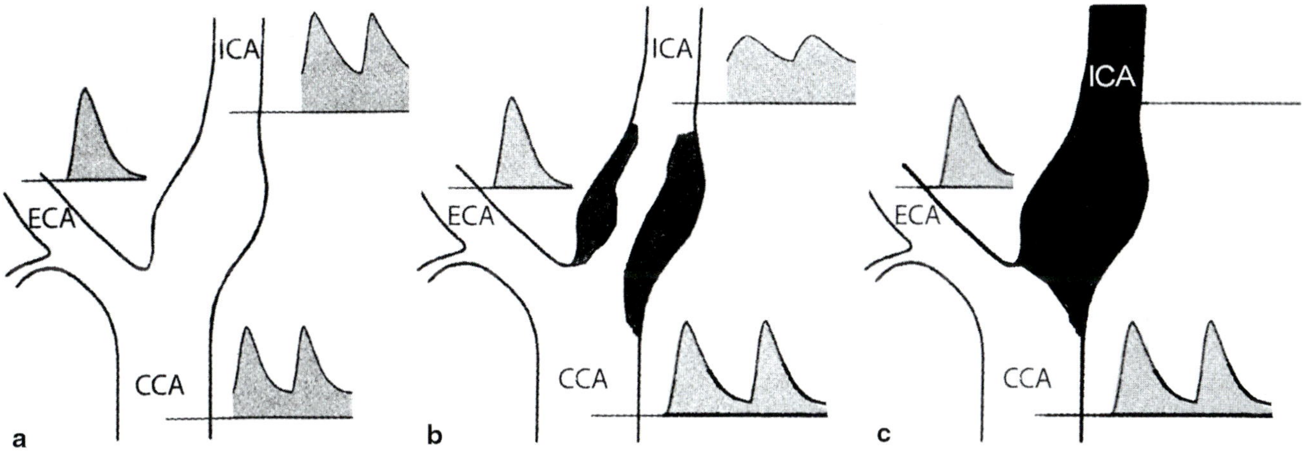


**Fig. 5.23 a–c** Diagrams of Doppler waveforms obtained from the carotid territory. **a** Normal flow patterns – internal carotid artery (ICA): low-resistance flow with little pulsatility and large diastolic flow component; external carotid artery (ECA): pulsatility higher than in the ICA – the ECA mainly supplies skin and muscles and therefore faces a higher peripheral resistance – but lower than in the extremity arteries, due to the ECA's supply to the glands; common carotid artery (CCA): mixed type, pulsatility intermediate between that of the ICA and ECA, which it supplies. **b** High-grade stenosis or occlusion of the ICA – flow in the CCA becomes increasingly pulsatile, approaching the flow pattern of the ECA (as the influence of the ICA on the CCA waveform recedes or disappears). If the ECA additionally assumes collateral function and supplies the brain via the supratrochlear artery, the ECA waveform will become less pulsatile (low peripheral resistance). ICA waveform distal to high-grade ICA stenosis: characteristic postocclusive flow with delayed systolic upslope and decreased pulsatility, i.e., larger diastolic flow component and reduced peak systolic velocity (PSV). The difference is less striking than the change from a triphasic to monophasic flow in extremity arteries as the ICA is a low-resistance artery and already has normal monophasic flow. **c** ICA occlusion – with the CCA exclusively supplying the ECA, its flow becomes more pulsatile with the waveform becoming similar to that of the ECA. The ECA may partially supply the brain (e.g., via the supratrochlear artery), reflected in a less pulsatile flow pattern with a larger diastolic component

ultrasound has a positive predictive value of 92.5–96.7% for demonstrating ICA occlusion (Kirsch et al. 1994).

B-mode imaging is fairly unreliable for detecting occlusions, especially recent ones, when intraluminal structures are still absent. Older occlusions, on the other hand, may be identified in the B-mode by the presence of internal echoes from the connective tissue in the shrunken lumen. However, an occluded artery is often difficult to distinguish from the surrounding tissue.

When the ICA is occluded, even greater care is required so as not to confuse the ICA and the ECA. In ICA occlusion, the *ECA supplies the brain* (Fig. A5.13) through the

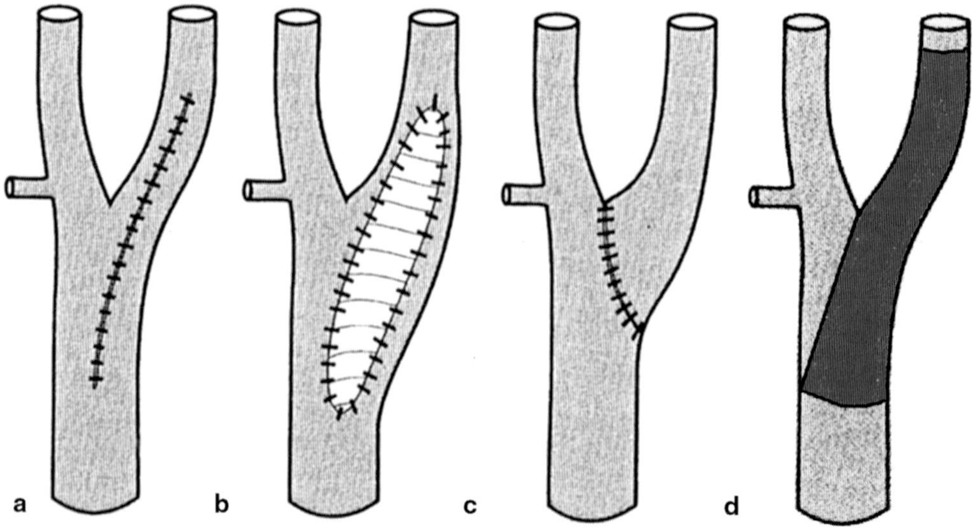


**Fig. 5.24 a–c** Techniques of vessel repair in internal carotid artery (ICA) stenosis. **a** Thromboendarterectomy (TEA) with direct closure. **b** TEA with patch angioplasty. **c** Eversion TEA. **d** Carotid percutaneous transluminal angioplasty (PTA) with stenting (CAS)

supratrochlear artery, *resulting in a less pulsatile Doppler waveform with a larger diastolic component* (Fig. 5.23). In this setting, the identity of the ECA must be confirmed by rhythmically tapping the temporal artery and looking for the transmitted pulsation (Fig. 5.22). This is important to avoid making a mistake and localizing the occlusion to the ECA.

In the rare cases of *CCA occlusion*, the ICA may be refilled by branches of the ECA serving as collaterals (e.g., superior thyroid artery; see Fig. A5.13). The corresponding Doppler spectrum will show retrograde flow in the proximal ECA with reversal to forward flow, but with a highly postocclusive character in the ICA.

Bypass grafting is indicated only in multi-vessel disease with reduced global perfusion, which may manifest as borderzone infarction.

#### 5.6.1.4 Postoperative Follow-Up

Three operative techniques are available for thromboendarterectomy (TEA) of the ICA (Fig. 5.24a–c).

In patients with a wide bulb, TEA can be performed with direct closure of the arteriotomy. A possible complication of this procedure is the inadvertent creation of a relative stenosis compared with the distal ICA lumen by pulling the vessel wall too tight.

This complication can be prevented by interposing a synthetic or venous patch after TEA to compensate for the relative narrowing that may be created by primary closure. Too wide a patch, on the other hand, will lead to ectasia or even aneurysm with development of turbulent flow.

In eversion TEA, the ICA is transected at its origin, and the outer layer over the stenosing plaque-intima cylinder is everted and dissected along the vessel until the intima appears fairly normal again. At this point the cylinder is transected, and the outer wall layer is re-inserted into the CCA.

Each of these three surgical techniques is prone to specific complications that must be carefully ruled out by postoperative ultrasonography. The complications of the primary closure technique, apart from relative narrowing as an early complication, include an intimal step or an intimal flap. Elongation of the ICA can lead to kinking with development of a stenosis unless the excessive segment is resected.

Patch angioplasty is susceptible to thrombotic deposits, but these may resolve, even after days, following treatment with heparin and platelet aggregation inhibitors (see Fig. A5.28). However, such thrombotic deposits can also induce TIAs or early occlusion. Aneurysmal dilatation due to excessive correction gives rise to turbulent flow. Suture aneurysms primarily occur in association with infection and after insertion of a synthetic patch. At the junction of the patch with the distal ICA, detachment of the intima can lead to the same complications as direct closure. Use of a venous patch can give rise to the formation of a true aneurysm due to the physiologically weaker venous wall. In the further course, recurrent stenosis may occur. Hence, in the follow-up examination, special attention must be paid to the suture area in order to identify stenosing intimal hyperplasia.

Suture line complications are rare in eversion TEA, while step formation at the transition to the native intima is somewhat more common (Table 5.15).

The specific complications of the individual operative techniques as outlined above must be borne in mind when performing the mandatory postoperative duplex scan. Postoperative sonography is impaired by scattering through edema, which may affect both B-mode imaging and spectral Doppler measurement. The use of a lower-frequency transducer (5 or even 3.5 MHz) yields B-mode scans with a poorer resolution but often facilitates both the identification of the target artery within the edematous tissue and spectral Doppler sampling for exclusion of early occlusion, residual stenosis, or stenosing thrombotic deposits.

**Table 5.15** Sonographic evaluation of postoperative complications after surgical internal carotid artery (ICA) reconstruction

| Operative technique | Operative site | Distal ICA |
|---|---|---|
| TEA with direct closure | Relative stenosis, recurrent or residual stenosis | Intimal step, intimal flap, intimal dissection, kinking, intimal hyperplasia |
| TEA with patch angioplasty | Thrombotic deposits in the patch area without/with hemodynamically significant stenosis, suture aneurysm, ectasia, infection, recurrent stenosis | Intimal step, intimal flap, intimal dissection, kinking, intimal hyperplasia |
| Eversion TEA | Retraction of suture line, suture aneurysm, recurrent stenosis | Intimal flap, intimal step, intimal dissection, intimal hyperplasia |

*TEA* thromboendarterectomy

The development of recurrent stenosis has been investigated in numerous studies with postoperative follow-up by duplex ultrasound. However, the studies do not address the problem of the diagnostic accuracy of ultrasound in detecting the above-described early complications under the poorer postoperative insonation conditions outlined here. Experience suggests that perioperative TIAs are often associated with thrombotic deposits on the synthetic patch, which are in part identified by a hemodynamic stenosis signal in the Doppler waveform and may completely resolve within days to weeks under heparin and antiplatelet therapy. No studies have investigated these complications in terms of their possible impact on patch selection. Unfortunately, B-mode ultrasound can only identify intimal flaps and dissection a few days later when swelling has subsided. Intimal flaps are then often seen as hyperechoic floating structures in the bloodstream.

The flow pattern in a *suture aneurysm* is the same as in a false aneurysm. To-and-fro flow can be demonstrated by color duplex or spectral Doppler (see Sect. 2.1.6.3 and Figs. A5.20 and A5.21).

If semiclosed endarterectomy of the ECA is performed, intimal flaps or dissection may give rise to stenosis or occlusion. These are rarely significant functionally and clinically but must be considered in the differential diagnosis when examining the ICA.

The *intimal step* remaining at the proximal end of the TEA in the CCA is often highly conspicuous on the B-mode scan but has no clinical or functional relevance as the step flattens out in the direction of the flowing blood.

*Recurrent stenosis* within the first 12 months of surgery is due to neointimal proliferation (unless the operation has been technically inadequate). Recurrence seen after 2 years is attributable to the progression of atherosclerosis. Based on the follow-up data from more than 160 studies including over 62,000 patients, the average restenosis rate is 6% (range, 0–50%). The incidence of symptomatic recurrent carotid stenosis is 2% with the recurrence rate being markedly higher after TEA with direct closure than after patch angioplasty (12% versus 5%). Taken together, about 20% of all stenoses seen after carotid TEA are accounted for by residual stenoses, 50% develop within 2 years, and 30% occur later. Long-term follow-up of 380 patients for 16 years revealed restenosis rates of 5.8%, 9.9%, 13.9%, and 23.4% after 1, 3, 5, and 10 years, respectively; however, only 2.1% of patients were found to have high-grade recurrent stenosis (>80%) (Mattos et al. 1993; Roth et al. 1999). Evaluation of the unoperated side and identification of progressive atherosclerosis appear to be more relevant than postoperative surveillance of the side where the TEA was performed. It has been proposed that routine duplex surveillance in the first 6 months is not required when an intraoperative completion study has confirmed the technical adequacy of the repair (Pross et al. 2001; Fig. 5.25). However, experience seems to indicate that it is common to detect thrombotic deposits in patients with postoperative TIAs despite normal intraoperative findings, especially when a synthetic patch has been used.

*Following carotid artery stenting* (*CAS*), ultrasound depicts the stent as a mesh-like structure. A diagnostic spectral waveform is difficult to obtain from the stented vessel segment by (color) duplex scanning during the first days, presumably because the stent is not yet incorporated. After this initial period, the scanning conditions are the same as before stent placement. Stents are also prone to thrombotic deposits, which may cause stenosis but will recede after initiation of treatment with heparin and platelet antiaggregators. The distal stent end is especially prone to recurrent stenosis (Fig. 5.26; see Figs. A5.32 and A5.33).

Recent studies suggest that revised ultrasound velocity criteria are needed for postprocedural surveillance of patients after CAS (Stanziale et al. 2005). *Loss of compliance* of the *arterial wall* after stent insertion results in *higher normal velocities*; thus, it has been proposed that a PSV of up to 150 or 180 cm/s should be considered normal in a stented carotid artery segment (Chahwan et al. 2007; Lal et al. 2004).

Another duplex ultrasound study of stented carotid arteries identified a PSV of 225 cm/s to be the optimal cutoff for diagnosing hemodynamically significant in-stent restenosis (>50% stenosis). This threshold had 90% sensitivity, 97% specificity, and 96% diagnostic accuracy (Stanziale et al. 2005).

The need for revised velocity criteria in stented carotid arteries was also confirmed by a recent ROC analysis (AbuRahma et al. 2008) performed to identify optimal duplex ultrasound velocity criteria for detecting different degrees of in-stent carotid restenosis. In this study, a PSV threshold of 154 cm/s for >30% stenosis showed 99% sensitivity and 89%

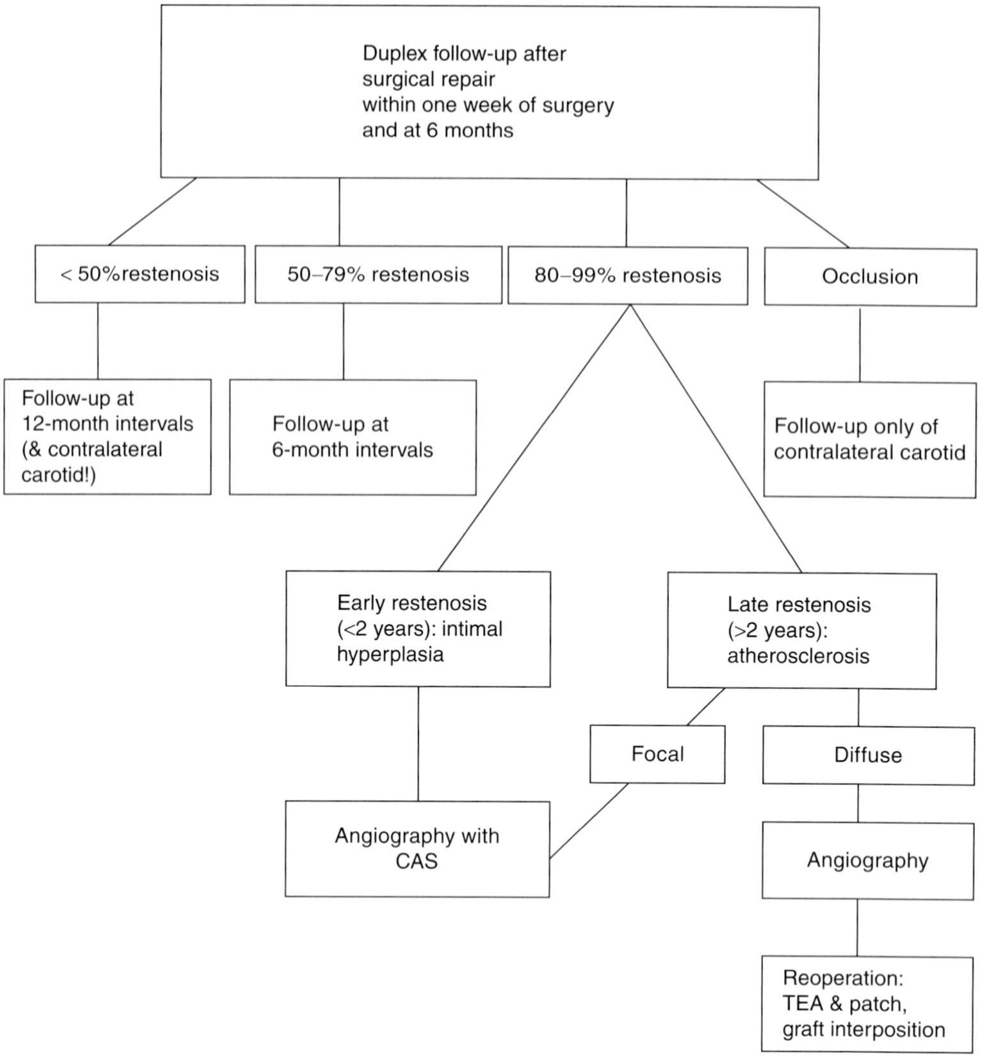


**Fig. 5.25** Diagnostic algorithm in patients after surgical repair

specificity. The optimal threshold velocity for >50% stenosis was 224 cm/s, which had 99% sensitivity, 90% specificity, 99% positive predictive value, 90% negative predictive value, and 98% overall accuracy. The ideal cutoff for >80% stenosis was 325 cm/s with 100% sensitivity, 99% specificity, and 99% accuracy. The diagnostic accuracy of PSV was compared with that of EDV and also with that of the ratio of PSV in the stented ICA to the PSV in the CCA. This comparison showed that PSV provided the most reliable criterion for sonographic stenosis grading in 144 patients in comparison with angiography. Nineteen of the patients had >50% in-stent restenosis. Large PSV ranges were found for different categories of stenosis (defined by angiography): range of 142–256 cm/s with a mean PSV of 178/s for 30–50% stenosis ($n$ = 38); 201–408 cm/s with a mean PSV of 278 cm/s for 50–80% stenosis ($n$ = 11); and 58–613 cm/s with a mean of 403 cm/s for 80–99% stenosis ($n$ = 8). These data indicate that the velocities encountered are affected not only by a loss of compliance in a stented carotid segment but also by systemic factors; therefore, *detection of a 100% increase in PSV* appears to be a *valid criterion for in-stent restenosis as well*, especially when the stenosis is located in the distal third of the artery (Fig. A5.33).

What has just been outlined regarding higher velocity thresholds for grading in-stent restenosis is based on the *NASCET criteria*. While these are generally used in the USA, *stenosis grading according to the ECST criteria* (*local stenosis grade*) is more common in Europe and is also used in this textbook. The fact that these two systems for stenosis grading exist may also explain the use of a PSV cutoff of 180 cm/s for identifying 50% stenosis in an untreated ICA (Brajesh Lal 2008) with a PSV of 250 cm/s defining >80% stenosis.

When the local grading system (based on ECST criteria) presented above is used (Sect. 5.5.1.1), the *velocity thresholds are lower* since the ECST criteria assign a higher degree to the same stenosis than do the NASCET criteria (distal degree).

Moreover, most studies use CT angiography as the standard of reference for color duplex imaging rather than angiography in two or three planes, neglecting the inherent methodological limitations of CT, especially in the carotid

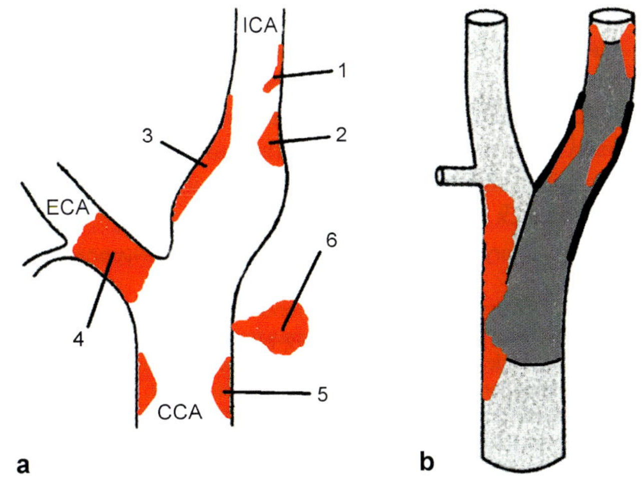


**Fig. 5.26** **a** Diagram of possible duplex ultrasound findings after carotid thromboendarterectomy (TEA): outcome of surgery (early and late complications) and progressive atherosclerosis. *1* Intimal flap; *2* recurrent stenosis due to plaque; *3* neointimal proliferation with recurrent stenosis; *4* postoperative external carotid artery (ECA) occlusion; *5* damage from clamping; step; plaque progression at proximal end of TEA; *6* suture aneurysm. **b** Early and late complications after stenting: neointimal hyperplasia, recurrent plaque with stenosis, stenosis of the ECA, whose origin is covered by the stent. Stent dislocation (not shown, see Fig. A5.35)

bifurcation. This introduces an additional inaccuracy into the ROC analysis of velocity thresholds. Most investigators use catheter angiography only in patients undergoing repeat interventional dilatation for higher-grade stenosis; as a result, the gold standard is available only for these cases.

A minor limitation of the ultrasound studies of carotid in-stent restenosis is the small number of cases investigated. Although some investigators present duplex ultrasound follow-up findings from over 100 patients who underwent CAS (Table 5.16), the ROC analyses they performed were only in subsets of 10–20 patients who underwent angiography because they had restenosis of at least 50% and were being considered as candidates for possible reintervention.

As a rule of thumb, it can be assumed that the velocity threshold (PSV) for grading stenosis in a stented artery should be 20–30% higher than in a native artery with the exact cutoff varying, depending on the design and the structure of the stent used.

A promising new approach that might overcome the poor quantification of hemodynamic parameters in stented carotid arteries is B-flow ultrasound (without or with speckle reduction imaging (SRI)). This technique is similar to angiography in that luminal narrowing is evaluated morphologically, the advantage being that the method is not angle-dependent (distinguishing it from hemodynamic assessment by spectral Doppler). The excellent discrimination between the blood and the vessel wall enables very accurate visualization of vascular stenosis in terms of surface morphology and degree of luminal narrowing in superficial arteries that can be examined with a high-frequency transducer (see Fig. A5.36). When used for stenosis grading in the stented ICA, two or three planes (as in angiography) are required (anterior and posterolateral orientation). As with other ultrasound modalities, B-flow imaging is subject to artifacts and is degraded by scattering and acoustic shadowing; this limits stenosis grading due to inadequate evaluation of the vessel wall. The option of SRI can eliminate artifacts to some extent, making individual tissues appear more uniform and hence improving the differentiation between the vessel lumen and the wall even further. Preliminary results (Jung et al. 2007) underscore the advantages of using B-flow ultrasound in patients who have undergone carotid artery stenting. Examples illustrate the advantages of this new technique in visualizing and quantifying restenosis in these patients. Data on the performance of B-flow ultrasound in comparison with other ultrasound techniques and with the gold standard, angiography, in a larger population of patients with restenosis after stent placement are not yet available.

As recurrent stenosis at the proximal stent end (i.e., the junction between the CCA and the stent) is less common than in-stent restenosis or stenosis at the distal stent end, abrupt doubling of PSV in a continuous Doppler measurement, an established ultrasound velocity criterion for diagnosing peripheral artery stenosis, can be used for the diagnosis of hemodynamically relevant in-stent restenosis (50% stenosis) in the carotid system as well.

**Table 5.16** Duplex ultrasound criteria for in-stent stenosis after carotid artery stenting (CAS)

| Author, year | No. | PSV (cm/s) | | | ICA/CCA ratio | | |
|---|---|---|---|---|---|---|---|
| | | >50% | >70% | >80% | >50% | >70% | >80% |
| Abu Rahma 2008 | 144/19 | 224 | | 325 | 3.4 | | 4.5 |
| Lal 2008 | 189/29 | 220 | | 340 | 2.7 | | 4.1 |
| Stanziale et al. 2005 | 118/19 | 225 | 350 | | 2.5 | 4.75 | |
| Chi 2007 | 13 | 240 | 450 | | 2.45 | 4.3 | |
| Wei Zhou 2008 | 237/22 | | 300 | | | 4 | |
| Bae Ju Kwon 2007 | | 200 | | | 2.5 | | |

*PSV* peak systolic velocity, *ICA* internal carotid artery, *CCA* common carotid artery

No.: number of patients with CAS examined with duplex ultrasound versus angiography or CT angiography/patients with >50% stenosis relevance if no neurologic symptoms or clinical complaints are present.

## 5.7 Diagnosis of Brain Death

An elevated intracranial pressure associated with trauma, hemorrhage, or edema is reflected in signs of increased peripheral resistance in proximal arterial segments. In the Doppler spectrum of the ICA, increasing intracranial pressure is indicated by a corresponding decrease in the diastolic flow component or even to-and-fro flow with a systolic forward and diastolic backward component (Figs. 5.30 and A5.46). However, the correlation between intracranial pressure and the pulsatility index varies as it is affected by individual factors and autoregulatory processes as well as the underlying disease. Therefore, no reproducible absolute values of intracranial pressure can be derived from the Doppler waveform or the pulsatility index.

Nevertheless, *interpretation of the Doppler waveform* will yield information on *relevant elevations of intracranial pressure*. When intracranial pressure exceeds diastolic blood pressure, the diastolic flow component disappears or becomes retrograde (to-and-fro flow) (Fig. A5.46), suggesting cessation of cerebral blood flow (Hassler et al. 1991). In Germany, since the end of 1991, *transcranial Doppler sonography* has been an accepted diagnostic modality for shortening the waiting time for diagnosing cerebral circulatory arrest. If, for technical reasons, the typical changes in the Doppler waveform cannot be demonstrated in the basal cerebral arteries, cerebral circulatory arrest can be diagnosed by using duplex sonography to demonstrate these changes in the flow profile (Fig. 5.30) of the extracranial segments of the internal carotid or vertebral arteries. In doing so, care must be taken to clearly identify the arteries supplying the brain and to differentiate them from other segments such as the ECA.

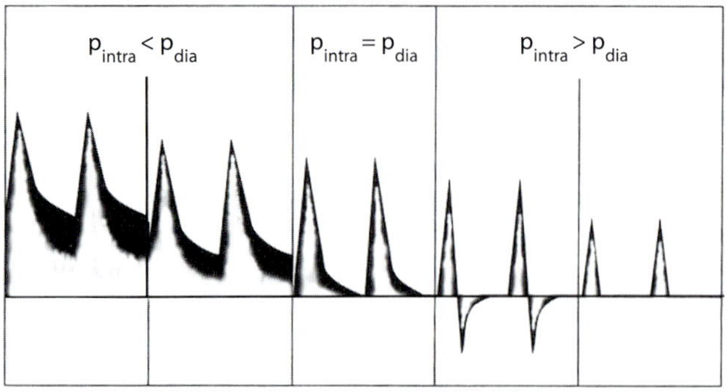


**Fig. 5.30** Effects of increasing intracranial pressure on pulsatility in the extracranial cerebral arteries. The diagrams of the Doppler waveforms from left to right reflect the decreasing diastolic component ($P_{dia}$ = diastolic blood pressure) with increasing intracranial pressure ($P_{intra}$) (according to Widder 1995)

## 5.8 Rare (Nonatherosclerotic) Vascular Diseases of the Carotid Territory

### 5.8.1 Dissection

Arterial dissection is the spontaneous or traumatic separation of the arterial wall layers caused by blood surging in through a tear in the intima. Elevation of the intima results in the creation of a false lumen alongside the true arterial lumen. If blood dissects between the media and adventitia, the latter is elevated, giving rise to a pseudoaneurysm. A blind-ending false lumen becomes thrombosed and compresses the true lumen, causing high-grade stenosis or occlusion in severe cases. When there is a second tear at the distal end, the blood can re-enter the true lumen and flow through both lumina. Dissection may cause various complications with manifestations ranging from headache to hemisymptoms. Seventy percent of patients with ICA dissection have no or only mild neurologic deficits, while 25% present with severe neurologic symptoms. Spontaneous resolution is common, especially when the false lumen becomes thrombosed and subsequent shrinkage of the thrombus causes the compression of the true lumen to recede.

There are three underlying causes of carotid dissection with different symptoms, treatments, and prognoses:

- *Spontaneous dissection*
- *Traumatic dissection* (blunt trauma or iatrogenic after puncture)
- *Aortic dissection* (Stanford type A) with subaortic extension

Common carotid dissection resulting from aortic dissection begins in the proximal portion, from where it can progress into the carotid bifurcation. In patients with suspected CCA dissection, the artery is examined in the transverse plane, starting as far anteriorly as possible using a convex or curved transducer. Spontaneous dissection of the CCA is very rare but may occur in patients with Marfan's syndrome (Harrer et al. 2006).

Traumatic and spontaneous dissection typically affects the ICA including the portion near the skull base, which is why the ultrasound examination must focus on these segments.

Carotid dissection as the cause of cerebral infarction primarily affects adolescents. It is typically due to trauma and rarely occurs spontaneously, commonly affecing carotid segments prone to injury from bony structures such as the skull base or the transverse foramina. Following an acute phase with a relatively high risk of embolization and occlusion, dissection has a good prognosis due to spontaneous recanalization over time.

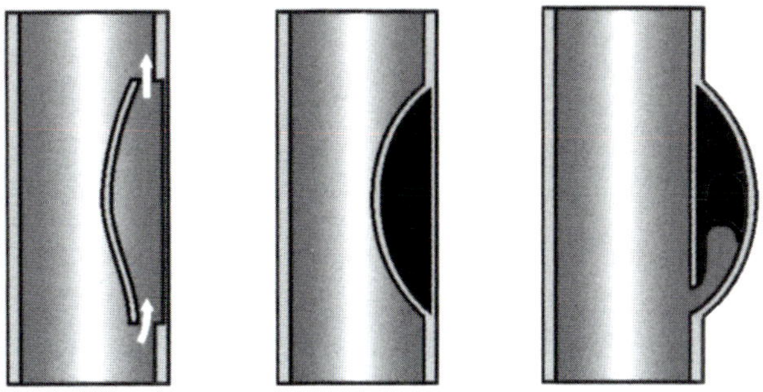

**Fig. 5.31** Diagrams of the sonomorphologic findings in different forms of dissection. The *first drawing* shows intimal dissection with the sites of entry and re-entry. The *second drawing* illustrates the situation in internal dissection with narrowing of the true lumen due to thrombosis of the false lumen. The *third drawing* presents the situation in external dissection, which is characterized by intramural hemorrhage between the media and adventitia with spindle-shaped or saccular dilatation but with little or no compression of the true lumen; this may lead to the formation of a pseudoaneurysm

The location and superficial course of the carotid artery allow good visualization of the sonomorphologic features of carotid dissection in the B-mode (Fig. 5.31):

- An *intraluminal intimal flap* separating the true and *false lumen*; the flap can often be seen flapping back and forth with pulsation.
- If *thrombosis of the false lumen* is present, in the case of dissection with an intimal tear, the thrombotic material will appear as a hypoechoic eccentric structure narrowing the true lumen over a variable length. The thrombotic false lumen typically has a somewhat higher echogenicity than the adjacent patent lumen.
- In the case of *external dissection*, intramural hemorrhage with thrombosis will result in an aneurysmal dilatation with low echogenicity of content and a visibly elevated adventitia.
- In the case of an *intimal tear*, the intima will be visualized as a *flapping structure* of higher echogenicity within the arterial lumen. In an older dissection, the intimal flap may assume the appearance of a circumscribed wall deposit in an otherwise normal-appearing artery. Short dissection can be iatrogenic – the result of inadvertent injury to the opposite arterial wall with the needle during catheterization and may cause short stenosis due to a structure protruding into the lumen and difficult to distinguish from plaque-like deposits.

Spectral Doppler findings obtained in a patent false lumen are highly variable, depending on the individual constellation and the site of sampling relative to the entry and re-entry points. There may be to-and-fro flow or even retrograde flow. The *flow signal* from the true carotid artery lumen may be *obscured by the more intense signal from the moving intimal flap.*

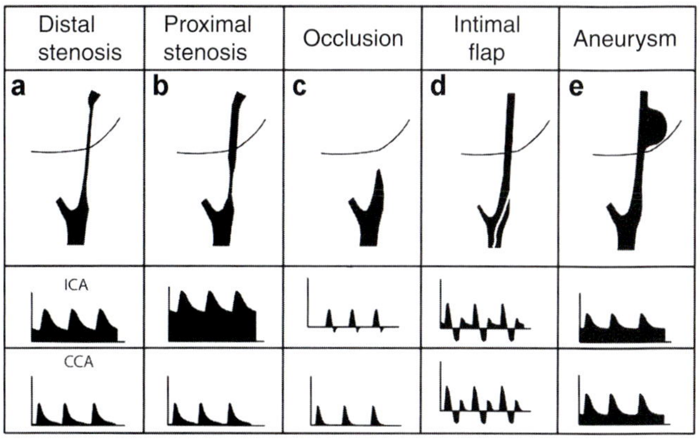


**Fig. 5.32 a–e** Diagrams of different flow profiles in the internal carotid artery in dissection, depending on localization, extent, thrombosis, and sites of entry and re-entry (from Widder 1995). **a** Long ICA dissection with varying flow velocities in the patent segment. **b** Short dissection with circumscribed flow acceleration at the site of luminal narrowing, which may be difficult to differentiate from atherosclerotic lesions or fibromuscular dysplasia. **c** Dissection-induced occlusion of the ICA with thump pattern (to-and-fro sign) in the patent segment. Flow in the CCA assumes the character of the ECA. **d** If the true and false dissection lumina are patent, flow profiles vary widely with the sites of entry and re-entry. The waveform from the true lumen depends on the degree of flow obstruction caused by the dissection. Fluttering of the intimal flap leads to a multiphasic waveform. **e** Distal formation of a pseudoaneurysm (typically beneath base of skull) cannot be detected by ultrasound because proximal flow is normal

*Thrombosis of the false lumen* is usually identified by a slightly higher echo level compared with the patent lumen. The Doppler waveform varies widely with the extent and type of dissection. In case of dissection-induced occlusion with a patent origin, a knocking waveform (thump pattern) is obtained and the profile of the CCA assumes the character of the ECA. Dissection with luminal narrowing is characterized by a waveform with a higher Doppler shift frequency and an increased angle-corrected flow velocity in the residual lumen over a long stretch of the ICA. With only minimal luminal narrowing, the spectral Doppler tracing from the ICA and CCA appears fairly normal (Fig. 5.32a–e).

Dissection of the CCA may also result from blunt trauma to the neck or hyperextension of the cervical spine. Additionally, it may be iatrogenic, the result of puncture of the cervical veins, or it may be secondary, the result of an aortic dissection extending into the CCA (type I according to De Bakey). Rarely, common carotid dissection extends into the ICA with patency of long stretches of the true and false lumen. In this form there may be forward flow in both lumina but to-and-fro flow or retrograde flow in the false lumen as well, depending on the site of re-entry (see. Fig. A5.24).

A study evaluating the usefulness of different duplex criteria in 23 patients with ICA dissection confirmed by MRI/MR angiography or conventional angiography revealed a detection rate of only 47.8% when morphologic critera alone were used (intramural hematoma, double lumen). Additional use of hemodynamic criteria (hemodynamic evidence of distal stenosis or occlusion) increased the detection rate to 73.9%. Sonographic follow-up after 3–6 weeks established a correct diagnosis in 91.3% of cases (hemodynamic signs of distal stenosis or occlusion with signs of resolution). Using *both morphologic and hemodynamic criteria*, duplex ultrasound is *highly sensitive in detecting dissection*; however, in some cases a sonographic follow-up examination is necessary for a definitive diagnosis (Arning 2005).

Dissection causing high-grade stenosis of the patent lumen can be diagnosed with 96% sensitivity using ultrasound with determination of hemodynamic parameters (Benninger et al. 2006).

### 5.8.2 Inflammatory Vascular Diseases (Takayasu's Arteritis)

Primary and secondary forms of vascular inflammation occur. *Secondary vasculitis* is associated with autoimmune diseases (collagen disease, systemic rheumatic disease), infections, and malignancies. These typically affect smaller vessels, and therefore rarely involve the large arteries supplying the brain.

More relevant for the major arteries supplying the brain is *primary vasculitis.Takayasu's arteritis*, also known as pulseless disease, is the most important primary vasculitis that may affect the extracranial cerebral arteries. It is a *giant cell* arteritis, predominantly of the aorta and its major branches, with the CCA and the subclavian artery as the extracranial cerebral arteries most frequently affected. As with all other forms of vasculitis, inflammatory thickening of the arterial wall (media) causes various degrees of luminal narrowing.

The etiology of Takayasu's arteritis is unknown but an immunologic basis is likely. It predominantly occurs in younger women. General symptoms include weakness, headache, fever, and weight loss. These symptoms as well as unspecific signs of inflammation are present before vascular stenosis or occlusion occurs, and an ultrasound examination of the subclavian artery and the CCA should be performed in cases of suspected Takayasu's arteritis. If the suspicion is confirmed by sonography, cortisone therapy is initiated to prevent vascular complications.

The *ultrasound appearance* of Takayasu's arteritis is characterized by *circumferential, homogeneous, and hypoechoic thickening of the arterial wall*, which primarily affects the media but may also involve the intima (*the so-called macaroni sign*). This concentric wall thickening distinguishes vasculitis from dissection with thrombosis of the false lumen, which instead causes eccentric narrowing of the true lumen (see Fig. A5.26). The appearance is also distinct from that of atherosclerotic lesions, which primarily damage the intima. Atherosclerotic lesions tend to exhibit focal variation, be more hyperechoic, and have irregular surfaces. While atherosclerosis can cause concentric luminal narrowing in patients with lipid metabolism disorders or diabetes, atherosclerotic lesions are primarily seen in the carotid bulb and the ICA. Conversely, Takayasu's arteritis affects the CCA and very rarely extends beyond the carotid bifurcation. Arteritis may also cause dilatation of the proximal aortic branches. Ultrasonography allows early diagnosis of the disease (Taniguchi et al. 1997) and is the method of choice for follow-up (Park et al 2000, Fukudome et al 1998), especially for documenting the regression of inflammatory wall thickening in patients on immunosuppressive treatment. The Doppler spectrum will show a continuously but only

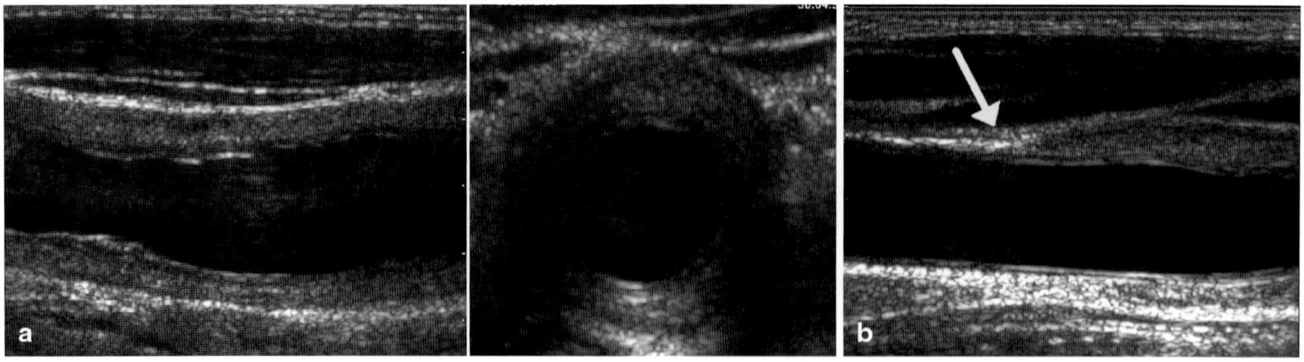


**Fig. 5.33** **a** Longitudinal and transverse images of circumferential wall thickening in Takayasu's arteritis. The longitudinal view (*left*) nicely illustrates that the hypoechoic inflammatory thickening predominantly involves the media; there is additional atherosclerotic thickening of the intima. **b** Inflammatory wall lesions in Takayasu's arteritis predominantly involve arterial segments close to the aorta, in particular the subclavian artery and the CCA, while the ICA is not involved. The image shows the transition from the thickened wall of the CCA to the carotid bifurcation, which is free of arteritis (*arrow*). In the left part of the image, the thickness of the artery wall is normal (courtesy of K. Amendt)

*moderately increased flow velocity, depending on the degree of concentric narrowing*. Ultrasound has a markedly higher accuracy than angiography, in particular in early disease. In advanced disease, there may be vascular occlusion due to inflammatory wall thickening. Medical therapy with the administration of anti-inflammatory and immunosuppressive agents is the treatment of choice. Bypass surgery is discouraged, even in occlusion, as the patency rate is poor.

The ECA can be involved in Takayasu's (with occlusion being quite common), but the ICA is not. Involvement of the latter suggests *Horton's giant cell arteritis*, also known as temporal arteritis or cranial arteritis.

Horton's disease of the extracranial cerebral arteries has a prevalence of 0.75% in individuals older than 50, and continues to become more prevalent with age. This form of giant cell arteritis also affects medium-sized and large arteries, predominantly the arteries of the abdomen and extremities as well as the supra-aortic arteries.

Although historically referred to as temporal arteritis, the disease can also involve the extracranial cerebral arteries (like Takayasu's arteritis) as well as the subclavian and axillary arteries. The ophthalmic artery can also be involved – which is much feared, as this can lead to blindness. A decisive difference, however, is that Horton's disease is an immunovasculitis of older people (>50 years). Thickening of the temporal artery, if involved, points to the diagnosis. Histologic workup of a segment of the diseased temporal artery was long considered the most important diagnostic measure, but comparative studies using ultrasound and clinical follow-up investigations show the accuracy of the histologic examination to be poorer than expected. In the sonographic examination, the main branch of the superficial temporal artery is identified in the transverse plane and traced upward until it divides. Thickening of the temporal artery may be segmental rather than continuous, and the thickened segments are evaluated in longitudinal and transverse planes in the B-mode. Temporal arteritis, like any form of vasculitis, causes circumferential wall thickening (halo or macaroni sign) with a wall thickness of 0.5–1.5 mm (Schmidt et al. 1997, 2003; Stammler et al. 2000). Blood flow velocity is decreased, and wall pulsation is absent or lower in the diseased temporal artery than on the contralateral side. These parameters have a high predictive value (Schmidt 2002, 2003), but their absence does not definitively rule out the disease, which may also spare the temporal artery. Thickening of the vessel wall recedes under immunosuppressive treatment, which correlates with a drop in laboratory inflammatory parameters.

*High-resolution ultrasound* of the *temporal artery* (if involved) has 97% specificity (Schmidt et al. 2005), and if the sonographic examination provides definitive evidence of vasculitis, treatment can be started without obtaining a biopsy (guidelines of the German Association of Scientific Medical Societies, AWMF guidelines). A biopsy is only required when ultrasound findings are inconclusive or normal but clinical signs suggest arteritis. A biopsy should be obtained from a sonographically suspicious wall segment to preclude false-negative results (as involvement is segmental).

## 5.8.3 Fibromuscular Dysplasia

Fibromuscular dysplasia is a rare vascular disease of unknown etiology that typically involves the renal arteries. It is a disease of medium-sized arteries and can therefore also affect the carotid territory, causing TIAs or even stroke. About 30% of patients with fibromuscular dysplasia have a cerebral aneurysm. In the vast majority of cases, the disease causes *luminal narrowing*. This is the result of *hyperplasia of the fibrous and smooth muscle tissue* in the media and must be differentiated from degenerative and inflammatory vascular conditions. The disease is characterized by multiple stenoses alternating with normal or dilated arterial segments, producing a *beaded appearance* on angiograms and high-resolution ultrasound images (the so-called *string-of-beads sign*). Flow in the residual lumen can be detected by color duplex or power Doppler imaging. The sonomorphologic appearance allows differentiation from atherosclerotic lesions, aided by the fact that fibromuscular dysplasia typically occurs in young women without atherosclerotic lesions in other vascular territories.

Carotid fibromuscular dysplasia is rarely diagnosed with duplex ultrasound as the first imaging test because the string-of-beads appearance caused by alternating luminal narrowing and dilatation usually spares the proximal 3–5 cm of the ICA. When fibromuscular dysplasia is suspected, the examiner must *follow the ICA as far cranially as possible* using a curved array transducer and lowering both the transmit frequency and the pulse repetition frequency toward the skull base. In general, ultrasound can only detect advanced disease with hemodynamically relevant stenosis located not too far cranially. Recent ultrasound studies report a prevalence of 0.05–0.14% (Labropoulos et al. 2007; Arning 2004) compared with 0.61% in catheter angiography studies (Sandok 1983).

## 5.8.4 Aneurysm

Aneurysm of the ICA is rare and may occur secondary to atherosclerotic or inflammatory vascular disease.

A true aneurysm is an aneurysm involving all three arterial wall layers and can be congenital, typically in patients with connective tissue disease, or it can be acquired. Mycotic or inflammatory aneurysm is caused by a localized infection of the arterial wall in the setting of inflammatory conditions of the head or neck region or in individuals in whom hematogenous spread has occurred, for example, in endocarditis.

True aneurysms of the extracranial cerebral arteries are very rare with reported rates of 0.4% (Painter et al. 1985) to 5.5% (Liapis et al. 1994). They are accounted for by atherosclerosis in 32% of cases, thrombosis in 17%, and dissection in 37% (Moreau et al. 1994). Before the era of antibiotic treatment, most true aneurysms were mycotic aneurysms developing secondary to tuberculosis and syphilis (Konstantinidis et al. 1998). In one study, however, only 5% of mycotic aneurysms involved the extracranial carotid arteries (Brown et al. 1995). Mycotic aneurysms have become very rare and are usually caused by staphylococci or streptococci, or less commonly by salmonella infections.

An aneurysm of the extracranial cerebral arteries becomes apparent as a *pulsating tumor of the neck*. B-mode ultrasound depicts the focal dilatation of the artery (saccular or spindle-shaped), and color duplex imaging allows evaluation of the patent lumen and demonstration of thrombotic deposits.

The definition of aneurysm that applies to the extracranial carotid and vertebral arteries (abrupt doubling of the lumen diameter) cannot readily be applied to the carotid bulb because of the wide normal variation in diameter seen here. Normal diameter variation must be differentiated from true aneurysmal dilatation, which is usually assumed when the external diameter reaches 14–15 mm. More clinically relevant, however, is the identification of thrombotic deposits in dilated arterial segments, which can give rise to embolism and cause cerebral infarction.

A spontaneous stroke rate of up to 50% has been reported for untreated carotid aneurysm (Valentine 2003), suggesting that even smaller aneurysms should be operated on. Other complications may result from local compression of adjacent structures such as the internal jugular vein, the trachea, the esophagus on the left side, and occasionally of a cerebral nerve (Numenthaler 1986). Rare cases of carotid aneurysms rupturing have also been reported. Color duplex ultrasound (or MR angiography) is the method of choice, enabling precise evaluation of the diameter and extent of the aneurysm as well as differentiation of thrombotic deposits (which is not possible with angiography) (see Fig. A5.22).

*Suture aneurysm* is a *false aneurysm* that may be noted as a pulsatile mass of the neck or may be detected at sonographic follow-up after carotid TEA. Color duplex ultrasound differentiates flow within the aneurysm from thrombotic material, and the characteristic "steam engine sound" caused by a high-frequency systolic signal, and retrograde flow throughout diastole can be heard in the aneurysm neck when Doppler interrogation is performed (see Fig. A5.21). The indication for surgical revision can be established without preoperative angiography.

## 5.8.5 Arteriovenous Fistula

An AV fistula is usually a sequela of trauma or iatrogenic manipulation (puncture, central venous catheter) and is conspicuous as a *mosaic of colors* due to *perivascular tissue vibration*. Spectral Doppler interrogation will not always demonstrate the fistula directly, which is why the diagnosis relies on the spectral Doppler demonstration of *high flow velocity* in the feeding artery, *especially during diastole*, and arterialized flow in the vein. The Doppler waveform obtained within the fistula depends on the flow volume but resembles the pattern in a stenosis with high systolic and diastolic flow velocities. AV fistulas in the carotid system primarily involve the CCA and the internal jugular vein because they lie close together. The *fistula flow volume* can be estimated by calculating the flow volume on the ipsilateral side by multiplying the mean flow velocity with the cross-sectional area of the CCA proximal to the fistula and then subtracting the CCA flow volume of the contralateral side.

## 5.8.6 Idiopathic Carotidynia

Idiopathic carotidynia was first mentioned in 1927 and has been recognized as a distinct clinical entity by the International Headache Society (IHS) since 1988. It is a neck pain syndrome presenting with severe unilateral pain of the upper neck region and responding well to treatment with nonsteroidal anti-inflammatory drugs. Ultrasound demonstrates echolucent thickening of the vessel wall, usually with moderate luminal narrowing but with the main part of the thickening extending outward, resembling the appearance in dissection or vasculitis. MRI has been reported to reveal no evidence of intramural hematoma but enhancement after administration of contrast medium, suggesting an inflammatory wall lesion (Burton et al. 2000; Arning 2004). Carotidynia *involves the carotid bifurcation with the distal CCA and the proximal ICA*, distinguishing the condition from dissection. As the thickened wall does not constrict the lumen, no hemodynamic signs of stenosis can be detected. Carotidynia is an example of a well-established clinical entity that required the advent of state-of-the-art imaging to identify morphologic changes (high-resolution ultrasound and MRI). The symptoms resolve spontaneously with follow-up imaging after 4 weeks demonstrating a return to almost normal wall thickness.

## 5.8.7 Vasospasm

Vasospasms can be induced by mechanical manipulation or medication used to treat vasculitis, or they can occur during episodes of migraine. They can cause cerebral or ocular ischemia, but the stenosis caused by spasm is usually so short that only a few reports describe it being visualized by ultrasound (Janzarik et al. 2007; Mosso et al. 2007); however, it is assumed that most instances of vasospasms go undetected. Treatment is with calcium antagonists. Color duplex imaging will show *a narrow lumen with stenotic flow, returning to normal* within hours. No morphologic wall changes are apparent; recurrent vasospasms usually affect the same arterial segment.

### 5.8.8 Compression by Tumor, Carotid Body Tumor

Compression of a carotid segment by cervical *tumors* or *lymph node metastases* is rare and more commonly affects the internal jugular vein. Carotid body tumors are highly vascularized masses located at the carotid bifurcation, where they cause the typical saddle deformity (splaying of the internal and external carotid branches by the tumor mass) on ultrasound. In the color duplex mode, multiple small tumor vessels are demonstrated.

The tumors arise from the 3–4 mm carotid body, a structure in the bifurcation that functions as a chemoreceptor and regulates $PO_2$, $PCO_2$, and the pH value. They are primarily supplied with blood from external carotid branches and rarely also from the thyrocervical trunk. Only 10–20% of carotid body tumors are diagnosed before surgery (Geiger et al. 1991). They are assumed to develop from paraganglial tissue, probably a residue of the neural crest. Hence, there may be multiple tumors and rarely also parajugular or paravagal tumors as well as tumors at the aortic arch. Histologically, adenomatous and angiomatous subtypes can be distinguished. The latter is very highly vascularized with an impressive appearance on color duplex imaging. Tumor growth in the area of the carotid bifurcation can encase or compress the arteries. *Color duplex evaluation* of the *localization* and *vascularization* of the tumor *contributes to the preoperative differentiation*, and the information on tumor extension facilitates radical surgical removal.

Color duplex imaging is also the method of choice for *monitoring* the *outcome of tumor embolization* in elderly or multimorbid patients in whom surgical resection should be avoided. Serial ultrasound allows evaluation of tumor growth and tumor vascularization.

## 5.9 Diagnostic Role of Duplex Ultrasound in Evaluating the Extracranial Cerebral Arteries

As a noninvasive diagnostic test, *duplex ultrasound* is the method of choice for confirming or ruling out suspected obstructive lesions of the carotid system. In the step-by-step diagnostic workup, it follows after a history has been obtained and a physical examination performed. The formerly widely used CW Doppler technique is less expensive, easy to perform, and has an accuracy of over 90% in detecting therapeutically relevant higher-grade stenosis (Keller 1990; Neuerburg-Heusler 1984). It is a suitable screening modality for patients with a reasonable suspicion of carotid stenosis if abnormal findings are subsequently verified by duplex imaging. However, *anatomic anomalies* and sudden changes in the angle of insonation due to *kinking* or *coiling* of the carotid artery may give rise to false-positive findings, and low-grade stenosis escapes detection by CW Doppler.

Duplex ultrasonography is noninvasive and has a sensitivity and specificity of over 90% in quantifying ICA stenosis, making it the diagnostic test of choice (Table 5.17). This is all the more so since *angiography*, the traditional gold standard, *has its limitations* as well. Its *accuracy*, determined by comparing the image interpretations performed by two independent radiologists, is *88–93%*, which is similar to the comparison of duplex ultrasound and angiography. This agreement is surprising since duplex ultrasound is based on hemodynamic evaluation while angiography is a morphologic method. One drawback to angiography is the fact that 3D plaques protruding into the vessel lumen are reduced to the two film dimensions, which impairs the reliability of stenosis measurement – despite mandatory assessment in two or three planes.

Duplex sonography is also the method of choice in all patients with nonatherosclerotic vascular conditions (inflammatory disease, dissection, aneurysm) because B-mode scanning depicts not only the luminal narrowing but wall changes and perivascular structures as well.

The *complications of angiography* include a stroke rate of 1–3% (Waugh et al. 1992), which is almost as high as the rate of complications experienced centers achieve with surgical management by carotid TEA. For this reason, the indication for carotid TEA is increasingly based on duplex ultrasound alone. In addition to the preoperative localization and quantification of

**Table 5.17** Role of duplex ultrasound in carotid artery surgery

| Decision to be made | Duplex criteria |
|---|---|
| Indication for surgery | • Degree of stenosis<br>• Plaque morphology<br>• Nonatherosclerotic vascular narrowing/disease<br>• Tandem stenosis |
| Timing of operation | • Early surgery, risk of occlusion/reischemia |
| Type of surgery/anesthesia | • Kinking: shortening of ICA<br>• Site of plaque/plaque length: general versus local anesthesia |
| Technical success | • Degree of residual/recurrent stenosis<br>• Postoperatively/stenting<br>• Complications of surgery |
| Outcome | • Recurrent stenosis, follow-up |

*ICA* internal carotid artery

carotid stenosis, sonography is also preferred for the postoperative follow-up after carotid TEA or PTA and stent placement.

In patients with higher-grade carotid stenosis (>70%), the stenosis grade alone establishes the indication for surgery and, if the sonographic examination allows confident grading, no further stenosis quantification or B-mode evaluation of *plaque morphology* is necessary (see Fig. 5.10). Sonomorphologic plaque evaluation *only has a role* in stage II carotid stenosis and stenosis of 60–70% or in stage I disease with higher-grade stenosis in patients in whom concomitant disease or advanced age demands caution in the consideration of surgical measures.

Many studies have been performed to investigate sonographic properties of plaques (e.g., echogenicity, surface, and contour) and to identify features that might allow prediction of the risk of embolism, but no consistent picture has emerged, and results are even contradictory. Furthermore, published data are not easily comparable because investigators use different study designs, descriptive criteria, and classification systems. Nevertheless, a few general conclusions regarding plaque morphology and echogenicity appear to be generally accepted. For one, the risk of stroke increases with plaque thickness, which is why the same stenosis grade is associated with a greater risk of embolism when caused by an eccentric plaque than when caused by a concentric plaque. This is because an eccentric plaque protruding into the blood stream is more susceptible to rupture of its cap. Such a plaque is often identified by characteristic longitudinal pulsation in the direction of blood flow in real-time B-mode ultrasound. An irregular surface seen on B-mode scans suggests atheromatous rather than fibrous plaque. Plaque with high lipid content is assumed to be echolucent and has a risk of embolism which is up to three times higher. In evaluating plaque echogenicity, however, the examiner must always bear in mind the inherent technical limitations of ultrasound resulting from the fact that a sonographic B-mode image is generated from echoes reflected off boundaries between tissues of different acoustic impedance. This means that low echogenicity merely indicates that a tissue is homogeneous but allows no conclusions to be drawn regarding other tissue properties such as elasticity. Moreover, evaluation of echogenicity is subjective and also depends on the equipment and settings used. To overcome these limitations, a *standardized measure of plaque echogenicity*, the gray-scale median (GSM), has been proposed. While this standardized analysis showed good interobserver correlation, agreement between sonomorphologic plaque classification and histopathologic examination of eversion TEA specimens was still poor.

Although plaque composition is heterogeneous, the individual components do not necessarily differ in acoustic impedance. Despite these limitations, sonographic plaque analysis can contribute additional information for estimating the risk of stroke. Rapidly progressive stenosis is four times more likely to cause TIAs and cerebral infarction than less progressive stenosis of a similar degree (Widder et al. 1992). Heterogeneous, mostly echolucent plaque is more likely to progress. A more definitive association between the sonomorphologic plaque type and stroke, however, cannot be established. Moreover, one also has to be aware that similar plaques may develop differently. Plaques considered harmless on the basis of their sonomorphologic and macroscopic appearance may rapidly turn into vulnerable, high-risk plaques, when intralesional hemorrhage occurs, for instance.

Given these limitations, there has been *controversy* regarding *the potential of ultrasound to predict the risk of carotid plaque embolism* ever since it was first used to characterize arterial plaque in the 1980s. Data from studies, including recent ones, comparing sonographic plaque evaluation with histologic findings in endarterectomy specimens are inconsistent, with some authors reporting high correlation between *histopathologic results* and *sonomorphologic appearance* and others finding poor or no correlation (Ratiff et al. 1985; Droste et al. 1997; Biasi et al. 1999; Widder et al. 1990; Schulte-Altedorneburg et al. 2000; Denzel et al. 2003; Gonçalves et al. 2004). Nevertheless, there appears to be some agreement that echolucent plaque has a high lipid content while echogenic plaque tends to be fibrous.

Investigators attempting to correlate sonomorphologic criteria with clinical symptoms (i.e., clinical proof of embolism) also present contradictory evidence. This also applies to studies performed using high-resolution ultrasound (Freilly 1992; Tegos et al. 2000; Cave et al. 1995; Meairs et al. 1999; Pedro et al. 2000; Widder et al. 1999).

The surface composition and structure also affect the risk of embolism. Some investigators report good sonographic evaluability of the plaque surface, in particular with regard to the detection of bowl-shaped recesses, which are assumed to represent ulcerations (Wijeyaratne et al. 2003; Schminke et al. 2000). However, discouraging results have been reported once again by studies comparing ultrasound with the histologic examination of endarterectomy specimens, in most cases finding poor or no correlation (sensitivities of 33–67% and specificities of 31–84%; Bluth et al. 1988; O'Leary et al. 1987; Sitzer et al. 1996).

In view of the methodological problems and contradictory study results, a word of caution is in order here regarding the sonomorphologic evaluation of plaque for predicting the risk of embolism. Further studies, in particular prospective ones, are needed for a definitive appraisal of the diagnostic value of the different sonographic criteria currently in use. Reproducible criteria are still lacking, and not even computer-based analysis provides satisfactory results. The retrospective analysis of patients operated on still does not allow any conclusions to be drawn regarding plaque ulceration and its prognosis. Prospective studies on the natural history and the search for criteria to differentiate ulcerations from stable niches would be desirable but face numerous methodological problems. In view of the known risk of embolism of plaque ulceration in high-grade stenosis, their follow-up observation in randomized studies is precluded for ethical reasons.

A new approach, intended to improve tissue differentiation and the identification of vulnerable plaque, processes the frequency of returning echoes rather than the amplitude as is done in conventional B-mode ultasound. This is known as virtual histology, and the rationale behind it is that all tissue types, and hence different plaque components (lipid, fibrous, necrotic, and hemorrhage), reflect echoes with slightly different frequencies. However, the differences are so small that practical experience with this method is only available with the high-resolution probes used in IVUS. The virtual histology information is color coded and superimposed on the B-mode image.

Initial CW Doppler imaging, as it used to be advocated by some investigators, is no longer necessary since a color duplex examination performed with adequate instrument settings enables continuous hemodynamic evaluation. Supplementary *transcranial ultrasonography*, on the other hand, provides useful additional information on intracranial arterial anomalies and stenosis. Severe intracranial flow obstruction such as occlusion of the middle cerebral artery may occasionally be suggested by indirect signs revealed by continuous Doppler interrogation or in the spectral waveform. However, the extracranial Doppler waveform may appear normal when there is good collateralization of high-grade stenosis of the main branch of the middle cerebral artery.

Duplex or color duplex ultrasound is highly reliable in evaluating the carotid bifurcation, the preferred site of carotid stenosis. Angiography does not yield any additional information in this area. The hemodynamic assessment by *duplex ultrasound* is *superior* in *grading carotid stenosis* compared with angiography, which merely depicts the perfused lumen in relation to the adjacent vessel segment. Only ultrasound provides information on *plaque morphology* (see Sect. 5.6.1.1 and Fig. 5.21). In the NASCET study, there was poor agreement between angiography and intraoperative findings with regard to the evaluation of plaque surface properties such as ulceration.

The *advantages of angiography* include the continuous visualization of the target vascular anatomy as well as better documentation of the findings. Whether angiography is necessary prior to carotid surgery following a duplex scan performed by an experienced examiner depends on the communication between the surgeon and the sonographer and the technical quality of the ultrasound examination. Another advantage of angiography is the detection of stenosis near the aortic arch and the base of the skull as well as intracranially. If the sonographic findings in these carotid segments are inconclusive, angiography should be performed.

If no angiography is performed prior to carotid TEA, the sonographer must work very conscientiously. This means reliable identification of the ICA and ECA. A high gain is required to differentiate between subtotal and total occlusion. In particular if the examination is impaired by calcified plaques, the sonographer must attempt to depict flow signals in the artery up to the base of the skull. However, a control angiography should be done in such cases and also if stenosis grading is impaired by heavy calcification.

Angiography or intra-arterial DSA is indicated only in those cases where the sonographic examination is inconclusive or where ultrasound examination of the extracranial cerebral arteries reveals indirect evidence of intracranial vascular pathology. Alternatively, a transcranial duplex examination can be performed.

In addition to angiography and color duplex ultrasound, the extracranial and intracranial cerebral arteries can be examined by *CT* or *MR angiography*. Unlike conventional angiography, which is a 2D projection technique, CT and MR angiography yield 3D datasets of blood flow in a specific body region, which can then be reconstructed in multiple planes for vascular evaluation.

A helical CT angiogram depicts the target vessels in relation to surrounding structures and is obtained after injection of iodine-based X-ray contrast medium. Arterial evaluation may be limited by adjacent structures of similar attenuation or bones and by premature opacification of veins. Bones may degrade the visualization of the carotid bulb, while superimposed veins may limit adequate arterial evaluation in the area of the carotid bifurcation. Time-consuming image postprocessing is required to ensure adequate evaluation in these cases. CT angiograms have high spatial resolution and are highly sensitive in detecting small flow volumes and slow flow, for example, distal to subtotal occlusion, but provide little information on blood flow direction or other hemodynamic parameters.

As with CT angiography, MRI also allows 3D reconstruction for the depiction of target vessels in relation to surrounding structures. Nearby bones do not limit evaluation and a contrast agent is not generally required but will markedly improve image quality and depiction of vessels with slow-flowing blood.

The signal intensity of blood on MR images is determined by various factors including the MR pulse sequence, the course of the vessel relative to the imaging plane, slice thickness, and flow profile. The depiction of flowing blood by MRI is a complex phenomenon, as the signal intensity of blood depends on numerous factors. Two basic phenomena are time-of-flight and phase-contrast effects, which are exploited by different MR techniques to highlight arteries and/or veins. Time-of-flight MR angiography can be manipulated to selectively image either the arteries or veins. To selectively highlight the arteries, the venous signal is suppressed. This is accomplished by application of a saturation band to flip longitudinal magnetization into the transverse plane, thereby suppressing venous enhancement in the imaging volume that would result from the inflow effect. The phase-contrast technique obtains information on the vascular system from deliberately induced flow-related phase shifts.These phase shifts depend on the speed of flowing protons and can be measured to calculate blood flow velocity. In-flow and phase-contrast MR angiography only use flow effects for vascular imaging. Contrast-agent-based MR techniques exploit the selective shortening of the T1 relaxation time of flowing blood (from 1,200 to 50 ms) during intravascular passage of the contrast agent to generate image contrast between vessels and stationary tissues. The use of special phased-array coils markedly improves the signal-to-noise ratio while at the same time shortening image acquisition time and increasing spatial resolution, thereby improving the differentiation between peripheral arteries and veins.

*MR angiography* differs from CT angiography in that blood flow itself rather than the contrast-enhanced blood is visualized in the image, and arteries and veins are differentiated using different pulse sequences and imaging techniques. Vessels are most accurately depicted on MR angiograms when blood flow is laminar. Vortexing and turbulent flow in a stenotic segment may impair quantitative assessment and lead to *overestimation of the degree of stenosis*, especially when the time-of-flight technique is used. These flow phenomena may also lead to misinterpretations in bifurcations and at the origins of branches. Injection of a contrast agent is necessary to visualize very slow flow. The combination of conventional MR imaging and MR angiography is an ideal imaging tool for a comprehensive evaluation of intracranial perfusion and parenchymal changes, providing diagnostic information to supplement color duplex imaging (extracranial cerebral arteries and stenosis quantification in the carotid bifurcation) in patients considered for CEA.

With the methodological limitations outlined above, CT angiography is primarily suited to evaluating the anterior and posterior arteries near the base of the skull as well as the origins of arteries arising from the aortic arch. MR angiography, on the other hand, enables good evaluation of the entire intracranial arterial territory including the carotid bulb. Color duplex imaging, however, performed with a high-frequency transducer remains the most suitable imaging tool for assessing the extracranial arteries supplying the brain, including the detection of pathology and stenosis grading. This is suggested by studies comparing different imaging modalities with the traditional gold standard, i.e., angiography performed in two or three planes.

With a sensitivity of 92% and a specificity of 74%, contrast-enhanced MR angiography has a poorer accuracy in identifying stenosis requiring surgical management than duplex ultrasound and is less accurate in determining the degree of stenosis as well. The two modalities are supplementary, with duplex ultrasound enabling adequate evaluation of the extracranial carotid system and MR angiography providing information on the intracranial vessels as well as on the supra-aortic origins of arterial branches. Together, the two modalities enable comprehensive diagnostic evaluation prior to surgical repair of carotid stenosis.

The indication for surgical management in patients with subclavian steal syndrome can be established if the clinical suspicion is confirmed by duplex imaging, but only angiography will enable exact identification of collateral pathways.

If initial treatment of carotid stenosis is conservative (e.g., antiplatelet therapy), follow-up ultrasonography should be performed every 6 months for timely surgery in case of rapid progression. In patients who undergo carotid TEA, a follow-up ultrasound examination is performed immediately after surgery and then at 6-month to 1-year intervals, depending on the results. A focus of follow-up is on identification of recurrent stenosis and complications such as suture aneurysm.

## 5.10 Atlas: Extracranial Cerebral Arteries

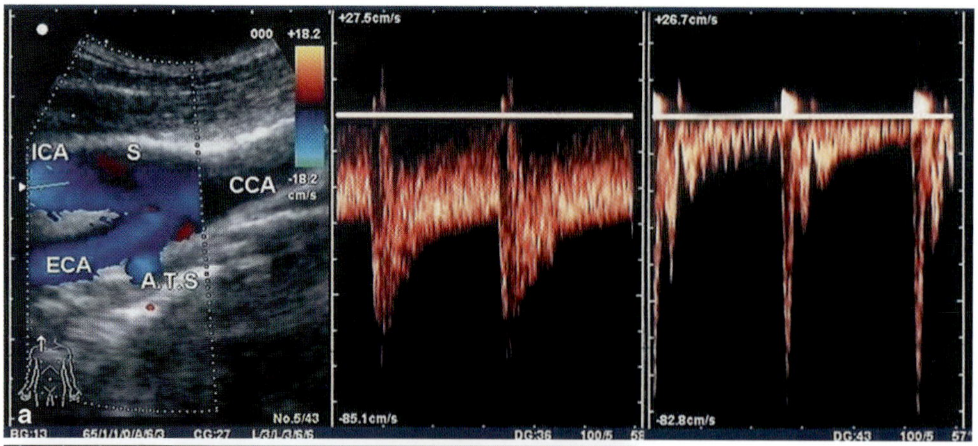


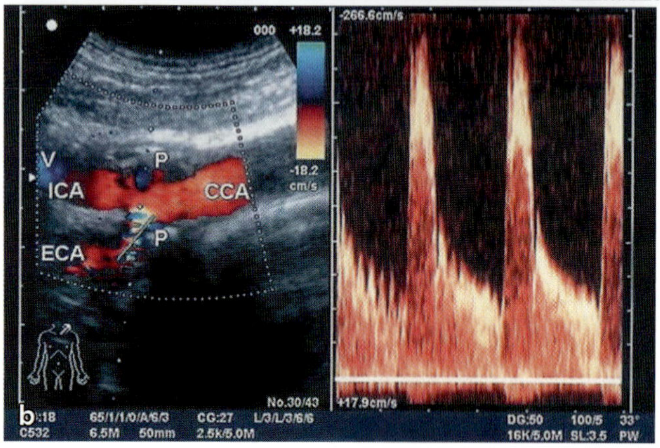


**Fig. A5.1 a, b**

**Carotid bifurcation (differentiation of ICA/ECA)**

**a** Longitudinal view of the carotid bifurcation obtained with the transducer in the posterolateral position. The internal carotid artery (*ICA*) is closer to the transducer. The color change in the bulb indicates retrograde flow components due to flow separation (*S*) (see Fig. 1.43b). The Doppler waveform of the ICA is characterized by a fairly large end-diastolic flow component. The external carotid artery (*ECA*) is identified further away from the transducer with flow separation at its origin (*red*) and the superior thyroid artery (*A.T.S*) arising from it. Doppler waveforms recorded in the ICA (*middle section*) and ECA (*right section*). The waveform of the ECA differs from that of the ICA in that it has a more pulsatile flow and shows the pulsation transmitted from the temporal artery upon rhythmical tapping anterior to the ear (*left part* of the ECA waveform).

**ECA stenosis**

**b** Stenosis of the ECA is associated with less pulsatile flow in the stenotic segment, which may make it difficult to correctly assign the stenosis to the ICA or ECA. The temporal tap sign enables reliable differentiation between the two arteries even if the ECA waveform is altered by stenosis. (Inverted color encoding of flow direction compared to **a**).

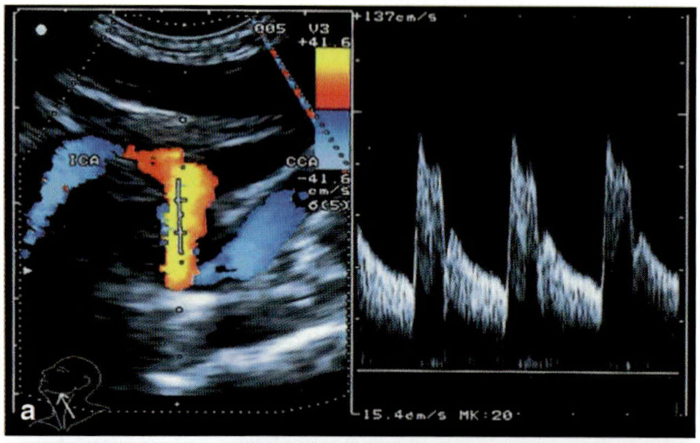


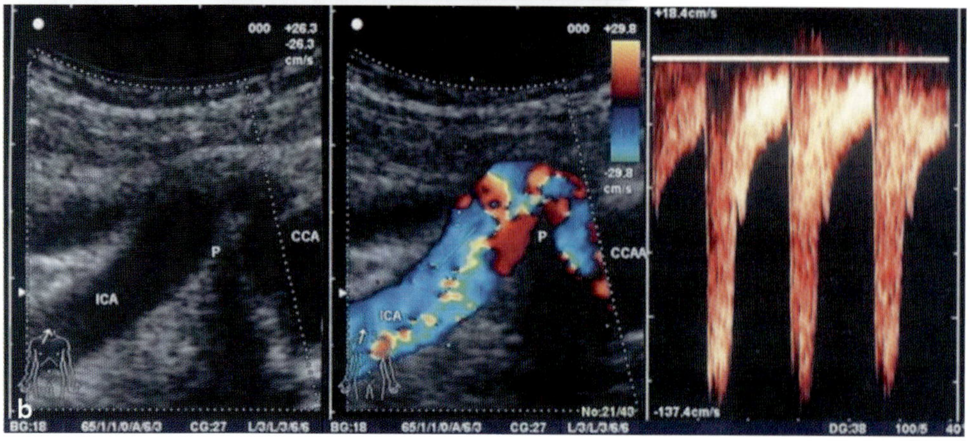


**Fig. A5.2 a, b**

**a** Elongation of the internal carotid artery (*ICA*) may lead to kinks or coils (see Fig. 5.1b). The resulting tortuosity of the ICA can give rise to different angles of insonation, which in turn can lead to localized increases in the Doppler shift frequency, which must not be misinterpreted as indicating stenosis. The corresponding color duplex image will show color aliasing in vessel segments insonated at a small angle. Depending on the insonation angle used, kinks or coils in the course of the ICA may be depicted as flow reversal (change in color coding). The color scan (*left section*) depicts the junction of the CCA with the ICA on the *right* and the distal ICA on the *left*. The Doppler waveform obtained after careful adjustment of the insonation angle shows laminar flow with a PSV of 95 cm/s, confirming that aliasing in the color mode is due to a small insonation angle. The color change from red to blue is caused by the change in flow direction relative to the transducer.

**Kinking with stenosis**

**b** Stenosis due to ICA kinking is rare. Such a stenosis may be caused by sclerotic wall changes with plaque (*P*) at the site of the kink. Here, a PSV of 145 cm/s indicates a stenosis of approximately 60%.

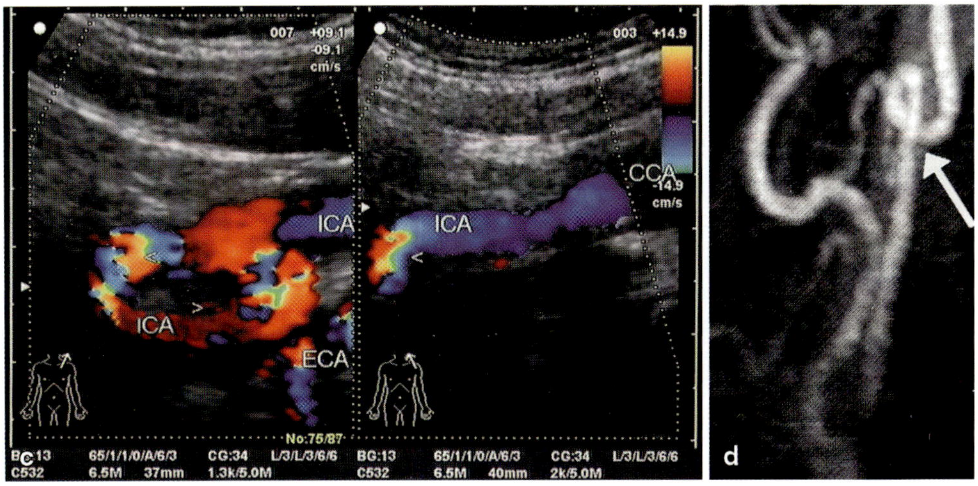


**Fig. A5.2 (continued)**
**Coiling**

**c** Coiling of the tortuous ICA is seen on color duplex images as a change in color coding, which indicates a change in flow direction relative to the transducer. The *right section* shows the proximal, straight segment of the ICA (first 2.5 cm) with the *arrowhead* indicating the transition to the coiled segment. The *left section* depicts the coiled segment and the transition from the straight portion (change from *blue*, flow away from transducer, to *red*, flow toward transducer). A coiled ICA segment is often not visualized in a single plane, but in most cases flexible transducer positioning will allow full evaluation. One segment is imaged at a 90° Doppler angle, resulting in the artifactual absence of flow. Color aliasing is due to use of a low pulse repetition frequency.
**d** Angiogram showing the coiled distal segment of the extracranial ICA.

**Fig. A5.3 a–c**
**Measurement of IMT**

**a** 38-year-old man with a history of hyperlipidemia, in whom an intima–media thickness (IMT) of 0.8 mm was measured in the far wall 2 cm proximal to the bifurcation (indicated by *plus signs*). An IMT of 0.8 mm is abnormal for the patient's age but would be normal for an individual over 60 (see Fig. 5.5 in Sect. 5.2.1).
**b** In another patient, measurement in the far wall of the CCA just before the bifurcation shows thickening of the intima–media complex to 0.9 mm and a plaque with a maximum thickness of 3.2 mm and an irregular surface to the right of it.

**Measurement of IMT – plaque**

**c** The thickness of the intima–media complex is measured in the wall away from the transducer, where the interface between the perfused lumen and the intima produces a sharp reflection due to the intervening flowing blood. The intima and media are indistinct with the second bright reflection occurring at the boundary between the adventitia and the connective tissue. The layer between these two reflections, which is measured, is the intima–media complex. The example shows abnormal IMT (1.1 mm) and a plaque with a thickness of 2.2 mm in the middle of the scan.

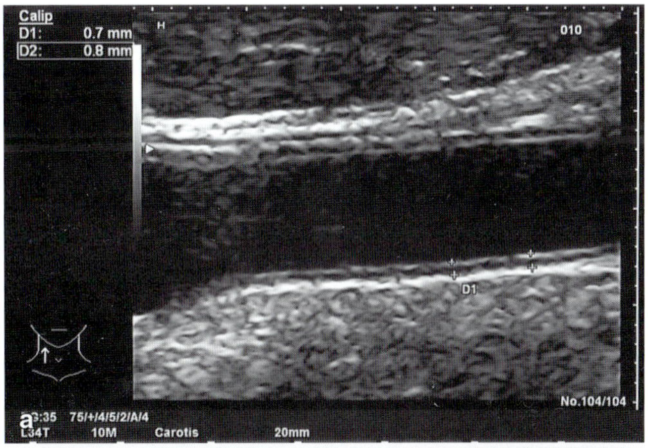


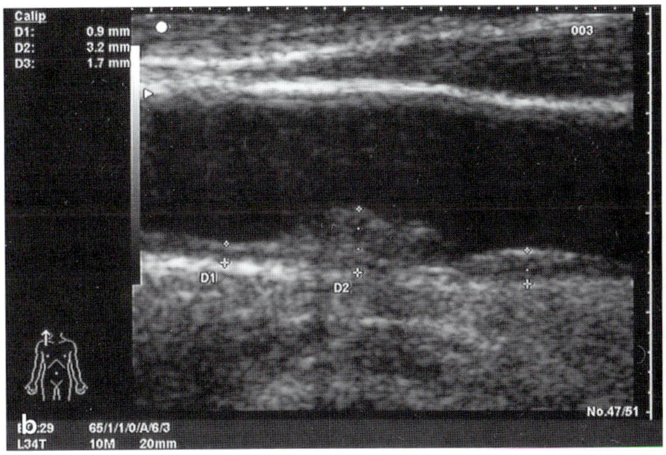


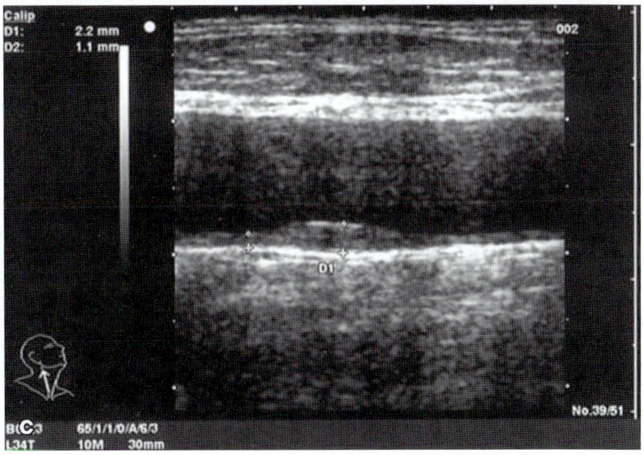

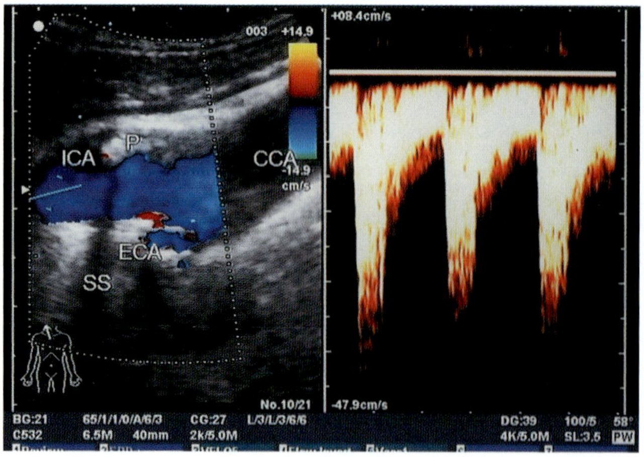


**Fig. A5.4**
**Plaque – grade of stenosis**
Internal carotid artery (*ICA*) evaluated from a posterior approach. The ICA origin is close to the transducer. The plaque (*P*) is not hemodynamically relevant as suggested by a PSV of 50 cm/s. However, both the color scan and the Doppler waveform show flow disturbance with backward flow.

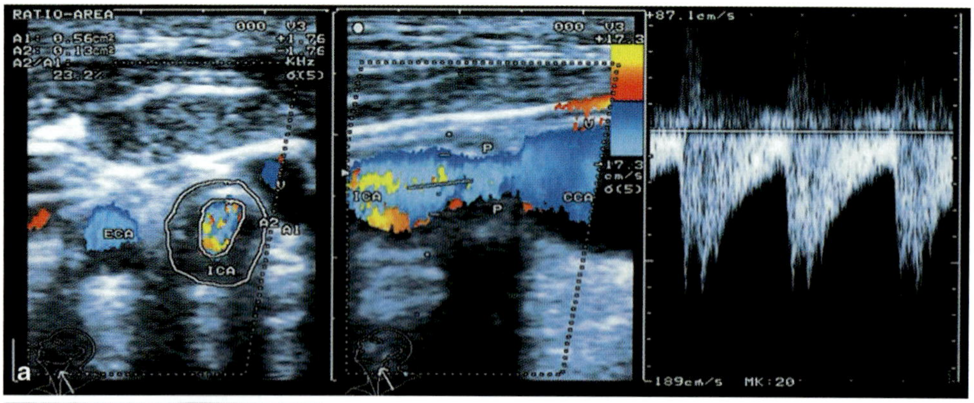


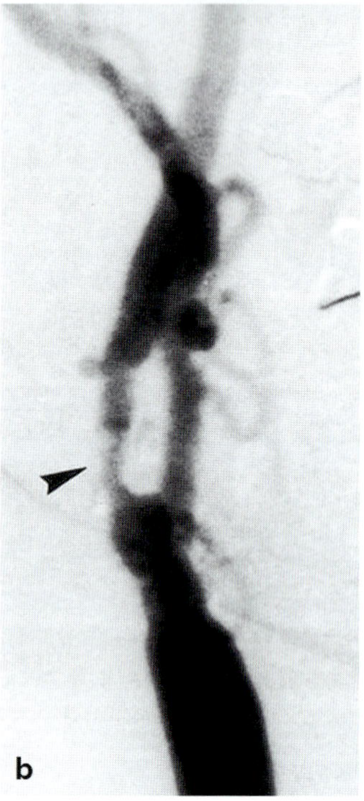


**Fig. A5.5 a, b**
**Stenosis with mild hemodynamic significance**
**a** A circular plaque in the internal carotid artery (*ICA*) reduces the cross-sectional area by 75% (*left section*). To achieve complete color filling of the perfused lumen in the transverse plane, a low pulse repetition frequency is employed, which produces aliasing. The *middle section* depicts central flow acceleration in lighter coloring (*yellow*) and eddy currents as a change in color coding (*red*) (see end of Sect. 1.2.3).
The hemodynamic stenosis grade with a PSV of 128 cm/s and spectral broadening correlates with the cross-sectional area reduction measured. (A 65–83% cross-sectional area reduction corresponds to a 40–60% diameter reduction (according to ECST criteria), suggesting a stenosis which is just becoming hemodynamically significant.) This is shown here for illustration only, and measurement of the cross-sectional area reduction from a transverse image should not be used for stenosis grading. All relevant stenoses are graded hemodynamically from spectral Doppler measurement performed in longitudinal orientation (see Sect. 5.6.1.2).
**b** Angiogram: Moderate stenosis of the ICA origin.

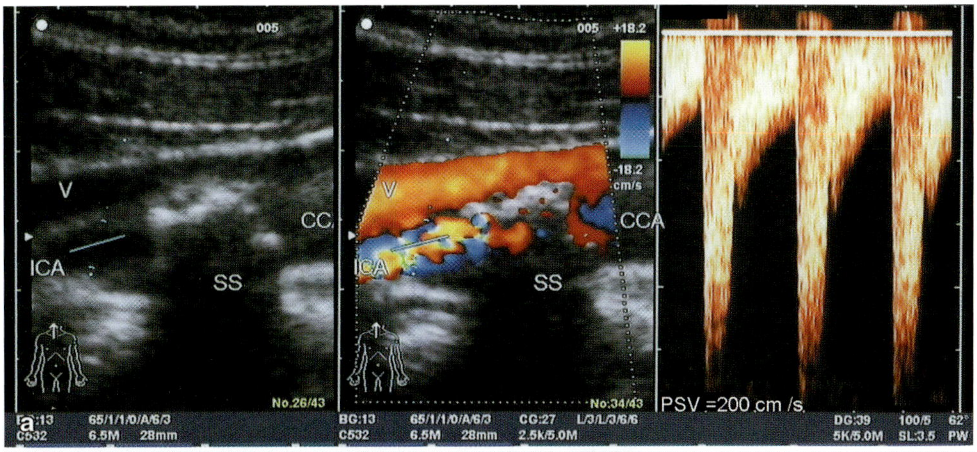


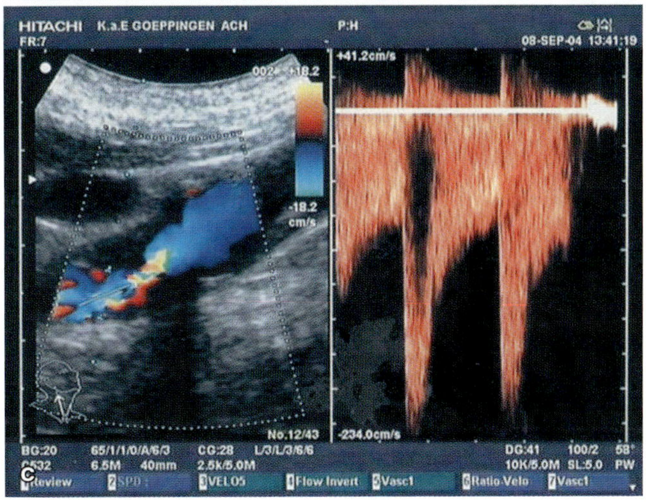


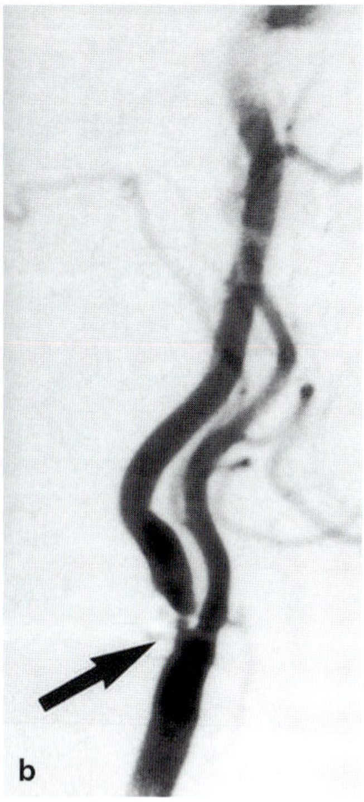


**Fig. A5.6 a–c**
**Moderate ICA origin stenosis**
**a** The extent of luminal narrowing caused by the plaque at the internal carotid artery (*ICA*) origin cannot be evaluated in the gray-scale mode due to calcification with posterior acoustic shadowing (*SS*). Color flow imaging is also impaired. Distal to the acoustic shadow, there is an eccentric jet with aliasing (*yellow*) and turbulent flow. Flow velocity is increased to 200 cm/s in systole and 70 cm/s at end diastole, indicating 60–80% stenosis (according to ECST criteria) by diameter reduction, which corresponds to an 84–95% cross-sectional area reduction. In this case, it was not possible to depict flow by moving the transducer and thus avoiding the calcification. Instead, a high gain was used to obtain a Doppler spectrum from the area of acoustic shadowing for hemodynamic quantification of the stenosis by determining PSV at the site of the plaque.
**b** Angiogram: 60–80% diameter reduction.
**c** Example of a plaque causing a similar degree of stenosis as in **a** but with better visualization of the stenosis due to the plaque not being calcified. Echolucency suggests a vulnerable plaque, but the surface is smooth. The plaque causes moderate to severe stenosis of the carotid bulb (aliasing, spectral Doppler flow velocity of 225 cm/s in systole and 80 cm/s at end diastole). The B-mode scan (*left section*) depicts the CCA on the *right* and the ICA on the *left*, both with flow coded in blue.

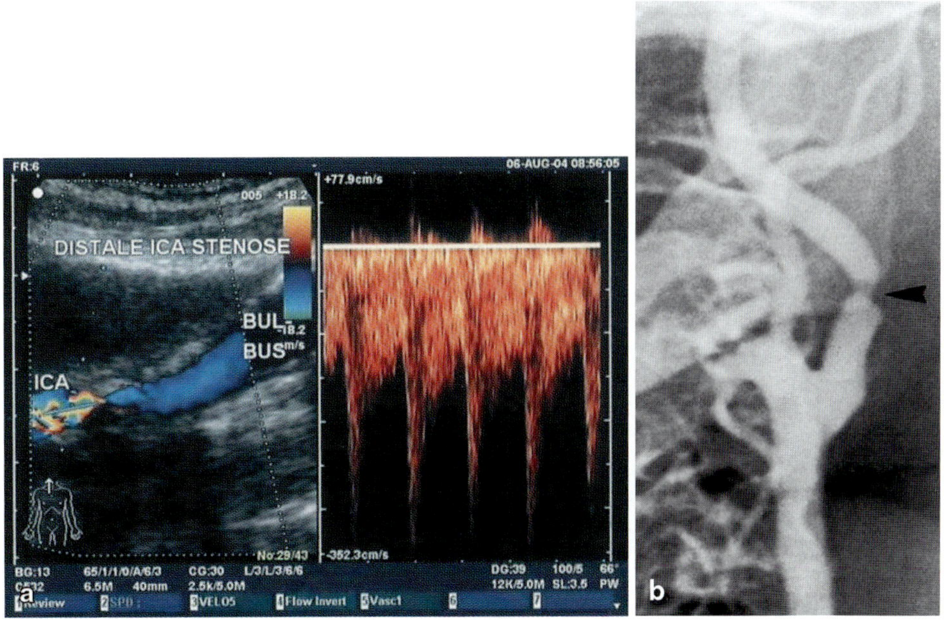


**Fig. A5.7 a, b**
**Distal ICA stenosis**
**a** From a posterolateral transducer position, stenosis is depicted in the ICA about 2.5 cm from the origin of the ECA. In the color duplex mode, stenosis is suggested by aliasing; the plaque is echolucent. A PSV of 380 cm/s suggests a diameter reduction of > 80%. More distal evaluation of the ICA is precluded by acoustic shadowing and scattering produced by connective tissue structures at the base of the skull. In the postoperative evaluation after carotid TEA, it is important to exclude stenosis at the distal patch end.
**b** Angiogram: Filling defect (*arrowhead*) just below the skull base and normal origin of the ICA.

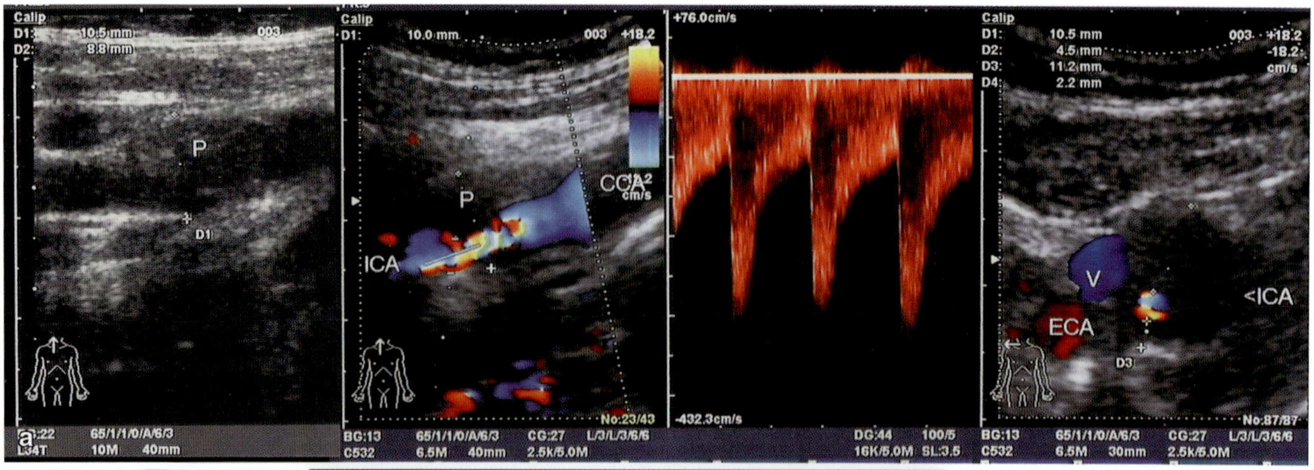


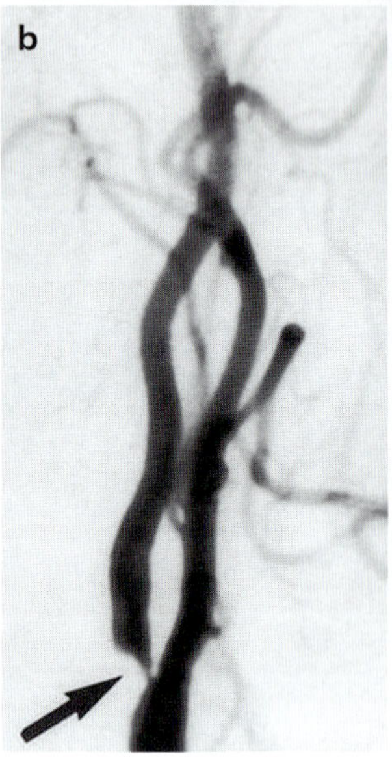


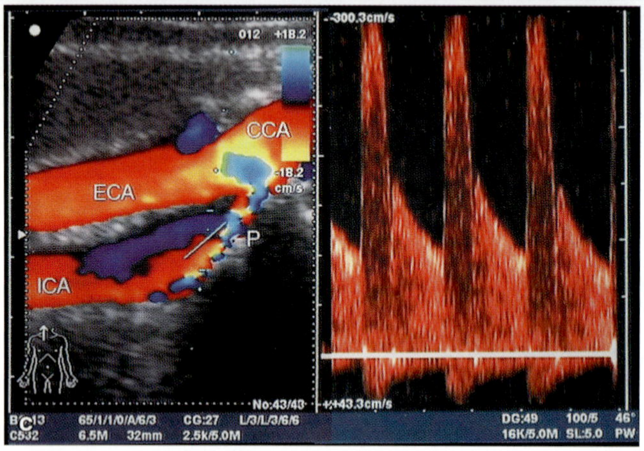


**Fig. A5.8 a–c**

**High-grade ICA origin stenosis**

**a** Echolucent, smooth plaque (*P*) at the origin of the ICA (*leftmost scan*). There is aliasing in the longitudinal color flow image with a PSV of 3 m/s in the corresponding Doppler waveform, suggesting high-grade stenosis of the ICA. Blue indicates normal flow direction toward the brain (away from transducer); red indicates turbulent flow with retrograde components.

The transverse view (*rightmost scan*) displays the sonomorphologic appearance of the echolucent, eccentric plaque in the carotid bulb (*ICA, indicated by plus signs*) and the resulting high-grade luminal narrowing. The external carotid artery (*ECA*) and jugular vein (*V*) are seen lateral to the ICA.

Accurate stenosis grading is not possible from transverse views (see Fig. A5.5 and Sect. 1.2.3); a rough estimate is that the diameter reduction is >80%.

**b** Angiogram: Eccentric high-grade stenosis (*arrow*) of the ICA.

**c** Eccentric high-grade ICA stenosis, which, unlike the stenosis in **a**, is caused by a calcified plaque (*P*) with acoustic shadowing (PSV of 380 cm/s). Blood flow directions with inverted color coding (red for arteries, despite flow away from transducer – as preferred by some examiners). Turbulent flow components (see Fig. 5.17a).

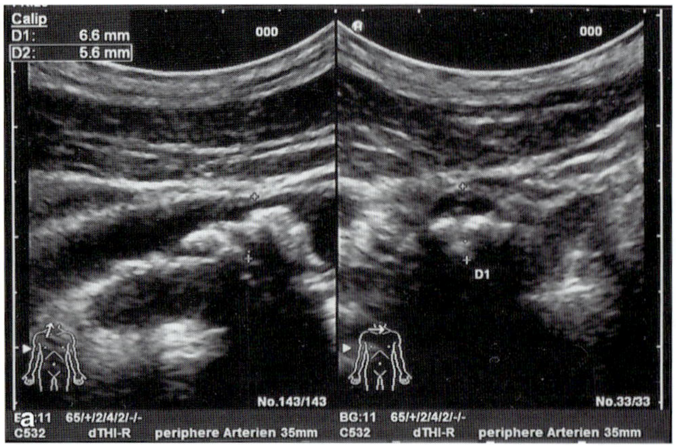


**Fig. A5.9 a–i**

**Evaluation of plaque morphology**

(see Figs. 5.12 – 5.14 in Sect. 5.6.1.1)

**a** Echogenic, calcified plaque (acoustic shadowing) with a relatively regular and smooth surface (longitudinal section on the *left* and transverse section on the *right*). The small indentation in the center of the plaque is a surface irregularity rather than an ulcer. The appearance in the longitudinal plane suggests a higher stenosis grade than the transverse view (70% diameter reduction) (see Fig. 5.12). Type III/IV plaque according to Gray-Weale classification but very eccentric. Compare echolucent, eccentric plaque causing the same degree of stenosis in Fig. 5.16.

**Fig. A5.9 (continued)**

**Concentric plaque**

**b** Plaque (*P*) in the near wall of higher echogenicity than a second plaque in the far wall; the more echolucent plaque appears smooth on the B-mode scan, and only the color flow image reveals eccentric protrusion into the lumen.

**c** Predominantly echolucent plaques with irregular surfaces in the near wall (*P*). The radius and shape of the surface defect suggest an irregular plaque contour rather than ulceration.

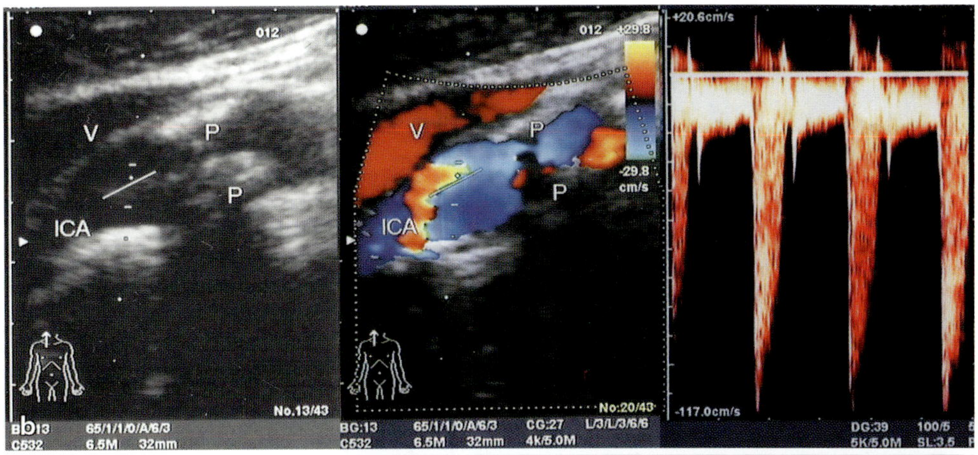


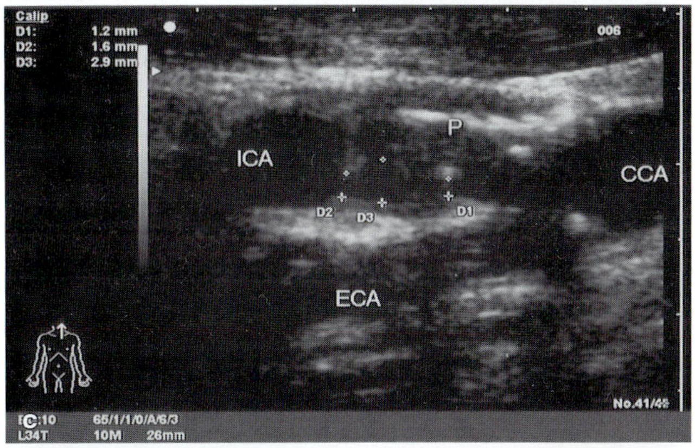


**d** Long, predominantly echolucent plaques in the ICA with eccentric portions and a smooth surface. The long plaque in this example is similar in appearance to dissection with thrombosis of the false lumen. Doppler waveform with a PSV of 330 cm/s suggests high-grade stenosis; smooth fibrotic plaques were seen intraoperatively. A uniformly fibrotic plaque is echolucent when it contains few boundaries between tissues of different acoustic impedance. Note that the angle correction cursor is parallel to the direction of flow and not to the wall.

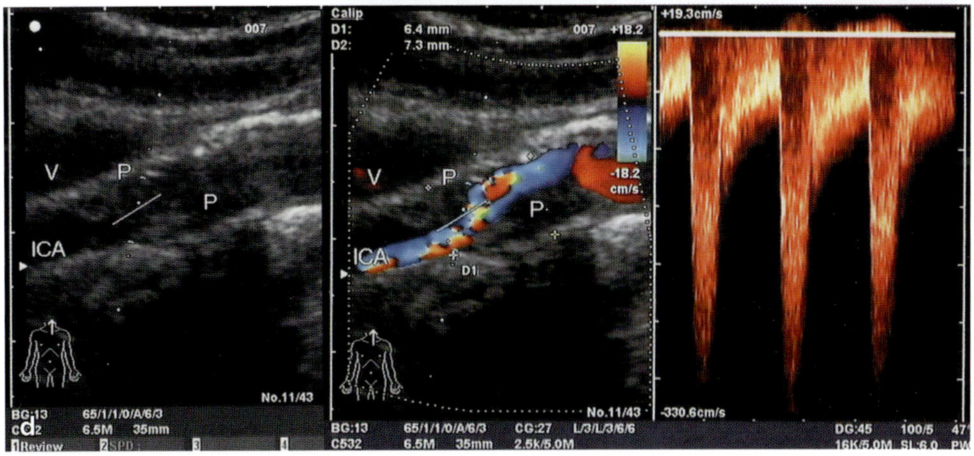


**g** Very short eccentric plaque with a very echolucent portion in the near wall and a more echogenic, partially calcified portion in the far wall (*P*). A short plaque like this (causing 60% stenosis) may escape detection unless the extracranial carotid territory is continuously imaged in the color duplex mode with spectral Doppler recording from the CCA to the base of the skull. Aliasing in the color duplex image and PSV of 150 cm/s in the stenotic jet.

**Carotid artery thrombus**

**h** Echolucent plaque must be differentiated from thrombus at the origin of the ICA (a very rare condition), which can be done applying sonomorphologic criteria: the thrombus adheres to the wall of the carotid bulb and is surrounded by a ring of blood that becomes increasingly wider and more complete toward the periphery; a circle of blood surrounds the thrombus tip (best seen on the transverse view). The thrombus (*X*) is echolucent and homogeneous, so that the color mode is necessary to reliably differentiate it from the surrounding blood.

**i** It is noteworthy that the degree of stenosis caused by carotid thrombus (as in peripheral arteries) is much lower in terms of hemodynamic effects measured by spectral Doppler (PSV < 1m/s) than would be expected based on its B-mode appearance. The patient suffered a stroke while being prepared for emergency surgery, and thrombolytic treatment was performed, which dissolved the carotid thrombus, except for a 4-mm residue on the bulb wall (not shown).

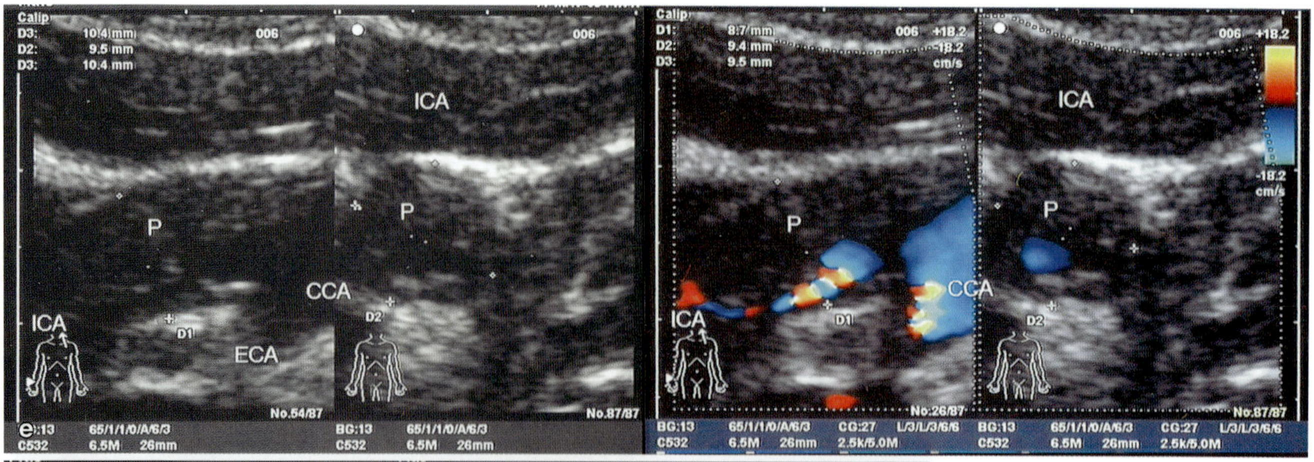


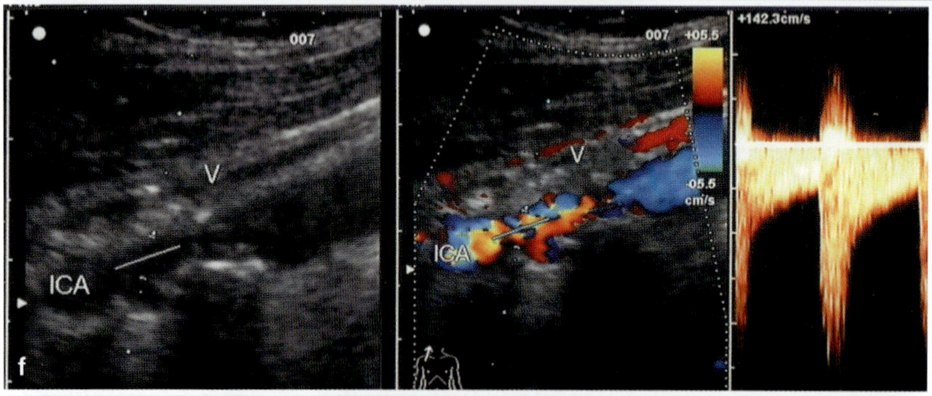


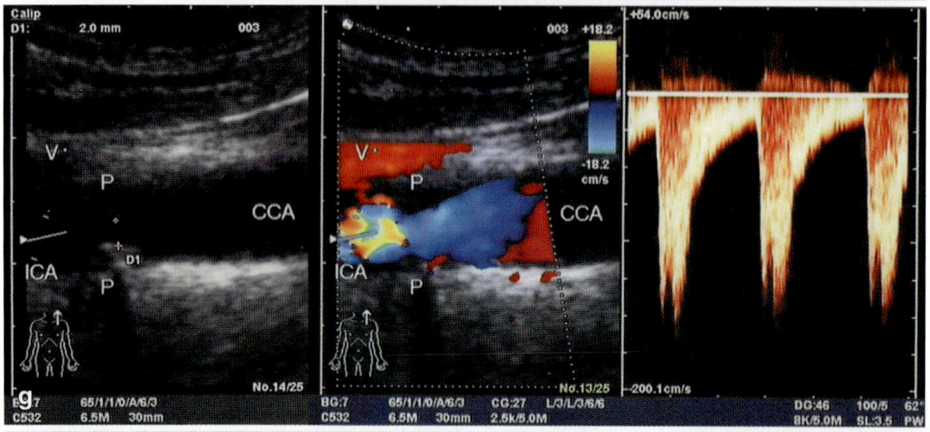


**Fig. A5.9 (continued)**

**e** An echolucent plaque may be difficult to differentiate from surrounding blood in B-mode imaging (longitudinal and transverse images *on the left*); demonstration of flow in the residual lumen in the color duplex mode is necessary and reveals high-grade stenosis caused by an eccentric plaque (longitudinal and transverse images *on the right*). If no gray-scale median (GSM) measurement is performed, the echogenicity of the plaque can be evaluated by comparing it with that of the sternocleidomastoid muscle, which is closer to the transducer. Low echogenicity corresponds to a GSM <30.

**f** Bowl-shaped defect at the distal end of a partially calcified plaque with echolucent noncalcified portions; the sharp demarcation of the defect with a bright boundary is more in keeping with a harmless defect niche rather than a fresh ulceration (and was confirmed intraoperatively). The PSV of 2.5 m/s corresponds to >70% stenosis. Mix of red and blue within the defect indicates eddy currents.

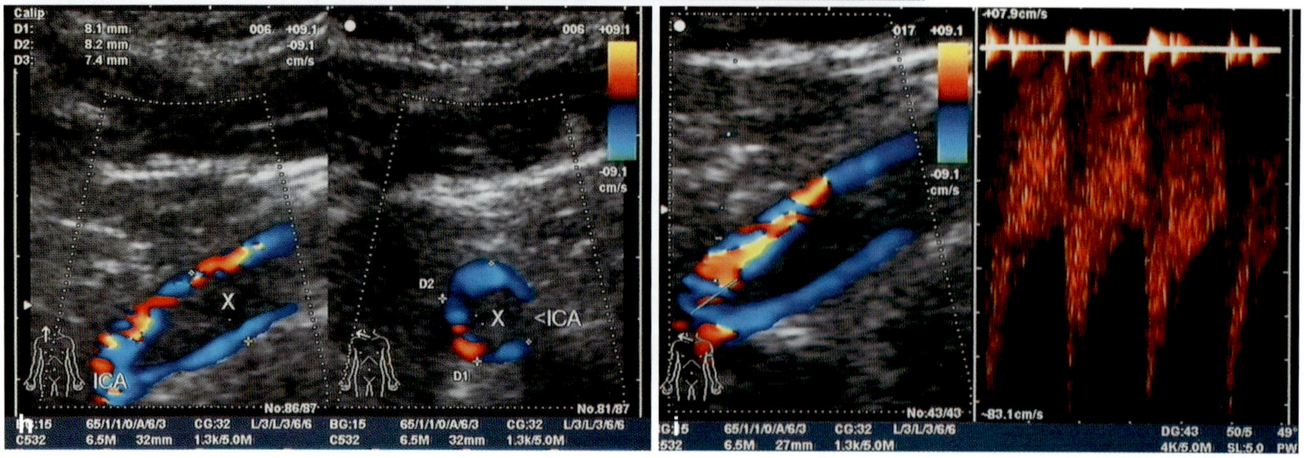

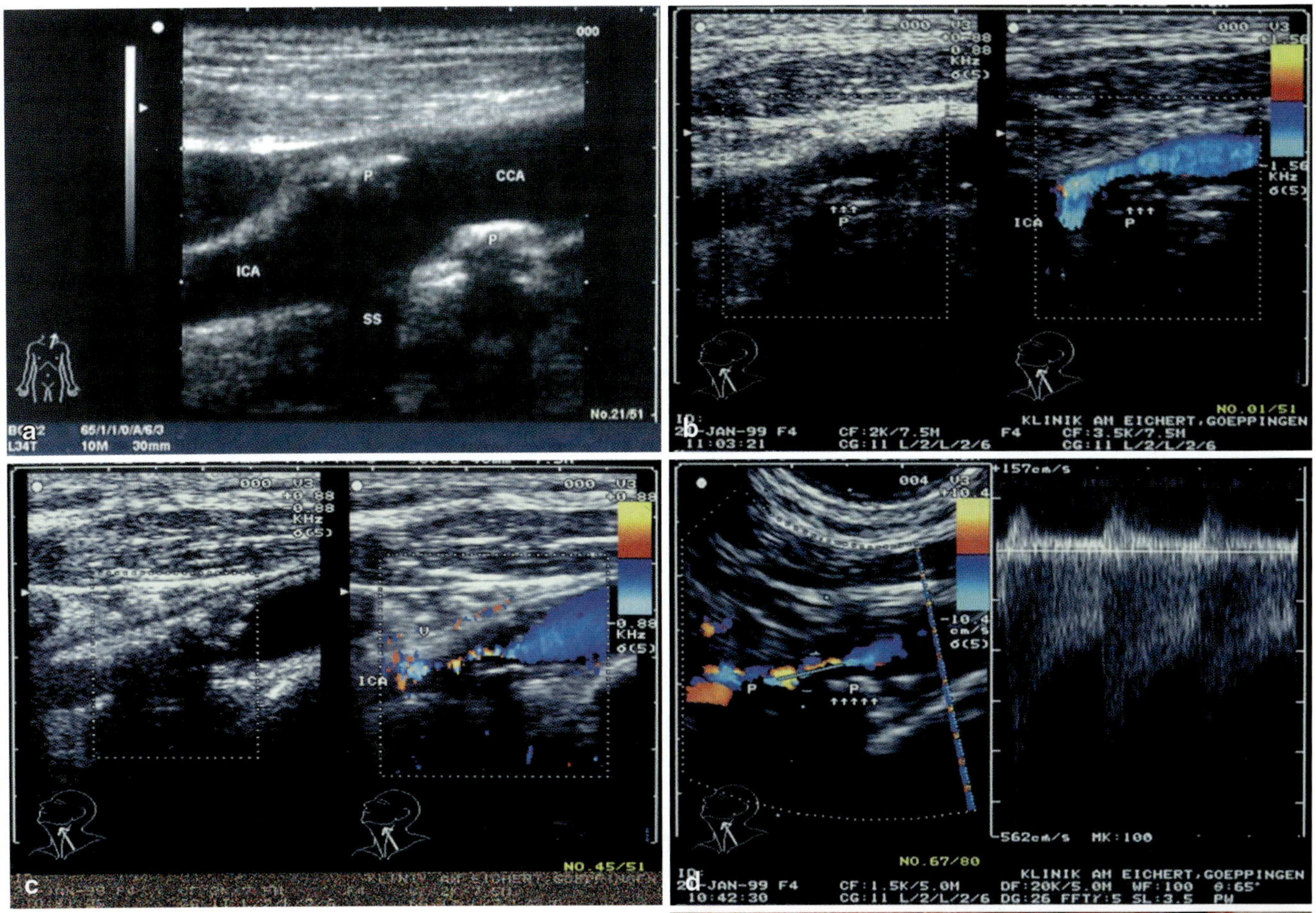


**Fig. A5.10 a–e**
**Plaque morphology – surface structure**
(see Sect. 5.6.1.1, Fig. 5.13b)

**a** A fairly homogeneous and echogenic plaque with smooth contours (*P*; far wall) is depicted in the more proximal portion of the carotid bulb. This appearance suggests a fibrous plaque with a low risk of embolism. More distally, an inhomogeneous and mostly echolucent plaque with a surface poorly delineated from the lumen is depicted. Portions of the plaque are only suggested by bright spots within the lumen. This appearance often correlates with an atheromatous plaque.
**b** B-mode scan (*left section*) depicting a plaque with 40–50% diameter reduction in the sagittal plane. Fairly homogeneous plaque with a regular surface (type I). Color duplex imaging confirms the smooth delineation of the plaque from the vessel lumen and the nearly normal laminar flow near the wall.
**c** Sagittal B-mode scan showing inhomogeneous plaques with acoustic shadowing and ill-defined contours. Bright spots suggest that the plaque extends almost to the center of the artery. Color duplex imaging is necessary for adequate evaluation, showing a very small residual lumen between the two plaques on the far and near wall. Flow acceleration is indicated by aliasing. Type IV plaque: echolucent, inhomogeneous, surface not delineated from vessel lumen.

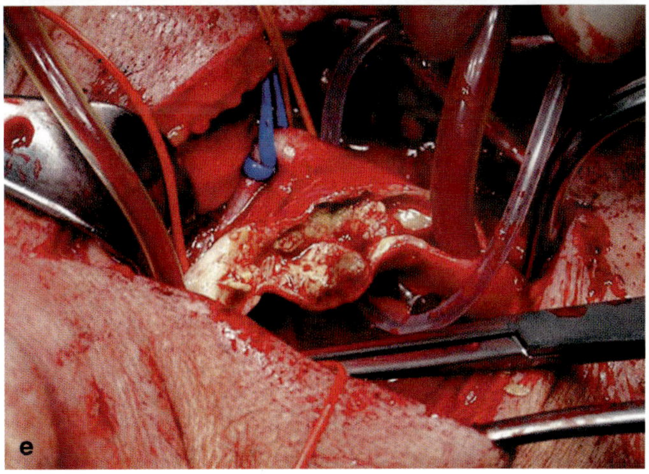


**d** Spectral Doppler measurement with the sample volume placed in the stenotic jet confirms high-grade stenosis with a PSV >4 m/s.
**e** Intraoperative confirmation of high-grade stenosis with a long plaque, predominantly of the atheromatous type (corresponding to **c** and **d**).

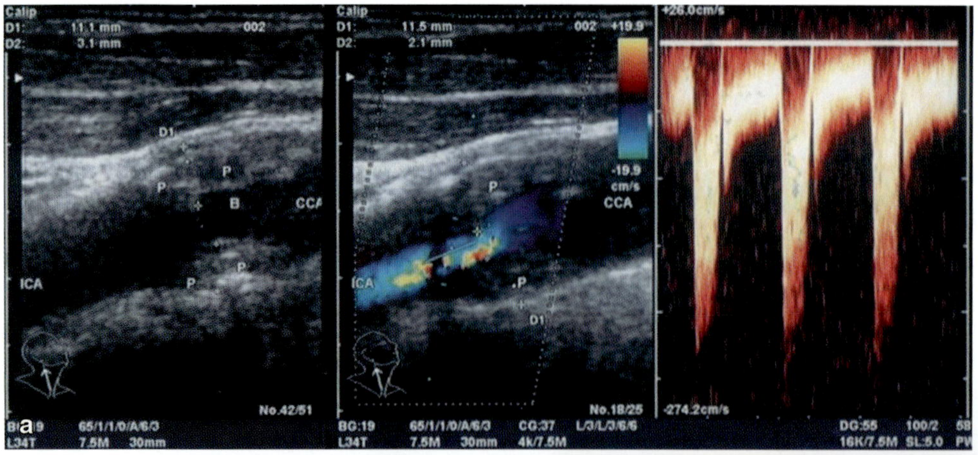


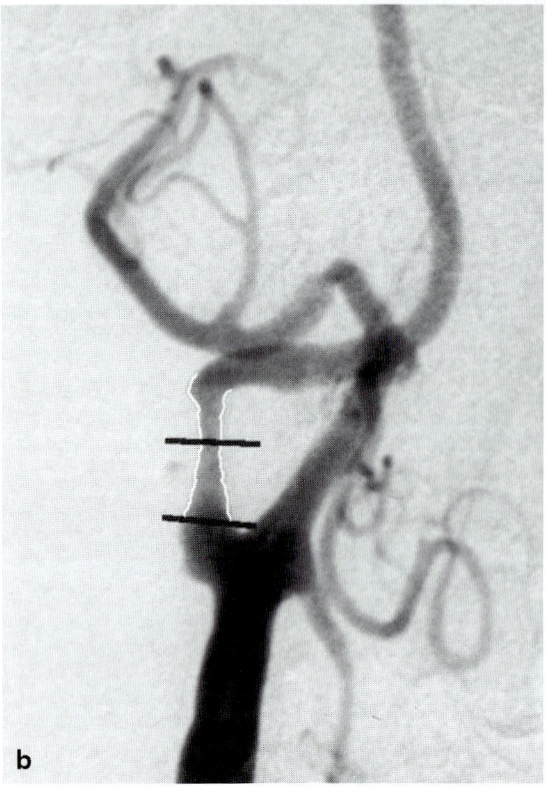


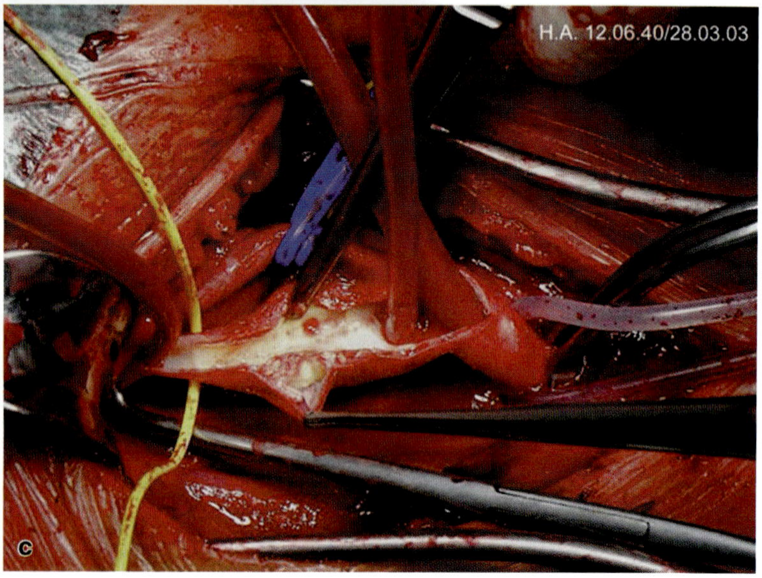


**Fig. A5.11 a–c**

**Plaque morphology – long concentric carotid stenosis (smooth, regular surface)**

**a** Gray-scale scan depicting a concentric, fairly homogeneous and smoothly marginated plaque in the center with a just barely visible, extremely echolucent extension in the cranial direction. Only the color duplex scan enables differentiation of the echolucent wall deposit and perfused lumen in the distal portion of the plaque. Spectral Doppler yields a PSV of 230 cm/s.

**b** Angiogram demonstrating a long concentric, smooth stenosis.

**c** Intraoperative photograph showing mostly fibrous plaque with a smooth surface.

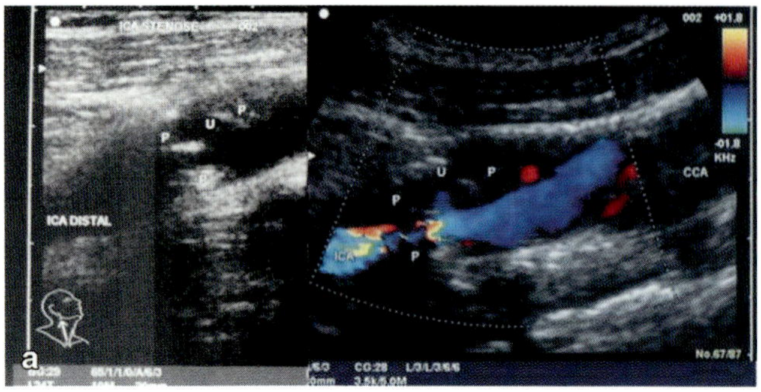


**Fig. A5.12 a–e**

**Plaque morphology – high-grade stenosis with ulceration**

**a** B-mode scan (*left section*) depicts an inhomogeneous plaque with a poorly demarcated surface and hyperechoic spots extending far into the lumen. The anechoic gap may indicate ulceration. The color duplex scan (*right section*) shows an eccentric, long, and fairly echolucent plaque beginning far proximally in the bulb and causing high-grade stenosis that extends to the distal end of the bulb between a plaque close to the transducer and one away from it (*P*). The anechoic bowl-shaped defect in the echolucent plaque in the B-mode scan with flow in the color mode (*U*) suggests ulceration.

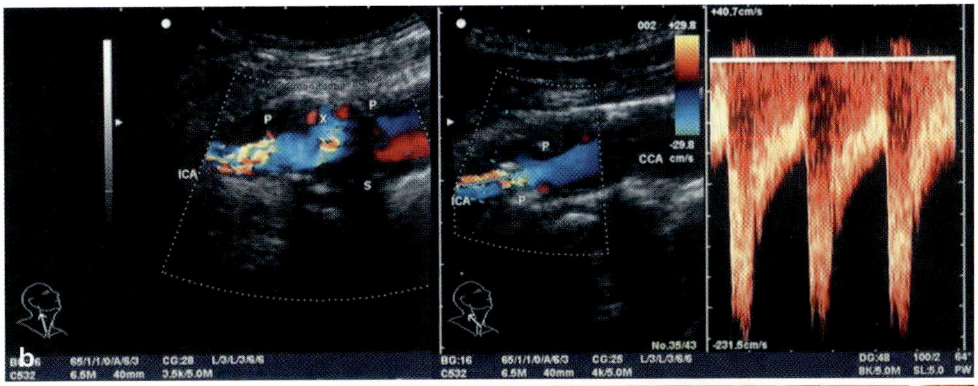


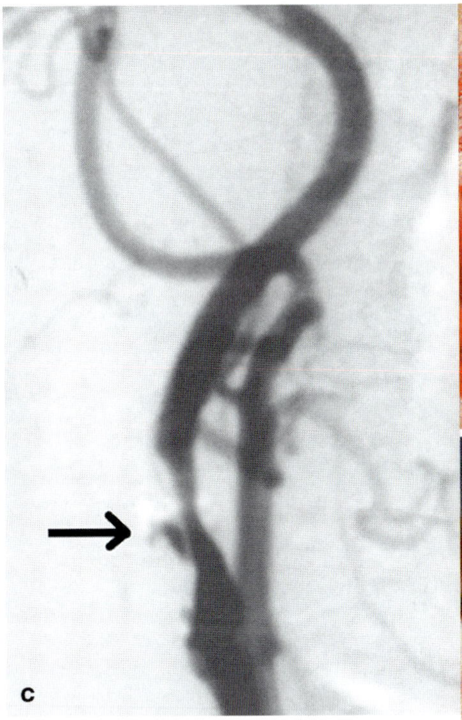


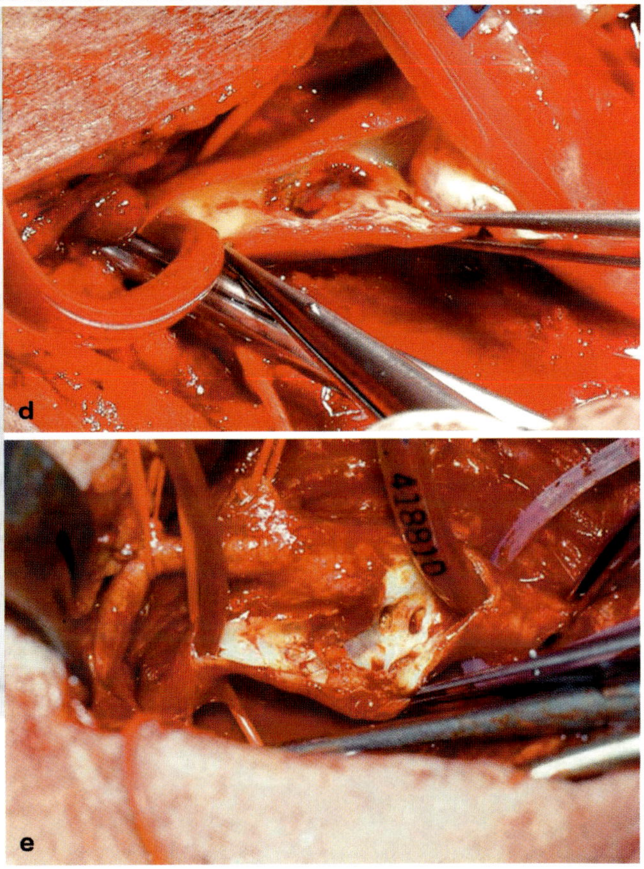


**Fig. A5.12 (continued)**

**b** The concentric plaque causing the high-grade stenosis begins directly distal to the ulceration (see *U* in **a**). Ulceration often occurs in the proximal portion of a highly stenotic plaque protruding far into the lumen. The arriving pulse wave (often depicted as longitudinal pulsatile plaque movement by gray-scale scanning) may cause rupture of the vulnerable plaque cap. In the example, the PSV in the stenotic jet (aliasing) is 220 cm/s.

**c** Angiography with a filling defect confirming the plaque contour demonstrated by ultrasound and also the ulceration.

**d** Intraoperatively, the atheromatous plaque and adjacent ulceration are confirmed at the sites already identified by ultrasonography and angiography (corresponding to **a**–**c**).

**e** Different case with the intraoperative finding of a washed-out niche, which is difficult to differentiate from ulceration by ultrasound and angiography.

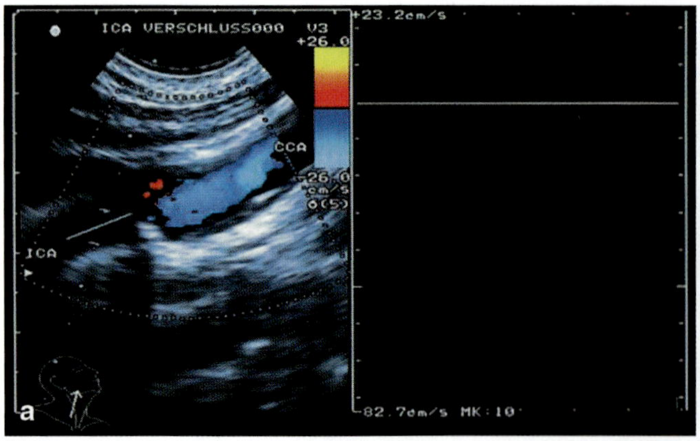


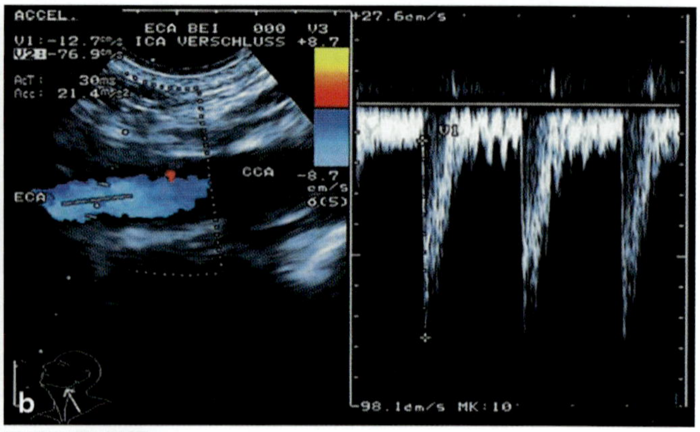


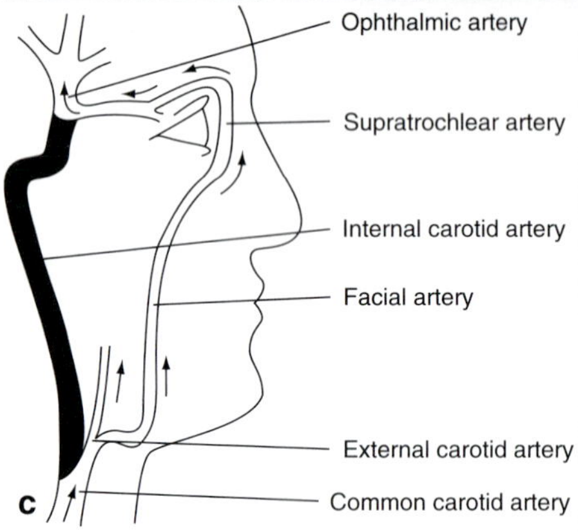


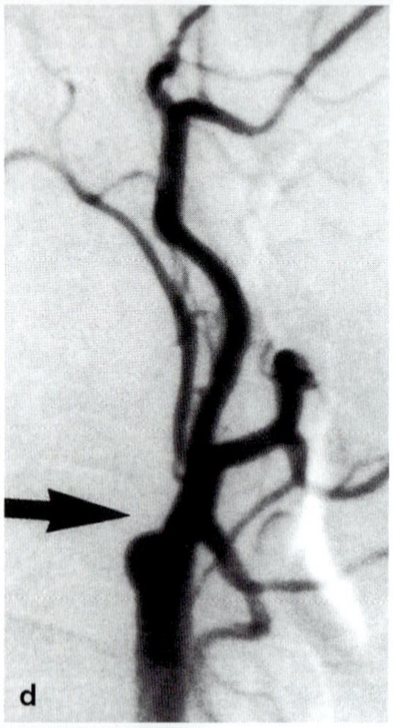


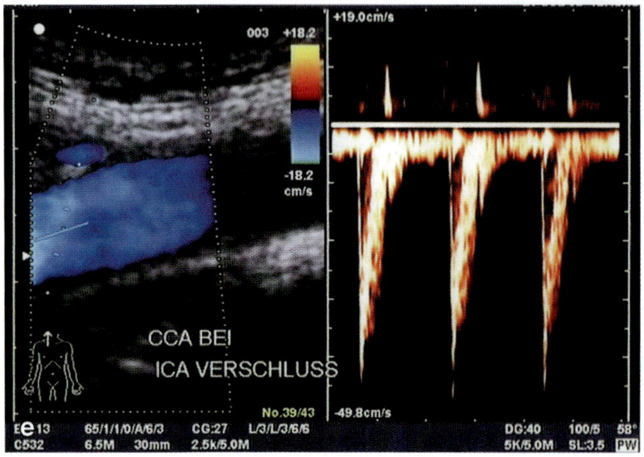


**Fig. A5.13 a–e**
**ICA occlusion**
**a** Neither spectral Doppler nor color duplex depicts flow signals in occlusion of the ICA. Calcified plaque with acoustic shadowing at origin. The *left section* shows the patent CCA on the *right*. To differentiate occlusion from subtotal occlusion, the ICA must be scanned for depiction of flow signals with a high gain to the level of the mandibular angle.
**b** When the ECA provides collateral flow via the supratrochlear artery, its spectral waveform shows a larger diastolic flow component. To avoid confusion with the ICA in this setting, the identity of the ECA should be confirmed by rhythmical tapping of the temporal artery (branch of ECA) and transmission of the pulsation to the Doppler spectrum (here seen in the middle cardiac cycle represented in the waveform).
**c** Diagram of collateral flow through the ECA and supratrochlear artery (with retrograde flow in CW Doppler examination) in ICA occlusion.
**d** Angiogram: ICA occlusion (*arrow*).
**e** Flow in the CCA becomes more pulsatile when the ICA is occluded with a Doppler waveform resembling that from the ECA, the only artery supplied by the CCA in this setting.

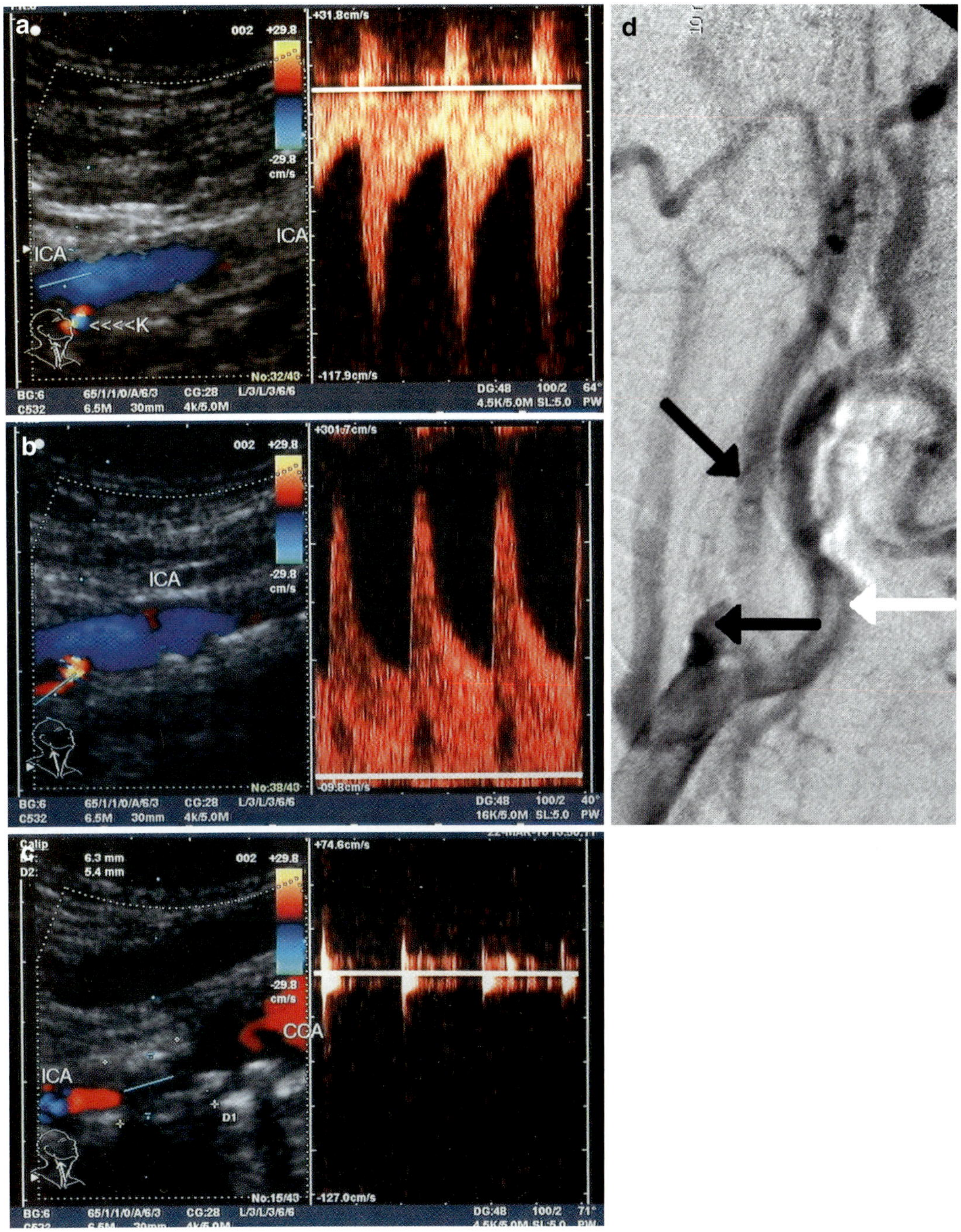


**Fig. A5.14 a–e**

**Proximal ICA occlusion with resupply of the distal extracranial segment through persistent primitive communications with the vertebrobasilar system and the ECA**

**a** Absence of flow signals in the proximal ICA with flow in the distal extracranial segment (color duplex and spectral Doppler) may suggest technical limitations or acoustic shadowing due to calcified plaque. This constellation has been referred to as pseudo-occlusion (residual blood flow that is difficult to detect in high-grade stenosis), but is rarely encountered when state-of-the-art ultrasound equipment is used. Another, though very rare, explanation for this observation is illustrated in the example shown: the proximal ICA is occluded, and the distal extracranial arterial segment is resupplied through persistent embryonic communications between the ICA and the vertebrobasilar system (proatlantic artery, primitive hypoglossal artery, primitive acoustic artery, primitive trigeminal artery) or the ECA (ascending pharyngeal artery). The blood flow in the resupplied distal ICA depends on the original diameter of these primitive arteries, and their widening when they assume collateral function in chronic ICA occlusion. This results in a variable blood flow velocity in the distal ICA. However, the PSV of 100 cm/s measured in the example presented here, with the sample volume placed in the distal ICA, is unusually high.

**b** Sample volume placed in a dilated persistent arterial connection between the ICA and vertebrobasilar system, which functions as a collateral and shows high flow (encoded in red, toward transducer).

**c** Color duplex imaging with a low pulse repetition frequency shows the length of the occlusion near the origin of the ICA (2 cm in length); this is confirmed by spectral Doppler measurement (signal from wall motion rather than flowing blood).

**d** Angiogram confirms occlusion of the ICA near its origin and resupply of the postocclusive segment through persistent embryonic arterial communications.The angiogram does not allow one to identify which of the primitive arteries provides collateral flow here.

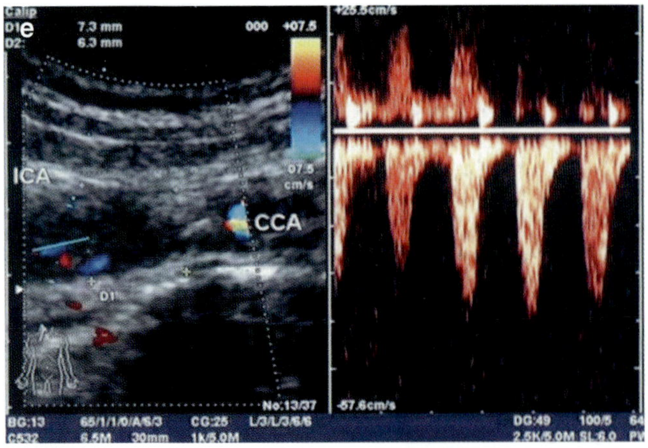


**Fig. A5.14 (continued)**

**Signs of recanalization in ICA occlusion**

**e** In the setting of ICA occlusion, the examiner must search for flow signals using a low pulse repetition frequency. Recanalization is uncommon and must be differentiated from pseudo-occlusion. In subtotal occlusion (which may be identified by isolated high-frequency flow signals when scanning with a high gain), there will be very slow flow in the poststenotic segment filling most of the lumen. In occlusion with recanalization (as shown in this example), flow signals indicating a thin, meandering current are depicted throughout the otherwise occluded extracranial ICA. In contrast to stenotic narrowing, recanalization is characterized by low flow velocities (30 cm/s in the example with atypical ICA flow signal due to changed resistance). Partial corkscrew-like flow is depicted in the color scan as an apparent change in flow direction (*blue–red*). This is reflected in the spectral waveform by a change in flow direction when the transducer is slightly angled or moved (flow partly toward the transducer in the left part of the spectrum, away from it on the right). The meander-like course of blood flow in the recanalized ICA precludes depiction of the flow signals in the bulb in a single scan plane.

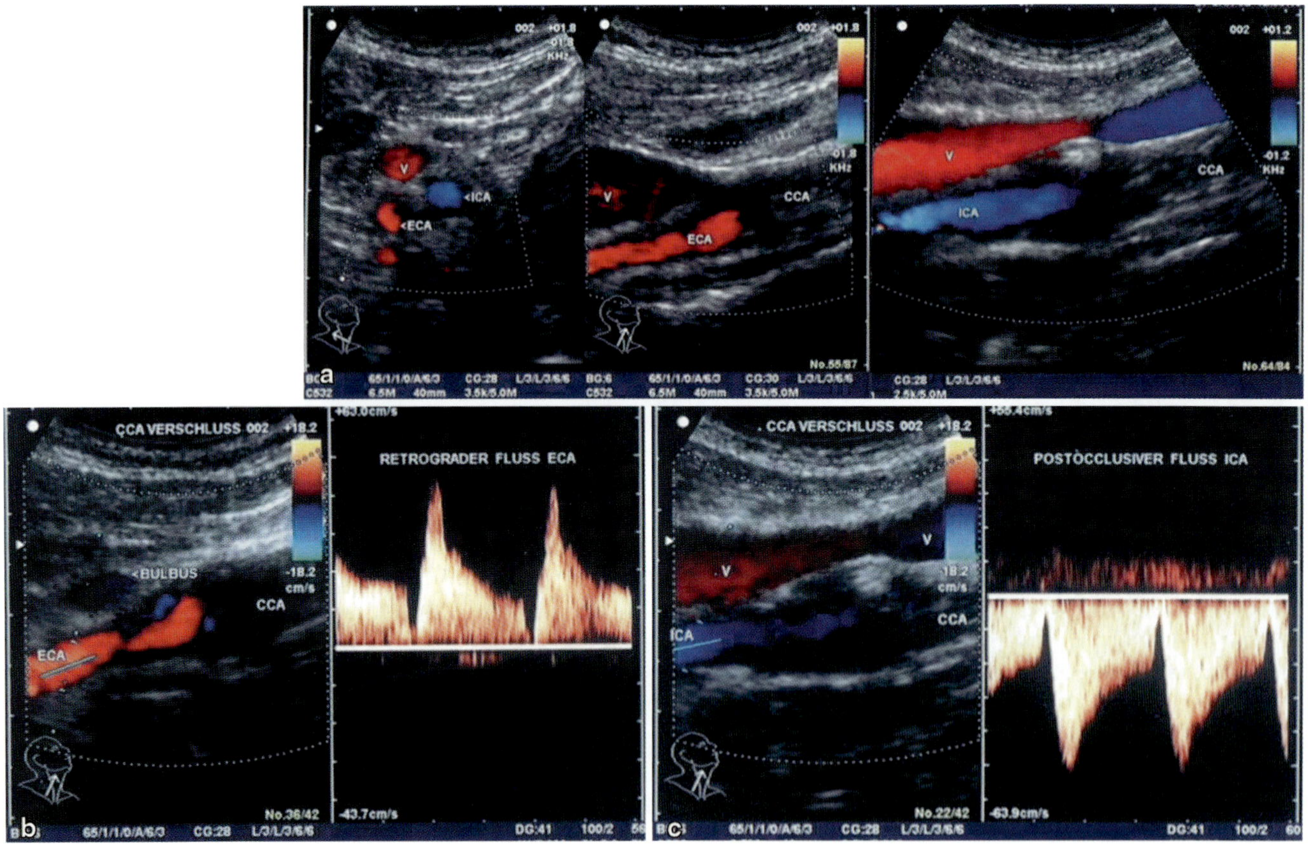


**Fig. A5.15 a–c**

**CCA occlusion/collaterals**

**a** In the setting of CCA occlusion with a patent bifurcation, the ICA will be refilled through branches of the ECA, primarily the superior thyroid artery, which in turn is supplied through branches of the thyrocervical trunk.

**b** Distal ECA branches may likewise contribute to the supply of the ICA. In this case there is retrograde flow along long segments of the ECA (displayed in *red*, toward the heart, same flow direction as in accompanying internal jugular vein).

**c** ICA with forward flow (displayed in *blue*, away from transducer). The postocclusive waveforms of the ECA and ICA show a delayed systolic increase and damped arterial flow profile.

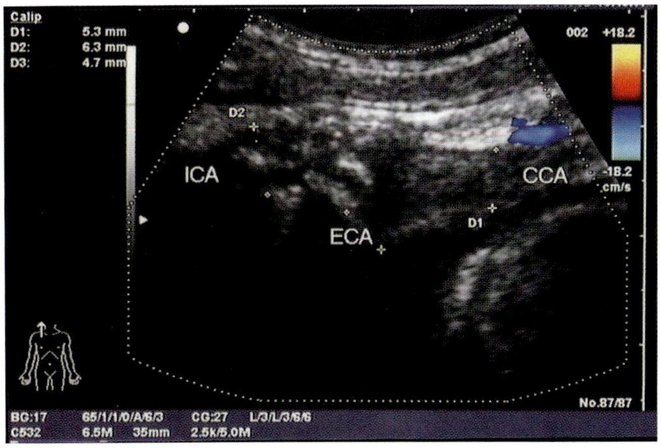


**Fig. A5.16**
**Complete extracranial carotid occlusion**
When the CCA is occluded, the examiner must visualize the bifurcation with the ICA and ECA to identify any flow signals. Evaluation may be difficult when extensive plaques with acoustic shadowing are present. The occluded segment appears very heterogeneous and contains areas of higher echogenicity, making it difficult to delineate the arterial lumen from the surrounding connective tissue. The arteries are indicated by *plus signs* (CCA: D1; ICA: D2; ECA: D3). Only venous flow (*blue*) is depicted in the scan.

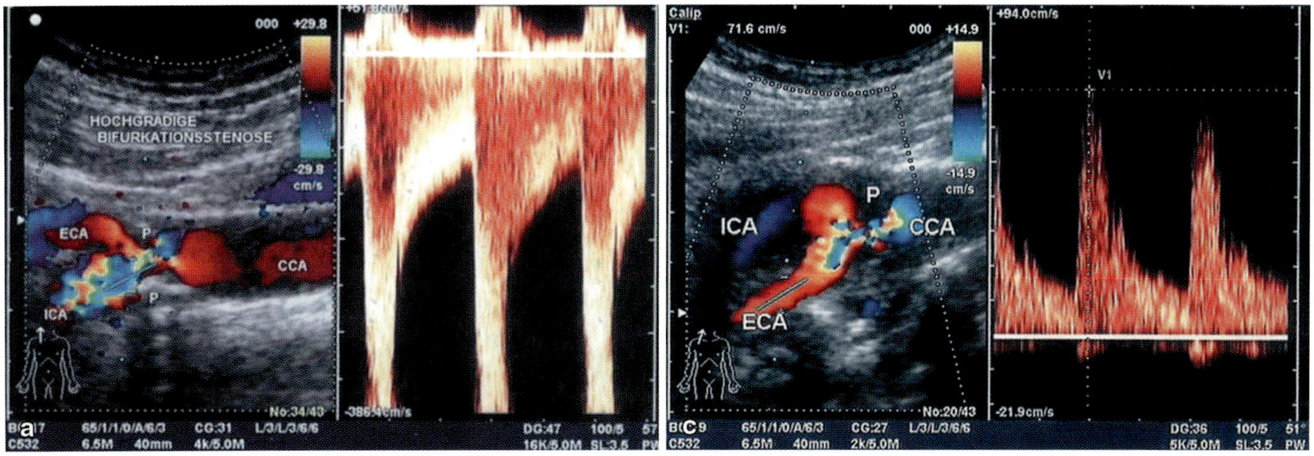


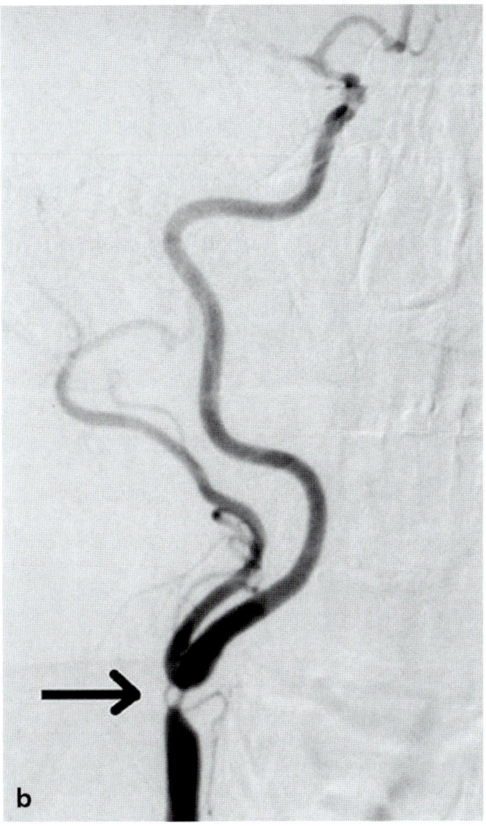


**Fig. A5.17 a–c**
**CCA stenosis**
**a** Preferred sites of CCA stenoses are the origin proximally and the area of the bifurcation distally. Concentric plaques (*P*) cause a high grade-stenosis just before the CCA divides into the ICA and ECA. The stenosis is indicated by aliasing and a PSV of over 4 m/s as determined by spectral Doppler.
**b** Angiogram of the high-grade stenosis of the distal CCA just before the bifurcation (same patient as in **a**).
**c** With increasing stenosis of the distal CCA, connecting vessels entering the ECA, e.g., via the superior thyroid artery, serve as collaterals. In the example, a high-grade stenosis of the CCA (*P*) is indicated by aliasing. There is retrograde flow in the *ECA* (displayed in red, toward transducer) with refilling of the *ICA*. The Doppler waveform from the ECA depicts backward flow to the heart (toward transducer). The large diastolic component reflects the fact that the ECA supplies the brain indirectly via the ICA in this setting. (Posterior transducer position as opposed to anterior position in **a**).

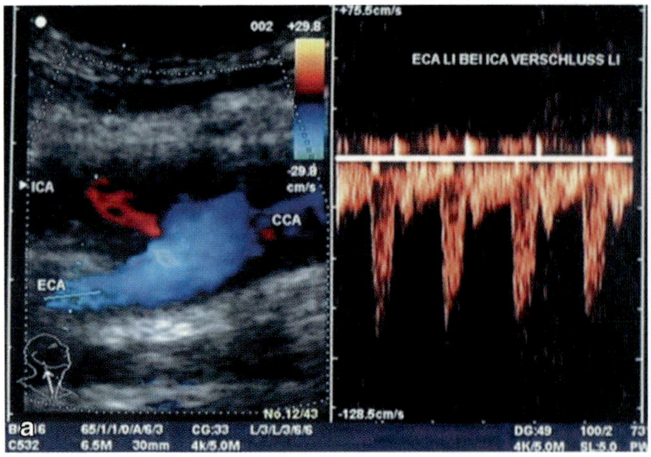


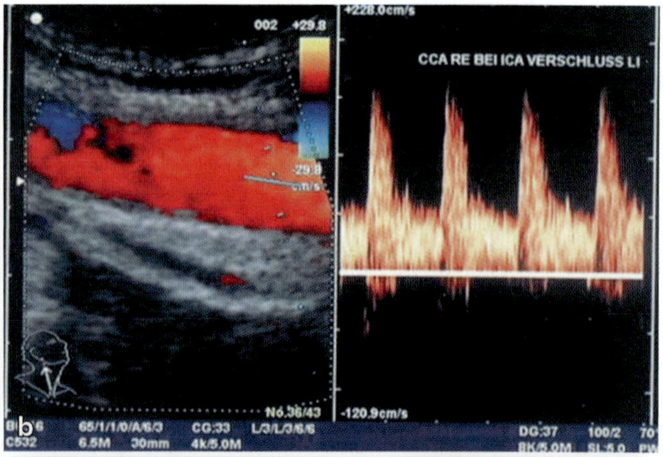


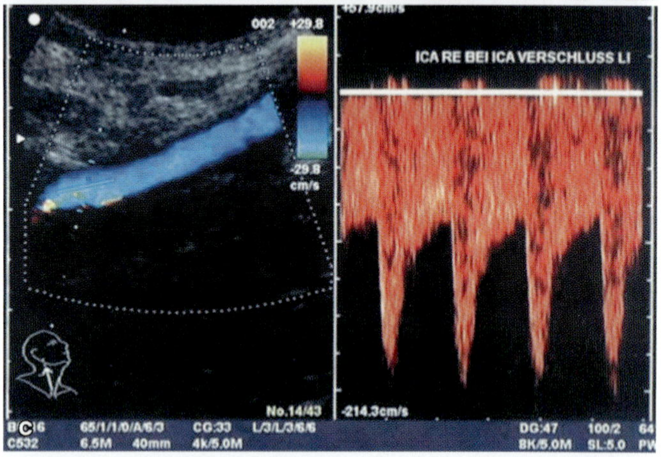


**Fig. A5.18 a–c**
**ICA occlusion - compensatory flow increase in collateral pathways**
The absent flow in ICA occlusion is compensated for by higher flow rates in the collateral arteries. The resulting higher flow velocities must not be misinterpreted as indicating stenosis. Faster flow is detected over long stretches of the collaterals, while no stenotic structures are identified.
**a** The ipsilateral ECA can become a collateral, assuming a flow profile resembling that of the ICA (to-and-fro flow – knocking waveform in the bulb).
**b** Occasionally, there may be an increased compensatory flow in the contralateral CCA as well (flow velocity increased to 150 cm/s over long stretch).
**c** PSV of 200 cm/s along long segment of contralateral ICA. The flow velocity is influenced by collateral flow in other vessels, and the increase is rarely as pronounced as in this example.

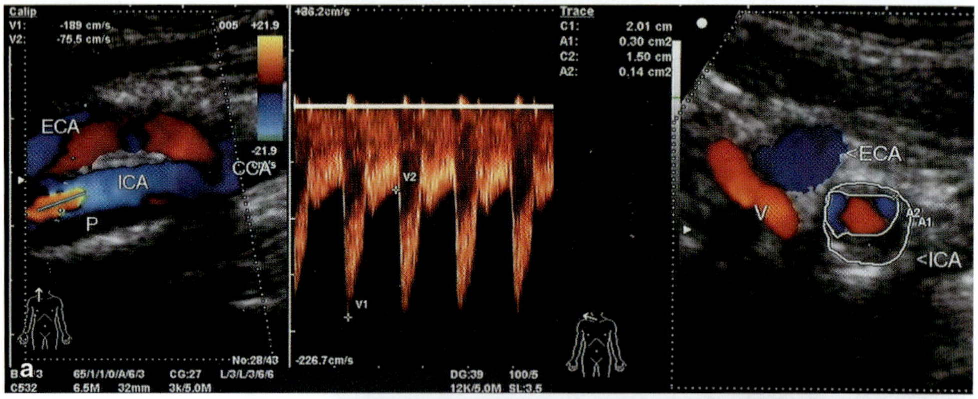


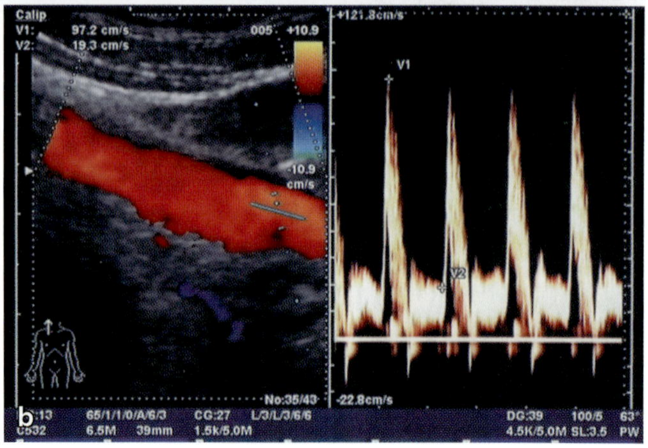


**Fig. A5.19 a, b**
**Correct interpretation of PSV measurement and stenosis grading in contralateral ICA occlusion**
**a** Long echolucent plaque (*P*; longitudinal image on the *left*) of the ICA, causing <50% luminal narrowing, while the PSV of 189 cm/s suggests 60–70% stenosis (according to ECST criteria). The maximum diameter reduction determined in the transverse plane (*right* image) of the ICA is just below 50% (beware of inherent limitations using this method), corresponding to a 50% cross-sectional area reduction (as the plaque is predominantly eccentric), which is not hemodynamically relevant. Cross-sectional area of patent lumen: 0.14 $cm^2$; total cross-sectional area of ICA: 0.3 $cm^2$.
**b** To determine whether the increased PSV in the ICA is due to increased flow, secondary to contralateral ICA occlusion, flow velocity in the CCA is measured. This reveals that the PSV has increased to 97.2 cm/s, a sign that there is collateral flow compensating for an occluded contralateral ICA. Consequently the PSV in the ICA cannot be used for calculating the stenosis grade in this artery without making allowance for this additional collateral flow.

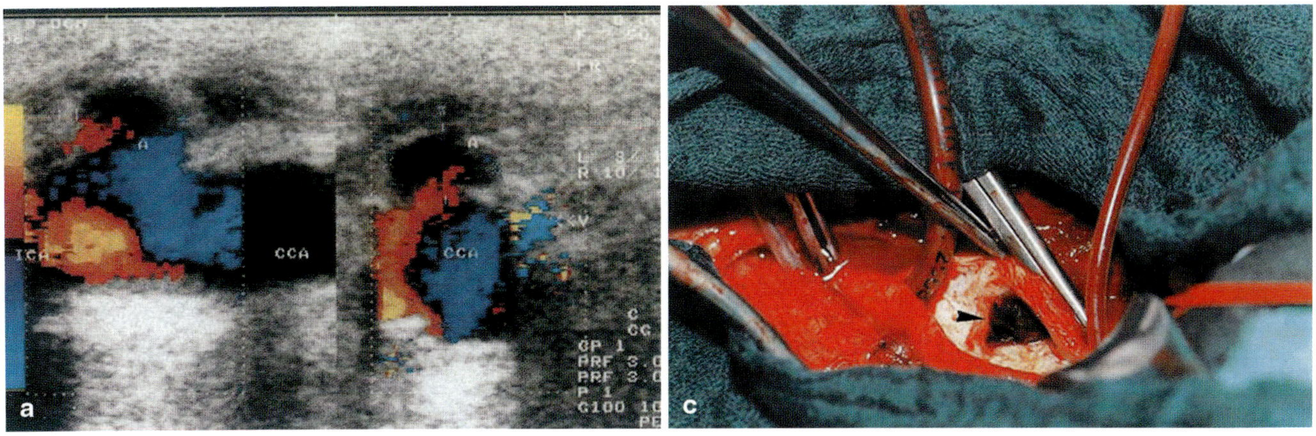


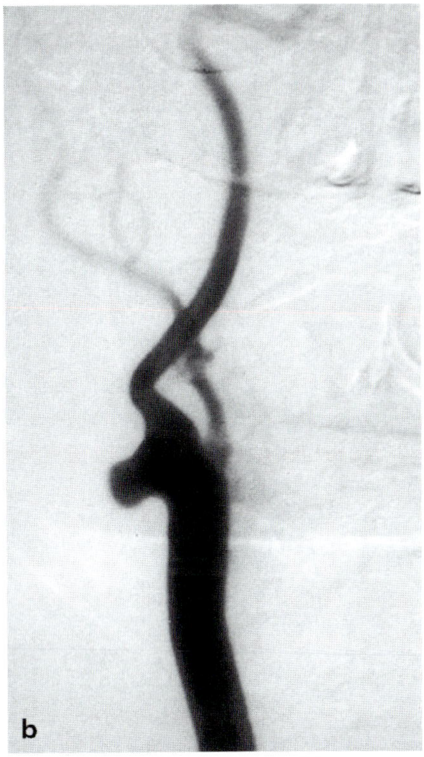


**Fig. A5.20 a–c**
**Suture aneurysm**
**a** Pulsatile mass of the neck 3 years after carotid TEA. Color-coded duplex sonography demonstrates a circumscribed outpouching with flow signals in the patch area at the origin of the ICA. Incomplete color filling of the pouch suggests partial thrombosis. No demonstration of stenosis.
**b** Control angiography confirms saccular aneurysm of the bulb.
**c** Intraoperatively, a suture aneurysm covered by connective tissue structures is seen with thrombotic deposits in the aneurysmal sac (*arrowhead*).

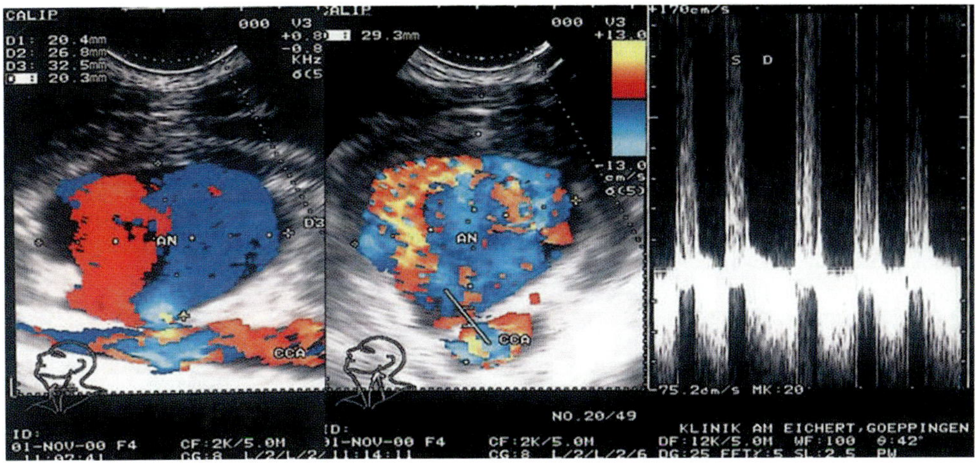


**Fig. A5.21**
**Complications after carotid TEA**
Virtually all false aneurysms of the carotid system are due to trauma or occur in the form of suture aneurysms after carotid TEA, particularly in patients who have undergone synthetic patch angioplasty. Suture aneurysms often indicate infection of the patch. Color duplex scanning shows a conspicuous mushroom-like structure protruding from the vessel, which can be palpated as a pulsating tumor in most cases. The color coding varies with the presence of thrombosis. The spectral waveform of the aneurysmal neck shows the typical to-and-fro sign indicating high-frequency systolic flow into the aneurysm and flow into the carotid lumen throughout diastole (*left section*: longitudinal scan; *middle section*: vessel cross section; *right section*: Doppler waveform).

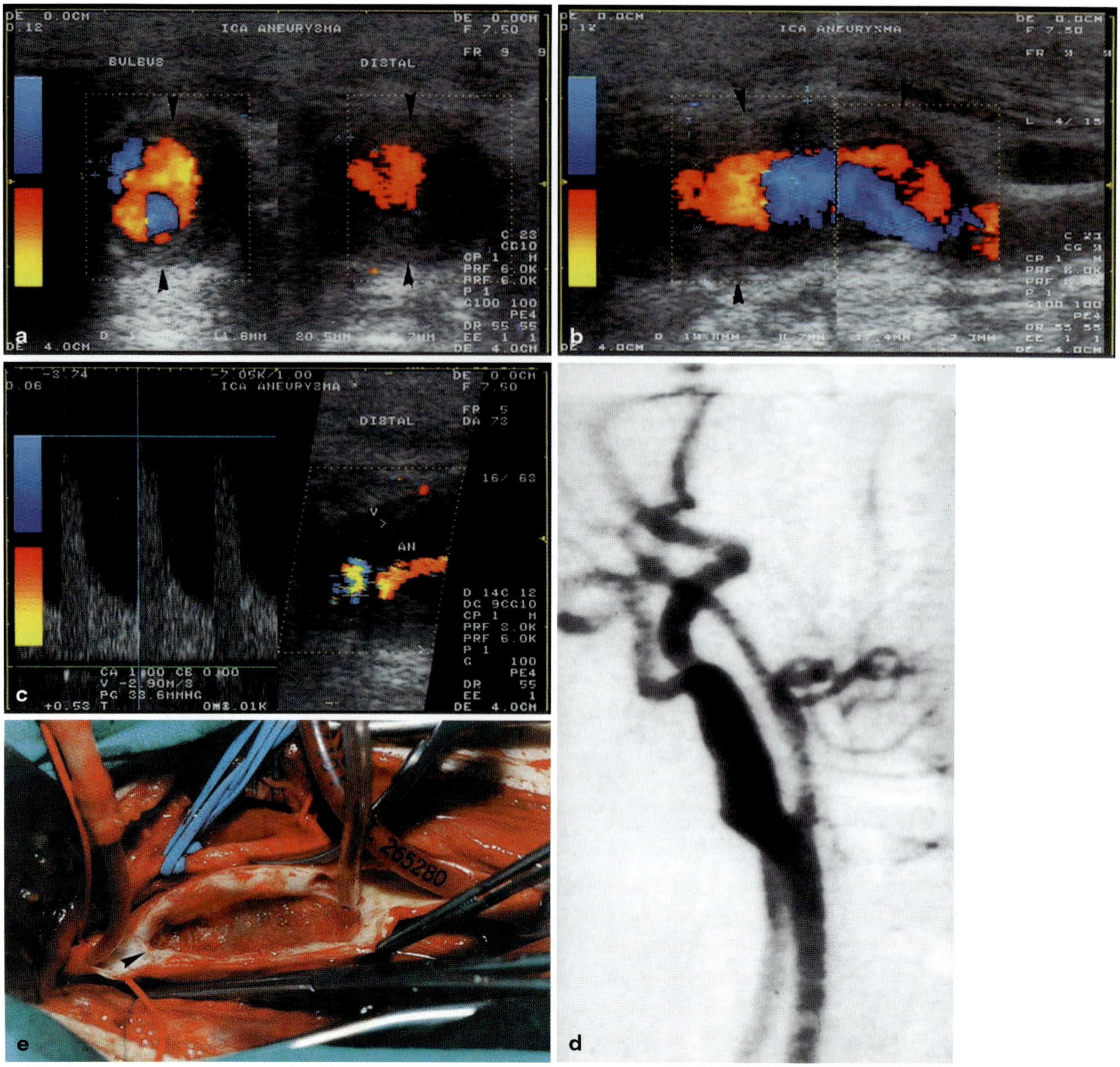


**Fig. A5.22 a–f**

**ICA aneurysm (true)**

**a** Transverse section showing the patent lumen of the ICA surrounded by hypoechoic aneurysmal thrombotic deposits (*arrowheads*) both in the bulb area (*left*) and in the distal segment (*right*). The aneurysm has a diameter of 2 cm.

**b** The longitudinal scan depicts the shape of the ICA aneurysm (CCA on the *right* and ICA on the *left*). The longitudinal scan also clearly demonstrates the size of the aneurysm (*arrowheads*) with hypoechoic mural thrombosis in relation to the color-coded flow in the patent lumen. The color change indicates eddy currents.

**c** At the distal end of the aneurysm there is flow acceleration with turbulence and aliasing (inverted display). Flow velocity is increased to 2.9 m/s as determined by spectral Doppler (inverted display depicting flow away from transducer above the baseline).

**d** Angiogram: Ectasia of the ICA. The mural thrombi make the aneurysm appear smaller than it actually is, and angiography does not provide information on the hemodynamic significance of the stenosis at the distal end of the aneurysm; all that is seen is less pronounced opacification due to luminal narrowing in the anteroposterior projection.

**e** Intraoperative site confirming the spindle-shaped aneurysm in the proximal ICA with mural thrombosis and fibrotic stenosis at its distal end (*arrowhead*). CCA with shunt on the *right* and distal ICA on the *left* with the aneurysm of the proximal ICA in between. The aneurysm is thrombosed and shows fibrous luminal narrowing at its distal end. ECA encircled with blue vessel loop.

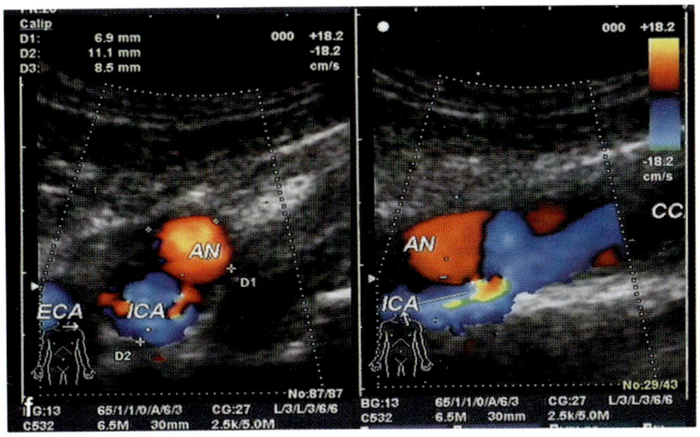


**Fig. A5.22 (continued)**
**Mycotic aneurysm of ICA**
**f** Transverse (*left*) and longitudinal (*right*) images of a saccular aneurysmal dilatation (*AN*) of the proximal ICA. The transverse view nicely depicts the aneurysm with flow toward the transducer (*red*). Flow direction in the ECA is normal (*blue*, toward the head).

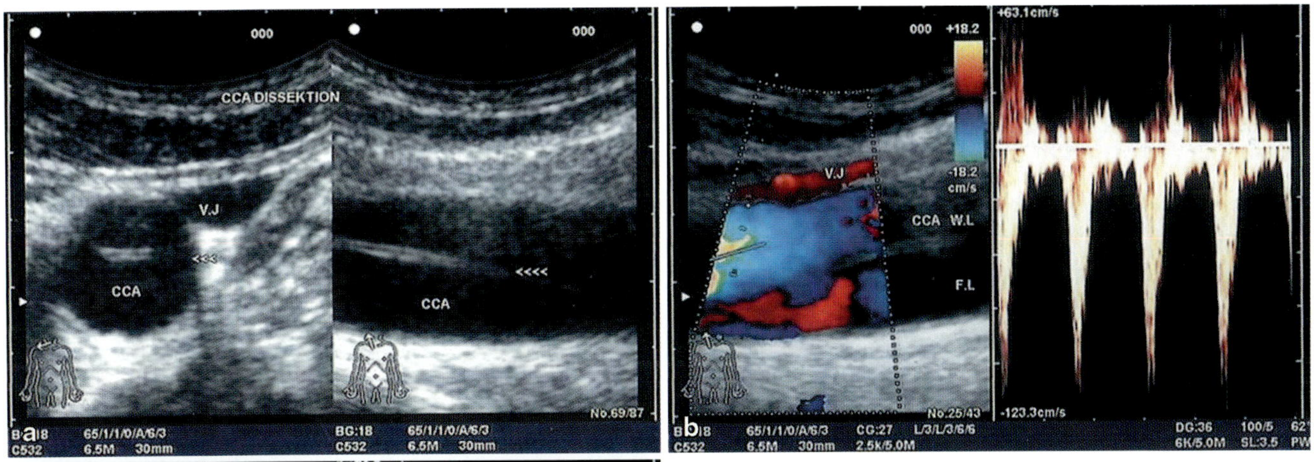


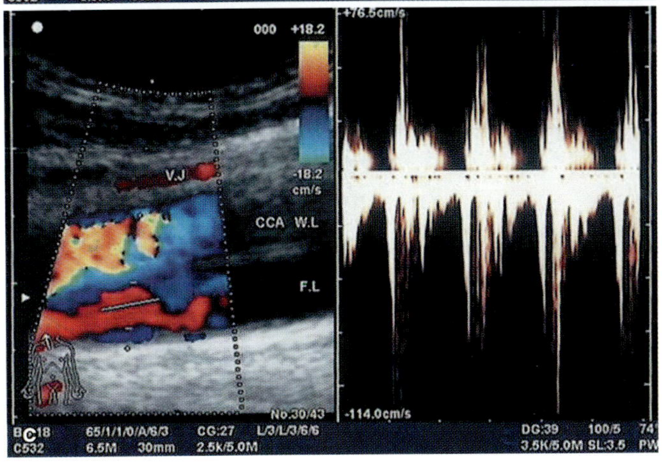


**Fig. A5.23 a–c**
**Dissection of CCA**
In type I aortic dissection according to De Bakey, the dissection extends into the CCA.
**a** The B-mode scan depicts the dissection in longitudinal and transverse planes as an intraluminal flap oscillating during the cardiac cycle.
**b** In the color mode, the true lumen of the CCA (*CCA W.L*) can be differentiated from the false lumen (*F.L*) as it shows flow in the cranial direction. (*V.J*, jugular vein).
**c** The flow direction demonstrated in the false lumen varies with the placement of the sample volume relative to the re-entry site. Forward flow is demonstrated if the re-entry is distal to the volume and to-and-fro flow, as in the example, if it is proximal.
The scan shows flow in the true lumen displayed in blue (toward the periphery, away from transducer) and aliasing, while there is to-and-fro flow in the false lumen (here displayed in red, toward transducer).

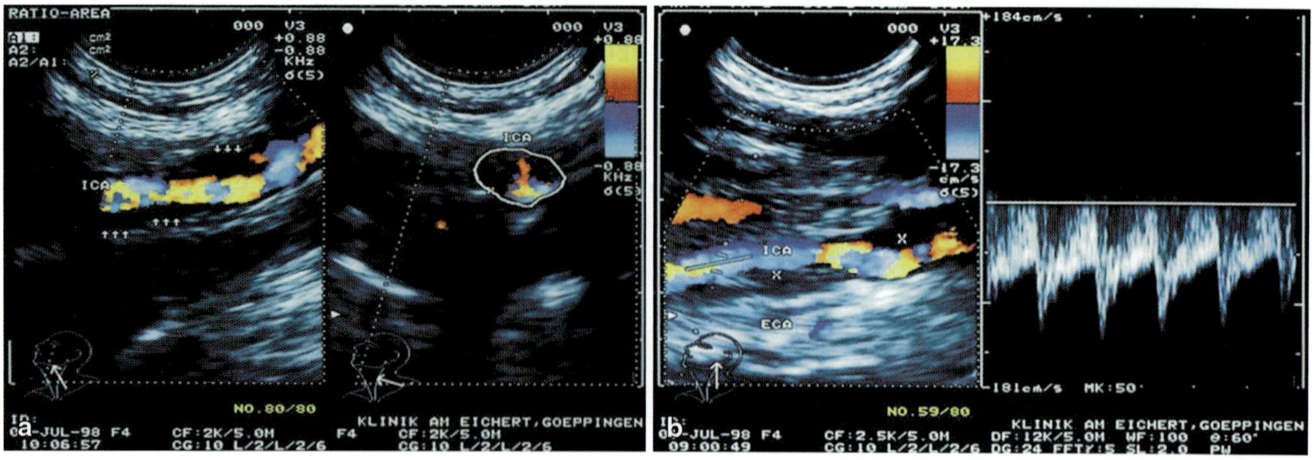


**Fig. A5.24 a, b**
**Posttraumatic dissection of ICA**
**a** Long posttraumatic dissection of the ICA extending from the bifurcation to the base of the skull with thrombosis of the false lumen. In contrast to atherosclerotic changes, thrombotic dissection is characterized by a homogeneous and hypoechoic sonomorphologic appearance. The thrombotic false lumen is long and tortuous and partly attaches to the vessel wall. Occasionally, as in the example, the entry site can be identified by the depiction of pulsatile flow signals in the color duplex mode (*right section*).
**b** As shown by spectral Doppler, the luminal narrowing due to dissection does not cause a higher-grade stenosis.

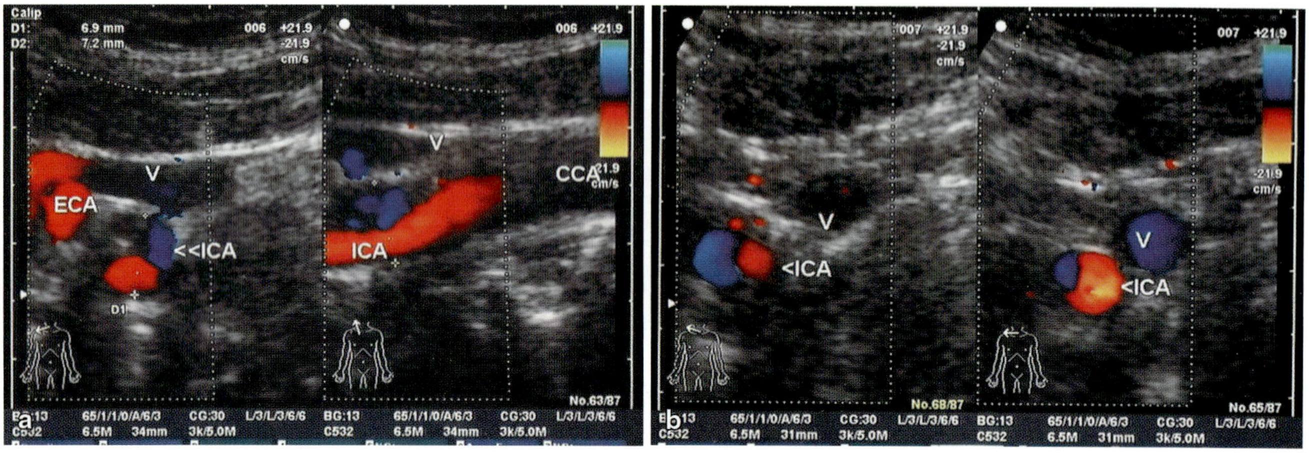


**Fig. A5.25 a, b**
**Posttraumatic dissection of ICA with patent true and false lumina**
**a** The dissection starts in the bulb of the ICA. The longitudinal scan on the *right* (*red*-coded true lumen, *blue*-coded false lumen) depicts the dissection with a less than 50% reduction in diameter. The transverse scan on the *left* shows the dissection membrane and partial thrombosis of the false lumen. *V* internal jugular vein, *ECA* external carotid artery.
**b** The dissection extends to the intracranial segment of the *ICA*. The variable diameter of the false lumen (*blue*) is seen in the middle portion of the extracranial ICA (*right section*) and its distal portion (*left section* with visible dissection membrane).

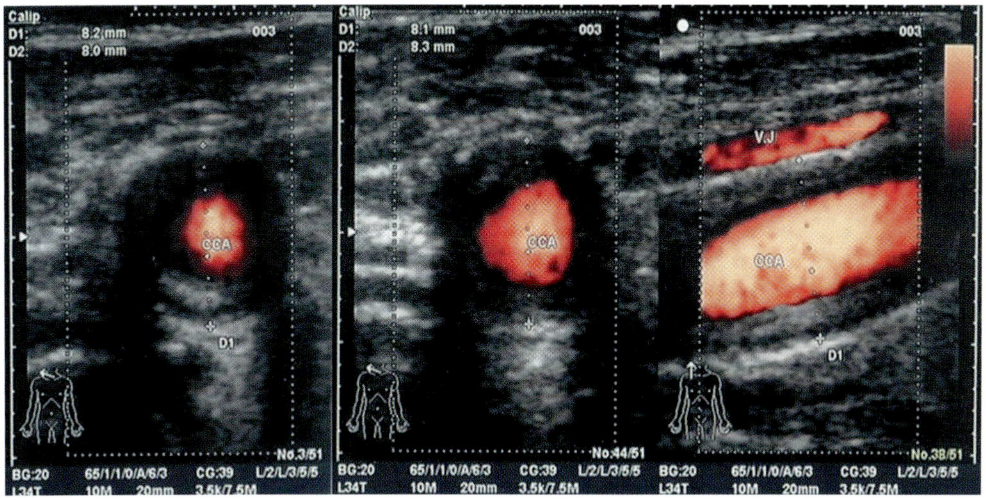


**Fig. A5.26**
**Takayasu's arteritis**
Concentric wall thickening is pathognomonic of arteritis. Takayasu's arteritis predominantly affects the subclavian artery and CCA. The most example shows the CCA in the power Doppler mode. The transverse scan (*leftmost section*) demonstrates concentric diameter reduction greater than 50% involving a long vessel segment. The transverse scan in the *middle* and the longitudinal scan on the *right* obtained after 2 weeks of cortisone treatment show slightly reduced but persistent concentric wall thickening of the CCA.

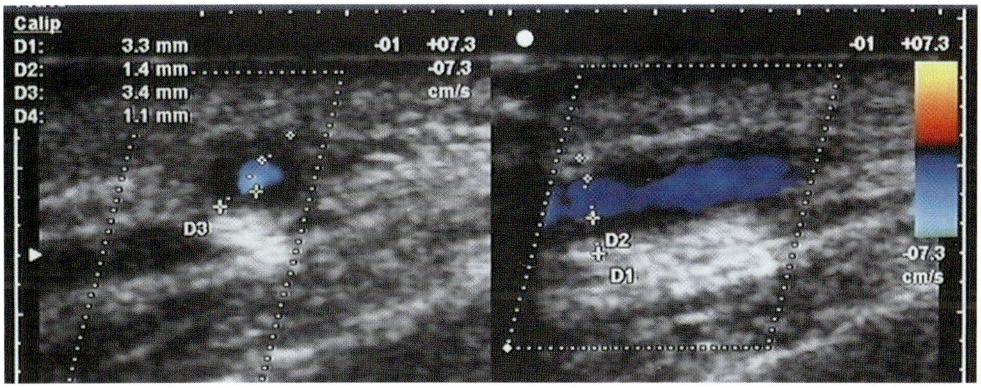


**Fig. A5.27**
**Temporal arteritis**
Transverse (*left*) and longitudinal (*right*) images of the temporal artery showing luminal narrowing (size reduction from 3 to 1 mm; normal lumen size indicated by *plus signs*) due to inflammatory concentric wall thickening.

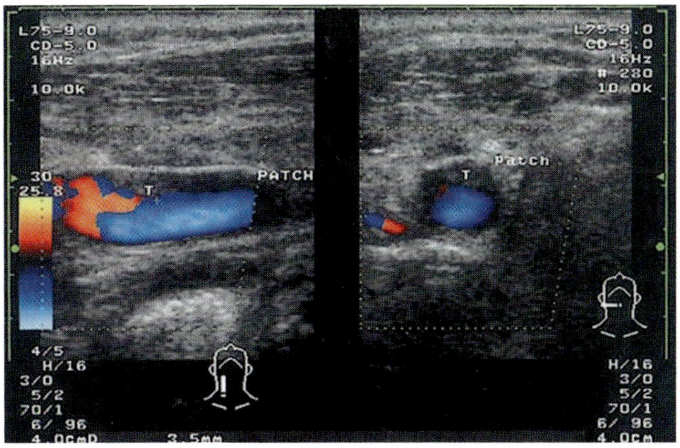


**Fig. A5.28**
**Postoperative follow-up**
Postoperatively, there may be luminal narrowing due to thrombotic deposits, in particular when a synthetic patch has been interposed. Thrombotic deposits protruding far into the lumen with hemodynamically significant narrowing are a source of embolism. The patch itself is seen as a hyperechoic interface (wall near transducer) with a hypoechoic deposit on the luminal side (*T*).

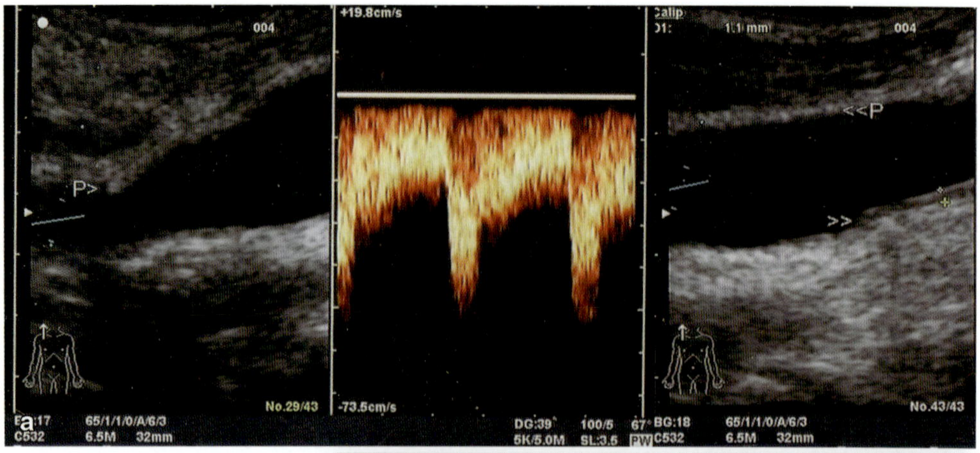


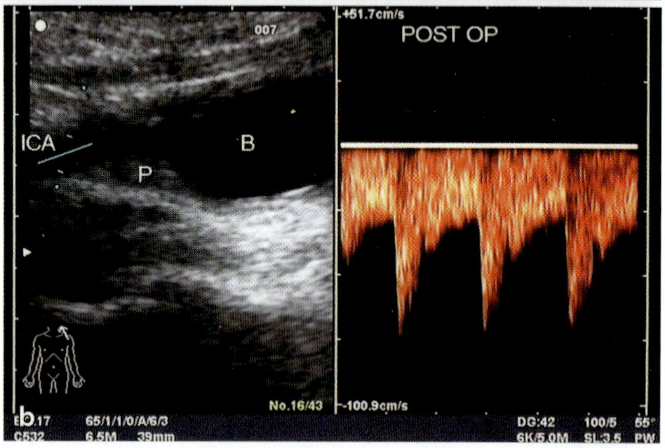


**Fig. A5.29 a, b**

**a** Early postoperative sonomorphologic appearance of the vessel wall after TEA with patch angioplasty (Dacron patch). The image on the *left* depicts the transition from the patch (*P*) to the native ICA with the sample volume for Doppler measurement. The corresponding Doppler waveform indicates normal flow velocities. The image on the *right* shows the proximal end of TEA and the patch (*P*) with a step in the far wall at the transition (indicated by *double arrowheads*). The caliber mismatch is unproblematic, causing no flow obstruction because the larger diameter is downstream. Intima–media thickness (IMT) in the CCA is increased to 1.1 mm (*plus sign*).
**b** At 6-month follow-up after TEA with patch closure, there is evidence of early neointimal formation at the transition from the patched segment to the distal ICA. There is good evaluation of this segment (*P*) using gray-scale imaging, which shows no hemodynamiclly relevant luminal narrowing (PSV of 90 cm/s).

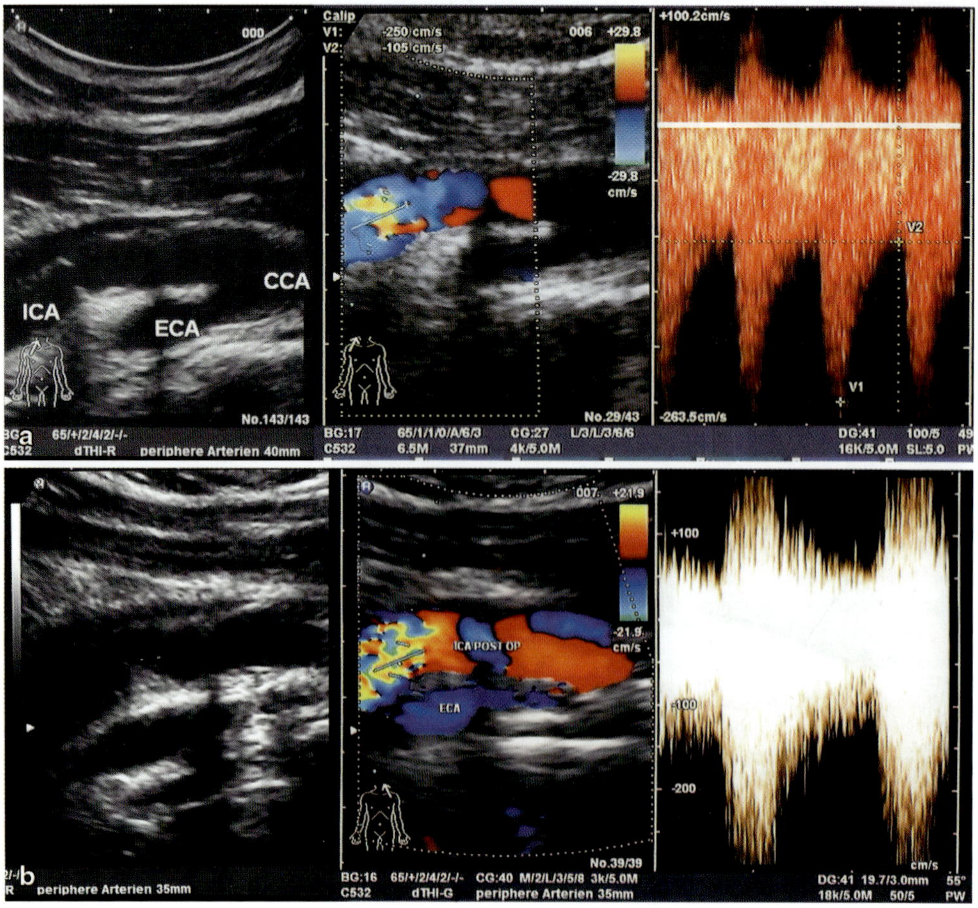


**Fig. A5.30 a–c**
**Recurrent stenosis**

**a** B-mode image after carotid TEA (eversion) shows an intimal flap within the lumen (to the right of the *ICA* label) with aliasing in the color duplex mode (color spillover obscuring the flap). The Doppler waveform (with a PSV of 250 cm/s) suggests >70% stenosis.
**b** Within a few weeks, a high-grade stenosis developed at the site of the intimal flap due to thrombotic deposits and neointima (PSV >300 cm/s with very turbulent flow in the stenotic segment).

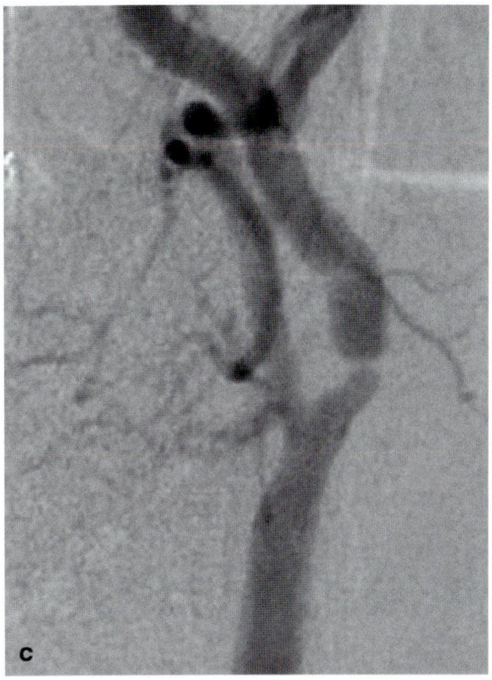


**Fig. A5.30 (continued)**

**c** Angiogram confirming high-grade stenosis after carotid TEA with subsequent intimal flap and thrombotic deposits.

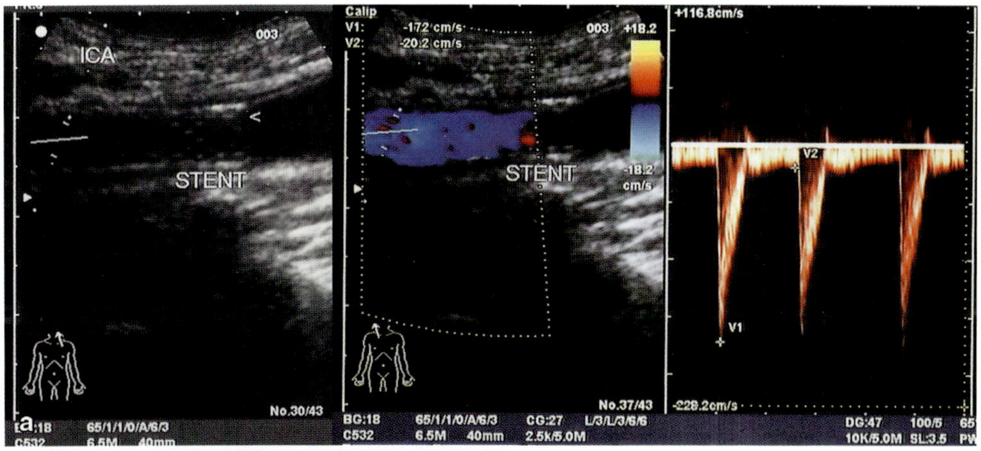


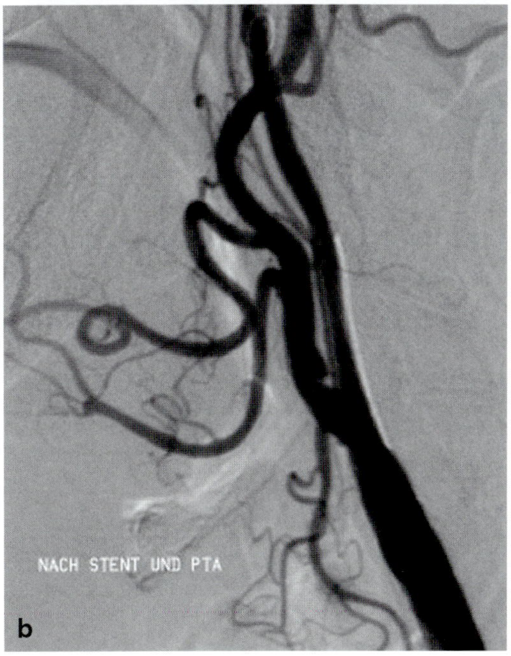


**Fig. A5.31 a, b**

**Change in pulsatility after stenting**

**a** The rigidity of the vessel wall after stenting (mesh-like structure; gray-scale image on the *left*) of the ICA results in more pulsatile flow with a higher PSV than in the native artery even when no luminal narrowing is present. The color flow image shows normal flow without aliasing, indicating absence of recurrent stenosis. Doppler waveform shows a very pulsatile flow pattern with an end-diastolic flow velocity of 20 cm/s and a PSV of 172 cm/s. The PSV indicates normal flow within the stented segment, but would indicate 60–70% stenosis in a native ICA (according to ECST criteria).

**b** Angiogram without signs of residual or recurrent stenosis in the stented ICA. The patent lumen within the stent is smaller than that of the native artery (which explains the higher PSV measured by Doppler).

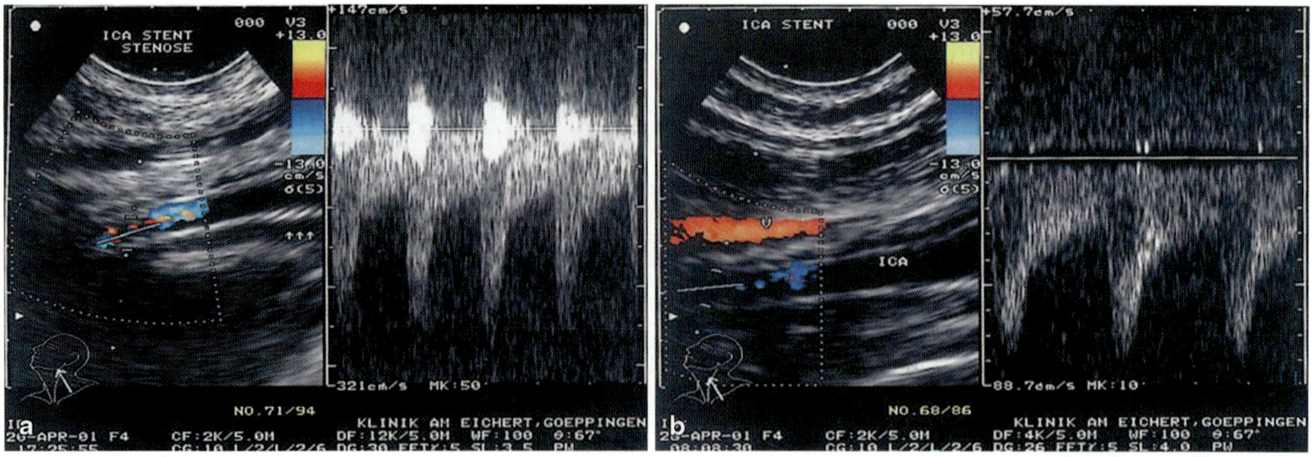


**Fig. A5.32 a, b**
**Stenosis after carotid stenting**
**a** Sonographic visualization, in particular in the color duplex mode, may be impaired during the first days after stenting due to poor tissue coupling. Visualization can be improved by use of a lower-frequency transducer. Thrombotic deposits within the stent will lead to flow acceleration in the spectral display if there is hemodynamically significant luminal narrowing (250 cm/s PSV, corresponding to >70% stenosis).
**b** Such thrombotic deposits will recede after some days of heparin and aspirin therapy. In the example shown, the PSV measured at identical sites decreased from 250 cm/s before treatment (**a**) to 80 cm/s after 5 days of treatment.

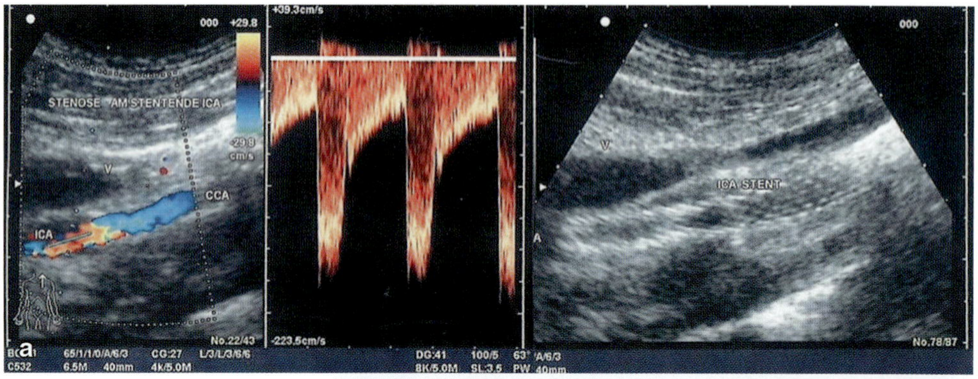


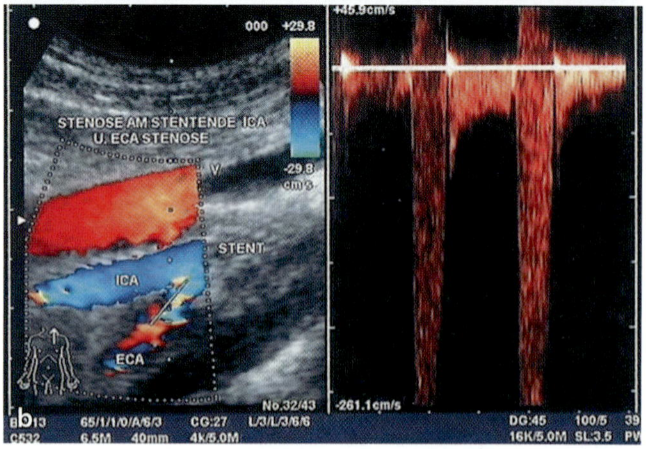


**Fig. A5.33 a–d**
**Carotid artery stent – stenosis grading**
In the follow-up of carotid artery stent implantation, the sonographer must in particular check the junction between stent and normal vessel for residual or recurrent stenosis and the stent position.
**a** The B-mode scan depicts the stent as a mesh-like structure in the vessel lumen. In the bulb the stent does not fit tightly to the vessel contour because the plaque has been pressed into the wall. B-mode scanning is highly unreliable in depicting thrombotic deposits or neointima shortly after stent placement. Evaluation for stenosis is done by color duplex or spectral Doppler. In the example, there is recurrent stenosis of 50–60% with a PSV of 190 cm/s (according to ECST criteria). The localized aliasing and the abrupt increase in flow velocity from 90 cm/s in the stent (prestenotic) to 190 cm/s at the stent end suggest that the increase is due to stenosis rather than to the normal loss of wall elasticity in a stented vessel segment.
**b** The stent extends over the origin of the ECA. The ECA is patent, but the Doppler waveform indicates high-grade stenosis.

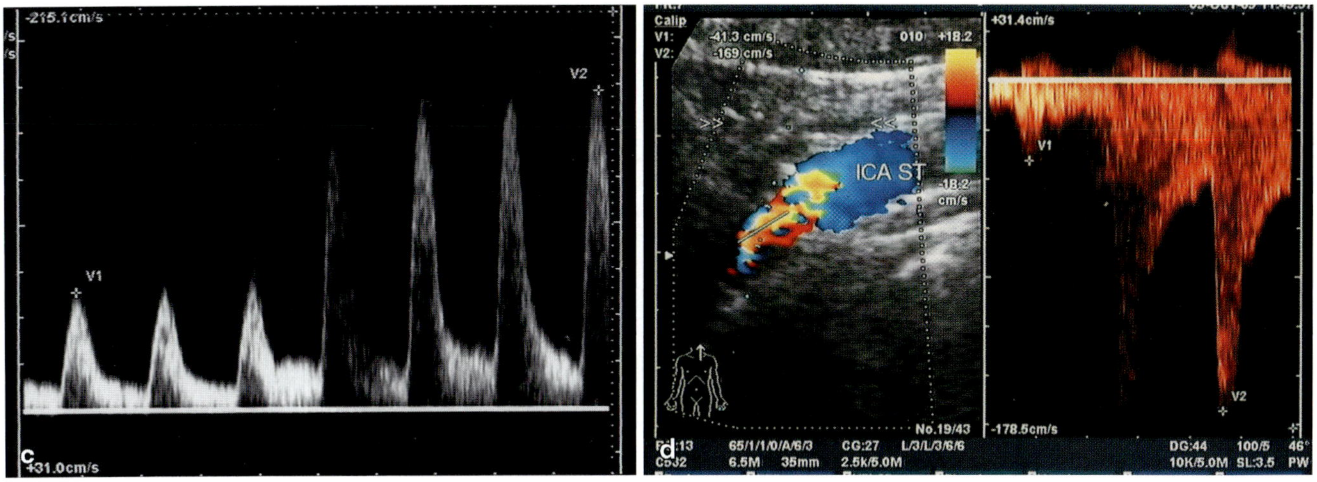


**Fig. A5.33 (continued)**

**Stenosis grading after stenting using PSV ratio**

**c** The severity of a recurrent stenosis within a stent can be estimated by continuous spectral Doppler recording (taking into account that the normal criteria do not apply because stenting reduces the lumen and wall elasticity, resulting in more pulsatile flow). The stent creates a tube of uniform caliber, which is why an abrupt increase in PSV is a reliable criterion of stenosis in this setting. In the example, the in-stent increase in PSV from 64.6 to 172 cm/s suggests >60% stenosis although the results of recent studies indicate that an absolute PSV of 172 cm/s alone does not constitute a criterion of hemodynamically relevant in-stent stenosis. The Doppler waveform is inverted. Note the higher pulsatility due to rigidity of the stented wall.

**d** Color duplex image suggests stenosis (aliasing) in the distal portion of an internal carotid stent (*ICA ST*). Based on current data, the PSV of 176 cm/s is not indicative of relevant in-stent stenosis, but when taking the increase from a prestenotic PSV of 54 cm/s into account, this does indeed suggest 60% stenosis. The Doppler spectrum was continuously recorded moving the transducer along the ICA, from proximal to distal, and includes the prestenotic signal and the sudden increase in PSV within the stenosis (from *left* to *right*). The segment along which the waveform was recorded is indicated by arrows in the B-mode image.

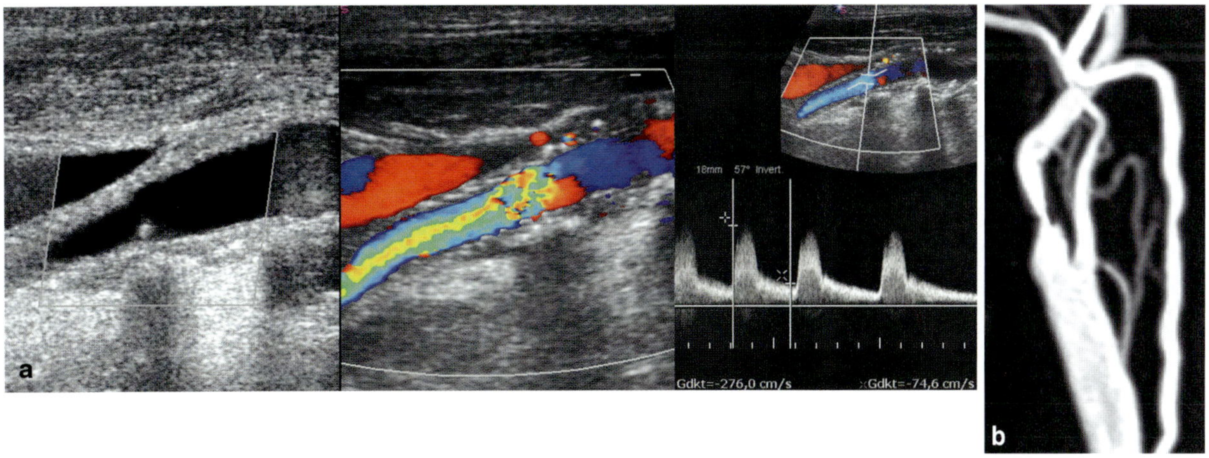


**Fig. A5.34 a–d**

**High-grade recurrent stenosis after ICA stenting**

**a** B-mode image (*leftmost* section) shows echogenic luminal narrowing within a stent. There is aliasing in the corresponding color flow image with a PSV of 280 cm/s measured by spectral Doppler. This corresponds to >70% stenosis (courtesy of K. Pfister).

**b** Angiogram showing short luminal narrowing, confirming the high-grade in-stent restenosis already seen at ultrasound.

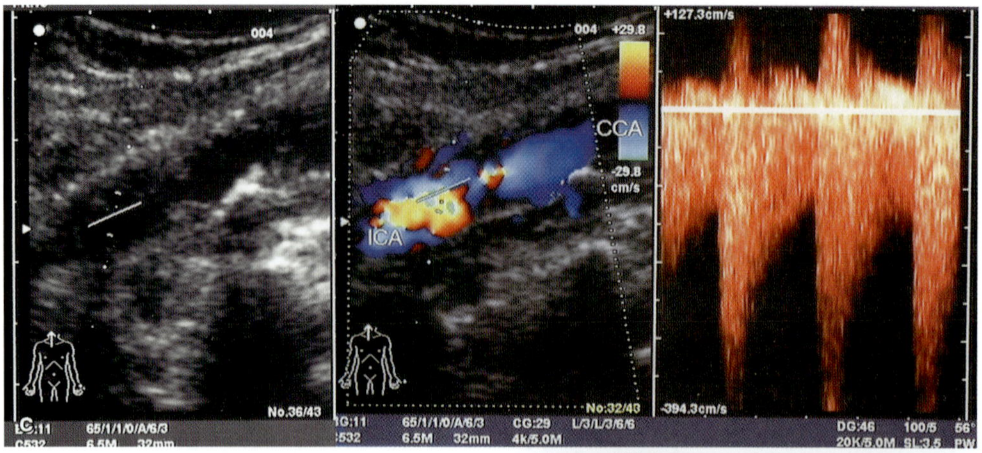


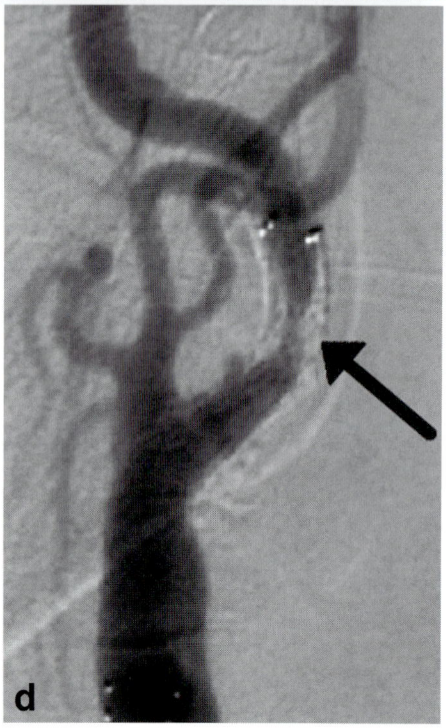


**Fig. A5.34 (continued)**

**High-grade recurrent stenosis 2 years after ICA stenting**

**c** High-grade in-stent restenosis 2 years after carotid percutaneous transluminal angioplasty (PTA) with stenting. Mixed echogenic and echolucent plaque with an intrastenotic PSV of almost 4 m/s.

**d** Angiogram: High-grade in-stent restenosis of the ICA.

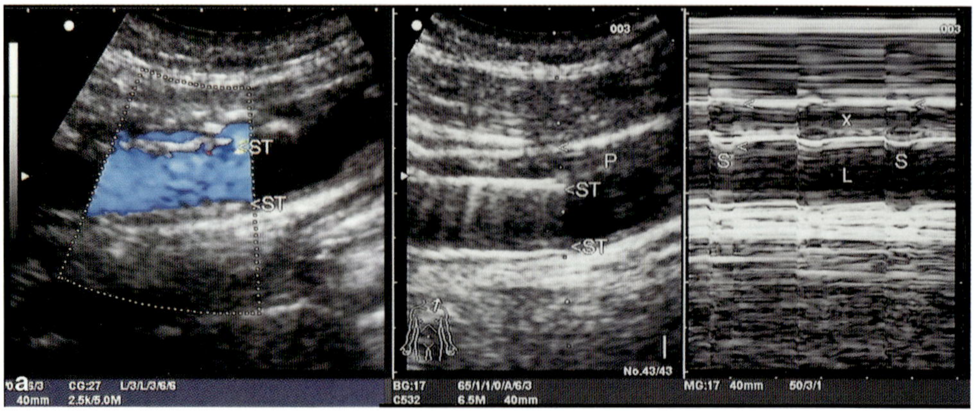


**Fig. A5.35 a–d**

**Stent dislocation**

**a** Longitudinal image (*left*) and time-motion image (*right*) showing dislocated stent (*ST*) in the ICA. There appears to be a lumen between the stent and the vessel wall, as indicated by flow signals. Note that the stent is subject to pulsatility effects (in the time–motion display), showing paradoxical stent motion as it is compressed by the flowing blood and pressure in the false lumen (*X*) during systole (*S*).

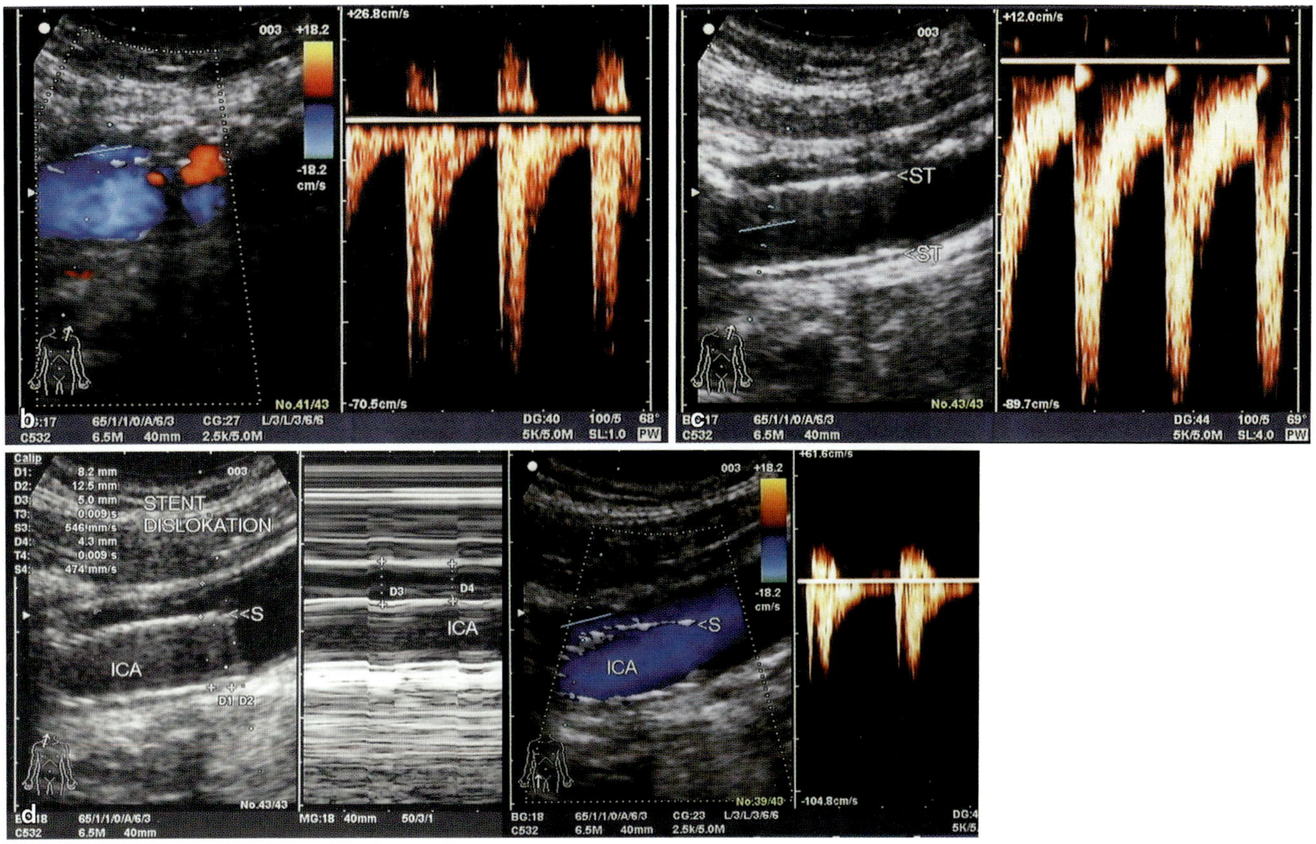


**Fig. A5.35 (continued)**

**b** The Doppler waveform from the false lumen between the stent and the wall demonstrates flow along the stent toward the head with a PSV of 60 cm/s.

**c** The stent lumen is not narrowed; PSV of 90 cm/s.

**d** The patient initially refused a repeat intervention and the stent became even more dislodged with an increase in the size of the lumen between the stent and the arterial wall (with demonstration of flow by spectral Doppler). This situation increases the risk of stent occlusion.

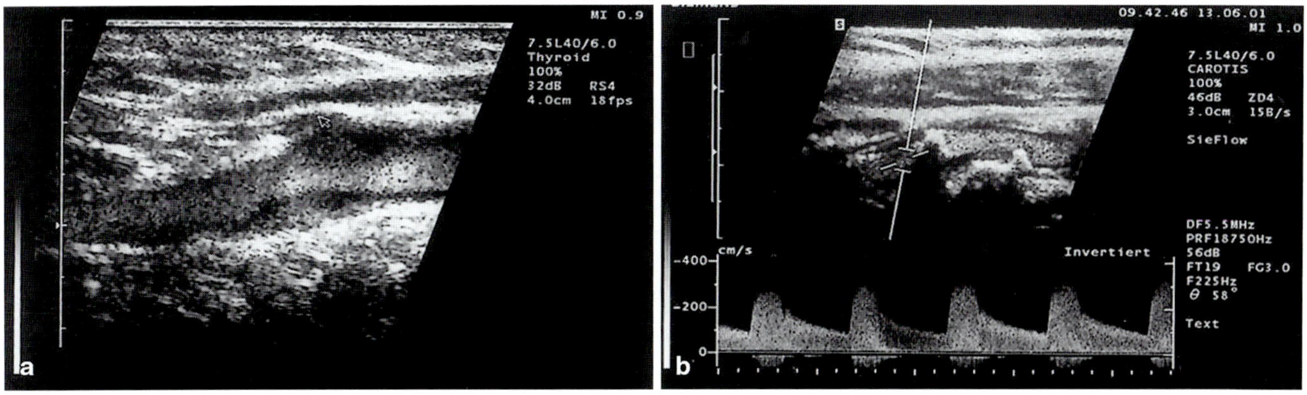


**Fig. A5.36 a–f**

**New ultrasound techniques: B-flow mode, 3D ultrasound**

**a** B-mode flow imaging analyzes the amplitude signal of the reflecting particles in the interval between two pulses. The movement of reflecting particles is encoded in terms of flow direction, velocity, and number. Vessel narrowing is depicted with higher signal intensity as a result of the larger number of reflecting particles and faster flow in segments with a reduced cross-sectional area. In addition to hemodynamic parameters, the B-flow mode also provides morphologic images with a high resolution of plaques and the vessel wall, enabling good differentiation of the plaque surface and surface irregularities (ulceration) from the patent lumen. In the example, a plaque near the transducer and another farther removed from it are depicted in the bulb. The luminal narrowing resulting from these plaques is indicated by the higher signal intensity of the perfused lumen in this area.

**b** Advantages of B-flow imaging are its little angle dependence and the good morphologic discrimination between vessel wall and patent lumen. Its major drawback is its susceptibility to artifacts induced by the highly pulsatile wall motion that occurs in the presence of high-grade stenosis caused by plaque. Moreover, like all ultrasound techniques, the B-flow mode is impaired by signal scattering and acoustic shadowing due to calcified structures. This is why B-flow imaging has not replaced the hemodynamic spectral Doppler technique, especially not in the presence of pathology. The figure illustrates how the B-flow image is degraded by acoustic shadowing from calcified plaque. The higher signal in the vessel lumen indicates the stenotic jet. Still, the Doppler spectrum continues to be the most reliable method for quantifying high-grade stenosis.

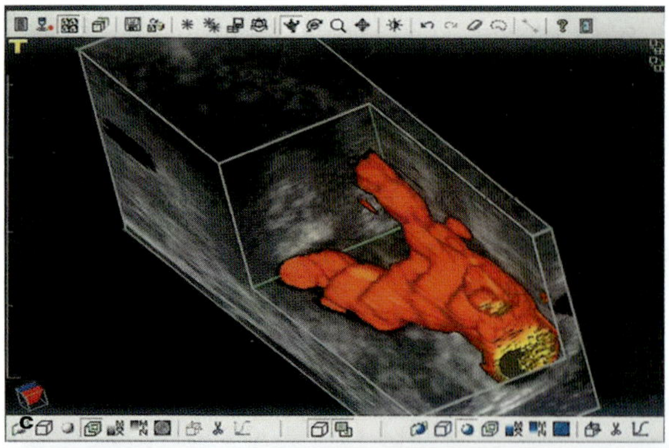

**Fig. A5.36 (continued)**

**c** Three-dimensional displays can provide a good overview of vascular anatomy and relationships in the presence of atypical variants or in case of elongation. At its current state of development, however, this technique contributes little to stenosis grading and evaluation of plaque morphology. Due to artifacts caused by vessel pulsation and atherosclerotic plaques, the 3D displays have no advantage over 2D displays in answering relevant angiologic and vascular surgical questions. The example depicts the CCA on the right with the superior thyroid artery above and the ICA (*bottom*) and ECA (*top*) on the left.

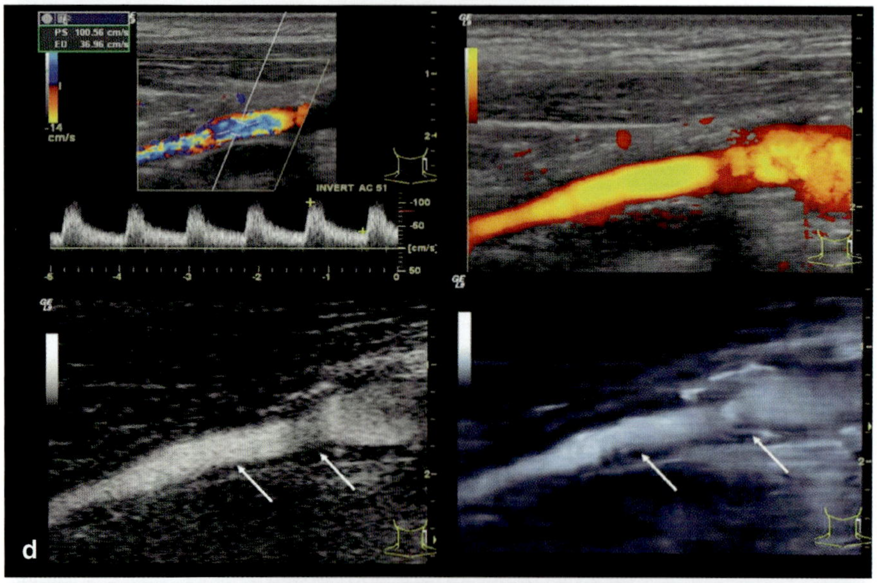


**B-flow imaging for stenosis grading after carotid artery stenting**

(Images **d**-**f** courtesy of M. Jung)

**d** High-grade ICA stenosis before (*top*) and after (*bottom*) stent placement. As described in Fig. 5.14, B-flow imaging (*top right*) is superior in visualizing a plaque ulcer or recess compared with the color-duplex mode (*top left*: 3D reconstruction). The images after stenting were obtained using the B-flow mode (*bottom left*) and the B-flow mode with speckle reduction imaging (SRI) (*bottom right*).

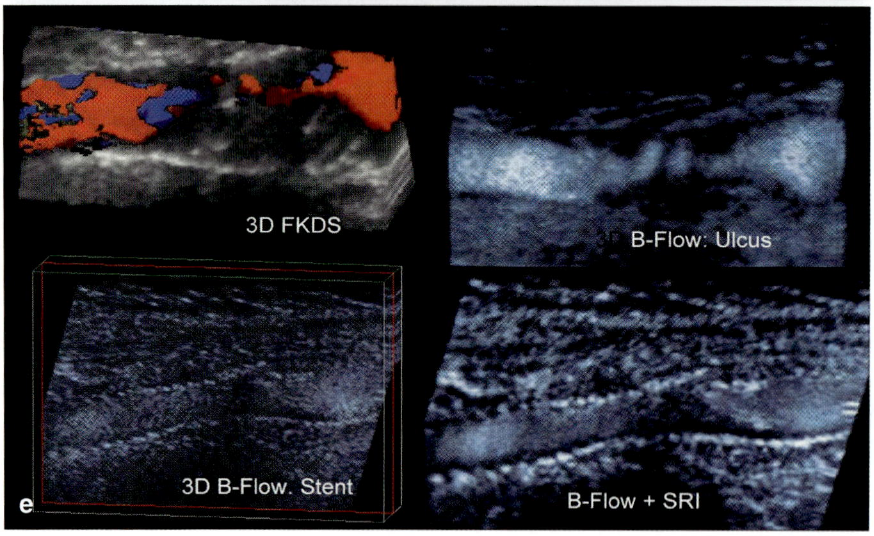


**e** In-stent intimal hyperplasia. Color duplex imaging with spectral Doppler analysis (*top left*) reveals no hemodynamically relevant stenosis; the other images were obtained using the power mode (*top right*), B-flow mode (*bottom left*), and B-flow mode with SRI (*bottom right*). B-flow imaging improves the visualization of small wall deposits or intimal hyperplasia compared with Doppler techniques (color duplex or power mode). The very accurate differentiation of wall structures from flowing blood on B-flow images allows morphologic stenosis grading.

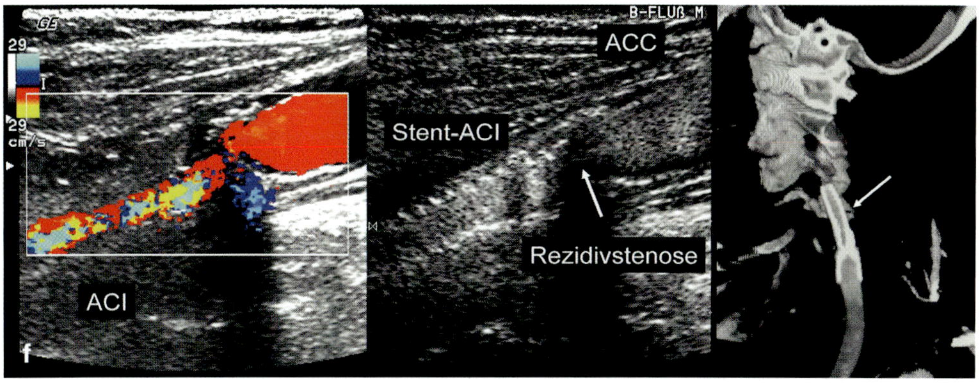


**Fig. A5.36 (continued)**
**f** Recurrent stenosis after carotid percutaneous transluminal angioplasty (PTA) with stenting on color duplex image (*left section*) and B-flow image (*middle section*). The example illustrates how B-flow imaging can be used to estimate the stenosis grade, but it also shows the limitations resulting from scattering and acoustic shadowing (calcification), to which B-flow imaging, like all ultrasound techniques, is subject. Computed tomography (CT) angiogram (*right section*) showing the stent in the ICA (*arrow*) with restenosis.

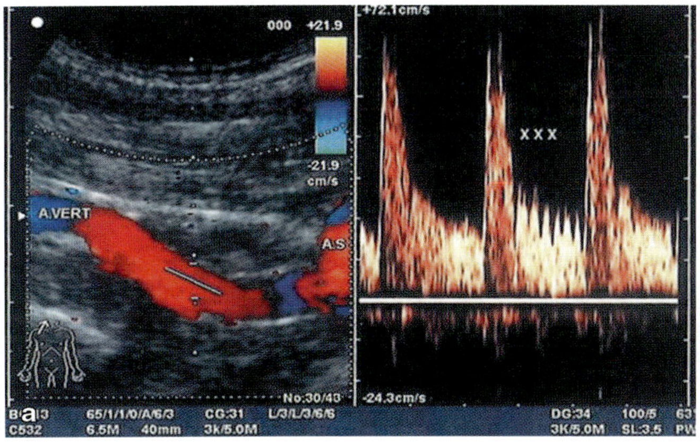


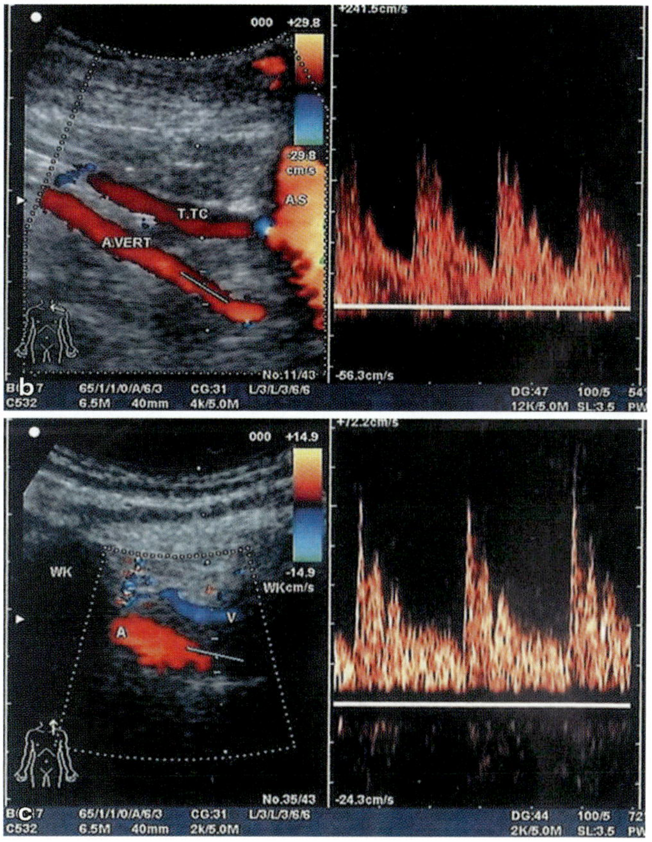


**Fig. A5.37 a–g**
**Vertebral artery**
**a** Origin of the vertebral artery from the subclavian artery interrogated with the transducer in the supraclavicular position. The flow profile is similar to that of the ICA. The vertebral artery is differentiated by the depiction in the Doppler waveform of a pulsation transmitted when tapping the mastoid area.
**b** The thyrocervical trunk must not be mistaken for the vertebral artery. It supplies the thyroid and therefore has a similar spectral waveform and comes more easily into view, in particular when the insonation conditions are poor, because its origin from the subclavian artery is closer to the transducer than that of the vertebral artery. In the example, the vertebral artery has a relatively high PSV due to hypoplasia of the contralateral artery.
**c** Vertebral artery coded in *red* between two transverse processes (*WK*). The vein (*V*) is depicted closer to the transducer with flow in *blue*.

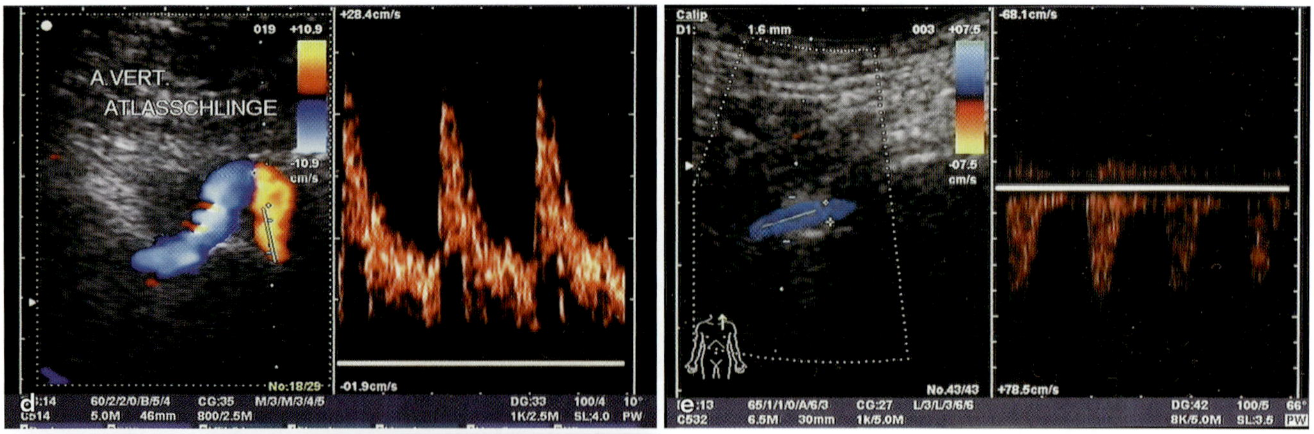


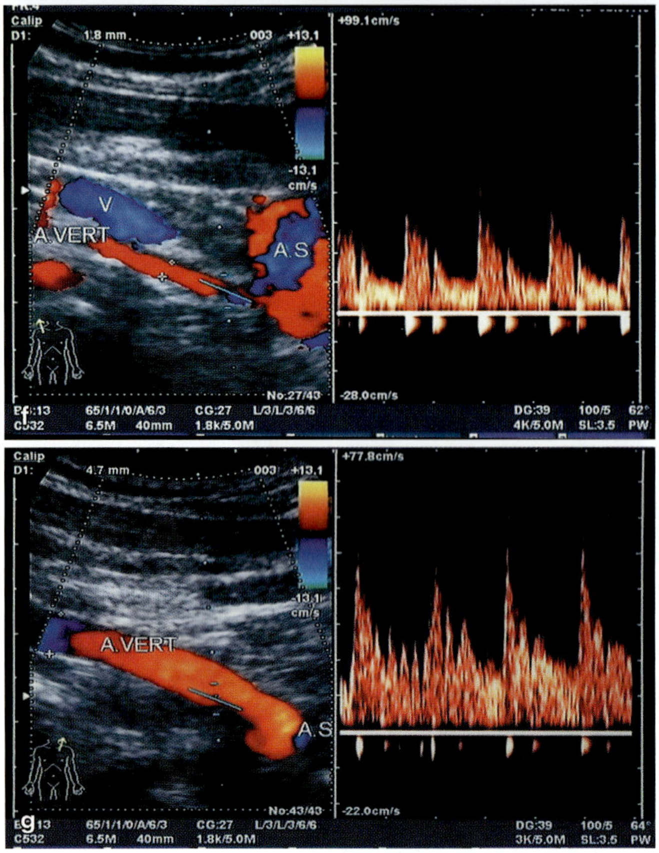


**Fig. A5.37 (continued)**

**d** Imaging of the vertebral artery by interrogation of the atlas loop (transducer placed below the mastoid and directed toward the contralateral eye) stems from the use of CW Doppler ultrasound and has become less important in duplex ultrasound. However, this approach is useful to sample the vertebral artery Doppler spectrum during functional tests performed to diagnose postural compression of the vertebral artery by a vertebral body. In this setting, evaluation of the atlas loop enables follow-up of postocclusive spectral changes in a fairly fixed position during movements of the neck. As with CW Doppler, changes in flow direction are reflected in the Doppler waveform. In the example, there is flow toward the transducer in the proximal portion of the atlas loop. In the distal atlas loop, flow is away from the transducer.

**Hypoplastic vertebral artery**

**e** Hypoplastic vertebral artery with a size of 1.6 mm and spectral Doppler measurement showing reduced flow velocity (PSV <40 cm/s), in particular during diastole (<5 cm/s). A compensatory increase in flow is measured in the contralateral vertebral artery (PSV of 90 cm/s and diameter of 4 mm; not shown).

**Hypoplastic vertebral artery**

**f** Hypoplastic vertebral artery (*A. VERT*) depicted at the origin from the subclavian artery (*A.S*); the diameter is 1.9 mm.

**g** The contralateral vertebral artery, also shown at the origin from the subclavian artery, is hyperplastic (diameter of 4.7 mm). The Doppler waveform from the hypoplastic vertebral artery shows more pulsatile flow. Individuals with a very hypoplastic vertebral artery have an increased risk of brain stem infarction. Moreover, as in the patient shown, certain rotational movements of the head can cause transient brain stem syndrome resulting from intermittent compression of the contralateral hyperplastic artery on its course through the foramina. This can cause reproducible episodes of vertigo with certain head positions, which will disappear after a few seconds when the head is returned to its normal position.

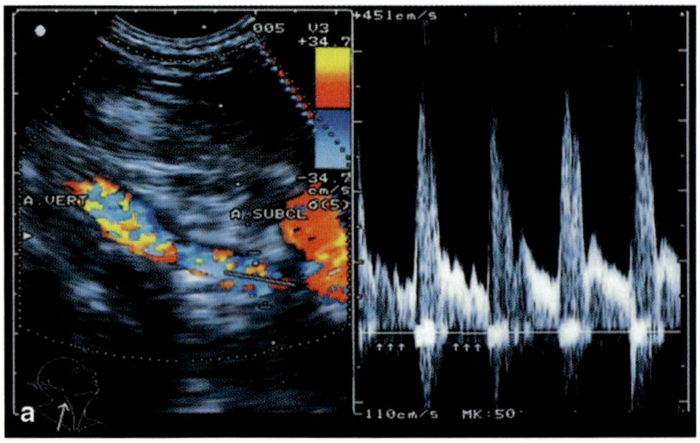


**Fig. A5.38 a, b**

**Stenosis at origin of vertebral artery**

**a** The color duplex scan shows aliasing due to turbulence at the origin of the vertebral artery (coursing *leftward* in the display) from the subclavian artery (*A.SUBCL*). At the origin, color coding is absent from the lumen due to plaque with acoustic shadowing. The spectral waveform documents the stenosis with a PSV of 320 cm/s and an EDV of 80 cm/s. The vertebral artery is distinguished from the thyrocervical trunk, which has a similar flow profile, by the transmission of the pulsation from rhythmical tapping of the distal vertebral artery below the mastoid (indicated by *arrows* in the spectral display).

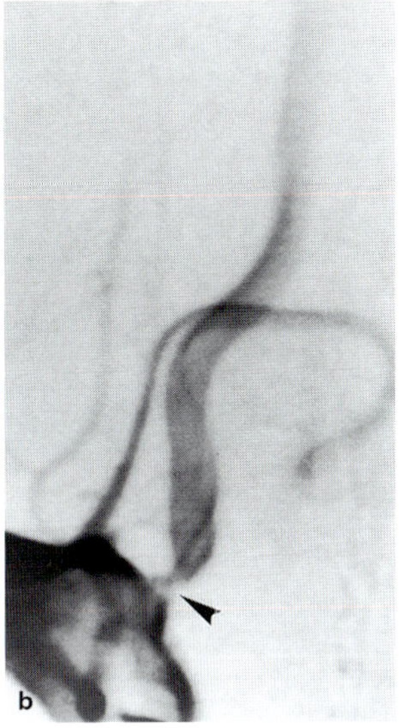


**Fig. A5.38 (continued)**
**b** Angiogram: Stenosis (*arrowhead*) at origin of vertebral artery.

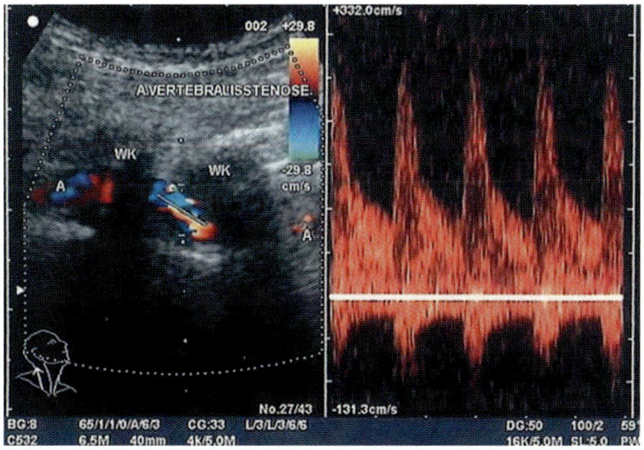


**Fig. A5.39**
**Distal vertebral artery stenosis**
Atherosclerotic vertebral artery stenoses typically occur at the origin from the subclavian artery. A more distal stenosis (in the V2 segment between C4 and C5 in the example presented) is more likely due to other causes such as constriction of the passageway through the transverse processes by exostosis or dissection. An increase in flow velocity (here 250 cm/s) indicates stenosis only if it is localized. Increased flow throughout the vertebral artery suggests a compensatory increase in perfusion due to hypoplasia of the contralateral branch or atherosclerotic occlusion of other arteries supplying the brain.

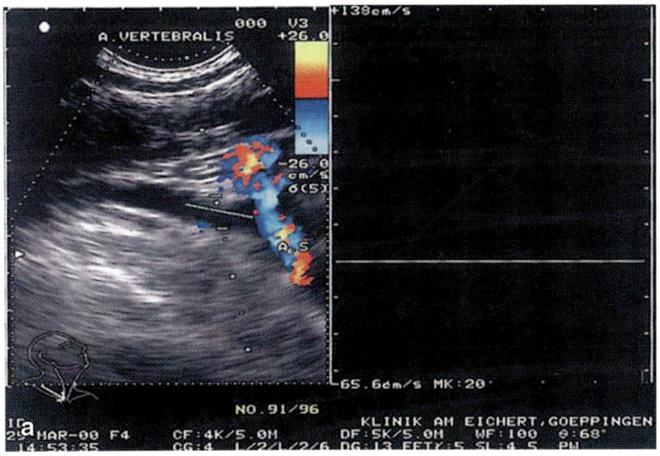


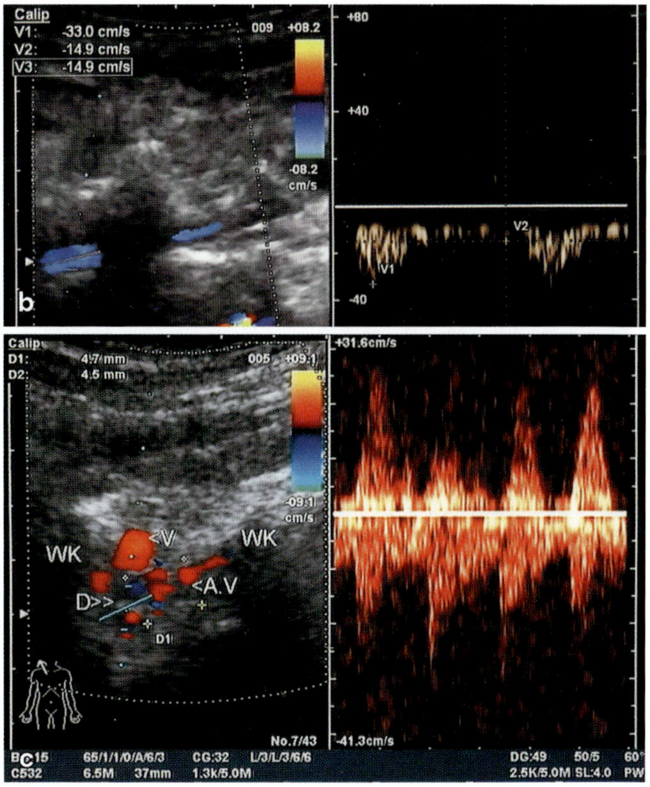


**Fig. A5.40 a–c**
**Vertebral artery occlusion**
**a** Both color duplex and spectral Doppler fail to depict flow signals in a tubular structure arising from the subclavian artery. The course of the structure corresponds to that of the vertebral artery, and the appearance is consistent with vertebral artery occlusion.
**b** Thin vertebral artery (2.2 mm) with refilling through spinal vessels just before the atlas loop and postocclusive flow (delayed systolic rise and PSV of 32 cm/s).

**Vertebral artery dissection**
**c** Following a failed endovascular intervention, an intimal flap (*D*) is visible in the vertebral artery between the transverse processes. At the site of sampling, flow in the true and false lumina is in opposite directions (coded in *red* and *blue*; above and below the baseline in the Doppler spectrum). *V* vertebral vein.

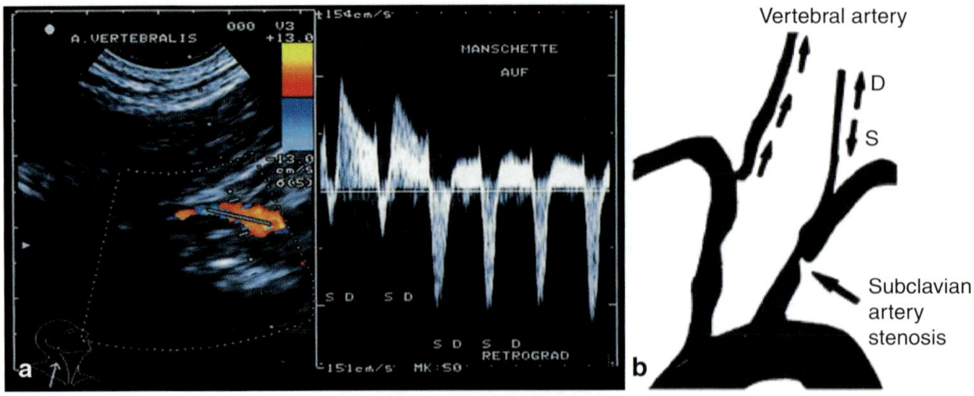


**Fig. A5.41 a, b**

**Subclavian steal syndrome – to-and-fro flow in the vertebral artery**

**a** The steal phenomenon in the vertebral artery varies with the grade of the subclavian stenosis, the collateral pathways, and the role of the vertebral artery in the collateral supply. The corresponding changes can be seen in the Doppler spectrum sampled after provoked ischemia. To-and-fro flow may be preserved and only change from primarily cranial flow to primarily central flow (toward subclavian artery). During compression of the ipsilateral arm, there is high diastolic flow in the cranial direction with only little retrograde flow in systole. Upon deflation of the arm cuff, there is a change in to-and-fro flow with a large retrograde systolic flow component (*S*) and only little antegrade flow in diastole (*D*).

**b** Diagram of to-and-fro flow in the vertebral artery in ipsilateral subclavian artery stenosis.

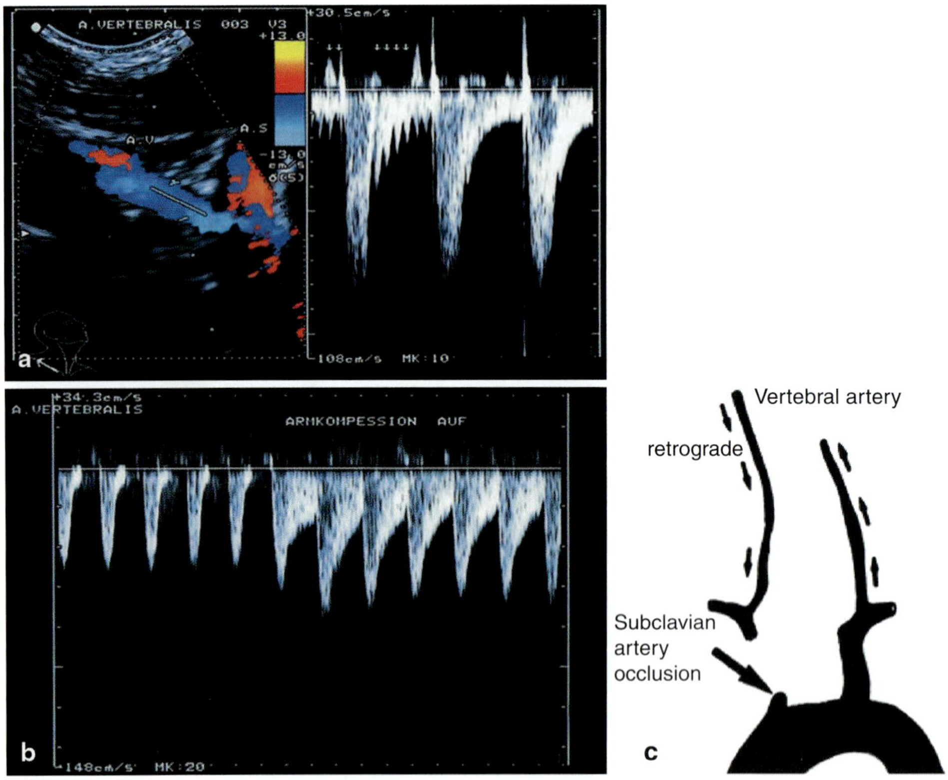


**Fig. A5.42 a–c**

**Subclavian steal syndrome – retrograde flow in the vertebral artery**

**a** In severe subclavian steal syndrome, retrograde flow from the ipsilateral vertebral artery (*A.V*) into the subclavian artery (*A.S*) will already be seen at rest (displayed in blue). This is verified by the spectral Doppler tracing with confirmation of the identity of the vertebral artery by transmission of pulsation upon tapping in the mastoid region. The subclavian artery is occluded proximal to the site of entry of the vertebral artery.

**b** Following ischemia upon release of the arm cuff, there is a marked increase in retrograde flow, in particular in diastole.

**c** Diagram of retrograde flow in the vertebral artery in ipsilateral subclavian artery occlusion with complete subclavian steal syndrome.

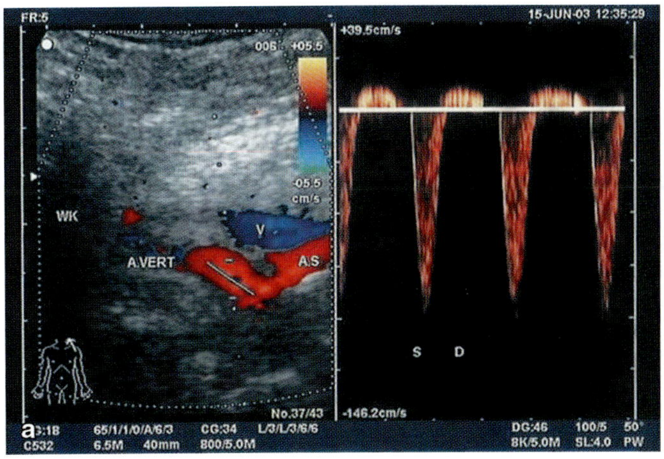


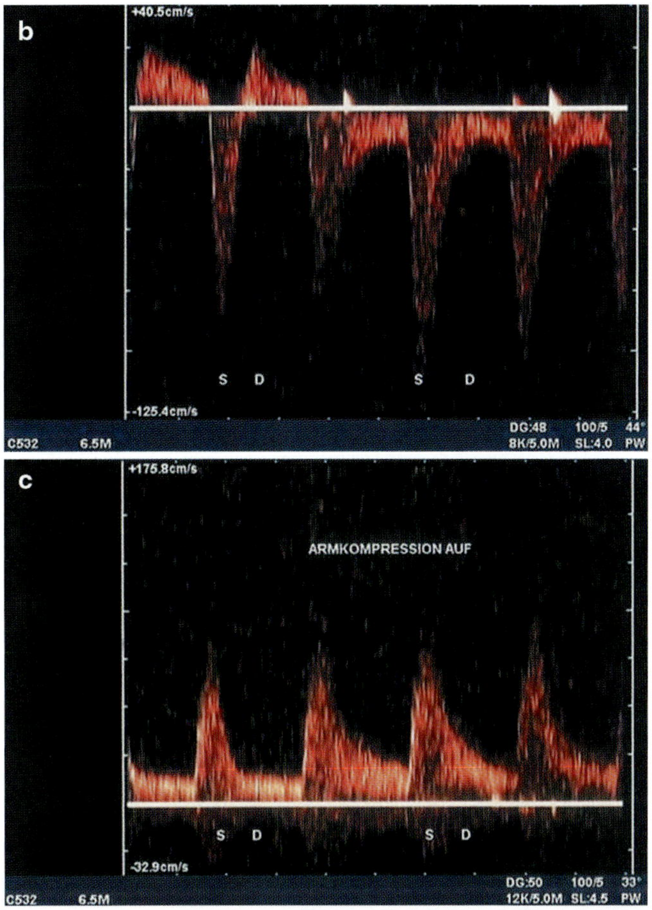


**Fig. A5.43 a–d**

**Subclavian steal syndrome with vertebrovertebral crossover**

**a** Depiction of the origin of the vertebral artery in central subclavian artery occlusion. The spectral waveform recorded at the origin of the vertebral artery (*A.VERT*) from the subclavian artery (*A.S*) demonstrates to-and-fro flow with a retrograde systolic component (away from transducer, toward heart) and an antegrade diastolic component (toward transducer, toward brain). The passage of the artery through the transverse process (*WK*) is shown at the left margin.

**b** In the provocative test, compression of the ipsilateral brachial artery with reduction of blood flow into the arm arteries leads to an increase in antegrade diastolic flow in the ipsilateral vertebral artery compared to rest (see **a**). Ischemia induced by release of the cuff (middle of the spectrum) results in a change from to-and-fro flow to a constant reversed flow from the vertebral artery into the subclavian artery (away from transducer).

**c** An increase in systolic and diastolic flow velocity (*S* = systole, *D* = diastole) in the vertebral artery on the opposite side of subclavian artery occlusion upon release of the cuff around the brachial artery on the side of occlusion proves vertebrovertebral crossover in the subclavian steal syndrome. In the example shown, the increase in velocity is not very pronounced, suggesting that there are other collateral routes to bypass the occluded subclavian artery.

**d** Angiogram with depiction of contrast medium crossover in occlusion of the left subclavian artery. The temporal course of the contrast medium passage shows flow from the right subclavian artery (*left*) into the right vertebral artery (*middle*) and into the left vertebral artery (*right*).

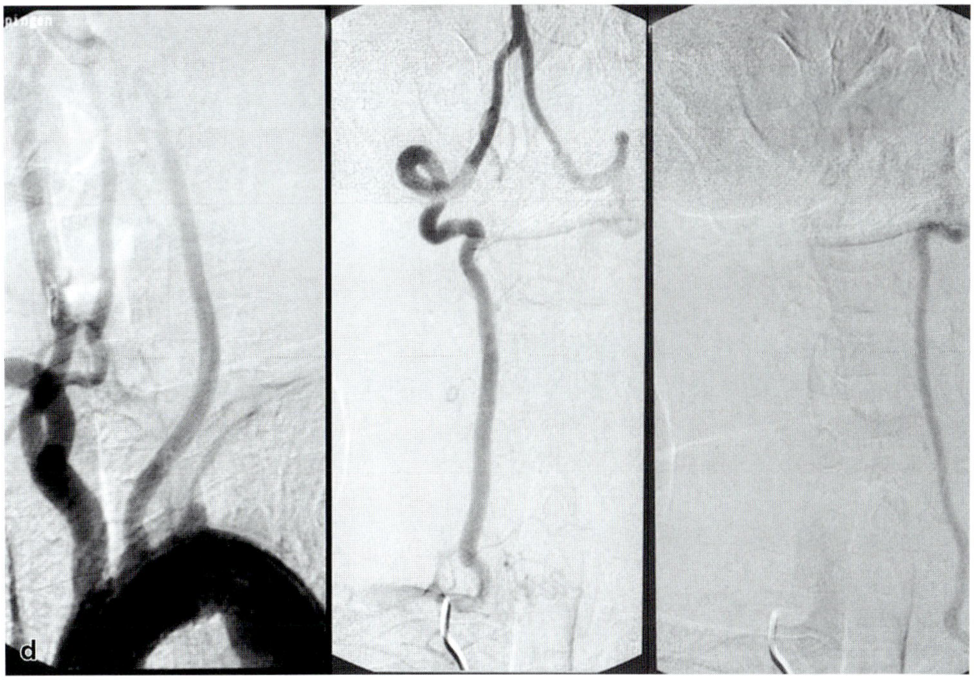

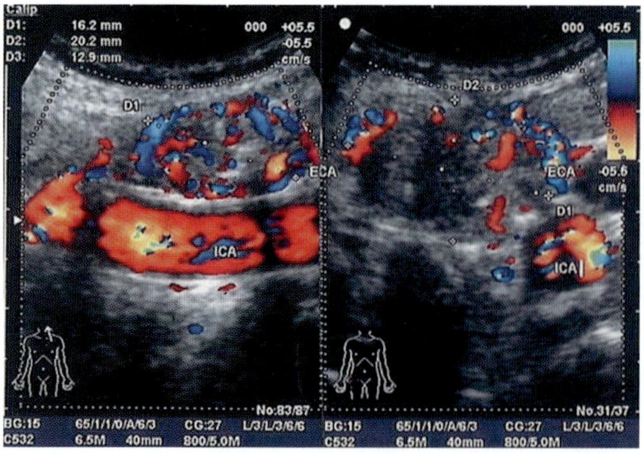


**Fig. A5.44**

**Carotid body tumor**

A carotid body tumor typically lies in the carotid bifurcation, causing the characteristic saddle deformity resulting from splaying of the two arteries. The tumor is highly vascularized, chiefly supplied by branches of the ECA, and usually confined to the space between the ICA and ECA. Occasionally, a carotid body tumor may also extend laterally (as in the example shown) or encase the vessels. Carotid body tumors in atypical locations must be differentiated from the more common lymphomas; this is accomplished by the demonstration of the good vascularization of carotid body tumors using duplex ultrasound. The longitudinal scan on the *left* obtained in anterior orientation depicts the well-perfused carotid body tumor anterior to the *ICA*. The transverse scan (*right*) shows the tumor in atypical location somewhat lateral to the ECA and ICA. The ECA is slightly compressed by the tumor (aliasing). Histologic confirmation of the tumor after surgical removal.

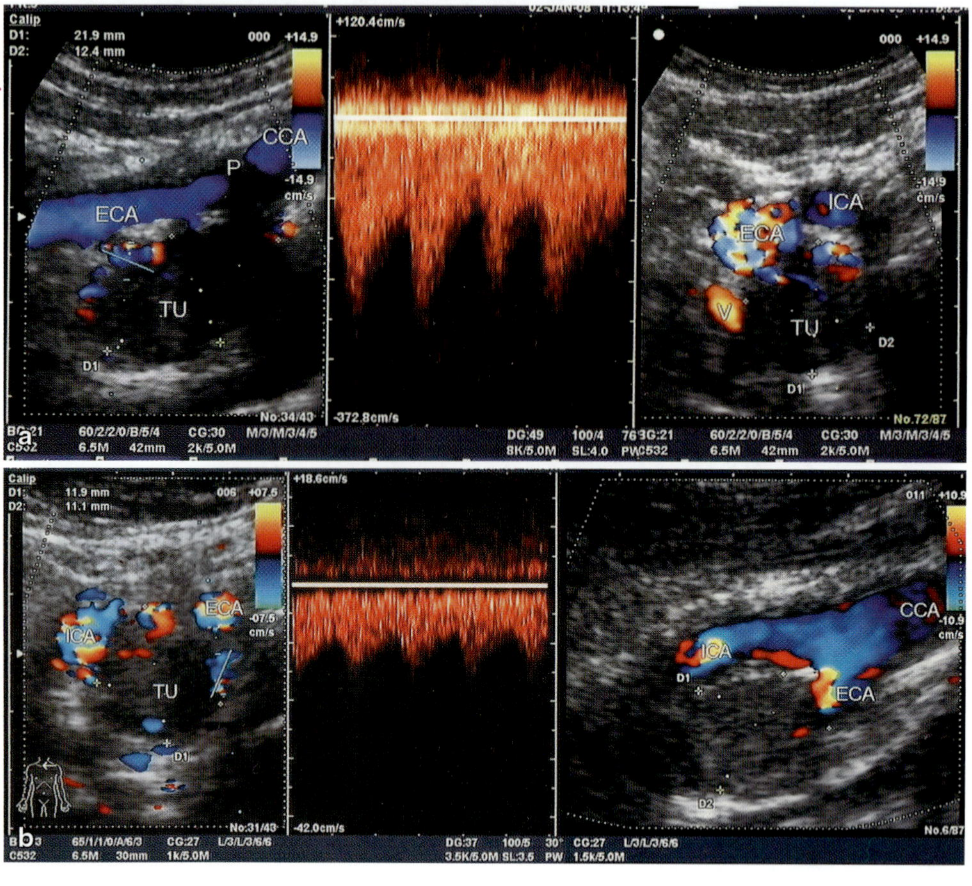


**Fig. A5.45 a, b**

**Carotid body tumor after embolization**

**a** Color duplex imaging (longitudinal scan on the *left*, transverse scan on the *right*) following transcranial embolization of a carotid body tumor (indicated by *plus signs*) supplied by the ECA in an over 90-year-old woman confirms successful treatment: there are no flow signals in the center of the tumor (with use of a low pulse repetition frequency) and only some residual flow in the capsule and outer margin. The Doppler waveform confirms marginal arterial tumor vascularization with low pulsatility and a large diastolic component.

**b** The histologic subtype of a carotid body tumor determines the degree of vascularization. The adenomatous tumor in this example is less vascularized than angiomatous carotid body tumors. The transverse view on the *left* shows the tumor located between and partially encasing the ECA and ICA and extending posteriorly. The image obtained after embolization treatment (longitudinal view on the *right*) shows absence of flow in the tumor.

**Fig. A5.46 a, b**

**Carotid body tumor – less vascularized**

**a** Longitudinal image of a carotid body tumor supplied by a branch arising from the ECA. The Doppler waveform from this branch shows a large diastolic flow component. The tumor is less vascularized than the one shown in Fig. A5.44.

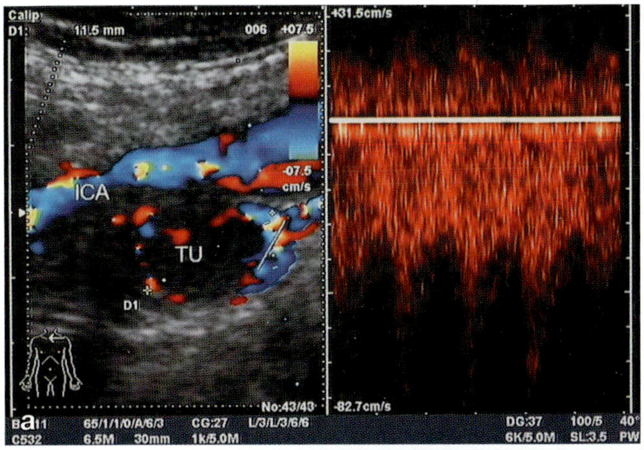


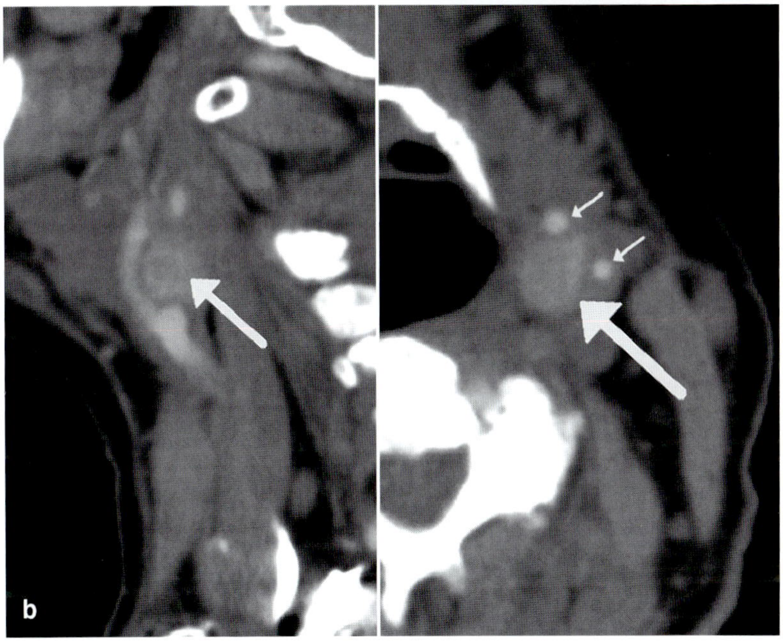


**b** Longitudinal and axial computed tomography images of the glomus tumor (*large arrow*); *small arrows* indicate the ICA and ECA.

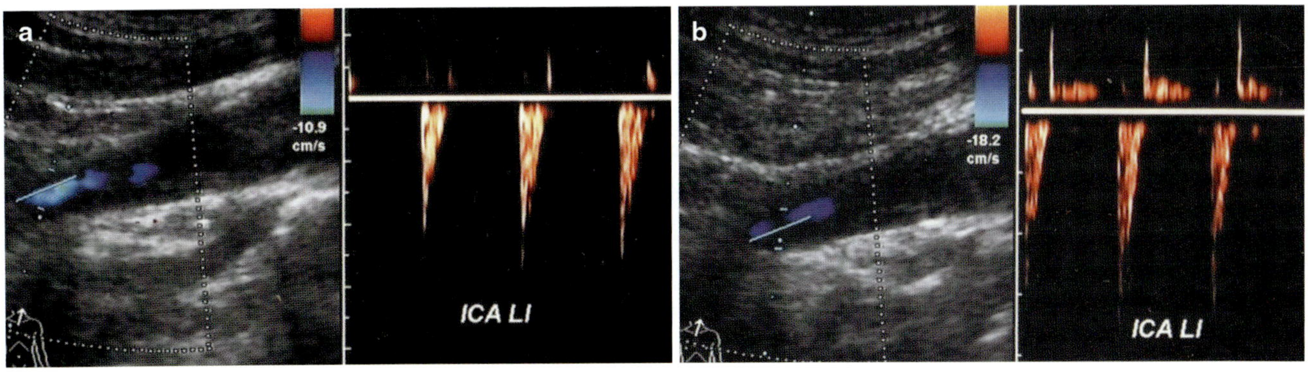


**Fig. A5.47 a, b**

**Diagnosis of brain death**

The diastolic flow velocity determines the pulsatility of blood flow, and this in turn is governed by vessel wall elasticity and especially by the degree of peripheral resistance. In the carotid system, peripheral resistance can increase when there is an increase in intracranial pressure, and diastolic flow velocity reflects diastolic blood pressure as a function of intracranial pressure. An increase in intracranial pressure therefore causes a decrease in diastolic flow velocity, and the end-diastolic flow component disappears when intracranial pressure matches the diastolic pressure. This can result in a flow signal resembling postocclusive flow (as shown in **a**): high pulsatility, no diastolic flow, and markedly reduced PSV (20 cm/s in the example, knocking waveform). The pressure situation can lead to to-and-fro flow with markedly reduced orthrograde flow velocity (PSV of 30 cm/s in **b**) and retrograde diastolic flow. To-and-fro flow or a waveform showing only early diastolic peaks indicates cerebral circulatory arrest.

# 6 Visceral and Retroperitoneal Vessels

## 6.1 Abdominal Aorta, Visceral and Renal Arteries

### 6.1.1 Vascular Anatomy

#### 6.1.1.1 Aorta

The *abdominal aorta* begins at the level of the diaphragm, crossing it via the aortic hiatus at the vertebral level of T12, and descends in front of or slightly to the left of the vertebral column. The diameter of the aorta decreases on its downward course from 25 to 20 mm. A diameter of up to 30 mm as a result of age-related dilatation is considered normal. An abrupt increase in diameter to more than 1.5 times that of the normal proximal segment is regarded as evidence of an aneurysm. The abdominal aorta divides into the two *common iliac arteries* at the L4/L5 level. The intestine is supplied with blood from the three large visceral branches arising from the anterior aspect of the aorta. Their pattern of supply is complex and has numerous variants. The lumbar arteries originate from the lateral aspect, and the two renal arteries course in a retroperitoneal direction. The arteries arising from the aorta from superior to inferior are described in detail below (Fig. 6.1).

#### 6.1.1.2 Visceral Arteries

Just below the aortic aperture of the diaphragm, the aorta gives off the *celiac trunk*, which, after 2–3 cm, divides into its two main branches, the common hepatic and splenic arteries. The *common hepatic artery* courses between the head of the pancreas and the lower edge of the liver into the hepatoduodenal ligament, where it gives off the gastroduodenal artery, an important collateral that connects to the superior mesenteric artery. It then continues to the liver as the *proper hepatic artery*. The *splenic artery* is very tortuous on part of its course along the upper border of the pancreas to the splenic hilum and supplies not only the spleen but also the body and tail of the pancreas as well as the greater curvature of the stomach.

About 0.5–2 cm below the celiac trunk lies the origin of the *superior mesenteric artery* at the L1/L2 level. It arises anteriorly at an acute angle of 15–30° relative to the aorta, and its proximal segment runs parallel to the aorta between the pancreas and renal vein. After about 4–5 cm, it gives off the inferior pancreaticoduodenal and middle colic arteries, which supply the transverse colon. The distal superior mesenteric artery divides into the jejunal, ileal, and ileocolic arteries supplying the small intestine.

Many anatomic variants exist. A normal celiac trunk is present in about 70% of the population. Alternatively, the hepatic or splenic artery may arise from the superior mesenteric artery, which has hemodynamic consequences for the affected vessels. A rare variant is a common origin of the hepatic, splenic, and superior mesenteric arteries from the aorta. In about 24% of cases, the liver is supplied by the superior mesenteric artery. There is good collateralization

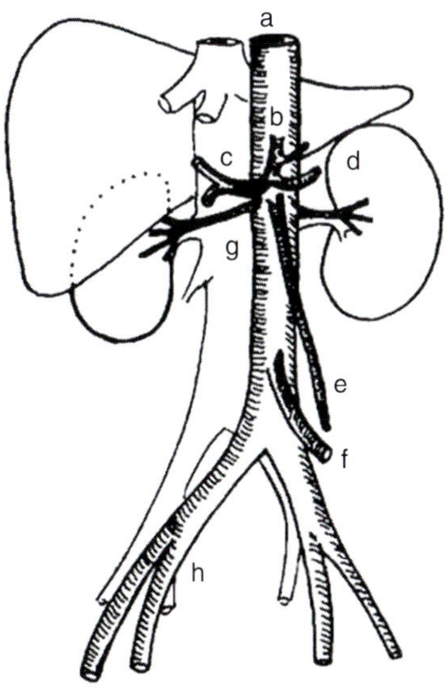


**Fig. 6.1** Diagram of major visceral and retroperitoneal arteries that are of significance for duplex sonography and vascular surgery: *a* aorta, *b* celiac trunk, *c* hepatic artery, *d* splenic artery, *e* superior mesenteric artery, *f* inferior mesenteric artery, *g* renal artery, *h* iliac arteries (common, internal, external)

W. Schäberle, *Ultrasonography in Vascular Diagnosis*,
DOI: 10.1007/978-3-642-02509-9_6, © Springer-Verlag Berlin Heidelberg 2011

should be performed *after a period of fasting* since the flow velocity is markedly increased after a meal.

Scatterers such as bowel gas can be pushed aside by applying pressure with the transducer. A small Doppler angle is achieved by moving and tilting the transducer, so that the sample volume in the target vessel comes to lie at the lateral edge of the B-mode scanning field displayed on the monitor.

The *respiratory mobility* of the intra-abdominal organs including the intra-abdominal and retroperitoneal vessels makes it necessary to scan the vessels during maximum inspiratory breath-holding or during shallow breathing with little respiratory motion. Proper breathing must be practiced with the patient before the examination.

#### 6.1.2.3 Renal Arteries

One way of identifying the renal arteries at their origins just below the easily visualized superior mesenteric artery is to localize the latter in transverse orientation and to then move the transducer 1–2 cm downward and look for the renal arteries as they arise from the aorta to the left and right (Fig. 6.5). A second landmark is the left renal vein (hypoechoic, broader band), which overcrosses the aorta, coursing between the aorta and the superior mesentering artery, before entering the vena cava. The left renal artery typically arises some millimeters below the right renal artery and both do not usually take a strictly horizontal course but move slightly downward. The right renal artery first courses anteriorly in a slightly curved fashion and then arches underneath the vena cava (Fig. 6.6a–c). The origins and proximal 3 cm of the renal arteries can be visualized and evaluated in over 90% of cases, while visualization of the middle third is often incomplete due to overlying bowel gas, especially on the left. The middle segment of the renal artery is easier to scan on the right, where the vena cava can be used as an acoustic window. *Interfering bowel gas can be displaced by pressing the transducer against the bowel.* The transducer is then moved to the right or left to achieve an optimal angle (as illustrated for the superior mesenteric artery in Fig. 6.2).

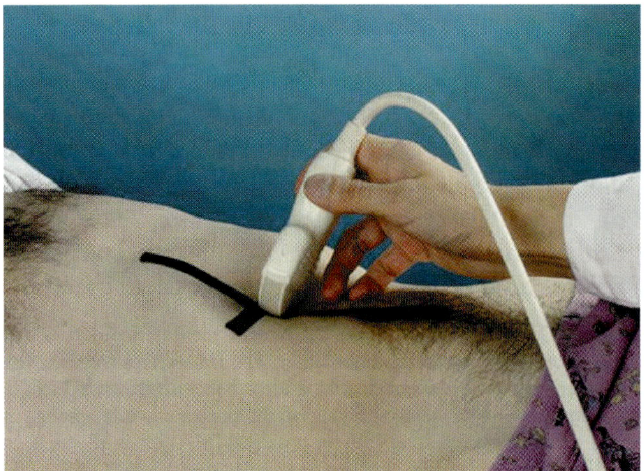

**Fig. 6.5** Epigastric transducer position (transverse upper abdominal section) for evaluation of the renal arteries

The *distal third can be scanned continuously from the flank* starting at the renal hilum and following the course of the artery proximally (Fig. 6.6). This transducer position also allows good Doppler measurement with a relatively small angle. The individual segmental arteries are identified in the color mode and evaluated for origin stenosis by having the patient exhale or by having the patient inhale and hold their breath. If renal artery infarction is suspected, the renal parenchyma is evaluated for perfusion defects seen on color duplex scans as wedge-shaped areas without color coding (high but artifact-free gain, low PRF).

A transplant kidney can often be imaged with a higher frequency than the abdomen in general, in particular when examining slender patients (4–7 MHz). Both the transplant renal artery and the vein should be continuously imaged by color duplex sonography with spectral Doppler sampling for stenosis quantification at the arterial and venous anastomoses. Additional Doppler measurements are obtained at sites of aliasing along the length of the transplant vessels (unless aliasing is due to the use of an inadequate PRF). As for all Doppler measurements, it is important to achieve an angle of insonation of less than 60° for measuring flow velocity, which may be difficult because the vessels are often tortuous. In the B-mode, the examiner assesses the echogenicity and contour of the parenchyma and looks for perirenal fluid collections, which can compress the vein or obstruct the flow of urine from the transplant, causing dilatation of the renal pelvis and calyces.

### 6.1.3 Normal Findings

#### 6.1.3.1 Aorta

The aorta changes in diameter and pulsatility from proximal to distal. The flow profile shows continuous diastolic flow immediately below the diaphragm and becomes triphasic, as in peripheral arteries, below the origins of the renal arteries. The diameter of the aorta decreases from a mean of 25 mm to 15–20 mm just above the bifurcation. Progressive age-related dilatation with diameters of up to 30 mm of the infrarenal portion is considered normal, in particular in individuals with atherosclerotic vascular lesions (Rieger and Schoop 1998).

#### 6.1.3.2 Visceral Arteries

Supplying parenchymal organs (liver and spleen) with a low peripheral resistance, the celiac trunk and hepatic and splenic

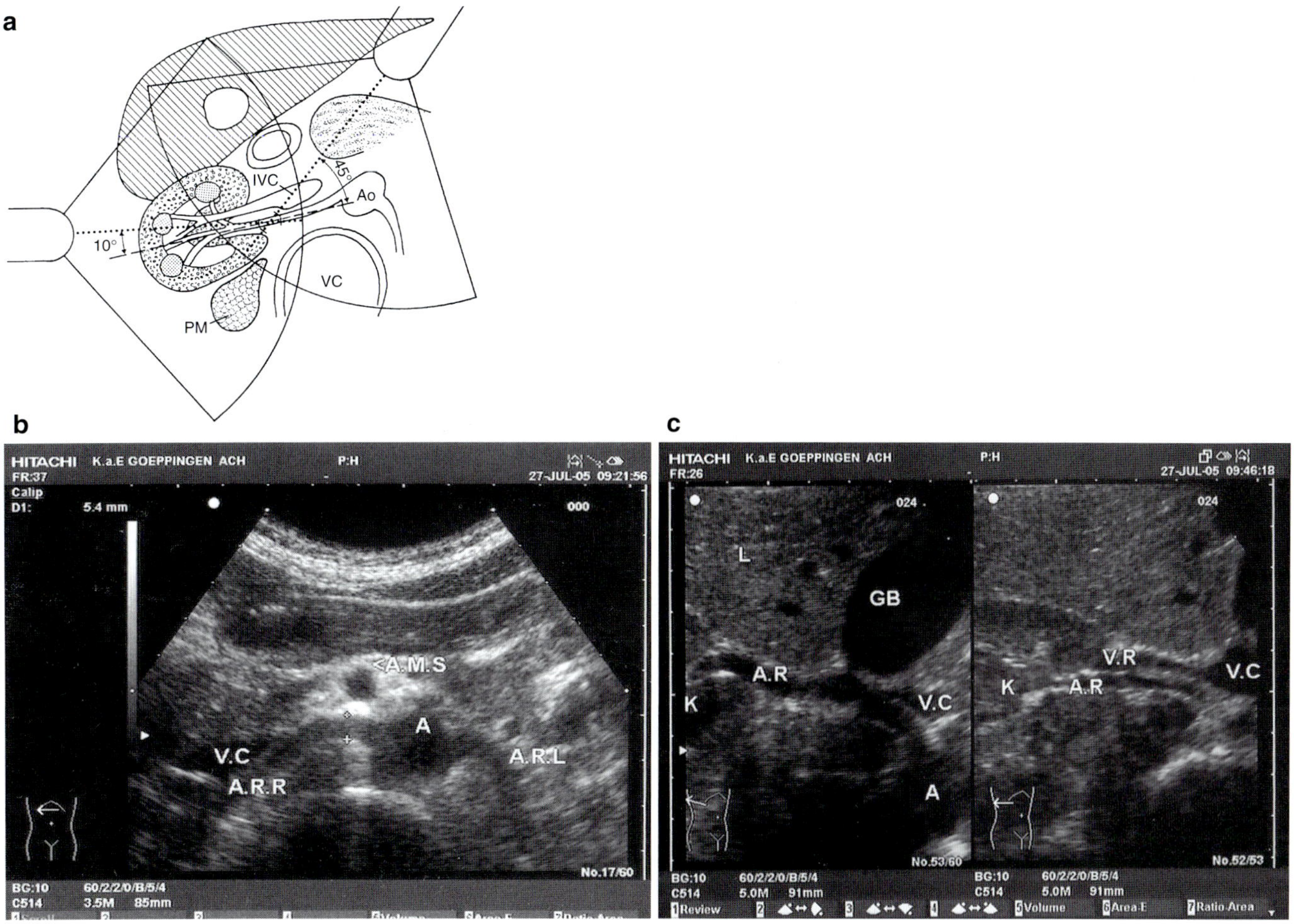


**Fig. 6.6** **a** Diagram of the right renal artery as it arises from the aorta (*Ao*) and of the renal vein entering the inferior vena cava (*IVC*). Anterior transducer position with a Doppler angle of 45°; lateral position with a Doppler angle of 10° (*VC* vertebral column, *PM* psoas muscle). **b** Sonoanatomy of the renal arteries. The renal artery origins from the aorta at the L1 level are seen in the epigastrium 1–2 cm below the superior mesenteric artery (*A.M.S*) orgin. It is rare, but when both arteries are visualized in the same plane, from the aorta to the renal hilum, the appearance suggests spread wings. Often, however, it is necessary to direct the transducer face toward the respective side in order to visualize a long segment of the renal artery; this is because the vascular bundle suspending the kidney is pulled slightly downward at its distal end. The right renal artery (*A.R.R*) passes under the vena cava (*V.C*) on its way to the renal hilum, while the left renal artery (*A.R.L*) courses posteroinferior to the pancreatic head in the retroperitoneum to reach the left renal hilum. **c** Visualization of the left renal artery is often impaired by air in the colon. On the right side, the liver and vena cava can be used as an acoustic window to achieve complete evaluation of the renal artery from its origin to the renal hilum. The *left section* shows the right renal artery (*A.R*) from the renal hilum (*K*) to the aorta (*A*), deep to the liver (*L*), gallbladder (*GB*), and vena cava (*V.C.*)

arteries have Doppler waveforms with a monophasic flow profile and a relatively large diastolic component, resulting in a low pulsatility index of 0.6–0.8. Studies demonstrate wide interindividual variation in systolic and end-diastolic velocities and in vessel diameters (Table 6.1). This situation makes it difficult to define cutoff values for identifying hemodynamically significant stenosis.

The Doppler waveform of the superior mesenteric artery reflects a mixed type combining the high pulsatility of peripheral arteries and the low pulsatility of arteries supplying parenchymal organs. At the same time, the superior mesenteric artery waveform is a good example of demand-adjusted changes in pulsatility. After a meal, there is an increase in peak systolic velocity (PSV), but above all in diastolic velocity, due to the reduced peripheral resistance, which is the main factor responsible for a lower postprandial pulsatility index.

In a study of normal vessels in 30 volunteers performed by the author as early as 1987, measurements in the superior mesenteric artery yielded a PSV of 134 ± 22.8 cm/s and a peak diastolic velocity of 20.8 ± 4.4 cm/s. The mean flow velocity was 23.4 ± 5.6 cm/s. During digestion-related hyperemia 1 h after eating, the PSV was increased to 196 ± 25 cm/s and peak diastolic velocity to 47.5 ± 8.3 cm/s. The mean postprandial flow velocity was 46 ± 7.4 cm/s. In the time-motion display, the diameter of the superior mesenteric artery varied on average between 8.04 mm during systole and 7.4 mm during diastole (Fig. 6.7). Based on these values, mean blood volume increased by 126%, from 639 mL/min preprandially to 1,447 mL/min postprandially.

Apart from mechanical, metabolic, and neural mechanisms, mesenteric artery perfusion is affected by *vasoactive substances*, in particular gastrointestinal hormones like

**Table 6.1** Normal values for the visceral arteries

| | $V_{max}$ (cm/s) | $V_{diast}$ (cm/s) | $V_{mean}$ (cm/s) | RI | Diameter (mm) |
|---|---|---|---|---|---|
| Celiac trunk (Bowersox et al. 1991; Jäger et al. 1992; Moneta et al. 1988) | 100–237 | 23–58 | 45–55 | 0.66–0.82 | 6–10 |
| Splenic artery (Nakamura et al. 1989; Sato et al. 1987) | 70–110 | | 15–40 | | 4–8 |
| Hepatic artery (Jäger et al. 1992; Nakamura et al. 1989; Sato et al. 1987) | 70–120 | | 20–40 | | 4–10 |
| Superior mesenteric artery[a] (Jäger et al. 1986; Sato et al. 1988; Sabba et al. 1991; Bowersox et al. 1991; Schäberle and Seitz 1991) | 124–218 | 5–30 | 15–35 | 0.75–0.9 | 5–8 |

[a]Values including preprandial and postprandial measurements

*Duplex parameters*: $V_{max}$ peak systolic velocity (PSV), $V_{diast}$ diastolic velocity, *RI* Pourcelot index, $V_{mean}$ mean flow velocity

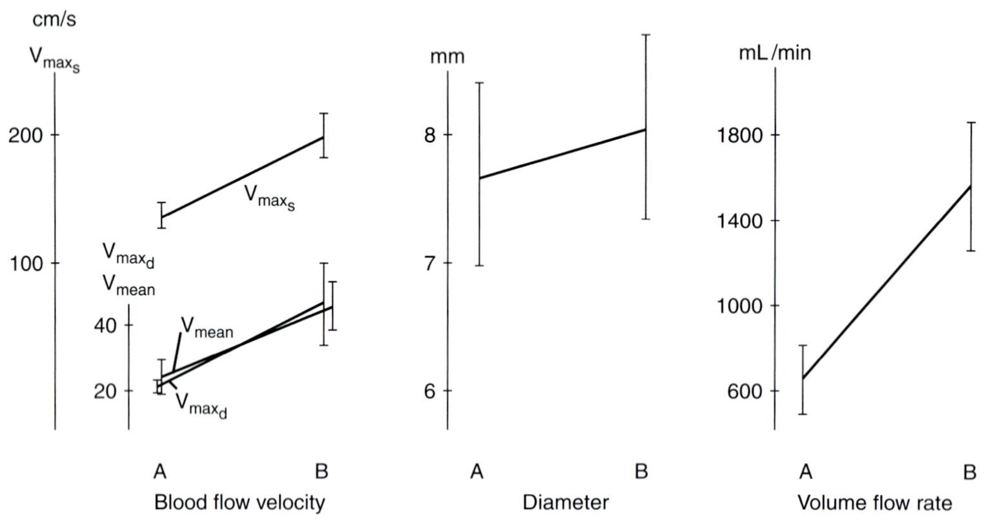


**Fig. 6.7** Preprandial (*A*) and postprandial (*B*) mean values (±15) of peak systolic velocity ($V_{max}$s), late diastolic velocity ($V_{max}$d), and mean blood flow velocity ($V_{mean}$) in the superior mesenteric artery in 30 healthy subjects. Also shown are the corresponding preprandial and postprandial vessel diameters and volume flow rates

gastrin, secretin, and glucagon as well as other vasoactive hormones like catecholamines, histamine, and bradykinin. The effect on mesenteric perfusion of drugs like nitrates, ergotamine, narcotic agents, or calcium antagonists can be demonstrated by flow measurements performed with duplex ultrasound. In the above-quoted study, it was also shown that mean blood flow velocity after nifedipine administration increased from 23.9 to 41.8 cm/s, corresponding to a 75% increase. The Pourcelot index decreased from 0.84 to 0.77, which is comparable to the postprandial decrease. These observations suggest that the increased mesenteric perfusion is primarily attributable to a reduced peripheral resistance in the mesenteric territory (see Fig. A6.4).

In patients with *anatomic variants* (hepatic artery arising from the superior mesenteric artery), higher threshold velocities must be defined (see Fig. A6.5). Moreover, blood flow in the superior mesenteric artery is also altered when the artery acts as a collateral in stenosis or occlusive disease of the celiac trunk.

Diagnostic evaluation of the inferior mesenteric artery is of little clinical significance. Systolic and diastolic flow velocities are comparable to those in the superior mesenteric artery but flow tends to be somewhat more pulsatile. Angle-corrected measurements in 20 subjects performed by our group yielded an average PSV of 109 cm/s and a peak diastolic velocity of 10 cm/s with wide interindividual variation. The flow profile was found to resemble that of the superior mesenteric artery but typically with a small diastolic component.

### 6.1.3.3 Renal Arteries

The renal arteries supply low-resistance parenchymal organs and therefore have a flow profile with little pulsatility and a large diastolic component. Measurements performed in 102 renal arteries without abnormalities on control angiography yielded a mean *peak systolic velocity* of 84.7 ± 13.9 cm/s and an end-diastolic velocity of 31.2 ± 7.8 cm/s. The Pourcelot index was 0.66 ± 0.07 (findings by our group, 1988). Visualization of the renal arteries for exclusion of stenosis by duplex scanning is possible in 85–90% of cases; the proximal third as the preferred site of atherosclerotic stenoses can be evaluated in over 90% of cases. The flow velocities reported in the literature not only vary widely from one study to the next but also within the studies. The range is <60–140 cm/s for PSV and 20–65 cm/s for end-diastolic velocity with a Pourcelot index of 0.6–0.8. Reported diameters range from 5 to 8 mm (Karasch et al. 1993; Hoffmann et al. 1991; Schäberle et al. 1992).

The peak systolic and diastolic velocities and the Pourcelot index are affected by vessel elasticity and peripheral resistance. Moreover, they are influenced by systemic blood pressure. In diabetics, medial sclerosis with reduced wall compliance and parenchymal changes result in lower diastolic flow and a higher Pourcelot index. PSV is slightly higher than in subjects with normal vessels.

## 6.1.4 Interpretation and Documentation

The ultrasound examination and the subsequent documentation of relevant findings of the visceral and retroperitoneal vessels depend on the clinical question to be answered. If an aortic aneurysm is diagnosed, the size is documented in the transverse plane; the report must also provide information on the localization and extent (infrarenal, relation to iliac arteries) as well as on the presence of thrombosis. Blood flow in the aorta should be documented by a longitudinally sampled Doppler waveform, though a color duplex scan may also be sufficient. In patients with suspected visceral artery stenosis, a Doppler spectrum is obtained from the origin of the respective vessel, the preferred site of visceral stenosis, and the vessel is scanned over a long stretch. The findings in patients evaluated for renal artery stenosis are documented by means of angle-corrected Doppler waveforms of the origins of both renal arteries. An angle-corrected waveform from the aorta is required as well. In fibromuscular dysplasia, either the findings in the middle thirds of both renal arteries are documented or Doppler waveforms from the origins and distal thirds (at the renal hilum) with comparison of resistance indices.

## 6.1.5 Clinical Role of Duplex Ultrasound

### 6.1.5.1 Aorta

Most ultrasound examinations of the retroperitoneal and intra-abdominal arteries are performed in patients with suspected aneurysm of the abdominal aorta. Morphologically, an aneurysm is characterized by concentric or eccentric luminal dilatation. The vast majority of aneurysms begins infrarenally (95%) and shows variable degrees of thrombosis. The underlying causes in Europe are as follows (in order of descending frequency):

- Atherosclerosis, 70–90%
- Idiopathic medial necrosis (Bollinger 1979), 8–10%
- Microbial infection including syphilis (which accounted for 20–30% of aneurysms in the 1960s), 4–5%
- Nonbacterial inflammatory aneurysms (arteritis), 3–5%
- Inflammation, 3–5%, and
- Congenital, 1–3%

Aneurysms predominantly occur in the elderly or in patients with multiple morbidity. Aneurysms with mural thrombosis tend to grow more slowly and are more stable. An aneurysm develops on the basis of rarefaction and fragmentation of the elastic membranes and atrophy of the muscular media of the vessel wall. Although the dilatative process primarily involves the media rather than the intimal and subintimal layers (the site of atherosclerotic lesions), *aneurysms in the European population are typically associated with atherosclerosis*. In other regions such as Africa, nonbacterial inflammatory conditions such as arteritis or bacterial infections account for a much larger proportion of aneurysms (20–30%), with a greater number of younger persons being affected. The sonographic evaluation of the vessel wall can provide information on the pathogenesis, on the basis of which specific therapeutic management can be initiated.

Ultrasound studies performed in Europe and the USA revealed a prevalence of 2.4% in the normal population aged 65–74 (>4 cm; Collin et al. 1988) and a prevalence of 4.9% (>3 cm) in an investigation of 1,800 subjects over 50 years of age without concomitant vascular disease (Akkersdijk et al. 1991). The prevalence increases to 10–14% in individuals with arterial occlusive disease or hypertension (Galland et al. 1991; Twomey et al. 1984).

Most aneurysms of the abdominal aorta are incidentally detected in patients undergoing an ultrasound examination for an unrelated problem (Allenberg et al. 1997). The therapeutic management of aortic aneurysms is guided by the risk of rupture, which increases with the diameter. When surgical repair is contemplated, the risk of rupture of the untreated aneurysm must be weighed against the risks of intraoperative and postoperative complications, which are considerable because aneurysms typically occur in older and multimorbid patients. The mortality risk is less than 5% in patients undergoing elective surgical resection and increases to over 50–60% in emergency surgery for a ruptured aneurysm. About 50% of patients die of aneurysmal rupture before they reach the hospital.

Several follow-up studies of subjects with aneurysms (Limet 1991; Nevitt et al. 1989; Zöllner et al. 1991) demonstrated a markedly higher risk of rupture for aneurysms greater than 5 cm, which is why a size of 5 cm evolved as the cutoff value for elective surgical resection of abdominal aortic aneurysm. Based on the UK Small Aneurysm Trial (1998), which compared the natural history and the risk of surgery, the cutoff diameter for surgical management was even elevated to 5.5 cm. Patients with an aneurysm smaller than 5.5 cm require close surveillance, and elective surgery is recommended if there is rapid growth of the aneurysm (>5 mm in 6 months), embolization to the periphery, pain, or the shape is very saccular. Other complications associated with aneurysm include compression of surrounding structures (veins, bowel) and fistulization.

*Prevention of rupture is the guiding principle of management.* Therefore, identification of factors contributing to the stabilization of an aneurysm is of crucial importance. Morphology seems to play a role (saccular aneurysms are more susceptible to rupture than spindle-shaped ones) as does thrombosis. Aneurysms with a thrombotic lining tend to grow more slowly, whereas growth appears to be accelerated by local pressure peaks associated with turbulent flow, which primarily occurs in saccular aneurysms.

The clinical usefulness as well as cost-effectiveness of B-mode scanning as the only *screening procedure* was impressively demonstrated in the recent English Multicentre Aneurysm Screening Study (MASS 2002). In this study, a total of 67,900 men aged 65–74 were screened for aortic aneurysms greater than 3 cm from 1997 to 1999. The results of this study suggest that a single ultrasound examination at the age of 65 with assignment of aneurysms to follow-up or repair reduces the risk of rupture by 53%.

*Dissecting aneurysm* (see Fig. 5.31 for aneurysm of the carotid artery) is caused by an intimal tear allowing blood to enter between the intima and media with formation of a false lumen along a segment of variable length where blood separates the intima from the media. The blood enters the false lumen at the upper point of entry and drains into the normal lumen at the distal end of the dissection (re-entry). Most aortic dissections originate in the thoracic aorta and from there may extend as far as the abdominal aorta. The distal extent serves to define different types of dissection, as in the De Bakey or Stanford classification. Transabdominal sonographic techniques are useful only for evaluating the abdominal aorta. Color duplex ultrasound is a valuable imaging modality, providing relevant information for therapeutic decision making including distal extent and involvement of visceral and renal arteries. The duplex examination will identify extension into aortic branches (renal and visceral arteries) or intermittent obstruction of the origins of these arteries by the intimal flap (Fig. A6.46).

Hence, the ultrasound examination provides *all the relevant information the surgeon needs* before open surgical aneurysm repair with patch placement: site of the aneurysm (suprarenal/infrarenal abdominal aorta), involvement of common or internal iliac artery, aneurysm extent, and presence of atherosclerotic disease in the pelvic arteries and femoral bifurcation.

Moreover, the sonographic characterization of an abdominal aortic aneurysm also provides important clues for identifying aneurysms amenable to stenting rather than open surgery. In general, the following sonomorphologic features preclude an endovascular procedure (unless a Y-stent graft is used): a short and conical proximal aneurysm neck, proximal kink >60°, mural thrombosis at the renal artery origins, accessory lower pole renal arteries, severe kinking of an iliac artery, and aneurysm extension into the internal iliac artery.

Once the decision has been made to eliminate an aortic aneurysm by stenting, a CT scan is necessary to make the measurements required for selecting an adequate stent.

In the *postinterventional surveillance* of patients treated with a stent, color duplex imaging can contribute to the early identification of stent migration and endoleaks (types I, II, III) and thus help prevent complications. Identification of an endoleak may be improved by administration of an ultrasound contrast agent.

*Arteritis* can also involve the aorta. Inflammation (e.g., giant cell arteritis, inflammatory aortic aneurysm) is suggested by thickening of the wall as opposed to thickening around the wall as is the case with retroperitoneal fibrosis (Ormond's disease). To make the distinction, the sonographer must carefully evaluate the proximal aortic branches and their origins as well as their relationship to the thickened area (see Figs. 6.19, A6.42, and A6.43).

Autopsy studies show that atherosclerosis, which is well known to be a systemic disease, begins earlier in the abdominal aorta than in other arteries, with atherosclerotic lesions appearing in the aorta about 5–10 years earlier than in the carotid or coronary arteries.

This is why the aortic arch is increasingly the focus of the search for the source of embolism in patients with embolic cerebral infarction, after atherosclerotic lesions of the carotid system have been ruled out. However, transesophageal echocardiography (TEE) is needed to reliably identify and characterize aortic plaque as a potential source of embolism.

Stenosis of the abdominal aorta can also be evaluated by ultrasound as it will typically be found in the distal aorta and the aortic bifurcation. Bilateral intermittent claudication is the chief symptom as more severe ischemia is prevented by good collateral pathways. The rare condition of aortic thrombosis may originate from atherosclerotic plaques. Predisposing factors are clotting disorders, paraneoplasia, and oral contraceptives. Aortic thrombus does not cause occlusion, but rather grows into the lumen in a cone-shaped manner, typically first becoming clinically apparent when it causes embolism of a peripheral artery.

Leriche's syndrome is an occlusive disease of the aortic bifurcation with occlusion of both common iliac arteries and of the aorta up to the origin of the inferior mesenteric artery or even to the renal artery origins. In patients with this syndrome, blood to the legs is mainly supplied by collaterals arising from the mesenteric and epigastric arteries.

### 6.1.5.2 Visceral Arteries

Symptomatic mesenteric artery stenosis (abdominal angina) is rare. The prognosis of *acute mesenteric artery occlusion* is crucially dependent on a prompt diagnosis upon hospitalization. It worsens rapidly within hours due to intestinal necrosis and short bowel syndrome even if patients are successfully operated on and survive the acute stage after bowel resection (Table 6.2).

The clinical symptoms are unspecific at the initial stage – with severe abdominal pain (Table 6.3) but no other clinicopathologic parameters; the progression is also rapid, occurring within the course of a few hours. For these reasons, timely diagnosis and embolectomy are essential. The abdominal ultrasound examination performed in most patients with unspecific abdominal pain should include (color) duplex imaging of the superior mesenteric artery. If the diagnosis of occlusion can be made with confidence by the spectral Doppler findings, immediate laparotomy can be performed without confirmatory angiography.

**Table 6.2** Acute occlusion syndrome of the superior mesenteric artery. Mortality rates in relation to delay between onset of abdominal symptoms and surgery (From Walter et al. 1992)

| Time from initial symptoms to surgery (h) | (n) | Mortality |
|---|---|---|
| 0–12 | 11 | 4 (36.3%) |
| 12–24 | 16 | 10 (62.5%) |
| >24 | 19 | 18 (94.7%) |
| Total | 46 | 32 (69.9%) |

*Visceral artery aneurysm* is uncommon but important to diagnose because of a high risk of rupture (in particular aneurysms of the splenic and hepatic arteries). Visceral aneurysm is often discovered incidentally in patients undergoing an ultrasound examination for evaluation of abdominal complaints. On color duplex scanning, a visceral aneurysm is distinguished from other lesions, in particular pseudocysts of the pancreas, at first glance.

In patients with clinical signs and symptoms of *abdominal angina* (postprandial pain, weight loss), the sonographic duplex examination must include not only the superior mesenteric artery but also the celiac trunk and possibly the inferior mesenteric artery as well. There is good collateralization of mesenteric occlusion through the Riolan anastomosis from the inferior mesenteric artery as well as through the gastroduodenal artery, pancreaticoduodenal artery, and hepatic artery (celiac trunk) (Fig. 6.13). This is why high-grade stenosis or occlusion of the superior mesenteric artery typically becomes clinically relevant only if there is concomitant stenosis or occlusion of a further visceral artery (celiac trunk or inferior mesenteric artery). Therefore, spectral Doppler recording must also include these arteries to search for stenosis, contributing to the differential diagnosis of abdominal pain and initiating proper treatment.

Stenosis of the celiac trunk may be caused by atherosclerosis (at origin) and rarely by fibromuscular dysplasia. An important cause of stenosis due to external compression is the *median arcuate ligament syndrome*, or celiac artery compression syndrome. The significance of intermittent stenosis caused by the median arcuate ligament is still controversial.

**Table 6.3** Acute occlusion syndrome of the superior mesenteric artery

| | Initial stage 1–6 h | Silent interval 7–12 (24) h | Terminal stage >(12) 24–48 h |
|---|---|---|---|
| Clinical presentation | Initial triad: | Receding pain | Paralytic ileus |
| | 1. Severe abdominal pain without local or generalized signs of peritonitis, clinically normal abdomen | Mild local changes, deteriorating general state, onset of intestinal paralysis | Peritonitis |
| | 2. Signs of shock in about 20% of cases | | Protracted shock |
| | 3. Diarrhea (anoxic) | | |
| Laboratory findings | | → Progressive leukocytosis → | |
| | | → Increase of serum lactate level → | |
| | | → Increasing CK and LDH levels → | |
| | | → Progressive acidosis → | |
| Plain radiography | Negative | Typically negative | Increased air content, fluid levels |
| B-mode ultrasound | Negative | Negative | Thickened bowel loops, air inclusions, (sub)total ileus of small intestine |
| Revascularization possible | +++ | ++ | (+) |
| Bowel resection necessary | – | (+) | ++ |
| Prognosis | | → Deteriorating → | |

The most important criterion for ligamentous compression is the respiratory variation in the degree of stenosis. The accompanying pain is most likely due to mechanical irritation of the celiac plexus. A reduction in perfusion (see Figs. 6.11 and 6.12) resulting from intermittent compression seems unlikely as there is good collateralization of the visceral vessels. Intermittent compression may, however, damage the vessel wall, thus leading to secondary stenosis. This is also confirmed by a study investigating the outcome of surgery, suggesting that a benefit can only be expected in patients with fixed celiac trunk stenosis (in both inspiration and expiration) and demonstration of a steal phenomenon by sonography and angiography (Walter et al. 1999). Surgery is indicated only in patients with epigastric pain and typical manifestations of abdominal angina such as postprandial symptoms and weight loss.

#### 6.1.5.3 Renal Arteries

For adequate treatment of high blood pressure, it is necessary to differentiate essential hypertension from secondary hypertension and, among the patients with secondary hypertension, identify those with renovascular hypertension, which is amenable to causal treatment (Table 6.4).

The average incidence of *renovascular hypertension* is 1–4% in an unselected population (van Bockel et al. 1989; Foster et al. 1973; Olbricht et al. 1991), but incidences as low as 0.18% and as high as 20% have also been reported (Arlart and Ingrisch 1984; Tucker and Lebbarthe 1977). These discrepancies are due to the use of different screening methods and the investigation of different groups of normal subjects and patients (presence of vascular risk factors and concomitant diseases, specific patient subsets). *Atherosclerotic stenosis* virtually always occurs *at the origins from the aorta* or, very rarely, at the branchings into segmental arteries; it predominantly affects older men with other occlusive vascular conditions. In contrast, stenosis due to *fibromuscular dysplasia* nearly exclusively involves the *middle thirds* of the renal arteries and primarily occurs in young women (Table 6.5). Hence, selective duplex scanning of the respective renal artery segments according to the suspected cause of stenosis can be performed.

**Table 6.4** Indications for ultrasonography of the renal arteries

- Workup of hypertension (atherosclerotic stenosis, fibromuscular dysplasia)
- Differentiation of <50% stenosis, higher-grade stenosis, and occlusion
- Follow-up after repair (operation, PTA with/without stenting)
- Suspected renal infarction
- Aortic aneurysm (spatial relationship to renal artery origins)
- Aortic dissection (possible involvement of renal arteries)
- Transplant kidney (anastomotic stenosis, rejection)

**Table 6.5** Fibromuscular and atherosclerotic renal artery stenosis

| | Fibromuscular stenosis | Atherosclerotic stenosis |
|---|---|---|
| Proportion | <10% | >90% |
| Age | Typically <40 years | Typically >40 years |
| Sex | Predominantly women | Predominantly men |
| Preferred site | Typically middle and rarely distal third | Origin or proximal third |
| Poststenotic dilatation | Frequent | Rare |
| Repair method of choice | PTA, surgical reconstruction | PTA, surgical reconstruction |

In searching for *renal artery stenosis* as a possible cause of hypertension in older patients with atherosclerotic lesions, the sonographer must therefore pay special attention to the origins of the renal arteries, where the stenosis is likely to be located in over 95% of this patient group. In younger patients, in whom a suspected stenosis may also be due to fibromuscular hyperplasia, the entire course of the renal arteries is carefully evaluated, in particular the middle third (see Table 6.5). Numerous studies have confirmed the validity of duplex sonography in identifying renal artery stenosis (Tables 6.7 and 6.8).

Until a few years ago, angiography was the method of choice for examining the renal arteries, but in recent years there has been a lively discussion of the relative merits of the different diagnostic tests. The ACC/AHA 2005 Practice Guidelines for the Management of Patients with Peripheral Arterial Disease (Hirsch et al. 2006) now favor duplex ultrasound scanning for diagnosing renal artery stenosis, followed by computed tomography (in patients without renal insufficiency) and magnetic resonance imaging; intra-arterial digital subtraction angiography (DSA) however remains the gold standard. These guidelines do not recommend scintigraphic methods or laboratory tests (including captopril challenge test). Controversy also exists about the degree of stenosis (in terms of diameter reduction) above which renal artery stenosis is able to induce hypertension and hence should be treated. As in other vascular territories, *hemodynamic relevance* (defined as >50% stenosis) *does not automatically imply clinical relevance*. Instead, clinical relevance of stenosis is defined by the diameter reduction that results in a relevant decrease in perfusion in the target organ. In the leg arteries, for instance, demand increases with activity, and hence a stenosis may cause problems only during activity but not at rest. For the carotid territory, a 60–70% diameter reduction is assumed to cause relevant perfusion reduction, and this threshold was also adopted for the renal arteries. However, simply applying these thresholds does not take into account that the renal arteries, unlike the extracranial carotid arteries, have no collaterals. More recent studies suggest that even stenosis causing only a *50% diameter reduction* produces a *marked increase in intra-arterial pressure gradients* (Staub et al. 2007). In this study, the mean systolic pressure gradient for angiographic stenosis of 50% was 24 mmHg. Other investigators found a significant *upregulation of renin* even for a 10% transstenotic pressure gradient (De Bruyne et al. 2006; Hirsch et al. 2006). Besides the need for a generally accepted threshold for clinically relevant renal artery stenosis, the other issue to be resolved is the degree of stenosis above which an attempt at revascularization is justified (percutaneous transluminal angioplasty (PTA) and stent/surgery). In the past, when surgery was the only treatment option, a higher cutoff was used because of the higher rate of morbidity compared with PTA. Catheter dilatation with stenting can be used more generously given the low complication rate and high success rate (internal quality assurance). It must be noted, however, that although studies show dilatation of renal artery stenosis to be slightly superior to medical treatment in terms of lowering arterial blood pressure and improving renal function, there is no evidence-based proof for this superiority (Balk et al. 2006; Jaarsveld et al. 2003). The disagreement about the stenosis threshold that justifies interventional or surgical treatment is also at the root of the controversy regarding the sonographic blood flow velocity cutoff and the diagnostic criteria to be used (direct or indirect): the largest group of authors advocate higher cutoff velocities, recommending PTA mainly for patients with higher-grade stenosis and severe renal dysfunction. Conversely, lower cutoff velocities are used by proponents of early PTA (typically to prevent fixed hypertension or parenchymal damage). The advocates of early PTA cannot make use of indirect sonographic criteria for diagnosis as these criteria yield reliable results only for higher-grade stenosis. Note also that PTA has no effect on essential hypertension in patients with secondary atherosclerotic wall lesions and renal artery stenosis.

After renal artery stenosis has been diagnosed by duplex ultrasound, no further diagnostic tests are needed prior to angiography with simultaneous PTA. In patients having undergone PTA with or without stenting, ultrasound is also the *follow-up method* of choice for identifying residual or recurrent stenosis.

Color duplex ultrasound should be routinely used after *kidney transplant* and can help to prevent graft loss by timely detection of early postoperative vascular complications (Aschwanden et al. 2006; Urbancic et al. 2001). The sonographic parameters determined in this examination, in particular the resistance index (RI), also serve as baseline for subsequent follow-up examinations. In the further posttransplant course, a color duplex scan should be performed whenever a deterioration of graft function or an increase in arterial blood pressure is noted.

### 6.1.6 Measurement Parameters, Diagnostic Criteria, and Role of Ultrasound

#### 6.1.6.1 Renal Arteries

A skilled examiner (more than 500 renal artery examinations) using a state-of-the-art high-end ultrasound machine should be able to identify and evaluate the renal arteries for stenosis in about 90% of cases. However, accessory renal arteries are more difficult to detect (40–70%) (Krumme et al. 1996). In the hands of an experienced examiner, sonographic evaluation of the renal arteries takes 10–20 min, depending on the acoustic window and clinical question to be answered (atherosclerotic stenosis: renal artery origins; fibromuscular dysplasia: middle thirds).

A wide range of different duplex scanning techniques and parameters have been proposed to differentiate normal findings and low-grade renal artery stenosis from hemodynamically significant higher-grade stenosis. This situation shows that all methods have their specific limitations, which one tries to overcome by using different approaches. The fact that the poor visualization of the proximal and middle thirds of the renal arteries precludes velocity measurement for direct demonstration of stenosis has prompted some investigators (Bönhof et al. 1990; Schwerk et al. 1994) to measure and compare peripheral resistance indices in both renal arteries. This is done by spectral Doppler sampling in the distal thirds of the arteries with scanning from the flank (Table 6.6).

In normal, unobstructed renal arteries, the *resistance index* (Pourcelot index, see Fig. 1.29) is roughly the same on both sides on condition that there is no unilateral renal parenchymal damage, which would lead to more pulsatile flow and thus affect the Pourcelot index as well. Distal to high-grade stenosis, flow is characterized by a delayed systolic upstroke and lower PSV, while diastolic flow is increased, resulting in a lower Pourcelot index. A Pourcelot index <0.5

**Table 6.6** Different sonographic criteria used for diagnosing renal artery stenosis

| |
|---|
| **Direct stenosis criteria** (measured at the origin or along the course of the renal artery) |
| • PSV >140 cm/s (Schäberle 1988; duplex) sensitivity 86%, specificity 83% |
| • PSV >180 cm/s (Karasch 1993; color duplex) sensitivity 92%, specificity 90% |
| • Renal-aortic ratio (RAR) >3.5 (Taylor 1988; duplex) sensitivity 84%, specificity 97% |
| **Indirect stenosis criteria** (measured in the distal renal artery/hilum) |
| • Unilateral decrease in RI (Pourcelot) >5–10% (Bönhof 1990; Schwerk 1994) |
| • Acceleration time >0.07 s (>0.1 s) measured in renal hilum (or in segmental artery) |

in the distal renal artery at the hilum suggests postocclusive changes due to upstream flow obstruction (see Fig. 6.8).

However, this method has poor sensitivity in patients with more pulsatile flow and higher Pourcelot indices resulting from a *loss of vascular compliance* due to atherosclerosis or medial sclerosis. In these patients, not even high-grade stenosis is associated with a Pourcelot index <0.5. On the other hand, *unilateral reduction of the Pourcelot index* of over 0.05 (or over 10%) compared with the contralateral side was found to have 82% sensitivity and 92% specificity for identifying >70% renal artery stenosis using angiography as the gold standard (Schwerk et al. 1994). This method takes into account elasticity losses as well as systemic factors such as the effects of hypercirculation or hypertension, which may be a source of error in flow velocity measurements for stenosis quantification.

These *indirect methods* are limited by the fact that they will miss bilateral renal artery stenosis. Moreover, the results are influenced by the presence of parenchymal damage, which affects the RI. Although renal damage with loss of parenchyma and renal atrophy can be identified by B-mode imaging and thus taken into account in the measurements, some uncertainty will remain. Parenchymal damage can also be caused by long-standing renal artery stenosis. In these cases, a high resistance index is an indicator of parenchymal damage and can be used to identify those *patients who will not benefit from renal artery recanalization* due to the extent of kidney damage that has already occurred at the time of diagnosis. This is assumed to be the case if the ipsilateral intrarenal Pourcelot index is >0.8–0.85 (Radermacher et al. 2000).

The *resistance index* can be regarded as *a sensitive indicator of an early, relevant loss of kidney function*. Especially in patients with signs of a hepatorenal syndrome, higher-grade liver cirrhosis is associated with vasoconstriction in the renal cortex, which is reflected in an increased RI. In these patients with advanced liver cirrhosis, the RI is increased before laboratory tests show any signs of impaired kidney function (Götzberger et al. 2008).

Ipsilateral comparison of Pourcelot indices at the renal artery origin and at its distal end near the renal hilum is a useful criterion for identifying stenosis due to fibromuscular fibrosis if overlying bowel gas or obesity precludes proper assessment of the middle third of the artery. Obstruction by high-grade stenosis will lead to more pulsatile flow with a higher Pourcelot index upstream and a lower index downstream due to a decreased PSV and a corresponding increase in diastolic flow velocity (Fig. 6.8a, b).

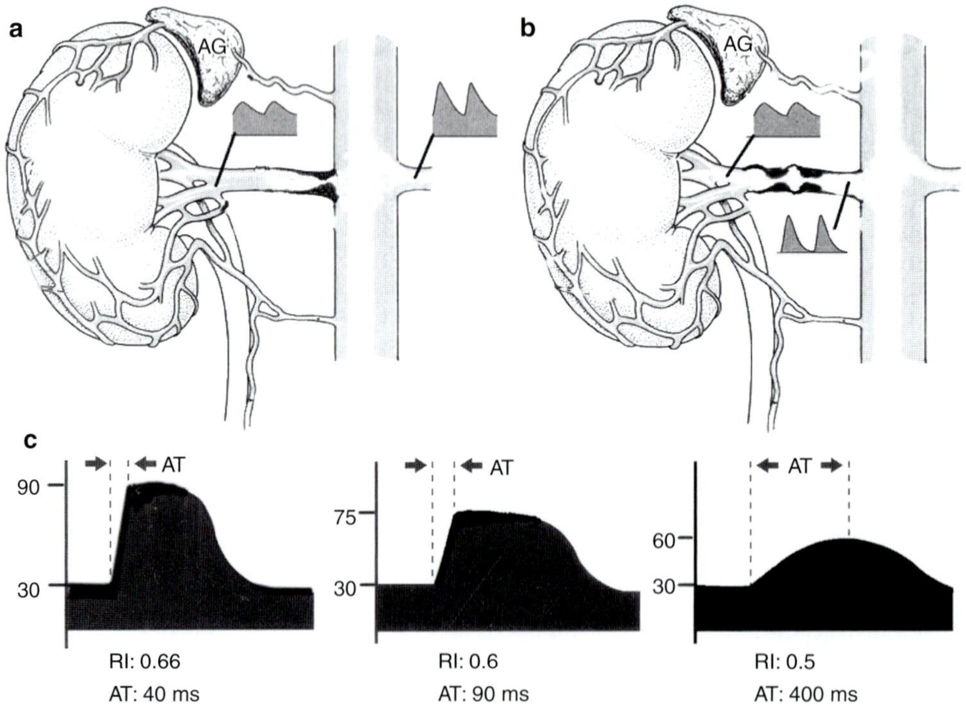


**Fig. 6.8** **a** Atherosclerotic renal artery stenosis (at origin). Changes in postocclusive waveform: less pulsatile flow with delayed systolic upstroke, slower peak systolic velocity (PSV), and corresponding increase in diastolic flow. Lower resistance index (RI, Pourcelot index) as compared with nonstenosed, contralateral artery (see Figs. A6.23 and A6.24). **b** Fibromuscular dysplasia of renal artery (see Fig. A6.25). Prestenotic and poststenotic waveforms with changes resulting from a stenosis in the middle third. Flow is more pulsatile upstream of the stenosis and becomes less pulsatile downstream with a markedly larger diastolic component and decreased RI (*AG* adrenal gland). **c** Poststenotic renal artery Doppler waveforms obtained at renal hilum (evaluation for indirect stenosis criteria). Diagrams illustrating the changes seen with increasing stenosis severity (*from left to right*): normal to mild stenosis, moderate stenosis (60–70%), high-grade stenosis. The poststenotic decrease in pressure is associated with a decrease in PSV, resulting in a lower RI (Pourcelot index). With increasing stenosis severity, the systolic upstroke (i.e., time to peak or acceleration time, *AT*) is delayed

To account for the *effects of hypertension* (false-positive results) on the angle-corrected spectral waveform, it has been proposed that the *PSV* in the aorta be related to that at the origin of the renal artery. The resulting *renal-aortic ratio* (*RAR*) is normally <3.5, with a ratio >3.5 indicating hemodynamically significant renal artery stenosis (60–99%). The RAR was found to have a surprisingly high sensitivity of 83–100% and a specificity of 73–97% (Frauchiger et al. 1995; Taylor et al. 1988; Miralles et al. 1996). Other investigators report poorer results (see Table 6.8).

As with stenosis of the carotid system or peripheral arteries, it is also possible, with a high degree of accuracy, to confirm renal artery stenosis directly on the basis of an increase in PSV above a specified threshold as determined from the angle-corrected Doppler waveform.

*Threshold velocities defined by receiver operating characteristic (ROC) curve analysis* always represent a compromise between sensitivity and specificity. A lower threshold increases sensitivity but decreases specificity, while a higher threshold improves specificity but at the cost of sensitivity. Moreover, the ideal threshold determined varies with the concomitant diseases present in the study population. For instance, a higher cutoff value will be identified in a population with a larger proportion of diabetics with medial sclerosis and more pulsatile flow or in a population including subjects with high blood pressure values during the examination.

In a study by our group (1989), conventional duplex ultrasound already had a sensitivity of 86% and specificity of 83% compared with intra-arterial catheter angiography/X-ray densitometry (Fig. A6.22). Forty-four of 170 renal arteries were found to have hemodynamically significant stenosis (>50%). The relatively high prevalence of renal artery stenoses in this population was due to a selection bias (special angiology consultation service). In a ROC curve analysis in that early study, we established a threshold of 140 cm/s to differentiate normal findings and low-grade stenosis from hemodynamically significant stenosis (Fig. 6.9).

In later studies (Karasch et al. 1993), in particular those placing the Doppler angle correction cursor using color duplex, which is especially important when interrogating the more curved right renal artery, *the cutoff value for significant stenosis was corrected to a PSV of 180 cm/s* (Table 6.7).

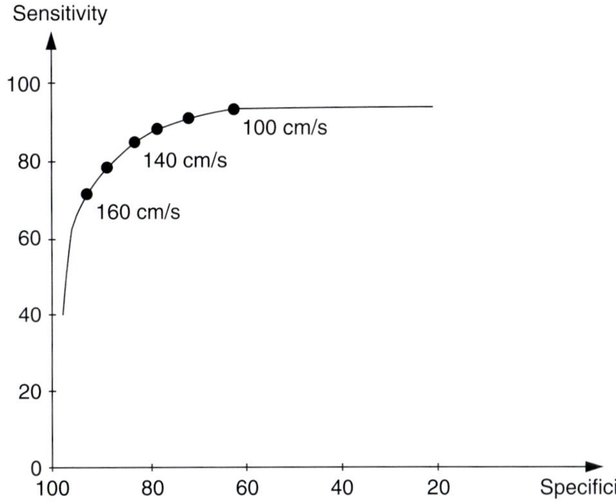


**Fig. 6.9** Receiver-operating characteristic (ROC) curve for identifying the optimal cutoff value (peak systolic velocity) for differentiating a normal renal artery or low-grade stenosis from hemodynamically significant stenosis (>50%). A threshold velocity of 140 cm/s yields a sensitivity of 86% and a specificity of 83% compared with 75% and 93%, respectively, for a velocity of 160 cm/s (investigated in 170 renal arteries, including 44 with significant stenosis)

**Table 6.7** Studies investigating the sensitivity and specificity of (color) duplex ultrasound in identifying hemodynamically significant renal artery stenosis using angiography as the gold standard

| Author | Total No. of renal arteries/No. of stenoses | Method/stenosis criteria | Sensitivity (%) | Specificity (%) | Reference angiography |
|---|---|---|---|---|---|
| Duplex imaging | | | | | |
| Avasthi 1984 | 52/26 | PSV > 100 cm/s | 89 | 73 | IA |
| Kohler 1986 | 43/? | RAR > 3.5 | 91 | 95 | ? |
| Ferretti 1988 | 104/27 | PSV >100 cm/s | 100 | 92 | conv. |
| Taylor 1988 | 58/14 | RAR > 3.5 | 84 | 97 | ? |
| Strandness 1990 | 58/14 | RAR > 3.5 | 84 | 97 | ? |
| Hoffmann 1991 | 85/64 | PSV >180 cm/s | 95 | 90 | ? |
| Schäberle 1988 | 91/44 | PSV >140 cm/s | 86 | 83 | IA, conv. |
| Color duplex imaging | | | | | |
| Breitenseher 1992 | 41/8 | PSV >120 cm/s | 17 | 89 | IA |
| Karasch 1993 | 277/109 | PSV >180 cm/s | 92.7 | 89.8 | conv., IA, IV |
| Spies 1995 | 268/42 | – | 93 | 92 | IA |
| Souza de Oliveira 2003 | 50 | PSV >150 cm/s | 83.3 | 89.5 | IA |

*IA* intra-arterial DSA, *IV* intravenous DSA, *conv.* conventional angiography

**Table 6.8** Studies investigating the accuracy of ultrasound in identifying hemodynamically relevant renal artery stenosis. Combination of different stenosis criteria (direct/indirect) for improving the diagnostic accuracy of ultrasound

| Author | Number | Method/stenosis criteria | Sensitivity (%) | Specificity (%) | Reference method |
|---|---|---|---|---|---|
| Zeller et al. 2001 | 69 (>70% stenosis) | RAR > 3.5 | 100 | 60 | Angiography |
| | | ΔRI > 0.5 | 77.5 | 99 | Angiography |
| | | RAR > 3.5 and ΔRI > 0.05 | 76 | 97 | Angiography |
| Krumme et al. 1996 | 135 (>50% stenosis) | PSV > 200 cm/s and ΔRI > 0.05 | 89 | 92 | Angiography |
| Hong et al. 1999 | 58 (60% stenosis) | PSV > 200 cm/s | 91 | 75 | Angiography |
| | | RAR > 3.5 | 72 | 92 | Angiography |
| | | AT > 100 ms | 50 | 86 | |
| | | Conclusion: use of a combination of criteria is recommended | | | |
| Motew et al. 2000 | 41 (> 60% stenosis) | PSV > 180 cm/s | 94 | 88 | Angiography |
| | | AT > 58 ms | 58 | 96 | Angiography |
| | | Conclusion: use of a combination of criteria is recommended | | | |
| Ripolles et al. 2001 | 60 (>75% stenosis) | AT > 80 ms | 89 | 99 | Angiography |
| | Age <50 | AT > 80 ms | 100 | 100 | Angiography |
| | Age >50 | AT > 80 ms | 75 | 97 | Angiography |
| | Age <50 | ΔRI > 0.05 | 90 | 93 | Angiography |
| | Age >50 | ΔRI > 0.05 | 0 | 100 | Angiography |
| | | Conclusion: ΔRI and AT are only reliable in patients younger than 50 years | | | |
| Radermacher et al. 1999 | 226 (>50% stenosis) | PSV > 180 cm/s and hilar PSV < 25% of intrastenotic PSV AT >70 ms | 96 | 98 | Angiography |
| Conkbayir et al. 2002 | 50 (>60% stenosis) | PSV > 180 cm/s | 89 | 88 | Angiography |
| | | RAR > 3.0 | 86 | 97 | Angiography |
| | | AT > 70 ms | 48 | 93 | Angiography |
| | | PSV > 180 cm/s or RAR > 3.0 | 92 | 88 | Angiography |
| | | PSV > 180 cm/s or RAR > 3.0 or AT >70 ms | 87 | 86 | Angiography |
| | | Conclusion: use of a combination of criteria is recommended | | | |
| Staub et al. 2007 | 49 (>50% stenosis) | PSV > 200 | 92 | 81 | Angiographic stenosis grade, intra-arterial pressure across stenosis |
| | | RAR > 3.0 | 83 | 91 | Angiography |
| | | ΔRI > 0.05 | 31 | 97 | Angiography |
| | 49 (>70% stenosis) | PSV > 250 cm/s | 89 | 70 | Angiography |
| | | RAR > 3.5 | 84 | 72 | Angiography |
| | | ΔRI > 0.05 | 42 | 91 | Angiographic stenosis grade, intra-arterial pressure across stenosis |
| | | Conclusion: PSV is recommended, may be combined with RAR (and ΔRI) to improve specificity | | | |

*AT* acceleration time, *ΔRI* side-to-side difference in intrarenal resistance indices, *RAR* renal-aortic ratio, *PSV* peak systolic velocity

The study of Karasch et al. (1993) included 185 renal arteries examined by color duplex imaging, which was found to have a sensitivity of 92% and a specificity of 91% using intra-arterial angiography and digital subtraction angiography (DSA) as reference methods.

Other investigators (Berland et al. 1990; Breitenseher et al. 1992; Desberg et al. 1990) used a PSV cutoff of 100–120 cm/s to differentiate low-grade from hemodynamically significant stenosis (based on the threshold used for the internal carotid artery). In light of more recent study, the results of these studies are no longer valid.

Staub et al. (2007) recently investigated different sonographic criteria routinely used for identifying significant renal artery stenosis, showing that using a PSV of 200 cm/s as a cutoff (identified by ROC analysis) had 92% sensitivity and 81% specificity in identifying stenosis with 50% diameter reduction. The renal-aortic velocity ratio (RAR) >2.5 was found to have a sensitivity of 92% and a specificity of 79% with an RAR of 3 decreasing sensitivity to 83% but raising sensitivity to 91%. The side-to-side difference in intrarenal resistance indices (deltaRI) of >0.05 had high specificity of 97% for detecting stenosis with at least 50% diameter reduction, while the sensitivity was only 31%.

The *poststenotic pressure drop with decreased perfusion* after higher-grade stenosis induces the affected kidney (renin-angiotensin system) *to counterregulate the seemingly low systemic blood pressure*. It was long assumed that poststenotic perfusion is decreased only if the stenosis grade is above 60–70%, which is reflected in a PSV of at least 200–220 cm/s. In inconclusive cases, stenosis of this magnitude can be diagnosed by additionally taking into account indirect criteria such as (audible) turbulence. For those who argue that interventional or surgical treatment should be reserved for higher-grade renal artery stenosis, the definition of a precise velocity cutoff is thus less important in identifying patients who should be treated. On the other hand, the above-quoted study of Staub et al. shows that 50% stenosis is already hemodynamically relevant, as suggested by an invasively measured pressure gradient of over 20 mmHg (Staub et al. 2007). This study impressively illustrates the problems encountered in defining cutoff points for the sonographic criteria used to diagnose renal artery stenosis (PSV, RAR, deltaRI). A high sensitivity is achieved at the cost of specificity, and vice versa. An ideal velocity cutoff value for the renal arteries should detect all stenoses causing at least 70% diameter reduction, that is, it should have a high sensitivity combined with a high negative predictive value; this is because the majority of investigators advocate intervention for these patients (Zeller et al. 2003). In those cases where renal artery stenosis can be treated by PTA with stenting, the diagnostic test should also reliably identify lower-grade stenosis (50%); the rationale here is that it has been shown that 50% stenosis is already associated with a poststenotic pressure drop. In these patients, PTA is an option if a benefit is expected based on the patient's clinical presentation and the effectiveness of other blood-pressure-lowering treatments. Hence, in this subset of patients, in whom PTA is contemplated as a realistic and beneficial treatment option, a lower cutoff point can be used even when it comes at the cost of a certain number of possibly unnecessary angiographies being performed, that is, in those patients who proceed to angiography with PTA standby based on the sonographic results. However, note that in patients with borderline renal artery stenosis, there is as yet no adequate evidence-based data available to prove any benefits of PTA over drug treatment of hypertension.

In diagnosing greater than 70% renal artery stenosis using a PSV of 200 cm/s, Olin et al. (1995) demonstrated a sensitivity and specificity of 98% and Krumme et al. (1996) a sensitivity of 71% and a specificity of 96%. Surprisingly good results were reported by Baxter et al. (1996), who found a sensitivity of 100% and a specificity of 95% using a PSV of 250 cm/s as a cutoff value.

Another indirect criterion, apart from a reduced poststenotic RI at the renal hilum, is a *delayed systolic rise (prolonged acceleration time)* or a *reduced acceleration index* (Kliewer et al. 1997; Stavros et al. 1994; Postman et al. 1996; Nazzal et al. 1997; Patriquin et al. 1992). An acceleration index of less than 300 cm/s and an acceleration time of more than 0.07 s are abnormal and indicate greater than 60% stenosis (Baxter et al. 1996; Kliewer et al. 1997; Stavros et al. 1992; Isaacson et al. 1995; Nazzal et al. 1997; Martin et al. 1991) with sensitivities of 78–89% and specificities of 92–98%. However, these *indirect criteria are not helpful in identifying moderate stenosis* (<70–80%) as the poststenotic blood flow abnormalities must reach a certain level before they are reliably reflected in changes in the indirect criteria that can determined by renal hilar ultrasound (not least of all because the parameters are influenced by renal parenchymal changes). Some investigators therefore propose an acceleration time cutoff of 0.1–0.12 s, which has been found to be more specific (Dawson 1996; Baxter et al. 1996; Isaacson et al. 1995). Comparison of direct duplex criteria of renal artery stenosis (intrastenotic PSV) and indirect criteria (poststenotic RI, acceleration time) with angiography showed better correlation for direct criteria, while the indirect criteria were found to be generally less reliable (van der Hulst et al. 1996). The unreliability is most likely attributable to renal parenchymal disease and loss of vascular compliance: many renal diseases and the chronic renal changes they cause lead to an increase in the vascular RI; these include both acute and chronic renal diseases, glomerulonephritis, pyelonephritis, urinary tract obstruction, and obstruction of the renal veins. Acceleration time is also affected by different factors such as arterial wall compliance and disturbed microcirculation in different conditions associated with renal parenchymal damage (in particular diabetic nephropathy). These factors affecting the indirect sonographic criteria in patients with underlying renal disease explain the discrepant results with part of the studies showing satisfactory to good correlation of indirect criteria with angiography, in particular in patients with at least 70% renal artery stenosis (Baxter et al. 1996; Dawson 1996; Stavros et al. 1992; Patriquin et al. 1992; Nazzal et al. 1997; Schwerk et al. 1994) and other investigators reporting poor diagnostic accuracy and very disappointing correlations (Kliewer et al. 1994, 1997; Postman et al. 1996; Isaacson et al. 1995; van der Hulst et al. 1996).

An interesting observation on the usefulness of indirect sonographic criteria such as acceleration time or side-to-side

### 6.1.6.2 Visceral Arteries

**Celiac Trunk** Stenosis of the celiac trunk is rare and may be caused by atherosclerosis or fibromuscular dysplasia. The rare median arcuate ligament syndrome is characterized by intermittent compression of the celiac trunk by the diaphragm during respiration. Atherosclerotic stenosis does not become clinically apparent unless several visceral arteries are obstructed.

Compression of the celiac trunk by the median arcuate ligament of the diaphragm (Fig. 6.11) can lead to luminal narrowing identifiable by angiography and ultrasound. It is controversial whether the nonspecific abdominal symptoms are due to hemodynamic disturbances or mechanical irritation of the celiac plexus. However, a primary vascular component of the compression syndrome is unlikely, given the rich collateral pathways.

The *median arcuate ligament syndrome* has a characteristic angiographic appearance, namely a concave constriction of the origin of the celiac trunk from above. On color duplex scanning, the constriction produces a stenosis signal in the celiac trunk that varies with the respiratory movement of the diaphragm. The stenosis increases with expiration and decreases with inspiration and may completely disappear in deep inspiration.

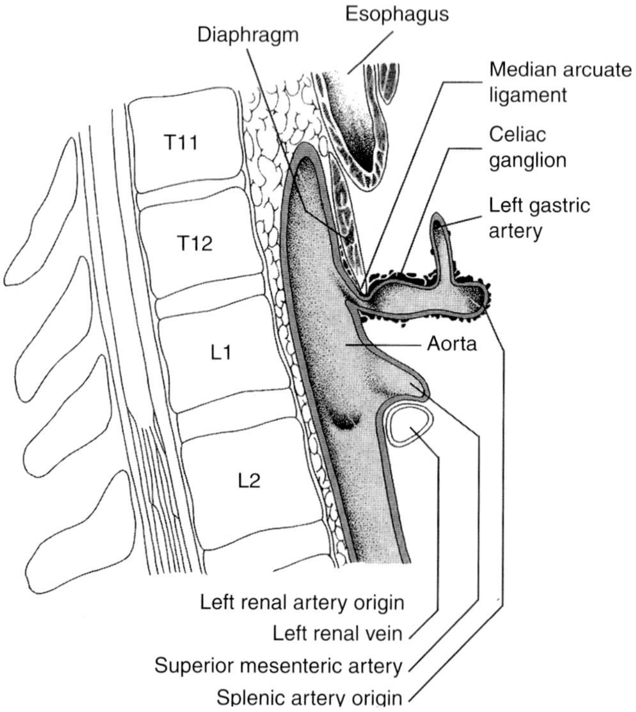


**Fig. 6.11** Topographic relationships between the median arcuate ligament, aorta, celiac trunk, superior mesenteric artery, and celiac ganglion. Mechanism of compression of the proximal celiac trunk by the arcuate ligament (from Schwilden 1987)

Collateral pathways and the release of blood circulation during inspiration ensure adequate perfusion. Still, the intermittent compression can damage the vessel wall and trigger deposition of thrombotic material, resulting in a so-called fixed stenosis and poststenotic dilatation, as in compression syndromes of other body regions. Upper abdominal pain is most likely due to the pressure exerted by the arcuate ligament and diaphragm on the vegetative nerves encircling the celiac artery.

As with other compression syndromes that are confirmed by a function test, duplex sonography is the method of choice for diagnosing the median arcuate ligament syndrome. The examination is performed during both *inspiration and expiration* to confirm *intermittent compression* of the celiac trunk by the ligament.

*Median arcuate ligament division* is indicated if a *fixed stenosis* is present, which is suggested by peak systolic velocities (PSV) exceeding 280 cm/s during both inspiration and expiration. Surgery is promising and likely to eliminate the compression-related symptoms in patients in whom a steal effect has been demonstrated by mesentericography or celiacography and in whom the clinical symptoms are due to this effect (abdominal angina with epigastric and postprandial pain and weight loss) and not to compression of the hypogastric plexus (pain). Other possible causes such as atherosclerotic stenosis of the mesenteric arteries, tumor compression, or chronic pancreatitis must be ruled out (Figs. 6.12, A6.7, and A6.8).

PSV exceeding 220–250 cm/s (fasting) indicates hemodynamically significant stenosis (>50%). However, in the celiac trunk, only higher-grade stenosis (>70%) with a PVS of more than 280–300 cm/s will affect perfusion in a

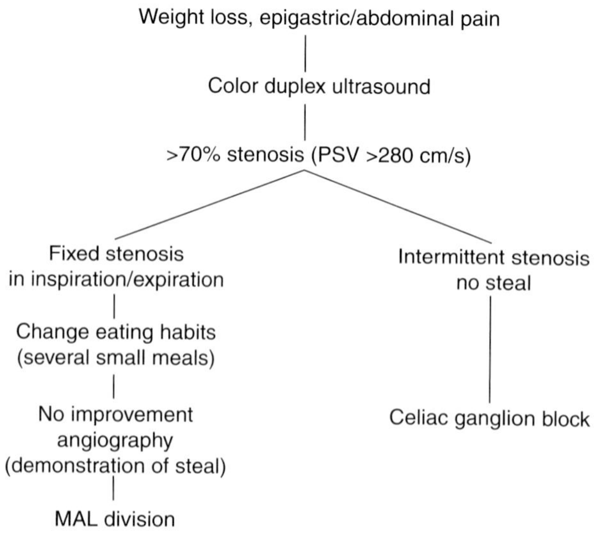


**Fig. 6.12** Diagnostic and therapeutic decision algorithm in median arcuate ligament (MAL) syndrome

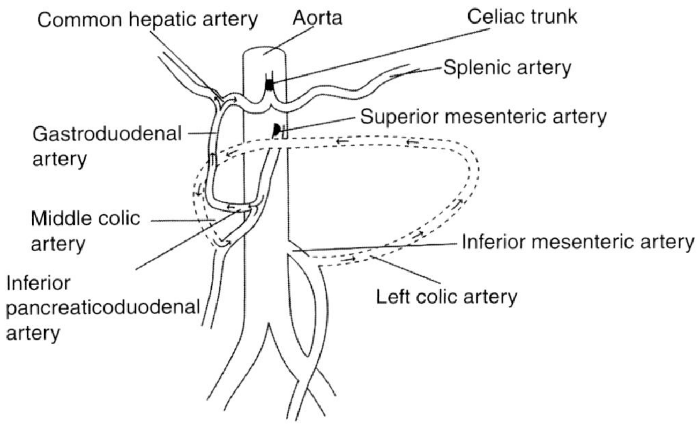


**Fig. 6.13** Diagram of collateral pathways in occlusion of the celiac trunk and/or superior mesenteric artery: the Riolan anastomosis (*dashed lines*) between the superior mesenteric artery and the middle colic artery is the main collateral pathway in occlusion of the superior mesenteric artery. Collaterals between the celiac trunk and superior mesenteric artery include the pancreaticoduodenal artery and the gastroduodenal artery, which joins the hepatic artery

significant manner (see Moneta et al. 1991). In case of *occlusion of the celiac artery*, backward flow can be demonstrated in the splenic or hepatic artery, depending on whether the occlusion is bridged by *collaterals* coursing to the splenic hilum or via the gastroduodenal artery. Flow velocities in the celiac trunk territory should be measured in the resting expiratory position to obtain a representative measurement.

Aneurysms of the hepatic and splenic arteries are rare and are conspicuous on B-mode scans as hypoechoic to anechoic round structures. Mural thrombosis is seen as echogenic layering. Aneurysms are differentiated from tumors or pseudocysts of the pancreas by the demonstration of flow in the color flow mode. Precise localization of a hepatic artery aneurysm is crucial for therapeutic decision making: an aneurysm proximal to the origin of the gastroduodenal artery from the common hepatic can be ligated without reconstruction because the liver will be supplied with blood via the gastroduodenal artery; for elimination of a more distal aneurysm (proper hepatic artery), however, vascular repair is necessary. Surgery is planned on the basis of the sonographically determined site of the aneurysm and its relationship to the origin of the gastroduodenal artery.

**Superior Mesenteric Artery** The *wide variation in flow volumes and systolic and end-diastolic velocities* in the mesenteric artery, *according to demand*, makes it necessary to examine patients in the fasting state in order to obtain standardized measurements and apply threshold velocities.

Mesenteric blood flow increases after eating (dilatation and increased flow velocity) and is also *affected by other physiologic and disease states as well as by pharmacologic agents*. Decreases in blood flow velocity and volume are observed after physical exertion and under the influence of vasopressin. An increase in mesenteric PSV and blood flow volume can be observed after glucagon administration and in individuals with severe hyperthyroidism or during acute episodes of inflammatory bowel disease with extensive involvement of bowel segments (Derko 2001).

*Quantification of blood flow* relies on the determination of the mean flow velocity and precise measurement of the vessel diameter. Diameter measurements in the superior mesenteric artery by our group demonstrated variations of about 10% between systole and diastole with ensuing differences in the cross-sectional area of up to 35%. Therefore, precise blood flow measurement makes it necessary to measure systolic and diastolic diameters separately and to calculate a mean vessel diameter according to the following formula, representing the two diameters according to their relative weight:

Mean vessel diameter / radius (R)

$$R = 1/3 \times (2 \times R_{\text{diastolic}} + R_{\text{systolic}})$$

For vessels up to 12 mm in diameter, the diameter can be measured most precisely using the leading-edge method (see Fig. 1.28) and scanning with a low transmit power. This method results in slight overestimation of the diameter but, for diameters of up to 10 mm, the overestimation is smaller than the underestimation that would result from using the inner-wall-to-inner-wall method. Also, the method enables systematization of the measurement error, which is important for serial measurements.

*Color duplex imaging* facilitates the identification of the mesenteric and renal arteries. Once the vessel has been brought into view, the Doppler spectrum is recorded for hemodynamic evaluation. Under good insonation conditions, color flow imaging will suggest a stenosis, but verification by spectral Doppler is necessary. Depending on the clinical question to be answered, spectral Doppler tracings should be sampled at the vessel origins, the preferred sites of atherosclerotic stenosis of the visceral arteries.

Atherosclerotic stenosis of visceral branches usually occurs at the origins from the aorta. Involvement of the peripheral branches is only seen in diabetics with generalized medial sclerosis. If there is high-grade atherosclerotic stenosis of only one of the three visceral artery origins, compensatory dilatation of the preformed *collateral pathways* will ensure adequate perfusion in most cases.

In general, *chronic intestinal ischemia* manifests as abdominal angina only if there is occlusion or stenosis of more than one visceral artery or in case of poor collateralization. The typical symptom is postprandial pain. *Calcified plaques* suggest a stenosis in the B-mode scan, but definitive evidence is provided only by flow acceleration with turbulence or the absence of flow signals in case of occlusion.

In *chronic mesenteric artery occlusion*, ultrasound will often depict the main *collateral vessels* (gastroduodenal, splenic, and inferior mesenteric arteries), but angiography provides a better overview and enables more reliable evaluation of collateral pathways.

Chronic mesenteric artery occlusion is due to atherosclerosis and is associated with extensive collateralization through the celiac trunk (primarily the pancreaticoduodenal artery) and the inferior mesenteric artery (Riolan anastomosis). Since patients with chronic mesenteric occlusion are usually thin, the dilated gastroduodenal artery is visualized by duplex ultrasound at the pancreatic head. The superior mesenteric artery is filled distally and shows postocclusive flow with a delayed and reduced systolic rise and a decreased Pourcelot index.

A peak systolic velocity (PSV) of over 200 cm/s in fasting patients suggests *stenosis of the origin of the superior mesenteric artery*. The increase in PSV is proportional to the degree of stenosis: >280 cm/s corresponds to >75% stenosis (Fig. 6.14). In a prospective study of over 100 patients with angiographic stenosis >70% in the superior mesenteric artery, duplex ultrasound with a cutoff value of 275 cm/s had a sensitivity of 92% and a specificity of 96% (Moneta et al. 1991, 1993). In the celiac trunk, sensitivity was 87% with 80% specificity when using a threshold of 200 cm/s for angiographic stenosis >70% (Moneta et al. 1993). These data only apply to fasting subjects with normal vascular anatomy.

Further studies are listed in Table 6.9. Note that other investigators use similar cutoff velocities (PSV and end-diastolic velocity) as Moneta et al. but to identify 50% stenosis.

Acute mesenteric occlusion due to embolism is easily and reliably demonstrated by (color) duplex imaging as the absence of flow if the occlusion is located near the origin of the mesenteric artery from the aorta. Peripheral mesenteric artery occlusions, on the other hand, pose a diagnostic problem. If there is extensive infarction of the small intestine but the main mesenteric artery is patent, the embolus is typically lodged at the divisions into jejunal branches or further distally at the origins of the ileocolic and right colic arteries. If there are patent branches such as the middle colic artery or proximal segments of the jejunal branches, the trunk of the mesenteric artery is patent as well. The overall reduction in blood flow and peripheral dilatation in the territory of the patent branches, which provide collateral flow via the arcades, is reflected in the corresponding spectral Doppler waveforms (see Figs. A6.11 and A6.12). The PSV is reduced, and the lower peripheral resistance results in a larger diastolic component with a decreased Pourcelot index (Table 6.10).

The waveform changes become more conspicuous with the number of occluded branches, which in turn increases the more proximal an embolus is located. However, the number of vessels involved has little clinical relevance and does not affect the patient's prognosis because the loss is

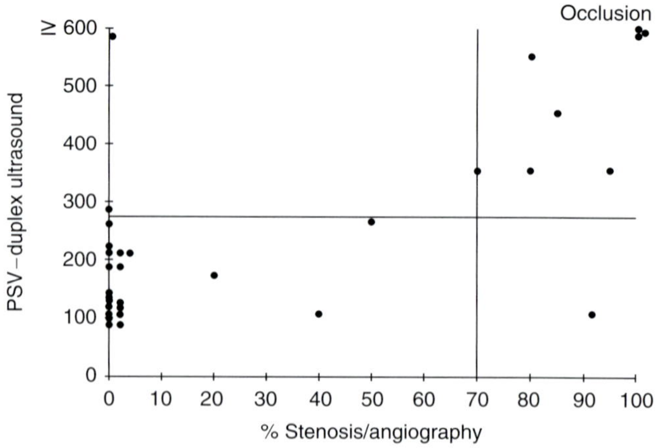


**Fig. 6.14** Duplex ultrasound versus angiography in the diagnostic evaluation of stenosis at the origin of the superior mesenteric artery (from Moneta et al. 1993). The horizontal line represents the peak systolic velocity (PSV) of 275 cm used as a threshold for angiographic 70% stenosis. The dots in the top right corner represent the mesenteric artery occlusions correctly diagnosed by duplex ultrasound. In the retrospective analysis, duplex ultrasound had 89% sensitivity and 92% specificity compared with angiography (*n* = 34)

**Table 6.9** Role of duplex sonography in diagnosing stenosis at the origins of the superior mesenteric artery and celiac trunk: stenosis criteria (velocity cutoff values) and accuracy compared with angiography

| Author | Stenosis (%) | Method (cm/s) | Superior mesenteric artery | | Celiac trunk | |
|---|---|---|---|---|---|---|
| | | | Sensitivity (%) | Specificity (%) | Sensitivity (%) | Specificity (%) |
| Bowersox 1991 | 50 | PSV > 300 | 63 | 100 | | |
| | 50 | EDV > 45 | 100 | 92 | | |
| Perko 1997, 2001 | 50 | PSV > 275 | 93 | 80 | | |
| | | EDV > 50 | 100 | 100 | | |
| Perko 1997, 2001 | 50 | PSV > 200 | | | 94 | 94 |
| Moneta 1991, 1993 | 70 | PSV > 275 | 92 | 96 | | |
| Moneta 1993 | 70 | PSV > 200 | | | 87 | 80 |

*Method*: stenosis grading based on duplex ultrasound measurement of flow velocity across the stenosis. *PSV* peak systolic velocity, *EDV* end-diastolic velocity

**Table 6.10** Duplex ultrasound criteria for pathology of the superior mesenteric artery

| |
|---|
| **Stenosis** |
| • PSV > 280 cm/s (fasting) |
| **Proximal occlusion** |
| • Absence of flow signals |
| **Distal occlusion** |
| • Under good insonation conditions, absence of flow signal in occluded mesenteric branch |
| • Indirect signs of hemodynamically significant changes in waveform of proximal vessel segment: |
| – Reduced PSV |
| – Reduced RI |
| – Thump pattern immediately before occlusion |

*PSV* peak systolic velocity, *RI* resistance index

compensated for by collateral flow through the patent branches and the arcades. In the worst case, short ischemic segments will persist and undergo necrosis over time, becoming clinically manifest as peritonitis (Fig. 6.15). Surgical resection of these short segments is indicated and has a good prognosis.

In summary, *more central* and extensive, and hence prognostically relevant, *mesenteric occlusions cause marked changes in the shape of the Doppler waveform* recorded from the *main trunk of the superior mesenteric artery* (see Fig. A6.12). Consequently, these spectral Doppler changes in conjunction with the above-described decreases in the Pourcelot index and PSV should prompt a careful evaluation of the individual mesenteric branches distally in longitudinal and transverse planes using color duplex ultrasound to identify flow (Fig. 6.16). In the absence of flow, either emergency angiography is required or, if the ultrasound diagnosis can be made with confidence, immediate laparotomy for embolectomy of the affected branches can be performed.

The ultrasound examination is technically difficult in advanced occlusion due to overlying air, pain, and inability of the patient to cooperate. In these cases, however, the clinical symptoms will lead to surgery with ultrasound contributing only little to therapeutic decision making. The prognosis is poor. In early occlusion, when the clinical symptoms are not yet severe enough to warrant emergency surgery (see Table 6.3), the insonation conditions are adequate and, in most cases, allow evaluation of the main mesenteric artery and proximal divisions.

Color duplex ultrasound is not the method of choice; it is not specific and is highly examiner-dependent. In the past, invasive catheter angiography was used for this purpose, now having been replaced by contrast-enhanced high-resolution multislice CT.

Nevertheless, the above-described spectral waveform changes are useful in identifying patients with embolic occlusion of the distal mesenteric trunk and several jejunal branches, which can ultimately cause necrosis of considerable portions of the bowel. These waveform changes and the ability to evaluate the individual jejunal branch origins and the distal mesenteric artery for patency (Fig. 6.16) make color duplex ultrasound a practical imaging test that can easily be integrated into the sonographic examination performed in all patients presenting with an acute abdomen. In an analysis by the author of 71 patients for whom there was a strong clinical suspicion of mesenteric artery occlusion and who were examined by duplex ultrasound on presentation, occlusion was confirmed in 19 patients (27%). These patients proceeded to surgery with embolectomy (which confirmed the ultrasound findings). Four patients (6%) had peripheral occlusion of individual jejunal branches or nonocclusive ischemic disease without changes in the Doppler spectrum recorded in the main mesenteric artery, but in these cases only very short bowel segments had to be resected. In 37 patients (52%), acute embolic mesenteric artery occlusion was ruled out by ultrasound; these findings were confirmed by the patients' clinical course or by operations performed for other underlying causes of acute abdomen. Eleven patients in whom ultrasound or Doppler measurement was technically inadequate underwent angiography, or contrast-enhanced CT.

To confirm or exclude ischemia as the cause of sonographically demonstrated thickening of bowel loops, the bowel wall and adjacent mesentery should be examined for the presence of flow using a high-resolution scanner operated with a high but artifact-free gain and low pulse repetition frequency. Color duplex detection of flow with confirmation by spectral Doppler measurement rules out ischemia. A large diastolic component in the Doppler waveform is indicative of an underlying inflammatory condition (see Fig. A6.14d).

The RI (Pourcelot index) of the superior mesenteric artery is an unspecific parameter because it is also decreased in patients with peritonitis or sepsis. In this setting, the decision for surgical revision can be made on clinical grounds in most cases and only inconclusive cases require a B-mode scan or other radiologic imaging test. The spectral waveform in sepsis or peritonitis differs from that in distal mesenteric occlusion (decreased PSV) in that the systolic rise is still relatively high and may be close to normal.

The indirect sonographic criteria cannot be quantified and, if present, should prompt further diagnostic testing (angiography) or, if warranted in conjunction with the clinical presentation, laparotomy. *Hypotension* and *tachycardia* as in septic shock or generalized *peritonitis* also cause marked hemodynamic changes in the mesenteric vessels and abnormal Doppler waveforms. Thus, the spectral waveform must always be interpreted in conjunction with the clinical presentation. However, a decreased Pourcelot index with a simultaneous decrease in peak systolic and mean flow velocities in the Doppler spectrum from the proximal superior

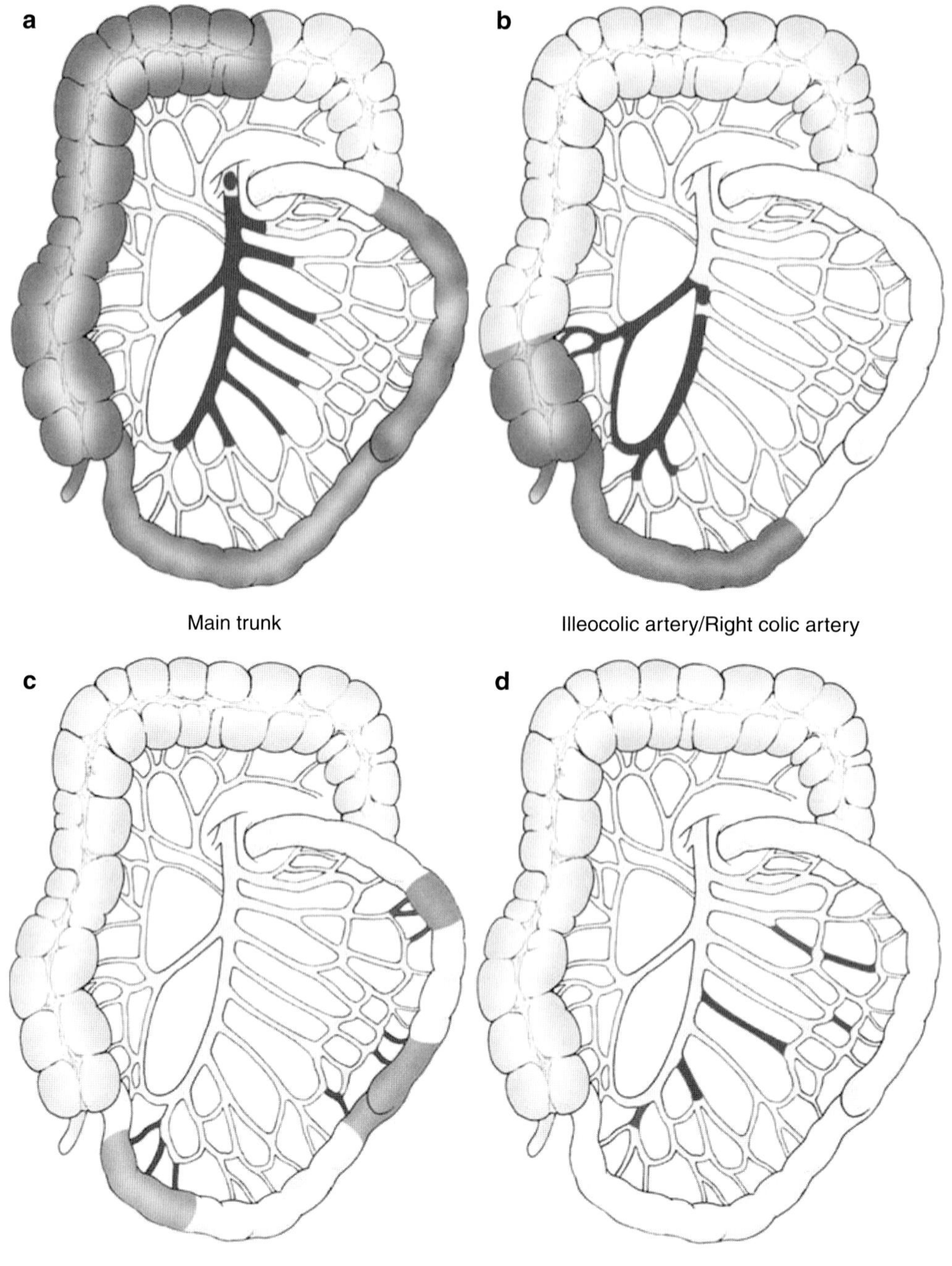


**Fig. 6.15 a–d** Acute mesenteric artery occlusion. The extent of intestinal necrosis varies with the level of occlusion. Occlusion of individual jejunal branches only will not lead to acute intestinal ischemia as the arcades ensure collateral flow from patent jejunal branches (**d**). The vasa recta are involved in non-occlusive intestinal ischemia (**c**). Proximal occlusions in which the mesenteric trunk is still patent are associated with necrosis of long intestinal segments and have a poor prognosis. The Doppler waveform in the patent mesenteric artery shows abnormal changes (**a**, **b**). The remaining patent branches dilate to provide maximum blood supply via the arcades, resulting in less pulsatile, low-resistance flow. Nevertheless, overall flow through the patent mesenteric trunk is reduced (decreased peak systolic velocity)

mesenteric artery always indicates peripheral occlusion of several mesenteric branches (Table 6.11).

If the hepatic artery arises from the superior mesenteric artery, the Doppler waveform obtained upstream of the origin will show a large diastolic component with a decrease in the Pourcelot index, because the flow pattern in this case is affected by the supply of a parenchymal organ. This must be borne in mind in interpreting the findings.

In *acute intestinal ischemia*, the hemodynamic changes must be interpreted in light of the B-mode findings. Rapidly progressive wall edema of the small intestine is identified by the typical bull's eye sign and leads to intestinal wall necrosis with cessation of peristalsis. Progressive intestinal wall thickening and free fluid around affected bowel loops are identified sonographically as are air bubbles in the intestinal wall and portal vein that characterize the late stage.

*Nonocclusive intestinal ischemia* has a poor prognosis and frequently occurs in patients with considerable comorbidity. The examiner must be aware of this condition as a differential diagnosis of proximal mesenteric occlusion. Circulatory

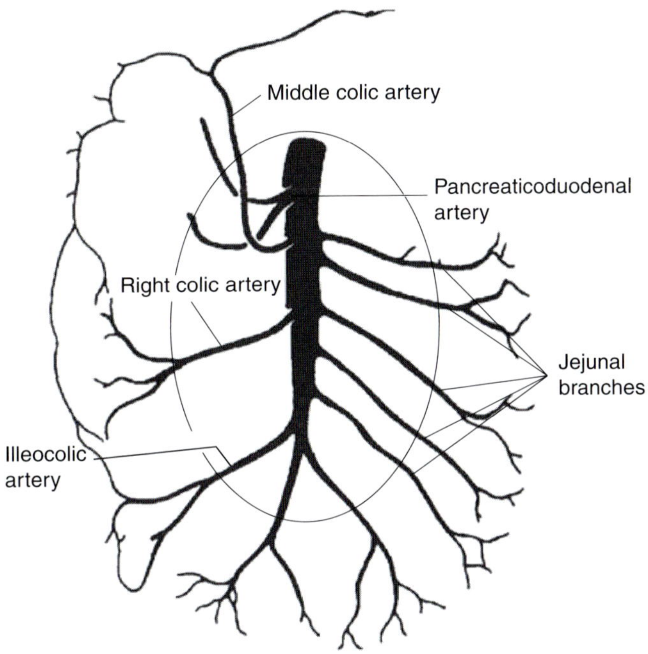


**Fig. 6.16** Divisions of the superior mesenteric artery with side branches. Under optimal conditions, color duplex scanning visualizes the main trunk, the division into jejunal branches, the right colic artery, and ileocolic artery (visible area outlined) (according to Kubale 1994)

**Table 6.11** Normal and abnormal changes of the spectral waveform from the superior mesenteric artery in relation to peripheral resistance

| |
|---|
| **Decreased RI (Pourcelot) with absolute or relative increase in diastolic flow component** |
| • With increase in mean flow velocity: |
| – Postprandial |
| – Drug-induced |
| – Inflammatory |
| – Tumor-related |
| • With decrease in mean flow velocity: |
| – Distal mesenteric artery occlusion (dilatation of collaterals) |
| **Increased RI (Pourcelot) with relative or absolute decrease in diastolic flow component** |
| • Diabetes mellitus (medial sclerosis) |
| • Acute extensive mesenteric vein thrombosis |

insufficiency, sepsis, and diabetes mellitus are pathogenetic factors. Ultrasonography has no role in the diagnosis since only the smaller, distal mesenteric branches are affected, while the superior mesenteric artery and the proximal segments of the main branches are patent. The diagnosis is confirmed angiographically before therapy with intra-arterial vasodilators is initiated.

An increased pulsatility of the mesenteric artery may be due to reduced wall elasticity in diabetes mellitus or indicate disturbed peripheral venous return, as in extensive mesenteric vein thrombosis.

The *diagnostic value of color duplex ultrasound* in evaluating infarction of the liver, spleen, or kidneys due to acute peripheral artery occlusion depends on the insonation conditions. The extent of infarction varies with the site of occlusion and blood supply through collateral routes. B-mode ultrasound shows poorly delineated, inhomogeneous, and hypoechoic areas, but not earlier than 1–3 days after the acute event (Seitz and Rettenmaier 1994). There are some case reports describing the use of color duplex imaging in patients with renal or splenic infarction, but ultrasound is most beneficial in guiding interventional procedures such as abscess drainage in superinfection of necrotic areas.

*Aneurysms of the visceral arteries* are very uncommon but tend to rupture. They are typically detected incidentally in patients undergoing B-mode ultrasound for diagnostic workup of abdominal symptoms (which may be due to pressure). They are differentiated from pseudocysts or other cystic tumorous lesions of the upper abdomen by their characteristic color duplex appearance. Locating the aneurysm to the splenic, superior mesenteric, or hepatic artery is important for the planning of the surgical procedure. As with all other vascular territories, the diagnostic evaluation of aneurysms falls in the domain of color duplex ultrasound: the flexibility in choosing the orientation of the scan plane enables reliable diameter measurement, identification of thrombotic wall deposits, and assessment of the patent residual lumen.

*Aneurysm* of the superior mesenteric artery is rare. Even less common are aneurysms of the gastroduodenal, pancreaticoduodenal, and inferior mesenteric arteries. They are typically mycotic aneurysms (staphylococci, salmonellae). They are located and measured as in other vascular territories. In planning the surgical procedure, it is crucial that their course and relationship to other vessels be determined exactly. Visceral artery aneurysm appears to be more common in patients with ectopic variants.

### 6.1.6.3 Aorta

Bilateral *intermittent claudication* may be caused by stenosis of the distal aorta. Therefore, the aorta should be scanned if the spectral display of the iliac artery shows poststenotic changes. High-grade stenosis of the aorta is indicated in the color duplex scan by a mosaic pattern resulting from perivascular vibration, as in an arteriovenous fistula. Severe atherosclerosis with calcified plaques impairs the detection of the high PSV in the stenosis jet in the duplex mode. On the other hand, extensive plaque with acoustic shadowing and poor delineation of the lumen in the B-mode scan often suggests high-grade stenosis of the aorta. The Doppler waveform sampled distal to the high-grade stenosis will show the typical postocclusive flow profile with a delayed systolic rise and large diastolic component.

*Aortic stenosis* occurs chiefly in the distal infrarenal segment including the bifurcation. In this setting, collateral supply with refilling of the iliac system is mainly ensured by the inferior mesenteric artery, which will become dilated and show high systolic (often over 200 cm/s) and end-diastolic flow velocities (see Fig. A6.51).

In acute obstruction of the terminal aorta (Leriche's syndrome), the lumen is discriminated in the B-mode scan by virtue of its being filled with hypoechoic material. If the occlusion is due to atherosclerosis, on the other hand, the aorta is difficult to differentiate from surrounding tissue. Flow is absent in both cases. In patients with poor visualization, chronic occlusion can be differentiated from high-grade stenosis by the absence of the mosaic pattern caused by perivascular vibration that is typical of stenosis.

*Thrombosis of the aorta* is visualized as a hypoechoic cone-like structure in the lumen. The tail of the thrombus is typically surrounded by flowing blood on all sides. Signs of luminal narrowing are seen on duplex scanning only when there is nearly complete occlusion. Most patients with aortic thrombosis present with embolism, often in both legs. In these cases, the circular flow signal seen around the hypoechoic thrombus on color flow images differentiates aortic thrombosis from an embolizing aneurysm (see Fig. A6.52).

The diagnosis of an *aortic aneurysm*, including its localization and characterization, falls in the domain of color duplex ultrasound (Table 6.12).

While angiography only depicts the residual lumen of an aneurysm, ultrasound enables *differentiation of mural thrombosis from the residual lumen*. The variable scanning plane of ultrasound relative to the course of the aneurysm facilitates determination of the true diameter compared with CT. Moreover, the use of standardized transverse sections as in CT may lead to overestimation of the aneurysmal diameter as a result of its elliptical profile in transverse orientation when there is concomitant elongation of the distal aorta (Figs. 6.17 and 6.18), which is often associated with atherosclerotic aneurysm (see Fig. A6.32). Ultrasonography does not have this methodological limitation as the transducer can be rotated to avoid oblique sectioning of the aortic lumen, thereby enabling more accurate diameter measurement. When interpreting the results of studies comparing the accuracy of CT and ultrasound in aneurysm size measurement, one must take into account the methodological limitations of CT in this context.

One study comparing maximum abdominal aortic aneurysm diameter measurement by ultrasound and CT in 83 patients found that standard axial CT yielded significantly larger diameters (4 mm on average) compared with orthogonal ultrasound (D1 versus D2 in Fig. 6.17). This difference between the two methods disappeared when reformatted orthogonal CT images were used for diameter measurement (Sprouse et al. 2004).

Intraobserver as well as interobserver studies show good reproducibility of the results if these aspects are taken into

**Table 6.12** Therapeutically relevant color duplex findings in aneurysm of the abdominal aorta

| |
|---|
| • Diameter of aneurysm (indication for surgery) |
| • Shape of aneurysm (saccular, spindle-shaped) |
| • Partial thrombosis |
| • Involvement of (common, internal) iliac arteries |
| • Infrarenal – suprarenal |
| Additional criteria if endovascular treatment is contemplated: |
| • Distance of proximal end of aneurysm to renal artery |
| • Degree of angulation in case of elongation of infrarenal aorta |
| • Conic neck of aneurysm |
| • Lumen of common femoral artery (large enough for stent insertion?) |

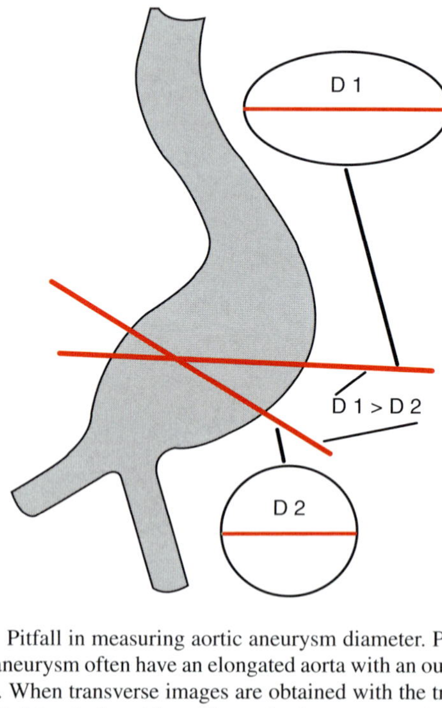


**Fig. 6.17** Pitfall in measuring aortic aneurysm diameter. Patients with an aortic aneurysm often have an elongated aorta with an outward curve to the left. When transverse images are obtained with the transducer in the normal abdominal position, this can lead to an overestimation of the diameter because the aorta is being measured in an ellipsoid scan plane (*D1*). Overestimation of aneurysm size by measurement in the wrong scan plane can also lead to overestimation of the rupture risk. The correct diameter of the aneurysm is measured by rotating the transducer clockwise until a round image of the aorta comes into view (*D2*)

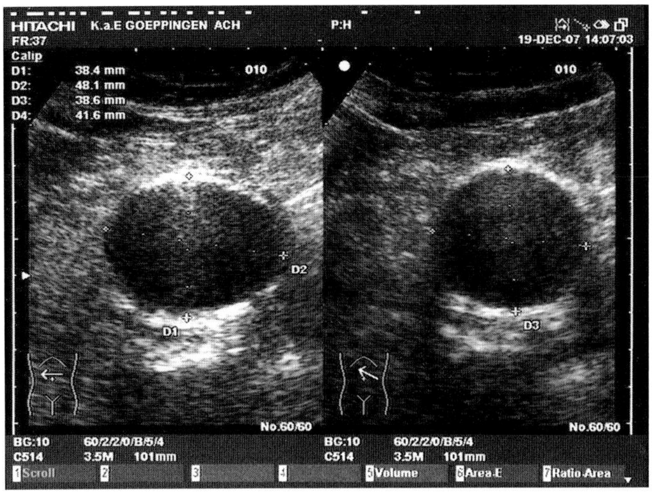


**Fig. 6.18** Diameter measurement in an elongated aorta. The largest diameter measured in a transverse scan plane is 48 mm (corresponding to a mid-abdominal axial CT scan). Rotating the transducer (see body marker) to obtain a correct scan plane (round cross section of the aorta) results in a diameter of 42 mm being measured

consideration and aneurysm diameter is measured using the method described in Fig. 6.17. A standardized measurement protocol is especially important for capturing the true development of aneurysm size in serial examinations (Sun et al. 2006; AbuRahma et al. 2006; Collins et al. 2007; Stavropoulous et al. 2007).

*Embolizing aortic aneurysm and saccular aneurysm require surgical management* irrespective of their size. Saccular aneurysms tend to exhibit turbulent flow on color duplex scanning, while laminar flow is more likely in smaller, spindle-shaped aneurysms. The local pressure peaks occurring in turbulent flow are associated with more rapid growth and a higher risk of rupture. In scanning patients with embolic occlusion of the leg arteries, thrombi in an aortic aneurysm should be ruled out as a source of embolism (see Fig. A6.31).

Rupture of an aneurysm is suggested by flank and back pain (occasionally abdominal pain), a palpable pulsating tumor, and shock. Gray-scale imaging depicts a hypoechoic structure of variable extent and comprising inhomogeneous or layered portions around the aorta in the retroperitoneum. Color duplex demonstrates paravascular flow signals at the site of the leak in case of a contained perforation. Leakage must be differentiated from other hypoechoic periaortal structures such as retroperitoneal fibrosis, horseshoe kidneys, or lymphoma, which may occur in conjunction with aneurysm (see Fig. A6.42). An examiner performing ultrasonography on an emergency basis must pay special attention to such accompanying conditions as they have important implications.

If there is perforation into the duodenum, the intestine may appear fluid-filled. Fistular connections to the vena cava can be demonstrated by color duplex.

*Aortic dissection* is only amenable to percutaneous ultrasound diagnosis if the intimal flap extends into the abdominal aorta. In the abdominal aorta, sonography is a valid method for assessing the extent of dissection, and spectral Doppler recording is helpful in identifying extension into branch vessels. The spectral Doppler examination also identifies intermittent obstruction of a branch origin by the intimal flap, moving synchronously with the heart.

In a dissecting aneurysm, splitting of the arterial wall with tearing of the intima is suggested on gray-scale ultrasonography by the presence of a flap in the vessel lumen; this flap can be identified by its hyperechoic reflection and typical undulating motion. Dissection is confirmed in the color mode by differing flow velocities and directions in the true and false lumens. The power mode will help demonstrate slow flow in the false lumen and differentiate it from partial thrombosis. The color-coded flow directions contribute to the identification of the entry and reentry sites. Knowledge of the relationships of the origins of the visceral and renal arteries to the true and false lumens is crucial for proper therapeutic management. As the dissected flap may extend into the groin, the iliac artery must be included in the examination (see Figs. A6.46 and A6.47).

Spectral Doppler measurement at the sites of aortic branch origins for identifying extension of the dissection into renal or visceral arteries as well as intermittent occlusion or stenosis of a branch artery by the intimal flap is important for the therapeutic approach. If there is extension into an aortic branch, the Doppler waveform will depict flow in the true and false lumens and may also contain signals produced by oscillation of the intimal flap. In the case of narrowing or intermittent occlusion of a branch origin by the intimal flap, spectral Doppler will reveal signs of stenosis or systolic deceleration (decreased systolic velocity or even zero flow) (Fig. A6.46).

Conversely, with isolated dissection of the abdominal aorta, which is rare (Knabe et al. 2001), the proximal origin of an aortic dissection is typically not detectable by transcutaneous sonography (except with the transducer directed retrosternally from a jugular position). In these cases, transesophageal echocardiography (Link 1999), computed tomography, or magnetic resonance imaging is required for diagnosis. The diagnostic accuracies in aortic dissection are 70% for ultrasound (Nienaber et al. 1993), 98% for transesophageal echocardiography (Sommer et al. 1996), and 100% for MRI (Silverman 2000) and CT.

*Follow-up* after surgical resection of an aneurysm or patch angioplasty of aortic stenosis is a domain of duplex ultrasonography, which allows early identification of complications such as suture aneurysm, anastomotic stenosis, recurrent stenosis, or abscess. Hypoechoic structures around a prosthesis, particularly

**Table 6.13** Therapeutically relevant sonographic criteria in the follow-up after treatment of aneurysms of the abdominal aorta

| |
|---|
| **After surgical repair:** |
| • Recurrent aneurysm – suture aneurysm |
| • Anastomotic stenosis (occlusion of Y-limb) |
| • Abscess |
| **After endovascular stent placement:** |
| • Endoleaks (types I, II, and III) |
| • Growth or shrinkage of the stented aneurysm |
| • Stent migration |
| • Occlusion |

at the sites of anastomosis, must be checked for flow signals in the color mode to differentiate suture aneurysm from postoperative hematoma and abscess. Suture aneurysms are false aneurysms and hence are characterized by to-and-fro flow (steam engine sound) in the spectral waveform (recorded at the site of the wall perforation; see Fig. A6.49). An abscess suspected on clinical grounds can be confirmed by ultrasound-guided fine-needle aspiration biopsy.

As outlined above, the sonographic examination allows adequate identification of suture aneurysm or anastomosis stenosis in the *follow-up* after surgical repair with vessel prosthesis insertion (Table 6.13). When ultrasound is used in the routine monitoring of patients who underwent endovascular treatment with stenting, a systematic approach is recommended to reliably identify endoleaks and other complications:

- *Gray-scale* examination in *transverse orientation* to evaluate the upper stent end and the relationship to the renal artery origins, followed by measurement of the largest diameter of the *residual aneurysm sac* as described (Figs. 6.18 and A6.34a).
- Transverse color duplex (low pulse repetition frequency) of the aneurysm site and the stent from the renal artery origins to the bifurcation, focusing on the origins of the lumbar arteries and the inferior mesenteric artery (Figs. A6.35 and A6.37).
- *Color duplex examination* of the stent ends in the longitudinal plane with spectral Doppler measurement to confirm patency, demonstrate stenosis, and identify type I endoleaks using an acute Doppler angle at the anchoring sites (Fig. A6.36).
- If *color duplex imaging reveals flow* in the aneurysm sac, this must be confirmed by spectral Doppler interrogation. This is especially important in the early postinterventional phase before complete thrombosis of the aneurysm has occurred and movement of the stent can mimic flow signals (pseudoendoleak). The Doppler spectrum differentiates pseudoleaks from true endoleaks by demonstrating to-and-fro flow. This flow pattern is characteristic of endoleaks as they are hemodynamically similar to false aneurysm, and it can also help to identify mirror artifacts by spectral Doppler in addition to insonation from different directions. The demonstration of to-and-fro flow is especially important in *differentiating small endoleaks from artifacts* (Fig. A6.37).

Whether the ultrasound examination allows adequate evaluation of a stented aneurysm depends very much on the individual acoustic window. Moreover, ultrasound also has inherent limitations, which is why there is *disagreement* regarding its role in monitoring patients with aneurysm stents and in particular its *reliability in identifying endoleaks* (Ashoke et al. 2005; Stanford et al. 2006). Some authors showed color duplex ultrasound to be sufficient to rule out endoleaks with reported sensitivities of 77–96% and specificities of 90–94% (D'Audiffret et al. 2001; Golzarian et al. 2002; Sato et al. 1998; Sun et al. 2006; AbuRahma et al. 2006; Collins et al. 2007; Stavropoulous et al. 2007); however, the patient populations investigated were small. The difficulty in detecting an endoleak is that it requires not only a good acoustic window and adequate machine settings (low pulse repetition frequency) but also a great deal of care in setting the Doppler angle correction cursor relative to the leak jet so as not to miss a type II endoleak. Moreover, motion and mirror artifacts in the stented aneurysm sac can mimic flow in the color duplex examination, which must be differentiated from an endoleak by spectral Doppler measurement.

The *Doppler waveform* obtained in the endoleak jet provides *hemodynamic information* (Figs. A6.34 to A6.39).

1. Low flow velocity (<30 cm/s), to-and-fro flow, and a small area of flow in the aneurysm sac with shrinkage – together these characteristics suggest a low risk of imminent rupture, and the patient can be observed (follow-up after 3 months).
2. Direct evidence of an endoleak or of flow into the aneurysm sac.
3. To-and-fro flow at the site of the endoleak or unidirectional flow into the aneurysm sac (dangerous endoleak with high risk of rupture). These spectral Doppler findings should prompt a search for a patent artery (lumbar, inferior mesenteric) with flow out of the perfused aneurysm sac.

There is agreement that perfusion of an aneurysm sac via thin side branches such as the lumbar arteries (type II endoleak) with small flow volumes can simply be observed and will often thrombose spontaneously. Nevertheless, there is an ongoing controversy about the criteria that differentiate this type of endoleak from those requiring interventional closure (Liewald et al. 2001; Parry et al. 2002; White et al. 2000). The PSV measured in the jet at the junction of the lumbar artery and the aneurysm sac can provide important clues as slow flow is a *good predictor of spontaneous thrombosis*, initially justifying watchful waiting without an intervention (Arko 2003).

Duplex sonography has sufficient diagnostic accuracy for the detection of type I and III endoleaks, which are characterized by higher blood flow velocities and are easier to locate.

Several investigators have shown that *ultrasound contrast agents* can improve the identification of endoleaks, in particular the more difficult to detect type II (Böhm et al. 2000; Henao 2006; McWilliams et al. 2002; Bendick et al. 2003; Heilberger et al. 1997; Clevert et al. 2008; Sarlon et al. 2009; Giannoni et al. 2007).

Recent studies (Jung 2008; Pfister et al. 2009) found real-time dynamic contrast-enhanced ultrasound to even be superior to CT angiography in detecting small type II endoleaks with very little flow. In a study investigating contrast-enhanced ultrasound using contrast harmonic imaging (CHI; see Sect. 1.1.5) in 50 patients with suspected endoleaks following endovascular treatment for abdominal aortic aneurysm, CHI and CT angiography concordantly detected endoleaks in 30 cases and found no endoleak in 20 cases. In one case, a combined type I/II endoleak was misclassified as type II. In one patient, a type II endoleak was initially only detected by CHI and later confirmed by follow-up CTA. Thus, CHI had 99% sensitivity, 93% specificity, 99% negative predictive value, and 95% positive predictive value. Time intensity curve (TIC) analysis was used in this study to evaluate perfusion of the aneurysm (Pfister et al. 2009). Standardized analysis of signal enhancement after administration of an ultrasound contrast agent can be used to compare enhancement in the stent and in the aneurysm sac; the intensity of enhancement varies with the amount of blood and can thus serve to quantify the endoleak, identifying at risk aneurysms that warrant treatment. While these new methods have the potential to improve the detection of endoleaks, the established sonographic techniques allow reliable assessment as well when used by a skilled and experienced examiner with a realistic sense of the limits of ultrasonography in general and a clear understanding of the quality of the examination in the individual patient.

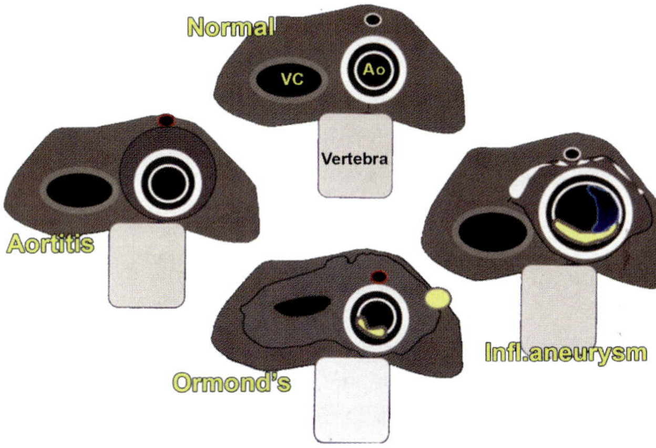


**Fig. 6.19** Evaluation of the course of the proximal inferior mesenteric artery as it arises from the aorta in order to differentiate between aortic wall thickening and retroperitoneal perivascular pathology (see Figs. A6.42 and A6.43). The *topmost* diagram illustrates the normal anatomy of the aorta (*Ao*) and vena cava (*VC*). The diagram labeled "Aortitis" illustrates the situation in giant cell arteritis: the inferior mesenteric artery arises from the aorta and then courses outside the hypoechoic, concentrically thickened wall of the aorta. In this condition, the mesenteric artery takes the shortest path possible through the abnormal tissue (inflammatory aortic wall). The diagram labeled "Infl. aneurysm" illustrates the findings in inflammatory abdominal aortic aneurysm. In this condition, the inferior mesenteric artery also passes through the aneurysmatically dilated and thickened abdominal wall, taking the shortest path possible, to then descend toward the left lower abdomen outside the thickened structure. If atherosclerotic lesions are present, these can help in differentiating inflammatory wall thickening from thrombotic deposits in the aneurysm sac: atherosclerotic lesions are found on the thickened, hypoechoic wall, and thrombotic deposits are found on top of the atherosclerotic plaques. The *bottom* diagram (labeled "Ormond's") illustrates the findings in retroperitoneal fibrosis. The proximal segment of the inferior mesenteric artery arising from the nonthickened aortic wall courses through the hypoechoic abnormal tissue before emerging out of it and passing to the left lower abdomen. In retroperitoneal fibrosis, the abnormal tissue pushes a longer proximal segment of the inferior mesenteric artery against the wall of the aorta. In patients with more advanced retroperitoneal fibrosis, which also encases the vena cava or even the ureter, the diagnosis and differentiation from other conditions are easier than in early disease with fibrotic tissue surrounding only the aorta. Therefore, the abnormal course of the proximal inferior mesenteric artery is the decisive criterion for making the differential diagnosis (diagrams courtesy of K. Amendt)

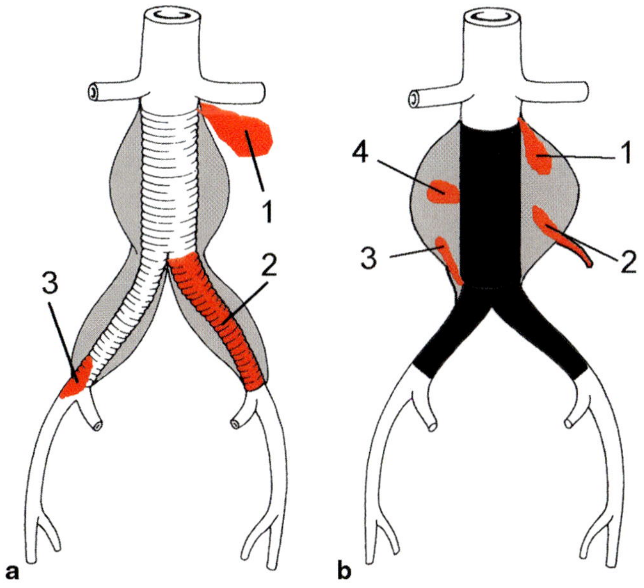


**Fig. 6.20 a, b** Complications of surgical and interventional aortic aneurysm repair. **a** Following surgical Y-prosthesis implantation: *1* suture aneurysm (typically at the upper anaostomosis or, if the prosthesis extends into the femoral artery, at the lower anastomosis), *2* occlusion of a prosthesis limb, *3* anastomosis stenosis (distal anastomosis). **b** Types of endoleaks following endovascular stenting: *1* type I endoleak, at the junction between the stent and the host artery, *2* type II endoleak, from patent lumbar arteries, *3* type III endoleak, separation of modular components, *4* type IV endoleak, device failure (due to material fatigue)

When these conditions are met, a meticulous examination performed using high-end equipment with optimally adjusted scan parameters (low pulse repetition frequency) allows one to confidently rule out an endoleak requiring treatment.

The aortic diameter can also provide diagnostic clues. A decrease in aortic diameter can be assumed to suggest adequate stent graft anchorage without endoleak (Giannoni et al. 1998; Thompson et al. 1998). Conversely, even a small increase in aneurysm diameter may be a sign of an endoleak and should prompt an intense search including the use of supplementary imaging modalities if required.

Duplex ultrasound is far inferior to radiologic imaging modalities in detecting stent fracture and stent migration, except when there is concomitant endoleak.

Patients having undergone endovascular treatment for aortic aneurysm require routine surveillance of the stented aneurysm at 6-month intervals with close monitoring by noninvasive duplex sonography (Fig. 6.20) if a type II endoleak not requiring reintervention is present (perfusion of the sac via patent lateral branches, lumbar arteries, or the superior mesenteric artery). Type I (anastomotic seals) and type III endoleaks are more readily identified sonographically due to faster blood flow. In general, computed tomography is the gold standard for the planning and follow-up of stent placement. Stent fracture and migration are reliably detected by conventional radiography.

*Aortic wall thickening* must be *differentiated from perivascular, retroperitoneal conditions* by B-mode imaging of the aorta with special attention paid to aortic branches. Retroperitoneal lymphomas are circumscribed round lesions but may become confluent. Retroperitoneal fibrosis is typically depicted as a more hypoechoic structure covering the aorta from anteriorly and involving the vena cava. It tapers off laterally and may also surround and compress the ureter.

The *course of the inferior mesenteric artery* can help to differentiate wall thickening of the aorta and retroperitoneal fibrosis (see Fig. 6.19). In retroperitoneal fibrosis, the proximal inferior mesenteric artery is pushed against the aorta and thus courses between the aortic wall and the hypoechoic fibrosis for several centimeters before piercing through the hypoechoic fibrotic cap (see Fig. A6.42). In *inflammation of the aorta* (e.g., giant cell arteritis), on the other hand, the inferior mesenteric artery pierces the hypoechoic layer around the patent lumen of the aorta directly at its origin to then descend into the left lower abdomen outside the hypoechoic thickening. Inflammatory wall thickening in aortitis and in inflammatory aortic aneurysm is circumferential and spares the vena cava. The flexible selection of scanning planes with the ultrasound scanner facilitates determination of the relationships between the aorta, hypoechoic thickening, perivascular structures, and other vessels, thereby contributing to the differential diagnosis.

## 6.2 Visceral and Retroperitoneal Veins

### 6.2.1 Vascular Anatomy

#### 6.2.1.1 Vena Cava

The inferior vena cava ascends parallel to the course of the aorta and to the right of the vertebral column. It is a capacitance vessel with an elliptical cross section and may vary in its anteroposterior diameter from 0.5 to 2.5 cm during the respiratory cycle. It is formed by the junction of the common iliac veins at the L4/L5 level slightly below the aortic bifurcation.

Variants and anomalies of the inferior vena cava are present in 1.5–4.0% of the population and are typically detected incidentally. They result from errors in the complex embryonic development with a disturbance in the transition from the symmetric venous system to the predominantly right-sided, asymmetric secondary system. The congenital anomalies were classified by Chuang et al. as early as 1974 (Table 6.14).

Vena cava anomalies typically involve the segment below the renal vein entries. Duplication (0.2–3%) and transposition (0.2–0.5%) are the most common anomalies. In both cases, the inferior vena cava, which runs to the left of the aorta, crosses the aorta together with or as part of the left renal vein to then continue on its course on the right side of the aorta.

#### 6.2.1.2 Renal Veins

The *right renal vein* courses anterior to the renal artery, and after 3–4 cm it empties into the vena cava at the level of

**Table 6.14** Congenital anomalies of the inferior vena cava (classification by Chuang et al. 1974)

| | Segment | |
|---|---|---|
| I | Postrenal | |
| | Type A: | Persistence of right posterior cardinal vein (retro- or circumaortic ureter) |
| | Type B: | Persistence of right supracardinal vein (normal inferior vena cava) |
| | Type C: | Persistence of left supracardinal vein (left-sided inferior vena cava) |
| | Type BC: | Persistence of both supracardinal veins (duplicated inferior vena cava) |
| II | Renal | Persistence of renal venous ring (circumaortic ureter) |
| III | Prerenal or hepatic | Absence of hepatic segment (azygos or hemiazygos vein continuation) |

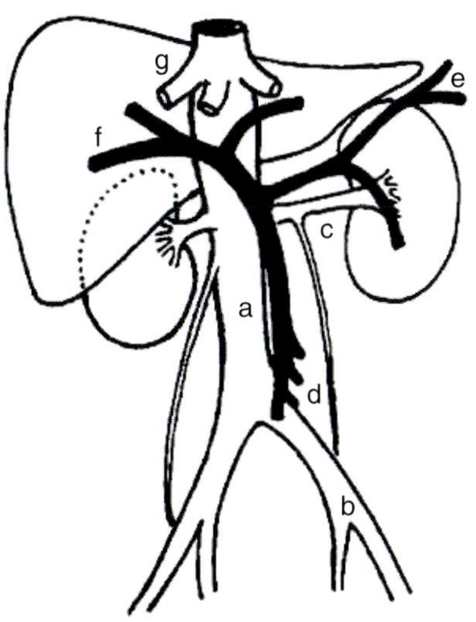


**Fig. 6.21** Diagram of the large retroperitoneal and intra-abdominal veins (*black*) (*a* vena cava, *b* (common, external, and internal) iliac veins, *c* renal vein, *d* superior mesenteric vein, *e* splenic vein, *f* portal vein, *g* hepatic veins)

the L1 vertebra. The *left renal vein* runs anterior and somewhat superior to the renal artery to then curve between the aorta and superior mesenteric artery and posteroinferior to the head of the pancreas on its way to the inferior vena cava. The left renal artery receives the ovarian or spermatic vein, which empty directly into the vena cava on the right (Fig. 6.21).

*Variants* include duplication of the renal veins and atypical terminations, for example, in the common iliac vein. Disturbed embryonic development may result in an atypical retroaortic course of the left renal vein.

### 6.2.1.3 Portal Venous System and Hepatic Veins

The *portal vein* is a thick trunk, normally 6–8 cm long, and has a transverse, anteroposterior elliptical diameter of 8–12 mm, rarely up to 16 mm, with wide physiologic variation. It passes through the hepatoduodenal ligament behind the proper hepatic artery and ascends to the porta hepatis, taking an oblique lateral course. Intrahepatically, it follows the branches of the hepatic artery and the bile ducts. The branching pattern of the portal vein provides the basis for the anatomic segmentation of the liver. The branches of the hepatic veins are intersegmental vessels.

The *splenic vein* passes behind the pancreas from the hilum of the spleen to the head of the pancreas, where it joins the superior mesenteric vein to form the portal vein (Fig. 6.22). The confluence of the two veins (see Fig. 6.21) is situated to the left of and behind the pancreatic head, somewhat lateral and inferior to the origin of the mesenteric artery.

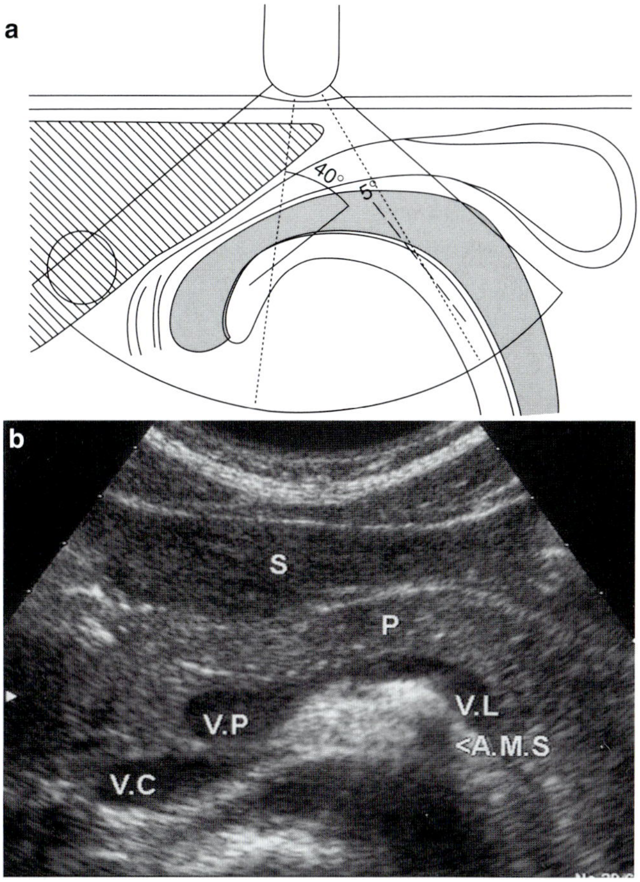


**Fig. 6.22** **a** Transverse diagram showing the course of the splenic vein from right to left, where it joins the superior mesenteric vein to form the portal vein, below the head of the pancreas. **b** Course of the splenic vein (*V.L*), posterior to the pancreas (*P*) and anterior to the superior mesenteric artery (*A.M.S*), terminating in the portal vein (*V.P*)

The *superior mesenteric vein* courses to the right of the superior mesenteric artery and anterior to the aorta.

Blood is drained from the liver through the segmental branches into the three major veins, which empty directly into the inferior vena cava.

## 6.2.2 Examination Technique

### 6.2.2.1 Vena Cava

The aorta can serve as a landmark when examining the vena cava. When assessing the lower extremity vessels for thrombosis or when assessing patients with pulmonary embolism, the course of the vena cava is followed after evaluation of the pelvic and leg veins. Evaluation in transverse orientation is followed by longitudinal scanning for spectral Doppler measurement. The confluence of the iliac veins at the level of the umbilicus can be identified using the aortic bifurcation at about the same level or slightly above for guidance.

In pelvic vein thrombosis, possible extension of the thrombus into the vena cava must be identified and the end of the thrombus evaluated for the presence of surrounding flow signals. The respiratory variation in vena cava diameter is determined in transverse or longitudinal images. Diameter variation is absent in thrombosis.

Compression ultrasound is unreliable in the abdomen (where compressibility is limited by anatomy), which is why (color) duplex ultrasonography is necessary to exclude thrombosis. Only in slender patients is it possible to reliably compress the vena cava. The flow velocity determined from the Doppler spectrum is likewise affected by respiratory phasicity and increasing cardiac pulsatility toward the heart. Below the kidneys, the liver can be used as an acoustic window with the portal vein being located anterior to the vena cava at the hilum of the liver.

#### 6.2.2.2 Renal Veins

The renal veins are scanned from the flank beginning at the renal hilum and following their course proximally (see Fig. 6.6). While the shorter right vein can be visualized from this position throughout its course to the inferior vena cava, the left vein (Fig. 6.23a, b) must be scanned from a medial approach in transverse orientation to visualize the segment between the aorta and superior mesenteric artery and its termination.

The superior mesenteric artery can serve as a landmark to avoid confusion of the left renal vein with the slightly more anterocranial splenic vein. The left renal vein runs posterior to the superior mesenteric artery, that is, between the latter and the aorta; the splenic vein courses anterior to it and then crosses over the superior mesenteric artery to enter the portal vein.

When searching for renal vein thrombosis or intravascular tumor extension in the B-mode, the examiner must look for the absence of respiratory diameter fluctuations and the presence of echogenic intraluminal material dilating the vein. The absence of flow in the spectral Doppler recording is diagnostic of thrombosis.

#### 6.2.2.3 Portal Vein and Superior Mesenteric Vein

The course of the portal vein from below the pancreatic head to the liver hilum is most easily accessible from a subcostal approach in an oblique plane through the right upper abdomen (Fig. 6.24). If visualization is impaired by overlying bowel gas, the liver hilum can be identified and the portal vein followed distally from an intercostal position (right flank) using the liver as an acoustic window. Portal vein thrombosis is suggested by the presence of echogenic material in the lumen and absent respiratory diameter variation in the B-mode image, and is subsequently confirmed by (color) duplex imaging.

*Suspected portal hypertension* is easily demonstrated sonographically by the identification of portal vein collaterals. The evaluation for collateral circulation comprises the following steps:

- Flow in the portal vein: to-and-fro flow, retrograde flow, or reduced flow velocity (unreliable sign; challenge test may be necessary to demonstrate reduced increase in portal vein flow).

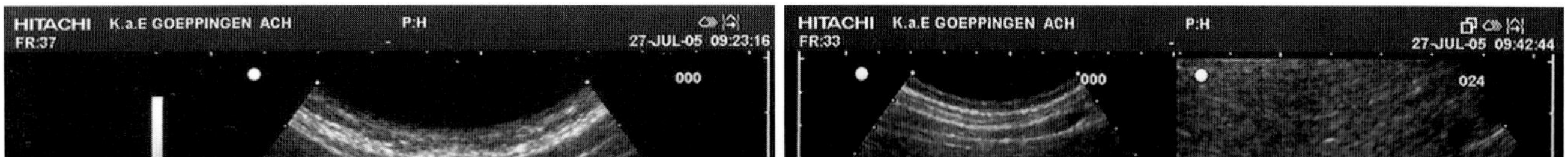


**Fig. 6.23** **a** The left renal vein (*V.R.*) courses from the left renal hilum, between the aorta (*A*) and superior mesenteric artery (*A.M.S*), to the vena cava (*V.C*). Transverse epigastric image. **b** With the transducer on the right flank (*left image*), the right renal vein (*V.R.*) can easily be evaluated from the hilum (*K*) to the vena cava (*V.C*). With a slightly more medial transducer position (*right image*), the liver can be used as an acoustic window, and the renal vein can be seen deep to the liver (*L*) from the renal hilum (*K*) to the vena cava (*V.C*), coursing anterior to the renal artery (*A.R.*)

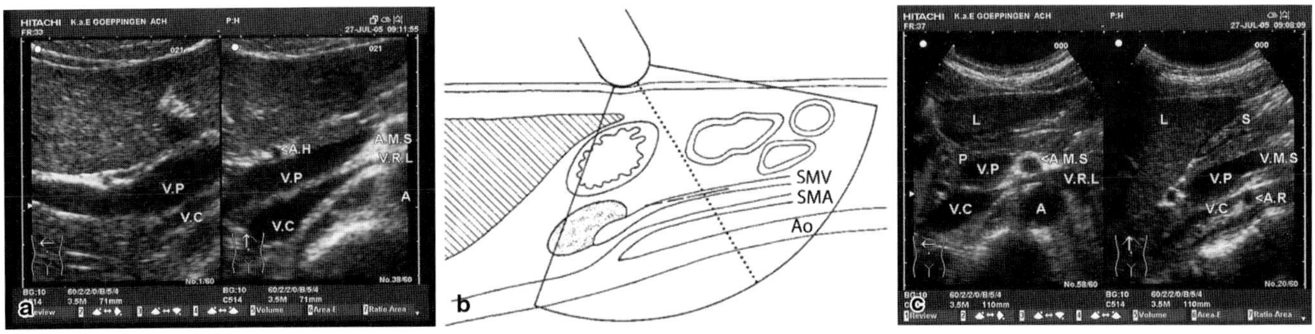


**Fig. 6.24** **a** In the subcostal oblique image (right upper abdomen), the portal vein formed by the confluence of the superior mesenteric and splenic veins posterior to the pancreatic head (*right* image) is seen anterior to the vena cava (*V.C*) on its course through the hepatoduodenal ligament to the liver hilum (left margin of *left* image). **b** Diagram of the course of the superior mesenteric vein (*SMV*; *SMA* superior mesenteric artery, *Ao* aorta). **c** The transverse upper abdominal image (*left* section) shows the vena cava (*V.C*) to the left of the aorta (landmark); the portal vein (*V.P*) courses inside the hepatoduodenal ligament deep to the liver (*L*). The longitudinal image (*right*) shows the superior mesenteric vein (*V.M.S.*) coursing anterior to the vena cava (*V.C*, landmark) and terminating in the portal vein (*V.P.*) posterior to the pancreatic head

- Flow direction in the splenic vein.
- Visualization of the falsiform ligament and demonstration of recanalization of the umbilical vein with hepatofugal flow (Cruveilhier–Baumgarten syndrome).
- Dilated left gastric vein (in longitudinal plane, arising from portal vein at the site of mesenteric vein termination).
- Varicose dilatation in the gallbladder wall.
- Identification of dilated gastroepiploic veins posterolateral to the left hepatic lobe and of splenorenal and splenogastric collaterals at the upper and lower poles of the kidney.

The superior mesenteric vein courses to the right of the superior mesenteric artery with a position to the left suggesting malrotation. The criteria for diagnosing thrombosis are the same as in the portal vein. If mesenteric vein thrombosis is suspected, the vein should be traced to the level of the jejunal branches, the ileocolic vein, and right colic vein. This is most easily accomplished in transverse orientation with slight angulation of the transducer in the color duplex mode.

## 6.2.3 Clinical Role of Duplex Ultrasound

### 6.2.3.1 Renal Veins

Acute renal vein thrombosis may present without symptoms or with pain and hematuria. The presentation depends on the extent of retroperitoneal collateralization, and based on this variability, it is assumed that renal vein thrombosis actually occurs more frequently than it is diagnosed. Thromboembolic complications are rare. Renal vein thrombosis can occur in patients with renal diseases such as nephrotic syndrome and glomerulonephritis, external compression of the renal vein by tumor or retroperitoneal lymphoma, and systemic conditions including clotting disorders and intra-abdominal inflammatory conditions such as acute pancreatitis or sepsis. The extent to which *renal function* is compromised (from normal to acute renal failure) varies with the *degree of thrombotic obstruction* and the *extent of retroperitoneal collateral flow* (through capsular and suprarenal veins). This is why a duplex ultrasound examination of the renal vein is indicated not only in patients with clinical signs and symptoms but also in patients with a retroperitoneal tumor, so that prompt anticoagulation treatment can be initiated.

Evaluation of the renal vein is mandatory before surgery for a retroperitoneal tumor and especially before nephrectomy for *renal cell carcinoma*. The presence of *tumor thrombus* in the vein is especially important for the surgical approach in renal cell carcinoma; it has been shown that 20–40% of patients with a large renal tumor have tumor thrombus in the renal vein, among them 5–10% with extension of the thrombus into the vena cava (Goncharenko et al. 1979; Levine 1990).

### 6.2.3.2 Portal Venous System

In patients with chronic hepatic dysfunction, the examination focuses on parenchymal damage, ranging all the way to cirrhosis. It also includes damage of the vascular system, in particular *portal hypertension*, which typically arises secondary to cirrhosis. Portal hypertension can result from obstruction at different levels: sinusoidal obstruction (typical in cirrhosis); presinusoidal obstruction in schistosomiasis, Wilson's disease, and myeloproliferative diseases; prehaptic thrombosis or occlusion of the portal vein; postsinusoidal hepatic vein occlusion; compression of the vena cava; and severe right ventricular insufficiency (Fig. A6.1b). In patients with cirrhosis, evaluation of portal hypertension is important for the explanation of the clinical findings and for the prognosis. Gray-scale ultrasound criteria are highly specific for liver cirrhosis, but sensitivity is very poor compared with liver biopsy and intraoperative findings. However, if portal hypertension is present and other underlying causes can be ruled out, then even with an inconclusive gray-scale examination, cirrhosis can be assumed. Criteria on gray-scale images include macroscopic and microscopic nodules, primarily identifiable as irregularities in the liver surface, and an increase in attenuation and echogenicity. Shrinkage of the right hepatic lobe with enlargement of the caudate lobe (caudate to right lobe ratio >0.65) was found to be 90–100% specific for liver cirrhosis but with a poor sensitivity of only 43% (Harbin et al. 1980; Giorgio et al. 1986). Splenomegaly is common in portal hypertension but is also not very specific. Duplex imaging allows adequate evaluation of blood flow in the portal vein and in the splenic and mesenteric veins in 93–95% of patients (Patriquin 1987; Yeh et al. 1996). The ultrasound criteria described below do not allow reliable exclusion of portal hypertension or cirrhosis.

## 6.2.4 Normal Findings

### 6.2.4.1 Vena Cava and Renal Veins

The *diameter* of the vena cava *varies* from 0.5 to 2.5 cm *with respiration.* The respiratory variations are reflected in the Doppler frequency spectrum recorded longitudinally. Additional cardiac (atrial) phasicity results in a *W-shaped flow profile* determined by the *pressure changes in the right atrium.* The first peak reflects the tricuspid valve movement during systole, followed by a decrease in flow velocity with increasing atrial filling and a second flow acceleration upon opening of the tricuspid valve, which produces the second peak. During atrial contraction, flow again becomes faster, sometimes with a short retrograde component.

The renal veins, in particular the right one, also show respiratory blood flow fluctuation, which is reflected in the Doppler waveform by an increase in flow velocity during inspiration and a decrease during expiration. The right renal vein also shows the atrial modulation. The left renal vein typically exhibits pulsatile variation due to brief compression of the vein during systole in the narrow passageway between the aorta and superior mesenteric artery. The left renal vein occasionally takes an atypical retroaortic course and rarely multiple branches are present on the left (4% versus about 20% on the right). If the left renal vein is not depicted between the aorta and superior mesenteric artery, the examiner must look for it behind the aorta at about the level of the origin of the left renal artery.

The major branches of the hepatic venous system and the right renal vein have the same flow character as the vena cava (cardiac (atrial) pulsatility and respiratory phasicity).

### 6.2.4.2 Portal Venous System

The normal portal vein is depicted by gray-scale sonography with an anechoic, smoothly delineated lumen below the liver and shows less marked respiratory caliber variation than the vena cava (usually 8–13 mm, larger caliber during deep inspiration). The (color) duplex mode depicts flow toward the liver with respiratory variation. The flow velocity is 10–25 cm/s (Seitz and Kubale 1988; Moriyasu et al. 1986; Giyanani et al. 1989) and may exceed 35 cm/s postprandially. Altogether, *flow velocities* in the portal system are characterized by *wide interindividual variation* and also increase after a meal as in the mesenteric circulation (two- to threefold increase in flow volume in the superior mesenteric artery and vein).

## 6.2.5 Documentation

The extent of documentation of the findings in the retroperitoneal veins as in the portal venous system depends on the clinical question to be answered. In addition to the B-scan findings and the Doppler waveform of the vein of interest, the perivenous findings should be documented as well. Specifically, vein compression and the extent of thrombotic changes must be recorded as well as collateral pathways in case of disturbed venous drainage, for example, retroperitoneal and splenorenal shunts in renal vein thrombosis and gastric or umbilical shunts in portal hypertension (liver cirrhosis, Cruveilhier–Baumgarten syndrome).

## 6.2.6 Abnormal Ultrasound Findings, Measurement Parameters, and Diagnostic Role

### 6.2.6.1 Vena Cava

The complex embryonic development of the venous system gives rise to numerous variants and malformations, all of which are rare. The vena cava can show the whole range of anomalies from aplasia to duplication.

Rare variants and atypical courses of the individual vessels are identified and differentiated from retroperitoneal lymph nodes by *color duplex scanning*. Compression of the vena cava is most commonly due to retroperitoneal lymph nodes or tumors, aortic aneurysm, or retroperitoneal fibrosis. Rare venous leiomyomas or leiomyosarcomas may also arise from the smooth muscle layer of the vena cava.

Since compression sonography is of limited use in demonstrating *thrombosis* of the retroperitoneal and visceral veins, *duplex imaging* and above all *color-coded duplex imaging* come in handy. However, criteria for *thrombosis* of the intra-abdominal or retroperitoneal veins are present on the *gray-scale scan* as well.

Thrombus may be visualized directly as a hyperechoic structure. In addition, thrombosis is suggested if the respiratory caliber variation typical of the larger retroperitoneal veins, in particular the vena cava, is lost. This finding is unspecific, and dilatation of the vena cava with reduced or absent caliber fluctuation may also occur in right ventricular failure. Thrombosis should always be ruled out by *color duplex imaging* with a low pulse repetition frequency and high gain or by obtaining a *Doppler waveform*. Apart from thrombosis, venous drainage may be obstructed by tumor compression, tumor infiltration, or intravascular tumor growth.

The clinical symptoms of thrombosis depend on its site, temporal course, and collateralization. In particular, in pelvic vein and *vena cava thrombosis*, there is a risk of embolism. Thrombosis of the vena cava is typically caused by an ascending thrombus from the pelvic and leg veins or extension of a tumor thrombus, for instance through the renal vein in renal cell carcinoma. Fresh ascending thrombus is typically hypoechoic and may be difficult to identify on gray-scale ultrasound in obese patients. In the further course, the thrombus undergoes hyalinization and becomes inhomogeneous. Thrombus organization with invasion of cells from the vessel wall and retraction of fibrin fibers results in increasing echogenicity and poorer delineation from the wall. Very old thrombi may undergo partial mural calcification.

Due to extensive collateralization, even occlusion of the vena cava may occasionally cause only a few clinical symptoms. Venous return occurs predominantly through the paravertebral plexus, the ascending lumbar and azygos venous systems, the superficial veins of the abdominal wall, and the portal collateral route. Color duplex scanning enables good evaluation of the collateral pathways in thrombosis of the vena cava or pelvic vein, though the findings have no clinical relevance in most cases.

Disturbed drainage of the pelvic and leg veins demonstrated by spectral Doppler (continuous flow without respiratory phasicity) may be due to central vena cava thrombosis caused by thrombus or tumor extension from the renal veins or compression of the vena cava by a retroperitoneal tumor or aortic aneurysm. Therefore, the examiner must carefully look for these possible causes. If the spectral waveform from the vena cava or pelvic veins shows pulsatile flow, a thorough search must be undertaken for an AV fistula, which may be caused by trauma, idiopathically, perforating aneurysm, or iatrogenically after surgery or puncture.

Tricuspid insufficiency or right ventricular failure affects caval blood flow, causing dilatation and abnormal changes in the shape of the Doppler waveform. Regurgitation into the right atrium in tricuspid insufficiency extends into the proximal vena cava, where it becomes apparent in the waveform by a reflux component during systole.

*Vena cava thrombosis* is typically due to *ascending thrombosis* of the pelvic and leg veins but may also be caused by *local obstruction* (external tumor compression or infiltration) as well as by *thrombus or tumor extension from renal veins* or thrombus extension from hepatic veins (Budd–Chiari syndrome).

### 6.2.6.2 Renal Veins

Just as in the inferior vena cava, *thrombosis* and central *tumor thrombus* of the renal veins can be identified sonographically under adequate insonation conditions. Therefore, preoperative *color duplex imaging* of the renal veins is sufficient prior to tumor nephrectomy. Venography has a similar diagnostic yield only if it is performed as venacavography with compression or provocative maneuvers. For this reason, contrast-enhanced CT is the primary alternative imaging modality in the routine clinical setting.

The nephrotic syndrome associated with glomerulonephritis is the most common cause of renal vein thrombosis. Predisposing factors are antithrombin III deficiency, sepsis, pregnancy, oral contraceptives, corticoid therapy, collagen diseases, and amyloidosis. Secondary renal vein thrombosis can be caused by obstruction due to retroperitoneal tumors, aortic aneurysm, or caval thrombosis as well as intravenous extension of renal tumors.

Unspecific signs of renal vein thrombosis, detectable by gray-scale ultrasound, are enlargement of the kidney and a reduced echogenicity of the renal parenchyma. In patients with an adequate acoustic window, the thrombus will be identified in the dilated vein as a hypoechoic and partially inhomogeneous structure within the lumen (Fig. A6.67). Thrombosis is confirmed by the absence of flow in color duplex scanning or in the Doppler waveform. Thrombus with partial occlusion may be seen as a defect in the color coding with a decrease or complete loss of cardiac pulsatility and respiratory phasicity of flow in the Doppler waveform. If there is good venous drainage through capsular veins and the suprarenal vein, venous flow may be detectable in the renal hilum even if the occlusion of the renal vein is complete.

*An indirect sign of acute renal vein thrombosis* in *the arterial waveform* is a marked reduction or even transient reversal of diastolic flow. This is due to a reflex vasoconstriction, and the flow pattern resembles that in rejection of a kidney transplant.

Apart from an increased peripheral resistance reflected in the waveform from the renal artery, acute renal vein thrombosis can also cause kidney enlargement. The *magnitude of these changes* depends on the *extent of collateral pathways* of the thrombosed renal vein, which may involve splenorenal shunts or retroperitoneal routes such as venous connections to the adrenal gland. Recanalization after acute renal vein thrombosis is seen on color duplex as meander-like flow in an otherwise dilated and echogenic lumen.

Markedly slower flow in the renal vein with loss of cardiac pulsatility and respiratory phasicity may also be seen in case of obstruction of the proximal inferior vena cava by a tumor or thrombosis or in right ventricular failure (acute: pulmonary embolism, chronic: tricuspid insufficiency).

Prior to tumor nephrectomy, sonographic evaluation of the renal veins is necessary to plan the surgical procedure according to the *stage of venous tumor extension*:

- *Stage I* is characterized by a button-like protrusion of tumor from the renal vein into the vena cava (see Fig. A6.68).
- In *stage II,* the tumor extends farther into the vena cava but the upper margin is still below the level of the hepatic vein termination.
- In *stage III,* the tumor extends to the level of the hepatic veins.
- In *stage IV,* there is tumor extension into the atrium.

Adequate sonographic evaluation of venous thrombus or tumor thrombus is impaired by superimposed bowel gas and by flow phenomena, especially when examining obese patients, resulting in adequate duplex evaluation of the renal veins in only 50–80% cases (Schwerk et al. 1994; Didier et al. 1987; Dubbins et al. 1986; London et al. 1989). When there is a good acoustic window, ultrasound has 95–100% accuracy in detecting tumor thrombus. With the poor overall visualization of the renal veins by ultrasound, contrast-enhanced computed tomography and magnetic resonance imaging are more accurate in preoperatively defining the extent of tumor thrombus in the renal vein and beyond and in planning the surgical resection. Therefore, these two imaging modalities should be used liberally as supplements to duplex ultrasound.

### 6.2.6.3 Superior Mesenteric Vein

*Mesenteric vein thrombosis* is a rare cause of intestinal necrosis. Therefore, duplex imaging performed to rule out mesenteric artery occlusion in patients presenting with the respective clinical symptoms should also include the mesenteric vein to exclude thrombosis there as well. Edematous *thickening of the intestinal walls* is a conspicuous finding and will already be seen on the gray-scale scan.

Causes include hematologic diseases, clotting disorders, abscess or sepsis, and tumor occlusion. Apart from acute thrombosis presenting with acute symptoms of intestinal necrosis, there may be chronic thrombosis with unspecific findings such as fever, leukocytosis, or thrombocytosis.

An abnormal course of the superior mesenteric vein suggests malrotation, which is confirmed if the superior mesenteric vein lies to the left of the superior mesenteric artery. If the vein is anterior to the artery, malrotation is present in one third of the cases. These indirect signs of malrotation are easily detected by duplex ultrasound.

The severity of the clinical symptoms associated with mesenteric vein thrombosis depends on the extent and site of the thrombus and collateralization. *Partial thrombosis of the superior mesenteric vein*, for instance, may be *fairly asymptomatic* and present with the clinical signs of enteritis only. Conversely, patients with extensive central mesenteric vein thrombosis may present with an acute abdomen due to intestinal necrosis. Early diagnosis with initiation of anticoagulation therapy is essential for preventing thrombosis progression. Therefore, the mesenteric vein should be evaluated in all patients with thickening of the intestinal wall on B-mode images. The criteria for thrombosis of the superior mesenteric vein are the same as in other vascular territories: *dilatation of the vein*, absence of respiratory diameter variation, possibly depiction of the thrombus as an echogenic intraluminal structure on the B-mode image, and *absence of flow signals* or only residual flow signals near the wall surrounding a central thrombus on *(color) duplex imaging* (see Figs. A6.58 and A6.59). Apart from intestinal wall thickening, another indirect sonographic sign of extensive mesenteric vein thrombosis is a more pulsatile flow in the arterial waveform in extensive mesenteric vein thrombosis (Table 6.15).

Abnormalities of the splenic vein (such as thrombosis) are negligible, both clinically and in terms of their therapeutic consequences, because extensive collateral routes exist. Nevertheless, the splenic vein may be involved in patients with portal hypertension from liver cirrhosis and is therefore included in the examination in these cases.

**Table 6.15** Mesenteric vein thrombosis

| |
|---|
| **Risk factors** |
| • Portal hypertension |
| • Sepsis |
| • Diverticulitis |
| • Paraneoplastic syndrome |
| • Autoimmune disease |
| • Clotting disorder |
| **Clinical presentation** |
| • From unspecific symptoms to acute abdomen (depending on the extent of collateralization) |
| **Duplex ultrasound findings** |
| • Hyperechoic thrombus |
| • Dilated vein |
| • No intraluminal flow signals |
| • Diastolic flow in the superior mesenteric artery may be reduced |
| • Thickening of bowel loops in gray-scale image |

### 6.2.6.4 Portal and Hepatic Veins

Portal vein thrombosis is diagnosed using the same sonomorphologic criteria as for other sites: dilatation of the lumen, absent respiratory diameter variation, and no flow signals or only residual flow signals around the thrombus on (color) duplex images. Acute portal vein thrombus tends to be hypoechoic and is clearly delineated from perivascular structures, whereas older thrombi contain more inhomogeneous and hyperechoic portions and their contours become blurred, resulting in poorer sonographic discrimination of the thrombotic vein. The Doppler spectrum sampled with an adequate angle will show absence of flows or more higher-frequency signals without respiratory phasicity around a central thrombus surrounded by flowing blood. Recanalization or *cavernous transformation* is associated with meander-like, tortuous flow in the portal vein, appearing as antegrade and retrograde flow components.

Acute portal vein thrombosis, like acute proximal mesenteric vein thrombosis, is characterized by the sudden onset of clinical symptoms. Collateralization in portal vein thrombosis is more extensive if the superior mesenteric vein is not involved; the collateral vessels in this case include the splenic vein and gastric veins, which can be depicted by duplex ultrasound.

The causes of portal vein thrombosis include liver cirrhosis, paraneoplasia, clotting disorders, and sepsis (Table 6.16). Similar to portal hypertension in liver cirrhosis, acute portal vein thrombosis is associated with widening of the veins serving as collaterals (splenic vein, esophagogastric vessels) and ascites, which can be identified sonographically as secondary signs.

The formation of collateral pathways in the liver hilum and partial recanalization of the thrombotic portal vein, so-called cavernous transformation, results in a worm-like meshwork of tortuous tubular structures, among which the former portal vein is no longer identifiable by color duplex imaging.

**Table 6.16** Etiology of thrombotic changes in the portal venous system

| |
|---|
| • Acute pancreatitis |
| • Chronic pancreatitis (may be associated with pseudocyst) |
| • Cancer (hepatocellular carcinoma, metastasis, pancreatic carcinoma) |
| • Idiopathic |
| • Abdominal infections |
| • Collagen diseases |
| • Myeloproliferative syndrome |
| • Trauma |
| • Status post splenectomy |
| • Pregnancy |
| • Drug treatment |
| • Liver disease, cirrhosis, thrombocytosis |
| • Antiphospholipid antibody syndrome |
| • Deficiency of AT3, protein C, protein S |

As with older thrombosis in other vessels, chronic portal thrombosis leads to shrinkage of the initially widened vessel with depiction of an inhomogeneous, more hyperechoic thrombus within the lumen. In contradistinction to acute portal thrombosis, the chronic form is associated with relatively unspecific, subacute clinical manifestations, in particular when it occurs secondary to liver cirrhosis.

Ultrasound *detects portal vein thrombosis* with a *sensitivity of 89–100%* and a *specificity of 95–100%*, which is comparable to its accuracy in the detection of deep leg vein thrombosis (Zwiebel 2000). Nevertheless, ultrasonography is limited by the fact that very slow to-and-fro flow in severe portal hypertension may escape detection.

Duplex ultrasound performed for *portal hypertension* provides valid information on portal venous flow in 93–95% of patients (Patriquin 1987; Yeh 1996; Seitz and Kubale 1988). The main criteria are:

- The portal vein diameter measured at gray-scale ultrasound
- The evaluation of *portal venous flow direction, flow profile,* and *velocity*
- The identification of portal caval shunts/*collateral pathways*

*Portal hypertension* can be caused by obstruction of the portal venous system at different levels (see Fig. A6.1b) and is associated with complex circulatory changes. The pressure in the portal vein is typically 2–4 mmHg above that in the inferior vena cava and hypertension is defined as a pressure increase to over 11 mmHg that persists for an extended period. *Indirect criteria* are dilatation of the superior mesenteric artery with *loss of respiratory phasicity*, an enlarged ventricular coronary vein diameter (>4 mm), and patency of the umbilical vein with high flow.

Portal hypertension secondary to liver cirrhosis is associated with *widening* of the portal vein and its distal tributaries and of the collaterals (portocaval, gastroesophageal, splenorenal, umbilical), which may already be noted at B-mode ultrasound (Table 6.17). Moreover, *respiratory diameter variation is lost* in the portal vein.

A portal vein diameter of more than 13 mm is a highly reliable predictor of portal hypertension with 95–100% specificity, while its sensitivity is only 45–50% (Bolondi et al. 1982), which is attributable to the wide variation in the normal portal vein diameter. Another criterion is the loss of respiratory variation in the portal vein diameter (best appreciated in the time-motion mode): an increase of 70% in the diameter of the superior mesenteric and splenic veins during deep inspiration has been reported to be normal, while an increase below this level has been found to have over 80% sensitivity and specificity for portal hypertension (Bolondi 1984). These two criteria – dilatation and pressure-related loss of the respiratory diameter variation – can be supplemented by a third sonographic criterion when

**Table 6.17** Ultrasound findings in portal hypertension

| | |
|---|---|
| **B-mode** | • Ascites, splenomegaly<br>• Possibly cirrhotic changes of hepatic vessel architecture and parenchymal structure<br>• Signs of congestion of the gallbladder and stomach walls<br>• Dilated portal vein with change from elliptical to circular profile<br>• Portocaval collaterals<br>• Portal vein thrombus (echogenic) |
| **Duplex** | • Reversed flow in portal vein<br>• Slower blood flow (diminished flow rate)<br>• Loss of respiratory phasicity<br>• Reduced increase in flow velocity after a test meal<br>• Portocaval collaterals |
| **Color duplex** | • Portocaval collaterals<br>• Stagnating/reversed blood flow<br>• Portal vein thrombosis |

evaluating the mesenteric, splenic, and portal veins: the increase in intravascular pressure causes these veins to lose their normal oval cross section and become round.

Atresia and hypoplasia of the portal vein are rare, as are anatomic variants and malformations. In individuals with a congenital extrahepatic portocaval shunt, portal venous blood from the mesentery and spleen drains directly into the inferior vena cava. As a result of this direct communication of the splenic and superior mesenteric veins with the vena cava (gray-scale scan), the cardiac pulsatility of venous return in the vena cava is transmitted to the mesenteric vein and reflected in the Doppler waveform from the latter.

*Aneurysm* of the *portal vein* is also rare and must be differentiated from pseudocysts of the pancreas, choledochal cysts, and liver cysts by the demonstration of flow in the color duplex mode.

The spectral waveform in portal hypertension shows a reduced mean flow velocity and loss of respiratory phasicity (see Table 6.17).

Preserved cardiac pulsatility in the portal vein can suggest portal hypertension secondary to severe right ventricular insufficiency, based on the assumption that marked atrial pulsations can be transmitted into the portal vein through the hepatic sinusoids. This abnormal cardiac pulsatility must be differentiated from normal respiratory phasicity (Moriyasu et al. 1986).

The *main diagnostic role of color-coded duplex ultrasonography* lies in the follow-up of portal hypertension with early identification of thrombosis as a possible complication. Moreover, it provides useful diagnostic information in presinusoidal, extrahepatic portal hypertension. The most common causes are primary or secondary tumor thrombosis, inflammatory diseases like pancreatitis, and slow flow due to liver cirrhosis. The presentation of portal vein thrombosis varies with the temporal course and collateralization, ranging from unspecific abdominal symptoms to an acute abdomen.

*Depending on the severity of portal hypertension*, spectral Doppler will demonstrate *antegrade flow with reduced velocity*, *to-and-fro flow*, or *flow reversal* when pressure is increased to over 30 mmHg. The normal cardiac pulsatility of the liver veins is lost in liver cirrhosis.

The *flow direction in the portal vein* is determined not only by the severity of cirrhosis and the magnitude of intraportal blood pressure but also by the *direction of collateral drainage* (Table 6.18 and Fig. A6.1b).

When the blood is chiefly drained through splenorenal shunts, flow in the portal vein is backward (hepatofugal), while *normal, hepatopetal flow* is preserved in the presence of a *patent umbilical vein* (Cruveilhier–Baumgarten syndrome). In this case, there may also be retrograde flow in the right portal vein branch with normal flow direction in the left portal branch), which feeds the recanalized umbilical vein.

*Difficulties in determining venous blood flow* mainly arise from the wide diameter fluctuations of veins, which are difficult to quantify. This also holds true for the portal vein with is extreme variation in diameter between inspiration and expiration. Because of these problems, determination of mean blood flow velocity in the portal vein is a more suitable parameter for discriminating between healthy individuals and patients with portal hypertension. Note, however, that the mean flow velocity is also influenced by the extent of collateralization. For instance, *pronounced flow in the patent and widened umbilical vein* (Cruveilhier–Baumgarten syndrome) *may mimic normal perfusion of the liver* with a fairly normal flow velocity in the portal vein because the blood drains through the umbilical vein, circumventing the sinusoids (see Fig. A6.1b).

Although the variable collateralization leads to a wide variation in mean portal flow velocities, both in intraindividual and interindividual comparison, significant differences are identified between healthy subjects and patients with portal hypertension when the mean values of larger study

**Table 6.18** Portocaval collaterals

| |
|---|
| **Shunts draining toward center:** |
| • Esophageal varices, gastric corpus and fundus varices (ventricular coronary vein – azygos vein, short gastric veins – azygos vein) |
| • Gastrosplenic shunts |
| • Portorenal and splenorenal collaterals |
| • Capsular veins of liver and spleen, diaphragmatic veins |
| **Shunts draining toward periphery:** |
| • Paraumbilical veins (Cruveilhier–Baumgarten syndrome) |
| • Splenolumbar shunts |
| • Mesenteric veins (superior and inferior mesenteric veins, ovarian vein, spermatic vein, rectal plexus) |

populations are compared. Several such studies demonstrated a statistically significant decrease from 15 cm/s in healthy subjects to half that value in patients with cirrhosis (Seitz and Kubale 1988). Though maximum venous flow velocity is decreased to 7–15 cm/s (mean 10 cm/s) in patients with cirrhosis, there is wide interindividual variation and overlap with the flow velocities in normal subjects, which may lead to misinterpretation in individual cases. The extent to which mean blood flow velocity is reduced also depends on collateralization.

In summary, *portal hypertension* is unlikely if maximum venous flow velocities ($V_{max}$) of over 20 cm/s are measured. It is probable if the velocity is 10–20 cm/s, and it is very likely *if flow velocity in the portal vein is below 10 cm/s.*

Duplex imaging also demonstrates a decrease in portal blood flow associated with the intake of beta-receptor blockers or somatostatins as well as an increase after eating or after glucagon challenge. *The increase in portal flow velocity after a test meal is less pronounced in patients with cirrhosis.*

Various tests have been proposed that promise more reliable discrimination of cirrhosis-induced portal hypertension from normal portal blood flow. Apart from the less marked increase in postprandial flow, drugs like beta-receptor blockers or nifedipine also have a less pronounced effect on the flow velocity in the portal vein in cirrhosis. Gaiani et al. (1989) compared 11 patients with liver cirrhosis and healthy controls 60 min after a test meal and found a markedly lower increase in diameter of 3% in the patients compared with 14% in controls; they also found a difference in flow velocity, which was 3.2% versus 24%. The flow volume after the test meal increased by only 8.5% in the patients, as opposed to 59% in the controls. Such clearcut results were not always confirmed by other study groups.

Another approach combines the increased vascular cross-sectional area and the decreased flow velocity in the portal vein in portal hypertension in a so-called *congestion index* ($cm^2/cm/s = cm \times s$). Normal individuals have an index <0.07 cm × s versus >0.1 cm × s in portal hypertension associated with liver cirrhosis (Moriyasu et al. 1985; Siringo et al. 1994) (see Fig. A6.61). Further studies are necessary to show whether glucagon-induced changes in portal venous flow can be used to estimate the hemodynamic reserve and whether measurement of portal flow velocity after propanolol administration may enable reliable identification of those patients requiring treatment for portal hypertension; this parameter however also shows wide interindividual variation.

A highly sensitive *direct sign* of portal hypertension is the presence of collateral pathways, seen either as enlargement of the short gastric veins or coronary vein with venous drainage to the esophageal plexus or as a patent umbilical vein (Cruveilhier–Baumgarten syndrome). Other collaterals including gastrorenal and splenorenal anastomoses and peripancreatic veins are less amenable to sonographic evaluation. When a systematic search is performed, 65–90% of the relevant portocaval collaterals can be identified by duplex imaging (Lafortune et al. 1987; Takayasu et al. 1984; Subramanyam et al. 1983). The left gastric vein with a normal diameter of less than 4 mm is usually well visualized, making it of great diagnostic importance in duplex ultrasound. A diameter of over 7 mm and hepatofugal flow indicate portal hypertension (Lafortune et al. 1984; Morin et al. 1992). The demonstration of hepatofugal flow in the recanalized umbilical vein, beginning in the round ligament, was found to have sensitivities and specificities of up to 100% (Gibson et al. 1989; Mostbeck et al. 1989). Occasionally, flow can be detected in the round ligament in individuals without portal hypertension; however, in these cases, blood flow velocity does not exceed 5 cm/s (Casarella et al. 1995; Lafortune et al. 1984, 1987). It is also helpful to identify and evaluate collaterals at the esophagogastric junction; these varices can be differentiated from enlarged lymph nodes by the demonstration of flow in the color duplex mode. Sonographic follow-up evaluation of the collateral pathways can also help in evaluating the outcome of treatment.

Quantitative determination of blood flow is not necessary in the routine diagnostic workup of portal hypertension since there is no close correlation between portal blood flow and portal hypertension due to the highly variable and ramified collateral system.

On the one hand, the decreased portal blood flow in portal hypertension due to cirrhosis can lead to a compensatory increase in arterial perfusion that is detectable sonographically. The higher perfusion can result in an enlargement of the cross-sectional areas of the hepatic veins both within and outside the liver.

On the other hand, progressive cirrhosis is associated with an increased resistance in the peripheral hepatic artery branches, resulting in a more pulsatile flow profile with increased resistance indices of 0.8 to 0.9.

Endoscopic obliteration of esophageal varices and the TIPSS procedure (transjugular intrahepatic portosystemic shunt with stenting) have led to a decrease in portocaval and splenorenal shunt operations. An important preoperative question to be answered is whether the portal, mesenteric, and splenic veins are patent. There is good evaluability of these veins by ultrasonography, which is why color duplex imaging has evolved into the method of choice. In the postoperative follow-up, ultrasound enables direct evaluation of shunt patency. The cardiac fluctuation of blood flow in the vena cava is transmitted to the anastomosed portal vein through the shunt.

When a distal splenorenal shunt (Warren shunt) is created, the relief of the portal vein leads to flow reversal in the splenic vein (hepatofugal flow). In the TIPSS procedure, a short circuit is established between the hepatic vein and portal vein under ultrasound guidance. Color duplex ultrasound can help in identifying a short puncture tract and is also useful for postinterventional surveillance of stent patency.

*In-stent stenosis* and shunt thrombosis are common, resulting in poor 1-year patency rates of 35–66% (Nazarian et al. 1994; Sterling and Darcy 1997; Kerlan et al. 1995). Patency can be improved by sonographic surveillance with timely revision. A sonographic examination should be performed within 24 h of shunt creation to determine stent location and flow velocity and to confirm the technical success of the procedure, especially at the junctions between the stent ends and the native vein. Further examinations should follow at 3-month intervals, with the following parameters being of particular importance:

- PSV in the shunt, which should be at least 50–60 cm/s (Chong et al. 1993; Foshager et al. 1995; Dodd et al. 1995; Feldstein et al. 1996). Normal peak shunt velocity ranges between 80 and 120 cm/s (Kanterman et al. 1997).
- Continuous color duplex imaging of the shunt should not reveal any mural thrombus: color-coded flow throughout the shunt lumen (with an adequate pulse repetition frequency) without defects and without aliasing.
- Stent ends should extend just as far as necessary into the vena cava and portal vein.
- Doppler spectrum, which should reveal largely continuous flow and at most slight cardiac pulsatility.
- Hepatopedal flow in the portal vein with a return to normal flow velocity.

Shunt stenosis is suggested by the abrupt doubling of flow velocity. A flow velocity of less than 50 cm/s indicates inadequate shunt flow and should prompt a thorough search for shunt stenosis or other causes (Bodner et al. 2000; Murphy et al. 1998; Kanterman et al. 1997; Dodd et al. 1995). Stenosis is common at the stent ends but may occur anywhere along the course of the shunt. A drop in intrashunt blood flow also reduces flow velocity in the portal vein, and in the case of severe stenosis produces sonographic and clinical signs of portal hypertension.

When the shunt is occluded, there is no flow in color duplex imaging or in the Doppler waveform; the findings in the portal vein correspond to those of portal hypertension obtained before creation of the shunt.

Like the vena cava, the *hepatic veins* are subject to both respiratory phasicity and cardiac pulsatility. Besides prandial fluctuations, the flow velocity is determined by the pressure in the thorax, right atrium, and abdomen. *Cardiac pulsatility* gives rise to a triphasic, W-shaped waveform reflecting the venous pressure variations during the cardiac cycle. The first velocity peak directed toward the vena cava occurs during systole and atrial filling. As the intra-atrial pressure increases, hepatofugal flow decreases in the hepatic veins and in the vena cava. Opening of the tricuspid valve leads to increased flow into the right ventricle and a second flow velocity peak in the hepatic veins and vena cava. During atrial contraction, there may be zero flow or retrograde, hepatopedal flow.

Another factor affecting the Doppler waveform of the hepatic veins is the consistency of the liver parenchyma. As elasticity is lost and the parenchyma becomes more rigid with progressive cirrhotic transformation, the waveform of the hepatic veins is increasingly flattened. This change is a diagnostic indicator of increasing tissue rigidity (see Fig. A6.56d, e) and also a prognostic factor with progressive phase loss of the waveform (Bolondi et al. 1991; Ohta et al. 1994). Cardiac pulsatility of blood flow is preserved in 30–50% of patients with cirrhosis. Typically, however, the waveform becomes flattened with progressive parenchymal damage, and a band-like curve in the hepatic veins is associated with a life expectancy of less than 2 years. A markedly abnormal flow pattern in the hepatic veins was found to have a diagnostic accuracy in Child A cirrhosis (chronic hepatitis C) of 77% (Colli et al. 1994) and a specificity of 78%.

Since parenchymal damage also occurs in other liver conditions associated with severe fatty degeneration, flattening of the waveform in the hepatic veins is not a specific indicator of cirrhosis. Moreover, there may be physiologic flattening in advanced pregnancy.

Budd–Chiari syndrome results from compromised hepatic venous outflow due to postsinusoidal obstruction. The obstruction may be caused by a mass (tumor, cyst, abscess), a liver vein thrombosis, or a congenital anomaly with a connective tissue membrane in the termination of the middle and/or left hepatic veins. In acute liver vein thrombosis, the vein is dilated with higher-echogenicity material in the lumen and absence of flow at color duplex sonography. In the chronic stage, there may be recanalization of an obstructed vein with sonographic demonstration of membranes and the identification of venovenous and portosystemic shunts in the color duplex mode.

Duplex imaging is a valid modality for the routine diagnostic evaluation of portal hypertension including initial diagnosis, hemodynamic evaluation of the portal vein, and follow-up. Radiologic modalities are only needed to examine patients with poor insonation conditions (massive ascites, meteorism) and to answer specific diagnostic questions. The flexibility in choosing sonographic imaging planes enables hemodynamic assessment as well as precise determination of topographic relationships. This is an advantage of ultrasound over angiographic procedures as well as over magnetic resonance imaging, especially with regard to the determination of blood flow volumes and flow direction.

# 6.3 Atlas: Visceral and Retroperitoneal Vessels

**Fig. A6.1 (continued)**

**Portal venous system/portal hypertension**

**b** Portal hypertension – causes and collateral circulation (from Droste 1989)

Portal hypertension (>15 cm $H_2O$ in portal venous system)

Causes

*A* Prehepatic obstruction: thrombosis of portal or splenic vein, tumor of adjacent organ (e.g., pancreas, stomach, duodenum, gallbladder)

*B* Intrahepatic obstruction:

$B_1$ Presinusoidal: schistosomiasis, Wilson's disease, myeloproliferative diseases (intrasinusoidal: chronic hepatitis, fatty liver)

$B_2$ Postsinusoidal obstruction: liver cirrhosis (cause of portal hypertension in 90% of cases), cytostatic treatment, etc.

*C* Posthepatic obstruction: hepatic vein occlusion (Budd-Chiari syndrome), compression of inferior vena cava, constrictive pericarditis

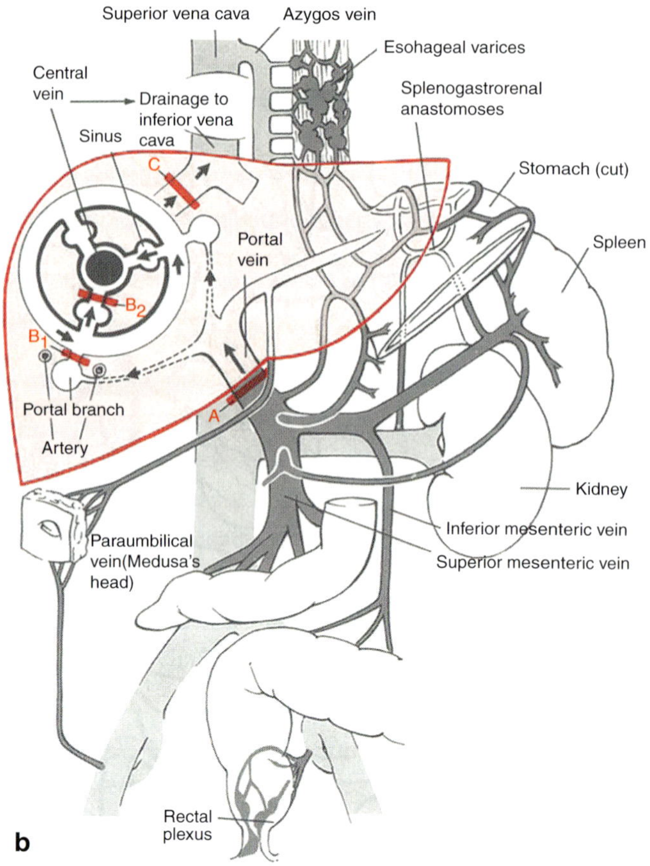


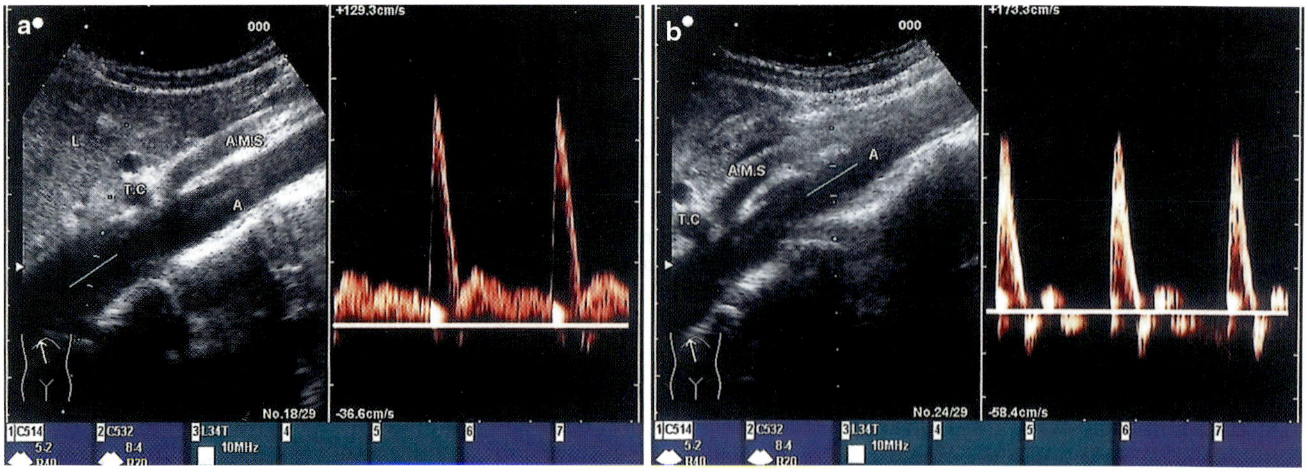


**Fig. A6.2 a, b**

**Flow profile in the aorta**

**a** Proximal to the origins of the visceral arteries (*T.C* celiac trunk, *A.M.S* superior mesenteric artery), the flow profile in the aorta is predominantly determined by the supply to parenchymal organs: a dip in early diastole is followed by constant diastolic flow. Flow in the aorta is of a mixed type resulting from blood flow to parenchymal organs (monophasic flow profile – low peripheral resistance) as well as flow to limbs (triphasic profile – high peripheral resistance).

**b** Distal to the origins of the visceral and renal arteries, the aorta has a triphasic flow profile (supply to limbs).

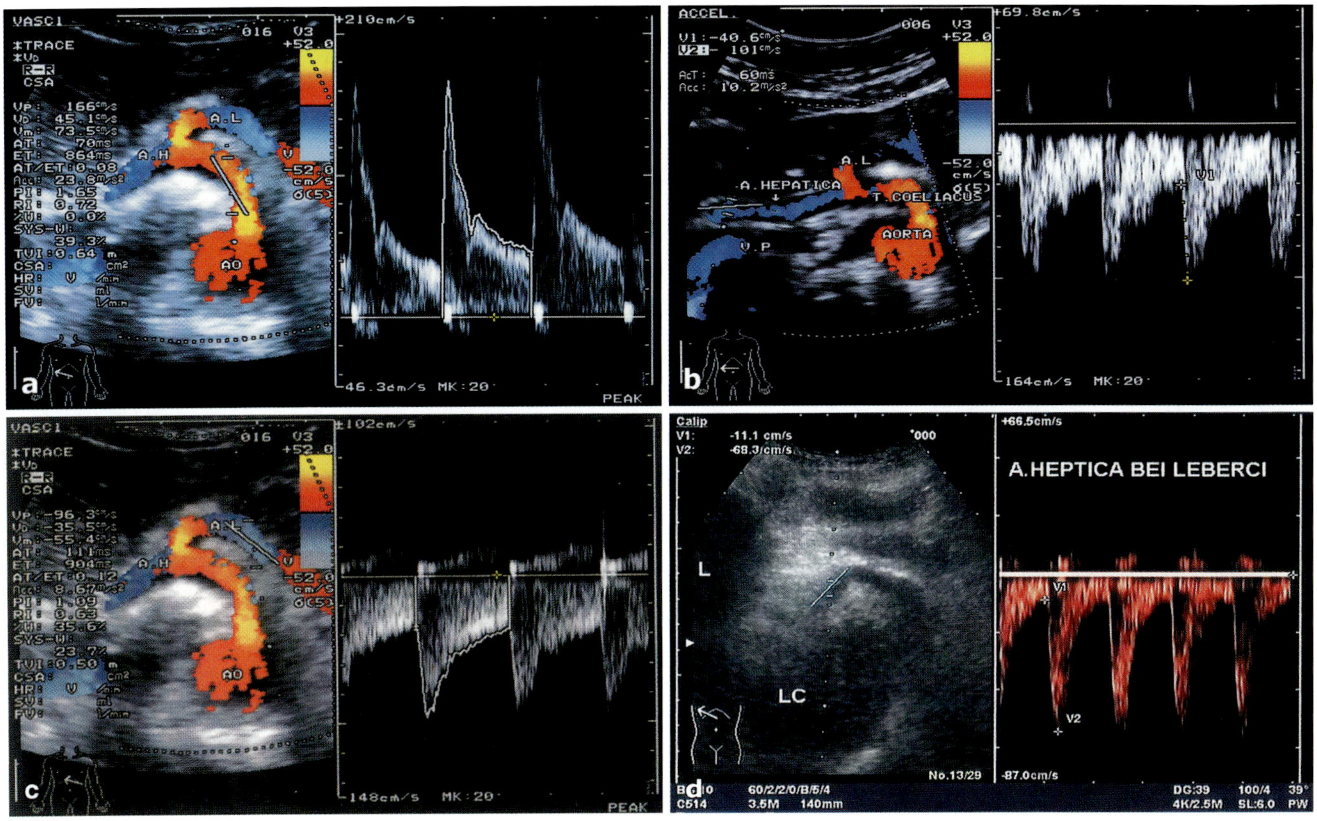


**Fig. A6.3 a–d**
**Celiac trunk**
**a** Transverse view of the celiac trunk showing its origin from the aorta (*AO*) and division into the hepatic artery (*A.H*) and splenic artery (*A.L*). The bifurcation is said to resemble a palm leaf or gull's wings.
Supplying parenchymal organs (spleen, liver), the celiac trunk, hepatic artery, and splenic artery show monophasic flow with a relatively large diastolic component, comparable to flow in the internal carotid artery. The aorta displayed in red gives off the celiac trunk anteriorly, likewise with flow depicted in red. The lighter color coding is not due to stenosis but to the angle of insonation.
This is confirmed by the normal Doppler waveform with a PSV of 165 cm/s and a peak end-diastolic velocity of 45 cm/s. Flow is synchronous with the cardiac cycle, exhibiting a low-frequency signal with a high amplitude due to wall motion in early systole.
**b** The hepatic artery courses to the hilum of the liver along the posterior aspect of the lower liver margin. The artery is coded in blue, indicating flow away from the transducer. The high diastolic flow is due to the low peripheral resistance of the liver. The splenic artery (*A.L*) first appears coursing in an anterior direction (toward transducer, coded red) and then turns posteriorly (blue) toward the splenic hilum.
**c** Splenic artery with typical Doppler waveform.

**Hepatic artery in liver cirrhosis**
**d** Liver cirrhosis is associated with parenchymal transformation, resulting in an increase in flow resistance in the hepatic artery. This is reflected in an increased Pourcelot index, which correlates with the extent of parenchymal change. In the case presented, the index is markedly increased to 0.83 (same patient as in Fig. A6.61a–c). There is marked enlargement and hypoechogenicity of the caudate lobe (*LC*) as a sign of severe cirrhosis.

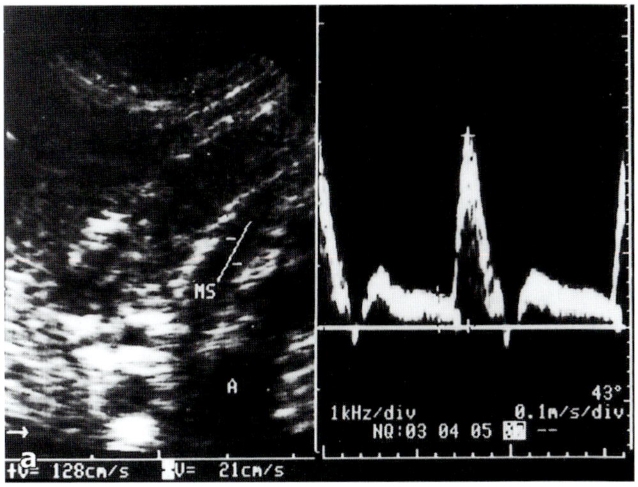


**Fig. A6.4 a–c**
**Mesenteric blood flow**
**a** Superior mesenteric artery with typical Doppler waveform of the mixed type. The end-diastolic flow component is intermediate between that of a peripheral artery and that of an artery supplying a parenchymal organ.
24-year-old fasting subject: normal blood flow in the superior mesenteric artery shortly after its origin with a PSV of 128 cm/s, peak end-diastolic velocity of 21 cm/s, and pulsatile flow. Gray-scale scan depicts the superior mesenteric artery (*MS*) arising from the aorta (*A*) at an acute angle.

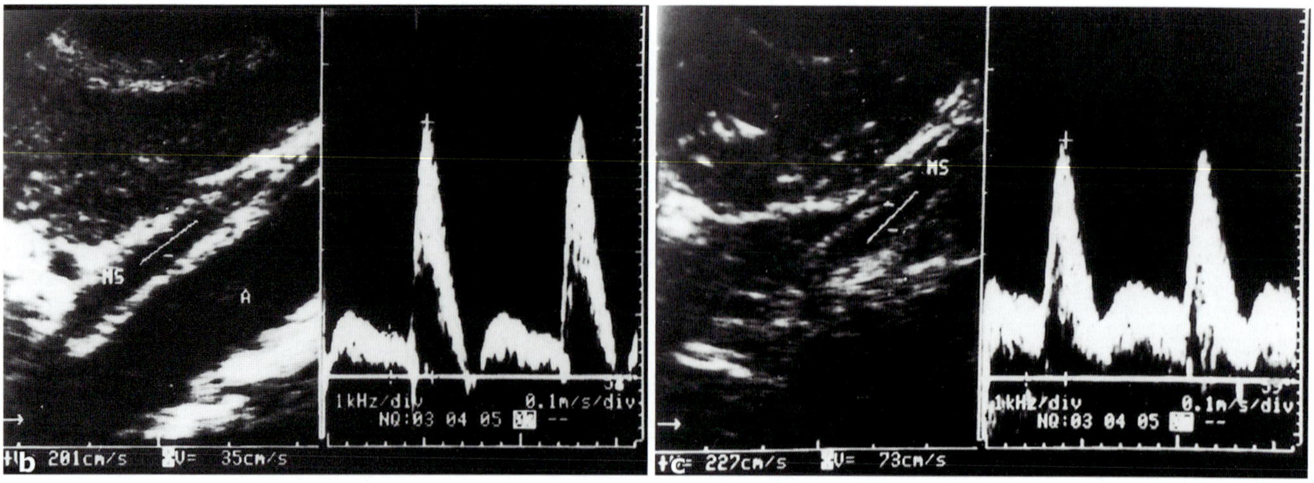


**Fig. A6.4 (continued)**

**b** Same subject as in **a**. Following administration of 20 mg of nifedipine, PSV increases to 201 cm/s, end-diastolic velocity to 35 cm/s.

**c** Postprandial increase in mesenteric blood flow (PSV of 227 cm/s, end-diastolic velocity of 73 cm/s). Assuming a threshold of 200 cm/s for >50% stenosis, the increased flow velocity observed after nifedipine administration and after eating would indicate a 50–60% stenosis in a fasting patient.

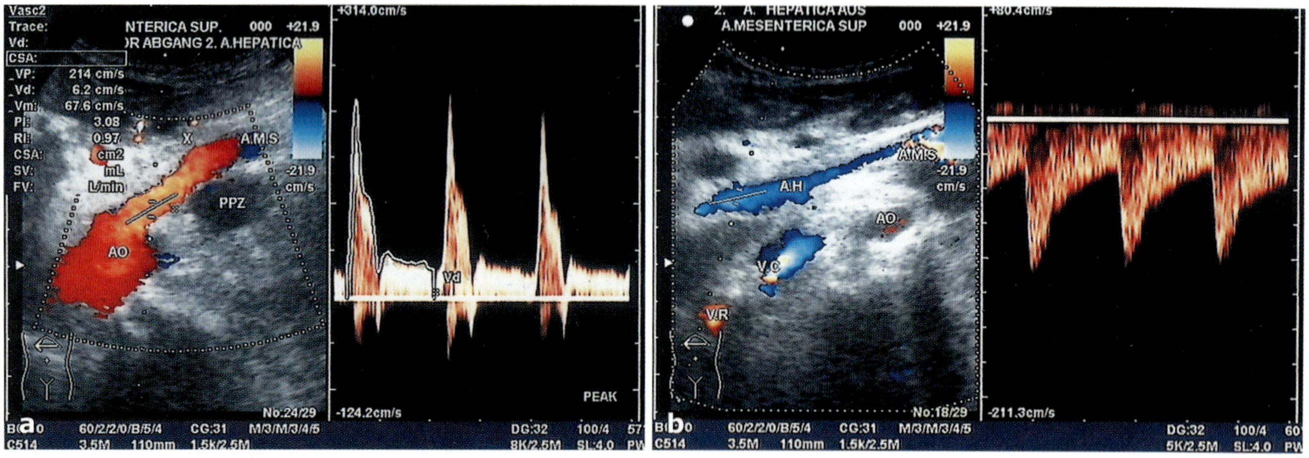


**Fig. A6.5 a–d**

**Waveform patterns of anatomic variants**

**a** Flow in a vessel as reflected in the Doppler waveform is determined by the organs it supplies. If the hepatic artery arises from the superior mesenteric artery, PSV is high even in the absence of stenosis (fasting velocity of 214 cm/s in the case presented). The patient has chronic pancreatitis with visualization of a pancreatic pseudocyst (*PPZ*) between the aorta and the superior mesenteric artery. The cyst is hypoechoic on the B-scan and can be differentiated from an aneurysm in the color duplex mode.

**b** In this rare case of duplication of the hepatic artery, the proximal superior mesenteric artery (transverse section through upper abdomen) gives off a hepatic artery (*A.H*) that courses to the right hepatic lobe along the anterior aspect of the vena cava (*V.C*). The left lobe is supplied by a normal hepatic artery arising from the celiac trunk.

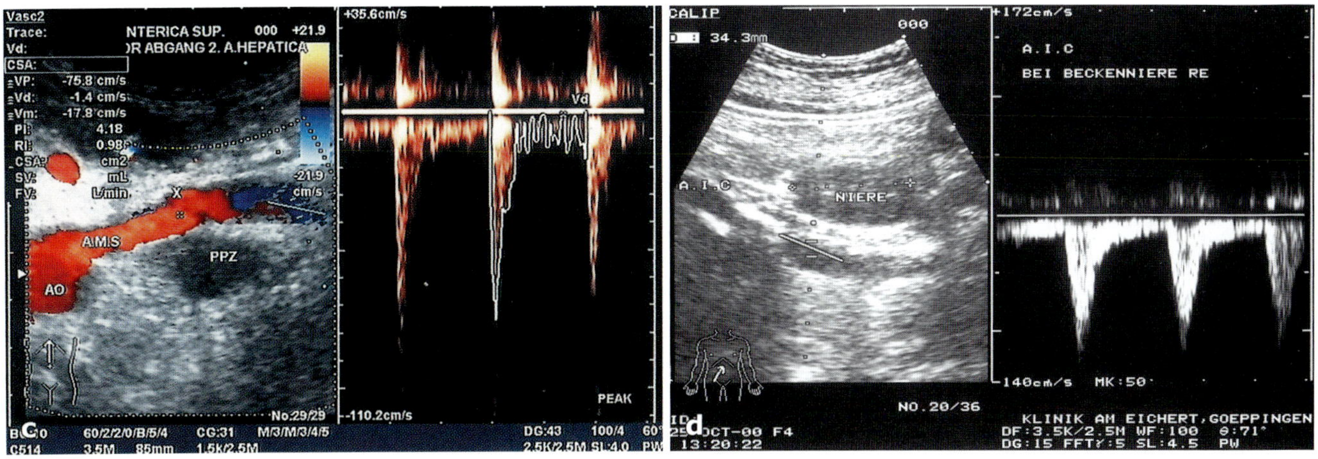


**Fig. A6.5 (continued)**

**c** Distal to the origin of the ectopic hepatic artery (at the level of the pancreatic pseudocyst, which serves as a landmark), the superior mesenteric artery shows flow with a smaller diastolic component and a reduced PSV. Proximal to the hepatic artery origin, flow is of the mixed type due to supply of two organs (liver and bowel). The examiner must be aware of these variants in interpreting changes in the Doppler waveform.

**d** In individuals with pelvic kidneys, as shown here (or in a transplant kidney anastomosed to the iliac artery), the normal Doppler waveform of the iliac artery proximal to the renal artery origin is monophasic rather than triphasic. The scan depicts part of the pelvic kidney above the common iliac artery. The monophasic waveform is due to blood supply to both the peripheral arteries and the renal artery and does not suggest postocclusive flow despite the presence of plaque proximal to the sample volume. Distal to the renal artery origin, the external iliac artery exhibits triphasic flow.

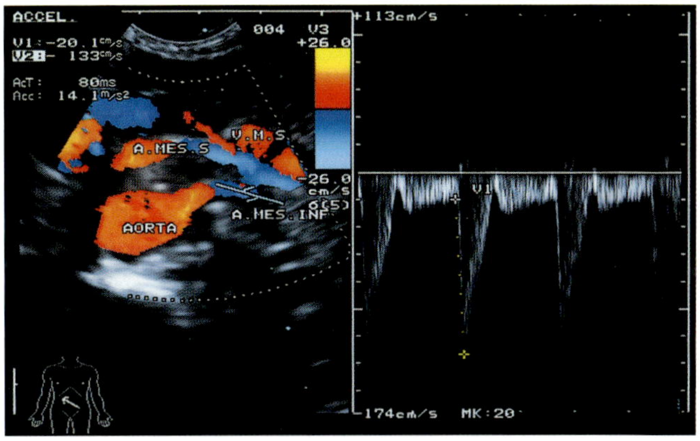


**Fig. A6.6**
**Inferior mesenteric artery**
Origin of the inferior mesenteric artery from the aorta. The Doppler waveform resembles that of the superior mesenteric artery but may occasionally show a smaller diastolic flow component or even end-diastolic zero flow. Anteriorly, a jejunal branch is depicted (blue, away from transducer) distal to the division of the superior mesenteric artery. The superior mesenteric artery dividing into the ileocolic and right colic arteries is seen anterior to the aorta with flow in the same direction coded in red. Directly anterior to the origin of the jejunal artery (displayed in blue), the jejunal vein with flow in the opposite direction (red) courses parallel to the artery and empties into the superior mesenteric vein (*V.M.S*).

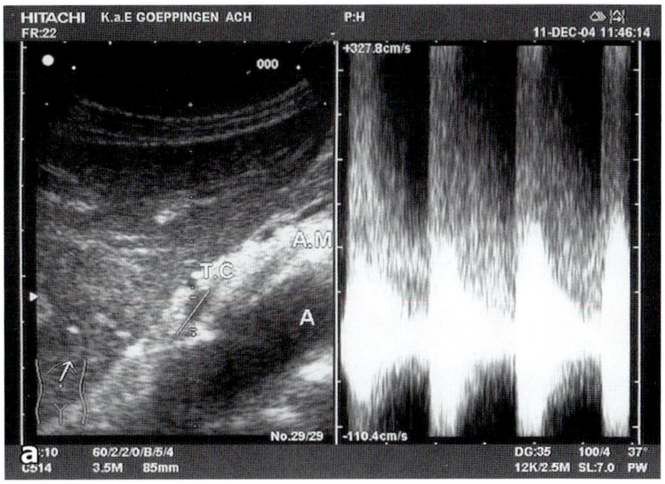


**Fig. A6.7 a–c**
**Stenosis of celiac trunk**
**a** The sample volume is placed in the celiac trunk (*T.C*) at its origin from the aorta. The Doppler waveform demonstrates high-grade stenosis with a PSV of over 4 m/s during both inspiration and expiration. Differentiation of the cause of celiac trunk stenosis (compression syndrome, atherosclerosis, fibromuscular dysplasia) is often difficult using B-mode ultrasound alone: wall assessment is impaired by the scan depth, the tangential angle of insonation relative to the celiac artery wall, and the presence of scatterers. The constant stenosis grade suggested by the unchanged PSV rules out median arcuate ligament syndrome; the demonstration of hyperechoic plaques in this case suggests that the stenosis is due to atherosclerosis.

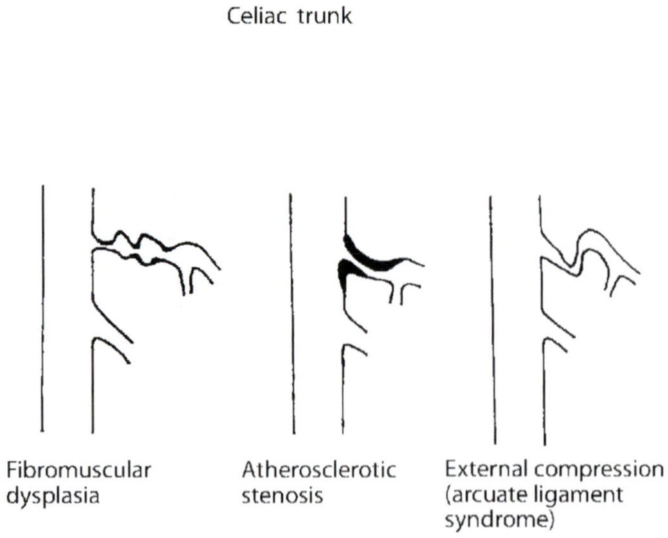


b

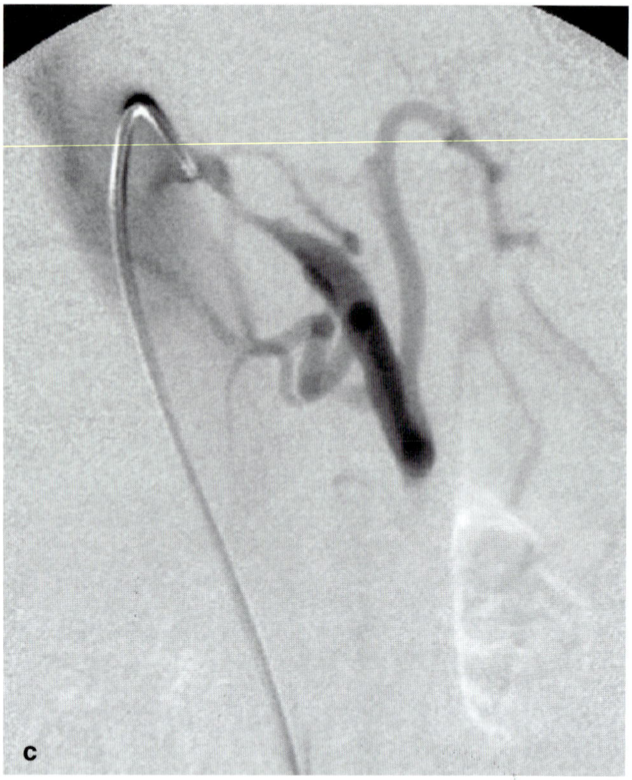


**Fig. A6.7 (continued)**
**b** Diagram of different types of celiac trunk stenosis and the underlying pathologies.
**c** Angiogram showing the stenosis at the origin of the celiac trunk.

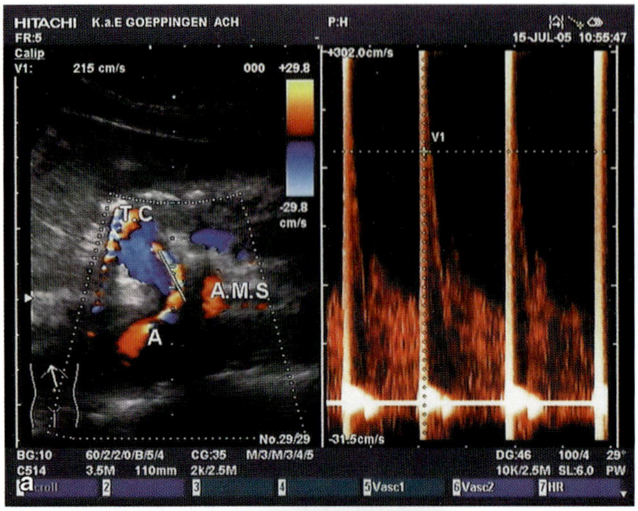


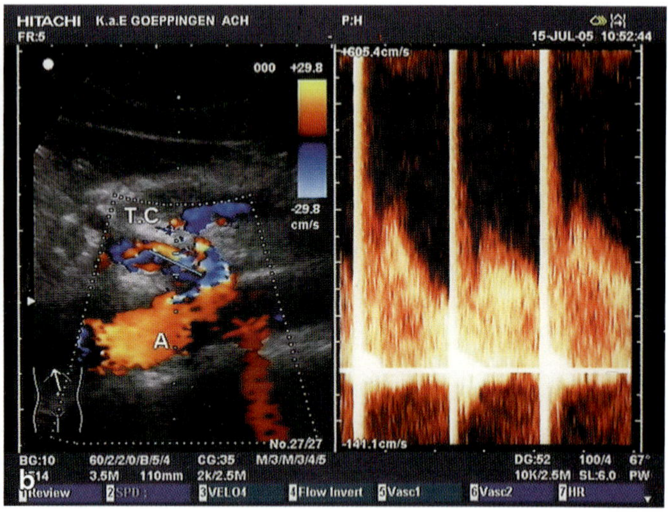


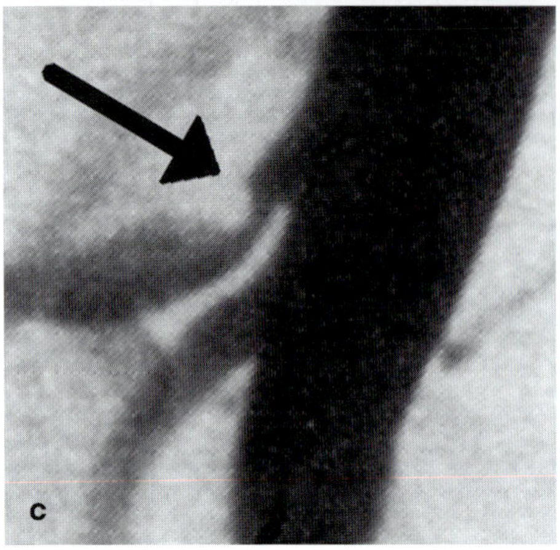


**Fig. A6.8 a–e**
**Median arcuate ligament syndrome**
**a**, **b** There is aliasing at the origin of the celiac trunk (*T.C*) from the aorta (*A*), which is above the origin of the superior mesenteric artery (*A.M.S*). With the sample volume placed just anterior to the origin, the spectral Doppler measurement yields peak velocities of 215 cm/s during systole and 90 cm/s at end diastole. Respiratory downward movement of the diaphragm displaces and compresses the celiac trunk, visible as a sharp bend in the proximal celiac segment in the color duplex image (in **b**). The corresponding spectral Doppler measurement (*right*) reveals peak velocities of 6 m/s during systole and 150 cm/s at end diastole, consistent with marked compression of the celiac trunk. PSV is difficult to measure in the proximal portions of aortic branches because of superimposed high amplitudes from pulsatile wall motion in early diastole, which cannot be eliminated from the spectrum by any wall filter.
**c** Angiogram confirming downward displacement and compression of the proximal celiac trunk by the median arcuate ligament.

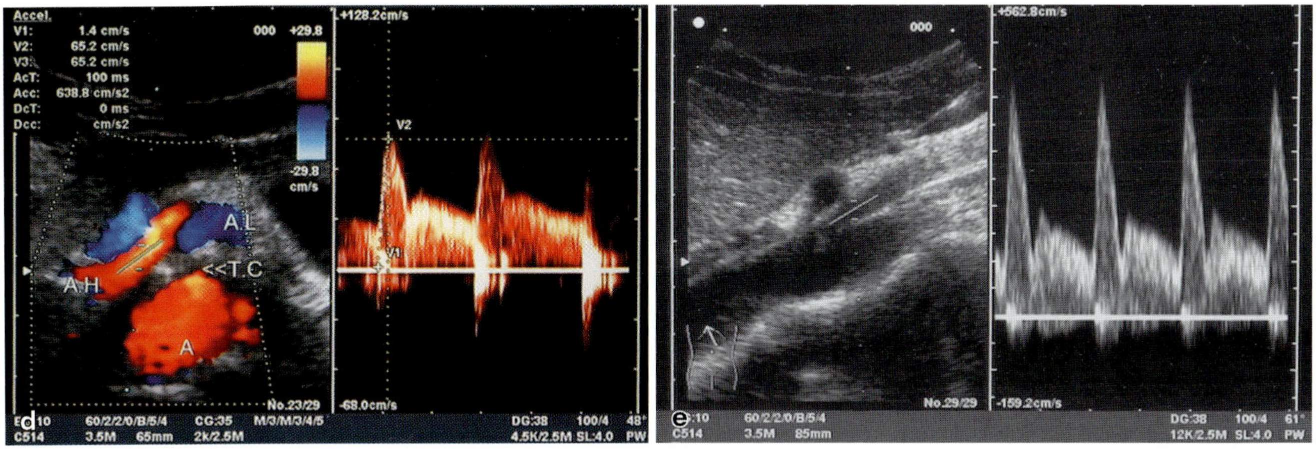


**Fig. A6.8 (continued)**

**Celiac trunk occlusion**

**d** Occlusion of the celiac trunk (*T.C*) is associated with retrograde blood flow in the hepatic artery (*A.H*) (red, flow toward transducer). The hepatic artery is refilled by the gastroduodenal artery and also supplies the splenic artery (*A.L*). The Doppler waveform is characteristic of an artery supplying a parenchymal organ and confirms retrograde flow in the hepatic artery. Blood flow direction in the splenic artery is normal.

**e** In occlusion of the celiac trunk, the liver and spleen are supplied by collaterals such as the pancreaticoduodenal and gastroduodenal arteries. The superior mesenteric artery (no stenosis) supplying these collaterals shows high blood flow velocity at its origin (average PSV of up to 4 m/s with 150 cm/s at end diastole) with a flow profile similar to that of arteries supplying parenchymal organs.

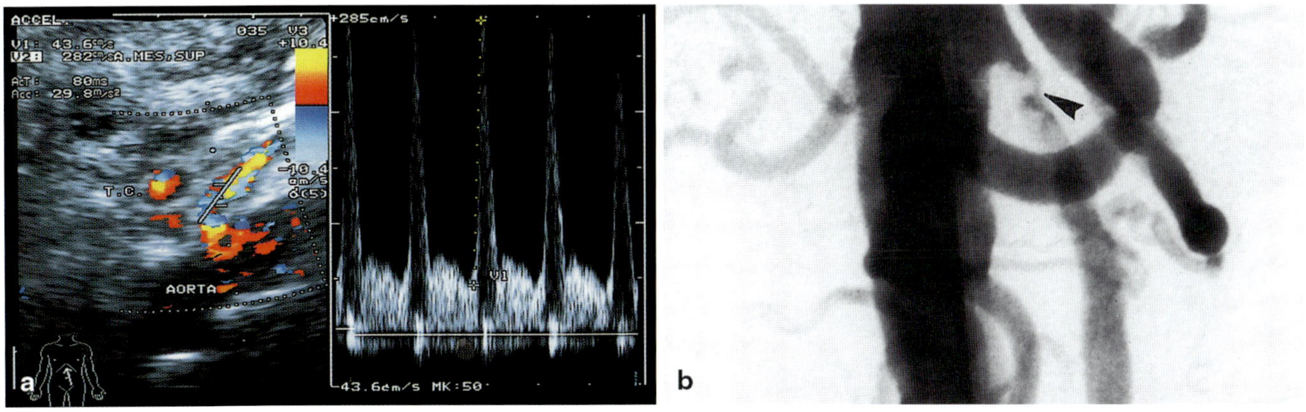


**Fig. A6.9 a, b**

**Mesenteric artery stenosis**

**a** The superior mesenteric artery arises from the aorta at an acute angle and courses anteriorly. The origin appears somewhat narrowed and shows aliasing with the color coding changing from red, to yellow, to blue as a sign of marked flow acceleration in the color duplex image. The Doppler waveform depicts turbulent flow with an increased velocity of 285 cm/s in systole and 43 cm/s at end diastole. Obtained in the fasting patient, the PSV suggests a stenosis of about 70%. The color duplex scan additionally visualizes the celiac trunk (*T.C.*) to the left of the mesenteric artery origin.

**b** Angiogram: Moderate stenosis at the origin of the superior mesenteric artery (*arrowhead*). Large-caliber inferior mesenteric artery acting as a collateral.

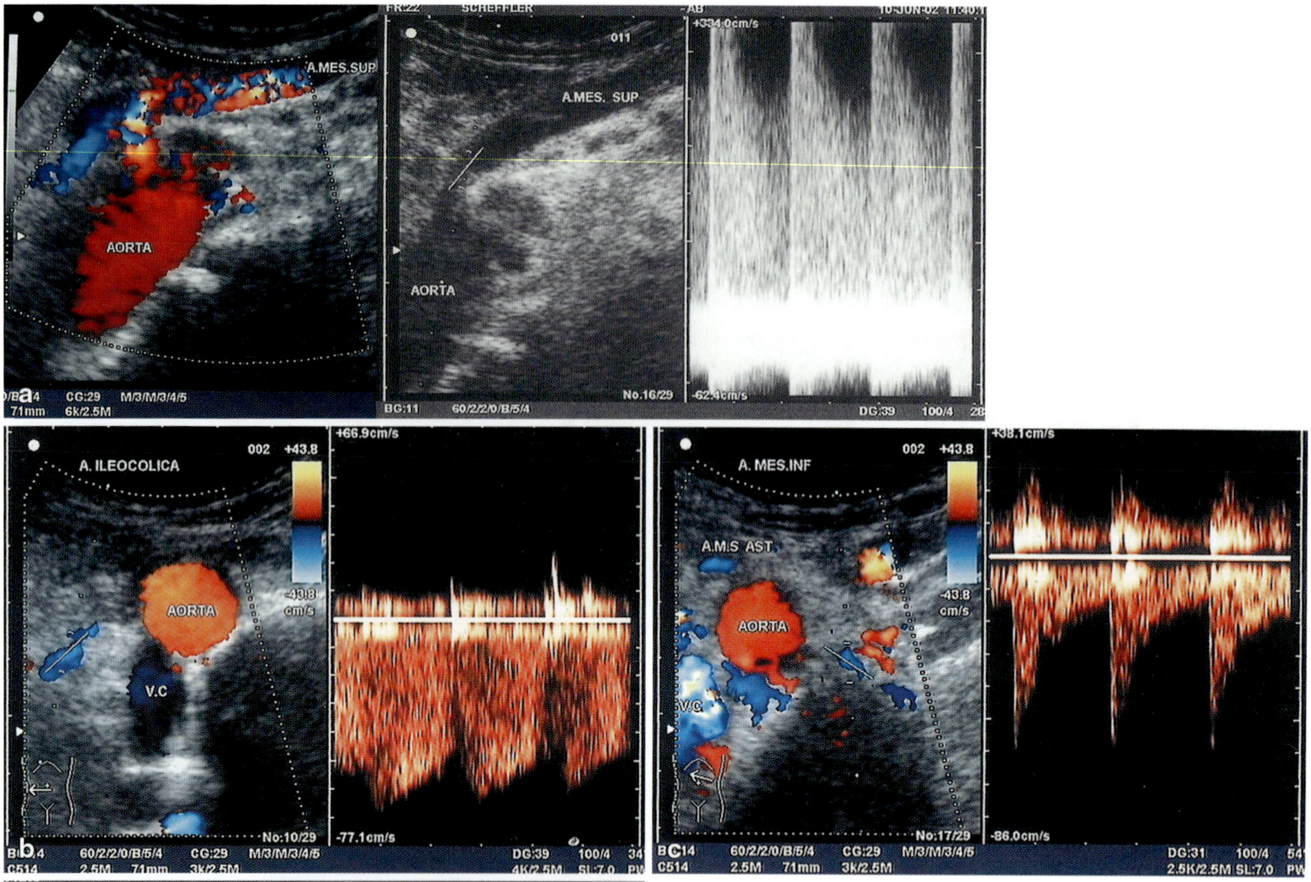


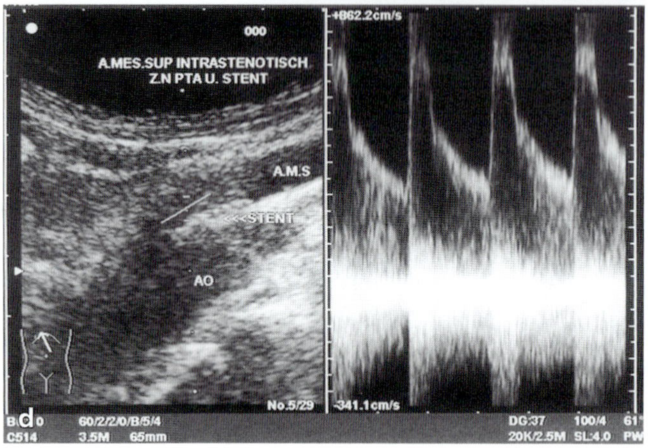


**Fig. A6.10 a–d**

**High-grade mesenteric artery stenosis**

**a** Aliasing in the color mode suggests high-grade stenosis at the origin of the superior mesenteric artery. The scanning conditions are usually good in the very thin patients presenting with suspected abdominal angina, but the arched course of the superior mesenteric artery at its origin may impair adequate angulation of the Doppler beam (*left section*). In inspiration, this segment of the superior mesenteric artery is straightened, which facilitates adjustment of the angle correction cursor parallel to the vessel wall and reduces the angle setting error (compare *middle* and *left sections*).

**b** The Doppler spectrum from the distal mesenteric branches (such as the ileocolic artery) shows the typical features of postocclusive flow with a markedly reduced pulsatility and an almost venous profile.

**c** The inferior mesenteric artery acts as a collateral via the Riolan anastomosis and hence shows an increased flow velocity, in particular in diastole.

**d** Sonographic follow-up after 2 months identifies the stent as a mesh-like structure in the wall area of the superior mesenteric artery. The Doppler waveform is characterized by a high-frequency signal with an angle-corrected PSV of over 8 m/s indicating high-grade restenosis.

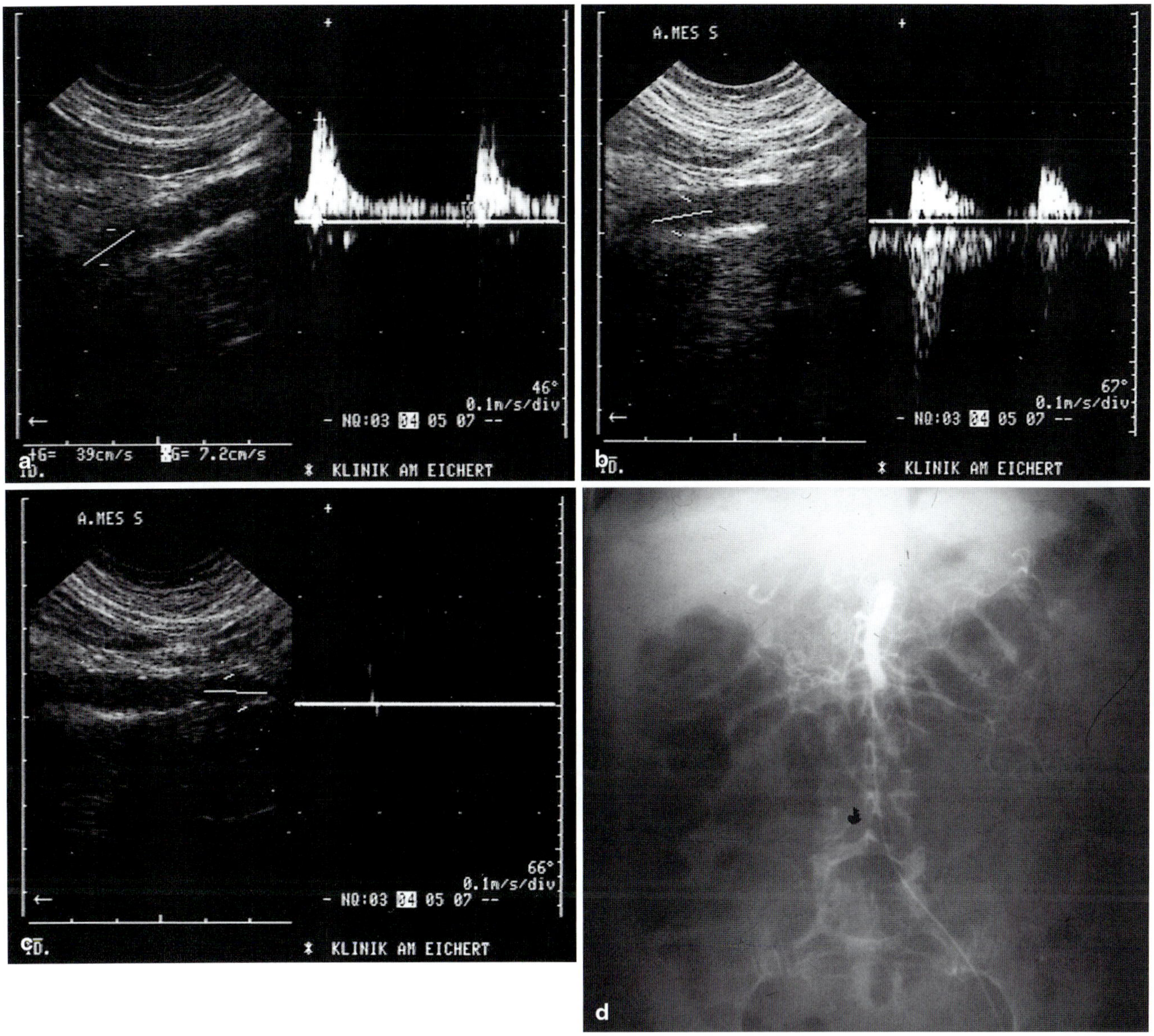


**Fig. A6.11 a–d**

**Acute mesenteric artery occlusion**

**a** Patient presenting with acute abdomen: abnormal Doppler waveform from the origin of the superior mesenteric artery with a decreased PSV of 39 cm/s. The end-diastolic peak velocity is 7.2 cm/s; the Pourcelot index is reduced.

**b** Continuous scanning of the superior mesenteric artery starting at its origin yields a flow profile more and more resembling a thump pattern with a decreasing flow velocity and absence of end-diastolic flow close to the occlusion. Flow in the mesenteric artery is toward the transducer and displayed above the baseline. The frequencies displayed below the baseline are from the middle colic artery, which arises near the sample volume.

**c** More distally, the superior mesenteric artery is occluded with zero flow in the Doppler spectrum despite a high gain.

**d** Angiogram showing patency of the trunk of the superior mesenteric artery and occlusion distal to the origin of the middle colic artery.

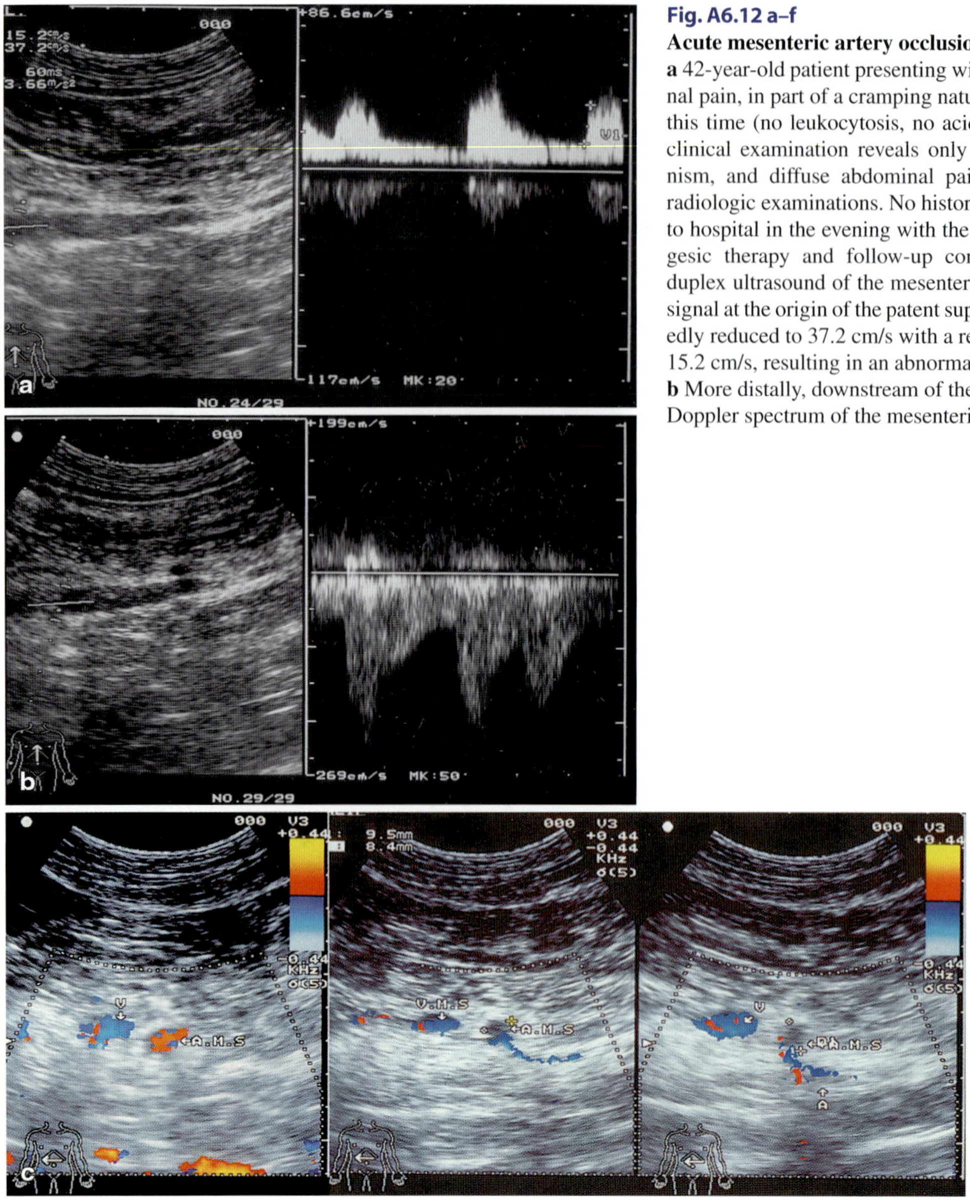


**Fig. A6.12 a–f**

**Acute mesenteric artery occlusion**

**a** 42-year-old patient presenting with a 3-hour history of severe abdominal pain, in part of a cramping nature. No abnormal laboratory values at this time (no leukocytosis, no acidosis, no elevated lactate level). The clinical examination reveals only mild tenderness, absence of peritonism, and diffuse abdominal pain. Normal B-mode ultrasound and radiologic examinations. No history of cardiac disease. Patient admitted to hospital in the evening with the tentative diagnosis of enteritis; analgesic therapy and follow-up contemplated. Additionally performed duplex ultrasound of the mesenteric arteries demonstrates an abnormal signal at the origin of the patent superior mesenteric artery. PSV is markedly reduced to 37.2 cm/s with a relatively large diastolic component of 15.2 cm/s, resulting in an abnormal resistance index of 0.59.

**b** More distally, downstream of the origin of the middle colic artery, the Doppler spectrum of the mesenteric artery shows a thump pattern.

**c** Color duplex scanning demonstrates a patent superior mesenteric artery to the level of the origins of the first jejunal branches. A proximal jejunal branch also shows color-coded flow signals. In the remainder of the superior mesenteric artery, neither color duplex nor spectral Doppler depicts flow signals. Emergency embolectomy with complete revascularization was performed without the necessity for bowel resection.

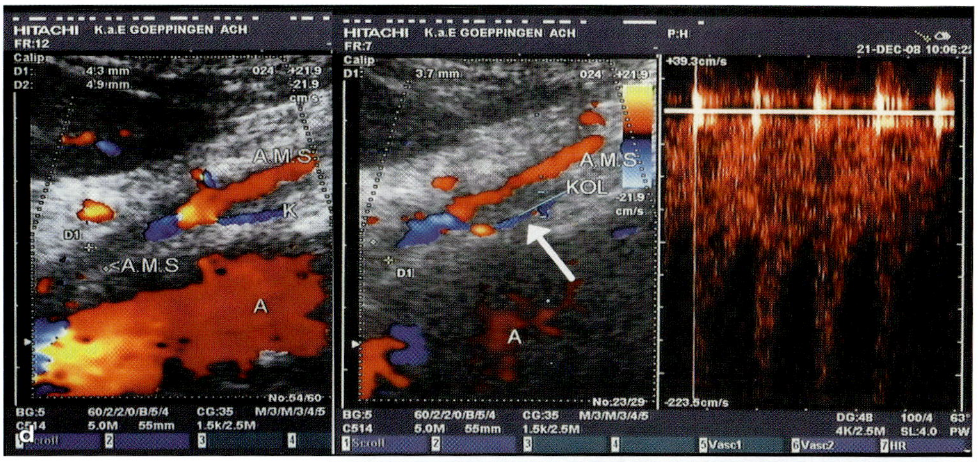


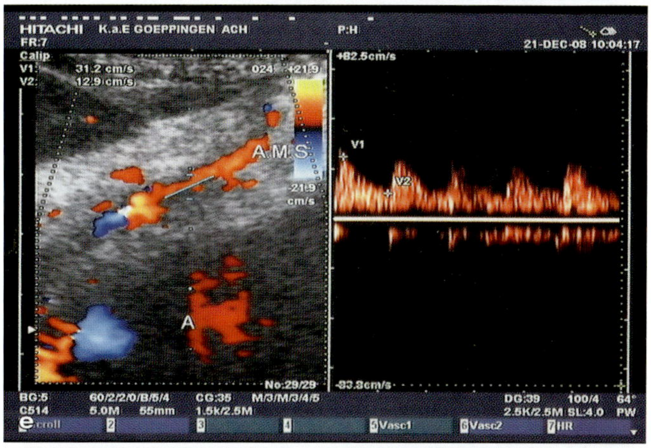


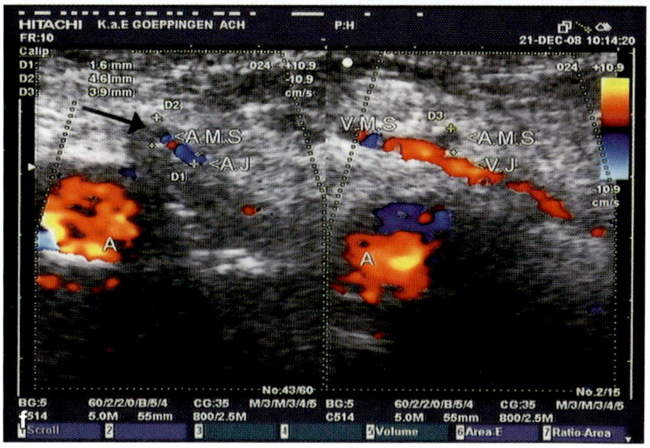


**Fig. A6.12 (continued)**

**Mesenteric artery occlusion - acute versus chronic**

**d** A slim 82-year-old woman with a history of intermittent abdominal pain was hospitalized for severe abdominal pain. On admission, she had a regular heart rate of 95 beats/min but a history of embolectomy of the leg in the year before. The color duplex examination reveals occlusion of the proximal superior mesenteric artery segment from its origin to the level of the pancreaticoduodenal artery origin (*K*). The superior mesenteric artery (*A.M.S*) is refilled by the gastroduodenal artery (arising from the hepatic artery) and the pancreaticoduodenal artery (*K*). With the gastroduodenal artery assuming collateral function (*KOL* in the second image), flow in this artery is high and clearly visualized. The spectral Doppler recording with the sample volume in this artery reveals a PSV of 220 cm/s and a peak end-diastolic velocity of 100 cm/s. These color duplex findings are consistent with chronic occlusion.

**e** With the sample volume placed in the superior mesenteric artery (*A.M.S.*), very low flow velocities of 31 cm/s during systole and 12 cm/s at end diastole are measured, suggesting poor collateralization or poor peripheral outflow.

**f** With the transducer in transverse orientation on the upper abdomen, the mesenteric artery trunk is examined in the color duplex mode with a low pulse repetition frequency, evaluating the jejunal origins for patency. Following the mesenteric artery downward, three jejunal branches are identified (*A.J*; *left* image) before the artery (*A.M.S*) first becomes partially occluded and then, more distally, completely occluded (no flow; *right* image). A jejunal vein branch (*V.J*) entering the mesenteric vein (*V.M.S*) is seen between the mesenteric artery anteriorly and the aorta (*A*) posteriorly. Taking these additional findings into account, the overall situation suggests acute embolic occlusion rather than chronic occlusion - despite the conclusion suggested by the findings described in **d**.

Intraoperatively, a short embolic occlusion at the mesenteric artery origin from the aorta was seen and a second occlusion of the distal segment with some patent jejunal branches arising from the patent portion between these occlusions. The patent portion was supplied by the pancreaticoduodenal artery (as demonstrated by the sonographic examination).

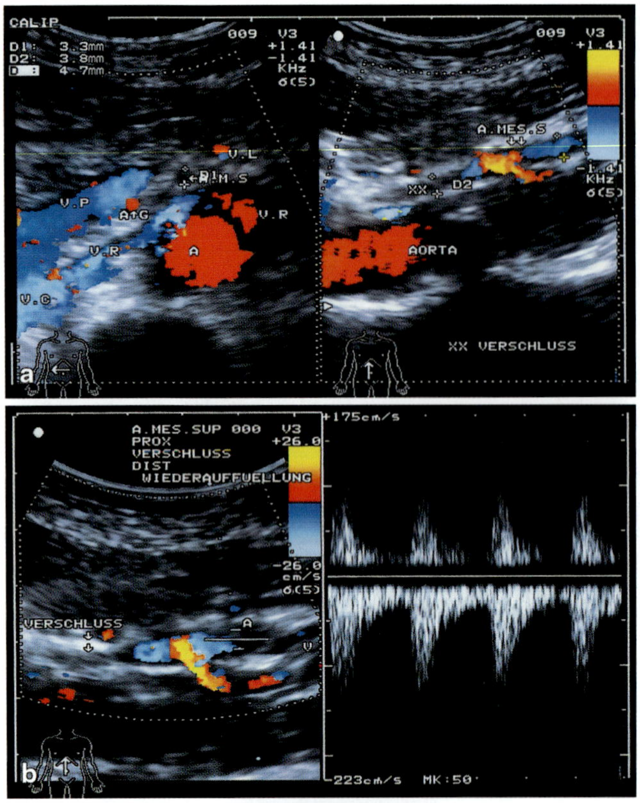


**Fig. A6.13 a–c**

**Chronic mesenteric artery occlusion**

**a** 50-year-old patient with symptoms of abdominal angina caused by proximal occlusion of the superior mesenteric artery with refilling through the gastroduodenal and pancreaticoduodenal arteries about 4 cm distal to its origin, as demonstrated by color duplex ultrasound. Transverse scan (*left section*) depicting the renal vein (*V.R*) and superior mesenteric artery (*A.M.S*) anterior to the aorta (*A*). Color duplex fails to demonstrate flow in the occluded superior mesenteric artery (3.3 mm). More anteriorly, the splenic vein (*V.L*) is seen; the renal vein (*V.R*), including its termination in the vena cava (*V.C*), is depicted longitudinally (flow coded in blue), to the left of the aorta sectioned obliquely. Anteriorly, the portal vein (*V.P*) is shown with blue-coded flow. Between the renal and portal veins, the cross section of the red gastroduodenal artery is seen at its junction with the pancreaticoduodenal artery. It is depicted beneath the lower margin of the portal vein and marked (*A↑G*). In transverse orientation, this collateral pathway can be followed in its entire length including refilling of the superior mesenteric artery.
The longitudinal scan (*right section*) depicts the superior mesenteric artery (*A.MES.S*) anterior to the obliquely sectioned aorta (red). Color signals are absent from the superior mesenteric artery segment in the left half of the scan (*XX*), where it is merely seen as a hypoechoic, tubular structure. Along its course to the right of the scan, it is refilled by the pancreaticoduodenal artery from posterolaterally (displayed in red, toward transducer). There is short backward flow in the unobstructed segment.
**b** The Doppler spectrum shows rather high flow in the postocclusive segment of the superior mesenteric artery (flow toward the periphery coded in blue in the color duplex image) with a postprandial PSV of 120 cm/s and an end-diastolic velocity of 30 cm/s, suggesting good collateral flow through the gastropancreaticoduodenal artery (coded red in the color image). Proximal to the entry of this collateral, the occluded segment of the superior mesenteric artery is depicted as a hypoechoic, tubular structure. Around the site of entry of the collateral, flow is highly turbulent. The postocclusive waveform shows a slightly delayed systolic rise, reduced pulsatility, and a larger end-diastolic component.

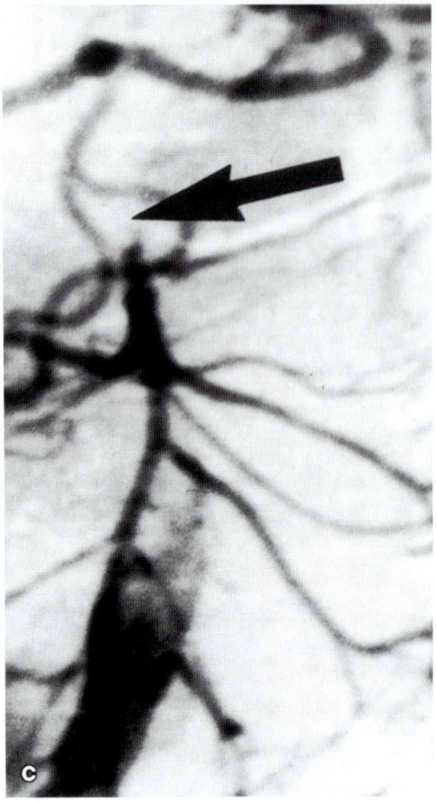

**c** Angiogram: Occlusion of the superior mesenteric artery at its origin (*arrow*) with refilling through the gastroduodenal and pancreaticoduodenal arteries. There is interference from the superimposed aorta at the lower margin. As a result of delayed contrast medium passage through the collateral pathways, the contrast medium has already disappeared from the aorta at the level of the celiac trunk and origin of the superior mesenteric artery by the time the refilled superior mesenteric artery becomes opacified.

**Fig. A6.14**
**Inflammatory bowel disease**
Acute abdomen with wall thickening of bowel loops on B-mode ultrasonography. The demonstration of flow in the bowel wall in the color duplex mode differentiates inflammatory thickening of the wall from thickening due to acute ischemia or mesenteric vein thrombosis, which exhibits the characteristic bull's eye sign. The inflammatory origin is also underlined by the large diastolic flow component in the Doppler waveform.

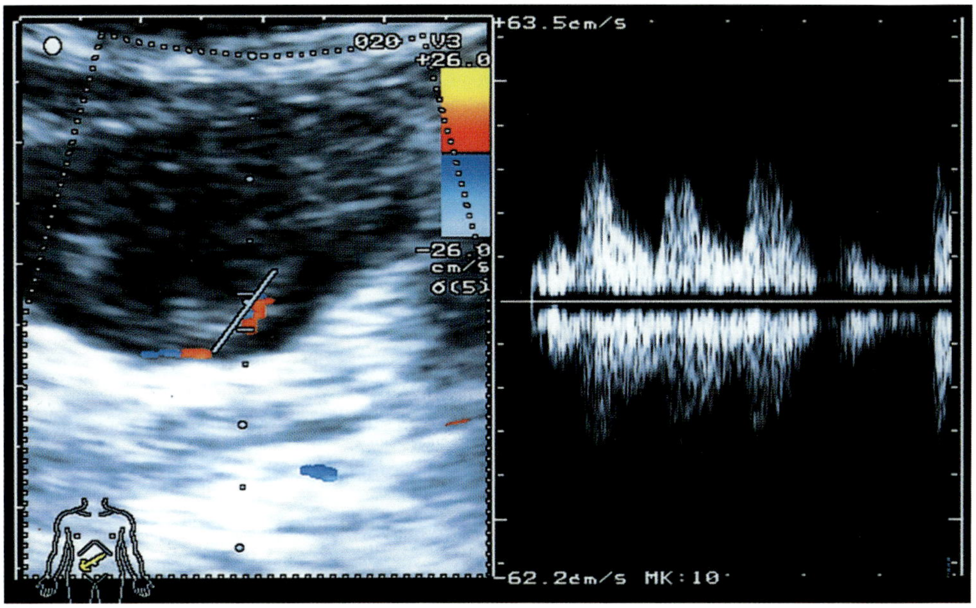


**Fig. A6.15 a, b**
**Superior mesenteric artery aneurysm**
**a** The *left section* shows the red-coded superior mesenteric artery (*A.M.S*) with an aneurysmal dilatation (*AN*) approximately 2 cm distal to its origin from the aorta. The color box covers only the central part of the scan, which is why flow is absent from the vessel segments depicted at the margins. In oblique orientation (*right section*), the extent of the aneurysm can be evaluated; again only the medial segment is displayed with color coding due to the small color box. The hepatic artery (*A.H*) arises from the aneurysm (*AN*) and shows turbulent flow at its origin. In this case, the hepatic artery is involved because of its abnormal origin from the superior mesenteric artery. This information is important for planning the surgical procedure because it means that the hepatic artery must be repaired as well. The aneurysm has a transverse diameter of 22 mm and a longitudinal diameter of 34 mm.

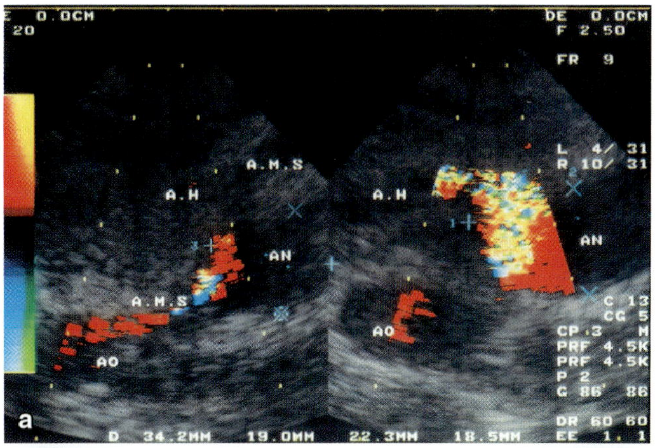


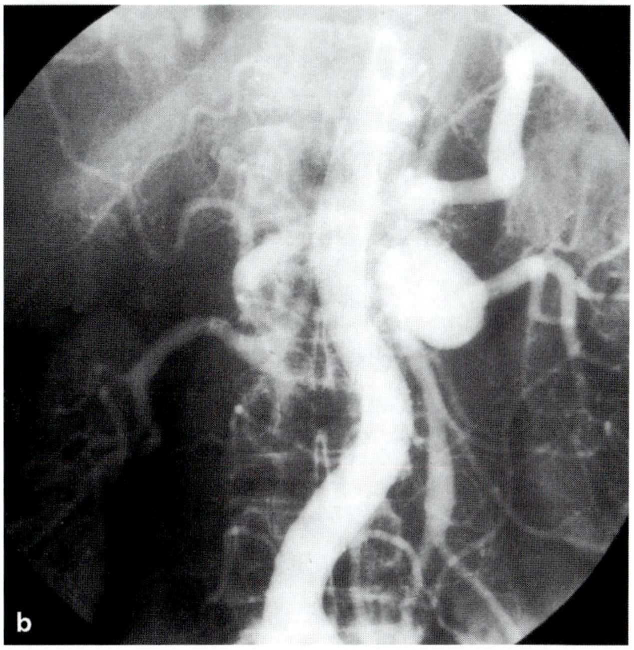


**b** Angiogram: Aneurysm at the origin of the superior mesenteric artery. The overlying aorta obscures the origin of the hepatic artery from the aneurysm. The splenic artery arises from the aorta.

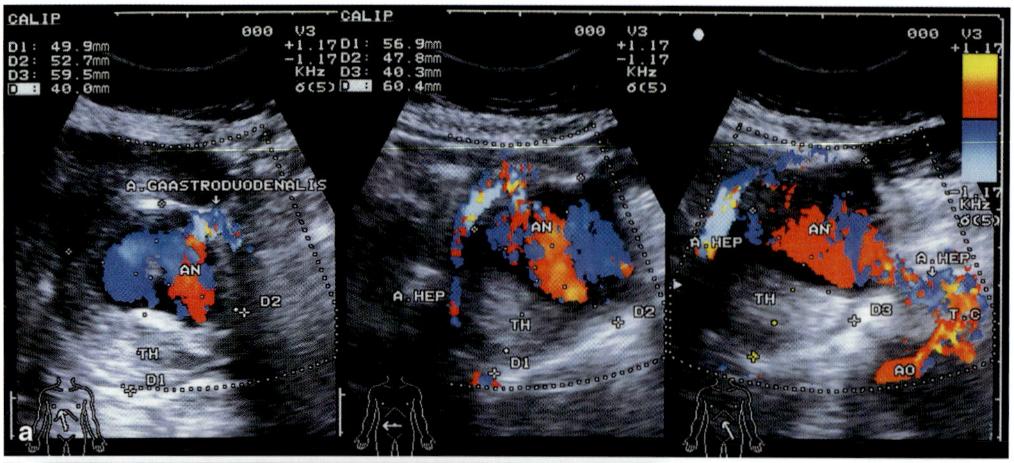


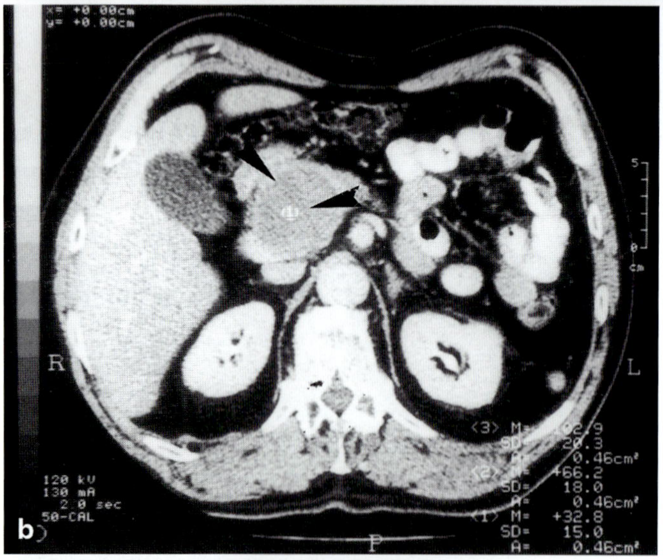


**Fig. A6.16 a–c**

**Hepatic artery aneurysm**

**a** A structure of mixed echogenicity measuring 6 × 5 cm and showing flow signals in the color duplex mode is depicted in the portal hilum. Stagnation thrombus (*TH*) is seen in the posterior portion of the aneurysm. For the surgical procedure, it is important to precisely locate the vessels entering and arising from the aneurysm (*AN*), in particular the gastroduodenal artery, which is shown to arise from the anteroinferior aspect of the aneurysm (blue, *left section*). The *middle section* depicts the elongated proper hepatic artery (*A.HEP*) curving around the aneurysm. The *right section* shows the common hepatic artery (*A.HEP*) emptying into the aneurysm and arising from the celiac trunk (*T.C*) on the *right side* of the scan.

As the hepatic artery aneurysm also involves the gastroduodenal artery, reconstruction of the hepatic artery is necessary after resection of the aneurysm. If the aneurysm were localized proximal to the gastroduodenal artery, the latter would ensure arterial supply of the liver.

**b** Upper abdominal CT showing subhepatic mass (*arrowhead*): hepatic artery aneurysm with partial thrombosis (*arrowhead*).

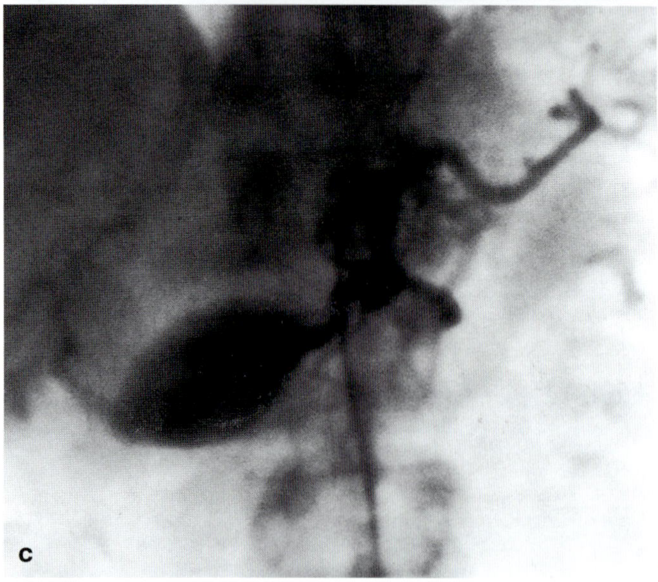


**c** Angiogram depicting hepatic artery aneurysm (*center*).

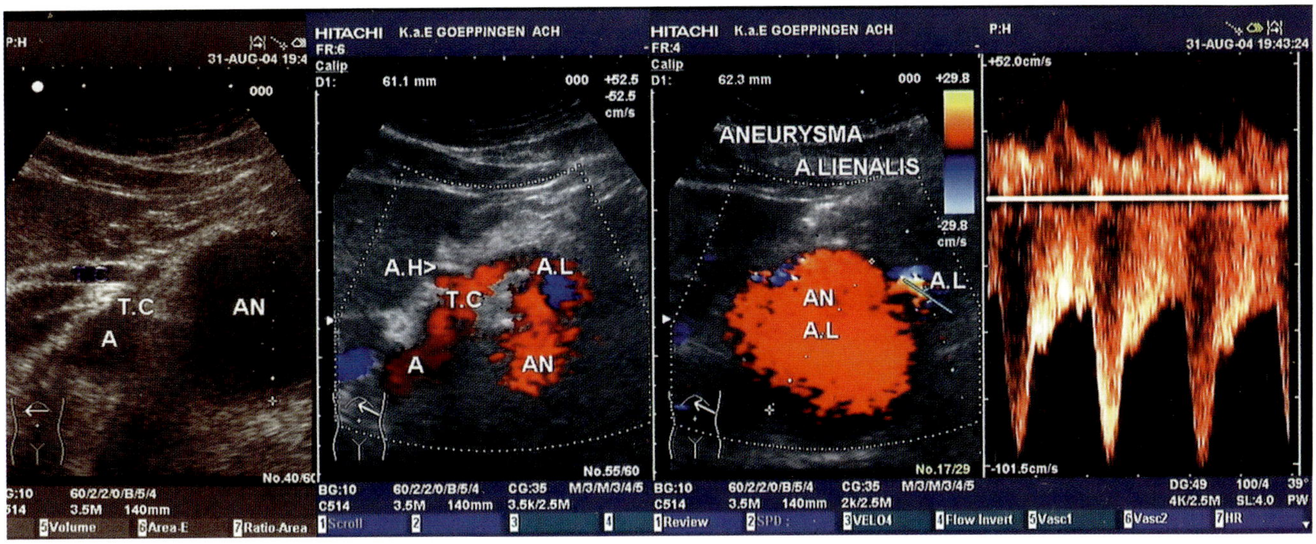


**Fig. A6.17**

**Splenic artery aneurysm**

The B-mode scan depicts an anechoic cystic lesion in the omental bursa (*leftmost*). The diagnosis of an aneurysm is suggested by the color coding in the duplex mode (*left center*). The junction of the aneurysm with the vessel is seen upon rotation of the transducer; in this example the splenic artery (*A.L*) shortly after its origin from the celiac trunk (*T.C*; *A* aorta, *A.H* hepatic artery). Moving the transducer laterally to the left (*right center*), the distal splenic artery (*A.L* with sample volume) can be traced along its course from the aneurysm (*A.N A.L*) to the splenic hilum. The vascular relationships of the aneurysm are thus determined sonographically prior to surgery. The Doppler waveform (*rightmost*) shows the typical low-resistance flow of the splenic artery.

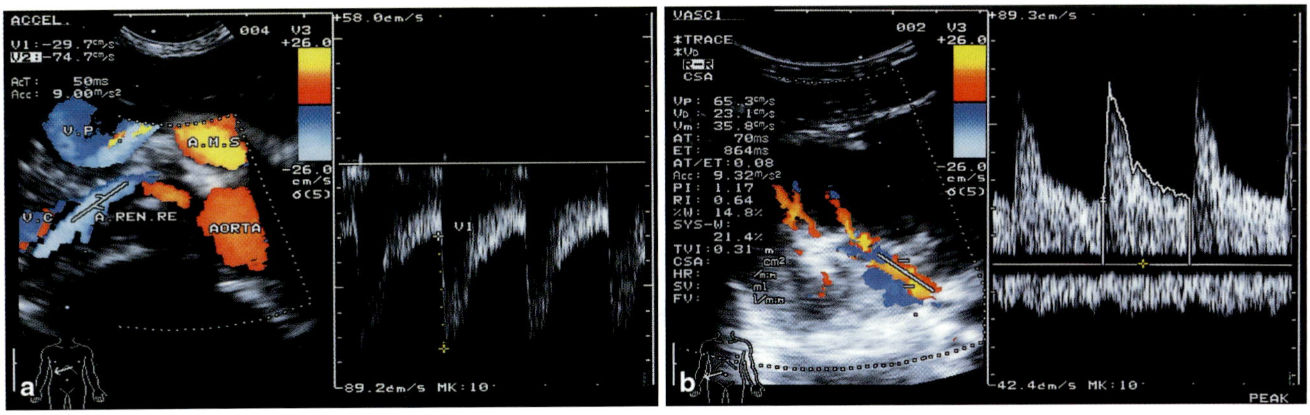


**Fig. A6.18 a–e**

**Course of the renal arteries**

**a** Adequate diagnostic evaluation for renal artery stenosis is crucially dependent on the meticulous visualization of the course of the renal artery. The transverse upper abdominal view on the *left* shows the right renal artery (*A.REN.RE*) following an arched course after arising from the aorta (proximal segment with flow in red toward transducer and distal segment with flow in blue away from transducer) below the vena cava (*V.C*). Anteriorly, the superior mesenteric artery (red, *A.M.S*) and portal vein (blue, *V.P*) are seen. A Pourcelot index of 0.6 is calculated for the origin of the renal artery from a PSV of 74.7 cm/s and an end-diastolic velocity of 29.7 cm/s.

**b** Renal artery at the renal hilum in transverse orientation scanned from the flank with the beam striking the vessel at an adequate angle. The waveform and the Pourcelot index are the same at the hilum as at the origin, suggesting that no hemodynamically significant stenosis is present along the course of the renal artery between these two sampling sites.

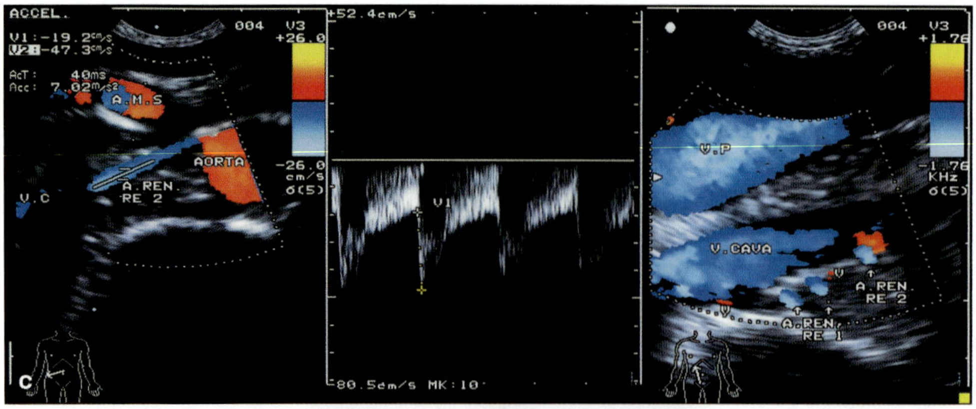


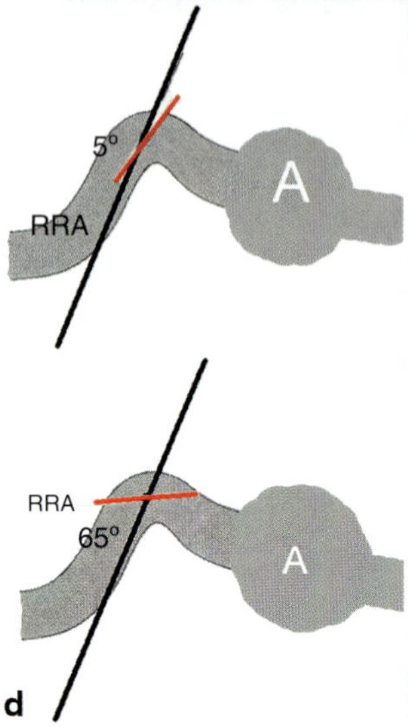


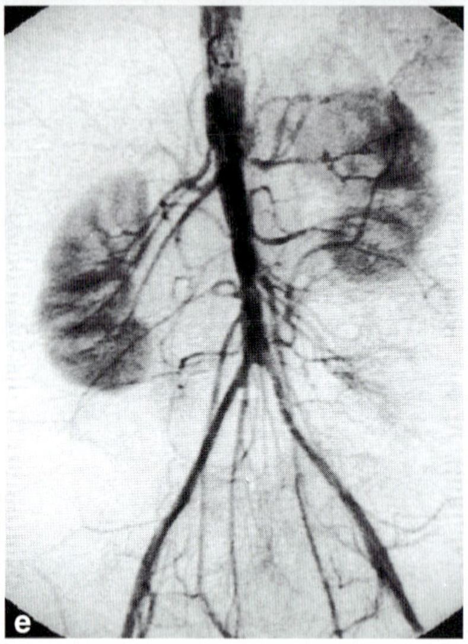


**Fig. A6.18 (continued)**

**c** Since 25% of all kidneys are supplied by a paired renal artery and hypertension may be caused by stenosis at the origin of the second branch, the sonographer must always look for a second renal artery branch by moving the transducer posteriorly in transverse orientation. Here, a second renal artery coded in blue arises from the aorta 1 cm from the origin of the first one. The characteristic renal artery waveform confirms the identity of the second artery.

In longitudinal orientation (*rightmost section*), the renal arteries can be identified posterior to the vena cava with blood flow in the renal arteries and in the vena cava depicted in blue. Three renal artery branches with blood flow coded in blue are seen below the vena cava; this is due to early division of the inferior branch of the paired renal artery on this side.

**d** Diagram illustrating the difficulties in placing the angle correction cursor parallel to the direction of flow in a tortuous or curved renal artery segment, which is not uncommon at the origin of the right renal artery (*RRA*). These pitfalls must be borne in mind to ensure correct grading of atherosclerotic renal artery stenosis, which tends to occur at the origin (see Fig. 1.23b).

**e** Angiogram: Two renal arteries arise from the aorta on the right with early division of the inferior branch.

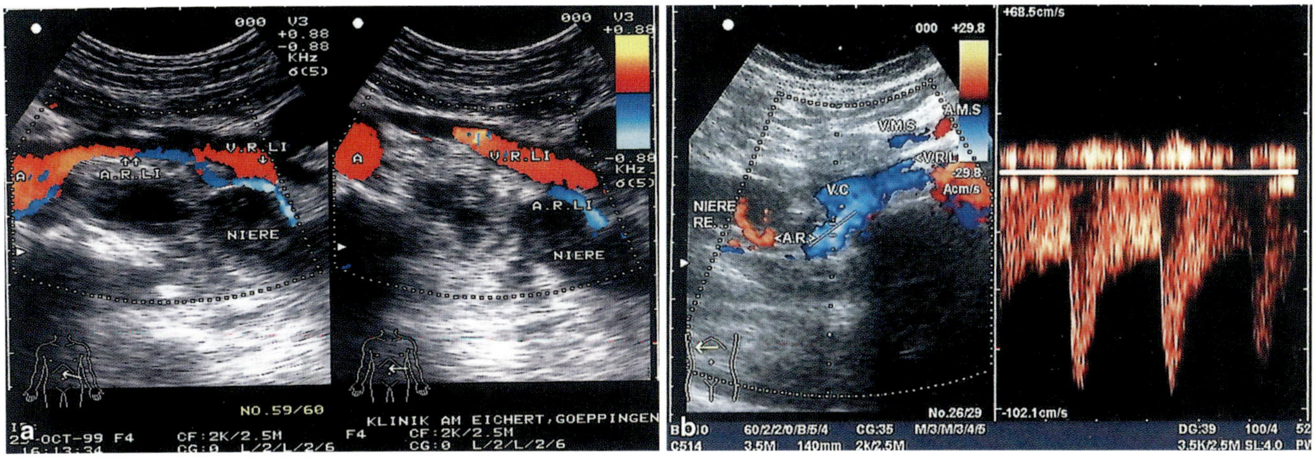


**Fig. A6.19 a, b**

**Sonoanatomy of the renal arteries**

**a** The left renal artery usually has a length of 5–6 cm, from the aorta to the renal hilum. Scattering by bowel gas makes it difficult to scan the entire length of the left artery in a single plane. The *left section* depicts the renal artery with flow in red (toward transducer) at its origin and in blue at the renal hilum (away from transducer). The change in the color coding does not indicate an actual change in flow direction but only a change relative to the transducer. The *right section* depicts the left renal vein (flow in red, toward transducer) anterior to the artery along its course to the vena cava anterior to the aorta.

**b** The image shows the right renal artery undercrossing the vena cava. Its proximal and middle thirds are depicted with flow coded in blue (*A.R*). The vena cava (*V.C*, blue) is seen anterior to it and the aorta is sectioned transversely (*A*, red) at the right margin. Anteriorly, the superior mesenteric artery (*A.M.S*) and vein (*V.M.S*) are seen. Between the aorta and the superior mesenteric artery, there is a short stretch of the left renal vein (*V.R.L*, blue). The distal third of the right renal artery is coded in red (flow toward transducer) at the renal hilum (*NIERE RE*). The Doppler spectrum was obtained from the middle third (posterior to the vena cava), the preferred site of stenosis in fibromuscular dysplasia.

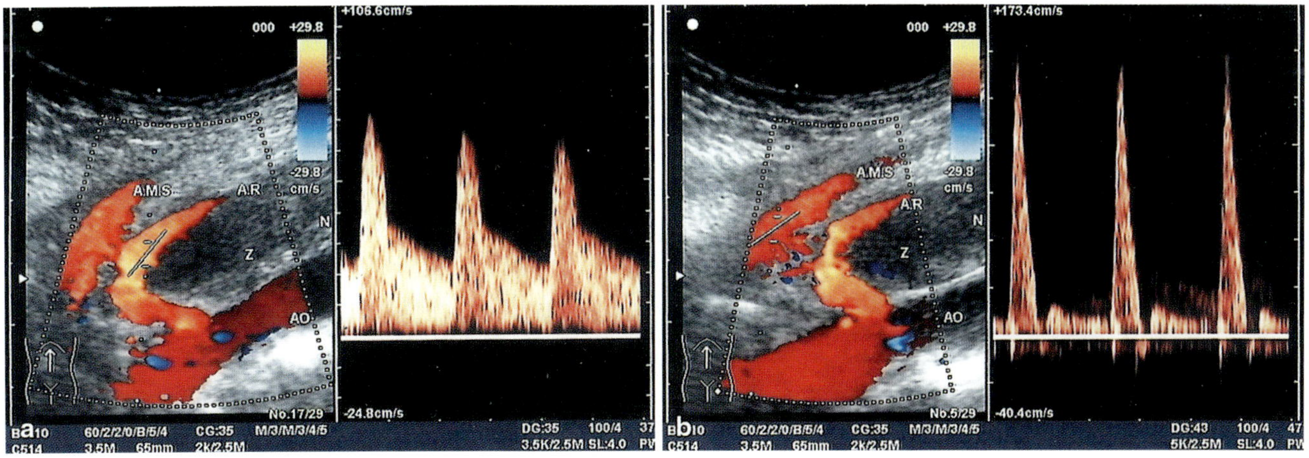


**Fig. A6.20 a–c**

**Horseshoe kidney**

**a** Horseshoe kidneys have atypical arteries and veins. Besides additional lower pole vessels, a fifth renal artery supplying the renal bridge crossing over the aorta may be present as in the example shown. The young woman had an infected renal cyst (*Z*) in the preaortic bridge of the horseshoe kidney. Pus was drained from the cyst under ultrasound guidance. In inconclusive cases, the Doppler waveform can help to establish the identity of a vessel. Here, the inferior of the two vessels, coursing over the cyst (*Z*) and renal parenchyma, displays the typical waveform of a renal artery and is thus identified as a supernumerary fifth renal artery (*A.R*).

**b** The artery coursing more superiorly (*A.M.S*) does not show the low-resistance flow typical of renal arteries but a mixed type characteristic of mesenteric arteries.

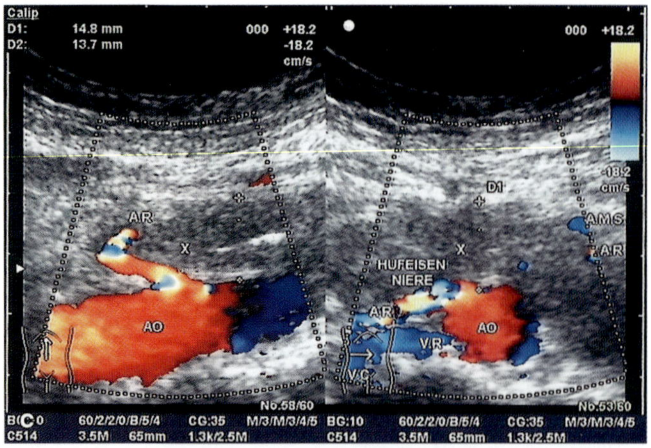


**Fig. A6.20 (continued)**

**c** Closer inspection of the vascular supply (transverse scan on the *right*, longitudinal scan on the *left*) shows the right lower polar artery (*A.R*) with flow in blue. This artery follows an atypical course, anterior to the vena cava on its way to the lower pole, after arising from the aorta (*AO*). A retroaortic renal vein with flow coded in blue (*V.R*) passes from the left lower pole into the vena cava (*V.C*). There is aliasing in the renal artery due to the low pulse repetition frequency used to detect slow venous (and arterial) flow. The longitudinal scan on the *left* again shows the fifth renal artery (*A.R*) coursing to the renal parenchyma in front of the aorta (*AO*) after drainage of the infected cyst (site indicated by the *X* in the kidney).

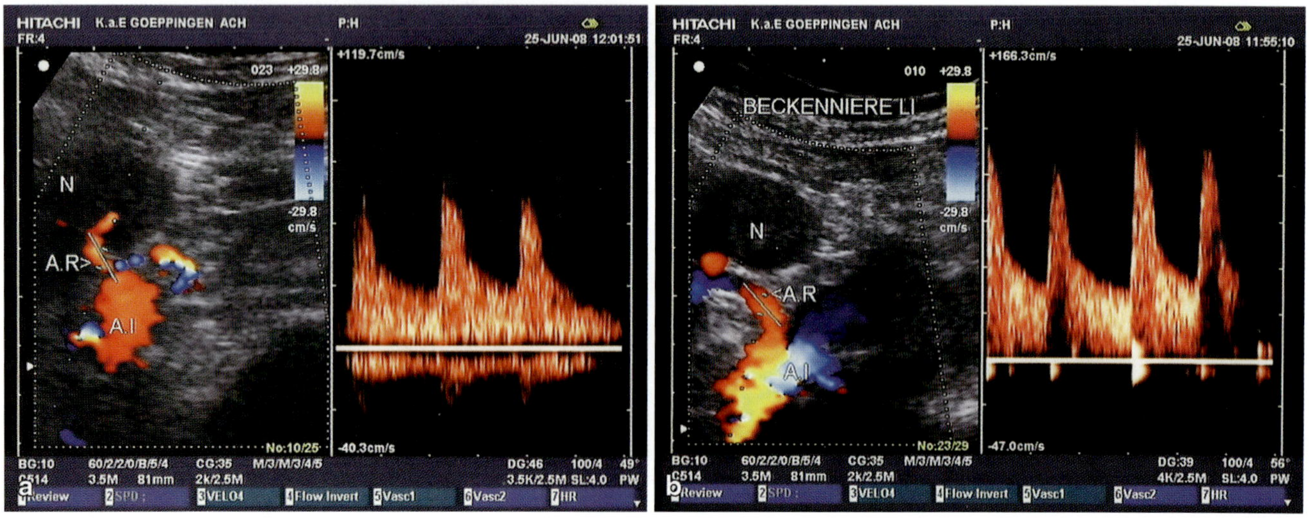


**Fig. A6.21 a, b**

**Pelvic kidney**

**a, b** If a kidney cannot be identified in its usual location in the flank, this should prompt a search for a pelvic kidney. The arterial supply of an ectopic pelvic kidney is highly variable with one or more renal arteries arising from the aorta or from the iliac artery. Also, the examiner must bear in mind that two polar arteries may be present and that stenosis in either of them can be the cause of hypertension. In the example, two polar arteries arising from the common iliac artery are identified; the two origins can be differentiated by moving and slightly rotating the transducer (lower pole artery in **a** and upper pole artery in **b**); stenosis in either artery is ruled out as flow velocity is below 120 cm/s.

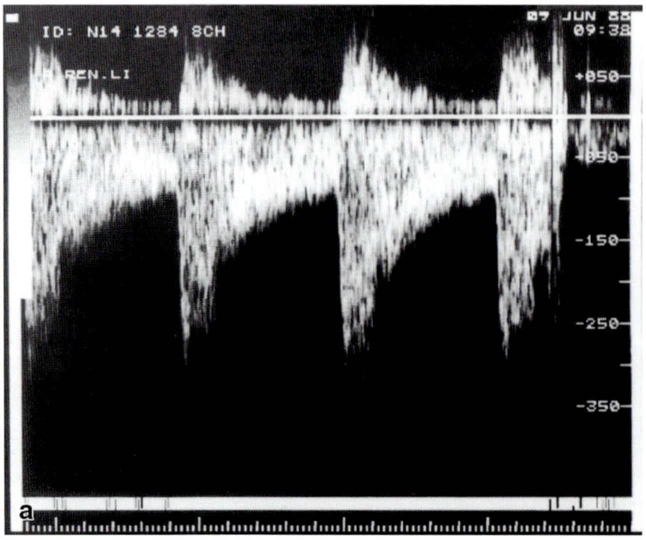


**Fig. A6.22 a–d**

**Renal artery stenosis – PTA**

**a** Doppler waveform obtained in the presence of moderate to severe stenosis at the origin of the renal artery with marked turbulence and a peak velocity of 310 cm/s during systole and 100 cm/s at end diastole.

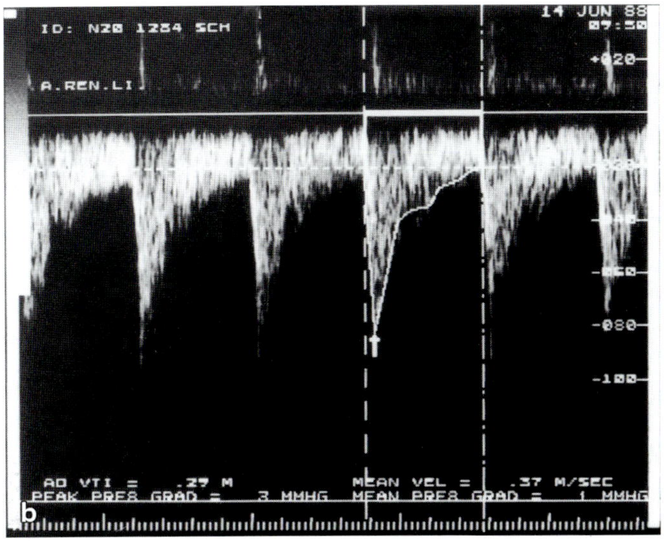


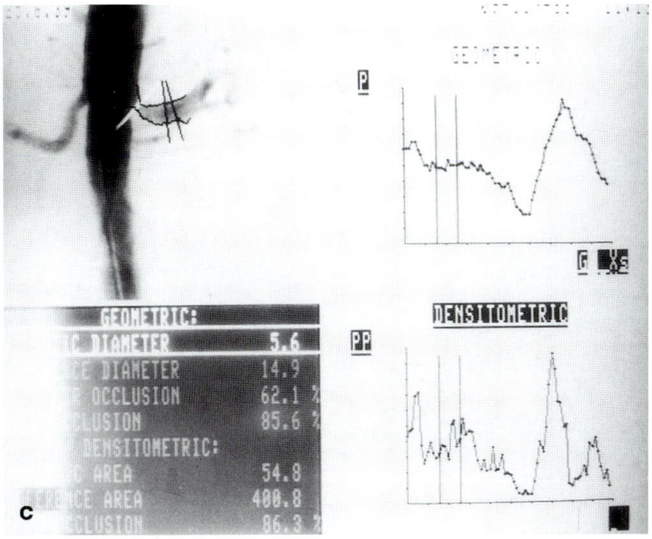


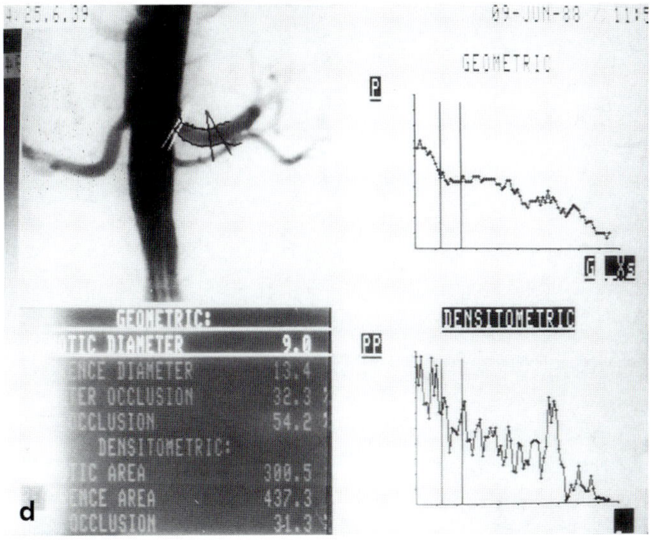


**Fig. A6.22 (continued)**
**b** Doppler waveform from the same renal artery as in **a** after percutaneous transluminal angioplasty (PTA). Return to normal blood flow velocities (PSV of 80 cm/s).
**c** X-ray densitometry (same patient as in **a**, before PTA): Measurement demonstrates a stenosis at the origin of the left renal artery with an area reduction of 86.3%.
**d** X-ray densitometry (same patient as before, after PTA; corresponding Doppler waveform in **b**): Residual stenosis with a 31.3% area reduction, which is hemodynamically nonsignificant. There is spectral broadening in the corresponding Doppler waveform but no accelerated flow.

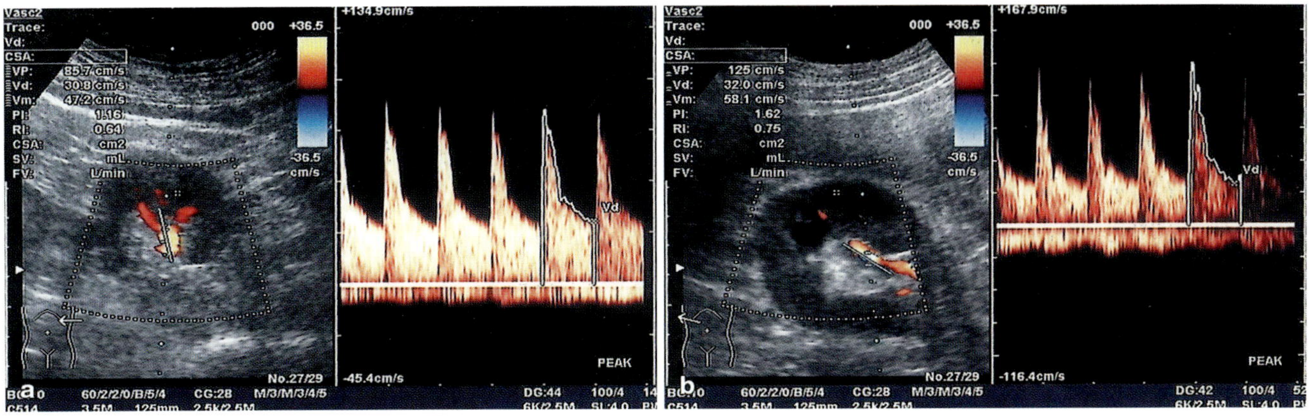


**Fig. A6.23 a–h**
**Renal artery stenosis – indirect criteria**
**a** The waveform from the renal hilum on the left yields a PSV of 85.7 cm/s and an end-diastolic peak velocity of 47.2 cm/s, from which a resistance index (RI; Pourcelot index) of 0.64 is calculated.
**b** The corresponding values in the right renal artery are: PSV of 125 cm/s, peak end-diastolic velocity of 58.1 cm/s, and a resulting RI of 0.75. The 10% RI difference is consistent with the diagnosis of renal artery stenosis and indicates postocclusive flow in the artery with the lower RI.

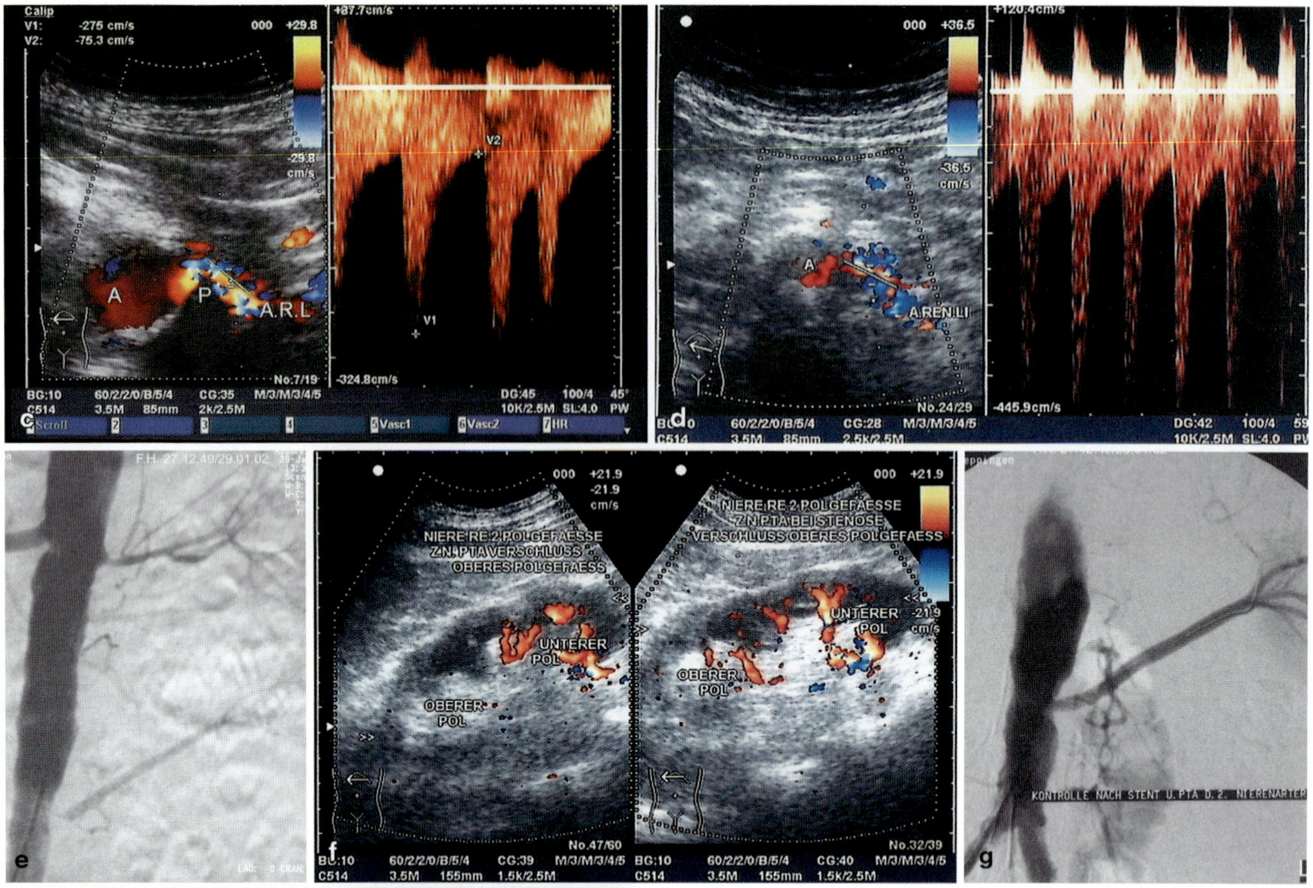


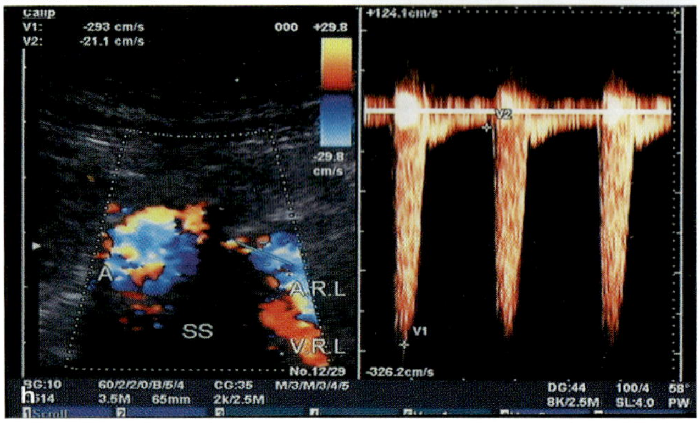


**Fig. A6.23 (continued)**

**High-grade renal artery stenosis - PTA**

**c** Patient with two left renal arteries and a PSV of 275 cm/s in the upper pole artery, consistent with high-grade stenosis caused by atherosclerotic plaque (*P*) at the origin of the artery. There is aliasing in the renal artery (A.R.L); the red flow signals anteriorly indicate the left renal vein (flow toward transducer).

**d** The Doppler waveform from the origin of the left renal artery confirms high-grade stenosis with a PSV over 4 m/s with the color duplex scan showing pronounced perivascular vibration (audible bruit on auscultation).

**e** Subsequent angiography with PTA confirms high-grade stenosis of both polar arteries on the left.

**f** Flank pain after PTA prompted a duplex ultrasound examination. In the lower pole of the kidney, both arterial and venous flow signals are obtained from the hilum to the periphery. The upper portion shows rarefied perfusion in the pole (capsular vessels) and no arterial flow at the hilum, consistent with occlusion of the upper pole artery after PTA.

**g** Angiogram confirms occlusion of the upper pole artery and normal flow in the lower pole artery.

**Renal artery stenosis – indication for PTA**

**h** High-grade stenosis of the left renal artery (*A.R.L*) with a PSV of 293 cm/s and an end-diastolic velocity of 21 cm/s, from which an RI of 0.9 is calculated. An RI of >0.8 indicates parenchymal damage and fixed hypertension, so that PTA is no longer a promising option. Stenotic plaque at the origin of the renal artery from the aorta (*A*) causes acoustic shadowing (*SS*). Retroaortic course of the left renal vein (*V.R.L*).

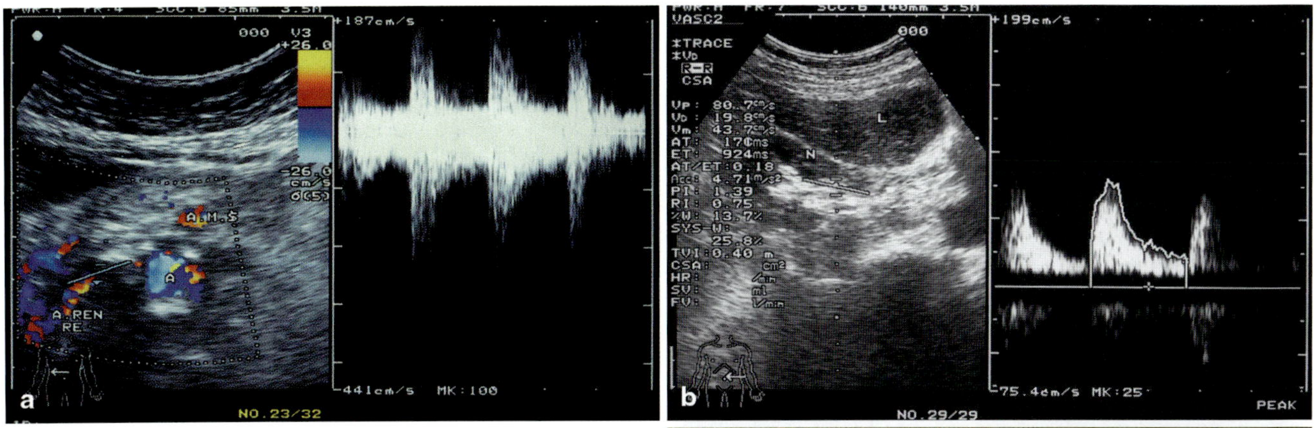


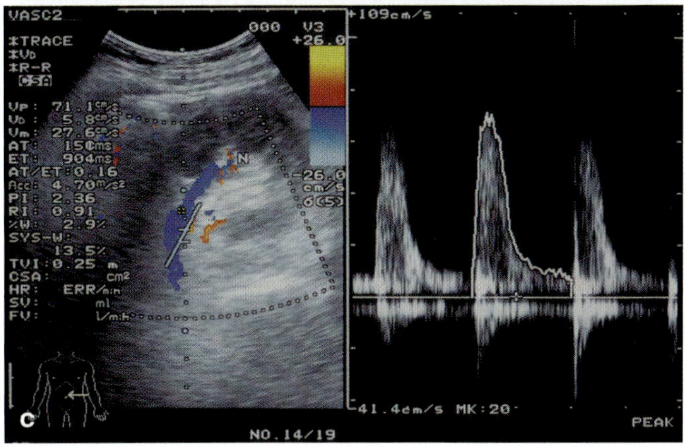


**Fig. A6.24 a–c**
**Renal artery stenosis in diabetes mellitus**

**a** In a patient with a long history of insulin-dependent diabetes mellitus and macro- and microangiopathy, the Doppler waveform obtained at the origin of the right renal artery shows turbulence and accelerated flow indicative of renal artery stenosis. Duplex imaging provides no adequate information for estimating the degree of stenosis due to plaque with acoustic shadowing at the origin. In interpreting the PSV of just over 2 m/s somewhat distal to the stenosis, one has to take into account possible hypertensive episodes during spectral Doppler sampling as well as the known higher pulsatility of blood flow with higher PSVs in diabetics. In the case presented, for instance, the resistance index (Pourcelot index) is calculated from the Doppler spectra of both distal (hilar) renal artery segments for confirmation of the hemodynamic significance of the stenosis.

**b** The spectrum from the right hilum yields a PSV of 80.7 cm/s and an end-diastolic peak velocity of 19.8 cm/s with a Pourcelot index of 0.75. The gray-scale scan depicts the liver (*L*) above the kidney.

**c** The spectrum from the left hilum shows more pulsatile flow with a PSV of 71.1 cm/s and an end-diastolic peak velocity of 5.8 cm/s; the Pourcelot index is 0.91. Compared with the spectrum recorded on the left side, the waveform of the right renal artery appears to be unusually normal, which is due to the fact that the effects of diabetes and stenosis cancel each other. The waveform from the left, which is too pulsatile for a renal artery, is attributable to medial sclerosis in long-standing diabetes mellitus and changes in the renal parenchyma. The much lower Pourcelot index of the right renal artery (over 10% in side-to-side comparison) is abnormal and indicates a hemodynamically significant proximal stenosis.

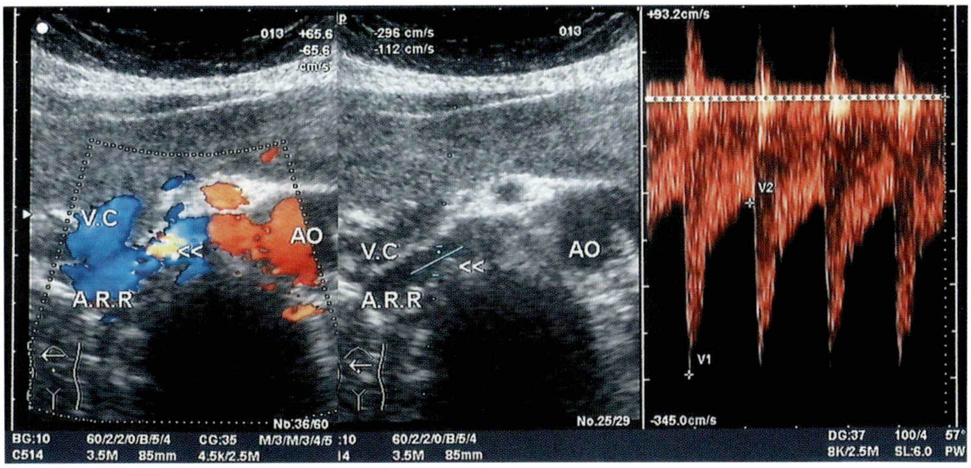


**Fig. A6.25**
**Fibromuscular dysplasia**

32-year-old patient with difficult-to-control hypertension. Color duplex imaging shows aliasing in the middle third (*arrow*) of the right renal artery (*A.R.R*) behind the vena cava (*V.C*) and 3 cm distal to its origin from the aorta (*AO*) - a sign of circumscribed flow acceleration. The Doppler spectrum yields a PSV of 3 m/s in the segment of the angiographically confirmed stenosis, which is due to fibromuscular dysplasia.

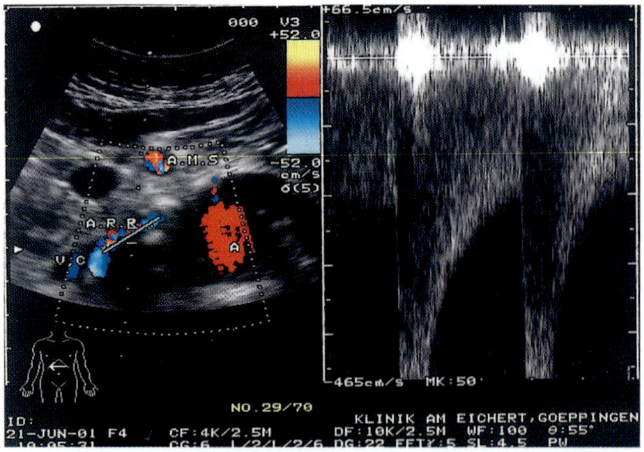


**Fig. A6.26**

**Suprarenal aortic aneurysm with renal artery stenosis**

Sonographic evaluation of the renal artery is indicated to evaluate the relationship of its origin to an aortic aneurysm. Here, scanning of the upper abdomen in transverse orientation shows the right renal artery arising from an aortic aneurysm with partial thrombosis and a diameter of 4.5 cm at the level of the origin of the renal artery (hypoechoic, concentric thrombus also at the renal artery origin). In addition, there is high-grade stenosis of the renal artery with a PSV exceeding 5 m/s.

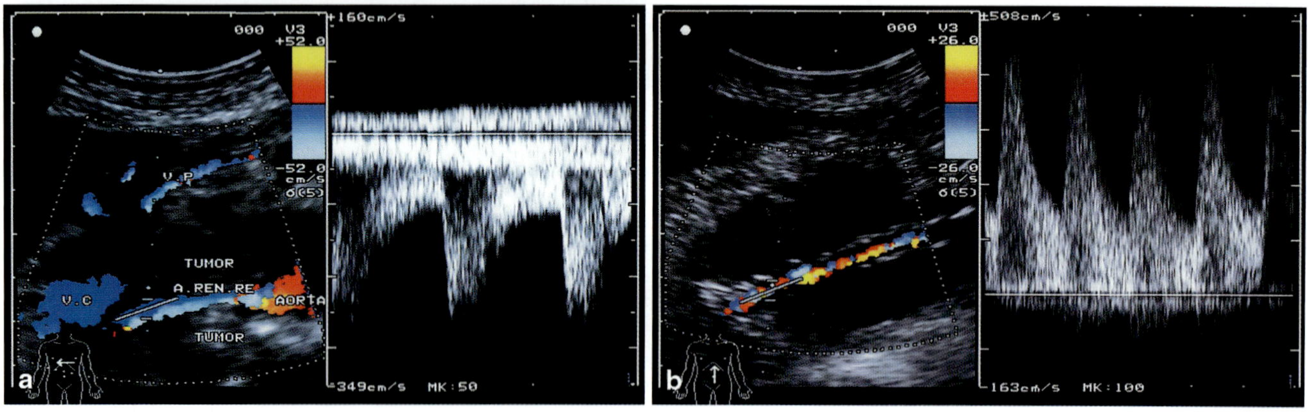


**Fig. A6.27 a, b**

**Vessel compression by tumor**

**a** A sarcoma (confirmed by ultrasound-guided core biopsy) splays the vena cava (*V.C*) and aorta in the retroperitoneum. A long segment of the renal artery (*A.REN.RE*) running through the tumor is moderately constricted (PSV of 250 cm/s measured by spectral Doppler). The vessels are located by color duplex imaging to avoid inadvertent vascular damage by subsequent ultrasound-guided core biopsy. Anteriorly, the portal vein (*V.P*) is also compressed by the tumor.

**b** The superior mesenteric artery encased by the tumor (sarcoma) at its root is also constricted along an extended segment (PSV of 450 cm/s).

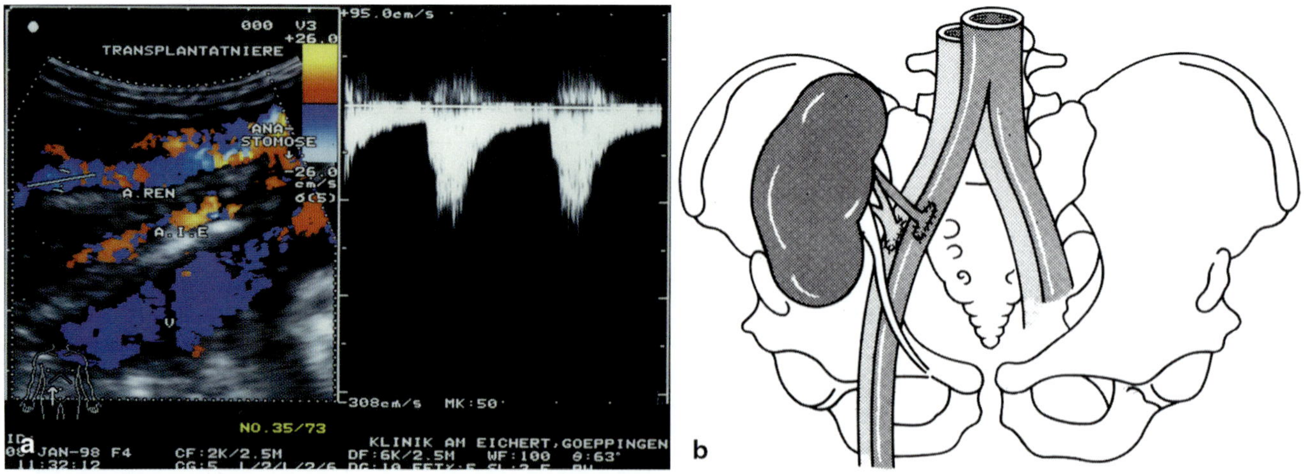


**Fig. A6.28 a, b**

**Transplant kidney**

**a** Color duplex scan depicting the artery of the transplant kidney, anastomosed to the iliac artery, with flow coded in blue (flow away from transducer), while the iliac artery is shown with flow in red (toward transducer). The Doppler waveform has a large diastolic component and the typical pattern of low-resistance flow indicating a functioning graft without rejection.

**b** Diagram of the connections of the renal transplant vessels to the iliac vessels.

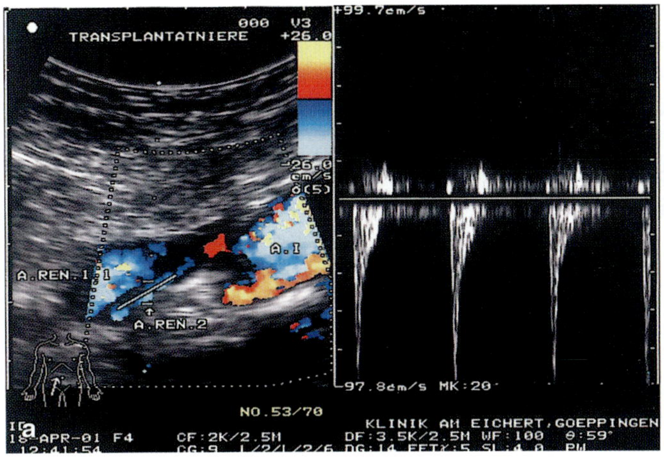


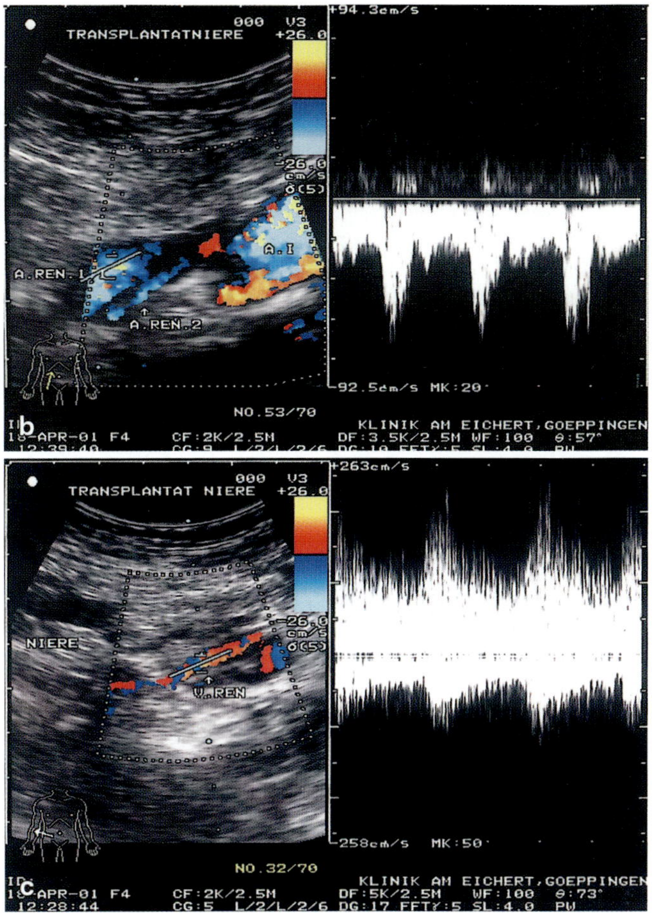


**Fig. A6.29 a–c**

**Transplant kidney – rejection – fistula**

Analysis of the Doppler waveform from the renal artery is an integral component of the diagnostic evaluation of kidney graft function and rejection. The renal artery of a transplant kidney anastomosed to the iliac artery is often more easily accessible to sonographic evaluation than the native renal artery.

**a** The two renal arteries supplying the kidney are depicted at their origins from the iliac artery (*A.I.*). A highly pulsatile waveform comparable to that of an extremity artery is obtained from the origin of the second renal artery (*A.REN.2*). This flow profile indicates rejection.

**b** Surprisingly, the other transplant artery, inserted above the first one, has a monophasic waveform with the low-resistance flow typical of normal kidney function.

**c** The Doppler waveform of the renal vein (flow toward transducer in the direction of the iliac vein) depicts a pulsatile flow profile with marked turbulence, which is typical of venous flow downstream of an AV fistula. Repeated biopsies performed in this patient for evaluation of transplant rejection led to the formation of a fistula, which explains why the artery (*A. REN.1*) supplying the fistula shows low-resistance flow despite rejection (as documented in the second artery, labeled as *A.REN.2*).

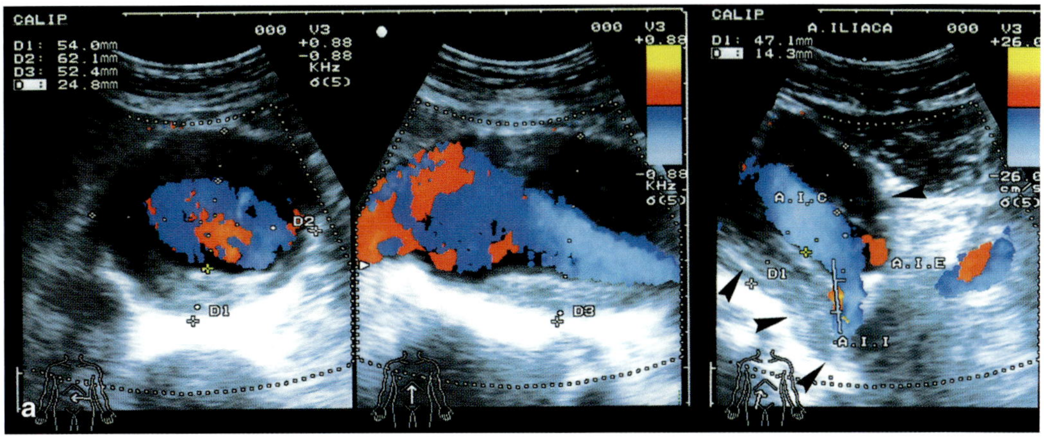


**Fig. A6.30 a–c**

**Aortic and iliac artery aneurysm**

**a** Partially thrombosed aortic aneurysm with infrarenal extension shown in transverse orientation (*left section*) and longitudinally (*middle section*). Evaluation of the perfused lumen is improved in the color duplex mode. The total diameter of the aneurysm is 6.2 cm. The mural thrombosis lining the lumen appears hypoechoic around the patent lumen. The aneurysm (*right section*, *arrowheads*) extends into the common iliac artery (*A.I.C*) and the proximal internal iliac artery (*A.I.I*). The elongated external iliac artery (*A.I.E*) leaves the scanning plane. The aneurysm has a total diameter of 47 mm with a patent lumen of 14 mm.

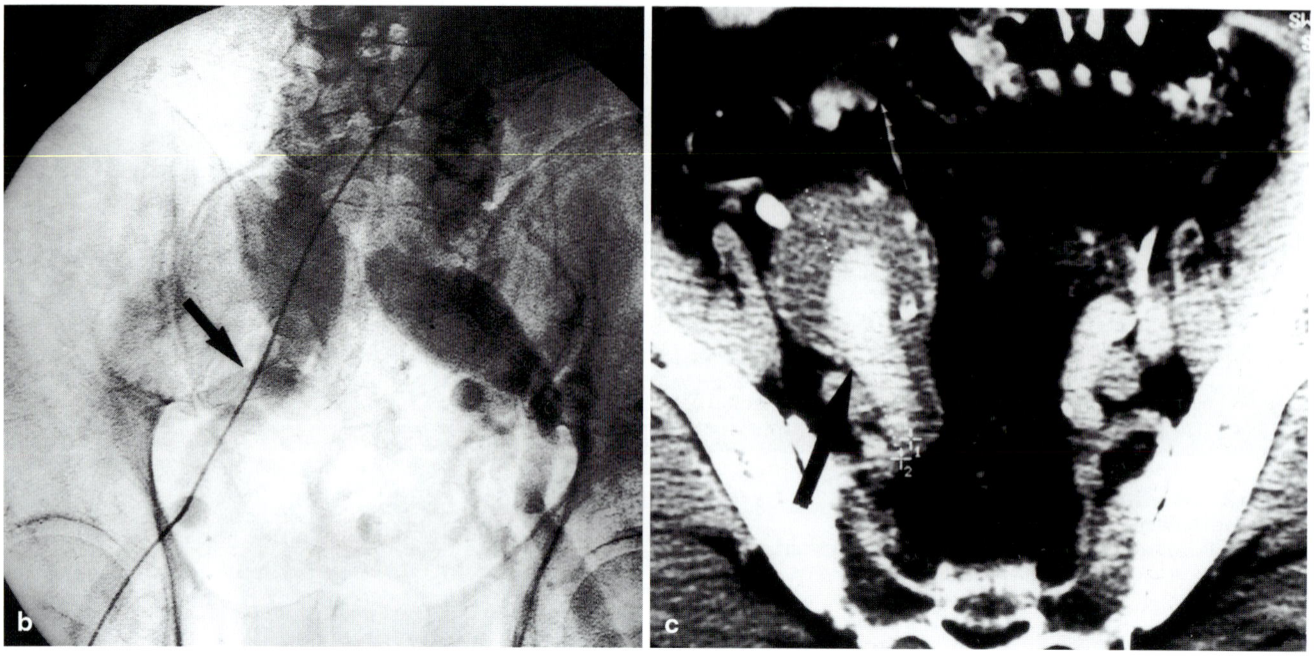


**Fig. A6.30 (continued)**

**b** Angiogram showing aneurysmal dilatation of the aorta and of the common iliac arteries. Due to mural thrombosis, the origin of the internal iliac artery on the right (*arrow*) seems not to be dilated.

**c** Contrast-enhanced CT: Aneurysm on the right (*arrow*) extending into the proximal internal iliac artery with mural thrombosis surrounding the perfused lumen. The internal iliac artery arises from the posterior aspect of the common iliac artery (see **a**).

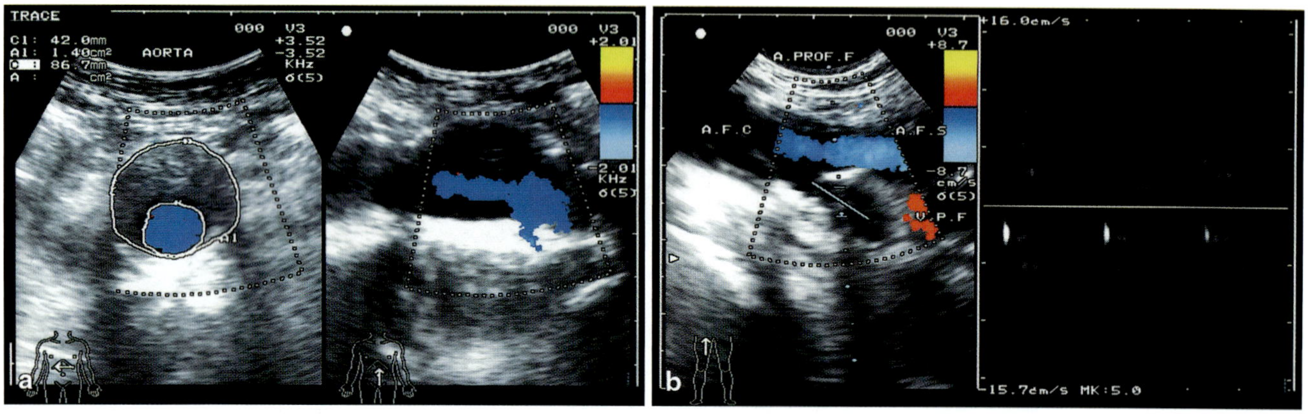


**Fig. A6.31 a, b**

**Aortic aneurysm with arterial embolism**

**a** While the risk of rupture correlates with the diameter of the aneurysm, the risk of embolism associated with the presence of thrombosis in an aneurysm is independent of its size. The saccular aneurysm shown has a size of only 4 cm with mural thrombosis reducing the size of the lumen to that of the normal vessel, especially in the saccular portion; nevertheless, this aneurysm was the source of distal emboli (see **b**). This is an indication for surgery irrespective of aneurysm size. Angiography shows no abnormalities as the perfused lumen of the aneurysm corresponds to that of the normal width of the aorta. The white outline in the left scan indicates the extent of the aneurysm; the longitudinal scan on the right depicts the saccular anterior outpouching and the thrombotic lining.

**b** Isolated occlusion of the profunda femoris artery with a patent superficial femoral artery (*A.F.S*) and common femoral artery (*A.F.C*) is typically due to embolism rather than atherosclerosis. Neither color duplex nor the Doppler waveform demonstrates flow in the profunda femoris. The gray-scale mode shows not only a posterior plaque with acoustic shadowing but also hypoechoic thrombotic material extending from the profunda femoris artery (sample volume) into the bifurcation.

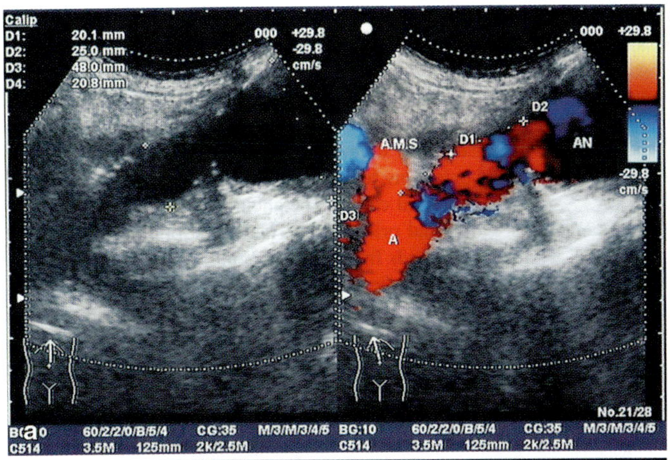


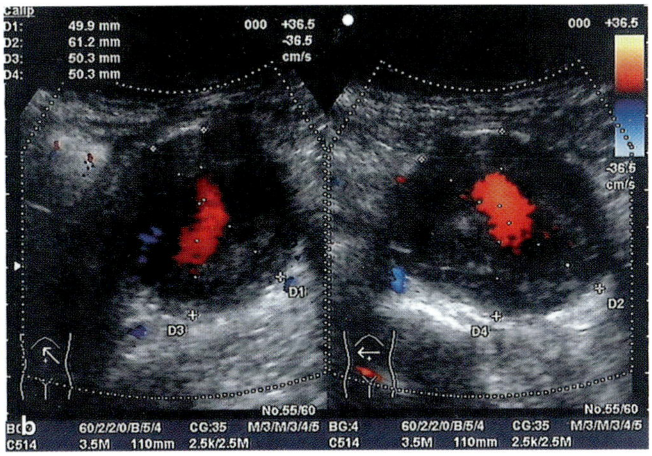


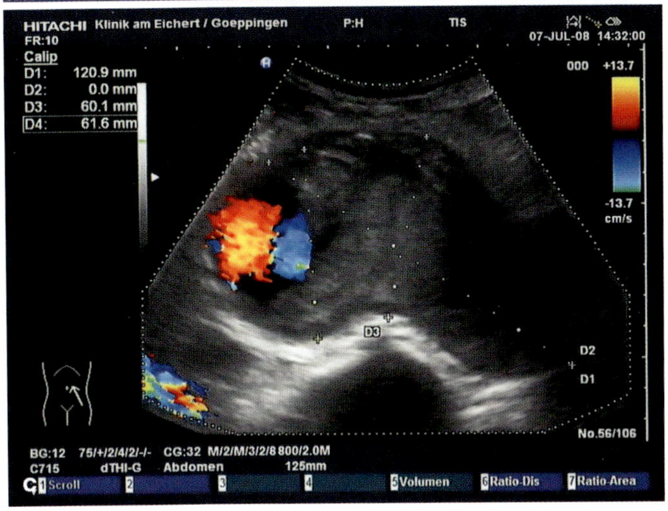


**Fig. A6.32 a–c**
**Aortic aneurysm**

**a** The therapeutic management of an aortic aneurysm is mainly dictated by its diameter, involvement of the iliac artery, presence of thrombosis, and infrarenal extent, including the distance to the renal artery origins, which is important when stenting is contemplated. Since the renal artery origins are best seen transversely, and the segment between the origins and the end of the aneurysm longitudinally, it is helpful to first identify the superior mesenteric artery in the longitudinal view and then use it as a guiding structure. The renal arteries arise 1–2 cm distal to the origin of the mesenteric artery. The segment between the end of the aneurysm and the superior mesenteric artery origin can thus be measured in longitudinal orientation. This value minus 2 cm is the distance between the renal artery origin and the aneurysm.

**b** Whether stenting is an option depends on the diameter of the aneurysm. Since a dilated aorta is also elongated (mostly with a left lateral convexity), imaging modalities with data acquisition in standardized transverse sections such as CT will lead to overestimation when the longest diameter of the elliptical lumen is measured (*right section*). To eliminate this source of error and perform a precise and reproducible cross-sectional diameter measurement (with little inter- and intraobserver variation) in repeat examinations, the sonographer must first identify the site of the largest diameter of the aneurysm. Next, the transducer must be rotated in this position to identify the smallest transverse diameter (typically the luminal profile will change from elliptical to circular), thus avoiding an oblique cross-sectional scan plane through the aorta, which would lead to an overestimation of the diameter. In the example, the measured diameter is thus reduced from 61 mm in the elliptical configuration (transverse mid-abdominal scan) to the true value of 50 mm (*left section*).

**Contained perforation of aortic aneurysm**

**c** Infrarenal, partially thrombosed abdominal aortic aneurysm measuring 6 cm ($D_3 + D_4$). The transverse lower abdominal scan reveals a contained perforation with complete thrombosis of the spilled blood at the time of the examination. The contour of the thrombosed aneurysm (*arrow*) is distinct from the clotted perivascular blood. The site of perforation is indicated by the contour disruption anterolaterally. The blood that escaped through the perforation into the psoas muscle has a total extent of 12 cm ($D_1$).

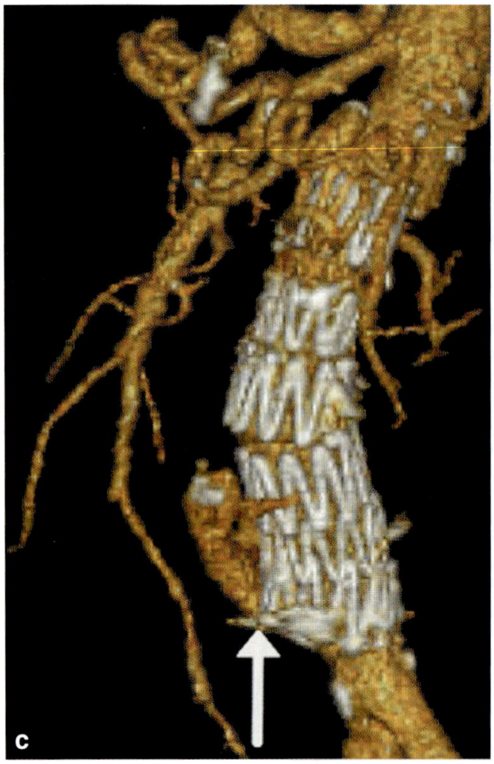


**Fig. A6.35 (continued)**
**c** 3D CT angiogram confirms the type I endoleak (*arrow*).

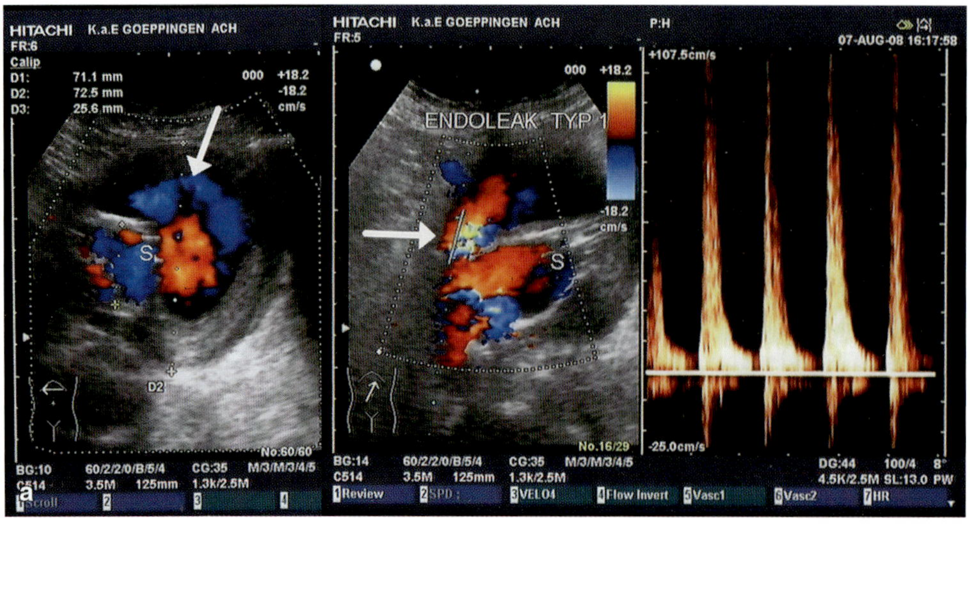


**Fig. A6.36 a, b**
**Type I endoleak after aortic stent**
**a** The color flow image shows flow within the stent (*S*) but also large color-coded areas indicating flow within the hypoechoic aneurysm sac. Blood enters the sac through a leak at the anastomotic seal below the renal artery origins (*middle section*), which is a type I endoleak (*arrow*). The Doppler spectrum from the site of the leak shows high-frequency monophasic flow with a PSV of over 1 m/s. The to-and-fro flow characteristic of endoleaks and false aneurysms (identical hemodynamic situation) is absent here. Unidirectional flow into an aneurysm through an endoleak will lead to rupture within a short time if there is no adequate drainage, underscoring the importance of searching for an outflow in such situation.

**Fig. A6.36 (continued)**

**b** Here, blood leaves the aneurysm through the inferior mesenteric artery (*A.M.I*; coded in blue, away from transducer, indicated by *arrow*) visualized along the hypoechoic aneurysm sac with the stent (*S*) and the two limbs (*R* and *L*). Farther to the left, the origin of the inferior mesenteric artery is seen with flow coded in red. The corresponding Doppler waveform from the inferior mesenteric artery reveals a rather large diastolic flow component.

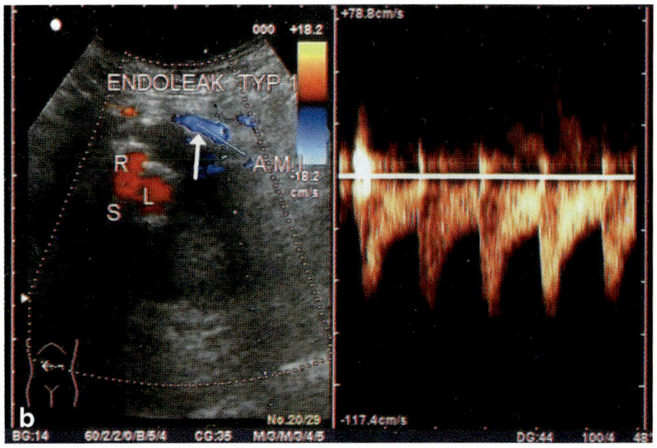


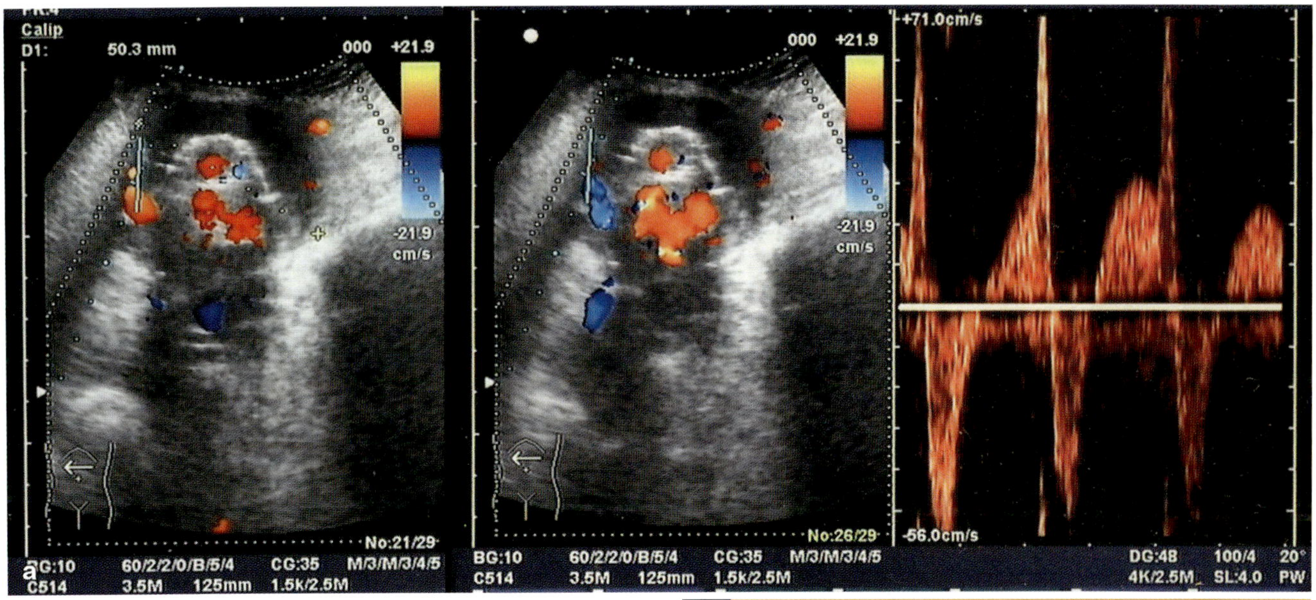


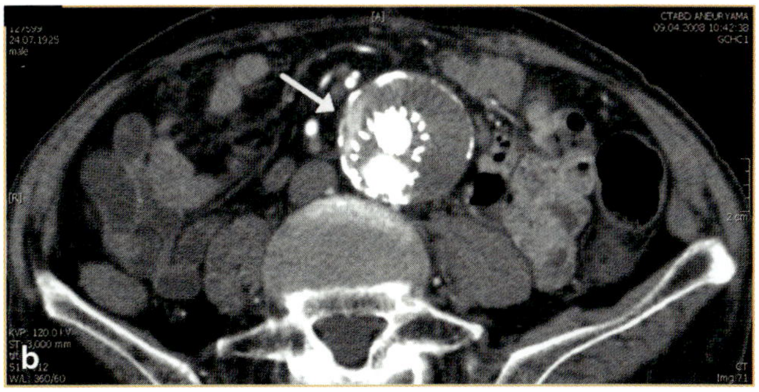


**Fig. A6.37 a, b**

**Type II endoleak**

**a** When the color flow image shows flow in the aneurysm sac, true flow must be differentiated from artifacts (migration artifact, mirror artifact, and artifact from pulsatile stent movement in the thrombosed aneurysm sac, especially shortly after stenting). Artifacts can be identified by insonation from different directions and spectral Doppler evaluation. Similar to false aneurysms in terms of hemodynamics, endoleak jets can be expected to display to-and-fro flow from the lumbar artery perfusing the aneurysm sac (flow into the sac during systole and back into the lumbar artery during diastole).

**b** Contrast-enhanced CT scan demonstrates blood flow into the aneurysm sac from a lumbar artery (type II endoleak).

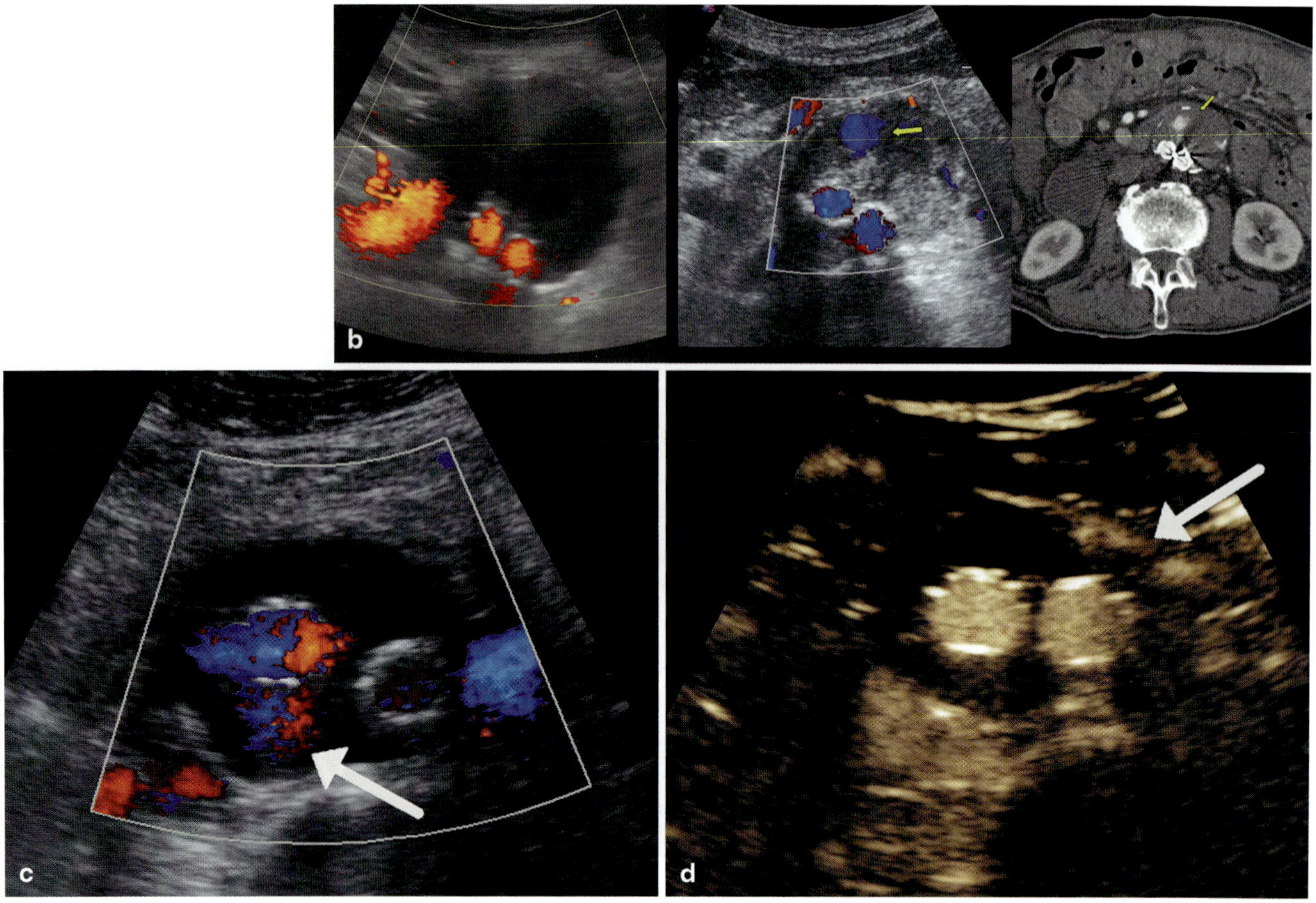


**Fig. A6.40 (continued)**
**Type II endoleak not detected with color duplex, only with contrast-enhanced ultrasound**
(images courtesy of K. Pfister)
**b** *Left image*: Color duplex examination following aortic aneurysm stenting fails to identify endoleak despite adequate machine settings. The two stent limbs are depicted in the posterior portion of the aneurysm sac with the patent vena cava laterally (power mode display, which is less angle-dependent and improves the detection of slow-flowing blood; see Table 1.8).
*Middle image*: Following administration of an ultrasound contrast agent, the signal enhancement reveals flow (*arrow*) in the aneurysm sac, consistent with a type II endoleak.
*Right image*: Contrast-enhanced computed tomography image confirms the endoleak (*arrow*).
**Type II endoleak – mirror artifact**
**c** Color duplex examination following aortic aneurysm stenting demonstrates no endoleak. The colored areas posterior to the right stent limb (*arrow*) suggest flow but are actually mirror artifacts produced by strongly reflecting surfaces.
**d** Contrast harmonic imaging: Following injection of ultrasound contrast medium, a contrast cloud appears in the anterolateral aspect of the aneurysm sac, suggesting a small endoleak.

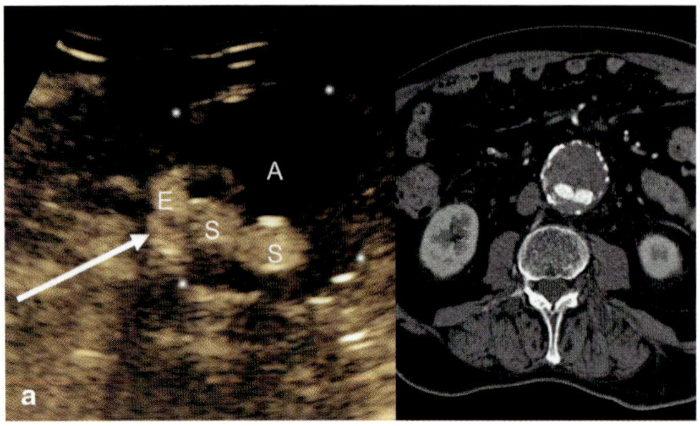


**Fig. A6.41 a–c**
**Type II endoleak not detected by CT, only by contrast-enhanced ultrasound**
(images courtesy of K. Pfister)
**a** *Right image*: Contrast-enhanced CT is inconclusive regarding the presence of an endoleak in the posterolateral aspect of the aneurysm sac on contrast-enhanced CT.
*Left image*: Contrast-enhanced ultrasound examination (contrast harmonic imaging, see Sect. 1.1.5) after aortic stenting (transverse orientation): there is contrast-enhanced flow (*arrow*) in the posterolateral portion of the aneurysm sac; the extent of the aneurysm is indicated by white dots. Same plane as the corresponding CT scan.

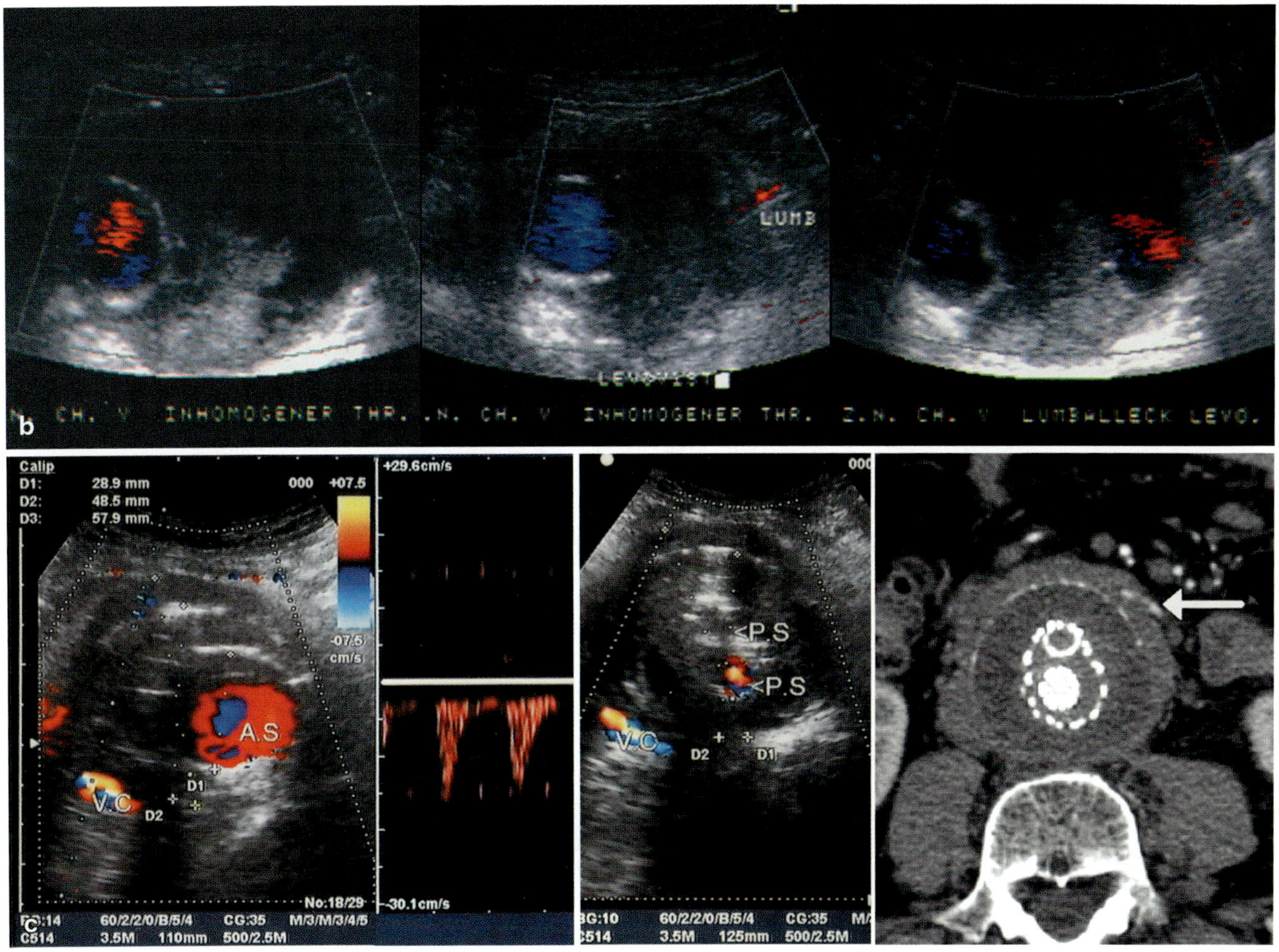


**Fig. A6.41 (continued)**

**Endoleak visualized with contrast-enhanced ultrasound** (images courtesy of P. Heilberger)

**b** Patent stent in a hypoechoic aneurysm with an increase in size at follow-up. No flow signals are obtained from the aneurysmal sac without contrast medium (*left section*).

One minute after administration of an echo enhancer (Levovist), leakage is visualized on the left side of the aneurysm with red-coded flow from a lumbar artery (*middle section*).

The scan on the *right* shows good visualization of the type II endoleak due to retrograde flow in the patent lumbar artery (depicted in red at the right margin) 3 min after echo enhancer administration.

**Follow-up of aortic aneurysm stenting – complication versus retroperitoneal fibrosis**

**c** Six-month sonographic follow-up after stent implantation identifies a margin of low echogenicity around the aneurysm sac. The residual diameter of the aneurysm sac in this plane (D2) is 48 mm (marked with *plus signs*). The 1-cm margin around the stented aneurysm is most conspicuous anteriorly; neither duplex ultrasound nor CT (*rightmost image*) demonstrates an endoleak. The Doppler waveform obtained from isolated color-coded flow signals within the hypoechoic margin (see sample volume) indicates normal intravascular blood flow and no to-and-fro flow (differential diagnosis: contained perforation). New onset retroperitoneal fibrosis is suspected (differential diagnosis: perigraft reaction) and confirmed by ultrasound-guided biopsy. *A.S* = main stent body; *V.C* = vena cava.

The *center right image* (obtained at a slightly lower level) shows occlusion of the more anterior stent limb adjacent to the 1-cm hypoechoic margin surrounding the aneurysm sac (indicated by *plus signs*); this is a complication occurring after aneurysm stenting. *P.S* stent limb, *V.C* vena cava.

*Rightmost image*: Contrast-enhanced CT shows enhancing tissue around the aneurysm sac (interpreted to indicate inflammatory hyperemia). Retroperitoneal fibrosis also explains why shrinkage of the aneurysm sac after stenting is minimal, even though neither contrast-enhanced CT nor color duplex imaging reveals an endoleak. Another CT finding consistent with retroperitoneal fibrosis is the demonstration of perfused arteries (*arrow*) coursing partially within the thickened margin and then being pushed back toward the aorta by perivascular fibrosis (see Figs. 6.19 and A6.42).

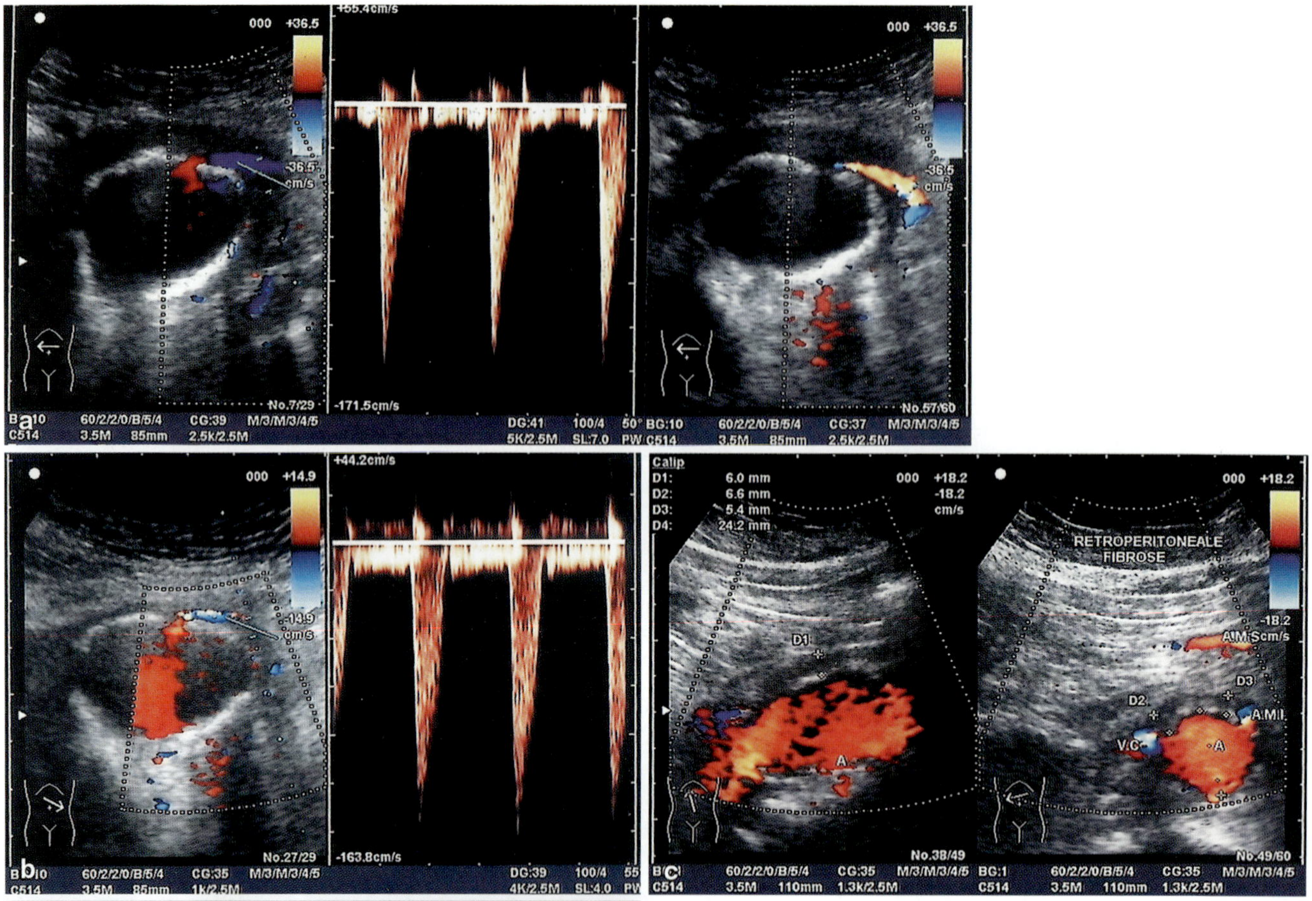


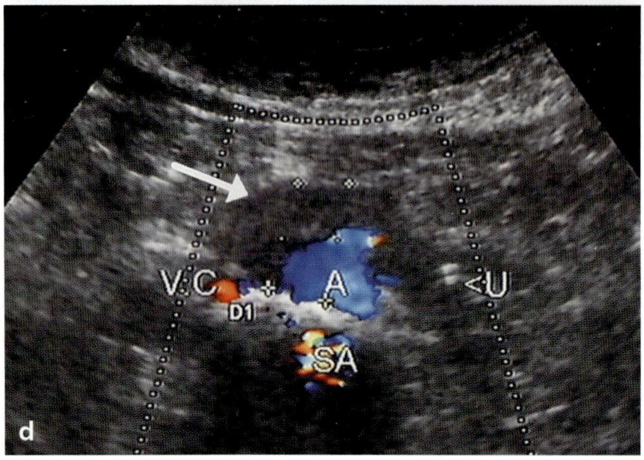


## Fig. A6.42 a–d

**Retroperitoneal fibrosis, aortitis, inflammatory aortic aneurysm**

**a** Retroperitoneal fibrosis (Ormond's disease) may be visualized as a hypoechoic cap-like structure anterior to the aorta. This condition differs from aortitis and inflammatory aortic aneurysm in that the process also involves the vena cava, which is ensheathed by fibrotic tissue. The origin and proximal course of the inferior mesenteric artery are evaluated to establish the differential diagnosis. In the presence of retroperitoneal fibrosis, the mesenteric artery, after arising from the left lateral aspect of the aorta, is pushed against the aortic wall, where it runs for some centimeters before piercing through the hypoechoic fibrotic layer and continuing its intra-abdominal course in the mesentery. The scan on the *left* depicts the inferior mesenteric artery arising from the aortic wall and the hypoechoic cap above. The scan on the *right* obtained in slightly oblique orientation documents the course of the inferior mesenteric artery with flow displayed in blue (away from transducer). It is pushed against the aortic wall by the hypoechoic structure. The *middle section* shows the corresponding Doppler waveform.

**b** After a few months of cortisone treatment, the hypoechoic layers around the aorta and vena cava have markedly decreased in thickness, from 1 cm (see **a**) to 4 mm, in the B-mode image; the fibrotic tissue however still pushes the inferior mesenteric artery against the aorta and its course remains unchanged. The Doppler spectrum sampled in the inferior mesenteric artery is presented on the *right*. Unchanged aneurysmal dilatation of the aorta with a diameter of 3.2 cm. While the oblique scan shown here was obtained for optimal evaluation of the superior mesenteric artery, this is not suitable for diameter measurement of the aorta.

**Retroperitoneal fibrosis – vena cava compression**

**c** Mild residual retroperitoneal fibrosis after treatment. The images document the involvement of the vena cava, which is important for the differential diagnosis. A hypoechoic layer (marked with *plus signs*, 6 mm thick) covers the aorta and vena cava (*V.C*, with aliasing due to compression) at the level of the inferior mesenteric artery origin (*A.M.I*).

**Retroperitoneal fibrosis – ureteral compression**

**d** Severe retroperitoneal fibrosis (marked with *plus signs*, 2 cm in thickness) forms a sort of shell around the aorta (*A*) and vena cava (*V.C*). There is compression of the vena cava and involvement of the ureter (*U*) with urinary congestion and displacement toward the aorta. The bright echoes from within the ureter are from a double-J stent (*SA* mirror artifact).

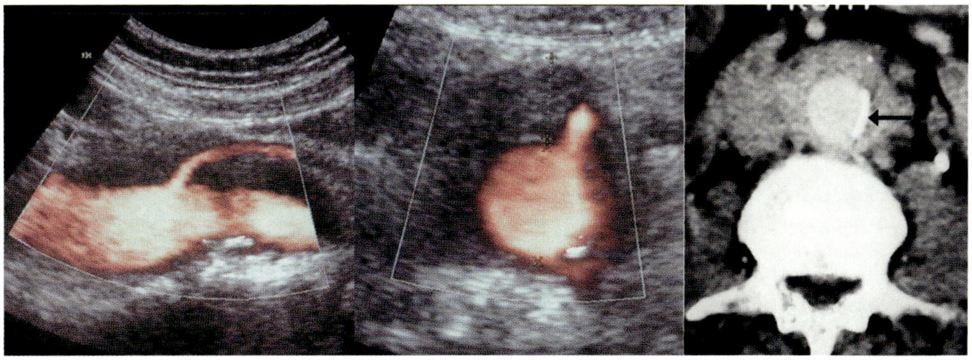

**Fig. A6.43**
**Aortitis (giant cell arteritis)**
Aortic wall thickening in giant cell arteritis (longitudinal scan on the *left*, transverse scan on the *right*; power mode). The course of the inferior mesenteric artery at its origin confirms disease of the aortic wall as the artery pierces the thickened wall directly at its origin without first being pressed to the aortic wall as is the case with fibrosis. The CT scan (rightmost image) shows circumferential wall thickening due to giant cell arteritis. Both ultrasound and CT additionally show plaque on the thickened wall in the lumen (courtesy of K. Amendt).

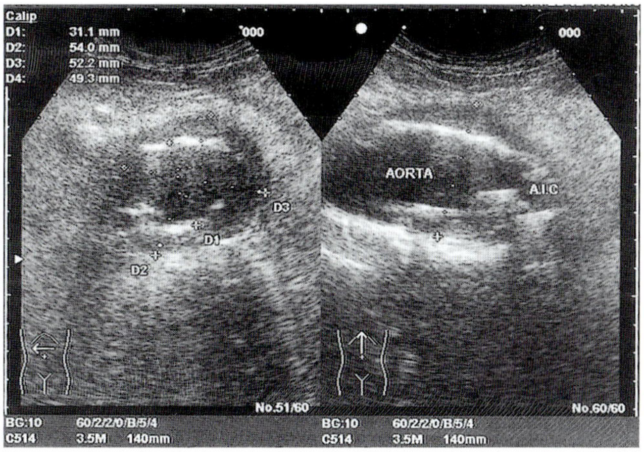


**Fig. A6.44**
**Inflammatory abdominal aortic aneurysm**
Transverse scan (*left*) and longitudinal scan (*right*) showing the typical appearance of an inflammatory aortic aneurysm. The aneurysm has a luminal diameter of 3.5 cm and exhibits atherosclerotic wall changes. The circumferential hypoechoic layer (1 cm) around the aneurysm confirms the inflammatory origin of the aneurysm.

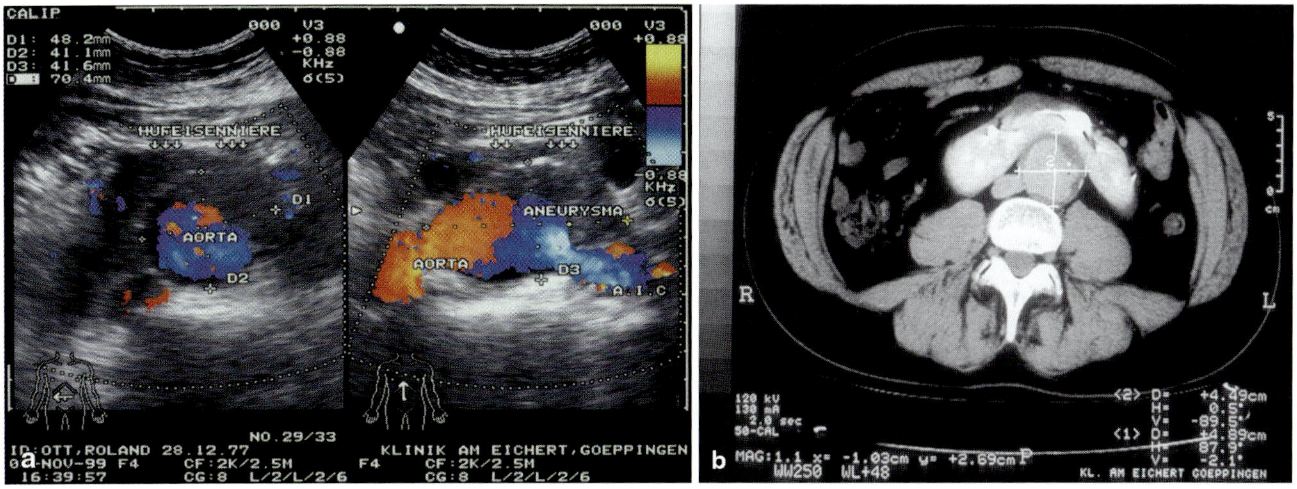


**Fig. A6.45 a, b**
**Horseshoe kidney in conjunction with abdominal aortic aneurysm**
**a** A horseshoe kidney is seen as a hypoechoic cap-like structure extending over the distal aorta. In the presence of a concomitant aortic aneurysm as in this case, the abnormal kidney must be sonomorphologically differentiated from the aortic wall as well as from other retroperitoneal structures or contained rupture of the aneurysm.
**b** CT confirming the abdominal aortic aneurysm and the horseshoe kidney.

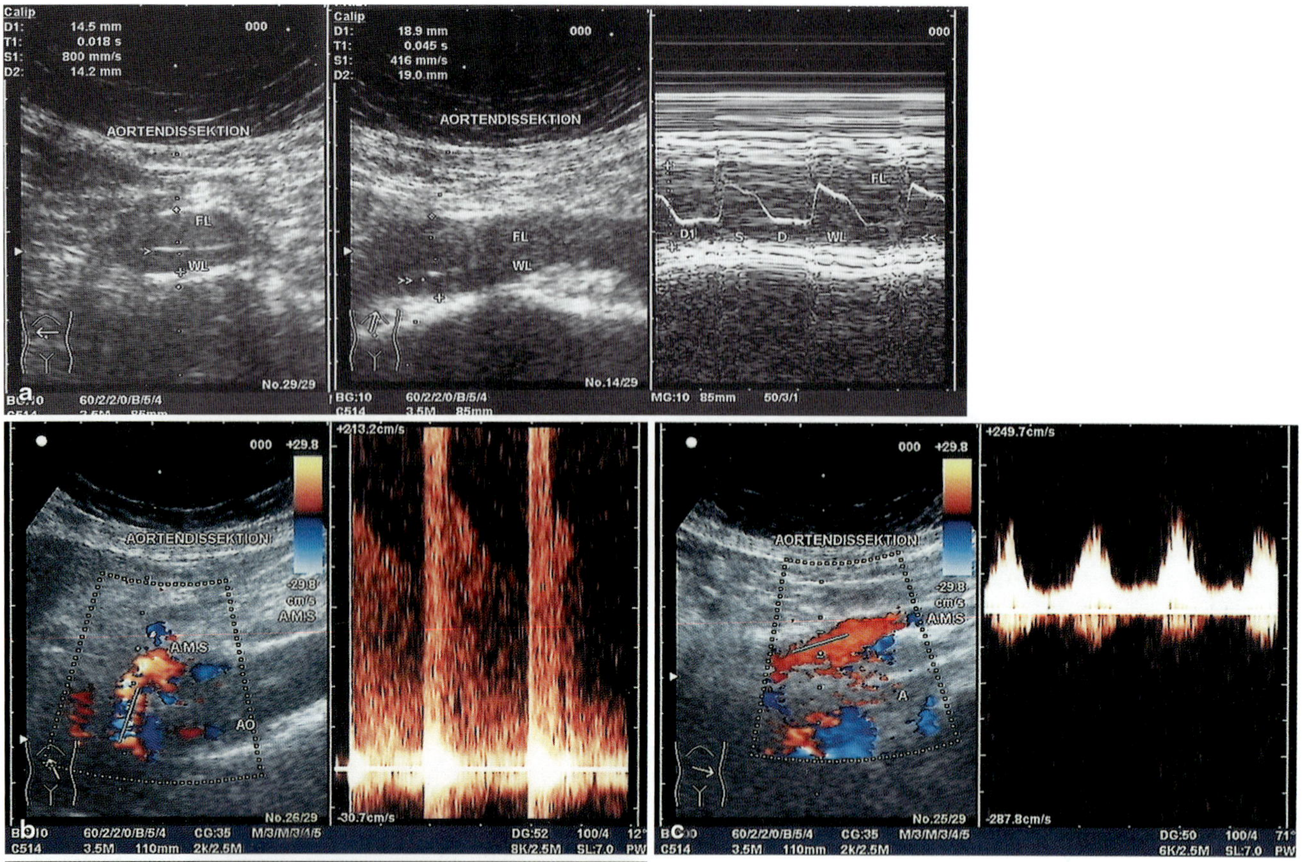


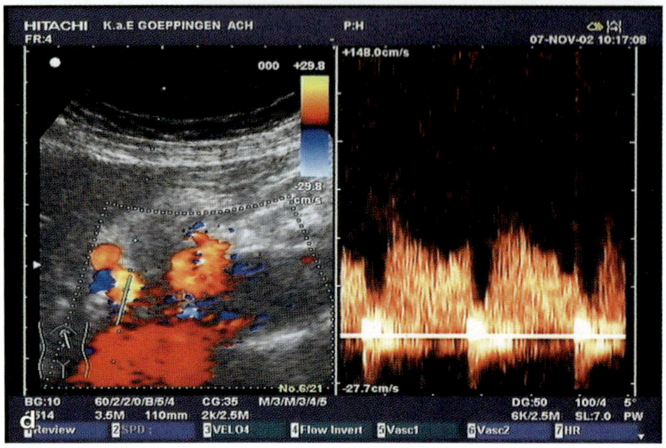


## Fig. A6.46 a–h

**Aortic dissection complicated by aortic branch involvement**

**a** Aortic dissection can be demonstrated by B-mode ultrasound when the intimal flap is insonated at a right angle (transverse view on the *left*, longitudinal view in the *middle*). The time-motion mode on the *right* shows the systolic-diastolic flap movement in the lumen. Scanning at a perpendicular angle enables differentiation of the true (*WL*) and false lumen (*FL*). The false lumen is compressed by the systolic pressure and expands again in diastole.

**b** The natural course and therapeutic measures in aortic dissection depend on the extent and involvement of aortic branches. Involvement of the superior mesenteric artery is associated with high-grade stenosis at the origin. Morphologically, the course of the intimal tear is difficult to identify. When the false lumen is located on the anterior side as in the case presented (see **a**), the superior mesenteric artery arises from the true lumen, and its origin is compressed by the false lumen or an intimal flap, resulting in flow obstruction with a typical stenosis waveform and a PSV of over 3 m/s (interpolated due to aliasing).

**c** Poststenotic Doppler waveform with the typical delay in systolic upstroke, turbulent flow, and a larger diastolic component. The color flow scan shows the aorta (*A*) with the flap deep to the superior mesenteric artery.

**d** The Doppler waveform from the celiac trunk of the patient shows systolic deceleration with near-zero flow. This decrease in systolic flow velocity is due to intermittent closure of the celiac artery origin by the aortic intimal flap; normal orthograde flow during diastole occurs because of pressure reversal pushing the flap back into the lumen. This dangerous situation with imminent arterial occlusion cannot be adequately visualized by any of the merely morphologic imaging modalities and can only be identified on the basis of the hemodynamic information provided by spectral Doppler measurement.

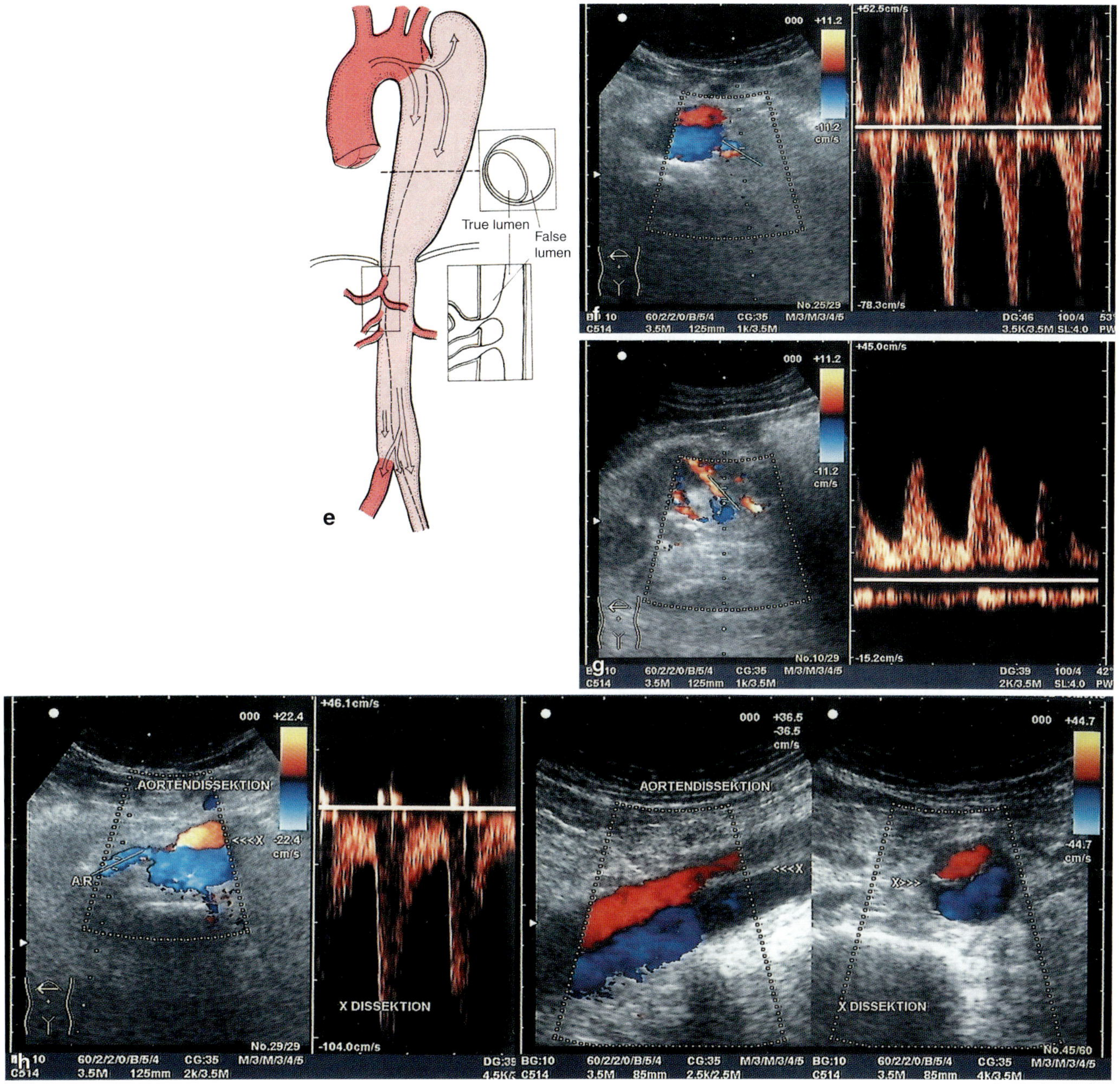


**Fig. A6.46 (continued)**

**e** Diagram of type III aortic dissection according to De Bakey (examples in **b** and **c**). The mesenteric arteries arising from the true lumen are compressed by the false lumen or the intimal flap (From Heberer and van Dongen 1993).

**f** If dissection involves the origin of a renal artery, there may be superimposition of the Doppler frequency spectra from the true and false lumina or – depending on the reentry site or the position of the sample volume in the dissected segment – to-and-fro flow as in the left renal artery shown.

**g** The waveform from a segmental artery in the left renal hilum demonstrates the typical postocclusive flow pattern with a reduced systolic upstroke and low PSV due to flow obstruction by the dissection.

**h** The right renal artery is not involved in the dissection and has a typical monophasic waveform with a PSV of 1 m/s. The further infrarenal course of the aortic dissection is shown in longitudinal (*right center*) and transverse planes (*rightmost*). The change in color coding may be due to the position of the reentry site or physiologic flow reversal (early diastolic reflux).

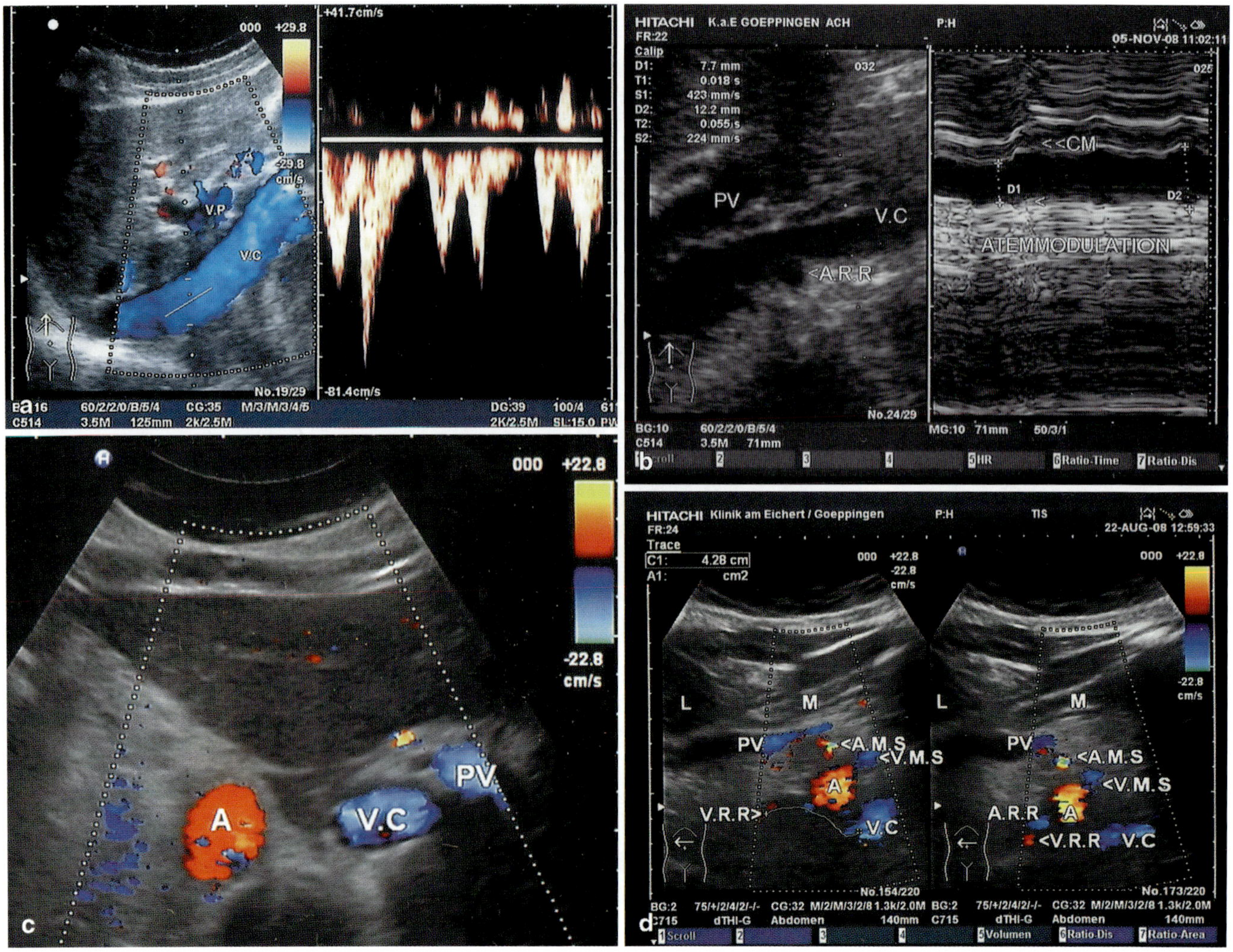


**Fig. A6.54 a–d**
**Vena cava**

**a** The cross-sectional area and flow velocity in the vena cava (*V.C*) vary with respiration. Flow is markedly faster during inspiration. In addition, the flow is subject to cardiac (atrial) pulsatility. The Doppler waveform typically shows two peaks, one during systole and the other upon opening of the atrioventricular valves (W-shaped waveform). There is marked reduction, cessation, or even a short reversal of flow during atrial contraction.

**b** The usual oval cross section of the vena cava can show diameter variation due to changes in intravascular pressure during the respiratory cycle (W-shaped waveform); cardiac pulsatility can also lead to variation (*arrow* in *left* image). This variation is nicely depicted in the time-motion display (*right* image).

**Situs inversus**

**c** There are some extreme anatomic variants of the vena cava; these are rare and include absence, doubling with one vena cava on either side of the aorta, and a single vena cava to the left of the aorta as in the case shown (here, in accordance with ultrasound convention, the left-lying vena cava is displayed to the right of the aorta). In complete situs inversus, the liver is located in the left upper abdomen, and the portal vein (*PV*) also ascends toward the liver hilum on the left side.

**d** Transverse upper abdominal image of a vena cava completely ascending to the left of the aorta in an individual with intestinal malrotation but without situs inversus. The portal vein (*PV*) arises from the superior mesenteric vein (*V.M.S*) on the left and crosses the superior mesenteric artery (*A.M.S*) on its way toward the liver hilum. The vena cava (*V.C*, displayed to the right of the aorta, by ultrasound convention) courses on the left side of the aorta (*A*) with the right renal vein (*V.R.R*) taking a retroaortic course (only intermittently visible in this image, course is marked in with white line). *M* stomach, *L* liver.

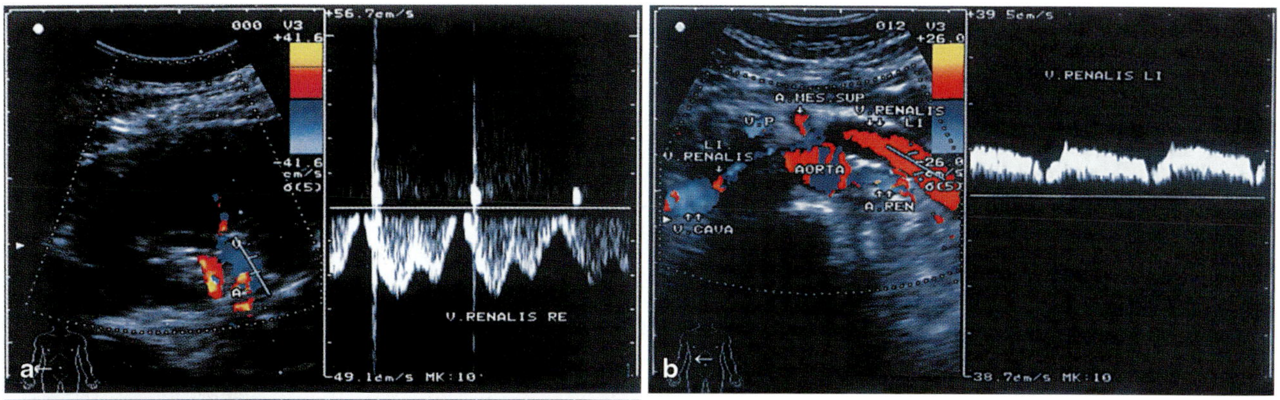


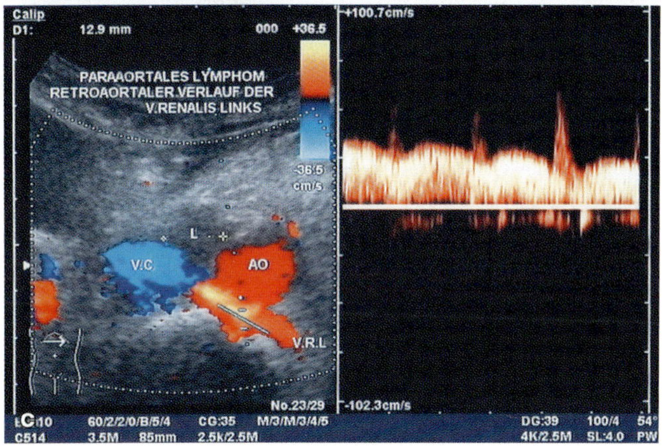


**Fig. A6.55 a–c**

**Right renal vein**

**a** Cardiac pulsatility and respiratory phasicity of blood flow are transmitted as far as the right renal vein at the hilum (vein: blue, segmental artery: red).

**Left renal vein**

**b** Cardiac pulsatility is typically lost in the left renal vein due to the narrow passage between the superior mesenteric artery and aorta. Instead, its flow variation is determined by the aortic pulse. Posterior to the red-coded renal vein, the renal artery is depicted in blue. The blue-coded renal vein has a rather large caliber in front of the narrow passage and then continues as a relatively thin vessel to the vena cava.

**Retroaortic left renal vein**

**c** If the left renal vein (*V.R.L*; red, flow toward transducer) is not identified between the aorta and superior mesenteric artery, the examiner must try and locate its entry into the vena cava (*V.C*) posterior to the aorta (*AO*). Identification of a retroaortic left renal vein is important prior to resection of an aortic aneurysm but is often an incidental finding, as in the case presented, where the atypical entry was identified in a patient in whom vascular sonography was performed prior to ultrasound-guided biopsy of a lymphoma (*L*).

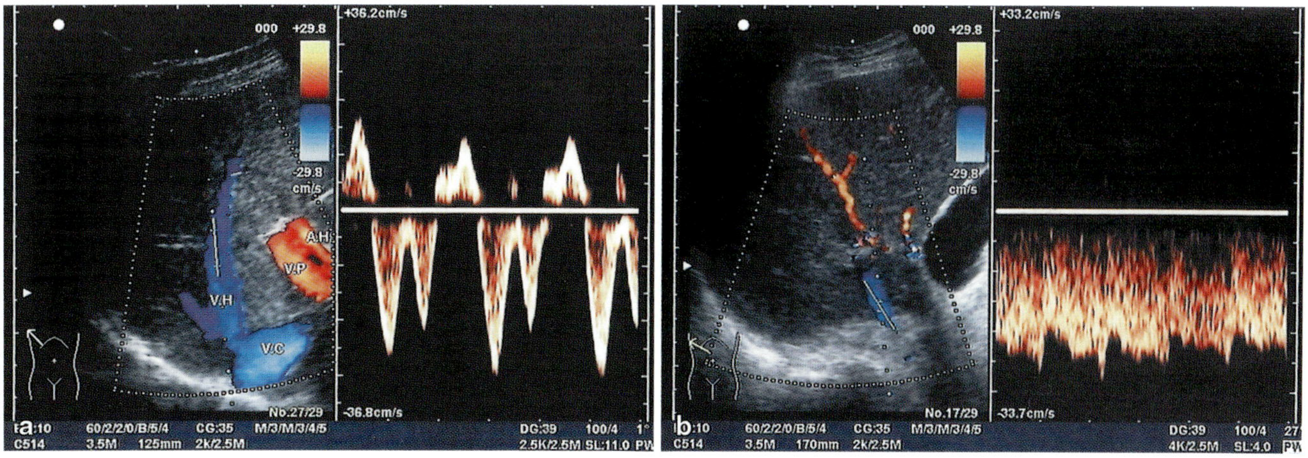


**Fig. A6.56 a–c**

**Doppler waveform of hepatic veins**

**a** W-shaped Doppler waveform with a first hepatofugal flow peak in systole, a second hepatofugal peak upon opening of the atrioventricular valves, and hepatopedal flow during atrial contraction. An abnormal spectrum resembling a sinus wave with to-and-fro flow in the extreme case is seen in patients with right ventricular failure.

**Abnormal waveform of hepatic veins in liver cirrhosis**

**b** The right hepatic vein in a patient with Child A liver cirrhosis scanned from the intercostal approach shows only residual cardiac pulsatility. The associated loss of parenchymal elasticity primarily prevents the decrease in flow velocity during atrial contraction, resulting in an increasingly band-like spectrum from the entry into the vena cava to peripheral branches (intermediate hepatic vein in blue, portal vein branch in red).

**Fig. A6.56 (continued)**

**Waveform of hepatic vein in liver cirrhosis**

**c** Nearly continuous flow in the hepatic vein in liver cirrhosis (transducer in flank position) with reduced cardiac pulsatility. Though the curve is flattened due to rigidity of the liver tissue, some residual cardiac pulsatility is still present, and the curve is not yet as flat as in **b** (*A* ascites).

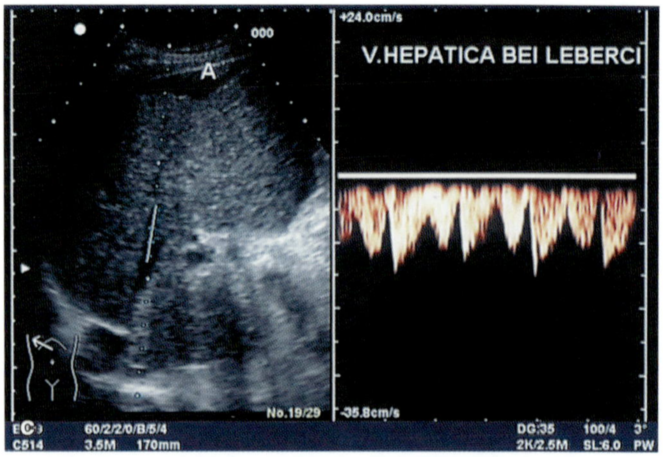


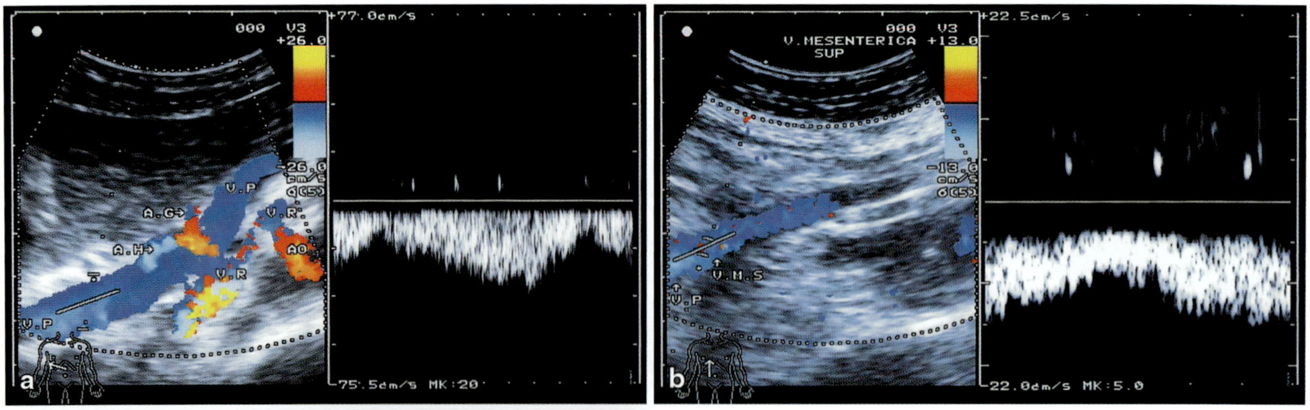


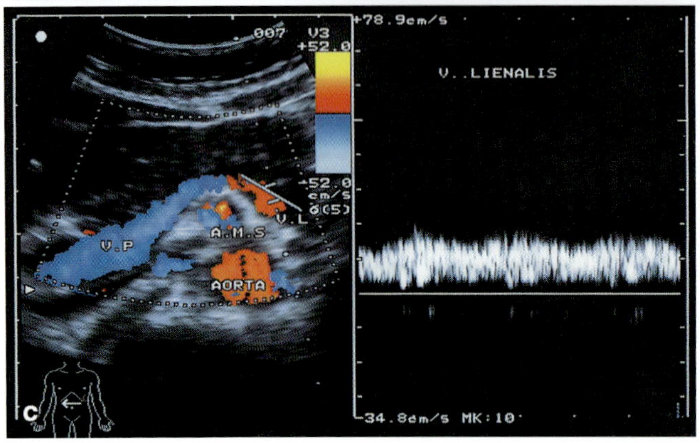


**Fig. A6.57 a–c**

**Portal vein and its tributaries**

**a** The Doppler spectrum from the portal vein is characterized by relatively wide variation in flow velocity, but flow is typically hepatocentral and slower during inspiration.

**b** The respiratory variation in blood flow velocity continues into the superior mesenteric vein, which is depicted to the right of the superior mesenteric artery.

**c** The splenic vein (*V.L*) is depicted at the lower edge of the pancreas with flow in red. It crosses over the root of the superior mesenteric artery (*A.M.S*) to enter (displayed in blue) the portal vein (*V.P*). The respiratory variation in flow velocity may continue into the splenic vein.

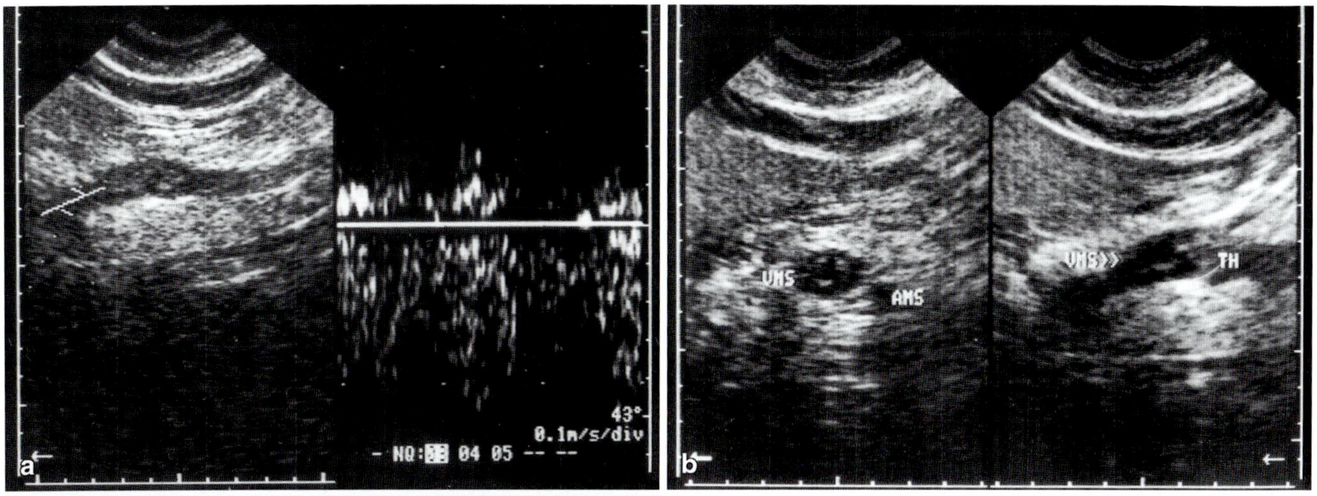


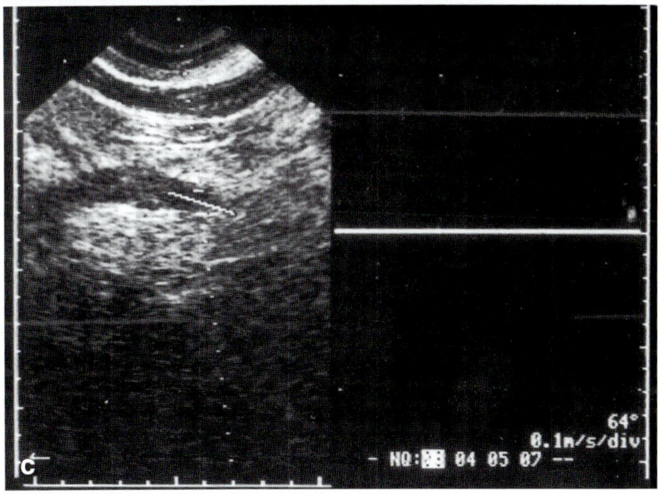


**Fig. A6.58 a–c**
**Mesenteric vein thrombosis**
**a** The clinical manifestations of mesenteric vein thrombosis range in severity from nonspecific symptoms to an acute abdomen with intestinal necrosis. In the example, flow is still present at the junction of the superior mesenteric vein with the portal vein depicted near the left margin of the gray-scale scan.
**b** The longitudinal and transverse images demonstrate occluding thrombosis (*TH*) of the superior mesenteric vein (*VMS*) extending into the portal vein. The transverse scan (*left section*) depicts the superior mesenteric artery (*AMS*) adjacent to the occluded vein.
**c** The Doppler frequency spectrum obtained from the mesenteric vein shows zero flow (sample volume as placed in the longitudinal image). In this patient, mesenteric vein thrombosis developed after postoperative sepsis; patency of the vein was restored with systemic thrombolytic treatment accompanied by the massive transfusion of packed red blood cells. Since the patient had the clinical picture of ischemia, treatment was necessary despite the preceding operation.

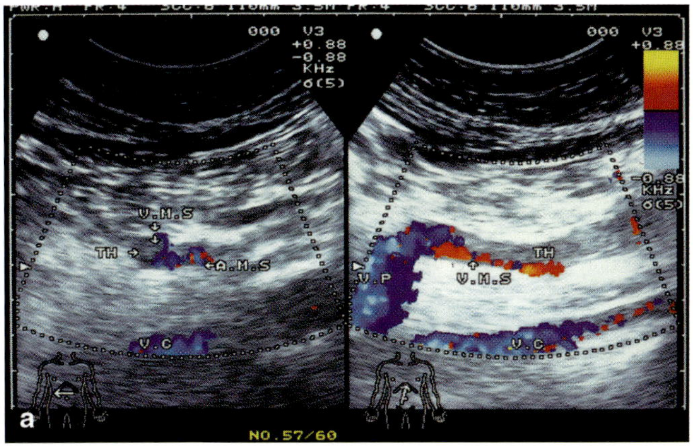


**Fig. A6.59 a, b**
**Mesenteric vein thrombosis – surrounded by flowing blood**
**a** The extent of thrombosis and collateralization determine whether the clinical manifestation will be mild with flu-like symptoms or severe with an acute abdomen due to intestinal necrosis. A 38-year-old patient with diffuse abdominal pain was treated conservatively for several days. Sonography was performed to exclude appendicitis and pancreatitis. Closer evaluation of the superior mesenteric vein by color duplex imaging revealed thrombosis of individual jejunal vein branches with protrusion of a thrombus into the trunk of the superior mesenteric vein. Mural mesenteric vein thrombosis causes obstruction of flow. Prompt initiation of full heparinization is necessary to prevent further appositional thrombus growth and subsequent intestinal necrosis.

**Fig. A6.59 (continued)**
**b** Digital subtraction angiogram confirms partial mesenteric vein thrombosis.

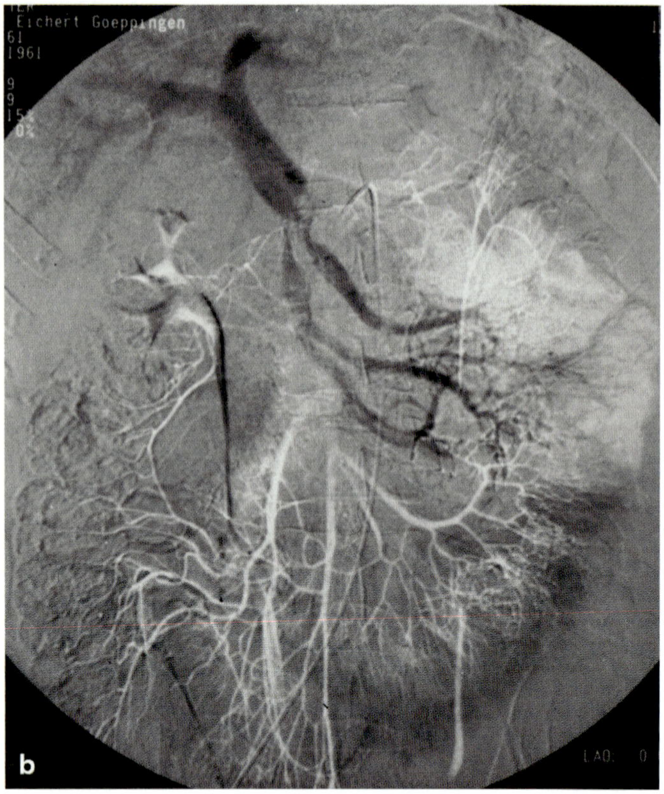


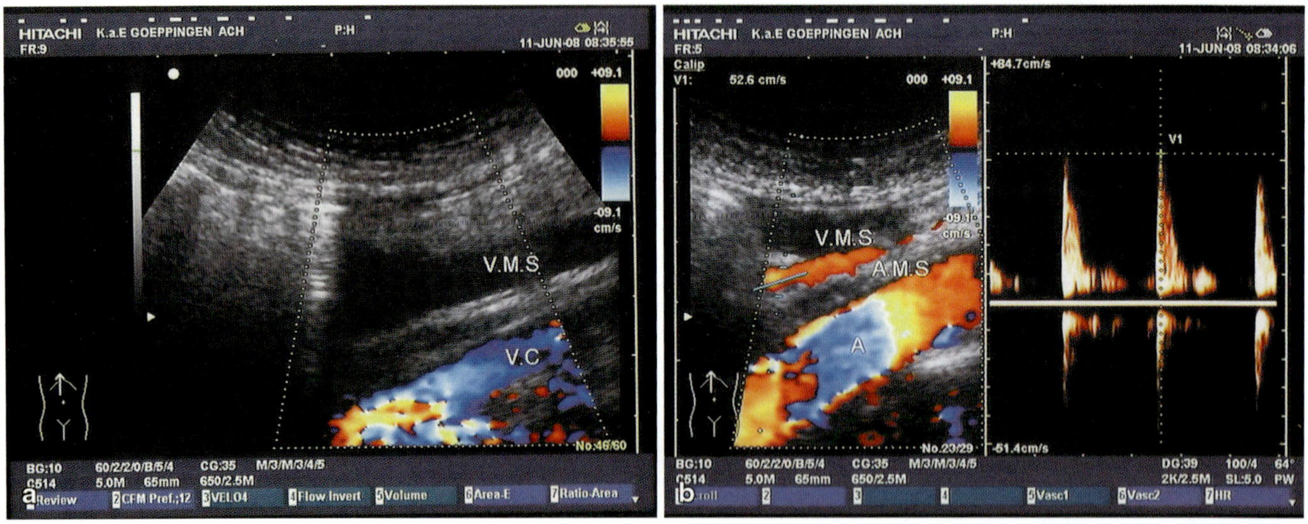


**Fig. A6.60 a–d**
**Thrombosis of portal veins and superior mesenteric vein**
**a** Complete thrombosis of the superior mesenteric vein (*V.M.S*) is indicated by the absence of flow signals despite a low pulse repetition frequency (indicated by aliasing in the vena cava, *V.C*). The vena cava is depicted posterior to the superior mesenteric vein.
**b** The superior mesenteric artery (*A.M.S*, red) comes into view when the transducer is moved to the left side. There is aliasing in the aorta posteriorly (*A*). With this transducer position, the superior mesenteric artery appears deep to the confluence of the superior mesenteric vein (*VMS*) and the splenic vein. Occluding thrombosis of the superior mesenteric vein affects flow in the artery, giving rise to a preocclusive thump pattern. PSV is markedly reduced (50 cm/s) and, along with the loss of diastolic flow, suggests high outflow resistance. Surprisingly, the 17-year-old woman had only mild diffuse abdominal pain (of the enteritic type) and mild meteorism, but no signs of peritonitis; there was mild leukocytosis without acidosis, and lactate levels were normal. The clinical symptoms persisted for 6 weeks before the diagnosis was made. Pre-existing portal vein thrombosis in an abnormal vein with severe ectasia led to the formation of collateral pathways, mainly via the inferior mesenteric vein, which is why the acute mesenteric vein thrombosis did not cause intestinal necrosis.

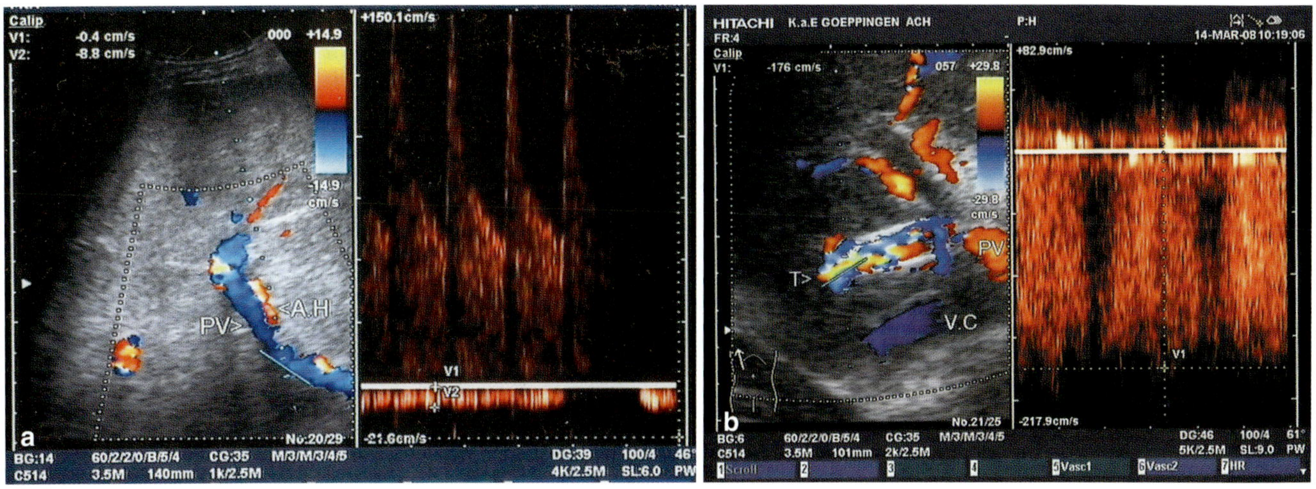


**Fig. A6.60 (continued)**

**c** There is a marked compensatory increase in blood flow in the hepatic artery with peak velocities of about 2 m/s during systole and 90 cm/s during diastole. The thrombosed portal vein (*PV*) is indicated by *plus signs*.

**d** Good collateral drainage through small veins is confirmed by the depiction of flow in the hepatoduodenal ligament (continuous high-frequency flow with a velocity of 50 cm/s) and around the gallbladder.

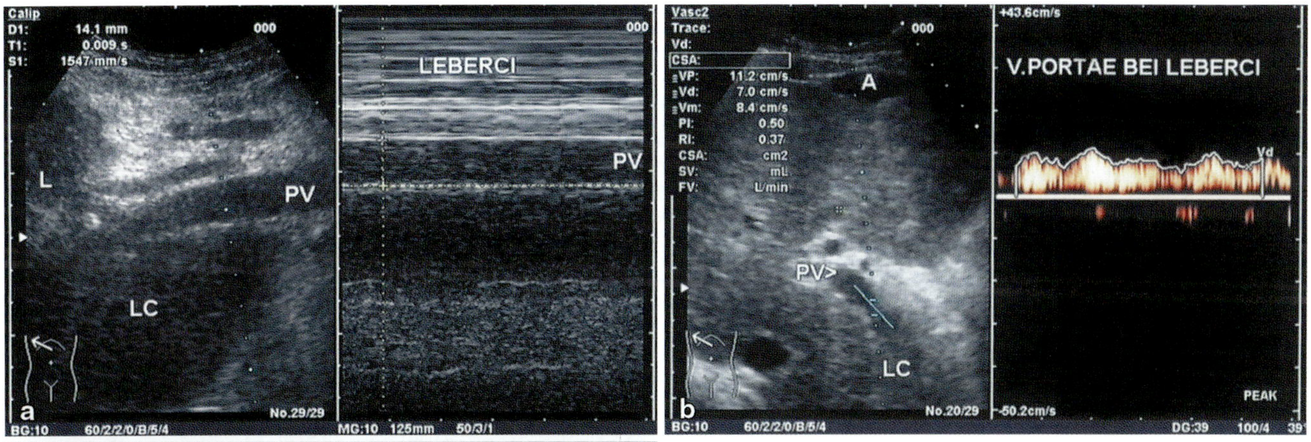


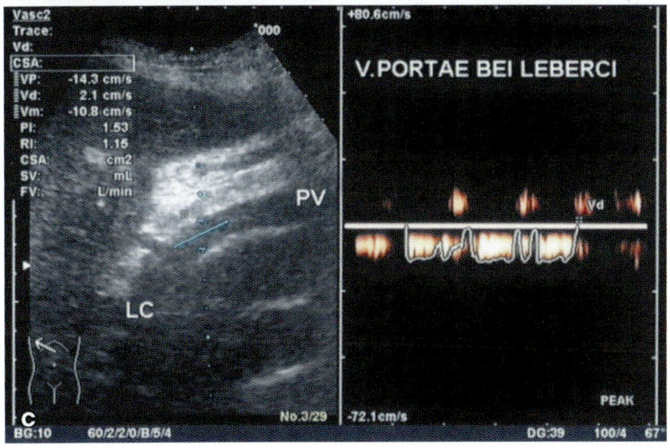


**Fig. A6.61 a–c**

**Portal hypertension**

**a** Loss of respiratory diameter variation in gray-scale ultrasound is a sign of portal hypertension. In the example, the time-motion mode demonstrates a constant diameter of 14 mm of the portal vein (*PV*).

**b** The patient presented has portal hypertension due to Child C liver cirrhosis. The peak flow velocity is markedly reduced to 11.2 cm/s with a mean flow velocity of 8.4 cm/s (intercostal transducer position). Other signs of liver cirrhosis depicted by ultrasound are perihepatic ascites (*A*) and the enlarged caudate lobe (*LC*). The congestion index is markedly increased to 0.2 cm × s.

**c** A reduced increase in postprandial flow velocity (mean flow velocity of 10.8 cm/s) is another sign of portal hypertension.

With an unchanged diameter of 14 mm, the postprandial flow velocity increases by only 20% (versus a mean increase >60% in normal individuals).

For didactic purposes, views of the portal vein from two transducer positions (**c**: subcostal; **b**: intercostal, from the flank) are shown with identical sample volumes in the vein. The intercostal approach permits smaller Doppler angles (in the example 38° versus 67°), thus yielding more accurate flow velocity measurements. Serial examinations should be performed with identical transducer positions.

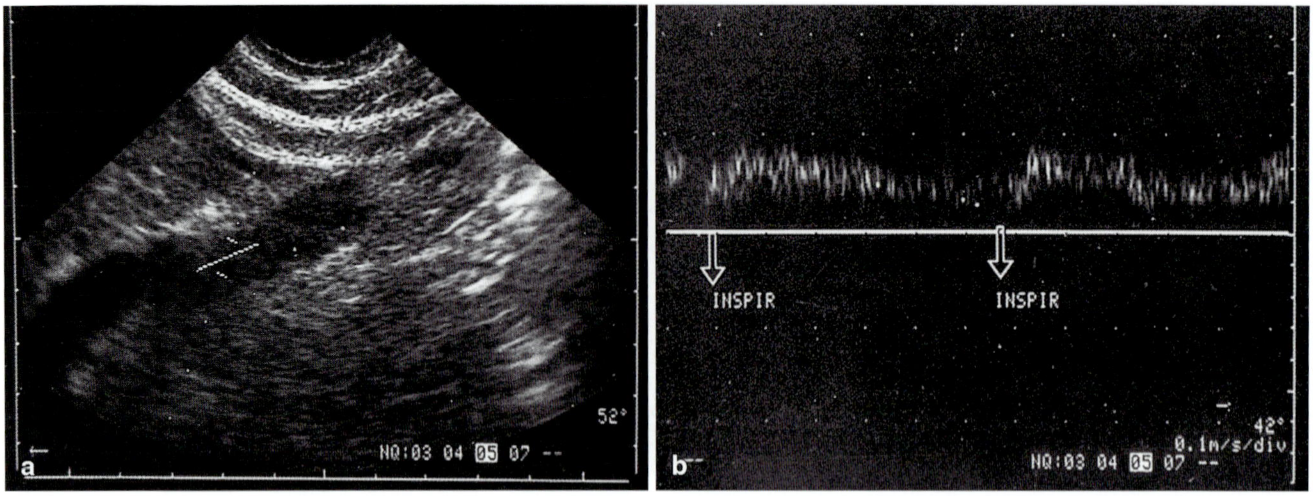


**Fig. A6.62 a, b**
**Collateral pathways in portal hypertension**
Collateral flow in portal hypertension may occur chiefly through the patent umbilical vein (Cruveilhier–Baumgarten syndrome). In the example, the umbilical vein is seen coursing toward the abdominal wall along the lower edge of the liver. Its diameter is markedly increased to over 1.5 cm. Flow in the umbilical vein is hepatofugal. The faster flow during inspiration suggests a communication with the portal vein and a high shunt volume.

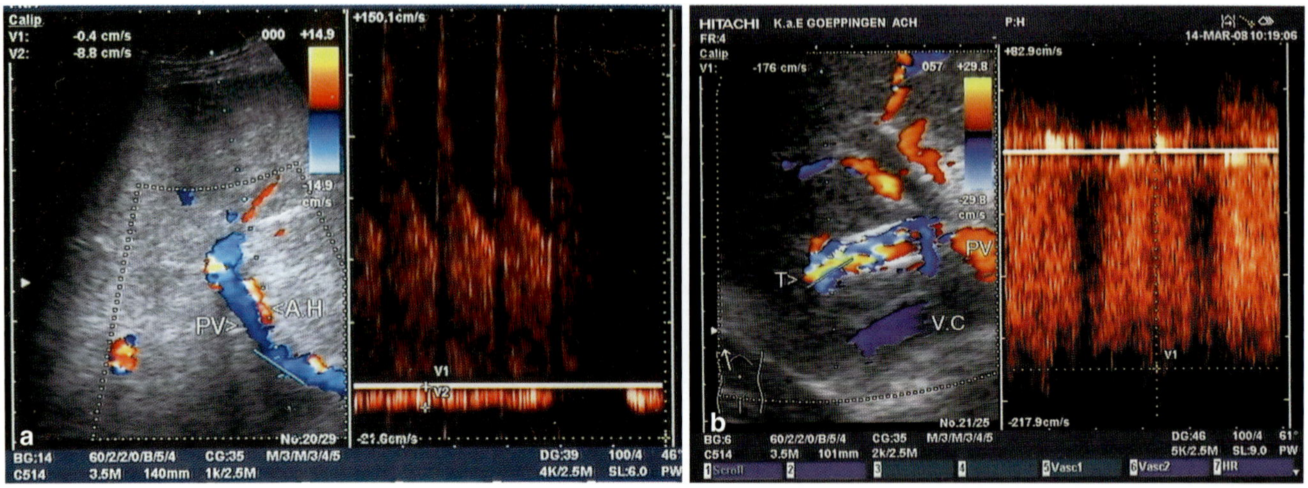


**Fig. A6.63**
**Hepatofugal flow in the portal vein**
**a** Intercostal view through a metastatic liver in a patient with cirrhosis. There is retrograde (hepatofugal) flow in the portal vein (*PV*; coded in blue, away from transducer) with high-volume flow in the opposite direction in the hepatic artery (*A.H*). The Doppler waveform from the portal vein confirms retrograde flow with a maximum velocity of 9 cm/s and also shows the high orthograde flow in the hepatic artery in the liver hilum (toward transducer) with a maximum flow velocity of 150.1 cm/s. This is because the sample volume in the portal vein includes a portion of the hepatic artery.

**TIPSS**
**b** Following creation of a transjugular intrahepatic portosystemic shunt (TIPSS), duplex follow-up reveals high flow within the shunt with a maximum velocity of 170 cm/s and little respiratory phasicity.

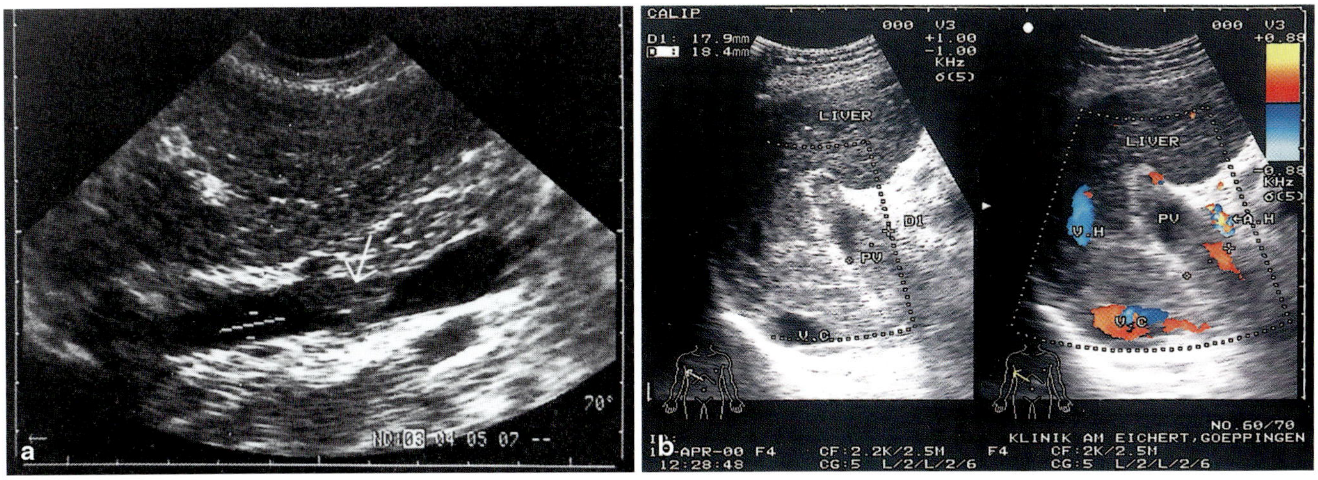


**Fig. A6.64 a, b**

**Portal vein thrombosis**

**a** Nonocclusive portal vein thrombus (*arrow*) following splenectomy due to hypersplenism with portal hypertension.

**b** Like thrombosis in other veins, portal vein thrombosis may already be suggested in the B-mode by a dilatation of the vessel and the presence of hyperechoic deposits in the lumen. Color duplex scanning from the intercostal approach using the liver as an acoustic window depicts flow signals (red, toward transducer). There is nearly complete thrombosis with residual flow close to the walls. The vena cava (*V.C*) is depicted posterior to the portal vein. There is aliasing in the hepatic artery (*A.H*) due to the low pulse repetition frequency (*V.H* hepatic vein).

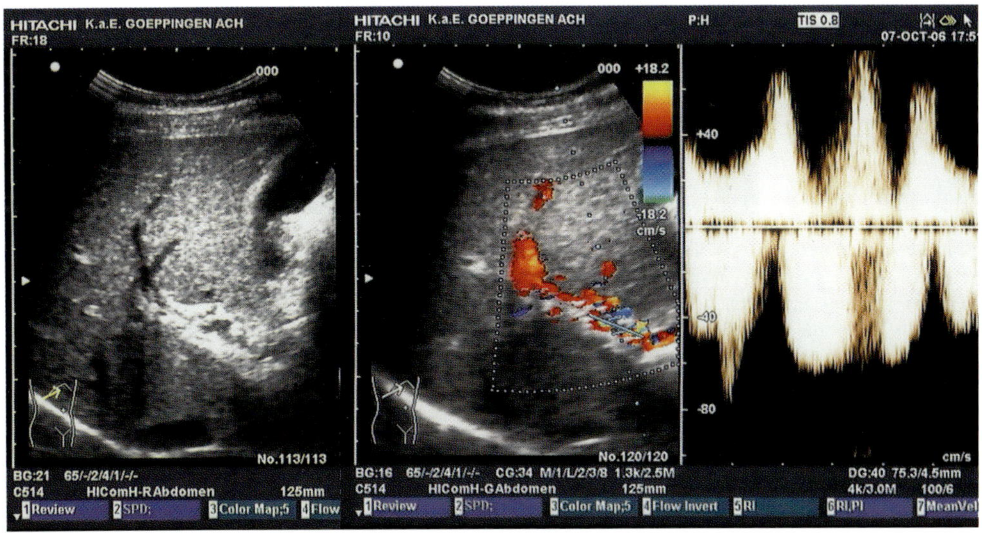


**Fig. A6.65**

**Cavernous transformation of the portal vein**

Cavernous transformation: failure to identify a normal-caliber portal vein in the liver hilum; instead, there are numerous hyperechoic tubular structures with thickened walls occupying the portal vein bed, consistent with cavernous transformation secondary to portal vein thrombosis. This diagnosis is confirmed by the mosaic of colors seen in these channels in the color flow image. This color pattern indicates different flow directions (relative to the transducer) in the tortuous venous channels.The Doppler waveform obtained from the transformed portal vein bed confirms blood flow toward and away from the transducer.

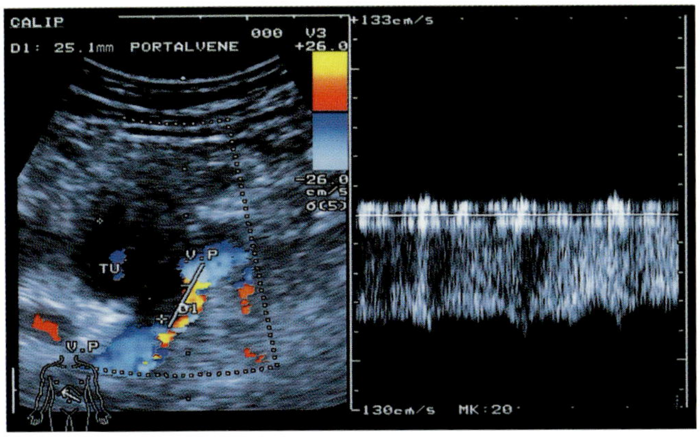


**Fig. A6.66**

**Tumor compression (portal vein)**

Carcinoma of the head of the pancreas with infiltration of the portal vein (*V.P*). The wall is poorly delineated and the lumen is compressed. Stenosis is suggested by aliasing in the color duplex scan and in the Doppler waveform. There is an intratumoral vessel (*TU*) with flow coded in blue.

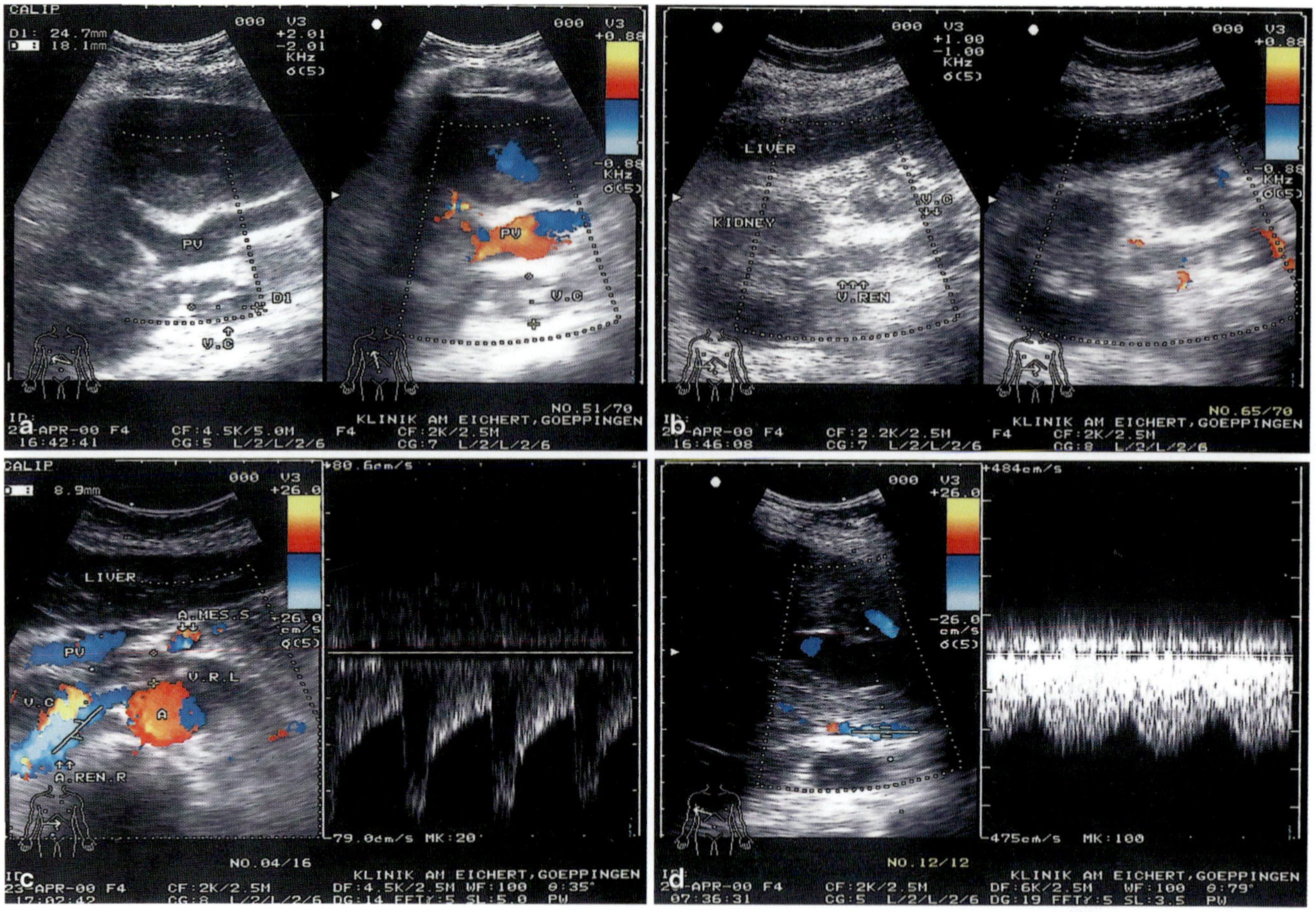


**Fig. A6.67 a–e**

**Vena cava thrombosis**

Following laser-induced thermoablation of a large paracaval liver metastasis (presenting with upper abdominal pain on the right), the patient developed subcapsular liver hemorrhage and complete thrombosis of the vena cava as a result of the local heat effect.

**a** B-scan (*left*) and color duplex scan (*right*) depicting the portal vein and vena cava (*V.C*) posterior to the liver hilum. The portal vein exhibits color-coded flow, while the vena cava (*V.C*) is depicted with hyperechoic internal echoes but without flow signals.

**b** The transverse view through the upper abdomen at a lower level shows residual flow near the walls displayed in red in the vena cava (right margin of *right section*) and no other flow signals. The scans also depict the renal vein (*V.REN*) joining the vena cava. The flow signals next to the renal vein indicate venous return through retroperitoneal collateral veins, chiefly the suprarenal vein.

**c** A tubular, hypoechoic structure without flow signals is depicted in the typical location of the left renal vein (*V.R.L*) between the superior mesenteric artery (*A.MES.S*) and the aorta (*A*), indicating thrombosis of this vein. The vein is markedly widened, and hyperechoic internal structures are depicted within the hypoechoic lumen. A vena cava thrombus next to the site of entry of the left renal vein is depicted with flow near the wall. Despite complete renal vein thrombosis, the right renal artery (*A.REN.R*) has a normal flow profile (*right section*). This arterial flow profile and the uneventful clinical course (no increase in creatinine or urea despite complete bilateral renal vein thrombosis) confirm good collateral function of the sonographically depicted retroperitoneal veins, in particular the suprarenal vein and the capsular veins.

**d** Example of a retroperitoneal collateral vein with the typical venous flow signal depicted between the liver and the thrombosed vena cava (hypoechoic tubular structure posterior to the sample volume) in an oblique abdominal view.

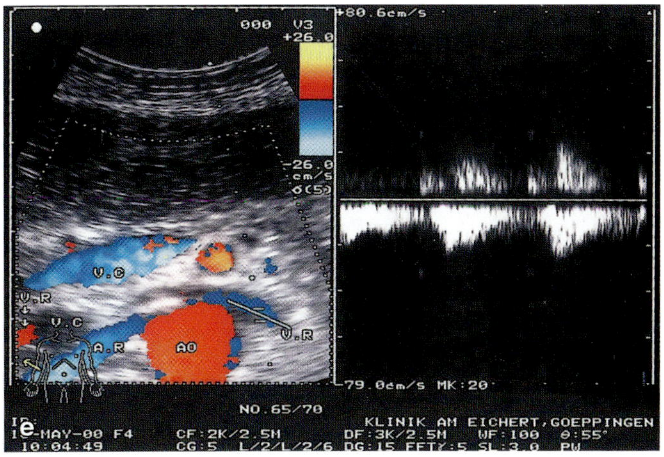


**Fig. A6.67 (continued)**

**e** Color duplex examination 18 days later already demonstrates spontaneous recanalization of both renal veins with flow signals in the formerly completely thrombosed renal vein (*V.R*; see **c**) between the aorta and superior mesenteric artery, also confirmed by the Doppler spectrum. There is also flow in the right renal vein (*V.R*/*arrow*) depicted in red near the left margin.

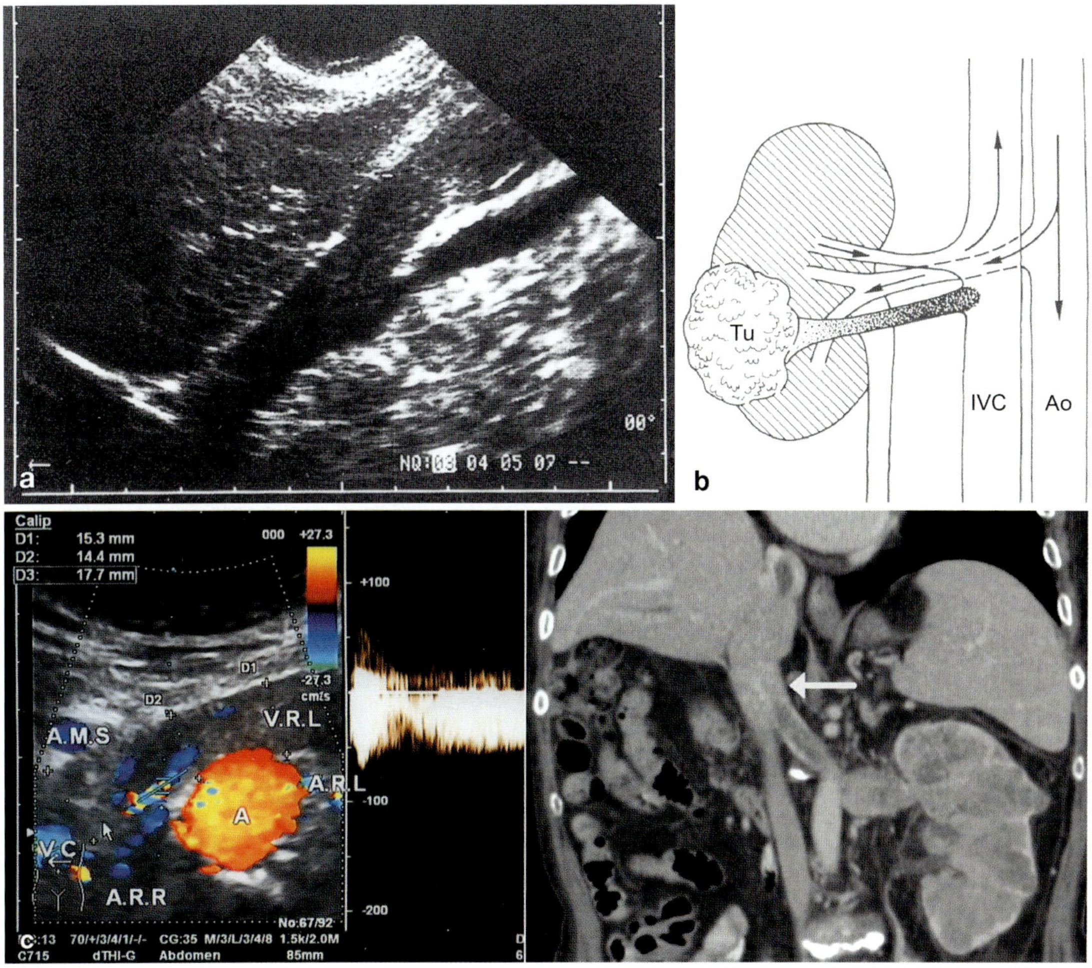


**Fig. A6.68 a–d**

**Renal vein thrombus**

**a** The right renal vein coursing below the liver is occluded by a thrombus protruding into the vena cava (stage I, renal cell carcinoma).

**b** Diagram of renal vein thrombosis due to tumor. Two renal veins are present, and the drawing shows growth of a thrombus from a renal cell carcinoma (*Tu*) into the vena cava (*IVC*) through one of the veins.

**Tumor thrombus ascending in vena cava**

**c** Renal cell carcinoma of the left kidney with nearly complete occlusion of the left renal vein (*V.R.L*) by tumor thrombus. The sample volume is placed in the residual patent lumen next to the thrombus. The thrombus extends into the vena cava (*V.C*, indicated by arrow). (*A.R.L* = left renal artery, *A.R.R* = right renal artery, *A.M.S* = superior mesenteric artery; *A* = aorta). The CT scan confirms the tumor thrombus in the left renal vein (*arrow*) with extension far into the vena cava (retrohepatic).

**Fig. A6.68 (continued)**

**d** Transverse and longitudinal images showing floating tumor thrombus extending from the renal vein into the vena cava (with corresponding time-motion display). The liver can be used as an acoustic window using a lateral intercostal approach, which is why acoustic shadowing from the ribs (*R*) is present (*V.C* = vena cava; *TH* = tumor thrombus, indicated by arrowhead; *L* = liver).

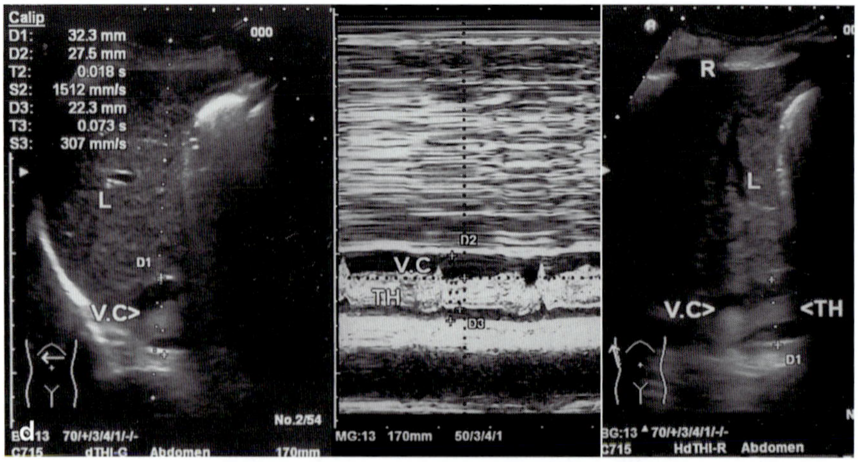


**Fig. A6.69**

**Vena cava compression**

Transverse and longitudinal images showing anterior displacement and compression of the vena cava (*V.C*) by a tumor (*T*). The site of the tumor and its relationship to the vena cava are consistent with a retroperitoneal connective tissue tumor (sarcoma in the case presented) and rule out lymphoma. The latter would be found lateral to the vena cava or aorta or anterior to the vena cava.

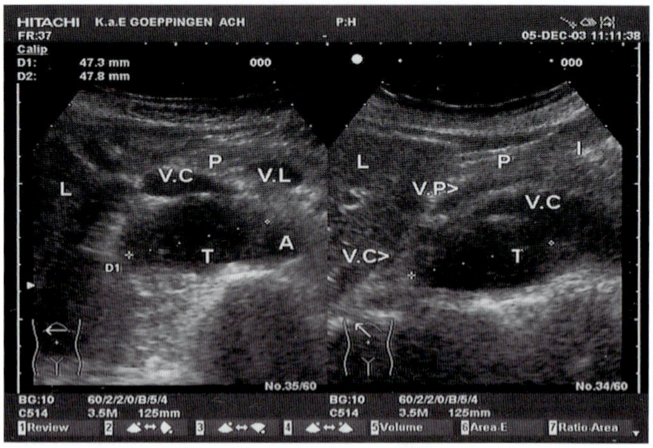


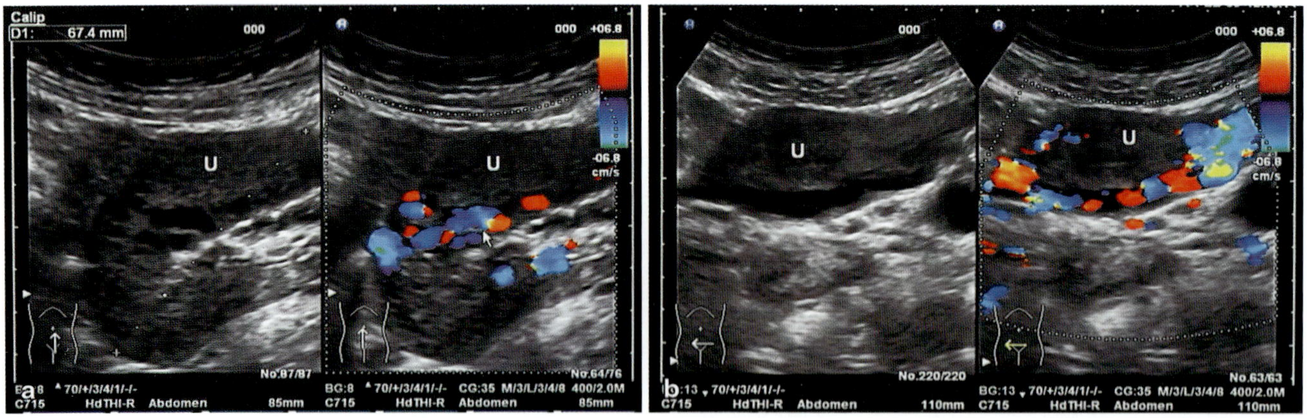


**Fig. A6.70 a–g**

**Varicose ovarian vein in nutcracker syndrome**

**a**, **b** Patient with lower abdominal pain and a sensation of pressure increasing in the evening. The ultrasound examination (B-mode images on the left in **a** and **b**) reveals anechoic tubular structures surrounding the uterus, which correspond to a perfused varicose venous network that communicates with the ovarian vein in the color duplex mode (transverse images in **b** and longitudinal images in **a**). These changes are due to disturbed venous drainage in the left renal vein, which is entrapped between the superior mesenteric artery and the aorta (nutcracker syndrome). As a consequence, venous blood from the kidney flows into the left ovarian vein.

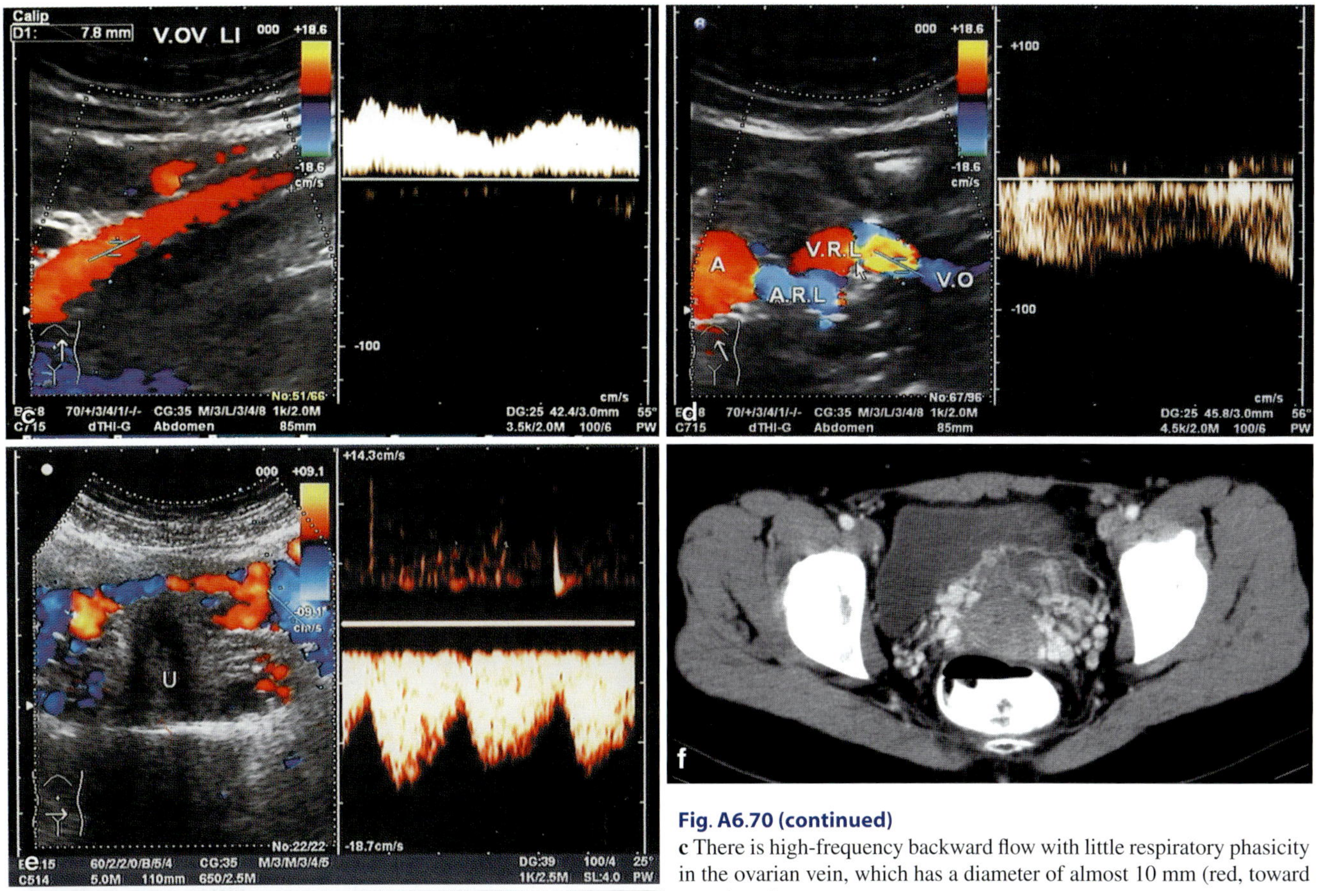


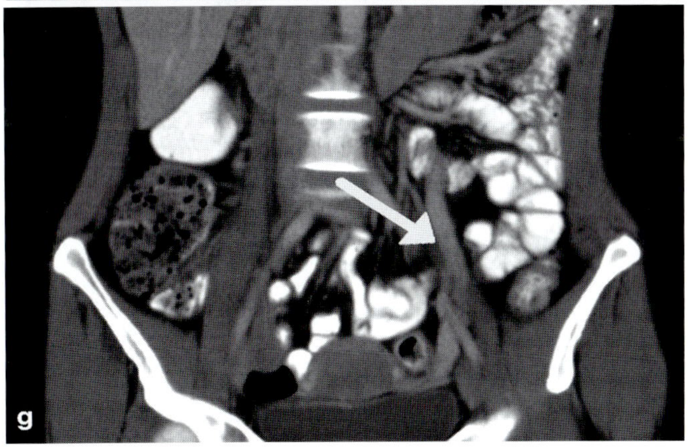


**Fig. A6.70 (continued)**

**c** There is high-frequency backward flow with little respiratory phasicity in the ovarian vein, which has a diameter of almost 10 mm (red, toward transducer).

**d** The ovarian vein (*V.O*, coded in blue) enters the left renal vein (*V.R.L*, red) anterior to the renal artery (*A.R.L*, blue). The Doppler waveform shows continuous retrograde high-frequency flow (into the periphery) in the ovarian vein, while the blood flow direction in the renal vein proximal to the site of entry of the ovarian vein is normal. These flow directions reveal that the high retrograde flow in the dilated ovarian vein is responsible for the varicose degeneration of the periovarian and periuterine veins; laparoscopic ligation of the ovarian vein is indicated.

**e** There is venous flow in all dilated veins.

**f** Axial CT scan showing dilated varicose venous networks surrounding the uterus and ovary.

**g** CT scan showing the dilated ovarian vein between the left renal vein and the ovary but with no other anomalies. CT provides no hemodynamic information.

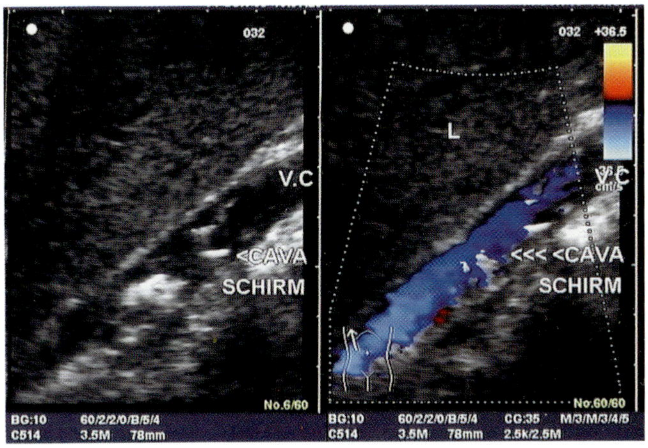


**Fig. A6.71**

**Vena cava umbrella**

Color duplex ultrasound is ideal for confirming proper placement of a vena cava umbrella and patency of the vein.

# Penile and Scrotal Vessels 7

Erectile dysfunction is the lack of copulative power due to failure to initiate an erection or to maintain an erection. The causes of erectile dysfunction include psychologic, neurophysiologic, endocrinologic, and vasculogenic factors. The vascular mechanisms involved in the process of erection are an increased arterial inflow and reduced venous outflow. The incidence of erectile dysfunction due to vascular causes increases with age. About 10% of men suffer from erectile dysfunction.

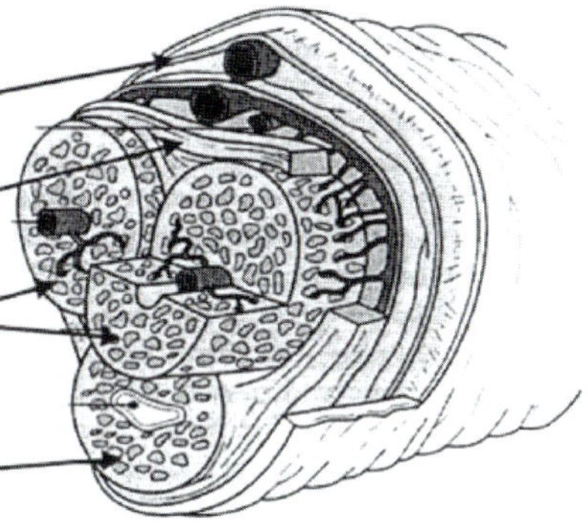


**Fig. 7.1** Anatomy of the erectile bodies of the penis

## 7.1 Vascular Anatomy

### 7.1.1 Penile Vessels

The penis is supplied with blood by the internal pudendal artery, a branch of the internal iliac artery. The internal pudendal artery gives off scrotal branches and continues as the common penile artery, which divides into four terminal branches. These are the urethral or spongiosal artery, the dorsal penile artery (chiefly supplying the skin and glans penis), the bulbourethral artery, and the deep penile artery, which gives off branches to the corpora cavernosa (Fig. 7.1). These branches course centrally in the cavernous bodies, which become rigid and enlarged when arterial inflow through these branches is increased. The other arteries play no relevant role in the erectile process.

Blood from the distal and middle portions of the corpora cavernosa is drained through the emissary and circumflex veins into the periprostatic plexus and from the proximal portion through the deep penile vein into the internal pudendal vein. The superficial dorsal vein of the penis drains primarily into the saphenofemoral junction through subcutaneous veins and the external pudendal veins. The blood from the deep penile vein flows out into the internal pudendal vein and the internal iliac vein (Fig. 7.2).

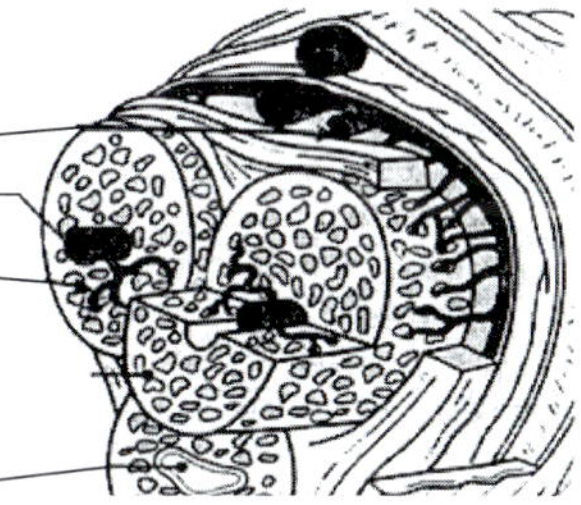


**Fig. 7.2** Arterial supply to the penis with arterial branches and venous drainage

In the flaccid state, the arterioles (helicine arteries) and the sinusoids in the cavernous spaces are contracted. There is only little blood flow due to high peripheral resistance. Conversely, the draining venules (emissary veins) are widely open; they drain the blood over a short stretch via the tunica albuginea into the circumflex vein and dorsal vein (Fig. 7.3a, b).

### 7.1.2 Scrotal Vessels

The paired testicular artery arises from the abdominal aorta and courses from the retroperitoneum to the internal inguinal

W. Schäberle, *Ultrasonography in Vascular Diagnosis*,
DOI: 10.1007/978-3-642-02509-9_7, © Springer-Verlag Berlin Heidelberg 2011

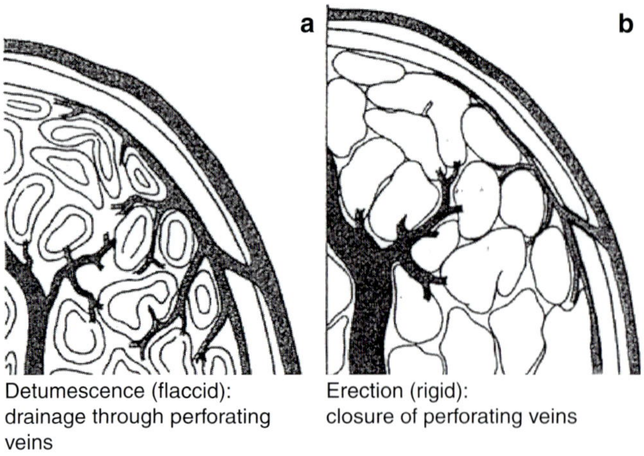


**Fig. 7.3** **a** Mechanism of penile erection: in the flaccid state (detumescence), the helicine arteries and sinusoids are contracted, while the subtonic emissary veins are open. **b** The initial event in penile erection is relaxation of the smooth muscle of the helicine arteries, enabling increasing arterial inflow into the cavernous spaces. At the same time, venous outflow is increasingly blocked by compression of the emissary veins. Adequate closure of cavernosal outflow is crucial for maintaining an erection

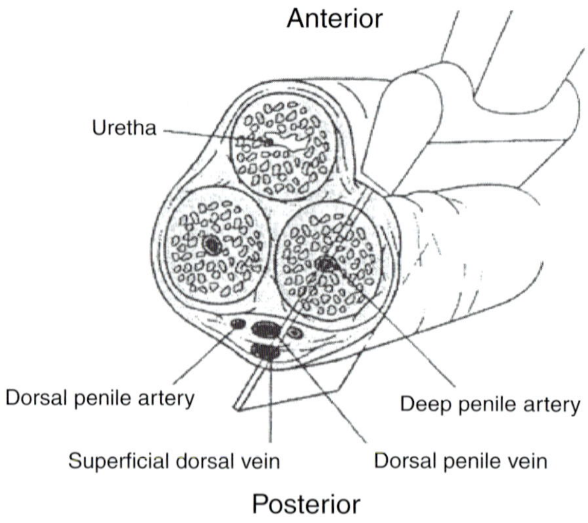


**Fig. 7.4** Transducer position for evaluating arterial and venous perfusion. The penis rests on the patient's lower abdomen (as in erection)

ring, from where it descends through the inguinal canal to the testis, surrounded by the veins of the pampiniform plexus and accompanied by the ductus deferens. The testicular veins drain the blood from the scrotum and testis via the pampiniform plexus. After having emerged from the internal inguinal ring, the right testicular vein courses to the inferior vena cava, the left testicular vein to the left renal vein.

## 7.2 Examination Technique

### 7.2.1 *Erectile Dysfunction*

The superficial course of the penile vessels enables their examination with a high-resolution, high-frequency transducer (7–10 MHz). The pulse repetition frequency and the wall filter should be set to detect slow flow.

With the patient in the supine position and the flaccid penis resting on the lower abdomen, the transducer is placed on the corpus carvernosum from a dorsal approach near the base in transverse orientation (Fig. 7.4). In this position, the B-mode examination is performed with special attention given to the thickness of the tunical albuginea and the penile septum. The color mode is then switched on, and the deep penile artery is identified on both sides with slight angulation of the transducer. In the longitudinal plane, a Doppler spectrum is obtained from both arteries to calculate angle-corrected flow velocities. To depict the slow flow velocities, a low pulse repetition frequency and wall filter are necessary, while the gain must be increased without allowing artifacts to occur.

Before scanning the penile vessels after administration of a vasoactive agent, written consent must be obtained from the patient after information about the examination and its possible side effects. For pharmacological induction of erection, prostaglandin (PGE1) or papaverine is injected into the left and right corpus cavernosum using a very fine needle (Fig. 7.5). To avoid side effects due to overdosage, the examination should be performed on 2 days with increasing dosages, beginning with 40 mg papaverine or 10 μg PGE1. If no adequate erection is achieved, 80 mg papaverine or 20 μg PGE1 should be injected into the right or left corpus cavernosum. Immediate outflow of the injected agent is prevented by short venous compression at the root of the penis. The erect penis is examined 4–5 min after injection by color duplex scanning in the same way as in the flaccid state. Following identification of the deep artery in transverse orientation, a spectral Doppler tracing is obtained from the proximal third of the artery near the base in longitudinal orientation with angle-corrected measurement of peak systolic and minimum end-diastolic flow velocities. The flow velocity measurements are repeated every 2–3 min until full erection is achieved. Finally, venous outflow in the deep femoral vein can be measured (Table 7.1).

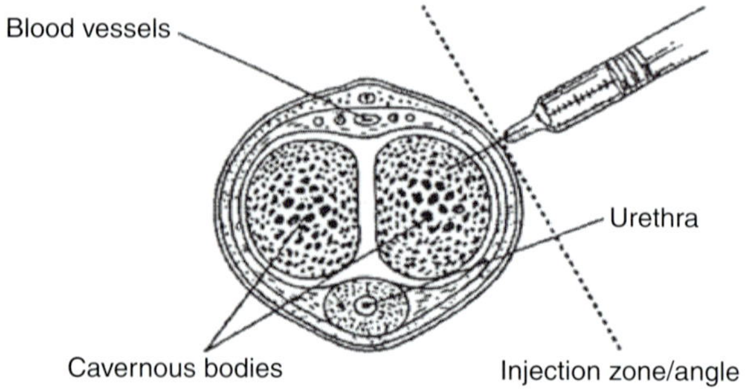


**Fig. 7.5** Technique of intracavernous PGE1 or papaverine injection

**Table 7.1** Pathophysiology of erectile dysfunction

| | |
|---|---|
| 50–70% organic, chiefly vasculogenic | |
| **Arterial** | Stenosis, occlusion (risk factors: atherosclerosis, diabetes mellitus, hypertension, hypercholesterinemia, smoking) |
| **Venous** | Venous leakage |
| Sinusoidal scar | Peyronie's disease, plaques – pain – detumescence |

The pharmacologically induced erection should subside within 4–6 h. Immediate treatment is required in case of longer persistence or priapism, the most dreaded complication of PGE1 injection (Stief et al. 2000). This complication has an incidence of 1–4% and chiefly occurs if the dose is too high and in young patients with psychogenic erectile dysfunction (Wagner and Kaplan 1993). Treatment consists in injecting 5–10 mg Effortil in 5 ml saline solution with a thin needle. If there is progression to priapism, the intracavernous blood must be drained by aspiration.

### 7.2.2 Scrotal Vessels

The principal vessels involved in scrotal and testicular perfusion, the testicular artery and vein, course close to the surface in the inguinal canal and are thus accessible to duplex scanning with a high-resolution transducer. The spermatic artery passes through the abdominal wall together with the ductus deferens at the inner inguinal ring and can be traced sonographically in the B-mode in the inguinal canal, through which it descends to the scrotum as part of the spermatic cord. The testicular artery and vein are identified in the transverse plane in the color duplex mode to then obtain spectral Doppler tracings in longitudinal orientation. In addition, in patients evaluated for the presence of varicocele, the diameters of the veins are determined in longitudinal and transverse planes and the vessels are tracked downward (pampiniform plexus).

## 7.3 Normal Findings

### 7.3.1 Penile Vessels

B-mode images depict the corpus cavernosum as a roundish structure of a fairly homogeneous texture and low echogenicity that is surrounded by the more echogenic tunica albuginea. The septum separating the corpora is also echogenic.

In the flaccid state, the small-caliber deep artery of the penis may be difficult to identify in the B-mode. The Doppler spectrum recorded after identification of the artery in the color duplex mode shows highly pulsatile flow due to high peripheral resistance. Intracavernous injection of PGE1 induces dilatation of the deep artery (seen in the B-mode) with an increase in systolic and diastolic flow velocities (Herbener et al. 1994; Mueller and Lue 1988; Quam et al. 1989). Normal arterial inflow results in a peak systolic velocity (PSV) of more than 30–35 cm/s 5–15 min after injection. The initially low peripheral resistance during the early phase of erection is associated with high diastolic flow (>5–10 cm/s). The physiologic closure of venous drainage during erection leads to increased peripheral arterial resistance in the fully erect penis. This results in a marked decrease in end-diastolic flow, usually to zero (pulsatile flow), 5–25 min after injection. A value of 5 cm/s or above indicates venous insufficiency of the cavernous bodies.

Blood flow in the deep artery of the penis during tumescence is determined by arteriolar vasodilation and the changed pressure in the erectile tissue. Increasing filling of the cavernous spaces by inflowing blood results in a pressure increase in the corpora cavernosa as long as there is proper venous closure. The ensuing increase in arterial flow resistance, in turn, leads to reduced diastolic flow (more pulsatile flow).

### 7.3.2 Scrotal Vessels

The testicular artery has a flow profile typical of a parenchyma-supplying vessel but with a rather small diastolic component. In a study of 30 men, a PSV of 14 cm/s (7.5–27.7 cm/s) and an end-diastolic velocity of 1.9 cm/s (0–4.7 cm/s) were found in the distal testicular artery. The resistance index (Pourcelot) was 0.84 (0.63–1; Middleton et al. 1989).

In the Valsalva test the testicular vein and pampiniform plexus veins exhibit zero flow after a brief reflux. Flow velocity varies with respiration; the normal diameter is less than 2–3 mm (Cvitanic et al. 1993).

## 7.4 Documentation

Transverse B-mode images of the penis depicting both cavernous bodies should be documented. The findings in the deep artery of the penis are documented together with the corresponding angle-corrected time-velocity spectra obtained in longitudinal orientation in the flaccid state and at different stages of erection (at intervals of 2–3 min after PGE1 or papaverine injection). The spectra should document the increase in PSV and end-diastolic velocity following injection as well as the subsequent decrease after full erection has been achieved.

The findings in the testicular artery are recorded at the level of the external inguinal ring in longitudinal orientation together with the spectral waveform. In the same way, the findings in the pampiniform plexus veins are documented with the patient breathing spontaneously and while performing a Valsalva maneuver. The diameter of the veins is measured and recorded in the transverse plane.

## 7.5 Clinical Role of Duplex Ultrasound

### 7.5.1 *Erectile Dysfunction*

Accurate data on the prevalence of erectile dysfunction is difficult to obtain due to underreporting because of embarrassment or because the problem is not considered worthy of medical attention. It is well established that the incidence increases with age (Kinsey et al. 1948). Erectile problems are reported by 2% of men up to 40 years of age, 7% of men aged 41–50, 15% of men aged 51–60, and 75% of those over 70. In a population of 100 men (mean age, 37 years), 7% reported disturbed initiation of erection and another 9%, disturbed maintenance of erection (Frank et al. 1978). Until the end of the 1980s, it was assumed that erectile dysfunction was of psychogenic origin in 90% of cases (Borst 1987).

Various diagnostic modalities such as cavernosometry, cavernosography, CW Doppler, duplex ultrasound, and arteriography are available. They are used to measure intracavernosal pressure or perfusion parameters during erection or to detect arterial flow obstructions and demonstrate relevant organic abnormalities in 50–70% of men presenting with erectile dysfunction (Stief et al. 1988; Tamura et al. 1993; Whitehead et al. 1990). This leaves a percentage of only 30–50% with predominantly psychogenic dysfunction.

Apart from vasculogenic causes, organic erectile dysfunction may be of neurogenic, endocrine, or drug-induced origin. Other causes include lesions of the cavernous bodies such as penile induration (Peyronie's disease). As with diabetes mellitus, vascular changes (macro- and microangiopathy) can aggravate a condition of primarily neurogenic origin (neuropathy of peripheral and autonomous nervous system). Vasculogenic erectile dysfunction may be caused by venous insufficiency of the corpora cavernosa with an incompetent veno-occlusive mechanism or by reduced arterial inflow due to stenosis or occlusion of the internal iliac artery or distal branches such as the pudendal artery (Table 7.2).

Prior to any examination, a detailed history must be obtained to identify possible nonvascular causes of the patient's erectile problems. This includes information on medications taken by the patient as well as a sociopsychological interview pertaining to his social situation with special emphasis on the duration, severity, and character of the erectile dysfunction. Also important is a history of prior diseases and trauma including operations, especially of the true pelvis, in order to obtain clues as to whether the erectile dysfunction is primarily organic or psychogenic in nature. Information must also be obtained on vascular risk factors, and the subsequent clinical examination should be performed focusing on vascular disorders.

**Table 7.2** Sonographic criteria

| | |
|---|---|
| **B-mode scan** (Longitudinal/ transverse) | • Detumescence/erection<br>• Corpus cavernosum, homogeneous echotexture of normal tissue (scars)<br>• Atherosclerotic changes/plaques |
| **Doppler** | • Spectral analysis<br>• PSV<br>• End-diastolic velocity<br>• Resistance index |
| **Course of erection** | • Changes in Doppler waveform (chiefly diastolic component) and resistance index during: detumescence – tumescence – erection – full erection |
| **Pharmacologically induced erection** | • Intracavernous injection: 10 μg PGE1 or 30–60 mg papaverine<br>• Spectral Doppler recordings every 3–5 min<br>Both deep penile arteries and deep dorsal vein<br>Maximum tumescence after 8–20 min |

Identification of the underlying mechanism is of outmost importance for successful treatment of erectile dysfunction. In patients in whom a nonpsychogenic cause has been identified, therapeutic management is initiated according to the algorithm presented in Fig. 7.7 and on the basis of the history, clinical findings, and invasive and noninvasive diagnostic tests. The therapeutic measures may include pharmacologic treatment and, in case of vasculogenic erectile dysfunction, venous resection or arterial revascularization, depending on the duplex sonographic findings.

### 7.5.2 *Acute Scrotum*

In patients presenting with acute scrotum, testicular torsion and inflammatory conditions such as epididymitis must be differentiated, but this may be difficult on the basis of the clinical presentation and history alone. If either of these conditions is suspected, imminent loss of testicular function due to necrosis requires rapid intervention and may necessitate surgical exposure of the testis in inconclusive cases. Complete testicular torsion with arterial obstruction is distinguished from incomplete torsion. In the latter, only venous return is affected, but the testis is also at risk. A noninvasive duplex examination must therefore always include both arterial inflow and venous outflow.

### 7.5.3 Varicocele

About 10–15% of sexually mature men suffer from varicocele, among them 5% with a severe form. Varicocele is a varicose condition of the veins of the pampiniform plexus that can be caused by valve incompetence of the spermatic vein. In addition, vascular anatomy also plays a role in the pathogenesis since the majority of varicoceles occur on the left side. While the right spermatic vein empties directly into the vena cava, the left vein opens into the left renal vein. The higher hydrostatic pressure, compression by the inferior mesenteric artery, and an atypical course are among the factors that may explain why varicocele is more common on the left side. In a study of 45 patients with left-sided varicocele, 25% were found to have a retroaortic course of the renal vein, while a periaortic course was present in 31 cases (Justich 1982; impaired drainage through the renal vein).

Although the exact mechanism of how a varicocele affects the spermiogram is not yet fully understood (hyperthermia, endocrine regulation, hypoxia/adrenal reflux), it is implicated as a cause of infertility as 30–50% of infertile men have a varicocele.

Surgery is indicated only in men with severe varicocele and manifest symptoms, unilateral testicular hypotrophy, or changes in the spermiogram with compromised spermiogenesis.

## 7.6 Abnormal Findings: Role of Duplex Ultrasound Parameters

### 7.6.1 Erectile Dysfunction

B-mode sonography depicts fibrosis in the homogeneously hypoechoic corpus cavernosum as hyperechoic strands. Fibrosis and calcifications of the tunica albuginea (Peyronie's disease) are reliably identified by thickening and an increased echogenicity. Extensive obstructive disease of the upstream arteries (both iliac arteries or pudendal artery) affect the flow character in the deep penile artery (postocclusive flow profile with decreased pulsatility) and thus suggest a vasculogenic cause of erectile dysfunction already in the flaccid state.

Injection of PGE1 or papaverine usually induces an increase in blood flow within 5–10 min (early tumescence phase) by lowering arterial resistance. As a result, PSV increases to more than 30–35 cm/s and end-diastolic velocity increases as well, while the Pourcelot index decreases. A pharmacologically induced PSV of less than 30 cm/s in the deep penile artery suggests inadequate arterial inflow. Studies demonstrated that a PSV below 25 cm/s after intracavernous injection was due to vascular obstruction, which was confirmed as the cause of erectile dysfunction in 88–100% of the patients by control angiography (Benson et al. 1993; Desai and Gilbert 1991; Quam et al. 1989). In another study of 42 patients with clinical signs of vasculogenic impotence, color duplex sonography was found to have a sensitivity of 82% and a specificity of 88% compared with selective penile DSA (Brandstetter et al. 1993). A delayed systolic upstroke in the Doppler waveform is another sign of proximal flow obstruction.

**Table 7.3** Erectile dysfunction (arterial – venous)

| | |
|---|---|
| **Erectile dysfunction** | |
| Arterial incompetence: | |
| PSV | <25 cm/s: severe arterial incompetence<br>Acceleration time increased to over 120 ms<br>25–30 cm/s: moderate arterial incompetence |
| Note: Limited diagnostic accuracy in nervous patients and in psychogenic erectile dysfunction | |
| **Venous erectile dysfunction** | |
| Clinical presentation: poor rigidity despite normal arterial function | |
| Duplex ultrasound: | Persistent diastolic flow >5 cm/s<br>Resistance index (RI) <1<br>Relevant flow in deep dorsal vein |
| Note: Examination of venous competence only in patients with normal arterial function | |
| Duplex scan provides only a preliminary diagnosis of venous incompetence, which needs to be confirmed by cavernosometry/cavernosography | |

If arterial inflow appears normal, the failure to achieve or maintain an erection may be due to dysfunction of the veno-occlusive mechanism (Table 7.3). In the intact state, the rise in intracavernous pressure leads to an increase in peripheral resistance with a decrease in the diastolic flow component. Conversely, if there is venous cavernous insufficiency, increased venous outflow results in persistent end-diastolic flow. An end-diastolic flow velocity of more than 5 cm/s indicates venous leakage with a sensitivity of over 90% (Quam et al. 1989). The magnitude of the end-diastolic flow velocity following maximum penile tumescence after intracavernous injection of a vasoactive agent correlates with venous outflow resistance and thus with the severity of venous leakage. Duplex sonography is a highly accurate tool for determining hemodynamic parameters during pharmacologically induced erection in the diagnostic evaluation and differentiation of vasculogenic causes of erectile dysfunction.

Based on the findings of the duplex scan, further invasive diagnostic tests may be indicated according to the algorithm for step-by-step diagnostic management presented in Fig. 7.6.

Treatment depends on whether initial intracavernous self-injection (ICSI) is successful or not (Fig. 7.7). Duplex measurements assist in choosing the most suitable therapy.

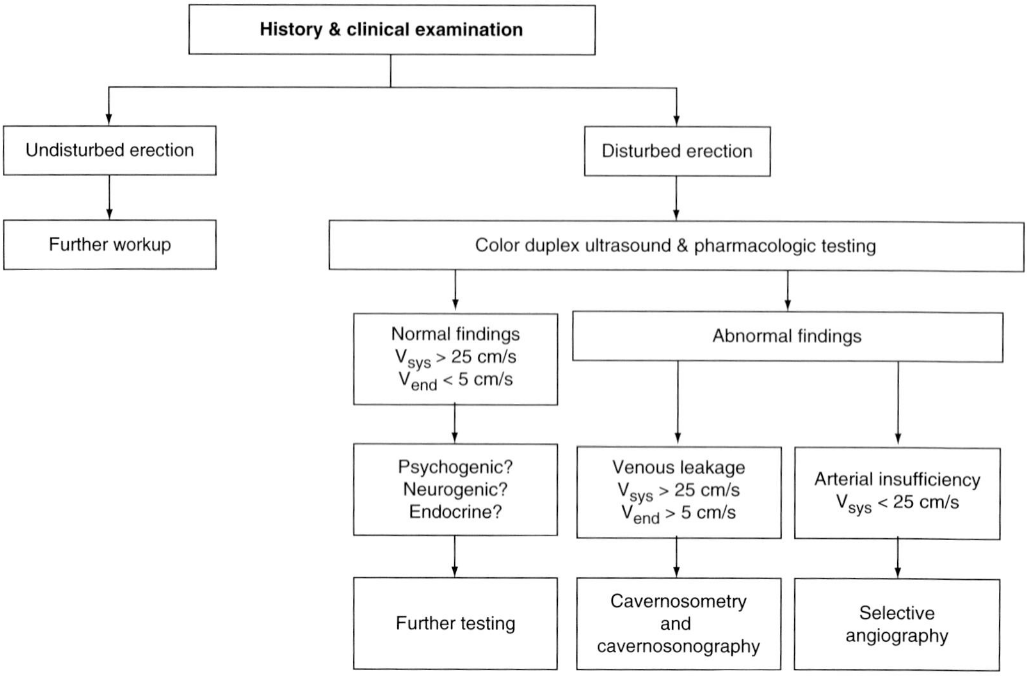


**Fig. 7.6** Algorithm for the diagnostic management of patients with erectile dysfunction. Systolic velocity ($V_{sys}$) and end-diastolic velocity ($V_{end}$) are measured by color duplex ultrasound

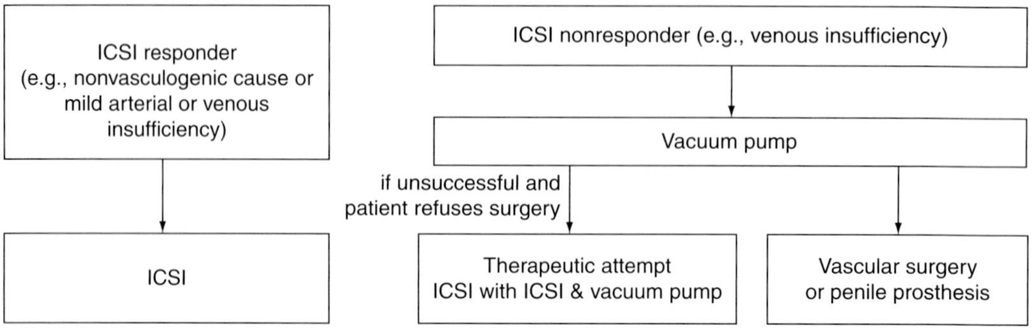


**Fig. 7.7** Algorithm for the therapeutic management of patients with erectile dysfunction (*ICSI* intracavernous self-injection)

## *7.6.2 Acute Scrotum*

In prepubertal boys, arterial flow is more difficult to detect due to the small testicular volume and slow blood flow; the scanner controls must be set to depict low frequency shifts, which means a low pulse repetition frequency and an adjusted gain. The normal testis appears uniform and is surrounded by a more echogenic capsule (tunica albuginea). Flow in the intratesticular segments of the testicular artery is monophasic with antegrade diastolic flow (low-resistance flow).When Doppler measurement is performed between the external inguinal ring and the testis, care must be taken not to confuse the testicular artery with one of the supratesticular arteries, which supply the testicular coverings and cremaster and therefore lack a diastolic flow component (higher diastolic resistance). In testicular torsion, the extent to which perfusion is compromised depends on the duration and severity. In incomplete torsion, the Doppler waveform is abnormal, but often shows some residual perfusion. In testicular torsion, spectral Doppler evaluation of the testicular artery is supplemented by color duplex imaging (low pulse repetition frequency) of the affected testis to evaluate arterial and venous flow. Spectral Doppler measurement of arterial and venous perfusion is necessary to rule out incomplete testicular torsion (acceleration time, flow velocity, side-to-side comparison).

Venous spectral Doppler measurement is necessary to confirm unimpaired venous drainage and rule out incomplete testicular torsion (which is characterized by unimpaired arterial inflow with impaired venous drainage).

In patients presenting with acute scrotum, duplex ultrasound has a specificity of nearly 100% for ruling out acute ischemia (Fitzgerald and Foley 1991). Several studies performed in small study populations found sensitivities of 86–100% and specificities of 100% in differentiating acute testicular or epididymal inflammation with hyperemia from testicular torsion with signs of acute ischemia (De Wire et al. 1992; Lerner et al. 1990; Middleton et al. 1990; Ralls et al. 1990). Inflammation is associated with low peripheral resistance, resulting in high diastolic flow. Conversely, testicular torsion is characterized by the absence of arterial flow in the twisted vessel segment and highly pulsatile flow without a diastolic flow component and a reduced PSV (compared to contralateral side) with a "knocking" waveform proximal to the lesion (in the inguinal ligament). The studies reported in the literature rarely address the problem of incomplete testicular torsion with impaired venous drainage only. To exclude testicular damage resulting from venous compromise, the vein must be evaluated by color duplex imaging from the scrotal compartment to the level of the inguinal ligament. The normal waveform shows flow with respiratory phasicity.

Following spontaneous or manual detorsion after an ischemic interval, the compensatory increase in testicular perfusion can be detected by spectral Doppler measurement. Appendiceal torsion also presents with acute onset of pain; ultrasound demonstrates the twisted appendage as a mass adjacent to the echogenic testis or epididymis. Color duplex imaging demonstrates increased perfusion in adjacent testicular and epiddidymal tissue. Perfusion is also increased in epididymitis, and the increase can be demonstrated by a color duplex examination and comparison with the contralateral epididymis.

In a study of 31 patients, a cutoff value of 15 cm/s for PSV was found to have an accuracy of 90% for identifying orchitis and of 93% for epididymitis (Brown et al. 1995). Another parameter used was the ratio of PSVs on both sides, with a ratio of greater than 1.9 being defined as indicating epididymitis or orchitis.

Taken together, published studies suggest that ultrasound is a highly valid diagnostic modality for the evaluation of testicular and penile perfusion and its abnormal changes. These reports however remain to be confirmed in larger patient populations. In unclear cases surgical exposure of the testis is still necessary.

### 7.6.3 Varicocele

Varicoceles are conspicuous as a convolution of veins in the scrotum. Dilatation of the veins of the pampiniform plexus with a diameter of over 3 mm is considered abnormal. The Valsalva test performed in the standing patient will induce continuous backward flow in the abnormally dilatated veins, which can be demonstrated both by color duplex sonography and in the Doppler waveform (Fitzgerald and Foley 1991). In a study of 63 infertile men, color duplex was found to be highly accurate in the diagnosis of varicoceles with a sensitivity of 97% and a specificity of 94% compared to venography of the spermatic vein (Trum et al. 1996). However, the clinical significance of a sonographically diagnosed varicocele must not be overestimated. In an investigation of 26 fertile men, 42% were found to have dilatated pampiniform plexus veins with diameters of more than 2–3 mm and signs of reflux (Cvitanic et al. 1993; Table 7.4).

**Table 7.4** Sonographic criteria of varicocele

| |
|---|
| • Testicular size (>2 mL difference) |
| • Plexus veins (>3 mm diameter) |
| • (Color) duplex ultrasound: reflux during normal respiration in the standing patient |

## 7.7 Atlas: Penile and Scrotal Vessels

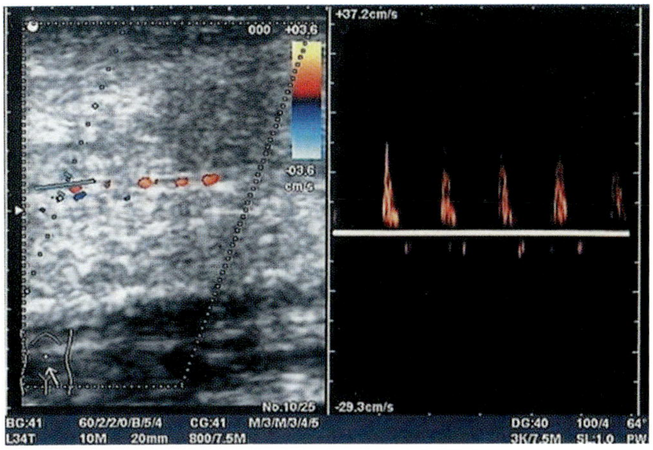


**Fig. A7.1**
**Detumescence**
In the flaccid state (detumescence), the deep artery of the penis demonstrates high-resistance flow with pulsatile systolic peaks but no significant diastolic flow.

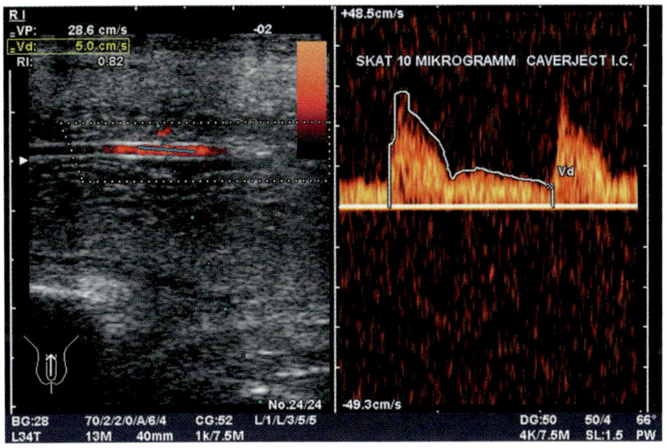


**Fig. A7.2**
**Doppler waveform following prostaglandin injection**
Markedly increased blood flow, especially in diastole, in the deep penile artery 10–15 min following injection of 10 μg PGE1 and relaxation of the smooth muscle of the sinusoids (low-resistance arterial inflow) (courtesy of F. Trinkler).

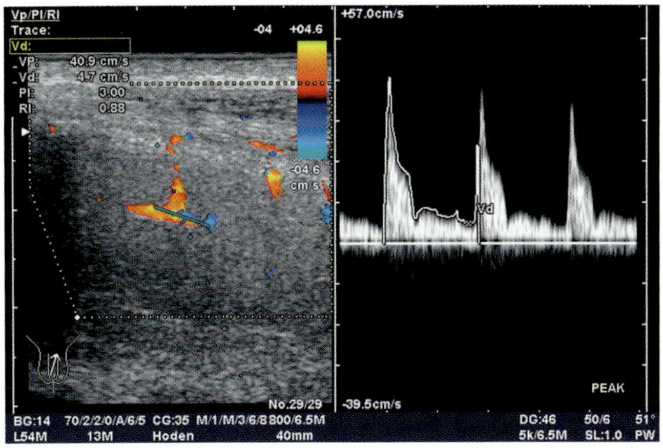


**Fig. A7.3**
**Doppler waveform – onset of erection**
With increasing erection brought on by continuous high arterial inflow, the sinusoids become filled, causing a build-up of counterpressure in the corpus cavernosum. As a result, peripheral resistance increases, and flow becomes more pulsatile. The diastolic flow component decreases and approaches zero in the further course. A peak systolic velocity >30 cm/s indicates normal arterial blood supply. In the example, PSV is 40 cm/s (courtesy of F. Trinkler).

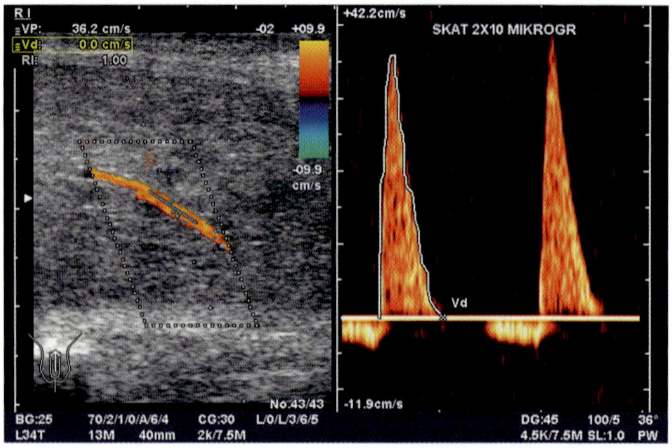


**Fig. A7.4**
**Doppler waveform – full erection**
Flow decreases again due to the high intracavernous pressure in full erection with absence of flow or retrograde flow during diastole. This is associated with a decrease in PSV. No flow is detected in the deep vein of the penis (courtesy of F. Trinkler).

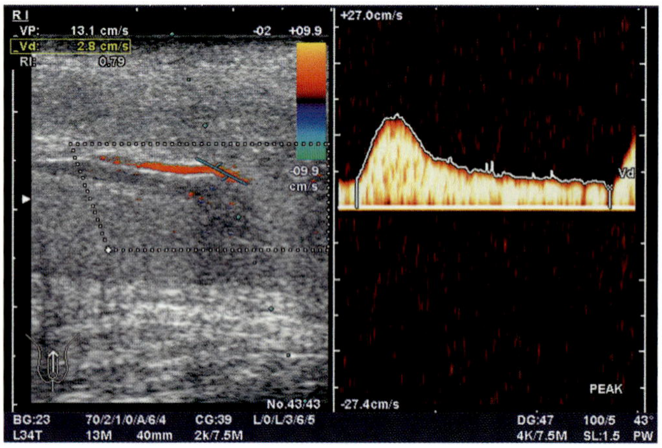


**Fig. A7.5**
**Doppler waveform – arterial insufficiency**
Inadequate arterial inflow is characterized by a smaller increase in PSV in the deep penile artery. In severe insufficiency, PSV drops below 25 cm/s. In the case presented, intracavernous injection of 10 µg PGE1 does not induce erection, and there is no adequate increase in flow during systole after a reasonable delay (5–15 min). A PSV of only 12 cm/s, a delayed systolic upsurge (prolonged acceleration time), and a larger diastolic flow component are typical signs of postocclusive flow. Here, the postocclusive flow is caused by upstream atherosclerotic stenoses. Because of the patient's high-grade arterial insufficiency, it is not possible to reliably determine whether there is concomitant venous leakage.

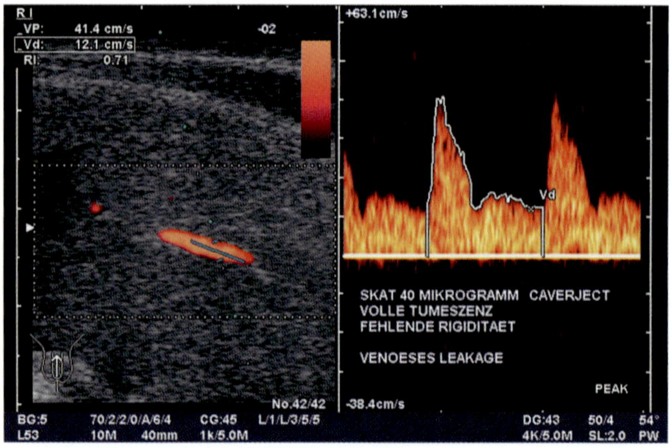


**Fig. A7.6**
**Doppler waveform – venous leakage**
In patients with venous leakage, rigidity is inadequate although full tumescence is achieved. Following intracavernous PGE1 injection, the Doppler spectrum from the deep penile artery demonstrates an adequate systolic increase with a flow velocity of 41 cm/s but no reduction during diastole. The high diastolic flow velocity indicates low peripheral resistance to venous outflow.

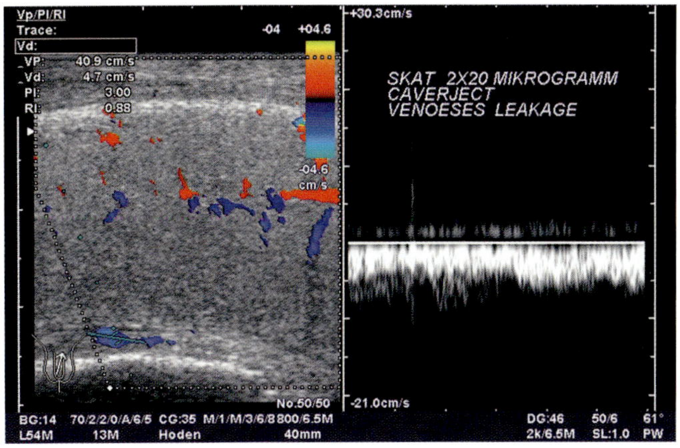


**Fig. A7.7**
**Venous insufficiency of corpora cavernosa**
Under normal conditions, no venous flow signal is obtained from the deep dorsal vein in the phase of full tumescence. In the patient presented, venous leakage is suggested by the demonstration of venous flow with a velocity of 10–20 cm/s.

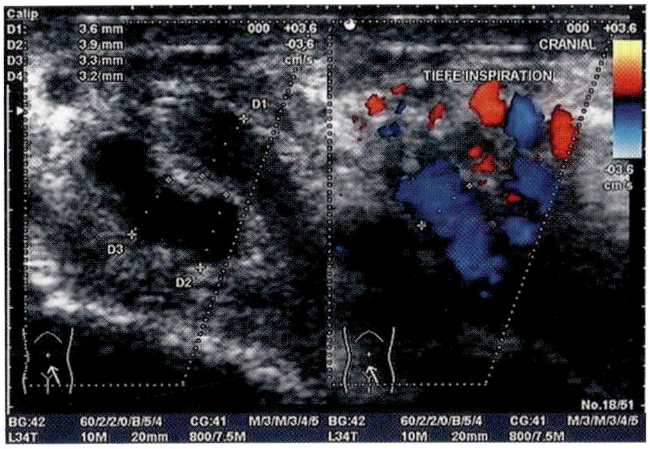


**Fig. A7.8**
**Varicocele**
Varicocele is identified by duplex sonography as dilatation of the veins of the pampiniform plexus to over 3 mm (*left section*), along with backward flow toward the testes during deep inspiration or Valsalva's maneuver (color duplex scan, *right section*).

# References


Abou-Zamzam AM Jr, Moneta GL, Edwards JM, Yeager RA, Taylor LM Jr, Porter JM (2000) Is a single preoperative duplex scan sufficient for planning bilateral carotid endarterectomy? J Vasc Surg 31:282–288

AbuRahma AF, Abu-Halimah S, Bensenhaver J, Dean LS, Keiffer T, Emmett M, Flaherty S (2008) Optimal carotid duplex velocity criteria for defining the severity of carotid in-stent restenosis. J Vasc Surg 48 589–594

AbuRahma AF, Burns W, Mullins DA (1999) Carotid artery dissection: a challenging diagnosis. W V Med J 95:17–19

AbuRahma AF, Covelli MA, Robinson PA, Holt SM (1999) The role of carotid duplex ultrasound in evaluating plaque morphology: potential use in selecting patients for carotid stenting. J Endovasc Surg 6:59–65

AbuRahma AF (2006) Fate of endoleaks detected by CT angiography and missed by color duplex ultrasound in endovascular grafts for abdominal aortic aneurysms. J Endovasc Ther 13:490–495

AbuRahma AF, Robinson P, Decanio R (1989) Prospective clinicopathologic study of carotid intraplaque hemorrhage. Am Surg 55:169–173

AbuRahma AF, Richmond BK, Robinson PA et al (1995) Effect of contralateral severe stenosis or carotid occlusion on duplex criteria of ipsilateral stenoses: comparative study of various duplex parameters. J Vasc Surg 22:751–762

AbuRahma AF, Robinson PA, Killmer SM, Kioschos JM, Roberts MD (1996) A critical analysis of cerebral computed tomography scanning before elective carotid endarterectomy and its correlation to carotid stenosis. Surgery 119:248–251

AbuRahma AF, Kyer PD, Robinson PA, Hannay RS (1998) The correlation of ultrasonic carotid plaque morphology and carotid plaque hemorrhage: clinical implications. Surgery 124:721–728

AbuRahma AF, Robinson PA, Stickler DL et al (1998) Proposed new duplex classification for threshold stenoses used in various symptomatic and asymptomatic carotid endarterectomy trials. Ann Vasc Surg 12:349–358

AbuRahma AF, Wulu JT Jr, Crotty B (2002) Carotid plaque ultrasonic heterogeneity and severity of stenosis. Stroke 33:1772–1775

ACAS (1995) Asymptomatic carotid atherosclerosis study. JAMA 273:1459–1461

Agrawal SK, Pinheiro L, Roubin GS et al (1992) Nonsurgical closure of femoral pseudoaneurysms complicating cardiac catheterization and percutaneous transluminal coronary angioplasty. J Am Coll Cardiol 20:610–615

Ahmad S, Blagg CR, Scribner BH (1998) Center and home chronic hemodialysis. In: Schrier RW, Gottschalk CW (eds) Diseases of the kidney, 4th edn. pp 3281–3322

Aitken AGF, Godden DJ (1987) Real-time ultrasound diagnosis of deep venous thrombosis: a comparison with venography. Clin Radiol 38:309–313

Akbari CM, LoGerfo FW (1999) Diabetes and peripheral vascular disease. J Vasc Surg 30:373–384

Akkersdijk GJM, Puylaert JBCM, Vries A (1991) Abdominal aortic aneurysm as an incidental finding in abdominal ultrasonography. Br J Surg 78:1261–1263

Alanen A, Kormano M (1985) Correlation of the echogenicity and structure of clotted blood. J Ultrasound Med 4:421–425

Albrecht T, Urbank A, Mahler M et al (1998) Prolongation and optimization of Doppler enhancement with a microbubble US contrast agent by using continuous infusion: preliminary experience. Radiology 207:339–347

Albrecht T, Hoffmann CW, Schettler S et al (2000) B-mode enhancement at phase-inversion US with air-based microbubble contrast agent: initial experience in humans. Radiology 216:273–278

Albrechtson LL, Olson QCG (1976) Thrombotic side effects of lower limb phlebography. Lancet II:723

Aldridge SC, Comerota AJ (1993) Popliteal venous aneurysm: report of two cases and review of the world literature. J Vasc Surg 18:708–715

Alexandrov AV, Brodie DS, McLean A et al (1997) Correlation of peak systolic velocity and angiographic measurement of carotid stenosis revisited. Stroke 28:339–342

Alexandrov AV, Vital D, Brodie DS, Hamilton Paul Grotta JC (1997) Grading carotid stenosis with ultrasound. Stroke 28:1208–1210

Allard L, Cloutier G, Durand L-G, Roederer GO, Langlois YE (1994) Limitation of ultrasonic duplex scanning for diagnosing lower limb arterial stenoses in the presence of adjacent segment disease. J Vasc Surg 19:650–657

Allenberg JR et al (1995) Endovascular reconstruction of infrarenal abdominal aortic aneurysm. Chirurg 66:870–877

Allenberg JR, Kallinowski F, Schumacher H (1997) Stand der Chirurgie des infrarenalen Aortenaneurysmas: Prävalenz und Versorgungssituation. Dt Ärztebl 94:A2830–A2834

Allon M, Robbin ML (2002) Increasing arteriovenous fistulas in hemodialysis patients: problems and solutions. Kidney Int 62:1109–1124

Alri M, Herba MJ, Reinhold C, Leclerc J, Ye S, Illescas FF, Bret PM (1996) Accuracy of sonography in the evaluation of calf deep vein thrombosis in both postoperative surveillance and symptomatic patients. AJR 166:1361–1367

Alson MD, Lang EV, Kaufman JA (1997) Pedal arterial imaging. Vasc Interv Radiol 8:9–18

Aly S, Sommerville K, Adiseshiah M (1998) Comparison of duplex imaging and angiography in evaluation of lower limb arteries. Br J Surg 85:1099–1102

Anderson IC, Baltaxe HA, Wolf GL (1979) Inability to show clot. One limitation of ultrasonography of the abdominal aorta. Radiology 132:693

Anjaria PD, Vaidya PN, Vahia VN et al (1973) Venous aneurysms. J Postgrad Med 20:142–144

Äppel RG, Bleyer AJ, Reavis S, Hansen KJ (1995) Renovascular disease in older patients beginning renal replacement therapy. Kidney Int 48:171–176

Appelman PT, de Jong TE, Lampman LE (1987) Deep venous thrombosis of the leg. Ultrasound findings. Radiology 163:743–746



W. Schäberle, *Ultrasonography in Vascular Diagnosis*,
DOI: 10.1007/978-3-642-02509-9, © Springer-Verlag Berlin Heidelberg 2011

Aprin H, Schwartz GB, Valderamma E (1987) Traumatic venous aneurysm. Clin Orthop 217:243–246

Araki CT, Back TL, Padberg FT, Thompson PN et al (1993) Refinements in the ultrasonic detection of popliteal vein reflux. J Vasc Surg 18:742–748

Arko FR (2003) Intrasac flow velocites predict sealing of type II endoleaks after endovascular abdominal aortic aneurysm repair. J Vasc Surg 37:8–15

Arlart IP, Ingrisch H (1984) Renovaskuläre Hypertonie. Radiologische Diagnostik und Therapie. Thieme Stuttgart

Arning C (2001) Nonatherosclerotic disease of the cervical arteries: role of ultrasonography for diagnosis. VASA 30:160–167

Arning C (2004) Die Karotidynie im Ultraschallbild: Mythos, Syndrom oder Krankheitsbild? Nervenarzt 75:1200–1203

Arning C (2005) Ultrasonographic criteria for diagnosing a dissection of the internal carotid artery. Ultraschall Med 26:24–28

Arning C, Grzyska U (2004) Color Doppler imaging of cervicocephalic fibromuscular dysplasia. Cardiovasc Ultrasound 2:7

Arning C, Oelze A, Lachenmayer L (1995) A rare cause of stroke: aortic dissection. Aktuell Neurol 22:189–192

Arning C, Hammer E, Kortmann H, Hahm H, Muller-Jensen A, Lachenmayer L (2003) Quantifizierung von A. carotis interna-Stenosen: Welche Ultraschallkriterien sind geeignet? Ultraschall Med 24:233–238

Arning C, von Reutern GM, Stiegler H, Görtler M (2010) Ultraschallkriterien zur Graduierung von Stenosen der A. carotis interna -Revision der DEGUM-Kriterien und Transfer in NASCET-Stenosierungsgrade [Revision of DEGUM ultrasound criteria for grading internal carotid artery stenoses and transfer to NASCET measurement]. Ultraschall in Med 31:251–257

Arnold JA, Modaresi KB, Thomas N, Taylor PR, Padayachee TS (1999) Carotid plaque characterization by duplex scanning: observer error may undermine current clinical trials. Stroke 30:61–65

Ascer E, Lorenso E, Pollina RM, Gennaro M (1995) Preliminary results of a nonoperative approach to saphenofemoral junction thrombophlebitis. J Vasc Surg 22:616–621

Ascer E, Pollina RM, Gennaro M, Lorensen E (1995) Noninvasive predictors of patency for infrapopliteaf PTFE bypasses with combined arteriovenous fistula and vein interposition technique. Am J Surg 170:103–105

Ascer E, Mazzariol F, Hingorani A, Dalles-Cunba S, Gade P (1999) The use of duplex ultrasound arterial mapping as an alternative to conventional arteriography for primary and secondary infrapopliteal bypasses. Am J Surg 178:162–165

Aschwanden M, Hess P, Labs KH et al (2003) Dialysis access-associated steal syndrome: the intraoperative use of duplex ultrasound scan. J Vasc Surg 37:211–213

Aschwanden M, Thalhammer C, Schaub S, Wolff T, Steiger J, Jaeger KA (2006) Renal vein thrombosis after renal transplantation - early diagnosis by duplex sonography prevented fatal outcome. Nephrol Dial Tranplant 21:825–826

Ashoke R, Brown LC, Rodway A, Choke E, Thompson MM, Greenhalgh RM, Powell JT (2005) Color duplex ultrasonography is insensitive for the detection of endoleak after aortic endografting: a systematic review. J Endovasc Ther 12:297–305

Atri M, Herba MJ, Reinhold C, Leclerc J et al (1996) Accuracy of sonography in the evaluation of calf deep vein thrombosis in both postoperative surveillance and symptomatic patients. Am J Roentgenol 166:1361–1367

Aube C, Oberti F, Korali et al (1999) Ultrasonographic diagnosis of hepatic fibrosis or cirrhosis. J Hepatol 30:472–478

Aune S, Pedersen OM, Trippestad A (1998) Surveillance of above-knee prosthetic femoropopliteal bypass. Eur J Vasc Endovasc Surg 16:509–512

Avasthi PS, Voyles WF, Er G (1984) Noninvasive diagnosis of renal artery stenosis by echo-Doppler velocimetry. Kidney Int 25:824–829

Avenarius JKA, Breek JC, Lampmann LEH, Lohle PNM, van Berge Henegouwen DP, Hamming JF (2002) The additional value of angiography after colour-coded duplex on decision making in patients with critical limb ischaemia. A prospective study. Eur J Vasc Endovasc Surg 23:393–397

AWMF-Leitlinien. Arteriitis cranialis. http://www.dgn.org/180.0.html

Baker SR, Burnard KG, Sommerville KM, Thomas ML, Wilson NM (1993) Comparison of venous reflux assessed by duplex scanning and descending phlebography in chronic venous disease. Lancet 341:400–403

Baker WH, Stoney RJ (1972) Acquired popliteal entrapment syndrome. Arch Surg 105:780–781

Baldassarre D, Amato M, Bondioli A, Sirtori CR, Tremoli E (2000) Carotid artery intima-media thickness measured by ultrasonography in normal clinical practice correlates well with atherosclerosis risk factors. Stroke 31:2426–2430

Balk E, Raman G, Chung M et al (2006) Effectiveness of management strategies for renal artery stenosis: a systematic review. Ann Intern Med 145:901–912

Bandyk DF, Cato RF, Towne JB (1985) A low flow velocity predicts failure of femoropopliteal and femorotibial bypass grafts. Surgery 98:799–809

Bandyk DF, Scabrook GR, Moldenauer P et al (1988) Hemodynamics of vein graft stenoses. J Vasc Surg 8:688–695

Bandyk DF, Schmitt DD, Seabrook GR (1989) Monitoring functional patency of in situ saphenous vein bypasses: the impact of a surveillance protocol and elective revision. J Vasc Surg 9:286–296

Bärlin E, Schäberle W, Junge H, Seitz K, Rettenmaier G (1988) In-vitro-Untersuchung zur Meßgenauigkeit der mittleren Blutflußgeschwindigkeit bei Duplex-Geräten. Ultraschall Klin Prax Suppl 1:69 (Abstr)

Barnes RW (1985) Doppler ultrasonic diagnosis of venous disease. In: Bernstein EF (ed) Noninvasive diagnostic techniques in vascular disease. Mosby, St. Louis, p 344

Barnes RW (1991) Noninvasive diagnostic assessment of peripheral vascular disease. Circulation 83(Suppl 2):120–127

Barnes RW, Wu KK, Hoak JC (1975) Fallibility of the clinical diagnosis of venous thrombosis. J Am Med Assoc 234:605

Barnes RW, Nix ML, Barnes CL, Lavender RC et al (1989) Perioperative asymptomatic venous thrombosis: role of duplex scanning versus venography. J Vasc Surg 9:251–260

Barrelier MT (1993) Superficial venous thromboses of the legs. Phlebologie 46:633–639

Bartels E (1992) Farbkodierte Dopplersonographie der Vertebralarterien. Vergleich mit der konventionellen Duplexsonographie. Ultraschall Med 13:59–66

Bartels E, Flügel KA (1993) Advantages of color Doppler imaging for the evaluation of the vertebral arteries. J Neuroimaging 3:229–233

Barwegen MGMH, van Dongen RJAM (1987) Neurovaskuläre Kompressionssyndrome an der oberen Thoraxapertur und ihre vaskulären Komplikationen. In: Heberer G, Van Dongen RJAM Hrsg (eds) Gefäßchirurgie. Springer, Berlin/Heidelberg/New York/Tokyo, pp S 571–S 584, Kirschnersche allgemeine und spezielle Operationslehre, Bd 11

Bassiouny HS, Sakaguchi Y, Mikucki SA et al (1977) Juxtalumenal location of plaque necrosis and neoformation in symptomatic carotid stenosis. J Vasc Surg 26:585–594

Baumgartner I, Maier SE, Koch M et al (1993) Magnetresonanzarteriographie, Duplexsonographie und konventionelle Arteriographie zur Beurteilung der peripheren arteriellen Verschlußkrankheit. Fortschr Röntgenstr 159:167–173

Baxter GM (2002) Imaging and renal transplantation. Imaging14: 285–298

Baxter GM, McKechnie S, Duffy P (1990) Colour Doppler ultrasound in deep venous thrombosis: a comparison with venography. Clin Radiol 42:32–36

Baxter GM, Aitchison F, Sheppard D et al (1996) Colour Doppler ultrasound in renal artery stenosis: intrarenal waveform analysis. Br J Radiol 69:810–815

Bay WH, Henry ML, Lazarus JM et al (1998) Predicting hemodialysis access failure with color flow Doppler ultrasound. Am J Nephrol 18:296–304

Beach KW (1992) 1975–2000: a quarter century of ultrasound technology. Ultrasound Med Biol 18:377–388

Becker D, Strobel D, Hahn EG (2000) Tissue harmonic imaging und contrast harmonic imaging. Verbesserung der Diagnose von Lebermetastasen. Internist 41:17–23

Becker HM, Kortmann H (1987) Nahtaneurysmen. In: Heberer G, Van Dongen RJAM (eds) Gefäßchirurgie. Springer, Berlin/Heidelberg/New York/Tokyo (Kirschnersche allgemeine und spezielle Operationslehre, Bd 11), pp 178–184

Beebe HG, Salles-Cunha SX, Scissons RP et al (1999) Carotid arterial ultrasound scan imaging: a direct approach to stenosis measurement. J Vasc Surg 29:838–844

Beil PRF, Brennan J (1991) Vein graft surveillance by duplex scanning and pressure measurement. In: Greenhalgh RM (ed) The maintenance of arterial reconstruction. Saunders WB, London, p 135

Belcaro G (1992) Evaluation of recurrent deep venous thrombosis using colour duplex scanning in comparison with venography. Vasa 21:22–26

Belcaro G, Nicolaides AN, Veller M (1995) Venous disorders. A manual of diagnosis and treatment. Saunders, London/Philadelphia/Toronto, p 99

Belkin M, Raftery KB, Mackey WC, McLaughlin RL, Umphrey SE, Kunkemueller A (1994) A prospective study of the determinants of vein graft flow velocity: implications for graft surveillance. J Vasc Surg 19:259–267

Bendick PJ et al (2003) Efficacy of ultrasound scan contrast agents in the noninvasive follow-up of aortic stent grafts. J Vasc Surg 37:381–385

Benninger DH, Baumgartner RW (2006) Ultrasound diagnosis of cervical artery dissection. Front Neurol Neurosci 21:70–84

Benninger DH, Georgiadis D, Gandjour J et al (2006) Accuracy of color duplex ultrasound diagnosis of spontaneous carotid dissection causing ischemia. Stroke 37:377–381

Benson CB, Aruny JE, Vickers MA (1993) Correlation of duplex sonography with arteriography in patients with erectile dysfunction. Am J Roentgenol (AJR) 160:71–73

Benvegna S, Cassina I, Giuntini G, Rusignuolo F, Talarico F, Florena M (1990) Atherothrombotic microembolism of the lower extremities (the blue toe syndrome) from atherosclerotic non-aneurysmal aortic plaques. J Cardiovasc Surg 31:87–91

Bergqvist D, Bjorck M, Ljungman C (2006) Popliteal venous aneurysm – a systematic review. World J Surg 30:273–279

Berguer R, Hwang NHC (1974) Critical arterial stenosis. A theoretical and experimental solution. Ann Surg 180:39–50

Berland LL, Koslin DB, Routh WD, Keller FS (1990) Renal artery stenosis: prospective evaluation of diagnosis with color duplex US compared with angiography – work in progress. Radiology 174: 421–423

Bernardi E, Prandoni P, Lensing AWA, Agnelli G, Guazzaloca G, Scannapieco C, Piovella F, Verlato F, Tomasi C, Moia M, Scarano L, Girolami A (1998) D-Dimer testing as an adjunct to ultrasonography: prospective cohort study. BMJ 317:1037–1040

Bettmann AM, Paulin S (1977) Leg phlebography: the incidence, nature and modifications of undesirable side effects. Radiol Diagn 122:101

Bettmann M, Salzman E, Rosenthal D et al (1980) Reduction of venous thrombosis complicating phlebography. AJR 134:1169–1172

Beven EG (1991) Thoracic outlet syndromes. In: Young JR, Graor RA, Olin JW, Bartholomew JR (eds) Peripheral vascular diseases. Mosby, St. Louis, pp 497–509

Biasi GM, Sampaolo A, Mingazzini P et al (1999) Computer analysis of ultrasonic plaque echolucency in identifying high risk carotid bifurcation lesions. Eur J Vasc Endovasc Surg 17:476–479

Biasi GM, Froio A, Diethrich EB, Deleo G, Galimberti S, Mingazzini P, Nicolaides AN, Griffin M, Raithel D, Reid DB, Valsecchi MG (2004) Carotid plaque echolucency increases the risk of stroke in carotid stenting: the Imaging in Carotid Angioplasty and Risk of Stroke (ICAROS) study. Circulation 110:756–762

Bicknell CD, Cheshire NJ (2003) The relationship between carotid atherosclerotic plaque morphology and the embolic risk during endovascular therapy. Eur J Vasc Endovasc Surg 26:17–21

Biemans RGM (1987) Kompressionssyndrom der Arteria poplitea. In: Heberer G, Van Dongen RJAM (eds) Gefäßchirurgie. Springer, Berlin/Heidelberg/New York/Tokyo, pp S 593–S 599, Kirschnersche allgemeine und spezielle Operationslehre, Bd 11

Biland L, Lemgo E, Widmer LK (1987) Zur Epidemiologie der venösen Thromboembolie. Internist 28:285

Blackshear WM, Phillips DJ, Chikos PM, Harley JD et al (1980) Carotid artery velocity patterns in normal and stenotic vessels. Stroke 11:67–71

Blättler W (1993) Komplikationen der Thrombophlebitis superficialis. Schweiz Med Wochenschr 123:223–228

Blättler W, Bulling B, Hertel T, Rabe E (1996) Leitlinien zur Diagnostik und Therapie der Thrombophlebitis. Phlebologie 25:197–198

Blättler W, Linder C, Blättler IK et al (1996) Die ambulante Behandlung der akuten tiefen Beinvenenthrombose (TVT): eine randomisierte prospektive Studie. Schweiz med Wochenschr 126 [Suppl]:74/1

Blättler W, Partsch H, Hertel T (1996) Leitlinien zur Diagnostik und Therapie der tiefen Bein- und Beckenvenenthrombose. Phlebologie 25:199–203

Bluth EL, Kay D, Merritt CRB, Sullivan M, Farr G, Mills NL, Foreman M, Sloan K, Schlater M, Stewart J (1986) Sonographic characterization of carotid plaque: detection of hemorrhage. AJR 146:1061–1065

Bluth EI, McVay LV, Merrit CRB, Sullivan MA (1988) The identification of ulcerative plaque with high resolution duplex sonographic carotid scanning. J Ultrasound Med 7:73–76

Bluth EL, Stavros AT, Marich KW, Wetzner SM, Aufrichtig D, Baker JD (1988) Carotid duplex sonography: a multicenter recommendation for standardized imaging and Doppler criteria. Radiographics 8:487–506

Bock RW, Gray-Weale AC, Mock PA et al (1993) The natural history of asymptomatic carotid artery disease. J Vasc Surg 17:160–171

Bodily K, Buttorff J, Nordesgaard A, Osborne R jr (1996) Aortoiliac reconstruction without angiography. Am J Surg 171:505–507

Bodner G, Peer S, Fries D et al (2000) Color and pulsed Doppler ultrasound findings in normally functioning transjugular intrahepatic portosystemic shunts. Eur J Ultrasound 12:131–136

Böhm B et al (2000) Der Wert der farbkodierten kontrastmittelunterstützten Duplexsonographie bei der Erfassung von Endoleckagen nach Aortenstentimplantation im mittleren Nachuntersuchungszeitraum von 15 Monaten. Gefäßchirurgie 5:225–231

Bollinger A (1979) Funktionelle Angiologie. Thieme, Stuttgart

Bollinger A, Franzeck UK (1982) Diagnose der tiefen Becken- und Beinvenenthrombose. Schweiz Med Wochenschr 112:550–556

Bolondi L, Gamrolfi L, Arienti V et al (1982) Ultrasonography in the diagnosis of portal hypertension: diminished response of portal vessels to respiration. Radiology 142:167–172

Bolondi L, Mazziotti A, Arienti V et al (1984) Ultrasonographic study of portal venous system in portal hypertension and after portosystemic shunt operations. Surgery 95:261–269

Bolondi L, Bassi SL, Gaiani S et al (1991) Liver cirrhosis: changes of Doppler waveform of hepatic veins. Radiology 178:513–516

Bolondi L, Gaiani S, Gebel M (1998) Portohepatic vascular pathology and liver disease: diagnosis and monitoring. Eur J Ultrasound 7(suppl 3):41–52

Bommer WJ, Miller L (1982) Real-time two dimensional color flow Doppler: enhanced Doppler flow imaging in the diagnosis of cardiovascular disease (abstr). Am J Cardiol 49:944

Bond MG, Wilmoth SK, Enevold GL et al (1989) Detection and monitoring of asymptomatic atherosclerosis in clinical trials. Am J Med 86:33–36

Bönhof J, Meairs SP, Wetzler H (1990) Duplex- und Farbdopplersonographische Kriterien von Nierenarterienstenosen. Ultraschall Klin Prax 5:187 (Abstr)

Bonnefous O, Pasque P (1986) Time domain formulation of pulse-Doppler ultrasound and blood velocity estimation by cross correlation. Academic, pp 73–85 (Ultrasonic Imaging 8)

Bork-Wölwer L, Wuppermann TH (1991) Verbesserung der nichtinvasiven Diagnostik der V. saphena magna- und der Vena saphena parva-Insuffizienz durch die Duplex-Sonographie. Vasa 20:343–347

Börner N (1991) Diagnostische und prognostische Bedeutung der Echomorphologie tiefer Becken-Bein-Venenthrombosen. Habilitationsschrift Mainz

Börner N, Todt M, Schuter CJ, Meyer J (1987) Sonographie in der Diagnostik venöser Thromben. Klin Wochenschr 65(Suppl 9):37

Boström A, Ljungman C, Hellberg A et al (2002) Duplex scanning as the sole preoperative imaging method for infrainguinal arterial surgery. Eur J Vasc Endovasc Surg 23:140–145

Bots ML, de Jong PT, Hofmann A et al (1997) Left, right, near or far wall common carotid intima-media thickness measurements: associations with cardiovascular disease and lower extremity arterial atherosclerosis. J Clin Epidemiol 50:801–807

Bouchet C, Magne JL, Lacaze R, Lebrun D, Franco A (1986) L'anévrisme de la veine poplitée: une cause rare d'embolie pulmonaire à répétition. J Mal Vasc 11:190–193

Bouhoutsos J, Martin P (1974) Popliteal aneurysm, a review of 116 cases. Br J Surg 61:469–475

Bouhoutsos J, Daskalakis E (1981) Muscular abnormalities affecting the popliteal vessels. Br J Surg 68:501–506

Bounameaux H (2002) Integrated diagnostic approach to suspected deep vein thrombosis and pulmonary embolism. VASA 31:15–21

Bounameaux H, De Moerloose P, Reber G (1994) Plasma measurement of D-dimer as diagnostic aid in suspected venous thromboembolism: an overview. Thromb Haemost 71:1–6

Bowersox JC, Zwolak RM, Walsh DB, Schneider JR, Musson A, LaBombard FE, Cronenwett JL (1991) Duplex ultrasonography in the diagnosis of celiac and mesenteric artery occlusive disease. Eur J Vasc Endovasc Surg 14:780–788

Brandl R, Orend KH, Becker HM (1993) Rekonstruktionsprinzipien bei peripherer arterieller Verschlußkrankheit der unteren Extremitäten. Dt Ärztebl 37:1616–1620

Brandstetter K, Schwarzer JU, Bautz W, Pickl U, Lenz M (1993) Vergleich der Farbduplexsonographie mit der selektiven penilen DSA bei der Abklärung der erektilen Dysfunktion. Fortschr Röntgenstr 158:405–409

Braun B, Scheffler P, Kiehl R, Wenzel E (1986) A duplex system for evaluation of venous function. In: Maurer PC, Becker HM, Heidrich H, Hoffmann G, Kriessmann A, Müller-Wiefel H, Prätorius C (eds) What is new in angiology? Trends and controversis, proceedings. Zuckschwerdt, München

Braunwald E (1984) Heart disease. A textbook of cardiovascular medicine. WB Saunders, Philadelphia

Breitenseher M, Kainberger F, Hübsch P, Trattnig S, Baldt M, Barton P, Karnel F (1992) Screening von Nierenarterienstenosen. Fortschr Röntgenstr 156:228–231

Brescia MJ, Cimino JE, Appel K, Burwich BJ (1966) Chronic hemodialysis using venipuncture and a surgically created arteriovenous fistula. New Engl J Med 275:1089–1092

Brittinger WD, Twittenhoff W-D, Walker G, Konrad N (1966) Revaskularisation des Dialyseshunts. Nieren- und Hochdruckkrankheiten 25:4–9

Brophy DP, Sheiman RG, Amatulle P, Akbari CM (2000) Iatrogenic femoral pseudoaneurysms: thrombin injection after failed US-guided compression. Radiology 214:278–282

Browman MW, Cooperberg PL, Harrison PB et al (1995) Duplex ultrasonography criteria for internal carotid stenosis of more than 70% diameter: angiographic correlation and receiver operating characteristic curve analysis. Can Assoc Radiol J 46:291–295

Brown JM, Hammers LW, Barton JW, Holland CK, Scoutt LM, Pellerito JS, Taylor KJ (1995) Quantitative Doppler assessment of acute scrotal inflammation. Radiology 197:427–431

Bruno S, Remuzzi G, Ruggenenti P (2004) Transplant renal artery stenosis. J Am Soc Nephrol 15:134–141

Bunk A, Buchcik R, Konopke R et al (2000) Farbdoppler, Power-Doppler, Echokontrastmittel. Verbesserung der perioperativen Diagnostik? Internist 41:29–36

Burckhardt CB (1993) Signalverarbeitung in Ultraschallabbildung, Doppler und Dopplerabbildung. Ultraschall Med 14:220–224

Burton BS, Syms MJ, Petermann GW et al (2000) MR imaging of patients with carotidynia. Am J Neuroradiol 21:766–769

Bushong SC, Archer BR (1991) Diagnostic ultrasound. Physics, biology and instrumentation. Mosby Year Book, St. Louis

Busuttil SJ, Franklin DP, Youkey JR et al (1996) Carotid duplex overestimation of stenosis due to severe contralateral disease. Am J Surg 172:144–148

Buth J, Disselhoff B, Sommeling C, Stam L (1991) Color-flow duplex criteria for grading stenosis in infrainguinal vein grafts. J Vasc Surg 14:716–726

Calligaro KD, Musser DJ, Chen AY, Dougherty MJ, McAffee-Bennett S, Doerr KJ et al (1996) Duplex ultrasonography to diagnose failing arterial prosthetic grafts. Surgery 120:455–459

Calligaro KD, Syrek JR, Dougherty MJ, Rua I, McAffee-Bennett S, Doerr KJ et al (1998) Selective use of duplex ultrasound to replace preoperative arteriography for failing arterial vein grafts. J Vasc Surg 27:89–95

Caplan LR (2008) Dissections of brain-supplying arteries. Nat Clin Pract Neurol 4:34–42

Cardella JF, Young AT, Smith TP et al (1988) Lower extremity venous thrombosis. Comparison of venography, impedance plethysmography and intravenous manometry. Radiology 168:109

Carpenter JP, Holland GA, Baum RA, Owen RS, Carpenter JT, Cope C (1993) Magnetic resonance venography for the detection of deep venous thrombosis: comparison with contrast venography and duplex Doppler ultrasonography. J Vasc Surg 18:734–741

Carpenter JP, Lexa FJ, Davis JT (1995) Determination of sixty percent or greater carotid artery stenosis by duplex Doppler ultrasonography. J Vasc Surg 22:697–703

Carpenter JP, Lexa FJ, Davis JT (1996) Determination of duplex Doppler ultrasound criteria appropriate to the North American Symptomatic Carotid Endarterectomy Trial. Stroke 27:695–699

Carr S, Farb A, Pearce WH, Virmani R, Yao JST (1996) Atherosclerotic plaque rupture in symptomatic carotid artery stenosis. J Vasc Surg 23:755–766

CASANOVA Study Group (1991) Carotid surgery versus medical therapy in asymptomatic carotid stenosis. Stroke 22:1229–1235

Casarella WJ (1995) Transjugular intrahepatic portosystemic shunt: a defining achievement in vascular and interventional radiology. Radiology 196:305

Cassar K, Engeset J (2005) Cystic adventitial disease: a trap for the unwary. Eur J Vasc Endovasc Surg 29:93–96

Cave EM, Pugh ND, Wilson RJ et al (1995) Carotid artery duplex scanning: does plaque echogenicity correlate with patient symptoms? Eur J Vasc Endovasc Surg 10:77–81

Chahlaoui J, Julien M, Nadeau P, Bruneau L, Roy P, Sylvestre J (1981) Popliteal venous aneurysm: a source of pulmonary embolism. AJR 136:415–416

Chahwan S, Miller MT, Pigott JP, Whalen RC, Jones L, Comerota AJ (2007) Carotid artery velocity characteristics after carotid artery angioplasty and stenting. J Vasc Surg 45:523–526

Chakfé N, Beaufigeau M, Geny B, Suret-Canale MA, Vix J, Groos N, Edah-Tally S, Steinmetz E, Kretz JG (1997) Extra-popliteal localization of adventitial cysts. Review of the literature. J Mal Vasc 22:79–85

Chalmers RTA, Hoballah JJ, Kresowik TF, Sharp WJ, Synn AY, Miller E, Corson JD (1994) The impact of color duplex surveillance on the outcome of lower limb bypass with segments of arm veins. J Vasc Surg 19:279–288

Chang BB, Leather RP, Kaufman JL, Kupinski AM, Leopold PW, Shah DM (1990) Haemodynamic characteristics of failing infrainguinal in situ vein bypass. J Vasc Surg 12:596–600

Chang YJ, Lin SK, Ryo SJ, Wai YY (1995) Common carotid artery occlusion: evaluation with duplex sonography. Am J Neuroradiol 16:1099–1105

Chawla Y, Santa N, Dhiman RK et al (1998) Portal hemodynamics by duplex Doppler sonography in different grades of cirrhosis. Dig Dis Sci 43:354–357

Chengelis DL, Bendick PJ, Glover PJL, Brown OW, Ranaval TJ (1996) Progression of superficial venous thrombosis to deep vein thrombosis. J Vasc Surg 24:742–748

Chong WK, Malisch TW, Mazer MJ et al (1993) Transjugular intrahepatic portosystemic shunts: US assessment with maximum flow velocity. Radiology 189:789–793

Chuang VP, Mena CE, Hoskins PA (1974) Congenital anomalies of the inferior vena cava. Review of embryogenesis and presentation of a simplified classification. Br J Radiol 47:206–213

Clark TW, Cohen RA, Kwak A et al (2007) Salvage of nonmaturing native fistulas by using angioplasty. Radiology 242:286–292

Claudon M, Cosgrove D, Albrecht T et al (2008) Guidelines and good clinical practice recommendations for contrast enhanced ultrasound (CEUS) – Update 2008. Ultraschall Med 29:28–44

Claudon M, Plouin PF, Baxter GM et al. for the Levovist Renal Artery Stenosis Study Group (2000) Renal arteries in patients at risk of renal arterial stenosis: multicenter evaluation of the echo-enhancer SHU 508 A at color and spectral Doppler US. Radiology 214:739–746

Clevert DA, Jung EM, Reiser M, Rupp N. Verbesserte diagnostische Sicherheit durch den Ultraschall B-Flow bei Gefäßdissektion. Institut für klinische Radiologie; München (DE); Radiologie; Passau (DE); Radiologie, München (DE)

Clevert DA, Minfaifar N, Weckbach S, Kopp R, Meimarakis G, Clevert DA, Reiser M (2008) Color duplex ultrasound and contrast-enhanced ultrasound in comparison to MS-CT in the detection of endoleak following endovascular aneurysm repair. Clin Hemorheol Microcirc 39:121–132

Clifford TA, Back T, Padberg FT, Thompson PN, Duran WN, Hobson RW (1993) Refinements in the ultrasonic detection of popliteal vein reflux. J Vasc Surg 18:742–748

Coelho JCU, Sigel B, Ryva JC, Machi J, Renigers SA (1982) B-mode sonography of blood clots. J Clin Ultrasound 10:323–327

Cogo A, Lensing AWA, Koopman MMW, Piovella F, Siragusa S, Wells PS, Villata S, Büller HR, Turpie AGG, Prandoni P (1998) Compression ultrasonography for diagnostic management of patients with clinically suspected deep vein thrombosis: a prospective cohort study. BMJ 316:17–20

Cogo A, Lensing WA, Prandoni P, Hirsh J (1993) Distribution of thrombosis in patients with symptomatic deep vein thrombosis. Arch Intern Med 153:2777–2780

Coleridge-Smith P, Labropoulos N, Partsch H, Myers K, Nicolaides A, Cavezzi A (2006) Duplex ultrasound investigation of the veins in chronic venous disease of the lower limbs–UIP consensus document. Part I. Basic principles. Eur J Vasc Endovasc Surg 31:83–92

Colli A, Cocciolo M, Riva C et al (1994) Abnormalities of Doppler waveform of the hepatic veins in patients with chronic liver disease: correlation with histologic findings. AJR 162:833

Collin J, Auranja H, Sutton GLJ, Lindsell D, Oxford D (1988) Screening programme for abdominal aortic aneurysm in men aged 65 to 74 years. Lancet II:613–615

Collins JT, Boros MJ, Combs K (2007) Ultrasound surveillance of endovascular aneurysm repair: a safe modality versus computed tomography. Ann Vasc Surg 21:671–675

Comerota AJ, Katz ML, Greenwald LL, Leefmans E, Czeredarczuk M, White JV (1990) Venous duplex imaging: should it replace hemodynamic tests for deep venous thrombosis? J Vasc Surg 11:53–59

Conkbayir I, Yucesoy C, Edguer T et al (2003) Doppler sonography in renal artery stenosis. An evaluation of intrarenal and extrarenal imaging parameters. Clin Imaging 27:256–260

Connell J (1978) Popliteal vein entrapment. Br J Surg 65:351

Cornus J, Pearson SD, Polak JF (1999) Deep venous thrombosis: complete lower extremity venous US examination in patients without known risk factors – outcome study. Radiology 211:637–641

Correas J-M, Hélénon O, Pourcelot L, Moreau J-F (1997) Ultrasound contrast agents. Acta Radiol 38(suppl 412):101–112

Cosgrove D (1997) Echo enhancers and ultrasound imaging. Eur J Radiol 26:64–76

Cosgrove DO, Arger PH (1982) Intravenous echoes due to laminar flow – experimental observations. Am J Radiol 139:953–956

Cosgrove DO, Blomley MJK, Jayaram V et al (1998) Echo-enhancing (contrast) agents. Ultrasound Q 14:66–75

Cossman DV, Ellison JE, Wagner WH, Carroll RM et al (1989) Comparison of contrast angiography to arterial mapping with color-flow duplex imaging in the lower extremities. J Vasc Surg 10:522–529

Crawford ES, Hess KR (1989) Abdominal aortic surgery. N Engl J Med 321:1040–1042

Creutzig A, von der Lieth H, Majewski A, Caspary L, Oestmann J, Alexander K (1988) Vascular complications of the compression syndrome of the anterior thoracic aperture (thoracic outlet syndrome). Med Klin (Munich) 83:133–136

Cronan J, Leen V (1989) Recurrent deep venous thrombosis: limitations of US. Radiology 170:739–742

Cronan JJ (1993) Venous thromboembolic disease: the role of US. Radiology 186:619–630

Cronan JJ (1996) Deep venous thromosis: one leg or both legs? Radiology 200:323–324

Cronan JJ (1997) Controversies in venous ultrasound. Semin Ultrasound CT MR 18:33–38

Cronan JJ, Dorfman GS, Gusmark J (1988) Lower-extremity deep venous thrombosis: further experience with and refinements of ultrasound assessment. Radiology 168:101

Cronan JJ, Dorfman GS, Scola FH, Schepps B, Alexander J (1987) Deep venous thrombosis. US assessment using vein compression. Radiology 162:191–194

Cronan JJ, Froehlich J, Dorfman GS, Scola FH, Schepps B (1988) Serial compression ultrasonography in a patient population at high risk for deep vein thrombosis. Radiology 169(Suppl):321 (Abstr)

Cronenwett JL, Krupski WC, Rutherford RB (2000) Abdominal aortic and iliac aneurysms. In: Rutherford RB (ed) Vascular surgery, 5th edn. Saunders, Philadelphia, pp 1246–1280

Cvitanic OA, Cronan JJ, Sigman M, Landau ST (1993) Varicoceles: postoperative prevalence – a prospective study with color Doppler US. Radiology 187:711–714

Czerny M, Trubel W, Claeys L et al (1997) Acute mesenteric ischemia. Zentralbl Chir 122:538–544

D'Audiffret A (2001) Follow-up evaluation of endoluminally treated abdominal aortic aneurysms with duplex ultrasonography: validation with computed tomography. J Vasc Surg 33:42–50

Daher A, Jones V, da Silva AF (2001) The role of popliteal vein incompetence in the diagnosis of saphenous-popliteal reflux using continuous wave Doppler. Eur J Vasc Endovasc Surg 21:350–352

Dahl JR, Freed TA, Burke MF (1976) Popliteal vein aneurysm with recurrent pulmonary thromboemboli. JAMA 236:2531–2532

Dasbach G, Schmitz I, Niehoff L, Edelmann M, Müller K-M (1998) Arteriosklerose der Karotisregion. Gefässchirurgie 3:151–157

Dauzat MM, Laroche J-R, Charras CC, Blin B, Domingo-Faye MM (1986) Real-time B-mode ultrasonography for better specificity in the noninvasive diagnosis of deep vein thrombosis. J Ultrasound Med 5:625–631

Davidson JT, Callis JT (1993) Arterial reconstruction of vessels in the foot and ankle. Ann Surg 217:699–708

Davies AH, Hawdon AJ, Sydes MR, Thompson SG (2005) Is duplex surveillance of value after leg vein bypass grafting? Principal results of the Vein Graft Surveillance Randomised Trial (VGST). Circulation 112:1985–1991

Davies KN, Humphrey PR (1993) Complications of cerebral angiography in patients with symptomatic carotid territory ischaemia screened by carotid ultrasound. J Neurol Neurosurg Psychiatry 56:967–972

Davis SM, Donnan GA (2003) Is carotid angiography necessary? Editors disagree. Stroke 34:1819

Dawson DL (1996) Noninvasive assessment of renal artery stenosis. Semin Vasc Surg 9:172–181

De Bakey ME, Henly WS, Cooley DA et al (1965) Surgical management of dissecting aneurysms of the aorta. J Thorac Cardiovasc Surg 49:130–148

De Bray JM, Baud JM, Dauzat M on behalf of the Consensus Conference (1997) Consensus concerning the morphology and the risk of carotid plaques. Cerebrovasc Dis 7:289–296

De Bray JM, Glatt B (1994) Quantification of atheromatous stenosis in the extracranial internal carotid. Artery. Cerebrovasc Dis 5:414–426

De Bruyne, Manoharan G, Pijls NHJ et al (2006) Assessment of renal artery stenosis severity by pressure gradient measurement. J Am Coll Cardiol 48:1851–1855

De Morais RH, Muglia VF, Mamere VF et al (2003) Duplex Doppler sonography of transplant renal artery stenosis. J Clin Ultrasound 31:135–141

De Smet AA, Kitslaar PJ (1990) A duplex criterion for aorto-iliac stenosis. Eur J Vasc Surg 4:275–278

De Smet AEA, Ermers EJM, Kitslaar PJ (1996) Duplex velocity characteristics of aortoiliac stenoses. J Vasc Surg 23:628–636

De Valois JC, van Schaik CC, Verzijlbergen F, van Ramshorst B, Eikelboom BC, Meuwissen OJA Th (1990) Contrast venography: from gold standard to "golden backup" in clinically suspected deep vein thrombosis. Eur J Radiol 11:131–137

DeWire DM, Begun FP, Lawson RK, Fitzgerald S, Foley WD (1992) Color Doppler ultrasonography in the evaluation of the acute scrotum. J Urol 147:89–91

DEGUM, Arbeitskreis Gefäßdiagnostik (1986) Richtlinien für die Durchführung dopplersonographischer Untersuchungen der Becken- und Beinvenen. Deutsche Ges Angiologie, Mitt 3. Demeter, Gräfelfing

Delcker A, Diener HC (1992) Die verschiedenen Ultraschallmethoden zur Untersuchung der A. vertebralis: Eine vergleichende Wertung. Ultraschall Med 13:213–220

Denzel C, Balzer K, Merhof D, Lang W et al (2009) 3D cross sectional view to investigate the morphology of internal carotid artery plaques. Is 3D ultrasound superior to 2D ultrasound? Ultraschall Med 29:291–296

Denzel C, Balzer K, Muller KM, Fellner F, Fellner C, Lang W (2003) Relative value of normalized sonographic in vitro analysis of arteriosclerotic plaques of internal carotid artery. Stroke 34:1901–1906

Denzel C, Fellner F, Wutke R, Bazler K, Muller KM, Lang W (2003) Ultrasonographic analysis of arteriosclerotic plaques in the internal carotid artery. Eur J Ultrasound 16:161–167

Derkx FH, Schalekamp MA (1994) Renal artery stenosis and hypertension. Lancet 344:237–239

Desai KM, Gilbert HG (1991) Noninvasive investigation of penile artery function. In: Kirby RS, Carson CC, Webster GD (eds) Impotence: diagnosis and management of male erectile dysfunction. Butterworth-Heinemann, Oxford, pp 81–91

Desberg AL, Paushter DM, Lammert GK et al (1990) Renal artery stenosis: evaluation with color Doppler flow imaging. Radiology 177:749–753

Diamaria G, Zittoun R, Reynes M (1981) Anévrisme veineux poplité révélé par une embolie pulmonaire. Ann Cardiol Angéiol 30:337–338

Dion JE, Gates PC, Fox AJ, Barnett HJM, Blom RJ (1987) Clinical events following neuroangiography: a prospective study. Stroke 18:997–1004

Disselhof B, Buth J, Jakimowicz J (1989) Early detection of stenosis of femoro-distal grafts. A surveillance study using colour-duplex scanning. Eur J Vasc Surg 3:43–48

Do DD, Zehender T, Mahler F (1993) Farbkodierte Duplexsonographie bei iatrogenen Aneurysmata spuria in der Leiste. Dtsch Med Wochenschr 118:656–660

Dodd GD III, Zajko AB, Orons PD et al (1995) Detection of transjugular intrahepatic portosystemic shunt dysfunction: value of duplex Doppler sonography. Am J Roentgenol 164:1119–1124

Doelman C, Duijm LE, Liem YS et al (2005) Stenosis detection in failing hemodialysis access fistulas and grafts: comparison of color Doppler ultrasonography, contrast-enhanced magnetic resonance angiography and digital subtraction angiography. J Vasc Surg 42:739–746

Donald IP, Edwards RD (1982) Fatal outcome from popliteal venous aneurysm associated with pulmonary embolism. Br J Radiol 55:930–931

Dorweiler B, Neufang A, Kreitner KF et al (2002) Magnetic resonance angiography unmasks reliable target vessels for pedal bypass grafting in patients with diabetes mellitus. J Vasc Surg 35:766–772

Dougherty MJ, Calligaro KD, DeLaurentis DA (1998) The natural history of "failing" arterial grafts in a duplex surveillance protocol. Ann Vasc Surg 12:255–259

Droste DW, Karl M, Bohle RM, Kaps M (1997) Comparison of ultrasonic and histopathological features of carotid artery stenosis. Neutral Res 19:380–384

Droste, v. Planta (1989) Memorix, 2. korrigierte Auflage: VCH Verlagsgesellschaft, Weinheim (Bundesrepublik Deutschland)

Dubbins PA (1986) Renal artery stenosis: duplex Doppler evaluation. Br J Radiol 59:225–229

Duerschmied D, Rossknecht A, Olsson L et al (2005) Contrast ultrasound perfusion imaging of lower extremities in peripheral arterial disease – a novel diagnostic method. Ultraschall Med 26:S39

Dunant JH, Eugenidis N (1973) Cystic degeneration of the popliteal artery. Vasa 2:156–159

Dyet JF, Nicholson AA, Eitles OF (2000) Vascular imaging and intervention in peripheral arteries in the diabetic patient. Diabetes Metab Res Rev 16(Suppl 1):16–22

Ebrahim S, Papacosta O, Whincup P et al (1999) Carotid plaque, intima media thickness, cardiovascular risk factors, and prevalent cardiovascular disease in men and women: The British Regional Heart Study. Stroke 30:841–850

ECST Collaborative Group (1996) Endarterectomy for moderate symptomatic carotid stenosis: interim results from the MRC European Carotid Surgery Trial. Lancet 347:1591–1593

ECST European Carotid Surgery Trialists Collaborative Group (1991) MRC European Carotid Surgery Trial: interim results for symptomatic patients with severe (70–90%) or with mild (0–29%) carotid stenosis. Lancet 337:1235–1243

Edmondson HT, Crowe JA (1972) Popliteal artery and venous entrapment. Am Surg 38:657–659

Edwards JM, Coldwell DM, Goldman ML, Strandness DE Jr (1991) The role of duplex scanning in the selection of patients for transluminal angioplasty. J Vasc Surg 13:69–74

Edwards JM, Moneta GL, Papanicolaou G et al (1995) Prospective validation of a new duplex ultrasound criteria for 70%–99% internal carotid stenosis. JEMU 16:3–7

Effeney DJ, Fiedmann MB, Gooding GAW (1984) Iliofemoral venous thrombosis: real-time ultrasound diagnosis, normal criteria, and clinical application. Radiology 150:787–792

Egbring J, Görg C (2007) Die asymptomatische Lungenembolie: Sollte jeder Patient mtit tiefer Beinvenenthrombose am Thorax geschallt werden? Ultraschall Med 28:375–379

Eiberg JP, Jensen F, Gronvall Rasmussen JB, Schroeder TV (2001) Screening for aortoiliac lesions by visual interpretation of the common femoral Doppler waveform. Eur J Vasc Endovasc Surg 22:331–336

Eiberg JP, Madycki G, Hansen MA, Christiansen S, Gronvall Rasmussen JB, Schroeder TV (2002) Ultrasound imaging of infrainguinal arterial disease has a high interobserver agreement. Eur J Vasc Endovasc Surg 24:293–299

Eichlisberger R, Jäger K (1989) Beeinflussung der venösen Hämodynamik durch Venenpharmaka. Nachweis mittels Duplexsonographie. Was gibt es Neues in der Phlebologie (Kurzfassung)

El-Barghouty N, Geroulkas G, Nicolaides A, Androulakis A, Bahal V (1995) Computer-assisted carotid plaque characterization. Eur J Vasc Endovasc Surg 9:389–393

El-Barghouty N, Nicolaides A, Bahal V, Geroulakos G, Androulakis A (1996) The identification of the high risk carotid plaque. Eur J Vasc Endovasc Surg 11:470–478

El-Barghouty NM, Levine T, Ladva S, Flanagan A, Nicolaides A (1996) Histological verification of computerised carotid plaque characterization. Eur J Vasc Endovasc Surg 11:414–416

Elias A, Le Croff G, Bouvier JL, Benichou A, Serradimigni A (1987) Value of real-time B-mode ultrasound imaging in the diagnosis of deep vein thrombosis of the lower limbs. Inter Angio 6:175–182

Eliasziw M, Rankin RN, Fox AJ, Hynes RB, Barnett HJ (1995) Accuracy and prognostic consequences of ultrasonography in identifying severe carotid artery stenosis. North American Symptomatic Carotid Endarterectomy Trial (NASCET) Group. Stroke 26:1747–1752

Elkind MS, Cheng J, Boden-Albala B, Paik MC, Sacco RL (2001) Elevated white blood cell count and carotid plaque thickness: Northern Manhattan Stroke Study. Stroke 32:842–849

Elsman BH, Legemate DA, van der Heyden FW et al (1996) The use of color-coded duplex scanning in the selection of patients with lower extremity arterial disease for percutaneous transluminal angioplasty: a prospective study. Cardiovasc Intervent Radiol 19:313–316

Elsman BH, Legemate DA, de Vos HJ, Mali WP, Eikelboom BC (1997) Hyperaemic colour duplex scanning for the detection of aortoiliac stenoses. A comparative study with intra-arterial pressure measurement. Eur J Vasc Endovasc Surg 14:462–467

Engberding R (1990) Untersuchungstechniken in der Echokardiographie. Springer, Berlin/Heidelberg/New York, pp S 183–S 193

Erickson SJ, Mewissen MW, Foley WD, Lawson TL, Middleton WD, Quiroz FA, Macrander SJ, Lipchik EO (1989) Stenosis of the internal carotid artery: assessment using color imaging compared with angiography. Am J Roentgenol 152:1299–1305

Evers EJ, Wuppermann Th (1995) Ultraschalldiagnostik bei postthrombotischem Syndrom. Eine vergleichende Untersuchung mittels Farbduplex, CW-Doppler, und B-Bildsonographie. Ultrachall Med 16:259–263

Evers EJ, Wuppermann Th (1997) Die Charakterisierung des postthrombotischen Refluxes mittels farbkodierter Duplexsonographie. VASA 26:190–193

Evers EJ, Wuppermann Th (1997) Langzeitverlaufbeobachtung der venösen Hämodynamik bei postthrombotischem Syndrom mittels Farbdopplersonographie. Ultraschall 18:S1, 54

Falk E (1992) Why do plaques rupture? Circulation 86:III-30–III-42

Falk RL, Smith DF (1987) Thrombosis of upper extremity thoracic inlet veins: diagnosis with duplex Doppler sonography. Am J Radiol 149:677–682

Faught WE, Mattos MA, van Bemmelen PS et al (1994) Color-flow duplex scanning of carotid arteries: new velocity criteria based on receiver operator characteristic analysis for threshold stenoses used in the symptomatic and asymptomatic carotid trials. J Vasc Surg 19:818–828

Fay JJ (1985) Anévrisme veineux poplité. Thèse Médicine, n° 78 Lille

Federman J, Anderson ST, Rosengarten DS, Pitt A (1977) Pulmonary embolism secondary to anomalies of deep venous system of the leg. Br Heart J 39:547–552

Feigenbaum H (ed) (1986) Echocardiography. Lea & Febiger, Philadelphia, pp 1–49

Feinstein SB, Cheirif J, ten Cate FJ et al (1990) Safety and efficacy of a new transpulmonary ultrasound contrast agent: initial multicenter clinical results. J Am Coll Cardiol 16:316–324

Feldstein VA, Patel MD, LaBerge JM (1996) Transjugular intrahepatic portosystemic shunts: accuracy of Doppler US in determination of patency and detection of stenoses. Radiology 201:141–147

Fell G, Phillips DJ, Chikos PM, Harley JD, Thiele BL, Strandness DE (1981) Ultrasonic duplex scanning for disease of the carotid artery. Circulation 64:1191–1195

Fellmeth BD, Roberts AC, Bookstein JJ et al (1991) Post-angiographic femoral artery injuries: nonsurgical repair with US-guided compression. Radiology 178:671–675

Fellner F, Janka R, Fellner C et al (1999) Post occlusion visualization of peripheral arteries with "floating table" MR angiography. Magn Reson Imaging 17:1235–1237

Ferrer EJM, Samso J, Serrando R et al (2000) Use of ultrasound in the diagnosis of carotid artery occlusion. J Vasc Surg 31:736–741

Ferretti G, Salomone A, Castagno PL (1988) Renovascular hypertension: a non-invasive duplex scanning screening. Int Angiol 7:219–223

Fillinger MF, Baker RJ Jr, Zwolak RM et al (1966) Carotid duplex criteria for 60% or greater angiographic stenosis: variation according to equipment. J Vase Surg 24:856–864

Finkenzeller T, Tacke J, Clevert DA, Jung W, Kubala R, Schreyer A, Feuerbach S, Jung EM (2008) Quantification of extracranial ICA stenoses with vessel ultrasound by CCDS and B-flow in comparison to 64-slice multidector CTA, contrast-enhanced MRA and DSA. Ultraschall Med 29:294–301

Fisher AJ, Paulson EK, Kliewer MA et al (1998) Doppler sonography of the portal vein and hepatic artery: measurement of a prandial effect in healthy subjects. Radiology 207:711–715

Fitzgerald SW, Foley WD (1991) Genitourinary system. In: Lanzer P, Yoganathan AP (eds) Vascular imaging by color Doppler and magnetic resonance. Springer, Berlin

Fitzgerald SW, Erickson S, DeWire DM, Foley WD, Lawson TL, Begun FP, Lawson RK (1992) Color Doppler sonography in the evaluation of the adult acute scrotum. J Ultrasound Med 11:543–548

Flanigan DP, Burnham SJ, Goodreau JJ et al (1979) Summary of cases of adventitial cystic disease of the popliteal artery. Ann Surg 189:165

Fletcher JP, Kershaw LZ, Barker DS (1990) Ultrasound diagnosis of lower limb deep venous thrombosis. Med J Aust 153:453–455

Flückiger F, Steiner H, Rabl H, Waltner F (1991) Zystische Adventitia-Degeneration der Arteria poplitea: Sonographische Sicherung der Diagnose. Ultraschall Med 12:84–87

Fobbe F (1993) Periphere Venen. In: Wolf KJ, Fobbe F (Hrsg) Farbkodierte Duplexsonographie. Thieme, Stuttgart/New York, pp 114–129

Fobbe F, Koennecke H-C, El Bedewi M, Heidt P, Boese-Landgraf J, Wolf K-J (1989) Diagnostik der tiefen Beinvenenthrombose mit farbkodierter Duplexsonographie. Fortschr Röntgenstr 151:569

Fobbe F, Ruhnke-Trautmann M, van Gemmeren D, Hartmann CA, Kania U, Wolf K-J (1991) Altersbestimmung venöser Thromben im Ultraschall. Fortschr Röntgenstr 155:344–348

Fobbe F, Wolf K-J (1988) Erste klinische Erfahrungen mit der Angiodynographie. Fortschr Röntgenstr 148:259–264

Foley WD (1991) Color Doppler flow imaging. Andover Medical Publishers, Boston

Foley WD, Middleton WD, Lawson TL, Erickson S, Quiroz FA, Macrander S (1989) Color Doppler ultrasound imaging of lower-extremity venous disease. Am Roentgenol 152:371–376

Föllinger O (1982) Laplace- and Fourier-transformation. AEG-Telefunken

Forsberg F, Liu J-B, Burns PN, Merton DA, Goldberg BB (1994) Artifacts in ultrasonic contrast agent studies. J Ultrasound Med 13:357–365

Forster S, Embree PM, O'Brien WD (1990) Flow velocity profile via time-domain correlation: error analysis and computer simulation. IEEE Trans Ultrason Ferroelectr Freq Control 37:164–175

Foshager MC, Ferral H, Nazarian GK et al (1995) Duplex sonography after transjugular intrahepatic portosystemic shunts (TIPS): normal hemodynamic findings and efficacy in predicting shunt patency and stenosis. Am J Roentgenol 165:1–7

Foster JH, Dean RH, Pinkterton JA, Rhamy RK (1973) Ten years experience with the surgical management of renovascular hypertension. Ann Surg 177:755–760

Francois GF, Jausseran JM, Giuly J (1988) Anévrisme veineux poplité. Presse Méd 17:755

Frank E, Anderson C, Rubinstein D (1978) Frequency of sexual dysfunction in "normal" couples. N Engl J Med 299:111–115

Fraser JD, Anderson DR (1999) Deep venous thrombosis: recent advances and optimal investigation with US. Radiology 211:9–24

Frauchiger B, Holtz D, Eichlisberger R, Jäger KA (1995) Duplexsonographie zur Abklärung der renovaskulären Hypertonie und bei Durchblutungsstörungen der Transplantatniere. In: Jäger KA, Eichlisberger R (eds) Sono-Kurs – Ein konzentrierter Refresherkurs über die gesamte Ultraschalldiagnostik. Karger, Basel, pp 114–127

Frauchiger B, Schmid HP, Roedel C, Moosmann P, Staub D (2001) Comparison of carotid arterial resistive indices with intima-media thickness as sonographic markers of atherosclerosis. Stroke 32:836–841

Frederick MG, Hertzberg BS, Kliewer MA et al (1996) Can the US examination for lower extremity deep venous thrombosis be abbreviated? A prospective study of 755 examinations. Radiology 199:45–47

Frimann-Dahl J (1935) Postoperative Röntgenuntersuchungen. Acta Chir Scand 76(Suppl):36

Frühwald, Blackwell (1992) Atlas der farbkodierten Dopplersonographie. Springer, Wien New York, p S 101

Fukudome Y, Abe I, Onaka U, Fujii K, Ohya Y, Fukuhara M, Kaseda S, Esaki M, Fujishima M (1998) Regression of carotid wall thickening after corticosteroid therapy in Takayasu´s arteritis evaluated by B-mode ultrasonography: report of 2 cases. J Rheumatol 25: 2029–2032

Furness PN, Philpott CM, Chorbadijan MT et al (2003) Protocol biopsy of the stable renal transplant: a multicenter study of methods and complication rates. Transplantation 76:969–973

Fürst G, Kuhn FP, Trappe RP, Modder U (1990) Diagnostik der tiefen Beinvenenthrombose. Farb-Doppler-Sonographie versus Phlebographie. Fortschr Geb Röntgenstr Neuen Bildgeb Verfahr 152:151–158

Fürst G, Saleh A, Wenserski F et al (1999) Reliability and validity of noninvasive imaging of internal carotid artery pseudo-occlusion. Stroke 30:1450–1455

Gaiani S, Bolondi L, Li Bassi S et al (1989) Effect of meal on portal hemodynamics in healthy humans and in patients with chronic liver disease. Hepatology 9:815–819

Gaitini D (1990) Late changes in veins after deep venous thrombosis: ultrasonic findings. Fortschr Röntgenstr 153:68–72

Gaitini D, Kaftori JK, Pery M, Weich YL, Markel A (1988) High-resolution real-time-ultrasonography in the diagnosis of deep vein thrombosis. Fortschr Roentgenstr 149:26

Galen SR, Gambino RS (1975) Beyond normality. The predictive value and efficiency of medical diagnoses. Wiley, New York, pp 10–14

Gallacher JJ, Hageman JH (1985) Popliteal vein aneurysm causing pulmonary embolus. Arch Surg 120:1173–1175

Galland RB, Simmons MJ, Torrie EPH (1991) Prevalence of abdominal aortic aneurysm in patients with peripheral vascular disease. Br J Surg 78:1259–1260

Garovic VD, Textor SC (2005) Renovascular hypertension and ischemic nephropathy. Circulation 112:1362–1374

Gartenschlager M, Klose KJ, Schmidt JA (1996) Diagnose flottierender venöser Thromben mittels Phlebo-Spiral-CT. Fortschr Geb Röntgenstr Neuen Bildgeb Verfahr 164:376–361

Garth KE, Carroll BA, Sommer FG, Oppenheimer DA (1983) Duplex ultrasound scanning of the carotid arteries with velocity spectrum analysis. Radiology 147:823–827

Geelkerken RH, Delahunt TA, Schultze Kool LJ et al (1996) Pitfalls in the diagnosis of origin stenosis of die coeliac and superior mesenteric arteries with transabdominal color duplex examination. Ultrasound Med Biol 22:695–700

Geiger A, Hammel D, Scheld HH, Böcker W (1991) Fallbericht: Ungewöhnlich großer Tumor eines Glomus caroticum. Angio 13:279–283

Gentile AT, Mills JL, Gooden MA, Westerband A, Cui H, Bermann SS et al (1997) Identification of predictors of lower extremity vein graft stenosis. Am J Surg 174:218–221

Geroulakos G, Domjan J, Nicolaides A, Stevens J et al (1994) Ultrasonic carotid artery plaque structure and risk of cerebral infarction on computed tomography. J Vasc Surg 20:263–266

Geroulakos G, Ramaswami G, Nicolaides A, James K, Labropoulos N, Belcaro G, Holloway M (1993) Characterization of symptomatic and asymptomatic carotid plaques using high-resolution real-time ultrasonography. Br J Surg 80:1274–1277

Gerson L, Martin H (1981) Ectasie veineuse d'origine traumatique. Angiologie 33:275–278

Giannoni MF, Bilotta F, Fiorani L, Fiorani P (1998) Regarding "reduction in aortic aneurysm size: early results after endovascular graft replacement". Letter: J Vasc Surg 27:981

Giannoni MF, Fanelli F, citone M, Cristina Acconcia M, Speziale F, Gossetti B (2007) Contrast ultrasound imaging: the best method to detect type II endoleak during endovascular aneurysm repair follow-up. Interact Cardiovasc Thorac Surg 6:359–362

Gibson MH, Mills JG, Johnson GE, Downs AR (1977) Popliteal entrapment syndrome. Ann Surg 185:341–348

Gibson RN, Gibson PR, Donlan JD, Clunie DA (1989) Identification of a patent paraumbilical vein by using Doppler sonography: importance in the diagnosis of portal hypertension. AJR Am J Roentgenol 153:513–516

Giorgio A, Amoroso P, Lettieri G et al (1986) Cirrhosis: value of caudate to right lobe ratio in diagnosis with US. Radiology 161: 443–445

Giustacchini P, Pisanti F, Citterio F et al (2002) Renal vein thrombosis after renal transplantation: an important cause of graft loss. Transplant Proc 34:2126–2127

Giyanani VL, Krebs CA, Nall LA, Eisenberg RL, Parvey HR (1989) Diagnosis of abdominal aortic dissection by image-directed Doppler sonography. J Clin Ultrasound 17:445–448

Glickerman DJ, Obregon RG, Schmiedl UP, Harrison SD, Macaulay SE, Simon HE, Kohler TR (1996) Cardiac-gated MR angiography of the entire lower extremity: a prospective comparison with conventional angiography. AJR 167:445–451

Golledge J, Beattie D, Greenhalgh R, Davies A (2009) Have the results of infrainguinal bypass improved with the widespread utilization of postoperative surveillance? Eur J Vasc Endovasc Surg 11:388–392

Golledge J, Iannos J, Walsh JA, Burnett JR, Foreman RK (2001) Critical assessment of the outcome of infrainguinal vein bypass. Ann Surg 234:697–701

Golzarian J et al (2002) Evaluation of abdominal aortic aneurysm after endoluminal treatment: comparison of color Doppler sonography and biphasic helical CT. Am J Rad 178:623–628

Gonçalves I, Lindholm MW, Pedro LM et al (2004) Elastin and calcium rather than collagen or lipid content are associated with echogenicity of human carotid plaques. Stroke 35:2795–2800

Gonçalves I, Moses J, Pedro LM, Dias N, Fernandes e Fernandes J, Nilsson J, Ares MPS (2003) Echolucency of carotid plaques correlates with plaque cellularity. Eur J Vasc Endovasc Surg 26:32–38

Gooding GAW, Perez S, Rapp JH, Drupski WC (1991) Lower-extremity vascular grafts placed for peripheral vascular disease. Prospective evaluation with duplex Doppler sonography. Radiology 180: 379–386

Goodman LR, Lipchick RJ (1996) Diagnosis of acute pulmonary embolism: time for a new approach. Radiology 199:25–27

Gorenstein A, Katz S, Schiller M (1987) Congenital aneurysms of the deep veins of the lower extremities. J Vasc Surg 5:765–768

Görtler M, Niethammer R, Widder B (1994) Differentiating subtotal carotid artery stenoses from occlusions by colour-coded duplex sonography. J Neurol 241:301–305

Gottlieb RH, Voci S, Syed L et al (2003) Randomized prospective study comparing routine versus selective use of sonography of the complete calf in patients with suspected deep venous thrombosis. AJR 180:241–245

Götzberger M, Kaiser C, Landauer N, Dieterle C, Heldwein W, Schiemann U (2008) Intrarenal resistance index for the assessment of early renal function impairment in patients with liver cirrhosis. Eur J Med Res 13: 383–387

Gramiak R, Shah PM (1968) Echocardiography of the aortic root. Invest Radiol 3:356–366

Grant EG, Benson CB, Moneta GL et al (2003) Carotid artery stenosis: gray-scale and Doppler US diagnosis – Society of Radiologists in Ultrasound Consensus Conference. Radiology 229:340–346

Grant EG, Duerinckx AJ, El Saden S et al (1999) Doppler sonographic parameters for the detection of carotid stenosis. Am J Roentgenol 172:1123–1129

Grant EG, Duerinckx AJ, El Saden SM et al (2000) Ability to use duplex US to quantify internal carotid stenoses: fact or fiction? Radiology 214:247–252

Grant EG, Perrella RR (1990) Wishing won't make it so: duplex Doppler sonography in die evaluation of renal transplant dysfunction. AJR 155:538–539

Grassbaugh JA, Nelson PR, Rzucidlo EM, Schermerhorn ML, Fillinger MF, Powell RJ et al (2003) Blinded comparison of preoperative duplex ultrasound and contrast arteriography for planning revascularisation at the level of the tibia. J Vasc Surg 37:1186–1190

Grassi CJ, Polak JF (1990) Axillary and subclavian venous thrombosis: follow-up evaluation with color Doppler flow US and venography. Radiology 175:651–654

Gray-Weale AC, Graham JC, Burnett JR et al (1988) Carotid artery atheroma: comparison of preoperative B-mode ultrasound appearance with carotid endarterectomy specimen. J Cardiovasc Surg 29:115–123

Gray-Weale AC, Graham JC, Burnett JR et al (1988) Comparison of preoperative B-mode ultrasound appearance with carotid endarterectomy specimen pathology. J Cardiovasc Surg 29:676–681

Green RM, McNamara J, Ouriel K, DeWeese JA (1990) Comparison of infrainguinal graft surveillance techniques. J Vasc Surg 11: 207–214

Greenwood LH, Yrizarry JM, Hallett JW (1982) Peripheral venous aneurysms with recurrent pulmonary embolism: report of a case and review of the literature. Cardiovasc Intervent Radiol 5:43–45

Greiner L (2005) Quantifizierung des Nichtquantifizierbaren oder: Maß und Zahl sind Schall und Rauch. Ultraschall Med 26:183–184

Griewig B, Morganstern C, Driesner F et al (1996) Cerebrovascular disease assessed by color flow and power Doppler ultrasonography. Comparison with digital subtraction angiography in internal carotid artery stenosis. Stroke 27:95–100

Grigg MJ, Nicolaides AN, Wolfe JHN (1988) Detection and grading of femorodistal vein graft stenoses: duplex velocity measurements compared with angiography. J Vasc Surg 8:661

Grigg MJ, Wolfe JHN, Tovar A, Nicolaides AN (1988) The reliability of duplex derived haemodynamic measurements in the assessment of femoro-distal grafts. Eur J Vasc Surg 2:177–181

Grogan J, Castilla M, Lozanski L et al (2005) Frequency of critical stenosis in primary arteriovenous fistulae before hemodialysis access: should duplex ultrasound surveillance be the standard of care? J Vasc Surg 41:1000–1006

Grønholdt ML, Nordestgaard BG, Schroeder TV, Vorstrup S, Sillesen H (2001) Ultrasonic echolucent carotid plaques predict future strokes. Circulation 104:68–73

Gross CM, Kramer J, Weingartner O et al (2001) Determination of renal arterial stenosis severity: comparison of pressur gradient and vessel diameter. Radiology 220:751–756

Grosser S, Kreymann G, Guthoff A, Taube C, Raedler A, Tilsner V, Greten H (1990) Farbkodierte Duplexsonographie bei Phlebothrombosen. Dtsch Med Wochenschr 115:1939–1944

Grosser S, Kreymann G, Kühns A (1991) Duplex-sonographisch quantifiziertes Shuntvolumen und dessen klinische Relevanz. Angio Arch 22:74–77

Gruss JD, Geissler C (1997) Aneurysms of the subclavian artery in thoracic outlet syndrome. Zentralbl Chir 122:730–734

Gubler FM, Laan R, van der Veen F (1996) The value of palpation, varicoscreen contact thermography and colour Doppler ultrasound in the diagnosis of varicocele. Hum Reprod 11:1232–1235

Haas SB, Tribus CB, Insall JN, Becker MW, Windsor RE (1992) The significance of calf thrombi after total knee arthroplasty. J Bone Joint Surg Br 74:799–802

Habscheid W (1988) Einsatz der Real-Time-Sonographie zur Diagnostik der tiefen Beinvenenthrombose auf einer internistischen Intensivstation. Intensivmed Notfallmed 25:326

Habscheid W (2006) Sonographie der Beinvenenthrombose. Ultraschall Med 27:512–532

Habscheid W, Landwehr P (1990) Diagnostik der akuten tiefen Beinvenenthrombose mit der Kompressionssonographie. Ultraschall Med 11:268–273

Habscheid W, Wilhelm T (1988) Diagnostik der tiefen Beinvenenthrombose durch Real-time-Sonographie. Dtsch Med Wochenschr 113:586–591

Habscheid W, Becker W, Höhmann M (1989) Diagnostik der tiefen Beinvenenthrombose. Dtsch Med Wochenschr 114:837

Habscheid W, Höhmann M, Klein S (1990) Kompressionssonographie als Verfahren zur Diagnose der akuten tiefen Beinvenenthrombose. Med Klin 85:6–12

Hach W (1981) Spezielle Diagnostik der primären Varikose. Demeter, Gräfelfing

Hach W, Girth E, Lechner W (1977) Einteilung der Stammvarikose der V. saphena magna in 4 Stadien. Phlebol Proktol 6:116–123

Hach W, Hach-Wunderle V (1994) Die Rezirkulationskreise der primären Varikose. Springer, Berlin/Heidelberg

Hach W, Hach-Wunderle V (1996) Phlebographie der Bein- und Beckenvenen. Schnetztor, Konstanz, pp 89–120

Hach W, Hach-Wunderle V (1998) Diagnostik der tiefen Bein- und Beckenthrombosen durch Phlebographie und Duplex-Sonographie. Hämostaselogie 18:11–17

Haerten R (1998) Power-Doppler-Verfahren. In: Bogdahn U, Becker G, Schlachetzki F (eds) Echosignalverstärker und transkranielle Farbduplex-Sonographie. Blackwell Wissenschafts-Verlag, Berlin Wien, pp 93–99

Haerten R, Kim J (1993) Verfahren der Farbdoppler-Sonographie – Ein Methodenvergleich. Ultraschall Med 14:225–230

Haimov H, Giron F, Jacobsen JH (1979) The expanded polytetrafluoroethylene graft. Three years` experience with 362 grafts. Arch Surg 114:673–677

Haire WD, Lynch TG, Lund GB, Lieberman RP, Edney JA (1991) Limitations of magnetic resonance imaging and ultrasound-directed (duplex) scanning in the diagnosis of subclavian vein thrombosis. J Vasc Surg 13:391–397

Hallam MJ, Reid JM, Cooperberg PL (1989) Color-flow Doppler and conventional duplex scanning of the carotid bifurcation: prospective double-blind correlative study. Am J Roentgenol 152:1101–1105

Hamann H, Cyba-Altunbay S, Schäfer H, Vollmar JF (1986) Asymptomatic carotid artery stenosis. Surgery versus medical treatment (Casanova Study). In: Maurer PC, Becker HM, Heidrich H, Hoffmann G, Kriessmann A, Müller-Wiefel H (eds) What is new in angiology? Zuckschwerdt, München, p 305

Hamulyak K, Lensing AWA, van der Meer J, Smid WM, van Ooy A, Hoek JA (1995) Subcutaneous low molecular-weight heparin or oral anticoagulants for the prevention of deep-vein thrombosis in elective hip and knee replacement? Thromb Haemost 76:1428–1431

Handa N, Matsumoto M, Maeda H, Hougaku H, Kamada T (1995) Ischemic stroke events and carotid atherosclerosis. Results of the Osaka Follow-up Study for Ultrasonograhic Assessment of Carotid Atherosclerois (the OSACA Study). Stroke 26:1781–1786

Hankey GJ, Warlow CP, Sellar RJ (1990) Cerebral angiographic risk in mild cerebrovascular disease. Stroke 21:209–222

Hansen LG, Boris P (1986) Aneurysma der Vena femoralis und poplitea. Radiologe 26:210

Harbin WP, Robert NH, Ferrucci JT (1980) Diagnosis of cirrhosis based on regional changes in hepatic morphology. Radiology 135:273–283

Harloff A, Strecker C, Reinhard M, Kollum M, Handke M, Olschewski M, Weiller C, Hetzel A (2006) Combined measurement of carotid stiffness and intima-media thickness improves prediction of complex aortic plaques in patients with ischemic stroke. Stroke 37:2708–2712

Harnoss B-M, Keller F, Häring R, Distler A, Maurer PC (Hrsg.) (1991) Der Dialyseshunt als chirurgische und nephrologische Aufgabe. Angio Archiv 22

Harns PL, de Cossart L, Moody P, Douglas H (1988) How can we detect and manage fibrous strictures within new grafts? Are these more problematical with reversed than in situ bypass? In: Greenhalgh RM (ed) Limb salvage and amputation for vascular disease. Saunder WB, London, p 221

Harolds JA, Friedman MH (1977) Venous aneurysms. South Med J 70:719–721

Harrer JU, Sasse A, Klötzsch C (2006) Intimal flap in a conmon carotid artery in a patient with Marfan's syndrome. Ultraschall Med 27:487–488

Hata J, Kamada T, Haruma K, Kusunoki H (2005) Evaluation of bowel ischemia with contrast-enhanced US: initial experience. Radiology 236:712–715

Hatle L, Angelson B (1985) Doppler ultrasound in cardiology. Lea & Febiger, Philadelphia, pp 22–26

Hatsukami TS, Prornozich JF, Zierler RE et al (1992) Color Doppler imaging of infrainguinal arterial occlusive disease. J Vasc Surg 16:527–533

Hauser M, Brunner U (1994) Neue pathophysiologische und funktionelle Gesichtspunkte zur Insuffizienz der Vena saphena parva. Vasa 22:338–341

Haaverstadt R, Fougner R, Myhre HO (1995) Venous haemodynamics and the occurrence of leg oedema in patients with popliteal aneurysm. Eur J Vasc Endovasc Surg 9:204–210

Haaverstadt R, Johnsen H, Saether OD, Myhre HO (1995) Lymph drainage and the development of post-reconstructive leg oedema is not influenced by the type of inguinal incision. A prospective randomised study in patients undergoing femoropopliteal bypass surgery. Euro Vasc Endovasc Surg 10:316–322

Heberer G, Van Dongen RJAM eds (1993) Gefäßchirurgie. Springer, Berlin/Heidelberg/New York/Tokyo, Kirschnersche allgemeine und spezielle Operationslehre, Bd 11

Hecking C, Aschwanden M, Dickenmann M et al (2006) Efficient haemodialysis despite complete central venous thrombosis. Vasa 35:243–244

Hedblad B, Wikstrand J, Janzon L et al (2001) Low-dose metroprolol CR/XL and fluvastatin slow progression of carotid intima-media thickness: main results from the Beta-Blocker Cholesterol-Lowering Asymptomatic Plaque Study (BCAPS). Circulation 103:1721–1726

Heilberger P et al (1997) Postoperative color flow duplex scanning in aortic endografting. J Endovasc Surg 4:262

Heine GH, Gerhart MK, Girndt M et al (2006) (2006) Intrarenale Widerstandsindices und Subklinische Artherosklerose als Prognosemarker bei nierentransplantierten Menschen. Ultraschall Med 27:S62

Heinrich U (1993) In-vitro-Untersuchungen an Gefäßstenosen mit der farbkodierten Duplexsonographie unter besonderer Berücksichtigung der Stenosegradbestimmung. Med Diss, Universität Würzburg

Henao EA, Hodge MD, Felkai DD, McCollum CH, Noon GP, Lin PH, Lumsden AB, Bush RL (2006) Contrast-enhanced duplex surveillance after endovascular abdominal aortic aneurysm repair: improved efficacy using a continuous infusion technique. J Vasc Surg 43:259–264

Henderson RD, Steinman DA, Eliasziw M, Barnett HJ (2000) Effect of contralateral carotid artery stenosis on carotid ultrasound velocity measurements. Stroke 31:2636–2640

Hendrickx PH, Roth U, Brassel F et al (1990) Phantomuntersuchungen zur Wertigkeit der farbkodierten Doppler-Sonographie bei der arteriellen Verschlußkrankheit der unteren Extremitäten. Fortschr Röntgenstr 152(1):1–5

Hendrickx PH, Roth U, von der Lieth H (1991) Wertigkeit der Angiodynographie zur Verlaufskontrolle chirurgischer Gefäßprothesen. Ultraschall Med 12:188–192

Hennerici M, Meairs S (2000) Imaging arterial wall disease. Cerebrovasc Dis 10 (Suppl 5):9–20

Hennerici M, Neuerburg-Heusler D (1988) Gefäßdiagnostik mit Ultraschall. Thieme, Stuttgart

Hennerici M, Steinke W, Rautenberg W (1989) High-resistance Doppler flow pattern in extracranial carotid dissection. Arch Neurol 46:670–672

Henricksen JH, Moller S, Schifter et al (1999) Increased arterial compliance in decompensated cirrhosis. J Hepatol 31:712–718

Herbener TE, Seftel AD, Nehra A, Goldstein I (1994) Penile ultrasound. Semin Urol 12:320–332

Herzog P, Anastasiu M, Wollbrink W, Herrmann W, Holtermüller K-H (1991) Real-time Sonographie bei tiefer Becken- und Beinvenenthrombose. Med Klinik 86:132–137

Hill SL, Donato AT (1994) Ability of the carotid duplex scan to predict stenosis, symptoms and plaque structure. Surgery 116: 914–920

Hirsch AT, Haskal ZJ, Hertzer NR et al (2006) ACC/AHA 2005 practice guidelines for the management of patients with peripheral arterial disease (lower extremities, renal, mesenteric, and abdominal aortic): a collaborative report from the American Association for Vascular Surgery/Society for Vascular Surgery, Society for Cardiovascular Angiography and Interventions, Society for Vascular Medicine and Biology, Society of Interventional Radiology, and the ACC/AHA Task Force on Practice Guidelines (Writing Committee to Develop Guidelines for the Management of Patients With Peripheral Arterial Disease). Circulation 113:e463–e654

Hirschl M, Bernt R (1990) Normalwerte, Reproduzierbarkeit und Aussagekraft duplexsonographischer Kriterien in der Venenfunktionsdiagnostik. Ultraschall Klin Prax 5:81–84

Hoballah JJ, Nazzal MM, Ryan SM, Martinasevic M, Jacobovicz C, Rossley N et al (1997) Is color duplex surveillance of infrainguinal polyetrafluoroethylene grafts worthwhile? Am J Surg 174: 131–135

Hoffmann U, Edwards JM, Carter S et al (1991) Role of duplex scanning for the detection of atherosclerotic renal artery disease. Kidney Int 39:1232–1239
Hofmann W, Forstner R, Sattlegger P, Ugurluoglu A, Magometschnigg H (2001) Die bildgebende Diagnostik pedaler Anschlussgefäße, Gefäßchirurgie. Springer, Originalarbeit, 6:98–102
Hofmann WJ, Walter J, Ugurluoglu A, Czerny M, Forstner R, Magometschnigg H (2004) Preoperative high-frequency duplex scanning of potential pedal target vessels. J Vasc Surg 39:169–175
Hollenbeck M, Hilbert N, Meusel F, Grabensee B (1994) Increasing sensitivity and specificity of Doppler sonographic detection of renal transplant rejection with serial investigation technique. Clin Investig 72:609–615
Hollerweger A, Macheiner P, Rettenbacher T, Gritzmann N (2000) Sonographische Diagnose von Muskelvenenthrombosen des Unterschenkels und deren Bedeutung als Emboliequelle. Ultraschall Med 21:66–72
Homma S, Ishii T, Tsugane S, Hirose N (1997) Different effects of hypertension and hypercholesterolemia on the natural history of aortic atherosclerosis by the stage of intimal lesion. Atherosclerosis 125:85–95
Homma S, Hirose N, Inagaki T, Suzuki M, Wakida Y (1999) Lifestyle, familial history and social background of Japanese centenarians. In: Tauchi H, Sato T, Watanabe T (eds) Japanese centenarians – medical research for the final stages of human aging. Editorial and Publishing Office of Japanese Centenarians, Aichi, Japan, pp 20–35
Homma S, Hirose N, Ishida H, Ishii T, Araki G (2001) Carotid plaque and intima-media thickness assessed by B-mode sonography in subjects ranging from young adults to centenarians. Stroke 32:830–835
Hood DB, Mattos MA, Mansour A et al (1996) Prospective evaluation of new duplex criteria to identify 70% internal carotid artery Stenosis. J Vasc Surg 23:254–261
Horrow MM, Stassi J, Shurman A, Brody JD, Kirby CL, Rosenberg HK (2000) The limitations of carotid sonography: interpretive and technology-related errors. Am J Roentgenol 174:189–194
Hua HT, Hood DB, Jensen CC et al (2000) The use of color flow duplex scanning to detect significant renal artery stenosis. Ann Vasc Surg 14:118–124
Huber TS, Ozaki CK, Flynn TC et al (2002) Prospective validation of an algorithm to maximize native arteriovenous fistulae for chronic hemodialysis access. J Vasc Surg 36:452–459
Hull R, Hirsh J, Sackett D et al (1981) Clinical validity of a negative venogram in patients with clinically suspected venous thrombosis. Circulation 64:622–625
Hull R, Raskop G, Leclerc J, Jay R, Hirsh J (1984) The diagnosis of clinically suspected venous thrombosis. Clin Chest Med 5: 439–456
Hust MH, Schuler A (1992) Farb-Doppler-gesteuerte Kompressionstherapie eines großen Aneurysma spurium der Arteria femoralis nach Linksherzkatheterisierung. Dtsch Med Wochenschr 117:1675–1678
Hust MH, Schuler A, Claußnitzer R, Wild K, Metzler B (1993) Farbdopplergesteuerte Kompressionstherapie. Dtsch Ärztebl 90:B2536–B2541
Huston J, James E, Brown RD Jr et al (2000) Redefined duplex ultrasonographic criteria for the diagnosis of carotid artery stenosis. Mayo Clin Proc 75:1133–1140
Iacob M, Ifrim S, Tanasescu C et al (2005) Duplex ultrasound evaluation of patients with carotid stenosis treated with stent implantation. Ultraschall Med 26:S104
Idu MM, Buth J, Hop WC, Cuypers P, van dc Pavoordt ED, Tordoir JM (1999) Factors influencing the development of vein-graft stenosis and their significance for clinical management. Eur J Vasc Endovasc Surg 17:15–21
Idu MM, Buth J, Hop WC, Cuypers P, van de Pavoordt ED, Tordoir JM (1998) Vein graft surveillance: is graft revision without angiography justified and what criteria should be used? J Vasc Surg 27: 399–411; discussion 412–413
Ihlberg L, Luther M, Tierala E, Lepäntalo M (1998) The utility of duplex scanning in infrainguinal vein graft surveillance: results from a randomised controlled study. Eur J Vasc Endovasc Surg 16:19–27
Ihnat DM, Mills JL, Dawson DL, Hughes JD, Hagino RT, DeMaioribus CA, Gentile AT, Westerband A (1999) The correlation of early flow disturbances with the development of infrainguinal graft stenosis: a 10-year study of 341 autogenous vein grafts. J Vasc Surg 30:8–15
Ikeda M, Fujimori Y, Tankawa H, Iwata H (1984) Compression syndrome of the popliteal vein and artery caused by popliteal cyst. Angiology 35:245–251
Insua JA, Young JR, Humphries AW (1970) Popliteal artery entrapmant syndrome. Arch Surg 101:771–775
Irninger W (1963) Histologische Altersbestimmung von Thrombosen und Embolien. Virchows Archiv Path Anat 336:220–237
Isaacson JA, Neumyer MM (1995) Direct and indirect renal arterial duplex and Doppler colour flow evaluations. J Vasc Technol 193:309–316
Isaacson JA, Zierler RE, Spittell PC, Strandness DE (1995) Noninvasive screening for renal artery stenosis: comparison of renal artery and renal hilar duplex scanning. J Vasc Technol 19:105–110
Ishikawa K (1987) Cystic adventitial disease of the popliteal artery and of other stem vessels in the extremities. Jap J Surg 17:221–229
Iwai T, Sato S, Yamada T et al (1987) Popliteal vein entrapment caused by the third head of the gastrocnemius muscle. Br J Surg 74: 1006–1008
Jaarsveld BC, Krijnen P, Pieterman H et al (2003) The effect of balloon angioplasty on hypertension in atherosclerotic renal-artery stenosis. N Engl J Med 342:1007–1014
Jack CR, Sharma R, Vemuri RB (1984) Popliteal venous aneurysm as a source of pulmonary emboli in a male. Angiology 35:55–57
Jacob A, Stock KW, Proske M, Steinbrich W (1996) Lower extremity angiography: improved image quality and outflow vessel detection with bilaterally antegrade selective digital subtraction angiography. A blinded prospective intraindividual comparison with aortic flush digital subtraction angiography. Invest Radiol 31:184–193
Jacobs NM, Grant EG, Schellinger D et al (1985) Duplex carotid sonography: criteria for stenosis, accuracy, and pitfalls. Radiology 154: 385–391
Jäger K (1987) Apparative Untersuchungen zur Diagnose der tiefen Venenthrombose. Internist 28:299–307
Jäger K (1989) Neuere diagnostische Methoden zur nichtinvasiven Lokalisation und hämodynamischen Beurteilung arterieller Obstruktionen. Internist 30:397–405
Jäger K, Bollinger A (1986) Blood flow velocity and diameter of the popliteal vein. Phlebology 85:260–263
Jäger K, Bollinger A, Siegenthaler W (1986) Duplex-Sonographie in der Gefäßdiagnostik. Dtsch Med Wochenschr 111:1608–1613
Jäger K, Bollinger A, Valli C, Ammann R (1986) Measurement of mesenteric blood flow by duplex scanning. J Vasc Surg 3:462–469
Jäger K, Eichlisberger R, Frauchiger B (1993) Stellenwert der bildgebenden Sonographie für die Diagnostik der Venenthrombose. Hamostaseologie 13:116–124
Jäger K, Martin PDL, RL HC, Roederer GO, Langlois YE, Strandness DE (1985) Noninvasive mapping of lower limb arterial lesions. Ultrasound Med Biol 11:515–521
Jäger K, Ricketts HJ, Strandness DE (1985) Duplex scanning for the evaluation of lower limb arterial disease. In: Bernstein EF (ed) Noninvasive diagnostic techniques in vascular disease. Mosby, St. Louis
Jäger K, Seifert H, Bollinger A (1989) M-mode echovenography. A new technique for the evaluation of venous wall and venous valve motion. Cardiovasc Res 23:25–30
Janzarik WG, Ringleb PA, Reinhard M et al (2007) Recurrent carotid artery vasospasms. Report of 2 cases. Stroke 37:2170–2173
Jennersjo CM, Fragerberg ICH, Karlander SG et al (2005) Normal D-dimer concentration is a common finding in symptomatic

outpatients with distal vein thrombosis. Blood Coagul Fibrinolysis 16:517–523

Johnson BF, Manzo RA, Bergelin RO, Strandness DE (1995) Relationship between changes in the deep venous system and the development of the postthrombotic syndrome after an acute episode of lower limb deep vein thrombosis: a one- to six-year follow-up. J Vasc Surg 21:307–313

Johnston DC, Chapman KM, Goldstein LB (2001) Low rate of complications of cerebral angiography in routine clinical practice. Neurology 57:2012–2014

Johnston KW, Rutherford RB, Tilson MD et al (1991) Suggested standards for reporting on arterial aneurysms. Ad Hoc Committee on Reporting Standards, Society for Vascular Surgery and North American Chapter, International Society for Cardiovascular Surgery. J Vasc Surg 13:459

Jongbloets LM, Lensing AW, Koopman MM, Buller HR, Cate JW (1994) Limitations of compression ultrasound for the detection of symptomless postoperative deep vein thrombosis. Lancet 343:114–144

Jorgensen JO, Hanel KC, Morgan AM, Hunt JM (1993) The incidence of deep venous thrombosis in patients with superficial thrombophlebitis of the lower limbs. J Vasc Surg 18:70–73

Juenemann KP, Persson-Juenenemann C, Alken P (1990) Pathophysiology of erectile dysfunction. Semin Urol 8:80

Jung EM, Kubale R, Clevert DA, Weskott HP, Prantl L, Herold T, Renz M, Rupp N, Tacke J (2007) B-flow and B-flow with 3D and SRI postprocessing before intervention and monitoring after stenting of the internal carotid artery. Clin Hemorheol Microcirc 36:35–46

Jung EM, Kubale R, Ritter G, Gallegos MT, Jungius KP, Rupp N, Clevert DA (2007) Diagnostics and characterisation of preocclusive stenoses and occlusions of the internal carotid artery with B-flow. Eur Radiol 17:439–447

Jung M (1990) Diagnosis of deep leg vein thrombosis: can we dispense with phlebography? Vasa [Suppl]: 30

Justich E (1982) The compression syndrome of the left renal vein. Rofo 136:404–412

Jutley RS, Cadle I, Cross KS (2001) Preoperative assessment of primary varicose veins: a duplex study of venous incompetence. Eur J Vasc Endovasc Surg 21:370–373

Kakkar VV (1972) The diagnosis of deep vein thrombosis using the 125I-fibrinogen test. Arch Surg 104:152–159

Kakkar VV, Howe CT, Flanc C et al (1969) Natural history of postoperative deep venous thrombosis. Lancet 2:230–232

Kakkos SK, Stevens JM, Nicolaides AN et al (2007) Texture analysis of ultrasonic images of symptomatic carotid plaques can identify those plaques associated with ipsilateral embolic brain infarction. Eur J Vasc Endovasc Surg 33:422–429

Kanterman RY, Vesely TM, Pilgram TK et al (1995) Dialysis access graft: anatomic location of venous stenosis and result of angioplasty. Radiology 195:135–139

Kanterman RY, Darcy MD, Middleton WD et al (1997) Doppler sonography findings associated with transjugular intrahepatic portosystemic shunt malfunction. Am J Roentgenol 168:467–472

Kaps M, Seidel G (1999) Echokontrastverstärkung in der neurologischen Ultraschalldiagnostik. Dt Ärztebl 96:A 276–A 280

Karacagil S, Almgren B, Bergström R, Bowald S, Eriksson I (1989) Postoperative predictive value of a new method of intraoperative angiographic runoff assessment in femoropopliteal bypass grafting. J Vasc Surg 10:400–407

Karacagil S, Löfberg AM, Almgren B et al (1994) Duplex ultrasound scanning for diagnosis of aortoiliac and femoropopliteal arterial disease. Vasa 23:325–329

Karacagil S, Löfberg AM, Granbo A, Lörelius LE, Bergqvist D (1996) Value of duplex scanning in evaluation of crural and foot arteries in limbs with severe lower limb ischaemia – a prospective comparison with angiography. Eur J Vasc Endovasc Surg 12:300–303

Karasch T, Neuerburg-Heusler D, Strauss A, Rieger H (1994) Farbduplexsonographische Kriterien zur Diagnose arteriosklerotischer Nierenarterienstenosen und -verschlüsse. In: Keller E, Krumme B Hrsg (eds) Farbkodierte Duplexsonographie in der Nephrologie. Springer, Berlin/Heidelberg/New York/Tokyo

Karasch T, Rieser R, Grün B, Strauss AL, Neuerburg-Heusler D, Roth F-J, Rieger H (1993) Bestimmung der Verschlusslänge in Extremitätenarterien. Farbduplexsonographie versus Angiographie. Ultraschall Med 14:247–254

Karasch T, Strauss AL, Grün B, Worringer M, Neuerburg-Heusler D, Roth F-J, Rieger H (1993) Farbcodierte Duplexsonographie in der Diagnostik von Nierenarterienstenosen. Dtsch Med Wochenschr 118:1429–1436

Karasch Th, Strauss AL, Worringer M, Neuerburg-Heusler D, Roth FJ, Rieger H (1993) Vergleich der Farbduplexsonographie mit verschiedenen angiographischen Verfahren in der Diagnostik arteriosklerotischer Nierenarterienstenosen und -verschlüsse. Ultraschall Klin Prax: 8 180

Kasai C, Namekawa K, Koyano R, Omoto R (1985) Real-time two-dimensional blood flow imaging using an autocorrelation technique. IEEE Trans Son Ultrason 32:458–463

Kathrein H (1991) Duplexsonographie von Dialyseshunts. Springer, Berlin/Heidelberg/New York/Tokyo

Kathrein H, König P, Dittrich P, Judmeier G (1988) Nichtinvasisve Beurteilung von Cimino-Brescia-Fisteln und PTFE-Shunts mit der Duplexsonographie. VASA Suppl 26:39–41

Katsamouris AN, Giannoukas AD, Tsetis D, Kostas T, Petinarakis I, Gourtsoyiannis N (2001) Can ultrasound replace arteriography in the management of chronic arterial occlusive disease of the lower limb? Eur J Vasc Endovasc Surg 21:155–159

Katz ML, Johnson M, Pomajzl MJ et al (1983) The sensitivity of real time B-mode carotid imaging in detection of ulcerated plaque. Bruit 8:13–16

Kearon C, Julian JA, Newnan TE, Ginsberg JS (1998) Noninvasive diagnosis of deep vein thrombosis. McMaster Diagnostic Imaging Practice Guidelines Initiative. Ann Intern Med 128:663–677

Keleinouridis V, Eckstein MR, Dembner AG, Waitman AG, Athanasoulis CA (1985) The normal leg venogram: significance in suspected vein thrombosis. Int Angiol 4:369–371

Keller F, Harnoss B-M, Czerlinsky H (1991) Umfrageergebnisse zur Technik der Ciminofistel. Angio Archiv 22:7–9

Keller F, Loewe HJ, Bauknecht KJ, Schwarz A, Offermann G (1988) Kumulative Funktionsraten von orthotopen Dialysefisteln und Interponaten. Dtsch Med Wschr 113:332–336

Kerlan RK Jr, LaBerge JM, Gordon RL, Ring EJ (1995) Transjugular intrahepatic portosystemic shunts: current status. Am J Roentgenol 164:1059–1066

Kerr ThM, Canley JJ, Johnson JR, Lutter KS, Riechmann GC, Cranley RD, True MA, Sampson M (1990) Analysis of 1084 consecutive lower extremities involved with acute venous thrombosis diagnosed by dublex scanning. Surgery 108:520–527

Kerr ThM, Lutter KS, Moeller DM, Hasselfeld KA, Roedersheimer LR, McKenna PJ, Winkler JL, Spirtoff K, Sampson MG, Cranley JJ (1990) Upper extremity venous thrombosis diagnosed by dublex scanning. Am J Surg 160:202–206

Kessler C, von Maravic M, Brückmann H, Kömpf D (1995) Ultrasound for the assessment of the embolic risk of carotid plaques. Acta Neurol Scand 92:231–234

Kiews P-W (1991) Color velocity imaging – Ein Vergleich der Verfahren zur farbkodierten Sonographie. Roentgenstrahlen (Philips Medizin Systeme) 65:1–6

Kiews P-W (1993) Physik und Technik der farblodierten Duplexsonographie (FKDS) In: Wolf KJ, Fobbe F (eds) Farbkodierte Duplexsonographie. Thieme, Stuttgart/New York, pp 248–295

Killewich LA, Bedford GR, Beach KW, Strandness DE Jr (1989) Diagnosis of deep venous thrombosis. A prospective study comparing duplex scanning to contrast venography. Circulation 79: 810–814

Killewich LA, Fisher C, Bartlett ST (1990) Surveillance of in situ infrainguinal bypass grafts: conventional vs color flow duplex ultrasonography. J Cardiovasc Surg 31:662

Killewich LA, Nunnelee JD, Auer AI (1993) Value of lower extremity venous duplex examination in the diagnosis of pulmonary embolism. J Vasc Surg 17:934–1030

Killewich LA, Bedford GR, Beach KW, Standness DE (1998) Spontaneous lysis of deep venous thrombi: rate and outcome. J Vasc Surg 9:89–97

Kimura K, Yasaka M, Moriyasu H, Tsuchiya T, Yamaguchi T (1994) Ultrasonographic evaluation of vertebral artery to detect vertebrobasilar axis occlusion. Stroke 25:1006–1009

Kinney EV, Bandyk DF, Mewissen MW et al (1991) Monitoring functional patency of percutaneous transluminal angioplasty. Arch Surg 126:743–747

Kirsch JD, Wagner LR, James EM, Charboneau JW, Nichols DA, Meyer FB, Hallett JW (1994) Carotid artery occlusion: positive predictive value of duplex sonography compared with angiography. J Vasc Surg 19:642–649

Kitamura A, Iso H, Imano H, Ohira T, Okada T, Sato S, Kiyama M, Tanigawa T, Yamagishi K, Shimamoto T (2004) Carotid intima-media thickness and plaque characteristics as a risk factor for stroke in Japanese elderly men. Stroke 35:2788–2794

Kliewer MA, Tupler RH, Hertzberg BS et al (1994) Doppler evaluation of renal artery stenosis: interobserver agreement in the interpretation of waveform morphology. Am J Roentgenol 162: 1371–1376

Kliewer MA, Herzberg BS, Keogan MT et al (1997) Early systole in the healthy kidney: variability of Doppler US waveform parameters. Radiology 205:109–113

Koelemay MJ, Legemate DA, van Gurp J, Ponson AE, Reekers JA, Jacobs MJ (1997) Colour duplex scanning and pulse-generated runoff for assessment of popliteal and cruropedal arteries before peripheral bypass surgery. Br J Surg 84:1115–1119

Koelemay MJW, Legemate DA, van Gurp JA, de Vos H, Balm R, Jacobs MJHM (2001) Interobserver variation of colour duplex scanning of the popliteal, tibial and pedal arteries. Eur J Vasc Endovasc Surg 21:160–164

Koennecke HC, Fobbe F, Hamed MM, Wolf KJ (1989) Diagnostik arterieller Gefäßerkrankungen der unteren Extremitäten mit der farbkodierten Duplexsonographie. Fortschr Röntgenstr 151:42–46

Kohler TR (1990) Evaluation of lower limb vessels. Clin Diagn Ultrasound 26

Kohler TR, Nance DR, Cramer MM et al (1987) Duplex scanning for diagnosis of aortoiliac and femoropopliteal disease: a prospective study. Circulation 76:1074–1080

Kohler TR, Strandness DE Jr (1986) Noninvasive testing for the evaluation of chronic venous disease. World J Surg 106:903–910

Koksoy C, Kuzu A, Kutlay J, Erden I, Ozcan H, Ergin K (1995) The diagnostic value of colour Doppler ultrasound in central venous catheter related thrombosis. Clin Radiol 50:687

Koopman MMW, Prandoni P, Piovella F et al (1996) Treatment of venous thrombosis with intravenous unfractionated heparin administered in the hospital as compared with subcutaneous low-molecular-weight heparin administered at home. N Engl J Med 334: 677–681

Koppenhagen K, Fobbe F (1993) Diagnostik der tiefen Beinvenenthrombose. Hämostaseologie Suppl 13:12–14

Korten E, Toonder IM, Schrama YC et al (2007) Dialysis fistulae patency and preoperative diameter ultrasound measurement. Eur J Vasc Endovasc Sur 33:467–471

Krause U, Kock HJ, Kröger K, Albrecht K, Rudofsky G (1998) Prevention of deep venous thrombosis associated with superficial thrombophlebitis of the leg by early saphenous vein ligation. VASA 27:34–38

Kreitner KF, Kalden P, Neufang A et al (2000) Diabetes and peripheral arterial occlusive disease: prospective comparison of contrast enhanced three-dimensional MR angiography with conventional digital subtraction angiography. Am J Roentgenol 174:171–179

Kremkau FW (1990) Doppler Ultrasound: Principles and Instruments. Saunders, Philadelphia

Kriessmann A, Bollinger A (Hrsg) (1978) Ultraschall-Doppler-Diagnostik in der Angiologie. Thieme, Stuttgart

Kriessmann A, Bollinger A, Keller H (Hrsg) (1982) Praxis der Doppler-Sonographie. Thieme, Stuttgart

Krings W, Adolph J, Diederich S, Urhahne S, Vassallo P, Peters PE (1990) Diagnostik der tiefen Becken- und Beinvenenthrombose mit hochauflösender real-time und CW-Doppler-Sonographie. Radiologe 30:525–531

Krishnabhakdi S, Espinola-Klein C, Kurz G, Neufang A, Schmidt W, Oelert H (2001) Sonographisches Venenmapping – Stellenwert in der distalen Bypasschirurgie. Ultraschall Med 22 Suppl: 60

Kroegel C, Reissig A (2003) Principle mechanisms underlying venous thromboembolism: epidemiology, risk factors, pathophysiology and pathogenesis. Respiration 70:7–30

Kroegel C (2003) Advances in the diagnosis and treatment of pulmonary embolism. Pulmonary embolism – how can you mend a broken clot? Respiration 70:4–6

Krumme B, Blum U, Schwertfeger E et al (1996) Diagnosis of renovascular disease by intra- and extrarenal Doppler scanning. Kidney Int 50:1288–1292

Krumme B, Lehnert T, Wollschläger H, Keller E (1995) Möglichkeiten und Grenzen der farbduplexgesteuerten Kompressionstherapie von punktionsbedingten Gefäßläsionen in der Leiste. Fortschr Röntgenstr 163:158–162

Krysiewicz S, Mellinger BC (1989) The role of imaging in the diagnostic evaluation of impotence. Am J Roentgenol 153:1133–1139

Kubale R (1987) Renovaskuläre Erkrankungen. In: Seitz K, Kubale R. Duplex Sonographie der abdominellen und retroperitonealen Gefäße. Weinheim

Kubale R (1993) Abdominelle Venen, portalvenöses System und Leber. In: Wolf K-J, Fobbe F (eds) Farbkodierte Duplexsonographie. Thieme, Stuttgart, pp 158–184

Kubale R (1994) Abdominelle und retroperitoneale Gefäße. In: Rettenmaier G, Seitz KH (eds) Sonographische Differentialdiagnostik. Chapman & Hall, Weinheim, S. 865ff

Kuzniec S, Kauffman P, Molnar LJ et al (1998) Diagnosis of limb and neck arterial trauma using duplex ultrasonography. Cardiovasc Surg 6:358–366

Labropoulos N, Volteas SK, Giannoukas AD, Toulpoupakis E, Delis K, Nicolaides AN (1996) Asymptomatic popliteal vein aneurysms. Vasc Surg 6:453–458

Labropoulos N, Leon LR Jr, Gonzalez-Fajardo JA et al (2007) Nonatherosclerotic pathology of the neck vessels: prevalence and flow patterns. Vasc Endovascular Surg 41:417–427

Labs K-H (1991) Die Aussagekraft der Dopplerspektralanalyse für die praktische Diagnostik. VASA Suppl 32:72–84

Lafortune M, Marleau D, Breton G et al (1984) Portal venous system measurements in portal hypertension. Radiology 151:27–30

Lafortune M, Patriquin H, Pomier G et al (1987) Hemodynamic changes in portal circulation after portosystemic shunts: use of duplex sonography in 43 patients. Am J Roentgenol 149:701–706

Lagerstedt CL, Olsson CC, Fagher BO, Öqvist BW, Albrechtsson U (1985) Need for long-term anticoagulant treatment in symptomatic calf-vein rhrombosis. Lancet 2:515–518

Lal BK, Hobson RW 2nd, Goldstein J, Chakhtoura EY, Duran WN (2004) Carotid artery stenting: is there a need to revise ultrasound velocity criteria? J Vasc Surg 39:58–66

Lal BK, Hobson RW 2nd, Hameed M, Pappas PJ, Padberg FT Jr, Jamil Z, Duran WN (2006) Noninvasive identification of the unstable carotid plaque. Ann Vasc Surg 20:167–174

Lal BK, Hobson RW 2nd, Tofighi B, Kapadia I, Cuadra S, Jamil Z (2008) Duplex ultrasound velocity criteria for the stented carotid artery. J Vasc Surg 47:63–73

Landry GJ, Moneta GL, Taylor LM, McLafferty RB, Edwards JM, Yeager RA et al (1999) Duplex scanning alone is not sufficient

imaging before secondary procedures after lower extremity reversed vein bypass graft. J Vasc Surg 29(2):270–280
Landwehr P, Lackner K (1990) Farbkodierte Duplexsonographie vor und nach PTA der Arterien der unteren Extremität. Fortschr Röntgenstr 152:35–41
Landwehr P, Tschammler A, Höhmann M (1990) Gefäßdiagnostik mit der farbkodierten Duplexsonographie. Dtsch Med Wochenschr 115:343–351
Lang W (2002) Arterielle Gefäßdiagnostik beim diabetischen Fußsyndrom. Gefäßchirurgie 7:122–127
Lange P, Houe T, Helgstrand UJV (2001) The efficacy of ultrasound-guided compression of iatrogenic femoral pseudo-aneurysms. Eur J Vasc Endovasc Surg 21:248–250
Lange SF, Trampisch HJ, Pittrow D, Darius H, Mahn M, Allenberg JR, Tepohl G, Haberl RL, Diehm C, for the getABI Study Group (2007) Profound influence of different methods for determination of the ankle brachial index on the prevalence estimate of peripheral arterial disease. BMC Public Health 7:147
Langerstedt CL, Olsson CG, Fagher BO et al (1985) Need for long-term anticoagulant treatment in symptomatic calf vein thrombosis. Lancet 2:515–518
Langholz J (1998) Investigation of peripheral arterial disease: the expanding role of echo-enhanced color flow Doppler and duplex sonography. Eur J Ultrasound 7(Suppl 3):53–61
Langholz J, Heidrich H (1991) Sonographische Diagnose der tiefen Becken-/Beinvenenthrombose: Ist die farbkodierte Duplexsonographie „überflüssig"? Ultraschall Med 12:176–181
Langholz J, Maul R, Heidrich H (1994) Was ist eine "hypoplastische" A. vertebralis? Überlegungen auf der Basis duplexsonographischer Untersuchungen. Vasa Suppl 43:50 (Abstr)
Langholz J, Schlief R, Heidrich H (1992) Verbesserung der farbcodierten Duplexsonographie durch Kontrastmittelgabe bei schwer untersuchbaren Regionen der peripher-arteriellen Becken-Bein-Strombahn. Ultaschall Med 13:234–238
Langholz JP (1997) Ultrasound contrast agents in peripheral vascular disease. In: Nanda NC, Schlief R, Goldberg BB (eds) Advances in echo imaging using contrast enhancement, 2nd edn. Kluwer, Dordrecht, pp 543–560
Langsfeld M, Gray-Weale AC, Lusby RJ (1989) The role of plaque morphology and diameter reduction in the development of new symptoms in asymptomatic carotid arteries. J Vasc Surg 9:548–557
Langsfeld M, Hershey FB, Thorpe L et al (1987) Duplex B-mode imaging for the diagnosis of deep venous thrombosis. Arch Surg 122: 587–591
Lanzer P, Yoganathan AP (1991) Vascular imaging by color Doppler and magnetic resonance. Springer, Berlin/Heidelberg/New York/Tokyo
Larch E (1993) Farbcodierte Duplexsonographie zur Beurteilung der Unterschenkelarterien bei peripherer arterieller Verschlußkrankheit. Vasa Suppl 41:14
Larch E, Minar E, Ahmadi R, Schnurer G, Schneider B, Stumpflen A, Ehringer K (1997) Value of color duplex sonography for evaluation of tibioperoneal arteries in patients with femoropopliteal obstruction: a prospective comparison with anterograde intraarterial digital subtraction angiography. J Vasc Surg 25:629–636
Laub G (1999) Principles of contrast-enhanced MR angiography. Basic and clinical applications. Magn Reson Imag Clin North Am 7:783–795
Lausen M, Jensen R, Wille-Jorgensen P et al (1995) Colour Doppler flow imaging ultrasonography versus venography as screening method for asymptomatic postoperative deep venous thrombosis. Eur J Radiol 20:200–204
Lee HM, Wang Y, Sostman HD et al (1998) Distal lower extremity arteries: evaluation with two-dimensional MR digital subtraction angiography. Radiology 207:505–512
Lee VS, Hertzberg BS, Workman MJ, Smith TP et al (2000) Variability of Doppler US measurements along the common carotid artery: effects on estimates of internal carotid arterial stenoses in patients with angiographically proved disease. Radiology 214:387–392
Legemate DA, Teeuwen C, Hoeneveld H, Eikelboom BC (1991) Value of duplex scanning compared with angiography and pressure measurement in the assessment of aortoiliac arterial lesions. Br J Surg 78:1003–1008
Leiner T, de Haan MW, Nelemans PJ et al (2005) Contemporary imaging techniques for the diagnosis of renal artery stenosis. Eur Radiol 15:2219–2229
Lensing AWA, Doris CI, McGrath FP, Cogo A, Sabine MJ, Ginsberg J et al (1997) A comparison of compression ultrasound with color Doppler ultrasound for the diagnosis of symptomless postoperative deep vein thrombosis. Arch Intern Med 157:765–768
Lensing AWA, Prandoni P, Brandjes D et al (1989) Detection of deep vein thrombosis by real-time B-mode ultrasonography. New Engl J Med 320:342–345
Lepore T, Savran J, van de Water J, Harrower H, Yablonski M (1978) Screening for lower extremity deep venus thrombosis. Am J Surg 135:529–534
Lerner RM, Medorach RA, Hulbert WC, Rabinowitz R (1990) Color Doppler US in the evaluation of acute scrotal disease. Radiology 176:355–358
Leu HJ (1973) Histologische Altersbestimmung von arteriellen und venösen Thromben und Embolie. Vasa 2:265–273
Leu HJ, Bollinger A, Pouliadis G, Brunner U, Soyka P (1977) Pathologie, Klinik, Radiologie und Chirurgie der zystischen Adventitia-Degeneration peripherer Blutgefäße. Vasa 6:94–99
Leung A, Hampson SJ, Singh MP et al (1983) Ultrasonic diagnosis of bilateral congenital internal jugular venous aneurysms. Br J Radiol 56:588–591
Levine M, Gent M, Hirsh J et al (1996) A comparison of low-molecular-weight heparin administered primarily at home with unfractionated heparin administered in the hospital for proximal deep vein thrombosis. N Engl J Med 334:677–681
Levy MM, Baum RA, Carpenter JP (1998) Endovascular surgery solely based on noninvasive preprocedural imaging. J Vasc Surg 28: 995–1003
Lewis BD, James EM, Charboneau JW et al (1989) Current applications of color Doppler imaging in the abdomen and extremities. Radiographics 9:599–631
Lewis BD, James EM, Welch TJ, Joyce JW, Hallett JW, Weaver AL (1994) Diagnosis of acute deep venous thrombosis of the lower extremities: prospective evaluation of colour Doppler flow imaging versus venography. Radiology 192:651–655
Lewis SC, Wardlaw JM (2002) Which Doppler velocity is best for assessing suitability for carotid endarterectomy? Eur J Ultrasound 15:9–20
Li R, Cai J, Tegeler C et al (1996) Reproducibility of extracranial carotid atherosclerotic lesions assessed by B-mode ultrasound: the ARIC Study. Ultrasound Med Biol 22:791–799
Liapis CD, Gugulakis A, Misiakos E, Verkokos C, Dousaitou B, Sechas M (1994) Surgical treatment of extracranial carotid aneurysms. Int Angiol 13:290–295
Liapis CD, Kakisis JD, Kostakis AG (2001) Carotid stenosis: factors affecting symptomatology. Stroke 32:2782–2786
Libby P (2002) The fire within. Sci Am 286:46–55
Liewald F et al (2001) Influence of treatment of type II leaks on the aneurysm surface area. Eur J Vasc Endovasc Surg 21:339–343
Ligush J, Reavis SW, Preisser JS, Hansen KJ (1998) Duplex ultrasound scanning defines operative strategies for patients with limb-threatening ischemia. J Vasc Surg 28:482–491
Limberg B (1991) Duplexsonographische Diagnose der portalen Hypertension bei Leberzirrhose. Einfluss einer standardisierten Testmahlzeit auf die portale Hämodynamik. Dtsch Med Wochenschr 116:1384–1387

Limet R, Sakalihassan N, Albert A (1991) Determination of the expansion rate and incidence of rupture of abdominal aortic aneurysms. J Vasc Surg 14:540–548

Lindholt JS, Vammen S, Juul S, Henneberg EW, Fasting H (1999) The validity of ultrasonographic scanning as screening method for abdominal aortic aneurysm. Eur J Vasc Endovasc Surg 17:472–475

Lindquist R (1977) Ultrasound as a complementary diagnostic method in deep vein thrombosis of the leg. Acta Med Scand 201:435

Lippert H, Pabst R (1985) Arterial variations in man. Bergmann, München

Lockhart ME, Robbin MI, Allon M (2004) Preoperative sonographic radial artery evaluation and correlation with subsequent radiocephalic fistula outcome. J Ultrasound Med 23:161–168

Lockhart ME, Robbin ML (2001) Hemodialysis access ultrasound. Ultrasound Q 17:157–167

Lohr JM, Kerr TM, Lutter KS, Cranley RD (1991) Lower extremity calf thrombosis: to treat or not to treat? J Vasc Surg 14:618–623

London GM, Safar ME (1989) Abnormalities of intrarenal hemodynamics in essential and renovascular hypertension in man. Presse Med 18:1459–1460

London GM, Safar ME (1989) Renal hemodynamics in patients with sustained essential hypertension and in patients with unilateral stenosis of the renal artery. Am J Hypertens 2:244–252

Londrey GL, Hodgson KJ, Spadone DP, Ramsey DE, Barkmeier LD, Summer DS (1990) Initial experience with colour-flow duplex scanning of infrainguinal bypass grafts. J Vasc Surg 12:284–290

Longo JM, Bilbao JI, Rousseau HP et al (1992) Color Doppler guidance in transjugular placement of intrahepatic portosystemic shunts. Radiology 184:281–283

Lopez JA, Espeland MA, Jarow JP (1992) Interpretation and quantification of penile blood flow studies using duplex ultrasonography. J Urol 146:1271

Lovelace TD, Moneta GL, Abou-Zamzam AM Jr et al (2001) Optimizing duplex follow-up in patients with an asymptomatic internal carotid artery stenosis of less than 60%. J Vasc Surg 33:56–61

Lovett JK, Gallagher PJ, Hands LJ, Walton J, Rothwell PM (2004) Histological correlates of carotid plaque surface morphology on lumen contrast imaging. Circulation 110:2190–2197

Ludwig D, Schwarting K, Korbel CM, Bruning A, Schiefer B, Stange EF (1998) The postprandial portal flow is related to the severity of portal hypertension and liver cirrhosis. J Hepatol 28:631–638

Ludwig M (1991) Quantitative Flußmessungen an Venen. Ultraschalldreiländertreffen, Lausanne

Ludwig M, Stumpe KO (1994) Karotisultraschall in der Früherkennung der Atherosklerose. Dtsch Ärztebl 91:745–746

Ludwig M, Mv P-K, Stumpe KO (2003) Intima media thickness of the carotid arteries: early pointer to arteriosclerosis and therapeutic endpoint. Ultraschall Med 24:162–174

Lue TF (1991) Physiology of penile erection. In: Jonas U, Thon WF, Stief CG (eds) Erectile dysfunction. Springer, Berlin/Heidelberg/New York, pp 44–65

Lujan S, Criado E, Puras E, Izquierdo LM (2002) Duplex scanning or arteriography for preoperative planning of lower limb revascularisation. Eur J Vasc Endovasc Surg 24:31–36

Lundell A, Lindblad B, Bergqvist D, Hansen F (1995) Femoropopliteal-crural graft patency is improved by an intensive surveillance program: a prospective randomized study. J Vasc Surg 21:26–34

Lusby RJ (1993) Plaque characterisation: does it identify high risk groups? In: Bernstein EF, Callow AD, Nicolaides AN, Shifrin EG (eds) Cerebral revascularisation. Med-Orion, London/Los Angeles/Nicosia, pp 93–107

Luska G, Risch U, Pellengahr M, von Boetticher H (1990) Farbcodierte dopplersonographische Untersuchungen zur Morphologie und Hämodynamik der Arterien des Beckens und der Beine bei gesunden Probanden. Fortschr Röntgenstr 153:246–251

Lutter KS, Kerr ThM, Roedersheimer LR, Lohr JM, Sampson MSG, Cranley JJ (1991) Superficial thrombophlebitis diagnosed by duplex scanning. Surgery 110:42–46

MacMathuna P, Vlavianos P, Westaby D, Williams R (1992) Pathophysiology of portal hypertension. Dig Dis (Suppl 1): 3–15

Magnusson MB, Nelzen O, Risberg B, Sivertsson R (2001) A colour Doppler ultrasound study of venous reflux in patients with chronic leg ulcers. Eur J Vasc Endovasc Surg 21:353–360

Makris SA, Karkos CD, Awad S, London NJ (2006) An "all-comers" venous duplex scan policy for patients with lower limb varicose veins attending a one-stop vascular clinic: is it justified? Eur J Vasc Endovasc Surg 32:718–724

Mannami T, Baba S, Ogata J (2000) Potential of carotid enlargement as a useful indicator affected by high blood pressure in a large general population of a Japanese city: the Suita study. Stroke 31: 2958–2965

Manthey J, Munderloh KH, Mautner JP, Köhl M, Fröhlich G (1994) Popliteal venous aneurysm with pulmonary and paradoxical embolization. Vasa 23:264–267

Marin J, Gosselin J, Khayat A (1978) A propos d'un cas d'anévrisme de la veine poplitée avec embolie pulmonaire. Phlebologie 31: 433–438

Markel A, Manzo RA, Bergelin RO, Strandness DE (1992) Valvular reflux after deep vein thrombosis: incidence and time of occurrence. J Vasc Surg 15:377–384

Markel A, Monzo R, Bergelin RO, Strandness DE (1992) Pattern and distribution of thrombi in acute venous thrombosis. Arch Surg 127:305–309

Marshall M (1990) Die Duplex-Sonographie bei phlebologischen Fragestellungen in Praxis und Klinik. Ultraschall Klin Prax 5:51–56

Marshall M (1990) Sklerosierungsreaktion großer Varizen im hochauflösenden Ultraschallbild. Phlebol Proktol 19:205–214

Martin RL, Nanra RS, Wlodarczyk J et al (1991) Renal hilar Doppler analysis in the detection of renal artery stenosis. J Vasc Tech 15:173–180

Mathiesen EB, Bonaa KH, Joakimsen O (2001) Echolucent plaques are associated with high risk of ischemic cerebrovascular events in carotid stenosis: the Tromso Study. Circulation 103:2171–2175

Mathis G, Bitschnau R, Gehmacher O et al (1999) Chest ultrasound in diagnosis of pulmonary embolism in comparison to helical CT. Ultraschall Med 20:54–59

Mathis G, Blank W, Reißig A et al (2005) Thoracic ultrasound for diagnosing pulmonary embolism. A prospective multicenter study of 352 patients. Chest 128:1531–1538

Matsagas MI, Vasdekis SN, Gugulakis AG et al (2000) Computer-assisted ultrasonographic analysis of carotid plaques in relation to cerebrovascular symptoms, cerebral infarction, and histology. Ann Vasc Surg 14:130–137

Mattos MA, Londrey GL, Leutz DW et al (1992) Color-flow duplex scanning for the surveillance and diagnosis of acute deep venous thrombosis. J Vasc Surg 15:366–376

Mattos MA, van Bemmelen PS, Barkmeier ID et al (1993) Routine surveillance after carotid endarterectomy: does it affect clinical management? J Vasc Surg 17:819–830

May R, Mignon G (1976) Spindelförmiges Aneurysma der Vena fibularis. Fortschr Röntgenstr 25:563–564

May R, Nissl R (1968) Aneurysma der Vena poplitea. Fortschr Röntgenstr 108:402–403

May R, Nissl R (1973) Die Phlebographie der unteren Extremität. Thieme, Stuttgart/New York

Mazzariol F, Ascher E, Salles-Cunha S, Grade P, Hingorani A (1999) Values and limitations of duplex ultrasonography as the sole imaging method of preoperative evaluation for popliteal and infrapopliteal bypasses. Ann Vasc Surg 13:1–10

Mazzariol F, Hingorani A, Gunduz Y, Yorkovich W, Salles-Cunha S (2000) Lower-extremity revascularisation without preoperative contrast arteriography in 185 cases: lessons learned with duplex ultrasound arterial mapping. Eur J Vasc Endovasc Surg 19:509–515

McLachan MSF, Thomson JG, Taylor DW, Kelly ME, Sackett DL (1979) Observer variation in the interpretation of lower limb venograms. Am J Radiol 132:227–229

McWilliams RG et al (2002) Detection of endoleak with enhanced ultrasound imaging: comparison with biphasic computed tomography. J Endovasc Ther 9:170–179

Meairs S, Hennerici M (1999) Four-dimensional ultrasonographic characterization of plaque surface motion in patients with symptomatic and asymptomatic carotid artery stenosis. Stroke 30: 1807–1813

Meerbaum S (1997) Microbubble fluid dynamics of echocontrast. In: Nanda NC, Schlief R, Goldberg BB (eds) Advances in echo imaging using contrast enhancement, 2nd edn. Kluwer, Dordrecht, pp 11–38

Meissner MH, Caps MT, Zierler BK, Bergelin RO, Manzo RA, Strandness DE (2000) Deep venous thrombosis and superficial venous reflux. J Vasc Surg 32:48–56

Melany ML, Grant EG, Farooki S et al (1999) Effect of US contrast agents on spectral velocities: in vivo evaluation. Radiology 211:427–431

Meltzer RS, Tickner EG, Sahines TP, Popp RL (1980) The source of ultrasound contrast effect. J Clin Ultrasound 8:121–127

Mendes RR, Farber MA, Marston WA et al (2002) Prediction of wrist arteriovenous fistula maturation with preoperative vein mapping with ultrasonography. J Vasc Surg 36:460–463

Mercer KG, Scott DJA, Turton EPL, Berridge DC, Weston MJ (1999) Can intra-operative flow measurements identify grafts for intensive duplex surveillance? St James University Hospital NHS Trust, Leeds

Merkus JW, Zeebregts CJ, Hoitsman AJ et al (1993) High incidence of arteriovenous fistula after biopsy of kidney allografts. Br J Surg 80:310–312

Merritt CRB, Bluth EI (1992) Ultrasound identification of plaque composition. In: Labs KH et al (eds) Diagnostic vascular ultrasound. Arnold, London, pp 213–223

Metz V, Braunsteiner A, Grabenwöger F et al (1988) Farbcodierte Doppler-Sonographie der Becken-Bein-Arterien: Überprüfung der Wertigkeit der Methode im Vergleich zur Angiographie. Fortschr Röntgenstr 149:314–316

Metz V, Dock W, Grabenwöger F et al (1992) Wertigkeit unterschiedlicher klinischer und bildgebender Verfahren für die postoperative Verlaufskontrolle extraanatomischer Bypasses. Fortschr Röntgenstr 154:172–175

Mewissen MW, Kinney EV, Bandyk DF et al (1992) The role of duplex scanning versus angiography in predicting outcome after balloon angioplasty in the femoropopliteal artery. J Vasc Surg 15:860–864

Meyer JL, Khalil RM, Obuchowski NA, Baus LK (1997) Common carotid artery: variability of Doppler US velocity measurements. Radiology 204:339–341

Meyer P, Rudofsky G, Nobbe N (1986) Das Histogramm des Okklusionsmaterials – ein neuer Prognoseparameter der thrombolytischen Therapie bei tiefer Beinvenenthrombose. In: Hansmann M, Koischwitz D, Lutz H, Trier HG (eds) Ultraschalldiagnostik 86. Springer, Berlin/Heidelberg/New York/Tokyo, p 138

Meyer P, Strobel M (2008) Intima-Media Dicke-Messung. Medizinische Verlagsgesellschaft, Berlin, pp 9–21

Mickley V (2006) Central vein obstruction in vascular access. Eur J Vasc Endovasc Surg 32:439–444

Middleton WD, Foley WD, Lawson TL (1988) Color flow Doppler imaging of carotid artery abnormalities. Am J Roentgenol 150:419–425

Middleton WD, Kellman GM, Melson GL, Madrazo BL (1989) Postbiopsy renal transplant arteriovenous fistulas: color Doppler US characteristics. Radiology 171:253–257

Middleton WD, Melson GL (1989) Testicular ischemia: color Doppler sonographic findings in five patients. AJR Am J Roentgenol 152:1237–1239

Middleton WD, Thorne DA, Melson GL (1989) Color Doppler ultrasound of the normal testis. AJR Am J Roentgenol 152:293–297

Middleton WD, Siegel BA, Melson GL, Yates CK, Andriole GL (1990) Acute scrotal disorders: prospective comparison of color Doppler US and testicular scintigraphy. Radiology 177:177–181

Miller N, Satin R, Tousignant L, Sheiner NM (1996) A prospective study comparing duplex scan and venography for diagnosis of lower-extremity deep vein thrombosis. Cardiovasc Surg 4:505–508

Millner R (ed) (1987) Ultraschalltechnik. Physik, Weinheim, pp 41–46

Mills JL, Bandyk DF, Gathan V, Esses GE (1995) The origin ot infrainguinal vein graft stenosis: a prospective study based on duplex surveillance. J Vasc Surg 21:16–22

Mills JL, Fujitani RM, Taylor SM (1993) The characteristics and anatomic distribution of lesions that cause reversed vein graft failure: a five-year prospective study. J Vasc Surg 17:195–206

Mills JL, Harris EJ, Taylor LM, Beckett WC (1990) The importance of routine surveillance of distal bypass grafts with duplex scanning: a study of 379 reversed vein grafts. J Vasc Surg 12:379–386

Miralles M, Cairols M, Cotillas J et al (1996) Value of Doppler parameters in the diagnosis of renal artery stenosis. J Vasc Surg 23: 428–435

Mitchell DG (1990) Color Doppler imaging: principles, limitations and artifacts. Radiology 177:1–10

Mödder U (1992) Farbcodierte Duplexsonographie (Angiodynographie). Fortschr Röntgenstr 157:204–209

Mohan CR, Hoballah JJ, Schueppert MT, Sharp WJ, Kresowik TF, Miller EV et al (1995) Should all in situ saphenous vein bypasses undergo permanent duplex surveillance? Arch Surg 130: 483–487

Mohan IV, Laheij RJF, Harris PL (2001) Risk factors for endoleak and the evidence for stent-graft oversizing in patients undergoing endovascular aneurysm repair. Eur J Vasc Endovasc Surg 21:344–349

Moll R, Habscheid W, Landwehr P (1991) Häufigkeit des Aneurysma spurium der Arteria femoralis nach Herzkatheteruntersuchung und PTA. Fortschr Röntgenstr 154:23–27

Moneta GL, Edwards JM, Chitwood RW et al (1993) Correlation of North American Symptomatic Carotid Endarterectomy Trial (NASCET): Angiographic definition of 70% to 99% internal carotid artery stenosis with duplex scanning. J Vasc Surg 17:152–157, discussion 157–159

Moneta GL, Edwards JM, Papanicolaou G, Hatsukami T, Taylor LM, Strandness DE, Porter JM (1995) Screening for asymptomatic internal carotid artery stenosis: duplex criteria for discriminating 60% to 99% stenosis. J Vasc Surg 21:989–994

Moneta GL, Lee RW, Yeager RA et al (1993) Mesenteric duplex scanning: a blinded prospective study. J Vasc Surg 17:79–86

Moneta GL, Taylor DC, Helton WS, Mulholland MW, Strandness DE (1988) Duplex ultrasound measurement of postprandial intestinal blood flow: effect of meal composition. Gastroenterology 95: 1294–1301

Moneta GL, Yeager RA, Antonovic R et al (1992) Accuracy of lower extremity arterial duplex mapping. J Vasc Surg 15:275–284

Moneta GL, Yeager RA, Dalman R et al (1991) Duplex ultrasound criteria for the diagnosis of splanchnic artery stenosis or occlusion. J Vasc Surg 14:511–520

Moneta GL, Yeager RA, Lee RW, Porter JM (1993) Noninvasive localisation of arterial occlusive disease: a comparison of segmental Doppler pressures and arterial duplex mapping. J Vasc Surg 17:578–582

Monreal M, Montserrat E, Salvador R, Bechini J, Donoso L, McCallegos J, Foz M (1989) Real-time ultrasound for diagnosis of symptomatic venous thrombosis and for screening of patients at risk. Correlation with ascending conventional venography. Angiology 39:527

Montefusco von Kleist CM, Bakal C, Sprayregen S, Rhodes BA, Veith FJ (1993) Comparison of duplex ultrasonography and ascending contrast venography in the diagnosis of venous thrombosis. Angiology 44:169–175

Monzer MAY, Wiese JA, Shamma AR (1987) The to-and-fro-sign: duplex Doppler evidence of femoral artery pseudoaneurysm. Am J Roentgenol 150:632–634

Moore WS (2003) For severe carotid stenosis found on ultrasound, further arterial evaluation is unnecessary. Stroke 34:1816–1817

Moreau P, Albat B, Thévenet A (1994) Surgical treatment of extracranial internal carotid artery aneurysm. Ann Vasc Surg 8:409–416

Morin C, Lafortune M, Pomier G et al (1992) Patent paraumbilical vein: anatomic and hemodynamic variants and their clinical importance. Radiology 185:253–256

Moriyasu F, Ban N, Nishida O et al (1986) Clinical application of an ultrasonic duplex system in the quantitative measurement of portal blood flow. J Clin Ultrasound 14:579–588

Moriyasu F, Ban N, Nishida O et al (1986) Portal hemodynamics in patients with hepatocellular carcinoma. Radiology 161:707–711

Moriyasu F, Nishida 0, Ban N, Nakamura T, Sakai M, Miyake T, Uchino H (1986) "Congestion index" of the portal vein. Am J Roentgenol 146:735–739

Moser KM, LeMoine JR (1981) Is embolic risk conditioned by location of deep venous thrombosis? Ann Intern Med 94:439–444

Mosso M, Jung HH, Baumgartner RW (2007) Recurrent spontaneous vasospasm of cervical carotid ophthalmic and retinal arteries causing repeated retinal infarcts: a case report. Cerebrovasc Dis 24:381–384

Mostbeck GH, Wittich GR, Herold C, Vergesslich KA, Walter RM, Frotz S, Sommer G (1989) Hemodynamic significance of the paraumbilical vein in portal hypertension: assessment with duplex US. Radiology 170:339–342

Motew SJ, Cherr GS, Craven TE et al (2000) Renal duplex sonography: main renal artery versus hilar analysis. J Vasc Surg 32:462–471

Moucka J, Jäger K (1990) Rationelle Abklärung bei peripherer arterieller Verschlußkrankheit. Schweiz Rundsch Med Prax 79:1553–1559

Mueller SC, Lue TF (1988) Evaluation of vasculogenic impotence. Urol Clin North Am 15:65–76

Muhm M, Polterauer P, Gstottner W, Temmel A et al (1997) Diagnostic and therapeutic approaches to carotid body tumors: review of 24 patients. Arch Surg 132:279–284

Müller-Schwefe CH, von Klinggräf G, Riepe G, Schröder A (1991) Das inflammatorische Bauchaortenaneurysma. Ultraschall Med 12:158–163

Müller-Wiefel H (1974) Untersuchungen zur Hämodynamik in den Venen der unteren Extremität. Physiologische, phathophysiologische und klinische Aspekte. Schattauer, Stuttgart

Mulligan SA, Matsuda T, Tanzer P et al (1991) Peripheral arterial occlusive disease: prospective comparison of MR angiography and color duplex US with conventional angiography. Radiology 178:695–700

Multicentre Aneurysm Screening Study Group (2002) Multicentre aneurysm screening study (MASS): cost effectiveness analysis of screening for abdominal aortic aneurysms based on four year results from randomised controlled trial. BMJ 325:1135, and Lancet 360: 1531–1539

Munda R, First MR, Alexander JW, Linnemann CC, Fidler JP, Kittur D (1983) Polytetrafluoroethylene graft survival in hemodialysis. J Am Med Ass 249:219–222

Murphy TP, Beechman RP, Kim HM et al (1998) Long-term follow-up after TIPS: use of Doppler velocity criteria for detecting elevation of portosystemic gradient. J Vasc Interv Radiol 9:275–281

Murphy TP, Cronan JJ (1990) Evolution of deep venous thrombosis: a prospective evaluation with US. Radiology 177:543–548

Naidich JB, Feinberg AW, Karp-Harmann H, Ka Tyma CG, Stein HL (1988) Contrast venography. Reassessment of its role. Radiology 168:97

Naidich JB, Torre JR, Pellerito JS et al (1996) Suspected deep venous thrombosis: is US of both legs necessary? Radiology 200:429–431

NASCET – North American Symptomatic Carotid Endarterectomy Trial Collaborators (1991) Beneficial effect of carotid endarterectomy in symptomatic patients with high-grade carotid stenosis. N Engl J Med 325:445–453

Nazarian GK, Gerral H, Casta eda-Zú iga WR et al (1994) Development of stenoses in transjugular intrahepatic portosystemic shunts. Radiology 192:231–234

Nazzal MM, Hoballah JJ, Miller EV et al (1997) Renal hilar Doppler analysis is of value in the management of patients with renovascular disease. Am J Surg 174:164–168

Nchimi A, Biquet JF, Brisbois D et al (2003) Duplex ultrasound as first-line screening test for patients suspected of renal artery stenosis: prospective evaluation in high-risk group. Eur Radiol 13:1413–1419

Neale ML, Chambers JL, Kelly AT et al (1994) Reappraisal of duplex criteria to assess significant carotid stenosis with special reference to reports from the North American Symptomatic Carotid Endarterectomy Trial and the European Carotid Surgery Trial. J Vasc Surg 20:642–649

Neglen P, Raju S (1992) A comparison between descending phlebography and duplex Doppler investigation in the evaluation of reflux in chronic venous insufficiency: a challenge to phlebography as the gold standard. J Vasc Surg 16:687–693

Neuerburg-Heusler D (1984) Dopplersonographische Diagnostik der extrakraniellen Verschlußkrankheit. Vasa Suppl 12:59–70

Neuerburg-Heusler D, Hennerici M (1995) Gefäßdiagnostik mit Ultraschall. Doppler- und farbcodierte Duplexsonographie. Thieme, Stuttgart

Neuerburg-Heusler D, Karasch TH (1991) Farbkodierte Duplex-Sonographie – Erweiterung der gefäßdiagnostischen Möglichkeiten? In: Maurer PC, Dörrler J, von Sommogy ST (eds) Gefäßchirurgie im Fortschritt. Thieme, Stuttgart

Neville RF, Abularrage CJ, White PW et al (2004) Venous hypertension associated with arteriovenous hemodialysis access. Semin Vasc Surg 17:50–56

Nevitt MP, Ballard DJ, Hallett JW Jr (1989) Prognosis of abdominal aortic aneuryms. A population-based study. N Engl J Med 321: 1009–1014

Nichols WN, O'Rourke MF (eds) (1990) McDonald's blood flow in arteries. Theoretical, experimental and clinical principles, 3rd edn. Edward Arnold, London, pp 54–76

Nicolaides AN (1995) Asymptomatic carotid stenosis and risk of stroke: identification of a high risk group (ACSRS): a natural history study. Int Angiol 14:21–23

Nielsen TG, Sillesn H, Schroeder TV (1995) Simple hyperaemia test as a screening method in a postoperative surveillance of infrainguinal in situ vein bypasses. Eur J Vasc Endovasc Surg 10:298–303

Nielsen TG, von Jessen F, Sillesen H (1993) Doppler spectral characteristics of infrainguinal vein bypasses. Eur J Vasc Surg 7:610–615

Noon GP, Jose LZ, Graig M et al (1984) Popliteal vein pseudoaneurysm: a case report. Surgery 96:942–945

Noppeney T, Noppeney J, Winkler M (2007) Das venöse Poplitealaneurysma. Gefäßchirurgie 12:187–190

Norgren L, Hiatt WR, Dormandy JA, Nehler MR, Harris KA, Fowkes FGR (2007) Inter-Society Consensus for the Management of Peripheral Arterial Disease (TASC II). Eur J Vasc Endovasc Surg 33:S1–S75

Norris CS, Greenfield LJ, Hermann JB (1985) Free-floating iliofemoral thrombus. Arch Surg 120:806–808

Numata K, Tanaka K, Kiba T et al (1999) Hepatic artery resistance after mixed-meal ingestion in healthy subjects and patients with chronic liver disease. J Clin Ultrasound 27:239–248

O'Donnell TF Jr, Erdoes L, Mackey WC et al (1985) Correlation of B-mode ultrasound imaging and arteriography with the pathologic findings of carotid endarterectomy. Arch Surg 120:443–449

O'Leary D, Glagov S, Zarins C, Giddens D (1991) Carotid artery disease. In: Rifkin MD, Charboneau JW, Laing FC (eds) Ultrasound

1991: Special Course Syllabus, 77th Scientific Assembly and Annual Meeting. RSNA, Oak Park, IL, pp 189–200
O'Leary DH, Holen J, Ricotta JJ et al (1987) Carotid bifurcation disease: prediction of ulceration with B-mode ultrasound. Radiology 162:523–525
O'Leary DH, Polak JF, Kronmal RA, Manolio TA, Burke GL, Wolfson SK Jr (1999) Carotid-artery intima and media thickness as a risk factor for myocardial infaction and stroke in older adults. Cardiovascular Health Study Collaborative Research Group. N Engl J Med 340:14–22
Oates CP, Pickard RS, Powell PH, Murthy LNS, Whittingham TAW (1995) The use of duplex ultrasound in the assessment of arterial supply to the penis in vasculogenic impotence. J Urol 153:354–357
Ohta M, Hashizume M, Tomikawa M et al (1994) Analysis of hepatic vein waveform by Doppler ultrasonography in 100 patients with portal hypertension. AJG 89:170–175
Olbricht CJ, Wanke B, Haubitz M, Koch KM (1991) Captopril spürt Stenosen der Nierenarterien auf. Med Trib 16:27
Olin JW, Piedmonte MR, Young JR et al (1995) The utility of duplex ultrasound scanning in the diagnosis of renal artery stenosis. Ann Intern Med 122:833–838
Olojugba DH, McCarthy MJ, Naylor AR, Bell PR, London NJ (1998) At what peak velocity ratio should duplex detected infrainguinal vein graft stenoses be revised? Eur J Vasc Endovasc Surg 15: 258–260
Omoloja AA, Racadio JM, McEnery PT (2002) Post-biopsy renal arteriovenous fistula. Pediatr Transplant 6:82–85
Orlic P, Vukas D, Drescik I et al (2003) Vascular complications after 725 kidney transplantations during 3 decades. Transplant Proc 35:1381–1384
Osman Y, Shokeir A, Ali-El-Dein B et al (2003) Vascular complications after live donor renal transplantation: study of risk factors and effects on graft and patient survival. J Urol 169:859–862
Ota H, Takase K, Rikimaru H, Tsuboi M, Yamada T, Sato A, Higano S, Ishibashi T, Takahashi S (2005) Quantitative vascular measurements in arterial occlusive disease. Radiographics 25:1141–1158
Owen RS, Carpenter JP, Baum RA, Perloff U, Cope C (1992) Magnetic resonance imaging of angiographically occult runoff vessels in peripheral arterial occlusive disease. N Engl J Med 326: 1577–1581
Owen WJ, Mc Coll I (1980) Venous aneurysm of the axilla simulating a soft tissue tumour. Br J Surg 67:577–578
Pages S et al (2001) Comparison of color duplex ultrasound and computed tomography scan for surveillance after aortic endografting. Ann Vasc Surg 15:155–162
Painter TA, Hertzer NR, Beven EG, O'Hara PJ (1985) Extracranial carotid aneurysms: report of six cases and review of the literature. J Vasc Surg 2:312–318
Papanicolaou G, Beach KW, Zierler RE, Strandness DE (1995) The relationship between arm-ankle pressure difference and peak systolic velocity in patients with stenotic lower extremity vein grafts. Ann Vasc Surg 9:554–560
Parent FN et al (2002) The incidence and natural history of type I and II endoleak: a 5-year follow-up assessment with color duplex ultrasound scan. J Vasc Surg 35:595–597
Park AE, McCarthy WJ, Pearce WH, Matsumura JS, Yao JST (1998) Carotid plaque morphology correlates with presenting symptomatology. J Vasc Surg 27:872–879
Park SH, Chung JW, Lee JW, Han MH, Park JH (2001) Carotid artery involvement in Takayasu´s ateritis: evaluation of the activity by ultrasonography. J Ultrasound Med 20:371–378
Parmar J, Aslam M, Standfield N (2007) Pre-operative radial arterial diameter predicts early failure of arteriovenous fistula (AVF) for haemodialysis. Eur J Vasc Endovasc Surg 33:113–115
Parry DJ et al (2002) Type II endoleaks: predictable, preventable, and sometimes treatable? J Vasc Surg 36:105–110
Partsch H (1976) "A-sounds" or "S-sounds" for Doppler ultrasonic evaluation of pelvic vein thrombosis. Vasa 5:16–19
Partsch H (1996) Diagnose und Therapie der tiefen Beinvenenthrombose. VASA 46 [Suppl]
Partsch H, Lofferer O (1972) Die Doppler-Detektoruntersuchung als Suchtest und Verlaufskontrolle von Bein-Beckenvenenthrombosen. Wien Klin Wochenschr 84:760–763
Partsch H, Mostbeck A (1979) Die Früherkennung der tiefen Unterschenkelvenenthrombophlebitis. Vasa 8:237–246
Partsch H, Oburger K, Mostbeck A et al (1992) Frequency of pulmonary embolism in ambulant patients with pelvic vein thrombosis: a prospective study. J Vasc Surg 16:715–722
Passman MA, Moneta GL, Nehler MR, Taylor LM et al (1995) Do normal early color-flow duplex surveillance examination results of infrainguinal vein grafts preclude the need for late graft revision? J Vasc Surg 22:476–484
Patel KR, Semel L, Oauss RH (1988) Extended reconstruction rate for limb salvage with intraoperative prereconstruction angiography. J Vasc Surg 7:531–537
Patel U, Khaw KK, Hughes NC (2003) Dopple ultrasound for detection of renal transplant artery stenosis – threshold peak systolic velocity needs to be higher in a low-risk or surveillance population. Clin Radiol 58:772–777
Patriquin H, LaFortune M, Burns PN, Dauzat M (1987) Duplex Doppler examination in portal hypertension: technique and anatomy. Am J Roentgenol 149:71–76
Patriquin HB, LaFortune M, Jéquier J-C et al (1992) Stenosis of the renal artery: assessment of slowed systole in the downstream circulation with Doppler sonography. Radiology 184:470–485
Paty PS, Kaufman JL, Koslow AR, Chang BB, Leather RP, Shah DM (1992) Adventitial cystic disease of the femoral vein: a case report and review of the literature. J Vasc Surg 15:214–217
Pedersen OM, Aslaksen A, Vik-Mo H, Bassoe AM (1991) Compression ultrasonography in hospitalized patients with suspected deep venous thrombosis. Arch Intern Med 151:2217–2220
Pedro LM, Pedro MM, Gonçalves I, Carneiro TF, Balsinha C, Fernandes e Fernandes R, Fernandes e Fernandes (2000) Computer-assisted carotid plaque analysis: characteristics of plaques associated with cerebrovascular symptoms and cerebral infarction. Eur J Vasc Endovasc Surg 19:118–123
Pedro LM, Fernandes e Fernandes J, Pedro MM (2002) Ultrasonographic risk score of carotid plaques. Eur J Vasc Endovasc Surg 24: 492–498
Pennestri F, Loperfido SMP et al (1984) Assessment of tricuspid regurgitation by pulsed Doppler ultrasonography of the hepatic veins. J Cardiol 54:363–368
Perkins JM, Galland RB, Simmons MJ, Magee TR (2000) Carotid duplex imaging: variation and validation. Br J Surg 87:320–322
Perko MJ (2001) Duplex ultrasound for assessment of superior mesenteric artery blood flow. Eur J Vasc Endovasc Surg 21:106–117
Perko MJ, Just S, Schroeder TV (1997) Importance of diastolic velocities in the detection of celiac and mesenteric artery diesease by duplex ultrasound. Eur J Vasc Endovasc Surg 26:288–293
Perlin SJ (1992) Pulmonary embolism during compression US of the lower extremity. Radiology 184:165–166
Perrier A, Bounameaux H (2001) Cost-effective diagnosis of deep vein thrombosis and pulmonary embolism. Thromb Haemost 86:475–487
Perrier A, Desmarais S, Miron MJ, de Moerloose P, Lepage R, Slosman D, Didier D, Unger PF, Patenaude JV, Bounameaux H (1999) Noninvasive diagnosis of venous thromboembolism in outpatients. Lancet 353:190–195
Persson AV, Jones C, Zide R, Er J (1989) Use of the triplex scanner in diagnosis of deep venous thrombosis. Arch Surg 124:593
Persson BG, Donner M, Peterson B, Eklof B, Wintzell K (1980) Aneurysm of the popliteal vein as a cause of pulmonary embolism. Acta Med Scand 208:407–410

Peterson BG, Longo GM, Kibbe MR, Matsumura JS, Blackburn D, Astleford P, Eskandari MK (2005) Duplex ultrasound remains a reliable test even after carotid stenting. Ann Vasc Surg 19:793–797

Petrick J (1996) Spektralanalyse von Ultraschall-Doppler-Signalen mit und ohne Ultraschallkontrastmittel an Modellen von Blutgefäßen. PhD Thesis, Berlin 1995. Köster, Berlin

Petrick J, Schlief R, Zomack M, Langholz J, Urbank A (1992) Pulsatiles Strömungsmodell mit elastischen Gefäßen für duplexsonographische Untersuchungen. Ultraschall Med 13:277–282

Petros JA, Andriole GL, Middleton WD, Picus DA (1991) Correlation of testicular color Doppler ultrasonography, physical examination and venography in the detection of left varicoceles in men with infertility. J Urol 145:785–788

Pezzullo JA, Perkins AB, Cronan JJ (1996) Symptomatic deep vein thrombosis: diagnosis with limited compression US. Radiology 198:67–70

Pfeil W, Jacksch R, Kotzerke M, Störk T, Selbach J. Behandlung von Pseudoaneurysmen der Femoralarterie mit Thrombininjektion – Das deutsche Multicenter-Register. Cardioangiologisches Centrum Bethanien, Frankfurt/M (DE); St. Vincenzkrankenhaus Essen (DE); Hegau Klinikum; Singen (DE); Karl Olga Krankenhaus; Stuttgart (DE); Caritas Krankenhaus; Bad Mergentheim (DE)

Pfister K et al. (2009) Detection and characterization of endoleaks following endovascular treatment of abdominal aortic aneurysms using Contrast Harmonic Imaging (CHI) with quantitative perfusion analysis (TCI) compared to CT angiography (CTA). Ultraschall Med (in press)

Planken RN, Keuter XH, Kessels AG et al (2006) Forearm cephalic vein cross-sectional area changes at incremental congestion pressures: towards a standardized and reproducible vein mapping protocol. J Vasc Surg 44:353–358

Ploenes C, Strauss A (1999) Die Wertigkeit der Farbduplexsonographie in der Diagnose von Nierenarterienstenosen und -verschlüssen. VASA Suppl 55:47

Podhaisky H, Hänsgen K, Seifert H, Taute BM (1996) Parameter und Einflußfaktoren der sonographischen Untersuchung des peripheren arteriellen Gefäßsystems. Herz/Kreisl 28:129–133

Polak JF (1992) Peripheral vascular sonography: a practical guide. Williams & Wilkins, Baltimore

Polak JF, Bajakian RL, O'Leary DH, Anderson MR, Donaldson MC, Jolesz FA (1992) Detection of internal carotid artery stenosis: comparison of MR angiography, color Doppler sonography, and arteriography. Radiology 182:35–40

Polak JF, Culter S, O'Leary D (1989) Deep veins of the calf: assessment with color Doppler flow imaging. Radiology 171:481–485

Polak JF, Cutter SS, O'Leary DH (1988) Doppler color flow imaging of the calf veins. Description of method and preliminary results. Radiology 169(Suppl):318 (Abstr)

Polak JF, Dobkin GR, O'Leary DH, Wang A-M, Cutler SS (1989) Internal carotid artery stenosis: accuracy and reproducibility of color-Doppler-assisted duplex imaging. Radiology 173:793–798

Polak JF, Donaldson MC, Dobkin GR et al (1990) Early detection of saphenous vein arterial bypass graft stenosis by color-assisted duplex sonography: a prospective study. AJR 154:857–861

Polak JF, Magruder CD, Whittemore AD et al (1990) Pulsatile masses surrounding vascular prostheses: real-time US color flow imaging. Radiology 170:363–366

Polak JF, Mitchell MI, Mannick JA, O'Leary DH, Donaldson MC, Whittemore AD (1990) Determination of the extent of lower-extremity peripheral arterial disease with color-assisted duplex sonography. Am J Roentgenol 155:1085–1089

Polak JF, O'Leary DH, Kronmal RA et al (1993) Sonographic evaluation of carotid artery atherosclerosis in elderly: relationship of disease severity to stroke and TIA. Radiology 188:363–370

Polak JF, Shemanski L, O'Leary DH, Lefkowitz D et al (1998) Hypoechoic plaque at US of the carotid artery: an independent risk factor for incident stroke in adults aged 65 years or older. Radiology 208:649–654

Polak KF, Karmel MI, Meyerowitz JA (1991) Accuracy of color Doppler flow mapping for evaluation of the severity of femoropopliteal arterial disease, a prospective study. J Vasc Interv Radiol 2:471–476

Poli A, Tremoli E, Colombo A et al (1988) Ultrasonographic measurement of the common carotid arterial wall thickness in hypercholesterolemic patients. Atherosclerosis 70:253–261

Pomposelli FB Jr, Marcaccio EJ, Gibbons GW et al (1995) Dorsalis pedis arterial bypass: durable limb salvage for foot ischemia in patients with diabetes mellitus. J Vasc Surg 21:375–384

Porst H (1987) Erektile Impotenz. Enke, Stuttgart, S73–77

Porter TR (1998) Transient response imaging. In: Bogdahn U, Becker G, Schlachetzki F (eds) Echosignalverstärker und transkranielle Farbduplex-Sonographie. Blackwell, Berlin Wien, pp 192–203

Postman CT, Bijlstra PJ, Rosenbusch G, Thien T (1996) Pattern recognition of loss of early systolic peak by Doppler ultrasound has a low sensitivity for the detection of renal artery stenosis. J Hum Hypertens 10:181–184

Prabhakaran S, Rundek T, Ramas R, Elkind MS, Paik MC, Boden-Albala B, Sacco RL (2006) Carotid plaque surface irregularity predicts ischemic stroke: the Northern Manhattan Study. Stroke 37: 2696–2701

Prandoni P (1993) Symptomatic deep vein thrombosis and the incidence of postthrombotic syndrome. Int Union of Angiology, Beaune, Oct. 6

Prandoni P, Cogo A, Bernardi E, Villalta S et al (1993) A simple ultrasound approach for detection of recurrent proximal vein thrombosis. Circulation 88:1730–1735

Prandoni P, Polistena P, Bernardi E et al (1997) Upper-extremity deep vein thrombosis: risk factors, diagnosis and complications. Arch Intern Med 157:357–362

Pross C, Shortsleeve CM, Baker JD et al (2001) Carotid endarterectomy with normal findings from a completion study: is there need for early duplex scan? J Vasc Surg 33:963–967

Prountjos P, Bastounis E, Hadjinikolaou L, Felekuras E, Balas P (1991) Superficial venous thrombosis of the lower extremities co-existing with deep venous thrombosis. A phlebographic study on 57 cases. Int Angiol 10:63–65

Quam JP, King BF, James EM, Lewis RW, Brakke DM et al (1989) Duplex and color Doppler sonographic evaluation of vasculogenic impotence. Amer J Roentgenol 153:1141–1147

Quandalle P, Sandement A, Chambon JP, Wurtz A (1989) L'anévrisme des veines profondes des membres inférieurs. J Chir 126(11): 586–590

Radermacher J, Chavan A, Bleck J, Vitzthum A et al (2000) Use of Doppler ultrasonography to predict the outcome of therapy for renal-artery stenosis. NEJM 334:410–417

Radermacher J, Chavan A, Schäffer J et al (2000) Detection of significant renal artery stenosis with color Doppler sonography: combining extrarenal and intrarenal approaches to minimize technical failure. Clin Nephrol 53:333–343

Radermacher J, Mengel M, Ellis S, Stuht S, Hiss M, Schwarz A, Eisenberger U, Burg M, Luft FC, Gwinner W, Haller H (2003) The renal arterial resistance index and renal allograft survival. N Engl J Med 349:115–124

Ragavendra BN, St H, Hilton S, Subramanyam BR, Rosen RJ, Lam SL (1986) Deep venous thrombosis: detection by probe compression of veins. J Ultrasound Med 5:89–95

Raghavendra BN, Rosen J, Lam S, Riles T, Horii SC (1984) Deep venous thrombosis. Detection by high-resolution real-time ultrasonography. Radiology 152:789

Rajfer J, Rosciszweski A, Mehringer M (1988) Prevalence of corporeal venous leakage in impotent men. J Urol 140:69

Ralls PW, Jensen MC, Lee KP, Mayekawa DS, Johnson MB, Halls JM (1990) Color Doppler sonography in acute epididymitis and orchitis. J Clin Ultrasound 18:383–386

Ranke C, Creutzig A, Alexander K (1992) Duplex scanning of the peripheral arteries: correlation of the peak velocity ratio with angiographic diameter reduction. Ultrasound Med Biol 18: 433–440

Ranke C, Creutzig A, Becker H, Trappe HJ (1999) Standardization of carotid ultrasound: a hemodynamic method to normalize for interindividual and interequipment variability. Stroke 30: 402–406

Ranke C, Hendrickx PH, Brasel F et al (1990) Duplexsonographie: Genauigkeit, Reproduzierbarkeit und Fehlermöglichkeiten. Dtsch Med Wochenschr 115:528–533

Ranke C, Trappe JH (1997) Blood flow velocity measurements for carotid stenosis estimation: interobserver variation and interequipment variability. Vasa 26:210–214

Rao AB, Koeller KK, Adair CF (1999) Paragangliomas of the head and neck: Radiologic-pathologic correlation. Radiographics 19: 1605–1632

Ratiff DA, Gallagher PJ, Hames TK et al (1985) Characterization of carotid artery disease: comparison of duplex scanning with histology. Ultrasound Med Biol 11:835–840

Redekop G, Marotta T, Weill A (2001) Treatment of traumatic aneurysms and arteriovenous fistulas of the skull base by using endovascular stents. J Neurosurg 95:412–419

Reilly LM (1990) Carotid intraplaque hemorrhage: noninvasive detection and clinical significance. In: Bernstein EF (ed) Noninvasive diagnostic techniques in vascular disease. Mosby, St Louis, pp 99–107

Reilly LM (1993) Importance of carotid plaque morphology. In: Bernstein EF (ed) Vascular diagnosis, 4th edn. Mosby Year Book, St. Louis, pp 333–340

Reilly LM, Lusby RJ, Hughes L et al (1983) Carotid plaque histology using real-time ultrasonography: clinical and therapeutic implications. Am J Surg 146:188–193

Reiter M, Horvat R, Puchner S, Rinner W, Polterauer P, Lammer J, Minar E, Bucek RA (2007) Plaque imaging of the internal carotid artery – correlation of B-flow imaging with histopathology. AJNR Am J Neuroradiol 28:122–126

Reuther GR, Wanjura D, Bauer H (1989) Acute renal vein thrombosis in renal allografts: detection with duplex Doppler US. Radiology 170:557–558

Rich NM (1982) Popliteal entrapment and adventitial cystic disease. Surg Clin North Am 62:449–465

Rich NM, Hughes CW (1967) Popliteal artery and vein entrapment. Am J Surg 113:696–698

Rich NM, Collins GJ, Mc Donald PT, Kozloff L, Clagett GP, Collins JT (1979) Popliteal vascular entrapment. Arch Surg 114: 1377–1384

RichterG,BöhmS,GörgDH,SchwerkWB(1992)Verlaufsbeobachtungen zur Echogenität venöser Gerinnungsthromben. Ultraschall Klin Prax 7:69–73

Richtlinien für die Durchführung Doppler- und duplexsonographischer Untersuchungen peripherer Arterien und Venen, extrakranieller hirnversorgender Halsarterien und intrakranieller Arterien des Arbeitskreises Gefäßdiagnostik der DEGUM (1991) Mitteil Angiol 1: 10–17

Ricotta JJ (1990) Plaque characterization by B-mode scan. Surg Clin North Am 70:191–199

Rieger H (1985) Durchblutungsstörungen. Adam Pharma Verlag GmbH, Essen

Rieger H, Schoop W (eds) (1998) Klinische Angiologie. Springer, Berlin/Heidelberg/New York, pp 627–666

Riehl J, Clasen W, Schmidt H, Kierdorf, H, Sieberth HG (1989) Altersabhängige Veränderungen der renalen Hämodynamik: Untersuchungen mittels Duplex-Sonographie. Ultraschall Klin Prax [Suppl] 1: 129

Rigsby Ch, Burns PN, Weltin GG, Chen B, Bia M, Taylor KJW (1987) Doppler signal quantitation in renal allografts: comparison in normal and rejecting transplants with pathologic correlation. Radiology 162:39–42

Ripolles T, Aliaga R, Morote V et al (2001) Utility of intrarenal Doppler ultrasound in the diagnosis of renal artery stenosis. Eur J Radiol 40:54–63

Robbin ML, Chamberlain NE, Lockhart ME et al (2002) Hemodialysis arteriovenous fistula maturity: US evaluation. Radiology 225: 59–64

Robbin ML, Oser RF, Allon M et al (1998) Hemodialysis access graft stenosis: US detection. Radiology 208:655–661

Rofsky NM, Adelman MA (2000) MR angiography in the evaluation of atherosclerotic peripheral vascular disease. Radiology 214:325–338

Rollins DL, Semrow CM, Friedell ML, Buchbinder D (1987) Use of ultrasonic venography in the evaluation of venous valve function. Am J Surg 154:189–191

Roobottom CA, Dubbins PA (1995) Significant disease of the celiac and superior mesenteric arteries in asymptomatic patients: predictive value of Doppler sonography. AJR 161:985–988

Roobottorn CA, Hunter JD, Weston MJ, Dubbins PA (1995) Hepatic venous Doppler waveforms: changes in pregnancies. J Clin Ultrasound 23:477–482

Rooijens PP, TordoirJH ST et al (2004) Radiocephalic wrist arteriovenous fistula for hemodialysis: meta-analysis indicates a high primary failure rate. Eur J Vasc Endovasc Surg 28:583–589

Rose SC, Zwiebel WJ, Miller FJ (1994) Distribution of acute lower extremity deep venous thrombosis in symptomatic and asymptomatic patients: imaging implications. J Ultra Med 13:243–250

Rose SC, Zwiebel WJ, Nelson BD et al (1990) Symptomatic lower extremity deep venous thrombosis: accuracy, limitations and role of color duplex flow imaging in diagnosis. Radiology 175:639–644

Rosen MP, Sheiman RG, Weintraub J, McArdle C (1996) Compression sonography in patients with indeterminate or low-probability lung scans: lack of usefulness in the absence of both symptoms of deep vein thrombosis and thromboembolic risk factor. AJR 166: 285–289

Rosfors S, Eriksson M, Höglund N, Johansson G (1993) Duplex ultrasound in patients with suspected aorto-iliac occlusive disease. Eur J Vasc Surg 7:513–517

Rosner NM, Doris PE (1988) Diagnosis of femoropopliteal venous thrombosis: comparison of duplex sonography and plethysmography. Am J Roentgenol 150:623–627

Ross B, Sherriff L, Walton L (1984) Diagnosis of deep-vein thrombosis: comparison of clinical evaluation, ultrasound, plathysmography, and venoscan with X-ray venogram. Lancet 716–719

Ross GJ, Violi L, Barber LW et al (1988) Popliteal venous aneurysm. Radiology 168:721–722

Rosvall M, Janzon L, Berglund G, Engstrom G, Hedblad B (2005) Incident coronary events and case fatality in relation to common carotid intima-media thickness. J Intern Med 257:430–437

Roth SM, Back MR, Bandyk DF et al (1999) A rational algorithm for duplex scan surveillance after carotid endarterectomy. J Vasc Surg 31:838–839

Rothwell PM (1999) Carotid endarterectomy for recently symptomatic carotid stenosis: consistent results from two large randomized controlled trials. Eur Heart J 20:1055–1057

Rothwell PM, Warlow CP (1999) Interpretation of operative risk of individual surgeons. European Carotid Surgery Trialists' Collaborative Group. Lancet 353:1325

Rothwell PM, Warlow CP (1999) Prediction of benefit from carotid endarterectomy in individual patients: a risk-modelling study. European Carotid Surgery Trialists' Collaborative Group. Lancet 353:2105–2110

Rothwell PM (2003) For severe carotid stenosis found on ultrasound, further arterial evaluation prior to carotid endarterectomy is unnecessary: the argument against. Stroke 34:1817–1819

Rothwell PM, Gibson R, Warlow CP (2000) Interrelation between plaque surface morphology and degree of stenosis on carotid angiograms and the risk of ischemic stroke in patients with symptomatic

carotid stenosis. On behalf of the European Carotid Surgery Trialists Collaborative Group. Stroke 31:615–621
Rothwell PM, Gibson RJ, Slattery J et al (1994) Equivalence of measurements of carotid stenosis: a comparison of three methods on 1001 angiograms. Stroke 25:2435–2439
Rothwell PM, Warlow CP, on behalf of the European Carotid Surgery Trialists' Collaborative Group (2000) Low risk of ischemic stroke in patients with reduced internal carotid artery lumen diameter distal to severe symptomatic carotid stenosis: cerebral protection due to low poststenotic flow? Stroke 31:622–630
Rotter W (1981) Gefäßveränderungen bei frischer und älterer Thrombose. In: Vinazzer H (ed) Thrombose und Embolie. Anästhesie Intensivmed 134:171–177
Roy-Chaudhury P, Kelly BS, Miller MA et al (2001) Venous neointimal hyperplasia in polytetraflouroethylene dialysis grafts. Kidney Int 59:2325–2334
Rubin JM, Bude RO, Carson PL et al (1994) Power Doppler US: a potentially useful alternative to mean frequency-based color Doppler US. Radiology 190:853–856
Rubin BG, Beak BI, Reilly JM (1995) Fusiform aneurysms of the popliteal vein. In: The American venous forum 7th annual meeting, 1995 Feb. 23–25, Fort Lauderdale, Fla., Abstract Book p 39
Rutherford RB, Baker D, Ernst C et al (1997) Recommended standards for reports dealing with lower extremity ischaemia: revised version. J Vasc Surg 26:517–538
Rutherford RB, Flanigan DP, Gupta SK, Johnston KW, Karmody A, Whittemore AD (1986) Suggested standards for reports dealing with lower extremity ischemia. J Vasc Surg 4:80–94
Saal JG, Dürk H (1994) Rationale Labordiagnostik in klinischer Immunologie und Rheumatologie. Internist 35:633–639
Sabba C, Ferraioli G, Genecin P et al (1991) Evaluation of postprandial hyperemia in superior mesenteric artery and portal vein in healthy and cirrhotic humans: an operator-blind echo-Doppler study. Hepatology 13:114–118
Sacks D, Robinson ML, Marinelli DL, Perlmutter GS (1990) Evaluation of the peripheral arteries with duplex after angioplasty. Radiology 176:39–44
Sakaguchi M, Kitagawa K, Nagai Y et al (2003) Equivalence of plaque score and intima-media thickness of carotid ultrasonography for predicting severe coronary artery lesion. Ultrasound Med Biol 29:367–371
Sandler DA, Duncan JS, Ward P et al (1984) Diagnosis of deep vein-thrombosis. Comparison of clinical evaluation, ultrasound, plethysmography and venoscan with x-ray venogram. Lancet II:716
Sandok BA (1983) Fibromuscular dysplasia of the internal carotid artery. Neurol Clin 1:17–26
Sanford RM, Bown MJ, Fishwick G, Murphy F, Naylor M, Sensier Y, Sharpe R, Walker J, Hartshorn T, London NJ, Sayers RD (2006) Duplex ultrasound scanning is reliable in the detection of endoleak following endovascular aneurysm repair. Eur J Vasc Encovasc Surg 32:537–541
Sarap MD, Wheeler WE (1988) Venous aneurysms. J Vasc Surg 8: 182–183
Sarkar R, Ro KM, Obrand DI, Ahn SS (1998) Lower extremity vascular reconstruction and endovascular surgery without preoperative angiography. Am J Surg 176:203–207
Sarlon G, Lapierre F, Sarlon E, Bartoli MA, Magnan PE, Branchereau A (2009) Endovascular aneurysm repair follow-up by unenhanced and contrast-enhanced duplex ultrasound. J Mal Vasc 34: 121–132
Sato DT et al (1998) Endoleak after aortic stent graft repair: diagnosis by color duplex ultrasound scan versus computed tomography scan. J Vasc Surg 28:657–663
Scarpato R, Gembarowicz R, Farber S, O'Donnell TF Jr, Kelly JJ, Callow AD, Deterling RA (1981) Intraoperative prereconstruction arteriography. Arch Surg 116:1053–1055
Schäberle W (1992) Diagnosis of venous aneurysm by ultrasound examination. In: Eurodop 92, Br Med Ultrasound Soc, London, p 157 (Abstr)
Schäberle W (1993) Ultraschall in der Venendiagnostik. Springer, Heidelberg
Schäberle W, Bärlin E, Seitz K (1988) Meßmethodische Probleme zur Durchmesser- bzw. Querschnittsbestimmung für die duplexsonographische Stromzeitvolumenmessung. Ultraschall Klin Prax Suppl 1:68 (Abstr)
Schäberle W, Eisele R (1990) Duplexsonographische Kriterien eines distalen Mesenterialarterienverschlusses. Ultraschall Klin Prax Suppl 5:197
Schäberle W, Eisele R (1991) Das sonographische Bild einer zystischen Adventitiadegeneration. VASA 33:207–208
Schäberle W, Eisele R (1991) Duplexsonographie versus Phlebographie in der Diagnostik der tiefen Beinvenenthrombose. Phlebologie 5:45
Schäberle W, Eisele R (1992) Duplexsonographische Diagnostik der Fossa poplitea zur Differenzierung seltener arterieller und venöser Gefäßerkrankungen. Ultraschall Klin Prax 7:148
Schäberle W, Eisele R (1996) Die sonographische Diagnostik, Verlaufsform und Therapie der cystischen Adventitiadegeneration. Ultraschall Med 17:131–137
Schäberle W, Eisele R (2001) Das venöse Aneurysma tiefer Beinvenen – Wertigkeit der Duplexsonographie in Operationsindikation und Therapieplanung. Ultraschall Med 22(Suppl):60–61
Schäberle W, Neuerburg Heusler D (1989) Wertigkeit der Duplexsonographie in der Diagnostik der Nierenarterienstenose. Klin Wochenschr 67(Suppl XVI):177
Schäberle W, Schock D, Eisele R (1992) Der isolierte Popliteal-arterienverschluß: Wertigkeit der Duplexsonographie in Differentialdiagnose und Therapieplanung. VASA Suppl 35:69–70
Schäberle W, Seitz K (1988) Meßmethodische Probleme der Durchmesser und Querschnittsbestimmung für die duplexsonographische Stromzeitvolumenmesseung. Ultraschall Klin Prax Suppl 1:68
Schäberle W, Seitz K (1991) Duplexsonographische Blutflußmessung in der Arteria mesenterica superior. Ultraschall Med 12:277–282
Schäberle W, Strauss A, Neuerburg-Heusler D, Roth FJ (1992) Wertigkeit der Duplexsonographie in der Diagnostik der Nierenarterienstenose und ihre Eignung in der Verlaufskontrolle nach Angioplastie (PTA). Ultraschall Med 13:271–276
Schäberle W, Ulrich Ch, Meyer H, Eisele R (1995) Großes Aneurysma spurium in der A. poplitea nach arthroskopischer Meniskektomie. Arthroskopie 8:41–43
Scharnke W, Claußnitzer R, Wild K, Schuler A (1995) Aneurysma der V. poplitea als Ursache von Lungenembolien. Diagnose mittels FDS. Ultraschall Med 16:114
Schatz IJ, Fine G (1962) Venous aneurysms. N Engl J Med 266:1310–1312
Schellong S (1993) Vaskulitiden. In: Alexander K (ed) Gefäßkrankheiten. Urban & Schwarzenberg, München/Wien/Baltimore
Schellong SM, Schwarz T, Halbritter K et al (2003) Complete compression ultrasonography of the leg veins as a single test for the diagnosis of deep vein thrombosis. Thromb Haemost 89:228–234
Schiff MJ, Feinberg W, Naidisch JB (1987) Noninvasive venous examinations as a screening test for pulmonary embolism. Arch Intern Med 147:505
Schild H, Berg S, Weber W, Schmied W, Steegmuller KW (1992) The venous aneurysm. Akt Radiol 2:75–80
Schindler JM, Kaiser M, Gerber A, Vuilliomenet A, Popovic A, Bertel O (1990) Colour coded duplex sonography in suspected deep vein thrombosis of the leg. BMJ 301:1369–1370
Schmidberger H, Hackl A, Ludwig G (1979) Aneurysma der Vena poplitea. Fortschr Röntgenstr 131:553–554
Schmidt JA, Bierbrauer A, Wichert P (1993) Vaskulitiden großer Arterien. Internist 34:615–621

Schmidt WA (2006) Takayasu and temporal arteritis. Front Neurol Neurosci 21:96–104

Schmidt WA, Blockmans D (2005) Use of ultrasonography and positron emission tomography in the diagnosis and assessment of large-vessel vasculitis. Curr Opin Rheumatol 17:9–15

Schmidt WA, Gromnica-Ihle E (2002) Incidence of temporal arteritis in patients with polymyalgia rheumatica: a prospective study using color Doppler ultrasonography of the temporal arteries. Rheumatology (Oxford) 41:46–52

Schmidt WA, Kraft HE, Vorpahl K, Völker L, Gromnica-Ihle EJ (1997) Color duplex ultrasonography in the diagnosis of temporal arteritis. N Engl J Med 337:1336–1342

Schminke U, Motsch L, Hilker L, Kessler C (2000) Three-dimensional ultrasound observation of carotid artery plaque ulceration. Stroke 31:1651–1655

Schmitt HE (1974) Möglichkeiten und Grenzen der antegraden Phlebographie. Vasa 3:440–445

Schmitt HE (1977) Aszendierende Phlebographie bei tiefer Beinvenenthrombose. Huber, Bern

Schneider JR, Walsh DB, McDaniel MD, Zwolak RM, Besso SR, Cronenwett JL (1993) Pedal bypass versus tibial bypass with autogenous vein: a comparison of outcome, and hemodynamic results. J Vasc Surg 17:1029–1038

Schneider PA, Ogawa DY (1998) Is routine preoperative aortoiliac arteriography necessary in the treatment of lower extremity ischemia? J Vasc Surg 28:28–36

Schneider PA, Ogawa DY, Rush MP (1999) Lower extremity revascularization without contrast arteriography: a prospective study of operation based upon duplex mapping. Cardiovasc Surg 7: 699–703

Schöllhorn J (1985) Cystic adventitia degeneration as a cause of dynamic stenosis of the popliteal artery: a case report. Angiology 36:809–814

Scholz H (1998) Der adäquate Gefäßzugang für die Hämodialyse. Bard, Impra, München

Schönhofer B, Bechthold H, Renner R et al (1992) Sonographische Befunde bei Varikophlebitis der Vena saphena magna. Dtsch Med Wochenschr 117:51–55

Schönhofer B, Bechthold H, Renner R, Bundschu HD (1991) Asymptomatische Lungenembolie bei Varikophlebitis der V. saphena magna: Sonographische Dokumentation von Wachstum und Ablösung eines Appositionsthrombus. Phlebologie 20:48

Schönhofer B, Bundschu HD, Wolf K, Grehn S (1992) Farbkodierte Duplexsonographie im Vergleich zur Phlebographie bei tiefer Bein- und Beckenvenenthrombose. Med Klin 87:172–178

Schoop W (1988) Praktische Angiologie. Thieme, Stuttgart/New York, pp 127

Schopohl J, Haen E, Ullrich T, Gärtner R (2000) Sildenafil (Viagra). Dt Ärztebl 97:A-311–A-315

Schraverus P, Dulieu J, Milleux P, Coulier B (1997) Cystic adventitial disease of the popliteal vein: report of a case. Acta Chir Belg 97:90–92

Schröder JO, Euler HH (1993) Systemischer Lupus erythematodes. Internist 34:351–361

Schroeder WB, Bealer JF (1992) Venous duplex ultrasonography causing acute pulmonary embolism: a brief report. J Vasc Surg 15: 1082–1083

Schuler A, Claußnitzer R, Dinkelacker S, Blank B, Braun B (1995) Farbdopplersonographie bei ascendierenden Thrombophlebitiden der unteren Extremität. Ultraschall Med 16:76

Schuler A, Wild K, Hofstätter F, Claussnitzer R, Blank W, Braun B (1993) Seltene Aneurysmalokalisationen – Farbdopplersonographische (FDS) Diagnostik. Ultraschall Klin Prax 8:201

Schulte-Altedorneburg G, Droste DW, Haas N et al (2000) Preoperative B-mode ultrasound plaque appearance compared with carotid endarterectomy specimen histology. Acta Neurol Scand 101:188–194

Schürmann R, Balzer TH, Schlief R (1992) Diagnostisches Potenzial von Ultraschall-Kontrastmitteln in der Gefäßdiagnostik. Ultraschall Med 13:234–238

Schwartz RA, Kerns DB, Mitchell DG (1991) Color Doppler ultrasound imaging in iatrogenic arterial injuries. Am J Surg 162:4–8

Schwarz A, Gwinner W, Hiss M et al (2005) Safety and adequacy of renal transplant protocol biopsies. Am J Transplant 5: 1992–1996

Schweizer J, Oehmichen F, Brandt HG, Altman E (1993) Farbkodierte Duplexsonographie und kontrastmittelverstärkte Duplexsonographie bei tiefer Beinvenenthrombose. Vasa 22:22–25

Schwerk WB, Restrepo IK, Stellwaag M et al (1994) Renal artery stenosis: grading with image-directed Doppler US. Evaluation of renal resistive index. Radiology 190:785–790

Scoble JE (1996) The epidemiology and clinical manifestations of atherosclerotic renal disease. In: Novick A, Scoble J, Hamilton G (eds) Renal vascular disease. WB Saunders, London/Philadelphia/ Toronto, pp 303–314

Seidel G, Cangür H, Meyer-Wiethe K et al (2006) On the ability of ultrasound parametric perfusion imaging to predict the area of infarction in acute ischemic stroke. Ultraschall Med 27:543–548

Seino Y, Fujimori H, Shimai S, Tanaka K, Takano T, Hayakawa H, Niimi Y (1994) Popliteal venous aneurysm with pulmonary embolism. Intern Med 33:779–782

Seitz K, Kubale R (1988) Duplexsonographie der abdominellen und retroperitonealen Gefäße. VCH, Weinheim

Sensier Y, Bell PR, London NJ (1998) The ability of qualitative assessment of the common femoral Doppler waveform to screen for significant aortoiliac disease. Eur J Vasc Endovasc Surg 15: 357–364

Sensier Y, Fishwick G, Owen R, Pemberton M, Bell PR, London NJ (1998) A comparison between color duplex ultrasonography and arteriography for imaging infrapopliteal lesions. Eur J Vasc Endovasc Surg 15:44–50

Sessa C, Nicolini P, Perrin M et al (2000) Management of symptomatic and asymptomatic popliteal venous aneurysm: a retrospective analysis of 25 patients and review of the literature. J Vasc Surg 32:902–912

Sheiman RG, McArdle CR (1995) Bilateral lower extremity US in the patient with unilateral symptoms of deep venous thrombosis: assessment of need. Radiology 194:171–173

Sheiman RG, McArdle CR (1999) Clinically suspected pulmonary embolism: use of bilateral lower extremity US as the initial examination—a prospective study. Radiology 212:75–78

Sheiman RG, Weintraub JL, McArdle CR (1995) Bilateral lower extremity US in the patient with bilateral symptoms of deep venous thrombosis: assessment of need. Radiology 196:379–381

Sidhu PS, Allan PL, Cattin F et al (2006) Diagnostic efficacy of SonoVue, a second generation contrast agent, in the assessment of extracranial carotid or peripheral arteries using colour and spectral Doppler ultrasound: a multicentre study. Br J Radiol 79:44–51

Silva MB Jr, Hobson RW, Pappas PJ et al (1998) A strategy for increasing use of autogenous hemodialysis access procedures: impact of preoperative noninvasive evaluation. J Vasc Surg 27:302–307

Silverstein MD, Heit JA, Mohr DN, Petterson TM, O'Fallon WM, Melton LJ (1998) Trends in the incidence of deep vein thrombosis and pulmonary embolism. Arch Intern Med 158:585–593

Silvestri M, Villain P, Boursier JL, Elias A (1987) L'anévrisme veineux poplité: une cause rare d'embolie pulmonaire. Presse Méd 16:2127

Siringo S, Bolondi L, Gaiani S et al (1994) The relationship of endoscopy, portal Doppler US flowmetry, and clinical and biochemical tests in cirrhosis. J Hepatol 20:11

Sitzer M, Fürst G, Fischer H, Siebler M et al (1993) Between-method correlation in quantifying internal carotid stenosis. Stroke 24: 1513–1518

Sitzer M, Muller W, Siebler M, Hort W et al (1995) Plaque ulceration and lumen thrombus are the main sources of cerebral microemboli in high-grade internal carotid stenosis. Stroke 26:1231–1233

Sitzer M, Wolfram M, Jörg R et al (1996) Color flow Doppler-assisted duplex imaging fails to detect ulceration in high-grade internal carotid artery Stenosis. J Vasc Surg 24:461–465

Skillman JJ, Kent KC, Porter DH, Kim D (1990) Simultaneous occurrence of superficial and deep thrombophlebitis in the lower extremity. J Vasc Surg 11:818–824

Smets D, Debing E, De Raeve H, Van den Brande P (1997) Venous aneurysm four years after greater saphenous vein stripping. Acta Chir Belg 97:194–195

Soulez G, Therasse E, Robillard P et al (1999) The value of internal carotid systolic velocity ratio for assessing carotid artery stenosis with Doppler sonography. Am J Roentgenol 172:207–212

Souza de Oliveira IR, Widmann A, Molnar LJ et al (2002) Colour Doppler ultrasound: a new index improves the diagnosis of renal artery stenosis. Ultrasound Med Biol 36:41–47

Sperschneider H, Stein G (1996) Update Nephrologie – Teil III. Nierenarterienstenose Rationelle Diagnostik. Med Klin 91: 517–520

Spies KP, Fobbe F, El-Bedewi M, Wolf KJ, Distler A, Schulte KL (1995) Color-coded duplex sonography for noninvasive diagnosis and grading of renal artery stenosis. Am J Hypertens 8:1222–1231

Sprouse LR 2nd, Meier GH 3rd, Parent FN, DeMasi RJ, Glickman MH, Barber GA (2004) Is ultrasound more accurate than axial computed tomography for determination of maximal abdominal aortic aneurysm diameter? Eur J Vasc Endovasc Surg 28:28–35

Stammler F, Ysermann M, Mohr W, Kuhn C, Goethe S (2000) Value of color-coded duplex ultrasound in patients with polymyalgia rheumatica without signs of temporal arteritis. Dtsch Med Wochenschr 125:1250–1256

Stanziale SF, Wholey MH, Boules TN, Selzer F, Makaroun MS (2005) Determining in-stent stenosis of carotid arteries by duplex ultrasound criteria. J Endovasc Ther 12:346–353

Staub D, Canevascini R, Huegli R et al (2007) Best duplex-sonographic criteria for the assessment of renal artery stenosis – correlation with intra-arterial pressure gradient. Ultraschall Med 28:45–51

Stavropoulos SW, Charagundla SR (2007) Imaging techniques for detection and management of endoleaks after endovascular aortic aneurysm repair. Radiology 243:641–655

Stavros AT, Parker SH, Yakes WF et al (1992) Segmental stenosis of the renal artery: pattern recognition of tardus and parvus abnormalities with duplex sonography. Radiology 184:487–492

Stavros T, Harshfield D (1994) Renal Doppler, renal artery stenosis, and renovascular hypertension: direct and indirect duplex sonographic abnormalities in patients with renal artery stenosis. Ultrasound Quart 12:217–263

Steckmeier B, Spengel FA, Küffer G et al (1989) Richtlinien zur Diagnostik und Therapie des Kompressionssyndroms der A. poplitea. Angio 11:105–113

Stein PD, Hull RD, Saltzman HA, Pineo G (1993) Strategy for diagnosis of patients with suspected acute pulmonary embolism. Chest 103:1553–1559

Steinke W, Els T, Hennerici M (1992) Comparison of flow disturbances in small carotid atheroma using a multi-gate pulsed Doppler system and Doppler color flow imaging. Ultrasound Med Biol 18:11–18

Steinke W, Hennerici M, Rautenberg W, Mohr JP (1992) Symptomatic and asymptomatic high-grade carotid stenoses in Doppler color-flow imaging. Neurology 42:131–138

Steinke W, Kloetzsch C, Hennerici M (1990) Carotid artery disease assessed by color Doppler flow imaging: correlation with standard Doppler sonography and angiography. Am J Roentgenol 154:1061–1068

Sterling KM, Darcy MD (1997) Stenosis of transjugular intrahepatic portosystemic shunts: presentation and management. Am J Roentgenol 168:239–244

Sterpetti AV, Hunter WJ, Schultz RD (1991) Importance of ulceration of carotid plaque in determining symptoms of cerebral ischemia. J Cardiovasc Surg 32:154–158

Sterpetti AV, Schultz RD, Feldhaus RJ, Davenport KL, Richardson M, Farina C, Hunter WJ (1988) Ultrasonographic features of carotid plaque and the risk of subsequent neurologic deficits. Surgery 104:652–660

Stewart MT, Moritz MW, Smith RB 3rd (1986) The natural history of carotid dysplasia. J Vasc Surg 3:305–310

Stief CG, Bähren W, Gall H, Scherb W (1988) Functional evaluation of penile hemodynamics. J Urol 139:734–737

Stief CG, Truss MC, Becker AJ, Kuczyk M, Jonas U (2000) Pharmakologische Therapiemöglichkeiten der Erektionsstörung. Dt Ärztebl 97:A-457–A-460

Stief CG, Wetterauer U (1988) Erectile responses to intracavernous papaverine and phentolamine: comparison of single and combined delivery. J Urol 140:1415–1416

Stiegler H, Rotter G, Standel R, Mosavi S, v. Kooten HJ, Weichenhain B, Baumann G (1994) Wertigkeit der Farbduplexsonographie in der Diagnose insuffizienter Vv. Perforantes. VASA 23:109–114

Stiegler H, Rotter G, Standl R et al (1993) Wertigkeit der Duplexsonographie in der Diagnose insuffizienter Vv. perforantes (Abstract). VASA 41(suppl):15

Stiegler H, Weichhain B, Chatzopulos D, Mathies R, Standl R, Mehnert H (1991) Untersuchungen zur Häufigkeit und Symptomatologie der Lungenembolie in Abhängigkeit von der Lokalisation der tiefen Beinvenenthrombose. VASA 20:119–124

Stierli P, Aeberhard P, Livers M (1992) The role of colour flow duplex screening in infrainguinal vein grafts. Eur J Vasc Surg 6:293–298

Strandness DE (1977) Thrombosis detection by ultrasound, plethysmography, and phlebography. Sem Nucl Med 7:213–218

Strandness DE (1990) Duplex scanning in diagnosis of renovascular hypertension. Surg Clin N Amer 70:109–117

Strauss A (2002) Duplexsonographie der Arterien und Venen. Radiologe 42:235–248

Strauss AL (2001) Farbduplexsonographie der Arterien und Venen. Springer, Berlin/Heidelberg/New York

Strauss AL, Beller KD (1996) Duplexsonographie mit Echokontrastmittel. Ultraschall Med 17:260–265

Strauss AL, Beller KD (1997) Arterial parameters under echo contrast enhancement. Eur J Ultrasound 5:31–38

Strauss AL, Ludwig M, Stein Hj, Horz R, Kopp H, Spengel F, Weber W für den Arbeitskreis Gefäßdiagnostik der Deutschen Gesellschaft für Ultraschall in der Medizin (DEGUM), Sektion Ultraschall der Deutschen Gesellschaft für Angiologie (DGA) und Kassenärztliche Vereinigung Koblenz (1999) Empfehlungen zur Qualitätssicherung in der Ultraschalldiagnostik der Gefäße (1996) VASA 28:135–139

Strauss AL, Roth FJ, Rieger H (1991) Duplexsonographische Bestimmung des Druckabfalls über Iliakaarterienstenosen. Med Klin 86:498–502

Strauss AL, Roth F-J, Rieger H (1993) Noninvasive assessment of pressure gradients across iliac artery stenoses: duplex and catheter correlative study. J Ultrasound Med 12:17–22

Strauss AL, Schäberle W, Rieger H, Neuerburg-Heusler D, Roth F-J, Schoop W (1989) Duplexsonographische Untersuchungen der A. profunda femoris. Z Kardiol 78:567–572

Strauss AL, Schäberle W, Rieger H, Roth F-J (1991) Use of duplex scan in the diagnosis of arteria profunda femoris stenosis. J Vasc Surg 13:698–704

Strauss AL, Scheffler A, Rieger H (1990) Dopplersonographische Bestimmung des Druckabfalls über peripheren Modell-arterienstenosen. VASA 19:207–211

Strauss AL, Weber G, Karasch T et al (1995) Quantifizierung hämodynamisch wirksamer Arterienstenosen mit Farbduplexsonographie. In: Ludwig M et al (eds) Doppler-/Duplex-Sonographie in der Intensivmedizin. Kagerer Kommunikation, Bonn, p 18

Streifler JY, Benavente OR, Fox AJ for the NASCET Group (1991) The accuracy of angiographic detection of carotid plaque ulceration. Results from the NASCET. Stroke 22:149

Stritecky-Kähler T (1994) Chirurgie der Krampfadern/Tomas Stritecky-Kähler. Thieme, Stuttgart/New York

Strotham G, Blebea J, Fowl RJ, Rosenthal GR (1995) Contralateral duplex scanning for deep venous thrombosis is unnecessary in patients with symptoms. J Vasc Surg 22:543–547

Sturzenegger M, Mattle HP, Rivoir A, Baumgartner RW (1995) Ultrasound findings in carotid artery dissection: analysis of 43 patients. Neurology 45:691–698

Subramanyam BR, Balthazar EJ, Madamba MR et al (1983) Sonography of porto-systemic collaterals in portal hypertension. Radiology 146:161–166

Sullivan DE, Peter DJ, Cranley JJ (1984) Real-time B-mode venous ultrasound. J Vasc Surg 1:465

Sumner OS, Lamberth A (1979) Reliability of Doppler ultrasound in the diagnosis of acute venous thrombosis both above and below the knee. Am J Surg 138:205

Sun Y, Cheng-Huai L, Chien-Jung L et al (2002) Carotid atherosclerosis, intima media thickness and risk factors – an analysis of 1781 asymptomatic subjects in Taiwan. Atherosclerosis 164:89–94

Sun Z (2006) Diagnostic value of color duplex ultrasonography in the follow-up of endovascular repair of abdominal aortic aneurysm. J Vasc Interv Radiol 17:759–764

Sutton-Tyrrell K, Wolfson SK, Thompson T, Kelsoey SF (1992) Measurement variability in Duplex scan assessment of carotid atherosclerosis. Stroke 23:215–220

Sztajzel R, Momjian-Mayor I, Comelli M et al (2006) Correlation of cerebrovascular symptoms and microembolic signals with the stratified gray-scale median analysis and color mapping of the carotid plaque. Stroke 37:824–829

Takayasu K, Moriyama N, Shima Y et al (1984) Sonographic detection of large spontaneous spleno-renal shunts and its clinical significance. Br J Radiol 57:565–570

Talbot SR (1982) Use of real-time imaging in identifying deep venous obstruction: a preliminary report. Bruit 6:41–42

Tamura M, Iriguchi H, Miyamoto T, Kimura K, Kawanishi Y, Numata A, Yuasa M, Hashimoto H, Terao N, Kagawa S (1993) Postoperative results of ligation of crura penis for impotence with corporal veno-occlusive insufficiency. Nippon Hinokika Gakkai Zasshi 84: 473–478

Tamura M, Iriguchi H, Miyamoto T, Kimura K, Kawanishi Y, Numata A, Yuasa M, Kagawa S (1993) Transition of diagnosis and treatment for impotence. Nippon Hinyokika Gakkai Zasshi 84:1397–1403

Taniguchi N, Itoh K, Honda M, Obayashi T, Nakamura M, Kawai F, Irie T (1997) Comparative ultrasonographic and angiographic study of carotid arterial lesions in Takayasu's arteritis. Angiology 48:9–20

Tartarini G, Bertoli D, Baglini R, Balbarini A, Mariani M (1988) Diagnosis of aortic dissection by color-coded Doppler. J Nucl Med 32:127–130

Taylor DC, Kettler MD, Moneta GL et al (1988) Duplex ultrasound in the diagnosis of renal artery stenosis: a prospective evaluation. J Vasc Surg 7:363–369

Taylor KJW, Burns PN, Wells PNT (1995) Clinical applications of Doppler ultrasound. Raven, New York, pp 1–33

Taylor PR, Tyrell MR, Crofton M et al (1992) Colour flow imaging in the detection of femoro-distal graft and native artery stenosis: improved criteria. Eur J Vasc Surg 6:232–236

Taylor PR, Wolfe JHN, Tyrrell MR, Mansfield AO, Nicolaides AN, Houston RE (1990) Graft stenosis: justification for 1-year surveillance. Br J Surg 77:1125–1128

Tegos TJ, Sabetai MM, Nicolaides AN et al (2000) Comparability of the ultrasonic tissue characteristics of carotid plaques. J Ultrasound Med 7:399–407

Tegos TJ, Sohail M, Sabetai MM et al (2000) Echomorphologic and histopathologic characteristics of unstable carotid plaques. AJNR Am J Neuroradiol 21:1937–1944

Teichgräber U, Gebel M, Benter T, Manns MP (1997) Duplexsonographische Charakterisierung des Lebervenenflusses bei Gesunden. Ultraschall Med 18:267–271

Tellis VA, Kohlberg WI, Bhat DJ, Driscoll B, Veith FJ (1979) Expanded polytetrafluoroethylene graft fistula for chronic hemodialysis. Ann Surg 189:101–105

Tessler FN, Gehring BJ, Gomes AS, Perrella RR, Ragavendra N, Busuttil RW, Grant EG (1991) Diagnosis of portal vein thrombosis: value of color Doppler imaging. Am J Roentgenol 157:293–296

Thalhammer C, Aschwander M, Mayr M et al (2006) Duplex sonography after living donor kidney transplantation: new insights in the early postoperativ phase. Ultraschall Med 27:141–145

Theodotou BC, Whaley R, Mahaley MS (1987) Complications following transfemoral cerebral angiography for cerebral ischemia: report of 159 angiograms and correlation with surgical risk. Surg Neurol 28:90–92

Thetter O (1987) Das Kompressionssyndrom des Arcus tendineus M. solei. In: Heberer G, Van Dongen RJAM (eds) Gefäßchirurgie. Springer, Berlin/Heidelberg/New York, pp 600–602, Kirschnersche allgemeine und spezielle Operationslehre, Bd 11

Thomas CHR, Röder H-U, Möser GH (1989) Diagnostische Wertigkeit der Angiodynographie bei pulsierenden Raumforderungen. Fortschr Röntgenstr 150:454–457

Thomas ML, Mc Donald LM (1978) Complications of ascending phlebography of the leg. Br Med J 2:317

Thompson MM et al (1998) Comparison of computed tomography and duplex imaging in assessing aortic morphology following endovascular aneurysm repair. Br J Surg 85:346–350

Thompson MM, Boyle JR, Hartshorn T et al (1998) Comparison of computed tomography and duplex imaging in assessing aortic morphology following endovascular aneurysm repair. Br J Surg 85:346–350

Tordoir JHM, De Bruin HG, Hoeneveld H, Eikelboom BC, Kitslaar PJ (1989) Duplex ultrasound scanning in the assessment of arteriovenous fistulas created for hemodialysis access: comparison with digital subtraction angiography. J Vasc Surg 10:122–128

Tordoir J, Canaud B, Haage P et al (2007) EBPG on vascular access. Nephrol Dial Transplant 22:ii88–ii117

Tranquet T, Bescos JM, Reparaz B (1989) Noninvasive methods in the diagnosis of isolated superior mesenteric vein thrombosis: US and CT. Gastrointest Radiol 14:321–325

Tratting S, Hübsch P, Frühwald F (1992) Farbkodierte Dopplersonographie der Vertebralarterien. In: Frühwald F, Blackwell DE (eds) Atlas der farbkodierten Doppersonographie. Gefäße und Weichteile des Halses und der oberen Extremität. Springer, Wien/New York, p 138

Tratting S, Hübsch P, Schuster H, Pötzleitner D (1990) Color-coded Doppler imaging of normal vertebral arteries. Stroke 21:1222–1225

Tratting S, Hübsch P, Schwaighofer B, Karnel F, Eilenberger M (1991) Vaskuläre Raumforderungen der Arteria carotis – Nachweis mit der farbkodierten Dopplersonographie. Ultraschall Med 12:70–73

Tratting S, Plötzleitner D, Hübsch P, Kaha K, Matula CH, Magometschnigg H (1992) Nicht-invasive Verlaufskontrolle mittels farbkodierter Doppler-Sonographie nach operativen Eingriffen an den extrakraniellen hirnversorgenden Arterien. Fortschr Röntgenstr 156:224–227

Tratting S, Schwaighofer B, Hübsch P, Schwarz M, Kainberger F (1991) Color-coded Doppler sonography of vertebral arteries. J Ultrasound Med 10:221–226

Treiman GS, Lawrence PF, Bhirangi K, Gazak CE (1999) Effect of outflow level and maximum graft diameter on velocity parameters of reversed vein grafts. J Vasc Surg 30:16–25

Trum JW, Gubler FM, Laan R, van der Veen F (1996) The value of palpation, varicoscreen contact thermography and color Doppler ultrasound in the diagnosis of varicocele. Hum Reprod 11:1232–1235

Tucker RM, Lebbarthe DR (1977) Frequency of surgical treatment for hypertension in adults in the Mayo Clinic from 1973–1975. Mayo Clin Proc 52:542–555

Tucker RM, Strog CG, Brennan LA, Sheps SG (1978) Renovascular hypertension: relationship of surgical curability to renin-angiotensin activity. Mayo Clin Proc 53:373–377

Turmel-Rodrigues L, Pengloan J, Baudin S et al (2000) Treatment of stenosis and thrombosis in haemodialysis fistulas and grafts by interventional radiology. Nephrol Dial Transplant 15: 2029–2036

Turmel-Rodrigues L, Mouton A, Birmele B et al (2001) Salvage of immature forearm fistulas for haemodialysis by interventional radiology. Nephrol Dial Transplant 16:2365–2371

Turton EP, Scott DJ, Richards SP, Weston MJ, Berridge DC, Kent PJ, Kester RC (1999) Duplex-derived evidence of reflux after varicose vein surgery: neoreflux or neovascularisation? Eur J Vasc Endovasc Surg 17:230–233

Twomey A, Twomey EM, Wilinks RA (1984) Unrecognized aneurysm disease in male hypertensive patients. Br J Surg 71:307–308

Ubbink DT, Legemate DA, Llull JB (2002) Color-flow duplex scanning of the leg arteries by use of a new echo-enhancing agent. J Vasc Surg 35:392–396

UK Small Aneurysm Trial Participants (1998) Results for randomised controlled trial of early elective surgery or ultrasonographic surveillance for small abdominal aortic aneurysms. Lancet 352:1649–1660

Umemura A, Yamada K (2000) B-mode flow imaging of the carotid artery. Stroke 32:2055–2057

Umscheid T et al (2001) Nachtuntersuchungen bei Patienten nach endovaskulärer Aortenprothesen-Durchführung, Anforderungen und Probleme. Gefäßchirurgie 6:185–193

Urbancic A, Buturovic-Ponkivar J (2001) Emergency intra- and perioperative Doppler after kidney transplantation – a guide for immediate surgical intervention. Transplant Proc 33:3320–3321

Uthoff H, Staub D, Meyerhans A et al (2008) Intima-media thickness and carotid resistive index: progression over 6 years and predictive value for cardiovascular events. Ultraschall Med 29:604–610

Valentine RJ (2003) Asymptomatic internal carotid artery aneurysm. J Vasc Surg 37:210

Valji K, Bookstein JJ (1993) Diagnosis of arteriogenic impotence: efficacy of duplex sonography as a screening tool. Am J Roentgenol 160:65–69

Valois de JC, van Schaik CC, Verzijlbergen F et al (1990) Contrast venography: from gold standard to "golden backup" in clinically suspected deep vein thrombosis. Eur J Radiol 11:131–137

Van Bemmelen PS, Bedford G, Beach K, Strandness DE (1989) Quantitative segmental evaluation of venous valvular reflux with duplex ultrasound scanning. J Vasc Surg 10:425–431

Van Bockel JH, van Schilfgaarde R, Van Brummelen P, Terpsta JL (1989) Renovascular hypertension. Surg Gynecol Obstet 169:467–478

Van Damme H, Demoulin JC, Zicot M, Creemers E, Trotteur G, Limet R (1992) Pathological aspects of carotid plaques. J Cardiovasc Surg 33:46–53

Van Damme H, Sakalihasan N, Limet R (1999) Fibromuscular dysplasia of the internal carotid artery: personal experience with 13 cases and literature review. Acta Chir Belg 99:163–168

Van der Hulst VP, van Baalen J, Kool LS et al (1996) Renal artery stenosis: endovascular flow wire study for validation of Doppler US. Radiology 200:165–168

Van der Zaag ES, Legemate DA, Nguyen T, Balm R, Jacobs MJ (1998) Aortoiliac reconstructive surgery based upon the results of duplex scanning. Eur J Vasc Endovasc Surg 16:383–389

Van Gemmeren D, Fobbe F, Ruhnke-Trautmann M (1991) Diagnostik tiefer Beinvenenthrombosen mit der farbkodierten Duplexsonographie und sonographische Altersbestimmung der Thrombose. Z Kardiol 80:523–528

Van Ramshorst B, Legemate DA, Verzijlbergen JF, Hoeneveld H, Eikelboom BC, de Valois JC, Meuwissen OJ (1991) Duplex scanning in the diagnosis of acute deep vein thrombosis of the lower extremity. Eur J Vasc Surg 5:255–260

Van Ramshorst B, van Bemmelen PS, Hoeneveld H, Eikelboom BC (1994) The development of valvular incompetence after deep vein thrombosis: a follow-up study with duplex scanning. J Vasc Surg 20:1059–1066

Vandendriessche M, Thiery L (1986) Aneurisma van de vena poplitea. Acta Chir Belg 86:37–40

Vasbinder GB, Nelemans PJ, Kessels AG et al (2001) Diagnostic test for renal artery stenosis in patients suspected of having renovascular hypertension: a meta-analysis. Ann Inter Med 135:401–411

Vascular Access 2006 Work Group (2006) Clinical practice guidelines for vascular access. Am J Kidney Dis 48:S176–S247

Vasdekis SN, Clarke GH, Nikolaides A (1989) Quantification of venous reflux by means of duplex scanning. J Vasc Surg 10:670–677

Vaskedis SN, Clarke GH, Hobbs JT, Nicolaides AN (1989) Evaluation of non-invasive and invasive methods in the assessment of short saphenous termination. Br J Surg 76:929–932

Vermeulen EGJ, Umans U, Rijbroek A, Rauwerda JA (2000) Percutaneous duplex-guided thrombin injection for treatment of iatrogenic femoral artery pseudoaneurysms. Eur J Vasc Endovasc Surg 20:302–304

Vicente DC, Kazmers A (1999) Acute mesenteric ischemia. Curr Opin Cardiol 14:453–458

Visser K, Idu MM, Buth J, Engel GL, Hunink MG (2001) Duplex scan surveillance during the first year after infrainguinal autologous vein bypass grafting surgery: costs and clinical outcomes compared with other surveillance programs. J Vasc Surg 33:123–130

Vitovec J (1976) Aneurysma venae poplitae. Cesk Radiol 30:328–340

Vogel P, Laing FD, Jeffrey RB jr, Wing VW (1987) Deep venous thrombosis of the lower extremity: US evaluation. Radiology 163:747–751

Voiculescu A, Hollenbeck M, Müller B et al (2005) Reversed diastolic flow after renal transplantation. A sign for a poor prognosis. Ultraschall Med 26:S73

Vollmar J (1982) Rekonstruktive Chirurgie der Arterien. Thieme, Stuttgart, pp 137–202

Von Behren P (1998) Harmonic imaging. In: Bogdahn U, Becker G, Schlachetzki F (eds) Echosignalverstärker und transkranielle Farbduplex-Sonographie. Blackwell Wiss.-Verlag, Berlin/Wien, pp 183–191

Vosshenrich R, Kopka L, Castillo E, Bottcher U, Graessner J, Grabbe E (1998) Electrocardiograph-triggered two-dimensional time-of-flight versus optimized contrast-enhanced three-dimensional MR angiography of the peripheral arteries. Magn Reson Imag 16: 887–892

Wagner G, Kaplan HS (1993) The new injection treatment for impotence. Brunner/Mazel, New York, pp 85–93

Waibel P, Fahrländer J, Ludin H (1966) Die chirurgische Behandlung der Angina abdominalis. Bedeutung der Diagnose in der Prophylaxe des arteriellen Mesenterialinfarktes. Schweiz med Wschr 96:10

Wain RA, Berdejo GL, Delvalle WN et al (1999) Can duplex scan arterial mapping replace contrast arteriography as the test of choice before infrainguinal revascularization? J Vasc Surg 29:100–107

Wallner B, Kratzsch G, Friedrich JM, Roth J (1989) Die intraarterielle DSA der Handarterien in der Diagnostik entzündlicher Bindegewebserkrankungen. RöFo 151:565–568

Walter P, Kubale R, Gross G, Beraneck E (1999) Das Kompressionssyndrom des Truncus coeliacus. Gefäßchirurgie 4:28–33

Walter P, Lindemann W, Koch P, Feifel G (1992) Der akute Mesenterialinfarkt. Klinikarzt 21:457–463

Ward B, Baker AC, Humphrey VF (1997) Nonlinear propagation applied to the improvement of resolution in diagnostic medical ultrasound. J Acoust Soc Am 101:143–154

Wardlaw JM, Chappell FM, Best JJK, Wartolowska K, Berry E, on behalf of the NHS Research and Development Health Technology Assessment Carotid Stenosis Imaging Group (2006) Non-invasive imaging compared with intra-arterial angiography in the diagnosis of symptomatic carotid stenosis: a meta-analysis. Lancet 367:1503–1512

Watanabe F, Yoshida T, Akizuki M, Mukai M (1993) Development of multiple subcutaneous nodules in a patient with rheumatic arthritis during methotrexate therapy. Ryumachi 33:74–79

Waugh J, Sacharias N (1992) Arteriographic complications in the DSA era. Radiology 182:243–246

Weber G, Strauss AL, Rieger H, Scheffler A, Eisenhoffer J (1992) Validation of Doppler measurement of pressure gradients across peripheral model arterial stenosis. J Vasc Surg 16:10–16

Weber J, May R (1990) Funktionelle Phlebologie. Thieme, Stuttgart

Wein AJ, van Arsdalen K, Hanno PhM, Levin RM (1991) Anatomy of male sexual function. In: Jonas U, Thon WF, Stief CG (eds) Erectile dysfunction. Springer, Berlin/Heidelberg/New York, pp 3–15

Weingarten MS, Branas CC, Czeredarzuk M, Schmidt JD, Wolferth CC (1993) Distribution and quantification of venous reflux in lower extremity chronic venous stasis disease with duplex scanning. J Vasc Surg 18:753–759

Weinmann EE, Chayen D, Kobzantzev ZV, Zaretsky M, Bass A (2002) Treatment of postcatheterisation false aneurysms: ultrasound-guided compression vs ultrasound-guided thrombin injection. Eur J Vasc Endovasc Surg 23:68–72

Weinmann S, Sandbichler P, Flora G (1986) Aneurysma der Vena poplitea als Quelle der Lungenembolie. Langenbecks Arch Chir 367: 107–112

Welch HJ, Young CM, Semegran AB, Iafrati MD, Mackey WC, O'Donnell TF (1996) Duplex assessment of venous reflux and chronic venous insufficiency: The significance of deep venous reflux. J Vasc Surg 24:755–762

Welger D, Müller JH (1988) Assoziierte thrombotische Prozesse des oberflächlichen, perforierenden und intramuskulären Venensystems bei Patienten mit akuter Phlebothrombose der unteren Extremitäten. Z Ges Inn Med Grenzgeb 43:15–18

Wells DH (1992) Physical and technical aspects of colour flow ultrasound. In: Labs KH et al (eds) Diagnostic vascular ultrasound. Arnold, London, pp 145–152

Wells PhS, Lensing AWA, Davidson BL, Prins MH, Hirsh J (1995) Accuracy of ultrasound for the diagnosis of deep venous thrombosis in asymptomatic patients after orthopedic surgery. Ann Intern Med 122:47–53

Wells PS, Anderson DR, Bormanis J, Guy F, Mitchel M, Gray L, Clement C, Robinson KS, Lewandowski B (1997) Value of assessment of pretest probability of deep-vein thrombosis in clinical management. Lancet 350:1795–1798

Wells PS, Hirsh J, Anderson DR, Lensing AW, Foster G et al (1995) Accuracy of clinical assessment of deep vein thrombosis. Lancet 345:1326–1330

Wermke W, Gassmann B (1998) Tumordiagnostik der Leber mit Echosignalverstärker. Springer

Weskot H, Özden N, Von Leitner E (2005) Stellenwert der kontrastverstärkten Sonograpie der ambulanten Diagnostik: Zuverlässigkeit und Ökonomie. Ultraschall Med 26:S73

Wesley S, Moore MD (2003) For severe carotid stenosis found on ultrasound, further arterial evaluation is unnecessary. Stroke 34:1816–1817

Westerband A, Mils JK, Kistler S, Berman SS, Hunter GC, Marek JM (1997) Prospective validation of threshold criteria for intervention in infrainguinal vein undergoing duplex surveillance. Ann Vasc Surg 11:44–48

Wheeler AHB (1985) Diagnosis of deep vein thrombosis. Am J Surg 150:7–14

Whelan TJ (1984) Popliteal artery entrapment. In: Rutherford RB (ed) Vascular surgery. Saunders, Philadelphia

White GH et al (2000) How should endotension be defined? History of a concept and evolution of a new term. J Endovasc Ther 7:435–438

Whitehead ED, Klyde BJ, Zussman S, Salkin P (1990) Diagnostic evaluation of impotence. Postgrad Med 88:123–26, 129–136

Whyman MR, Ruckley CV, Fowkes FG (1993) A prospective study of the natural history of femoropopliteal artery stenosis using duplex ultrasound. Eur J Vasc Surg 7:444–447

Widder B (1999) Doppler- und Duplexsonographie der hirnversorgenden Arterien. Springer, Berlin/Heidelberg/New York/Tokyo

Widder B, Görtler M (2004) Doppler- und Duplexsonographie der hirnversorgenden Arterien. Springer, Heidelberg

Widder B, von Reutern GM, Neuerburg-Heusler D (1986) Morphologische und dopplersonographische Kriterien zur Bestimmung von Stenosierungsgraden an der A. carotis interna. Ultraschall Med 7:70–75

Widder B, Berger G, Bressmer H et al (1988) Reproduzierbarkeit sonographischer Kriterien zur morphologischen Beurteilung von Karotisstenosen. Ultraschall Klin Prax 5(Suppl 1)

Widder B, Arnolds B, Drews S (1990) Terminologie der Ultraschallgefäßdiagnostik. Ultraschall Med 11:214–218

Widder B, Berger G, Hackspacher J, Horz R, Nippe A, Paulat K, Schäfer H, Wieller C, Willeit J (1990) Reproduzierbarkeit sonographischer Kriterien zur Charakterisierung von Karotisstenosen. Ultraschall Med 11:56–61

Widder B, Paulat K, Hackspacher J et al (1990) Morphological characterization of carotid artery stenosis by ultrasound duplex scanning. Ultrasound Med Biol 16:349–354

Widder B, Kleiser B, Hackspacher J, Reuchlin G, Dürr A (1992) Sonomorphological prediction of progressive carotid artery stenoses. In: Oka M et al (eds) Recent advances in neurosonology. Elsevier, Amsterdam, pp 425–429

Wijeyaratne SM, Jarvis S, Stead LA et al (2003) A new method for characterizing carotid plaque: multiple cross-sectional view echomorphology. J Vasc Surg 37:778–784

Wilkinson R (1996) Epidemiology and clinical manifestation. In: Novick A, Scoble J, Hamilton G (eds) Renal vascular disease. Saunders, London/Philadelphia/Toronto, pp 171–184

Wilson YG, Davies AH, Currie IC, Morgan M, McGrath C, Baird RN (1996) Vein graft stenosis: incidence and intervention. Eur J Vasc Endovasc Surg 11:164–169

Wixon CL, Mills JL, Westerband A, Hughes JD, Ihnat DM (2000) An economic appraisal of lower extremity bypass graft maintenance. J Vasc Surg 32:1–12

Wolf KJ, Fobbe F (1993) Farbkodierte Duplexsonographie. Thieme, Stuttgart

Wolf Y et al (2000) Duplex ultrasound scanning versus computed tomographic angiography for postoperative evaluation of endovascular abdominal aortic aneurysm repair. J Vasc Surg 32:1142–1148

Wölfle KD, Bruijnen H, Mayer B, Loeprecht H (1994) Verlaufskontrolle infrainguinaler Bypassoperationen: Bedeutung von systolischer Spitzengeschwindigkeit und Arm-Knöchel-Index für die Bewertung femorodistaler Rekonstruktionen. VASA 23:349–356

Wölfle KD, Bruijnen H, Limmer S, Loeprecht H (1999) Autologe Distal-origin-Bypasses zur Überbrückung infrapoplitealer Verschlussprozesse bei Diabetikern mit kritischer Fußischämie. Gefässchirurgie 4:220–228

Wölfle KD, Neudert S, Mayer B, Storm G, Bruijnen H, Loeprecht H (1992) Duplexsonographische Überwachung von infrainguinalen arteriellen Rekonstruktionen: Kann dadurch ein drohender Bypassverschluss erkannt werden? Zentrbl Chir 117:540–546

Wolverson MK, Bashiti HM, Peterson GJ (1983) Ultrasonic tissue characterization of atheromatous plaques using a high resolution real time scanner. Ultrasound Med Biol 6:669–709

Wong JK, Duncan JL, Nichols DM (2003) Whole-leg duplex mapping for varicose veins: observations on patterns of reflux in recurrent and primary legs, with clinical correlation. Eur J Vasc Endovasc Surg 25:267–275

Woodcock JP, Fitzgerald DE, Labs KH et al (1992) Consensus on problem areas in diagnostic vascular ultrasound. In: Labs KH et al (eds) Diagnostic vascular ultrasound. Arnold, London, pp 321–325

Woods VL, Zvaifler NJ (1985) Pathogenesis of systemic lupus erythematosus. In: Kelly WN, Harris ED, Ruddy SR, Siedge CB (eds) Textbook of rheumatology, 3rd edn. Saunders, Philadelphia/London/Toronto, pp 1077–1100

Wuppermann T (1986) Varizen, Ulcus cruris und Thrombose. Springer, Berlin/Heidelberg/New York/Tokyo

Wuppermann T, Exler U, Mellmann J, Kestilä M (1981) Noninvasive quantitative measurement of regurgitation insufficiency of the superior saphenous vein by Doppler-ultrasound: a comparison with clinical examination and phlebography. Vasa 10:24–27

Wuppermann T, Knospe E, Reiss HD, Mellmann J (1981) Diagnostik insuffizienter Venae perforantes: Trefferquoten verschiedener Untersuchungsmethoden bei der primären Varikosis. In: May R, Partsch H, Staubesand J (eds) Venae perforantes. Urban & Schwarzenberg, München

Yao JST, Flinn WR, McCarty WJ, Bergan JJ (1986) The role of noninvasive testing in the evaluation of chronic venous problems. World J Surg 10:91–918

Yao JST, Van Bellen B, Flinn WR, Bergan JJ (1982) Aneurysms of the venous system. In: Bergan JJ, Yao JST (eds) Aneurysms, diagnosis and treatment. Grune & Stratton, New York, pp 515–529

Yeh H-C, Stancato-Pasik A, Ramos R, Rabinowitz JG (1996) Paraumbilical venous collateral circulations: colour Doppler ultrasound features. J Clin Ultrasound 24:359–366

Yurcel EK, Fisher JS, Egglin TK, Geller SC, Waltman AC (1991) Isolated calf venous thrombosis: diagnosis with compression ultrasound. Radiology 179:443–446

Yurdakul M, Tola M, Cumhur T (2006) B-flow imaging for assessment of 70% to 99% internal carotid artery stenosis based on residual lumen diameter. J Ultrasound Med 25:211–215

Zachrisson BE, Norback B (1974) Phlebographic diagnosis of a soleus vein aneurysm. Vasa 3:308

Zannetti S, De Rango P, Parente B, Parlani G, Verzini F, Maselli A, Nardelli L (2000) Role of duplex scan in endoleak detection after endoluminal abdominal aortic aneurysm repair. Eur J Vasc Endovasc Surg 19:531–535

Zanow J, Petzold K, Petzold M et al (2006) Flow reduction in high-flow arteriovenous access using intraoperative flow monitoring. J Vasc Surg 44:1273–1278

Zeller T, Frank U, Müller C et al (2003) Predictors of improved renal function after percutaneous stent-supported angioplasty of severe atherosclerotic ostial renal artery stenosis. Circulation 108:2244–2249

Zeller T, Frank U, Späth M et al (2001) Fabduplexonographische Darstellbarkeit von Nierenarterien und Erkennung hämodynamisch relevanter Nierenarterienstenosen. Ultraschall Med 22: 116–121

Zierler RE (1999) Vascular surgery without arteriography: use of Duplex ultrasound. Cardiovasc Surg 7:74–82

Zöllner N, Zoller WG, Spengel FA, Weigold B, Schewe CK (1991) The spontaneous course of small abdominal aortic aneurysms. Aneurysmal growth rates and life expectancy. Klin Wochensch 69:633–639

Zuber M, Koch B, Gause A, Pfreundschuh M (1996) Diagnose- und Klassifikationskriterien in der Rheumatologie: Kollagenosen. Dtsch Med Wochenschr 121:913–919

Zwiebel WJ (1987) Spectrum analysis in carotid Doppler sonography. Ultrasound Med Biol 13:625–636

Zwiebel WJ (1997) Doppler parameters for carotid stenosis. Semin Ultrasound CT MR 18:66

Zwiebel WJ (2000) Vascular disorders of the liver. In: Zwiebel WJ (ed) Introduction to vascular ultrasonography. Saunders, Philadelphia/London/Toronto, pp 431–454

Zwolak RM (2000) Arterial duplex scanning. In: Rutherford RB (ed) Vascular surgery, 5th edn. Saunders, Philadelphia/London/Sydney, pp 192–214

Zwolak RM, Fillinger MF, Walsh DB, LaBombard FE, Musson A, Darling CE, Cronenwett JL (1998) Mesenteric and celiac duplex scanning: a validity study. J Vasc Surg 27:1078–1108

# Subject Index



W. Schäberle, *Ultrasonography in Vascular Diagnosis*,
DOI: 10.1007/978-3-642-02509-9, © Springer-Verlag Berlin Heidelberg 2011